International Table of Atomic Weights (1983)

Based on relative atomic mass of $^{12}C = 12$.

The following values apply to elements as they exist in materials of terrestrial origin and to certain artificial elements. When used with due regard to footnotes, they are reliable to ± 1 in the last digit, or ± 3 when followed by an asterisk (*). Value in parentheses is the mass number of the isotope of longest half-life.

	Symbol	Atomic number	Atomic weight
Actinium	Ac	89	227.028
Aluminum	Al	13	26.9815[a]
Americium	Am	95	(243)
Antimony	Sb	51	121.75*
Argon	Ar	18	39.948[b,c,d,g]*
Arsenic	As	33	74.9216[a]
Astatine	At	85	(210)
Barium	Ba	56	137.33[g]
Berkelium	Bk	97	(247)
Beryllium	Be	4	9.01218[a]
Bismuth	Bi	83	208.980[a]
Boron	B	5	10.81[c,d,e]
Bromine	Br	35	79.904[c]
Cadmium	Cd	48	112.41[g]
Calcium	Ca	20	40.08[g]
Californium	Cf	98	(251)
Carbon	C	6	12.011[b,d]
Cerium	Ce	58	140.12
Cesium	Cs	55	132.905[a]
Chlorine	Cl	17	35.453[c]
Chromium	Cr	24	51.996[c]
Cobalt	Co	27	58.9332[a]
Copper	Cu	29	63.546*[c,d]
Curium	Cm	96	(247)
Dysprosium	Dy	66	162.50
Einsteinium	Es	99	(252)
Erbium	Er	68	167.26*
Europium	Eu	63	151.96[g]
Fermium	Fm	100	(257)
Fluorine	F	9	18.9984[a]
Francium	Fr	87	(223)
Gadolinium	Gd	64	157.25*[g]
Gallium	Ga	31	69.72
Germanium	Ge	32	72.59*
Gold	Au	79	196.967[g]
Hafnium	Hf	72	178.49*
Helium	He	2	4.00260[b,c,g]
Holmium	Ho	67	164.930[a]
Hydrogen	H	1	1.0079[b,d]
Indium	In	49	114.82[g]
Iodine	I	53	126.905[a]
Iridium	Ir	77	192.22*
Iron	Fe	26	55.847*
Krypton	Kr	36	83.80[c,d]
Lanthanum	La	57	138.906*[b,g]
Lawrencium	Lr	103	(260)
Lead	Pb	82	207.2[d,g]
Lithium	Li	3	6.941*[c,d,e,g]
Lutetium	Lu	71	174.967*
Magnesium	Mg	12	24.305[c,g]
Manganese	Mn	25	54.9380[a]
Mendelevium	Md	101	(258)
Mercury	Hg	80	200.59*
Molybdenum	Mo	42	95.94
Neodymium	Nd	60	144.24*[g]
Neon	Ne	10	20.179*[c,e]
Neptunium	Np	93	237.048[f]
Nickel	Ni	28	58.69
Niobium	Nb	41	92.9064[a]
Nitrogen	N	7	14.0067[b,c]
Nobelium	No	102	(259)
Osmium	Os	76	190.2[g]
Oxygen	O	8	15.9994[b,c,d]
Palladium	Pd	46	106.42[g]
Phosphorus	P	15	30.9738[a]
Platinum	Pt	78	195.08*
Plutonium	Pu	94	(244)
Polonium	Po	84	(209)
Potassium	K	19	39.0983*
Praseodymium	Pr	59	140.908[a]
Promethium	Pm	61	(145)
Protactinium	Pa	91	231.036[f]
Radium	Ra	88	226.025[f,g]
Radon	Rn	86	(222)
Rhenium	Re	75	186.207[c]
Rhodium	Rh	45	102.906[a]
Rubidium	Rb	37	85.4678*[c,g]
Ruthenium	Ru	44	101.07*[g]
Samarium	Sm	62	150.36[g]
Scandium	Sc	21	44.9559[a]
Selenium	Se	34	78.96*
Silicon	Si	14	28.0855*
Silver	Ag	47	107.868[c,g]
Sodium	Na	11	22.9898
Strontium	Sr	38	87.62[g]
Sulfur	S	16	32.06[d]
Tantalum	Ta	73	180.948*[b]
Technetium	Tc	43	(98)
Tellurium	Te	52	127.60*[g]
Terbium	Tb	65	158.925[a]
Thallium	Tl	81	204.383*
Thorium	Th	90	232.038[f,g]
Thulium	Tm	69	168.934[a]
Tin	Sn	50	118.71*
Titanium	Ti	22	47.88*
Tungsten	W	74	183.85*
Unnilpentium	Unp	105	(262)
Unnilhexium	Unh	106	(263)
Unnilquadium	Unq	104	(261)
Uranium	U	92	238.029[b,c,e,g]
Vanadium	V	23	50.9415*[b,c]
Xenon	Xe	54	131.29[a,e,g]
Ytterbium	Yb	70	173.04*
Yttrium	Y	39	88.9059[a]
Zinc	Zn	30	65.39
Zirconium	Zr	40	91.224[g]

[a]Element with only one stable nuclide.
[b]Element with one predominant isotope (about 99 to 100% abundance).
[c]Element for which the atomic weight is based on calibrated measurements.
[d]Element for which known variation in isotopic abundance in terrestrial samples limits the precision of the atomic weight given.
[e]Element for which users are cautioned against the possibility of large variations in atomic weight due to inadvertent or undisclosed artificial isotopic separation in commercially available materials.
[f]Most commonly available long-lived isotope.
[g]In some geological specimens this element has an anomalous isotopic composition, corresponding to an atomic weight significantly different from that given.

CHEMISTRY & CHEMICAL REACTIVITY

Thermite reaction, Fe_2O_3 + Al (Charles D. Winters)

CHEMISTRY & CHEMICAL REACTIVITY

JOHN C. KOTZ
SUNY Distinguished Teaching Professor
State University of New York
College at Oneonta

KEITH F. PURCELL
Professor of Chemistry
Kansas State University

SAUNDERS COLLEGE PUBLISHING
Philadelphia New York Chicago
San Francisco Montreal Toronto
London Sydney Tokyo Mexico City
Rio de Janeiro Madrid

Address orders to:
383 Madison Avenue
New York, NY 10017

Address editorial correspondence to:
210 West Washington Square
Philadelphia, PA 19105

Text Typeface: Times Roman
Compositor: General Graphic Services
Acquisitions Editor: John Vondeling
Developmental Editor: Jay Freedman
Project Editor: Carol Field
Copy Editor: Charlotte Nelson
Art Director: Carol C. Bleistine
Text Designer: Edward A. Butler
Cover Designer: Lawrence R. Didona
Layout Artist: Dorothy Chattin
Text Artwork: J & R Technical Services
Production Manager: Tim Frelick
Assistant Production Manager: JoAnn Melody

Cover Credit: Fireworks/© COMSTOCK, Inc.

Library of Congress Cataloging-in-Publication Data

Kotz, John C.
Chemistry and chemical reactivity.

Includes index.

1. Chemistry. I. Purcell, Keith F., 1932–
II. Title.
QD31.2.K68 1987 540 86-26084

ISBN 0-03-058349-7

CHEMISTRY & CHEMICAL REACTIVITY 0-03-058349-7

Library of Congress catalog card number 86-26084.

890 9876543

CBS COLLEGE PUBLISHING
Saunders College Publishing
Holt, Rinehart and Winston
The Dryden Press

Preface

The title of this book, CHEMISTRY & CHEMICAL REACTIVITY, was chosen to convey its principal themes: a broad overview of the principles of chemistry and the reactivity of chemical elements and compounds. While attempting to provide a firm foundation in these areas, it is our hope also to convey a sense of chemistry as a field that not only has a lively history but also one that is currently dynamic, with important new developments on the horizon. We also hope to provide some insight into the chemical aspects of the world around us. For example, what is the role and importance of carbon dioxide in our environment, why is iron important in our economy and why does it rust, and what is it that gives the color and boom to fireworks? By tackling the principles leading to answers to these questions, you can come to a better general understanding of nature and to an appreciation for some of the consumer products coming from the chemical industry. Indeed, one of the objectives of this book is to provide the tools needed to function as a chemically literate citizen. Learning something of the chemical world is as important as understanding some basic mathematics and biology, and just as important as having an appreciation for fine music and literature.

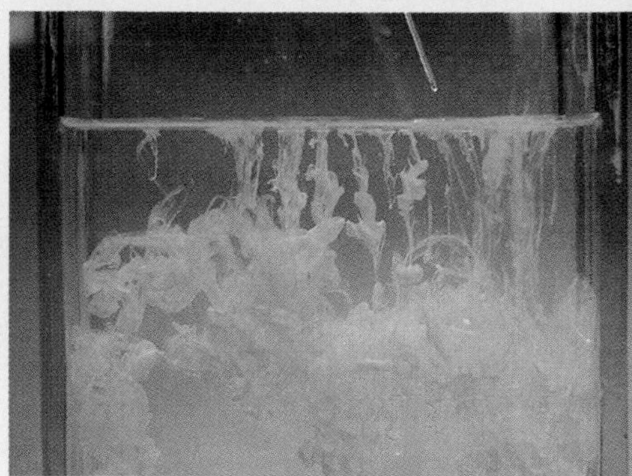
Charles D. Winters

Above all, you should realize that the authors of this book became chemists because, simply put, it is fun to discover new compounds and find new ways to apply chemical principles. We hope to have conveyed in this book that sense of enjoyment as well as our awe at what is known about chemistry and, just as important, what is not known!

AUDIENCE

CHEMISTRY & CHEMICAL REACTIVITY is a textbook for introductory courses in chemistry for students interested in further study in science, whether that science is biology, chemistry, engineering, geology, physics, or related subjects. Our assumption is that students beginning this course will have had a basic foundation in algebra and some in general science. Although undeniably helpful, a previous exposure to chemistry is neither assumed nor required.

PHILOSOPHY AND APPROACH

When this book was planned, we had two major, but not independent, goals. The first was to construct a book that students would enjoy reading and that would offer, at a reasonable level of rigor, chemistry and chemical principles in a format and organization typical of college and university courses today. Further, we wanted to convey the utility and importance of chemistry by introducing the properties of the elements, their compounds, and their reactions as early as possible and by focusing the discussion as much as possible on these subjects.

A glance at the introductory chemistry texts currently available shows that there is a generally common order of treatment of chemical principles used by educators. With a few minor changes we have followed that order as well. However, that is not to say that the chapters cannot be used in some other order. For example, Chapter 6 (Gases) may be coupled with Chapter 11 (Liquids and Solids). The introduction to oxidation-reduction reactions in Chapter 3 may be moved easily to Chapter 19, or, as is done in the course of one of the authors, it can be taught as a part of the discussion of stoichiometry in Chapter 4. Further, one of the authors of this text regularly teaches the material on equilibria involving insoluble solids (Chapter 17) before acid-base equilibria (Chapters 15 and 16), and he introduces kinetics (Chapter 13) and thermodynamics (Chapter 18) as a unit, after all of the material on equilibria. Although chapters are loosely organized into groups with common themes, every attempt has been made to make the chapters as independent as possible.

The order of topics in the text was also devised to introduce as early as possible the background required for the laboratory experiments usually done in General Chemistry. For this reason, chapters on common reaction types (acid-base and oxidation-reduction reactions, Chapter 3), stoichiometry (Chapter 4), thermochemistry (Chapter 5), and gases (Chapter 6) begin the book.

The American Chemical Society has recently urged educators to put "chemistry" back into introductory chemistry courses. As inorganic chemists, we agree wholeheartedly. Therefore, we have tried to describe the elements, their compounds, and their reactions as early and as often as possible in three ways. First, there are numerous color photographs of reactions occurring, of the elements and of common compounds, and of common laboratory operations and industrial processes. Further, we have tried to bring material on the properties of elements and compounds as early as possible into the Exercises and Study Questions, and new principles are introduced using realistic chemical situations. Additionally, there are Special Sections on the nature of some important elements (The Essential Elements, Gold, Mercury, Platinum), photographic essays on the chemistry of some of the metals (Silver, Lead, Nickel, and Iron), and on other interesting aspects of chemistry (The Origin of the Elements, the Chemistry of Space, the Chemistry of Carbon Dioxide, and the Chemistry of Fireworks). Finally, Part 5 is devoted to a more systematic study of descriptive chemistry, largely from the point of view of chemicals important in our economy and in the world around us.

ORGANIZATION

CHEMISTRY & CHEMICAL REACTIVITY is organized into five parts, each containing several chapters.

Part 1 The Basic Tools of Chemistry

Certain basic ideas and methods form the fabric of chemistry, and these are introduced in Part 1. Chapter 1 defines some important terms and is a review of units and mathematical methods. Chapter 2 introduces some basic ideas of atoms and molecules, and introduces you to one of the most important organizational devices of chemistry, the periodic table. In Chapter 3 you will begin to learn some principles of chemical reactivity, and in Chapter 4 you will be introduced to some of the numerical methods used by chemists to extract quantitative information from chemical reactions. Chapter 5 is the first introduction to the energy involved in chemical processes, and Chapter 6 is an introduction to the properties of gases and the use of mathematical models to predict and understand the physical properties of matter.

Part 2 The Structure of Atoms and Molecules
The first goal of this section is to outline, in Chapters 7 and 8, the current theories of the arrangement of electrons in atoms and some of the historical developments that led to these ideas. With these ideas, we can understand why atoms and their ions have different chemical and physical properties. So that these properties can be recalled, and predictions made, the discussion is tied closely to the arrangement of elements in the periodic table. In Chapter 9 we take up for the first time how the electrons of atoms in a molecule may lead to chemical bonding and the properties of these bonds. In addition, we can show how to derive the three-dimensional structure of simple molecules. Finally, Chapter 10 considers come of the major theories of chemical bonding in more detail.

Part 3 States of Matter
Intermolecular forces, which lead to the liquid and solid states of matter, are described in Chapter 11, with particular attention given to liquid and solid water. Chapter 11 also considers the solid state, an area of chemistry that is currently undergoing a renaissance. In Chapter 12 we take up the properties of solutions, intimate mixtures of gases, liquids, and solids.

Part 4 The Control of Chemical Reactions
The first chapter in this section, Chapter 13, examines the important question of the rates of chemical processes and the factors controlling these rates. With this in mind, we move to Chapters 14 through 17, a group of chapters that considers chemical reactions at equilibrium. After an introduction to equilibrium in Chapter 14, we are especially interested in reactions involving acids and bases in water (Chapters 15 and 16) and in reactions leading to insoluble salts (Chapter 17). To tie together the discussion of chemical equilibria we have further explored the science of thermodynamics in Chapter 18. As a final topic in this section we explore in Chapter 19 a major class of chemical reactions, those involving the transfer of electrons.

Part 5 The Chemistry of the Elements and Their Compounds
Although the chemistry of the various elements has been described throughout the book to this point, Part 5 considers this topic in a more systematic way. Chapters 20 and 21 are devoted to the elements of Groups 1A to 4A, groups dominated by metals. Chapter 22 explores Groups 5A to 8A, where the elements are generally nonmetals. Carbon and its compounds are of such importance in industry, in our economy, and in ourselves and our environment, that Chapters 23 and 24 are devoted almost exclusively to this topic. The transition elements have a chemistry somewhat different than the other groups, so this is taken up separately in Chapter 25. Finally, to conclude this broad survey of chemistry, Chapter 26 surveys the behavior of the radioactive elements.

LEARNING AIDS

Much effort has gone into the illustrations for this book. We hope that the **full color drawings** will help in leading to an understanding of the principles illustrated. The **color photographs** were carefully planned to illustrate common elements, common compounds, and reactions or processes in progress. Not only do we hope these will be of some interest to the reader but that they convey some sense of the fact that much of chemistry is about reactions that are vigorous and colorful.

Color has also been used in many chemical equations to highlight the element or group that is changing. The colors of the spectrum are also often used in tables to point up the succession of values from low values (in red) to high (in blue).

Our students tell us that clearly worked **Examples** are one of the most useful learning aids in a book, so you will find nearly 200 such examples. Following each example, there is usually one or more **Exercises**. Each of the nearly 300 exercises further illustrates the preceding worked example, and its answer, worked out in brief, is found in Appendix M.

Marginal notes have also been used frequently to highlight very important points, to add a point of interest, or to alert you to places where the topic was discussed earlier in the book or where it will be discussed again in later chapters.

A **summary** of the main ideas of a chapter is given at the end of each chapter; important terms are usually defined again briefly in this summary.

At the close of each chapter you will find a number of **Study Questions**. The first few questions are usually called *Review Questions* and are designed simply to prompt you to re-read portions of the chapter if the answer is not quickly obvious. Next, questions on each major section of the chapter are given, with the questions roughly in order of increasing difficulty. At the end of the study questions you will find a few *General Questions,* for which you will have to decide which principles to apply. The answers to the questions designated by a blue number are given in Appendix N.

At the back of the text are a number of **Appendices**, some of which have been mentioned already. These cover a review of mathematical methods (A), naming of compounds (B), and tables of conversion factors (D) and important constants (C–K). Finally, a glossary of terms is included in the combined **Index/Glossary** at the very back of the book.

Inside the covers to the book there is a periodic table and a listing of the names of the elements (front) and commonly used constants and conversion factors (back).

SUPPLEMENTS

A number of supplements have been designed to accompany this text. All have been written by persons with many years of experience in chemical education.

The **Study Guide** (by Harry Pence of SUNY–Oneonta) that accompanies this text has been designed around the key objectives of the book.

Each chapter includes a list of the main concepts, important terms, questions testing mastery of each objective, a test evaluating overall mastery of the chapter, and a set of comprehensive questions. There is also a chapter on the use of chemical equivalents and normality. Answers are provided for all questions and tests.

The **Laboratory Manual** (by Charles W.J. Scaife of Union College and O.T. Beachley Jr. of SUNY–Buffalo) is the result of many years of collaboration on the development of laboratory experiments for general and introductory inorganic chemistry. All experiments have been thoroughly tested and attention has been given to cost and safety.

Detailed answers to designated Study Questions are also found in the **Student Solutions Manual** (by Alton J. Banks of Southwest Texas State University).

An **Instructor's Manual** (by J. Kotz and K. Purcell) is available. This manual lists the important objectives for each chapter and gives suggestions for organization of the course, as well as alternative organizations. In addition, there are suggested classroom demonstrations (with detailed instructions in some cases), and there are worked-out solutions to the questions not designated by a blue number in the text.

A **Student Lecture Outline** (by Ronald Ragsdale of the University of Utah) is available for students to help in organizing the material in the text, and a **Problems Book** has also been prepared (by Ronald Ragsdale). The latter provides additional practice problems and sample examination questions.

Several computer-based aids are available. First, there is a **Computerized Test Bank** (by C. Beeson of San Diego State University), which enables one to generate multiple-choice quizzes and examinations. The **Wilkie Computer Chemistry Package** (by C. Wilkie of Marquette University) aids students in reviewing important concepts and is available in both Apple II/e and IBM formats. Lastly, **COMPress: Introduction to General Chemistry** (by Stanley Smith of the University of Illinois, Urbana, and Ruth Chabay of the University of Wisconsin) provides a graphics-oriented computer tutorial for students with no previous chemistry background. It is also available in Apple and IBM formats.

How to Study Chemistry (by Irwin Becker of Villanova University) is a booklet that gives suggestions on how to study college-level general chemistry. It is free with this text.

Audio Tapes (by B. Shakhashiri) enable the student to study and learn chemistry at his or her own pace.

Finally, **100 color overhead transparencies** of important illustrations are available. The illustrations chosen are those most often used in the classroom.

ACKNOWLEDGMENTS

Writing a book of this size can take several years of almost continuous work and can be a difficult and (at times) painful process. However, we have had the support and encouragement of family and of some wonderful friends, colleagues, and students.

The talented editorial staff of Saunders College Publishing has been enormously helpful throughout this project. Their warm good humor, friendship, and dedication to the project have made this an enjoyable experience.

Much of the credit for the successful completion of this project goes to our Publisher, John Vondeling, and our Development Editor, Jay Freedman. We have worked with John for many years and thank him for his friendship, his support and confidence, and his knowledge of good restaurants. Jay Freedman, who is also a good friend, is unfortunately no longer in publishing, but he was without doubt the best in the business. It was he who molded and guided this book into its final form, and we shall be forever in his debt.

Our Project Editor Carol Field has been patient and helpful throughout many crises. Her attention to detail has made this, the first full-color chemistry book produced at Saunders, a success. Carol Bleistine, Art Director, has had an unfailing sense of the correct colors, layout, and illustrations in guiding the design. The Manager of Editing, Design, and Production, Tim Frelick, and Kate Pachuta, Editorial Assistant, have kept the project well organized.

The color photography for this book is largely the work of Charles D. Winters (of SUNY–Oneonta), a talented professional. Charlie was able to translate the sometimes vague thoughts of the authors into stunning photographs. It was an immense pleasure working with him and listening to his endless supply of tall tales. (We also thank him for supplying some "chemicals"; the honey in Figure 11.19 is from the bees that he keeps at his home in the Catskill Mountains.) Finally, Professor Joseph Tausta (of SUNY–Oneonta) supplied some ideas for photographs and helped in setting up the reactions.

Photos not done by Charles D. Winters had to be located at companies and universities around the world. Amy Leary, Developmental Assistant at Saunders, very pleasantly took care of this time-consuming task.

We also want to thank Susan Hughson (SUNY–Oneonta) for checking all the Exercise and Study Question answers and for proofreading text. Further, we thank Professor W. Lawrence Armstrong (SUNY–Oneonta) for his help with the chapters on organic chemistry and biochemistry, and, finally, we acknowledge the use of another Saunders book, "Chemistry: A Modern Introduction" (by F. Brescia, S. Mehlman, F.C. Pellegrini, and S. Stambler) in preparing Chapter 26.

The most important acknowledgment is last. We thank, with deepest gratitude, the person who did the most in keeping this project moving forward, Katie Kotz. The book simply could not have been completed without her love, confidence, and support.

JCK and KFP
September 1986

Reviewers

We believe that the success of a book is due in no small measure to the quality of the reviewers, and our reviewers were extraordinarily helpful and insightful. We wish to acknowledge with gratitude the efforts of all those listed below. Our special thanks go to Professors Beachley, Pence, Post, Rochow, and Titus, who carefully read several drafts of the manuscript and made many invaluable comments.

Bruce Ault, University of Cincinnati
Alton Banks, Southwest Texas State University
O.T. Beachley, SUNY–Buffalo
Jon M. Bellama, University of Maryland
James M. Burlitch, Cornell University
Geoffrey Davies, Northeastern University
Glen Dirreen, University of Wisconsin
John M. DeKorte, Northern Arizona University
Darrell Eyman, University of Iowa
Lawrence Hall, Vanderbilt University
James D. Heinrich, Southwestern College
Forrest C. Hentz, North Carolina State
Marc Kasner, Montclair State College
Philip Keller, University of Arizona
Herbert C. Moser, Kansas State University
John Parson, Ohio State University
Lee G. Pedersen, University of North Carolina
Harry E. Pence, SUNY Oneonta
Charles Perrino, California State University (Hayward)
Elroy Post, University of Wisconsin, Oshkosh
Ronald Ragsdale, University of Utah
Eugene Rochow, Harvard University
Steven Russo, Indiana University
George H. Schenk, Wayne State University
Peter Sheridan, Colgate University
Kenneth Spitzer, Washington State University
Donald D. Titus, Temple University
Charles A. Trapp, University of Louisville
Trina Valencich, California State University (Los Angeles)

A NOTE ON SAFETY

TO STUDENTS

The photographs that appear in this book were chosen to illustrate the discussion in the text and to show how chemical reactions actually occur. Some of the reactions illustrated involve very corrosive chemicals and occur very rapidly and with the evolution of a great deal of energy. They are unsafe unless carried out under appropriate conditions by trained chemists wearing protective clothing. Under *no* circumstances should you attempt these reactions yourself.

TO INSTRUCTORS

Many of the reactions illustrated in this book can be done safely in a lecture room. However, some can only be done in a laboratory hood using protective clothing. Further information on the reactions and on safety can be found in the ''Instructor's Manual'' accompanying this text and in ''Chemical Demonstrations,'' Volumes 1 and 2, by B. Shakhashiri (University of Wisconsin Press, 1983 and 1985).

Contents Overview

Charles D. Winters

Contents

Charles D. Winters

Charles D. Winters

Charles D. Winters

Charles D. Winters

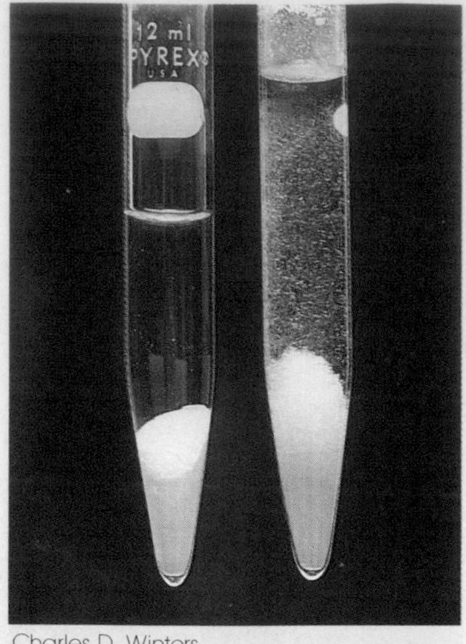

Charles D. Winters

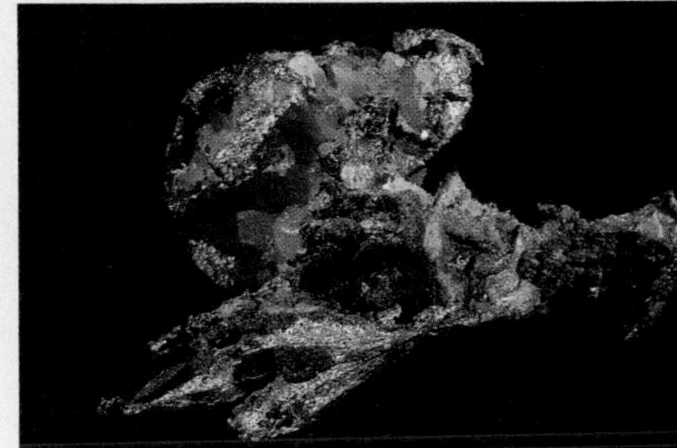

Royal Ontario Museum, Toronto

Charles D. Winters

PART ONE

THE BASIC TOOLS OF CHEMISTRY

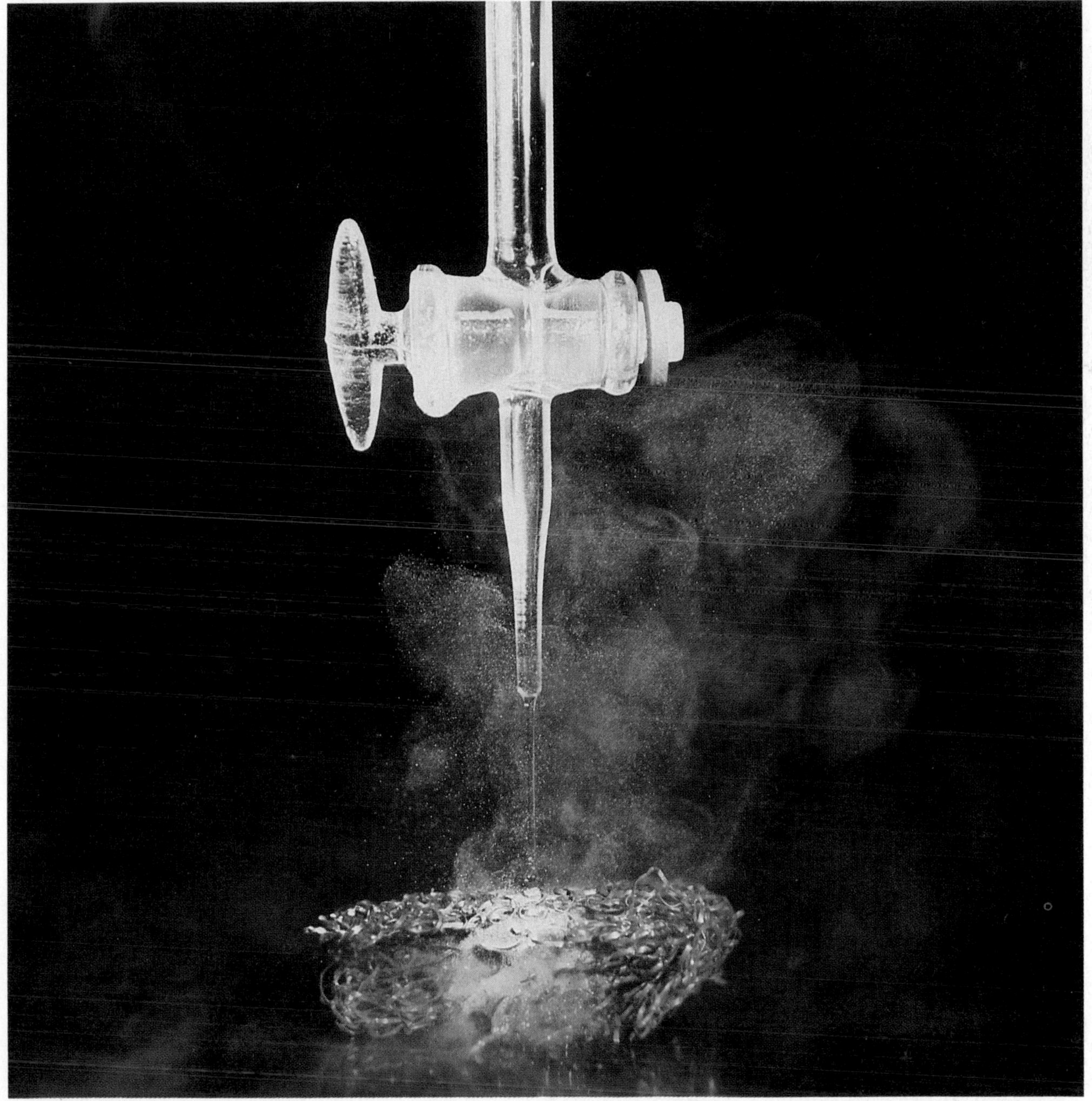

Copper wool reacting with nitric acid.

PART ONE: PREFACE

Chemistry is the study of the chemical elements and their compounds. You are just beginning a tour of the properties of elements such as helium, lithium, phosphorus, and one with a name that is hard to pronounce—molybdenum. Compounds such as water, ammonia, and poisonous, foul-smelling hydrogen sulfide are on the tour as well. We are going to ask why the chemical elements have the properties they have and how they were formed. We also want to know why compounds behave as they do, why molecules have particular shapes and what the forces are that hold them together, why diamonds are colorless, enormously hard objects, and why water is such a strange substance. It is these and hundreds of other questions that make up the fabric of this book and your introductory course in chemistry.

Even to begin to answer some of the questions we have posed, we must lay a firm foundation of principles. Thus, Chapter 1 defines some important terms and is a review of units and numerical methods. Chapter 2 introduces some basic ideas of atoms and molecules. This is followed by a first look at chemical reactivity and its patterns in Chapter 3. There you will learn some properties of ionic compounds and how to predict the course of some chemical reactions. In Chapter 4 you will see the methods of extracting quantitative information from chemical reactions, and in Chapter 5 you will take up some aspects of the energy involved in chemical processes. Finally, in Chapter 6 you will study gases and will see how simple mathematical models can be used to predict the behavior of one of the phases of matter. With these principles in place, you can go on in confidence to the remaining sections of the text.

1

Some Useful Ideas and Tools of Chemistry

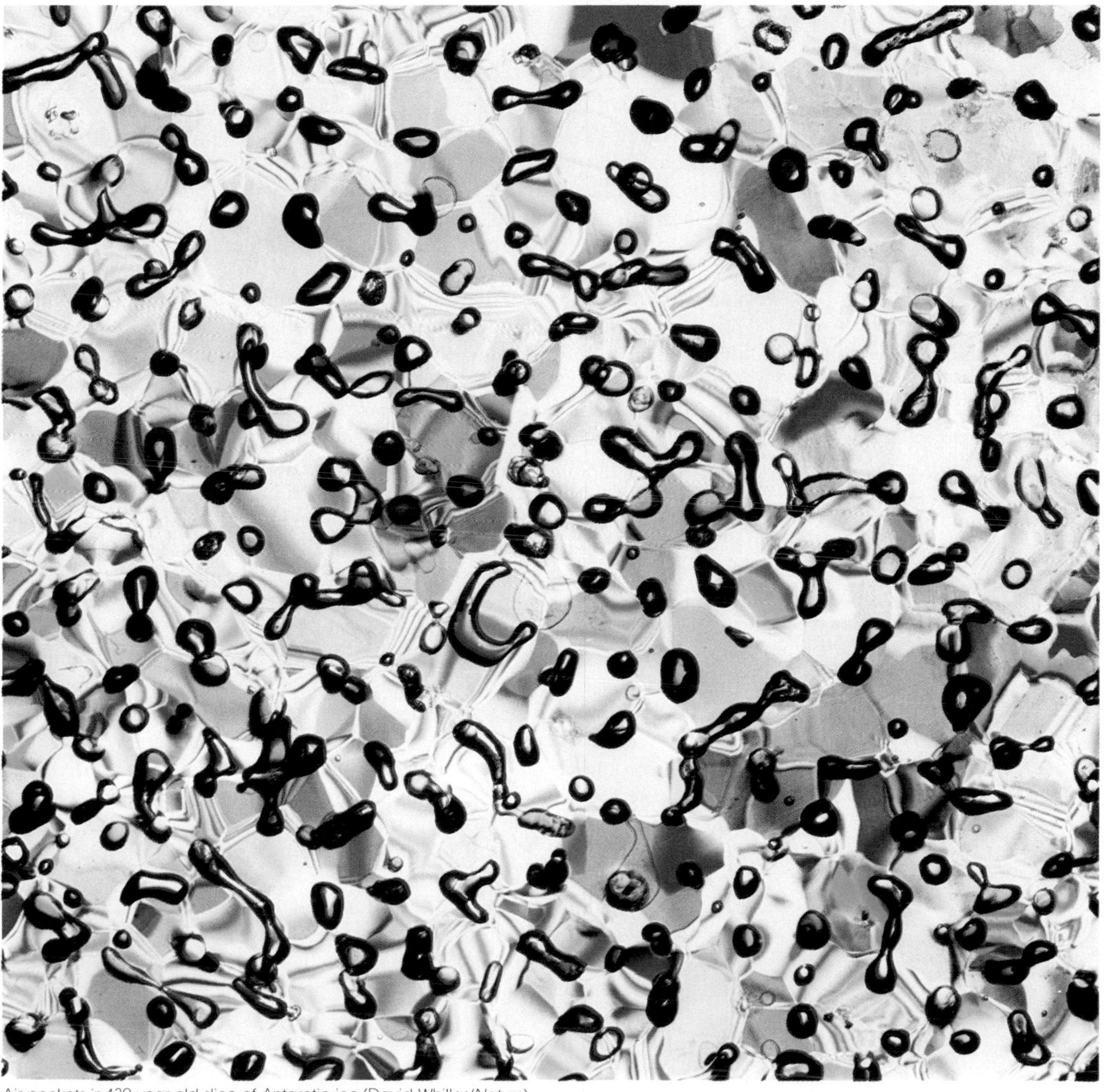

Air pockets in 130-year-old slice of Antarctic ice (David Whillas/*Nature*)

CHAPTER OUTLINE

1.1 THE STATE OF CHEMISTRY TODAY

Pauling's ideas of chemical bonding, and other approaches, are described in Chapters 9 and 10.

The American chemist Linus Pauling was awarded the Nobel Prize in Chemistry in 1954 for developing new ideas on the bonding of atoms, and in 1962 he received the Prize for Peace for his contributions to international relations. He has stated, "Every aspect of the world today—even politics and international relations—is affected by chemistry." One of our objectives in this book is to show the truth of this statement.

In chemistry, the word **synthesis** means that a compound has been made by combining other compounds or groups. Nature, as well as chemists, is adept at synthesis.

The effects of chemistry—for better or worse—are all around you and within you. Your body functions by a complicated series of chemical processes, which are only partly understood. There are thousands of man-made chemicals to treat the symptoms or even the causes of illness, to eliminate a headache, or even to stop the spread of a cancer. You walk on carpets of synthetic fibers, and these same synthetic fibers are in many kinds of clothing. You also use household chemicals to clean, polish, bleach, and unclog. To feed the expanding population of our earth, agriculture must be expanded and made more efficient. The discovery in the early part of this century of a process to manufacture ammonia cheaply has had an enormous impact on the world, since ammonia and its salts are necessary for making fertilizers. Since then, chemists have developed a host of herbicides, pesticides, and fungicides, without which large-scale production of foodstuffs would simply not be possible.

Fritz Haber (1868–1934) was the developer of the Haber process for ammonia synthesis from nitrogen of the atmosphere and hydrogen (Chapter 22).

However, there can be another side to the creativity of chemists. Ammonia is also used to make explosives, and one of the men responsible for the ammonia process also developed chemical warfare agents first used in World War I. Pesticides and herbicides have greatly benefited agriculture, and countless people in Third World countries are alive because these chemicals have led to a more abundant food supply. On the other hand, the release of a chemical used in an intermediate step of pesticide synthesis killed or sickened hundreds of people in Bhopal, India, in 1985.

The chemistry community has always been mindful of its responsibilities to society. As more and more is known about chemicals and their interactions in the environment, it has become easier to make a decision on whether to introduce a new product or how to dispose of a waste material.

Linus Pauling is correct that chemistry plays a role in international affairs. The fact that many metals and minerals are found in commercially usable forms only in certain countries has a great impact on the world's

Ammonia is applied to a field. The nitrogen contained in the ammonia is essential to plant growth. (Farmland Industries, Inc.)

economy. For example, the U.S. battery industry depends on antimony, the steel industry needs manganese and chromium, and the automotive and chemical industries require platinum. Deposits of these metals that can be mined economically are located mainly in countries that have had poor relations with the United States, and a change in international relations could lead to loss of these and other metals. In an effort to reduce this dependence on imports, chemists are at work on developing alternative materials. For example, ceramics are beginning to be used in some applications where metals had previously served.

Opportunities for the coming generation of chemists are practically limitless. The depletion of our supplies of petroleum and of crucial metals and minerals forces us to search for new sources of energy and new materials. As we come to understand the biochemistry of plants and animals better and learn to manipulate biochemical processes, we have the opportunity to control or eradicate certain diseases and generally to enhance the quality of life.

But, whether you are interested in a career in chemistry or not, why should *you* study chemistry today? Our primary objective is to show you that chemistry has not only a rich and interesting history but an enormously exciting future as well. To meet your responsibilities to yourself and society, you need to know how scientists define and answer questions about natural phenomena and use that knowledge to the benefit of humanity. Furthermore, study of chemistry is centrally important not only to all of science but also in helping to explain and understand yourself and the world about you. When you breathe oxygen, what happens to it? How do the batteries in your radio or calculator work? What are some solutions to the energy problems of the world? How can one create new materials to replace metals? Why does ice float on water, and why can you skate on the ice? How might life itself have begun on our planet?

It has been said, "Chemistry is the central subject in a liberal arts curriculum. It stands between the traditional humanities on the one hand and modern physics on the other hand."* It focuses on you and the world around you. It is the result of generations of intensely creative human

Many new uses are being found for ceramics, materials that often contain the element silicon. They are strong and resistant to corrosion. Their electrical insulating qualities make them ideal for the base of microelectronic chips. Further, because of their light weight, strength, and ability to withstand high temperatures, they are increasingly being used for such things as these turbine rotors in turbo charged engines. (GTE Laboratories, Inc.)

*H.A. Bent, *Chemical and Engineering News*, March 12, 1984, p. 44.

thought and experiment, and it contains the seeds of the continuing progress of humankind.

1.2 THE METHODS OF SCIENCE

Before you begin your study of chemistry, it is useful to have some insight into the way any scientific study is done. Although it may not seem so now, it is easy to formulate an idea, a topic for study. The goal is to state a problem that is narrow enough in scope so that there is some realistic possibility of coming to a useful conclusion. It is reasonable to ask, for example, how cancers begin in a human body, but it is decidedly *not* reasonable to expect an answer any time soon.

Having posed a reasonable question, you study previous experimental work done in the field, so that you will have some notion of possible answers. After forming an **hypothesis**—a tentative explanation or prediction of experimental observations—you perform experiments designed to yield results that should eliminate erroneous explanations. This requires collection of both qualitative and quantitative information or data. **Qualitative** information consists of non-numerical observations about the problem, while **quantitative** information means numerical data. With the results of your experiments, you then revise and extend your original hypothesis and continue to test it with additional experiments. After having done a large number of experiments, and after continually checking to ensure that your results are *reproducible*, a pattern of behavior or results will begin to emerge. At this point, you may be able to summarize your observations in the form of a **law**, a concise verbal or mathematical statement of a relation that is always the same under the same conditions.

Qualitative observations would include the color of a chemical and its physical state.

Once you have performed enough reproducible experiments that you have been led to a "law of nature," you should formulate a theory to explain the law. A **theory** is a unifying principle that explains a body of facts and the laws based on them. Excellent examples of theories are those developed to account for chemical bonding (Chapters 9 and 10). It is a fact that atoms are held together or bonded to one another. But how and why? There are theories currently used in chemistry to answer such questions, but are they correct? Do they have limits? If so, what are they? Can the theories be improved or are completely new theories necessary? Laws are the facts of nature and do not change. Theories are inventions of the human mind; theories can change as new facts are uncovered.

People outside of science usually have the idea that science is an intensely logical field. They picture a white-coated chemist moving logically from hypothesis to conclusion without human emotion or foibles. Nothing could be further from the truth! Often, scientific results and understanding arise quite by accident. A wonderful example is the development of a new cancer chemotherapy agent called "cisplatin." In 1964 Barnett Rosenberg and his colleagues at the University of Michigan were studying the effects of electricity on the growth of bacteria. To test this effect, they inserted platinum electrodes (conductors of electricity) into a bacteria culture. Within a few hours they observed that all of the

DR. BARNETT ROSENBERG AND "CISPLATIN"

In 1964 Dr. Barnett Rosenberg, Professor of Biophysics at Michigan State University, discovered the cancer-combating drug "cisplatin." It is especially effective against certain types of cancers. In his left hand he holds a mouse that will die of cancer in a few days. In his right hand is a mouse that although infected with cancer has received cisplatin and will survive.

Barros Research Institute

bacteria had ceased growing. After considerable research, they were able to show that a platinum-containing substance, which came from the platinum of the electrodes and other chemicals in the system, was responsible. This is the point at which *creativity and insight* intervened. They reasoned that, since cancer involves unrestrained cell growth, perhaps this platinum-containing substance would be a cancer chemotherapy agent. Indeed it was, and the Food and Drug Administration approved the material for marketing in 1979.

1.3 SOME DEFINITIONS

One of the problems with studying chemistry is that there are many new terms that you must learn to use, and there are a few that are necessary from the beginning (Figure 1.1).

Matter, anything that has mass and occupies space, consists not only of things you can see and touch but also such things as air, which

FIGURE 1.1

The components of matter and the relation between mixtures and substances.

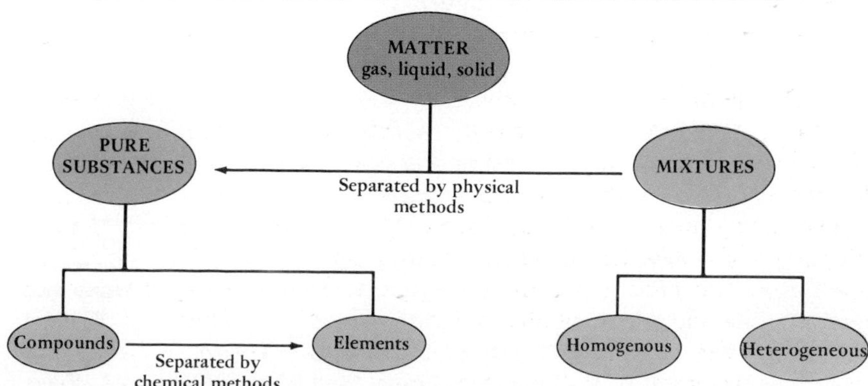

A few kinds of matter have properties of more than one phase, such as "liquid crystals" and glasses.

you cannot see. Matter can exist in three **phases**: solids, liquids, and gases. Some kinds of matter can exist in all three phases, even simultaneously under special conditions.

Solids consist of particles arranged into a definite, rigid shape that does not change much with temperature (Figure 1.2). A **liquid** also has a definite volume, but, unlike a solid, a liquid usually does not have a shape of its own (Figure 1.3). The particles making up a liquid may flow to assume the shape of the container. A liquid changes volume only to a small extent with change in temperature, although often to a greater extent than a solid.

In contrast with a solid or liquid, the particles of a **gas** completely fill any container. If the container is not rigid, changes in temperature can lead to large changes in the volume of a gas. Finally, gases can be expanded and compressed over enormous ranges of volume, suggesting that the particles are much more widely separated from one another in a gas than in a liquid or solid (Figure 1.4).

Atoms and Elements. Although these terms are described in more detail in Chapter 2, we need them to continue here. An **atom** is the smallest particle of an element that retains the chemical properties of the element. *All* matter is composed of different kinds of atoms, and matter that is composed of only one kind of atom is an **element**. The names and symbols of the 108 known elements are shown inside the front cover of the book. Bromine and iodine (Figure 1.4) and iron and gold are but a few of these elements.

FIGURE 1.2
Crystals of azurite, a copper-containing mineral. (Brian Parker, Tom Stack & Associates)

FIGURE 1.3
Solid "ice cubes" have a definite shape and do not fill the container evenly. After melting, liquid water assumes the shape of the container. (Charles D. Winters)

Molecules and Compounds. A **compound** is composed of two or more atoms chemically combined in definite proportions. A **molecule** is the smallest unit of a compound that retains the chemical characteristics of the compound. Most of the research done in chemistry involves the study of the transformation of one compound into another, so you will spend much of your time examining molecules: their shapes, the forces holding them together, and their chemical and physical properties.

Substances and Mixtures. A **substance** is a form of matter that has a definite composition (every sample of a substance contains the same kinds of atoms in the same proportions) and distinct properties. Table salt or sugar, the oxygen and nitrogen of the air, or diamonds (pure carbon) are substances. Each of these differs from the others by its composition, its taste or smell, or its ability to dissolve in water. *A substance is pure by definition.*

A **mixture** is a combination of two or more substances in which each substance retains its identity. A cup of coffee, with sugar and milk, is a mixture of many substances, as is a soft drink, a piece of cement, or a coin.

Mixtures can be **homogeneous** or **heterogeneous**. A homogeneous mixture has the same composition throughout the mixture. If you stir table salt into pure water, the salt dissolves and the mixture is a homogeneous solution (see Figure 1.5). A **solution** is a homogeneous mixture of two or more substances. The material dissolved is called the **solute** and the medium in which it is dissolved is the **solvent**. In a salt water solution, salt is the solute and water is the solvent. A mixture of solid grains of a compound salt and grains of sand is **heterogeneous**, since particles of each

FIGURE 1.4
Bromine (left) is a deep brown-orange liquid, but it is volatile enough that some is in the gas phase. Iodine (right) is a violet solid, but some sublimes to give a violet vapor. (Charles D. Winters)

FIGURE 1.5
A homogeneous solution. A green compound, nickel nitrate (the solute), is stirred into water (the solvent), where it dissolves to form a homogeneous solution. (Charles D. Winters)

(a)

(b)

(c)

(d)

FIGURE 1.6

Separating a copper compound from sand. The heterogeneous mixture (a) is placed in water where the copper compound dissolves (b). The mixture is filtered, leaving the sand in the filter (c); the copper-containing solution passes through the filter. Evaporation of the water leaves blue crystals of the copper compound (d). (Charles D. Winters)

component of the mixture remain separate and can be observed as individual substances (Figure 1.6).

In either a homogeneous or a heterogeneous mixture, the components can be separated into pure substances by *physical means*, that is, without changing the specific atom ratios within the particles. For example, the components of the solution in Figure 1.5 could be separated by evaporating the water to leave the nickel compound as a green solid, and, if the experiment is done properly, pure liquid water can be condensed from the gas phase. Components of a heterogeneous mixture can also be separated physically, as illustrated in Figure 1.6.

The components of a mixture can also be separated by *chemical means*, but this involves changing the atom ratios within the particles. This is often done in **chemical analysis**, where the components of a mixture are transformed into new substances that can in turn be observed or separated by physical means. This is discussed in more detail in Chapters 3 and 4.

Physical and Chemical Properties. Your friends recognize you by your physical appearance: height, weight, hair color, and so on. The same is true of chemical substances. Each substance has a set of **physical properties**, properties that can be measured and observed without changing the atom ratios within the substance. Such properties include color, the temperature at which a substance melts or boils, density, and physical state (Figure 1.7). Elemental bromine and iodine (Figure 1.4) clearly differ from one another in their color and physical state.

FIGURE 1.7

A physical change. A flower is placed in liquefied nitrogen (which boils at −196°C). The flower freezes and crumbles to pieces when touched. (Charles D. Winters)

Chemical properties, on the other hand, are properties that matter exhibits when it undergoes a change in atom ratios within the particles. When gasoline burns in your automobile engine or metals rust and corrode, their chemical composition changes (Figure 1.8).

The physical properties of substances can be classified further as extensive and intensive. **Extensive** properties are those that depend on the amount of matter present. Thus, the volume and mass of a sample are extensive properties, as they both are directly proportional to the amount of matter in the sample. In contrast, **intensive** properties do not depend on the amount of sample; they are the same, no matter what the sample size. The temperature at which a sample melts and the color of a material are both intensive properties. Water is colorless and freezes at 0°C (32°F), whether you have a spoonful or a ton of it.

No two substances have the same combination of chemical and physical properties under the same conditions, so we can use these differences to identify substances. Many of the physical properties of oxygen and nitrogen are very similar; for example, both are colorless gases at room temperature. However, a burning match will go out if it is put into a flask of nitrogen, but it will burn brightly in oxygen. Thus, the two gases have different chemical properties. Also, it is not difficult to tell the difference between two nonmetals such as bromine and iodine (Figure 1.4).

FIGURE 1.8

A chemical change. A copper compound is dissolved in water to give a homogeneous blue solution. On adding a piece of aluminum foil (top), a chemical change occurs. As described in Chapters 3 and 19, copper metal is formed from the dissolved copper compound, and elemental aluminum metal is transformed into an aluminum compound (bottom). This is evidenced by the coating of copper on the surface of the ball of foil, and by the observation that the blue color of the dissolved copper compound diminishes with time. In addition, a physical change, an increase in temperature, is observed as reaction occurs. (Charles D. Winters)

1.4 UNITS OF MEASUREMENT

Scientists have traditionally used the **metric system** for recording and reporting their measurements. This is a decimal system, in which all of the units are expressed as powers of ten times some basic unit.

To establish a uniform set of units, the General Conference of Weights and Measurements in 1960 recommended a single base unit to be used for each measured quantity. The resulting system is called the *Système International d'Unités* (International System of Units), abbreviated **SI**. The seven SI base units are listed in Table 1.1. Larger and smaller quantities are expressed by using the appropriate prefix (Table 1.2) with the base unit. For instance, highway distances are given in *kilo*meters, where 1 kilometer (km) is exactly 1000 or 10^3 meters (m). Objects in the laboratory are often measured in *centi*meters (cm) or *milli*meters (mm). The prefix *centi-* means that 1 centimeter is 1/100 of a meter (1 cm $= 1 \times 10^{-2}$ m), and 1 millimeter is 1/1000 of a meter (1 mm =

TABLE 1.1 SI Base Units

PHYSICAL QUANTITY	NAME OF UNIT	ABBREVIATION
Mass	kilogram	kg
Length	meter	m
Time	second	s
Temperature	kelvin	K
Amount of substance	mole	mol
Electric current	ampere	A
Luminous intensity	candela	cd

TABLE 1.2 Selected Prefixes Used in the Metric System

PREFIX	ABBREVIATION	MEANING	EXAMPLE
Mega-	M	10^6	1 megaton = 1×10^6 tons
Kilo-	k	10^3	1 kilogram (kg) = 1×10^3 grams
Deci-	d	10^{-1}	1 decimeter (dm) = 0.1 m
Centi-	c	10^{-2}	1 centimeter (cm) = 0.01 m
Milli-	m	10^{-3}	1 millimeter (mm) = 0.001 m
Micro-	μ*	10^{-6}	1 micrometer (μm) = 1×10^{-6} m
Nano-	n	10^{-9}	1 nanometer (nm) = 1×10^{-9} m
Pico-	p†	10^{-12}	1 picometer (pm) = 1×10^{-12} m

*This is the Greek letter mu (pronounced "mew").
†This prefix is pronounced "peako."

The quantity 1000 can be expressed in *scientific* or *exponential notation*, where $1000 = 1 \times 10^3$. The quantity 1/10 or 0.1 is expressed as 1×10^{-1}. This notation will be used throughout the book and is explained in Appendix A.

1×10^{-3} m). On a smaller scale still, dimensions of molecules are often given in *nano*meters (nm) (1 nm = 1×10^{-9} m) or *pico*meters (pm) (1 pm = 1×10^{-12} m).

SI units for all other physical quantities are derived from the seven base units, and several of these are listed in Table 1.3. Some factors that allow you to convert between SI and non-SI units are given inside the back cover of the book.

LENGTH

The original "standard meter" was a bar made of platinum and iridium kept at Sèvres, France. In 1983, however, the standard was redefined as the distance traveled by light in a vacuum in 1/299,792,500 seconds.

The **meter** is the standard unit of length. One meter is equivalent to 3.28 feet or 39.37 inches. It is a convenient unit in human terms. A human leg is roughly a meter long, and you can hold a meter stick lengthwise comfortably between your outstretched hands.

Just as distances in the nondecimal English system of inches and feet are broken into smaller units, the meter is subdivided into 100 centimeters, 1000 millimeters, a million micrometers, a billion (10^9) nanometers, and so on. Centimeters and millimeters are most convenient for measuring objects around the laboratory, and nanometers or picometers are used for dimensions at the molecular level.

For dimensions at the molecular level, many scientists use the **Ångstrom** unit where 1 Å is equivalent to 10^{-10} meters.

EXAMPLE 1.1

CONVERTING CENTIMETERS TO METERS

In track and field competition, an excellent distance for the pole vault is 585 cm. (a) What is this distance in meters? (b) What is this distance in feet?

TABLE 1.3 Some Commonly Used Units Derived from SI Base Units

QUANTITY	UNIT NAME	SYMBOL	DEFINITION
Area	square meter	m^2	
Volume	cubic meter	m^3	
Density	kilogram per cubic meter	kg/m^3	
Force	newton	N	$kg \cdot m/s^2$
Pressure	pascal	Pa	N/m^2
Energy	joule	J	$kg \cdot m^2/s^2$
Electric charge	coulomb	C	$A \cdot s$
Electric potential difference	volt	V	$J/(A \cdot s)$

Solution (a) To solve this problem, you need to know the relation between centimeters and meters. That is, 1 m is the same as 100 cm (Table 1.2). Therefore,

$$585\ \text{cm}\left(\frac{1\ \text{m}}{100\ \text{cm}}\right) = 5.85\ \text{m}$$

To convert a measurement in one unit to another unit, we multiply by a "factor" (in this case, 1 m/100 cm) that relates equal quantities in the two units. This factor is just another name for the number 1, so the *quantity* is not changed. The factor is written so that the units in the denominator cancel with the original units, leaving the desired units. Here units of centimeters cancel, and we are left with units of meters.

(b) The table inside the back cover of the book lists only a relation between centimeters and inches, so we first convert 585 cm to inches

$$585\ \text{cm}\left(\frac{1\ \text{in}}{2.54\ \text{cm}}\right) = 230.\ \text{in}$$

and then convert inches to feet.

$$230.\ \text{in}\left(\frac{1\ \text{ft}}{12\ \text{in}}\right) = 19.2\ \text{ft}$$

EXAMPLE 1.2

DISTANCES ON THE MOLECULAR LEVEL

The distance between the oxygen atom and a hydrogen atom in a water molecule is 95.7 pm. What is this distance in meters? In nanometers?

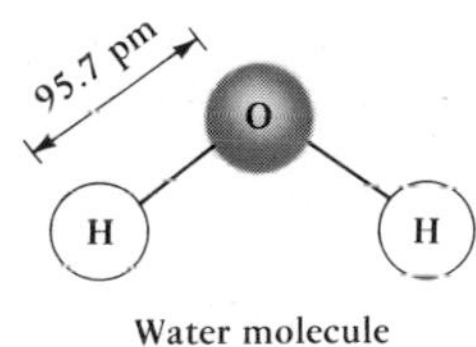

Water molecule

Solution In each case, you must first know the relation between the picometer (pm) and the desired unit. In this case, 1 pm is exactly 1×10^{-12} m (see Table 1.2). Thus, you multiply the distance in pm by the factor (10^{-12} m/1 pm) so that units of pm cancel, leaving an answer in meters.

$$95.7\ \text{pm}\left(\frac{10^{-12}\ \text{m}}{1\ \text{pm}}\right) = 95.7 \times 10^{-12}\ \text{m} = 9.57 \times 10^{-11}\ \text{m}$$

To relate picometers and nanometers, you must know that 1 nm = 10^{-9} m (Table 1.2). Therefore, you can take the distance in meters, which you just found above, and multiply by the factor 1 nm/10^{-9} m.

$$9.57 \times 10^{-11}\ \text{m}\left(\frac{1\ \text{nm}}{10^{-9}\ \text{m}}\right) = 9.57 \times 10^{-2}\ \text{nm}$$

Examples 1.1 and 1.2 were solved by the technique of "dimensional analysis," an approach to problem solving described in Section 1.6.

EXERCISE 1.1 INTERCONVERTING UNITS OF LENGTH

A standard U.S. postage stamp is 2.5 cm long. What is this length in meters? In millimeters? In inches?

Answers to all of the Exercises in the text are given in Appendix M.

AREA AND VOLUME

The units of area and volume are derived from the base unit of length, that is, areas can be given in square meters (m^2) and volumes in cubic meters (m^3). Unfortunately, volumes in cubic meters are not very convenient for everyday use. For example, the volume of a common labo-

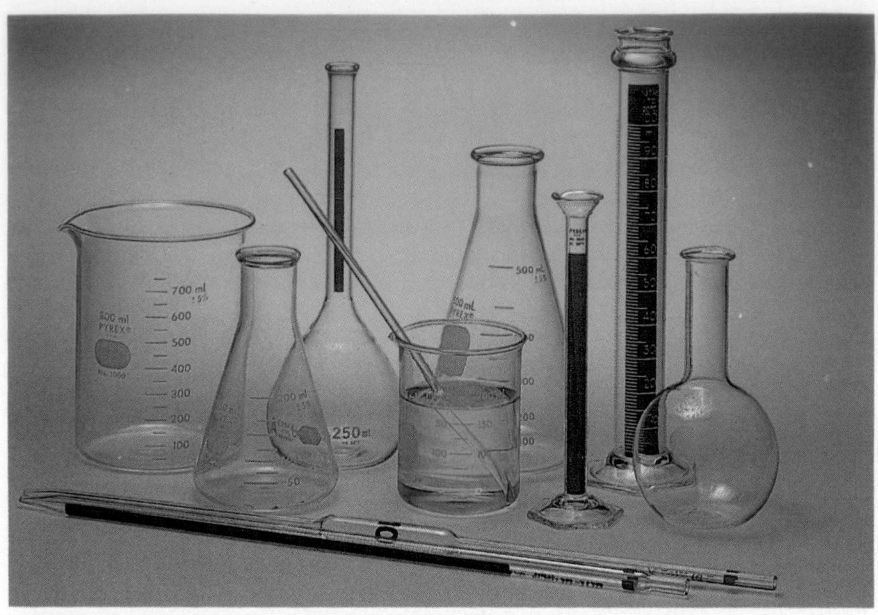

FIGURE 1.9
Some common volumetric glassware. Two pipets are in the foreground, and the 250-mL flask with a very long neck (rear) is a volumetric flask; if it is filled to a mark on the neck, the volume of liquid is known precisely. The beaker contains a solution of green nickel nitrate. (Charles D. Winters)

ratory beaker is 0.0006 m^3 (Figure 1.9), and a soccer ball has a volume of about 0.005 m^3. Chemists most often work with volumes of chemicals in the range of 0.001 m^3 (1 cubic decimeter) or less. A cube having sides equal to 10 cm (0.1 m or 1 dm) has a volume of 10 cm × 10 cm × 10 cm = 1000 cm^3 (or 0.001 m^3 or 1 dm^3). This volume is called a **liter**, symbolized by **L**.

$$1000\ \text{cm}^3 = 1\ \text{dm}^3 = 1\ \text{liter or } 1\ \text{L}$$

Thus, the 0.0006 m^3 beaker mentioned above has a volume of 600 cm^3 or 0.6 L.

Notice that the *entire* conversion factor (100 cm/1 m) must be cubed, in order to cancel units of m^3.

$$0.0006\ \text{m}^3 \left(\frac{100\ \text{cm}}{1\ \text{m}}\right)^3 = 0.0006\ \cancel{\text{m}^3} \left(\frac{1\ 000\ 000\ \text{cm}^3}{1\ \cancel{\text{m}^3}}\right) = 600\ \text{cm}^3$$

Expressing the calculation in scientific notation,

$$6 \times 10^{-4}\ \text{m}^3 \left(\frac{100\ \text{cm}}{1\ \text{m}}\right)^3 = 6 \times 10^{-4}\ \cancel{\text{m}^3} \left(\frac{1.00 \times 10^6\ \text{cm}^3}{1\ \cancel{\text{m}^3}}\right)$$
$$= 6 \times 10^2\ \text{cm}^3$$

and the volume in liters is

$$600\ \cancel{\text{cm}^3} \left(\frac{1\ \text{L}}{1000\ \cancel{\text{cm}^3}}\right) = 0.6\ \text{L}$$

The liter is a convenient unit to use in the laboratory, as is the cubic centimeter or "cc."

Finally, notice that, since there are 1000 cm^3 in a liter, one cubic centimeter is one thousandth of a liter or a **milliliter**, **mL**.

$$1 \text{ cm}^3 = 0.001 \text{ liter} = 1 \text{ milliliter (mL)}$$

Chemists often use the terms milliliter and cubic centimeter interchangeably.

EXERCISE 1.2 VOLUME

(a) A standard wine bottle has a volume of 750 mL. How many liters does this represent? (b) One U.S. gallon is equivalent to 3.785 liters. How many liters are there in a 2.0-quart container of dish detergent? (4 quarts = 1 gallon)

MASS AND WEIGHT

The terms *mass* and *weight* are used interchangeably in everyday speech, but scientists give different meanings to the two words. The **mass** of a body is a measure of the quantity of matter in that body. The **weight** of a body, on the other hand, is the response of that body to the force of gravity and so depends both on the amount of material *and* on the force of gravity. This difference is easily understood. An astronaut weighs 150 pounds or so on the earth, but, on the moon he weighs so little that he is able to leap tall rocks in a single bound! However, on earth or on the moon, the astronaut has the same mass.

Since weight varies with the force of gravity, and that force varies with the distance from the center of the earth, it must therefore be true that weight depends on the altitude of your laboratory. How then can you "weigh out" the same amount of matter in Miami and in Denver, cities that differ in altitude by nearly a mile? The answer is to compare the weights in both places with objects of known mass, called "standards," using a laboratory balance (Figure 1.10).

FIGURE 1.10
A laboratory balance. A sample of sulfur is being weighed. (Charles D. Winters)

The SI base unit of mass is the **kilogram (kg)**. Smaller masses are expressed in grams (1000 g = 1 kg) or milligrams (10^6 mg = 1 kg or 10^3 mg = 1 g). Again, the prefixes of Table 1.2 apply.*

The English system based on the pound refers to *weight*, not mass. Under standard gravity (at sea level), however, a mass of 1 kg has a weight of 2.2046 pounds, about the weight of a quart of milk. Alternatively, the mass of an object weighing 1 pound (at standard gravity) is 0.4536 kg or 453.6 g.

EXAMPLE 1.3

MASS IN KILOGRAMS AND GRAMS

A U.S. penny has a mass of 2.65 g. Express this mass in kilograms and milligrams.

Solution

(a) $$2.65 \not{g} \left(\frac{1 \text{ kg}}{1000 \not{g}}\right) = 0.00265 \text{ kg}$$

(b) $$2.65 \not{g} \left(\frac{1000 \text{ mg}}{1 \not{g}}\right) = 2.65 \times 10^3 \text{ mg}$$

*The original metric system used the gram as the base unit of mass. A later conference changed the official mass unit to the kilogram.

EXERCISE 1.3 MASS

One pound is equivalent to 453.6 g at standard gravity. (a) How many kilograms are equivalent to 3.00 pounds? (b) How many milligrams are equivalent to 0.500 pounds? (c) How many pounds are equivalent to 4.00 kilograms?

DENSITY

Density is the ratio of the mass of an object to its volume.

$$\text{Density} = \frac{\text{mass}}{\text{volume}}$$

If any two of these quantities are known for a specific sample of matter, they can be used to calculate the third quantity. That is, algebraic manipulation of the definition gives

$$\text{Volume} = \frac{\text{mass}}{\text{density}} \qquad \text{and} \qquad \text{Mass} = \text{density} \times \text{volume}$$

Gold has a density of 19.3 g/cm^3. That is, 1.00 cm^3 of metallic gold has a mass of 19.3 g. Water, on the other hand, is much less dense, having a density of 1.00 g/cm^3 at 25°C.*

EXAMPLE 1.4

DENSITY

Suppose you have 250. cm^3 (or milliliters) of ethyl alcohol (the common ''alcohol'' in alcoholic beverages). If the density of ethyl alcohol is 0.789 g/cm^3 (at 20°C), how many grams of alcohol does this represent?

Solution Because we know the density and volume of the sample, we will use the definition above in the form ''mass = density × volume.'' Note that the volume is multiplied by a factor having units of mass over volume, and the units of cm^3 cancel.

$$250.\ \cancel{\text{cm}^3}\left(\frac{0.789\ \text{g}}{1\ \cancel{\text{cm}^3}}\right) = 197\ \text{g}$$

EXERCISE 1.4 DENSITY

The density of dry air is 1.12×10^{-3} g/cm^3. What volume of air, in cubic centimeters, will have a mass of 1.00 kg?

TIME

Unlike the units we have discussed thus far, the fundamental unit of time—the second, symbolized by ''s''—is not a metric unit. Instead, time units were established by the Sumerians (who lived in what is now Iraq) 40 or more centuries ago! Prior to 1964 the second was defined as 1/86,400 of

*Because the volume of a liquid is more strongly affected by temperature than that of most solids, it is usual to indicate the temperature at which a liquid's density is measured.

a mean solar day. Since there are irregularities in the solar day, however, the second is now defined in terms of certain radiation produced by an atom of cesium.

In 1 second, light can travel 2.99792500 10^8 m or about $7\frac{1}{2}$ times around the earth at the equator! But in one nanosecond (10^{-9} s), the light travels only 0.299 m or about 30 centimeters, that is, about the length of this book!

TEMPERATURE

In the summertime it is a great pleasure to go for a swim in the local pool, in a lake, or in the ocean. The cooling sensation comes from the feeling that heat has flowed from you to the water, because the water is cooler than your body. **Temperature** is a property of an object that determines the direction of heat flow when that object contacts another. The greater the temperature difference, the greater the tendency for heat to flow away from a hotter object to a cooler one. The number that represents the temperature difference depends on the size of the degree on your measuring device, a thermometer. That is, it depends on the scale of measurement.

FAHRENHEIT AND CELSIUS TEMPERATURE SCALES In the United States we commonly use the **Fahrenheit** scale, but the **Celsius** scale is used in most other countries and in science. Both scales are based on the properties of water. The Celsius scale is defined by the freezing point of pure water as 0°C and the boiling point as 100°C.* The size of the degree on the Fahrenheit scale is equally arbitrary. Fahrenheit defined 0°F as the freezing point of a solution in which he had dissolved the maximum possible amount of salt (because this was the lowest temperature he could reproduce reliably), and he intended 100°F to be the normal human body temperature (but this turned out to be 98.6°F). Today, the reference points are set at 32°F (the freezing point of pure water) and 212°F (the boiling point of pure water). The temperature difference between these points is 180 Fahrenheit degrees. This means that 100 Celsius degrees cover the same range as 180 Fahrenheit degrees (Figure 1.11). That is, the Celsius

Daniel Fahrenheit (1686–1736), a German instrument maker, was the first to use the mercury-in-glass thermometer. Anders Celsius (1701–1744) was a Swedish astronomer.

*To be entirely correct, we must specify that water boils at 100°C and freezes at 0°C only when the pressure of the surrounding atmosphere is 1 standard atmosphere. We shall discuss pressure and its effect on boiling point in Chapter 11.

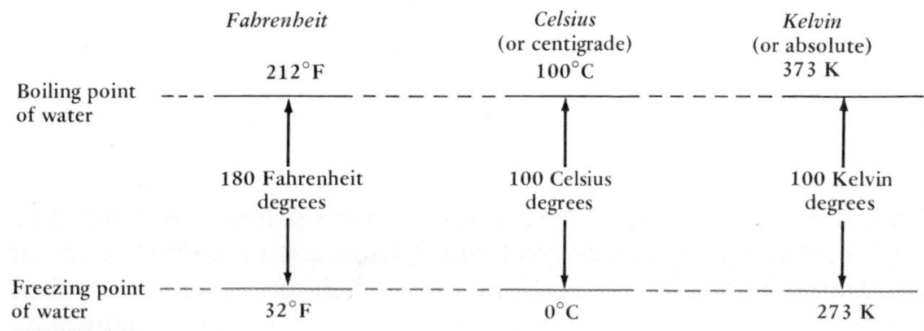

FIGURE 1.11

A comparison of Fahrenheit, Celsius, and kelvin temperature scales. The reference or starting point for the kelvin scale is *absolute zero* (0 K = −273.15°C), the lowest temperature theoretically or experimentally obtainable.

degree is almost twice as large as the Fahrenheit degree; it takes only 5°C to cover the same range as 9°F.

$$\frac{100°\text{C}}{180°\text{F}} = \frac{5°\text{C}}{9°\text{F}}$$

This relationship is used to convert a temperature on one scale to a temperature on the other scale.

$$°\text{C} = \frac{5°\text{C}}{9°\text{F}}(°\text{F} - 32°\text{F}) \qquad \text{or} \qquad °\text{F} = \left(\frac{9°\text{F}}{5°\text{C}}\right) °\text{C} + 32°\text{F}$$

EXAMPLE 1.5

CONVERTING A FAHRENHEIT TEMPERATURE TO A CELSIUS TEMPERATURE

Your normal body temperature is 98.6°F. Show that this corresponds to 37.0°C.

Solution

$$\text{Body temperature in °C} = \frac{5°\text{C}}{9°\text{F}}(98.6°\text{F} - 32.0°\text{F})$$
$$= [(5/9)(66.6)]°\text{C} = 37.0°\text{C}$$

EXAMPLE 1.6

CONVERTING A CELSIUS TEMPERATURE TO A FAHRENHEIT TEMPERATURE

The temperature in Lisbon, Portugal, is often 17°C at about noon. What is this temperature in °F?

Solution

$$°\text{F} = \left(\frac{9°\text{F}}{5°\text{C}}\right) 17°\text{C} + 32°\text{F} = 63°\text{F}$$

EXERCISE 1.5 TEMPERATURE CONVERSION
The temperature in Arizona often reaches 110°F. What is this temperature in °C?

Laboratory work is almost always done using Celsius thermometers. To calibrate your senses on this scale, you should know that water freezes at 0°C, a comfortable room temperature is about 22°C, your body temperature is 37°C, and the hottest water you could put your hand into is about 60°C.

Notice that the degree symbol (°) is not used with Kelvin temperatures, and the unit is called a *kelvin* (not capitalized).

THE KELVIN SCALE Winter temperatures in many places can easily drop below 0°F, that is, to temperatures given by negative numbers. In the laboratory, even colder temperatures can be achieved quite easily, and the temperatures are given by even more negative numbers. However, there is a limit to how low the temperature can go. It can be proved that you cannot reach a temperature lower than −273.15°C (or −459.67°F) or **absolute zero** (Chapter 18).

William Thomson, known as Lord Kelvin (1824–1907), first suggested a temperature scale that did not use negative numbers. Kelvin's scale, now adopted as the SI standard, uses the same size degree as the Celsius scale, but it begins at absolute zero. Thus, the freezing point of water is reached at 273.15 degrees above the starting point; that is, 0°C is the same as 273.15 **kelvins** or 273.15 K. Temperatures on the Celsius scale are readily converted to the kelvin scale, and vice versa, using the relation

$$T\ (\text{K}) = t\ (^\circ\text{C}) + 273.15$$

EXERCISE 1.6 TEMPERATURE CONVERSIONS

Carry out the following temperature conversions: (a) 25°C to K. (b) Liquified nitrogen boils at 77 K. What is this temperature in °C? In °F?

1.5 HANDLING NUMBERS

PRECISION AND ACCURACY

The **precision** of a measurement refers to the agreement of repeated measurements of a value with one another. **Accuracy** is the agreement between the measured quantity and the accepted value. A highly precise number may be inaccurate because of an error that is the same for each measurement. As an example, consider the data in Table 1.4. Three students were asked to determine the mass of a piece of metal whose mass is known to be 0.520 g.

The data for Student A are neither very precise nor accurate; the individual values differ widely from one another, and the average value is not accurate. Student B was able to determine the mass of the metal more precisely; the three values deviate but little from one another, but the average mass is still not accurate. In contrast, the data for student C is both precise and accurate.

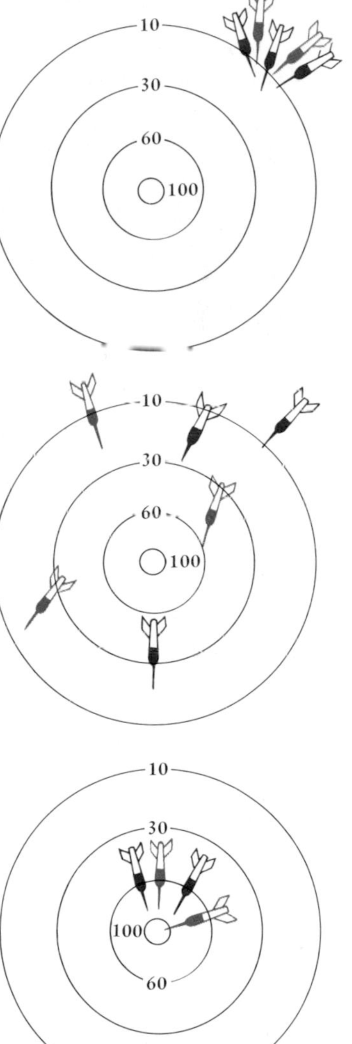

FIGURE 1.12
Precision and accuracy. Top: poor accuracy, good precision. Middle: poor accuracy, poor precision. Bottom: good accuracy, good precision.

SIGNIFICANT FIGURES

Suppose you wish to find the percentage of mass lost (as water) by a popcorn kernel when popped. The mass before heating was found to be 0.123 g, and the mass after heating was 0.108 g. The typical analytical balance, however, is accurate to within 1 mg or 0.001 g. This means that the mass may have been as low as 0.122 g or as high as 0.124 g. The

TABLE 1.4 Data to Illustrate Precision and Accuracy

	MEASUREMENT (g)			
	1	2	3	AVERAGE (g)
Student A	0.521	0.515	0.509	0.515
Student B	0.516	0.515	0.514	0.515
Student C	0.521	0.520	0.520	0.520

implied error is 1 part in 123 or about 0.8%. Similarly, writing 0.108 g implies the mass is between 0.107 or 0.109 g. Both numbers you observed had *three significant figures*; all three digits are important and were experimentally determined.

The % mass loss can be calculated as

$$\% \text{ Mass lost} = \frac{\text{mass of water lost}}{\text{mass of kernel}} \times 100 = \frac{(0.123 - 0.108)\text{g}}{0.123 \text{ g}} \times 100$$

Handling numbers in scientific notation and other mathematical operations are reviewed in Appendix A in the back of the book.

When performing this calculation on an electronic calculator, you would observe 12.19512195. An answer with this many digits implies that you made the original measurement far more precisely than was actually the case. The 10-digit number implies that the final digit is 4, 5, or 6 and that all preceding digits are correct. That is, it implies you are sure of the result to 1 part in 1.2 billion. This is certainly *not* the case! Instead, the proper result is 12%, a number with two significant figures.

The following guidelines regarding significant figures are helpful, but above all else, use common sense!

GUIDELINES FOR DETERMINING SIGNIFICANT FIGURES

Rule 1. To determine the number of significant figures in a measurement, read the number from left to right and count all of the digits, starting with the first digit that is not zero.

EXAMPLE	NUMBER OF SIGNIFICANT FIGURES
1.23 g	3
0.00123 g	3; the zeros to the left of 1 simply locate the decimal point. To avoid confusion, write numbers of this type in scientific notation; thus, $0.00123 = 1.23 \times 10^{-3}$.
2.0 g and 0.020 g	Both of these numbers have two significant digits. When a number is greater than 1, all zeros to the right of the decimal point are significant. For a number less than 1, all zeros to the right of the first significant digit are significant.
100 g	In numbers that do not contain a decimal point, "trailing" zeros may or may not be significant. If only the last digit is uncertain there are three significant figures, so the number should be written in scientific notation as 1.00×10^2. Alternatively, this idea can be conveyed by following the number with a decimal point (100.), a practice followed in this book. It is also possible that there is an error of 10 in the number, in which case there are only two significant figures, and the number should be written as 1.0×10^2. Finally 1×10^2 would imply only one significant figure.
100 cm/meter	Infinite number of significant figures, because this is a defined quantity.
$\pi = 3.1415926 \ldots$	The value for π is a known to a large number of significant figures; you may choose the number of digits appropriate to your calculation.

Zeros to the *left* of the first non-zero digit of a number are *not* significant; they serve only to place the decimal. Zeros to the *right* of the last non-zero digit are significant when the number contains a decimal point. Thus, 10 has 1 significant figure, whereas 10. has 2, 10.0 has 3, and 10.50 has 4 significant figures.

Rule 2. When adding or subtracting, the number of decimal places in the answer should be equal to the number of decimal places in the number with the fewest places.

0.12	2 significant figures	2 decimal places
1.6	2 significant figures	1 decimal place
10.967	5 significant figures	3 decimal places
12.696		

The answer above should be reported as 12.7, a number with 1 decimal place, because 1.6 has only 1 decimal place.

Rule 3. In multiplication and division, the number of significant figures in the answer should be the same as that in the quantity with the fewest significant figures.

$$\frac{0.01208}{0.0236} = 0.512 \text{ or, in exponential notation, } 5.12 \times 10^{-1}$$

Since 0.0236 has only three significant figures, the answer must be limited to three significant figures.

Rule 4. When a number is rounded off (that is, the number of significant figures is reduced), the last digit retained is increased by 1 only if the first digit to be dropped is 5 or greater.*

FULL NUMBER	NUMBER ROUNDED TO THREE SIGNIFICANT FIGURES
12.696	12.7
16.249	16.2
18.35	18.4
18.351	18.4

One last word regarding significant figures and calculations. In working problems in chemistry on a pocket calculator, you should carry through the calculation all of the digits allowed by the calculator and round off only at the end of the problem. Rounding off in the middle can introduce significant errors. If your answers do not quite agree with those in the back of the book, this may be the source of the disagreement.

EXERCISE 1.7 SIGNIFICANT FIGURES

(a) How many significant figures are there in 12.63 and in 0.063? (b) What is the sum of 12.63 and 0.063? (c) What is the product of 12.63 and 0.063?

1.6 PROBLEM SOLVING BY DIMENSIONAL ANALYSIS

Dimensional analysis is a systematic way of solving numerical problems. Simply put, every number in a problem must have some units associated with it; density, for example, is given as mass per unit volume (g/cm^3). When the numbers in a calculation are manipulated in the correct way, their units must cancel out to leave the final answer in the appropriate units. If you set up the problem incorrectly, the units will not cancel properly, and you will know immediately that you have made a mistake.

Dimensional analysis has already been used in the examples in this chapter. However, since the method is such a valuable tool in chemistry,

---

*A modification of this rule is sometimes used to reduce the accumulation of roundoff errors: If the first digit dropped is 5 and there are no following digits or all following digits are zeros (*e.g.*, 18.35 or 18.3500), then the last digit retained is increased by 1 only if it is odd. Thus, 18.35 and 18.45 are both rounded to 18.4.

it is worthwhile to illustrate it further using some more familiar items and the type of calculation you have often done in your head. Let us say you need 30 cans of soft drinks for a party, and the cost is $2.15 per six-pack of cans. What is the total cost? The strategy in solving the problem is first to convert the units "cans" to the unit "six-pack," since the cost is given in units of "dollars per six-pack."

Step 1. Find the number of six-packs required. Convert units of "cans" to units of "six-packs."

$$\text{Number of six-packs} = 30\ \cancel{\text{cans}} \left(\frac{1\ \text{six-pack}}{6\ \cancel{\text{cans}}}\right) = 5\ \text{six-packs}$$

Notice that if you had multiplied 30 times 6, the units would have come out to be (cans^2/six-pack), which is clearly nonsense.

Step 2. Find the total cost. Convert units of "six-packs" into units of "$."

$$\text{Total price} = 5\ \cancel{\text{six-packs}} \left(\frac{\$2.15}{\cancel{\text{six-pack}}}\right) = \$10.75$$

Now that you have worked through the example, notice that in each step you always multiplied the starting unit by a factor that gave the answer in the desired unit. Such factors as (1 six-pack/6 cans) or ($2.15/six-pack) are called **conversion factors**. Conversion factors are multipliers that relate the desired unit to the starting unit.

$$\text{Conversion factor} = \frac{\text{desired unit}}{\text{starting unit}}$$

The numerator of the factor must be equal or equivalent to the denominator. Thus, 1 six-pack is equivalent to 6 cans, and $2.15 is the amount you must pay for 1 six-pack. Other "conversion factors" are 453.6 g/pound, 100 cm/m, or 5°C/9°F.

EXAMPLE 1.7

DIMENSIONAL ANALYSIS: DENSITY AND VOLUME

If a frying pan needs a Teflon coating that is 1.00 mm thick, and the area covered is 36.0 square inches, how many pounds of Teflon are required to coat the pan? Teflon has a density of 0.805 g/cm^3.

Teflon-coated pans.

Solution The first step of the problem is to calculate the total volume of Teflon needed. Once this is done, you can convert the volume to a mass by using the density.

Step 1. Volume of Teflon required.

The volume of the Teflon can be calculated from the product of the thickness and the area. Unfortunately, these dimensions are in different units and cannot be directly multiplied until both are in the same unit. You could convert the thickness to inches, so that the product of thickness (inches) and area (inches^2) gives the volume in cubic inches (inches^3). Alternatively, you could convert the area to square millimeters so that volume is obtained in cubic millimeters. However, we choose to do neither. Thinking ahead to the next step in the problem, you will want to use the density in units of g/cm^3. Therefore, it would be most useful to obtain the volume of Teflon in units of cm^3 so that it is compatible with density. Thus,

$$\text{Thickness} = 1.00 \text{ mm} \left(\frac{1 \text{ cm}}{10 \text{ mm}}\right) = 0.100 \text{ cm}$$

$$\text{Area} = 36.0 \text{ in}^2 \left(\frac{2.54 \text{ cm}}{\text{in}}\right)^2 = 232 \text{ cm}^2$$

The factor 2.54 cm/in is given in the table at the back of the book. Since you need cm^2/in^2, you simply square the conversion factor *and* its units (= 6.45 cm^2/in^2).

With the area and thickness in the same units, you can now calculate the volume.

$$\text{Volume of Teflon coating} = \text{thickness} \times \text{area} = (0.100 \text{ cm})(232 \text{ cm}^2) = 23.2 \text{ cm}^3$$

Step 2. Weight of Teflon required.

In solving any problem, you should keep firmly in mind the answer for which you are aiming and, just as importantly, the units of that answer. In this case, you want the weight of Teflon in pounds. As you saw in Example 1.4, the density will allow you first to convert the volume of Teflon to a mass in grams.

$$23.2 \text{ cm}^3 \left(\frac{0.805 \text{ g}}{\text{cm}^3}\right) = 18.7 \text{ g}$$

Now that you have the mass in grams, it is a simple matter to convert it to the weight in pounds, since you know (from the table at the back of the book) that 1 pound is equivalent to 454 g.

$$18.7 \text{ g} \left(\frac{1 \text{ pound}}{454 \text{ g}}\right) = 0.0412 \text{ pounds} = 4.12 \times 10^{-2} \text{ pounds}$$

Notice that the final answer has three significant figures.

EXERCISE 1.8 MASS, VOLUME, AND DENSITY

Mercury is a metal, but unlike almost all other metals, it is a liquid with a density of 13.6 g/cm^3 at room temperature. It is also poisonous and should be treated with great care and respect.

(a) If you have 100. mL of mercury, how many grams do you have? How many pounds?

(b) If you spill 100. mL of mercury on the floor, and it spreads out into a puddle that is 2.0 mm thick, what area of the floor is covered? Calculate the area in square centimeters and square inches. (Incidentally, mercury is also expensive; 100 mL would cost approximately $90)

SUMMARY

In science, an investigation of a problem usually follows a well worn pathway (Section 1.2). After reviewing work done before, the scientist constructs a **hypothesis**, a tentative explanation or prediction of the observed facts. This hypothesis is revised as further experimental work is done, and the final result can be a **law**, a concise verbal or mathematical statement of a relation that is always the same under the same conditions. To explain the law, we attempt to devise a **theory**, a principle that explains a body of facts and the laws based on them.

Matter is anything having mass and occupying space (Section 1.3). It can exist in three **states** or **phases**: solid, liquid, and gas. A **substance** is a form of matter that has a definite composition and distinct properties. A **mixture** is a combination of two or more substances in which each retains its identity. Mixtures can be **homogeneous**, that is, have the same composition throughout the mixture. Such mixtures can be called **solutions**, where a **solute** is dissolved in a **solvent**. In contrast, in a **heterogeneous** mixture the components can still be observed as individual substances.

Pure substances have **physical and chemical properties**. Physical properties, such as color and density, can be measured and observed without changing the substance's chemical composition. A chemical property, though, requires that the substance undergoes a chemical change, such as reaction with oxygen.

Chemists perform both **qualitative** and **quantitative** experiments, the latter involving numerical information. Units are attached to this information (Section 1.4). The standard unit of length is the **meter (m)**, a unit subdivided into centimeters (1 m = 100 cm), millimeters (1 m = 1000 mm), and so on. Volume measurements are made in cubic meters or in smaller units such as cubic centimeters (1000 cm^3 = 1 liter). The **kilogram (kg)** is the standard unit of mass, but grams (1 kg = 1000 g) and milligrams (1 kg = 10^6 mg) are more often used in the laboratory. The standard unit of time is the **second (s)**, while temperatures are measured in **degrees Celsius (°C)** or in **kelvins (K)**. The latter has the same degree size as the Celsius scale, but measurement begins at **absolute zero**, 273.15 degrees below zero on the Celsius scale.

In handling numerical information you should recognize the difference between the **precision** and **accuracy** of data (Section 1.5). The former is the agreement between repeated determinations of a given value, while the latter is the agreement between the measured value and the accepted value. **Significant figures** are also important, and guidelines for their use are given in Section 1.5.

STUDY QUESTIONS

In solving many of the problems below, you will need to consult Table 1.2, Appendix D, or the back endsheets for units and conversion factors. Questions for which answers are given in Appendix N are indicated by a colored number.

GENERAL QUESTIONS

1. In each case, tell whether the underlined property listed is a physical or chemical property:
   (a) The normal color of bromine is red-orange.
   (b) Iron is transformed into rust in the presence of air and water.
   (c) Dynamite can explode when it interacts with oxygen.
   (d) The density of uranium metal is 19.07 g/cm^3.
   (e) Aluminum metal, the "foil" you use in the kitchen, melts at 660°C.
2. In what phase do you normally find each of the following?
   (a) rust
   (b) the oxygen you breathe
   (c) limestone
   (d) gasoline
3. Small chips of iron are mixed with sand. Is it a homogeneous or heterogeneous mixture? Suggest a way to separate the iron and sand.
4. Decide whether each statement reflects a law or a theory:
   (a) The beginning of the universe occurred as a "big bang."
   (b) In all chemical processes, matter is never lost; it is conserved.
   (c) Various plants and animals have evolved over the history of the planet.

5. In each case, point out which is qualitative information and which is quantitative:
   (a) A purple solid has a mass of 1.25 g.
   (b) A 0.025 g piece of silvery magnesium floats on oil.
   (c) 25 mL of a blue copper sulfate solution react exactly with 25 mL of a colorless ammonia solution.
6. Decide whether each of the underlined items is an intensive or an extensive property.
   (a) The melting point of sodium metal is 98°C.
   (b) A chemical experiment requires 250 mL of water.
   (c) The bromine in Figure 1.4 is a red-orange vapor and liquid.
   (d) The density of gold is 19.3 g/cm^3.

LENGTH, AREA, VOLUME, AND MASS

7. The average lead pencil, new and unused, is 19 cm long. What is its length in millimeters? In meters? In inches?
8. An excellent height in the pole vault is 18 feet, 11.5 inches. What is this height in meters?
9. In track and field, the so-called metric mile is 1600. meters. What fraction of a mile does this represent?
10. The world record for the 100.-meter dash is 9.95 seconds. What is the average speed of the runner in miles per hour?
11. The classic marathon is 26.2 miles. How many kilometers is this equivalent to?
12. Road signs in Germany advise motorists to drive no faster than 130. km/hour. What is this speed in miles per hour?
13. The distance from Paris (France) to Amsterdam (The Netherlands) is about 5.0×10^2 km. How many miles does this represent?
14. The maximum speed limit in the United States is 55 miles per hour. What is this speed in kilometers/hour? In feet/second?
15. A standard U.S. postage stamp is 2.5 cm long and 2.1 cm wide. What is the area of the sample in cm^2? In m^2? In inches^2?
16. The separation between carbon atoms in a diamond is 0.154 nm. (a) What is their separation in meters? (b) Another unit frequently used in science is the Ångstrom, where 1 Å = 10^{-10} m. What is the carbon atom separation in Ångstrom units?
17. As discussed in Chapter 7, light and other forms of radiation can be described as waves, the distance between crests of a wave being the wavelength. If a radiowave has a wavelength of 13 cm, what is its wavelength in meters? In inches? In feet?
18. A Saab automobile has a luggage compartment of dimensions 100. cm × 100. cm × 150. cm. What is the volume of the compartment in liters? In cubic meters?
19. A popular brand of backpack has an upper compartment that is 10 inches × 15.5 inches × 15 inches and a bottom compartment that is 7 inches × 15.5 inches × 9 inches. What is the capacity of the pack in inches^3? In cm^3? In liters?
20. The maximum weight a person can carry comfortably in a backpack is about 60. pounds. What is the mass of this load in grams? In kilograms?
21. A popular tent for backpacking purposes weighs 3 pounds, 11 ounces. What is this mass in grams? In kilograms?
22. A new U.S. quarter has a mass of 5.63 g. What is its mass in kilograms? In milligrams?
23. A sleeping bag for use in very cold climates is filled with 3 pounds of goose down. How many grams of down are used? How many kilograms?
24. If you weigh 160. pounds, what is your weight in grams? In kilograms?
25. A typical laboratory beaker has a volume of 800. mL. What is its volume in cm^3? In liters? In m^3? In dm^3 (where dm = decimeter)?
26. A Volkswagen engine has a displacement of 120. cubic inches. What is this volume in liters?
27. A camping stove has a fuel capacity of 0.60 pints. The storage bottle for the extra fuel is advertised as having a capacity of 0.30 L. When the stove has used all of its fuel, does the storage bottle contain enough fuel to refill the stove completely?
28. The smallest repeating unit of a crystal of common salt is a cube with an edge length of 0.563 nm. (a) What is this length in Ångstrom units? (See Study Question 16.) (b) What is the volume of this cube in nm^3? In cm^3?
29. The mass of a gemstone is often measured in "carats" where 1 carat = 0.200 g. If the annual, worldwide production of diamonds is 12.5 million carats, how many grams does this represent?
30. One 2.0-ounce serving of macaroni provides 8.0 g of protein, 42 g of carbohydrate, and 1.0 g of fat. Calculate the mass percentage of each of the three substances in the 2.0 ounces of macaroni.

DENSITY

31. The density of carbon in the form of diamonds is 3.51 g/cm^3. What is this density in kg/m^3?
32. Common sugar has a density of 1.587 g/cm^3. What would be the volume (in cm^3) of one pound of sugar?
33. Water has a density at 25°C of 0.997 g/cm^3. If you have 1.00 quart of water, what is its mass in grams? In pounds?
34. A chemist needs 2.00 g of a liquid compound. (a) What volume of the compound is necessary if the density of the liquid is 0.718 g/cm^3? (b) If the compound costs \$2.41 per milliliter, what is the cost of the reagent?

SPECIAL SECTION: GOLD!

FIGURE 1.13
Crystallized gold. (Allen B. Smith, Tom Stack & Associates)

A Chinese book from about AD 100 says, "Gold is noncorruptible, and therefore the most valuable of things. Men feeding on it attain longevity. The gold dust, having entered the five internal organs, spreads foggily. Vaporizing and permeating, it reaches the four limbs. Thereupon, the complexion becomes rejuvenated, hoary hair regains its blackness, and new teeth grow where fallen used to be."

1 metric ton = 1000 kg or 2204.6 pounds.

Gold! The mere thought of it has driven people to perform centuries of misguided experiments, to murder, to exploit one another, and to explore the earth (Figure 1.13). Alexander the Great invaded Persia for it, the Portuguese set sail into the unknown, the Spanish explored South and Central America, and the settlement of the United States and Canada was spurred. The ancient alchemists made countless, fruitless attempts to synthesize gold, partly because the Greek philosopher Aristotle (384–322 BC) believed all things tend to reach perfection. Since gold was a model of perfection, it was a simple extension of this idea that all other metals try to reach the perfect state. Thus, you find "recipes" in the writings of the alchemists such as "Take mercury, fix it with the body of Italian antimony. Cast the white earth so prepared on copper. . . . Add yellow electron and you will have gold." It was simple! Unfortunately, whatever this recipe means in modern chemistry is not known.

The Chinese, responsible for so many things in our civilization, tried to prepare gold, and it has been suggested that the word "chemistry" comes from a Chinese word meaning "gold-making juice." Gold making was important in Chinese culture not only for the lure of the metal itself but also because eating gold was said to guarantee immortality.

Gold has many uses other than its obvious one in jewelry. For example, it is a prime method of payment in international commerce. Thousands of tons of bullion or gold bars lie in vaults in the Federal Reserve Bank in New York and at Fort Knox, Kentucky. In all, about 30,000 metric tons of gold bullion are stored in the U.S.

Because gold is malleable, a good electrical conductor, and does not corrode, it is used in electrical circuits of television sets, computers, and calculators. But its greatest use in the United States is still in jewelry. In fact, more than 2.5 million school class rings containing gold are made every year.

Eighty thousand metric tons of gold are estimated to have been removed from the Earth in the last 6000 years. This is a staggering amount considering that the concentration of gold in the earth's crust is low (only about 0.004 g per million grams of crust or 0.004 parts per million) (Figure 1.14). Thus, extraordinary methods have to be used to extract it from the ground. In ancient times gold was recovered from river sand by washing the water over a sheep's fleece, a practice that was probably the origin of the "Golden Fleece" of Greek mythology. More recently, river sand has been "panned" for gold, a method that uses the fact that gold's density (19.3 g/cm^3) is much higher than that of sand (2.5 g/cm^3).

As traditional sources of gold are played out, it is increasingly taken from deep mines. The mines of South Africa, some nearly 11,000 feet

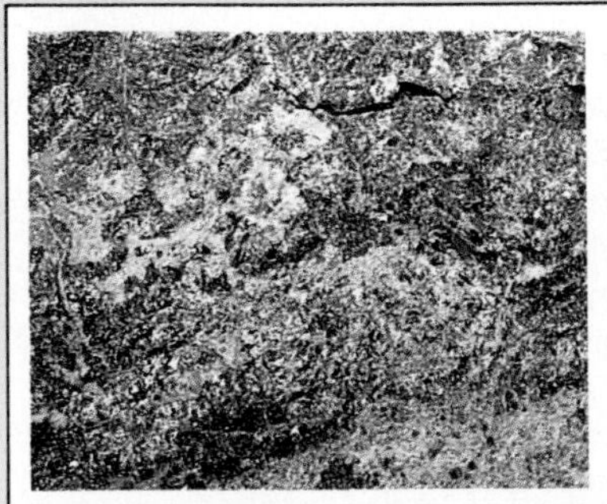

FIGURE 1.14
A vein of high grade gold ore. (Brian Parker, Tom Stack & Associates)

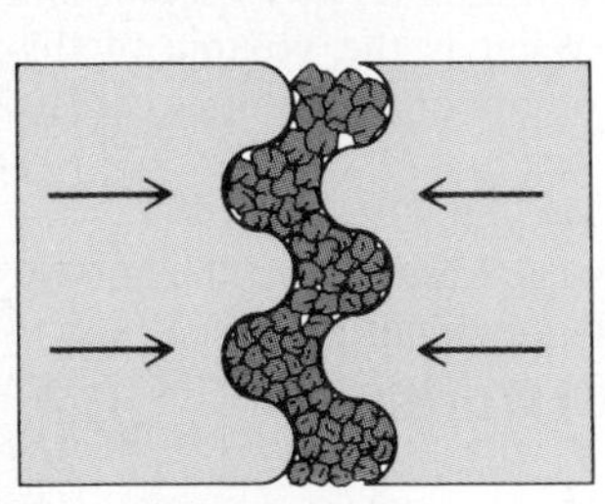

1. Like a mighty nutcracker, steel jaws shatter ore into softball-size fragments. Hand sorters discard pieces lacking gold.

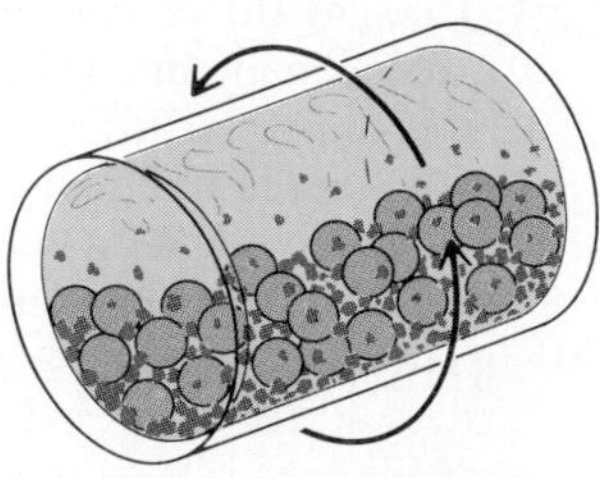

2. After further crushing, the ore mixes with water and enters a revolving cylinder, to be pulverized by tumbling steel balls or bars.

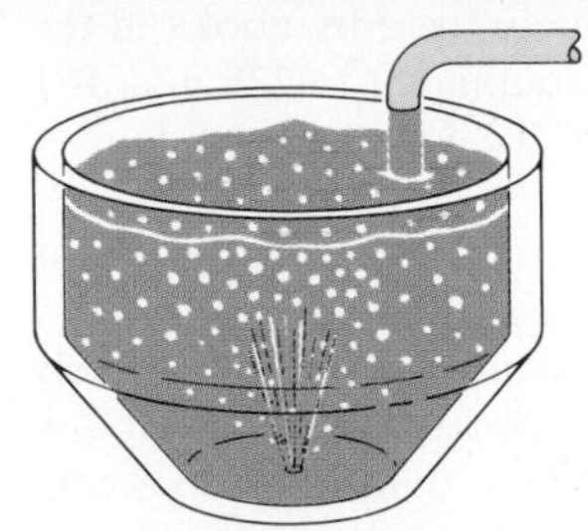

3. Air jets and mechanical arms in agitator tanks mix cyanide into powdered ore and water, called slime. This releases gold from rock.

4. The gold-cyanide solution and slime funnel into vast tanks where the rock particles slowly sink. The clarified solution is fed into filtration units.

5. Gold-cyanide solution is filtered to strain out any remaining rock particles, and then is deaerated.

6. Zinc dust added to the solution separates the cyanide from the gold, which emerges as an impure powder.

7. The gold is melted with fluxes such as borax. As the metal cools in the bottom of a conical mold, the fluxes combine with impurities and float as slag.

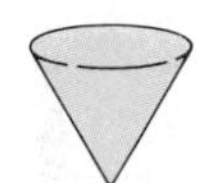

8. The final product is a "button" of 90% gold and about 10% silver. Further refining yields 99.6% gold.

FIGURE 1.15
A scheme for recovering gold from rock. About 2½ tons of rock must be processed to yield an ounce of gold and a tiny amount of silver.

deep, produce almost two thirds of the world's gold. After gold-bearing rock is brought up from the mine, it is processed by a method in use for many years (Figure 1.15).

At the end of the refining process, the gold is 99.95% pure and is called "24 carat" gold. Usually jewelry is not made of 24 carat gold, since it would be too soft. Instead, copper and silver are added to make it harder (and cheaper). In the United States, jewelry is usually made of 14 carat gold that contains 14/24 or 58.33% gold. We shall describe more of the chemistry of gold and the other metals in the course of this book. However, as you probably can guess gold is not a very reactive element. Unlike silver, which tarnishes when exposed to sulfur and its compounds, gold is the only metal that does not react with sulfur. Further, it is not affected by many acids. Only *aqua regia*, or "kingly water," a 3:1 mixture of concentrated hydrochloric and nitric acids, will dissolve gold.

A "carat" is a way of expressing the proportion of gold in a sample, a carat being 1/24 part by weight.

35. The "cup" is a volume widely used by cooks in the United States. One cup is equivalent to 225 mL. If 1 cup of olive oil has a mass of 205 g, what is the density of the oil?
36. At 25°C the density of water is 0.997 g/cm^3, whereas the density of ice at −10°C is 0.917 g/cm^3. (a) If a soft-drink can (volume = 250. mL) is filled completely with pure water and then frozen at −10°C, what volume will the solid occupy? (b) Could the ice be contained within the can?
37. Liquid sodium can be used to cool nuclear reactors. (a) If liquid sodium has a density of 0.93 g/cm^3, how many grams of sodium would be required to fill a container with a volume of 15 liters? (b) How many pounds of sodium would be required?
38. Peanut oil has a density of 0.92 g/cm^3. If a recipe calls for 1 cup of peanut oil (1 cup = 225 mL), how many grams of peanut oil are you using?
39. Silver has a density of 10.5 g/cm^3. If you have a silver coin with a mass of 6.0 g (about the same as a U.S. quarter), what is the volume of the silver in the coin?

TEMPERATURE

40. Many laboratories use 25°C as a standard temperature. What is this temperature in °F? In K?
41. Make the following temperature conversions:

	°F	°C	K
(a)	57	_____	_____
(b)	_____	37	_____
(c)	−40.	_____	_____
(d)	_____	_____	77
(e)	_____	60.	_____
(f)	1000.	_____	_____

42. The temperature on the surface of the sun is 5.50×10^3°C. What is this temperature in °F? In K?
43. Oxygen freezes to a solid at −218°C. What is this temperature in °F? In K?
44. Solid gallium has a melting point of 29.8°C. If you hold this metal in your hand, what will be its physical state? That is, will it be a solid or liquid? (Prove your answer with appropriate calculations.)
45. Titanium is used in industrial applications where a high melting point is important. Its melting point is 3294°F. What is this temperature in °C? In K?
46. Helium, an element produced in the sun, has a melting point of −272.3°C and a boiling point of −268.6°C. Express these temperatures in °F and in kelvins.

SIGNIFICANT FIGURES

47. What is the average mass of three objects whose individual masses are 10.3 g, 9.334 g, and 9.25 g?
48. What is the volume, in cm^3, of a backpack whose dimensions are 22.86 cm × 38.0 cm × 76 cm?
49. A 2-quart saucepan has a radius (r) of 83.5 mm and a height (h) of 9.5 cm. What is the volume of this saucepan in cm^3? (The volume of a cylindrical object is given by $\pi r^2 h$.)
50. What is the quotient of 546/760.0?
51. Solve the equation below for n and report the answer to the correct number of significant figures.

$$(11.2/760.0)(123.4) = n(0.0821)(298.3)$$

USING UNITS

52. In 1983, as in most years, sulfuric acid was the chemical produced in greatest amount by industry. In that year, 69.45 billion pounds of the acid were produced. How many tons does that represent? How many kilograms? How many grams? (1 ton is exactly 2000 pounds)
53. One source of sulfuric acid is elemental sulfur. In 1981, 12.1 million metric tons of yellow solid sulfur were produced. (1 metric ton = 1000 kg) How many kilograms of sulfur does that represent? How many pounds?
54. A sheet of $8\frac{1}{2} \times 11$ inch paper is 0.10 mm thick. It can be considered as a rectangular solid for which the volume is length × width × thickness. What is the volume of the sheet in cm^3? If a sheet has a mass of 4.58 g, what is its density?
55. An ancient gold coin is 2.2 cm in diameter and 3.0 mm thick. It is a cylinder for which volume = (π)(radius)^2(thickness) where radius = $\frac{1}{2}$ diameter. If the density of gold is 19.3 g/cm^3, what is the mass of the coin in grams? Assume a price of gold of $410 per ounce. How much is the coin worth? (1 ounce = 28.4 g)
56. Heat is commonly measured in units of calories (cal), but more recently scientists have accepted the SI unit, the joule (Chapter 5). If there are 4.184 joules per calorie, how many joules does a dieter expend if the food eaten provides 9.0×10^5 cal/day? [When discussing food, we use 1000 calories (small calories) = 1 Calorie with a capital "C." Thus, 900,000 "small calories" is equal to 900 Calories, a small food intake.]
57. The aluminum in a package containing 75 square feet of kitchen foil weighs approximately 12 ounces. Aluminum has a density of 2.70 g/cm^3. What is the approximate thickness of the aluminum foil in millimeters? (1 ounce = 28.4 g, and volume = area × thickness)
58. A tanker has spilled 1.00×10^4 L of oil into the ocean and the oil film covers the sea with a layer 3.0×10^2 nm thick. What is the area covered by the oil film in square meters? In square miles? (Volume = area × thickness)

2

Atoms, Molecules, and Ions

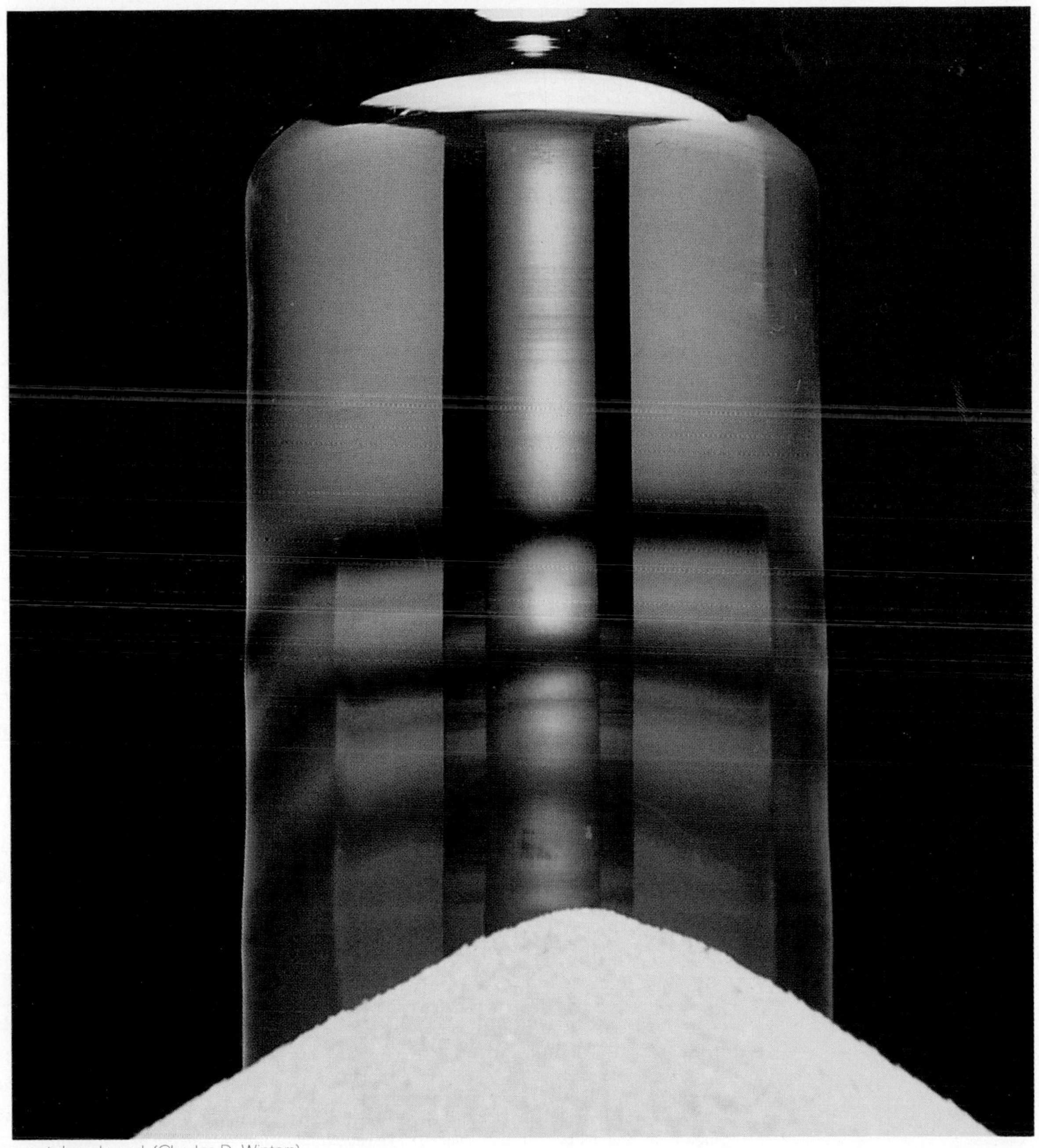

Silicon crystal and sand. (Charles D. Winters)

CHAPTER OUTLINE

In 1803 John Dalton first proposed the basic ideas or hypotheses of an **atomic theory of matter**. He said this:

1. All matter is made of **atoms**. These indivisible and indestructible objects are the ultimate chemical particles.*
2. All the atoms of a given element are identical, in both weight and chemical properties. However, atoms of different elements have different weights and different chemical properties.
3. **Compounds** are formed by the combination of different atoms in the ratio of small whole numbers. (For example, water is composed of hydrogen and oxygen in a ratio of 2 atoms of hydrogen to 1 of oxygen.) As described later in this chapter, we know now that **molecules** are the ultimate chemical particles of compounds.
4. A **chemical reaction** involves only the combination, separation, or rearrangement of atoms; atoms are neither created or destroyed in the course of ordinary chemical reactions.

John Dalton's hypotheses have been completely accepted into our culture, and we assume that you are aware of at least the first of his ideas. You are probably also aware of the last of his ideas, although under a different name. This is really just a statement of the **law of conservation of matter**, that is, matter can neither be created nor destroyed. It had been established somewhat earlier by the great French chemist Antoine Lavoisier (1743–1794). After describing some fundamental properties of atoms and molecules in this chapter, we shall take up Lavoisier and his role in chemistry when chemical reactions are first described in Chapter 3.

2.1 ELEMENTS

The concept of elements originated with the Greek philosopher Aristotle (384–323 BC) who said,

*It is only fair to point out that there is still some controversy about what Dalton really meant by the word "atom." See R. Mierzecki, *Journal of Chemical Education* 58:1006, 1981.

TABLE 2.1 Derivation of Element Names and Symbols

ELEMENT	SYMBOL	DATE OF DISCOVERY	DISCOVERER (COUNTRY)*	DERIVATION OF NAME OR SYMBOL
Berkelium	Bk	1950	G.T. Seaborg, S.G. Thompson, A. Ghiorso (U.S.)	Berkeley, California (site of Seaborg's laboratory)
Copper	Cu	Ancient		Latin, *cuprum*, copper. Derived from *Cyprium*, the Island of Cyprus, the main source of copper in the ancient world.
Einsteinium	Es	1952	A. Ghiorso (U.S.)	Albert Einstein
Iron	Fe	Ancient		Latin, *ferrum*, iron
Lead	Pb	Ancient		Latin, *plumbum*, lead, meaning heavy
Oxygen	O	1774	J. Priestley (G.B.) K.W. Scheele (Swed.)	French, *oxygene*, generator of acid, derived from the Greek, *oxy* and *genes* meaning acid forming (oxygen was thought to be part of all acids).
Silver	Ag	Ancient		Latin, *argentum*, silver
Tin	Sn	Ancient		Latin, *stannum*, tin
Tungsten	W	1783	J.J. and F. de Elhuyer (Sp.)	Name from Swedish, *tung sten*, meaning heavy stone; symbol from wolframite, a mineral.

*U.S., United States; G.B., Great Britain; Swed., Sweden; Sp., Spain.

Everything is either an element or composed of elements. . . .

A modern definition, however, might be that an **element** is a form of matter that cannot be broken down into simpler forms by ordinary means.*

Chemists have discovered and studied 108 elements so far. Each of them has been given a *name* and a *symbol*, listed in the tables at the front of the book. Some of the elements (gold, silver, iron, mercury, and tin, for example) were known in relatively pure form to the Greeks and Romans and to the alchemists of ancient China, the Arab world, and medieval Europe. Many others were discovered in the 18th and 19th centuries, but two elements (107 and 108) were discovered only in the 1980s. Many elements have names and symbols with Latin or Greek roots, but more recently discovered elements have been named for their place of discovery or for a person or place of significance in the history of science. Some examples of the origins of element names are given in Table 2.1.

The table at the front of the book, in which each element is enclosed in a box, is called the **periodic table**.

Be sure to notice that only the first letter of an element symbol is capitalized. For example, copper is Cu and not CU, a notation that would mean a combination of carbon and uranium.

*We know now that atoms are composed of electrons, protons, neutrons, and other subatomic particles. However, an atom cannot be broken down completely into these particles by ordinary means, that is, by the forms of energy that are available in the usual chemistry laboratory.

JOHN DALTON (1766–1844)

John Dalton was born on about the fifth of September, 1766, in the village of Eaglesfield in Cumberland, England. His family was quite poor, and his formal schooling ended at age 11. However, he was clearly a bright young man, and with the help of influential patrons, he began a teaching career at the age of 12. Shortly thereafter, he made his first attempts at scientific investigation, observations of the weather. This was a study that was to last his lifetime. In fact, he made over 200,000 observations of weather conditions by the end of his life. In 1793 Dalton moved to Manchester, England, where he took up a post as a tutor at the New College, but he left there in 1799 to pursue scientific inquiry on a full-time basis. It was not long after this, on October 21, 1803, that he read the paper introducing his "Chemical Atomic Theory" to the Literary and Philosophical Society of Manchester. This was followed by lectures in London and in other cities in England and Scotland, and his reputation as a scientist rapidly increased. He was first proposed for membership in the top scientific society in Britain, the Royal Society, in 1810, and many other honors followed over his lifetime.

2.2 ATOMS

An **atom** is the smallest particle of an element that retains the chemical properties of the element. The word "atom" comes from the Greek meaning "incapable of being cut." If a piece of sodium, for example, is cut in half and each of these halves is in turn cut in half, and so on, the Greek philosophers reasoned that eventually it would no longer be possible to cut the last piece in half. Therefore, they reasoned that matter was made up of very small, uncuttable pieces, which we now call atoms.

Atoms, we now know, are not indivisible; if they are split, however, they lose their chemical identity. They are composed of three primary types of particles: **electrons, protons,** and **neutrons**. The **nucleus** or core of the atom is made up of protons with a positive electrical charge and neutrons with no charge. The negatively charged electrons are found in the space about the nucleus; for an atom, which must have no net electrical charge, the number of negatively charged electrons equals the number of positively charged protons in the nucleus.

Atoms are extremely small; the radius of the typical atom is between 30 and 150 pm (or about 0.000000000030 m $= 3.0 \times 10^{-11}$ m). Actually, as discussed in Chapter 7, the atom is mostly empty space, since the radius of the nucleus is only about 0.1 pm! To give you some feeling for these sizes, look at it this way:

(a) One teaspoon of water (about 1 cm^3) contains about three times as many atoms as the Atlantic Ocean contains teaspoons of water.

(b) If, after the flood of about 3000 BC, Noah had started to string hydrogen atoms on a thread at a rate of one atom a second for 8 hours a day, the chain would only be about 1.6 meters (about 5 feet) long today!

All atoms of a given element have the same number of protons in the nucleus, and, because the atom is electrically uncharged, the number of electrons surrounding the nucleus is the same as the number of protons.

TABLE 2.2 Properties of Subatomic Particles

PARTICLE	MASS		ELECTRICAL	
	GRAMS	ATOMIC MASS UNITS	CHARGE	SYMBOL
Electron	9.1083×10^{-28}	0.0005483	-1	$_{-1}^{0}e$
Proton	1.6726×10^{-24}	1.007276	$+1$	$_{1}^{1}H$
Neutron	1.6750×10^{-24}	1.008665	0	$_{0}^{1}n$

This number is called the **atomic number**, and it is given the symbol **Z**. In the periodic table at the front of the book, the atomic number of each element is the number above the element's symbol. Magnesium, for example, has a nucleus containing 12 protons, so its atomic number is 12; uranium has 92 protons in its nucleus and Z = 92.

EXERCISE 2.1 ATOMIC NUMBERS, NAMES, AND SYMBOLS OF THE ELEMENTS

(a) Using the tables in the front of the book, give the atomic numbers of the following elements: Li, Ar, Cr, Ag, and Ra. (b) Give the name of each element in part (a). (c) What are the symbols for beryllium, chlorine, manganese, antimony, platinum, and plutonium?

1 amu = $\frac{1}{12}$ the mass of a carbon atom having 6 protons and 6 neutrons.

Thousands of experiments have led to the establishment of a scale of *relative* atomic masses based on the **atomic mass unit (amu)**. This standard is based on the mass of a carbon atom that has six protons and six neutrons in its nucleus. Such an atom is defined to have a mass of exactly 12 atomic mass units or 12 amu. The mass of every other known element is established relative to the mass of carbon. Thus, for example, experiments have shown that an oxygen atom is, on the average, 1.33 times heavier than a carbon atom, so oxygen has a mass of 1.33 × 12.0 amu or 16.0 amu.

Masses of the basic atomic particles in amu (Table 2.2) have been determined experimentally. Notice that the proton and neutron have masses very close to 1 amu, while the electron is only about 1/1900 times as heavy.

With this definition of the basic unit of mass, the mass of any atom for which the nuclear composition is known can be estimated. The proton and neutron have masses so close to 1 amu that the difference can usually be ignored and even a large number of electrons will not greatly affect the mass of the atom. Therefore, the mass of an atom is estimated simply by adding up the number of protons and neutrons. The result is called the **mass number** of that particular atom and is given the symbol **A**. For example, a fluorine atom has 9 protons and 10 neutrons, so its mass number is A = 19. The most common atom of iron has 26 protons and 30 neutrons, so it has a mass number of 56. An atom of a particular composition is fully symbolized by the notation

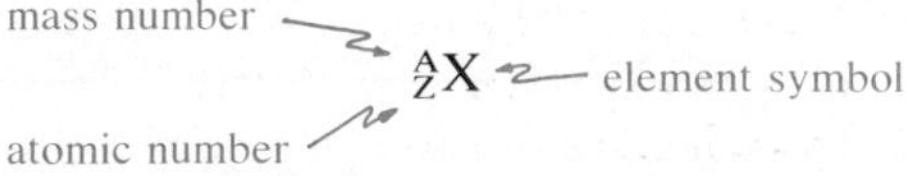

(where the subscript indicating the atomic number is optional, because the element symbol tells you what the atomic number must be). For example, the fluorine atom described above would have the symbol $^{19}_{9}F$ or just ^{19}F. In words, you would say "fluorine-19."

EXAMPLE 2.1

CALCULATING THE MASS NUMBER OF AN ATOM

What is the mass number of a tin atom if that atom has 69 neutrons?

Solution First, you must know the atomic number of tin, since this will give the number of protons in the atom. Looking in the tables at the front of the book, you find that tin (symbol = Sn) has an atomic number of 50. Therefore, the mass number, A, is

$$A = \text{number of protons} + \text{number of neutrons} = 50 + 69 = 119$$

FIGURE 2.1

(a) The mass spectrometer, an instrument for determining the mass of atoms, molecules, and molecular fragments. (b) Representation of a "time-of-flight" mass spectrometer. The time required for positive ions (such as H^+, O^+, C^+, and so on) to reach the ion detector is measured; the heavier an ion the longer it takes to travel down the tube. The ion detector is connected to an instrument that measures the number of ions reaching the detector in a given time. The time-of-flight tube is vacuum sealed, so the process occurs at a very low pressure. (c) The ion detector in a mass spectrometer also measures the relative number of atoms, that is, the *relative abundance* of each istotope. A *mass spectrum,* such as that below for antimony, shows the relative abundance of atoms as a function of mass number (a, Nicolet)

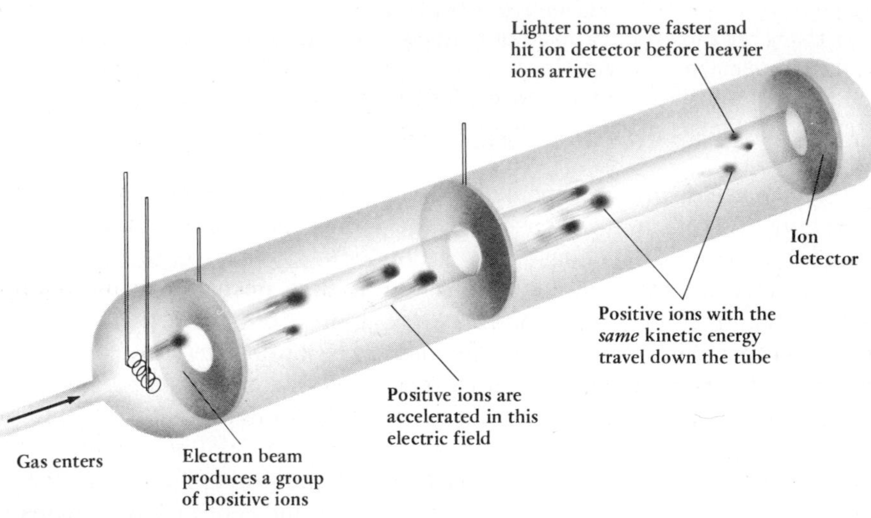
Lighter ions move faster and hit ion detector before heavier ions arrive
Ion detector
Positive ions with the *same* kinetic energy travel down the tube
Positive ions are accelerated in this electric field
Gas enters
Electron beam produces a group of positive ions

(b)

Nicolet

(a)

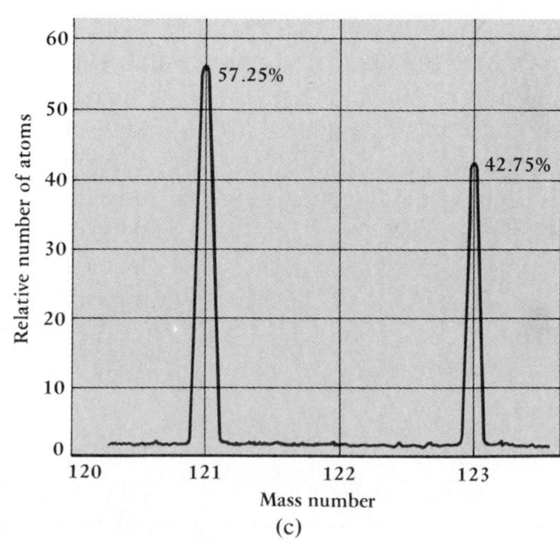
60
50
40
30
20
10
0
57.25%
42.75%
Relative number of atoms
120
121
122
123
Mass number

(c)

EXERCISE 2.2 CALCULATING THE MASS NUMBER OF AN ATOM

Give the mass number of each of the following atoms: (a) magnesium with 12 neutrons; (b) copper with 34 neutrons; and (c) mercury with 120 neutrons.

EXERCISE 2.3 ATOMIC SYMBOLS

(a) Give the appropriate symbol for the iron-57 and tin-119 atoms. (b) How many protons, neutrons, and electrons are contained in an atom of $^{195}_{78}Pt$, platinum-195?

The *actual masses* of atoms have been determined experimentally, although sophisticated and expensive instruments are required (Figure 2.1). It is always observed that while the actual mass is almost the same as the mass number the actual mass is not an integral number. For example, the actual mass of an iron atom with 32 neutrons is 57.933272 amu.

An obvious thing to do is to check the experimental masses by adding up all the exact masses of the protons, neutrons, and electrons of an atom to see whether you obtain the experimental mass. However, you would find the sum of the exact particle masses is not quite equal to the actual mass. This is not a mistake. Rather, the difference (sometimes called the *mass defect*) is related to the forces binding the particles of the nucleus to one another; more will be said about this in Chapter 26.

2.3 ISOTOPES

For many years minerals containing boron have been mined in Death Valley, California. If you were to examine the boron atoms of some borax mined in that region, you would find that, while all boron atoms have 5 nuclear protons, some of the atoms have 5 neutrons while others have 6. That is, you would find a collection of $^{10}_{5}B$ and $^{11}_{5}B$ atoms. These are *isotopes* of boron. **Isotopes** are atoms having the same atomic number Z but a different mass number A, because the atoms have a different number of neutrons.

If two atoms differ in the number of *protons* they contain, they are different *elements*; if they differ only in the number of *neutrons*, they are *isotopes*. It is the number of protons and extranuclear electrons that determine the chemistry of the atom.

Most elements have at least two stable isotopes (Table 2.3). There are very few elements with only one isotope (aluminum, fluorine and

TABLE 2.3 Exact Masses of the Stable Isotopes of Some Elements

ELEMENT	SYMBOL	ATOMIC WEIGHT	MASS NUMBER	EXACT MASS	PERCENT ABUNDANCE
Hydrogen	H	1.00797	1	1.007825	99.9855
			2	2.014102	0.0145
Boron	B	10.811	10	10.012939	19.91
			11	11.009305	80.09
Oxygen	O	15.9994	16	15.994914	99.759
			17	16.999134	0.0374
			18	17.999160	0.2039
Magnesium	Mg	24.312	24	23.985045	78.80
			25	24.985840	10.15
			26	25.982591	11.05
Uranium	U	238.03	234	234.040875	0.0056
			235	235.043900	0.7205
			238	238.050734	99.274

phosphorus, for example), and there are others with many isotopes (tin has nine stable isotopes). Generally, we refer to a particular isotope by giving its mass number (for example, uranium-238, ^{238}U), but some isotopes are so important that they have special names. Hydrogen atoms all have one proton: When that is the only nuclear particle, the element is called simply "hydrogen"; when one neutron is added to give $^{2}_{1}H$, it is called "deuterium" or heavy hydrogen; and adding yet another neutron gives radioactive hydrogen, $^{3}_{1}H$, or "tritium."

2.4 ATOMIC WEIGHT

Boron atoms have two different isotopic masses, 10.0129 and 11.0093. This means that the average mass of a collection of boron atoms will be neither 10 nor 11 but somewhere in between, the actual value depending on the proportion of each kind of isotope.

The percentage of atoms of a natural sample of the pure element represented by a particular isotope has been called its **percent abundance**.

$$\text{Percent abundance} = \frac{\text{number of atoms of a given isotope}}{\text{total number of isotopic atoms}} \times 100$$

If these percent abundances can be determined for each isotope, then the average mass can be determined. Such determinations can be done using a device called a mass spectrometer (Figure 2.1), and some results are given in Table 2.3. From this table you can see that the boron isotopes ^{10}B and ^{11}B have percent abundances of 19.91% and 80.09%, respectively. This means that, if you could count out 10,000 boron atoms from an "average" natural sample, 1,991 of them will have a mass of 10.0129 amu and 8,009 amu of them will have a mass of 11.0093 amu. The *average mass* of a representative sample of atoms is called the **atomic weight**. For boron, the atomic weight is 10.81 amu, as is shown by the calculation in Example 2.2.

Chemists usually use the term **atomic weight** of an element rather than "atomic mass." Although the quantity is more properly a "mass" than a "weight," the term "atomic weight" is so commonly used that it has become accepted.

EXAMPLE 2.2

CALCULATING AVERAGE ATOMIC MASS FROM RELATIVE ABUNDANCES

A natural sample of boron consists of two isotopes. One has an exact mass of 10.0129 and its percent abundance is 19.91%; the other isotope, of mass 11.0093, has a percent abundance of 80.09%. Calculate the atomic weight.

Solution The average value for a series of numbers is always found by summing the products of "value" times "fractional abundance."

The fractional abundance is the percent abundance expressed as a number between 0 and 1.

$$\begin{aligned}\text{Average atomic mass} &= (\text{fractional abundance of isotope 1})(\text{mass of isotope 1}) \\ &\quad + (\text{fractional abundance of isotope 2})(\text{mass of isotope 2}) + \ldots\end{aligned}$$

For the boron sample, the calculation is as follows:

$$\begin{aligned}\text{Average atomic mass of boron} &= \text{atomic weight} \\ &= (0.1991)(10.0129) + (0.8009)(11.0093) = 10.81\end{aligned}$$

EXERCISE 2.4 CALCULATING ATOMIC WEIGHT

Verify that the atomic weight of chlorine is 35.45, given the following information:

^{35}Cl, exact mass = 34.96885, percent abundance = 75.77%

^{37}Cl, exact mass = 36.96590, percent abundance = 24.23%

The atomic weight of each element has been determined, and it is these masses that appear in the tables in the front of the book. In the periodic table, each element's box contains the atomic number, the element symbol, and the atomic weight.

If the masses are known for the isotopes of a given element, it is also possible to compute the percent abundances by "reversing" the procedure outlined in Example 2.2. The next example shows how this is done.

Copper

atomic number →	29
atomic symbol →	Cu
atomic weight →	63.546

EXAMPLE 2.3

CALCULATING ISOTOPE ABUNDANCES

Calculate the percent abundances of copper-63 and copper-65 if the atomic weight is 63.54 amu and the exact masses of the isotopes are ^{63}Cu = 62.93 amu and ^{65}Cu = 64.93 amu. Assume no other isotopes exist.

Solution This problem has two unknown quantities, the percent abundances of ^{63}Cu and ^{65}Cu. Recall from algebra that to solve for two unknowns, two equations are required. The key here is to realize that the sum of the fractional isotope abundances (each less than 1) must equal 1.

$$^{63}\text{Cu fractional abundance} + {}^{65}\text{Cu fractional abundance} = 1$$

Both of these abundances are unknown. However, if one of them is assigned the value x, then the other is $1 - x$. That is,

if	^{63}Cu fractional abundance $= x$
then	$x +$ ^{65}Cu fractional abundance $= 1$
and so	^{65}Cu fractional abundance $= 1 - x$

The expression "^{65}Cu abundance $= 1 - x$" is the first of the two equations you need. The second is the same type of expression used in Example 2.2.

$$\text{Average atomic mass of Cu} = ({}^{63}\text{Cu fractional abundance})(62.93) + ({}^{65}\text{Cu fractional abundance})(64.93)$$

These two equations can now be solved in the usual way. Substitute the expressions for the Cu isotope abundances into the second equation, and solve for x.

$$\begin{aligned} \text{Atomic weight} &= (x)(62.93) + (1 - x)(64.93) \\ 63.54 &= (x)(62.93) + (1 - x)(64.93) \\ 63.54 &= 62.93x + 64.93 - 64.93x \\ 63.54 - 64.93 &= (62.93 - 64.93)x \\ x &= \frac{63.54 - 64.93}{62.93 - 64.93} = 0.695 \end{aligned}$$

We find that $x = 0.695$. This means that the ^{63}Cu fractional abundance is 0.695 or that the percentage abundance is 69.5%. It then follows that the fractional abundance of ^{65}Cu must be $1 - x = 0.305$, so the percent abundance is 30.5%.

Native copper deposits. (Harold L. Levin)

AVOGADRO AND HIS NUMBER

Lorenzo Romano Amadeo Carlo Avogadro (1766–1856), an Italian, was educated as a lawyer and practiced the profession for many years. However, in about 1800 he turned as well to science and was the first professor in Italy in mathematical physics. It was in 1811 that he first suggested the hypothesis, which we know now to be a law, that "equal volumes of gases under the same conditions have equal numbers of molecules." From this eventually came the concept of the mole. Avogadro did not see the acceptance of his ideas in his own lifetime, and it was not until another Italian, Cannizzaro, discussed them at the great conference in Karlsruhe, Germany, in 1860 that scientists were convinced.

One of the great difficulties presented by Avogadro's number is comprehending its size. It may help to write it out in full as

$$6.022 \times 10^{23} = 602{,}200{,}000{,}000{,}000{,}000{,}000{,}000$$

or as

$$602{,}200 \times 1 \text{ million} \times 1 \text{ million} \times 1 \text{ million}$$

But, think of it this way: If you have Avogadro's number of popcorn kernels, and pour them over the continental United States, the country would be covered to a depth of 9 miles! Or, if you divide one mole of pennies equally among every man, woman, and child in the United States, one person could pay off the national debt (currently about \$1.4 trillion or 1.4×10^{12}) and still have 20 trillion dollars or so left for incidental expenses.

How is Avogadro's number determined? It is clearly not possible to count all of the atoms in a mole. (If a computer counted 10 million atoms per second, it would take about 2 billion years to count all of the atoms in a mole.) There are in fact at least four or five experimental ways to do it, one of them being the measurement of the dimensions of the smallest repeating unit in a solid element (see Study Question 46 and Chapter 11).

2.5 THE MOLE

One fascinating part of chemistry is the discovery of some new substance when one element is mixed with another. But chemistry is also a quantitative science. When two chemicals are mixed, you would like to know how many atoms of each are used. This means that there must be some method of counting atoms, no matter how small they are. The solution to this problem is to have a convenient unit of matter that contains a known number of particles in a known mass. The "chemical counting unit" that has come into use is the **mole**.

The word "mole" was apparently introduced in about 1896 by Wilhelm Ostwald, who derived the term from the Latin word *moles* meaning a "heap" or "pile." The mole, whose symbol is "mol," is the SI base unit for measuring "amount of substance" (Table 1.1); it is defined as follows:

> A **mole** is the amount of substance that contains as many elementary particles (atoms, molecules, or other particles) as there are atoms in 0.0120 kg (12.0 g) of carbon-12 isotope.

The key to understanding the concept of *the mole* is that it *always contains the same number of particles, no matter what the substance*. But how many particles? Many, many experiments over the years have established that number as

$$1 \text{ mole} = 6.022045 \times 10^{23} \text{ particles}$$

This number, usually rounded off to 6.022×10^{23}, is known as **Avogadro's Number** in honor of Amadeo Avogadro, an Italian lawyer and physicist (1766–1856) who conceived the basic idea (see *Avogadro and His Number*).

The mole is the chemist's six-pack or dozen. Many objects in our everyday life come in similar counting units. Shoes, socks, and gloves are sold by the pair, soft drinks by the six-pack, and eggs and donuts by the dozen. Atoms are used by the mole. Thus, the mole is just a counting unit.

The atomic mass scale is a relative scale. Since the ^{12}C atom is chosen as the standard, then one can say, as we did in Section 2.2, that an oxygen atom is 16.0/12.0 or 1.33 times more massive than a carbon atom. Similarly, a fluorine atom is 19.0/12.0 or 1.58 times more massive than a carbon atom. Since *a mole always has the same number of particles*, this means that a mole of oxygen atoms is always 1.33 times more massive than a mole of carbon atoms, and a mole of fluorine atoms is 1.58 times more massive than a mole of carbon atoms. As is shown in the following box, the outcome is that the mass of 1.00 mole of O atoms is 16.0 g. Similarly, the mass of 1.00 mole of F atoms is 19.0 g.

RELATIVE ATOMIC MASSES

Since the ratio of O and C atom masses in amu is

$$\frac{\text{Mass of one O atom}}{\text{Mass of one C atom}} = \frac{16.0 \text{ amu}}{12.0 \text{ amu}} = 1.33$$

the ratio of the masses of 1 mole of O and C must also be 1.33.

$$\frac{\text{Mass of 1.00 mole of O atoms}}{\text{Mass of 1.00 mole of C atoms}} = \frac{\text{mass of } 6.022 \times 10^{23} \text{ atoms of O}}{\text{mass of } 6.022 \times 10^{23} \text{ atoms of C}} = 1.33$$

Since 1.00 mole of carbon is defined as 12.0 g, this means that 1 mole of O must be 16.0 g.

$$\frac{\text{Mass of 1.00 mole of O}}{\text{Mass of 1.00 mole of C}} = 1.33 = \frac{16.0 \text{ g/mol}}{12.0 \text{ g/mol}}$$

The mass of 1 mole of atoms of one kind (6.022×10^{23} atoms) is called the **molar mass**, and it is numerically equal to the atomic weight in atomic mass units (amu). Thus, the molar mass of oxygen (O) is 16.0 g/mol, while that of fluorine (F) is 19.0 g/mol and that of lead (Pb) is 207.2 g/mol. The dimensional units of molar mass are "g/mol."

$$\text{Molar mass of O} = \text{mass of 1 mole of O atoms} = 6.022 \times 10^{23}\ \text{O atoms} = 16.0\ \text{g/mol}$$

$$\text{Molar mass of F} = \text{mass of 1 mole of F atoms} = 6.022 \times 10^{23}\ \text{F atoms} = 19.0\ \text{g/mol}$$

$$\text{Molar mass of Pb} = \text{mass of 1 mole of Pb atoms} = 6.022 \times 10^{23}\ \text{Pb atoms} = 207.2\ \text{g/mol}$$

Figure 2.2 shows 1 mole quantities of some common elements. Although each of these "piles of atoms" has a different volume and a different mass, each contains 6.022×10^{23} atoms.

EXERCISE 2.5 ATOMIC WEIGHT AND MOLAR MASS

(a) What is the atomic weight of manganese? What is the mass of 1.00 mole of this element? (b) What is the mass of 6.022×10^{23} atoms of silicon? (c) How many atoms are there in 196.97 g of gold?

According to John Dalton, atoms combine with one another in the ratio of small whole numbers. A modern interpretation of this observation is that atoms combine with one another 1 mole for 1 mole, or 1 mole for 2 moles, or 2 moles for 3 moles, and so on. This is a matter taken up in Chapters 3 and 4. The point we want to make here is that the *mole concept is the cornerstone of quantitative chemistry*. To deal with this, you must learn to make the conversions "moles → mass" or "mass → moles" easily. Dimensional analysis (Section 1.6) tells you that this may be done in the following way:

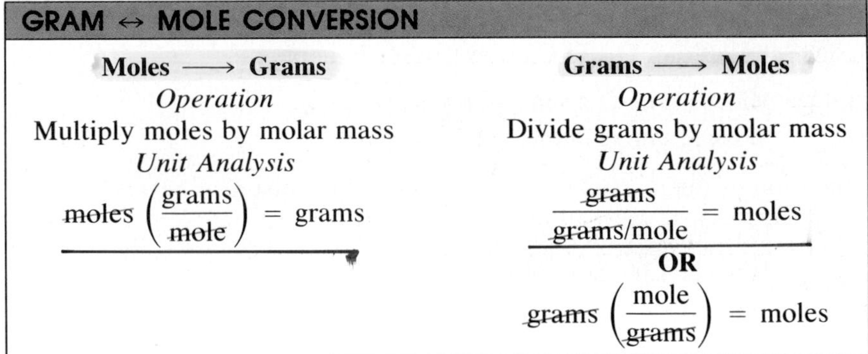

GRAM ↔ MOLE CONVERSION

Moles ⟶ Grams	**Grams ⟶ Moles**
Operation	*Operation*
Multiply moles by molar mass	Divide grams by molar mass
Unit Analysis	*Unit Analysis*
$\cancel{\text{moles}}\left(\dfrac{\text{grams}}{\cancel{\text{mole}}}\right) = \text{grams}$	$\dfrac{\cancel{\text{grams}}}{\cancel{\text{grams}}/\text{mole}} = \text{moles}$
	OR
	$\cancel{\text{grams}}\left(\dfrac{\text{mole}}{\cancel{\text{grams}}}\right) = \text{moles}$

EXAMPLE 2.4

MOLES TO GRAMS

How many grams are contained in 2.50 moles of sulfur (S)?

Solution For a conversion between mass and moles, you will always need the molar mass. The tables in the front of the book give 32.06 g/mol for S. Thus, the number of grams of sulfur in 2.50 moles is

$$2.50\ \cancel{\text{mol}}\ \text{sulfur}\left(\frac{32.06\ \text{g}}{1\ \cancel{\text{mol}}}\right) = 80.2\ \text{g sulfur}$$

FIGURE 2.2
One mole quantities of some common elements. The two "copper-colored" metal sheets at the rear together represent 1 mole of elemental copper (Cu, 63.54 g). On the left, in the middle row, is a ball of 1 mole of aluminum foil (Al, 27.0 g), and on the right, in the middle, is a 1 mole pile of magnesium chips (Mg, 24.3 g). On the left in the front is a pile of yellow sulfur (32.1 g), and on the right in the front is 1 mole of sodium lumps (23.0 g). In the middle of the photo is a 25 mL Erlenmeyer flask containing 1 mole of liquid mercury (Hg, 200.6 g). (Richard Roese)

EXAMPLE 2.5

GRAMS TO MOLES

How many moles are represented by 1.00 pound of silicon, an element used in semiconductors? (1.00 pound = 454 g)

Solution The molar mass of silicon is 28.1 g/mol, so the number of moles of silicon is

$$454\,\cancel{\text{g}}\text{ Si}\left(\frac{1\text{ mol}}{28.1\,\cancel{\text{g}}}\right) = 16.2\text{ mol silicon}$$

EXERCISE 2.6 GRAM/MOLE CONVERSIONS

(a) How many grams are contained in 2.5 moles of aluminum? (b) How many moles are represented by 1.00 pound of lead?

Once the number of moles represented by a given mass of element is known, you can then determine the number of atoms contained in the sample.

EXAMPLE 2.6

THE NUMBER OF ATOMS IN A SAMPLE

How many atoms of silicon are there in 1.00 pound of silicon?

Solution In Example 2.5 you found that 1.00 pound of silicon (454 g) represents 16.2 moles of silicon. Therefore, the number of atoms of silicon is

$$16.2\,\cancel{\text{mol}}\text{ Si}\left(\frac{6.022\times10^{23}\text{ atoms Si}}{1\,\cancel{\text{mol}}\text{ Si}}\right) = 9.76\times10^{24}\text{ atoms Si}$$

EXERCISE 2.7 THE NUMBER OF ATOMS IN A SAMPLE

How many atoms of uranium are there in 50.0 g of the element?

Now that you are more familiar with the mole concept and with Avogadro's number, it is interesting to see what one atom weighs. The following example shows you that it is indeed an extremely small mass, just as you have suspected.

EXAMPLE 2.7

THE MASS OF AN ATOM

Calculate the mass of 1 atom of mercury.

Solution You know that 1 mole of mercury has a mass of 200.6 g and that this represents 6.022×10^{23} atoms. Therefore, you know the relation between mass and number of atoms.

$$\left(\frac{200.6 \text{ g Hg}}{1 \cancel{\text{mol}} \text{ Hg}}\right)\left(\frac{1 \cancel{\text{mol}} \text{ Hg}}{6.022 \times 10^{23} \text{ atoms Hg}}\right) = \frac{3.33 \times 10^{-22} \text{ g}}{1 \text{ atom of Hg}}$$

The mass of one atom is indeed incredibly small!

EXERCISE 2.8 THE MASS OF AN ATOM

(a) Calculate the mass of 1 hydrogen atom. (b) The mass of 1 mercury atom was found in Example 2.7. What is the ratio (mass of 1 Hg atom/mass of 1 H atom)? Did you need to calculate the mass of 1 atom of each of these elements to find this ratio? Why or why not?

2.6 THE PERIODIC TABLE OF THE ELEMENTS

The periodic table of elements given in Figure 2.3 and in the front of the book contains a wealth of information. Since we shall refer to it often, we want to introduce you to some of its main features and terminology. The historical development of the periodic table and periodic law are described in more detail in Chapter 7, and much of this textbook is devoted to examining the chemical and physical properties of the elements and their interrelationships.

The elements are arranged in the table in order of increasing atomic number in such a way that elements having similar chemical and physical properties lie in vertical columns called **groups**. The groups are numbered 1 through 8, and each has a letter A or B. The horizontal rows of the table are called **periods**, and they are numbered beginning with 1 for the period containing only H and He.

Photographs of some of the elements appear in Figure 2.4

The table can be divided into several regions. There are **metals** (violet, blue, or green in Figure 2.3), **nonmetals** (in red), and elements that fit neither of these categories and so are called **metalloids** (in yellow). Although we shall carefully describe the difference between metals and nonmetals later, you should be familiar with it from your everyday experience. Automobiles are made of iron (Fe) and aluminum (Al), both metals. Oxygen and nitrogen are the gases you breathe and are nonmetals. All the nonmetals in the periodic table lie to the right of the heavy zigzag line (between Al and Si, Ge and As, Sb and Te, and Po and At). Most

Alkali and alkaline earth metals

Groups →

Periods ↓

Transition metals

Metalloids

Main group metals

Nonmetals

1A	2A	3B	4B	5B	6B	7B	8B	8B	8B	1B	2B	3A	4A	5A	6A	7A	8A
																1 H	2 He
3 Li	4 Be											5 B	6 C	7 N	8 O	9 F	10 Ne
11 Na	12 Mg											13 Al	14 Si	15 P	16 S	17 Cl	18 Ar
19 K	20 Ca	21 Sc	22 Ti	23 V	24 Cr	25 Mn	26 Fe	27 Co	28 Ni	29 Cu	30 Zn	31 Ga	32 Ge	33 As	34 Se	35 Br	36 Kr
37 Rb	38 Sr	39 Y	40 Zr	41 Nb	42 Mo	43 Tc	44 Ru	45 Rh	46 Pd	47 Ag	48 Cd	49 In	50 Sn	51 Sb	52 Te	53 I	54 Xe
55 Cs	56 Ba	57 *La	72 Hf	73 Ta	74 W	75 Re	76 Os	77 Ir	78 Pt	79 Au	80 Hg	81 Tl	82 Pb	83 Bi	84 Po	85 At	86 Rn
87 Fr	88 Ra	89 †Ac	104 Rf	105 Ha													

*Lanthanide series	58 Ce	59 Pr	60 Nd	61 Pm	62 Sm	63 Eu	64 Gd	65 Tb	66 Dy	67 Ho	68 Er	69 Tm	70 Yb	71 Lu
†Actinide series	90 Th	91 Pa	92 U	93 Np	94 Pu	95 Am	96 Cm	97 Bk	98 Cf	99 Es	100 Fm	101 Md	102 No	103 Lw

FIGURE 2.3
The periodic table. Metals are shown in blue or green, nonmetals in red, and metalloids in yellow.

FIGURE 2.1 SOME OF THE ELEMENTS OF THE PERIODIC TABLE

(Charles D. Winters)

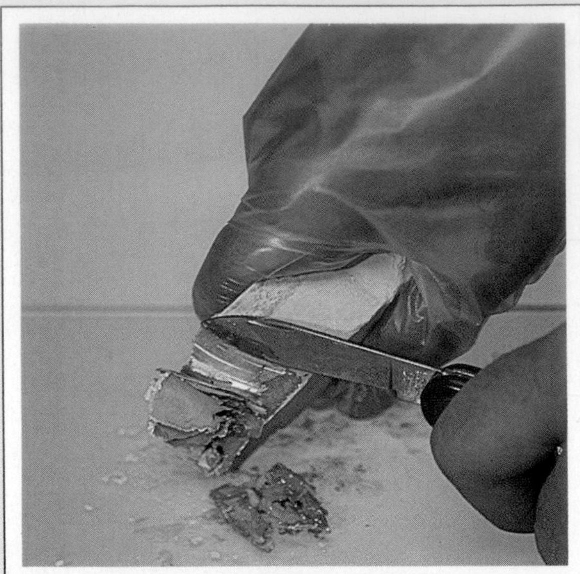

Group 1A: Sodium (Na).

Group 2A: Magnesium (Mg) (left) and calcium (Ca) (right).

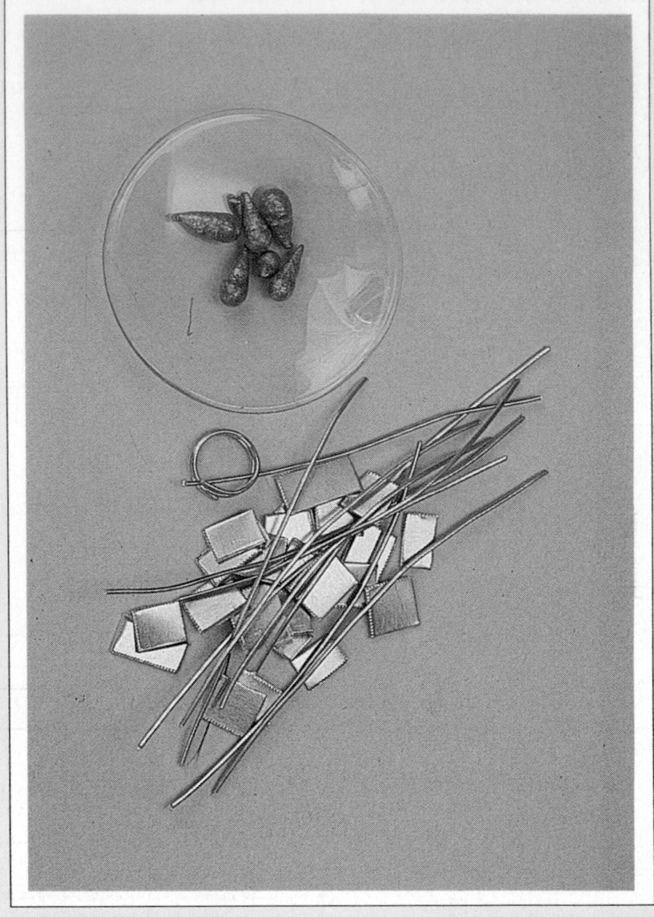

Group 3A: Indium (In) (top) and aluminum (Al) (bottom).

Some fourth period transition metals (left to right) Ti, V, Cr, Mn, Fe, Co, Ni, Cu.

Group 2B: Zinc (Zn) (left) and mercury (Hg) (right).

Group 4A: Carbon (C) (bottom), silicon (Si) (left middle), tin (Sn) (right middle), lead (Pb) (top).

Group 5A: Nitrogen (N_2) [liquid N_2].

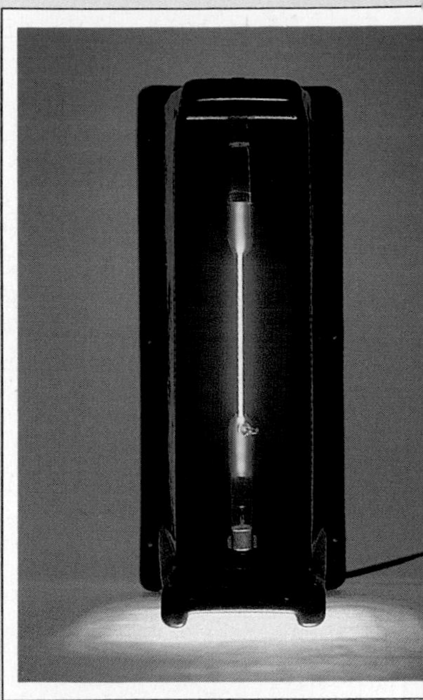

Group 8A: Neon (Ne).

Group 5A: White phosphorus (P_4).

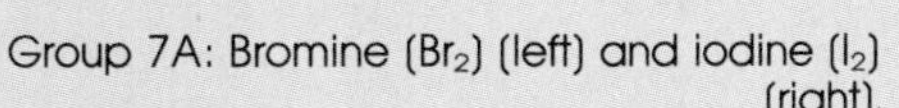

Group 7A: Bromine (Br_2) (left) and iodine (I_2) (right). ►

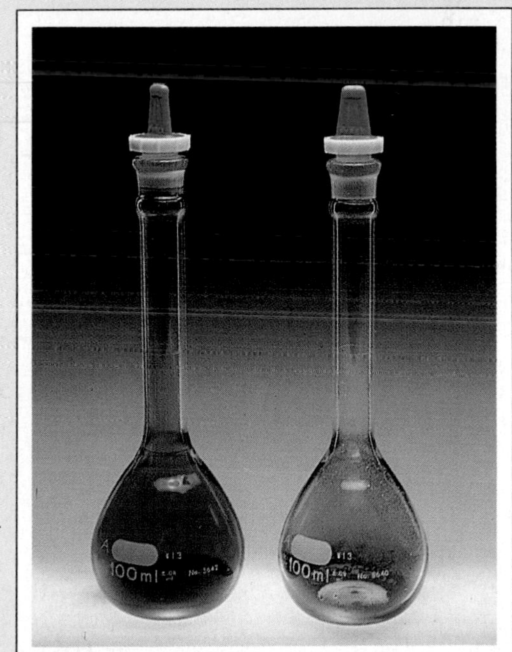

Group 5A: Arsenic (As) (left), antimony (Sb) (right), bismuth (Bi) (top).

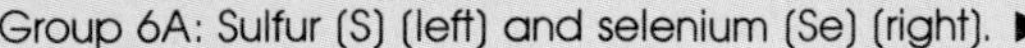

Group 6A: Sulfur (S) (left) and selenium (Se) (right). ►

of the elements lying immediately next to this line have some properties that can be thought of as metallic but others that are considered characteristic of nonmetals. Such elements (B, Si, Ge, As, Sb, and Te) are called metalloids.

It is the property of being neither metal nor nonmetal that gives the **metalloids** the important electrical properties that make them semiconductors, the heart of integrated circuits.

Many of the groups of the periodic table or collections of groups have meaningful names, and you should learn them.

GROUP 1A: ALKALI METALS

The ancient Arab alchemists studied the properties of many natural substances and found that the ashes, called *al-qali*, of certain plants gave water solutions that felt slippery on the hands. It is now known that these substances contain compounds of Group 1A elements. Since these elements are also metals, they have come to be called the "alkali metals" (Figure 2.4).

The lightest element of the alkali metals, lithium, is not common, but a trace of lithium is important to your mental health; lithium deficiency results in manic depressive behavior. In contrast with lithium, sodium is the seventh most abundant element in the earth's crust and potassium is the eighth most abundant. The heaviest element, francium, is radioactive and has only a fleeting existence in nature.

GROUP 2A: ALKALINE EARTH METALS

These elements can also form substances that are alkaline in water solution. The lightest element, beryllium, is very low in abundance, and its compounds can be carcinogenic. Again like Group 1A, the next two elements, magnesium and calcium, are much more abundant; they are the sixth and fifth most abundant elements in the earth's crust, respectively (Figure 2.5). Calcium is particularly well known, since it occurs in vast

(a)

(b)

FIGURE 2.5

Minerals containing alkaline earth metals. (a) Calcite, calcium carbonate. (b) Barite, barium sulfate. (a, Brian Parker; b, Allen B. Smith, Tom Stack & Associates)

deposits of limestone. The basic chemical constituent of limestone, calcium carbonate, is also the chief substance in corals, sea shells, and chalk. Radium, the heaviest element, is radioactive.

GROUPS 3B THROUGH 8B AND 1B AND 2B: TRANSITION ELEMENTS OR TRANSITION METALS

All the elements in these groups are shown in green in Figure 2.3. All are metals and range from some of the most abundant elements (iron is the second most abundant metal, after aluminum, and the fourth most abundant element) to some of the least abundant. All have unique properties. Silver, gold, and platinum are coveted for their beauty, as well as their economic value.

Included here are the "inner transition elements." The *lanthanide elements* fit into the table between lanthanum (La) and hafnium (Hf). The *actinide elements,* as their name implies, then fit between actinium (Ac) and element 104.

Gold was described in the Special Section at the end of Chapter 1, and platinum is described at the end of Chapter 13.

GROUP 3A

This group has no special name, but it contains aluminum, the most abundant metal in the earth's crust (8.3%). Aluminum is exceeded only in abundance by the nonmetals O (45.5%) and Si (25.7%). These three elements are found combined in clays and other common minerals.

GROUP 4A

The lightest element of this group is carbon, an element found uncombined as graphite and diamonds (Figure 2.6). In combined form, it is the basis of the substances of which we are all made. It is 17th in abundance in the earth's crust and is found in the crust mainly in the form of carbonates (Figure 2.5).

Silicon is the second most abundant element in the earth's crust, occurring mainly in combination with oxygen. Together, silicon and oxy-

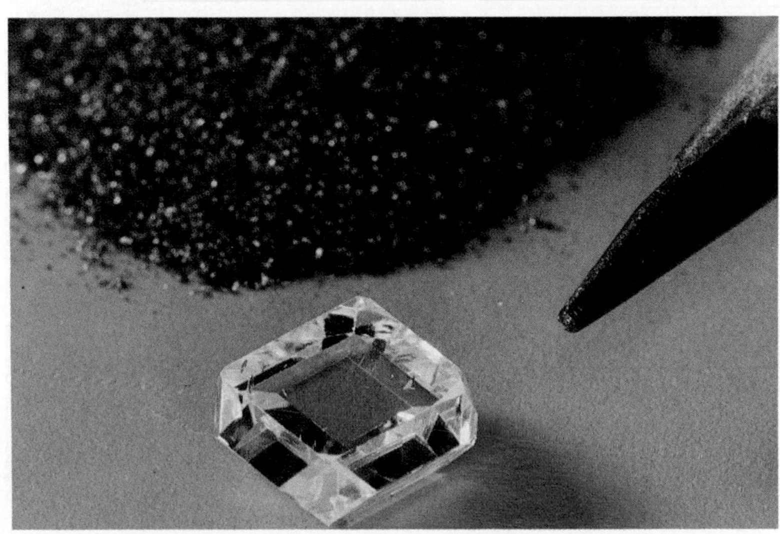

FIGURE 2.6

An artificial diamond, a pure form of carbon, is seen at the bottom. Graphite, another form of carbon, is at the top. It is one constituent of pencil lead. (General Electric)

FIGURE 2.7

Amethyst, an impure form of silicon dioxide. (Allen B. Smith, Tom Stack & Associates)

gen comprise four out of five of all atoms available near the surface of the earth! You know it in the form of silicon-oxygen materials such as sand, quartz, clay, and mica (Figure 2.7).

As in the previous groups, the heaviest elements, germanium, tin, and lead, are the least abundant. However, tin and lead have been known for centuries, since they are easily obtained from their ores. Tin, when mixed with copper, makes bronze, a material used for utensils and weapons for centuries, and the Romans used lead extensively for water pipes and plumbing.

GROUP 5A

Once again, there is no common name for this group. The lightest element, nitrogen, comprises about three fourths of the atmosphere, but in the earth's crust it is only about as abundant as gallium. Phosphorus has a fascinating chemistry and is essential to life as a key element in teeth and bones (Figure 2.8).

FIGURE 2.8

Apatite, a mineral containing calcium, fluorine, and phosphorus. It is also a major constituent of teeth and bones. (Bill Tronca, Tom Stack & Associates)

GROUP 6A

Oxygen is the lightest element of this group and is the key to life as we know it on this planet. Sulfur has been known since ancient times as brimstone, the "burning stone" of the Devil. The element has enormous economic importance largely because it is the basis of sulfuric acid, the chemical produced in largest amount by industry.

GROUP 7A: HALOGENS

The name of this group of elements comes from the Greek words *hals* for "salt" and *genes* for "forming." That is, the elements are salt forming, and the best known of these is common table salt, sodium chloride. There are many others, however, because these elements react with all the metals and nearly all the other elements of the periodic table.

It is a common error to misspell the name of the element F. The name is spelled f-l-u-o . . . and not f-l-o-u. . . . If the name were spelled "flourine" it would be pronounced "flower-ine."

You may know fluorine as part of the compounds added to toothpaste and to many municipal water supplies and as an important constituent of Teflon.

Chlorine is the most abundant of the elements of this group and has been known since earliest times in the form of common sea salt. The remaining elements, bromine, iodine, and astatine, are much lower in abundance. Even so, iodine is essential to a hormone that regulates growth, the reason that table salt is "iodized." Astatine, like many of the heaviest elements, is radioactive and occurs in negligible amounts in nature.

GROUP 8A: THE RARE GASES

As their name implies, these elements, helium, neon, argon, krypton, xenon, and radon, are all gases and all are low in their abundance on the earth. They are often called "inert" gases or "noble" gases, although krypton and xenon have recently been found to be reactive toward certain other elements.

2.7 COMPOUNDS, MOLECULES, AND MOLECULAR FORMULAS

John Dalton said that **compounds** form by the combination of atoms in the ratio of small whole numbers. Now we know that the smallest unit of a compound that retains the chemical characteristics of the compound is a **molecule**. The composition of a molecule can be represented by a **molecular formula**, which expresses the number of atoms of each type within one molecule of the compound.

When compounds are formed directly from elements or from other compounds, one striking feature is that the characteristics of the constituent elements are lost. This, and the concept of the molecular formula, is illustrated by the reaction in Figure 2.9. Red phosphorus reacts violently with bromine, a foul-smelling, red-orange liquid, to give phosphorus tribromide, a colorless liquid. The formula of this compound, PBr_3, conveys the fact that there are four atoms per formula unit: one atom of phosphorus and three atoms of bromine. The subscript to the right of the element's symbol indicates the number of atoms of that element in the molecule. If the subscript is omitted, it is understood to be one, as for P in PBr_3. Similarly, in the water molecule, H_2O, there is 1 atom of oxygen for every 2 atoms of hydrogen.

Consider the molecular formula of a compound such as ammonium phosphate, a common fertilizer. We could write its formula with the elements in alphabetic order as $H_{12}N_3O_4P$, but you will more commonly see it as $(NH_4)_3PO_4$. This latter version shows you immediately that the compound is composed of two different types of "chemical units," NH_4 and PO_4, and that they are present in the ratio 3 NH_4 to 1 PO_4.* There are many common "units" of atoms in chemistry, and we shall always keep the atoms of those units together when writing formulas. Section 2.9 considers this is more detail.

The formula of ethyl alcohol is written below in two different ways.

C_2H_6O	CH_3CH_2OH
molecular formula	**modified molecular formula**

*In Section 2.9 you will see that these "units" are in fact the ions NH_4^+ and PO_4^{3-}.

(a)

(b)

FIGURE 2.9

The reaction of phosphorus and bromine. (a) Solid red phosphorus (in the evaporating dish) and liquid bromine, Br_2 (in the graduated cylinder). (b) When the bromine is poured onto the phosphorus, a vigorous reaction produces phosphorus tribromide, PBr_3. (Charles D. Winters)

In the **molecular formula**, symbols of the elements are written in alphabetical order, each with a subscript indicating the total number of atoms of that type in the molecule.* In the **modified molecular formula**, the atoms are written in groups. Chemists often write formulas in this manner to show the sequence of atoms in the molecule or to emphasize the chemically important or "functional group" (OH in ethyl alcohol).

EXERCISE 2.9 WRITING FORMULAS

Write molecular formulas for the following compounds: (a) A molecule of hydrazine, a rocket fuel, has two atoms of nitrogen and four of hydrogen. (b) A molecule of oxalic acid, a compound found in many plants, has two carbon atoms, two hydrogen atoms, and four oxygen atoms. (c) A molecule of ferrocene has ten atoms of carbon, ten atoms of hydrogen, and one atom of iron.

Nonmetals

3A	4A	5A	6A	7A	8A
				1 H 1.00797	2 He 4.0026
5 B 10.811	6 C 12.01115	7 N 14.0067	8 O 15.9994	9 F 18.9984	10 Ne 20.183
13 Al 26.9815	14 Si 28.086	15 P 30.9738	16 S 32.064	17 Cl 35.453	18 Ar 39.948
31 Ga 69.72	32 Ge 72.59	33 As 74.9216	34 Se 78.96	35 Br 79.909	36 Kr 83.80
49 In 114.82	50 Sn 118.69	51 Sb 121.75	52 Te 127.60	53 I 126.9044	54 Xe 131.30
81 Tl 204.37	82 Pb 207.19	83 Bi 208.980	84 Po (210)	85 At (210)	86 Rn (222)

FIGURE 2.10
Elements in blue exist as diatomic molecules in the elemental form, while phosphorus consists of P_4 molecules (yellow), sulfur as S_8 molecules (red), and carbon (graphite and diamond) as networks of carbon atoms (gray).

2.8 ELEMENTS AND MOLECULES

All but the heaviest of the elements have been isolated in large amounts in pure form. As you can see in Figures 2.3 and 2.4, most elements are metals and most of these are solids. In the solid state, metals consist of atoms packed together as closely as possible. On the other hand, nonmetals are often gases, liquids, or solids consisting of discrete atoms or molecules. The rare gases, for example, are found as uncombined atoms in nature. In contrast, hydrogen, nitrogen, oxygen, and the halogens are all *diatomic* or "two-atom" molecules in their elemental form (Figure 2.10).

Phosphorus, sulfur, and carbon are also interesting elements to consider. Elemental phosphorus can be found as a tetratomic or "four-atom" molecule P_4, while elemental sulfur exists as the S_8 molecule. Finally, pure elemental carbon is found in two forms: diamond and graphite. In both of these, there are extended networks of carbon atoms, as you shall see in Chapter 11.

2.9 IONS

A **cation** is a positively charged ion. The word is pronounced cat-ion.

Atoms of almost all the elements can gain or lose electrons in ordinary chemical reactions to form **ions**. Indeed, a characteristic of *metals* is that metal atoms easily lose electrons to form ions with a positive electrical charge, ions commonly called **cations**. In the reaction below, a lithium atom, which has three protons and three electrons, loses one of its electrons to form the Li^+ cation. The superscript $^+$ indicates a positive electric charge on the ion and means that there is one more (positive) proton in the atomic nucleus than there are (negative) electrons about the nucleus.

*In the chemical literature a convention you will often see is that, for molecules containing C, H, and other elements, C is written first in the molecular formula and then H; the other elements follow in alphabetical order.

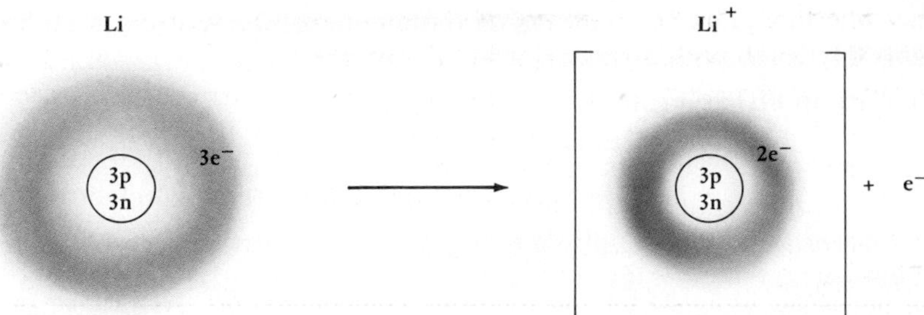

When more than one electron is lost, the number is written with the charge sign; for example, when calcium ion is formed by loss of two electrons, the symbol is Ca^{2+}.

In contrast with metals, *nonmetals* frequently gain electrons to give ions with a negative electrical charge. Such ions are commonly called **anions**. In the reaction illustrated below, the fluorine atom with 9 protons and 9 electrons has gained an electron to give F^-, an anion having 9 protons and 10 electrons. There are several useful guidelines for predicting the likely number of electrons lost by a metal or gained by a nonmetal.

An **anion** is a negatively charged ion. The word is pronounced **ann**-ion.

An electron is represented by e^-.

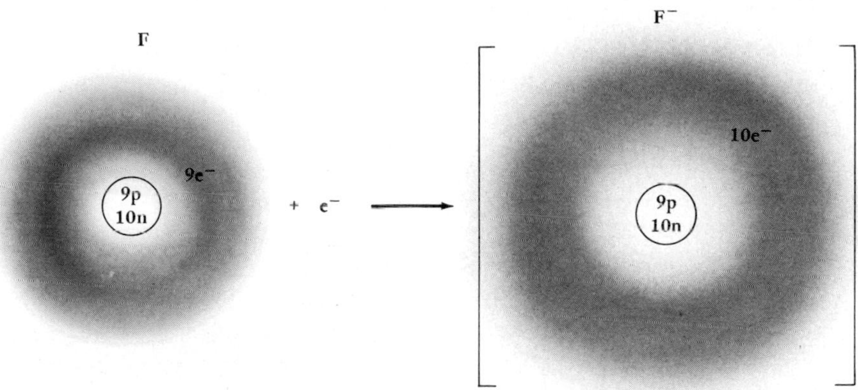

CHARGES OF MONATOMIC IONS

1. For the metals of Groups 1A through 3A, for 1B and 2B, and for the metals of Group 4A, the number of electrons lost is equal to their group number. Therefore, these metals form positive ions whose charge is equal to the group number of the metal.

   Na (Group 1A)(11 protons, 11 electrons) $\longrightarrow$ e^- + Na^+ (11 protons, 10 electrons)

   Ca (Group 2A)(20 protons, 20 electrons) $\longrightarrow$ $2e^-$ + Ca^{2+} (20 protons, 18 electrons)

   Al (Group 3A)(13 protons, 13 electrons) $\longrightarrow$ $3e^-$ + Al^{3+} (13 protons, 10 electrons)

2. The transition metals of Groups 3B through 8B also form cations, but it is more difficult to predict the number of electrons lost. For

the time being, however, a useful rule of thumb for these metals is that they form ions with a +2 or +3 charge.

Fe (Group 8B)(26 protons, 26 electrons)

$$\longrightarrow 2e^- + Fe^{2+} \text{ (26 protons, 24 electrons)}$$

$$\longrightarrow 3e^- + Fe^{3+} \text{ (26 protons, 23 electrons)}$$

3. The maximum number of electrons gained by a nonmetal atom is equal to 8 minus the group number of the element. For example, nitrogen is in Group 5A and so can gain 8 − 5 = 3 electrons to form a −3 ion.

$$\text{N (Group 5A)(7 protons, 7 electrons)} + 3e^- \longrightarrow N^{3-} \text{ (7 protons, 10 electrons)}$$

The ions O^{2-}, S^{2-}, F^-, Cl^-, Br^-, and I^- in particular are common in naturally occurring compounds and are listed in Table 2.4 with other anions and their names. A negative ion's name is formed from the root of the element's name plus the suffix *-ide*.

4. Hydrogen can either gain or lose electrons, depending on the other elements it encounters.

$$\text{H (1 proton, 1 electron)} \xrightarrow{-e^-} H^+ \text{ (1 proton, 0 electrons)}$$

$$\text{H (1 proton, 1 electron)} \xrightarrow{+e^-} H^- \text{ (1 proton, 2 electrons)}$$

H^+ is called simply the "hydrogen ion," while H^- is named "hydride."

5. The rare gases do not lose or gain electrons, except in unusual cases.

EXERCISE 2.10 PREDICTING ION CHARGES

Using the guidelines above, predict possible charges for the ions formed from the following elements:

(a) K (b) Se (c) Be (d) V (e) Co (f) Cs

The ions we have described so far are **monatomic**, they consist of one atom bearing a charge. If the ion is positive, it has the name of the metal from which it is derived. In cases where a metal may have more than one positive charge, we indicate the charge with a Roman numeral. For example,

ION	ION NAME
Na^+	sodium ion
Ca^{2+}	calcium ion
Fe^{2+}	iron(II) ion
Fe^{3+}	iron(III) ion

Common monatomic negative ions are named in Table 2.4.

Many other common ions in chemistry are groups of atoms bound together. Ions such as SO_4^{2-}, OCl^-, NH_4^+, $C_2H_3O_2^-$, and MnO_4^- are called **polyatomic** or many-atom ions (Table 2.5). In a polyatomic ion the superscript is the charge on the *group* of atoms. Thus, a carbonate ion, CO_3^{2-}, has a −2 charge on a group of four atoms.

The naming of ions and compounds is discussed in more detail in Appendix B.

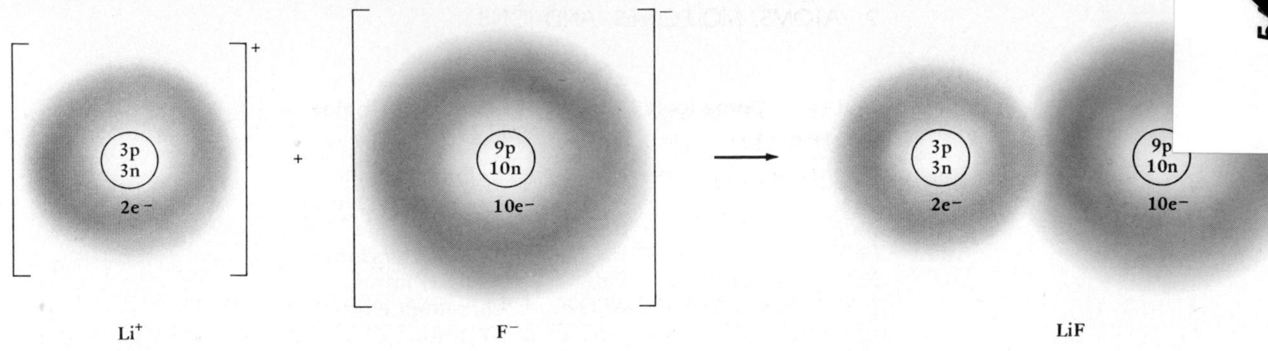

COMPOUNDS FORMED FROM IONS

Opposite electrical charges attract, so a positive ion (cation) and negative ion (anion) attract one another to form an **ionic compound**. When the number of positive charges carried by the cation (or cations) in a compound is equal to the number of negative charges carried by the anion (or anions), then an electrically neutral or uncharged compound is formed. For example, a calcium ion bearing a +2 charge can interact with anions of −1, −2, or −3 charge in the following ways.

ION COMBINATION	PRODUCT	OVERALL CHARGE ON PRODUCT
Ca^{2+} + 2 Cl^-	$CaCl_2$	0 = (+2) + (2 × −1)
Ca^{2+} + CO_3^{2-}	$CaCO_3$	0 = (+2) + (−2)
3 Ca^{2+} + 2 PO_4^{3-}	$Ca_3(PO_4)_2$	0 = (3 × +2) + (2 × −3)

TABLE 2.4 Anions of Nonmetals

4A	5A	6A	7A
C^{4-} carbide	N^{3-} nitride	O^{2-} oxide	F^- fluoride
	P^{3-} phosphide	S^{2-} sulfide	Cl^- chloride
		Se^{2-} selenide	Br^- bromide
		Te^{2-} telluride	I^- iodide

TABLE 2.5 Names and Composition of Some Common Polyatomic Ions

CATIONS: Positive Ions			
NH_4^+	ammonium		
ANIONS: Negative Ions			
OH^-	hydroxide	NO_2^-	nitrite
CN^-	cyanide	NO_3^-	nitrate
$C_2H_3O_2^-$	acetate		
		SO_3^{2-}	sulfite
CO_3^{2-}	carbonate	SO_4^{2-}	sulfate
HCO_3^-	hydrogen carbonate (or bicarbonate)	HSO_4^-	hydrogen sulfate (or bisulfate)
PO_4^{3-}	phosphate		
MnO_4^-	permanganate		
$Cr_2O_7^{2-}$	dichromate		
ClO_4^-	perchlorate		

TABLE 2.6 Some Ionic Compounds and Their Names

COMPOUND			NAME
$Na^+ + Br^-$	$\longrightarrow$	NaBr	Sodium bromide
$NH_4^+ + Cl^-$	$\longrightarrow$	NH_4Cl	Ammonium chloride
$Ba^{2+} + 2\ NO_3^-$	$\longrightarrow$	$Ba(NO_3)_2$	Barium nitrate
$Fe^{2+} + 2\ Cl^-$	$\longrightarrow$	$FeCl_2$	Iron(II) chloride
$Fe^{3+} + 3\ NO_3^-$	$\longrightarrow$	$Fe(NO_3)_3$	Iron(III) nitrate
$2\ Al^{3+} + 3\ O^{2-}$	$\longrightarrow$	Al_2O_3	Aluminum oxide

Some other examples of compounds formed by the combination of positive and negative ions are given in Table 2.6, and each is given with its proper name. Notice that the name of each ionic compound is simply a combination of the appropriate positive and negative ion names; the positive ion is named first and is shown first in the formula. Further details on naming compounds are given in Appendix B.

As you will see in the course of this book, a compound formed of ions often has unique properties. Many of these arise from the fact that there is no such thing as an "ionic molecule," in the sense that there is a single unit of NaCl or NH_4Cl as there is a molecule of H_2 or S_8. Rather, ionic compounds commonly exist as extended three-dimensional networks of positive and negative ions; it is impossible to say a certain anion/cation pair belong together to form one molecule. This is illustrated by the model of the structure of common table salt, NaCl, in Figure 2.11. Each Na^+ ion is surrounded by six Cl^- ions, and each Cl^- is in turn surrounded by six Na^+ ions.

It is important that you be able to recognize the ions and their relative numbers in the formula of an ionic compound. The following example illustrates how to do this.

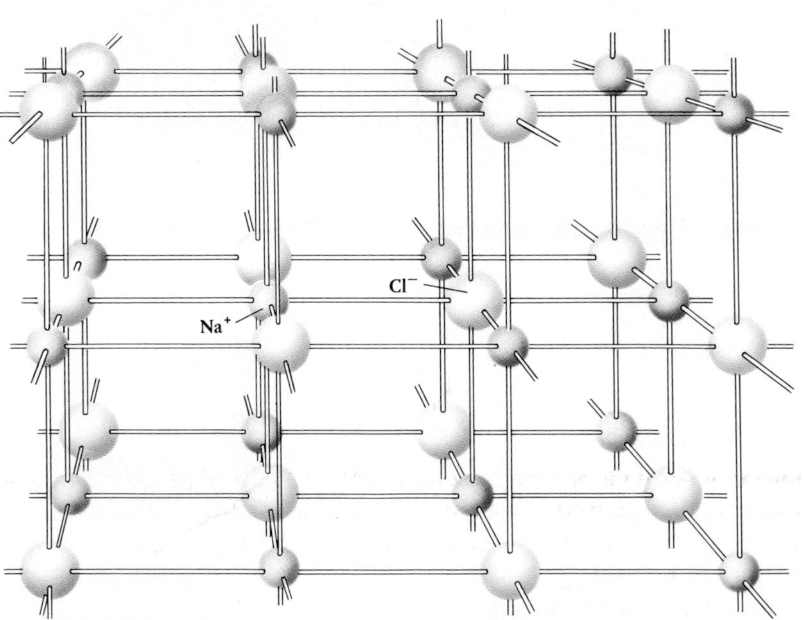

FIGURE 2.11

A model of a crystal of sodium chloride. The lines between ions are not chemical bonds but are simply reference lines showing the relationship of Na^+ (red) and Cl^- (green) in space.

EXAMPLE 2.8

IONIC COMPOUNDS

In each of the following ionic compounds, tell what ions are present, give their relative number and give the name of the compound:

(a) $MgBr_2$ (b) Li_2CO_3 (c) $KMnO_4$

Solution (a) $MgBr_2$ is composed of one magnesium ion, Mg^{2+}, and two bromide ions, Br^-. Its name is magnesium bromide.

$$Mg^{2+} + 2\ Br^- \longrightarrow MgBr_2$$

When a halogen, such as bromine, is associated only with a metal, you can assume the halogen is an anion with a single negative charge. Magnesium is an element of Group 2A and *always* has a charge of +2 in its compounds.

(b) Li_2CO_3 is composed of two lithium ions, Li^+, and one carbonate ion, CO_3^{2-}. Its name is lithium carbonate.

$$2\ Li^+ + CO_3^{2-} \longrightarrow Li_2CO_3$$

Here you must learn to recognize a carbonate ion and recall its formula and charge. To help remember that carbonate is a 2− anion, you can see that lithium is a Group 1A ion and so *always* has a charge of +1 in its compounds.

(c) $KMnO_4$ is composed of one potassium ion, K^+, and one permanganate ion, MnO_4^-. Its name is potassium permanganate.

$$K^+ + MnO_4^- \longrightarrow KMnO_4$$

Potassium is a Group 1A element and so *always* has a charge of +1 in its compounds. You should learn to recognize the permanganate ion, since it is commonly used in the laboratory.

EXERCISE 2.11 IONIC COMPOUNDS

In each of the following ionic compounds, tell what ions are present, give their relative number, and give the name of the compound:

(a) NaF (b) $NaC_2H_3O_2$ (c) $Cu(NO_3)_2$

EXERCISE 2.12 IONIC COMPOUNDS

Give the formula and name of the ionic compounds formed in each of the following cases: (a) compound from ammonium and nitrate ions; (b) compound from cobalt(II) and sulfate ions; and (c) compound from nickel(II) and cyanide ions.

2.10 MOLAR MASS, MOLECULAR WEIGHT, AND FORMULA WEIGHT

The molecular formula of a compound tells you the type of atoms in the molecule and the quantity of each. For example, in one molecule of ammonia, NH_3, there is one atom of N and three atoms of H. If 100 atoms of N and 300 atoms of H are combined, then 100 molecules of NH_3 would form. Now suppose you combine

Avogadro's number of N atoms or 6.022×10^{23} atoms of N or 1 mole of N or 14.01 g of N	+	$3 \times$ Avogadro's number of H atoms or 18.066×10^{23} of H atoms or 3 moles of H atoms or 3.03 g of H atoms
equivalent quantities		equivalent quantities

↓

are equivalent to

Avogadro's number of molecules of NH_3
or
6.022×10^{23} molecules of NH_3
or
1 mole of NH_3
or
17.04 g of NH_3

Since Avogadro's number of particles is a **mole** (Section 2.5), the mass of 1.000 mole or Avogadro's number of ammonia molecules is 17.04 g. This mass is called the **molar mass**. For all substances, the molar mass in grams is numerically equivalent to the **molecular weight**, the sum of the atomic weights of all of the atoms of the molecule. The molecular weight is the average mass of a molecule of the substance, expressed in atomic mass units.

COMPOUND	MOLECULAR WEIGHT (amu)	MOLAR MASS (g/mol)	MASS OF ONE MOLECULE (g)
NH_3	17.04	17.04	2.830×10^{-23}
H_2O	18.02	18.02	2.992×10^{-23}
CH_2Cl_2	84.93	84.93	1.410×10^{-23}

Be sure to notice in the table above that the molecular weight is expressed in amu. To find the mass of one molecule in grams, you must divide the molar mass by Avogadro's number.

$$\left(\frac{17.04 \text{ g NH}_3}{1 \cancel{\text{mol}}}\right)\left(\frac{1 \cancel{\text{mol}}}{6.022 \times 10^{23} \text{ molecules}}\right) = \frac{2.830 \times 10^{-23} \text{ g}}{1 \text{ NH}_3 \text{ molecule}}$$

For ionic compounds such as NaCl that do not exist as individual molecules, it is more appropriate to refer to the sum of atomic masses as the **formula weight**. Thus, the formula weight of NaCl is 58.44 amu (22.99 amu for Na plus 35.45 amu for Cl).

Figure 2.12 is a photograph of 1 mole quantities of a range of common compounds. To find the molecular or formula weight of each of these compounds, and thus its molar mass, you need only to add up the atomic weights of each element in one molecule or one formula unit.

FIGURE 2.12
One mole quantities of a range of compounds. The white compound is NaCl (MW = 58.44 g/mol); the blue compound is $CuSO_4 \cdot 5\ H_2O$ (MW = 249.68 g/mol); the deep red compound is $CoCl_2 \cdot 6\ H_2O$ (MW = 237.95 g/mol); the green compound is $NiCl_2 \cdot 6\ H_2O$ (MW = 237.70 g/mol); and the orange compound is $K_2Cr_2O_7$ (MW = 294.19 g/mol). (Richard Roese)

EXAMPLE 2.9

MOLAR MASS

Calculate the molar mass of (a) ethyl alcohol, C_2H_6O; (b) ammonium sulfate, $(NH_4)_2SO_4$; and (c) barium chloride dihydrate, $BaCl_2 \cdot 2\ H_2O$.

Solution (a) Ethyl alcohol, C_2H_6O.

$$\text{2 moles of C per mole of alcohol} = 2\ \cancel{\text{mol C}}\left(\frac{12.01\text{ g C}}{1\ \cancel{\text{mol C}}}\right) = 24.02\text{ g C}$$

$$\text{6 moles of H per mole of alcohol} = 6\ \cancel{\text{mol H}}\left(\frac{1.01\text{ g H}}{1\ \cancel{\text{mol H}}}\right) = 6.06\text{ g H}$$

$$\text{1 mole of O per mole of alcohol} = 1\ \cancel{\text{mol O}}\left(\frac{16.00\text{ g O}}{1\ \cancel{\text{mol O}}}\right) = \underline{16.00\text{ g O}}$$

$$\text{Molar mass} = 46.08\text{ g/mol}$$

(b) Ammonium sulfate, $(NH_4)_2SO_4$. This compound contains two ammonium ions. Therefore, in one formula unit there are 2 nitrogen atoms and 8 hydrogen atoms.

$$\text{2 moles of N per mole sulfate} = 2\ \cancel{\text{mol N}}\left(\frac{14.01\text{ g N}}{1\ \cancel{\text{mol N}}}\right) = 28.02\text{ g N}$$

$$\text{8 moles of H per mole sulfate} = 8\ \cancel{\text{mol H}}\left(\frac{1.01\text{ g H}}{1\ \cancel{\text{mol H}}}\right) = 8.08\text{ g H}$$

$$\text{1 mole of S per mole sulfate} = 1\ \cancel{\text{mol S}}\left(\frac{32.06\text{ g S}}{1\ \cancel{\text{mol S}}}\right) = 32.06\text{ g S}$$

$$\text{4 moles of O per mole sulfate} = 4\ \cancel{\text{mol O}}\left(\frac{16.00\text{ g O}}{1\ \cancel{\text{mol O}}}\right) = \underline{64.00\text{ g O}}$$

$$\text{Molar mass} = 132.2\text{ g/mol}$$

(c) A compound having water as an integral part of the solid is called a "hydrate." Thus, the formula $BaCl_2 \cdot 2\ H_2O$ tells us there are *two* molecules of water associated with every unit of $BaCl_2$. This means that each formula unit contains one Ba^{2+} ion, 2 Cl^- ions, *four* H atoms, and *two* O atoms.

$$1 \text{ mole of } Ba^{2+} \text{ ions} = 1 \cancel{\text{mol } Ba^{2+}} \left(\frac{137.3 \text{ g } Ba^{2+}}{1 \cancel{\text{mol } Ba^{2+}}}\right) = 137.3 \text{ g } Ba^{2+}$$

$$2 \text{ moles of } Cl^- \text{ ions} = 2 \cancel{\text{mol } Cl^-} \left(\frac{35.45 \text{ g } Cl^-}{1 \cancel{\text{mol } Cl^-}}\right) = 70.90 \text{ g } Cl^-$$

$$4 \text{ moles of H atoms} = 4 \cancel{\text{mol H}} \left(\frac{1.008 \text{ g H}}{1 \cancel{\text{mol H}}}\right) = 4.032 \text{ g H}$$

$$2 \text{ moles of O atoms} = 2 \cancel{\text{mol O}} \left(\frac{16.00 \text{ g O}}{1 \cancel{\text{mol O}}}\right) = \underline{32.00 \text{ g O}}$$

$$\text{Molar mass} = 244.2 \text{ g/mol}$$

As a final point, notice in (c) that the mass of Ba^{2+} ions (and of Cl^- ions) is the same as the atomic weight. Loss or gain of electrons leads to a loss or gain of an insignificant amount of weight.

EXERCISE 2.13 MOLECULAR WEIGHT AND MOLAR MASS
Calculate the molecular weight and molar mass of (a) limestone, $CaCO_3$, and (b) caffeine, $C_8H_{10}N_4O_2$.

For a summary of calculations with moles see page 40.

In Chapter 3 we shall begin to deal with combinations of molecules. Here you will see that they combine by one molecule reacting with another, or two molecules with one other, and so on. This means of course that the ratio of combining moles will also be one to one, or two to one, and so on. Because we have no instrument capable of measuring directly amounts of compounds in mole units, but only in mass units, it is important for you to be able to convert "mass of compound → moles of compound" and "moles → mass." This is done in exactly the same way you learned when dealing with moles of elements.

EXAMPLE 2.10

MASS ↔ MOLES CONVERSIONS

Express 23.2 g of ethyl alcohol in moles, and find the number of grams of ammonium sulfate represented by 3.50 moles.

Solution We use the molar masses found in Example 2.9.

$$\text{(a)} \quad 23.2 \cancel{\text{g}} \; C_2H_6O \left(\frac{1 \text{ mol}}{46.08 \cancel{\text{g}}}\right) = 0.503 \text{ moles } C_2H_6O$$

$$\text{(b)} \quad 3.50 \cancel{\text{mol}} \; (NH_4)_2SO_4 \left(\frac{132.2 \text{ g}}{1 \cancel{\text{mol}}}\right) = 463 \cancel{\text{g}} \; (NH_4)_2SO_4$$

EXERCISE 2.14 MASS ↔ MOLES CONVERSION
In each of the conversions, see Exercise 2.13 for the molar mass of the compound. (a) How many moles are represented by 1.00×10^3 g of limestone? (b) How many grams of caffeine are necessary to have 2.50×10^{-3} moles of the compound?

One molar mass of any substance is the mass of material that contains Avogadro's number of particles. Thus, each of the compounds pictured in Figure 2.12 represents 6.022×10^{23} molecules or formula units. With this knowledge, you can determine the number of molecules in any sample from its mass or even determine the mass of one molecule.

EXAMPLE 2.11

MOLES → MOLECULES

In Example 2.10, 23.2 g of ethyl alcohol was found to be equivalent to 0.503 moles. How many molecules does this mass represent? What is the mass of 1 ethyl alcohol molecule?

Solution

(a) $$0.503\ \cancel{\text{mol}}\ C_2H_6O \left(\frac{6.022 \times 10^{23}\ \text{molecules}}{1\ \cancel{\text{mol}}}\right) = 3.03 \times 10^{23}\ \text{molecules}\ C_2H_6O$$

(b) To calculate the mass of 1 molecule, the mass of a mole is divided by the number of molecules in a mole.

$$\left(\frac{46.1\ \text{g}}{1\ \cancel{\text{mol}}}\right)\left(\frac{1\ \cancel{\text{mol}}}{6.022 \times 10^{23}\ \text{molecules}}\right) = 7.66 \times 10^{-23}\ \text{g/molecule}$$

Again notice that the problem is set up so that units of "mol" cancel and leave units of g/molecule in the answer.

EXERCISE 2.15 MOLES → MOLECULES

A can of artificially sweetened soft drink may contain 70 mg of aspartame, or NutraSweet, $C_{14}H_{18}N_2O_5$. (a) Calculate the number of moles of aspartame in the soft drink. (b) Calculate the number of molecules of aspartame in the soft drink. (c) Calculate the mass of 1 molecule of aspartame.

2.11 DETERMINATION OF THE FORMULA OF A COMPOUND

Although it is often possible to predict the formulas of simple ionic compounds, there are millions of molecules of greater complexity. Given a sample of a compound, how do you determine its formula? The answer lies in chemical analysis, a major branch of chemistry that deals with the determination of molecular formula and structure, among other things.

EMPIRICAL AND MOLECULAR FORMULAS

The **law of constant composition** states that any sample of a pure compound always consists of the same elements combined in the same proportions by mass. Every molecule of ammonia, for example, always has the formula NH_3. That is, 1 molecule contains 1 atom of N and 3 atoms of H, or 1 mole of pure ammonia always contains 1 mole of N and 3 moles of H. It follows then that 1 molar mass, 17.03 g of ammonia, always contains

14.01 g of N and 3.02 g of H. Thus, you see so far that there are at least two equivalent ways to express molecular composition: (a) in terms of the number of atoms of each type per molecule or (b) in terms of atom masses per mole.

Equivalent Ways of Expressing Molecular Composition:

(a) a formula giving number of atoms of each type per molecule
(b) mass of each element per mole of compound
(c) mass of each element per 100 g of compound (**percent composition**)

There is a third way to express molecular formulas, a way closely related to (b) above. Composition can be given by the mass of each element in the compound relative to the total mass of the compound. That is, the formula can be expressed as **percent composition**. For ammonia this is

$$\text{Percent by weight N in NH}_3 = \frac{14.01 \text{ g N}}{17.03 \text{ g NH}_3} \times 100$$

$$= 82.27\% \text{ or } \frac{82.27 \text{ g N}}{100.0 \text{ g NH}_3}$$

$$\text{Percent by weight H in NH}_3 = \frac{3.02 \text{ g}}{17.03 \text{ g NH}_3} \times 100$$

$$= 17.73\% \text{ or } \frac{17.73 \text{ g H}}{100 \text{ g NH}_3}$$

EXERCISE 2.16 PERCENT COMPOSITION

Express the composition of each of the following compounds in terms of the mass of each element in 1.00 mole of the compound and the percent by weight of each element:

(a) NaCl, sodium chloride
(b) CH_4, methane
(c) $(NH_4)_2SO_4$, ammonium sulfate
(d) $C_8H_{10}N_4O_2$, caffeine

Now consider the *reverse* of the process above: use relative masses or percent composition to find a molecular formula. That is, if you know the identity of the elements in a sample and can determine by chemical analysis the mass of each element in a given mass of compound (the percent composition), you can then calculate the relative number of moles of each element in one mole of compound and then the relative number of atoms of each type in a molecule.

To see how to use percent composition data to find the relative number of atoms of each type in a molecule, consider hydrazine, a compound closely related to ammonia. A sample of pure hydrazine consists of 87.42% by weight N and 12.58% by weight H. If you have a 100.0 g sample of hydrazine, the percent composition tells you that the sample contains 87.42 g of N and 12.58 g of H. Therefore, the number of moles of each element in the 100.0 g sample is

$$87.42 \text{ g N} \left(\frac{1 \text{ mol N}}{14.01 \text{ g N}}\right) = 6.240 \text{ moles N}$$

$$12.58 \text{ g H} \left(\frac{1 \text{ mol H}}{1.008 \text{ g H}}\right) = 12.48 \text{ moles H}$$

What is important now is to use this information to find the *ratio* of the number of moles of each element. For hydrazine the ratio is 2 to 1,

$$\frac{12.48 \text{ moles H}}{6.240 \text{ moles N}} = \frac{2.00 \text{ moles H}}{1.00 \text{ moles N}}$$

showing that there are 2 moles of H atoms for every mole of N atoms in the sample, or in one molecule there are 2 atoms of H for every atom of N.

Percent composition data allow you to arrive at the atom ratios in the molecule. However, remember that a molecular formula must convey *two* pieces of information: (a) the *relative numbers* of each type of atom in a molecule (the atom ratios) and (b) the *total number* of atoms in the molecule. For hydrazine you know that there are twice as many H atoms as N atoms. This means that the molecular formula could be NH_2, but N_2H_4, N_3H_6, and so forth are also possible. *Percent composition data give only the simplest possible ratio of the elements.* Thus, a formula such as NH_2, where the atom ratio is the simplest possible, is called the **empirical formula**. In contrast, the **molecular formula** shows the true number of atoms per molecule; it is some integer multiple of the empirical formula.

The empirical formula of a compound is based on the simplest possible ratio of the elements. To define the molecular formula you must know the molecular weight *from experiment*. For hydrazine you must determine experimentally that the molar mass of the compound is 32.0 g/mol; that is, it is equivalent to the mass of two moles of the hypothetical molecule NH_2 (which will have the molar mass 16.0 g/mol). Thus, the molecular formula of hydrazine is N_2H_4.

EXAMPLE 2.12

EMPIRICAL AND MOLECULAR FORMULAS

A reactive compound contains 13.2% B and 86.8% Cl by weight. (a) What is the empirical formula of the compound? (b) If the molar mass of the compound is known from a separate experiment to be 164 g/mol, what is the molecular formula?

Solution To determine empirical and molecular formulas, there are always *two basic steps*: (1) determine the empirical formula (atom ratios) from percent composition data and (2) determine the molecular formula (total atoms per molecule) from the empirical formula and molar mass.

Step 1. Empirical Formula (Atom Ratios). The percent composition indicates that there are in 100.0 g of the compound 13.2 g of B and 86.8 g of Cl. Therefore, the first step is to express these masses in moles.

$$13.2 \cancel{\text{g B}} \left(\frac{1 \text{ mol B}}{10.8 \cancel{\text{g B}}}\right) = 1.22 \text{ moles B}$$

$$86.8 \cancel{\text{g Cl}} \left(\frac{1 \text{ mol Cl}}{35.5 \cancel{\text{g Cl}}}\right) = 2.45 \text{ moles Cl}$$

The ratio of the number of moles of each element in a given mass is the same as the ratio of number of atoms in a molecule. Here, that ratio is 2 Cl atoms to 1 B atom.

$$\frac{2.45 \text{ moles Cl}}{1.22 \text{ moles B}} = \frac{2.00 \text{ moles Cl}}{1.00 \text{ moles B}} = \frac{2 \text{ atoms Cl}}{1 \text{ atom B}}$$

Therefore, the *empirical formula* is BCl_2.

Step 2. Molecular Formula (Total Atoms). Any molecular formula having a 1:2 boron-to-chlorine ratio is possible, so the molecular formula could be the same as the empirical formula (BCl_2), twice the empirical formula (B_2Cl_4), three times the empirical formula (B_3Cl_6), and so on. You can find the number of empirical formula units in one molecular unit by dividing the molar mass (164 g/mol) by the mass of the empirical unit.

$$\frac{164 \text{ g/mol of compound}}{81.8 \text{ g/mol of } BCl_2} = 2.00 \text{ mol } BCl_2 \text{ per mol compound}$$

Thus, the *molecular formula* must be B_2Cl_4.

EXAMPLE 2.13

EMPIRICAL AND MOLECULAR FORMULAS

One oxide of iron is called "magnetite." It is composed of 72.4% Fe and 27.6% oxygen. What is the empirical formula of magnetite?

Solution First calculate the number of moles of each element in a 100 g sample.

$$\text{Moles of Fe} = 72.4 \cancel{\text{g Fe}} \left(\frac{1 \text{ mol Fe}}{55.85 \cancel{\text{g Fe}}}\right) = 1.30 \text{ moles Fe}$$

$$\text{Moles of O} = 27.6 \cancel{\text{g O}} \left(\frac{1 \text{ mol O}}{16.0 \cancel{\text{g O}}}\right) = 1.73 \text{ moles O}$$

Second, find the ratio of moles of O to moles of Fe.

$$\frac{\text{Moles of O}}{\text{Moles of Fe}} = \frac{1.73 \text{ moles O}}{1.30 \text{ moles Fe}} = \frac{1.33}{1.00}$$

The mole ratio is 1.33/1.00 or $1\frac{1}{3}$ to 1. Since the ratio of atoms in a formula is given in whole numbers, you would express the O to Fe ratio here as 4 to 3. That is, the empirical formula of magnetite is Fe_3O_4.

EXERCISE 2.17 EMPIRICAL AND MOLECULAR FORMULAS

Boron hydrides, compounds containing only boron and hydrogen, form a large class of compounds. One consists of 78.3% B and 21.7% H; its molar mass is 27.6 g/mol. What are the empirical and molecular formulas of this compound?

EMPIRICAL FORMULAS BY COMBUSTION ANALYSIS

The empirical formula of a compound can be determined if the percent composition of the compound is known. But where do percent composition data come from? Various methods of analysis are used, but one method that works well for compounds that burn in oxygen is *analysis by combustion*. Compounds containing C and H (and other elements) burn in oxygen to give CO_2 and H_2O (and other oxides). The CO_2 and H_2O can be isolated separately and their masses determined as illustrated in Figure 2.13. According to the **law of conservation of matter**, *all* the

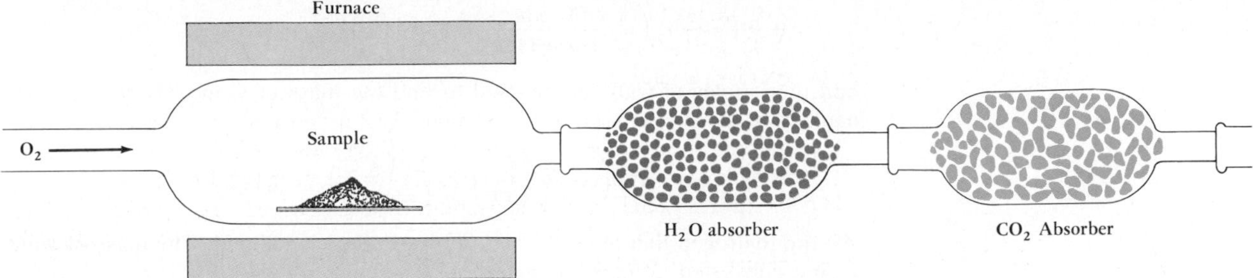

FIGURE 2.13
If a compound containing C and H is burned in oxygen, CO_2 and H_2O are formed and the mass of each can be determined. The H_2O is absorbed by magnesium perchlorate, and the CO_2 is absorbed by finely divided sodium hydroxide supported on asbestos. The mass of each absorbent before and after combustion will give the mass of the CO_2 and H_2O trapped by its absorbent. Only a few milligrams of compound are needed for analysis.

carbon that was in the compound will be found after combustion as CO_2. Thus, we have the relationships

$$\left.\begin{array}{l}\text{1 g of C in a compound}\\ \text{1 mol of C in a compound}\\ \text{12.01 g of C in a compound}\end{array}\right\} \xrightarrow{\text{burn in pure oxygen}} \left\{\begin{array}{l}\text{1 g of C in } CO_2\\ \text{1 mol of C in 1 mol of } CO_2\\ \text{12.01 g of C in 44.01 g of } CO_2\end{array}\right.$$

For every mole of CO_2 isolated, a mole of C, or 12.01 g, must have been contained in the original compound. Similarly, for the hydrogen in the combusted material, you have

$$\left.\begin{array}{l}\text{1 g of H in a compound}\\ \text{2 mol H in a compound}\\ \text{2.0 g of H in a compound}\end{array}\right\} \xrightarrow{\text{burn in pure oxygen}} \left\{\begin{array}{l}\text{1 g of H in } H_2O\\ \text{2 mol of H in 1 mol of } H_2O\\ \text{2.0 g of H in 18.02 g of } H_2O\end{array}\right.$$

Here it is important to notice that one mole of water as a combustion product means that the original compound contained *two* moles of H atoms.

Based on these relationships, the amount of C and H in a compound can be determined from the masses of water and carbon dioxide formed in the combustion.

EXAMPLE 2.14

DETERMINING THE EMPIRICAL FORMULA OF A COMBUSTIBLE COMPOUND

Oxalic acid is used in the paint, cosmetics, and ceramics industries and is found in many plants and vegetables. It contains the elements C, H, and O. If 0.513 g of the acid is burned in oxygen, 0.501 g of CO_2 and 0.103 g of H_2O result. What is the empirical formula of oxalic acid? If the molar mass of the acid is 90.04 g/mol, what is the molecular formula?

Solution 1.000 mole of C in the oxalic acid will lead to 1.000 mole of CO_2. Thus, the mass of CO_2 is first used to find the number of moles of C isolated as CO_2,

$$0.501 \cancel{\text{g } CO_2} \left(\frac{1 \text{ mol } CO_2}{44.01 \cancel{\text{g } CO_2}}\right) = 0.0114 \text{ mol } CO_2$$

and then moles of CO_2 can be used to find the mass of C in CO_2 *and* in oxalic acid.

$$0.0114 \cancel{\text{mol } CO_2} \left(\frac{1 \cancel{\text{mol C}}}{1 \cancel{\text{mol } CO_2}}\right)\left(\frac{12.01 \text{ g C}}{1 \cancel{\text{mol C}}}\right) = 0.137 \text{ g C}$$

All the hydrogen that appears in H_2O was in the oxalic acid. The mass of water is first converted to moles of water

$$0.103 \cancel{\text{g } H_2O} \left(\frac{1 \text{ mol } H_2O}{18.02 \cancel{\text{g } H_2O}}\right) = 0.00572 \text{ mol } H_2O$$

and then moles of water are related first to moles of H in the combusted compound,

$$0.00572 \cancel{\text{mol } H_2O} \left(\frac{2 \text{ mol H}}{1 \cancel{\text{mol } H_2O}}\right) = 0.0114 \text{ mol H}$$

and then to the mass of H in water *and* in oxalic acid.

$$0.0114 \cancel{\text{mol H}} \left(\frac{1.008 \text{ g H}}{1 \cancel{\text{mol H}}}\right) = 0.0115 \text{ g H}$$

The calculations reveal that, in 0.513 g of oxalic acid, there are 0.137 g of C and 0.0115 g of H; the remaining mass, 0.365 g, must be oxygen. To find the formula of oxalic acid, we need only to find the number of moles of each element and then find the mole ratios.

$$0.137 \cancel{\text{g C}} \left(\frac{1.00 \text{ mol C}}{12.01 \cancel{\text{g C}}}\right) = 0.0114 \text{ mol C}$$

$$0.0115 \cancel{\text{g H}} \left(\frac{1.00 \text{ mol H}}{1.01 \cancel{\text{g H}}}\right) = 0.0114 \text{ mol H}$$

$$0.365 \cancel{\text{g O}} \left(\frac{1.00 \text{ mol O}}{16.0 \cancel{\text{g O}}}\right) = 0.0228 \text{ mol O}$$

To find the mole ratio of elements, divide the number of moles of each element by the smallest number of moles.

$$\frac{0.0114 \text{ mol H}}{0.0114 \text{ mol C}} = \frac{1.00 \text{ mol H}}{1.00 \text{ mol C}} \qquad \frac{0.0228 \text{ mol O}}{0.0114 \text{ mol C}} = \frac{2.00 \text{ mol O}}{1.00 \text{ mol C}}$$

For every C atom in the molecule, there is one H atom and two O atoms. That is, the *empirical formula* of oxalic acid is CHO_2.

To determine the molecular formula, the experimental molar mass and the molar mass of one empirical formula unit are compared.

$$\frac{90.04 \text{ g/mol oxalic acid}}{45.02 \text{ g/mol of } CH_2O} = \frac{2.000 \text{ mol } CH_2O}{1.000 \text{ mol oxalic acid}}$$

Thus, the *molecular formula* of oxalic acid is twice the empirical formula, that is, $C_2H_2O_4$.

EXERCISE 2.18 FORMULA DETERMINATION FROM COMBUSTION

Compounds consisting of carbon, hydrogen, and metals are called "organometallic" compounds. One of the best known is called "ferrocene," a molecule containing C, H, and Fe. If 0.652 g of ferrocene is burned in oxygen, 1.542 g of CO_2 and 0.315 g of H_2O are isolated. (The iron is converted to Fe_2O_3.) What is the empirical formula of ferrocene and what is its empirical formula weight?

DETERMINING THE FORMULA OF A HYDRATED COMPOUND

As a final example of the determination of molecular formulas, consider the problem of **hydrated salts**. Many of the compounds in Figure 2.12 are hydrated ionic compounds, that is, molecules of water are associated with the ions of the compound. The beautiful green nickel(II) salt in Figure 2.12, for example, has a formula of $NiCl_2 \cdot 6\ H_2O$. The "dot" between $NiCl_2$ and 6 H_2O indicates that six moles of water are associated with every mole of $NiCl_2$, so the molar mass of the compound is 237.7 g/mol.

Hundreds of hydrated salts are known. Since there is no simple way to predict how much water will be involved, it must be determined experimentally. Such an experiment usually involves heating the hydrate so that all the water is released from the crystal and driven away (Figure 2.14). Only the **anhydrous** salt, a material "without water," is left. The difference in mass of the hydrated and anhydrous material is the mass of water lost by the hydrated compound. As outlined in the following example, you can then calculate the ratio of moles of water to moles of anhydrous salt.

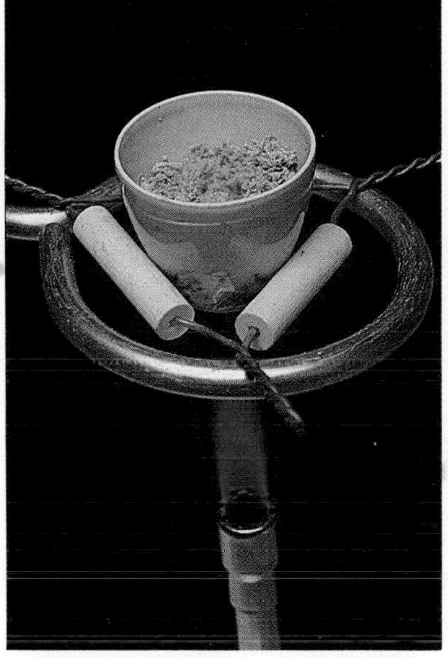

FIGURE 2.14
A crucible containing a hydrated salt (e.g., $CuSO_4 \cdot 5\ H_2O$) is being heated to drive off all the water to leave the anhydrous salt ($CuSO_4$). (Charles D. Winters)

EXAMPLE 2.15

DETERMINING THE FORMULA OF A HYDRATED SALT

Hydrated copper(II) sulfate is the deep blue solid in Figure 2.12. Suppose you know only that it has a formula of $CuSO_4 \cdot x\ H_2O$, but you do not yet know the value of x. To find x, you weigh 1.023 g of the blue solid and heat it in a porcelain crucible (Figure 2.14). (Heating is continued until the mass no longer decreases, so you are certain that all the water has been driven off.) The mass of anhydrous, white $CuSO_4$ is 0.654 g. How many moles of water are there per mole of $CuSO_4$? That is, what is the value of x?

Solution (a) Calculate the mass of water released.

$$1.023\ \text{g hydrate} - 0.654\ \text{g anhydrous salt} = 0.369\ \text{g water lost}$$

(b) Calculate the moles of water driven off.

$$0.369\ \cancel{\text{g}} \left(\frac{1\ \text{mol}}{18.0\ \cancel{\text{g}}}\right) = 0.0205\ \text{mole}\ H_2O$$

(c) All the white, anhydrous solid left is $CuSO_4$. Express the quantity of this salt, 0.654 g, in moles.

$$0.654\ \cancel{\text{g}} \left(\frac{1\ \text{mol}}{159.6\ \cancel{\text{g}}}\right) = 0.00410\ \text{mole}\ CuSO_4$$

(d) Find the ratio of moles of water to moles of $CuSO_4$.

$$\frac{\text{moles}\ H_2O}{\text{moles}\ CuSO_4} = \frac{0.0205\ \text{mole}\ H_2O}{0.00410\ \text{mole}\ CuSO_4} = \frac{5.00}{1.00}$$

The water-to-$CuSO_4$ ratio is 5:1, so the empirical formula of the blue hydrated salt is $CuSO_4 \cdot 5\ H_2O$.

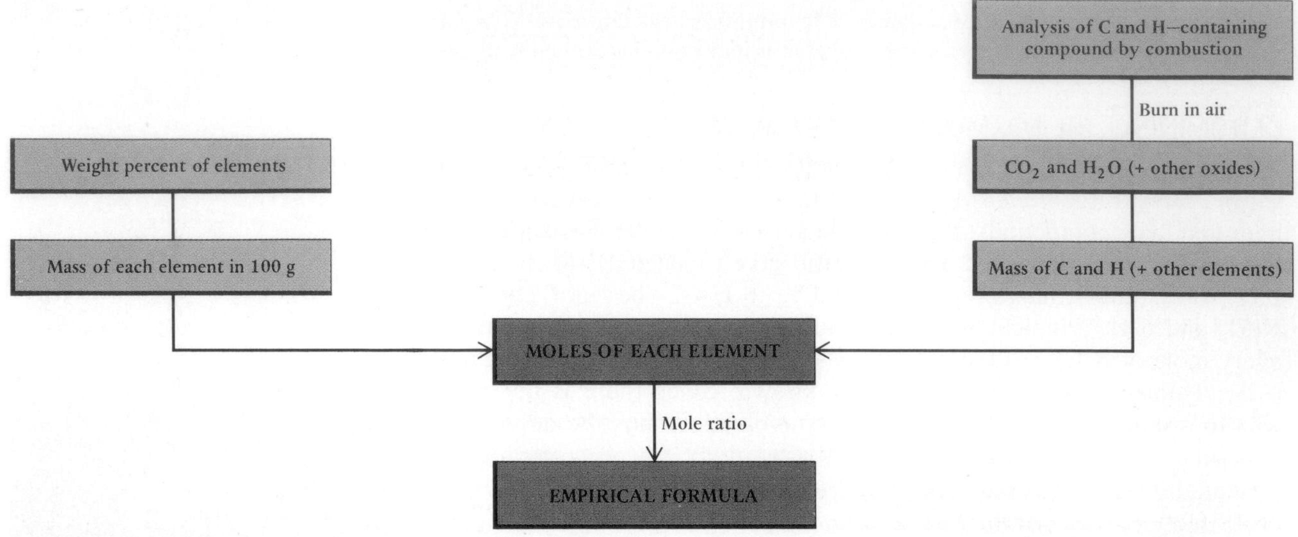

FIGURE 2.15
Various approaches to finding the empirical formula of a compound.

EXERCISE 2.19 DETERMINING THE FORMULA OF A HYDRATED SALT

Borax powder has the formula $Na_2B_4O_7 \cdot x\ H_2O$. Determine the molar mass of the hydrated compound from the following data: (a) weight of borax powder taken for analysis equals 2.145 g and (b) weight of white powder, $Na_2B_4O_7$, remaining after heating the original sample sufficiently to drive off all the water equals 1.130 g.

You have now seen several examples of the determination of the empirical formula of a compound; be sure to notice that they all follow the same pattern. The various approaches are illustrated in Figure 2.15.

2.12 CHEMICAL ANALYSIS OF MIXTURES

Besides using chemical analysis to determine compound formulas, it is also possible to devise ways to find the quantity of an element or compound in a mixture. As an example of a chemical analysis, suppose you wish to determine the amount of barium in a sample that contains an ionic compound of barium (Ba^{2+}) and some impurity. A procedure for doing this is outlined in Figure 2.16. In the figure you see that a weighed sample containing an unknown amount of barium is dissolved in water. Sulfuric acid, H_2SO_4, is added to the solution, and the Ba^{2+} ion in the water and the SO_4^{2-} ion from the added acid combine to produce $BaSO_4$, an ionic compound that is *not* soluble in water. If sufficient acid is added, all the Ba^{2+} that was in the unknown sample is now found as $BaSO_4$. Insoluble white $BaSO_4$ can be isolated by filtration, and, after drying, its mass can be determined. Since you know that barium sulfate always contains 1 mole of Ba^{2+} per mole of $BaSO_4$ (or 137.3 g Ba per 233.4 g $BaSO_4$), the number of moles (or grams) of Ba isolated as $BaSO_4$ can be determined readily.

(a)

(b)

(c)

FIGURE 2.16

(a) A sample containing an unknown amount of Ba^{2+} is dissolved in water. (b) Excess sulfuric acid is added, and the SO_4^{2-} ion from the acid and the Ba^{2+} from the sample combine to form a precipitate of insoluble $BaSO_4$. (c) The $BaSO_4$ is collected in a filter. After drying and weighing the $BaSO_4$ precipitate, the amount of Ba^{2+} in the solid can be calculated. (Charles D. Winters)

EXAMPLE 2.16

CHEMICAL ANALYSIS

Suppose you have an old sterling silver spoon and wish to determine how much silver it contains. (Sterling silver is a mixture or alloy of silver and copper.) A piece of spoon weighing 1.175 g is taken for analysis by a procedure outlined in the margin. The piece is placed in nitric acid, which dissolves the metals by converting all the silver metal to Ag^+ (and all the Cu to Cu^{2+}). Excess HCl is added to the acid solution of Ag^+ and Cu^{2+} ions, and the Ag^+ and Cl^- (from HCl) combine to produce the insoluble ionic compound AgCl. After filtering the solution to isolate the solid AgCl from the solution and drying the solid, the solid is found to have a mass of 1.449 g. What is the weight percent of Ag in the sterling silver spoon?

Analysis of Ag in the Presence of Cu

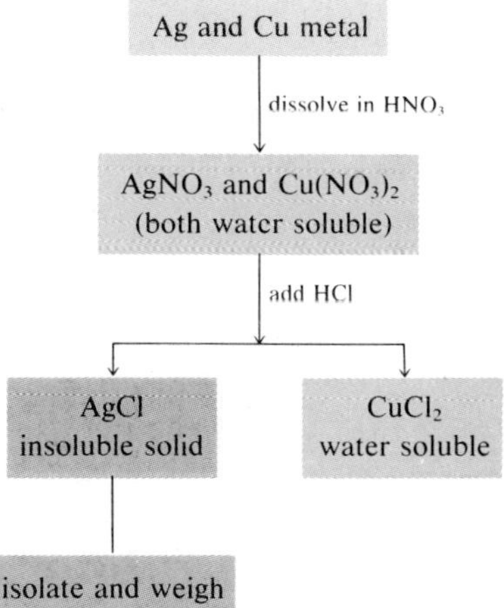

Solution The isolated AgCl contains all the Ag that was in the 1.175 g piece of spoon. Since one mole of Ag in the spoon must lead to one mole of AgCl, the first step is to find moles of AgCl isolated.

Step 1. Calculate moles of AgCl isolated.

$$1.449 \cancel{\text{g AgCl}} \text{ isolated} \left(\frac{1 \text{ mol AgCl}}{143.3 \cancel{\text{g AgCl}}}\right) = 0.01011 \text{ mol AgCl}$$

Step 2. Calculate moles Ag in AgCl and in the spoon sample.

$$0.01011 \cancel{\text{mol AgCl}} \left(\frac{1 \text{ mol Ag}}{1 \cancel{\text{mol AgCl}}}\right) = 0.01011 \text{ mol Ag}$$

Step 3. Calculate the mass of Ag in the spoon sample.

$$0.01011 \cancel{\text{mol Ag}} \left(\frac{107.9 \text{ g Ag}}{1 \cancel{\text{mol Ag}}}\right) = 1.091 \text{ g Ag}$$

Step 4. Calculate the weight percent of silver in the spoon sample.

$$\text{Weight percent Ag in spoon} = \frac{1.091 \text{ g Ag}}{1.175 \text{ g spoon}} \times 100 = 92.86\% \text{ Ag}$$

EXERCISE 2.20 CHEMICAL ANALYSIS

Refer to Figure 2.16 for the procedure for a barium analysis. Suppose you have a solid that consists of some $BaCl_2$ contaminated with NaCl. To analyze the mixture, you must find a way to separate the Ba^{2+} from the Na^+ and then isolate the Ba^{2+} ion in the form of a compound of known formula. Therefore, you take 1.023 g of the solid mixture, dissolve it in water, and add H_2SO_4 to form insoluble $BaSO_4$ and leave NaCl in solution. If the $BaSO_4$ has a mass of 0.560 g after isolating and drying, calculate (a) the weight percentage of barium in the sample and (b) the grams of $BaCl_2$ in the original mixture.

SUMMARY

An **element** is a form of matter that cannot be broken down into simpler forms by ordinary means (Section 2.1). Each of the known 108 elements has a *name* and *symbol* and is listed in the **periodic table** in the front of the book.

An **atom** is the smallest particle of an element that retains the chemical properties of the element (Section 2.2). Each atom is composed of many subatomic particles, the most important for chemistry being the **protons** and **neutrons** in the nucleus and **electrons** in the space about the nucleus. The **atomic number (Z)** of an element corresponds to the number of protons in the nucleus (*and* the number of electrons outside the nucleus). Masses of atoms are given on a relative scale based on the **atomic mass unit (amu)**, where 1 amu is exactly $\frac{1}{12}$ the mass of a carbon atom having 6 protons and 6 neutrons. The **mass number (A)** of a particular atom is the sum of the number of protons and neutrons in the nucleus. The composition of an atom, X, can be given by the symbol ${}^{A}_{Z}X$, although the subscript Z is optional. **Isotopes** are atoms having the same atomic number but different mass numbers (Section 2.3). Virtually every element has more than one stable isotope. The weighted average of the isotopic masses, expressed in atomic mass units, is the **atomic weight** of the element.

The **mole** (Section 2.5) is the amount of substance that contains as many elementary particles (which equal **Avogadro's number, 6.022 × 10²³**) as there are atoms in exactly 12 g of carbon. The mass of 1 mole in grams is called the **molar mass**; it is numerically equivalent to the atomic weight expressed in atomic mass units. The mole is the cornerstone of quantitative chemistry, so it is *extremely important* that you know how to do the conversion "mass → moles" or "moles → mass."

Mass ⟶ Moles

$$\frac{\text{Mass in grams}}{\text{Molar mass (g/mol)}} = \text{moles}$$

Moles ⟶ Mass

$$\text{Moles} \times \text{molar mass (g/mol)} = \text{mass in grams}$$

The number of atoms in a sample can be determined by first converting the sample mass to moles and then doing the calculation

Moles × Avogadro's number (6.022×10^{23} atoms/mol) = atoms

If the elements are arranged in order of increasing atomic number, as in the **periodic table** in the front of the book, one finds a regular repetition of properties (Section 2.6). Each vertical column is a **group** of elements having similar properties. (The horizontal rows of elements are called **periods**.) Each group of the table is given a number and letter designation. Groups labeled A are sometimes called **main group elements**, while B groups are the **transition elements** or transition metals.

Most of the known elements are **metals**. In the periodic table, the metallic elements are those to the left, while the **nonmetals** are at the right. A zig-zag line beginning at boron, and running to Po and At, roughly separates the metals and nonmetals. The elements along this line share properties of both types of elements and are often called **metalloids**.

Elements of Group 1A are commonly called the **alkali metals**, those of Group 2A are the **alkaline earth metals**, Group 7A are the **halogens**, and Group 8A elements are called the **rare gases**.

A **compound** is a combination of atoms in the ratio of small, whole numbers, and a **molecule** is the smallest particle of a compound that has the same atom ratio as the entire compound and retains the chemical characteristics of the compound (Section 2.7). The atomic composition of a compound can be expressed in terms of a **molecular** or **structural formula**.

Nitrogen, oxygen, and the halogens are diatomic or "two-atom" molecules (N_2, O_2, F_2, etc.) in their elemental form (Section 2.8). Elemental phosphorus consists of P_4 molecules, and elemental sulfur is composed chiefly of S_8 molecules. Carbon has two well-known forms, diamond and graphite.

Many compounds are composed of ions (Section 2.9). Metal atoms commonly lose one or more electrons to give positive ions or **cations**, while nonmetal atoms can gain electrons to form **anions** with a negative electrical charge. For a metal in a group other than the transition series, the cation positive charge is often equal to the number of the group in which it is found. The negative charge on a single-atom, or **monatomic**, nonmetal anion is given by (8 minus the group number). Besides monatomic ions, there are many **polyatomic** ions; some are listed in Table 2.5.

Ionic compounds are formed by the attraction of positive and negative ions (Section 2.9). Be sure to know how to name and formulate ionic compounds.

The **molar mass** of a compound is the weight in grams of 1 mole (Avogadro's number of molecules). It is numerically equivalent to the **molecular weight,** the sum of the atomic weights of all of the atoms of the molecule. The molecular weight is the average mass of a molecule of the substance, expressed in atomic mass units. For ionic compounds, which do not consist of individual molecules, we often refer to the sum of atomic weights for atoms in the formula as the **formula weight**.

Molecular composition can be expressed in terms of *either* the relative number of atoms *or* relative masses of atoms of each type. Thus, one can express the composition of a compound in terms of **percent com-**

(Summary continues on page 72)

SPECIAL SECTION: THE ESSENTIAL ELEMENTS

As our knowledge of biochemistry—the chemistry of living systems—increases, we learn that more and more elements are essential to life. The simplest definition of an **essential element** is that it is required for the maintenance of life. Its absence will result in death or serious malfunction, and no other element can wholly take its place. Of the 108 known elements, only 29 or so are said to be "essential," and these can be divided roughly into two categories: (a) 11 "bulk elements" and (b) 18 "trace elements."

Vitamins are essential biological organic nutrients, and the essential elements are their inorganic counterparts. Unlike the vitamins, essential elements cannot be made by living organisms but must be present in the environment. Therefore, it is not surprising that the relative quantities of essential elements in the human body (Table 2.7) reflect in many ways the relative abundances of the known elements in the earth's oceans and, to a lesser extent, in the crust. (Element abundances are described in detail in the special section in Chapter 8 (Figure 8.19).) Because of this, there has been some speculation that life began in or close to water. Iron is the most abundant of the transition metals in the cosmos, and it is the dominant transition metal in the biochemistry of vertebrates.

Four elements (O, C, H, and N, Table 2.7) account for most of the mass of the human body. Indeed, about 99% of the structure of the body is built of elements up to atomic number 21. Exceptions are Li, Be, Al, and the rare gases. It is important to point out, however, that lithium is used effectively to treat acute mania and depression, while an excess of aluminum has been found in the nerve cells of brains of persons who died of Alzheimer's disease.

The remaining 1% of the essential elements come largely from the first series of transition metals. In fact, their importance is out of all proportion to their abundance in the body, as iron, zinc, and copper are among the most important of the essential elements.

Much of the 3 or 4 grams of iron in the body are used in hemoglobin, the substance responsible for carrying oxygen to the cells of the body. Iron deficiency is marked by fatigue, infections, mouth inflammation, and

TABLE 2.7 Relative Amounts of Essential Elements in the Human Body

ELEMENT	PERCENT BY MASS (G OF ELEMENT IN 100 G SAMPLE)
Oxygen	65
Carbon	18
Hydrogen	10
Nitrogen	3
Calcium	1.5
Phosphorus	1.2
Potassium, sulfur, chlorine	0.2
Sodium	0.1
Magnesium	0.05
Iron, cobalt, copper, zinc, iodine	< 0.05
Selenium, fluorine	< 0.01

other signs.* The average human body contains about 2 grams or so of zinc, and a deficiency of zinc will be noticed by loss of appetite, failure to grow, skin changes, and impaired healing of wounds. There is also some evidence that zinc deficiency in pregnant animals can result in birth defects.

The average adult has about 75 mg of copper, with about one third of that in the liver and brain, about one third in the muscles, and the rest in other tissues. It is involved in many biochemical functions, so a deficiency shows up in a variety of ways: anemia, degeneration of the nervous system, impaired immunity, and defects in the color and structure of hair, among others.

The elements of Figure 2.17 are essential to health and can be obtained from many common foods (Table 2.8). However, an overabundance can be detrimental. For example, the optimum amount of selenium is thought to be in the range of 50 to 200 micrograms/day, while 200 to 1000 micrograms is considered toxic, and amounts in excess of this are lethal. Indeed, at the moment there is considerable controversy over reports that birds and other wildlife are being killed or losing reproductive capacity because of selenium poisoning in wildlife refuges in the western United States.†

A complete understanding of the functioning of the trace elements is a challenging area of current research for bioinorganic chemists, biochemists, and nutritionists.

---

*Dietary deficiencies of iron are common, but overdoses of iron supplements can be toxic. This is why chewable mineral supplements in bottles have "child-proof" caps.

†You can read about the selenium problem in *Science*, July 12, 1985, p 144.

TABLE 2.8 Sources of Some Essential Elements

ELEMENT	SOURCE	AMOUNT OF ELEMENT AVAILABLE (MG ELEMENT/100 G OF EDIBLE PORTION)
Iron	Brewer's yeast	17.3
	Wheat bran	14.9
	Beef liver	8.8
	Eggs	2.3
Zinc	Brazil nuts	4.2
	Beef liver	3.9
	Egg yolk	3.5
	Chicken	2.6
Copper	Oysters	13.7
	Brazil nuts	2.3
	Beef liver	1.1
	Eggs	0.2
Calcium	Swiss cheese	925
	Whole milk	118
	Broccoli	103
Selenium	Butter	0.15
	Cider vinegar	0.09

position (Section 2.11). If the percent by weight of each element in a compound is known from experiment (Sections 2.11 and 2.12), then the **empirical formula** or simplest possible ratio of elements is known. The empirical formula is converted to the **molecular formula** with the *experimentally determined* molecular weight.

STUDY QUESTIONS

ELEMENTS AND ATOMS

1. Give the name of each of the following elements:
(a) C (c) Cl (e) Mg
(b) Na (d) P (f) Ge
2. Give the name of each of the following elements:
(a) Mn (c) K (e) As (g) W
(b) F (d) Hg (f) Xe (h) Ra
3. Give the symbol for each of the following elements:
(a) lithium (d) silicon
(b) titanium (e) gold
(c) iron (f) lead
4. Give the symbol for each of the following elements:
(a) silver (e) tin
(b) aluminum (f) barium
(c) plutonium (g) krypton
(d) cadmium (h) ruthenium
5. Element 25 is found in the form of oxide "nodules" at the bottom of the sea. What is the name and symbol of this element?
6. Give the mass number of each of the following atoms: (a) beryllium with 5 neutrons, (b) titanium with 26 neutrons, and (c) gallium with 39 neutrons.
7. Give the complete symbol ($^{A}_{Z}X$) for each of the atoms in Study Question 6.
8. Give the complete symbol ($^{A}_{Z}X$) for each of the following atoms: (a) sodium with 12 neutrons, (b) argon with 21 neutrons, and (c) lead with 126 neutrons.
9. How many electrons, protons, and neutrons are there in (a) calcium-40, ^{40}Ca; (b) tin-119, ^{119}Sn; and (c) plutonium-244, ^{244}Pu?
10. Fill in the columns of blanks in the table (one column per element).

Symbol	$^{45}_{21}Sc$	$^{33}_{16}S$	___	___
Number of protons	___	___	8	___
Number of neutrons	___	___	9	31
Number of electrons in the neutral atom	___	___	___	25

11. For ruthenium-101, (a) how many protons are in the nucleus, (b) how many neutrons are in the nucleus, and (c) how many electrons are arranged about the nucleus?

ISOTOPES AND ATOMIC WEIGHT

12. Of the isotopes given below, which are isotopes of the same element?
(a) $^{20}_{10}X$ (b) $^{22}_{11}X$ (c) $^{22}_{10}X$ (d) $^{19}_{9}X$ (e) $^{21}_{10}X$
13. A natural sample of gallium consists of two isotopes of masses 68.95 and 70.95 with abundances of 60.16% and 39.84%, respectively. What is the average atomic weight of gallium?
14. Bromine, like chlorine, has two stable isotopes. For bromine the isotopic masses are 78.9183 (relative abundance, 50.69%) and 80.9163 (relative abundance, 49.31%). Calculate the average atomic weight of bromine.
15. Silicon is found in nature combined with oxygen to give sand, quartz, agate, and similar materials. The element has three stable isotopes.

Exact Mass	Relative Abundance (%)
27.97693	92.23
28.97649	4.67
29.97376	3.10

Calculate the average atomic weight of silicon from the data above.

16. The metallic element chromium has four stable isotopes.

Exact Mass	Relative Abundance (%)
49.9461	4.35
51.9405	83.79
52.9407	9.50
53.9389	2.36

Calculate the average atomic weight of chromium.

17. The element with an atomic weight of 72.6 was the basis of some early transistors. State its name and symbol.
18. The volcanic eruption of Mt. St. Helens produced a considerable quantity of a radioactive element in the gaseous state. The element has an atomic number of 86. What are the symbol, name, and atomic weight of this element?
19. About 42 billion pounds of element 7 were produced industrially in the United States in 1983. What are the name, symbol, and atomic weight of this element?
20. The element with an atomic weight of 78.96 is important to your health if ingested in very small amounts (90 micrograms/day); an excess causes loss of hair. What are the name, symbol, and atomic number of this element?

21. Antimony, one of the elements known to the ancient alchemists, has two stable isotopes: ^{121}Sb (mass, 120.90) and ^{123}Sb (mass, 122.90). Calculate the relative abundances of the two isotopes.
22. Magnesium is commonly extracted from seawater. Magnesium-24 is its most abundant isotope (78.99%); its exact mass is 23.985. If the atomic weight of magnesium is 24.312, what are the relative abundances of magnesium-25 (mass, 24.986) and magnesium-26 (mass, 25.983)?

MOLES

23. Calculate the number of moles represented by each of the following:
    (a) 127.08 g of Cu
    (b) 20.0 g of calcium
    (c) 16.75 g of Al
    (d) 0.012 g of potassium
    (e) 5.0 mg of americium
24. A chunk of sodium metal, Na, if thrown into a bucket of water, produces a dangerously violent explosion from the reaction of sodium with water. If 50.4 g of sodium is used, how many moles of sodium does that represent?
25. Black gunpowder contains several chemicals, among them sulfur and carbon. A typical powder is about 10.09% S (by mass) and 14.29% C. If you have 1.00 pound (454 g) of gunpowder, how many grams of sulfur and how many grams of carbon are present? How many moles of each?
26. An average sample of coal contains about 3.0% by mass sulfur. How many moles of sulfur are there in 1.0 ton of coal (1 T = 2000 lb; 1 lb = 454 g)?
27. Drinking water can contain many different chemicals in trace amounts. If a water sample contains 0.0180 ppm Hg (ppm = parts per million; 1 ppm = 1 g of Hg per 10^6 g of water), how many grams of mercury are you ingesting when you drink one cup of water (225 mL)? How many moles of mercury are you ingesting? (1.00 mL H_2O = 1.00 g)
28. Calculate the number of grams in each of the following:
    (a) 0.10 mole of iron
    (b) 2.31 moles of Si
    (c) 0.0023 mole of carbon
    (d) 0.54 mole of sodium
    (e) 6.03 moles of gold
29. The aluminum foil in a package of kitchen foil weighs approximately 12 ounces. How many moles of aluminum will you get when you buy a package of this foil? (1 ounce = 28.4 g)
30. Gems and precious stones are measured in carats, a weight unit equivalent to 200. mg. If you have a 2.3 carat diamond in a ring, how many moles of carbon do you have? (A diamond is a very pure form of carbon.)
31. Chemicals are sometimes sold by the mole. For example, you can buy 0.038 mole of silver for one dollar. How many grams of silver will you get for your dollar?
32. The international markets in precious metals operate in the weight unit "troy ounce" (where 1 troy ounce is equivalent to 31.1 g). Platinum sells for $325 per troy ounce. (a) How many moles are there in one troy ounce? (b) If you have $5000 to spend, how many grams and how many moles of platinum can be purchased?
33. Gold prices flucuate, depending on the international situation. If gold currently sells for $338.70 per troy ounce, how much must you spend to purchase 1.00 mole of gold? (1 troy ounce is equivalent to 31.1 g)
34. A piece of platinum foil 0.254 mm thick and 50.0 mm × 50.0 mm sells for $525.00. How many moles of platinum are you buying if the density of platinum is 21.45 g/cm^3?
35. Scrap iron sells for $114.00 per ton. The density of iron is 7.86 g/cm^3. (a) How many moles of iron can you buy for $1000.00? (b) If $1000.00 worth of iron is delivered to your home in the form of a cube, how long is one side of the cube? (1 T = 9.08×10^5 g)
36. Analysis of water from the Atlantic Ocean indicates, among other things, that it contains 45 tons of silver and 9.0 pounds of gold per cubic mile. (a) How many grams of silver and gold are there in one cubic mile? (1 T = 2000 lb; 1 lb = 454 g) (b) How many moles of silver and gold are there per cubic mile? (c) How many grams of silver and gold are there per liter of ocean water?
37. The Statue of Liberty in New York harbor is made of 2.00×10^5 pounds of copper sheets bolted to an iron framework. How many grams and how many moles of copper does this represent? (1 lb = 454 g)

AVOGADRO'S NUMBER

38. For each of the quantities in Question 28, calculate the number of atoms of the element.
39. For each of the quantities in Question 23, calculate the number of atoms of each element.
40. What is the mass of one copper atom? Of one atom of tungsten (W)?
41. The average mass of one gold atom in a sample of naturally occurring gold is 3.2702×10^{-22} g. What is the molar mass of gold?
42. One of the events of a track and field meet is the shotput. If the shot is made of iron and if it contains 7.83×10^{25} atoms of iron, could you lift it easily? (Calculate the weight of the shot in pounds and compare it with your lifting ability. One pound = 454 g.)

43. A 5-cent coin, the nickel, actually contains mostly copper. The metal used is 75% copper and 25% nickel by weight. If a 5-cent coin weighs 5.10 g, how many grams of copper and nickel does it contain? How many moles of each? How many atoms of each?

44. Assume for the sake of this problem that a standard soft-drink or beer can measures 2 × 2 × 5 inches. If you stack one mole of these cans over the United States, how deeply would the cans be stacked? (The land area of the continental United States is 3.6×10^6 miles^2.)

45. If all of the people in the United States, 250 million people, were put to work counting the atoms in a mole of gold and if each person could count one atom per second day and night for 365 days a year, how many years would it take to finish the count?

46. The structure of a pure solid can be viewed as spherical atoms stacked neatly into a regular array, and the solid can often be divided into tiny cubes each containing a given number of atoms. For example, a silicon crystal consists of cubes each containing 8 spherical silicon atoms, and each cube is 543.1 picometers on a side. Use this information, along with the average mass of silicon (28.0854 g/mol) and its density (2.328994 g/cm^3), to determine Avogadro's number.

PERIODIC TABLE

47. (a) Name all of the elements of one group of the periodic table. (b) Give the symbols for all of the elements of one period of the periodic table.

48. For each of the elements in Question 2, tell whether it is a metal, metalloid, or nonmetal, give the number of the group and period in which it is found, and give the name of the group if possible. (For example, Ar is argon; it is a nonmetal and is found in Group 8A in period 3; the group is called the "rare gases.")

49. Arsenopyrite is an important mineral. It is composed of Fe, As, and S. Which element is the metal in this mineral? Which one is the nonmetal? Which one is the metalloid?

WRITING FORMULAS AND NAMING COMPOUNDS

(See Appendix B for details of naming compounds.)

50. Write the molecular formula of each of the following compounds: (a) Anatase, a common titanium-containing mineral, has a titanium atom and two oxygen atoms per formula unit. (b) One of a series of compounds called the boron hydrides has four boron atoms and ten hydrogen atoms per molecule. (c) Aluminum trimethyl has one aluminum atom, three carbon atoms, and nine hydrogen atoms per formula unit. (d) Vitamin C or ascorbic acid has six carbon atoms, eight hydrogen atoms, and six oxygen atoms per molecule.

51. Give formulas for the following ionic compounds. Be sure to write the formula by giving the cation first and then the anion.

(a) potassium iodide	(e) magnesium sulfide
(b) calcium oxide	(f) sodium carbonate
(c) sodium bromide	(g) ammonium nitrate
(d) lithium oxide	(h) cesium hydroxide

52. Give formulas for the following ionic compounds.

(a) ammonium sulfate	(e) potassium cyanide
(b) aluminum carbonate	(f) chromium(III) oxide
(c) calcium phosphate	(g) tin(II) sulfide
(d) gallium(III) oxide	(h) iron(III) hydroxide

53. Give formulas for each of the ionic compounds below.

(a) sodium acetate	(e) titanium(IV) bromide
(b) silver perchlorate	(f) uranium(IV) fluoride
(c) potassium chlorate	(g) calcium hypochlorite
(d) calcium hydrogen phosphate	(h) sodium nitrite

54. Name each of the following ionic compounds and give the formula, charge, and number of each ion that makes up the compound.

(a) $NaHCO_3$	(e) $NaCN$
(b) $Ca_3(PO_4)_2$	(f) $CuSO_4$
(c) NH_4Br	(g) $KMnO_4$
(d) $KClO_4$	(h) ZnO

55. Name each of the following ionic compounds and give the formula, charge, and number of each ion that makes up the compound.

(a) $MgCO_3$	(e) Na_3PO_4
(b) $Fe(ClO_4)_3$	(f) CoO and Co_2O_3
(c) $(NH_4)_2SO_3$	(g) $CrCl_3$
(d) $BeSO_4$	(h) $SnCl_2$ and $SnCl_4$

56. Write the formulas of all of the compounds that can be made by combining each of the cations with each of the anions below. Name each compound formed.

Cations	**Anions**
Na^+	CO_3^{2-}
Sr^{2+}	I^-
NH_4^+	NO_3^-

57. Name all of the compounds in the following chemical equation:

$$Cd(NO_3)_2 + (NH_4)_2S \longrightarrow 2\ NH_4NO_3 + CdS$$

MOLECULAR WEIGHT, MOLES, AND AVOGADRO'S NUMBER

58. Calculate the molecular weight of each of the following compounds:

(a) Fe_2O_3, iron(III) oxide

(b) BF_3, boron trifluoride

(c) N_2O, dinitrogen oxide (laughing gas)
(d) $MnCl_2 \cdot 4\ H_2O$, manganese(II) chloride tetrahydrate
(e) $C_6H_8O_6$, ascorbic acid or vitamin C

59. Calculate the molecular weight of each of the following compounds:
(a) $B_{10}H_{14}$, a boron hydride once considered as a rocket fuel
(b) $C_6H_2(CH_3)(NO_2)_3$, TNT, an explosive
(c) $PtCl_2(NH_3)_2$, a new cancer chemotherapy agent called "cisplatin"
(d) $CH_3CH_2CH_2CH_2SH$, has a skunk-like odor
(e) $C_{20}H_{24}N_2O_2$, quinine, used as an antimalarial drug

60. How many moles are represented by 1.00 g of each of the following compounds?
(a) CH_3OH, methyl alcohol
(b) Cl_2CO, phosgene, a poisonous gas
(c) NH_4NO_3, ammonium nitrate
(d) $MgSO_4 \cdot 7\ H_2O$, magnesium sulfate heptahydrate (epsom salt)
(e) $AgC_2H_3O_2$, silver acetate

61. Assume you have 0.250 g of each of the following compounds. How many moles of each are represented?
(a) $C_7H_5NO_3S$, saccharin, an artificial sweetener
(b) $C_{13}H_{20}N_2O_2$, procaine, a "pain killer" used by dentists
(c) $C_{20}H_{14}O_4$, phenolphthalein, a dye

62. Tin(II) fluoride is used in a well-known brand of toothpaste to prevent tooth decay. (a) How many moles of SnF_2 are there in 0.050 g of SnF_2? (b) How many F^- ions and how many Sn^{2+} ions are present in 0.050 g?

63. Strychnine, $C_{21}H_{22}N_2O_2$, is a powerful poison and has been used to eradicate rats. If a can of rat poison contains 0.75 g of strychnine, how many molecules of the compound are present?

64. Vinyl chloride, C_2H_3Cl, is used to make polyvinylchloride (PVC), a plastic from which many useful items are made. If you have one ton (2.00×10^3 lb) of vinyl chloride, how many moles and how many molecules of the compound are present? (1.00 lb = 454 g)

65. Arrange the following in order of increasing mass:
(a) 3.0×10^{23} molecules of C_4H_{10}
(b) 1 penny (about 3 g)
(c) 6.0×10^{23} molecules of CO
(d) 1.0 mole of B_2H_6
(e) 1 molecule of N_2

66. A package of baking soda contains 2.00 pounds (908 g) of $NaHCO_3$. (a) What is the name of the compound $NaHCO_3$? (b) How many moles of $NaHCO_3$ are there in the package? (c) How many moles and how many grams of oxygen atoms are in the package?

67. Monosodium glutamate, MSG, is a common food additive; its formula is

$$HOOCCH_2CH_2CH(NH_2)COONa.$$

(a) Write the molecular formula for MSG, and calculate its molecular weight. (b) How many moles of MSG are there in 2.00 g (about 1 teaspoonful) of MSG? How many molecules?

68. The most important beryllium-containing mineral is beryl, which occurs mostly as large blue-green crystals with the formula $Be_3Al_2(SiO_3)_6$. (a) What is the formula weight of beryl? (b) How many moles of beryl are there in a 0.25-g crystal? (c) How many grams of beryl must you have in order to have 10. g of beryllium?

69. Compounds containing carbon, hydrogen, and a metal are often called "organometallic." One of the best known of these is ferrocene, $(C_5H_5)_2Fe$, an orange solid. If you buy the compound from a chemical supply house for $110/kg, how much are you paying per mole?

70. An Alka-Seltzer tablet contains 324 mg of aspirin ($C_9H_8O_4$), 1904 mg of $NaHCO_3$, and 1000. mg of citric acid ($C_6H_8O_7$). (The last two compounds react with each other to provide the "fizz," bubbles of CO_2, when the tablet is put in water.) (a) Calculate the number of moles of each substance in the tablet. (b) If you take one tablet, how many molecules of aspirin are you consuming?

71. Some types of freon are used as the propellant in spray cans of paint, hair spray, and other consumer products. However, the use of freons is being curtailed, because there is some suspicion that they may cause environmental damage. If there are 250 g of the freon CCl_2F_2 in a spray can, how many molecules are you releasing to the air when you empty the can?

72. Benzene, C_6H_6, is an important industrial chemical and is produced in large amounts in spite of the fact that it is implicated as a cause of leukemia. In 1983, 9.48×10^9 pounds of benzene were produced. How many moles does this represent? (1.00 lb = 454 g)

73. DDT, $C_{14}H_9Cl_5$, is an insecticide and belongs to a class of compounds called "chlorinated hydrocarbons." Although DDT was enormously successful in controlling insects, it has also caused considerable environmental damage by interfering with the reproduction of birds. If you use 1.00 pound of DDT on several acres of farmland, how many molecules of the compound are you spreading over the fields? How many grams of chlorine (as Cl) are contained in 1.00 pound of DDT? (1.00 lb = 454 g)

74. Pepto-Bismol, which helps provide soothing relief for upset stomach, contains 300 mg of bismuth subsalicylate, $C_7H_5BiO_4$, per tablet. If you take two tab-

lets for your stomach distress, how many moles of the "active ingredient" are you taking? How many grams of Bi are you consuming in the two tablets?

75. A chemical commonly called "dioxin" has been very much in the news in the past few years. (It is the by-product of herbicide manufacture and is thought to be quite toxic.) Its formula is $C_{12}H_4Cl_4O_2$. If you have a sample of dirt (1 ounce or 28.4 g) that contains 1.0 $\times 10^{-4}$% dioxin, how many moles of dioxin are in the dirt?

76. Vitamin C, ascorbic acid, has the formula $C_6H_8O_6$. (a) The recommended daily dose of vitamin C is 60. milligrams. How many moles are you consuming if you ingest 60 milligrams of the vitamin? (b) A typical tablet contains 1.00 gram of vitamin C. How many moles of vitamin C does this represent? (c) When you consume 1.00 gram of vitamin C, how many oxygen atoms are you eating?

77. Rotenone, $C_{23}H_{22}O_6$, is the active component in many garden insecticides. If a can of insecticide contains 0.500 pound (227 g) of powdered insecticide and the insecticide is 5.0% by weight rotenone, how many moles of rotenone are present? How many molecules?

78. Quinine, isolated from the bark of the cinchona tree, is used in the form of its salt with HCl as an important antimalarial drug. The formula of the drug is $C_{20}H_{24}N_2O_2 \cdot 2\ HCl$. The usual dose of the drug is 2.0 g per day. How many moles of the drug would you be taking at this dose level?

PERCENT COMPOSITION

79. Calculate the molecular weight of these compounds and the weight percent of each element.
    (a) C_2H_6, ethane, a hydrocarbon fuel
    (b) $C_2H_4O_2$, acetic acid, an important ingredient in vinegar
    (c) $C_2H_3O_5N$, peroxyacetyl nitrate, an objectionable compound in photochemical smog
    (d) $C_4H_{10}O_3NPS$, acephate, an insecticide

80. Acrylonitrile, H_2CCHCN, is the basis of many important plastics and fibers. (a) Calculate the molecular weight. (b) Calculate the weight percent of each element in the compound.

81. The formula of DDT is $C_{14}H_9Cl_5$ (see Question 73). Calculate the molecular weight of the compound and the weight percent of each element.

82. Hexachlorophene, $C_{13}H_6Cl_6O_2$, is a germicide in soaps. Calculate the weight percentage of each element in the compound.

83. A close chemical relative of DDT is methoxychlor ($C_{16}H_{15}Cl_3O_2$), a compound used in garden insecticides. (a) Calculate the weight percentage of each element in the compound. (b) If the estimated fatal dose for a human is 7.5 g/kg of body weight, what is your fatal dose? (Remember that 1.00 lb = 0.454 kg.)

EMPIRICAL AND MOLECULAR FORMULAS

84. The empirical formula of maleic acid is CHO. Its molar mass is 116.1 g/mol. What is its molecular formula?

85. Acetylene is a colorless gas that is used as a fuel in welding torches, among other things. It is 92.26% C and 7.74% H. Its molar mass is 26.02 g/mol. Calculate the empirical and molecular formulas.

86. There is a large family of boron–hydrogen compounds called boron hydrides. All have the formula B_xH_y and almost all react with air and burn or explode. One member of this family contains 88.5% B; the remainder is hydrogen. Which of the following is its empirical formula: BH_3, B_4H_{10}, B_5H_7, B_5H_{11}, B_6H_{12}.

87. Nitrogen and oxygen form an extensive series of oxides of general formula N_xO_y; at least seven are known. One of them is a blue solid that comes apart, reversibly, in the gas phase. It contains 36.84% N. What is the empirical formula of this oxide?

88. Acetic acid is the important ingredient in vinegar. It is composed of carbon (40.0%), hydrogen (6.71%), and oxygen (53.29%). Its molar mass is 60.0 g/mol. Determine the empirical and molecular formulas of the acid.

89. What is the molecular formula of a substance that contains in 1.00 mole of sample 1.00 mole of S, 24.1 $\times 10^{23}$ atoms of F, and 71.0 grams of Cl?

90. An analysis of nicotine, a poisonous compound found in tobacco leaves, shows that it is 74.0% C, 8.65% H, and 17.35% N. Its molar mass is 162 g/mol. What are the empirical and molecular formulas of nicotine?

91. Cacodyl, a compound containing arsenic, was reported in 1842 by the German chemist Bunsen. It has an almost intolerable garlic-like odor. Its molar mass is 210 g/mol, and it is 22.88% C, 5.76% H, and 71.36% As. Determine its empirical and molecular formulas.

92. The action of bacteria on meat and fish produces a poisonous compound called cadaverine. As its name and origin imply, it stinks! It is 58.77% C, 13.81% H, and 27.40% N. Its molar mass is 102.2 g/mol. Determine the molecular formula of cadaverine.

93. Vanillin is a common flavoring agent. It has a molar mass of 152 g/mol and is 63.15% C and 5.30% H; the remainder is oxygen. Determine the molecular formula of vanillin.

94. Fluorocarbonyl hypofluorite was recently isolated, and analysis showed it to be 14.6% C, 39.0% O, and 46.3% F. If the molar mass of the compound is 82 g/mol, determine the empirical and molecular formulas of the compound.

95. Naphthalene, best known in the form of "moth balls," is composed only of carbon (93.75%) and hydrogen (6.25%). If the molar mass of the compound is 128 g/mol, what is the molecular formula of naphthalene?

96. A major oil company has used a gasoline additive called MMT to boost the octane rating of its gasoline. What is the empirical formula of MMT if it is 49.5% C, 3.2% H, 22.0% O, and 25.2% Mn?

97. The hemoglobin from the red corpuscles of most mammals contains about 0.33% iron. Physical measurements indicate that hemoglobin is a very large molecule with a molar mass of about 6.8×10^4 g/mol. How many moles of Fe are there in one mole of hemoglobin? How many iron atoms are there in one molecule of hemoglobin?

98. Metal carbonyls are molecular compounds formed by metals and carbon monoxide, CO. Although they can be quite stable at room temperature, they usually decompose to the metal and CO when heated in a vacuum. Assume that you heat 1.400 g of $Mn_x(CO)_y$ and obtain 0.394 g of pure manganese. The CO is lost in the course of the experiment. If the molar mass of the compound is 390 g/mol, what are the empirical and molecular formulas of the compound?

99. Butane, which contains only C and H, is a commonly used fuel in camping stoves. To determine the formula of butane, assume you burn 0.580 g of it and obtain 1.760 g of CO_2 and 0.900 g of H_2O. What is the empirical formula of butane?

100. Anthracene, which contains only carbon and hydrogen, is an important source of dyes. Determine its empirical and molecular formulas from the following data: (a) When 2.50 mg is burned in pure oxygen, 8.64 mg of CO_2 and 1.26 mg of H_2O are isolated. (b) The experimental molar mass of the compound is 178 g/mol.

101. Vitamin C is a compound containing the elements C, H, and O. Determine the empirical formula of vitamin C from the following data: 4.00 mg of the solid vitamin is burned in oxygen to give 6.00 mg of CO_2 and 1.632 mg of H_2O.

102. The formula of a boron hydride, a compound containing B and H, can be determined by burning a sample in oxygen and collecting and weighing the combustion products, B_2O_3 and H_2O. Assume that a sample of a particular boron hydride is burned and 1.740 g of B_2O_3 and 0.810 g of H_2O are obtained. What is the empirical formula of the compound?

103. Silicon and hydrogen form a series of interesting compounds, Si_xH_y. To find the formula of one of them, you take a 6.22 g sample of the compound and burn it in oxygen. On doing so, all of the Si is converted to 12.02 g of SiO_2 and all of the H to 5.40 g of H_2O. What is the empirical formula of the silicon compound? If the experimental molar mass is 62.2 g/mol, what is its molecular formula?

104. Metal carbonyls, compounds containing a metal bonded to carbon monoxide, constitute a large class of compounds. Iron, for example, forms three such compounds of general formula $Fe_x(CO)_y$. The empirical formula of one of them can be determined from the following data: A 1.256 g sample of a black solid, $Fe_x(CO)_y$, is burned in air to give 0.598 g of Fe_2O_3 and 1.317 g of CO_2. If the molar mass of the compound is 504 g/mol, what is the molecular formula of the compound?

105. Vinyl chloride is used in the plastics industry, but it has recently been implicated as a cancer-causing agent. The compound contains C, H, and Cl; if you burn it in oxygen, you obtain CO_2, H_2O, and HCl. Assume that you burn 3.125 g of vinyl chloride and obtain 4.400 g of CO_2 and 1.825 g of HCl plus some water. From the weights of CO_2 and HCl, you can determine the weight percentage of C and Cl in vinyl chloride. What is the empirical formula of the compound? If its molar mass is 62.3 g/mol, what is its molecular formula?

106. Ruthenium chemistry is quite interesting, and a good starting material for such studies is $RuCl_3 \cdot x\ H_2O$. If you heat 1.056 g of the hydrated salt and find that only 0.838 g of $RuCl_3$ remains when all of the water has been driven off, what is the value of x?

107. The "alum" used in cooking is potassium aluminum sulfate hydrate, $KAl(SO_4)_2 \cdot x\ H_2O$. To find the value of x, you can heat a sample of the compound to drive off all of the water and leave only $KAl(SO_4)_2$. Assume that you heat 4.74 g of the hydrated compound and that it loses 2.16 g of water. What is the value of x?

108. If "epsom salt," $MgSO_4 \cdot x\ H_2O$, is heated to 250°C, all of the water of hydration is lost. On heating a 1.687 g sample of the hydrate, 0.824 g of $MgSO_4$ remains. How many molecules of water are there per formula unit of $MgSO_4$?

109. Copper sulfate as commonly used in the laboratory is the hydrated compound $CuSO_4 \cdot 5\ H_2O$ (see Figure 2.14). If you heat the solid to at least 150°C, all of the water of hydration is lost. On heating 10.5 g of $CuSO_4 \cdot 5\ H_2O$ to this temperature, how many grams of water would be lost and how many grams of anhydrous $CuSO_4$ would remain?

CHEMICAL ANALYSIS

110. You can perform a quantitative analysis for barium in a sample by isolating the Ba^{2+} ion in the sample in the form of insoluble $BaSO_4$ (Figure 2.16). If 8.23 g of $BaSO_4$ can be obtained from 20.0 g of a barium-containing ore, what is the weight percentage of barium in the ore?

111. The amount of calcium present in milk can be determined by adding oxalate ion, $C_2O_4^{2-}$ (in the form of its water-soluble sodium salt, $Na_2C_2O_4$); the insoluble compound CaC_2O_4 is precipitated. Suppose you take a 75.0-g sample of milk and isolate 0.288 g of CaC_2O_4 from it. What is the weight percentage of calcium in the milk?

112. The magnets you buy in novelty stores are often made of "Alnico," a mixture of aluminum, nickel, and cobalt (hence the name Al-ni-co). You can analyze a sample of Alnico for its nickel content by dissolving a piece in nitric acid (to form Ni^{2+} in solution), and then, under suitable conditions, adding a compound called dimethylglyoxime to precipitate the nickel in the form of a red compound whose formula is $Ni(C_4H_7N_2O_2)_2$. Suppose you analyze a 0.4734 g piece of magnet, and you obtain 0.4659 g of $Ni(C_4H_7N_2O_2)_2$. What is the weight percent of Ni in the Alnico magnet?

113. A 4.22-g sample of $CaCl_2$ and NaCl was dissolved in water, and the solution was treated with sodium carbonate to precipitate the calcium as $CaCO_3$. After isolating the solid $CaCO_3$, it was heated to drive off CO_2 and form 0.959 g of CaO. What is the weight percent of $CaCl_2$ in the original 4.22-g sample?

114. A common laboratory experiment for introductory chemistry courses involves the analysis of a mixture of $BaCl_2$ and $BaCl_2 \cdot 2\,H_2O$. You are given a white powder that is a mixture of these compounds and are asked to determine the weight percentage of each. You find that 1.292 g of the mixture has a mass of only 1.187 g after heating to drive off all of the water present. (a) What is the weight percentage of water in the mixture? (b) What is the weight percentage of $BaCl_2 \cdot 2\,H_2O$ in the mixture? (c) What is the weight percentage of Ba in the mixture?

115. A simple salt containing only the magnesium ion, chloride ion, and water is analyzed for all of its components. A 1.072-g sample of the compound is first heated to release all of the water of hydration, and 0.503 g of a salt containing only magnesium ion and chloride ion is obtained. This 0.503-g sample is analyzed for magnesium ion by adding aqueous $(NH_4)_3PO_4$ to precipitate the magnesium as $MgNH_4PO_4$; 0.725 g of this solid is obtained. The 0.503-g sample is also analyzed for chloride ion by adding aqueous $AgNO_3$ to precipitate the chloride as AgCl; 1.514 g is isolated. What is the complete, empirical formula of the hydrated salt? From your knowledge of the probable charges on magnesium and chloride ions, is this the formula you would expect?

GENERAL QUESTIONS

116. A compound with a formula M_3N contains 0.673 g of N per gram of the metal M. What is the atomic weight of M? What element is M?

117. A compound containing only a metal and oxygen, MO, can be decomposed to the elements (M and O_2) by heating. If 4.386 g of the compound form 4.063 g of M on heating, what is the atomic weight of M? What is the probable identity of M?

118. The formula of a compound containing Cu and S can be determined by heating together a weighed amount of copper and an excess of sulfur in a crucible. After the reaction is complete, any unreacted sulfur can be disposed of by heating vigorously to turn the excess sulfur to a vapor. The following data were obtained in the laboratory:

Mass of crucible	19.732 g
Mass of crucible plus Cu	27.304 g
Mass of crucible plus the Cu–S compound	29.214 g

What is the simplest formula of the Cu–S compound?

119. A 1.00-g sample of europium chloride, $EuCl_x$, is dissolved in water and $AgNO_3$ is added. All of the chlorine precipitates in the form of 1.28 g of AgCl. Calculate the atomic weight of europium from these data.

120. You are given a sample of pure potassium chlorate, $KClO_3$, with a mass of 4.008 g. After heating to drive off all of the O, the remaining solid, KCl, was found to have a mass of 2.438 g. This residue was dissolved in water and treated with $AgNO_3$. This led to the formation of a precipitate of AgCl, which contained all of the Cl in original sample. The AgCl had a mass of 4.687 g. After further treatment of the AgCl, it was found to contain 3.531 g of Ag. Use these data to calculate the atomic weights of Ag, Cl, and K relative to O = 15.999?

121. Equal weights of Zn metal and iodine, I_2, are mixed together, and the iodine is converted completely to ZnI_2. What fraction by weight of the original zinc remains unreacted?

122. A 1.000-g mixture of copper(I) oxide, Cu_2O, and copper(II) oxide, CuO, was reduced quantitatively to give 0.839 g of metallic copper. What is the weight of copper (I) oxide in the original 1.000-g sample?

123. A mixture of KBr and NaBr weighing 0.560 g was dissolved in water and then treated with $AgNO_3$. All the bromide ion from the original sample was recovered in the form of 0.970 g of AgBr. What is the fraction by weight of KBr in the original sample?

3

An Introduction to Chemical Reactions

Reaction of aluminum and bromine to give aluminum bromide (Charles D. Winters)

CHAPTER OUTLINE

Chemistry in general, and this course in particular, has two fundamental objectives: (1) to understand the structure of matter and the forces that hold it together and (2) to understand the reactions of molecules, ions, and atoms, and the forces controlling those reactions. In the previous chapters some aspects of atoms, simple ions, molecules, and molecular ions are described in preliminary form, so you are now acquainted with the basics of the first objective of the course. This chapter turns your attention to the concept of the chemical reaction and lays the groundwork for later chapters.

3.1 CHEMICAL EQUATIONS

Take a piece of thin ribbon of magnesium metal and put it into a bunsen burner flame. It will burst into flame, give off a brilliant white light, and then flicker out to leave a pile of white powder as its epitaph (Figure 3.1). The process or **chemical reaction** that has occurred is the combination of solid Mg with gaseous O_2 in the air to give solid magnesium oxide, MgO. We can depict this by the following **chemical equation** where (s) and (g) show the physical states of the **reactants** (the materials put into the reaction) and the **product**.*

$$2\ Mg(s) + O_2(g) \longrightarrow 2\ MgO(s)$$

In addition to showing the reactants and products in their physical states, a **balanced chemical equation** shows the relative amounts of products and reactants.

In the 18th century, Antoine Lavoisier showed that there is *conservation of matter* in any chemical process: matter can neither be created nor destroyed. This means 2 atoms of Mg and 1 molecule of O_2 (that is, two atoms of O) must produce 2 "molecules" of MgO,† a compound containing one atom each of Mg and O. Thus, the balanced equation tells you that 2 atoms of Mg were used to react with 2 atoms of O and that

FIGURE 3.1

A piece of magnesium ribbon burns in air to give the white solid magnesium oxide (MgO) as the product. (Charles D. Winters)

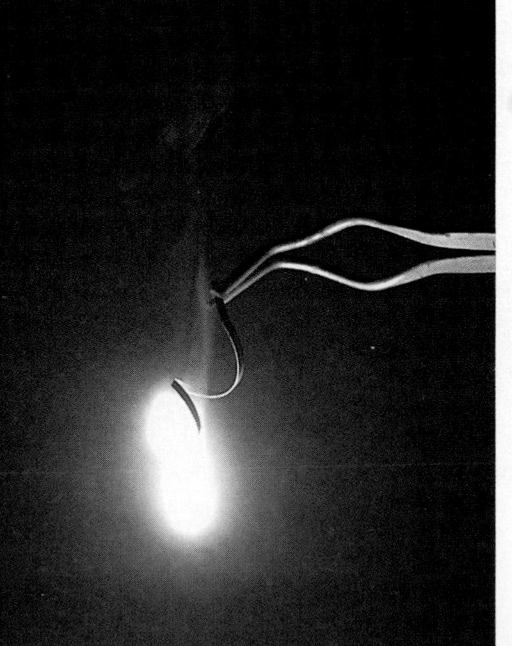

*You might be curious as to the origin of the light given off in the burning of magnesium. As described in Chapters 5 and 18, reactions almost always involve the *net* input or production of energy. In the case of Mg and O_2, there is a net production of energy, some of the energy appearing in the form of visible light and the rest as heat.

†MgO is actually an ionic compound with the same structure as NaCl (Figure 2.11). For the sake of simplicity, however, we consider one MgO unit a "molecule."

ANTOINE LAURENT LAVOISIER (1743–1794)

On Monday, August 1, 1774, the Englishman Joseph Priestley (1733–1804) first isolated oxygen from the decomposition of mercury(II) oxide, HgO. Priestley did not at first recognize the importance of the discovery, but he mentioned it to Lavoisier in October, 1774. One of Lavoisier's contributions to science was his recognition of the importance of exact scientific measurements and of carefully planned experiments, and he applied these methods to the study of oxygen. From this work he came to believe the gas entered the composition of all acids and so named it "oxygen," from Greek words meaning "to form an acid." In addition, his experiments suggested to him that oxygen combined with carbon in the body, a reaction that was the source of the heat in a living organism. Although he did not understand the details of the process, this was a step in the development of biochemistry.

Lavoisier was a prodigious scientist and introduced principles of naming chemical substances that are still in use today. Further, he wrote a textbook in which he applied for the first time the principle of the conservation of matter to chemistry and used the idea to write early versions of chemical equations.

As Lavoisier was an aristocrat, he came under suspicion during the Reign of Terror of the French Revolution, and his career was cut short on May 8, 1794, by the guillotine.

they appear combined in the product. Of course, the balanced equation also tells you that 1000 atoms of Mg react with 500 molecules of O_2 or 1000 atoms of O (500 molecules × 2 atoms/molecule = 1000 atoms of O) and produce 1000 "molecules" of MgO. To carry this argument further, it is also true that 6.022×10^{23} molecules of O_2 (1 mole) react with $2 \times 6.022 \times 10^{23}$ (12.04×10^{23}) atoms of Mg (2 moles) to give 12.04×10^{23} molecules of MgO (2 moles). As demanded by the conservation of matter, the total number of atoms used as reactants was 24.08×10^{23}, and the total number of atoms appearing in the product is 24.08×10^{23}.

2 Mg(s)	+ O_2(g)	$\longrightarrow$	2 MgO(s)
2 atoms Mg	+ 1 molecule O_2	$\longrightarrow$	2 "molecules" MgO
2 moles Mg	+ 1 mole O_2	$\longrightarrow$	2 moles MgO
12.04×10^{23} atoms Mg	+ 6.02×10^{23} molecules O_2	$\longrightarrow$	12.04×10^{23} "molecules" MgO
12.04×10^{23} atoms Mg	+ 12.04×10^{23} atoms O	$\longrightarrow$	24.08×10^{23} total atoms Mg and O

From Chapter 2 you know that 1 mole of Mg is equivalent to 24.13 g and that 1 mole of O_2 amounts to 32.00 g. The balanced equation tells you that, if one mole of O_2 is consumed, then 2 moles of Mg must also be used (48.62 g), so the total mass of reactants must be 80.62 g.

1 mole O_2	=	32.00 g
+ 2 moles Mg	=	+ 48.62 g
Total mass of reactants	=	80.62 g

The principle of the conservation of matter demands that the same mass, 80.62 g of MgO, results from the reaction. Of course the balanced equation shows that this is the case.

$$2\ \cancel{\text{mol MgO}}\left(\frac{40.31\text{ g MgO}}{1\ \cancel{\text{mol MgO}}}\right) = 80.62\text{ g MgO}$$

The word "stoichiometry," pronounced "stoy-key-AHM-uh-tree," is derived from the Greek words *stoicheion* (meaning "element") and *metron* (meaning "measure").

The relationship between the masses of chemical reactants and products is called **stoichiometry**, and the coefficients (or multiplying numbers) in a balanced equation are the **stoichiometric coefficients**. We shall take up the important role of stoichiometry in Chapter 4. However, you should first recognize that *an equation must be balanced before useful quantitative information can be obtained.*

3.2 BALANCING CHEMICAL EQUATIONS

Balancing a chemical equation ensures that the same number of atoms of each element involved appears on each side of the equation. All chemical equations can be balanced by inspection, although some will involve more inspection than others.

One general class of chemical reactions involves the *reaction of metals with halogens* such as Cl_2, Br_2, and I_2 to give metal halides. For example, zinc reacts with iodine to give zinc iodide (Figure 3.2),

$$\text{Zn(s)} + \text{I}_2\text{(s)} \longrightarrow \text{ZnI}_2\text{(s)}$$

and aluminum reacts vigorously with liquid bromine to give aluminum bromide Al_2Br_6 (Figure 3.3).

$$\text{Al(s)} + \text{Br}_2\text{(l)} \xrightarrow{\text{(unbalanced equation)}} \text{Al}_2\text{Br}_6\text{(s)}$$

The Zn/I_2 equation is balanced as written above, since there are 1 Zn and 2 I atoms on each side. For the Al/Br_2 equation, however, we balance the Al atoms by placing a 2 in front of Al on the left,

$$2\text{ Al(s)} + \text{Br}_2(\ell) \xrightarrow{\text{(unbalanced equation)}} \text{Al}_2\text{Br}_6\text{(s)}$$

FIGURE 3.2

A metal reacting with a halogen. Zinc dust (Zn) [on the left in (a)] reacts with iodine (I_2) [on the right in (a)] to give zinc iodide (ZnI) (b). The heat of the reaction is great enough that excess iodine evaporates as a purple vapor (b). (Charles D. Winters)

(a)

(b)

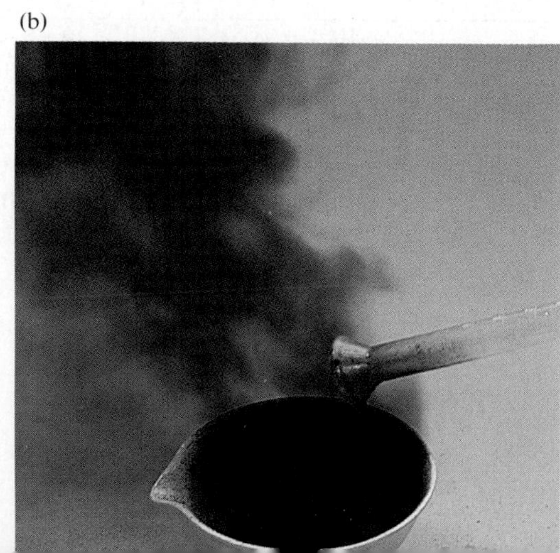

and then a 3 in front of the Br_2 on the left to balance the 6 Br atoms appearing on the right side of the equation.

$$2\ Al(s) + 3\ Br_2(\ell) \xrightarrow{\text{balanced equation}} Al_2Br_6(s)$$

Reactions with oxygen, O_2, are another major class of chemical reaction (and are described in Section 3.4). A **combustion reaction**, the reaction of a compound with O_2 to form products in which all elements are combined with oxygen, is one such reaction. Such reactions are typical of compounds containing C and H, and an excellent example is the combustion of propane, C_3H_8, a hydrocarbon commonly used in industry and homes as LP gas.

$$C_3H_8(g) + O_2(g) \xrightarrow{\text{(unbalanced equation)}} CO_2(g) + H_2O(g)$$

Compounds having only C and H (or C, H, and O) always give just CO_2 and H_2O on complete combustion. Although there are several systematic ways to balance such an equation, our advice is usually to leave the oxygen balance to the last step. Generally, it is best to balance the C atoms first, followed by H, and then O.

Step 1. *Balance the C atoms.* There are 3 carbon atoms in the reactants, so there must be 3 in the products. Therefore, we need 3 CO_2 molecules on the right side.

$$C_3H_8 + O_2 \longrightarrow 3\ CO_2 + H_2O$$

Step 2. *Balance the H atoms.* There are 8 H atoms in the reactants. Each molecule of water has 2 H atoms, so 4 molecules of water will have the required 8 H atoms.

$$C_3H_8 + O_2 \longrightarrow 3\ CO_2 + 4\ H_2O$$

Step 3. *Balance the oxygen atoms.* There are 10 oxygen atoms on the right side ($3 \times 2 = 6$ in CO_2 and $4 \times 1 = 4$ in water). Therefore, you need 5 O_2 molecules to supply the required 10 oxygen atoms.

$$C_3H_8 + 5\ O_2 \longrightarrow 3\ CO_2 + 4\ H_2O$$

Step 4. *Verify that each element is balanced.* The reaction involves 3 C atoms, 8 H atoms, and 10 O atoms on each side.

EXAMPLE 3.1

BALANCING THE EQUATION FOR A COMBUSTION REACTION

Write a balanced equation for the combustion of butane, C_4H_{10}.

Solution As is always the case for compounds containing only C, H, and O, the products will be CO_2 and H_2O if the reaction goes to completion. Therefore, the unbalanced equation is

$$C_4H_{10}(g) + O_2 \longrightarrow CO_2(g) + H_2O(g)$$

Step 1. *Balance the C atoms.* 4 C atoms in butane require the production of 4 CO_2 molecules.

$$C_4H_{10} + O_2 \longrightarrow 4\ CO_2 + H_2O$$

(a)

(b)

(c)

FIGURE 3.3
Bromine (Br_2), an orange-brown liquid and aluminum metal (a) react so vigorously that the aluminum becomes molten and glows white hot (b). The vapor in (b) consists of vaporized Br_2 and some of the product, Al_2Br_6. At the end of the reaction the beaker is coated with aluminum bromide and products of its reaction with atmospheric water. (Charles D. Winters)

Step 2. *Balance the H atoms.* There are 10 H atoms on the left, so 5 molecules of H_2O are required on the right.

$$C_4H_{10} + O_2 \longrightarrow 4\ CO_2 + 5\ H_2O$$

Step 3. *Balance the O atoms.* As the reaction stands after step 2, there are 13 O atoms on the right ($4 \times 2 = 8$ for CO_2 plus $5 \times 1 = 5$ for H_2O) and 2 on the left. That is, there is an odd number on the right and an even number on the left. There are two equally valid ways to balance the oxygen.

Solution (*1*). To have 13 atoms of oxygen on the left side, use a stoichiometric coefficient of 13/2; you can do this because

$$\left(\frac{13}{2}\right)\left(\frac{2 \text{ atoms of O}}{\text{molecule of } O_2}\right) = 13 \text{ oxygen atoms}$$

Therefore, the balanced equation will be

$$C_4H_{10} + 13/2\ O_2 \longrightarrow 4\ CO_2 + 5\ H_2O$$

Solution (2). Taking the equation as it stands after step 2, multiply each coefficient by 2 so that there is an even number of oxygen atoms on the right side (i.e., 26).

$$2\ C_4H_{10} + (\text{—}\ O_2) \longrightarrow 8\ CO_2 + 10\ H_2O$$

Now the O_2 on the left can be balanced by multiplying by a whole number instead of a fraction.

$$2\ C_4H_{10} + 13\ O_2 \longrightarrow 8\ CO_2 + 10\ H_2O$$

While you will find it convenient at times to use fractional coefficients, generally equations are written with whole-number coefficients.

Step 4. As *verification*, notice that there are 8 C atoms, 20 H atoms, and 26 O atoms on each side of the equation from solution (2) above.

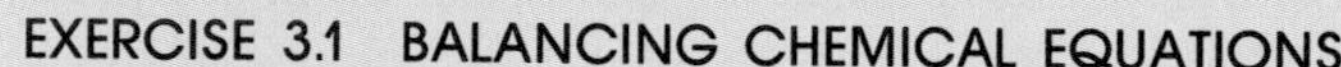

EXERCISE 3.1 BALANCING CHEMICAL EQUATIONS

Balance the chemical equations for the following:

(a) the oxidation of iron

$$Fe(s) + O_2(g) \longrightarrow Fe_2O_3(s)$$

(b) the combustion of methane

$$CH_4(g) + O_2(g) \longrightarrow CO_2(g) + H_2O(g)$$

(c) the combustion of B_4H_{10} in O_2 to give $B_2O_3(s)$ and $H_2O(g)$

(d) the reaction of CO with H_2 to give methyl alcohol, CH_3OH (a reaction that serves as the basis of a process for making synthetic fuel from coal)

(e) the combustion of octane, C_8H_{18}

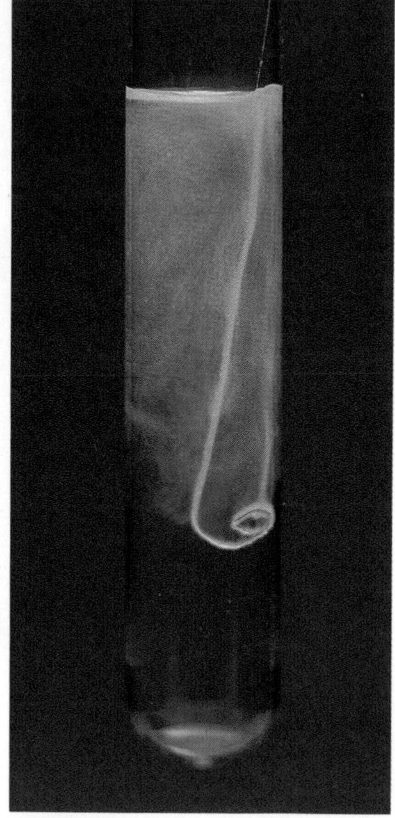

FIGURE 3.4

A ribbon of magnesium metal reacts with aqueous HCl to give H_2 gas and aqueous $MgCl_2$. (Charles D. Winters)

3.3 REACTIONS OF IONIC COMPOUNDS IN WATER

A large number of reactions proceed in water, that is, in *aqueous solution.* For example, if you put a magnesium ribbon into aqueous hydrochloric acid, the mixture will bubble furiously as H_2 gas is given off according to the balanced equation (Figure 3.4),

$$Mg(s) + 2\ HCl(aq) \longrightarrow MgCl_2(aq) + H_2(g)$$

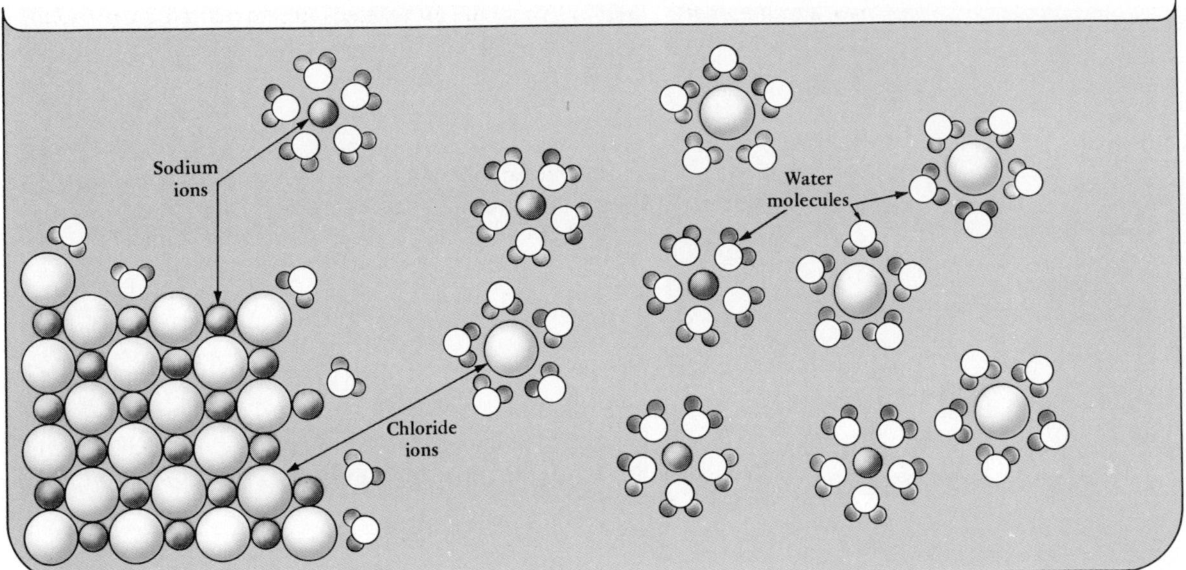

FIGURE 3.5
A model for the process of dissolving NaCl in water.

where (aq) indicates HCl and $MgCl_2$ are in *aq*ueous solution. One of the products, $MgCl_2$, is an ionic compound, and HCl forms H^+ and Cl^- ions in water. It is important to understand some special aspects of reactions involving ions, so we wish first to describe the properties of ionic compounds in water and then turn to their reactions.

SOME PROPERTIES OF IONIC COMPOUNDS

ELECTROLYTES Years ago you made the "experimental" observation that common table salt, NaCl, dissolves in water to form a solution. The dissolving process on a molecular scale is illustrated in Figure 3.5. When a crystal of Na^+ and Cl^- ions is placed in water, water molecules are attracted to the ions (by forces described in Chapter 11). This water-ion attraction cloaks each ion on the surface of the crystal with water molecules, and the ions are pulled into the bulk water. The ions, now sheathed in water molecules, are free to move about. Under normal conditions, the movement of ions in solution is random, and the Na^+ and Cl^- ions are dispersed uniformly throughout the solution. However, if you place two **electrodes** or electrically conducting plates in the solution, and connect the plates to a battery, the plates take on an electrical charge, one positive and one negative. (The purpose of the battery is to remove electrons from one electrode and move them to the other.) Since the ions are also electrically charged, cations will migrate through the solution to the negative plate and anions to the positive plate (Figure 3.6). This movement of ions in solution constitutes an **electric current**. As the cations collect near the negatively charged electrode and anions around the positive one, the battery finds it easier to move electrons from one electrode to the other.

When a compound dissolves in the water to give a **solution**, the compound is the **solute**, the water is the **solvent**. (See Section 1.3.)

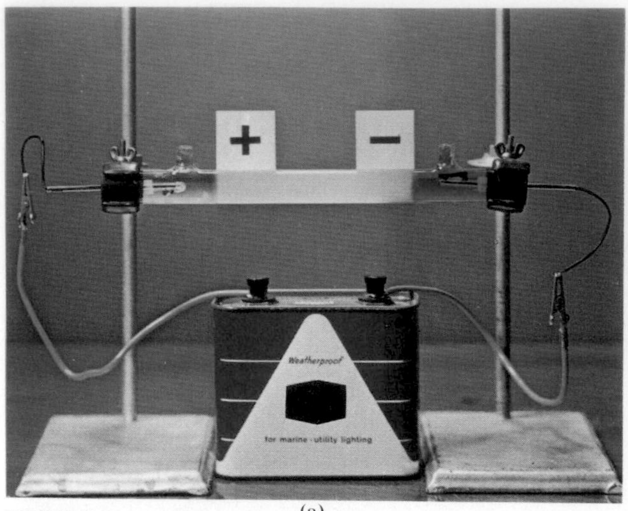

(a)

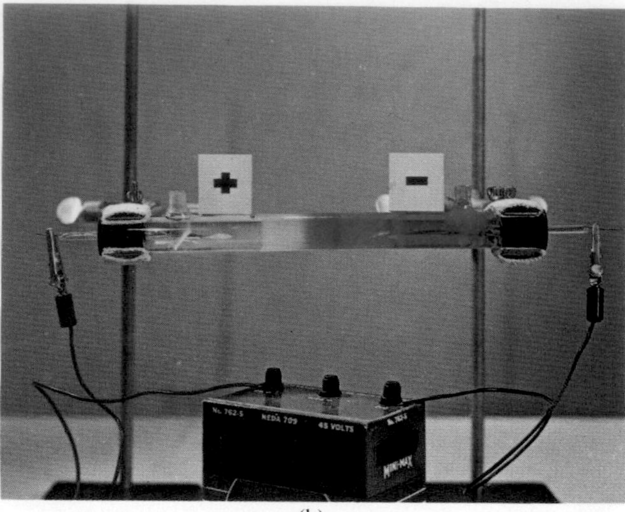

(b)

FIGURE 3.6
Solutions of blue Cu^{2+} ions and yellow CrO_4^{2-} (chromate) ions are separated by a gel (a). When attached to a battery, electrodes in the apparatus become electrically charged, and the ions migrate toward the oppositely charged electrode (b). (The anion associated with Cu^{2+} and the cation associated with CrO_4^- also migrate to the appropriate electrode. However, they are colorless and so are not visible.) (From H. C. Metcalfe)

Therefore, simultaneous with ion movement in solution, electrons move in the external circuit. If a light bulb is inserted into the circuit as in Figure 3.7, the bulb will light, showing that the circuit is complete. Ionic compounds that behave in this manner when placed in solution are called **electrolytes**.

Electrolytes can be classified roughly as *strong* or *weak*. NaCl and other ionic compounds that dissolve in water (see below) come apart or *dissociate completely* into ions in water and are good conductors of electricity; they are strong electrolytes. Conversely, there are some substances, such as acetic acid (see Figure 3.7 and the section on acids and bases), that provide only a few ions and so conduct electricity only poorly; they are weak electrolytes.

The words anion, cation, and electrolyte originated with Michael Faraday. See his biography in the box.

There are also substances that dissolve in water but do not allow the solution to conduct electricity; they are called **non-electrolytes**. There are many compounds that are non-electrolytes, among them pure water, ethyl alcohol, common sugar, and antifreeze.

SOLUBILITY OF IONIC COMPOUNDS IN WATER Not all ionic compounds will dissolve in or are soluble in water. There are in fact many that do not, and there are still others that dissolve only to a small extent. Fortunately, we can make some general statements about which types of ions can lead to soluble ionic compounds.

$NaNO_3$ dissolves to the extent of 92 g per 100 mL of water at 25°C.

Ionic compounds that dissolve in water can be assumed to be electrolytes.

Table 3.1 is a set of broad guidelines that can help you predict the likelihood that an ionic compound will be soluble in water.* For example, $NaNO_3$ contains both an alkali metal cation, Na^+, and the nitrate anion, NO_3^-; therefore, the compound is predicted to be soluble. Further, since it dissociates into its ions in water, $NaNO_3$ is a strong electrolyte.

*You will occasionally run across an ionic compound that does not follow these guidelines.

TABLE 3.1 General Guidelines for Solubility in Water

SOLUBLE COMPOUNDS	EXCEPTIONS
Almost all sodium (Na^+), potassium (K^+), and ammonium (NH_4^+) salts	
All chlorides (Cl^-), bromides (Br^-), iodides (I^-)	Ag^+, Hg_2^{2+}, Pb^{2+}
All fluorides (F^-)	Mg^{2+}, Ca^{2+}, Sr^{2+}, Ba^{2+}, Pb^{2+}
All nitrates (NO_3^-), chlorates (ClO_3^-), perchlorates (ClO_4^-), and acetates ($C_2H_3O_2^-$)	Acetates of Ag^+ and Hg_2^{2+} are only moderately soluble
All sulfates (SO_4^{2-})	Sr^{2+}, Ba^{2+}, Pb^{2+}; (Ca^{2+} and Ag^+ moderately soluble)
POORLY SOLUBLE SALTS	**EXCEPTIONS**
All carbonates (CO_3^{2-}), phosphates (PO_4^{3-}), oxalates ($C_2O_4^{2-}$), and chromates (CrO_4^{2-})	Na^+, K^+, NH_4^+
All sulfides (S^{2-})	Alkali and alkaline earth metal ions and NH_4^+
All hydroxides (OH^-) and oxides (O^{2-})	Alkali metals (those of Ca, Sr, and Ba only moderately soluble)

$$NaNO_3 \text{ in water} \longrightarrow Na^+(aq) + NO_3^-(aq)$$
strong electrolyte

On the other hand, CuS is quite *in*soluble, just like almost all sulfides except those where the cation is an alkali metal ion or the ammonium ion. Other examples of common salts are given in Figure 3.8. Be sure to notice that, *as long as the compound contains one of the ions leading to solubility* (Table 3.1), *it is at least moderately soluble in water.*

(a)

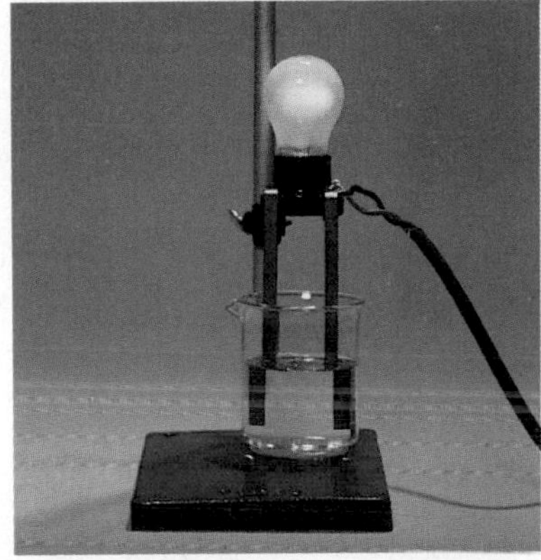
(b)

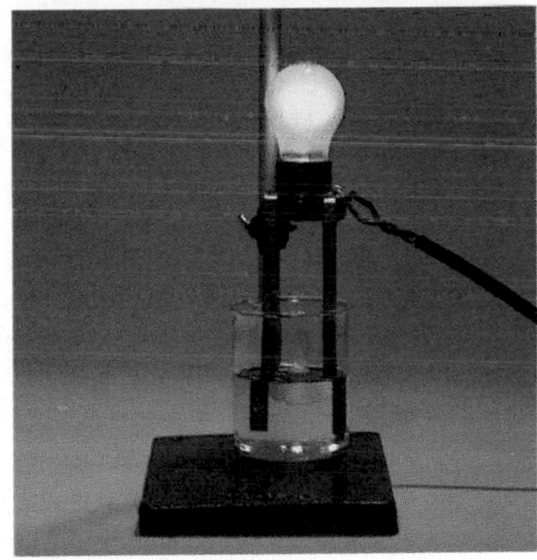
(c)

FIGURE 3.7 When an electrolyte is dissolved in water in the beaker, the circuit is completed, and the light bulb glows. The greater the number of ions in solution, the brighter the bulb will glow. (a) Pure water is a nonelectrolyte. (b) Dilute acetic acid, a weak electrolyte, supplies only a few ions to the solution. (c) Dilute HCl, hydrochloric acid, is a strong electrolyte. (John King)

EXAMPLE 3.2

SOLUBILITY GUIDELINES

For each of the following ionic compounds, predict whether it is likely to be water soluble. If it is soluble, tell what ions exist in solution.

(a) KCl (b) $MgCO_3$ (c) NiO (d) CaI_2

Solution You must first recognize the cation and anion involved and then decide the probable water solubility. As long as one ion is listed in Table 3.1 as leading to solubility, the compound is likely to be soluble.

(a) KCl is composed of K^+ and Cl^-. According to Table 3.1, the presence of either of these ions means that the compound is likely to be soluble in water. Indeed, its actual solubility is about 35 g in 100 mL of water at 20°C. Thus, a solution of KCl actually consists of K^+ and Cl^- ions, and KCl is an electrolyte.

$$KCl \text{ in water} \longrightarrow K^+(aq) + Cl^-(aq)$$

(b) Magnesium carbonate is composed of the Mg^{2+} cation and the CO_3^{2-} anion. Mg^{2+} is in the alkaline earth group, a group that usually does not form soluble compounds. The carbonate ion usually gives *in*soluble compounds (Table 3.1), unless combined with something like Na^+ or NH_4^+. Therefore, $MgCO_3$ is

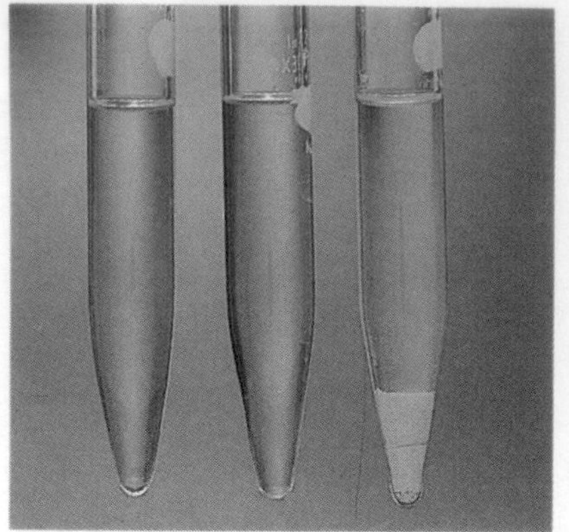

(a) $Ba(NO_3)_2$, $BaCl_2$, $BaSO_4$

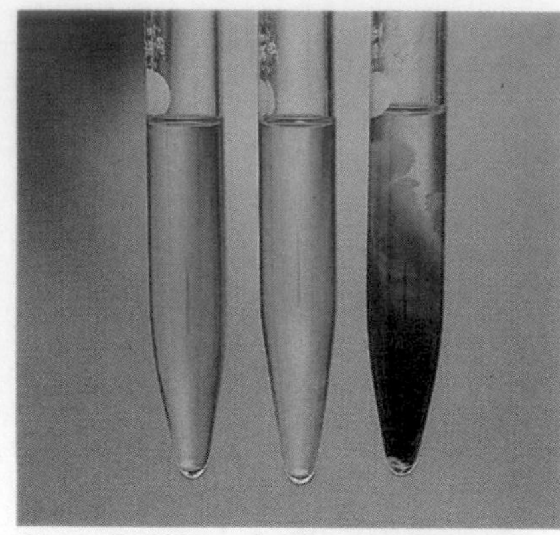

(b) $Cu(NO_3)_2$, $CuSO_4$, $Cu(OH)_2$

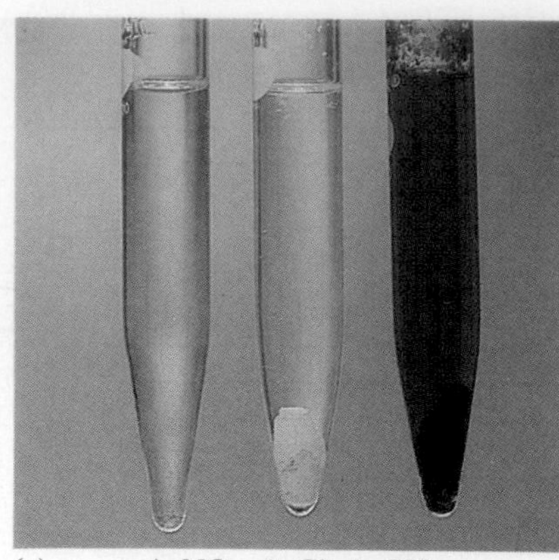

(c) $AgNO_3$, AgCl, AgOH

(d) $(NH_4)_2S$, CdS, Sb_2S_3, PbS

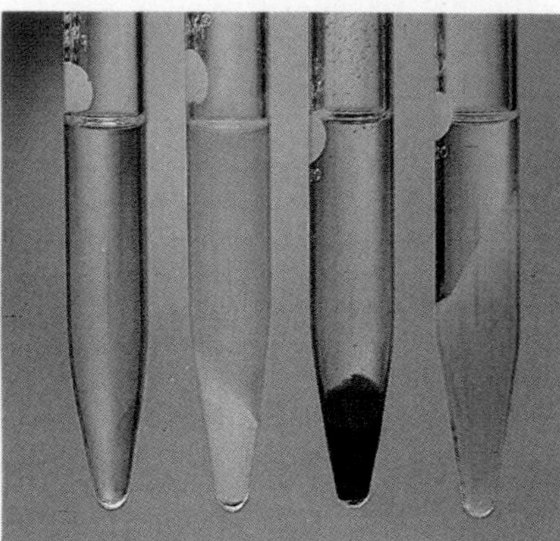

(e) NaOH, $Ca(OH)_2$, $Fe(OH)_3$, $Ni(OH)_2$

FIGURE 3.8

With certain exceptions, salts of the common anions Cl^- and NO_3^- are soluble in water. Although many sulfates are water soluble as well, there are some that are not. Finally, the sulfide ion, S^{2-}, almost invariably gives insoluble salts. Compare these results with Table 3.1. (Charles D. Winters)

not predicted to be soluble in water. (The actual solubility of $MgCO_3 \cdot 3\ H_2O$ is less than 0.2 g per 100 mL of water.)

(c) Nickel(II) oxide is composed of Ni^{2+} and O^{2-}. Again, Table 3.1 suggests that oxides are soluble only when O^{2-} is combined with an alkali metal ion; Ni^{2+} is a transition metal ion, so NiO is insoluble.

(d) Calcium iodide is composed of Ca^{2+} and I^- ions. According to Table 3.1 almost all iodides are soluble in water, so CaI_2 is a water-soluble electrolyte.

$$CaI_2 \text{ in water} \longrightarrow Ca^{2+}(aq) + 2\ I^-(aq)$$

Notice that the compound gives *two* I^- ions on dissolving in water. (A common misconception is that I_2 or some ion such as I_2^{2-} is found in solution. All halide-containing compounds produce F^-, Cl^-, Br^-, or I^- when dissolved in water.)

EXERCISE 3.2 SOLUBILITY GUIDELINES

Tell whether each compound below is likely to be soluble in water. If the compound dissolves in water, tell what ions exist in aqueous solution.
(a) NaBr (b) $BaSO_4$ (c) K_2CO_3 (d) AgCl (e) $(NH_4)_2SO_4$

EXERCISE 3.3 SOLUBILITY GUIDELINES

Write formulas for (a) a soluble ionic compound containing the nitrate ion, (b) an insoluble compound containing the sulfide ion, and (c) a soluble compound containing Ni^{2+}.

ANIONS, CATIONS, AND MICHAEL FARADAY (1791–1867)

Many of the terms we have been using, such as anion, cation, electrode, and electrolyte, originated with Michael Faraday, one of the most influential men in the history of chemistry. Faraday was apprenticed to a bookbinder in London (England) when he was only 13. This suited him, however, as he enjoyed reading the books sent to the shop for binding. One of these chanced to be a small book on chemistry, and his appetite for science was whetted. He soon began performing some experiments on electricity, and, in 1812, a patron of the shop invited Faraday to accompany him to a lecture at the Royal Institution by one of the most famous chemists of the day, Sir Humphrey Davy. Faraday was so intrigued by Davy's lecture that he wrote to ask Davy for a position as an assistant. Faraday was accepted and began work in 1813. His work was so fruitful and Faraday was so talented that he was made the Director of the Laboratory of the Royal Institution about 12 years later.

ACIDS AND BASES

Acids and bases represent a major class of chemical substances. For the moment, we shall define a **base** as a compound that ionizes in water to provide hydroxide ion, OH^-, and a cation.

Acids and bases and their reactions are described in Section 3.4 and in Chapters 15 and 16.

$$\underset{\text{strong electrolyte}}{\overset{\text{base}}{NaOH(aq)}} \longrightarrow Na^+(aq) + OH^-(aq)$$

An **acid** is a compound that ionizes in water to provide a hydrogen ion, H^+, and an anion.

$$\underset{\text{strong electrolyte}}{\overset{\text{acid}}{HCl(aq)}} \longrightarrow H^+(aq) + Cl^-(aq)$$

A few of the most common acids and bases are given in Table 3.2 and in Figure 3.9. *Be sure to familiarize yourself with these compounds and their names.* While all the compounds listed in Table 3.2 are electrolytes, some are *strong* electrolytes and others are *weak* electrolytes. HCl, HNO_3, $HClO_4$, NaOH, and KOH, for example, are strong electrolytes. When placed in water, they are essentially 100% ionized. One mole of HCl will produce 1.0 mole of H^+ ions and 1.0 mole of Cl^- ions.

See Appendix B for further comments on the names of common acids and bases.

Acids and bases that are strong or weak electrolytes are also called "strong" or "weak" acids or bases.

TABLE 3.2 Common Acids and Bases

ACIDS		BASES	
HCl	Hydrochloric acid	NaOH	Sodium hydroxide
HNO_3	Nitric acid	KOH	Potassium hydroxide
$HClO_4$	Perchloric acid	$Ca(OH)_2$	Calcium hydroxide
H_2SO_4	Sulfuric acid	NH_3	Ammonia
H_3PO_4	Phosphoric acid		
$HC_2H_3O_2$	Acetic acid		
H_2CO_3	Carbonic acid		

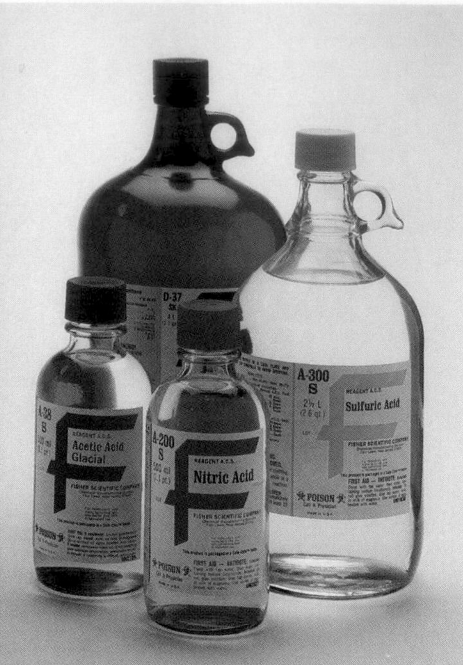

FIGURE 3.9
Some common acids: nitric, HNO_3; sulfuric, H_2SO_4; and acetic, $HC_2H_3O_2$. (Fisher Scientific Company)

Acetic acid is a typical weak electrolyte (Figure 3.7). When placed in water, only about 1% of the molecules ionize to form H^+ ion and acetate ion, $C_2H_3O_2^-$.

$$\underset{\text{weak electrolyte}}{\overset{\text{acid}}{HC_2H_3O_2(aq)}} \longrightarrow H^+(aq) + C_2H_3O_2^-(aq)$$

Ammonia, a base, is also a weak electrolyte because it produces very little OH^- ion on reaction with water.

$$\underset{\text{weak electrolyte}}{\overset{\text{base}}{NH_3(aq)}} + H_2O(\ell) \longrightarrow NH_4^+(aq) + OH^-(aq)$$

Nonetheless it is convenient to think of aqueous ammonia as a solution of ammonium hydroxide, NH_4OH.

Some acids are capable of providing more than one H^+ ion per molecule, and sulfuric acid is an example.

$$H_2SO_4(aq) \longrightarrow H^+(aq) + HSO_4^-(aq)$$
$$HSO_4^-(aq) \longrightarrow H^+(aq) + SO_4^{2-}(aq)$$

The loss of the first proton to give HSO_4^- is essentially complete, so the acid is a strong electrolyte.

NET IONIC EQUATIONS

Let us return to the reaction of magnesium with aqueous HCl (Figure 3.4).

$$Mg(s) + 2\ HCl(aq) \longrightarrow MgCl_2(aq) + H_2(g)$$

Hydrochloric acid, HCl(aq), and magnesium chloride are strong electrolytes in water. Therefore, we can rewrite the Mg(s)/HCl(aq) equation as

$$Mg(s) + 2\ H^+(aq) + 2\ Cl^-(aq) \longrightarrow Mg^{2+}(aq) + 2\ Cl^-(aq) + H_2(g)$$

Now Cl^- ions appear both on the reactant and product sides of the equation in *exactly* the same form. Such ions are often called **spectator ions** because they are not involved in the net process. This means that no stoichiometric information is lost if the equation is written without them, so we can simplify the equation to

$$Mg(s) + 2\ H^+(aq) \longrightarrow Mg^{2+}(aq) + H_2(g)$$

A **net ionic equation** involves only atoms whose chemical state changes in the course of the reaction.

The equation that results from leaving out the spectator ions is called a **net ionic equation**. Such equations make it easier to see the essential part of the complete reaction.

This is not to imply that Cl^- is totally unimportant in the Mg/HCl reaction. Indeed, a "naked" H^+ or Mg^{2+} ion simply cannot exist in solution; a negative ion of some kind *must be present* to counterbalance the positive ion charge (so the balancing ion is often called a "counter ion"). Finally, having left the negative counter ions out of the net ionic equation implies that *any* anion will do, as long as it forms water-soluble compounds; thus, we could have used HNO_3 or HBr as the source of H^+.

As a final point concerning net ionic equations, you should recognize that there is *conservation of charge* as well as mass in a balanced chemical equation. Thus, in the Mg/HCl net ionic equation there is a +2 electrical charge on each side of the equation. On the left side of the equation there are two H^+ ions for a total electrical charge of +2 (the Mg atom has no electrical charge). This is balanced by a magnesium ion on the right side with an electrical charge of +2.

EXAMPLE 3.3

WRITING AND BALANCING NET IONIC EQUATIONS

Write a balanced, net ionic equation for the reaction of $AgNO_3$ with $CaCl_2$ to give AgCl and $Ca(NO_3)_2$.

Solution

Step 1. Write the complete, balanced equation.

$$2\ AgNO_3 + CaCl_2 \longrightarrow 2\ AgCl + Ca(NO_3)_2$$

Step 2. Decide on the solubility of each compound from Table 3.1. One general guideline was that nitrates are almost always soluble, so $AgNO_3$ and $Ca(NO_3)_2$ are water soluble. Further, with a few exceptions (e.g., AgCl), chlorides are water soluble. Therefore, we can write

$$2\ AgNO_3(aq) + CaCl_2(aq) \longrightarrow 2\ AgCl(s) + Ca(NO_3)_2(aq)$$

Step 3. At this stage in our development of chemistry, we shall assume that all soluble ionic compounds are electrolytes. Therefore,

$$AgNO_3(aq) \longrightarrow Ag^+(aq) + NO_3^-(aq)$$

$$CaCl_2(aq) \longrightarrow Ca^{2+}(aq) + 2\ Cl^-(aq)$$

$$Ca(NO_3)_2(aq) \longrightarrow Ca^{2+}(aq) + 2\ NO_3^-(aq)$$

This results in the following complete ionic equation.

$$2\ Ag^+(aq) + 2\ NO_3^-(aq) + Ca^{2+}(aq) + 2\ Cl^-(aq) \longrightarrow 2\ AgCl(s) + Ca^{2+}(aq) + 2\ NO_3^-(aq)$$

Step 4. There are two spectator ions in the complete ionic equation (Ca^{2+} and NO_3^-), so these can be eliminated to give the net ionic equation.

$$2\ Ag^+(aq) + 2\ Cl^-(aq) \longrightarrow 2\ AgCl(s)$$

To finish the job, realize that each species in the net equation is preceded by a coefficient of 2. Therefore, the equation can be simplified by dividing through by 2.

$$Ag^+(aq) + Cl^-(aq) \longrightarrow AgCl(s)$$

Step 5. Finally, notice that the sum of ion charges is the same on both sides of the equation. On the left, +1 and −1 give zero; on the right the electrical charge on AgCl is also zero.

EXAMPLE 3.4

WRITING AND BALANCING NET IONIC EQUATIONS

Write a balanced, net ionic equation for the reaction of calcium chloride with sodium carbonate to give sodium chloride and calcium carbonate.

Solution Using the guidelines of Chapter 2, we first decide on the formulas and then write the *unbalanced* equation.

$$CaCl_2 + Na_2CO_3 \longrightarrow CaCO_3 + NaCl$$

It then follows that the complete balanced equation is

$$CaCl_2 + Na_2CO_3 \longrightarrow CaCO_3 + 2\ NaCl$$

Now, to write the net ionic equation, find the spectator ions and eliminate them. Sodium salts are usually water soluble, as are most chlorides. Unless they are associated with an alkali metal, carbonates are not often soluble. This means that Na_2CO_3 is soluble (and also an electrolyte), while $CaCO_3$ is insoluble. Therefore, a complete ionic equation can be written as

$$Ca^{2+}(aq) + 2\ Cl^-(aq) + 2\ Na^+(aq) + CO_3^{2-}(aq) \longrightarrow CaCO_3(s) + 2\ Na^+(aq) + 2\ Cl^-(aq)$$

From this, you can see that the spectator ions are Na^+ and Cl^-, so the balanced net ionic equation can be written as

$$Ca^{2+}(aq) + CO_3^{2-}(aq) \longrightarrow CaCO_3(s)$$

This is the net ionic equation for the formation of limestone, a mineral of enormous importance on our earth.

EXERCISE 3.4 WRITING NET IONIC EQUATIONS

Balance each of the following equations and write net ionic equations.

(a) $BaCl_2 + Na_2SO_4 \longrightarrow BaSO_4 + NaCl$

(b) $(NH_4)_2S + Cd(NO_3)_2 \longrightarrow CdS + NH_4NO_3$

(c) Lead nitrate reacts with potassium chloride to give lead chloride and potassium nitrate.

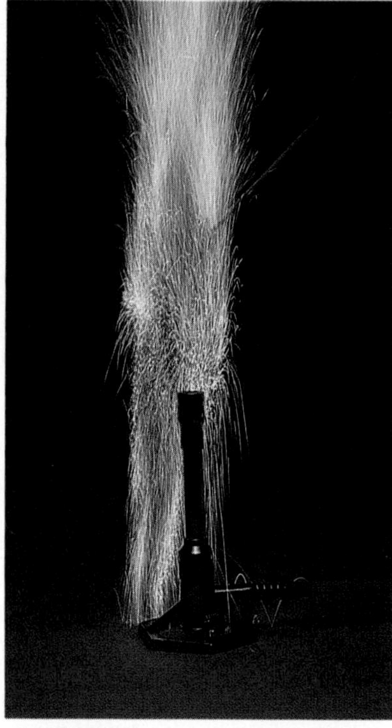

FIGURE 3.10
Iron burns vigorously in oxygen to give a mixture of iron(II) oxide (FeO) and iron(III) oxide (Fe_2O_3). (Charles D. Winters)

3.4 SOME COMMON TYPES OF CHEMICAL REACTIONS

Elements and compounds can react with one another to produce a bewildering array of new compounds. But how can one predict what reactions can occur and what the products will be? Fortunately, chemists have developed some guidelines that allow predictions to be made about a wide range of reactions. For example, you know from earlier in this chapter that any compound containing only C, H, and O will always react with an excess of O_2 to produce only CO_2 and H_2O in a *combustion reaction* and that magnesium will react with oxygen to form an oxide.

It is important to realize at the outset that there are a relatively few common types of chemical reactions. You will see these throughout the book, so it is useful to introduce a few of them briefly at this point.

REACTIONS OF ELEMENTS AND COMPOUNDS WITH OXYGEN

The reaction of hot magnesium metal and oxygen in the air to give magnesium oxide (commonly known as "magnesia") is an example (Figure 3.1) of the fact that all of the elements in the periodic table, with the exception of the rare gases, form binary or "two-element" *oxides*, M_xO_y.

You can predict the formulas of the products of metal + O_2 reactions from the fact that the products must be electrically neutral. The oxide ion is a dianion, O^{2-}, and you can predict a reasonable charge for metals using the guidelines of Chapter 2. By balancing positive and negative ion charges, you can predict that aluminum (Al, Group 3A), for example, will react with O_2 to give Al_2O_3 (2 Al^{3+} and 3 O^{2-}), a compound known as *alumina* or *corundum*.

Alumina, Al_2O_3, is extraordinarily nonreactive and hard, and, because it forms a coating on aluminum, it protects aluminum window frames, furniture, and airplanes from corrosion.

$$4\ Al(s) + 3\ O_2(g) \longrightarrow \underset{\text{aluminum oxide}}{2\ Al_2O_3(s)}$$

Metals of the transition series can have a variety of positive charges, but, as discussed in Chapter 2, they most often have charges of +2 or +3. Thus, when iron reacts with oxygen it forms both FeO and Fe_2O_3 (Figure 3.10).

$$2\ Fe(s) + O_2(g) \longrightarrow \underset{\text{iron(II) oxide}}{2\ FeO(s)}$$

$$4\ Fe(s) + 3\ O_2(g) \longrightarrow \underset{\text{iron(III) oxide}}{2\ Fe_2O_3(s)}$$

Almost all nonmetals (He, Ne, Ar, and possibly Kr are exceptions) also combine with O_2, and the compounds formed by carbon are good examples. Both CO and CO_2 are gases under normal conditions.

Caution! Nonmetal/O_2 reaction products cannot always be predicted from your knowledge of the periodic table at this point. You should know the few reactions given here.

$$2\ C(s) + O_2(g) \longrightarrow \underset{\text{carbon monoxide}}{2\ CO(g)}$$

$$\downarrow \tfrac{1}{2}\ O_2(g)$$

$$C(s) + O_2(g) \longrightarrow \underset{\text{carbon dioxide}}{CO_2(g)}$$

Carbon dioxide is the nontoxic gas in soft drinks, beer, and champagne. CO, in contrast, is toxic. This molecule can combine with hemoglobin, the molecule in your blood system that picks up oxygen in the lungs and carries it to the cells in your body. Because CO reacts even more strongly with hemoglobin than does O_2, the carbon monoxide prevents the hemoglobin from absorbing O_2.

The fact that carbon forms two compounds with O_2 is illustrative of a general feature of the reactions of all nonmetals with O_2: C, N, P, and S all form several binary oxides, as illustrated in the reactions that follow.

Elemental nitrogen, N_2, and O_2 are the major components of the air we breathe. In an automobile engine, small amounts of these gases combine at high temperatures to give the simple compound NO, nitrogen oxide.

$$N_2(g) + O_2(g) \xrightarrow{\text{energy}} \underset{\substack{\text{nitrogen oxide} \\ \text{(colorless gas)}}}{2\ NO(g)}$$

This is also a product when a lightning discharge passes through air during a thunderstorm. However, just as CO can be converted to CO_2 under the right conditions, NO can react with excess O_2 to give NO_2 (Figure 3.11).

FIGURE 3.11

Nitrogen dioxide, NO_2, a brown gas. It can be formed when oxygen and nitrogen combine during fuel combustion in automobile engines, and so NO_2 can be seen in polluted air. Here it is a product of the reaction of copper metal and nitric acid (HNO_3). The other product of the reaction is copper nitrate, the compound responsible for the blue-green color of the solution. See Exercise 3.15. (Charles D. Winters)

FIGURE 3.12
Sulfur burns in oxygen to give sulfur dioxide and trioxide. (Charles D. Winters)

$$2\ NO(g) + O_2(g) \longrightarrow 2\ NO_2(g)$$

nitrogen dioxide (brown gas)

Both NO and NO_2 are toxic and are important links in the chemical chain leading to the production of smog from air, sunlight, and automobile exhaust (Chapter 22).

Phosphorus, in the same periodic group as N, is exceedingly toxic; only 50 milligrams is a fatal dose for an adult human. The element in its most common form is a yellowish-white solid, P_4. The slight amount of P_4 in the vapor above the solid at room temperature reacts with O_2 and gives off a phosphorescent glow, a fact that led to the name of the element. At least six different products are possible when P_4 is burned in air, but the one product you should be aware of now is the white solid P_4O_{10}.

$$P_4(s) + 5\ O_2(g) \longrightarrow P_4O_{10}(s)$$

tetraphosphorus decoxide (white solid)

Sulfur, the Group 6A neighbor of oxygen, forms two oxides, SO_2 and SO_3, in reactions that have great environmental importance: they are the beginning of the production of acid rain (Figure 3.12).

$$S_8(s) + 8\ O_2(g) \longrightarrow 8\ SO_2(g)$$

sulfur dioxide

$$\downarrow 4\ O_2(g)$$

$$8\ SO_3(g)$$

sulfur trioxide

EXERCISE 3.5 FORMING BINARY COMPOUNDS
Predict the products and balance equations for the following reactions that produce simple binary compounds.
(a) $Li(s) + O_2(g) \longrightarrow$
(b) $V(s) + O_2(g) \longrightarrow$
(c) $Ga(\ell) + O_2(g) \longrightarrow$

FIGURE 3.13
Adding a drop of KI to a solution of $Pb(NO_3)_2$ leads to the formation of a precipitate of insoluble PbI_2 and leaves water-soluble KNO_3 in solution. (Charles D. Winters)

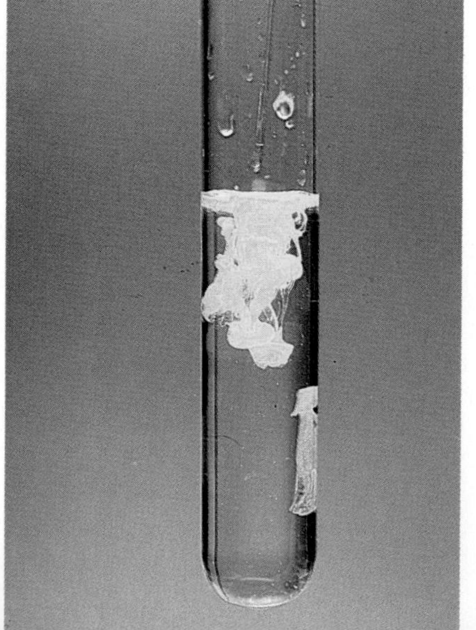

EXERCISE 3.6 REACTIONS TO FORM BINARY COMPOUNDS
Write a balanced equation for the preparation of each of the following binary compounds. Give the systematic name for each binary product.
(a) ____ + ____ $\longrightarrow GeO_2(s)$
(b) ____ + ____ $\longrightarrow Cr_2O_3(s)$

PRECIPITATION AND ACID–BASE REACTIONS

Exchange reactions, also called **metathesis** or **double-replacement** reactions, proceed by the interchange of reactant cation/anion partners. Two major categories of such reactions are **precipitation reactions** (Figure 3.13)

$$Pb(NO_3)_2(aq) + 2\ KI(aq) \longrightarrow 2\ KNO_3(aq) + PbI_2(s)$$

and **acid–base reactions**.

$$HNO_3(aq) + KOH(aq) \longrightarrow KNO_3(aq) + HOH(\ell)\ [= H_2O]$$

Mixing just any two compounds together does not ensure they will react. For example, mixing aqueous solutions of NaCl and KNO_3 will not produce $NaNO_3$ and KCl; the reactants will just sit in the flask and get wet. Why then do exchange reactions occur and how can you predict which ones will work? Among other things, you can predict an exchange reaction if (a) a water-insoluble product results from two soluble reactants or (b) a stable molecule such as water is formed. The former is the basis of precipitation reactions, and the latter is the reason for the effectiveness of acid–base reactions.

FIGURE 3.14
A drop of a water-soluble sulfide salt is placed in a solution of water-soluble $CuCl_2$, and black, insoluble CuS is formed. (Charles D. Winters)

PRECIPITATION REACTIONS Precipitation reactions produce an insoluble salt or **precipitate** from soluble reactants. Table 3.1 lists ions that are likely to lead to insoluble compounds. There are many positive/negative ion combinations that can give insoluble materials, so many precipitation reactions are possible. For example, copper(II) sulfide is easily precipitated by the reaction of a water-soluble copper(II) salt with a water-soluble sulfide salt (Figure 3.14).

Complete Reaction

$$CuCl_2(aq) + Na_2S(aq) \longrightarrow CuS(s) + 2\ NaCl(aq)$$

Net Ionic Equation

$$Cu^{2+}(aq) + S^{2-}(aq) \longrightarrow CuS(s)$$

If a soluble copper(II) salt in nature comes in contact with a source of sulfide ions (say from a volcano or a natural gas pocket), CuS precipitates.

Precipitation reactions are often responsible for the deposits in the earth of metal ion salts called "ores."

EXERCISE 3.7 PRECIPITATION REACTIONS

Complete and balance the following equations for precipitation reactions. Indicate whether each substance is soluble or insoluble in water. Write the net ionic equations.

(a) $AgNO_3 + LiCl \longrightarrow$

(b) $NiCl_2 + Na_2S \longrightarrow$

ACID–BASE REACTIONS Exchange reactions between an *acid* and a *base* produce a **salt** and *water*. Such a reaction is often called a **neutralization**, since it tends to make the solution "neutral," that is, neither acidic nor basic. One example of a neutralization is the combination of the base NaOH with hydrochloric acid, HCl, to give sodium chloride and water.

A salt is an ionic compound whose cation comes from a base and whose anion comes from an acid.

NaOH(aq)	+	HCl(aq)	$\longrightarrow$	NaCl(aq)	+	HOH (ℓ)
sodium hydroxide "lye"		hydrochloric acid "muriatic acid"		sodium chloride "sea salt"		water

Table 3.2 lists some common acids and bases. In every case, reaction of one of these acids with one of the bases produces a salt and water.

Hydrochloric acid and nitric acid in Table 3.2 are strong electrolytes in water and so are often called "strong acids." This means the complete ionic equation for the reaction of HCl(aq) and NaOH(aq), for example, should be written as

$$Na^+(aq) + OH^-(aq) + H^+(aq) + Cl^-(aq) \longrightarrow Na^+(aq) + Cl^-(aq) + HOH\ (\ell)$$

and, since Na^+ and Cl^- ions appear on both sides of the equation, the *net ionic reaction* is simply the combination of H^+ and OH^- to give water.

$$H^+(aq) + OH^-(aq) \longrightarrow HOH(\ell)$$

This will always be the net ionic equation for the neutralization reaction of any strong acid with any strong base. The other ions of the base (the cation) and acid (the anion) remain unchanged. If the water is evaporated, however, the cation and anion form a solid salt. In the example above, NaCl would be isolated, while NaOH and nitric acid, HNO_3, would give sodium nitrate, $NaNO_3$.

$$NaOH(aq) + HNO_3(aq) \longrightarrow NaNO_3(aq) + HOH(\ell)$$

EXAMPLE 3.5

ACID–BASE REACTIONS

Write the balanced equation for the reaction of aqueous ammonia and nitric acid.

Solution At this stage in your understanding of chemistry it is convenient to treat aqueous ammonia as a solution of ammonium hydroxide, NH_4OH. Therefore, we can write

$$\underset{\text{base}}{NH_4OH(aq)} + \underset{\text{acid}}{HNO_3(aq)} \longrightarrow \underset{\text{salt}}{NH_4NO_3(aq)} + \underset{\text{water}}{HOH(\ell)}$$

EXERCISE 3.8 ACID–BASE REACTIONS

Write the balanced equation for the reaction of sulfuric acid, H_2SO_4, and KOH. Assume that *both* H^+ ions from the acid combine with OH^- from KOH.

PREPARATION OF COMPOUNDS USING EXCHANGE REACTIONS Many chemists have as their main occupation the preparation of new materials. Exchange reactions represent one way to do this, but more information is needed if you want to use them for preparative purposes.

An exchange reaction may give you the desired product, but then you are faced with the problem of isolating the compound. For example, in the case of the reaction of NaOH with HCl in aqueous solution, the final solution contains only NaCl in water. To obtain the sodium chloride as a solid, you could simply evaporate the water from the solution, leaving the salt as a white, crystalline solid.

Alternatively, the desired salt may be obtained using a precipitation reaction. The compound you want could be either the insoluble salt or the soluble species left in aqueous solution. For example, you could prepare insoluble barium sulfate by the procedure illustrated in Figure 2.16. Sulfuric acid was poured into a solution of water-soluble $BaCl_2$ to give insoluble $BaSO_4$ and soluble HCl.

$$BaCl_2(aq) + H_2SO_4(aq) \longrightarrow BaSO_4(s) + 2\,HCl(aq)$$

Filtration was then used to separate the insoluble and soluble species. Insoluble barium sulfate was trapped in the paper filter, and the solution containing the soluble hydrochloric acid passed through.

EXAMPLE 3.6

PREPARATION OF A COMPOUND BY A PRECIPITATION REACTION

Prepare calcium carbonate, $CaCO_3$, by a precipitation reaction.

Solution

Step 1. Is the compound soluble or insoluble in water? According to Table 3.1, carbonates are generally insoluble, except when the cation is from Group 1A (e.g., Na^+, K^+). Therefore, a precipitation reaction is appropriate.

Step 2. Choose water-soluble salts as reactants, one containing Ca^{2+} and the other containing CO_3^{2-}. Nitrates are generally soluble, so calcium nitrate, $Ca(NO_3)_2$, is a good choice. As a source of carbonate ion, sodium carbonate, Na_2CO_3, is reasonable, since sodium salts are generally soluble.

Step 3. Write the balanced equation.

Complete equation: $Ca(NO_3)_2(aq) + Na_2CO_3(aq) \longrightarrow CaCO_3(s) + 2\ NaNO_3(aq)$

Net ionic equation: $Ca^{2+}(aq) + CO_3^{2-}(aq) \longrightarrow CaCO_3(s)$

The insoluble product could be isolated by filtration, $CaCO_3$ being held by the filter and $NaNO_3$ remaining in the water.

EXAMPLE 3.7

PREPARING A COMPOUND BY AN ACID–BASE REACTION

Prepare potassium acetate, $KC_2H_3O_2$, by an acid–base reaction.

Solution

Step 1. Is the compound soluble or insoluble in water? According to Table 3.1, acetates are generally soluble. Therefore, an acid–base reaction, where one product is the desired salt and the other is water, is appropriate.

Step 2. Acetic acid (Table 3.2) is the source of acetate ion, $C_2H_3O_2^-$, and potassium hydroxide can supply potassium ion, K^+.

Step 3. Write the balanced equation.

$$HC_2H_3O_2(aq) + KOH(aq) \longrightarrow KC_2H_3O_2(aq) + HOH\ (\ell)$$

A precipitation reaction could be used as an alternative, but the product other than $KC_2H_3O_2$ must then be insoluble. For example, mixing the water-soluble salts barium acetate and potassium sulfate gives the desired soluble salt potassium acetate and insoluble barium sulfate.

$$Ba(C_2H_3O_2)_2(aq) + K_2SO_4(aq) \longrightarrow 2\ KC_2H_3O_2(aq) + BaSO_4(s)$$

EXERCISE 3.9 WRITING EXCHANGE REACTIONS

Complete and balance the following equations for exchange reactions. Tell whether each is an acid–base or a precipitation reaction.

(a) $H_2SO_4 + CsOH \longrightarrow$
(b) $MgCl_2 + Na_2CO_3 \longrightarrow$
(c) $HNO_3 + Ca(OH)_2 \longrightarrow$
(d) $CdCl_2 + Na_2S \longrightarrow$

FIGURE 3.15

A bed of calcium carbonate (limestone), Verde River, Arizona. (James Cowlin)

EXERCISE 3.10 PREPARING IONIC COMPOUNDS BY EXCHANGE REACTIONS

Write a balanced equation to show how you would prepare each of the following salts by an acid–base or precipitation reaction.

(a) NaCl (b) KNO_3 (c) $Ba(NO_3)_2$ (d) FeS

REACTIONS OF METAL CARBONATES

Calcium is the fifth most abundant element in the earth's crust and is the third most abundant metal (after Al and Fe). Virtually all the calcium is in the form of calcium carbonate, $CaCO_3$, coming from the fossilized remains of earlier marine life (Figure 3.15). Limestone, a form of calcium carbonate, is enormously important in our economy, as is sodium carbonate, Na_2CO_3. The chemistry of these and other metal carbonates is important, and it ties together the chemistry of metal oxides and the acid–base reactions described above.

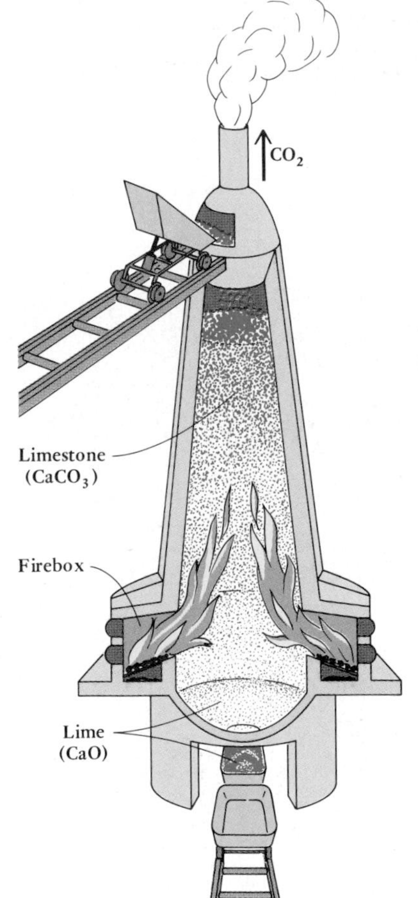

FIGURE 3.16

Calcium oxide or "quicklime" (CaO) is prepared by decomposing limestone in a process called "calcination" by heating in a kiln to 800°C to 1000°C. The process is energy intensive, requiring up to $\frac{1}{3}$ ton of a coal to produce 1 ton of CaO. Nonetheless, in 1984, 32.2 billion pounds of CaO were produced in the United States. As such, CaO is the fourth most widely used chemical. Almost half of it is used in the steel industry to remove impurities from the iron, and much of the rest is converted to calcium hydroxide or "slaked lime."

$$CaO(s) + H_2O(\ell) \longrightarrow Ca(OH)_2(s)$$

This slightly soluble hydroxide is used as a base in water treatment, for removal of polluting SO_2 from power plant emissions,

$$Ca(OH)_2(aq) + SO_2(g) \longrightarrow CaSO_3(s) + H_2O(\ell)$$

in paper making, and in the manufacture of bleach, $Ca(OCl)_2$.

$$2\ Ca(OH)_2(aq) + 2\ Cl_2(g) \longrightarrow Ca(OCl)_2(s) + CaCl_2(s) + 2\ H_2O(\ell)$$

A characteristic reaction of metal carbonates is that when heated, they decompose to form CO_2 and the corresponding metal oxide.

$$\underset{\text{limestone}}{CaCO_3(s)} \longrightarrow \underset{\text{lime}}{CaO(s)} + CO_2(g)$$

Thus, limestone produces calcium oxide, a compound commonly called "lime" (Figure 3.16). This of course represents a very convenient way to make a metal oxide, assuming the appropriate carbonate is available.

Another characteristic reaction of metal carbonates is their exchange reaction with acids.

$$CaCO_3(s) + 2\ HCl(aq) \longrightarrow CaCl_2(aq) + H_2CO_3(aq)$$
$$\downarrow$$
$$H_2O(\ell) + CO_2(g)$$

A salt and H_2CO_3, carbonic acid, are always the products. Carbonic acid, however, is unstable and rapidly forms water and CO_2 gas. If the reaction is done in an open beaker (Figure 3.17), the gas bubbles out of the solution. In fact, *it is gas formation that represents the driving force for the exchange*.

FIGURE 3.17

A piece of blackboard chalk, which is mostly $CaCO_3$, reacts rapidly with acid (HCl) to give a salt ($CaCl_2$), water, and CO_2 gas. (Charles D. Winters)

EXERCISE 3.11 METAL CARBONATE CHEMISTRY

Write a balanced equation to show the result of heating magnesium carbonate. Write a balanced equation for the reaction of sodium carbonate with nitric acid.

3.5 OXIDATION–REDUCTION REACTIONS

Reactions of metals and nonmetals with O_2 fit into a broad class of reactions called *oxidation–reduction* reactions. In addition to the metal oxidations, there are many other processes, some with very complicated equations, and balancing such equations requires a systematic approach. Since you will encounter many such reactions in your laboratory work, techniques of balancing these equations are the subject of this section.

The terms oxidation and reduction come from reactions known to chemists for centuries. Earliest man learned how to change metal oxides and sulfides to the metal, that is, how to reduce them. Mercury sulfide, known as cinnabar or vermilion (a mineral used since prehistoric times), can be *reduced* to liquid mercury simply by heating in air.

$$HgS(s) + O_2(g) \longrightarrow Hg(\ell) + SO_2(g)$$

Cassiterite or tin(IV) oxide, SnO_2, was found in Britain centuries ago. The metal is a major component of the alloy of copper and tin called bronze,* an important material in Roman times. Tin ore is very easily reduced to the metal by heating with carbon.

If mercury sulfide, the red solid, is heated in air, liquid mercury metal is formed. (Charles D. Winters)

*An **alloy** is an intimate mixture of two or more metals. Bronze, for example, usually contains about 20% tin and 80% copper.

reduced to

$$SnO_2(s) + \underset{\text{reducing agent}}{2\,C(s)} \longrightarrow Sn(s) + 2\,CO(g)$$

In this process carbon is the agent that brings about the reduction of tin ore to tin metal, so carbon is called the *reducing agent*.

Oxidation is the opposite of reduction. This too is a process known for centuries, and Figure 3.1 is an excellent example. Here oxygen is the agent of oxidation or *oxidizing agent*.

oxidized to

$$Mg(s) + \underset{\text{oxidizing agent}}{\tfrac{1}{2}\,O_2(g)} \longrightarrow MgO(s)$$

Notice in the reduction of SnO_2 with carbon that the carbon combines with oxygen to give CO, that is, the carbon is oxidized.

These experimental observations point to several fundamental conclusions concerning oxidation–reduction reactions: (a) If one substance is oxidized, another must be reduced. (b) The reducing agent is itself oxidized, and the oxidizing agent is reduced. Such conclusions become more obvious when described in terms of modern chemical theory. *Oxidation and reduction reactions are those involving transfer of electrons.* When a substance **accepts electrons**, it is said to be **reduced**. This is because there is a *reduction* in the real or apparent electrical charge on an atom of the substance. In the reaction below, Cu^{2+} is reduced to uncharged Cu(s) by accepting electrons from zinc metal. Since zinc metal is the "agent" that supplies electrons and causes the Cu^{2+} ion to be reduced, Zn is called the **reducing agent** (Figure 3.18).

Oxidation–reduction reactions are often referred to as **redox** reactions.

Cu^{2+} is reduced to Cu; Cu^{2+} accepts electrons and is the oxidizing agent.

$+2e^-$ from Zn

$$Cu^{2+}(aq) + Zn(s) \longrightarrow Cu(s) + Zn^{2+}(aq)$$

$-2e^-$ to Cu

Be sure to notice that the Cu^{2+}/Zn reaction is written as a net ionic equation. All redox reactions in this section are also net equations. (See Section 3.3.)

Zn is oxidized to Zn^{2+}; Zn donates electrons and is the reducing agent.

Zinc metal releases electrons on going to Zn^{2+}; since its electrical charge has increased, it is said to have been *oxidized*. In order for this to happen, there must be something available to take the electrons offered by the zinc. In this case, Cu^{2+} is the electron acceptor and so it the "agent" that causes Zn metal to be oxidized; thus, Cu^{2+} is the **oxidizing agent**. In every oxidation–reduction reaction something is reduced (and so is itself

Reducing agent: donates electron(s) and is oxidized. **Oxidizing agent**: accepts electron(s) and is reduced.

the oxidizing agent) and something is oxidized (and so is the reducing agent).*

OXIDATION NUMBERS

How can you tell an oxidation–reduction reaction when you see one? The answer is to look for a change in the *oxidation number* of an element in the course of the reaction. The **oxidation number** of an element in a compound is defined as the charge an atom has, or appears to have, when the electrons of the compound are counted according to a certain set of rules. In Chapter 9, we shall examine the basis of the counting rules, but for the moment it is quite easy to learn some simple guidelines for assigning oxidation numbers. This will allow you to determine whether a given reaction is an oxidation–reduction process.

The guidelines for determining the oxidation number of an element in a compound are as follows:

At this point the *only* reason to use oxidation numbers is to decide when a redox reaction has occurred.

1. **In free elements, each atom has an oxidation number of 0.** The oxidation number of Zn in metallic zinc is 0, the apparent charge or oxidation number of each I atom in I_2 is 0, and that of each S atom in S_8 is 0.
2. **For ions consisting of a single atom, the oxidation number is equal to the charge on the ion.** The oxidation number of Cl^- is -1 and of Zn^{2+} is $+2$. Recall from Chapter 2 that elements of Periodic Groups 1A through 3A form ions with a positive charge equal to the group number. The charge on the simple ion of aluminum, Al^{3+}, and its oxidation number, are $+3$, for example.
3. **The oxidation number of H is $+1$ and of O is -2 in most compounds.** Although this statement applies to many, many compounds, there are a few important exceptions. When H forms a binary compound with a metal, the metal forms a positive ion and the H becomes a -1 anion, H^- or hydride. Thus, in NaH the oxidation number of Na is $+1$ and that of H is -1. Another exception to the rule is that oxygen can have an oxidation number of -1 in a class of compounds called *peroxides*, compounds based on the O_2^{2-} ion. For example, in H_2O_2, hydrogen peroxide, H has an oxidation number of $+1$ and that of O is -1.
4. **The algebraic sum of the oxidation numbers in a neutral compound must be zero; in an polyatomic ion, the sum must be equal to the ion charge.** Examples of this rule are the compounds above and the following:

   (a) Li_2O is a neutral compound. If O has an oxidation number of -2, then that of Li must be $+1$ (as predicted by its position in Group 1A).

   $Li_2O = (2Li^+)(O^{2-}) \qquad \text{Charge} = 0 = 2(+1) + (-2)$

   (b) H_3PO_4 also has an overall charge of 0. If the oxygen has an oxidation number of -2 and that of H is $+1$, then P must have an oxidation number of $+5$.

FIGURE 3.18

A clean strip of zinc (in front of the beaker at the left) is placed in a dilute solution of $CuSO_4$. With time, Zn reduces Cu^{2+}, so copper metal coats the zinc strip (in the beaker at the right). At the same time Zn^{2+} ions enter solution. Notice that the blue color of aqueous Cu^{2+} ions has greatly diminished in the beaker at the right. (The reaction between Cu^{2+} and Al in Figure 1.8 is similar to the Cu^{2+}/Zn reaction.) (Charles D. Winters)

*To help you remember the language of oxidation–reduction reactions you might recall the phrase "LEO the GERman." LEO stands for "Loses Electrons, Oxidized" and GER stands for "Gains Electrons, Reduced."

$H_3PO_4 = (3\ H^+)(P^{5+})(4\ O^{2-})$

$\text{Charge} = 0 = 3(+1) + (+5) + 4(-2)$

(c) The permanganate ion, MnO_4^-, is a commonly used ion having an overall charge of -1. Since O has an oxidation number of -2, this means that Mn has an apparent charge or oxidation number of $+7$.*

$MnO_4^- = [(Mn^{7+})(4\ O^{2-})]^-$ $\quad$ $\text{Charge} = -1 = (+7) + 4(-2)$

(d) Dichromate ions are also widely used in the laboratory, and the $Cr_2O_7^{2-}$ ion has an overall charge of -2. Since O has an oxidation number of -2, then Cr must have an oxidation number of $+6$.*

$Cr_2O_7^{2-} = [(2\ Cr^{6+})(7\ O^{2-})]^{2-}$ $\quad$ $\text{Charge} = -2 = 2(+6) + 7(-2)$

EXERCISE 3.12 OXIDATION NUMBERS

Determine the oxidation number of each element in the following ions or compounds:

(a) MgO (d) HS^- (g) $NaCl$
(b) PH_3 (e) ClO_3^- (h) HNO_3
(c) AlH_3 (f) $S_2O_3^{2-}$ (i) ZnS

Using the concept of oxidation number, you can now see why a reaction such as

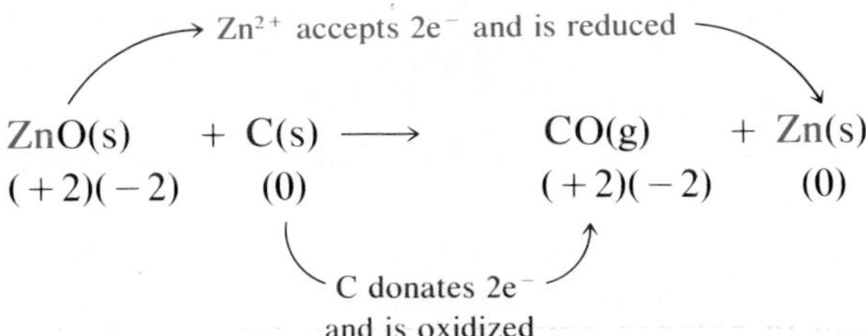

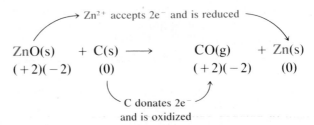

is said to be an oxidation–reduction reaction. The zinc starts with an oxidation number of $+2$ in ZnO and ends as the metal, oxidation number 0. Thus, the oxidation number of Zn has decreased from $+2$ to 0: zinc has been reduced. Carbon begins as pure, elemental carbon, oxidation number 0, and it ends with an oxidation number of $+2$ in CO. The oxidation number of C has increased from 0 to $+2$, and it is said to have been oxidized.

The reactions of Figures 3.2 and 3.3 are also oxidation–reduction processes.

Zn donates $2e^-$ and is oxidized

$$Zn(s) + I_2(s) \longrightarrow ZnI_2(s)$$
$$(0) \qquad (0) \qquad\qquad (+2)(-1)$$

I_2 accepts $2e^-$ and is reduced to I^-

*Except in simple ions, the oxidation number does *not* represent the real electrical charge on that atom, but oxidation numbers are useful for keeping track of electrons in reactions.

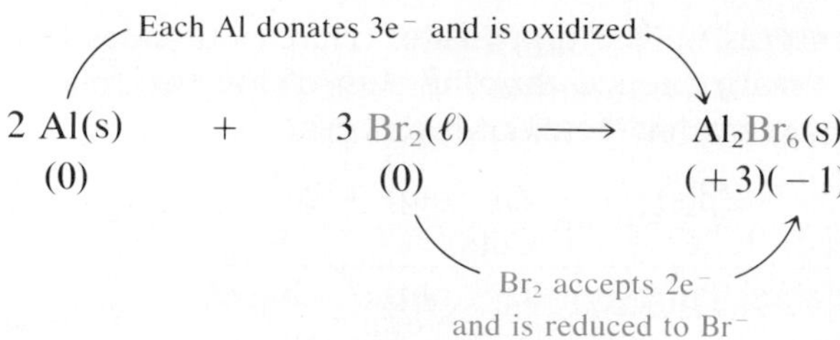
Each Al donates 3e⁻ and is oxidized
2 Al(s) + 3 Br₂(ℓ) ⟶ Al₂Br₆(s)
(0) (0) (+3)(−1)
Br₂ accepts 2e⁻
and is reduced to Br⁻

In both of these reactions we find the halogen combined only with a metal. When this is the case, we always consider the halogen has formed a halide ion (I^- or Br^-) with an oxidation number of -1. Thus, zinc must have an oxidation number of $+2$ and aluminum of $+3$, just as predicted from their position in the periodic table. In both cases, the metal donates electrons and is the reducing agent, and the halogen molecules, X_2, accept electrons and are oxidizing agents. Except in extremely rare circumstances, *metals are reducing agents*, and reactions of metals with halogens always involve halogens acting as oxidizing agents.

EXERCISE 3.13 RECOGNIZING OXIDATION–REDUCTION REACTIONS

For each of the following reactions, tell whether it is an oxidation–reduction reaction. If so, tell which substance is oxidized and which one is reduced.

(a) $S(s) + O_2(g) \longrightarrow SO_2(g)$

(b) $Fe(s) + Cl_2(g) \longrightarrow FeCl_2(s)$

(c) $NaOH(aq) + HCl(aq) \longrightarrow NaCl(aq) + H_2O(\ell)$

(d) $3\ ZnS(s) + 8\ HNO_3(aq) \longrightarrow 3\ ZnSO_4(aq) + 8\ NO(g) + 4\ H_2O(\ell)$

BALANCING OXIDATION–REDUCTION EQUATIONS

The systematic way to balance equations for oxidation–reduction reactions can be illustrated with the reaction of aqueous copper ions and metallic zinc (see Figure 3.18).

$$Cu^{2+}(aq) + Zn(s) \longrightarrow Cu(s) + Zn^{2+}(aq)$$

For zinc to change from metallic zinc to zinc ion, the metal must lose two electrons, and a balanced equation can be written depicting this.

Zn loses electrons, is oxidized, and is the reducing agent

$$Zn(s) \longrightarrow Zn^{2+}(aq) + 2e^-$$

The same is true for the transformation of copper ion to copper metal, except that the Cu^{2+} ion must gain electrons.

Cu^{2+} gains electrons, is reduced, and is the oxidizing agent

$$Cu^{2+}(aq) + 2e^- \longrightarrow Cu(s)$$

Since each of these equations represents a portion of the total reaction, they are often called **half reactions**. Notice that *each half reaction is balanced for mass and charge*. There is a **mass balance** since there is one atom of each kind on each side of the equation. There is a **charge balance** because the algebraic sum of charges on one side of the equation equals the sum of the charges on the other side. (Here both sides have a net charge of zero.) Finally, notice that the sum of the two half reactions equals the total equation for the overall process.

$$\begin{array}{rcl} Zn(s) & \longrightarrow & Zn^{2+}(aq) + 2e^- \\ + \; Cu^{2+}(aq) + 2e^- & \longrightarrow & Cu(s) \\ \hline Zn(s) + Cu^{2+}(aq) & \longrightarrow & Zn^{2+}(aq) + Cu(s) \end{array}$$

$2e^-$ appears on the right in the first equation and on the left in the second; like a spectator ion, it is not written in the final equation.

Summing the half reactions this way conveys the idea that the electrons produced by zinc are consumed by the copper(II) ions.

This example illustrates the general approach to balancing oxidation–reduction reactions. (a) First break the process into half reactions, one involving the donation of electrons (the reducing agent) and the other involving the acceptance of electrons (the oxidizing agent). (b) Each of these half reactions is then balanced, first for mass and then for charge. (c) After making sure that the reducing half reaction gives off as many electrons as the oxidizing half reaction requires, the half reactions are added to give the net, balanced equation. Examples 3.8 through 3.10 introduce you to situations you are likely to encounter.

FIGURE 3.19

The oxidation of copper metal by silver ion. A clean piece of copper screen is placed in a solution of silver nitrate, $AgNO_3$. With time, the copper reduces Ag^+ to Ag, silver metal crystals, and the copper is oxidized to Cu^{2+}. The blue color of the solution is due to the presence of aqueous copper(II) ion. (Charles D. Winters)

EXAMPLE 3.8

BALANCING AN OXIDATION–REDUCTION EQUATION

Balance the equation for the reaction of Ag^+ with copper (Figure 3.19).

$$Ag^+(aq) + Cu(s) \longrightarrow Cu^{2+}(aq) + Ag(s)$$

Solution

Step 1. *Recognize the reaction as an oxidation–reduction process.* To recognize that the reaction is an oxidation–reduction process, you need only to *look for a change in the oxidation numbers of the reactants*; you do *not* need to know what that change is. In this case, there is a change in the oxidation numbers of Ag^+ (from +1 to 0) and Cu (from 0 to +2).

Step 2. *Break the process into half reactions.*

Reduction (decrease in Ag oxidation number) $\quad Ag^+(aq) \longrightarrow Ag(s)$

Oxidation (increase in Cu oxidation number) $\quad Cu(s) \longrightarrow Cu^{2+}(aq)$

Step 3. *Achieve a mass balance by balancing the half reactions so that the same number of atoms of each kind appear on each side.* Both half reactions here are already mass balanced.

Step 4. *Achieve a charge balance by adding negative electrons to the side having an excess of positive charge.*

$$Ag^+(aq) + e^- \longrightarrow Ag(s)$$

(Ag^+ acquires electrons and is the oxidizing agent)

$$Cu(s) \longrightarrow Cu^{2+}(aq) + 2e^-$$

(Cu donates electrons and is the reducing agent)

Step 5. *Multiply the half reactions by appropriate factors so that the reducing agent donates as many electrons as the oxidizing agent consumes.* Here one atom of copper produces two electrons, whereas one ion of Ag^+ acquires only one electron. Therefore, two Ag^+ ions are required to consume the two electrons produced by a Cu atom, and we multiply the Ag^+/Ag half reaction by 2.

$$2\ Ag^+(aq) + 2e^- \longrightarrow 2\ Ag(s)$$

Notice that the half reaction is still balanced for mass and charge (0 on both sides of the equation).

Step 6. *Add the half reactions together to achieve the final, balanced overall equation.*

$$\begin{array}{rcl} 2\ Ag^+(aq) + 2e^- & \longrightarrow & 2\ Ag(s) \\ Cu(s) & \longrightarrow & Cu^{2+}(aq) + 2e^- \\ \hline 2\ Ag^+(aq) + Cu(s) & \longrightarrow & Cu^{2+}(aq) + 2\ Ag(s) \end{array}$$

Step 7. *Check the final result to ensure that there is a mass and charge balance.* Here there are two silver atoms or ions and one copper atom or ion on each side, and the total charge on each side is +2. The equation is balanced.

EXERCISE 3.14 BALANCING AN EQUATION FOR AN OXIDATION–REDUCTION REACTION

Balance the equation

$$Cr^{2+}(aq) + I_2(aq) \longrightarrow Cr^{3+}(aq) + I^-(aq)$$

Show the balanced half reactions and the balanced overall equation. Identify (a) the oxidizing agent, (b) the reducing agent, (c) the substance oxidized, and (d) the substance reduced.

A problem you will often confront when balancing equations for reactions in aqueous solution is that water and its ions, H^+ and OH^-, can enter into the reaction. In acidic conditions, H^+ may be a reactant or product, and in basic conditions it is possible for OH^- to behave similarly. Under other circumstances, the *pair* of species H^+ and H_2O must be used to balance equations for reactions in acid solution. Similarly, the *pair* OH^- and H_2O may be needed to balance an equation for a reaction in basic solution.* Examples 3.9 and 3.10 show how to determine when H^+, OH^-, H_2O/H^+, or H_2O/OH^- are needed and how to place them in the equation.

EXAMPLE 3.9

BALANCING EQUATIONS FOR OXIDATION–REDUCTION REACTIONS IN ACID SOLUTION

Balance the equation for the reaction of permanganate ion with oxalic acid in *acid solution* (Figure 3.20).

$$MnO_4^-(aq) + H_2C_2O_4(aq) \longrightarrow Mn^{2+}(aq) + CO_2(g)$$

Solution

Step 1. *Recognize the reaction as an oxidation–reduction process.* The oxidation numbers of Mn in MnO_4^- and C in $H_2C_2O_4$ both change.

Step 2. *Break the process into half reactions.*

$$H_2C_2O_4(aq) \longrightarrow CO_2(g)$$

$$MnO_4^-(aq) \longrightarrow Mn^{2+}(aq)$$

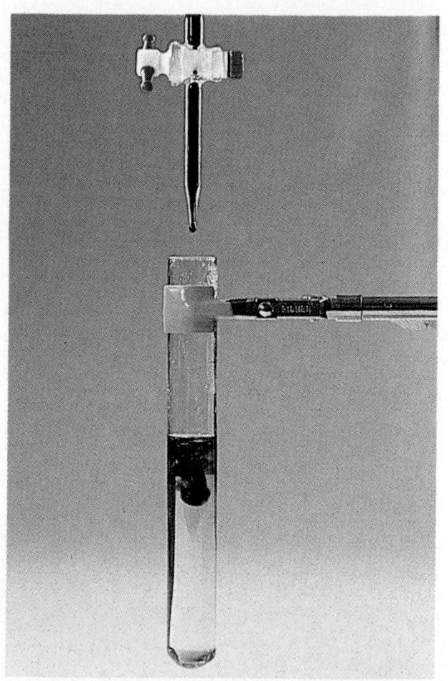

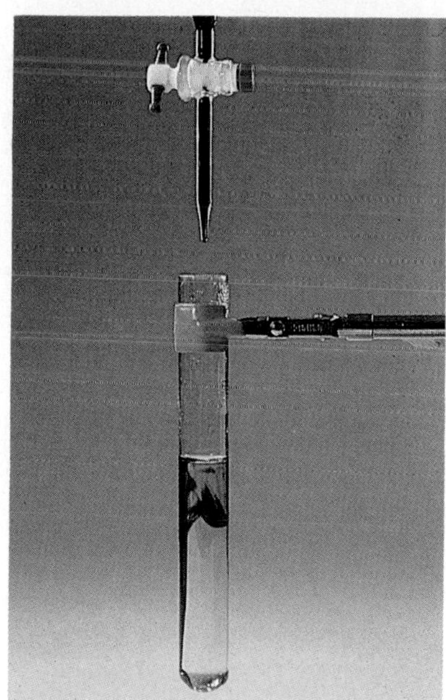

FIGURE 3.20
Potassium permanganate, $KMnO_4$, is an oxidizing agent, owing to the ability of the MnO_4^- ion to accept electrons. Here a drop of deep purple $KMnO_4$ solution is placed in a solution of colorless oxalic acid, $H_2C_2O_4$, a reducing agent that supplies the electrons to reduce the MnO_4^- ion (top). The purple color of $KMnO_4$ fades as the reaction proceeds, since the purple MnO_4^- ion is converted to the nearly colorless Mn^{2+} ion (bottom). (Charles D. Winters)

*You will *never* observe a case where both H^+ and OH^- are used together in the same reaction; they react to produce H_2O.

Step 3. *Balance each half reaction for mass.* Begin by balancing all atoms except H and O; these latter atoms are always the last to be balanced. In the first equation there are 2 C atoms on the left. To have 2 C atoms on the right, place a 2 in front of CO_2.

$$H_2C_2O_4(aq) \longrightarrow 2\ CO_2(g)$$

Now notice that O has also been balanced (4 O atoms on each side), leaving only H to be balanced by adding H in some form to the right side of the equation. Since we are told that the reaction occurs in an acid solution, a mass balance can be achieved if the hydrogen appears in the form of the H^+ ion. Thus, the final, mass-balanced equation is

$$H_2C_2O_4(aq) \longrightarrow 2\ CO_2(g) + 2\ H^+(aq)$$

Turning to the permanganate half reaction, you see that Mn is already balanced.

$$MnO_4^-(aq) \longrightarrow Mn^{2+}(aq) + \text{(need 4 O atoms)}$$

However, an oxygen-containing species must be added to the right side. In acid solution, that species is water. Thus, to balance the four O atoms in MnO_4^- on the left, add 4 H_2O molecules to the right side.

$$MnO_4^-(aq) \longrightarrow Mn^{2+}(aq) + 4\ H_2O(\ell)$$

Now there are 8 unbalanced H atoms on the right. To balance these, use the fact that H^+ ion exists in acid solution.

$$8\ H^+(aq) + MnO_4^-(aq) \longrightarrow Mn^{2+}(aq) + 4\ H_2O(\ell)$$

The equation is now balanced for mass.

To balance oxygen in acid solution, add H_2O to the oxygen-deficient side and then 2 H^+ for every H_2O to the other side of the equation.

Be sure to notice that you can *never* add just O^{2-} to balance oxygen in an equation. In acid solution O^{2-} is converted to H_2O according to the following mass- and charge-balanced equation.

$$O^{2-} + 2\ H^+ \longrightarrow H_2O$$

Indeed, this equation suggests another way to look at the half reaction above: the MnO_4^- ion loses 4 O^{2-} ions on being transformed into Mn^{2+}, but these four O^{2-} ions react with H^+ in the acid solution to produce 4 H_2O molecules.

Step 4. *Balance the half reactions for charge.* The mass-balanced $H_2C_2O_4$ equation has a charge of 0 on the left side but +2 on the right side. Therefore, 2 e^- are added to the more positive right side.

$$H_2C_2O_4(aq) \longrightarrow 2\ CO_2(g) + 2\ H^+(aq) + 2\ e^-$$

The mass-balanced MnO_4^- half reaction has a charge of +7 on the left and +2 on the right. Therefore, $5e^-$ are added to the more positive left side.

$$MnO_4^-(aq) + 8\ H^+(aq) + 5e^- \longrightarrow Mn^{2+}(aq) + 4\ H_2O(\ell)$$

You can now identify the oxidizing and reducing agents. The MnO_4^- ion consumes electrons and so is the oxidizing agent, and $H_2C_2O_4$ supplies electrons and so is the reducing agent.

Step 5. *Multiply the half reactions by appropriate factors so that the reducing agent donates as many electrons as the oxidizing agent consumes.* The $H_2C_2O_4$ half reaction should be multiplied by 5, and the MnO_4^- half reaction should be multiplied by 2, so that each equation involves 10 electrons.

$$5[H_2C_2O_4(aq) \longrightarrow 2\ CO_2(g) + 2\ H^+(aq) + 2\ e^-]$$

$$2[MnO_4^-(aq) + 8\ H^+(aq) + 5e^- \longrightarrow Mn^{2+}(aq) + 4\ H_2O(\ell)]$$

Step 6. *Add the half reactions to give the overall equation.*

$$
\begin{array}{rcl}
5\ H_2C_2O_4(aq) & \longrightarrow & 10\ CO_2(g) + 10\ H^+(aq) + 10\ e^- \\
2\ MnO_4^-(aq) + 16\ H^+(aq) + 10e^- & \longrightarrow & 2\ Mn^{2+}(aq) + 8\ H_2O(\ell) \\
\hline
2\ MnO_4^-(aq) + 6\ H^+(aq) + 5\ H_2C_2O_4(aq) & \longrightarrow & 2\ Mn^{2+}(aq) + 10\ CO_2(g) + 8\ H_2O(\ell)
\end{array}
$$

Step 7. *Check the final result to ensure there is a mass and charge balance.* (a) Mass balance: Both sides of the equation have 2 Mn, 28 O, 10 C, and 16 H atoms. (b) Charge balance: Each side has a net charge of +4.

EXERCISE 3.15 BALANCING AN EQUATION FOR AN OXIDATION–REDUCTION REACTION IN ACID SOLUTION

The reaction of Cu and HNO_3 was shown in the photograph on page 93.

$$Cu(s) + NO_3^-(aq) \longrightarrow NO_2(g) + Cu^{2+}(aq)$$

Show the balanced half reactions and balanced overall equation. Identify (a) the oxidizing agent, (b) the reducing agent, (c) the substance reduced, and (d) the substance oxidized.

In Example 3.9, we were concerned with a reaction in acid solution. In acid solution we can only work with H^+ or the H^+/H_2O pair to achieve balanced equations. Conversely, in basic solution, we can only use OH^- or the OH^-/H_2O pair to balance an equation. The following example illustrates how to balance an equation in basic solution.

In basic solution OH^- (paired with H_2O) can be used to balance oxygen.

EXAMPLE 3.10

BALANCING AN EQUATION FOR AN OXIDATION–REDUCTION REACTION IN BASIC SOLUTION

In the following equation, Bi^{3+} is reduced to bismuth metal by a tin(II) compound in basic solution. Balance the equation.

$$Bi(OH)_3(s) + SnO_2^{2-}(aq) \longrightarrow Bi(s) + SnO_3^{2-}(aq)$$

Solution

Step 1. *Verify that the reaction is an oxidation–reduction process.* The oxidation numbers of both Bi and Sn change.

Step 2. *Break the process into half reactions.*

$$Bi(OH)_3(s) \longrightarrow Bi(s)$$

$$SnO_2^{2-}(aq) \longrightarrow SnO_3^{2-}$$

Step 3. *Balance each half reaction for mass.* The bismuth half reaction involves the simple loss of OH^- ions as $Bi(OH)_3$ is changed to Bi(s). In basic solution OH^- alone can be a product or reactant, so OH^- is a product of the half reaction,

$$Bi(OH)_3(s) \longrightarrow Bi(s) + 3\ OH^-(aq)$$

and it is balanced for mass.

In the tin-containing half reaction, the Sn atoms are balanced, but the left side of the reaction is deficient in oxygen.

$$\text{(need 1 O atom)} + SnO_2^{2-}(aq) \longrightarrow SnO_3^{2-}(aq)$$

In basic solution, there are two oxygen-containing species, OH^- and H_2O. However, OH^- is oxygen rich, since it can be thought of as capable of supplying oxide ion (and also water because of the requirements of mass and charge balance) according to the balanced equation

$$2\ OH^- \longrightarrow O^{2-} + H_2O$$

To balance oxygen in a redox process in basic solution, add OH^- to the oxygen-deficient side and water to the other according to the balanced equation $2\ OH^- \rightarrow O^{2-} + H_2O$.

Therefore, OH^- is added to the oxygen-deficient side of the half reaction.

$$?\ OH^-(aq) + SnO_2^{2-}(aq) \longrightarrow SnO_3^{2-}(aq)$$

To decide on the number of OH^- ions to be added, notice above that 2 OH^- ions must be used to supply one O^{2-} (with 1 H_2O being produced at the same time). Therefore, you must add 2 OH^- to transform the SnO_2^{2-} ion into SnO_3^{2-}.

$$2\ OH^-(aq) + SnO_2^{2-}(aq) \longrightarrow SnO_3^{2-}(aq) + H_2O(\ell)$$

Step 4. *Balance the half reactions for charge.*

$$Bi(OH)_3(s) + 3e^- \longrightarrow Bi(s) + 3\ OH^-(aq)$$

$$SnO_2^{2-}(aq) + 2\ OH^-(aq) \longrightarrow SnO_3^{2-}(aq) + H_2O(\ell) + 2e^-$$

Now you see that $Bi(OH)_3$ is the oxidizing agent (electron acceptor) and SnO_2^{2-} is the reducing agent (electron donor).

Step 5. *Multiply each half reaction by an appropriate factor to balance the number of electrons given off and taken on.*

$$2\ [Bi(OH)_3(s) + 3e^- \longrightarrow Bi(s) + 3\ OH^-(aq)]$$

$$3\ [SnO_2^{2-}(aq) + 2\ OH^-(aq) \longrightarrow SnO_3^{2-}(aq) + H_2O(\ell) + 2e^-]$$

Step 6. *Add the half reactions to give the overall equation.*

$$\begin{array}{r}
2\ Bi(OH)_3(s) + 6e^- \longrightarrow 2\ Bi(s) + 6\ OH^-(aq) \\
3\ SnO_2^{2-}(aq) + 6\ OH^-(aq) \longrightarrow 3\ SnO_3^{2-}(aq) + 3\ H_2O + 6e^- \\
\hline
2\ Bi(OH)_3(s) + 3\ SnO_2^{2-}(aq) \longrightarrow 2\ Bi(s) + 3\ SnO_3^{2-}(aq) + 3\ H_2O(\ell)
\end{array}$$

Step 7. *Check the final result to insure there is a mass and charge balance.* (a) Mass balance: both sides of the equation have 2 Bi, 3 Sn, 6 H, and 12 O atoms. (b) Charge balance: both sides of the equation have a net charge of −6.

EXERCISE 3.16 BALANCING AN EQUATION FOR AN OXIDATION–REDUCTION REACTION IN BASIC SOLUTION

Balance the following equation for a reaction occurring in basic solution.

$$Cr(s) + ClO_4^-(aq) \longrightarrow Cr(OH)_3(s) + ClO_3^-(aq)$$

Show each balanced half reaction, and identify (a) the oxidizing agent, (b) the reducing agent, (c) substance oxidized, and (d) the substance reduced.

The examples above illustrate the fundamental principles for balancing equations involving electron transfer. The steps you go through are always the same, but there are many variations in detail, especially in the mass balance step 2. Only a few of these variations can be illustrated here, so the best way to learn how to balance such equations is to practice.

EXERCISE 3.17 BALANCING EQUATIONS FOR HALF REACTIONS

Balance for mass and charge the equations for the following half reactions. For each reaction, tell whether the substance on the left is an oxidizing or reducing agent and whether the overall process is an oxidation or reduction.

(a) $S_2O_3^{2-}(aq) \longrightarrow S_4O_6^{2-}(aq)$ (neutral solution)
(b) $NO_3^-(aq) \longrightarrow NO(g)$ (acid solution)

(c) $C_2H_5OH(aq) \longrightarrow C_2H_4O(aq)$ (acid solution)
(d) $PH_3(aq) \longrightarrow H_3PO_4(aq)$ (acid solution)
(e) $ClO^-(aq) \longrightarrow Cl^-(aq)$ (basic solution)

EXERCISE 3.18 BALANCING EQUATIONS FOR OXIDATION–REDUCTION REACTIONS

Using the stepwise procedure outlined above, balance for mass and charge equations for the following oxidation–reduction reactions. In each case, identify the oxidizing and reducing agents.

(a) $S_2O_3^{2-}(aq) + I_2(aq) \longrightarrow S_4O_6^{2-}(aq) + I^-(aq)$ (neutral solution)
(b) $Cu(s) + NO_3^-(aq) \longrightarrow Cu^{2+}(aq) + NO(aq)$ (acid solution)
(c) $Fe^{2+}(aq) + MnO_4^-(aq) \longrightarrow Fe^{3+}(aq) + Mn^{2+}(aq)$ (acid solution)
(d) $Cr(OH)_3(s) + ClO^-(aq) \longrightarrow Cl_2(aq) + CrO_4^{2-}(aq)$ (basic solution)

SUMMARY

In any chemical change, matter is conserved. Although the atoms involved are rearranged into different species in the course of a reaction and the number of molecules may change, the total number of atoms of each kind in the reactants and products must be the same. Thus, a **balanced chemical equation** shows the relative amounts of products and reactants, the amounts being indicated by **stoichiometric coefficients** (Sections 3.1 and 3.2).

Properties of ionic compounds are explored further, since their reactions are important (Section 3.3). Ionic compounds that contain certain ions can dissolve in water to a significant extent (Table 3.1). In doing so, they break up or dissociate into their ions. The resulting solution conducts electricity, so the dissolved ionic compounds are called **electrolytes**. Common acids and bases are listed in Table 3.2. **Acids** ionize to provide H^+ ions in water, while **bases** provide OH^- ions.

Some ions do not enter directly into reactions involving ionic compounds and so are called **spectator ions**. These can be deleted from the complete equation for the reaction to give a **net ionic equation** (Section 3.3).

Throughout this book, you will encounter (a) reactions of compounds or elements with oxygen, (b) **exchange** or **metathesis reactions** (**acid–base** and **precipitation reactions**), and (c) **oxidation–reduction reactions**, among others (Section 3.4). **Combustion** reactions involve the combination of a compound of C and H or of C, H, and O with O_2 to give CO_2 and H_2O. **Metal carbonates** decompose on heating to give metal oxide and CO_2; with acid they give a metal salt, water, and CO_2.

Oxidation-reduction reactions involve the exchange of electrons between compounds (Section 3.5). A compound is said to be **reduced** if the **oxidation number** of one of its atoms is reduced by acquiring electrons from another species, the **reducing agent**. Conversely, a compound is **oxidized** if the oxidation number of one of its atoms is increased because that compound has transferred one or more electrons to another species, the **oxidizing agent**. Equations for oxidation-reduction reactions can be balanced if the equation is divided into **half reactions**, one for the species being reduced and another for the species being oxidized.

SPECIAL SECTION: MERCURY

Mercury, a metal, is a liquid at room temperature. It boils at 356.6°C and freezes at −38.87°C. It has a density of 13.594 g/cm³. See page 99.

Mercury, or *quicksilver,* is about as abundant in the Earth's crust as silver. Since mercury-containing ore is an obvious red mineral that occurs in concentrated deposits, it is one of the oldest known elements. Its principal ore is cinnabar or natural vermilion, the common names of mercury(II) sulfide (see page 99). Prehistoric man mixed powdered vermilion with a little water and used it as a pigment to draw on the walls of caves. Ancient peoples also found that heating the red rock in a fire gave liquid quicksilver.

$$HgS(s) + O_2(g) \longrightarrow Hg(liq) + SO_2(g)$$

Samples of the metal have been found in graves of the 16th century BC, and Aristotle, in the 4th century BC, was first to call it "liquid silver."

Ancient Hindu wise men thought mercury was an aphrodisiac. It has been widely used for its medicinal properties, and it was even prescribed for severe constipation; drink a liter of mercury, and nothing stands in its way!

The ability of mercury to form **amalgams** or intimate mixtures with other metals such as gold and silver has been known for centuries. If a liquid amalgam of gold and mercury is rubbed onto a copper object, the object becomes covered with gold. Since mercury is volatile, heating drives away the mercury and leaves a gold coating. The same idea has been used to extract metallic gold from natural sources. Mixing gold-bearing soil with mercury gives a liquid amalgam that is easily separated from dirt. Heating this amalgam then drives off the mercury and leaves gold behind. It was for this reason that the Spanish fleet carried casks of mercury to the New World and returned with silver and gold.

There may be as many as 3000 uses for the metal in our modern society. Dentists use an amalgam with silver and tin to fill cavities; a drop of mercury goes into every fluorescent lamp; and the liquid metal is used in thermometers and barometers and in electrical equipment of many kinds.

Compounds of mercury also find many uses. Mercury oxide, for example, is used in batteries as the oxidizing agent (where Zn is the reducing agent, $Zn + 2\ OH^- \rightarrow ZnO + H_2O + 2e^-$). (See Chapter 19.) By arranging the compounds properly, the electrons provided by Zn can be passed through a wire to the HgO ($HgO + H_2O + 2e^- \rightarrow Hg + 2\ OH^-$), a passage of "current" that can be tapped to power your calculator. Many mercury compounds, such as mercurochrome, are antiseptics, inhibiting the growth of bacteria and mildew.

Three mercury salts: mercury(II) iodide (HgI_2) (left), mercury(II) chloride ($HgCl_2$) (middle), and mercury(II) oxide (HgO) (right). See also page 99.

But mercury and its salts have a darker side, and none has expressed it better than Alfred Stock (1876–1946), a German chemist. Stock said that "mercury is a strong poison, particularly dangerous because of its liquid form and noticeable volatility even at room temperature. . . . [He] found from personal experience . . . that protracted stay in an atmosphere charged with only 1/100 of the amount of mercury required for its saturation, [can] induce chronic mercury poisoning. [It] reveals itself [first] as an affection of the nerves, causing headaches, numbness, mental lassitude, depression, and the loss of memory."

A well-known example of mercury poisoning is "hatter's disease." Years ago felt was made by dipping furs into vats containing mercury(II) nitrate. Workers in these ill-ventilated shops breathed mercury-laden fumes and soon developed tremors, lost teeth, had difficulty in walking, and lost mental ability. Legend has it that the Mad Hatter in Lewis Carroll's *Alice in Wonderland* was patterned after a victim of this affliction.

STUDY QUESTIONS

REVIEW QUESTIONS

1. Find in the chapter one example of each of the following reaction types and write the balanced equation for the reaction: (a) combustion; (b) reaction of O_2 with a metal; (c) reaction of O_2 with a nonmetal.
2. Find two examples each of precipitation reactions and acid-base reactions in the chapter. Write balanced equations for these reactions.
3. What is an electrolyte? How can we differentiate between a weak and a strong electrolyte? Give an example of each.
4. Name the spectator ions in each of the following reactions:
   (a) Calcium carbonate and hydrochloric acid.
   $$CaCO_3(s) + 2\,H^+(aq) + 2\,Cl^-(aq) \longrightarrow CO_2(g) + Ca^{2+}(aq) + 2\,Cl^-(aq) + H_2O(\ell)$$
   (b) Nitric acid and magnesium hydroxide.
   $$2\,H^+(aq) + 2\,NO_3^-(aq) + Mg(OH)_2(s) \longrightarrow 2\,H_2O(\ell) + Mg^{2+}(aq) + 2\,NO_3^-(aq)$$
5. Name the water-insoluble product in each reaction.
   (a) $CuCl_2(aq) + H_2S(aq) \longrightarrow CuS + 2\,HCl$
   (b) $CaCl_2(aq) + K_2CO_3(aq) \longrightarrow 2\,KCl + CaCO_3$
   (c) $AgNO_3(aq) + NaI(aq) \longrightarrow AgI + NaNO_3$

BALANCING EQUATIONS

6. Balance the following equations.
   (a) $Al(s) + O_2(g) \longrightarrow Al_2O_3(s)$
   (b) $N_2(g) + H_2(g) \longrightarrow NH_3(g)$
   (c) $C_6H_6(\ell) + O_2(g) \longrightarrow H_2O(g) + CO_2(g)$
7. Balance the following equations.
   (a) $Al(s) + Cl_2(g) \longrightarrow AlCl_3(s)$
   (b) $SiO_2(s) + C(s) \longrightarrow Si(s) + CO(g)$
   (c) $Fe(s) + H_2O(g) \longrightarrow Fe_3O_4(s) + H_2(g)$
8. Balance the following equations.
   (a) $UO_2(s) + HF(\ell) \longrightarrow UF_4(s) + H_2O(\ell)$
   (b) $B_2O_3(s) + HF(\ell) \longrightarrow BF_3(g) + H_2O(\ell)$
   (c) $BF_3(g) + H_2O(\ell) \longrightarrow HF(\ell) + H_3BO_3(s)$
9. Balance the following equations.
   (a) $MgO(s) + Fe(s) \longrightarrow Fe_2O_3(s) + Mg(s)$
   (b) $H_3BO_3(s) \longrightarrow B_2O_3(s) + H_2O(g)$
   (c) $NaNO_3(s) + H_2SO_4(\ell) \longrightarrow Na_2SO_4(s) + HNO_3(g)$
10. Balance the following equations.
   (a) $Na_2O_2(s) + H_2O(\ell) \longrightarrow NaOH(aq) + H_2O_2(aq)$
   (b) $PH_3(g) + O_2(g) \longrightarrow P_4O_{10}(s) + H_2O(g)$
   (c) $C_2H_3Cl(\ell) + O_2(g) \longrightarrow CO_2(g) + H_2O(g) + HCl(g)$
11. Balance the following equations.
   (a) $CaF_2(s) + H_2SO_4(\ell) \longrightarrow CaSO_4(s) + HF(g)$
   (b) $N_2O(g) \longrightarrow N_2(g) + O_2(g)$
   (c) $NH_4NO_3(s) \longrightarrow N_2O(g) + H_2O(g)$
12. Balance the following equations.
   (a) Reaction to produce hydrazine, N_2H_4.
   $$H_2NCl(aq) + NH_3(g) \longrightarrow NH_4Cl(aq) + N_2H_4(aq)$$
   (b) Reaction of fuel used in moon lander and space shuttle.
   $$(CH_3)_2N_2H_2(\ell) + N_2O_4(\ell) \longrightarrow N_2(g) + H_2O(g) + CO_2(g)$$
   (c) Reaction of calcium carbide to produce acetylene.
   $$CaC_2(s) + H_2O(\ell) \longrightarrow Ca(OH)_2(s) + C_2H_2(g)$$
13. Balance the following equations.
   (a) Reaction of calcium cyanamide to produce ammonia.
   $$CaCN_2(s) + H_2O(\ell) \longrightarrow CaCO_3(s) + NH_3(g)$$
   (b) Reaction to produce diborane, B_2H_6.
   $$NaBH_4(s) + H_2SO_4(aq) \longrightarrow Na_2SO_4(aq) + B_2H_6(g) + H_2(g)$$
   (c) Reaction to rid water of hydrogen sulfide, H_2S, a foul-smelling compound.
   $$H_2S(aq) + Cl_2(aq) \longrightarrow HCl(aq) + S(s)$$

PROPERTIES OF IONIC COMPOUNDS

14. Which compound or compounds in each of the following groups is (are) expected to be soluble in water?
   (a) CuO, $CuCl_2$, and $CuCO_3$
   (b) $AgCl$, Ag_3PO_4, and $AgNO_3$
   (c) KCl, K_2CO_3, and $KMnO_4$
   (d) $NaCl$, $AgCl$, and $BaCl_2$
15. Give the formula for
   (a) a soluble compound containing the acetate ion.
   (b) an insoluble sulfide.
   (c) a soluble hydroxide.
   (d) an insoluble chloride.
16. Each compound below is water soluble. What ions are produced in water?
   (a) NaI (c) $KHSO_4$
   (b) K_2SO_4 (d) $NaCN$
17. Each compound below is water soluble. What ions are produced in water?
   (a) KOH (c) $NaNO_3$
   (b) K_2CO_3 (d) $(NH_4)_2SO_4$
18. Tell whether each of the following is water soluble or not. If soluble, tell what ions are produced.
   (a) $BaCl_2$ (c) $Pb(NO_3)_2$
   (b) $Cr(NO_3)_2$ (d) $BaSO_4$
19. Tell whether each of the following is water soluble or not. If soluble, tell what ions are produced.
   (a) Na_2CO_3 (c) CdS
   (b) $CuSO_4$ (d) $CaBr_2$

20. If you wanted to make a water-soluble compound containing Cu^{2+}, name four anions you could choose to combine with the copper ion. Conversely, if you wanted an insoluble Cu^{2+} compound, what two anions would you choose?
21. If you wanted to make a water-soluble compound containing Ca^{2+}, name two anions you could use. Conversely, if you wanted an insoluble Ca^{2+} compound, what two anions would you choose?
22. Tell whether each of the following is water soluble or not. If soluble, tell what ions are produced.
(a) $NiCO_3$ (d) $Cu(ClO_4)_2$
(b) $Zn(C_2H_3O_2)_2$ (e) $K_2Cr_2O_7$
(c) KCN (f) FeS

ACIDS AND BASES

23. Write a balanced equation to show how nitric acid ionizes in water.
24. Write a balanced equation to show how calcium hydroxide ionizes in water.

NET IONIC EQUATIONS

25. Balance each of the following equations, and then write the net ionic equation.
(a) $Zn(s) + HCl(aq) \longrightarrow H_2(g) + ZnCl_2(aq)$
(b) $Mg(OH)_2(s) + HCl(aq) \longrightarrow MgCl_2(aq) + H_2O(\ell)$
(c) $HNO_3(aq) + Ca(OH)_2(aq) \longrightarrow Ca(NO_3)_2(aq) + H_2O(\ell)$
(d) $HCl(aq) + MnO_2(s) \longrightarrow MnCl_2(aq) + Cl_2(g) + H_2O(\ell)$
26. Balance each of the following equations, and then write the net ionic equation.
(a) $(NH_4)_2S(aq) + Cu(NO_3)_2(aq) \longrightarrow CuS(s) + NH_4NO_3(aq)$
(b) $Pb(NO_3)_2(aq) + HCl(aq) \longrightarrow PbCl_2(s) + HNO_3(aq)$
(c) $BaCO_3(s) + HCl(aq) \longrightarrow BaCl_2(aq) + H_2O(\ell) + CO_2(g)$
27. Balance each of the following equations, and then write the net ionic equation. Refer to Tables 3.1 and 3.2 for information on acids and bases and on solubility. Show phases for all reactants and products.
(a) $Ba(OH)_2 + HNO_3 \longrightarrow Ba(NO_3)_2 + H_2O$
(b) $BaCl_2 + Na_2CO_3 \longrightarrow BaCO_3 + NaCl$
(c) $Na_2S + Ni(NO_3)_2 \longrightarrow NiS + NaNO_3$
(d) $ZnCl_2 + KOH \longrightarrow KCl + Zn(OH)_2$

CLASS OF CHEMICAL REACTIONS

28. Complete and balance the following equations involving oxygen.
(a) $Mg(s) + O_2(g) \longrightarrow$
(b) $Ca(s) + O_2(g) \longrightarrow$
(c) $In(s) + O_2(g) \longrightarrow$
29. Complete and balance the following equations involving oxygen.
(a) $Ti(s) + O_2(g) \longrightarrow$ titanium(IV) oxide
(b) $S_8(s) + O_2(g) \longrightarrow$ sulfur dioxide
(c) $Se(s) + O_2(g) \longrightarrow$ selenium dioxide
30. Complete and balance equations for the following combustion reactions.
(a) $CH_4(g) + O_2(g) \longrightarrow$
(b) $C_8H_{18}(\ell) + O_2(g) \longrightarrow$
(c) $C_2H_5OH(\ell) + O_2(g) \longrightarrow$
31. Complete and balance equations for the following combustion reactions.
(a) $C_{10}H_{10}Fe(s) + O_2(g) \longrightarrow Fe_2O_3(s) + \cdots$
(b) $B_5H_9(\ell) + O_2(g) \longrightarrow B_2O_3(s) + \cdots$
(c) $Si_2H_6(g) + O_2(g) \longrightarrow SiO_2(s) + \cdots$
32. Write a balanced equation for the formation of
(a) carbon monoxide
(b) nickel(II) oxide
(c) chromium(III) oxide
33. Write a balanced equation for the formation of
(a) copper(I) oxide
(b) arsenic(III) oxide
(c) zinc oxide
34. Complete and balance equations for the following exchange reactions. Tell whether each is an acid–base or a precipitation reaction.
(a) $HCl(aq) + KOH(aq) \longrightarrow$
(b) $AgNO_3(aq) + KCl(aq) \longrightarrow$
(c) $H_2SO_4(aq) + NaOH(aq) \longrightarrow$
35. Complete and balance equations for the following exchange reactions. Tell whether each is an acid–base or a precipitation reaction.
(a) $Cu(NO_3)_2(aq) + Na_2S(aq) \longrightarrow$
(b) $Pb(NO_3)_2(aq) + KCl(aq) \longrightarrow$
(c) $LiOH(aq) + HClO_4(aq) \longrightarrow$
36. Complete and balance equations for the following exchange reactions. Tell whether each is an acid–base or a precipitation reaction.
(a) $Al(OH)_3(s) + HNO_3(aq) \longrightarrow$
(b) $BaCl_2(aq) + H_2SO_4(aq) \longrightarrow$
(c) $MnCl_2(aq) + (NH_4)_2S(aq) \longrightarrow$
37. Complete and balance equations for the following exchange reactions. Tell whether each is an acid–base or a precipitation reaction.
(a) $HNO_3(aq) + Ca(OH)_2(s) \longrightarrow$
(b) $HCl(aq) + Fe(OH)_2(s) \longrightarrow$
(c) $SrCl_2(aq) + Na_2CO_3(aq) \longrightarrow$
38. Write a balanced equation showing how you could prepare each of the following salts by an acid–base reaction.
(a) $NaNO_3$
(b) KCl
(c) K_3PO_4
(d) Cs_2SO_4

39. Write a balanced equation showing how you could prepare each of the following salts by an acid–base reaction.
(a) $NaC_2H_3O_2$
(b) NH_4NO_3
(c) $MgSO_4$
(d) Na_3PO_4

40. Write a balanced equation showing how you would prepare each of the following insoluble salts by a precipitation reaction.
(a) AgBr (d) $BaSO_4$
(b) $CaCO_3$ (e) ZnS
(c) NiS

41. Write a balanced equation showing how you would prepare each of the following compounds by an exchange reaction.
(a) KNO_3 (d) $BaCl_2$
(b) $SrSO_4$ (e) $(NH_4)_2SO_4$
(c) ZnS

42. Write a balanced equation showing at least one way to prepare each of the following compounds.
(a) Na_2SO_4 (c) $MgCl_2$
(b) Li_2O (d) SnO_2
(c) CdS

43. Selecting only from the list of reagents given below, show how you would prepare (a) NaCl, (b) $PbCl_2$, (c) $BaSO_4$, and (d) AgI

Available reagents

KCl	$BaCl_2$	Na_2SO_4	$Pb(NO_3)_2$	$AgNO_3$
HCl	AgCl	$NaNO_3$	KI	$PbCO_3$
NH_4Cl	NaOH	PbI_2	$BaCO_3$	HNO_3

44. Complete and balance the following equations:
(a) $CaCO_3(s) + HNO_3(aq) \longrightarrow$
(b) $SrCO_3(s) + \text{heat} \longrightarrow$
(c) $K_2CO_3(aq) + H_2SO_4(aq) \longrightarrow$
(d) $CuCO_3(s) + HClO_4(aq) \longrightarrow$

45. Using the reactions of metal carbonates, give a balanced equation for the preparation of (a) NaCl, (b) barium nitrate, and (c) BaO.

46. Give balanced equations for two methods of preparing lithium oxide, one from the metal and one from the appropriate carbonate.

47. Using the reactions of metal carbonates, give a balanced equation for the preparation of (a) $LiClO_4$, (b) nickel(II) chloride, and (c) copper(II) oxide.

OXIDATION NUMBERS

48. Determine the oxidation number of each element in the following ions or compounds.
(a) BrO^- (d) IO_3^-
(b) $C_2O_4^{2-}$ (e) $HClO_4$
(c) I_2

49. Determine the oxidation number of each element in the following ions or compounds.
(a) PF_5 (d) H_3AsO_4
(b) SCl_2 (e) $C_6H_{12}O_6$
(c) SO_4^{2-}

50. Determine the oxidation number of each element in the following ions or compounds.
(a) XeO_4^{2-} (c) C_3H_8
(b) N_2H_4 (d) C_7H_6O

OXIDATION–REDUCTION REACTIONS

51. In each of the following reactions, tell which reactant is oxidized and which is reduced.
(a) $2\ Mg(s) + O_2(g) \longrightarrow 2\ MgO(s)$
(b) $C_2H_4(g) + 3\ O_2(g) \longrightarrow 2\ CO_2(g) + 2\ H_2O(g)$
(c) $Si(s) + 2\ Cl_2(g) \longrightarrow SiCl_4(\ell)$

52. In each of the following reactions, tell which reactant is oxidized and which is reduced.
(a) $Ca(s) + 2\ HCl(aq) \longrightarrow CaCl_2(aq) + H_2(g)$
(b) $Cr_2O_7^{2-}(aq) + 3\ Sn^{2+}(aq) + 14\ H^+(aq) \longrightarrow 2\ Cr^{3+}(aq) + 3\ Sn^{4+}(aq) + 7\ H_2O(\ell)$
(c) $FeS(s) + 3\ NO_3^-(aq) + 4\ H^+(aq) \longrightarrow 3\ NO(g) + SO_4^{2-}(aq) + Fe^{3+}(aq) + 2\ H_2O(\ell)$

HALF REACTIONS

Balance equations for the half reactions in Questions 53 through 56. Tell whether the reactant is an oxidizing or reducing agent and if the overall process is an oxidation or reduction. Unless noted otherwise, all are carried out in acid solution, meaning that H^+ or H^+/H_2O *may* be used to balance the equation.

53. (a) $Cr(s) \longrightarrow Cr^{3+}(aq)$
(b) $Fe^{3+}(aq) \longrightarrow Fe^{2+}(aq)$
(c) $AsH_3(g) \longrightarrow As(s)$
(d) $VO_3^-(aq) \longrightarrow V^{2+}(aq)$

54. (a) $Br_2(aq) \longrightarrow Br^-(aq)$
(b) $H_2S(aq) \longrightarrow S(s)$
(c) $VO^{2+}(aq) \longrightarrow V^{3+}(aq)$
(d) $U^{4+}(aq) \longrightarrow UO_2^+(aq)$

55. (a) $Cr_2O_7^{2-}(aq) \longrightarrow Cr^{3+}(aq)$
(b) $N_2H_5^+(aq) \longrightarrow N_2$
(c) $CH_3CHO(aq) \longrightarrow CH_3COOH(aq)$
(d) $Bi^{3+}(aq) \longrightarrow HBiO_3(aq)$

56. (a) $HOI(aq) \longrightarrow I^-(aq)$
(b) $NO(g) \longrightarrow HNO_2(aq)$
(c) $C_6H_5CH_3(aq) \longrightarrow C_6H_5COOH(aq)$
(d) $SO_4^{2-}(aq) \longrightarrow H_2SO_3(aq)$

57. The half reactions here are in basic solution. You *may* need to use OH^- or OH^-/H_2O to balance the equation.
(a) $Sn(s) \longrightarrow Sn(OH)_4^{2-}(aq)$
(b) $MnO_4^-(aq) \longrightarrow MnO_2(s)$
(c) $ClO^-(aq) \longrightarrow Cl^-(aq)$

58. The half reactions here are in basic solution. You *may* need to use OH^- or OH^-/H_2O to balance the equation.
(a) $CrO_2^-(aq) \longrightarrow CrO_4^{2-}(aq)$
(b) $Br_2(aq) \longrightarrow BrO_3^-(aq)$
(c) $Ni(OH)_2(s) \longrightarrow NiO_2(s)$

BALANCING OXIDATION–REDUCTION EQUATIONS

Use the half-reaction method to balance the equations in Questions 59 through 66.

59. Reaction (a) is in neutral solution. Reactions (b) through (d) are in acid solution, so you *may* need to use H^+ or H^+/H_2O to balance the equation.
(a) $Cl_2(aq) + Br^-(aq) \longrightarrow Br_2(aq) + Cl^-(aq)$
(b) $Sn(s) + H^+(aq) \longrightarrow Sn^{2+}(aq) + H_2(g)$
(c) $Al(s) + Sn^{4+}(aq) \longrightarrow Al^{3+}(aq) + Sn^{2+}(aq)$
(d) $Zn(s) + VO^{2+}(aq) \longrightarrow Zn^{2+}(aq) + V^{3+}(aq)$

60. The reactions here are in acid solution. You *may* need to use H^+ or H^+/H_2O to balance the equation.
(a) $Hg^{2+}(aq) + Cu(s) \longrightarrow Cu^{2+}(aq) + Hg(\ell)$
(b) $MnO_2(s) + Cl^-(aq) \longrightarrow Mn^{2+}(aq) + Cl_2(g)$
(c) $I^-(aq) + Br_2(aq) \longrightarrow IO_3^-(aq) + Br^-(aq)$
(d) $Zn(s) + NO_3^-(aq) \longrightarrow Zn^{2+}(aq) + N_2O(g)$

61. The reactions here are in acid solution. You *may* need to use H^+ or H^+/H_2O to balance the equation.
(a) $Ag^+(aq) + HCHO(aq) \longrightarrow Ag(s) + HCOOH(aq)$
(b) $MnO_4^-(aq) + C_2H_5OH(aq) \longrightarrow Mn^{2+}(aq) + C_2H_4O(aq)$
(c) $H_2S(aq) + Cr_2O_7^{2-}(aq) \longrightarrow S(s) + Cr^{3+}(aq)$
(d) $Zn(s) + VO_3^-(aq) \longrightarrow V^{2+}(aq) + Zn^{2+}(aq)$

62. The reactions here are in acid solution. You *may* need to use H^+ or H^+/H_2O to balance the equation. In (d) $C_6H_8O_6$ is vitamin C.
(a) $MnO_4^-(aq) + HSO_3^-(aq) \longrightarrow Mn^{2+}(aq) + SO_4^{2-}(aq)$
(b) $Cr_2O_7^{2-}(aq) + Fe^{2+}(aq) \longrightarrow Cr^{3+}(aq) + Fe^{3+}(aq)$
(c) $Ag(s) + NO_3^-(aq) \longrightarrow NO_2(g) + Ag^+(aq)$
(d) $Br_2(aq) + C_6H_8O_6(aq) \longrightarrow Br^-(aq) + C_6H_6O_6(aq)$

63. The reactions here are in acid solution. You *may* need to use H^+ or H^+/H_2O to balance the equation.
(a) $IO_3^-(aq) + HSO_3^-(aq) \longrightarrow I^-(aq) + SO_4^{2-}(aq)$
(b) $OsO_4(s) + C_4H_8(OH)_2(aq) \longrightarrow Os^{4+}(aq) + C_4H_8O_2(aq)$
(c) $U^{4+}(aq) + MnO_4^-(aq) \longrightarrow Mn^{2+}(aq) + UO_2^+(aq)$
(d) $CuS(s) + NO_3^-(aq) \longrightarrow Cu^{2+}(aq) + S(s) + NO(g)$

64. The reactions here are in acid solution. You *may* need to use H^+ or H^+/H_2O to balance the equation.
(a) $Sn^{2+}(aq) + IO_4^-(aq) \longrightarrow I^-(aq) + Sn^{4+}(aq)$
(b) $I_2(aq) + S_4O_6^{2-}(aq) \longrightarrow H_2SO_3(aq) + I^-(aq)$
(c) $Zn(s) + H_2AsO_4^-(aq) \longrightarrow AsH_3(g) + Zn^{2+}(aq)$
(d) $PbS(s) + NO_3^-(aq) \longrightarrow Pb^{2+}(aq) + SO_4^{2-}(aq) + NO(g)$

65. The reactions here are in *basic* solution. You *may* need to add OH^- or OH^-/H_2O to balance the equation.
(a) $Zn(s) + ClO^-(aq) \longrightarrow Zn(OH)_2(s) + Cl^-(aq)$
(b) The following reaction is a **disproportionation**: One substance, Br_2, functions both as the reducing and oxidizing agent.
$Br_2(aq) \longrightarrow Br^-(aq) + BrO_3^-(aq)$
(c) $ClO^-(aq) + CrO_2^-(aq) \longrightarrow Cl^-(aq) + CrO_4^{2-}(aq)$

66. The reactions here are in *basic* solution. You *may* need to add OH^- or OH^-/H_2O to balance the equation.
(a) $Fe(OH)_2(s) + CrO_4^{2-}(aq) \longrightarrow Fe_2O_3(s) + Cr(OH)_4^-(aq)$
(b) $PbO_2(s) + Cl^-(aq) \longrightarrow ClO^-(aq) + Pb(OH)_3^-(aq)$
(c) $Al(s) + OH^-(aq) \longrightarrow Al(OH)_4^-(aq) + H_2(g)$
(d) $CN^-(aq) + CrO_4^{2-}(aq) \longrightarrow CNO^-(aq) + Cr(OH)_4^-(aq)$

4

Stoichiometry: Quantitative Information from Chemical Equations

Reaction of red phosphorus and bromine to give phosphorus tribromide (Charles D. Winters)

CHAPTER OUTLINE

4.1 WEIGHT RELATIONS IN CHEMICAL REACTIONS
4.2 REACTIONS WHERE ONE REAGENT IS PRESENT IN LIMITED SUPPLY
4.3 PERCENT YIELD
4.4 REACTIONS IN SOLUTION

A major aspect of chemistry is the study of the reactions of elements and compounds. Among the many things about chemical reactions we would like to know are the nature of the products when several compounds or elements react with one another. If a product turns out to have commercial potential, the efficiency of the reaction is then of interest. Can the product be made in sufficient quantity and purity that it can be sold for a reasonable price? Questions such as these mean that reactions must be studied quantitatively. **Stoichiometry**, the subject of this chapter, is the study of the quantitative relations between amounts of reactants and products.

4.1 WEIGHT RELATIONS IN CHEMICAL REACTIONS

The **conservation of matter** is the guiding principle of chemical stoichiometry. In the preceding chapter you saw that burning a piece of magnesium wire in the oxygen of air will give magnesium oxide, MgO (Figure 3.1). If you use 2.00 moles of magnesium wire (48.6 g), then the following balanced equation shows you that 1.00 moles or 32.0 g of O_2 must be used—no more or less—and that 2.00 moles or 80.6 g of MgO are produced.

$$2\,\text{Mg(s)} + \text{O}_2\text{(g)} \longrightarrow 2\,\text{MgO(s)}$$

2 Mg(s)	O_2(g)	2 MgO(s)
2 atoms	1 molecule	2 molecules
2.00 moles	1.00 mole	2.00 moles
48.6 g	32.0 g	80.6 g

The balanced equation for burning magnesium applies to the reaction of Mg with O_2 no matter how much Mg is used. If 0.020 mole of Mg is used, then 0.010 mole of O_2 will be required and 0.020 mole of MgO will form. You can confirm by experiment that 0.81 g of reactants would be consumed, and 0.81 g of MgO must be produced.

Following this line of reasoning further, suppose you have a piece of magnesium having a mass of only 0.145 g. How much O_2 must there be to consume the magnesium completely? The following steps lead to the answer:

Step 1. Write a balanced equation for the reaction. This is

$$\text{Mg(s)} + \text{O}_2\text{(g)} \longrightarrow 2\,\text{MgO(s)}$$

Step 2. Convert the mass of magnesium to moles. This must be done because the balanced equation uses mole units, not masses.

$$0.145 \text{ g Mg} \left(\frac{1 \text{ mol of Mg}}{24.3 \text{ g Mg}}\right) = 5.97 \times 10^{-3} \text{ mol Mg}$$

Step 3. Relate the number of moles of the given reactant (Mg) to the number of moles of the other reactant (O_2) (or product) whose amount you wish to know.

$$5.97 \times 10^{-3} \text{ mol Mg} \left(\frac{1 \text{ mol } O_2 \text{ required}}{2 \text{ mol Mg available}}\right) = 2.98 \times 10^{-3} \text{ mol } O_2$$

required to use up completely the available Mg

Here you multiply the number of moles of available reagent by a **stoichiometric factor**, a mole-ratio factor relating moles of the desired compound to moles of the available reagent. The *stoichiometric factor comes directly from the balanced chemical equation*, hence the reason for balancing equations for chemical reactions before proceeding with calculations.

Step 4. Knowing the moles of O_2 required to use up all the available Mg, you can calculate the mass of O_2 required. This is done simply by reversing the operation carried out in step 2. That is,

$$2.98 \times 10^{-3} \text{ mol } O_2 \left(\frac{32.0 \text{ g } O_2}{1 \text{ mol } O_2}\right) = 9.54 \times 10^{-2} \text{ g } O_2$$

Since the object of this example is to find the mass of O_2 required, the problem is complete.

At this point you may also wish to know about the amount of MgO produced in the reaction of 0.145 g of magnesium with oxygen. Because of the principle of the conservation of matter, you can answer this simply by adding up the masses of Mg and O_2 used (giving 0.240 g of MgO produced). Alternatively, you could repeat steps 3 and 4 above, but with the appropriate stoichiometric factor and molar mass.

Step 3′. Relate number of moles of Mg to the number of moles of MgO produced.

$$5.97 \times 10^{-3} \text{ mol Mg} \left(\frac{1 \text{ mol MgO produced}}{1 \text{ mol Mg used}}\right) = 5.97 \times 10^{-3} \text{ mol MgO produced}$$

Step 4′. Convert the moles MgO produced to grams.

$$5.97 \times 10^{-3} \text{ mol MgO} \left(\frac{40.3 \text{ g MgO}}{1 \text{ mol MgO}}\right) = 0.240 \text{ g MgO}$$

Using this alternative may seem a bit silly here, since only one product is formed, but simply adding the masses of reactants to obtain the product mass works only if the reaction gives a single product. When more than one product results from a reaction, it is necessary to apply steps 3 and 4 for each product.

The example above was meant to show you that there is a definite series of steps to follow in carrying out a stoichiometric calculation. To help you see this more clearly, consider the scheme outlined in Figure 4.1.

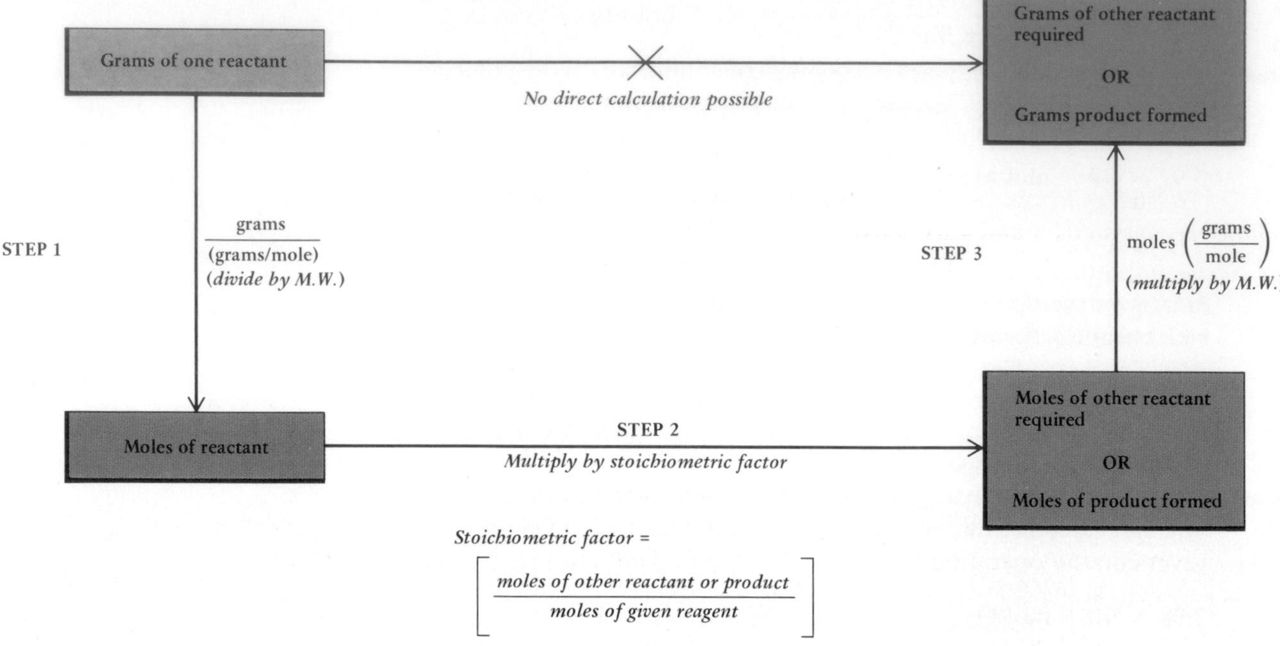

FIGURE 4.1

A scheme outlining the stoichiometric relation between the mass of one reactant and the mass of another reactant or product.

EXERCISE 4.1 WEIGHT RELATIONS IN REACTIONS

Sulfur dioxide, which forms when sulfur burns in air (Figure 3.12), is a major air pollutant.

$$S_8(s) + 8\ O_2(g) \longrightarrow 8\ SO_2(g)$$

If one pound (454 g) of sulfur is burned to give SO_2, how many grams of O_2 are required and how many grams of SO_2 are formed?

Propane, C_3H_8, can be used as a fuel in your home or car, because it is easily liquefied and transported. You learned in the preceding chapter that this and other hydrocarbons combust in O_2 to give CO_2 and H_2O.* If you burn 1.00 pound or 454 g of propane as fuel for your car, how much oxygen would be required to burn this amount of fuel completely and how much H_2O and CO_2 would be produced? Before beginning to do any calculation, first write a balanced chemical equation.

Step 1. Balance the equation.

$$C_3H_8(g) + 5\ O_2(g) \longrightarrow 3\ CO_2(g) + 4\ H_2O(g)$$

This having been done, you can now proceed to the calculation.

Step 2. Convert the mass of compound available (propane) to moles.

$$454\ \cancel{g\ C_3H_8} \left(\frac{1\ mol\ C_3H_8}{44.1\ \cancel{g\ C_3H_8}}\right) = 10.3\ mol\ C_3H_8$$

Step 3. Relate the moles of propane available to moles of O_2 required.

*See Example 3.1.

$$10.3\ \cancel{\text{mol } C_3H_8}\left(\frac{5 \text{ mol } O_2 \text{ required}}{1\ \cancel{\text{mol } C_3H_8} \text{ available}}\right) = 51.5 \text{ mol } O_2 \text{ required}$$

Step 4. Convert moles of O_2 required to grams (or any other desired unit).

$$51.5\ \cancel{\text{mol } O_2} \text{ required}\left(\frac{32.0 \text{ g } O_2}{1\ \cancel{\text{mol } O_2}}\right) = 1.65 \times 10^3 \text{ g } O_2 \text{ required}$$

To find out how much CO_2 can be produced, for example, you simply repeat steps 3 and 4 as above.

Step 3′. Relate the moles of C_3H_8 available to moles CO_2 produced.

$$10.3\ \cancel{\text{mol } C_3H_8} \text{ available}\left(\frac{3 \text{ mol } CO_2 \text{ produced}}{1\ \cancel{\text{mol } C_3H_8} \text{ available}}\right)$$
$$= 30.9 \text{ mol } CO_2 \text{ produced}$$

Step 4′. Convert moles CO_2 produced to grams.

$$30.9\ \cancel{\text{mol } CO_2}\left(\frac{44.0 \text{ g } CO_2}{1\ \cancel{\text{mol } CO_2}}\right) = 1.36 \times 10^3 \text{ g } CO_2$$

How can you now find out how much H_2O was formed? Go through steps 3 and 4 again? Of course! But it would be easier to recognize that the total mass of reactants (454 g C_3H_8 + 1650 g O_2 = 2.10×10^3 g) must be the same as the total mass of products (1.36×10^3 g CO_2 + mass of water). Therefore, the mass of water must be 2.10×10^3 g − 1.36×10^3 g = 740. g.

All stoichiometry problems fit the mold of the questions outlined above. There will be some differences from one situation to another (for example, you may not be given the mass of available reagent to be converted to moles, step 2, but will have to obtain it from its density and volume), but don't let that obscure the following simple procedure (Figure 4.1):

1. Balance the equation.
2. Convert the amount of available reagent to moles.
3. Convert moles of available reagent to moles of other reagent.
4. Convert moles of other reagent to grams.

EXERCISE 4.2 WEIGHT RELATIONS IN CHEMICAL REACTIONS

Iron reacts with oxygen to give iron(III) oxide, Fe_2O_3. If an ordinary iron nail (assumed to be pure iron) has a mass of 1.05 g, how many grams of Fe_2O_3 would it produce if the nail turned completely to this anhydrous oxide in the process of rusting? How many grams of oxygen, O_2, are required for the reaction?

4.2 REACTIONS WHERE ONE REAGENT IS PRESENT IN LIMITED SUPPLY

The objective of the examples in the preceding section was to find the exact amount of material required to consume completely a given quantity of some other reagent. However, in carrying out a reaction, a chemist, or nature for that matter, rarely supplies exact, stoichiometric amounts of reactants. As an example, consider the following reaction, the main product of which is a member of a family of compounds called "freons."

Freon is the name given to a family of compounds containing C and F, and frequently another element such as Cl, Br, or H. They are used typically as the coolant in refrigerators and air conditioners.

$$3\ CCl_4(\ell) + 2\ SbF_3(s) \longrightarrow \underset{\text{freon-12}}{3\ CCl_2F_2(\ell)} + 2\ SbCl_3(s)$$

When this reaction (and countless others employed in the chemical industry) is actually carried out, the cheaper and more abundant reagent is usually used in a greater amount than is stoichiometrically necessary according to the balanced equation. This is done to ensure that all of the more expensive reagent, SbF_3 in this case, is consumed completely.

From the discussion above, you see that it is reasonable to react the expensive chemical SbF_3 with a much greater amount of less expensive CCl_4 than is called for by the balanced equation. Thus, on completion of the freon synthesis, all the SbF_3 in principle will have been converted to product, and some CCl_4, which was originally present in stoichiometric excess, will remain. How much freon is formed? It depends only on the amount of SbF_3 you had at the start, not on the amount of CCl_4, since the latter was present in a stoichiometric excess. Thus, we call SbF_3 the **limiting reagent** since the amount present limits the amount of product formed.

The **limiting reagent** limits the amount of product formed in a reaction. All stoichiometry calculations are based on the total consumption of limiting reagent.

As an analogy to the chemical "limiting reagent" situation, consider what happens if you try to make some salami sandwiches. Assume you have 15 loaves of bread and 30 slices of salami. If each sandwich requires 3 slices of salami, you can make only 10 sandwiches. Since this will use only 20 slices of bread, you will still have a lot of bread left over! The bread is the "excess reagent" and the salami is the "limiting reagent," since the amount of salami available limits the number of sandwiches that can be made. Furthermore, although salami is bought by the pound and bread by the loaf, the combining units are *slices*, and your calculation had to be done in those units, just as stoichiometry problems are done in units of moles. The "mole" conversion factor here is (3 slices salami/2 slices bread).

EXAMPLE 4.1

A REACTION WITH ONE REAGENT IN LIMITED SUPPLY

Assume that you combine 150. g of CCl_4 with 100. g of SbF_3 to give freon-12. (a) On a mole basis, which reagent is in excess and which is the limiting reagent? (b) How many grams of the freon can be formed? (c) How many grams of which reagent are left after all the limiting reagent has been consumed and the maximum amount of freon has been formed?

Solution The first question you should ask when confronted by a problem of this type is if the reactants are present in the correct stoichiometric ratio or if any of the reactants is in limited supply (Figure 4.2). Since the stoichiometric ratio is a *mole ratio*, you must first find the moles available of each reagent. This will then allow you to say which of the reactants is present in short supply.

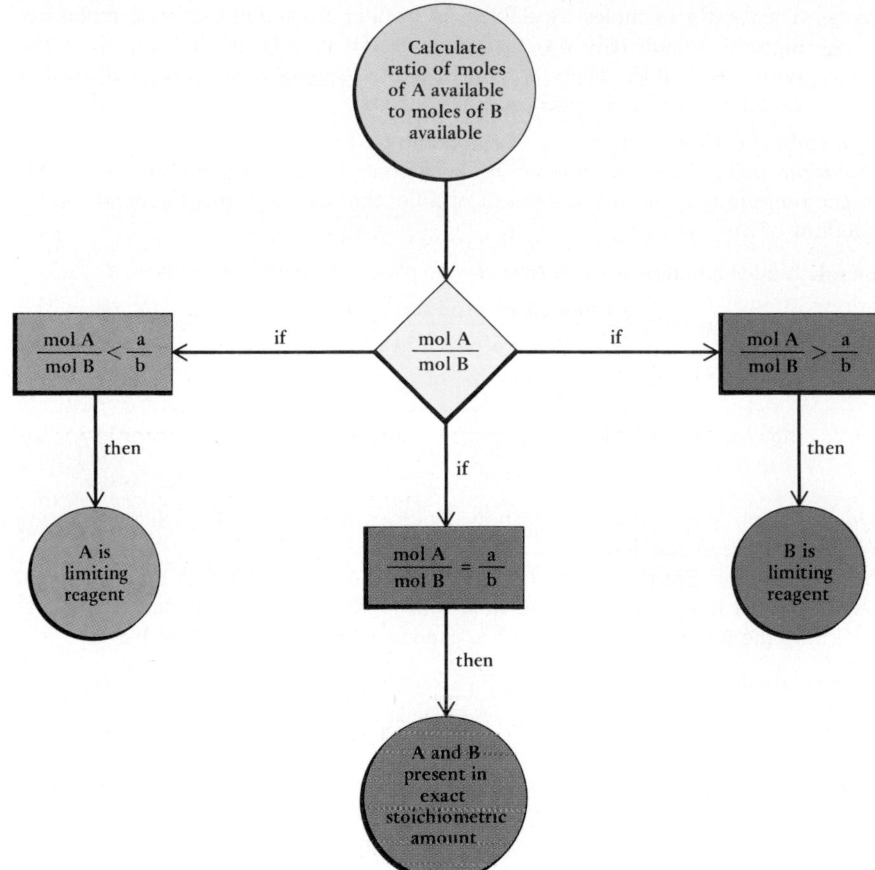

FIGURE 4.2

A scheme for solving limiting reagent problems involving a reaction between X and Y of the type $aX + bY \longrightarrow$ Products, where a and b are the stoichiometric coefficients of X and Y, respectively, in the balanced equation. Moles of reactants actually available are symbolized by A (= moles X) and B (= moles Y).

Set-up Steps:

Step 1. Write the balanced equation for the reaction (see text).

$$3\ CCl_4(\ell) + 2\ SbF_3(\ell) \longrightarrow 3\ CCl_2F_2(\ell) + 2\ SbCl_3(s)$$

Step 2. Calculate the number of moles available of each reactant.

$$150.\ \text{g}\ \cancel{CCl_4}\left(\frac{1\ \text{mol}\ CCl_4}{154\ \text{g}\ \cancel{CCl_4}}\right) = 0.974\ \text{mol}\ CCl_4$$

$$100.\ \text{g}\ \cancel{SbF_3}\left(\frac{1\ \text{mol}\ SbF_3}{179\ \text{g}\ \cancel{SbF_3}}\right) = 0.559\ \text{mol of}\ SbF_3$$

Question (a): On a mole basis, which of the reagents is limiting?
Solution to (a):

Step 3. From the balanced equation (step 1), calculate the required mole ratio of reactants. Then calculate the ratio of moles of reagent actually available (from step 2).

$$\text{Ratio of moles required} = \frac{2\ \text{moles}\ SbF_3}{3\ \text{moles}\ CCl_4} = 0.67$$

$$\text{Ratio of moles available} = \frac{0.559\ \text{mole}\ SbF_3}{0.974\ \text{mole}\ CCl_4} = 0.574$$

The "ratio of moles available" is smaller than the "ratio of moles required," which tells you that there is not enough of the reagent in the numerator, SbF_3. Therefore, SbF_3 is the *limiting reagent*, and all further calculations will be based on this reagent.

Question (b): How many grams of freon can be formed?
Solution to (b): This question can be answered, since you know now that SbF_3 is the *limiting reagent*; the amount of product that can be formed depends on the amount of SbF_3 present.

Step 4. Find the moles of freon that can be produced from 0.559 mole of SbF_3.

$$0.559 \cancel{\text{mol } SbF_3} \left(\frac{3 \text{ mol freon produced}}{2 \cancel{\text{mol } SbF_3} \text{ required}}\right) = 0.839 \text{ mole freon}$$

Step 5. Convert moles of freon to grams of freon.

$$0.839 \cancel{\text{mol } CCl_2F_2} \left(\frac{121 \text{ g } CCl_2F_2}{1 \cancel{\text{mol } CCl_2F_2}}\right) = 101 \text{ g } CCl_2F_2$$

amount of freon formed if all SbF_3 is used

Question (c): Which reagent and how much of it remains after all the limiting reagent has been consumed?
Solution to (c): From question (a) above you know that CCl_4 is the reagent in excess. As a first step in finding how many grams of CCl_4 remain, you must calculate the number of moles of CCl_4 used in consuming all the SbF_3.

Step 6. Calculate moles of CCl_4 required to react with 0.559 mole of SbF_3.

$$0.559 \cancel{\text{mol } SbF_3} \left(\frac{3 \text{ mol } CCl_4 \text{ required}}{2 \cancel{\text{mol } SbF_3} \text{ available}}\right) = 0.839 \text{ mol } CCl_4 \text{ required}$$

Step 7. Calculate moles of CCl_4 remaining after reaction. You know from step 2 above that 0.974 mole of CCl_4 was available. Therefore, 0.135 mole of CCl_4 will remain on completion of the reaction.

0.974 mol of CCl_4 available at the beginning
− 0.839 mol of CCl_4 required to consume 0.559 mole of SbF_3

0.135 mol of CCl_4 in excess over that needed to consume the SbF_3

Step 8. Convert moles of excess CCl_4 to the mass in grams.

$$0.135 \cancel{\text{mol } CCl_4} \left(\frac{154 \text{ g } CCl_4}{1 \cancel{\text{mol } CCl_4}}\right) = 20.8 \text{ g of excess } CCl_4$$

All three questions posed at the beginning have now been answered.

EXERCISE 4.3 LIMITING REAGENTS

Aluminum chloride, $AlCl_3$, is an inexpensive reagent used in many industrial processes. It is made by treating scrap aluminum with chlorine according to the following balanced equation.

$$2 \text{ Al(s)} + 3 \text{ Cl}_2\text{(g)} \longrightarrow 2 \text{ AlCl}_3\text{(s)}$$

If you start with 2.70 g of Al and 4.05 g of Cl_2, which reagent is limiting? How many grams of $AlCl_3$ can be produced? How many grams of which reactant, Al or Cl_2, will remain when the reaction is completed?

Reactions where one reagent is in excess and the other limits the amount of product formed represent the way chemical reactions are usually carried out. It is generally obvious to the person doing the reaction which reagent is in excess, since the experiment is designed that way in the beginning.

Another useful point to consider before going on is the following. The freon problem above was constructed to combat the common misconception that chemicals combine gram for gram. The mass ratio of CCl_4 to SbF_3 is 3 to 2, exactly the same as the stoichiometric mole ratio from the balanced equation. After working through the problem, however, you see that there are in fact too many *grams* of CCl_4. Chemicals react in the ratio of small whole numbers of *moles* or molecules, *not* weight units such as grams, pounds, or tons.

4.3 PERCENT YIELD

The quantity of product you should isolate from a chemical reaction is the **theoretical yield**. However, if you have worked in the chemistry laboratory, you must know how difficult it is to account for every drop of liquid and every crumb of solid. Thus, the **actual yield** of the compound may be less than the theoretical yield. To judge the efficiency of a chemical reaction and the techniques used to obtain the desired compound in pure form, chemists often compare the actual and theoretical yields by calculating their ratio and call the result the **percent yield** (Figure 4.3).

$$\text{Percent yield} = \frac{\text{actual yield}}{\text{theoretical yield}} \times 100$$

EXAMPLE 4.2

PERCENT YIELD

Consider the reaction of salicylic acid and acetic anhydride to produce aspirin.

$$\underset{\text{salicylic acid}}{2\ C_7H_6O_3(s)} + \underset{\text{acetic anhydride}}{C_4H_6O_3(\ell)} \longrightarrow \underset{\text{aspirin}}{2\ C_9H_8O_4(s)} + H_2O(\ell)$$

Assume you have 14.43 g of the acid and that it is the limiting reagent. Enough acetic anhydride is present to consume the acid completely. If you obtain 6.26 g of pure aspirin from the reaction, what is the percent yield of aspirin?

Solution This problem differs from the stoichiometry problems done thus far in that there is one additional step. You should first follow the usual steps to convert grams of salicylic acid to grams of aspirin expected, that is, the theoretical yield. The percent yield is then the ratio of the actual yield, 6.26 g, to the theoretical yield.

Step 1. The balanced chemical equation has been written above.

Step 2. You know that salicylic acid is the limiting reagent here. Calculate the moles of this acid available.

$$14.43\ \cancel{\text{g salicylic acid}} \left(\frac{1\ \text{mol}\ C_7H_6O_3}{138\ \cancel{\text{g}\ C_7H_6O_3}}\right) = 0.105\ \text{mol salicylic acid}$$

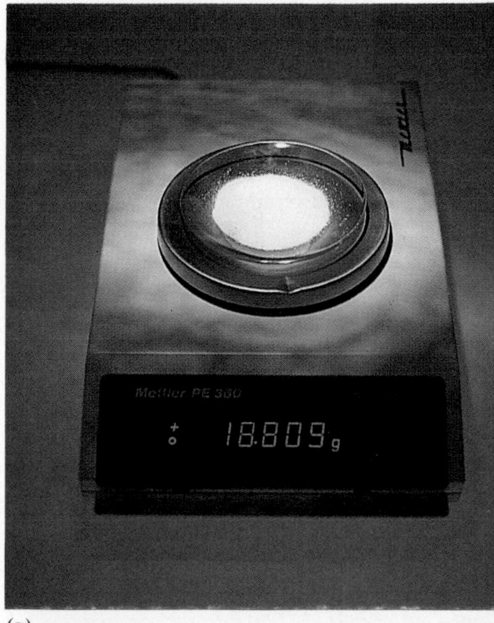

(a)

(b)

FIGURE 4.3
The reaction described in Example 4.2 should give 18.81 g of aspirin, a white solid (the theoretical yield) (a). Instead, only 6.26 g of the product is actually obtained (b). The percent yield is (6.26 g/18.81 g)100 = 33.3%. (Charles D. Winters)

Step 3. Using the stoichiometric factor from the balanced equation, relate moles of salicylic acid available to moles of aspirin that could be produced if everything worked perfectly.

$$0.105 \cancel{\text{mol acid}} \left(\frac{2 \text{ mol aspirin produced}}{2 \cancel{\text{mol acid}} \text{ available}}\right) = 0.105 \text{ mol aspirin}$$

Step 4. Calculate the grams of aspirin that could be obtained if everything worked perfectly. This is the *theoretical yield*.

$$0.105 \cancel{\text{mol aspirin}} \left(\frac{180 \text{ g } C_9H_8O_4}{1 \cancel{\text{mol } C_9H_8O_4}}\right) = 18.8 \text{ aspirin}$$

Step 5. Calculate the *percent yield* if, as stated in the problem, the *actual yield* was 6.26 g.

$$\text{Percent yield} = \frac{6.26 \text{ g aspirin actually obtained}}{18.8 \text{ g aspirin theoretically obtainable}} \times 100 = 33.3\%$$

EXERCISE 4.4 PERCENT YIELD

Professor H.C. Brown of Purdue University received the Nobel Prize in Chemistry in 1979 for his work on the chemistry of diborane, B_2H_6. This gas can be prepared by the following reaction (which is carried out in a nonaqueous solvent):

$$3 \text{ NaBH}_4 + 4 \text{ BF}_3 \longrightarrow 3 \text{ NaBF}_4 + 2 \text{ B}_2\text{H}_6\text{(g)}$$

If you begin with 18.9 g of $NaBH_4$ (and excess BF_3), and you isolate 7.50 g of B_2H_6 gas, what is the percent yield of B_2H_6?

4.4 REACTIONS IN SOLUTION

Many chemical reactions are carried out with pure liquids, solids, and gases. The synthesis of ammonia, for example, is done by reacting N_2 and H_2 directly in the gas phase, and sulfuric acid is made by oxidizing solid sulfur with O_2, the resulting SO_3 gas being dissolved in water only in the final step. Perhaps the majority of reactions, however, are done with the chemicals dissolved in some medium such as water. Sodium hydroxide and chlorine, for example, are made by passing electricity through a concentrated aqueous solution of sodium chloride. Other examples are the reactions occurring at this moment within your body where all the reactants are bathed in water. To work quantitatively with reactions in solution, we shall continue to use units of moles, but the amounts of reactants and products are given in "volume of solution" units rather than mass. The concept of *solution concentration* is required to be able to convert "volume of solution" units to mole units.

SOLUTION CONCENTRATION: MOLARITY

You are already familiar with the concept of concentration. For example, we know that there are about 3,000,000 people in Kansas and that the state has a land area of 82,276 square miles. Therefore, the average concentration of people is about 36 people per square mile. In just the same way we can say, for instance, that there is a certain amount of chemical, called the **solute**, dissolved in a given amount of **solvent** or in a given quantity of solution. One efficient way to define solution concentrations is in moles of solute per liter of solution, that is, in **molarity** units.

The solvent in a solution is the substance whose phase does not change on adding the solute, the substance dissolved. Water is liquid before and after adding NaCl, but solid NaCl is dispersed in the solution as ions.

$$\text{Molarity } (M) = \frac{\text{moles of compound}}{\text{liters of solution}}$$

For example, if you dissolve 342 g, or 1.00 mole, of sugar ($C_{12}H_{22}O_{11}$) in enough water to give a solution whose total volume is 1.00 liter, the concentration will be 1.00 mole per liter or 1.00 molar. We often abbreviate this as 1.00 M, where the capital M stands for "moles per liter." Another notation you will often see is

$$[C_{12}H_{22}O_{11}] = 1.00\ M$$

Placing the formula of the compound in square brackets implies that the concentration of the solute in *moles of compound per liter of solution* is being specified.

The terms "moles per liter" and "molar" are used interchangeably.

It is important to notice in expressing solution concentration in molarity that reference is always to *liters of solution* and not to liters of solvent. If you add one liter of water to one mole of a solid compound to make a one molar solution, the final solution volume may not be one liter (Figure 4.4). Therefore, when making solutions, *add the solvent to the already-dissolved solute until the desired solution volume is reached.*

EXAMPLE 4.3

SOLUTION MOLARITY

Potassium permanganate, $KMnO_4$, which was used at one time as a germicide in the treatment of burns, is a common laboratory chemical. It is a shiny, purple-black solid that dissolves in water to give a beautiful purple solution. Assume you dissolve 0.395 g of $KMnO_4$ in enough water to give 250. mL of solution (Figure 4.5). What is the molar concentration of $KMnO_4$?

Solution As is almost always the case, the first step in a problem is to convert material masses into moles.

$$0.395 \cancel{\text{ g KMnO}_4} \left(\frac{1 \text{ mol KMnO}_4}{158 \cancel{\text{ g KMnO}_4}}\right) = 0.00250 \text{ mol KMnO}_4$$

Now that the moles of dissolved material and the volume of solution are known, the solution concentration can be determined by dividing the moles of $KMnO_4$ by the volume of solution.

$$\text{Molarity of KMnO}_4 = \frac{0.00250 \text{ mol KMnO}_4}{0.250 \text{ L}} = 0.0100\ M$$

EXERCISE 4.5 SOLUTION MOLARITY

Sodium bicarbonate, $NaHCO_3$, is used in baking powder formulations, in fire extinguishers, in the manufacture of plastics and ceramics, and so on. If you have 26.3 g of the compound and dissolve it in enough water to make 200. mL of solution, what is its molar concentration?

FIGURE 4.4

To make a 0.100 M solution of $CuSO_4$, 24.9 g or 0.100 mol of $CuSO_4 \cdot 5\ H_2O$ (the blue crystalline solid) was placed in a volumetric flask. In this photo, exactly 1.00 L of water was measured out and slowly added to the 1.00 L volumetric flask containing the solid copper compound. When enough water had been added that the *solution volume* was exactly 1.00 L, approximately 8 mL of water (the quantity in the small graduated cylinder) remained. This emphasizes that molar concentrations are defined as moles per liter of *solution* and not per liter of water (or other solvent). (Charles D. Winters)

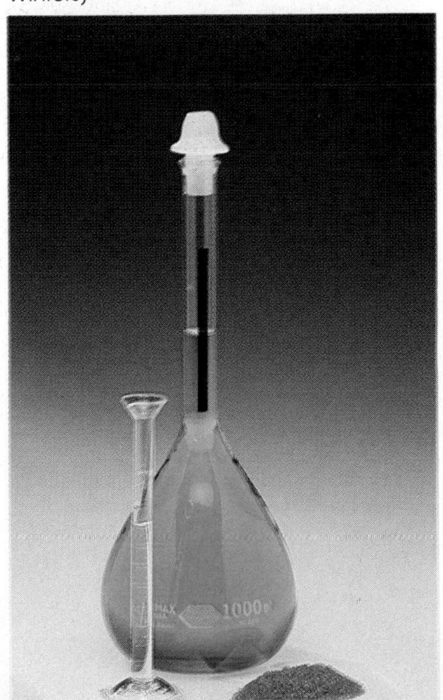

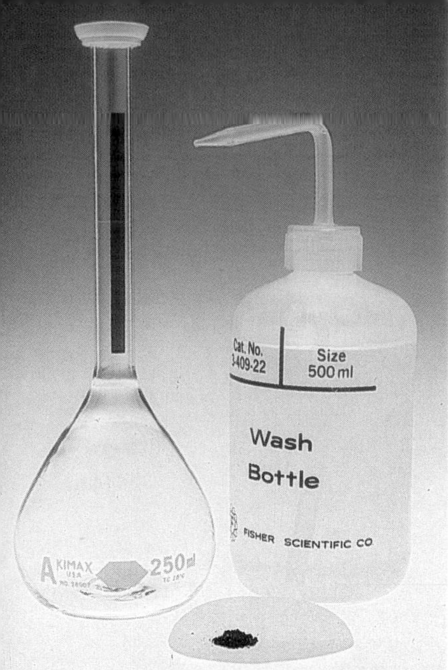

(a)

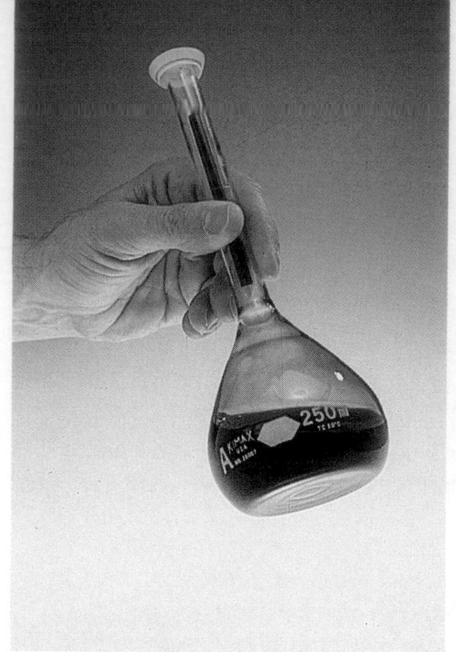

(b)

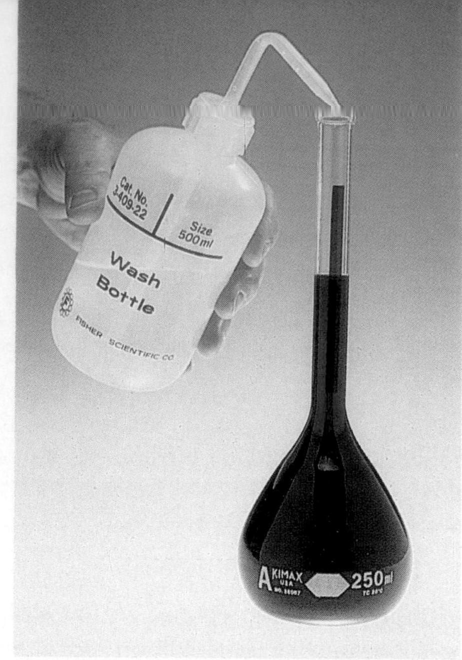

(c)

FIGURE 4.5

A 0.0100 *M* solution of $KMnO_4$ is made by adding enough water to 0.395 g of $KMnO_4$ to make 0.250 L of solution (a). To ensure the correct solution volume, the $KMnO_4$ is placed in a volumetric flask and dissolved in a small amount of water (b). After dissolving is complete, sufficient water is added to fill the flask to the mark; the flask contains 0.250 L of solution (c). (Charles D. Winters)

The calculation of $KMnO_4$ concentration in Example 4.3 showed that $[KMnO_4] = 0.0100\ M$. It is important to recognize that $KMnO_4$ is a strong electrolyte in water; that is, it dissociates completely into its ions, K^+ and MnO_4^- (Figure 4.6a).

$$KMnO_4 \text{ in water} \longrightarrow K^+(aq) + MnO_4^-(aq)$$

Sometimes we are interested only in the concentration of a particular ion in solution. This can be determined from the concentration of the compound and its ionization reaction. For $KMnO_4$, the coefficients in the

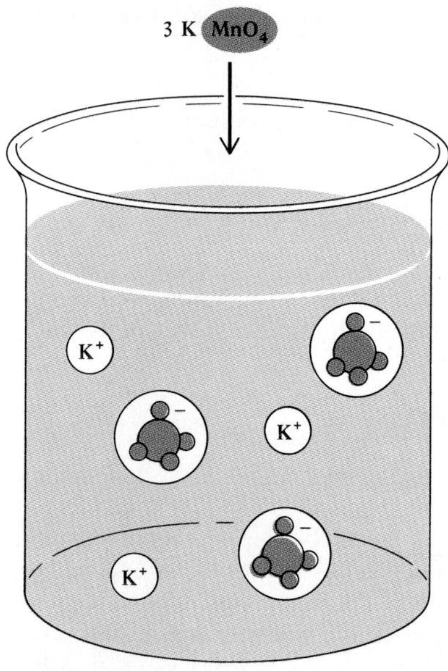

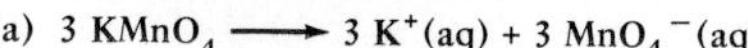

(a) $3\ KMnO_4 \longrightarrow 3\ K^+(aq) + 3\ MnO_4^-(aq)$

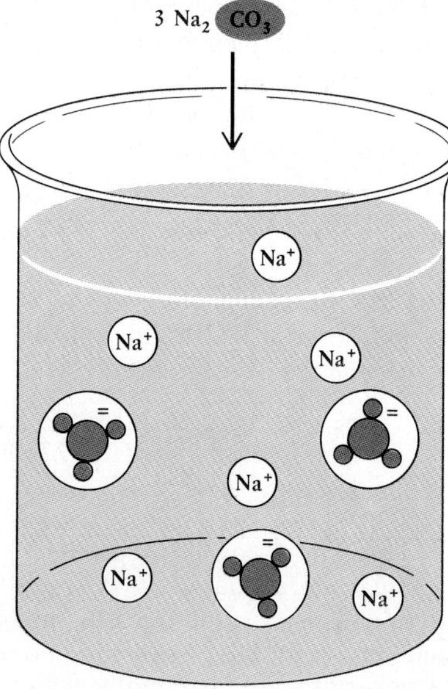

(b) $3\ Na_2CO_3 \longrightarrow 6\ Na^+(aq) + 3\ CO_3^{2-}(aq)$

FIGURE 4.6

(a) When $KMnO_4$ is dissolved in water, one mole of K^+ ions and one mole of MnO_4^- ions form for every mole of $KMnO_4$ dissolved. (b) Dissolving one mole of Na_2CO_3, however, produces two moles of Na^+ and one mole of CO_3^{2-}.

ionization reaction tell you that 1 mole of $KMnO_4$ dissociates into 1 mole of K^+ and 1 mole of MnO_4^-. Accordingly, 0.0100 *M* $KMnO_4$ gives a concentration of K^+ in the solution of 0.0100 *M*, the same as the concentration of MnO_4^-. Thus, the total concentration of ions is 0.0200 *M*, while the original concentration of the $KMnO_4$ solute was 0.0100 *M*.

This same idea holds for any other substance that produces ions when dissolved in a solvent such as water (Figure 4.6b). For example,

$$Na_2CO_3 \text{ in water} = 2\ Na^+(aq) + CO_3^{2-}(aq)$$

Thus, if one mole of Na_2CO_3 were dissolved in enough water to make 1 liter of solution, the concentration of the sodium ion would be $[Na^+] = 2\ M$ and that of the carbonate ion would be $[CO_3^{2-}] = 1\ M$; total ion concentration is 3 *M*. These results come from the conversion of $[Na_2CO_3] = 1.0\ M$ to $[Na^+]$ and $[CO_3^{2-}]$ by the stoichiometric factors (2 mol Na^+/1 mol Na_2CO_3) and (1 mol CO_3^{2-}/1 mol Na_2CO_3), respectively.*

EXERCISE 4.6 ION CONCENTRATIONS IN SOLUTION

Both of the following are strong electrolytes, dissociating completely into constituent ions when dissolved in water. For each case, write a balanced equation for the ionization of the compound and then state the molar concentration of each ion in solution and the *total* concentration of all ions together.

(a) 1 *M* HCl (b) 0.5 *M* $(NH_4)_2SO_4$

A situation a chemist often faces is the opposite of the one above. That is, a given volume of solution of known concentration must be prepared. The problem is to find out what mass of solute to use.

Suppose, for example, you wish to prepare 2.0 L of a 1.5 *M* solution of Na_2CO_3. You are given a bottle of solid Na_2CO_3, some distilled water, and a 2.0 L volumetric flask.* To make the solution, weigh out the necessary quantity of Na_2CO_3 as accurately as possible, place the solid in the volumetric flask, and then add some water to dissolve the solid; after the solid is dissolved completely, more water is added to bring the solution volume to 2.0 liters. You will have a solution of the desired concentration and in the volume specified.

EXAMPLE 4.4

MAKING SOLUTIONS OF KNOWN CONCENTRATION

How many grams of Na_2CO_3 are required to make 2.0 L of 1.5 *M* Na_2CO_3?

Solution As usual, you must first calculate the moles of substance required.

$$2.0\ \cancel{\text{L of solution}} \left(\frac{1.5 \text{ mol } Na_2CO_3}{1.0\ \cancel{\text{L solution}}}\right) = 3.0 \text{ mol } Na_2CO_3 \text{ required}$$

---

*Chemists often use the expression "compound XY is ___ *M* in solution." However, if XY is an ionic compound, it will form X and Y ions on dissolving; XY does not exist as such in solution. Thus, the expression "compound XY is ___ *M*" is a way of saying that ___ moles of *XY* units are dissolved in enough water to make one liter of solution.

†A volumetric flask is a special flask with a line marked on its neck (Figure 4.5); if it is filled with solution to this line, you have exactly the volume of solution specified.

(a)

(b)

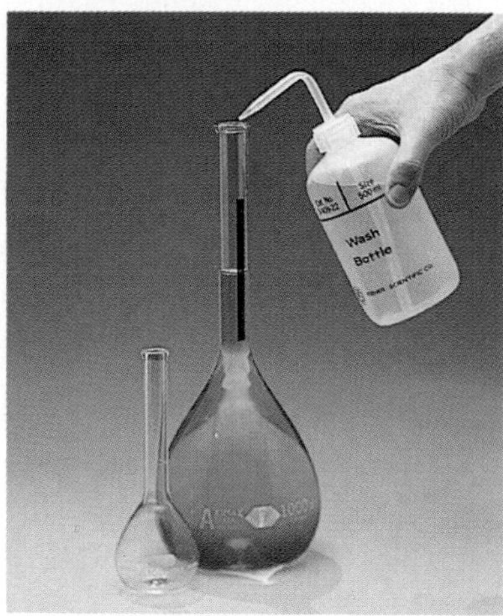

(c)

FIGURE 4.7

Making a solution by dilution. (a) A 100. mL volumetric flask is filled to the mark with 0.100 *M* $K_2Cr_2O_7$. (b) This is transferred to a 1.00 L volumetric flask. (c) The 1.00 L volumetric flask is filled to the mark with distilled water. The concentration of the now diluted $K_2Cr_2O_7$ is 0.0100 *M*. (Charles D. Winters)

Now that you know moles of Na_2CO_3 required, this is converted to grams.

$$3.0 \cancel{\text{mol Na}_2\text{CO}_3} \left(\frac{106 \text{ g Na}_2\text{CO}_3}{1 \cancel{\text{mol Na}_2\text{CO}_3}}\right) = 3.2 \times 10^2 \text{ g Na}_2\text{CO}_3$$

EXERCISE 4.7 MAKING SOLUTIONS OF KNOWN CONCENTRATION

An experiment in your laboratory requires 500. mL of a 0.0200 *M* solution of $KMnO_4$. You are given a bottle of solid $KMnO_4$, some distilled water, and a 500. mL volumetric flask. Describe how you would go about making up the required solution.

Before going on, you should be aware that there are other ways of defining the concentration of chemicals in solution. For the moment, molarity is the most useful for our purposes, but other units will be introduced later in this text as needed (Chapter 12).

PREPARING A SOLUTION BY DILUTION OF A MORE CONCENTRATED SOLUTION

In the first part of this section the preparation of solutions was described. The solid reagent was weighed and water or other solvent was added to bring the total solution volume to the desired amount. Another common method for preparing solutions of specific concentration is to *begin with a more concentrated one and add water to make it more dilute*. Many of the solutions prepared for your use in your laboratory course are probably made this way. It is often more efficient to store a few liters of a concentrated solution and then add water to make it into many more liters of a dilute solution. The following example illustrates the technique of preparing a dilute solution from a more concentrated one.

EXAMPLE 4.5

PREPARING SOLUTIONS BY DILUTION

You need 1.00 L of a 0.0100 *M* $K_2Cr_2O_7$ (potassium dichromate) solution. You have available some 0.100 *M* $K_2Cr_2O_7$, some volumetric flasks, and distilled water (Figure 4.7). How much of the more concentrated solution do you need and how much water must be added to give finally 1.00 L of 0.0100 *M* $K_2Cr_2O_7$?

Solution As in stoichometry problems, there is no *direct* way from the beginning (volume units) to the end (volume units) of a "dilution problem." The way, as always, lies through moles. The final solution must contain a certain number of moles of $K_2Cr_2O_7$ per liter. Therefore, you must transfer into the flask that number of moles of $K_2Cr_2O_7$ from the more concentrated solution. That is, *the number of moles of solute in the final solution must equal the number of moles taken from the more concentrated solution*. Dilution with water will then give you the correct number of moles in the desired volume of the final solution.

Step 1. Calculate the number of moles of $K_2Cr_2O_7$ that the final solution must contain.

$$1.00 \cancel{\text{L}} \text{ of solution} \left(\frac{0.0100 \text{ mol K}_2\text{Cr}_2\text{O}_7}{\cancel{\text{L}}}\right) = 0.0100 \text{ mol K}_2\text{Cr}_2\text{O}_7$$

must be contained in the final, dilute solution

Since the number of moles of $K_2Cr_2O_7$ in the final solution must be the same as the number taken from the more concentrated solution, you will need to transfer 0.0100 mole of $K_2Cr_2O_7$ into the volumetric flask from the more concentrated solution.

Step 2. Calculate the volume of 0.100 M $K_2Cr_2O_7$ solution that will give the required 0.0100 mole of $K_2Cr_2O_7$.

$$0.0100\ \cancel{\text{mol } K_2Cr_2O_7}\text{ required}\left(\frac{1.00\text{ L of more concentrated solution}}{0.100\ \cancel{\text{mol } K_2Cr_2O_7}}\right)$$
$$= 0.100\text{ L of } 0.100\ M\ K_2Cr_2O_7\text{ required}$$

Step 3. Make up the final, more dilute solution. 0.100 liters or 100. mL of 0.100 M $K_2Cr_2O_7$ is measured out with a 100. mL volumetric flask and transferred completely to the 1.00 L volumetric flask. Adding approximately 900 mL of distilled water, enough to bring the total volume to the mark on the 1.00 L flask, will give 1.00 liter of 0.0100 M $K_2Cr_2O_7$.

Let us take another look at the example above. The central point of the calculation was that the number of moles of $K_2Cr_2O_7$ in the final, dilute solution had to be equal to the number of moles of $K_2Cr_2O_7$ taken from the more concentrated solution. This number of moles could be calculated in either of the following ways:

(a) moles of $K_2Cr_2O_7$ in the final, dilute solution = 0.0100 mole = M_dV_d
(b) moles of $K_2Cr_2O_7$ taken from the concentrated solution = 0.0100 mole = M_cV_c

where M and V are the molarity and volume, respectively, of the dilute (d) and concentrated (c) solutions. Since both MV products are equal to the same number of moles, we can say that

moles of reagent taken from concentrated solution =
moles of the reagent in dilute solution

$$M_cV_c = M_dV_d$$

This handy expression holds for all cases where a more concentrated solution is used to make a more dilute one. You can use it to find, for example, the molarity of the dilute solution, M_d, from values of M_c, V_c, and V_d.

EXERCISE 4.8 PREPARING SOLUTIONS BY DILUTION
An experiment calls for you to use 300. mL of 1.00 M NaOH, but you are given only a large bottle of 3.00 M NaOH. Tell how you would make up the 1.00 M NaOH in the desired volume.

SOLUTION STOICHIOMETRY

It is important to be able to derive quantitative relationships for reactions in solution, since many reactions in fact occur that way. Having now defined the concept of "concentration" and a convenient unit of concentration, you are ready to study some examples where quantitative relationships are involved.

EXAMPLE 4.6

SOLUTION STOICHIOMETRY

Metallic magnesium reacts with aqueous HCl to give magnesium chloride and hydrogen (see Figure 3.4). This reaction, or similar ones using other metals such as zinc, may be used as a convenient preparation of H_2 in the laboratory. Suppose you wanted to prepare 0.0061 moles of H_2. How many milliliters of 3.0 *M* HCl would you need to use with a handful (that is, an excess) of magnesium?

Solution

Step 1. The first step in a stoichiometry problem is to write a balanced chemical equation to depict the reaction involved.

$$Mg(s) + 2\ HCl(aq) \longrightarrow MgCl_2(aq) + H_2(g)$$

This tells you that 1 mole of H_2 is produced for every 2 moles of HCl used. This is the stoichiometric mole ratio factor you need to find the number of moles of HCl required.

Step 2. Number of moles of HCl required.

$$0.0061\ \cancel{\text{mol } H_2}\ \text{desired}\left(\frac{2\ \text{mol HCl required}}{1\ \cancel{\text{mol } H_2}\ \text{desired}}\right) = 0.012\ \text{mol HCl required}$$

Step 3. With the number of moles of HCl required now known, this can be converted to liters of HCl required with another conversion factor, the solution concentration.

$$0.012\ \cancel{\text{mol HCl}}\left(\frac{1.0\ \text{L solution}}{3.0\ \cancel{\text{mol HCl}}}\right) = 0.0041\ \text{L solution}$$

Since an answer in units of "milliliters" is requested, you can then convert 0.0041 L to mL. That is, 0.0041 L = 4.1 mL.

EXAMPLE 4.7

SOLUTION STOICHIOMETRY

A common type of reaction is that between a metal carbonate and an aqueous acid to give a salt and gaseous CO_2 (Chapter 3).

$$\begin{array}{ccccccccc} \text{metal carbonate} & + & \text{acid} & \longrightarrow & \text{salt} & + & \text{carbon dioxide} & + & \text{water} \\ Na_2CO_3(aq) & + & 2\ HCl(aq) & \longrightarrow & 2\ NaCl(aq) & + & CO_2(g) & + & H_2O(\ell) \end{array}$$

Suppose you have 100. mL of 1.50 *M* Na_2CO_3 (Example 4.4). How many milliliters of 0.750 *M* HCl would you have to add to the sodium carbonate solution to consume the Na_2CO_3 completely?

Solution

Step 1. The balanced equation for the reaction is given above.

Step 2. Calculate the number of moles of Na_2CO_3 to be used.

$$0.100\ \cancel{\text{L } Na_2CO_3}\left(\frac{1.50\ \text{mol } Na_2CO_3}{1.00\ \cancel{\text{L } Na_2CO_3}}\right) = 0.150\ \text{mol } Na_2CO_3$$

Step 3. Relate the number of moles of Na_2CO_3 available to the number of moles of HCl required. The required stoichiometric mole ratio factor is obtained from the balanced equation.

$$0.150\ \cancel{\text{mol } Na_2CO_3}\left(\frac{2\ \text{mol HCl required}}{1\ \cancel{\text{mol } Na_2CO_3}\ \text{available}}\right) = 0.300\ \text{mol HCl}$$

Step 4. Calculate the number of liters of HCl required using moles of HCl required and the HCl concentration.

$$0.300 \text{ mol HCl required} \left(\frac{1.00 \text{ L solution}}{0.750 \text{ mol HCl}}\right) = 0.400 \text{ L solution}$$

The final step tells you that 0.400 L or 400. mL of solution is required.

You should be sure to notice that the steps in the examples above are similar to those used in the general stoichiometry scheme in Figure 4.1. The chief difference between them is that grams and moles are interconverted in Figure 4.1, whereas volume and moles are interconverted in Examples 4.6 and 4.7. That is, these examples follow part b of Figure 4.8 instead of part a.

After looking at the scheme in Figure 4.8, it should be clear that parts A and B can be combined diagonally across the diagram. For example, you could use the volume of one reactant to calculate the grams of other material needed or grams of product formed. [Here you would move in the following manner: volume of one reactant—(multiply by mol/L) → moles of that reactant—(multiply by stoichiometric factor) → moles of other reactant—(multiply by g/mol) → grams of other reactant.] The exercises that follow illustrate this point.

EXERCISE 4.9 SOLUTION STOICHIOMETRY

In Example 4.7 above, you found that 100. mL of 1.50 *M* Na_2CO_3 will react completely with 400. mL of 0.750 *M* HCl. How many grams of NaCl and CO_2 are produced in this reaction? What is the molar concentration of NaCl in the solution remaining at the end of the reaction? (Neglect the volume of water produced by the reaction.)

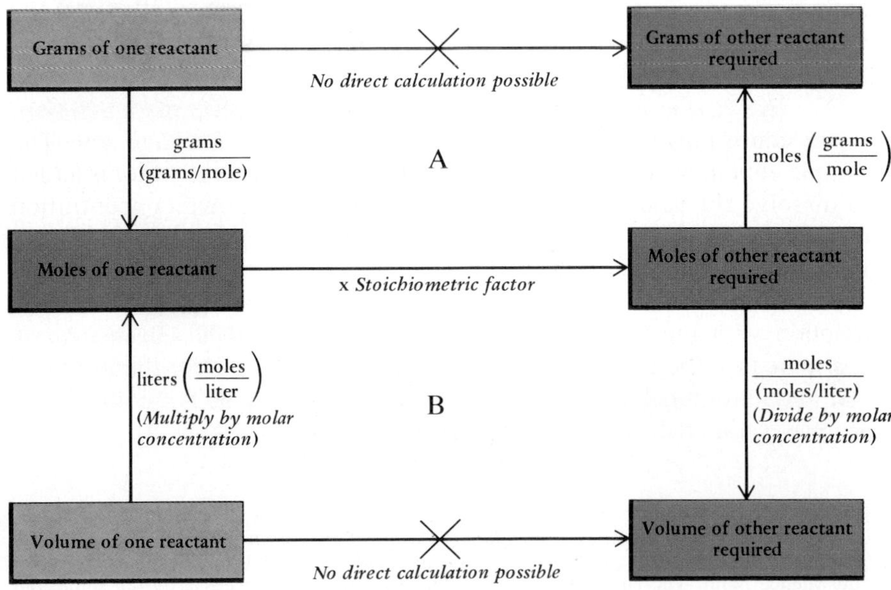

FIGURE 4.8
A scheme outlining stoichiometric relations for reactions in solution.

EXERCISE 4.10 SOLUTION STOICHIOMETRY
If you have 25.0 mL of 0.750 *M* HCl, how many grams of Na_2CO_3 are required to react completely with the acid? How many grams of NaCl will be produced?

TITRATIONS

Your study of stoichiometry so far should have convinced you that

(a) if you know the precise reaction occurring between the two reactants,
(b) if the reaction is complete and rapid,
(c) and if you know the exact quantity of one of the reactants,

then you can always obtain the exact amount of any of the other substances in the reaction. This is the essence of any technique of **quantitative chemical analysis**, the determination of the *amount* of a given constituent of a mixture.

Qualitative analysis is the determination only of the identity of the constituents of a mixture. **Quantitative analysis** is the determination of the quantity of a constituent.

Suppose, for example, you want to analyze a type of common clover for the quantity of oxalic acid, $H_2C_2O_4$, in the leaves. We know that this acid reacts with the base sodium hydroxide in aqueous solution according to the balanced equation

$$H_2C_2O_4(aq) + 2\ NaOH(aq) \longrightarrow Na_2C_2O_4(aq) + 2\ H_2O(\ell)$$

Assume you can carry out the reaction in such a way that you know when the sodium hydroxide being added is *exactly* the amount needed to react with *all* the oxalic acid present in solution. If you know the concentration of the sodium hydroxide and if you measure the exact volume of the base you have used to this point, you can then tell precisely how much oxalic acid was present in a given mass of clover leaves.

Having defined the analytical strategy, there are two points that must be explained. (a) How can you carry out the reaction so that you know *exactly* when the NaOH has reacted with all the oxalic acid and how much has been used? (b) How can you determine exactly how much oxalic acid is present in the sample?

The answer to the first of these questions is to carry out a **titration**, a procedure illustrated in the series of photographs in Figure 4.9. The sample containing oxalic acid is placed in a flask, and some water is added to dissolve the acid. Sodium hydroxide of exactly known concentration is placed in a buret, a measuring cylinder most commonly of 50.0 mL volume and calibrated in 0.1 mL divisions. As the sodium hydroxide is added slowly to the acid solution in the flask, the acid is consumed by reaction with the base. The point at which the acid has been *exactly* consumed by the base is called the **equivalence point**; here the moles of OH^- added is equal to the moles of H^+ originally present. The equivalence point is estimated by adding an **indicator** to the solution.* An indicator

The **equivalence point** in an acid–base titration is the point at which moles H^+ = moles OH^-.

*Actually the indicator detects the *endpoint* of the titration. An indicator is chosen that turns color at a hydrogen ion concentration as close as possible to that existing at the equivalence point. The endpoint and equivalence point are not necessarily the same, but their difference in practice is negligible. See Chapter 17.

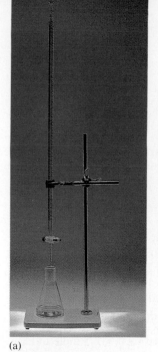

(a)

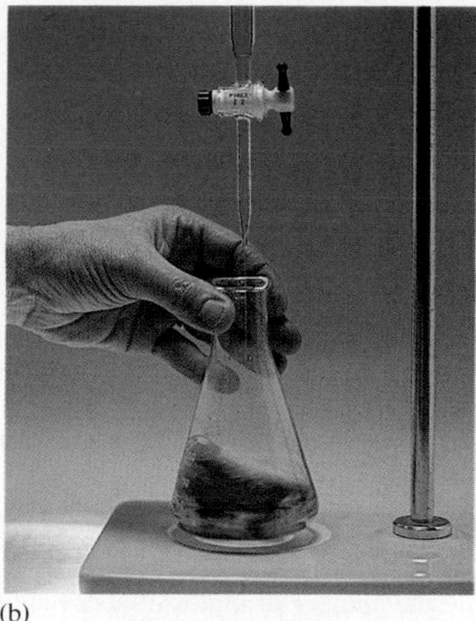

(b)

(c)

FIGURE 4.9

Titration of an acid in aqueous solution with a base. (a) The buret, a volumetric measuring device calibrated in divisions of 0.1 mL, is filled with base of known concentration. (b) Base is added slowly to the solution from the buret. (c) A change in color of an indicator dye signals the equivalence point, the point at which moles H^+ = moles OH^-. (Charles D. Winters)

is a dye that changes color as a solution changes from having an excess of acid to having an excess of base or vice versa. The juice of red cabbage leaves, for example, is an excellent indicator of acidity, turning red in the presence of acids and blue in bases (Figure 4.10).

When the equivalence point has been estimated in a titration, the volume of base used from the beginning of the titration can be determined by reading the calibrated buret (Figure 4.9). If you know the concentration of the base in units of moles/liter, you can then calculate the exact number of moles of base used from

$$\text{Moles of base added} = \text{molarity of base} \left(\frac{\text{mol}}{\text{L}}\right) \times \text{volume of base (L)}$$

Since the balanced equation for the acid–base reaction gives you the stoichiometric factor, you can use this mole-ratio conversion factor to convert moles of base added to the exact number of moles of acid that were present in the original sample.

FIGURE 4.10

The juice of red cabbage leaves is an indicator of the hydrogen ion concentration in a solution. The solution at the left is acidic (0.10 *M* HCl), the solution in the middle is neutral (equal concentrations of H^+ and OH^-), and the solution at the right is basic (0.10 *M* NaOH). (Charles D. Winters)

EXAMPLE 4.8

AN ACID–BASE TITRATION

Suppose you have 1.034 g of clover leaves and you extract the oxalic acid from them into a small amount of water. This solution of oxalic acid is found to require 34.47 mL of 0.100 *M* NaOH for titration to the equivalence point. What is the weight percent of oxalic acid in the leaves?

Solution

Step 1. Write the balanced equation.

$$H_2C_2O_4(aq) + 2\ NaOH(aq) \longrightarrow Na_2C_2O_4(aq) + 2\ H_2O$$

Step 2. Calculate the moles of base used. (Notice that milliliters of base have been changed first into liters of base.)

$$0.03447\ \cancel{\text{L NaOH}}\ \text{solution}\left(\frac{0.100\ \text{mol NaOH}}{\cancel{\text{L NaOH}}}\right) = 0.00345\ \text{mol NaOH}$$

Step 3. Calculate the number of moles of acid required by 0.00345 mole of base.

$$0.00345\ \cancel{\text{mol of NaOH}}\left(\frac{1\ \text{mol acid}}{\cancel{2\ \text{mol NaOH}}}\right) = 0.00172\ \text{mole of acid required}$$

Step 4. Calculate the number of grams of acid in solution.

$$0.00172\ \cancel{\text{mol}\ H_2C_2O_4}\left(\frac{90.0\ \text{g}\ H_2C_2O_4}{1.00\ \cancel{\text{mol}\ H_2C_2O_4}}\right) = 0.155\ \text{g}\ H_2C_2O_4$$

Step 5. Calculate the weight percentage of $H_2C_2O_4$ in the 1.034 g of leaves.

$$\text{Weight percentage} = \frac{0.155\ \text{g}\ H_2C_2O_4}{1.034\ \text{g of leaves}} \times 100 = 15.0\%$$

EXERCISE 4.11 AN ACID–BASE TITRATION

Vinegar contains acetic acid, $HC_2H_3O_2$. You can determine the mass of acetic acid in a vinegar sample by titrating with sodium hydroxide of known concentration. The reaction that occurs is

$$HC_2H_3O_2(aq) + NaOH(aq) \longrightarrow NaC_2H_3O_2(aq) + H_2O(\ell)$$

If you find that a 25.00 mL sample of vinegar requires 28.33 mL of a 0.953 *M* solution of NaOH for titration to the equivalence point, how many grams of acetic acid are there in the vinegar sample? What is the molar concentration of the acetic acid in the vinegar?

Standardization is the procedure by which the accurate concentration of an acid, base, or other reagent is determined.

In the example and exercise above, the exact concentration of the base, NaOH, was known. The procedure by which the exact concentration of an acid, a base, or other reagent is determined is called **standardization**, and there are two general approaches illustrated by the following example and exercise.*

*You might think that you could make up a solution of NaOH simply by weighing out some solid NaOH as accurately as possible and dissolving it in water in a volumetric flask. Practical considerations get in the way of this seemingly simple approach. Unfortunately, NaOH, which is sold as little pellets, rapidly picks up water and CO_2 from the air and so cannot be weighed accurately.

EXAMPLE 4.9

STANDARDIZATION OF A REAGENT

An acid such as HCl can be standardized by using it to titrate a base such as Na_2CO_3. Sodium carbonate is a solid that can be obtained in pure form and can be weighed accurately. If you find that 0.250 g of Na_2CO_3 requires 25.76 mL of HCl for titration to the equivalence point, what is the exact molar concentration of the HCl?

Solution

Step 1. Write a balanced equation for the reaction.

$$Na_2CO_3(aq) + 2\ HCl(aq) \longrightarrow 2\ NaCl(aq) + CO_2(g) + H_2O(\ell)$$

Step 2. Calculate the number of moles of Na_2CO_3 used.

$$0.250\ \cancel{g\ Na_2CO_3}\left(\frac{1.00\ mol\ Na_2CO_3}{106\ \cancel{g\ Na_2CO_3}}\right) = 2.36 \times 10^{-3}\ mol\ Na_2CO_3$$

Step 3. Convert the moles of Na_2CO_3 available to the moles of HCl required for complete reaction, that is, for titration to the equivalence point.

$$2.36 \times 10^{-3}\ \cancel{mol\ Na_2CO_3}\left(\frac{2\ mol\ HCl\ required}{1\ \cancel{mol\ Na_2CO_3}}\right) = 4.72 \times 10^{-3}\ mol\ HCl$$

Step 4. In the two steps above you found that 2.36×10^{-3} mole of Na_2CO_3 requires 4.72×10^{-3} mole of HCl for complete titration. Therefore, there must have been 4.72×10^{-3} mole of HCl in the 25.76 mL of HCl solution. This means the concentration of HCl is

$$[HCl] = \frac{4.72 \times 10^{-3}\ mol\ HCl}{0.02576\ L} = 0.183\ M$$

EXERCISE 4.12 STANDARDIZATION OF A BASE

Hydrochloric acid, HCl, can be purchased from chemical supply houses in solutions that are exactly 0.100 *M*, so these solutions can be used to standardize the solution of a base. If you titrate to the equivalence point 25.00 mL of a sodium hydroxide solution with 32.56 mL of 0.100 *M* HCl, what is the concentration of the base?

Acid–base titrations are extremely useful for quantitative chemical analysis, and many study questions are given at the end of this chapter to illustrate their scope. You should not get the impression, however, that a titration will involve only an acid reacting with a base. In Chapter 3 some common types of chemical reactions were described, among them acid–base reactions and electron transfer or oxidation–reduction reactions. The latter type also lends itself well to chemical analysis by titration, because many of these reactions are also 100% complete, and there are methods for estimating the equivalence point. Before describing the use of electron transfer reactions in analysis, however, there is one fact that should be clear. *Before using any reaction as an analytical method, you must know the balanced chemical equation.* For acid–base reactions, this is usually straightforward. One mole of acid reacts with one, two, or three moles of base, or vice versa. In oxidation–reduction titrations, however, the equations can be more complicated, and the special techniques described in Section 3.4 are required for balancing them.

An example of an oxidation–reduction titration used in chemical analysis is outlined below and an exercise follows. Before working these problems, you should be aware of the one major difference between acid–base and oxidation–reduction titrations. In the latter case, *the equivalence point of an oxidation–reduction titration is reached when moles of oxidizing agent and reducing agent have been mixed in the correct stoichiometric ratio.*

EXAMPLE 4.10

AN OXIDATION–REDUCTION TITRATION

Vitamin C is the simple compound $C_6H_8O_6$. Besides being an acid, it is also able to transfer electrons to a molecule that can accept electrons, and it is the electron transfer reaction that is used in the analysis of vitamin C. One method for determining the amount of vitamin C in a sample is to titrate it with a solution of bromine (Br_2), an element capable of accepting electrons (i.e., acting as an oxidizing agent).

$$C_6H_8O_6(aq) + Br_2(aq) \longrightarrow 2\ HBr(aq) + C_6H_6O_6(aq)$$

Suppose you find that a 1.00 g "chewable vitamin C tablet" requires 27.85 mL of 0.102 *M* Br_2 solution for titration to the equivalence point. How many grams of vitamin C are contained in the tablet?

Solution

Step 1. Write a balanced equation for the reaction involved to find the required stoichiometric mole ratio conversion factor. In this case, the equation has been written above.

Step 2. Calculate the moles of Br_2 used in the titration.

$$0.02785\ \cancel{\text{L Br}_2}\ \text{solution}\left(\frac{0.102\ \text{mol Br}_2}{\cancel{\text{L Br}_2}}\right) = 2.84 \times 10^{-3}\ \text{mol Br}_2$$

Step 3. Relate the number of moles of Br_2 to moles of $C_6H_8O_6$. Since the stoichiometric factor is 1 mole of Br_2 to 1 mole of vitamin C, the number of moles of vitamin C titrated must have been 2.84×10^{-3}.

Step 4. Calculate the number of grams of vitamin C.

$$2.84 \times 10^{-3}\ \cancel{\text{mol C}_6\text{H}_8\text{O}_6}\left(\frac{176\ \text{g C}_6\text{H}_8\text{O}_6}{1.00\ \cancel{\text{mol C}_6\text{H}_8\text{O}_6}}\right)$$

$$= 0.500\ \text{g or}\ 500\ \text{mg C}_6\text{H}_8\text{O}_6$$

EXERCISE 4.13 OXIDATION–REDUCTION IN ANALYSIS

It is possible to determine the iron content of a mineral or ore by converting the iron to the iron(II) ion, Fe^{2+} and then titrating this ion with potassium permanganate, $KMnO_4$. The balanced, net ionic equation for the reaction is

$$\underset{\text{(purple)}}{MnO_4^-(aq)} + 8\ H^+(aq) + \underset{\text{(colorless)}}{5\ Fe^{2+}(aq)} \longrightarrow \underset{\text{(colorless)}}{Mn^{2+}(aq)} + \underset{\text{(colorless)}}{5\ Fe^{3+}(aq)} + 4\ H_2O(\ell)$$

This is an excellent analytical reaction, because it is easy to detect when all the iron(II) has reacted. The MnO_4^- ion is deep purple. On reaction with Fe^{2+}, however, the color disappears, because the Mn^{2+} ion is colorless. Thus, as $KMnO_4$ is added from a buret, the purple color disappears when the solutions mix. When all the Fe^{2+} has been consumed, any additional $KMnO_4$ will give the solution a permanent purple color. Therefore, one carries out this titration until the initially colorless, Fe^{2+}-containing solution just turns a faint purple color (Figure 4.11).

Assume that you find 1.026 g of an iron-containing ore requires 24.34 mL of 0.0200 *M* $KMnO_4$. How many grams of iron does the ore sample contain? What is the weight percentage of iron in the ore?

The analysis of materials by the method of acid–base or oxidation–reduction titration is a powerful tool, and it is widely used. You will doubtless be exposed to this technique in the laboratory, and you may wish to learn a slightly different approach for treating the information. This approach uses **chemical equivalents** and defines solution concentration in terms of **normality**. For further information, see the Study Guide that accompanies this text.

SUMMARY

Stoichiometry is the study of mass relations in chemical reactions, and its guiding principle is the **conservation of matter**. To relate the mass of one reactant to the mass required of another reactant or of a product, four steps are always followed (Section 4.1): (1) Balance the equation for the reaction. (2) Calculate moles of the reactant. (3) Relate moles of product expected to moles of reactant used with a **stoichiometric factor**. (This factor comes from the balanced equation and is always moles of required reactant or product divided by moles of known reactant). (4) Calculate mass of product from moles of product.

Most reactions are done with one reactant present in a smaller amount than required by stoichiometry. This reactant controls the amount of product formed and is called the **limiting reagent** (Section 4.2). You

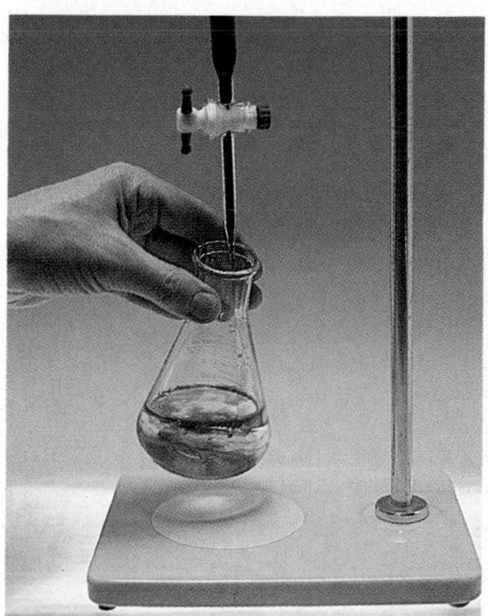

FIGURE 4.11

Titration of a solution containing Fe^{2+} (a reducing agent) (flask at left) with $KMnO_4$ (an oxidizing agent) (in the buret). As $KMnO_4$ is added to the solution the iron(II) is oxidized and the deep purple MnO_4^- ion is reduced. The products (Fe^{3+} and Mn^{2+}) are nearly colorless. Just past the equivalence point, when a *slight* excess of $KMnO_4$ has been added, the solution takes on a faint purple color. The appearance of this color indicates that the titration is completed. (Charles D. Winters)

must often decide on the limiting reagent before you can proceed to calculate the mass of product expected.

The **percent yield** of a reaction is used to express the efficiency of chemical reactions (Section 4.3). This is the ratio of the amount of product actually isolated to the amount of product expected if the reaction had worked according to theory (called the **theoretical yield**).

$$\text{Percent yield} = \frac{\text{actual yield}}{\text{theoretical yield}} \times 100$$

Many important chemical reactions occur in solution, so the concentration of material (the **solute**) in the reaction medium (the **solvent**) must be defined (Section 4.4). A very convenient unit of concentration is **molarity**.

$$\text{Molarity } (M) = \frac{\text{moles solute}}{\text{liters of solution}}$$

In another often-used form, this equation is moles solute $= M \times V$.

Reagent solutions are often prepared by diluting a more concentrated solution of that reagent (Section 4.4). The number of moles of material in a solution remains the same after being diluted. Therefore, the product of molarity and volume of a concentrated solution must be equal to the product of molarity and volume of the diluted solution (c, concentrated; d, diluted).

$$M_c V_c = M_d V_d$$

This allows you to find the new concentration, for example, when diluting a given amount of solution of known concentration to a new volume.

A **titration** is a way to carry out a reaction in a very precise manner and to use it for analysis (Section 4.4). In an acid–base titration, acid of known concentration is added to a base of unknown concentration (or vice versa) until the number of moles of H^+ added is exactly equal to the number of moles of OH^- present. This is called the **equivalence point**. If the volume of acid is known exactly, then moles of acid are known exactly, and, if the balanced equation for the overall reaction is known, then moles of base are known. Oxidation–reduction titrations are identical, except that the equivalence point is reached when the amounts of reducing and oxidizing agents are in the correct stoichiometric ratio.

STUDY QUESTIONS

GENERAL STOICHIOMETRY PROBLEMS

1. Silver bromide is an important component of photographic film. It can be made by the reaction

   $AgNO_3(aq) + NaBr(aq) \longrightarrow AgBr(s) + NaNO_3(aq)$

   If you have 2.6 g of sodium bromide, how many grams of silver nitrate are required for complete reaction? How many grams of sodium nitrate and of silver bromide will you obtain?

2. Laughing gas, N_2O, is made by the careful thermal decomposition of ammonium nitrate.

   $NH_4NO_3(s) \longrightarrow N_2O(g) + 2\ H_2O(g)$

   If you begin with 1.00×10^3 g (about 2 lb) of ammonium nitrate, what is the theoretical yield of laughing gas?

3. Iron oxidizes in hot, dry air to produce Fe_3O_4. If 5.58 g of iron reacts with O_2 to give this oxide, how many grams of O_2 are required?

4. The final step in the manufacture of pure platinum (for use in automobile catalytic converters and other purposes) is the reaction

$$3\ (NH_4)_2PtCl_6(s) \xrightarrow{\text{heat}} 3\ Pt(s) + 2\ NH_4Cl(g) + 2\ N_2(g) + 16\ HCl(g)$$

(a) If you heat 22.2 g of $(NH_4)_2PtCl_6$, what is the theoretical yield of Pt metal?
(b) How much HCl is produced when you heat 22.2 g of $(NH_4)_2PtCl_6$?

5. Passing an electrical current into brine, a solution of sodium chloride in water, gives hydrogen gas, chlorine gas, and sodium hydroxide according to the equation

$$2\ NaCl(aq) + 2\ H_2O(\ell) \xrightarrow{\text{electricity}} H_2(g) + Cl_2(g) + 2\ NaOH(aq)$$

Assume you have 1.00×10^3 g of NaCl in solution. Calculate the mass of each of the products expected from the reaction above.

6. Hydrogen chloride, HCl, can be made conveniently in the laboratory by the reaction of NaCl, common table salt, with sulfuric acid. As described in Chapter 3, this is an exchange or metathesis reaction, giving HCl(g) and Na_2SO_4 as products.
(a) Write a balanced equation depicting this reaction.
(b) If 20. g of NaCl are used, how many grams of H_2SO_4 are required for complete reaction?
(c) If 20. g of NaCl are used, and only 5.6 g of HCl are isolated, what is the percent yield of HCl?
(d) If only 5.6 g of HCl are obtained, how many grams of Na_2SO_4 must also have been formed?

7. The overall equation for the reduction of iron ore in a blast furnace is

$$Fe_2O_3(s) + 3\ CO(g) \longrightarrow 2\ Fe(s) + 3\ CO_2(g)$$

(a) How many moles of CO are required to give 2.4 moles of Fe? How many grams of CO are required?
(b) How many grams of CO are required to consume completely 2.0 moles of Fe_2O_3?
(c) To produce 27.9 kg of Fe, how many kilograms of Fe_2O_3 are required?
(d) If each pound (454 g) of iron(III) oxide, Fe_2O_3, produces one-half pound of Fe, what is the percent yield of iron?

8. Hydrofluoric acid, HF, is never sold in glass bottles, because glass is composed of calcium and sodium silicates that react with HF. The reaction can be depicted, for simplicity, as the interaction of silicon dioxide and the acid.

$$SiO_2(s) + 4\ HF(aq) \longrightarrow SiF_4(g) + 2\ H_2O(\ell)$$

(a) How many grams of HF would you need to react completely with 2.0 moles of SiO_2?
(b) How many grams of SiF_4 would be produced from 300. g of pure silica, SiO_2? (Assume more HF is present than required.)
(c) Calculate the percent yield if 1.28 g of SiO_2 give 1.08 g of SiF_4 as product.

9. On April 16, 1947, the S.S. Grandchamp blew up in the harbor of Texas City, Texas, and the explosion set off a chain reaction of explosions and fires that eventually killed 570 people. The original blast was from the explosive decomposition of ammonium nitrate, a compound used as a fertilizer, to give nitrogen, oxygen, and water.

$$2\ NH_4NO_3(s) \longrightarrow 2\ N_2(g) + O_2(g) + 4\ H_2O(\ell)$$

(a) If a shipload of ammonium nitrate (3.00×10^4 T) (1 T = 2000 lb; 1 lb = 454 g) explodes, how many tons of each of the products is formed?
(b) If the explosion is done on a small scale, and 45.0 g of H_2O was isolated, what was minimum amount of ammonium nitrate present in the beginning?
(c) Assume the decomposition of ammonium nitrate is not complete. Only 1.0 g of O_2 are isolated from 8.0 g of NH_4NO_3. What is the percent yield of O_2?

10. $Pt(NH_3)_2Cl_2$, called "cisplatin," has recently been found to be effective in treating certain types of cancers. It is synthesized by the following reaction:

$$K_2PtCl_4(aq) + 2\ NH_3(aq) \longrightarrow 2\ KCl(aq) + Pt(NH_3)_2Cl_2(aq)$$

If you begin with 1.00 g of K_2PtCl_4, how many grams of NH_3 must be used for complete reaction? How many grams of $Pt(NH_3)_2Cl_2$ will be produced?

11. Ammonia is used throughout the world as a fertilizer and in making nitrogenous plastics and fibers. It is usually manufactured in the Haber process, the direct reaction of N_2 with H_2 (in the presence of other compounds that accelerate the reaction).

$$N_2(g) + 3\ H_2(g) \longrightarrow 2\ NH_3(g)$$

If you use 280. pounds of N_2, how many pounds of H_2 are required for complete reaction? How many pounds of NH_3 can be produced? (1.00 lb = 454 g)

12. Ammonia can be made by treating calcium cyanamide with water.

$$CaCN_2(s) + 3\ H_2O(\ell) \longrightarrow CaCO_3(s) + 2\ NH_3(g)$$

How many moles of water would be required to react with $CaCN_2$ to produce 68.0 g of NH_3? How many grams of $CaCN_2$ are used in the process?

13. Oil paintings in which the "white lead" has been blackened by reaction with H_2S (from air pollution or

from the glaze over the painting itself), forming black lead sulfide, may be cleaned by washing with hydrogen peroxide, H_2O_2. The reaction for the cleaning process is

$$PbS(\text{black solid}) + 4\ H_2O_2(aq) \longrightarrow PbSO_4(s) + 4\ H_2O(\ell)$$

(a) How many grams of H_2O_2 must be used to clean off 0.24 g of PbS?

(b) If 0.072 g of H_2O form in the reaction, how many grams of $PbSO_4$ must also have been formed?

14. Most metal carbonates give the appropriate metal oxide and CO_2 when heated. For example,

$$CaCO_3(s) \longrightarrow CaO(s) + CO_2(g)$$

Limestone is mostly calcium carbonate, $CaCO_3$, but other minerals are usually present as well. Assume a 1.605 g sample of limestone is heated and decomposed completely to CaO, 0.657 g of CO_2, and an inert residue. What is the weight percentage of $CaCO_3$ in the limestone sample?

15. An oxide of scandium, with a mass of 1.423 g, is chemically reduced with H_2 to give H_2O and 0.929 g of Sc metal. What is the formula of the scandium oxide? How much water is formed?

16. Some solid CaO in a test tube picks up water from the atmosphere and is thereby changed completely to $Ca(OH)_2$.

$$CaO(s) + H_2O(\ell) \longrightarrow Ca(OH)_2(s)$$

The initial mass of CaO and test tube is 10.860 g. After picking up water to give calcium hydroxide, the total mass of tube and $Ca(OH)_2$ is 11.149 g. What is the mass of the test tube?

17. It has been proposed that magnesium oxide can be used to remove sulfur dioxide from the flue gas of factories and power plants (where SO_2 comes from burning fossil fuels). The magnesium oxide reacts with SO_2 and O_2 according to the equation

$$MgO(s) + SO_2(g) + \tfrac{1}{2}\ O_2(g) \longrightarrow MgSO_4(s)$$

The MgO comes from $MgCO_3$, a naturally occurring mineral, according to the equation

$$MgCO_3(s) + \text{heat} \longrightarrow MgO(s) + CO_2(g)$$

To remove 20. million tons of SO_2 from factory and plant gases each year, how many tons of $MgCO_3$ must be mined? How much $MgSO_4$ will be produced (and where can we put it)?

LIMITING REAGENT PROBLEMS

18. Dinitrogen tetrafluoride, N_2F_4, can be produced by the reaction of NH_3 with F_2.

$$2\ NH_3(g) + 5\ F_2(g) \longrightarrow N_2F_4(g) + 6\ HF(g)$$

If 4.00 g of NH_3 and 14.0 g of F_2 are allowed to react, which reagent is in excess and by how much? How many grams of N_2F_4 and HF are produced?

19. Disulfur dichloride, S_2Cl_2, is used to vulcanize rubber. It can be made by treating molten sulfur with gaseous chlorine.

$$S_8(s) + 4\ Cl_2(g) \longrightarrow 4\ S_2Cl_2(g)$$

If you begin with 32.0 g of sulfur and 71.0 g of Cl_2, how many grams of S_2Cl_2 can be produced? What quantity of which starting material will remain after the maximum amount of S_2Cl_2 has been formed?

20. The *Starship Enterprise* on *Star Trek* really used B_5H_9 and O_2 as a fuel. The two react according to the following balanced equation.

$$2\ B_5H_9(\ell) + 12\ O_2(g) \longrightarrow 5\ B_2O_3(s) + 9\ H_2O(g)$$

(a) If one fuel tank holds 126 kg of B_5H_9, and the other fuel tank holds 192 kg of liquid O_2, which fuel tank will be emptied first?

(b) When one fuel tank is emptied, how much remains in the other tank?

(c) When the reaction has gone as far as possible, how much water has been formed?

21. Methyl alcohol, CH_3OH, is a clean-burning, easily handled fuel. It can be made by the direct reaction of CO and H_2 (obtained from coal and water).

$$CO(g) + 2\ H_2(g) \xrightarrow{\text{catalyst}} CH_3OH(\ell)$$

Assume you start with 12.0 g of H_2 and 74.5 g of CO. (a) Which of the reactants is in excess? What mass (in grams) of this reagent is left after reaction is complete? (b) How many grams of methyl alcohol can be obtained theoretically?

22. In Example 4.2, aspirin was synthesized from salicylic acid and acetic anhydride. If you begin with 1.0 kg each of the two reactants, which is the limiting reagent? What is the theoretical yield of aspirin under these conditions?

23. Phosphoric acid is made in enormous quantities every year by treating phosphate rock with acid. For example,

$$\underset{\text{apatite}}{Ca_5(PO_4)_3F(s)} + 5\ H_2SO_4 + 10\ H_2O \longrightarrow 3\ H_3PO_4 + \underset{\text{gypsum}}{5\ CaSO_4 \cdot 2\ H_2O} + HF$$

(a) If you begin with 1.00 ton of apatite, how many grams of phosphoric acid can be obtained? That is, what is the *theoretical yield* of H_3PO_4? (1 T = 908,000 g; Molar mass apatite = 504 g/mol)

(b) If the amount of acid actually obtained is 2.50×10^5 g, what is the percent yield of phosphoric acid?

(c) If the ton of apatite had been treated with 1 ton of sulfuric acid, which is the limiting reagent? (Prove your answer with appropriate calculations.)

CONCENTRATION OF SOLUTIONS

24. If 2.60 g of NaBr is dissolved in enough water to make 160. mL of solution, what is the molar concentration of NaBr? How many milliliters of 0.10 *M* NaBr would you need to supply 2.60 g of NaBr?
25. For each of the following solutions, tell how many grams of solute would be necessary for its preparation.
    (a) 0.10 L of 0.10 *M* $AgNO_3$
    (b) 5.0 mL of 0.05 *M* NaCN
    (c) 0.10 L of 0.10 *M* $BaCl_2$
    (d) 250 mL of 0.0014 *M* $KMnO_4$
26. For each of the following solutions, tell how many grams of solute would be necessary for its preparation.
    (a) 0.25 L of 0.050 *M* $Na_2C_2O_4$
    (b) 0.125 L of 0.15 *M* $K_2Cr_2O_7$
    (c) 0.50 L of 0.10 *M* $KHCO_3$
27. What would be the molar concentrations of the solute in each of the following solutions?
    (a) 0.50 L containing 5.6 g of $NaClO_4$
    (b) 0.10 L containing 2.3 g of KNO_3
    (c) 0.25 L containing 1.5 g of C_4H_8O
    (d) 50. mL containing 0.55 g of NaOH
    (e) 1.55 liters containing 153 g of Na_2CO_3
28. For each of the following solutions, give the concentration of each ion in solution.
    (a) 1.0 *M* NaBr
    (b) 0.50 *M* Na_3PO_4
    (c) 0.10 *M* $ZnCl_2$
    (d) 0.10 *M* $KClO_4$
29. If you dilute 50.0 mL of 0.300 M HCl to 300. mL, what is the HCl concentration in the diluted solution?
30. If you require 500. mL of 0.15 M NaOH, how many milliliters of 1.00 M NaOH must you dilute?
31. Sucrose, common table sugar, has the formula $C_{12}H_{22}O_{11}$. If you add one spoonful of sugar (3.4 g) to 250 mL of coffee, what is the molar concentration of the sugar?
32. To make 100. mL of a solution that is 0.25 *M* in chloride ion, how many grams of $MgCl_2$ would you need to dissolve?
33. You are a laboratory assistant for your chemistry course. There are 250 students in the course, and each needs to use 25 mL of a 0.15 *M* solution of copper sulfate. How many liters of copper sulfate solution should you prepare? How many grams of $CuSO_4 \cdot 5\,H_2O$ must you dissolve to make this solution?
34. If you need 100. mL of 0.15 M $CuSO_4$, how many milliliters of 0.50 M $CuSO_4$ must you dilute? How many milliliters of water must be added? How many grams of $CuSO_4$ does the dilute solution contain?

REACTIONS IN SOLUTION: GENERAL STOICHIOMETRY

35. An aqueous solution of NaCl gives $H_2(g)$, $Cl_2(g)$, and NaOH when an electrical current is passed through the solution (Question 5). If you begin with 10. liters of 0.15 *M* NaCl, how many grams of H_2 can be formed? How many grams of NaOH are formed?
36. H_2S is a foul-smelling compound that can be a component of polluted air. As noted in Question 13 it reacts with Pb^{2+} ion in paint to give black lead sulfide, PbS. This same reaction can be done in aqueous solution.

    $$H_2S(aq) + Pb(NO_3)_2(aq) \longrightarrow PbS(s) + 2\,HNO_3(aq)$$

    If you have 50. mL of 0.10 *M* H_2S, and add lead nitrate in excess, how many moles of PbS can be formed? How many grams of $Pb(NO_3)_2$ will be used?
37. Chlorine, $Cl_2(g)$, can be made by treating HCl or a chloride salt in acid solution with an strong oxidizing agent, MnO_2 for example. The balanced, net ionic equation for the reaction is

    $$4\,H^+(aq) + 2\,Cl^-(aq) + MnO_2(s) \longrightarrow Mn^{2+}(aq) + 2\,H_2O(\ell) + Cl_2(g)$$

    If you have 125 mL of a 0.100 *M* solution of HCl and an excess of MnO_2, how many moles of Cl_2 can be formed? How many grams of Cl_2 can be formed? How many grams of MnO_2 would be necessary to complete the reaction?
38. Diborane, B_2H_6, can be produced by the following reaction.

    $$2\,NaBH_4(aq) + H_2SO_4(aq) \longrightarrow 2\,H_2(g) + Na_2SO_4(aq) + B_2H_6(g)$$

    How many grams of $NaBH_4$ and how many milliliters of 1.00 *M* H_2SO_4 should be used to prepare 4.14 g of B_2H_6? (Assume a 100% yield of B_2H_6.)
39. If H_2S is dissolved in drinking water, the water has an unpleasant odor, so water treatment plants take steps to remove the H_2S. This can be done by passing chlorine gas through the water and then filtering the precipitated sulfur.

    $$H_2S(aq) + Cl_2(g) \longrightarrow 2\,HCl(aq) + S(s)$$

    If your drinking water contains 10. parts per million (10 g/million g H_2O) of H_2S, how much Cl_2 must you use to remove all the H_2S from the water of a city which uses 10^6 metric tons (1 metric ton = 1000 kg) of water per day?
40. One use of hydrazine, N_2H_4, is as a scavenger of dissolved oxygen in hot water heating systems. (The O_2 leads to corrosion and must be removed.)

    $$N_2H_4(aq) + O_2(aq) \longrightarrow N_2(g) + 2\,H_2O(\ell)$$

    (a) If you dissolve one pound of hydrazine (454 g) in

a 400. L heating system, what is the molarity of the hydrazine?

(b) How many moles of O_2 can be removed by the added hydrazine?

41. Hydrazine, N_2H_4, is a base like ammonia, and so can react with an acid such as sulfuric acid.

$$2\ N_2H_4(aq) + H_2SO_4(aq) \longrightarrow 2\ N_2H_5^+(aq) + SO_4^{2-}(aq)$$

If you dissolve one pound (454 g) of hydrazine in 400. liters of water, what is the molarity of the solution? How many liters of concentrated sulfuric acid (18.0 *M*) would be required to neutralize the hydrazine? If one bottle of concentrated acid contains 2.50 liters of acid, how many bottles of sulfuric acid would be required for neutralization?

ACID–BASE TITRATIONS

42. How many milliliters of 0.250 *M* HCl would be required to neutralize completely 2.50 g of NaOH?

43. How many milliliters of 0.250 *M* HCl would be required to neutralize completely 36.5 mL of 0.100 *M* NaOH?

44. You dissolve 2.50 g of $Ba(OH)_2$ in enough water to make 250. mL of solution. How many moles of HCl would be required to neutralize completely this solution? If the HCl solution is 0.110 *M*, how many milliliters would be required to neutralize the $Ba(OH)_2$ solution?

45. Sodium carbonate, Na_2CO_3, is a good compound to use to standardize acid solutions.

$$Na_2CO_3(aq) + 2\ HCl(aq) \longrightarrow 2\ NaCl(aq) + H_2O(\ell) + CO_2(g)$$

If 42.43 mL of HCl solution is used to titrate 0.251 g of Na_2CO_3 to the equivalence point, what is the molar concentration of the acid?

46. Potassium acid phthalate, $KHC_8H_4O_4$, is used to standardize solutions of bases. The acidic anion reacts with bases according to the net ionic equation

$$HC_8H_4O_4^-(aq) + OH^-(aq) \longrightarrow H_2O(\ell) + C_8H_4O_4^{2-}(aq)$$

If a 0.902 g sample of potassium acid phthalate is dissolved in water and titrated to the equivalence point with 39.45 mL of NaOH, what is the molar concentration of the NaOH?

47. If apple cider is allowed to ferment, the result is vinegar, the distinctive ingredient of which is acetic acid, $HC_2H_3O_2$. The acid reacts with NaOH according to the balanced equation

$$HC_2H_3O_2(aq) + NaOH(aq) \longrightarrow NaC_2H_3O_2(aq) + H_2O(\ell)$$

If 100. mL of 1.00 *M* NaOH is required to neutralize completely a 500. mL sample of vinegar, how many grams of acetic acid are there in the vinegar sample?

48. Malic acid is a naturally occurring acid found in fruits, especially in apples. It reacts with NaOH according to the equation

$$C_4H_6O_5(aq) + 2\ NaOH(aq) \longrightarrow Na_2C_4H_4O_5(aq) + 2\ H_2O(\ell)$$

How many milliliters of 0.520 *M* NaOH are required to react completely with 1.34 g of malic acid?

49. A soft drink contains an unknown amount of citric acid, $C_6H_8O_7$. If 100. mL of the soft drink require 36.51 mL of 0.0102 *M* NaOH to neutralize completely the citric acid, how many grams of citric acid does the soft drink contain per 100 mL? What is the molar concentration of the citric acid? The reaction of citric acid and NaOH is

$$C_6H_8O_7(aq) + 2\ NaOH(aq) \longrightarrow Na_2C_6H_6O_7(aq) + 2\ H_2O(\ell)$$

50. Succinic acid, $C_4H_6O_4$, occurs in fungi and lichens, but it can be made in an industrial process. It is used in the manufacture of paints, perfumes, dyes, and so on. If 0.885 g of succinic acid is dissolved in enough water to make 100. mL of solution, what is the molarity of the solution? If the acid reacts with KOH to give the salt $K_2C_4H_4O_4$, write a balanced equation to show the reaction. How many grams of KOH are required to react completely with 0.885 g of succinic acid?

51. The labels on chemical bottles sometimes deteriorate or fall off if the chemical is stored for a long time in a chemical stockroom. Suppose you are the laboratory assistant for your course and you find a bottle in the stockroom that says on it only "potassium bi . . . ," the end of the second word having been obliterated. You know that it could be either of two acids: potassium biphthalate, $KC_8H_5O_4$, or potassium bitartrate, $KC_4H_5O_6$ (cream of tartar). Each acid loses only one mole of H^+ per mole of acid on reaction with a base such as NaOH. You find by titration that 1.021 g of the unknown acid requires 24.32 mL of 0.206 *M* NaOH. What acid is in the bottle?

OXIDATION–REDUCTION TITRATIONS

52. A commonly used reagent to supply Fe^{2+} in aqueous solution is $Fe(NH_4)_2(SO_4)_2 \cdot 6\ H_2O$ (MW = 392.2). If you use 1.050 g of this compound, and titrate the Fe^{2+} ions in the solution to the equivalence point with 32.45 mL of aqueous $KMnO_4$, what is the molar concentration of the $KMnO_4$? (See Exercise 4.13 for the balanced equation for the reaction involved.)

53. Iodine, I_2, reacts with the thiosulfate ion according to the equation

$$2\ S_2O_3^{2-}(aq) + I_2(aq) \longrightarrow S_4O_6^{2-}(aq) + 2\ I^-(aq)$$

If you dissolve 6.50 g of I_2 in enough water to make 500. mL of solution, how many milliliters of 0.0500 *M* $Na_2S_2O_3$ solution are needed for complete reaction?

54. Oxalic acid is found in many plants and vegetables. Like vitamin C (see Example 4.10), oxalic acid is both an acid and a reducing agent. It can react with an oxidizing agent such as $KMnO_4$ according to the balanced equation

$$2\ MnO_4^-(aq) + 5\ H_2C_2O_4(aq) + 6\ H^+(aq) \longrightarrow 2\ Mn^{2+}(aq) + 10\ CO_2(g) + 8\ H_2O(\ell)$$

How many milliliters of 0.0200 *M* $KMnO_4$ would be required to react completely with 0.892 g of oxalic acid, $H_2C_2O_4$?

55. Oxalic acid is a naturally occurring reducing agent and reacts with the oxidizing agent $KMnO_4$ according to the equation in Question 54. Suppose a 15.67 g sample of green leaves requires 23.45 mL of 0.0250 *M* $KMnO_4$ for titration to the equivalence point. What is the weight percent of oxalic acid in the leaves?

56. Sodium thiosulfate, $Na_2S_2O_3$, is used as a "fixer" in black and white photography. Assume you have a bottle of sodium thiosulfate and want to determine its purity. You can titrate the thiosulfate ion with I_2 according to the equation in Question 53. If you use 40.21 mL of 0.246 *M* I_2 in the titration, what is the weight percent of $Na_2S_2O_3$ in a 3.232 g sample of impure $Na_2S_2O_3$?

57. The amount of calcium in blood is often determined by adding oxalate ion to the blood sample to precipitate solid calcium oxalate according to the equation

$$Ca^{2+}(\text{in blood}) + C_2O_4^{2-}(aq) \longrightarrow CaC_2O_4(s)$$

The solid calcium oxalate is then separated from the blood, dissolved in acid, and titrated with $KMnO_4$ according to the equation in Question 54. Find the number of milligrams of Ca^{2+} in 100. mL of blood using the following information: The Ca^{2+} in a 5.00 mL sample of blood is precipitated as CaC_2O_4, and the calcium oxalate is titrated to the equivalence point with 11.63 mL of 0.00100 *M* $KMnO_4$.

58. The metal content of ores can sometimes be determined by oxidation–reduction titration. Suppose you have a sample that contains chromium, probably in the form of Cr_2O_3. To analyze the ore, you convert the chromium to the ion $Cr_2O_7^{2-}$, and then titrate the solution of this ion with Fe^{2+} in acid. The *unbalanced* equation for the titration is

$$Fe^{2+}(aq) + Cr_2O_7^{2-}(aq) \longrightarrow Fe^{3+}(aq) + Cr^{3+}(aq)$$

If you take a 1.70 g sample of chromium-bearing ore and find that it requires 40.2 mL of a 0.200 *M* Fe^{2+} solution for complete reaction, what is the weight percent of chromium in the sample?

59. The iron in an iron ore is brought into solution as Fe^{2+} and then titrated with standardized $Ce(SO_4)_2$ according to the equation

$$Fe^{2+}(aq) + Ce^{4+}(aq) \longrightarrow Ce^{3+}(aq) + Fe^{3+}(aq)$$

Calculate the weight percent of iron in an iron ore if a 15.45 g sample of the ore requires 42.34 mL of 0.113 *M* $Ce(SO_4)_2$ for titration to the equivalence point.

60. The lead content of a sample can be estimated by converting the lead to PbO_2 and dissolving the PbO_2 in an acid solution of KI. This liberates I_2 according to the *unbalanced* equation

$$PbO_2(s) + H^+(aq) + I^-(aq) \longrightarrow Pb^{2+}(aq) + I_2(aq)$$

The liberated I_2 is then titrated with $Na_2S_2O_3$ (see Question 53), and the amount of liberated I_2 is related to the amount of lead in the sample. If 0.576 g of lead-containing ore requires 35.23 mL of 0.0500 *M* $Na_2S_2O_3$ for titration of the liberated I_2, calculate the weight percentage of lead in the ore.

61. A small amount of gold in samples can be estimated by first converting the gold to $AuCl_3$ and then treating this with an excess of I^-. The reaction that occurs is

$$AuCl_3(aq) + 3\ KI(aq) \longrightarrow AuI(s) + I_2(aq) + 3\ KCl(aq)$$

The liberated I_2 is then titrated with $Na_2S_2O_3$ (according to the equation in Question 53), and the amount of I_2 is related to the amount of gold. If a 1.050 g sample of gold ore, which is 0.0300% Au by weight, reacts with KI, how many milliliters of 1.00×10^{-4} *M* $Na_2S_2O_3$ are required to titrate the liberated I_2?

62. A standard method of analyzing for copper is to use the following sequence of reactions:

    1. Dissolve the copper metal in nitric acid to give Cu^{2+}.
    2. Reduce the Cu^{2+} with I^- to liberate I_2.

    $$2\ Cu^{2+}(aq) + 4\ I^-(aq) \longrightarrow 2\ CuI(s) + I_2(aq)$$

    3. Titrate the liberated I_2 with standardized $Na_2S_2O_3$ according to the equation in Question 53.

    A 5-cent coin, the nickel, is actually composed mostly of copper (75.0% Cu and 25.0% Ni). How many milliliters of 0.500 *M* $Na_2S_2O_3$ must be used to consume the I_2 liberated by a 5-cent coin weighing 5.10 g?

GENERAL QUESTIONS

63. Balance the following equation and answer the questions that follow.

$$MnO_2(s) + I^-(aq) + H^+(aq) \longrightarrow I_2(aq) + Mn^{2+}(aq) + 2\ H_2O(\ell)$$

The amount of I_2 liberated in the reaction above can be quantitatively determined by titrating the I_2 with thiosulfate, $S_2O_3^{2-}$.

$$I_2(aq) + 2\ S_2O_3^{2-}(aq) \longrightarrow S_4O_6^{2-}(aq) + 2\ I^-(aq)$$

(a) If you use 25.0 mL of 0.100 *M* $S_2O_3^{2-}$ solution in this titration, how many grams of MnO_2 (MW = 86.9) must have been used to produce the I_2 titrated?
(b) In the reaction above, you used 0.100 *M* $S_2O_3^{2-}$. If you need 500. mL of this solution for your laboratory work, how many milliliters of 0.400 *M* $S_2O_3^{2-}$ would you have to dilute to make the required 500. mL?
(c) To make up 500. mL of 0.400 *M* $Na_2S_2O_3$ (Molar mass = 158 g/mol), how many grams of the solid sodium salt would be required?

64. One reason for the widespread use of platinum is its relative chemical inertness. It will dissolve, however, in "aqua regia," a mixture of nitric and hydrochloric acid.

$$3\ Pt(s) + 4\ HNO_3(aq) + 18\ HCl(aq) \longrightarrow 3\ H_2PtCl_6(aq) + 4\ NO(g) + 8\ H_2O(\ell)$$

(a) If you have 10.0 g of Pt, how many grams of chloroplatinic acid, H_2PtCl_6 (*MW* = 410), can be produced?
(b) How many grams of nitrogen oxide, NO, would be produced from 10.0 g of Pt?
(c) How many milliliters of 10.0 *M* nitric acid would be required for complete reaction with 10.0 g of Pt?
(d) If you have 10.0 g of Pt and 180. mL of 5.00 *M* HCl (plus excess HNO_3), which is the limiting reagent?

5
Thermochemistry: Energy and Chemistry

Thermite reaction [iron(III) oxide and aluminum] (Charles D. Winters)

CHAPTER OUTLINE

Energy! We think so much about its sources and uses. Where does the energy come from to move our modern society and how can we use it more efficiently? How can we provide energy to developing countries? Are there new sources of energy that are possible now or in the future? These and others are important questions to all of us. The contents of this chapter, therefore, are central to much of the chemistry that follows, because in it we begin to build a framework in which these questions can be considered and begin to focus on the specific question of energy from chemical processes.

5.1 ENERGY AND ITS FORMS

Energy is usually defined as the capacity to do work and transfer heat. **Work** is most often defined as the product of force (f) times the distance (d) over which the force acts.

$$\text{Work} = w = \text{force} \times \text{distance} = f \times d \tag{5.1}$$

This is an equation that you experience all the time. If you climb a mountain (Figure 5.1), you apply a force against the force of gravity to carry yourself and a pack up the mountain. Similarly, you perform work when you carry this textbook up stairs. Of course you can do this work because you have the energy or capacity to do so, the energy having been provided by the food you have eaten. This chemical source is only one of the many forms that energy may take, some of which are relevant to us as chemists and we will discuss them briefly.

FIGURE 5.1
Work! Carrying yourself and a pack up a mountain involves a force acting over a distance. (Spencer Swanger, Tom Stack & Associates)

Potential energy is possessed by an object by virtue of its position and is associated with attractions and repulsions. If you stand at the end of a diving board (Figure 5.2), you have considerable potential energy due to your height above the surface of the water. When you dive, your initial potential energy is converted progressively into **kinetic energy**. A body has kinetic energy because of its motion. This form of energy depends on the mass (m) and velocity (v) of the falling or moving object. That is,

$$\text{Kinetic energy} = KE = \tfrac{1}{2}mv^2 \tag{5.2}$$

During the dive your mass is constant, but you move faster and faster as you fall, so your kinetic energy increases, and it does so at the expense

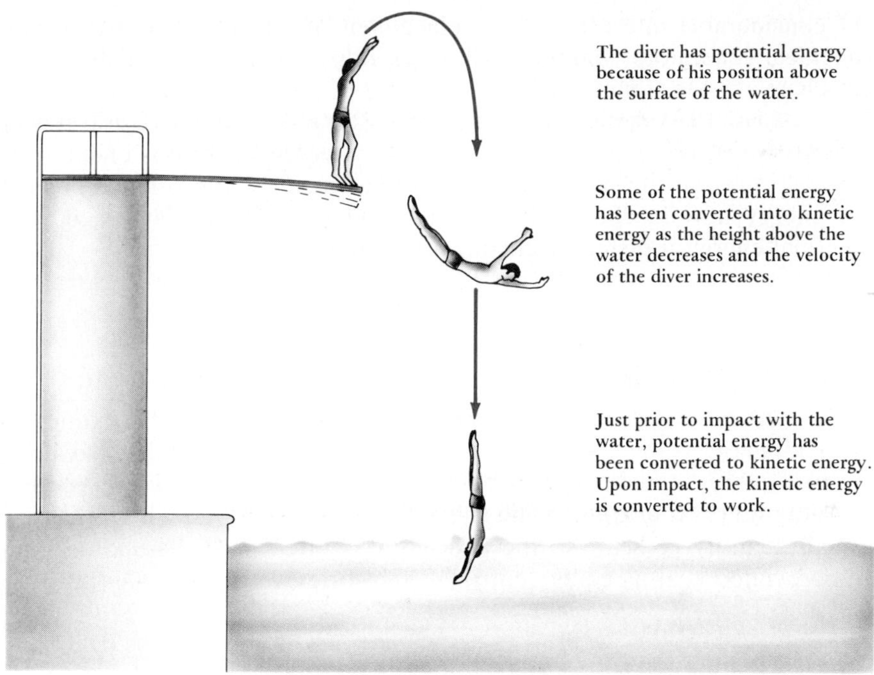

The diver has potential energy because of his position above the surface of the water.
Some of the potential energy has been converted into kinetic energy as the height above the water decreases and the velocity of the diver increases.
Just prior to impact with the water, potential energy has been converted to kinetic energy. Upon impact, the kinetic energy is converted to work.

FIGURE 5.2
The interconversion of potential energy, kinetic energy, and work.

of potential energy. Potential energy is progressively lost as you approach the water.

At the moment of impact of your body with the water, your velocity is abruptly changed, so much of your kinetic energy is lost. However, because of the **principle of the conservation of energy**, the total energy (potential + kinetic) must not change. Your potential energy has been converted into kinetic energy throughout the fall, and the kinetic energy is converted into work on impact with the water; that is, your impact has rearranged the water, forcing it aside to make room for your body.

Heat or **thermal energy** is one form of molecular energy. If you rub your hands together on a cold day, friction converts some of the energy of motion into heat. Thermal energy is usually associated with the motions of atoms or molecules in a solid, liquid, or gas. The more rapid these motions, the higher the thermal energy level of the material. A measure of ''thermal energy level'' can be made with a thermometer; the higher the temperature, the greater the motions of the atoms or molecules.* One objective of this chapter is to explore the idea of heat or thermal energy.

Radiant and **electrical energy** are two additional forms of molecular and atomic energy. Radiant energy is the type supplied by a radiator in a room or by the sun to the earth. Electrical energy can be generated by mechanical devices or chemical reactions. Chemical sources are currently

Solar energy is a topic much discussed as one way to alleviate the world's energy crisis.

*We have been careful to differentiate between energy (or thermal content) and temperature. The energy content of a cup of hot coffee is much less than that of a bath tub full of hot water at the same temperature. Your cup of hot coffee may provide only enough energy to melt one ice cube before coming to room temperature, but the tub of hot water could melt a whole icebox full of ice before it comes to room temperature. The subject of heat or thermal content is taken up in the next section.

of considerable interest in the replacement of oil- and coal-driven mechanical generators, and we shall discuss this aspect of chemistry, electrochemistry, in Chapter 19.

Chemical reactions consume or release energy in the forms of heat (thermal energy), light (radiant energy), or electricity. In this chapter we want to present the basic ideas of **thermodynamics**, the science of heat or energy flow in chemical reactions, and we will be concerned with thermal energy transfer by chemical reactions.

5.2 UNITS OF ENERGY

All forms of energy can be converted to heat, so some of the units in which energy is measured are really those that measure heat. A **calorie** is the amount of heat required to raise the temperature of 1.00 g of pure liquid water 1.0 degree Celsius, from 14.5°C to 15.5°C. This is a very small quantity of heat, so the **kilocalorie** is often used. The kilocalorie (abbreviated kcal) is equivalent to 1000 calories.

$$1 \text{ kilocalorie} = 1 \text{ kcal} = 1000 \text{ calories} \quad (5.3)$$

Most of us were raised to think in calories, probably because we hear about dieting or read breakfast cereal boxes. However, the calorie used with regard to food is Calorie with a capital *C*, a unit equivalent to the kilocalorie. Thus, a breakfast cereal that gives you 100 Calories really provides 100 kcal or 100×10^3 calories (with a small *c*).

1 calorie = 4.184 joules
The "joule" is named for James P. Joule (1818–1889). Joule was a student of John Dalton (Chapter 2) and the son of a brewer in Manchester, England.

In this textbook we shall use the **joule**, the SI unit of energy. One calorie is equivalent to 4.184 joules. The joule is used because it is directly derived from the units used to calculate energy. If a 2.0-kg object (about 4 pounds) is moving with a velocity of 1.0 meter per second (roughly 2 mph), the kinetic energy is

$$\text{Kinetic energy} = \tfrac{1}{2}\,mv^2 = \tfrac{1}{2}\,(2.0 \text{ kg})\left(\frac{1.0 \text{ m}}{\text{s}}\right)^2 = 1.0\,\frac{\text{kg}\cdot\text{m}^2}{\text{s}^2} = 1.0 \text{ joule}$$

To give you some feeling for joules, if you drop a six-pack of soft drink cans on your foot, the kinetic energy at the moment of impact is around a calorie or two, that is, 4 to 10 joules.

EXAMPLE 5.1

KINETIC ENERGY CALCULATION

A baseball, which weighs about $\frac{1}{4}$ pound (114 g), can be thrown at a speed of 100 miles per hour (44.7 m/s). (a) Calculate the kinetic energy of the baseball in joules. (b) Convert the answer in joules to kilocalories.

Solution In part (a) you need Equation 5.2.

$$\begin{aligned} KE &= \tfrac{1}{2}\,mv^2 \\ &= (\tfrac{1}{2})\,(114 \text{ g})\left(\frac{1.00 \text{ kg}}{1000 \text{ g}}\right)\left(\frac{44.7 \text{ m}}{\text{s}}\right)^2 = 114 \text{ kg}\cdot\text{m}^2/\text{s}^2 \\ &= 114 \text{ J} \end{aligned}$$

Part (b) asks you to make a unit conversion.

$$KE = 114 \cancel{J} \left(\frac{1.000 \text{ cal}}{4.184 \cancel{J}}\right) = 27.2 \text{ cal}$$

$$KE = 27.2 \cancel{\text{cal}} \left(\frac{1 \text{ kcal}}{1000 \cancel{\text{cal}}}\right) = 0.0272 \text{ kcal}$$

EXERCISE 5.1 KINETIC ENERGY CALCULATION

A certain electron with a mass of 9.11×10^{-28} g has a kinetic energy of 1.60×10^{-18} J. What is the velocity of the electron in meters/second? In miles/hour? (Make sure you use the numbers in the correct units. Recall that units of energy are $kg \cdot m^2/s^2 = J$.)

EXERCISE 5.2 ENERGY UNIT CONVERSIONS

A hot dog provides 150 Calories. What is this energy in joules?

5.3 SPECIFIC HEAT AND HEAT CAPACITY

One calorie is the amount of heat energy needed to raise the temperature of 1.00 g of water (at 14.5°C) 1.00 degree Celsius. This quantity of heat (1.00 cal or 4.184 joules) is called the **specific heat** of water; it is "specific" in the sense that the heat is specific to a unit mass of matter and for a unit change in temperature. The specific heat of a substance can be determined experimentally by determining the quantity of heat gained or lost by a definite mass of the substance as the temperature rises or falls.

$$\text{Specific heat} = \frac{\text{heat lost or gained}}{(\text{g of material})(\text{degrees temperature change})} \tag{5.4}$$

As an example, it has been found that 60.5 joules are required to raise the temperature of 25.0 g of ethylene glycol (an organic compound used in automobile antifreeze) 1.00 degree Celsius (or 1.00 K). This means that the specific heat is

$$\text{Specific heat of ethylene glycol} = \frac{60.5 \text{ joules}}{(25.0 \text{ g})(1.00 \text{ K})} = 2.42 \frac{\text{J}}{\text{g} \cdot \text{K}}$$

The specific heat has been determined for many substances, and a few are listed in Table 5.1. You will notice that water has a high specific heat, and it is in fact one of the highest known. This is important in our everyday lives, since a high specific heat means that many joules must be absorbed by a large body of water to raise its temperature just a degree or so. Conversely, many joules must be lost before the temperature of the water drops by more than a degree.

The specific heat of a substance allows you to calculate the heat lost or gained in some thermal process by measuring the mass of the substance and the temperature change. Conversely, you can calculate the temperature change that should occur when adding or taking away a given

TABLE 5.1 Specific Heat Values for some Elements, Compounds, and Common Solids

SUBSTANCE		SPECIFIC HEAT (J/g · K)
Elements		
Al	Aluminum	0.902
C	Graphite	0.720
Cu	Copper	0.385
Au	Gold	0.128
Fe	Iron	0.451
Compounds		
$NH_3(\ell)$	Ammonia	4.70
$CCl_4(\ell)$	Carbon tetrachloride	0.861
$C_2H_5OH(\ell)$	Ethyl alcohol	2.46
$(CH_2OH)_2(\ell)$	Ethylene glycol (antifreeze)	2.42
H_2O	Liquid water	4.18
	Ice	2.06
$CCl_2F_2(g)$	Freon-12	0.598
Common Solids		
Common steel		0.45
Cement		0.88
Glass		0.84
Granite		0.79
Wood		1.76

amount of heat. Equation 5.4 allows you to do this, but it may be more convenient to rewrite it in the following form:

The symbol "q" is used to designate the heat gained or lost.

Heat lost or gained = (specific heat)(mass)(degrees temperature change)

$$q = (\text{J/g} \cdot \text{K})(\text{mass in g})(\Delta T \text{ in kelvins}) \qquad (5.5)$$
$$= (\text{sp. ht.})(m)(\Delta T)$$

In this expression the temperature change, ΔT, is always calculated as

$$\Delta T = \text{final temperature} - \text{initial temperature}$$

and so heat is lost when $q < 0$ and heat is gained when $q > 0$.*

EXAMPLE 5.2

USE OF SPECIFIC HEAT

How many joules of energy must a cup of coffee lose for its temperature to drop from 60.0°C to the temperature of your body (37.0°C)? Assume that the cup holds 250. mL of coffee (with a density of 1.00 g/mL, the coffee weighs 250. g), and that the specific heat of coffee is the same as water.

---

*Anytime you take the difference between two quantities in thermochemistry, you should *always subtract the initial quantity from the final quantity*. A natural consequence of this convention is that the algebraic sign of the calculated result indicates heat gain (+; $q > 0$) or loss (−; $q < 0$) by the substance being studied; this is an important point, as you will see in Example 5.2 and in the next section.)

Solution The heat change can be obtained from Equation 5.5.

$$\text{Heat change} = \left(\frac{4.184\ \text{J}}{\text{g} \cdot \text{K}}\right)(250.\ \text{g})\ (\underset{\text{final temp.}}{37.0°\text{C}} - \underset{\text{initial temp.}}{60.0°\text{C}})$$

$$= -24.1 \times 10^3\ \text{J} = -24.1\ \text{kJ}$$

Notice that the final answer has a *negative sign*. This denotes that *energy is being lost* by the water as the temperature of the water declines. (Conversely, a positive sign would have indicated that heat energy was gained.)

EXAMPLE 5.3

HEAT LOST BY WATER ON COOLING

The volume of the ocean that is 1.0 mile square and has a depth of 10. feet contains 7.9×10^{12} liters of water. How many joules of energy are lost when the water temperature drops by 1.0 degree Celsius? Assume the density of ocean water is 1.0 g/mL and that the specific heat of sea water is the same as that of pure water.

Solution Since the specific heat equation is written in grams, you must first convert the volume to grams.

$$7.9 \times 10^{12}\ \cancel{\text{L}} \left(\frac{1000\ \cancel{\text{mL}}}{1.0\ \cancel{\text{L}}}\right)\left(\frac{1.0\ \text{g}}{\cancel{\text{mL}}}\right) = 7.9 \times 10^{15}\ \text{g}$$

Knowing the mass of the piece of ocean being studied, you can now calculate the heat lost.

Note that 1 degree on both the kelvin and Celsius scales has the same magnitude, and so these units cancel in Example 5.3.

$$\text{Heat change} = \left(\frac{4.184\ \text{J}}{\text{g} \cdot \text{K}}\right)(7.9 \times 10^{15}\text{g})(-1.0°\text{C})$$

(Notice that the temperature change is negative!)

$$\text{Heat change} = \text{heat lost by water} = -3.3 \times 10^{16}\ \text{J} = -3.3 \times 10^{13}\ \text{kJ}$$

EXERCISE 5.3 USING SPECIFIC HEAT

The cup of coffee in Example 5.2 lost 24.1 kJ when cooled from 60.0°C to 37.0°C. If this same amount of heat is used to warm a piece of aluminum weighing 250. g, what would the final temperature of the aluminum be if its initial temperature was 37.0°C? (The specific heat of Al is 0.902 J/g · K.)

Before leaving this topic, it is useful to point out again that the higher the specific heat of a substance the greater its capacity to absorb heat energy before changing its temperature by one degree Celsius. If you worked Exercise 5.3, you saw that a given amount of heat will raise the temperature of a piece of aluminum considerably more than an amount of water of equal mass.

The **molar heat capacity** is an alternative way to express the capacity of a substance to absorb heat and to compare substances. That is,

$$\text{Molar heat capacity (J/mol} \cdot \text{K)} = \text{(specific heat)(molar mass)} \qquad (5.6)$$

Thus, the specific heat of water is 4.184 J/g · K, while the molar heat capacity is 75.3 J/mol · K, or 18.0 times the specific heat of 4.184 J/g · K.

5.4 ENERGY CHANGES IN CHEMICAL PROCESSES: ENTHALPY

If carbon dioxide gas is cooled to $-78°C$, the compound can be obtained in a solid form, a form commonly called "dry ice." The process is reversible in that CO_2 molecules in dry ice can return to the gas phase by absorbing heat energy.

The double arrows in the $CO_2(s)/CO_2(g)$ equation, $\rightleftharpoons$, are meant to convey the notion that the chemical processes are reversible—solid can be converted to gas or gas to solid.

$$CO_2(\text{solid}, -78°C) + \text{heat} \rightleftharpoons CO_2(\text{gas}, 25°C)$$

It is this latter process that we want to describe in the proper language of thermodynamics.

The physical process being described is given above as a chemical equation where heat is a "reactant" when solid CO_2 is converted to a gas, or heat is a "product" when gas is changed to solid. The CO_2 sample is called the **system**. Since one can only observe the *transfer* of heat from one place to another, there must be two substances between which the transfer occurs. That is, the system must have **surroundings**.

The material being studied is the **system**, and the rest of the universe is the **surroundings**.

The system is the substance in whose energy content we are primarily interested, and everything outside this is the surroundings. The system may be contained within an actual physical boundary, in a beaker or within a cell in your body, for instance. Alternatively, the boundary can be purely imaginary; for example, you could study the solar system within its surrounding, the galaxy.

In the process $CO_2(s) \rightarrow CO_2(g)$, heat must flow into the system from the surroundings since energy is needed for the regular array of molecules making up the solid to be disrupted and converted into dispersed CO_2 molecules in the gas phase at a higher temperature. Heat flow into a system from its surroundings is called an **endothermic process**, in contrast with an **exothermic process** where heat flows from the system to the surroundings.*

Direction of Heat Flow
System $\longrightarrow$ Surroundings
$q < 0$, *Exo*thermic (*exo* = "out")
Surroundings $\longrightarrow$ System
$q > 0$, *Endo*thermic (*endo* = "in")

According to the **principle of the conservation of energy**, all the heat energy flowing into a system from its surroundings must go somewhere; no energy can be lost. For $CO_2(s) \rightarrow CO_2(g)$, the heat input has three effects: (a) it provides the energy necessary to overcome the forces holding the molecules together in the solid state at $-78°C$, allowing them to break away from one another and become a gas; (b) it warms the gas from $-78°C$ to room temperature; and (c) it allows the gas to expand to a larger volume (if the pressure is kept constant). If, as shown in Figure 5.3a, the CO_2 is not held in a rigid container, the volume of the system increases, but it can only do so if the system pushes aside the atmosphere (just as the diver in Figure 5.2 pushed aside the water). Thus, the system does work on its surroundings. The energy associated with steps (a) and (b) causes internal changes in the system; we represent this by ΔE (called

Energy diagram showing relation of ΔE, q, w, and ΔH for $CO_2(s) \rightarrow CO_2(g)$

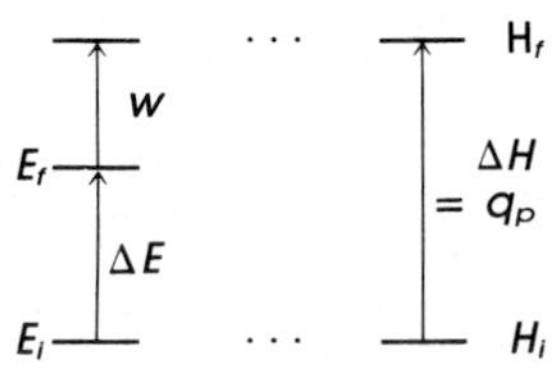

*There is a simple experiment you can do to illustrate an exothermic and an endothermic process. Hold a large rubber band (the system) against your upper lip (the surroundings). Keeping the band on your lip, quickly stretch it to its limit. You will feel the temperature of the band rise noticeably; stretching is exothermic. On relaxing the rubber band, you will notice that its temperature drops; contraction is endothermic.

(a)

(b)

FIGURE 5.3

A system absorbing heat at constant pressure or constant volume. (a) When heat is absorbed, the solid CO_2 at $-78°C$ is transformed into a gas that then warms to room temperature and fills and expands the balloon. The pressure of the system is constant, so q_p = heat absorbed at constant pressure = ΔH. (b) When heat is absorbed the solid CO_2 at $-78°C$ is transformed into a gas that then warms to room temperature and fills the sealed container. The pressure in the container increases. However, since there is no "mechanical" connection with the surroundings, no work is done by the system on its surroundings. Therefore, q_v = heat absorbed at constant volume = ΔE. (Charles D. Winters)

"delta E"), which stands for "change in internal energy." Step (c) is associated with work on the surroundings; this is called "w." Since energy must be conserved, the heat energy put into the system must be equal to the total energy required for the first two processes (ΔE) *plus* the work done by the system (w) in the third process. This idea can be expressed by the equation

$$q = \Delta E + w$$

an equation that is often written in the following fashion.

$$\Delta E = q - w \tag{5.7}$$

This is a statement of the **first law of thermodynamics**, or the law of the conservation of energy. All the energy flowing into the system must be used to increase the internal energy of the system and, if possible, to allow the system to do some work on its surroundings.

The system just described did work on the surroundings because the CO_2 was allowed to expand; the pressure within the system remained constant and the same as the atmosphere. Therefore, heat was added to the system at a constant pressure. To remind ourselves of this we write $q_p = \Delta E + w$, where q_p means "heat added at constant pressure." The alternative to this situation is that where CO_2 is held in a rigid container (Figure 5.3b); the gas is not allowed to expand into the surroundings, and its pressure is no longer constant. The volume of the system, however, is constant. No work can be done under these circumstances, so all the heat added at constant volume (q_v) must be used to increase the internal energy of the system ($q_v = \Delta E$ since $w = 0$).

In plants and animals, as well as in the laboratory, reactions usually occur at constant pressure, so the proper form of the first law of thermodynamics under these conditions is $q_p = \Delta E + w$. The heat added to (or lost by) a system at constant pressure is given the special name **enthalpy** change, symbolized by $\boldsymbol{\Delta H}$ ("delta H"). Thus,

$$\Delta H = q_p = \text{heat added to or lost by a system at constant pressure}$$
$$= \Delta E + w$$

and

In equation 5.7 we follow the convention that w is positive when work is done *by* the system. Alternatively, the 1st law is often written $\Delta E = q + w$, where w is defined as positive when work is done *on* the system. These two forms of the 1st law are equivalent.

The word "enthalpy" is pronounced "en-THAL-pee."

$$\Delta H = \text{(heat content of the system at the end of a process)} - \text{(heat content of the system at the start of the process)}$$
$$= H_{\text{final}} - H_{\text{initial}} \quad (5.8)$$

Positive enthalpy change $+\Delta H$ = endothermic

Negative enthalpy change $-\Delta H$ = exothermic

From these equations, you can see that, if the heat content of the final system is greater than that of the initial system (the case for solid CO_2 to gaseous CO_2), the heat content has increased (Figure 5.4), and the process has a positive ΔH; the process is said to be **endo**thermic. Conversely, if the heat content of the final system is less than that of the initial system, heat has flowed out of the system, and ΔH is negative; the process is **exo**thermic.

Enthalpy changes can be determined for any chemical process. For any chemical reaction, the products represent the "final system" and the reactants the "initial system," so Equation 5.8 can be rewritten as Equation 5.9.

$$\Delta H = H_{\text{products}} - H_{\text{reactants}} \quad (5.9)$$

The heat put into the water to decompose it to the elements is "stored" in H_2 and O_2 as chemical potential energy that can be released by formation of water.

For example, the left side of Figure 5.5 shows that the heat content of the products of decomposition of water is greater than that of the reactants, so there must be a net heat input of 242 kJ *per mole of water decomposed*; that is, the enthalpy change for

$$H_2O(g) + 242 \text{ kJ} \longrightarrow \tfrac{1}{2} O_2(g) + H_2(g)$$

In thermochemistry, the value of ΔH is for the case when the coefficients of the balanced equation are read in moles.

the *endothermic* decomposition of water is $\Delta H = +242$ kJ/mol. If there were two moles of H_2O, however, twice as much heat energy or 484 kJ would be required for complete reaction. *The change in energy or enthalpy for a reaction is directly proportional to the quantity of material present.*

EXERCISE 5.4 HEAT ENERGY CALCULATION

How much heat energy would be required to decompose 12.5 g of H_2O vapor to the elements?

The decomposition of water is a reversible process. Water can be decomposed to its elements, or the elements can combine to form water. The amount of heat energy involved in the combination reaction

$$H_2(g) + \tfrac{1}{2} O_2(g) \longrightarrow H_2O(g) + 242 \text{ kJ}$$
$$\Delta H = -242 \text{ kJ}$$

H_2 combustion fuels the Space Shuttle and is considered seriously as fuel for cars and homes, because so much energy is provided per gram. (See Chapter 20.)

is the same as for the decomposition reaction except that the combination reaction is *exothermic*. That is, $\Delta H = -242$ kJ per mole of water formed. The fact that the reaction for the formation of water gives off exactly the

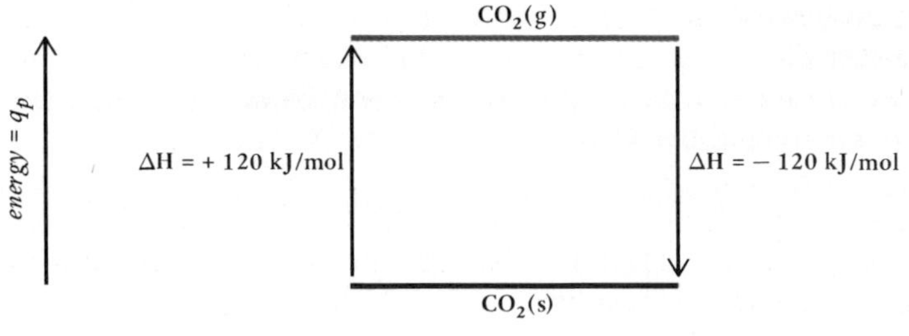

FIGURE 5.4

Potential energy diagram for the interconversion of $CO_2(s)$ and $CO_2(g)$.

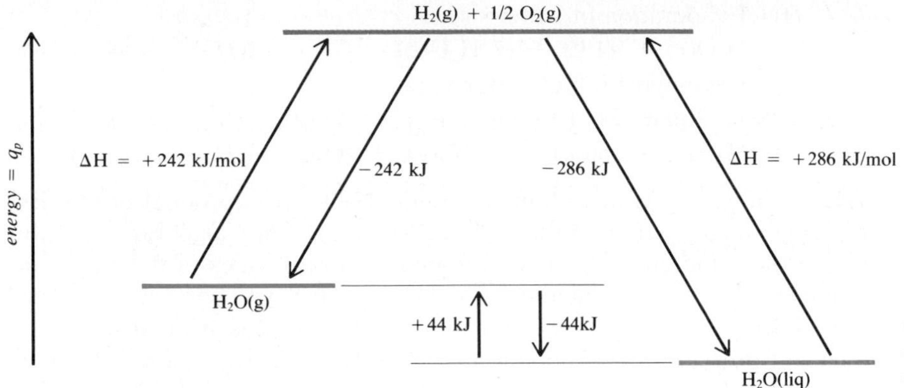

FIGURE 5.5
Potential energy diagram for the interconversion of $H_2O(g)$, $H_2O(\ell)$, and its elements.

same amount of energy as needed for the decomposition of water is mandated by the principle of the conservation of energy.

The amount of energy obtained or needed in a chemical reaction depends on the physical state (solid, liquid, or gas) in which the reactants and products are found. You have seen that the transformation of solid CO_2 to gaseous CO_2 requires an expenditure of energy. Therefore, it should come as no surprise that the decomposition of *liquid* water to H_2 and O_2

$$H_2O(\ell) + 286\ \text{kJ} \longrightarrow H_2(g) + \tfrac{1}{2}\ O_2(g)$$

$$\Delta H = +286\ \text{kJ}$$

requires a different amount of energy than the decomposition of water vapor. The energy relationship between water in the liquid or vapor state and its elements is shown in Figure 5.5. This should make it clear that energy is required to convert liquid water into water vapor, and it is this fact that leads to the requirement for a greater amount of energy to decompose liquid water to its elements than to decompose water vapor. It is also for this reason that your skin cools when you leave a swimming pool on a hot day; the liquid water (the system) absorbs heat from your skin (the surroundings) as the water evaporates, and your skin temperature drops.

5.5 HESS'S LAW

One consequence of the first law of thermodynamics is **Hess's law of constant heat of summation**. Just as chemical reactions adhere to the law of mass conservation, so must they obey the law of energy conservation. This is the basis of Hess's law that states, if a reaction is the sum of two or more other reactions, ΔH for the overall process must be the sum of the ΔH's of the constituent reactions. As an example, the decomposition of *liquid* water into its elements $H_2(g)$ and $O_2(g)$ can be thought of as the sum of two other processes: (1) the transformation of liquid water to water vapor and (2) the decomposition of water vapor to the elements.

(1) Evaporation of water.
$H_2O(\ell) + 44\ kJ \longrightarrow H_2O(g) \qquad \Delta H = +44\ kJ$

(2) Decomposition of water vapor.
$H_2O(g) + 242\ kJ \longrightarrow H_2(g) + \frac{1}{2}\,O_2(g) \qquad \Delta H = +242\ kJ$

(1) + (2) $H_2O(\ell) + 286\ kJ \longrightarrow H_2(g) + \frac{1}{2}\,O_2(g) \qquad \Delta H = +286\ kJ$

When adding the two reactions together, $H_2O(g)$ is a product of the first reaction and a reactant in the second. Thus, as in any algebraic equation where the same quantity or term appears on both sides of the equation, you can cancel out the quantity "$H_2O(g)$." Also, you can see that there are 44 kJ needed to vaporize a mole of water (this heat absorbed at constant pressure, q_p, is commonly called the **enthalpy** or **heat of vaporization**), and another 242 kJ are needed to decompose a mole of water vapor. Therefore, 286 kJ are needed in all, so the overall enthalpy change is +286 kJ per mole of water (Figure 5.5).

It is often important to know the quantity of heat energy involved in a chemical process, so knowledge of ΔH values is useful. Such values can be determined experimentally for many reactions, but it is not a simple task; besides, it would be impossible to measure values for every conceivable reaction. Fortunately, Hess's law allows us to take ΔH values for simple reactions, which have been determined experimentally, and combine those values to obtain enthalpy values for new or more complex reactions.

EXAMPLE 5.4

HESS'S LAW: ENTHALPY CHANGE FOR THE FORMATION OF A COMPOUND

Suppose you wish to know the enthalpy change for the formation of $PCl_5(s)$ from its elements.

$$\frac{1}{4}\,P_4(s) + \frac{5}{2}\,Cl_2(g) \longrightarrow PCl_5(s) \qquad \Delta H = ?$$

This value is difficult to obtain experimentally, but ΔH values can be determined from the following reactions.

$$\frac{1}{4}\,P_4(s) + \frac{3}{2}\,Cl_2(g) \longrightarrow PCl_3(\ell) \qquad \Delta H_1 = -320\ kJ$$

$$PCl_3(\ell) + Cl_2(g) \longrightarrow PCl_5(s) \qquad \Delta H_2 = -124\ kJ$$

Solution The solution to this problem is to add together, just as they are written above, the equation for the formation of $PCl_3(\ell)$ from the elements and the equation for the further reaction of $PCl_3(\ell)$ to give $PCl_5(s)$. The compound PCl_3 is a product of the first reaction and a reactant in the second. Therefore, PCl_3 "cancels out" when the equations are added, and the final equation obtained is the one desired. Since the final equation can be obtained by the simple addition of the two equations involving PCl_3, ΔH for the formation of PCl_5 from its elements is simply the sum of ΔH_1 and ΔH_2. That is,

$$\Delta H(\text{formation of } PCl_5 \text{ from } Cl_2 \text{ and } P_4) = \Delta H_1 + \Delta H_2 = -444\ kJ$$

EXAMPLE 5.5

HESS'S LAW IN A COMBUSTION REACTION

Calculate the quantity of heat given off when benzene, C_6H_6, burns in oxygen to give water and carbon dioxide.

$$C_6H_6(\ell) + \tfrac{15}{2}\ O_2(g) \longrightarrow 6\ CO_2(g) + 3\ H_2O(\ell)$$

This is a combustion reaction (Chapter 3), and the heat evolved in such a process is called the **enthalpy of combustion** (or often simply the heat of combustion).

The following information can be found in reference books (or in Table 5.2 or Appendix L in this book):

1.	$6\ C(\text{graphite}) + 3\ H_2(g) \longrightarrow C_6H_6(\ell)$	$\Delta H_1 = +49.0$ kJ
2.	$C(\text{graphite}) + O_2(g) \longrightarrow CO_2(g)$	$\Delta H_2 = -393.5$ kJ
3.	$H_2(g) + \tfrac{1}{2}\ O_2(g) \longrightarrow H_2O(\ell)$	$\Delta H_3 = -285.8$ kJ

Solution You cannot just add together Equations 1 through 3 above, as they are now written, and obtain the desired reaction. C_6H_6 should be a reactant, but Equation 1 is for its formation. Therefore, you will have to reverse the first equation *and* reverse the sign of ΔH_1: If the formation of C_6H_6 is endothermic, its decomposition must be exothermic.

1′. $C_6H_6(\ell) \longrightarrow 6\ C(\text{graphite}) + 3\ H_2(g) \qquad \Delta H_1' = -49.0$ kJ

Both Equations 2 and 3 are in the correct direction as written: CO_2 and H_2O are products as required in the combustion reaction. However, each has only 1 mole of product (CO_2 and H_2O), whereas 6 moles of CO_2 and 3 moles of H_2O are needed as products. To conserve mass, you must multiply Equations 2 and 3 by 6 and 3, respectively, and their ΔH values must be multiplied by the same factors.

2′. $6\ C(\text{graphite}) + 6\ O_2(g) \longrightarrow 6\ CO_2(g) \qquad \Delta H_2' = 6 \times \Delta H_2 = -2361.0$ kJ

3′. $3\ H_2(g) + \tfrac{3}{2}\ O_2(g) \longrightarrow 3\ H_2O(\ell) \qquad \Delta H_3' = 3 \times \Delta H_3 = -857.4$ kJ

It is now possible to add the three equations above (1′ + 2′ + 3′) to obtain the desired equation for benzene combustion. Notice that 6 moles of C(graphite) appear in Equation 1′ as product and in Equation 2′ as reactant; therefore, C(graphite) cancels out. The number of moles of O_2 required, $\tfrac{15}{2}$, is achieved by adding 6 moles from Equation 2′ and $\tfrac{3}{2}$ moles from Equation 3′.

The three equations will now add correctly, and you can simply add the $\Delta H'$ values for the three reactions.

$$\Delta H_{\text{total}} \text{ for } C_6H_6 \text{ combustion} = \Delta H_1' + \Delta H_2' + \Delta H_3' = -3267.4 \text{ kJ}$$

EXERCISE 5.5 USING HESS'S LAW

Calculate the heat evolved in the combustion of ethyl alcohol. That is, calculate its enthalpy of combustion.

$$C_2H_5OH(\ell) + 3\ O_2(g) \longrightarrow 2\ CO_2(g) + 3\ H_2O(\ell) \qquad \Delta H = ?$$

Use information from Example 5.5 and the following equation:

$$2\ C(\text{graphite}) + 3\ H_2(g) + \tfrac{1}{2}\ O_2(g) \longrightarrow C_2H_5OH(\ell) \qquad \Delta H = -277.7 \text{ kJ}$$

5.6 STATE FUNCTIONS

The object of Example 5.4 was to find the the enthalpy of the reaction when elemental phosphorus reacted with chlorine to give PCl_5. In principle, this reaction can be done in two ways: (1) direct combination of P_4 and Cl_2 to give PCl_5 or (2), in stages, where PCl_3 is an intermediate along the way (Figure 5.6). The enthalpy change could be measured for both pathways, and each pathway would evolve the *same* quantity of heat. This is always true of any reaction. *No matter how you go from starting materials to final products, the net amount of heat evolved or required will always be the same. The enthalpy change does not depend on the path you choose to follow*. Thus, the enthalpy change is called a **state function**, a quantity whose changed value is determined *only* by its initial and final values.

There are many commonly measured quantities that are state functions; among them are potential and kinetic energy, pressure, volume, temperature, and the size of your bank account. You could have arrived at your current bank balance, say $25, by having simply deposited $25. Alternatively, you could have deposited $100 and then withdrawn $75. The volume of a balloon is a state function. You can blow up a balloon to a large volume and then let some air out to arrive at the desired volume. As an alternative, you could blow it up in stages, adding tiny amounts of air to arrive at the same volume as before. The final volume does not depend in any way on how you got there. In both of these examples, there are an infinite number of ways to arrive at the final state, but the final value depends only on the size of the balance or the balloon and not on the path taken from the initial to the final state.

Since enthalpy is a state function, in principle one can specify the enthalpy of the reactants ($H_{\text{initial}} = H_{\text{reactants}}$) and that of the product ($H_{\text{final system}} = H_{\text{product}}$). The difference between these enthalpies is the change for the system.

$$\Delta H_{\text{reaction}} = H_{\text{final}} - H_{\text{initial}} = H_{\text{product}} - H_{\text{reactants}}$$

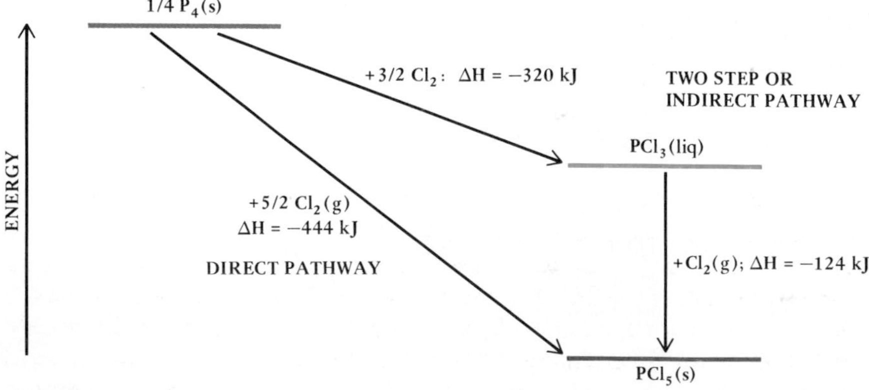

FIGURE 5.6
Two of the possible pathways from elemental phosphorus and gaseous chlorine to $PCl_5(s)$.

Since the reaction starts and finishes at the same place no matter which pathway is chosen, $\Delta H_{reaction}$ must always be independent of pathway. However (and this is a big "however"), unlike volume, temperature, pressure, energy, or a bank balance, it is not possible to determine the absolute enthalpy of a compound. *Only the change in enthalpy in a reaction can be determined experimentally.* The heat evolved or required in a chemical process is only a reflection of the *difference* in heat content of the reactants and products. One can only determine that the heat content of the products is greater or less than that of the reactants by a certain amount, and the potential energy diagram for the decomposition of water to its elements (Figure 5.5) illustrates this idea.

5.7 STANDARD CONDITIONS AND STANDARD ENTHALPIES OF FORMATION

To define precisely the conditions of an experiment, a chemist often states them in terms of **standard conditions**: a pressure of 1 atmosphere and a given temperature, most often 25°C (298 K). The physical state in which an element or compound is found with these conditions is called its **standard state.** Thus, the standard state for H_2 is the gaseous state, whereas that for NaCl is the solid state. For an element such as carbon that can exist in either of two solid states at 1 atm pressure, graphite rather than diamond has been chosen as the standard state.

When reactions occur with all the reactants and products in their standard states, we say that the heat evolved or required is the **standard enthalpy of reaction, $\Delta H°$**, where the superscript ° indicates standard conditions. All the reactions we have discussed to this point have met these conditions, so all the ΔH values should have ° attached.

It is common to use the term "heat of reaction" interchangeably with "enthalpy of reaction." Understand that it is only the heat at constant pressure that is equivalent to the enthalpy change.

The heat or enthalpy of a reaction for the formation of one mole of a compound from its elements, with everything at standard conditions, is called the **molar enthalpy of formation**, ΔH_f°, the subscript $_f$ indicating a heat or enthalpy of formation. Some of the reactions already discussed define standard heats of formation,

$$H_2(g) + \tfrac{1}{2}\,O_2(g) \longrightarrow H_2O(\ell) \qquad \Delta H_f^\circ = -285.8 \text{ kJ/mol}$$

$$\tfrac{1}{4}\,P_4(s) + \tfrac{3}{2}\,Cl_2(g) \longrightarrow PCl_3(\ell) \qquad \Delta H_f^\circ = -320 \text{ kJ/mol}$$

and Table 5.2 and Appendix L list those for many other compounds. Notice there are no heats of formation for *elements* such as C(graphite) or O_2 in these tables. *Standard enthalpies of formation for the elements are **defined** as zero* at 298 K. The reason for this is that we must have a point of reference for our scale of enthalpies. As mentioned previously, only the *change* in enthalpy in a reaction relative to some starting point can be determined. Just as time and latitude or longitude are measured relative to some reference point on the globe, so are enthalpies of reaction measured with the elements at standard conditions as the reference point.

TABLE 5.2 Selected Standard Molar Enthalpies of Formation at 298 K

SUBSTANCE		STANDARD MOLAR HEAT OF FORMATION (kJ/mol)
$Al_2O_3(s)$	Aluminum oxide	−1675.7
$BaCO_3(s)$	Barium carbonate	−1216.3
$CaCO_3(s)$	Calcium carbonate	−1206.9
$CaO(s)$	Calcium oxide	−635.1
$CCl_4(\ell)$	Carbon tetrachloride	−135.4
$CH_4(g)$	Methane	−74.81
$C_2H_5OH(\ell)$	Ethyl alcohol	−277.7
$CO(g)$	Carbon monoxide	−110.5
$CO_2(g)$	Carbon dioxide	−393.5
$C_2H_2(g)$	Acetylene	+226.7
$C_2H_4(g)$	Ethylene	+52.3
$C_2H_6(g)$	Ethane	−84.7
$C_3H_8(g)$	Propane	−103.8
n-$C_4H_{10}(g)$	Butane	−888.0
$CuSO_4(s)$	Copper(II) sulfate	−771.4
$H_2O(g)$	Water vapor	−241.8
$H_2O(\ell)$	Liquid water	−285.8
$HF(g)$	Hydrogen fluoride	−271.1
$HCl(g)$	Hydrogen chloride	−92.3
$HBr(g)$	Hydrogen bromide	−36.4
$HI(g)$	Hydrogen iodide	+26.48
$KF(s)$	Potassium fluoride	−562.3
$KCl(s)$	Potassium chloride	−436.7
$KBr(s)$	Potassium bromide	−380.7
$MgO(s)$	Magnesium oxide	−601.7
$MgSO_4(s)$	Magnesium sulfate	−1170.6
$Mg(OH)_2(s)$	Magnesium hydroxide	−924.5
$NaF(s)$	Sodium fluoride	−573.6
$NaCl(s)$	Sodium chloride	−411.1
$NaBr(s)$	Sodium bromide	−361.1
$NaI(s)$	Sodium iodide	−287.8
$NH_3(g)$	Ammonia	−46.11
$NO(g)$	Nitrogen oxide	+90.3
$NO_2(g)$	Nitrogen dioxide	+33.2
$PCl_3(\ell)$	Phosphorus trichloride	−319.7
$PCl_5(s)$	Phosphorus pentachloride	−443.5
$SiO_2(s)$	Silicon dioxide (quartz)	−910.9
$SnCl_2(s)$	Tin(II) chloride	−325.1
$SnCl_4(\ell)$	Tin(IV) chloride	−511.3
$SO_2(g)$	Sulfur dioxide	−296.8
$SO_3(g)$	Sulfur trioxide	−395.7

EXAMPLE 5.6

WRITING EQUATIONS TO DEFINE ENTHALPIES OF FORMATION

The standard molar enthalpy of formation of gaseous ammonia is −46.11 kJ/mol. Write a balanced equation for which the enthalpy of reaction is −46.11 kJ.

Solution The equation you wish to write must show the formation of 1 mole of $NH_3(g)$ from the elements in their standard states; both N_2 and H_2 are gases at 25°C and 1 atmosphere pressure. Therefore, the correct equation is

$$\tfrac{1}{2}\,N_2(g) + \tfrac{3}{2}\,H_2(g) \longrightarrow NH_3(g) \qquad \Delta H^\circ_{rxn} = \Delta H^\circ_f = -46.11\ \text{kJ}$$

(where the subscript "rxn" is an abbreviation for "reaction"). Another way to write this equation would have been to include the heat evolved as a product. That is,

$$\tfrac{1}{2}\,N_2(g) + \tfrac{3}{2}\,H_2(g) \longrightarrow NH_3(g) + 46.11\ \text{kJ}$$

EXERCISE 5.6 WRITING EQUATIONS TO DEFINE ENTHALPIES OF FORMATION

(a) The molar enthalpy of formation of AgCl(s) is −127.1 kJ/mol. Write a balanced equation for which the enthalpy of reaction is −127.1 kJ. (b) The molar enthalpy of formation of methyl alcohol, $CH_3OH(\ell)$, is −238.7 kJ/mol. Write a balanced equation for which the heat of reaction is −238.7 kJ.

The standard heats of formation in Table 5.2 and Appendix L are just a few of the many that have been determined by various direct and indirect methods. Much of this work has been done by the U.S. National Bureau of Standards, and the values are collected in extensive tables published by the Bureau. Example 5.7 further illustrates the use of these enthalpies of formation and of Hess's law to find the enthalpy of a reaction.

EXAMPLE 5.7

USING ENTHALPIES OF FORMATION

Calculate the heat required to decompose limestone (calcium carbonate) to lime (calcium oxide) and carbon dioxide, with all substances at standard conditions.

$$CaCO_3(s) \longrightarrow CaO(s) + CO_2(g)$$

The following information is given:

$$\begin{aligned}
\Delta H^\circ_f\,[CaCO_3(s)] &= -1206.9\ \text{kJ/mol}\\
\Delta H^\circ_f\,[CaO(s)] &= -635.1\ \text{kJ/mol}\\
\Delta H^\circ_f\,[CO_2(g)] &= -393.5\ \text{kJ/mol}
\end{aligned}$$

Solution First, write the equations that correspond to the enthalpies of formation. It should then be more clear how to add them together to obtain the desired equation.

$$\begin{aligned}
&Ca(s) + C(\text{graphite}) + \tfrac{3}{2}\,O_2(g) \longrightarrow CaCO_3(s) && \Delta H^\circ_f = -1206.9\ \text{kJ}\\
&Ca(s) + \tfrac{1}{2}\,O_2(g) \longrightarrow CaO(s) && \Delta H^\circ_f = -635.1\ \text{kJ}\\
&C(\text{graphite}) + O_2(g) \longrightarrow CO_2(g) && \Delta H^\circ_f = -393.5\ \text{kJ}
\end{aligned}$$

In the first reaction, $CaCO_3(s)$ is the product, but you need to have it on the left side of the final equation as a reactant. Therefore, this equation must be reversed *and* the sign of ΔH° must also be reversed from negative to positive. On the other hand, CaO(s) and $CO_2(g)$ must appear as products in the final equation, so the second and third equations must have the same direction and sign of ΔH° given above.

$$\begin{aligned}
&CaCO_3(s) \longrightarrow Ca(s) + C(\text{graphite}) + \tfrac{3}{2}\,O_2(g) && \Delta H^\circ = +1206.9\ \text{kJ}\\
&Ca(s) + \tfrac{1}{2}\,O_2(g) \longrightarrow CaO(s) && \Delta H^\circ_f = -635.1\ \text{kJ}\\
&\underline{C(\text{graphite}) + O_2(g) \longrightarrow CO_2(g)} && \underline{\Delta H^\circ_f = -393.5\ \text{kJ}}\\
&CaCO_3(s) \longrightarrow CaO(s) + CO_2(g) && \Delta H^\circ = +178.3\ \text{kJ}
\end{aligned}$$

After the equation for the formation of $CaCO_3$ has been reversed, the other two equations can be added to it as they are; O_2, Ca, and C(graphite) cancel out, and the desired equation is obtained. The decomposition of limestone, $CaCO_3$, is endothermic; that is, 178.3 kJ are required per mole of the carbonate decomposed.

You can find the heat or enthalpy of any reaction if you can find a set of reactions that when added together will give the desired reaction and if the enthalpies of the reactions to be added are known. However, there is one other useful conclusion from Examples 5.5 and 5.7. You may have noticed that the mathematics of the problem really amount to the expression

$$\Delta H^\circ_{rxn} = \Sigma\, a\, \Delta H^\circ_f\,(\text{products}) - \Sigma\, b\, \Delta H^\circ_f\,(\text{reactants}) \qquad (5.10)$$

where Σ (the Greek letter *sigma*) means to "sum up" and *a* and *b* are the moles involved in the balanced equation. This expression simply tells you to add up the heats of formation of the products (multiplying each by its stoichiometric coefficient) and subtract from this sum the sum of the heats of formation of the reactants (each having been multiplied by its coefficient). Whenever you have the heats of formation of each compound in a reaction, the heat of the reaction can be calculated from Equation 5.10. However, recognize that this is just a *shortcut* around writing down the actual reactions for compound formation.

EXAMPLE 5.8

USING ENTHALPIES OF FORMATION

Calculate the enthalpy of combustion of benzene (Example 5.5) using Equation 5.10. The balanced equation is

$$C_6H_6(\ell) + \tfrac{15}{2}\, O_2(g) \longrightarrow 6\, CO_2(g) + 3\, H_2O(\ell)$$

and the data available to you are

$$\Delta H^\circ_f\,[C_6H_6(\ell)] = +49.0 \text{ kJ/mol}$$

$$\Delta H^\circ_f\,[CO_2(g)] = -393.5 \text{ kJ/mol}$$

$$\Delta H^\circ_f\,[H_2O(\ell)] = -285.8 \text{ kJ/mol}$$

$\Delta H^\circ_f\,[O_2(g)] = 0$ (Recall that the heats of formation of the elements in their standard states are zero by definition.)

Solution

$$\Delta H^\circ_{rxn} = \{6\,\Delta H^\circ_f[CO_2(g)] + 3\,\Delta H^\circ_f[H_2O(\ell)]\} - \{\Delta H^\circ_f[C_6H_6(\ell)] + \frac{15}{2}\,\Delta H^\circ_f[O_2(g)]\}$$

$$\Delta H^\circ_{rxn} = \{6 \text{ mol}(-393.5 \text{ kJ/mol}) + 3 \text{ mol}(-285.8 \text{ kJ/mol})\}$$

$$- \{1 \text{ mol}(+49.0 \text{ kJ/mol}) + \frac{15}{2} \text{ mol}\,(0 \text{ kJ/mol})\}$$

$$\Delta H^\circ_{rxn} = -3267.4 \text{ kJ}$$

EXERCISE 5.7 USING ENTHALPIES OF FORMATION

Propane, C_3H_8, is widely used as a fuel. Calculate the heat given off when 1.00 mole of propane burns in air (that is, reacts with O_2) to give the usual products of combustion. (This reaction heat is often called the **enthalpy** or **heat of combustion**.) Appropriate enthalpies of formation can be obtained from Table 5.2. Assume water is produced as a liquid, $H_2O(\ell)$.

5.8 MEASURING HEATS OF REACTION: CALORIMETRY

The heat of a reaction can be measured in many ways, but perhaps the easiest to describe at this point is the **combustion calorimeter** (Figure 5.7). The sample, a solid or liquid that is combustible, is placed in the sample dish that is encased in the "bomb." (The bomb is a cylinder about the size of a large fruit juice can with heavy steel walls and ends.) The bomb is then placed in a container filled with water, the walls of this container being well insulated. After filling the bomb with an atmosphere of pure O_2, the mixture of O_2 and sample is ignited, usually by an electrical spark. If the experiment has been set up properly, the sample burns completely in the pure oxygen atmosphere to give the products of combustion. The heat so generated warms up the bomb and the water around it. In this configuration, the oxygen and the compound represent the *system* and the bomb and water around it are the *surroundings*. From the principle of the conservation of energy and the fact that the apparatus is a rigid container (so no work is done by the system), one can write the expression

Heat flowing from the system = heat flowing into the surroundings

Heat evolved by the reaction = heat absorbed by the surroundings

$$-q_{\text{reaction}} = q_{\text{bomb}} + q_{\text{water}}$$

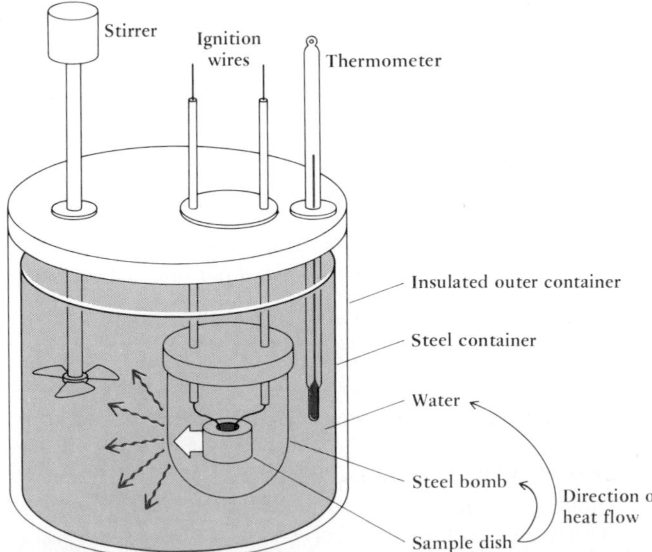

FIGURE 5.7

A combustion calorimeter. A combustible sample is burned in pure O_2 in a steel "bomb." The heat generated by the reaction flows to the bomb *and* to the water surrounding the bomb and warms both to the same temperature. By measuring the temperature increase, the heat evolved by the reaction can be determined.

If you measure the temperature change of the water (which also tells you the temperature change of the bomb), and if you know the heat capacities of the bomb and the water, you can calculate the heat given off by the reaction.

The heat flowing to the bomb from the reaction is calculated from the heat capacity of the bomb (C_{bomb}, units of joules/K) and the temperature change, T.

$$q_{bomb} = C_{bomb} \times \Delta T$$

The heat from the reaction taken up by the water can be calculated from the specific heat of the water, its mass, and the temperature change.

$$q_{water} = (4.184 \text{ J/g} \cdot \text{K})(m_{water})(\Delta T)$$

EXAMPLE 5.9

DETERMINING THE HEAT OF A REACTION USING A CALORIMETER

Octane, C_8H_{18}, a primary constituent of gasoline, burns in air to give the usual products of combustion.

$$C_8H_{18}(\ell) + \frac{25}{2} O_2(g) \longrightarrow 8\ CO_2(g) + 9\ H_2O(\ell)$$

A 1.00-g sample of octane is burned in a calorimeter that contains 1.20 kg of water. The temperature of the water and the bomb rises from 25.00°C to 33.20°C. If the heat capacity of the bomb, C_{bomb}, is 837 J/K calculate (a) the heat given off in the combustion of the 1.00-g sample of C_8H_{18} and (b) the heat given off per mole of C_8H_{18}.

Solution The heat flowing from the reaction into the bomb is calculated from

$$\begin{aligned} q_{bomb} &= (C_{bomb})(\Delta T) = (837 \text{ J/K})(33.20°\text{C} - 25.00°\text{C}) \\ &= +6.86 \times 10^3 \text{ J} \end{aligned}$$

The heat flowing from the reaction into the water of the calorimeter is

$$\begin{aligned} q_{water} &= (4.184 \text{ J/g} \cdot \text{K})(m_{water})(\Delta T) \\ &= (4.184 \text{ J/g} \cdot \text{K})(1.20 \times 10^3 \text{ g})(33.2°\text{C} - 25.0°\text{C}) \\ &= +41.2 \times 10^3 \text{ J} \end{aligned}$$

The total heat *given off* by the reaction is equal to the sum of the heats flowing into the bomb and the water, *with the sign reversed*. Thus,

$$\text{Total heat given off by 1.00 g of octane} = -48.1 \times 10^3 \text{ J} = -48.1 \text{ kJ}$$

The calculation shows that 48.1 kJ are evolved per gram of octane burned. The molar mass of octane is 114 g. Therefore, *the heat evolved per mole of octane* is

$$q_{reaction} (\text{kJ/mol}) = (-48.1 \text{ kJ/g})(114 \text{ g/mol}) = -5.48 \times 10^3 \text{ kJ/mol}$$

EXERCISE 5.8 HEAT OF COMBUSTION

A 1.00-g sample of ordinary table sugar (sucrose, $C_{12}H_{22}O_{11}$) is burned in a combustion calorimeter. The temperature of the 1.50×10^3 g of water in the calorimeter rises from 25.00°C to 27.32°C. If the heat capacity of the bomb is 837 J/°C and the specific heat of water is 4.184 J/g · K, calculate (a) the heat evolved per gram of sucrose and (b) the heat evolved per mole of sucrose.

The combustion calorimeter method can give the heat evolved per mole of combustible material. However, you should recognize that the heat that was measured is that evolved at constant volume, q_v. Earlier in this chapter, this heat was designated as being equivalent to ΔE, the change in internal energy. Recall that ΔH, the change in enthalpy, is the heat evolved or required at constant pressure. Fortunately, ΔE and ΔH are related in a relatively simple way, so ΔH can be obtained from the calorimeter experiment. To be able to do this now is not important for you, so we shall skip a discussion of this conversion. However, you should at least be aware of this problem.

SUMMARY

Energy, the capacity to do work and transfer heat, takes many forms. This chapter presents the basic ideas of the science of energy flow in chemical processes, that is, of chemical **thermodynamics**. The focus is on the heat energy involved in chemical processes.

Calories (cal), or the SI unit the **joule** (J) (1 calorie = 4.184 J), are the usual units of energy (Section 5.2). The **specific heat** of a substance is the heat required to increase the temperature of 1.00 g of the substance by 1.00°C (Section 5.3). Therefore, the heat lost or gained by a substance can be calculated from

$$\begin{aligned} q &= \text{heat lost or gained} \\ &= \text{(specific heat)(g of material)(temperature change)} \\ &= (\text{J/g} \cdot \text{K})(\text{g})(\Delta T \text{ in kelvins}) \end{aligned}$$

where ΔT is always defined as final temperature minus initial temperature. Thus, the quantity of heat has a negative sign if the temperature has declined. Conversely, the sign is positive if the temperature has increased. The **molar heat capacity** of a substance is the product of its specific heat and molar mass; thus, its units are J/mol · K.

One can only observe the *transfer* of heat from one place to another. In the language of thermodynamics, it is transferred between a **system** (the part of the universe under study) and its **surroundings** (the rest of the universe)(Section 5.4). The process is said to be **exothermic** if heat flow direction is system → surroundings; it is **endothermic** if the flow is in the opposite sense. Since a law of nature states that, like mass, energy must be conserved, the heat energy (q) flowing into a system can be used to change the internal energy of the system (ΔE) and allow the system to perform work (w) on its surroundings. This statement can be written in equation form as

$$q = \Delta E + w \tag{5.7}$$

and is called the **first law of thermodynamics**. When the heat transfer occurs at constant pressure (q_p), then q_p is defined as the **enthalpy** change, ΔH. If the process is exothermic, then ΔH is negative; if endothermic, ΔH is positive. The value of ΔH is specific to the process and depends on the quantity of material involved.

Hess's law states that, if a reaction is the sum of two or more other reactions, then ΔH for the overall process is the sum of the ΔH's of the constituent reactions (Section 5.5).

ΔH for the process is a **state function**, a quantity whose changed value is determined only by its initial and final values (Section 5.6). The change in the function does not depend on the path followed in the change. Besides enthalpy, potential and kinetic energy, pressure, volume, and temperature are examples of state functions.

When an element or compound is in its most stable form under **standard conditions** (pressure of 1 atmosphere and, usually, a temperature of 25°C), it is said to be in its **standard state** (Section 5.7). The enthalpy change of a reaction when all of the reactants and products are in their standard states is the **standard enthalpy change** (symbolized by $\Delta H°$). Specifically, the enthalpy for the formation of a compound from its elements, everything in their standard states, is called the **enthalpy of formation** (symbolized by ΔH_f°). (ΔH_f° of an element is 0 by definition.) Many such values have been determined and are given in Table 5.2 and Appendix L. Using Hess's law, it is possible to calculate ΔH_f° for a new compound or to determine $\Delta H°$ for a reaction.

The heat given off by a chemical reaction, particularly a combustion, can be determined in a **calorimeter** (Section 5.8).

STUDY QUESTIONS

REVIEW QUESTIONS

1. Name two laws stated in this chapter and provide a statement of each.

2. Define specific heat and molar heat capacity and state the difference between them.

3. Based on your experience, when ice melts to liquid water, is the process exothermic or endothermic? When liquid water freezes to ice at 0°C, is this exothermic or endothermic?

4. What is the first law of thermodynamics? Explain in words and use a mathematical expression.

5. For each of the following, define the system and the surroundings and give the direction of heat flow:
(a) Propane is burning in a bunsen burner in the laboratory.
(b) Water drops, sitting on your skin after a dip in the ocean, evaporate.
(c) Two chemicals are mixed in a flask sitting on a laboratory bench. A reaction occurs and heat is evolved.

6. The temperature of the room in which you are sitting is 25°C. Why is this temperature said to be a state function?

7. Is the distance you travel between your home town and your college or university a state function? Why or why not?

8. Your house is made of wood and glass. Assuming an equal amount of sunshine falls on a wooden wall and a piece of glass, which will warm faster? Explain briefly.

ENERGY UNITS

9. Convert the following energies to joules or Calories as specified.
(a) A 2-inch thick piece of two-layer chocolate cake with frosting provides 1670 kilojoules (1.67×10^6 J). What is this in Calories?
(b) If you are on a diet that calls for eating no more than 1200 Calories per day, how many joules will this be?
(c) A typical breakfast cereal provides 110 Calories per 1 ounce serving. How many joules of energy are thereby provided?

10. To change 1 mole of water from ice to liquid water at 0°C, 1436 calories are required. (a) How many joules are required per mole of water? (b) How many joules are required per gram? (c) If liquid water is changed to ice at 0°C, how much heat is evolved?

11. To melt lead at 327°C requires a heat input of 1224 calories per mole. If you wish to melt 1.00 pound (454 g) of lead, how many joules are required?

12. A 1.00-ton satellite is moving with a velocity of 2.50×10^4 mph. What is its kinetic energy in kJ? (1 T = 2000 lb and 1 mile = 1.61 km)

SPECIFIC HEAT AND HEAT CAPACITY

13. Assume that 15,000 joules of heat are given off when the temperature of 250 g of lemonade drops from 25°C to 10°C. If this heat is used to melt ice, and if 334 joules are required to melt 1.0 gram of ice, how many grams of ice can be melted with 15,000 joules of heat energy?

14. Ethylene glycol, $(CH_2OH)_2$, is often used as an antifreeze in your car. (a) Which requires more heat energy to warm from 25.0°C to 100.°C, pure water or an equal mass of pure ethylene glycol? Calculate the amount of heat energy required for 1.00 kg of each. (b) Assume that the density of water and glycol are both 1.00 g/cm³. If you have a 5.00-quart cooling system in your automobile, compare water and glycol as to the quantity of heat energy (in joules) the liquid in the system can absorb on raising the temperature from 25.0°C to 100.°C (1 qt = 1.10 L).

15. Using the data in Table 5.1, calculate the molar heat capacity of (a) aluminum, (b) iron, and (c) $CCl_4(\ell)$.

16. For each compound below, calculate the specific heat from the molar heat capacity (as listed in tables published by the U.S. National Bureau of Standards).
(a) $C_2H_2(g)$, acetylene (43.93 J/mol · K)
(b) $CF_3CCl_3(g)$, one of a class of compounds called "Freons" (120.5 J/mol · K).
(c) $CO(NH_2)_2(s)$, urea, a fertilizer (93.14 J/mol · K).

ENTHALPY

17. Energy is stored in the body in the form of adenosine triphosphate, ATP. It forms on reaction of adenosine diphosphate, ADP, with phosphoric acid.

$$\text{ADP} + H_3PO_4 + 38\text{ kJ} \longrightarrow \text{ATP} + H_2O$$

Is the reaction endothermic or exothermic?

18. When iron reacts with O_2 to form $Fe_2O_3(s)$, 824 kJ are evolved per mole of the oxide formed. How many kJ of heat energy are evolved when 27.9 g of Fe are reacted with excess O_2 to produce $Fe_2O_3(s)$?

19. "Gasohol," a mixture of gasoline and ethyl alcohol, is a possible automobile fuel. The alcohol produces energy in a combustion reaction with O_2.

$$C_2H_5OH(\ell) + 3\,O_2(g) \longrightarrow 2\,CO_2(g) + 3\,H_2O(\ell)$$

If 0.115 g of alcohol evolves 3.42 kJ when burned at constant pressure, what is the *molar* enthalpy (or heat) of combustion for ethyl alcohol?

20. A laboratory volcano can be made from ammonium dichromate. When ignited, the compound decomposes in a fiery display.

$$(NH_4)_2Cr_2O_7(s) \longrightarrow N_2(g) + 4\,H_2O(g) + Cr_2O_3(s)$$

If the decomposition produces 315 kJ per mole of ammonium dichromate, how much heat energy would be produced by 28.4 g (1 oz) of the solid?

21. On the sideline at every sporting event, there is often a box of "chemical cold packs" in case of sprains. These cold packs usually have two pouches, one containing solid ammonium nitrate and another containing water (often dyed for no particular reason). When the contents of the pouches mix, the ammonium nitrate dissolves in the water,

$$NH_4NO_3(s) \longrightarrow NH_4^+(aq) + NO_3^-(aq)$$

but the dissolving process is very endothermic. The reaction absorbs approximately 326 J/g of ammonium nitrate. To lower the temperature of 220. g of water (the typical amount contained in a "cold pack") by 20.0°C, how many grams of ammonium nitrate would you need to dissolve? (This assumes the ammonium nitrate dissolves completely and that the water is perfectly insulated from its surroundings; no heat flows in from the surroundings.)

22. The thermite reaction, the reaction between aluminum and iron(III) oxide,

$$2\,Al(s) + Fe_2O_3(s) \longrightarrow Al_2O_3(s) + 2\,Fe(s)$$

produces a tremendous amount of heat, 824 kJ per mole of Al_2O_3 produced. If you begin with 3.0 g of Fe_2O_3 and 2.1 g of iron is produced, is the heat produced sufficient to raise the temperature of the iron to its melting point, 1530°C? (Sp. ht. Al_2O_3 = 0.775 J/g · K.)

23. White phosphorus, P_4, ignites in the air to produce heat, light, and P_4O_{10}. The heat evolved is 3010 kJ per mole of P_4O_{10} formed.

$$P_4(s) + 5\,O_2(g) \longrightarrow P_4O_{10}(s)$$

This is an important reaction, since the phosphorus oxide is then treated with water to give phosphoric acid to be consumed in making detergents, toothpaste, soft drinks, and on and on (see Chapter 22). About 500×10^3 tons of elemental phosphorus are made annually in the United States. If you oxidize just 1 ton (9.08×10^5 g) to the oxide, how much heat is evolved (in kJ)?

HESS'S LAW

24. Lead has been known and used since ancient times. To obtain the metal, the ore, PbS (galena), is first heated in air to form PbO,

$$PbS(s) + \tfrac{3}{2}\,O_2(g) \longrightarrow PbO(s) + SO_2(g) \qquad \Delta H = -413.7\text{ kJ}$$

and the oxide is reduced with carbon.

$$PbO(s) + C(s) \longrightarrow Pb(s) + CO(g) \qquad \Delta H = +106.8\text{ kJ}$$

To obtain the lead from 1.00 kg of pure galena, how much heat energy is required or evolved?

25. Three very important reactions to the semiconductor industry are
(a) the reduction of silicon dioxide to crude silicon

$$SiO_2(s) + 2\,C(s) \longrightarrow Si(s) + 2\,CO(g) \qquad \Delta H = +689.9\text{ kJ}$$

(b) the formation of silicon tetrachloride

$$Si(s) + 2\,Cl_2(g) \longrightarrow SiCl_4(g) \qquad \Delta H = -657.0\text{ kJ}$$

(c) the reduction of silicon tetrachloride to pure silicon with magnesium

$$SiCl_4(g) + 2\,Mg(s) \longrightarrow 2\,MgCl_2(s) + Si(s) \qquad \Delta H = -625.6\text{ kJ}$$

What is the overall energy involved in changing 1.00 mole of sand, SiO_2, into very pure silicon?

26. The oxidation of sulfur to its oxides, followed by the absorption of SO_3 into water, produces sulfuric acid. The acid is the most heavily produced chemical in modern industrial countries, the typical plant producing about 750 tons per day (1 T = 9.08×10^5 g). If the following reactions are involved, calculate the amount of heat developed on producing 750. tons of the acid from elemental sulfur.

$$S(s) + O_2(g) \longrightarrow SO_2(g) \qquad \Delta H = -297.0\text{ kJ}$$

$$SO_2(g) + \tfrac{1}{2}\,O_2(g) \longrightarrow SO_3(g) \qquad \Delta H = -9.8\text{ kJ}$$

$$SO_3(g) + H_2O\text{ (in 98\% }H_2SO_4) \longrightarrow H_2SO_4(\ell) \qquad \Delta H = -130.0\text{ kJ}$$

You will observe that the amount of heat produced is enormous, so much so that the heat is used to operate the plant and generate electricity to be sold to local utilities.

27. Using the following reactions, calculate a value for the energy required to break a mole of C—N bonds. That is, calculate ΔH for C—N(g) $\longrightarrow$ C(g) + N(g).

(a) Energy to break C—N bond and produce graphite and N_2 molecules

$$C\text{—}N(g) \longrightarrow C(\text{graphite}) + \tfrac{1}{2}\,N_2(g) \qquad \Delta H = -416\text{ kJ}$$

(b) Energy involved when gaseous C atoms condense to form graphite

$$C(g) \longrightarrow C(\text{graphite}) \qquad \Delta H = -717\text{ kJ}$$

(c) Energy evolved when N atoms form N_2 molecules

$$2\,N(g) \longrightarrow N_2(g) \qquad \Delta H = -946\text{ kJ}$$

STANDARD ENTHALPIES OF FORMATION

28. The molar enthalpy of formation of glucose, $C_6H_{12}O_6$, is −1260 kJ/mol. (a) Is the formation of glucose from its elements endothermic or exothermic? (b) Write a balanced equation depicting the formation of glucose from its elements and for which the enthalpy of reaction is −1260 kJ.

29. For each compound below, write a balanced equation depicting its formation (1 mole) from its elements. Look up the standard molar enthalpy of formation in Table 5.2 or Appendix L.

(a) $Al_2O_3(s)$ (d) $NH_4NO_3(s)$
(b) $TiCl_4(\ell)$ (e) $COCl_2(g)$
(c) $Mg(OH)_2(s)$

30. In photosynthesis, the sun's energy brings about the combination of CO_2 and H_2O to form O_2 and a carbon-containing compound such as a sugar or hydrocarbon. In its simplest form, the reaction would be

$$CO_2(g) + 2\,H_2O(\ell) \longrightarrow 2\,O_2(g) + CH_4(g)$$

Using the enthalpies of formation in Table 5.2, (a) decide if the reaction is exothermic or endothermic and (b) calculate the enthalpy of reaction.

31. Calcium oxide reacts with CO_2 to give calcium carbonate. Using enthalpies of formation in Table 5.2, calculate the heat evolved when 1.00 mole of CaO forms 1.00 mole of $CaCO_3$.

32. An important reaction in the production of sulfuric acid is

$$SO_2(g) + \tfrac{1}{2}\,O_2(g) \longrightarrow SO_3(g)$$

It is also a key reaction in the formation of acid rain, beginning with the air pollutant SO_2. Using the data of Table 5.2, calculate the enthalpy change for the reaction.

33. Refer to Question 24 above. Show how the enthalpies of the two reactions were obtained from $\Delta H°$ data.

34. Acetylene is used as a fuel in the "oxy-acetylene" torch for welding. Calculate the heat energy involved (at constant pressure) when 1.00 g of acetylene, $C_2H_2(g)$, burns in air to give $CO_2(g)$ and $H_2O(g)$. Use the data available in Table 5.2.

35. Using the data in Table 5.2 (and Appendix L if necessary), calculate enthalpies of reaction for the following processes:

(a) A method of obtaining titanium metal

$$TiCl_4(\ell) + 2\,Mg(s) \longrightarrow Ti(s) + 2\,MgCl_2(s)$$

(b) The Romans used CaO as mortar in stone structures. The CaO was mixed with water to give $Ca(OH)_2$, and this slowly reacted with CO_2 in the air to give limestone.

$$Ca(OH)_2(s) + CO_2(g) \longrightarrow CaCO_3(s) + H_2O(g)$$

(c) The first step in the production of nitric acid from ammonia involves the oxidation of NH_3.

$$4\,NH_3(g) + 5\,O_2(g) \longrightarrow 4\,NO(g) + 6\,H_2O(g)$$

36. Refer to Question 25 above. Show how each of the reaction enthalpy changes was obtained from $\Delta H°$ data.

37. Nitroglycerin is a powerful explosive, giving four different gases when it is detonated.

$$2\,C_3H_5(NO_3)_3(\ell) \longrightarrow 3\,N_2(g) + \tfrac{1}{2}\,O_2(g) + 6\,CO_2(g) + 5\,H_2O(g)$$

(a) Given that the enthalpy of formation of nitroglycerin is −364 kJ/mol and the other enthalpies of formation in Table 5.2, calculate the energy (heat) liberated when 1.00 mole of nitroglycerin is detonated.

(b) If 10.0 pounds of nitroglycerin is detonated, how many kilojoules of energy are released?

38. Suppose you measured the enthalpies of the following two reactions in the laboratory.

$$2\ C(graphite) + 2\ H_2(g) \longrightarrow C_2H_4(g) \qquad \Delta H_f^\circ = +52.3\ kJ$$

$$C_2H_4Cl_2(g) \longrightarrow Cl_2(g) + C_2H_4(g) \qquad \Delta H_{rxn}^\circ = +116\ kJ$$

Calculate the standard molar enthalpy of formation of $C_2H_4Cl_2(g)$.

39. Hydrazine and dimethylhydrazine both react spontaneously with O_2 and can be used as rocket fuels.

$$N_2H_4(\ell) + O_2(g) \longrightarrow N_2(g) + 2\ H_2O(g)$$

hydrazine

$$N_2H_2(CH_3)_2(\ell) + 4\ O_2(g) \longrightarrow 2\ CO_2(g) + 4\ H_2O(g) + N_2(g)$$

dimethylhydrazine

The molar enthalpy of formation of liquid hydrazine is +50.6 kJ/mol and that of liquid dimethylhydrazine is +42.0 kJ/mol. By doing appropriate calculations, decide whether hydrazine or dimethylhydrazine gives more heat *per gram* when reacted with O_2. (Other enthalpy of formation data can be obtained from Table 5.2.)

40. The reaction that occurs when a typical fat, glyceral trioleate, is metabolized in the body is

$$C_{57}H_{104}O_6(s) + 80\ O_2(g) \longrightarrow 57\ CO_2(g) + 52\ H_2O(\ell)$$

(a) 37.8 kJ is evolved when 1.00 g of this fat (MW = 884) is metabolized. Using data in Table 5.2, calculate the molar enthalpy of formation of fat in kJ/mol.

(b) How many kilojoules of energy must be evolved in the form of heat if you want to get rid of 1.00 pound (454 g) of this fat by combustion? How many calories is this?

41. The combustion of diborane, B_2H_6, proceeds according to the equation

$$B_2H_6(g) + 3\ O_2(g) \longrightarrow B_2O_3(s) + 3\ H_2O(g)$$

and 1941 kJ of energy is liberated per mole of $B_2H_6(g)$ consumed. Calculate the molar enthalpy of formation of $B_2H_6(g)$ using this information, the data in Table 5.2, and the fact that ΔH_f° for $B_2O_3(s)$ is −1273 kJ/mol.

42. Oxygen difluoride is a colorless, very poisonous gas that reacts rapidly with water vapor to produce O_2, HF, and heat.

$$OF_2(g) + H_2O(g) \longrightarrow O_2(g) + 2\ HF(g) + 318\ kJ$$

Using the data of Table 5.2 and the information above, calculate the molar enthalpy of formation of OF_2.

43. Given the following equations, calculate the molar heat of formation of $MnO_2(s)$.

$$MnO_2(s) \longrightarrow MnO(s) + \tfrac{1}{2}\ O_2(g) \qquad \Delta H = +132\ kJ$$

$$MnO_2(s) + Mn(s) \longrightarrow 2\ MnO(s) \qquad \Delta H = -240\ kJ$$

CALORIMETRY

44. How many kilojoules of heat are evolved by a reaction in a calorimeter where the temperature of the bomb and the water increase from 19.50°C to 22.83°C? The bomb has a heat capacity of 650. J/°C, and the calorimeter contains 320. g of water.

45. Sulfur (2.56 g) was burned in a calorimeter with excess $O_2(g)$. The temperature increased from 21.25°C to 26.72°C. The bomb had a heat capacity of 923 J/°C, and the calorimeter contained 815 g of water. Calculate the heat evolved, per mole of SO_2 formed, in the course of the reaction $S(s) + O_2(g) \rightarrow SO_2(g)$.

GENERAL PROBLEMS

46. Given the data below and the fact that the standard heat of formation of formic acid, HCOOH, is −379 kJ, calculate ΔH_1, the heat involved in the formation of formaldehyde, H_2CO.

$$C(s) + \tfrac{1}{2}\ O_2(g) + H_2(g) \longrightarrow H_2CO(g) \qquad \Delta H_1^\circ = ?$$

$$\tfrac{1}{2}\ O_2(g) + H_2CO(g) \longrightarrow HCOOH(g) \qquad \Delta H_2^\circ = -270\ kJ$$

47. Given the following information, calculate the molar heat or enthalpy of formation of liquid hydrazine, N_2H_4.

$$N_2H_4(\ell) + O_2(g) \longrightarrow N_2(g) + 2\ H_2O(g) \qquad \Delta H^\circ = -534\ kJ$$

$$H_2(g) + \tfrac{1}{2}\ O_2(g) \longrightarrow H_2O(g) \qquad \Delta H^\circ = -242\ kJ$$

48. Some years ago Texas City, Texas, was devastated by the explosion of a shipload of ammonium nitrate, a compound intended to be used as a fertilizer. When heated, ammonium nitrate decomposes, exothermically, to N_2O and water.

1. $NH_4NO_3(s) \longrightarrow N_2O(g) + 2\ H_2O(g)$

If the heat of this exothermic reaction is contained, higher temperatures are generated, at which point ammonium nitrate can decompose explosively to N_2, H_2O, and O_2.

2. $2\ NH_4NO_3(s) \longrightarrow 2\ N_2(g) + 4\ H_2O(g) + O_2(g)$

If oxidizable materials are present, fires can break out, as was the case at Texas City.

Using enthalpies of formation in Appendix L, answer the following questions. (a) If the heat of formation of $N_2O(g)$ is 82.0 kJ/mol, how much heat is evolved at constant pressure by Reaction 1 under standard conditions? (b) If 8.00 kg of ammonium nitrate explode (Reaction 2), how much heat is evolved at constant pressure under standard conditions?

49. During World War II the Germans added powdered aluminum (from ground-up airplanes) to ammonium nitrate to give powerful bombs. The oxygen released in the decomposition of NH_4NO_3 (Question 48 and chapter-opening photograph for Chapter 13) combined with the aluminum to give Al_2O_3 in an exothermic process.

$$2\ Al(s) + 3\ NH_4NO_3(s) \longrightarrow 3\ N_2(g) + 6\ H_2O(g) + Al_2O_3(s)$$

If you mixed 8.00 kg of ammonium nitrate with a stoichiometric excess of powdered aluminum, how much heat is evolved?

50. A piece of iron rod (400. g) is heated in a bunsen flame and is then plunged into an insulated beaker containing 1.00 L of water (1000 g). The original temperature of the water was 20.0°C. After the iron rod and the water have reached the same temperature, it is 32.8°C. What was the original temperature of the iron rod?

51. A 10.0 g piece of aluminum is heated to 300.°C in a bunsen flame and then dropped into 250. mL of water (density = 1.00 g/cm³). The aluminum transfers heat to the water, the former cooling down and the water warming up from 25.0°C to the final temperature; the two finally come to the same temperature. Use the principle of the conservation of energy to predict the final temperature of the water with the aluminum immersed in it.

52. Boron–hydrogen compounds, or boron hydrides, react with oxygen, some violently so, to give water and boric oxide, B_2O_3. For example, pentaborane, B_5H_9, reacts as

$$2\ B_5H_9(\ell) + 12\ O_2(g) \longrightarrow 9\ H_2O(g) + 5\ B_2O_3(s) + 8703\ \text{kJ}$$

Calculate the standard molar heat of formation of $B_5H_9(\ell)$ using data in Table 5.2 (or Appendix L) and $\Delta H_f^\circ\ [B_2O_3(s)] = -1273$ kJ/mol.

53. Uranium is used as fuel in nuclear power plants. The uranium-235 isotope is the fissionable isotope. Since natural uranium contains only a small amount of this isotope, the uranium must be enriched in uranium-235 before it can be used (Chapter 26). To do this, uranium oxide is first converted to a gaseous uranium compound, UF_6, and isotope separation is done by a gaseous diffusion technique (Chapter 6). Some key reactions are:

$$UO_2(s) + 4\ HF(g) \longrightarrow UF_4(s) + 2\ H_2O(g)$$

$$UF_4(s) + F_2(g) \longrightarrow UF_6(g)$$

Recently a ship sank in the English Channel while carrying about 225 tons of UF_6 to the Soviet Union for enrichment of the uranium. How much heat energy would be involved in producing 225 tons of $UF_6(g)$ from UO_2? (1 T = 9.08×10^5 g) Some necessary enthalpies of formation are: $\Delta H_f^\circ\ [UO_2(s)] = -1085$ kJ/mol; $\Delta H_f^\circ\ [UF_4(s)] = -1914$ kJ/mol; and $\Delta H_f^\circ\ [UF_6(g)] = -2147$ kJ/mol.

6

Gases and Their Behavior

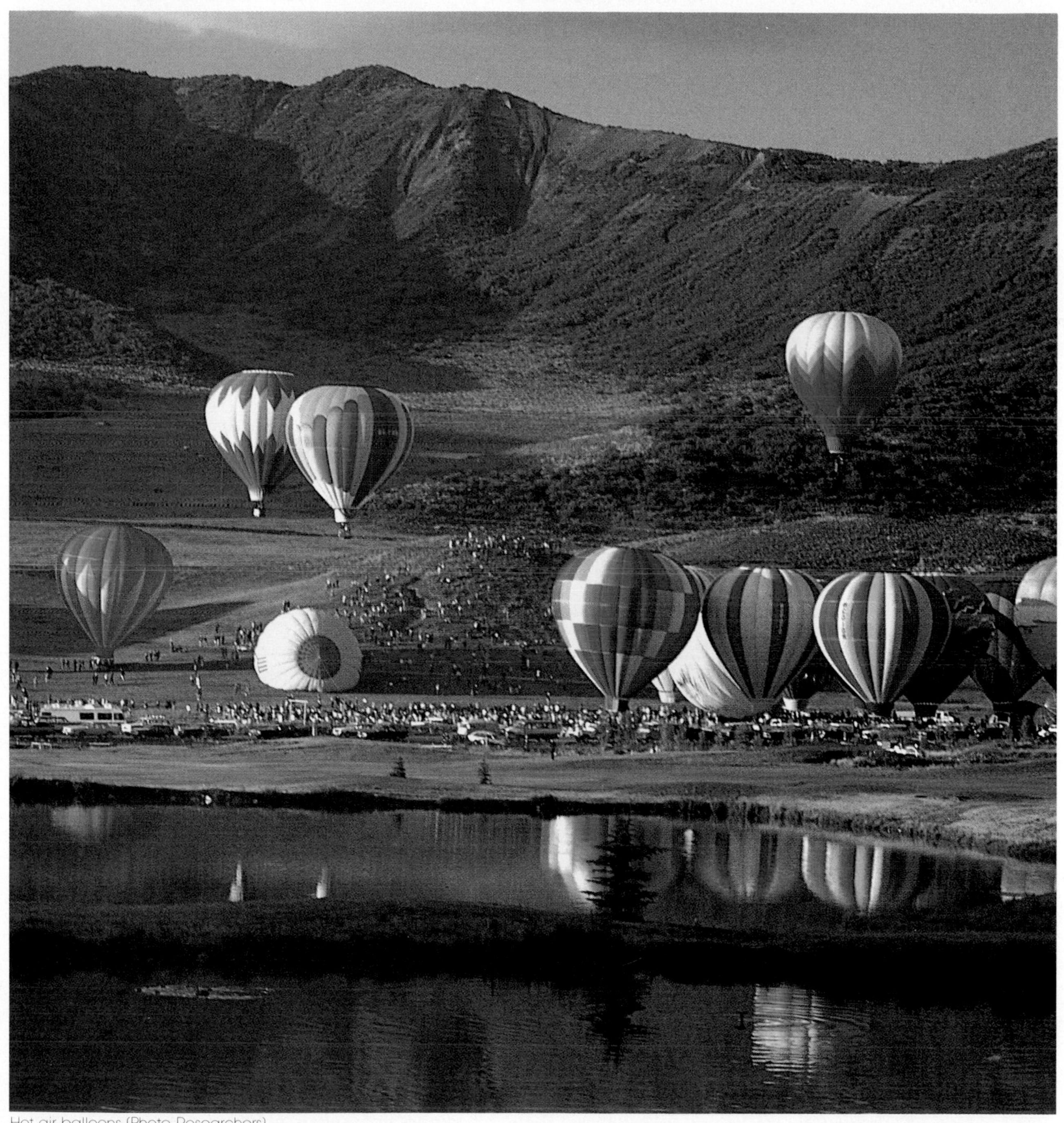

Hot air balloons (Photo Researchers)

CHAPTER OUTLINE

We all live at the bottom of a sea of gas molecules. Molecular oxygen and nitrogen make up almost 99% of the earth's atmosphere, but other gases are important as well. Carbon dioxide, for example, is taken in by plants and converted to sugars and other substances essential for life, and a blanket of ozone gas, O_3, surrounds the earth at high altitudes and protects us from dangerous radiation from the sun.

In this chapter, the fundamentals of the physical behavior of gases are discussed for at least three important reasons. First, our gaseous atmosphere provides one means of transferring energy and material throughout the globe, and it is the source of life-giving chemicals. Second, gas behavior is well understood and can be expressed in terms of simple mathematical models. The modeling of nature is one of the great endeavors of science, and a study of gas behavior will introduce you to this approach. Third, many common elements and compounds are in the gaseous state under normal conditions of pressure and temperature (Table 6.1). Further, many common liquids can be vaporized, and the properties of these vapors are important.

A **gas** is a substance that is normally in the gaseous state at ordinary pressures and temperatures. A **vapor** is the gaseous form of a substance that is normally a liquid or solid at ordinary temperatures and pressures. Thus, we often speak of helium gas and water vapor.

6.1 THE PROPERTIES OF GASES

To describe the bulk physical properties of a gas, the chemist has learned that only four quantities are needed to define the **state of the gas**: (a) the *quantity* of gas, n (in moles); (b) the *temperature* of the gas, T, in kelvins; (c) the *volume* of gas, V, in liters; and (d) the gas *pressure*, P. Before seeing how they are related, we must examine the concept of pressure and its units.

TABLE 6.1 Some Common Gaseous Elements and Compounds

(at 1 atm pressure and 25°C)

He	CO
Ne	CO_2
Ar	CH_4
Kr	C_2H_4
Xe	C_3H_8
H_2	HF
O_2	HCl
O_3 (ozone)	HBr
N_2	HI
F_2	NO
Cl_2	NO_2
	H_2S
	HCN

GAS PRESSURE

The pressure of a gas is a measure of the force that it exerts on its container. **Force** is the physical quantity that causes a change in the motion of a mass if it is free to move. For instance, gravity is the force that the earth exerts on all objects near it, causing the same acceleration (change in velocity per unit time) for all objects. The gravitational force is also known as the **weight** of the object.

Newton's first law states that force = mass × acceleration. The SI units of mass and acceleration are kg and m/s^2, so the units of force are kg · m/s². This derived unit is given the name **newton (N)**. In the English system, force and weight are measured in pounds, so 1 pound = 4.448 newtons.

Pressure is defined as the force exerted on an object divided by the area over which the force is exerted.

A **manometer** can be used to measure the pressure of an enclosed gas sample (see boxed material).

$$\text{Pressure} = \frac{\text{force}}{\text{area}}$$

This book, for example, weighs a little over 4 pounds and has an area of 85 in.^{2}, so it exerts a pressure of about 0.05 pounds/in.^{2} when it lies flat on a surface; in SI units, the weight is about 20 newtons and the area is approximately 550 cm^2, so the pressure is about 0.03 N/cm^2.

The air around us exerts a pressure on everything it touches, and we can measure that pressure with a **barometer** (Figure 6.1). To see how we use a barometer to measure pressure, consider the pressure that the mercury in the column exerts on the mercury in the beaker. This pressure is, by definition, the weight of the mercury column divided by the cross-sectional area of the tube. Since weight is proportional to mass, we can write the pressure as

$$\text{Pressure} = \frac{\text{weight of mercury column}}{\text{cross-section area of column}} \propto \frac{\text{mass of mercury column}}{\text{area of column}}$$

The mass of an object is equal to its volume times its density (see Chapter 1), so the pressure P exerted by the mercury column is

$$P \propto \frac{\text{volume of mercury column} \times \text{density of Hg}}{\text{area of column}}$$

$$P \propto \frac{(\text{height of column} \times \text{area of column}) \times \text{density of Hg}}{\text{area of column}}$$

$$P \propto \text{height of column} \times \text{density of Hg}$$

This means that the pressure exerted by the column of mercury in a barometer such as that in Figure 6.1 depends *only* on the height of the mercury column, and not on its area. Since this pressure is equal to the atmospheric pressure outside the beaker, it is natural to measure atmospheric pressure (or the pressure of any other gas) by measuring the height of the column of mercury it can support. Thus, a common unit of pressure is the **millimeter of mercury (mm Hg)**, also called the **torr** in honor of Evangelista Torricelli (1608–1647), who invented the barometer in 1643.

At sea level, the pressure of the atmosphere can support a column of mercury about 760 mm high (depending on weather conditions), so the **standard atmosphere (atm)** is defined as

$$1 \text{ standard atmosphere} = 1 \text{ atm} = 760 \text{ mm Hg (exactly)}$$

The SI unit of pressure is the **pascal (Pa)**, named for the French mathematician and philosopher Blaise Pascal (1623–1662). It is the only pressure unit that is defined directly in terms of force per unit area.

$$1 \text{ pascal (Pa)} = 1 \text{ newton/meter}^2$$

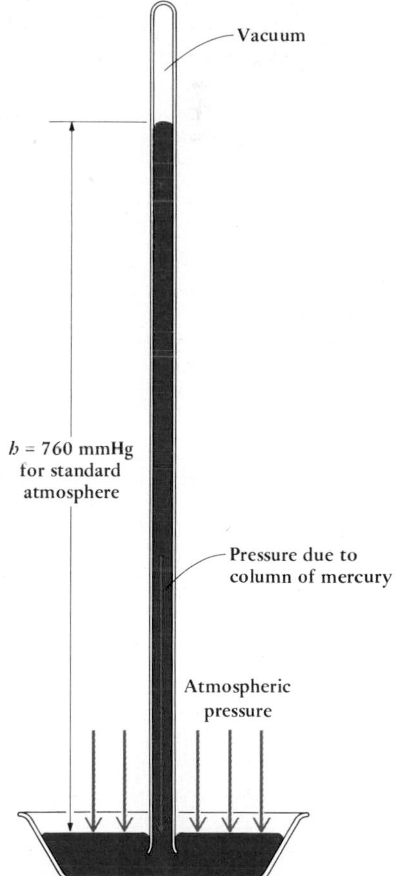

FIGURE 6.1

A mercury barometer. The pressure of the atmosphere on the surface of the mercury in the beaker is transmitted through the liquid and acts to support the weight of the mercury in the tube.

PRESSURE MEASUREMENT WITH A MANOMETER

Gas pressures are often measured in the laboratory with a U-tube manometer, which is a mercury-filled, U-shaped glass tube. The closed side of the manometer has been evacuated so that no gas remains to press on the mercury surface on this side. The other side of the manometer is open to the gas whose pressure is to be measured. When the gas presses on the mercury in the open side, the gas pressure in mm Hg is read directly as the difference in the mercury levels in the closed and open sides.

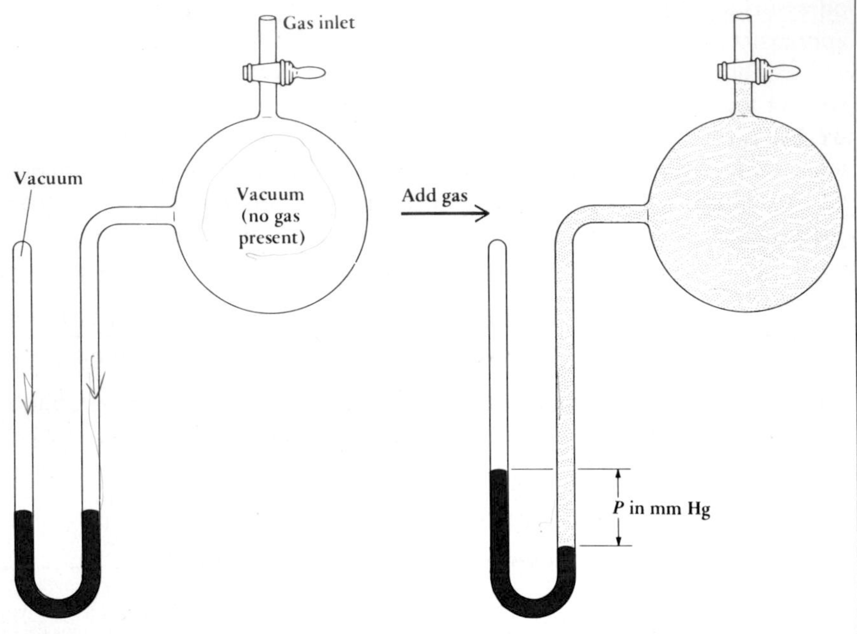

This is a very small unit compared to ordinary pressures, so the kilopascal (kPa) is more often used. The relationship among the units we have listed is

$$1 \text{ atm} = 760. \text{ mm Hg} = 760. \text{ torr} = 101.325 \times 10^3 \text{ Pa} = 101.325 \text{ kPa}$$

EXAMPLE 6.1

PRESSURE UNIT CONVERSIONS

Convert a pressure of 745 mm Hg into its corresponding value in units of torr, atm, and Pa.

Solution The torr and mm Hg units are equivalent; thus 745 mm Hg = 745 torr.

The relation between mm Hg and atm is 1 atm = 760. mm Hg. Notice that the given pressure is less than 760. mm, so the atm equivalent is less than 1 atm.

$$745 \cancel{\text{mm Hg}} \left(\frac{1.00 \text{ atm}}{760. \cancel{\text{mm Hg}}}\right) = 0.980 \text{ atm}$$

The relation between mm Hg and Pa is 101.3 kPa = 760.0 mm Hg. Therefore,

$$745 \cancel{\text{mm Hg}} \left(\frac{101.3 \text{ kPa}}{760. \cancel{\text{mm Hg}}}\right) = 99.3 \text{ kPa}$$

EXERCISE 6.1 PRESSURE UNIT CONVERSION
Rank the following pressures in decreasing order of magnitude (largest first, smallest last): 75 kPa, 300. torr, 0.60 atm, and 350. mm Hg.

6.2 THE GAS LAWS

Experimentation in the 17th and 18th centuries led to three gas laws that provide the basis for our understanding of gas behavior.

BOYLE'S LAW

Robert Boyle (1627–1691) has sometimes been called the "father of chemistry" because he studied a broad range of subjects related to chemistry. He is best known, however, for making one of the first quantitative studies of gas behavior, which he called "the Spring of Air." Boyle observed that the volume of a confined gas is inversely proportional to the pressure exerted on the gas. Since all gases behave in this manner, we know this as **Boyle's law**.

Boyle's law can be demonstrated by the experiment in Figure 6.2. A hypodermic syringe is filled with air and sealed. When pressure is

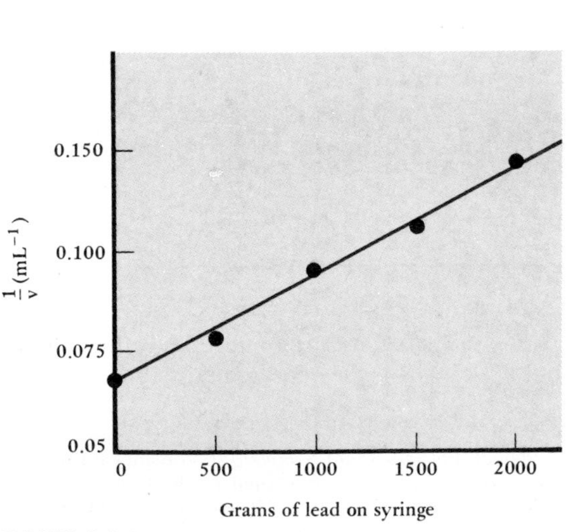

(a)

(b)

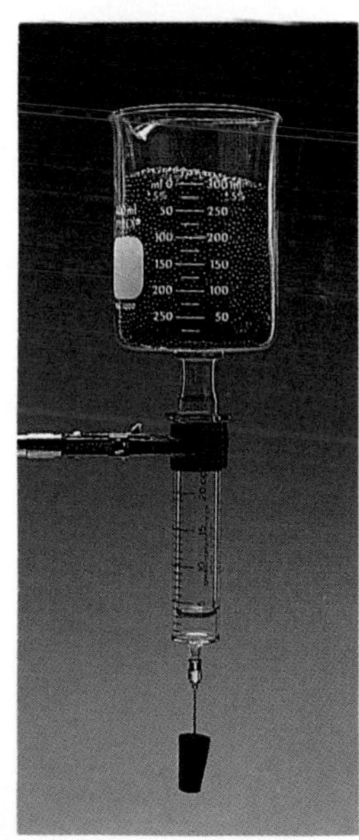

FIGURE 6.2
Boyle's Law. A syringe, containing some air, was sealed, and then lead shot was added to the beaker on top of the barrel. As the mass of lead increased (from photo a to photo b), and thereby increased the pressure on the gas in the syringe, the air was compressed. A plot of (1/volume of air) *vs.* mass of lead shows there is a linear relation between 1/V and P (as measured by the mass of lead). (See D. Davenport, *Journal of Chemical Education*, **1962,** *39,* 252.) (Charles D. Winters)

ROBERT BOYLE (1627–1691)

Robert Boyle was born in Ireland, in a home that still stands, as the 14th and last child of the first Earl of Cork. He first published his studies of gases in 1660, and a book, *The Sceptical Chymist,* was published in 1680. The subtitle of this book, which had a profound effect on those interested in science in that day, was "Chymico–Physical Doubts and Paradoxes, Touching the Experiments whereby Vulgar Spagirists are wont to Endeavour to Evince their Salt, Sulfur, and Mercury, to be the True Principles of Things." Not everyone applauded all of Boyle's work, however. Isaac Newton, a young man when Boyle's career was at its peak, once remarked about one of Boyle's papers that "I question not but that the great wisdom of the noble author will sway him to high silence till he shall be resolved of what consequence the thing may be . . . [by] the judgment of some other that thoroughly understands what he speaks about."

applied to the movable plunger of the syringe, the volume of air in the sealed apparatus decreases. When the pressure is plotted as a function of $1/V$, a straight-line relation is observed. This tells you that the pressure and volume are *inversely proportional.*

$$P \propto \frac{1}{V}$$

The symbol $\propto$ means "proportional to."

When two quantities are proportional to each other, they can be equated if a *proportionality constant*, C_b, is introduced.* Thus,

$$P = C_b \frac{1}{V} \qquad \text{or} \qquad PV = C_b \tag{6.1}$$

The last relation, $PV = C_b$, leads to a useful statement of Boyle's law: For a given quantity of gas at a given temperature, the product of pressure and volume is a constant:

$$P_1V_1 = P_2V_2 = P_3V_3 = P_nV_n = C_b$$

where C_b is determined by the quantity of gas (in moles) and the temperature. This means that if the PV product is known for one set of conditions (P_1 and V_1), then the volume expected (V_2) for a new pressure (P_2) can be calculated, for example.

EXAMPLE 6.2

BOYLE'S LAW

A sample of gaseous CO_2 has a pressure of 55 mm Hg in a 125-mL flask. If this sample is transferred to a 650-mL flask, what is the expected pressure of the gas?

---

*If you are not acquainted with the idea of a *proportionality constant*, consider it this way. The amount of money you earn for a job is directly proportional to the time you spend working.

Money earned $\propto$ time spent working

This can be made into an equality if the proportionality constant "money earned/time" is introduced.

Money earned = (money earned/time)(time spent working)

This is the approach we use in solving problems throughout this textbook.

Solution It is useful to make a table of the information provided.

ORIGINAL CONDITIONS	FINAL CONDITIONS
$P_1 = 55$ mmHg	$P_2 = ?$
$V_1 = 125$ mL	$V_2 = 650$ mL

Since you know that $P_1V_1 = P_2V_2$, then

$$P_2 = \frac{P_1V_1}{V_2} = \frac{(55 \text{ mm Hg})(125 \cancel{\text{mL}})}{(650 \cancel{\text{mL}})} = 11 \text{ mm Hg}$$

As expected, when the gas was placed in a larger volume, the pressure that it exerted was lower.

EXERCISE 6.2 BOYLE'S LAW

If the CO_2 sample in Example 6.2 is placed in another flask and the pressure of the gas was found to be 78 mm Hg, what is the volume of the container?

CHARLES'S LAW

If a given amount of gas is held at a constant pressure, then its volume will be directly proportional to the absolute temperature ($V \propto T$). This means that $V = C_c \times T$, or that

$$C_c = \frac{V}{T} \tag{6.2}$$

This statement, which came from experiments by the Frenchmen Jacques Charles, is illustrated in an extreme manner in Figure 6.3.

To illustrate Charles's law, consider some experimental data for the volume of a gas as a function of temperature (Figure 6.4). A sample of a gas is heated and cooled, and its volume (at constant pressure) is plotted as a function of temperature. In other words, the amount of gas (in moles) and the pressure are being held constant, and V and T are the variables. Once again a straight-line relation is observed. The volume is found to double when T, the kelvin temperature, is doubled. The *direct proportionality* of V and T means that

$$\frac{V_1}{T_1} = \frac{V_2}{T_2} = \frac{V_3}{T_3} = \frac{V_n}{T_n}$$

and so the volume of gas at some temperature can be calculated if its volume is known at one other temperature. For example, using the data for 366 K in Figure 6.4, we can calculate the volume, V_2, the gas would have at 0.0°C ($T_2 = 273$ K). Since you know that $V_1/T_1 = V_2/T_2$, then

$$V_2 = \frac{V_1}{T_1}T_2 = \left(\frac{300.\ \text{mL}}{366\ \cancel{\text{K}}}\right)(273\ \cancel{\text{K}}) = 224 \text{ mL}$$

FIGURE 6.3

A dramatic illustration of Charles's law. Some air-filled balloons are placed in liquid nitrogen (at 77 K). The volume of the gas is dramatically reduced at this temperature. After all the balloons have been placed in the liquid nitrogen, they are poured out again and the balloons reinflate to their original volume when warmed back to room temperature. (Charles D. Winters)

Volume–Temperature Relations
(for 0.010 mole of a gas at 1.0 atm pressure)

Volume, V (mL)	Temperature, t (°C)	Temperature, T (K)	V/T (mL/K)
100	−151	122	0.820
150	−90	183	0.820
200	−29	244	0.820
250	32	305	0.820
300	93	366	0.820
400	214	487	0.821

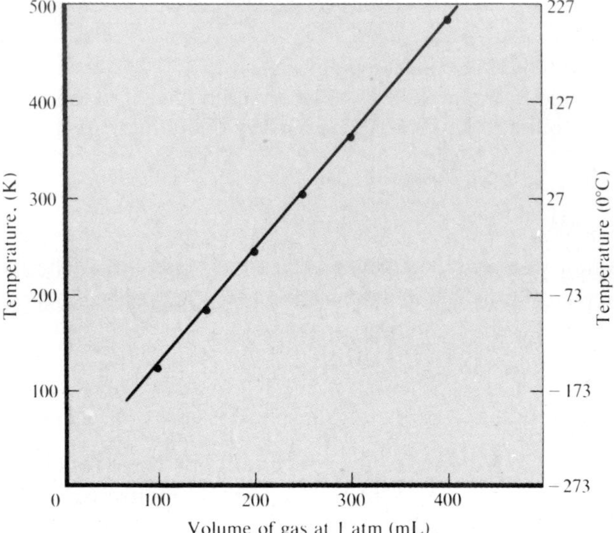

FIGURE 6.4
Charles's law. For a given quantity of gas at a constant pressure, the volume of a gas and its kelvin temperature are directly proportional.

Calculations based on Charles's law are further illustrated by the following example and exercise. Be sure to notice that *the temperature, T, must be expressed in kelvins.*

EXAMPLE 6.3

CHARLES'S LAW

Suppose you have a sample of CO_2 in a gas-tight syringe (see Figure 6.2). The gas volume is 25.0 mL at room temperature (20°C). What is the final volume of the gas if you hold the syringe in your hand to raise its temperature to 37°C?

Solution To organize the information, construct a table. Notice that the temperature has been converted to kelvins.

ORIGINAL CONDITIONS	FINAL CONDITIONS
$V_1 = 25.0$ mL	$V_2 = ?$
$T_1 = 20 + 273 = 293$ K	$T_2 = 37 + 273 = 310.$ K

Since you know that $V_1/T_1 = V_2/T_2$, you can rearrange the equation to solve for V_2 and substitute the information given.

$$V_2 = \frac{V_1}{T_1} T_2 = \left(\frac{25.0 \text{ mL}}{293 \text{ K}}\right)(310. \text{ K}) = 26.5 \text{ mL}$$

As expected, the volume of the gas increased as the temperature increased.

JACQUES ALEXANDRE CESAR CHARLES (1746–1823)

The French chemist Charles was most famous in his lifetime for his experiments in ballooning. The first such flights were actually made by the Montgolfier brothers in June, 1783, using a large spherical balloon made of linen and paper and filled with hot air. In August, 1783, however, a different group, supervised by Jacques Charles, tried a different approach. Exploiting his recent discoveries in the study of gases, Charles decided to inflate the balloon with hydrogen. Since hydrogen would easily escape a paper bag (see Section 6.7), Charles made a bag of silk coated with a rubber solution. Inflating the bag to its final diameter took several days and required nearly 500 pounds of acid and 1000 pounds of iron to generate the hydrogen gas. A huge crowd watched the ascent on August 27, 1783. The balloon stayed aloft for almost 45 minutes and traveled about 15 miles, but, when it landed in a village, the people were so terrified that they tore it to shreds.

(The Bettmann Archive)

EXERCISE 6.3 CHARLES'S LAW

A balloon is inflated with helium to a volume of 4.5 L at room temperature (25°C). If you take the balloon outside on a cold day (− 10°C), what is the new volume of the balloon?

Sir William Thomson (1824–1907), known as Lord Kelvin, realized the significance of Charles's work. Kelvin recognized that, if the line in a V versus T plot (such as Figure 6.4) is continued to the point where the gas has zero volume, the temperature would be *absolute zero*. Since all gases will liquify at temperatures above absolute zero, however, it is not possible to measure gas volumes at very low temperatures to verify this directly by experiment. Instead, one can define the V versus T relation for a gas sample at a variety of pressures. As illustrated in Figure 6.5, each of these V versus T lines should *extrapolate* to absolute zero. By performing many such experiments, the absolute zero of temperature is

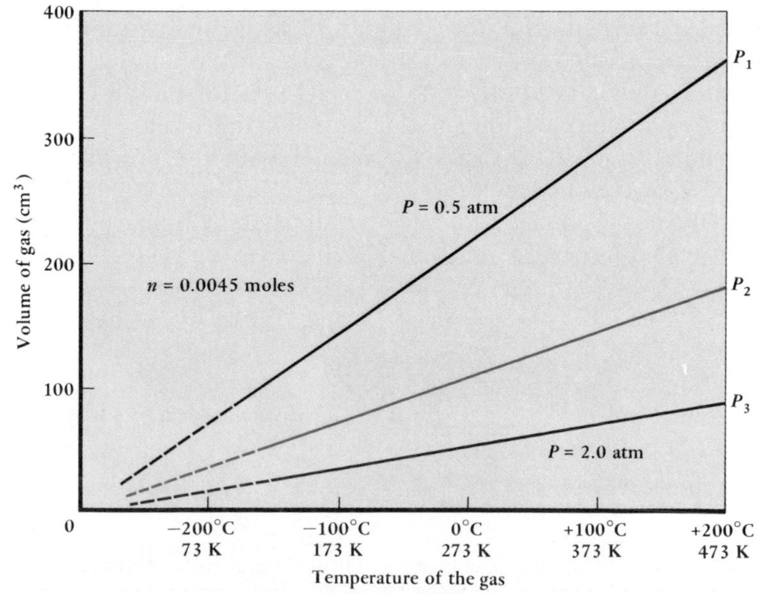

FIGURE 6.5

Experimental determination of absolute zero. Each solid line represents the variation in volume of a gas sample with temperature at a given pressure. If the pressure at which the measurements are made is changed, the slope of the V vs. T line is different. In this figure, the pressures increase from P_1 to P_3. All gases liquify at a sufficiently low temperature, so the lines are extrapolated below the liquifaction temperature (the dashed portion). When extrapolated, all lines reach zero volume at the same temperature, −273.15°C.

JOSEPH LOUIS GAY-LUSSAC (1778–1850)

The Frenchman Gay-Lussac was a chemist, but he contributed to physics, meteorology, and physiology. Although we chiefly remember him for his experiments on gases, he established the elementary nature of sulfur, was the first to prepare sodium and potassium by chemical means, and was the first to isolate boron. He was also, like Jacques Charles, a balloonist. In fact, in 1804, Gay-Lussac ascended to 23,000 feet in a balloon to sample the air; it was an altitude record that stood for many years. In another high-altitude flight, Gay-Lussac had to throw overboard several items to lighten the balloon and gain height. One item he decided to sacrifice was an old white kitchen chair that he had used for a seat. The chair landed near a peasant girl who was minding sheep near a French village. After considerable debate, the local citizenry and priest decided that the incident was a miracle, but wondered why God apparently owned such shabby furniture.

found to be $-273.15°C$, and Kelvin took this as the starting point of the *absolute temperature scale*.

GAY-LUSSAC AND AVOGADRO'S LAW

At the beginning of the 19th century John Dalton published his classic work, *A New System of Chemical Philosophy*, in which he developed his atomic theory (Chapter 2). In spite of the basic correctness of his ideas, Dalton clung to the notion that the majority of compounds were composed of two atoms. Thus, he believed the formulas of water and ammonia, for example, must be HO and HN, respectively. However, in 1809 Joseph Gay-Lussac showed that two volumes of hydrogen gas were required to combine with one volume of oxygen gas to form two volumes of water vapor, all gases being at the same temperature and pressure. Based on these results and similar ones for other gas-phase reactions, Gay-Lussac established the law of combining volumes: Volumes of gases, at the same temperature and pressure, will combine with one another in the ratio of small whole numbers. Such an idea did not agree with Dalton's notion that water had the formula HO, for example. If, as Dalton assumed, the combining gases consisted of atoms, H and O, then the only way they could react in a volume ratio of 2:1 to give HO was to split an O atom in half, a clear violation of Dalton's notion of indivisible atoms. The only possible explanation must be that oxygen and hydrogen are diatomic molecules and give H_2O molecules.

By volumes of gas, we mean 1 L of oxygen combines with 2 L of hydrogen, or 125 mL of oxygen combines with 250 mL of hydrogen, and so on.

Dalton did not accept Gay-Lussac's results, but the Italian physicist and lawyer, Amadeo Avogadro, did, and the latter extended Gay-Lussac's

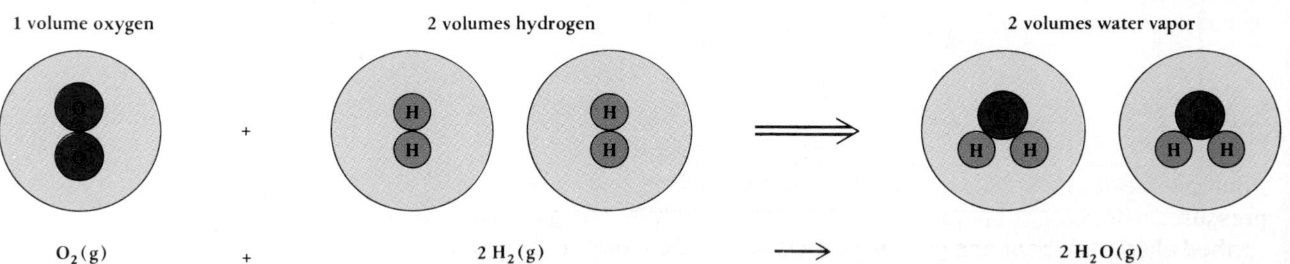

law to its logical conclusion. In 1811 Avogadro published his **hypothesis** that equal volumes of gases under the same conditions of temperature and pressure have equal numbers of molecules. Thus, two volumes of H_2 must have twice the number of molecules as one volume of O_2, and they should produce one volume of H_2O.

There is more on Avogadro in Chapter 2; also, see the brief biography of Gay-Lussac in the boxed text.

Avogadro's law follows directly from Avogadro's hypothesis: The volume of a gas, at a given temperature and pressure, is directly proportional to the quantity of gas. Thus, V is proportional to n ($V \propto n$), the number of moles of gas, so

$$V = C_a n \qquad (6.3)$$

where C_a is the proportionality constant. This is a useful relationship, since it directly connects the properties of gases to chemical stoichiometry as you have already seen above.

EXAMPLE 6.4

AVOGADRO'S LAW

Ammonia can be made directly from the elements according to the equation

$$N_2(g) + 3\ H_2(g) \longrightarrow 2\ NH_3(g)$$

If one begins with 12.6 L of $H_2(g)$ at a given temperature and pressure, what volume of $N_2(g)$ is required for complete reaction (at the same temperature and pressure)? What is the theoretical yield of NH_3 in liters?

Solution Since gas volume is proportional to the number of moles of a gas, we can use volumes instead of moles and treat this just as we did stoichiometry problems in Chapter 4.

(a) Calculate liters of N_2 required by multiplying the number of liters of H_2 available by a stoichiometric factor in liters.

$$12.6\ \text{L}\ H_2(g)\ \text{available}\left(\frac{1\ \text{L}\ N_2(g)\ \text{required}}{3\ \text{L}\ H_2(g)\ \text{available}}\right) = 4.20\ \text{L}\ N_2\ \text{required}$$

(b) Calculate liters of gaseous ammonia, NH_3, produced.

$$12.6\ \text{L}\ H_2(g)\ \text{available}\left(\frac{2\ \text{L}\ NH_3(g)\ \text{produced}}{3\ \text{L}\ H_2(g)\ \text{available}}\right) = 8.40\ \text{L}\ NH_3\ \text{produced}$$

EXERCISE 6.4 AVOGADRO'S LAW

Methane burns in oxygen to give the usual products, CO_2 and H_2O.

$$CH_4(g) + 2\ O_2(g) \longrightarrow 2\ H_2O(g) + CO_2(g)$$

If 22.4 L of gaseous CH_4 are burned, what volume of O_2 is required for complete combustion? What volumes of H_2O and CO_2 are produced? Assume all gases are measured at the same temperature and pressure.

6.3 THE IDEAL GAS LAW

Four interrelated quantities can be used to describe the state of a gas: pressure, volume, temperature, and quantity. The three gas laws described above tell how one gas property is affected as another is changed,

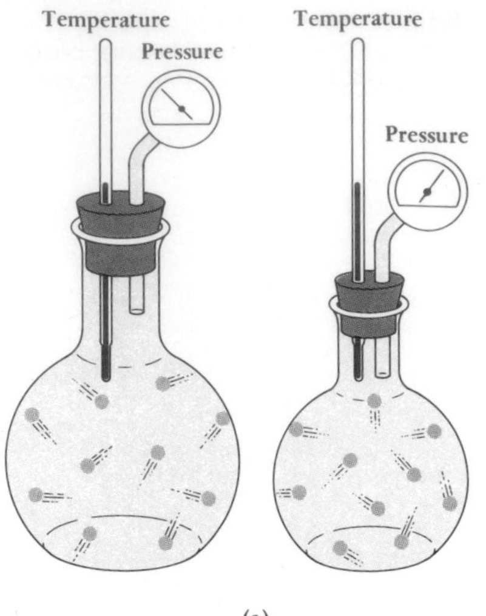

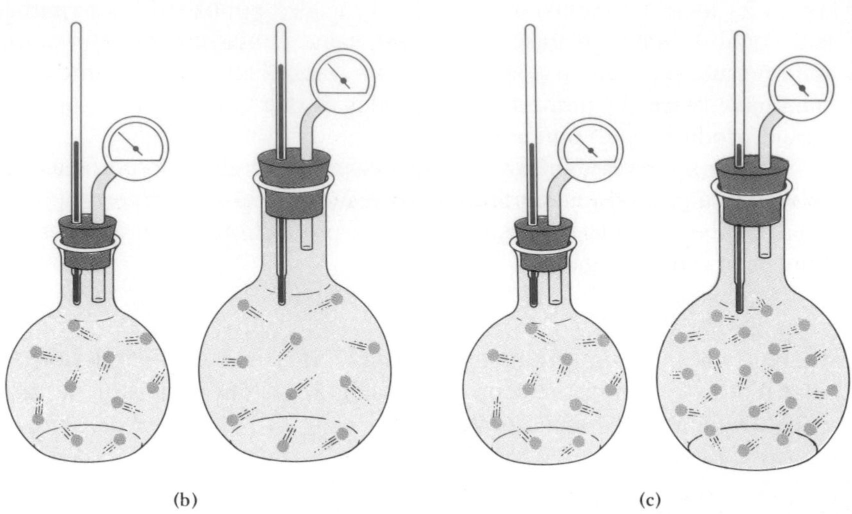

(a) (b) (c)

FIGURE 6.6

A summary of the gas laws. (a) Boyle's law. At a constant temperature, the pressure exerted by a given quantity of gas increases as the volume of the gas decreases. (b) Charles's law. At a constant pressure, the volume a given quantity of gas occupies increases as the temperature of the gas increases. (c) Avogadro's law. At a constant temperature and pressure, the volume occupied by a gas depends directly on the quantity of gas.

assuming the other properties remain fixed (Figure 6.6). In summary, these laws are

BOYLE'S LAW	CHARLES'S LAW	AVOGADRO'S LAW
$V \propto \frac{1}{P}$	$V \propto T$	$V \propto n$
(constant T, n)	(constant P, n)	(constant T, P)

If all three laws are combined the result is

$$V \propto \frac{nT}{P}$$

The proportionality constant is the product $C_aC_bC_c$, which we now label R and call the **gas constant**:

$$V = R\,\frac{nT}{P}$$

or

$$PV = nRT \tag{6.4}$$

This last equation is called the **ideal gas law**, since it describes the state of a so-called ideal gas. As you will learn in Section 6.8, there is no such thing as an "ideal gas." However, real gases at pressures around an atmosphere or less and temperatures around room temperature behave close enough to ideality that $PV = nRT$ is an adequate description.

To use the equation $PV = nRT$, we need a value for R. Many experiments show that under conditions of **standard temperature and pressure (STP)** (a gas temperature of 0°C or 273.15 K and a pressure of 1.0000 atm) 1.0000 mole of gas occupies 22.414 liters, a quantity called the **standard molar volume**. Substituting these values into the ideal gas law, we can solve for R.

$$R = \frac{PV}{nT}$$

$$= \frac{(1.0000 \text{ atm})(22.414 \text{ L})}{(1.0000 \text{ mol})(273.15 \text{ K})} = 0.082058 \frac{\text{L} \cdot \text{atm}}{\text{K} \cdot \text{mol}}$$

Although R was calculated to 5 significant figures, we shall generally round off the value to three significant figures (0.0821).

EXAMPLE 6.5

THE IDEAL GAS LAW

The helium in a 1.5-L flask at 25°C exerts a pressure of 425 mm Hg. How many moles of helium are there in the flask?

Solution First, write down the information provided.

$V = 1.5$ L $\quad P = 425$ mm Hg $\quad T = 25$°C $\quad n = ?$

To use the ideal gas law, the pressure must be expressed in atmospheres and the temperature in kelvins. Therefore,

$$P = 425 \text{ mm Hg} \left(\frac{1 \text{ atm}}{760. \text{ mm Hg}}\right) = 0.559 \text{ atm}$$

$$T = 25 + 273 = 298 \text{ K}$$

Now rearrange the ideal gas law to solve for the number of moles, n.

$$n = \frac{PV}{RT}$$

and substitute into this expression.

$$n = \frac{(0.559 \text{ atm})(1.5 \text{ L})}{(0.0821 \text{ L} \cdot \text{atm/K} \cdot \text{mol})(298 \text{ K})} = 0.034 \text{ mole He}$$

EXERCISE 6.5 THE IDEAL GAS LAW

The balloon used by Charles in his historic flight in 1783 was filled with about 1300 moles of H_2. If the temperature of the gas was 20.°C, and its pressure was 750 mm Hg, what was the volume of the balloon? (To have some idea of the size of this balloon, you may find it interesting to calculate its radius or diameter from the volume. Recall that the radius of a spherical object and its volume are related by $V = \frac{4}{3}(\pi r^3)$.)

The ideal gas law is useful for calculating one of the four properties of a gas when the other three are known. However, there are many times when one needs to know what happens to the state of a *given amount of gas* when there is a *change* in one, two, or even three of the parameters P, V, and T. Since R is a universal constant for gases, this means that a gas exerting a pressure P_1 in a volume V_1 at a temperature T_1 must obey the equation

$$R = \frac{P_1V_1}{nT_1}$$

If the quantity of the gas, n, remains the same, but the other conditions change to new values P_2, V_2, and T_2, we write the same expression for R.

$$R = \frac{P_2V_2}{nT_2}$$

Under either set of conditions the quotient PV/nT is equal to R. Therefore,

$$\frac{P_1V_1}{nT_1} = \frac{P_2V_2}{nT_2} \tag{6.5}$$

This is sometimes known as the **general gas law**.

EXAMPLE 6.6

THE GENERAL GAS LAW

You have a sample of a gas in a small container (12.5 mL). The pressure of the gas is 685 mm Hg at room temperature (22.0°C). If you hold the container in your hand (so the temperature climbs to 37.0°C), what is the new pressure?

Solution Here the quantity of gas is constant, as is the volume in which it is contained. Therefore, we are interested in the change in pressure as the temperature increases. We shall begin by setting out the information given in a table. Notice that pressures and temperatures have been changed to units of atmospheres and kelvins, respectively.

INITIAL CONDITIONS	FINAL CONDITIONS
V_1 = 12.5 mL	V_2 = 12.5 mL
P_1 = 685 mm Hg	P_2 = ?
T_1 = 22.0°C (295 K)	T_2 = 37.0°C (310. K)
$n_1 = n_2$	

From the text above, the general gas law is

$$\frac{P_1V_1}{n_1T_1} = \frac{P_2V_2}{n_2T_2}$$

Since $n_1 = n_2$ and $V_1 = V_2$, these quantities can be canceled from the equation,

$$\frac{P_1}{T_1} = \frac{P_2}{T_2}$$

and the resulting equation can then be solved for P_2.

$$P_2 = P_1\frac{T_2}{T_1} = 685 \text{ mm Hg}\left(\frac{310.\text{ K}}{295\text{ K}}\right) = 720.\text{ mm Hg}$$

The pressure has increased, as expected from the increase in temperature.

The answer to Example 6.6 points up a useful general conclusion: *For a given quantity of gas in a constant volume, the pressure is directly proportional to the absolute temperature.* When you drive an automobile for some distance, for instance, friction warms the air in the tires, and, since the tire volume is nearly constant, the pressure increases.

EXAMPLE 6.7

THE GENERAL GAS LAW

A gas occupying a volume of 2.00 L at 756 mm Hg pressure and 273 K is transferred to a 4.00-L container where its pressure is 189 mm Hg. What is the temperature of the gas after the transfer?

Solution In this problem a given quantity of gas under one set of conditions is allowed to expand into a container of larger volume. As expected from Boyle's law, the pressure has dropped. From Charles's and Gay-Lussac's law, you might have expected the gas temperature to increase with increasing volume. However, remember that V and T are directly proportional *only* if the pressure is constant, and this is not true here. Let us use the general gas law to find out what has happened to the gas temperature.

INITIAL VALUES	FINAL VALUES
V_1 = 2.00 L	V_2 = 4.00 L
P_1 = 756 mm Hg (0.995 atm)	P_2 = 189 mm Hg (0.249 atm)
T_1 = 273 K	T_2 = ?
$n_1 = n_2$	

The general gas law states that $P_1V_1/n_1T_1 = P_2V_2/n_2T_2$. Here n_1 and n_2 are equal and cancel out of the expression. Solving for the unknown quantity, T_2, we have

$$T_2 = T_1 \times \frac{P_2}{P_1} \times \frac{V_2}{V_1} = 273 \text{ K} \left(\frac{0.249 \text{ atm}}{0.995 \text{ atm}}\right) \left(\frac{4.00 \text{ L}}{2.00 \text{ L}}\right) = 137 \text{ K}$$

Notice that the final temperature is less than the initial temperature. Is this reasonable? The volume of gas increased by a factor of 2, which in itself should indicate an increase in T by a factor of 2. However, the pressure decreased by a factor of 4, which could have been caused by a decrease in T by a factor of 4. Since the pressure change is larger than the volume change, you expect T to decrease.

The cooling of a gas during expansion is common. In fact, this is the principle by which air and other gases are cooled until they liquefy.

EXERCISE 6.6 THE GENERAL GAS LAW

You have a 20.-L cylinder of helium at a pressure of 150 atm at 30.°C. How many balloons can you fill, each to a volume of 5.0 L, on a day when the atmospheric pressure is 755 mm Hg and the temperature is 22°C?

GAS DENSITY CALCULATIONS

The density of a gas at a given temperature and pressure (Figure 6.7) is a useful quantity. Let us see how this is related to the ideal gas law. By rearranging the expression $PV = nRT$, we have

$$\frac{n}{V} = \frac{P}{RT}$$

Since the number of moles (n) of any compound is given by its mass (m) divided by its molar mass (M), we can rewrite the equation above as

$$\frac{m}{MV} = \frac{P}{RT}$$

FIGURE 6.7

Gas density. The two balloons are filled with nearly equal quantities of gas at the same temperature and pressure. The blue balloon contains low-density helium (d = 0.18 g/L), while the red balloon contains argon, a higher-density gas (d = 1.8 g/L). In comparison, the density of dry air at 1 atm pressure and 25°C is about 1.2 g/L. (Charles D. Winters)

This expression can again be rearranged since *density* (d) is mass/volume (m/V). Therefore,

$$d = \frac{m}{V} = \frac{PM}{RT} \tag{6.6}$$

We now have an expression relating the density of a gas to its molar mass at a given temperature and pressure.

EXAMPLE 6.8

GAS DENSITY AND MOLAR MASS

The density of an unknown gas at STP is 1.429 g/L. Calculate its molar mass.

Solution List the information given in a short table, recalling that "STP" is shorthand for 1 atm and 0°C.

$$\begin{aligned} d &= 1.429 \text{ g/L} \\ P &= \text{standard pressure} = 1.000 \text{ atm} \\ T &= \text{standard temperature} = 273.15 \text{ K} \\ R &= 0.08206 \text{ L} \cdot \text{atm/K} \cdot \text{mol} \\ M &= ? \end{aligned}$$

The equation for gas density is rearranged to solve for the molar mass (M).

$$\begin{aligned} M &= \frac{dRT}{P} \\ &= \frac{(1.429 \text{ g/L})(0.08206 \text{ L} \cdot \text{atm/K} \cdot \text{mol})(273.15 \text{ K})}{(1.000 \text{ atm})} \\ &= 32.03 \text{ g/mol} \end{aligned}$$

EXERCISE 6.7 GAS DENSITY CALCULATION

Calculate the density of air at 22.0°C and 745 mm Hg if its molar mass (average) is 29.0 g/mol.

This discussion of gas density has some important practical implications. From the equation $d = M(P/RT)$, we know that gas density is directly proportional to the molar mass (for gases at the same T and P). Air, with an average molecular weight of about 29, has a density of about 1.2 g/L (at 1 atm and 25°C). This means that gases or vapors more dense than about 1.2 g/L, such as CO_2, SO_2, and gasoline, will settle along the ground if released into the atmosphere. Conversely, gases such as H_2, He, CO, CH_4 (methane), and NH_3 (ammonia) will rise if released into the atmosphere. The release of methyl isocyanate (H_3CNCO) into the air on December 3, 1984, in Bhopal, India, killed several hundred people and injured many thousands because the toxic vapors were heavier than air and settled close to ground where they could be inhaled.

CALCULATING THE MOLAR MASS OF A GAS FROM *P*, *V*, AND *T* DATA

Either the ideal gas law or the equation for gas density can be used to determine the molar mass of a gas.

EXAMPLE 6.9

CALCULATING THE MOLAR MASS OF A GAS FROM *P*, *V*, AND *T* DATA

A 0.100-g sample of a compound of empirical formula CH_2F_2 occupies 0.0470 L at 298 K and 755 mm Hg. What is the molar mass of the compound?

Solution This problem is typical of the laboratory measurement of the molar mass of a gas. The experiment is performed when one has synthesized a new gaseous compound and does not yet know its molar mass. Begin by organizing the data.

$$V = 0.0470 \text{ L} \qquad P = 755 \text{ mm Hg} \qquad T = 298 \text{ K} \qquad n = ?$$

(a) We can find the molar mass by first using the ideal gas law to calculate n, the number of moles equivalent to 0.100 g of the gas.

$$n = \frac{PV}{RT}$$

When substituting into the rearranged gas equation, you must make sure the units of P, V, and T are compatible with the units of R. Here this means P is converted to atmospheres.

$$P = 755 \text{ mm Hg} \left(\frac{1 \text{ atm}}{760 \text{ mm Hg}}\right) = 0.993 \text{ atm}$$

Therefore,

$$n = \frac{(0.993 \text{ atm})(0.0470 \text{ L})}{(0.08206 \text{ L} \cdot \text{atm/K} \cdot \text{mol})(298 \text{ K})} = 0.00191 \text{ mole}$$

Since you know the number of grams in the sample and now know the number of moles to which it is equivalent, you can calculate the molar mass.

$$\text{Molar mass} = \frac{0.100 \text{ g}}{0.00191 \text{ mol}} = 52.4 \text{ g/mol}$$

In this particular case the experimental molar mass is equal to the empirical formula weight.

(b) The gas density equation can also be used to solve for the molar mass. The density of the gas here is 0.100 g in 0.0470 L or

$$d = \frac{0.100 \text{ g}}{0.0470 \text{ L}} = 2.13 \text{ g/L}$$

If the equation $d = MRT/P$ is solved for molar mass (M), and then density, pressure, and temperature are substituted into the expression, we have

$$M = \frac{dRT}{P} = \frac{(2.13 \text{ g/L})(0.08206 \text{ L} \cdot \text{atm/K} \cdot \text{mol})(298 \text{ K})}{0.993 \text{ atm}} = 52.4 \text{ g/mol}$$

EXERCISE 6.8 MOLAR MASS FROM *P*, *V*, AND *T* DATA

A 0.105-g sample of a gaseous compound has a pressure of 560. mm Hg in a volume of 125 mL at 23.0°C. What is its molar mass?

6.4 THE GAS LAWS AND CHEMICAL REACTIONS

When introducing Avogadro's law we found that gas volumes could be used instead of units of moles in chemical stoichiometry calculations. This section explores this aspect of gas behavior in more detail.

EXAMPLE 6.10

GAS LAWS AND STOICHIOMETRY

Calculate the volume of H_2 (measured at 0.0°C and 1.0 atm pressure) produced by 12 g of zinc reacting with excess sulfuric acid (Figure 6.8).

$$\mathrm{Zn(s) + H_2SO_4(aq) \longrightarrow ZnSO_4(aq) + H_2(g)}$$

Solution One mole of H_2 gas is produced for every mole of zinc used. Therefore, we first find the moles of zinc consumed.

$$12 \text{ g Zn}\left(\frac{1 \text{ mol Zn}}{65.38 \text{ g Zn}}\right) = 0.18 \text{ mol Zn}$$

Since there is a 1:1 relation between moles of Zn consumed and moles of H_2 produced, 0.18 mole of H_2 is formed. To find the volume of H_2 formed, you can solve the ideal gas law for volume and substitute the given data.

$$V = \frac{nRT}{P} = \frac{(0.18 \text{ mol H}_2)(0.082 \text{ L} \cdot \text{atm/K} \cdot \text{mol})(273 \text{ K})}{1.0 \text{ atm}} = 4.1 \text{ L}$$

Alternatively, you could recognize that the conditions of the experiment are STP, for which the standard molar volume (the volume of one mole) is 22.4 L. Using this as a conversion factor, we arrive at the same answer.

$$0.18 \text{ mol H}_2\left(\frac{22.4 \text{ L}}{1 \text{ mol H}_2}\right) = 4.0 \text{ L}$$

EXAMPLE 6.11

GAS LAWS AND REACTIONS

Gaseous ammonia is synthesized from nitrogen and hydrogen by the reaction

$$\mathrm{N_2(g) + 3\ H_2(g) \xrightarrow[500°C]{\text{iron catalyst}} 2\ NH_3(g)}$$

If 125 L of gaseous H_2 at 25.0°C and 542 mm Hg are combined with excess N_2, how many liters of NH_3 measured at 500°C and 1.00 atm can be produced?

Solution One approach to the problem is to convert the volume of H_2 gas to the volume it would have if measured at the same conditions as the product, NH_3. Then Avogadro's law can be used to relate the volume of H_2 available to the volume of NH_3 produced.

(a) Volume of H_2 when the pressure is 1.00 atm and the temperature is 500°C.

INITIAL CONDITIONS	FINAL CONDITIONS
V_1 = 125 L	V_2 = ?
P_1 = 542 mm Hg (0.713 atm)	P_2 = 1.00 atm
T_1 = 25.0°C (298 K)	T_2 = 500°C (773 K)
$n_1 = n_2$	

FIGURE 6.8
The reaction of zinc metal with sulfuric acid to give hydrogen gas. (Charles D. Winters)

To solve for V_2, rearrange the general gas law, $P_1V_1/T_1 = P_2V_2/T_2$.

$$V_2 = V_1 \times \frac{T_2}{T_1} \times \frac{P_1}{P_2} = 125\ \text{L}\left(\frac{773\ \text{K}}{298\ \text{K}}\right)\left(\frac{0.713\ \text{atm}}{1.00\ \text{atm}}\right) = 231\ \text{L}$$

(b) Since the volume of H_2 is now known under the same conditions at which the product volume is measured, Avogadro's law can be used. The volume of NH_3 produced at the conditions given is

$$231\ \text{L H}_2\left(\frac{2\ \text{L NH}_3}{3\ \text{L H}_2}\right) = 154\ \text{L of NH}_3$$

EXAMPLE 6.12

THE GAS LAWS AND STOICHIOMETRY

Some commercial drain cleaners contain sodium hydroxide and powdered aluminum (Figure 6.9). When the mixture is poured into a drain full of water and dirt, the reaction that occurs is:

$$2\ \text{Al(s)} + 2\ \text{NaOH(aq)} + 6\ \text{H}_2\text{O}(\ell) \longrightarrow 2\ \text{NaAl(OH)}_4\text{(aq)} + 3\ \text{H}_2\text{(g)}$$

The heat generated by the reaction helps to melt any grease, and the gas being generated stirs up the particles and helps to unclog the drain. If you use 5.6 g of aluminum powder and excess NaOH, how many liters of gaseous H_2 measured at 742 mm Hg and 22.0°C are produced?

Solution The first step is to calculate moles of Al available and then to find moles of H_2 generated. Once the quantity of H_2 theoretically obtainable is known, then its volume can be calculated.

(a) Moles of Al available are

$$5.6\ \text{g Al}\left(\frac{1.0\ \text{mol Al}}{27\ \text{g Al}}\right) = 0.21\ \text{mol Al}$$

Moles of H_2 produced by the aluminum are

$$0.21\ \text{mol Al}\left(\frac{3\ \text{mol H}_2}{2\ \text{mol Al}}\right) = 0.31\ \text{mol H}_2\ \text{expected}$$

(b) From $PV = nRT$, solve for V and then substitute the conditions of the measurement. Remember that the temperature must be converted to kelvins (22.0°C = 295 K) and the pressure must be in atmospheres (742 mm Hg = 0.976 atm).

$$V = \frac{nRT}{P} = \frac{(0.31\ \text{mol})(0.0821\ \text{L} \cdot \text{atm/K} \cdot \text{mol})(295\ \text{K})}{(0.976\ \text{atm})}$$

$$= 7.7\ \text{L H}_2$$

EXERCISE 6.9 GAS LAWS AND STOICHIOMETRY

Gaseous oxygen reacts with aqueous hydrazine to produce water and gaseous nitrogen according to the balanced equation

$$\text{N}_2\text{H}_4\text{(aq)} + \text{O}_2\text{(g)} \longrightarrow 2\ \text{H}_2\text{O}(\ell) + \text{N}_2\text{(g)}$$

If a solution contains 180 g of N_2H_4, what is the maximum volume of O_2 that will react with the hydrazine if the O_2 is measured at a barometric pressure of 750 mm Hg and room temperature (21°C)?

FIGURE 6.9

The reaction of sodium hydroxide with powdered aluminum to produce H_2 gas and sodium aluminate, $NaAl(OH)_4$. The reaction is used by popular brands of drain cleaners. (Charles D. Winters)

6.5 GAS MIXTURES AND PARTIAL PRESSURES

The air you breathe is a blend of oxygen, nitrogen, carbon dioxide, water vapor, and small amounts of other gases. It is logical that the ideal gas law applies to this gas mixture as well as to pure gases. There is, after all, no quantity in the ideal gas law that depends in any way on the chemical constitution of the gas molecules. Only by doing experiments dealing with the chemical properties of a gas can you tell a mixture of gases from a pure gas.

In applying the ideal gas law to gas mixtures, the number of moles of gas present in the sample is taken as the total of all the moles of the components; that is,

$$n_{\text{total}} = n_{\text{A}} + n_{\text{B}} + \cdots \tag{6.7}$$

It is the total number of moles of gas, n_{total}, that is used as n in the gas law. For gas law calculations, a mixture of 0.5 mole of O_2 and 0.5 mole of N_2 is indistinguishable from 1.0 mole of either gas, for example.

The pressure exerted by each component in a mixture is related to the number of moles of that component. Suppose you fill a balloon of volume V with a mixture of nitrogen and oxygen to a pressure P at a temperature T (Figure 6.10). The gas law applied to this mixture is

$$PV = n_t RT \tag{6.8}$$

where n_t represents moles of N_2 plus moles of O_2. The equation can be rewritten in the form

$$PV = (n_{\text{N}_2} + n_{\text{O}_2})RT$$

where n_{N_2} = moles of nitrogen and n_{O_2} = moles of oxygen. Expanding the right side of this equation and rearranging, we get

$$PV = n_{\text{N}_2}RT + n_{\text{O}_2}RT$$

and

$$P = \frac{n_{\text{N}_2}RT}{V} + \frac{n_{\text{O}_2}RT}{V}$$

Now, $n_{\text{N}_2}RT/V$ is the pressure (P_{N_2}) that would be exerted at a temperature T in a volume V by n_{N_2} moles of nitrogen if there were only molecules of nitrogen present. Similarly, $n_{\text{O}_2}RT/V$ is the pressure that would be exerted by the O_2 molecules alone in the same volume V. Thus,

$$P_{\text{total}} = P_{\text{N}_2} + P_{\text{O}_2}$$

The quantities P_{N_2} and P_{O_2} are called the **partial pressures** of the two components. John Dalton was the first to formulate the idea that *the total pressure exerted by a mixture of gases is the sum of the partial pressures of the individual gases in the mixture*. In his honor this statement is now called **Dalton's law of partial pressures**.

Dalton is an important figure in the history of chemistry. See Section 2.1.

To use Dalton's law in working with gas mixtures, we shall write an expression for the ratio of the partial pressure of one gas A (P_{A}) in the mixture to the total pressure of the mixture (P_{total}). Since $P = nRT/V$, we can write

$$\frac{P_A}{P_{total}} = \frac{n_ART/V}{n_{total}RT/V} = \frac{n_A}{n_{total}}$$

Notice that the ratio of the partial pressure of gas A to the total pressure of the gas mixture is the same as the ratio of the moles of gas A to the total moles of gas. This ratio, n_A/n_{total}, is called the **mole fraction** of A, and it can be given the symbol X_A.

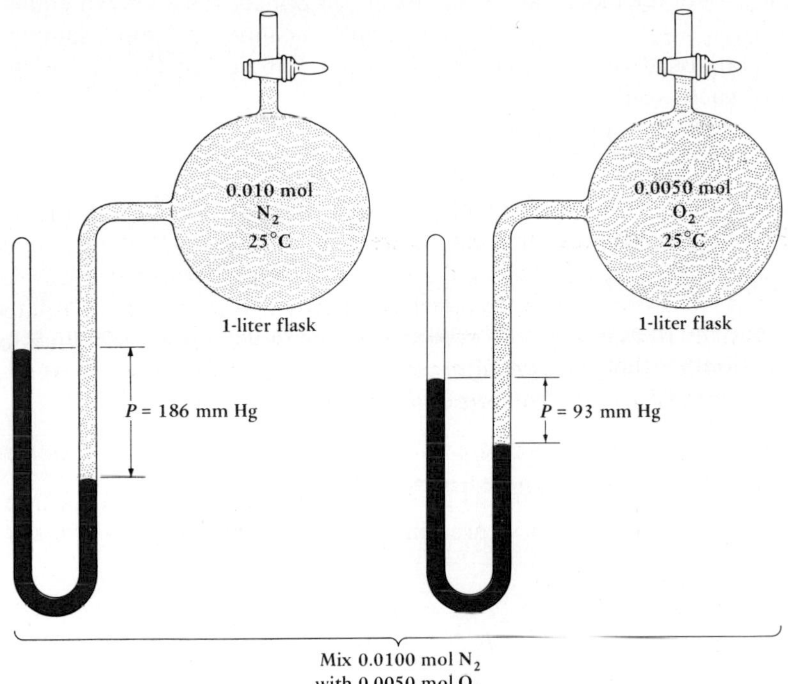

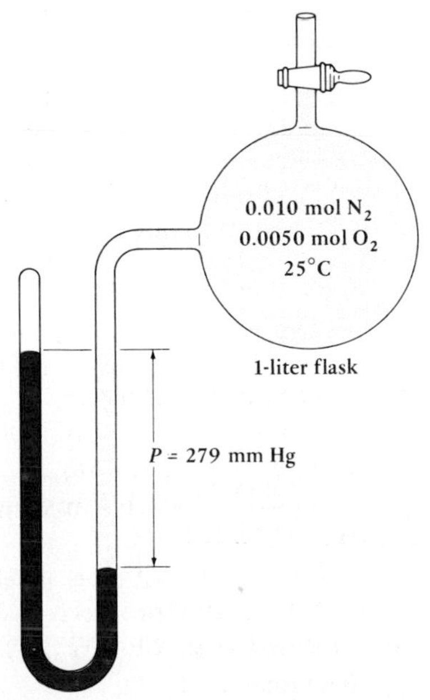

FIGURE 6.10

Dalton's law. (a) 0.010 mole of N_2 in a 1.0-L flask at 25°C exerts a pressure of 186 mm Hg. (b) 0.0050 mole of O_2 in a 1.0-L flask at 25°C exerts a pressure of 93 mm Hg. (c) The 0.010 mole sample of N_2 and 0.0050 mole sample of O_2 are mixed in the same 1.0-L flask at 25°C. The total pressure, 279 mm Hg, is the total of the pressures each gas would have if it were alone in the flask.

$$\frac{P_A}{P_{total}} = \frac{n_A}{n_{total}} = \text{mole fraction A} = X_A \quad (6.9)$$

In general, the mole fraction of any component A of a mixture is

$$\text{Mole fraction of component A} = X_A = \frac{\text{moles of A}}{\text{moles A} + \text{moles B} + \text{moles C} + \cdots} \quad (6.10)$$

where the sum of the mole fractions of each component must *always* equal 1 ($X_A + X_B + X_C + \cdots = 1$). As an example, assume a 5.0-mole sample of air is composed of 3.9 moles of N_2 and 1.1 moles of O_2, the mole fraction of each would be

$$X_{N_2} = \frac{3.9 \text{ mol } N_2}{3.9 \text{ mol } N_2 + 1.1 \text{ mol } O_2} = 0.78$$

$$X_{O_2} = \frac{1.1 \text{ mol } N_2}{3.9 \text{ mol } N_2 + 1.1 \text{ mol } O_2} = 0.22$$

where $X_{N_2} + X_{O_2} = 1$.

Applying the concept of the mole fraction to gas laws, you can see from equation 6.9 that *the partial pressure of a gas in a mixture is equal to the product of its mole fraction and the total pressure.*

$$P_A = X_A P_{total} \quad (6.11)$$

Partial pressure of gas A = (mole fraction of A)(total pressure)

Applied to a sample of air at a pressure of say 742 mm Hg, this means that

$$\text{Partial pressure of } N_2 = P_{N_2} = (0.78)(742 \text{ mm Hg}) = 579 \text{ mm Hg}$$
$$\text{Partial pressure of } O_2 = P_{O_2} = (0.22)(742 \text{ mm Hg}) = 163 \text{ mm Hg}$$

EXAMPLE 6.13

PARTIAL PRESSURES OF GASES

Since fluorocarbons are no longer thought to be environmentally safe as aerosol propellants, chemists have had to devise substitutes. Two gases that have been used are propane (C_3H_8) and isobutane (C_4H_{10}). Suppose you have prepared a mixture of 22 g of propane and 11 g of isobutane. This mixture is then forced into a can until the total pressure is 1.5 atm. What are the partial pressures of propane and isobutane?

Solution The partial pressure of each gas in a mixture is equal to the mole fraction of that gas times the total pressure. The total pressure is known, and other data are given that allow mole fractions to be calculated.

Step 1. Calculate mole fractions.

$$\text{Moles } C_3H_8 = 22 \text{ g } C_3H_8 \left(\frac{1.0 \text{ mol } C_3H_8}{44 \text{ g } C_3H_8}\right) = 0.50 \text{ mole}$$

$$\text{Moles } C_4H_{10} = 11 \text{ g } C_4H_{10} \left(\frac{1.0 \text{ mol } C_4H_{10}}{58 \text{ g } C_4H_{10}}\right) = 0.19 \text{ mole}$$

$$\text{Mole fraction } C_3H_8 = \frac{0.50 \text{ mol } C_3H_8}{0.69 \text{ total moles}} = 0.72$$

Since

$X_{C_3H_8} + X_{C_4H_{10}} = 0.72 + X_{C_4H_{10}} = 1$

we now know that

$X_{C_4H_{10}} = 0.28$

Step 2. Calculate partial pressures.

Partial pressure of $C_4H_{10} = 0.28\ P_{total} = 0.28\ (1.5\ atm) = 0.42\ atm$

Since

$P_{C_3H_8} + P_{C_4H_{10}} = 1.5\ atm$

this means

$P_{C_3H_8} = 1.5\ atm - 0.42\ atm = 1.1\ atm$

One practical application of Dalton's law is a laboratory experiment where you generate a gas such as N_2 or H_2 and collect it by displacing water from a container (Figure 6.11). Here the total gas pressure in the collecting flask is the sum of the pressures of the N_2 or H_2 *and* the water vapor. Many liquids such as water evaporate to some extent, and the vapor exerts a pressure that depends on the temperature of the system. Appendix F provides the vapor pressures of water at different temperatures, and the concept of vapor pressure is described in more detail in Chapter 11.

EXAMPLE 6.14

PARTIAL PRESSURES: COLLECTING A GAS OVER WATER

Small quantities of H_2 gas can be prepared in the laboratory by the following reaction.

$$Fe(s) + 2\ HCl(aq) \longrightarrow Fe^{2+}(aq) + 2\ Cl^-(aq) + H_2(g)$$

Assume that you carried out this experiment and collected 0.500 L of H_2 gas as illustrated in Figure 6.11. The temperature of the gas mixture was 26.0°C, and

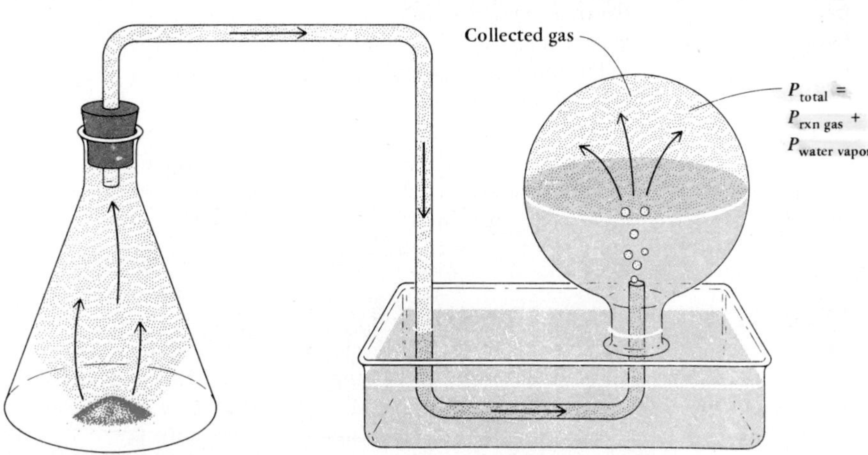

FIGURE 6.11
Collecting a gas over water. The pressure of the generated gas forces some of the water from the collection flask. Thus, when the pressure of the collected gas is measured (P_{total}), it is the sum of the pressure of the gas from the reaction ($P_{rxn\ gas}$) *plus* the pressure generated by water vapor from the evaporation of water in the collection flask ($P_{water\ vapor}$). See Example 6.14.

the total pressure of gas in the flask was 745 mm Hg. How many total moles of gas (hydrogen + water vapor) were there in the flask, and how many moles of hydrogen did you prepare?

Solution

Step 1. Calculate the total number of moles of gas.

P = total pressure = 745 mm = 0.980 atm
V = 0.500 L
T = 26.0°C = 299 K
n = total moles of gas = moles H_2 + moles H_2O vapor = ?

$$n = \frac{PV}{RT} = \frac{(0.980 \text{ atm})(0.500 \text{ L})}{(0.08206 \text{ L} \cdot \text{atm/K} \cdot \text{mol})(299 \text{ K})} = 0.0200 \text{ total mole of gas}$$

Step 2. Calculate moles of $H_2(g)$. There are two gases present in the receiver: H_2 and H_2O vapor from the evaporation of the liquid water in the trough in Figure 6.11. Therefore, 0.0200 mole is the sum of the number of moles of each of these gases. Likewise, the total pressure is the sum of the partial pressures of H_2 and H_2O.

$$P_{\text{total}} = P_{H_2} + P_{H_2O}$$

To find the moles of H_2, you must know the partial pressure of H_2.

$$P_{H_2} = P_{\text{total}} - P_{H_2O}$$

The partial pressure of water vapor over liquid water can be obtained from tables in a handbook (see Appendix F in this text). Here P_{H_2O} is 25.0 mm Hg, so P_{H_2} = 720. mm Hg.

Now you can use the ratio of P_{H_2} to the total pressure to find the mole fraction of H_2 in 0.0200 total mole of gas.

$$\text{Mole fraction } H_2 = \frac{P_{H_2}}{P_{\text{total}}} = \frac{720. \text{ mm Hg}}{745 \text{ mm Hg}} = 0.966$$

Moles H_2 = X_{H_2} (total moles) = (0.966)(0.0200 mol) = 0.0193 mol H_2

Alternatively, you could use the partial pressure of H_2 (in atm = 0.947) in the gas law equation to find n_{H_2}.

$$n_{H_2} = \frac{P_{H_2}V}{RT} = \frac{(0.947 \text{ atm})(0.500 \text{ L})}{(0.08206 \text{ L} \cdot \text{atm/K} \cdot \text{mol})(299 \text{ K})} = 0.0193 \text{ mol } H_2$$

EXERCISE 6.10 PARTIAL PRESSURES

One theory of the origin of life is that simple molecules such as H_2, O_2, N_2, and CO_2 were combined to form more complicated molecules, the energy being supplied by atmospheric lightning. In an experiment to see whether N_2 and H_2 will combine in the presence of an electrical discharge to form ammonia (NH_3), a 1.0 L flask was filled with N_2 at STP and a 3.0 L flask was filled with H_2 at STP. These two flasks were then connected through a valve and the valve was opened, allowing the gases to mix in the two flasks. What are the partial pressures of the N_2 and H_2 in the gas mixture?

EXERCISE 6.11 PARTIAL PRESSURES

In an experiment similar to that in Figure 6.11, 352 mL of gaseous nitrogen were collected in a flask over water at a temperature of 22.0°C. The total pressure of the gas in the collection flask was 742 mm Hg. How many grams of N_2 were collected? (See Appendix F for water vapor pressure.)

6.6 KINETIC MOLECULAR THEORY OF GASES

The macroscopic behavior of gases, as summarized by the ideal gas law, gives us little information about how the *molecules* of a gas behave. This is the question we wish to take up now, and the treatment we use, *the kinetic molecular theory of gases*, was developed many years ago by scientists such as Daniel Bernoulli, Rudolph Clausius, James Clerk Maxwell, and Max Boltzmann.

Our concept of a gas is that of particles in motion. If your friend is wearing a perfume or your pizza smells good, how do you know it? To put such lovely effects into scientific terms, molecules of the perfume or food odor enter the gas phase and drift through space until they reach the cells of your body that react to odors. The same thing happens in the laboratory when bottles of aqueous ammonia and hydrochloric acid sit side by side and form ammonium chloride in the air (Figure 6.12).

FIGURE 6.12
Illustration of the movement of molecules in the gas phase. Open beakers of aqueous ammonia and HCl are placed side by side. When molecules of NH_3 and HCl escape from the beakers and encounter one another, you observe the formation of ammonium chloride (NH_4Cl) as a white cloud. (Charles D. Winters)

If you changed the temperature of the environment of the bottles in Figure 6.12 and then compared the times needed for a cloud of ammonium chloride to begin to form, you would find that the time would be longer at lower temperatures. Obviously, molecules in the gas phase move more slowly at lower temperature: molecular speeds depend on temperature. In fact, it can be shown that the *average kinetic energy*, $\overline{KE}$, *of a collection of gas molecules depends only on the temperature.*

$$\overline{KE} \propto T \qquad (6.12)$$

The horizontal bar over KE indicates an average value. The kinetic energy of any *one* molecule is given by

$$KE = \tfrac{1}{2}mu^2 \qquad (6.13)$$

where u is the speed of that molecule. If you were to measure the speeds of the trillions of molecules in a gas sample, however, you would find that some have a speed u_1, some have a different speed u_2, and so on. The *average* speed $\overline{u}$ of the molecules is

$$\overline{u} = \frac{u_1 + u_2 + \cdots}{N} \qquad (6.14)$$

where N is the total number of molecules. This means the *average* kinetic energy of many molecules, $\overline{KE}$, must be related to the *average of the squares* of their speeds, $\overline{u^2}$, by the equation

The horizontal bar over KE and u indicates an average value.

$$\overline{KE} = \tfrac{1}{2}m\overline{u^2} \qquad (6.15)$$

Now, since KE is proportional to T *and* to $m\overline{u^2}$, this means $\frac{1}{2}m\overline{u^2} \propto T$, and we can say that

$$\tfrac{1}{2}m\overline{u^2} = CT \qquad (6.16)$$

where C is the proportionality constant. We shall return to this useful relation in a moment.

$\overline{u^2}$ is called the mean square speed.

Another observation you have made about gases is that they are readily compressible. For example, you can "pump up" a tire by forcing or compressing gas into it. By contrast, it is difficult to squeeze a solid much if at all, and liquids can only be slightly compressed. This must mean that the distance between gas particles (atoms or molecules) is very large in relation to the actual size of the particles, whereas the opposite is true in solids and liquids. Finally, we noted earlier that gases occupy completely the volume available to them, in contrast to liquids and solids. This is evidence that the forces of attraction between gas molecules are weak, and the molecules are easily separated and free to move.

The principal tenets of the **kinetic molecular theory of gases** are a summary of the preceding ideas.

1. Gases consist of molecules whose separation is much larger than the molecules themselves.
2. The molecules of a gas are in continual, random, and rapid motion.
3. The average energy of the gas molecules is determined by the temperature of the gas.
4. The molecules collide with each other and with the walls of their container, but they do so without net loss of energy.*

A gas that strictly obeys all the points above is said to be an "ideal" gas. In practice, as you will see below, there is no such thing as an ideal gas, but some do come close to this behavior. If you assume an ideal gas, though, it is possible to derive the ideal gas law from the statements outlined above.

KINETIC MOLECULAR THEORY AND THE GAS LAWS

The gas laws, which came from experiment, can be explained by the kinetic molecular theory. The starting place is to write equations describing how pressure arises from the collisions of gas molecules with the container walls. The force developed by these collisions depends on how many collisions there are per unit time and the average force exerted in each collision. When the temperature of a gas is increased, the average force of a collision on the walls of its container increases because the average kinetic energy of the molecules increases. Also, since the speed increases with temperature, there are more collisions per second. Thus, the collective force per unit area is greater, and the pressure increases. Mathematically, $P \propto T$ when n and V are constant.

Increasing the number of molecules of gas at a fixed temperature and volume does not change the average collision force, but it does increase the number of collisions occurring per second. Thus, the pressure increases, and so we can say that $P \propto n$ when V and T are constant.

If the pressure is not allowed to increase when either the number of moles of gas or temperature is increased, the volume of the container (the area over which the collisions can take place) must increase. This is expressed by the relation $V \propto nT$ when P is constant, a statement that is a combination of *Avogadro's law* and *Charles's law*.

---

*Scientists call such collisions "perfectly elastic"; colliding billiard balls come close to having elastic collisions, while basketballs have inelastic collisions. In a perfectly elastic collision of a fast and a slow molecule, the fast one will slow down and the slow one will speed up, but the sum of the kinetic energies of the two molecules will be the same after the collision as before.

Finally, if n and T are fixed, and the volume of the container is decreased, you know from *Boyle's law* that the pressure of the gas will increase. Although the number of molecules is constant, and the average impact force is constant at constant T, the number of collisions per second increases, and the pressure increases. This means that $P \propto 1/V$ when n and T are constant.

DISTRIBUTION OF MOLECULAR SPEEDS

The speed of molecules in the gas phase and the relative number having that speed can be measured experimentally. If the number of molecules having a particular speed is plotted *versus* that speed, the result is a **Boltzmann distribution curve** illustrated in Figure 6.13. Here you see that some molecules are fast (have a high kinetic energy), while others are slow (have a low kinetic energy). However, there is a most common speed, the speed at which the maximum in the distribution curve is observed.

From Equation 6.16 you know that average of the square of molecular speed ($\overline{u^2}$) and temperature are directly related, so the shape of the speed distribution curve must change with temperature. For example, the maximum shifts to a higher average speed at a higher temperature; the number of molecules with low speeds decreases, while the number with higher speeds increases. Even though the curve for the higher temperature is flatter and broader than that at lower temperature, the areas under the two curves are the same because the number of molecules in the sample is fixed.

One last point concerns molecular mass, average speed, and temperature. The average energy, $\overline{KE}$, is fixed by the temperature, so the product ($\frac{1}{2}m\overline{u^2}$) (Equation 6.16) is also fixed. Since two gases with different molecular masses must still have the same average kinetic energy at the

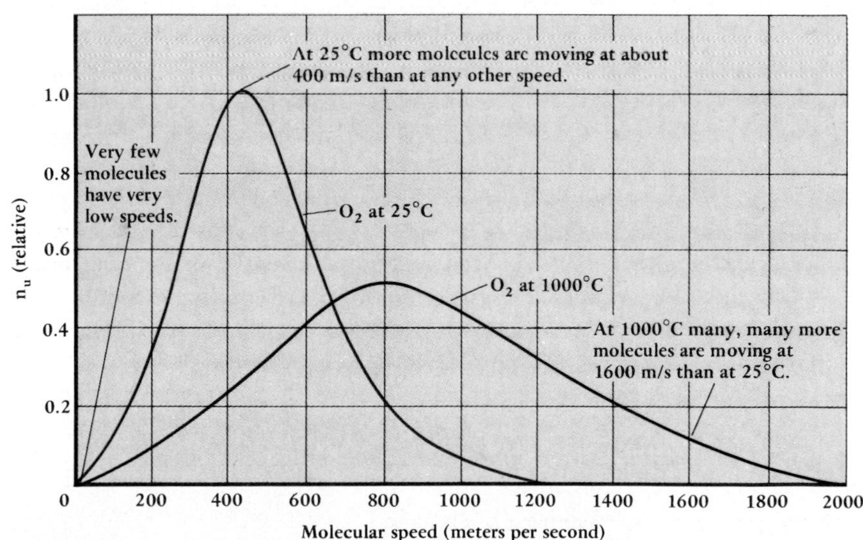

FIGURE 6.13
A plot of the relative number of molecules with a given speed versus that speed. The curve is called a Boltzmann distribution of molecular speeds. The purple curve shows the effect on the distribution of increasing the temperature of the gas.

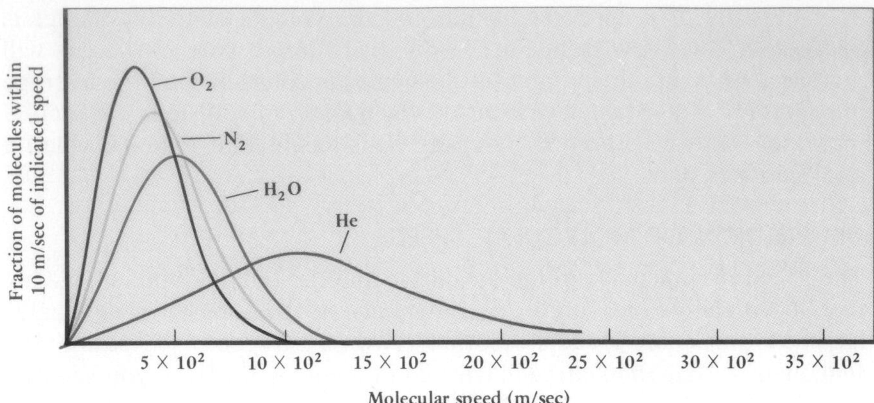

FIGURE 6.14
Effect of molecular mass on the Boltzmann distribution curve at a given temperature. Notice the similarity of the effect of increasing mass in this figure to the effect of temperature in Figure 6.13.

same temperature, the heavier gas molecules must have a lower average speed (Figure 6.14). James Maxwell (1831–1879) expressed this relation in a quantitative form when he showed that the root of the mean squared speed ($\sqrt{\overline{u^2}}$, called the **root-mean-square** or **rms speed**), the temperature (T), and the molar mass (M) are related by the equation (called "Maxwell's equation")

$$\sqrt{\overline{u^2}} = \sqrt{\frac{3RT}{M}} \tag{6.17}$$

(Here R is the gas constant expressed in appropriate units, 8.314 J/K · mol.)

EXAMPLE 6.15

MOLECULAR SPEED

Calculate the rms speed of oxygen molecules at 25°C.

Solution We must use Equation 6.17 with M in units of kg/mol. The reason for this is that R is in units of J/K · mol, and 1 J = 1 kg · m²/s². Thus, the molar mass of O_2 is 32.0×10^{-3} kg/mol.

$$\sqrt{\overline{u^2}} = \sqrt{\frac{3RT}{M}} = \sqrt{\frac{3(8.314 \text{ J/K} \cdot \text{mol})(298 \text{ K})}{(32.0 \times 10^{-3} \text{ kg/mol})}}$$
$$= \sqrt{2.32 \times 10^5 \text{ J/kg}}$$

To obtain the answer in m/s, we use the relation 1 J = 1 kg · m^2/s^2. This means we have

$$\sqrt{\overline{u^2}} = \sqrt{2.32 \times 10^5 \text{ kg} \cdot \text{m}^2/\text{kg} \cdot \text{s}^2}$$
$$= \sqrt{2.32 \times 10^5 \text{ m}^2/\text{s}^2}$$
$$= 482 \text{ m/s}$$

This speed is equivalent to about 990 miles per hour!

EXERCISE 6.12 MOLECULAR SPEEDS
Calculate the rms speed of helium atoms at 25°C.

One application of the gas laws arises in deep sea diving. To explore the floor of the oceans, scientists have built diving ships to operate at

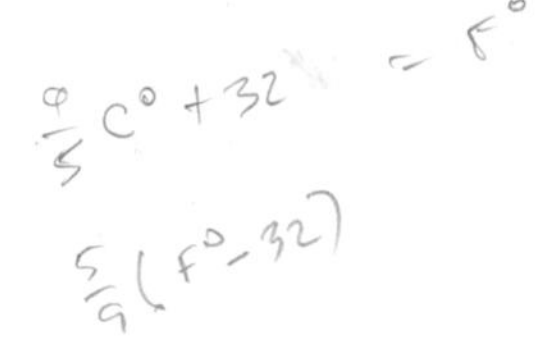

great depths. These ships are filled with an atmosphere of oxygen and helium at high pressures, but the oxygen is at a lower mole fraction than in normal sea-level air. It has been found that our bodies function best when the partial pressure of oxygen is about 0.21 atm. If the pressure of gas one breathes is 2 atmospheres, for example, and the composition of the gas were that of normal air, this would make the partial pressure of oxygen about 0.4 atm. Such high oxygen partial pressures are toxic, so the gas divers breathe must have a lower mole fraction of oxygen.

Not only is the fraction of oxygen reduced in diving gases, but the nitrogen of normal air is replaced by helium. Nitrogen is more soluble in blood and body fluids at high pressures and leads to *nitrogen narcosis*, a condition similar to alcohol intoxication. Helium is less soluble and is thus more suitable to dilute the oxygen. The problem with helium is that divers inside a diving chamber feel decidedly chilly, even when the inside temperature is a normally comfortable 70°C. The reason for this can be explained by the kinetic molecular theory. Helium atoms move with an average velocity about 2.6 times that of nitrogen molecules at the same temperature (see Exercise 6.12). When gas molecules strike the human body, they carry away some of the heat generated by metabolism. Since He atoms move so much more rapidly, they collide more frequently with the body and are thus more efficient in carrying away heat energy.

EXERCISE 6.13 KINETIC THEORY OF GASES

Consider increasing one gas parameter (P, V, n, or T) at a time. Which one has the same effect on the average molecular speed as increasing the mass of the gas molecules?

6.7 GRAHAM'S LAWS OF DIFFUSION AND EFFUSION

Gaseous **diffusion** is the gradual mixing of the molecules of two or more gases owing to their molecular motions, and Figures 6.12 and 6.15 are excellent illustrations of this property of gases. **Effusion** is closely related to diffusion. In Figure 6.16 gas molecules are held in one compartment, but they gradually escape to an empty compartment through small openings in the barrier.

Thomas Graham (1805–1869), a Scottish chemist, studied the effusion of gases and found experimentally that the rates of effusion of two gases were inversely proportional to the square roots of their molar masses at the same temperature and pressure.

$$\frac{\text{Rate of effusion of gas 1}}{\text{Rate of effusion of gas 2}} = \sqrt{\frac{M \text{ of gas 2}}{M \text{ of gas 1}}} \tag{6.18}$$

This is known as **Graham's law**, an equation explained readily by Maxwell's equation (Equation 6.17). Clearly, the rate of effusion, the amount of material moving from one place to another in a given amount of time, will depend on the root-mean-square speed. Therefore, the ratio of these speeds is the same as the ratio of effusion rates.

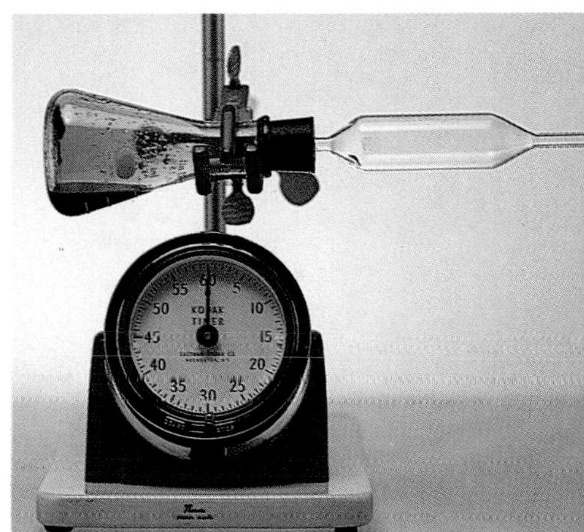

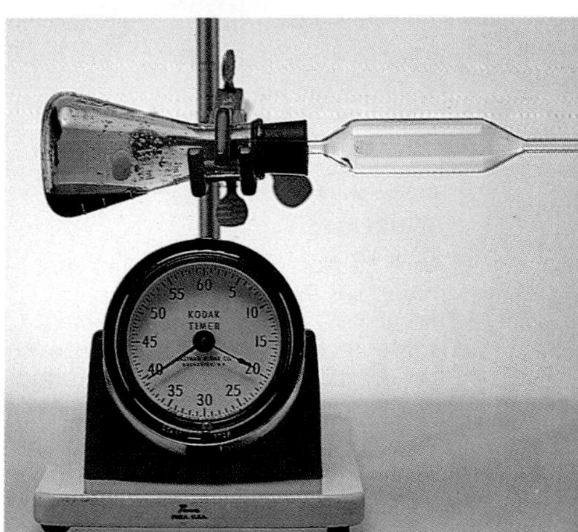

FIGURE 6.15
Brown NO_2 gas diffuses out of the flask in which it was generated and into the attached tube in a matter of minutes. (The NO_2 was made by reacting copper with nitric acid. You may notice green-blue crystals of another reaction product, copper(II) nitrate, on the inside of the flask.) (Charles D. Winters)

$$\frac{\text{Rate of effusion of 1}}{\text{Rate of effusion of 2}} = \frac{\sqrt{\overline{u^2}} \text{ of gas 1}}{\sqrt{\overline{u^2}} \text{ of gas 2}} = \sqrt{\frac{3RT/(M \text{ of gas 1})}{3RT/(M \text{ of gas 2})}} \quad (6.19)$$

Canceling out like terms, we have the simple expression developed by Graham.

Be sure to notice in Graham's law that there is an *inverse* relation between speed and molecular weight.

$$\frac{\text{Rate of effusion of 1}}{\text{Rate of effusion of 2}} = \sqrt{\frac{M_2}{M_1}} \quad (6.20)$$

In Figure 6.16 you see that a greater relative number of H_2 molecules effuse in a given time than N_2 molecules. Graham's law now allows a quantitative comparison of their relative rates.

$$\frac{\text{Rate of effusion of } H_2}{\text{Rate of effusion of } N_2} = \sqrt{\frac{M \text{ of } N_2}{M \text{ of } H_2}} = \sqrt{\frac{28.0 \text{ g/mol}}{2.02 \text{ g/mol}}} = \frac{3.72}{1}$$

This calculation tells you that H_2 molecules will effuse through the barrier 3.72 times faster than the N_2 molecules.

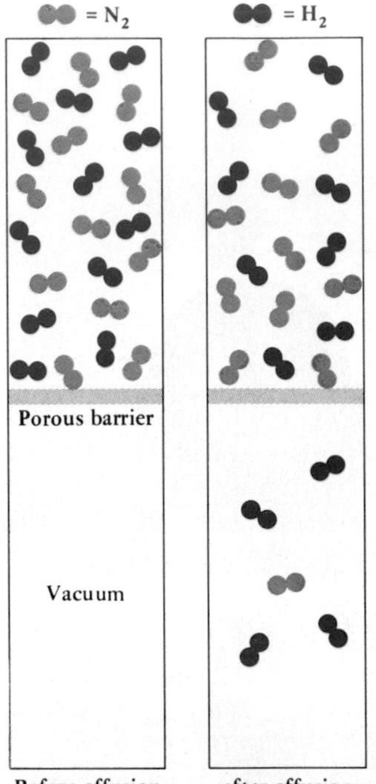

FIGURE 6.16

An illustration of the effusion of gas molecules through the pores of a membrane or other porous barrier. Lighter molecules with greater speed move through the channels more rapidly than heavier, slower molecules.

EXAMPLE 6.16

GRAHAM'S LAW OF EFFUSION

Tetrafluoroethylene, C_2F_4, effuses through a barrier at the rate of 4.6×10^{-6} moles per hour. An unknown gas, consisting only of nitrogen, oxygen, and fluorine atoms, effuses at the rate of 6.5×10^{-6} moles per hour under the same conditions. What is the molar mass of the unknown gas? Suggest a formula for the unknown gas.

Solution Graham's law tells us a lighter molecule effuses more rapidly than a heavier one. Since the unknown gas effuses more rapidly than C_2F_4 (M = 100. g/mol), the unknown must have a molar mass less than 100.

$$\frac{\text{Rate of unknown}}{\text{Rate of } C_2F_4} = \sqrt{\frac{M \text{ of } C_2F_4}{M \text{ of unknown}}}$$

$$\frac{6.5 \times 10^{-6} \text{ mol/hr}}{4.6 \times 10^{-6} \text{ mol/hr}} = \sqrt{\frac{100. \text{ g/mol}}{M}}$$

$$1.41 = \sqrt{\frac{100. \text{ g/mol}}{M}}$$

To solve for the unknown molar mass, square both sides of the equation,

$$2.00 = \frac{100. \text{ g/mol}}{M}$$

and then rearrange to obtain M.

$$M = 50.0 \text{ g/mol}$$

This molecule is known to contain only N, O, and F. If the atomic weights of these elements are added, one obtains 49 g/mol. Therefore, a likely formula for the unknown is simply NOF, a compound called nitrosyl fluoride.

EXERCISE 6.14 GRAHAM'S LAW

A pure sample of methane, CH_4, is found to effuse through a porous barrier in 1.50 minutes. Under the same conditions, an equal mass of an unknown gas effuses through the barrier in 4.73 minutes. What is the molar mass of the unknown gas?

Graham's law allows us to explain the earlier observation concerning the balloon used by Charles in 1783. In the brief biography of Charles, we mentioned that Charles filled the balloon with gaseous hydrogen, but he had to go to extraordinary lengths to keep the gas in the balloon. Unlike the first hot air balloons that were made of paper, Charles had to use silk and coat it with rubber. The reason for this is simple: at the same temperature, the lighter-weight H_2 molecules have a higher average speed than N_2 or O_2 molecules of air, so H_2 molecules would diffuse rapidly through paper, a more porous material than silk coated with rubber.

For much the same reason that Charles used special materials in his hydrogen-filled balloon, you should be cautious when buying a helium-filled, rubber balloon at a carnival. Lightweight helium atoms have a higher average speed than the heavier N_2 or O_2 molecules of air, so helium can rapidly diffuse through the rubber skin of a balloon and the balloon deflates more rapidly than a comparable air-filled balloon.

GRAHAM'S LAW AND ISOTOPE SEPARATION

In the race to construct atomic explosives during World War II, considerable quantities of uranium enriched in the ^{235}U isotope were needed. One technique that was developed uses gas effusion, a method still in use today (Figure 6.17). To separate the ^{235}U isotope from the more abundant ^{238}U isotope, all the uranium was converted to UF_6, one of the few metal compounds that can be a gas at room temperature. Then, just as N_2 and H_2 were separated by diffusion in Figure 6.16, the lighter molecules of $^{235}UF_6$ are separated by diffusion from the heavier molecules of $^{238}UF_6$.

FIGURE 6.17

The gaseous diffusion plant for separation of uranium isotopes at Oak Ridge National Laboratory. Interior view of process equipment shows arrangement of piping, compressors, motors, and large tank-shaped "diffusers." Uranium in the form of a gaseous compound (uranium hexafluoride) is pumped into the diffuser vessels where, at each stage, the uranium is slightly enriched in the fissionable U-235 isotope. For use as fuel in nuclear power plants, the uranium is enriched to approximately 3% U-235 (compared to .7% for uranium found in the natural state). (Oak Ridge National Laboratory)

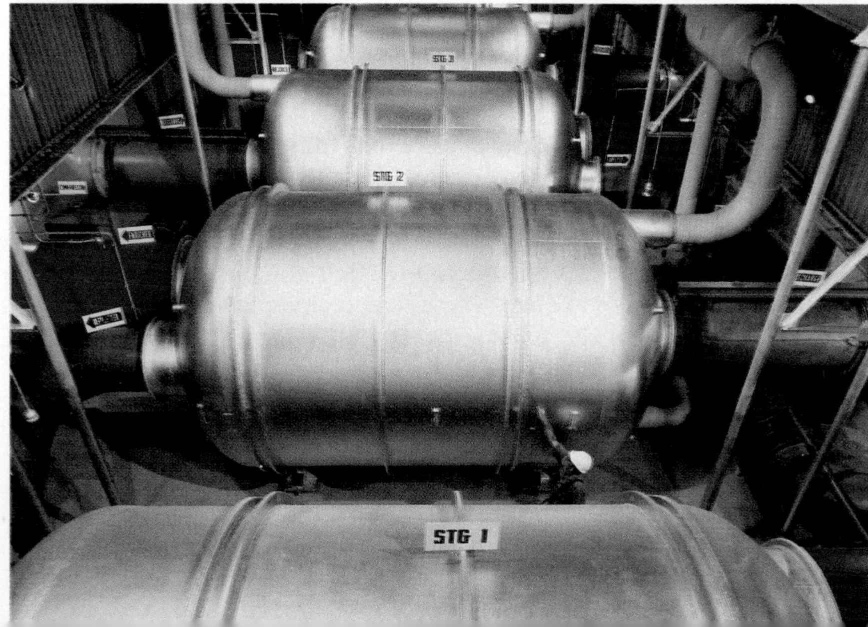

Graham's law allows us to answer some intriguing questions about isotope separation by gaseous effusion. For example, what is the maximum separation of uranium isotopes that can be expected after a single effusion step? The solution is to use Graham's law to calculate the ratio of the effusion rates of the two isotopes.

$$\frac{\text{Rate of } ^{235}\text{UF}_6}{\text{Rate of } ^{238}\text{UF}_6} = \sqrt{\frac{352}{349}} = 1.0043 = \text{enrichment factor}$$

Since the lighter $^{235}UF_6$ is the faster molecule, more molecules of $^{235}UF_6$ than of $^{238}UF_6$ will pass through the effusion barrier per second. There will be 1.0043 times as many $^{235}UF_6$ molecules as $^{238}UF_6$ molecules after one passage through the porous barrier (the rest of the $^{238}UF_6$ molecules remain behind the barrier and are removed from the system). Clearly, many effusion steps are required to give a reasonable degree of isotope separation.

6.8 NON-IDEAL GASES

If you are working with a gas at approximately room temperature and at pressures of 1 atmosphere or less, the "ideal" gas law is remarkably successful in relating the pressure, volume, temperature, and mole quantities of the gas. However, at much higher pressures or at low temperatures, serious deviations from the ideal gas law sometimes set in. The origins of these deviations are easy to understand in terms of breakdowns in the assumptions of the kinetic theory of gases.

The most easily visualized problem arises from the assumption of the kinetic molecular theory that gas molecules occupy a negligibly small volume relative to the container volume. At STP, the average molar volume of an ideal gas is 22.4 L/mol (22.4×10^{-3} m^3/mol). This means that the volume available to one atom of a gas such as helium is

$$V = \frac{22.4 \times 10^{-3} \text{ m}^3/\text{mol}}{6.02 \times 10^{23} \text{ molec/mol}} = 3.72 \times 10^{-26} \text{ m}^3/\text{molec}$$

If we assume this volume is a spherical space, we can calculate that the radius of this imaginary sphere is about 2000 picometers.* Since we know from other experiments that the actual radius of a helium atom is only 50 pm, this means that there is much free space in a sample of gaseous helium at STP. However, if the pressure of the helium is greatly increased, say to 500 atm at 0°C, the volume of free space allowed for the helium atom is severely reduced (to a sphere only about 260 pm in radius). Interactions between helium atoms are more likely, and the ideal gas law is no longer applicable. The effect of the reduction in free volume at high pressures is that pressure changes will lead to smaller volume changes than predicted by the ideal gas law. The volume occupied by a gas at high pressures,

---

$$^*r = \sqrt[3]{\frac{3V}{4\pi}} = \sqrt[3]{\frac{3(3.72 \times 10^{-26} \text{ m}^3)}{4(3.14)}} = 2.07 \times 10^{-9} \text{ m or } 2070 \text{ pm}$$

$V_{observed}$, will be *larger* than that expected on the basis of the ideal gas law, V_{ideal}.

The second assumption that can break down at high pressure or low temperature is that molecules do not attract one another. In fact they do, and these forces between molecules (*intermolecular forces*) can sometimes be large enough to cause measurable deviations from ideality. Intermolecular forces are important because they cause molecules to "stick together" momentarily and thus reduce the number of particles present in the gas phase to less than the number that you calculate from the mass of gas. This will cause the pressure × volume product of the gas to be less than expected from the ideal gas law: for a flexible container under atmospheric pressure the actual volume will be smaller than expected, and, for a rigid container, the pressure will be smaller than expected.

Intermolecular forces are described more fully in Chapter 11. It is just these forces that allow a gas to be converted to a liquid or solid at a sufficiently low temperature or high pressure.

To summarize: (a) The finite molecular size of real gas molecules causes the observed gas volume to be greater than ideal at high pressures. (b) Attractions between gas molecules tend to make the observed gas volume smaller than ideal. The two effects oppose one another, with the former being important only at higher pressures. Figure 6.18 shows the competition between the negative effect of molecular association on molar volume at the low end of the high pressure range and the positive effect of molecular volume at the higher end of the range. The minimum in the curve marks the point at which the effects caused by molecular volume have become important, and the curve crosses the dotted line $V_{observed}/V_{ideal} = 1$ at the pressure at which the molecular volume effect exactly cancels the association effect.

The Dutch physicist Johannes van der Waals (1837–1923) studied the breakdown of the ideal gas law equation and developed another equation to describe gases at high pressures or gases in which effects of in-

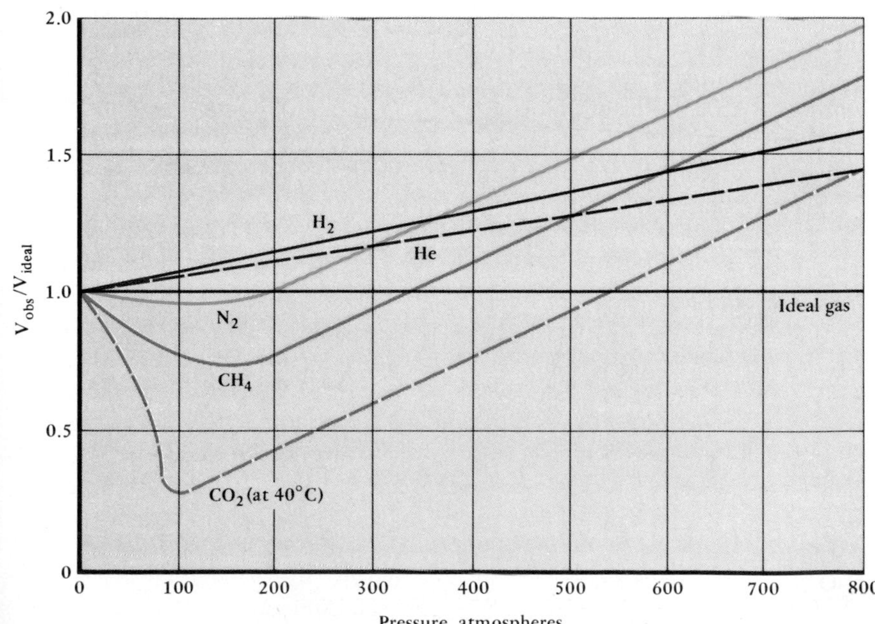

FIGURE 6.18
The effect of pressure on the molar volumes of gases. V_{obs} is the experimental molar volume at 25°C. If this volume were equal to V_{ideal} (from $PV = nRT$), then $V_{obs}/V_{ideal} = 1$.

termolecular forces can be appreciable. Under such circumstances, the so-called **van der Waals equation** makes better predictions than the ideal gas law.

The van der Waals Equation

$$[P_{obs} + a\left(\frac{n}{V}\right)^2][V_{obs} - bn] = nRT$$

correction for intermolecular forces (the $a(n/V)^2$ term)

correction for molecular volume (the bn term)

P_{obs} = observed pressure

V_{obs} = observed volume

a and b = van der Waals constants

Although this equation might appear complicated at first glance, the terms in square brackets are simply those of the ideal gas law, each corrected for the effects discussed above.

The pressure correction term, $a(n/V)^2$, allows for intermolecular forces. The term (n/V) is the concentration of the gas in the absence of association due to intermolecular forces. Since the actual gas pressure is lower than the ideal pressure (from $PV = nRT$) owing to intermolecular forces, this means that $a(n/V)^2$ is *added* to the observed pressure, P_{obs}, to make this portion of the equation equal to P in the $PV = nRT$ equation. The constant a typically has values in the range 0.01 to 10 atm(L/mol)^2, but the values have no simple relation to other molecular properties.

The "bn" term in van der Waals's equation corrects the container volume to a smaller value, the volume actually available to the gas molecules. Here n is the number of moles of gas and b is an experimental quantity that gives the correction (per mole) for the molecular volume. Typical values of b range from 0.01 to 0.1 L/mol, roughly increasing with increasing molecular size (Table 6.2).

As an example of the importance of these corrections, consider a sample of 8.00 moles of chlorine gas, Cl_2, in a 4.00-L tank at 27.0°C. The ideal gas law would lead you to expect a pressure of 49.2 atm. The real pressure, obtained from the van der Waals equation, however, is only 29.5 atm, 20 atm less than the ideal pressure!

TABLE 6.2 van der Waals Constants

SUBSTANCE	a (atm · L^2/mol^2)	b (L/mol)
He	0.034	0.0237
Ar	1.34	0.0322
H_2	0.244	0.0266
N_2	1.39	0.0391
O_2	1.36	0.0318
Cl_2	6.49	0.0562
CO_2	3.59	0.0427
CH_4	2.25	0.0428
H_2O	5.46	0.0305

EXERCISE 6.15 VAN DER WAALS'S EQUATION

Using both the ideal gas law and van der Waals's equation, calculate the pressure expected for 10.0 mole of helium gas in a 1.00-L container at 25°C.

SUMMARY

To describe the state of a gas one must specify the quantity (n) in moles, the temperature (T) in kelvins, the volume (V) in liters, and the pressure (P) in atmospheres. Pressure measurement is described in Section 6.1, as are other units of pressure, especially mm Hg and pascals. One standard atmosphere is equivalent to 760 mm Hg. Conditions of **STP**, standard temperature and pressure, are 273 K and 1 atm.

The experimental basis of the gas laws is outlined in Section 6.2. Boyle's law specifies that the pressure of a gas is inversely proportional to its volume ($P \propto 1/V$) at constant T and n. Charles's law and Gay-Lussac's law tell us that gas volume is directly proportional to temperature ($V \propto T$) at constant n and P. Finally, Avogadro's law specifies that gas volume is directly proportional to its quantity ($V \propto n$) at constant T and P.

If the three gas laws are combined into one statement, the **ideal gas law** ($PV = nRT$) results (Section 6.3). Since PV/nT is always equal to R, the **general gas law** can be derived from the ideal gas law:

$$\frac{P_1V_1}{n_1T_1} = \frac{P_2V_2}{n_2T_2}$$

It applies to a given quantity of gas undergoing a change in conditions

Dalton's law of partial pressures specifies that the total pressure of a mixture of gases is the sum of the **partial pressures** of the individual gases in the mixture (Section 6.5).

$$P_{\text{total}} = P_{\text{A}} + P_{\text{B}} + P_{\text{C}} + \cdots$$

The partial pressure of a gas in a mixture (P_{A}) is given by its **mole fraction** (X_{A}) times the total pressure of the mixture (P_{total}).

$$P_{\text{A}} = X_{\text{A}}P_{\text{total}}$$

The mole fraction of a component (A) of a mixture is defined as the number of moles of A divided by the total moles of all components.

The gas laws can be understood in terms of the **kinetic molecular theory**, a theory of gas behavior at the molecular level (Section 6.6). One important tenet of the theory is that the *average* kinetic energy of gas molecules ($\overline{KE}$) is proportional to the temperature of the gas.

$$\overline{KE} = \tfrac{1}{2}\, m\overline{u^2} \propto T$$

Here $\overline{u^2}$ is the average of the squares of the molecular speeds. Since $\overline{KE}$ is determined by temperature, heavier molecules (large m) must move with a slower average speed ($\overline{u}$) than lighter molecules (small m) at a given temperature.

Graham's law of effusion states that the rates of effusion of two gases (r_1 and r_2) are inversely proportional to the square roots of their molar masses [$(M)_1$ and $(M)_2$] at the same temperature and pressure.

$$\frac{r_1}{r_2} = \sqrt{\frac{(M)_2}{(M)_1}}$$

Kinetic molecular theory assumes, among other things, that gas molecules have no volume and that they do not interact with one another by **intermolecular forces**. These assumptions lead to a simple model of *ideal gas* behavior as expressed by $PV = nRT$. However, neither assumption is completely correct because a *real gas* has a more complex behavior than an ideal gas. One description of the behavior of real gases is found in the **van der Waals equation**.

STUDY QUESTIONS

REVIEW QUESTIONS

1. Name the three gas laws that interrelate P, V, and T. Explain the relationships in words and in equations.

2. What conditions are represented by STP? What is the volume of a mole of ideal gas under these conditions?

3. Show how to calculate the molar mass of a gas from P, V, and T measurements and other information.

4. Explain how to determine the density of a gas from P, V, and T data and other information.

5. State Avogadro's law. Relate your discussion to the formation of water vapor from its elements.

$$2\ H_2(g) + O_2(g) \longrightarrow 2\ H_2O(g)$$

For example, if 2 moles of H_2 are used at STP, how many liters of O_2 are required at STP and how many liters of H_2O are produced at STP?

6. State Dalton's law. If the air you breathe is 78% N_2 and 22% O_2 (on a mole basis), what is the mole fraction of O_2? What is the partial pressure of O_2 when atmospheric pressure is 745 mm Hg?

7. What are the basic assumptions of the kinetic molecular theory? Which of these assumptions are most nearly correct and which are violated to some extent by a real gas?

8. Explain Boyle's law on the basis of kinetic molecular theory.

9. In van der Waals's equation, what properties of a real gas are accounted for by the constants a and b?

10. State Graham's law in words and in equation form. If gas A effuses four times more rapidly than gas B at 25°C, does this ratio of rates of effusion increase, decrease, or remain the same as the temperature of the gas increases?

UNITS OF MEASUREMENT

11. Gas temperatures must be expressed in kelvins when doing calculations. Convert each of the following temperatures to kelvins.
(a) room temperature, 22°C
(b) 98.6°C
(c) −78°C
(d) 215°C

12. Gas pressure can be expressed in units of mm Hg, torr, atmospheres, and pascals (although mm Hg and atm are used exclusively in this book). Do the following unit conversions.
(a) 740 mm Hg to atm
(b) 1.25 atm to mm Hg
(c) 0.025 atm to mm Hg
(d) 16 mm Hg to atm
(e) 542 mm Hg to torr
(f) 45 torr to mm Hg
(g) 740 mm Hg to kPa
(h) 0.50 atm to kPa

13. Make the following conversions of pressure units.
(a) 781 mm Hg to atm
(b) 356 mm Hg to kPa
(c) 35 torr to atm
(d) 210 kPa to mm Hg
(e) 0.015 kPa to atm
(f) 95.0 kPa to mm Hg

14. Rank the following pressures in increasing order (smallest first): 150 kPa, 742 mm Hg, 0.89 atm, and 650 torr.

PRESSURE MEASUREMENT

15. Refer to page 174 and answer the following question. If the mercury levels in a U-tube manometer are as illustrated on the next page, express the gas pressure in mm Hg, atm, and torr, and kPa.

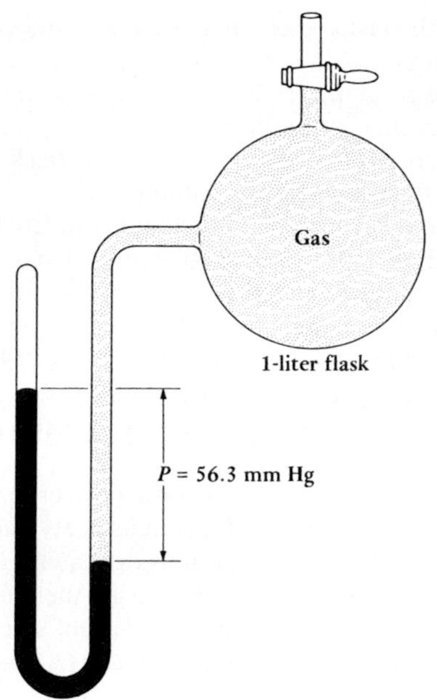

16. Refer to page 174 and answer the following question. If the pressure of a gas is 95.0 kPa, what is the difference (in mm Hg) in the mercury levels in the U-tube manometer?

THE GAS LAWS

17. A sample of HCl gas is placed in a 256-mL flask where it exerts a pressure of 67.5 mm Hg. What is the pressure of this gas sample if it is transferred to a 135-mL flask at the same temperature?
18. A sample of HCl has a pressure of 67.5 mm Hg in a 256-mL flask. If the sample is transferred to a new flask where it exerts a pressure of 23.6 mm Hg at the same temperature, what is the volume of this new flask?
19. A 25.0-mL sample of gas is enclosed in a gas-tight syringe (Figure 6.2) at 22°C. If the syringe is immersed in an ice bath (0°C), what is the new gas volume?
20. A bicycle tire is inflated to a pressure of 55 lb/in.^{2} at 15°C. If the tire is heated to 35°C, what is the pressure in the tire? (For simplicity, assume the tire volume cannot change.)
21. H_2O can be made by combining gaseous O_2 and H_2. If you begin with 1.0 L of H_2(g) at 380 mm Hg and 25°C, how many liters of O_2(g) would you need for complete reaction if the O_2 gas is also measured at 380 mm Hg and 25°C?
22. Gaseous silane, SiH_4, ignites spontaneously in air according to the reaction

$$SiH_4(g) + 2\ O_2(g) \longrightarrow SiO_2(s) + 2\ H_2O(g)$$

If 5.2 L of SiH_4 are treated with O_2, how many liters of O_2 are required for complete reaction? How many liters of H_2O vapor are produced? Assume all gases are measured at the same temperature and pressure.

THE IDEAL GAS LAW

23. A 1.00-g sample of water is allowed to vaporize completely in a 10.0-L container. What is the pressure of the water vapor at a temperature of 150°C?
24. 30.0 kg of helium are placed in a balloon. What is the volume of the balloon if the final pressure is 1.20 atm and the temperature is 22°C?
25. To find the volume of a flask, it was first evacuated so that it contained no gas at all. Next, 4.4 g of CO_2 was introduced into the flask. On warming to 27°C, the gas exerted a pressure of 730 mm Hg. What is the volume of the flask?
26. Which of the following gas samples contains the largest number of gas molecules? Which contains the smallest number?
    (a) 1.0 L of H_2 at STP
    (b) 1.0 L of Ne at STP
    (c) 1.0 L of H_2 at 27°C and 760 mm Hg
    (d) 1.0 L of CO_2 at 0°C and 800 mm Hg
27. A sample of H_2 exerts a pressure of 3.0 atm at 25°C in an explosion-proof container. What is the pressure of H_2 on heating to 110°C?
28. A sample of gaseous helium in a 1.50-L flask at 273 K exerted a pressure of 1.00 atm. After heating the flask to 346 K, what is the pressure of the gas?
29. If you have a 150.-L tank of gaseous CO and the gas exerts a pressure of 41.8 mm Hg at 25°C, how many moles of CO are there in the tank?
30. How many grams of helium are required to fill a 5.0-L balloon to a pressure of 1.1 atm at 25°C?
31. A sample of H_2 gas occupies 615 mL at 27.0°C and 575 mm Hg. When the gas is cooled, its volume is reduced to 455 mL and its pressure is reduced to 385 mm Hg. What is the new temperature of the gas?
32. A sample of gas occupies 754 mL at 22°C and a pressure of 165 mm Hg. What is its volume if the temperature is raised to 42°C and the pressure is raised to 265 mm Hg?
33. Assume that one of the cylinders of an automobile engine has a volume of 400. cm^3. The engine takes in air at a pressure of 1.00 atm and a temperature of 27°C and compresses it to a volume of 50.0 cm^3 at 77°C. What is the final pressure of the gas in the cylinder? (The ratio of before and after volumes, in this case 400:50 or 8:1, is called the compression ratio.)

34. Assume that a helium-filled balloon needs to displace at least 1.00×10^5 L of air. You fill the balloon with helium to a volume of 1.05×10^5 L on the ground where the pressure is 745 mm Hg and the temperature is 20.0°C. When the balloon ascends to a height of 2 miles where the pressure is only 600. mm Hg and the temperature is −33°C, does the balloon still displace the required 1.00×10^5 L of air?
35. 1.00 g of a gaseous compound occupies 0.820 L at 1.00 atm and 3.0°C. Which of the following is the correct formula: (a) B_4H_{10}, (b) B_2H_6, (c) B_5H_9?
36. A newly discovered gas has a density of 2.39 g/L at 23.0°C and 715 mm Hg. What is the molar mass of this gas?
37. A 0.982-g sample of an unknown gas exerts a pressure of 700. mm Hg in a 450.-mL container at 23°C. What is the molar mass of the gas?
38. Forty miles above the earth's surface the temperature is 250 K and the pressure is only 0.20 mm Hg. What is the density of air (M = 29.0 g/mol) at this altitude?
39. A common liquid ether [diethyl ether, $(C_2H_5)_2O$] vaporizes easily at room temperature. If the vapor has a pressure of 233 mm Hg in a flask at 25.0°C, what is the density of the vapor?
40. Chloroform is a common liquid used in the laboratory. It vaporizes readily. If the pressure of chloroform is 195 mm Hg in a flask at 25.0°C and the density of the vapor is 1.25 g/L, what is the molar mass of chloroform?
41. Acetaldehyde is a common liquid that vaporizes readily. A pressure of 331 mm Hg is observed in a 125-mL flask at 0.0°C, and the density of the vapor is 0.855 g/L. What is the molar mass of acetaldehyde?
42. If 12.0 g of O_2 is required to inflate a balloon to a certain size at 27°C, what mass of O_2 is required to inflate it to the same size (and pressure) at 127°C?
43. The pressure gauges on two cylinders of O_2 of equal volume register the same pressure. However, one cylinder (A) is inside the chemistry laboratory (room temperature = 23°C) and the other cylinder (B) is outside in the snow (temperature = −10°C). Which cylinder, A or B, contains the greater mass of O_2?
44. Two identical flasks are filled with different gases; both flasks are at the same temperature, and both gases exert the same pressure. The mass of gas A is 0.34 g, while that of gas B is 0.48 g. It is known that gas B is ozone, O_3. Which of the following is gas A: (a) O_2, (b) SO_2, or (c) H_2S?
45. If equal masses of O_2 and N_2 are placed in separate containers of equal volume at the same temperature, which of the following statements is true? If false, tell why it is false.
    (a) Both flasks contain the same number of molecules.
    (b) The pressure in the N_2-containing flask is greater than that in the flask containing O_2.
    (c) There are more molecules in the flask containing O_2 than in the one containing N_2.
46. You have two gas-filled balloons, one containing He and the other H_2. The H_2 balloon is twice the size of the He balloon. The pressure of gas in the H_2 balloon is 1 atm while that in the He balloon is 2 atm. The H_2 balloon is outside in the snow (30°F) while the He balloon is inside a warm building (70°F).
    (a) Which balloon contains the greater number of molecules?
    (b) Which balloon contains the greater mass of gas?
47. A 75-L steel tank at 70°F contains acetylene at a pressure of 2000 psi (psi = pounds per square inch). How many grams of acetylene (C_2H_2) are there in the tank?
48. Suppose you have two pressure-proof steel cylinders of equal volume, one containing CO (A) and the other acetylene, C_2H_2 (B).
    (a) If you have 1 kg of each compound, in which cylinder is the pressure greater at 25°C?
    (b) Now suppose cylinder A has a total pressure of 10 atm at 25°C, while B has a pressure of 9 atm measured at 0°C. Which cylinder contains the greater number of molecules?

GAS LAWS AND STOICHIOMETRY

49. If 1.0×10^3 g of uranium are converted to gaseous UF_6, what pressure of UF_6 would be observed at 32°C in a chamber that has a volume of 3.0×10^2 L?
50. $Ni(CO)_4$ can be made by reacting finely divided nickel with gaseous CO. If you have CO in a 1.50-L flask at a pressure of 418 mm Hg at 25.0°C, what is the maximum number of grams of $Ni(CO)_4$ that can be made?
51. If the boron hydride B_4H_{10} is treated with pure oxygen it combusts to give B_2O_3 and H_2O.

$$2\ B_4H_{10}(g) + 11\ O_2(g) \longrightarrow 4\ B_2O_3(s) + 10\ H_2O(g)$$

    If a 0.050-g sample of the boron hydride burns completely in O_2, what will be the pressure of gaseous water in a 4.25-L flask at 30.0°C?
52. Hydrazine reacts with O_2 according to the equation

$$N_2H_4(g) + O_2(g) \longrightarrow N_2(g) + 2\ H_2O(g)$$

    Assume the O_2 to combust the hydrazine is in a 450-L tank at 26°C. If you wish to combust completely a 10.-kg sample of hydrazine, to what pressure should you fill the O_2 tank in order to have sufficient oxygen?
53. Butane can be used as a fuel in an automobile engine. It burns in O_2 according to the equation

$$2\ C_4H_{10}(g) + 13\ O_2(g) \longrightarrow 8\ CO_2(g) + 10\ H_2O(g)$$

If one cylinder of the engine is filled with butane to a pressure of 1.5 atm at 500°C, it requires 2.45 g of O_2 for complete combustion. How many grams of O_2 would be required if the cylinder were filled with butane at a total pressure of 2.0 atm at 500°C?

54. Hydrogen can be made in the "water gas reaction."

$$C(s) + H_2O(g) \longrightarrow H_2(g) + CO(g)$$

If you begin with 250 L of gaseous water at 120°C and 2.0 atm pressure, how many grams of H_2 can be made?

GAS MIXTURES

55. What is the pressure in atmospheres of a gas mixture that contains 1.0 g of H_2 and 8.0 g of Ar in a 3.0-L container at 27°C?

56. A 3.0-L bulb containing He at 145 mm Hg is connected by a valve to a 2.0-L bulb containing Ar at 355 mm Hg. (See the figure below.) Calculate the partial pressure of each gas and the total pressure after the valve between the flasks is opened.

57. A cyclopropane–oxygen mixture (C_3H_6 + O_2) can be used as an anesthetic. Assume that a tank containing such a mixture has the following partial pressures: $P(\text{cyclo})$ = 170 mm Hg and $P(O_2)$ = 570 mm Hg.

(a) What is the ratio of the number of moles of cyclopropane to the number of moles of O_2?

(b) What are the mole fractions of O_2 and C_3H_6?

(c) If the tank contains 160 g of O_2, how many grams of C_3H_6 are present?

58. Gaseous CO exerts a pressure of 45.6 mm Hg in a 56.0-L tank at 22.0°C. If the gas is released into a room with a volume of 2.70×10^4 L, what is the partial pressure of CO (in mm Hg) in the room at 22°C?

59. A collapsed balloon is filled with He to a volume of 10. L at a pressure of 1.0 atm. Oxygen (O_2) is then added so that the final volume of the balloon is 30. L with a total pressure of 1.0 atm. The temperature, constant throughout, is equal to 20°C.

(a) How many grams of He does the balloon contain?

(b) What is the final partial pressure of He in the balloon?

(c) What is the partial pressure of O_2 in the balloon?

(d) What are the mole fractions of each gas?

60. The boron hydride B_2H_6 combusts in O_2.

$$B_2H_6(g) + 3\ O_2(g) \longrightarrow B_2O_3(s) + 3\ H_2O(g)$$

If the reactant gases were mixed in the correct stoichiometric ratio and then placed in a flask at 25°C so that the total pressure was 10. mm Hg, what were the partial pressures of B_2H_6 and O_2 before combustion?

61. The most effective rocket fuels are lightweight liquids that react to produce many molecules of gaseous products. One of the best fuels is dimethylhydrazine. When mixed with dinitrogen tetroxide, N_2O_4, it powered the *Lunar Lander* in the missions to the moon. The two components react according to the equation

$$(CH_3)_2N_2H_2(\ell) + 2\ N_2O_4(\ell) \longrightarrow 3\ N_2(g) + 4\ H_2O(g) + 2\ CO_2(g)$$

If 2.5 moles of dimethylhydrazine react completely with N_2O_4 and if the product gases are collected at 27°C in a 250-L tank, what is the total pressure in the tank?

62. Explosives are effective if they produce a large number of gaseous molecules as products. Nitroglycerin, for example, detonates according to the equation

$$2\ C_3H_5N_3O_9(s) \longrightarrow 6\ CO_2(g) + 3\ N_2(g) + 5\ H_2O(g) + \tfrac{1}{2}\ O_2(g)$$

If 1.0 g of nitroglycerin explodes, calculate the volume the product gases would occupy if their total pressure is 1.0 atm at 5.0×10^2°C. Compare with the volume the product gases would occupy at 25°C and 1.0 atm pressure.

63. A miniature laboratory volcano can be made with ammonium dichromate. When ignited it decomposes in a fiery display.

$$(NH_4)_2Cr_2O_7(s) \longrightarrow N_2(g) + 4\ H_2O(g) + Cr_2O_3(s)$$

If 5.0 g of ammonium dichromate are used and if the gases from this reaction are trapped in a 2.0-L flask at 25°C, what is the total pressure of the gas in the flask? What are the partial pressures of N_2 and H_2O?

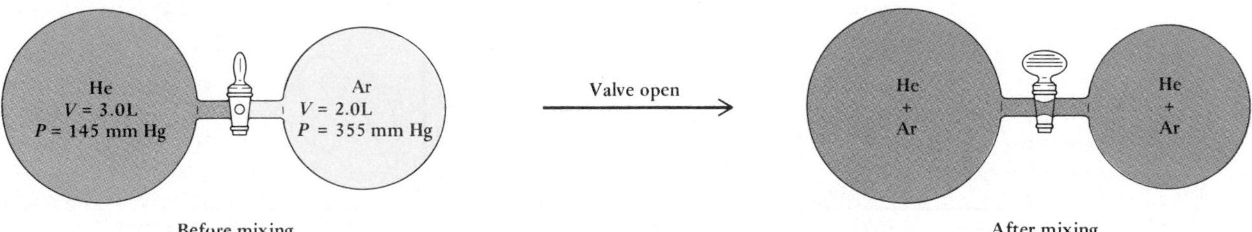

64. The density of air at 20. km above the earth's surface is 92 g/m^3. The pressure is 42 mm Hg and the temperature is −63°C. Assuming that the atmosphere contains only O_2 and N_2, calculate (a) the average molar mass of the atmosphere and (b) the mole fraction of each gas.

65. If potassium chlorate is heated, it decomposes to KCl and O_2.

$$2\ KClO_3(s) \longrightarrow 2\ KCl(s) + 3\ O_2(g)$$

Assume that you have heated 1.56-g $KClO_3$ and collected the evolved O_2 over water (see Figure 6.7). If the temperature of the gas is 21.0°C, what volume should it occupy if the *total* pressure of the gas in the collection flask is equal to the barometric pressure (742 mm Hg)? (See Appendix F for water vapor pressure data.)

66. You are given 1.56 g of a mixture of $KClO_3$ and KCl. When heated, the $KClO_3$ decomposes to KCl and O_2

$$2\ KClO_3(s) \longrightarrow 2\ KCl(s) + 3\ O_2(g)$$

and 327 mL of O_2 are collected over water at 19°C. The *total* pressure of the gas in the collection flask is 735 mm Hg. What is the weight percentage of $KClO_3$ in the sample? (See Appendix F for water vapor pressure data.)

KINETIC MOLECULAR THEORY

67. You are given two flasks of equal volume. Flask A contains H_2 at 0°C and 1 atm pressure. Flask B contains CO_2 gas at 0°C and 2 atm pressure. Compare these two gases with respect to each of the following:
(a) Average kinetic energy per molecule.
(b) Average molecular velocity.
(c) Number of molecules.

★68. Equal masses of gaseous N_2 and Ar are placed in separate flasks of equal volume, both gases being at the same temperature. Tell whether each of the following statements is true or false. Briefly explain your answer in each case.
(a) The pressure is greater in the Ar flask.
(b) Ar atoms have a greater average velocity than the N_2 molecules.
(c) The molecules of N_2 collide more frequently with the walls of the flask than do the atoms of Ar.

69. Which graph at the bottom of the page would best represent the distribution of molecular speeds for the gases acetylene (C_2H_2) and N_2? Both gases are in the same flask with a total pressure of 750 mm Hg. The partial pressure of N_2 is 500 mm Hg.

70. Calculate the rms speed of xenon atoms in a sample of the gas at 25°C and compare it with the rms speed for helium atoms as calculated in Exercise 6.12.

71. Calculate the rms speed for CCl_4 molecules at 25°C. What is the ratio of its speed to that of H_2 at the same temperature?

72. Place the following gases in order of increasing average molecular speed at 25°C: (a) Kr, (b) CH_4, (c) N_2, and (d) CH_2Cl_2.

GRAHAM'S LAW

★73. Rank the following gases in order of increasing rate of effusion: (a) He, (b) Xe, (c) CO, and (d) C_2H_6.

74. A gas whose molar mass you wish to know effuses through an opening at a rate only one third as great as the effusion rate of helium. What is the molar mass of the unknown gas?

75. The rms speed of an O_2 molecule is 482 m/s at 298 K. What is the rms speed of a xenon atom at that same temperature?

76. A sample of uranium fluoride is found to effuse at the rate of 12.7 mg/hr. Under comparable conditions, gaseous I_2 effuses at the rate of 15.0 mg/hr. What is the molar mass of the uranium fluoride?

NON-IDEAL GASES

★77. Based on their respective van der Waals constants, is CH_4 or N_2 expected to behave more nearly like an ideal gas at a pressure of 50 atm?

78. In the text it is stated that the pressure of 8.00 moles of Cl_2 in a 4.00-L tank at 27.0°C should be 29.5 atm, if calculated using the van der Waals equation. Verify

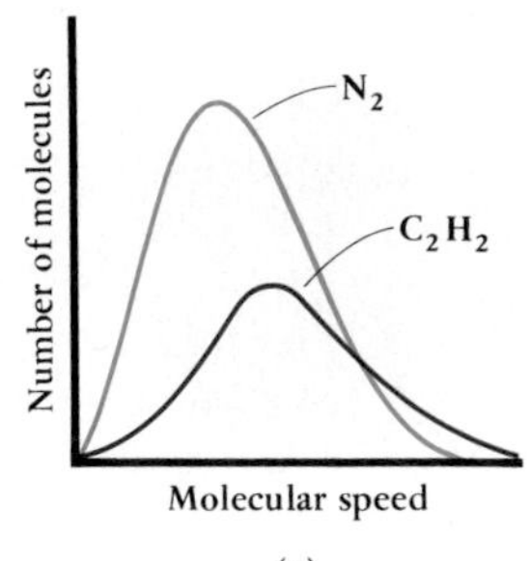

(a)

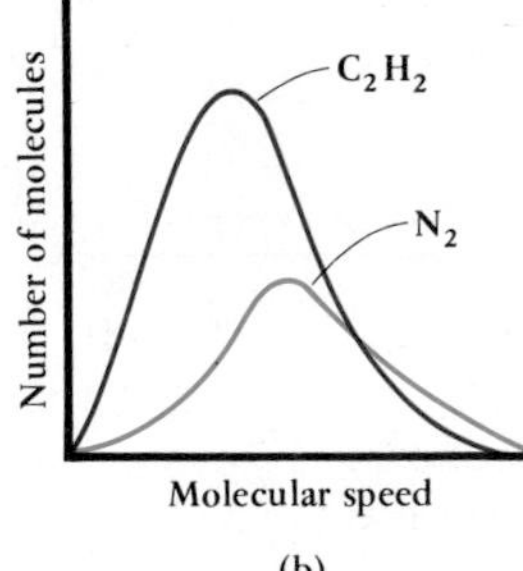

(b)

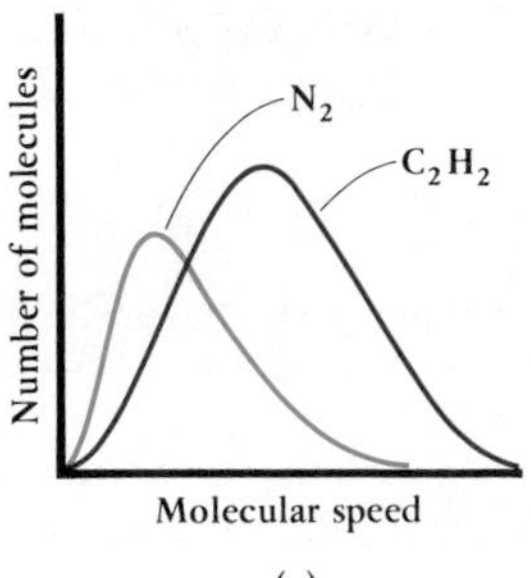

(c)

this result and compare it with the pressure expected from the ideal gas law.

79. Using both the ideal gas law and the van der Waals equation, calculate the volume expected for 1.0 mole of CH_4 at 150 atm and at 350 atm. Assume that the temperature is 20.0°C. Compare your calculations with Figure 6.18.

GENERAL QUESTIONS

*80. 1.0 L of a gaseous compound of hydrogen, carbon, and nitrogen gave upon combustion 2.0 L of CO_2, 3.5 L of H_2O vapor, and 0.50 L of N_2 at STP. What is the empirical formula of the compound?

81. You are given a solid mixture of $NaNO_2$ and NaCl and are asked to analyze it for the amount of $NaNO_2$ present. To do so you cause it to react with sulfamic acid, HSO_3NH_2, in water according to the equation

$$NaNO_2(aq) + HSO_3NH_2(aq) \longrightarrow NaHSO_4(aq) + H_2O(\ell) + N_2(g)$$

What is the weight percentage of $NaNO_2$ in 1.012 g of the solid mixture if reaction with sulfamic acid produces 325 mL of N_2 gas? The gas was collected over water at a temperature of 21.0°C and with the barometric pressure equal to 731.0 mm Hg.

82. Hydrazine can be used to remove oxygen from hot-water heating systems by the following reaction:

$$N_2H_4(aq) + O_2(g) \longrightarrow N_2(g) + 2\ H_2O(\ell)$$

What is the maximum number of liters of N_2 that could be produced by 250 L of 0.010 M N_2H_4? The N_2 gas is collected at 25°C and its pressure is 730 mm Hg.

83. Octane, a component of gasoline, reacts with O_2 according to the equation

$$2\ C_8H_{18}(g) + 25\ O_2(g) \longrightarrow 16\ CO_2(g) + 18\ H_2O(g)$$

Assume the volume of one cylinder of an automobile engine is 500. cm^3. If air enters the engine at 50.°C and 1.0 atm of pressure, how many grams of octane should the fuel injection system send to the cylinder in order to completely consume all of the O_2 in the air? Assume that the mole fraction of O_2 in the air is 0.20.

84. The *Starship Enterprise* is not really fueled by dilithium crystals! Rather, it uses a mixture of diborane, B_2H_6, and oxygen. The two react according to the equation

$$B_2H_6(g) + 3\ O_2(g) \longrightarrow B_2O_3(s) + 3\ H_2O(g)$$

(a) If all of the gases involved in the reaction above are at the same temperature, what are their relative rms speeds?

(b) B_2H_6 and O_2 are contained in separate fuel tanks of equal volume and at the same temperature. The pressure in the O_2 tank is 45 atm. If the B_2H_6 tank is to contain an amount of B_2H_6 that will exactly react with the O_2, what must the pressure be in the B_2H_6 tank (in atm)?

85. A xenon fluoride can be prepared by heating a mixture of Xe and F_2 to a high temperature in a pressure-proof container made of nickel. Assume that xenon gas was added to a 0.25-L container until its pressure was 0.12 atm at 0.0°C. Fluorine gas was then added until the total pressure was 0.72 atm at 0.0°C. After the reaction was complete, the xenon had been consumed completely and the pressure of the F_2 remaining in the container was 0.36 atm at 0°C. What is the empirical formula of compound prepared from Xe and F_2?

86. The atmosphere is a mixture of gases with a total pressure equal to the barometric pressure. Assume for this problem that this pressure is 740 mm Hg. Further assume that the gases of the air have the following partial pressures: $P(N_2)$ = 557 mm Hg, $P(CO_2)$ = 23 mm Hg; $P(H_2O)$ = 40 mm Hg; $P(O_2)$ = ?

(a) What is the partial pressure of O_2?

(b) List the gases in order of increasing mole fraction in a sample of air.

(c) List the gases in order of increasing average molecular speed.

(d) The van der Waals a and b constants for each gas are given in Table 6.2. Which of the gases would be expected to behave *most* ideally?

87. The partial pressure of O_2 in air is 158 mm Hg at 25°C and a total pressure of 760. mm Hg. The air that is expired from your lungs after breathing has a partial pressure of O_2 of 115 mm Hg under the same conditions. How many moles of O_2 are absorbed by your lungs from 1.0 L of air?

88. A sample of He has a volume of 10.00 mL at 0°C but a volume of 13.66 mL at 100°C. (The pressure of the gas sample at both temperatures is the same.) Use this to prove that 0°C = 273 K.

89. Assume you have synthesized a new compound of boron and hydrogen. Its empirical and molecular formulas can be determined from the following information.

(a) To determine its empirical formula, you burn it in pure O_2 to give B_2O_3 and H_2O. The water from the combustion of 0.492 g of B_xH_y has a mass of 0.540 g.

(b) 0.0631 g of the boron hydride was quickly evaporated into a flask with a volume of 120. mL; the gas exerted a pressure of 98.6 mm Hg at 23°C.

90. Methylthallium, $Tl(CH_3)_x$, has been prepared. To find its formula (the value of x), you decompose the compound with HCl. The hydrogen of the acid combines with the CH_3 to produce CH_4 gas. Assume you decompose 0.102 g of $Tl(CH_3)_x$ and find that the isolated CH_4 gas has a pressure of 90.6 mm Hg in a volume of 256 mL at 30.0°C. What is the correct formula of methylthallium?

91. A mixture of aluminum and zinc weighing 1.67 g was completely dissolved in acid. In the process the mixture evolved 1.69 L of H_2, measured at 273 K and 1.00 atm pressure. What is the mass of aluminum in the mixture? (The aluminum reacts with acid according to the equation $Al(s) + 3\ H^+(aq) \rightarrow Al^{3+}(aq) + \frac{3}{2}\ H_2(g)$. For the zinc reaction, see Example 6.10.)

PART TWO

THE STRUCTURE OF ATOMS AND MOLECULES

Red light is emitted by excited hydrogen gas (Trifid Nebula, Sagittarius) (Anglo-Australian Telescope Board)

PART TWO: PREFACE

To understand the physical and chemical properties of the elements and their compounds, it is important to understand the underlying structure of atoms. Thus, our first goal in this section is to outline the current theories of the arrangement of electrons in atoms and some of the important historical developments which led to these notions (Chapters 7 and 8). With these ideas, you can then understand why atoms and their ions vary in size and in their ability to lose or gain electrons. So that these variations in properties can be remembered, and predictions made, we shall tie this discussion closely to the arrangement of elements in the periodic table (Chapter 8).

With a firm understanding of atomic structure, it is possible to see how atoms can join to form molecules. In Chapter 9 we shall first consider how the electrons of the atoms of the molecule are divided into various functions of bonding electrons and nonbonding or lone pair electrons. Having made such decisions, we can then show how one can derive the three-dimensional structure of simple molecules. Once the structure is known, various properties of molecules can then be explained, a topic pursued throughout the book. Chapter 10 considers in more detail the major theories of molecular bonding and extends these ideas to bonding in metals.

7

The Structure of The Atom

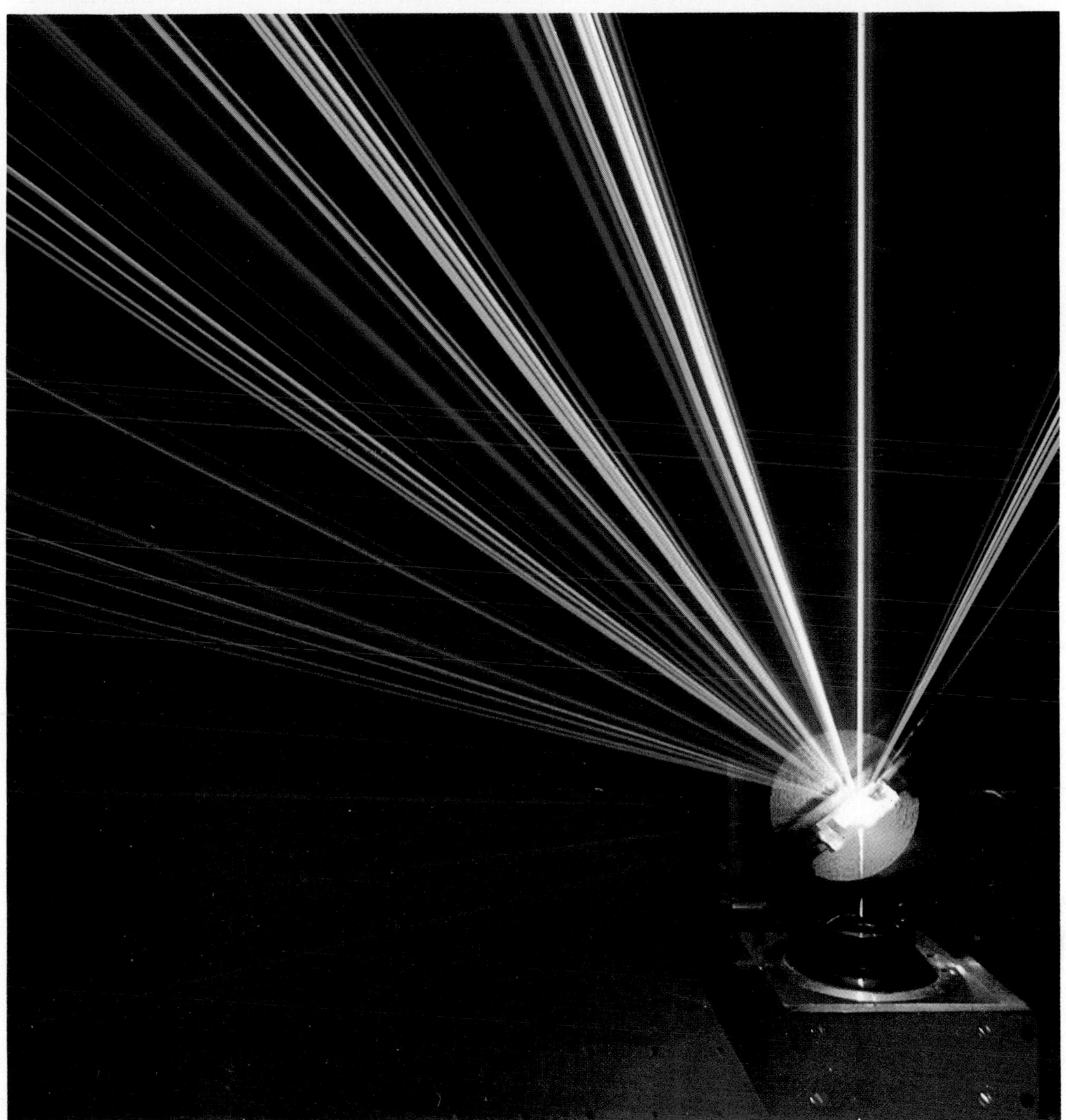

The separation of light emitted by selenium (Fritz Goro)

CHAPTER OUTLINE

Those of us who are chemists were attracted to this field for many reasons. Perhaps most important was the fact that we were interested in why elements or compounds behave as they do, why and how they interact with one another, why some are colored, why some are liquids and others solids or gases, or why some are poisons in our bodies while others are beneficial.

The fundamental concept of the atom is basic to an understanding of the important questions in chemistry and their answers. In this chapter and the next we shall explore the basic ideas of atomic structure and begin to see the relation to practical, everyday chemistry. This chapter first leads you through a brief view of the development of atomic theory, a period of history replete with marvelous characters who are fascinating not only for the brilliance of their contributions to mankind but also for being interesting people. Then we shall describe some ideas from physics that are useful to an understanding of atomic structure, and finally we lay out some of the principles of atomic structure.

As you read these chapters, pay attention to the changing notions of atomic structure in this century and be aware that our current ideas are simply the best theories we have at present.

7.1 HISTORICAL PERSPECTIVE: RUTHERFORD AND THE NUCLEAR ATOM

We have come a long way from John Dalton's revolutionary hypothesis that all matter is composed of atoms to our modern ideas of atomic structure, and the road leads directly through the work of Michael Faraday (1791–1867).

Faraday began his brilliant career as an assistant to Sir Humphrey Davy (1778–1829), one of the most influential scientists in England in his time. Among other things, Davy was interested in the effects of electricity on chemical substances in solution. Davy's curiosity about this subject infected Faraday, and the latter spent the remainder of his life studying electrical phenomena. In 1832 Faraday found that when he passed electricity through a chemical substance the mass of material affected was proportional to the duration and strength of the current. From these experiments he was able to deduce that ions were electrically charged and that their charges were all multiples of some smallest charge. In 1891, another English scientist, E. Johnstone Stoney, proposed the name **electron** for this natural unit of electricity, and the same name was applied a few years later to an isolated particle having a negative charge.

In Thomson's time beams of electrons were called cathode rays, because the electrons originated at the negative pole or "cathode" of an electrical discharge. This name has stayed with us, and TV tubes are often called CRT's, cathode ray tubes.

At the close of the 19th century, J.J. Thomson (1856–1909) was the Cavendish Professor of Experimental Philosophy at Cambridge University in England (Figure 7.1); this was the most important position in physics in the English-speaking world at the time, and it was in his laboratory that sophisticated experiments on beams of electrons were carried out first. The most important outcome of Thomson's experiments was a measurement of a fundamental property of the electron, the ratio of its charge (e) to its mass (m). The currently accepted value is

$$\text{Electron charge/mass} = e/m = 1.76 \times 10^8 \text{ coulombs/gram*}$$

The design of Thomson's experiment allowed only the measurement of the *ratio* of charge and mass. To find out one or the other of these quantities, a separate experiment had to be performed. This experiment was done in 1909 by Robert Millikan of the University of Chicago (Figure 7.2), and the value for the electron charge was found to be 1.6022×10^{-19} coulombs. Combined with Thomson's charge/mass ratio, this leads to an electron mass of 9.1095×10^{-28} g.

The two decades from 1895 to 1915 were an exciting time in physics because many important discoveries were made. Our modern view of the atom was shaped in that period, and the next important development after Thomson's and Millikan's experiments came from the laboratory of Ernest Rutherford (1871–1937) of the University of Manchester in England.

Thomson and his students had developed a picture of the atom that has often been called the "plum pudding" model. They assumed the atom was a positive sphere of matter in which electrons were sprinkled about randomly, like the raisins in a plum pudding.

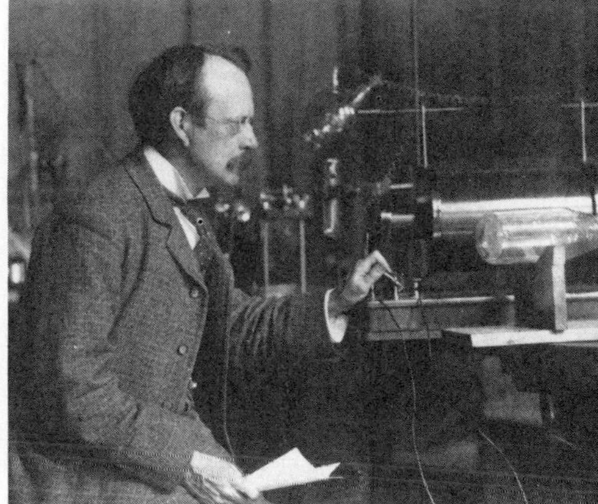

FIGURE 7.1
Sir J.J. Thomson (1856–1909) in his laboratory at Cambridge University, England. Thomson received the Nobel Prize in 1906 for his work on the properties of the electron. (University of Cambridge, Cavendish Laboratory)

*Since only the magnitude of e/m is important here, we have neglected the negative sign of the charge. The coulomb is the SI unit for the *quantity* of electricity. It is abbreviated C.

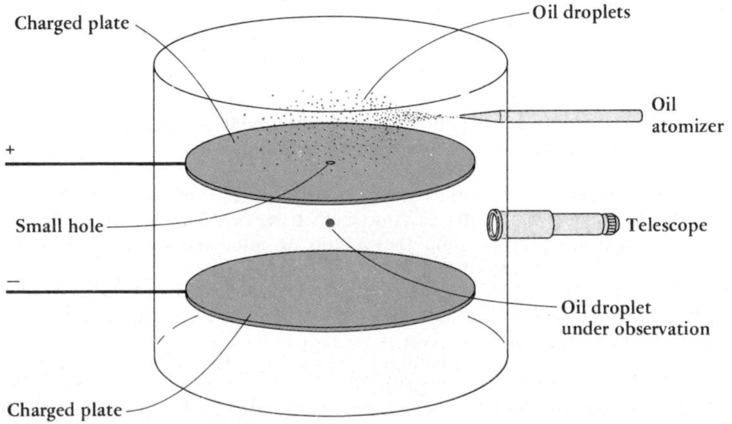

FIGURE 7.2
Millikan oil drop experiment. A fine mist of oil droplets is introduced into the chamber. The gas molecules inside the chamber are ionized (split into electrons and positive ions) by a beam of x-rays (not shown). The electrons adhere to the oil droplets, some droplets having one electron, some two electrons, and so forth. These negatively charged oil droplets fall under the force of gravity into the region between electrically charged plates. By carefully adjusting the voltage on the plates, the force of gravity is exactly counterbalanced by the attraction of the negative oil drop to the upper, positively charged plate. Analysis of these forces leads to a value for the charge on an electron.

ERNEST RUTHERFORD (1871–1937)

Lord Rutherford was born and raised in New Zealand but went to Britain in 1895 to work with J.J. Thomson at Cambridge University. In 1899 he moved to McGill University in Canada where he pursued research on radioactivity. Among his other work at McGill, he discovered that uranium gave out two kinds of emissions that he named alpha and beta. He soon proved that the alpha particles were positively charged, and later he deduced that they consisted of a helium atom without the two electrons. He also showed that a beta particle is an electron. For this work, he received the Nobel Prize in Chemistry in 1908.

In 1907 Rutherford moved to Manchester University in England and it was there that he performed the experiment that uncovered the true nature of the atom. In 1919 he moved back to Cambridge and assumed the position formerly held by J.J. Thomson. Not only was Rutherford responsible for very important work in physics, but he guided the work of ten future recipients of the Nobel Prize.

One large question that remained for Thomson was the number of electrons in the atom, and his original estimate was that the number was about 1000 times the atomic weight of the atom! To test this, Thomson fired a beam of electrons at a very thin metal foil. If his model were correct, the high energy electron beam would travel through the foil and emerge from the other side. However, while moving through the foil, the electrons of the beam would encounter electrons of the atoms, and the negative charges would repel. A deflection of the beam from its straight path should be observed, with the size of the deflection related to the number of electrons in the atom. Thomson did indeed observe a deflection, but it was much smaller than he expected, and he was forced to revise his estimate of the number of electrons, but not his "pudding" model of the atom.

Ernest Rutherford had been Thomson's student a few years before. He believed Thomson's atomic model and, in about 1910 or 1911, he decided to test it further.* In his study of radioactivity, Rutherford had earlier discovered a positively charged, massive particle that he called the alpha (α) particle.†

Rutherford believed that, if Thomson's model were correct, very little deflection of a beam of more massive alpha particles should occur if they were fired at a thin gold foil target. So, with this idea in mind,

*The evolution of events leading up to Rutherford's experiment is outlined because it illustrates for you how science really works. An investigator has an idea about the structure or chemistry of a substance and wants to test that idea by the best means available. The methods chosen depend on the background of the investigator and on the availability of equipment or techniques. This could mean, of course, that the best approach may not be used; the results may not be definitive, and they could be interpreted easily to fit the preconceived idea. This was presumably the case with Thomson's experiment. Rutherford happened to choose a more appropriate method, because techniques using alpha particles had been developed in his laboratory. Rutherford also had the genius and good fortune to interpret his results properly.

†We now know that the alpha particle is a helium atom nucleus. The two electrons of the helium atom have been stripped away, leaving the two protons and two neutrons of the nucleus. Because of this, the particle has an electrical charge of +2.

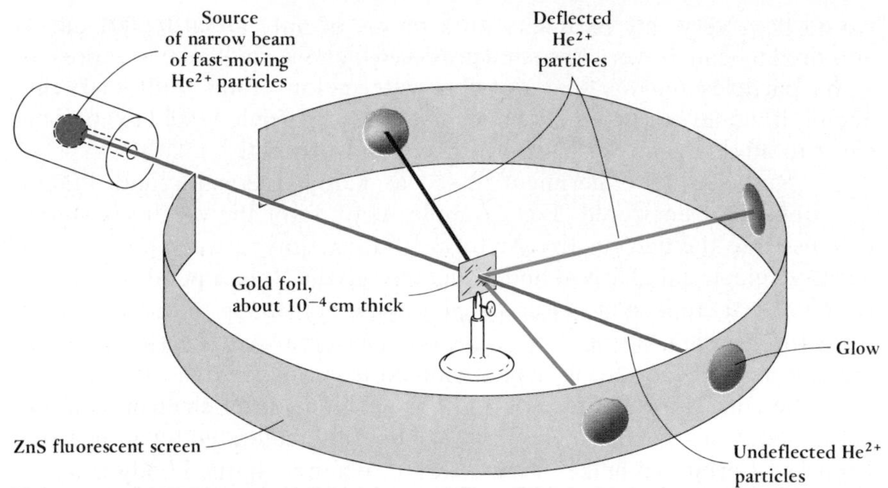

FIGURE 7.3

The experimental arrangement for the Rutherford experiment. A beam of alpha particles is directed at a very thin (about 6×10^{-5} cm) piece of gold foil. A luminescent screen surrounding the foil detects particles passing through the foil (blue) or deflected by encounters with atomic nuclei (green and red).

Rutherford's students Geiger and Marsden carried out their now famous experiment (Figure 7.3). While they generally did find that alpha particles were deflected to only a small extent, they were astonished to find that a *very few* particles were deflected nearly backwards (Figure 7.4)! Rutherford described this unexpected result by saying "It was about as credible as if you had fired a 15-inch (artillery) shell at a piece of tissue paper and it came back and hit you." The only way to account for this was to discard Thomson's model and assume that the positive charge of the atom is concentrated in a very small volume of space, which Rutherford called the **nucleus**, and that the electrons of the atom occupy the space about the nucleus. From their results, Rutherford, Geiger, and Marsden were able to calculate that the charge on the gold nucleus is in the range 100 ± 20 (actual value, 79) and that the nucleus had a radius of about 10^{-12} cm (actual value is closer to 10^{-13} cm). The radius of the atom was found to be about 100,000 times that of the nucleus.

Hans Geiger invented a device to measure or count radioactive disintegrations, a device we still use to this day and which bears his name. See Chapter 26.

The picture of the atom that emerged from Rutherford's experiment was extraordinary. To give you some feeling for the result, imagine enlarging the gold foil, Rutherford's target, about one billion times. On this scale a gold atom would be about the size of a watermelon, but the nucleus

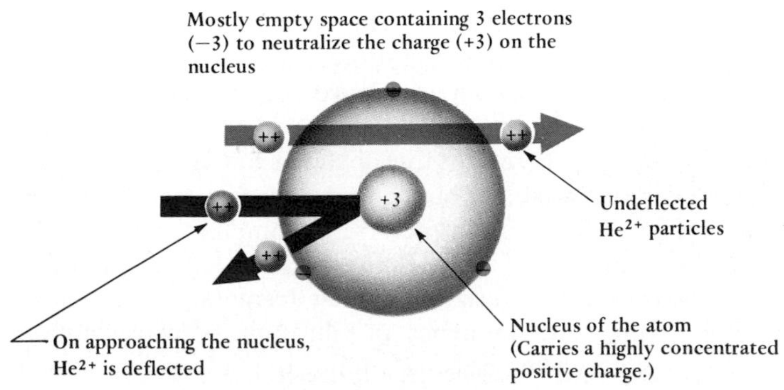

FIGURE 7.4

A helium ion, He^{2+} (an alpha particle), encountering an atom (Li in this case). In the Rutherford model, the atom is mostly empty space with a dense nucleus containing most of the atom's mass and all of its positive charge. Most of the alpha particles pass unimpeded through the foil because they do not come near an atomic nucleus. A few, however, closely approach the positively charged nucleus of the atom, and, since such He^{2+} ions are subjected to a strong unbalanced force, they are deflected through large angles. (The drawing is not to scale.)

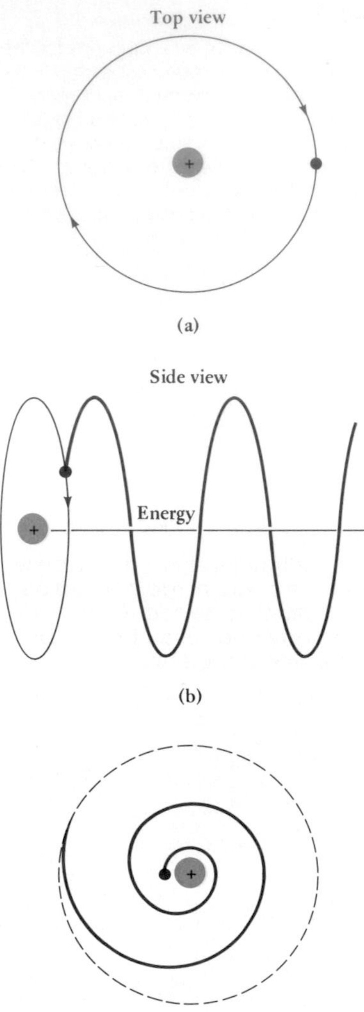

FIGURE 7.5

A problem with the Rutherford atom according to classical physics. Classical physics tells us that a charge (an electron) moving in an electrical field (the positive nucleus), (a), would radiate energy, (b). If this were the case the electron would gradually slow down, spiral in toward the nucleus, and eventually be trapped. The fact that this does not occur, (c), means that electrons in atoms do not obey the laws of classical physics.

would be a very tiny particle with a radius of only about 0.0001 cm. If you fired a beam of very fine sand grains (which we shall take to represent alpha particles on this scale) at the watermelon atoms with such tiny nuclei, it would be extremely unusual if a sand particle would even come close to any nucleus, let alone strike it almost directly.

Rutherford's experiment raised as many interesting and difficult questions as it answered. For example, why don't the electrons simply collapse into the nucleus (by the force of attraction between positive and negative electrical charges) and ultimately give a "plum pudding" atom? Given that there is some force keeping the electrons in their space apart from the nucleus, an obvious view is to assume that the electrons are moving in orbits about the nucleus, much as the earth and other planets orbit the sun. But this presents another problem. If an electron is circulating about the nucleus in an orbit, it must be accelerated by a force that keeps it in orbit. To make matters worse, when a charged body is accelerated it loses energy by radiation. This would mean the circling electron would lose energy and spiral into the nucleus, collapsing the atom (Figure 7.5). (An analogy to this is the Skylab space station, which gradually lost its orbital energy and crashed into the earth's atmosphere.) Because this does not occur, it must mean electrons in atoms do not obey the laws of classical physics. A way out of the dilemma outlined above was offered in 1913 by a young Danish physicist, Niels Bohr. Although Bohr's explanation sent science in the right direction, he quickly recognized its limitations, and he and others eventually developed a more complete and useful model of atomic structure, **quantum mechanics**. Quantum mechanics is the approach currently used in science, and we shall outline that model after we have described the concepts on which it is based.

7.2 ELECTROMAGNETIC RADIATION

We all are familiar with water waves, and you may also know that radiation such as light can be described as wave motion. Indeed, the Englishman James Maxwell developed an elegant mathematical theory in 1864 to describe all forms of radiation in terms of oscillating or wave-like electric and magnetic fields in space (Figure 7.6). Hence, radiation such as light, microwaves, television and radio signals, and x-rays has come to be called more precisely **electromagnetic radiation**. This view is important to our understanding of how electrons behave in atoms.

Two typical waves are shown in Figure 7.7. (These are called sinusoidal waves because of the mathematical function describing them.) The **wavelength** of a wave is given by the distance between successive crests or high points in the wave (or between successive troughs or low points). This distance can be given in meters, nanometers, or whatever unit is convenient. The symbol for wavelength is the Greek letter λ (lambda).

Waves can also be characterized by their **frequency**, symbolized by the Greek letter ν (nu). The frequency of vibration, or wave motion, is the number of complete waves passing a point in some given amount

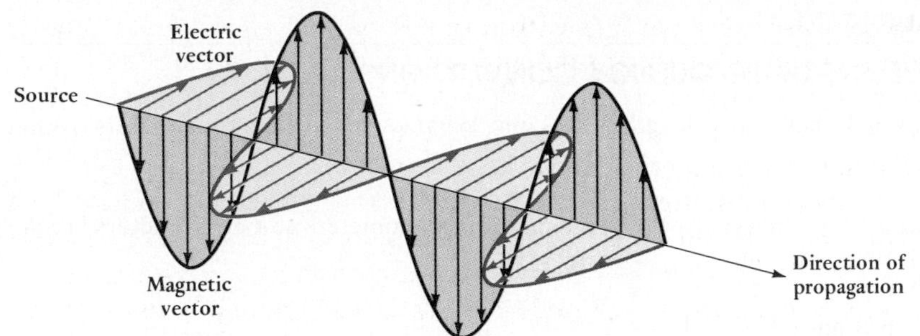

FIGURE 7.6

Electromagnetic radiation. In the 1860's James Maxwell developed the currently accepted theory that all forms of radiation are propagated through space as vibrating electric and magnetic fields, the fields being at right angles with one another. Each of the fields is described by a sine wave. Such oscillating fields emanate from vibrating charges in a source such as a light bulb or radio antenna.

of time. We usually refer to the frequency as the number of cycles or "peak to peak" events per second, a unit often written s^{-1} (standing for 1/second) and now called the **hertz**.

If you have ever enjoyed water sports, you are familiar with the height of waves. In more scientific terms, the maximum height of a wave, as measured from the axis of propagation, is called its **amplitude**. In Figure 7.7, notice that the (sinusoidal) wave has zero amplitude at the beginning, at the end, and midway between the ends of each wavelength. All the points of zero amplitude are called **nodes**; they always occur at intervals of $\lambda/2$ for a sinusoidal wave.

Finally, the **velocity** of a moving wave is important. As an analogy, consider cars in a traffic jam traveling bumper to bumper. If each car is 16 feet long, and if a car passes you every four seconds (that is, the frequency is 1 per 4 seconds or $\frac{1}{4}\ s^{-1}$), then the traffic is "moving" at the speed of (16 feet)($\frac{1}{4}\ s^{-1}$) or 4 feet per second. This multiplication of length times frequency to give velocity holds for any periodic motion, including a wave.

$$\begin{array}{ccccc} \text{Wavelength(m)} & \times & \text{frequency}(s^{-1}) & = & \text{velocity(m/s)} \\ (\lambda) & & (\nu) & = & \text{v} \end{array} \tag{7.1}$$

For visible light and other electromagnetic radiation, the velocity is the well known constant $\boldsymbol{c = 2.99795 \times 10^8}$ **m/s**. This means that, if you know the wavelength of a light wave, you can readily calculate its frequency and vice versa.

The velocity of light is commonly symbolized by *c*.

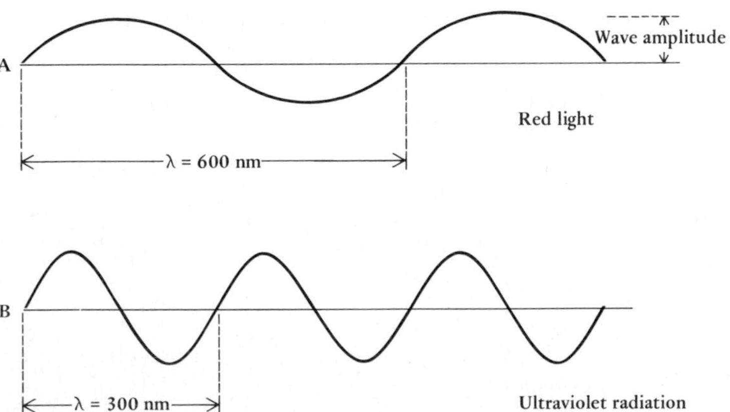

FIGURE 7.7

Two typical waves. The wavelength of red light, wave A, is twice that of wave B (a particular radiation in the ultraviolet region); that is $\lambda_A = 2\lambda_B$. Since both waves are traveling at the same velocity, the ultraviolet light completes two cycles or vibrations in the time the red light completes one. Thus, the frequency of the red light is one-half that of the ultraviolet light ($\nu_A = \frac{1}{2}\nu_B$).

EXAMPLE 7.1

WAVELENGTH–FREQUENCY CONVERSIONS

Orange light has a wavelength of 620 nm. What is its wavelength in meters? What is its frequency?

Solution To convert from a wavelength in nanometers to one in meters, recall that 1 nm is 10^{-9} m.

$$620 \text{ nm} \left(\frac{10^{-9} \text{ m}}{\text{nm}}\right) = 6.2 \times 10^{-7} \text{ m}$$

Frequency can now be obtained from the equation $c = \lambda\nu = 2.998 \times 10^8$ m/s.

$$\nu = c\,\frac{1}{\lambda} = \left(\frac{2.998 \times 10^8 \text{ m}}{\text{s}}\right)\left(\frac{1}{6.2 \times 10^{-7} \text{ m}}\right)$$
$$= 4.8 \times 10^{14} \text{ s}^{-1}$$

EXERCISE 7.1 WAVELENGTH–FREQUENCY CONVERSIONS

Your favorite FM radio station has a frequency of 104.6 MHz (MHz = megahertz = 10^6 s^{-1}). What is the wavelength (in meters) of the radiation emitted by this station?

You can remember the colors of visible light, in order of decreasing wavelength, by the famous name ROY G BIV, standing for red-orange-yellow, green, blue-indigo-violet.

We are all bathed constantly in electromagnetic radiation, such as the radiation you can see, visible light. As you know, visible light really consists of a **spectrum** of colors, ranging from red light at the long wavelength end to violet light at the short wavelength end of the spectrum (Figure 7.8). Visible light is only a small portion of the total electromagnetic spectrum. Ultraviolet radiation, the radiation that leads to sunburn, has wavelengths shorter than visible light; x-rays and γ-rays, the latter emitted in the process of radioactive disintegration of some atoms, are found at shorter wavelengths still. At longer wavelengths than visible light, infrared radiation, the type of radiation that is sensed as heat from a fire, is encountered first. Longer still is the wavelength of the radiation in a microwave oven and that used in television and radio transmission.

EXERCISE 7.2 RADIATION, WAVELENGTH, AND FREQUENCY

Which color in the visible spectrum has the highest frequency? Which has the lowest frequency? Is the radiation used in a microwave oven higher or lower in frequency than that of your favorite FM radio station? (See Figure 7.8.)

The wave motion we have described so far is that of *traveling* waves. There is another type of wave motion, called **standing** or **stationary** waves, that is also relevant to modern atomic theory. If you tie down a string at both ends, as you would the string of a guitar or violin, and pluck the string, it will vibrate as a standing wave (Figure 7.9). There are several important points to note about standing waves: (1) A standing wave is characterized by having fixed ends where the amplitude is zero; the wave does not appear to travel along the line. (2) There are always two or more places where the vibrating string never moves; that is, the wave amplitude is zero at these points, which are called **nodes**. As with traveling waves, the distance between consecutive nodes is always $\lambda/2$. (3) In the first of the vibrations illustrated in Figure 7.9, the distance between the ends of

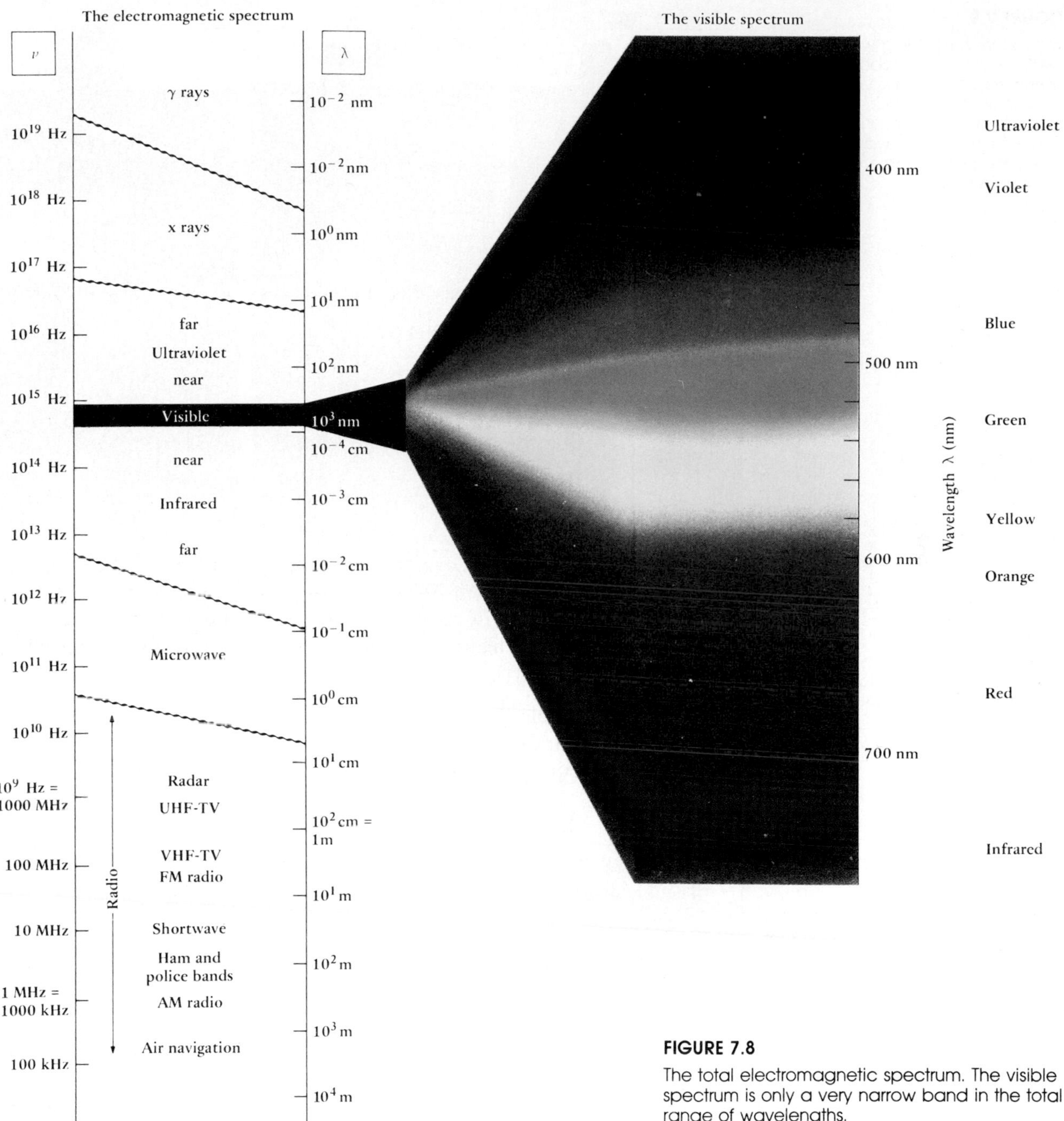

FIGURE 7.8

The total electromagnetic spectrum. The visible spectrum is only a very narrow band in the total range of wavelengths.

the string, a, is $\lambda/2$; in the second vibration the string length equals one complete wavelength, or $2(\lambda/2)$; in the third, the string length is $(3/2)\lambda$, or $3(\lambda/2)$. Could the distance between the ends of a standing wave vibration ever be $(3/4)\ \lambda$, or $3/2(\lambda/2)$? Certainly not! For standing wave vibrations, both those you can see and those on an atomic scale, only certain wave-

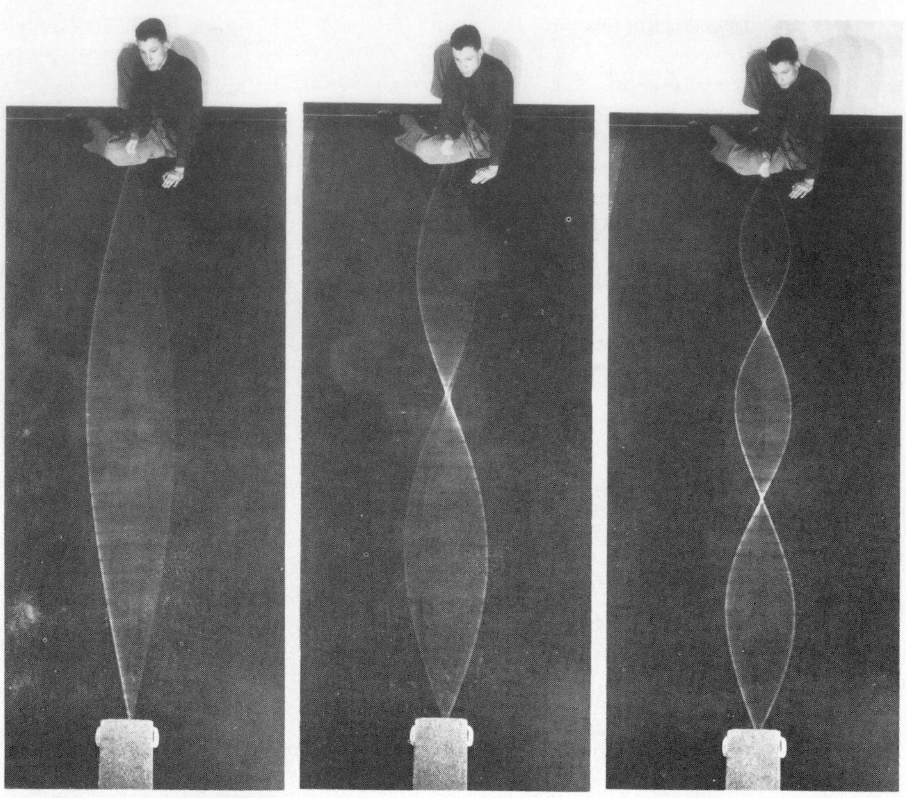

FIGURE 7.9
Illustration of standing waves (which occur in a rubber tube stretched between two points). In the first wave, the end-to-end distance is $\frac{1}{2}\lambda$, in the second wave it is λ, and in the third wave it is $\frac{3}{2}\lambda$. (Education Development Center, Newton, Mass.)

lengths are possible. Because the standing wave ends must be nodes, the only allowed vibrations are those where $a = n(\lambda/2)$ [where a is the distance from one end or "boundary" to the other, and n is an integer (1, 2, 3, . . .)]. This is an example of *quantization* in nature, a concept we turn to next.

EXERCISE 7.3 STANDING WAVES

The line drawn below is 10 cm long. Using this line,

(a) draw a standing wave with 1 node between the ends. What is the wavelength of this wave? (b) Draw a standing wave with 2 nodes between the ends. What is its wavelength? (c) If the wavelength of the standing wave is 5 cm, how many waves will fit within the boundaries? How many nodes will there be?

7.3 THE QUANTIZATION OF ENERGY

A **spectrum** is a plot of the intensity of emitted or absorbed radiation against the wavelength of the radiation.

If you have ever looked at the heating elements on an electric stove, you know that a heated object can glow red. The spectrum of the light given off by a glowing object is represented by Figure 7.10. Most of the light is emitted with a wavelength in the red region of the spectrum, but lesser amounts are found in other regions as well. If the object is heated to a

MAX PLANCK (1858–1947)

Max Karl Ernst Ludwig Planck was raised in Munich, Germany, where his father was an important professor at the University. When still in his teens Planck decided to become a physicist in spite of the advice of the head of the physics department at Munich that "The important discoveries [in physics] have been made. It is hardly worth entering physics anymore." Planck worked for a time at the University in Munich, but he later went to Berlin and studied thermodynamics (see Chapter 18) and wrote his thesis on the Second Law. This deep interest in thermodynamics led him eventually to the problem of the "ultraviolet catastrophe" and to his revolutionary hypothesis. The discovery was announced two weeks before Christmas in 1900 and he was awarded the Nobel Prize in 1918. Einstein said that it was a longing to find harmony and order in nature, a "hunger in his soul," that spurred Planck on.

still higher temperature, it will glow yellow and finally "white hot." The spectrum of the emitted light is similar to that at the lower temperature in Figure 7.10, but the wavelength of the maximum is shifted more to the yellow and then to the blue or violet. At the close of the 19th century, physicists were trying to understand these spectra and their changes with temperature. Several theories had been developed, but all predicted a continuous increase in light intensity as the wavelength of the emitted light became shorter, that is, as radiation in the ultraviolet was considered. This was obviously not in agreement with experiment, and the problem was given the marvelous name of the "ultraviolet catastrophe."

The temperature of the stars is given in terms of the color of their glow; there are red giant stars, yellow stars, and white dwarfs.

In 1900 Max Planck (1858–1947) solved the problem of the ultraviolet catastrophe with an explanation that contained within it the seeds of a revolution in scientific thought. Planck made what was for that time an incredible assumption: There is a minimum amount of energy that can be gained or lost by an atom, and all energy gained or lost must be some integer multiple, n, of that minimum. Symbolically,

$$\text{Energy} = nh\nu$$

where h is a proportionality constant, now called **Planck's constant**, with the value 6.6262×10^{-34} joule·seconds. The frequency ν is the lowest

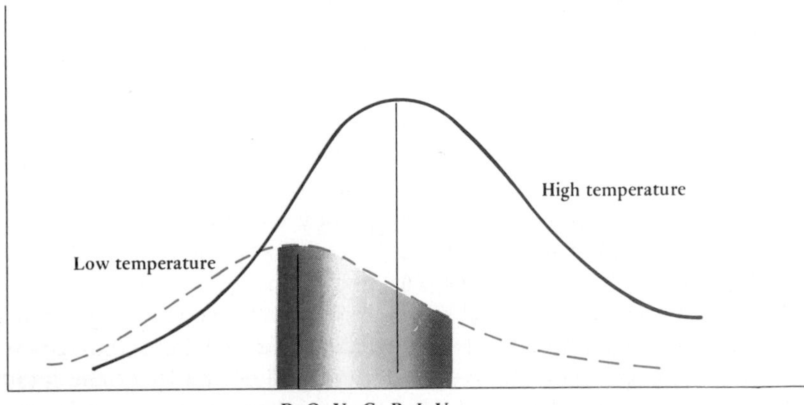

FIGURE 7.10

The spectrum of radiation given off by a heated body. The dotted line is the spectrum for a "red hot" object, since the wavelength of maximum intensity occurs at the wavelength of red light. As the temperature of the object increases, the color of the body becomes more orange and then yellow, as the wavelength of the highest intensity radiation moves toward higher frequencies (toward the ultraviolet). At very high temperatures, the object will be "white hot" when there is a comparable intensity of radiation at all wavelengths in the visible spectrum.

frequency that can be absorbed or emitted by the atom, and the minimum energy change, $h\nu$, is called a **quantum**.

Combining his new idea with existing theories of energy distribution in solid bodies, Planck was able to calculate a spectrum for a glowing body that reproduced the experimental spectrum. More importantly, experiments in other areas of physics in the years after Planck's original work showed that Planck's law applies to all phenomena on the atomic and molecular scale. The quantum revolution was born!

EXAMPLE 7.2

USING PLANCK'S LAW

In Example 7.1 you found that orange light has a frequency of $4.84 \times 10^{14}\ s^{-1}$. What is the energy of one quantum of orange light?

Solution Since the frequency of the light is given, we can multiply this by Planck's constant to obtain the energy of one quantum.

$$E = h\nu = (6.626 \times 10^{-34}\ J \cdot s)(4.84 \times 10^{14}\ s^{-1})$$
$$= 3.21 \times 10^{-19}\ J$$

7.4 MORE HISTORICAL PERSPECTIVE: ALBERT EINSTEIN AND NIELS BOHR

The modern quantum mechanical view of atomic structure came about through a series of brilliant insights by fascinating men. We have previously described a few of the early developments, but there are still more, and the people involved are among the giants of science.

ALBERT EINSTEIN AND THE PHOTOELECTRIC EFFECT

In 1900 Albert Einstein (1879–1955) was working in the patent office in Bern, Switzerland. He liked his job of reading patents for technical devices, partly because he found he could finish his day's work in a few hours and then spend the remainder of his time thinking about physics, his real interest. One thing that caught his eye were some experiments done in Germany on the *photoelectric effect*, experiments that resisted explanation by classical ideas.

The **photoelectric effect** occurs when light bombards the surface of a metal and electrons are ejected (Figure 7.11). Experiments had showed electrons would be ejected from the surface only if light of at least a minimum frequency was used. If the frequency was lower than the minimum required, no effect was observed, regardless of the intensity of the light. If the frequency of the light was above the minimum, however, increasing the light intensity caused more and more electrons to be ejected.

All these observations, Einstein decided, could be explained by combining Planck's idea of energy quanta with the notion that light could be described not only as having wave-like properties but also as having

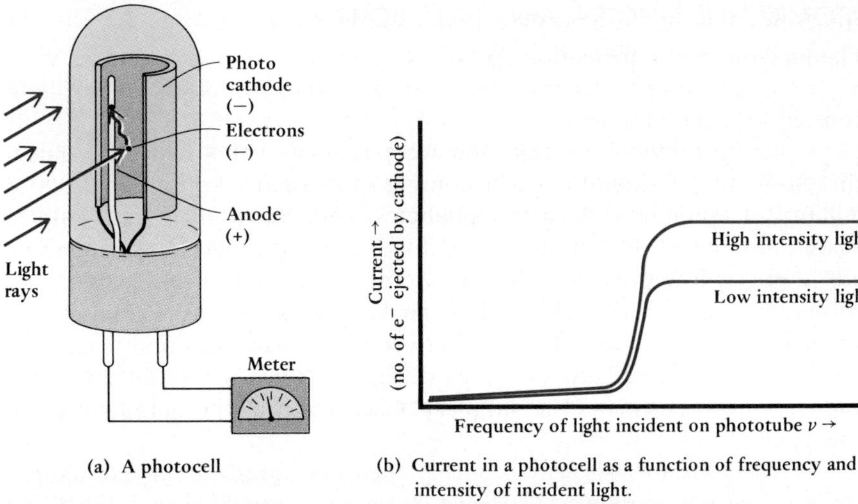

(a) A photocell

(b) Current in a photocell as a function of frequency and intensity of incident light.

FIGURE 7.11

The photoelectric effect. (a) A photocell operates by the photoelectric effect. The chief component of the cell is a photosensitive cathode. This is a material, usually a metal, that ejects electrons if light of sufficient energy falls on it. The ejected electrons move to the anode and a current flows in the cell. Such a device can be used as a switch in electrical circuits. (b) As the frequency of the incident light is increased, no current is observed until the critical frequency is passed. Light of this frequency and higher has enough energy to dislodge an electron from the surface of the photocathode. If higher intensity light is used, the only effect is to cause more electrons to be released from the surface; the onset of current is observed at the same frequency as with lower intensity light. When light of higher frequency than the minimum is used, the excess energy simply makes the electron escape the atom with more speed.

particle-like properties. Einstein assumed these "particles," now called **photons**, carry the energy given by Planck's law; that is, the energy of each photon in a stream of photons is proportional to the frequency of its wave.

$$E = h\nu \tag{7.2}$$

With Einstein's revolutionary postulate, the photoelectric effect can be explained. It is easy to imagine that a high-energy particle would have to bump into an atom to cause the atom to lose an electron. Furthermore, it is reasonable to accept the idea that an electron can be torn away from the atom only if some minimum amount of energy is used. If light is particle-like, as Einstein said, then the greater the intensity of light, the more photons there are. It then follows that the atoms of a metal surface will not lose electrons when the metal is bombarded by millions of photons if no individual photon has enough energy to remove an electron from an atom. Once the critical minimum energy (that is, minimum light frequency) is exceeded, the energy content of each photon is sufficient to cause one metal atom to lose an electron. Given this minimum energy, more photons dislodge more electrons. Thus, the connection is made between light intensity and number of electrons ejected after the minimum energy is exceeded.

(a)

(b)

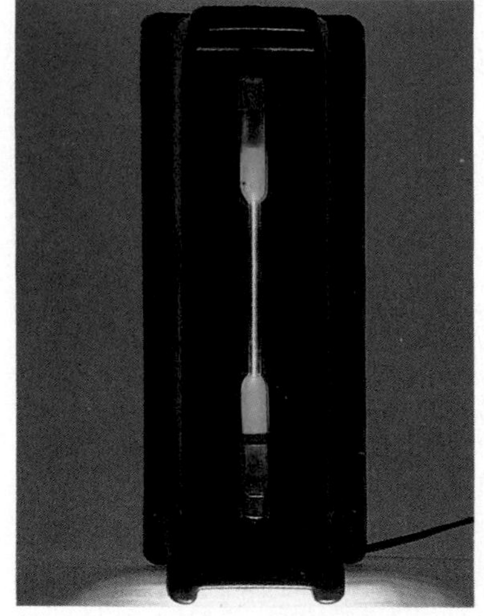

(c)

FIGURE 7.12
Light emission from excited gases: (a) Ne, (b) Ar, and (c) Hg. (Charles D. Winters)

ATOMIC LINE SPECTRA AND NIELS BOHR

The final piece of information that played a major role in our modern view of atomic structure was the observation of the properties of the light emitted by excited atoms.

In the previous section you were reminded that a heated object can glow (emit radiation). Light coming from such a source consists of millions of minutely different frequencies, and the result is that you see a continuous light. In a similar way, if you put sufficient electrical energy into a gas at low pressure, the gas atoms or molecules are excited and will glow (Figure 7.12). This is familiar to you as "neon" lights used in advertising signs. The difference between a glowing gas and a glowing piece of metal is that the gas gives off light of only a few different frequencies. In fact, this is the simplest observation of the quantization of energy in atoms.

The emission of particular frequencies of light from an excited gas can be observed with a **spectroscope** or **spectrometer** (Figure 7.13). Light from a glowing gas is passed through a narrow slit, thereby isolating a thin beam or line of light. The beam is then passed through some device (a prism or, in modern instruments, a diffraction grating) that separates the light into its various wavelengths, and we observe it as a **line spectrum**. Glowing neon, for example, would appear red to your eye, but when you pass the light through a prism, the light is actually seen to be composed of a few different colors; the wavelength appropriate to red predominates, however, so this is the color you observe.

In the 19th century, a Swiss music teacher, Johann Balmer (1825–1898), worked out a mathematical relation that accounted for the three lines of longest wavelength in the visible emission spectrum of hydrogen atoms. His results were rewritten by Johannes Rydberg (1854–1919), who found the wavelength of the lines could be predicted from the equation

$$\frac{1}{\lambda} = R\left(\frac{1}{4} - \frac{1}{n^2}\right) \qquad (n > 2) \qquad (7.3)$$

This is called the **Rydberg equation**: n is an integer associated with each line, and R, now called the **Rydberg constant**, has the value 1.0968×10^7 m^{-1}. If n is 3, the wavelength of the red line in the hydrogen spectrum is obtained; if $n = 4$, the wavelength of the green line is obtained; and if $n = 5$, the blue line. This group of visible lines (and others for which $n = 6, 7, 8$, and so on) are now known as the **Balmer series** of lines.

Niels Bohr, a young Danish physicist who worked for a time in Rutherford's laboratory, provided the first connection between line spectra and the quantum ideas of Planck and Einstein.* Bohr ignored problems of classical physics such as that illustrated by Figure 7.5, and he introduced the idea that the single electron of a hydrogen atom could occupy only certain *quantized energy states* or **stationary states** as he called them. Combining this quantization postulate with the laws of motion from clas-

*If you are interested in this period of history and Bohr's contributions, read the article by B. L. Haendler in the *Journal of Chemical Education*, 59:372, 1982.

NIELS BOHR (1885–1962)

Niels Bohr was born in Copenhagen, Denmark. He earned a Ph.D. in physics in Copenhagen in 1911 and then went to work first with J.J. Thomson in Cambridge, England, and later with Rutherford in Manchester. It was there that he began to develop the ideas that a few years later led to the publication of his theory of the structure of the atom and his explanation of atomic spectra. (For this work, he received the Nobel Prize in 1922.) After being with Rutherford for a very short time, Bohr returned to Copenhagen, where he eventually became director of the Institute for Theoretical Physics. There many young physicists carried on his work, and seven of these men later received Nobel Prizes for their studies in chemistry and physics, among them Werner Heisenberg, Wolfgang Pauli, and Linus Pauling.

sical mechanics, Bohr showed that the energy of the single electron of the hydrogen atom, in the nth stationary state, is given by the simple equation

$$\text{Energy} = E_n = -\frac{Rhc}{n^2} \qquad (7.4)$$

where R is a proportionality constant; h is Planck's constant; c is the velocity of light; and n is an integer having values of 1, 2, 3, and so on, each value characterizing a different stationary state of the atom.

Bohr was able to express R in terms of other known physical quantities and to calculate its value as $R = 1.0968 \times 10^7\ \text{m}^{-1}$, exactly the value of the experimental Rydberg constant! From this relation he could determine the energies of the quantized states of the electron in the H atom (Figure 7.14) and could derive Equation 7.3, which had previously been only an experimentally determined relation.

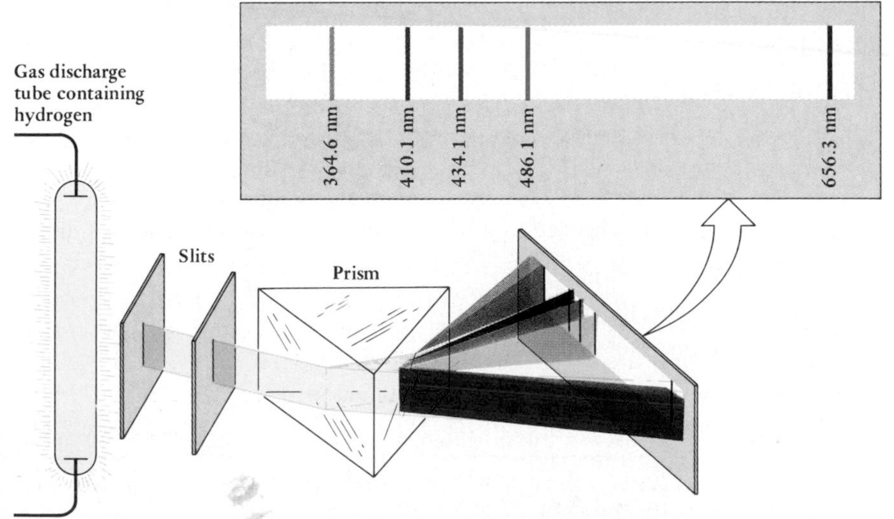

FIGURE 7.13
The line spectrum of excited H atoms. The emitted light is passed through a series of slits to isolate a narrow beam of light, and this beam is then passed through some device such as a prism that separates the light into its component wavelengths. A photographic plate or an instrument detects the separate wavelengths as individual lines. Hence, the name "line spectrum" for the light emitted by a glowing gas.

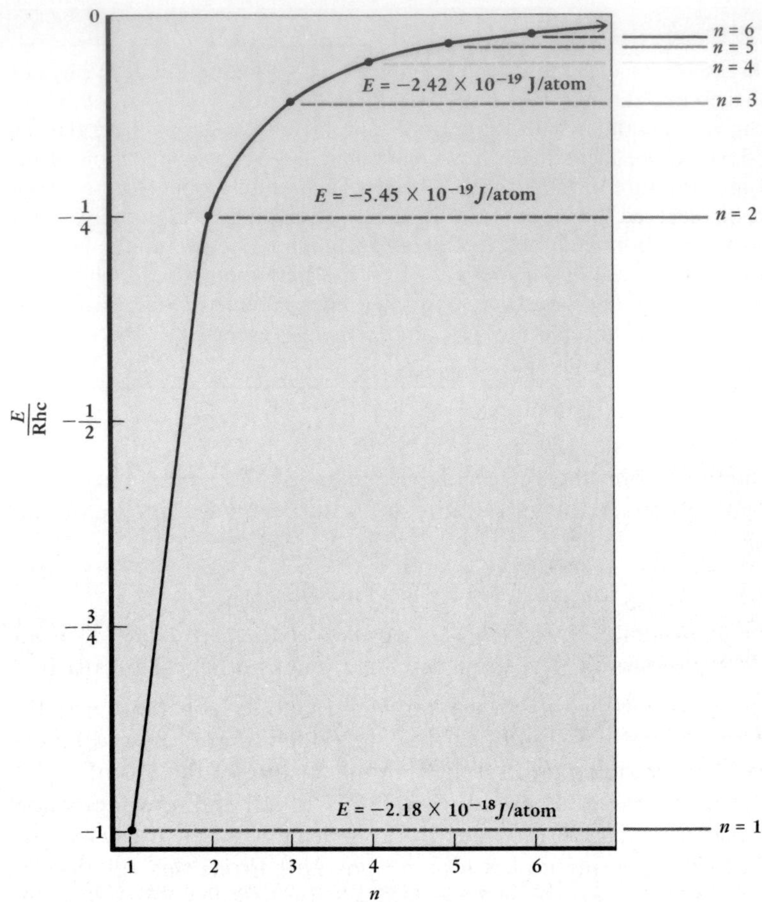

FIGURE 7.14

A graph of $E = -Rhc/n^2$ versus n. Energies are given in J/atom. Notice that the differences between successive energy states become smaller as n becomes larger.

EXAMPLE 7.3

ELECTRON ENERGIES

What are the energies of the $n = 1$ and $n = 2$ states of the hydrogen atom in (a) J/atom and (b) kJ/mol?

Solution (a) The values needed to use Equation 7.4 to solve for energies are

$$R = 1.0968 \times 10^7 \text{ m}^{-1}$$
$$h = 6.6262 \times 10^{-34} \text{ J} \cdot \text{s}$$
$$c = 2.9979 \times 10^8 \text{ m/s}$$

Thus, when $n = 1$, the energy of a single H atom is

$$E_1 = -\frac{Rhc}{n^2} = -\frac{(1.0968 \times 10^7 \text{ m}^{-1})(6.6262 \times 10^{-34} \text{ J} \cdot \text{s})(2.9979 \times 10^8 \text{ m/s})}{1^2}$$
$$= -2.1788 \times 10^{-18} \text{ J/atom}$$

and when $n = 2$ it is

$$E_2 = \frac{E_1}{2^2} = -\frac{(-2.1788 \times 10^{-18}\ \text{J/atom})}{4} = -5.4469 \times 10^{-19}\ \text{J/atom}$$

(b) The conversion from J/atom to kJ/mol is accomplished using Avogadro's number and the relation 1 kJ = 1000 J.

$$E_1 = \left(\frac{-2.179 \times 10^{-18}\ \text{J}}{\text{atom}}\right)\left(\frac{6.022 \times 10^{23}\ \text{atoms}}{\text{mol}}\right)\left(\frac{1\ \text{kJ}}{1000\ \text{J}}\right)$$

$$= -1312\ \text{kJ/mol}$$

Similarly, $E_2 = -328.0$ kJ/mol

Notice that the calculated energies are both negative with E_1 more negative than E_2. This arises because Bohr's equation reflects the fact that the force of attraction between oppositely charged bodies is given by Coulomb's law: attractive energy = − (charge on A)(charge on B)/(distance between A and B). The stronger the attractive force, the more negative the energy. Coulomb's law also tells us that the closer the negatively charged electron is to the positive nucleus, the stronger the force of attraction. As discussed in Section 7.7, an electron with $n = 1$ is closer to the nucleus on average than one with $n = 2$.

A major assumption of Bohr's theory was that an electron in an atom would remain in its lowest energy state unless otherwise disturbed (an assumption shared with the quantum mechanical approach in use today). It is the consequence of changing from one energy state to another that is important in explaining line spectra; only in this way can energy be absorbed or evolved. As you saw in Example 7.3 (and as illustrated in Figure 7.14), an electron with $n = 1$ has the most negative energy and is the most strongly attracted to the nucleus. If the electron of the hydrogen atom has $n = 1$, the atom is said to be in its **ground state**, the lowest energy stationary state.

An electron with $n = 2$ is less strongly attracted to the nucleus than one with $n = 1$, and the energy of an $n = 2$ electron is less negative (Example 7.3). Therefore, to disturb the electron in the $n = 1$ state and move it to the $n = 2$ state, the atom must absorb energy. We say that the electron must be **excited** (Figure 7.15). If the hydrogen atom has its electron in any state but $n = 1$ (the ground state), the atom is said to be in an **excited state**. Using Bohr's equations we can calculate the amount

Macayo Neon Sign. (James Cowlin) When excited, neon gas emits visible light most prominently in the red region of the spectrum. In doing so, the atoms return to the ground state.

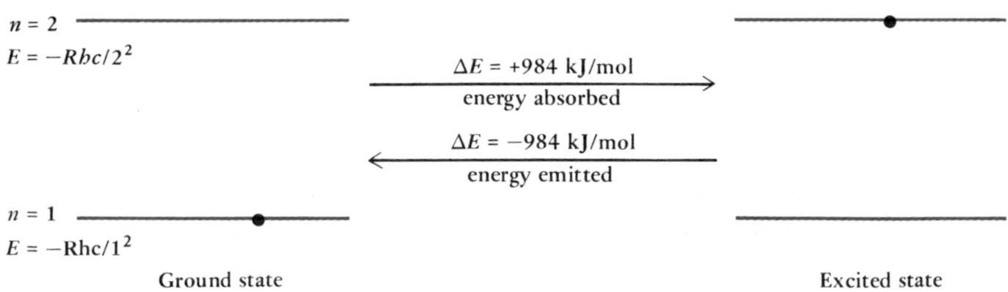

FIGURE 7.15

Absorption ($\Delta E = +$) and emission ($\Delta E = -$) of energy by the electron in the H atom moving from the $n = 1$ state (the ground state) to the $n = 2$ state (the excited state) and back again.

of energy needed to carry the H atom from the ground state to its first excited state ($n = 2$). As you learned in Chapter 5, the difference in energy between the two states is

$$\begin{aligned}\Delta E &= E_{\text{final state}} - E_{\text{initial state}} \\ &= E_2 - E_1 = \left(-\frac{Rhc}{2^2}\right) - \left(-\frac{Rhc}{1^2}\right) \\ &= \frac{3}{4}Rhc \\ &= 1.6341 \times 10^{-18}\ \text{J/atom or 984 kJ/mol of H atoms}\end{aligned}$$

where we have used the energies of the $n = 1$ and $n = 2$ states as calculated in Example 7.3. The amount of energy that must be *absorbed* by the atom so that an electron can move from the first to the second energy state is $0.75Rhc$, no more and no less. If $0.8Rhc$ of energy is provided, $0.05Rhc$ will not be used. If $0.7Rhc$ is provided, no transition between states is possible. The energy difference between energy states is quantized.

Moving an electron from a state of low n to one of higher n is an endothermic process; energy is absorbed, and the sign of ΔE must be positive. The opposite process, an electron "falling" from a level of higher n to one of lower n will therefore emit energy. For example, for a transition from $n = 2$ to $n = 1$,

$$\begin{aligned}\Delta E &= E_{\text{final state}} - E_{\text{initial state}} \\ &= E_1 - E_2 = \left(-\frac{Rhc}{1^2}\right) - \left(-\frac{Rhc}{2^2}\right) \\ &= -\frac{3}{4}Rhc\end{aligned}$$

The process is exothermic; that is, 984 kJ must be *evolved* or *emitted* per mole of H atoms.

Depending on how energy is added to a collection of H atoms in the gas phase, some atoms have their electrons excited from the $n = 1$ state to $n = 2$ or 3 or higher states. After absorbing energy, these electrons naturally move back down to lower levels (not necessarily directly to $n = 1$) and *emit* energy in the process. This is the source of the numerous lines observed in the *emission spectrum* of H atoms, and the same basic explanation holds for other elements. For hydrogen, the series of lines having energies in the ultraviolet region (called the **Lyman series**, Figure 7.16) comes from electrons moving from states with n greater than 1 to that with $n = 1$.

For excited hydrogen atoms, the series of lines that have energies in the visible region arise from electrons moving from states with $n = 3$ or greater to the state with $n = 2$. This is the series studied by Balmer, and the equation originally written by Balmer and modified by Rydberg can now be derived from Bohr's theory. Since the final state for the series of visible lines is $n = 2$, we can write

$$\Delta E = E_{\text{final}} - E_{\text{initial}}$$

$$\Delta E_{\text{Balmer}} = \left(-\frac{Rhc}{2^2}\right) - \left(-\frac{Rhc}{n_i^2}\right)$$

$$\Delta E_{\text{Balmer}} = -Rhc\left(\frac{1}{2^2} - \frac{1}{n_i^2}\right) \qquad \text{where } n_i = 3, 4, \ldots \qquad (7.5)$$

Now, we can combine Planck's law (Equation 7.2) with Equation 7.1 to find the relation between the energy of light radiated, ΔE_r, and its wavelength.

$$\Delta E_r = h\nu = \frac{hc}{\lambda}$$

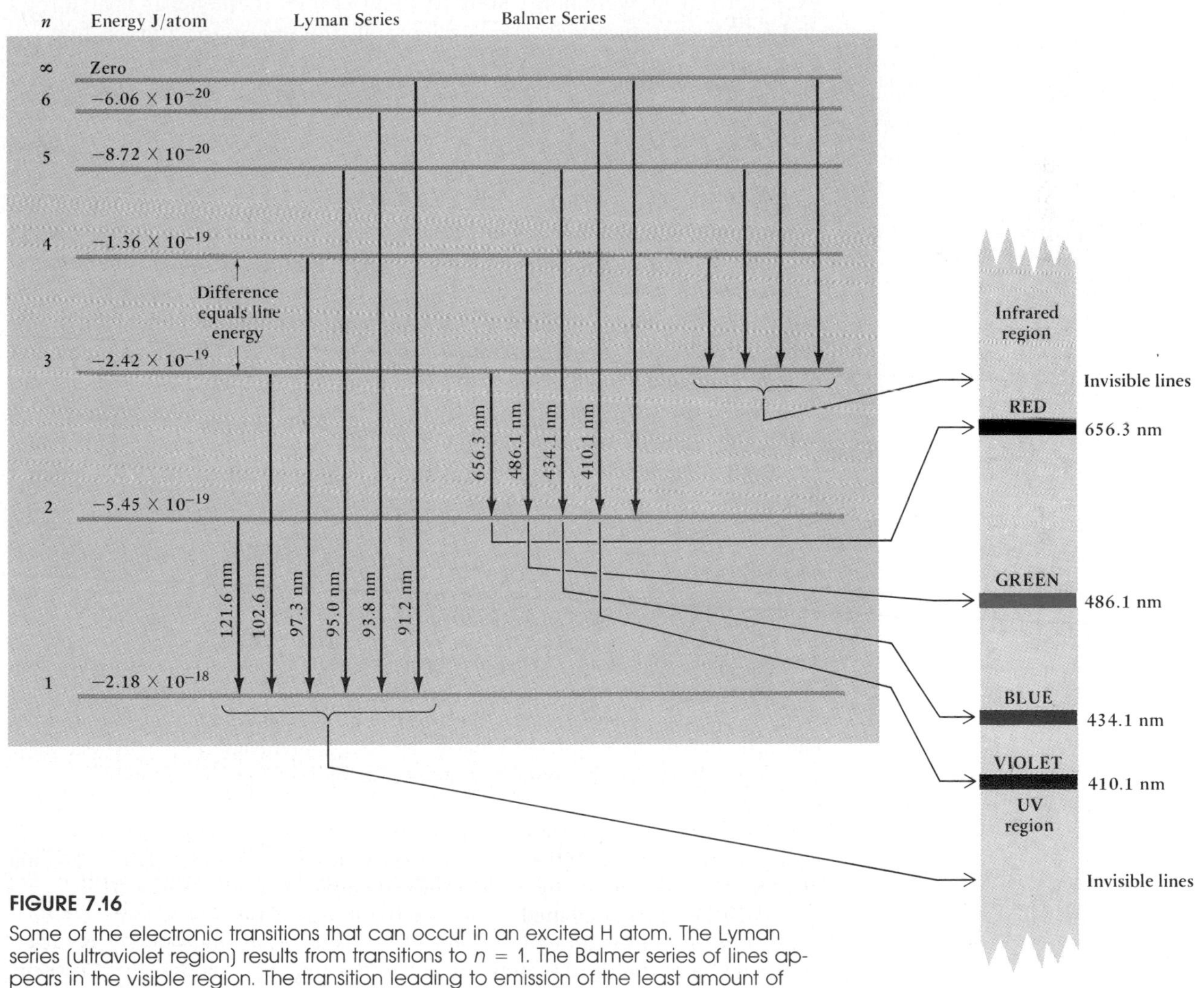

FIGURE 7.16

Some of the electronic transitions that can occur in an excited H atom. The Lyman series (ultraviolet region) results from transitions to $n = 1$. The Balmer series of lines appears in the visible region. The transition leading to emission of the least amount of energy in this series is $n = 3$ to $n = 2$ (a line in the red region, $\lambda = 656.3$ nm).

Bohr's theory said that this energy is exactly equal to the difference in energy between the two states of the hydrogen atom. Equating ΔE_r with ΔE_{Balmer} and dividing both sides by hc gives

$$\frac{1}{\lambda} = -R\left(\frac{1}{4} - \frac{1}{n_i^2}\right) \qquad \text{where } n_i = 3, 4, \ldots$$

If the minus sign, which signals the exothermic nature of the process, is removed, this is exactly Rydberg's equation (Equation 7.3). Bohr had succeeded in deriving an experimental relation from theoretical considerations, a real triumph for his theory.

In summary, we now recognize that the origin of atomic line spectra is the movement of electrons between quantized energy states. If an electron moves from a higher energy state to a lower one, energy is emitted and an *emission line* is observed. On the other hand, if an electron is excited from a lower energy state to a higher one, then energy is absorbed and an *absorption line* is seen. In general, the energy of a given line is

$$\Delta E = E_{\text{final}} - E_{\text{initial}}$$

$$\Delta E = -Rhc\left(\frac{1}{n_{\text{final}}^2} - \frac{1}{n_{\text{initial}}^2}\right) \tag{7.6}$$

When Bohr's paper describing his ideas was first published in 1913, Einstein declared it to be "one of the great discoveries."

Bohr's theory was an important advance in physics and chemistry, because it introduced the concept of energy quantization for phenomena on the atomic scale, a concept that is still an important part of modern physics and chemistry. But it soon became apparent that there were major defects in the theory. It explained only the spectrum of the H atom and of other systems having only one electron (such as He^+), for example. Further, another part of Bohr's theory, the idea that the electron moves about the nucleus like the planets about the sun, is incorrect. However, great improvements were made in atomic theory in the early 1920s, events in which Bohr was an enthusiastic and prominent participant.

EXAMPLE 7.4

ENERGIES OF ATOMIC SPECTRAL LINES

The Paschen series of lines in the hydrogen spectrum occurs in the infrared region. The electrons that produce them are moving from higher states to the $n = 3$ state (Figure 7.16). (a) Give n_{initial} for the transition that would account for the least energetic line of the series. (b) What is the energy involved in the least energetic transition? What are the frequency and wavelength of the emitted light?

Solution (a) The least energetic transition is that involving the smallest possible change in state energy. If the electron is to end with $n = 3$, then the least amount of energy will be emitted by an electron moving from $n = 4$ to $n = 3$.

(b) The energy emitted by an electron moving from $n = 4$ to $n = 3$ is

$$\begin{aligned}\Delta E &= -Rhc\left(\frac{1}{3^2} - \frac{1}{4^2}\right)\\ &= -0.04861\ Rhc\end{aligned}$$

In Example 7.3 we found that $Rhc = 1312$ kJ/mol, so the $n = 4$ to $n = 3$ transition involves an energy change of

$$\Delta E = -0.04861\ (1312\ \text{kJ/mol}) = -63.78\ \text{kJ/mol}$$

The frequency is calculated from Planck's law.

$$\nu = \frac{\Delta E}{h} = \frac{(63.78\ \text{kJ/mol})(10^3\ \text{J/kJ})}{(6.022 \times 10^{23}\ \text{electrons/mol})(6.626 \times 10^{-34}\ \text{J}\cdot\text{s})}$$
$$= 1.598 \times 10^{14}\ \text{s}^{-1}$$

Once the frequency has been obtained, the wavelength can be calculated from the relation $\lambda\nu = c = 2.998 \times 10^8$ m/s.

$$\lambda = \left(\frac{2.998 \times 10^8\ \text{m}}{\text{s}}\right)\left(\frac{1\ \text{s}}{1.598 \times 10^{14}}\right) = 1.876 \times 10^{-6}\ \text{m}$$
$$= 1.876 \times 10^{-6}\ \text{m}\ (10^9\ \text{nm/m}) = 1876\ \text{nm}$$

The experimental value is 1875.1 nm, extremely good agreement between experiment and theory for a theory as simple as Bohr's.

7.5 THE WAVE PROPERTIES OF THE ELECTRON

Einstein used the photoelectric effect to demonstrate that light, which is usually thought of as having wave properties, can also be thought of in terms of particles or photons. This fact was pondered by Louis Victor de Broglie (1892–1977). He reasoned that if light can be considered as having both wave and particle properties, why doesn't matter behave similarly? That is, could an object such as an electron, which we have considered a particle to this point, also exhibit wave properties in some circumstances? Specifically, in 1925, de Broglie proposed that a free electron of mass m moving with a velocity v should have an associated wavelength given by the equation

$$\lambda = h/mv \tag{7.7}$$

This idea was revolutionary, and experimental proof was not long in coming. Davisson and Germer, working at the Bell Telephone Laboratories, found that a beam of electrons was diffracted like light waves by the atoms in a thin sheet of metal foil and that de Broglie's relation was followed quantitatively. Since diffraction is an effect readily explained by the wave properties of light, it followed that electrons also can be described by equations of waves under some circumstances.

De Broglie's equation suggests that *any* particle of mass m traveling with velocity v has an associated wavelength. Yet, if λ is to be large enough to measure, the product of m and v must be *very* small because h is so small. For example, a 114-g baseball traveling at 110 mph has a large mv product (5.61 kg·m/s^{-1}) and, therefore, the incredibly small wavelength of 1.2×10^{-34} m! Unfortunately, this is such a tiny value that it cannot be measured with any instrument now available, so the wave nature of a moving baseball cannot be confirmed experimentally.

Thus, it is only possible to observe wave-like properties for particles of extremely small mass, for example, electrons, protons or neutrons.

EXAMPLE 7.5

USING DE BROGLIE'S EQUATION

Calculate the wavelength associated with an electron of $m = 9.11 \times 10^{-28}$ g traveling at 40.0% of the velocity of light.

Solution First, consider the units involved. Wavelength comes from h/mv where h is Planck's constant in units of joule · sec. As discussed in Example 6.1, Joules are equivalent to kg · m²/s². Therefore, the mass in de Broglie's equation will have to be used in kilograms and the velocity in m/s.

$$\text{Wavelength of a free electron} = \lambda = \frac{h}{mv}$$

The velocity of the electron is 40.0% of 2.998×10^8 m/s or

$$v = (0.400)(2.998 \times 10^8 \text{ m/s}) = 1.20 \times 10^8 \text{ m/s}$$

and the mass of the electron in kilograms is 9.11×10^{-31} kg.

Substituting these values into the equation for wavelength, we have

$$\lambda = \frac{(6.6262 \times 10^{-34} \text{ kg} \cdot \text{m}^2/\text{s}^2)(\text{s})}{(9.11 \times 10^{-31} \text{ kg})(1.20 \times 10^8 \text{ m/s})}$$
$$= 6.07 \times 10^{-12} \text{ m}$$

In nanometers, the wavelength is $(6.07 \times 10^{-12} \text{ m})(1.00 \times 10^9 \text{ nm/m}) = 0.00607$ nm. This wavelength is about 1/20 of the diameter of an H atom.

EXERCISE 7.4 DE BROGLIE'S EQUATION

Calculate the wavelength associated with a neutron having a mass of 1.675×10^{-24} g and an energy of 6.21×10^{-21} J. (Remember that the kinetic energy of a particle is $E = mv^2/2$.)

Erwin Schrödinger (1887–1961) was born in Vienna and studied at the University there. Following service in World War I as an artillery officer, he became a professor of physics at various universities; in 1928 he succeeded Max Planck as professor of theoretical physics at the University of Berlin. He shared the Nobel Prize in physics (with Paul Dirac) in 1933. (AIP Niels Bohr Library)

7.6 SCHRÖDINGER, HEISENBERG, AND THE DAWN OF WAVE MECHANICS

After World War I, Niels Bohr assembled a group of physicists in Copenhagen, a group that set out to derive a comprehensive theory for the behavior of electrons in atoms from the viewpoint of the electron as a particle. Independently, Erwin Schrödinger (1887–1961), an Austrian, worked toward the same goal, but he used de Broglie's hypothesis that an electron in an atom could be described by equations appropriate to wave motion. Both Bohr and Schrödinger were successful, but theoreticians today primarily use Schrödinger's concept, because it has been found to be more convenient. In any event, the general theoretical approach to understanding atomic behavior, developed by Bohr, Schrödinger, and their associates, has come to be called **quantum mechanics** or **wave mechanics**.

Before you can appreciate Schrödinger's solution, there is another point to be emphasized. After de Broglie's suggestion that an electron can be described as having wave properties, a great debate raged in physics. How can an electron be described as *both* a particle and as a wave? When J.J. Thomson measured the charge/mass ratio of an electron early in this century, he did an experiment that showed the particle-like nature of the electron. On the other hand, Davisson and Germer observed its wave-like properties. One must conclude that *the electron has dual properties*. The result of a given experiment can be described *either* by the physics of waves *or* of particles. However, there is no single experiment that can be done to *simultaneously* show the electron to be a wave and a particle!

What does the **wave–particle duality** have to do with electrons in atoms? Werner Heisenberg (1901–1976) and Max Born (1882–1970) provided the answer. Heisenberg proposed the so-called **uncertainty principle**, which can be expressed mathematically as

$$\Delta x \cdot \Delta(mv) > h \tag{7.8}$$

This simple expression means that any uncertainty in the position of the electron (Δx) times any uncertainty in its momentum (Δmv) must be greater than Planck's constant. In other words, if you try to define exactly the momentum of some particle ($\Delta mv \rightarrow 0$), be it a car or an electron, you will have to forego knowledge of its position ($\Delta x \rightarrow \infty$) at the instant you measure its momentum. As outlined in the following example, the mass of a moving car is large, so even a small uncertainty in its velocity leads to such a small error in position that we know the position of the car accurately. On the other hand, an electron has so little mass that you will be unsure about its position in space if you know its momentum with any reasonable accuracy.

The momentum of a body (mass × velocity = mv) is related to its kinetic energy ($mv^2/2$). $mv^2/2 = (mv)^2/2m$ = (momentum)$^2/2m$

EXAMPLE 7.6

HEISENBERG'S UNCERTAINTY PRINCIPLE

(a) What is the uncertainty in the position of an electron ($m = 9.11 \times 10^{-28}$ g) moving at 40.0% of the velocity of light with a velocity uncertainty of 0.100%.

(b) What is the uncertainty in the position of an automobile ($m = 1.00 \times 10^3$ kg) moving at 60.0 ± 0.100 mph (26.8 ± 0.0450 m/s)?

Solution (a) Position uncertainty for electron = $\Delta x > h/\Delta mv$

$$\Delta x > \frac{(6.6262 \times 10^{-34}\ \text{kg} \cdot \text{m}^2/\text{s}^2)(\text{s})}{(9.11 \times 10^{-31}\ \text{kg})(2.998 \times 10^8\ \text{m/s})(0.400)(0.00100)}$$
$$> 6.07 \times 10^{-9}\ \text{m}$$

The uncertainty in the electron position amounts to about 6 nm, a very large distance considering that distances on the atomic and molecular scale are measured in nanometers. For example, the diameter of an H atom is about 0.1 nm.

(b) Position uncertainty for an automobile = $\Delta x > h/\Delta mv$.

$$\Delta x > \frac{(6.6262 \times 10^{-34}\ \text{kg} \cdot \text{m}^2/\text{s}^2)(\text{s})}{(1.00 \times 10^3\ \text{kg})(0.0450\ \text{m/s})} = 1.47 \times 10^{-35}\ \text{m}$$

This is such a small uncertainty that current instruments cannot measure it. We thus know the car's position very accurately.

Werner Heisenberg (1901–1976) earned a PhD in theoretical physics at the University of Munich (Germany) in 1923 and then studied with Max Born and later Niels Bohr. He received the Nobel Prize in physics in 1932. (AIP Niels Bohr Library)

The uncertainty principle suggests that we can know the energy of the electron (a quantity directly related to mv), but this knowledge will come at the expense of knowing exactly where the electron is located. Therefore, Max Born proposed that the results of quantum mechanics should be interpreted as follows: if we choose to know the energy of an electron in an atom with little uncertainty, we shall have to accept a correspondingly large uncertainty in its position in the space about the atomic nucleus. What we *can do* is calculate the **probability** of finding that electron within a given region of space, and we turn to this viewpoint in the next section.

7.7 THE WAVE MECHANICAL VIEW OF THE ATOM

Erwin Schrödinger combined de Broglie's equation with classical equations for wave motion. From these he derived a new equation, often called the wave equation or **Schrödinger equation**, to describe the behavior of an electron in the hydrogen atom. This equation is complex and mathematically difficult to solve except in simple cases. The mathematics are of no concern to us here, but the solutions are chemically important.

WAVE FUNCTIONS

Solving Schrödinger's equation leads to a set of functions called **wave functions**, symbolized by the Greek letter ψ (psi). There are several important points to be made about these wave functions, which characterize the electron as a matter–wave.

1. Only certain vibrations, called standing waves (Figure 7.9), will occur in a vibrating string. The same situation occurs in an atom; although electron motion is not as simple as that of a vibrating string, there still are *only certain allowed wave functions*.
2. Each wave function, ψ_n, corresponds to an allowed energy ($-Rhc/n^2$) for the electron in the atom. This is like Bohr's result; for each integer n there is an atomic state characterized by its own wave function ψ_n and energy E_n.
3. Points 1 and 2 amount to saying that *the energy of the electron is quantized.* The concept of quantization enters Schrödinger's theory naturally with the basic assumption of an electron matter wave. This is in contrast with Bohr's theory where quantization is imposed as a postulate at the start.
4. Each wave function ψ_n itself has no readily interpretable physical meaning. Instead, *the square of* ψ_n (when calculated for the electron at a point *P*) *gives the intensity of the electron wave or the probability of finding the electron* at the point *P* in space about the nucleus. The theory does not predict the position of the "particle." Furthermore, since Schrödinger's theory precisely defines the energy of the electron, Heisenberg's uncertainty principle tells us this must result in a large uncertainty in electron position. This is why we can only define the *probability* of the electron being at a certain point in space when in a given energy state.

5. The wave for each of the allowed energy states is called an **orbital**. We shall say more about orbitals in a moment.
6. To solve Schrödinger's equation for an electron in a three-dimensional world, three integer numbers—the quantum numbers, n, ℓ, and m_ℓ—must be introduced. These quantum numbers may have only certain combinations of values, as outlined below. We shall use these combinations to define the energy states and orbitals available to the electron.
7. The amplitude of the electron wave at a point depends on the distance of the point from the nucleus. In some cases, there is also dependence on the "latitude" and "longitude" of the point.

THE QUANTUM NUMBERS AND ATOMIC ORBITALS

In a three-dimensional world, three parameters are required to describe the location of an object in space (Figure 7.17). For the quantized atomic electron, this requirement leads to the existence of **three quantum numbers** $\boldsymbol{n}$**,** $\boldsymbol{\ell}$**, and** $\boldsymbol{m_\ell}$. For the case of the hydrogen atom, the first of these three quantum numbers alone is sufficient to describe the energy of the electron, but all three are needed to define the probability of finding that electron in a given region of space. Before looking into the meanings of the three quantum numbers, it is important to say that they are all integers but that their values cannot be randomly selected. The value of n limits the possible values of ℓ, which in turn limit the values of m_ℓ.

Think of the three quantum numbers as an orbital "zipcode." A zipcode defines your location, in part, by state, county, and city. n, ℓ, and m_ℓ define an orbital by giving the electron shell, the subshell, and the orbital within that subshell.

To determine electron energies in multielectron atoms, a fourth quantum number, **m_s, the spin quantum number**, is required. This is discussed in the next chapter where the distribution of electrons in real atoms is described.

n, the principal quantum number $= 1, 2, 3, \cdots$

The principal quantum number n can have any integer value from 1 to infinity. As the name implies, it is the most important quantum number because its value determines the total energy of the electron. For the hydrogen atom the quantum mechanical relation between the electron energy E and the quantum number n is the same as that given in Bohr's equation (7.4)($E = -Rhc/n^2$). Since all the quantities in this equation are constants except n, this means that *the energy of the electron in the H atom varies only with the value of n*.

The value of n also gives one a measure of the most probable distance of the electron from the nucleus: the greater the value of n, the more probable it is that the electron is found further from the nucleus. That is, $\boldsymbol{n}$ is a measure of the **orbital radial size or diameter**.

As you will see later, in atoms having more than one electron, two or more electrons may have the same n value. These electrons are then said to be in the same **electron shell**, the shells being numbered according to their major quantum number.*

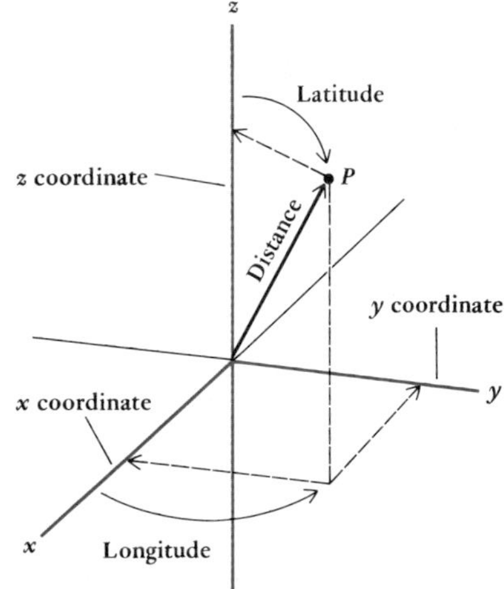

FIGURE 7.17
The cartesian coordinate system. The position of a point P in space can be specified by giving the x, y, and z coordinates.

ℓ, the angular momentum quantum number $= 0, 1, 2, \cdots (n - 1)$

The angular momentum quantum number is related to the **shape of electron orbitals**, and the *number of values* of ℓ for a given value of n will

*An earlier notation used letters for the major electron shells: K, L, M, N, and so on, corresponding to $n = 1, 2, 3, 4$, and so on.

ℓ is related to orbital shape and describes the types of subshells within a shell. The number of values of ℓ (number of subshells) in a shell = n for that shell.

tell you how many different *orbital types* or electron **subshells** there are in a particular electron shell. The integer values that ℓ may have are limited by the value of n: ℓ may be an integer from 0 up to and including $n - 1$. In other words, if n is 1, then there is only one ℓ value possible; ℓ can only be 0, and there can only be one type of orbital or subshell in the $n = 1$ electron shell. In constrast, when $n = 4$, ℓ can have four values of 0, 1, 2, and 3. Because there are four values of ℓ, there are four orbital types or four subshells within the fourth major quantum shell.

The values of the ℓ quantum number are usually coded by a letter according to the scheme below.

The orbital labels come from the early days of atomic spectroscopy when lines in emission spectra were called *s*harp, *p*rincipal, *d*iffuse, and *f*undamental.

VALUE OF ℓ	CORRESPONDING ORBITAL LABEL
0	*s*
1	*p*
2	*d*
3	*f*

Thus, for example, a subshell with a label of $\ell = 1$ is called a "*p* subshell," and an orbital found in that subshell is called a "*p* orbital."

m_ℓ, magnetic quantum number $= 0, \pm 1, \pm 2, \pm 3, \cdots \pm \ell$

m_ℓ is related to the spatial orientation of an orbital in a given subshell. The number of m_ℓ values = the number of orbitals in a subshell.

The first quantum number (n) locates the electron in a particular electron shell, and the second (ℓ) places it in a particular subshell of orbitals within the shell. The third quantum number (m_ℓ) then specifies in which orbital within the subshell the electron is located.

The integer values that m_ℓ may have are limited by ℓ; m_ℓ values can range from $+\ell$ to $-\ell$ with 0 included. (For example, when $\ell = 2$, m has the five values $+2$, $+1$, 0, -1, -2.) The number of values of m_ℓ for a given ℓ tell you how many orbitals of a given type there are in that subshell.

The quantum number rules can give us chemically useful information. Let us expand on them and see where this leads.

If $n = 1$, the rules tell us that ℓ can only be 0 and so m_ℓ must also have a value of 0. This means that, in the electron shell closest to the nucleus, there is only one type of orbital or one subshell (ℓ has only one value), and there is only one orbital of this type (m_ℓ has only one value). This orbital is labeled "1*s*," the "1" conveying the value of n and "*s*" telling you that ℓ is 0. When $\ell = 0$*, an s orbital is indicated, and there can be only one s orbital in a given shell of electrons.*

Electron orbitals are labeled by first giving the value of n and then the value of ℓ in the form of its letter code. For $n = 1$ and $\ell = 0$, the label is 1*s*.

When $n = 2$, ℓ can now have two values: 0 and 1, so there are two types of orbitals in the second shell. One of these is the 2*s* orbital ($n = 2$ and $\ell = 0$), and the other is the 2*p* orbital ($n = 2$ and $\ell = 1$). Since the values of m_ℓ can range from $+1$ to -1 including 0 when $\ell = 1$, this means that there must be three *p* orbitals. *When* $\ell = 1$*, p orbitals are indicated, and there are always three of them.* (The opposite is true as well: for *p* orbitals, ℓ is always 1.)

For a given n, there will be n values of ℓ and n orbital types; this means there is a total of n^2 orbitals in the nth shell.

When an electron is characterized by $n = 3$, there are three orbital types (the three values of ℓ are 0, 1, and 2). Since you see ℓ values of 0 and 1 again, you know that two of the orbital types in the $n = 3$ shell will be 3*s* (one orbital) and 3*p* (three orbitals). The third orbital type is *d*,

TABLE 7.1 Summary of the Quantum Numbers, Their Interrelationships, and the Orbital Information Conveyed

PRINCIPAL QUANTUM NUMBER	ANGULAR MOMENTUM	MAGNETIC QUANTUM NUMBER	NUMBER AND TYPE OF ORBITALS IN THE SUBSHELL	
Symbol = n Values = 1, 2, 3, . . . (Orbital size, Energy)	Symbol = ℓ Value = 0 . . . $n - 1$ (Orbital Shape)	Symbol = m_ℓ Values = $-\ell$ . . . 0 . . . $+\ell$ (Orbital Orientation)	Number = number of values of m_ℓ = $2\ell + 1$ (Orbitals in a Shell = n^2)	
1	0	0	1 s orbital	(1 orbital in the $n = 1$ shell)
2	0	0	1 $2s$ orbital	(4 orbitals of 2 types in the $n = 2$ shell)
	1	+1, 0, −1	3 $2p$ orbitals	
3	0	0	1 $3s$ orbital	(9 orbitals of 3 types in the $n = 3$ shell)
	1	+1, 0, −1	3 $3p$ orbitals	
	2	+2, +1, 0, −1, −2	5 $3d$ orbitals	
4	0	0	1 $4s$ orbital	(16 orbitals of 4 types in the $n = 4$ shell)
	1	+1, 0, −1	3 $4p$ orbitals	
	2	+2, +1, 0, −1, −2	5 $4d$ orbitals	
	3	+3, +2, +1, 0, −1, −2, −3	7 $4f$ orbitals	

indicated by $\ell = 2$. Since m_ℓ can be +2, +1, 0, −1, and −2 when $\ell = 2$, there are five *d* orbitals (no more and no less). Thus, when $\ell = 2$ you can count on the fact that *there are five orbitals labeled d*.

Besides *s*, *p*, and *d* electron orbitals, many elements have *f* electron orbitals, that is, orbitals for which $\ell = 3$. There will be seven of them as given by the seven possible values of m_ℓ when $\ell = 3$ (+3, +2, +1, 0, −1, −2, −3).*

The three quantum numbers introduced so far and their allowed values are summarized in Table 7.1. To understand atomic structure fully, you should be thoroughly familiar with these rules.

EXERCISE 7.5 USING THE QUANTUM NUMBERS

Without looking at Table 7.1, complete the following statements:

(a) When $n = 2$, the values of ℓ can be _____ and _____.

(b) When $\ell = 1$, the values of m_ℓ can be _____ , _____ , and _____ , and the subshell has the letter label _____.

(c) When $\ell = 2$, this is called a _____ subshell.

(d) When a subshell is labeled *s*, the value of ℓ is _____ and m_ℓ has the value _____.

(e) When a subshell is labeled *p*, there are _____ orbitals within the subshell.

(f) When a subshell is labeled *f*, there are _____ values of m_ℓ and there are _____ orbitals within the subshell.

THE SHAPES OF ATOMIC ORBITALS

The chemistry of an element and of the compounds it forms is determined by which orbitals are filled, which are empty, and the shapes of those orbitals. Now that we know something about the types and number of orbitals available, we want to turn to the question of orbital shape.

*No known elements have electrons in orbitals with an ℓ value greater than 3 in the ground state. If they existed, however, $\ell = 4$ would correspond to *g* orbitals, $\ell = 5$ to *h* orbitals, and so on down the alphabet.

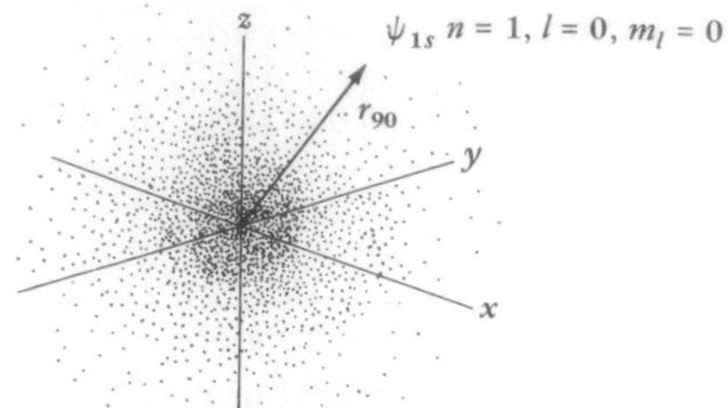

(a) Dot picture of an electron with a 1*s* atomic orbital. Each dot represents the position of the electron at a different instant in time. Note that the dots cluster closest to the nucleus. r_{90} is the radius within which the electron is found 90% of the time.

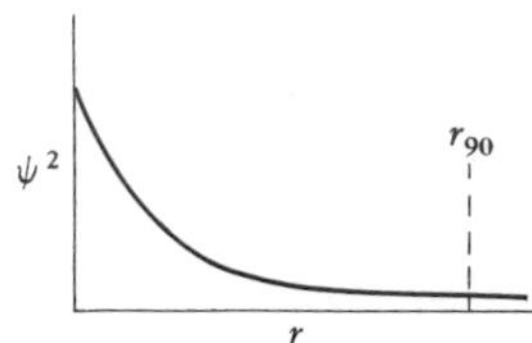

(b) A plot of the probability density as a function of distance from the nucleus for a one-electron atom with the electron in the 1*s* orbital.

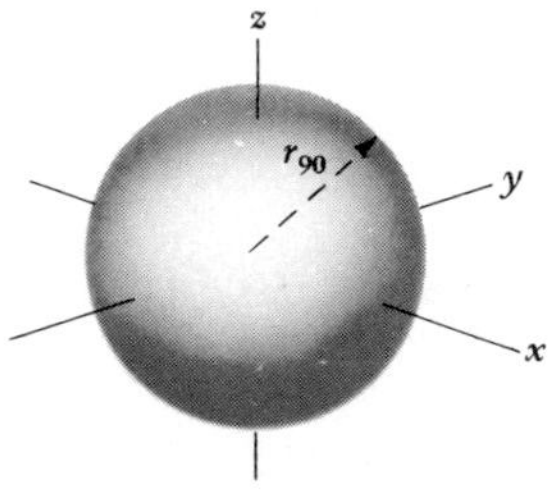

(c) The surface of the sphere within which the electron is found 90% of the time in a 1*s* orbital.

FIGURE 7.18

Different views of a 1*s* ($n = 1$ and $\ell = 0$) orbital for the electron in the hydrogen atom. (a) Dot picture of a 1*s* electron orbital. Each dot represents a possible position for the electron. The higher the density of dots, the higher the probability of the electron in that region. Note that the dots cluster closest to the nucleus. r_{90} is the radius within which the electron is found 90% of the time. (b) A plot of the electron probability density as a function of distance from the nucleus. (c) The sphere within which is found 90% of the 1*s* electron orbital.

When an electron has a value of $\ell = 0$, we say the electron occupies an *s* orbital. But what does this mean? What is an *s* orbital? What does it look like? To answer this question, we shall have to investigate the wave function for an electron with $n = 1$ and $\ell = 0$, that is, the 1*s* orbital. If you could photograph the electron with that orbital at one-second intervals for a few thousand seconds, the composite picture would resemble the drawing in Figure 7.18a, a picture that is meant to be like that of a photograph of a standing wave (Figure 7.9) taken at frequent intervals. The fact that there is a greater density of "dots" close to the nucleus indicates that the electron is most often found near the nucleus (and less likely to be found farther away). Putting this in the language of quantum mechanics, we say that the greatest probability of finding the electron is in a tiny volume of space around the nucleus, while it is less probable that it is farther away. The "thinning" of the electron cloud at increasing distance, shown by the decreasing density of dots in Figure 7.18a, is illustrated further in Figure 7.18b. Here we have plotted the square of the electron wave function for the electron in a 1*s* orbital as a function of the distance of the electron from the nucleus. As noted in point 4 on page 238, the value of ψ^2 at a point *P* gives the electron probability at that point. The units of ψ^2 at each point *P* are 1/volume, so the numbers on this plot are related to probability of finding the electron in each cubic nanometer, for example. Hence, ψ^2 is called the **probability density**. For the 1*s* orbital ψ^2 is very high for points in the tiny volume of space immediately surrounding the nucleus, but it drops off rapidly as the distance from the nucleus increases. Notice that the probability approaches but never reaches zero, even at very large distances.

For the 1*s* orbital, Figure 7.18a shows that the electron is most likely found within a sphere with the nucleus at the center. No matter which direction you proceed from the nucleus, the probability of finding the electron along a line in that direction drops off as depicted by Figure 7.18b. Indeed, *all s orbitals are spherical in shape*.

The visual image of Figure 7.18a is that of a cloud whose density is small at large distances from the center; there is no sharp boundary beyond which the electron is never found. However, *s* and other orbitals are often depicted as having a sharp boundary (Figure 7.18c), simply because it is easier to draw such pictures. To arrive at this picture we have drawn a sphere about the nucleus in such a way that there is a 90% chance of finding the electron somewhere inside. There are many misconceptions about pictures of this type. Be sure to understand that this surface is not real. The nucleus is surrounded by an "electron cloud" and not by an inpenetrable surface "containing" the electron. The electron is not distributed evenly throughout the volume enclosed by the surface but is most likely found nearer the nucleus.

An electron in a 2*s* orbital is also represented by a spherical electron cloud, but there are important differences from the 1*s* orbital. First, the 2*s* cloud is larger than the 1*s* cloud: the point of maximum probability for the 2*s* electron is found slightly farther from the nucleus than that of the 1*s* electron. The second difference is seen in Figure 7.19. When moving along a line away from the nucleus, the probability of finding the 2*s*

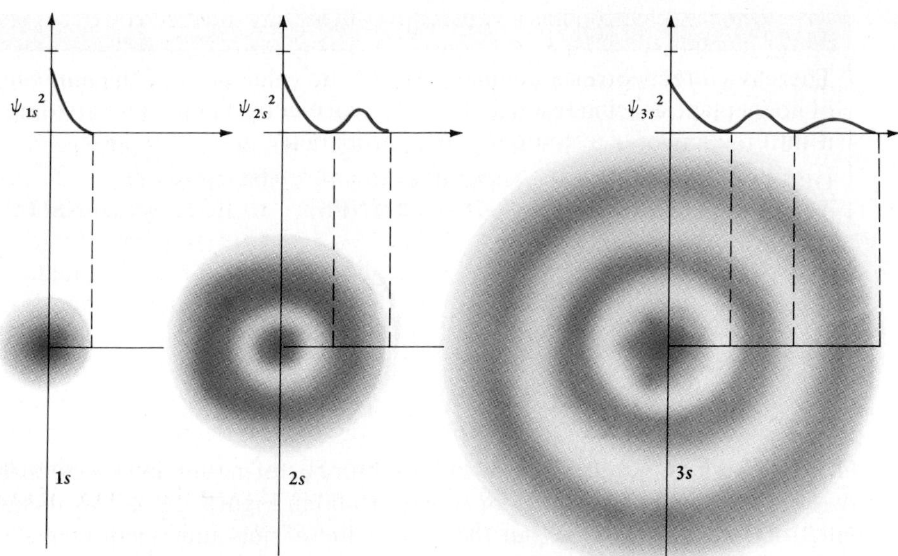

FIGURE 7.19

Probability density plots and electron cloud pictures comparing the 1*s*, 2*s*, and 3*s* atomic orbitals for the H atom. The electron cloud is shown in a plane containing the nucleus. All three orbitals are spherical in three-dimensional space, but the 2*s* orbital has 1 spherical node and the 3*s* orbital has 2 spherical nodes. For a given orbital type, the number of spherical nodes increases as the energy increases.

electron drops off rapidly and actually reaches zero at 0.1058 nm. That is, there is a **node** in the electron wave at this distance.* Beyond this distance, the probability rises again, passes through a maximum at 0.2116 nm, and then trails off at still larger distances. This happens no matter in which direction you move away from the nucleus. If you could stand on the nucleus and look into space in any direction, you would always see a node at 0.1058 nm. Therefore, the probability of finding the 2*s* electron is always interrupted at 0.1058 nm, and all the points at which this happens describe the surface of a sphere. Since there is *no* probability of finding an electron on this surface, it is called a **spherical node**. The number of spherical nodes for any type of orbital is always given by

$$\text{Number of spherical nodes} = n - \ell - 1 \tag{7.9}$$

For example, if you consider *s* orbitals, for which ℓ is always 0, you will find the following numbers of spherical nodes:

ORBITAL	NUMBER OF SPHERICAL NODES $(n - \ell - 1)$
1*s*	0
2*s*	1
3*s*	2
4*s*	3

*There are two points of interest in this regard. First, recall that a standing wave (Figure 7.10) has at least two nodes and perhaps more. Second, a question our students often ask is how the electron moves across the node. Unfortunately, this question has no meaning. If you think of the electron obeying the physics of waves, it is no longer fair to think about it as a particle moving about in space. The standing waves in Figure 7.10 have nodes, but this does not mean that the string does not exist there; it simply means that there is no amplitude at the node. The same is true of electron waves.

RELATIONSHIP BETWEEN ℓ AND NUMBER OF NODAL PLANES

There is a direct correspondence between the value of ℓ and the *number* of nodal planes slicing through the spherical electron cloud and dividing it into regions of electron density or probability.

ORBITAL TYPE	VALUE OF ℓ	NUMBER OF NODAL PLANES	REGIONS OF ELECTRON DENSITY
s	0	0	1
p	1	1	2
d	2	2	4
f	3	3	8

A spherical node is a spherical surface on which there is no probability of finding the electron. The number of such surfaces in an atom is $n - \ell - 1$.

Thus, as the value of n increases, the number of nodes in the electron wave also increases (as seen for the 3*s* orbital in Figure 7.19). This means all *s* orbitals are spheres, but there is an increasing number of layers of electron density as n increases. As an analogy, think of an onion that has layer on layer of white or pink matter separated by a thin space.

Atomic orbitals for which $\ell = 1$ are called *p* orbitals. All *p* orbitals have one plane in which an electron is never found, that is, a **nodal plane** (Figure 7.20). The electron can never be in the nodal plane that slices through the nucleus; the regions of electron probability lie on either side of this plane. This means that, unlike the *s* orbital, there is no likelihood of finding the electron at the nucleus, and so a plot of ψ^2 versus distance (Figure 7.20a) starts at zero probability at the nucleus, rises to a maximum at 0.1058 nm (for a 2*p* orbital), and then drops off at still greater distances. If you enclose within a boundary surface 90% of the electron density, the view in Figure 7.20b is appropriate: the electron cloud fills a space that resembles a weight lifter's "dumbbell," and so this is the term by which chemists often describe *p* orbital shapes.

According to Table 7.1, when $\ell = 1$, then m_ℓ can only be $+1$, 0, and -1. That is, there are three types of $\ell = 1$ or *p* orbitals. Since there are three mutually perpendicular directions in space (x, y, and z), the *p* orbitals are commonly visualized as lying along these directions, and they are labeled according to the axis along which they lie (p_x, p_y, and p_z) (Figure 7.20b).

You have just seen above that *s* orbitals (for which $\ell = 0$) are spheres with no nodal planes, while *p* orbitals (for which $\ell = 1$) have a single nodal plane. It follows that *d orbitals, for which $\ell = 2$, have two nodal planes that slice the electron cloud into four sections.* In an *x-y-z* world, there are three mutually perpendicular planes: *xy*, *xz*, and *yz* (Figure 7.21). Three of the five possible *d* orbitals lie in these planes. For example, the d_{xy} orbital in Figure 7.21a lies in the *xy* plane, and the electron cloud has been sliced into four regions of electron probability by two nodal planes, the *xz* and *yz* planes. Once again, there is no probability of finding the electron in a *d* orbital on one of these nodal surfaces or at the nucleus, but only in the regions between the planes.

Of the two remaining *d* orbitals, the $d_{x^2-y^2}$ orbital is easier to visualize. It lies in the *xy* plane, but, unlike the d_{xy} orbital, the regions of

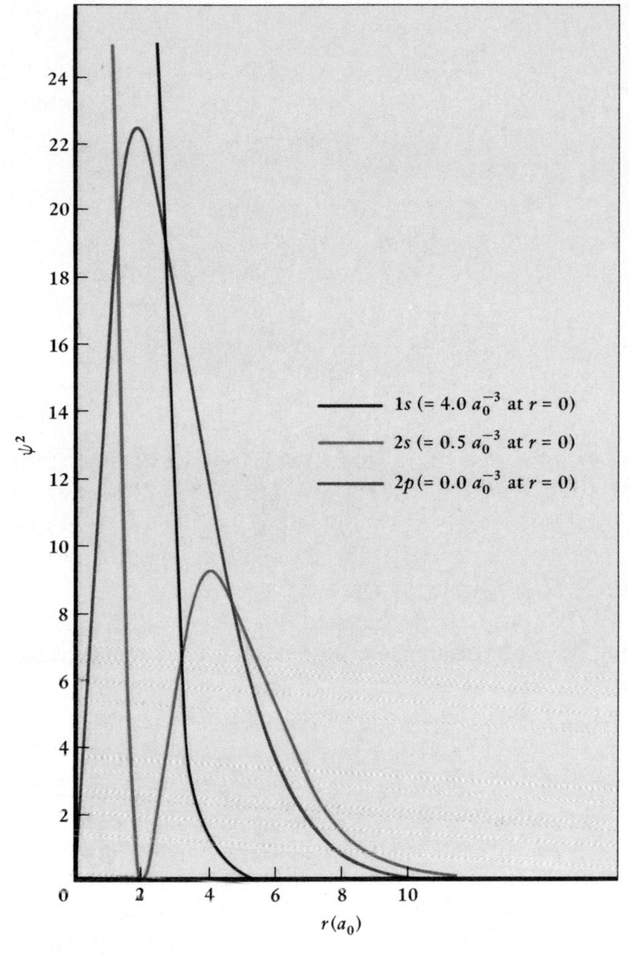

(a)

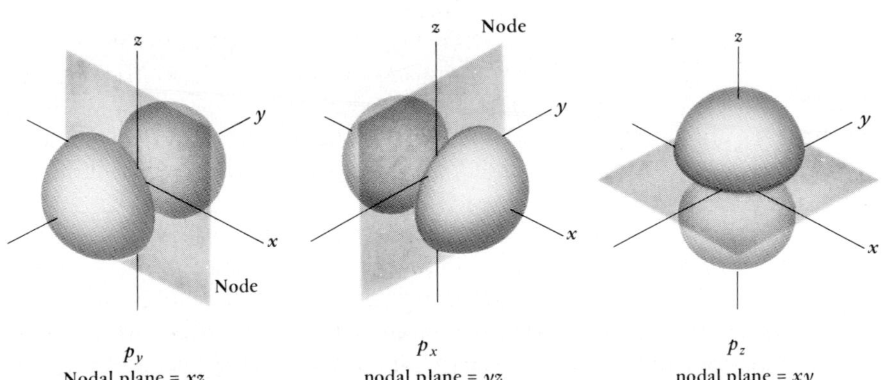

(b)

FIGURE 7.20

(a) Probability density plots for 1*s*, 2*s*, and 2*p* electron orbitals. The horizontal axis represents the distance from the nucleus in units of a_0 (atomic units), where a_0 = 0.0529 nm. Units of the vertical axis, which gives the probability or electron wave intensity, are $10^{-3}/a_0^3$. Note that the 1*s* electron wave is without a node, whereas the 2*s* electron wave has a node at $2a_0$ while the 2*p* electron wave has a node at the nucleus. (The 1*s* and 2*s* probabilities very close to the nucleus are not infinite but are indeed quite large.) (b) The 2*p* orbitals for a one-electron atom. The subscript letter on the orbital notation indicates the cartesian axis along which the orbital lies. The plane passing through the nucleus is the *planar node*; all *p* orbitals have one planar node because ℓ = 1.

electron density lie *along* the *x* and *y* axes. Thus, the cutting nodal planes bisect the angles made by the *x* and *y* axes.

The final *d* orbital, d_{z^2}, has two main regions of electron density along the *z* axis, but there is also a "belt" or "donut" of electron density in the *xy* plane. This orbital also has two nodal surfaces, but the surfaces are not flat. Think of an ice cream cone sitting with its tip at the nucleus.

FIGURE 7.21

The *d* atomic orbitals. (a) The *xz* and *yz* nodal planes of the d_{xy} orbital. The subscript *xy* on the orbital designation means that the orbital lies in the *xy* plane. (b) All *d* orbitals have $\ell = 2$, so all have two planar nodes. (The "planes" for the d_{z^2} orbital are actually conical in shape.)

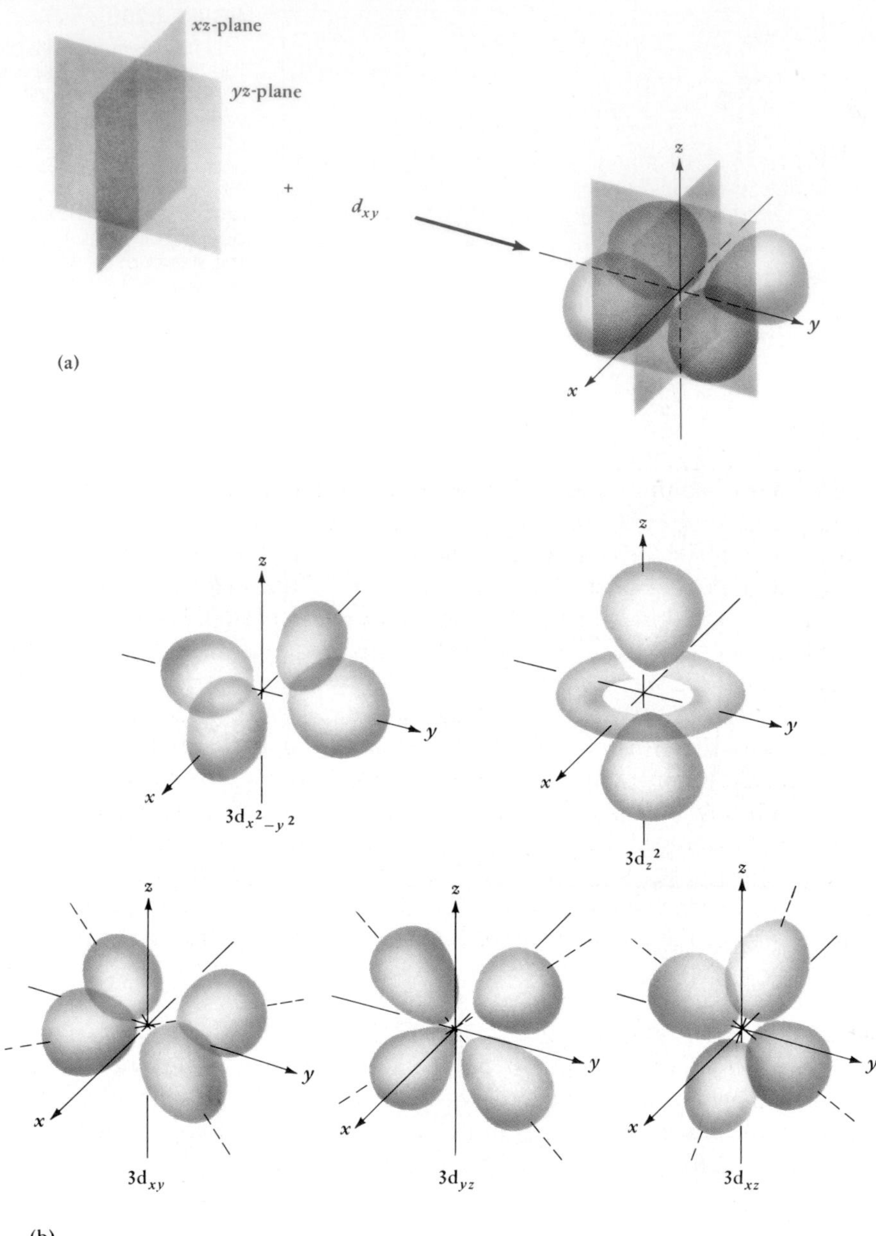

One of the electron clouds along the *z* axis will sit inside the cone. If you use another cone pointing the opposite direction from the first cone, again with its tip at the nucleus, another region of electron density will fit inside this second cone. The region outside both cones defines the remaining, donut-shaped region of electron density or probability.

The set of seven *f* orbitals all have $\ell = 3$, meaning that there are three nodal planes slicing through the nucleus and carving up the electron cloud into eight regions of probability. This makes these orbitals less easily visualized than the ones we have seen so far, but one example of an *f* orbital is illustrated in Figure 7.22.

EXERCISE 7.6 ORBITAL SHAPES

(a) What are the n and ℓ values for each of the following orbitals: $6s$, $4p$, $5d$, and $4f$?

(b) Look carefully at the plot of ψ^2 against distance from the nucleus for a $2p$ orbital in Figure 7.20. How many planar nodes does this orbital have? How many spherical nodes?

(c) Now draw a plot such as the one in Figure 7.20(a) for a $3p$ orbital. Make sure you take into account the proper number of planar and spherical nodes.

FIGURE 7.22
One of the seven possible atomic f electron orbitals. Notice the presence of 3 nodal planes (xy, xz, yz), as required by an orbital with an ℓ value of 3. (Charles D. Winters)

SUMMARY

The modern quantum mechanical view of the atom has been developed in this century through some important experiments (Section 7.1). **Thomson** first measured the charge/mass ratio (e/m) of the electron shortly after 1900. **Millikan** was able to derive the mass of the electron (9.1095×10^{-28}g) after experimentally determining the charge (1.6022×10^{-19} coulombs) in his "oil drop" experiment. **Rutherford**, attempting to verify predictions made by Thomson, carried out an experiment that showed for the first time that most of the mass of the atom is concentrated in the nucleus, and the electrons occupy the space outside the nucleus.

In the same period in which the experiments above were performed, **Einstein** postulated an explanation for the **photoelectric effect** (Section 7.4) that showed that electromagnetic radiation can be thought of as consisting of massless particles called **photons**. These photons have an energy content given by **Planck's law** (Section 7.3), a statement of the equivalency of the energy content and the frequency of radiation (ν)

$$E = h\nu \qquad (7.2)$$

where h is called Planck's constant (6.6262×10^{-34} J · s).

Before all of the experiments mentioned above, it was known that all **electromagnetic radiation** could be described by the physics of waves, where the product of wavelength (λ) and frequency (ν) are equal to the velocity of light ($c = 2.998 \times 10^8$ m/s)(Section 7.2)

$$\lambda\nu = c \qquad (7.1)$$

De Broglie later proposed (Section 7.5) that an electron and other particles of comparable mass could also be described by the physics of waves. The length of the wave appropriate for an object of mass m moving with velocity v is

$$\lambda = h/mv \qquad (7.7)$$

The first atomic model that provided some explanation for experimental observations was that of **Bohr** (Section 7.4) who postulated that the electron in the hydrogen atom could occupy only certain **stationary states**, each with an energy given by

$$E = -Rhc/n^2 \qquad (7.4)$$

where R is the **Rydberg constant**, h Planck's constant, c the velocity of light, and n an integer equal to or greater than 1. Bohr postulated that the electron could move to another state, the amount of energy absorbed or emitted being equal to the difference in energy between the two states. If the electron moves from a particular state to one of lower n, energy is emitted as electromagnetic radiation of discrete wavelength. (In the case of excited H atoms, some radiation, the **Balmer lines**, are seen in the visible region.)

The Bohr model proved incapable of extension to atoms heavier than H, and a new model called **quantum** or **wave mechanics** was developed by Bohr, **Schrödinger**, and their associates (Sections 7.6 and 7.7). This model describes the electron of the H atom by a wave. From this model has come a set of matter wave functions or orbitals for the electron in the atom. Each wave function corresponds to an allowed energy of the electron in the atom; that is, the electron can take on only certain quantized energies in the atom. The energy of the electron in the H atom in a given state is known and is given by the same equation as that derived by Bohr (Equation 7.4). The position of the electron is not known with certainty, however; *only the probability of the electron being in a given region of space can be calculated.* (This is the interpretation of the quantum mechanical model by **Heisenberg** and is the reason for his postulate called the **uncertainty principle**.)

The electron wave for each allowed energy state is called an **orbital**. Each orbital is described by **three quantum numbers: n, ℓ, and m_ℓ.**

QUANTUM NUMBER	NAME	VALUES	ORBITAL PROPERTY DESCRIBED
n	Principal	1,2,3 . . .	Orbital size Label for an electron shell
ℓ	Angular	$0,1,2 \ldots, n-1$	Orbital shape Label for electron subshell s orbital for $\ell = 0$ p orbital for $\ell = 1$ d orbital for $\ell = 2$ f orbital for $\ell = 3$
m_ℓ	Magnetic	$+\ell \ldots 0 \ldots -\ell$	Orbital orientation

All s orbitals are spherical in shape. All p orbitals have a "dumb bell" shape, since all p orbitals have one **nodal plane** (because all have $\ell = 1$, and ℓ = number of nodal planes) slicing through the nucleus and dividing the region of electron density into two halves. All d orbitals have four regions of electron density, since all d orbitals have two nodal surfaces (all have $\ell = 2$).

STUDY QUESTIONS

REVIEW QUESTIONS

1. Our modern view of atomic structure was developed through many experiments. Name at least three of these experiments, the outcome of the experiment, and the person most associated with that experiment.

2. There are several important mathematical relations in this chapter. Give the equation for each of the following:
(a) The relationship between wavelength, frequency, and velocity of radiation.

(b) The relation between energy and frequency of radiation.
(c) The energy of an electron in a given energy state of the H atom.

3. State Planck's law in words and in a mathematical relation.
4. Name the colors of visible light beginning with that of highest energy.
5. Draw a picture of a standing wave and use this to define the terms wavelength, amplitude, and node.
6. What is a photon? Explain how the photoelectric effect implies the existence of photons.
7. In what region of the electromagnetic spectrum is the Lyman series of lines found? The Balmer series?
8. What were two major assumptions of Bohr's theory of atomic structure?
9. Bohr pictured the electrons of the atom as being located in definite orbits about the nucleus, just as the planets orbit the sun. Criticize this model in view of the quantum mechanical view.
10. What is the "wave–particle duality? What are its implications in our modern view of atomic structure?
11. What is Heisenberg's uncertainty principle? Explain how it applies to our modern view of atomic structure.
12. How do we interpret the physical meaning of the square of the wave function, ψ^2? What are the units of ψ^2?
13. What are the three quantum numbers used to describe an orbital? What property of an orbital is described by each quantum number? Specify the rules that govern the values of each quantum number.
14. On what does the amplitude of the electron wave in an atom depend?
15. Give the number of planar nodes for each orbital type: *s*, *p*, *d*, and *f*.
16. What is the maximum number of *s* orbitals that are found in a given electron shell? The maximum number of *p* orbitals? Of *d* orbitals? Of *f* orbitals?
17. Match the values of ℓ below with orbital type (*s*, *p*, *d*, or *f*).

ℓ VALUE	ORBITAL TYPE
3	_____
0	_____
1	_____
2	_____

18. Draw a picture of the 90% boundary surface of an *s* orbital and the p_x orbital. Be sure your drawing shows why the *p* orbital is labeled p_x and not p_y, for example.

IMPORTANT EXPERIMENTS

19. In Millikan's oil drop experiment, some droplets captured one electron, some two, others three and so on. If you performed the experiment, and you observed the following charges on individual oil droplets, what would you conclude is the largest possible value of the fundamental unit of charge on one electron?

OIL DROP OBSERVED	OBSERVED CHARGE ON DROP (UNITS OF 10^{-19} COULOMBS)
a	−3.2
b	−4.8
c	−6.4
d	−2.4
e	−1.6

20. Thomson's experiments on the nature of the electron were done in an instrument that allowed him to pass a beam of electrons through magnetic and electric fields. A modern instrument that uses a similar technique is a mass spectrometer (Figure 2.1). In this instrument one or more electrons are stripped away from a molecule, and the now positively charged ion is passed through a magnetic field. Ions of different charge/mass ratio travel paths of different curvature through the magnetic field, a feature that allows the ions to be identified. Careful examination of the ions produced by a sample allows a chemist to determine what molecules are present in an unknown sample. Calculate the charge/mass ratio in units of coulombs/amu for each of the following molecular ions: (a) CO^+, (b) O_2^+, and (c) $C_6H_6^+$.
21. If the gold target in Rutherford's experiment (Figure 7.4) was 6.0×10^{-5}-cm thick, about how many atoms thick was the target? Assume that the atoms of gold line up just touching one another from one side of the foil to the other and that they are hard spheres. (The radius of a gold atom is 0.144 nm.)

ELECTROMAGNETIC RADIATION

22. The U.S. Navy has proposed a system, called ELF (for "extra long frequency"), for communicating with submerged submarines. The system uses radio waves with a frequency of 76 s^{-1}. What is the wavelength of this radiation in meters? In miles? (1 mile = 1.61 km)
23. The colors of the visible spectrum, with each wavelength corresponding to a different color, are given in Figure 7.8. What colors of light involve less energy than yellow light? Which color of visible light has photons of greater energy: green or violet?
24. Green light has a wavelength of approximately 500. nm. What is the frequency of this light? What is the energy in joules of one photon of green light? What is the energy in joules of 1.00 mole of photons of green light?

25. Perform the following conversions: (a) Red light has a wavelength of about 600. nm. What is its frequency? Calculate the energy of 1 photon of red light. (b) 200. nm light is located in what region of the electromagnetic spectrum? What is its frequency? What is the energy of a photon of 200. nm radiation? (c) Assume a microwave oven operates at a frequency of $1.00 \times 10^{11}\ s^{-1}$. What is the wavelength of this radiation in meters? What is its energy per photon?

26. The yellow light emitted by the sodium lamps in streetlights has wavelengths of 589.6 nm and 589.0 nm. What is the frequency of the 589.0 nm light? What is the energy of 1 mole of photons of yellow light with a wavelength of 589.0 nm?

27. An AM radio station at 600. KHz broadcasts with a frequency of $6.00 \times 10^5\ s^{-1}$ (1 KHz = 1 kilohertz = $1000\ s^{-1}$). What is the wavelength of this signal in meters?

✓28. Place the following types of radiation in order of increasing energy per photon:
(a) radar signals
(b) radiation from a microwave oven
(c) gamma rays from a nuclear reaction
(d) red light from a neon sign
(e) ultraviolet radiation from a sun lamp

29. Radiation in the microwave region of the electromagnetic spectrum is not very energetic. However, it is this radiation that cooks your food in a microwave oven. If you bombard a cherry pie with 1 mole of photons with a wavelength of 0.50 cm, how much energy is your pie absorbing?

30. When *Voyager 2* encountered the planet Saturn in 1981, the planet was approximately 1 billion miles (10^9 miles) from the earth. How long did it take for the television picture signal to reach the Earth from Saturn? (1 mile = 1.61 km).

PHOTOELECTRICITY

✓31. Which of the following best describes the importance of the photoelectric experiment as explained by Einstein?
(a) The experiment showed that light is electromagnetic radiation.
(b) The intensity of a light beam is related to its frequency.
(c) Light can be thought of as consisting of massless particles whose energy is given by Planck's equation, $E = h\nu$.

32. To cause a sodium atom on a metal surface to lose an electron, an energy of 220. kJ/mol is required. Calculate the longest possible wavelength of light that can ionize a sodium atom. What is the region of the electromagnetic spectrum in which this radiation is found?

33. Assume you are an engineer designing a space probe to land on a distant planet. You wish to use a switch that works by the photoelectric effect. That is, light falling on the surface of a metal causes the metal to release some electrons, which flow to a positively charged pole and close the electrical circuit. The metal you wish to use in your device requires 6.7×10^{-19} J (per atom) to remove an electron. You know that the atmosphere of the planet on which your device must work filters out all wavelengths of light less than 540 nm. Will your device work on the planet in question? Why or why not?

ATOMIC SPECTRA AND THE BOHR ATOM

34. Light is given off by a sodium- or mercury-containing streetlight when the atoms are excited in some way. The light you see arises for which of the following reasons:
(a) electrons moving from a given quantum level to one of higher n.
(b) electrons being removed from the atom, thereby creating a metal cation.
(c) electrons moving from a given quantum level to one of lower n.
(d) electrons whizzing about the nucleus in an absolute frenzy.

35. Consider only the following quantum levels for the H atom.

________ $n = 4$
________ $n = 3$

________ $n = 2$

________ $n = 1$

The emission spectrum of an excited H atom will consist of transitions between these levels.
(a) How many emission lines are possible, considering only the four quantum levels above?
(b) Photons of the highest energy will be emitted in a transition from the level with n = _____ to the level with n = _____.
(c) The emission line having the longest wavelength corresponds to a transition from the level with n = _____ to the level with n = _____.

36. If energy is absorbed by a hydrogen atom in its ground state, the atom is excited to a higher energy state. For example, the excitation of an electron from the orbital with $n = 1$ to an orbital with $n = 4$ requires radiation with a wavelength of 97.2 nm. Which of the following transitions would require radiation of longer wavelength than this?
(a) $n = 2$ to $n = 4$
(b) $n = 1$ to $n = 3$
(c) $n = 1$ to $n = 5$
(d) $n = 3$ to $n = 5$

✓37. Calculate the energy of the electron in the $n = 4$ orbital of the H atom using Bohr's energy equation (Equation 7.4). (Answer in kJ/mol.)

38. The energy emitted when an electron moves from an excited orbital to a lower energy orbital in any atom can be observed as electromagnetic radiation.
    (a) Which involves the emission of the greater energy in the H atom, an electron changing from $n = 4$ to $n = 1$ or an electron changing from $n = 5$ to $n = 2$? Explain fully.
    (b) Calculate the wavelength of light emitted when an electron changes from $n = 4$ to $n = 1$ in the H atom. In what region of the spectrum is this radiation found?

39. If sufficient energy is absorbed by an atom, an electron can be lost by the atom and a positive ion formed. The amount of energy required is called the *ionization energy*. In the H atom, the ionization energy is that required to change the electron from $n = 1$ to $n =$ infinity. Knowing that the energy of a quantum level for an atom is given by $E = -Rhc/n^2$, calculate the ionization energy for the H atom.

40. Excited H atoms give off many emission lines. One series of lines, called the Pfund series, occurs in the infrared region. It results when an electron changes from higher orbitals to orbitals with $n = 5$. Calculate the wavelength and frequency of the lowest energy line of this series.

41. A line in the Balmer series of emission lines of excited H atoms has a wavelength of 410.17 nm. This series originates from electrons changing from high energy orbitals to the $n = 2$ orbitals. To account for the 410.17 nm line, what is the value of n for the initial orbitals?

De BROGLIE AND MATTER WAVES

42. When an electron moves with a velocity of 3.0×10^8 cm/s, what is its de Broglie wavelength?

✓43. If an electron has an energy of 100. V (1.00 electron volt $= 1.60 \times 10^{-19}$ J), what is its wavelength? (Remember that $E = mv^2/2$.)

44. A beam of neutrons ($m = 1.67492 \times 10^{-27}$ kg) has a velocity of 0.40 m/s. Calculate the wavelength of the neutron wave.

QUANTUM MECHANICS

45. Complete the following table:

QUANTUM NUMBER	ATOMIC PROPERTY DETERMINED BY QUANTUM NUMBER
____	Orbital size
____	Relative orbital orientation
____	Orbital shape

✓46. Answer the following questions:
    (a) When $n = 3$, what are the possible values of ℓ?
    (b) When ℓ is 3, what are the possible values of m_ℓ? What type of orbital corresponds to an ℓ of 3?
    (c) For a 4s orbital, what are the possible values of n, ℓ, and m_ℓ?
    (d) For a 5f orbital, what are the possible values of n, ℓ, and m_ℓ?
    (e) When $n = 4$, $\ell = 2$, and $m_\ell = -2$, to what orbital type does this refer? (Give the orbital label; e.g., 1s.)
    (f) How many orbitals are there in the $n = 6$ electron shell? How many subshells? What are the letter labels of the subshells?
    (g) When a subshell is labeled g, how many orbitals are there in the subshell? What are the values of m_ℓ?

47. A possible excited state of the H atom has the electron in a 4p orbital. List all possible sets of quantum numbers n, ℓ, and m_ℓ for this electron.

48. How many subshells are there in the electron shell with principal quantum number $n = 5$?

✓49. Explain briefly why each of the following is *not* a possible set of quantum numbers for an electron in an atom.
    (a) $n = 2$, $\ell = 2$, $m_\ell = 0$
    (b) $n = 3$, $\ell = 0$, $m_\ell = -2$
    (c) $n = 6$, $\ell = 0$, $m_\ell = 1$

50. What is the maximum number of orbitals that can be identified by each of the following sets of quantum numbers?
    (a) $n = 4$, $\ell = 2$
    (b) $n = 5$
    (c) $n = 2$, $\ell = 2$
    (d) $n = 3$, $\ell = 1$, $m_\ell = -1$
    (e) $n = 3$, $\ell = 0$, $m_\ell = +1$
    (f) $n = 5$, $\ell = 1$
    (g) $n = 7$, $\ell = 5$

51. How many spherical and planar nodes does each of the following orbitals possess?

	SPHERICAL NODES	PLANAR NODES
2s	____	____
5d	____	____
5f	____	____

52. Consider a 4s orbital. How many spherical nodes does it have? How many planar nodes? Draw a graph for the 4s orbital showing how the electron probability (ψ^2, electron density) changes as the distance from the nucleus increases.

53. Question 52 above considers the 4s orbital. Compare the 4s and 4p orbitals by drawing a graph showing how the electron probabilities (ψ^2 curves) for the 4s and 4p orbitals change as the distance from the nucleus increases.

54. The square of the wave function for a $2p$ orbital is plotted against distance from the nucleus in Figure 7.20a. What is the approximate distance from the nucleus at which the maximum probability density is reached?

55. State which of the following orbitals cannot exist according to the quantum theory: $2s$, $2d$, $3p$, $3f$, $4f$, and $5g$. Briefly explain your answers.

56. Rank the following orbitals *in the H atom* in order of increasing energy: $3s$, $2s$, $2p$, $4s$, $3p$, $1s$, and $3d$.

57. Write the complete set of quantum numbers (n, ℓ, and m_ℓ) that quantum theory allows for each of the following: (a) $2p$, (b) $3d$, and (c) $4f$.

58. From memory, sketch shapes of the electron clouds for each of the following atomic orbitals: (a) $1s$, (b) $2p_x$, (c) $3d_{xy}$.

59. A given orbital is labeled by the magnetic quantum number $m_\ell = -1$. This could *not* be a
(a) g orbital (d) p orbital
(b) f orbital (e) s orbital
(c) d orbital
Briefly explain your answer.

60. Consider the $5f$ orbitals. These orbitals have _____ spherical nodes and _____ planar nodes.

General Questions

61. If 5% of the energy supplied to a light bulb is radiated as visible light, how many photons are emitted per second by a 100. W bulb? Assume the wavelength of all the visible light is 560 nm. (1 W = 1 J/s)

62. In Question 42, you calculated the de Broglie wavelength of an electron moving with a velocity of 3.0×10^8 cm/s. If the error in the velocity measurement is 0.010%, what is the error in the measured position of the electron?

63. In what way does Bohr's model of the atom violate the uncertainty principle?

64. Pictured below are "contour maps" of two atomic orbitals. The curved lines connect all points in the plane of the contours where there is equal probability of finding an electron. (Such diagrams are similar to topographical maps of land areas.)
(a) Give the n and ℓ values for the orbital picture labeled (a). (Hint: How many spherical and how many planar nodes does the orbital possess?) How should this orbital be labeled (e.g., $5f$)?
(b) For the orbital labeled (b), give its values of n and ℓ and its letter designation (e.g., $1s$, $6f$, etc.).

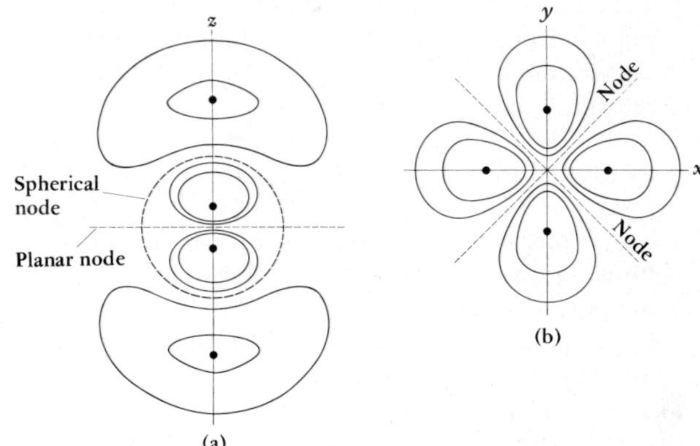

8

Atomic Electron Configurations and Chemical Periodicity

Some Group 4A elements: (left to right) tin (Sn), silicon (Si), and lead (Pb) (Charles D. Winters)

CHAPTER OUTLINE

The modern view of the atom was developed in the previous chapter. We know now that electrons are arranged in shells in the space around the nucleus; each of these shells is distinguished by the quantum number n and consists of one or more subshells. Each subshell is distinguished by the quantum number ℓ and contains one or more orbitals, and there are strict mathematical rules governing the number and type of orbitals within the atom. The picture that emerges is a beautiful one that reflects the order of the natural world around us, but it is exactly true only for the case of the H atom. To be useful, the model must be applicable to atoms with more than one electron. Hydrogen is the most abundant element in the universe, oxygen and silicon are most abundant on our planet, and many other elements are important to the maintainence of life (Chapter 2). Since chemical properties depend on the number and arrangement of electrons in an atom, one objective of this chapter is to develop a workable picture of the electronic structure of elements other than hydrogen.

Another objective of this chapter is to explore some of the physical properties of atoms: the ease with which atoms lose or gain electrons to form ions and the concept of atomic and ionic radii. These properties are directly related to atomic electron arrangement and thus to the chemistry of the elements and their compounds.

FIGURE 8.1
A compass points to the geographical north pole of the Earth. (Charles D. Winters)

8.1 ELECTRON SPIN

Three quantum numbers (n, ℓ, and m_ℓ) allow us to define the orbital for an electron. To describe completely an electron in an atom with many electrons, however, we still need one more quantum number, the **electron spin quantum number, m_s**. In approximately 1920, theoretical chemists realized that, since electrons interact with a magnetic field, there must be one more concept to explain the behavior of electrons in atoms. It was soon verified experimentally that the electron behaves as though it has a spin. To understand this property and its relation to atomic structure, we should understand something of the general phenomenon of magnetism.

MAGNETISM

The needle of a compass at a given location on the earth will always point in a given direction, no matter how the compass is located (Figure 8.1). The needle is a natural *magnet* and consists of *magnetite*, a magnetic

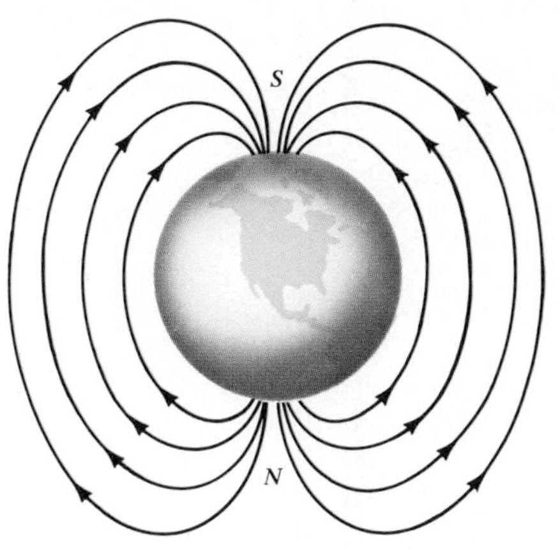

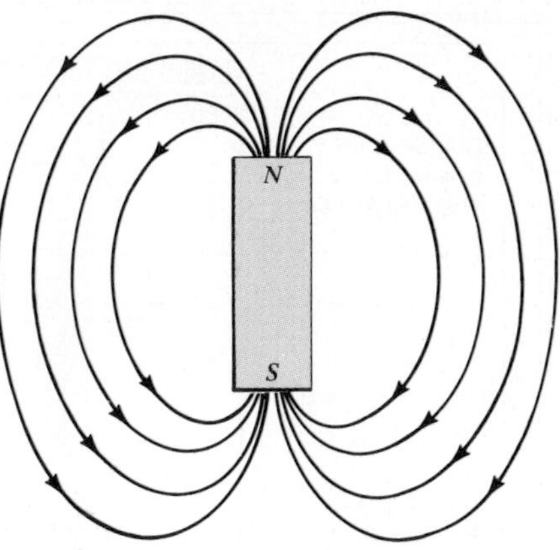

FIGURE 8.2

The magnetic field of the earth and of a bar magnet. The magnetic field comes out of one end, arbitrarily called the "north pole," N, and loops toward and enters the "south pole," S. The *geographic north pole* of the earth, named before the introduction of the term "magnetic pole," *is the magnetic south pole.*

oxide of iron, Fe_3O_4. In 1600, William Gilbert concluded that the earth is also a large spherical magnet giving rise to a magnetic field that surrounds the planet (Figure 8.2). The compass needle is "drawn" or "attracted" into this magnetic field, one end of the needle pointing approximately to the earth's geographic north pole. Thus, we say the end of the compass needle pointing north is the magnet's "magnetic north pole" or simply its "north pole," designated N. The other end of the needle is its "south pole," designated S.

Similar magnetic poles (N–N or S–S) repel one another, while opposite poles (N–S) attract (Figure 8.3). Since the magnetic north pole of the compass needle points to the earth's geographic north pole, this means the latter is the earth's magnetic south pole.

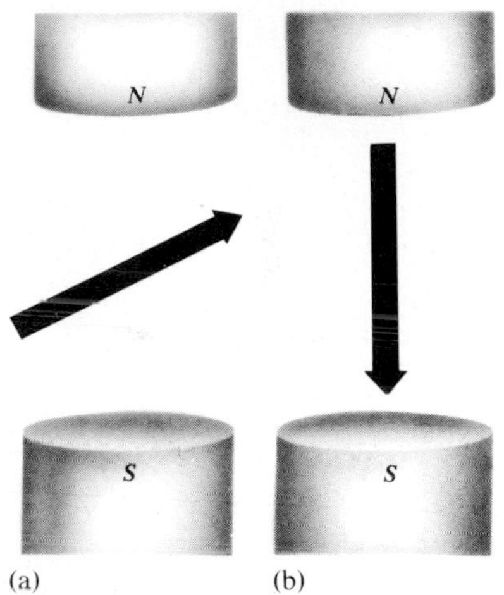

FIGURE 8.3

A magnetic needle in a strong magnetic field. The repulsion of similar poles (a) and attraction of opposite poles (b) cause the needle to line up as shown, the N pole of the needle facing the S pole of the magnet.

Most substances are not magnetic and cannot be attracted by a magnetic field. They are said to be **diamagnetic**; such substances are in fact repelled by a magnetic field. There are, however, natural or synthetic magnets, such as an alloy called "Alnico" (for Al, Ni, and Co), that can attract pieces of iron, steel, nickel, and other materials. Substances that are attracted to a magnetic field are termed **paramagnetic**, and the magnitude of the effect can be determined with an apparatus such as that illustrated in Figure 8.4.

PARAMAGNETISM AND UNPAIRED ELECTRONS

The underlying cause of paramagnetism is the presence of one or more unpaired electrons. It has been verified experimentally that an electron in an atom behaves as though it has a spin. This spin is much like that of the earth spinning on its axis, and, since the electron is electrically charged, the spinning charge generates a magnetic field with north and south magnetic poles (Figure 8.5); that is, the electron acts like a tiny bar magnet.

Even paramagnetic materials exhibit diamagnetism. The net magnetic effect is a balance of the two. *Ferromagnetism*, a very strong magnetic effect, is observed when clusters of atoms cooperatively adopt the same alignment of their unpaired electrons.

The properties of magnets are related to the number of *unpaired electrons* in the atoms of which the magnet is composed. Thus, as expected, hydrogen atoms are paramagnetic to the extent of one unpaired electron. Helium atoms, which have two electrons, are *not* paramagnetic, however. The explanation for this experimental observation rests on two

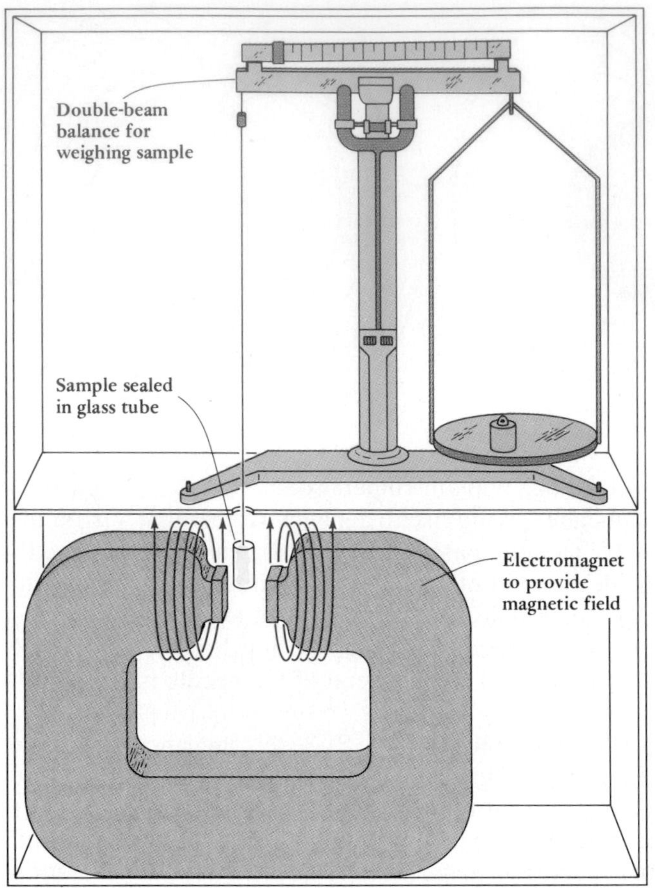

FIGURE 8.4
A magnetic balance used to measure the magnetic properties of a sample. The sample is first weighed with the electromagnet turned off. The magnet is then turned on and the sample reweighed. If paramagnetic, the sample is drawn into the magnetic field, and the apparent weight increases.

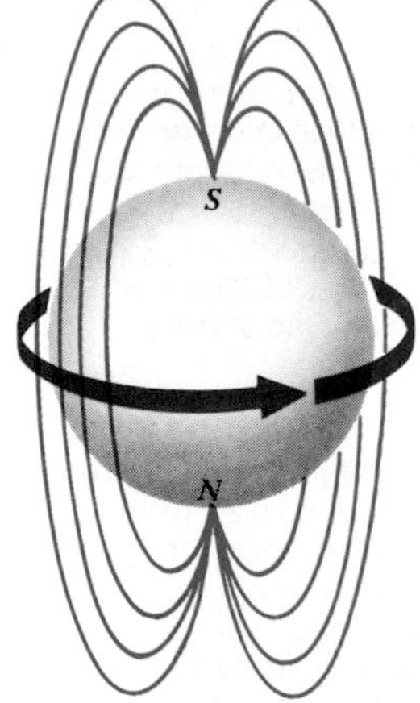

FIGURE 8.5
Electron spin. In atoms and molecules, the electron in motion generates a magnetic field whose properties are identical to the field set up by a bar magnet or by the earth. (Of course, the strengths of these magnetic fields differ considerably.)

hypotheses: (1) the two electrons are assigned to the same orbital and (2) electron spin is quantized (Figure 8.6). The quantization of electron spin means that there are *only two possible orientations of an electron in a magnetic field*, one associated with a *spin quantum number*, m_s, of $+\frac{1}{2}$ and the other with an m_s value of $-\frac{1}{2}$. To account for the lack of paramagnetism of helium, we must assume the two electrons assigned to the same orbital have opposite spin directions; we say they are **paired**. This means the paramagnetism of one electron, $\overset{N}{\underset{S}{\uparrow}}$, is canceled by the paramagnetism of the second electron, $\overset{S}{\underset{N}{\downarrow}}$, of opposite spin. The implications of this observation are enormous and open the way to explain the electron configurations of atoms with more than one electron.

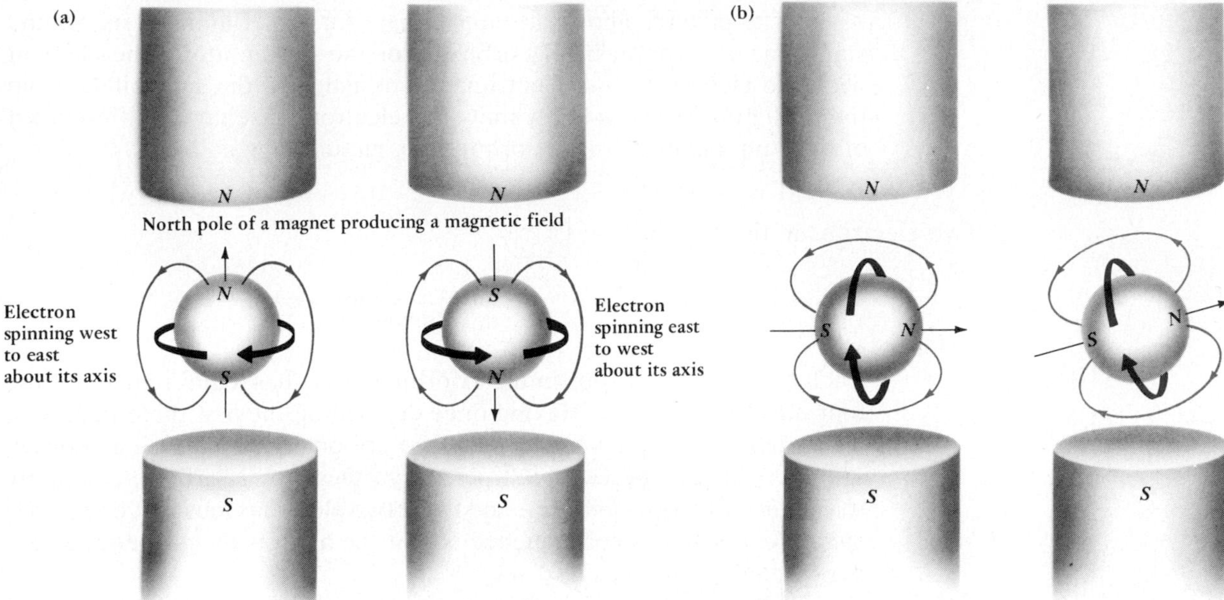

FIGURE 8.6

Quantization of electron spin. (a) The two assumed spins of the electron, a "micromagnet," relative to a magnetic field. Any other position, such as those illustrated in (b), is forbidden. Therefore, we can say the spin of an electron is quantized.

8.2 THE PAULI EXCLUSION PRINCIPLE

To make the quantum theory consistent with experiment, Wolfgang Pauli stated in 1925 the **Pauli exclusion principle**: No two electrons in an atom can have the same set of four quantum numbers (n, ℓ, m_ℓ, and m_s). This principle leads to yet another important conclusion, proved below, that *no atomic orbital can be assigned to (or "contain") more than two electrons.*

Orbitals are not literally things or boxes in which electrons are placed. "Orbital" and "wave function" are synonymous, so it is not conceptually correct to talk about electrons *in* or *occupying* orbitals, although it is commonly done.

If we consider the 1s orbital of the H atom, this orbital is defined by the set of quantum numbers $n = 1$, $\ell = 0$, and $m_\ell = 0$. No other set of numbers can be used. If an electron has this orbital, the electron spin direction must also be specified as sketched below. The orbital is shown as a "box," and the electron spin in one direction is depicted by an arrow.

Electron in 1s orbital = [↑] (1s) Quantum Number Set

$$n = 1, \ell = 0, m_\ell = 0, m_s = +\tfrac{1}{2}$$

If there is only one electron with a given orbital, you can picture the electron as an arrow pointing either up or down. Thus, an equally valid combination of quantum numbers and "box" diagram would be

Electron in 1s orbital = [↓] (1s) Quantum Number Set

$$n = 1, \ell = 0, m_\ell = 0, m_s = -\tfrac{1}{2}$$

The spin and magnetic field arrows for an electron point in opposite directions. From now on the arrow for an electron refers to its spin.

The pictures above are appropriate for the H atom in its ground state: one electron in the 1*s* orbital. For the helium atom, the element with two electrons, both electrons are assigned to the 1*s* orbital. From the Pauli principle, you know that each electron must have a different set of quantum numbers, so the orbital box picture now is

Two electrons in the 1*s* orbital = [↑↓]

this electron has $n = 1, \ell = 0, m_\ell = 0, m_s = -\frac{1}{2}$
this electron has $n = 1, \ell = 0, m_\ell = 0, m_s = +\frac{1}{2}$

Each of the two electrons in the 1*s* orbital of He has a different set of the four quantum numbers. Having once decided on the first three numbers, which tell you this is a 1*s* orbital, there are only two remaining choices, $m_s = \pm\frac{1}{2}$. Thus, the 1*s* orbital, and any other atomic orbital, can apply to *no more than two electrons*, and these two electrons must have *opposite spin directions*. The consequence is that the atom is diamagnetic, as experimentally observed.

The $n = 1$ electron shell in any atom can accommodate no more than two electrons. But what about the $n = 2$ shell? There can be one *s* orbital and three *p* orbitals in the $n = 2$ electron shell (Table 7.1). Since each orbital can apply to two electrons and no more, the 2*s* orbital is

TABLE 8.1 Number of Electrons Accommodated in the Various Possible Electron Shells and Subshells

ELECTRON SHELL	SUBSHELLS AVAILABLE (NUMBER AVAILABLE $= n$)	NUMBER OF ELECTRONS POSSIBLE IN THE SUBSHELL $[2(2\ell + 1)]$	TOTAL ELECTRONS POSSIBLE FOR THE nTH ELECTRON SHELL $(2n^2)$
1	*s*	2	2
2	*s*	2	8
	p	6	
3	*s*	2	18
	p	6	
	d	10	
4	*s*	2	32
	p	6	
	d	10	
	f	14	
5	*s*	2	50
	p	6	
	d	10	
	f	14	
	*g**	18	
6	*s*	2	72
	p	6	
	d	10	
	*f**	14	
	*g**	18	
	*h**	22	

*These orbitals are not used in the ground state of any known element.

SUMMARY OF QUANTUM NUMBERS

You can think of the four quantum numbers used to describe an electron in an atom as an "electronic zipcode."

n: shell to which electron is assigned (1, 2, 3, . . .)
ℓ: the subshell within the shell (0, 1, 2, . . . , $n - 1$)
m_ℓ: orbital within the subshell ($-\ell$, . . . , 0, . . . , $+\ell$)
m_s: electron spin direction ($\pm \frac{1}{2}$)

assigned to two electrons and the three $2p$ orbitals to six electrons, for a total of eight electrons possible with $n = 2$. As seen in Table 8.1, this analysis is carried farther for the shells normally observed in real elements.

The number of orbitals in the nth electron shell is n^2 and the number of electrons in the shell is $2n^2$.

8.3 ORBITAL ENERGIES AND ELECTRON ASSIGNMENT

A goal of this chapter is an understanding of the distribution of electrons in atoms with many electrons. The heavier elements can be "built" by assigning electrons to a succession of orbitals, no more than two electrons to an orbital. Generally, electrons are assigned to orbitals of successively higher energies because this will make the total energy of all the electrons as low as possible. With elements having more than 18 electrons, however, we find there are small differences in orbital energies, which complicates matters slightly, and so some heavier elements do not always follow a neat set of rules.

ORDER OF ORBITAL ENERGIES AND ASSIGNMENTS

Quantum theory leads to the conclusion that in the H atom the energy of an electron depends only on the value of n ($E = -Rhc/n^2$, Equation 7.4). For heavier atoms, however, the situation is more complex. For atoms of more than one electron, the actual order of orbital energies is given in Figure 8.7. Here you see that *orbital energies of many-electron atoms depend on both n and ℓ.* The orbitals with $n = 3$, for example, do not all have the same energy; rather, they are in the order $3s < 3p < 3d$.

The orbital energy order in Figure 8.7, and the determination of the actual electron configurations of the elements, lead to *two general rules for the order of assignment of electrons to orbitals.*

1. Orbital assignments follow a sequence of increasing $n + \ell$.
2. For two orbitals of the same $n + \ell$, electrons are assigned first to orbitals of lower n.

These rules mean electrons are *usually* assigned in order of increasing orbital energy. However, there are exceptions. For example, as you will see in Section 8.4, electrons are assigned to a $4s$ orbital ($n + \ell = 4$) before being assigned to $3d$ orbitals ($n + \ell = 5$). This order of assignment is observed in spite of the fact that the energies of these orbitals are in the order $3d \leq 4s$ (Figure 8.7).

The reason for the generally observed ordering of orbital assignments rests in part on the concept of **effective nuclear charge (Z^*)**. As

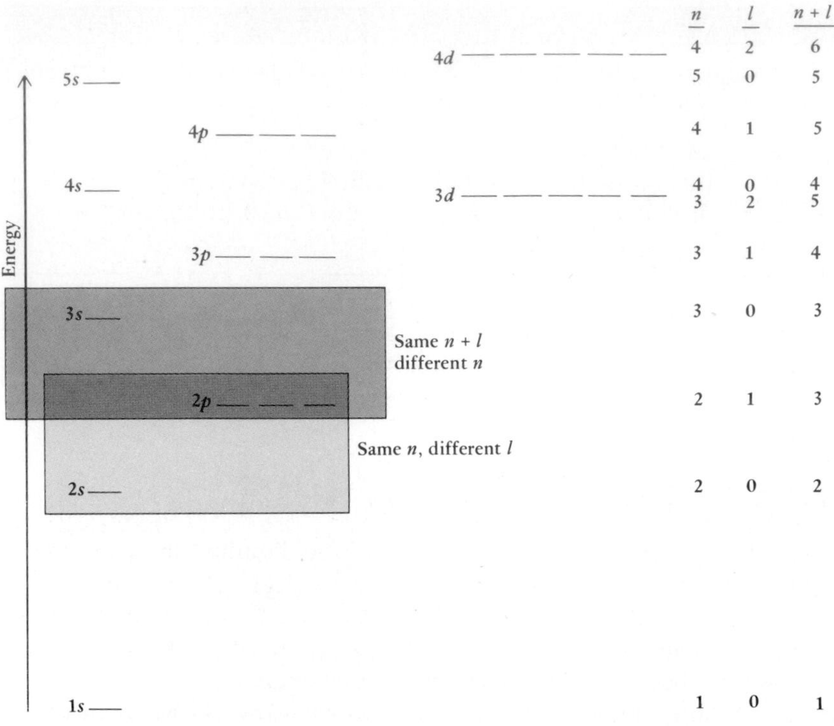

FIGURE 8.7
Orbital energies for a many-electron atom generally increase as the sum ($n + \ell$) increases. That is, orbital energies depend not only on the value of n (as is true for the H atom) but also on the angular momentum quantum number ℓ. If two orbitals have the same ($n + \ell$) value, the orbital with the lower n generally lies lower in energy (compare $2s < 3p$ and $3p < 4s$).

explained in more detail in the section that follows, Z^* for an electron is the charge on the nucleus of a multielectron atom as modified by the presence of electrons between that electron and the nucleus. In an atom with only one electron, such as the H atom, the 2*s* and 2*p* orbitals have the same energy. However, for an atom with three electrons, such as lithium, the presence of the 1*s* electrons causes the 2*s* orbital to lie lower in energy than the 2*p* orbitals.

EXERCISE 8.1 ORDER OF ORBITAL ASSIGNMENTS

Using the "$n + \ell$" rules above, predict the following orbital assignments for a many-electron atom.

(a) Which is assigned first, 3*s* or 3*d*?
(b) Which is assigned first, 5*d* or 6*s*?
(c) Which is assigned first, 4*f* or 5*s*?

SHIELDED AND EFFECTIVE NUCLEAR CHARGE

You can memorize rules on the succession of orbital assignments, but it is important to understand why they work. You know from quantum theory that the greater the value of n the further the electron is from the nucleus. Additionally, the attraction between a positive charge (the nucleus) and a negative charge (the electron) falls off as the distance increases (Coulomb's law). This explains why an electron in a 3*s* orbital has a higher (or less negative) energy than one in a 2*s* orbital. But why, in a many-electron atom, should an electron in a 2*s* orbital have a lower energy than an electron in a 2*p* orbital? This question can be answered in part by referring to Figure 7.20a and Figure 8.8.

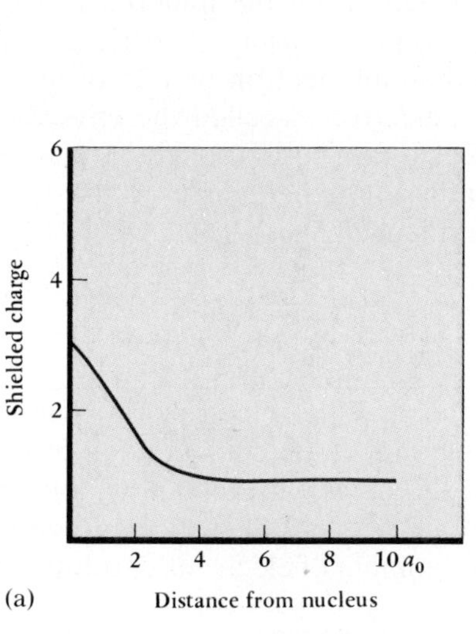

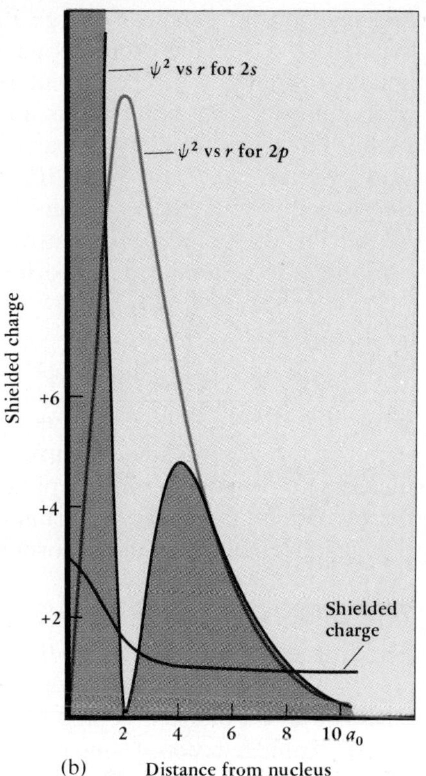

FIGURE 8.8

Shielded nuclear charge. (a) The shielded nuclear charge is the positive charge an electron "feels" at a given distance from the nucleus. The shielded charge is shown for the third electron in the Li atom. The nucleus has three protons, and two electrons have been assigned to the 1*s* orbital. When the third electron is placed near the nucleus, it feels a charge of +3. However, as the electron moves away from the nucleus, the positive nuclear charge is screened or shielded by the negative 1*s* electrons, and the third electron feels a lower nuclear charge. When the third electron is outside the region occupied by the 1*s* electrons, the shielded charge felt by the third electron is +1. (b) The 2*s* orbital allows the third electron of Li to spend more time where the shielded charge is large than does the 2*p* orbital.

In Figures 8.8b and 7.20a you should notice that the probability plot for the 2*s* orbital lies almost completely within the region of probability for the 2*p* orbital. Chemists say that the 2*s* and 2*p* orbitals **penetrate** each other. What distinguishes these orbitals is that the 2*s* orbital has a much higher probability than the 2*p* orbital very close to the nucleus.

The third electron in lithium occupies the 2*s* orbital, rather than 2*p*, because the 2*s*–2*p* probability difference near the nucleus leads to a lower energy for a 2*s* than for a 2*p* electron. Lithium, with a nucleus containing three protons, has two electrons in the 1*s* orbital. When the third electron is almost at the nucleus (regardless of its orbital type), Figure 8.8a shows you that it will "feel" a charge of +3. However, the positive charge "felt" by this third electron drops off rapidly as it moves away from the nucleus, and the charge finally reaches +1. The reason for this drop in charge is that the negative charges of the two electrons in the 1*s* orbital **shield** or **screen** the effect of the positive nuclear charge, the screening effect becoming greater as the third electron moves through and eventually outside the region occupied by the 1*s* electrons. The charge felt by any electron at various distances from the nucleus in a real atom is called the **shielded nuclear charge**.

Now turn to Figure 8.8b. We know any electron is found with different probabilities at different distances from the nucleus. Most importantly, the electron experiences a different shielded nuclear charge at each distance from the nucleus. Figure 8.8b shows that the 2*s* orbital has a higher probability than the 2*p* orbital at any point within the region of space where the nuclear charge is high. The energy of an electron described by a specific orbital is an average of its energies at all points in space, the energy at each point being weighted by the importance (or probability) of that point. Thus, an electron in a 2*s* orbital experiences a higher *average* shielded nuclear charge than an electron in a 2*p* orbital. The *average shielded charge* felt by the electron is called the *effective nuclear charge, Z**. This means a 2*s* electron is more strongly attracted to the nucleus than a 2*p* electron. Identical arguments can be used to establish the general order of orbital energies in Figure 8.7.

The **effective nuclear charge**, Z^*, is the *average* shielded nuclear charge felt by an electron in a multielectron atom.

8.4 ELECTRON CONFIGURATIONS OF THE ELEMENTS

Now it is possible to describe the arrangement of electrons in many-electron atoms. It is important for you to connect the configuration of an element to its position in the periodic table, since this will allow you to organize a large number of chemical facts. However, the device in Figure 8.9 can also help to decide the order of assignment of electrons to orbitals.

ELECTRON CONFIGURATIONS OF THE MAIN GROUP ELEMENTS

The first element in the periodic table, *hydrogen*, has one electron with a 1*s* orbital. One way to depict this is with the **orbital box diagram** used earlier, but an alternative and even more frequently used method is the **spectroscopic notation**. Using the latter method, the electron configuration of H is $1s^1$, or "one ess one."

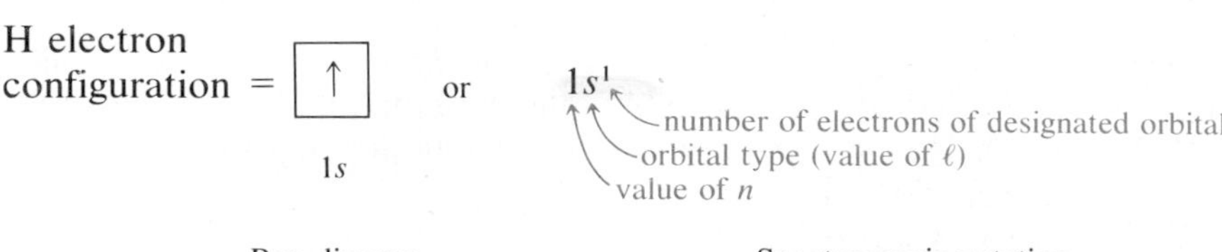

We can use the same methods to describe the configurations of the other elements. Configurations of the first ten elements are illustrated in Table 8.2.

Following hydrogen and helium, *lithium* (Group 1A), with three electrons, is the first element in the second period of the periodic table. The first two electrons must be assigned to the 1*s* orbital, so the third electron must use the $n = 2$ shell. According to the energy level diagram in Figure 8.7, that electron must be assigned to the 2*s* orbital. The spectroscopic notation, $1s^22s^1$, is read as "one ess two, two ess one."

The position of Li in the periodic table tells you its configuration immediately. All the elements of Group 1A (and 1B) have one electron assigned to an *s* orbital of the *n*th shell, where *n* is the number of the period in which the element is found (Figure 8.10). For example, potassium is the first element in the $n = 4$ row, so potassium has the electron configuration of the element preceding it in the table (Ar) *plus* a final electron assigned to the 4*s* orbital. Copper, in Group 1B, will also have one electron assigned to the 4*s* orbital, plus 28 other electrons assigned to other orbitals.

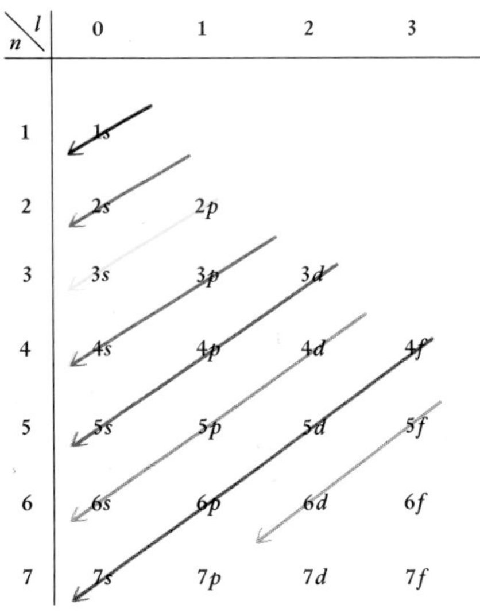

FIGURE 8.9

A device for remembering the orbital filling order in atoms. Write all subshells of a given *n* on the same line with all subshells of a given ℓ in the same column. Draw parallel arrows diagonally from upper right to lower left. The arrows are read from top to bottom, and tail to head. The order is thus 1*s*, 2*s*, 2*p*, 3*s*, 3*p*, 4*s*, 3*d*, 4*p*, 5*s*, 4*d*, . . . and so on.

TABLE 8.2 Electron Configurations of Elements with $Z = 1$ to 10

	1s	2s	2p		
ℓ	0	0	1		
m_ℓ	0	0	+1	0	−1
H $1s^1$	↑				
He $1s^2$	↑↓				
Li $1s^22s^1$	↑↓	↑			
Be $1s^22s^2$	↑↓	↑↓			
B $1s^22s^22p^1$	↑↓	↑↓	↑		
C $1s^22s^22p^2$	↑↓	↑↓	↑	↑	
N $1s^22s^22p^3$	↑↓	↑↓	↑	↑	↑
O $1s^22s^22p^4$	↑↓	↑↓	↑↓	↑	↑
F $1s^22s^22p^5$	↑↓	↑↓	↑↓	↑↓	↑
Ne $1s^22s^22p^6$	↑↓	↑↓	↑↓	↑↓	↑↓

Beryllium, in Group 2A, will have two electrons assigned to the $1s$ orbital plus two additional electrons. It can be seen in Figure 8.7 that the $2s$ orbital is still appropriate, so the configuration of Be is $1s^22s^2$. All elements of Group 2A have electron configurations of [electrons of preceding rare gas + ns^2], where n is the period in which the element is found in the periodic table.

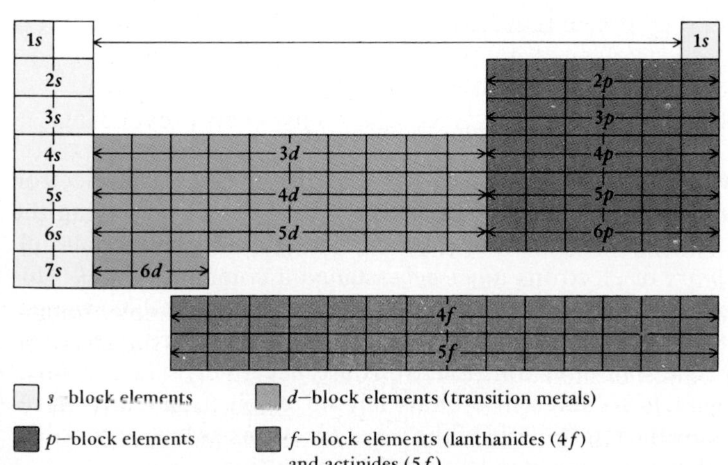

FIGURE 8.10
Electron configurations and the periodic table. The outermost electrons of an element are assigned to the indicated orbitals.

At *boron* (Group 3A) you first encounter an element in the block of elements on the right side of the periodic table. Since the 1*s* and 2*s* orbitals are filled in a boron atom, the fifth electron must be assigned to a 2*p* orbital. In fact, *all the elements from Group 3A through Group 8A are characterized by electrons assigned to p orbitals*, so these elements are sometimes called the ***p* block elements**. All have the general configuration of ns^2np^x where x = group number − 2. Additional further comments need to be made about several of them.

The three *p* orbitals of any shell have the same energy (and are thus said to be *degenerate*), so it is arbitrary into which orbital you place the electron. This means there are six possible, equivalent sets of quantum numbers for the 2*p* electron of boron.

n	ℓ	m_ℓ	m_s		m_ℓ = +1	0	−1
2	1	+1	+½	=	↑		
2	1	0	+½	=		↑	
2	1	−1	+½	=			↑
2	1	+1	−½	=	↓		
2	1	0	−½	=		↓	
2	1	−1	−½	=			↓

The electron can be assigned to any 2*p* orbital and, once assigned, it can have either spin direction.

Carbon (Group 4A) is the second element in the *p* block, so there is a second electron assigned to the 2*p* orbitals. This electron *must* be assigned to either of the two remaining *p* orbitals, and it *must* have the same spin direction as the first *p* electron.

	+1	0	−1		+1	0	−1		+1	0	−1	
	↑	↑		or	↑		↑	or		↑	↑	$= 2p^2$
m_ℓ =	+1	0	−1		+1	0	−1		+1	0	−1	
ℓ =		1				1				1		

C atom ground state configurations

In general, when electrons are assigned to *p*, *d*, or *f* orbitals, each electron is assigned a different orbital of the subshell, each electron having the same spin as the previous one; this proceeds until the subshell is half full, after which *pairs* of electrons must be assigned a common orbital. This procedure follows **Hund's rule**, which states that the most stable arrangement of electrons is that with the maximum number of unpaired electrons, all with the same spin direction. Electrons are negatively charged particles, so assignment to different orbitals minimizes electron–electron repulsions, making the total energy of the set of electrons as low as possible. Giving them the same spin also lessens their repulsions.

As a final point, notice that carbon is the second element in the *p* block of elements, so there must be two *p* orbital electrons (besides the *s* electrons already present). Since C is in the second period of the table, the *p* orbitals involved are $2p$. Thus, you can immediately write the carbon electron configuration by reference to the periodic table: starting at H and moving left to right across the successive periods you write $1s^2$ to reach the end of period 1, and then $2s^2$ and finally $2p^2$ to bring the electron count to six. Carbon is in Group 4A of the periodic table, because there are four electrons in the $n = 2$ shell.

Following nitrogen (Group 5A), *oxygen* (Group 6A) has a total of six electrons in its outer shell: two of the electrons are assigned to the $2s$ orbital, and, as it is the fourth element in the *p* block, the other four electrons are assigned to $2p$ orbitals. With nitrogen the $2p$ subshell was half full, one electron per orbital. Therefore, for oxygen the fourth $2p$ electron must pair up with one already present. Again, it makes no difference which orbital this electron is assigned (the $2p$ orbitals all have the same energy), but it *must* have a spin opposite to the electron already assigned to the orbital (Table 8.2).

Like all the other elements in Group 8A, *neon* is a rare gas. All Group 8A elements (except helium) have eight electrons in the shell of highest *n* value, so all have the configuration ns^2np^6 where *n* is the period in which the element is found. That is, all the elements have filled *ns* and *np* subshells. As you will see, this correlates with their nearly complete chemical inertness.

The next element after neon is *sodium*, and a new period is begun. Since Na is the first element with $n = 3$, the added electron must be assigned to the $3s$ orbital. (Remember that all elements in Group 1A have the ns^1 configuration.) Thus, the complete configuration of Na would be that of neon plus one $3s$ electron.

Na: $1s^22s^22p^63s^1$ or $[Ne]3s^1$

↑ **rare gas notation** for abbreviated configurations

We have written the electron configuration in two ways, one an abbreviated form called the **rare gas notation**. The arrangement preceding the $3s$ electron is that of the rare gas neon, so, instead of writing out "$1s^22s^22p^6$," we represent the completed electron shells by placing the symbol of the corresponding rare gas in brackets.

EXAMPLE 8.1

ELECTRON CONFIGURATIONS

Using the spectroscopic notation, give the electron configuration of silicon.

Solution Silicon, element 14, is the fourth element in the third period ($n = 3$), and it is in the *p* block. Therefore, the last four electrons placed in the atom have the configuration $3s^23p^2$. These are preceded by the completed shells $n = 1$ and $n = 2$, the electron arrangement for Ne. Therefore, the complete configuration of Si is

Si configuration: $1s^22s^22p^63s^23p^2$ or $[Ne]3s^23p^2$

EXAMPLE 8.2

ELECTRON CONFIGURATIONS AND QUANTUM NUMBERS

Write a set of quantum numbers for each of the electrons in $n = 4$ for Ca.

Solution Calcium is the second element in the fourth period. Therefore, there are two electrons in $n = 4$, and, since Ca is in the s block of elements, these are $4s$ electrons. The element is preceded by the rare gas Ar, so the electron configuration of Ca is $[Ar]4s^2$. Using a box notation, the configuration is [Ar] [⇅], and possible sets of quantum numbers are

	n	ℓ	m_ℓ	m_s
For ↑	4	0	0	$+\frac{1}{2}$
For ↓	4	0	0	$-\frac{1}{2}$

The fact that the electrons are assigned to an s orbital makes $\ell = 0$, which in turn means m_ℓ must be 0.

EXAMPLE 8.3

ELECTRON CONFIGURATIONS AND QUANTUM NUMBERS

Write an acceptable set of quantum numbers that describe each electron in a silicon atom.

Solution The electron configuration of Si is given in Example 8.1. In box notation, its configuration would be

Electron	1, 2	3, 4	5–10			11, 12	13	14	
	[⇅]	[⇅]	[⇅]	[⇅]	[⇅]	[⇅]	[↑]	[↑]	[ ]
Orbital	$1s$	$2s$	$2p$	$2p$	$2p$	$3s$	$3p$	$3p$	$3p$

and an acceptable set of quantum numbers would be

ELECTRON	n	ℓ	m_ℓ	m_s
1, 2	1	0	0	$\pm\frac{1}{2}$
3, 4	2	0	0	$\pm\frac{1}{2}$
5–10	2	1	+1	$\pm\frac{1}{2}$
	2	1	0	$\pm\frac{1}{2}$
	2	1	−1	$\pm\frac{1}{2}$
11, 12	3	0	0	$\pm\frac{1}{2}$
13	3	1	+1*	$+\frac{1}{2}$‡
14	3	1	0†	$+\frac{1}{2}$§

*The value of m_ℓ can be any of the set, not *necessarily* +1.
†Once $m_\ell = +1$ is used for electron 13, then 14 must have a different m_ℓ (0 or −1).
‡Either m_s $+\frac{1}{2}$ or $-\frac{1}{2}$ can be used.
§Once electron 13 is assigned $m_s = +\frac{1}{2}$, then electron 14 must have the same value of m_s.

EXERCISE 8.2 ORBITAL BOX DIAGRAMS, SPECTROSCOPIC NOTATION, QUANTUM NUMBERS

(a) What element has the configuration $1s^2 2s^2 2p^6 3s^2 3p^5$? (b) Using the spectroscopic notation and a box diagram, show the electron configuration of sulfur. (c) Write one possible set of quantum numbers for the $3p$ electron of aluminum.

ELECTRON CONFIGURATIONS OF THE TRANSITION ELEMENTS

When $n = 3$, ℓ can be 0, 1, or 2, corresponding to *s*, *p*, and *d* orbitals, respectively. The 3*s* and 3*p* subshells are filled at argon, and the periodic table indicates that the next element is potassium, the first element of the fourth period. This means, though, that potassium must have the configuration $1s^22s^22p^63s^23p^64s^1$ ($[Ar]4s^1$), a configuration given by the $(n + \ell)$ rule. A valid question here is why the 4*s* orbital is used before the 3*d* orbitals.* The reason for this, a touchy balance of electron repulsions, is complex and well beyond the scope of this course. We accept the complications of nature here and rely on the "$n + \ell$" rule.

After electrons have been assigned to the 4*s* orbital, the 3*d* orbitals are those next utilized. Accordingly, *scandium* must have the configuration $[Ar]3d^14s^2$ and *titanium* follows with $[Ar]3d^24s^2$ and *vanadium* with $[Ar]3d^34s^2$. Notice in the orbital box diagrams in Table 8.3 that one electron is assigned to each of the five possible *d* orbitals and that all have the same spin direction.

On arriving at *chromium*, we come to what some might call an anomaly in the order of orbital assignment. For complex reasons, among them the minimization of electron–electron repulsions, *chromium has one electron assigned to each of the six available 4s and 3d orbitals*. Particularly in the heavier transition elements (and this includes the lanthanides and actinides) there are many so-called "anomalies." The origin of all of them is that the energies of all electrons must be considered, not just the outer few, and so electron arrangements are determined by many interrelated factors; the net result cannot always be predicted by broad generalizations.

It is shown in Table 8.4 that heavier elements do not always precisely follow the $n + \ell$ rule. You should remember the specific cases of Cr and Cu and be aware that others exist.

The four elements (Mn, Fe, Co, Ni) following chromium follow the $n + \ell$ guideline. On coming to *copper*, however, we arrive at the point where the energy of the 3*d* orbitals is so much lower than that of 4*s* orbitals that copper has all its 3*d* orbitals filled with electron pairs, and the 4*s* orbital has only a single electron. (Hence, copper is an element in Group 1B.) With *zinc* at the end of the transition series, the 4*s* and 3*d* orbitals are filled (Zn = $[Ar]3d^{10}4s^2$), thus preparing the way to begin filling the 4*p* orbitals in the *p* block of elements beginning with gallium, Ga.

The fifth period ($n = 5$) follows the pattern of the fourth period. The sixth period, however, presents us with another interesting problem, the inner transition series or **lanthanides**. As their name implies, the lanthanide elements begin with *lanthanum*, La. As the first element in the transition series in the sixth period, La has the electron configuration $[Xe]5d^16s^2$.† The next element (Ce, *cerium*) is set out in a separate row at the bottom of the table, and it is with the elements of this row that *f*

---

*Theoreticians have found that the 4*s* orbital energy is *slightly greater* than the 3*d* orbital energy for all elements K to Cu. However, there is a difference in how the core electrons repel each other when the outer 3*d* electron is changed to a 4*s* electron, and this overcomes the effect of the small 4*s*–3*d* energy gap and causes the 4*s* orbital to be favored over 3*d* for the outer electron at K and Ca.

†The guidelines on page 259 would lead you to predict that the 4*f* orbitals should begin filling before the 5*d* set. As explained above, such minor deviations from the general guidelines occasionally occur, particularly with the heavier metals.

TABLE 8.3 Orbital Box Diagrams for the Elements Ca Through Zn

		3d					4s	
		+2	+1	0	−1	−2	0	$= m_\ell$
Ca	[Ar] $4s^2$						↑↓	
Sc	[Ar] $3d^1 4s^2$	↑					↑↓	
Ti	[Ar] $3d^2 4s^2$	↑	↑				↑↓	
V	[Ar] $3d^3 4s^2$	↑	↑	↑			↑↓	
Cr*	[Ar] $3d^5 4s^1$	↑	↑	↑	↑	↑	↑	
Mn	[Ar] $3d^5 4s^2$	↑	↑	↑	↑	↑	↑↓	
Fe	[Ar] $3d^6 4s^2$	↑↓	↑	↑	↑	↑	↑↓	
Co	[Ar] $3d^7 4s^2$	↑↓	↑↓	↑	↑	↑	↑↓	
Ni	[Ar] $3d^8 4s^2$	↑↓	↑↓	↑↓	↑	↑	↑↓	
Cu*	[Ar] $3d^{10} 4s^1$	↑↓	↑↓	↑↓	↑↓	↑↓	↑	
Zn	[Ar] $3d^{10} 4s^2$	↑↓	↑↓	↑↓	↑↓	↑↓	↑↓	

*These configurations do not follow the "$n + \ell$" rule.

orbitals begin to come into use; the configuration of Ce is [Xe]$4f^1 5d^1 6s^2$. As we move across the lanthanide series, the pattern continues (with some variations; see Table 8.4): electrons are added to the seven *f* orbitals, singly at first and then in pairs, until they are all assigned. As expected, the configuration of the last of these elements (Lu, *lutetium*) is [Xe]$4f^{14} 5d^1 6s^2$. At this point, we return to the main body of the table with hafnium (Hf), where filling of 5*d* orbitals is resumed.

The seventh period begins with the radioactive elements *francium* (Fr) and *radium* (Ra), where electrons are assigned to the 7*s* orbital. The element *actinium* (Ac) heads up the inner transition series, the **actinide elements**, which stretch from *thorium* (Th) through *lawrencium* (Lr). Of the transition elements of the seventh period, where electrons are being assigned to 6*d* orbitals, there is evidence for elements 104 through 107 and 109.

When you complete this section you should be able to depict the electron configuration of any element using the periodic table as a guide.

EXAMPLE 8.4

ELECTRON CONFIGURATIONS OF TRANSITION ELEMENTS

Using the spectroscopic notation, give electron configurations for palladium (Pd) and osmium (Os).

Solution Proceeding along the periodic table, we come to the rare gas krypton, Kr, before arriving at Pd (element 46) in the fifth period. Following the 36 electrons of Kr there are 2 electrons in $5s$ orbitals (for Rb and Sr) and 8 electrons in the $(n - 1)d$ or $4d$ orbitals. Therefore, the palladium configuration is $[Kr]4d^8 5s^2$.

Osmium is a sixth period element, one that follows the lanthanide series. After xenon, Xe, the next 2 electrons are assigned to the $6s$ orbital and then 1 electron is assigned to a $5d$ orbital at La, lanthanum. Next, 14 electrons are assigned to $4f$ orbitals on crossing the lanthanide series from Ce to Yb. Filling the $5d$ orbitals then resumes at Hf, and an additional 5 electrons are assigned to these orbitals (for a total of six $5d$ electrons) to reach Os. Thus, the Os configuration is $[Xe]4f^{14}5d^6 6s^2$.

EXERCISE 8.3 ELECTRON CONFIGURATIONS

Using a periodic table, and without looking at Table 8.4, write electron configurations for the following elements: (a) S, (b) Zn, (c) Zr, (d) In, (e) Pb, and (f) U. Use the spectroscopic notation. When you have finished, check your answers in Table 8.4.

8.5 THE PERIODIC TABLE

The arrangement of elements in the periodic table is clearly connected to atomic structure. However, the concept of the periodic table has been known for the last 100 years, and it was developed well before our more recent knowledge of electron configurations. The history of the development of the table is replete with interesting characters, and it is worth pausing to consider at least one, Dmitrii Mendeleev (1834–1907).

In the 19th century many chemists tried to find relationships between atomic weights and the properties of the elements, but these attempts largely failed because atomic weights were not known for all the elements, and many measured values were inaccurate. However, at a conference in 1860 in Germany, Stanislao Cannizzaro (1826–1910) described his method of determining accurate and unambiguous atomic weights, and this began to set things right. A young Russian chemist attending the conference, Mendeleev, doubtless heard Cannizzaro's paper and was clearly influenced by it. He was to become the founder of the concept of chemical periodicity.

Partly because of inaccurate atomic weights, the state of chemistry was so chaotic early in the 19th century that the great German chemist Friedrich Wölher remarked in 1835 that chemistry reminded him of "a monstrous and boundless thicket . . . into which one may well dread to enter."

As happens so often in science, Cannizzaro built on the work of others, the Frenchman Charles Gay-Lussac (1778–1850) and fellow Italian Amadeo Avogadro (1776–1856).

Mendeleev's ideas on chemical periodicity began with his work on a book on inorganic chemistry. To help organize the material for this book, he had a file of note cards, one for each element. On each of these cards he wrote the atomic weight, then known more accurately due to Cannizzaro's work, and some properties of the element. When he arranged these cards in order of increasing atomic weight of the element, he saw there was a repetition of properties every eight or eighteen elements. Thus was born the concept of chemical periodicity and Mendeleev's periodic table (Figure 8.11).

Although other chemists before Mendeleev had considered a relation between atomic weight and atomic properties, we especially recognize Mendeleev because he had the genius to realize that (a) there were many elements yet to be discovered and (b) the characteristics of elements could be predicted from their atomic weights (and their position in his

TABLE 8.4 The Electronic Configurations of the Atoms of the Elements

ELEMENT	ATOMIC NUMBER	POPULATIONS OF SUBSHELLS 1*s*	2*s*	2*p*	3*s*	3*p*	3*d*	4*s*	4*p*	4*d*	4*f*	5*s*
H	1	1										
He	2	2										
Li	3	2	1									
Be	4	2	2									
B	5	2	2	1								
C	6	2	2	2								
N	7	2	2	3								
O	8	2	2	4								
F	9	2	2	5								
Ne	10	2	2	6								
Na	11				1							
Mg	12				2							
Al	13	Neon core			2	1						
Si	14				2	2						
P	15				2	3						
S	16				2	4						
Cl	17				2	5						
Ar	18	2	2	6	2	6						
K	19							1				
Ca	20							2				
Sc	21						1	2				
Ti	22						2	2				
V	23						3	2				
Cr	24						5	1				
Mn	25	Argon core					5	2				
Fe	26						6	2				
Co	27						7	2				
Ni	28						8	2				
Cu	29						10	1				
Zn	30						10	2				
Ga	31						10	2	1			
Ge	32						10	2	2			
As	33						10	2	3			
Se	34						10	2	4			
Br	35						10	2	5			
Kr	36	2	2	6	2	6	10	2	6			
Rb	37											1
Sr	38											2
Y	39									1		2
Zr	40									2		2
Nb	41	Krypton core								4		1
Mo	42									5		1
Tc	43									5		2
Ru	44									7		1
Rh	45									8		1
Pd	46									10		
Ag	47									10		1
Cd	48									10		2

ELEMENT	ATOMIC NUMBER		POPULATIONS OF SUBSHELLS									
			4d	*4f*	*5s*	*5p*	*5d*	*5f*	*6s*	*6p*	*6d*	*7s*
In	49		10		2	1						
Sn	50		10		2	2						
Sb	51		10		2	3						
Te	52		10		2	4						
I	53		10		2	5						
Xe	54		10		2	6						
Cs	55		10		2	6			1			
Ba	56		10		2	6			2			
La	57		10		2	6	1		2			
Ce	58		10	1	2	6	1		2			
Pr	59		10	3	2	6			2			
Nd	60		10	4	2	6			2			
Pm	61		10	5	2	6			2			
Sm	62		10	6	2	6			2			
Eu	63		10	7	2	6			2			
Gd	64		10	7	2	6	1		2			
Tb	65		10	9	2	6			2			
Dy	66		10	10	2	6			2			
Ho	67		10	11	2	6			2			
Er	68		10	12	2	6			2			
Tm	69		10	13	2	6			2			
Yb	70		10	14	2	6			2			
Lu	71		10	14	2	6	1		2			
Hf	72		10	14	2	6	2		2			
Ta	73		10	14	2	6	3		2			
W	74		10	14	2	6	4		2			
Re	75	Krypton core	10	14	2	6	5		2			
Os	76		10	14	2	6	6		2			
Ir	77		10	14	2	6	7		2			
Pt	78		10	14	2	6	9		1			
Au	79		10	14	2	6	10		1			
Hg	80		10	14	2	6	10		2			
Tl	81		10	14	2	6	10		2	1		
Pb	82		10	14	2	6	10		2	2		
Bi	83		10	14	2	6	10		2	3		
Po	84		10	14	2	6	10		2	4		
At	85		10	14	2	6	10		2	5		
Rn	86		10	14	2	6	10		2	6		
Fr	87		10	14	2	6	10		2	6		1
Ra	88		10	14	2	6	10		2	6		2
Ac	89		10	14	2	6	10		2	6	1	2
Th	90		10	14	2	6	10		2	6	2	2
Pa	91		10	14	2	6	10	2	2	6	1	2
U	92		10	14	2	6	10	3	2	6	1	2
Np	93		10	14	2	6	10	4	2	6	1	2
Pu	94		10	14	2	6	10	6	2	6		2
Am	95		10	14	2	6	10	7	2	6		2
Cm	96		10	14	2	6	10	7	2	6	1	2
Bk	97		10	14	2	6	10	9	2	6		2
Cf	98		10	14	2	6	10	10	2	6		2
Es	99		10	14	2	6	10	11	2	6		2
Fm	100		10	14	2	6	10	12	2	6		2
Md	101		10	14	2	6	10	13	2	6		2
No	102		10	14	2	6	10	14	2	6		2
Lr	103		10	14	2	6	10	14	2	6	1	2
Unq	104		10	14	2	6	10	14	2	6	2	2
Unp	105		10	14	2	6	10	14	2	6	3	2
Unh	106		10	14	2	6	10	14	2	6	4	2

TABELLE II

REIHEN	GRUPPE I. — R^2O	GRUPPE II. — RO	GRUPPE III. — R^2O^3	GRUPPE IV. RH^4 RO^2	GRUPPE V. RH^3 R^2O^5	GRUPPE VI. RH^2 RO^3	GRUPPE VII. RH R^2O^7	GRUPPE VIII. — RO^4
1	H = 1							
2	Li = 7	Be = 9,4	B = 11	C = 12	N = 14	O = 16	F = 19	
3	Na = 23	Mg = 24	Al = 27,3	Si = 28	P = 31	S = 32	Cl = 35,5	
4	K = 39	Ca = 40	— = 44	Ti = 48	V = 51	Cr = 52	Mn = 55	Fe = 56, Co = 59, Ni = 59, Cu = 63.
5	(Cu = 63)	Zn = 65	— = 68	— = 72	As = 75	Se = 78	Br = 80	
6	Rb = 85	Sr = 87	?Yt = 88	Zr = 90	Nb = 94	Mo = 96	— = 100	Ru = 104, Rh = 104, Pd = 106, Ag = 108.
7	(Ag = 108)	Cd = 112	In = 113	Sn = 118	Sb = 122	Te = 125	J = 127	
8	Cs = 133	Ba = 137	?Di = 138	?Ce = 140	—	—	—	— — — —
9	(—)	—		—	—	—	—	
10	—	—	?Er = 178	?La = 180	Ta = 182	W = 184	—	Os = 195, Ir = 197, Pt = 198, Au = 199.
11	(Au = 199)	Hg = 200	Tl = 204	Pb = 207	Bi = 208	—	—	
12	—	—	—	Th = 231	—	U = 240	—	— — — —

FIGURE 8.11

Dmitrii Mendeleev's periodic table of 1872. The spaces marked by blank lines represent elements that Mendeleev recognized as unknown in 1872, and places were left for them in the table.

table). As you can see from Figure 8.11, if Mendeleev had simply placed the known elements in order of increasing atomic weight, titanium (Ti), for example, would be in Group 3 with B and Al, elements with which Ti has little chemical similarity. Instead, Mendeleev placed Ti in Group 4 because of its similarity to Si, Sn, and especially Zr. He thus left a place in Group 3 and predicted that an element would be found to fit this position. Indeed, each of the dashes in Mendeleev's table indicated an undiscovered element. As brilliant as these conclusions were at the time, his prediction of the properties of the undiscovered elements was just as important. For example, he left a place in the table for an element he called "eka-silicon," an element that should have an atomic weight of approximately 72. Based on the properties of Si and Sn, he predicted the properties of this new element. As you can see from Table 8.5, his predicted properties nearly match the properties of germanium (Ge), an element not discovered until 1886. This prediction and others convinced chemists at the time of the value and correctness of Mendeleev's idea.*

In spite of Mendeleev's great achievement, problems arose when new elements were discovered and more accurate atomic weights were determined. Using the table in the front of this book, examine the following pairs of elements: Ar and K, Te and I, and Co and Ni. In each case, the first named element has a *greater* atomic weight than the second named element. Arranging them in order of increasing weight would, for example, move potassium, a reactive metal, into the group with the rare gases,

*The German Lothar Meyer (1830–1895) is also given credit for uncovering the concept of chemical periodicity, because he pointed out the periodicity of the physical properties of elements with increasing atomic weight. Mendeleev's ideas have been considered more important, however, because he recognized their predictive value.

DMITRII IVANOVICH MENDELEEV (1834–1907)

Mendeleev and the German Lothar Meyer (1830–1895) are honored for their discovery of the concept of chemical periodicity.

Mendeleev was born in Tobolsk, Siberia, but was educated in St. Petersburg (Leningrad) where he lived virtually all his life. He taught at St. Petersburg University and there wrote books and published his concept of periodicity in 1871.

It is interesting that Mendeleev did little else with chemical periodicity after his initial articles. He went on to other interests, among them studying the natural resources of Russia and their commercial applications. In 1876 he visited America to study the fledgling oil industry. He was apparently much impressed with the industry but not with the country. He found America uninterested in science, and he felt it carried on the worst features of European civilization.

By the end of the 19th century, political unrest was growing in Russia, and Mendeleev lost his position at the University. He was appointed Chief of the Chamber of Weights and Measures for Russia, however, and established an inspection system for guaranteeing the honesty of weights and measures used in Russian commerce.

All pictures of Mendeleev show him with long hair. He made it a rule to cut his hair only once a year in the spring, whether he had to appear at an important occasion or not.

elements that are nearly chemically inert. The fault lies with Mendeleev's assumption that the properties of the elements are periodic functions of their atomic weight.

H.G.J. Moseley, a young scientist working with Ernest Rutherford in 1913, found that the wavelengths of x-rays emitted by a particular element are related in a precise, mathematical way to the *atomic number* of the element. He quickly realized that other atomic properties may be similarly related to atomic number and not, as Mendeleev had believed, to atomic weight. Indeed, if the elements are arranged in order of increasing atomic number, the chemical periodicity observed by Mendeleev is preserved, and the defects in the Mendeleev table are corrected. It was therefore Moseley who discovered the real **law of chemical periodicity**: The properties of the elements are periodic functions of their atomic numbers.

TABLE 8.5 Properties That Mendeleev Predicted for the Element "Eka-silicon" and Those Observed for the Element Germanium

PROPERTY	MENDELEEV'S PREDICTION FOR EKA-SILICON, 1871	OBSERVED FOR GERMANIUM, 1886
Atomic weight	72.0	72.3
Specific gravity of element	5.5	5.469
Color	Dark gray	Grayish white
Specific gravity of the oxide	4.7	4.703
Boiling point of chloride	100°C	86°C
Boiling point of ethyl derivative	160°C	160°C

MODERN PROBLEMS WITH THE PERIODIC TABLE

The International Union of Pure and Applied Chemistry (IUPAC) and the American Chemical Society (ACS) have recently decided to adopt a new version of the periodic table (see inside front cover). Here the groups are numbered 1 through 18, thereby eliminating the problem of calling some groups A and others B. The table has not been accepted universally and has in fact generated some controversy within the international community of chemists. It may be several years before the issue is settled.

The IUPAC has also recommended three-letter symbols for the elements of atomic number 104 and higher. One reason for this is to remove the competition among the discoverers of these elements in devising names. Element 104 has the symbol Unq, where U stands for *un* or 1, n stands for *nil* or 0, and q stands for *quad* or four. Following this line of reasoning, 106 is Unh, where the h stands for *hexa*. Again, this system has not been universally accepted, so you will see tables with symbols such as Rf (rutherfordium) for element 104.

8.6 ATOMIC PROPERTIES AND PERIODIC TRENDS

The periodic table is organized into information about the physical and chemical properties of the elements. For example, in this chapter we have used the periodic table as a guide in defining the electron configurations of the elements. In Chapter 2 we described the division of the table into a section of metallic elements and another section of nonmetals. Within these broad classes there are groups of elements, each having a characteristic chemistry. For example, there are the alkali metals (Group 1A) and alkaline earth metals (Group 2A). In both cases, these metals can react to produce H_2 and a metal hydroxide when exposed to water.

$$\text{Group 1A} \quad Na(s) + H_2O(\ell) \longrightarrow NaOH(aq) + \tfrac{1}{2} H_2(g)$$

$$\text{Group 2A} \quad Mg(s) + 2\, H_2O(\ell) \longrightarrow Mg(OH)_2(s) + H_2(g)$$

Except for $Be(OH)_2$, the metal hydroxides provide basic solutions in water, a feature that is the source of the name of these groups. In all their chemical activity, it is a characteristic of these metals that they typically lose electrons to form $+1$ (Group 1A) or $+2$ (Group 2A) cations.

An objective of this section is to begin to show how atomic electron configurations are related to some physical properties of the elements and why those properties change in a reasonably predictable manner when moving down groups and across periods. Based on this knowledge, you will be able eventually to organize and predict new chemical and physical properties of elements and compounds.

ATOMIC SIZE

Electron orbitals have no sharp boundary beyond which the electron never strays. But how then do we define the size or radius of an atom? It is a

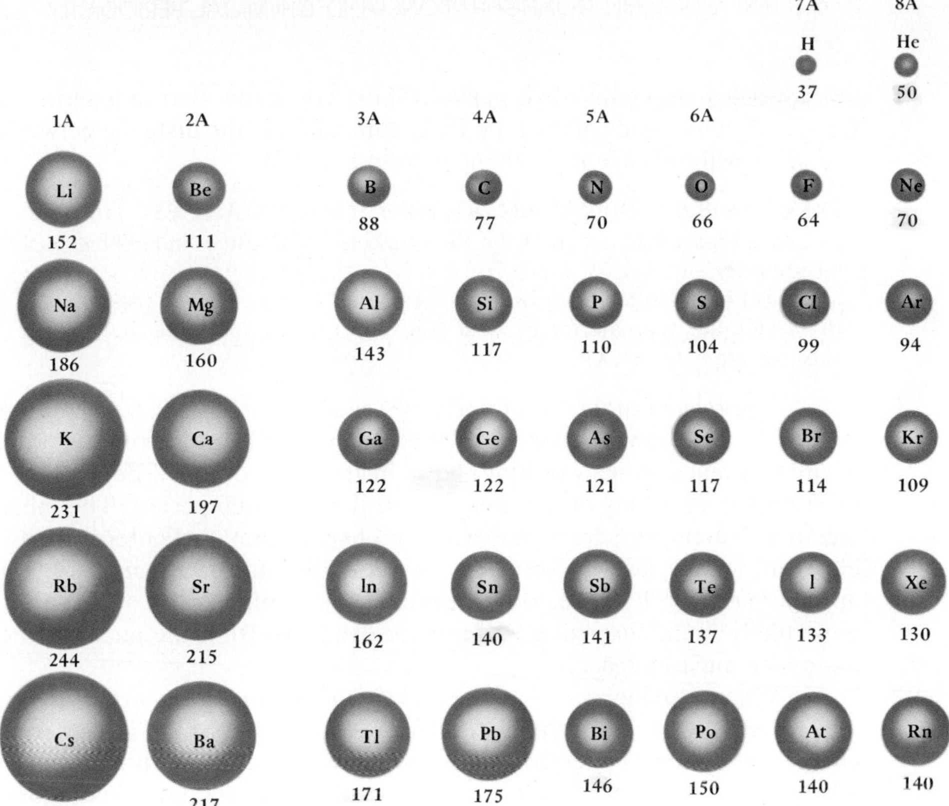

FIGURE 8.12
Atomic radii (in picometers, 1 pm $= 10^{-12}$ m) for the main group elements.

problem not yet solved to everyone's satisfaction, and the fact that new tables of atomic radii are still being compiled suggests that there are fundamental difficulties involved and reminds us that we should treat any set of radii with extreme caution.

For atoms that form simple diatomic molecules, Cl_2 for example, the radius of the atom can be defined experimentally by first finding the distance between the centers of the two atoms. One half of this distance is *assumed* to be a good estimate of the atom radius.*

In Cl_2 and H_2 molecules, for example, the atom–atom distances are 198 pm and 74 pm, respectively. This means that the Cl radius is 99 pm and that for H is 37 pm. A test of the reasonableness of these atomic radius estimates is whether we can use them to estimate the distance between H and Cl in HCl. As you can see, the estimated and experimental values are in reasonable agreement. This means this approach can be used to measure other atomic radii. For example, the radii of O, C, and S can now be estimated by measuring the O—H, C—Cl, and H—S distances in H_2O, CCl_4, and H_2S. Using this technique, and others, a reasonable set of atom radii has been determined for most elements, and some are illustrated in Figure 8.12. For *the main group elements, atomic radii increase going down a group in the periodic table*

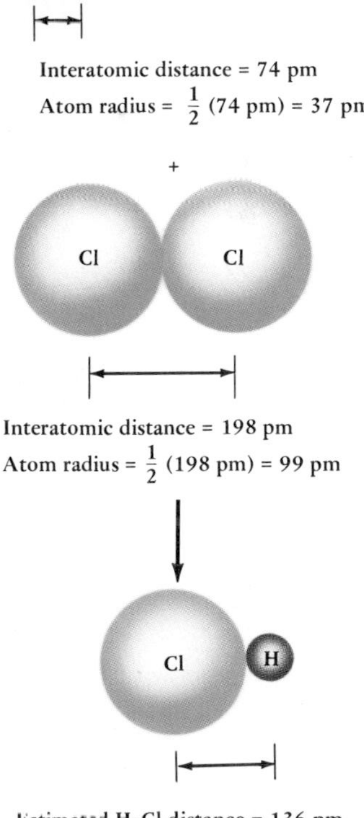

*For molecules where one atom is bonded to the other by a single pair of electrons, this method actually gives what is called the *single bond covalent radius*.

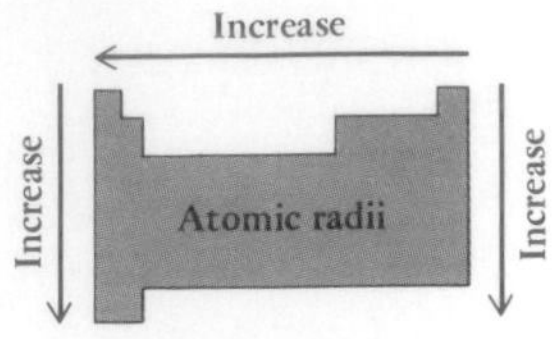

General trends in atomic radii of A Group elements with position in periodic table

and decrease going across a period. Thus, you know that in a series of Group 4A compounds of the type ECl_4, for example, the distance between E and Cl will increase as E changes from C to Pb.

EXERCISE 8.4 ESTIMATING ATOM–ATOM DISTANCES

(a) Using Figure 8.12, estimate the H—O and H—S distances in H_2O and H_2S, respectively.

(b) If the interatomic distance in Br_2 is 228 pm, what is the radius of Br? Using this estimate for Br, and that for Cl above, estimate the distance between atoms in BrCl.

The reason *atomic radii increase on descending a periodic group* is clear. Going down Group 1A, for example, the last electron added is always assigned to an *s* orbital and is in the electron shell beyond that used by the electrons of the elements in the previous period. The inner electrons shield or screen the nuclear charge from the outermost ns^1 electron, so the last electron feels an effective nuclear charge, Z^*, of approximately +1. Since, on descending the group, the ns^1 electron is most likely found at greater and greater distances from the nucleus, the atom size must increase.

When moving *across a period of main group elements, the size decreases because the effective nuclear charge increases.* The nuclear charge felt by the $2s^1$ electron of Li was shown graphically in Figure 8.8. When you add another 2*s* electron when going to beryllium, this second electron enters the same region of space as the first 2*s* electron. Both of these electrons have a significant probability of being close to the nucleus, and both experience the greater nuclear charge in Be (+4) than in Li (+3). Therefore, the two electrons are attracted strongly to the nucleus, and the 1*s* and 2*s* orbitals, and thus the atom, shrink.

The nuclear charge of boron is +5. The two 2*s* electrons are not completely effective in screening the nuclear charge, however, and the 2*p* electron is attracted strongly; again, the core (1*s*) and valence orbitals (2*s* and 2*p*) shrink still more from beryllium to boron (Figure 8.12).

Continuing across the period starting with boron, more protons are added to the nucleus. Again, the 2*s* and 2*p* electrons do not completely screen each other from the nuclear charge, so the Z^* values of the elements increase, and the atoms shrink still more. Notice, though, that the shrinkage is smaller after boron. As more and more electrons occupy the 2*p* orbitals, the size of the atom reflects the balancing of two effects: (1) the increase in nuclear charge across the periodic table should lead to decreasing atomic size, but (2) as more electrons are added, increasing repulsions between these electrons lead to electron cloud expansion. The net result is that the atom radius contracts slowly when moving across the periodic table from boron to neon.

The sizes of transition metal atoms change little proceeding from vanadium across the series, since the size is largely determined by the 4*s* orbital, which bears at least one electron in each element. The similar sizes have a profound effect on transition metal chemistry (Chapter 25).

IONIZATION ENERGY

The ionization energy of an atom is the energy required to remove an electron from an atom or ion in the gas phase.

$$\text{Atom in ground state(g)} + \text{energy} \longrightarrow \text{Atom}^+\text{(g)} + e^-$$
$$\Delta E = \text{ionization energy } (IE)$$

The process of ionization involves moving an electron from a given electron shell to a position outside the atom, that is, to n = infinity (Figure 8.13). Energy is always required, so the process is endothermic and the sign of the ionization energy is always positive.

Each atom can have a series of ionization energies, since more than one electron can always be removed (except for H). For example, the first three ionization energies of Mg(g) are

$$\underset{1s^22s^22p^63s^2}{\text{Mg(g)}} \longrightarrow \underset{1s^22s^22p^63s^1}{\text{Mg}^+\text{(g)}} + e^- \qquad IE(1) = 738 \text{ kJ/mol}$$

$$\underset{1s^22s^22p^63s^1}{\text{Mg}^+\text{(g)}} \longrightarrow \underset{1s^22s^22p^63s^0}{\text{Mg}^{2+}\text{(g)}} + e^- \qquad IE(2) = 1450 \text{ kJ/mol}$$

$$\underset{1s^22s^22p^6}{\text{Mg}^{2+}\text{(g)}} \longrightarrow \underset{1s^22s^22p^5}{\text{Mg}^{3+}\text{(g)}} + e^- \qquad IE(3) = 7734 \text{ kJ/mol}$$

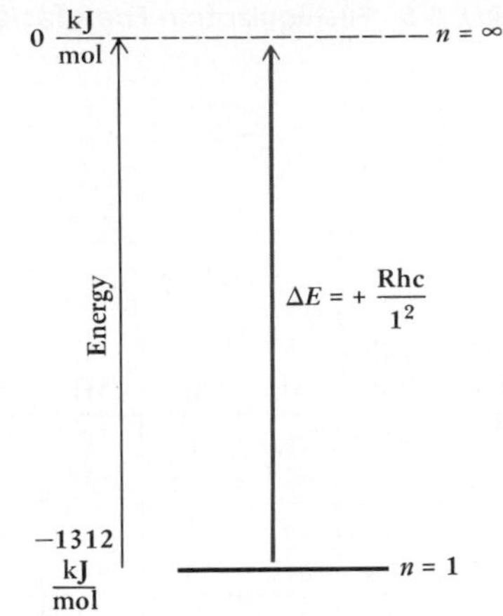

FIGURE 8.13

Ionization energy for an electron in $n = 1$ for the H atom. $IE = \Delta E = E_{\text{final}} - E_{\text{initial}} = 0 - (-Rhc/n^2)$. When $n = 1$, $E = Rhc$ or 1312 kJ/mol.

Notice that removing each subsequent electron requires more and more energy. In the Mg atom, there are 12 protons and 12 electrons, but in the Mg^+ ion there are 12 protons and only 11 electrons. One of the 3s electrons is no longer there to screen its partner; the remaining 3s electron experiences a higher Z^* and is thus held more tightly in the ion than in the atom (which makes it more difficult to remove than the first electron). The third ionization energy is higher for two reasons: (1) There are now only 10 electrons in Mg^{2+}, so Z^* experienced by the remaining electrons has gone up. (2) The third electron removed is a 2p type, which has a much larger Z^* than the 3s type removed previously.

The observation that the third IE of Mg is much higher than the first two is a confirmation of the shell structure of atomic electrons.

For main group or A-type elements, *first ionization energies generally decrease down a periodic group and increase across a period* (Figure 8.14 and Table 8.6). The ionization energy decrease going down the

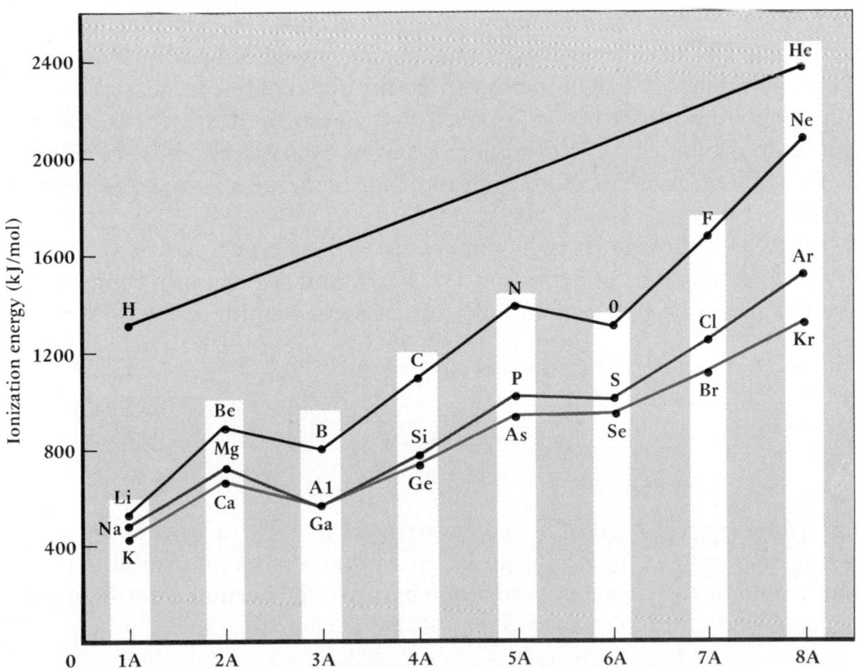

FIGURE 8.14

First ionization energies of the elements of the main group elements of the first four periods.

TABLE 8.6 First Ionization Energies of the Elements (kJ/mol)*

1A																	8A
H 1312	2A											3A	4A	5A	6A	7A	He 2374
Li 521	Be 897											B 801	C 1090	N 1400	O 1312	F 1680	Ne 2084
Na 496	Mg 738	3B	4B	5B	6B	7B	8B			1B	2B	Al 577	Si 782	P 1013	S 1004	Cl 1254	Ar 1525
K 415	Ca 589	Sc 627	Ti 656	V 647	Cr 656	Mn 714	Fe 762	Co 762	Ni 733	Cu 743	Zn 907	Ga 579	Ge 762	As 946	Se 946	Br 1139	Kr 1351
Rb 405	Sr 550	Y 618	Zr 618	Nb 666	Mo 685	Tc 704	Ru 714	Rh 724	Pd 801	Ag 733	Cd 869	In 560	Sn 704	Sb 830	Te 869	I 1013	Xe 1168
Cs 376	Ba 502	La 540	Hf 676	Ta 762	W 772	Re 762	Ds 840	Ir 869	Pt 869	Au 888	Hg 1004	Tl 589	Pb 714	Bi 704	Po 811	At 897	Rn 1033

*The energies represent the energy necessary to remove an electron from the element in the gas phase.

table occurs for the same reason that the size increases in this direction: the first-removed electron is farther and farther from the nucleus and so less and less energy is required for its removal. Additionally, recall that the energy of an electron depends on n [$E \propto -(1/n^2)$, Equation 7.4]. As n becomes larger, E is less negative, so ΔE (the ionization energy) is smaller.

Increase

First IE

General trends in first ionization energies of A Group elements with position in periodic table. Exceptions occur at Groups 3A and 4A.

There is a general increase in ionization energy when moving across a period of the periodic table due to an ever increasing effective nuclear charge. The trend, however, is not smooth (Figure 8.14), and its peaks and valleys give us further insight into atomic structure. First, in spite of the general trend, the ionization energy for boron is less than that for beryllium. The reason for this is that the $2s$ orbital of beryllium is lower in energy (higher Z^*) than the boron $2p$ orbital. Therefore, less energy is required to remove the boron $2p$ electron than the $2s$ electron of beryllium. The $2s/2p$ orbital energy difference at boron establishes a new beginning for the general trend of increasing ionization energies across the second period.

When reaching oxygen, however, the general trend is again offset. Oxygen has the configuration $1s^2 2s^2 2p^4$, and the electron removed to give O^+(g) is one of a pair of electrons assigned to the same $2p$ orbital.

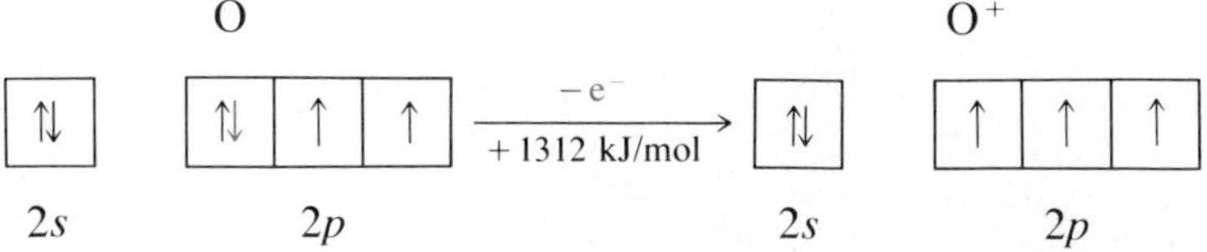

Electron–electron repulsion causes oxygen to have a lower ionization energy than expected. Z^* for an electron that shares an orbital is lower than would be the case if it were alone because the partner electron exerts a higher force of repulsion than an electron assigned to a different orbital. To remove one of a pair of electrons with the same orbital is thus easier than would otherwise be the case.

After oxygen, the trend to higher ionization energies is again resumed due to the continued increase in Z^* and the fact that all succeeding atoms have in common higher electron–electron repulsion.

For the rare gases, the elements that end the periods, the ionization energies reach a maximum for the period. The effective nuclear charge has become large enough that the electrons are firmly held. This is the reason the rare gases do not readily form ions or compounds.

This analysis of the general trends and specific details of ionization energies is a valuable lesson in chemistry: the chemical and physical properties of the elements and the compounds they form often depend on several factors, and it is not always simple to predict which will be most important for any given element.

ELECTRON AFFINITY

Some atoms have an affinity for electrons and can acquire one or more to form negative ions. This occurs because the effective nuclear charge is often sufficient for the added electron(s) to be attracted to the nucleus. A measure of the **electron affinity** of an element is the energy involved when an electron is brought from an infinite distance away up to a gaseous atom and absorbed by it to form a gaseous ion (Table 8.7).

$$A(g) + e^- \longrightarrow A^-(g)$$

$$\Delta E = \text{electron affinity } (EA)$$

When a stable anion is formed, energy is released and the sign of the energy change is negative. The greater the electron affinity the more

TABLE 8.7 Electron Affinity Values for Some Elements (kJ/mol)*

1A	2A	2B	3A	4A	5A	6A	7A	8A
H −72.4								He (+21)
Li −59.8	Be (+241)		B −23.2	C −122	N 0	O −142	F −322	Ne (+29)
Na −53.1	Mg (+232)		Al −44.4	Si −120	P −74.3	S −201	Cl −348	Ar (+35)
K −48.3	Ca (+156)	Cu −123	Ga (−35.7)	Ge −116	As −77.2	Se −195	Br −324	Kr (+39)
Rb −47.3	Sr (+120)	Ag −125	In (−33.8)	Sn −121	Sb −101	Te −190	I −295	Xe (+41)
Cs −45.4	Ba (+52)	Au −223	Tl (−48.3)	Pb −101	Bi −101	Po (−174)	At (−270)	Rn (+41)

*Estimated values are given in parentheses.

negative the value of *EA*. Fluorine, for example, has an electron affinity of −322 kJ/mol.

$$F(g) + e^- \longrightarrow F^-(g) + 322 \text{ kJ/mol}$$

$$\underset{2s}{\boxed{\uparrow\downarrow}}\ \underset{2p}{\boxed{\uparrow\downarrow}}\ \underset{2p}{\boxed{\uparrow\downarrow}}\ \underset{2p}{\boxed{\uparrow\ \ }} + e^- \longrightarrow \underset{2s}{\boxed{\uparrow\downarrow}}\ \underset{2p}{\boxed{\uparrow\downarrow}}\ \underset{2p}{\boxed{\uparrow\downarrow}}\ \underset{2p}{\boxed{\uparrow\downarrow}}$$

In contrast, iodine has an electron affinity of only −295 kJ/mol.

If an unstable ion is formed, the process is endothermic and electron affinity is a positive quantity. Helium has an EA of +21 kJ/mol, for example, because the electron must be added to a much higher energy orbital.

$$He(g) + e^- + 21 \text{ kJ/mol} \longrightarrow He^-(g)$$

$$\underset{1s}{\boxed{\uparrow\downarrow}}\ \underset{2s}{\boxed{\ \ \ }} + e^- \longrightarrow \underset{1s}{\boxed{\uparrow\downarrow}}\ \underset{2s}{\boxed{\uparrow\ \ }}$$

The periodic trends in electron affinity are closely related to those for ionization energy and size (Figure 8.15). There is a general increase

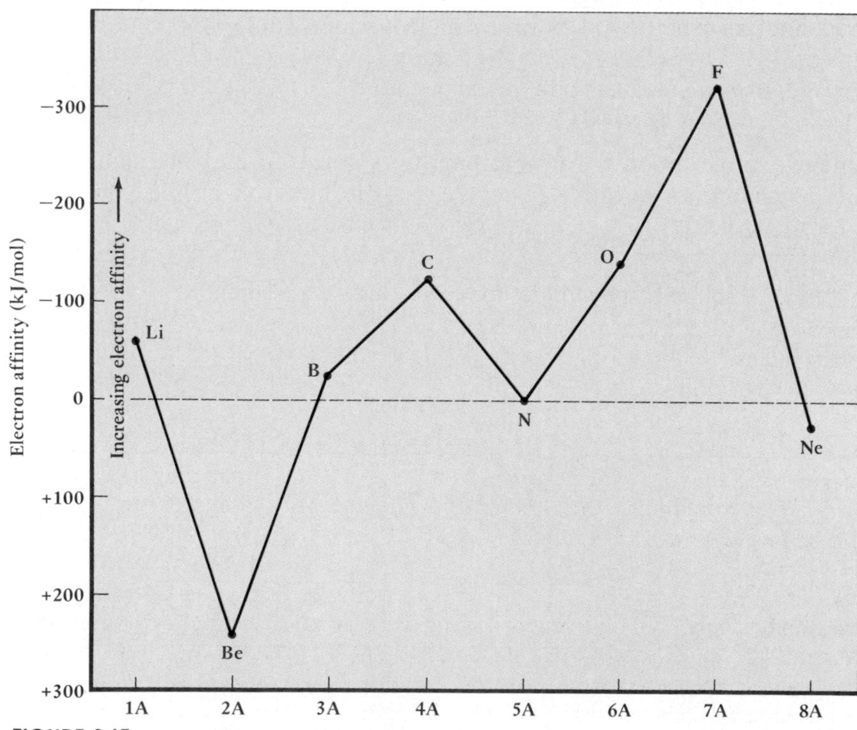

FIGURE 8.15

Electron affinities for the second period elements. Electron affinity is defined as the energy involved in the addition of a mole of electrons to a mole of gaseous atoms of the element. The more negative the energy the greater the electron affinity.

across a period due to the general increase in Z^*, but there is evidence again for the competing effects of electron–electron repulsions and changes in nuclear charge. For example, just as electron–electron repulsions cause the ionization energy of oxygen to be lower than expected, this same effect means nitrogen has almost no affinity for an electron.

$$N(g) + e^- \xrightarrow[\text{or required}]{\text{no energy evolved}} N^-(g)$$

↑↓	↑	↑	↑	+ e⁻ ⟶	↑↓	↑↓	↑	↑
2s	2p	2p	2p		2s	2p	2p	2p

When descending a periodic group, we expect the electron affinity to decrease for the same reason that atom size increases and ionization energy decreases. A glance at Table 8.7 shows this is *roughly* true, but there are numerous exceptions. Consider the halogens, for example. The electron affinity does indeed drop as expected from Cl to I, but fluorine has a lower electron affinity than chlorine! The reason stems from unusually large electron–electron repulsions in fluorine. Adding an electron to the nine already present in fluorine means that ten electrons must be packed in a relatively small volume, leading to severe repulsions between electrons. Chlorine, bromine, and iodine have larger atomic volumes than fluorine, so adding an electron does not lead to such severe electron–electron repulsions as in fluorine. The electron affinity of fluorine is thus anomalously low.

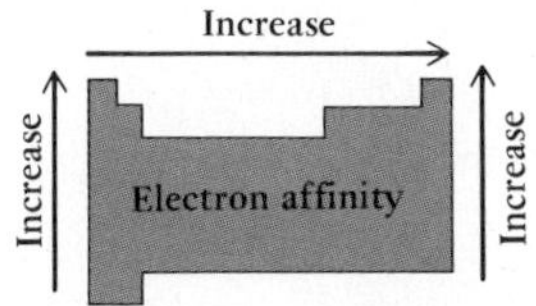

General trends in electron affinities of A Group elements with position in periodic table. Exceptions occur at Groups 2A and 5A.

EXAMPLE 8.5

PERIODIC TRENDS

Compare the three elements N, O, and P. (a) Place them in order of increasing atomic radius. (b) Which has the largest ionization energy? (c) Which has the larger electron affinity, N or O?

Solution (a) *Atomic size*. Atomic radius declines moving across a period, so nitrogen must have a larger radius than oxygen. However, radius increases down a periodic group. Since N and P are in the same group (Group 5A), P must be larger than N. Therefore, in order of increasing size, the elements are $O < N < P$.

(b) *Ionization Energy (IE)*. Ionization energy generally increases across a period and decreases down a group; a large decrease in *IE* occurs from the second to the third period elements. This should make $IE(P) < IE(N) < IE(O)$. However, recall that electron–electron repulsions modify this trend, and reverse the positions of O and N. Thus, $IE(P) < IE(O) < IE(N)$.

(c) *Electron affinity (EA)*. Electron affinity, like ionization energy, generally increases across a period and decreases down a group (although the trends are not as clear as trends in ionization energy). Thus, $EA(N) < EA(O)$.

EXERCISE 8.5 PERIODIC TRENDS

Compare the three elements Al, C, and Si.

(a) Place the three elements in order of increasing atomic radius.

(b) Rank the elements in order of increasing ionization energy. (Try to do this without looking at Table 8.6; then compare your estimates with the table.)

(c) Which element, Al or Si, is expected to have the larger electron affinity (more negative value for the energy involved when the atom acquires an electron)?

8.7 IONS

Elements may gain or lose electrons to form ions. The electron configurations and properties of these ions are important in determining the chemistry of compounds they form. For example, ion size can be a controlling feature in interesting and vital processes. The fact that iron(II) and iron(III) and not some other ions are used in hemoglobin, the oxygen-carrying molecule in blood, is determined partly by their size.

ION ELECTRON CONFIGURATIONS

When an element loses one or more electrons to form a positive ion, the electrons lost are those of highest n. If all the electrons available for removal have the same n, then the electrons lost are those of highest "$n + \ell$." For example, carbon has four electrons with $n = 2$, but only two of these need to be removed to form the C^{2+} ion. Since the $2p$ electrons have $n + \ell = 3$, whereas the $2s$ electrons have a value of only $n + \ell = 2$, it is the $2p$ electrons that are removed to form C^{2+}. To form a C^{4+} ion, all the $n = 2$ electrons are removed.

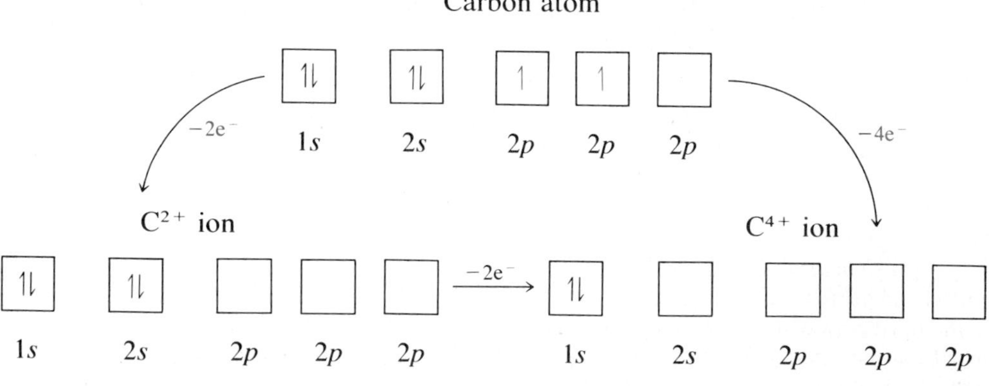

The transition metals all form positive ions, and the "highest n first" rule applies here as well. For example, Fe^{2+} is formed by loss of the two $4s$ electrons, rather than any $3d$ electrons.

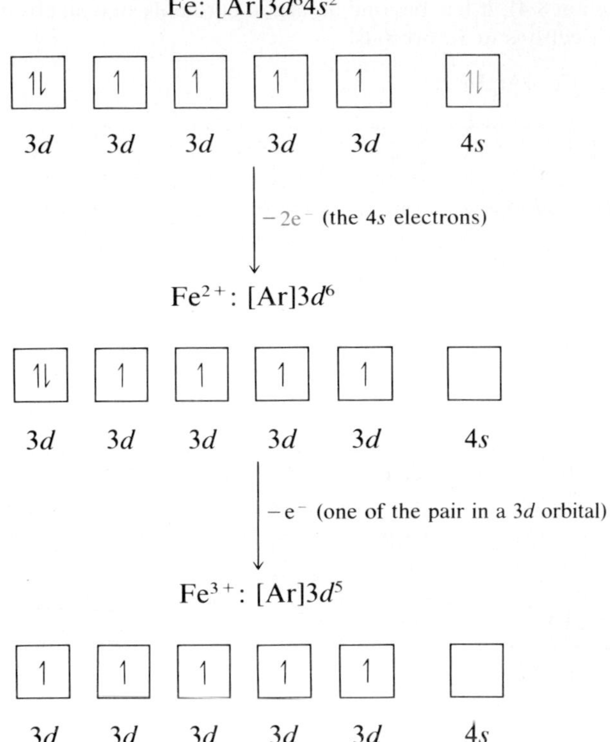

Only after the $4s$ electrons are removed to form Fe^{2+} are $3d$ electrons affected. Notice that Fe^{3+} is formed from Fe^{2+} by removal of one electron of a pair in a $3d$ orbital. Properties of transition metal compounds are often determined by the number of d electrons, and the number of these electrons that are unpaired. Therefore, you must know how to derive the correct electron configuration of these metal ions.

A configuration with all $3d$ electrons unpaired is the lowest energy configuration because electron–electron repulsions have been minimized.

Atoms and ions with unpaired electrons are *paramagnetic*, that is, they are capable of being attracted into a magnetic field (Section 8.1). Paramagnetism is important here because it allows us to prove that transition metal ions must be formed by loss of ns electrons before $(n - 1)d$ electrons. For example, the Fe^{2+} ion above is paramagnetic to the extent of 4 unpaired electrons, and the Fe^{3+} ion has five unpaired electrons. If three $3d$ electrons were removed to form Fe^{3+}, for example, the ion would still be paramagnetic but only to the extent of three unpaired electrons.

EXAMPLE 8.6

CONFIGURATIONS OF TRANSITION METAL IONS

Give the electron configurations for copper, Cu, and for its +1 and +2 ions. Is the metal or either of its ions paramagnetic? If so, how many unpaired electrons does each have?

Solution Copper is in Group 1B in the fourth period transition metal series. Recalling that the metal has a configuration different from that predicted by the

"$n + \ell$" rule (Section 8.4), it has beyond an argon core only one electron in the 4*s* orbital and ten electrons in 3*d* orbitals.

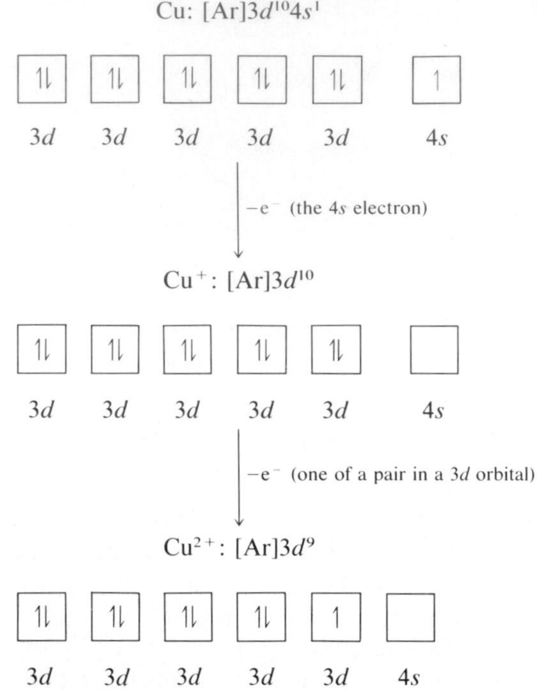

Copper metal and Cu^{2+} each have one unpaired electron and so should be paramagnetic. In contrast, Cu^+ has no unpaired electrons so the ion and its compounds are diamagnetic.

EXERCISE 8.6 METAL ION CONFIGURATIONS

Depict electron configurations for V^{2+}, V^{3+}, and Co^{3+}. Use orbital box diagrams and show only the *ns* and $(n - 1)d$ electrons. Are any of the ions paramagnetic? If so, give the number of unpaired electrons.

ION SIZES

It is clearly shown in Figure 8.16 that the periodic trends in the sizes of a few common ions are the same as those for neutral atoms: positive or negative ions of the same group increase in size when descending the group. But pause for a moment and compare Figure 8.16 with Figure 8.12. When an electron is removed from an atom to form a cation, the size shrinks considerably; *the radius of a cation is always smaller than the atom from which it is derived*. For example, the radius of Li is 152 pm whereas that of Li^+ is only 60 pm. This is understandable, because the electron removed (a 2*s*) is more radially extended than the 1*s* electrons that remain. In addition, the effective nuclear charge experienced by the remaining 1*s* electrons has increased because there are now fewer electrons (2 assigned to the 1*s* orbital) to repel each other. Thus, the 1*s* orbital contracts toward the nucleus, and the Li^+ ion is smaller than the Li atom.

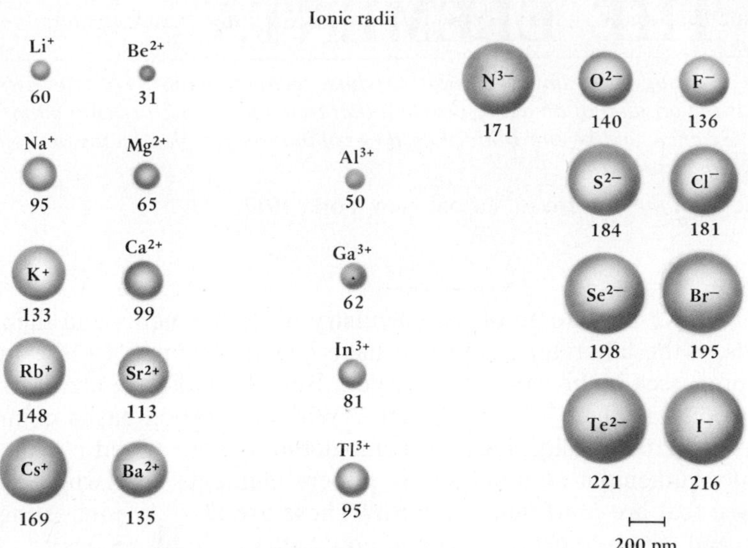

FIGURE 8.16
Relative sizes of some common ions (radii in pm).

The same general argument holds for all the positive ions of Group 1A. However, since the size of the neutral atoms of Group 1A increases down the group, the sizes of their respective positive ions also increase.

The same trend in ion sizes is seen in Group 2A as well. However, here there is an even more dramatic decrease in size on forming the +2 ion. In this case two electrons have been removed; for example, Be has 4 protons and 4 electrons, while Be^{2+} has only 2 electrons to be attracted to 4 protons. Because of the higher nuclear charge in Be^{2+} ($1s^2$) than in Li^+ ($1s^2$) the $1s$ orbital of Be^{2+} is more contracted than the $1s$ orbital of Li^+.

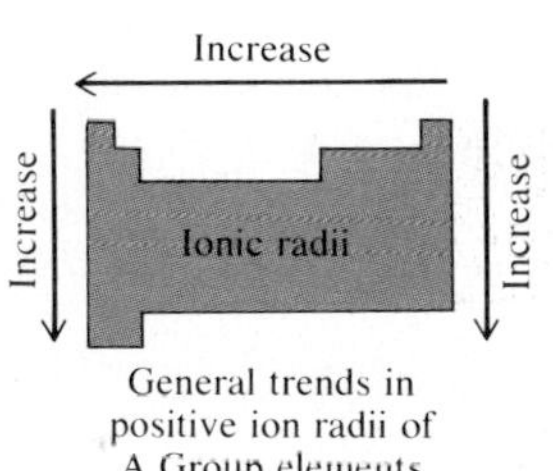

General trends in positive ion radii of A Group elements with position in periodic table

It is clear from Figure 8.17 that *anions are always larger than the atoms from which they are derived.* Here the argument is the opposite of that used to explain positive ion radii. The F atom, for example, has 9 protons and 9 electrons. On gaining an electron, the anion still has 9 protons, but there are now 10 electrons present. The fluoride ion is much larger than the fluorine atom because *increased electron–electron repulsions have caused the atom to swell when an electron is added.*

(Text continues on page 289)

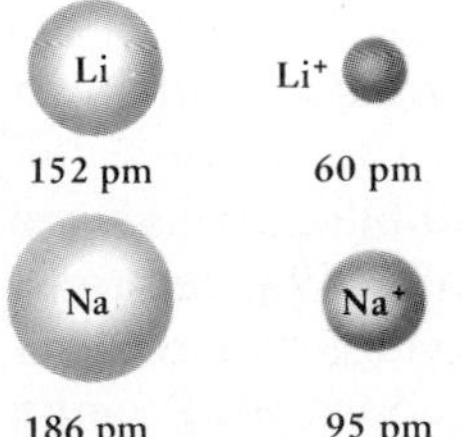

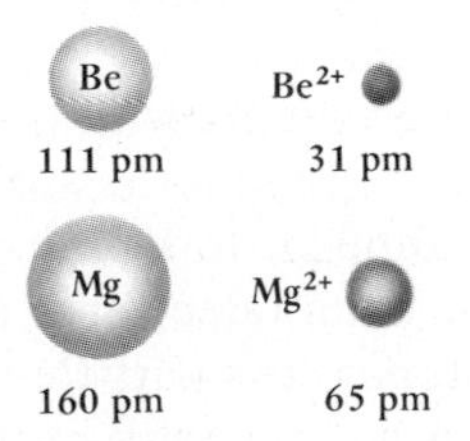

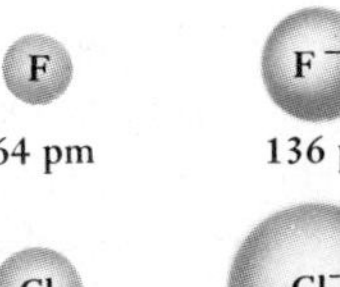

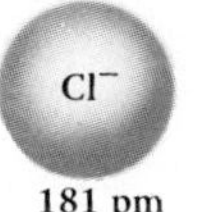

FIGURE 8.17
Atom and ion sizes. The nuclear charge stays constant when an ion is formed, so cations are smaller and anions larger than the atoms from which they are derived.

SPECIAL SECTION: THE ORIGIN OF THE ELEMENTS

This great saga of cosmic evolution, to whose truth the majority of scientists subscribe, is the product of an act of creation that took place about twenty billion years ago. Science, unlike the Bible, has no explanation for the occurrence of that extraordinary event.

R. Jastrow, *Until the Sun Dies*, Norton, New York, 1977.

By now you have seen some of the chemistry of the elements and their compounds in the laboratory, and you have begun to connect these to chemical processes in the world around you. But where did the elements originate? Why are there so few of them? Why are carbon-based compounds so numerous? Why does the hemoglobin of your blood contain iron and not ruthenium or uranium? Are there elements somewhere in the universe that are not found on earth? These are all very interesting questions, and we are most fortunate in now having some answers!

The composition of a number of stars in the universe and of the planets and moons of our solar system has been determined. It is a great relief to find that there is no evidence for an element found in some distant star that does not exist on earth. Further, although there are variations from star to star and planet to planet, the relative abundances of the elements throughout the universe are approximately the same. These are very important observations, since they mean that the element-forming processes are general throughout the universe.

The relative abundances of a few the elements in the universe are illustrated in Figure 8.18. The most striking features of this figure are those that follow and are the facts that must be taken into account in any theory of the origin of the elements.

(a) 90% to 95% of the atoms in the universe are hydrogen atoms.
(b) 5% to 10% of all atoms are helium.
(c) All of the other elements taken together make up only about 1% of the universe, even on a weight basis.
(d) Lithium, beryllium, and boron are mysteriously rare.
(e) Elements of even atomic number are more abundant than those with odd atomic number.
(f) There is a general decline in abundance from oxygen to lead. However, there is a very pronounced maximum in relative abundance around iron.
(g) There are no stable elements with mass numbers greater than about 210.

The universe is thought to have begun about 20 billion years ago as a subatomic particle soup (electrons, positrons or positive electrons, protons, neutrons, and massless particles such as neutrinos and antineutrinos and photons in a ball recently estimated to be about 10^{-28} cm in diameter). The "big bang" theory hypothesizes that an explosion of unimaginable fury began the series of events that has led ultimately to the universe and to life on earth.

Within seconds of the big bang, the explosion of the super-dense ball of particles, the synthesis of the elements began. Free neutrons are

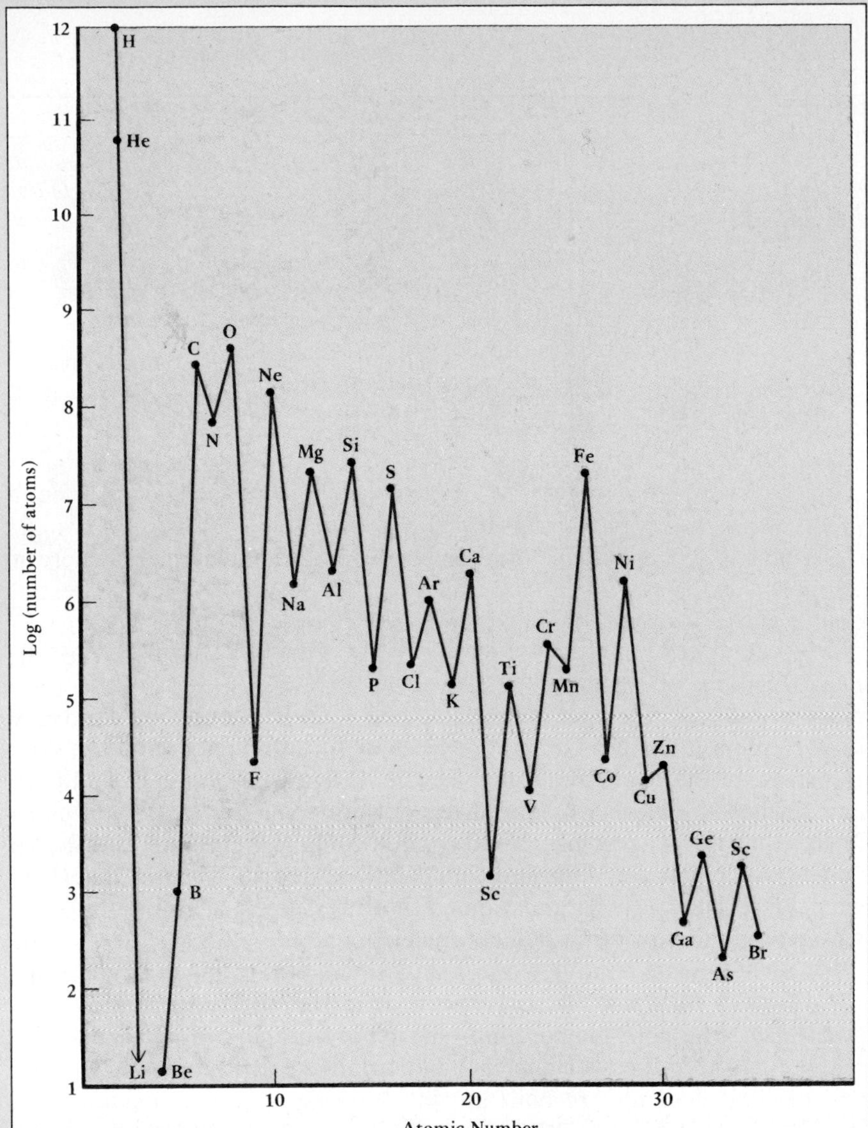

FIGURE 8.18

The cosmic abundances of the lighter elements as a function of atomic number. Abundances are expressed as numbers of atoms per 10^{12} atoms of H and are plotted on a (base-10) logarithmic scale. (Data taken from G.O. Abell, *Exploration of the Universe,* 4th ed. Saunders College Publishing, Philadelphia, 1982, p. 706.)

known to decay to give protons, electrons, and a large amount of energy.

$$\text{Neutron} \longrightarrow \text{proton} + \text{electron} + \text{energy}$$

Since a hydrogen atom has a nucleus consisting of a single proton, this is the origin of the hydrogen in our universe. To continue the process of element formation, hydrogen nuclei fused in the high temperature furnace of the big bang and continue to do so in stars, to give helium nuclei, the next most important building block of the elements.

At the temperatures found in a star or at the time of the "big bang," all electrons are stripped from hydrogen, helium, or other atoms, leaving only the bare nucleus.

In the seconds, minutes, and hours following the moment of Creation, the gas-filled, chaotic universe rapidly expanded and began to cool. After perhaps a billion years, it had cooled to the point that the hydrogen and helium began to coalesce into galaxies and, within the galaxies, into stars, a process that continues even today. The primordial stars contracted under the force of gravitational attraction, and the temperature began to climb, finally reaching about 10 million degrees, the critical temperature for "hydrogen burning" or fusion to form more helium. This exothermic

FIGURE 8.19

Hydrogen fusion continues for 99% of the lifetime of a star. When the hydrogen is essentially depleted, the star collapses until its core reaches a temperature in excess of 100 million degrees Celsius. At this temperature, helium nuclei begin to fuse to produce carbon nuclei. The process is thought to proceed through an unstable $^{8}_{4}Be$ isotope as an intermediate.

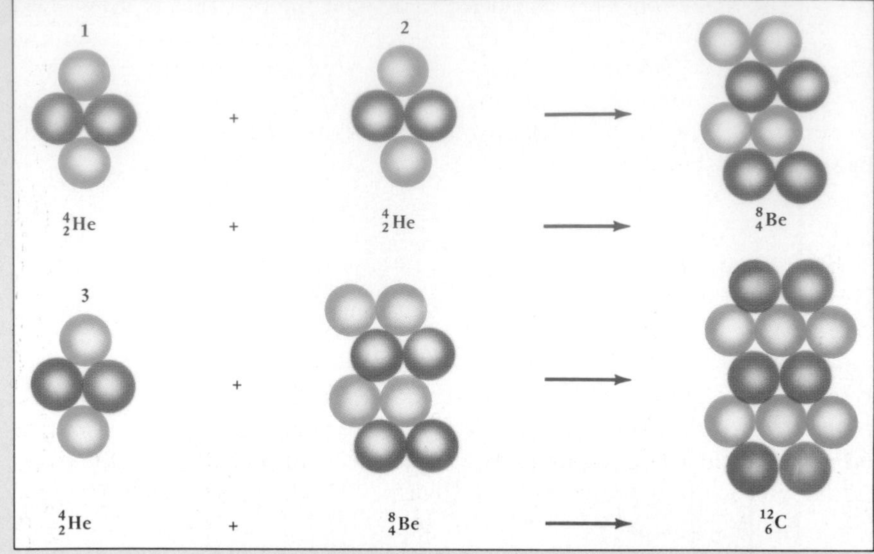

reaction raises the temperature of the star, and helium nuclei can begin to coalesce at a hundred million degrees or so into nuclei of still heavier elements. In the "helium-burning" cycle, it is the carbon nucleus that is first formed (Figure 8.19). The direct formation of carbon from helium means that the process has circumvented lithium, beryllium, and boron and explains their very low abundance in the universe. The fact that these elements exist, however, must mean that they are formed in a different way than by the fusion of helium nuclei.

Our sun is a relatively small star, since ones of much larger mass, say 10 to 30 solar masses, have been observed. Because of their very great mass, the core temperatures must become very great, and fusion reactions to form heavier nuclei than carbon are now possible. Heavier and heavier elements are built up, in a process involving helium, $^{4}_{2}He$, similar to that in Figure 8.19. Since helium is an element of "even mass," this explains the observation that elements of even mass are more abundant than those of odd mass.

Elements heavier than iron form by neutron capture. As mentioned, a neutron can be transformed into a proton and an electron. This means that when a nucleus captures a stray neutron and the neutron becomes a proton, the atom is now an atom of next higher atomic number. See Chapter 26.

Energy is evolved when light nuclei coalesce or fuse to form nuclei lighter than iron. However, iron formation is the end of this exothermic series! To continue the fusion process requires energy. Therefore, when a massive star has essentially consumed its fuel and iron has been formed, the core collapses to a super-dense state *in less than a second*. The shock waves from this collapse expand outward through the envelope of gases surrounding the core, heating the gases and providing the energy to create heavy nuclei. This outward blast is seen as a gigantic explosion called a supernova, many of which have been observed. The explosion sends debris rocketing into space, where over additional millions of years it coalesces with hydrogen, helium, and other elements to form new stars. Our sun is just such a "second-generation" star, and our earth and the other planets of the solar system were formed from the debris of the explosions of massive stars. It is for this reason that there is a wide variety of elements found on the earth and other planets.

The oxide ion, O^{2-}, is **isoelectronic** with the F^-, that is, they both have the same number of electrons (10). The oxide ion, however, is much larger than the fluoride ion because the oxide ion has only 8 protons available to attract the 10 electrons, whereas F^- has 9 protons. It is useful to compare the sizes of isoelectronic ions across the periodic table. For example, consider the isoelectronic series O^{2-}, F^-, Na^+, and Mg^{2+}.

ION	O^{2-}	F^-	Na^+	Mg^{2+}
Ionic radius (pm)	140	136	95	65
Number of nuclear protons	8	9	11	12
Number of electrons	10	10	10	10

All the ions above have a total of ten electrons. However, the O^{2-} or oxide ion has only eight protons in its nucleus to attract these electrons, while F^- has nine, Na^+ has eleven, and Mg^{2+} has twelve. As the *proton/electron ratio* increases in a series of isoelectronic ions so does the effective nuclear charge for each kind of electron present, and the ionic size decreases. As you can see in Figure 8.16, this is generally true for all isoelectronic series of ions.

As a general observation in this section, notice the importance of nuclear charge and electron–electron repulsions in determining periodic trends.

EXERCISE 8.7 ION SIZES

What is the trend in sizes of the ions N^{3-}, O^{2-}, and F^-? Briefly explain why the trend exists.

SOME COMMON MONATOMIC IONS

You learned in Chapter 2 that many metals form positive ions with a charge equal to the group number (Table 8.8). In contrast, nonmetals can form negative ions with a charge equal to the group number minus 8. For both metals and nonmetals this charge was equated with the oxidation number of the ion. From the discussion of ionization energy, electron affinity, and ion electron configurations, you can now see the reasons for these observations.

The first four ionization energies for sodium, magnesium, and aluminum are listed in Table 8.9. A modest amount of energy (496 kJ/mol) is required to form Na^+(g), while almost 10 times that much is needed to form Na^{2+}(g) because the second electron lost must be a firmly held n = 2 electron. Since energies of this magnitude are not available in normal chemical reactions, the Group 1A element sodium commonly forms only a +1 ion. Similarly, it is energetically too expensive to form Mg^{3+}(g) or Al^{4+}(g), so these elements form only Mg^{2+}(g) and Al^{3+}(g) in normal

TABLE 8.8 Common Monatomic Ions of the Main Group Elements*

	GROUP AND ION						
PERIOD	1A	2A	3A	4A	5A	6A	7A
1	Li^+	Be^{2+}	B	C	N^{3-}	O^{2-}	F^-
2	Na^+	Mg^{2+}	Al^{3+}	Si	P	S^{2-}	Cl^-
3	K^+	Ca^{2+}	Ga^{3+}	Ge	As	Se^{2-}	Br^-
4	Rb^+	Sr^{2+}	In^{3+}	Sn^{2+}	Sb	Te^{2-}	I^-
5	Cs^+	Ba^{2+}	Tl^+, Tl^{3+}	Pb^{2+}	Bi^{3+}		

*Shaded elements with no charge do not commonly form compounds having monatomic ions.

TABLE 8.9 Ionization Energies of Na, Mg, and Al

	IONIZATION ENERGY (kJ/mol)			
ELEMENT	FIRST	SECOND	THIRD	FOURTH
Na	496	4562	6912	9543
Mg	738	1450	7733	10540
Al	577	1816	2744	11577

reactions. This is the reason monatomic ion charges are equal to the group number, and the ions have **rare gas configurations**, electron configurations that are the same as that of the rare gas (Ne) closest to the element in the periodic table.

The heavier elements of Group 3A, gallium for example, also form +3 ions. Here the ion cannot have the same configuration as the rare gas argon, because too many electrons would have to be lost. Instead, only the 4*s* and 4*p* electrons are lost to leave a **pseudo–rare gas configuration**.

$$\text{Ga [Ar]}3d^{10}4s^24p^1 \longrightarrow \text{Ga}^{3+}\ \text{[Ar]}3d^{10} + 3e^-$$

Boron, the lightest element of Group 3A, does not act as a +3 ion in any of its compounds. In Table 8.6 you see that the first ionization energy is 30% higher than that of aluminum, largely because boron has a much higher effective nuclear charge than aluminum (as reflected by the boron atom's small size). Therefore, boron typically forms compounds with covalent bonds, the topic of the next chapter.

In Group 3A we also see for the first time an element having two possible charges or oxidation numbers. Thallium forms compounds with oxidation numbers of +1 (Tl^+) or +3 (Tl^{3+}), and there is some evidence for this in gallium and indium chemistry.

$$\text{Tl [Xe]}4f^{14}5d^{10}6s^26p^1 \longrightarrow \text{Tl}^+\ \text{[Xe]}4f^{14}5d^{10}6s^2 + e^-$$

Similarly, in Group 4A, tin and lead form ions (Sn^{2+} and Pb^{2+}) with a charge lower than the group number. In both Groups 3A and 4A, one or two *p* electrons have been removed to give the ion of lower oxidation number. The pair of *ns* electrons that remains has been called an "inert pair," and the occurrence of an oxidation number 2 less than the group number has been called the **inert pair effect**.* The phenomenon is quite general among the *p* block elements in periods containing transition metals.

Ionization energies and electron affinities increase across a period. Therefore, especially for carbon, silicon, and germanium of Group 4A, appreciable energies are required to form cations, while their electron affinity is not sufficient to lead to negative ions. Like boron, these elements typically form only covalently bonded compounds. Due to the decline in ionization energy down a group, the heavier elements of Group 4A will

*"Inert pair effect" is a misleading name, since the energy required to remove these electrons is not appreciably greater than that for the lighter elements. (The third *IE* for Al is 2744 kJ/mol, whereas that for Tl is 2877 kJ/mol.) Instead, the *ns* pair only appears to be inert, because the energy required for its removal is not compensated by the energy released on forming two additional bonds.

form compounds containing M^{2+}, as the "inert pair effect" is again observed.

In Group 5A the electron affinity of nitrogen is sufficient that N^{3-} ions are observed, but phosphorus, arsenic, and antimony are intermediate in ionization energy and electron affinity and generally do not form simple positive or negative ions. Only bismuth forms a +3 ion.

The elements of Groups 6A and 7A all have sufficiently high electron affinities that they form simple negative ions isoelectronic with the next higher rare gas. For example, the oxide ion O^{2-} has the same electron configuration as Ne ($1s^22s^22p^6$), and Cl^- is isoelectronic with Ar ($1s^22s^22p^63s^23p^6$). Even hydrogen can be considered a pseudohalogen, since it readily adds an electron to form the hydride ion, H^- ($1s^2$), isoelectronic with helium.

Stable, monatomic positive and negative ions have rare or pseudorare gas electron configurations.

In Chapter 2 you learned that transition metals typically form +2 and +3 ions. This is reasonable, since their ionization energies are in the same range as main group metals. To achieve higher charges requires much more energy and is not commonly observed. Finally, with the exception of scandium (Sc) and possibly titanium (Ti), the metal ions do not achieve a rare configuration, since this would require the loss of too many electrons and an impossibly high energy cost.

EXERCISE 8.8 ION CONFIGURATIONS

Write electron configurations for F^-, S^{2-}, and In^+. Do these ions have a rare gas or pseudo–rare gas configuration?

SUMMARY

In the previous chapter three quantum numbers (n, ℓ, and m_ℓ) were used to describe an atomic orbital. To describe an electron completely, however, a fourth quantum number, the **electron spin quantum number m_s**, is needed (Section 8.1). m_s has only two values: $\pm\frac{1}{2}$. Each electron in an atom can now be "addressed" specifically, since *no two electrons in the same atom can have the same set of four quantum numbers* (the **Pauli exclusion principle**; Section 8.2). This leads to the further conclusion that *no atomic orbital can be assigned more than two electrons* and that the two electrons in an orbital must have opposite spins (different values of m_s). Atoms, ions, or compounds having unpaired electrons are **paramagnetic**, that is, they have the property of being attracted to a magnetic field. (**Diamagnetic** atoms, ions, or compounds are repelled from a magnetic field.)

A major goal of the chapter is an understanding of the distribution of electrons in atoms and the connection to the periodic table (Section 8.4). Electrons are generally assigned to orbitals of an atom in order of increasing orbital energy (Section 8.3). In the H atom the orbital energies increase with increasing n, but, in a many-electron atom, the orbital energies depend on both n and ℓ: (1) assignments of electrons to orbitals follow the sequence of increasing ($n + \ell$); (2) for two orbitals with the same $n + \ell$, the one with lower n is assigned first (Figure 8.3). (There are minor exceptions to these guidelines, especially in the heavy transition metals.) The underlying reason for these guidelines is that electron–electron repulsions offset nuclear-electron attraction, the net result of which is interpreted in terms of how atomic orbitals **penetrate** one another and lead to differences in **effective nuclear charge** (**Z^***) for different orbitals.

In placing electrons in atomic orbitals, the Pauli exclusion principle and **Hund's rule** are followed: The most stable arrangement of electrons in a set of orbitals all having the same energy (three *p* orbitals, five *d* orbitals, and so on) is that with the maximum number of unpaired electrons, all electrons having the same spin direction.

Using a periodic table as a guide, electron configurations of the elements can be depicted by an **orbital box notation** or a **spectroscopic notation**. (In both cases, configurations can be abbreviated with the **rare gas notation**.)

Dmitrii Mendeleev founded the modern concept of periodicity (Section 8.5). Mendeleev realized that many elements remained undiscovered in the 19th century and predicted their properties. Moseley recognized the true **law of chemical periodicity**: The properties of the elements are periodic functions of their atomic numbers.

Important physical properties of the elements are: **size**, **ionization energy** (*IE*, the energy required to remove an electron from an atom), and **electron affinity** (*EA*, the energy involved when a gaseous atom absorbs an electron to form a negative ion of the element; Section 8.6). Their general periodic trends are

1. Atomic size: decreases across a period and increases down a group.
2. Ionization energy: increases across a period and decreases down a group.
3. Electron affinity: increases across a period and decreases down a group.

Ion size is also important (Section 8.7). Negative ions are larger than the elements from which they are derived, while positive ions are smaller. Trends in ion sizes parallel those of the neutral atoms. Ion charges can be rationalized in terms of ionization energy and electron affinity (Section 8.7).

STUDY QUESTIONS

REVIEW QUESTIONS

1. Give the four quantum numbers, specify their allowed values, and tell what property of the electron they describe.
2. What is the Pauli exclusion principle?
3. What is the "$n + \ell$" guideline? Use it to tell to which orbital an electron is assigned first, 5*s* or 4*d*.
4. What is the shielded charge felt by an electron? How is this related to effective nuclear charge?
5. Give an example of orbital penetration and its effect.
6. Using lithium as an example, show the two methods of depicting electron configurations (orbital box diagram and spectroscopic notation).
7. What is Hund's rule? Give an example of the use of this rule.
8. What is the rare gas notation? Write an electron configuration using this notation.
9. Name at least three people connected with the historical development of the periodic table.
10. What was Mendeleev's major contribution to the development of the concept of periodicity?
11. Tell what happens to atomic size, ionization energy, and electron affinity when proceeding across a period and down a group.
12. Explain how the size of atoms change and why they change when proceeding across a period of the periodic table.

13. Explain how the ionization energy of atoms changes and why the change occurs when proceeding down a group of the periodic table.

14. Explain why the sizes of transition metal atoms are nearly identical across a period.

15. Write electron configurations to show the first two ionization processes for potassium. Explain why the second ionization energy is much higher than the first.

16. Explain why the ionization energies of Si, P, and S are in the order Si $<$ P $>$ S.

17. Why is the radius of Li^+ so much smaller than the radius of Li? Why is the radius of F^- so much larger than the radius of F?

18. What is the origin of the paramagnetic effect? Is Li paramagnetic? What about Li^+? Is either of these diamagnetic?

19. Name three common ions that are isoelectronic with Ne.

20. Which ions in the following list are likely to be formed: K^{2+}, Cs^+, Al^{4+}, F^{2-}, and Se^{2-}? Do any of these ions have a rare gas configuration?

WRITING ELECTRON CONFIGURATIONS

21. Write electron configurations for the following elements using both the spectroscopic notation and orbital box diagrams: (a) Mg, (b) Al, (c) Cl.

22. Using the spectroscopic notation, give the electron configuration of vanadium, V. (The name of the element was derived from *vanadis*, a Scandinavian goddess.) Compare your answer with Table 8.2.

23. Using the spectroscopic notation, write the electron configurations of chromium and iron.

24. Germanium was an element that had not been discovered when Mendeleev formulated his ideas of chemical periodicity. He predicted its existence, however, and it was indeed found in 1886 by Winkler. Depict its electron configuration using the spectroscopic notation.

25. Using orbital box diagrams (and the rare gas notation), compare the electron configurations of the first four elements of the halogen group.

26. Using orbital box diagrams, depict the electron configurations of the following ions: (a) Na^+, (b) Al^{3+}, and (c) Cl^-.

27. Using the spectroscopic notation, write electron configurations for the following:

(a) Strontium, Sr, named for a town in Scotland.

(b) Zirconium, Zr. This metal is exceptionally resistant to corrosion and so has important industrial applications. Moon rocks show a surprisingly high zirconium content compared with rocks on the earth.

(c) Rhodium, Rh, used in jewelry and in catalysts in industry.

(d) Tin, Sn. A metal used in the ancient world. Alloys of tin (solder, bronze, and pewter) are important.

28. The lanthanides, or rare earths, are now only "medium rare." All can be purchased for a reasonable price. Depict electron configurations for the following elements:

(a) Europium, Eu. It is the most expensive of the rare earth elements; one gram can be purchased for \$50 to \$100.

(b) Ytterbium, Yb. It is less expensive than Eu, as Yb costs only about \$15 per gram. It was named for the village of Ytterby in Sweden where a mineral source of the element was found.

29. Use the rare gas and spectroscopic notations to show electron configurations for the following metals of the third transition series.

(a) Tungsten, W. The element finds extensive use in the filaments of electric lamps and television tubes. It has the highest melting point of all the elements.

(b) Platinum, Pt, used by pre-Colombian Indians in jewelry. It does not oxidize in air, no matter how high the temperature. Therefore, it is used to coat missile nose cones and in jet engine fuel nozzles.

30. The actinide americium (Am) is a man-made, radioactive element that has found use in home smoke detectors. Using the rare gas notation and spectroscopic notation, depict its electron configuration.

31. Depict electron configurations for the following elements of the actinide series of elements. Use the rare gas and spectroscopic notations.

(a) Plutonium, Pu. It is one of the most dangerous radiological poisons known because of the high rate at which it evolves alpha particles. It is best known as the explosive material in nuclear weapons and as a by-product of nuclear power plant operation.

(b) Einsteinium, Es. The element was named for the famous physicist Albert Einstein.

32. Among the last elements of the periodic table are those with atomic numbers 104 through 109. Element 104 was originally given the name rutherfordium, Rf, to honor the physicist Rutherford (Chapter 7). Depict its electron configuration using the spectroscopic and rare gas notations.

33. Element 109 is the latest to be discovered. It was produced in August, 1982, by a team at Germany's Institute for Heavy Ion Research. Depict its electron configuration using the spectroscopic and rare gas notations. Name another element found in the same group as 109.

34. Using orbital box diagrams and the rare gas notation, depict the electron configurations of: (a) Ti, (b) Ti^{2+}, and (c) Ti^{4+}.

35. Element 25 can be found at the bottom of the sea in the form of oxide "nodules."
    (a) Depict the electron configuration of this element using the spectroscopic notation and an orbital box diagram.
    (b) Using an orbital box diagram, show the electrons beyond those of the preceding rare gas for the +2 ion.

36. Pt^{2+} is the central ion in a new reagent for treating cancer, $(NH_3)_2PtCl_2$. Depict the electron configuration of Pt^{2+} using the rare gas notation and an orbital box diagram.

37. Cobalt commonly exists as +2 and +3 ions. Using orbital box diagrams and the rare gas notation, show electron configurations of these ions.

38. The most common oxidation state of indium, In, is +3. Using the spectroscopic and rare gas notations, give the electron configuration of the ion.

39. The rare earths or lanthanides commonly exist as +3 ions. Using an orbital box diagram, and the rare gas notation, show the electron configurations of the following:
    (a) Sm and Sm^{3+} (samarium)
    (b) Ho and Ho^{3+} (holmium)

40. Suppose a new element with atomic number 113 has been discovered.
    (a) Predict its electron configuration using the rare gas and spectroscopic notations.
    (b) Name another element in the periodic group in which this new element would be placed.

ELECTRON CONFIGURATIONS AND QUANTUM NUMBERS

41. Depict the electron configuration for Be using the orbital box method. Give a complete set of four quantum numbers for each of the electrons in this atom.

42. Depict the electron configuration for silicon using the orbital box and rare gas notations. Give a complete set of quantum numbers for each electron beyond those of the preceding rare gas.

43. Using an orbital box diagram and the rare gas notation, show the electron configuration of Ti. Give one possible set of four quantum numbers for each of the electrons beyond those of the preceding rare gas.

44. Using an orbital box diagram, show the electron configuration of gallium, Ga, beyond those of the preceding rare gas. Give a set of quantum numbers for the last electron added.

45. A neutral atom has two electrons of $n = 1$, 8 electrons with $n = 2$, 8 electrons with $n = 3$, and 1 electron with $n = 4$. Assuming this element is in its ground state, supply the following information: (a) atomic number; (b) total number of *s* electrons; (c) total number of *p* electrons; and (d) total number of *d* electrons.

46. What is the maximum number of electrons that can be identified with each of the following sets of quantum numbers? In some cases, the answer may be "none." In such cases, explain why this is the correct answer.
    (a) $n = 2$ and $\ell = 1$.
    (b) $n = 5$
    (c) $n = 4$ and $\ell = 4$.
    (d) $n = 3$, $\ell = 1$, and $m_\ell = -1$
    (e) $n = 4$ and $\ell = 2$
    (f) $n = 3$, $\ell = 0$, $m_\ell = +1$

47. What is maximum number of electrons that can be associated with the following sets of quantum numbers? In one case, the answer is "none." Explain why this is true.
    (a) $n = 4$, $\ell = 3$
    (b) $n = 6$, $\ell = 1$, $m_\ell = -1$
    (c) $n = 3$, $\ell = 3$, $m_\ell = -3$
    (d) $n = 2$, $\ell = 1$, $m_\ell = 1$, $m_s = +\frac{1}{2}$

48. Explain briefly why each of the following is *not* a possible set of quantum numbers for an electron in an oxygen atom.
    (a) $n = 2$, $\ell = 2$, $m_\ell = 0$, $m_s = +\frac{1}{2}$
    (b) $n = 2$, $\ell = 1$, $m_\ell = -1$, $m_s = 0$

49. Which of the following is *not* an allowable set of quantum numbers? Explain your answer briefly.

	n	ℓ	m_ℓ	m_s
(a)	2	0	0	$-\frac{1}{2}$
(b)	1	1	0	$+\frac{1}{2}$
(c)	2	1	−1	$-\frac{1}{2}$
(d)	6	5	5	$-\frac{1}{2}$

50. No known elements have electrons assigned to 7*p* orbitals in the ground state. If these orbitals are utilized by electrons in a yet to be discovered element, how many 7*p* electrons can there be? 6

51. In the fictitious element with atomic number 120, what is the last orbital assigned?

52. How many 7*s* electrons are there in element 104? 2

53. How many *complete* electron shells are there in element 71?

54. For which element is the last electron placed in the third ($n = 3$) quantum shell? Ar

55. How many *p* orbital electron pairs are there in selenium, Se?

56. Which of the following sets of quantum numbers could describe the outermost electrons in the ground state of an atom?

ELECTRON	n	ℓ	m_ℓ	m_s
(a) First	1	1	1	$+\frac{1}{2}$
Second	1	1	1	$-\frac{1}{2}$
(b) First	2	0	0	$+\frac{1}{2}$
Second	2	0	0	$-\frac{1}{2}$
Third	2	1	−1	$+\frac{1}{2}$
Fourth	2	1	−1	$+\frac{1}{2}$
(c) First	2	0	0	$+\frac{1}{2}$
Second	2	0	0	$-\frac{1}{2}$

57. Consider the sets of quantum numbers in Question 56. Which set(s) does(do) not obey the quantum number rules?

58. A possible excited state for the H atom has an electron in a 4*p* orbital. List all possible sets of quantum numbers (n, ℓ, m_ℓ, and m_s) for this electron.

59. Write down a complete set of quantum numbers for each of the electrons beyond the nearest rare gas for scandium (Sc).

PERIODIC PROPERTIES

60. Use the data in Figure 8.12 to estimate E—Cl bond distances when E is a Group 4A element.

61. Estimate the Xe—F bond distance in XeF_2 from the information in Figure 8.12. (Known Xe—F distances are in the range of 190 pm.)

62. Arrange the following elements in order of increasing size: Al, B, C, K, and Na. (Try doing it without looking at Figure 8.12, then check yourself by looking up the necessary atomic radii.)

63. Select the atom or ion in each pair that has the larger radius.
(a) Cl or Cl^-
(b) Al or N
(c) In or Sn

64. Place the following elements and ions in order of decreasing size: Ar, K^+, Cl^-, S^{2-}, and Ca^{2+}.

65. The ionization energies for the removal of the first electron in Si, P, S, and Cl are as listed below. Briefly rationalize this trend.

ELEMENT	FIRST IONIZATION ENERGY (kJ/mol)
Si	780
P	1060
S	1005
Cl	1255

66. Rank the following ionization energies in order from the smallest value to the largest value. Briefly explain your answer.
(a) first *IE* of Be
(b) first *IE* of Li
(c) second *IE* of Be
(d) second *IE* of Na
(e) first *IE* of K

67. Rank the following in order of increasing ionization energy: Zn, Ca, Ca^{2+}, and Cl^-. Briefly explain your answer.

68. Predict which of the following elements would have the greatest difference between the first and second ionization energy: Si, Na, P, and Mg. Briefly explain your answer.

69. In general as you move across the periodic table, the electron affinity of the elements increases. One exception to this trend, however, is that there is a large decrease in electron affinity when going from Group 4A elements to those in Group 5A. Refer to orbital box diagrams to explain why this decrease is plausible and indeed expected.

70. Explain why the first ionization energy of Mg is greater than that of Na, whereas the second ionization energy of Mg is lower than the second ionization energy of Na.

71. Which of the following groups of elements is arranged correctly in order of increasing ionization energy?
(a) C < Si < Li < Ne
(b) Ne < Si < C < Li
(c) Li < Si < C < Ne
(d) Ne < C < Si < Li

72. The following are isoelectronic species: Cl^-, Ar, and K^+. Rank them in order of increasing (a) size, (b) ionization energy, and (c) electron affinity.

73. Compare the elements Li, K, C, and N.
(a) Which has the largest atomic radius?
(b) Which has the largest electron affinity?
(c) Place the elements in order of increasing ionization energy.

74. Compare the elements B, Al, C, and Si.
(a) Which has the most metallic character?
(b) Which has the largest atomic radius?
(c) Which has the largest electron affinity?
(d) Place the three elements B, Al, and C in order of increasing first ionization energy.

75. Periodic trends. Explain each answer briefly.
(a) Place the following elements in order of increasing ionization energy: F, O, and S.
(b) Which has the largest ionization energy: O, S, or Se?
(c) Which has the largest electron affinity: Se, Cl, or Br?
(d) Which has the largest radius: O^{2-}, F^-, or F?

76. Periodic trends. Explain each answer briefly.
(a) Rank the following in order of increasing atomic radius: O, S, and F.
(b) Which has the largest ionization energy: P, Si, S, or Se?
(c) Place the following in order of increasing radius: Ne, O^{2-}, N^{3-}, or F^-.
(d) Place the following in order of increasing ionization energy: Cs, Sr, Ba.

77. How many unpaired electrons does Ti^{2+} have? The Co^{3+} ion? Is either of these paramagnetic?
78. Which of the following atoms or ions is paramagnetic and which is diamagnetic: Al, Al^{3+}, Ba, V, V^{3+}?
79. Are any of the +2 ions of the elements Ti through Zn diamagnetic? Which +2 ion has the greatest number of unpaired electrons?
80. Two elements in the first transition series (Sc through Zn) have four unpaired electrons in their +2 ions. What elements fit this description?
81. Two elements in the second transition series (Y through Cd) have four unpaired electrons in their +3 ions. What elements fit this description?
82. What periodic groups of elements can form +2 ions? (Consider only main group elements.)
83. What periodic groups of elements can form +3 ions? (Consider only main group elements.)
84. Which of the following ions are unlikely and why: Cs^+, In^{4+}, Fe^{6+}, Te^{2-}, Sn^{5+}, and I^-?

GENERAL

85. Name the element corresponding to each characteristic below:
    (a) The element with the electron configuration $1s^22s^22p^63s^23p^4$.
    (b) The element in the alkaline earth group that has the largest atomic radius.
    (c) The element in Group 5A that has the largest ionization energy.
    (d) The element whose +2 ion has the configuration $[Kr]4d^6$.
    (e) The element with the lowest electron affinity in Group 7A.
    (f) The element whose electron configuration is $[Ar]3d^{10}4s^1$.
    (g) The element whose electron configuration is $[Xe]4f^75d^16s^2$.

86. Answer the questions below about the elements A and B, which have the electron configurations shown.

    A = . . . $4p^65s^1$ B = . . . $3p^63d^{10}4s^24p^4$

    (a) Is element A a metal, nonmetal, or metalloid?
    (b) Which element would have the greater ionization energy?
    (c) Which element should have the greater electron affinity?
    (d) Which element has larger atoms?

87. Answer the following questions about the elements with the electron configurations below.

    A = . . . $3p^64s^2$ B = . . . $3p^63d^{10}4s^24p^5$

    (a) Is element A a metal, metalloid, or nonmetal?
    (b) Is element B a metal, metalloid, or nonmetal?
    (c) Which element is expected to have the larger ionization energy?
    (d) Which element should have the greater electron affinity?
    (e) Which element should be the smaller of the two?

88. Suppose a new element, tentatively given the symbol Et, has just been discovered. Its atomic number is 113.
    (a) Depict the electron configuration of the element using the spectroscopic notation.
    (b) Name another element you would expect to find in the same group as Et.
    (c) Give the formulas for the compounds of Et with O and Cl.

89. Complete the following table of properties for the three as yet "undiscovered" elements: J, Ok, and Zt. All have atomic numbers greater than 105.

SYMBOL	VALENCE ELECTRON CONFIGURATION	GROUP	POSSIBLE CHARGE ON ION	METAL, METALLOID, OR NONMETAL
J	ns^2np^6	____	____	____
Ok	$ns^2(n-1)d^1$	____	____	____
Zt	ns^2np^2	____	____	____

9

Basic Concepts of Bonding and Molecular Structure

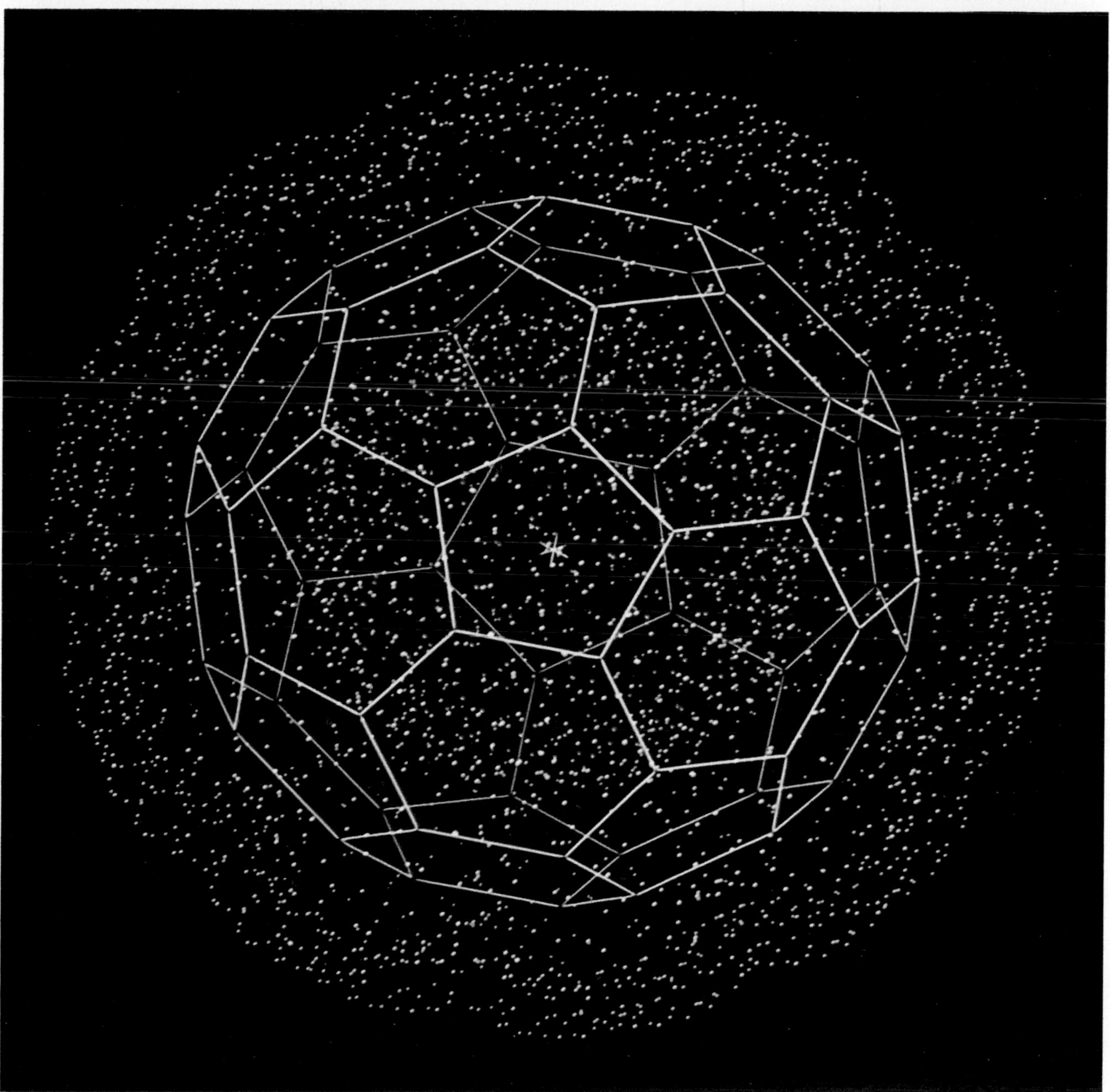

Computer drawn structure of C_{60} molecule with La in the center. (John C. Spurlino and Florentine A. Quiocho, Rice University/*CE News*)

CHAPTER OUTLINE

In Chapters 7 and 8 you developed an understanding of what we think electrons do in atoms. It is now time to investigate what happens when two or more atoms come close enough to form a molecule and how this behavior explains the three-dimensional shape of molecules.

9.1 VALENCE ELECTRONS

The outermost electrons of an atom, the ones most affected by the approach of another atom, are called **valence electrons**. The rare gas core electrons and the filled *d* shell electrons of the Group 3A elements are not greatly affected by reactions with other atoms, so we focus our attention on the behavior of the outer *ns* and *np* electrons (and *d* electrons in unfilled subshells of the transition metals). This means the valence electrons for a few typical elements are

ELEMENT	CORE ELECTRONS	VALENCE ELECTRONS	PERIODIC GROUP
Na	$1s^22s^22p^6$	$3s^1$	1A
Si	$1s^22s^22p^6$	$3s^23p^2$	4A
Ti	$1s^22s^22p^63s^23p^6$	$4s^23d^2$	4B
As	$1s^22s^22p^63s^23p^63d^{10}$	$4s^24p^3$	5A

Here you see that the *number of valence electrons of each element is equal to the group number*. The fact that every element in a given group has the same number of valence electrons accounts for the similarity of chemical properties among members of the group.

A useful device for keeping track of the valence electrons of *main group elements* is the **Lewis electron dot symbol**, first suggested by G.N. Lewis, about 1916. In this notation, the nucleus and core electrons are represented by the atomic symbol. The valence electrons, represented by dots, are then placed around the symbol one at a time until they are used up or until all four sides are occupied; any remaining electrons are paired with the ones already there. The Lewis symbols for the first two periods are

The two *ns* electrons of the Group 2A through 8A atoms are paired in the atom's electron configurations, but the Lewis symbol shows them as unpaired. This will be useful later.

1A	2A	3A	4A	5A	6A	7A	8A
ns^1	ns^2	ns^2np^1	ns^2np^2	ns^2np^3	ns^2np^4	ns^2np^5	ns^2np^6
· Li	· Be ·	· B ·	· C ·	· N ·	: O ·	: F :	: Ne :
· Na	· Mg ·	· Al ·	· Si ·	· P ·	: S ·	: Cl :	: Ar :

GILBERT NEWTON LEWIS 1875–1946

In a paper published in the *Journal of the American Chemical Society* in 1916, Gilbert N. Lewis introduced the theory of the shared electron pair chemical bond and revolutionized chemistry. It is to honor this contribution that we often refer to "electron dot" structures as Lewis structures. However, he also made major contributions to other fields such as thermodynamics, isotope studies, and the interaction of light with substances. Of particular interest in this text is the extension of his theory of bonding to a *generalized theory of acids and bases*. This theory, often referred to as Lewis acid–base theory, will be described in Chapter 15.

G.N. Lewis was born in Massachusetts but raised in Nebraska. After earning his B.A. and Ph.D. at Harvard University, he began his academic career. In 1912 he was appointed Chairman of the Chemistry Department at the University of California, Berkeley, and he remained there for the rest of his life. Lewis felt that a chemistry department should both teach and advance fundamental chemistry, and he was not only a productive researcher but also a teacher who profoundly affected his students. Among his ideas was the large use of problem sets in teaching, an idea that we see in great use today.

The Lewis symbol emphasizes the rare gas configuration, ns^2np^6, as a stable, low-energy state. In fact, the bonding behavior of the main group elements can often be considered the result of gaining, losing, or sharing valence electrons to achieve the same configuration as the nearest rare gas. Since all rare gases (except He) have eight valence electrons, this observation is called the **octet rule**. It is only a guideline, so there are many exceptions, but it provides an easy way to predict the results of the most common reactions.

9.2 CHEMICAL BOND FORMATION

In Chapters 2 and 8 we discussed the formation of ions and ionic compounds. In those cases where metals from the left side of the periodic table interact with nonmetals from the far right side, the result is usually the complete transfer of one or more electrons from one atom to another and the creation of an **ionic bond**.

$$\text{Na}\cdot + :\ddot{\underset{\cdot}{\text{Cl}}}: \longrightarrow \text{Na}^+ + :\ddot{\underset{\cdot\cdot}{\text{Cl}}}:^- \qquad \text{(ionic compound)}$$

Guidelines for predicting which bonds are ionic will be described in Section 9.4. Primarily, covalent bonding is described in this chapter; more will be said about ionic bonding in Chapter 11.

When the elements lie closer together in the periodic table, however, electrons are more often *shared between atoms* and the result is a **covalent bond**.

$$:\ddot{\underset{\cdot\cdot}{\text{I}}}\cdot + \cdot\ddot{\underset{\cdot\cdot}{\text{Cl}}}: \longrightarrow :\ddot{\underset{\cdot\cdot}{\text{I}}}:\ddot{\underset{\cdot\cdot}{\text{Cl}}}: \qquad \text{(covalent compound)}$$

Guidelines for predicting which bonds are ionic will be described in Section 9.4 Primarily, covalent bonding is described in this chapter; more will be said about ionic bonding in Chapter 11.

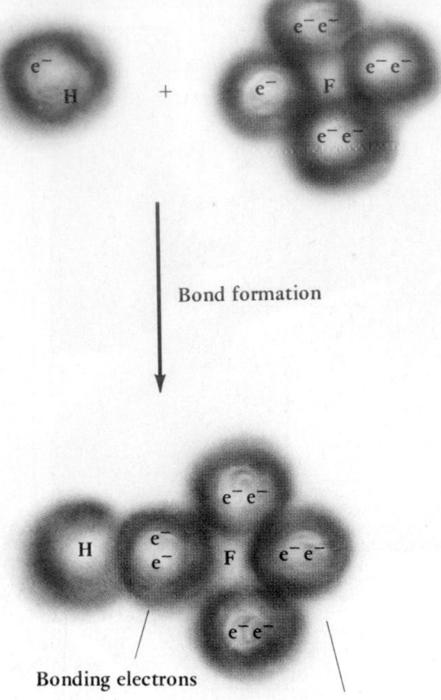

FIGURE 9.1

The formation of a covalent bond between an H atom and a F atom. One pair of electrons (one electron from each atom) flows into the internuclear region and is attracted to both the H and the F nuclei. It is this mutual attraction of 2 (or sometimes 4 or 6) electrons by two nuclei that leads to a force of attraction between two atoms and assists in bond formation.

Covalent bond formation can be approached on two levels. The simpler goes no further than recognizing the repulsion between like electrical charges and the attraction between opposite charges. When two atoms approach each other closely enough for their electron clouds to interpenetrate, the electrons and nuclei repel each other; at the same time, however, each atom's electrons attract the other atom's nucleus. If the total attractive force is greater than the total repulsive force, a covalent bond is formed (Figure 9.1).

The second view of bonding, based on quantum mechanics, is more adaptable to mathematical analysis, but a bit harder to visualize. Here we imagine combining an atomic orbital (that is, a wave function) from each of the two atoms to form a **bond orbital**. Like an atomic orbital, a bond orbital can describe either one or two electrons; if there are two electrons (a **bond pair**) in a bond orbital, they must be paired with opposite spins. All the valence electrons of the atoms not involved in bonding are described by **lone-pair orbitals**, which are concentrated outside the bond region.

The more the attraction between the bonding electrons and the nuclei exceeds the repulsion between the nuclei and between lone electron pairs, the stronger the bond will be between the atoms. Of course, a stronger bond means a more stable molecule with a lower potential energy. The diagram in Figure 9.2 shows the energy changes as a pair of electrons initially associated with separate H atoms becomes a bonding pair in H_2. The energy of the electrons in a bond orbital, where the electrons are attracted by two nuclei, is lower than their energy in valence atomic orbitals.* Thus, in general, electrons fall to a lower potential energy when they become bonding electrons, and this energy is given off in the form of heat and/or light.

*There is also a quantum mechanical effect related to the size of the bond region compared to the size of the atomic orbital; because the electron is free to move in a larger space, its kinetic energy is lower. This effect is quite important in explaining certain types of bonds, but we shall not explore it further.

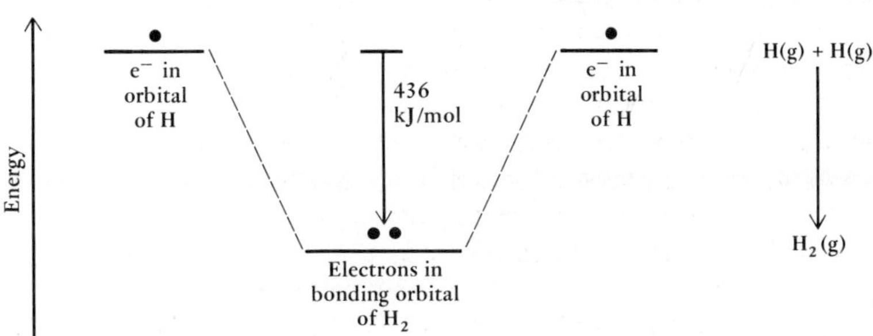

FIGURE 9.2

Energy charges occurring in the course of bond formation between two H atoms.

9.3 NUMBER OF BOND PAIRS AROUND AN ATOM

THE OCTET RULE

The view of covalent bonding just described implies that each unpaired valence electron in the Lewis structure of an isolated atom is available for sharing with another atom to form one bond. Thus, to predict the number of bonds an atom can form, just count unpaired electrons. From the table of Lewis structures in Section 9.1 you can see that the number of unpaired electrons on an atom of Groups 4A through 8A is just 8 minus the group number. (The number of unpaired electrons is equal to the group number for Groups 1A to 3A, but these elements, except boron, usually form ionic rather than covalent compounds.) For example, oxygen in Group 6A has $8 - 6 = 2$ unpaired electrons and forms 2 bonds.

Table 9.1 shows Lewis formulas for a number of covalent compounds. As is customary, bonding pairs are represented by dashes instead of a pair of dots, to distinguish them from lone pairs. Notice that atoms with four or more valence electrons have four pairs of valence electrons (total of bond pairs and lone pairs) around that atom, as predicted by the Lewis rule.

TABLE 9.1 Examples of Lewis Structures by Periodic Group*

	1A	2A	3A	4A	5A	6A	7A	8
Lewis structure	H—H	H—Be—H	H—B(—H)—H	H—C(—H)(—H)—H	H—N̤(—H)—II	H—Ö(—H):	H—F̤̈:	:N̤̈e:
Chemical formula	H_2	BeH_2	BH_3†	CH_4	NH_3	H_2O	HF	Ne
Bond pairs	1	2	3	4	3	2	1	0
Lone pairs	0	0	0	0	1	2	3	4
Total pairs of electrons at central atom	1	2	3	4	4	4	4	4
Lewis structure			:F̤̈—B(—F̤̈:)—F̤̈:	H₂C=CH₂ (H—C(—H)=C(—H)—H)	:N≡N:	:C̤̈l—S̤(—C̤̈l:):	:C̤̈l—F̤̈:	
Chemical formula			BF_3	C_2H_4	N_2	SCl_2	ClF	
Bond pair			3	4	3	2 1	1 1	
Lone pairs			0	0	1	2 3	3 3	
Total pairs of electrons at central atom			3	4	4	4 4	4 4	

*In each case, the number of bond pairs of electrons, lone pairs of electrons, and total pairs of electrons around atoms other than H are given.

†BH_3 is unstable. The simplest, isolable B—H containing molecule is B_2H_6.

EXERCISE 9.1 OCTET RULE

Which of the following combinations of lone pairs and bond pairs around a single atom A are consistent with the octet rule?

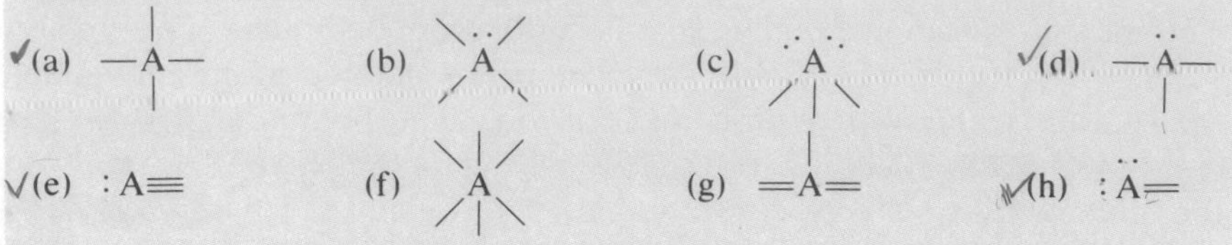

(a) —A— (b) A (c) A (d) —A—

(e) :A≡ (f) A (g) =A= (h) :A=

SINGLE AND MULTIPLE BONDS Some of the examples in Table 9.1 have a single pair of electrons (a **single** bond) between atoms. Single bonds are also called **sigma bonds**, symbolized by the Greek letter σ. Other structures in Table 9.1 indicate two or three electron pairs (a multiple bond) between the same pair of atoms. In a **double** or **triple** bond, one of the bonds is a sigma bond, but the second (and third if present) is a **pi bond**, denoted by the Greek letter π. *Multiple bonds are most often formed by C, N, O, and S atoms.*

Sigma and pi bonds differ in shape and strength, which makes them behave differently in chemical reactions (Chapter 10).

EXERCISE 9.2 SIGMA AND PI BONDS

For each Lewis structure in Table 9.1, give the number of sigma and pi bonds around each atom that is not hydrogen. For example, C in C_2H_4 (ethylene) has three sigma bonds and one pi bond.

COORDINATE COVALENT BONDS In all of the compounds shown so far, each atom contributes one unpaired electron to a bond pair, as in

$$H\cdot + \cdot H \longrightarrow H:H$$

Some elements, such as nitrogen and phosphorus, tend to share a lone pair with another atom that is short of electrons, leading to the formation of a **coordinate covalent bond**:

$$H^+ + :NH_3 \longrightarrow [NH_4]^+$$

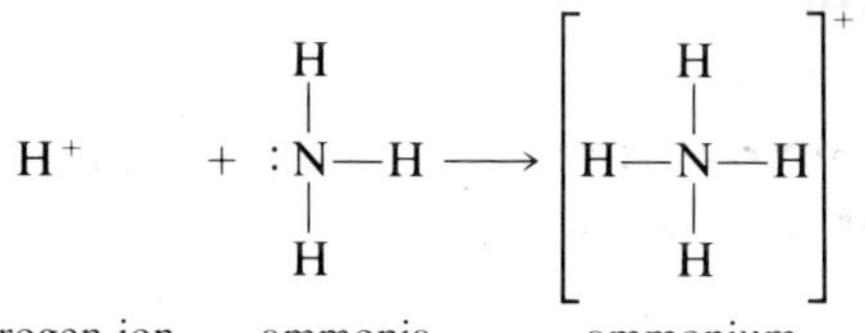

hydrogen ion (no electrons) + ammonia molecule ⟶ ammonium ion

Once such a bond is formed, it is the same as any other bond; in the ammonium ion, for instance, all four bonds are identical.

EXCEPTIONS TO THE OCTET RULE

FEWER THAN EIGHT VALENCE ELECTRONS In Table 9.1, several compounds have fewer than four pairs in the valence shell. Hydrogen, of course, can accommodate at most two electrons in its shell, so it shares only two electrons with another atom. In BeH_2 there are only four valence electrons about Be, and there are only six about boron in BH_3 and BF_3. These facts

TABLE 9.2 Examples of Lewis Structures in Which the Central Atom Has an Expanded Valence*

	GROUP 4A	GROUP 5A	GROUP 6A	GROUP 7A	GROUP 8
Molecules with five total pairs		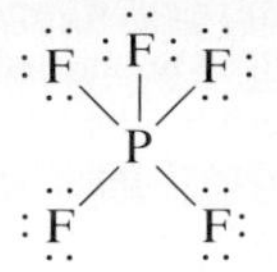 P (with 5 F)	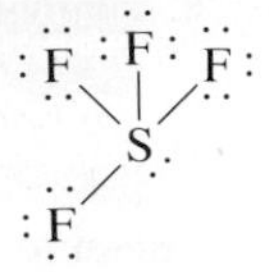 S (with 4 F)	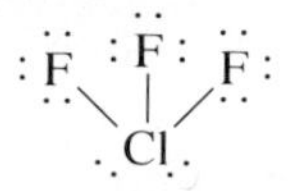 Cl (with 3 F)	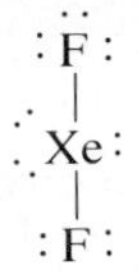 Xe (with 2 F)
Bond pairs		5	4	3	2
Lone pairs		0	1	2	3
Molecules with six total pairs	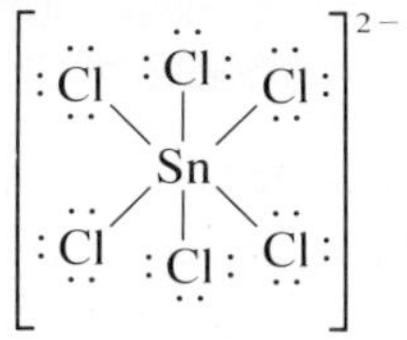 $[SnCl_6]^{2-}$	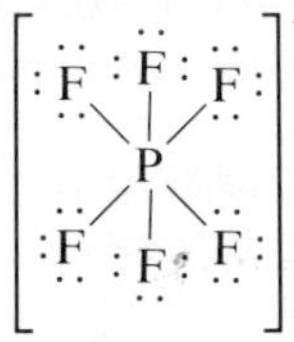 $[PF_6]^{-}$	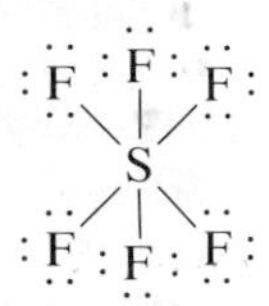 S (with 6 F)	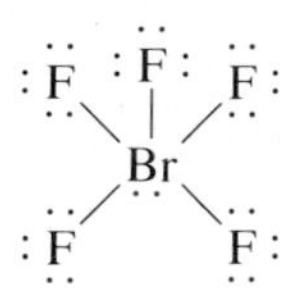 Br (with 5 F)	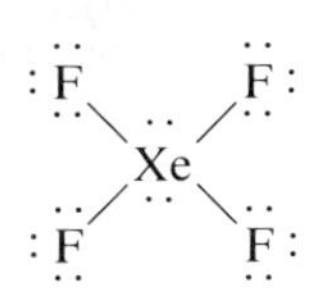 Xe (with 4 F)
Bond pairs	6	6	6	5	4
Lone pairs	0	0	0	1	2

*In each case, the number of bond pairs and lone pairs about the central atoms are given.

account for the ready formation of coordinate covalent bonds such as that in $H_3N—BF_3$; the N atom lone pair is shared with boron, and both N and B now have an octet of electrons.

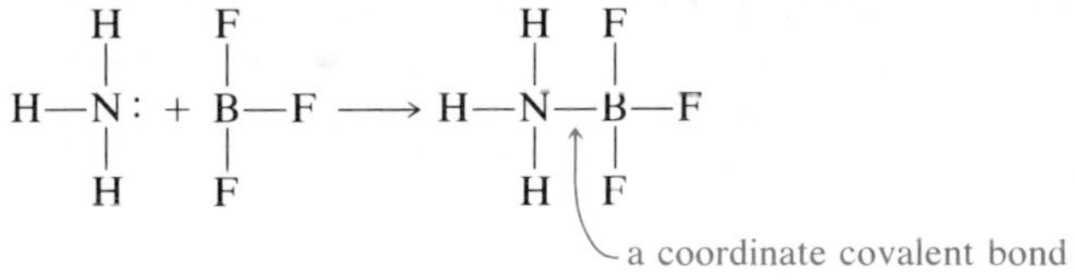

$$H_3N: + BF_3 \longrightarrow H_3N{\rightarrow}BF_3$$

a coordinate covalent bond

EXPANDED VALENCE Elements of the third or higher periods can be surrounded by more than four valence pairs in certain compounds and are said to display **expanded valence**. The number of bonds formed depends on a balance between the ability of the nucleus to attract electrons and the repulsion between the pairs. Examples of such compounds are shown in Table 9.2.

ODD-ELECTRON COMPOUNDS A *few* compounds contain an odd number of electrons, and thus cannot obey the octet rule. The molecule NO, for example, has 5 valence electrons from N and 6 from O, for a total of 11 valence electrons. Other examples are NO_2 and ClO_2.

According to the usual rules, O_2 should have the Lewis structure $:\ddot{O}{=}\ddot{O}:$. However, experiment shows the molecule has two *unpaired* electrons. This result cannot be explained by Lewis theory (see Section 10.2).

DRAWING LEWIS STRUCTURES

It is very important that you learn to draw Lewis structures. Among other things, they can help us predict the molecular structure, that is, the way the atoms are positioned in space relative to each other. The following are guidelines for drawing Lewis structures, and some examples follow.

GUIDELINES FOR DRAWING LEWIS STRUCTURES

1. To predict the arrangement of atoms within the molecule, use the following rules:

A better criterion for the central atom is that it should be the atom of lowest electronegativity. The concept of electronegativity is introduced in Section 9.4.

Each negative charge on an ion represents an additional valence electron, and each positive charge represents the loss of an electron.

  a. H is always an end or *terminal* atom. It is connected to only one other atom, never to two or more (see Table 9.1).*
  b. Simple compounds containing an element bonded to oxygen or halogen have these latter atoms as terminal atoms. (In cases where the halogen and oxygen are bonded together, the oxygen is usually the terminal atom.)
  c. The atom of lowest electron affinity in the molecule or ion is the central atom.
2. Having decided on the general arrangement of atoms, count the number of valence electrons in the molecule by adding up the group numbers of the elements. For ions, also *add* to the sum of group numbers the ion charge for a negative ion or *subtract* the ion charge for a positive ion. The number of valence electron pairs to be used in the Lewis structure is *half* the total number of electrons.
3. Place one pair of electrons (a sigma bond) between each pair of bonded atoms.
4. Subtract from the total number of valence electron pairs the number of bonds you drew in step 3. This is the pool of electron pairs from which you form the lone pairs, as well as any pi bond pairs if necessary.
5. Place lone pairs about each terminal atom (except H) to satisfy the octet rule.
  a. If pairs are still left at this point, assign them to the central atom. If the central atom is from the third or higher period, it can accommodate more than four electron pairs.
  b. If the central atom is not yet surrounded by four electron pairs, convert one or more terminal atom lone pairs to pi bond pairs. In this way the pairs are still associated with the terminal atom, and they are also a part of the central atom. Not all elements form pi bonds. You can use the general rule that *only C, N, O, and S form a pi bond to another atom of the same element or with another atom of this group of four elements.* That is, there can be bonds such as C=C, C=N, S=O, and so on.

EXAMPLE 9.1

DRAWING LEWIS STRUCTURES

Draw Lewis structures for ammonia, NH_3; formaldehyde, H_2CO; the hypochlorite ion, OCl^-; and the nitronium ion, NO_2^+.

Solution for Ammonia (a) The N atom is clearly the central atom.

(b) The total number of valence electrons is the sum of the group numbers: 5 (for N) + 3 (1 for each H) = 8. There are 4 valence pairs.

(c) Form a single bond between each pair of atoms, using 3 pairs of electrons.

```
H—N—H
  |
  H
```

(d) After sigma bond formation, one pair of electrons remains, and this becomes a lone pair on the central atom.

```
  ..
H—N—H
  |
  H
```

*There are some exceptions to this statement, especially in the series of B/H-containing compounds called boron hydrides. You will not encounter them until Chapter 21, however.

Each H atom now has a share in one pair of electrons as required, while the central N atom has a share in four electron pairs. Three of the nitrogen pairs are shared sigma bonding electron pairs, while the fourth is a lone electron pair.

Solution for Formaldehyde (a) Carbon, the atom of lowest electron affinity, is the central atom.

(b) There is a total of 12 valence electrons or six pairs: 2 (1 for each H) + 4 (for C) + 6 (for O) = 12.

(c) Form a single bond between each pair of atoms, using 3 of the 6 pairs.

H—C—H
|
O

(d) The 6 electrons that remain are placed around the O atom as lone pairs.

H—C—H
|
:Ö:

(e) The C atom is deficient by one pair. The solution is to make one of the lone pairs on the O atom into a bonding pair, making a pi bond between O and C. Now both C and O have a share in four pairs of electrons.

H—C—H ⟶ H—C—H
| ‖
:Ö: :O:

The carbon has a share in three sigma bond pairs and one pi bond pair, while the O atom has two lone pairs and shares one sigma and one pi bond pair.

Solution for OCl^- Hypochlorite ion has 14 valence electrons [6 (for O) + 7 (for Cl) + 1 (for ion charge) = 14] or 7 valence electron pairs. After forming the O—Cl bond, and distributing the remaining 6 pairs around the "terminal" atoms, both atoms have a share in four electron pairs as required.

[:Ö—C̈l:]$^-$

Solution for NO_2^+ (a) The N atom has a lower electron affinity than O and so is the central atom.

(b) The valence electron total is 16 e^- or 8 pairs: 12 (6 for each O atom) + 5 (for N) − 1 (for ion positive charge) = 16.

(c) After forming two N—O bonds, the six remaining pairs of electrons are distributed on the terminal atoms until each such atom has a share in a total of four electron pairs.

[:Ö—N—Ö:]$^+$

(d) After step c, the N atom has a deficiency of two electron pairs. Thus, one lone pair of electrons from each O atom is converted to a pi bonding electron pair.

[:Ö=N=Ö:]$^+$

Each atom in the ion now has a share in four electron pairs. Nitrogen shares two sigma and two pi pairs, and each oxygen has two lone pairs and shares one sigma and one pi bond pair.

EXERCISE 9.3 DRAWING LEWIS STRUCTURES
Sketch the Lewis structure for each of the following molecules or ions. Give the number of sigma and pi bonds in which each atom is involved for (a) carbon monoxide, CO; (b) the ion NO^+; (c) hydrogen cyanide, HCN; and (d) hydrogen sulfide, H_2S.

EXAMPLE 9.2

DRAWING LEWIS STRUCTURES FOR MOLECULES WITH A CENTRAL ATOM OF EXPANDED VALENCE

Sketch the Lewis structure for xenon difluoride, XeF_2.

Solution (a) Xenon, the atom of lowest electron affinity, is the central atom.

(b) There is a total of 22 valence electrons or 11 pairs of electrons: 14 (for 2 F atoms) + 8 (for Xe) = 22.

(c) Form two Xe—F single bonds and then distribute the remaining electron pairs as lone pairs on the terminal atoms.

:F̤̈—Xe—F̤̈:

Each fluorine atom can accommodate only four pairs so the three pairs remaining are placed on xenon, an atom that can accommodate more than a total of four bonding and lone pairs.

:F̤̈—:Ẍ̤e:—F̤̈:

Thus, XeF_2 is a stable molecule with a total of 5 electron pairs (2 sigma bond pairs and 3 lone pairs) surrounding the central Xe atom.

EXERCISE 9.4 DRAWING LEWIS STRUCTURES WHERE THE CENTRAL ATOM HAS AN EXPANDED VALENCE
Sketch the Lewis structure for each of the following molecules: (a) SF_4, sulfur tetrafluoride; (b) ClF_3, chlorine trifluoride; and (c) PCl_5, phosphorus pentachloride.

RESONANCE

O 127.8 pm
O 116.8° O
Ozone, O_3

Ozone, O_3, protects the earth and its inhabitants from intense ultraviolet radiation from the sun. The compound is an unstable, blue, diamagnetic gas with a characteristic odor. Its structure is pictured here. As you will see in the next section, the number of bonding electron pairs between two atoms is important in determining bond length and strength. Ozone has equal O—O bond lengths, implying that there is an equal number of bond pairs on each side of the central O atom. However, using the guidelines for drawing Lewis structures, you might come to a different conclusion. A possible dot structure would be

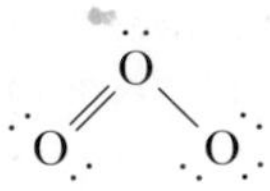

The structure has a double bond on one side of the central O atom and a single bond on the other side. If this were the electronic structure of O_3, one bond would be shorter (O=O) than the other (O—O). This is not the case, and one way to reconcile this with the experimental observation of equal O—O bond lengths is to use alternate dot structures called **resonance structures**. When drawing the O_3 dot structure, the final step is to convert a lone pair to a pi bond pair. In O_3 this can be done in either of two ways, and two equivalent structures or resonance structures are produced.

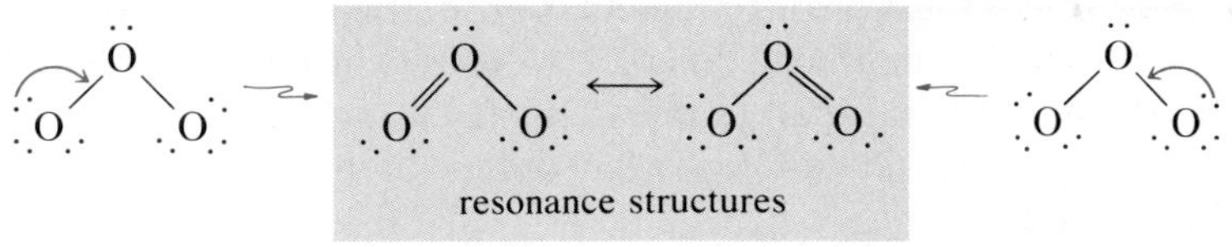

resonance structures

It is conventional to connect resonance structures by a double-headed arrow, ↔.

These resonance structures are meant to convey the idea that two pi bonding electrons are spread permanently over the entire structure of the ion. Therefore, the separate resonance structures might be represented by the composite picture in the margin.

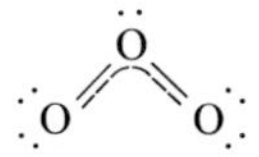

Composite picture of the O_3 resonance structures

The carbonate ion, CO_3^{2-}, provides another good example of resonance structures. Here there are *three* equivalent structures.

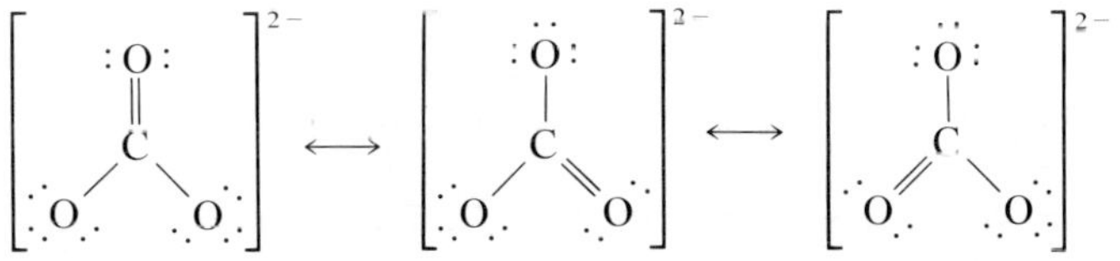

You can see that resonance structures arise whenever there is a question about which of two or three atoms contribute lone pairs to achieve an octet of electrons about a central atom by multiple bond formation.

When drawing resonance structures there are several important things to notice. First, only electron pairs are moved, not atoms. Second, simply moving a lone pair from one position around an atom to another position on that same atom does not mean you have a different resonance structure. Resonance structures always differ in the number of bond pairs between two atoms.

$$[:\ddot{N}=C=\ddot{O}:]^- \longleftrightarrow [:\ddot{N}=C=\underset{..}{O}:]^-$$

These are the same structure. They are *not* resonance structures because the number of NC and CO bond pairs is not changed.

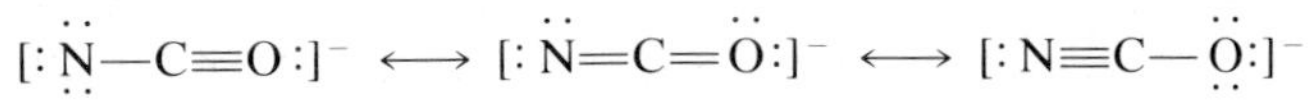

$$[:\underset{..}{\ddot{N}}-C\equiv O:]^- \longleftrightarrow [:\ddot{N}=C=\ddot{O}:]^- \longleftrightarrow [:N\equiv C-\underset{..}{\ddot{O}}:]^-$$

These are resonance structures. Lone pairs and bond pairs have been interconverted. The number of NC and CO bond pairs changes.

EXERCISE 9.5 DRAWING RESONANCE STRUCTURES
Draw resonance structures for (a) SO_2 and (b) NO_3^-.

9.4 BOND PROPERTIES

BOND ORDER

The **bond order** is the number of bonding electron pairs shared by two atoms in a molecule. Various molecular properties can be understood by this concept, including the distance between two atoms (bond length) and the energy required to separate the atoms from each other (bond energy). In this text you will encounter bond orders of 1 through 3, as well as fractional bond orders.

BOND ORDER = 1 The bond order is 1 when there is only a sigma bond between the two bonded atoms. Examples are the single bonds in the following molecules.

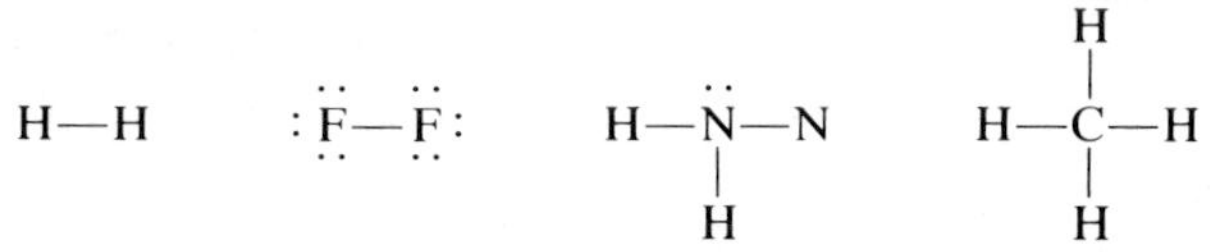

BOND ORDER = 2 The order is 2 when there are two shared pairs between two atoms. One of these pairs forms a sigma bond and the other pair forms a pi bond. Examples are the C=O bonds in CO_2 and the C=C bond in ethylene, C_2H_4.

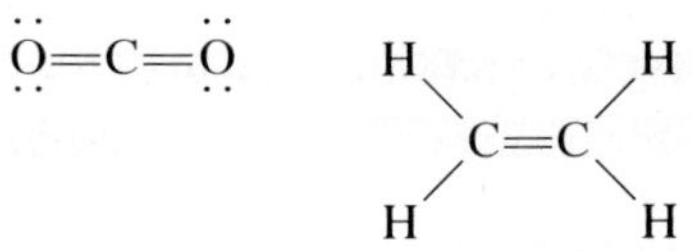

There can never be more than one sigma bond pair between a pair of atoms. All additional bond pairs are pi pairs.

BOND ORDER = 3 An order of 3 occurs when two atoms are connected by one sigma bond and two pi bonds. Examples are the carbon–carbon bond in acetylene (C_2H_2), the carbon–oxygen bond in carbon monoxide (CO), and the carbon–nitrogen bond in the cyanide ion (CN^-).

:C≡O: H—C≡C—H [:C≡N:]$^-$

Fractional bond orders occur for molecules or ions having resonance structures. The oxygen–oxygen bonds of O_3 have an *average* bond order of $\frac{3}{2}$, for example. A resonance structure of O_3 has one O—O bond and one O=O bond, for a total of three shared, bonding pairs to account for two oxygen–oxygen links. Using the definition of bond order below, the order is seen to be $\frac{3}{2}$.

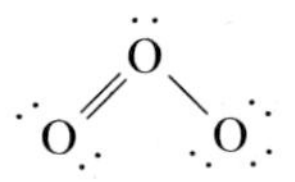

$$\text{Bond order} = \frac{\text{number of shared pairs linking X and Y}}{\text{number of X–Y links}} = \frac{3}{2}$$

BOND LENGTH

As described in Chapter 8, the most important factor determining **bond length**, the distance between two bonded atoms, is the sizes of the atoms themselves. For given elements, the order of the bond then determines the final value of the distance.

Recall from Figure 8.12 that atom sizes vary in a fairly smooth way with the position of the element in the periodic table. When you compare bonds of the same order, the bond length will be greater for the larger atoms. Thus, bonds involving carbon and another element would increase in length along the series

$$\underrightarrow{\text{C—N} < \text{C—C} < \text{C—P}}$$
Increase in bond distance

Similarly, a C═O bond will be shorter than a C═S bond, and a C≡N bond will be shorter than a C≡C bond. Each of these trends can be predicted from the relative sizes shown in the margin, and some common bond lengths are given in Table 9.3.

Relative Atom Sizes for Groups 4A, 5A, and 6A.

The effect of bond order is evident when you compare bonds *between the same two atoms*. For example, the bonds become shorter as the bond order increases in the series C—O, C═O, and C≡O.

Bond	C—O	C═O	C≡O
Bond Order	1	2	3
Bond Length (pm)	143	122	113

Adding a pi bond to the sigma bond in C—O shortens the bond by only 21 pm on going to C═O, rather than reducing it by half as you might have expected. The second pi bond results in a 9 pm reduction in bond length from C═O to C≡O.

TABLE 9.3 Some Approximate Single and Multiple Bond Lengths*

SINGLE BOND LENGTHS

	Group										
	1A	4A	5A	6A	7A	4A	5A	6A	7A	7A	7A
	H	C	N	O	F	Si	P	S	Cl	Br	I
H	58	110	98	94	92	145	138	132	127	142	161
C		154	147	143	141	194	187	181	176	191	210
N			140	136	134	187	180	174	169	184	203
O				132	130	183	176	170	165	180	199
F					128	181	174	168	163	178	197
Si						234	227	221	216	231	250
P							220	214	209	224	243
S								208	203	218	237
Cl									198	213	232
Br										228	247
I											266

MULTIPLE BOND LENGTHS

Bond	Length	Bond	Length
C═C	134	C≡C	121
C═N	127	C≡N	115
C═O	122	C≡O	113
N═O	115	N≡O	108

*In picometers (pm); 1 pm = 10^{-12} m.

The carbonate ion, CO_3^{2-}, which has three equivalent resonance structures, has a C—O bond order of 4/3. Not surprisingly, this leads to a bond distance of 129 pm, a distance intermediate between the C—O single bond and C=O double bond.

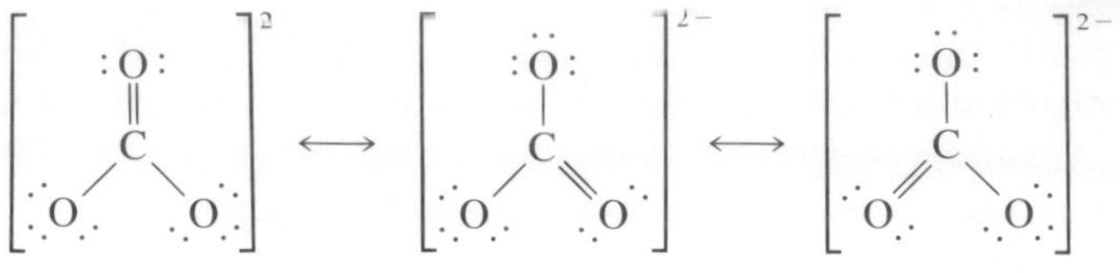

C—O bond order = 4/3; C—O bond length = 129 pm

EXERCISE 9.6 BOND DISTANCES AND BOND ORDER

(a) Give the bond order of each of the following bonds and arrange them in order of decreasing bond distance: C=N, C≡N, and C—N.

(b) Draw resonance structures for NO_2^-. What is the NO bond order in this ion? Consult Table 9.3 for N—O and N=O bond lengths. Compare these with the NO bond length in NO_2^- (124 pm). Account for any difference you observe.

Table 9.3 lists *average* values of bond lengths. Variations in neighboring parts of a molecule can affect the length of a particular bond. For example, the common C—H bond has a length of 105.9 pm in acetylene, H—C≡C—H, but a length of 109.3 pm in methane, CH_4. Be aware that there can be a variation of as much as 10% from the average values listed in Table 9.3.

BOND ENERGY

The greater the number of bonding electron pairs between a pair of atoms, the shorter the bond. This implies that atoms are held together more tightly when there are multiple bonds, and so it should not be surprising that there is a relation between bond order and the energy required to separate atoms.

Suppose you wish to separate, by means of chemical reactions, the carbon atoms in ethane ($H_3C—CH_3$), ethylene ($H_2C=CH_2$), and acetylene (HC≡CH) for which the bond orders are 1, 2, and 3, respectively. For the same reason that the ethane C—C bond is the longest of the series, and the acetylene C≡C bond the shortest, the separation will require the least energy for ethane and the most energy for acetylene.

$$\text{Molecule} + \text{energy} \xrightleftharpoons[\text{energy released}]{\text{energy supplied}} \text{molecular fragments}$$

$$H_3C—CH_3(g) + 347\ \text{kJ} \longrightarrow H_3C(g) + CH_3(g) \qquad \Delta H = +347\ \text{kJ/mol}$$

$$H_2C=CH_2(g) + 611\ \text{kJ} \longrightarrow H_2C(g) + CH_2(g) \qquad \Delta H = +611\ \text{kJ/mol}$$

$$HC{\equiv}CH(g) + 837\ \text{kJ} \longrightarrow HC(g) + CH(g) \qquad \Delta H = +837\ \text{kJ/mol}$$

TABLE 9.4 Some Average Single and Multiple Bond Energies*

SINGLE BONDS

	H	C	N	O	F	Si	P	S	Cl	Br	I
H	436	414	389	464	569	293	318	339	431	368	297
C		347	293	351	439	289	264	259	330	276	238
N			159	201	272		209		201	243?	
O				138	184	368	351		205		201
F					159	540	490	285	255	197?	
Si						176	213	226	360	289	213
P							213	230	331	272	213
S								213	251	213	
Cl									243	218	209
Br										192	180
I											151

MULTIPLE BONDS

N=N	418	C=C	611
N≡N	946	C≡C	837
C=N	615	C=O (in O=C=O)	803
C≡N	891	C=O (as in H_2C=O)	745
O=O (in O_2)	498	C≡O	1075

*In kJ/mol.

The energy that must be supplied to a gaseous molecule to separate two of its atoms is called the **bond dissociation energy** (or **bond energy** for short) and is given the symbol D. As D represents energy supplied to the molecule from its surroundings, D has a positive value, and *the process of breaking bonds in a molecule is always endothermic*.

The amount of energy supplied to break the carbon–carbon bonds in the molecules above must be the same as the amount of energy released when the same bonds form. *The formation of bonds from atoms in the gas phase is always exothermic*. This means, for example, that ΔH for the formation of $H_3C—CH_3$ from two $CH_3(g)$ fragments is −347 kJ/mol.

Some experimental bond energies are tabulated in Table 9.4, and you should be aware of several important points: (1) The energies listed are all *positive*; they are energies required to *break* the bond in question. If you need the energy evolved when the *bond is formed*, the magnitude is the same, but the sign is *negative*. (2) The energies of Table 9.4 are *average bond energies*. The energy of a C—H bond, for example, is given as 414 kJ/mol. However, this energy may vary by as much as 30 to 40 kJ/mol from molecule to molecule, for the same reason that bond lengths vary from one molecule to another. (3) The values in Table 9.4 are defined in terms of *gaseous* atoms or molecular fragments. If a reactant or product is in the solid or liquid state, you must first provide energy to convert it to a gas before using these values. (4) Finally, notice once again the connection between bond energy and bond order.

CALCULATING REACTION ENERGIES FROM BOND ENERGIES In reactions between molecules, bonds in the reactants are broken and new connections are established. If the total energy released when new bonds are

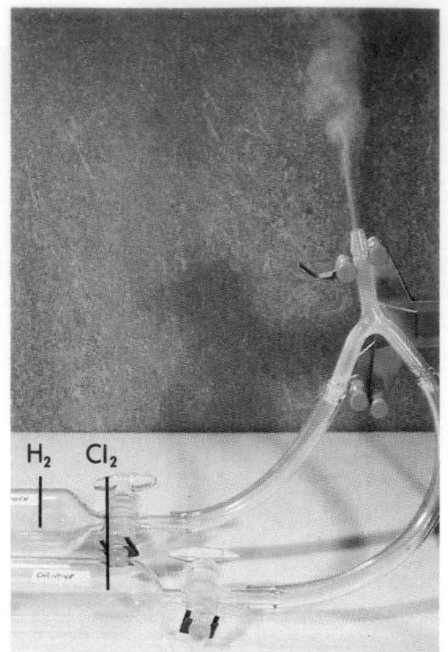

FIGURE 9.3
Hydrogen burns in chlorine to form hydrogen chloride, a colorless gas. Hydrogen chloride combines with moisture in the air to form the fog visible in the picture.

formed exceeds the energy required to break the original bonds, the overall reaction is exothermic. If the opposite is true, then the overall reaction is endothermic.

The following example shows you how to estimate, using the data in Table 9.4, whether the reaction is exothermic or endothermic.

EXAMPLE 9.3

USING BOND ENERGIES TO ESTIMATE HEATS OF FORMATION

Estimate the heat of formation of HCl(g) using the bond energies of Table 9.4 (Figure 9.3).

$$\text{H—H(g)} + \text{Cl—Cl(g)} \longrightarrow 2\ \text{H—Cl(g)} \quad \tfrac{1}{2}\,\Delta H^\circ_{rxn} = \Delta H^\circ_f[\text{HCl(g)}]$$

Solution The bond energies for the reactants and products are defined as follows (with values from Table 9.4):

$$\text{H—H(g)} \longrightarrow 2\ \text{H(g)} \qquad \Delta H^\circ = 436\ \text{kJ/mol}$$
$$\text{Cl—Cl(g)} \longrightarrow 2\ \text{Cl(g)} \qquad \Delta H^\circ = 243\ \text{kJ/mol}$$
$$\text{H—Cl(g)} \longrightarrow \text{H(g)} + \text{Cl(g)} \qquad \Delta H^\circ = 431\ \text{kJ/mol}$$

Recognize that 1 mole of H—H bonds and 1 mole of Cl—Cl bonds must be broken (for a total of 2 moles of bonds broken). If no new bonds were subsequently formed, 679 kJ of energy (436 kJ + 243 kJ) would have to be supplied to separate the mole of H_2 molecules and the mole of Cl_2 molecules into their atoms. However, there is a subsequent reaction of the H and Cl atoms to form 2 moles of H—Cl bonds, and this reaction *releases* 862 kJ of energy (2 moles of bonds × 431 kJ per mole of bonds). To summarize

$$\begin{array}{ll} \text{H—H(g)} \longrightarrow 2\ \text{H(g)} & \Delta H^\circ = +436\ \text{kJ} \\ \text{Cl—Cl(g)} \longrightarrow 2\ \text{Cl(g)} & \Delta H^\circ = +243\ \text{kJ} \\ 2\ \text{H(g)} + 2\ \text{Cl(g)} \longrightarrow 2\ \text{H—Cl(g)} & \Delta H^\circ = -862\ \text{kJ} \\ \hline \text{H—H(g)} + \text{Cl—Cl(g)} \longrightarrow 2\ \text{H—Cl(g)} & \Delta H^\circ_{rxn} = -183\ \text{kJ} \end{array}$$

Thus,

$$\Delta H^\circ_f[\text{HCl(g)}] = \tfrac{1}{2}\,\Delta H^\circ_{rxn} = -91.5\ \text{kJ/mol}$$

From the result above, you would predict that the reaction is exothermic. That is, 91.5 kJ are released per mole of HCl formed. Since the equation written above represents the formation of HCl in its gaseous state from the elements H_2 and Cl_2 (in their standard states), we have found that −91.5 kJ is the standard heat of formation of HCl(g), a result that agrees well with the experimental value of −92.3 kJ/mol.

As you saw in Chapter 5, there are often shortcuts in thermodynamic calculations. When using bond energies to find the heat or enthalpy of a reaction you can use the equation

$$\Delta H^\circ_{reaction} = \Sigma mD\ \text{(bonds broken)} - \Sigma nD\ \text{(bonds made)}$$

where D is the bond dissociation energy, and m and n are moles of bonds broken and made, respectively. This equation tells you to multiply the bond energy for each bond broken (all of these energies are endothermic, so the sign is +) by the number of bonds of that type, and add up all of these products. You then subtract from this total the sum of energies of bonds formed (because the sign of energies for bonds made is −). Applying this to the formation of HCl(g)

$$H_2(g) + Cl_2(g) \longrightarrow 2\ HCl(g)$$

you would write

$$\Delta H^\circ_{reaction} = \overbrace{(1\text{ mole H—H bonds} + 1\text{ mole Cl—Cl bonds})}^{\text{bonds broken}} - \overbrace{(2\text{ moles H—Cl bonds})}^{\text{bonds formed}}$$
$$= [1\text{ mol}(435\text{ kJ/mol}) + 1\text{ mol}(243\text{ kJ/mol})] - [2\text{ mol}(431\text{ kJ/mol})]$$
$$= -\ 184\text{ kJ}$$

EXAMPLE 9.4

USING BOND ENERGIES TO CALCULATE ENTHALPIES OF FORMATION

Estimate the heat of formation of hydrazine, N_2H_4, using the bond energies of Table 9.4.

$$N{\equiv}N(g) + 2\ H{-}H(g) \longrightarrow H_2\ddot{N}{-}\ddot{N}H_2\ (g)$$

Solution Here 1 mole of N≡N triple bonds and 2 moles of H—H single bonds are broken, and energy is required. On the other hand, energy is released when 1 mole of N—N single bonds and 4 moles of H—N bonds are made in forming the product.

$$\Delta H^\circ_{reaction} = \overbrace{(D_{N\equiv N} + 2\ D_{H-H})}^{\text{bonds broken}} - \overbrace{(D_{N-N} + 4\ D_{N-H})}^{\text{bonds made}}$$
$$= [946\text{ kJ} + 2\text{ mol}(435\text{ kJ/mol})] - [159\text{ kJ} + 4\text{ mol}(389\text{ kJ/mol})]$$
$$= +\ 101\text{ kJ}$$

Since the equation as written depicts the formation of one mole of hydrazine from its elements, the enthalpy of reaction we have calculated is the molar enthalpy of formation of hydrazine; that is, $\Delta H^\circ_f[N_2H_4(g)] = +101$ kJ/mol. The value determined using average bond energies is in good agreement with the value of +95.4 kJ/mol from the U.S. National Bureau of Standards.

EXAMPLE 9.5

USING BOND ENERGIES TO ESTIMATE ENTHALPIES OF FORMATION

Acetylene provides considerable heat when it is burned, making it a suitable fuel for welding metal (Figure 9.4). Estimate the enthalpy of formation of acetylene, H—C≡C—H, using the bond energies of Table 9.4 and the enthalpy of vaporization of graphite (the latter being the standard state of carbon).

$$C(\text{graphite}) \longrightarrow C(g) \qquad \Delta H^\circ_{vaporization} = +717\text{ kJ/mol}$$

FIGURE 9.4

The synthesis and combustion of acetylene, C_2H_2. The compound, a colorless gas, is produced by combining calcium carbide, CaC_2, and water. (Charles D. Winters)

$$CaC_2(s) + 2\ H_2O(\ell) \longrightarrow C_2H_2\ (g) + Ca(OH)_2(s)$$

Solution The equation for the formation of acetylene from its elements is

$$2\text{C(graphite)} + \text{H—H(g)} \longrightarrow \text{H—C}\equiv\text{C—H(g)}$$

Recall that bond energies can be used only when all of the atoms or molecules are in the gas phase. In this problem, therefore, you must include the energy necessary to convert carbon from its standard state, the black solid graphite, into carbon atoms in the gas phase. The series of reactions in this case is the following:

2 [C(graphite) → C(g)]	$2\,[\Delta H^\circ = +717$ kJ/mol]
H—H(g) → 2 H(g)	$\Delta H^\circ = D_{H—H} = +436$ kJ/mol
2 C(g) + 2 H(g) → H—C≡C—H	$\Delta H^\circ = 2\,(-D_{C—H}) + (-D_{C\equiv C})$
	$= 2$ mol$(-414$ kJ/mol) + (1 mol) $(-837$ kJ/mol)
	$= -1665$ kJ
2 C(graphite) + H_2(g) → H—C≡C—H(g)	$\Delta H^\circ = +205$ kJ/mol

The equation above represents the formation of 1 mole of acetylene from the elements, so $\Delta H^\circ_{rxn} = \Delta H^\circ_f = +205$ kJ/mol. (Table 5.2 shows $\Delta H^\circ_f = +226.7$ kJ/mol. The deviation between the experimental and calculated values arises from the use of average bond energies.)

A chemist often wishes to know the energy of a bond in a particular compound, since the strength of that bond is an important factor in determining the ease of reactions in which the bond is replaced. The problem that follows is an example of the method that is used to *estimate* the energy of a particular bond.

EXAMPLE 9.6

ESTIMATING THE ENERGY OF A BOND

Suppose the C=C double bond energy were not available, and you wished to estimate it. You need a reaction in which all of the bond energies are known except the one of interest, the C=C bond energy. A combustion reaction is a good choice here, because energies of bonds with oxygen are readily available. Using an apparatus similar to that in Figure 5.7, you can determine the heat of combustion of ethylene, $H_2C{=}CH_2$ to be -1323 kJ/mol of C_2H_4, and you can then calculate $D_{C=C}$.

Solution The balanced equation for the combustion of ethylene is

$$\text{H}_2\text{C=CH}_2\text{(g)} + 3\ \text{O=O(g)} \longrightarrow 2\ \text{O=C=O(g)} + 2\ \text{H—O—H(g)}$$

$$\Delta H^\circ_{combust} = -1323 \text{ kJ/mol}$$

$$\begin{aligned}
\Delta H^\circ_{combust} &= \text{sum of energies to break bonds} \\
&\quad - \text{sum of energies of bonds made} \\
&= [D_{C=C} + 4\,D_{C—H} + 3\,D_{O=O}] - [4\,D_{C=O} + 4\,D_{O—H}] \\
-1323 \text{ kJ} &= [D_{C=C} + 4 \text{ mol}(414 \text{ kJ/mol}) + 3 \text{ mol}(498 \text{ kJ/mol})] - \\
&\quad [4 \text{ mol}(803 \text{ kJ/mol}) + 4 \text{ mol}(464 \text{ kJ/mol})] \\
-1323 \text{ kJ} &= D_{C=C} + (-1918 \text{ kJ}) \\
D_{C=C} &= 595 \text{ kJ/mol}
\end{aligned}$$

This estimate of the bond energy is in good agreement with the accepted value of $D_{C=C}$ for C_2H_4, which is 611 kJ/mol.

EXERCISE 9.7 ESTIMATING ENTHALPY OF FORMATION

Estimate the enthalpy of formation of tetrafluorohydrazine, N_2F_4, from the data in Table 9.4.

$$\text{N}{\equiv}\text{N(g)} + 2\ \text{F}{-}\text{F(g)} \longrightarrow \text{F}_2\ddot{\text{N}}{-}\ddot{\text{N}}\text{F}_2\text{(g)} \qquad \Delta H^\circ_f = ?$$

EXERCISE 9.8 ESTIMATING ENTHALPY OF FORMATION

Using the enthalpy of vaporization of graphite [C(s) → C(g)] in Example 9.5 and bond energies from Table 9.4, estimate the enthalpy of formation of carbon tetrachloride, $CCl_4(g)$.

EXERCISE 9.9 ESTIMATING BOND ENERGIES

The enthalpy of formation of formic acid, HCO_2H, is −378.6 kJ/mol.

$$\text{C(graphite)} + \text{O}{=}\text{O(g)} + \text{H}{-}\text{H(g)} \longrightarrow \text{H}{-}\underset{\underset{\text{O}}{\|}}{\text{C}}{-}\text{O}{-}\text{H(g)}$$

Estimate the bond energy of the C=O bond in the formic acid molecule. Use the enthalpy of vaporization of C(graphite) in Example 9.5 and the data in Table 9.4.

BOND POLARITY AND ELECTRONEGATIVITY

In Chapter 8 you learned that not all atoms hold onto their valence electrons with equal strength. The elements all have different values of ionization energy and electron affinity. More to the point, the valence electrons of each different element experience a different **effective nuclear charge, Z^***. If two different elements form a bond, the one with higher Z^* will attract the shared pair more strongly than the other. Only when two atoms of the same kind form a bond can we presume that the bond pair is shared equally between the two atoms.

See page 260 for a description of Z^*.

Consider a bond between H and Cl. The valence electrons of H and Cl are subject to different values of Z^* so their shared sigma bond pair will, on the average, spend more time near the atom of higher Z^* (Cl) and less time near the atom of lower Z^* (H). The atom around which the sigma pair spends most of its time, Cl, has a slight excess of negative charge, so H must have a slight deficiency of electrons and take on an equal, but positive, charge. Another way to say this is that the bond has *polarity*. Such a bond is said to be a **polar covalent** bond. We usually indicate polarity by using the symbols $\delta+$ and $\delta-$, which indicate *partial* + and − charges. Some polar bonds in common molecules are

When covalently bonded atoms have different values of Z^*, the bond is **polar covalent**.

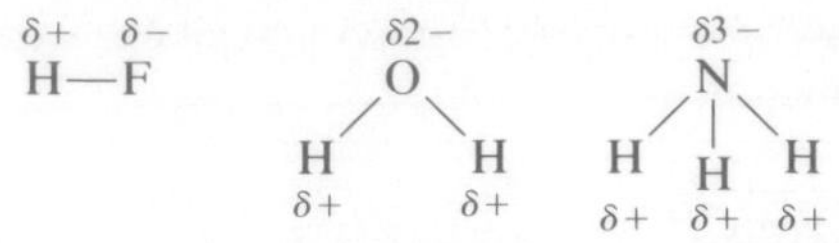

$\overset{\delta+}{H}-\overset{\delta-}{F}$ $\quad$ $\overset{\delta+}{H}-\overset{\delta 2-}{O}-\overset{\delta+}{H}$ $\quad$ $\overset{\delta 3-}{N}(\overset{\delta+}{H})_3$

When atoms of very different Z^* are bonded, the bond is **ionic**.

In the extreme case of very different Z^* values for the bonded atoms, the sigma pair spends essentially all of its time about the atom of higher Z^*. The sigma bond pair then becomes a lone pair at the atom of higher Z^*, and the bond is so polar that we refer to it as **ionic**; a cation is now associated with an anion. In contrast, if there is *equal* sharing of the covalent bond pair, the bond is **nonpolar**.

How can we predict whether a bond is nonpolar, polar, or ionic? Of these classes, the only clearly defined one is nonpolar. The difference between polar covalent and ionic is only a matter of degree. The following guidelines come from experimental observation:

(a) If the two bonded atoms are identical, the bonding electrons are equally shared and the bond is nonpolar. Examples include O_2, N_2, and H_2.

(b) If the two atoms are different but occur in the same region of the periodic table, there is only a moderate difference in their attraction for the bond electrons and the bond is polar covalent. Examples of such bonds are O—F, C—N, C—O, and N—O.

(c) Metals have low values of Z^*, whereas oxygen and the halogens (especially F and Cl) have high Z^* values. In general, bonds between metals (with oxidation number $\leq +3$) and oxygen or the halogens are highly polar or ionic.

What is the physical basis of these guidelines? In the last chapter you learned that the ionization energy of an atom is the energy necessary to remove valence electrons from an atom; thus, it is a measure of the tendency of an atom to attract its electrons. We cannot assume, though, that atom ionization energies have any direct meaning for atoms in molecules, since they apply strictly only to individual atoms in the gas phase. However, it is clear that some concept related to the ability of atoms to attract shared bond pairs in molecules would be useful. Linus Pauling advanced the necessary idea in the 1930s, and his concept of **electronegativity**, χ, has been used since as a measure of the ability of an atom in a molecule to attract electrons to itself. Using the principles of thermodynamics, Pauling defined a value for the electronegativity of each element (Table 9.5), and others have since defined similar scales based on various physical and chemical properties of atoms in molecules.

The trends in electronegativity are clearly related to those for the ionization energies and electron affinities for free atoms.

As you examine Table 9.5 you should notice several special features and periodic trends.

(a) The element with the largest electronegativity (4.0) is fluorine in the upper right corner of the table. The element in the bottom left corner of the table, francium, has the smallest value. In general, electronegativity increases diagonally up and to the right in the table, a trend clearly related to the trends in ionization energy and electron affinity.

(b) The greatest variation in electronegativity across any of the periods occurs in the second period (Li, . . ., F).

TABLE 9.5 Electronegativity Values and Their Periodic Relationships

Key: <1.0; 1.0–1.4; 1.5–1.9; 2.0–2.4; 2.5–2.9; 3.0–4.0

1 H 2.1																
3 Li 1.0	4 Be 1.5											5 B 2.0	6 C 2.5	7 N 3.0	8 O 3.5	9 F 4.0
11 Na 1.0	12 Mg 1.2											13 Al 1.5	14 Si 1.8	15 P 2.1	16 S 2.5	17 Cl 3.0
19 K 0.9	20 Ca 1.0	21 Sc 1.3	22 Ti 1.4	23 V 1.5	24 Cr 1.6	25 Mn 1.6	26 Fe 1.7	27 Co 1.7	28 Ni 1.8	29 Cu 1.8	30 Zn 1.6	31 Ga 1.7	32 Ge 1.9	33 As 2.1	34 Se 2.4	35 Br 2.8
37 Rb 0.9	38 Sr 1.0	39 Y 1.2	40 Zr 1.3	41 Nb 1.5	42 Mo 1.6	43 Tc 1.7	44 Ru 1.8	45 Rh 1.8	46 Pd 1.8	47 Ag 1.6	48 Cd 1.6	49 In 1.6	50 Sn 1.8	51 Sb 1.9	52 Te 2.1	53 I 2.5
55 Cs 0.8	56 Ba 1.0	57 La 1.1	72 Hf 1.3	73 Ta 1.4	74 W 1.5	75 Re 1.7	76 Os 1.9	77 Ir 1.9	78 Pt 1.8	79 Au 1.9	80 Hg 1.7	81 Tl 1.6	82 Pb 1.7	83 Bi 1.8	84 Po 1.9	85 At 2.1
87 Fr [illegible]	88 Ra 1.0	89 Ac 1.1														

58 Ce 1.1	59 Pr 1.1	60 Nd 1.1	61 Pm 1.1	62 Sm 1.1	63 Eu 1.1	64 Gd 1.1	65 Tb 1.1	66 Dy 1.1	67 Ho 1.1	68 Er 1.1	69 Tm 1.1	70 Yb 1.0	71 Lu 1.2
90 Th 1.2	91 Pa 1.3	92 U 1.5	93 Np 1.3	94 Pu 1.3	95 Am 1.3	96 Cm 1.3	97 Bk 1.3	98 Cf 1.3	99 Es 1.3	100 Fm 1.3	101 Md 1.3	102 No 1.3	103 Lr 1.5

(c) The "*s* block" elements, Groups 1A and 2A, have somewhat similar values of electronegativity, and those for the "p block" elements (Groups 3A through 7A) are rather similar. However, there is considerable difference between the electronegativity values of the two different blocks of elements.

(d) No values are given in Table 9.5 for the rare gases. The reason for this is that only xenon (and perhaps krypton) forms compounds, and then only under special circumstances.

The periodic trends mean that the most electronegative atoms are in the upper right corner of the table, while the least electronegative are in the lower left corner. If atoms from these two opposite regions of the periodic table form a chemical compound, their large electronegativity difference means that the more electronegative atom of the pair will more strongly attract the bonding electrons, and the bond will be polar. If the polarization is extreme, the bond is ionic.

EXAMPLE 9.7

ESTIMATING BOND POLARITIES

For each of the following bond pairs, tell which is the more polar and indicate the negative and positive poles.

(a) Cs—F and Li—I (b) C—S and P—P (c) C—O and C—S

Solution (a) Cs and F are in the lower left and upper right corners, respectively, of the periodic table, and they form one of the most ionic bonds. Li lies above Cs in the table (Li is more electronegative than Cs), and I is below F (I is less electronegative than F); this means that the Li—I bond is less ionic or less polar than Cs—F.

Element	Cs	Li	F	I
Electronegativity	0.8	1.0	4.0	2.5

For the Cs—F bond, the difference in electronegativity, $\Delta\chi$, is 4.0 − 0.8 = 3.2, whereas it is 2.5 − 1.0 = 1.5 for Li—I. In both bonds, the halogen is the more electronegative and so is the more negative atom.

$$\overset{\delta+}{\text{Cs}}\text{—}\overset{\delta-}{\text{F}} \qquad \overset{\delta+}{\text{Li}}\text{—}\overset{\delta-}{\text{I}}$$
$$\Delta\chi = 3.2 \qquad \Delta\chi = 1.5$$

(b) The P—P bond is nonpolar (or purely covalent), because the bond is between two atoms of the same kind. C is in the second period, while S is in the third period but closer to the right side of the table. It is no surprise to find, therefore, that C and S both have the same electronegativity (2.5), and their bond is also predicted to be nonpolar.

(c) O lies above S in the periodic table, so O is more electronegative than S. This means the C—O bond must be more polar than the C—S bond, which was predicted in part b to be nonpolar. For the C—O bond, O is the more negative atom of the dipolar bond: $\overset{\delta+}{\text{C}}\text{—}\overset{\delta-}{\text{O}}$.

EXERCISE 9.10 POLAR BONDS

For each of the following pairs of bonds, decide which is the more polar. For each polar bond, indicate the negative and positive poles. First make your prediction from the relative atom positions in the periodic table; then check your prediction by calculating $\Delta\chi$.

(a) H—F and H—I (b) B—C and B—F (c) C—Si and C—S

OXIDATION NUMBERS

Oxidation numbers, which we introduced in Chapter 3 as an aid in considering oxidation–reduction reactions, are related to electronegativity. With the background of the present chapter, we can now understand the basis of these numbers.

The electrical charge on a free atom is zero. If the atom is bound to another in a molecule, however, it is impossible to say what its charge may be, since some valence electrons are shared with other atoms. It is possible, though, to define the limiting case to determine at least the sign and maximum value of the charge on an atom involved in a bond. This limiting situation arises if we agree that all the bond pair electrons belong to the more electronegative atom in a bond, which amounts to assuming that all bonds are ionic. The charge on the atom calculated in this "ionic limit" is called the **oxidation number**.

Oxidation number = group number of element
− number of electrons surrounding the element in the ionic limit

Here you see that the oxidation number is given by the number of electrons acquired by the atom in excess of its valence electrons (negative oxidation number) or the number released by the atom (positive oxidation number) in the ionic limit.

EXAMPLE 9.8

OXIDATION NUMBERS

Determine the oxidation number of F in HF and S in SCl_2.

Solution In HF the F atom is the more electronegative atom, so F is surrounded by 8 electrons in the ionic limit. Thus, its oxidation number is $7 - 8 = -1$.

H—F: → (H)(:F:) equivalent to $H^+ F^-$

ionic limit

For SCl_2, the S is less electronegative than Cl (see Table 9.5). Therefore, in the ionic limit the S atom has only 4 electrons. This means its oxidation number is $6 - 4 = +2$.

:Cl—S—Cl: → (:Cl:)(S)(:Cl:) equivalent to $Cl^- S^{2+} Cl^-$

ionic limit

Since electrons *must* be evenly shared in a homonuclear diatomic molecule such as H_2, the oxidation number of each element must be 0.

Oxidation numbers of the elements in compounds and polyatomic ions containing both oxygen and a halogen can pose problems. As an example, the Lewis dot structure for the perchlorate ion is

$$\left[\begin{array}{ccc} & :\ddot{O}: & \\ & | & \\ :\ddot{O}- & Cl & -\ddot{O}: \\ & | & \\ & :\ddot{O}: & \end{array}\right]^-$$

Since each O atom is more electronegative than the Cl, in the ionic limit the bonding electrons cluster about the O atoms, a feature that can be represented as

$$\left[\begin{array}{ccc} & (:\ddot{O}:)^{2-} & \\ (:\ddot{O}:)^{2-} & (Cl)^{7+} & (:\ddot{O}:)^{2-} \\ & (:\ddot{O}:)^{2-} & \end{array}\right]^-$$

Thus, Cl is assigned the extremely high oxidation number of $+7$. Experiments by atomic physicists with such ions show they will voraciously attract electrons. It is inconceivable that any species under normal conditions can really contain an ion in such a highly charged state. Indeed, any charge in excess of about ± 2 on an atom in a chemical compound is improbable. This tells you that *oxidation numbers usually do not represent the real charge on an atom.* They are simply a convention that indicates the ionic limiting value. However, you will find them useful for two purposes: (1) to keep track of electrons in molecules and in chemical reactions (as in Chapter 3) and (2) to indicate the more positive and more negative atoms in a molecule. The latter is often of use in understanding how molecules react to form new bonds.

EXERCISE 9.11 OXIDATION NUMBERS FROM LEWIS STRUCTURES

For each atom in each of the following formulas, calculate the oxidation number from the information in the Lewis structure.

(a) SF_4 (d) PF_3
(b) CO_3^{2-} (e) SO_3
(c) ClO_3^-

In (e) is it necessary to average the values from the three possible resonance structures?

FORMAL CHARGES ON ATOMS

The ionic limit used to assign oxidation numbers is unrealistic for covalent bonds. An alternate view, which assumes each bond pair is shared equally by two atoms (the "nonpolar limit"), often provides a more reasonable estimate of charges. These are called **formal atom charges**, and they are given by the expression

$$\text{Formal atom charge} = \text{group number} - \text{number of lone pair electrons} - \tfrac{1}{2}(\text{number of bonding electrons})$$

Here we count the lone pair electrons about an atom and one half of the bonding electrons that it shares. To illustrate, reexamine the case of ClO_4^-. Begin again with the Lewis structure.

$$\left[\begin{array}{ccccc} & & :\ddot{O}: & & \\ & & | & & \\ :\ddot{O} & — & Cl & — & \ddot{O}: \\ & & | & & \\ & & :\underset{..}{O}: & & \end{array}\right]^-$$

Now, the formal charge is -1 for each oxygen (the oxidation number was -2),

$$\text{Formal charge for oxygen} = 6 - 6 - \tfrac{1}{2}(2) = -1$$

and the formal charge for chlorine is $+3$ (the oxidation number was $+7$).

$$\text{Formal charge for chlorine} = 7 - 0 - \tfrac{1}{2}(8) = +3$$

Notice that the signs of the formal charges and oxidation numbers are the same (as is most often the case), but the magnitude of the charge is more realistic. Also notice that *the sum of the formal charges must add up to the ion charge*.

Two fundamental ideas in chemistry apply to atom formal charge. (1) Atoms in molecules (or ions) should have formal charges as close to zero as possible (the **principle of electroneutrality**). (2) A molecule (or ion) is most stable when any negative charge resides on the most electronegative atom. Since some molecules or ions can have several resonance structures, the structure best satisfying these principles is the most important. For example, let us use atom formal charges as an aid in deciding which of two resonance structures is more reasonable for CO_2.

	A	B
Formal charges	0 0 0	+1 0 −1
Resonance structures	$:\ddot{O}{=}C{=}\ddot{O}:$	$:O{\equiv}C{-}\ddot{\underset{\cdot\cdot}{O}}:$

For structure A, each atom has a formal charge of 0, a favorable situation. In B, however, one oxygen atom has a charge of +1. This is an unfavorable state for the very electronegative oxygen atom, so resonance structure B is of little importance.

EXAMPLE 9.9

ATOM FORMAL CHARGES AND RESONANCE STRUCTURES

There are three possible resonances structures for the cyanate ion, NCO^-. Using atom formal charges, decide which is the most reasonable of these structures.

Solution Formal charges have been calculated for each of the NCO^- resonance structures given on page 307 and are listed above the appropriate atom in each structure.

	A	B	C
Formal charges	−2 0 +1	−1 0 0	0 0 −1
Resonance structures	$[:\ddot{\underset{\cdot\cdot}{N}}{-}C{\equiv}O:]^-$	$[:\ddot{N}{=}C{=}\ddot{O}:]^-$	$[:N{\equiv}C{-}\ddot{\underset{\cdot\cdot}{O}}:]^-$

As an example of the calculation of formal charge, consider structure A.

$$\begin{aligned}\text{Formal charge for N} &= 5 - 6 - \tfrac{1}{2}(2) = -2\\ \text{Formal charge for C} &= 4 - 0 - \tfrac{1}{2}(8) = 0\\ \text{Formal charge for O} &= 6 - 2 - \tfrac{1}{2}(6) = \underline{+1}\\ \text{Sum of formal charges} &= -1 = \text{charge on ion}\end{aligned}$$

In structure A, the O atom has a formal charge of +1, a very unfavorable situation for this electronegative atom. Therefore, A contributes little to the overall electronic structure of the cyanate ion. Structure B places a −1 charge on N and 0 on oxygen, not an unfavorable situation. However, structure C may be favored slightly because the negative atom in the structure is the most electronegative one in the ion.

EXERCISE 9.12 FORMAL CHARGES

The Lewis structure of H_2BF is depicted by A below. To satisfy the electron deficiency of the boron atom, one could write the resonance structure B. Calculate the formal charge on each atom in structure B and comment on the possibility that this resonance structure has any importance.

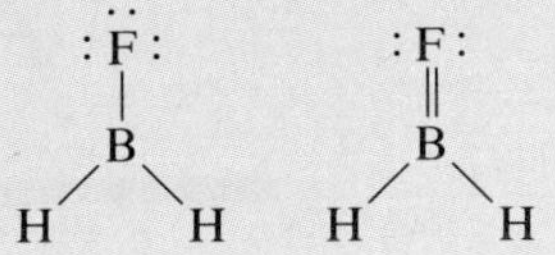

A: $:\ddot{F}:$ single-bonded to B, which is bonded to two H atoms.

B: $:F:$ double-bonded to B, which is bonded to two H atoms.

9.5 MOLECULAR SHAPE

Lewis structures only show how many bond pairs and lone pairs surround a given atom. However, all molecules are three dimensional, and most molecules have their atoms in more than one plane in space. It is often important to know the way molecules fill space, because the structure partly determines the chemical functioning of the molecule. Pharmaceutical companies, for example, use knowledge of molecular shape to design drugs that will fit into the site in the body where pain is to be relieved or disease attacked.

To convey a sense of three dimensionality for a molecule drawn on a piece of flat paper, we use sketches such as the "ball-and-stick" model of methane, or we can draw structures in perspective using "wedges" for bonds that emerge from or recede into the plane of the drawing.

CORRELATION BETWEEN STRUCTURE AND VALENCE ELECTRON PAIRS: THE VSEPR MODEL

A sampling of perspective sketches and ball-and-stick models of molecules for which we have already drawn Lewis structures is shown in Table 9.6. Although you do not yet know how to predict the structures in this table, there is an easy way to do this. Notice how the molecular shape changes with the number of sigma bonds plus lone pairs about the central atom.

SIGMA BONDS + LONE PAIRS ON CENTRAL ATOM	STRUCTURE OF MOLECULE IN TABLE 9.6
2	Linear
3	Trigonal planar
4	Tetrahedral (or pyramidal)
5	Trigonal bipyramidal
6	Octahedral

The repulsion is partly due to the electrostatic force between negatively charged electrons.

The idea that will allow us to predict the molecular structure is that each lone pair or bond group (sigma + pi pairs) repels all other lone pairs and bond pair groups. Because the pairs try to avoid one another, they move as far apart as possible, and, since all of the pairs are "tied" to the same central atom nucleus, they can only orient themselves so as to make the angles between themselves as large as possible. This is the

(a)

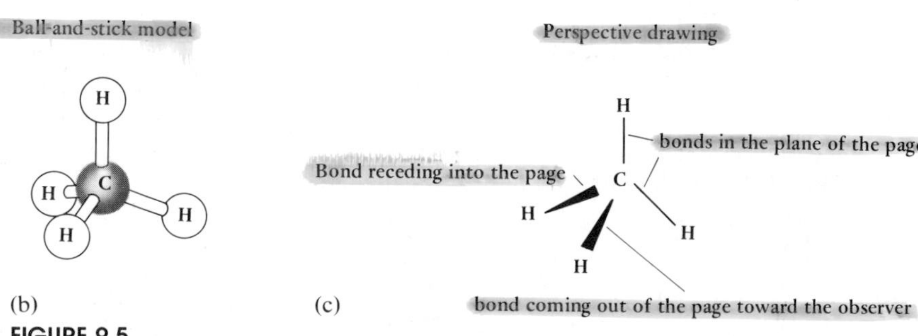

FIGURE 9.5

The structure of methane, CH_4, to show the ways molecular structures will be illustrated. (a) Photograph and (b) drawing of a ball-and-stick model. (c) Perspective drawing. (a, Charles D. Winters)

TABLE 9.6 Perspective Molecular Sketches

MOLECULE	GEOMETRY	PERSPECTIVE SKETCH	BALL-AND-STICK MODEL
CO_2	Linear	O=C=O 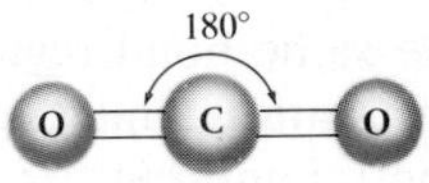 	180°
CO_3^{2-}	Trigonal planar 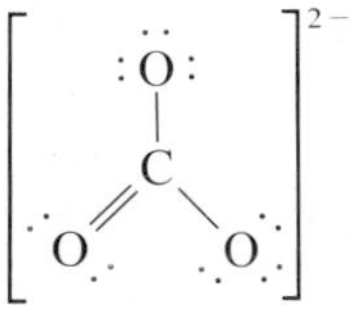 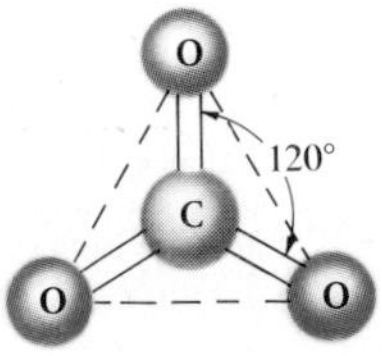 		120°
NH_3	Pyramidal  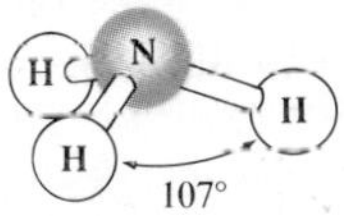 		107°
CH_4	Tetrahedral 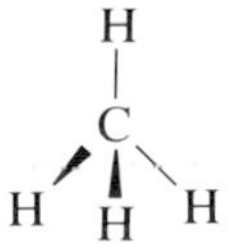 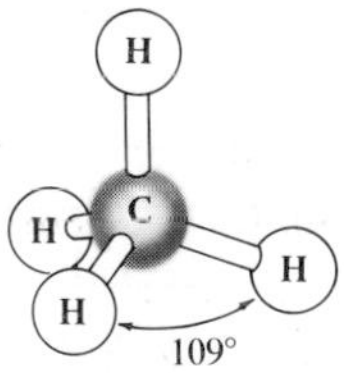 		109°
PCl_5	Trigonal bipyramidal 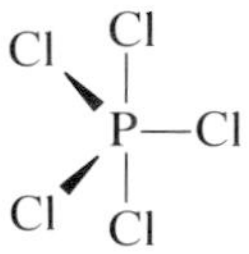 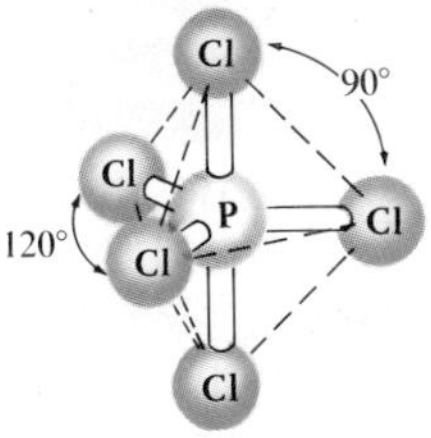 		90°, 120°
SF_6	Octahedral  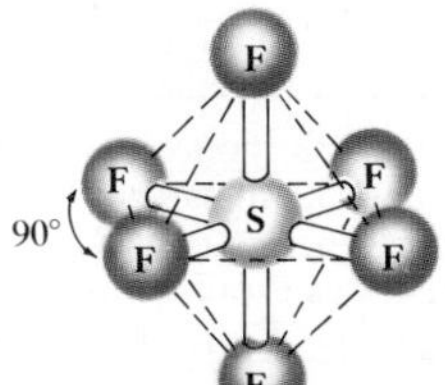 		90°

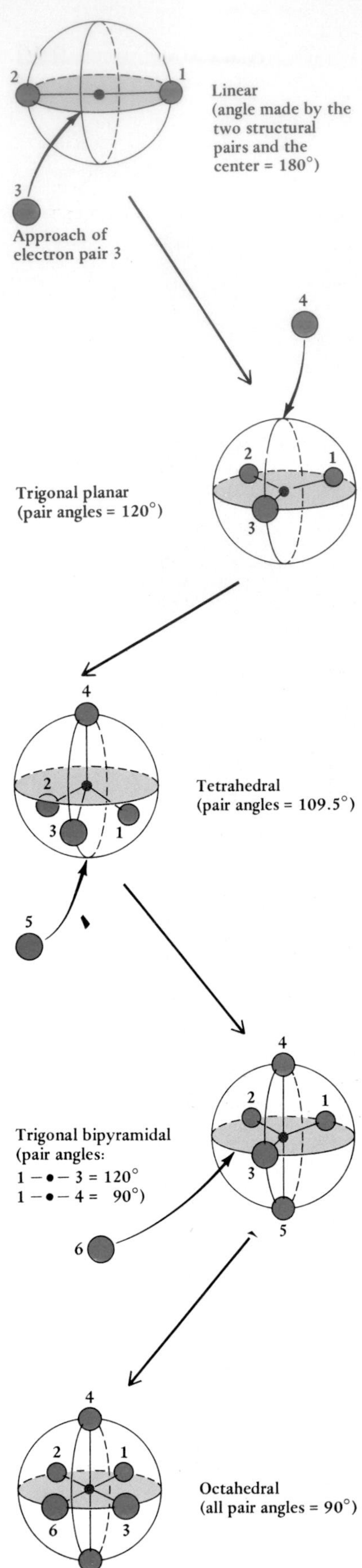

essence of the ***V**alence **S**hell **E**lectron **P**air **R**epulsion* model (**VSEPR** for short) for predicting molecular structures.*

To show you how to use the VSEPR model, we shall first designate as **structural pairs** the sigma and lone pairs about an atom. (The pi pairs are deliberately excluded as structural pairs, because each pi pair occupies the same bond region as a sigma pair.) To visualize the structural pair orientations, imagine them situated on the surface of a sphere with the central atom at the center.

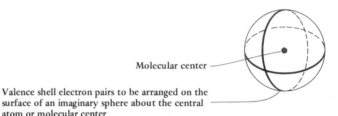

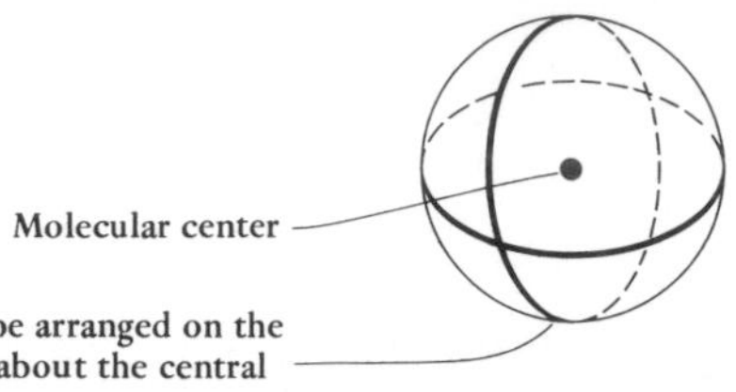

If there are only *two structural pairs* about the nucleus, they can best avoid each other if they are 180° apart. This means that the two pairs and the central atom are in a line; the arrangement is *linear*. If we add a third structural pair midway between the first two pairs, you see that the newly added pair will push the first two back until the angles between them are 120°, and they all lie in the same plane. We say this arrangement is *trigonal planar*. If a fourth structural pair is added, it will take a position above or below the triangular plane (where it finds the most space) and will push back the previous three pairs until all of the angles are equal, this time 109.5°, and the orientation of pairs is *tetrahedral*. If a fifth pair enters the valence shell, it will do so opposite any of the first four pairs and will push its three nearest neighbors back to a trigonal planar arrangement. The orientation of the five pairs is now *trigonal bipyramidal* (if you connect the electron pairs with imaginary lines the structure looks like two triangular pyramids that share a triangular face).

Finally, if a sixth structural pair is added to the five of the trigonal bipyramid, this new pair will find the most room in the triangular plane and enter there, pushing back its two nearest neighbors so that all of the angles are 90°, and the electron pair orientation is *octahedral*.

To summarize, what we have found so far are the orientations about the central atom of 2 to 6 structural pairs (Figure 9.6). The arrangement for a particular number of pairs is called its **structural-pair geometry**. Some or all of the structural pairs around an atom in a molecule will be bond pairs, and it only remains to mark the bond pairs with the atomic

*The VSEPR model was devised by Ronald J. Gillespie (1924–) and Ronald S. Nyholm (1917–1971). Gillespie was born in England and Nyholm in Australia, but both received the Ph.D. in chemistry at University College, London. Nyholm made many contributions to chemistry as Professor of Chemistry at University College, London, until his untimely death. Gillespie has been a Professor of Chemistry at McMaster University (Canada) since 1960.

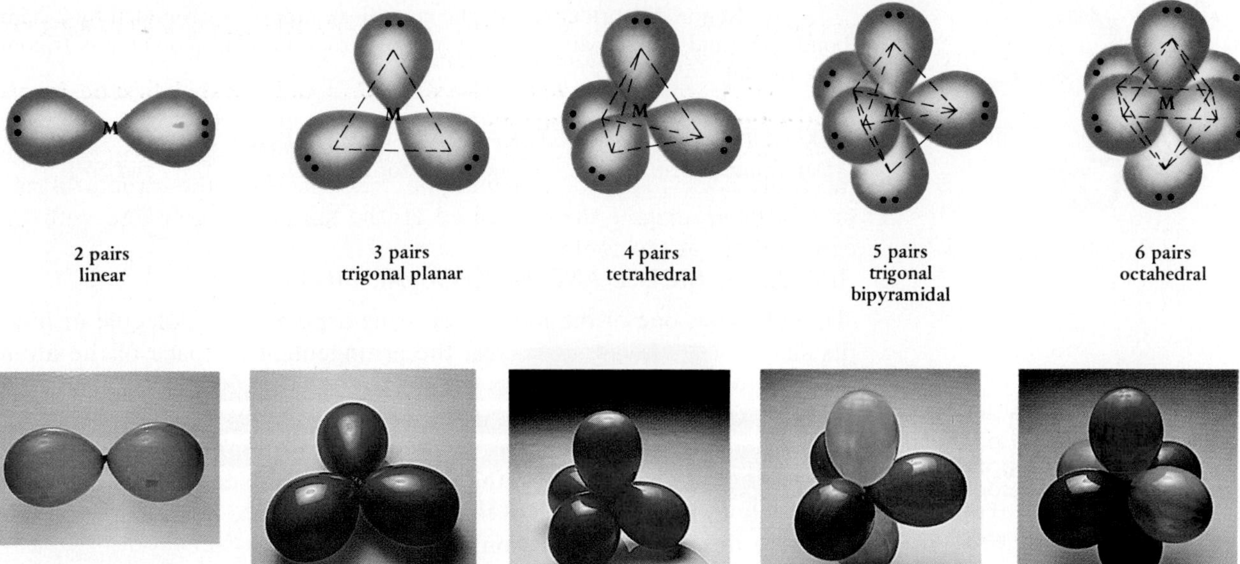

FIGURE 9.6
Structural pair geometries for 2 to 6 structural pairs. The balloon models are especially convenient. If one ties together several balloons of similar size and shape they will naturally assume the geometries illustrated. (Charles D. Winters)

symbols of the terminal atoms to arrive at the full, three-dimensional molecular geometry.

EXAMPLE 9.10

STRUCTURAL-PAIR GEOMETRIES

Give the structural-pair geometry for (a) NH_3, (b) H_2CO, and (c) XeF_2.

Solution One *must* first draw the Lewis structure for the molecule or ion. In this case, Lewis structures have already been presented in Examples 9.1 and 9.2 and are reproduced below.

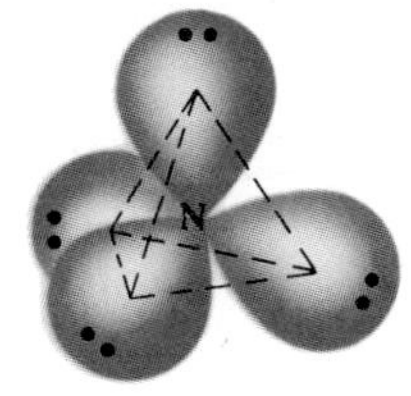

NH_3 structural pair geometry

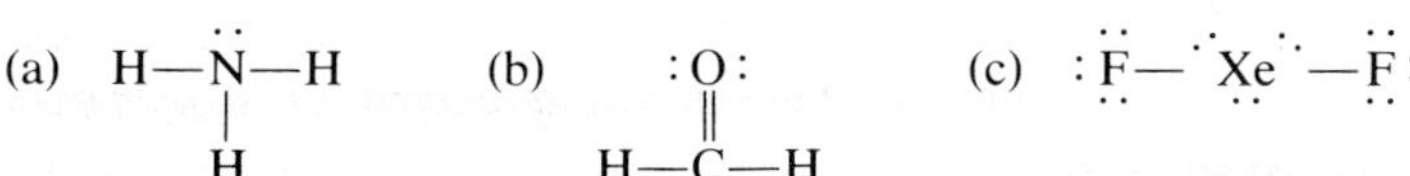

(a) Ammonia, NH_3: Four pairs of electrons (3 sigma bond pairs and 1 lone pair) surround the central N atom. Therefore, the structural pair geometry around N is tetrahedral.

(b) Formaldehyde, H_2CO: The C atom is surrounded by 3 sigma bonding pairs. (There is also one pi bonding pair, but this is ignored for purposes of deciding on geometry.) Therefore, the structural pair geometry is planar trigonal.

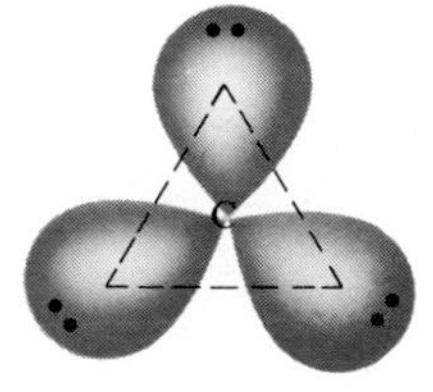

H_2CO structural pair geometry

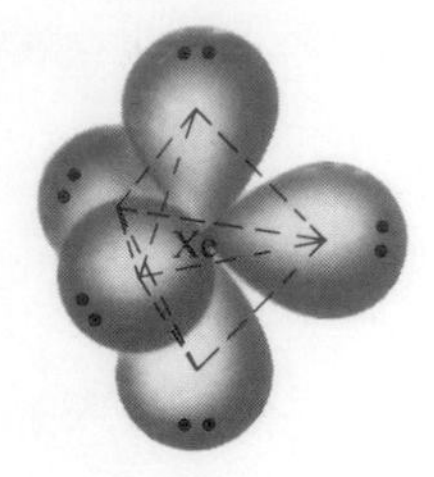

XeF_2 structural pair geometry

(c) Xenon difluoride, XeF_2: The central Xe atom is surrounded by 2 sigma bond pairs and 3 lone pairs. Therefore, the structural-pair geometry is trigonal bipyramidal.

EXERCISE 9.13 STRUCTURAL-PAIR GEOMETRIES
Determine the structural pair geometry for (a) H_2O, (b) NO_2^+, and SF_4.

THE VSEPR MODEL AND MOLECULAR SHAPE

The **structural-pair geometry** of an atom in a molecule is the location of the lone pairs and sigma bond pairs surrounding that atom. The **molecular geometry** of a molecule or ion is defined by the location of the bond pairs only.

To a chemist, one of the most interesting aspects of a molecule or ion is its shape or **molecular geometry**, the arrangement in space of the atoms bonded to a central atom. This may or may not be the same as the shape described by the structural pairs, the *structural-pair geometry* of the ion or molecule, depending on whether there are any lone pairs around the central atom. To derive the three-dimensional molecular geometry, we need only to identify which of the structural pairs are bond pairs and then describe the bond pair locations.

MOLECULAR SHAPES FOR CENTRAL ATOMS WITH NORMAL VALENCE When the central atom of a molecule or ion obeys the octet rule, there can be no more than four structural pairs, and the complete range of molecular shapes that can arise from various combinations of bond and lone pairs is illustrated in Table 9.7. These are by far the most commonly occurring structural types for normal valence compounds, and you should familiarize yourself with them thoroughly.

The VSEPR model does not apply to molecules having a transition metal atom or ion as the central atom.

You must recognize that the *name given to the shape of a molecule is the word that best describes the relative positions of the atoms*. The name does *not* always describe the location of the structural pairs. For example, the structure of water is said to be "bent," because this describes the relative *atom* locations. Water has a bent configuration because the four structural pairs are at the corners of a tetrahedron. However, only two of these pairs are used to bind H atoms, and the two bond pairs determine the molecular geometry.

The NO_2^- ion is also described as bent. The ion has three structural pairs at the corners of a triangle, but only two of these pairs are sigma bond pairs. The net effect is that the three atoms appear in a bent configuration.

Finally, look at the examples having pi bonds: CO_2, NO_2^-, and CO_3^{2-}. Notice that, in counting structural pairs, we have lumped pi and sigma pairs together. Thus, the C atom in CO_2 is surrounded by two structural-pair groups, and a linear structure results. The N in NO_2^- and the C in CO_3^{2-} are the centers of three groups of structural pairs. Since all three structural pairs in CO_3^{2-} are used in bonding, the ion shape is said to be trigonal planar.

You can describe the shape of virtually any molecule by thinking through the following steps: (1) Sketch the Lewis dot structure. (2) De-

TABLE 9.7 Structural Pair Geometries and Molecular Shapes for Molecules with Two, Three, and Four Structural Pairs about the Central Atom

NUMBER OF STRUCTURAL PAIRS	STRUCTURAL PAIR GEOMETRY	BOND PAIRS	LONE PAIRS	MOLECULAR SHAPE	EXAMPLE
2	Linear	2	0	Linear (180°)	O=C=O
3	Trigonal planar	3	0	Trigonal planar (120°)	$[CO_3]^{2-}$
		2	1	Bent	$[NO_2]^-$
4	Tetrahedral	4	0	Tetrahedral (109°)	CH_4
		3	1	Pyramidal	NH_3
		2	2	Bent	H_2O

scribe the structural-pair geometry. (3) Decide which structural pairs are sigma bond pairs (or bond pair groups), and then describe their location.

EXAMPLE 9.11

FINDING THE SHAPES OF MOLECULES

What are the molecular shapes of PH_3, SO_2, and NO_3^-?

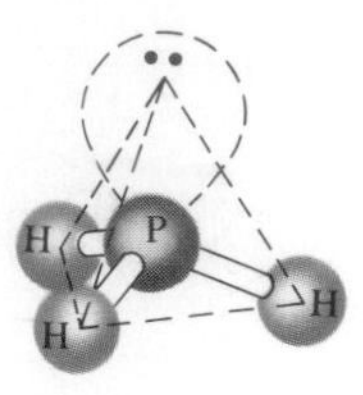

PH_3 structure

Solution (a) The Lewis structure of phosphine, PH_3,

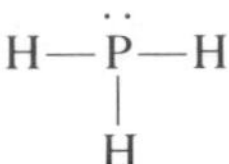

reveals that the central P atom has four structural pairs, so the structural-pair geometry is tetrahedral. Since three of the four structural pairs are used to bond terminal atoms, the central P atom and the three H atoms form a pyramidal molecular shape like NH_3.

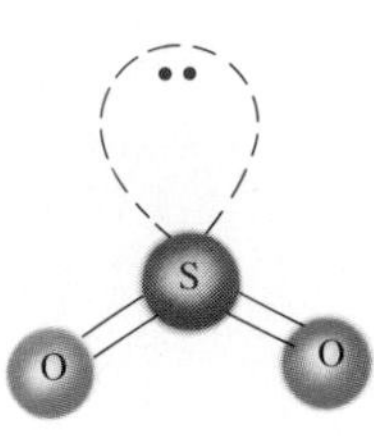

SO_2 structure

(b) Either of the two resonance structures of SO_2 shows that there are three structural-pair groups around the sulfur atom. (Again, remember that pi bond pairs do not count as structural pairs.) This means the structural-pair ge-

ometry is trigonal planar. However, as only two structural pairs are used in sigma bonding, the molecular shape is bent, just like the NO_2^- ion.

(c) The NO_3^- ion has the same number of valence electrons as the CO_3^{2-} ion in Table 9.7. Thus, like the carbonate ion, the nitrate ion is trigonal planar.

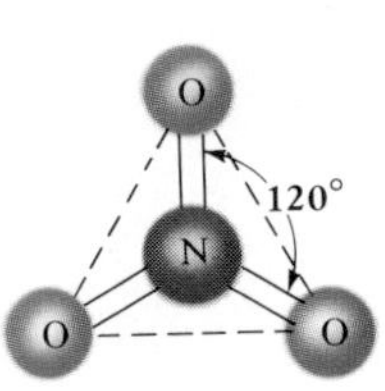

NO_3^- structure

EXERCISE 9.14 DETERMINING MOLECULAR SHAPES

Determine the structural-pair geometry and molecular shape for (a) carbon disulfide, CS_2; (b) hydrogen sulfide, H_2S; and (c) phosphate ion, PO_4^{3-}.

BOND ANGLES IN MOLECULES AND IONS Not only can you determine the overall shape of a molecule or ion from the VSEPR model, but finer details can also be predicted and explained. For example, methane (CH_4), ammonia (NH_3), and water (H_2O) are based on a tetrahedral arrangement of structural pairs. In methane, each electron pair is used to bind an H atom, and the angle between neighboring H atoms, the **bond angle**, is 109.5°. Notice that, as you proceed from methane to ammonia and then to water, the number of lone pairs increases and the bond angle decreases.

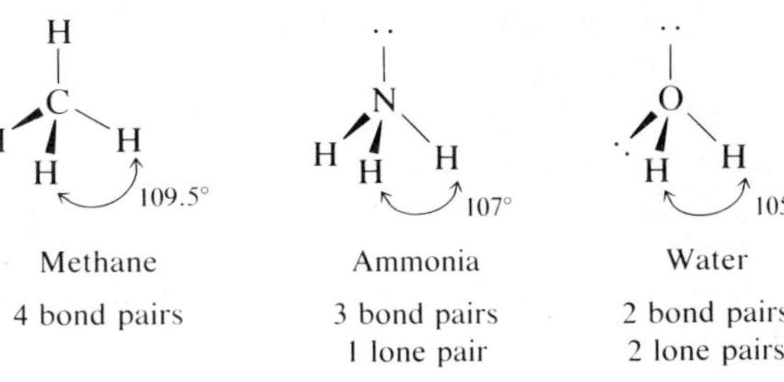

Methane: 4 bond pairs

Ammonia: 3 bond pairs, 1 lone pair

Water: 2 bond pairs, 2 lone pairs

The reason for this is that lone pairs require more space on the surface of the imaginary sphere about the central atom. Bond pairs are drawn into the bond region by strong attractive forces of *two* nuclei and are, therefore, relatively compact in their spatial needs; they are "skinny." For a lone pair, however, there is only one nucleus attracting the electron pair, and this nuclear charge is not so effective in overcoming the normal repulsive forces between two negative electrons; as a result, lone pairs are "fat."

MOLECULAR SHAPES FOR CENTRAL ATOMS WITH EXPANDED VALENCE For a central atom with five or six structural pairs (the trigonal bipyramid or octahedron), there are quite a few molecular structure possibilities. Fortunately, not all these possibilities occur in nature as stable structures, because there seems to be a limit of three lone pairs about the central atom. The molecular structures that have been observed are illustrated in Table 9.8.

Let's look first at the entries in Table 9.8 for the case of five structural pairs. The angles in the triangular plane are all 120°, while the angles between any of these pairs and an upper or lower pair are only 90°. (Because the positions in the trigonal plane lie in the equator of the imaginary sphere around the central atom, they are called the **equatorial** positions. The north and south poles are called **axial** positions.) Each equatorial position thus lies further from all other positions than does an axial position. This means the "fat" lone pairs occupy an equatorial position rather than an axial position. In XeF_2, the three lone pairs all occupy equatorial positions, forcing the bonding pairs into axial positions, so the molecular geometry is linear. In ClF_3, the two lone pairs are equatorial; two bond pairs are axial and the third is allowed into the equatorial plane, so the molecular geometry is T-shaped.

The trigonal bipyramid

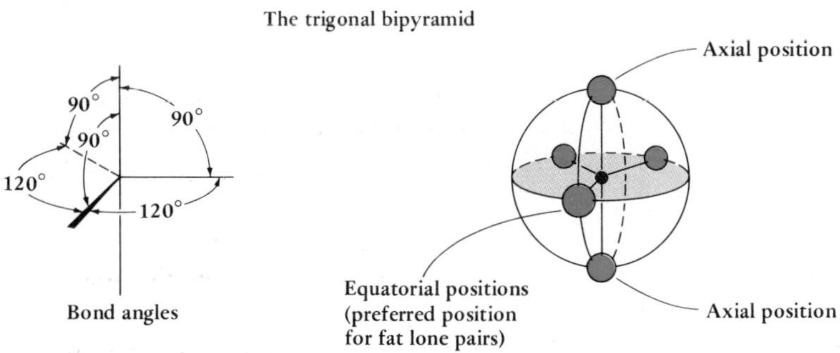

Unlike the trigonal bipyramid, all of the angles in the octahedron are 90°. There are no distinct axial and equatorial positions; all positions are the same. Therefore, if a molecule has one lone pair, as in BrF_5, it makes no difference where you place it. On the other hand, if the molecule or ion has two lone pairs, as in ICl_4^-, each of the lone pairs needs as much room as possible. This is best achieved by placing the lone pairs on opposite sides of the center. In this way, the pairs flanking each "fat" lone pair are "skinny" bond pairs.

The octahedron

TABLE 9.8 Structural Pair Geometries and Molecular Shapes for Molecules with Five or Six Valence Shell Electron Pairs

STRUCTURAL PAIR GEOMETRY	NUMBER OF BOND PAIRS	NUMBER OF LONE PAIRS	MOLECULAR SHAPE	DESCRIPTION	EXAMPLE (LONE PAIRS NOT SHOWN)
Trigonal bipyramid	5	0		Trigonal bipyramidal	PF_5
	4	1		Unsymmetrical tetrahedron	SF_4
	3	2		T-Shaped	ClF_3
	2	3		Linear	XeF_2
Octahedral	6	0		Octahedral	SF_6
	5	1		Square pyramidal	BrF_5
	4	2		Square planar	XeF_4

EXAMPLE 9.12

VSEPR AND MOLECULAR STRUCTURE

Draw the Lewis dot structure for $OXeF_4$ and then describe the structural-pair geometry and the molecular shape.

Solution The molecule has a total of 42 valence electrons: 8 for Xe + 6 for O + 4 × 7 for F. This leads to a dot structure with a total of 6 electron pairs about the central Xe atom, one of these pairs being a lone pair.

```
 :F̤  :Ö:  F̤:
    \  |  /
      Xe
    /  ¨  \
 :F̤        F̤:
```

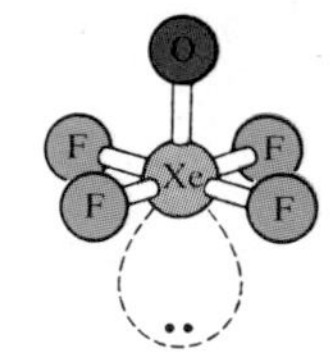

The six total pairs about the Xe atom mean that the structural-pair geometry is octahedral. Since only five of the structural pairs are sigma bond pairs, this means the molecular shape is a square pyramid. Although we will not worry for the moment about the exact placement of the O and F atoms, they could in principle be placed so that either the O or F is across the molecule from the lone pair. It is known in this case that the real geometry is that shown.

EXERCISE 9.15 MOLECULAR AND IONIC STRUCTURE
Draw the Lewis structure for ICl_2^- and then decide on the structural-pair and molecular geometry of the ion.

EXERCISE 9.16 BOND ANGLES
What are the approximate values of the S—C—S and H—S—H angles in CS_2 and H_2S, respectively?

9.6 MOLECULAR POLARITY

The adjective "polar" was used in Section 9.4 to describe the situation of separated positive and negative charges in a bond. However, because most molecules have at least some polar bonds, molecules can themselves be polar. In a polar molecule, there is an accumulation of electron density toward one side of the molecule, so that one end of the molecule bears a slight negative charge, $\delta-$; the other end has a slight positive charge of equal value, $\delta+$ (Figure 9.7). The experimental measure of this separation of charge is the molecule's **dipole moment**, which is defined as the product of the size of the charge (δ) and the distance of separation. The units of dipole moment are therefore coulomb · meters; a convenient unit is the debye (D), defined as 1 D = 3.34×10^{-30} C · m. Experimental dipole moments of a few molecules are given in the table in the margin.

Polar molecules will align themselves with an electric field as in Figure 9.7b, and they will also align themselves with each other. This interaction of polar molecules is an extraordinarily important effect in water and other substances, as you will see in Chapter 11 in particular.

(*Text continues on page 334*)

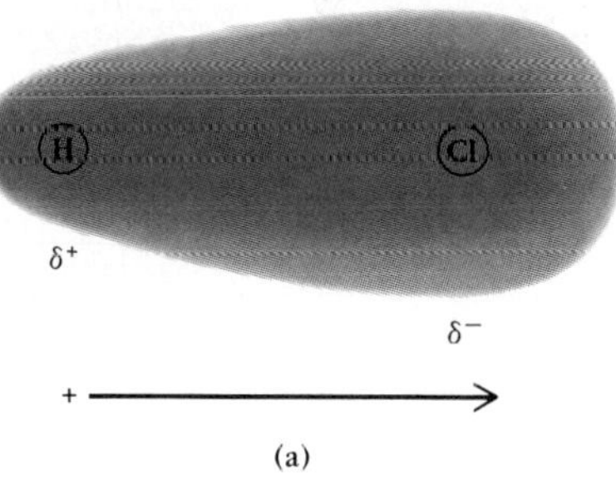

(a)

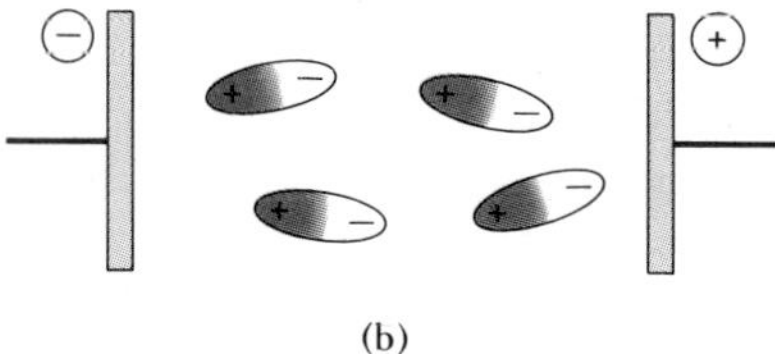

(b)

FIGURE 9.7

Polar molecules. (a) In a polar molecule the valence electron density has shifted slightly to one side of the molecule. To show the direction of charge transfer, that is, the direction of molecular polarity, we often use an arrow with a positive "tail," +→. (b) A molecule with a dipole moment is called a polar or dipolar molecule. When placed in an electric field the positive end of the molecule tends to align itself with the negative side of the field. This affects the capacitance of the plates (their ability to hold a charge) and provides a way experimentally to measure the magnitude of the dipole.

SPECIAL SECTION: CHEMISTRY OF INTERSTELLAR SPACE

The way in which elements are formed in the stars is outlined in the special section for Chapter 8.

The elements were generated in the fires and explosions of the stars, and molecules and molecular ions built from these elements are now found in interstellar space. Here we want to catalog a few of these molecules (Table 9.9), some surprisingly large, and describe some current ideas of their formation.

TABLE 9.9 Some Interstellar Molecules

DIATOMIC	POLYATOMIC
H_2	H_2O, water
CH	HCN, hydrogen cyanide
OH	H_2S, hydrogen sulfide
C_2	C_2H_5OH, ethyl alcohol
CN	NH_3, ammonia
CO	C_2H_2, acetylene
NO	H_2CO, formaldehyde
SiO	CH_3CN, acetonitrile

About 90% of the matter in our galaxy, the Milky Way, is held in stars, while the remaining 10% is in interstellar space. The interstellar material is mostly hydrogen, but other elements are present as well (Figure 9.8). The distribution of this material is not uniform, however. Most space is empty, with fewer than 0.1 particles/cm^3, while the interstellar matter concentrates in clouds. Interstellar clouds are often classed as either "diffuse" or "dense." Diffuse clouds can be penetrated by galactic starlight, radiation that is usually in the approximate wavelength range 100 to 200 nm. Since the clouds largely contain atomic hydrogen, and since H atoms can be ionized to H^+ by radiation with a wavelength of 91.2 nm, many of the H atoms are ionized. In fact, this energy divides the chemistry of these clouds into two groups of species. Elements and molecules more easily ionized than H atoms exist as ions (and so this includes other elements such as C), while those with ionization energies greater than that of H atoms exist as neutral species (such as N). The resulting chemistry in diffuse clouds is quite simple, and the only available species are mostly diatomic molecules or ions such as H_2, HD, OH, CO, CH, N_2, O_2, CN, and so on.

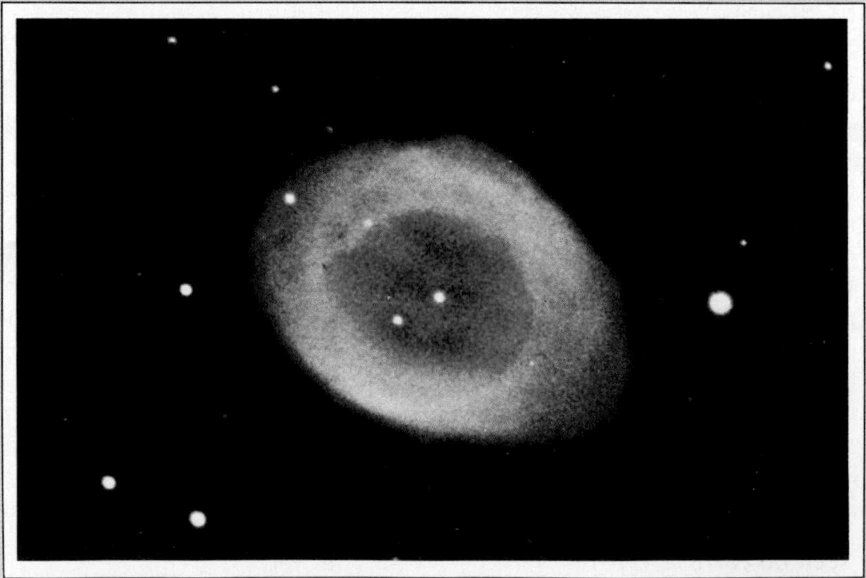

FIGURE 9.8
The Ring Nebula, a planetary nebula in Lyra. Red hydrogen emission is visible around the outer edge while green radiation from oxygen shows in the center. (Palomar Observatory)

"Dense" interstellar clouds are opaque to ultraviolet starlight, so light-induced processes do not occur. Rather, it is thought that chemical reactions are brought on by cosmic radiation. For example, H_2 is ionized by such radiation

$$\text{Cosmic ray} + H_2 \longrightarrow H_2^+ + e^-$$

to produce the hydrogen molecule–ion, which then reacts with more H_2 to give H_3^+ and H atoms.

$$H_2^+ + H_2 \longrightarrow H_3^+ + H$$

The H_3^+ ion is the key to the production of many other molecules by the reaction

$$H_3^+ + A \longrightarrow AH^+ + H_2$$

$$A \text{ is CO, } N_2\text{, O, N, } C_2$$

$$AH^+ \text{ is } HCO^+\text{, } N_2H^+\text{, } OH^+\text{, } NH^+\text{, and } C_2H^+$$

For example, if A is an oxygen atom, the OH^+ ion can be generated. The ions produced in such reactions above are extremely reactive and can react with H_2, the most abundant species in the cloud. For example, the OH^+ ion would give the ionized water molecule.

$$OH^+ + H_2 \longrightarrow H_2O^+ + H$$

Reactions such as those outlined above can account for many of the species in Table 9.9, but more complex processes must be responsible for such interesting species as HC_9N, which presumably has the structure H—C≡C—C≡C—C≡C—C≡C—C≡N.

FIGURE 9.9
Comet West, a bright comet visible in 1976. (Martin Grossman)

The recent visit of Halley's comet to the solar system has greatly renewed interest in the chemistry of comets (Figure 9.9). Comets are particularly interesting because they may preserve a unique record of conditions at the time of the formation of the planets of the solar system.

It is currently thought that there is a large, spherical cloud of comets (called the *Oort cloud* for the Dutch astronomer who proposed the idea) surrounding the solar system. One of these comets occasionally comes near enough for observation.

At the center of the head of a comet is the nucleus, which is a few kilometers across at most. The "dirty snowball" hypothesis is the generally accepted theory of the composition of this nucleus. According to this model, the comet consists of a mixture of various ices, such as H_2O, CO_2, NH_3, and CH_4, as well as small dirt particles.

The "dirty snowball" theory was proposed in 1950 by F.L. Whipple of the Harvard and Smithsonian Observatories. Neutral water molecules have been observed in Halley's comet.

Regardless of the nature of the nucleus, it is the *coma* of a comet (about 10^5 km in diameter) that is generally observed. As a comet approaches the sun, its ices sublime (go directly from solid to gas). The resulting gases expand into the vacuum of space and, in the process, drag along small particles of dust. The coma shines partly because the gas and dust reflect sunlight toward us and partly because the gases are excited enough by sunlight to radiate light themselves. It is the light given off by the excited atoms and molecules that helps astrometers unravel the chemistry of the comet. For example, the spectrum of light given off by comet West (Figure 9.9) has established the existence of CN, C_2, C_3, NH_2, Na, and O. In addition, many simple neutral species (CH, CO, OH, CH_3CN, HCN, NH_3, etc.), ions (CO^+, OH^+, H_2O^+, etc.), and atoms (K, Ca, V, Mn, Fe, Co, Mo, and Cu) have been discovered in other comets.

It is reassuring that the molecules found in space are largely those seen on earth, a fact that has implications for all the universe.

Some Dipole Moments

MOLECULE	MOMENT (IN DEBYE UNITS)
H_2	0
H_2O	1.94
NH_3	1.46
CH_4	0
CH_3Cl	1.86

To predict when a simple molecule will be polar, we need to consider if the molecule has polar bonds and how these bonds are positioned relative to one another. Let us consider first a linear triatomic molecule such as carbon dioxide, CO_2. Here each C—O bond is polar, with the oxygen atom the negative end of the bond dipole.

$$\overset{\delta-}{O}\overset{\longleftarrow\!\!+}{—}\overset{\delta 2+}{C}\overset{+\!\!\longrightarrow}{—}\overset{\delta-}{O} \qquad \text{no net dipole}$$

Each bond dipole is a vector, since it has direction and magnitude. Like forces, bond dipole vectors can be added and subtracted. In the CO_2 molecule, the dipole vectors have the same magnitude, but they are in the opposite direction. Therefore, there is no net dipole moment for the molecule. (This is the same as if two people of equal strength are pulling at opposite ends of a rope.)

If a triatomic molecule is bent, it will probably have a dipole moment. Water is an excellent example of this. Here both the O—H bonds are polar, and both bond dipoles point toward O. There is thus a net movement of charge away from the H atoms, toward the O atom. One side of the molecule has a negative charge (O), and the other side is positive (the H's). The bond dipoles add together to give a non-zero value.

2δ− O; H δ+, H δ+ — net dipole = 1.94 D

Molecules of the type below are nonpolar when all of the terminal atoms are the same.

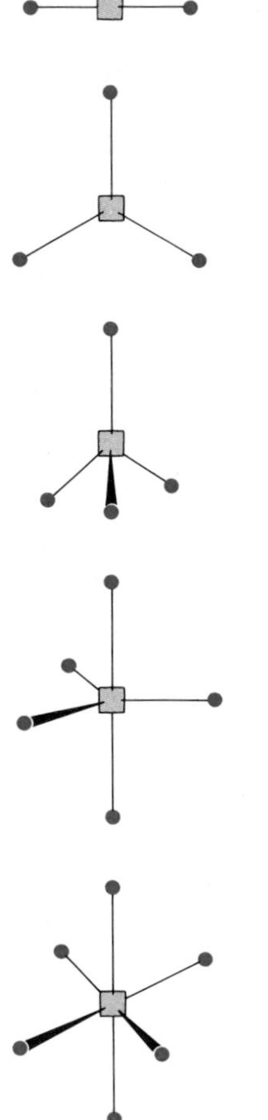

In general, simple molecules of the type CT_n, where terminal atoms T are symmetrically placed about the center C, are nonpolar. In contrast, bent and pyramidal molecules are usually dipolar.

A simple tetrahedral molecule such as CCl_4, carbon tetrachloride, is nonpolar, because the bond dipole in one direction is canceled by the bond dipoles in the opposite direction. However, if one of the Cl atoms is replaced by H to give $CHCl_3$, chloroform, the molecule is polar.

Cl δ−; C 4δ+; Cl δ−, Cl δ−, Cl δ− — no net dipole

H δ+; C 2δ+; Cl δ−, Cl δ−, Cl δ− — net dipole

The electronegativity of H (2.1) is less than C (2.5), so the bond dipoles now all point in the same direction. Thus, $CHCl_3$ has a net dipole with the negative end toward the Cl atoms. That is, negative charge accumulates on the Cl's and away from the H atom.

EXAMPLE 9.13

MOLECULAR POLARITY

Are nitrogen trifluoride (NF_3), dichloromethane (CH_2Cl_2), and boron trifluoride (BF_3) polar or nonpolar? If polar, indicate the direction of polarity.

Solution NF_3 has the same pyramidal molecular structure as ammonia (NH_3). Since F is more electronegative than N, each bond dipole points to fluorine. Thus, there is a net build up of negative charge on the fluorine side of the molecule, so there is a net dipole moment in that direction.

$\ddot{N}$ $3\delta+$, F F F, $\delta-$ $\delta-$ $\delta-$ — net dipole

In CH_2Cl_2, a tetrahedral molecule, Cl is more electronegative than C, while H is less electronegative. Therefore, there is a movement of negative charge away from H toward Cl, and CH_2Cl_2 has a net dipole with the negative end at the Cl atoms.

H $\delta+$, C, H $\delta+$, Cl $\delta-$, Cl $\delta-$ — H, Cl, C, H, Cl — net dipole

As predicted for a molecule with three structure pairs around the central atom, BF_3 is trigonal planar.

$\delta-$ F, B $3\delta+$, $\delta-$ F, F $\delta-$ — no net dipole

Since F is more electronegative than B, the B—F bonds are polar with F being the negative end. The molecule is nonpolar, though, since the three terminal atoms are identical; the dipole of one B—F is offset by the dipoles of the opposite B—F bonds.

EXERCISE 9.17 MOLECULAR POLARITY

For each of the following molecules, decide whether the molecule is polar and which side is positive and which is negative: (a) $BFCl_2$, (b) NH_2Cl, and (c) SCl_2.

Peter Debye (1884–1966) was born and educated in Europe but became Professor of Chemistry at Cornell University in 1940. He was noted for his work on x-ray diffraction, electrolyte solutions, and the properties of polar molecules. He received the Nobel Prize in 1936.

SUMMARY

The **valence electrons** of an atom are important in chemical bonding. For main group elements, the number of valence electrons is the same as its group number in the periodic table (Section 9.1). Valence electrons can be depicted by a **Lewis electron dot symbol**. In forming bonds, main group elements gain, lose, or share electrons to achieve the configuration of the nearest rare gas; this is the basis of the **octet rule**.

In **covalent bonding**, 2, 4, or 6 electrons are shared between atoms (Section 9.2). (Compounds where the bonding electrons reside nearly or completely on one atom of the bonded pair are called **ionic**.) Valence electrons not used in bond formation are called **unshared** or **lone electrons**. Bonding electrons generally occur in pairs if at all possible. In covalent molecules these electrons reside in a **bonding orbital**.

For elements of the second period in particular, the number of bonding and lone pair electrons surrounding an atom does not exceed 8 (four pairs). (This observation is the basis of the octet rule.) Atoms in this condition are said to be in their **normal valency state**. The elements C, N, O, and F are *always* surrounded by four pairs of electrons in their molecules and ions. (Be and B may be surrounded by only 2 or 3 pairs, respectively.) However, many of the other elements of the periodic table can display an **expanded valence** where the central atom is surrounded by 5, 6, or even 7 electron pairs.

In forming bonds with another element, an element always forms at least one bond called a **sigma bond** (σ). In some cases, C, O, S, and N may form two or three bonds to another element of the same kind (or to another in this grouping). The second (and third) bonds are called **pi bonds** (π). **Coordinate covalent bonds** occur when one element of a pair supplies the bonding electron pair. **Odd-electron molecules**, which have an uneven number of electrons, are observed but are rare.

The distribution of electrons between bond and lone pairs in molecules can be depicted by **Lewis dot structures** (Section 9.3).

When more than one acceptable Lewis structure can be written for a molecule, the possible structures are called **resonance structures**. Resonance structures differ by the number of bond pairs between a given pair of atoms.

Some properties of chemical bonds are important (Section 9.4). The **bond order** is the number of bonding electron pairs between a given pair of atoms. The **bond length** depends on the sizes of the bonded atoms and on bond order. The **bond dissociation energy (*D*)** is the energy that must be supplied to break a covalent bond. The process of breaking bonds is always endothermic; conversely, when a bond is formed, the process is always exothermic. The bond energy for a particular atom pair depends on the type of atoms bonded and on the bond order. Bond energies can be used to estimate reaction enthalpies.

If two different elements form a bond, the element with the greater effective nuclear charge will attract the shared pair more strongly than the other atom (Section 9.4). The bond will then be **polar**, the atom at one end of the bond having a slight positive charge ($\delta+$) and the atom at the other end having a slight negative charge ($\delta-$). A measure of the ability of an atom in a molecule to attract electrons to itself is the **electronegativity**, χ.

Assuming that all the electrons in a bond belong to the more electronegative element, the **oxidation number** of a bonded atom can be calculated. If the atom charge is calculated assuming that the bonding electrons are shared equally, however, the result is the **formal charge** of the atom (Section 9.4).

A major goal of this chapter is to devise a method to predict the shape or **molecular geometry** for simple molecules and ions. This can be done using the **Valence Shell Electron Pair Repulsion** or **VSEPR** theory (Section 9.5). The sigma bonding and lone electron pairs (**structural pairs**) surrounding an atom are oriented in space so as to make the angles between them as large as possible. The geometry described by these pairs is called the **structural-pair geometry**. The location of the structural pairs that are sigma bonding pairs defines the **molecular geometry**. The angle between bonded atoms depends on the presence and location of any lone pairs.

If a molecule has polar bonds, there can be an accumulation of negative charge in one region of the molecule (Section 9.6). The molecule as a whole is then said to be **polar** or **dipolar** and to have a **dipole moment**.

STUDY QUESTIONS

REVIEW QUESTIONS

1. Give the number of valence electrons for Li, Sc, Zn, Si, and Cl.
2. Explain the difference between an ionic bond and a covalent bond.
3. Refer to Table 9.1 and answer the following questions:
   (a) What molecules do not obey the octet rule?
   (b) How many lone pairs and how many bond pairs are there in ammonia (NH_3) and HF?
   (c) How many sigma bonds and how many pi bonds are there in N_2 and in C_2H_4?
4. Boron compounds often do not obey the octet rule. Illustrate this with BCl_3. Show how the molecule can obey the octet rule by forming a coordinate covalent bond with ammonia (NH_3).
5. Refer to Table 9.2 and answer the following questions:
   (a) Do any molecules having expanded valence have a second period element as the central atom?
   (b) What is the maximum number of bond and lone pairs that surround the central atom in any of these molecules?
6. Which of the following are odd-electron molecules: NO_2, SCl_2, NH_3, and NO_3?
7. Draw the resonance structures of SO_2. Explain how the average number of S—O bonds is 1.5.
8. Consider the following structures for the formate ion, HCO_2^-. Designate which two are resonance structures and which is equivalent to one of the resonance structures. What is the average number of C—O bonds?

(a) 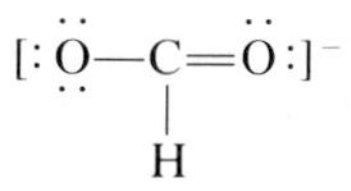

(b) 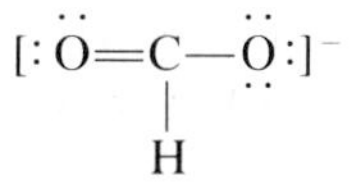

(c) 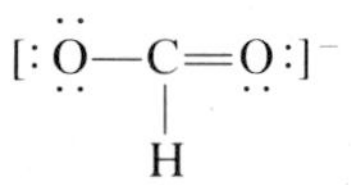

9. Give the bond orders of the bonds in acetylene, H—C≡C—H. How many sigma bonds are there and how many pi bonds?
10. Explain why the C—O bond order in the carbonate ion, CO_3^{2-}, is 1.33.
11. Consider a series of molecules in which carbon is bonded to second period elements: C—O, C—F, C—N, C—C, and C—B. Place these bonds in order of increasing bond length.
12. Define bond dissociation energy. When a C—H bond is broken, is the enthalpy of reaction assigned a negative or a positive sign? Explain briefly.
13. What is the relation between bond order, bond length, and bond energy for a series of related bonds, say C—N bonds?
14. If you wished to calculate the enthalpy of the reaction

$$O{=}O(g) + H{-}H(g) \longrightarrow 2\ H{-}O{-}H(g)$$

what bond energies would you need? Outline the calculation, being careful to show correct algebraic signs.

15. Define and give an example of a polar covalent bond. Give an example of a nonpolar bond.
16. Define electronegativity. Describe the difference between electronegativity and electron affinity.
17. Describe the trends in electronegativity in the periodic table.
18. Describe the difference between oxidation number and formal charge. Which is often the more realistic description of the charge on an atom in a molecule?
19. What is the principle of electroneutrality? How does it apply to the possible resonance structures of CO_2?
20. What is the VSEPR theory? What is the physical basis of the theory?
21. What is the difference between the structural-pair geometry and the molecular geometry of a molecule? Use the water molecule as an example of your discussion.
22. Designate the structural-pair geometry for each case of 2 to 6 electron pairs around a central atom.
23. What molecular geometries are possible for each of the following:

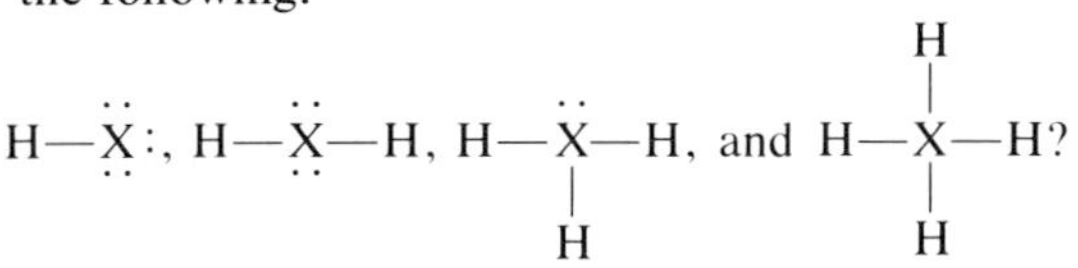

Give the H—X—H bond angle for each of the last three.

24. If you have three structural pairs around a central atom, how can you have a trigonal planar molecule? A bent molecule? What bond angles are predicted in each case?
25. Draw a trigonal bipyramid of electron pairs. Designate the axial and the equatorial pairs. Are there similarly axial and equatorial pairs in an octahedron?

26. Ammonia, NH_3, has a molecular structure similar to NF_3 in Example 9.13. Does NH_3 have a dipole moment? If so, what is the direction of the net dipole in NH_3?

VALENCE ELECTRONS

27. Give the number of valence electrons and the periodic group number for each of the following atoms:
    (a) N (f) C
    (b) B (g) Cl
    (c) S (h) P
    (d) Na (i) Ne
    (e) Mg
28. Give the number of valence electrons for an atom in Groups 1A, 3A, 3B, and 4A.

THE OCTET RULE AND EXPANDED VALENCES

29. Is it possible to have pi bonds between two atoms without there being a sigma bond?
30. For each of the A groups in the periodic table, give the number of bonds an element is expected to form if it obeys the octet rule.
31. Which of the following elements could have expanded valences? That is, which can form compounds with five or six valence pairs?
    (a) C (d) F (g) Se
    (b) P (e) Cl (h) Sn
    (c) O (f) B

LEWIS STRUCTURES

32. Which of the following atoms would you expect to occur widely as terminal atoms in molecule or ions: H, C, B, P, S, F, Al, O, Cl?
33. Draw Lewis structures for the following molecules or ions:
    (a) H_2CO (f) BF_4^-
    (b) HOCl (g) NO_2^+
    (c) NF_3 (h) PO_4^{3-}
    (d) SO_3^{2-}
    (e) ClO_3^-
34. Draw Lewis structures for the following molecules:
    (a) Formic acid, HCOOH. The atomic arrangement is

    H—C(—O)(—O—H) .

    (b) Acetonitrile, $H_3C—CN$.
    (c) Tetrafluoroethylene (the molecule from which Teflon is built), F_2CCF_2.
    (d) Methyl alcohol, $H_3C—OH$.
35. The following molecules or ions have two or more resonance structures. Show all of the resonance structures for each species.
    (a) SO_2 and SO_3
    (b) Nitric acid, HNO_3. The arrangement of atoms is

    H—O—N(—O)(—O) .

    (c) Nitromethane, $H_3C—NO_2$.
    (d) Formamide, $CHONH_2$. The atom arrangement is

    O and H on C; C—N; N bonded to H and H .

36. Draw Lewis structures (and resonance structures where appropriate) for the following molecules and ions. What similarities and differences are there in this series?
    (a) CO_2
    (b) N_3^-
    (c) OCN^-
37. Draw Lewis structures for the following molecules or ions.
    (a) BrF_3
    (b) I_3^-
    (c) XeO_2F_2 (both O and F are terminal atoms)

BOND PROPERTIES

38. Give the number of sigma bonds and number of pi bonds for the following molecules. Tell the bond order between each atom pair. (These are all molecules for which you sketched Lewis structures in the questions above.)
    (a) H_2CO
    (b) SO_3^{2-}
    (c) NO_2^+
    (d) CN^-
    (e) acetonitrile, H_3CCN
    (f) SO_2 and SO_3
39. Compare the S—O and P—O bond lengths in SO_3^{2-} and PO_4^{3-}. Which is the shorter bond?
40. What are the orders of the N—O bonds in NO_2^- and NO_2^+? The N—O bond length in one of these ions is 110 pm and in the other ion 124 pm. Which bond length corresponds with which ion? Account for these different bond lengths.
41. In each pair of bonds below, decide which will be the shorter:
    (a) B—Cl and Ga—Cl
    (b) C—O and Sn—O
    (c) P—S and P—O
    (d) Si—N and P—O
42. Which bond will require more energy to break, the CO bond in formaldehyde (H_2CO) or the CO bond in carbon monoxide (CO)?

BOND ENERGIES AND ENTHALPIES OF REACTION

43. Using the bond energies of Table 9.4, estimate the standard heat of formation of each of the following molecules:
(a) $NH_3(g)$
(b) $H_2O(g)$
(c) H—O—O—H(g), hydrogen peroxide

44. Using the methods of Example 9.5, calculate the standard heat of formation of ethylene, $H_2C{=}CH_2$ (g). Compare your result with the heat of formation of acetylene that was obtained in Example 9.5. Why is ΔH_f° for ethylene so much less positive than ΔH_f° for acetylene?

45. The standard heat of formation, ΔH_f°, of H_2S is −20.63 kJ/mol, and the heat of formation of S(g) [from S(s)] is 278.8 kJ/mol. Using this information and the bond energies of Table 9.4, estimate the H—S bond energy. Is the H—S bond energy greater or less than the H—O bond energy? From this comparison, what could you conclude about the trend in H—*X* bond energies as the atomic weight of *X* increases?

46. Using the following information:

ΔH_f° [HI(g)] = 26.48 kJ/mol
ΔH° for $I_2(s) \longrightarrow I_2(g)$ is 62.44 kJ/mol

calculate the bond energy of the H—I bond in hydrogen iodide, HI(g). Compare your calculated value with the bond energies of the other hydrogen halides (HF, HCl, HBr) in Table 9.4. Based on this comparison, what would you conclude about the trend in bond energies in H—*X* as the atomic weight of *X* increases?

47. What is meant by the phrases "combustion reaction" and "heat of combustion"? Write an equation for the reaction of O_2 with CH_4 to illustrate these concepts.

48. The equation for the combustion of methyl alcohol is

$$H_3COH(g) + \tfrac{3}{2} O_2(g) \longrightarrow CO_2(g) + 2\ H_2O(g)$$

Using the bond energies in Table 9.4, estimate the heat of this reaction, that is, the heat of combustion of methyl alcohol.

49. The bunsen burners in your laboratory are fueled by either natural gas, which is primarily methane (CH_4), or by propane, $H_3C—CH_2—CH_3$. Using the bond energies in Table 9.4, calculate the heat of combustion of each of these substances, $CH_4(g)$ and $C_3H_8(g)$. Which provides the greater amount of heat *per gram*?

50. Using the bond energies of Table 9.4, and the heat of vaporization of C(graphite) (+717 kJ/mol) (see Example 9.5), calculate the standard heat of formation of one of the basic building blocks of life, the amino acid glycine.

glycine = H—C(H)(NH$_2$)—C(=O)—O—H

51. In Chapter 2, the standard state of phosphorus was described as consisting of P_4 molecules. If the enthalpy of the following reaction is +1259 kJ per mole of P_4,

$$P_4(s) \longrightarrow 4\ P(g) \qquad \Delta H^\circ = +1259 \text{ kJ/mol of } P_4$$

calculate the standard heat of formation of phosphine, $PH_3(g)$. (The necessary bond energies are given in Table 9.4.)

52. The compound oxygen difluoride is quite unstable, giving oxygen and HF on reaction with water.

$$OF_2(g) + H_2O(g) \longrightarrow O{=}O(g) + 2\ HF(g) \qquad \Delta H^\circ = -318 \text{ kJ}$$

The enthalpy of this reaction is −318 kJ. Using bond energies, calculate the bond dissociation energy of the O—F bond in OF_2.

53. Bond energy calculations can only give approximate values for heats of reaction or formation. The reason for this is that the bond energies are only average values and are not necessarily specific to the compound under consideration. To illustrate this point, carry out the indicated calculations on the following reaction:

$H_3C—C(=O)—CH_3(g) \rightleftharpoons H_3C—C(—O—H){=}CH_2(g)$

acetone ⇌ 2-hydroxypropene

(a) The standard heat of formation of acetone is −216 kJ/mol and that of 2-hydroxypropene is −117 kJ/mol. Use these values to calculate the heat of the reaction above. (The two compounds above are *isomers* of each other. They each have the same formula, but the atoms are connected in different arrangements.)
(b) Using the bond energies in Table 9.4, estimate the heat of the reaction. Compare your estimate with the calculation in part (a).

54. Phosgene, Cl_2CO, is a highly toxic gas that has been used as a war gas in World War I. It is an insidious poison as it is not immediately toxic, even when lethal concentrations are inhaled. Using the bond energies of Table 9.4, estimate the heat of the reaction of carbon monoxide and chlorine to produce phosgene.

$$CO(g) + Cl_2(g) \longrightarrow Cl_2CO(g)$$

55. Hydrogenation reactions, the addition of H_2 to a molecule, are widely used in industry to transform one

compound into another. For example, the molecule propene (called an "olefin" because of the C=C double bond) is converted to propane (called an "alkane" because there are only C—H bonds and C—C single bonds) by addition of H_2.

$$H_3C{-}CH{=}CH{-}H(g) + H_2(g) \longrightarrow H_3C{-}CH_2{-}CH_2{-}H(g)$$

Using the bond energies of Table 9.4, estimate the heat of hydrogenation of propene.

56. Acetone can be converted into isopropyl alcohol, rubbing alcohol, by a hydrogenation reaction.

$$\underset{\text{acetone}}{H_3C{-}C({=}O){-}CH_3(g)} + H_2(g) \longrightarrow \underset{\text{isopropyl alcohol}}{H_3C{-}CH(OH){-}CH_3}$$

Estimate the heat of this reaction using the bond energies of Table 9.4.

57. Urea is widely used as a fertilizer because of its high nitrogen content, so better methods for its production are always being sought. Using Table 9.4, estimate the heat of a reaction to make urea.

$$2\ NH_3(g) + CO(g) \longrightarrow \underset{\text{urea}}{H_2N{-}C({=}O){-}NH_2(g)} + H_2(g)$$

BOND POLARITY AND ELECTRONEGATIVITY

58. Which of the following statements are correct and which are incorrect? For ones that are not true, change them to make them correct.
   (a) Of two bonded atoms, the one closer to the upper right corner of the periodic table is the more electronegative.
   (b) In the SO bond, S is the more electronegative atom.
   (c) In the H—F bond, H is the more electronegative atom.
   (d) In the CO bond, the oxygen is the more electronegative atom.
   (e) In the H—B bond, both atoms have almost the same electronegativity.

59. In each pair of bonds, indicate the more polar bond and use an arrow to show the direction of polarity in each bond.
   (a) C—O and C—N (c) P—H and P—N
   (b) N—O and P—S (d) B—H and B—I

60. Which of the compounds below would have the most polar bonds and which the least polar bonds?
   (a) CH_4 (b) NH_3 (c) H_2O (d) HF

61. Place the following compounds in order of increasing degree of ionic character in the X—F bond.
   (a) BeF_2 (c) HF (e) NF_3
   (b) F_2 (d) OF_2

OXIDATION NUMBERS AND FORMAL CHARGES

62. Using Lewis structures and relative atom electronegativities, give the oxidation number of each element in each of the following molecules or ions.
   (a) H_2O (e) ClO^-
   (b) H_2O_2 (f) SO_2
   (c) CN^- (g) SO_3
   (d) HPO_3^{2-} (h) N_2O

63. Using Lewis structures and relative atom electronegativities, give the oxidation number of each element in each of the following molecules or ions.
   (a) H_2S (e) XeF_2
   (b) CH_4 and H_2CCH_2 (f) ICl_2^+
   (c) H_2CO (g) OF_2
   (d) ClF_3

64. Calculate the formal charge on each atom in each of the following molecules and ions. In each case, compare the formal charge with the oxidation number.
   (a) H_2O (e) HOF
   (b) CH_4 (f) XeF_2
   (c) NO_2^+ (g) ICl_2^-
   (d) OCl^-

65. Two resonance structures are possible for SO_2 and HCO_2^-. What is the formal charge on each element in the molecule and ion, respectively? Are the formal charges the same in each resonance structure? If not, what is the *average* formal charge on each atom?

66. Three resonance structures are possible for dinitrogen oxide, N_2O.
   (a) Draw the three resonance structures.
   (b) Calculate the formal charge on each atom in each resonance structure.
   (c) Based on formal charges, decide on the most reasonable resonance structure.

67. The resonance structures of the NCO^- ion were given in Example 9.9, and the most likely resonance structure was determined. If NCO^- were to react with H^+, which end of the ion is the H^+ ion most likely to "attack" (N or O)? (The H^+ ion, being positive, will "attack" the more negative end of the NCO^- ion.) Having decided to which end of NCO^- the H^+ would become attached, draw the Lewis structure of cyanic acid, HNCO, with the H atom attached to the correct terminal atom of the NCO grouping.

68. The cyanate ion is NCO^- with the least electronegative atom (atom of lowest electron affinity) in the

center. The very unstable fulminate ion, CNO^-, has the same formula, but now N is in the center.
(a) Draw the three possible resonance structures of CNO^-.
(b) Calculate the formal charge on each atom in each resonance structure.
(c) On the basis of formal charges, decide on the resonance structure with the most reasonable distribution of charge.
(d) Mercury fulminate is so unstable it is used in blasting caps. Can you offer an explanation for this instability? (*Hint*: Are the formal charges in any resonance structure reasonable in view of the relative electronegativities of the atoms?)

MOLECULAR SHAPE

69. Is the following statement true or false? If false, rewrite it to make it correct. "Only bond pairs around an atom are considered structural-pairs."

70. "Sigma and pi bond pairs between two atoms are considered as one structural-pair." If the preceding statement is false, rewrite it to make it correct.

71. For each molecule in Table 9.6 tell how many structural-pairs there are about the central atom.

72. Is it always true that the name given to the structural-pair shape is also the correct name for the molecular shape? Explain your answer briefly.

73. Draw the Lewis structure for each of the following molecules or ions and then (1) give the name of the central atom structural-pair shape, (2) give the name of the molecular (or ionic) shape, and (3) comment on similarities and differences in each series of molecules and ions.
(a) Three atoms: CO_2, NO_2^-, SO_2, O_3, and ClO_2^-.
(b) Four atom species with oxygen: BO_3^{3-}, CO_3^{2-}, NO_3^-, SO_3^{2-}, ClO_3^-.
(c) Fluorine terminal atoms: BF_3, CF_4, NF_3, OF_2, HF.

74. Describe the shape or structure you would expect for the following molecules or ions:

(a) BH_3	(e) ClF_2^+
(b) NH_2Cl	(f) SF_2
(c) $SnCl_3^-$	(g) BeF_2
(d) $SnCl_4$	

75. The following are examples of molecules or ions for which the central atom has expanded valence. (1) Give the name of the central atom structural-pair shape and (2) give the name of the molecular shape. (a) ClF_3, (b) ClF_4^-, (c) ClF_5, (d) SF_4, (e) PF_5, (f) PF_6^-, (g) SiF_6^{2-}, (h) XeF_4.

76. Given that the spatial requirements of lone pairs are much greater than that of bond pairs, explain why
(a) XeF_2 has a linear molecular structure and not a bent one.
(b) ClF_3 has a T-shaped structure and not a planar, triangular structure.

77. Which would have the greater O—N—O bond angle, NO_2^- or NO_2^+? Explain your answer briefly.

78. Which compound has the larger bond angle, NF_3 or OF_2? Explain briefly.

79. Give approximate values for the indicated bond angles in the following molecules.
(a) SO_2, the O—S—O angle
(b) BF_3, the F—B—F angle
(c) H—Ö—N(=Ö)—Ö: (angles 1 and 2 indicated)
(d) H—C(H)(H)—C≡N: (angles 1 and 2 indicated)
(e) H—C(H)(H)—Ö—H (angles 1 and 2 indicated)
(f) $SnCl_3^-$, the Cl—Sn—Cl angle
(g) H₂N̈—C(=Ö:)—N̈H₂ (angles 1 and 2 indicated)
(h) SeF_4
(i) SOF_4, all angles (the O is in an equatorial position)
(j) $XeOF_4$, all angles (the O and F atoms are all terminal atoms)

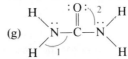

80. Compare the F—Cl—F angles in ClF_2^+ and ClF_2^-. From Lewis structures determine the approximate bond angle in each ion. Explain which ion has the greater angle and why.

81. The structure of ClF_3 is given in Table 9.8. Experiment shows that the F_{axial}—Cl—$F_{equatorial}$ angle is 87.5°. Explain briefly why this angle should be less than the expected 90°.

MOLECULAR POLARITY

82. Which of the following molecules should be polar? For each polar molecule, indicate the direction of polarity; that is, indicate the negative and positive ends.
(a) N_2O (b) CS_2 (c) PCl_3 (d) XeF_2

83. Which of the following molecules should have the largest polarity? That is, in which is there the greatest difference in charge between the ends of the molecule?
(a) H_2S (b) H_2O (c) Cl_2O

84. Which of the following molecules should have a dipole moment: (a) FBH_2, (b) PH_3, (c) XeF_4? In each case show the direction of the net dipole.

85. The compound $C_2H_2Cl_2$ can exist in two forms. Does either of these have a net dipole moment? If yes, give the direction of the net dipole.

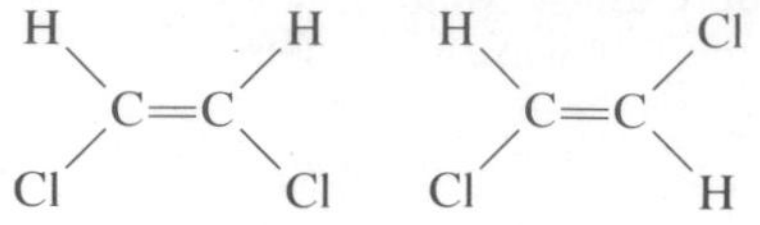

cis trans

GENERAL QUESTIONS

86. One of the most objectionable compounds in photochemical smog is peroxyacetyl nitrate, PAN. It has the structure below. (a) How many sigma bonds are there in the molecule? (b) How many pi bonds are there in the molecule? (c) Give the approximate values of the angles labeled 1–4.

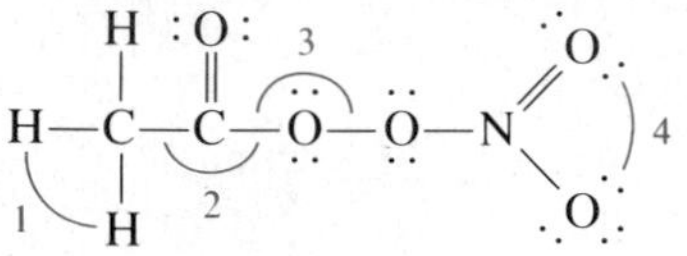

87. In 1962 Watson and Crick received the Nobel Prize for their simple but elegant model for the "heredity molecule" DNA. The key to their structure (the famous double helix) was an understanding of the geometry and bonding capabilities of the nitrogen-containing bases thymine, adenine, guanine, and cytosine. The thymine molecule has the structure below. (a) There are three carbon-to-carbon bonds in the molecule. Which is the shortest? (b) How many sigma bonds are there in thymine? How many pi bonds? (c) Give the approximate values of the indicated bond angles. (d) Which are the most polar bonds in the molecule?

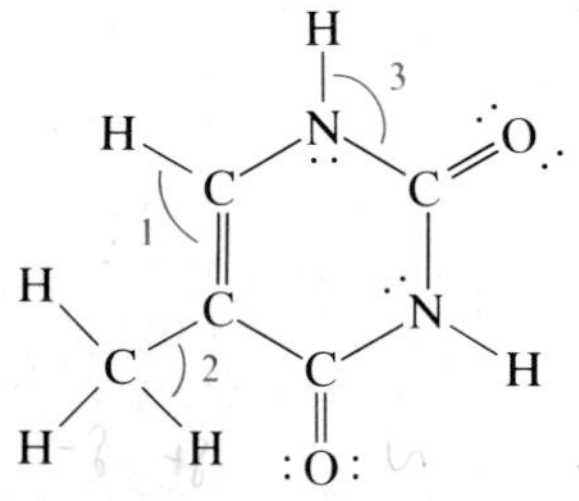

88. Histidine is one of the basic amino acids important to life. Its structure is

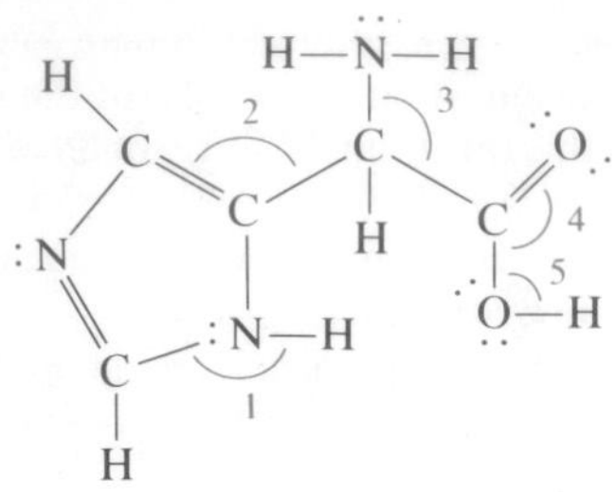

(a) How many sigma bonds are there in histidine? How many pi bonds?
(b) Which carbon–carbon bond is the shortest of the three in the molecule?
(c) Give approximate values of the indicated bond angles.
(d) Which of the carbon–carbon bonds in the molecule would be the strongest? That is, which should have the highest bond energy?

89. This problem concerns the addition of H_2 to benzene to form cyclohexane.

$$C_6H_6(g) + 3\,H_2(g) \longrightarrow C_6H_{12}(g)$$

benzene cyclohexane

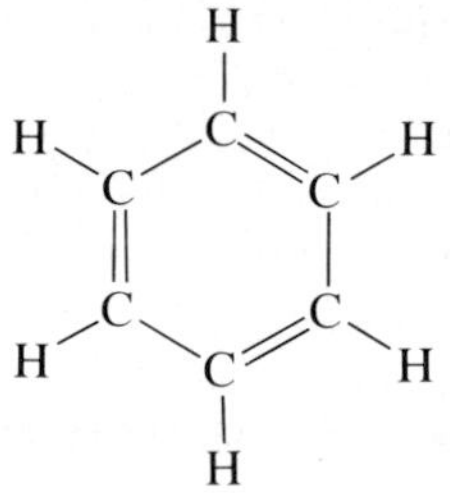

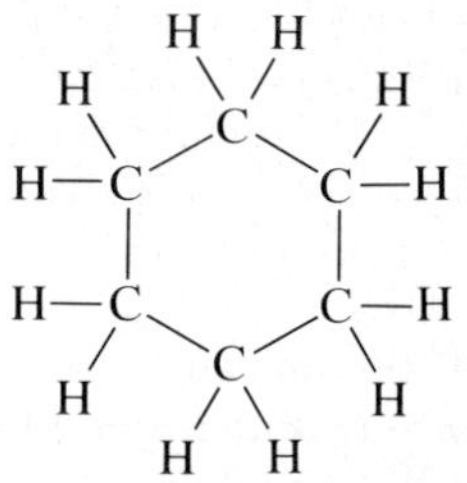

(a) Calculate the heat of the reaction knowing that $\Delta H_f^\circ[C_6H_6(g)] = +82.8$ kJ/mol and $\Delta H_f^\circ[C_6H_{12}(g)] = -123.1$ kJ/mol.
(b) Benzene has two resonance structures (Chapter 23), and this makes the C—C bond order 1.5. However, assume for the moment that the C_6 ring consists of three C=C bonds and three C—C bonds, and calculate the heat of the hydrogenation reaction using bond energies.
(c) There should be a difference between the actual heat of hydrogenation calculated in (a) and the estimated value in (b). This difference occurs because we assumed resonance in (a) but not in (b). What does this tell you about the effect of resonance on the "stability" of a compound?

10

Further Concepts of Chemical Bonding: Orbital Hybridization, Molecular Orbitals, and Metallic Bonding

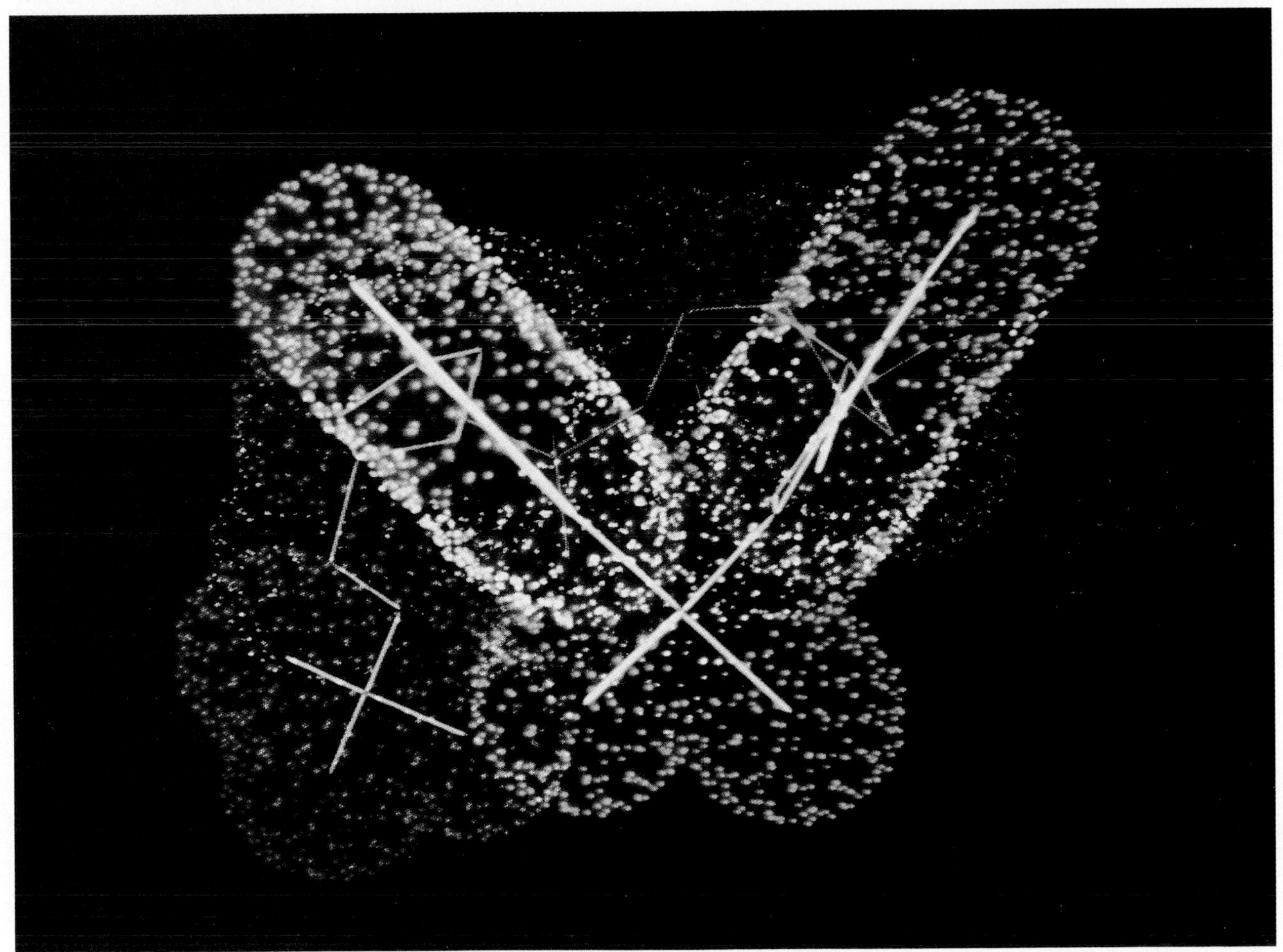

Computer-drawn structure of cis-platin, $(NH_3)_2PtCl_2$ (Dr. Stephen Lippard, MIT)

CHAPTER OUTLINE

The preceding chapter outlined the methods that chemists use to represent bonds between atoms. This chapter will introduce, still in an elementary way, some more ideas on how atoms use their orbitals in bond formation.

There are two commonly used approaches to chemical bonding: **valence bond (VB) theory** and **molecular orbital (MO) theory**. The former was first developed by Linus Pauling, while the latter came from work by Robert Mullikan. Mullikan's approach is to combine pure atomic orbitals on each atom to derive *molecular orbitals* that are spread or **delocalized** over the molecule. Only after developing the molecular orbitals are the pairs of electrons of the molecule assigned to these orbitals; thus, the molecular electron pairs are more or less uniformly distributed over the molecule. In contrast, Pauling's approach is more closely tied to Lewis's idea of electron pair bonds, where each electron pair is confined to the space between two bonded atoms and of lone pairs of electrons localized on a particular atom. Only bonding and nonbonding (lone pair) orbitals are considered.

Linus Pauling received the Nobel Prize in chemistry for this work in 1954, and Mullikan was awarded the Nobel Prize in 1966.

Although the VB and MO theories may seem quite different at an elementary level, they lead ultimately to equivalent results, as a more advanced treatment would show.

10.1 VALENCE BOND THEORY

The description of covalent bonding in Chapter 9 can now be given in the terminology of valence bond theory. According to this theory, two atoms form a bond when both of the following conditions occur:

1. There is **orbital overlap** between the two atoms. If two H atoms approach each other closely enough, their 1*s* orbitals can partially occupy the same region of space (Figure 10.1).
2. A maximum of two electrons, of opposite spin, is present in the overlapping orbitals.

Usually one electron of a bond pair comes from each atom. If one of the two atoms supplies both electrons, the result is called a coordinate covalent bond.

Due to orbital overlap, the pair of electrons is found within a region influenced by both nuclei. This means that both electrons are mutually attracted to both atomic nuclei, and this, among other factors, leads to bonding.

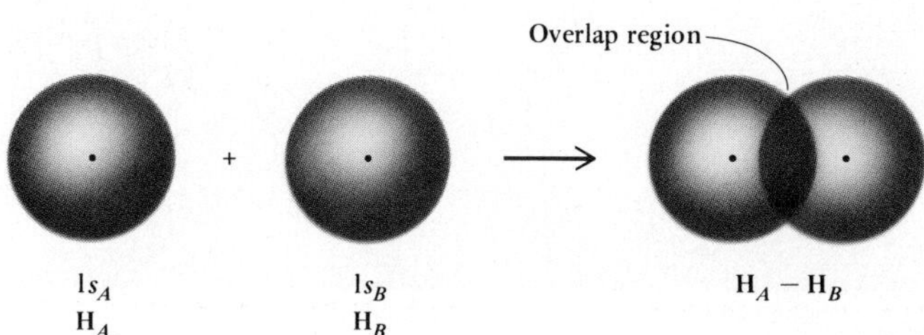

FIGURE 10.1
The overlap of 1*s* orbitals from two H atoms to form the sigma bond of H_2.

LINUS PAULING (1901–)

Linus Pauling was born in Portland, Oregon in 1901, the son of a druggist. He earned a B.Sc. degree in chemical engineering from Oregon State College in 1922 and completed his Ph.D. in chemistry at the California Institute of Technology in 1925. Before joining Cal Tech as a faculty member, he traveled to Europe where he worked briefly with Erwin Schrödinger and Niels Bohr. In chemistry he is best known for his work on chemical bonding. Indeed, his book *The Nature of the Chemical Bond* has influenced a generation of scientists, and it was for this work that he was awarded the Nobel Prize in Chemistry in 1954. However, shortly after World War II, Pauling and his wife began a crusade to limit nuclear weapons, a crusade that came to fruition in the limited test ban treaty of 1963. For this effort, Pauling was awarded the 1963 Nobel Prize for Peace. Never before had any person received two unshared Nobel Prizes.

As the extent of overlap between two orbitals increases, the strength of the bond increases. This is seen in Figure 10.2 as a drop in energy as two H atoms, originally far apart, come closer and closer together. However, the figure also shows that, as the atoms come very close to one another, the energy increases rapidly, due to the repulsion of one positive nucleus by the other. Thus, there is an optimum distance, the observed bond distance, at which the total energy is at a minimum; here there is a balance of attractive and repulsive forces.

The overlap of two *s* orbitals, one from each of two atoms (Figure 10.1), leads to a **sigma bond**: the electron density of a sigma bond is greatest along the axis of the bond. Sigma bonds can also form by the overlap of an *s* orbital with a *p* orbital or by the head-to-head overlap of two *p* orbitals (Figure 10.3).

Since *s* orbitals are spherical, two H atoms could approach one another from any direction, and a sigma bond would form. Other types of orbital overlap (*s*/*p* and *p*/*p*) are directional, however. The criterion of **maximum overlap** is part of the valence bond theory. In forming a bond, two atoms are arranged to give the greatest orbital overlap possible, since this leads to the strongest possible sigma bond. This means two *p* orbitals should overlap directly along the axis of the bond to form the strongest possible sigma bond (Figure 10.4).

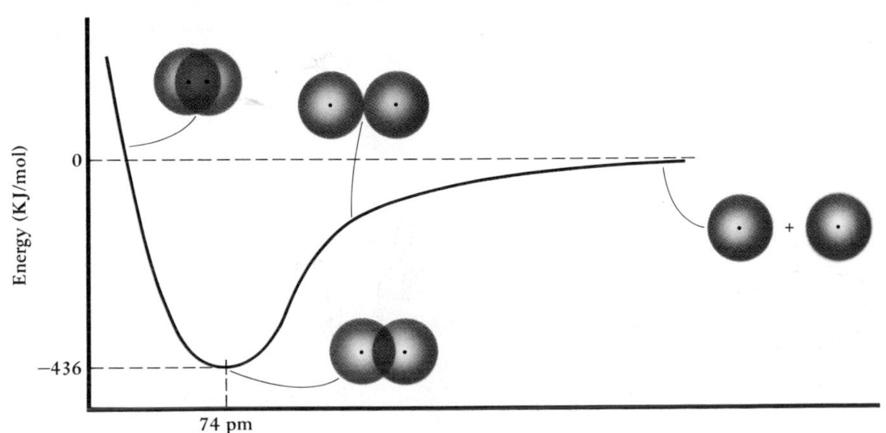

FIGURE 10.2
Total potential energy change in the course of H—H bond formation.

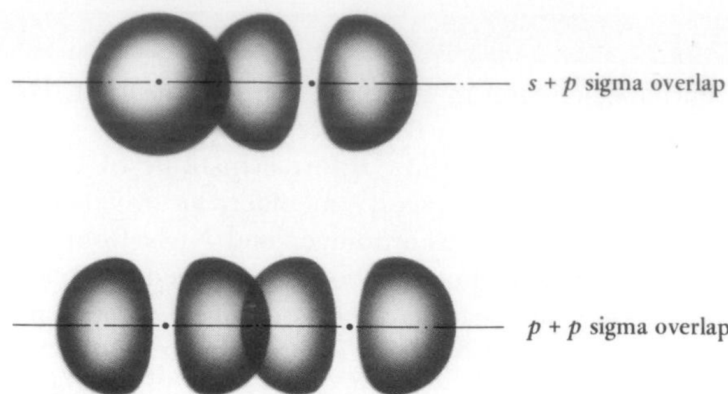

FIGURE 10.3
Sigma bond formation. (a) The *s* orbital overlaps a *p* orbital. (b) Two *p* orbitals overlap head to head.

HYBRID ORBITALS

An isolated carbon atom has two unpaired electrons, and so might be expected to form only two bonds.

Carbon Electron Configuration

$$[\mathrm{He}]\ 2s^2 2p_x^{\ 1} 2p_y^{\ 1} \qquad \text{or} \qquad [\mathrm{He}]\ \underset{2s}{\underline{\uparrow\downarrow}}\ \underset{2p}{\underline{\uparrow}}\ \underset{2p}{\underline{\uparrow}}\ \underset{2p}{\underline{\ \ }}$$

However, there are four C—H bonds in methane, CH_4. Furthermore, VSEPR theory tells us, and experiment confirms it, that the structural-pair geometry of the C atom in CH_4 is tetrahedral. There must be *four equivalent* bonding electron pairs around the C atom. The three *p* orbitals around an isolated atom lie at 90° to one another (Figure 7.20). Therefore, if sigma bonds were formed in some manner using pure *s* and *p* orbitals,

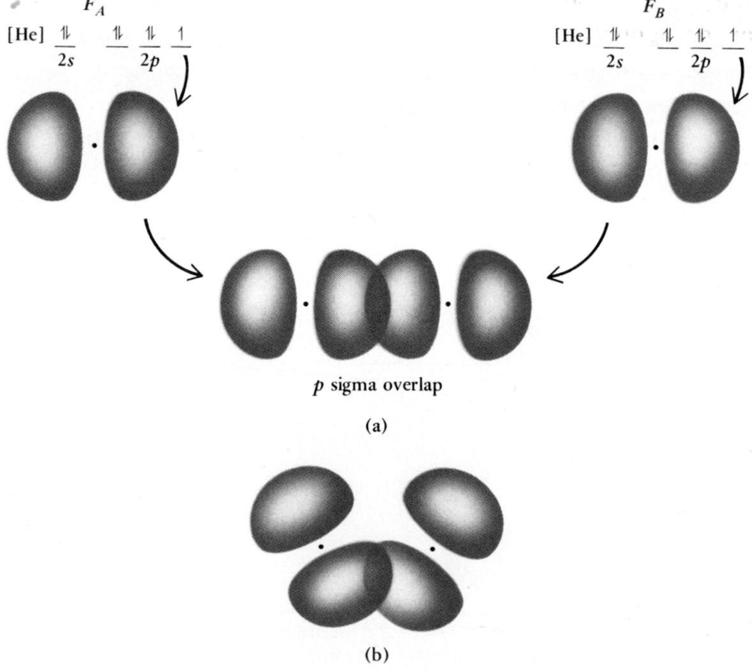

FIGURE 10.4
Sigma bond formation in F_2. (a) Maximum overlap of atomic orbitals occurs when the orbitals interact along the axis of the orbital. (b) If the orbitals do not overlap head to head, there is a smaller overlap and a weaker bond.

H — C(H)(H) — H

Electron dot structure

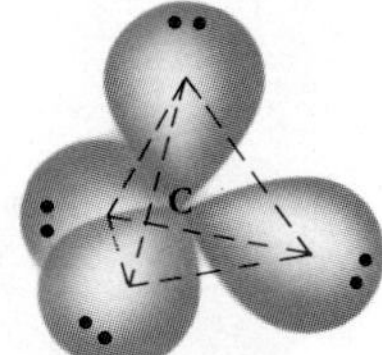

Structural-pair geometry

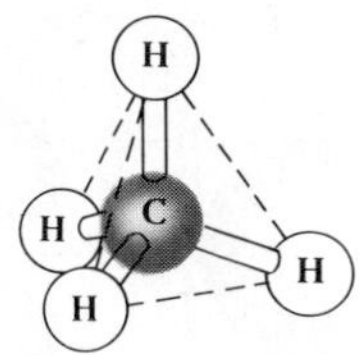

Molecular geometry

the bonds would neither be equivalent nor would they be arranged correctly in space. Some other scheme is required to account for C—H bonds at an angle of 109°.

Pauling proposed **orbital hybridization** as a way to explain the formation of bonds by the maximum overlap of atomic orbitals and yet accommodate the use of *s* and *p* orbitals. In order for the four C—H bonds of methane to have their maximum strength, there must be maximum orbital overlap between the carbon orbitals and the H-atom *s* orbitals at the corners of a tetrahedron. Thus, Pauling suggested that the approach of the H atoms to the isolated C atom causes distortion of the four carbon *s* and *p* orbitals. These orbitals **hybridize** or combine in some manner to provide *four equivalent hybrid orbitals that point to the corners of a tetrahedron* (Figure 10.5).* We label *each hybrid orbital* as $\mathbf{sp^3}$, since the orbitals are the result of the combination of one *s* and three *p* orbitals on one atom. Each hybrid orbital combines the properties of its *s* and *p* orbital parents.

You may be familiar with hybridization in agriculture, where two varieties of a plant are mixed to provide a new variation that reflects some characteristics of its parents.

Be sure to notice that *four* atomic orbitals produce *four* hybrid orbitals. There is *always* a 1:1 correspondence between the number of atomic orbitals used and the number of hybrid orbitals produced.

The theory of orbital hybridization is an attempt to explain how in CH_4, for example, there can be four equivalent bonds directed to the corners of a tetrahedron. Another outcome of hybrid orbital theory is that hybrid orbitals are more extended in space than any of the atomic orbitals from which they are formed. This important observation means that greater overlap can be achieved between C and H in CH_4, for instance, and stronger bonds result than without hybridized orbitals.

The four sp^3 hybrid orbitals have the same shape, but they differ in their direction in space. Each also has the same energy, which is the weighted average of the parent *s* and *p* orbital energies. Four sigma bonds are to be formed by carbon, so each of the four valence electrons of carbon is assigned, according to Pauli's principle and Hund's rule, to a separate hybrid orbital.

This is the reason the four valence electrons of C were shown as unpaired on page 298.

energy

2p ↿ ↿ —

2s ⥮

isolated C atom

→

↿ ↿ ↿ ↿

four sp^3 hybrid orbitals in CH_4

*Pure atomic orbital wave functions can be combined mathematically to give hybrid orbital wave functions. That fact alone is sufficient here; the mathematics need not concern us.

FIGURE 10.5
Hybrid orbital formation. (a) The four *s* and *p* atomic orbitals are combined or hybridized to form four new sp^3 hybrid orbitals. These orbitals are equivalent and are directed to the corners of a tetrahedron (angle 109.5°). (b) The shape of one of the four hybrid orbitals. The small "back" lobe of each orbital was omitted in (a) for clarity. (c) The C—H *sigma bond* bond in CH_4 formed from a carbon sp^3 hybrid orbital and a hydrogen 1*s* orbital.

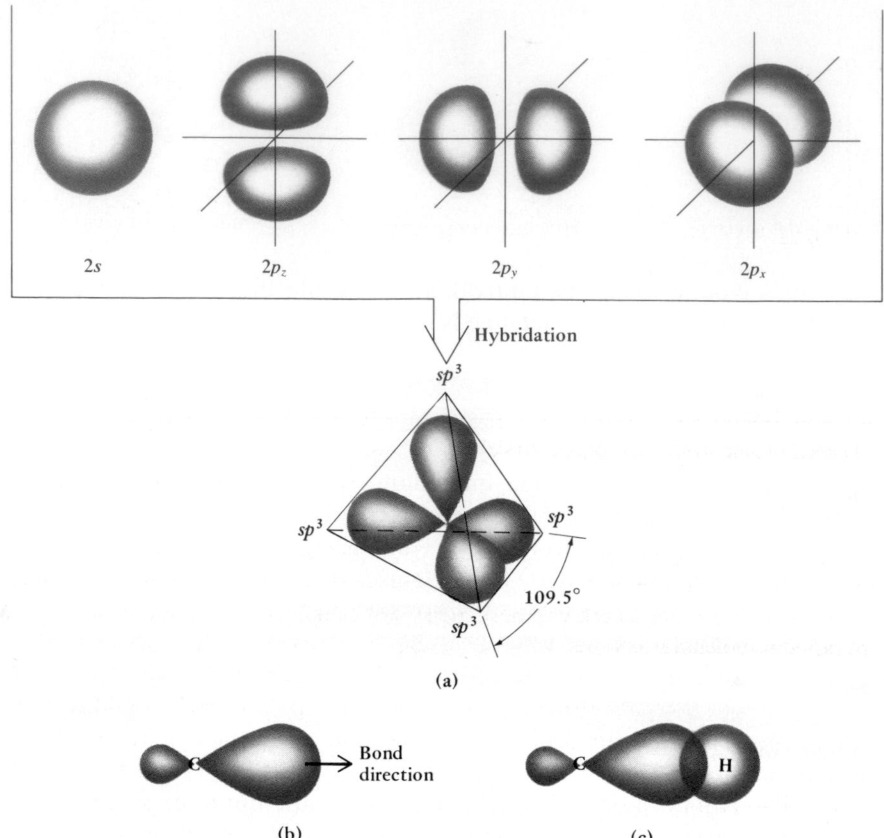

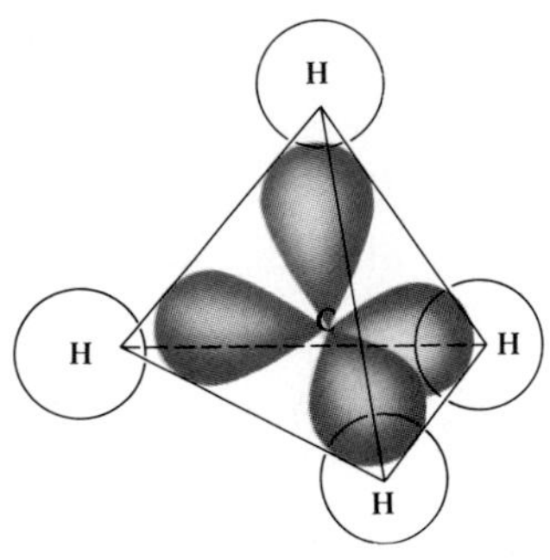

Overlap of each half-filled sp^3 hybrid orbital with a half-filled hydrogen 1*s* orbital gives *four equivalent C—H bonds arranged tetrahedrally*, as required by experimental evidence.

Hybrid orbitals can also be used to explain bonding and structure for such common molecules as H_2O and NH_3. An isolated O atom has two unpaired valence electrons as required for two bonds, but these electrons are in orbitals 90° apart.

Oxygen Electron Configuration

$1s^2 2s^2 2p^4$ or [He] $\underset{2s}{\underline{\uparrow\downarrow}}\ \underset{2p}{\underline{\uparrow\downarrow}}\ \underset{2p}{\underline{\uparrow}}\ \underset{2p}{\underline{\uparrow}}$

However, we know that the water molecule is based on an approximate tetrahedron of structural pairs: the two bond pairs are 105° apart, and the lone pairs occupy the other corners of the tetrahedron. If we allow the four *s* and *p* orbitals of oxygen to distort or hybridize on approach of the H atoms, four sp^3 hybrid orbitals are created. Two of these orbitals are occupied by unpaired electrons, and lead to the O—H sigma bonds. The other two orbitals contain pairs of electrons and so are the lone pairs of the water molecule.

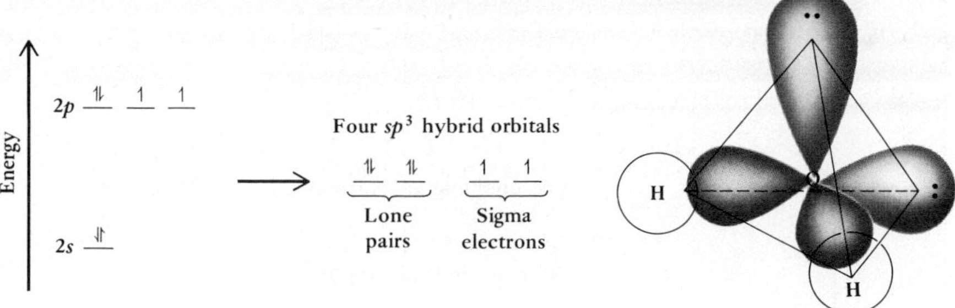

EXAMPLE 10.1

HYBRID ORBITALS IN BONDING

Describe the bonding in ammonia, NH_3, using orbital hybridization.

Solution The electron dot structure of ammonia shows that there are four structural pairs (one lone pair and three bond pairs), which must be at the corners of a tetrahedron. Since three of these pairs are bond pairs, the molecule has a pyramidal molecular geometry.

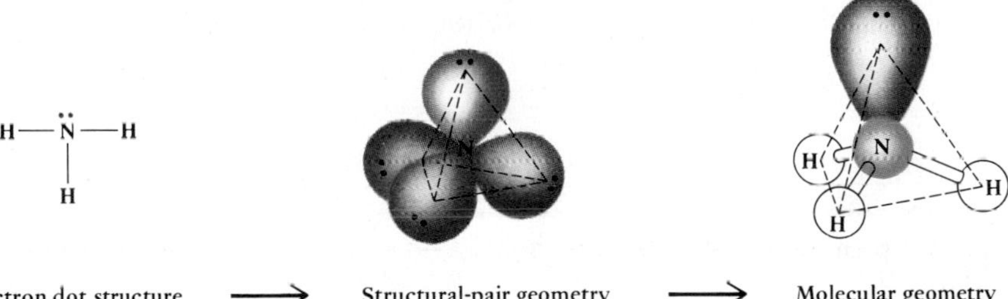

To explain bonding in NH_3 and an H—N—H angle of approximately 107°, we invoke orbital hybridization. The *s* and *p* orbitals of nitrogen combine to give four sp^3 orbitals, one containing a lone pair of electrons and each of the other three having one unpaired electron.

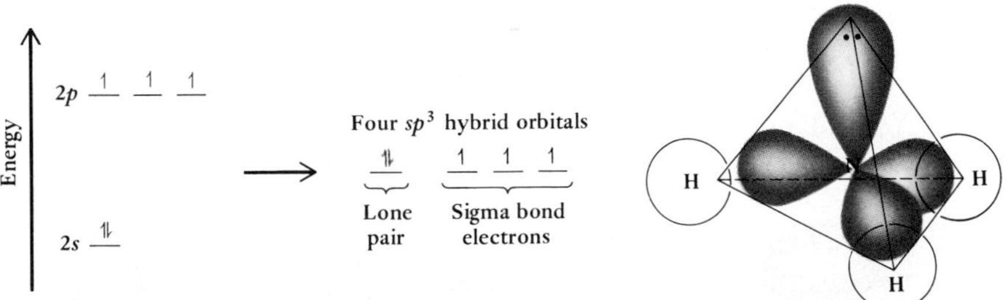

Thus, there are three bond orbitals and one lone pair orbital at the corners of a tetrahedron.

Other combinations of atomic orbitals also lead to hybrid atomic orbitals. For example, an s orbital and two p orbitals on the same atom can combine to form three sp^2 hybrid orbitals that are directed to the corners of a planar triangle (Figure 10.6).

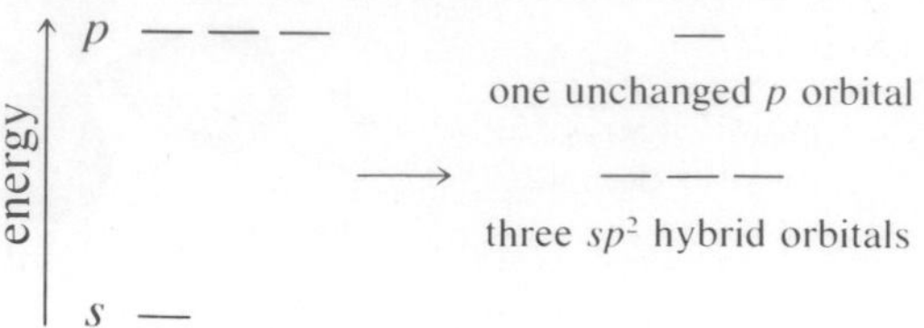

Similarly, one s and one p orbital may hybridize to form two sp hybrid atomic orbitals. Here the hybrid orbitals are directed away from one another with an angle of 180° (Figure 10.6).

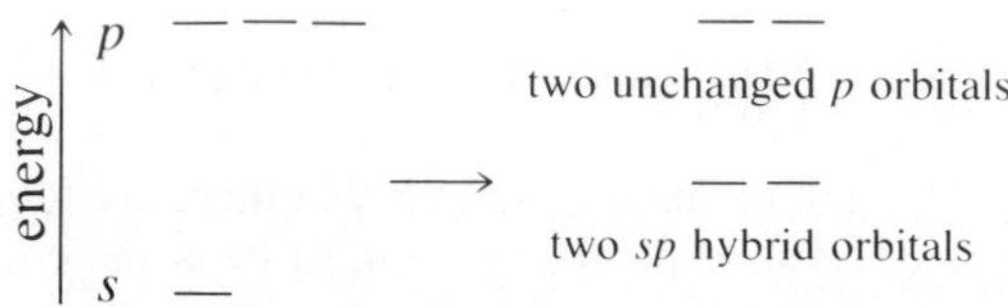

In all hybridization schemes, *the number of hybrid orbitals produced is equal to the number of pure atomic orbitals used in the combination*. Thus, in both sp^2 and sp hybridization, one or two pure p orbitals, re-

FIGURE 10.6
Formation of sp^2 and sp hybrid orbitals.

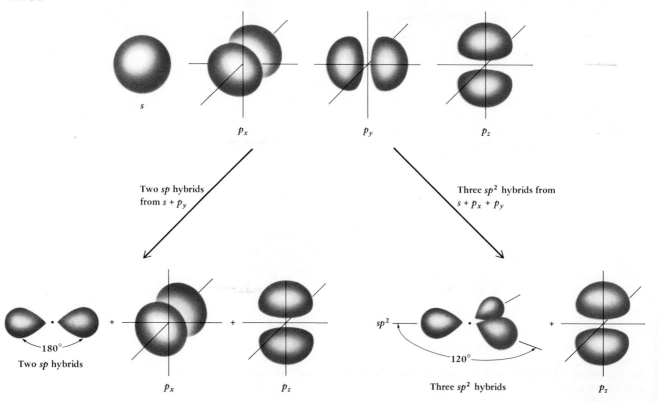

spectively, remain after hybridization. These can remain empty or can be utilized in pi bond formation, as described below.

EXAMPLE 10.2

HYBRIDIZATION AND BONDING

Describe the bonding in BF_3 using orbital hybridization.

Solution The structural-pair and molecular geometry of BF_3 are both trigonal planar.

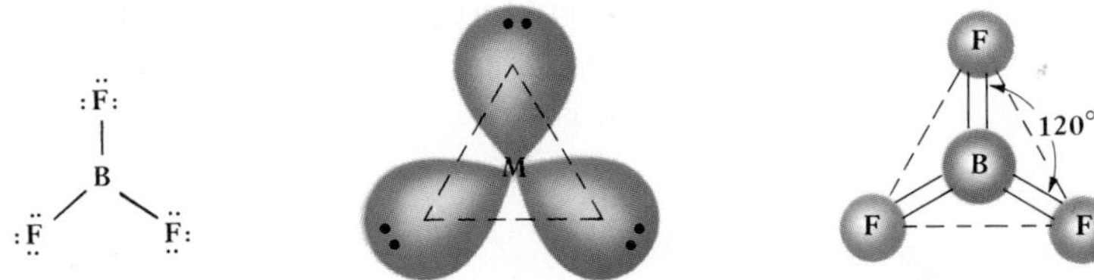

Three bonds, at the corners of a triangle, must be formed. Thus, three boron atom orbitals (the *s* orbital and the two *p* orbitals in the molecular plane) are hybridized to form sp^2 orbitals,

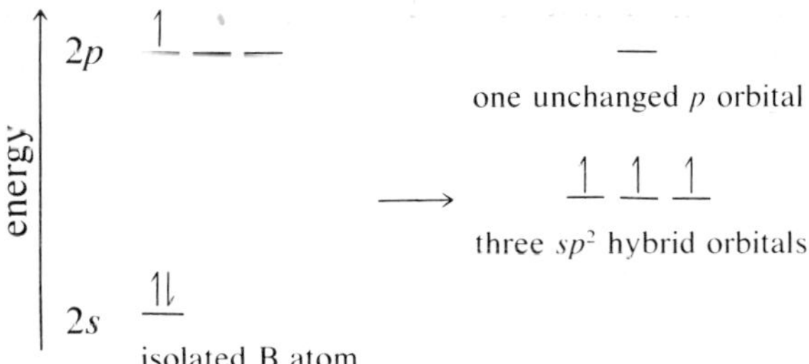

and each hybrid orbital contains an unpaired electron for bond formation with an F atom.

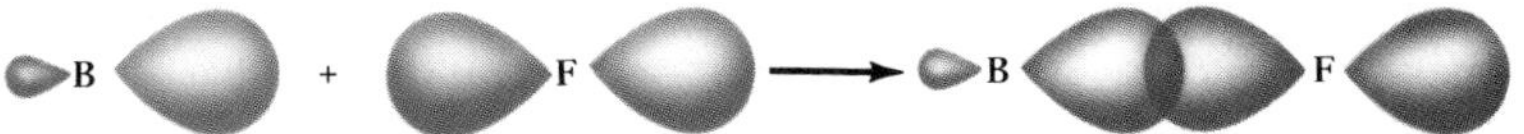

Here the boron *p* orbital not used in hybridization remains unfilled. This fact means that such molecules are reactive, as they tend to acquire an electron pair for the empty orbital from another molecule if possible.

EXAMPLE 10.3

RECOGNIZING HYBRID ORBITALS

What hybrid orbital sets are required for the central atoms in (a) PCl_3 and (b) BeH_2?

Solution (a) For PCl_3 the electron dot structure and VSEPR theory tell us that the P atom must be surrounded by four electron pairs approximately at the corners of a tetrahedron.

:Cl—P—Cl:
|
:Cl:

Four structural pairs arranged tetrahedrally require the P atom to be sp^3 hybridized. Phosphorus, like nitrogen, is in Group 5A. Therefore, the bonding in PCl_3 resembles that in NH_3 (Example 10.1). The phosphorus lone pair is assigned to one sp^3 hybrid orbital, and the P—Cl bonds are formed by overlap of a Cl half-filled p orbital with a half-filled P sp^3 orbital.

(b) BeH_2 is a linear molecule.

Linear geometry and two sigma bonds are achieved by forming sp hybrid orbitals on Be (from the s orbital and the p orbital that lies along the HBeH axis).

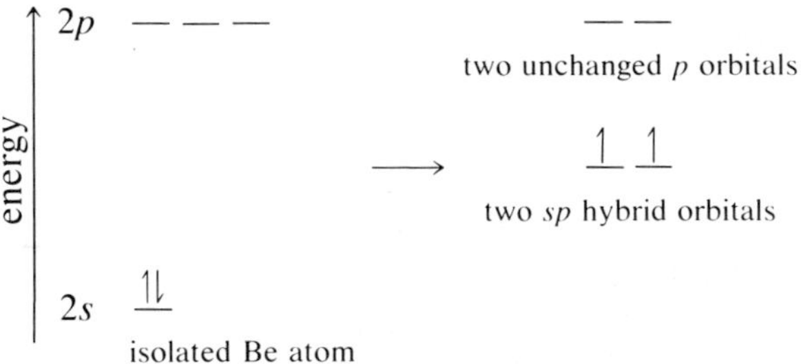

The unhybridized p orbitals of the Be atom are not utilized in a simple BeH_2 molecule.

EXERCISE 10.1 HYBRID ORBITALS AND BONDING

Describe the bonding in SCl_2 using hybrid orbitals.

HYBRID ORBITALS FOR ATOMS OF EXPANDED VALENCE

Elements of the third and higher periods can exhibit expanded valence in molecules such as PF_5 and SF_6. Here the central atom's valence shell is expanded to five or six pairs, so hybrid theory requires that five or six atomic orbitals be hybridized. Since there are only four atomic orbitals of the s and p type in a valence shell, the extra one or two orbitals required come from a d subshell of the shell containing the s and p valence orbitals. The hybrid orbital sets for five or six valence shell electron pairs are given in Table 10.1 and Figure 10.7, along with those sets already described.

d orbitals are energetically accessible for hybrid formation only for elements of the third and higher periods. Hence, only these elements exhibit expanded valence.

TABLE 10.1 Hybrid Orbital Sets for Two to Six Electron Pairs

HYBRID ORBITAL SET	NUMBER OF HYBRID ORBITALS (NUMBER OF SIGMA AND LONE ELECTRON PAIRS)	GEOMETRY	EXAMPLE
sp	2	Linear	Be in BeF_2
sp^2	3	Trigonal planar	B in BF_3
sp^3	4	Tetrahedral	C in CH_4
sp^3d	5	Trigonal bipyramidal	P in PF_5
sp^3d^2	6	Octahedral	S in SF_6

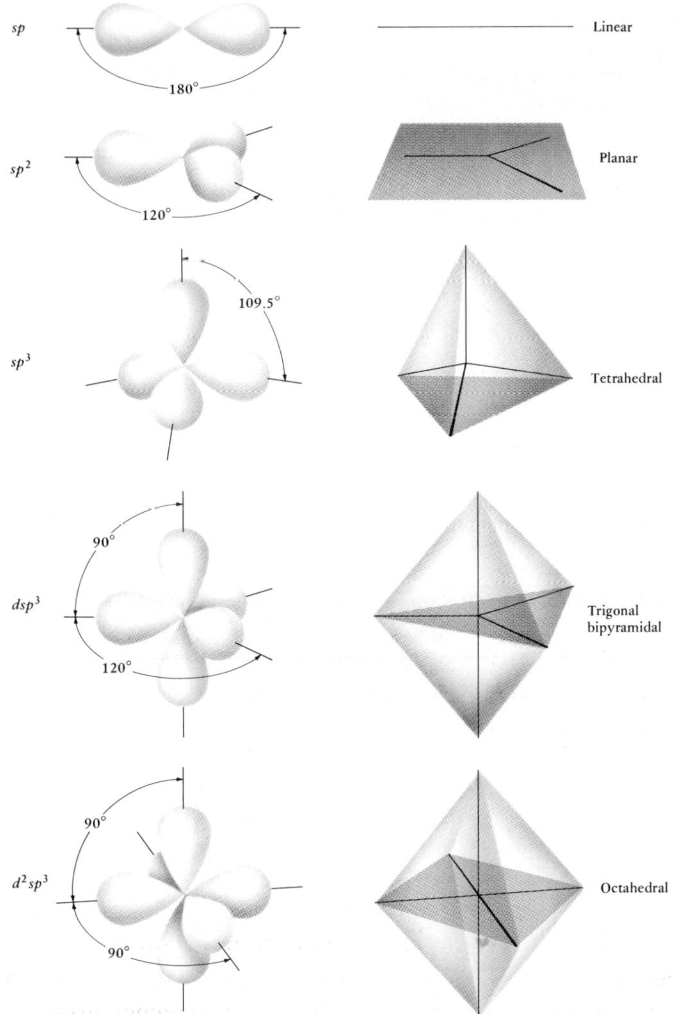

FIGURE 10.7
The geometry of the hybrid orbital sets for two to six structural electron pairs.

In forming a hybrid orbital set, the *s* orbital is always used, plus as many *p* orbitals (and *d* orbitals) as are required to give the necessary number of sigma bonding and lone pair orbitals.

EXAMPLE 10.4

HYBRIDS IN EXPANDED-VALENCE ATOMS

Describe the bonding in PF_5 using hybrid orbitals.

Solution The structural-pair geometry of PF_5 is trigonal bipyramidal.

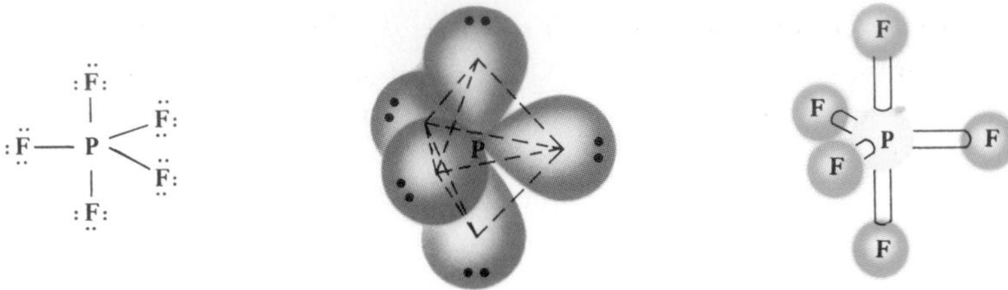

According to Figure 10.7, the hybrid scheme sp^3d is required.

energy

3d — — — — —   — — — —
four unhybridized d orbitals

3p $\underline{1}\ \underline{1}\ \underline{1}$ ⟶ $\underline{1}\ \underline{1}\ \underline{1}\ \underline{1}\ \underline{1}$
five sp^3d hybridized orbitals for sigma bonds

3s $\underline{1\!\downarrow}$

isolated P atom

EXERCISE 10.2 HYBRIDS IN EXPANDED-VALENCE ATOMS

Describe the bonding in XeF_4 using hybrid orbitals. Remember to consider first the electron dot structure and then the structural-pair geometry.

MULTIPLE BONDING

There are two types of chemical bonds: **sigma** (σ) bonds and **pi** (π) bonds. According to valence bond theory, sigma bonds arise from the overlap of atomic orbitals so that the electrons of the bond lie along the bond axis. *Pi bonds come from the sideways overlap of p atomic orbitals* (Figure 10.8). This means the overlap region is above and below the internuclear axis.

Pi bonds never occur alone without the bonded atoms also being joined by a sigma bond. Thus, a double bond consists of a sigma bond

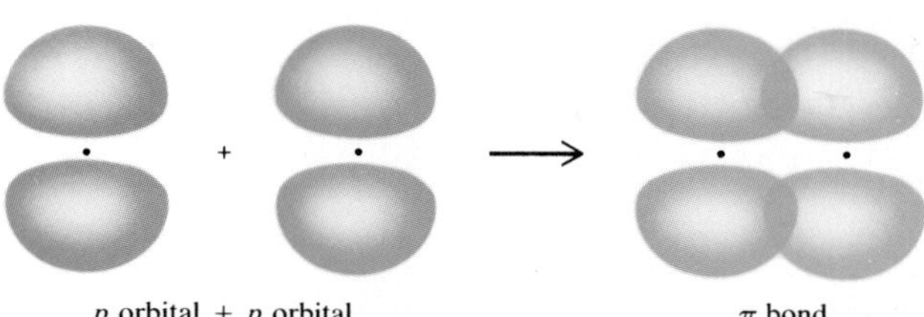

FIGURE 10.8
Formation of a pi bond by sideways overlap of p orbitals.

and a pi bond, while a triple bond consists of a sigma bond and two pi bonds.

pi bond → C=C ← sigma bond   two pi bonds → C≡C ← one sigma bond

Since a pi bond is formed from pure p atomic orbitals, one on each of two atoms, *a pi bond may form only if unhybridized p orbitals remain* on the bonded atoms, in the correct orientation, after accounting for sigma bond formation. Therefore, only atoms having sp or sp^2 hybridization can be involved in multiple bonding.

One of the simplest examples of pi bonding is found in ethylene, C_2H_4.

$H_2C=CH_2$ (H—C—H angle 120°)

Each carbon atom has three sigma-bonding electron pairs, so VSEPR theory predicts that the geometry about the carbon atoms in C_2H_4 is trigonal planar. To rationalize the sigma bonding of carbon to two hydrogen atoms and another carbon in a trigonal plane, the carbon atom must be sp^2 hybridized. Thus, each carbon atom has three sp^2 hybrid orbitals in a plane, and an unhybridized p orbital perpendicular to that plane (Figure 10.6). Each of these four orbitals accounts for one unpaired electron.

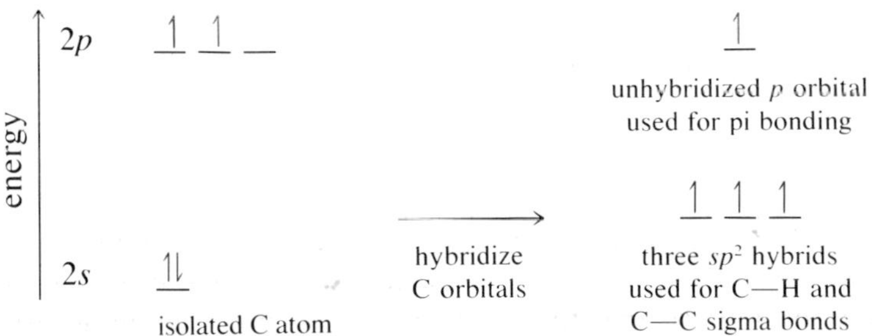

If two such C atoms approach one another, head-to-head overlap of one sp^2 hybrid orbital on each atom will give a sigma bond (Figure 10.9), and

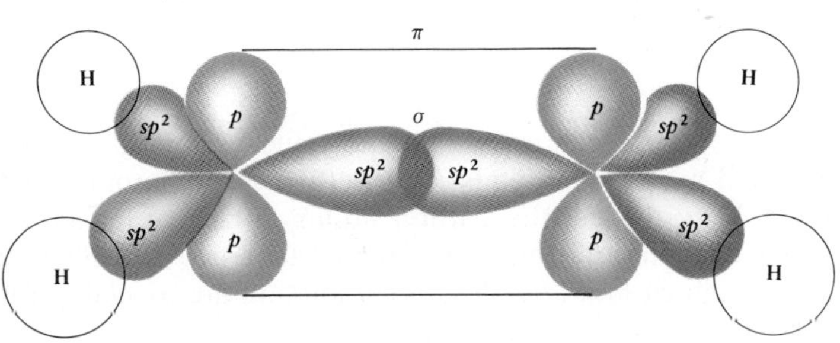

FIGURE 10.9

Bonding in ethylene, C_2H_4. A C—C sigma bond comes from head-to-head overlap of carbon sp^2 hybrid orbitals, and a pi bond comes from sideways overlap of carbon p orbitals.

A pi bond can form only if the triangular planes of sigma orbitals of each carbon are in the same plane. If one end is twisted relative to the other, the *p* orbitals will not lie side by side. Hence, all atoms around an atom with one pi bond must be in the same plane.

sideways overlap of the half-filled, pure *p* orbitals will form a pi bond. Thus, the C=C double bond consists of one sigma bond and one pi bond. To complete the bonding picture in ethylene, each of the two remaining half-filled sp^2 hybrid orbitals on the C atoms overlaps with a half-filled hydrogen 1*s* orbital to form a C—H sigma bond.

In Chapter 9 we pointed out that carbon can form multiple bonds with oxygen, sulfur, and nitrogen. To understand how such bonds are possible, consider formaldehyde, H_2CO.

H
C=O — sp^2
H
sp^2

This trigonal planar molecule is based on an sp^2 hybridized carbon atom. Thus, just as for the carbon atoms in ethylene, an unhybridized *p* orbital, perpendicular to the molecular plane, is available for pi bonding, this time with oxygen. If we also consider the oxygen atom as sp^2 hybridized,

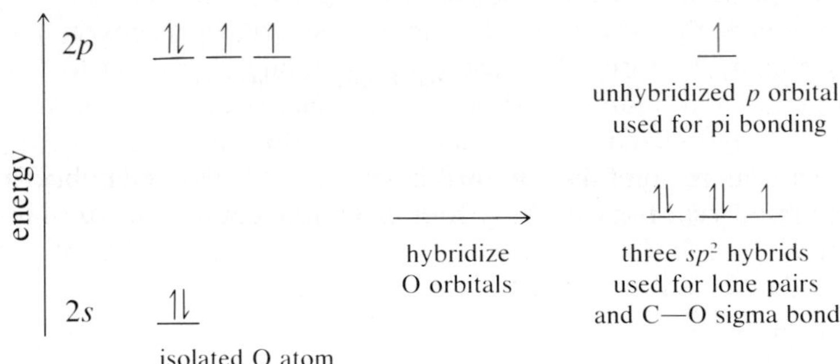

these hybrid orbitals can account for the C—O sigma bond and the oxygen lone pairs. The unhybridized oxygen *p* orbital, describing an unpaired electron, is then available for pi bond formation with the carbon *p* orbital.

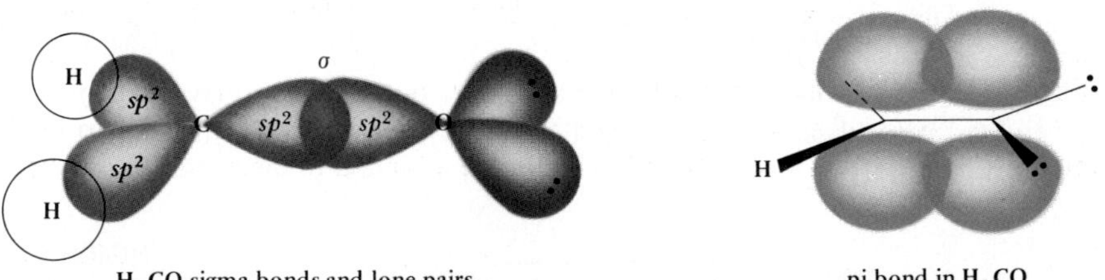

H_2CO sigma bonds and lone pairs **pi bond in H_2CO**

Acetylene, H—C≡C—H, is a simple molecule with a triple bond. Here one sigma bond and two pi bonds join the two carbon atoms. The structural-pair geometry around the carbon atoms is clearly linear, so each carbon atom is *sp* hybridized. This means that *two* half-filled *p* orbitals remain on each carbon after hybridization and are available for pi bond formation.

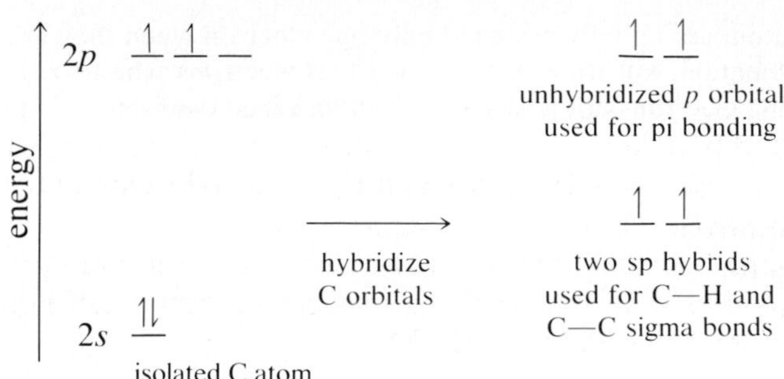

The orbitals about each C atom can be pictured as

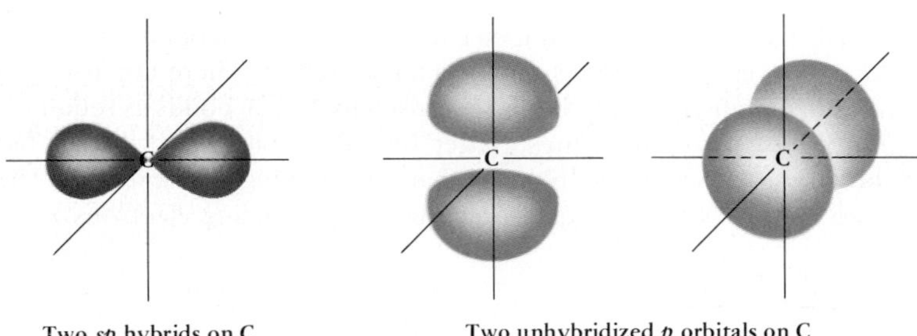

where the two *sp* hybrids are 180° apart, and the two unhybridized *p* orbitals are 90° apart and lie in planes perpendicular to the *sp* axis. The *sp* hybrids are used for C—C and C—H sigma bond formation, and the pure *p* orbitals overlap to produce *two* carbon–carbon pi bonds (Figure 10.10).

EXAMPLE 10.5

THE C≡O TRIPLE BOND IN CARBON MONOXIDE

Describe the bonding in carbon monoxide, CO, using orbital hybridization.

Solution The electron dot structure, :C≡O:, tells us that both the carbon and the oxygen must be *sp* hybridized. Further, each atom must use one *sp* hybrid orbital for sigma bond formation and one hybrid orbital for a lone pair.

To "bookkeep" electrons here properly, there is another useful rule of thumb to be followed when hybrid orbitals are assigned to electrons: the hybrid

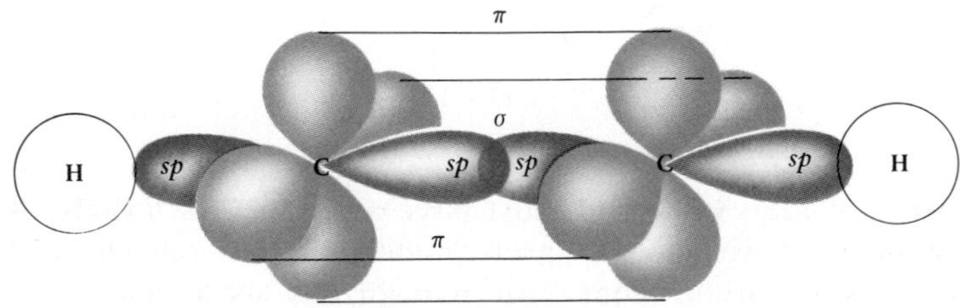

FIGURE 10.10

Bonding in acetylene, H—C≡C—H. A C—C sigma bond comes from head-to-head overlap of carbon *sp* hybrid orbitals, and two pi bonds come from sideways overlap of carbon *p* orbitals.

orbitals of an atom are initially assigned only one electron when they are used in sigma bond formation, but they are assigned two electrons when used for lone pairs. Remaining electrons are assigned to unhybridized orbitals.

	sp hybridized C	*sp* hybridized O
Unhybridized *p* orbitals for pi bonding	1 _	1l 1
Two *sp* hybrids	1l (lone pair) 1 (sigma bond)	1l (lone pair) 1 (sigma bond)

Now there is a half-filled *sp* hybrid orbital on each atom for sigma bond formation, as well as a lone pair on each atom assigned to an *sp* hybrid. There are also two pairs of electrons in unhybridized *p* orbitals to be used for two pi bonds as required. (Be aware that the way we show these latter four electrons to be distributed between C and O is irrelevant. The point is that after having accounted for the sigma bonds, we are left with the required number of pi bonding electrons.)

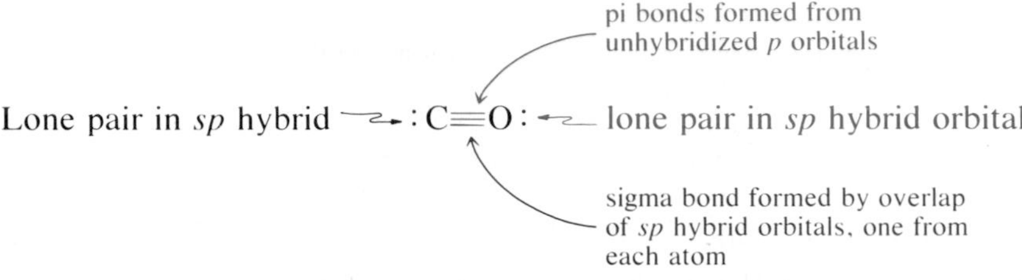

The orbital picture for CO is identical with HC≡CH in Figure 10.10, except that the C—H bond pairs are lone pairs.

EXERCISE 10.3 TRIPLE BONDS

Describe the bonding in the nitrogen molecule, N_2.

There are thousands of carbon-based molecules that have multiple bonds, and you will encounter many of them in this book and in further study of chemistry. It is therefore valuable to examine the bonding in a somewhat more complex case. Consider as an example acetic acid, $H_3C—COOH$, the important ingredient in vinegar. Its electron dot structure is shown here.

109° H, 120°, O ← sp^2
H—C—C ← sp^2, 109°
H, ↖ sp^3, O—H
↖ sp^3

The carbon atom of the CH_3 group must have a tetrahedral structural-pair geometry, so the C—H angle is approximately 109°. In valence bond terms, this means this carbon atom is sp^3 hybridized. The other carbon atom has a trigonal planar structural-pair geometry, with substituents at

120°C. This carbon must be sp^2 hybridized. Finally, the oxygen atom of the C—O—H grouping is surrounded by two bond pairs and two lone pairs and so has a tetrahedral structural-pair geometry. Thus, the C—O—H angle is approximately 109° and the O atom must be sp^3 hybridized. Finally, we account for the C=O link exactly as in the H_2CO molecule described earlier. Both the C atom and the O atom are taken as sp^2 hybridized, so an unhybridized p orbital remains on each atom to form the carbon–oxygen pi bond.

EXERCISE 10.4 BONDING AND HYBRIDIZATION

Analyze the bonding in acetonitrile, $H_3C—C≡N$:. Give values for the H—C—H, H—C—C, and C—C—N angles, and indicate the hybridization of both carbon atoms and the nitrogen atom.

PI RESONANCE STRUCTURES AND HYBRID ORBITALS

Pi resonance structures involve an electron pair that is used alternately to form a pi bond between two atoms and a lone pair on one of these atoms. As an example, consider ozone, O_3, which has two resonance structures.

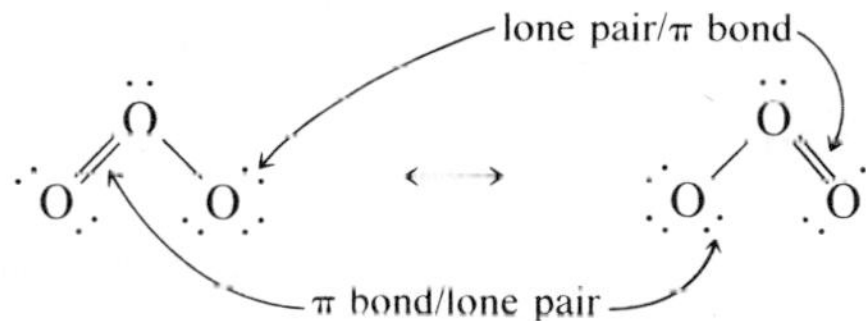

To visualize the bonding in ozone and to account for resonance structures, assume that all three O atoms are sp^2 hybridized. The central atom uses sp^2 hybrid orbitals to form two sigma bonds and to accommodate a lone pair. The terminal atoms use their sp^2 hybrids to form only one sigma bond and to accommodate two lone pairs.

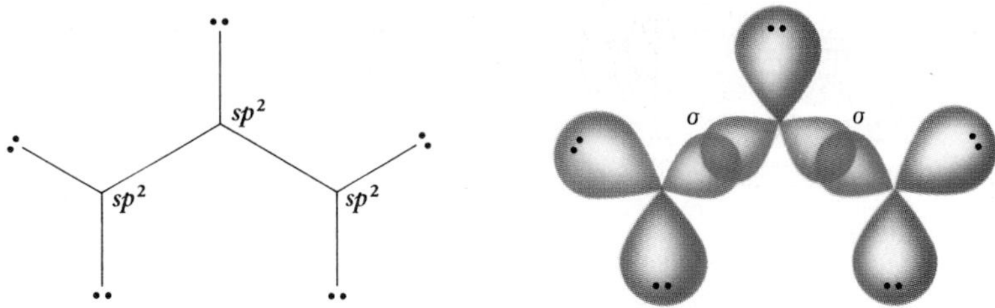

Top view of ozone showing sigma bonds and lone pairs in oxygen sp^2 hybrid orbitals

Since O_3 has nine valence electron pairs to be accommodated, the sigma framework and lone pairs illustrated above account for seven of these pairs. The pi bonds in ozone arise from the two remaining electron pairs.

Since we assume each oxygen atom is sp^2 hybridized, a p orbital *perpendicular to the O_3 plane* must remain on each atom.

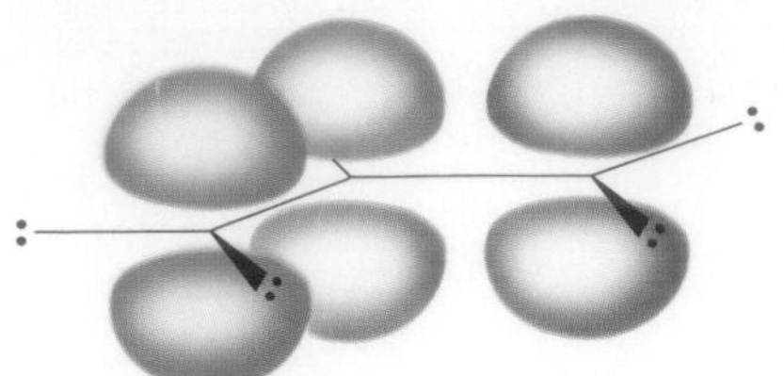

The oxygen p orbitals remaining after sigma bonds and lone pairs are accommodated in the O_3 plane

sp^2 hybridized O

Energy

p orbital used for out-of-plane pi bonds

Three sp^2 hybrid orbitals for in-plane sigma bonds and lone pairs

According to hybrid orbital theory, one *pair* of these p orbitals can form a pi bond using one of the remaining electron pairs. The other p orbital is then left to accommodate the final electron pair as a lone pair.

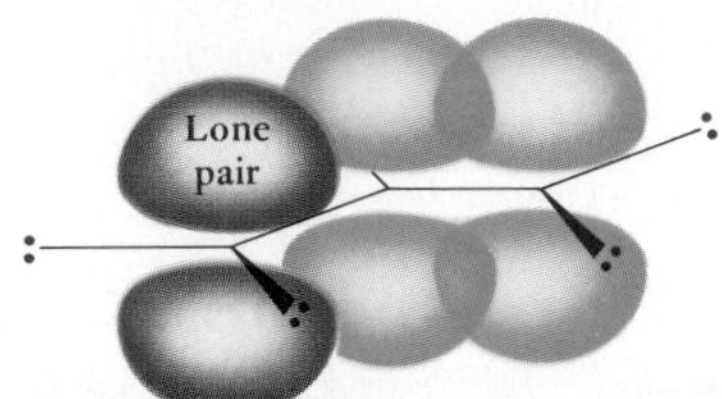

Formation of a pi bond between one pair of O atoms

It is obvious, however, that the O—O pi bond could form using the central atom p orbital with a p orbital on either of the terminal *O* atoms. The lone pair is then assigned to the p orbital unused in the multiple bond. Hence, there are two possible arrangements, which we have called pi resonance structures.

EXERCISE 10.5 RESONANCE STRUCTURES AND HYBRIDIZATION

Describe the resonance structures of NO_2^- in terms of hybrid orbitals.

10.2 MOLECULAR ORBITAL THEORY

Molecular orbital (MO) theory is an alternative way to view electron orbitals in molecules. In contrast to the localized bond and lone pair orbitals of valence bond theory, pure s and p atomic orbitals of the atoms in the molecule combine to produce orbitals that are spread or delocalized over several atoms or even over the entire molecule. The new orbitals are called *molecular orbitals*, and they can have different energies. Just as with orbitals in atoms, molecular orbitals are assigned to electrons according to the Pauli principle and Hund's rule.

One reason for learning about the MO concept is that it correctly predicts the electronic structures of certain molecules that do not follow the electron-pairing assumptions of the Lewis approach. The most common example is the O_2 molecule. Using the rules of Chapter 9, you would draw the electron dot structure of the molecule as $:\ddot{O}=\ddot{O}:$, with all electrons paired. However, experiments clearly show that the O_2 molecule is *paramagnetic* and that is has exactly two unpaired electrons per molecule. It is sufficiently magnetic that solid O_2 clings to the poles of a magnet (Figure 10.11). The molecular orbital approach can account for the paramagnetism of O_2 more easily than can valence bond theory. To

Paramagnetism arises from the presence of unpaired electrons. See page 254 for a more complete discussion.

FIGURE 10.11

Solid O_2 is paramagnetic and so clings to the poles of a magnet. The compound was cooled to a very low temperature to cause it to solidify. Notice that the solid is light blue. (Masterton, Slowinski, and Stanitski, *Chemical Principles*, 6th ed., Saunders College Publishing, 1986)

see how MO theory can apply to O_2 and other small diatomic molecules, we shall first describe *four principles* of the theory.

PRINCIPLES OF MOLECULAR ORBITAL THEORY

According to valence bond theory, the number of hybrid orbitals produced on an atom is *always* the same as the number of atomic orbitals combined on that atom. The same principle holds in molecular orbital theory, except that *only orbitals on two different atoms are combined*. The **first principle** of molecular orbital theory is that the number of molecular orbitals produced is *always* equal to the number of atomic orbitals brought by the combining atoms. To see the consequences of this, consider first the H_2 molecule.

BONDING AND ANTIBONDING MOLECULAR ORBITALS IN H_2 When the 1*s* orbitals of two atoms overlap, two molecular orbitals result. The principles of molecular orbital theory tell us that, in one of the resulting molecular orbitals, the 1*s* regions of electron density *add* together to lead to an increased probability that electrons are found in the bond region (Figure 10.12).* Thus, electrons in such an orbital attract both nuclei. Since the

*Orbitals are electron waves. Therefore, a way to view molecular orbital formation is to assume that two electron waves, one from each atom of the bonded pair, interfere with one another. The interference can be constructive (to give a bonding MO) or destructive (to give an antibonding MO).

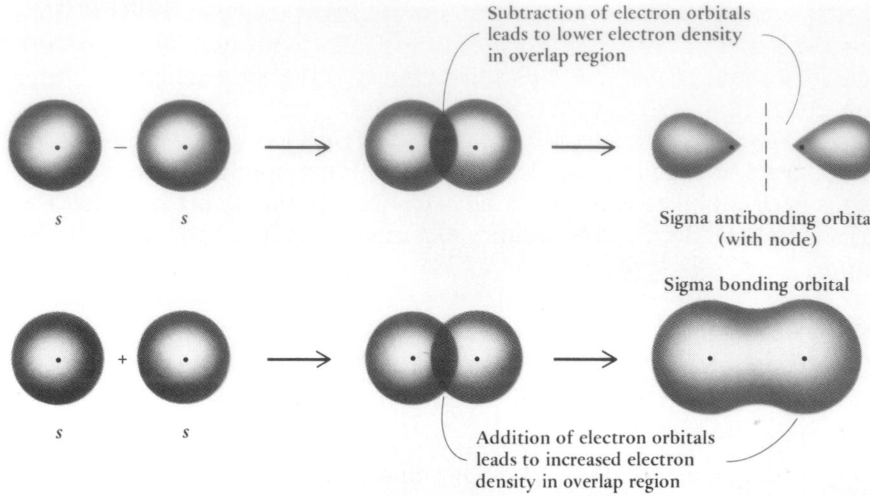

FIGURE 10.12

Formation of bonding and antibonding molecular orbitals from two *s* atomic orbitals. Notice the presence of a node in the antibonding orbital.

atoms are thereby bound together, the molecular orbital is called a **bonding molecular orbital**. Moreover, it is a sigma orbital, since the region of electron probability lies directly along the bond axis. We label this molecular orbital σ_{1s}.

Since two combining atomic orbitals *must* produce two molecular orbitals, the other combination is constructed by *subtracting* one orbital from the other (Figure 10.12). When this happens there is reduced electron probability *between* the nuclei for the molecular orbital. This is called an **antibonding molecular orbital**. Since it is also a sigma orbital, it is labeled σ^*_{1s}, where the asterisk conveys the notion of an antibonding orbital.

Antibonding orbitals have nodes and higher energies than the average energy of the atomic orbitals from which they were created. Their upward energy displacement is *slightly* greater than the downward displacement of the bonding orbitals.

A **second principle** of molecular orbital theory is that the bonding molecular orbital is lower in energy than the parent orbitals, and the antibonding orbital is higher in energy (Figure 10.13). The average energy of the molecular orbitals is slightly higher than the average energy of the parent atomic orbitals.

A **third principle** of molecular orbital theory is that the electrons of the molecule are placed in orbitals of successively higher energy; the Pauli principle and Hund's rule are obeyed. Thus, electrons occupy the lowest energy orbitals available, and they do so with spins paired. Since the energy of the electrons in the bonding orbital of H_2 is lower than that of either parent 1*s* electron, the H_2 molecule is stable. We write the electron configuration of H_2 as $(\sigma_{1s})^2$.

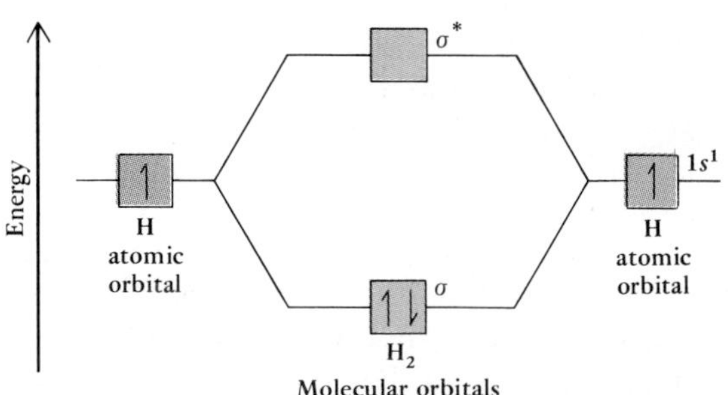

FIGURE 10.13

Energy level diagram for the molecular orbitals from two 1*s* atomic orbitals. The two electrons of H_2 are placed in the σ_{1s} MO.

Next consider putting two helium atoms together to form He_2. Since both He atoms have 1*s* valence orbitals, they combine to produce the same kind of molecular orbitals as in H_2. The four helium electrons are assigned to these orbitals according to the scheme shown in Figure 10.14. The pair of electrons in σ_{1s} stabilizes He_2. However, the two electrons in σ^*_{1s} destabilize the He_2 molecule a little more than the two electrons in σ_{1s} stabilize He_2. Thus, molecular orbital theory predicts that He_2 has no net stability, and laboratory experiments indeed show that two He atoms have little tendency to combine.

Actually the upward energy displacement of σ^*_{1s} is slightly greater than the downward displacement of σ_{1s}. This means the net energy change for four electrons is positive and ensures that He_2 is less stable than two separate He atoms.

BOND ORDER Bond order was defined in Chapter 9 as the net number of bonding electron pairs linking a pair of atoms. This same concept can be applied directly to molecular orbital theory, but now we define bond order as

Bond order = *net* number of bonding electron pairs
= number of electron pairs in bonding molecular orbitals − number of electron pairs in antibonding molecular orbitals

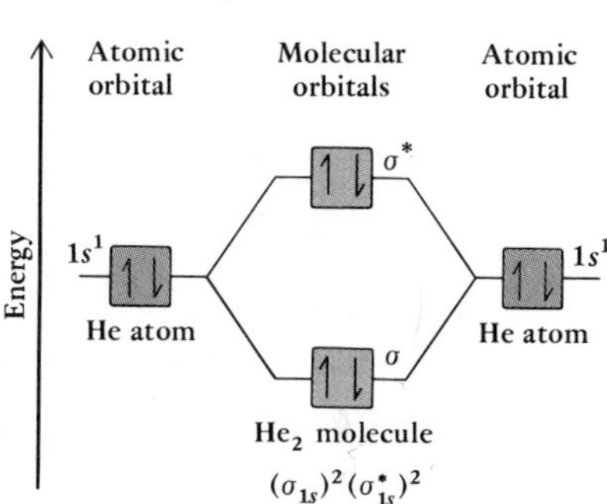

FIGURE 10.14
Energy level diagram for the hypothetical He_2 molecule.

In the H_2 molecule, there is one electron pair in a bonding orbital, so H_2 has bond order of 1. In contrast, the effect of the σ_{1s} pair in He_2 is cancelled by the effect of the σ^*_{1s} pair, so the bond order is 0.

Fractional bond orders are also possible. For example, even though He_2 does not exist, the He_2^+ ion has been detected. Its molecular orbital electron configuration would be $(\sigma_{1s})^2(\sigma^*_{1s})^1$. Here there is one electron pair in a bonding molecular orbital, but one-half pair in an antibonding orbital. Therefore, the net bond order is $\frac{1}{2}$.

EXERCISE 10.6 MOLECULAR ORBITALS AND BOND ORDER
Write the configuration of the H_2^- ion in molecular orbital terms. What is the bond order of the ion? Do you expect it to exist?

MOLECULAR ORBITALS OF Li_2 and Be_2 A **fourth principle** of molecular orbital theory is that atomic orbitals combine most effectively with orbitals of the same type and similar energy. This principle comes into operation when we move past He_2 to Li_2. A lithium atom has two valence orbitals of the *s* type (1*s* and 2*s*), so a 1*s*/2*s* overlap is theoretically possible. However, because the 1*s* and 2*s* orbitals are quite different in energy, the 1*s*/2*s* interaction cannot make an important contribution. Thus, the molecular orbitals can be considered to come only from 1*s*/1*s* and 2*s*/2*s* interactions (Figure 10.15).

The electron configuration of dilithium is

$$Li_2 \qquad (\sigma_{1s})^2(\sigma^*_{1s})^2(\sigma_{2s})^2$$

Bond order = 2 bonding pairs − 1 antibonding pair = 1 net bonding pair

The electrons of σ_{1s} and σ^*_{1s} make no net contribution to the bonding; all the bonding in Li_2 is due to the electron pair assigned to the σ_{2s} orbital. This is of course exactly what we observed in drawing electron dot structures: electrons of filled subshells are not considered valence electrons. In molecular orbital terms, whenever an atom has completed *s*, *p*, or *d*

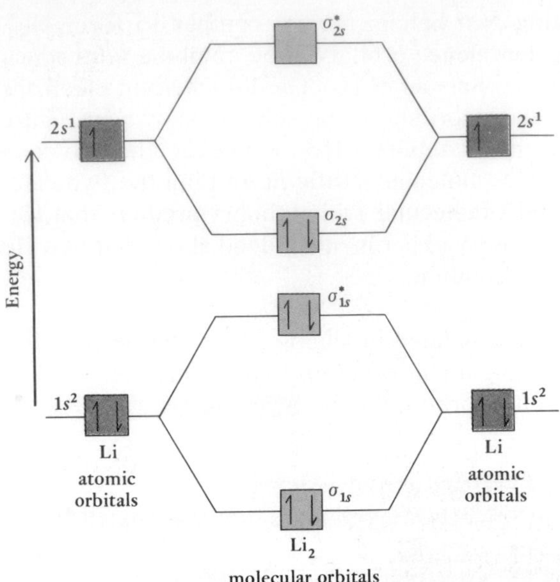

FIGURE 10.15
Energy level diagram for the combination of two atoms with 1*s* and 2*s* atomic orbitals. The electron configuration shown is for Li_2.

subshells, these are called *core electrons*, and they are assigned to bonding and antibonding molecular orbitals that offset one another. Thus, these electrons no longer need to be considered.

A diberyllium molecule, Be_2, is not stable. Neglecting the 1*s* core electrons, its electron configuration would be

$$Be_2 \qquad \text{(core electrons)}\ (\sigma_{2s})^2(\sigma^*_{2s})^2$$

Bond order = 1 bonding pair − 1 antibonding pair = 0 net bonding pairs

and you see that there are no net bonding electron pairs.

EXERCISE 10.7 MOLECULAR ORBITALS IN DIATOMIC MOLECULES

Be_2 does not exist. What about the Be_2^+ ion? Describe its electron configuration in molecular orbital terms. What is the net bond order? Do you expect the ion to be stable?

MOLECULAR ORBITALS FOR HOMONUCLEAR, DIATOMIC MOLECULES

When two identical atoms form a molecule, this is called a **homonuclear, diatomic molecule**. With many of the principles of molecular orbital theory in place, we are ready to account for bonding in such molecules formed by second period or *p* block elements, and important examples are N_2, O_2, and F_2. First, however, we need to see what types of molecular orbitals form when elements have both *s* and *p* orbitals.

MOLECULAR ORBITALS FROM ATOMIC *p* ORBITALS Three types of interactions are possible for two elements that both have *s* and *p* orbitals (Figure 10.16). Sigma bonding and antibonding molecular orbitals are formed by *s* orbitals interacting in the usual manner. Similarly, it is pos-

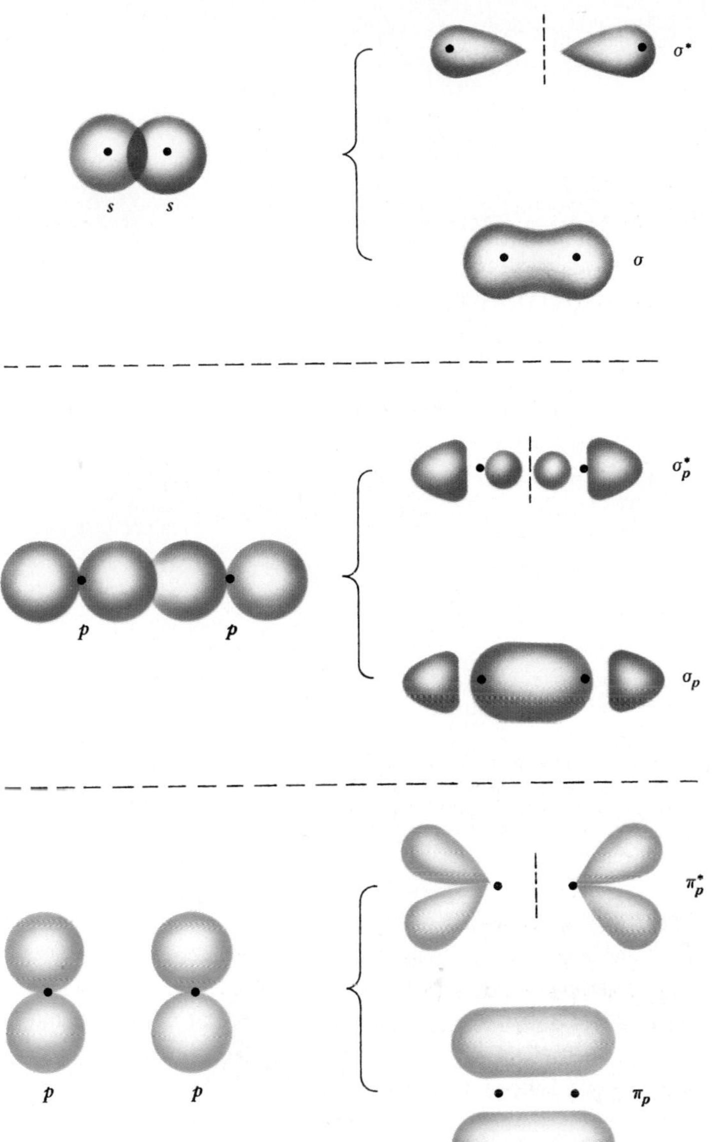

FIGURE 10.16
The bonding and antibonding molecular orbitals that come from *s* and *p* orbital overlap. The drawings represent only the general shapes of the orbitals.

sible for a *p* orbital on one atom to interact with a *p* orbital on the other atom in a head-to-head fashion to produce a sigma bonding and antibonding molecular orbital pair. And finally, each atom has *two p* orbitals in a plane perpendicular to the axis of the head-to-head *p* orbital interaction. These *p* orbitals can interact sideways to give *pi* bonding and antibonding molecular orbitals. Thus, two *p* orbitals on each of two atoms produce two pi bond orbitals and two pi antibonding orbitals.

MOLECULAR ORBITAL ENERGY DIAGRAMS FOR *p* BLOCK ELEMENTS The molecular orbital energy level diagram in Figure 10.17 gives the configurations for all diatomic molecules from B_2 through F_2, where the valence

FIGURE 10.17
A molecular orbital energy level diagram for X_2, a diatomic molecule of second period elements.

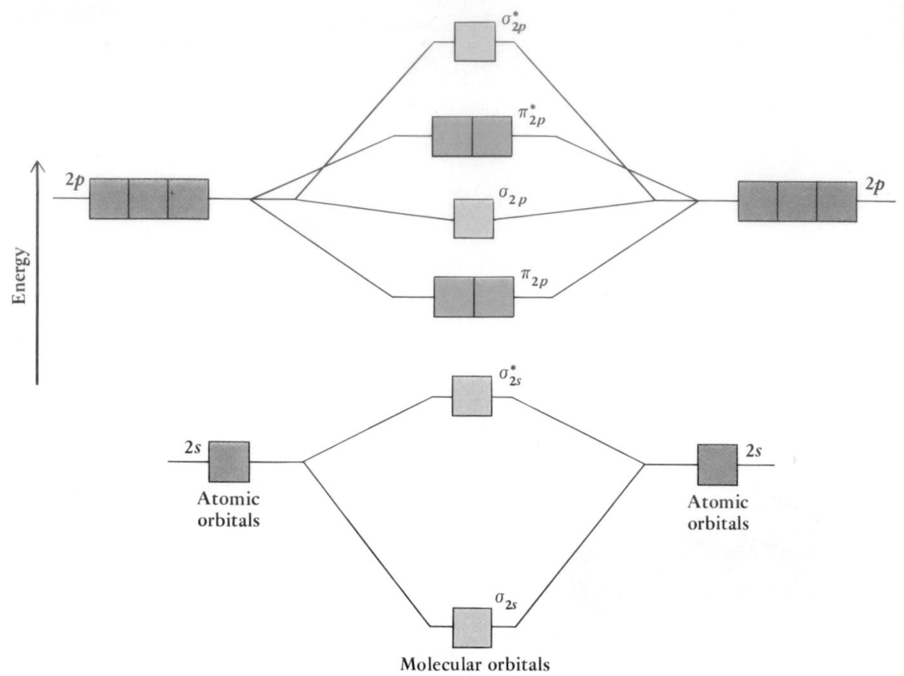

electrons of the isolated atoms occupy 2*s* and 2*p* atomic orbitals. Notice that the *p* orbitals are of two kinds: one *p* orbital on each atom gives a sigma bonding orbital and a sigma antibonding orbital, while the other two *p* orbitals on each atom give two pi bonding orbitals and two pi antibonding orbitals. The diagram shows that the bonding and antibonding sigma orbitals from 2*s* interactions are lower in energy than the MOs from 2*p* interactions. The reason is that 2*s* orbitals have a lower energy than 2*p* orbitals in the separated atoms. The separation of bonding and antibonding orbitals is greater for σ_{2p} than for π_{2p}. This happens because *p* orbitals overlap to a greater extent when they are oriented head to head than when placed side by side. The greater the orbital overlap, the greater the stabilization of the bonding MO and the greater the *de*stabilization of the antibonding MO.

A complication in the order of energy for the molecular orbitals of second period elements arises from the possibility of 2*s* and 2*p* interactions. Mixing of *s* and *p* orbitals causes the σ_{2s} and σ^*_{2s} orbitals to be lower in energy than otherwise expected, and the σ_{2p} and σ^*_{2p} orbitals to be pushed up in energy. This is the reason the energy lowering and raising for the σ_{2s} and σ^*_{2s} orbitals (and for the σ_{2p} and σ^*_{2p} orbitals) is not symmetric in Figure 10.17.

Mixing of *s* and *p* orbitals is important only for B_2, C_2, and N_2. Thus, Figure 10.17 applies strictly only to these molecules. For O_2 and F_2, σ_{2p} is lower in energy than π_{2p}. Nonetheless, the order given in Figure 10.17 leads to the correct bond order and correctly predicts the paramagnetism of O_2, conclusions of great importance.

ELECTRON CONFIGURATIONS FOR HOMONUCLEAR, DIATOMIC MOLECULES Molecular orbital electron configurations are given for the diatomic molecules B_2 through F_2 in Table 10.2. We find there is an excellent correlation between the electron configurations and the bond orders, bond lengths, and bond dissociation energies shown at the bottom of the table.

TABLE 10.2 Molecular Orbital Occupations and Physical Data for Homonuclear, Diatomic Molecules of Second Period Elements

	B_2	C_2	N_2	O_2	F_2
σ^*_{2p}					
π^*_{2p}				↑ ↑	↑↓ ↑↓
σ_{2p}			↑↓	↑↓	↑↓
π_{2p}	↑ ↑	↑↓ ↑↓	↑↓ ↑↓	↑↓ ↑↓	↑↓ ↑↓
σ^*_{2s}	↑↓	↑↓	↑↓	↑↓	↑↓
σ_{2s}	↑↓	↑↓	↑↓	↑↓	↑↓
Bond order	One	Two	Three	Two	One
Bond-dissociation energy (kJ/mol)	290	620	941	495	155
Bond distance (pm)	159	131	110	121	143

B_2 and C_2 are not ordinary molecules; C_2, for example, has been observed only in the vapor phase over solid carbon at high temperatures. It is, however, worth noticing that the higher predicted bond order for C_2 than for B_2 agrees well with the higher bond dissociation energy and shorter bond length of C_2.

We know from experiment (and have also predicted from the electron dot structure) that N_2 is a diamagnetic molecule with a short, strong triple bond. The molecular orbital picture is certainly in agreement, predicting a bond order of 3.

The molecular orbital electron configuration for O_2 clearly shows that the bond order is two. Hund's rule requires two unpaired electrons, exactly as determined by experiment. Thus, a simple molecular orbital picture leads to a reasonable view of the bonding in paramagnetic O_2, a point on which simple valence bond theory failed.

O_2 has four bonding pairs (one σ_{2p}, two π_{2p}, and one σ_{2s}. Their bonding effect is partly cancelled by two antibonding pairs (one σ^*_{2s} and two π^*_{2p} half pairs).

Finally, molecular orbital theory predicts the bond order of F_2 to be one, and the molecule does indeed have the weakest bond of the series in Table 10.2.

Molecular orbital theory can be applied not only to diatomic molecules but also to some diatomic ions. The following example and exercise explore this use of MO theory.

EXAMPLE 10.6

MOLECULAR ORBITAL THEORY AND IONS

When potassium reacts with O_2, potassium superoxide, KO_2, is produced. The superoxide ion is O_2^-. Write the molecular orbital electron configuration of the ion and predict its bond order and magnetic character.

Solution Using the energy level diagram in Figure 10.17, the configuration of the ion is

$$O_2^- \qquad \text{(core electrons)}\ (\sigma_{2s})^2(\sigma_{2s}^*)^2(\pi_{2p})^4(\sigma_{2p})^2(\pi_{2p}^*)^3$$

It is predicted to be paramagnetic to the extent of one unpaired electron, a prediction confirmed by experiment. The bond order is $1\frac{1}{2}$, since there are 4 bond pairs and $2\frac{1}{2}$ antibonding pairs. This is a smaller bond order than in O_2, so we predict the O—O bond in O_2^- should be longer than in O_2. In fact, the superoxide ion has an O—O bond length of 132 pm, while that in O_2 is 121 pm.

EXERCISE 10.8 MOLECULAR ORBITAL THEORY AND IONS

The cations O_2^+ and N_2^+ are important components of the earth's upper atmosphere. Write the electron configuration of O_2^+. Predict its bond order and magnetic character.

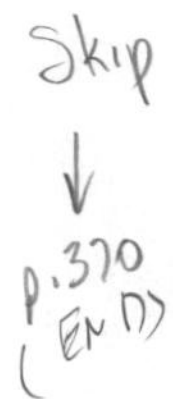

MOLECULAR ORBITALS OF HETERONUCLEAR, DIATOMIC MOLECULES

If the two atoms of a diatomic molecule are different, this is called a **heteronuclear** molecule. Important examples include CO and NO. With some modification, the molecular orbital energy level diagram in Figure 10.17 can be applied to them, and it is useful to look at one such case.

Because of the difference in effective nuclear charge (Section 8.3), atomic orbitals having the same set of quantum numbers in atoms of different elements have slightly different energies. Therefore, when a molecular orbital is formed by overlap of two nonequivalent atomic orbitals (such as the 2*s* orbitals of C and O), molecular orbital theory predicts that the bonding molecular orbital has a larger contribution from the atom of greater electronegativity. Conversely, the antibonding partner has a greater contribution from the atom of lower electronegativity. This principle is illustrated by the energy level diagram for CO in Figure 10.18. It is the same as Figure 10.17, except that the MO energy levels have shifted as a reflection of the shifts in the 2*s* and 2*p* energies of C and O.

There are 10 valence electrons in CO, so the molecular orbital scheme is

$$:C{\equiv}O: \qquad \text{(core electrons)}\ (\sigma_{2s})^2(\sigma_{2p})^2(\pi_{2p})^4(\sigma_{2s}^*)^2$$

The bond order derived from the MO configuration is three, just as given by the dot structure, and the molecule is diamagnetic as confirmed by experiment. Finally, the electrons of highest energy are those in a molecular orbital closest in energy to the 2*s* orbital of the C. These electrons can be identified as the lone pair on the carbon atom, and it is this pair that CO uses in its chemical reactions. In fact, it is in this way that CO can react with the iron in the hemoglobin of your blood system and can thereby prevent the hemoglobin from interacting with O_2 as it usually does.

DELOCALIZED MOLECULAR ORBITALS

We previously used the concept of *resonance* to describe the electronic structure of ozone, O_3 (Section 10.1). One advantage of molecular theory

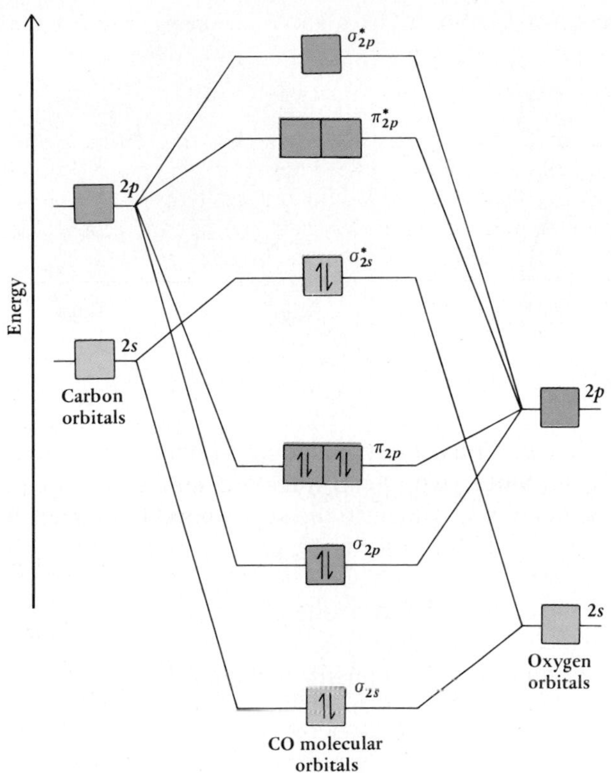

FIGURE 10.18
An approximate molecular orbital energy level diagram for CO. The ten valence electrons of the molecule fill the molecular orbitals through σ^*_{2s}. The energy levels of the isolated O atom are lower than those of the C atom, as indicated by oxygen's greater electronegativity.

is that it does not introduce such an artificial notion as resonance. Instead, O_3 is described in terms of *delocalized pi molecular orbitals*.

In the previous section we considered the "skeleton" of the molecule to be the sigma and lone pairs around the three atoms in the O_3 plane. There you saw that there must then be three p atomic orbitals, one on each O atom, perpendicular to that plane. To simplify the molecular orbital description of O_3, we also consider two separate sets of molecular orbitals: (a) the sigma and lone pairs and (b) the three perpendicular p orbitals that form the pi bonds.

Here we want to consider *only the pi molecular orbitals* perpendicular to the O_3 plane. According to the principles of molecular orbital theory, the three O atom p atomic orbitals can form three pi molecular orbitals (Figure 10.19). One of these molecular orbitals is bonding, one is nonbonding (has an energy the same as the parent p orbitals), and the third is antibonding.

When treating O_3 bonding in hybrid orbitals terms, we concluded that there are two electron pairs to be assigned to the pi orbitals. The same is true in molecular orbital terms. Therefore, the three pi molecular orbitals accommodate two pi electron pairs. One of these occupies the pi bonding orbital, which is delocalized over the three O atoms and has a lower energy than the parent p orbitals. The second pair occupies the nonbonding orbital, which, as the name implies, neither adds to nor detracts from bonding between O atoms. The antibonding orbital, with a

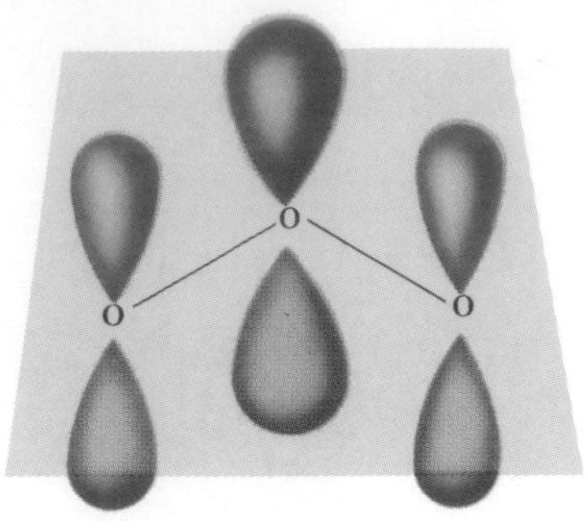

Antibonding π orbital

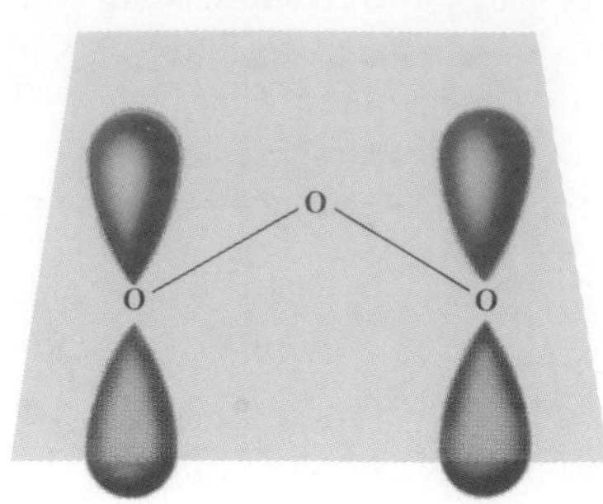

Nonbonding π orbital

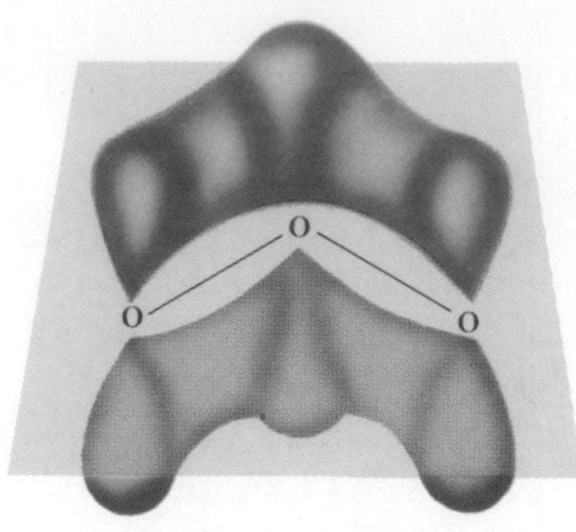

Bonding π orbital

FIGURE 10.19
The three pi molecular orbital of ozone, O_3.

higher energy, is unoccupied. Thus, the pi bonding in ozone is accomplished with only one pair. Since two O—O links are served by this pi pair, the order of each pi bond is $\frac{1}{2}$. Added to the sigma bond linking each pair of oxygens, the net O—O bond order is $1\frac{1}{2}$, exactly as given by the resonance structures.

This approach to the pi bonding in ozone can be applied to many other molecules with delocalized pi electrons. We shall examine a few more cases in the chapter on organic chemistry (Chapter 23).

SUMMARY

Two commonly used theories of chemical bonding are **valence bond theory** and **molecular orbital theory**. Valence bond theory views bonding as arising from the **overlap** of atomic orbitals on two atoms to give a bonding orbital with electrons localized between the bonded atoms (Section 10.1). In a sigma bond (σ), orbitals (s with s, p with p, or s with p) overlap head to head, and the bonding electron density is concentrated along the axis of the bond. If p orbitals remain after sigma bond formation, one or two pi (π) bonds may form by the sideways overlap of p orbitals; pi bonding electron density lies on either side of the bond axis.

To achieve maximum overlap of atomic orbitals in sigma bonding orbitals, the concept of **orbital hybridization** was introduced (Section 10.1). For example, atomic s and p orbitals may combine to produce four equivalent orbitals, each labeled sp^3, at the corners of a tetrahedron. Other hybrid orbital sets are sp (2 hybrid orbitals, linear geometry), sp^2 (3 hybrid orbitals, trigonal planar), sp^3d (5 hybrid orbitals, trigonal bipyramidal), and sp^3d^2 (6 hybrid orbitals, octahedral). The number of hybrid orbitals formed on an atom is equal to the number of atomic orbitals that are combined on that atom.

In molecular orbital theory, pure s and p atomic orbitals of the atoms in the molecule combine to produce **molecular orbitals** that are generally delocalized over many atoms (Section 10.2). The number of molecular orbitals formed is always equal to the number of atomic orbitals brought by the combining atoms. Some of these molecular orbitals will be **bonding** (and lie lower in energy than the energy of the parent atomic orbitals in the isolated atoms), while others will be **antibonding** and lie higher in energy. The electrons of the molecule are assigned to these orbitals, beginning with those of lowest energy, according to the Pauli principle and Hund's rule.

The bond order in a diatomic molecule is the net number of bonding electron pairs linking the atoms. Thus, bond order is given by the difference between the number of bonding electron pairs and the number of antibonding electron pairs.

The bonding in **homonuclear**, diatomic molecules of second row elements and some properties of these molecules can be rationalized using the molecular orbital scheme in Figure 10.17. Such a scheme needs only slight modification to serve for **heteronuclear**, diatomic molecules.

In addition to diatomic molecules, molecular orbital theory can be applied to molecules of more atoms, and it is especially useful for those with delocalized pi electrons.

The **band theory** of metallic bonding is molecular orbital theory applied to metallic solids (Special Section). There are so many atoms in a crystal that a large number of molecular orbitals is formed, the orbitals forming bands of energy levels. Depending on the separation of the filled **valence band** and partially filled or empty **conduction band**, a solid may be an electrical **conductor**, **semiconductor**, or **insulator**.

STUDY QUESTIONS

REVIEW QUESTIONS

1. What is one important difference between molecular orbital and valence bond theory?
2. Inspect Figure 10.2 and tell why the energy drops to a minimum and then rapidly increases as the H—H distance decreases.
3. What is the most important difference between a pi bond and a sigma bond?
4. What is the maximum number of hybrid orbitals a carbon atom may form? What is the minimum number? Explain briefly.
5. What is the maximum number of hybrid orbitals a third period element, such as sulfur, may form? Explain briefly.
6. What are the angles between regions of electron density in *sp*, sp^2, and sp^3 hybrid orbital sets?
7. For each of the following structural-pair geometries, tell what hybrid orbital set is used: tetrahedral, linear, trigonal planar, octahedral, trigonal bipyramidal.
8. If an atom is *sp* hybridized, how many pure *p* orbitals remain on the atom? How many pi bonds can the atom form?
9. Give an example of a molecule with a central atom of expanded valence. Tell what hybrid orbitals are used by this atom.
10. Describe four principles of molecular orbital theory.
11. Draw contour diagrams of the bonding and antibonding molecular orbitals of H_2 and tell how they differ.
12. Is the following statement true or false? If false, change it to give a true statement. If two atomic orbitals produce bonding and antibonding molecular orbitals, the antibonding orbital is always lower in energy.
13. What is meant by the order of a bond? How are bond order, bond length, and bond dissociation energy related?
14. How is molecular orbital theory related to the bonding in metals?
15. Explain briefly the theory of bonding in metals. What is the name usually applied to this theory?

HYBRID ORBITALS

16. Draw the electron dot structure of H_2S. What is its structural-pair and molecular geometry? Describe the bonding in the molecule using hybrid orbitals.
17. Tell what hybrid orbital set is used by the underlined atom in each of the following molecules or ions:
    (a) $\underline{B}Cl_3$ (c) $H\underline{C}Cl_3$
    (b) $\underline{N}O_2^+$ (d) $H_2\underline{C}O$
18. Tell what hybrid orbital set is used by the underlined atom in each of the following molecules or ions:
    (a) $\underline{C}S_2$ (c) $\underline{N}O_3^-$
    (b) $\underline{N}O_2^-$ (d) $\underline{B}H_4^-$

19. Give the hybrid orbital set used by sulfur in each of the following molecules or ions:
(a) SO_2 (b) SO_3 (c) SO_4^{2-}

20. N_2O, dinitrogen oxide, is commonly called "laughing gas." It is used as an anesthetic and as a propellant in cans of whipped cream. (a) Sketch the electron dot structure of the molecule (the atoms of which are in the order NNO). (b) What are the structural-pair and molecular geometries? (c) What is the hybridization of the central nitrogen atom?

21. Give the hybrid orbital set used by the underlined atom in each of the following molecules or ions:
(a) $\underline{S}F_6$ (c) $\underline{I}Cl_2^-$
(b) $\underline{S}F_4$ (d) $O\underline{Xe}F_4$

22. Give the electron dot structure and structural-pair and molecular geometries for XeF_2. Describe the bonding in the molecule using hybrid orbitals.

23. Give the electron dot structure and structural-pair and molecular geometries for ICl_4^-. Describe the bonding in the molecule using hybrid orbitals.

24. Give the electron dot structures of ClF_2^+ and ClF_2^-. What are the structural-pair and molecular geometries of each ion? What hybrid orbital set is used by Cl in each ion?

25. Give the hybrid orbital set used by each underlined atom in the following molecules:

(a) $H_3\underline{C}—\underline{O}—H$

(b) $H_3C—\underline{C}(H)=\underline{C}H_2$

(c) $H_3C—\underline{C}H_2—\underline{C}(=O)—H$

26. What is the hybridization of the carbon atom in CO_2? Give a complete description of the sigma and pi bonding in this molecule.

27. What is the hybridization of the carbon atom in the NCO^- ion? Give a complete description of the sigma and pi bonding in this ion.

28. What is the hybridization of the carbon atom in phosgene, Cl_2CO? Give a complete description of the sigma and pi bonding in this molecule.

29. What is the hybridization of the sulfur atom in sulfur dioxide, SO_2? Give a complete description of the sigma and pi bonding in this molecule. In particular, show how many electron pairs participate in the resonance structures.

30. What is the hybridization of the nitrogen atom in the nitrate ion, NO_3^-? Describe the orbitals involved in the formation of an N=O bond.

31. Sulfur trioxide, SO_3, is formed as an intermediate in the manufacture of sulfuric acid. What is the hybridization used by sulfur in SO_3? How many resonance structures are there? Explain how hybrid orbitals and pure *p* orbitals are involved in forming one of the resonance structures.

MOLECULAR ORBITAL THEORY

32. Hydrogen, H_2, can be ionized to give H_2^+. Write the electron configuration of the ion in molecular orbital terms. What is its bond order? Is the H—H bond stronger or weaker in H_2^+ than in H_2?

33. Write molecular orbital electron configurations for Li_2^+ and Li_2^-. Compare the Li—Li bond order in Li_2 with that in its ions, Li_2^+ and Li_2^-.

34. Which ones of the homonuclear, diatomic molecules of the second row elements (from Li_2 to Ne_2) are paramagnetic? Which ones have a bond order of one? Which ones have bond orders of two? What diatomic molecule has the highest bond order?

35. Calcium carbide, CaC_2, contains the acetylide ion, C_2^{2-}. Sketch the molecular orbital energy level diagram for the ion. How many net sigma bonds and pi bonds does the ion have? What is the C—C bond order? How has the bond order changed on adding electrons to C_2 to obtain C_2^{2-}?

36. Which of the following molecules or molecule–ions should be paramagnetic? Assume the molecular orbital diagram in Figure 10.17 applies to all of them.
(a) NO (d) Ne_2^+
(b) OF^- (e) CN
(c) O_2^{2-}

37. Oxygen, O_2, can acquire one or two electrons to give O_2^- (superoxide ion) or O_2^{2-} (peroxide ion). Write molecular orbital electron configurations for the ions and then compare them with O_2 on the following basis: (a) magnetic character, (b) net number of sigma and pi bonds, (c) bond order, and (d) O—O bond length.

38. Nitrogen oxide, NO, is a colorless gas with an extensive chemistry. (a) Is NO diamagnetic or paramagnetic? If paramagnetic, how many unpaired electrons does it have? (b) Assume the molecular orbital diagram for CO applies to NO. What is the highest energy molecular orbital assigned to an electron? (c) What is the N—O bond order? (d) If NO is ionized to form NO^+, does the N—O bond become stronger or weaker than in NO? Explain briefly.

39. Write the electron configuration of the CO^+ ion. Is the ion paramagnetic? What is the net number of sigma and pi bonds? What is the bond order? Do you expect the C—O bond to be stronger or weaker in CO^+ than in CO?

METALS

40. What is the important difference between an insulator, a conductor, and a semiconductor?

41. Explain how a metal such as lithium can conduct electricity.

GENERAL QUESTIONS

42. The organic compound below is a member of a class known as oximes.

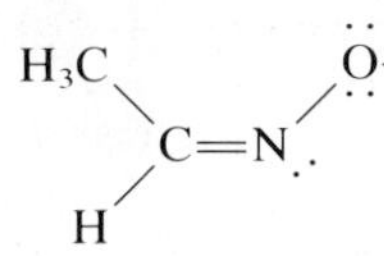

(a) What are the hybridizations of the two carbon atoms? (b) What is the approximate C—N—O bond angle?

43. Acrolein has a pungent odor and irritates eyes and mucous membranes. It is a component of photochemical smog.

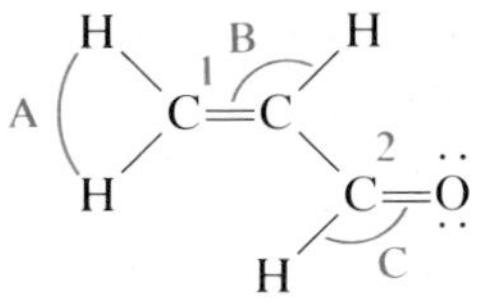

(a) What are the hybridizations of carbon atoms 1 and 2? What are the approximate values of the angles marked A, B, and C?

44. The compound sketched below is acetylsalicylic acid, better known by its common name, aspirin.

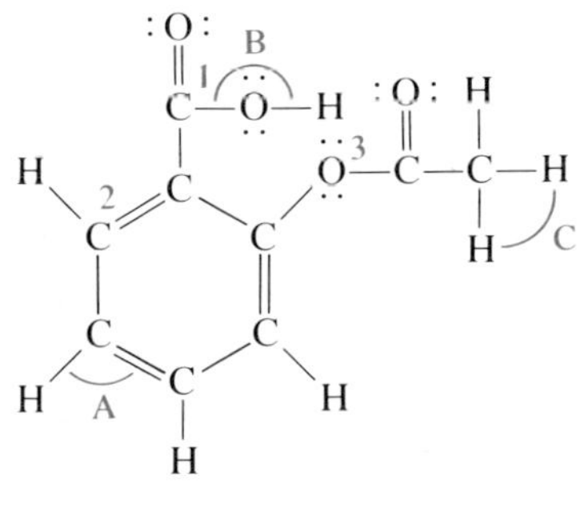

(a) How many pi bonds are there in aspirin? How many sigma bonds? (b) What are the approximate values of the angles marked A, B, and C? (c) What hybrid orbitals are used by carbon atoms 1 and 2 and oxygen atom 3?

45. Tabun, the compound sketched below, is a chemical warfare agent.

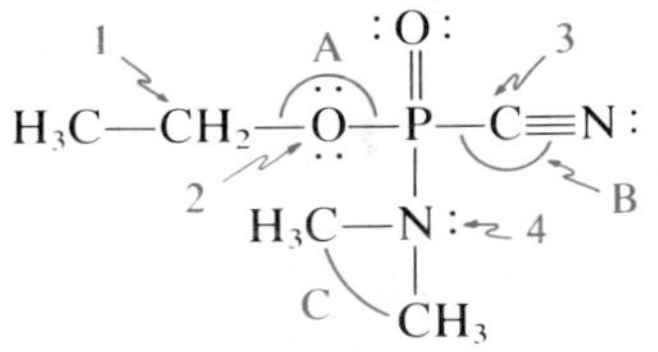

(a) How many sigma and pi bonds are there in the molecule? (b) Give the hybridization of each atom 1 through 4. (c) Which C—N bond is the shortest in the molecule? (d) Which P—O bond is shorter? (e) Give the approximate values of the bond angles A through C.

46. Histamine is found in normal body tissues and in blood. It has the structure below.

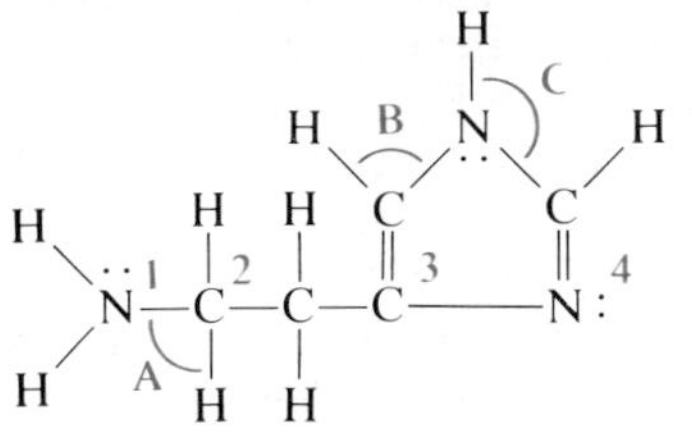

(a) How many sigma and pi bonds does the molecule have? (b) Give the hybridizations of atoms 1 through 4. (c) What are the approximate values of the bond angles A through C?

47. Boron trifluoride, BF_3, can accept a pair of electrons from another molecule to form a coordinate covalent bond (see Chapter 9), as in the following reaction with ammonia.

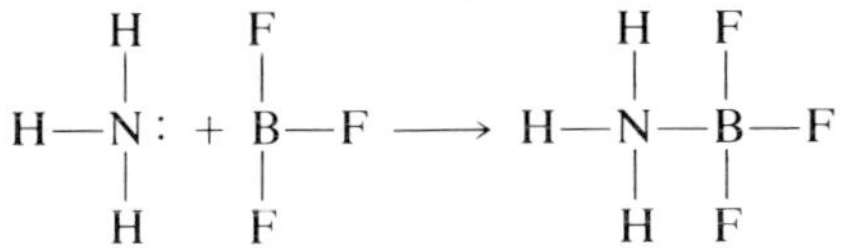

$$H_3N: + BF_3 \longrightarrow H_3N{-}BF_3$$

(a) What is the geometry about the boron atom in BF_3? In $H_3N—BF_3$? (b) What is the hybridization of boron in the two compounds in the reaction?

48. The simple valence bond picture of O_2 does not agree with the molecular orbital view. Compare these two theories with regard to the peroxide ion, O_2^{2-}. Do they lead to the same magnetic character and bond order?

49. Nitrogen, N_2, can ionize to form N_2^+ or absorb an electron to give N_2^-. Compare these species with regard to (a) their magnetic character, (b) net number of pi bonds, (c) bond order, (d) bond length, and (d) bond strength.

50. Bonding in ozone was described in Section 10.2. If ozone forms O_3^-, will the O—O bond lengths change? What happens to the O—O bond length if O_3 ionizes to O_3^+? Explain briefly. (Electrons are added to or removed from the pi molecular orbitals in ozone and other molecules with delocalized pi molecular orbitals.)

51. The ammonium ion is important in chemistry. Discuss any changes in hybridization and bond angles which

may occur in the combination of ammonia and a proton.

$H^+ + NH_3 \longrightarrow NH_4^+$

52. Antimony pentafluoride reacts with HF according to the equation

$2\ HF + SbF_5 \longrightarrow [H_2F]^+[SbF_6]^-$

What is the hybridization of Sb in the reactant and product? Draw an electron dot structure of H_2F^+. Is the ion linear or bent in structure? What is the hybridization of F in H_2F^+?

53. In many chemical reactions atom hybridization changes. In each reaction below, tell what change, if any, occurs in the underlined atoms.
(a) $H_2\underline{C}{=}\underline{C}H_2 + H_2 \rightarrow H_3C{-}CH_3$
(b) $\underline{P}(CH_3)_3 + I_2 \rightarrow P(I)_2(CH_3)_3$
(c) $\underline{Xe}F_2 + F_2 \rightarrow XeF_4$
(d) $\underline{Sn}Cl_4 + 2\ Cl^- \rightarrow SnCl_6^{2-}$

54. Iodine and oxygen form a complex series of periodates, among them the ions IO_4^- and IO_5^{3-}. Draw electron dot structures of these ions and give their structural-pair and molecular geometries. What is the hybridization of the iodine atom in these ions?

55. Xenon is the only rare gas element that forms well characterized compounds. Two xenon–oxygen compounds are XeO_3 and XeO_4. Draw electron dot structures of each of these compounds and give their structural-pair and molecular geometries. What are the hybrid orbital sets used by xenon in these two oxides?

56. The molecular orbital diagram for a simple tetrahedral species such as methane (CH_4) or ammonium ion (NH_4^+) is

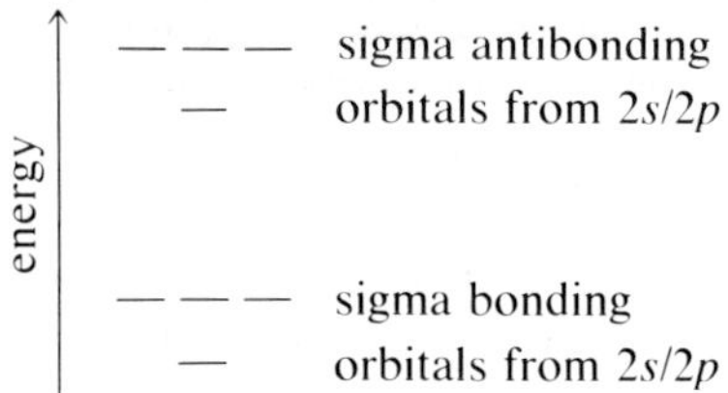

In methane, for example, the four sigma bonding orbitals are fully occupied by electrons for a bond order of one for each C—H bond. What is the orbital occupation for the ammonium ion? If ammonium ion absorbs an electron to give neutral NH_4, are the N—H bonds shortened or lengthened? What is the N—H bond order in NH_4?

57. Whereas elemental nitrogen is a diatomic molecule with an N≡N triple bond, the next element in Group 5A, phosphorus, exists in several forms, among them white or yellow P_4. The four P atoms are arranged at the corners of a tetrahedron.

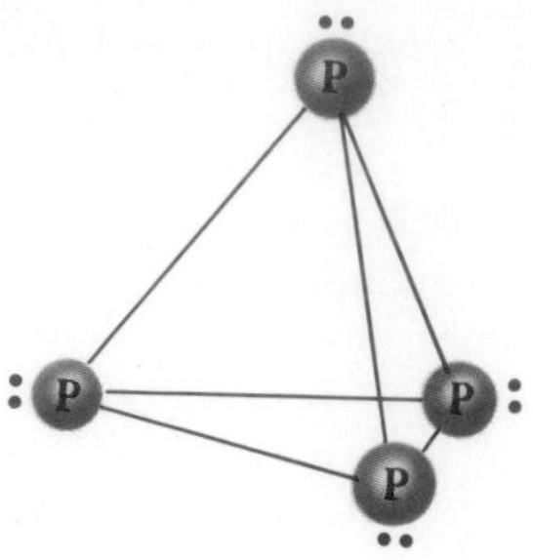

Describe the bonding in this molecule in terms of hybrid orbital theory.

58. The elements of the second period from boron to oxygen form compounds of the type $X_2E{-}EX_2$ where X can be H or a halogen. Sketch possible molecular structures for B_2F_4, C_2H_4, N_2H_4, and O_2H_2. Give the hybridizations of E in each molecule and specify approximate X—E—E bond angles. What is the major difference between the carbon-containing compound and the other molecules?

59. Rings of elements are very commonly observed in chemistry. In each ring below, give the hybridization of the indicated atom(s). Comment in each case on whether you think the ring is totally flat or if it may be puckered.

(a) C in benzene

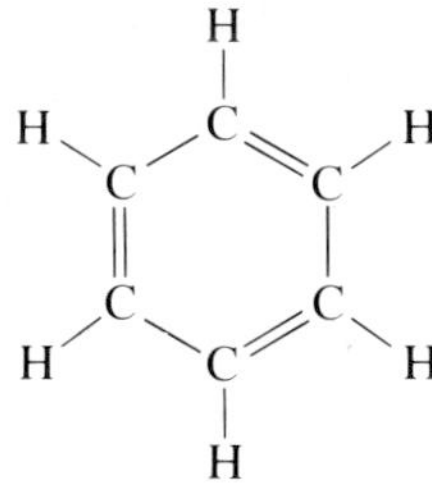

(b) S in elemental sulfur

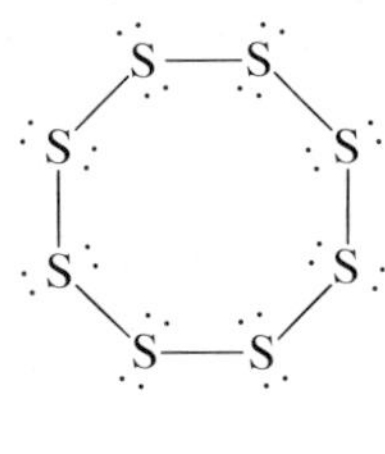

(c) B and N in borazine

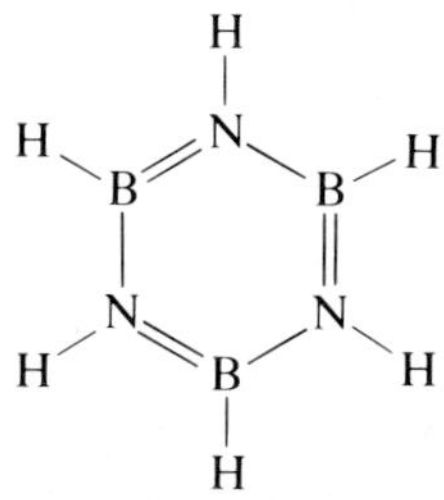

(d) Si and O in a silicate

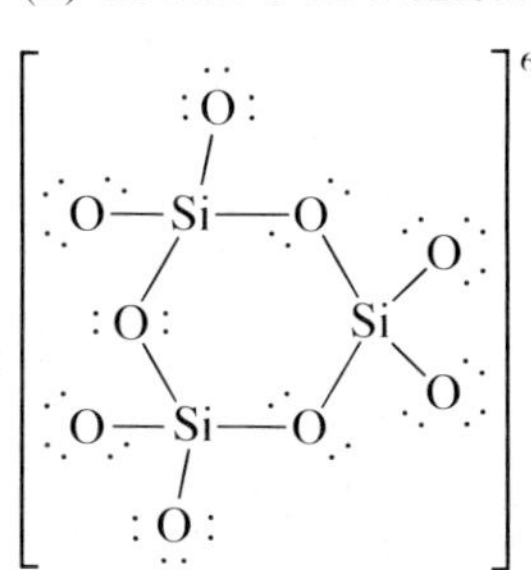

60. The sulfamate ion, $H_2N{-}SO_3^-$, can be thought of as forming from the amide ion, NH_2^-, and sulfur trioxide, SO_3. (a) Sketch a structure for the sulfamate ion. Include estimates of bond angles. (b) What changes in hybridization do you expect for N and S in the course of the reaction, $NH_2^- + SO_3 \rightarrow H_2N{-}SO_3^-$?

SPECIAL SECTION: METALS AND SEMICONDUCTORS

The molecular orbital model is useful for describing properties of metals. Metal crystals can be viewed as "supermolecules" held together by delocalized bonds formed from the atomic orbitals of all of the atoms in the crystal. Since there is a large number of atoms in a crystal, there is an even larger number of atomic orbitals to form molecular orbitals. There are so many molecular orbitals formed from each kind of atomic orbital, in fact, that the energy spacing between levels is very small, and we can speak of a "band" of orbitals over a range of energies for each orbital type (Figure 10.20). Electrons filling the bottom of a band lead to bonding, while those filling the top lead to antibonding. The band is composed of as many levels as there are contributing atomic orbitals, and each level can hold two electrons of opposite spin. This is the basic idea of the **band theory** of metallic bonding.

A one-gram crystal of lithium contains about 8.7×10^{22} atoms or 26×10^{22} atomic orbitals that can form 26×10^{22} molecular orbitals.

At 0 kelvin the continuous band of energy levels for each orbital type is filled from the lowest level up to an energy called the **Fermi level** for each orbital type (Figure 10.21). Where the band is not filled completely, any input of energy, no matter how small, can cause an electron to move from a lower, filled level to a higher energy, empty level. Since the band of energy levels in a metal is essentially continuous, it can absorb radiation of nearly any wavelength, and, because of this ability to absorb light of all colors, an unpolished metal would be expected to be black. Instead, a polished metal surface is reflective. When light causes an electron in a metal to move to a higher energy state, the electron can immediately reemit a photon of the same energy by returning to the original energy level. It is because of this rapid and efficient reemission of light that metal surfaces are reflective and appear lustrous.

Thermal energy can promote electrons into the empty levels above the Fermi level. Since the molecular orbitals are delocalized over the entire metal, this accounts for electrical conductivity.

Variations in metal color (e.g., gold and silver) arise from differences in the number of states available at particular energies above the Fermi level.

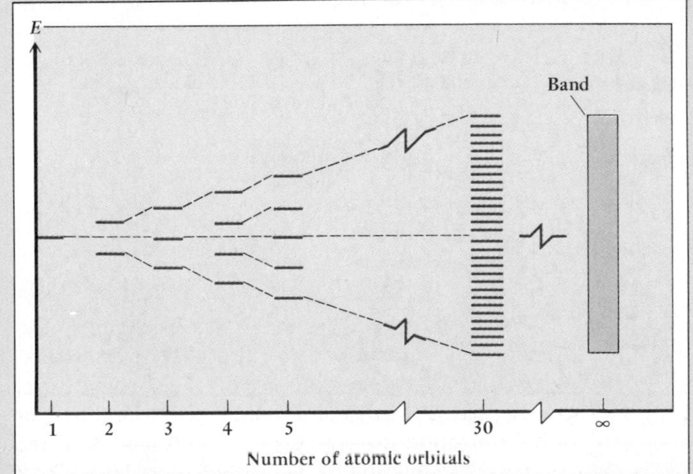

FIGURE 10.20

Bands of molecular orbitals in a metal crystal. As more and more atoms with the same valence orbitals are added, the number of molecular orbitals grows, until they merge into a band of orbitals.

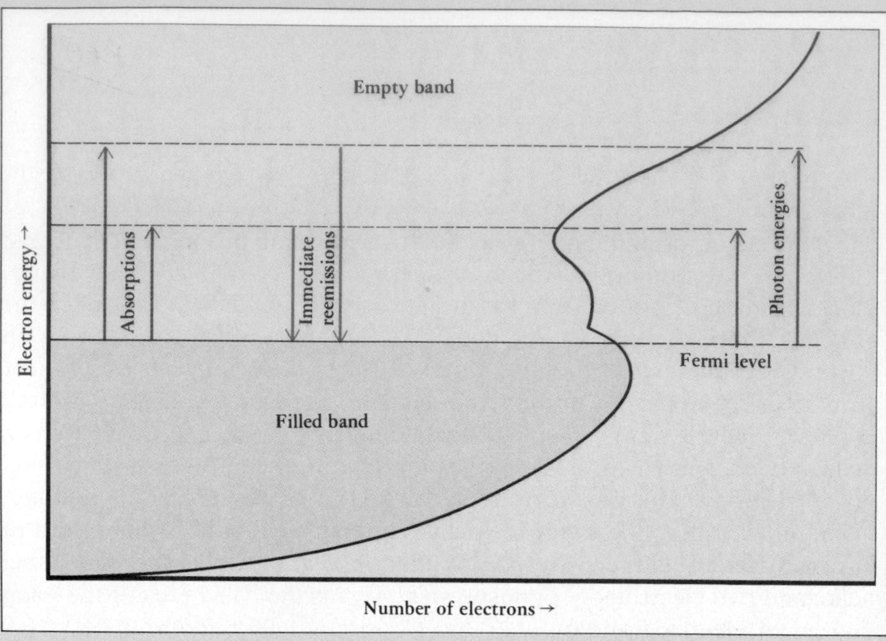

FIGURE 10.21

Electronic structure of a metal. There is an essentially continuous band of allowed energy levels that is filled up to an energy called the Fermi level. All higher energy levels are empty and can accept electrons, which can be promoted to these levels either by absorption of thermal energy or light energy. For example, light of almost any energy can be absorbed. Efficient and rapid reemission of this light causes metal surfaces to be lustrous. (Redrawn from K. Nassau, *Scientific American*, October 1980, p. 124.)

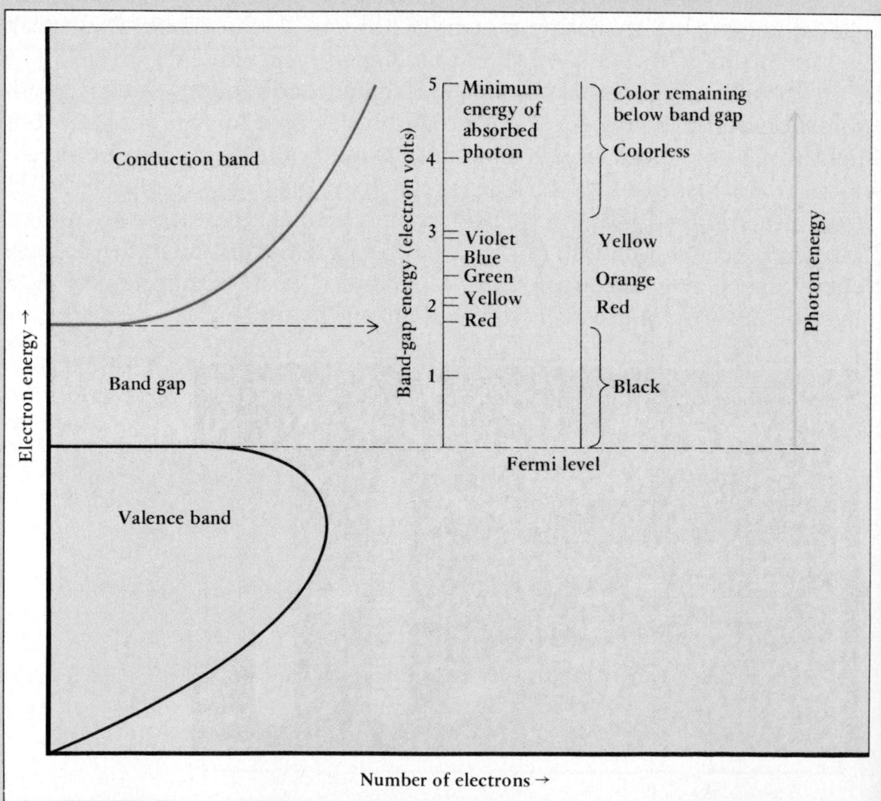

FIGURE 10.22

The band structure of a semiconductor. There is an energy gap, E_g, separating the filled valence band from the empty conduction band. Thus, there is a minimum energy (corresponding to red light in this example) that must be absorbed to move the electron across the band gap. (Energies are given here in electron volts: 1 eV = 96.35 kJ/mol). If the energy required to promote an electron is less than that of red light, the solid will appear black in white light. If the energy required is that of yellow light, for example, then the solid will appear to be red in white light. (Redrawn from K. Nassau, *Scientific American*, October 1980, p. 124.)

Semiconductors are also characterized by wide, continuous bands of electronic levels. What distinguishes metals and semiconductors, however, is that two energy bands are slightly separated in a semiconductor. The lower energy band is a **valence band**, which is completely filled with electrons in the ground state. The second band of levels makes up the **conduction band**; in the ground state the latter band is completely empty. Separating the valence and conduction band is a gap with no levels. The magnitude of this band gap, which has an energy of E_g, determines many of the properties of semiconductors.

The color of a pure semiconductor is determined by E_g. Unlike metals, electrons in the valence band cannot absorb light of any wavelength and be promoted to the conduction band. Instead, there is a minimum energy required, equal to E_g, to lift the electron from the top of the valence band to the bottom of the conduction band. If the band gap is smaller than the lowest energy of visible light, then all wavelengths of visible light are absorbed and the unpolished material is black (Figure 10.22). In other cases, such as silicon, the band gap is small, and the reemission of light is rapid and efficient, so the solid can have a metal-like luster.

A larger energy gap can lead to colored materials, such as yellow CdS (Figure 10.23) or red HgS (where E_g = 200 kJ/mol). In CdS, for example, the band gap is smaller than the energy of blue and violet light, so light of these energies is required to promote electrons from the valence to the conduction band. This means that yellow light is reflected, and this is the color of light seen by the eye.

For a more complete discussion of light absorption, transmission, and color see Chapter 25.

Finally, there are materials in which the band gap is so large that they are colorless and are electrically nonconducting. Diamonds, with a band gap of about 520 kJ/mol, are in this category.

FIGURE 10.23

Colors of semiconducting materials. Pure cadmium sulfide, CdS, at the left has E_g = 205 kJ/mol. The result is that violet light is absorbed, and the complementary color, yellow, is seen. In pure cadmium selenide, CdSe, the band gap energy is only 150 kJ/mol, so all wavelengths of visible light are absorbed, and the material appears black. (K. Nassau)

PART THREE
STATES OF MATTER

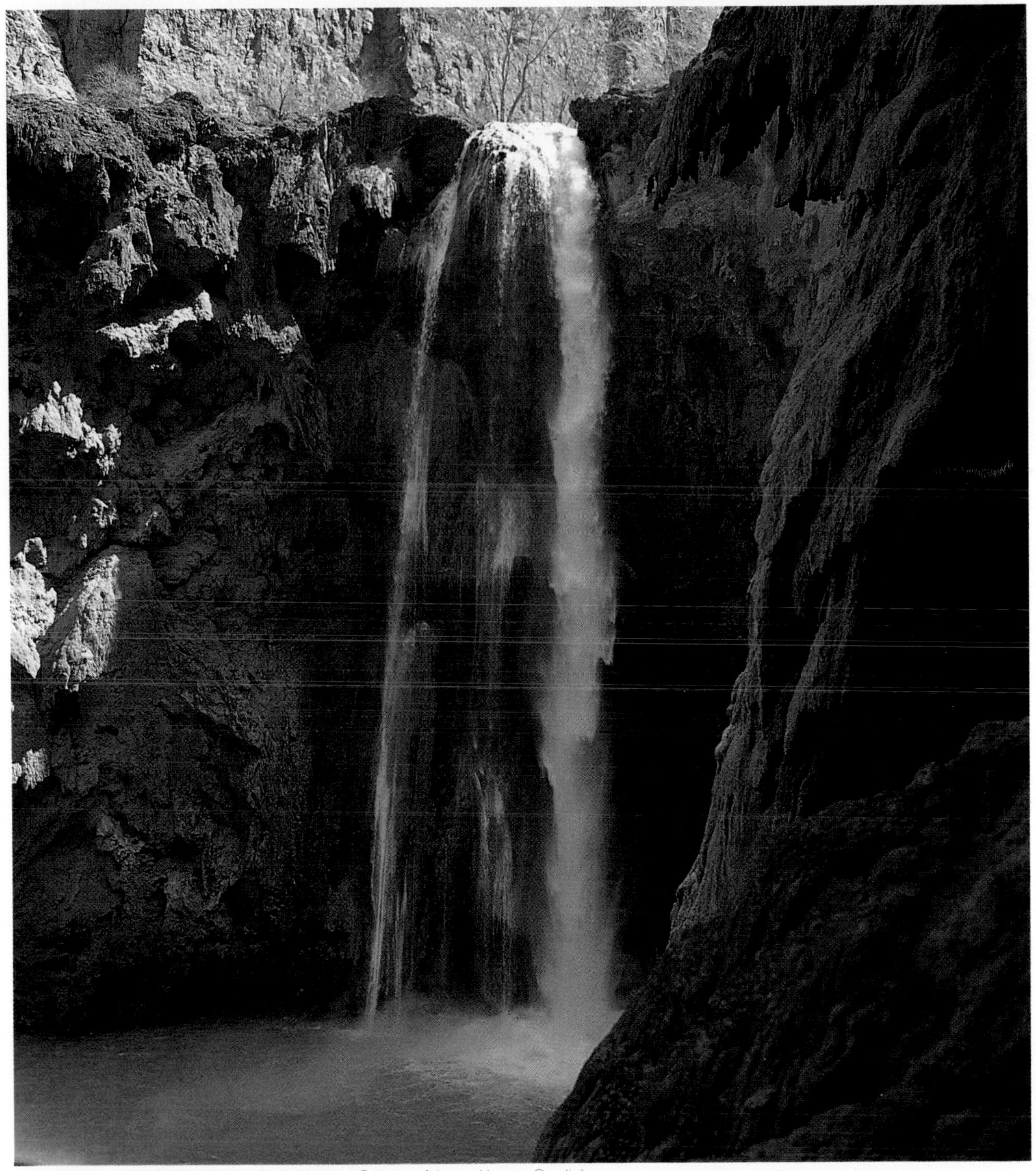

Solids, liquids, and solutions, Moony Falls, Havasu Canyon, Arizona (James Cowlin)

PART THREE: PREFACE

The chemical elements and their compounds can exist as gases, liquids, and solids. The study of gases (Chapter 6) is useful because we see that their behavior can be explained and predicted on the basis of simple mathematical models. When the forces between gaseous atoms or molecules become significant, liquids and solids can form (Chapter 11). Intermolecular forces are examined in Chapter 11, with particular attention given to those between water molecules in liquid and solid water. In contrast with liquids, solids often have beautifully regular structures, and we can relate their structures to their molecular formulas. Finally, gases, liquids, and solids can form intimate mixtures or solutions (Chapter 12). Examples are the carbon dioxide that gives the bubbles to champagne, the salt in the sea, or mixtures of elements that we use as semiconductors or transistors.

11

Intermolecular Forces, Liquids, and Solids

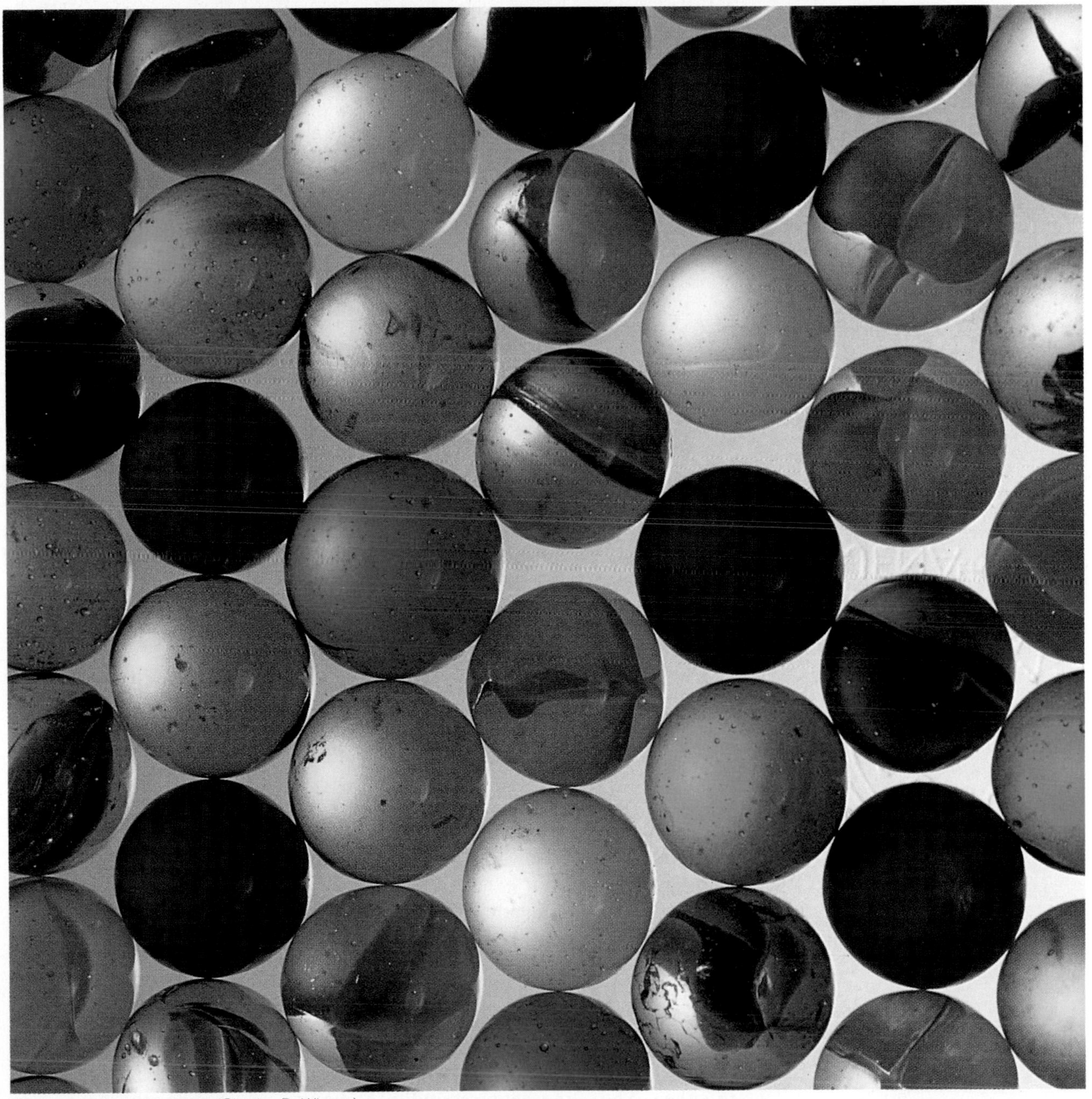

Efficient packing of spheres (Charles D. Winters)

CHAPTER OUTLINE

You are surrounded by gases, liquids, and solids: the gases of the atmosphere, liquid water in oceans and lakes, and solid earth. Of the 108 known elements, the vast majority are solids; only a few are gases (H_2, N_2, O_2, the halogens, and the rare gases), and fewer still are liquids (Hg and Br_2). The behavior of gases was explored in Chapter 6, so we turn now to the other phases of matter. You will find this a useful chapter because it will explain, among other things, why your body cools when you sweat, how bodies of water can influence local climate, why diamonds are hard but graphite is slippery, and something of the internal structure of beautifully crystalline solids.

11.1 PHASES OF MATTER AND THE KINETIC MOLECULAR THEORY

The kinetic molecular theory of gases assumes that gas molecules or atoms are widely separated, that there are no intermolecular forces, that the molecules are in continual, random, and rapid motion, and that their average kinetic energy is determined only by the gas temperature. It is on the first two points that gases differ from liquids and solids.

22.4 L/mol is the molar volume of an ideal gas. However, the actual molar volume of the polar molecule N_2O is slightly different.

Gases, unlike liquids or solids, can be compressed, because there is so much space between gas molecules. The "laughing gas" (dinitrogen oxide, N_2O) that a dentist uses as an anesthetic occupies 22.27 liters per mole under standard conditions. If one mole of laughing gas were cooled to about −90°C, it would turn into a liquid and would occupy only 35.1 cm^3, a reduction in volume by a factor of about 635! Since the number of molecules has not changed, the molecules in liquid N_2O must be much closer together than those in gaseous laughing gas. Clearly, there is a tremendous amount of empty space in the gas phase compared with the liquid.

In contrast with the change in volume when converting from a gas to a liquid, there is no dramatic change in volume when a liquid is converted to a solid. For example, one mole of liquid oxygen (O_2) occupies 28.1 cm^3 (at its boiling point, −182.7°C), whereas one mole of solid oxygen takes up only slightly less volume, 22.7 cm^3 at −253°C. This means that molecules of solid are packed together only a bit more tightly than the molecules in the liquid phase. These phase differences for gas, liquid, and solid NO_2 are illustrated in Figure 11.1.

Each molecule of a compound in the liquid or solid phase is closer to a neighbor than in the gas phase. In fact, they cannot come much closer together. Thus, the **compressibility** of liquids and solids, the change in volume with change in pressure, is small compared with gases. The air/fuel

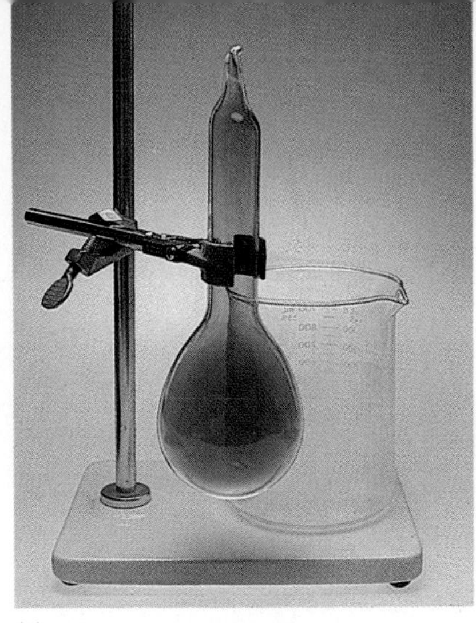

(a)

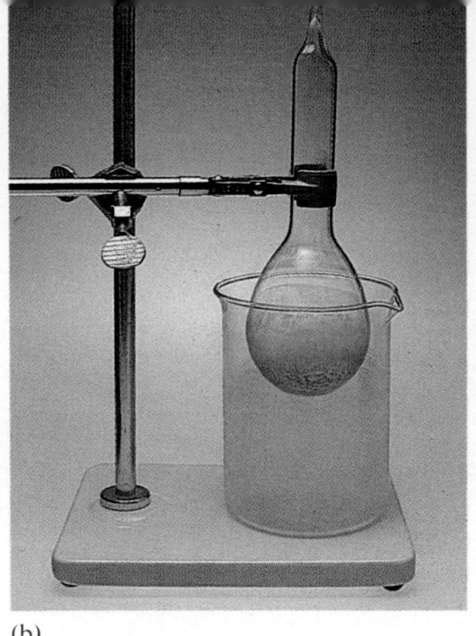

(b)

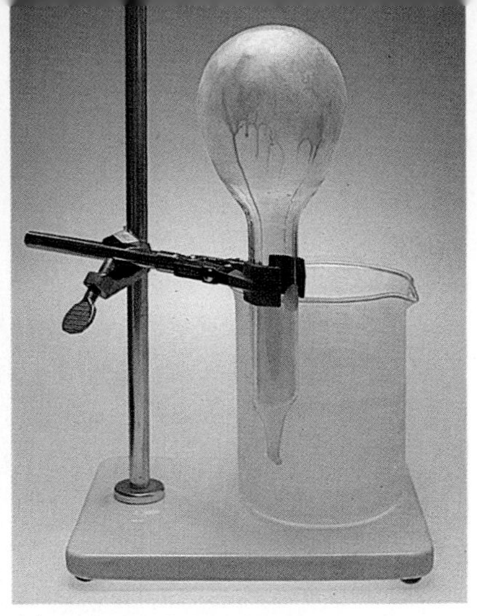

(c)

FIGURE 11.1

Three phases of matter for NO_2. (a) The brown gas is frozen to a solid (b) in liquid nitrogen. (c) When warmed, the solid melts to liquid NO_2. (Charles D. Winters)

mixture in your car's engine is routinely compressed by a factor of about 10. In contrast, the volume of liquid water changes only by 0.005% per atmosphere of pressure applied to it.

The kinetic molecular theory of gases assumes there are only weak forces attracting molecules to one another, so gases can expand infinitely to fill their containers uniformly and completely. In contrast, there are strong **intermolecular forces** in liquids and solids (Section 11.2), and these attractive forces allow the formation of condensed phases. Furthermore, the intermolecular forces in a liquid or solid mean that the volume of a given mass in one of these phases is fixed (the two quantities are related at a given temperature by the density). In addition, liquids (and solids if they are powdered) can be poured from one container to another, while the forces between molecules keep the mass of substance together.

In all phases of matter, molecules are in constant motion, the difference being that the motions are more restricted in liquids and solids. The average energy of motion of molecules depends only on the temperature: the higher the temperature the higher the average energy. Furthermore, just as for molecules in the gas phase, liquid phase molecules have a distribution of kinetic energies: the higher the temperature, the greater the relative number of molecules with high kinetic energy and the greater the spread of energies (Figure 11.2).

Molecules of a liquid are in constant, rapid motion. They translate (move from place to place), vibrate, and rotate. In the liquid phase, however, the distance traveled by a molecule before collision with another one is more restricted than in a gas. When the substance becomes a solid, molecular motion (translation and usually rotation) are severely restricted, and molecular motion is reduced to vibrations.

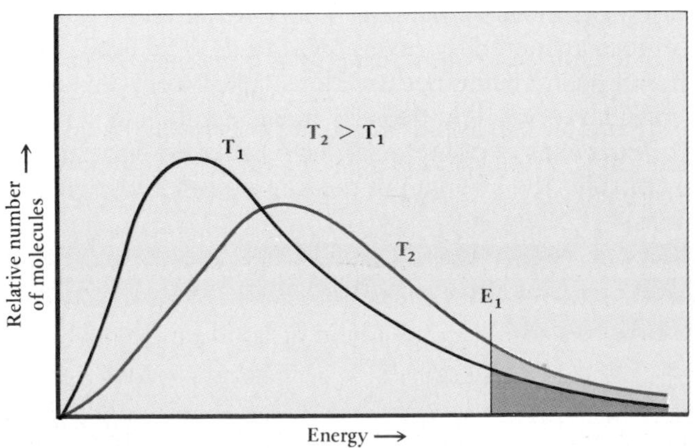

FIGURE 11.2

The distribution of molecular energies in the liquid phase. T_2 is greater than T_1, so the proportion of molecules having an energy greater than E_1 is larger at T_2 than at T_1.

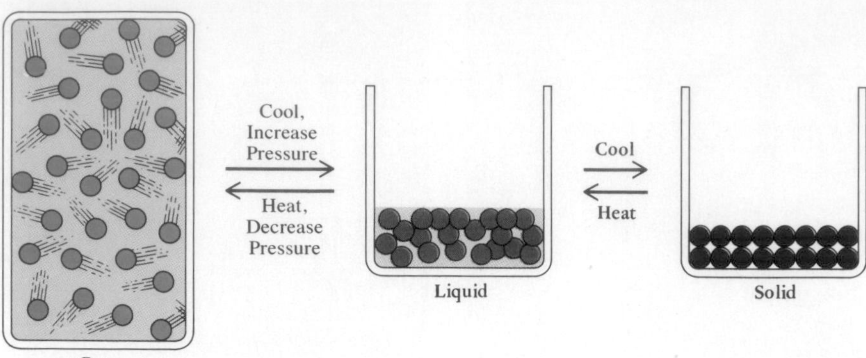

FIGURE 11.3
The three states of matter—gas, liquid, and solid—from the point of view of the kinetic molecular theory.

The view of gases, liquids, and solids we have developed to this point is summarized in Figure 11.3. Molecules move freely in the gas phase. Liquids form when the intermolecular forces are strong enough to bind molecules together, and motion is more restricted. Still, a liquid is like a very dense gas in that the molecules can move rapidly and randomly. As the liquid becomes a solid, however, molecular motion is even more restricted, intermolecular forces are strong, and the molecules "pack" themselves into a regular array.

11.2 INTERMOLECULAR FORCES

Without **intermolecular forces** (the interactions between molecules, between ions, or between ions and molecules), all substances would be gases. Before studying liquids and solids, therefore, it is important to describe the forces that lead to their formation.

The various types of intermolecular forces are listed in Table 11.1. They range from relatively strong ion–ion forces to very weak forces between induced dipoles. The weaker forces, those not involving ions, are collectively known as **van der Waals forces**, named for the same chemical physicist who developed an equation to describe the behavior of real gases (Chapter 6).

ION–ION INTERACTIONS

NaCl is a solid and not a gas because the Na^+ ion is attracted by several Cl^- ions, which are in turn attracted by several Na^+ ions. The final result is a solid, regular array of positive and negative ions called a **crystal lattice** (Figure 11.4). We consider crystal lattices in more detail later in the chapter. However, for purposes of comparison with other intermolecular forces, it is useful to consider the strength of the interaction between two oppositely charged ions.

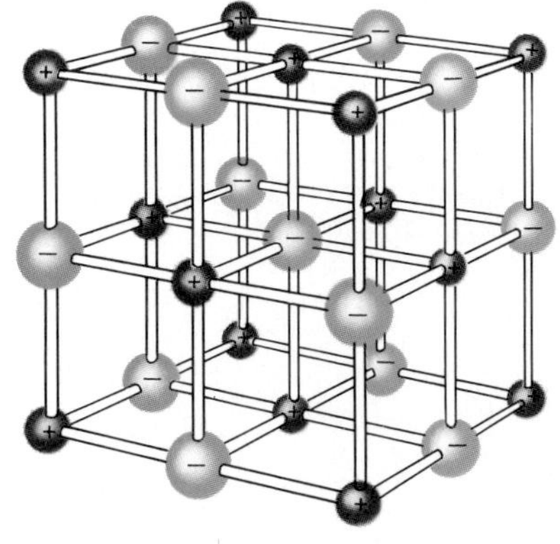

FIGURE 11.4
The crystal lattice of an alkali halide salt such as LiF, NaCl, or KBr.

One measure of the force of attraction between two oppositely charged ions is the **enthalpy of dissociation** of an ionic salt in the gas phase to its ions, also in the gas phase.

$$MX(g) + \text{energy} \longrightarrow M^+(g) + X^-(g) \qquad \text{Energy} = \Delta H_{\text{dissociation}}$$

$$NaCl(g) + 548 \text{ kJ} \longrightarrow Na^+(g) + Cl^-(g) \qquad \Delta H_{\text{diss}} = +548 \text{ kJ/mol}$$

TABLE 11.1 Summary of Intermolecular Forces

TYPE OF INTERACTION	PRINCIPAL FACTORS RESPONSIBLE FOR THE INTERACTION	DISTANCE (d) DEPENDENCE OF INTERACTION ENERGY	MAGNITUDE OF THE INTERACTION
Ion–ion	Charge on the ions	$1/d$	400–4000 kJ/mol
Ion–dipole	Ion charge; dipole moment	$1/d^2$	40–600 kJ/mol
Dipole–dipole	Dipole moment	$1/d^3$	5–25 kJ/mol
Hydrogen bonding	Dipole moment	—	10–40 kJ/mol
Dipole-induced dipole	Dipole moment; polarizability	$1/d^6$	2–10 kJ/mol
Induced dipole–induced dipole	Polarizability	$1/d^6$	0.05–40 kJ/mol

The energy expressed as ΔH_{diss} can be calculated from **Coulomb's law**, which states that the energy of interaction is the product of the ion charges divided by the distance between the ions

$$\text{Coulombic energy per mole} = \Delta H_{diss} \propto \frac{n_+ n_-}{d} \qquad (11.1)$$

Here n_+ is the number of positive charges on the cation (for example, $n_+ = 2$ for Mg^{2+}) and n_- is the number of negative charges for the anion ($n_- = 3$ for PO_4^{3-}), and d is the distance between the ion centers in the crystal lattice.* The important aspect of this equation is that the force of attraction between ions depends directly on ion charges (the higher the charges the greater the attraction) and inversely on the distance between them (the greater the distance the weaker the attraction).

Ion–ion attractions have a profound effect on such properties of ionic solids as their melting points and solubilities in water, topics to be described later. Here, however, it is useful to see the periodic trends in such interactions. The energies for the dissociation of alkali metal halides

---

*The exact equation for calculating the coulomb energy is

$$\text{Coulombic energy per mole} = \frac{N(n_+e)(n_-e)}{4\pi\epsilon d} \qquad (11.1)$$

N is Avogadro's number, e is the charge on the electron (1.60×10^{-19} coulomb), d is the distance between the ions (in meters), and ϵ is a constant (8.85×10^{-12} coulomb^2/J · m). For NaCl, where $d = 2.36 \times 10^{-10}$ m, you would find coulomb energy is 587 kJ/mol, in reasonable agreement with experiment.

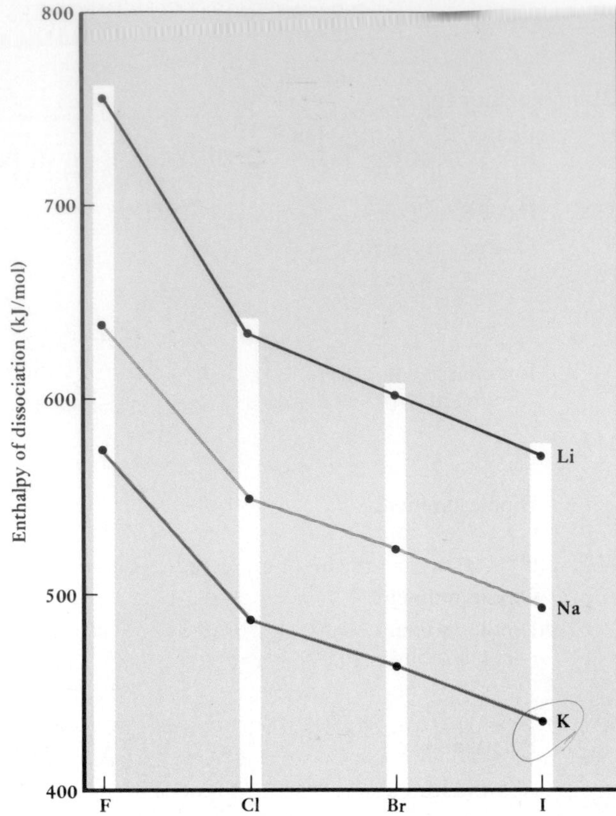

FIGURE 11.5
Enthalpies of dissociation for the alkali metal halide MX(g) to give the gaseous ions $M^+(g)$ and $X^-(g)$.

are plotted in Figure 11.5. Notice, for example, that ΔH_{diss} for NaCl is less than that for LiCl, but larger than that for KCl, in agreement with Coulomb's law. More energy is required to dissociate salts of smaller ions.

EXERCISE 11.1 ION–ION FORCES

The temperature required to melt an ionic solid depends in large part on the coulombic forces between ions: the greater the forces, the higher the melting point. In each pair, which ionic solid should have the higher melting point: (a) LiF or KI and (b) MgO or NaCl?

ION–DIPOLE FORCES

The energy involved in the attraction between a positive or negative ion and a dipolar molecule is less than that found for ion–ion attraction but greater than any of the other intermolecular forces we shall describe.

Water is an excellent example of a polar molecule, a molecule with positive and negative electrical poles.

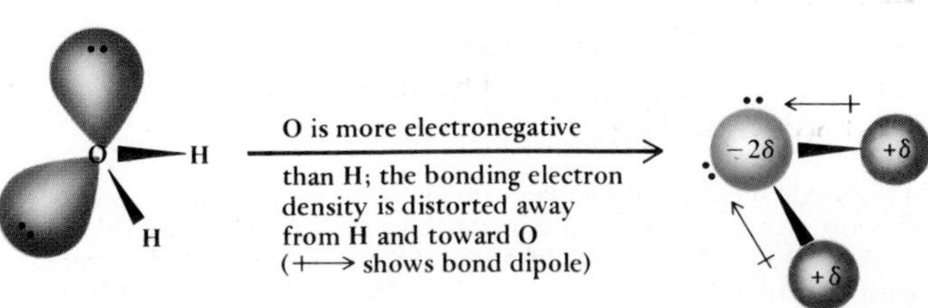

Thus, if a water molecule encounters an ion, there will be a force of attraction between the ion and the molecule.

As you learned in Chapter 9, when attraction occurs, the attracting particles release energy on bond formation.

$$Na^+(g) + x\ H_2O(g) \longrightarrow [Na(H_2O)_x]^+(aq) + 397\ kJ/mol$$
$$(x \text{ probably} = 6)$$

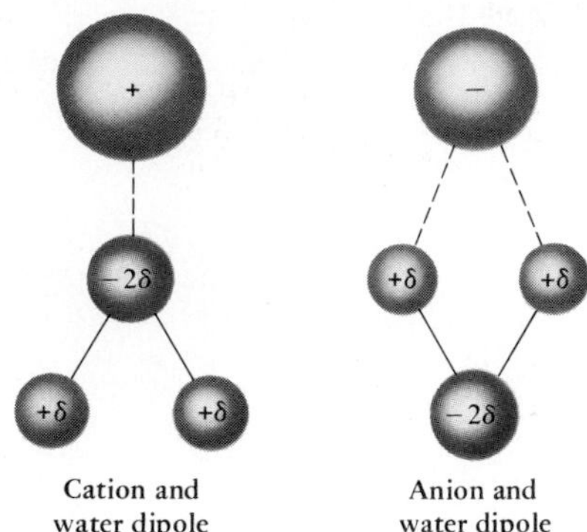

A metal ion binding several water molecules is said to be **hydrated**, and the energy released is the **heat** or **energy of hydration**. If some polar molecule other than water is involved as the solvent, the more general term **solvation** is used, and we would talk about **solvation energy**.

Hydration is important in our watery world. The oceans, lakes and rivers consist of many substances dissolved in the water (e.g., Na^+, Ca^{2+}, Cl^-, CO_2, HCO_3^-, CO_3^{2-}, O_2). The interaction between the water and these substances not only causes them to dissolve but also greatly influences the properties of the water.

Neutral molecules, as well as ions, can also be hydrated or solvated.

As defined in Chapter 3, acids are a source of hydrogen ions, H^+, in water. Not surprisingly, there is a strong ion–dipole interaction between H^+ and water. The H—Cl covalent bond in molecular HCl is moderately strong. However, it is readily broken when HCl gas is bubbled into water, because the H—Cl dipole interacts with water so strongly that the bond is broken, and water molecules cluster about the resultant H^+ and Cl^- ions. The H^+ ion is extremely small, so water molecules are strongly bound to H^+ (its heat of hydration is 1080 kJ/mol), and ions of the general formula $[(H_2O)_nH]^+$ with $n \leq 4$ are found in solution. We usually symbolize this by indicating that the hydronium ion, H_3O^+ (or $[(H_2O)H]^+$), is formed when an acid such as HCl, HBr, HI, HNO_3, or H_2SO_4 is placed in water.

Interaction of water and HCl dipoles

2δ− 2δ
H—O · · · · H—Cl · · · H—O
δ+ δ+ δ− δ+
H H
δ+ δ+

$$HCl(g) + (n + 1)\ H_2O \longrightarrow H_3O^+(aq) + [Cl(H_2O)_n]^-(aq)$$

The importance of ion–dipole attractions is not limited to solutions. You will notice that many solids in the laboratory have formulas such as $BaCl_2 \cdot 2\ H_2O$ or $CoCl_2 \cdot 6\ H_2O$. The formula of the cobalt(II) salt, for example, is better written as $[Co(H_2O)_4Cl_2)] \cdot 2\ H_2O$, because four of the six water molecules are associated with the Co^{2+} ion by ion–dipole attraction (Figure 11.6).

There are two general observations we can make about hydrated salts. First, water is most commonly associated with *positive* ions. Second, because of Coulomb's law, small ions or those of high charge most strongly attract water dipoles. Thus, small ions such as Li^+ and Na^+ often form hydrated salts, while salts of K^+, Rb^+, Cs^+, and NH_4^+ are usually anhydrous (without water of hydration). The effect of ion charge is nicely illustrated by the series of salts of Cs^+ (CsCl, anhydrous), Ba^{2+} ($BaCl_2$, with one or two water molecules of hydration), and La^{3+} ($LaCl_3$, with 6 or 7 water molecules of hydration). The higher the ion charge the more strongly water dipoles interact.

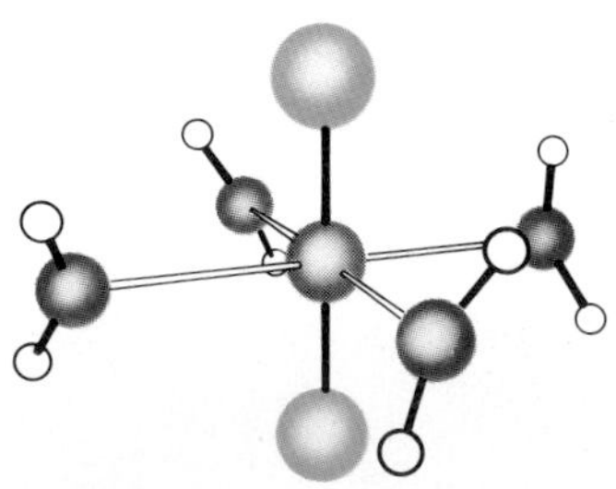

FIGURE 11.6

The structure of $Co(H_2O)_4Cl_2$. The four water molecules are bound to Co^{2+} by ion–dipole attractions. (Green = Cl; red = O; gray = Co)

EXERCISE 11.2 HYDRATION ENERGY

Using Coulomb's law, explain why the energy of hydration of Na^+ is 397 kJ/mol, while that of Cs^+ is 255 kJ/mol, and that of Mg^{2+} is 1908 kJ/mol.

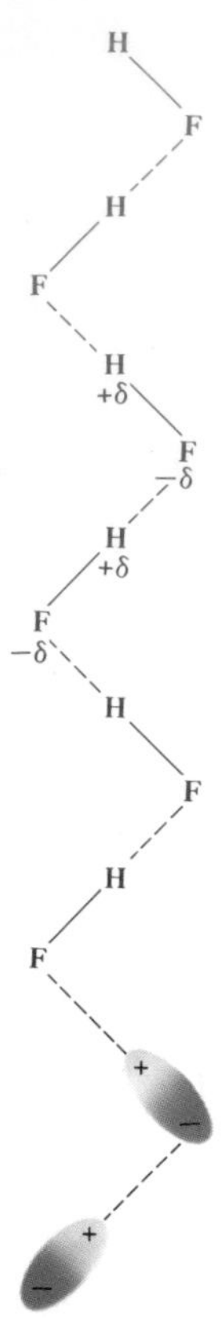

Polar HF molecules form zig-zag chains in the solid state, each HF interacting with two neighbors.

DIPOLE–DIPOLE FORCES

If we represent dipolar molecules as cigar-shaped objects, a series of them can interact as follows.

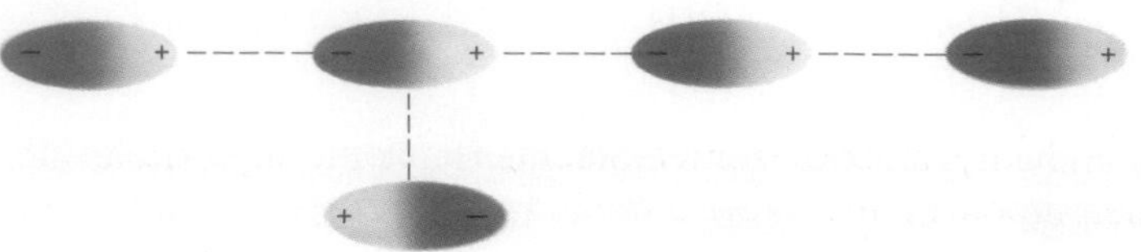

Energy is released when two dipoles attract one another. In part, this is why you must cool a polar gas to convert it to a liquid: the heat evolved on interaction must be removed. Conversely, energy is required to separate two interacting dipoles, and this is part of the reason that water or any other dipolar molecule in the liquid state needs to be heated to convert its molecules from the liquid to the gas phase.

Two different polar molecules can associate as well. For example, water can be dissolved readily in ethyl alcohol (C_2H_5OH) because there is a strong interaction between these two kinds of dipolar molecules, and considerable heat is evolved when they are mixed.

$\delta+$ H (upper) and $\delta+$ H (lower) bonded to $\overset{2\delta-}{O} \cdots \overset{\delta+}{H}\text{—}\overset{\delta-}{O}\text{—}CH_2\text{—}CH_3$

In contrast, water does not dissolve gasoline to any appreciable extent, because the hydrocarbon molecules (such as $H_3CCH_2CH_2CH_2CH_2CH_3$) that constitute gasoline are not sufficiently polar. Water–hydrocarbon attractions are so weak that they cannot disrupt the stronger water–water attractions (Figure 11.7).

HYDROGEN BONDING

Hydrogen bonding is a special form of dipole–dipole attraction, and enhances dipole–dipole attractions. When hydrogen is attached to a very electronegative atom X, the interaction between the H—X bond dipole and another dipole is significantly greater than expected for ordinary dipole–dipole interactions. This interaction, called **hydrogen bonding**, is of great practical significance.

The electronegativities of N (3.0), O (3.5), and F (4.0) are among the highest of all the elements, while that of H (2.1) is considerably less. Therefore, covalent bonds between H and N, O, or F are extremely polar. Thus, if an atom of an adjacent molecule (or even one in the same molecule) is the negative end of a bond dipole, a strong attraction can occur. This is especially true if the neighboring atom is small, has a lone pair of electrons, and can closely approach the H atom. The small, highly electronegative atoms N, O, and F are therefore involved in strong hydrogen bonding, while larger atoms of similar electronegativity [Cl (3.0) and S (2.8)] have a much smaller tendency to form hydrogen bonds.

To illustrate the role of hydrogen bonding in determining the physical properties of a compound, consider two polar, organic molecules,

FIGURE 11.7

Oil on water. Since the two do not mix, oil, the less dense liquid, floats on the water. If the oil layer is only a thin film, it diffracts light, and you see a rainbow of colors. (Charles D. Winters)

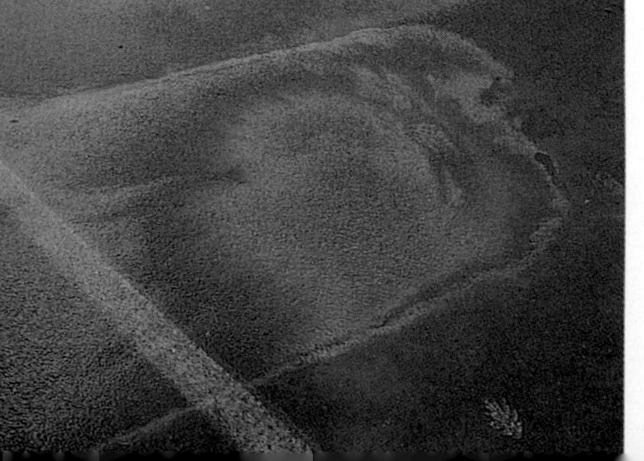

TABLE 11.2 A Comparison of the Physical Properties of Ethyl Alcohol and Dimethyl Ether

COMPOUND		DIPOLE MOMENT (D)	MELTING POINT (°C)	BOILING POINT (°C)
C_2H_5OH	Ethyl alcohol	1.7	−115	78
$H_3C—O—CH_3$	Dimethyl ether	1.3	−141	−25

dimethyl ether and ethyl alcohol. Both molecules have the same molecular formula and their dipole moments are comparable. However, their melting and boiling points are vastly different (Table 11.2). A higher melting point and boiling point for one of two similar molecules means that stronger intermolecular forces are involved for the compound requiring the higher temperatures. Methyl alcohol requires higher temperatures because alcohol molecules may interact through hydrogen bonding, whereas hydrogen bonding is not important for the ether.

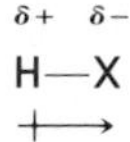

The polarity of a bond between H and O, N, or F.

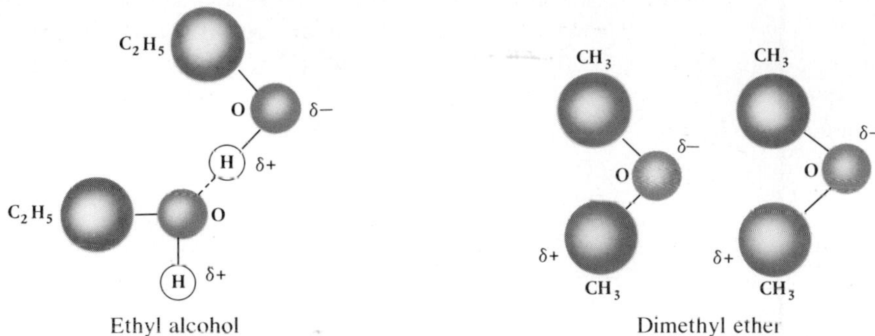

Ethyl alcohol Dimethyl ether

The consequences of hydrogen bonding for several series of structurally analogous molecules are listed in Figure 11.8. The hydrogen compounds of the Group 4A elements (CH_4, SiH_4, GeH_4, and SnH_4) are all nonpolar, tetrahedral molecules, and their boiling points are expected to increase with molecular mass (largely due to increased intermolecular dispersion forces as explained below). It is certainly true for the compounds of Group 4A and for the higher molecular weight compounds of the other groups. However, the boiling points of NH_3, H_2O, and HF clearly do not follow the expected trends within their groups. It is of course just these compounds for which hydrogen bonding between molecules is expected to be most important. Water, for example, might be expected to boil at around −90°C, the temperature you find if you just extrapolate the line for H_2S, H_2Se, and H_2Te. However, hydrogen bonding has raised the boiling point almost 200°C over this!

The most important forms of H-bonding

F—H...F	F—H...O	F—H...N
O—H...F	O—H...O	O—H...N
N—H...F	N—H...O	N—H...N

INTERACTIONS INVOLVING INDUCED DIPOLES — skip to p. 392

PERMANENT DIPOLE-INDUCED DIPOLE FORCES A nonpolar molecule such as O_2 or CO_2 can dissolve to a small extent in water because polar molecules such as water can **induce** a dipole in molecules without permanent dipole moments.

Picture a polar water molecule approaching a nonpolar molecule such as O_2 (Figure 11.9). The electron cloud of an isolated oxygen molecule is distributed, on average, symmetrically between the bonded O—O atoms. However, as the negative end of the water molecule approaches,

Dipole-induced dipole forces are important because this is the effect that allows gases such as O_2 and CO_2 to dissolve in water, and their solutions are of concern in our economy and our environment. Dissolved CO_2 can produce hard water; see Chapter 17.

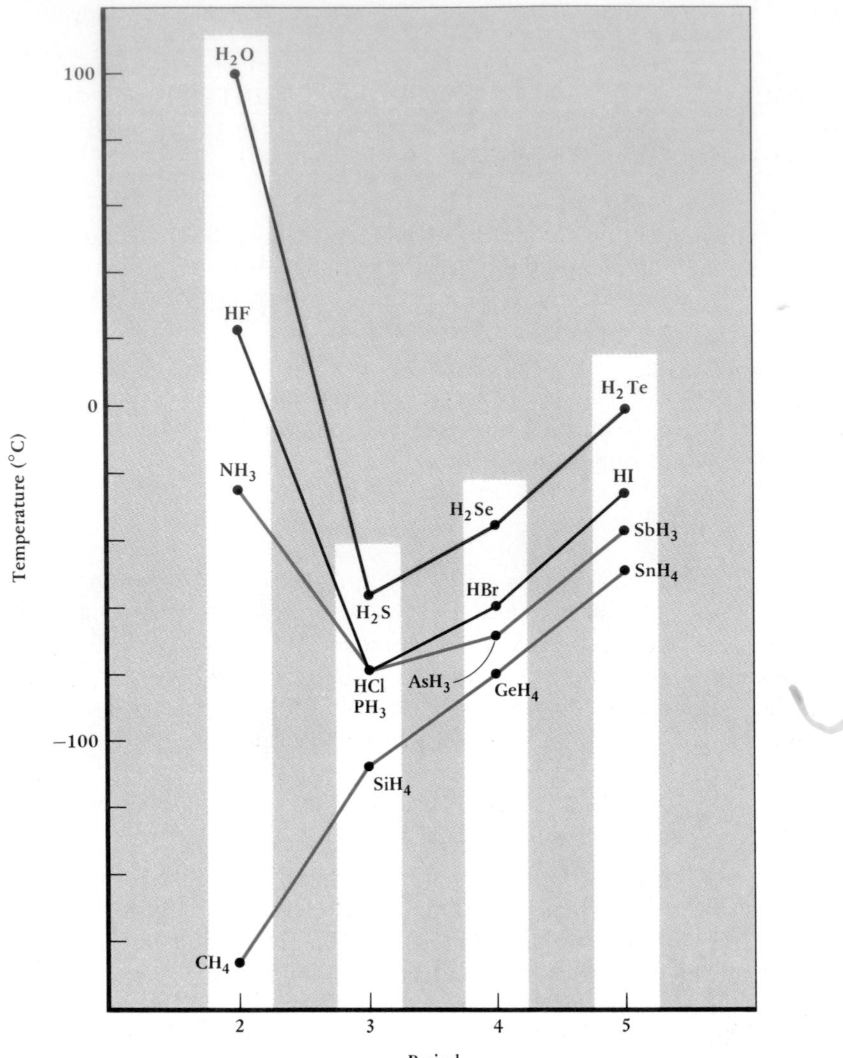

FIGURE 11.8

The boiling points of some nonmetal hydrides. Lines connect molecules containing atoms from the same group of the periodic table. Notice the effect of hydrogen bonding on the boiling points of H_2O, HF, and NH_3.

the O_2 electron cloud is distorted by repulsion between the cloud and the negative end of the water dipole. That is, a dipole has been *induced* in the otherwise nonpolar O_2 molecule, and water and O_2 can now be attracted to one another.

The amount by which the electron cloud of an atom (such as Ne or Ar) or a symmetrical molecule without a permanent dipole (such as O_2, N_2, I_2, or CCl_4) can be distorted and a dipole induced depends on the polarizability of the atom or molecule. The **polarizability** of an atom or molecule is a measure of the extent to which the electron cloud of the molecule can be distorted by an external electric charge. Atoms or molecules with large, extended charge clouds, such as I_2, can be polarized more readily than those with small, compact clouds such as He, H_2, or F_2.

A table of the solubilities of the atmospheric gases in seawater (Table 11.3) illustrates dipole-induced dipole interactions. Here you see

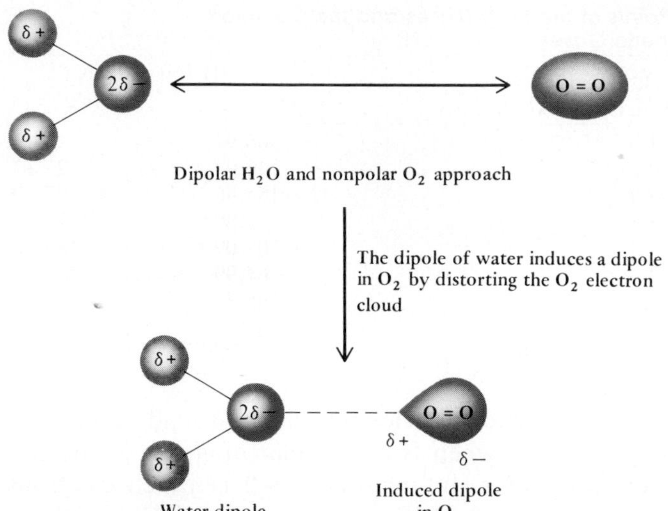
δ+
2δ−
δ+
O = O
Dipolar H_2O and nonpolar O_2 approach
The dipole of water induces a dipole in O_2 by distorting the O_2 electron cloud
δ+
2δ−
δ+
O = O
δ+
δ−
Water dipole
Induced dipole in O_2

FIGURE 11.9
Dipolar water inducing a dipole in nonpolar O_2.

a clear trend to higher solubility with increasing atomic or molecular mass of the nonpolar gas. The greater the mass, the greater the number of electrons in the electron cloud to be polarized. Since the magnitude of the induced dipole is directly related to polarizability, this means there should be a reasonably direct relation between the mass of a nonpolar molecule and the strength of its interaction with a permanent dipole.

INDUCED DIPOLE-INDUCED DIPOLE FORCES The weakest of all intermolecular forces is between two induced dipoles. Such forces are often called **London forces** or **dispersion forces**. In spite of their extreme weakness, they are important. Nonpolar molecules of I_2 must attract one another in some manner because I_2 is a solid at room temperature. Similarly, N_2, O_2, and the rare gases can all be liquefied at low temperatures. In Table 11.4 you see a trend to higher boiling points of nonpolar gases with increasing atomic or molecular weight, in part because dispersion forces generally become stronger with increasing mass.

To understand how two nonpolar molecules can attract one another, remember that the electrons in atoms and molecules are in a state

TABLE 11.3 The Solubility of the Rare Gases and the Common Atmospheric Gases in Seawater

		SOLUBILITY (moles/m³)	
GAS	MOLECULAR WEIGHT	0°C	24°C
He	4	0.36	0.31
Ne	20.2	0.42	0.36
Ar	39.9	1.7	1.0
Kr	83.8	3.2	1.9
Xe	131.3	6.1	3.1
N_2	28.0	0.80	0.54
O_2	32.0	1.9	1.1
CO_2	44.0	65	32

Data taken from W.S. Broecker, *Chemical Oceanography*, Harcourt Brace Jovanovich, Inc., New York, 1974, p. 117.

TABLE 11.4 Boiling Points of the Rare Gases and the Common Atmospheric Gases

GAS	ATOMIC OR MOLECULAR MASS	BOILING POINT °C	BOILING POINT K
He	4.0	−268.97	4.18
Ne	20.2	−246.02	27.13
Ar	39.9	−185.86	87.29
Kr	83.8	−152.89	120.26
Xe	131.3	−107.09	166.06
Rn	222	−64.99	208.16
N_2	28.0	−195.8	77.4
O_2	32.0	−187.97	85.18

of constant motion. On average, the electron cloud around an atom is spherical in shape. However, when two nonpolar atoms or molecules approach each other, attractions and repulsions between their electrons and nuclei can lead to distortions in their electron clouds. That is, dipoles can be induced momentarily in neighboring atoms or molecules, and these induced dipoles can lead to intermolecular attraction (Figure 11.10).

Two nonpolar atoms or molecules (Time average shape is spherical)

Momentary attractions and repulsions between nuclei and electrons in neighboring molecules lead to induced dipoles.

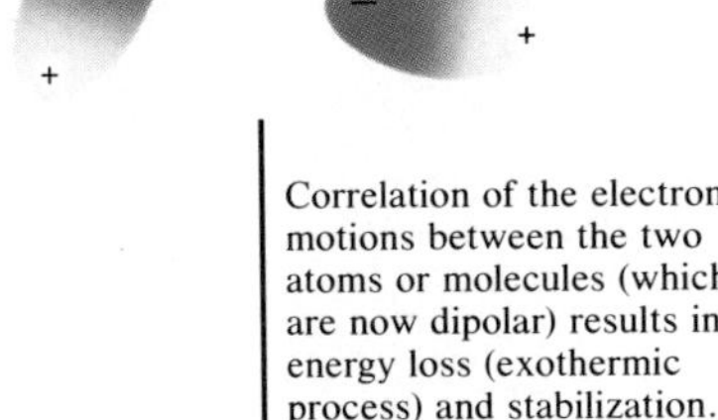

Correlation of the electron motions between the two atoms or molecules (which are now dipolar) results in energy loss (exothermic process) and stabilization.

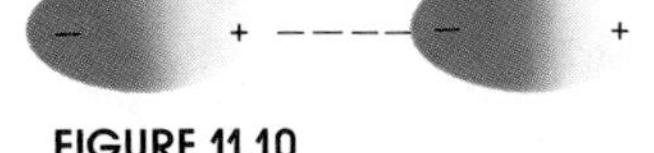

FIGURE 11.10
Interaction between two induced dipoles.

EXAMPLE 11.1

INTERMOLECULAR FORCES

Decide what type of intermolecular force is involved in each case and place them in order of increasing strength of interaction: (a) CH_4 . . . CH_4; (b) H_2O . . . CH_3OH; and (c) Li^+ . . . H_2O.

Solution (a) Both of the molecules have perfect tetrahedral geometry, and neither has a net dipole moment. Therefore, they can only interact through the weakest of all intermolecular forces, induced dipole/induced dipole.

(b) Both water and CH_3OH (methyl alcohol, Table 11.2) have a dipole moment. Further, both have an —O—H bond. Therefore, they can interact through hydrogen bonding, a specific type of dipole–dipole force.

```
δ+ 2δ−      δ+  δ−
H—O · · · · H—O
  |           |
  H           CH₃
  δ+
```

(c) The Li^+ . . . H_2O pair involves the relatively strong interaction of an ion with a dipole.

In order of increasing strength, the interactions are (a) < (b) < (c).

EXERCISE 11.3 INTERMOLECULAR FORCES

Decide what type of intermolecular force is involved in (a) N_2 . . . N_2 and (b) H_2O . . . CO_2. Which is stronger?

11.3 PROPERTIES OF LIQUIDS

Water is the most common liquid you encounter, but it has some unique properties. Here we want to describe some properties common to all liquids, before turning to those unique to water in Section 11.5.

ENTHALPY OF VAPORIZATION

A liquid can have a regular structure in a small region, but, because intermolecular forces are weak, there can be no long-range order. It is possible that some molecules at or near the surface have enough kinetic energy to overcome the intermolecular attractive forces and allow them to move from the liquid phase into the gas phase (Figure 11.11). This is the molecular-level description of the process of **vaporization** or **evaporation**.

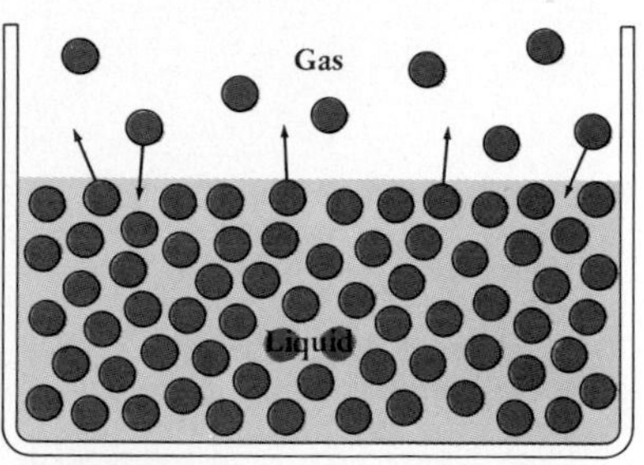

FIGURE 11.11
Molecules in liquid and gas phases. All the molecules in both phases are moving, although the distance traveled in the liquid before collision with another molecule is small. Some of these liquid phase molecules are moving with an energy (E_1 in Figure 11.2) great enough to break the intermolecular attractive forces in the liquid. Thus, they can escape to the gas phase or *evaporate*. At the same time, some molecules rebound into the surface after colliding with molecules of gases in the air or with other molecules of the same type.

$$\text{Liquid} + \text{heat} \xrightarrow{\text{vaporization}} \text{gas} \qquad \text{Heat required} = \Delta H_{\text{vaporization}}$$

To give escaping molecules enough energy to overcome intermolecular forces, energy must be supplied (usually as heat). Thus, vaporization is always an endothermic process, and the **enthalpy of vaporization**, ΔH_{vap}, is always positive (Table 11.5).

If there are molecules of the liquid in the gas phase, it is possible that some have lost the high kinetic energy that enabled them to escape from the liquid and that the direction of motion of these molecules is back toward the liquid surface. Such molecules may reenter the liquid phase in the process of **condensation**.

$$\text{Gas} \xrightarrow{\text{condensation}} \text{liquid} + \text{heat} \qquad \text{Heat evolved} = -\Delta H_{\text{vap}}$$

Since new intermolecular bonds are made, heat is evolved. Condensation is always an exothermic process, and the heat evolved is equal but opposite in sign to the enthalpy of vaporization. For example, one mole of water vapor gives up 40.7 kJ on condensing to liquid water, which means $\Delta H_{\text{condensation}} = -\Delta H_{\text{vap}} = -40.7$ kJ/mol.

TABLE 11.5 Heats of Vaporization, Vapor Pressures, and Critical Points for Common Compounds

COMPOUND	HEAT OF VAPORIZATION (kJ/mol)	TEMPERATURE REQUIRED TO ACHIEVE A GIVEN VAPOR PRESSURE (°C)		CRITICAL TEMPERATURE (°C)	CRITICAL PRESSURE (atm)
		100 mm	760 mm*		
HF	25.2	−28.2	19.7	188	64
HCl	17.5	−114.0	−84.8	51.4	82.1
HBr	19.3	−97.7	−66.5	90.0	84.5
HI	21.2	−72.1	−35.1	150	81.9
CH_4 (methane)	8.9	−181.4	−161.5	−82.1	45.8
C_2H_6 (ethane)	15.7	−119.3	−88.6	32.2	48.2
C_3H_8 (propane)	19.0	−79.6	−42.1	96.8	42.0
C_4H_{10} (butane)	24.3	−54.1	−0.5	152.0	37.5
NH_3	25.1	−68.4	−33.6	132.5	112.5
H_2O	40.7	51.6	100.0	374.1	218.3
SO_2	26.8	−46.9	−10.0	157.4	78.6
He	0.08		−269.0	−267.9	2.3
Ne	1.8		−246.0	228.7	26.9
Ar	6.5		−185.9	−122	48.0
Xe	12.6		−107.1	16.6	58.0

*The temperature at which the vapor pressure is 760 mm is the **normal boiling point**.

A system of liquid and gas phase molecules in a closed container is said to be in a state of **dynamic equilibrium** when the *rate* (mass per unit time) at which molecules are vaporizing is equal to the rate at which molecules are condensing.

$$\text{Liquid} + \text{heat} \rightleftharpoons \text{gas}$$

The state of equilibrium is a concept used throughout chemistry and one we shall return to often.

At equilibrium, the opposing molecular transfer processes continue to operate, but the effect of one (evaporation) cancels the effect of the other (condensation); the mass of each is constant in time.

EXAMPLE 11.2

ENTHALPY OF VAPORIZATION

You put 1.00 liter of water (about 4 cupsful) in a pan at 25°C and it slowly evaporates. How much heat must have been absorbed by the water to vaporize?

Solution There are three pieces of information you need to solve this problem:

1. The enthalpy of vaporization of water: ΔH_{vap} = 40.7 kJ/mole.
2. The density of water: 0.997 g/cm^3 at 25°C. (This is needed, because ΔH_{vap} has units related to mass of water; you were given the volume of water.)
3. The molar mass of water: 18.0 g/mol.

Now, you know that 1.00 liter of water is 1.00 × 10 cm^3. Given the density of water at 25°C, this amount of water has a mass of 997 g and is equivalent to 55.4 moles. Therefore, the amount of heat required is

$$55.4 \text{ mol } H_2O \left(\frac{40.7 \text{ kJ}}{1 \text{ mol}}\right) = 2.25 \times 10^3 \text{ kJ}$$
$$= \text{heat required for evaporation}$$

2250 kJ is equivalent to about $\frac{1}{4}$ of your food intake on a normal day.

EXERCISE 11.4 ENTHALPY OF VAPORIZATION

The molar enthalpy of vaporization of wood alcohol, methanol (CH_3OH), is 37.6 kJ/mol. How much heat energy is required to evaporate 1.00 kg of this alcohol?

EXAMPLE 11.3

CALCULATING THE ENTHALPY OF VAPORIZATION

The standard enthalpy of formation of liquid grain alcohol, ethanol (C_2H_5OH), is −277.7 kJ/mol, and the standard enthalpy of formation of ethyl alcohol in the gas phase is −235.1 kJ/mol. What is the standard molar enthalpy of vaporization of liquid ethyl alcohol (at 25°C)?

Solution The enthalpy of formation of liquid ethyl alcohol implies the equation

$$2\ C(s) + 3\ H_2(g) + \tfrac{1}{2}\ O_2(g) \longrightarrow C_2H_5OH(\ell) \qquad \Delta H_1 = -277.7 \text{ kJ/mol}$$

and the enthalpy of formation of the alcohol vapor is given by

$$2\ C(s) + 3\ H_2(g) + \tfrac{1}{2}\ O_2(g) \longrightarrow C_2H_5OH(g) \qquad \Delta H_2 = -235.1 \text{ kJ/mol}$$

If you add the second equation to the reverse of the first

$$C_2H_5OH(\ell) \longrightarrow 2\,C(s) + 3\,H_2(g) + \tfrac{1}{2}\,O_2(g) \qquad -\Delta H_1 = +277.7\text{ kJ}$$

$$2\,C(s) + 3\,H_2(g) + \tfrac{1}{2}\,O_2(g) \longrightarrow C_2H_5OH(g) \qquad \Delta H_2 = -235.1\text{ kJ}$$

$$C_2H_5OH(\ell) \longrightarrow C_2H_5OH(g) \qquad \Delta H_{net} = +42.6\text{ kJ}$$

the resulting equation is that for the conversion of liquid to vapor, and you see that the heat required is 42.6 kJ per mole of alcohol at 25°C.

EXERCISE 11.5 ENTHALPY OF VAPORIZATION

The standard enthalpies of formation of liquid and gaseous carbon tetrachloride, CCl_4, are given in Appendix L. Use them to calculate the molar enthalpy of vaporization for $CCl_4(\ell)$ at 25°C.

In Example 11.2 you learned that an enormous amount of heat is required to convert liquid water to water vapor. This is important to your environment and to your physical well being. When you exercise vigorously, your body responds by sweating to rid itself of the excess heat. The sweat, which is mostly water, can be evaporated by the input of heat energy. The source of this energy is the heat generated by your muscles, so the evaporation of sweat removes the excess heat, and your body is cooled.

It is interesting to look at trends among the enthalpy of vaporization data in Table 11.5. You saw earlier that the boiling points of a related series of nonpolar liquids, say the rare gases, tend to increase with increasing atomic or molecular mass (Figure 11.8 and Table 11.4), a reflection of increasing intermolecular dispersion forces. Similarly, the boiling points and enthalpies of vaporization of the hydrogen halides, HX (where X = F, Cl, Br, I), increase with increasing strength of intermolecular bonding, particularly hydrogen bonding and dipole–dipole forces. Finally, be sure to notice the uniqueness of water; owing to extensive hydrogen bonding, it has an anomalously high enthalpy of vaporization.

EXERCISE 11.6 ENTHALPIES OF VAPORIZATION AND INTERMOLECULAR FORCES

Explain the relation between the enthalpies of vaporization, boiling points, and intermolecular forces for ammonia (NH_3) and ethane (C_2H_6). Necessary data are given in Table 11.5.

VAPOR PRESSURE AND BOILING POINT

EQUILIBRIUM VAPOR PRESSURE If you put water in an open beaker, eventually the water will evaporate completely. However, if you put some water in a sealed flask (Figure 11.12), the liquid will evaporate only until the rate of vaporization equals the rate of condensation, that is, until equilibrium is established. At this point, the net amount of water in each of the liquid and gas phases does not change. In contrast, the water in the open beaker can never come to equilibrium with the water molecules in the gas phase; air movement and diffusion quickly remove the water vapor from the vicinity of the surface of the liquid water.

When equilibrium is established, the pressure exerted by the water vapor is called the **equilibrium vapor pressure** (often, but not correctly, called just the vapor pressure). The equilibrium vapor pressure of any

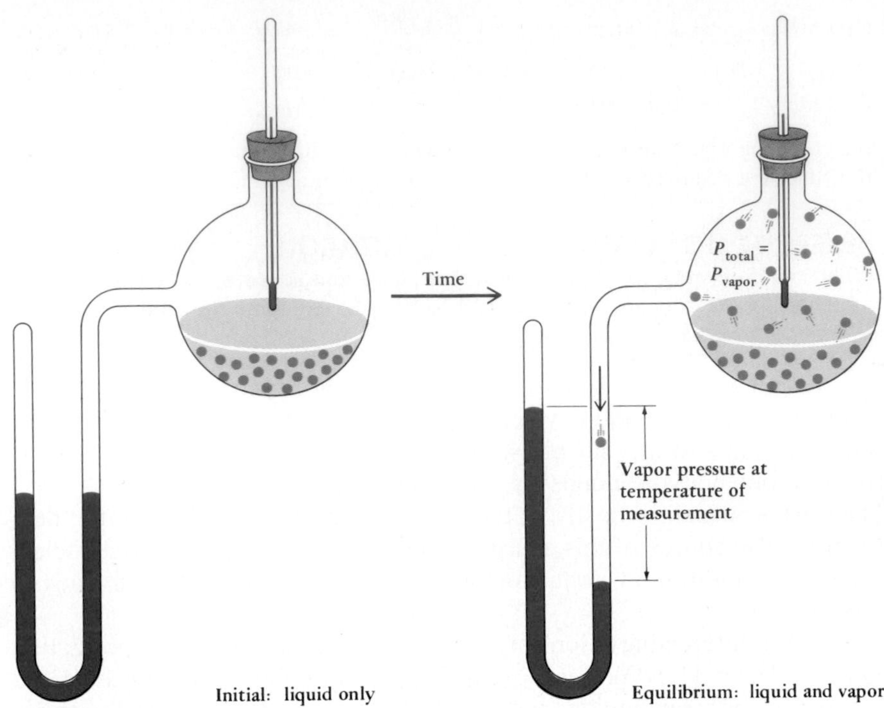

FIGURE 11.12

A volatile liquid is placed in a sealed, evacuated flask (left). In the beginning, no molecules of liquid are in the vapor phase. With time, some of the liquid evaporates, and the molecules of the vapor exert a pressure (right). If the pressure is measured when the liquid and vapor are in equilibrium, the pressure is called the **equilibrium vapor pressure.**

substance is a measure of the tendency of its molecules to escape from the liquid phase and enter the gas phase *at a given temperature*. We often also refer to this tendency as the **volatility** of the compound. The higher the equilibrium vapor pressure at a given temperature, the more **volatile** the compound.

Vapor pressure data for a variety of compounds are given in Table 11.5. Be sure to notice that the vapor pressure of a compound at a given temperature is also a reflection of its intermolecular forces, just as is its enthalpy of vaporization. As stronger and stronger intermolecular forces are involved, the substance is less volatile at a given temperature.

Since the average energy of a molecule in the liquid phase is a function of the temperature (Figure 11.2), molecules move more rapidly, and the rate of vaporization increases as the temperature rises. Because of this, the equilibrium vapor pressure must also increase with temperature (Figure 11.13). Any point along a curve for a compound in Figure 11.13 represents the pressure and temperature at which liquid and vapor are in equilibrium. For example, at 25°C the vapor pressure of water is 24 mm Hg, whereas it is 149 mm Hg at 60°C. This means that if water is placed in an evacuated flask that is maintained at 60°C, liquid will evaporate until the pressure exerted by the vapor is 149 mm. (If not enough water is placed in the flask, it is possible that all the liquid will evaporate before this equilibrium pressure is reached.)

At the conditions of T and P given by any point on a curve in Figure 11.13, liquid and vapor are in dynamic equilibrium. If a point *not* on the curve is considered, the system is *not* at equilibrium.

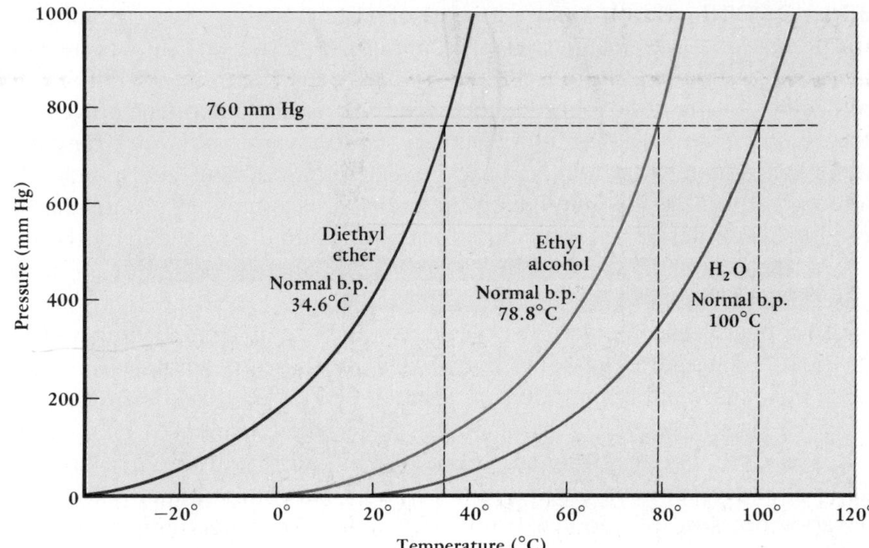

FIGURE 11.13
Vapor pressure curves for diethyl ether ($C_2H_5OC_2H_5$), ethyl alcohol (C_2H_5OH), and water. Each curve represents all the conditions of *T* and *P* where the two phases are in equilibrium. Under all conditions to the left of a curve, the compound exists as a liquid, and to the right of a curve it exists as a vapor.

EXAMPLE 11.4

VAPOR PRESSURE

If 1.00 liter of water is placed in a small room that has a volume of 2.30×10^4 L, will the water all evaporate at 25°C? (The density of water at 25°C is 0.997 g/cm³, the molar mass of water is 18.0 g/mol, and the vapor pressure of water at 25°C is 23.8 mm Hg.)

Solution One approach to solving this problem is to calculate the amount of water that must evaporate in order to exert a pressure of 23.8 mm Hg in a volume of 2.30×10^4 L at 25°C.

$$n \text{ moles} = \frac{PV}{RT} = \frac{\left(\dfrac{23.8 \text{ mm}}{760 \text{ mm/atm}}\right) 2.30 \times 10^4 \text{ L}}{(0.0821 \text{ L}\cdot\text{atm/K}\cdot\text{mol})(298 \text{ K})}$$

$$n = 29.4 \text{ mol } H_2O$$

$$29.4 \text{ mol } H_2O \left(\frac{18.0 \text{ g}}{1 \text{ mol}}\right) = 530. \text{ g}$$

$$530. \text{ g} \left(\frac{1 \text{ cm}^3}{0.997 \text{ g}}\right) = 532 \text{ cm}^3$$

The calculation shows that about half of the 1.00 L of water will evaporate in order to achieve an equilibrium water vapor pressure of 23.8 mm Hg at 25°C.

EXERCISE 11.7 VAPOR PRESSURE

If you seal 0.50 g of water in an evacuated 5.0-L flask and heat the whole assembly to 60°C, will the pressure be equal to or less than the equilibrium vapor pressure at this temperature? What if you used 2.0 g of water? Under either set of conditions, will there be liquid water left in the flask, or will all the water be in the gas phase?

BOILING POINT If you have a beaker of water open to the atmosphere, the weight of the atmosphere is pressing down on the surface. As heat is added to the water, more and more water can evaporate, pushing the molecules of the atmosphere aside. When the beaker is open, however, the total pressure at the liquid surface (water vapor plus air) remains indistinguishable from atmospheric pressure. If sufficient heat is added, a temperature is reached at which the vapor pressure equals atmospheric pressure; bubbles of vapor begin to form in the liquid, and the liquid *boils*.

Vapor bubbles in a boiling liquid are observed because, at the molecular level, there are large numbers of molecules throughout the liquid phase that have sufficient kinetic energy to break away from the liquid.

The **boiling point** of a liquid is the temperature at which the vapor pressure is equal to the external pressure. If the external pressure is 1 atmosphere, the temperature is designated the **normal boiling point.** Normal boiling points of some liquids are also listed in Table 11.5.

The normal boiling point of water at sea level is 100°C. But if you live in Salt Lake City, Utah, where the barometric pressure is less than 1 atmosphere (about 650 mm Hg), the boiling point is different. Looking at a vapor pressure curve for water, you find that water should boil (have a vapor pressure of about 650 mm Hg) when the water temperature is about 95°C. Cooks know that food has to be heated a bit longer in Salt Lake to achieve the same effect as in New York City at sea level.

To shorten cooking time in Denver, or anywhere else, use a pressure cooker. This is a sealed pot that allows water vapor to build up to pressures slightly greater than the external pressure. The boiling point of water increases, and foods cook faster.

It is evident from Figure 11.13 that there is not a linear relation between vapor pressure and temperature for any liquid. Rather, a curved relation is always exhibited, a behavior that can be predicted accurately by the **Clausius–Clapeyron equation**.

The equation was derived from work by the German physicist R. Clausius (1822–1888) and the Frenchman B.P.E. Clapeyron.

$$\ln P = -(\Delta H_{vap}/RT) + C \quad (11.2)$$

The term ln P is the natural logarithm of the vapor pressure, ΔH_{vap} is the enthalpy of vaporization, T is the kelvin temperature at which P is measured, R is the ideal gas constant in energy units (8.3143×10^{-3} kJ/K · mol), and C is a constant characteristic of the compound. Equation 11.2 is useful because it gives us an excellent way of determining the enthalpy of vaporization of a compound. We measure the vapor pressure at two different temperatures and calculate the difference between ln P_1 (P_1 is the vapor pressure at temperature T_1) and ln P_2 (P_2 is the vapor pressure at T_2).

$$\ln P_2 - \ln P_1 = \left(\frac{-\Delta H_{vap}}{RT_2} + C\right) - \left(\frac{-\Delta H_{vap}}{RT_1} + C\right)$$

When this equation is simplified, you can see that the heat of vaporization can be calculated readily from experimentally determined pressures and temperatures.

$$\ln \frac{P_2}{P_1} = \frac{\Delta H_{vap}}{R}\left[\frac{1}{T_1} - \frac{1}{T_2}\right] \quad (11.3)$$

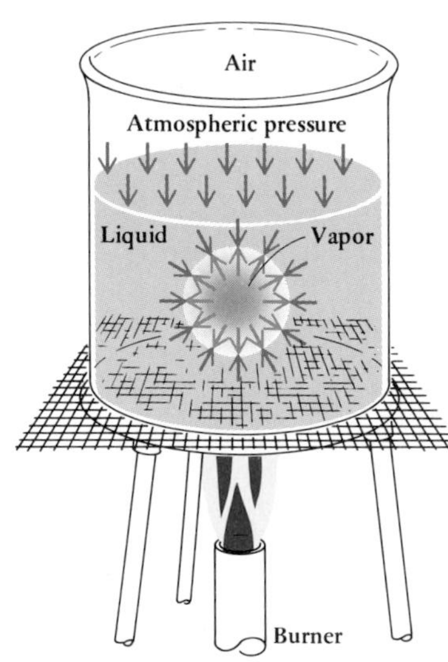

A liquid boils when its equilibrium vapor pressure is equal to the atmospheric pressure.

EXAMPLE 11.5

USING THE CLAUSIUS–CLAPEYRON EQUATION

The organic liquid, 1,2-ethanediol ($C_2H_6O_2$, commonly called ethylene glycol) can be used as an antifreeze in car radiators. Its vapor pressure at 100°C is 14.9 mm Hg while it is 49.1 mm Hg at 125°C. Calculate the enthalpy of vaporization of this compound.

Solution You are given $P_1 = 14.9$ mm Hg at $T_1 = 373$ K (100°C) and $P_2 = 49.1$ mm Hg at $T_2 = 398$ K (125°C). Substituting this information into the Clausius–Clapeyron equation (Equation 11.3), we find

$$\ln \frac{49.1 \text{ mm Hg}}{14.9 \text{ mm Hg}} = \frac{\Delta H_{vap}}{8.3143 \times 10^{-3} \text{ kJ/K} \cdot \text{mol}} \left(\frac{1}{373 \text{ K}} - \frac{1}{398 \text{ K}} \right)$$

$$1.19 = \frac{\Delta H_{vap}}{8.3143 \times 10^{-3} \text{ kJ/K} \cdot \text{mol}} (0.000168)$$

$$\Delta H_{vap} = 58.9 \text{ kJ/mol}$$

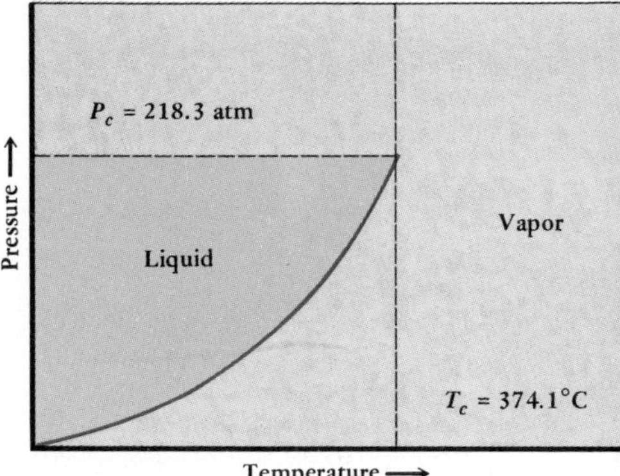

FIGURE 11.14

The vapor pressure curve for water near the critical temperature (T_c). This temperature is "critical" because, beyond this point, vapor cannot be converted to liquid, no matter how high the pressure exerted on the vapor.

EXERCISE 11.8 THE CLAUSIUS–CLAPEYRON EQUATION

Diethyl ether, $(C_2H_5)_2O$, has vapor pressures of 57 mm Hg and 534 mm Hg at −22.8°C and 25.0°C, respectively. Calculate its molar enthalpy of vaporization.

CRITICAL TEMPERATURE AND PRESSURE

The vapor pressure of a liquid will continue to increase with temperature up to the **critical point**, where the curve of vapor pressure versus temperature comes to an abrupt halt (Figure 11.14). The temperature at which this occurs is the **critical temperature**, T_c, and the corresponding vapor pressure is the **critical pressure**, P_c. The significance of this point is that, if the temperature exceeds T_c, all the liquid molecules have sufficient kinetic energy to separate from each other, regardless of the pressure, and a conventional liquid no longer exists. Instead the substance is often called a **supercritical fluid**.

Notice in Table 11.5 that values of T_c and P_c correlate well with other measures of intermolecular forces.

A knowledge of critical temperatures and pressures (see Table 11.5) is important in designing air conditioners, for example. These devices cool a room by using the room heat to vaporize a liquid. To make the process continuous, however, the vapor is condensed back to the liquid state by compressing the vapor to a pressure higher than P_c. Compounds called Freons are often used as the fluid in air conditioners and refrigerators because some of them can be liquefied at reasonable pressures. For example, Freon-11 (CCl_3F) has a T_c of 198°C and a P_c of 43.5 atm. This means CCl_3F vapor can be converted to a liquid by applying modest pressures at any temperature below 198°C, a temperature far above room temperature.

In the supercritical state, certain fluids like water and carbon dioxide take on unexpected properties, such as the ability to dissolve normally insoluble materials. This property is just beginning to be exploited in the oil, biochemicals, and food industries. Especially promising is supercritical CO_2. This fluid does not dissolve water or polar compounds such as sugars, but it does dissolve nonpolar oils, which constitute many flavorings. In West Germany a food company is already using supercritical CO_2 to extract the caffeine from coffee, for example.

SURFACE TENSION AND VISCOSITY

Molecules at the surface of a liquid behave differently than those in the interior because molecules in the center interact with molecules all around

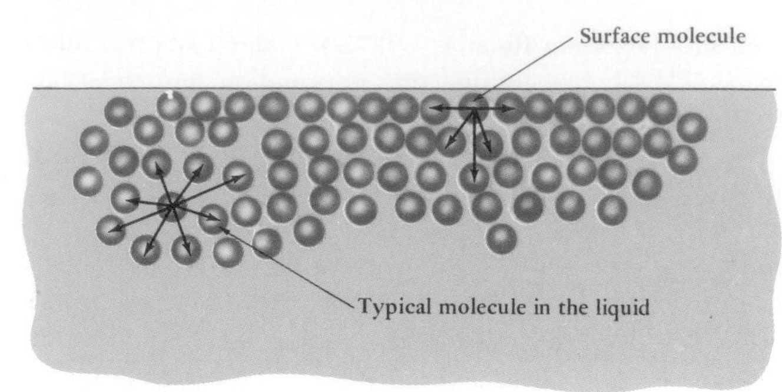

FIGURE 11.15
The difference in forces acting on a molecule within the liquid phase and on a molecule at the surface of the liquid.

them (Figure 11.15). In contrast, surface molecules are affected only by those below the surface layer. This leads to a net inward force of attraction on the surface molecules, contracting the surface and making it behave as though it had a skin. It is this that allows water bugs and autumn leaves to float on water and causes rain drops to be spheres and not little cubes.

The **surface tension** of a liquid is the energy required to break through the surface or to disrupt a liquid drop and spread the material out as a film. As you can see in Table 11.6, the surface tension of mercury is larger than water, so even relatively heavy objects can float on the liquid metal surface (Figure 11.16).

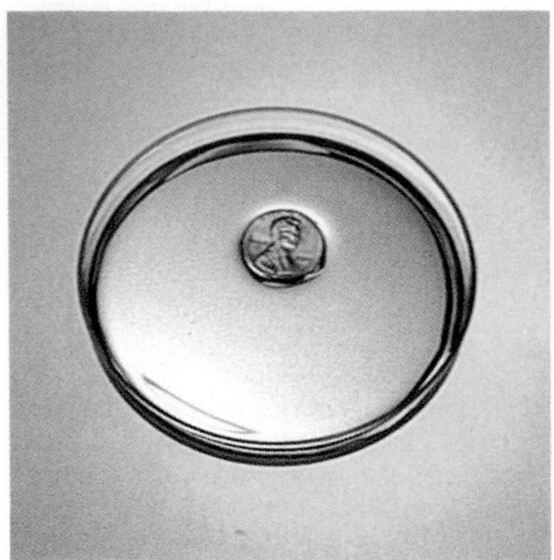

FIGURE 11.16
Mercury has a higher surface tension than most other common liquids at 25°C. (Charles D. Winters)

Capillary action is closely related to surface tension. When a small diameter glass tube is placed in water, the water rises in the tube, just as water rises in a piece of paper in water (Figure 11.17). Because there are polar Si—O bonds on the surface of the glass, polar water molecules are attracted by **adhesive** forces between the two different substances. These forces are sufficiently strong that they can compete with the **cohesive** forces between the water molecules themselves. Thus, the water molecules begin to climb the walls of the tube. However, the surface tension of the water (from cohesive forces) is great enough that the water level rises in the tube as well. The rise will continue until the various forces (adhesion between water and glass, cohesion between water molecules, and the force of gravity on the water column) are in equilibrium. It is these forces which lead to the characteristic concave or downward-

FIGURE 11.17
Capillary action. Polar water molecules are attracted to the O—H bonds in paper fibers, and water rises along the fibers. If a line of ink is placed in the path of the rising water, the different components of the ink are attracted differently to the water and are separated in a process called *chromatography*. (Charles D. Winters)

TABLE 11.6 Surface Tension and Viscosity of Some Liquids

COMPOUND NAME AND FORMULA		SURFACE TENSION J/m² at 20°C*
Water	H_2O	7.29×10^{-2}
Ethyl alcohol	H_3CCH_2OH	2.23×10^{-2}
Mercury	Hg	46×10^{-2}
		VISCOSITY (kg/m · sec)
Water	H_2O	1.00×10^{-3}
Ethyl alcohol	H_3CCH_2OH	1.20×10^{-3}
n-Heptyl alcohol	$H_3CCH_2CH_2CH_2CH_2CH_2CH_2OH$	8.53×10^{-3}
Olive oil		84×10^{-3}

*The energy in joules necessary to increase the surface area (in m²) by a given amount.

curving **meniscus** that you see for water in a drinking glass or in a laboratory test tube (Figure 11.18).

There are of course some liquids such as mercury for which cohesive forces (high surface tension) are much greater than adhesive forces. Mercury will not climb the walls of a glass capillary, and when it is placed in a tube it will have a convex or upward-curving meniscus (Figure 11.18).

Another important property of liquids is their **viscosity**, the resistance of liquids to flow (Figure 11.19). It is measured by determining the time a given mass takes to flow through a thin, vertical tube under the force of gravity. When you turn over a glassful of water, it empties quickly, but it takes much more time to empty a glassful of more viscous motor oil or honey. Although intermolecular forces play a role in determining the viscosity of a substance, there are clearly other factors. In Table 11.6 you will notice that viscosity increases as the length of the carbon chain of alcohols increases. These chains are floppy and become entangled with one another; the longer the chain the greater the tangling and the greater the viscosity.

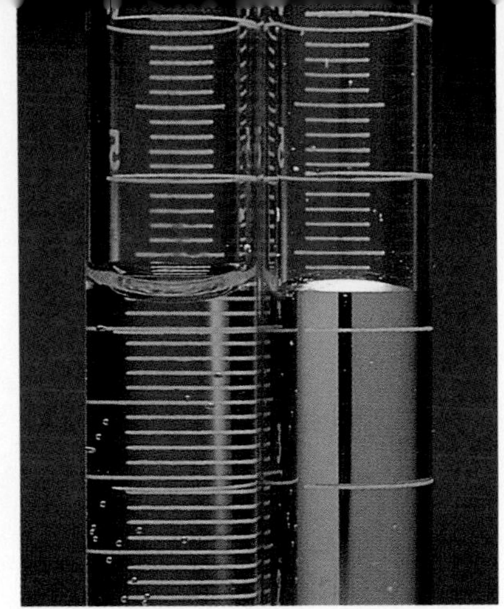

FIGURE 11.18

The meniscus formed by two different liquids. Water (left): The cohesive forces between molecules are partly overcome by the adhesive forces between water and glass, and water has a concave meniscus. Mercury (right): The cohesive forces within mercury are significantly greater than the adhesive forces between mercury and glass, and mercury has a convex meniscus. (Charles D. Winters)

11.4 SOLIDS

There are many kinds of solids in the world around us (Figure 11.20), and solid-state chemistry is one of the booming areas of science. In this section, we can give only a brief introduction to the principles of this fascinating subject.

AN OUTLINE OF SOLID STATE CHEMISTRY

In both gases and liquids, molecules continually move; they translate randomly as well as rotate and vibrate. Because of this, there can be no

FIGURE 11.19

Honey is a very viscous fluid. (Charles D. Winters)

FIGURE 11.20

Crystals of fluorite (CaF_2, left), calcite ($CaCO_3$, center), and copper sulfate ($CuSO_4 \cdot 5\ H_2O$, right). (Charles D. Winters)

TABLE 11.7 Types of Solids and Modes of Bonding

TYPE OF SOLID	EXAMPLES	TYPE OF BONDING OR INTERMOLECULAR FORCE
Ionic	NaCl and other metal halides	Ionic
Metals	Fe, Zn, Cu, Au, Ag, etc.	Metallic
Molecular solids	H_2O as ice; I_2, sulfur	Covalently bonded small molecules
		Dipole–dipole, hydrogen bonding, and induced dipole forces bind molecules together as solid
Network solids	Graphite, diamond, quartz	Extended network of covalently bonded atoms
Imperfect solids		
Disordered	AgI and other "super-ionics" (used in some batteries)	Ionic
Nonstoichio-metric	Semiconductors such as CdS	Ionic to metallic
Amorphous	Glass; organic polymers (polyethylene, polystyrene, etc.)	Covalently bonded networks; networks bound together by dipole–dipole or induced dipole forces

extensive, orderly array of molecules in the gas or liquid state, and it is on this point that you encounter a major difference between liquids and solids. In solids, the molecules, atoms, or ions cannot translate (although they vibrate and occasionally rotate), and there can be long-range order, a characteristic of the solid state.

Long-range order differentiates true solids from glasses, which are just supercooled liquids.

To organize solid state chemistry, we might classify different types of solids as in Table 11.7. This section is devoted chiefly to ionic solids, but a brief summary of other types of solids is given as well.

SOME PRINCIPLES OF SOLID STATE STRUCTURES

UNIT CELLS AND CRYSTAL LATTICES The beautiful regularity of a crystal of salt or of a metal suggests that it has an internal symmetry as well. Indeed, all crystal lattices are built of **unit cells**, the smallest, repeating, internal unit that has all the symmetry characteristics of the crystalline solid.

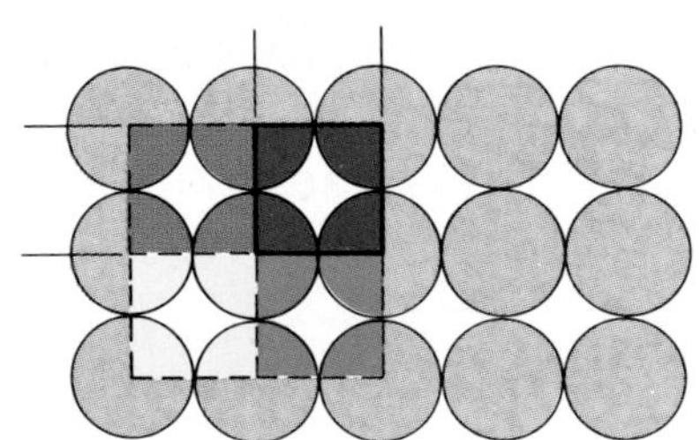

FIGURE 11.21
Repeating pattern of circles with a square unit cell. Each circle at the four corners contributes $\frac{1}{4}$ of its area to the area inside the square, for a net of one circle within a unit cell.

To understand the idea of a unit cell, look at the repeating pattern of circles in Figure 11.21. If lines are drawn from the center of one circle to the centers of neighboring circles, a square results, and a net of one circle is located within each square. This is one unit cell for the pattern of circles, because the entire pattern can be built of identical square unit cells, each containing the equivalent of one circle.

Solids can be built by piling up three-dimensional unit cells like building blocks, as the artist M.C. Escher illustrated in a well-known drawing (Figure 11.22). The points defining each unit cell, the *lattice*

FIGURE 11.22

"Cubic Space Division" by M.C. Escher. Cubic unit cells form a cubic crystal lattice. (Haags Gemeentemuseum)

points, represent equivalent environments for an ion, metal atom, or molecule. The total array of points for a crystal is called the **crystal lattice**.

To construct crystal lattices, nature uses seven three-dimensional unit cells that differ from one another in that their sides have different relative lengths and their edges meet at different angles (Figure 11.23). However, we shall be concerned here only with **cubic unit cells**, cells with equal edges and all angles equal to 90°. They are easily visualized and are commonly observed in nature. Within the cubic class, there are three

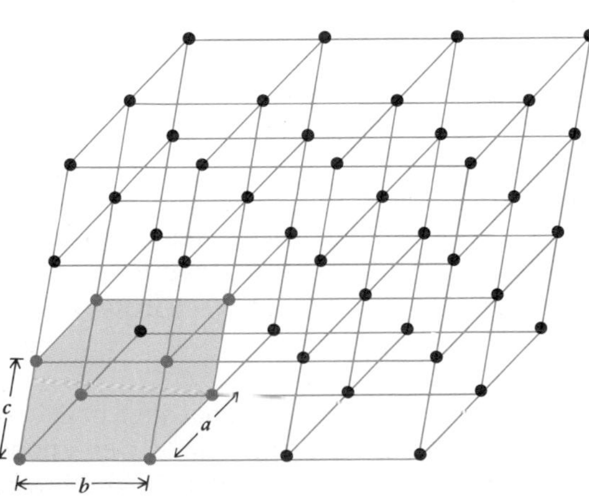

FIGURE 11.23

All units cells are parallelepipeds. They have six sides with parallel opposite edges (parallelograms). In a simple cubic unit cell, all edges are equal and all angles are 90°.

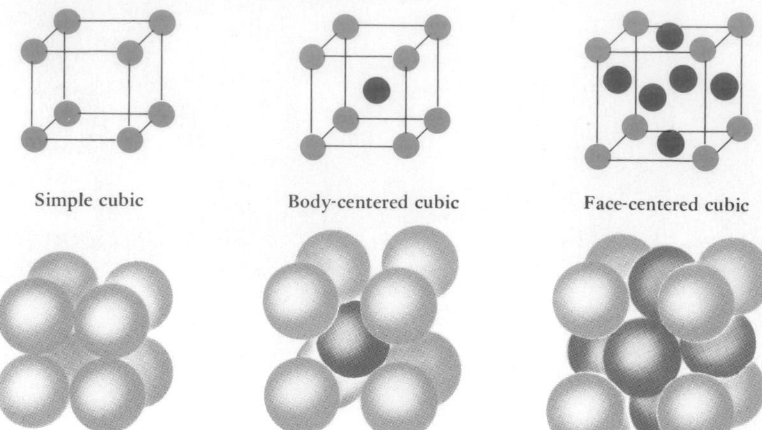

FIGURE 11.24

The three different type of cubic unit cells. The top row shows the lattice points of the three cells. The bottom row shows the same unit cells, but the atoms (or ions) defining the unit cell are considered as space-filling spheres in contact with one another. All the spheres represent identical atoms or ions; the colors are meant only to show the different locations: at the cube corners (green) or in the cube center or cube faces (red).

sc = simple cubic
bcc = body-centered cubic
fcc = face-centered cubic

subtypes: **simple cubic (sc)**, **body-centered cubic (bcc)**, and **face-centered cubic (fcc)** (Figure 11.24). All three have eight identical atoms or ions at the corners of a cube. However, the bc and fcc arrangements differ from simple cubic in having additional atoms, of the same type as the corner atoms, at other locations. The **bcc** structure is called "body centered" because it has one additional atom at the center of the cube or "body." The **fcc** arrangement is called "face centered" because it has, in the center of each of the six cube faces, an ion or atom of the same type as the corner ions or atoms.

The formula of an ionic solid is related to the number and type of holes filled by the cations in a stack of anions packed as efficiently as possible.

STRUCTURES AND FORMULAS OF IONIC SOLIDS The formula of a crystalline, *ionic* compound is directly related to its unit cell structure. To see this, let us view the ions of lattices as spherical objects packed together as efficiently as possible. It is shown in Figure 11.25 that the NaCl lattice, for example, actually consists of the larger Cl^- ions packed together as closely as possible, with the smaller Na^+ ions inserted into the holes between the Cl^- ions.

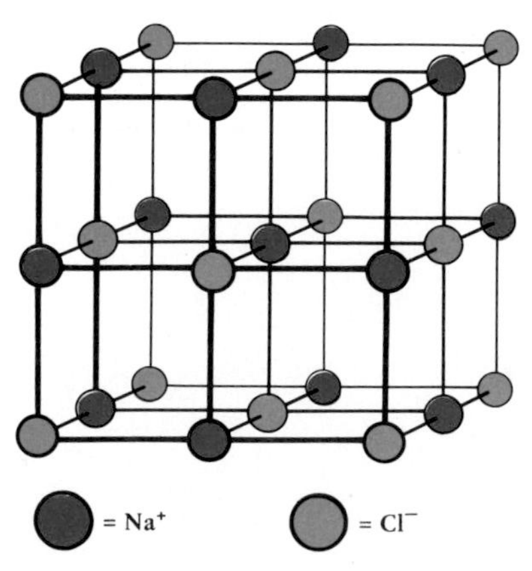

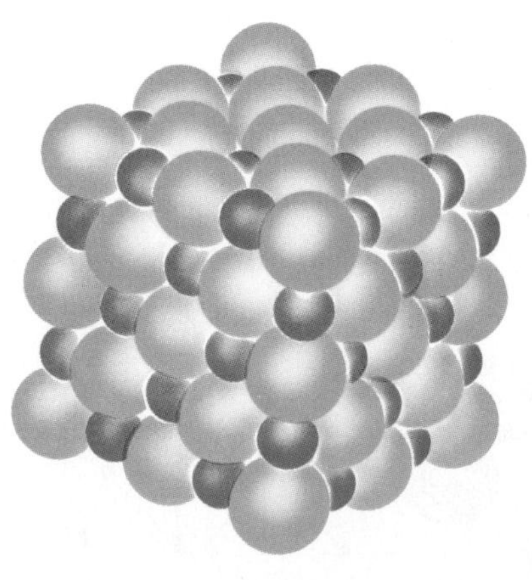

FIGURE 11.25

NaCl crystal lattice. (a) An expanded view of one unit cell. The lines only represent the connections between lattice points. They are not bonds. (b) A more extended view showing the ions as spheres, the smaller Na^+ ions packed into a lattice of larger Cl^- ions.

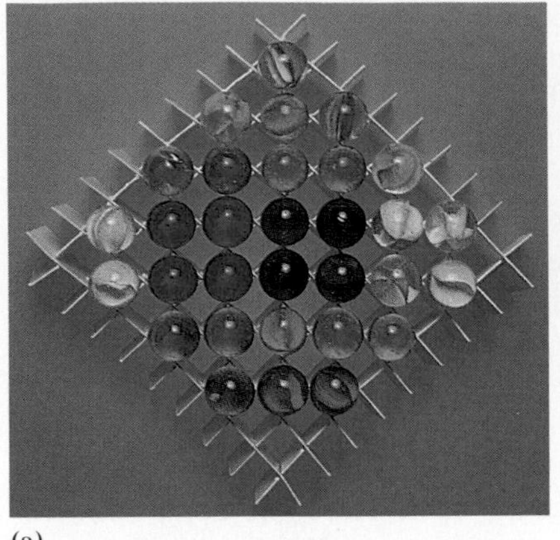
(a)

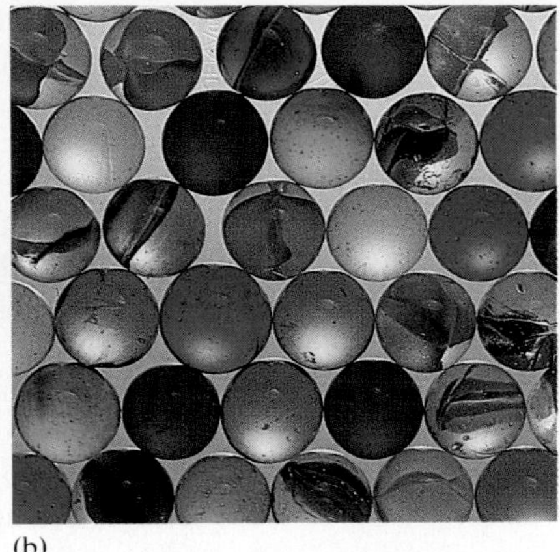
(b)

FIGURE 11.26

Packing of spheres in one layer. (a) Marbles are arranged in a square pattern, and each marble contacts four others. (b) Each marble contacts six others at the corners of a hexagon. It can be seen that (b) is a more efficient packing arrangement than (a). (Charles D. Winters)

In most crystal lattices, atoms or ions fill space as efficiently as possible. To see this, start by packing spherical ions or atoms in one layer (Figure 11.26). The spheres can be placed so that each directly contacts four others (Figure 11.26a), or so that each touches six others at the corners of a hexagon (Figure 11.26b). In either case, holes remain, but it is easy to prove that 11.26b is the more efficient packing scheme (Study Question 11-78).

To fill three dimensional space, more layers are added on top of the first. If you start with the square arrangement above (Figure 11.26a) for the first layer and simply stack the next layer of spherical ions or atoms directly on top of the first, the **simple cubic** unit cell is formed (Figure 11.24).

Close-packing arrangements make more efficient use of space than the simple cube (Figure 11.27). One of these arrangements is called **cubic**

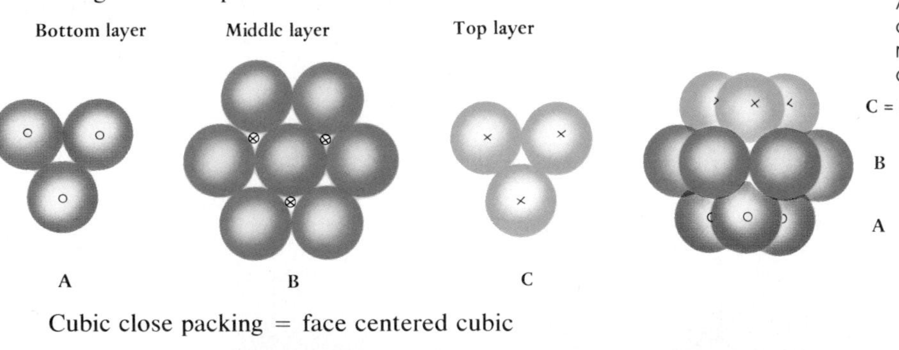

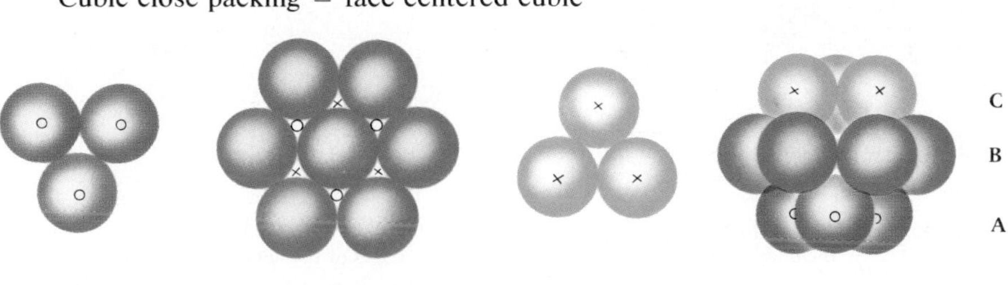

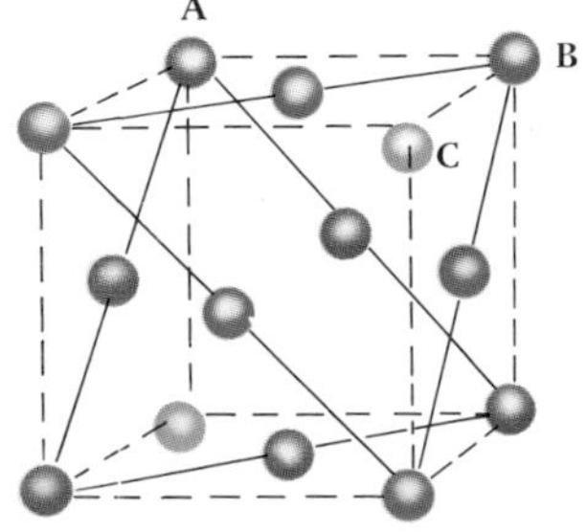

FIGURE 11.27

The two most efficient ways to pack atoms or ions in crystalline materials. In both arrangements, the packing within each layer is the hexagonal pattern shown in Figure 11.26b and described in the text. (a) In the *hexagonal close-packed (hcp)* arrangement, the layers placed above and below a given layer fit into the same depressions on either side of the middle layer. In a three-dimensional crystal, the layers repeat in their pattern in the manner ABABAB. . . . (b) In *cubic close packing (ccp),* the atoms of the bottom layer rest in depressions marked "o", and those of the top layer are oppositely oriented, resting in the depressions marked "x." In a crystal, the pattern is repeated ABCABCABC . . . By turning the whole crystal, you can see that the ccp arrangement is just the face-centered cubic structure.

A simple cube of spheres fills only 68% of the space inside a cube, whereas close-packing methods use 74% of available space.

close packing and gives the **face-centered cubic (fcc) lattice** described earlier. The other possible close packing arrangement, called **hexagonal close packing**, is also common in nature, but it is more difficult to visualize; the rest of the discussion will concentrate on simple and face-centered cubic arrangements.

The unit cell lattices in Figures 11.24 and 11.26 clearly have empty spaces or holes. These holes are the key to understanding the relation between lattice structure and stoichiometry, since the lattices of many ionic compounds are built by taking a simple cubic or face-centered cubic lattice of ions of one type and placing ions of opposite charge in the holes left in the lattice.

The hole in the simple cube is in the center of the cell, and this is the only type of space available (Figure 11.24). The ionic compound cesium chloride, CsCl, adopts just this structure, a simple cubic lattice of Cl^- ions, with Cs^+ ions in the center of each cube.

The structure of sodium chloride, NaCl, represents one of the most common ways of constructing an ionic solid. We can think of this salt as a face-centered cubic lattice of Cl^- ions with Na^+ in all the **octahedral lattice holes**. The holes are said to be octahedral because each Na^+ is surrounded at the corners of an octahedron by six Cl^- ions (Figure 11.28).

Both CsCl and NaCl can be viewed as lattices of cations (sc or fcc, respectively) with anions in the holes. However, it is more usual to form the unit cell from the larger anions and pack the smaller cations in the holes.

To understand the relation between the formulas of NaCl and CsCl, we need only find the maximum number of hole ions permitted per lattice ion. This is done in the next example and exercise.

EXAMPLE 11.6

FORMULA AND STRUCTURE OF SODIUM CHLORIDE

Show that the crystal structure of NaCl must lead to a formula with one cation and one anion.

Solution Formula and structure are related by counting the *net* number of ions that define the unit cell and then finding how many ions of opposite charge there are in lattice holes.

FIGURE 11.28

Octahedral holes in a face-centered cubic lattice. If a cation is placed in the center of a face-centered cubic anion lattice, it is surrounded by an octahedron of anions. In addition, there are equivalent octahedral holes in the center of each of the 12 cube edges. A cation in each edge hole will have four nearest-neighbor ions of opposite charge in the same cube and two in a neighboring cube. In all common crystals with face-centered anion lattices, *all* octahedral holes are filled with cations.

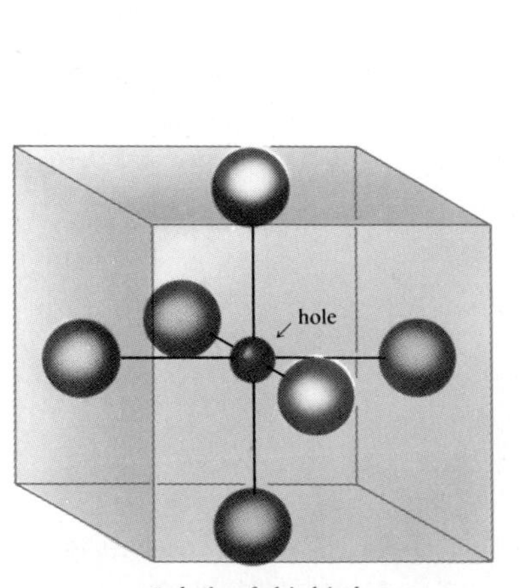

1 hole of this kind in the body center

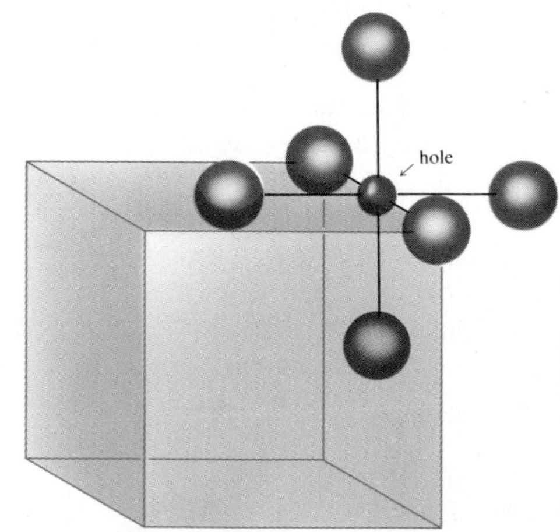

12 holes of this kind on the 12 edges of the cube (a net of 3 holes)

(a) Octahedral holes in a face-centered lattice

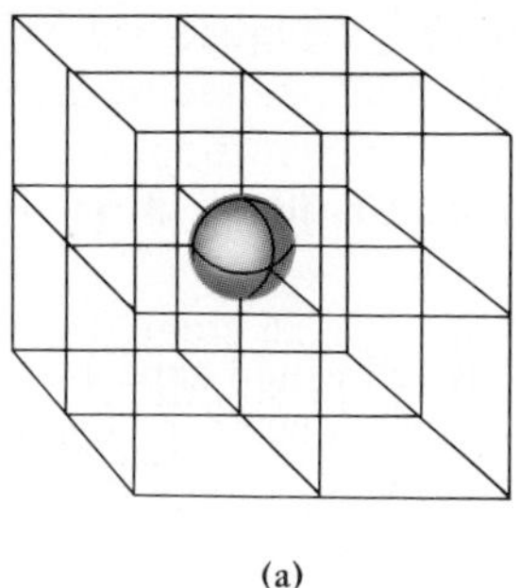

(a)

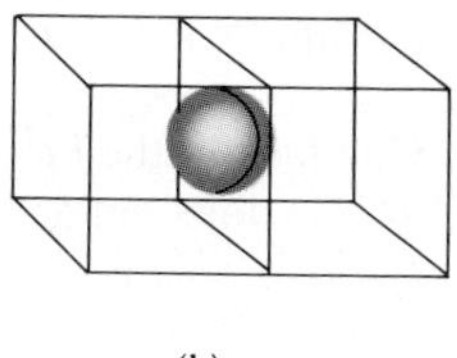

(b)

FIGURE 11.29

Ion sharing at cube corners and faces. (a) In the simple cubic or face-centered cubic lattices, each corner ion is shared equally among 8 cubes; $\frac{1}{8}$ of each ion is within any one cube. (b) In the face-centered lattice, each ion in a cube face is shared equally between 2 cubes; each ion of this type contributes $\frac{1}{2}$ of itself to a given cube.

We consider NaCl as a lattice with Cl^- ions defining a face-centered cube with Na^+ ions in octahedral holes (Figure 11.25). In any cubic cell there are 8 corner ions. In Figure 11.29 you see that a corner ion is shared with seven other cubes. Therefore, each corner ion contributes only $\frac{1}{8}$ of its volume to a particular unit cell. In addition, each of six face-centered Cl^- ions contributes one half of its volume to the unit cell, for a net of three more lattice anions. Thus, *the face-centered cubic lattice contains four net lattice ions*:

$$(8 \text{ cube corners})(\tfrac{1}{8} \text{ ion per corner}) + (6 \text{ faces})(\tfrac{1}{2} \text{ ion per face}) = 4 \text{ net lattice ions per fcc unit cell}$$

The Na^+ ions are placed in octahedral holes of the fcc Cl^- ion lattice. One octahedral hole at the cube center and one such hole in each of the 12 cube edges, each of the edge holes being shared among four cubes, are shown in Figures 11.25 and 11.29. Thus, the total number of octahedral holes in a face-centered cubic lattice is

$$(1 \text{ octahedral hole at cube center}) + (12 \text{ edges})(\tfrac{1}{4} \text{ octahedral hole per edge}) = 4 \text{ total octahedral holes}$$

Since the numbers of lattice ions and octahedral holes for face-centered lattices are both four, sodium chloride has a 1:1 ratio of Na^+ and Cl^- ions.

EXERCISE 11.9 IONIC STRUCTURE AND FORMULA

Cesium chloride, CsCl, has a cubic unit cell of Cl^- ions with Cs^+ ions in the cubic holes. Prove that the formula of the salt must be one Cs^+ to one Cl^-.

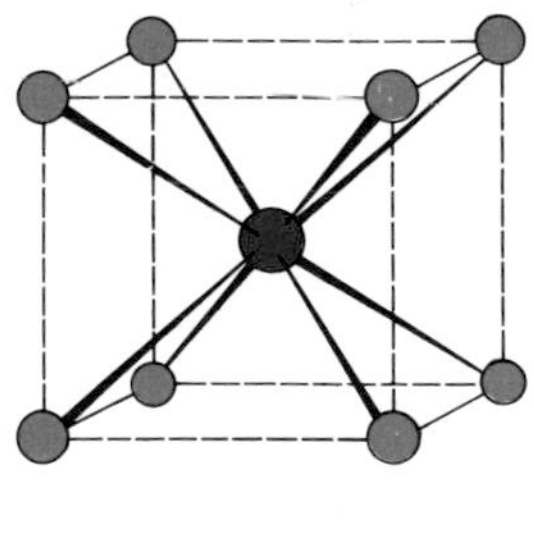

Cesium chloride (CsCl) crystal structure.

In summary, compounds of the type MX may be formed by M^{n+} ions occupying *all* of the cubic holes of a simple cubic X^{n-} lattice (Exercise 11.9), or M^{n+} ions in *all* of the octahedral holes in a face-centered cubic X^{n-} lattice (Figure 11.27). CsCl is a good example of the former, and NaCl is the outstanding example of the latter. Indeed, the sodium chloride or "rock salt" structure is adopted by many classes of compounds, most especially by all the alkali metal halides (except CsCl, CsBr, and CsI), all the oxides and sulfides of the alkaline earth metals, and all the oxides of formula MO of the first row transition metals.

ION SIZE AND CRYSTAL DENSITY Ions have different radii, depending on their position in the periodic table. Since crystals form by packing ions as closely as possible, the size of the unit cell and the density of the solid depend on the radii of the ions involved.

Although we have not described such a case, be aware that there are also 8 tetrahedral holes for cations in a fcc lattice of anions. If they are all filled, the salt has the formula M_2X (as in Na_2S).

EXAMPLE 11.7

LATTICE SIZE AND ION RADIUS

The radius of a Na^+ ion is 95 pm and the radius of a Cl^- ion is 181 pm. Calculate the volume of the NaCl unit cell in pm^3 and in cm^3.

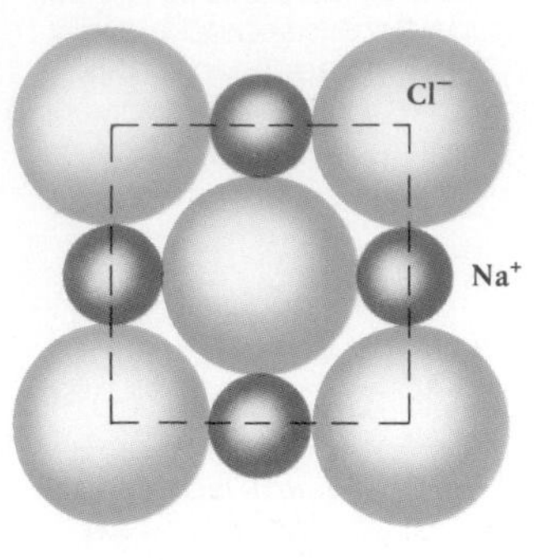

Solution One face of the face-centered NaCl unit cell would appear as shown at the left. The Cl^- ions define the lattice and the ions along each edge just touch one another. This means that one edge of the unit cell is equal to one Cl^- ion radius plus twice the radius of Na^+ plus another Cl^- ion radius, or

$$\text{NaCl unit cell edge} = 181 \text{ pm} + 2(95 \text{ pm}) + 181 \text{ pm} = 552 \text{ pm}$$

Since the crystal is cubic, the volume of the unit cell is the cube of the edge.

$$\text{Volume of unit cell} = (\text{edge})^3 = (552 \text{ pm})^3 = 1.68 \times 10^8 \text{ pm}^3$$

Converting this to cubic centimeters,

$$\text{Volume in cm}^3 = 1.68 \times 10^8 \text{ pm}^3(10^{-10} \text{ cm/pm})^3 = 1.68 \times 10^{-22} \text{ cm}^3$$

EXERCISE 11.10 UNIT CELL DIMENSIONS
KCl has the same crystal structure as NaCl. Calculate the volume of a KCl unit cell using the ion sizes given in Figure 8.16.

EXAMPLE 11.8

CRYSTAL PACKING, DIMENSIONS, AND DENSITY

The volume of a NaCl unit cell is 1.68×10^{-22} cm^3 (Example 11.7). What is the density of NaCl?

Solution The density of a material is its mass per unit volume. Since we know the volume of the NaCl unit cell, we need only to find its mass in order to find the density.

First, recognize that the unit cell contains 4 Na^+ and 4 Cl^- ions. You can obtain the mass of 1 Na^+ and 1 Cl^- from the formula weight of NaCl and Avogadro's number:

$$\left(\frac{58.5 \text{ g}}{1 \text{ mol}}\right)\left(\frac{1 \text{ mol}}{6.02 \times 10^{23} \text{ molecules}}\right) = 9.72 \times 10^{-23} \text{ g/molecule}$$

Since there are 4 NaCl "molecules" per unit cell, the cell has a mass of

$$\left(\frac{9.72 \times 10^{-23} \text{ g}}{\text{molecule}}\right)\left(\frac{4 \text{ NaCl molecules}}{\text{unit cell}}\right) = 3.89 \times 10^{-22} \text{ g/unit cell}$$

Now we have both the volume of the unit cell (from Example 11.7) and its mass. The density is then

$$\text{Density} = \left(\frac{3.89 \times 10^{-22} \text{ g}}{1 \text{ unit cell}}\right)\left(\frac{1 \text{ unit cell}}{1.68 \times 10^{-22} \text{ cm}^3}\right) = 2.31 \text{ g/cm}^3$$

The experimental density is 2.164 g/cm^3. The small discrepancy between the calculated and experimental values comes from the use of averaged Na^+ and Cl^- radii. (Using the experimental density and working the problem in reverse gives a unit cell edge distance of 562 pm, a value about 2% different from the value of 552 pm calculated using averaged ion radii.)

EXERCISE 11.11 SOLID DENSITY FROM UNIT CELL DIMENSIONS
Using the volume of a KCl unit cell that was found in Exercise 11.10, calculate the density of solid KCl.

METALS, MOLECULAR AND NETWORK SOLIDS, AND IMPERFECT SOLIDS

All but 26 of the chemical elements are **metals**, and all of those except mercury are solids under normal conditions. Structurally, the majority of metals have body-centered cubic or close-packed arrangements of atoms (Figure 11.24).* However, recall that a body-centered cubic arrangement of metal atoms does not represent as efficient a method of packing as the appropriately named close-packed structures. Because their packing efficiencies are similar, however, modest changes in temperature and pressure on a metal can sometimes induce changes between structures.

Bonding in metals was described in the special section for Chapter 10.

Covalently bonded molecules such as H_2O, I_2, CO_2, and many others condense to form solids if the intermolecular forces are sufficiently strong. Further, when such molecules are assembled into a solid, they generally adopt the usual body-centered cubic or close-packing forms. For example, iodine is a crystalline solid at room temperature with I_2 molecules at the lattice points in a face-centered cubic arrangement.

There are a number of solids composed of **networks of covalently bonded atoms**. Excellent examples are graphite and diamond, **allotropes** of carbon; that is, different forms of the same element under the same conditions of temperature and pressure (Figures 11.30 and 11.31).

Graphite is the classic example of a planar network solid. Here sp^2 hybridized carbon atoms form a planar network of six-membered rings. Just as you can draw resonance structures for the simple hydrocarbon benzene, so too can resonance structures be imagined for a layer of graphite.

Resonance structures of benzene, C_6H_6

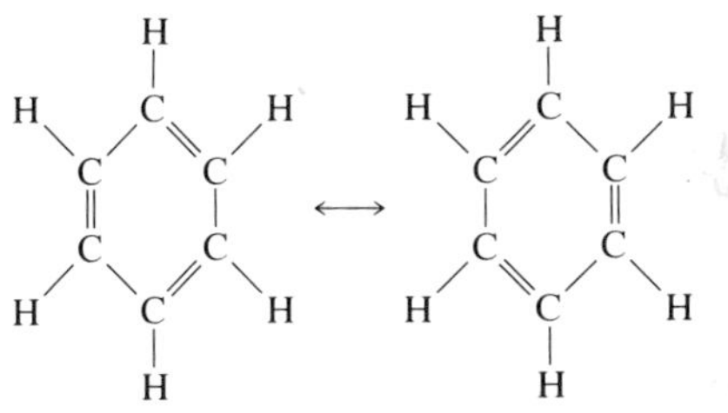

Graphite layer

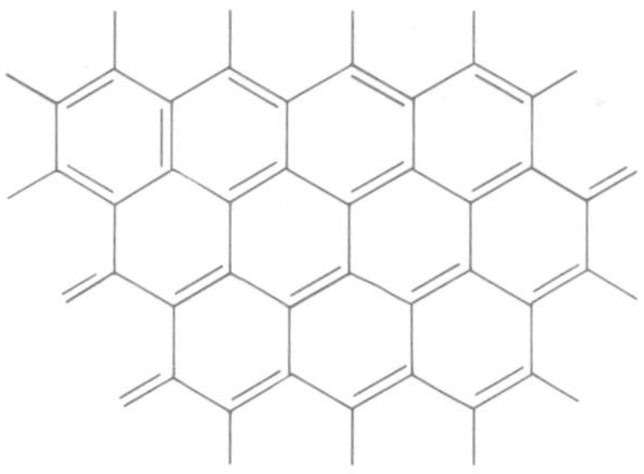

In fact, so many resonance structures can be drawn for this network of carbon atoms that the pi electrons of each ring are effectively delocalized over the entire sheet. This is why graphite is black (for a photon of almost any energy there is an electron available to absorb it) and why graphite will conduct electricity along the direction of the sheets.

---

*Ionic solids *never* have a body-centered cubic arrangement. CsCl has been said to have this structure, but it does not. Instead, it is more properly a simple cubic Cl^- lattice with Cs^+ in cubic holes.

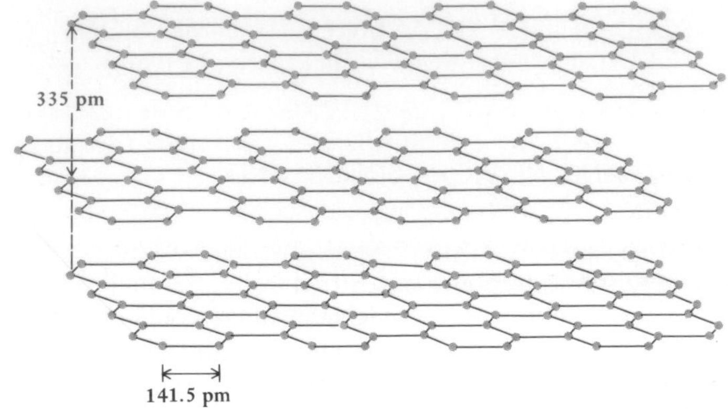

FIGURE 11.30
The layered structure of graphite.

The covalent bonds within the graphite sheets are quite strong, but, as indicated by the large distance between sheets (Figure 11.30), they are bonded to one another only by weak dispersion forces. Due to the weakness of intersheet bonding, the sheets slide past one another readily and allow graphite to act as a lubricant. Graphite is also used commercially as an electrical conductor.

Diamonds are also built of six-membered carbon rings. However, here each carbon is sp^3 hybridized and is bonded to four others to give a nonplanar three-dimensional network. One way to view the diamond lattice shows its relation to graphite: the C=C pi bonds of graphite are replaced with C—C sigma bonds between carbon atoms in adjacent sheets. There are two results of this cross-linking of graphite sheets: (1) Since there are no longer sheets to slide past one another, diamonds are not easily deformed; they are one of the hardest solids known and are used industrially as abrasives. (2) Diamonds are denser than graphite. Furthermore, diamonds are also very high melting, colorless when absolutely pure, and nonconductors of electricity. Finally, diamonds are much less chemically reactive than graphite.

DENSITY	(g/cm^3)
Graphite	2.22
Diamond	3.51

More chemistry of carbon is discussed in Chapters 21 and 23.

Silicates, compounds of silicon and oxygen, are network solids. These important compounds, which form the basis of many minerals, are described in Chapter 21.

"Imperfect" solids include the semiconductors and magnetic materials that are the heart of computers and allow you to record video and

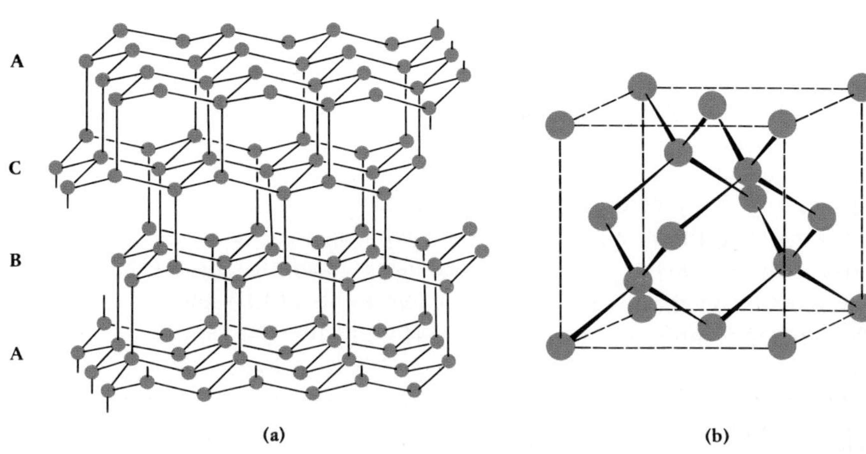

FIGURE 11.31
Two views of the diamond structure. (a) The three-dimensional network of C_6 rings with the basic structural element of tetrahedral, sp^3 hybridized carbon atoms, leading to puckered six-membered rings. (b) The unit cell of diamond. You can view it as a face-centered arrangement of C atoms with four other C atoms located in holes in the cube.

sound images. By definition, they are not regular arrays of ions or atoms, so it is unfortunately difficult to describe them without going into great detail. Nonetheless, it is useful to comment on one type of imperfect solid: stoichiometric compounds with imperfect lattices. **Stoichiometric solids** have formulas in agreement with the usual rules of oxidation numbers. The alkali metal halides are examples. If the lattice is imperfect, say by a pair of ions missing from their normal sites in a lattice (Figure 11.32), this is known as a **Schottky defect**. On the other hand, if an M^+ ion occupies a site other than a normal one, this is called a **Frenkel defect**. Either case can lead to extraordinary electrical, magnetic, and optical properties.

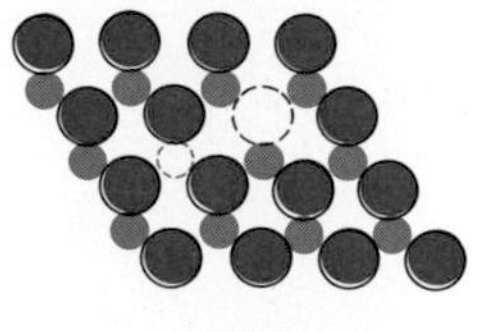
(a)

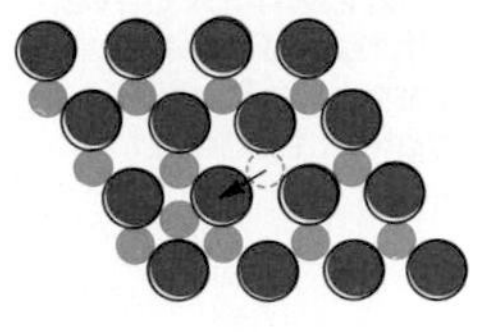
(b)

FIGURE 11.32
Lattice defects. (a) Schottky defects arise in a MX lattice, for example, when a unit of MX is missing from its expected position in the lattice. (b) A Frenkel defect occurs when an M^{n+} ion is found in a lattice hole other than the one expected.

Amorphous solids are solids without long-range order. Examples include some of our most important materials, among them glass and organic **polymers** such as polyethylene and polypropylene. The prefix *poly* in "polymer" means that many units (called mono*mers*) have been linked together. Thus, polyethylene is formed by linking ethylene monomers in a very long chain.

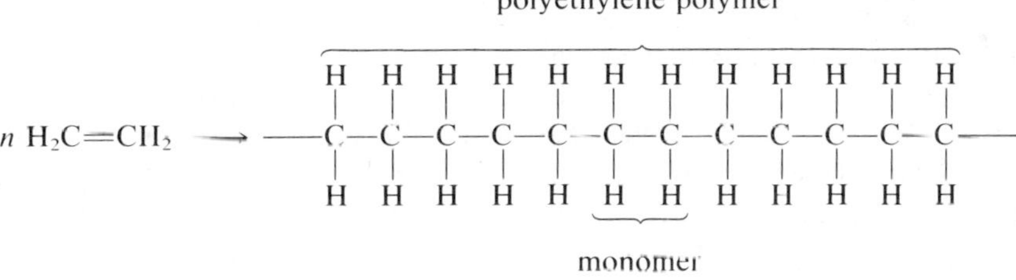

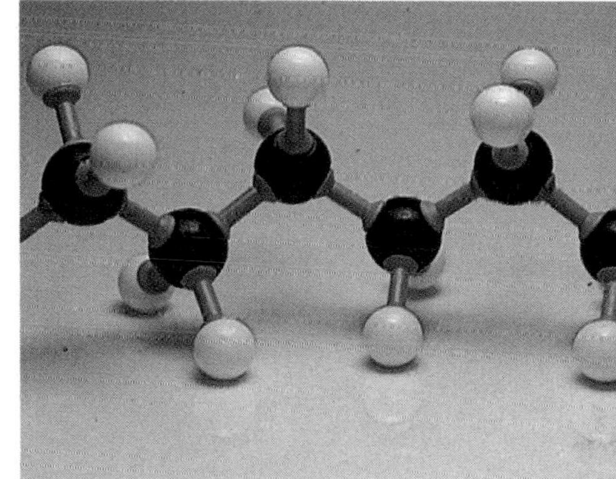
Portion of polyethylene polymer. (Charles D. Winters)

Interactions between polymer chains occur by weak dispersion forces and lead to solids that can be molded into the various shapes that you see for "plastic" objects. Polymers, both man-made and natural, are described in detail in Chapter 24.

THE PHYSICAL PROPERTIES OF SOLIDS

We know now that the outward shape of a crystalline solid is a reflection of its internal structure. But what about the temperature at which solids melt, their hardness, or solubility in water, properties certainly of interest to geologists and others? A few of these are explored below.

THERMOCHEMISTRY OF SOLID FORMATION: LATTICE ENERGY Sodium chloride melts at 800°C and potassium iodide at 680°C. These rather high temperatures must reflect very strong forces holding the solid intact, forces we can explore by understanding the thermochemistry of crystal lattice formation.

In Section 11.2 you learned about the energy of interaction between two ions in the gas phase and found that the energy depended on ion charge and on their "bond distance" in the resulting gaseous molecule. We are now interested in the additional factors that determine the enthalpy of formation of a *crystalline*, ionic solid. To analyze these factors, we shall calculate the energy involved in the formation of NaCl crystals from sodium and chlorine, each in its standard state.

$$Na(s) + \frac{1}{2} Cl_2(g) \longrightarrow NaCl(s) \qquad \Delta H_f^\circ = -411.0 \text{ kJ/mol}$$

This enthalpy can be calculated by breaking the overall process into a series of steps such as those outlined in Figure 11.33. According to Hess's Law, first introduced in Chapter 5, the sum of the energies for steps 1 through 5 in Figure 11.33 is equal to the enthalpy of formation of solid NaCl (Example 11.9). This approach to analyzing reaction energies is called a **Born–Haber cycle**, so named for Max Born and Fritz Haber, two German scientists prominent in thermodynamics earlier in this century.

Fritz Haber is best known for developing the currently used method for producing ammonia from H_2 and N_2. See Chapters 1 and 21.

You have seen all the energies listed in the Born–Haber cycle in Figure 11.33, except for that in step 5. The heat evolved when gaseous ions form a crystalline solid is called the **enthalpy of crystallization**, ΔH_{cryst}.

$$Na^+(g) + Cl^-(g) \longrightarrow NaCl(s) + \text{heat}$$
$$\text{Heat} = \text{enthalpy of crystallization} = \Delta H_{cryst}$$

The negative of this heat, that is, $-\Delta H_{cryst}$,

$$NaCl(s) + \text{heat} \longrightarrow Na^+(g) + Cl^-(g)$$
$$\text{Heat} = -\Delta H_{cryst} \cong \text{lattice energy}$$

is nearly equal to the **lattice energy**, the energy required for one mole of crystalline, ionic solid to be converted to widely separated, gaseous ions. Just like the dissociation energies of gaseous ionic molecules (Section 11.2), lattice energies depend directly on ion charge and inversely on the distance between ions in the lattice. And, just as in the case of the dissociation energies, there is a periodicity to lattice energies as illustrated in Figure 11.34. The lattice energy clearly drops as ion size (and so cation–anion separation) increases for the alkali metal halides. As we shall see below, this has a clear correlation with such observable physical properties of solids as their melting point.

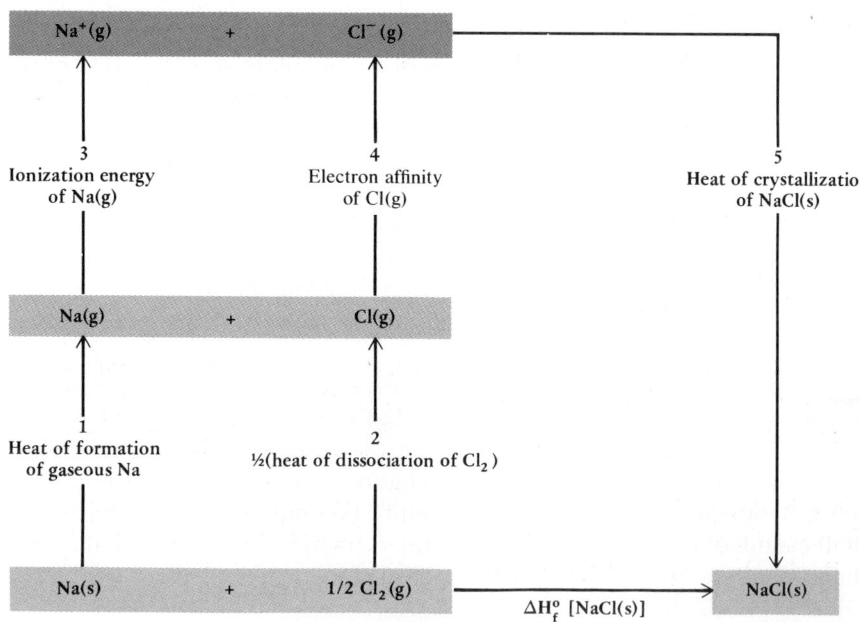

FIGURE 11.33

A Born–Haber cycle for the formation of the crystalline, ionic solid NaCl from its elements in their standard states. Energies associated with the various steps are given in Example 11.9.

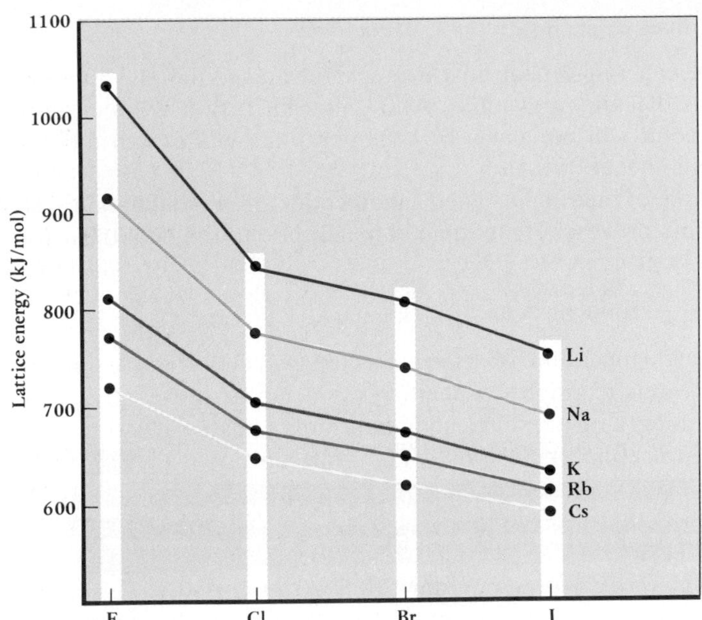

FIGURE 11.34
Trends in the lattice energies of the alkali metal halides.

EXAMPLE 11.9

USING A BORN–HABER CYCLE TO CALCULATE ΔH_f°[NaCl(s)].

Given the energies in steps 1 through 5 in the Born–Haber cycle for the formation of NaCl(s), calculate the heat of formation of this crystalline solid.

Step 1. Heat of formation of Na(g) = 108.4 kJ/mol.
Step 2. One-half the heat of dissociation of one mole of Cl_2 = $\frac{1}{2}$(243.3 kJ/mol).
Step 3. Ionization energy of Na(g) = 495.7 kJ/mol.
Step 4. Electron affinity of Cl(g) = −348.5 kJ/mol.
Step 5. Experimental ΔH_{cryst} = −779.0 kJ/mol

Solution According to Hess's Law (Chapter 5), the sum of the energies of steps 1 through 5 should be equal to the heat of formation of NaCl(s). Adding the energies above, we find that

$$\Delta H_f^\circ[\text{NaCl(s)}] = 108.4 \text{ kJ} + \tfrac{1}{2}(243.3 \text{ kJ}) + 495.7 \text{ kJ} - 348.5 \text{ kJ} - 779.0 \text{ kJ}$$
$$= -401.8 \text{ kJ/mol.}$$

The calculated value is in good agreement with the experimental value.

EXERCISE 11.12 BORN–HABER CYCLE CALCULATION

Calculate the enthalpy of formation of KCl(s) using the Born–Haber cycle approach. Use the heat of dissociation for Cl_2 and the electron affinity of Cl(g) in the example above. The ionization energy of K(g) is found in Chapter 8. The ΔH_{cryst} for KCl(s) is −708 kJ/mol, and the enthalpy of formation of K(g) is 90.14 kJ/mol.

SOME PHYSICAL PROPERTIES OF SOLIDS The **melting point** of a solid is the temperature at which the crystal lattice collapses and solid is converted to liquid. Just as is the case for liquid/vapor interconversion, **melting** requires energy input, an energy often called the **heat of fusion**.

$$\text{Solid} + \text{heat} \longrightarrow \text{liquid} \qquad \text{Heat} = \Delta H_{\text{fusion}}$$

Heats of fusion can range from just a few thousand joules per mole to many thousands of joules per mole. A low heat of fusion will certainly mean that the solid will melt at a low temperature, while high melting points reflect high heats of fusion.

The reverse of melting is called **solidification** or **crystallization**, and it is an exothermic process. Heat equal in magnitude to the heat of fusion is evolved.

$$\text{Liquid} \longrightarrow \text{solid} + \text{heat} \qquad \text{Heat} = -\Delta H_{\text{fusion}}$$

The melting temperature of a solid can convey a considerable amount of information. Some examples of melting point temperatures are given in Table 11.8, along with the corresponding molar heats of fusion, and there are several useful conclusions which you can draw. (1) Low molecular weight, nonpolar substances that form molecular solids are low melting. However, the melting point increases as the molecular weight increases. This happens because dispersion forces become stronger with increasing molecular weight (and number of electrons). As the dispersion forces increase, an ever increasing amount of energy must be supplied to the molecules to achieve enough kinetic energy to break down the intermolecular forces and allow the orderly structure of the solid to collapse. This increasing energy requirement is reflected in the increasing heat of fusion. (2) The same trend is observed for the hydrogen halide series, although there are dipole–dipole forces involved here. In addition, as the molecular weight increases, dispersion forces increase, and the melting point increases. (3) Next, it is clear that hydrogen bonding and dipolar forces have a profound effect in such compounds as water and benzoic acid (C_6H_5COOH). (4) Finally, extended ion–ion forces such as exist in

TABLE 11.8 Melting Points and Heats of Fusion of Some Solids

COMPOUND	MELTING POINT °C	HEAT OF FUSION J/mol	TYPE OF INTERMOLECULAR FORCES
O_2	−218	445	Dispersion forces; low molecular weight
HCl	−114	1990	All three have dipole–dipole forces; attraction enhanced by dispersion forces that increase with molecular weight.
HBr	−87	2406	
HI	−51	2871	
CF_4	−184	—	All three CX_4 molecules are nonpolar; dispersion forces increase with increasing molecular weight.
CCl_4	−23	3280	
CBr_4	90	—	
H_2O	0	6008	Hydrogen bonding; compare with H_2S
H_2S	−86	2395	Dipole–dipole forces, dispersion forces increase with molecular weight.
H_2Te	−49	6987	
C_6H_5COOH	122	17300	Hydrogen bonding; an organic acid
NaF	992	29288	All ionic solids have extended ion–ion interactions. Note that the general trend is the same as that in lattice energies.
NaCl	800	30208	
NaBr	747	25690	
KF	875	27196	Again, note that melting point declines as ion–ion distances increase.
RbF	833	17280	
CsF	705	10250	

ionic solids generally lead to melting points higher than 100°C. Be sure to notice the relation of the melting points of the alkali metal halides to their lattice energies.

The melting point of an impure solid (one consisting of an intimate mixture of two or more components) is lower than that of a pure compound (Chapter 12). Therefore, if you measure the melting point of a solid, you can assess its relative purity. For this reason, melting point determinations are a common laboratory operation.

One other important property of solids is the direct conversion of solids to gases by **sublimation** (Figure 11.35),

$$\text{Solid} + \text{heat} \longrightarrow \text{gas} \qquad \text{Heat} = \Delta H_{\text{subl}}$$

an endothermic process whose enthalpy is called the heat or **enthalpy of sublimation**. Water, which has a heat of sublimation of 51 kJ/mol, can be converted from solid ice to water vapor quite readily. One of the best examples of the use of this is in a frost-free refrigerator. During certain times, the freezer compartment is warmed slightly. Molecules of water free themselves from the surface of the ice (the vapor pressure of ice at 0°C is 4.60 mm), and they enter the gas phase where they are removed in a current of air blown through the freezer. Other common substances that sublime are I_2 and dry ice, solid CO_2.

FIGURE 11.35

Sublimation. The compound $(C_5H_5)_2Fe$ (called ferrocene) is a yellow-orange solid with a substantial vapor pressure when it is heated. The vapor condenses as yellow-orange crystals on the cold evaporating dish covering the beaker. (Charles D. Winters)

11.5 THE SPECIAL PROPERTIES OF LIQUID AND SOLID WATER

One of the most striking differences between our planet and others in the solar system is the existence of liquid water on the earth. Three fourths of the globe is covered with oceans, the polar regions are vast ice fields, and even the soil and rocks hold large amounts of water. Water has played a major role in the history of the people of the earth, and it is a significant factor in controlling the climate of our planet and the life on it. Ice floats on water, lakes freeze from the top down and not from the bottom up, very large amounts of energy are needed to change the temperature of a body of water to any considerable extent, and snow flakes have a six-sided geometry. All these features reflect the ability of water molecules to cling tenaciously to one another by hydrogen bonding, the source of the unique properties of water.

The oxygen atom of a water molecule is the center of a tetrahedron of two H atoms and two lone pairs of electrons, and each point of the tetrahedron can be involved in hydrogen bonding to a neighboring water molecule (Figure 11.36). In solid water (ice) the hydrogen bonds are 177 pm in length, compared with 99 pm for a normal O—H covalent bond. This difference of course reflects the weaker nature of the hydrogen bond.

Each of the water molecules hydrogen-bonded to a central water molecule can in turn hydrogen-bond to yet more water molecules. Although this could lead to a complex, random structure, Nature often exhibits a tendency to achieve great regularity, and ice is beautifully regular (Figure 11.36). It is an open cage structure in which the oxygen atoms of hydrogen-bonded water molecules are arranged at the corners of puckered, six-sided rings or hexagons. Each edge of a hexagonal ring consists of a normal O—H covalent bond and a hydrogen bond. It is interesting that snowflakes are always based on six-sided figures, a reflection of the internal molecular structure of ice.

You will see many more examples of six-sided rings in chemistry; the structure of diamond is very similar to that of ice (Figure 11.31).

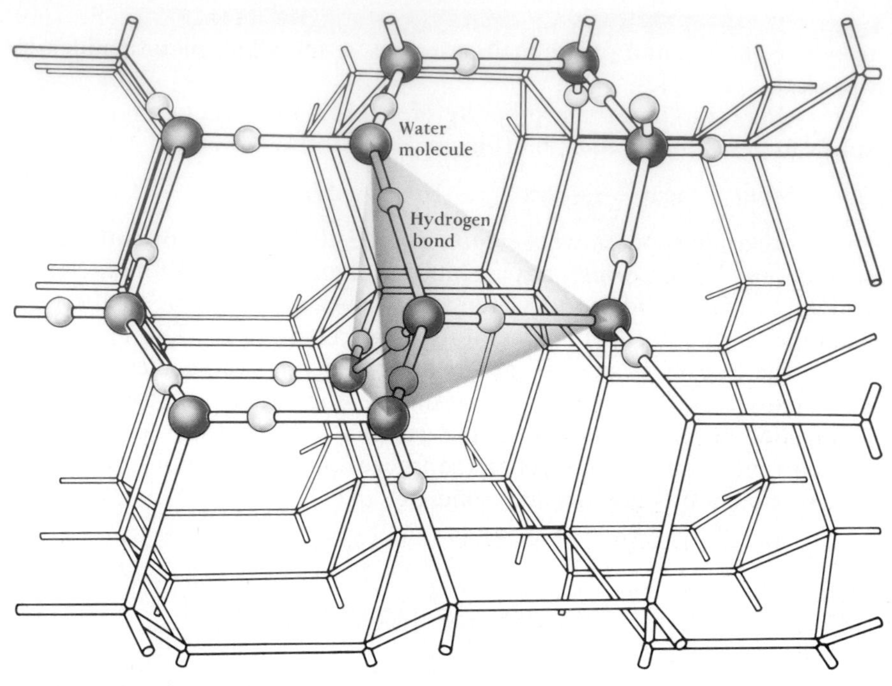

FIGURE 11.36

Ice. Each water molecule in ice is hydrogen bonded to four others. Oxygen atoms of hydrogen-bonded water molecules are at the corners of puckered, six-membered rings.

When ice melts to water the regular structure of ice collapses to some extent, but approximately 85% of the hydrogen bonds still remain. Although the structure of water is still a matter of great debate, it does seem certain that there are ice-like clumps of water molecules interspersed with individual water molecules. The major consequence of the open structure of ice compared with liquid water is that the liquid is more dense than ice. This density difference is small, but it is all it takes to make ice float on top of water.

At 4°C, liquid water has a density of 1.0 g/cm^3, while ice at 0°C has a density of 0.917 g/cm^3.

Not only are the densities of liquid water and ice different but the density of liquid water also changes with temperature (Figure 11.37). When water melts at 0°C, there is a great increase in density. When the now liquid water warms slightly, its density increases still more as the ice-like liquid structure breaks down further, and the liquid becomes more compact. The density reaches a maximum at 4°C, but it begins to decline again at higher temperatures as more and more hydrogen bonds are broken and the structure loosens.

The way that water density changes with temperature means that lakes will not freeze solidly from the bottom up in the winter. When lake water cools with the approach of winter, the density increases, the cooler water sinks, and the water and dissolved material turn over until all the water reaches 4°C, the maximum density. (This carries oxygen-rich water to the lake bottom to restore the oxygen used during the summer, and it brings nutrients to the top layers of the lake.) As the water cools further, it stays on the top of the lake, since water cooler than 4°C is less dense than 4°C water. With further heat loss, ice can then begin to form on the surface, floating there and protecting the underlying water and aquatic life from further heat loss.

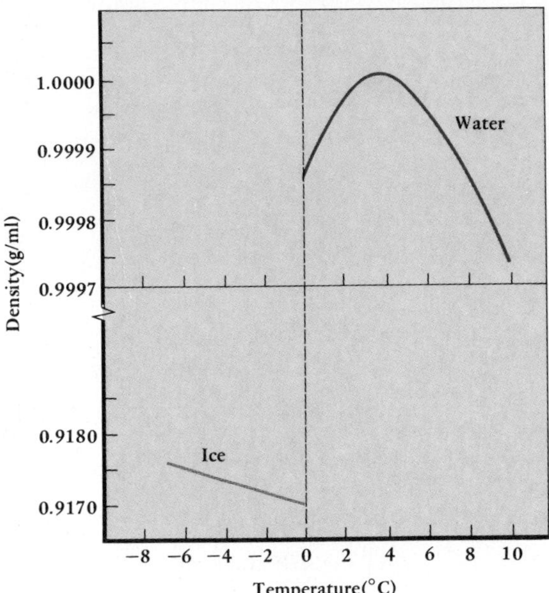

FIGURE 11.37
The temperature dependence of the density of ice and liquid water.

11.6 PHASE CHANGES

Now that we know something of the basic properties of liquids and solids, we want to put this information together in two ways: **heating/cooling curves** and **phase diagrams**. It is here that you will see why water has such an important effect on your climate and why you can skate on an ice-covered pond.

HEATING AND COOLING CURVES

Let us take 1.00 mole of water in the form of ice at −10°C. How much heat is necessary to warm the mole of water until it becomes steam at 110°C, and how much time is required if you add heat at the constant rate of 100. J/min? What happens is shown graphically in Figure 11.38.

The specific heat of ice is 2.09 J/g· K. This means that raising the temperature of 1.00 mole of ice by 10.0 degrees requires

$$(18.0 \text{ g})(2.09 \text{ J/g} \cdot \text{K})(10.0 \text{ K}) = 376 \text{ J}$$

Since you are adding heat at the rate of 100. J per minute, this means that 3.76 minutes are required, and the lowest segment of the graph in Figure 11.38 reflects this fact.

When the ice has warmed to 0°C, it can be melted to water at 0°C by further input of heat equal to the heat of fusion. Since the heat of fusion of ice is 6.01 kJ/mol, this quantity of heat is required for the melting process alone. The temperature of the system stays at 0°C until all of the solid has been converted to liquid, and the time required is 60.1 min.

$$\left(6010 \frac{\text{J}}{\text{mol}}\right)(1.00 \text{ mol})\left(\frac{1.00 \text{ min}}{100. \text{ J}}\right) = 60.1 \text{ min}$$

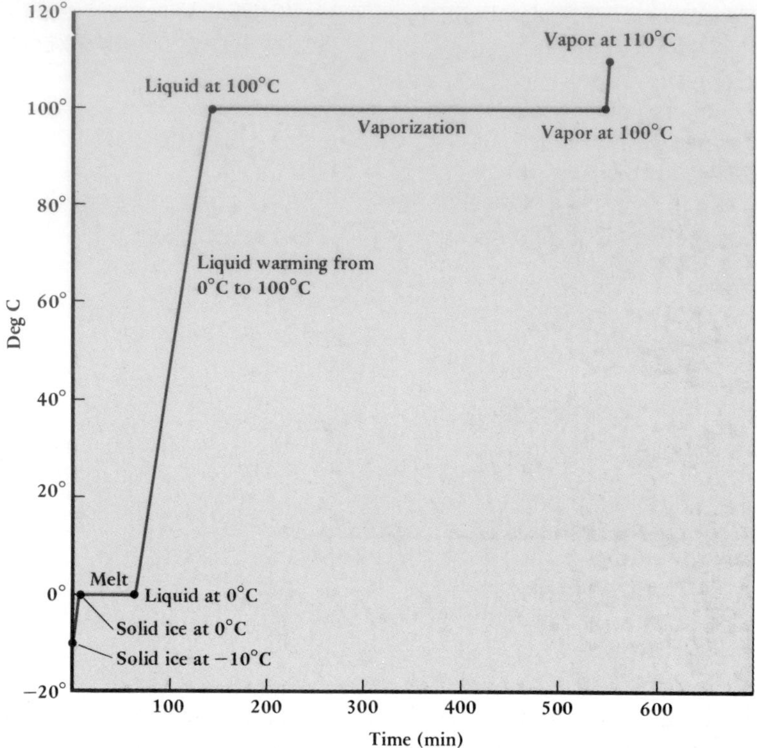

FIGURE 11.38

The heating/cooling curve for water. One mole of water is heated from −10°C (where it is in the form of ice) to +110°C (where it is steam). A constant heating rate of 100. J/min is assumed.

Now that liquid water has been formed, it can be warmed to the boiling point and evaporated to steam by further heat input. Using the specific heat of water (4.18 J/g · K), one mole of water requires 7520 joules for a 100. degree temperature rise (and 75.2 minutes are required). At the boiling point, a further 40.7 kJ (the molar heat of vaporization) are required to evaporate completely the mole of water at 100°C. Finally 2.0 J/g · K are needed to heat the steam from 100°C to the final temperature.

If you cool the 1.00 mole sample from steam to ice, the curve in Figure 11.38 would be identical. As long as the rates of heat input and removal are the same, heating and cooling are just the reverse of one another.

An important aspect of the curve in Figure 11.38 is its relation to the breaking of intermolecular bonds. When heating a solid or liquid, there is a rapid rise in temperature up to the point of a phase change. The rate at which the temperature rises is inversely related to the specific heat of the phase, which is smaller for solids and gases than for liquids. At the molecular level, a rise in temperature for a solid, liquid, or gas depends on how much of the added energy shows up as increased molecular kinetic energy, which in turn is largely determined by how easily and to what extent the molecules are separated. When heating a solid, the molecules do not separate much, and most of the added energy serves only to make them vibrate more violently. Thus, the temperature of a solid rises quickly for a given heat input. For gases, the molecules are already widely separated and added heat simply causes them to fly about more rapidly. Again, added heat causes the temperature of a gas to increase rapidly. When a liquid is heated, however, large separations between molecules must occur, meaning that more heat must be used to raise the temperature 1 degree.

In the plateau or phase-change regions of Figure 11.38, essentially all the energy added is used to separate molecules, and very little appears as a kinetic energy increase (or temperature increase).

The oceans and large lakes have an enormous effect on weather, due in large part to the high specific heat of water. In autumn, the temperature of the atmosphere drops. As the temperature of the ocean or lake water equalizes to that of the air, ocean or lake water gives up heat to the atmosphere and the air temperature drop is moderated. Further, since so much heat must be given off for every degree drop in water temperature, the decline in water temperature is gradual. Thus, the temperature of the ocean or large lake is generally higher than the average air temperature on land until late in the winter.

EXERCISE 11.13 HEATING AND COOLING

If the lakes and oceans of another planet are filled with ammonia, how many joules of energy are given off when 1.00 mole of liquid ammonia cools from −33.0°C (its boiling point) to −43.0°C? (Specific heat of ammonia is 4.70 J/g · K.) Compare this with the amount of heat given off by 1.00 mole of water cooling by 10.0°C.

PHASE DIAGRAMS

A **phase diagram** shows how the phases of a system are affected by changes in pressure and temperature. Each line in the diagram represents all of the conditions of T and P at which equilibrium exists between the two phases on either side of the line (Figure 11.39).

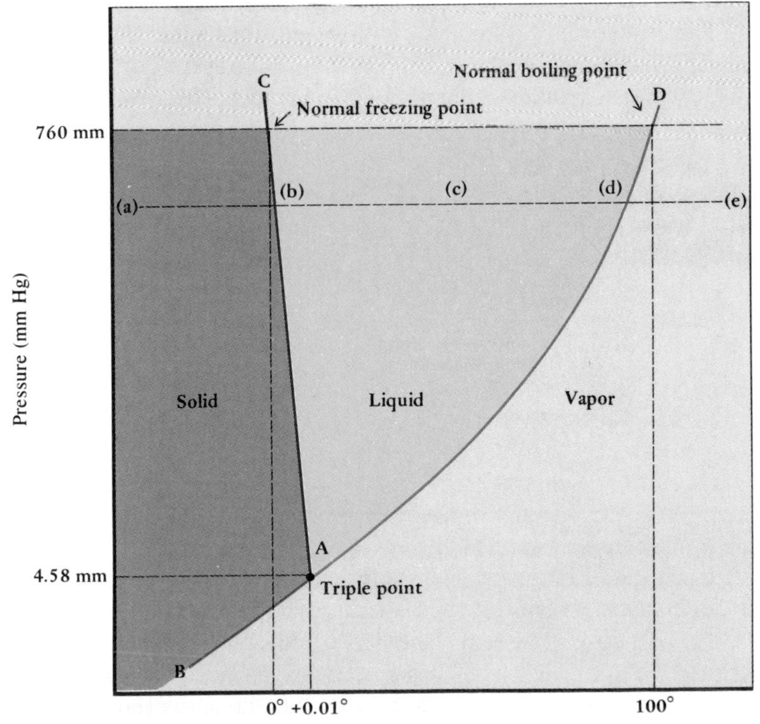

FIGURE 11.39

Phase diagram for water. (The scale is greatly exaggerated. See text for an explanation of the line *ae*.)

As described in Section 11.3, the equilibrium vapor pressure for water rises with temperature along the curved line *AD*, and the pressure reaches 760 mm Hg at 100°C, the normal boiling point. As the temperature of a liquid decreases, so does its vapor pressure until point *A*, the **triple point**, is reached (at $P = 4.58$ mm Hg and $T = 0.01$°C for water). Here, all three phases (solid, liquid, and vapor) are in equilibrium.

At temperatures and pressures below the triple point, ice can be in equilibrium with water vapor. The equilibrium vapor pressure of ice subliming to vapor at a given temperature is given by the line *AB*.

The line *AC* shows the conditions of pressure and temperature at which solid/liquid equilibrium exists. (Since no vapor pressure is involved here, the pressure referred to is the external pressure on the liquid.) Water is highly unusual, because this line has a negative slope. That is, the higher the external pressure, the lower the melting point. The change for water is approximately 0.01°C for each atmosphere increase in pressure.

Of all the thousands of substances studied, only water, bismuth, and antimony have solid/liquid equilibrium lines of negative slope.

The negative slope of the water solid/liquid equilibrium line can be explained based on our knowledge of the structure of water and ice. When pressure is increased on an object, common sense tells you the object will react by becoming smaller if possible. The material in the object will be forced into a smaller volume and thus become more dense. Since liquid water is denser than ice, due to the open lattice structure of ice, ice and water in equilibrium will respond to increased pressure by melting ice to form more water, since the same mass of water requires less volume. This is illustrated below using a solid/liquid equilibrium line of exaggerated slope. In addition, this diagram helps to explain why you can skate on ice. When you put skates on your feet, your weight is concentrated on a very small area. Therefore, your weight per unit area, the pressure you exert on the ice, is very large, and you compress the ice. Assuming the temperature remains constant, the ice changes to the denser liquid phase, and the film of water lubricates your sliding motion.*

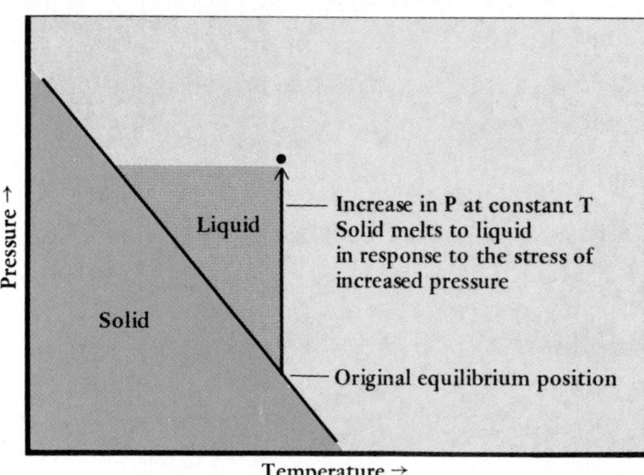

*Ice skates have very thin blades, so your weight is concentrated on a very small area of the ice. In this way, a person of average weight can easily exert 500 atm of pressure on the ice. Since the melting point of ice changes by about −0.01°C when the pressure is increased 1 atm, this means that a melting point change of −5.0°C (= −0.01°C/atm)(500 atm) can take place beneath the blade of a skate. This effect, combined with some frictional heating, helps you glide across the surface of the ice. (From this analysis you can also see why it is more difficult to skate on ice when the temperature of the air is quite cold (say 0°F).)

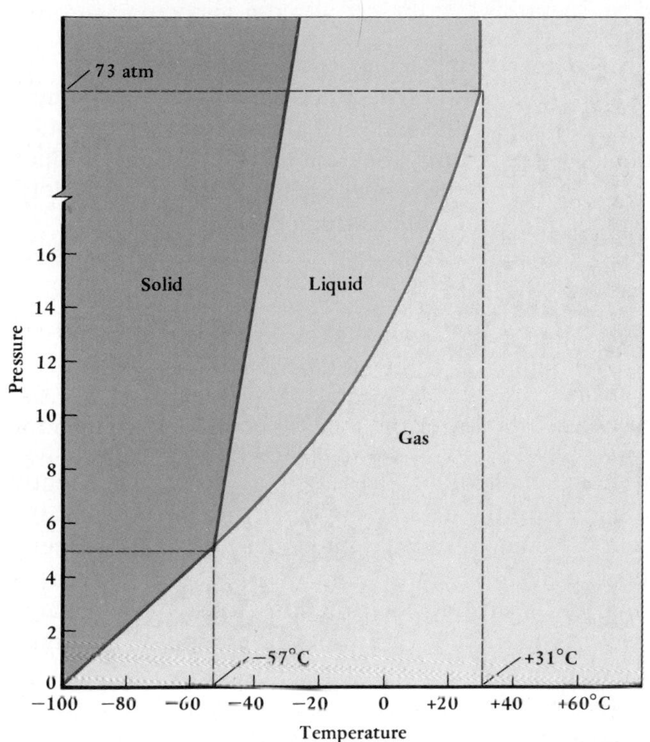

FIGURE 11.40
Phase diagram for CO_2.

Finally, let us correlate the water phase diagram (Figure 11.39) with the heating/cooling curve (Figure 11.38). Let us say that the atmospheric pressure is about 740 mm Hg and that you have an ice sample at − 10°C. As heat is added, you move along the horizontal dotted line in Figure 11.39, starting at point *a*. Solid alone exists at this point. When the line AC is reached, ice melts to water at *b* as heat is added. Once the ice is all melted, the water is now warmed, moving along the line from *b* to *d*. At all points between *b* and *d*, say *c*, the sample exists only as water; if vapor is present there is no equilibrium and vapor is condensing to liquid. When the point *d* is reached, liquid can come into equilibrium with vapor, and the equilibrium vapor pressure is 740 mm Hg. As more heat is added, the water evaporates and is converted completely to vapor at point *d*. The vapor is heated further to point *e*.

The basic features of the phase diagram for carbon dioxide (Figure 11.40) are the same as those of water, except the CO_2 solid/liquid equilibrium line has a positive slope. Solid CO_2 is more dense than the liquid, so the solid will sink to the bottom in a container of liquid CO_2. Also notice that the vapor pressure of solid CO_2 is 1 atm at −78°C. This means that solid CO_2 will sublime to CO_2 gas if you have a piece of it at room temperature. This is why solid CO_2 is called *dry ice*; it looks like water–ice, but it does not melt.

EXERCISE 11.14 PHASE DIAGRAMS

Suggest a set of conditions of temperature and pressure at which liquid CO_2 can exist.

SUMMARY

Because of **intermolecular forces**, molecules can be attracted to one another strongly enough that condensation occurs to give liquids and solids. Just as in gases, molecules in the liquid and solid phase retain some motion and have a distribution of kinetic energy (Section 11.1). Liquids and solids differ from gases, however, in that the former are much less **compressible**, and many solids have definite and regular structures.

Intermolecular forces range from very strong to very weak (Section 11.2).

Ion–ion > ion–dipole > dipole–dipole (including hydrogen bonding)
> dipole–induced dipole > induced dipole–induced dipole

Ion–ion forces, such as those holding together crystals of sodium chloride, can range from several hundred to several thousand kilojoules per mole. The **enthalpy of dissociation** of an ionic molecule [e.g., NaCl(g)] is directly proportional to the charges on the ions and inversely proportional to the distance between the ions (**Coulomb's law**). **Ion–dipole** forces cause metal ions to be **hydrated**. The negative end of the water dipole (oxygen) is attracted to the positive ion. As in ion–ion attractions, ions of larger charge or smaller size interact most strongly with water or other dipoles. Many molecules are polar, so the positive end of the dipole of one molecule is attracted to the negative end of the dipole of another molecule by **dipole–dipole forces**. A polar molecule can cause a distortion or **polarize** the electron cloud in a neighboring nonpolar molecule. Induced dipole interactions are exceedingly weak and depend on, among other things, the **polarizability** or "distortability" of the electron cloud of the nonpolar molecule.

Hydrogen bonding is a special form of dipole–dipole interaction. This occurs most strongly when H is attached to O, N, or F. Because of hydrogen bonding, water has many unique properties (Section 11.5).

Many properties of liquids are important (Section 11.3). All liquids can **evaporate** to the vapor phase, the energy required being the **enthalpy of vaporization**. Evaporation of liquid will continue at a given temperature until a state of **dynamic equilibrium** is attained when the rate of evaporation equals the rate at which vapor phase molecules **condense** or reenter the liquid phase. At equilibrium, the pressure exerted by the vapor phase molecules is called the **equilibrium vapor pressure**. More **volatile** compounds have higher equilibrium vapor pressures at a given temperature. If the vapor pressure equals the atmospheric or external pressure, the liquid **boils**. When the external pressure is 1 atmosphere, the temperature is the **normal boiling point**. The equilibrium vapor pressure of a liquid increases with temperature. The relationship between vapor pressure, temperature, and enthalpy of vaporization is expressed by the **Clausius–Clapeyron equation**.

Vapor pressure will continue to increase with temperature until the **critical temperature**, T_c, and **critical pressure**, P_c, are reached (Section 11.3). No liquid can exist above this temperature.

Since there are **cohesive forces** between identical liquid molecules, a liquid acts as though it has a skin, and the energy necessary to break through the surface is called the **surface tension**. It is the surface tension of water, together with **adhesive forces** between water and glass, that causes water to have a concave **meniscus** in a glass laboratory beaker. Cohesive intermolecular forces also account in part for the resistance to flow or **viscosity** of liquids.

Solids are characterized by long-range order (Section 11.4). The basic unit of a **crystal lattice** is called a **unit cell** that, on displacement in three dimensions, leads to the observed lattice. There are seven basic unit cells, but Section 11.4 is largely concerned with the **simple cube** and the **face-centered cube** for ionic compounds. The formula of an ionic compound is related to the number and type of holes that fill with cations in a simple cubic or face-centered cubic lattice of anions.

In addition to ionic solids, there are metals; solids formed by covalently bonded molecules (H_2O, I_2); network solids (graphite, diamond); imperfect and nonstoichiometric solids (semiconducting metal sulfides and oxides); and amorphous solids (glass, plastics and polymers) (Section 11.4).

The thermodynamics of the interaction of a metal and a halogen, for example, to form an ionic solid can be analyzed using a **Born–Haber cycle**. Among the energies involved is the **lattice energy**.

$$NaCl(s) + \text{energy} \longrightarrow Na^+(g) + Cl^-(g)$$
$$\text{Energy required} = \text{lattice energy}$$

The lattice energy is related (1) directly to the charge of the ions and (2) inversely to the distance between ions in the solid.

The heat required to melt a solid to give a liquid is called the **heat of fusion.** The heat or **enthalpy of sublimation** is the energy required to convert a solid directly to its vapor.

A **phase diagram** (Section 11.6) shows the relation between the three phases for a compound under various conditions of temperature and pressure.

STUDY QUESTIONS

REVIEW QUESTIONS

1. Name the types of forces that can be involved between molecules, between ions and molecules, and between ions and ions. Place them in order of decreasing strength.
2. On what does the enthalpy of dissociation of a gaseous ionic compound depend? Illustrate your answer with the compounds NaCl, NaI, and $MgCl_2$.
3. Explain how a water molecule can interact with a molecule such as CO_2. What intermolecular force is involved?
4. Hydrogen bonding is most important when H is attached to certain electronegative atoms. What are those atoms?
5. Explain how hydrogen bonding leads to the decline in density of water from 4°C to solid ice at 0°C.
6. Explain why the specific heat of water is so large compared with many other liquids. How can this affect climate?
7. Explain evaporation of liquids and condensation of vapor in terms of the kinetic molecular theory.
8. Define normal boiling point, critical pressure, and critical temperature.
9. Explain, in terms of intermolecular forces, the trends in boiling points of the molecules EH_3, where E is a Group 5A element (Figure 11.8).
10. Explain the difference between cohesive and adhesive forces. What is their role in explaining the concave meniscus observed for water in glass?
11. List the major types of solids with an example of each.

12. Explain the difference between the three types of cubic units cells. Which is never observed for ionic solids?
13. Consider a face-centered lattice of bromide ions. How many *net* Br^- are contained within this unit cell? Where are the octahedral "holes" in this lattice?
14. Define the lattice energy of an ionic solid. How is it related to the size of the ions and their charge?
15. Draw the phase diagram for water. Label the normal boiling point, melting point, triple point, and show what regions of pressure and temperature are appropriate to solid, liquid, and vapor.
16. Explain why the solid/liquid (S/L) equilibrium line in the water phase diagram has a negative slope. What is the requirement for a S/L line of positive slope?

INTERMOLECULAR FORCES

17. Answer each of the following questions by indicating the appropriate intermolecular force.
    (a) One example of a hydrated salt is $CuSO_4 \cdot 5\,H_2O$. What kind of force is responsible for binding water molecules to the copper ions of the copper sulfate?
    (b) What type of intermolecular forces must be overcome to: (1) melt ice; (2) melt solid I_2; (3) dissolve NaCl in water; or (4) convert NH_3 from liquid to vapor?
18. Rank the following in order of increasing strength of intermolecular forces in the pure substances, and indicate which ones are most likely to exist as gases at 25°C and 1 atm: KI, Ne, CH_4, CO, $MgSO_4$.
19. To melt an ionic solid, energy must be supplied to disrupt the forces between the ions so that the regular array of ions collapses, loses order, and becomes a liquid. If the distance between anion and cation in a crystalline solid is decreased (and the ion charges remain the same), is the melting point expected to increase or decrease? Why?
20. Considering intermolecular forces, which compound in the following pairs of ionic compounds should have the higher melting point: (a) NaCl or RbCl; (b) BaO or MgO; (c) NaCl or $MgCl_2$?
21. Which of the following compounds would be expected to form hydrogen bonds with water: (a) CH_3—O—CH_3, (b) CH_4, (c) HF, (d) HCOOH, (e) I_2, (f) CH_3OH?
22. Ethyl alcohol (H_3C—CH_2—OH) and water mix in all proportions. Draw molecular structures showing how water and alcohol interact. What type of intermolecular force is involved?
23. Acetone, $(H_3C)_2C{=}O$, is a common laboratory solvent. However, it is usually contaminated with water. Why does acetone absorb water so readily? Draw molecular structures showing how water and acetone can interact. What type of intermolecular force is involved in the interaction?
24. Tell which member of each of the following pairs of compounds you would expect to have the higher boiling point: (a) O_2 and N_2; (b) SO_2 and CO_2; (c) HF and HI; and (d) SiH_4 and GeH_4. In each case, tell what intermolecular forces are involved as well.
25. Explain why the boiling point of H_2S is lower than that of water.
26. Tell what type of intermolecular force is important in converting each of the following from a gas to a liquid: (a) CO_2; (b) NH_3; (c) $CHCl_3$; and (d) CCl_4.
27. Consider the following four compounds: (1) $MgCl_2$; (2) NH_3; (3) CH_4; and (4) CO. (a) Which of the compounds above is most likely to be a crystalline solid at room temperature? (b) Place the four substances in order of increasing boiling point.
28. In each of the following pairs of salts, which is more likely found with a hydrated cation: (a) Li_2SO_4 or Cs_2SO_4; (b) $CaCl_2$ or CsCl; (c) NH_4NO_3 or $Mg(NO_3)_2$; (d) RbCl or $NiCl_2$?
29. Rationalize the observation that 1-propanol (H_3C—CH_2—CH_2—OH) has a boiling point of 97.2°C, whereas a compound with the same empirical formula, methyl ethyl ether (H_3C—CH_2—O—CH_3) boils at 7.4°C.
30. Cooking oil floats on top of water. From this observation, what conclusions can you draw regarding the type of molecules found in cooking oil?

LIQUIDS

31. Answer the following questions with "increase," "not change," or "decrease."
    (a) If the intermolecular forces in a liquid increase, the normal boiling point of the liquid will _____.
    (b) If the intermolecular forces in a liquid increase, the vapor pressure of the liquid will ________.
    (c) If you increase the surface area of a liquid, the vapor pressure will ____________________.
32. Some physical properties for a number of compounds are listed in Table 11.5. Using the ideas of molecular polarizability and dispersion forces, explain the trend observed in the vaporization enthalpies and boiling points of the hydrocarbons methane through butane.
33. Sketch curves to compare the distribution of molecular energy for liquid H_2O at 25°C and at 90°C.
34. The liquid used in refrigerators and air conditioners is generally one or more of a family of compounds called "Freon." One compound belonging to this group is Freon-11, CCl_3F. If you fill an air conditioner with 1.00 kg of the compound, and if the enthalpy of vaporization of CCl_3F is 24.8 kJ/mol, how much heat is required to vaporize all of the Freon-11 at its boiling point of 23.8°C?
35. Freon-12, CCl_2F_2, is very commonly used as the refrigerant in home appliances. Its enthalpy of vaporization is 20. kJ/mol. How many joules of heat energy

are needed to vaporize 1.0 kg of Freon-12 at its boiling point?

36. Using the thermodynamic information in Appendix L, calculate the molar enthalpy of vaporization at 25°C for (a) $Br_2(\ell)$, (b) diamonds, (c) $CHCl_3$ (chloroform), and (d) $SnCl_4(\ell)$.

37. Acetonitrile, H_3CCN, a commonly used compound, is a liquid at room temperature and normal pressure. The standard enthalpy of formation of liquid acetonitrile is 31.38 kJ/mol. If the enthalpy of vaporization of the compound is 33.85 kJ/mol, what is the standard enthalpy of formation of $H_3CCN(g)$?

38. Using Figure 11.13, answer the following questions:
(a) What is the approximate equilibrium vapor pressure of water at 60°C. Compare your answer with the data in Appendix F.
(b) At what temperature will water have an equilibrium vapor pressure of 600 mm Hg?
(c) Compare the equilibrium vapor pressures of water and ethyl alcohol at 70°C.
(d) What is the equilibrium vapor pressure of diethyl ether at room temperature (approximately 20°C)?

39. You put some 60°C water in a plastic milk carton and seal the top very tightly. What will happen to the shape of the plastic milk carton when the water cools?

40. If you put a few drops of diethyl ether (Figure 11.13) on your hand, will it evaporate or remain as a liquid?

41. Assume you seal 1.0 g of diethyl ether (Figure 11.13) in an evacuated 100.-mL flask. (There are no molecules of any other gas or liquid in the flask.) If the flask is held at 30°C, what is the approximate gas pressure in the flask? If the flask is placed in an ice bath, will additional liquid ether evaporate or will some ether vapor condense to a liquid? What is the gas pressure at 0°C?

42. Some vapor pressure data for benzene, a common organic solvent, are given in the table below. Use them to answer the following questions.

TEMPERATURE (°C)	VAPOR PRESSURE (mm Hg)
7.6	40
26.1	100
60.6	400
80.1	760

(a) What is the normal boiling point of benzene?
(b) Plot these data, vapor pressure on the vertical axis against temperature on the horizontal axis. At what temperature does the liquid have an equilibrium vapor pressure of 250. mm Hg? At what temperature is it 650. mm Hg?
(c) Calculate the enthalpy of vaporization.

43. Equilibrium vapor pressure data for octane, C_8H_{18}, are given below. Use these data to calculate the enthalpy of vaporization and to estimate the normal boiling point of the liquid.

VAPOR PRESSURE (mm Hg)	TEMPERATURE (°C)
13.6	25
45.3	50
127.2	75
310.8	100

44. Vapor pressure data for liquid sodium are given below. Use these data to estimate the heat of vaporization of liquid sodium and its normal boiling point.

TEMPERATURE (°C)	VAPOR PRESSURE (mm Hg)
549	10
633	40
701	100
823	400

45. If you place 1.0 L of ethyl alcohol in a room that is 3.0 meters wide, 2.5 meters deep, and 2.5 meters high, will all of the alcohol evaporate? If some remains, how much will there be? The vapor pressure of ethyl alcohol at 25°C is 60. mm Hg. The density of ethyl alcohol (C_2H_5OH) at this temperature is 0.79 g/cm^3.

46. Freon-14, CF_4, has a critical temperature of −45.7°C and a critical pressure of 37 atm. Are there any conditions under which this compound can be a liquid at room temperature? Explain your answer briefly.

47. The polar liquid methyl alcohol, H_3COH, is placed in a glass tube. Will the meniscus be concave or convex?

48. Liquid ethylene glycol, $HO—C_2H_4—OH$, is one of the main ingredients in commercial antifreeze. Would you predict its viscosity to be greater or less than that of ethyl alcohol (see Table 11.6)?

SOLID STATE STRUCTURES

49. Can $CaCl_2$ have the NaCl structure? Briefly explain your answer.

50. Why is it not possible for a salt with the formula M_3X (for example, Na_3PO_4) to have a face-centered cubic lattice of X anions and M cations in octahedral holes?

51. RbCl crystallizes in the NaCl structure. (a) What is the volume of the unit cell? (You will need the ion radii in Figure 8.16 for this calculation.) (b) What is the density of the solid?

52. Calcium fluoride is the well-known mineral fluorite. It is known that each unit cell contains 4 Ca^{2+} ions and 8 F^- ions (because the F^- ions fill what are called "tetrahedral" holes in a face-centered Ca^{2+} lattice). The edge of a CaF_2 cubic unit cell has a length of 5.4×10^{-8} cm. What is the density of CaF_2 in g/cm^3?

53. LiH crystallizes with H^- ions defining either a simple cubic or a face-centered cubic lattice. The solid has

a density of 0.77 g/cm^3. If the edge of the unit cell is 4.086×10^{-8} cm, how many Li^+ and H^- ions are there in one unit cell? Which unit cell is appropriate for LiH?

54. If a simple cubic unit cell is formed so that the spherical atoms or ions just touch one another along each edge, calculate the percentage of empty space within the unit cell.

METALS AND OTHER SOLIDS

55. As noted in the text, pure metals crystallize in the same form as ionic compounds. One type of iron, for example, has a "body-centered cubic" lattice. The edge of the cube measures 2.87×10^{-8} cm.
(a) How many atoms of iron are there per unit cell?
(b) What is the volume of the unit cell?
(c) What is the density of this form of iron in g/cm^3?

56. Aluminum metal crystallizes in the face-centered cubic (cubic close packed) arrangement. If the atomic radius of aluminum is 143 pm, calculate (a) the length of a cell edge, (b) the volume of a unit cell, and (c) the density of solid aluminum.

57. A picture of the diamond unit cell is included in Figure 11.31. How many carbon atoms are there in one unit cell? If the density of diamond is 3.51 g/cm^3, calculate (a) the volume of the unit cell and (b) the length of an edge.

SOLIDS: PHYSICAL PROPERTIES

58. The ions of NaF and MgO are isoelectronic, and the internuclear distances are about the same (231 pm and 210 pm, respectively). Why then are the melting points of NaF and MgO so different (992°C and 2642°C, respectively)?

59. The lattice energies of the alkali metal halides are plotted in Figure 11.34. Explain the trend for MI, the metal iodides. Why do the lattice energies drop as the metal is changed from Li to Cs?

60. Would you expect the melting point of CsF to be greater or less than CsI? Compare KBr and RbBr on the same basis.

61. The standard enthalpy of formation of KCl(s) is −436 kJ/mol, and its lattice energy is 708 kJ/mol. The heat of sublimation of K(g) is 90.14 kJ/mol. From these data and some in Example 11.9, calculate the ionization energy of K(g).

62. Calculate the lattice energy of RbCl using a Born–Haber cycle and the following information (and additional information as needed from Example 11.9): (a) enthalpy of formation of RbCl(s) = −435.35 kJ/mol and (b) enthalpy of formation of Rb^+(g) from Rb(s) = 490.1 kJ/mol.

63. The enthalpy of formation of Zn(g) is 130 kJ/mol. What is the heat or enthalpy of sublimation of zinc?

64. Water has a very high heat of fusion (6.01 kJ/mol). Account for this heat in terms of the structure of solid water.

65. The molar heat of fusion of water is 6.01 kJ/mol. How much heat energy is required to melt one ice cube in a soft drink (to form liquid water at 0°C) if the cube has a mass of 15.0 g?

66. Account for the following trend in the heats of fusion of rubidium halides: RbCl (18.4 kJ/mol), RbBr (15.5 kJ/mol), and RbI (12.5 kJ/mol).

67. Benzene is an organic liquid that freezes at 5.5°C to beautiful, feather-like crystals. How much heat is evolved when 15.5 g of benzene freeze at 5.5°C? (The molar heat of fusion of C_6H_6 is 9.94 kJ/mol.) If the 15.5-g sample of benzene is remelted, again at 5.5°C, what quantity of heat is required to convert it to a liquid at this temperature?

68. During hailstorms in the midwest, very large hailstones can fall from the sky. (Some are the size of golf balls!) To preserve some very large stones, we once put them in the freezing compartment of a frost-free refrigerator. Unfortunately, after a day or two they had disappeared. Why?

HEATING/COOLING CURVES AND PHASE DIAGRAMS

69. Liquid ammonia, $NH_3(\ell)$, is still commonly used as a refrigerating liquid and as a heat transfer liquid. The specific heat of the liquid is 4.7 J/g · K and that of the vapor is 2.2 J/g · K. Its enthalpy of vaporization is 25.1 kJ/mol. If you heat 10. kg of liquid ammonia from −50.0°C to its boiling point of −33.6°C, and then on to 0°C, how much heat energy must you supply?

70. Your air conditioner may use Freon-12 (CCl_2F_2) as the heat transfer fluid. Its normal boiling point is −30°C, and the heat of vaporization is 165 J/g; and the gas and liquid have specific heats of 0.61 J/g · K and 0.97 J/g · K, respectively. How much heat must be evolved when 10.0 kg of Freon-12 is cooled from +40°C to −40°C?

71. A phase diagram of xenon is given on page 427. Use it to answer the following questions.
(a) In what phase is xenon found at room temperature and 1.0 atmosphere pressure?
(b) If the pressure exerted on the xenon is 0.75 atmosphere, and the temperature is −114°C, in what phase does the xenon exist?
(c) If you measured the vapor pressure of liquid xenon and found 380 mm Hg, what is the temperature of the liquid phase?
(d) What is the vapor pressure of the solid at −122°C?
(e) Which is the denser phase, solid or liquid? Briefly explain.

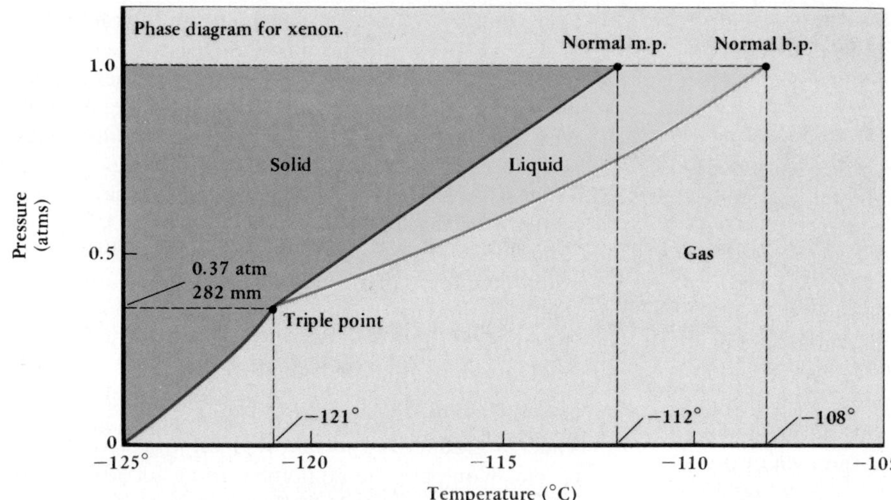

72. Construct an approximate phase diagram for O_2 from the following information: normal boiling point, 90.18 K; normal melting point, 54.8 K; and triple point, 54.34 K (and a pressure of 2 mm Hg). Very roughly estimate the vapor pressure of liquid O_2 at −196°C, the lowest temperature that is reached easily in the laboratory. Is the density of liquid O_2 greater than or less than that of solid O_2?

73. Assume you are on another planet that is covered with "dry ice," CO_2. Could you "ice" skate on the solid? (See Figure 11.40.)

74. A small lake might have a surface area of 1 square mile and an average depth of 100 feet. If the lake water cools from 25°C to 4°C, the temperature of maximum density, how many joules of heat must be lost? (Assume the density of water is 1.0 g/cm³.)

GENERAL QUESTIONS

75. The equilibrium vapor pressure of $CCl_4(\ell)$, carbon tetrachloride, is 113 mm Hg at 25°C. From this, and the enthalpies of formation of liquid and vapor CCl_4, estimate the normal boiling point of CCl_4.

76. Refer to the NaCl lattice and give the formula of a salt in which cations fill only half of the octahedral holes in an anion lattice. That is, define p and q in M_pX_q.

77. Some camping stoves contain liquid butane (C_4H_{10}). They work only when the outside temperature is warm enough to allow the butane to have a reasonable vapor pressure (and so they are not especially good for camping when the temperature is below approximately 0°F). The enthalpy of vaporization of butane is 24.3 kJ/mol. (a) Your fuel tank contains 190. g of liquid butane. How many joules of heat energy are required to vaporize all of the butane? (b) What is the vapor pressure of butane at 0°F? (You will need to use the Clausius–Clapeyron equation and data from Table 11.5 for this calculation.) Why doesn't this type of stove work well when the temperature of the air is low?

78. Two ways of packing spherical ions in one plane are shown in Figure 11.26. Prove that B is the more efficient method. A unit cell of A contains portions of four circles and one hole. In B, packing coverage can be calculated by looking at a triangle that contains portions of three circles and one hole. Show that A fills about 80% of available space while B fills closer to 90% of available space.

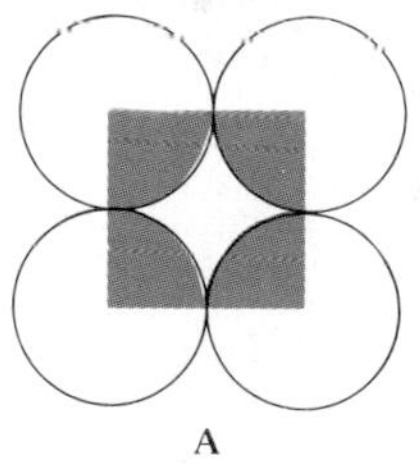

A

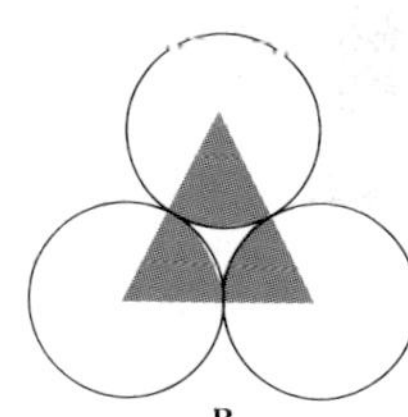

B

79. Rutile or TiO_2 crystallizes in a structure that is characteristic of many other ionic compounds. How many formula units of TiO_2 are there in the unit cell of rutile illustrated below? (The oxide ions marked by a dot are in the faces of the unit cell. The oxide ions marked by X are wholly *within* the unit cell.)

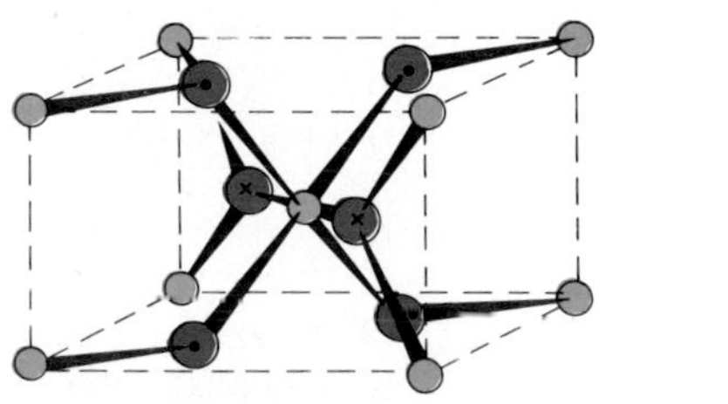

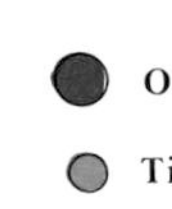

80. Cesium iodide, CsI, consists of a simple cubic lattice of I^- ions with Cs^+ ions in the cubic lattice holes. If the edge of the cube is 0.445 nm, what is the distance from the center of the Cs^+ in the center of the cube to the center of one of the I^- ions at the cube corners? An I^- ion has a radius of 0.216 nm. (You should recognize that the I^- ions at the unit cell corners do not contact one another, but they are in contact with the Cs^+ ion in the center of the unit cell.) What is the radius of a Cs^+ ion?
81. Mercury is the only metal that is a liquid at room temperature. It and many of its compounds are dangerous poisons if breathed, swallowed, or even absorbed through the skin. If the vapor pressure of liquid mercury is 1.0 mm at 126.2°C and if its normal boiling point is 357.0°C, calculate its equilibrium vapor pressure at room temperature (25°C).
82. Freon-113 ($C_2Cl_3F_3$) has a boiling point of 47.6°C and an enthalpy of vaporization of 146 J/g; the liquid has a specific heat of 0.91 J/g · K. How much heat is required to warm 1.00 kg of Freon-113 from room temperature (25°C) to its boiling point and then to vaporize the entire amount at 47.6°C?
83. In the course of laboratory experiments in organic chemistry, you may run into a situation where you want to remove a small amount of water from an organic solvent such as chloroform. You can do this by putting a few grams of an anhydrous salt into the flask containing the solvent. If the salt has a hydrated form, hydration occurs, and water is effectively removed from the liquid. If you were doing this experiment, which salt would you choose: $MgSO_4$ or $(NH_4)_2SO_4$?
84. Construct a plot of the boiling points of the rare gases (Table 11.4) vs. atomic mass. Suggest a reason for the fact that the rather smooth trend to higher boiling points is interrupted on going from Ar to Xe. That is, why are the boiling points of the heavier rare gases so much higher than expected?
85. Suggest a reason why the boiling point of O_2 is slightly higher than that of N_2.
86. Equation 11.1 (page 385) allows us to calculate the coulombic force of attraction between two ions. Use this equation to calculate the coulombic force between Mg^{2+} and O^{2-} in one mole of MgO(g). That is, calculate the enthalpy of the reaction

$$MgO(g) \longrightarrow Mg^{2+}(g) + O^{2-}(g)$$

The O^{2-} radius is 140 pm, and the Mg^{2+} radius is 65 pm. Compare the coulombic force for MgO with that of NaF in Figure 11.5. Recognizing that Na^+ and F^- are isoelectronic with Mg^{2+} and O^{2-}, respectively, comment on the effect of ion charge on the coulombic force.
87. Calculate enthalpies of dissociation for rubidium chloride (RbCl) and cesium chloride (CsCl), that is, the enthalpies for the reaction (see Equation 11.1)

$$MCl(g) \longrightarrow M^+(g) + Cl^-(g)$$

Use the experimental internuclear distances of 2.79×10^{-10} m and 3.32×10^{-10} m for RbCl and CsCl, respectively. Are their relative enthalpies of dissociation in the order expected? Comment on their values relative to those in Figure 11.5.

12

Solutions and Their Behavior

The ocean, a solution of many solutes in water (Louis Goldman, Photo Researchers, Inc.)

CHAPTER OUTLINE

A **solution** is a homogeneous mixture of two or more substances in a single phase. It is usual to think of the component present in largest amount as the **solvent** and the other component(s) as the **solute**. When you try to think of solutions, those that occur to you first probably involve water as the solvent: fruit juice, soft drinks, and beer. However, some consumer products (lubricating oils, gasoline, household cleaners) are solutions involving a liquid other than water as solvent.

Our objective in this chapter is to develop an understanding of gases, liquids, and solids dissolved in liquids. However, you should be aware that solutions do not have to involve liquid solvents. The air you breathe is a solution of nitrogen, oxygen, carbon dioxide, water vapor, and other gases. Glass objects are all around you. Although glass is variously referred to as an amorphous solid or as a super-cooled liquid, it is a solution of metal oxides (Na_2O and CaO, among others) in silicon dioxide.

Common sense tells you that adding a solute to a pure liquid will change the properties of the liquid, since the intermolecular forces will be changed or disrupted. Indeed, that is precisely why some solutions are made. For instance, by adding antifreeze to your car's coolant water you can prevent the radiator from boiling over in the summer and freezing up in the winter. As we will see shortly, the changes in freezing and boiling points from pure solvent to solution are called **colligative properties**. The magnitude of colligative properties—changes in freezing point, boiling point, and vapor pressure from solvent to solution, as well as solution osmotic pressure—ideally depend only on the number of solute particles per solvent molecule and not on the nature of the solute or solvent.

There are three major topics to be covered in this chapter. First, since colligative properties depend on the relative number of solvent and solute particles in solution, we need convenient ways of describing solution concentrations in these terms. Second, we are interested in how and why the solution process occurs. This will give us some insight into the third topic, the colligative properties themselves.

12.1 UNITS OF CONCENTRATION

The concentration unit that has served us so well to this point, molarity, is not going to work with colligative properties. Let's understand why.

Molarity is defined as the number of moles of material per liter of solution.

$$\text{Molarity } (M) \text{ of compound A in solution} = \frac{\text{moles of A}}{\text{liters of solution}}$$

For example, Figure 4.4 shows a 0.100 molar aqueous solution of $CuSO_4$ being made by adding enough water to 0.100 mole (25.0 g) of $CuSO_4 \cdot 5\,H_2O$ to make 1.00 L of solution. We did not pay attention to how much of the solvent (water) we actually added; we were concerned only with having the correct total solution volume. However, as Figure 4.4 made clear, if we had begun with 1.00 L of water, some water would have been unused when the solution volume was 1.00 L, because the solute takes up space. If 1.00 L of water had been added, the molarity would be less than 0.100 M. The only advantage in adding all the water would be so that we know how many molecules of water we added to a given number of Cu^{2+} and SO_4^{2-} ions. This is just the information we need to define the colligative properties of a solution.*

There are several methods of expressing concentration that allow us to define the number of solute particles per solvent molecule. These are mole fraction, molality, weight percent, and parts per million.

The **mole fraction** of a component A, X_A, of a solution is defined as the number of moles of A divided by the total number of moles of all of the components of the solution. For A in a solution containing A, B, and C, the mole fraction of A is

$$\text{Mole fraction of A} = X_A = \frac{\text{moles of A}}{\text{moles of A} + \text{moles of B} + \text{moles of C}} \qquad (12.1)$$

For example, for a solution that contains 1.0 mole (46 g) of alcohol, C_2H_5OH, in 9.0 moles (162 g) of water, the mole fraction of alcohol is 0.10 and that of water is 0.90.

$$X_{alc} = \frac{1.0\text{ mol alc}}{1.0\text{ mol alc} + 9.0\text{ mol } H_2O}$$

$$X_{alc} = 0.10$$

$$X_{water} = \frac{9.0\text{ mol } H_2O}{1.0\text{ mol alc} + 9.0\text{ mol } H_2O}$$

$$X_{water} = 0.90$$

Obviously, *the sum of the mole fractions of all components in the solution must equal exactly 1.*

$$X_{alc} + X_{water} = 1.0$$

The **molality** of a solution is defined as the number of moles of solute per kilogram of solvent.

---

*Another problem with using molarity is that, since the density of a liquid is temperature dependent, the volume of solution also depends on temperature. Therefore, molarity is temperature dependent. To be entirely correct you must report the temperature when you use solution concentrations in terms of molarity.

$$\text{Molality of A} = m_A = \frac{\text{moles of A}}{\text{kilograms of solvent}} \tag{12.2}$$

For the alcohol/water mixture mentioned above, 1.0 mole of alcohol is dissolved in 0.162 kg of water, so the molality of A is

$$\text{Molality of A} = \frac{1.0 \text{ mol A}}{0.162 \text{ kg solvent}} = 6.2\ m$$

(Notice that molality is denoted with a small italicized "*m*" while molarity is signaled with a capital italicized "*M*.")

Expressing a concentration as a **weight percentage*** is straightforward, since it is simply the mass of solute per 100 g of solution.

Weight % of A

$$= \frac{\text{grams of A}}{\text{grams of A + grams of any other solute + grams solvent}} \times 100 \tag{12.3}$$

Thus, our alcohol/water mixture is 46 g (1.0 mol) of alcohol in 162 g of water or in a total solution mass of 208 g.

$$\text{Weight \% alcohol} = \frac{46 \text{ g solute}}{(162 + 46) \text{ g solution}} \times 100$$
$$= 22\% \text{ alcohol by weight}$$

Finally, very dilute solutions often have concentrations expressed in **parts per million** (or per billion, per trillion, and so forth). It is generally understood to be a weight composition unit, so 1 part per million (1 ppm) is one gram of solute in a million grams of solution. For example, the Dead Sea in the Middle East has a bromide ion concentration of 4600 ppm or 4600 g of Br^- per million grams of sea water.

EXAMPLE 12.1

CALCULATING MOLE FRACTIONS, MOLALITY, AND WEIGHT PERCENT

If you dissolve 10.0 g of sugar, $C_{12}H_{22}O_{11}$, (about 1 heaping spoonful) in a cup of water (250. g), what are the mole fraction, molality, and weight percentage of sugar?

Solution 10.0 g of sugar (MW = 342.3) are equivalent to 0.0292 moles and 250. g of water represent 13.9 moles.

Mole Fraction:

$$X_{\text{sugar}} = \frac{0.0292 \text{ mol sugar}}{0.0292 \text{ moles sugar} + 13.9 \text{ moles } H_2O} = 0.00210$$

*Strictly speaking, this is a *mass* percentage; however, long usage has made "weight percentage" the standard term.

Molality:

$$m = \frac{0.0292 \text{ mol sugar}}{0.250 \text{ kg water}} = 0.117 \text{ molal in sugar}$$

Weight Percentage:

Weight percentage sugar

$$= \frac{10.0 \text{ g sugar}}{(10.0 \text{ g sugar}) + (250 \text{ g water})} \times 100 = 3.85\% \text{ sugar}$$

EXAMPLE 12.2

PARTS PER MILLION

If the Dead Sea contains 4600 ppm bromide ion, what is the molality of the ion?

Solution 4600 ppm represents 4600 g of Br^- in 1.0×10^6 g of water. This mass of bromide is equivalent to 58 moles, and one million grams of water is 1.0×10^3 kg.

$$\text{Molality of } Br^- = \frac{58 \text{ mol } Br^-}{1.0 \times 10^3 \text{ kg}} = 0.058\ m$$

EXERCISE 12.1 MOLE FRACTIONS AND MOLALITY

Assume you add 1.0×10^3 g of $C_2H_4(OH)_2$ (ethylene glycol) as an antifreeze to 4.0 kg of water in the radiator of your car. What are the mole fraction and molality of the ethylene glycol?

12.2 THE SOLUTION PROCESS

Both ammonium chloride and sodium hydroxide dissolve readily in water, but the water cools down when NH_4Cl dissolves while it warms up when NaOH dissolves. On the other hand, $CaCO_3$ dissolves in water only to a very small extent. Why do these things happen? What controls the solution process?

First, let's define some of the terms we have been using. When a solute is *dissolved* in a solvent, we mean that the attractive forces between solute and solvent particles are great enough to overcome the attractive forces (intermolecular forces) within the pure solute and solvent. As described in Chapter 11, when solutes are dissolved, they become **solvated** if solvent molecules are bonded firmly to solute molecules or ions (usually by dipole–dipole or ion–dipole forces). If water is the solvent, this is called more specifically **hydration**.

$$\mathrm{Li^+} \cdots \overset{\delta 2-}{\mathrm{O}}(\overset{\delta+}{\mathrm{H}})_2$$

Li^+ hydrated by water

Solubility is the maximum amount of material that will dissolve in a given amount of solvent at a given temperature to produce a stable solution. In Chapter 3 you learned some guidelines for the solubility of common ionic compounds, but we can now be somewhat more quantitative. For example, it is suggested in Table 3.1 that common nitrates and

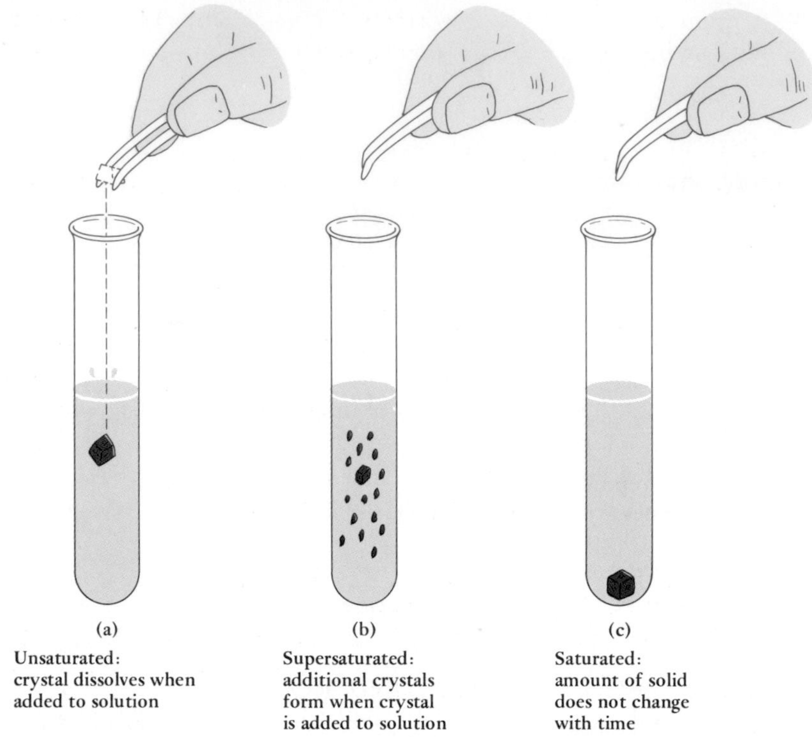
(a)
Unsaturated:
crystal dissolves when
added to solution
(b)
Supersaturated:
additional crystals
form when crystal
is added to solution
(c)
Saturated:
amount of solid
does not change
with time

FIGURE 12.1

The difference between unsaturated, saturated, and supersaturated solutions. In a, some solute is added to a solvent, and the solute dissolves. The solution was unsaturated with solute. In b, the solution was supersaturated. Adding extra solute led to the precipitation of some dissolved solute. The solution in c is saturated, since no more solute will dissolve; the concentration of solute does not change with time.

chlorides are usually soluble, except for the chlorides of Ag(I), Pb(II), and Hg(I). For example, we find that 952 g of silver nitrate will dissolve in 100 mL of water at 100°C, so we say this is a *soluble* salt. In contrast, only 0.00217 g of silver chloride dissolves in 100 mL of water at 100°C, and we say this is *insoluble*, even though it does have some degree of solubility.*

When the maximum amount of solute has been dissolved in a solvent and equilibrium is attained, the solution is said to be **saturated**. The concentration of dissolved solute in a saturated solution is a measure of the solubility of the solute. If the concentration of solute is less than the saturation amount, the solution is **unsaturated.** On the other hand, there are instances when a solution can temporarily contain more solute than the saturation amount, and these are called **supersaturated**. Figure 12.1 shows you what will happen when more solute is added to a solution in each of these conditions.

The concept of solubility is the basis of innumerable laboratory operations and industrial processes, and it can be the controlling factor

*When can you say properly that one substance is soluble and another is insoluble? There is no hard and fast rule about this, because there are degrees of solubility; nothing is totally without solubility at some temperature. Chemists frequently say something is not soluble, partly soluble, or soluble. Usually what they mean is that something is soluble if they can see that most of a solid has gone into solution. If only some has dissolved, it is partly soluble. If they cannot observe the dissolution of any solid, it is said to be insoluble or sparingly soluble.

in our environment and in living organisms. Thus, it is useful to explore the solubility or insolubility of common substances.

LIQUIDS DISSOLVING IN LIQUIDS

If two liquids mix to an appreciable degree to form a solution they are said to be **miscible**. In contrast, **immiscible** liquids will not mix to form a solution; they will exist in contact with each other as separate layers.

The nonpolar liquids octane (a component of gasoline) and benzene are miscible; they will mix in all proportions to form a homogeneous solution (Figure 12.2a). On the other hand, polar water and nonpolar octane are immiscible. You would observe that the less dense liquid, octane, simply floats as a layer on top of the more dense water layer (Figure 12.2b). Finally, polar ethyl alcohol molecules (C_2H_5OH) will dissolve in all proportions in water; beer, wine, and other alcoholic beverages contain amounts of alcohol ranging from a few percent to more than 50%. It is these observations that have led to the familiar rule of thumb: *like dissolves like*. That is, two or more nonpolar liquids frequently are miscible with each other, just as are two or more polar liquids. Liquids of different types will not mix to an appreciable degree.

What is the molecular basis for the "like dissolves like" guideline? Molecules of pure octane or pure benzene are held together in the liquid phase by induced-dipole/induced-dipole (dispersion) forces (Section 11.2). Similar forces can exist between an octane molecule and a benzene molecule, and the molecules are attracted to one another.

Crude oil consists of molecules resembling octane in Figure 12.2, so it is not miscible with water and is also less dense. Therefore, when an oil spill occurs in the environment, oil spreads across the water in a thin film, and a small amount of oil can cover a very large area (Figure 11.7). Since it forms a dark, thick coating on the water, it can affect aquatic life by preventing oxygen from mixing with the water and sunlight from penetrating the surface.

The rule of thumb "Like dissolves like" is useful and understandable. Polar solvents must be used to dissolve polar solutes, and nonpolar solutes are usually soluble in nonpolar solvents.

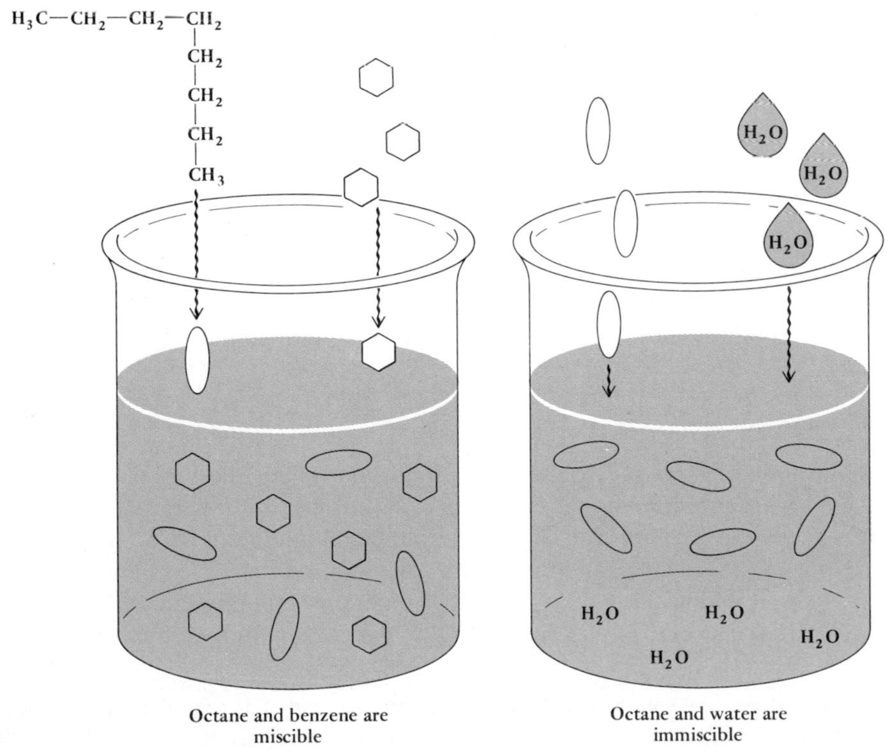

FIGURE 12.2

Miscibility. The two nonpolar molecules octane (C_8H_{18}, ⬭) and benzene (C_6H_6, ⬡) are miscible. However, weak intermolecular forces between water and octane cannot overcome the very strong forces between water molecules and allow octane to dissolve in water.

Weak

Weak

H—O---H—O---H—O---H—O---H—O---H—O---H—O

Strong

FIGURE 12.3

Intermolecular forces in water and octane. Water molecules interact strongly through hydrogen bonding, but there are only weak dispersion forces between octane molecules. Similarly, the octane/water forces are weak. Therefore, on mixing water and octane, the water/octane interaction is not strong enough to compensate for breaking hydrogen bonds in water, and water and octane are not miscible to any appreciable extent. (Octane is $H_3C—CH_2—CH_2—CH_2—CH_2—CH_2—CH_2—CH_3$, ⬭.)

In contrast to the benzene/octane solution, water and octane do not mix (Figure 12.3). The cause of this observation is analyzed in the following scheme.

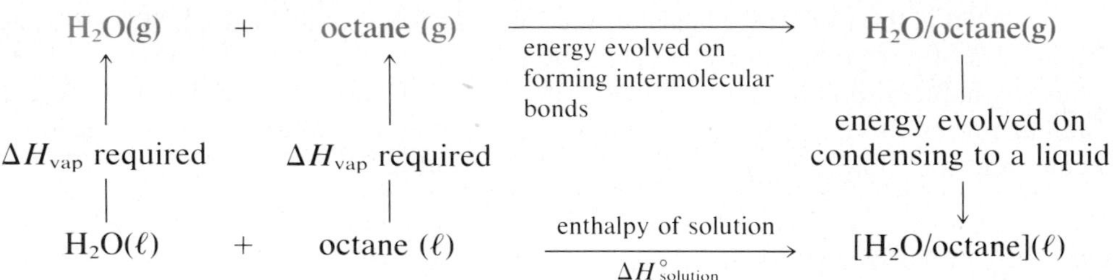

The **enthalpy of solution**, $\Delta H^\circ_{solution}$, is the net heat energy involved in the process of solution formation. It is seen here, according to Hess's law (Chapter 5), as the sum of the heats of vaporization and the energies evolved on forming intermolecular H_2O/octane bonds and on condensing the gaseous H_2O/octane to the liquid state. The energy required to break hydrogen bonds and vaporize water is enormous. Therefore, this overwhelms the energy evolved in the other steps, and the overall enthalpy change, the enthalpy of solution, must be positive; that is, the process must be highly endothermic.* Polar and nonpolar liquids do not usually mix.

SOLIDS DISSOLVING IN LIQUIDS

Solid iodine, I_2, is an example of a **molecular solid**, the molecules being held together by only moderately strong intermolecular dispersion forces. To dissolve this nonpolar solid in a nonpolar liquid, say CCl_4, the forces between I_2 molecules and between CCl_4 must be disrupted. The heat or enthalpy of solution is close to zero, however, since the energy required to pull apart solid I_2 and liquid CCl_4 is approximately returned by the

*Although heat of solution is important in determining solubility, another quantity, the *entropy* of solution, is also a contributing factor. This is considered in Chapter 18.

formation of new intermolecular bonds between I_2 and CCl_4. Thus, it is not surprising that I_2 is soluble in CCl_4 (Figure 12.4).

Sucrose, a sugar, is also a molecular solid, but the molecules interact in the solid phase through hydrogen bonds. Therefore, sugar is

Sucrose (cane sugar)

quite soluble in water. Hydrogen bonding between sugar and water is strong enough that energy is supplied to disrupt the sugar/sugar and water/water interactions. On the other hand, sugar is not soluble at all in CCl_4 or other nonpolar liquids. A nonpolar liquid does not interact sufficiently well with sugar to cause it to dissolve.

Network solids include graphite, diamond, and quartz, and your intuition tells you they do not dissolve readily in water. After all, where would all the beaches be if sand dissolved readily in water? Bonding in network solids is simply too strong to be broken down by an interaction with water dipoles.

Common **ionic solids** are not soluble at all in nonpolar solvents, and they vary greatly in their solubility in water (Figure 3.8). At 20°C, 100 g of water will dissolve 74.5 g of $CaCl_2$, but only 0.0014 g of limestone, $CaCO_3$, can be dissolved no matter how hard you try. It is useful to explore the reasons for the widely different solubilities of ionic solids.

(a)

(b)

FIGURE 12.4

Water, carbon tetrachloride (CCl_4), and iodine. Water and CCl_4 are not miscible, and the less dense water layer is found on top of the more dense CCl_4 layer. A small amount of iodine is dissolved in water to give a brown solution (top). However, the nonpolar molecule I_2 is more soluble in nonpolar CCl_4, as indicated by the fact that I_2 dissolves preferentially in CCl_4 to give a purple solution after shaking the mixture (bottom). (Charles D. Winters)

DISSOLVING IONIC SOLIDS Salt dissolves in water because strong ion–dipole forces lead to strong ion hydration and help to break down the cation–anion attraction in the crystal lattice. There are two ways to analyze the overall energy involved here. First, the heat or enthalpy of solution ($\Delta H_{solution}$) can be estimated from the lattice energy of NaCl (ΔH_1) and the heat of hydration of the gaseous ions to form aqueous hydrated ions (ΔH_2).

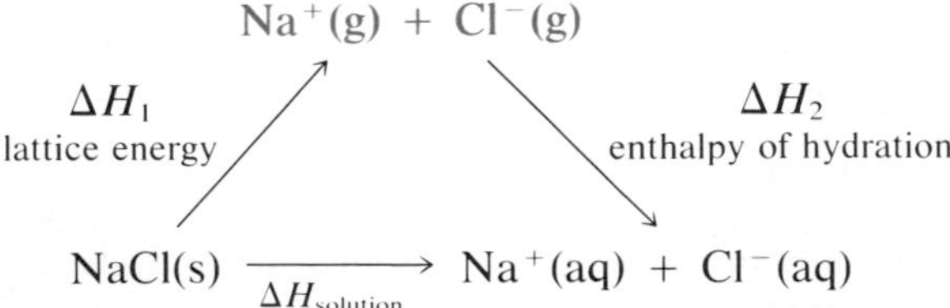

According to Hess's law, the heat of solution is the sum of ΔH_1 and ΔH_2. For NaCl these are

$$\Delta H_{solution} = \text{lattice energy} + \text{heat of hydration} = \Delta H_1 + \Delta H_2$$

$$\Delta H_{solution} = 774 \text{ kJ/mol} + (-760 \text{ kJ/mol}) = +14 \text{ kJ/mol}$$

TABLE 12.1 Water Solubility of Some Ionic Compounds

COMPOUND	LATTICE ENERGY kJ/mol	ENTHALPY OF HYDRATION kJ/mol	SOLUBILITY IN H_2O (g/100 cm^3)
NaCl	774	−760	35.7 (0°C)
LiF	1032	−1005	0.3 (18°C)
KF	813	−819	92.3 (18°C)
RbF	776	−792	130.6 (18°C)
$SrCl_2$	2110	−2161	53.8 (20°C)
AgCl	916	−851	8.9×10^{-5} (10°C)

and the solution process is only slightly endothermic. Sodium chloride is therefore water soluble. Although solubility cannot be analyzed with just a few simple factors, a major requirement is an exothermic or only slightly endothermic heat of solution. This means that solubility is favored when the energy required to break down the lattice (lattice energy) is smaller than or approximately equal to the energy given off when the ions are hydrated (enthalpy of hydration). This conclusion is illustrated by a few data in Table 12.1 where you will notice that the insoluble salt AgCl has a very endothermic enthalpy of solution.

EXERCISE 12.2 ENTHALPY OF SOLUTION

Using the data in Table 12.1, calculate the enthalpy of solution for AgCl and RbF. Comment on the relation between the enthalpy of solution and solubility for these two salts.

It is appropriate to try to *estimate* the enthalpy of solution from the lattice energy and heats of hydration of the gaseous ions, since this process can be visualized. One can "see" a crystal lattice "flying apart" into ions in the gas phase and then these ions plunging into water. Unfortunately, only a few of the necessary thermodynamic quantities are available, and so we are interested in a more general method of obtaining $\Delta H^\circ_{solution}$. Consider again the process of dissolving NaCl(s).

$$\text{NaCl(s)} \longrightarrow \text{NaCl(aq)}$$

It is evident that the enthalpy of solution can be estimated from

$$\Delta H^\circ_{solution} = \Delta H^\circ_f[\text{NaCl(aq)}] - \Delta H^\circ_f[\text{NaCl(s)}]$$

These values are available from the tables published by the U.S. National Bureau of Standards.

$$\Delta H^\circ_f[\text{NaCl(aq)}] = -407.1 \text{ kJ/mol} \qquad \Delta H^\circ_f[\text{NaCl(s)}] = -411.0 \text{ kJ/mol}$$

and they give a value of $\Delta H^\circ_{solution}$ of +3.9 kJ, an endothermic value just as was that estimated from the data in Table 12.1. Some data that you can use to calculate solution enthalpies for other compounds are given in Table 12.2.

EXERCISE 12.3 CALCULATING ENTHALPY OF SOLUTION

Using the data in Table 12.2, calculate the enthalpy of solution of ammonium nitrate.

TABLE 12.2 Data to Calculate Enthalpy of Solution

COMPOUND	ΔH_f°(s) (kJ/mol)	ΔH_f°(aq, 1 m) (kJ/mol)
LiF	−616.0	−611.1
KF	−526.3	−585.0
RbF	−557.7	−583.8
KCl	−436.7	−419.5
RbCl	−435.4	−418.3
NaOH	−425.6	−470.1
NH_4NO_3	−365.6	−339.9

In Exercise 12.3 you calculated the enthalpy of solution of ammonium nitrate in water, and you should have found it to be quite positive. The energy required to dissolve NH_4NO_3 comes from the surroundings, so the water cools dramatically when the solid solute is added (Figure 12.5). This is the basis of "chemical cold packs" that you may have seen in a hospital or at a sporting event. A typical cold pack consists of a sealed bag of water inside a bag of ammonium nitrate. When the bags are squeezed firmly, the inner pouch breaks and water and ammonium nitrate mix. The endothermic solution process causes the water temperature to drop 5°C or so.

Dissolving NH_4NO_3 is highly endothermic.

$$NH_4NO_3(s) + \text{heat} + H_2O(\ell) \longrightarrow NH_4NO_3(aq)$$

FACTORS AFFECTING SOLUBILITY: PRESSURE AND TEMPERATURE

Geologists and biochemists, among others, are interested in the solubility of gases such as CO_2 and O_2 in water, and all scientists need to know

(a)

(b)

FIGURE 12.5

An endothermic solution process. (a) Hydrated barium hydroxide [$Ba(OH)_2 \cdot 8\ H_2O$] and ammonium nitrate (NH_4NO_3) are mixed in a flask, and an acid–base reaction occurs.

$$Ba(OH)_2 \cdot 8\ H_2O(s) + 2\ NH_4NO_3(s) \longrightarrow Ba(NO_3)(s) + 2\ NH_3(aq) + 10\ H_2O(\ell)$$

Excess ammonium salt dissolves in the water produced in the reaction. (b) The dissolving process is so endothermic that if the flask is placed on a wet, wooden block the water freezes and the block is attached to the flask. (Charles D. Winters)

about the solubility of solids in liquids. Two external factors controlling such processes are temperature and gas pressure. Both can affect the solubility of gases in liquids, whereas only temperature is important in determining the solubility of solids in liquids.

Henry's Law holds strictly for gases that do not interact chemically with the solvent (such as NH_3 that forms NH_4^+ and OH^-).

PRESSURE EFFECTS: HENRY'S LAW The effect of the pressure of a gas on its solubility in a liquid is often dramatic. It is governed by **Henry's law**, Equation 12.4, which states that the amount of gas that dissolves in a given amount of liquid at a specified temperature depends directly on the partial pressure of the gas above the liquid.

$$\text{Molality of solute gas} = kP \text{ (at constant temperature)} \quad (12.4)$$

P = partial pressure of the gaseous solute

k = Henry's Law constant, a constant *characteristic of the solute and solvent*

Carbonated soft drinks are good examples of Henry's law. They are packed under pressure in a chamber filled with carbon dioxide gas, some of which dissolves in the drink. When you open the can, the partial pressure of CO_2 above the solution drops, so the concentration of CO_2 in solution also drops, and gas bubbles out of solution (Figure 12.6). The same can happen with gases dissolved in your blood if you are an underwater diver. Nitrogen is soluble in blood, so if you breathe a high pressure mixture of O_2 and N_2 deep under water, there can be an appreciable concentration of N_2 in your blood system. If you ascend too rapidly, the nitrogen is released as the pressure decreases and forms bubbles in the blood. This affliction, sometimes called the "bends," is painful and often fatal because blood is forced out of capillaries, and the brain, for example, can be deprived of oxygen. To partly circumvent this problem, deep sea divers sometimes use a helium/oxygen mixture instead of nitrogen/oxygen, because helium is not nearly as soluble in blood as nitrogen.

FIGURE 12.6
Illustration of Henry's Law. The greater the partial pressure of CO_2 over the soft drink in a can of soda, the greater the amount of CO_2 dissolved. There is more CO_2 dissolved in the closed can than in the can open to the atmosphere. (Charles D. Winters)

EXAMPLE 12.3

USING HENRY'S LAW

Oxygen has a Henry's law constant of 1.7×10^{-6} molal/mm Hg when dissolved in water at 25°C. What is the concentration of O_2 in water at 25°C when O_2 has a partial pressure of 150 mm Hg? (This is approximately its partial pressure in dry air.)

Solution From Henry's law, we have

$$\text{Molality of } O_2 = (1.7 \times 10^{-6} \text{ molal/mm Hg})(150 \text{ mm Hg}) = 2.6 \times 10^{-4}\ m$$

EXERCISE 12.4 USING HENRY'S LAW

Henry's law constant for CO_2 in water at 25°C is 4.44×10^{-5} molal/mm Hg. What is the concentration of CO_2 in water when the CO_2 partial pressure is one third of an atmosphere?

TEMPERATURE EFFECTS: LE CHATELIER'S PRINCIPLE You know from experience that much more gas is released from a can of soda that is opened when it is warm rather than when it is cold. Sometimes the gas bubbles out of solution so rapidly that soda gushes out of the can. This illustrates the fact that *temperature affects solubility*.

Gases dissolving in solvents are exothermic processes; that is, the enthalpy of solution is negative.

$$\text{Gas} + \text{liquid solvent} \longrightarrow \text{saturated solution} + \text{heat}$$

The process continues until a saturated solution is formed and *equilibrium* is achieved. At this point, the dissolving process continues, but it is counterbalanced by gas molecules coming out of solution with the consumption of heat.

$$\text{Saturated solution} + \text{heat} \longrightarrow \text{gas} + \text{liquid solvent}$$

At equilibrium, as many molecules come out of solution as dissolve in a given time. We depict this situation by showing the original process with arrows running in both directions.

$$\text{Gas} + \text{liquid solvent} \rightleftharpoons \text{saturated solution} + \text{heat}$$

Henri Le Chatelier (1850–1936) was a French chemist who developed his concept of effects on equilibrium in connection with the chemistry of cement.

To understand how temperature affects solubility, we introduce **Le Chatelier's principle**. This states that a change in any of the factors determining an equilibrium will cause the system to adjust in order to reduce or counteract the effect of change. For example, if we heat a solution of a gas in a liquid, the equilibrium position will shift in an attempt to use up some of the added heat. That is, the reaction

The factors that can determine an equilibrium are concentration, pressure, volume (for gases), and temperature. Only temperature effects are considered here. Chapters 14 to 17 consider all factors in more detail.

$$\text{Gas} + \text{liquid solvent} \rightleftharpoons \text{saturated solution} + \text{heat}$$

add heat
shift left

will shift to the left, back to free gas molecules and pure solvent. This is why a dissolved gas always becomes less soluble with increasing temperature (see Table 11.3).

When solids dissolve in liquids, the process can be endothermic, just as you saw with NaCl above,

$$\text{Heat} + \text{solid} + \text{liquid solvent} \longrightarrow \text{saturated solution}$$

Here, when equilibrium is reached, as many molecules come out of solution as dissolve in a given time. If adding heat is the change made, the reaction will be affected as the heat is used, and Le Chatelier's principle predicts that the solubility of such salts will increase with increasing temperature.

To understand Le Chatelier's principle, think of a reaction as a tube of toothpaste. If you squeeze on one side, the "reaction" pops up on the other. If you "squeeze" an endothermic process by adding heat, the reaction moves to give more of the material on the product side.

$$\text{Heat} + \text{solid} + \text{liquid solvent} \rightleftharpoons \text{saturated solution}$$

add heat
shift right

This is true for many common ionic compounds, as illustrated in Figure 12.7.

FIGURE 12.7
The temperature dependence of the solubility of some ionic compounds in water.

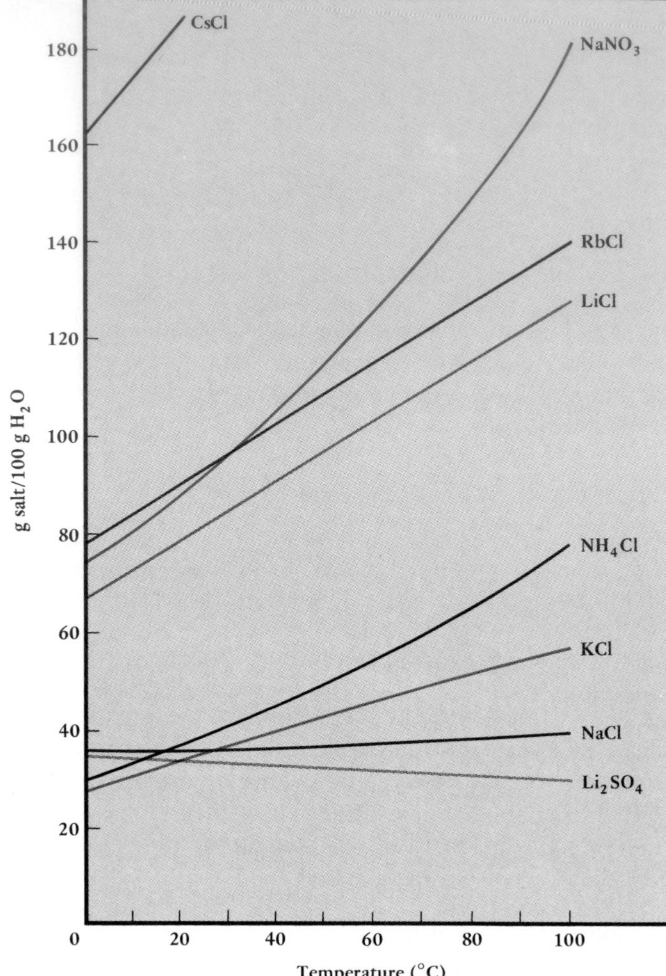

There are also some salts that have a negative enthalpy of solution (the process is *exo*thermic).

Solid + liquid solvent $\longrightarrow$ saturated solution + heat

and they become less soluble as the temperature increases; the solubility curve of Li_2SO_4 seen in Figure 12.7 is one example of this behavior.

In view of the discussion in the previous chapter of trends in lattice energy (see Figure 11.34), it is interesting to notice the trend in solubilities of the alkali metal chlorides in Figure 12.7. In general, as the lattice energy of the salts declines, their solubility increases. Thus, the solubilities increase in the order NaCl < KCl < RbCl < CsCl. Lithium chloride is clearly an exception. In fact, lithium salts often have properties that do not reflect the general trends in Group 1A.

12.3 COLLIGATIVE PROPERTIES

When sodium chloride is dissolved in water, the vapor pressure of the solution is different from that of pure water, as is its melting point, boiling point, and osmotic pressure. These changes between pure solvent and solution are called the **colligative properties** of the solution.

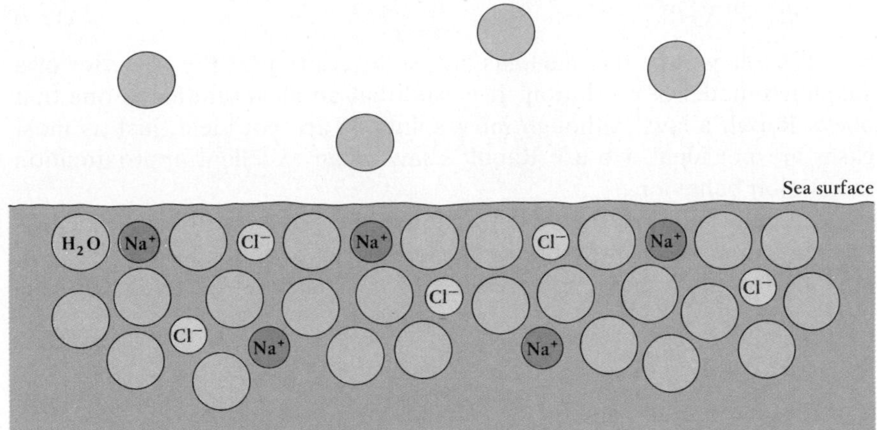

FIGURE 12.8

Seawater is an aqueous solution of sodium chloride and many other salts. The vapor pressure of water over an aqueous solution is not as large as the vapor pressure of water over pure water at the same temperature.

We shall begin this discussion by studying the effect of a nonvolatile solute (such as salt or sugar) on the vapor pressure of the solvent, and this will lead us to an understanding of other colligative properties.

CHANGES IN VAPOR PRESSURE: RAOULT'S LAW

At the surface of a solution (Figure 12.8) there are molecules of water as well as ions from the dissolved salts. Water molecules can leave the liquid and enter the gas phase, exerting a vapor pressure. However, there are not as many water molecules at the surface as in pure water, because some molecules have been displaced by dissolved ions. Therefore, not as many water molecules are available to leave the liquid surface, and the vapor pressure is lower than that of the pure water at a given temperature. From this analysis, it should make sense that the vapor pressure of the solvent, P_{solvent}, will be proportional to the relative number of solvent molecules in a solution, that is, to their mole fraction. If $P_{\text{solvent}} \propto X_{\text{solvent}}$, then we have

$$P_{\text{solvent}} = KX_{\text{solvent}} \tag{12.5}$$

(where K is a constant). This equation tells you that, if there are only half as many solvent molecules present at the surface of a solution as at the surface of the pure liquid, then the vapor pressure of the solvent in a solution will be only half as great as that of the pure solvent at the same temperature.

If we are dealing only with pure solvent, then we can write

$$P^{\circ}_{\text{solvent}} = KX_{\text{solvent}}$$

where $P^{\circ}_{\text{solvent}}$ is the vapor pressure of the pure solvent. Since X_{solvent} is 1 in pure solvent, this means $P^{\circ}_{\text{solvent}} = K$; the vapor pressure of pure solvent is a constant. Substituting for K in Equation 12.5, we arrive at an equation called **Raoult's law**.

$$P_{\text{solvent}} = X_{\text{solvent}}P^{\circ}_{\text{solvent}} \tag{12.6}$$

If the solution contains more than one volatile component, then Raoult's law can be written for any one such component, A, as

$$P_A = X_A P_A^\circ \tag{12.7}$$

Raoult's law is, like the ideal gas law, a description of the behavior of a simplified model of a solution. It is said that an **ideal solution** is one that obeys Raoult's law. Although most solutions are not ideal, just as most gases are not ideal, we use Raoult's law as an excellent approximation to solution behavior.

In any solution, the mole fraction of the solvent will always be less than 1, so the vapor pressure of the solvent over the solution ($P_{solvent}$), must be less than the vapor pressure of the pure solvent ($P^\circ_{solvent}$). This vapor pressure lowering, $\Delta P_{solvent}$, is given by

$$\Delta P_{solvent} = P^\circ_{solvent} - P_{solvent} \qquad \text{where } P_{solvent} < P^\circ_{solvent}$$

Substituting Raoult's law for the second term, we now have

$$\Delta P_{solvent} = P^\circ_{solvent} - (X_{solvent} P^\circ_{solvent})$$

or

$$\Delta P_{solvent} = (1 - X_{solvent}) P^\circ_{solvent}$$

In a solution that has only the solvent and one nonvolatile solute, the sum of the mole fractions of solvent and solute must be equal to 1.

$$X_{solvent} + X_{solute} = 1$$

Therefore, $1 - X_{solvent} = X_{solute}$, and the equation for $\Delta P_{solvent}$ can be rewritten as

$$\Delta P_{solvent} = X_{solute} P^\circ_{solvent} \tag{12.8}$$

That is, the *lowering of the vapor pressure of the solvent is proportional to the mole fraction (the relative number of particles) of solute*. This expression and Raoult's law from which it derived are two of the most useful ways to describe the physical properties of solutions.

EXAMPLE 12.4

USING RAOULT'S LAW

A large business in the northeastern United States is making maple syrup. In the spring the sugar maple trees are tapped for their sap. This sap is boiled to evaporate most of the water and leave a concentrated solution of sweet, tasty maple syrup. Although the components of this solution are complex, let us assume that the sap from the tree is just water and sucrose ($C_{12}H_{22}O_{11}$). If the water is evaporated from this solution until it contains 50.0% sucrose by mass, what is the vapor pressure of the water over the solution at 60°C? (The vapor pressure of pure water at 60°C is 149 mm Hg.) (Although it is a poor assumption here, we shall assume this is an ideal solution.)

Solution To calculate the new vapor pressure using Raoult's law, the mole fraction of water needs to be determined first. If the solution is 50.0% sugar by mass, 100. g of solution contains 50.0 g of sucrose and 50.0 g of water.

$$\text{Moles of sugar} = 50.0 \text{ g} \left(\frac{1 \text{ mol}}{342 \text{ g}}\right) = 0.146 \text{ mole}$$

$$\text{Moles of water} = 50.0\text{ g}\left(\frac{1\text{ mol}}{18.0\text{ g}}\right) = 2.78\text{ moles}$$

The mole fraction of water can now be determined.

$$X_{water} = \frac{2.78\text{ mol H}_2\text{O}}{2.78\text{ mol H}_2\text{O} + 0.146\text{ mol sugar}} = 0.950$$

Finally, the vapor pressure of the water over the maple syrup can be calculated from Raoult's law (Equation 12.6).

$$P_{water} = X_{water}\,P^{\circ}_{water} = (0.950)(149\text{ mm Hg}) = 142\text{ mm Hg}$$

$$\Delta P_{water} = 149\text{ mm Hg} - 142\text{ mm Hg} = 7\text{ mm Hg}$$

One of the uses of colligative properties in the chemistry laboratory is the determination of the molar mass of a compound. The following example shows you how this is done in one case.

EXAMPLE 12.5

CALCULATING A MOLAR MASS USING RAOULT'S LAW

There is an aluminum-containing compound of the type $[Al(CH_3)_3]_x$ where x is some whole number (1, 2, 3, . . .). The molar mass of the compound can be determined by a Raoult's law experiment. For example, pure benzene, used as the solvent in the experiment, has a vapor pressure of 95.00 mm Hg at 25°C. When 0.144 g of the aluminum compound is dissolved in 10.00 g of benzene, the vapor pressure is now 94.27 mm Hg. Assuming that the solution is ideal (and the aluminum-containing compound does not contribute to the vapor pressure), calculate the molar mass of $[Al(CH_3)_3]_x$ and determine x. (Benzene is C_6H_6.)

Solution Knowing the vapor pressure of the pure solvent and that of the solvent over the solution, you can calculate the mole fraction of the solvent.

$$P^{\circ}_{benzene} = 95.00\text{ mm Hg} \qquad P_{benzene} = 94.27\text{ mm Hg}$$

From Raoult's law you have: $X_{benzene} = P_{benzene}/P^{\circ}_{benzene}$

$$X_{benzene} = 0.9923$$

The mole fraction of benzene is defined by moles benzene/total moles. That is,

$$X_{benzene} = 0.9923 = \frac{\dfrac{10.00\text{ g}}{78.11\text{ g/mol benzene}}}{\text{moles Al compd} + \dfrac{10.00\text{ g}}{78.11\text{ g/mol benzene}}}$$

If this expression is solved for "moles of Al compd," you find

$$\text{Moles of Al compound} = 0.0009934$$

This means that the molar mass of the compound is

$$\frac{0.144\text{ g}}{0.0009934\text{ moles}} = 145.0\text{ g/mol}$$

Since $Al(CH_3)_3$ has a mass of 72.10 g per formula unit, the value of x must be 2, and the compound is $[Al(CH_3)_3]_2$.

EXERCISE 12.5 USING RAOULT'S LAW

The vapor pressure of pure water at 90°C is 526 mm Hg. If you add 10.0 g of sugar ($C_{12}H_{22}O_{11}$) to 225 mL (225 g) of water, what is the vapor pressure of the water? (In essence, you will determine the drop in vapor pressure that occurs when you add a heaping teaspoonful of sugar to a cup of tea or coffee.)

When solutions are very dilute, we can make a useful *approximation* to Equation 12.8. For a *dilute* solution (solute concentration $< 0.1\ m$), the change in vapor pressure when a solute is added to a solvent is proportional to the molality of the solution, and the proportionality constant K_{vap} is characteristic of the solvent.

$$\Delta P_{solvent} = K_{vap} m_{solute} \qquad \text{where } K_{vap} = P^\circ_{solvent}\left(\frac{M_{solvent}}{1000}\right) \qquad (12.9)$$

$$\text{and } M_{solvent} = \text{solvent molar mass}$$

This equation is useful because it is directly related to our next topics, the changes that occur in the boiling point and freezing point of a liquid when a solute is added.

EXAMPLE 12.6

CHANGES IN VAPOR PRESSURE

Using Equation 12.8, calculate the change in vapor pressure of the solvent benzene (C_6H_6) at 60°C when 0.200 mole of a nonvolatile solute is added to 100. g of benzene. The vapor pressure of benzene at 60°C is 400. mm Hg. Compare this calculation with the result from the more approximate form of Raoult's law given by Equation 12.9.

Solution Equation 12.8 gives the change in vapor pressure as $\Delta P_{solvent} = X_{solute}P^\circ_{solvent}$. To use this equation, you must first calculate X_{solute}. Since 100. g of benzene is equivalent to 1.28 moles,

$$X_{solute} = \frac{0.200 \text{ mol solute}}{0.200 \text{ mol solute} + 1.28 \text{ mol benzene}} = 0.135$$

this means

$$\Delta P_{solvent} = (0.135)(400.\text{ mm Hg}) = 54.1 \text{ mm Hg}$$

If Equation 12.9 is used instead to *estimate* $\Delta P_{solvent}$, you will find since

$$K_{vap} = P^\circ_{solvent}\,(M_{solvent}/1000) = 31.2 \text{ mm Hg/molal}$$

then

$$\Delta P_{solvent} = (31.2 \text{ mm Hg/molal})(2.00\ m) = 62.4 \text{ mm Hg}$$

The agreement between Equation 12.8 and its approximate form is rather poor for this relatively concentrated solution. (Try redoing the calculation for a more dilute solution (0.200 m in benzene) and comment on the agreement.)

EXERCISE 12.6 USING RAOULT'S LAW TO CALCULATE VAPOR PRESSURE CHANGES

Nitroglycerin, $C_3H_5(NO_3)_3$, is used in the manufacture of dynamite and in the treatment of certain coronary problems. If you add 0.454 g of the compound to 100. g of benzene at 25°C, what change in vapor pressure should be observed? (The vapor pressure of pure benzene at 25°C is 95 mm Hg.)

BOILING POINT ELEVATION

Raoult's law (Equation 12.6) tells you that the vapor pressure of a solution must be lower than that of the pure solvent. Assume that 0.200 mole of a nonvolatile solute is added to 100. g of the volatile solvent benzene (C_6H_6). (For a similar case, see Example 12.6). Using Raoult's law you should find that the vapor pressure of benzene at 60°C will drop from 400 mm Hg to 346 mm Hg. We have marked this point on the vapor pressure curve in Figure 12.9. Now, what happens to the vapor pressure when the temperature of the solution is raised another 10 degrees? $P^\circ_{benzene}$ becomes larger with increasing temperature, so $P_{benzene}$ must also become larger. The equilibrium vapor pressure of the solution will be defined by the lower curve in Figure 12.9.

Let us explore the consequences of the lowering of solvent vapor pressure due to a nonvolatile solute. Recall that the *normal boiling point* of a volatile liquid is the temperature at which its vapor pressure is equal to 1 atm or 760 mm Hg. Since the vapor pressure of a solvent is lowered

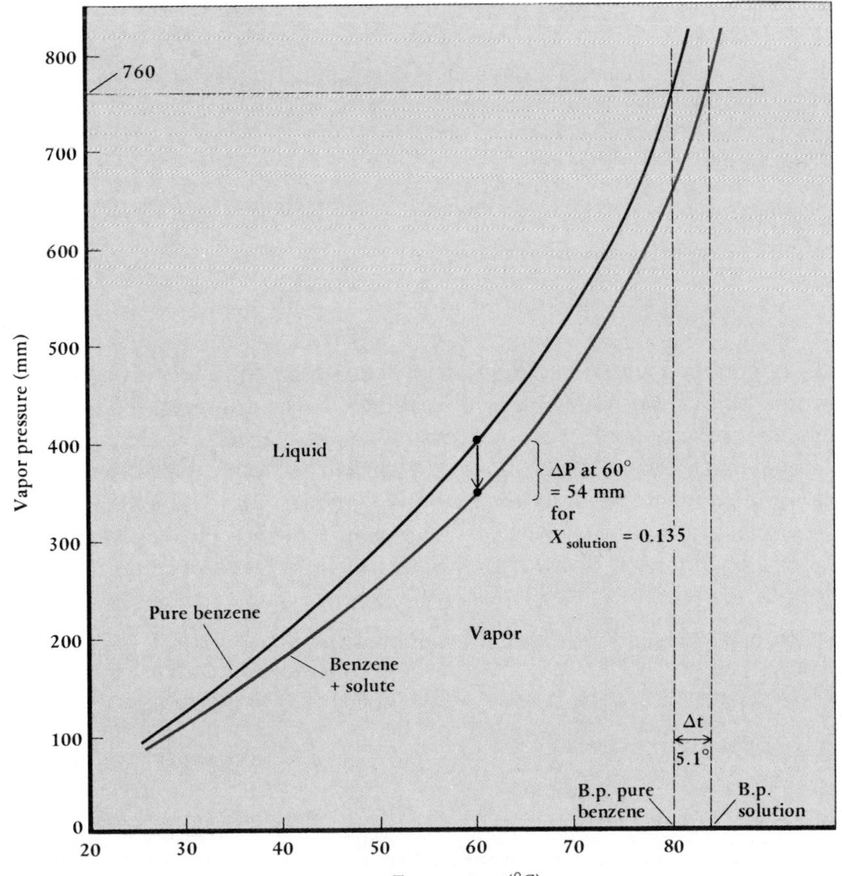

FIGURE 12.9

Lowering of the vapor pressure of a volatile solvent, benzene, by addition of a nonvolatile solute. See Example 12.6 for details. At 60°C the vapor pressure of the benzene was lowered by 54 mm Hg from 400 to 346 mm Hg. At the normal boiling point of benzene, 80.1°C, the vapor pressure is lowered from 760 to 657 mm Hg. To boil, the vapor pressure of the benzene must be 760 mm, so the solution boils 5.1°C higher than the pure solvent, that is, at 85.2°C.

on adding a solute, this must mean that *the boiling point of the solution is elevated relative to that of the pure solvent*. The change is shown graphically in Figure 12.9.

How large is the elevation in boiling point? You previously saw that the change in vapor pressure of the solvent in a dilute solution, ΔP, is given by a constant times the solution molality ($\Delta P = K_{vap}m$). Since the vapor pressure of a liquid and its boiling point are related, it makes sense that the *elevation* of the boiling point of a solvent and the *change* in its vapor pressure on adding a solute are related ($\Delta t_{bp} \propto \Delta P$). This must mean, then, that the change in boiling point of a liquid on adding a solute is proportional to the number of solute particles.

The proportionality constants K_{bp} and K_{vap} are not equal to each other.

$$\text{Elevation in boiling point} = \Delta t_{bp} = K_{bp}m_{solute} \quad (12.10)$$

where K_{bp}, the **boiling point elevation constant**, is characteristic of the solvent (Table 12.3).

EXAMPLE 12.7

BOILING POINT ELEVATION

In Example 12.5 you learned that the aluminum compound had a formula of $[Al(CH_3)_3]_2$. What is the boiling point of a solution of 0.144 g of this compound dissolved in 10.0 g of benzene?

Solution You cannot calculate the boiling point directly, only its change. Since we know K_{bp}, we only need to calculate the molality m. In this case you have 0.00100 mole of aluminum compound and 0.0100 kg of solvent, so the solution is 0.100 molal, and

$$\Delta t_{bp} = 2.53 \text{ deg/molal}(0.100 \text{ molal}) = 0.253°\text{C}$$

Since the boiling point is *raised* relative to that of the pure solvent, the boiling point of the solution is 80.1°C + 0.253°C = 80.4°C.

A common use of colligative properties in the laboratory is the determination of the molar mass of a solute. If the compound is soluble in a solvent of appreciable vapor pressure, or with a reasonable K_{bp}, then the molar mass can be determined, at least in principle. Unfortunately, there are disadvantages to both methods. To have an ideal solution, the solute concentration must be low (say under 0.10 m). However, as you learned in Example 12.6, the lowering of the vapor pressure can be quite

TABLE 12.3 Some Boiling Point Elevation and Freezing Point Depression Constants

SOLVENT	NORMAL Bp PURE SOLVENT °C	K_{bp} (deg/molal)	NORMAL Fp PURE SOLVENT °C	K_{fp} (deg/molal)
Water	100.0	+0.512	0.0	−1.86
Benzene	80.1	+2.53	5.5	−4.90
Camphor	—	—	179.75	−40.0
Naphthalene	—	—	80.2	−6.8

small for a dilute solution, and this means that the measurement of $\Delta P_{\text{solvent}}$ is difficult. It is easier to measure small changes in boiling point, but the disadvantage of determining molar mass in a boiling solvent is that the compound must be nonvolatile at the boiling point and must be stable to heat. Nonetheless, both methods are used.

EXAMPLE 12.8

DETERMINING MOLAR MASS BY ELEVATION IN BOILING POINT

Some beautiful blue crystals of azulene (0.640 g) were dissolved in 100. g of benzene. The boiling point of the solution was 80.23°C. Calculate the molar mass of azulene.

Solution From the elevation in boiling point (and K_{bp} from Table 12.3), you can first determine the molality of the solution.

$$m = \frac{\Delta t_{bp}}{K_{bp}} = \frac{(80.23 - 80.10)°\text{C}}{2.53\ \text{deg/molal}} = 0.051\ m$$

Knowing the molality, you can then calculate the molar mass.

$$\left(\frac{0.051\ \text{mol}}{1.00\ \text{kg solvent}}\right)(0.100\ \text{kg solvent}) = 0.0051\ \text{mol azulene}$$

$$\frac{0.640\ \text{g azulene}}{0.0051\ \text{mol}} = 125\ \text{g/mol}$$

Azulene, whose formula is $C_{10}H_8$, actually has a molar mass of 128.2 g/mol. As you see, we have experimentally *estimated* its molar mass with an error of about 2%. In real laboratory work, this would represent a very good determination.

EXERCISE 12.7 DETERMINING MOLAR MASS BY ELEVATION IN BOILING POINT

Determine the molar mass of oil of wintergreen (methyl salicylate) from the following data: 1.25 g of the oil are dissolved in 100. g of benzene. The boiling point of the solution is 80.31°C.

One common application of boiling-point elevation is the summer protection your car's engine receives from "all-season" antifreeze. Among other compounds, the main ingredient of commercial antifreeze is 1,2-ethanediol (HOC_2H_4OH), commonly called ethylene glycol. The car's radiator and cooling system are sealed to keep the coolant under pressure, so that it will not vaporize at normal engine temperatures. However, when the air temperature is high in the summer, the radiator could still "boil over" if it were not protected with antifreeze. By adding antifreeze, you make a solution that has a higher boiling point and can be used even under severe weather conditions.

FREEZING POINT DEPRESSION

Another consequence of dissolving a solute in a solvent is that the freezing point of the solution is lower than that of the pure solvent. For an ideal solution, the depression of the freezing point is given by an equation similar to that for the elevation of the boiling point:

$$\text{Freezing point depression} = \Delta t_{fp} = K_{fp} m_{solute} \qquad (12.11)$$

where K_{fp} is the **freezing point depression constant**. A few values of this constant are given in Table 12.3. Notice especially the constant for camphor. Although camphor is a solid at room temperature, when melted it is a good solvent for many compounds (except salts). Its large constant means that even a very dilute solution will exhibit a large change in freezing point. This makes camphor a good choice for reasonably accurate molar mass determinations.

EXAMPLE 12.9

USING FREEZING POINT DEPRESSION FOR THE DETERMINATION OF MOLAR MASS

Calculate the molar mass of azulene (see Example 12.8) from the following information: 0.640 g of azulene is dissolved in 100. g of camphor. The freezing point of the solution is 177.75°C.

Solution The freezing point of the solution is 2.00°C lower than that of pure camphor.

$$\begin{aligned}\Delta t_{fp} &= \text{fp of solution} - \text{fp of pure camphor}\\ &= 177.75°\text{C} - 179.75°\text{C} = -2.00°\text{C}\end{aligned}$$

This means the molality of the solution is

$$m = \frac{\Delta t_{fp}}{K_{fp}} = \frac{-2.00°\text{C}}{-40.0 \text{ deg/molal}} = 0.0500 \text{ molal}$$

and so the number of moles of azulene must be

$$\text{Moles azulene} = \left(\frac{0.0500 \text{ mol}}{1.00 \text{ kg}}\right) 0.100 \text{ kg} = 0.00500 \text{ mol azulene}$$

Thus,

$$\text{Molar mass of azulene} = \frac{0.640 \text{ g}}{0.00500 \text{ mol}} = 128 \text{ g/mol}$$

As a final point, compare the size of the freezing point depression of 0.640 g of azulene in camphor with the size of the boiling point elevation in benzene. You can appreciate that it is easier to measure a 2.00°C change than one of only 0.13°C.

EXERCISE 12.8 DETERMINING MOLAR MASS BY FREEZING POINT DEPRESSION

What is the molar mass of oil of wintergreen if 1.25 g of the oil, when dissolved in 100. g of benzene, produce a solution with a freezing point of 5.10°C? (Compare the freezing point depression here with the boiling point elevation in Exercise 12.7. Considering ease of measurement, which may be the better method to use in the laboratory?)

The practical aspects of depressing the freezing point are similar to those of the elevation of boiling point. The very name of the liquid that you add to the radiator in your car, antifreeze, indicates its purpose. The label on a container of antifreeze tells you, for example, to add 6 quarts of antifreeze to a 12-quart cooling system in order to lower the freezing point to −34°F and raise the boiling point to +226°F.

EXAMPLE 12.10

FREEZING POINT DEPRESSION

How many grams of ethylene glycol [$C_2H_4(OH)_2$] must be added to 5.50 kg of water to lower the freezing point of the water from 0.0° to −10.0°C? (This is approximately the situation in the radiator of your car.)

Solution The solution concentration and the amount by which the freezing point is depressed are related by $\Delta t_{fp} = K_{fp}m$. Therefore, the molality of the solution having a freezing point depression of −10.0° must be

$$m = \frac{-10.0°\text{C}}{-1.86 \text{ deg/molal}} = 5.38 \text{ molal} = \frac{5.38 \text{ mol glycol}}{1.00 \text{ kg H}_2\text{O}}$$

Since you have 5.50 kg of water, the amount of antifreeze that must be dissolved is

$$(5.50 \text{ kg})(5.38 \text{ mol/kg}) = 29.6 \text{ moles of C}_2\text{H}_4(\text{OH})_2$$

The molar mass of $C_2H_4(OH)_2$ is 62.1 g/mol, so the mass required is

$$29.6 \text{ moles} \left(\frac{62.1 \text{ g}}{1 \text{ mol}}\right) = 1840 \text{ g}$$

The density of ethylene glycol is 1.1 kg/L, so the volume of antifreeze to be added is 1.8 kg/(1.1 kg/L) = 1.7 L.

WHY IS A SOLUTION FREEZING POINT DEPRESSED?*

Imagine the freezing process this way. When the temperature drops to the freezing point of a pure solvent, freezing begins with a few molecules clustering together to form a tiny amount of solid. More molecules of the liquid move to the surface of the solid, and the solid grows. Recall that a quantity of heat, the heat of fusion, is evolved; as long as this heat is removed, solidification continues. However, if the heat is not removed, the opposing processes of freezing and melting can come into equilibrium; at this point, the number of molecules moving from solid to liquid is the same as the number moving from liquid to solid in a given time.

The heat of fusion of a liquid was discussed in Chapter 11.

But what happens in a freezing *solution*? Again, a few molecules of solvent cluster together to form some solid. More and more solvent molecules join them, and the solid phase, which is pure solid solvent, continues to grow as long as the heat of fusion is removed. At the same time some solvent molecules are returning to the liquid from the solid. The freezing and melting processes can come into equilibrium when the

*To keep the answer to our question simple, we shall consider the case where the solute does not form a solid solution with the solvent. This is generally true for water; very few substances form solid solutions with ice.

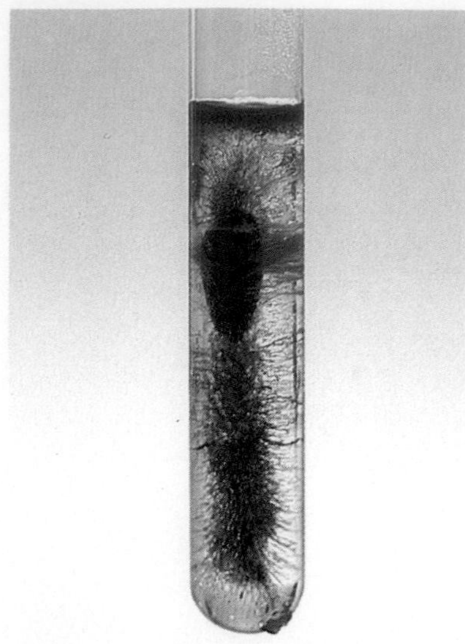

FIGURE 12.10

A purple dye was dissolved in water and the solution frozen slowly. Pure ice formed, and the solution became more concentrated, until equilibrium was reached. (Charles D. Winters)

number of molecules moving in each direction is the same in a given time. However, there is a problem. The liquid layer next to the solid contains solute molecules or ions, whereas the solid is pure solvent. [This is analogous to the situation at the solution/vapor interface in Figure 12.8.] Therefore, if the temperature is held at the normal freezing point of the pure solvent, the number of molecules of the solid (pure solvent) entering the liquid phase must be greater than the number of solvent molecules leaving the solution and depositing on the solid, in a given time. Why? For the same reason the vapor pressure of a solution is lower than that of the pure solvent: solute molecules have replaced some solvent molecules at the liquid/vapor or liquid/solid interface. Thus, in order to have the same number of solvent molecules moving in each direction (solid → liquid and liquid → solid) in a given time, the temperature must be lowered to slow down movement from solid to liquid. That is, the freezing temperature of a solution must be less than that of the pure liquid solvent.

As a solution freezes, solvent molecules are removed from the liquid phase and are deposited on the solid. This means that the concentration of the solute in the liquid solution increases, and the solution freezing point further declines as a result. That is, *a solution does not have a sharply defined freezing point*. However, we usually take the freezing point of a solution as the point at which solid solvent crystals first begin to appear.

When aqueous solutions freeze, the fact that the solid is pure ice while the solution is more concentrated than before (Figure 12.10) can be put to some practical use. One way to purify a liquid solvent in the laboratory is to freeze it slowly, saving the solid (pure solvent) and throwing away any remaining solution (now more concentrated with impurities). Early Americans (and some contemporary ones as well) knew that a drink called "apple jack" can be made this way. Fermenting apple cider produces a small amount of alcohol. If the fermented cider is cooled, some of the water freezes to pure ice, leaving a solution higher in alcohol content.

DISSOCIATION OF IONIC SOLUTES AND COLLIGATIVE PROPERTIES

If you live in the northern United States, you know that you can put salt on snowy and icy roads or sidewalks to melt the snow and ice. If local absorption of sunlight warms the ice, then a small amount of liquid water forms and dissolves some salt. The freezing point of the water is lowered and more liquid water forms. Salts such as NaCl or $CaCl_2$ are used because they are inexpensive and dissolve readily in water. In addition, they liberate their ions when dissolved. Since all colligative properties depend on the number of particles dissolved, dissolving 1 mole of NaCl, for example, should have the same effect as dissolving 2 moles of sugar, a nonionic, nondissociating compound.

EXAMPLE 12.11

FREEZING POINT AND IONIC SOLUTIONS

Assuming that NaCl dissociates completely into its ions when dissolved in water, how much sodium chloride must be dissolved in 5.50 kg of water to lower the freezing point from 0°C to −10°C? How much sugar (sucrose, $C_{12}H_{22}O_{11}$) will accomplish the same result?

Solution From Example 12.10, you know that you will need 5.38 moles of dissolved particles per kilogram of solvent or a total of 29.6 moles in 5.50 kg. However, there is a difference between ethylene glycol and common salt. The ethylene glycol molecule stays intact in solution, whereas NaCl gives Na^+ and Cl^- ions.

$$NaCl(s) + H_2O(\ell) \longrightarrow Na^+(aq) + Cl^-(aq)$$

That is, *if* dissociation is complete, one mole of NaCl gives two moles of dissolved ions. This means that the required 29.6 moles of particles in solution will be made up of equal numbers of Na^+ and Cl^- ions.

$$29.6 \text{ moles of particles required} = \text{moles } Na^+ + \text{moles } Cl^-$$

Therefore, the number of moles of *undissociated* NaCl is $\frac{1}{2}$ of 29.6 or 14.8, and the mass of NaCl required is

$$(14.8 \text{ moles NaCl required})(58.5 \text{ g/mol}) = 866 \text{ g (or about 1.9 pounds)}$$

Unlike NaCl, sucrose dissolves in water without dissociating into ions. One mole of solid sucrose gives one mole of dissolved sucrose.

$$C_{12}H_{22}O_{11}(s) + \text{water} \longrightarrow C_{12}H_{22}O_{11}(aq)$$

To lower the freezing point of water 10°C, we calculated above that 29.6 moles of dissolved particles are required. Therefore, 29.6 moles of sugar are needed. This amounts to 10.1 kg or about 22 pounds! (Why do we need a greater mass of sugar than of ethylene glycol for the same freezing point depression?)

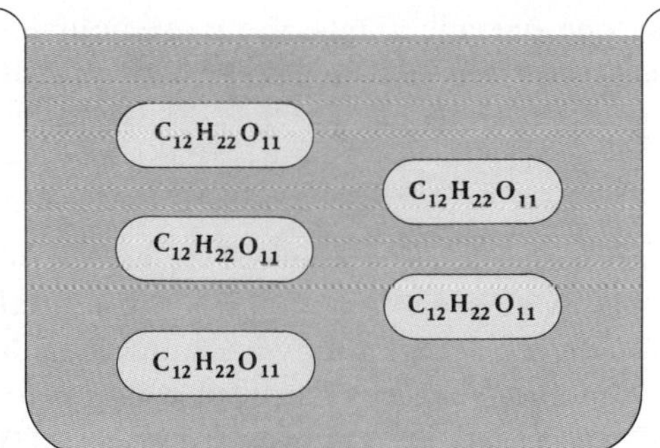

Sugar solution

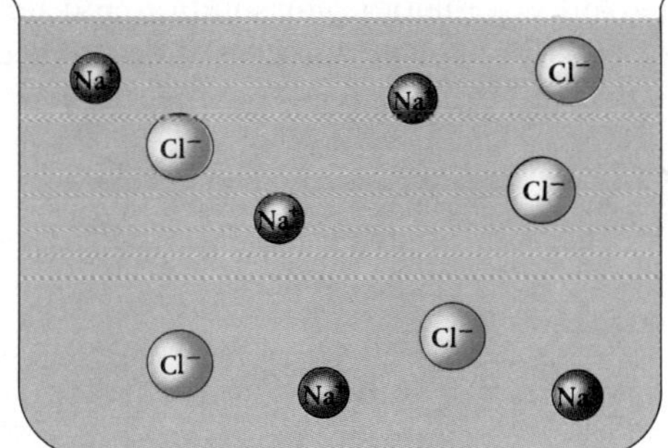

NaCl solution

Equal molar quantities of sugar and NaCl affect the freezing point (and boiling point) of a solution differently. For example, 5 moles of NaCl produce twice as many particles in solution as 5 moles of sugar.

EXERCISE 12.9 FREEZING AND BOILING POINTS OF SOLUTIONS OF IONIC COMPOUNDS

In Example 12.11, we found that 866 g of NaCl in 5.50 kg of water give a freezing point depression of 10°C. What is the boiling point of this solution?

EXERCISE 12.10 FREEZING AND BOILING POINTS OF SOLUTIONS OF IONIC COMPOUNDS

Calculate the freezing point of 500. g of water when (a) it contains 25.0 g of NaCl and (b) it contains 25.0 g of $CaCl_2$. Assume complete dissociation of both salts in water.

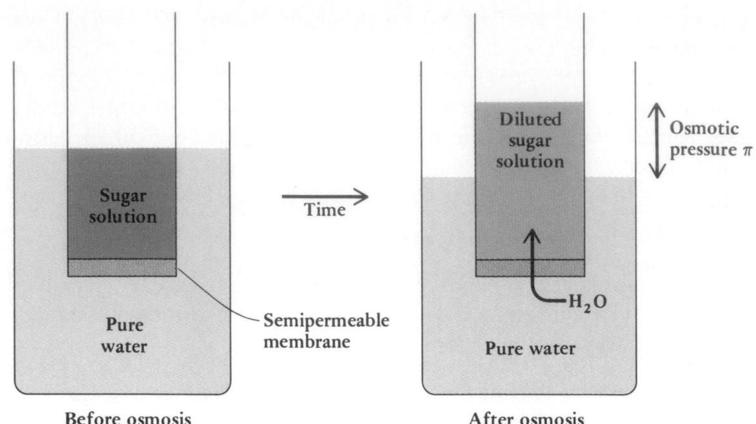

FIGURE 12.11

The process of osmosis. At the beginning of the experiment, a tube of aqueous sugar solution is placed in a beaker of pure water. The solution and pure water are separated by a membrane permeable only to water molecules and not sugar molecules. With time, water flows from a region of low solute concentration (pure water) to one of higher solute concentration (the sugar solution). Flow will continue until the *osmotic pressure* of water flowing into the solution is countered by the pressure of the column of solution in the tube above the water level in the beaker.

In the example and exercises above, we have *assumed* that the ionic salts *completely* dissociate in aqueous solution. Unfortunately, reality is not as simple: dissociation *may not* be complete. For example, suppose you dissolve 1.0 mole of KBr in 1.0 kg of water. The molality of both K^+ and Br^- is expected to be 1.0 *m*, giving a total molality of particles of 2.0 *m*. This is expected to give a freezing point of (2.0 molal) · (−1.86 deg/molal) or −3.72°C. Instead, the actual freezing point of the solution is closer to −3.29°C, a smaller freezing point depression than expected. The reason for this is that, at any instant, a few of the ions in solution form *ion-pairs*, species where the cation and anion "stick together" in solution and behave as one particle. Thus, there are actually fewer particles than if dissociation were complete, and the freezing point depression is slightly smaller than anticipated.

OSMOTIC PRESSURE

Osmosis can be demonstrated with a simple experiment. The beaker in Figure 12.11 contains pure water, and we assume there is a concentrated sugar solution in the tube. The liquids are separated by a **semipermeable membrane**, a thin sheet of material (such as vegetable tissue or cellophane) through which only certain types of molecules may pass; here, water

Ion-pairs may form in more concentrated solutions, so the freezing point depression and boiling point elevation are not as large as expected.

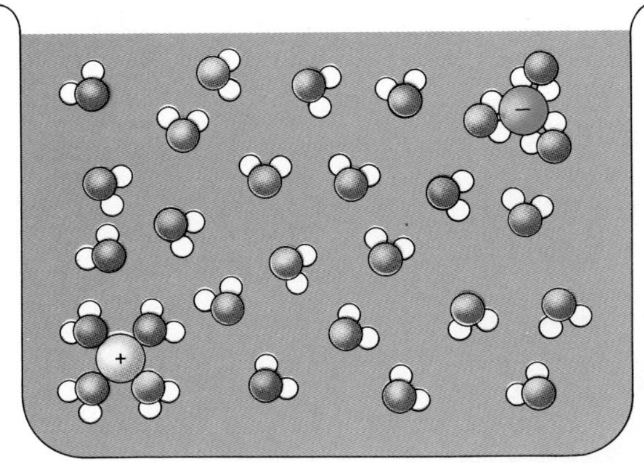
Dilute solution

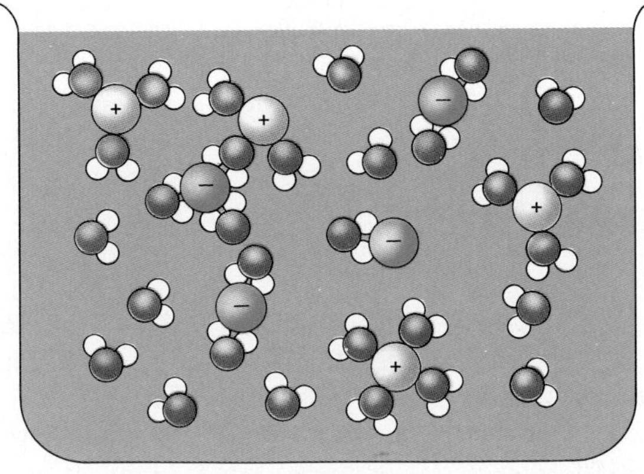
Concentrated solution

molecules can pass but larger sugar molecules cannot. When the experiment is begun, the liquid levels in the beaker and the tube are the same. With time, however, the level of the sugar solution inside the tube rises, the level of pure water in the beaker falls, and the sugar solution becomes steadily more dilute. After a while, the changes stop; an equilibrium state is reached.

Osmosis has occurred in the experiment in Figure 12.11. *Solvent molecules moved through the semipermeable membrane from a region of lower solute concentration to a region of higher solute concentration.* This is an important phenomenon, since it is one factor controlling the flow of material in and out of living cells.

From a molecular point of view, the semipermeable membrane does not present a barrier to the movement of water molecules, so they move through the membrane in both directions. However, on the sugar-solution side, there are not as many water molecules striking the membrane in a given time as on the pure-water side, because sugar molecules get in the way. Thus, in a given time, more water molecules pass through the membrane from pure water to sugar solution than in the opposite direction. In effect, water molecules tend to move from regions of high water concentration (low solute concentration) to regions of low water concentration (high solute concentration). The same is true for any solvent, as long as the membrane passes solvent molecules but not solute molecules.

The molecular-level description of osmosis should remind you of the discussion of the reason for vapor pressure lowering (Secton 12.3).

It may not be obvious why the system reaches equilibrium. After all, the solution in the tube can never reach zero sugar concentration, which would be required to equalize the number of water molecules moving through the membrane in each direction in a given time. The answer lies in the increasing mass of the sugar solution. As osmosis continues, water moves into the sugar solution, and the mass of the sugar solution will increase as the level of the solution in the tube rises higher and higher above the level of the water in the beaker. Eventually the pressure exerted by this mass of solution will counterbalance the pressure of the water moving through the membrane from the pure-water side, and there will be no further net movement of water. An equilibrium of forces is achieved. The force exerted by water molecules passing through the membrane when the system is at equilibrium is called the **osmotic pressure, Π**, and a measure of this pressure is the difference in height of the solution and the level of pure water in the beaker.

From experimental measurements on dilute solutions, it is known that osmotic pressure (in atmospheres) and concentration (M, in moles/liter) are related by the simple equation

$$\Pi = MRT \tag{12.12}$$

where R is the gas constant (0.0821 L · atm/K · mol) and T is the absolute temperature. This equation is analogous to the ideal gas law, with Π taking the place of P and M being equivalent to n/V.

According to the osmotic pressure equation, the pressure exerted by a 1 molar solution of particles at 25°C would be

$$\Pi = (1 \text{ mol/L})(0.0821 \text{ L} \cdot \text{atm/deg} \cdot \text{mol})(298 \text{ deg}) = 24.4 \text{ atm}$$

Since pressures on the order of 10^{-3} atm are easily measured, concentrations of about 10^{-4} M can be determined. This makes osmosis an ideal method for measuring the molar masses of very large molecules, many of which are of biological importance.

EXAMPLE 12.12

OSMOTIC PRESSURE AND MOLAR MASS

The osmotic pressure found for a solution of 5.0 g of horse hemoglobin in 1.0 L of water is 1.8×10^{-3} atm at 25°C. What is the molar mass of the hemoglobin?

Solution Solving Equation 12.12 for the solution concentration, we have

$$M = \frac{\Pi}{RT} = \frac{1.80 \times 10^{-3} \text{ atm}}{(0.082 \text{ L} \cdot \text{atm/K} \cdot \text{mol})(298 \text{ K})}$$
$$= 7.4 \times 10^{-5} \text{ mol/L}$$

The measured concentration is achieved by dissolving 5.0 g of hemoglobin per liter. Therefore,

$$\text{Molar mass of hemoglobin} = \frac{5.0 \text{ g/L}}{7.4 \times 10^{-5} \text{ mol/L}} = 68{,}000 \text{ g/mol}$$

EXERCISE 12.11 OSMOTIC PRESSURE AND MOLECULAR WEIGHT

Beta-carotene is the most important of the A vitamins. Its molar mass can be determined by measuring the osmotic pressure generated by a given mass of carotene dissolved in the solvent chloroform. Calculate the molar mass of carotene if 7.68 mg, dissolved in 10.0 mL of chloroform, give an osmotic pressure of 26.57 mm Hg at 25°C.

In the health professions, osmotic pressure is called **tonicity**, and an iso-osmotic solution is said to be **isotonic**. Solutions may also be **hypertonic** or **hypotonic**.

Osmosis has great practical significance, especially if you are a nurse or physician. Patients who become dehydrated through illness often need to be given water and nutrients intravenously. However, one cannot simply drip water into the patient's vein. Rather, the solution must have the same overall solute concentration as the patient's blood. That is, the intravenous solution must be *iso-osmotic*. If you used pure water, the inside of a blood cell would have a higher solute concentration (lower water concentration), and water would flow into the cell. This is called a *hypo-osmotic* situation, and the result would be bursting of the red blood cells (Figure 12.12). The opposite situation, *hyperosmotic*, would occur if the intravenous solution is more concentrated than the blood cell. In this case the cell would lose water and shrivel up. To combat this, a dehydrated patient is rehydrated in the hospital with a sterile saline solution that is about 0.9% sodium chloride, a solution that is iso-osmotic with the cells of the body.

Cell shriveling by osmosis is the same thing that happens when vegetables or meat are cured in brine. Brine is just a concentrated solution of NaCl. If you put a fresh cucumber or carrot into brine, water will flow out of its cells and into the brine, leaving behind a shriveled vegetable (Figure 12.13). With the proper spices, however, a cucumber becomes a delicious pickle!

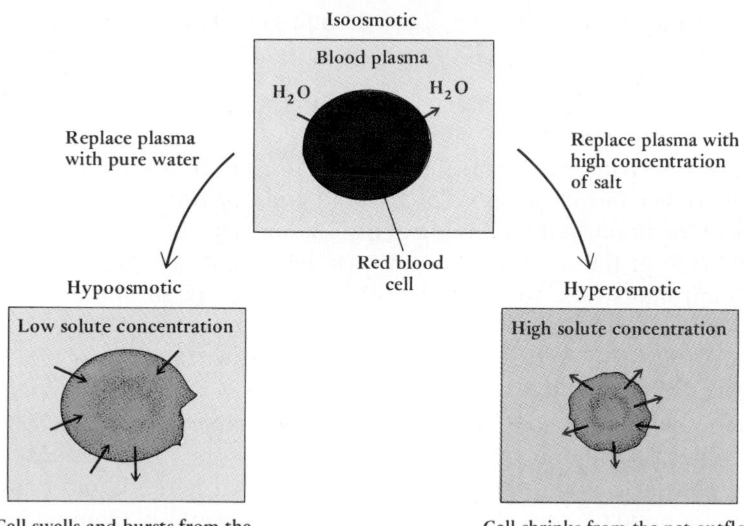

FIGURE 12.12

When a red blood cell is placed in a medium different from blood plasma, the cell can swell or shrink from the osmosis of water into or out of the cell.

12.4 IDEAL SOLUTIONS WITH TWO OR MORE VOLATILE COMPONENTS

To keep things simple to this point, we have described the behavior of solutions in which only the solvent is volatile. When the *solute* in an ideal solution is also volatile, both of the volatile species will obey Raoult's law. For the case of an ideal solution with two volatile components, A and B, the partial pressures of their vapors are

$$P_A = X_A P_A^\circ \qquad \text{and} \qquad P_B = X_B P_B^\circ$$

The *total* vapor pressure over such a solution will be the sum of these partial pressures,

$$P_{\text{total}} = P_A + P_B = X_A P_A^\circ + X_B P_B^\circ$$

The way this works for a solution made from 2 moles of toluene ($C_6H_5CH_3$, $X_{\text{Tol}} = 0.67$) and 1 mole of benzene (C_6H_6, $X_{\text{Ben}} = 0.33$) is shown in Figure 12.14. At 20°C, the vapor pressures of the pure solvents are

$$\text{Benzene: } P_{\text{Ben}}^\circ = 75 \text{ mm Hg} \qquad \text{Toluene: } P_{\text{Tol}}^\circ = 22 \text{ mm Hg}$$

Therefore, the partial pressures of benzene and toluene will be

$$P_{\text{Ben}} = (0.33)(75 \text{ mm Hg}) = 25 \text{ mm Hg}$$
$$P_{\text{Tol}} = (0.67)(22 \text{ mm Hg}) = 15 \text{ mm Hg}$$

and so the total pressure is

$$P_{\text{total}} = P_{\text{Tol}} + P_{\text{Ben}} = 40. \text{ mm Hg}$$

It is important to recognize that the vapor over the solution is richer in the more volatile component, benzene. The relative amount of benzene in the vapor is given by the ratio of its vapor pressure to the total pressure (Equation 6.9), and the same is true of toluene.

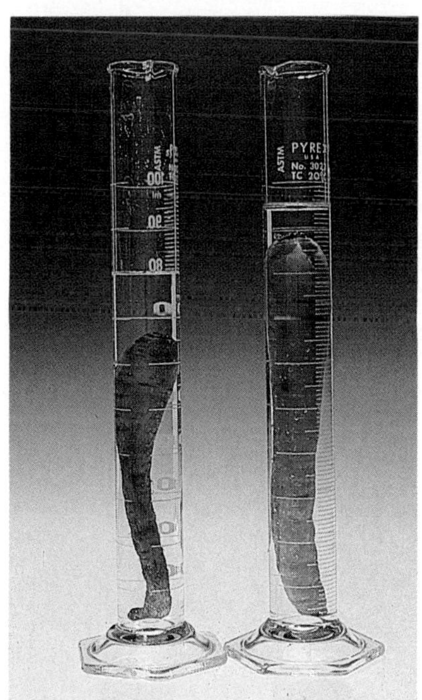

FIGURE 12.13

Osmosis. After some hours a carrot in a strong NaCl solution shows the effects of osmosis (left). (Water has flowed out of the carrot into the NaCl solution, leaving the carrot limp.) A carrot in pure water (right) is not appreciably affected. (Charles D. Winters)

The mole fractions of benzene and toluene in the gas phase are calculated using Dalton's law (Equation 6.9).

$$X_{Ben} \text{ in the vapor} = P_{Ben}/P_{total} = 25 \text{ mm Hg}/40. \text{ mm Hg} = 0.63$$
$$X_{Tol} \text{ in the vapor} = P_{Tol}/P_{total} = 15 \text{ mm Hg}/40. \text{ mm Hg} = 0.38$$

This tells you that although $\frac{1}{3}$ of the molecules in the solution are benzene, 63% of them in the vapor phase are benzene.

It is always true that, for ideal solutions in equilibrium with vapor, *the vapor is richer in the more volatile component of the solution*. This is the basis of the important technique of **distillation**. This is used routinely on a small scale in the laboratory, and on a huge scale in the chemical and petroleum industries, to separate volatile species from one another (Chapter 23).

A laboratory scale distilling apparatus is shown in Figure 12.15. To understand how this works, suppose you place in the flask the benzene/toluene solution we have been using as an example. When the flask is heated, vapor rises in the apparatus until a cooler zone (the condenser) is reached, and the vapor condenses to a liquid. What is the liquid composition at this point? It must be the same as the vapor phase from which it condensed: 63 mole percent benzene and 38 mole percent toluene. Thus, you have produced a liquid enriched in the more volatile component, and, if this enriched liquid is distilled yet another time, the condensed vapor will be even richer still in the more volatile component. Carrying out this

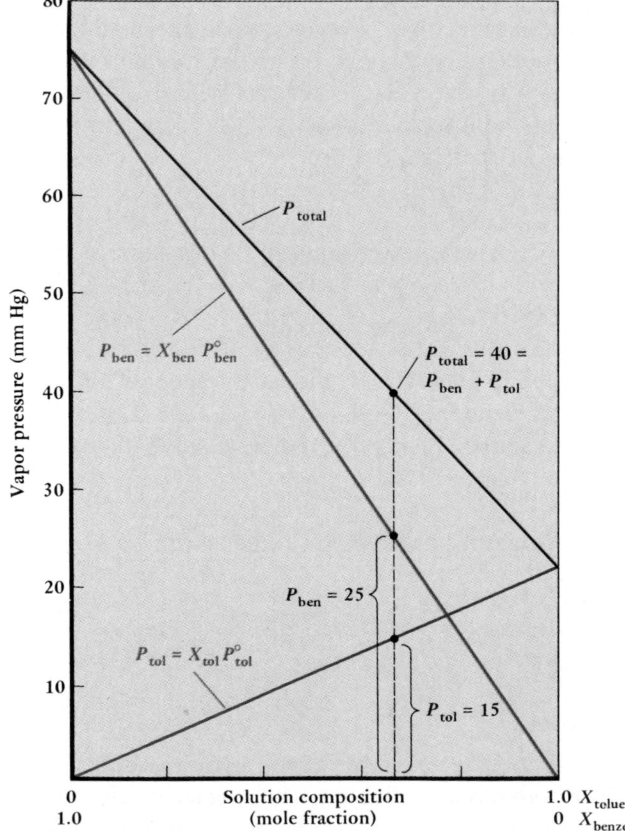

FIGURE 12.14

Variation in total vapor pressure for an ideal solution having two volatile components (benzene and toluene). When the mole fraction of benzene is $\frac{1}{3}$ and that of toluene is $\frac{2}{3}$, the total vapor pressure is 40. mm Hg. (Note that the horizontal axis is labeled with both mole fraction scales; this can be done because the solution contains only two volatile substances.)

operation many times in a specially designed *fractional distillation* apparatus can lead eventually to separation of the more-volatile from the less-volatile material.

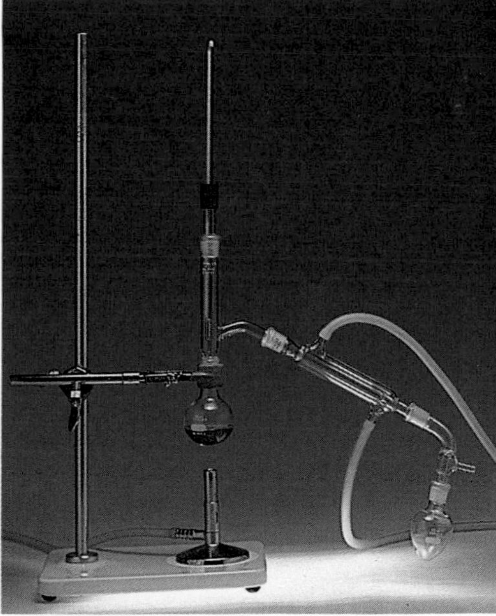

FIGURE 12.15

A simple laboratory apparatus for distillation of volatile liquids. A solution is placed in the round flask and heated gently. The more volatile component preferentially distills from the solution and can be collected in the flask at the end of the water-cooled condenser. (Charles D. Winters)

EXAMPLE 12.13

MIXTURES OF TWO VOLATILE COMPONENTS

Assume you have the liquid condensed in the distillation described in the text above. What is the vapor pressure of the mixture? What is the composition of the vapor phase?

Solution The mole fractions of benzene and toluene in the condensed vapor were 0.63 and 0.38, respectively. Therefore, the vapor pressures over this liquid are

$$P_{\text{Ben}} = (0.63)(75 \text{ mm Hg}) = 47 \text{ mm Hg}$$
$$P_{\text{Tol}} = (0.38)(22 \text{ mm Hg}) = 8.4 \text{ mm Hg}$$

This gives a total vapor pressure of 55 mm Hg.

From these data you can determine (from Equation 6.9) that the new vapor phase will consist of

$$X_{\text{Ben}} \text{ in the vapor} = 47 \text{ mm Hg}/55 \text{ mm Hg} = 0.85$$
$$X_{\text{Tol}} \text{ in the vapor} = 8.4 \text{ mm Hg}/55 \text{ mm Hg} = 0.15$$

Thus, if this new liquid mixture is distilled, the condensed vapor will be richer still (now 85%) in the more volatile component.

SUMMARY

Solutions are homogeneous mixtures of two or more substances in a single phase; the **solvent** is the component present in largest amount and the **solute** is the other component(s). This chapter is concerned with the **colligative properties** of solutions, properties that depend ideally on solute concentration and the nature of the solvent but not on the nature of the solute. These properties include changes in solution **vapor pressure**, **melting point** and **boiling point**, and the **osmotic pressure**.

In Section 12.1, the various units of solute concentration are defined and used: **mole fraction**, **molality**, **weight percent**, and **parts per million**. The first two in particular are used to define solution colligative properties.

Intermolecular forces provide one key to understanding the solution process (Secton 12.2), since they are closely related to the **enthalpy of solution**, the net heat energy involved in the process of solution formation. One criterion for a solute dissolving in a solvent is that the process be exothermic or only slightly endothermic. When the $\Delta H^\circ_{\text{solution}}$ is significantly positive, the energy evolved on forming intermolecular bonds between solute and solvent does not offset the energy required to break down solute–solute or solvent–solvent intermolecular bonding. This is the basis of the "like dissolves like" rule of thumb. For example, ionic solids are dissolved only in highly polar solvents. The process of dissolving ionic solids in particular can be analyzed quantitatively using (a) lattice energies and heats of hydration or (b) the standard enthalpies of formation of the crystalline solid and the aqueous, 1 molal solution.

Pressure and temperature affect solubility (Section 12.2). For a gas, the greater its partial pressure, the greater the solubility, a fact expressed by **Henry's law** (Equation 12.4). The effect of temperature depends in large part on the enthalpy of solution: solutes for which the solution process is endothermic generally increase in solubility with temperature (and vice versa).

It is possible to describe accurately the colligative properties of an **ideal solution** (Section 12.3), a solution obeying **Raoult's law**:

$$P_{\text{solvent}} = X_{\text{solvent}} P^{\circ}_{\text{solvent}} \tag{12.6}$$

Solvent vapor pressure of solution =
mole fraction of solvent × vapor pressure of pure solvent

This tells us that, when a solute is dissolved in a solvent, the vapor pressure of solvent declines by an amount $\Delta P_{\text{solvent}}$.

$$\Delta P_{\text{solvent}} = X_{\text{solute}} P^{\circ}_{\text{solvent}} \tag{12.8}$$

When a solute is added to a solvent, the boiling point of a solution is elevated relative to that of the pure solvent by an amount

$$\Delta t_{\text{bp}} = K_{\text{bp}} m_{\text{solute}} \tag{12.10}$$

and the freezing point is depressed by an amount

$$\Delta t_{\text{fp}} = K_{\text{fp}} m_{\text{solute}} \tag{12.11}$$

(where K_{bp} and K_{fp} are positive and negative constants, respectively, that depend on the solvent) (Section 12.3).

Osmosis is the movement of molecules through a *semipermeable membrane* from a region of low solute concentration to a region of high solute concentration. The **osmotic pressure**, Π, of a solution can be found from the equation

$$\text{Osmotic pressure} = \Pi = MRT \tag{12.12}$$

where M is the molarity of the solution, R is the gas constant, and T the kelvin temperature.

STUDY QUESTIONS

REVIEW QUESTIONS

1. Is the following statement true or false? If false, change it to read correctly. "Colligative properties depend on the nature of the solvent and solute and on the concentration of solute."
2. Name the four colligative properties described in this chapter. Write the mathematical expression that describes each of these.
3. Define molality and tell how it differs from molarity, another concentration unit.
4. How is the solubility of a gas or solid related to the algebraic sign of the enthalpy of solution?
5. Explain the relation between enthalpy of solution of a gas or solid and the temperature dependence of solubility.
6. How are the aqueous solubilities of NaCl, KCl, RbCl, and CsCl in Figure 12.7 related to their lattice energies?
7. Name three effects that govern the solubility of a gas in water.

8. If you dissolve equimolar amounts of NaCl and $CaCl_2$ in water, the calcium salt lowers the freezing point of the water almost 1.5 times as much as the NaCl. Why?
9. Explain why a cucumber shrivels up when you put it in a concentrated solution of salt.
10. Explain how you can separate two volatile liquids from one another by distillation.

CONCENTRATIONS

11. One of life's great pleasures is to make homemade ice cream in the summer. Fresh milk and cream, sugar, and fruit are churned in a bucket suspended in ice and water of which the freezing point has been lowered by adding rock salt. One manufacturer of home ice cream freezers recommends that you use 2.50 lbs (1130 g) of rock salt (NaCl) to 16.0 lbs of ice (7250 g) in a four-quart freezer. For the solution when this mixture melts calculate (a) the weight percentage of NaCl, (b) the mole fraction of NaCl, and (c) the molality of the solution.
12. The Salton Sea in California has a relatively high concentration of lithium ion, 1.9 ppm. What is the molality of Li^+ in this water? (See Example 12.2.)
13. Silver ion is found with an average concentration of 28 parts per billion in U.S. water supplies. (a) What is the molality of the silver ion? (b) If you wanted 1.0×10^2 g of silver and could recover it chemically from water supplies, how many liters of water would you have to treat? (Assume the density of water is 1.0 g/cm³.)
14. Fill in the blanks in the table below.

COMPOUND	MOLALITY	WEIGHT PERCENT	MOLE FRACTION
NaCl	0.25	____	____
C_2H_5OH	____	5.0%	____
$C_{12}H_{22}O_{11}$	0.10		____
NH_3	____	30.0	____
CH_3COOH (acetic acid)	____	0.05	____

15. Hydrochloric acid is sold as a concentrated aqueous solution. If the molarity of commercial HCl is 12.0 and its density is 1.18 g/cm³, calculate (a) the molality of the solution and (b) the weight percent of HCl in the solution.
16. Concentrated sulfuric acid has a density of 1.84 g/cm³ and is 95.0% by weight H_2SO_4. What is the molality of this acid? What is its molarity?
17. Water at 25°C has a density of 0.997 g/cm³. Calculate the molality and molarity of pure water at this temperature.
18. At 30°C, 300. g of silver nitrate can be dissolved in 100. g of water. Calculate (a) the molality of the solution, (b) the mole fraction of silver nitrate, and (c) the weight percentage of silver nitrate.
19. A 10.7 molal solution of NaOH has a density of 1.33 g/cm³ at 20°C. Calculate the (a) mole fraction of NaOH, (b) the weight percentage of NaOH, and (c) the molarity of the solution.
20. If you want to prepare a solution that is 0.200 molal in $NaNO_3$, how many grams of the salt would you add to 500. g of water? What is the mole fraction of $NaNO_3$ in the resulting solution?
21. If you dissolve 2.00 g of $CaCl_2$ in 750 g of water, what is the molality of $CaCl_2$? What is the total molality of the ions? (That is, what is the total concentration of Ca^{2+} and Cl^- ions?)
22. If you want a solution that is 0.100 molal in ions, how many grams of $MgCl_2$ must you dissolve in 150. g of water?

THE SOLUTION PROCESS

23. Give reasons for the following:
    (a) Octane is very miscible with CCl_4.
    (b) Methyl alcohol, H_3COH, dissolves in all proportions in water.
    (c) Sodium bromide is very poorly soluble in diethyl ether (C_2H_5—O—C_2H_5).
    (d) Carbon tetrachloride (CCl_4) has a density of 1.6 g/cm³. When poured into water, it does not dissolve; water floats on top of the layer of CCl_4.
24. Account for the fact that alcohols such as methyl (CH_3OH) and ethyl (C_2H_5OH) are quite miscible with water, while an alcohol with a long carbon chain such as octyl alcohol ($C_8H_{17}OH$) is poorly soluble in water.
25. Water and CCl_4 (carbon tetrachloride) are not miscible; when put into the same beaker, they form layers, the CCl_4 on the bottom (see Figure 12.4). If the water layer initially contains I_2, most of the halogen will transfer to the CCl_4 layer if you shake or stir the water–CCl_4 mixture. Why?
26. A student weighs 150. g of KBr and adds it to 100. g of water held at 100°C. Not all of the solid dissolves. After filtering the solution (while still hot), the student finds that the undissolved KBr amounts to 46.0 g. What is the solubility of KBr at 100°C in g solute/100 g solvent?
27. Potassium nitrate dissolves in water in the following amounts per 100. g of water: (a) 31.6 g at 20°C, (b) 85.5 g at 50°C, and (c) 202 g at 90°C. Is the enthalpy of solution of KNO_3 expected to be positive or negative? Calculate the heat of solution from the following information: $\Delta H_f^\circ[KNO_3(s)] = -494.6$ kJ/mol and $\Delta H_f^\circ[KNO_3(aq, 1\ m)] = -459.7$ kJ/mol.
28. Briefly rationalize the following trend in enthalpies of solution in water in terms of lattice energies (assuming this factor alone determines relative solubilities): (a) NaCl(s), 3.88 kJ/mol; (b) KCl(s), 17.2 kJ/mol; (c) RbCl(s), 17.3 kJ/mol; and (d) CsCl(s), 17.8 kJ/mol.

29. The ionic compound ammonium chloride could be used as a "chemical cold pack." When the solid is mixed with water, the solution becomes very cold. Calculate the enthalpy of formation of $NH_4Cl(aq)$ $[\Delta H°(NH_4Cl,\ 1m)]$

$$NH_4Cl(s) + 14.7\ kJ/mol \longrightarrow NH_4^+(aq) + Cl^-(aq)$$

if the standard enthalpy of formation of solid NH_4Cl is -314.4 kJ/mol, and the enthalpy of solution is $+14.7$ kJ/mol.

30. Some ionic compounds evolve heat when mixed with water. In fact, dissolving $CaCl_2$ is so exothermic that it is used in a commercial "hot pack." Calculate the enthalpy of the reaction

$$CaCl_2(s) \longrightarrow Ca^{2+}(aq) + 2Cl^-(aq)$$

from $\Delta H_f°[CaCl_2(s)] = -795$ kJ/mol and $\Delta H_f°[CaCl_2(aq)] = -877.89$ kJ/mol.

31. The standard enthalpy of formation of NaI(s) is -288 kJ/mol and its enthalpy of solution in water is -7.5 kJ/mol. Calculate the standard enthalpy of formation of NaI(aq).

32. The enthalpy of solution of NaI is -7.5 kJ/mol while that of KI is $+20.3$ kJ/mol. Thinking in terms of the contributions to enthalpy of solution from lattice energy and enthalpy of hydration, rationalize the fact that the enthalpy of solution of NaI can be negative whereas the majority of such salts have positive heats of solution.

HENRY'S LAW

33. Compare the concentration of CO_2 in water at 25°C when the partial pressure of the gas is one third of an atmosphere (see Exercise 12.4) and 1.0 atmosphere.

34. The partial pressure of O_2 in your lungs is 25 mm Hg to 40 mm Hg. How much O_2 can dissolve in water at 25°C if the O_2 partial pressure is 40. mm Hg? (Henry's law constant for O_2 at 25°C is 1.68×10^{-6} molal/mm Hg.)

35. The Henry's law constants for O_2 and N_2 are, respectively, 1.68×10^{-6} and 8.53×10^{-7} (in units of molal/mm Hg at 25°C). If dry air is 78.3 mole % N_2 and 21.0 mole % O_2, calculate the molality of each gas dissolved in water if the total gas pressure is 760 mm Hg.

VAPOR PRESSURE CHANGES

36. Ethylene glycol, $C_2H_4(OH)_2$, is a common ingredient of automotive antifreeze. If you mix 1.80×10^3 g of water and 620. g of ethylene glycol, what is the vapor pressure of water over the mixture at the normal boiling point of water, that is, at 100°C? (Assume ethylene glycol is not volatile.)

37. The vapor pressure of water is 17.5 mm Hg at 20°C. (a) What is the vapor pressure of the solution when 15.0 g of urea, $(NH_2)_2CO$, is dissolved in 0.500 kg of water? (b) Compare the vapor pressure in part (a) with that of a solution composed of 15.0 g of sugar ($C_{12}H_{22}O_{11}$) in 0.500 kg of water at 20°C. (c) What quantity of sugar would have to be dissolved in 0.500 kg of water to produce the same vapor pressure as that given by the solution in part (a)?

38. 10.0 g of a nonvolatile solute are dissolved in 100. g of benzene (C_6H_6). The vapor pressure of pure benzene at 30°C is 121.8 mm Hg, while that of the solution is 113.0 mm Hg at the same temperature. What is the molar mass of the solute?

39. A solution prepared from 20.0 g of a nonvolatile solute in 154 g of the solvent carbon tetrachloride (CCl_4) has a vapor pressure of 504 mm Hg at 65°C. (The vapor pressure of pure CCl_4 is 531 mm Hg at 65°C.) What is the approximate molar mass of the solute?

40. Do Example 12.6 using 0.0200 mol of nonvolatile solute in 100. g of benzene. Compare your result with that *estimated* using Equation 12.9. What does this tell you about the validity of Equation 12.9 for more dilute solutions?

BOILING POINT ELEVATION

41. Verify the result in Figure 12.9 that 0.200 mol of a nonvolatile solute in 100. g of benzene (C_6H_6) produces a solution whose boiling point is 85.2°C.

42. In Study Question 37 you calculated the vapor pressure change for a solution composed of 15.0 g of urea in 0.500 kg of water. What is the boiling point of this solution?

43. Chloroform, $CHCl_3$, has a normal boiling point of 61.7°C. If K_{bp} for chloroform is 3.63 deg/molal, what is the boiling point of a solution composed of 15.0 g of $CHCl_3$ and 0.515 g of the nonvolatile solute acenaphthalene, $C_{12}H_{10}$ (a component of coal tar)?

44. You add 0.255 g of an orange, crystalline compound whose empirical formula is $C_{10}H_8Fe$ to 11.12 g of benzene. The boiling point of the benzene rises from 80.10 to 80.26°C. What is the molecular formula of the compound?

45. Refer to Figure 12.7. When you have dissolved the maximum amount of NaCl possible in 100. g of water at 100°C, what will be the boiling point of this solution?

46. Hexachlorophene has been used in germicidal soap. What is its molar mass if 0.640 g of the compound, dissolved in 25.0 g of chloroform, produces a solution whose boiling point is 61.93°C? (The normal boiling for chloroform is 61.70°C, and its K_{bp} is 3.63 deg/molal.)

FREEZING POINT DEPRESSION

47. If you use only water and pure ethylene glycol, $C_2H_4(OH)_2$, in your car's cooling system, how many

grams of the glycol must be added to each quart of water to give protection to $-31°C$? (One quart of water has a mass of 946 g.)

48. Some ethylene glycol, $C_2H_4(OH)_2$, was added to your car's cooling system along with 5.0 kg of water. (a) If the freezing point of the water–glycol solution is $-15°C$, how many grams of $C_2H_4(OH)_2$ must have been added? (b) What is the boiling point of the solution?

49. If you have ever made homemade ice cream, you know that you cool the milk and cream by immersing the container in ice and a concentrated solution of rock salt (NaCl) in water. If you want to have a water–salt solution that freezes at $-10°C$, how many grams of NaCl will you have to add to 3.0 kg of water?

50. Erythritol is a compound that occurs naturally in algae and fungi. It is about twice as sweet as sucrose. A solution of 2.50 g of erythritol in 50.0 g of water freezes at $-0.773°C$. What is the molar mass of erythritol?

51. Suppose you have just made a compound containing only boron and fluorine. To determine its molar mass you dissolve 0.146 g in 10.0 g of benzene and find that the freezing point of the solution is 4.75°C. (The freezing point of pure benzene is 5.48°C and K_{fp} for benzene is -4.90 deg/molal). If the compound contains 22.1% boron, what are the empirical and molecular formulas of the boron–fluorine compound?

52. List the following aqueous solutions in order of increasing melting point: (a) 0.1 *m* sugar; (b) 0.1 *m* NaCl; (c) 0.08 *m* $CaCl_2$; (d) 0.04 *m* Na_2SO_4. (The last three are all assumed to dissociate completely in water.)

53. A solution prepared by dissolving 9.41 g of $NaHSO_3$ in 1.00 kg of water freezes at $-0.33°C$. From these data decide which of the following equations is the correct expression for the ionization of $NaHSO_3$ (assuming complete ionization in each case).
(a) $NaHSO_3(aq) \longrightarrow Na^+(aq) + HSO_3^-(aq)$
(b) $NaHSO_3(aq) \longrightarrow Na^+(aq) + H^+(aq) + SO_3^{2-}(aq)$

54. A compound is known to be a potassium salt, KX. If 4.00 g of the salt are dissolved in 100. g of water, the solution freezes at $-1.28°C$. Which of the elements of Group 7A is X?

OSMOTIC PRESSURE

55. Meats such as fish and pork can be preserved by "salting" them. In terms of the phenomenon of osmosis, explain what happens when meat is placed in a strong salt solution.

56. A line in the "Rime of the Ancient Mariner," a well known poem by Coleridge, is about a ship's crew stranded in a small boat in the ocean, out of food and drink. One of the crew says "Water, water everywhere and not a drop to drink." Why is the ocean water so unfit to drink?

57. Calculate the concentration of solute particles in human blood if the osmotic pressure is 7.53 atm at 37°C, the temperature of the body.

58. Cold-blooded marine animals and fish have blood that is iso-osmotic (isotonic) with sea water. If sea water freezes at $-2.30°C$, what is the osmotic pressure of the blood of marine life at 20°C?

TWO VOLATILE COMPONENTS IN A SOLUTION

59. A solution, assumed to be ideal, is made from 1.0 mole of toluene ($C_6H_5CH_3$) and 2.0 moles of benzene (C_6H_6). The vapor pressures of the pure solvents are 22 mm Hg and 75 mm Hg, respectively, at 20°C. What is the total vapor pressure of the mixture? (Does your calculation agree with Figure 12.14?) What is the relative amount of each compound in the vapor phase?

60. At 60°C the vapor pressure of pure benzene is 388 mm Hg and that of toluene is 137 mm Hg. Calculate the mole fraction of benzene in an ideal solution of benzene and toluene that has a total vapor pressure of 266 mm Hg at 60°C.

61. Calculate the total vapor pressure at 25°C of an ideal solution made by mixing two volatile liquids: 10.0 g of CH_2Cl_2 and 1.00 g of hexene (C_6H_{12}) at 25°C. At 25°C, the vapor pressure of pure CH_2Cl_2 is 435 mm Hg and that of pure C_6H_{12} is 166 mm Hg.

62. Hexane, C_6H_{14}, has a vapor pressure of 151 mm Hg at 25°C, whereas that of cyclohexane, C_6H_{12}, is 98 mm Hg at this temperature. (a) Using graph paper, construct a vapor pressure/composition diagram as in Figure 12.14. (b) Calculate the vapor pressure of an ideal solution having 25 g of hexane and 50 g of cyclohexane. (c) What are the relative amounts of hexane and cyclohexane in the vapor over the solution in part (b)?

63. If you mix 5.00 mL of methylene chloride (CH_2Cl_2, density = 1.34 g/cm³) and 10.0 mL of chloroform ($CHCl_3$, density = 1.49 g/cm³), what is the total pressure of the solution at 25°C? (The pure solvents have vapor pressures of 435 mm Hg and 195 mm Hg, respectively, at 25°C, and they are assumed to form an ideal solution.)

GENERAL QUESTIONS

64. Which of the following pairs would be most likely to form an ideal solution?
(a) C_3H_8 (propane) and H_2O
(b) C_3H_8 (propane) and CCl_4
(c) C_2H_5OH (ethyl alcohol) and CH_3OH (methyl alcohol)

65. Refer to Figure 12.7. If excess $NaNO_3$ is added to 100. g of water held at 60°C, how many grams will dissolve? When you cool the solution to 20°C, how many grams of $NaNO_3$ will precipitate from solution?

66. Enthalpies of hydration of individual ions can be calculated, and some are given below. (a) If the enthalpy change is calculated for $M^{n+}(g) \rightarrow M^{n+}(aq)$, explain briefly why they are always negative. (b) Explain the trends you observe in the following series of hydration enthalpies:
(a) Mg^{2+}, −1980 kJ/mol; Ca^{2+}, −1650 kJ/mol; Ba^{2+}, −1360 kJ/mol.
(b) Li^{+}, −545 kJ/mol; Mg^{2+}, −1980 kJ/mol; Al^{3+}, −4750 kJ/mol.

67. In the past some people used a mixture of ethyl alcohol and water in car radiators. If the freezing point of an alcohol–water mixture is −10.0°C, what is the vapor pressure of the mixture ($P_{alc} + P_{water}$) at 25°C? (Information you need to solve this problem is: (a) the vapor pressure of alcohol at 25°C = 60 mm Hg; (b) the vapor pressure of pure water at 25°C = 24 mm Hg. Assume an ideal solution.)

68. Which aqueous solution would have the greater water vapor pressure at 25°C: (a) 33.0 g of sugar (MW = 342) in 180. g of water or (b) 5.85 g of NaCl in 180. g of water? Which would have the higher boiling point?

69. Arrange the following aqueous solutions in order of increasing vapor pressure at 25°C: (a) 0.35 *m* $C_2H_4(OH)_2$ (nonvolatile solute); (b) 0.50 *m* sugar; (c) 0.20 *m* KBr (a strong electrolyte); (d) 0.20 *m* Na_2SO_4 (a strong electrolyte).

70. You have a small but seemingly pure sample of an unknown white powder containing only C, H, and O. To analyze the compound, you determine its empirical formula by analysis for C and H, and its molar mass by freezing point depression. (a) What is the empirical formula if a 0.0151-g sample gives 0.0111-g H_2O and 0.0449-g CO_2 on combustion? (b) If you dissolve 1.00 g of the powder in 100. g of camphor, you find that the melting point of the solution is 177.05°C. Pure camphor melts at 179.75°C. What is the molar mass of the white powder? (c) What is the true molecular formula of the white powder?

71. Using the solutions in Study Question 52, list them in order of increasing boiling point.

72. Calculate the boiling point of each of the aqueous solutions in Study Question 69.

73. If you added ethylene glycol to your car's radiator to lower the freezing point to −31°C, what would be the corresponding boiling point of the water in your radiator? (See Study Question 47.)

74. In chemical research we often send newly synthesized compounds to commercial laboratories for analysis. These laboratories determine the weight percent of C and H by burning the compound and collecting the evolved CO_2 and H_2O. They determine the molecular weight by measuring the osmotic pressure of a solution of the compound. Calculate the empirical and molecular formulas of a compound given the following information: (a) The compound contains 73.87% C and 8.21% H; the remainder is chromium. (b) At 25°C, the osmotic pressure of 5.00 mg of the unknown dissolved in 100. mL of chloroform solution is 3.17 mm Hg.

75. *Reverse osmosis* is being used to obtain fresh water from sea water by desalinization (salt removal). A membrane permeable to water but not salt separates compartments containing sea water and fresh water. The normal flow of water, due to osmosis, from the fresh water to the sea water compartment is reversed by applying a pressure to the sea water compartment. For purposes of this problem assume that sea water is 3.0 weight percent NaCl. What is the minimum pressure (above atmospheric pressure) that must be supplied to cause reverse osmosis to occur at 25°C?

PART FOUR

THE CONTROL OF CHEMICAL REACTIONS

A spiral wave in the Belousov-Zhabotinskii reaction (oxidation and decarboxylation of malonic acid by bromate ions catalyzed by ferroin). (Stefan C. Müller)

PART FOUR: PREFACE

New compounds are created by allowing elements and compounds to react with one another. To many scientists, this is the essence of chemistry. But, if a reaction occurs, how rapidly does it occur and through what sequence of steps does it proceed? This is the subject of Chapter 13, a study of the rates and mechanisms of reactions.

When a reaction occurs, it can reach a state of equilibrium. After defining chemical equilibria, and examining their general properties in Chapter 14, we shall proceed to study reactions in aqueous solutions in particular. We are especially interested in reactions involving acids and bases (Chapters 15 and 16), since biochemical and other natural systems are so highly dependent on acid-base reactions. Further, many environmental reactions involve the precipitation and dissolving of insoluble compounds, so this is the subject of Chapter 17.

To tie together the discussion of chemical equilibria we shall further explore the science of thermodynamics (Chapter 18), and you will learn how to predict if a given chemical reaction can occur.

As a final topic in this section, we shall describe in more detail another major class of reactions: oxidation-reduction processes (Chapter 19). Since such reactions can be caused by electricity, or can result in the flow of electrical current, we treat such reactions under the heading of "electrochemistry."

13

Chemical Kinetics: The Rates and Mechanisms of Chemical Reactions

Explosive decomposition of NH_4NO_3, catalyzed by Cl^- ions, in the presence of powdered zinc. (Charles D. Winters)

CHAPTER OUTLINE

So far we have been interested in what kinds of reactions occur. However, another large area of chemistry, called **chemical kinetics**, is the study of the speed or rate of reactions under various conditions and what this means in terms of the **mechanisms** of reactions, the sequence of events at the molecular level that control the speed and outcome of a reaction. Ultimately, chemists hope to learn how to control reactions and to devise better methods for making chemicals.

It is the purpose of this chapter to introduce you to the chemist's concepts and language for understanding the rates and mechanisms of chemical reactions. You will find that molecular structures and bond strengths are important to this story, as are reactant concentrations and the presence or absence of substances called catalysts.

13.1 KINETICS AND CONDITIONS

The first step in a kinetics study is to acquire experimental information on the reaction speed or rate under a variety of conditions. At the outset, you should be aware that the following conditions affect the speed of a chemical process:

a. *Concentrations of reactants.* The natural gas in a bunsen burner burns in air.

$$2\ C_2H_6(g) + 7\ O_2(g) \longrightarrow 6\ H_2O(\ell) + 4\ CO_2(g)$$

However, if the concentrations of the hydrocarbon and O_2 are increased by compressing them in a much smaller volume, as in the cylinder of an automobile engine, combustion occurs at explosive speed.

b. *Temperature.* An egg can be kept without spoiling in a refrigerator for some time. However, a few minutes in boiling water speeds up the reactions that congeal the white and yolk, and you have a hard-boiled egg.

c. *Catalyst.* Hydrogen peroxide, H_2O_2, can decompose to water and oxygen.

$$2\ H_2O_2(\ell) \longrightarrow O_2(g) + 2\ H_2O(\ell)$$

The peroxide is reasonably stable for many months, when stored in a cool place. However, a catalyst speeds up the reaction tremendously. An insect called the bombardier beetle uses the process as its defense mechanism. It uses a biological catalyst, an *enzyme*, to cause the reaction to occur with explosive speed (Figure 13.1).

d. *Physical state of the reactants*. The effect of the physical state of the reactants is shown in Figure 13.2. The oxidation of organic matter is violently fast when it is finely divided. This effect of particle size on reaction rate explains why we dissolve reactants in water or other solvents before carrying out a reaction. When reactants are placed in solution, all molecules are fully exposed to interactions with each other and the reaction proceeds more quickly.

After defining more precisely what is meant by a reaction rate, we shall study what happens to the concentrations of reactants and products during the time a reaction occurs and how temperature and catalysts affect the rate.

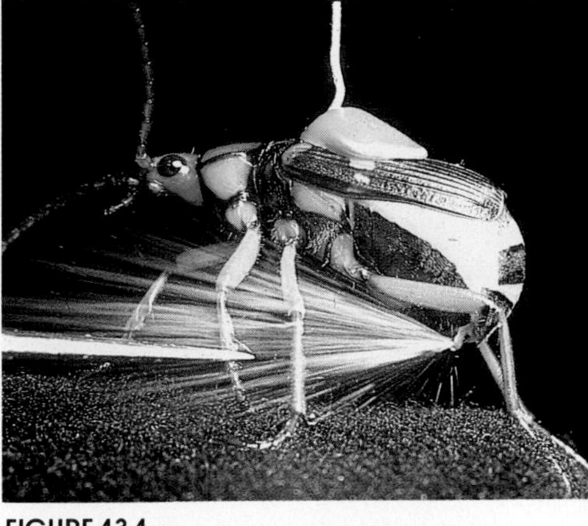

FIGURE 13.1

A bombardier beetle uses the catalyzed decomposition of hydrogen peroxide as a defense mechanism. The oxygen gas formed in the decomposition forces out water and other chemicals with explosive force. Since the reaction is very exothermic, the water comes out as steam. (Thomas Eisner with Daniel Aneshansley)

13.2 CHEMICAL REACTION RATES

The word "rate" implies that some measurable quantity is changing with time. For instance, a car's rate is found by measuring the change in its position, Δp, over a given time period, Δt. For example, if you travel 55 miles in 60 minutes, you are moving at a speed of 55 mph. When describing chemical reactions, the **reaction rate** is the change in concentration of a reagent per unit time. Consider the decomposition of hydrogen peroxide.

$$2\ H_2O_2(aq) \longrightarrow 2\ H_2O(\ell) + O_2(g)$$

The rate of this reaction can be found by measuring the quantity of O_2 evolved over a given time interval. If the amount of O_2 is measured in moles per liter, then the rate of reaction can be given in units of mol/(L · min) or mol/(L · s), and it is expressed by

The square brackets around O_2 imply that O_2 is measured in moles per liter.

$$\text{Rate of reaction}\left(\frac{\text{mol/L}}{\text{min}}\right) = \frac{\Delta[O_2]}{\Delta t}$$

Recall that the Greek letter Δ (delta) means a change in some quantity has been measured; it is always obtained by subtracting the initial value of concentration or time from the final value.

(a)

(b)

FIGURE 13.2

Lycopodium powder. (a) The spores of this common moss burn only with difficulty when piled in a dish. (b) If the finely divided powder is sprayed into a flame, however, combustion is rapid. (Charles D. Winters)

FIGURE 13.3
Concentration versus time for O_2 and H_2O_2 during the decomposition of hydrogen peroxide.

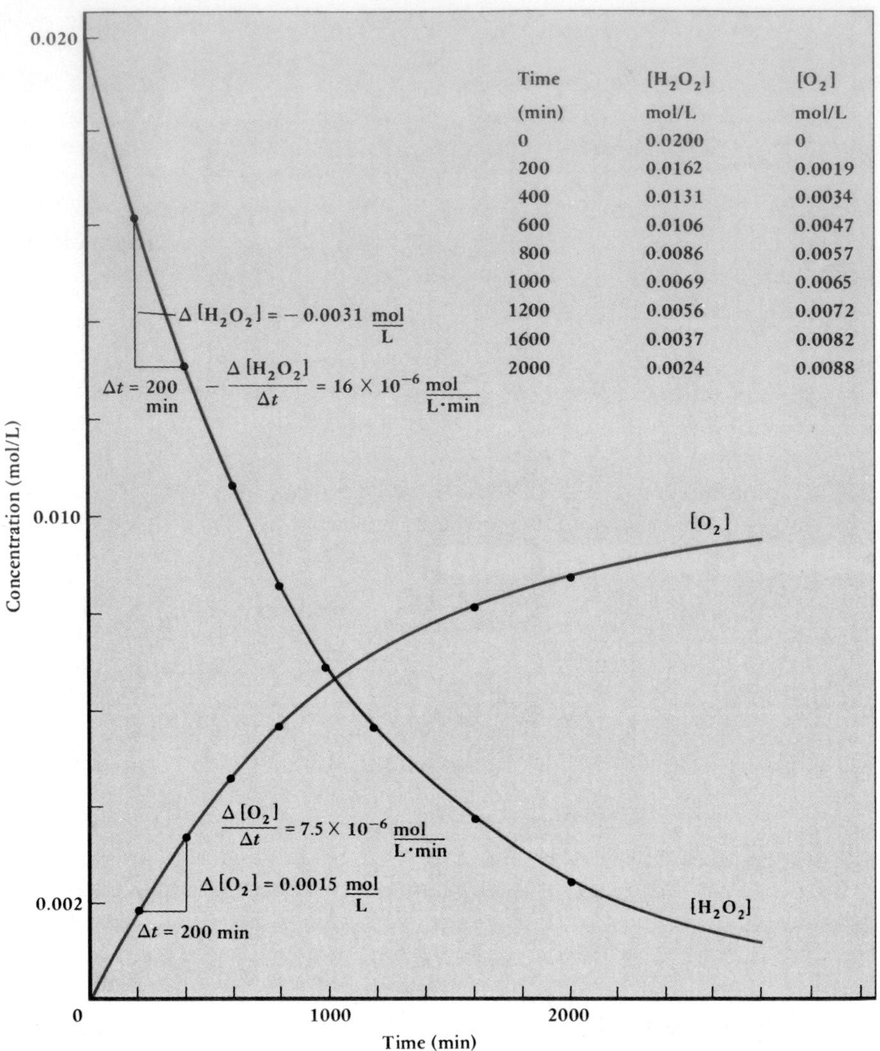

Time (min)	$[H_2O_2]$ mol/L	$[O_2]$ mol/L
0	0.0200	0
200	0.0162	0.0019
400	0.0131	0.0034
600	0.0106	0.0047
800	0.0086	0.0057
1000	0.0069	0.0065
1200	0.0056	0.0072
1600	0.0037	0.0082
2000	0.0024	0.0088

The increase in concentration of O_2 and the decrease in H_2O_2 during the decomposition of hydrogen peroxide is shown in Figure 13.3. Each point on these curves gives the concentration of that compound at a particular time. To obtain an *average reaction rate* over some period of time, we can find the change in concentration of O_2 or H_2O_2 during that time. For example, during the time interval from 200 to 400 minutes after the reaction starts, the rate of formation of O_2 is

$$\text{Rate of formation of } O_2 = \frac{\Delta[O_2]}{\Delta t} = \frac{(0.0034 - 0.0019)\text{ mol/L}}{(400 - 200)\text{ min}}$$
$$= 7.5 \times 10^{-6} \text{ mol/(L} \cdot \text{min)}$$

The rate of formation of a product, $\Delta[P]/\Delta t$, is always a positive quantity, while the rate of disappearance of a reactant, $\Delta[R]/\Delta t$, is always a negative quantity.

Since $[O_2]$ must increase with time during H_2O_2 decomposition, $\Delta[O_2]$ is a positive quantity and so is $\Delta[O_2]/\Delta t$. Conversely, the amount of H_2O_2 must decrease with time, so $\Delta[H_2O_2]/\Delta t$ must be a negative quantity.

$$\text{Rate of disappearance of } H_2O_2 = \frac{\Delta[H_2O_2]}{\Delta t} = \frac{(0.0131 - 0.0162) \text{ mol/L}}{(400 - 200) \text{ min}}$$

$$= -16 \times 10^{-6} \text{ mol/(L} \cdot \text{min)}$$

The value of an average rate as calculated above depends on which data points are selected. To get around this difficulty, we draw a short *tangent line* to the curve at the desired point, measure the changes in concentration and time from the tangent line, and use these values in the equation. The result is that the rate is the *slope* of the tangent line as illustrated in the margin. This can be a tedious process, so we often approximate the slope of the tangent line by the slope of a line drawn between adjacent data points; as the points are taken closer together, the approximation becomes better. By this method we can obtain what is effectively an *instantaneous rate* at a given concentration. If these rates are then plotted versus time, the curves shown in Figure 13.4 are obtained. Extrapolating the rate versus time curves to zero time then gives the *initial rate of reaction*, a useful experimental quantity, as you shall see in Section 13.4.

A line tangent to a curve.

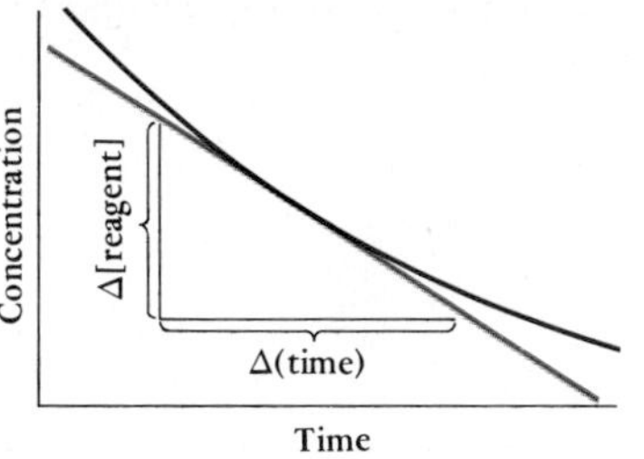

Another feature shown in Figures 13.3 and 13.4 is that the rate of O_2 formation is half the rate of peroxide decomposition.

$$\text{Rate of formation of } O_2 = \tfrac{1}{2}(\text{rate of decomposition of } H_2O_2)$$

$$\frac{\Delta[O_2]}{\Delta t} = \frac{1}{2}\left(-\frac{\Delta[H_2O_2]}{\Delta t}\right)$$

The stoichiometry of peroxide decomposition tells you that two moles of H_2O_2 must decompose to produce one mole of O_2. Therefore, O_2 can only form at one-half the rate that H_2O_2 decomposes.

We have been careful to talk about the "rate of decomposition of H_2O_2" and the "rate of formation of O_2." However, one often speaks simply of the "rate of the reaction." By convention, the rate of reaction is a positive number, so we change the sign of any rate measured in terms of the concentration of a reactant. In addition, to remove the effect of stoichiometry, the rate for each reactant or product is divided by the coefficient of that molecule in the balanced equation. For example, in the decomposition of hydrogen peroxide,

$$2\ H_2O_2(\ell) \longrightarrow O_2(g) + 2\ H_2O(\ell)$$

the reaction rate is given by any of the equal expressions.

$$\text{Reaction rate} = \frac{1}{2}\left(-\frac{\Delta[H_2O_2]}{\Delta t}\right) = +\frac{\Delta[O_2]}{\Delta t} = \frac{1}{2}\left(+\frac{\Delta[H_2O]}{\Delta t}\right)$$

When the "rate of reaction" is plotted versus time, the same curve is obtained, regardless of which concentration is being measured; it is shown as a dashed line for the peroxide decomposition in Figure 13.4.

FIGURE 13.4

Plot of the instantaneous rate of reaction for the disappearance of H_2O_2 and the appearance of O_2 in the decomposition of hydrogen peroxide. The green line gives the "rate of the reaction"; see the text for details. Notice that as the reaction progresses the rates of peroxide decomposition and oxygen formation both approach zero, since the concentration of peroxide decreases with time.

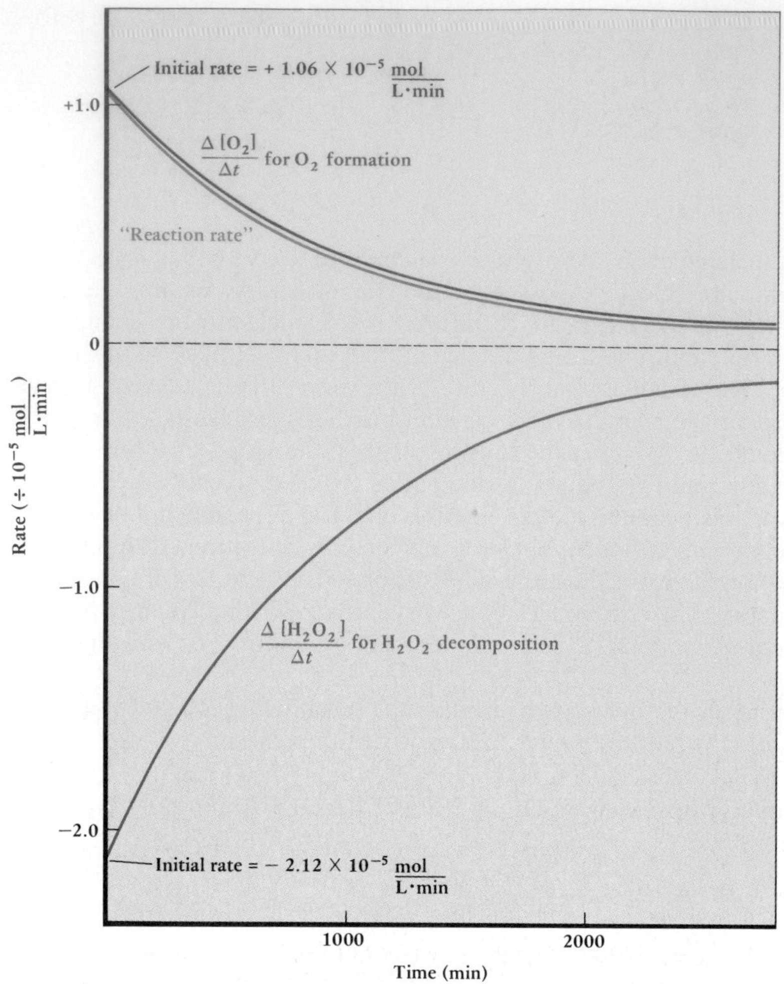

Data for Figure 13.4

$[H_2O_2]$ (mol/L)	TIME (min)	RATE OF DISAPPEARANCE OF H_2O_2 mol/(L · min) ÷ 10^{-5}	RATE OF FORMATION OF O_2 mol/(L · min) ÷ 10^{-5}
0.020	0	−2.12	+1.06
0.018	99	−1.91	0.95
0.016	211	−1.70	0.85
0.014	337	−1.48	0.74
0.012	482	−1.27	0.64
0.010	654	−1.06	0.53
0.008	864	−0.85	0.42
0.006	1136	−0.64	0.32
0.004	1518	−0.42	0.21
0.002	2172	−0.21	0.11
0.001	2826	−0.11	0.05

EXAMPLE 13.1

RELATIVE RATES AND STOICHIOMETRY

Give the relative rates for disappearance of reactants and formation of products for the following reaction.

$$4\ PH_3(g) \longrightarrow P_4(g) + 6\ H_2(g)$$

Solution To equate rates, we have to divide Δ(reagent)/Δt by the stoichiometric coefficient in the balanced equation. Therefore,

$$\text{Reaction rate} = -\frac{1}{4}\left(\frac{\Delta[PH_3]}{\Delta t}\right) = +\frac{\Delta[P_4]}{\Delta t} = +\frac{1}{6}\left(\frac{\Delta[H_2]}{\Delta t}\right)$$

$$-\tfrac{1}{4}(\text{rate of disappearance of } PH_3) = \text{rate of appearance of } P_4$$
$$= \tfrac{1}{6}(\text{rate of appearance of } H_2)$$

Since 4 moles of PH_3 disappear for every mole of P_4 formed, the rate of P_4 formation can only be one fourth of the rate of PH_3 disappearance. Similarly, P_4 must appear at only one sixth of the rate that H_2 appears.

EXERCISE 13.1 RELATIVE RATES AND STOICHIOMETRY
What are the relative rates (including sign) for each reactant and each product in the decomposition of nitrosyl chloride, NOCl?

$$2\ NOCl(g) \longrightarrow 2\ NO(g) + Cl_2(g)$$

13.3 MEASURING CONCENTRATIONS IN RATE STUDIES

In order to calculate reaction rates, a chemist has to measure the concentration of a reactant or product at various times during a reaction. For slow reactions, this can be done by removing a small sample of the reaction mixture and analyzing its composition by a method such as a titration, as described in Chapters 4 and 16. If a gas is consumed or produced (as in the decomposition of H_2O_2), the change in pressure can be followed as the reaction proceeds.

These classical "bench chemistry" methods are being replaced by instrumental techniques that rely on changes in physical properties such as light absorption (Figure 13.5). If it can be determined that one reactant or product absorbs light of a particular wavelength that is not absorbed by any of the other compounds in the reaction mixture, then the amount of light absorbed from a beam of that wavelength is generally proportional to the concentration of the absorbing molecule.* Such methods have the advantage of being able to measure concentrations continuously, without the delay involved in sampling. They can be automated, and many such instruments can send their data directly to a computer for continuous monitoring.

13.4 EFFECT OF CONCENTRATION ON REACTION RATE

The relation between reaction rate and the concentrations of reactants is given by a mathematical equation called a *rate law*. This section is con-

*This proportionality is called Beer's law, and it is obeyed reasonably well by many gas mixtures and dilute solutions.

FIGURE 13.5

Vacuum apparatus used to measure the rate of formation of solid UO_2 on the surface of liquid uranium at 1300 K. (The pressure in the apparatus is 10^{-10} mm Hg.) A vidicon camera, a special television camera, is used to observe the reaction and to record information to magnetic tape. These data are then used to determine the reaction rate. (Laboratory of W. Mclean, Lawrence Livermore National Laboratory. Courtesy of R. Jarnigan)

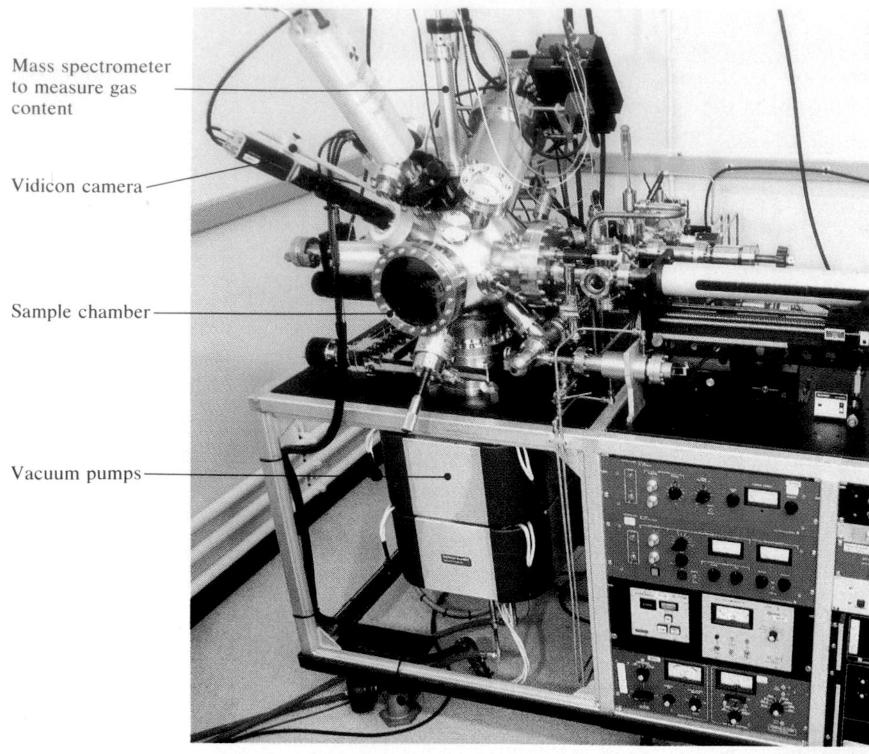

cerned with the form of the rate law and its derivation for some typical reactions.

THE RATE LAW

Reaction rates are proportional to reactant concentrations; the lower the concentration, the lower the rate. In some cases, rate and reactant concentration may be directly proportional, while in other cases, the rate may increase much faster than an increase in concentration. To find the exact relation between rate and concentration, *we must do some experiments* as outlined in the previous section, measuring the rate while varying reactant concentrations. Then, to show exactly how concentrations and reaction rate are related, we can write a **rate law**. For simpler reactions such as

$$a\,A + b\,B \xrightarrow{C} x\,X \qquad \text{where C is the catalyst}$$

the rate expression will always have the form

$$\text{Initial reaction rate} = k[A]_0^m[B]_0^n[C]^p$$

This equation expresses the fact that the initial rate of reaction is proportional to the reactant concentrations, each concentration being raised to some power. It is important to recognize that these exponents, m, n, and p here, are not necessarily the stoichiometric coefficients for the balanced chemical equation. The exponents, which can be zero or positive or negative whole numbers or fractions, must be determined by experiment. For example, experiment shows that the decomposition of H_2O_2 in the presence of iodide ion as catalyst has the rate law

$$\text{Initial rate} = k[H_2O_2]_0[I^-]_0$$

where the exponent on each concentration term is 1, even though the stoichiometric coefficient of H_2O_2 is 2 and I^- does not even appear in the balanced equation.

The symbol "k" is called the **rate constant** and is a proportionality constant relating rate and concentration. It must be evaluated for each reaction at each temperature.

Notice the word "initial" in the rate expression, and the 0 subscripts on the concentrations. The initial reaction rate is the instantaneous rate at the very beginning of the reaction,* as illustrated by Figure 13.4.

THE ORDER OF A REACTION

As you shall see in Section 13.7, a great deal of information is conveyed about reaction mechanisms by the "order" of a reaction. The **order** with respect to a particular reactant is the exponent of its concentration term in the rate law, and the **total reaction order** is the sum of the exponents on all the concentration terms.

An example of a **first order reaction** (a reaction with a total order of 1) is the change in the structure of a molecule. For example, azobenzene changes from one form (*cis*) to another (*trans*)

$$\underset{cis}{C_6H_5\!\!-\!\!N{=}N\!\!-\!\!C_6H_5} \longrightarrow \underset{trans}{C_6H_5\!\!-\!\!N{=}N\!\!-\!\!C_6H_5}$$

with an initial rate of 0.11 mol/(L · hr) when $[cis] = 2.2\ M$ at 77°C. The experimentally determined rate law for the process is

$$-\frac{\Delta[cis]_0}{\Delta t} = \frac{0.11\ \text{mol/L}}{\text{hr}} = k[cis]_0$$

This reaction is first order because the exponent of $[cis]$ in the rate law is 1. Experimentally, this means that the reaction rate changes in direct proportion with a change in $[cis]$. For example, if $[cis]$ is doubled, the rate will also double.

The decomposition of H_2O_2 catalyzed by iodide ion is a **second order reaction** overall, since the sum of the exponents is 2.

$$-\frac{\Delta[H_2O_2]_0}{\Delta t} = k[H_2O_2]_0[I^-]_0$$

---

*The initial rate and concentrations are used because, when the reaction has just begun, it is proceeding only in the forward direction. At the start there are no products present to allow the reverse reaction to replace reactants. See Section 13.7.

However, we would also say that the reaction is first order with respect to H_2O_2 and first order with respect to I^-, since the exponent on each concentration term is 1. This means that, if you double the concentration of either H_2O_2 or I^-, the reaction rate will double. If both concentrations are doubled, the rate would increase by a factor of four.

Referring to the I^- catalyzed decomposition of H_2O_2, what would happen to the initial rate if the initial concentrations of both H_2O_2 and I^- are halved?

First and second order reactions are quite common, but other orders are observed as well. One thoroughly studied reaction (whose mechanism is described in Section 13.7) is the oxidation of NO to give NO_2, a **third order** reaction.

$$2\ NO(g) + O_2(g) \longrightarrow 2\ NO_2(g)$$

$$-\frac{\Delta[NO]_0}{\Delta t} = k[NO]_0^2[O_2]_0$$

Fractional orders are also seen, as in the combination of H_2 and Br_2 to give HBr.

$$H_2(g) + Br_2(g) \longrightarrow 2\ HBr(g)$$

$$-\frac{\Delta[H_2]_0}{\Delta t} = k[H_2]_0[Br_2]_0^{1/2}$$

Reactions can also be zero order in a reactant. In such cases changes in reactant concentration have no effect on reaction rate.

Here the total reaction order is $\frac{3}{2}$.

DETERMINATION OF THE RATE LAW

We do not have any generally useful theory that will allow us to predict rate laws.

The rate law must be found from experiments that are designed to determine the rate as the initial concentration of one reagent at a time is varied. The general approach is to examine how the initial rate varies with the initial concentration of one reactant, while the other reactants have fixed initial concentrations.

To see how to determine a rate law, let us take a very simple reaction, A → B. We assume the rate depends only on the concentration of A in some manner. Thus, the rate law will be "initial rate $= k[A]_0^m$." Now, it is possible that the exponent m is 0, $\frac{1}{2}$, 1, 2, and so on (or even a negative number). The table below shows how the rate would depend on $[A]_0$ for each of these cases.

VALUE OF m (REACTION ORDER)	EFFECT ON INITIAL RATE WHEN $[A]_0$ IS DOUBLED FROM 1 *M* to 2 *M*
0	$[2]^0 = 1$, so no effect on rate
$\frac{1}{2}$	$[2]^{1/2} = 1.41$, so rate increases by 1.41
1	$[2]^1 = 2$, so rate increases by 2
2	$[2]^2 = 4$, so rate increases by 4

As an example of the determination of reaction order, let us analyze experimental data for the reaction

$$a\ A + b\ B \longrightarrow x\ X$$

$$\text{Initial rate} = -\frac{\Delta[A]}{\Delta t} = k[A]_0^m[B]_0^n$$

where we assume the initial rate does not depend in some way on [X], but only on [A] and [B]. It can be seen in the first three lines of data in

TABLE 13.1 Rate Data for the Reaction a A + b B → x X

EXPERIMENT NUMBER	INITIAL RATE mol/(L · hr)	INITIAL CONCENTRATIONS (mol/L) $[A]_0$	$[B]_0$
1	0.50×10^{-2}	0.50	0.20
2	0.50×10^{-2}	0.75	0.20
3	0.50×10^{-2}	1.00	0.20
4	1.00×10^{-2}	0.50	0.40
5	1.50×10^{-2}	0.50	0.60

Table 13.1 that $[A]_0$ was varied as $[B]_0$ was kept fixed. Even though $[A]_0$ was doubled from Experiment 1 to Experiment 3, the initial reaction rate was unchanged. The initial rate is independent of $[A]_0$, so m must be 0. The reaction is zero order in [A].

Now consider the dependence of initial rate on $[B]_0$. In Experiments 4 and 5 in Table 13.1, $[A]_0$ is fixed and $[B]_0$ varies. Here $[B]_0$ increases by a factor of 1.5, and the initial rate increases by the same factor. Thus, the change in rate is *directly* proportional to the change in $[B]_0$, so $n = 1$, and the reaction is first order in $[B]_0$. (Do experiments 1 and 4 agree with an order of 1 for $[B]_0$?)

In summary, the dependence of rate on the concentrations in Table 13.1 gives a rate expression of

$$\text{Initial rate} = k[A]_0^0[B]_0^1 = k[B]_0$$

Here the overall reaction order is 1, since $m + n = 0 + 1 = 1$.

In the special case where all reactants have initial concentrations of 1 M, we get

$$\text{Initial rate} = k[A]_0^m[B]_0^n \ldots = k$$

The value of the rate in this case is called the **specific rate**, and, since it is equal to k, the constant k is sometimes called the **specific rate**. None of the data sets in Table 13.1 gives the specific rate directly. However, knowing the rate law, k can be calculated from one of the sets. Using Experiment 3, for example, we have

$$0.50 \times 10^{-2} \text{ mol/(L} \cdot \text{hr)} = k(1.00 \text{ mol/L})^0(0.20 \text{ mol/L})^1$$

$$0.50 \times 10^{-2} \text{ mol/L} \cdot \text{hr} = k(0.20 \text{ mol/L})$$

and so

$$k = 2.5 \times 10^{-2}\text{/hr}$$

The values of the exponents on the concentration terms in a rate expression can often be found simply by inspecting the experimental data, but a more rigorous algebraic method must be used at times.

EXAMPLE 13.2

DETERMINATION OF REACTION ORDER

Derive the rate expression and the value of k for the decomposition of acetaldehyde

$$CH_3CHO(g) \longrightarrow CH_4(g) + CO(g)$$

using the data below. Experiment shows that CH_4 and CO are not involved in the rate law.

EXPERIMENT NUMBER	INITIAL CONCENTRATION OF CH_3CHO (mol/L)	INITIAL RATE OF DISAPPEARANCE OF CH_3CHO mol/(L·s)
1	0.10	0.020
2	0.20	0.080
3	0.30	0.180
4	0.40	0.320

Solution Here the rate law must be

$$\text{Initial rate} = -\frac{\Delta[CH_3CHO]_0}{\Delta t} = k[CH_3CHO]_0^m$$

where m, the reaction order, and k, the rate constant, are unknown. Inspecting the data above, you see that the initial rate increases by a factor of 4 when the initial concentration doubles. For example, the CH_3CHO concentration doubles from Experiment 1 to Experiment 2, and the rate increases by (0.080/0.020) = 4. Since the rate increases as the *square* of the concentration, the value of m must be 2, and the reaction order is 2. This means the rate is

$$\text{Initial rate} = k[CH_3CHO]_0^2$$

Now that the rate expression is known, we can substitute one set of experimental information into the expression and find k. Taking data from Experiment 1, we have

$$0.020 \text{ mol/(L} \cdot \text{s)} = k\,(0.10 \text{ mol/L})^2$$

and so

$$k = 2.0 \text{ L/mol} \cdot \text{s}$$

EXAMPLE 13.3

ALGEBRAIC METHOD FOR FINDING REACTION ORDERS

In this example we shall use the same data as in Example 13.2, but now we shall use an algebraic approach.

Solution The first step here is to take the ratio of rates for any two experiments, say 1 and 2.

$$\frac{\text{Rate 1}}{\text{Rate 2}} = \frac{0.020 \text{ mol/(L} \cdot \text{s)}}{0.080 \text{ mol/(L} \cdot \text{s)}} = \frac{k[CH_3CHO]_1^m}{k[CH_3CHO]_2^m} = \frac{k(0.10)^m}{k(0.20)^m}$$

This simplifies to

$$\frac{1}{4} = \left(\frac{1}{2}\right)^m$$

Although it may be obvious now that $m = 2$ (since $\frac{1}{2}$ squared is $\frac{1}{4}$), a general approach here is to take the logarithm of each side of the equation.

$$\log\left(\frac{\text{rate 1}}{\text{rate 2}}\right) = \log\left(\tfrac{1}{4}\right) = \log\left(\tfrac{1}{2}\right)^m$$

and so

$$\log\left(\tfrac{1}{4}\right) = m \log\left(\tfrac{1}{2}\right)$$

and

$$(-0.602) = m\,(-0.301)$$

which gives a value of $m = 2$.

Rules for working with logarithms are summarized in Appendix A.

EXERCISE 13.2 REACTION ORDER

In the following reaction, a Co—Cl bond is replaced by a Co—OH_2 bond.

$$[Co(NH_3)_5Cl]^{2+}(aq) + H_2O(\ell) \longrightarrow [Co(NH_3)_5H_2O]^{3+}(aq) + Cl^-(aq)$$
$$\text{Initial rate} = k\{[Co(NH_3)_5Cl]^{2+}\}_0^m$$

Using the data below, find the value of m in the rate expression and calculate the value of k.

EXPERIMENT	INITIAL CONCENTRATION OF $[Co(NH_3)_5Cl]^{2+}$ (mol/L)	INITIAL RATE mol/(L · min)
1	1.0×10^{-3}	1.3×10^{-7}
2	2.0×10^{-3}	2.6×10^{-7}
3	3.0×10^{-3}	3.9×10^{-7}
4	1.0×10^{-2}	1.3×10^{-6}

13.5 RELATIONSHIPS FOR CONCENTRATION AND TIME

Chemists often wish to know how long a reaction must proceed to reach a predetermined particular concentration of some reagent, or what the reactant and product concentrations will be after some time has elapsed. One way to do this is to collect experimental data and construct curves such as those in Figure 13.3. However, this is inconvenient and time consuming. A simpler approach is to determine the reaction order for each reagent by experiment and then derive an equation that relates concentration and time. This equation will then allow us to calculate a concentration at any time, or vice versa.

FIRST AND SECOND ORDER CONCENTRATION/TIME EQUATIONS

As an example of a first order process, let us return to a simple reaction such as the transformation of a *cis* compound to a *trans* compound (page 475).

$$A \longrightarrow B$$

$$-\frac{\Delta[A]}{\Delta t} = k[A]$$

Using the methods of calculus, this equation can be transformed into another very useful expression.

Equation 13.1 is valid for cases where the product B does not revert to the reactant A. This is assumed to be true in the examples in this chapter.

$$\ln\left(\frac{[A]}{[A]_0}\right) = -kt \qquad (13.1)$$

If $[A]_0$ is the initial concentration of the reactant (at $t = 0$), then Equation 13.1 allows us to find the concentration of A at some other time t (=[A]). Alternatively, it can be used to find the time elapsed until A reaches some predetermined concentration.

ln is a natural logarithm (see Appendix A). In base-10 logarithms, Equation 13.1 is

$$2.303 \log \frac{[A]}{[A]_0} = -kt$$

EXAMPLE 13.4

THE FIRST ORDER RATE EQUATION

The transformation of cyclopropane to propene is first order in the cyclo compound.

$$\underset{\text{cyclopropane}}{\text{cyclo-}(CH_2)_3} \longrightarrow \underset{\text{propene}}{CH_3{-}CH{=}CH_2} \qquad \text{Rate} = k[\text{cyclo}]$$

If the initial concentration of cyclopropane is 0.050 *M*, how many hours must elapse for the concentration of the compound to drop to 0.010 *M*? (The rate constant k is 5.4×10^{-2}/hr.)

Solution The first order rate equation applied to this reaction is

$$\ln \frac{[\text{cyclo}]}{[\text{cyclo}]_0} = -kt$$

where [cyclo], $[\text{cyclo}]_0$, and k are given.

$$\ln\left(\frac{0.010}{0.050}\right) = -(5.4 \times 10^{-2}/\text{hr})t$$

$$\frac{-\ln(0.20)}{5.4 \times 10^{-2}/\text{hr}} = t$$

$$\frac{-(-1.61)}{5.4 \times 10^{-2}/\text{hr}} = t$$

$$t = 30.\ \text{hours}$$

To derive the first order rate equation using calculus, we first write it in differential form.

$$-\frac{dA}{dt} = k[A]$$

After rearranging the equation we integrate it between the limits time = 0 to time = t.

$$\int_{t=0}^{t=t} \frac{dA}{[A]} = -k \int_{t=0}^{t=t} dt$$

which gives $\{\ln[A] - \ln[A]_0\} = -k(t - 0)$. When this is rearranged, Equation 13.1 is obtained.

EXAMPLE 13.5

USING THE FIRST ORDER RATE EQUATION

In dilute NaOH at 20°C, the decomposition of hydrogen peroxide is first order in H_2O_2 only,

$$2\ H_2O_2(aq) \longrightarrow 2\ H_2O(\ell) + O_2(g) \qquad \text{Rate} = k[H_2O_2]$$

and the rate constant is 1.06×10^{-3}/min. If the initial concentration of H_2O_2 is 0.020 *M*, what is the concentration after exactly 100 minutes?

Solution Here Equation 13.1 is written as

$$\ln\frac{[H_2O_2]}{[H_2O_2]_0} = -kt$$

where $[H_2O_2]_0$, k, and t are known. To calculate $[H_2O_2]$, the peroxide concentration after 100 minutes, it is convenient to rearrange the equation to

$$\ln[H_2O_2] - \ln[H_2O_2]_0 = -kt$$

Substituting into this expression, we have

$$\ln[H_2O_2] - \ln(0.020) = -(1.06 \times 10^{-3}/\text{min})(100\ \text{min})$$
$$\ln[H_2O_2] - (-3.91) = -(1.06 \times 10^{-1})$$
$$\ln[H_2O_2] = -3.91 - 0.106$$
$$\ln[H_2O_2] = -4.02$$

Taking the antilogarithm of -4.02, we find the concentration of peroxide after 100 minutes as

$$[H_2O_2] = 0.018\ M$$

EXERCISE 13.3 USING THE FIRST ORDER RATE EQUATION

The rate constant for the first order conversion of *cis* to *trans*-$C_6H_5—N{=}N—C_6H_5$ is 2.22/hr.

$$\underset{cis}{C_6H_5\diagdown N{=}N \diagup C_6H_5} \longrightarrow \underset{trans}{C_6H_5\diagdown N{=}N \diagdown C_6H_5}$$

If the initial concentration of *cis* is 0.010 M, what is the concentration of the compound after 19 minutes?

The time dependence of concentration is different for a second order reaction than for a first order reaction. For a general reaction that is *second order with respect to A and second order overall*, the rate law is

$$-\frac{\Delta[A]}{\Delta t} = k[A]^2$$

and the concentration/time equation is

$$\frac{1}{[A]} - \frac{1}{[A]_0} = kt \qquad (13.2)$$

Equation 13.2 may also be used for reactions that are 2nd order overall and 1st order with respect to each of two reactants A and B (Rate = $k[A][B] = k[A]^2 = k[B]^2$) when $[A]_0 = [B]_0$.

Taking the gas phase decomposition of HI as an example,

$$HI(g) \longrightarrow \tfrac{1}{2}H_2(g) + \tfrac{1}{2}I_2(g)$$

Rate of disappearance of HI $= k[HI]^2$

$$k = \frac{30.}{\text{mol/(L}\cdot\text{min)}}\ (\text{at } 443°C)$$

We can calculate that only 3.3 minutes are required for the concentration to fall from 0.010 M to 0.0050 M at 443°C.

EXERCISE 13.4 USING THE SECOND ORDER CONCENTRATION/TIME EQUATION

Using the rate constant given in the text above, calculate the concentration of HI after 10. minutes have elapsed for the reaction at 443°C.

HALF-LIFE AND REACTION RATE

The rate constant k is a good indicator of the speed of a chemical reaction, and it is a compact way of comparing one reaction with another. Another useful measure of reaction speed is the **reaction half-life, $t_{1/2}$**. As we shall now show, k and $t_{1/2}$ are related to one another in simple ways.

The half-life of a reaction is the time required for the concentration of one of the reactants to decrease to $\frac{1}{2}$ of its initial value. Specifically, for reactant A, $t_{1/2}$ is the time when

$$[A] = \tfrac{1}{2}[A]_0 \qquad \text{or} \qquad \frac{[A]}{[A]_0} = \tfrac{1}{2}$$

where $[A]_0$ is the initial concentration and [A] is the concentration after the reaction is half completed. To find $t_{1/2}$ we use the concentration/time equations we derived above (Equations 13.1 and 13.2).

HALF-LIFE OF A FIRST ORDER REACTION Taking the first order rate equation,

$$\ln \frac{[A]}{[A]_0} = -kt$$

and substituting the fact that $[A] = \frac{1}{2}[A]_0$ when $t = t_{1/2}$, we have

$$\ln \frac{1}{2} = -kt_{1/2}$$

or

$$\ln 2 = kt_{1/2}$$

Solving this for $t_{1/2}$, we find

$$t_{1/2} = \frac{\ln 2}{k} = \frac{0.693}{k}$$

To illustrate the concept of half-life for first order reactions, we have reproduced Figure 13.3, the plot of $[H_2O_2]$ versus time for the decomposition of the hydrogen peroxide (Figure 13.6). Since the rate constant k for this reaction is 1.06×10^{-3}/min, the half-life is

$$t_{1/2} = \frac{0.693}{k} = \frac{0.693}{1.06 \times 10^{-3}/\text{min}} = 654 \text{ min}$$

If the initial concentration of H_2O_2 is 0.020 M, the concentration is only 0.010 M after 654 minutes. Notice that the concentration drops again by half after another 654 minutes. That is, after two half-lives (1308 minutes) the concentration is only $(\frac{1}{2})(\frac{1}{2}) = \frac{1}{4}$ or 25% of the initial concentration. After three half-lives (1962 minutes), the concentration has now dropped to $(\frac{1}{2})(\frac{1}{2})(\frac{1}{2}) = \frac{1}{8}$ or 12.5% of the initial value; here $[H_2O_2] = 0.0025\ M$.

The concept of a half-life of a first order process is widely used, especially when referring to the rate of decay of radioactive isotopes as you will see in Chapter 26. Here we wish to apply it to some other first order reactions.

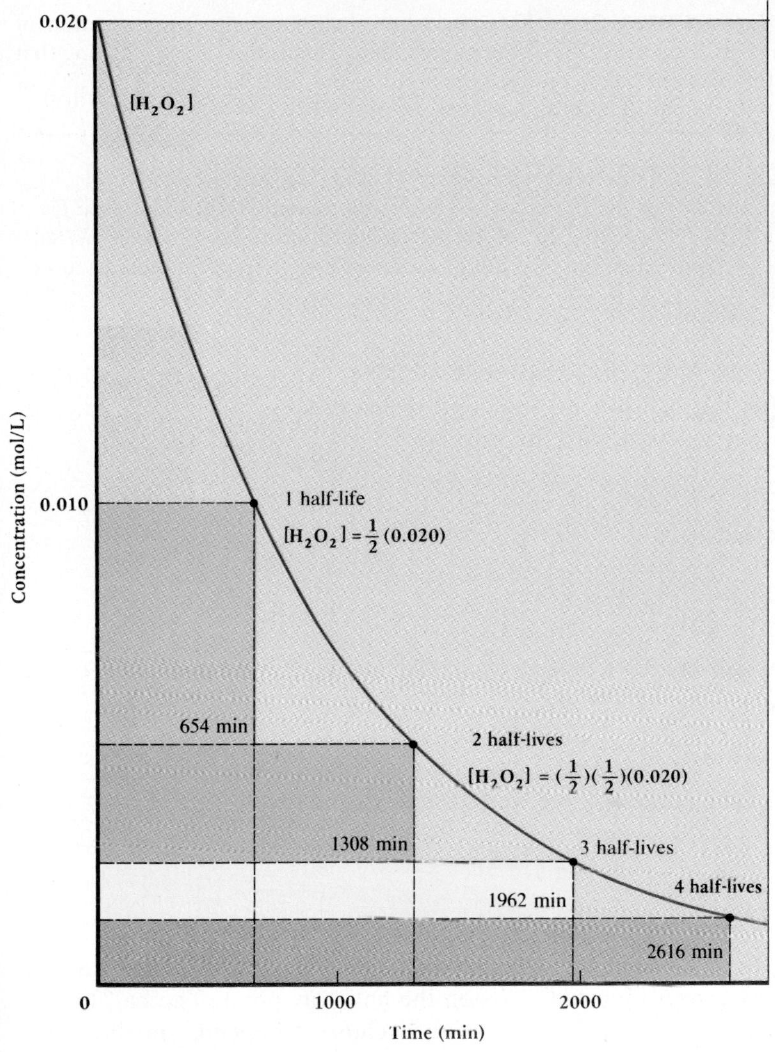

FIGURE 13.6

The concentration versus time curve for the disappearance of H_2O_2. At 654 minutes ($t_{1/2}$) the initial concentration of H_2O_2 has halved. During each successive interval of 654 minutes the concentration again halves.

EXAMPLE 13.6

HALF-LIFE

Molecules with coordinate covalent bonds, B:→A, may exchange one of the pieces, A or B, with an excess of that reagent. For example,

$$R_3N{:}\!\rightarrow GaR_3 + {}^*GaR_3 \longrightarrow R_3N{:}\!\rightarrow {}^*GaR_3 + GaR_3$$

where R is CH_3 in this case and the asterisk * helps us keep track of the GaR_3 molecule that is exchanging. The rate law is

$$\text{Initial rate} = k[R_3N{:}\!\rightarrow GaR_3]_0$$

and k is 31/s. Find the half-life of the reaction (when half of GaR_3 has been replaced by $^*GaR_3$). Calculate the time required for 87.5% of the GaR_3 to be replaced.

Solution The half-life of the exchange is

$$t_{1/2} = \frac{0.693}{31/\text{s}} = 0.022 \text{ s}$$

and tells us that the reaction is very rapid.

In Figure 13.6 you see that, after three half-lives, the concentration of reactant has dropped to 12.5% of the initial value. This is the same as saying that 87.5% of the reactant has been used. Therefore, the time required for 87.5% of the GaR_3 to be replaced is three half-lives or 3×0.022 s = 0.066 s.

EXERCISE 13.5 THE HALF-LIFE OF A FIRST ORDER REACTION
The rate constant for the transformation of cyclopropane to propene (see Example 13.4) is 5.40×10^{-2}/hr. What is the half-life of this reaction? What fraction of cyclopropane remains after 51.2 hours? What fraction remains after 18 hours?

HALF-LIFE FOR A SECOND ORDER REACTION An equation can also be derived for the half-life of a second order process. For the particular second order reaction with the rate law

$$\text{Rate} = k[\text{A}]^2$$

we know that

$$\frac{1}{[\text{A}]} - \frac{1}{[\text{A}]_0} = kt$$

Since one half-life has elapsed when $[\text{A}] = [\text{A}]_0/2$, we have

$$\frac{2}{[\text{A}]_0} - \frac{1}{[\text{A}]_0} = kt_{1/2}$$

Solving for the initial $t_{1/2}$, we find an interesting result.

$$t_{1/2} = \frac{1}{k[\text{A}]_0}$$

For both first and second order reactions $t_{1/2}$ is inversely proportional to k. This is expected, since the larger the value of k the faster the reaction and the less time is required to reach the half-way point. The half-life for a second order reaction is interesting because it depends on the initial concentration of reactant: this means that one can shorten the half-life by increasing the initial concentration $[\text{A}]_0$. In Section 13.6 you will see why this must be true.

GRAPHICAL METHOD FOR DISTINGUISHING FIRST AND SECOND ORDER REACTIONS

The algebraic method introduced in Section 13.4 for determining reaction order is the method to use when initial rates and reactant concentrations are available. In the laboratory, however, initial rates are not *directly* determined. Rather, we collect information on the concentrations of reactants and products as a function of time. Only after drawing a plot such as that in Figure 13.3 can you draw a plot of rate versus concentration (Figure 13.4). It is from this plot that you can then determine, by extrapolation, the initial rate of reaction for a given concentration. Then, armed with some initial rates as a function of initial concentration, you can derive the reaction order.

A more convenient method of determining reaction order is suggested by Equations 13.1 and 13.2, the rate equations for first and second order processes. Rearranging both of these equations slightly,

First order	Second order
$\ln[A] = -kt + \ln[A]_0$	$\dfrac{1}{[A]} = kt + \dfrac{1}{[A]_0}$
$\downarrow \quad \Downarrow \quad \downarrow$	$\downarrow \quad \Downarrow \quad \downarrow$
$y = ax + b$	$y = ax + b$

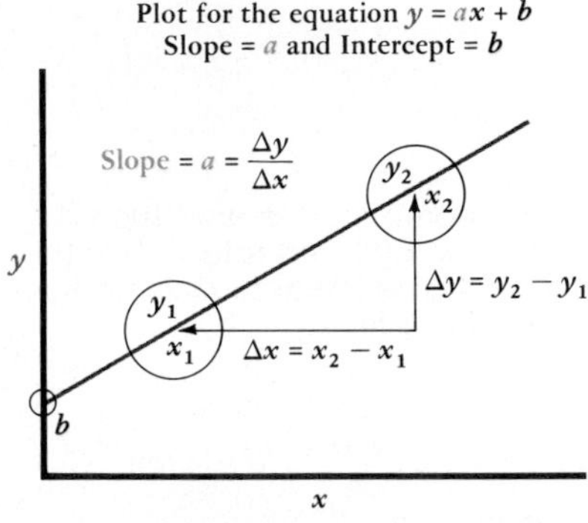

you see that each is an equation of a straight line, $y = ax + b$, where a is the *slope* of the line, and b is the *intercept* when $x = 0$.

A plot of concentration versus time for a first order reaction is a curved line (Figure 13.3). However, if the data are plotted as ln[reactant] versus time, a straight line should result. In Figure 13.7 we have plotted the data for the first order decomposition of H_2O_2 in this way and do indeed see a straight line. Notice that the slope of the line is negative, because of the negative sign in the equation ($\ln[A] \propto -t$). *Only for a first order reaction will you observe a straight line of negative slope if you plot the logarithm of concentration against time.* Not only can you pin-

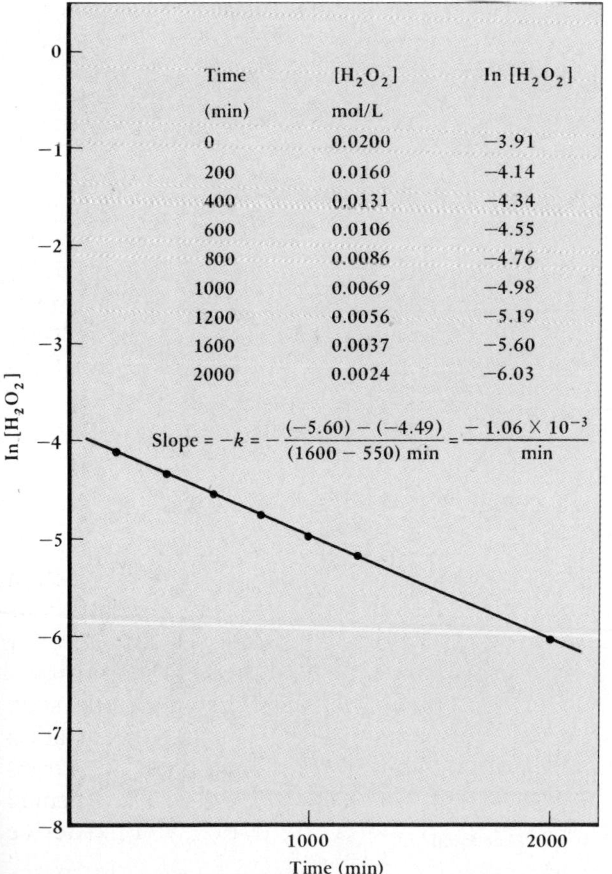

Time (min)	$[H_2O_2]$ mol/L	$\ln [H_2O_2]$
0	0.0200	−3.91
200	0.0160	−4.14
400	0.0131	−4.34
600	0.0106	−4.55
800	0.0086	−4.76
1000	0.0069	−4.98
1200	0.0056	−5.19
1600	0.0037	−5.60
2000	0.0024	−6.03

FIGURE 13.7

Plot of ln $[H_2O_2]$ versus time. The straight line plot of negative slope indicates a first order reaction. The negative of the slope gives the rate constant k.

point the reaction order this way but you can also find the rate constant from the slope of the plot ($k = -\text{slope}$).

The decomposition of NO_2 is a second order process.

$$NO_2(g) \longrightarrow NO(g) + \tfrac{1}{2}\,O_2(g)$$

$$\text{Initial rate} = k[NO_2]_0^2$$

and concentration versus time data have been plotted in Figure 13.8 in two ways: (a) $\ln[NO_2]$ versus time and (b) $1/[NO_2]$ versus time. Now you see that a plot of $1/[NO_2]$ gives a straight line as predicted for a second order process. Conversely, plotting the data as though it were a first order reaction clearly gives a curved line (in Figure 13.8a). These examples of two commonly observed rate laws illustrate a useful way to determine reaction order: A straight line is observed *only* when $\ln[A]$ is plotted against time for a first order process (Rate $= k[A]$) or when $1/[A]$ is plotted against time for a second order process of the type Rate $= k[A]^2$.

More complex rate laws such as rate $= k[A][B]$ and rate $= k[A]^2[B]$ will not give linear plots for $\ln[A]$ or $1/[A]$ versus time.

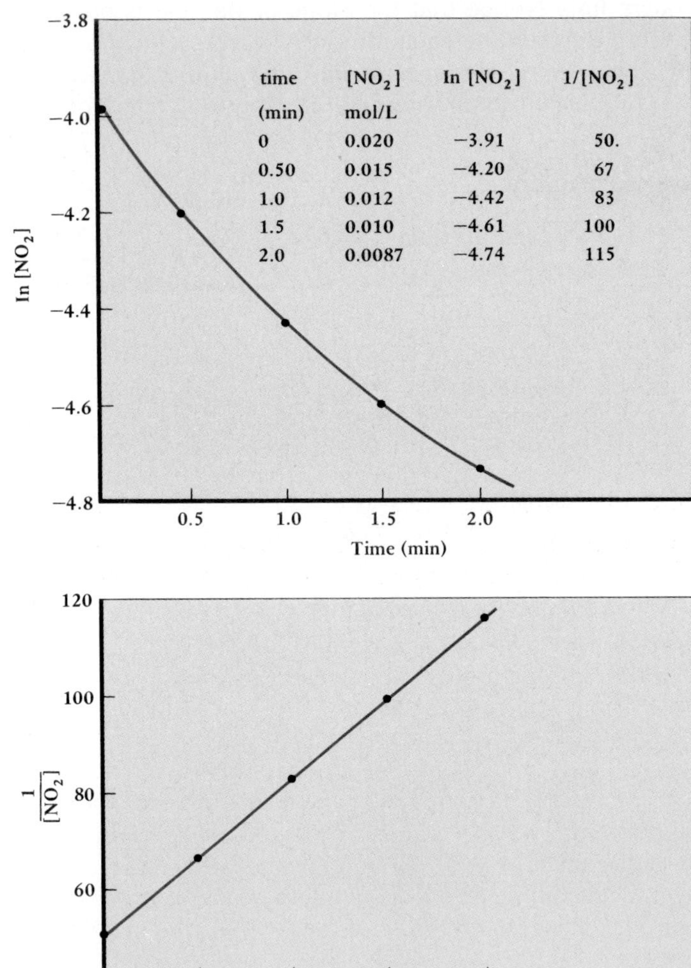

time (min)	$[NO_2]$ mol/L	$\ln[NO_2]$	$1/[NO_2]$
0	0.020	−3.91	50.
0.50	0.015	−4.20	67
1.0	0.012	−4.42	83
1.5	0.010	−4.61	100
2.0	0.0087	−4.74	115

FIGURE 13.8
Concentration versus time curves of NO_2 decomposition. (a) $\ln[NO_2]$ versus time. (b) $1/[NO_2]$ versus time. For a rate law of the form Rate $= k[A]^2$, a straight line is observed only when $1/[A]$ is plotted against time.

EXERCISE 13.6 DETERMINATION OF REACTION ORDER BY A GRAPHICAL METHOD

For the reaction of $(CH_3)_3CBr$ with OH^-,

$$(CH_3)_3CBr + OH^- \longrightarrow (CH_3)_3COH + Br^-$$

the following data were obtained in the laboratory.

TIME (seconds)	$[(CH_3)_3CBr]$ (mol/L)
0	0.10
30	0.074
60	0.055
90	0.041

Plot these data as $\ln[(CH_3)_3CBr]$ versus time and $1/[(CH_3)_3CBr]$ versus time. Is the reaction first order or second order? What is the value of the rate constant k?

13.6 RATE CHANGES WITH TEMPERATURE

It is generally observed that reactions occur faster at higher temperatures and slower at lower temperatures (Figure 13.9). The decomposition of dinitrogen pentoxide, for example,

$$2\ N_2O_5(g) \longrightarrow 4\ NO_2(g) + O_2(g)$$

has a rate constant of 3.46×10^{-5}/s at 25°C, while it is 1.35×10^{-4}/s at 35°C, an increase in rate by a factor of almost four for a 10 degree rise in temperature.

In this section we want to explore the theories of the way reactions occur in order to understand the cause of the temperature dependence of reaction rates.

FIGURE 13.9

Hydrolysis of a green cobalt (II) compound. The compound has two Co—Cl bonds, one of which is replaced by a bond between Co^{2+} and H_2O to give a red compound. The reaction is slow at room temperature, so the solution at the right has remained green. If the solution is placed in warm water, however, the rate increases and the solution soon turns pink. (Charles D. Winters)

TRANSITION STATE THEORY

Understanding why there are wide variations in reaction rates and why reactions are faster at higher temperatures begins with the idea that there is an energy barrier to all reactions. An analogy is found in the following scene. Suppose you have taken a basketball to the observation deck of the Eiffel Tower in Paris, France. Rolling the ball toward the edge does not mean the ball will fall to the ground, even though there is a great driving force for it to do so, because there is a barrier around the deck. Only if the ball has sufficient vibrational or ''bouncing energy'' is it able to get over the barrier and begin its plunge to the ground. If several bags of balls were emptied onto the observation deck, only those bouncing vigorously would get over the barrier. The number of balls falling from the top of the tower in some time (the rate) would depend on the number of balls (the concentration) and the fraction of the total that bounced with enough energy.

In chemical reactions, vibrating atoms are like the basketballs on the Eiffel Tower. Chemical reactions also encounter energy barriers to

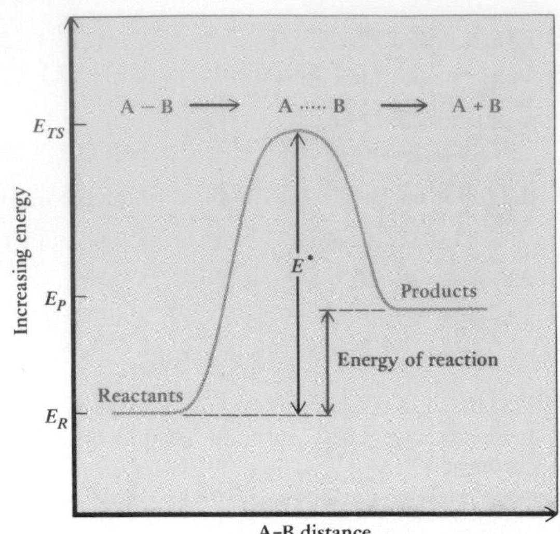

FIGURE 13.10
Energy diagram for a reaction step. Reactant energies, E_R, are given by the "foot" of the barrier at the left and product energies, E_P, are given by "foot" of the barrier at the right. The net reaction energy is given by $E_P - E_R$; the difference may have a positive (endothermic) or negative (exothermic) sign. The activation energy for the forward reaction E_f^* (R → P) is the difference between the transition state energy (E_{TS}) and the reactant energy, $E_f^* = E_{TS} - E_R$.

the separation of atoms or to the transfer of an atom between two molecular fragments. When the barriers to reaction are too high, the reaction cannot occur at a detectable rate because too few molecules are sufficiently energetic to cross the barrier.

Chemists illustrate the notion of a barrier with a **reaction energy diagram**. The way the energy changes as the bond between atoms A and B in a molecule is broken and the fragments separate is shown in Figure 13.10. The energy first increases as the bond is broken, and then it decreases as the electrons and other nuclei in the A- and B-containing fragments rearrange in space in order to keep the energy as low as possible. This is why we speak of the reaction passing over an energy barrier in the course of the reaction.

An example of a bond-breaking reaction that would have an energy diagram like that in Figure 13.10 is

$$R_3C{-}Cl \longrightarrow R_3C^+ + Cl^-$$

where R is the CH_3 or methyl group. In the course of the reaction the R—C—R bond angles increase, and the C—Cl bond lengthens. This is

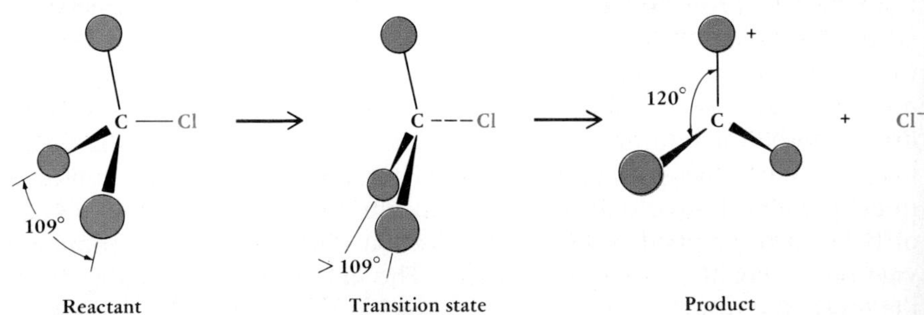

the reason the energy increases, finally reaching a maximum at the top of the barrier when the C—Cl bond is broken. Then, after the Cl^- ion has departed, the the R—C bonds shorten and strengthen and the energy of the system drops.

The energy at the maximum of the reaction energy diagram is called the **transition state energy**, E_{TS}, because the reacting system is in the *transition state* between reactants and products. A system in the transition state is called the **activated complex**, and the difference in energy between the reactant molecule or molecules and the transition state is called the **activation energy**, E_f^*. This is the energy that the reacting system must absorb from its environment in order to react. At any temperature above 0 K, all molecules have increasingly vibrating bonds. In the case of the R_3C—Cl reaction above, the C—Cl bond can break if the amplitude of the vibration of the bond becomes large enough, but this can occur only if the molecule absorbs enough energy.

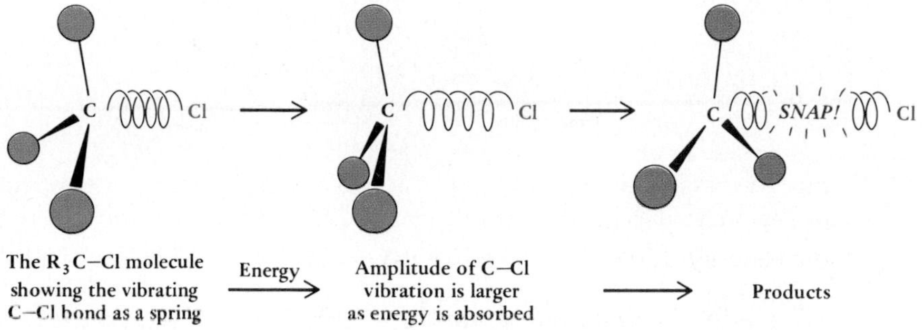

C Cl ⟶ C Cl ⟶ C SNAP! Cl

The R_3C–Cl molecule showing the vibrating C–Cl bond as a spring —Energy⟶ Amplitude of C–Cl vibration is larger as energy is absorbed ⟶ Products

In the case of $(CH_3)_3C$—Cl, the activation energy is about 100 kJ/mol, depending on the solvent in which the reaction occurs.

KINETICS AND NET ENERGY OF REACTION The theoretical approach to reaction rates we have been describing is called **transition state theory**. One aspect of this theory is the relationship between kinetics and thermodynamics. After the system passes over the energy barrier in Figure 13.10 the energy of the products, E_P, is greater than that of the reactants, E_R. Thus, a net amount of energy is consumed; the reaction is *endothermic*.

In contrast, the reaction illustrated in Figure 13.11 is *exothermic*. Here the reaction of 1 mole of H_2 and 1 mole of I_2 to give 2 moles of HI has an activation energy of 170 kJ. The energy of the products is lower than that of the reactants, so the reaction is exothermic by approximately 13 kJ ($\Delta E_{rxn} \simeq -13$ kJ).

When reactants have accumulated energy and are poised at the top of the activation energy barrier, they may progress to products or may fall back down the barrier to give reactants again. This implies that any reaction and its exact reverse must have the same activated complex, so the net energy of reaction ΔE is just the difference between the activation energy in the forward direction, E_f^*, and the activation energy for the reverse process, E_r^*. Since the net energy for $H_2 + I_2 \rightarrow 2\ HI$ ($= \Delta E$) is -13 kJ,

$$\Delta E = E_f^* - E_r^* = 170 - 183 = -13 \text{ kJ}$$

this means the net energy for the *reverse* reaction, the decomposition of HI to H_2 and I_2, is $+13$ kJ ($= \Delta E'$).

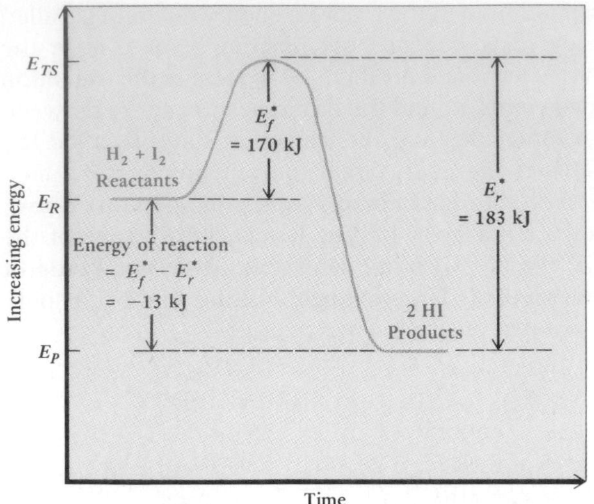

FIGURE 13.11
Reaction energy diagram for an exothermic process. Here 1 mole each of H_2 and I_2 produce 2 moles of HI with an activation energy (E_f^*) of 170 kJ. The reverse reaction, 2 moles of HI producing H_2 and I_2, has an activation energy (E_r^*) of 183 kJ. Therefore, the net energy change for $H_2 + I_2 \rightarrow 2\text{ HI}$ is $E_f^* - E_r^* = -13$ kJ.

$$\text{Net energy (for } 2\text{ HI} \longrightarrow H_2 + I_2) = \Delta E'$$

$$\Delta E' = E_f^{*\prime} - E_r^{*\prime} = 183 - 170 = +13 \text{ kJ}$$

COLLISION THEORY OF REACTION RATES

The activation energy largely determines the great variation observed in rates of different reactions at a given temperature. In addition, it lies at the heart of an explanation of why the rate of a single reaction can change with temperature. To do this, we turn to the **collision theory**. This theory assumes that a *minimum* condition for two molecules A and B to react is that their centers of mass must come within a certain critical distance of each other. It is relatively simple to calculate the number of collisions that must occur between two gas phase molecules A and B under given conditions. For example, if both gases are at 1 atm pressure and 0°C, it is found that the collision rate is on the order of 10^8 mol/(L · s). This corresponds (when $E_f^* = 0$) to an enormous reaction rate. If two reacting gases are mixed under these conditions, they would be almost entirely consumed in 10^{-7} seconds! There are in fact a few such reactions, and among them is the reaction of oxygen atoms with NO_2.

$$O(g) + NO_2(g) \longrightarrow O_2(g) + NO(g)$$

On the other hand, there are many common reactions that have rates in the range 10^{-2} mol/(L · s), reactions that are 10^{10} *slower* than predicted by assuming that a reaction occurs every time two colliding molecules react. Therefore, there must be other criteria determining reaction rate than the collision rate.

One of those other criteria is that reacting molecules must collide in an *effective* way. For example, when two HI molecules react to give H_2 and I_2

$$2\text{ HI}(g) \longrightarrow H_2(g) + I_2(g)$$

the activated complex is thought to have a trapezoidal geometry.

$$\begin{matrix} H & & H \\ | & + & | \\ I & & I \end{matrix} \longrightarrow \left[\begin{matrix} H \cdots H \\ \vdots \quad \vdots \\ I \cdots\cdots I \end{matrix} \right] \longrightarrow \begin{matrix} H-H \\ + \\ I-I \end{matrix}$$

activated complex

Any collision that does not come close to this favorable arrangement will not lead to reaction. Since chance dictates that there will be many more collisions that don't have the correct geometry than do, this cuts down our original estimate of reaction rate based simply on the number of collisions. For simple molecules, the reduction is about 10^{-1}, but for more complicated molecules it can be as much as 10^{-5}. This means that, even for complicated molecules, the favorable collision rate should be around 10^5 mol/(L · s). Unfortunately, this is still not low enough to agree with rates of most reactions, so there must be yet another factor.

The clue to further understanding is that reaction rates are very sensitive to temperature (Figure 13.9). The fundamental reason is that a rise in temperature causes molecules to move about more rapidly and leads *both* to more energetic collisions *and* to a greater number of collisions per second. The first of these is the more important.

Recall from the discussion of kinetic molecular theory in Chapter 6 that the average velocity of gas molecules and, even more importantly, the *distribution of molecular velocities*, are sensitive to temperature. As shown on the left in Figure 13.12, the number of molecules with high energy increases greatly with temperature, even though the average energy (velocity) increases only slightly. The significance of this is shown by comparing these energy distribution diagrams with an activation energy diagram (on the right in Figure 13.12). The number of molecules with an energy in excess of the activation energy is given by the shaded area in the distribution diagrams. As the temperature increases, the number of molecules with energies greater than the minimum required increases greatly. This is the factor that causes reaction rates to be so sensitive to temperature.

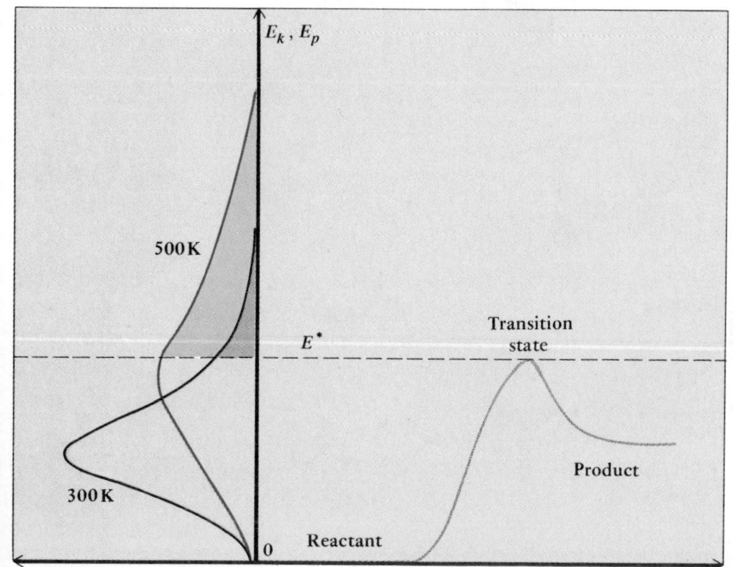

FIGURE 13.12

For reaction to occur, molecular kinetic energy (E_K) must be converted, by collisions between molecules, to potential energy (E_P). Furthermore, reactant molecules must have their initial potential energy increased by E^* (the difference between the transition state energy and the initial reactant energy). At 300 K only a small fraction of reactant molecules (pink area) have $E_P \geq E^*$, and the reaction rate is low. At 500 K, a much larger fraction of molecules (pink plus blue areas) now have $E_P \geq E^*$, and the reaction rate is higher.

EFFECT OF TEMPERATURE ON REACTION RATE: ARRHENIUS EQUATION

Svante Arrhenius (1859–1927) was a Swedish chemist who derived the relation between the rate constant and temperature from experiment.

Collision theory tells us that reaction rates depend on the energy and the number of collisions between reacting molecules, on whether the collisions have the correct geometry, and on the temperature. These requirements are summarized by the **Arrhenius equation**

$$k = \text{reaction rate constant} = Ae^{-E^*/RT} \tag{13.3}$$

where R is the gas constant with a value of 8.31×10^{-3} kJ/K · mol. A is called the *frequency factor* and has units of 1/time. It depends on the number of collisions occurring and the fraction of them that have the correct geometry. The factor $e^{-E^*/RT}$ is always less than one, and is interpreted as the *fraction of molecules having the minimum energy required for reaction*. As the table in the margin shows, the factor changes significantly with temperature, and so it reflects the area of the shaded region in the energy distribution curve in Figure 13.12.

TEMPERATURE	VALUE OF $e^{-E^*/RT}$ FOR $E^* = 40$ kJ
298 K	9.7×10^{-8}
400 K	5.9×10^{-6}
600 K	3.3×10^{-4}
800 K	2.4×10^{-3}

The Arrhenius equation is valuable primarily because it can be used to (a) calculate the value of the activation energy from the temperature dependence of the rate constant and (b) calculate the rate constant for a given temperature if the activation energy and A factor are known.

If we take the natural logarithm of each side of Equation 13.3, we have

$$\ln k = \ln A - \frac{E^*}{RT}$$

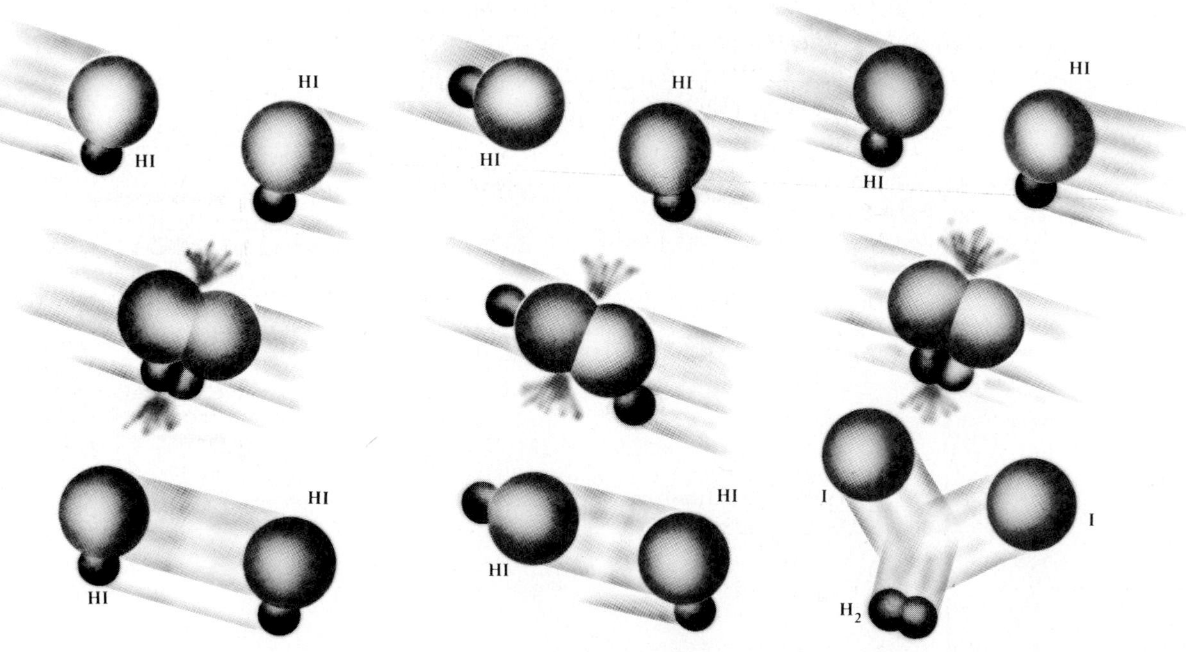

Collisions of HI molecules to form H_2 and 2 I (which subsequently form I_2 molecules). To be effective, colliding molecules must have sufficient energy and be in the correct orientation.

In slightly rearranged form it becomes

$$\begin{array}{ccccccc} \ln k & = & -\dfrac{E^*}{R}\left(\dfrac{1}{T}\right) & + & \ln A & \leftarrow & \text{Arrhenius equation} \\ \downarrow & & \downarrow \swarrow & & \downarrow & & \\ y & = & ax & + & b & \leftarrow & \text{equation for straight line} \end{array} \qquad (13.4)$$

The factor *A* in the Arrhenius equation can be calculated once E^* is known. We simply take *k* for one value of *T* and use Equation 13.3. However, an interpretation goes well beyond the level of this text. Be aware, though, that *A* is generally smaller as the reactants become larger and more complex.

Here you see that the equation is that of a straight line. Thus, if we plot the natural logarithm of *k* versus $1/T$, we find a downward sloping line with a slope of $(-E^*/R)$ (Figure 13.13). This means that E^* can be calculated if values of *k* are determined experimentally at several temperatures. The calculation is illustrated in Example 13.7.

EXAMPLE 13.7

DETERMINATION OF E^* FROM THE ARRHENIUS EQUATION

Using the experimental rate constant data below, calculate the activation energy E^* for the reaction

$$2\ N_2O(g) \longrightarrow 2\ N_2(g) + O_2(g)$$

TEMPERATURE (K)	k [L/(mol · s)]
1125	11.59
1053	1.67
1001	0.380
838	0.0011

Solution The first step is to find the reciprocal of the kelvin temperature and the natural logarithm of *k*.

$(1/T \times 10^4)$/K	ln *k*
8.889	2.450
9.497	0.513
9.990	−0.968
11.9	−6.81

These data are then plotted as illustrated in Figure 13.13. Choosing the blue points on the graph, the slope is

$$\text{Slope} = \frac{\Delta \ln k}{\Delta(1/T)} = \frac{2.0 - (-5.6)}{(9.0 - 11.5)10^{-4}/\text{K}} = -\frac{7.6}{2.5 \times 10^{-4}}\ \text{K}$$

$$= -3.0 \times 10^4\ \text{K}$$

The activation energy is then evaluated from

$$\text{Slope} = -\frac{E^*}{R}$$

$$-3.0 \times 10^4\ \text{K} = -\frac{E^*}{8.31 \times 10^{-3}\ \text{kJ/K} \cdot \text{mol}}$$

$$E^* = 250\ \text{kJ/mol}$$

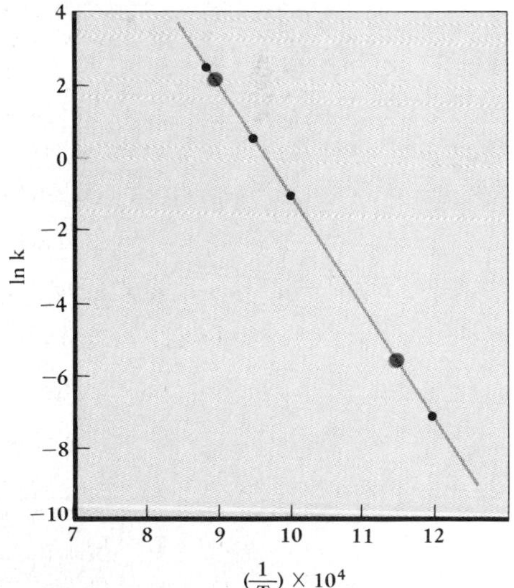

FIGURE 13.13
A plot of ln *k* versus $1/T$ for the reaction

$$2\ N_2O(g) \longrightarrow 2\ N_2(g) + O_2(g).$$

The slope of the line gives E^* as outlined in Example 13.7.

Equation 13.4 can be manipulated further to obtain it in a useful form. The slope of the line in Figure 13.13 is given by

$$\frac{\ln k_2 - \ln k_1}{(1/T_2) - (1/T_1)} = -\frac{E^*}{R}$$

If this is rearranged, we find

$$\ln \frac{k_2}{k_1} = -\frac{E^*}{R}\left[\frac{1}{T_2} - \frac{1}{T_1}\right] \tag{13.5}$$

This equation tells us that if k is known at two different temperatures (k_1 at T_1 and k_2 at T_2), then we can calculate E^*. Alternatively, it allows us readily to calculate a rate constant at some temperature if k is known at some other temperature and E^* has been determined previously.

EXAMPLE 13.8

CALCULATING E^* FROM THE TEMPERATURE DEPENDENCE OF k

Using values of k determined at two different temperatures, calculate the value of E^* for the decomposition of HI.

$$2\ HI(g) \longrightarrow H_2(g) + I_2(g)$$

$k = 2.15 \times 10^{-8}$ L/(mol · s) at 650 K and 2.39×10^{-7} L/(mol · s) at 700 K.

Solution Here we use Equation 13.5, since we have k_1 at T_1 and k_2 at T_2.

$$\ln \frac{2.39 \times 10^{-7}\ \text{L/(mol} \cdot \text{s)}}{2.15 \times 10^{-8}\ \text{L/(mol} \cdot \text{s)}} = -\frac{E^*}{8.31 \times 10^{-3}\ \text{kJ/K} \cdot \text{mol}}\left[\frac{1}{700\ \text{K}} - \frac{1}{650\ \text{K}}\right]$$

$$\ln (11.1) = -\frac{E^*}{8.31 \times 10^{-3}\ \text{kJ/K} \cdot \text{mol}}(-0.000110/\text{K})$$

$$E^* = 182\ \text{kJ/mol}$$

EXAMPLE 13.9

CALCULATING k FROM KNOWLEDGE OF E^*

The reaction of phosphine with diborane

$$PH_3(g) + B_2H_6(g) \longrightarrow H_3P{:}\!\rightarrow\! BH_3(g) + BH_3(g)$$

has an activation energy of 48.0 kJ/mol. If you have measured the rate of the reaction at room temperature, 298 K, at what temperature would you observe a doubling of the room-temperature rate?

Solution We know that k_1 was measured at $T_1 = 298$ K. The problem tells us that k_2, the rate constant at some new temperature T_2, is twice k_1, that is, $k_2 = 2k_1$. Substituting these facts into equation 13.5, we have

$$\ln\left(\frac{2k_1}{k_1}\right) = -\frac{48.0\ \text{kJ/mol}}{8.31 \times 10^{-3}\ \text{kJ/K} \cdot \text{mol}}\left[\frac{1}{T_2} - \frac{1}{298\ \text{K}}\right]$$

$$\ln 2 = -(5.78 \times 10^3\ \text{K})\left[\frac{1}{T_2} - 0.00336\right]$$

$$0.693 = -(5.78 \times 10^3\ \text{K})(1/T_2) + 19.4$$

$(-5.78 \times 10^3)(-0.00336)$

$$-18.7 = -(5.78 \times 10^3\ \text{K})(1/T_2)$$

$$T_2 = 309\ \text{K}$$

At the beginning of this section we noted that a chemist's "rule of thumb" is that rates double with every ten degree rise in temperature. This problem shows that this is correct for some reactions; here, doubling was observed for an 11 K temperature increase.

The rule of thumb that reaction rate doubles for every ten degree rise in temperature is reasonably correct for reactions with $E^* \cong 50$ kJ/mol.

EXERCISE 13.7 E^* FROM THE TEMPERATURE DEPENDENCE OF k

F atoms can be exchanged between molecules of $(CH_3)_3PF_2$ with a relatively rapid rate.

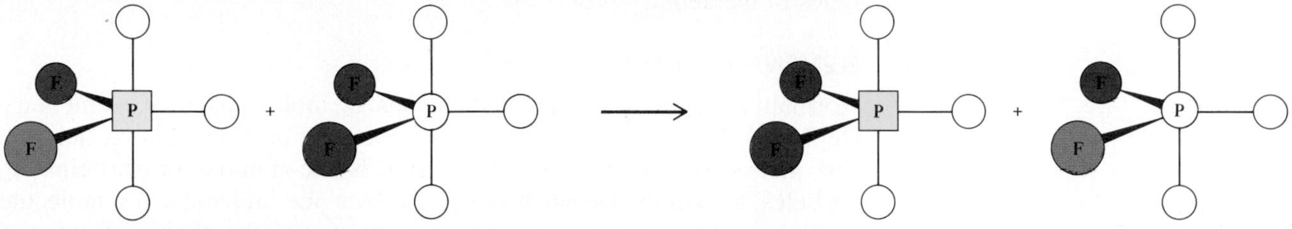

The rate constant (k_1) at 300. K is 5.076 L/(mol · s), while the constant at 330. K (k_2) is 9.839 L/(mol · s). Calculate E^*.

13.7 REACTION MECHANISMS

The rate law for a reaction can be determined by experiment, and, from that rate law, you can make an educated guess about the **reaction mechanism**, the sequence of bond-making and bond-breaking steps that occur during the conversion of reactants to products.

Rate laws are experimentally determined *laws* of nature, while a mechanism is a *theory* of the sequence of events that may be occurring at the molecular level.

In some reactions the conversion of reactants to products occurs in a single step. Nitrogen oxide and ozone react in this manner.

$$NO(g) + O_3(g) \longrightarrow NO_2(g) + O_2(g)$$

Most chemical reactions, however, do not occur in this simple manner. Instead, they involve a sequence of steps. For example, the important chemical hydrazine (N_2H_4) is made by the Raschig process. The overall reaction is

$$2\ NH_3(aq) + NaOCl(aq) \xrightarrow{\text{basic solution}} N_2H_4(aq) + NaCl(aq) + H_2O(\ell)$$

but the reaction actually occurs in a series of steps beginning with rapid formation of chloramine, NH_2Cl, by reaction of ammonia and hypochlorite ion.

Mechanism step 1:

$$NH_3(aq) + OCl^-(aq) \xrightarrow{\text{fast}} NH_2Cl(aq) + OH^-(aq)$$

The chloramine then proceeds to the final hydrazine product in several ways, among them the steps

Mechanism step 2:

$$NH_2Cl(aq) + NH_3(aq) \xrightarrow{\text{slow}} N_2H_5^+(aq) + Cl^-(aq)$$

Mechanism step 3:

$$N_2H_5^+(aq) + OH^-(aq) \xrightarrow{\text{fast}} N_2H_4(aq) + H_2O(\ell)$$

Each of the steps in the reaction sequence is called an **elementary process**, a simple event in which some chemical transformation occurs. The collection of elementary steps that lead to the overall reaction is the *reaction mechanism*. The mechanism of a reaction *must* be determined by experiment. To see how this is done, we shall first describe the three types of elementary processes.

ELEMENTARY STEPS

Normally, reactions occur by collisions of molecules, and elementary steps are classified according to the number of molecules that participate. The **molecularity** of an elementary step is the number of participating particles, so a **unimolecular** process involves one molecule. If a molecule has been given sufficient energy by collisions with other molecules, a bond may break to give stable or unstable products. For example.

$$Ni(CO)_4(g) \longrightarrow Ni(CO)_3(g) + CO(g)$$

A **bimolecular process** must involve two molecules. The molecules may be the same (as in the collision of two HI molecules on page 492),

$$A + A \longrightarrow \text{products}$$

or different.

$$A + B \longrightarrow \text{products}$$

The collision of an O_3 molecule with a molecule of NO to give NO_2 and O_2 is a bimolecular process, as is each step in the Raschig process shown above.

The collision of three molecules, a **termolecular process**, is not a highly probable event, unless one of the molecules involved is in very high concentration, like a solvent molecule. Most such processes involve the reaction of two particles, where the function of the third particle is to carry away the excess energy produced when a new chemical bond is formed by the first two particles. In the following termolecular reaction, N_2 is unchanged.

Elementary steps with molecularity greater than 3 are rare, since it is highly unlikely that four or more molecules of the right types will collide with the correct orientation.

$$O(g) + O_2(g) + N_2(g) \longrightarrow O_3(g) + \text{energetic } N_2(g)$$

If O and O_2 have sufficient energy to react to form an O_3 molecule, the ozone must have enough energy to dissociate back to reactants. However, in the activated complex formed by the three reactants, N_2 absorbs the energy evolved when the new O—O bond forms to give ozone.

The rate law and total order of a reaction cannot be predicted from its stoichiometry. In contrast, *the rate law of an elementary step is given by the product of the rate constant and the concentrations of the reactants* in the step. This means that we can write the rate law for any elementary steps, as in the examples below.

ELEMENTARY STEP	MOLECULARITY	RATE LAW
$A \longrightarrow$ product	Unimolecular	Rate = $k[A]$
$A + B \longrightarrow$ product	Bimolecular	Rate = $k[A][B]$
$A + A \longrightarrow$ product	Bimolecular	Rate = $k[A]^2$
$2A + B \longrightarrow$ product	Termolecular	Rate = $k[A]^2[B]$

This conclusion is obvious if you consider, for example, a bimolecular elementary step involving A reacting with B. Since, at the very least, A and B must collide with one another to react, the number of collisions is proportional to the concentration of each species. For example, assume the concentrations of both A and B are 2, and k is 1.

$$\text{Rate} = k[\text{A}][\text{B}] = (1)[2][2] = 4$$

This means the rate is 4. On the other hand, if the concentration of A is doubled, then

$$\text{Rate} = (1)[4][2] = 8$$

the rate is doubled to 8. This is even more evident if we illustrate the number of possible ways that collisions can occur between the reacting molecules. For 2 A's and 2 B's, four A–B collisions are possible. If the number of A molecules is doubled to 4, there are then 8 possible collisions.

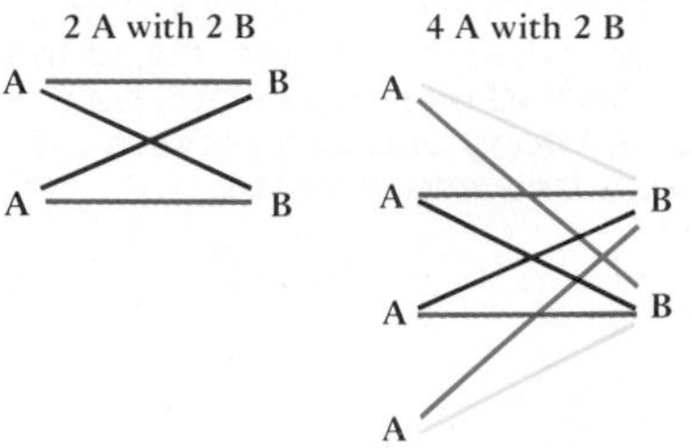

The same conclusion is evident for a unimolecular process. Here we simply have a collection of molecules rearranging or decomposing independently of one another. If the number of molecules involved is increased, the rate will increase proportionately.

The molecularity of an elementary process and its order are the same. A unimolecular process must be first order, a bimolecular process must be second order, and a termolecular process must be third order. However, the opposite is emphatically *not* true. If you discover experimentally that a reaction is first order, this does not mean it occurs in a single, unimolecular elementary step. Similarly, a second order rate law does not imply the reaction occurs in a single, bimolecular elementary step. An example of this is the decomposition of N_2O_5. Here the rate law is

$$2\,N_2O_5(g) \longrightarrow 4\,NO_2(g) + O_2(g)$$
$$\text{Rate} = k[N_2O_5]$$

but the mechanism involves a series of both unimolecular and bimolecular steps.

To see how the experimentally observed rate law is connected with a possible mechanism or sequence of elementary steps requires some chemical intuition and detective work. This is the subject of the next section.

EXERCISE 13.8 ELEMENTARY STEPS AND RATE LAWS

Write the rate law for each elementary reaction step given below.

(a) $Ni(CO)_4(g) \longrightarrow Ni(CO)_3(g) + CO(g)$
(b) $O_3(g) \longrightarrow O_2(g) + O(g)$
(c) $NO_2(g) + NO_3(g) \longrightarrow NO_2(g) + O_2(g) + NO(g)$

EXERCISE 13.9 ELEMENTARY STEPS AND MOLECULARITY

The reaction of carbon monoxide with chlorine to give phosgene, $COCl_2$, has the following mechanism:

Step 1. Fast	$Cl_2(g) \rightleftharpoons 2\,Cl(g)$
Step 2. Slow	$Cl(g) + CO(g) \longrightarrow COCl(g)$
Step 3. Fast	$COCl(g) + Cl_2(g) \longrightarrow COCl_2(g) + Cl(g)$
Overall reaction	$Cl_2(g) + CO(g) \longrightarrow COCl_2(g)$

What is the molecularity of elementary step 2? Write the rate law for the second step.

REACTION MECHANISMS AND RATE LAWS

The rate law is determined by experiment. The mechanism is a good guess (an hypothesis) about the way the reaction occurs. Several mechanisms can correspond to an experimental rate law, so the postulated mechanism can be quite wrong and can provoke disputes between scientists.

One of the great unknowns in chemistry today is the way in which many reactions occur, especially those of biological interest. Thus, the objective of many kinetic investigations is to unravel the mechanism of the reaction. Now that you understand the concept of elementary steps in a mechanism, we can begin to see the connection between the *experimental* rate law and the reaction mechanism, the *hypothesis* about the way the reaction occurs.

Imagine a two-step reaction for which the rates of both steps are known.

	ELEMENTARY STEP	RATE (reactions/second)	RELATIVE RATE
	$A + B \xrightarrow{k_1} X + I$	0.001	Slow (E* large)
	$A + I \xrightarrow{k_2} Y$	100	Fast (E* small)
Total reaction	$2\,A + B \longrightarrow X + Y$		

In the first reaction A and B come together and *slowly* produce one of the products (X) plus another reactive species (I). Almost as soon as I is formed, however, it is rapidly consumed by reaction with an additional molecule of A. The products X and Y are seen to be the result of two steps, and they can be produced *no faster than the rate of the slowest step*. The rate of the overall reaction is limited by, and is exactly equal to, the rate of this slow step. Therefore, the slowest elementary step of a sequence is called the **rate determining step**. Here the rate determining, elementary step is bimolecular and so has the rate law

$$\text{Rate} = k_1[\text{A}][\text{B}]$$

where k_1 is the rate constant of step 1. The overall reaction

$$2\text{A} + \text{B} \longrightarrow \text{X} + \text{Y}$$

is not an elementary process; yet it follows a second order rate law, because the slowest step in the mechanism is bimolecular.

Now let us turn to a few actual examples of mechanisms. Experiment shows that the reaction of nitrogen dioxide with fluorine

Overall reaction

$$2\,\text{NO}_2(\text{g}) + \text{F}_2(\text{g}) \longrightarrow 2\,\text{NO}_2\text{F}(\text{g})$$

has a second order rate law.

$$\text{Reaction rate} = -\frac{1}{2}\left(\frac{\Delta[\text{NO}_2]}{\Delta t}\right) = k[\text{NO}_2][\text{F}_2]$$

The experimental rate law immediately rules out the possibility that the reaction occurs in a single step, since a rate law for an elementary step identical with the overall stoichiometric equation would be

$$\text{Rate} = k[NO_2]^2[F_2]$$

and this rate law does *not* agree with experiment. Therefore, since experiment shows that 2 molecules of NO_2 cannot react directly with 1 molecule of F_2 to give the product, at least one other step must be involved. The simplest possibility is a mechanism similar to the symbolic one above. That is, NO_2 and F_2 first produce one molecule of product plus a fluorine atom, and the F atom then reacts with additional NO_2 to give one more molecule of product.

At this introductory level you cannot be expected to derive reaction mechanisms. However, given a mechanism, you can decide if it is in agreement with an experimental rate law.

ELEMENTARY STEPS AND RELATIVE RATES

Step 1. Slow $\quad NO_2(g) + F_2(g) \xrightarrow{k_1} NO_2F(g) + F(g)$

Step 2. Fast $\quad NO_2(g) + F(g) \xrightarrow{k_2} NO_2F(g)$

If we *assume* that the first, bimolecular step is rate determining, its rate law would be

$$\text{Rate} = k_1[NO_2][F_2]$$

exactly the rate law observed experimentally. The experimental rate constant k is then seen to be the same as k_1.

The F atom that is formed slowly in the NO_2/F_2 reaction is called an intermediate. A **reaction intermediate** is produced in one step of a reaction sequence, but it is consumed in a subsequent step. As a result it does not appear in the net stoichiometric equation. Reaction intermediates usually have a very fleeting existence, but they occasionally have long enough lifetimes to be observed.

Another reaction intermediate in this section is I in the A + B reaction at the beginning of the section.

EXERCISE 13.10 ELEMENTARY STEPS AND REACTION MECHANISM

Oxygen atom transfer from nitrogen dioxide to carbon monoxide is a simple reaction

$$NO_2(g) + CO(g) \longrightarrow NO(g) + CO_2(g)$$

and has the following rate law at temperatures less than 500 K.

$$\text{Rate} = k[NO_2]^2$$

Can this reaction occur in one bimolecular step whose stoichiometry is the same as the overall reaction? Explain your answer briefly.

EXERCISE 13.11 ELEMENTARY STEPS AND REACTION MECHANISM

The Raschig reaction produces N_2H_4 from NH_3 and OCl^- in aqueous solution. Three possible elementary steps were given at the beginning of this section. Which step of the three is the rate determining step? Write the rate law only for the rate determining elementary step.

MECHANISMS WITH A FAST INITIAL EQUILIBRIUM STEP

The reactions we have examined so far all had a rate determining elementary step followed by one or more fast steps. Many reactions proceed, however, by the reactants rapidly producing some intermediates that subsequently react in a slow step. There are unique aspects to these mechanisms, so we want to consider them as a separate topic.

The oxidation of nitrogen oxide by oxygen

$$2\ NO(g) + O_2 \longrightarrow 2\ NO_2(g)$$

$$\text{Rate} = k[NO]^2[O_2]$$

has an experimental rate law that is third order overall. This rate law would be correct for the reaction occurring in a single, termolecular elementary step. However, there is experimental evidence for a reaction intermediate, and so the process is thought to occur in two elementary steps. Unlike the previous examples, the second step is the rate determining step.

ELEMENTARY STEPS AND RELATIVE RATES

Step 1. Fast, at equilibrium $\quad NO(g) + O_2(g) \underset{k_{-1}}{\overset{k_1}{\rightleftharpoons}} OONO(g)$ (intermediate)

Step 2. Slow, rate determining $\quad NO(g) + OONO(g) \xrightarrow{k_2} 2\ NO_2(g)$

Overall reaction $\quad 2\ NO(g) + O_2(g) \longrightarrow 2\ NO_2(g)$

Therefore, the rate law predicted by the mechanism should be

$$\text{Rate} = k_2[NO][OONO]$$

This rate law cannot be compared directly with the experimental rate law since it contains the concentration of an unobserved intermediate, OONO. The experimental rate law should be written in terms of compounds appearing in the overall equation and should not include reaction intermediates. Therefore, we want to express the postulated rate law in a way that eliminates the intermediate. To do this, we must look carefully at the first reaction.

At the beginning of the reaction, NO and O_2 react rapidly and produce the intermediate OONO with a rate constant k_1.

$$\text{Rate of production of OONO} = k_1[NO][O_2]$$

Since the intermediate is consumed only very slowly in the second step, some OONO reverts to NO and O_2 before it can be consumed in the second step.

$$\text{Rate of reversion of OONO to NO and } O_2 = k_{-1}[OONO]$$

As NO and O_2 form OONO their concentration drops, so the rate of the forward reaction drops. At the same time, the concentration of OONO builds up, so the rate of the reverse reaction increases. Eventually the rates of the forward and reverse reactions become the same, and the system reaches a state of *dynamic equilibrium*. Because these two reactions are so much faster than the second elementary step of the sequence, equilibrium is established before any significant amount of OONO is consumed by NO to give NO_2. The state of equilibrium remains throughout the lifetime of the overall reaction.

The concept of equilibria in chemical reactions is discussed in detail in the next four chapters.

A state of equilibrium occurs when the forward and reverse rates in a chemical reaction are equal, so we can say

At equilibrium (forward rate = reverse rate)

$$k_1[NO][O_2] = k_{-1}[OONO]$$

Rearranging this equation, we find

$$\frac{k_1}{k_{-1}} = \frac{[OONO]}{[NO][O_2]}$$

Here the concentration of the product of the first reaction, divided by the product of the concentrations of the reactants, is equal to the ratio of rate constants. We call this ratio of constants k_1/k_{-1} the **equilibrium constant** and give it the symbol $\boldsymbol{K}$.

$$K = \frac{[OONO]}{[NO][O_2]}$$

As you shall see in the next four chapters, the equilibrium constant always has this same general form.

Now we are in a position to eliminate [OONO] from the rate law that came from the postulated sequence of steps for the NO_2/O_2 reaction. Rearranging the equilibrium constant expression, we find

$$[OONO] = K[NO][O_2]$$

If this is substituted into the rate law,

$$\text{Rate} = k_2\{K[NO][O_2]\}[NO]$$

which becomes

$$\text{Rate} = k_2K[NO]^2[O_2]$$

Since *both* k_2 and K are constants, their product is a constant k, so we have

$$\text{Rate} = k[NO]^2[O_2]$$

This is exactly the rate law that came from experiment, so we believe we have postulated a reasonable mechanism. However, it is not the *only* reasonable mechanism. As we see in the following example, at least one other mechanism is consistent with the same experimental rate law.

EXAMPLE 13.10

MECHANISMS WITH A FAST INITIAL STEP

The NO/O_2 reaction described in the text could also occur by the sequence of steps

ELEMENTARY STEPS AND RELATIVE RATES

Step 1. Fast, at equilibrium	$NO(g) + NO(g)$	$\underset{k_{-1}}{\overset{k_1}{\rightleftharpoons}}$	$N_2O_2(g)$ intermediate
Step 2. Slow, rate determining	$N_2O_2(g) + O_2(g)$	$\xrightarrow{k_2}$	$2\ NO_2(g)$
Overall reaction	$2\ NO(g) + O_2(g)$	$\longrightarrow$	$2\ NO_2(g)$

Show that this reaction also leads to the experimentally observed rate law.

Solution The rate law for the rate determining, elementary step is

$$\text{Rate} = k_2[N_2O_2][O_2]$$

The compound N_2O_2 is an intermediate, which is consumed in the slow step and cannot appear in the final derived rate law. Therefore, we must express the derived rate law in another way, and we can do this by recognizing that the intermediate is related to NO_2 by the equilibrium constant expression for the reactants and products in step 1. (Notice that the concentration of each molecule in the equilibrium constant expression is raised to a power equal to its stoichiometric coefficient in the balanced chemical equation for step 1. Here 2 NO molecules are reactants in step 1.)

$$K = \frac{[N_2O_2]}{[NO]^2}$$

If this expression is solved for $[N_2O_2]$, we have $[N_2O_2] = K[NO_2]^2$. When this is substituted into the derived rate law,

$$\text{Rate} = k_2\{K[NO_2]^2\}[O_2]$$

we have an equation identical with the experimental law, where $k_2K = k_{exp}$.

The NO/O_2 reaction has a rate law that suggests two possible mechanisms. The challenge that often faces the chemist interested in kinetics is to decide which mechanism is correct or if there may be a better choice. In this case further chemical investigation showed the presence of the species OONO as a short-lived intermediate, confirming the mechanism given on page 500.

EXERCISE 13.12 MECHANISMS WITH A FAST INITIAL STEP

The reaction of H_2 and Br_2 to give HBr has the following experimental rate law,

$$H_2(g) + Br_2(g) \longrightarrow 2\,HBr(g)$$
$$\text{Experimental rate law} = k[H_2][Br_2]^{1/2}$$

and is thought to occur by the by the mechanism

ELEMENTARY STEPS AND RELATIVE RATES

Step 1. Fast, equilibrium	$Br_2(g)$	$\underset{k_{-1}}{\overset{k_1}{\rightleftharpoons}}$	$2\,Br(g)$
Step 2. Slow	$Br(g) + H_2(g)$	$\xrightarrow{k_2}$	$HBr(g) + H(g)$
Step 3. Fast	$H(g) + Br_2(g)$	$\xrightarrow{k_3}$	$HBr(g) + Br$
Overall equation	$H_2(g) + Br_2(g)$	$\longrightarrow$	$2\,HBr(g)$

Here we assume a rapid, initial equilibrium step supplies the reaction intermediate, a Br atom that is required in the rate determining step. Using the method outlined in Example 13.10, show that the postulated mechanism leads to the experimental rate law.

To summarize our discussion of rate laws and reaction mechanisms, we can emphasize the following points:

a. The first step in a kinetics investigation is to perform experiments that define the effect of reactant concentrations on the rate of the reaction. This gives the experimental rate law.

b. A mechanism for the reaction is proposed on the basis of the experimental rate law, the principles of stoichiometry and molecular structure and bonding, general chemical experience, and imagination.

c. The reaction mechanism is used to derive a rate law. This rate law must contain only those species present in the overall chemical reaction and not reaction intermediates. If the derived and experimental rate laws are the same, the postulated mechanism *may* be a reasonable hypothesis of the reaction sequence. The measured rate constant is then seen to be the rate constant of an elementary step or a product of elementary rate constants and equilibrium constants.

d. If more than one mechanism can be proposed, all of which predict derived rate laws in agreement with the experimental rate law, then further experiments must be done to confirm the correctness of one of them (or the incorrectness of the others).

13.8 CATALYSIS

In the preceding sections you have seen that the initial rate of a reaction is given by an expression such as "initial rate $= k[R_1]_0[R_2]_0$" and that the rate constant k depends on the temperature, the size of the activation energy barrier, and the effectiveness of collisions between reactants. Raising the concentration of a reactant increases the rate, while raising the temperature increases the value of k. However, there are two other ways to speed up a reaction: (a) The activation energy barrier can be lowered by making bond breaking or formation easier in the rate determining step easier. (b) The effectiveness of the collisions can be increased by finding a way to improve the orientation of reacting molecules. Either or both of these objectives can be achieved by using a catalyst. A **catalyst** is any reagent that can increase the rate of a reaction while not being consumed by the reaction. That is, a catalyst can be used to ensure that the reactants come together in the correct orientation, it can make bond making or bond breaking occur more easily, or it can alter the mechanism by opening up another pathway having a lower activation energy.

Biochemical reactions use more than a thousand different catalysts called **enzymes**.

There are two general types of catalysts: homogeneous and heterogeneous. **Homogeneous catalysts** are reagents that are in the same phase as the reaction mixture. This usually means the reaction mixture is liquid, and the catalyst can dissolve in it. In contrast, **heterogeneous catalysts** are found in a different phase than the reactants. In this case the catalyst is usually a solid, while the reactants are dissolved in a solvent or are in the gas phase.

Almost 5 billion pounds of catalysts, worth about one billion dollars, are used in the United States every year. Over 90% are acids (HF, H_2SO_4, H_3PO_4). The remainder are heterogeneous catalysts, largely metals and metal oxides.

The decomposition of H_2O_2 to O_2 and H_2O is an example of the use of a homogeneous catalyst to lower an activation energy barrier (Figure 13.1). In the absence of a catalyst, the reaction proceeds only slowly with an activation energy of 76 kJ/mol. If a little iodide is added, though, the reaction occurs readily, in large part because the activation energy is only 57 kJ/mol (Figure 13.14). At the molecular level the function of the I^- ion is apparently to produce the reactive ion IO^- as an intermediate.

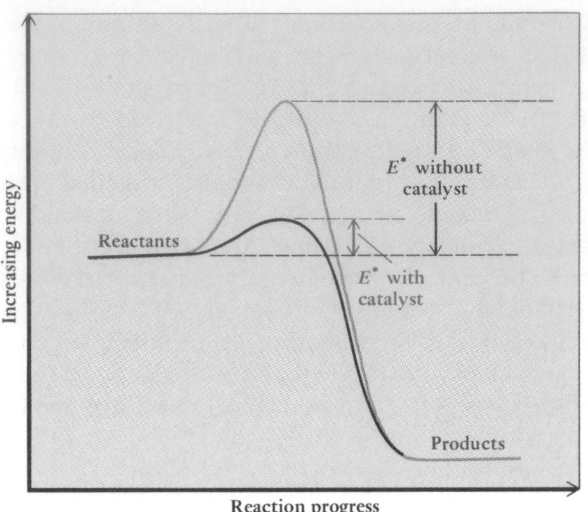

FIGURE 13.14
A catalyst can accelerate a reaction by lowering the activation energy E^*. With a smaller barrier to overcome, a greater fraction of reacting molecules has sufficient energy to surmount the barrier, so the reaction occurs more rapidly.

ELEMENTARY STEPS AND RELATIVE RATES

Step 1. Slow $\quad H_2O_2(aq) + I^-(aq) \xrightarrow{k_1} IO^-(aq) + H_2O(\ell)$

Step 2. Fast $\quad H_2O_2(aq) + IO^-(aq) \xrightarrow{k_2} H_2O(\ell) + O_2(g) + I^-(aq)$

Overall reaction $\quad 2\,H_2O_2(aq) \longrightarrow 2\,H_2O(\ell) + O_2(g)$

Here you see that I^- is a catalyst since it speeds up the reaction, but it is not required in the overall stoichiometry (Figure 13.15).

Acid-catalyzed reactions are other important examples, occurring not only in the chemistry laboratory but also in biological systems. Acid catalysis involves the transfer of a proton to the reacting molecule,

$$R + H^+ \longrightarrow HR^+$$

and this altered molecule now reacts more rapidly than before.

$$R \xrightarrow{\text{slow}} \text{products} \qquad \text{but} \qquad HR^+ \xrightarrow{\text{fast}} \text{products}$$

For example, an organic compound called an *ester* can react with water to replace the —OR group with —OH. The products are compounds called *acids* and *alcohols* (see Chapter 23).

FIGURE 13.15
Hydrogen peroxide decomposition can also be accelerated by the heterogeneous catalyst MnO_2. Here a 30% aqueous solution of H_2O_2 is dropped onto solid, black MnO_2, with rapid decomposition to give O_2 and H_2O; the latter is evolved as steam due to the high heat of reaction. (Charles D. Winters)

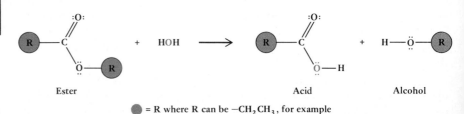

$$R{-}C({=}O){-}O{-}R + HOH \longrightarrow R{-}C({=}O){-}O{-}H + H{-}O{-}R$$

Ester $\qquad$ Acid $\qquad$ Alcohol

● = R where R can be $-CH_2CH_3$, for example

If a tiny quantity of acid is added, the H^+ ion attaches itself to the C=O group by using an oxygen lone pair of electrons to form a bond.

This species is now much more susceptible to reaction with water, and the ester quickly falls apart to products.

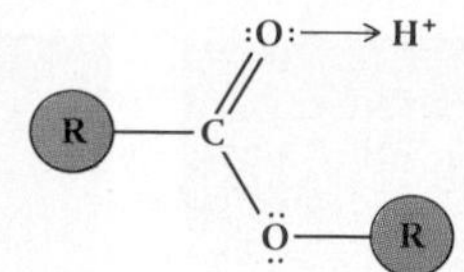

Heterogeneous catalysts are extremely important in the chemical industry. Chemical companies profit by combining small, inexpensive molecules to produce larger ones. The combination process is often one that can occur but only at such a slow rate that the process would not be commercially useful. Therefore, catalysts have to be invented to speed up the reaction, and heterogeneous ones are generally the most practical (Figure 13.16). Several important examples include the synthesis of ammonia from H_2 and N_2,

$$3\ H_2(g) + N_2(g) \xrightarrow{\text{Fe-containing catalyst/400°C/200 atm}} 2\ NH_3(g)$$

the synthesis of methanol from synthetic gas (CO and H_2),

$$2\ H_2(g) + CO(g) \xrightarrow{\text{aluminum and chromium oxides}} H_3COH(g)$$

and the oxidation of SO_2 in the production of sulfuric acid.

$$2\ SO_2(g) + O_2(g) \xrightarrow{V_2O_5} 2\ SO_3(g)$$
$$SO_3(g) + H_2O(\ell) \longrightarrow H_2SO_4(\ell)$$

FIGURE 13.16

Heterogeneous catalysts used by industry in petroleum refining and to make polymers, to oxidize, and to add hydrogen to other molecules. (Harshaw/Filtrol Partnership)

You are most familiar with heterogeneous catalysts through their use in the exhaust system of automobiles (Figure 13.17). Their purpose is to insure that the combustion of carbon monoxide, hydrocarbons, and nitrogen oxides is complete,

$$2\ CO(g) + O_2(g) \xrightarrow{\text{catalyst}} 2\ CO_2(g)$$

$$2\ C_8H_{18}(g) + 25\ O_2(g) \xrightarrow{\text{catalyst}} 16\ CO_2(g) + 18\ H_2O(g)$$

and to convert nitrogen and sulfur oxides to molecules less harmful to the environment. The air drawn into an automobile engine contains N_2 as well as O_2. At the high temperature of combustion, some N_2 reacts with O_2 to give NO. This is a serious air pollutant, so catalysts have been

FIGURE 13.17

The cylinders contain the catalyst used to control hydrocarbons, nitrogen oxides, and carbon monoxide in automotive exhaust systems. The catalyst system consists of metals of the platinum group such as platinum and rhodium. (Johnson Matthey)

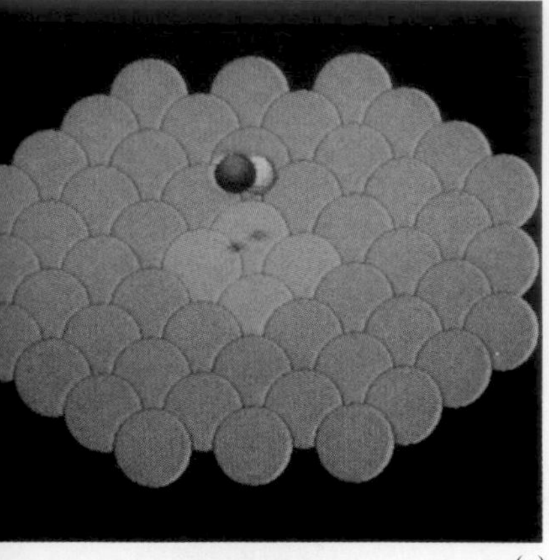
(a)

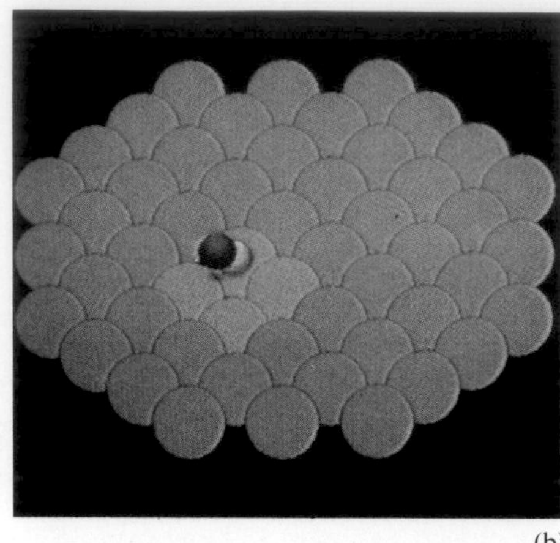
(b)

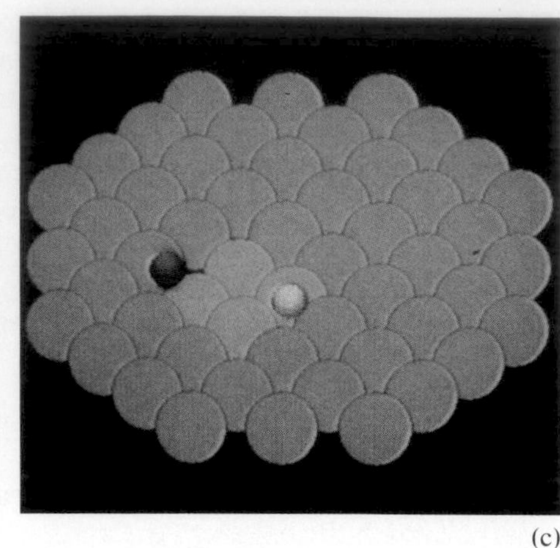
(c)

FIGURE 13.18
A computer simulation of the dissociation of NO into N (reddish) and O (white) atoms on a platinum surface. In (a) the NO approaches the surface. After 1.5×10^{-12} seconds it interacts with the surface (b) and then dissociates (c), 1.7×10^{-12} seconds after approaching the surface. (John Tully, AT&T Bell Laboratories)

developed that catalyze not only the oxidation of carbon compounds but also the decomposition of NO back to the elements.

$$2\ NO(g) \xrightarrow{\text{catalyst}} N_2(g) + O_2(g)$$

The role of the heterogeneous catalyst in the reactions above is probably to weaken the bonds in the reacting H_2, N_2, or O_2 molecules and to assist in product formation. For example, in Figure 13.18 you see a portion of a computer-generated movie of NO molecules dissociating into N and O atoms on the surface of a platinum metal catalyst.

SUMMARY

Chemical kinetics deals with the theory and measurement of chemical **reaction rates** (the change in concentrations of reactants and products with time) and chemical **mechanisms** (the sequence of events at the molecular level that control reaction speed and the outcome of reactions).

Reaction rates are affected by reactant concentrations, temperature, the presence of a catalyst, and the physical state of the reactants.

The rate of a reaction must be determined by experiment (Sections 13.2 and 3). The concentration of a reactant or product is measured as a function of time, and a plot similar to Figure 13.3 is prepared. The reaction rate at a particular time is the slope of the line tangent to the concentration/time curve. The rate of a reaction is highest at the beginning of the process, but it approaches zero with time. The rates of disappearance of reactants, or of appearance of products, are related by their stoichiometric coefficients. Thus, for a reaction such as A → 2 B,

$$-\text{Rate of disappearance of A} = -\frac{\Delta[\text{A}]}{\Delta t}$$

$$= \frac{1}{2}(\text{rate of appearance of B}) = +\frac{1}{2}\frac{\Delta[\text{B}]}{\Delta t}$$

The *rate of reaction* is always positive and is calculated by dividing the rate for each reactant or product by its stoichiometric coefficient.

The relation between reaction rate and reactant concentrations is given by a **rate law** (Section 13.4). For the general reaction

$$\text{a A} + \text{b B} \xrightarrow{\text{C}} \text{x X} \qquad \text{(where C is a catalyst)}$$

the rate law will always have the form

$$\text{Initial reaction rate} = k[\text{A}]_0^m[\text{B}]_0^n[\text{C}]^p$$

where $[\text{reagent}]_0$ is the initial concentration and k is the **rate constant**. Each exponent (m, n, and p) is called the **reaction order** with respect to A, B, or C, respectively. The sum of exponents is the **total order** of the reaction. The exponents are not necessarily the same as the stoichiometric coefficients in the balanced equation; they may have positive and negative values of whole or fractional numbers. They *must* be determined by experiment.

A first order reaction has a rate law of the form "initial rate = $k[\text{A}]_0$," while a second order reaction would have a rate law such as "initial rate = $k[\text{A}]_0^2$" or "initial rate = $k[\text{A}]_0[\text{B}]_0$." Other possibilities for these and reactions of other order are given in the text.

For a first order process of the form "initial rate = $k[\text{A}]_0$," the concentration of A at some time t (=[A]) is related to the concentration of A at time = 0 (=$[\text{A}]_0$) by (Section 13.5)

$$\ln\left(\frac{[\text{A}]}{[\text{A}]_0}\right) = kt \tag{13.1}$$

For the second order rate law "initial rate = $k[\text{A}]_0^2$," the relation is

$$\frac{1}{[\text{A}]} - \frac{1}{[\text{A}]_0} = kt \tag{13.2}$$

The **half-life, $t_{1/2}$**, of a reaction is the time required for the concentration of any of the reactants to decrease to $\frac{1}{2}$ of its initial value. For a first order process, $t_{1/2} = 0.693/k$.

The order of a reaction can be distinguished graphically. If a straight line is observed when ln[reactant] is plotted versus time, the reaction is first order in the reactant and the slope of the line is $-k$. For a second order reaction of the form "initial rate = $k[\text{A}]_0^2$," a plot of 1/[A] versus time is a straight line with a slope of k.

As bonds break and new bonds form, the energy of the reacting system increases and then decreases, creating an energy barrier to the reaction. Such energy barriers are illustrated by a **reaction energy diagram** (Section 13.6), where, as time progresses, the reaction moves from reactants to products through the **transition state.** A system in the transition state is called the **activated complex.** The energy at the top of the barrier is the **transition state energy**, and the difference between this energy and the energy of the reactants is the **activation energy, E^***, for the reaction. The activation energy largely determines the variation in rates observed for different reactions. The theory relating rates and energies is called the **transition state theory**.

(*Text continues on page 510*)

SPECIAL SECTION: PLATINUM

Platinum currently sells for \$300 to \$400 per troy ounce where 1 troy ounce = 31.103481 g.

Platinum, which is so widely used in heterogeneous and homogeneous catalysts, is a precious metal. Like gold, platinum is used to make fine jewelry, but it has many more practical uses than gold (Figure 13.19).

The people of South America invented ways of using platinum to make jewelry sometime in the first century AD, and the Spanish discovered these treasures and the deposits of platinum there in the 16th century. Indeed, the name of the metal comes from its Spanish name, *platina* or "little silver."

FIGURE 13.19

Because of its high melting point (1769°C), electrical conductivity, resistance to corrosion, and other chemical properties, platinum is widely used in industry. Here a platinum laboratory crucible is being heated, without damaging it, to a very high temperature in air. (Englehard Corporation)

The sands of the rivers of the west coast of South America are rich in gold. Platinum is often found along with the gold, but the early Spanish explorers often painstakingly picked out the grains of platinum to separate it from bits of gold. In some cases, they believed that the platinum was gold that had not been buried long enough to ripen into yellow gold, and they threw it back into the rivers to age!

Platinum did not find its way to Europe until the early 18th century. In 1736 a Spanish naval officer, de Ulloa, observed platinum in the gold mines of South America. On his way back to Spain in 1745, however, his ship was attacked by privateers and eventually captured by the British navy. He and his papers describing platinum were taken to London where members of the Royal Society became aware of his work and actually elected de Ulloa a member of the Society. In the meantime, a sample of the metal was brought to London and work began on the study of its properties.

Because of its color, beauty, and value, platinum was known as "white gold" by its early European discoverers.* However, it was soon apparent that it was much more difficult to work with than gold, because its melting point (1769°C) is much higher than those of other metals. Many attempts were made to melt platinum, but the French chemist Lavoisier was apparently the first to achieve the feat on a small scale in 1782. It was not until late in the 19th century that methods were devised so that larger amounts of platinum could be melted and thus worked into useful products.

Platinum, which is about as abundant as gold in the earth's crust, is usually found with other metals often called the "platinum group metals" (ruthenium and osmium, rhodium and iridium, and palladium and platinum). All of these metals are economically important, although not quite to the extent of platinum itself.

The annual worldwide production of precious metals is estimated to be about 10,000 tons of silver, 1,000 tons of gold, and only 70 to 80 tons of platinum and palladium combined.

Until about 1820 all of the platinum came from South America, but then deposits were discovered in Russia, which supplied the world for many years. Now the greatest amount is mined in South Africa in a

*The term "white gold" is now applied to an alloy of gold and palladium, the latter a metal in the same group as platinum.

FIGURE 13.20

The worker in the photo is installing a layer of platinum/rhodium wire gauze in a reactor for the production of nitric acid from ammonia. The most important end product is nitrate fertilizer. (Johnson Matthey)

geologic zone called the "Merensky Reef." Even in this platinum-rich zone, however, roughly 10 tons of rock must be removed to isolate a troy ounce of platinum.

The annual consumption of platinum in the United States is estimated to be about 1,300,000 troy ounces or about 45 tons. By far the greatest fraction of this use is in emission control catalysts for automotive use, and the next most important use is in the chemical industry. As mentioned in Chapter 1, the production of inexpensive fertilizers has been crucial to the world economy. Much of this fertilizer comes from the synthesis of ammonia and the conversion of this into nitrate fertilizers (see Chapter 22). The second step in this overall process is carried out in enormous chemical plants which use platinum/rhodium gauze as the catalyst (Figure 13.20).

Since platinum has a high melting point and is not easily corroded, it is used to make vessels for high temperature processes. For example, very high purity optical glass is made by melting the components in platinum crucibles, and fiberglass is produced by forcing molten glass through many small holes in a sheet of platinum (Figure 13.21).

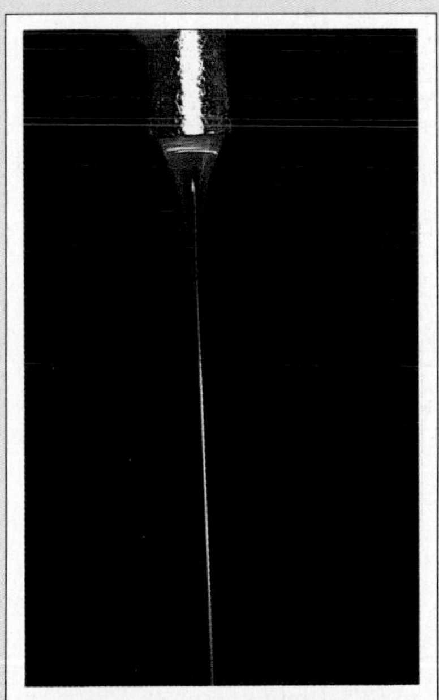

FIGURE 13.21

Glass fibers are formed by drawing molten glass through hundreds of small openings in a plate made of a platinum–gold alloy. (Owens-Corning Fiberglas Company)

All the uses described for platinum thus far rely on its properties as a metal. However, the chemistry of platinum and the metals to which it is most closely related have an extraordinarily rich and interesting chemistry. Many of their compounds act as catalysts for chemical reactions, but the newest and most exciting use of a platinum-containing compound is as the cancer chemotherapy agent *cisplatin*, $(H_3N)_2PtCl_2$, and its preparation was noted in Chapter 1.

Collision theory assumes that reacting molecules or atoms must collide in order to react. The theory supposes the reaction rate constant depends on the number of collisions occurring with the correct geometry and on the temperature. This relation is expressed in terms of the **Arrhenius equation**.

$$k = Ae^{-E^*/RT} \qquad (13.3)$$

(activation energy: E^*; A: frequency of collisions with correct geometry)

The factor $e^{-E^*/RT}$ is always less than 1 and is the fraction of molecules having the minimum energy required for reaction. The Arrhenius equation can be used, in rearranged form (Equation 13.5), to determine E^* if the values of k are known at two different temperatures.

A reaction mechanism consists of a series of **elementary steps**, each involving the reaction of one, two, or three molecules. The number of interacting particles gives the **molecularity** of a step (one particle is a *unimolecular* step, two a *bimolecular* step, and three a *termolecular* step.) The rate law for each elementary step is the product of the concentrations of the reactants in the step. The slowest elementary step in a mechanism is called the **rate determining step**. If the experimental rate law is the same as the rate law derived from the mechanism, then the postulated mechanism is a reasonable hypothesis.

Catalysts enhance the speed of a reaction by *lowering the energy barrier*, E^*, and by giving the reacting molecules the *correct geometry* for efficient reaction. A catalyst can appear as a reactant in the rate law, in contrast with its lack of appearance in the overall balanced equation. The catalyst can be **homogeneous**, if in the same phase as the other components of the reaction, or **heterogeneous**, if the catalyst and reaction components are in different phases.

STUDY QUESTIONS

REVIEW QUESTIONS

1. The rate law for a chemical reaction can be determined by which of the following:
   (a) Theoretical calculations.
   (b) Measuring the rate of the reaction as a function of the concentration of the reacting species.
   (c) Determining the equilibrium constant for the reaction.
   (d) Measuring the rate of the reaction as a function of temperature.
2. Name four conditions that determine the rate of a chemical reaction.
3. In Figure 13.4 the instantaneous rate of decomposition of H_2O_2 and of formation of O_2 are plotted as a function of time. Explain why the rate of disappearance of H_2O_2 is twice the rate of formation of O_2 at any given time.
4. Refer to Figure 13.4 and explain why the rate of disappearance of H_2O_2 decreases with time.
5. Using the rate law "initial rate $= k[A]_0^2[B]_0^0$," define the order of the reaction with respect to A and B and the total order of the reaction.
6. If a reaction has the experimental rate law "initial rate $= k[A]_0^2$," explain what happens to the rate when the concentration of A is tripled. When the concentration of A is halved?
7. A reaction has the experimental rate law "initial law $= k[A]_0^2[B]_0$." If the concentration of A is doubled, and the concentration of B is halved, what happens to the reaction rate?
8. Give the equation that allows us to find the concentration of reactant as a function of time for a first order reaction. Define each term in the equation.

9. Derive the expression for the half-life of a first order reaction ($t_{1/2} = 0.693/k$).
10. After five half-life periods for a first order reaction, what fraction of reactant remains?
11. If you plot 1/[reactant] versus time and observe a straight line, what is the order of the reaction? If ln[reactant] is plotted versus time, and a straight line of negative slope is observed, what is the order of the reaction?
12. Draw a reaction energy diagram for an exothermic process. Show the activation energies of the forward and reverse processes, the energy of the transition state, and how you can calculate the net energy of reaction.
13. Explain how collision theory accounts for the temperature dependence of reaction rates.
14. Write the Arrhenius equation and define each part of the equation.
15. What is the *mechanism* of a chemical reaction? In your discussion, define *elementary step* and *rate determining step*.
16. Define the term *molecularity* in terms of the collision theory.
17. Define the term *reaction intermediate*. Give an example using one of the reaction mechanisms described in the text.
18. Indicate whether each of the following statements is true or false. Change the wording of each false statement to make it true.
    (a) In stepwise reactions the rate determining step is the slow one.
    (b) It is possible to change the rate constant for a reaction by changing the temperature.
    (c) As a first order reaction proceeds at a constant temperature, the rate remains constant.
    (d) The rate constant for a reaction is independent of reactant concentrations.
19. What is a catalyst? What is its effect on the energy barrier for a reaction? On the frequency factor in the Arrhenius equation?
20. Explain the difference between a homogeneous and a heterogeneous catalyst. Give an example of each.

REACTION RATES

21. For each reaction below give the relative rates of formation of products and disappearance of reactants.
    (a) $2\ O_3(g) \longrightarrow 3\ O_2(g)$
    (b) $2\ HOF(g) \longrightarrow 2\ HF(g) + O_2(g)$
    (c) $N_2(g) + 3\ H_2(g) \longrightarrow 2\ NH_3(g)$
22. What are the relative rates of disappearance of reactants and formation of products for each reaction below?
    (a) $HC{\equiv}CH(g) + HCN(g) \longrightarrow H_2C{=}CHCN(g)$
    (b) $CH_4(g) + 2\ O_2(g) \longrightarrow CO_2(g) + 2\ H_2O(g)$
    (c) $2\ C_2H_6(g) + 7\ O_2(g) \longrightarrow 4\ CO_2(g) + 6\ H_2O(g)$
23. Experimental data are listed below for the hypothetical reaction A → 2 B.

TIME (seconds)	[A] (mol/L)
0.00	1.000
10.0	0.833
20.0	0.714
30.0	0.625
40.0	0.555

    (a) Plot these data, connect the points with a smooth line, and calculate the rate of disappearance of A for each 10-second interval from 0 to 30 seconds.
    (b) Calculate the rate of appearance of B for the time interval from 10 to 20 seconds. How is this related to the rate of disappearance of A in the same time interval?
24. A compound called phenyl acetate reacts with water according to the equation

$$CH_3{-}\overset{\overset{\displaystyle O}{\|}}{C}{-}O{-}C_6H_5(aq) + H_2O(\ell) \longrightarrow CH_3{-}\overset{\overset{\displaystyle O}{\|}}{C}{-}OH(aq) + C_6H_5{-}OH(aq)$$

phenyl acetate → acetic acid + phenol

and the following data were collected at 5°C.

TIME (min)	[PHENYL ACETATE] (mol/L)
0	0.55
0.25	0.42
0.50	0.31
0.75	0.23
1.00	0.17
1.25	0.12
1.50	0.085

    (a) Plot these data. Describe the shape of the curve observed. Compare this with Figure 13.3.
    (b) Calculate the rate of disappearance of phenyl acetate during the period 0.20 min to 0.40 min and then during the time period 1.2 min to 1.4 min. Compare the values of these rates and tell why the one is smaller than the other.
    (c) What is the rate of appearance of phenol during the time period 1.00 min to 1.25 min?
25. Two molecules of butadiene, C_4H_6, can join to form C_8H_{12}.

$C_4H_6(g) \longrightarrow \frac{1}{2} C_8H_{12}(g)$

Concentration versus time data have been collected at 500°C, and are listed below.

TIME (sec)	$[C_4H_6]$(mol/L)
195	1.62×10^{-2}
604	1.47×10^{-2}
1246	1.29×10^{-2}
2180	1.10×10^{-2}
4140	0.84×10^{-2}
4655	0.80×10^{-2}
6210	0.68×10^{-2}
8135	0.57×10^{-2}

(a) Plot these data and calculate the rate of disappearance of C_4H_6 over the time period 1000s to 2000s.
(b) Compare the rate of disappearance of C_4H_6 in the time 1000s to 2000s with the rate of appearance of C_8H_{12} in the same time.
(c) What is the reaction rate for the first 1000 seconds?

CONCENTRATION AND RATE LAWS

26. The reaction $CO(g) + NO_2(g) \rightarrow CO_2(g) + NO(g)$ is second order in NO_2 and zero order in CO at temperatures less than 500 K.
(a) Write the rate law for the reaction.
(b) How will the reaction rate change if the NO_2 concentration is halved?

27. For each of the following expressions, give the reaction order in terms of each reagent and for the overall reaction.
(a) initial rate $= k[A]_0[B]_0^2$
(b) initial rate $= k[A]_0[B]_0[C]^{-1}$
(c) initial rate $= k[A]_0^0$
(d) initial rate $= k[A]_0^3[B]_0^{-\frac{1}{2}}$

28. For a reaction of the type $A + B \rightarrow C$, it is experimentally observed that doubling the concentration of B causes the reaction rate to be increased fourfold, but doubling the concentration of A has no effect on the rate. The rate equation is therefore
(a) initial rate $= k[A]_0^2$
(b) initial rate $= k[A]_0[B]_0$
(c) initial rate $= k[B]_0^2$
(d) initial rate $= k[A]_0$

29. The transfer of an O atom from NO_2 to CO has been studied at 540 K

$$CO(g) + NO_2(g) \longrightarrow CO_2(g) + NO(g)$$

and the following data were collected. Use them to (a) write the rate law, (b) determine the reaction order, and (c) calculate the rate constant.

$[CO]_0$(mol/L)	$[NO_2]_0$(mol/L)	INITIAL RATE (mol/L · hr)
5.1×10^{-4}	0.35×10^{-4}	3.4×10^{-8}
5.1×10^{-4}	0.70×10^{-4}	6.8×10^{-8}
5.1×10^{-4}	0.18×10^{-4}	1.7×10^{-8}
1.0×10^{-3}	0.35×10^{-4}	6.8×10^{-8}
1.5×10^{-3}	0.35×10^{-4}	10.2×10^{-8}

30. A study of the reaction $2A + B \rightarrow C + D$ gave the following results:

	INITIAL CONCENTRATIONS (*M*)		
EXPT. NO.	$[A]_0$	$[B]_0$	INITIAL RATE (*M*/sec)
1	0.10	0.05	6.0×10^{-3}
2	0.20	0.05	1.2×10^{-2}
3	0.30	0.05	1.8×10^{-2}
4	0.20	0.15	1.1×10^{-1}

(a) What is the rate law for this reaction?
(b) What is the overall order of the reaction?
(c) Could this reaction occur in a single step?

31. The bromination of acetone is acid catalyzed.

$$\underset{\text{acetone}}{CH_3COCH_3} + Br_2 \xrightarrow{\text{acid catalyst}} CH_3COCH_2Br + H^+ + Br^-$$

The rate of disappearance of bromine was measured for several different initial concentrations of acetone, bromine, and H^+.

$[CH_3COCH_3]_0$	$[Br_2]_0$	$[H^+]_0$	INITIAL RATE OF DISAPPEARANCE OF Br_2 (*M*/sec)
0.30	0.05	0.05	5.7×10^{-5}
0.30	0.10	0.05	5.7×10^{-5}
0.30	0.05	0.10	12.0×10^{-5}
0.40	0.05	0.20	31.0×10^{-5}
0.40	0.05	0.05	7.6×10^{-5}

(a) Deduce the rate law for the reaction.
(b) What is the order with respect to each reactant? The overall order?
(c) What is the numerical value of k, the rate constant?

32. One of the major eye irritants in smog is formaldehyde, CH_2O, formed by reaction of ozone with ethylene.

$$C_2H_4(g) + O_3(g) \longrightarrow 2\,CH_2O(g) + \tfrac{1}{2}\,O_2(g)$$

(a) What are the relative rates of disappearance of O_3 and appearance of CH_2O?
(b) Using the data below, determine the rate law and rate constant.

(c) What is the reaction order in terms of O_3? In terms of C_2H_4?

$[O_3]_0$ (mol/L)	$[C_2H_4]_0$ (mol/L)	INITIAL RATE OF APPEARANCE OF CH_2O mol/(L · s)
0.50×10^{-7}	1.0×10^{-8}	1.0×10^{-12}
1.5×10^{-7}	1.0×10^{-8}	3.0×10^{-12}
1.0×10^{-7}	2.0×10^{-8}	4.0×10^{-12}

33. The data below were collected for the reaction A + B → D + E.

$[A]_0$ (mol/L)	$[B]_0$ (mol/L)	INITIAL RATE (mol/L · min)
0.10	0.20	2.0×10^2
0.10	0.10	5.0×10^1
0.30	0.10	1.5×10^2

(a) Write the rate law for the reaction and give the order with respect to each reactant.
(b) Calculate the rate constant *k*.
(c) What is the initial rate of reaction where $[A]_0 = [B]_0 = 0.20$?
(d) What concentration of B is required to give an initial rate of 6.0×10^2 mol/(L · min) when $[A]_0 = 0.30$ M?

34. One reaction which may occur in air polluted with nitrogen oxide is

$$2\,NO(g) + O_2(g) \longrightarrow 2\,NO_2(g)$$

Using the data below, answer the questions which follow.

EXPERIMENT	INITIAL CONCENTRATIONS $[NO]_0(M)$	$[O_2]_0(M)$	RATE OF APPEARANCE OF NO_2 (M/sec)
1	0.001	0.001	7×10^{-6}
2	0.001	0.002	14×10^{-6}
3	0.001	0.003	21×10^{-6}
4	0.002	0.003	84×10^{-6}
5	0.003	0.003	189×10^{-6}

(a) What is the order of reaction with respect to each reactant?
(b) What is the overall order?
(c) Write the rate law for the reaction.
(d) Calculate the rate of appearance of NO_2 when $[NO] = [O_2] = 0.005\,M$.

35. Nitryl fluoride, an explosive compound, can be made by treating nitrogen dioxide with fluorine.

$$2\,NO_2(g) + F_2(g) \rightarrow 2\,NO_2F(g)$$

Use the rate data below to answer the questions which follow.

EXPERIMENT	INITIAL CONCENTRATIONS (mol/L) $[NO_2]_0$	$[F_2]_0$	$[NO_2F]_0$	INITIAL RATE mol/(L · s)
1	0.001	0.005	0.001	2×10^{-4}
2	0.002	0.005	0.001	4×10^{-4}
3	0.006	0.002	0.001	4.8×10^{-4}
4	0.006	0.004	0.001	9.6×10^{-4}
5	0.001	0.001	0.001	4×10^{-5}
6	0.001	0.001	0.002	4×10^{-5}

(a) Write the rate law for the reaction.
(b) What is the order of reaction with respect to each component of the reaction?
(c) What is the numerical value of the rate constant *k*?

CONCENTRATION/TIME EQUATIONS AND REACTION HALF-LIFE

36. The transformation of cyclopropane to propene was described in Example 13.4. If the initial concentration of cyclopropane is 0.050 *M*, how many hours must elapse for the concentration to drop to 0.025 *M*? (The rate constant is 5.4×10^{-2}/hr.)

37. For a reaction with the rate law "initial rate = $k[A]_0$," and $k = 3.33 \times 10^{-6}$/hr, what is the half-life of the reaction?

38. The decomposition of N_2O_5 (to give NO_2 and O_2) follows the rate law "rate = $k[N_2O_5]$" where *k* is 5.0×10^{-4} per second at a particular temperature.
(a) What is the half-life of the reaction?
(b) How long does it take for the N_2O_5 concentration to drop to one tenth of its original value?

39. The decomposition of phosphine, PH_3, proceeds according to the equation

$$4\,PH_3(g) \longrightarrow P_4(g) + 6\,H_2(g)$$

It is found the reaction has the rate law "initial rate = $k[PH_3]$." If the half-life is 37.9 seconds, how much time is required for three fourths of the PH_3 to decompose?

40. Molecules of butadiene, C_4H_6, can couple to form C_8H_{12}. The rate law for the reaction is

$$\text{Rate} = k[C_4H_6]^2$$

If the rate constant is estimated to be 0.014 (L/mol · s), what is the half-life of the reaction (in seconds) when $[C_4H_6]_0 = 0.016\,M$? If the initial concentration of C_4H_6 is 0.016 *M*, how much time (in seconds) must elapse for the concentration to drop to 0.0016 *M*?

41. Mercury is eliminated from the body in a first order reaction with a half-life of 60 days. If a person eats a piece of tuna fish contaminated with 1 milligram of mercury, how much of this mercury will remain in the body after 6 months?

42. Bond breaking reactions are important to study, and one investigated recently can be written

$$M{-}SCH_3 \longrightarrow M + SCH_3$$

where M is a metal to which other groups are attached. One of the authors of this book recently studied the kinetics of the process and found it to be first order in $M{-}SCH_3$ with a half-life of 40. minutes at room temperature.
(a) What is the rate constant for the reaction?
(b) If you begin with a 0.0060 *M* solution of $M{-}SCH_3$, how much is left after 2.0 hours?
(c) Concentrations of $M{-}SCH_3$ as low as 10^{-4} *M* can be detected. If you begin with a concentration of $M{-}SCH_3$ of 0.0060 *M*, how long does it take to reach 1.0×10^{-4} *M*, the limit of the detection method?

43. The Cr^{3+} complex ion below changes its shape (it *isomerizes*) by a first order reaction. The half-life is found to be 1.07 hours.

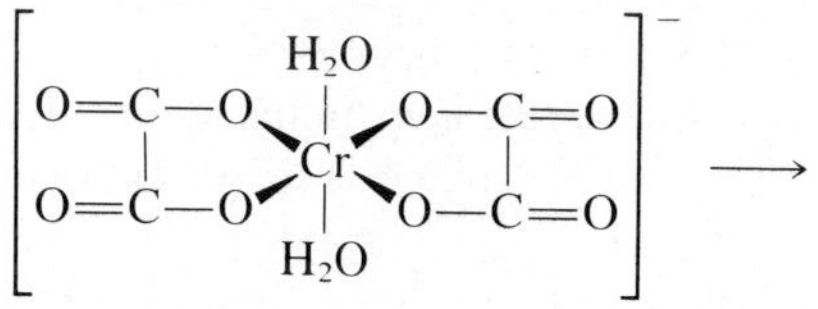

trans form

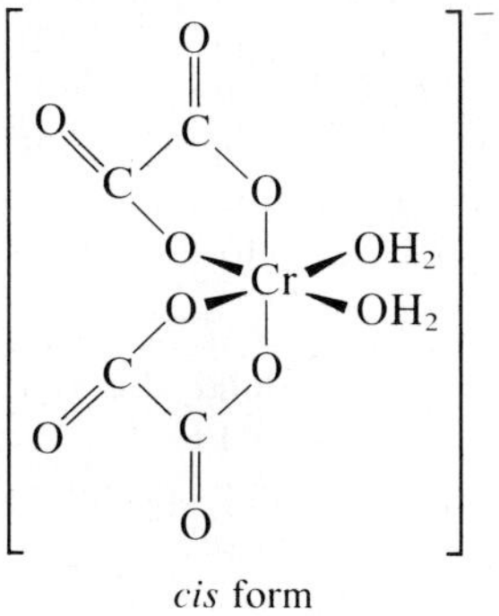

cis form

(a) What is the rate constant for the reaction?
(b) What fraction of the original form (the *trans* form) remains after 3.21 hours of reaction time?
(c) When the concentration of the *trans* form is 0.125 *M*, what is the rate of reaction in moles per liter per hour?
(d) How long will it take to reduce the concentration of the *trans* form to 1.00% of its original concentration?

44. It is desirable that pesticides eventually decompose in the environment to give harmless products. Let us assume the half-life for the decomposition of a certain pesticide is 10.2 years. If the present concentration of the pesticide in a lake is 3.1×10^{-5} g/mL, what was its concentration 4 years ago?

45. Although hypochlorous acid, HOCl, was one of the first compounds known of chlorine, the fluorine analog, HOF, was only recently isolated. Instability is its most prominent chemical property, and it decomposes by a first order reaction to HF and O_2 with a half-life of only 30. minutes at room temperature.

$$HOF(g) \longrightarrow HF(g) + \tfrac{1}{2} O_2(g)$$

If the partial pressure of HOF in a 1.0-L flask is initially 100. mm Hg at 25°C, what is the total pressure in the flask and the partial pressure of HOF after 30. minutes? After 45 minutes?

GRAPHICAL ANALYSIS OF RATE LAWS

46. The decomposition of N_2O_5 in the gas phase

$$2\ N_2O_5(g) \longrightarrow 4\ NO_2(g) + O_2(g)$$

has been thoroughly studied. Some data for the concentration of N_2O_5 as a function of time have been plotted below. What is the order of the reaction? What is the rate law for the process?

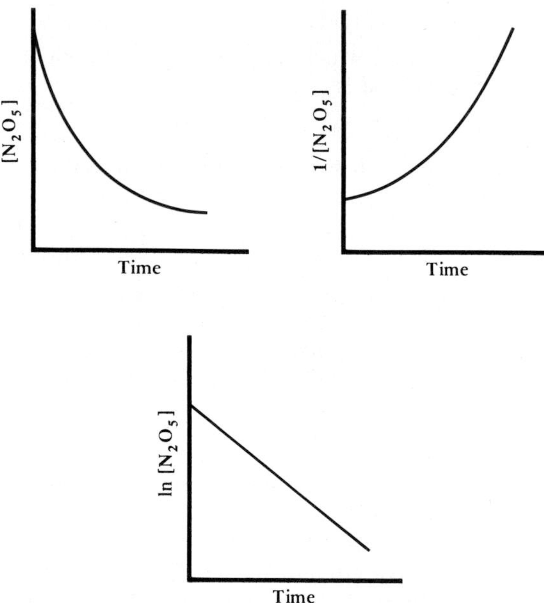

47. Common sugar, sucrose, reacts in dilute acid solution to give the simpler sugars glucose and fructose. Both of the simple sugars have the same formula $C_6H_{12}O_6$.

$$C_{12}H_{22}O_{11}(aq) + H_2O(\ell) \longrightarrow 2\ C_6H_{12}O_6(aq)$$

The rate of this reaction has been studied in acid solution, and the following data were obtained.

TIME (min)	$[C_{12}H_{22}O_{11}](M)$
0	0.316
39	0.274
80	0.238
140	0.190
210	0.146

(a) Plot the data above as ln[sucrose] versus time and 1/[sucrose] versus time. What is the order of the reaction?
(b) Write the rate law for the reaction and calculate the rate constant k.
(c) Estimate the concentration of sucrose after 175 minutes.

48. Data for the reaction of phenylacetate with water are given in Study Question 24. Plot these data as ln[phenylacetate] and 1/[phenylacetate] versus time. What is the order of the reaction with respect to phenylacetate? Determine the rate constant from your plot and calculate the half-life for the reaction.

49. Data for the reaction of an organic bromide (A) with an organic amine (B) were obtained in methyl alcohol when equal initial concentrations of reactants were used at 35°C.

$$\underset{A}{C_6H_5-\overset{\overset{\displaystyle O}{\|}}{C}-CH_2Br} + \underset{B}{C_5H_5N} \longrightarrow$$

$$\underset{\text{products}}{[C_6H_5-\overset{\overset{\displaystyle O}{\|}}{C}-CH_2NC_5H_5]^+ + Br^-}$$

TIME (min)	[A] (mol/L)
0	0.0385
200	0.0288
400	0.0230
600	0.0191
800	0.0163
1000	0.0143

(a) Plot these data as ln[A] and 1/[A] versus time. What is the order of the reaction?
(b) What is the rate constant?
(c) What is the half-life of the reaction?
(d) Based on your answer to part (a), the fact that the initial concentrations of A and B were equal ($[A]_0 = [B]_0$), and the marginal note on page 481, decide on the rate law for this reaction.

50. Data for the combination of two C_4H_6 molecules to give C_8H_{12} are given in Study Question 25.

$$C_4H_6(g) \longrightarrow \tfrac{1}{2}\, C_8H_{12}(g)$$

Plot these data as $\ln[C_4H_8]$ and $1/[C_4H_8]$ versus time. What is the order of the reaction? Determine the rate constant from your plot and calculate the initial half-life of the reaction.

KINETICS AND ENERGY

51. For the hypothetical reaction A + B → C + D, the activation energy is 32 kJ/mol. For the reverse reaction (C + D → A + B), the activation energy is 58 kJ/mol. Is the reaction A + B → C + D exothermic or endothermic?

52. Use the diagram below to answer the following questions.
(a) Is the reaction exothermic or endothermic?
(b) What is the approximate value of ΔE for the forward reaction?
(c) What is the activation energy in each direction?
(d) A catalyst is found that lowers the activation energy of the reaction by about 10 kJ/mol. How will this catalyst affect the rate of the *reverse* reaction?

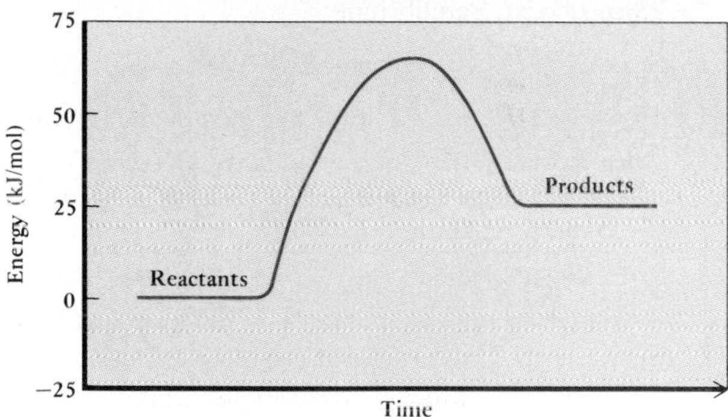

53. Calculate the activation energy (E^*) for the reaction

$$N_2O_5(g) \longrightarrow 2\ NO_2(g) + \tfrac{1}{2}\ O_2(g)$$

from the observed rate constants at the following two temperatures: (a) at 25°C, $k = 3.46 \times 10^{-5}$/s and (b) at 55°C, $k = 1.5 \times 10^{-3}$/s.

54. The decomposition of dinitrogen pentoxide

$$2\ N_2O_5(g) \longrightarrow 4\ NO_2(g) + O_2(g)$$

has the rate law "initial rate $= k[N_2O_5]_0$." It has been found experimentally that the decomposition is 20.% complete in 6.0 hours at 300. K and 20.% complete in 1.0 minute at 350 K. Calculate the rate constant and the half-life at 300. K.

55. If the rate constant for a reaction triples in value when the temperature rises from 300. K to 310. K, what is the activation energy of the reaction?

56. For a certain reaction, the constant A in the Arrhenius equation is 6.00×10^{12} mol/(L · s) and $E^* = 100.$ kJ/mol. What is the rate constant at 400. K?

57. The data below give the temperature dependence of the rate constant for the reaction $N_2O_5(g) \rightarrow$

$2\ NO_2(g) + \frac{1}{2} O_2(g)$. Plot these data in the appropriate way to derive the activation energy for the reaction.

T (K)	k (per second)
338	4.87×10^{-3}
328	1.50×10^{-3}
318	4.98×10^{-4}
308	1.35×10^{-4}
298	3.46×10^{-5}
273	7.87×10^{-7}

MECHANISMS

58. Write the rate law for each elementary step below. Give the molecularity of each step.
(a) $NO(g) + NO_3(g) \rightarrow 2\ NO_2(g)$
(b) $Cl(g) + H_2(g) \rightarrow HCl(g) + H(g)$
(c) $(CH_3)_3CBr(aq) \rightarrow (CH_3)_3C^+(aq) + Br^-(aq)$

59. The reaction between chloroform and chlorine gas proceeds in a series of elementary steps.

Step 1. Fast, equilibrium
$Cl_2(g) \rightleftharpoons 2\ Cl(g)$
Step 2. Slow
$CHCl_3(g) + Cl(g) \longrightarrow CCl_3(g) + HCl(g)$
Step 3. Fast
$CCl_3(g) + Cl(g) \longrightarrow CCl_4(g)$
Overall reaction
$CHCl_3(g) + Cl_2(g) \longrightarrow CCl_4(g) + HCl(g)$

(a) Which of the steps is the rate determining step?
(b) Write the rate law for the rate determining step.
(c) What is the molecularity of each step?

60. The ozone layer in the earth's upper atmosphere is important in shielding the earth from very harmful ultraviolet radiation. The ozone, O_3, decomposes according to the equation

$2\ O_3(g) \longrightarrow 3\ O_2(g)$

The mechanism of the reaction is thought to proceed through an initial fast equilibrium and a slow second step.

Step 1. Fast, equilibrium $O_3(g) \rightleftharpoons O_2(g) + O(g)$
Step 2. Slow $O_3(g) + O(g) \longrightarrow 2\ O_2(g)$

(a) Which of the steps is the rate determining step?
(b) Write the rate law for the rate determining step.
(c) What is the molecularity of each step?

61. The Raschig process for producing hydrazine was described in the introduction to Section 13.7. An alternative to steps 2 and 3 are

Alternate step 2: Slow
$NH_2Cl(aq) + OH^-(aq) \longrightarrow NHCl^-(aq) + H_2O(\ell)$

Alternate step 3: Fast
$NHCl^-(aq) + NH_3(aq) \longrightarrow N_2H_4(aq) + Cl^-(aq)$

(a) Which of these is the rate determining step? What is its molecularity?
(b) What is the rate law for the rate determining elementary step?

62. Assume that an A molecule reacts with two B molecules in a one-step process to give AB_2. That is, $A + 2B \longrightarrow AB_2$.
(a) Write the rate law for this reaction.
(b) If the initial rate of appearance of AB_2 is 2.0×10^{-5} mol/(L · s) when the initial concentrations of A and B are 0.30 M, calculate the rate constant for the reaction.

63. At temperatures less than 500K the reaction between carbon monoxide and nitrogen dioxide

$NO_2(g) + CO(g) \longrightarrow CO_2(g) + NO(g)$

follows the rate law "rate of appearance of $CO_2 = k[NO_2]^2$." Which of the three mechanisms suggested below best agrees with the experimentally observed rate law?

Mechanism 1: $NO_2 + CO \longrightarrow CO_2 + NO$ (single, elementary step)

Mechanism 2: $NO_2 + NO_2 \longrightarrow NO_3 + NO$ (slow)
$NO_3 + CO \longrightarrow NO_2 + CO_2$ (fast)

Mechanism 3: $NO_2 \longrightarrow NO + O$ (slow)
$CO + O \longrightarrow CO_2$ (fast)

64. The reaction $A + 2B \rightarrow D$ occurs by the following mechanism:

Step 1. $A + B \rightleftharpoons X$ very rapid equilibrium
Step 2. $X + C \longrightarrow Y$ slow
Step 3. $Y + B \longrightarrow D$ fast

Which of the rate laws below is correct?
(a) $R = k[C]$
(b) $R = k[D]$
(c) $R = k[A][B]$
(d) $R = k[A][B]^2[C]$
(e) $R = k[A][B][C]$

65. The mechanisms below have been proposed for the following reaction:

$2\ H_2(g) + 2\ NO(g) \longrightarrow N_2(g) + 2\ H_2O(g)$

The experimental rate law is "rate of appearance of $N_2 = k[NO]^2[H_2]$." For each mechanism, indicate whether or not it is consistent with the observed rate law.

Mechanism 1: $H_2 + NO \longrightarrow H_2O + N$ (slow)
$N + NO \longrightarrow N_2 + O$ (fast)
$O + H_2 \longrightarrow H_2O$ (fast)

Mechanism 2: $2\ NO \rightleftharpoons N_2O_2$ (fast)
$N_2O_2 + H_2 \longrightarrow H_2O + N_2O$ (slow)
$N_2O + H_2 \longrightarrow N_2 + H_2O$ (fast)

66. The mechanism for the reaction of chloroform with chlorine was given in Study Question 59. Show that this agrees with the experimental rate law "rate = $k[CHCl_3][Cl_2]^{1/2}$.

67. The rate law for the decomposition of ozone described in Study Question 60 is "rate = $k[O_3]^2/[O_2]$." Show how this can be derived from the elementary steps in Study Question 60.

CATALYSIS

68. Which of the following statements are true?
   (a) The concentration of a homogeneous catalyst may appear in the rate law.
   (b) A catalyst never affects the Arrhenius "frequency factor" *A*.
   (c) If a reaction is exothermic, a catalyst affects the quantity of heat evolved.
   (d) A catalyst can change the course of a reaction and allow different products to be produced.
   (e) A catalyst can cause a change in which elementary step in a reaction is rate determining.

69. Which of the following reactions appear to involve a catalyst? In those cases where a catalyst is present, tell whether it is homogeneous or heterogeneous.
   (a) $CH_3CO_2CH_3(aq) + H_2O(\ell) + H^+(aq) \rightarrow CH_3CO_2H(aq) + CH_3OH(aq) + H^+(aq)$
   (b) $2\ H_2(g) + O_2(g) \rightarrow 2\ H_2O(g)$
   (c) $2\ H_2(g) + O_2(g) + Pt(s) \rightarrow 2\ H_2O(g) + Pt(s)$
   (d) $NH_3(aq) + CH_3Cl(aq) + H_2O(\ell) \rightarrow Cl^-(aq) + NH_4^+(aq) + CH_3OH(aq)$

70. Hydrogenation reactions, processes wherein H is added to a molecule, are usually catalyzed. An excellent catalyst is a very finely divided metal suspended in the reaction solvent. Tell why finely divided rhodium, for example, is a much more efficient catalyst than a small block of the metal.

GENERAL QUESTIONS

71. The experimental rate law for the decomposition of $N_2O_5(g)$ is "initial rate = $k[N_2O_5]_0$." A postulated reaction mechanism is

Fast, equilibrium	$N_2O_5(g) \rightleftharpoons NO_2(g) + NO_3(g)$
Slow	$NO_2(g) + NO_3(g) \longrightarrow NO(g) + O_2(g) + NO_2(g)$
Fast	$NO(g) + NO_3(g) \longrightarrow 2\ NO_2(g)$

   (a) What is the overall reaction?
   (b) Show that the observed rate law is compatible with the postulated mechanism.
   (c) If $k = 5.0 \times 10^{-4}$/s, how much time must elapse for the concentration of N_2O_5 to drop to 10.% of its original value?

72. Sketch a reaction energy diagram for an exothermic reaction that proceeds through two steps, with the second step rate determining. If the first step is rate determining, how does this change your diagram?

73. A catalyst can allow a reaction to proceed along a different pathway with the lower activation energy. If the catalyst lowers the activation energy by 5 kJ/mol, how much faster is the reaction, if both occur at 298 K (and *A* does not change)?

74. Tell whether the following statements are true or false. If false, rewrite the sentence to make it correct.
   (a) The rate controlling elementary step in a reaction is the slowest step in the mechanism.
   (b) It is possible to change the rate constant by changing the temperature.
   (c) As a reaction proceeds at constant temperature, the rate remains constant.
   (d) A negative exponent for a reagent in the rate law means the reagent is a reactant in an elementary step prior to the rate determining step.
   (e) A reaction that is third order overall must involve more than one step.

14
Chemical Equilibria: General Concepts

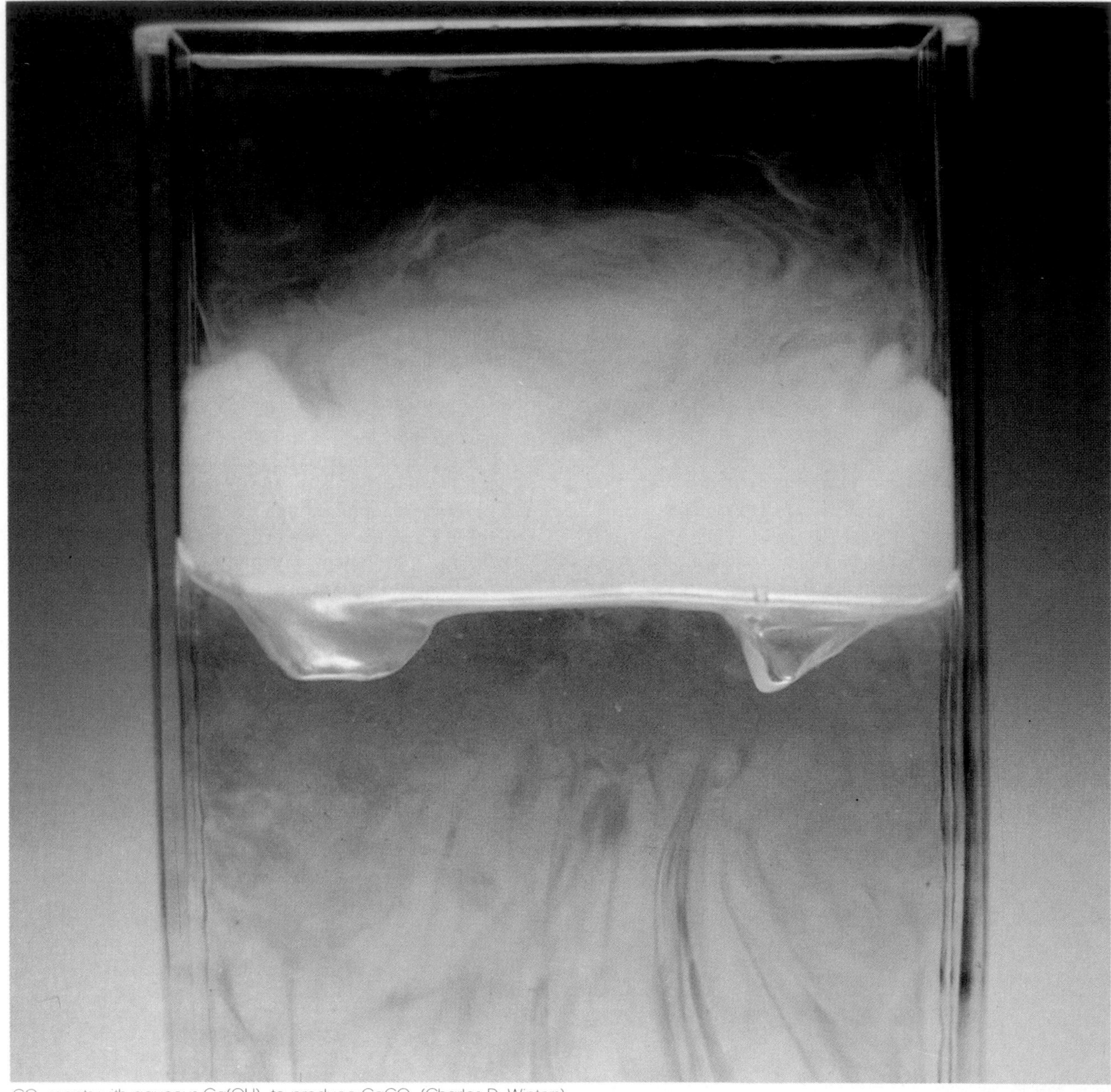

CO_2 reacts with aqueous $Ca(OH)_2$ to produce $CaCO_3$ (Charles D. Winters)

CHAPTER OUTLINE

Our goal in this chapter is to explore the consequences of the fact that all chemical reactions are *reversible* and that in a closed system a state of *equilibrium* can eventually be achieved between reactants and products. A major result of this exploration will be an ability to describe "chemical reactivity" in more quantitative terms. As you will see, the concentrations of reactant and product at equilibrium are a measure of the intrinsic tendency of chemical reactions to proceed from reactants to products.

14.1 THE NATURE OF THE EQUILIBRIUM STATE

We first introduced you to the idea of reaction reversibility in Chapter 11 when describing equilibria in the case of changes between phases. From these discussions you began to develop some ideas about the nature of the state of equilibrium, but now we must make these ideas more explicit.

As an example of a chemical equilibrium, consider the reaction that leads to the formation of limestone stalactites and stalagmites in some caves (Figure 14.1). These originate from the reaction of limestone, a form of calcium carbonate, with CO_2 and water.

An equilibrium between reactants and products can be depicted using a set of arrows of the type $\rightleftarrows$ or $\rightleftharpoons$.

$$CaCO_3(s) + CO_2(g) + H_2O(\ell) \rightleftharpoons Ca^{2+}(aq) + 2\,HCO_3^-(aq)$$

$CaCO_3$ is found in underground deposits, a leftover of ancient oceans. If water seeping through the ground contains dissolved CO_2, the reaction above can occur to give aqueous Ca^{2+} and HCO_3^- ions. This reaction illustrates, first of all, that all chemical reactions are **reversible,** just as

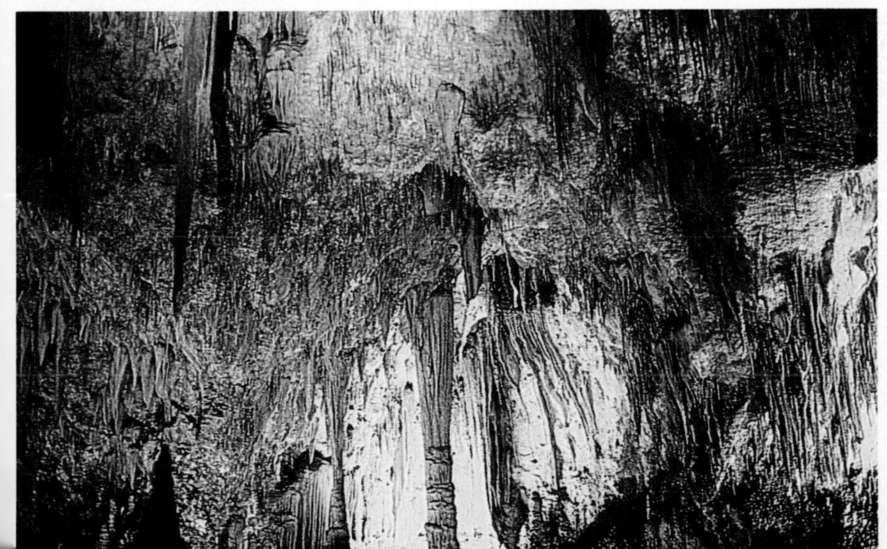

FIGURE 14.1
Calcium carbonate stalactites cling from the ceiling and stalagmites grow from the floor in a cave. (National Park Service, F.E. Mang, Jr.)

are phase changes. As a result, for reversible processes there will always be conditions of pressure and temperature for which reactants and products can coexist at equilibrium. To prove reaction reversibility, let us do an experiment with some soluble salts containing the Ca^{2+} and HCO_3^- ions mentioned above (say $CaCl_2$ and $NaHCO_3$). Putting them in an open beaker of water in the laboratory, you will soon see bubbles of CO_2 gas and solid $CaCO_3$ (Figure 14.2). If the CO_2 is swept away, all the dissolved Ca^{2+} and HCO_3^- ions will eventually disappear from solution, having been converted completely to calcium carbonate.

$$Ca^{2+}(aq) + 2\,HCO_3^-(aq) \longrightarrow CO_2(g) + H_2O(\ell) + CaCO_3(s)$$

Now do another experiment, but this time put some $CaCO_3(s)$ and some water in a flask under a high partial pressure of CO_2 gas just as nature does underground. Eventually some of the limestone will begin to dissolve to give $Ca^{2+}(aq)$ and $HCO_3^-(aq)$ according to the first equation above. From these experiments it is clear that the system containing limestone, water, carbon dioxide, and aqueous calcium bicarbonate is reversible, a fact that we convey by using a double set of arrows to connect the two sides of the balanced equation.

$$CaCO_3(s) + H_2O(\ell) + CO_2(g) \rightleftharpoons Ca^{2+}(aq) + 2\,HCO_3^-(aq)$$
$$Ca^{2+}(aq) + 2\,HCO_3^-(aq) \rightleftharpoons CO_2(g) + H_2O(\ell) + CaCO_3(s)$$

The next important feature of chemical equilibria is that they are **dynamic;** that is, both the forward and reverse reactions take place continually at equal rates while the system is at equilibrium. *Equilibrium does not involve the cessation of the reactions, but rather a balance between them.* A simple experiment can be used to illustrate the dynamic nature

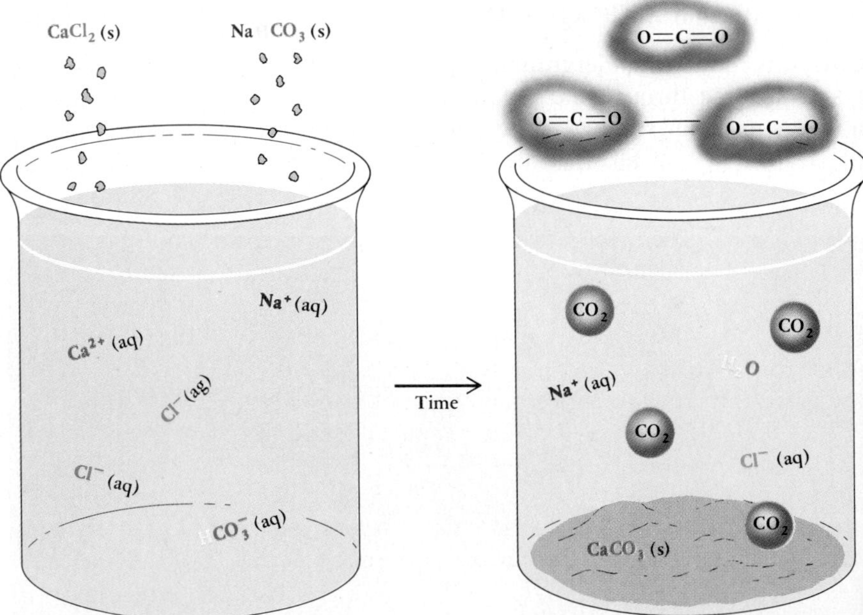

FIGURE 14.2
Some $CaCl_2$ and $NaHCO_3$ are added to a beaker of water. With time, the reaction

$$Ca(HCO_3)_2(aq) \rightleftharpoons CaCO_3(s) + CO_2(g) + H_2O(\ell)$$

occurs, evolving CO_2, precipitating $CaCO_3$, and leaving NaCl in solution.

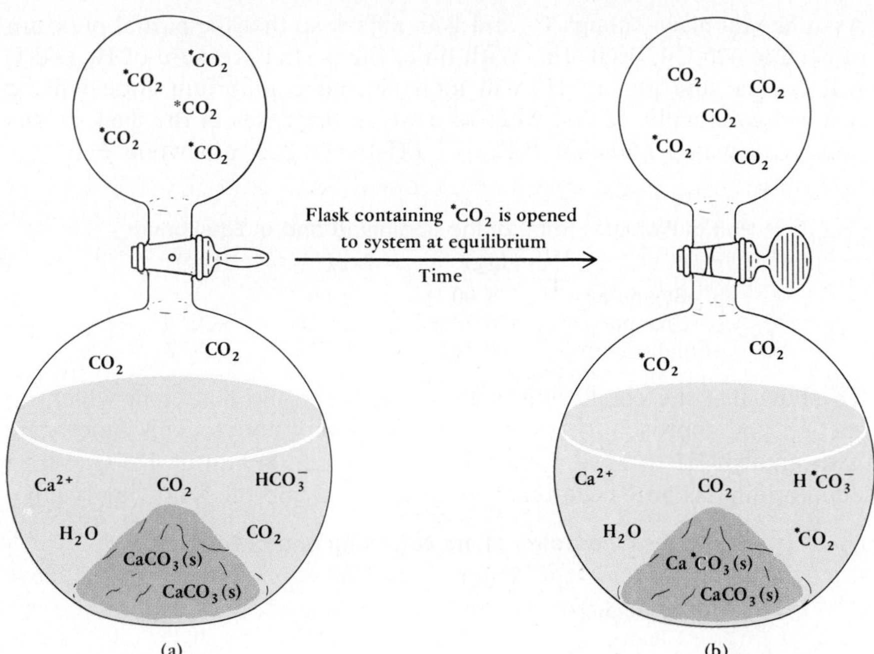

FIGURE 14.3
Illustration of the dynamic nature of an equilibrium. (a) Calcium carbonate, water, and CO_2 are allowed to come to equilibrium in a closed flask. (b) The flask is then opened to another flask containing CO_2 tagged with radioactive carbon (*C). The CO_2 pressure in the two flasks is the same. Because equilibria are dynamic and reversible, *C is mixed throughout the system.

of the calcium carbonate/CO_2 system. Place some $CaCO_3(s)$ in a sealed flask with some water; add some CO_2 to a partial pressure of about 1 atm. Until a state of equilibrium is achieved, the Ca^{2+} and HCO_3^- concentrations will increase, and the CO_2 pressure will decrease. No further net change is observed after this time, because the rate at which $CaCO_3$, H_2O, and CO_2 react to give $Ca(HCO_3)_2(aq)$ is exactly balanced by the rate at which the newly formed calcium bicarbonate reverts to calcium carbonate, water, and carbon dioxide. Now open the flask to a second flask also containing CO_2 gas *at the same pressure as the equilibrium pressure*, but ''label'' this CO_2 with radioactive ^{14}C (Figure 14.3). We have done nothing to disturb the equilibrium we have achieved, but a new form of CO_2 gas has been allowed to enter the reaction. You can see the results if you open the system after some time and sample the $CaCO_3(s)$ and $HCO_3^-(aq)$. Both would be found to contain some radioactive carbon! This equilibrium, like all others, is dynamic. The radioactive carbon dioxide has been distributed throughout the system by the ''forward'' and ''reverse'' reactions. In any system, these reactions are always occurring, whether the system is at equilibrium or not. When the system is approaching equilibrium, one reaction must be more rapid than the other, but, when equilibrium has been achieved, their rates must be the same.

The relation between forward and reverse reaction rates was discussed in Chapter 13.

Not only are equilibrium processes reversible and dynamic, but the *nature and properties of the equilibrium state will be the same, no matter what the direction of approach.* To see why this is true, we take the reaction of hydrogen and iodine to produce hydrogen iodide, a reaction that has been thoroughly studied.

$$H_2(g) + I_2(g) \rightleftharpoons 2\ HI(g)$$

Assume you place enough H_2 and I_2 in a flask so that the partial pressure of each at 425°C is 1.00 atm. With time, the partial pressure of H_2 and I_2 will decline and that of HI will increase; an equilibrium state will be reached eventually. If you were to analyze the gases in the flask at this point you would find that $P(H_2) = P(I_2) = 0.212$ atm while $P(HI) =$ 1.58 atm.

Partial Pressures (atm) at the Beginning and at Equilibrium

	$H_2(g)$	$I_2(g)$	HI(g)
Beginning	1.00	1.00	0
Change	−0.788	−0.788	+1.58
Equilibrium	0.212	0.212	1.58

To show that the equilibrium state is the same, no matter in which direction you approach it, reverse the experiment above. This time begin with enough HI to have a partial pressure of 2.00 atm at 425°C. When equilibrium has now been reached, you would find the following results:

Partial Pressures (atm) at the Beginning and at Equilibrium

	$H_2(g)$	$I_2(g)$	HI(g)
Beginning	0	0	2.00
Change	+0.212	+0.212	−0.42
Equilibrium	0.212	0.212	1.58

The experiment shows quite clearly that the partial pressures of the compounds in this reaction are the same, no matter in which direction you have approached the reaction.*

14.2 THE EQUILIBRIUM CONSTANT

Equilibria involving changes between phases are also reversible and dynamic and can be approached from either direction. However, there is a difference in the ways we describe the equilibrium position in a phase change and in a chemical reaction. To describe the equilibrium position of water evaporating at a particular temperature you need only give the equilibrium vapor pressure. In contrast, to describe the equilibrium position for chemical reactions it is often necessary to give equilibrium concentrations of several reagents. Fortunately, there is a simple way to relate the equilibrium concentrations of the reactants to those of the products for a chemical reaction at a given temperature. This is the **equilibrium constant expression,** a function of such great importance in chemistry that

*We must be careful in interpreting this third aspect of chemical equilibria. As an example, take the H_2/I_2 reaction above. If you begin with 1 mole of H_2 and 1 mole of I_2, equilibrium will be established at the same point as if you had begun with 2 moles of HI. However, if you had used 2 moles of hydrogen with 1 mole of iodine, a new equilibrium state with different concentrations of the gases would be reached. To approach this last equilibrium state from the opposite direction, you would have to mix 2 moles of HI with 1 mole of H_2. The statement that the equilibrium state has the same properties no matter what the direction of approach presupposes that the number of atoms of each element used is the same in the forward and reverse experiments.

it is described in detail as applied to general equilibria in this chapter and to equilibria in aqueous solutions in Chapters 15 through 17.

THE EQUILIBRIUM CONSTANT EXPRESSION

The notion that the equilibrium concentrations of reactants and products are interrelated in a simple manner is easy to prove by experiments such as those for the H_2/I_2 gas phase reaction in Table 14.1.

$$H_2(g) + I_2(g) \rightleftharpoons 2\ HI(g)$$

In the first four experiments, H_2 and I_2 were mixed, and HI was formed; when equilibrium had been achieved, the partial pressure of each component of the mixture was determined. The last two experiments were done in the reverse manner; HI was the reactant and H_2 and I_2 were the products. The important point of these experiments is that the equilibrium partial pressures of the reaction components are related through a simple numerical ratio. That is, the ratio

$$\frac{P_{HI}^2}{P_{I_2}P_{H_2}} = 55.17$$

is the same within experimental error for all six experiments!

Hundreds of other experiments have proved that the relationship found for the hydrogen iodide equilibrium arises from the general expression for an **equilibrium constant.** For the general reaction

$$aA + bB \rightleftharpoons cC + dD$$

experiments show that concentrations of reactants and products are related by the *equilibrium constant expression*

$$\text{Equilibrium constant} = K = \frac{[C]^c[D]^d}{[A]^a[B]^b} \qquad (14.1)$$

where the value of the constant K depends on the particular reaction *and* on the temperature. The equilibrium constant expression tells you that the *product concentrations appear in the numerator and reactant concentrations in the denominator. Each concentration is raised to the power of its stoichiometric coefficient in the balanced equation.*

The experimental equilibrium constant expression for the H_2/I_2 reaction is written in partial pressures, while Equation 14.1 is in concentration units. Expressions for K can be written in either units, since, for gases, the ideal gas law specifies that concentration = $(n/V) = P/RT$.

The basic form of the equilibrium expression can be predicted by thermodynamics. This is described in Chapter 18.

TABLE 14.1 Equilibrium Partial Pressures for the Reaction $H_2(g) + I_2(g) = 2\ HI(g)$

PARTIAL PRESSURES (atm)			
H_2	I_2	HI	$K = P_{HI}^2/P_{H_2}P_{I_2}$
0.1645	0.09783	0.9447	55.46
0.2583	0.04229	0.7763	55.17
0.1274	0.1339	0.9658	54.68
0.1034	0.1794	1.0129	55.31
0.02703	0.02745	0.2024	55.21
0.06443	0.06540	0.4821	55.16
			Average = 55.17

These data were taken from an article by Taylor and Crist, *J. Am. Chem. Soc.*, 63:1377, 1941. The temperature was 698.6 K.

The concentrations in the equilibrium constant expression are usually given in moles per liter (M). However, for a gas, moles per liter (n/V) is proportional to the partial pressure of the gas ($n/V = P/RT$), and K expressions for gases can be written in terms of gas partial pressures. We shall return to this point later.

USING EQUILIBRIUM CONSTANT EXPRESSIONS

There are several things you should know about equilibrium constant expressions so that you can apply them in practical situations.

PURE SOLIDS AND LIQUIDS IN EQUILIBRIUM EXPRESSIONS *The concentrations of pure solids, of pure liquids, and of solvents in dilute solutions never appear in equilibrium constant expressions.* For example, the oxidation of solid, yellow sulfur produces colorless sulfur dioxide (Figure 3.12).

$$S(s) + O_2(g) \rightleftharpoons SO_2(g)$$

Following the general principle that products appear in the numerator and reactants in the denominator, you would write

$$K' = \frac{[SO_2(g)]}{[S(s)][O_2(g)]}$$

However, since sulfur is a molecular solid and since the concentration of molecules within any solid is fixed, the sulfur concentration is not changed either by the reaction or by addition or removal of some solid. Further, it is an experimental fact that the equilibrium concentrations (or partial pressures) of O_2 and SO_2 are not changed by the amount of sulfur, as long as there is some solid sulfur present at equilibrium. Therefore, it is conventional to include the constant concentrations of any solid reactants and products in the equilibrium constant itself and write

$$K'[S(s)] = K = \frac{[SO_2(g)]}{[O_2(g)]} = 4.2 \times 10^{52} \text{ (at 25°C)}$$

This same principle applies to reactions in aqueous solutions such as

$$AgCl(s) \rightleftharpoons Ag^+(aq) + Cl^-(aq)$$

where only the concentrations of the silver and chloride ions in solution would be included in the equilibrium constant expression.

$$K = [Ag^+(aq)][Cl^-(aq)] = 1.8 \times 10^{-10} \text{ (at 25°C)}$$

The silver chloride concentration, [AgCl(s)], does not appear in the expression, since the amount of solid metal halide does not affect the equilibrium (if some AgCl is present).

The concentration of water, $[H_2O]$, is effectively constant for reactions occurring in aqueous solution. For example, ammonia in water functions as a weak base due to the interaction of NH_3 with water,

$$NH_3(aq) + H_2O(\ell) \rightleftharpoons NH_4^+(aq) + OH^-(aq)$$

$$K' = \frac{[NH_4^+][OH^-]}{[NH_3][H_2O]}$$

$$K = K'[H_2O] = \frac{[NH_4^+][OH^-]}{[NH_3]} = 1.8 \times 10^{-5}$$

and formic acid is a weak acid since it can transfer an H^+ ion to water to form the hydronium ion.

$$HCOOH(aq) + H_2O(\ell) \rightleftharpoons HCOO^-(aq) + H_3O^+(aq)$$

$$K' = \frac{[HCOO^-][H_3O^+]}{[HCOOH][H_2O]}$$

$$K = K'[H_2O] = \frac{[HCOO^-][H_3O^+]}{[HCOOH]} = 1.8 \times 10^{-4}$$

In neither case is water included in the final equilibrium constant expression because, by convention, $[H_2O]$ is included in the accepted value of K. Ammonia, formic acid, and their products are present in such small concentrations in dilute solutions that the concentration of water is essentially not affected by their presence.

Finally, when dissolved materials and gases are both present at equilibrium, their concentrations or partial pressures, respectively, appear in the equilibrium expression. The best example of this is a reaction we have already used: the dissolving of calcium carbonate in the presence of water and CO_2.

$$CaCO_3(s) + H_2O(\ell) + CO_2(g) \rightleftharpoons Ca^{2+}(aq) + 2\ HCO_3^-(aq)$$

$$K = \frac{[Ca^{2+}][HCO_3^-]^2}{P_{CO_2}}$$

CHANGING STOICHIOMETRIC COEFFICIENTS Another important aspect of equilibrium expressions is the consequence of changing stoichiometric coefficients. For example, oxidation of sulfur can give, ultimately, sulfur trioxide

$$S(s) + \tfrac{3}{2}\ O_2(g) \rightleftharpoons SO_3(g)$$

and the equilibrium constant expression for this reaction as written would be

$$K_1 = \frac{[SO_3]}{[O_2]^{3/2}} = 1.1 \times 10^{65} \text{ at } 25°C$$

However, you can write this equation equally well as

$$2\ S(s) + 3\ O_2(g) \rightleftharpoons 2\ SO_3(g)$$

and the equilibrium constant expression would now be

$$K_2 = \frac{[SO_3]^2}{[O_2]^3} = 1.2 \times 10^{130} \text{ at } 25°C$$

When you compare these two expressions, you find that

$$K_2 = 1.2 \times 10^{130} = (1.1 \times 10^{65})^2 = K_1^2$$

In general, when the stoichiometric coefficients of a balanced equation are multiplied by some factor, the equilibrium constant for the new equation (K_2) is the old equilibrium constant (K_1) raised to the power of the multiplication factor. Above, the new equation was obtained by multiplying the first equation by *two*. Therefore, K_2 is the *square* of K_1 in the example above.

Since the value of K depends on stoichiometric coefficients, you must specify the balanced equation when given a K value.

REVERSING EQUATIONS A closely related situation is the equilibrium constant expression for a reaction and its reverse. Compare the combination of hydrogen and oxygen to produce water

$$2\,H_2(g) + O_2(g) \rightleftharpoons 2\,H_2O(g) \quad K_1 = \frac{[H_2O]^2}{[H_2]^2\,[O_2]} = 3.4 \times 10^{81} \text{ at } 25°C$$

or the splitting of water into its elements.

$$2\,H_2O(g) \rightleftharpoons 2\,H_2(g) + O_2(g) \quad K_2 = \frac{[H_2]^2[O_2]}{[H_2O]^2} = 2.9 \times 10^{-82} \text{ at } 25°C$$

Comparing the two expressions above, we find that $K_2 = K_1^{-1} = 1/K_1$. In general, the equilibrium constants for a reaction and its reverse are the reciprocals of one another.

ADDING REACTIONS It is often necessary to add two equations together to obtain the equation for a net process. As an example, use the successive oxidation of sulfur to the dioxide and then to the trioxide (where K values are given for 25°C).

$$S(s) + O_2(g) \rightleftharpoons SO_2(g) \quad K_1 = \frac{[SO_2]}{[O_2]} = 4.2 \times 10^{52}$$

$$SO_2(g) + \tfrac{1}{2}\,O_2(g) \rightleftharpoons SO_3(g) \quad K_2 = \frac{[SO_3]}{[SO_2]\,[O_2]^{1/2}} = 2.6 \times 10^{12}$$

$$S(s) + \tfrac{3}{2}\,O_2(g) \rightleftharpoons SO_3(g) \quad K_{net} = \frac{[SO_3]}{[O_2]^{3/2}} = 1.1 \times 10^{65}$$

The calculation above indicates that $K_{net} = K_1K_2$. In general, when you add two or more equations to produce a net equation, the equilibrium constant for the net reaction is the *product* of the equilibrium constants of the added reactions. Conversely, when you subtract one equation (K_2) from another equation (K_1) to obtain a net equation, the equilibrium constant for the net equation is the *quotient* K_1/K_2.

K_c and K_p In some instances the value of K depends on whether the equilibrium constant is written in terms of concentrations (K_c) or gas pressures (K_p). For example, for the decomposition of limestone

$$CaCO_3(s) \rightleftharpoons CaO(s) + CO_2(g)$$

K_c is simply equal to $[CO_2]$. From Chapter 6 you know that the concentration of a gas is proportional to its partial pressure (at constant temperature).

$$\text{Concentration} = \frac{\text{moles}}{\text{L}} = \frac{n}{V_{\text{gas}}} \propto P_{\text{gas}}$$

This means that K for the limestone decomposition could also be written in terms of the partial pressure of CO_2.

$$K_p = P_{CO_2}$$

The important point here is that K_c is not numerically equal to K_p. (The exact connection is derived below in Example 14.1.) When you work with equilibrium constants, make sure you know how the concentrations of reactants and products were expressed.

K_c and K_p will be identical when the number of moles of gases (or moles of aqueous species) are the same on each side of the balanced equation. See Example 14.1.

THE MAGNITUDE OF K You have probably already noticed that the values of K extend over a tremendous range, from about 10^{-10} for dissolving silver chloride, to 10^{81} for the combustion of hydrogen, and to even higher values for the oxidation of sulfur. What is the practical significance of these values? The answer is straightforward: the larger the value of K, the more the products are favored at equilibrium at the specified temperature (that is, the greater the extent of reaction). AgCl is a very poorly soluble salt, as you learned in a general way in Chapter 3, so the numerator of the equilibrium constant expression is much smaller than the denominator. On the other hand, the combustion of hydrogen proceeds almost completely from reactants to products, so the numerator of K is large and the denominator is small.

EXERCISE 14.1 WRITING EQUILIBRIUM EXPRESSIONS

Write equilibrium constant expressions for the following chemical equations.

(a) $PCl_5(g) \rightleftharpoons PCl_3(g) + Cl_2(g)$

(b) $HCl(g) + LiH(s) \rightleftharpoons H_2(g) + LiCl(s)$

(c) $Cu(OH)_2(s) \rightleftharpoons Cu^{2+}(aq) + 2\ OH^-(aq)$

(d) $[Cu(NH_3)_4]^{2+}(aq) \rightleftharpoons Cu^{2+}(aq) + 4\ NH_3(aq)$

EXERCISE 14.2 WORKING WITH EQUILIBRIUM EXPRESSIONS

(a) If K_p for the dissociation of N_2O_4 is 6.75 at 25°C,

$N_2O_4(g) \rightleftharpoons 2\ NO_2(g)$

what is the value of K_p for the dissociation reaction at 25°C written in the following manner?

$2\ NO_2(g) \rightleftharpoons N_2O_4(g)$

Or if written as follows?

$\frac{1}{2}\ N_2O_4(g) \rightleftharpoons NO_2(g)$

(b) if K_p for the formation of phosgene ($COCl_2$) is 6.6×10^{11} at 25°C,

$CO(g) + Cl_2(g) \rightleftharpoons COCl_2(g)$

what is the value of K_p for its dissociation into CO(g) and $Cl_2(g)$ at the same temperature?

$$COCl_2(g) \rightleftharpoons CO(g) + Cl_2(g)$$

EXERCISE 14.3 *K* AND EXTENT OF REACTION

If solid AgCl and AgI were each placed in 1.0 L of water in separate beakers, in which beaker would $[Ag^+]$ be greater?

$$AgCl(s) \rightleftharpoons Ag^+(aq) + Cl^-(aq) \qquad K = 1.8 \times 10^{-10}$$
$$AgI(s) \rightleftharpoons Ag^+(aq) + I^-(aq) \qquad K = 1.5 \times 10^{-16}$$

EXAMPLE 14.1

THE RELATION BETWEEN K_c AND K_p

For the formation of H_2O from its elements

$$2\ H_2(g) + O_2(g) \rightleftharpoons 2\ H_2O(g)$$

the equilibrium constant expression can be written in terms of concentrations or partial pressures.

$$K_c = \frac{[H_2O]^2}{[H_2]^2[O_2]} \qquad K_p = \frac{P_{H_2O}{}^2}{P_{H_2}{}^2 P_{O_2}}$$

What is the relation between K_c and K_p?

Solution From the ideal gas law you know that $n/V = P/RT$. Therefore,

$$K_c = \frac{[P_{H_2O}/RT]^2}{[P_{H_2}/RT]^2[P_{O_2}/RT]} = \frac{P_{H_2O}{}^2}{P_{H_2}{}^2 P_{O_2}}(RT)$$

From the expression for K_p above, you can see that K_c is actually

$$K_c = K_p(RT)$$

That is, K_c is K_p multiplied by the factor RT, where R is the ideal gas constant (0.08206 L · atm/K · mol) and T is the absolute temperature. Since K_p is 1.39×10^{80}, this means that K_c is

$$K_c = K_p(RT) = (1.39 \times 10^{80})\ (0.08206)(298) = 3.40 \times 10^{81}$$

The expression derived in Example 14.1 comes from a universal equation, one which is usually written in the form

$$K_p = K_c(RT)^{\Delta n} \quad \text{or} \quad K_c = \frac{K_p}{(RT)^{\Delta n}} \tag{14.2}$$

where Δn is the change in the number of moles of gas from reactants to products. That is,

$$\Delta n = [\text{total moles of gaseous products} - \text{total moles of gaseous reactants}]$$

In the formation of water from its elements (Example 14.1), $\Delta n = 2 - 3 = -1$. Thus, $K_p = K_c(RT)^{-1}$, which is the same thing as $K_c = K_p(RT)^1$. This is exactly the expression we derived above for this particular case.

TABLE 14.2 Summary of Changes in Equilibrium Expressions That Occur When Changing Stoichiometric Coefficients, Reversing Equations, or Adding Equations

CHANGE MADE	EFFECT ON K
Stoichiometric coefficients in a balanced equation are changed by a factor of n.	$K_{new} = K_{old}^n$
Balanced chemical equation is reversed.	$K_{new} = 1/K_{old}$
Several balanced equations are added to obtain a net, balanced equation.	$K_{net} = K_1K_2K_3 \ldots$ K_{net} is the product of the K's of the added equations.

EXERCISE 14.4 K_p/K_c CONVERSIONS

The value of K_p for the following reaction is 6.0×10^5 at 25°C. What is K_c?

$$N_2(g) + 3\ H_2(g) \rightleftharpoons 2\ NH_3(g)$$

This section is an important one. You should now know the following:

1. how to write an equilibrium constant expression from the balanced equation, recognizing that pure solids and liquids do not appear in such expressions;
2. how the value of K changes as the stoichiometric coefficients are changed by some factor or as the equation is reversed;
3. how to find K for an equation that is the sum of other equations;
4. how the value of K depends on the way in which equilibrium concentrations are expressed and how to convert K in terms of concentrations (K_c) and K expressed in partial pressure (K_p);
5. and how the value of K reflects the extent to which a reaction proceeds at the temperature for which K is specified. Many of these points are summarized in Table 14.2.

In the following section we shall further illustrate the meaning of the equilibrium constant by describing how to interpret K graphically and by introducing the "reaction quotient." Then, in the final section, you will practice using equilibrium expressions quantitatively.

14.3 THE MEANING OF EQUILIBRIUM CONSTANTS

GRAPHICAL PRESENTATION OF K

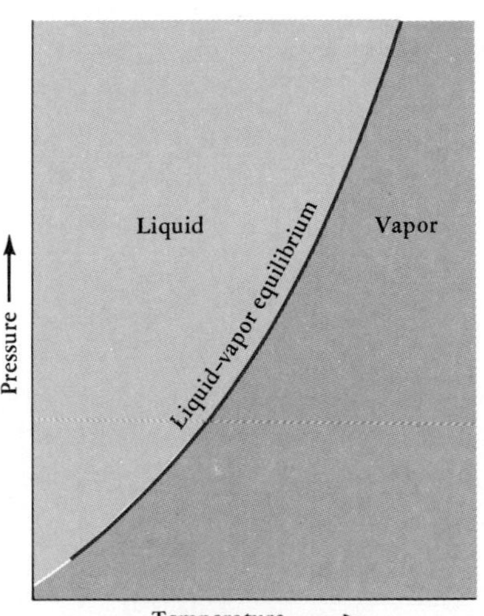

In Chapter 11 you first encountered phase diagrams, graphic representations of the equilibria involved in transformations between the phases of matter. There you saw that the liquid–vapor equilibrium for a substance is defined by a well known mathematical equation relating temperature, heat of vaporization, and equilibrium vapor pressure. The equation led to a curved line in a plot of temperature versus equilibrium vapor pressure, a line that defines all the conditions of T and P at which equilibrium can exist. At any point not on the line, equilibrium cannot exist.

There is only one equation defining liquid–vapor equilibria, but there can be many different forms of equilibrium constant expressions relating reactants and products. Nonetheless, any particular equilibrium constant expression involving only two concentrations can be represented graphically. To provide some insight into chemical equilibria and help you understand a later discussion of the consequences of a reaction *not* at equilibrium, we shall graph one type of equilibrium expression that is common to several different chemical reactions.

Let us consider reactions that have an equilibrium constant expression of the form $K = x/y$. The oxidation of zinc with copper(II) ion would have such an equilibrium constant expression

$$Zn(s) + Cu^{2+}(aq) \rightleftharpoons Zn^{2+}(aq) + Cu(s)$$

$$K = \frac{[Zn^{2+}]}{[Cu^{2+}]} = 2 \times 10^{37} \text{ at } 25°C$$

The *n*-butane → isobutane reaction is an example of an *isomerization.* The formula of the reactant and product are the same, but the atom-to-atom connections are different.

as would the transformation of *n*-butane to isobutane.

$$\underset{n\text{-butane}}{H_3C{-}CH_2{-}CH_2{-}CH_3} \rightleftharpoons \underset{\text{isobutane}}{H_3C{-}\overset{\overset{\displaystyle CH_3}{|}}{C}H{-}CH_3} \qquad K = \frac{[iso]}{[n]} = 2.5 \text{ at } 25°C$$

In each case the equilibrium constant expression is just the equation of a straight line, $y = ax + b$, where a is the slope of the line and b is the intercept when $x = 0$. Here $(1/K)$ is the slope, a, and the intercept, b, is zero.

$$K = x/y$$

$$y = \left(\frac{1}{K}\right) x$$

$$y = ax + b$$

A plot of concentrations that satisfy the expression $K = [iso]/[n] = x/y = 2.5$ is shown in Figure 14.4. *Any pair of concentrations* lying on this straight line, with a slope of (1/2.5), will satisfy the equilibrium condition.

It is useful to see how equilibrium is approached when you begin with a pure reactant. If you place 2.0 moles of *n*-butane in a 1.0-L flask, the *n*-butane will isomerize under appropriate conditions. As it does so, one mole of isobutane is formed for every mole of *n*-butane used. This means that the reacting system follows a line with a slope of -1 beginning at the initial concentration of *n*-butane (the dashed line marked *A* in Figure 14.4), and the net reaction terminates when the equilibrium line is met. This occurs when 1.43 moles of isobutane have been formed and 0.57 moles of *n*-butane remain.

The process is similar if you begin with pure isobutane. Each molecule of isobutane that disappears is replaced by one of *n*-butane; starting with 2.0 mol/L of isobutane, the system now follows line *B* in Figure 14.4, and the net reaction terminates on meeting the equilibrium line. Since you began with 2.0 mol/L of isobutane, equilibrium is achieved at the same position as if you had begun with 2.0 mol/L of *n*-butane; the position of equilibrium is the same no matter what the direction of approach.

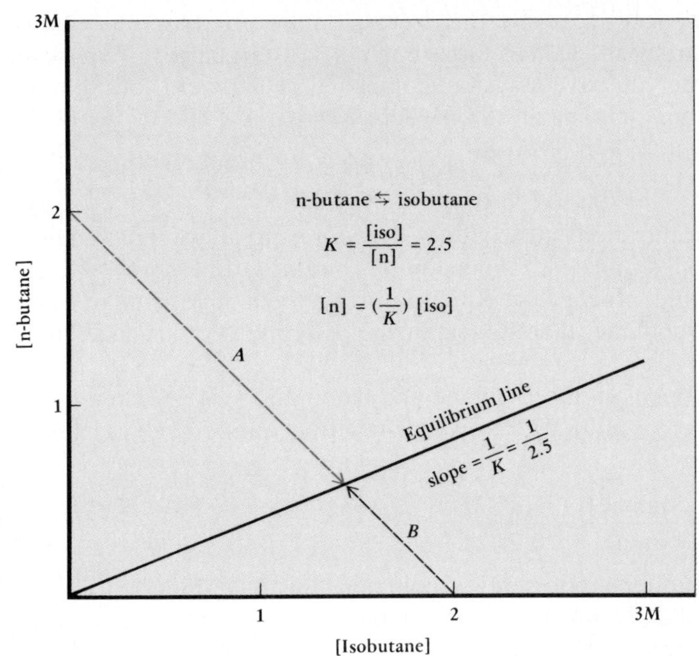

FIGURE 14.4

Graphical illustration of the *n*-butane/isobutane equilibrium. If one begins with 2.0 *M* *n*-butane, the approach to equilibrium is along the dashed line *A*, and equilibrium is achieved at $[n] = 0.57\ M$ and $[iso] = 1.43\ M$. Beginning with 2.0 *M* isobutane, one follows the dashed line *B* and arrives at the same equilibrium position.

EXERCISE 14.5 GRAPHICAL ILLUSTRATION OF *K*

Using Figure 14.4, estimate the equilibrium concentration of *n*-butane when [isobutane] = 2.0 *M*. Use this estimated concentration and [isobutane] = 2.0 *M* to estimate the value of *K*.

THE REACTION QUOTIENT

The graphical treatment of equilibrium constants is helpful for two reasons: (1) Such graphs illustrate the fact that there is an infinite number of possible sets of concentrations of reactants and products (all the points along the equilibrium line) that can exist at equilibrium at a given temperature. (2) Graphs provide a clear way of seeing, from the given concentrations, when a system is or is not at equilibrium. Unfortunately, though, graphic presentation of chemical equilibria is usually not done, because most equilibrium constant expressions contain three or more variables. Our experience with the simple equilibrium described above, however, suggests some generalizations that apply to all systems. Consider a general chemical equation of the type

$$aA + bB \rightleftharpoons cC + dD$$

For this equation we can write the algebraic expression

$$\frac{[C]^c[D]^d}{[A]^a[B]^b} = Q_c$$

where Q_c is called the **reaction quotient**. The expression for Q_c has the *appearance* of the equilibrium constant expression, but Q_c differs from

K_c in that the concentrations in the expression are *not necessarily* equilibrium concentrations. Taking the *n*-butane ⇌ isobutane system as an example, assume you have a system composed of 1.0 mol/L of *n*-butane and 1.0 mol/L of isobutane at 25°C (where $K = 2.5$)

$$Q_c = \frac{[\text{isobutane}]}{[n\text{-butane}]} = \frac{1.0\ M}{1.0\ M} = 1.0$$

This set of concentrations clearly does *not* represent a system at equilibrium, since $Q_c < K_c$. Instead, this point lies on line *A* in Figure 14.4, and the system would proceed to equilibrium by the conversion of some *n*-butane into isobutane, thereby increasing [*iso*] and decreasing [*n*] until $Q_c = K_c$.

What happens in the *n*-butane and isobutane system when $[n] = 0.25\ M$ and $[iso] = 1.75\ M$? Now the reaction quotient Q_c is greater than K_c.

$$Q_c = \frac{[\text{isobutane}]}{[n\text{-butane}]} = \frac{1.75\ M}{0.25\ M} = 7.0$$

The reaction quotient represents a point on line *B* in Figure 14.4; the system is again clearly not at equilibrium, but it proceeds to an equilibrium state by converting isobutane to *n*-butane. Only when $Q_c = K_c$ is the system finally at equilibrium.

Our analysis of the *n*-butane ⇄ isobutane system leads to some general conclusions that can be summarized as follows:

If $Q_c < K_c$, the system is not at equilibrium: The ratio of products to reactants is too small. Reactants must be further converted to products (thus increasing Q_c) to achieve equilibrium (when $Q_c = K_c$).

If $Q_c = K_c$, the system is at equilibrium.

If $Q_c > K_c$, the system is not at equilibrium: The ratio of products to reactants is too large. To reach equilibrium, products must be converted to reactants (thus decreasing the value of Q_c until $Q_c = K_c$).

The relationship of the reaction quotient, Q_c, and the equilibrium constant, K_c, allows us to decide if a system is at equilibrium. If it is not at equilibrium, we can also decide in which direction the reaction will proceed to reach equilibrium.

EXAMPLE 14.2

THE REACTION QUOTIENT

Consider the system $PCl_5(g) \rightleftarrows PCl_3(g) + Cl_2(g)$. At 250°C, the equilibrium constant K_c is 4.0×10^{-2}. If the concentrations of Cl_2 and PCl_3 are both 0.30 *M*, while the concentration of $PCl_5(g)$ is 3.0 *M*, is the system at equilibrium? Is Q_c larger than, equal to, or smaller than K_c? If the system is not at equilibrium, in which direction does the reaction proceed?

Solution The equilibrium constant expression for the reaction is

$$K_c = \frac{[PCl_3]\,[Cl_2]}{[PCl_5]} = 4.0 \times 10^{-2}$$

If the concentrations of reactants and products are substituted into the reaction quotient expression we have

$$Q_c = \frac{(0.30)(0.30)}{3.0} = 3.0 \times 10^{-2}$$

The value of Q_c is less than the value of K_c ($Q_c < K_c$), so the reaction is not at equilibrium. It proceeds to equilibrium by converting more PCl_5 to products, thus increasing the concentration terms in the numerator and decreasing that in the denominator until $Q_c = K_c$.

EXERCISE 14.6 THE REACTION QUOTIENT

At 2000 K the equilibrium constant, K_p, for the formation of NO(g)

$$N_2(g) + O_2(g) \rightleftharpoons 2\ NO(g)$$

is 4.0×10^{-4}. If, at 2000 K, the partial pressure of N_2 is 0.50 atm, that of O_2 is 0.25 atm, and that of NO is 4.2×10^{-3} atm, decide whether the system is at equilibrium. If not, predict which way the reaction will proceed.

14.4 SOME CALCULATIONS WITH THE EQUILIBRIUM CONSTANT

It is often necessary or helpful to know the concentrations of reactants and products for a reaction at equilibrium. As you have seen, these are connected through the equilibrium constant, a numerical value that can be determined by experimental measurement of equilibrium concentrations. This section is devoted to some examples of the calculation of K from experimental information and the use of these K values. For the moment we shall concentrate largely on simple gas phase reactions, leaving most of the discussion of reactions in aqueous solution to Chapters 15 through 17.

EXAMPLE 14.3

CALCULATION OF *K* FROM EXPERIMENTAL INFORMATION

2.0 moles of Br_2 are placed in a 2.0-L flask at 1756 K, a temperature at which the halogen molecule dissociates to Br atoms.

$$Br_2(g) \rightleftharpoons 2\ Br(g)$$

If $Br_2(g)$ is 1.0% dissociated at this temperature, calculate K_c.

Solution The equilibrium constant expression is

$$K_c = \frac{[Br]^2}{[Br_2]}$$

To determine the value of K_c, we need to know the *equilibrium values* of $[Br_2]$ and [Br].

(a) At equilibrium

$$\begin{aligned}\text{Amount of } Br_2 &= \text{initial amount} - \text{amount of } Br_2 \text{ dissociated}\\ &= \text{initial amount} - \text{(fraction dissociated)(initial amount)}\\ &= 2.0 \text{ moles} - (0.010)(2.0 \text{ moles})\\ &= 1.98 \text{ moles}\end{aligned}$$

Equilibrium concentration $[Br_2] = 1.98 \text{ mol}/2.0 \text{ L} = 0.99\ M$.

(b) Since the equilibrium amount of Br_2 is 1.98 moles, this means that 0.020 mole of Br_2 must have been converted to Br. Therefore, the amount of Br at equilibrium can be calculated.

$$\text{Amount of Br at equilibrium} = 0.020 \text{ mol } Br_2 \text{ dissociated}\left(\frac{2 \text{ mol Br}}{1 \text{ mol } Br_2}\right)$$

$$= 0.040 \text{ mol}$$

$$[Br] \text{ at equilibrium} = 0.040 \text{ mol}/2.0 \text{ L} = 0.020\ M$$

(c) Substitute the equilibrium concentrations into the equilibrium expression.

$$K_c = \frac{[Br]^2}{[Br_2]} = \frac{(0.020)^2}{(0.99)} = 4.0 \times 10^{-4}$$

EXAMPLE 14.4

CALCULATION OF *K* FROM EXPERIMENTAL INFORMATION

Calculate the equilibrium constant at 25°C for the reaction

$$2\ NOCl(g) \rightleftharpoons 2\ NO(g) + Cl_2(g)$$

In one experiment, 2.00 moles of NOCl were placed in a 1.00-L flask, and the concentration of NO after equilibrium was achieved was 0.66 moles/liter.

Solution After writing the expression for *K* from the balanced chemical equation,

$$K_c = \frac{[NO]^2[Cl_2]}{[NOCl]^2}$$

Always keep in mind that the algebraic expression for *K* demands equilibrium concentrations.

our next goal is to define the concentration of each reagent at equilibrium. To organize this problem, we have set out the data in a small table, and you will usually find it helpful to do the same.

	[NOCl]	[NO]	$[Cl_2]$
Concentration before reaction (M)	2.00	0	0
Concentration change when proceeding to equilibrium (M)	−0.66	+0.66	+(1/2)(0.66)
Concentration at equilibrium (*M*)	2.00 − 0.66 = 1.34	0 + 0.66	0 + 0.33

Let us first see how the equilibrium concentrations in the table were derived. The initial concentration of NOCl was 2.00 M, and the equilibrium value of [NO] was found experimentally to be 0.66 *M*. This is the starting point in our calculation of the equilibrium concentrations of the gases.

The balanced equation tells you that half as many moles of $Cl_2(g)$ are produced as moles of NO(g). Therefore, the equilibrium concentration of $Cl_2(g)$ must be half that of NO(g) or 0.33 M.

Similarly, the balanced equation shows that for every mole of NO(g) produced, one mole of NOCl(g) is used. Therefore, at equilibrium

Concentration of NOCl = concentration before reaction − concentration that disappeared on proceeding to equilibrium
= 2.00 M − 0.66 M = 1.34 M

With all the equilibrium amounts and concentrations known, the value of K_c at 25°C can be calculated.

$$K = \frac{[NO]^2[Cl_2]}{[NOCl]^2} = \frac{(0.66)^2(0.33)}{(1.34)^2} = 0.080$$

Once the equilibrium constant for a reaction is known at a given temperature, that value of K can be used to find equilibrium concentrations of reactants and products under other conditions (if the temperature does not change).

EXAMPLE 14.5

CALCULATION OF A CONCENTRATION FROM AN EQUILIBRIUM CONSTANT

The equilibrium constant for the dissociation of AgCl(s) to its ions in aqueous solution

$$AgCl(s) \rightleftharpoons Ag^+(aq) + Cl^-(aq)$$

is 1.8×10^{-10}. If the concentration of $Ag^+(aq)$ is 1.0×10^{-5} M, what is the concentration of Cl^- that can be in equilibrium with silver ion at that concentration?

Solution The equilibrium constant expression for this reaction is

$$K_c = [Ag^+][Cl^-]$$

Therefore, the equilibrium concentration of Cl^- must be

$$[Cl^-] = \frac{K_c}{[Ag^+]} = \frac{1.8 \times 10^{-10}}{1.0 \times 10^{-5}} = 1.8 \times 10^{-5}\ M$$

EXAMPLE 14.6

CALCULATIONS USING THE EQUILIBRIUM CONSTANT

The equilibrium constant K_c (= K_p) for

$$H_2(g) + I_2(g) \rightleftharpoons 2\ HI(g)$$

was determined using the experimental data of Table 14.1. If you place 1.00 mole each of $H_2(g)$ and $I_2(g)$ in a 0.500-L flask at 699 K, what are the equilibrium concentrations of $H_2(g)$, $I_2(g)$, and HI(g)?

Solution The first step is to write the balanced equation. Since this is done above, the next step is to write the equilibrium constant expression.

$$K = \frac{[HI]^2}{[H_2]\,[I_2]} = 55.17$$

Next, we define equilibrium concentrations and organize the information using a small table.

	$[H_2]$	$[I_2]$	$[HI]$
Original Concentrations (M)	2.00	2.00	0
Concentration change when proceeding to equilibrium (M)	$-x$	$-x$	$+2x$
Equilibrium Concentrations (M)	$2.00-x$	$2.00-x$	$2x$

The equilibrium concentrations of H_2 or I_2 will be the amount (in mol/L) that was present before the reaction *minus* the amount (in mol/L) that was transformed into HI when equilibrium was achieved. You do not know the amount (in mol/L) of H_2 or I_2 that disappeared, only that the same number of moles/liter of each must have been involved. We call this unknown amount x (mol/L). The balanced equation tells you that the concentration of HI at equilibrium is twice the concentration of H_2 or I_2 that disappeared, which can now be expressed as $2x$.

Now that the equilibrium concentrations of all three compounds are defined in terms of a single unknown, we can write the equilibrium constant expression and set it equal to the experimentally determined value.

$$K = 55.17 = \frac{[HI]^2}{[H_2]\,[I_2]} = \frac{(2x)^2}{(2.00-x)^2}$$

The unknown x can be found by taking the square root of both sides and solving.

$$\sqrt{K} = 7.428 = \frac{2x}{2.00-x}$$

$$7.428\,(2.00 - x) = 2x$$

$$14.9 - 7.428x = 2x$$

$$14.9 = 9.428x$$

$$x = 1.58$$

The concentration of H_2 or I_2 that disappears in the process of establishing equilibrium is therefore 1.58 M.

Now we return to the expressions for $[H_2]$, $[I_2]$, and $[HI]$ to obtain

$$[H_2]_{equi} = [I_2]_{equi} = 2.00 - x = 0.42\ M$$

$$[HI]_{equi} = 2x = 3.16\ M$$

It is important that you verify that these values are correct by substituting them back into the equilibrium expression to calculate K and see if the calculated K is the same as that originally given. In this case, $(3.16)^2/(0.42)^2 = 56$. The discrepancy with $K = 55.17$ is due to the fact that we know $[H_2]$ and $[I_2]$ to only two significant figures.

EXAMPLE 14.7

CALCULATIONS USING THE EQUILIBRIUM CONSTANT

When you find a bottle of concentrated nitric acid in the laboratory, it often has a brown gas, NO_2, hovering over the surface of the liquid. This simple molecule can bind to another NO_2 molecule and form a dimer, $O_2N—NO_2$, that is, a molecule having two NO_2 units. NO_2 and its dimer, N_2O_4, are in equilibrium in the gas phase.

$$N_2O_4(g) \rightleftharpoons 2\ NO_2(g) \quad K_p = 0.11 \text{ (at 25°C)}$$

If the total pressure of an equilibrium mixture of N_2O_4 and NO_2 is 1.5 atm, calculate the partial pressure of each gas.

Solution The equilibrium constant expression is

$$K = \frac{P_{NO_2}{}^2}{P_{N_2O_4}} = 0.11$$

This equation contains two unknowns, the pressures of the two different gases. To solve it, you will have to reduce the equation to only a single unknown quantity. This can be done as follows:

(a) The equilibrium total pressure is the sum of the partial pressures.

$$P_{total} = P_{NO_2} + P_{N_2O_4} = 1.5 \text{ atm}$$

and so the partial pressure of N_2O_4 at equilibrium is

$$P_{N_2O_4} = 1.5 \text{ atm} - P_{NO_2}$$

(b) From the step above, you can now write the equilibrium expression in terms of a single unknown, the partial pressure of NO_2.

$$K = \frac{P_{NO_2}{}^2}{1.5 - P_{NO_2}} = 0.11$$

Rearranging this expression you have

$$0.17 - 0.11\ P_{NO_2} = P_{NO_2}{}^2$$

or

$$P_{NO_2}{}^2 + 0.11\ P_{NO_2} - 0.17 = 0$$
$$ax^2 + bx + c = 0$$

The rearranged equilibrium expression is a quadratic equation of the general form $ax^2 + bx + c = 0$. Such equations can be solved by the formula given in Appendix A.

$$x = \frac{-b \pm \sqrt{b^2 - 4ac}}{2a}$$

In this case $a = 1$, $b = 0.11$, and $c = -0.17$. Substituting these values we have

$$x = P_{NO_2} = \frac{-(0.11) \pm \sqrt{(0.11)^2 - 4(1)(-0.17)}}{2(1)}$$

$$= P_{NO_2} = \frac{-(0.11) \pm 0.83}{2}$$

$$= 0.36 \text{ or } -0.47$$

If we use $+0.83$, then $x = 0.36$, or if we use -0.83, then $x = -0.47$. Since x stands for P_{NO_2}, a negative value is physically meaningless. Therefore, we now know that $x = 0.36$ must be the equilibrium partial pressure of NO_2. Since $P_{N_2O_4}$ is $1.5 - P_{NO_2}$, the equilibrium pressure of N_2O_4 is 1.1 atm.

As a final step in the problem, you should verify this answer by substituting back into the equilibrium expression. The quotient $(0.36)^2/(1.1)$ is 0.12, a value slightly different than the value of K due to rounding numbers in the course of the problem.

In later chapters, it is useful and important to be able to carry out simple equilibrium calculations, especially those involving aqueous solutions. The best way to learn how to do this is to practice!

EXERCISE 14.7 CALCULATING AN EQUILIBRIUM CONSTANT

Phosphorus pentachloride decomposes in the gas phase to chlorine and phosphorus trichloride vapor.

$$PCl_5(g) \rightleftharpoons PCl_3(g) + Cl_2(g)$$

Calculate the equilibrium constant for this reaction at 300°C knowing that the original pressure of $PCl_5(g)$ before it dissociated was 1.50 atm, and the equilibrium pressures of $PCl_3(g)$ and $Cl_2(g)$ are both 1.34 atm.

EXERCISE 14.8 CALCULATION OF AN EQUILIBRIUM CONCENTRATION FROM *K*

Again consider the equilibrium in Exercise 14.7 above. If you begin with enough PCl_5 so that its pressure is 3.00 atm before dissociation and if you now know that $K_p = 11.2$ at this temperature, what are the equilibrium pressures of the reactants and products at 300°C?

EXERCISE 14.9 CALCULATION FROM AN EQUILIBRIUM CONSTANT

Manganese sulfide is a poorly soluble salt, dissolving only to a slight extent in water.

$$MnS(s) \rightleftharpoons Mn^{2+}(aq) + S^{2-}(aq) \qquad K_c = 5.1 \times 10^{-15}$$

What is the maximum concentration of Mn^{2+} that can exist in water when $[S^{2-}] = 2.0 \times 10^{-9}$ *M*?

14.5 EXTERNAL FACTORS AFFECTING EQUILIBRIA

Chemical and physical equilibria are influenced by three factors:

INFLUENCE	EFFECT
1. Temperature changes	Value of *K* changes.
2. Concentration changes (addition or removal of reactant or, for a gas phase reaction, change in volume)	*K* is constant; position of equilibrium is changed (equilibrium quantities of both reactants *and* products change).
3. Addition of catalyst	*K* is constant; rate of approach to equilibrium is affected, but eventual equilibrium mixture is the same.

The first two influences will take some discussion, but the third has already been described in Chapter 13. The main point to recall, however, is that the *addition of a catalyst to a reaction does not influence the position of the equilibrium, only the rate of approach to equilibrium.*

EFFECT OF TEMPERATURE CHANGES ON EQUILIBRIA

In Chapters 11 and 12 you learned about temperature effects on phase equilibria and on the solubility of salts in water and other solvents. We noted, for example, that an increase in temperature leads to an increase in the vapor pressure of a liquid and an increase in the solubility of most salts. These observations are special cases of **Le Chatelier's principle.** This principle states that a change in any of the factors that determine the equilibrium conditions of a system will cause the system to change in such a manner as to reduce or counteract the effect of the change. This was applied specifically to the variation in solubility with temperature in Section 12.2. Now we wish to apply it to chemical reactions.

What is the effect of a temperature change on the equilibrium position in a chemical reaction? If you know the enthalpy change of the reaction (ΔH_{rxn}), you can make a qualitative prediction from Le Chatelier's principle. As an example, consider the reaction of N_2 with O_2 to give nitrogen oxide, NO.

$$N_2(g) + O_2(g) + 180.5\text{ kJ} \rightleftharpoons 2\text{ NO}(g)$$

$$K = \frac{[NO]^2}{[N_2][O_2]}$$

$$K\text{ (at 2000 K)} = 4.1 \times 10^{-4} \qquad K\text{ (at 2500 K)} = 36 \times 10^{-4}$$

The NO formed in automobile engines as a by-product of fuel combustion is exhausted with the other combustion products. NO can react further with O_2 and other oxidants in the atmosphere to give NO_2, the source of much of the air pollution in urban areas.

We are surrounded by nitrogen and oxygen, but you know they do not react appreciably at room temperature. The position of equilibrium lies almost completely to the left. However, if a mixture of N_2 and O_2 is heated above 700°C, as in an automobile engine, the equilibrium shifts toward NO, and a greater concentration of NO can exist in equilibrium with the reactants. The experimental equilibrium constants given above show that [NO] increases and $[N_2]$ and $[O_2]$ decrease as temperature increases.

The *activation energy*, E^*, for the reaction of N_2 with O_2 is undoubtedly very great, so the reaction rate is very small at ordinary temperatures. See Chapter 13.

To see why the value of K can change with temperature, consider the N_2/O_2 reaction further. The enthalpy of the reaction is +180.5 kJ, so we can consider heat as a "reactant." Since temperature determines the position of equilibrium, Le Chatelier's principle tells us that input of energy (as heat) will cause the equilibrium to shift in a direction that minimizes or counteracts this input. The way to counteract the energy input here is to consume some of the added heat by producing more NO (by consuming N_2 and O_2) and by increasing the temperature. Thus, an increase in temperature is associated with production of NO *and* an increase in K.

A more quantitative analysis of temperature effects on equilibria can be made using thermodynamics (Chapter 18).

Two molecules of the brown gas NO_2 can combine readily to form N_2O_4, and equilibrium is readily achieved in a closed sytem (Figure 14.5).

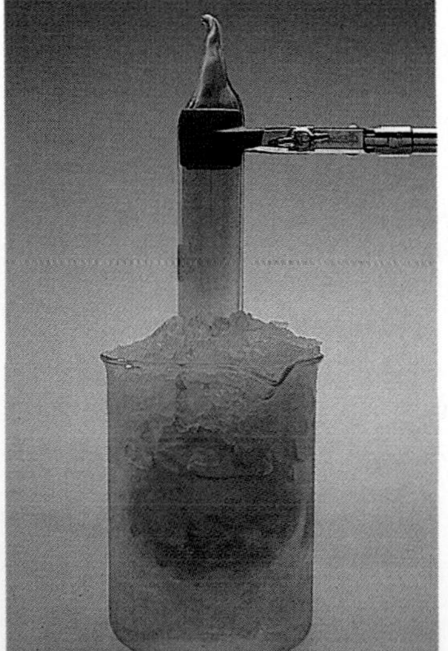

FIGURE 14.5

Effect of temperature on the NO_2/N_2O_4 equilibrium. The tubes in the photograph both contain a mixture of NO_2 and N_2O_4. As predicted by Le Chatelier's principle, the equilibrium favors colorless N_2O_4 at lower temperatures. This is clearly seen in the tube on the left, where the gas in the ice bath at 0°C is only slightly brown because there is only a small concentration of NO_2. At 50°C (the tube on the right), equilibrium is shifted toward NO_2, a brown gas. (Charles D. Winters)

$$2\ NO_2(g) \rightleftharpoons N_2O_4(g) + 57.2\ kJ$$
brown colorless

$$K = \frac{[N_2O_4]}{[NO_2]^2}$$

$$K\ (\text{at } 298\ K) = 8.8 \qquad K\ (\text{at } 273\ K) = 76$$

Here the reaction is exothermic, so heat can be considered a "reaction product." Thus, if heat is removed, as in Figure 14.5, some temperature decrease occurs. According to Le Chatelier's principle, this consumption of heat can be counteracted if the reaction produces more heat by the combination of NO_2 to give more N_2O_4. Thus, the equilibrium concentration of NO_2 declines, that of N_2O_4 increases, and the values of K above become larger as the temperature declines.

EXERCISE 14.10 TEMPERATURE EFFECTS ON EQUILIBRIA

Predict the effect of temperature changes on each of the following equilibria.

(a) Does the concentration of the ammonium ion increase or decrease with an increase in temperature?

$$NH_4NO_3(s) + 26.2\ kJ \rightleftharpoons NH_4^+(aq) + NO_3^-(aq)$$

(b) When the temperature declines, what happens to the concentration of $N_2(g)$?

$$2\ N(g) \rightleftharpoons N_2(g) + 946\ kJ$$

EFFECT OF ADDITION OR REMOVAL OF A REAGENT

If the temperature is constant, the value of K is also constant. Nonetheless, if the concentration of some reactant or product is changed from its equilibrium value, the reaction must shift to a new equilibrium position. To see how this can happen, we shall look at two simple examples. First, let us return to the isomerization of *n*-butane to isobutane for which we previously graphed the equilibrium condition (Figures 14.4 and 14.6).

$$n\text{-butane} \rightleftharpoons \text{isobutane}$$

$$K_c = 2.5 = \frac{[iso]}{[n]} \qquad (\text{at } 25°C)$$

Suppose that the equilibrium concentration of isobutane in a 1.0-L flask is 1.25 *M* and that of *n*-butane is 0.50 *M*. Now add 1.5 moles of *n*-butane to the flask. After equilibrium is reestablished, what are the new concentrations?

First, let us ask what Le Chatelier's principle would predict qualitatively. A good analogy to Le Chatelier's principle is a toothpaste tube: squeeze a toothpaste tube at one end and toothpaste pops out the other end. In this case, we added more *n*-butane than equilibrium allows, so some must be converted to isobutane.

The ratio $[iso]/[n]$, which we called the *reaction quotient* (Q) in Section 14.4, will be smaller than 2.5 immediately after adding extra *n*-butane.

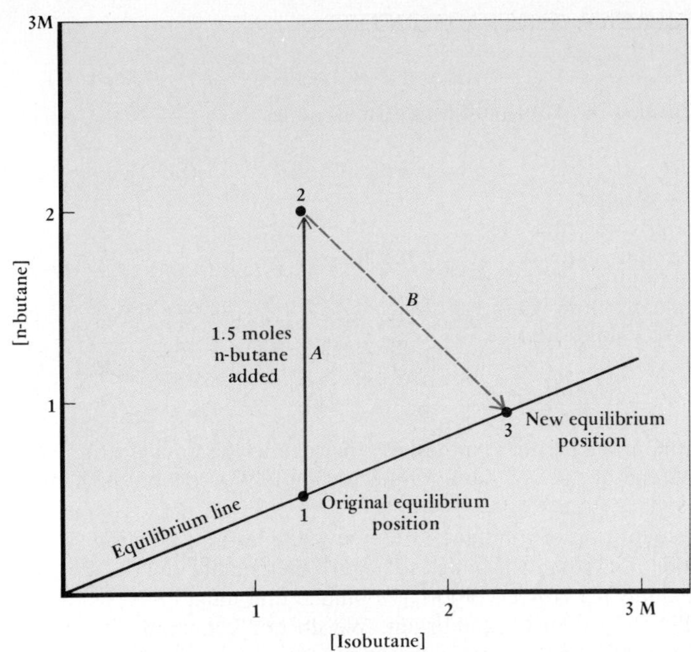

FIGURE 14.6

The effect of adding a reagent to a system at equilibrium. The system *n*-butane $\rightleftharpoons$ isobutane was at equilibrium at $[n] = 0.500\ M$ and $[iso] = 1.25\ M$ (point 1). If 1.50 mol/L of *n*-butane is added, the system moves to point 2. Equilibrium is eventually restored by moving along line *B* to a new equilibrium position (point 3) where $[n] = 0.93\ M$ and $[iso] = 2.32\ M$.

After adding excess n-butane to an equilibrium mixture

$$Q = \frac{[iso]}{[n]} = \frac{(1.25 \text{ moles present})}{(0.50 \text{ present} + 1.5 \text{ added})} < 2.5$$

As *n*-butane is converted to isobutane, the amount of isobutane increases and the amount of *n*-butane declines, until the ratio $[iso]/[n] = 2.5 = K$ is reestablished.

You can see how this happens by using the graph in Figure 14.6. Starting with $[iso] = 1.25$ and $[n] = 0.5$ (point 1), you add 1.5 moles/liter of *n*-butane and arrive (following line *A*) at point 2. This point is not on the equilibrium line and so does not represent an equilibrium condition. To arrive at equilibrium, the system must convert some *n*-butane to isobutane. Since the conversion is done by changing 1 molecule of *n*-butane into 1 molecule of isobutane, the path followed by the system is line *B*. This brings you to the new equilibrium point (3). From the graph you can see that point 3 is approximately at $[iso] = 2.32\ M$ and $[n] = 0.93\ M$, that is, *K* is once again 2.5. Notice that the effect of adding some *n*-butane is to make *both* $[n]$ *and* $[iso]$ higher at the new equilibrium point than at the old point.

Some equilibrium expressions are more complex than the simple one for the *n*-butane reaction, so you do not want to rely on graphical solutions. It is usually easier to take an algebraic approach.

EXAMPLE 14.8

AFFECTING AN EQUILIBRIUM BY CHANGING CONCENTRATIONS

In this example, we shall work out an algebraic solution to the *n*-butane case outlined above. Assume equilibrium has been established in a 1.00-L flask with $[n] = 0.500$ and $[iso] = 1.25$. Then 1.50 moles of *n*-butane are added. What are the new equilibrium concentrations of *n*- and isobutane?

Solution First organize the information in a small table.

COMPOUND	n-BUTANE	ISOBUTANE
Concentration at old equilibrium position (M)	0.500	1.25
Concentration immediately after adding n-butane	0.500 + 1.50	1.25
Change in concentration when proceeding to equilibrium	$-x$	$+x$
Concentration at new equilibrium position (M)	0.500 + 1.50 $-x$	1.25 + x

Included in the table are algebraic expressions that define the concentrations of n-butane and isobutane at the new equilibrium position. We arrived at them as follows:

(a) The concentration of n-butane at the new equilibrium position will be the old concentration *plus* what was added to disturb the equilibrium (1.50 moles/1.00 L = 1.50 M) *minus* the concentration of n-butane that must be converted to isobutane in order to reestablish equilibrium. We do not yet know how many moles/liter of n-butane disappear when going to the new equilibrium position, so we have designated it by x.

(b) The amount of isobutane at the new equilibrium position is the amount that was already present (1.25 M) *plus* the amount (x mol/L) that comes from the conversion of added n-butane.

Having defined algebraically $[n]$ and $[iso]$ at the new equilibrium position and remembering that K is a constant equal to 2.50 throughout, we can write

$$K = \frac{[iso]_{\text{new}}}{[n]_{\text{new}}} = 2.50 = \frac{1.25 + x}{0.500 + 1.50 - x}$$

$$2.50(0.500 + 1.50 - x) = 1.25 + x$$

$$3.75 = 3.5x$$

$$x = 1.07 \text{ moles}$$

You now know that the new equilibrium position is established at

$$[n] = 0.500 + 1.50 - x = 0.93 \text{ mol/L} \quad [iso] = 1.25 + x = 2.32 \text{ mol/L}$$

as suggested by the graphical analysis in Figure 14.6. Be sure to verify that $[n]/[iso] = 2.32/0.93 = 2.5$.

EXERCISE 14.11 CONCENTRATION CHANGES AND EQUILIBRIUM

Equilibrium exists between n-butane and isobutane when $[n] = 0.20\ M$ and $[iso] = 0.50\ M$. What are the equilibrium concentrations of n-butane and isobutane if 2.00 mol/L of isobutane are added to the original mixture? First estimate the new equilibrium concentrations from Figure 14.6 and then calculate the exact values.

What happens to equilibrium concentrations if you simply change the size of the container for a reaction involving gases? (This occurs, for example, when air and fuel vapor are compressed in an automobile engine.) At first you might conclude that nothing will happen. However, remember that concentration is moles/volume, so if the volume of a gas changes the concentration must also change, and the equilibrium position can be affected. As an example, again consider the equilibrium

$$2\ NO_2(g) \rightleftharpoons N_2O_4(g)$$

brown gas colorless gas

$$K_p = \frac{P_{N_2O_4}}{P_{NO_2}^{\,2}}$$

What happens to this equilibrium if the volume of the container holding the gases is reduced? According to Boyle's law, the immediate result is that the partial pressures of both gases will increase. However, the numerator in the K_p expression will only double if the volume is halved, but the denominator will increase by four (since P_{NO_2} is squared). It is clear from the K_p expression that the system can no longer be at equilibrium. To accommodate the stress of the volume increase, Le Chatelier's principle states that the equilibrium will shift in such a way that the volume decrease will be partly counteracted. The best way to do this is to use up some NO_2, since only one molecule of N_2O_4 replaces every two of NO_2 used (and so the system requires less space). Thus, the equilibrium shifts to the side of the reaction with the fewer number of gas molecules. This reduces the partial pressure of NO_2 and increases $P_{N_2O_4}$, tending to make the quotient $P_{N_2O_4}/P_{NO_2}^{\,2}$ equal once again to K_p. This seems to make sense, but let's analyze it algebraically.

Boyle's law: $P \propto 1/V$. See Chapter 6.

Since $P = (n/V)RT$, we substitute this for the partial pressures of each gas in the K_p expression.

$$K_p = \frac{P_{N_2O_4}}{P_{NO_2}^{\,2}} = \frac{(\text{moles } N_2O_4/V)RT}{[(\text{moles } NO_2/V)RT]^2}$$

Cancelling like terms in the numerator and denominator, you find that

$$K_p = \left(\frac{\text{moles } N_2O_4}{(\text{moles } NO_2)^2}\right)\left(\frac{V}{RT}\right)$$

Now we see that the equilibrium constant for the reaction is simply the ratio of moles multiplied by V/RT. This means that, if V is changed (and T is constant), the ratio of moles must also change to keep K a constant. If V decreases, then the amount of N_2O_4 must increase, which can be accomplished only by reducing the number of moles of NO_2.

The conclusions for the system NO_2/N_2O_4 can be generalized in the following way: For any reaction involving gases, the stress of a volume decrease will be counterbalanced by a shift in the equilibrium to the side of the reaction with the fewer number of molecules.

If the number of moles of gas is the same on both sides of the balanced equation, as in $N_2(g) + O_2(g) \rightleftharpoons 2\ NO(g)$, then changing the volume has no effect on the equilibrium.

EXERCISE 14.12 EFFECT OF CONCENTRATION CHANGES ON EQUILIBRIA

The formation of ammonia from its elements is an important industrial process.

$$3\ H_2(g) + N_2(g) \rightleftharpoons 2\ NH_3(g)$$

(a) Does the equilibrium shift to the left or right when extra $H_2(g)$ is added? When extra ammonia is added?

(b) What is the effect on the position of the equilibrium when the volume allowed to the system is increased? Does the equilibrium point shift left or right or is it unchanged?

SUMMARY

This chapter is an important one, because it has introduced you formally to the notion that chemical systems eventually come to equilibrium at a point decided by the temperature, pressure, and concentration of reactants and products. You have found that all chemical and phase equilibria share three characteristics: they are dynamic, they are reversible, and the equilibrium position can be approached from either direction (Section 14.1). The equilibrium position is described by a simple mathematical statement, the **equilibrium constant expression** (Section 14.2). For the general reaction of (aA + bB $\rightleftarrows$ cC + dD), one can write the following expression

$$Q = \frac{[C]^c\,[D]^d}{[A]^a\,[B]^b}$$

where $[A]^a$, and so forth, are concentrations of reactants and products, each raised to the power of its stoichiometric coefficient. Q has been called the **reaction quotient**. When the concentrations in the reaction quotient are those at equilibrium, Q is equal to K, the equilibrium constant. If $Q < K$ the system is not at equilibrium, and reactants are converted to products on proceeding to equilibrium. When $Q > K$, the system is again not at equilibrium, and products are converted to reactants on moving to a state of equilibrium.

Equilibrium constants are generally determined by experimental measurement of equilibrium concentrations, and their magnitudes can range from extremely small values (for example, 10^{-18} for FeS dissolving in water) to very large values (10^{81} for the combustion of hydrogen). Equilibrium constants are appropriate only for the temperature at which they are measured and for the chemical equation as written. For example, if the chemical equation is reversed or if the stoichiometric coefficients are changed in the balanced equation, the expression for K and value of K change accordingly (Section 14.2 and Table 14.2).

In the same way that phase equilibria are shown graphically, the equilibrium point can be plotted as a function of reactant and product concentration (Section 14.3). However, algebraic solutions rather than graphic ones are generally used, and one of your major goals in this chapter should be to learn to carry out calculations involving the use of the equilibrium constant expression (Section 14.4).

Once a system has reached equilibrium, it can be disturbed by external forces such as a change in the temperature or by addition or removal of a reagent. (The latter can be done by physically adding or removing a reactant or product, or it can be accomplished by changing the volume, for example, of a gas phase system.) The equilibrium system always responds according to **Le Chatelier's principle:** a change in any of the factors that determine the equilibrium point of a system will cause the system to change in such a manner as to offset the effect of the change.

STUDY QUESTIONS

REVIEW QUESTIONS

1. Name three important features of the equilibrium condition.
2. Tell whether each statement below is true or false. If false, change the wording to make it true.
   (a) The magnitude of the equilibrium constant is always independent of temperature.
   (b) When two chemical equations are added to give a net equation, the equilibrium constant for the net equation is the product of the equilibrium constants of the summed equations.
   (c) The equilibrium constant for a reaction has the same value as *K* for the reverse reaction.
   (d) For the equilibrium process $CaCO_3(s) \rightleftharpoons CaO(s) + CO_2(g)$, only the concentration of CO_2 appears in the equilibrium constant expression.
   (e) For the equilibrium process $CaCO_3(s) \rightleftharpoons CaO(s) + CO_2(g)$, the value of *K* is the same, whether the amount of CO_2 is expressed as moles/liter or as gas pressure.
3. Neither $PbCl_2$ nor PbF_2 is appreciably soluble in water. If solid $PbCl_2$ and solid PbF_2 are placed in water in separate beakers, in which beaker is the concentration of Pb^{2+} greater? Equilibrium constants for these solids dissolving in water are

   $$PbCl_2(s) \rightleftharpoons Pb^{2+}(aq) + 2\,Cl^-(aq) \quad K = 1.7 \times 10^{-5}$$
   $$PbF_2(s) \rightleftharpoons Pb^{2+}(aq) + 2\,F^-(aq) \quad K = 3.7 \times 10^{-8}$$

4. Some equilibrium constant expressions can be illustrated graphically (see Figure 14.4). Sketch a graph for an equilibrium expression of the form $K = xy$.
5. How does the *reaction quotient* different from an equilibrium constant?
6. If the reaction quotient is smaller than the equilibrium constant for a reaction such as $A \rightleftharpoons B$, does this mean that reactant A continues to disappear to form B, or does B form A, on moving to equilibrium?
7. Using Le Chatelier's principle, explain how increasing the temperature would affect the equilibrium $CaCO_3(s) + \text{heat} \rightleftharpoons CaO(s) + CO_2(g)$. If more $CaCO_3$ is added to a flask in which this equilibrium exists, how is the equilibrium affected? What if some additional CO_2 is placed in the flask?

WRITING EQUILIBRIUM CONSTANT EXPRESSIONS

8. Write equilibrium constant expressions, in terms of reactant and product concentrations, for the following reactions:
   (a) $2\,H_2O_2(g) \rightleftharpoons 2\,H_2O(g) + O_2(g)$
   (b) $PCl_3(g) + Cl_2(g) \rightleftharpoons PCl_5(g)$
   (c) the formation of carborundum from sand

   $$SiO_2(s) + 3\,C(s) \rightleftharpoons SiC(s) + 2\,CO(g)$$

   (d) $H_2(g) + S(s) \rightleftharpoons H_2S(g)$
   (e) formation of ozone from oxygen

   $$3\,O_2(g) \rightleftharpoons 2\,O_3(g)$$

   (f) $SiH_4(g) + 2\,O_2(g) \rightleftharpoons SiO_2(s) + 2\,H_2O(g)$
   (g) $MgO(s) + SO_2(g) + \frac{1}{2}\,O_2(g) \rightleftharpoons MgSO_4(s)$
   (h) the roasting of lead sulfide ore (commonly called galena)

   $$2\,PbS(s) + 3\,O_2(g) \rightleftharpoons 2\,PbO(s) + 2\,SO_2(g)$$

9. Write equilibrium constant expressions, in terms of reagent concentrations or pressures as appropriate, for each of the following reactions:
   (a) decomposition of thallium trichloride

   $$TlCl_3(s) \rightleftharpoons TlCl(s) + Cl_2(g)$$

   (b) dissociation of a copper-chloride complex

   $$CuCl_4^{2-}(aq) \rightleftharpoons Cu^{2+}(aq) + 4\,Cl^-(aq)$$

   (c) $2\,NO_2(g) \rightleftharpoons 2\,NO(g) + O_2(g)$
   (d) the water gas shift reaction

   $$CO(g) + H_2O(g) \rightleftharpoons CO_2(g) + H_2(g)$$

   (e) $4\,H^+(aq) + 2\,Cl^-(aq) + MnO_2(s) \rightleftharpoons Mn^{2+}(aq) + 2\,H_2O(aq) + Cl_2(g)$
10. The following data are taken from a classic study of the equilibrium reaction (at 986°C and 1 atm):

    $$H_2(g) + CO_2(g) \rightleftharpoons H_2O(g) + CO(g)$$

    Equilibrium Composition (mole percent)

CO_2	H_2	$CO = H_2O$
0.69	80.52	9.40
7.15	46.93	22.96
21.44	22.85	27.86
34.43	12.68	26.43

    Using these data, derive a value for the equilibrium constant for the reaction. At this temperature, are the products or reactants more favored? If the reaction is reversed ($H_2O + CO \rightleftarrows H_2 + CO_2$), what is the value of K for this reaction?

11. Consider the following two equilibria involving $SO_2(g)$ and their corresponding equilibrium constants.

$SO_2(g) + \frac{1}{2} O_2(g) \rightleftharpoons SO_3(g)$ K_1

$2\ SO_3(g) \rightleftharpoons 2\ SO_2(g) + O_2(g)$ K_2

Which of the following expressions relates K_1 to K_2?
(a) $K_2 = K_1^2$
(b) $K_2^2 = K_1$
(c) $K_2 = K_1$
(d) $K_2 = 1/K_1$
(e) $K_2 = 1/K_1^2$

12. Hydrogen can react with elemental sulfur to give that smelly, toxic gas H_2S according to the reaction

$H_2(g) + S(solid) \rightleftharpoons H_2S(g)$ $K = 7.6 \times 10^5$

(a) If the equilibrium constant for the reaction above is 7.6×10^5 at 25°C, determine the value of the equilibrium constant for the reaction written as

$\frac{1}{2} H_2(g) + \frac{1}{2} S(s) \rightleftharpoons \frac{1}{2} H_2S(g)$

(b) Is the value of the equilibrium constant written in terms of gas pressures different from that written in terms of reagent concentrations?

13. At 400°C, the equilibrium constant (K_p) for the Haber synthesis of ammonia is 1.7×10^{-4} for the reaction written as below.

$3\ H_2(g) + N_2(g) \rightleftharpoons 2\ NH_3(g)$

Calculate the value of K for the same reaction written as

$\frac{3}{2} H_2(g) + \frac{1}{2} N_2(g) \rightleftharpoons NH_3(g)$

14. Equilibrium constants for the following reactions have been determined at 823 K:

$CoO(s) + H_2(g) \rightleftharpoons Co(s) + H_2O(g)$ $K_1 = 67$

$CoO(s) + CO(g) \rightleftharpoons Co(s) + CO_2(g)$ $K_2 = 490$

Using this information, calculate K's (at the same temperature) for

$CO_2(g) + H_2(g) \rightleftharpoons CO(g) + H_2O(g)$ K_3

and the commerically important water gas shift reaction

$CO(g) + H_2O(g) \rightleftharpoons CO_2(g) + H_2(g)$ K_4

15. The equilibrium constant K_p for $N_2O_4(g) \rightleftarrows 2\ NO_2(g)$ is 0.11 at 25°C. Calculate the equilibrium constant for

$NO_2(g) \rightleftharpoons \frac{1}{2} N_2O_4(g)$

at the same temperature.

16. Lead sulfide can be converted to metallic lead in the following series of reactions:

$2\ PbS(s) + 3\ O_2(g) \rightleftharpoons 2\ PbO(s) + 2\ SO_2(g)$ K_1

$PbO(s) + C(s) \rightleftharpoons Pb(\ell) + CO(g)$ K_2

$PbO(s) + CO(g) \rightleftharpoons Pb(\ell) + CO_2(g)$ K_3

If the overall process is

$2\ PbS(s) + 3\ O_2(g) + C(s) \rightleftharpoons 2\ Pb(\ell) + CO_2(g) + 2\ SO_2(g)$

what is the equilibrium constant of the net reaction, K_{net}, in terms of known values of K_1, K_2, and K_3?

GRAPHICAL ILLUSTRATION OF EQUILIBRIUM CONSTANTS

17. The n-butane $\rightleftarrows$ isobutane equilibrium described in the text and by Figure 14.4 has a value of K of 2.5 at 25°C.
(a) If the concentration of n-butane is 1.0 M, and that of isobutane is 2.9 M, is the system at equilibrium? If not, which compound increases in concentration on proceeding to equilibrium?
(b) From Figure 14.4 determine the equilibrium concentration of isobutane when n-butane is 0.75 M. Verify your result by calculating it from the equilibrium expression.

18. The hydrocarbon cyclohexane, C_6H_{12}, can isomerize or change into methylcyclopentane, a compound of the same formula but different structure.

$C_6H_{12}(g) \rightleftharpoons C_5H_9CH_3(g)$
cyclohexane methylcyclopentane

The equilibrium constant is estimated to be 0.12 at 25°C. (Stevenson and Morgan, *J. Am. Chem. Soc.*, 70:2773, 1948.)
(a) Construct a plot of the equilibrium points as follows: plot the concentration of cyclohexane (*hex*) from 0 to 10 M on the vertical axis and that of methylcyclopentane (*pent*) from 0 to 1 M along the horizontal axis.
(b) What is the slope of the equilibrium line on your plot? What is the value of K_c?
(c) From the graph estimate the equilibrium concentration of "*pent*" when that of "*hex*" is 4.5 M. Verify the graphically determined value by a calculation from the equilibrium expression.

19. The equilibrium $AgCl(s) \rightleftarrows Ag^+(aq) + Cl^-(aq)$ has an equilibrium constant of 1.8×10^{-10}. Graph this equilibrium constant expression, which has the form $K = xy$. Using your graph, (a) estimate the equilibrium concentration of $Ag^+(aq)$ when $[Cl^-]$ is 0.75×10^{-5} M and (b) tell whether the system is at equilibrium when $[Ag^+] = 1.5 \times 10^{-5}$ M and $[Cl^-] = 2.5 \times 10^{-5}$ M.

CALCULATING AND USING EQUILIBRIUM CONSTANTS

20. At high temperature, hydrogen and carbon dioxide react to give water and carbon monoxide.

$H_2(g) + CO_2(g) \rightleftharpoons H_2O(g) + CO(g)$

(a) If it is found at 986°C that there are 11.2 atm each of CO and water vapor and 8.8 atm each of H_2 and CO_2 at equilibrium, calculate the equilibrium constant.

(b) If there were 8.8 moles of H_2 and CO_2 in a 1.0-L container at equilibrium, how many moles of CO(g) and $H_2O(g)$ would be present?

21. Carbon dioxide reacts with carbon to give carbon monoxide according to the equation

$C(s) + CO_2(g) \rightleftharpoons 2\ CO(g)$

At 700°C, a 2.0-L flask is found to contain at equilibrium 0.10 moles of CO, 0.20 moles of CO_2, and 0.40 moles of C. Calculate the equilibrium constant, K_c, for this reaction at the specified temperature.

22. At 1500°C water vapor is 10.% dissociated into $H_2(g)$ and $O_2(g)$. (That is, 10.% of the original water has been transformed into products, and 90.% remains as water.)

$H_2O(g) \rightleftharpoons H_2(g) + \frac{1}{2}\ O_2(g)$

(a) Assuming a water concentration of 2.0 *M* before dissociation, calculate the equilibrium constant in concentration units.

(b) If H_2O dissociates to give an equilibrium concentration of H_2 of 0.40 *M* at 1500°C, what was the original concentration of H_2O?

23. Initially, 0.84 moles of $PCl_5(g)$ and 0.18 moles of $PCl_3(g)$ are mixed in a 1.0-L container. It is later found that only 0.72 moles of PCl_5 are present when the system has reached equilibrium. Calculate the value of K_c for the reaction

$PCl_5(g) \rightleftharpoons PCl_3(g) + Cl_2(g)$

24. When 2.0 moles of HI(g) are placed in a 1.0-L container at 25°C and allowed to dissociate according to the equation

$2\ HI(g) \rightleftharpoons H_2(g) + I_2(g)$

it is found that 20.% of the HI has dissociated at equilibrium. Calculate K_c and K_p.

25. For the simple equilibrium system

$2\ NO_2(g) \rightleftharpoons N_2O_4(g)$

the equilibrium constant K_c is 8.8 at 25°C. If analysis of the system shows that 2.0×10^{-3} moles of NO_2 are present in a 10.-L flask along with 1.5×10^{-3} moles of N_2O_4, is the system at equilibrium? If it is not at equilibrium, does the concentration of NO_2 increase or decrease on proceeding to equilibrium?

26. Nitrosyl bromide, NOBr(g), decomposes according to the equation

$NOBr(g) \rightleftharpoons NO(g) + \frac{1}{2}\ Br_2(g)$

with an equilibrium constant of $K_p = 0.15$ at 350°C. If 1.0 atm of NOBr, 0.8 atm of NO, and 0.4 atm of Br_2 are mixed at 350°C, will any net reaction occur? If a net reaction is observed, will NO be formed or consumed?

27. The equilibrium constant K_c for the decomposition of $COBr_2$

$COBr_2(g) \rightleftharpoons CO(g) + Br_2(g)$

is 0.190 at 73°C. If the concentrations of both CO(g) and $Br_2(g)$ are 0.402 *M*, and the concentration of $COBr_2(g)$ is 0.950, is the system at equilibrium? If it is not at equilibrium, does $[COBr_2]$ increase or decrease on proceeding to equilibrium?

28. Nitrogen dioxide can be formed from nitrogen oxide and oxygen.

$2\ NO(g) + O_2(g) \rightleftharpoons 2\ NO_2(g)$

At 1000 K the equilibrium constant K_c is 1.20.

(a) If the system is analyzed and the following concentrations are found, is the system at equilibrium: $[O_2] = 1.25\ M$, $[NO] = 2.25\ M$, and $[NO_2] = 3.25\ M$?

(b) If the system is not at equilibrium, does the concentration of NO_2 increase or decrease on proceeding to equilibrium?

29. The hydrocarbon C_4H_{10} can exist in two forms, *n*-butane and isobutane. The value of K_c for the interconversion of the two forms is 2.5 at 25°C.

n-butane $\rightleftharpoons$ isobutane

If you place 0.017 moles of *n*-butane in a 0.50-L flask at 25°C and allow equilibrium to be established, what will be the equilibrium concentrations of the two forms of butane?

30. If some solid ZnS in placed in water, it is found that only a very small amount of the metal sulfide dissolves. What is the concentration of Zn^{2+} in the water? K_c for dissolving ZnS is 1.1×10^{-21}.

$ZnS(s) \rightleftharpoons Zn^{2+}(aq) + S^{2-}(aq)$

31. Barium sulfate is a poorly soluble salt.

$BaSO_4(s) \rightleftharpoons Ba^{2+}(aq) + SO_4^{2-}(aq)$

What is the maximum concentration of Ba^{2+} that can exist in equilibrium with $5.2 \times 10^{-3}\ M\ SO_4^{2-}$?

32. Zinc carbonate dissolves very poorly in water ($K_c = 1.5 \times 10^{-11}$).

$ZnCO_3(s) \rightleftharpoons Zn^{2+}(aq) + CO_3^{2-}(aq)$

If you place some solid $ZnCO_3$ in some water, what are the molar concentrations of Zn^{2+} and CO_3^{2-} when equilibrium has been achieved?

33. When solid ammonium carbamate sublimes, it dissociates completely into ammonia and carbon dioxide according the equation

$N_2H_6CO_2(s) \rightleftharpoons 2\ NH_3(g) + CO_2(g)$

At 25°C, experiment shows that the total pressure of the gases in equilibrium with the solid is 0.116. What is the equilibrium constant for the reaction?

34. At 633°C, the equilibrium constant K_c for the dissociation of ammonia to its elements is 6.56×10^{-3}.

$2\ NH_3(g) \rightleftharpoons N_2(g) + 3\ H_2(g)$

What is the value of K_p, the constant written in terms of reactant and product pressures?

35. Consider the following equilibrium process at 700°C.

$2\ H_2(g) + S_2(g) \rightleftharpoons 2\ H_2S(g)$

Analysis shows that, at equilibrium, there are 2.50 moles of $H_2(g)$, 1.35×10^{-5} moles of $S_2(g)$, and 8.70 moles of $H_2S(g)$ present in a 12.0-L flask. Calculate K_c and K_p.

36. Two moles of PCl_5 and two moles of Cl_2 are mixed and allowed to reach equilibrium according to the reaction

$PCl_5\ (g) \rightleftharpoons PCl_3(g) + Cl_2(g)$

If the equilibrium amount of $PCl_3(g)$ is x moles, which is the correct expression for the equilibrium constant?

(a) $(2-x)/x(2+x)$ (c) $(2+x)/x(2-x)$
(b) $(2+x)x/(2-x)$ (d) $(2-x)x/(2+x)$

37. Calcium carbonate decomposes at high temperature to form solid CaO and gaseous CO_2.

$CaCO_3(s) \rightleftharpoons CaO(s) + CO_2(g)$

If the equilibrium constant (K_c) at some temperature is 0.10, how many grams of CO_2 are formed when equilibrium is established at that temperature in a 1.0-L container?

38. Magnesium carbonate may be formed by reaction of MgO with $CO_2(g)$.

$MgO(s) + CO_2(g) \rightleftharpoons MgCO_3(s) \qquad K_c = 1.2$

How many moles of $CO_2(g)$ are present in a 2.0-L container when 2.0×10^2 g of $MgCO_3(s)$ have been produced at equilibrium?

39. The equilibrium constant K_p for the reaction

$2\ HCN(g) \rightleftharpoons H_2(g) + C_2N_2(g)$

is 4.00×10^{-4} at 500°C. What is the equilibrium partial pressure of $C_2N_2(g)$ at 500°C if pure HCN was originally present at 25.0 atm?

40. The equilibrium constant K_c for the reaction

$CO_2(g) + H_2(g) \rightleftharpoons CO(g) + H_2O(g)$

is 1.6 at 690°C. What is the concentration of each substance at equilibrium in a mixture prepared by originally placing 0.50 moles of $CO_2(g)$ and 0.50 moles of $H_2(g)$ in a 5.0-L container at 690°C?

41. At 633°C, 3.60 moles of ammonia are placed in a 2.00-L vessel and allowed to decompose to the elements.

$2\ NH_3(g) \rightleftharpoons N_2(g) + 3\ H_2(g)$

If K_c is 6.56×10^{-3} for this reaction at this temperature, calculate the equilibrium concentration of each reagent.

Le CHATELIER'S PRINCIPLE

42. Hydrogen, bromine, and HBr are in equilibrium in a 1-L container.

$H_2(g) + Br_2(g) \rightleftharpoons 2\ HBr(g) + 68\ kJ$

How will each of the following changes affect the indicated quantities? Write "increase," "decrease," or "no change."

CHANGE	$[Br_2]$	$[HBr]$	K_c
Some H_2 is added to the container.	______	______	______
The temperature of the gases in the container is increased.	______	______	______
Partial pressure of HBr is increased.	______	______	______
Volume of flask is increased.	______	______	______

43. K_p for the following reaction is 0.16 at 25°C. The enthalpy change for the reaction at standard conditions is −344 kJ.

$2\ NOBr(g) \rightleftharpoons 2\ NO(g) + Br_2(g)$

Predict the effect of the following changes on the position of the equilibrium; that is, state which way the equilibrium will shift (left, right, or no change) when each of the following changes is made.

(a) Addition of more $Br_2(g)$ ______________
(b) Removal of some NOBr(g) ______________
(c) Increase in the container volume ______________
(d) Decrease in temperature ______________

44. The oxidation of NO to NO_2 is exothermic.

$2\ NO(g) + O_2(g) \rightleftharpoons 2\ NO_2(g)$

If the system is at equilibrium and the temperature is increased, will the concentration of NO increase, remain the same, or decrease?

45. Consider the n-butane $\rightleftharpoons$ isobutane system ($K_c = 2.5$). The system is originally at equilibrium with $[n] = 1.0\ M$ and $[iso] = 2.5\ M$.

(a) If 0.50 mol/L of isobutane is suddenly added and then the system shifts to a new equilibrium position, what are the equilibrium concentrations of each gas?

(b) If 0.50 mol/L of *n*-butane is added, and the system shifts to a new equilibrium position, what are the equilibrium concentrations of each gas?

46. K_p for the decomposition of ammonium hydrogen sulfide is 0.11 at 25°C.

$$NH_4HS(s) \rightleftharpoons NH_3(g) + H_2S(g)$$

(a) When the pure salt decomposes in a flask, what are the equilibrium pressures of NH_3 and H_2S?
(b) If NH_4HS is placed in a flask already containing 0.50 atm of NH_3 and then the system is allowed to come to equilibrium, what are the equilibrium pressures of NH_3 and H_2S?
(c) Do your results in parts (a) and (b) conform with Le Chatelier's principle?

47. When solid ammonium carbamate sublimes, it dissociates completely into ammonia and carbon dioxide according to the equation ($K_p = 2.31 \times 10^{-4}$)

$$N_2H_6CO_2(s) \rightleftharpoons 2\ NH_3(g) + CO_2(g)$$

If 0.05 atm of CO_2 is added to the flask after equilibrium has been established at 25°C, will the final pressure of CO_2 be greater or less than 0.05 atm? Will the pressure of NH_3 be greater or less than $P(NH_3)$ in the absence of added CO_2?

48. The equilibrium constant K_c for the decomposition of $COBr_2$ is 0.190 at 73°C.

$$COBr_2(g) \rightleftharpoons CO(g) + Br_2(g)$$

(a) 1.50 moles of $COBr_2$ are placed in a 2.00-L flask and heated to 73°C. What are the equilibrium concentrations of each compound?
(b) If 1.00 mole of CO is added to the system at equilibrium, what are the concentrations of each compound after equilibrium is reestablished?

GENERAL PROBLEMS

49. The equilibrium constant K_p for the formation of $PCl_5(g)$ from $PCl_3(g)$ and $Cl_2(g)$ is 0.087 at 300°C.

$$PCl_3(g) + Cl_2(g) \rightleftharpoons PCl_5(g)$$

If you place 1.0 mole each of PCl_3 and Cl_2 in a 5.0-L flask and heat to 300°C, what is the maximum concentration of PCl_5 that can exist at equilibrium?

50. Two molecules of gaseous acetic acid can form a dimer through hydrogen bonds.

$$2\ H_3CCOOH \rightleftharpoons H_3C{-}C\begin{matrix} \diagup O{-}H\cdots O \diagdown \\ \diagdown O\cdots H{-}O \diagup \end{matrix}C{-}CH_3$$

The equilibrium constant K_p at 25°C has been determined to be 1.3×10^3 (where the pressures are measured in atmospheres). Assume that acetic acid is present initially at a partial pressure of 10. mm Hg and that no dimer is present initially.

(a) What percentage of acetic acid is converted to dimer?
(b) The energy of the hydrogen bonds is 132 kJ/mol. As the temperature goes up, in which direction does the equilibrium shift?

51. The equilibrium constant for the *n*-butane ⇄ isobutane isomerization reaction is 2.5 at 25°C. If you mix 1.75 moles of *n*-butane and 1.25 moles of isobutane, is the system at equilibrium? If not, when it proceeds to equilibrium which reagent increases in concentration? Calculate the concentrations of the two compounds when the system reaches equilibrium.

52. K_p at 2000 K for the formation of NO(g) is 4.0×10^{-4}.

$$N_2(g) + O_2(g) \rightleftharpoons 2\ NO(g)$$

(a) What is the value of K_c?
(b) If analysis shows that the concentrations of N_2 and O_2 are both 0.25 *M*, and that of NO is 0.0042 *M*, is the system at equilibrium?
(c) If the system is not at equilibrium, in which direction does the reaction proceed?
(d) When the system is at equilibrium, what are the equilibrium concentrations?

53. At 1800 K, oxygen dissociates very slightly into its atoms.

$$O_2(g) \rightleftharpoons 2\ O(g) \qquad K_p = 1.7 \times 10^{-8}$$

If you place 1.0 mole of O_2 in a 10.-L vessel and heat it to 1800 K, how many O atoms are present in the flask?

54. The equilibrium constants for the dissociation of three complexes of trimethylborane, $B(CH_3)_3$, have been determined.

$$L\text{-}B(CH_3)_3(g) \rightleftharpoons L(g) + B(CH_3)_3(g)$$

L	K_p (atm) at 100°C
$(CH_3)_3P$	0.128
$(CH_3)_3N$	0.472
H_3N	4.62

(a) If you begin an experiment by placing 0.010 mole of each complex in a flask, which would have the largest partial pressure of $B(CH_3)_3$ at 100°C?
(b) If 0.73 g (0.010 mole) of $H_3N{-}B(CH_3)_3$ is placed in a 1.0×10^2-mL flask and heated to 100°C, what is the partial pressure of each gas in the equilibrium mixture and what is the total pressure?
(c) What is the percent dissociation of $H_3N{-}B(CH_3)_3$?
(d) If you use 0.010 mole of the $(CH_3)_3N$ complex, calculate the equilibrium partial pressure of each gas at 100°C. What is the percent dissociation of the complex? Compare this with the answer to part (c) above and with your expectations in part (a).

55. Calcium fluoride is a poorly soluble salt.

$$CaF_2(s) \rightleftharpoons Ca^{2+}(aq) + 2\ F^-(aq)$$

If the equilibrium constant for this process is 3.9×10^{-11}, what are the equilibrium concentrations of $Ca^{2+}(aq)$ and $F^-(aq)$ when CaF_2 is placed in some water?

56. The concentrations given in Example 14.2 are not those at equilibrium. Given that the reaction begins with the concentrations in Example 14.2, what are the equilibrium concentrations of PCl_5, PCl_3, and Cl_2?

57. A reaction important in smog formation is

$$O_3(g) + NO(g) \rightleftharpoons O_2(g) + NO_2(g) \qquad K_c = 6.0 \times 10^{34}$$

(a) If the initial concentrations are $[O_3] = 1.0 \times 10^{-6}$ M, $[NO] = 1.0 \times 10^{-5}$ M, $[NO_2] = 2.5 \times 10^{-4}$ M, and $[O_2] = 8.2 \times 10^{-3}$ M, is the system at equilibrium? If not, in which direction does the reaction proceed?

(b) If the temperature is increased, as on a very warm day, will the concentrations of the products increase or decrease?

58. At high temperatures, carbon and carbon dioxide react to form carbon monoxide.

$$C(s) + CO_2(g) \rightleftharpoons 2\ CO(g)$$

(a) When equilibrium is established at 1000 K, the total pressure in the system is 4.70 atm. If $K_p = 1.72$, what are the partial pressures of CO and CO_2?

(b) If enough CO_2 had been placed in the reaction vessel before reaction so that $P(CO_2) = 1.11$ atm when $T = 1000$ K, what are the partial pressures of CO and CO_2 when equilibrium has been established?

15

Acids and Bases

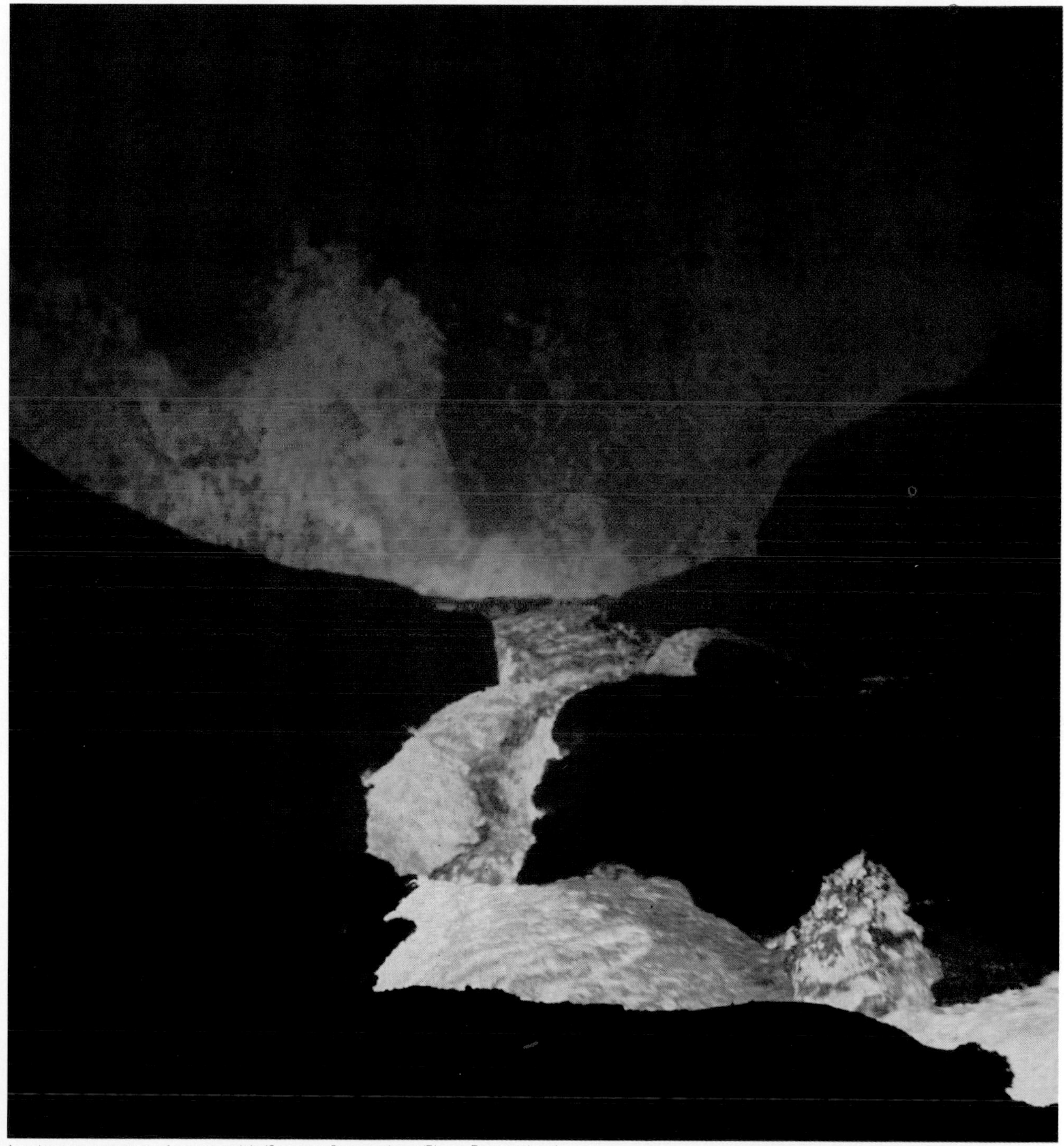

A volcano, a source of many acids (Soames Summerhays/Photo Researchers)

CHAPTER OUTLINE

It has been said that "the ocean is the result of a gigantic acid–base titration; acids that have leaked out of the interior of the earth are titrated with bases that have been set free by the weathering of primary rock."

Acids and bases provide the cornerstone of life on the earth. Basic carbonate and hydrogen carbonate ions from rocks and from CO_2 are present in most natural waters along with such other substances as borate, phosphate, arsenate, silicate, and ammonia. In addition, volcanoes and hot springs can give strongly acidic waters due to the presence of HCl and SO_2. This witch's brew allows acids and bases to interact with one another and with carbon dioxide. Biological activities such as photosynthesis and respiration influence the hydrogen ion content of natural waters through the use or production of CO_2, the most important acid-forming substance in nature.

With such reactions occurring in our environment, it is no wonder that acids and bases have been studied for hundreds of years. The terms acid, alkali, and salt are themselves hundreds of years old. Acid comes from the latin word *acidus,* meaning sour. It was probably originally applied to vinegar, but it was later the name given to other substances that had a sour taste. Alkali is derived from an Arabic word for the ashes that came from heating certain plants. Since potash, or potassium carbonate, is the primary product of this process, and since water solutions of potash feel soapy and taste bitter, the term alkali was later applied more generally to substances having these pıoperties. Finally, the word *salt,* which finds roots in many languages, probably originally meant sea salt or sodium chloride, but it now has a broader meaning.

With time the terms acid and alkali were applied more broadly. Robert Boyle, whose work with gases led to "Boyle's law," wrote in 1684 that alkalis give soapy solutions, restore vegetable colors reddened by acids, and react with acids to give what he called "indifferent salts." Acids, on the other hand, he characterized by their sour taste, their ability to be corrosive, to redden blue vegetable colors, and to lose all these properties when brought into contact with alkalis.

Here you see a very early, explicit mention of the use of acid–base indicators. The juice of the red cabbage is an excellent indicator of acidity (Figure 4.10).

By the 18th century, salts were recognized as the product of the interaction of an acid and an alkali, a concept that became a major chemical theory. About 1750, the Frenchman Rouelle contended that a neutral salt was formed by an acid reacting with any substance—a water-soluble alkali, an "earth," a metal, and so on—capable of serving as "a base for [the salt]." Thus, the term *base* entered the vocabulary of chemistry, and we now recognize reactions such as the following as salt producers.

$$2\ HCl(aq) + Ca(OH)_2(s)\ \text{(slaked lime)} \longrightarrow CaCl_2(aq) + 2\ H_2O(\ell)$$

$$2\ HCl(aq) + CaCO_3(s)\ \text{(limestone, an ``earth'')} \longrightarrow CaCl_2(aq) + H_2O(\ell) + CO_2(g)$$

$$2\ HCl(aq) + Ca(s)\ \text{(a metal)} \longrightarrow CaCl_2(aq) + H_2(g)$$

Due to their universal importance in chemistry, more general notions of classifying and explaining acid–base behavior and reactions have been developed over the past 100 years. The concept we have used thus far in this book came from Svante Arrhenius (Swedish, 1859–1927). He proposed that an acid is a substance giving the hydrogen ion (H^+) as one of the products of ionic dissociation in water and that a base gives a hydroxide ion (OH^-) as one of the dissociation products.

Arrhenius Definition

ACID	BASE
$H_xB \longrightarrow xH^+ + B^{x-}$	$M(OH)_y \longrightarrow M^{y+} + yOH^-$
$HCl(aq) \longrightarrow H^+(aq) + Cl^-(aq)$	$NaOH(aq) \longrightarrow Na^+(aq) + OH^-(aq)$

Unfortunately this concept is limited to water because it refers to ions (H^+) and (OH^-) derived from water. A truly general concept of acids and bases should be appropriate to other solvents. Still another limitation is that the base dissociates to produce OH^-. This is certainly true of lye (NaOH), slaked lime [$Ca(OH)_2$], and milk of magnesia [$Mg(OH)_2$]. However, there are many substances that are bases, such as ammonia (NH_3), but that do not have OH^- ions as part of their original formulation.

In this chapter we shall study only two of the more general acid–base concepts, those of J.N. Brønsted (Danish, 1879–1947) and G.N. Lewis (American, 1875–1946). However, since we shall be most concerned with acids and bases in water, we first want to say something more about the properties of this solvent.

Every year several books and many scientific papers are published on acid–base chemistry. One of the intellectual challenges of modern chemistry is to understand thoroughly the interaction of an acid and a base and to predict the strength of that interaction.

15.1 WATER, THE HYDRONIUM ION, AND AUTO-IONIZATION

The H^+ ion is a bare proton with a radius of less than 0.1 pm. This means that if this tiny particle with its highly concentrated charge is placed in water it will be attracted very strongly to the negative water dipole.

$$H^+(g) + H_2O(\ell) \longrightarrow [H_3O]^+(aq) + \text{Energy}$$

Hydrated proton
= hydronium ion

Thus, a proton in water can only exist associated with water molecules as a hydrated proton. By analogy with the ammonium ion, NH_4^+, we call the simplest proton–water complex, H_3O^+, the **hydronium ion.** You recall from Chapter 11, however, that there is extensive hydrogen bonding in water, and the hydrated proton is probably even better represented by other formulations such as

$H_5O_2^+$ and $H_7O_3^+$

It is also certain that the hydroxide ion is associated with water molecules through hydrogen bonding, perhaps in the form $H_3O_2^-$.

An acid such as HCl does not need to be present for the hydronium ion to exist in water. In fact, *two water molecules can interact with one another to produce a hydronium and a hydroxide ion* by proton transfer from one water molecule to another.

$$2\ H_2O(aq) \rightleftharpoons H_3O^+(aq) + OH^-(aq)$$

$$\mathrm{H{-}\ddot{O}{:}(H) + H{-}\ddot{O}{:}(H) \rightleftharpoons H{-}\ddot{O}{-}H^+(H) + {:}\ddot{O}{:}^-(H)}$$

That this so-called **auto-ionization** of water occurs was proved many years ago by Friedrich Kohlrausch (1840–1910). He found that, even after water is painstakingly purified, it still conducts electricity to a very small extent, since auto-ionization produces a very low concentration of H_3O^+ and OH^- ions even in the purest water. Water auto-ionization is the cornerstone of our concepts of aqueous acid–base behavior.

Since we are most interested in the properties of acids and bases in water, the properties of water and hydronium and hydroxide ions are important in all our discussions. It is understood that both the H^+ and OH^- ions must be associated with water, but, for simplicity, we shall generally write them simply as $H^+(aq)$ or $OH^-(aq)$.

15.2 THE BRØNSTED CONCEPT OF ACIDS AND BASES

In 1923 J.N.. Brønsted in Copenhagen and J.M. Lowry in Cambridge (England) independently suggested a new concept of acid and base behavior. They proposed that an **acid** is any substance that *can donate a proton* to any other substance.

$$\underset{\text{acid}}{HNO_3(aq)} + H_2O(\ell) \longrightarrow H_3O^+(aq) + NO_3^-(aq)$$

and a **base** is a substance that *can accept a proton* from any other substance.

$$\underset{\text{base}}{NH_3(aq)} + H_2O(\ell) \longrightarrow NH_4^+(aq) + OH^-(aq)$$

These statements have come to be known as the *Brønsted definition of acids and bases,* because Brønsted and his students substantiated and extended the idea.

There is a wide variety of Brønsted acids, and you are familiar with many of them. Acids such as HF, HCl, HNO_3, CH_3COOH, and HCN are all capable of donating one proton and so are called **monoprotic acids.** However, there are also many acids capable of donating two or more protons, and these are called **polyprotic acids.** Some common polyprotic acids include sulfuric acid, phosphoric acid, carbonic acid (H_2CO_3), hydrosulfuric acid (H_2S), and oxalic acid ($H_2C_2O_4$).

$$\begin{array}{l} O{=}C{-}O{-}H(aq) \\ \quad\ | \\ O{=}C{-}O{-}H \end{array} + H_2O(\ell) \longrightarrow \begin{array}{l} O{=}C{-}O^-(aq) \\ \quad\ | \\ O{=}C{-}O{-}H \end{array} + H_3O^+(aq)$$

oxalic acid — hydrogen oxalate

$$\begin{array}{l} O{=}C{-}O^-(aq) \\ \quad\ | \\ O{=}C{-}O{-}H \end{array} + H_2O(\ell) \longrightarrow \begin{array}{l} O{=}C{-}O^-(aq) \\ \quad\ | \\ O{=}C{-}O^- \end{array} + H_3O^+(aq)$$

hydrogen oxalate — oxalate

Just as there are acids that can donate more than one proton, there are bases that can accept more than one proton and so are called **polyprotic bases.** Such bases include the anions of the polyprotic acids listed above, that is, SO_4^{2-}, PO_4^{3-}, CO_3^{2-}, S^{2-}, or $C_2O_4^{2-}$. All behave to a greater or lesser extent like the sulfide ion.

$$S^{2-}(aq) + H^+(aq) \longrightarrow HS^-(aq)$$
$$HS^-(aq) + H^+(aq) \longrightarrow H_2S(aq)$$

base

There are also **amphiprotic** substances, molecules or ions that can behave as either a Brønsted acid or base. Water is an excellent example. For instance, with HCl, water acts as a base in accepting a proton from the acid.

$$\underset{\text{acid}}{HCl(aq)} + \underset{\text{base}}{H_2O(\ell)} \longrightarrow H_3O^+(aq) + Cl^-(aq)$$

However, water is an acid when donating a proton to ammonia.

$$\underset{\text{base}}{NH_3(aq)} + \underset{\text{acid}}{H_2O(\ell)} \longrightarrow NH_4^+(aq) + OH^-(aq)$$

Besides water, other amphiprotic substances you have encountered so far include

	$+H^+$	Amphiprotic Ions		$-H^+$	
H_2S	$\longleftarrow$	HS^-	(hydrogen sulfide)	$\longrightarrow$	S^{2-}
H_3PO_4	$\longleftarrow$	$H_2PO_4^-$	(dihydrogen phosphate)	$\longrightarrow$	HPO_4^{2-}
$H_2PO_4^-$	$\longleftarrow$	HPO_4^{2-}	(monohydrogen phosphate)	$\longrightarrow$	PO_4^{3-}
$H_2C_2O_4$	$\longleftarrow$	$HC_2O_4^-$	(hydrogen oxalate)	$\longrightarrow$	$C_2O_4^{2-}$
H_2CO_3	$\longleftarrow$	HCO_3^-	[hydrogen carbonate (bicarbonate)]	$\longrightarrow$	CO_3^{2-}

The amphiprotic nature of the HPO_4^{2-} and HCO_3^- ions in particular is especially important in biochemistry.

Finally, let us consider briefly the interaction between two substances, one considered an acid relative to water and the other considered a base relative to water.

$$\underset{\text{acid}}{HCl(aq)} + \underset{\text{base}}{NH_3(aq)} \longrightarrow NH_4^+(aq) + Cl^-(aq)$$

A proton has clearly been transferred from the acid to the base in a reaction that can be thought of as the sum of three processes:

$$
\begin{array}{llll}
HCl(aq) & + H_2O(\ell) & \longrightarrow H_3O^+(aq) & + Cl^-(aq) \\
NH_3(aq) & + H_2O(\ell) & \longrightarrow OH^-(aq) & + NH_4^+(aq) \\
H_3O^+(aq) & + OH^-(aq) & \longrightarrow 2\ H_2O(\ell) & \\
\hline
HCl(aq) & + NH_3(aq) & \longrightarrow NH_4^+(aq) & + Cl^-(aq)
\end{array}
$$

where water has functioned as the proton transfer agent. As you have previously seen when writing net equations, the principal aspect of acid–base reactions is the interaction of hydronium and hydroxide ions to give water, which is the opposite of water auto-ionization. Since the net result of such a reaction can be a solution with neither acidic nor basic characteristics, an acid–base reaction is often called a **neutralization** reaction. We will study this topic in detail in Chapter 16.

CONJUGATE ACID–BASE PAIRS

There is considerably more to the Brønsted notion than just the definitions. In each of the chemical equations above there has been a *transfer* of a proton. If an acid has donated a proton, there must be a substance that accepts that proton, a base. For example, the bicarbonate or hydrogen carbonate ion transfers a proton to the base water

$$\underset{\text{acid}}{HCO_3^-(aq)} + \underset{\text{base}}{H_2O(\ell)} \rightleftharpoons CO_3^{2-}(aq) + H_3O^+(aq)$$

and in the process the carbonate anion and hydronium ion have been produced. Since this reaction is reversible, the carbonate ion could act as a base, accept a proton from the proton donor H_3O^+, and re-form the bicarbonate ion. Thus, bicarbonate and carbonate ions are related to one another by the loss or gain of H^+, as are H_2O and H_3O^+. A pair of compounds or ions related by gain or loss of one H^+ is called a **conjugate acid–base pair.** Thus, CO_3^{2-} is the conjugate base of the acid HCO_3^-,

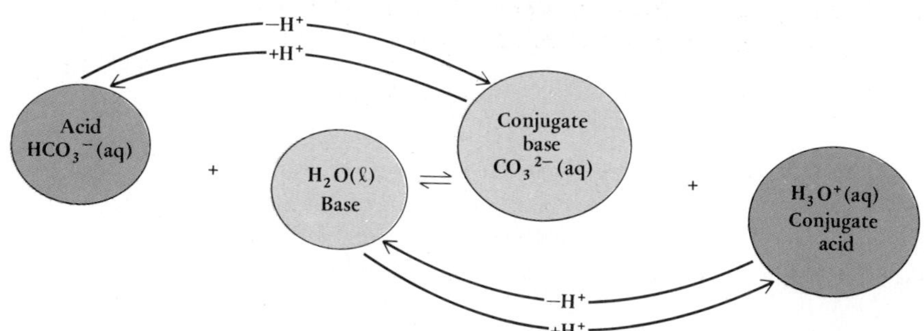

and HCO_3^- is the conjugate acid of the base CO_3^{2-}. *Every acid–base interaction involving H^+ transfer has two conjugate acid–base pairs.* To see that this is the case, look at the reactions above and those in Table 15.1.

EXERCISE 15.1 CONJUGATE ACIDS AND BASES

(a) In the following reaction, identify the acid on the left and its conjugate base on the right. Similarly, identify the base on the left and its conjugate acid on the right.

$$HCN(aq) + NH_3(aq) \rightleftharpoons NH_4^+(aq) + CN^-(aq)$$

(b) What is the conjugate base of H_2S?
(c) What is the conjugate acid of NO_3^-?

RELATIVE STRENGTHS OF ACIDS AND BASES

In water some acids are better proton donors than others, and some bases are better proton acceptors than others. For example, a dilute solution of hydrochloric acid consists largely of $H^+(aq)$ and $Cl^-(aq)$ ions; the acid is nearly 100% ionized, and so it is considered a *strong Brønsted acid.*

A strong Brønsted acid is a molecule or ion that has a strong tendency to transfer a proton to water. A strong Brønsted base is a molecule or ion that has a strong tendency to accept a proton from water, producing OH^-.

Strong acid (≈100% ionization)

$$0.1\ M\ HCl(aq) + H_2O(\ell) \longrightarrow 0.1\ M\ H_3O^+(aq) + 0.1\ M\ Cl^-(aq)$$

In contrast, acetic acid ionizes only to a very small extent. Acetic acid, and any acid where less than about 1% of the acid is ionized, is considered a *weak Brønsted acid.*

Weak acid (< ≈1% ionization)

$$\underset{\text{acetic acid}}{0.1\ M\ CH_3COOH(aq)} + H_2O(\ell) \rightleftharpoons 0.001\ M\ H_3O^+(aq) + \underset{\text{acetate ion}}{0.001\ M\ CH_3COO^-(aq)}$$

The oxide ion is a strong Brønsted base in aqueous solution because it has a strong tendency to accept a proton from water to produce OH^-.

Strong base ($[OH^-] \simeq$ initial concentration of base, O^{2-})

$$O^{2-}(aq) + H_2O(\ell) \longrightarrow 2\ OH^-(aq)$$

Aqueous ammonia or carbonate ion, however, produce only a very small concentration of OH^- and are weak Brønsted bases.

Weak base ($[OH^-] << $ initial concentration of base, NH_3)

$$NH_3(aq) + H_2O(\ell) \rightleftharpoons OH^-(aq) + NH_4^+(aq)$$

$$CO_3^{2-}(aq) + H_2O(\ell) \rightleftharpoons OH^-(aq) + HCO_3^-(aq)$$

According to the Brønsted model, an acid donates a proton and produces a conjugate base. It is an important corollary of this model that

TABLE 15.1 Some Acids and Bases

NAME	ACID 1		BASE 2		BASE 1		ACID 2*
Hydrogen chloride	HCl	+	H_2O	⇄	Cl^-	+	H_3O^+
Nitric acid	HNO_3	+	H_2O	⇄	NO_3^-	+	H_3O^+
Hydrogen carbonate	HCO_3^-	+	H_2O	⇄	CO_3^{2-}	+	H_3O^+
Acetic acid	H_3CCOOH	+	H_2O	⇄	H_3CCOO^-	+	H_3O^+
Hydrogen cyanide	HCN	+	H_2O	⇄	CN^-	+	H_3O^+
Hydrogen sulfide	H_2S	+	H_2O	⇄	HS^-	+	H_3O^+
Ammonia	H_2O	+	NH_3	⇄	OH^-	+	NH_4^+
Carbonate ion	H_2O	+	CO_3^{2-}	⇄	OH^-	+	HCO_3^-
Water	H_2O	+	H_2O	⇄	OH^-	+	H_3O^+

*Acid 1 and base 1 are a conjugate pair, as are base 2 and acid 2.

(a) the stronger the acid, the weaker its conjugate base,

and

(b) the stronger the base, the weaker its conjugate acid.

Thus, aqueous HCl is a strong acid because it has a strong tendency to donate a proton to water and produce the conjugate base Cl^-. At the same time, water, acting as a base, has accepted the proton to produce the conjugate acid H_3O^+. Since we know that the reaction has proceeded

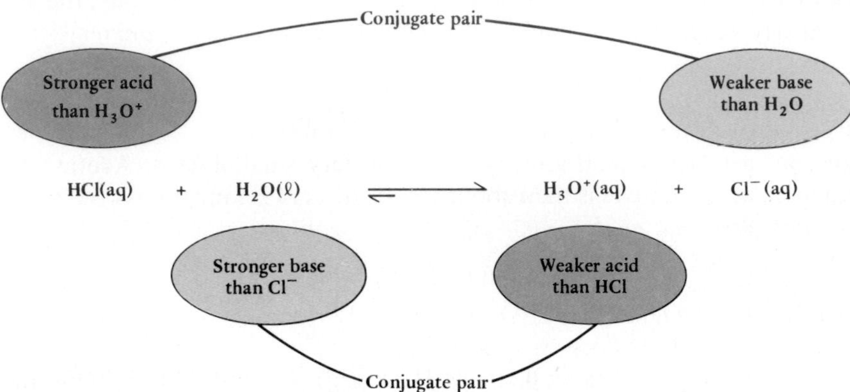

almost completely to the right, at equilibrium there is essentially no HCl left intact in solution. This must mean that the acid and base on the left are each stronger than the acid and base on the right. Of the two bases in solution, H_2O and Cl^-, water must be the stronger base and wins out in the competition for the proton. Of the two acids, HCl and H_3O^+, HCl is the better able to donate a proton. We denote this by using arrows of unequal length, showing the reaction largely proceeds to the right.

To save space and effort we often abbreviate acetic acid, CH_3COOH, as HOAc and the acetate ion, CH_3COO^-, as OAc^-.

Acetic acid, that important component of vinegar and bad wine, is the classic example of a weak acid. Acetic acid ionizes to a very small

Conjugate pair

Weaker acid than H_3O^+

Stronger base base than H_2O

$$HOAc(aq) + H_2O(\ell) \rightleftharpoons H_3O^+ + OAc^-(aq)$$

Weaker base than OAc^-

Stronger acid than HOAc

Conjugate pair

extent in water. Thus, of the two acids present in aqueous acetic acid, (HOAc and H_3O^+), it is the hydronium ion that is the stronger. Of the two bases (H_2O and OAc^-), the acetate ion must be the stronger. At equilibrium, the solution consists mostly of acetic acid with only a small concentration of acetate ion and hydronium ion.

This is the reason that you learned in Chapter 3 that the formation of water is the driving force of acid–base reactions.

Based on these two examples of the relative extent of acid–base reactions, another corollary of the Brønsted model is that

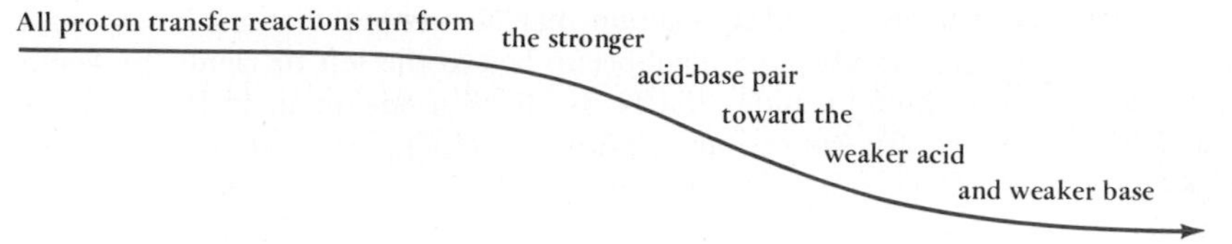

All proton transfer reactions run from
the stronger
acid-base pair
toward the
weaker acid
and weaker base

Acids and bases can be ordered depending on their relative abilities to donate or accept protons, and we have done so for a few acids and bases in Table 15.2. At the very top on the left are the strongest acids, those substances that so strongly donate protons that their conjugate bases are extremely weak. The opposite situation is true at the bottom of the table. For example, the H_2 molecule can be considered an acid in the sense that it is conceivable that it can give the conjugate base H^-, the hydride ion, on losing H^+. However, hydride is such a strong base that it will extract H^+ from water to give H_2, often with explosive violence (Figure 15.1).

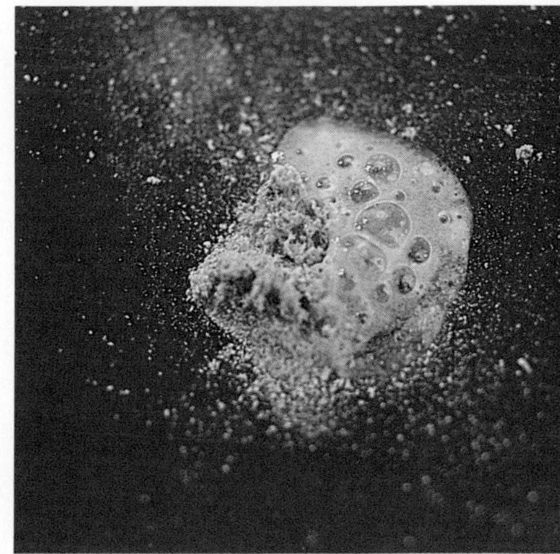

FIGURE 15.1

The basic properties of the hydride ion, H . Calcium hydride, CaH_2, is the source of the H^- ion. The ion is such a powerful proton acceptor that it reacts vigorously with the proton donor water to give H_2 (H^- + H^+). (Charles D. Winters)

WRITING ACID–BASE REACTIONS TO ACCOUNT FOR RELATIVE ACID–BASE STRENGTHS

The chart of acids and bases in Table 15.2 can be used to write acid–base reactions and to predict whether the equilibrium lies to the left or right. The examples that follow show you how to do this.

TABLE 15.2 Relative Strengths of Acids and Bases

CONJUGATE ACID		CONJUGATE BASE	
NAME	FORMULA	FORMULA	NAME
Perchloric acid	$HClO_4$	ClO_4^-	Perchlorate ion
Sulfuric acid	H_2SO_4	HSO_4^-	Hydrogen sulfate ion
Hydrochloric acid	HCl	Cl^-	Chloride ion
Nitric acid	HNO_3	NO_3^-	Nitrate ion
Hydronium ion	H_3O^+	H_2O	Water
Hydrogen sulfate ion	HSO_4^-	SO_4^{2-}	Sulfate ion
Phosphoric acid	H_3PO_4	$H_2PO_4^-$	Dihydrogen phosphate ion
Acetic acid	$HC_2H_3O_2$	$C_2H_3O_2^-$	Acetate ion
Hexaaquoaluminum ion	$[Al(H_2O)_6]^{3+}$	$[Al(H_2O)_5(OH)]^{2+}$	Hydroxopentaaquoaluminum ion
Carbonic acid	H_2CO_3	HCO_3^-	Hydrogen carbonate ion
Hydrogen sulfide	H_2S	HS^-	Hydrogen sulfide ion
Dihydrogen phosphate ion	$H_2PO_4^-$	HPO_4^{2-}	Hydrogen phosphate ion
Ammonium ion	NH_4^+	NH_3	Ammonia
Hydrogen cyanide	HCN	CN^-	Cyanide ion
Hydrogen carbonate ion	HCO_3^-	CO_3^{2-}	Carbonate ion
Phenol	C_6H_5OH	$C_6H_5O^-$	Phenoxide ion
Water	H_2O	OH^-	Hydroxide ion
Ethyl alcohol	C_2H_5OH	$C_2H_5O^-$	Ethoxide ion
Ammonia	NH_3	NH_2^-	Amide ion
Methylamine	CH_3NH_2	CH_3NH^-	Methylamide ion
Hydrogen	H_2	H^-	Hydride ion
Methane	CH_4	CH_3^-	Methide ion

↑ Increasing Acid Strength

Increasing Base Strength ↓

EXAMPLE 15.1

WRITING ACID–BASE REACTIONS

Write a balanced equation for the reaction that occurs between each acid–base pair in water and tell whether the equilibrium lies to the left or right. (a) Acetic acid, HOAc, and sodium cyanide, NaCN. Is HCN, a poisonous acid, formed to a significant extent? (b) Ammonium chloride, NH_4Cl, and sodium carbonate, Na_2CO_3.

Solution (a) It is shown in Table 15.2 that the conjugate base of acetic acid is the acetate ion, CH_3COO^- or OAc^-. The other reactant, sodium cyanide, is a water-soluble salt and dissociates to give Na^+ and the cyanide ion, CN^-. Sodium ion is not listed in Table 15.2 because it is neither a Brønsted acid or base in water (see Section 15.5). The CN^- ion, however, is a base with HCN as its conjugate acid. Therefore, we can write the reaction

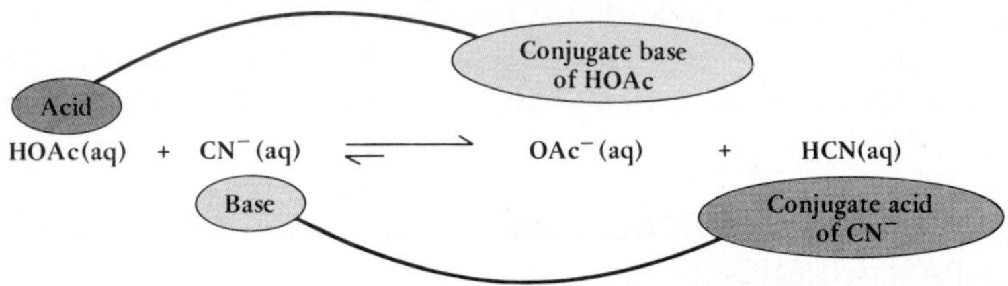

To decide on which side the equilibrium lies, left or right, we can compare the two acids (or the two bases) in the reaction. According to the chart, HCN is a weaker acid than HOAc, and OAc^- is a weaker base than CN^-. Since reactions run toward the weaker acid and base, this must mean that, at equilibrium, the reaction favors the HCN/OAc^- pair. This tells you that you certainly would not want inadvertently to mix acetic acid and a cyanide salt since HCN will be the product. The average fatal dose of HCN is 50 to 60 mg or about 0.002 moles.

(b) Ammonium chloride dissociates to give NH_4^+ and Cl^- in aqueous solution, ions that are both listed in Table 15.2. The ammonium ion is an acid according to Table 15.2,

$$\underset{\text{acid}}{NH_4^+(aq)} + \underset{\text{base}}{H_2O(\ell)} \rightleftharpoons \underset{\text{conjugate base}}{NH_3(aq)} + \underset{\text{conjugate acid}}{H_3O^+(aq)}$$

and, in principle, the chloride ion is a base in water.

$$\underset{\text{base}}{Cl^-(aq)} + \underset{\text{acid}}{H_2O(\ell)} \rightleftharpoons \underset{\text{conjugate acid}}{HCl(aq)} + \underset{\text{conjugate base}}{OH^-(aq)}$$

It is shown in Table 15.2 that OH^- is a *much* stronger base than Cl^-, and HCl is a *much* stronger acid than H_2O. Therefore, we can consider that this reaction does not occur. In general, *any anion X^- that is a weaker base than water does not react with water to give HX and OH^-*. Thus, for ammonium chloride we only need to worry about the NH_4^+ ion.

We can disregard the base properties of ClO_4^-, HSO_4^-, Cl^-, and NO_3^- in aqueous solution. Their solutions are neutral, since they are conjugate bases of strong acids.

Sodium carbonate, Na_2CO_3, dissociates in water to give Na^+ and CO_3^{2-}. As in (a) above, we can neglect any contribution from Na^+. Carbonate, however, is a base according to Table 15.2.

$$CO_3^{2-}(aq) + H_2O(\ell) \rightleftharpoons HCO_3^-(aq) + OH^-(aq)$$

base | acid | conjugate acid | conjugate base

Therefore, the ammonium and carbonate ions can interact according to the equation

$$NH_4^+(aq) + CO_3^{2-}(aq) \rightleftharpoons NH_3(aq) + HCO_3^-(aq)$$

acid | base | conjugate base | conjugate acid

We can see from Table 15.2 that NH_4^+ is a stronger acid than HCO_3^- and that CO_3^{2-} is a stronger base than NH_3. Therefore, some reaction does indeed occur between the ammonium ion and the carbonate ion in water.

EXERCISE 15.2 PREDICTING THE DIRECTION OF PROTON TRANSFER REACTIONS

For each reaction below, predict whether the equilibrium lies predominantly to the left or to the right.

(a) $HSO_4^-(aq) + NH_3(aq) \rightleftharpoons NH_4^+(aq) + SO_4^{2-}(aq)$

(b) $HCO_3^-(aq) + HS^-(aq) \rightleftharpoons H_2S(aq) + CO_3^{2-}(aq)$

EXERCISE 15.3 WRITING ACID–BASE REACTIONS

If ammonium chloride and sodium sulfate are mixed in water, write a balanced equation for the acid–base reaction that could, in principle, occur. Does the reaction occur to any significant extent?

15.3 WATER AND THE pH SCALE

Stronger Brønsted acids and bases will lead to larger concentrations of $H^+(aq)$ or $OH^-(aq)$ for a given number of moles of substance. If these concentrations could be measured quantitatively, we would have a way to compare acid and base strengths and to predict more accurately reaction direction and extent. In this section we want to outline one way to measure these concentrations.

THE WATER IONIZATION CONSTANT, K_w

Water auto-ionizes, transferring a proton from one water molecule to another and producing a hydronium ion and a hydroxide ion.

$$H_2O(\ell) \rightleftharpoons H^+(aq) + OH^-(aq)$$

Although it is clear that H^+ in aqueous solution exists in the form of ions such as H_3O^+, we shall generally use the symbol H^+ from here on to keep chemical and mathematical equations as simple as possible.

Since the hydroxide ion is a much stronger base than water and the hydronium ion is a much stronger acid than water (Table 15.2), the equilibrium lies far to the left side. In pure water at 25°C only about two in 10^9 molecules is in the ionic form at any instant. To express this idea more quantitatively, we can write an equilibrium constant.

$$K = \frac{[H^+][OH^-]}{[H_2O]}$$

However, in dilute aqueous solutions (say 0.1 *M* solute or less) the concentration of solvent water can be considered to be a constant (55.4 *M*), so we include that in the constant *K* and write the equilibrium constant instead as

$$K[H_2O] = K_w$$

$$K_w = \text{Water Ionization Constant} = [H^+][OH^-]$$
$$= 1.008 \times 10^{-14} \text{ at } 25°C$$

The equation $K_w = [H^+][OH^-]$ is valid in pure water and in *any* aqueous solution.

This expression, and the value of the water auto-ionization constant, $K_w = 1.0 \times 10^{-14}$, are important and should be committed to memory.

In pure water, the transfer of a proton between two water molecules leads to one H^+ and one OH^-. Since this is the only source of these ions, we know that $[H^+]$ must equal $[OH^-]$ in pure water. This means that

$$[H^+] = [OH^-] = \sqrt{K_w} = \sqrt{1.0 \times 10^{-14}}$$

or

$$[H^+] = [OH^-] = 1.0 \times 10^{-7}\ M$$

The hydronium ion and hydroxide ion concentrations in pure water are both 10^{-7} *M* at 25°C, and the water is said to be **neutral**. In an **acidic solution**, however, the concentration of hydronium ion must be greater than 10^{-7} *M*. Similarly, in a **basic solution**, the concentration of OH^- must be greater than 10^{-7} *M*. Of course, since the product of the H^+ and OH^- concentrations must be equal to 10^{-14}, this means that a basic solution is also characterized by a hydronium ion concentration less than 10^{-7} *M*.

EXAMPLE 15.2

ION CONCENTRATIONS IN A SOLUTION OF STRONG BASE

If 0.0010 mole of NaOH is added to enough water to make 1.0 L of solution, what are the hydroxide and hydronium ion concentrations?

Solution NaOH is 100% dissociated into ions in water, so the initial concentration of OH^- is 0.0010 *M*. Before adding NaOH, the water initially contained a small concentration of OH^- (10^{-7} *M*) from the auto-ionization of the water. However,

$$H_2O(\ell) \rightleftharpoons H^+(aq) + OH^-(aq)$$

we ignore this small initial amount of OH^- from water and imagine that ionization of water occurs only *after* excess OH^- has been added. Thus, the net OH^- concentration = 0.0010 *M* + OH^- from water.

	$[H^+]$	$[OH^-]$
Concentration before H_2O ionization (*M*)	0	0.0010
Change when proceeding to equilibrium	$+x$	$+x$
Concentration after equilibrium is established (*M*)	x	$(0.0010 + x)$

We can solve for x from the expression relating hydronium and hydroxide ion to K_w.

$$K_w = [H^+][OH^-] = (x)(0.0010 + x) = 1.0 \times 10^{-14}$$

$$x^2 + 0.0010x - 1.0 \times 10^{-14} = 0$$

This is a quadratic expression of the form $ax^2 + bx + c = 0$ (where $a = 1$, $b = 0.0010$ and $c = -10^{-14}$), and it can be solved exactly for x using the quadratic formula (Appendix A). However, it is possible to make an approximation that will allow us to avoid this. Le Chatelier's principle tells us that the equilibrium is disturbed by adding extra OH^-, so the water auto-ionization reaction will shift to the left to decrease the extra OH^- concentration. Thus, the value of x, the concentration of H^+ and OH^- coming from water, must be smaller than 10^{-7} M. If a number smaller than 0.0000001 M (that is, 10^{-7}) is added to 0.0010 M, the sum is effectively 0.0010. Therefore, x in the term $(0.0010 + x)$ can be neglected, and we have the following approximate, but very nearly correct, expression.

$$K_w = 1.0 \times 10^{-14} \cong (x)(0.0010)$$

$$x = [H^+] \text{ in presence of } 0.0010\ M \text{ NaOH} \cong 1.0 \times 10^{-11}\ M$$

As a final step, we can check our approximation.

$$[H^+][OH^-] = (1.0 \times 10^{-11})(0.0010 + 1.0 \times 10^{-11}) \cong 1.0 \times 10^{-14}$$

The calculated ion product and the given K agree, so the approximation is valid.

EXERCISE 15.4 HYDROXIDE ION CONCENTRATION IN THE SOLUTION OF A STRONG ACID

Gaseous HCl (0.073 g) is bubbled into 5.0×10^2 mL of water to make an aqueous hydrochloric acid solution. What are the concentrations of H^+ and OH^- in this solution?

THE pH SCALE

Rather than express hydronium and hydroxide ion concentrations as very small numbers or as exponentials, it is more convenient to use the **pH**. The pH of a solution is defined as the negative of the base-10 logarithm (log) of the hydronium ion concentration,

$$\text{pH} = -\log\,[H^+]$$

In general, pX = −log[X].

In a similar way, the **pOH** of a solution is defined as the negative of the base-10 logarithm of the hydroxide ion concentration.

$$\text{pOH} = -\log\,[OH^-]$$

An alternative and useful form of these definitions is

$$[H^+] = 10^{-\text{pH}} \qquad [OH^-] = 10^{-\text{pOH}}$$

As you will soon see, these equations make it easy to convert pH or pOH values into hydronium or hydroxide ion concentrations, respectively.

In pure water, the hydronium and hydroxide ion concentrations are both 1.0×10^{-7} M. Therefore,

$$\text{pH} = -\log(1.0 \times 10^{-7}) = -[\log(1.0) + \log(10^{-7})]$$

$$\text{pH} = -[(0.00) + (-7)]$$

$$\text{pH} = 7.00 = \text{pH of pure water}$$

In the same way, you can show that the pOH of pure water is also 7.00.

If we take the logarithms of both sides of the expression $K_w = [H^+][OH^-]$, we obtain another useful expression.

$$K_w = [H^+][OH^-] = 1.0 \times 10^{-14}$$

$$-\log([H^+][OH^-]) = -\log(1.0 \times 10^{-14})$$

$$(-\log[H^+]) + (-\log[OH^-]) = 14.00$$

$$\textbf{pH + pOH = 14.00}$$

The sum of the pH and the pOH of a solution is equal to 14.00 at 25°C.

EXAMPLE 15.3

CALCULATING pH

The basic solution in Example 15.2 had $[OH^-] = 1.0 \times 10^{-3}$ *M* and $[H^+] = 1.0 \times 10^{-11}$ *M*. Calculate the pH and pOH of this solution.

Solution Let us begin by converting the hydrogen ion concentration to pH. Since $[H^+] = 1.0 \times 10^{-11}$ *M*, this means that

$$pH = -\log[H^+] = -\log(1.0 \times 10^{-11}) = -(-11.00) = 11.00$$

The pOH can be obtained quickly from the fact that the sum of the pH and the pOH must equal 14.00. Since the pH is 11.00, this must mean that the pOH is 3.00. We can prove that, however, by calculating the pOH directly from $[OH^-]$.

$$pOH = -\log[OH^-] = -\log(1.0 \times 10^{-3}) = -(-3.00) = 3.00$$

For a review of logarithms and significant figures see Appendix A.

As a footnote to this problem, notice the use of significant figures and logarithms. As outlined further in Appendix A, the mantissa of a logarithm (the numbers to the right of the decimal) has as many significant figures as the number whose log was found. Here we found the log of 1.0×10^{-11}, a number with two significant figures. Therefore, the log (11.00), must have two numbers to the right of the decimal (both zeros in this case).

We will assume 25°C in all of our calculations, unless specifically stated otherwise.

A 0.0040 *M* solution of HCl has a pH of 2.40 (Exercise 15.4), while a 0.0010 *M* NaOH solution has a pH of 11.00 (Example 15.3). Since the pH of pure water is 7.00 (at 25°C), we can say in general that *solutions with pH less than 7.00 are acidic, while solutions with pH greater than 7.00 are basic* (at 25°C). The relation between acidity, basicity, and pH or pOH can be shown graphically as follows:

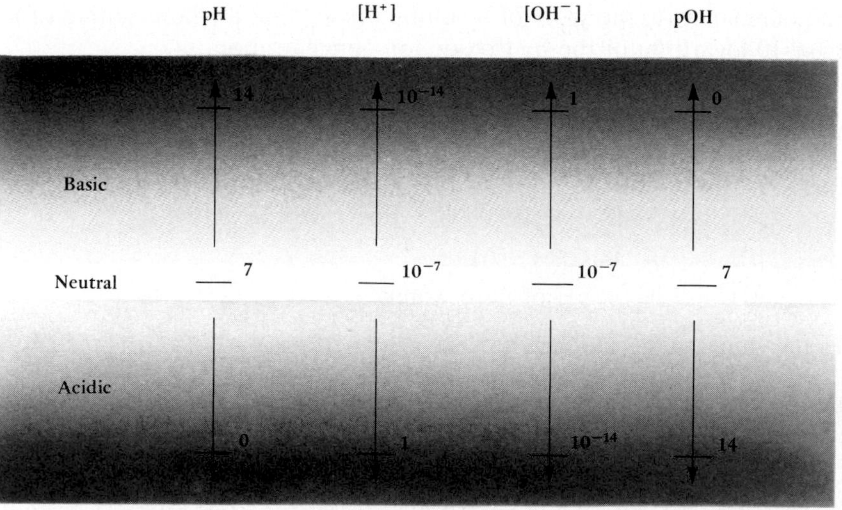

To give you a feeling for the pH scale, the approximate pH values for some common aqueous solutions are given in Figure 15.2.

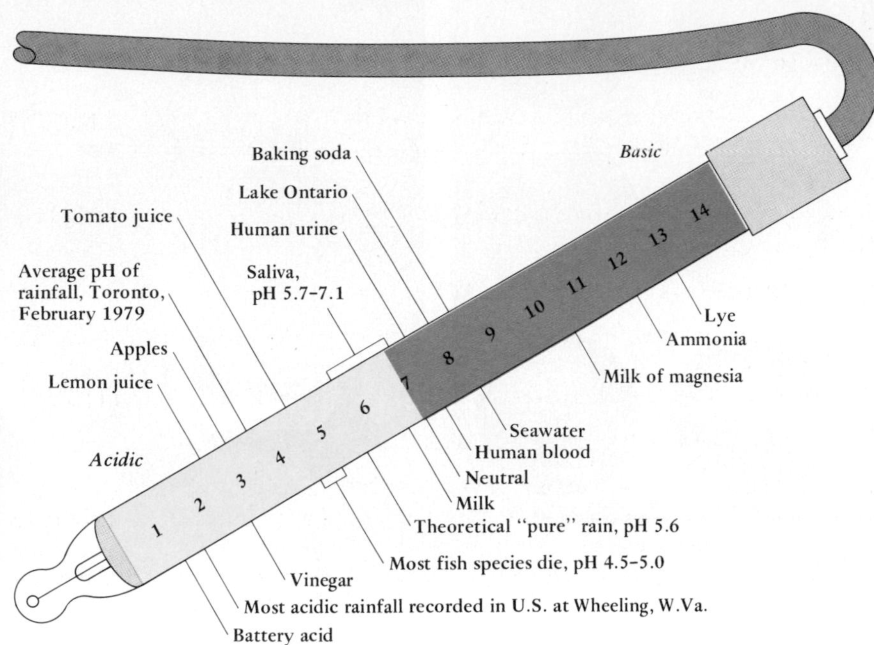

FIGURE 15.2

The pH of some common aqueous solutions. The scale is superimposed on the drawing of a pH electrode used in the measurement of pH by an instrumental method.

EXAMPLE 15.4

pH AND HYDRONIUM ION CONVERSIONS

(a) The $[H^+]$ of wine can be 1.6×10^{-3} *M*. Calculate its pH. (b) The $[H^+]$ of rain water in some instances has been found to be 5.0×10^{-5} *M*. Calculate its pH. (c) The pH of seawater is 8.30. Calculate $[H^+]$ and $[OH^-]$.

Solutions

(a) pH of the wine $= -\log[H^+] = -\log(1.6 \times 10^{-3}) = 2.80$

(b) pH of the rain $= -\log[H^+] = -\log(5.0 \times 10^{-5}) = 4.30$

(c) If the pH is 8.30, this means that

$$[H^+] = 10^{-pH} = 10^{-8.30} = 5.0 \times 10^{-9}\ M$$

There are two ways to find the $[OH^-]$. You can make use of the relation

$$K_w = [H_3O^+][OH^-] = 1.0 \times 10^{-14}$$

and calculate $[OH^-]$ from it.

$$(5.0 \times 10^{-9})[OH^-] = 1.0 \times 10^{-14}$$
$$[OH^-] = 2.0 \times 10^{-6}\ M$$

Alternatively, you can find the pOH from pH + pOH = 14, and then convert pOH to the $[OH^-]$. Try this to see if you do indeed obtain 2.0×10^{-6}.

EXERCISE 15.5 pH AND HYDRONIUM ION CONVERSION

The pH of blood is 7.30. What are the hydronium and hydroxide ion concentrations?

(a)

(b)

FIGURE 15.3

Indicators and pH. (a) A "universal indicator" in solutions of various pH values. Solutions in the top row have pH values of 1 through 4, those in the middle row have values 5 through 8, and those in the bottom row have pH's of 9 through 12. (b) Indicators in common products showing that vinegar is acidic with a pH of about 3, a carbonated beverage has a pH of 4 to 5, and a household cleaner is strongly basic. (Charles D. Winters)

pH INDICATORS

The pH of a solution may be determined chemically or instrumentally. The chemical method involves the use of an **indicator**, a substance that changes color in some known pH range (Figure 15.3). Recall that acids were originally defined by the fact they made certain vegetable dyes turn red, and indicators are still quite often large molecules derived from plants. These dyes or indicators can exist in conjugate acid and base forms, and they have different colors depending on the form in which they are found. Common litmus paper is impregnated with a natural plant juice that is red in solutions more acidic than approximately pH 5 but blue when the pH exceeds approximately 8.2 in a basic solution.

Although not derived from plants, phenolphthalein is a common acid–base indicator. In acid solution it is in its protonated form (H_2In) and is colorless. When the solution becomes basic in the pH interval 8.2

[Structure: Phenolphthalein (colorless) (aq) ⇌ $2\,H^+(aq)$ + conjugate base of phenolphthalein (red) (aq)]

Colorless

Phenolphthalein

Red

Conjugate base of phenolphthalein

$$H_2In(aq) \rightleftharpoons 2\,H^+(aq) + In^{2-}(aq)$$

to 9.8, two protons from the H_2In are lost in a reaction with the base, and the conjugate base (In^{2-}) is a beautiful pink (Figure 4.9).

As you can see in Table 15.3, there is a wide variety of indicators that can be used in many pH ranges. However, it can also be seen in

TABLE 15.3 Some Acid-Base Indicators

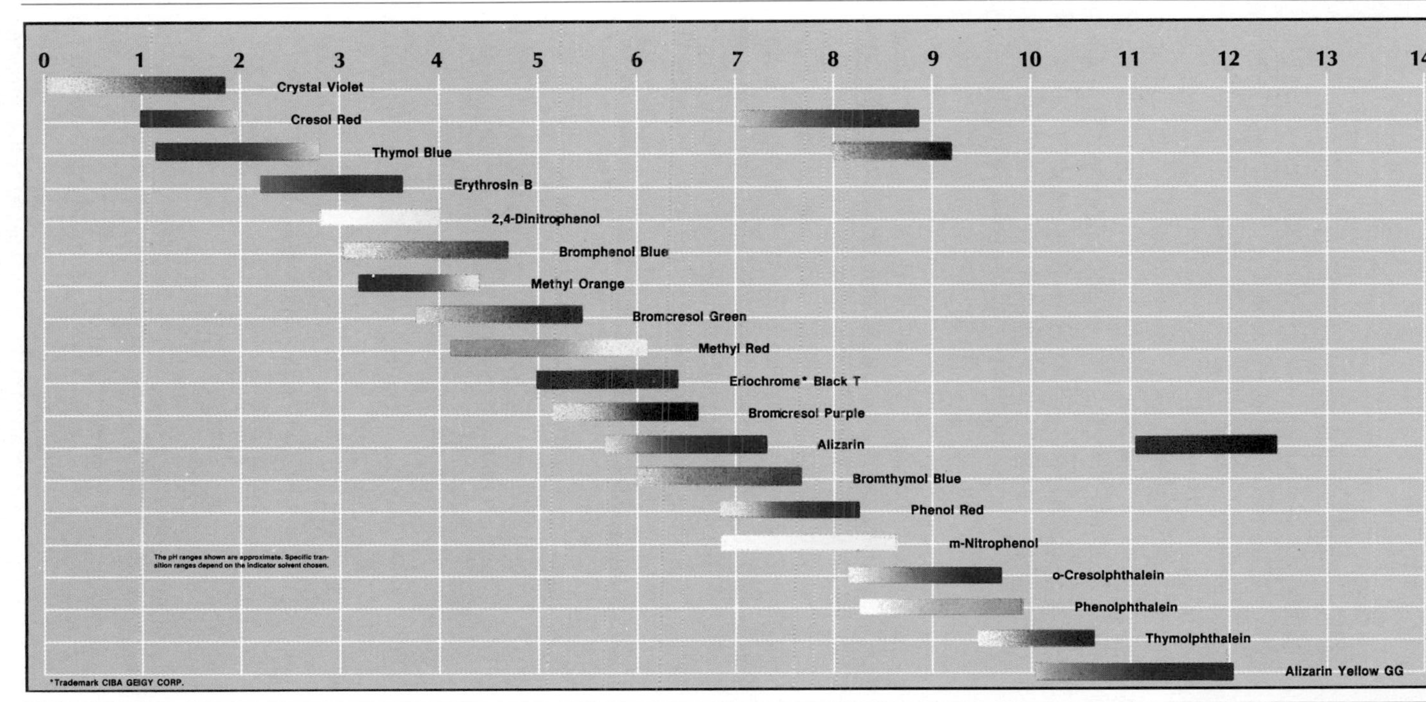

Hach Company

Table 15.3 that indicators change color over a *range* of pH values. If you wish a more accurate estimate of pH, a modern **pH meter** is far preferable (Figure 15.4).

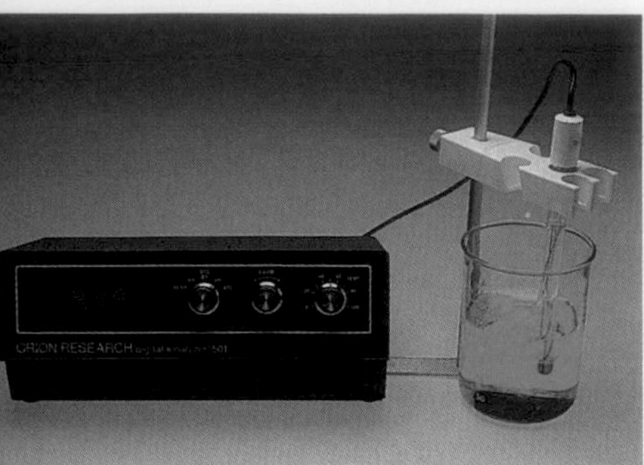

FIGURE 15.4

An electronic instrument for the measurement of pH. The blue solution is 0.10 *M* $CuSO_4$, and the pH of 4.5 illustrates the fact that many metal ions give weakly acidic solutions (Section 15.5). (Charles D. Winters)

15.4 STRONG ACIDS AND BASES

Some common acids are listed in Table 15.2 in descending order of their ability to donate a proton. *The hydronium ion is the strongest acid that can exist in water.* You will notice, however, that several acids are listed even higher in the table than H_3O^+. How can this be? When they are placed in water, they dissociate completely to H_3O^+ and their respective conjugate bases. Thus, for instance, HCl *is* a stronger acid than H_3O^+, but HCl does not exist as such *in water*; rather, H_3O^+ and Cl^- are present, and H_3O^+ is the strongest acid species in the solution. Acids that ionize 100% in solution are called **strong acids**, but they can be no stronger than H_3O^+ in water. This observation is called the **leveling effect**.

Some common strong acids:
- Hydrohalic acids: HCl, HBr, HI
- Nitric acid: HNO_3
- Sulfuric acid: H_2SO_4
- Perchloric acid: $HClO_4$

Some common strong bases:
- Group 1A hydroxides: LiOH, NaOH, KOH
- Group 2A hydroxides: $Mg(OH)_2$, $Ca(OH)_2$

Just as there are no acids that can be stronger in water than H_3O^+, no base can be stronger than OH^- in aqueous solution. In Table 15.2, **strong bases** are those species lower than OH^- in the right column: $C_2H_5O^-$, NH_2^-, CH_3NH^-, H^-, and CH_3^-, all of which react completely with water to produce OH^-. The strong bases that you commonly encounter, however, are hydroxides or oxides of Group 1A and 2A metals; with sodium hydroxide and potassium hydroxide perhaps most familiar to you. Although not nearly as soluble as NaOH in water, calcium hydroxide and magnesium hydroxide are among the cheapest bases known. Millions of tons of calcium oxide or *quicklime* are made annually by decomposing limestone, $CaCO_3$, at temperatures of 800°C to 1000°C (see Chapter 3).

$$CaCO_3(s) \longrightarrow CaO(s) + CO_2(g)$$

In a reaction characteristic of many metal oxides, CaO forms *slaked lime* or $Ca(OH)_2$ when mixed with water.

$$CaO(s) + H_2O(\ell) \longrightarrow Ca(OH)_2(s) \qquad \Delta H^\circ = -271 \text{ kJ/mol}$$

Over 10 million tons of NaOH or caustic soda are made every year, and even laboratory grade material costs less than \$10/kg. On the other hand, CsOH costs over \$600 per kilogram!

Clear, aqueous solutions of slaked lime are known as *limewater*. The poor solubility of $Ca(OH)_2$ and $Mg(OH)_2$ means that, if you mix a teaspoonful or so with water, it will be just a suspension of white powder. The suspension is basic, so $Mg(OH)_2$ is used as an antacid and is often called "milk of magnesia."

15.5 WEAK ACIDS AND BASES

Very few acids and bases strongly donate or accept protons, respectively. *The vast majority of acids and bases are weak.*

Perhaps the easiest way to tell if an acid or base is weak is to test the pH of an aqueous solution of the acid or base. A strong acid gives a pH value that is very close to the hydronium ion concentration expected on the basis of complete ionization. A 0.01 *M* solution of HCl has a pH of 2, while a 0.01 *M* solution of NaOH has pH of 12. On the other hand, a sample of vinegar, which is essentially a 0.8 *M* aqueous solution of the weak acid acetic acid, has a pH no less than 3 (Figure 15.5).

The relative strength of an acid or base can be expressed quantitatively with an equilibrium constant. For a general weak acid HA, for instance, we can write

$$HA(aq) \rightleftharpoons H^+(aq) + A^-(aq)$$

$$K_a = \frac{[H^+][A^-]}{[HA]}$$

where the equilibrium constant has a subscript "a" to indicate that it is for a weak acid. Similarly, we can write the equilibrium expression for a weak base (where *K* is labeled with a subscript *b*).

$$\text{Base(aq)} + H_2O(\ell) \rightleftharpoons \text{base } H^+(aq) + OH^-(aq)$$

$$K_b = \frac{[\text{base } H^+][OH^-]}{[\text{base}]}$$

FIGURE 15.5
Acids and bases (left to right): 0.10 *M* HCl, vinegar (acetic acid), 0.10 *M* NaOH. Indicator is thymol blue (see Table 15.3). (Charles D. Winters)

Now we can turn to two tasks. First, we want to see what species can be weak acids or bases and examine some typical values of *K*. Then, we can explore how one can determine K_a or K_b from experimental measurements and how to use the values in a practical way.

WEAK ACIDS

MOLECULAR ACIDS Molecular acids are neutral molecules that have one or more ionizable hydrogen atoms. A few are pictured below with their K_a values (and others are listed in Table 15.4). Notice that all of the K_a values are quite small, reflecting very little ionization of the acid in solution. Hydrocyanic and carbonic acids are quite weak, for example, while oxalic acid is only moderately weak, and its uses in the dye industry, as a paint remover, and so on, reflect this.

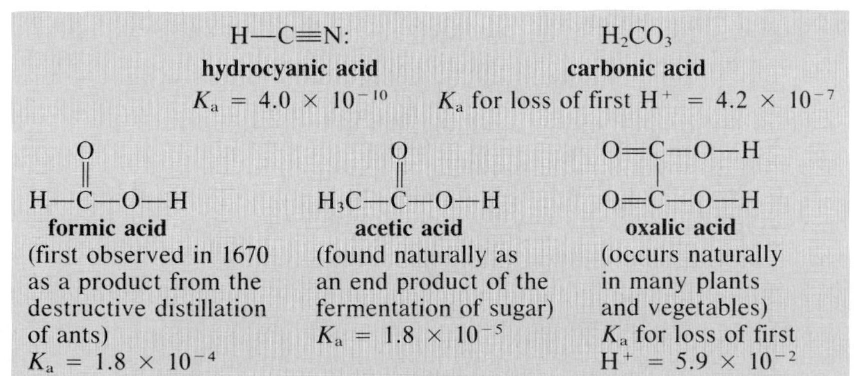

H—C≡N:
hydrocyanic acid
$K_a = 4.0 \times 10^{-10}$

H_2CO_3
carbonic acid
K_a for loss of first $H^+ = 4.2 \times 10^{-7}$

H—C(=O)—O—H
formic acid
(first observed in 1670 as a product from the destructive distillation of ants)
$K_a = 1.8 \times 10^{-4}$

H_3C—C(=O)—O—H
acetic acid
(found naturally as an end product of the fermentation of sugar)
$K_a = 1.8 \times 10^{-5}$

O=C—O—H
|
O=C—O—H
oxalic acid
(occurs naturally in many plants and vegetables)
K_a for loss of first $H^+ = 5.9 \times 10^{-2}$

ANIONS AS WEAK ACIDS Six anionic Brønsted acids are listed in Table 15.4. The dihydrogen phosphate anion, for example, is the acid in baking powder,

The reaction of $H_2PO_4^-$ and HCO_3^- to give CO_2 and HPO_4^{2-}. On the left is a solution of NaH_2PO_4 and on the right a solution of $NaHCO_3$. The bromcresol purple indicator shows the $H_2PO_4^-$ solution is acidic and the HCO_3^- solution is basic.

TABLE 15.4 Ionization Constants of Weak Acids and Bases

	ACID	K_a	BASE	K_b
Sulfurous acid	H_2SO_3	1.2×10^{-2}	HSO_3^-	8.3×10^{-13}
Hydrogen sulfate ion	HSO_4^-	1.2×10^{-2}	SO_4^{2-}	8.3×10^{-13}
Phosphoric acid	H_3PO_4	7.5×10^{-3}	$H_2PO_4^-$	1.3×10^{-12}
Hydrofluoric acid	HF	7.2×10^{-4}	F^-	1.4×10^{-11}
Nitrous acid	HNO_2	4.5×10^{-4}	NO_2^-	2.2×10^{-11}
Formic acid	$HCHO_2$	1.8×10^{-4}	CHO_2^-	5.6×10^{-11}
Benzoic acid	$HC_7H_5O_2$	6.3×10^{-5}	$C_7H_5O_2^-$	1.6×10^{-10}
Acetic acid*	$HC_2H_3O_2$	1.8×10^{-5}	$C_2H_3O_2^-$	5.6×10^{-10}
Propionic acid	$HC_3H_5O_2$	1.4×10^{-5}	$C_3H_5O_2^-$	7.1×10^{-10}
Carbonic acid	H_2CO_3	4.2×10^{-7}	HCO_3^-	2.4×10^{-8}
Hydrogen sulfide	H_2S	1×10^{-7}	HS^-	1×10^{-7}
Dihydrogen phosphate ion	$H_2PO_4^-$	6.2×10^{-8}	HPO_4^{2-}	1.6×10^{-7}
Hydrogen sulfite ion	HSO_3^-	6.2×10^{-8}	SO_3^{2-}	1.6×10^{-7}
Hypochlorous acid	HClO	3.5×10^{-8}	ClO^-	2.9×10^{-7}
Boric acid	H_3BO_3	7.3×10^{-10}	$H_2BO_3^-$	1.4×10^{-5}
Ammonium ion	NH_4^+	5.6×10^{-10}	NH_3	1.8×10^{-5}
Hydrocyanic acid	HCN	4.0×10^{-10}	CN^-	2.5×10^{-5}
Hydrogen carbonate ion	HCO_3^-	4.8×10^{-11}	CO_3^{2-}	2.1×10^{-4}
Hydrogen phosphate ion	HPO_4^{2-}	3.6×10^{-13}	PO_4^{3-}	0.028
Hydrogen sulfide ion	HS^-	1.3×10^{-13}	S^{2-}	0.077

↑ Increasing Acid Strength (upward); Increasing Basic Strength ↓ (downward)

For the acids: $HX(aq) \rightleftharpoons H^+(aq) + X^-(aq) \qquad K_a = \frac{[H^+][X^-]}{[HX]}$

For the bases: $X^-(aq) + H_2O \rightleftharpoons HX(aq) + OH^-(aq) \qquad K_b = \frac{[HX][OH^-]}{[X^-]}$

Notice that $K_aK_b = K_w$ for a given substance. See Box, page 581.

*The formula for acetic acid is often written as CH_3COOH and that for acetate ion as CH_3COO^-. Appropriate abbreviations are HOAc and OAc^-, respectively.

$$H_2PO_4^-(aq) \rightleftharpoons H^+(aq) + HPO_4^{2-}(aq) \qquad K_{a2} = 4.8 \times 10^{-11}$$

where its function is to provide H^+ to produce CO_2 from $NaHCO_3$.

$$H_2PO_4^-(aq) + HCO_3^-(aq) \longrightarrow CO_2(g) + H_2O(\ell) + HPO_4^{2-}(aq)$$

CATIONS AS WEAK ACIDS The ammonium ion is an excellent example of a cation acting as a weak Brønsted acid (Figure 15.6).

$$\underset{\text{acid}}{NH_4^+(aq)} + \underset{\text{base}}{H_2O(\ell)} \rightleftharpoons \underset{\text{conjugate base of } NH_4^+}{NH_3(aq)} + \underset{\text{conjugate acid of } H_2O}{H_3O^+(aq)}$$

$$K_a = 5.6 \times 10^{-10}$$

FIGURE 15.6

Weak acids (left to right): 0.10 *M* acetic acid, HSO_4^- (Na^+ salt), saturated CO_2 (which gives H_2CO_3), NH_4^+ (NO_3^- salt), and Cd^{2+} (NO_3^- salt). A universal indicator was used (see Figure 15.3). (Charles D. Winters)

It is clearly shown in Table 15.2 that the reactions of $H_2PO_4^-$ and NH_4^+ with water above would proceed to a stronger combination of acid and base. Therefore, these reactions occur only to a small extent, and the values of K_a are correspondingly small.

When the salt of a metal cation is placed in water, the metal ion becomes hydrated (Figure 15.6). This ion–water interaction is sufficiently strong that an aqueous metal ion is usually surrounded by six water molecules, so we represent a hydrated ion as $[M(H_2O)_6]^{n+}$. In some cases, particularly for ions such as Al^{3+} and transition metal ions with a charge of +2 and +3, the attraction between the ion and attached water molecules is so great that an O—H bond breaks in one or more of these water molecules, and the hydrogen appears in solution as H^+.

Metal ion hydration was described in Chapter 11.

$$[Fe^{3+} \longleftarrow :\ddot{O}(H)\text{—}H]^{3+}(aq) \longrightarrow [Fe^{3+} \longleftarrow :\ddot{O}\text{—}H^-]^{2+}(aq) + H^+(aq)$$

$$[Fe(H_2O)_6]^{3+}(aq) \rightleftharpoons [Fe(H_2O)_5(OH)]^{2+}(aq) + H^+(aq) \qquad K_a = 6.3 \times 10^{-3}$$

Such a reaction with water, in which an O—H bond is broken, is called a **hydrolysis reaction**, and such reactions have a definite impact on the chemistry of our environment.

Equilibrium constants have been determined for the hydrolysis of a number of cations, and a few values are given in Table 15.5. Notice that common ions such as Na^+, K^+, Mg^{2+}, and Ca^{2+} are not included; little if any hydrolysis of these hydrated ions takes place. Thus, for salts of these and other ions of Groups 1A and 2A, the metal ion does not contribute to the acidity of the solution. Since these are the important cations found in seawater, for example, the pH of the sea must be controlled by other species.

TABLE 15.5 Acid Ionization Constants for the Hydrolysis of Hydrated Metal Ions at 25°C

METAL ION	K_a
Al^{3+}	7.9×10^{-6}
Pb^{2+}	1.5×10^{-8}
Fe^{2+}	3.2×10^{-10}
Fe^{3+}	6.3×10^{-3}
Co^{2+}	1.3×10^{-9}
Ni^{2+}	2.5×10^{-11}
Cu^{2+}	1.6×10^{-7}

THE ACID IONIZATION CONSTANT, K_a The K_a values found in Table 15.4 and in a more extensive table in Appendix G were all determined experimentally. There are several ways to approach the problem, but the usual method is to determine the pH of the solution. The following example illustrates this method.

EXAMPLE 15.5

K_a FROM pH MEASUREMENT

Lactic acid is a monoprotic acid that occurs naturally in sour milk and arises from metabolism in the human body. When 0.10 mole of lactic acid, $HC_3H_5O_3$, is dissolved in enough water to make 1.0 L of solution, the pH of solution is found to be 2.43. What is the K_a of lactic acid?

Solution The equation for the equilibrium interaction of lactic acid with water is

$$\underset{\text{HLac}}{HC_3H_5O_3(aq)} \rightleftharpoons H^+(aq) + \underset{\text{Lac}^-}{C_3H_5O_3^-(aq)}$$

and the equilibrium constant expression is

$$K_a = \frac{[H^+][Lac^-]}{[HLac]}$$

The main piece of information we have to begin with is the pH. Converting this to $[H^+]$, you should find

$$[H^+] = 10^{-pH} = 10^{-2.43} = 3.7 \times 10^{-3}\ M$$

To solve for K_a we need to know equilibrium concentrations of each species. The following table is used to summarize the situation before and after equilibrium is attained.

	[HLac]	$[Lac^-]$	$[H^+]$
Before ionization (*M*)	0.10	0	10^{-7} (from the water)
Change when going to equilibrium	$-x$	$+x$	$+x$
After equilibrium (*M*)	$(0.10 - x)$	x	$(x + 10^{-7})$

In the table above, x is the concentration of lactate ion, Lac^-, *and* the concentration of hydronium ion, H^+, coming from lactic acid at equilibrium. Since we know from the pH that $[H^+]$ after equilibrium is established is $3.7 \times 10^{-3}\ M$, this must be equal to the H^+ produced by the acid *plus* that from water. However, Le Chatelier's principle informs us that the extra H^+ added to the water by lactic acid ($= x$) will cause the water auto-ionization equilibrium to shift toward water, thereby reducing $[H^+]$ from the water. Since total $[H^+] = 3.7 \times 10^{-3}\ M$, and since $[H^+]$ from water is less than 10^{-7}, the pH is almost totally a reflection of $[H^+]$ from lactic acid. Thus, without introducing a significant error, we can say that $[H^+]$ at equilibrium is simply x and that $x = 3.7 \times 10^{-3}\ M$. (For weak acids and bases, this approximation can *almost* always be made, but be careful and think through each case to be certain.) Now we know the equilibrium concentrations are

$$x = [H^+] = [Lac^-] = 3.7 \times 10^{-3}\ M$$

$$[HLac] = 0.10 - 0.0037 \cong 0.10\ M$$

and the K_a for lactic acid can be derived.

$$K_a = \frac{[H^+][Lac^-]}{[HLac]} = \frac{(3.7 \times 10^{-3})(3.7 \times 10^{-3})}{0.10}$$

$$K_a = 1.4 \times 10^{-4}$$

EXERCISE 15.6 K_a FROM pH

When 0.10 mol of propanoic acid is dissolved in sufficient water to give 1.0 L of solution, the pH of the solution is 2.94 after equilibrium has been established. Determine K_a for propanoic acid. The acid ionizes according to the balanced equation

$$H_3C-CH_2-COOH(aq) \rightleftharpoons H^+(aq) + H_3C-CH_2-COO^-(aq)$$

Values of K_a for many weak acids of all types have been measured. With these values, the pH of a solution of given concentration can be calculated, for example.

EXAMPLE 15.6

EQUILIBRIUM CONCENTRATIONS AND pH FROM K_a

Benzoic acid occurs free and in combined form in nature. For example, most berries contain up to 0.05% of the acid by weight.

$$\text{HC}_4\text{H}_4\text{C—COOH(aq)} \rightleftharpoons \text{HC}_4\text{H}_4\text{C—COO}^-\text{(aq)} + \text{H}^+\text{(aq)}$$

benzoic acid (HBz) — benzoate ion (Bz^-)

$$K_a = 6.3 \times 10^{-5}$$

If 0.020 mol of the acid is mixed with sufficient water to make 1.0 L of solution, calculate the pH of the solution.

Solution Organize the information in the problem in a table.

	[HBz]	$[H^+]$	$[Bz^-]$
Before ionization (M)	0.020	10^{-7} (from water)	0
Change when going to equilibrium	$-x$	$+x$	$+x$
After equilibrium is achieved (M)	$(0.020 - x)$	x	x

Every mole of benzoic acid that ionizes (x) produces 1 mole of benzoate ion (so $[Bz^-] = x$ at equilibrium) and 1 mole of hydronium ion (so $[H^+]$ from the acid ionization alone $= x$ at equilibrium). To simplify our work, without introducing a significant error, we shall say once again that $[H^+]$ is equal only to that coming from the weak acid; H^+ from water is neglected.

Substituting the concentrations of the acid and its ions into the K_a expression, we have

$$K_a = \frac{(x)(x)}{0.020 - x} = 6.3 \times 10^{-5}$$

You should recognize that this is a quadratic expression of the form $ax^2 + bx + c = 0$ (where $a = 1$, $b = K_a$, and $c = -0.020\ K_a$),

$$K_a(0.020 - x) = x^2$$

$$0.020\ K_a - K_a x = x^2$$

$$x^2 + K_a x - 0.020\ K_a = 0$$

an expression that can be solved with the quadratic formula (Appendix A). (Here x is found to be 0.0011 M when rounded to the correct number of significant figures). Although there are times (Example 15.7) when it is necessary to solve equilibrium expressions by the quadratic formula, much of the time this is not required. Equilibrium constants are not usually known to more than two significant figures. Unless the acid is relatively strong, and the original concentration of acid quite small, the quadratic equation and an approximate solution give identical answers to two significant figures. But how do you know when this will be the case? There is a useful rule of thumb described in the box below.

The approximate equilibrium expression for the acid HA

$$K_a = \frac{[H^+][A^-]}{[HA]_e} = \underbrace{\frac{[H^+]^2}{[HA]_0 - [H^+]}}_{\text{exact expression}} \cong \underbrace{\frac{[H^+]^2}{[HA]_0}}_{\text{approximate expression}}$$

$[HA]_e$ = equilibrium concentration of HA

$[HA]_0$ = initial concentration of HA

can be used when $[HA]_0 > 100\ K_a$; then $[HA]_e = [HA]_0 -$ amount of acid dissociated $\cong [HA]_0$. If $[HA]_0 < 100\ K_a$, then the exact expression form must be solved.

In the benzoic acid problem here, $100(K_a) = 6.3 \times 10^{-3}$. $[HA]_0$ (= 0.020 *M*) is greater than this, so the following approximation is valid.

$$K_a = \frac{[H^+]^2}{0.020} = 6.3 \times 10^{-5}$$

Solving this approximate expression, we find that $[H^+] = 1.1 \times 10^{-3}$ *M*, the *same* as the value from the quadratic equation. Now that $[H^+]$ is known, we also know that $[Bz^-] = 1.1 \times 10^{-3}$ *M* and $[HBz] \cong 0.10$ *M*. The last step is to calculate the pH, and here we find

$$pH = -\log[H^+] = -\log(1.1 \times 10^{-3}) = 2.96$$

As a final point notice that we are justified in not including the concentration of H^+ from water in our calculation. Benzoic acid supplies an H^+ concentration of 1.1×10^{-3} *M*, while that from water is less than 10^{-7} *M*.

EXAMPLE 15.7

pH FROM K_a

Formic acid was first obtained in 1670 as a product of the destructive distillation of ants, whose Latin genus name is *Formica*. It is a moderately weak monoprotic acid with a K_a of 1.8×10^{-4}.

$$HCOOH(aq) \rightleftharpoons H^+(aq) + HCOO^-(aq)$$

If you have a 0.0010 *M* solution of the acid, what is the pH of the solution? What is the concentration of the formic acid at equilibrium?

Solution The usual equilibrium-concentration table is

	[HCOOH]	$[HCOO^-]$	$[H^+]$
Before ionization (*M*)	0.0010	0	10^{-7}
Change on proceeding to equilibrium	$-x$	$+x$	$+x$
At equilibrium (*M*)	$(0.0010 - x)$	x	x

(Notice that we have already made the approximation, explained in Example 15.5, regarding the concentration of H^+ from water ionization at equilibrium.)

Substituting the equilibrium concentrations into the expression for K_a we have

$$K_a = \frac{[HCOO^-][H^+]}{[HCOOH]} = \frac{(x)(x)}{0.0010 - x} = 1.8 \times 10^{-4}$$

The approximation that $0.0010 - x \cong 0.0010$ *cannot* be made in this case, because $100(K_a) = 0.018$, a number larger than the initial concentration of the acid, $[HCOOH]_0$

(= 0.0010 *M*). Therefore, we must use the quadratic equation to solve the expression

$$ax^2 + bx + c = 0$$

$$x^2 + (1.8 \times 10^{-4})x + (-1.8 \times 10^{-4})(0.0010) = 0$$

Here $a = 1$, $b = 1.8 \times 10^{-4}$, and $c = -1.8 \times 10^{-7}$, and we apply the solution to a quadratic equation as follows:

$$x = \frac{-b \pm \sqrt{b^2 - 4ac}}{2a}$$

$$x = [H^+] = \frac{-(1.8 \times 10^{-4}) \pm \sqrt{(1.8 \times 10^{-4})^2 - 4(1)(-1.8 \times 10^{-7})}}{2(1)}$$

$$x = [H^+] = \frac{-1.8 \times 10^{-4} \pm 8.7 \times 10^{-4}}{2}$$

Since $[H^+]$ cannot have a negative value, the only valid root of the equation is $[(-1.8 + 8.7)10^{-4}]/2$ or

$$x = [H^+] = 3.4 \times 10^{-4}\ M$$

This means that

$$[HCOO^-] = [H^+] = x = 0.00034\ M$$

$$[HCOOH] = 0.0010 - x \cong 0.0007\ M$$

(If solved using the approximate expression $K = x^2/(0.0010)$, x would be equal to 0.00042 *M* or more than a 20% error.)

Finally, from the calculated $[H^+]$ we find that the pH is

$$pH = -\log(3.4 \times 10^{-4}) = 3.47$$

a pH indicating that the solution is acid, as expected.

EXERCISE 15.7 EQUILIBRIUM CONCENTRATIONS FROM K_a

What are the equilibrium concentrations of acetic acid, the acetate ion, and H^+ when a 0.10 *M* solution of acetic acid ($K_a = 1.8 \times 10^{-5}$) is allowed to come to equilibrium?

The goals of this section were (1) to see what types of substances could behave as weak acids and (2) to develop methods of treating their ionization quantitatively. The mathematical approach derived to this point can be summarized as follows. For the weak acid HA

$$HA(aq) \rightleftharpoons H^+(aq) + A^-(aq)$$

the equilibrium expression, in terms of the equilibrium concentrations of HA, H^+, and A^-, is

$$K_a = \frac{[H^+][A^-]}{[HA]}$$

If the original concentration of HA is $[HA]_0$, then the equilibrium concentration of HA must be $[HA]_0 - [H^+]$. Therefore, we can write K_a in a form that will allow us to solve for $[H^+]$ if we know $[HA]_0$ (and ignore the H^+ that arises from H_2O auto-ionization).

$$K_a = \frac{[H^+]^2}{[HA]_0 - [H^+]}$$

FIGURE 15.7
Weak bases (left to right): tap water, 0.10 *M* acetate ion (Na^+ salt), 0.10 *M* CO_3^{2-} (Na^+ salt), 0.30 *M* PO_4^{3-} (Na^+ salt), and 0.10 *M* NH_3. A universal indicator was used (see Figure 15.3).

To solve for $[H^+]$ generally requires the use of the quadratic equation. However, this can be avoided if we can make the approximation that $[HA]_0 - [H^+] \cong [HA]_0$. Such an approximation can be made if $[HA]_0$ is much larger than $[H^+]$, a situation that occurs when $[HA]_0$ is at least 100 times larger than K_a. Under these circumstances

$$[H^+] \cong \sqrt{[HA]_0 K_a}$$

WEAK BASES

A variety of weak bases are important in chemistry, and a few are listed in Table 15.4 in order of increasing base strength. Each is listed with its value of K_b at the far right of the table.

Approximately 15 to 20 million tons of ammonia are made annually by the direct combination of hydrogen with atmospheric nitrogen. The major uses of ammonia are in the manufacture of nitric acid and in the production of solid fertilizers such as urea ($H_2N—CO—NH_2$) and ammonium nitrate.

MOLECULAR BASES Ammonia, NH_3, which is perhaps the best known weak base, produces a very small amount of hydroxide ion when accepting a proton from water (Figure 15.7).

$$NH_3(aq) + H_2O(\ell) \rightleftharpoons NH_4^+(aq) + OH^-(aq) \qquad K_b = 1.8 \times 10^{-5}$$

The compound is also the parent of a large series of derivatives called *amines*, where one or more of the H atoms of NH_3 is replaced by some other substituent. For example, if the substituent is the methyl group, CH_3, we have

$H_3C—\ddot{N}—H$ (with H on N)	$H_3C—\ddot{N}—CH_3$ (with H on N)	$H_3C—\ddot{N}—CH_3$ (with CH_3 on N)
methylamine	dimethylamine	trimethylamine
$K_b = 5.0 \times 10^{-4}$	$K_b = 7.4 \times 10^{-4}$	$K_b = 7.4 \times 10^{-4}$

Pyridine, the compound in Example 15.8, is yet another example of a weak base, and others include nicotine ($C_{10}H_{14}N_2$), caffeine ($C_8H_{10}N_4O_2$), and the antimalarial compound quinine ($C_{20}H_{24}N_2O_2$).

EXAMPLE 15.8

THE pH OF A SOLUTION OF A WEAK BASE

Pyridine, a common weak base, was discovered in coal tar in 1846. Its ionization constant, K_b, is 1.5×10^{-9}.

$$\underset{\text{weak base = pyridine (py)}}{C_5H_5N\text{:}(aq)} + H_2O(\ell) \rightleftharpoons OH^-(aq) + \underset{\text{conjugate acid (pyH}^+\text{)}}{C_5H_5NH^+(aq)}$$

$$K_b = 1.5 \times 10^{-9}$$

If you have dissolved 0.010 moles of pyridine (py) in enough water to make 1.0 L of solution, what is the pH of the solution?

Solution The equilibrium expression for this weak base can be written as

$$K_b = 1.5 \times 10^{-9} = \frac{[pyH^+][OH^-]}{[py]}$$

and the concentrations of the base and ionic products are

	[py]	$[pyH^+]$	$[OH^-]$
Before ionization (*M*)	0.010	0	0
Change in concentration when moving to equilibrium	$-x$	$+x$	$+x$
At equilibrium (*M*)	$(0.010 - x)$	x	x

If the expressions for the equilibrium concentrations are substituted into the K_b expression, we have

$$K_b = 1.5 \times 10^{-9} = \frac{(x)(x)}{0.010 - x}$$

There are three points to be noticed here.

1. Hydroxide ion arises not only from the weak base but also from the autoionization of water. Since the latter source provides so little, we have made the reasonable approximation that $[OH^-]$ is simply equal to the hydroxide arising from the weak base. This is analogous to the approximation used with a weak acid (Example 15.5).
2. The expression for K_b is analogous to the general expression used for weak acids,

$$K_b = \frac{[OH^-]^2}{[base]_0 - [OH^-]}$$

where $[base]_0$ is the initial concentration of base.

3. The expression above can be solved exactly for $[OH^-]$ if the quadratic expression is used. However, the denominator $[base]_0 - [OH^-]$ can be reduced to $[base]_0$ if $[OH^-] << [base]_0$. The same "rule of thumb" given in Example 15.6 can be applied to a base: if $100(K_b)$ is less than $[base]_0$, then the denominator is just $[base]_0$. Here $100(K_b) = 1.5 \times 10^{-7}$, a value much smaller than $[base]_0$ (0.010 *M*), so the approximation is valid.

Making the approximations outlined above, we have

$$K_b = \frac{[OH]^2}{[base]_0} = \frac{[OH]^2}{0.010}$$

$$[OH^-] = 3.9 \times 10^{-6}\ M.$$

Now we see that our approximation was valid, since $0.010 - 0.0000039 \cong 0.010\ M$.

Based on the calculated OH^- concentration, the pOH is

$$pOH = -\log[OH^-] = -\log(3.9 \times 10^{-6}) = 5.41$$

The pH can be obtained from the fact that the sum of pOH and pH must be 14.00. Thus, the pH is 8.59, indicating only a very weakly basic solution.

ANIONS AS WEAK BASES You have already seen some examples of anions acting as Brønsted bases in aqueous solution (Figure 15.7). For example, ionization of the strong Brønsted acid HCl in water produces chloride ion as the conjugate base.

$$\underset{\text{acid}}{HCl(aq)} + H_2O(\ell) \rightleftharpoons H_3O^+(aq) + \underset{\text{conjugate base of HCl}}{Cl^-(aq)}$$

Similarly, the weak acid HCN gives a conjugate base, the cyanide ion, in water.

$$\underset{\text{acid}}{HCN(aq)} + H_2O(\ell) \rightleftharpoons H_3O^+(aq) + \underset{\text{conjugate base of HCN}}{CN^-(aq)}$$

Since both Cl^- and CN^- are Brønsted bases, aqueous solutions of their salts may be basic. However, in Example 15.1 we argued that Cl^- does not react with water to a measurable extent, since it is the conjugate base of a strong acid.

According to the information in Table 15.2, HCl is a much stronger acid than H_2O, so the reaction of Cl^- with H_2O would have to go from a weak (H_2O and Cl^-) to a strong (HCl and OH^-) combination of acid and base.

In contrast to Cl^-, the cyanide ion, CN^-, hydrolyzes in water to produce a measurable concentration of hydroxide ion,

$$\underset{\text{base}}{CN^-(aq)} + \underset{\text{acid}}{H_2O(\ell)} \rightleftharpoons \underset{\text{conjugate acid of } CN^-}{HCN(aq)} + \underset{\text{conjugate base of water}}{OH^-(aq)}$$

and the value of the equilibrium constant for the hydrolysis can be determined.

$$K_b = \frac{[HCN][OH^-]}{[CN^-]} = 2.5 \times 10^{-5}$$

Like Cl^-, cyanide ion is a weaker base than OH^-, so the equilibrium is predicted to lie to the left. However, according to Table 15.2, CN^- is closer in base strength to OH^- than is Cl^-. Therefore, we can guess that CN^- will react with water to a small extent, a guess confirmed by the small value of K_b (Table 15.4).

Based on the examples of chloride ion and cyanide ion, the following general statements can be made:

(a) When the salt containing the anion of a strong acid (HCl, HNO_3, $HClO_4$) is dissolved in water, hydrolysis of the anion to give OH^- does not occur to a measurable extent.

(b) The anion or conjugate base, A^-, of any weak acid HA will hydrolyze to produce a measurable concentration of hydroxide ion (Figure 15.7).

$$A^-(aq) + H_2O(\ell) \rightleftharpoons HA(aq) + OH^-(aq)$$

Values of K_b are given in Table 15.4 for the conjugate bases of common weak acids and the following example shows that such anionic bases can produce solutions with pH greater than 7.

The value of K_a for a weak acid and K_b for its conjugate base are related by the expression $K_aK_b = K_w$ (see box, page 581). This expression tells us that the stronger the acid the weaker the conjugate base, an observation that is confirmed in Tables 15.2 and 15.4.

EXAMPLE 15.9

pH OF THE SOLUTION OF AN ANIONIC BASE

If 0.010 mole of sodium acetate, CH_3COONa, is added to enough water to make 1.0 L of solution, what is the pH of the solution?

Solution According to the solubility guidelines (Table 3.1) sodium acetate is a very soluble salt. It is also an electrolyte, dissociating completely into sodium and acetate ions in solution. As noted previously, the sodium ion does not hydrolyze to a measurable extent. However, the acetate ion is the conjugate base of a weak acid, so it is this species in which we are interested. The equilibrium reaction of the acetate ion is

$$\underset{\text{acetate ion (OAc}^-\text{) weak base}}{CH_3COO^-(aq)} + H_2O(\ell) \rightleftharpoons \underset{\text{acetic acid (HOAc) conjugate acid}}{CH_3COOH(aq)} + OH^-(aq)$$

and the base ionization constant expression is (where K_b is given in Table 15.4)

$$K_b = \frac{[HOAc][OH^-]}{[OAc^-]} = 5.6 \times 10^{-10}$$

The concentrations of acetate ion, acetic acid, and hydroxide initially and at equilibrium are

	$[OAc^-]$	$[HOAc]$	$[OH^-]$
Before ionization (M)	0.010	0	10^{-7}
Change in concentration when moving to equilibrium	$-x$	$+x$	$+x$
At equilibrium (M)	$(0.010 - x)$	x	x

The approximations outlined in Example 15.6 are made in this case. Thus, the K_b expression reduces to

The approximation that $[\text{base}]_0 - [OH^-] \cong [\text{base}]_0$ is *almost* always valid for anionic weak bases, but you should think through every situation to make certain.

$$K_b = 5.6 \times 10^{-10} = \frac{[OH^-]^2}{0.010}$$

and $[OH^-]$ is found to be 2.4×10^{-6} M. Solving for $[H^+]$ we find

$$K_w = 1.0 \times 10^{-14} = [H^+][OH^-] = [H^+](2.4 \times 10^{-6})$$

$$[H^+] = 4.2 \times 10^{-9}\ M$$

This is indeed a small value, so the approximation that $0.010 - x \cong 0.010$ is valid.

When the hydrogen ion concentration is converted to pH, we find

$$pH = -\log[H^+] = -\log(4.2 \times 10^{-9}) = 8.38$$

EXERCISE 15.8 WEAK BASES AND pH

Calculate the pH of 0.010 M solutions of (a) NH_3 and (b) NaCN. The K_b values are listed in Table 15.4.

ACID–BASE PROPERTIES OF SALTS

Many of the compounds in nature and in the laboratory are salts, and many produce acidic or basic solutions upon hydrolysis in water (Figures 15.6 and 15.7). For example, you have already seen that (a) the ammonium ion and some metals ions (most notably ions such as Al^{3+}, Pb^{2+}, and transition metal ions of +2 and +3 charge) provide a measurable amount of hydronium ion in aqueous solution and (b) anions (A^-) of weak acids (HA) produce basic solutions. These conclusions are sufficiently general and of such importance that they are summarized in Table 15.6, and K_a and K_b values for conjugate acid–base pairs are given in Table 15.4.

K_b for the conjugate base of a weak acid is often called its **hydrolysis constant**, K_h. Its value is connected in a simple way to K_a for the weak acid. See the boxed material for an explanation.

Table 15.6 is valuable because you can use this information to predict, for example, that

1. A salt such as $NaNO_3$ gives a neutral solution since Na^+ does not hydrolyze to an appreciable extent and since NO_3^- is the conjugate base of a strong acid.
2. An aqueous solution of K_2S should be basic since S^{2-} is the conjugate base of the very weak acid HS^- (K_b for S^{2-} is 0.077 in Table 15.4), while K^+ does not hydrolyze appreciably.
3. An aqueous solution of $FeCl_2$ should be weakly acidic, since Fe^{2+} hydrolyzes to give an acidic solution (K_a for Fe^{2+} is 3.2×10^{-10} in Table 15.5), while Cl^- is the conjugate base of the strong acid HCl.

Some additional explanation is needed concerning salts of anions such as HCO_3^- and HS^-. Since they have an ionizable hydrogen, they can in principle act as acids.

$$\underset{\text{acid}}{HS^-(aq)} \rightleftharpoons S^{2-}(aq) + H^+(aq) \qquad K_a = 1.3 \times 10^{-13}$$

However, keep in mind that they are also the conjugate bases of weak acids.

$$\underset{\text{base}}{HS^-(aq)} + H_2O(\ell) \rightleftharpoons H_2S(aq) + OH^-(aq) \qquad K_b = 1 \text{ x } 10^{-7}$$

If you are curious about the pH of a solution of a salt where both the cation and the anion contribute to the acidity of the solution (for example, ammonium acetate), see Study Question 110.

Whether the solution is acidic or basic will depend on the relative size of K_a and K_b. In the case of hydrogen sulfide, K_b is much larger than K_a, so $[OH^-]$ in a solution of a hydrosulfate salt is much larger than $[H^+]$. Therefore, a solution of NaHS, for example, will be basic.

TABLE 15.6 Acid–Base Properties of Some Common Ions in Water Solution

	NEUTRAL		BASIC		ACIDIC
Anion	Cl^- Br^- I^-	NO_3^- ClO_4^- SO_4^{2-}	$C_2H_3O_2^-$ F^- CO_3^{2-} S^{2-} PO_4^{3-}	CN^- NO_2^- HCO_3^- HS^- HPO_4^{2-}	HSO_4^-
Cation	Li^+ Na^+ K^+	Mg^{2+} Ca^{2+} Ba^{2+}	None		Al^{3+} NH_4^+ Transition metal ions

DERVIATION OF HYDROLYSIS CONSTANTS, K_h, FOR ANIONS AND CATIONS

Acetate ion, OAc^-, is the conjugate base of acetic acid, so the equilibrium constant for reaction of OAc^- with water has been designated as K_b, a base ionization constant. However, this K_b is often called the *hydrolysis constant* for acetate ion and is given the symbol K_h. If K_a for a weak acid is known, then K_b ($= K_h$) for its conjugate base can be derived. To see how this is done, notice what is achieved when the ionization reaction of HOAc and the hydrolysis reaction of OAc^- are added together.

$$HOAc(aq) \rightleftharpoons OAc^-(aq) + H^+(aq) \qquad K_a = \frac{[OAc^-][H^+]}{[HOAc]} = 1.8 \times 10^{-5}$$

$$OAc^-(aq) + H_2O(\ell) \rightleftharpoons HOAc(aq) + OH^-(aq) \qquad K_b = \frac{[HOAc][OH^-]}{[OAc^-]} = 5.6 \times 10^{-10}$$

$$H_2O(\ell) \rightleftharpoons H^+(aq) + OH^-(aq) \qquad K_w = K_aK_b = [H^+][OH^-] = 1.0 \times 10^{-14}$$

The result is simply the equation for the auto-ionization of water, and we see that K_w is just the product of K_a for acetic acid and K_b for its conjugate base, the acetate ion. Indeed, it is this relation that gives the values of K_b ($= K_h$) for the anions listed in Table 15.4. For example, K_a for acetic acid is known to be 1.8×10^{-5}, so

$$K_b \text{ for acetate, the conjugate base of acetic acid} = \frac{K_w}{K_a} = \frac{1.0 \times 10^{-14}}{1.8 \times 10^{-5}}$$

$$K_b = K_h = 5.6 \times 10^{-10}$$

The relation $K_w = K_aK_b$ also holds for a weak base and its conjugate acid. For example, K_a ($= K_h$) for the NH_4^+ ion is

$$K_a \text{ for } NH_4^+ = K_h - \frac{K_w}{K_b \text{ for } NH_3} = \frac{1.0 \times 10^{-14}}{1.8 \times 10^{-5}} = 5.6 \times 10^{-10}$$

EXERCISE 15.9 PREDICTING THE pH OF SOLUTIONS

Predict for each salt in aqueous solution whether the pH will be greater than, less than, or equal to 7.

(a) NaCl
(b) $FeCl_3$
(c) NH_4NO_3
(d) $NaHCO_3$

POLYPROTIC ACIDS

A few of the more important of the acids capable of donating more than one proton are listed in the margin. In every case the loss of each successive proton becomes more difficult, as indicated by the K_a values of Table 15.4 (and Appendix G). For example, sulfuric acid is listed above H_3O^+ in Table 15.2 and so is a strong acid for the loss of one proton in water.

Polyprotic Acids
Phosphoric (H_3PO_4)
Sulfuric (H_2SO_4)
Hydrosulfuric (H_2S)
Oxalic ($H_2C_2O_4$)
Carbonic (H_2CO_3)

$$\underset{\text{strong acid}}{H_2SO_4(aq)} \xrightarrow{\approx 100\% \text{ ionization}} \underset{\text{conjugate base}}{HSO_4^-(aq)} + H^+(aq)$$

To lose the second proton, however, is more difficult as shown by the much smaller K_a value of the hydrogen sulfate ion (Figure 15.6).

$$HSO_4^-(aq) \xrightleftharpoons[]{\approx 3\% \text{ ionization for } 0.1\ M \text{ solution}} SO_4^{2-}(aq) + H^+(aq)$$

weak acid — conjugate base

Carbonic acid, an important acid in our environment, behaves somewhat differently than sulfuric acid. Here, *both* protons are lost with difficulty, the first more easily than the second.

$$H_2CO_3(aq) \rightleftharpoons HCO_3^-(aq) + H^+(aq) \qquad K_{a1} = 4.2 \times 10^{-7}$$

$$HCO_3^-(aq) \rightleftharpoons CO_3^{2-}(aq) + H^+(aq) \qquad K_{a2} = 4.8 \times 10^{-11}$$

One reason for the difference in K_a values is that a negative ion is produced when H^+ is lost by an acid. To lose another positive ion (H^+) from the conjugate base anion is increasingly difficult because the ion must accommodate a larger and larger negative charge.

Notice the interesting analogy here between increasing difficulty of proton loss for polyprotic acids and ionization energies of atoms, where each successive loss of an electron requires more energy.

For many acids, as for carbonic acid, each successive loss of a proton is about 10^4 to 10^6 times more difficult than the previous ionization step. This means that the first ionization step of a polyprotic acid produces up to about a million times more H^+ than the second step. Hydrosulfuric acid, the acid associated with the smell of rotten eggs, is an example.

$$H_2S(aq) \rightleftharpoons HS^-(aq) + H^+(aq) \qquad K_{a1} = 1.0 \times 10^{-7}$$

$$HS^-(aq) \rightleftharpoons S^{2-}(aq) + H^+(aq) \qquad K_{a2} = 1.3 \times 10^{-13}$$

For solutions of H_2S, and many other inorganic polyprotic acids, *the pH of the solution depends primarily on the hydronium ion generated in the first ionization step*; hydronium ion from the second step can be neglected.

EXAMPLE 15.10

pH OF A POLYPROTIC ACID SOLUTION

Hydrogen sulfide can dissolve in water to the extent of approximately 0.10 *M*. What is the pH of such a solution? What are the equilibrium concentrations of $[H_2S]$, $[HS^-]$, and $[S^{2-}]$?

Solution The aqueous equilibria of H_2S are written in the text above, and we said that the equilibrium value of $[H^+]$ depends *almost* completely on the hydronium ion generated in the first step. So, let us assume first that *only* the first step occurs,

$$H_2S(aq) \rightleftharpoons HS^-(aq) + H^+(aq)$$

and calculate the resulting concentrations.

	$[H_2S]$	$[HS^-]$	$[H^+]$
Before ionization (*M*)	0.10	0	10^{-7}
Change in concentration when moving to equilibrium	$-x$	$+x$	$+x$
At equilibrium (*M*)	$(0.10 - x)$	x	x

The concentration of H^+ (x) can be derived from the expression

$$K_{a1} = 1.0 \times 10^{-7} = \frac{[HS^-][H^+]}{[H_2S]} = \frac{(x)(x)}{0.10 - x}$$

From Example 15.6, you know that we can make the approximation that $0.10 - x \cong 0.10$ [since $100(K_{a1}) = 10^{-5}$ and $0.10 > 10^{-5}$]. Therefore,

$$[H^+] = [HS^-] = x = \sqrt{(0.10)K_{a1}} = 1.0 \times 10^{-4}\ M$$

(and we see the approximation is valid because $0.010 - 0.0001 \cong 0.010$). From this, we can obtain the pH,

$$pH = -\log(1.0 \times 10^{-4}) = 4.00$$

and the concentration of the acid H_2S.

$$[H_2S] = 0.10 - 0.0001 \cong 0.10\ M$$

Next, we turn to the problem of obtaining the concentration of the sulfide ion, S^{2-}, a product of the second ionization step.

$$HS^-(aq) \rightleftharpoons S^{2-}(aq) + H^+(aq)$$

	$[HS^-]$	$[S^{2-}]$	$[H^+]$
Before ionization (M)	1.0×10^{-4}	0	1.0×10^{-4}
Change in concentration when moving to equilibrium	$-y$	$+y$	$+y$
At equilibrium (M)	$(1.0 \times 10^{-4} - y)$	y	$(1.0 \times 10^{-4} + y)$

We have previously noted that the second ionization step occurs to a very small extent. This means that y, the amount of S^{2-} and H^+ produced in the second step, is *much* smaller than 10^{-4}. Here $[HS^-]$ before ionization $= 10^{-4}\ M$, and $100(K_{a2}) \cong 10^{-12}$. Thus, we can write the following approximate expression.

$$K_{a2} = \frac{[H^+][S^{2-}]}{[HS^-]} = 1.3 \times 10^{-13} = \frac{(1.0 \times 10^{-4})(y)}{1.0 \times 10^{-4}}$$

Since $[H^+]$ and $[HS^-]$ have nearly identical values, they cancel from the expression, and we find that $[S^{2-}]$ is simply equal to K_{a2}.

$$y = [S^{2-}] = K_{a2} = 1.3 \times 10^{-13}$$

EXERCISE 15.10 POLYPROTIC ACIDS

Suppose you have a 0.10 M solution of oxalic acid, $H_2C_2O_4$. What is the pH of the solution? What is the concentration of the oxalate ion, $C_2O_4^{2-}$?

Example 15.10 points to a general conclusion. For polyprotic acids, H_2A, where K_{a1} and K_{a2} are different by 10^3 or more, it is useful approximation to say that

(a) $[H^+]$ comes from the first ionization step and is given by $[H^+] \cong \sqrt{[H_2A]_0 K_{a1}}$

(b) $[HA^-] = [H^+]$

(c) $[A^{2-}] = K_{a2}$

15.6 EFFECTS OF STRUCTURE AND BONDING ON ACID STRENGTH

One of the most interesting aspects of chemistry is the correlation between molecular structure, bonding, and observed properties. Here it is useful to analyze the connection between the structure and bonding of some acids and their relative strengths.

When an acid HA dissociates in water, we can think of the process as the sum of a series of steps. (For simplicity, we shall ignore the solvation of the individual species, since this can have *about* the same effect in analogous cases.)

(a) *H—A bond breaking*. (The bond breaks *homolytically*; one bonding electron is retained by each partner.)

$$H—A \longrightarrow H\cdot + \cdot A$$

(b) *Loss of an electron by H to form H^+.*

$$H\cdot \longrightarrow H^+ + e^-$$

(c) *Gain of electron by A to form A^-.*

$$A\cdot + e^- \longrightarrow A:^-$$

Since we are only interested in *relative* acidities, we can simplify our analysis still further by ignoring the second step, one common to all acid dissociations. Thus, to compare the relative strengths of several acids we get some insight into the problem by comparing (a) the H—A bond strengths in the acids and (c) the relative electron affinities of the fragment A. In general, the more easily the H—A bond is broken and the greater the electron affinity of A, the greater the relative strength of the acid. These two effects can work together (a weak H—A bond and a high A electron affinity) to lead to a strong acid. However, it is often observed that they work in opposite directions (a weak H—A bond but a low electron affinity for A), so the balance between the effects controls acidity.

One of the first series of acids to consider is the binary hydrogen halides, HA. Their relative acid strengths are known to be in the order

— Increasing acid strength →

	HF <	HCl <	HBr <	HI
H—A bond strength (kJ/mol)	569	431	368	297
Electron affinity of A (kJ/mol)	−322	−348	−324	−295

Here it is evident that H—A bond strengths largely control the relative acidities of these compounds. The strongest acid, HI, has the weakest H—X bond. If electron affinity of A were the controlling factor, then HCl would have been the strongest acid.

Another example of the change in acidity for a series of analogous acids in one periodic group is H_2A where A is a Group 6A element.

— Increasing acid strength →

	H_2O <	H_2S <	H_2Se <	H_2Te
K_{a1} =	1×10^{-14}	1×10^{-7}	1.3×10^{-4}	2.3×10^{-3}

Since the H-A bond energies decrease in the order H—O > H—S > H—Se > H—Te, while element electron affinities increase in the series Te < S < O, it is apparent that H—A bond energies again control the acidity of this series. Thus, it would appear generally true that *analogous acids of the elements of a periodic group will increase in strength on descending the periodic table* and that *H—A bond energies are the controlling feature*.

Next, let us examine acids in which the proton is always bonded to the same element, but other changes are made in the molecule. Oxyacids fit this description, and two series of interest are

This generalization applies only when ionizable H is attached directly to the element A (as in HA = HF). For acids such as HOCl, where H is attached to O, and O is then bonded to a different atom, the opposite periodic trend is generally observed. That is, the values of K_a are in the order HOCl > HOBr > HOI.

—— Increasing acid strength ⟶

O=N—O—H (nitrous acid) < O—N(=O)—O—H (nitric acid)

Cl—O—H (hypochlorous acid) < O—Cl—O—H (chlorous acid) < O—Cl(—O)—O—H (chloric acid) < O—Cl(—O)(—O)—O—H (perchloric acid)

In general, *the greater the number of O atoms in an oxyacid, the greater the relative acid strength.* In all the acids, the bond broken is O—H. Although this bond strength can be modified by the other atoms in the molecule, the major effect in these series is the difference in the electron affinities of the A fragments. The greater the number of electronegative O atoms attached to N or Cl, the greater the electron affinity of the fragment. Thus, O_2N—O has a greater electron affinity than ON—O, and nitric acid is a much stronger acid than nitrous acid. In general, the greater the number of O atoms, the more readily the fragment A can accommodate the negative charge left when H^+ dissociates.

There is one other interesting question. Why can only relatively few substances behave as Brønsted acids, even though there are hundreds of compounds with bonds to hydrogen? Acetic acid, for example, loses the hydrogen of the OH group (and not that attached to carbon) as H^+,

$$H_3C—C(=O)—O—H \longrightarrow H_3C—C(=O)—O^- + H^+$$

and a negative charge is left on the oxygen atom of the acetate ion. Such cleavage can only occur if the oxygen atom can accommodate the negative charge; since oxygen has a high electronegativity, it can accept this charge relatively easily. In contrast, the C—H hydrogen of acetic acid (and many other molecules) is not dissociated as H^+ in the presence of water, in spite of the fact that the C—H bond dissociation energy is slightly less than the O—H dissociation energy. The carbon atom is not sufficiently electronegative to accommodate the negative charge left if the bond breaks as C—H ⟶ C:$^-$ + H^+.

EXERCISE 15.11 RELATIVE ACID STRENGTHS

In each pair of acids, tell which is the stronger and why.

(a) H_2SO_4 (sulfuric acid) and H_2SO_3 (sulfurous acid).

(b) H_3AsO_3 (arsenious acid) and H_3AsO_4 (arsenic acid).

Verify your answers using Appendix G.

15.7 THE LEWIS CONCEPT OF ACIDS AND BASES

G.N. Lewis is the same scientist who developed the first concepts of the electron pair bond. See page 299 for a short biography of this important American scientist.

Ideas developed by Gilbert N. Lewis in the 1930s have led to a comprehensive approach to understanding acid–base interactions. The **Lewis definition** can be stated very simply: An *acid* is a substance that can *accept a pair of electrons to form a new bond*, and a *base* is a substance that can *donate a pair of electrons to form a new bond*. This means that an acid–base reaction can occur when a base provides a pair of electrons to share with an acid. The result is often called an acid–base **adduct** or **complex**, and the bond formed between the pair is a **donor–acceptor** or **coordinate covalent** bond.

$$\underset{\text{acid}}{A} + \underset{\text{base}}{B:} \longrightarrow \underset{\text{adduct or complex}}{B{:}\!\rightarrow\! A}$$

A simple example of a Lewis acid–base reaction is one already mentioned many times, the formation of the hydronium ion from H^+ and water.

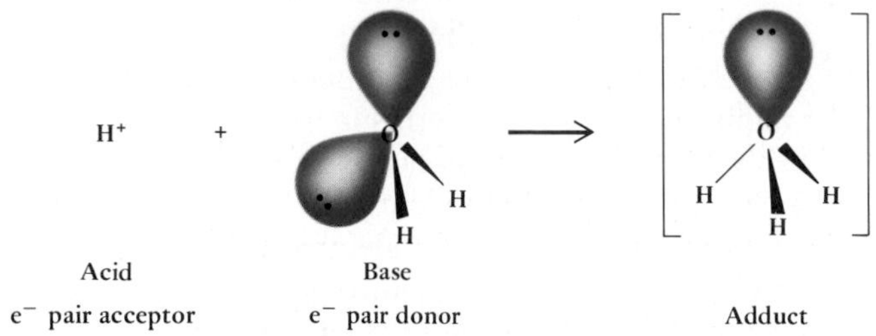

H^+ + H_2O ⟶ $[H_3O]^+$

Acid, e^- pair acceptor; Base, e^- pair donor; Adduct

The H^+ ion has no electrons in its valence or 1*s* orbital, while the water molecule has two unshared pairs of electrons (located in sp^3 hybrid orbitals). One of the pairs can be shared between H^+ and water, thus forming an O—H bond. A similar interaction occurs between H^+ and the base ammonia to form the ammonium ion.

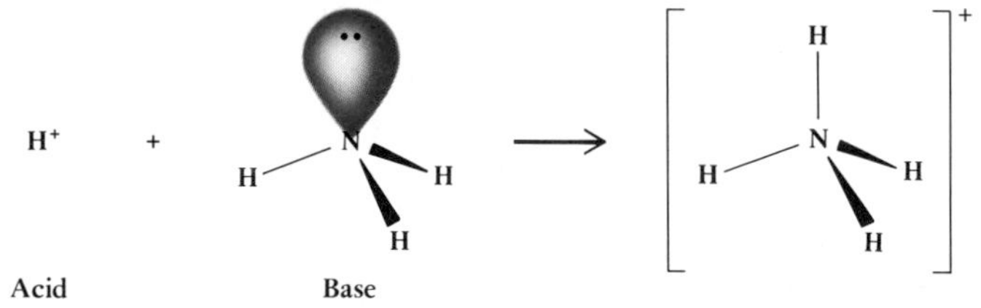

H^+ + NH_3 ⟶ $[NH_4]^+$

Acid; Base

Such reactions are very common. In general, they involve acids that are (a) cations or (b) neutral molecules with an available, empty valence orbital and bases that are (a) anions or (b) neutral molecules with a lone electron pair.

CATIONIC LEWIS ACIDS

Recall the previous discussion of the hydration of cations in Chapter 11.

All cations are potential Lewis acids. Not only do they attract electrons due to their positive charge, but all have at least one empty orbital. This empty orbital can overlap the orbital bearing the base electron pair and thereby form a two-electron chemical bond. Consider the beryllium ion, Be^{2+}, as an example. The Be^{2+} ion has no electrons in the valence shell

(n = 2), so there are four empty orbitals available: $2s$ and the three $2p$ orbitals. Water molecules are electron pair donors, Lewis bases, so Be^{2+}

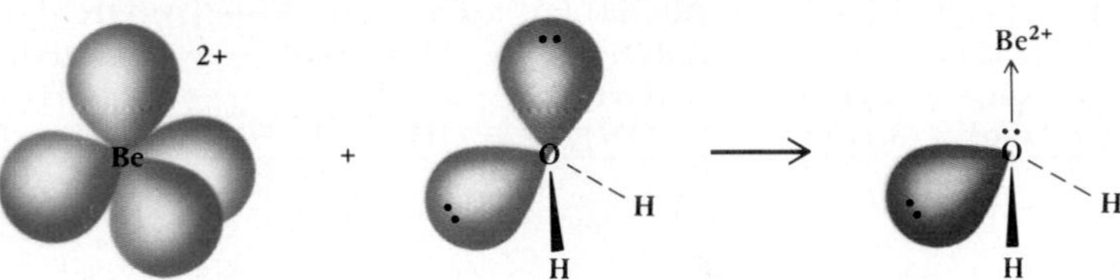

and H_2O form an acid–base adduct. Four empty orbitals on beryllium mean that up to four donor–acceptor bonds can form.

Most compounds of beryllium are toxic; breathing beryllium-containing dust or getting it on your skin can lead ultimately to cancer. When fluorescent lights were first developed the material coating the inside contained beryllium. With the realization of the health hazard of beryllium compounds, however, a substitute was found.

$$Be^{2+}(aq) + 4\ H_2O(\ell) \longrightarrow [Be(H_2O)_4]^{2+}(aq)$$

The **coordination number** of a metal ion in a complex ion is the number of donor atoms bonded to the central ion. The coordination number of a given cation depends on the charge and size of the metal ion, the nature of the base, and other factors; it is usually four or six. The hydrated beryllium ion has a coordination number of four.

The final product, Be^{2+} surrounded by four bond pairs (one each from four water molecules), is presumably tetrahedral in shape (as predicted by the ideas outlined in Chapters 9 and 10). The VSEPR theory suggests that four electron pairs about a central ion or atom will be located at the points of a tetrahedron, so we presume the beryllium ion is sp^3 hybridized.

Species such as $[Be(H_2O)_4]^{2+}$, $[Co(NH_3)_6]^{3+}$, $[CuCl_4]^{2-}$, and $Pt(NH_3)_2Cl_2$ are often called *complexes, coordination complexes*, or, if charged, *complex ions*. Particularly if it bears a charge, the formula of the complex is enclosed within square brackets (see Chapter 25). The Lewis base attached to the central metal ion is often called a **ligand**.

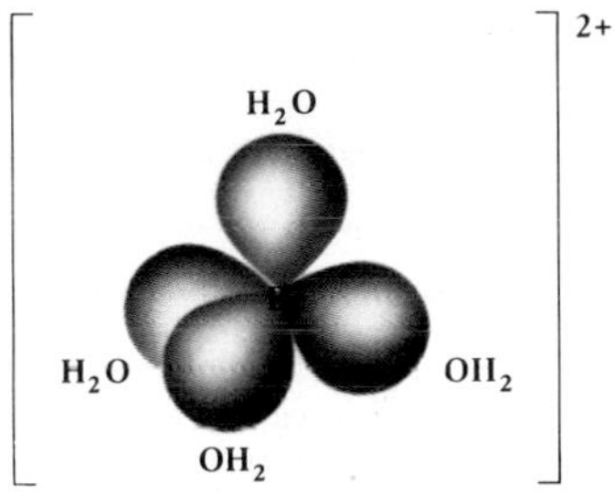

The very high effective nuclear charge (Z^*) of the Be^{2+} ion means that the electrons of the coordinate covalent bond are very strongly attracted to the cation. As a result, the O—H bonds are *polarized*, because the oxygen of water, in relinquishing electrons to Be^{2+}, demands more from its O—H bonds.

$$Be^{2+} \longleftarrow :\overset{\delta-}{\ddot{O}}\text{—}H^{\delta+} \quad (\text{O bonded also to } H^{\delta+})$$

The result is that a coordinated water molecule can more easily lose a proton than a normal water molecule, and the $[Be(H_2O)_4]^{2+}$ complex ion functions as a Brønsted acid or proton donor.

$$[Be(H_2O)_4]^{2+}(aq) \longrightarrow [Be(H_2O)_3(OH)]^{+}(aq) + H^{+}(aq)$$

$$[Be(H_2O)_3(OH)]^{+}(aq) \longrightarrow Be(H_2O)_2(OH)_2(aq) + H^{+}(aq)$$

For this same reason many other small, highly charged metal cations are Brønsted acids (see Table 15.5).

The hydroxide ion is an excellent Lewis base and so binds readily to metal cations to give metal hydroxides. An important feature of the chemistry of many metal hydroxides is that they are **amphoteric**. That is, they can behave as a Lewis acid and react with a Lewis base,

An **amphoteric** substance is a Brønsted base and a Lewis acid. An **amphiprotic** substance is a Brønsted acid and base (p. 555).

$$Zn(OH)_2(s) + 2\ OH^{-}(aq) \longrightarrow [Zn(OH)_4]^{2-}(aq)$$

TABLE 15.7 Some Common Amphoteric Metal Hydroxides

HYDROXIDE	REACTION AS A BASE	REACTION AS AN ACID
$Al(OH)_3$	$Al(OH)_3(s) + 3\ H^+ \longrightarrow Al^{3+}(aq) + 3\ H_2O$	$Al(OH)_3(s) + OH^-(aq) \longrightarrow [Al(OH)_4]^-(aq)$
$Zn(OH)_2$	$Zn(OH)_2(s) + 2\ H^+ \longrightarrow Zn^{2+}(aq) + 2\ H_2O$	$Zn(OH)_2(s) + 2\ OH^-(aq) \longrightarrow [Zn(OH)_4]^{2-}(aq)$
$Sn(OH)_4$	$Sn(OH)_4(s) + 4\ H^+ \longrightarrow Sn^{4+}(aq) + 4\ H_2O$	$Sn(OH)_4(s) + 2\ OH^-(aq) \longrightarrow [Sn(OH)_6]^{2-}(aq)$
$Cr(OH)_3$	$Cr(OH)_3(s) + 3\ H^+ \longrightarrow Cr^{3+}(aq) + 3\ H_2O$	$Cr(OH)_3(s) + OH^-(aq) \longrightarrow [Cr(OH)_4]^-(aq)$

or behave as a Brønsted base and react with a Brønsted acid (Table 15.7).

$$Zn(OH)_2(s) + 2\ H^+(aq) \longrightarrow Zn^{2+}(aq) + 2\ H_2O(\ell)$$

One of the best examples of amphoterism is aluminum hydroxide, $Al(OH)_3$ (Figure 15.8). Aluminum is widely used as a structural material, partly because of its light weight and partly because of its resistance to corrosion. Its corrosion resistance arises from a chemically inert and hard coating of oxidized metal, Al_2O_3. Nonetheless, the coating can be broken, and the metal can then be dissolved in base, producing hydrogen as a useful by-product.

$$2\ Al(s) + 2\ KOH(aq) + 6\ H_2O(\ell) \longrightarrow 2\ KAl(OH)_4(aq) + 3\ H_2(g)$$

The aluminate ion, $[Al(OH)_4]^-$, is a Brønsted base, so acidification of its solutions produces the insoluble metal hydroxide.

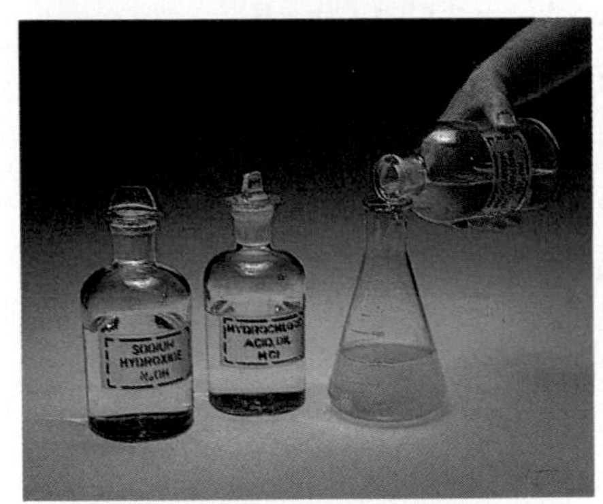

(a)

(b)

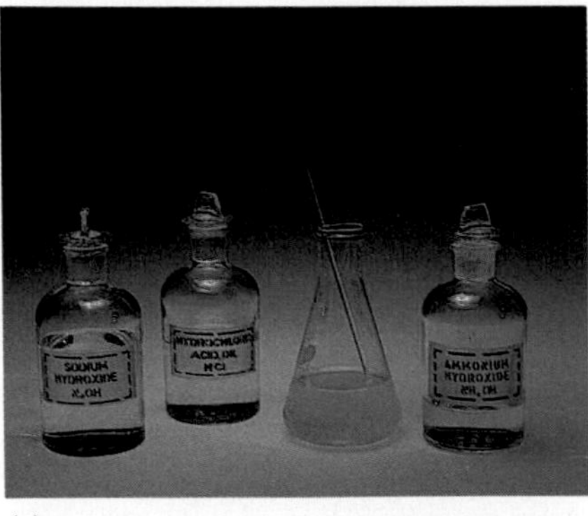

(c)

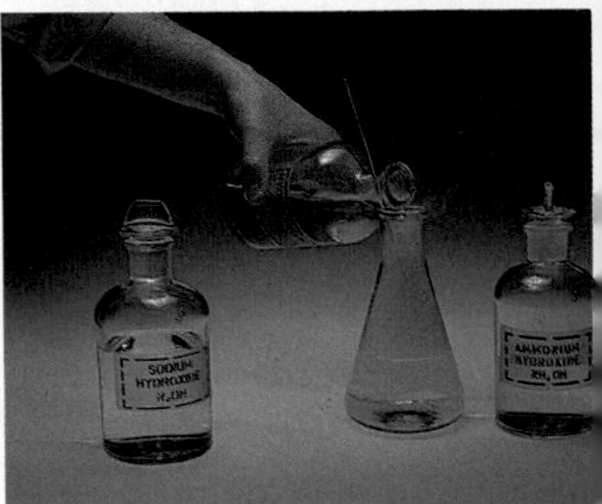

(d)

FIGURE 15.8
The amphoteric nature of $Al(OH)_3$. (a) Adding aqueous ammonia to a soluble salt of Al^{3+} leads to precipitation of insoluble $Al(OH)_3$. (b) Adding a strong base (NaOH) to $Al(OH)_3$ dissolves the precipitate. Here aluminum hydroxide acts as an acid toward OH^- and forms the soluble sodium salt of the complex ion $[Al(OH)_4]^-$. (c) If we begin again with freshly precipitated $Al(OH)_3$, it is dissolved (d) as strong acid (HCl) is added. In this case aluminum hydroxide acts as a base and forms a soluble aluminum salt and water. (Charles D. Winters)

Overall reaction:
$2\ KAl(OH)_4(aq) + H_2SO_4(aq) \longrightarrow 2\ Al(OH)_3(s) + 2\ H_2O(\ell) + K_2SO_4(aq)$
Net ionic equation:
$[Al(OH)_4]^-(aq) + H^+(aq) \longrightarrow Al(OH)_3(s) + H_2O(\ell)$

The metal hydroxide is amphoteric, and adding OH^- returns it to $Al(OH)_4^-$,

$$Al(OH)_3(s) + OH^-(aq) \longrightarrow Al(OH)_4^-(aq)$$

while it can react with acid as a Brønsted base.

$$2\ Al(OH)_3(s) + 3\ H_2SO_4(aq) \longrightarrow Al_2(SO_4)_3(aq) + 6\ H_2O(\ell)$$

Aluminum sulfate crystallizes from aqueous solution as $Al_2(SO_4)_3 \cdot 18\ H_2O$, a salt commonly called "paper maker's alum" (see Chapter 21).

Metal ions also form a large series of complexes with the Lewis base ammonia, $:NH_3$. For example, silver ion readily forms a water-soluble, colorless complex ion in liquid ammonia or aqueous ammonia. Indeed, this complex ion is so stable that the very insoluble compound AgCl can be dissolved in aqueous ammonia.

$$AgCl(s) + 2\ NH_3(aq) \longrightarrow [H_3N{:}\!\longrightarrow\!Ag\!\longleftarrow\!{:}NH_3]^+(aq) + Cl^-(aq)$$

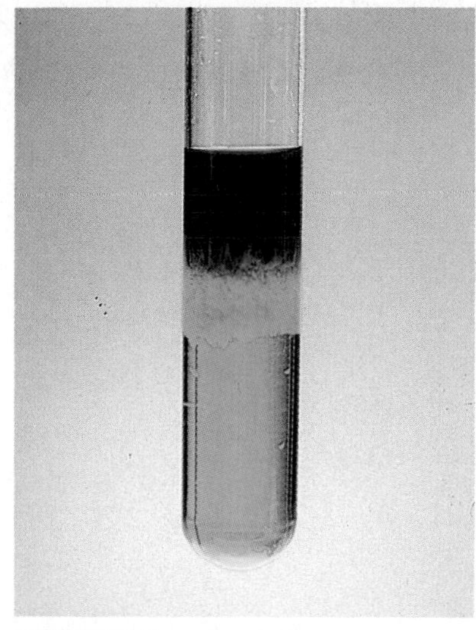

FIGURE 15.9
The Lewis acid–base complex $[Cu(NH_3)_4]^{2+}$. Here aqueous ammonia $[NH_3(aq)]$ was added to aqueous $CuSO_4$ (the light blue solution in the bottom of the test tube). The small concentration of OH^- in aqueous ammonia first formed insoluble blue-white $Cu(OH)_2$ (the solid in the middle of the tube). However, with additional NH_3 the deep blue, soluble complex ion $[Cu(NH_3)_4]^{2+}$ formed (the solution at the top of the tube). (Charles D. Winters)

Light blue aqueous copper(II) ions also react with ammonia to produce beautiful, deep blue complex ions with four ammonia molecules surrounding each metal ion (Figure 15.9).

$$Cu^{2+}(aq) + 4\ NH_3(aq) \longrightarrow \left[\begin{array}{c} NH_3 \\ | \\ H_3N{:}\!-\!Cu\!-\!{:}NH_3 \\ | \\ NH_3 \end{array}\right]^{2+}(aq)$$

NEUTRAL MOLECULES AS LEWIS ACIDS

We discussed in Chapter 9 that the early elements of second period elements do not obey the octet rule all the time. Boron compounds in particular often have only three pairs of electrons around the boron atom. On the other hand, elements of higher periods can often accommodate more than four pairs of electrons. Both of these features mean that such compounds frequently behave as Lewis acids.

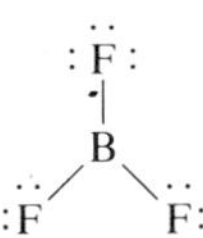

Electron dot structure of BF_3.

Boron trifluoride is produced in ton quantities from borax, fluorospar, and sulfuric acid, all cheap chemicals.

$$\underset{\text{borax}}{Na_2B_4O_7(s)} + \underset{\text{fluorospar}}{6\ CaF_2(s)} + 8\ H_2SO_4(aq) \longrightarrow$$
$$4\ BF_3(g) + 2\ NaHSO_4(aq) + 6\ CaSO_4(s) + 7\ H_2O(\ell)$$

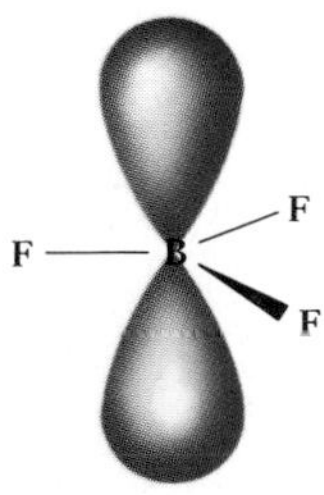

Boron p orbital is unused in sp^2 hybrid formation in a molecule such as BF_3.

BF_3, a colorless gas, is an excellent Lewis acid. The molecule is trigonal and planar with an sp^2 hybridized boron atom. This means that the boron has an unused p orbital perpendicular to the molecular plane, and it is this orbital that may be attacked by a Lewis base such as ammonia.

$BF_3(g) + NH_3(g) \longrightarrow$

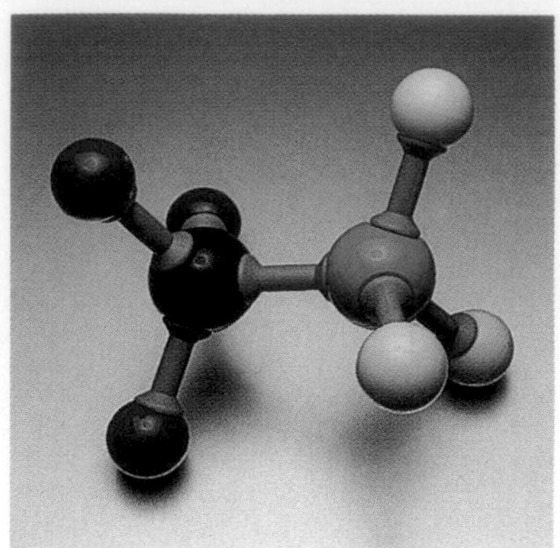

$F_3B \leftarrow NH_3$

Similarly, BF_3 and the other boron trihalides are readily attacked by the Lewis base water, presumably to give initially $H_2O\!:\!\rightarrow BX_3$. However, the reaction proceeds further, eventually giving the hydrolysis products HX and $B(OH)_3$, boric acid.

$$BX_3(g) + 3\ H_2O(\ell) \longrightarrow B(OH)_3(aq) + 3\ HX(aq)$$

Carbon dioxide can also act as a Lewis acid and react with the Lewis base OH^- in aqueous solution.

$$H{-}\ddot{\underset{..}{O}}\!:^- + \overset{\delta-}{\ddot{O}}{=}\overset{\delta+}{C}{=}\overset{\delta-}{\ddot{O}} \longrightarrow H{-}\ddot{\underset{..}{O}}\!:^- \rightharpoonup C^+\!\begin{matrix}\nearrow O\\ \searrow :\!O\!:^-\end{matrix} \longrightarrow \begin{matrix}H & & :\!O\!:\\ & :\!O{-}C & \\ & & :\!O\!:^-\end{matrix}$$

Because oxygen is electronegative, C—O bond electrons flow away from carbon and toward oxygen in CO_2. This causes the carbon atom to be slightly positive, and the negatively charged Lewis base OH^- can attack the carbon to give, ultimately, the bicarbonate ion. This is why the concentration of a solution of sodium hydroxide or potassium hydroxide can change during storage in air; if you have standardized a base, you must use it fairly soon, or CO_2 from the air can lower the effective OH^- ion concentration.

EXERCISE 15.12 LEWIS ACIDS AND BASES

Tell if each of the following is a Lewis acid or Lewis base.

(a) PH_3 (b) BCl_3 (c) H_2S (d) SF_4

THE EXTENT OF LEWIS ACID–BASE REACTIONS: FORMATION CONSTANTS

The extent to which reaction occurs between Ag^+ and ammonia, for example, can be expressed in terms of an equilibrium constant.

$$Ag^+(aq) + 2\ NH_3(aq) \rightleftharpoons [Ag(NH_3)_2]^+(aq)$$

$$K = \frac{\{[Ag(NH_3)_2]^+\}}{[Ag^+][NH_3]^2} = 1.6 \times 10^7$$

Since the formula of a complex ion is given in square brackets, we have enclosed the bracketed formula in { } to indicate the molar concentration of the complex ion.

The large constant for the formation of the silver–ammonia complex indicates that there is a very great tendency to form this ion. On the other hand, the smaller constant for lead(II) with chloride ion

$$Pb^{2+}(aq) + 4\ Cl^-(aq) \rightleftharpoons [PbCl_4]^{2-}(aq)$$

$$K = \frac{\{[PbCl_4]^{2-}\}}{[Pb^{2+}][Cl^-]^4} = 10$$

tells us that the concentration of the complex ion will not be large at equilibrium.

Each of the complexes you have seen so far involves more than one Lewis base molecule or ion (ammonia or Cl^-) combining with the Lewis acid, the metal ion. Formation of such complexes can be envisioned as a stepwise process, with Lewis base molecules being added successively to the central ion.

$$Cu^{2+}(aq) + NH_3(aq) \rightleftharpoons [Cu(NH_3)]^{2+}(aq) \qquad K_1 = 1.7 \times 10^4$$

$$[Cu(NH_3)]^{2+}(aq) + NH_3(aq) \rightleftharpoons [Cu(NH_3)_2]^{2+}(aq) \qquad K_2 = 3.2 \times 10^3$$

$$[Cu(NH_3)_2]^{2+}(aq) + NH_3(aq) \rightleftharpoons [Cu(NH_3)_3]^{2+}(aq) \qquad K_3 = 8.3 \times 10^2$$

$$[Cu(NH_3)_3]^{2+}(aq) + NH_3(aq) \rightleftharpoons [Cu(NH_3)_4]^{2+}(aq) \qquad K_4 = 1.5 \times 10^2$$

In some cases, such as the formation of the copper–ammonia complex, it is possible to measure the value of K for each successive addition. Since the overall reaction

$$Cu^{2+}(aq) + 4\ NH_3(aq) \rightleftharpoons [Cu(NH_3)_4]^{2+}(aq)$$

is the sum of four reaction steps, this means the overall equilibrium constant is the product of the stepwise constants.

Notice that the equilibrium constant for each successive NH_3 addition is smaller than the previous one. This is frequently observed.

$$K_{total} = \frac{\{[Cu(NH_3)_4]^{2+}\}}{[Cu^{2+}][NH_3]^4} = K_1 \times K_2 \times K_3 \times K_4 = 6.8 \times 10^{12}$$

Overall equilibrium constants for a variety of complex ions are given in Appendix J. Since these are constants for the formation of the complex ion, they are called **formation constants.***

The formation constant can be used to calculate the concentrations of the complex, metal ion, or free Lewis base at equilibrium. Of course such calculations are complicated by the fact that complexes are formed in steps, so there can exist in solution some of each of the intermediate ions or molecules. However, because K for each step is so large, the assumption usually made is that the complex ion containing the most Lewis base groups is dominant when [Lewis base] >> [metal ion].

EXAMPLE 15.11

USING FORMATION CONSTANTS

Suppose that 50. mL of 1.0 M NH_3 is mixed with 50. mL of 1.0×10^{-3} M Ag^+. Calculate the concentrations of NH_3, Ag^+, and $[Ag(NH_3)_2]^+$ at equilibrium.

$$Ag^+(aq) + 2\ NH_3(aq) \rightleftharpoons Ag(NH_3)_2^+(aq)$$

$$K = 1.6 \times 10^7 = \frac{\{[Ag(NH_3)_2]^+\}}{[Ag^+][NH_3]^2}$$

Solution When equal volumes of solutions are mixed, the concentrations of the dissolved species become half of their original values. Thus, the initial concentrations in the combined solutions are

$$[Ag^+] = \tfrac{1}{2}(1.0 \times 10^{-3}) = 5.0 \times 10^{-4}\ M$$

$$[NH_3] = \tfrac{1}{2}(1.0) = 0.50\ M$$

*An alternative name is **stability constant**, since it reflects the "stability" of the complex in solution. However, the same information is conveyed by **dissociation** or **instability constants**. These are for the dissociation of the complex ion as in

$$[Cu(NH_3)_4]^{2+}(aq) \rightleftharpoons Cu^{2+}(aq) + 4\ NH_3(aq)$$
$$K_{dissoc} = 1.5 \times 10^{-13}$$

Since this reaction is the *reverse* of the formation reaction, the equilibrium constant, K_{dissoc}, is the *reciprocal* of the formation constant.

Our strategy in solving this problem is as follows: The equilibrium constant for the formation of the silver–ammonia complex is very large, and the concentration of ammonia is high. Therefore, when ammonia and silver ion are mixed, you *imagine* that *all* of the Ag^+ ion forms $[Ag(NH_3)_2]^+$, and stoichiometry tell us that the *initial* concentration of the complex ion must be 5.0×10^{-4} *M*.

	$[Ag^+]$	$[NH_3]$	$[Ag(NH_3)_2{}^+]$
Concentration before mixing (*M*)	0.0005	0.50	0
Concentration after mixing (*M*) (complete reaction assumed)	0	$0.50 - 2(0.00050) \cong 0.50$	0.00050

Next, we imagine that the reaction reverses to a slight extent on proceeding to an equilibrium state, and we calculate the equilibrium concentrations.

	$[Ag^+]$	$[NH_3]$	$[Ag(NH_3)_2{}^+]$
Concentration before equilibrium (*M*)	0	0.50	0.00050
Change on going to equilibrium	$+x$	$+2x$	$-x$
Concentration at equilibrium (*M*)	x	$(0.50 + 2x)$	$(0.0005 - x)$

These concentrations can be substituted into the equilibrium expression for the chemical equation connecting the species, and x can be determined. Before doing so, however, you can make the assumption that x is extremely small since the formation constant of the complex ion is so great. This means that $[NH_3]$ is effectively 0.50 *M* and $[Ag(NH_3)_2]^+$ is approximately 0.00050 *M*.

$$K = 1.6 \times 10^7 = \frac{\{[Ag(NH_3)_2]^+\}}{[Ag^+][NH_3]^2} = \frac{(0.00050)}{x(0.50)^2}$$

$$x = [Ag^+] = 1.3 \times 10^{-10}\ M$$

As a final step, observe that the approximation was reasonable. The quantities $(0.50 + 2x)$ and $(0.0005 - x)$ are indeed effectively 0.50 and 0.0005, respectively.

EXERCISE 15.13 USING FORMATION CONSTANTS

Calculate the equilibrium concentrations of Cu^{2+}, NH_3, and $[Cu(NH_3)_4]^{2+}$ when 500. mL of 3.00 *M* NH_3 is mixed with 500. mL of 2.00×10^{-3} *M* $Cu(NO_3)_2$. $K_{formation}$ for $[Cu(NH_3)_4]^{2+} = 6.8 \times 10^{12}$.

SUMMARY

Interest in acids and bases began hundreds of years ago, and useful ideas to understand their behavior are still being developed. A cornerstone of current ideas of acid–base behavior in aqueous solution is that water undergoes **auto-ionization** to give the **hydronium ion**, H_3O^+, and hydroxide ion, OH^-, to a small extent even in pure water (Section 15.1).

$$2\ H_2O(\ell) \rightleftharpoons H_3O^+(aq) + OH^-(aq)$$

The acid/base definition probably most widely used by chemists to describe aqueous chemistry is that proposed by Brønsted: an aqueous acid is a proton donor, while a base is a proton acceptor (Section 15.2). Thus, HCO_3^- is an acid

$$\underset{\text{acid}}{HCO_3^-(aq)} + \underset{\text{base}}{H_2O(\ell)} \rightleftharpoons \underset{\text{conjugate base of } HCO_3^-}{CO_3^{2-}(aq)} + \underset{\text{conjugate acid of } H_2O}{H_3O^+(aq)}$$

and a base

$$\underset{\text{base}}{HCO_3^-(aq)} + \underset{\text{acid}}{H_2O(\ell)} \rightleftharpoons \underset{\text{conjugate acid of } HCO_3^-}{H_2CO_3(aq)} + \underset{\text{conjugate base of } H_2O}{OH^-(aq)}$$

(Because the same ion, HCO_3^-, can be both a Brønsted acid and base, it is said to be **amphiprotic**.) In each reaction, the acceptance or donation of a proton by HCO_3^- produces a **conjugate** partner. Conjugate acids and bases are related to one another by the gain or loss of a proton. Every acid–base reaction involving H^+ transfer has two conjugate acid–base pairs (Section 15.2).

Acids and bases vary considerably in their ability to act as proton donors and acceptors (Section 15.2). HCl is a strong acid and is nearly 100% ionized in dilute aqueous solutions. In contrast, HF is only weakly ionized. This means that HCl has a great tendency to donate a proton *and* that its conjugate base, Cl^-, only weakly accepts a proton to return to HCl. Conversely, since HF is a weak acid, its conjugate base, F^-, is relatively strong, readily accepting a proton from water to re-form HF. These acids illustrate two corollaries of the Brønsted concept: (1) the stronger an acid or base, the weaker its conjugate base or acid, respectively. (2) In general, acid–base reactions proceed in the direction favoring the weaker acid–base pair (Table 15.2).

The concentration of hydronium ion in solution can be expressed in terms of pH (Section 15.3).

$$pH = -\log[H^+]$$

The auto-ionization of pure water produces H^+ and OH^- in equal amounts, and the extent to which this reaction occurs is expressed by the equilibrium constant K_w where

$$K_w = [H^+][OH^-] = 1.008 \times 10^{-14}$$

For a neutral solution, $[H^+] = [OH^-] = 1.00 \times 10^{-7}\ M$. Thus, the pH is 7.00. For an acid solution, with an excess of H^+ over OH^-, $[H^+]$ is greater than $1.00 \times 10^{-7}\ M$, so the pH is less than 7. Conversely, for a basic solution, $[OH^-] > 10^{-7}\ M$, $[H^+] < 10^{-7}\ M$, and so the pH is > 7.

Strong acids (and bases) produce as many moles of H_3O^+ (or OH^-) in solution as moles of acid (or base) dissolved (Section 15.4). Although there are relatively few strong acids and bases, there are many weak ones (Section 15.5). In the case of weak acids and bases, ionization occurs to a small extent, and the concentration of acid or base is related to its ionization products by the equilibrium constant K_a or K_b.

ACID

$$HA(aq) \rightleftharpoons H^+(aq) + A^-(aq)$$

$$K_a = \frac{[H^+][A^-]}{[HA]}$$

BASE

$$B(aq) + H_2O(\ell) \rightleftharpoons BH^+(aq) + OH^-(aq)$$

$$K_b = \frac{[BH^+][OH^-]}{[B]}$$

K_a values for a large number of weak acids are typically around 1.0×10^{-5}, although some are much lower (Table 15.6).

Weak acids (Tables 15.2, 15.4, and 15.5) in aqueous solution can be **monoprotic** acids such as hydrocyanic acid (HCN) or acetic acid

(CH_3COOH), which can donate only one proton in solution. There are also **polyprotic** acids such as oxalic acid ($H_2C_2O_4$), which donate two or more protons. Cations such as NH_4^+, Fe^{3+}, or Al^{3+} can also act as acids in water. Weak bases include compounds such as ammonia (NH_3) or anions that are the conjugate bases of weak acids; examples of the latter include CN^-, F^-, and acetate (CH_3COO^-) ions.

The relative strengths of acids HA can be rationalized in terms of the strength of the H—A bond and the electron affinity of A (Section 15.6). In general, analogous acids of the elements in a given periodic group increase in strength on descending the periodic table. Further, in a series of oxyacids, the greater the number of O atoms attached to the central element, the stronger the acid.

Lewis theory is the most general of all current ideas of acid–base behavior (Section 15.6). This concept defines an **acid** as a substance that can accept a pair of electrons and a **base** as a substance that can donate a pair of electrons. An acid and a base can interact to form an **adduct** or **complex** with a **donor–acceptor** or **coordinate covalent** bond between the acid and base. Thus, the proton and any other cation are Lewis acids, as are many metal hydroxides. For example, the Lewis acid Zn^{2+} can form $Zn(OH)_2$ when reacting with the Lewis base OH^- ($Zn^{2+} \longleftarrow :\ddot{O}H^-$). The $Zn(OH)_2$ can also act as a Lewis acid in reacting with additional OH^- to form $Zn(OH)_4^{2-}$. However, $Zn(OH)_2$ is also a Brønsted base, since it can react with H^+ to give H_2O and Zn^{2+}. Such substances as $Zn(OH)_2$, which are Lewis acids and Brønsted bases, are said to be **amphoteric**.

The extent of the interaction between a Lewis acid and base is often expressed in terms of a **formation constant** (Section 15.7).

$$Ag^+(aq) + 2\ NH_3(aq) \rightleftharpoons Ag(NH_3)_2^+(aq)$$

$$K_{formation} = \frac{\{[Ag(NH_3)_2^+]\}}{[Ag^+][NH_3]^2}$$

The magnitude of the formation constant depends on the metal cation and the base.

STUDY QUESTIONS

REVIEW QUESTIONS

1. Outline the main ideas of the Brønsted and Lewis theories of acids and bases.
2. Show that water can be a Brønsted base and a Lewis base.
3. Write a balanced equation depicting the auto-ionization of water.
4. Write balanced chemical equations showing that phosphoric acid is a polyprotic acid.
5. Write a balanced chemical equation showing that water is an amphiprotic substance.
6. Designate the acid and the base on each side of the following equation. Designate the conjugate partner of each.

   $NH_4^+(aq) + CN^-(aq) \rightleftharpoons NH_3(aq) + HCN(aq)$
7. The behavior of the acid–base indicator phenolphthalein is described in Section 15.3. Using Le Chatelier's principle, explain why the dye exists as a colorless molecule in acidic solution, while it is deep red in basic solution.
8. Give two representative examples of (a) strong acids, (b) weak acids, and (c) weak bases.
9. If Ni^{2+} exists as $[Ni(H_2O)_6]^{2+}$ in aqueous solution, write a balanced equation to show how hydrolysis of the ion leads to an acidic solution. What is the equilibrium constant expression for this reaction?
10. Explain the relation between Tables 15.2 (Relative Strengths of Acids and Bases) and 15.6 (Acid–Base Properties of Some Common Ions).

11. Explain why hypochlorous acid (HOCl, $K_a = 3.2 \times 10^{-8}$) is a much weaker acid than perchloric acid ($HClO_4$, strong acid).
12. Define the term "acid–base adduct." Give an example.
13. Define the term "amphoteric." Give an example of an amphoteric metal hydroxide.
14. Illustrate the term "formation constant" using the reaction

$$Ag^+(aq) + 2\ NH_3(aq) \rightleftharpoons [Ag(NH_3)_2]^+(aq)$$

GENERAL ACID–BASE THEORY

15. In water, potassium carbonate can form K^+ and HCO_3^- ions. Write a balanced equation showing this process. Does K_2CO_3 give an acidic or basic solution?
16. Dissolving ammonium bromide in water gives an acidic solution. Write a balanced equation showing how this can occur.
17. Write balanced equations showing how the HPO_4^{2-} ion of sodium monohydrogen phosphate can be a Brønsted acid or a Brønsted base.
18. If a hydride ion (H^-) salt, NaH, is put in water, it reacts almost explosively to form $H_2(g)$. Write a balanced equation showing this. Is the resulting solution acidic or basic?
19. Liquid ammonia auto-ionizes just as water does.
    (a) Write a balanced equation for the auto-ionization process of liquid ammonia.
    (b) What is the conjugate acid of NH_3? The conjugate base?
    (c) If $NaNH_2$ is dissolved in liquid ammonia, is it an acid or a base?
20. A neutralization reaction occurs on mixing an acid and a base.
    (a) Write a balanced equation for the reaction occurring when equal volumes of 1.0 *M* NaOH and 1.0 *M* H_2SO_4 are mixed.
    (b) Write the balanced equation for the reaction occurring when equal volumes of 2.0 *M* NaOH and 1.0 *M* H_2SO_4 are mixed.
21. Write the formula and give the name of the conjugate base of each of the following acids:
    (a) HCN (d) HNO_2
    (b) HSO_4^- (e) NH_3
    (c) HF
22. Write the formula and give the name of the conjugate acid of each of the following bases:
    (a) NH_3 (d) Br^-
    (b) HCO_3^- (e) HSO_4^-
    (c) HS^-
23. For each substance listed below, tell whether you expect it to behave as a Brønsted acid or base, and give its conjugate partner. (See Tables 15.1 and 15.2.)
    (a) Br^- (c) H_3PO_4
    (b) $[Al(H_2O)_6]^{3+}$ (d) CH_3COO^-
24. For each substance listed below, tell whether you expect it to behave as a Brønsted acid or base, and give its conjugate partner. (See Tables 15.1 and 15.2.)
    (a) PO_4^{3-} (d) CO_3^{2-}
    (b) H^- (e) OH^-
    (c) NH_4^+
25. In each of the following acid–base reactions, identify the acid and base on the left, and their conjugate partners on the right.
    (a) $HCOOH(aq) + H_2O(\ell) \longrightarrow HCOO^-(aq) + H_3O^+(aq)$
    (b) $H_2S(aq) + NH_3(aq) \longrightarrow NH_4^+(aq) + HS^-(aq)$
    (c) $HSO_4^-(aq) + OH^-(aq) \longrightarrow SO_4^{2-}(aq) + H_2O(\ell)$
26. In each of the following acid–base reactions, identify the acid and base on the left, and their conjugate partners on the right.
    (a) $CH_3COOH(aq) + C_5H_5N(aq) \longrightarrow CH_3COO^-(aq) + C_5H_5NH^+(aq)$
    (b) $N_2H_4(aq) + HSO_4^-(aq) \longrightarrow N_2H_5^+(aq) + SO_4^{2-}(aq)$
    (c) $[Al(H_2O)_6]^{3+}(aq) + OH^-(aq) \longrightarrow [Al(H_2O)_5OH]^{2+}(aq) + H_2O(\ell)$
27. Ammonium chloride and sodium dihydrogen phosphate, NaH_2PO_4, are mixed in water. Using Table 15.1, write a balanced equation for the acid–base reaction that could, in principle, occur. Does the reaction occur to any significant extent?
28. Acetic acid, $HC_2H_3O_2$, and sodium hydrogen carbonate, $NaHCO_3$, are mixed in water. Using Table 15.2, write a balanced equation for the acid–base reaction that could, in principle, occur. Does the reaction occur to any significant extent?
29. Hydrogen sulfide, H_2S, and sodium acetate, $NaC_2H_3O_2$, are mixed in water. Using Table 15.2, write a balanced equation for the acid–base reaction that could, in principle, occur. Does the reaction occur to any significant extent?
30. Several acids are listed below with their respective equilibrium constants.

$$HF(aq) \rightleftharpoons H^+(aq) + F^-(aq) \qquad K_a = 7.2 \times 10^{-4}$$

$$HS^-(aq) \rightleftharpoons H^+(aq) + S^{2-}(aq) \qquad K_a = 1.3 \times 10^{-13}$$

$$HOAc(aq) \rightleftharpoons H^+(aq) + OAc^-(aq) \qquad K_a = 1.8 \times 10^{-5}$$

    (a) Which is the strongest acid? Which is the weakest?
    (b) What is the conjugate base of the acid HF?
    (c) Which acid has the weakest conjugate base?
    (d) Which acid has the strongest conjugate base?

31. Several acids are listed below with their respective equilibrium constants:

$C_6H_5OH(aq) \rightleftharpoons H^+(aq) + C_6H_5O^-(aq)$ $K_a = 1.3 \times 10^{-10}$

$HCOOH(aq) \rightleftharpoons H^+(aq) + HCOO^-(aq)$ $K_a = 1.8 \times 10^{-4}$

$HC_2O_4^-(aq) \rightleftharpoons H^+(aq) + C_2O_4^{2-}(aq)$ $K_a = 6.4 \times 10^{-5}$

(a) Which is the strongest acid? Which is the weakest?
(b) Which acid has the weakest conjugate base?
(c) Which acid has the strongest conjugate base?

32. Several bases are listed below with their respective K_b values:

$NH_3(aq) + H_2O(\ell) \rightleftharpoons NH_4^+(aq) + OH^-(aq)$ $K_b = 1.8 \times 10^{-5}$

$C_5H_5N(aq) + H_2O(\ell) \rightleftharpoons C_5H_5NH^+(aq) + OH^-(aq)$ $K_b = 1.5 \times 10^{-9}$

$N_2H_4(aq) + H_2O(\ell) \rightleftharpoons N_2H_5^+(aq) + OH^-(aq)$ $K_b = 8.5 \times 10^{-7}$

(a) Which is the strongest base? Which is the weakest?
(b) What is the conjugate acid of C_5H_5N?
(c) Which base has the strongest conjugate acid? Which has the weakest?

33. State which of the following compounds or ions has the strongest conjugate base and briefly explain your choice: (a) HSO_4^-, (b) $HC_2H_3O_2$, and (c) HClO.

34. Which of the following compounds has the strongest conjugate acid? Briefly explain your choice. (a) CN^-, (b) NH_3, (c) SO_4^{2-}

35. Using Table 15.2, suggest a reason for the fact that the oxide ion is a powerful Brønsted base in aqueous solution.

$O^{2-}(aq) + H_2O(\ell) \longrightarrow 2\ OH^-(aq)$

36. For each reaction below, predict whether the equilibrium lies predominantly to the left or to the right.
(a) $CO_3^{2-}(aq) + 2\ HCl(aq) \rightleftharpoons H_2CO_3(aq) + 2\ Cl^-(aq)$
(b) $NH_2^-(aq) + H_2O(\ell) \rightleftharpoons NH_3(aq) + OH^-(aq)$
(c) $CN^-(aq) + NH_3(aq) \rightleftharpoons HCN(aq) + NH_2^-(aq)$
(d) $SO_4^{2-}(aq) + HOAc(aq) \rightleftharpoons HSO_4^-(aq) + OAc^-(aq)$

37. For each reaction below, predict whether the equilibrium lies predominantly to the left or to the right.
(a) $NH_2^-(aq) + H_2O(\ell) \rightleftharpoons NH_3(aq) + OH^-(aq)$
(b) $C_6H_5OH(aq) + SO_4^{2-}(aq) \rightleftharpoons C_6H_5O^-(aq) + HSO_4^-(aq)$
(c) $NH_3(aq) + HCO_3^-(aq) \rightleftharpoons NH_4^+(aq) + CO_3^{2-}(aq)$

pH CALCULATIONS

38. A certain table wine has a pH of 3.40. What is the hydrogen ion concentration of the wine? Is it acidic or basic?

39. Make the following interconversions. In each case, tell whether the solution is acidic or basic.

	pH	$[H^+]$ *M*
(a)	1.00	____
(b)	10.50	____
(c)	____	1.3×10^{-5}
(d)	____	5.6×10^{-10}
(e)	____	6.7×10^{-8}
(f)	7.35	____
(g)	5.25	____
(h)	____	2.5×10^{-2}

40. Calculate the pOH and $[OH^-]$ for each solution in Study Question 39.

41. The base $Ca(OH)_2$ is almost insoluble in water; only 0.50 g can be dissolved in 1.0 L of water at 25°C. If the dissolved substance is completely dissociated into its constituent ions, what is the pH of the solution?

42. The ionization constant, K_w, for water at the temperature of your body (37°C) is 2.5×10^{-14}. What is the pH of a neutral solution at this temperature?

43. Suppose you have a 5.0×10^{-4} *M* solution of HCl. (a) If you add a drop of thymol blue indicator to the solution (Table 15.3), what color will the indicator be in solution? (b) If you add a drop of phenolphthalein, what will be the color of the solution?

44. Suppose you have a 3.2×10^{-3} *M* solution of NaOH. What color will the solution be when you add a drop of the indicator phenolphthalein? When you add a drop of alizarin? (See Table 15.3.)

GENERAL ACID–BASE EQUILIBRIA

45. A 2.5×10^{-3} *M* solution of an unknown acid has a pH of 3.80 at 25°C.
(a) What is the hydrogen ion concentration of the solution?
(b) Is the acid a strong acid, a moderately weak acid (K_a of about 10^{-5}), or a very weak acid (K_a of about 10^{-10})?

46. The ionization constant of a weak acid HA is 4.0×10^{-9}. Calculate the equilibrium concentrations of H^+, A^-, and HA in a 0.040 *M* solution of the acid.

47. What are the equilibrium concentrations of H^+, acetate ion, and acetic acid in a 0.20 *M* aqueous solution of acetic acid (HOAc)?

48. If you have a 0.025 *M* solution of HCN, what are the equilibrium concentrations of H^+, CN^-, and HCN? What is the pH of the solution?

49. Phenol, C_6H_5OH, is a weak organic acid.

$C_6H_5OH(aq) \rightleftharpoons C_6H_5O^-(aq) + H^+(aq)$

Although somewhat toxic to humans, it is widely used as a disinfectant and in the manufacture of plastics. If you dissolve 1.56 g of the acid in enough water to make 1.00 L of solution, what is the equilibrium hydrogen ion concentration? What is the pH of the solution?

50. Many organic acids have a foul odor, and propionic acid, C_2H_5COOH, is no exception. The acid, which ionizes in water according to the equation,

$$C_2H_5COOH(aq) \rightleftharpoons C_2H_5COO^-(aq) + H^+(aq) \qquad K_a = 1.8 \times 10^{-5}$$

is used in the manufacture of plastics. If you dissolve 2.35 g of the acid in enough water to make 1.00 L of solution, what is the equilibrium hydrogen ion concentration? What is the pH of the solution?

51. Barbituric acid, $C_4H_4N_2O_3$, has a K_a of 9.9×10^{-5}, while that of nicotinic acid is 1.4×10^{-5}. For a 0.010 M solution of each of these monoprotic acids, which will have the higher pH?

52. Place the following acids (a) in order of increasing strength and (b) in order of increasing pH assuming you have a 0.10 M solution of each acid.
(a) Valeric acid, $K_a = 1.5 \times 10^{-5}$
(b) Glutaric acid, $K_a = 3.4 \times 10^{-4}$
(c) Hypobromous acid, $K_a = 2.9 \times 10^{-9}$

53. Calculate the pH of a 0.025 M solution of the weakest acid in Study Question 52.

54. A fictitious acid HX has $K_a = 1.3 \times 10^{-3}$. After first deciding whether the quadratic formula is needed to solve the equilibrium expression, calculate $[H^+]$ and pH for a 0.010 M solution of the acid.

55. A hypothetical weak base MOH has $K_b = 5.0 \times 10^{-4}$ for the reaction

$$MOH(aq) \rightleftharpoons M^+(aq) + OH^-(aq)$$

Calculate the equilibrium concentrations of MOH, M^+, and OH^- in a 0.15 M solution of MOH.

56. What are equilibrium concentrations of NH_3, NH_4^+, and OH^- in a 0.15 M solution of aqueous ammonia? What is the pH of the solution?

57. The weak base methylamine, CH_3NH_2, has $K_b = 5.0 \times 10^{-4}$. It reacts with water according to the equation

$$CH_3NH_2(aq) + H_2O(\ell) \rightleftharpoons CH_3NH_3^+(aq) + OH^-(aq)$$

Calculate the equilibrium hydroxide ion concentration in a 0.25 M solution of the base. What are the pH and pOH of the solution?

58. Calculate the pH of a 0.12 M aqueous solution of the base aniline, $C_6H_5NH_2$. ($K_b = 4.2 \times 10^{-10}$.)

59. Hydroxylamine, NH_2OH, has a K_b of 6.6×10^{-9}. What are the pH and pOH of a 0.051 M solution of the base?

60. Ethanolamine is a base with a K_b of 2.8×10^{-5}. A closely related base, ethylamine, has a K_b of 5.6×10^{-4}. (a) If you have a 0.10 M solution of each, which has the higher pH? (b) Which of the two bases is stronger? Calculate the pH of its 0.10 M solution.

61. A 0.10 M solution of chloroacetic acid, $ClCH_2COOH$, has a pH of 1.95. Calculate K_a for the acid.

62. Saccharin, $C_7H_5NO_3S$, is an acid in aqueous solution. It has a dissociation constant K_a of 2.1×10^{-12}. Calculate the pH of a solution made by dissolving 3.0 g in enough water to make 1.0 L of solution. (This is close to a saturated solution of saccharin.)

63. Butyric acid, C_3H_7COOH, is found in rancid butter and has a well deserved, malodorous reputation. In a well ventilated room, 0.20 moles of the acid are dissolved in water to give 500. mL of a solution whose pH is found to be 2.60. Assuming that butyric acid is monoprotic (it has only one ionizable proton), calculate K_a for the acid.

64. After deciding whether to use the quadratic formula to solve the equilibrium expression, calculate the pH of a 0.0010 M aqueous solution of HF.

65. It was found that a solution of hydrofluoric acid, HF, had a pH of 2.30. Calculate the equilibrium concentrations of HF, F^-, and H^+, and calculate the amount of HF originally dissolved per liter.

66. If each of the salts listed below were dissolved in water to give a 0.10 M solution, which solution would have the highest pH? Which would have the lowest pH?
(a) Na_2S
(b) Na_3PO_4
(c) NaF
(d) $NaC_2H_3O_2$ (sodium acetate)

67. Which of the following common food additives would give a basic solution when dissolved in water?
(a) $NaNO_3$ (used as a meat preservative)
(b) $NaC_7H_5O_2$ (sodium benzoate; used as a soft drink preservative)
(c) Na_2HPO_4 (used as an emulsifier in the manufacture of pasteurized Kraft cheese).

68. Will $NaHSO_4$ give an acidic or basic solution in water? What about Na_2HPO_4?

69. For each salt below, predict whether an aqueous solution will have a pH less than, equal to, or greater than 7.
(a) $AlCl_3$ (e) NaH_2PO_4
(b) NH_4Br (f) $NaNO_3$
(c) Na_2S (g) $KClO_4$
(d) NaF

70. Calculate the equilibrium hydrogen ion concentration in a 0.20 M solution of the salt ammonium chloride, NH_4Cl.

71. Calculate the hydrogen ion concentration in a 0.015 *M* solution of the salt sodium acetate, $NaC_2H_3O_2$.

72. Sodium cyanide is the salt of the weak acid HCN. Calculate the concentration of H^+, OH^-, HCN, and Na^+ in a solution prepared by dissolving 0.22 moles of NaCN in 550 mL of pure water at 25°C.

73. The sodium salt of propionic acid, $NaC_3H_5O_2$, is used as an antifungus agent by veterinarians. Calculate the equilibrium concentration of H^+ and OH^-, and the pH, for a solution of 0.10 *M* $NaC_3H_5O_2$. (See Table 15.4 for the properties of the $C_3H_5O_2^-$ ion.)

74. About this time, you may be wishing you had an aspirin. Aspirin is an organic acid with a K_a of 3.27×10^{-4} for the reaction

$$HC_9H_7O_4(aq) \rightleftharpoons C_9H_7O_4^-(aq) + H^+(aq)$$

If you have two tablets, each having 0.325 g of aspirin (mixed with a neutral "binder" to hold the tablet together) and dissolve them in a glass of water (200. mL), what is the pH of the solution?

75. *m*-Nitrophenol, a weak acid, can be used as an indicator, since it is yellow at a pH above 8.6 and colorless at a pH below 6.8. If the pH of a 0.010 *M* solution of the compound is 3.44, calculate the K_a of the compound.

76. Lactic acid, $C_3H_6O_3$, occurs in sour milk as a result of lactic acid bacteria. What is the pH of a solution of 56 mg of lactic acid in 250 mL of water? K_a for lactic acid = 1.4×10^{-4}. (See Example 15.5.)

77. Chloroacetic acid, $ClCH_2COOH$, is a moderately weak acid ($K_a = 1.40 \times 10^{-3}$). Suppose you dissolve 94.5 mg of the acid in 100. mL of water. What is the pH of the solution?

POLYPROTIC ACIDS AND BASES

78. At room temperature and 1 atm pressure, a saturated H_2S solution has $[H_2S] = 0.10$ *M*, $[HS^-] = 1.0 \times 10^{-4}$ *M*, and $[S^{2-}] = 1.3 \times 10^{-13}$ *M*.
    (a) What is the pH of the solution?
    (b) Now suppose the solution is made basic (without any change in solution volume). Which of the following happens?
        (1) Only $[H_2S]$ will increase.
        (2) Both $[H_2S]$ and $[S^{2-}]$ will increase.
        (3) $[H_2S]$ decreases and $[S^{2-}]$ increases.
        (4) $[H^+]$ increases.
        (5) $[H_2S]$, $[HS^-]$, and $[S^{2-}]$ all remain constant.

79. Oxalic acid is a relatively weak acid capable of losing two protons. Calculate the equilibrium constant for the overall reaction from K_{a1} and K_{a2}.

$$H_2C_2O_4(aq) \rightleftharpoons C_2O_4^{2-}(aq) + 2\,H^+(aq)$$

80. Hydrazine, N_2H_4, can interact with water in two stages.

$$N_2H_4(aq) + H_2O(\ell) \rightleftharpoons N_2H_5^+ + OH^- \qquad K_{b1} = 8.5 \times 10^{-7}$$

$$N_2H_5^+(aq) + H_2O(\ell) \rightleftharpoons N_2H_6^{2+} + OH^- \qquad K_{b2} = 8.9 \times 10^{-16}$$

(a) What is the concentration of OH^-, $N_2H_5^+$, and $N_2H_6^{2+}$ in a 0.010 *M* aqueous solution of hydrazine?
(b) What is the pH of the 0.010 *M* solution of hydrazine?

81. Ethylenediamine, $H_2N—C_2H_4—NH_2$, can interact with water in two steps, giving OH^- in each step. (See Appendix H.) If you have a 0.15 *M* aqueous solution of the amine, calculate the concentration of OH^- and $[H_3N—C_2H_4—NH_3]^{2+}$.

82. Ascorbic acid (vitamin C, $C_6H_8O_6$) is a diprotic acid ($K_{a1} = 7.9 \times 10^{-5}$ and $K_{a2} = 1.6 \times 10^{-12}$). What is the pH of a solution that contains 5.0 mg of acid per mL of water?

83. H_2SO_3, sulfurous acid, is a weak acid capable of providing two H^+ ions.
    (a) What is the pH of a 0.45 *M* solution of H_2SO_3?
    (b) What is the equilibrium concentration of the sulfite ion, SO_3^{2-}, in the 0.45 *M* solution of H_2SO_3?

RELATIVE STRENGTHS OF ACIDS

84. Predict the relative strengths of the acids in each pair: (a) H_3PO_4 (phosphoric acid) and H_3PO_3 (phosphorous acid), (b) H_2SeO_3 (selenous acid) and H_2SeO_4 (selenic acid).

85. A marginal note on page 585 explains that oxyacids decrease in strength as the central element increases in atomic weight. This being the case, predict the order of K_a values for H_2SO_4 (sulfuric acid) and H_2SeO_4 (selenic acid).

86. Which of the following should be the strongest Brønsted acid and which should be the weakest: (a) CH_3OH, (b) ClOH, and (c) LiOH?

87. Of the three acids, $ClCH_2COOH$, CH_3COOH, and CCl_3COOH, which is the strongest and which is the weakest? Briefly explain.

LEWIS ACIDS AND BASES

88. For each substance listed below, tell whether it is a Lewis acid or base:
    (a) Mn^{2+}
    (b) $:NH_2(CH_3)$
    (c) hydroxylamine, H_2NOH
    (d) SO_2 in the reaction $SO_2(g) + BF_3(g) \longrightarrow O_2S{:}BF_3(s)$
    (e) $Zn(OH)_2$ in the reaction: $Zn(OH)_2(s) + 2\,OH^-(aq) \longrightarrow Zn(OH)_4^{2-}(aq)$

89. Silver cyanide can react in two ways:

$$AgCN(s) + CN^-(aq) \longrightarrow [Ag(CN)_2]^-(aq)$$

$$AgCN(s) + H^+(aq) \longrightarrow Ag^+(aq) + HCN(aq)$$

Which reaction shows AgCN as a Lewis acid? In which reaction is it behaving as a Brønsted base? Is AgCN amphiprotic or amphoteric?

90. Draw a Lewis dot structure of SbF_5, and tell whether it should be a Lewis acid or Lewis base.

91. Trimethylamine, $(H_3C)_3N:$, is a common reagent. It interacts readily with diborane, B_2H_6. The latter dissociates to BH_3, and this forms a complex with the amine, $(H_3C)_3N:BH_3$. Is the BH_3 fragment a Lewis acid or a Lewis base?

92. Carbon monoxide, $:C{\equiv}O:$, forms complexes with low-valent metals. For example, $Ni(CO)_4$ and $Fe(CO)_5$ are well known. CO also forms complexes with the iron of hemoglobin, preventing the hemoglobin from acting in its normal way, that is, taking up oxygen. Is CO a Lewis acid or Lewis base?

93. Draw a Lewis dot structure of ICl_3. Does it function as a Lewis acid or base in reacting with Cl^- to form ICl_4^-? What are the likely molecular structures of ICl_3 and ICl_4^-?

94. Is $Fe(CO)_5$ best classed as a Lewis acid, Lewis base, Brønsted acid, or Brønsted base in the following reaction (which occurs in a nonaqueous solvent)?

$$Fe(CO)_5 + H^+ \longrightarrow HFeCO_5^+$$

FORMATION CONSTANTS

95. If you have a 0.1 M solution of Cd^{2+}, which should you add, NH_3 or CN^-, to achieve the greatest lowering of the Cd^{2+} ion concentration? Formation constants for complex ions of Cd^{2+} are as follows: 1.0×10^7 for $[Cd(NH_3)_4]^{2+}$; 1.3×10^{17} for $[Cd(CN)_4]^{2-}$.

96. What is the concentration of $Cu^{2+}(aq)$ in a solution that was originally 0.050 M $Cu^{2+}(aq)$ and 3.0 M $NH_3(aq)$?

97. What are the concentrations of $[CdCl_4]^{2-}$, Cd^{2+}, and Cl^- in a solution that was originally 0.010 M in Cd^{2+} and 1.0 M in Cl^-?

98. Silver and gold often occur together in nature, and both are dissolved as cyanide complex ions when treated with oxygen and CN^-. If you have a 0.010 M aqueous solution of $[Ag(CN)_2]^-$, what is the concentration of silver ion, Ag^+, in solution? K_{form} for $[Ag(CN)_2]^-$ from Ag^+ and CN^- is 5.6×10^{18}.

GENERAL PROBLEMS

99. The dissociation of acetic acid is an endothermic process. What effect does an increase in temperature have on the value of K_a for the acid?

100. To what volume should 100. mL of any weak acid HA with a concentration 0.20 M be diluted in order to double the percentage dissociation?

101. Given the following solutions:

0.1 M NH_3	0.1 M NH_4Cl
0.1 M Na_2CO_3	0.1 M $NaC_2H_3O_2$
0.1 M NaCl	0.1 M $NH_4C_2H_3O_2$
0.1 M $HC_2H_3O_2$	

(a) Which of the solutions are acidic?
(b) Which of the solutions are basic?
(c) Which of the solutions is most acidic?

102. Arrange the following 0.1 M solutions in order of increasing pH.

(a) NaCl (d) HCl
(b) NH_3 (e) KOH
(c) $NaC_2H_3O_2$ (f) $HC_2H_3O_2$

103. Arrange the following 1.0 M solutions in order of increasing pH.

(a) NaBr (d) HBr
(b) NH_4Cl (e) NaOH
(c) NaCN (f) N_2H_4 (hydrazine)

104. Which of the following is a base according to the Arrhenius concept and which is better described as a base by the Brønsted concept?

(a) NaOH (d) KCN
(b) Na_2CO_3 (e) sodium ethoxide, NaC_2H_5O.
(c) $Ca(OH)_2$

105. Nicotinic acid, $C_6H_5NO_2$, is found in minute amounts in all living cells, but there are appreciable amounts in liver, yeast, milk, adrenal glands, white meat, and corn. Whole wheat flour contains about 60. micrograms per gram of flour. One gram of the acid dissolves in 60. mL of water and gives a pH of 2.70. What is the approximate value of K_a for the acid?

106. Nicotine, $C_{10}H_{14}N_2$, has two basic nitrogen atoms, and both can react with water to give a basic solution.

$$Nic(aq) + H_2O(\ell) \rightleftharpoons NicH^+(aq) + OH^-(aq)$$

$$NicH^+(aq) + H_2O(\ell) \rightleftharpoons NicH_2^+(aq) + OH^-(aq)$$

K_{b1} is 7.0×10^{-7} and K_{b2} is 1.1×10^{-10}. Calculate the approximate pH of the 0.020 M solution.

107. Sodium benzoate, $Na[C_6H_5COO]$, is soluble in water and is commonly used as a food preservative. If 9.0 g of the salt are dissolved in sufficient water to make exactly 500. mL of solution, what is the pH of the solution?

108. Not only can one determine formation constants for complex ions, but it is also possible to measure them for complexes of neutral Lewis acids. For example, K_{form} $(=K_p)$ for the dimethyl ether complex of BF_3, $(CH_3)_2O{:}{\rightarrow}BF_3$, is 5.8.

$$BF_3(g) + (CH_3)_2O(g) \rightleftharpoons (CH_3)_2O{:}{\rightarrow}BF_3(g)$$

If you place 1.0 g of the complex in a 1.0-L flask at 25°C, what is the total pressure in the flask? What are the partial pressures of the Lewis acid, the Lewis base, and the complex?

109. The thiosulfate complex of silver is important in developing photographs.

$$Ag^+(aq) + 2\ S_2O_3^{2-}(aq) \rightleftharpoons [Ag(S_2O_3)_2]^{3-} \qquad K = 2.0 \times 10^{13}$$

If 250 mL of 0.0010 *M* $AgNO_3$ is mixed with 250 mL of 5.0 *M* $Na_2S_2O_3$, calculate

(a) the original concentrations of Ag^+ and $S_2O_3^{2-}$;

(b) the concentrations of Ag^+ and $[Ag(S_2O_3)_2]^{3-}$ at equilibrium.

110. Let us consider a salt of a weak base and weak acid such as ammonium cyanide. Both NH_4^+ and CN^- ion hydrolyze in aqueous solution, but the net reaction can be considered as a proton transfer from NH_4^+ to CN^-.

$$NH_4^+(aq) + CN^-(aq) \rightleftharpoons NH_3(aq) + HCN(aq)$$

(a) Show that the equilibrium constant for this reaction, K_{total}, is

$$K_{total} = \frac{K_w}{K_a K_b}$$

where K_a is the dissociation constant for the weak acid HCN and K_b is the constant for the weak base NH_3.

(b) Prove that the hydrogen ion concentration in this solution must be given by

$$[H^+] = \sqrt{\frac{K_w K_a}{K_b}}$$

(c) What is the pH of a 0.15 *M* solution of ammonium cyanide?

16

Reactions Between Acids and Bases

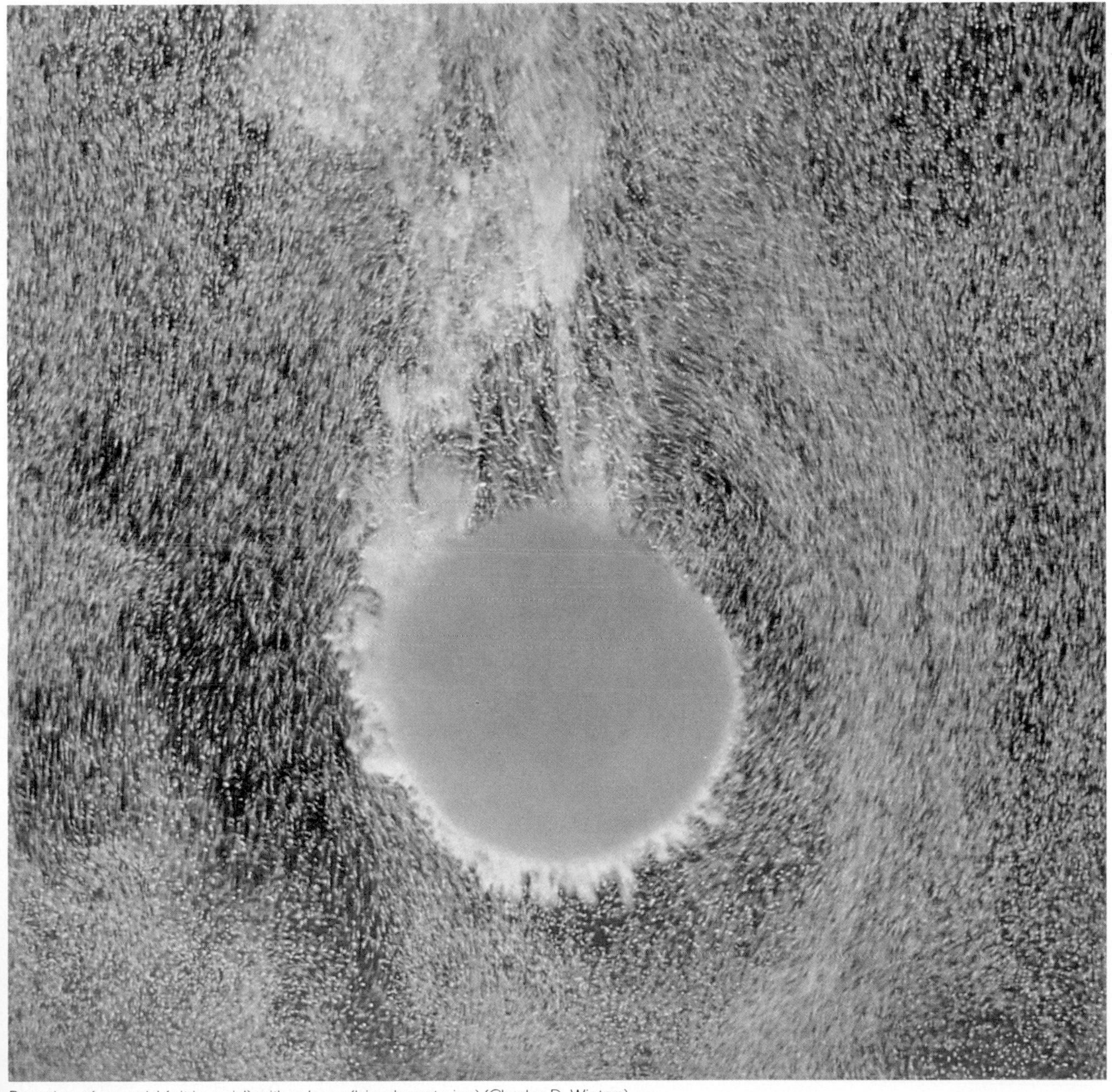

Reaction of an acid (citric acid) with a base (bicarbonate ion) (Charles D. Winters)

CHAPTER OUTLINE

In the previous chapter we began to explore the chemistry of acids and bases. Reactions of acids with bases are all around you and within you. The pH of the oceans and your blood is controlled by the chemistry of carbonic acid. Well over 50 million tons of fertilizers such as ammonium nitrate, made by the reaction of the base ammonia with nitric acid, are manufactured every year. The acid in "acid rain" comes from the interaction of airborne, nonmetal oxides such as CO_2, SO_2, and nitrogen oxides with water, and this acid rain falls to the earth where it is neutralized by reaction with minerals.

This chapter continues the exploration of acid–base chemistry with particular emphasis on the results of acid–base reactions, the control of such reactions, and some reactions of practical concern.

16.1 ACID–BASE REACTIONS

Using Table 15.2, the direction of an acid–base reaction can be predicted. For example, acetic acid ($HC_2H_3O_2$ or HOAc) should react to a significant extent with ammonia in water.

$$HC_2H_3O_2(aq) + NH_3(aq) \longrightarrow NH_4^+(aq) + C_2H_3O_2^-(aq)$$

Remember that reactions always proceed in the direction of the weaker acid–base pair, and you find here that NH_4^+ is a weaker acid than HOAc and OAc^- is a weaker base than NH_3. We can verify this because the overall reaction is the sum of three other reactions whose values of K are known.

As we did in Chapter 15, for simplicity we shall indicate hydrogen ion in aqueous solution by $H^+(aq)$.

Ionization of the acid

$$HOAc(aq) \rightleftharpoons H^+(aq) + OAc^-(aq)$$

$$K_a = 1.8 \times 10^{-5}$$

Ionization of the base

$$NH_3(aq) + H_2O(\ell) \rightleftharpoons NH_4^+(aq) + OH^-(aq)$$

$$K_b = 1.8 \times 10^{-5}$$

Union of hydronium ion and hydroxide ion

$$H^+(aq) + OH^-(aq) \rightleftharpoons H_2O(\ell)$$

$$K = 1/K_w = 1.0 \times 10^{14}$$

K for a net reaction is the product of the constants of the summed reactions.

Net reaction

$$HOAc(aq) + NH_3(aq) \rightleftharpoons NH_4^+(aq) + OAc^-(aq)$$

$$K_{net} = \frac{K_aK_b}{K_w} = 3.2 \times 10^4$$

As the acid and base create H_3O^+ and OH^-, they are "swept up" by the water-formation reaction with its extraordinarily high equilibrium constant of 10^{14}. Thus, Le Chatelier's principle causes both dissociation reactions to shift much further to the right than either would go if it occurred alone in solution, and the overall process has a large equilibrium constant.

The reaction of acetic acid and ammonia is a good example of a weak acid reacting with a weak base. However, this is only one of four possible types of acid–base reactions in aqueous solution.

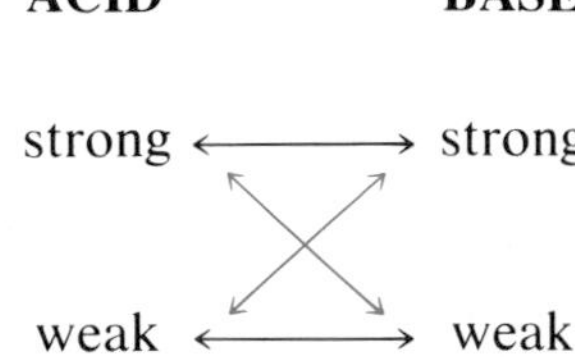

All of these combinations of acid and base will be considered in turn.

THE REACTION OF A STRONG ACID WITH A STRONG BASE

Strong acids and bases are effectively 100% ionized in solution (Section 15.4), so the equilibrium constant for their ionization is very large. Therefore, if we mix HCl with NaOH, the reactions occurring are

Ionization of the acid

$$HCl(aq) = H^+(aq) + Cl^-(aq) \qquad K_a > 1$$

Dissociation of the base

$$NaOH(aq) = Na^+(aq) + OH^-(aq) \qquad K_b > 1$$

Union of hydrogen ion and hydroxide ion

$$H^+(aq) + OH^-(aq) \longrightarrow H_2O(\ell) \qquad K = \frac{1}{K_w} = 1 \times 10^{14}$$

Net reaction

$$HCl(aq) + NaOH(aq) \longrightarrow NaCl(aq) + H_2O(\ell)$$

$$K_{net} = \frac{K_a K_b}{K_w} > 10^{14}$$

The enormous value of K shows the reaction is, for practical purposes, *quantitatively complete*. Thus, if equal numbers of moles of NaOH and HCl are mixed, the result is just a solution of NaCl in water. Since the constituents of NaCl, Na^+ and Cl^- ions, arise from a strong base and a strong acid, respectively, they produce a neutral aqueous solution (Table 15.6). Because of this, reactions of acids and bases have come to be called **neutralizations**.

THE REACTION OF A STRONG BASE WITH A WEAK ACID

Naturally occurring acids include oxalic acid (present as the potassium or calcium salt in plants of the *oxalis* family in particular), malic acid (in apples and many other fruits), and succinic acid (in fungi and lichens).

Nature is full of weak acids. Let us take as an example formic acid, HCOOH, and its reaction with the strong base sodium hydroxide. The overall reaction can again be considered the sum of three reactions.

Ionization of the acid

$$HCOOH(aq) \rightleftharpoons H^+(aq) + HCOO^-(aq) \qquad K_a = 1.8 \times 10^{-4}$$

Dissociation of the base

$$NaOH(aq) = Na^+(aq) + OH^-(aq) \qquad K_b > 1$$

Union of hydrogen ion and hydroxide ion

$$H^+(aq) + OH^-(aq) \longrightarrow H_2O(\ell) \qquad K = \frac{1}{K_w} = 1.0 \times 10^{14}$$

K for the net reaction of a weak acid with a strong base is the reciprocal of K_b of the conjugate base of the weak acid. Here $K_a = 1.8 \times 10^{10} = 1/(5.6 \times 10^{-11})$.

Net reaction

$$NaOH(aq) + HCOOH(aq) \longrightarrow Na^+(aq) + HCOO^-(aq) + H_2O(\ell)$$

$$K_{net} = \frac{K_aK_b}{K_w} > 1.8 \times 10^{10} = \frac{1}{K_b \text{ for } HCOO^-}$$

Once again, the equilibrium constant for the overall process is large, so the net reaction goes essentially to completion. Assuming that equal molar quantities of acid and base were mixed, the solution contains only sodium formate ($NaCHO_2$), a salt that is 100% dissociated in water. However, while the sodium ion is the cation of a strong acid and so gives a neutral solution, the formate ion is the anion of a weak acid and is therefore a base (Table 15.4). The solution will be basic.

$$HCOO^-(aq) + H_2O(\ell) \rightleftharpoons HCOOH(aq) + OH^-(aq)$$

$$K_b \text{ (from Table 15.4)} = 5.6 \times 10^{-11}$$

In general, mixing equal molar amounts of *a strong base with a weak acid will produce a salt wherein the anion is the conjugate base of the weak acid*. The *solution will be basic*, the pH depending on K_b of the anion. This point is made in Example 15.9 (page 579) and in the example that follows.

EXAMPLE 16.1

pH AT THE EQUIVALENCE POINT OF A STRONG BASE/WEAK ACID REACTION

In Chapter 4 you learned that the amount of an acid in solution, for example, can be determined by *titration* with a base (Figure 4.9). When the amount of H^+ in solution is equal to the amount of OH^-, this is the *equivalence point* of the titration.

Suppose 50. mL of 0.10 *M* NaOH are mixed with 50. mL of 0.10 *M* formic acid, HCOOH. What is the pH of the resulting solution? (Assume the solution volumes add to give 1.0×10^2 mL.)

Solution The reaction that occurs when these solutions are mixed is given in the text above. Here we are mixing 0.0050 moles (0.050 L × 0.10 *M*) of HCOOH with 0.0050 moles of NaOH, and the product is 0.0050 moles of $NaCHO_2$. Since we mixed 50. mL of each reagent, the salt is dissolved in 1.0×10^2 mL of water. The concentration of the salt, therefore, is 0.050 *M*.

With the salt concentration known, we can solve for the pH of the solution. As suggested above, the solution will be basic due to the *hydrolysis* of the conjugate base of formic acid according to the equation:

$$HCOO^-(aq) + H_2O(\ell) \rightleftharpoons HCOOH(aq) + OH^-(aq)$$

	$[HCOO^-]$	$[HCOOH]$	$[OH^-]$
Before hydrolysis (M)	0.050	0	0
Change when proceeding to equilibrium	$-x$	$+x$	$+x$
At equilibrium (M)	$(0.050 - x)$	x	x

The problem can be simplified if $x << 0.050$. Using the "rule of thumb" for simplifying problems (Example 15.6), we see that 100 times K_b is much less than the original $HCOO^-$ ion concentration (0.050 M), so $0.050 - x \cong 0.050\ M$. Therefore,

In Example 16.1 $[OH^-]$ is given as 0 before equilibrium is established. It is actually 10^{-7} M, but, as described in Chapter 15, this can generally be neglected in calculations of this type. The same simplification is made in other examples in this chapter.

$$K_b = \frac{[HCOOH][OH^-]}{[HCOO^-]} = \frac{(x)(x)}{0.050 - x} \cong \frac{x^2}{0.050} = 5.6 \times 10^{-11}$$

$$x = [OH^-] = 1.7 \times 10^{-6}\ M$$

and so

$$[H^+] = 6.0 \times 10^{-9}\ M$$

which gives

$$pH = 8.22$$

The solution is indeed basic as predicted by the fact that $HCOO^-$ is the conjugate base of a weak acid.

EXERCISE 16.1 pH WHEN AN ACID–BASE REACTION IS COMPLETED

How many milliliters of 0.100 M NaOH are required to react completely with 0.976 g of the weak, monoprotic acid benzoic acid (C_6H_5COOH)? What is the pH of the solution after reaction? (See Table 15.4 for the K_b of the benzoate ion, $C_6H_5COO^-$).

There are several weak acids with more than one proton. For these, reaction with OH^- occurs stepwise. Carbonic acid, which is explored more in the special section in Chapter 17, is an important example. If a strong base such as NaOH reacts with this acid, we find that the first reaction goes nearly to completion before the second begins.

Common polyprotic weak acids include H_3PO_4 and H_2SO_3.

$$H_2CO_3(aq) + OH^-(aq) \rightleftharpoons HCO_3^-(aq) + H_2O(\ell)$$

$$HCO_3^-(aq) + OH^-(aq) \rightleftharpoons CO_3^{2-}(aq) + H_2O(\ell)$$

If one mole of OH^- ions is added per mole of H_2CO_3, the product is one mole of the hydrogen carbonate ion, HCO_3^-. If a second mole of OH^- ions is subsequently added, the final product is the carbonate ion, CO_3^{2-}.

THE REACTION OF A STRONG ACID WITH A WEAK BASE

A good example of a reaction of this type is that between the strong acid hydrochloric acid and the weak base ammonia.

Ionization of the acid

$$HCl(aq) = H^+(aq) + Cl^-(aq) \qquad K_a > 1$$

Dissociation of the base

$$NH_3(aq) + H_2O(\ell) \rightleftharpoons NH_4^+(aq) + OH^-(aq) \qquad K_b = 1.8 \times 10^{-5}$$

Union of hydrogen ion and hydroxide ion

$$H^+(aq) + OH^-(aq) \longrightarrow H_2O(\ell) \qquad K = \frac{1}{K_w} = 10^{14}$$

K for the net reaction of a strong acid with a weak base is the reciprocal of K_a for the conjugate acid of the weak base. Here $K_a = 5.6 \times 10^{-10} = 1/(1.8 \times 10^9)$.

Net reaction

$$HCl(aq) + NH_3(aq) \longrightarrow NH_4^+(aq) + Cl^-(aq)$$

$$K_{net} = \frac{K_a K_b}{K_w} > 1.8 \times 10^9 = \frac{1}{K_a \text{ for } NH_4^+}$$

Just as in the case of the other "strong–weak" combination above, the overall equilibrium constant is quite large, and the reaction proceeds essentially to completion. After reaction of equal molar amounts of acid and base, the solution contains the salt ammonium chloride. Is the solution neutral, acidic, or basic? You can tell by looking at each ion of the salt. The Cl^- ion is the conjugate base of a strong acid and so gives neutral solutions (Table 15.6). However, the ammonium ion is the conjugate acid of the weak base ammonia, so it produces an acidic solution.

$$NH_4^+(aq) + H_2O(\ell) \rightleftharpoons NH_3(aq) + H_3O^+(aq)$$

$$K_a \text{ for } NH_4^+ = 5.6 \times 10^{-10} \text{ (Table 15.4)}$$

In general, *a strong acid and a weak base will give an acidic solution* when equal molar quantities are mixed.

EXAMPLE 16.2

pH AT THE EQUIVALENCE POINT OF A STRONG ACID/WEAK BASE TITRATION

Suppose you titrate 1.00×10^2 mL of 0.10 M HCl with 50. mL of 0.20 M NH_3. What is the pH of the solution at the equivalence point?

Solution The balanced equation for the HCl/NH_3 reaction is given in the text above. As explained, the ammonium ion of the product is a weak Brønsted acid. This means the pH of the solution depends on K_a for NH_4^+ and on its concentration. The latter is found as follows:

$$\left.\begin{array}{l}(0.100 \text{ L})(0.10\ M \text{ HCl}) = 0.010 \text{ mol HCl used} \\ (0.050 \text{ L})(0.20\ M\ NH_3) = 0.010 \text{ mol } NH_3 \text{ used}\end{array}\right\} \longrightarrow 0.010 \text{ mol } NH_4Cl$$

The reactants were present initially in equal amounts (0.010 mole), so 0.010 mole of NH_4Cl must have been produced. Since the solution volume after titration is the total of the two reacting solutions, 1.50×10^2 mL, the product ion concentrations are

$$[NH_4^+] = [Cl^-] = 0.010 \text{ mol}/0.150 \text{ L} = 0.067\ M$$

Now we can solve for the hydrogen ion concentration of the solution using the equilibrium constant for ammonium ion ionization.

	$[NH_4^+]$	$[NH_3]$	$[H^+]$
Before ionization (M)	0.067	0	0
Change when proceeding to equilibrium	$-x$	$+x$	$+x$
At equilibrium (M)	$0.067 - x$	x	x

Once again, we can simplify the problem (Example 15.6): 100 times K_a for NH_4^+ is 5.6×10^{-8}, a number much less than the original concentration of NH_4^+ (0.067 M). Therefore, we can assume that x is very small relative to the original ammonium ion concentration, so $0.067 - x \cong 0.067\ M$.

$$K_a = 5.6 \times 10^{-10} = \frac{[NH_3][H^+]}{[NH_4^+]} = \frac{(x)(x)}{(0.067 - x)} \cong \frac{x^2}{0.067}$$

$$x = [NH_3] = [H^+] = 6.1 \times 10^{-6}\ M$$

$$pH = 5.21$$

The pH of 5.21 indicates that the solution is acidic, as was anticipated for the salt of a weak base.

FIGURE 16.1

A commercial remedy for excess stomach acid. The bubbles are carbon dioxide, CO_2, from the reaction between a Brønsted acid (citric acid, $C_6H_8O_7$) and a Brønsted base, HCO_3^- (from sodium bicarbonate). (Charles D. Winters)

EXERCISE 16.2 pH AFTER STRONG ACID/WEAK BASE REACTION

Aniline, $C_6H_5NH_2$, is a weak organic base first discoverd in 1826 when the deep blue dye indigo was heated. If you mix 50. mL of 0.20 M HCl with 0.93 g of aniline, are the acid and base completely consumed? What is the pH of the resulting solution? (K_a of $C_6H_5NH_3^+$ is 2.4×10^{-5})

$$HCl(aq) + C_6H_5NH_2(aq) \longrightarrow C_6H_5NH_3^+(aq) + Cl^-(aq)$$

There are some weak bases, especially anions, that are capable of reacting with more than one proton. The carbonate ion is an important example.

$$H^+(aq) + CO_3^{2-}(aq) \rightleftharpoons HCO_3^-(aq) \qquad K = 1/K_a \text{ for } HCO_3^-$$

$$H^+(aq) + HCO_3^-(aq) \rightleftharpoons H_2CO_3(aq) \qquad K = 1/K_a \text{ for } H_2CO_3$$

$$H_2CO_3(aq) \rightleftharpoons H_2O(\ell) + CO_2(g)$$

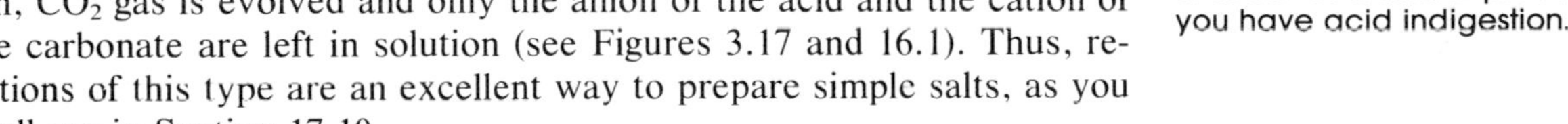

If enough acid is added to a solution of carbonate or hydrogen carbonate ion, CO_2 gas is evolved and only the anion of the acid and the cation of the carbonate are left in solution (see Figures 3.17 and 16.1). Thus, reactions of this type are an excellent way to prepare simple salts, as you shall see in Section 17.10.

This is why you take "bicarbonate of soda" or a similar product when you have acid indigestion.

A SUMMARY OF ACID–BASE REACTIONS

When writing an acid–base reaction, you must pay attention to whether the reactants are strong or weak. When one reactant is weak, the pH of the solution after mixing equal molar amounts of acid and base is controlled by the conjugate base of the weak acid or conjugate acid of the weak base. This means the equilibrium constant for the acid–base reaction is the reciprocal of the K for ionization of the conjugate base or acid. These relationships are summarized in Table 16.1.

Not included in Table 16.1 is one final case, the reaction of a weak acid with a weak base. For example, if formic acid, a weak acid, is mixed with ammonia, a weak base, the following reaction occurs.

$$HCOOH(aq) + NH_3(aq) \rightleftharpoons NH_4^+(aq) + HCOO^-(aq)$$

TABLE 16.1 Characteristics of Acid–Base Reactions

TYPE	NET IONIC EQUATION	K	SPECIES PRESENT AFTER EQUAL MOLAR AMOUNTS ARE MIXED (SOLUTION pH)
Strong acid–strong base HCl + NaOH	$H^+(aq) + OH^-(aq) \rightleftharpoons H_2O$	1.0×10^{14}	Na^+, Cl^- neutral, pH = 7
Weak acid–strong base HCOOH + NaOH	$HCOOH(aq) + OH^-(aq) \rightleftharpoons HCOO^-(aq) + H_2O$	1.8×10^{10}	Na^+, $HCOO^-$ basic, pH > 7
Strong acid–weak base HCl + NH_3	$H^+(aq) + NH_3(aq) \rightleftharpoons NH_4^+(aq)$	1.8×10^{9}	Cl^-, NH_4^+ acidic, pH < 7

If equal molar quantities of acid and base were mixed, the resulting solution will contain only ammonium formate, NH_4CHO_2. Is this solution acidic or basic? Ammonium ion is the conjugate acid of a weak base, and so should contribute to the solution acidity.

$$K_a \text{ for } NH_4^+ = 5.6 \times 10^{-10}$$

$$NH_4^+(aq) \rightleftharpoons NH_3(aq) + H^+(aq)$$

On the other hand, formate ($HCOO^-$) is the conjugate base of a weak acid; it should make the solution basic.

$$K_b \text{ for } HCOO^- = 5.6 \times 10^{-11}$$

$$HCOO^-(aq) + H_2O(\ell) \rightleftharpoons HCOOH(aq) + OH^-(aq)$$

It is difficult to analyze this situation quantitatively, so we shall make only a *qualitative* prediction. To do this, consider the ionization constants for the ions in solution, and you see that $K_a > K_b$. This tells us that more H^+ is produced than OH^-, so the solution is expected to be slightly acidic; the pH will be less than 7. *In general, when a weak acid and a weak base react in equal molar amounts in solution, a salt will be produced; the pH of the solution will depend on the relative K values of the conjugate base and acid.*

EXERCISE 16.3 pH OF A SOLUTION OF A SALT OF A WEAK ACID AND WEAK BASE

Suppose you mix 50 mL of 0.10 *M* acetic acid and 50 mL of 0.10 *M* pyridine (C_5H_5N). (K_a for the pyridinium ion, $C_5H_5NH^+$ is 6.7×10^{-6}.) Is the solution acidic or basic?

16.2 THE COMMON ION EFFECT AND BUFFER SOLUTIONS

One of the objectives of this chapter is to follow the course of events during an acid–base titration. Before we can do this, though, we must explore the nature of solutions of acids and bases that contain another solute, in particular an ion that is common to the acid or base.

THE COMMON ION EFFECT

Lactic acid, $CH_3CH(OH)COOH$, is a weak acid found in sour milk, apples and other fruit, beer and wine, and several plants.

$$\underset{\text{lactic acid}}{CH_3CH(OH)COOH(aq)} \rightleftharpoons H^+(aq) + \underset{\text{lactate ion}}{CH_3CH(OH)COO^-(aq)}$$

$$K_a = 1.3 \times 10^{-4}$$

Suppose you are analyzing a sample containing lactic acid by titrating the sample with a solution of the strong base sodium hydroxide.

$$\underset{\text{lactic acid}}{CH_3CH(OH)COOH(aq)} + OH^-(aq) \rightleftharpoons \underset{\text{lactate ion}}{CH_3CH(OH)COO^-(aq)} + H_2O(\ell)$$

FIGURE 16.2
The common ion effect. An acid (0.25 *M* HOAc, left) is mixed with a base (0.10 *M* NaOAc, sodium acetate, right). The indicator (bromphenol blue, Table 15.3) shows the resulting solution (center) has a lower H^+ concentration (a pH of about 5) than the HOAc solution. (Charles D. Winters)

At the equivalence point in the titration, the acid will have been consumed and converted completely to its conjugate base, the lactate ion. However, *before* the equivalence point, some lactate ion is present along with unreacted lactic acid. Thus, *before the equivalence point, the weak acid is present along with some amount of its conjugate base*. If we halt the titration at some intermediate stage before the equivalence point and allow equilibrium to be established, it is observed that the ionization of the remaining lactic acid is affected by the lactate ion. This is called the **common ion effect**, since an ion (here the lactate ion) "common" to the dissociation of the acid is present in an amount greater than that produced by simple acid dissociation. In an acid ionization, the presence of some conjugate base in solution will, according to Le Chatelier's principle, limit the extent to which the acid ionizes and thus affect the pH of the solution (Figure 16.2).

EXAMPLE 16.3

THE COMMON ION EFFECT

What is the pH of 0.25 *M* aqueous acetic acid? What is the pH of the solution after adding enough sodium acetate to make the solution 0.10 *M* in the salt? (Ignore any change in volume on adding the solid sodium acetate.)

Solution As a first step, let us determine the pH of the 0.10 *M* HOAc solution using the techniques outlined in Chapter 15.

(a) Write a balanced equation to remind yourself of the equilibrium in question.

$$HOAc(aq) \rightleftharpoons H^+(aq) + OAc^-(aq)$$

and then set up a table of concentrations.

	[HOAc]	$[H^+]$	$[OAc^-]$
Concentrations before ionization (*M*)	0.25	0	0
Change when proceeding to equilibrium	$-x$	$+x$	$+x$
Concentrations at equilibrium (*M*)	$0.25 - x$	x	x

$$K_a = 1.8 \times 10^{-5} = \frac{[H^+][OAc^-]}{[HOAc]} = \frac{(x)(x)}{0.25 - x} \cong \frac{x^2}{0.25}$$

The hydrogen ion concentration can be found from the approximate expression $K_a = x^2/(0.25)$. Here 100 times K_a is 1.8×10^{-3}, a number much smaller than the original concentration of acid, and so the approximation that $(0.25 - x)\ M \cong 0.25\ M$ is valid. Therefore, $x = [H^+] = [OAc^-] \cong 2.1 \times 10^{-3}\ M$. This means the pH of the solution is 2.67.

(b) Now we can tackle the problem of the added "common ion." Sodium acetate is a very soluble salt that is 100% dissociated into its ions. The Na^+ ion is the conjugate acid of a strong base and so is neutral in solution (Table 15.6). On the other hand, OAc^- is the conjugate base of the weak acid HOAc, so it should contribute OH^- to the solution,

$$OAc^-(aq) + H_2O(\ell) \rightleftharpoons HOAc(aq) + OH^-(aq)$$

and thus reduce $[H^+]$ below $2.1 \times 10^{-3}\ M$, the concentration just found for the solution containing only acid.

As outlined in a box below, we shall approach this situation in the following way: *Imagine* that the solution first contains HOAc and that this does not ionize to any extent. Then add OAc^- as its sodium salt. Now imagine that the HOAc ionizes in the presence of OAc^- to give H^+ and *more* OAc^-, both in an amount y. (Without added OAc^-, ionization of the acid gave these ions in an amount x, but because of Le Chatelier's principle $y < x$). Thus, in our summary table, we begin with two species in solution, and the concentration of acetic acid declines by y and the concentrations of H^+ and OAc^- increase by y.

	[HOAc]	$[H^+]$	$[OAc^-]$
Concentrations before acid ionization (M)	0.25	0	0.10
Change when proceeding to equilibrium	$-y$	$+y$	$+y$
Concentrations at equilibrium (M)	$0.25 - y$	y	$0.10 + y$

These concentrations can now be substituted into the usual equilibrium constant expression.

$$K_a = \frac{[H^+][OAc^+]}{[HOAc]} = \frac{(y)(0.10 + y)}{(0.25 - y)} = 1.8 \times 10^{-5}$$

To solve this requires the quadratic formula, unless we can make some approximations. What is the relative magnitude of y? The amount of OAc^- formed in the pure acetic acid solution was only 0.0021 M, but now it is going to be even less. Therefore, the term $(0.10 + y)\ M \cong 0.10\ M$ and $(0.25 - y)\ M \cong 0.25\ M$. Now we have a much simpler expression to solve for y.

$$K_a = 1.8 \times 10^{-5} = \frac{(0.10)(y)}{0.25}$$

$$y = [H^+] = 4.5 \times 10^{-5}\ M$$

$$pH = 4.35$$

Now compare the *before* and *after* situations. The pH before adding sodium acetate was 2.67, but, because of the basic nature of the acetate ion, the acidity has declined (the pH has increased to 4.35) after adding sodium acetate.

EXERCISE 16.4 THE COMMON ION EFFECT

Assume you have a 0.30 M solution of formic acid (HCOOH), and add enough sodium formate (NaCHOO) to make the solution 0.10 M in the salt. Calculate the pH of the formic acid solution before and after adding sodium formate.

$$\underset{\text{formic acid}}{HCOOH(aq)} \rightleftharpoons H^+(aq) + \underset{\text{formate ion}}{HCOO^-(aq)}$$

A NOTE ON SOLVING COMMON ION PROBLEMS

In these chapters on equilibrium calculations you will see some cases where a system at equilibrium such as

$$A + B \rightleftharpoons X + Y$$

$$K = \frac{[X][Y]}{[A][B]}$$

is stressed by the addition of an extra amount of reactant (A or B) or product (X or Y). Le Chatelier's principle tells us that the reactant and product concentrations will shift to new values so that the ratio of product to reactant concentrations remains equal to K. We are often interested in knowing what these new concentrations are, and the most straightforward way to find them is to *ignore the equilibrium that existed before adding extra reagent.* Therefore, for purposes of the calculation, the initial concentrations of reactants (or products) will be those originally in the solution *plus* the added reagent. Only then do we imagine that equilibrium is established. The equilibrium concentration of a particular species will then be (amount originally present) *plus* (amount of extra material added) *plus* or *minus* (material appearing or disappearing owing to the equilibrium reaction).

BUFFER SOLUTIONS

The normal pH of human blood is 7.4. Experiment clearly shows that the addition of a small quantity of strong acid or base to blood, say 0.01 mole to a liter of blood, leads to a change in pH of only about 0.1 pH units (Figure 16.3). In comparison, if you add 0.01 mole of HCl to 1.0 L of water, the pH drops from 7 to 2, while addition of 0.01 mole of NaOH increases the pH from 7 to 12. Blood, and many other body fluids, are said to be *buffered;* their pH is resistant to change upon addition of strong acid or base.

In general, two species are required in a **buffer** (buffered solution), one capable of reacting with added OH^- ions (an acid) and another that can consume added H^+ ions (a base). An additional requirement is that the two species must not react with one another. This means a buffer is usually prepared from roughly equal quantities of (a) a weak acid and its conjugate base (acetic acid and acetate ion, for example) or (b) a weak base and its conjugate acid (ammonia and ammonium ion, for example). Some systems commonly used in the laboratory are given in Table 16.2.

To see how a buffer solution works, let us consider an acetic acid/acetate buffer. The acetic acid, the weak acid, is needed to consume any added hydroxide ion.

$$HOAc(aq) + OH^-(aq) \rightleftharpoons OAc^-(aq) + H_2O(\ell) \qquad K = 1.8 \times 10^9$$

The equilibrium constant for the reaction is very large because OH^- is a *much* stronger base than OAc^-, as we know from Table 15.2. This means

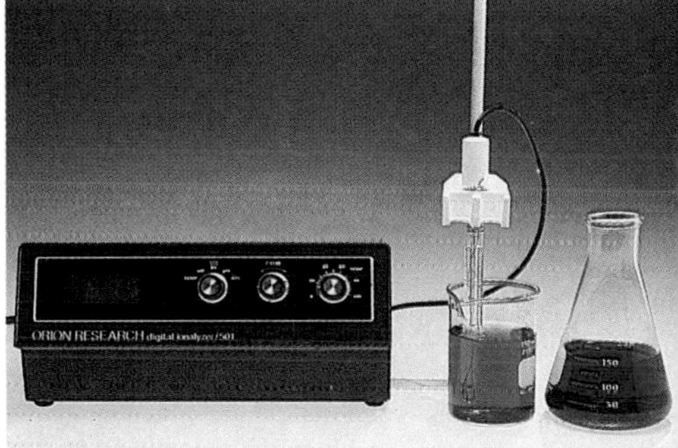

(a)

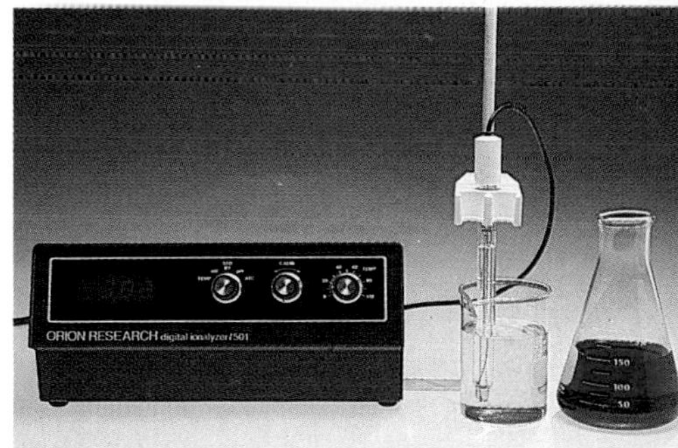

(b)

FIGURE 16.3

Buffer solutions. (a) The pH electrode is indicating the pH of water that contains a trace of acid (and bromcresol green indicator; see Table 15.3). The solution at the right is a buffer having a pH the same as human blood (7.4). (b) When 5 ml of 0.10 *M* HCl is added to each solution, the pH of the water drops several units, while the pH of the buffer stays constant. (Charles D. Winters)

TABLE 16.2 Some Commonly Used Buffer Systems

WEAK ACID	CONJUGATE BASE	ACID K_a	pK_a	USEFUL pH RANGE
Phthalic acid $C_6H_4(COOH)_2$	Hydrogen phthalate $C_6H_4(COOH)COO^-$	1.3×10^{-3}	2.89	1.9–3.9
Acetic acid $HC_2H_3O_2$	Acetate $C_2H_3O_2^-$	1.8×10^{-5}	4.75	3.7–5.8
Dihydrogen phosphate $H_2PO_4^-$	Hydrogen phosphate HPO_4^{2-}	6.2×10^{-8}	7.21	6.2–8.2
Hydrogen phosphate HPO_4^{2-}	Phosphate PO_4^{3-}	4.8×10^{-13}	12.32	11.3–13.3

that any OH^- entering the solution from an outside source is consumed completely. Similarly, any hydrogen ion added to the solution can be consumed completely by the acetate ion present in the buffer.

$$OAc^-(aq) + H^+(aq) \longrightarrow HOAc(aq) \qquad K = 5.6 \times 10^4$$

The equilibrium constant for this reaction is also quite large because H^+ is a *much* stronger acid than HOAc, again as suggested by the data in Table 15.2.

Because of its composition a buffer solution is just a special case of the "common ion effect."

Now that we have established that a buffer should effectively remove small amounts of added acid or base, work through the following example, which shows more quantitatively how the pH of a solution can be maintained.

EXAMPLE 16.4

A BUFFER SOLUTION

Calculate the pH change that occurs when 1.0 mL of 1.0 *M* HCl is added (a) to 1.0 L of pure water and (b) to 1.0 L of acetic acid/sodium acetate buffer (where [HOAc] = 0.70 *M* and [OAc^-] = 0.60 *M*).

Notice that, when a strong acid is added to water, we neglect the amount of H^+ from the water in calculating the final H^+ ion concentration. (See Example 15.2.)

Solution (a) *Adding Acid to Pure Water*. 1.0 mL of 1.0 *M* HCl represents 0.0010 moles of acid. If this is added to 1.0 L of pure water, the hydrogen ion concentration is 1.0×10^{-3} *M*. This means the [H^+] concentration is raised from 10^{-7} to 10^{-3} *M*, so the pH falls from 7 to 3.

(b) *pH of Acetic Acid/Acetate Buffer Solution*. We shall first determine the pH of the buffer solution before any HCl is added. The balanced equation connecting the species in solution is

$$HOAc(aq) \rightleftharpoons H^+(aq) + OAc^-(aq)$$

and the table of concentrations is

	[HOAc]	[H^+]	[OAc^-]
Concentrations before ionization (*M*)	0.70	0	0.60
Change when proceeding to equilibrium	$-x$	$+x$	$+x$
Concentrations at equilibrium (*M*)	$0.70 - x$	x	$0.60 + x$

Before writing out the equilibrium constant expression, you can see that x is going to be much less than 0.70 or 0.60, since Le Chatelier's principle predicts that the equilibrium will lie far to the left because of the added acetate ion. Therefore,

$$K_a = \frac{[H^+][OAc^-]}{[HOAc]}$$

or

$$[H^+] = \frac{[HOAc]}{[OAc^-]} K_a \cong \frac{(0.70)}{(0.60)} (1.8 \times 10^{-5}) = 2.1 \times 10^{-5}\ M$$

$$pH = 4.68$$

Adding HCl to the Buffer Solution. When 1.0 mL of 1.0 M HCl is added to 1.0 L of the buffer, the following reaction occurs,

$$H^+(aq) + OAc^-(aq) \longrightarrow HOAc(aq)$$

and the concentration of each ion or molecule in this equation can be described as follows.

	$[H^+]$ (FROM HCl)	$[OAc^-]$ (FROM BUFFER)	[HOAc] (FROM BUFFER)
Mol/L before reaction	0.0010	0.600	0.700
Change when reaction is complete	−0.0010	−0.001	+0.001
Mol/L after *complete* reaction	0	0.599	0.701

Since the added HCl reacts *completely* with acetate to produce acetic acid, the buffer once again contains only the weak acid and its salt. Therefore, we only need to consider the equilibrium

$$HOAc(aq) \rightleftharpoons H^+(aq) + OAc^-(aq)$$

where HOAc ionizes in the presence of its common ion OAc^-.

	[HOAc]	$[H^+]$	$[OAc^-]$
Concentrations before ionization (M)	0.701	0	0.599
Change when proceeding to equilibrium	$-y$	$+y$	$+y$
Concentrations after equilibrium (M)	$0.701 - y$	y	$0.599 + y$

As usual, we make the approximation that y, the amount of H^+ formed by dissociating HOAc in the presence of OAc^-, is small compared with 0.701 or 0.599. Therefore, we can write

$$[H^+] = \frac{[HOAc]}{[OAc^-]} K_a = \frac{0.701}{0.599} (1.8 \times 10^{-5}) = 2.1 \times 10^{-5}\ M$$

$$pH = 4.68$$

Within the number of significant figures allowed, *the pH does not change in the buffer solution* after adding HCl, even though it changed by 4 units when 1 mL of 1.0 M HCl was added to 1.0 L of pure water. Buffer solutions do indeed "buffer." In this case the acetate ion consumed the added hydrogen ion and thus resisted a change in pH.

EXERCISE 16.5 BUFFER SOLUTIONS

Calculate the pH of 0.500 L of a buffer solution composed of 0.50 M formic acid (HCO_2H) and 0.70 M sodium formate ($NaHCO_2$) before and after adding 10. mL of 1.0 M HCl.

GENERAL EXPRESSIONS FOR BUFFER SOLUTIONS In Example 16.4, we solved for the hydrogen ion concentration of the acetic acid/acetate buffer solution by rearranging the K_a expression to give

$$[H^+] = \frac{[HOAc]}{[OAc^-]} K_a$$

This result can be generalized for any buffer. For a buffer solution based on a *weak acid and its conjugate base*, the $[H^+]$ is always given by

$$\mathbf{[H^+] = \frac{[acid]}{[conjugate\ base]} K_a}$$

When the usual expression for K_b is rearranged, the hydroxide ion concentration in a buffer composed of *a weak base and its conjugate acid* (e.g., NH_3 and NH_4^+) is

$$\mathbf{[OH^-] = \frac{[base]}{[conjugate\ acid]} K_b}$$

The equations above are often written in a slightly different manner for convenience in use. If we take the negative logarithm of each side of the equation for $[H^+]$, for example, we have

$$-\log[H^+] = \left\{-\log \frac{[acid]}{[conjugate\ base]}\right\} + \{-\log K_a\}$$

You know that $-\log[H^+]$ is defined as pH, and from Chapter 15 you recall that the definition can be extended to other quantities. Thus, $-\log K_a$ is equivalent to pK_a. Furthermore, since $\{-\log[acid]/[conjugate\ base]\}$ is equivalent to $\{+\log[conjugate\ base]/[acid]\}$, we can now rewrite the equation above as

$$\mathbf{pH = pK_a + \log \frac{[conjugate\ base]}{[acid]}}$$

The Henderson–Hasselbalch equation is only valid when the ratio (acid/conjugate base) is no larger than 10 and no smaller than 0.10.

This equation is known as the **Henderson–Hasselbalch equation**. It is quite useful, since many texts list acid and base dissociation constants in terms of pK values.

EXAMPLE 16.5

USING THE HENDERSON–HASSELBALCH EQUATION

2.00 g of benzoic acid (C_6H_5COOH) and 2.00 g of sodium benzoate (C_6H_5COONa) are dissolved in enough water to make 1.0 L of solution. What is the pK_a of the acid? What is the pH of the solution? (See Example 15.6, p. 572.)

Solution (a) Our first objective is to find the pK_a, that is, $-\log K_a$, for benzoic acid. Appendix G gives the K_a of benzoic acid as 6.3×10^{-5}. Thus,

$$-\log (6.3 \times 10^{-5}) = pK_a = 4.20$$

(b) The next objective is to calculate the pH of the solution. To do this, you first need the concentration of the acid (benzoic acid) and its conjugate base (benzoate ion).

$$2.00 \text{ g benzoic acid} \left(\frac{1 \text{ mol}}{122 \text{ g}}\right) = 0.0164 \text{ moles benzoic acid}$$

$$2.00 \text{ g sodium benzoate} \left(\frac{1 \text{ mol}}{144 \text{ g}}\right) = 0.0139 \text{ moles sodium benzoate}$$

Since the solution volume is 1.0 L, we know that [benzoic acid] = 0.0164 *M* and [sodium benzoate] = 0.0139 *M*. Therefore, the Henderson–Hasselbalch equation can be written

$$pH = 4.20 + \log \frac{0.0139\ M}{0.0164\ M}$$

$$pH = 4.20 + \log(0.848) = 4.13$$

EXERCISE 16.6 USING THE HENDERSON–HASSELBALCH EQUATION

K_a for HCO_3^- is 5.61×10^{-11}. What is the pK_a of the acid? Suppose that you mix 15.0 g of $NaHCO_3$ with 18.0 g of Na_2CO_3 in enough water to make 1.00 L of solution. Use the Henderson–Hasselbalch equation to calculate the pH of the solution.

PREPARING BUFFER SOLUTIONS There are two obvious requirements for a buffer solution: (1) it should have the capacity to control the pH after the addition of reasonable amounts of acid and base and (2) it should control the pH at the desired value.

By *buffer capacity*, we mean that there must be a large enough concentration of acetic acid in an acetic acid/acetate buffer, for example, to consume all of the hydrogen ion that may be added with only a very small change in solution pH (see Example 16.4).* The requirement is usually met by using 0.10 *M* to 1.0 *M* solutions of reagents. However, any buffer eventually loses its capacity if enough strong acid or base is added.

Now to the second point. The equation for hydrogen ion concentration of an acid buffer

$$[H^+] = \frac{[\text{acid}]}{[\text{conjugate base}]} K_a$$

shows that there are two factors that can be controlled: (1) K_a of the acid and (2) the acid/conjugate base ratio. Usually an acid is selected whose K_a is close to the desired $[H^+]$, and then the exact value of $[H^+]$ desired is achieved by adjusting the acid/conjugate base ratio. The following example illustrates this.

EXAMPLE 16.6

PREPARING A BUFFER SOLUTION

Suppose you wish to prepare a buffer solution to maintain the pH at 4.30. A list of possible acids (and their conjugate bases) is

---

*For an excellent illustration of buffer capacity see C.J. Donahue, M.G. Panek, *J. Chem. Educ.* 62:337, 1985.

ACID	CONJUGATE BASE	K_a
HSO_4^-	SO_4^{2-}	1.2×10^{-2}
HOAc	OAc^-	1.8×10^{-5}
HCN	CN^-	4.0×10^{-10}

Which combination should be selected and what should the ratio of acid to base be?

Solution The desired hydrogen ion concentration is $5.0 \times 10^{-5}\ M$,

$$pH = 4.30, \text{ so } [H^+] = 10^{-pH} = 10^{-4.30} = 5.0 \times 10^{-5}\ M$$

so the correct choice of acid is now evident. Of the acids given, only acetic acid (HOAc) has a K_a value close to that of the desired $[H^+]$. Therefore, you need only adjust the ratio $[HOAc]/[OAc^-]$ to achieve the exact, desired hydrogen ion concentration.

$$[H^+] = 5.0 \times 10^{-5} = \frac{[HOAc]}{[OAc^-]}(1.8 \times 10^{-5})$$

To satisfy the expression above, you must have a ratio $[HOAc]/[OAc^-]$ of 2.8/1. Therefore, if you add 0.28 moles of acetic acid and 0.10 moles of sodium acetate to enough water to make one liter of solution, the solution will constitute a buffer capable of controlling the pH at 4.30.

EXERCISE 16.7 PREPARING A BUFFER SOLUTION

Using an acetic acid/sodium acetate buffer solution, what ratio of acid to conjugate base will you need to maintain the pH at 5.00? Explain how you would make up such a solution.

The last example brings up one more important point concerning buffer solutions. The hydrogen ion concentration depends on the ratio [acid]/[conjugate base] (or hydroxide ion concentration depends on [base]/[conjugate acid]). However, although we have written these ratios in terms of reagent concentrations, it is not concentrations that are important; rather, what is important is *the **relative** number of moles of acid and conjugate base*. Since both reagents are dissolved in the same solution, their concentrations depend on the same solution volume. In Example 16.6, the ratio 2.8/1 for acetic acid and sodium acetate implied that 2.8 moles of acetic acid and 1.0 mole of sodium acetate were dissolved per liter.

$$\frac{[HOAc]}{[OAc^-]} = \frac{2.8 \text{ mol/L}}{1.0 \text{ mol/L}} = \frac{2.8 \text{ mol}}{1.0 \text{ mol}}$$

Alternatively, 1.4 moles of acetic acid and 0.50 moles of sodium acetate, or any amounts that give a ratio of 2.8/1, could have been used. Further, since the actual concentration is not important, the acid and its conjugate could have been dissolved in any reasonable amount of water. The ratio of acid to conjugate base is maintained, no matter what the solution volume may be. This is the reason that commercially available buffer solutions are sold as concentrated solutions. To use them, you only need to take a small amount and dilute it with pure water to a convenient volume (Figure 16.4).

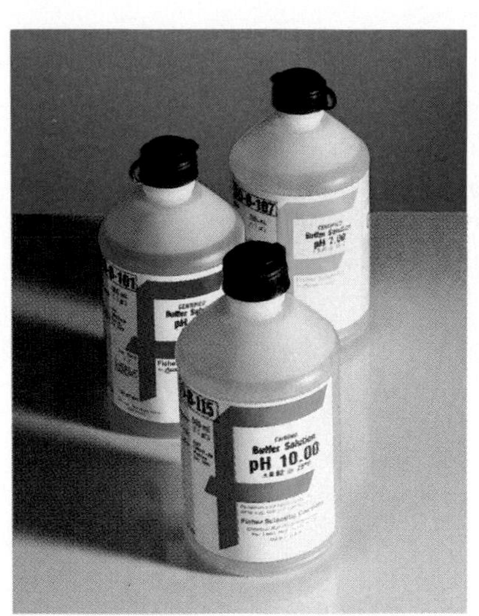

FIGURE 16.4
Commercial buffer solution. To use the solution, take a small volume and dilute it with pure water to a larger volume. The ratio [acid]/[conjugate base] or [base]/[conjugate acid] remains constant. Thus, the pH of the buffer remains constant no matter what the volume of solution. (Fisher Scientific Company)

16.3 ACID–BASE TITRATION CURVES

In Chapter 4 we described the stoichiometry involved in acid–base titrations, a method for the accurate analysis of the amount of acid or base in a sample. In the current chapter you have learned something more about such titrations. For example, you know now that the pH at the equivalence point is 7, and the solution is truly neutral, only when a strong acid is titrated with a strong base or vice versa. If one of the substances being titrated is weak, then the pH at the equivalence point is not 7. We found that

(a) Weak acid + strong base $\longrightarrow$ pH > 7 at equivalence point due to hydrolysis of conjugate base of weak acid

(b) Strong acid + weak base $\longrightarrow$ pH < 7 at equivalence point due to hydrolysis of conjugate acid of weak base

Now we wish to describe the *change in pH* of a solution of an acid or base as it is being titrated with a base or acid. To illustrate this, the following examples explore two of the most common situations: (a) the titration of a strong acid with a strong base and (b) the titration of a weak acid with a strong base.

EXAMPLE 16.7

THE TITRATION OF A STRONG ACID WITH A STRONG BASE

You have 50.0 mL of a 0.100 *M* solution of HCl. Calculate the pH of the solution as 0.100 *M* NaOH is slowly added. Plot the results with pH on the vertical axis and milliliters of added base on the horizontal axis.

Solution (a) HCl is a strong acid, so its concentration is equivalent to the H^+ concentration. Since [HCl] $= 1.00 \times 10^{-1}$ *M* in the original solution, the pH is 1.000.

(b) The number of moles of HCl originally present is

$$\text{Moles H}^+ = (0.0500\text{ L})(0.100\ M) = 5.00 \times 10^{-3}\text{ moles}$$

(c) Now add the first increment of NaOH, 10.0 mL. After 10.0 mL of NaOH have been added, this means 1.00×10^{-3} moles of base have been added.

$$(0.0100\text{ L NaOH})(0.100\text{ moles/L}) = 1.00 \times 10^{-3}\text{ moles of NaOH added}$$

Since the net reaction in the titration of a strong acid with a strong base is

$$H^+(aq) + OH^-(aq) \longrightarrow H_2O(\ell)$$

the added NaOH is completely consumed by reaction with 1.0×10^{-3} moles of HCl, and the moles of HCl remaining are

0.00500 mol HCl originally present
− 0.00100 mol HCl consumed by addition of 0.00100 mol of added NaOH

0.00400 mol HCl remaining after adding 10.0 mL of NaOH

Remembering that the total solution volume now is 60.0 mL (50.0 mL of HCl originally + 10.0 mL of NaOH added), the H^+ concentration now is

$$[H^+] = \frac{4.00 \times 10^{-3} \text{ moles}}{0.0600 \text{ L}} = 6.67 \times 10^{-2}\ M$$

$$pH = 1.176$$

(d) In general, the hydrogen ion concentration up to the equivalence point can be calculated, just as it was after the first addition, from the expression

$$[H^+] = \frac{\text{original moles acid} - \text{total moles of base added}}{\text{volume acid} + \text{total volume base added}}$$

For instance, after 45.0 mL of NaOH have been added, the acid concentration is

$$[H^+] = \frac{5.00 \times 10^{-3} \text{ moles HCl} - (0.0450 \text{ L})(0.100\ M)}{0.0500 \text{ L} + 0.0450 \text{ L}} = 5.26 \times 10^{-3}\ M$$

and the pH is 2.279.

(e) When the volume of added base has reached 49.0 mL,

$$[H^+] = \frac{5.00 \times 10^{-3} \text{ moles HCl} - (0.0490 \text{ L})(0.100\ M)}{0.0500 \text{ L} + 0.0490 \text{ L}} = 1.01 \times 10^{-3}\ M$$

$$pH = 2.996$$

the pH indicates the solution is still quite acidic.

(f) Finally, when 50.0 mL of base have been added, the equivalence point is reached, since 5.00×10^{-3} moles of base have now been added. All of the acid has been consumed by base, and only a solution of the "neutral ions" Na^+ and Cl^- remains. The pH is therefore 7.00 (see page 603), since in pure water or in water containing only a neutral salt such as NaCl,

$$[H^+] = [OH^-] = \sqrt{K_w} = 1.00 \times 10^{-7}$$

(g) When the equivalence point has been passed, no acid remains; NaOH is added to water containing NaCl, a neutral salt. Therefore, we calculate the pH from the concentration of the OH^- ion. For example, after only 1.0 mL of base has been added beyond the equivalence point (total volume of base added = 51.0 mL), the pH is 11.0.

$$[OH^-] = \frac{\text{moles excess base}}{\text{total volume}} = \frac{(1.0 \times 10^{-3} \text{ L})(0.100\ M)}{0.050 \text{ L acid} + 0.051 \text{ L base}} = 9.9 \times 10^{-4}\ M$$

$$[H^+] = \frac{K_w}{[OH^-]} = \frac{1.00 \times 10^{-14}}{9.9 \times 10^{-4}} = 1.0 \times 10^{-11}\ M$$

$$pH = 11.00$$

The results of the calculations outlined above are summarized in the table and curve shown in Figure 16.5.

EXERCISE 16.8 TITRATION OF A STRONG ACID WITH A STRONG BASE

For the titration outlined in Example 16.7, verify the pH calculated for the addition of (a) 40.0 mL of NaOH and (b) 60.0 mL of NaOH. What is the pH after 49.9 mL of NaOH are added?

The pH at each point in the titration of the strong acid HCl with the strong base NaOH is illustrated in Figure 16.5. As calculated in Example 16.7, the pH rises slowly until one is very close to the equivalence

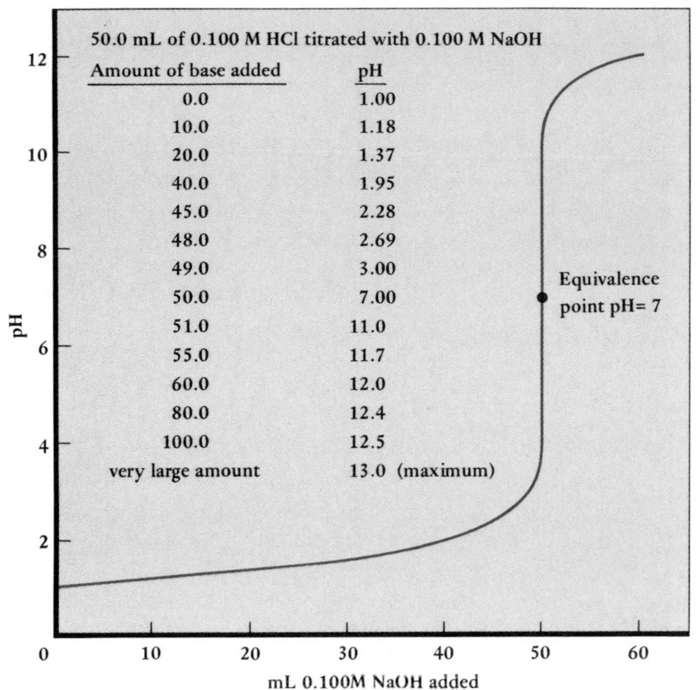

50.0 mL of 0.100 M HCl titrated with 0.100 M NaOH
Amount of base added | pH
0.0 | 1.00
10.0 | 1.18
20.0 | 1.37
40.0 | 1.95
45.0 | 2.28
48.0 | 2.69
49.0 | 3.00
50.0 | 7.00
51.0 | 11.0
55.0 | 11.7
60.0 | 12.0
80.0 | 12.4
100.0 | 12.5
very large amount | 13.0 (maximum)
Equivalence point pH= 7
pH
12
10
8
6
4
2
0
10
20
30
40
50
60
mL 0.100M NaOH added

FIGURE 16.5

The change in pH as a strong acid is titrated with a strong base. In this case 50.0 mL of 0.100 *M* HCl is titrated with 0.100 *M* NaOH.

point. Then the pH increases very rapidly, rising 7 units (the $[H^+]$ increases by a factor of 10 million!) when only a very small amount of base (perhaps a drop or two) is added (Figure 16.6). After the equivalence point is passed, only a small further rise in pH is seen.

The *titration of a weak acid with a strong base* is somewhat different from the strong acid/strong base titration just described. To illustrate, let us look carefully at the titration curve for 50.0 mL of 0.100 *M* acetic acid with 0.100 *M* NaOH (Figure 16.7).

$$HOAc(aq) + NaOH(aq) \longrightarrow NaOAc(aq) + H_2O(\ell)$$

FIGURE 16.6

Titration of a strong acid (HCl) with a strong base (NaOH) at the end point. The pH changed abruptly from acid (a) to base (b) when one drop of 1 *M* NaOH was added to the solution. (The indicator is bromcresol green; see Table 15.3). (Charles D. Winters)

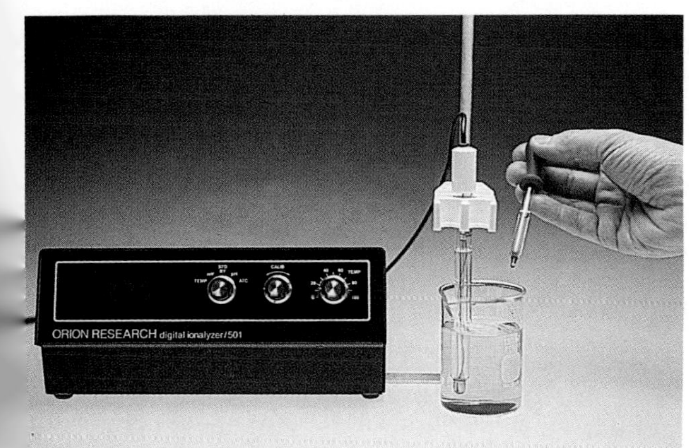

(a)

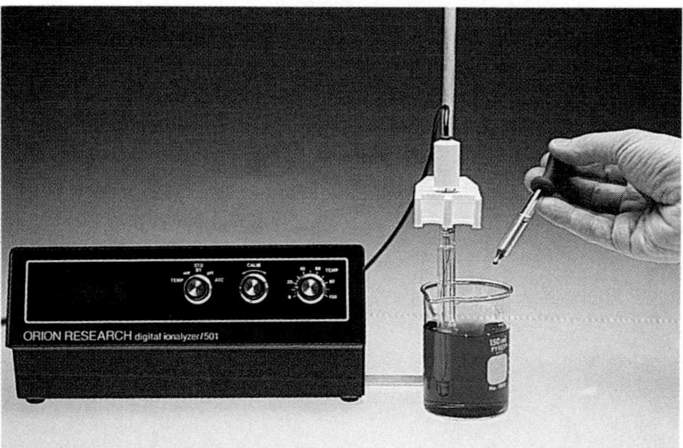

(b)

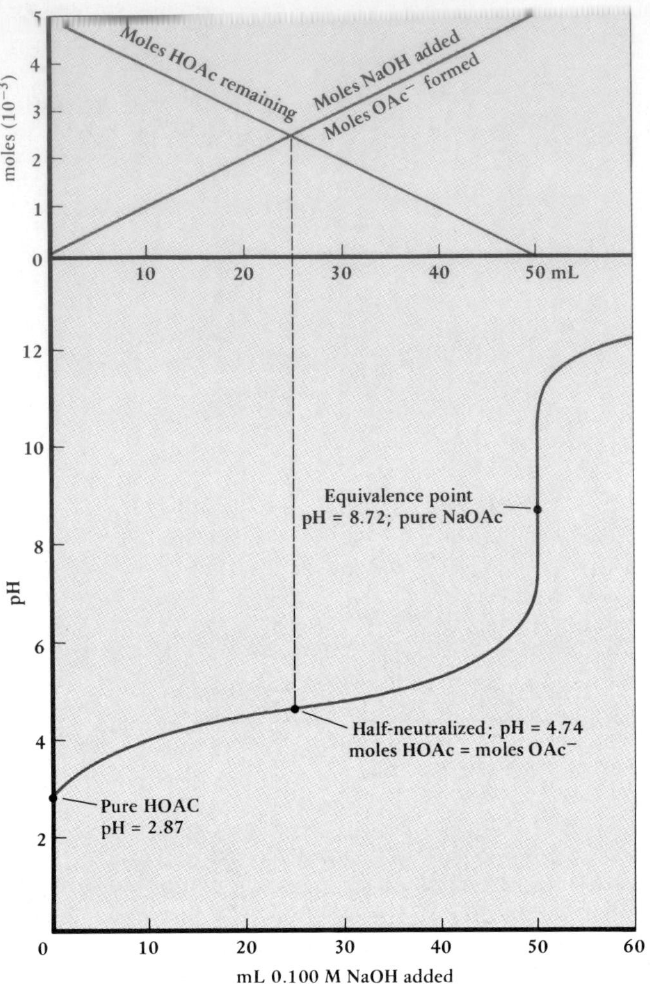

FIGURE 16.7

The change in pH during the titration of a weak acid with a strong base. (50.0 mL of 0.100 *M* HOAc is titrated with 0.100 *M* NaOH.) (Top) Moles of acid remaining or moles of conjugate base formed as a function of milliliters of added base. The point at which these lines cross is the half-neutralization point. (Bottom) The change in pH of the solution as a function of milliliters of added base.

There are four points or regions on this curve that are especially interesting: (1) the pH before titration begins; (2) the pH when the acid is half-neutralized; (3) the pH at the equivalence point; and (4) the pH when base is added beyond the equivalence point. The pH before any base has been added (2.87) is found in the usual way (Example 15.6) for the ionization of a weak acid. At the equivalence point of the titration, the solution consists simply of sodium acetate, and so the pH (8.72) is controlled by acetate ion, the conjugate base of acetic acid (see Example 16.1).

Now let us determine the pH at the half-neutralization point in the HOAc/NaOH titration and the pH after the equivalence point. As sodium hydroxide is added to the acetic acid, the hydroxide is consumed by a reaction producing sodium acetate. This means that at every point between the beginning of the titration (where only pure, weak acid is present) and the equivalence point (where only sodium acetate is present) the solution contains acetic acid *and* its salt, sodium acetate. These are of course the components of a buffer solution, and the hydrogen ion concentration can be found from the following expression when the mole ratio is greater than 0.1.

$$[H^+] = \frac{\text{moles acetic acid}}{\text{moles acetate ion}} K_a$$

But what happens when exactly half of the acid has been consumed by base? If the initial number of moles of acid were M_0, then at the *half-neutralization point*,

$$\text{Moles acid remaining} = \text{moles acid consumed} = \tfrac{1}{2} M_0$$

$$\begin{aligned}\text{Moles acetate ion formed} &= \text{initial moles acid} - \text{moles acid consumed} \\ &= M_0 - \tfrac{1}{2} M_0 = \tfrac{1}{2} M_0\end{aligned}$$

Therefore, since both the moles of acid remaining and moles of acetate formed are $\frac{1}{2} M_0$,

$$[H^+] \text{ at half-neutralization point} = \frac{\text{acid remaining}}{\text{acetate ion formed}} K_a$$

$$[H^+] \text{ at half-neutralization point} = K_a$$

At the half-neutralization point it is always a good approximation that

(a) in the titration of a weak acid with a strong base, $[H^+] = K_a$ of the weak acid (or pH = pK_a),
(b) in the titration of a weak base with a strong acid, $[OH^-] = K_b$ of the weak base (or pOH = pK_b).

A common way to determine the K_a of a weak acid or the K_b of a weak base is to titrate it with a strong base or acid, respectively, and determine the pH at the half-neutralization point.

Thus, in the particular case of the titration of HOAc with strong base, $[H^+] = 1.8 \times 10^{-5}$ *M* at the half-neutralization point, and so the pH = 4.74.

As a final point concerning the titration of a weak acid with a strong base, we wish to know the pH *after* the equivalence point. In the case of the HOAc/NaOH titration, the solution would consist of the salt sodium acetate and any NaOH that was added *in excess* after passing the equivalence point. Both acetate ion and hydroxide ion are bases, so the total concentration of OH^- would be the sum of that from excess NaOH plus that produced by hydrolysis of acetate ion.

$$OAc^-(aq) + H_2O(\ell) \rightleftharpoons HOAc(aq) + OH^-(aq)$$

However, the OH^- concentration from the hydrolysis reaction is very small compared with that from excess NaOH, so, after the equivalence point

$$[OH^-] = \frac{\text{moles excess } OH^- \text{ from NaOH}}{(\text{total volume, L})}$$

EXERCISE 16.9 TITRATION OF A WEAK ACID WITH A STRONG BASE

The titration of 50.0 mL of 0.100 *M* acetic acid with 0.100 *M* NaOH is described in the text above. What is the pH of the solution when 20.% of the acid has been neutralized? After 55.0 mL of NaOH has been added?

16.4 ACID–BASE INDICATORS

The goal of a titration is to add a standard solution in an amount that is stoichiometrically equivalent to the substance with which it reacts. This condition is achieved at the **equivalence point**. As you may now realize from laboratory experience, the equivalence point is only of theoretical concern. Usually we can only *estimate* the location of the equivalence point by some change in the solution, as for example a change in the color of a dye or **indicator**. This change occurs at the **end point** of the titration, and it is hoped that the difference between the end and equivalence points is negligible.

An indicator (HIn) is usually an organic dye that is itself a weak acid.

$$\mathrm{HIn(aq)} \rightleftharpoons \mathrm{H^+(aq)} + \mathrm{In^-(aq)}$$

$$K_a \text{ of the indicator} = \frac{[\mathrm{H^+}][\mathrm{In^-}]}{[\mathrm{HIn}]}$$

The acid form of the compound (HIn) has one color, while the conjugate base (In^-) has another color (Table 15.3). According to Le Chatelier's principle, addition of H^+ or OH^- to an indicator will cause one color or the other to appear, depending on whether HIn or In^- is the predominant species.

If the K_a expression for the indicator is rearranged

$$[\mathrm{H^+}] = \frac{[\mathrm{HIn}]}{[\mathrm{In^-}]} K_a$$

it is more evident that a tiny amount of an indicator can in fact reveal the pH of a solution by changing the $[HIn]/[In^-]$ ratio. If a drop or two of some dilute indicator solution is added in an acid–base titration, the ratio $[HIn]/[In^-]$ is controlled by the hydrogen ion concentration of the solution. As the hydrogen ion concentration of the solution changes, the ratio $[HIn]/[In^-]$ must shift to maintain the equality in the equation above. Therefore, when the hydrogen ion concentration is high, [HIn] must be large and $[In^-]$ small. The indicator color will be that of HIn. In the case of the common indicator phenolphthalein, the dye is colorless at high $[H^+]$ or low pH, whereas the indicator thymol blue is red under these circumstances (Table 15.3 and page 569). When the pH increases, and $[H^+]$ declines, the ratio $[HIn]/[In^-]$ must also decrease. Thus, the concentration of In^- increases, and the color of In^- becomes predominant. Phenolphthalein, for example, is red above pH 10 whereas thymol blue is yellow between pH 3 and 8 and purple above pH 9.

In principle, the color of the indicator changes when $[H^+] = K_a$ of the indicator (because $[HIn] = [In^-]$ at this point), and so you might think that you could accurately determine $[H^+]$ by carefully observing the color change. In practice, your eyes are not quite that good. Usually, you see the color of HIn when $[HIn]/[In^-]$ is about 10/1, and the color of In^- when $[HIn]/[In^-]$ is about 1/10. This means the color change is observed over a $[H^+]$ interval of about 100, corresponding to 2 pH units. This is not really a problem, however, as you can see in Figures 16.5 and

16.7; at the equivalence point of these titrations the pH changes by as many as 7 units.

As Table 15.3 shows, there is a variety of indicators available, each changing color in a different pH range. If you are analyzing a weak acid or base by titration, you will have to choose an indicator that changes color in a range that includes the pH to be observed at the equivalence point. This means that an indicator that changes color in the pH range 7 ± 2 should be used for a strong acid/strong base titration. On the other hand, the pH at the equivalence point in the titration of a weak acid with a strong base is greater than 7. Therefore, you should use an indicator that changes color at a pH of about 8.

EXERCISE 16.10 INDICATORS

Use Table 15.3 to decide which indicator would be best to use in the titration of NH_3 with HCl in Example 16.2.

SUMMARY

Four types of acid–base reactions were considered in Section 16.1, and the general results when mixing equal amounts of acid and base are

Strong acid + strong base ⟶ neutral solution

$$HCl + NaOH \longrightarrow H_2O + NaCl$$

Weak acid + strong base ⟶ basic solution

$$HOAc + NaOH \longrightarrow H_2O + NaOAc \text{ (a basic salt)}$$

Strong acid + weak base ⟶ acidic solution

$$HCl + NH_3 \longrightarrow NH_4Cl \text{ (an acidic salt)}$$

Weak acid + weak base ⟶

pH depends on relative K_a and K_b values

$$HOAc + NH_3 \longrightarrow NH_4OAc$$

Acetic acid ionizes to acetate and hydrogen ions. If either of these is already present or is added after equilibrium is achieved, the acetic acid will ionize to a smaller extent, as predicted by Le Chatelier's principle. The general effect is often referred to as the **common ion effect** (Section 16.2). The specific case of a *weak acid* ionizing in the presence of a significant concentration of its *conjugate base* (HOAc) with OAc^-) or a *weak base* ionizing in the presence of its *conjugate acid* (NH_3 with NH_4^+) can give a buffer solution. A **buffer solution** will resist a change in pH when a strong acid or base is added. The hydrogen ion concentration of an acid buffer (say $HOAc/OAc^-$) is

$$[H^+] = \frac{[\text{acid}]}{[\text{conjugate base}]} K_a$$

If we take the negative logarithm of each side, the result is the **Henderson–Hasselbalch equation**.

$$pH = pK_a + \log \frac{[\text{conjugate base}]}{[\text{acid}]}$$

Either expression shows that the pH of the buffer depends primarily on the K_a of the acid and secondarily on the ratio of the amount of acid to amount of conjugate base. (Similar expressions can be written for a buffer based on a weak base and its conjugate acid.)

The titration of a strong acid with a strong base follows the course outlined in Figure 16.5 (Section 16.3). A very large, vertical rise in the pH is seen at the equivalence point and the pH at the equivalence point is 7. If a weak acid is titrated with a strong base, however, the pH at the equivalence point is greater than 7 (Figure 16.7). Further, at the point at which the acid has been half-neutralized, $[H^+]$ is equal to the K_a of the weak acid (or pH = pK_a).

Acid–base indicators (In) (Section 16.4) are themselves generally weak acids in which the acid form (HIn) has one color and the conjugate base (the anion In^-) has quite a different color. In solutions of low pH, HIn will predominate, while In^- will be the predominant species in high pH solutions. Thus, the color changes with change in pH.

STUDY QUESTIONS

REVIEW QUESTIONS

1. What is the equivalence point of an acid–base titration? (If you are not certain, review the subject of titrations in Chapter 4.)
2. Tell if the pH is equal to 7, less than 7, or greater than 7 when equal molar amounts of (a) a weak base reacts with a strong acid, (b) a strong base reacts with a strong acid, and (c) a strong base reacts with a weak acid.
3. Sketch the general shape of the titration curve when (a) a strong base is titrated with a strong acid and (b) a weak base is titrated with a strong acid. In each case indicate if the pH at the equivalence point is equal to 7, less than 7, or greater than 7.
4. Briefly describe how a buffer solution can control the pH of a solution when excess strong acid is added. Use NH_3/NH_4Cl as an example.
5. Briefly explain the difference between the equivalence point of a titration and its end point.

ACID–BASE REACTIONS AND TITRATIONS

6. Calculate the $[H^+]$ and the pH at the equivalence point in the titration of 22.0 mL of 0.10 *M* acetic acid, CH_3COOH, with 22.0 mL of 0.10 *M* NaOH.
7. Calculate the hydrogen ion concentration and the pH at the equivalence point in a titration of 50.0 mL of 0.40 *M* NH_3 with 0.40 *M* HCl.
8. For each titration below, tell whether the pH at the equivalence point will be less than, equal to, or greater than 7.
   (a) 0.10 *M* acetic acid, CH_3COOH, with 0.10 *M* KOH
   (b) 0.015 *M* NH_3 with 0.015 *M* HCl
   (c) 0.0020 *M* HNO_3 with 0.0020 *M* NaOH
   (d) 0.45 *M* H_2SO_4 with 0.45 *M* NaOH (both hydrogen ions of sulfuric acid are titrated)
   (e) 0.10 *M* formic acid, HCOOH, with 0.10 *M* ammonia, NH_3
9. What is the pH of the solution when 20.0 mL of 0.15 *M* HCl is mixed with
   (a) 20.0 mL of 0.15 *M* NH_3
   (b) 30.0 mL of 0.15 *M* NaOH
10. Phenol, a weak organic acid, C_6H_5OH, has many uses, and about 2.5 billion pounds are produced annually in the United States. Assume you dissolve 0.500 g of phenol in 1.0×10^2 mL of water and titrate it with 0.100 *M* NaOH.

    Complete equation

    $$C_6H_5OH(aq) + NaOH(aq) \longrightarrow NaC_6H_5O(aq) + H_2O(\ell)$$

    Net equation

    $$C_6H_5OH(aq) + OH^-(aq) \longrightarrow C_6H_5O^-(aq) + H_2O(\ell)$$

    What are the concentrations of all of the species at the equivalence point: Na^+, H^+, OH^-, $C_6H_5O^-$? What is the pH at the equivalence point?
11. Calculate the pH at the equivalence point in a titration of 25.0 mL of 0.12 *M* formic acid, HCOOH, with 25.0 mL of 0.12 *M* NaOH.

THE COMMON ION EFFECT

12. What is the pH of a 0.100 *M* solution of lactic acid, $C_3H_6O_3$ ($K_a = 1.35 \times 10^{-4}$)? If you add the conjugate base of the acid, lactate ion ($C_3H_5O_3^-$), will the pH of the solution increase or decrease? Answer this question by assuming that you add 2.75 g of sodium lactate ($NaC_3H_5O_3$) to 5.00×10^2 mL of the lactic acid solution. What is the pH of this new solution? (Lactic acid is found in sour milk, in sauerkraut, and in muscles after activity.)
13. What is the pH of the solution when the following amounts of NH_4Cl are added to 250 mL of 0.12 *M* NH_3? (a) 2.20 g (b) 5.65 g
14. What is the pH of a 0.15 *M* acetic acid solution? How many grams of sodium acetate, $NaC_2H_3O_2$, must you add to 1.50 L of the 0.15 *M* acetic acid in order to achieve a pH of 5.40?
15. What is the pH of the solution if you add 1.0. mL of 1.0 *M* HCl to 1.0 L of 0.10 *M* acetic acid? What is the pH if 10. mL of 1.0 *M* HCl is added?
16. Will the pH of the second solution increase, decrease, or stay the same when
    (a) 10. mL of 0.10 *M* HCl is added to 100. mL of 0.10 *M* NaOH
    (b) 10. g of NH_4Cl is added to 1.0 L of 0.10 *M* NH_3
    (c) 20. g of NaCl is added to 1.0 L of 0.10 *M* sodium acetate
    (d) 1.0 mL of 1.0 *M* HCl is added to 1.0 L of 0.10 *M* pyridine, a weak base

BUFFER SOLUTIONS

17. To study many natural processes in the laboratory you must maintain the system at a carefully controlled pH by using a buffer. Which of the following combinations would be the best choice to maintain the pH at approximately 7?
    (a) H_3PO_4/NaH_2PO_4 (b) NaH_2PO_4/Na_2HPO_4
    (c) Na_2HPO_4/Na_3PO_4
18. How many grams of ammonium chloride would have to be added to 5.0×10^2 mL of 0.10 *M* NH_3 solution to give a pH of 9.00?
19. How many grams of sodium acetate, $NaC_2H_3O_2$, must be added to 1.0 L of 0.15 *M* acetic acid to give a pH of 4.5?
20. If a buffer solution is made of 12.2 g of benzoic acid (C_6H_5COOH) and 7.20 g of sodium benzoate (C_6H_5COONa) in 5.00×10^2 mL of solution, what is the pH of the buffer? If the buffer is diluted to 1.50 L, what is the pH of the new solution?
21. A buffer solution is prepared from 5.0 g of NH_4NO_3 and 0.10 L of 0.10 *M* NH_3. What is the pH of the solution?
22. Calculate the pH that will result when 0.010 mole of HCl is added to 1.0 L of a balanced acetic acid buffer where originally [HOAc] = [OAc^-] = 0.20 *M*?
23. A 1.00-L solution contains 0.300 mole of $NaNO_2$ and 0.400 mole of HNO_2. How many grams of $NaNO_2$ must be added to this solution if a pH of 4.00 is desired?
24. What is the pH of a buffer solution made of 100. mL of 0.100 *M* NH_3 and 100. mL of 0.100 *M* NH_4Cl? If the solution is diluted to 2.00 liters, what is the pH of the new solution?
25. A buffer solution is prepared by adding 0.500 mole of ammonium chloride to 500. mL of 2.00 *M* solution of ammonia and then diluting the solution to 2.00 liters.
    (a) What is the pH of the 2.00-L solution?
    (b) If 0.0100 mole of gaseous HCl is bubbled into the 2.00 liters of buffer solution, now what is the pH of the solution?
26. If you dissolve 0.0100 mole of NaOH in a 2.00-L solution that originally had [$H_2PO_4^-$] = [HPO_4^{2-}] = 0.100 *M*, calculate the resulting pH.
27. *Buffer capacity* is defined as the number of moles of a strong base or strong acid (of the type HA or MOH) that are required to change the pH of one liter of the buffer solution by one unit. What is the buffer capacity of a solution that is 0.10 *M* in HOAc and 0.10 *M* in NaOAc?
28. How many moles of HCl must be added to 1.00 L of a buffer made from 0.150 *M* NH_3 and 10.0 g of NH_4Cl to decrease the pH by one unit?
29. A common buffer solution is made from potassium hydrogen phthalate, $C_6H_4(COOH)COOK$ ($K_a = 3.9 \times 10^{-6}$), and its conjugate base $C_6H_4(COO)_2^{2-}$. The dianion is made from the monoanion by adding some strong base such as NaOH.

$$C_6H_4(COOH)COOK(aq) + OH^-(aq) \longrightarrow K^+(aq) + C_6H_4(COO)_2^{2-}(aq) + H_2O(\ell)$$

    What is the pH of a buffer solution made by mixing 43.0 mL of 0.10 *M* NaOH and 50.0 mL of 0.10 *M* $C_6H_4(COOH)COOK$?
30. The pH of human blood is controlled by several buffer systems, among them the reaction

$$H_2PO_4^-(aq) + H_2O(\ell) \rightleftharpoons H_3O^+(aq) + HPO_4^{2-}(aq)$$

    If K_a for $H_2PO_4^-$ is 6.2×10^{-8}, calculate the ratio of $H_2PO_4^-/HPO_4^{2-}$ in normal blood having a pH of 7.40.

TITRATION CURVES AND INDICATORS

31. Sketch the curve for the titration of 0.10 *M* NaOH with 0.10 *M* HCl. Indicate the pH at the beginning of the titration and at the equivalence point.
32. Calculate the pH of a solution originally containing 20.0 mL of 0.11 *M* NH_3. Next calculate the pH after the addition of 5.00, 11.0, 15.0, 20.0, 22.0, and 25.0 mL of 0.10 *M* HCl. Plot the titration curve. What indicator in Table 15.3 would be best to use to detect the equivalence point?

33. Construct a rough plot of pH versus volume of base added for the titration of 50. mL of 0.050 *M* HCN with 0.075 *M* NaOH.
(a) How many milliliters of base are required to neutralize the HCN?
(b) What is the pH before any base is added?
(c) What is the pH at the half-neutralization point?
(d) What is the pH when 95% of the required NaOH has been added?
(e) What is the pH at the equivalence point?
(f) What indicator would be most suitable to use to detect the equivalence point (Table 15.3)?
(g) What is the pH when 105% of the required NaOH has been added?

34. Assume you dissolve 0.221 g of trimethylamine, $(CH_3)_3N$, in 50.0 mL of water and titrate it with 0.100 *M* HCl.
(a) What is the concentration of the amine solution in the beginning?
(b) What is the pH of the solution before HCl is added?
(c) What is the pH at the half-neutralization point?
(d) How many milliliters of acid are required to neutralize the amine?
(e) What is the pH at the equivalence point?
(f) Construct a rough titration curve for the titration and decide on a suitable indicator (Table 15.3).

35. Using Table 15.3, suggest an indicator to use in each of the following titrations:
(a) HCl and the weak base pyridine
(b) NaOH and formic acid
(c) Formic acid and ammonia
(d) Hydrazine (N_2H_4) and HCl
(e) Sodium carbonate titrated to HCO_3^- with HCl

36. Without doing lengthy calculations, sketch a titration curve for the titration of 10 mL of 0.10 *M* CO_3^{2-} with 0.10 *M* HCl.

GENERAL QUESTIONS

37. Suppose you add 12.5 mL of 4.15 *M* acetic acid to 25.0 mL of 1.00 *M* NaOH. Calculate the hydrogen ion concentration and pH of the resulting solution.

38. Calculate the concentrations of NH_4^+, OH^-, NH_3, and Na^+ in a solution that was originally 0.040 *M* NaOH and 0.20 *M* in NH_3.

39. What is the pH of a solution made by mixing 20.0 mL of 1.00 *M* HF and 0.400 g of NaOH?

40. During active exercise, lactic acid ($K_a = 1.4 \times 10^{-4}$) is produced in the muscle tissues. At the pH of the body (pH = 7.4), which form will be primarily present: undissociated lactic acid [$CH_3CH(OH)COOH$] or the lactate ion [$CH_3CH(OH)COO^-$]?

41. Arrange the following solutions in order of increasing pH (all reagents are 0.1 *M*).
(a) NaCl
(b) NH_3
(c) HOAc/NaOAc
(d) HCl
(e) KOH
(f) HOAc

42. The chemical name for aspirin is acetylsalicylic acid. It is believed that the analgesic and other desirable properties of aspirin are due not to the aspirin itself but to the simpler compound salicylic acid, $C_6H_4(OH)COOH$, that results from the breakdown of aspirin in the stomach. If the pH of gastric juice remains constant at 2.0, calculate the percentage of salicylic acid that will be present in the stomach in the form of the salicylate ion, $C_6H_4(OH)COO^-$. K_a for salicylic acid is 1.1×10^{-3}.

43. 50.0 mL of 0.0500 *M* HF is titrated with 0.100 *M* NaOH. Calculate the pH at the equivalence point.

44. Two acids, each approximately 10^{-2} *M* in concentration, are titrated separately with a strong base. The acids show the following pH values at the equivalence point: HA, pH = 9.5 and HB, pH = 8.5.
(a) Which acid, HA or HB, is the stronger acid?
(b) Which of the conjugate bases, A^- or B^-, is the stronger base?

45. A 4.0×10^{-1} *M* solution of an acid HA has a pH of 2.40. What is the pH of an equimolar solution of HA and the Na^+ salt of its conjugate base, NaA?

46. The weak acid benzoic acid can be used to inhibit biological growth. The table below shows that the minimum toxic concentration of benzoic acid depends on the pH of the solution.

pH	MINIMUM BENZOIC ACID REQUIRED (*M*)
3.5	1.2×10^{-3}
4.0	1.6×10^{-3}
4.5	3.0×10^{-3}
5.0	7.3×10^{-3}
5.5	21.0×10^{-3}
6.0	64.0×10^{-3}

Interpret the pH dependence in terms of the chemistry of benzoic acid. Which is the more toxic species: benzoic acid (BzOH) or the benzoate ion (BzO^-)?

47. Show that pOH = pK_b at the half-neutralization point in the titration of a weak base with a strong acid (e.g., NH_3 with HCl).

48. The hydrochloride of the amino acid glycine undergoes two acid dissociation reactions in aqueous solution. (The Cl^- is not shown in the net ionic equation).

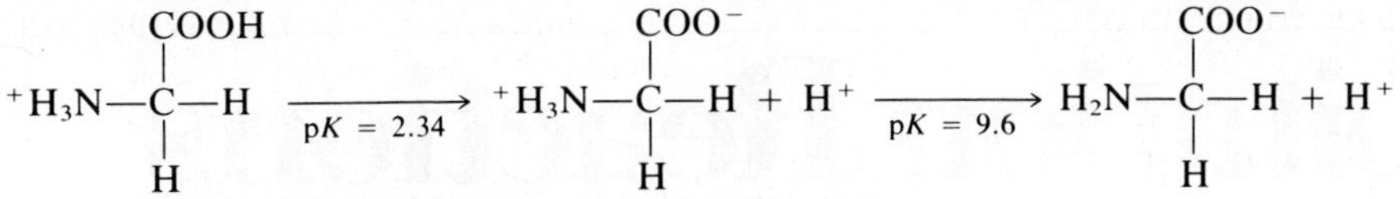

$$^{+}H_3N{-}\overset{\displaystyle COOH}{\underset{\displaystyle H}{C}}{-}H \xrightarrow[pK\,=\,2.34]{} {}^{+}H_3N{-}\overset{\displaystyle COO^-}{\underset{\displaystyle H}{C}}{-}H + H^+ \xrightarrow[pK\,=\,9.6]{} H_2N{-}\overset{\displaystyle COO^-}{\underset{\displaystyle H}{C}}{-}H + H^+$$

(a) What is the pH of a 0.0010 *M* glycine hydrochloride solution?

(b) Sketch a titration curve for the titration of 50. mL of 0.1 *M* glycine hydrochloride with 0.1 *M* NaOH. Indicate the pH of the equivalence point in the titration.

49. The compound procaine hydrochloride is sold under the brand name Novocaine, among others.

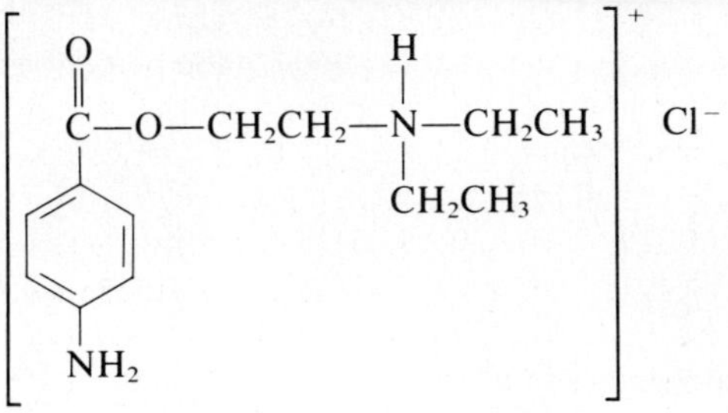

$$\left[H_2N{-}C_6H_4{-}\overset{\displaystyle O}{\overset{\|}{C}}{-}O{-}CH_2CH_2{-}\overset{\displaystyle H}{\underset{\displaystyle CH_2CH_3}{N}}{-}CH_2CH_3 \right]^+ Cl^-$$

The *Merck Index of Chemicals and Drugs* states that the pH of a 0.1 *M* aqueous solution is 6. What is the K_a of procaine hydrochloride as a weak acid?

50. When 16.0 mL of 0.20 *M* benzoic acid is mixed with 32.0 mL of 0.10 *M* NH_3, what is the pH of the resulting solution? See Study Question 110 in Chapter 15 for the method of calculating the H^+ ion concentration.

17 Precipitation Reactions

A precipitation reaction [$Pb(NO_3)_2$ + KI] (Charles D. Winters)

CHAPTER OUTLINE

Beginning in Chapter 3, we introduced you to several types of reactions, among them **exchange reactions**. We divided these further into *precipitation reactions* and *acid–base reactions*. The previous two chapters were devoted to acids and bases and their reactions, and this one considers precipitations.

17.1 SOLUBILITY OF SALTS

A precipitation reaction is an exchange reaction in which one of the products is an insoluble compound,

$$MA(aq) + BX(aq) \longrightarrow MX(s) + BA(aq)$$

$$CaCl_2(aq) + Na_2CO_3(aq) \longrightarrow CaCO_3(s) + 2\ NaCl(aq)$$

that is, a compound having a solubility of less than about 0.01 mole of dissolved material per liter of solution. If you stir calcium carbonate (as the mineral calcite) into pure water, only about 6 mg will dissolve per liter at 25°C. No wonder that sea shells, which are mostly calcium carbonate, do not dissolve appreciably in the sea. On the other hand, you know that "sea salt," NaCl, is very soluble in water.

How do you know when to predict an insoluble compound as the product of a reaction? In Chapter 3 we described some guidelines for making such predictions (Table 3.1), and we can now point out some intriguing applications of these guidelines. First, we should expect seawater to be a solution of many soluble salts. When seawater is evaporated, you would find that the mass percentage of anions in the dry residue is 55.4% chloride ion, approximately 7.8% sulfate ion, and 0.013 to 0.19% bromide ion. As you can see in Table 3.1, these ions generally give soluble or moderately soluble salts with the cations found in seawater: sodium (30.5 to 30.9%), potassium (0.9 to 1.2%), calcium (about 1.2%), and magnesium (about 3.7%).

A moderately soluble compound will have a solubility somewhat greater than about 0.01 mole per liter, while soluble compounds will dissolve to the extent of at least 0.1 mole per liter.

Sea shells do dissolve in very deep water in the ocean. Calcite is unusual in that it is about twice as soluble at 2°C as it is at 25°C (it has an exothermic enthalpy of solution). Thus, cold, deep ocean water promotes solubility. Further, there is more dissolved CO_2 in the deep ocean, so the following reaction occurs to convert $CaCO_3(s)$ to the more soluble $Ca(HCO_3)_2$.

$$CaCO_3(s) + H_2O(\ell) + CO_2(aq) \longrightarrow Ca^{2+}(aq) + 2\ HCO_3^-(aq)$$

The solubility guidelines of Table 3.1 are general *guidelines* and *not laws* of nature. There will be exceptions.

TABLE 17.1 Some Common Minerals

CHEMICAL NAME	FORMULA	COMMON NAMES (AND USES)
Calcium carbonate	$CaCO_3$	Calcite, aragonite, iceland spar (source of lime, CaO)
—	$CuCO_3 \cdot Cu(OH)_2$	Azurite (one source of copper)
Mercury sulfide	HgS	Cinnabar (source of mercury)
Zinc sulfide	ZnS	Zinc blende (one source of zinc)
Lead(II) sulfide	PbS	Galena (one source of lead)
—	FeS_2	Iron pyrite, fool's gold
—	$Ca_{10}(PO_4)_6(OH)_2$	Apatite (source of phosphate for phosphoric acid and fertilizers)
Calcium fluoride	CaF_2	Fluorite or fluorospar (source of HF and other inorganic fluorides)
Calcium sulfate	$CaSO_4 \cdot 2\ H_2O$	Gypsum
—	Fe_3O_4	Magnetite (one source of iron)
Titanium(IV) oxide	TiO_2	Rutile (for paint pigments)
Aluminum oxide	Al_2O_3	Bauxite (source of Al)

Second, our economy depends on various metals and chemicals produced from minerals found in the earth (Figure 17.1). These minerals are often concentrated in certain geographic areas and are usually insoluble salts. (If they were soluble, they would have been dissolved by ground water.) A few examples are listed in Table 17.1. Notice that all these minerals fit the guidelines for insolubility in Table 3.1 [including the oxides, which are related directly to hydroxides (by dehydration) or carbonates (by loss of CO_2)].

Third, many economically important metal ores are sulfides, carbonates, or oxides, so methods have to be found to recover metals from insoluble materials. The refining of nickel is an example of the use of precipitation reactions. Nickel oxide is found in very low concentrations in iron-bearing rock. To recover the nickel, it is leached or extracted from the rock with sulfuric acid to give soluble nickel sulfate and water, an exchange reaction that owes its "driving force" to the formation of water. In another exchange reaction, the nickel ion can then be precipitated with sulfide ion to give black nickel(II) sulfide.

Leaching from rock:

$$NiO(s) + H_2SO_4(aq) \longrightarrow NiSO_4(aq) + H_2O(\ell)$$

Precipitation as sulfide:

$$NiSO_4(aq) + H_2S(aq) \longrightarrow H_2SO_4(aq) + NiS(s)$$

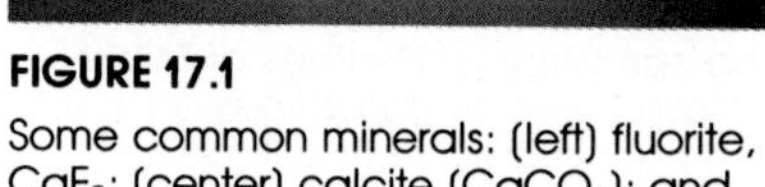

FIGURE 17.1
Some common minerals: (left) fluorite, CaF_2; (center) calcite ($CaCO_3$); and (right) iron pyrite or fool's gold (FeS_2). (Charles D. Winters)

EXERCISE 17.1 SOLUBILITY OF SALTS
Using the guidelines in Table 3.1, predict whether each of the following will be soluble or insoluble in water.

(a) AgBr (d) KF
(b) K_2CrO_4 (e) $NH_4C_2H_3O_2$
(c) $SrCO_3$ (f) $Ba(NO_3)_2$

EXERCISE 17.2 PRECIPITATION REACTIONS

Which of the following mixtures of salts should lead to a precipitate when 0.1 *M* solutions are mixed? Give the formula of the precipitate.

(a) $NaCl(aq) + AgNO_3(aq)$
(b) $KI(aq) + Pb(NO_3)_2(aq)$
(c) $CuCl_2(aq) + Mg(NO_3)_2(aq)$

17.2 THE SOLUBILITY PRODUCT

Silver bromide, AgBr, is used in photographic film. The water-insoluble salt can be made by adding a water-soluble silver salt ($AgNO_3$) to an aqueous solution of a bromide-containing salt (KBr) (see Special Section: Silver Chemistry).

See Special Section: Silver Chemistry, page 663.

$$Ag^+(aq) + Br^-(aq) \longrightarrow AgBr(s)$$

If some of the precipitated AgBr is placed in pure water, some salt will eventually dissolve and an equilibrium will be established.

$$AgBr(s) \rightleftharpoons Ag^+(aq)(5.7 \times 10^{-7}\ M) + Br^-(aq)(5.7 \times 10^{-7}\ M)$$

When the AgBr has dissolved to the greatest extent possible, the solution is said to be *saturated*. At this point, an experiment would show that the concentrations of the silver and bromide ions in solution are each about 5.7×10^{-7} moles per liter at 25°C. The extent to which an insoluble salt dissolves can be expressed in terms of the equilibrium constant for dissolving process. In this case,

Recall from Chapter 14 that the "concentration" of solids does not appear in an equilibrium constant expression.

$$K = [Ag^+(aq)][Br^-(aq)]$$

When silver bromide dissolves in water, one mole of Ag^+ ions and one mole of Br^- ions are produced for every mole of AgBr dissolved.* Therefore, the solubility of silver bromide can be determined if the concentration of either the silver ion *or* the bromide ion is measured. This means that the equilibrium expression above tells us that the *product* of the two measures of the *solubility* is a *constant*. Hence, this constant has come to be called the **solubility product constant**, and it is often designated by K_{sp}.

$$K_{sp} = [Ag^+][Br^-]$$

*Dissolving ionic solids is actually a complex process. It almost always involves more than just the simple equilibria shown here. As a result, you should realize that simple K_{sp} calculations can be quite inaccurate. In particular, the calculated solubility can be incorrect, especially for salts having ions with charges larger than + or − 1. It has been noted that "the solubility product [is more properly] a description, not of solubility, but simply of the ionic concentrations in the saturated solution" (L. Meites, J.S.F. Pode, and H.C. Thomas, *J. Chem. Educ.* 43:667, 1966).

TABLE 17.2 K_{sp} Values of Some Poorly Soluble Salts

COMPOUND	K_{sp} AT 25°C
$CaCO_3$	3.8×10^{-9}
$SrCO_3$	9.4×10^{-10}
$BaCO_3$	8.1×10^{-9}
CaF_2	3.9×10^{-11}
BaF_2	1.7×10^{-6}
CdS	3.6×10^{-29}
PbS	8.4×10^{-28}
CuS	8.7×10^{-36}
HgS	3.0×10^{-53}
CuCl	1.9×10^{-7}
AgCl	1.8×10^{-10}
AuCl	2.0×10^{-13}

Since the concentrations of both Ag^+ and Br^- are 5.7×10^{-7} *M* when silver bromide is in equilibrium with its ions, this means that K_{sp} is

$$K_{sp} = [5.7 \times 10^{-7}][5.7 \times 10^{-7}] = 3.3 \times 10^{-13} \text{ (at 25°C)}$$

In general, K_{sp} is always of the form

$$A_xB_y(s) \rightleftharpoons x\ A^{y+}(aq) + y\ B^{x-}(aq)$$

$$K_{sp} = [A^{y+}]^x[B^{x-}]^y$$

This means that you would write K_{sp} expressions as follows.

$$CaF_2(s) \rightleftharpoons Ca^{2+}(aq) + 2\ F^-(aq) \qquad K_{sp} = [Ca^{2+}][F^-]^2 = 3.9 \times 10^{-11}$$

$$Ag_2SO_4(s) \rightleftharpoons 2\ Ag^+(aq) + SO_4^{2-}(aq) \qquad K_{sp} = [Ag^+]^2[SO_4^{2-}] = 1.7 \times 10^{-5}$$

$$Bi_2S_3(s) \rightleftharpoons 2\ Bi^{3+}(aq) + 3\ S^{2-}(aq) \qquad K_{sp} = [Bi^{3+}]^2[S^{2-}]^3 = 1.6 \times 10^{-72}$$

The numerical values of K_{sp} for a few salts are given in Table 17.2, but many more values are collected in Appendix I. Notice that all are given for a temperature of 25°C.

As with any equilibrium constant, K_{sp} values change with temperature. Whether the K_{sp} will increase or decrease will depend on whether the enthalpy of solution of the solid is endothermic or exothermic. See the marginal note above concerning the solubility of calcite.

EXERCISE 17.3 WRITING K_{sp} EXPRESSIONS

Write K_{sp} expressions for the following insoluble salts and look up numerical values for the constant in Appendix I.

(a) CdS
(b) BiI_3
(c) Ag_2CO_3

17.3 DETERMINING K_{sp} FROM EXPERIMENTAL MEASUREMENTS

In practice, solubility product constants are determined by careful laboratory measurements using various chemical and spectroscopic methods. In general, we shall not go into these methods, but we shall simply assume that the given ion concentrations in fact have been measured experimentally.

EXAMPLE 17.1

K_{sp} FROM SOLUBILITY MEASUREMENTS

The solubility of silver iodide, AgI, is 1.22×10^{-8} mol/L. Calculate K_{sp} for AgI. (See Special Section: Silver Chemistry.)

Solution When silver iodide dissolves, the reaction is represented by the equation

$$AgI(s) \rightleftharpoons Ag^+(aq) + I^-(aq)$$

and so the equilibrium expression will be

$$K_{sp} = [Ag^+][I^-]$$

The question above says that the "solubility" of silver iodide is 1.22×10^{-8} mol/L. Since each mole of AgI that dissolves will give rise to one mole of Ag^+ and one mole of I^- in solution, this means that the concentration of each ion is 1.22×10^{-8} *M*. K_{sp} is the product of these concentrations.

$$K_{sp} = [1.22 \times 10^{-8}][1.22 \times 10^{-8}] = 1.49 \times 10^{-16}$$

EXAMPLE 17.2

K_{sp} FROM SOLUBILITY MEASUREMENTS

Lead(II) chloride dissolves to a slight extent in water according to the equation

$$PbCl_2(s) \rightleftharpoons Pb^{2+}(aq) + 2\ Cl^-(aq)$$

Calculate the K_{sp} if the lead ion concentration has been found to be 1.62×10^{-2} mol/L.

Solution If $PbCl_2$ is placed in water and allowed to dissolve, the Cl^- concentration must be *twice* the Pb^{2+} concentration.

$$\text{If } [Pb^{2+}] = 1.62 \times 10^{-2}\ M, \text{ then } [Cl^-] = 2[Pb^{2+}] = 3.24 \times 10^{-2}\ M$$

To calculate K_{sp} we simply substitute these values into the appropriate expression.

$$K_{sp} = [Pb^{2+}][Cl^-]^2 = [1.62 \times 10^{-2}][3.24 \times 10^{-2}]^2 = 1.70 \times 10^{-5}$$

EXERCISE 17.4 K_{sp} FROM SOLUBILITY MEASUREMENTS

Calculate the K_{sp} value for each of the following salts dissolved in water if the silver ion concentration in the saturated solution is known for each. Verify your answer using the table of values in Appendix I.

(a) AgSCN dissolves to give Ag^+ and SCN^- ions. ($[Ag^+]$ at equilibrium = 1.0×10^{-6} mol/L.)

(b) Ag_2S dissolves to give Ag^+ and S^{2-}. ($[Ag^+]$ at equilibrium = 5.8×10^{-17} mol/L.)

17.4 CALCULATING SALT SOLUBILITY FROM K_{sp}

Through a variety of laboratory measurements, the K_{sp} values for many insoluble salts have been determined. These are an invaluable aid, since we can estimate from them the solubility of the solid salt and thereby can tell the extent to which a solid salt will dissolve or if a solid will precipitate if solutions of its anion and cation are mixed. We turn now to methods to estimate the solubility of a salt from its K_{sp}. In Section 17.7, we shall see how to use these predictions to plan the separation of ions that are mixed in solution.

EXAMPLE 17.3

SOLUBILITY FROM K_{sp}

The K_{sp} of $CaCO_3$ (as the mineral calcite) is 3.8×10^{-9} at 25°C. Calculate the solubility of calcium carbonate in pure water in (a) moles per liter and (b) grams per liter.

Solution The equation for the solubility of $CaCO_3$ is

$$CaCO_3(s) \rightleftharpoons Ca^{2+}(aq) + CO_3^{2-}(aq)$$

and the K_{sp} expression is

$$K_{sp} = [Ca^{2+}][CO_3^{2-}]$$

Equilibria involving the carbonate ion are more complex than can be presented here. The solubility of $CaCO_3$ is best defined only in terms of $[Ca^{2+}]$, since hydrolysis of CO_3^{2-} actually makes $[CO_3^{2-}]$ *slightly* less than $[Ca^{2+}]$. (See Section 17.10 and the special section in this chapter.)

When 1 mole of calcium carbonate dissolves, 1 mole of Ca^{2+} and 1 mole of CO_3^{2-} ions are produced. Thus, the solubility of $CaCO_3$ can be measured by determining the concentration of either Ca^{2+} or CO_3^{2-}. Let us denote the *solubility* of $CaCO_3$ (in mol/L) by the symbol S. This means that both $[Ca^{2+}]$ and $[CO_3^{2-}] = S$ at equilibrium.

	$[Ca^{2+}]$	$[CO_3^{2-}]$
Initial concentration	0	0
Equilibrium concentration (M)	S	S

Since K_{sp} is the product of the calcium and carbonate ion concentrations, it is the square of the solubility.

$$K_{sp} = (S)(S) = S^2$$

Given that the K_{sp} of $CaCO_3$ in pure water is 3.8×10^{-9}, the value of S is

$$S = \sqrt{3.8 \times 10^{-9}} = 6.2 \times 10^{-5}\ M$$

The solubility of $CaCO_3$ in pure water is 6.2×10^{-5} mol/L. To find its solubility in g/L, we need only to multiply by the molar mass of $CaCO_3$.

$$\text{Solubility in g/L} = (6.2 \times 10^{-5}\ \text{mol/L})(1.0 \times 10^2\ \text{g/mol}) = 0.0062\ \text{g/L}$$

EXAMPLE 17.4

K_{sp} AND SOLUBILITY

Knowing that the K_{sp} of MgF_2 is 6.4×10^{-9}, calculate the solubility of the salt in (a) moles per liter and (b) grams per liter.

Solution Once again we shall begin by writing an equation for dissolving the salt

$$MgF_2(s) \rightleftharpoons Mg^{2+}(aq) + 2\ F^-(aq)$$

and the K_{sp} expression

$$K_{sp} = [Mg^{2+}][F^-]^2$$

Our next problem is to define the salt solubility in terms that will allow us to solve the K_{sp} expression for this value. It is evident that, if one mole of magnesium fluoride dissolves, *one* mole of Mg^{2+} and *two* moles of F^- will appear in solution. From this we conclude that MgF_2 solubili'y is equivalent to the concentration of Mg^{2+}, $[Mg^{2+}]$. Thus, if we denote the solubility of MgF_2 by S, then $[Mg^{2+}] = S$ and $[F^-] = 2S$.

	$[Mg^{2+}]$	$[F^-]$
Initial concentrations	0	0
Equilibrium concentrations (M)	S	$2S$

We can now substitute into the K_{sp} expression as follows:

$$K_{sp} = [Mg^{2+}](F^-]^2 = (S)(2S)^2 = 4S^3$$

Solving for the solubility S, we have

$$S = \sqrt[3]{\frac{K_{sp}}{4}} = \sqrt[3]{\frac{6.4 \times 10^{-9}}{4}} = 1.2 \times 10^{-3}\ M$$

We now know that the solubility of MgF_2 is 0.0012 mol/L. To find the solubility in g/L, we need only multiply by the gram molecular weight.

$$(0.0012 \text{ mol/L})(62.3 \text{ g/mol}) = 0.075 \text{ g/L of } MgF_2$$

EXERCISE 17.5 SALT SOLUBILITY FROM K_{sp}

Using the K_{sp} values of Appendix I, calculate the solubility in moles per liter for (a) CdS and (b) $Mg(OH)_2$.

The relative solubilities of salts can often be deduced by comparing values of solubility product constants. However, you must be very careful when doing this. For example, the K_{sp} for AgCl is 1.8×10^{-10} and that for Ag_2CrO_4 is 9.0×10^{-12}. In spite of the fact that silver chromate has a numerically smaller K_{sp} value, it is approximately 10 times more soluble than silver chloride. Solving for their solubilities, you would find $S_{chloride} = 1.34 \times 10^{-5}$ mol/L, whereas $S_{chromate} = 1.3 \times 10^{-4}$ mol/L. Valid comparisons of solubility on the basis of K_{sp} values can be made only for salts having the same ion ratio (C^+/A^- or A^-/C^+). Thus, you can directly compare solubilities of 1:1 salts, for example, by comparing K_{sp} values. The K_{sp} values for the silver halides are in the order

See Special Section: Silver Chemistry, page 663.

$$\text{AgCl } (K_{sp} = 1.8 \times 10^{-10}) > \text{AgBr } (K_{sp} = 3.3 \times 10^{-13}) > \text{AgI } (K_{sp} = 1.5 \times 10^{-16})$$

and their solubilities are

$$S(\text{AgCl}) > S(\text{AgBr}) > S(\text{AgI})$$

Similarly, you could compare 1:2 salts such as PbX_2 where the solubility order is $S_{Cl} > S_{Br} > S_I$, the same order as the K_{sp} values.

$$PbCl_2\ (K_{sp} = 1.7 \times 10^{-5}) > PbBr_2\ (K_{sp} = 6.3 \times 10^{-6}) > PbI_2\ (K_{sp} = 8.7 \times 10^{-9})$$

$$S\ (PbCl_2) > S\ (PbBr_2) > S\ (PbI_2)$$

EXERCISE 17.6 COMPARING SALT SOLUBILITIES

Using K_{sp} values, tell which salt in each pair is more soluble in water.

(a) AgCl and AgCN
(b) ZnS and PbS
(c) $MgCO_3$ and $CaCO_3$
(d) $Mg(OH)_2$ and $Ca(OH)_2$
(e) HgS and FeS

An interesting example of the practical consequences of relative solubilities of insoluble salts occurs in the sea. One group of marine animals, the pteropods, forms the mineral aragonite, one type of $CaCO_3$. Most other marine organisms, however, form calcite, another type of calcium carbonate. Measurements at 2°C and at ocean depths of several thousand meters show that aragonite is about 1.5 times more soluble than calcite. This suggests why pteropod shells are not found at all beyond depths of a few hundred meters in the Pacific Ocean, for example, whereas

Solitary coral, a form of calcium carbonate. (Charles Seaborn)

substantial calcite deposits are found to depths of 3500 meters. At greater depths the temperature is lower and the concentration of dissolved CO_2 is greater. The solubility of $CaCO_3$ in either form *increases* with a drop in temperature and the reaction

$$CaCO_3(s) + H_2O(\ell) + CO_2(aq) \rightleftharpoons Ca^{2+}(aq) + 2\ HCO_3^-(aq)$$

occurs to a greater extent. Since this is a dynamic equilibrium, as the more soluble form (aragonite) dissolves, the less soluble form (calcite) will be reprecipitated.

17.5 PRECIPITATION OF INSOLUBLE SALTS

Metal-bearing ores often contain the metal in the form of an insoluble salt (Figure 17.1), and, to complicate matters, the ores often contain several such metal salts. Virtually all industrial methods for separating metals from their ores involve dissolving the metal salts to obtain the metal ion or ions in solution. The solution is then usually concentrated in some manner, and a precipitating agent is added to precipitate selectively only one type of metal ion as an insoluble salt. In the case of nickel, the ion can be precipitated as insoluble sulfide, NiS, or as nickel carbonate.

$$Ni^{2+}(aq) + S^{2-}(aq) \rightleftharpoons NiS(s) \qquad K_{sp} = 3.0 \times 10^{-21}$$

$$Ni^{2+}(aq) + CO_3^{2-}(aq) \rightleftharpoons NiCO_3(s) \qquad K_{sp} = 6.6 \times 10^{-9}$$

The final step in obtaining the metal itself is to reduce the ion to the metal either chemically or electrochemically (Chapter 19).

Our goal in this section is to work out methods to determine whether a precipitate will form under a given set of conditions. For example, if Ag^+ ion and Cl^- ion are present at given concentrations, will AgCl precipitate? A useful way to answer this is to use a graph of K_{sp}.

K_{sp} AND THE REACTION QUOTIENT, Q

Silver chloride, like silver bromide, is used in photographic films. It dissolves to a very small extent in water and has a correspondingly small value of K_{sp}.

$$AgCl(s) \rightleftharpoons Ag^+(aq) + Cl^-(aq)$$

$$K_{sp} = [Ag^+][Cl^-] = 1.8 \times 10^{-10}$$

The K_{sp} expression is simple. Mathematically, it is just "constant equals product of two ion concentrations" or "$K = xy$." If you plot the concentrations of silver and chloride ions that satisfy this expression, you obtain the curve shown in Figure 17.2. This curve represents all of the combinations of silver ion and chloride ion concentrations where equilibrium exists with solid silver chloride. When the product of ion concentrations is equal to K_{sp}, the solution is said to be **saturated**, so the curve represents all of the conditions at which the solution is saturated with silver chloride.

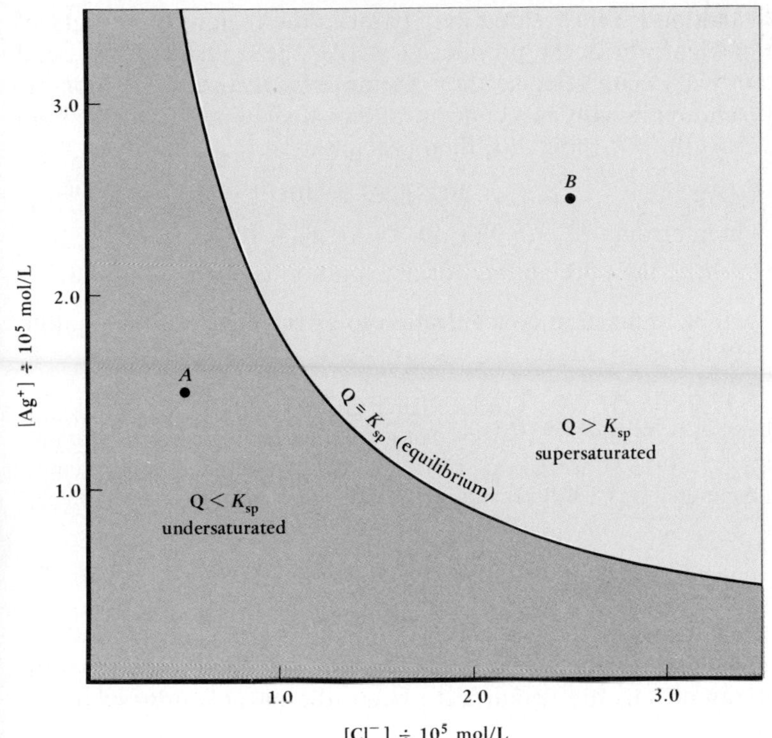

FIGURE 17.2

The solubility product constant for AgCl ($K_{sp} = 1.8 \times 10^{-10}$). The curve represents all combinations of $[Ag^+]$ and $[Cl^-]$ at which equilibrium can exist. If a solution is represented by a point to the left of the line, it is undersaturated ($Q < K_{sp}$). To the right of the line solutions are supersaturated ($Q > K_{sp}$).

Will AgCl precipitate from a solution in which the silver and chloride ion concentrations are represented by a point to the left of the curve, say point A? To answer this we need only to calculate Q, the *reaction quotient*. For the general process

See Section 14.3 for discussion of the reaction quotient, Q.

$$A_xB_y(s) \rightleftharpoons x\ A^{y+}(aq) + y\ B^{x-}(aq)$$

$Q = [A^{y+}]^x[B^{x-}]^y$, where $[A^{y+}]$ and $[B^{x-}]$ are the actual ion concentrations and not necessarily those at equilibrium. You can decide

the system is at **equilibrium** if $\boldsymbol{Q = K_{sp}}$

the solution is **undersaturated** and precipitation does not occur if $\boldsymbol{Q < K_{sp}}$

the solution is **supersaturated** and precipitation occurs if $\boldsymbol{Q > K_{sp}}$

For example, at point A in Figure 17.2, $[Ag^+] = 1.5 \times 10^{-5}\ M$ and $[Cl^-]$ is $0.5 \times 10^{-5}\ M$. There can be no AgCl precipitate because $Q < K_{sp}$.

Q (at point A) $= [Ag^+][Cl^-] = [1.5 \times 10^{-5}][0.5 \times 10^{-5}] = 0.75 \times 10^{-10}$

Q is less than K_{sp}, $(0.75 \times 10^{-10} < 1.8 \times 10^{-10})$

If $Q < K_{sp}$, then no precipitate forms; solution is undersaturated.

If the initial ion concentrations are represented by *any* point to the left of the curve in Figure 17.2 ($Q < K_{sp}$), all silver and chloride ions are in solution, and the solution is *undersaturated*. To proceed to equilibrium, additional Ag^+ and/or Cl^- ions must be added to give a set of concentrations defined by the equilibrium line, that is, until $Q = K_{sp}$.

If a solution is represented by a point in the region to the right of the curve in Figure 17.2, the product $Q = [Ag^+][Cl^-]$ exceeds K_{sp} and precipitation will occur because the solution is *supersaturated*. For example, if a solution initially has concentrations of silver and chloride both equal to 2.5×10^{-5} *M* (point *B*), then precipitation must occur.

$Q = [Ag^+](Cl^-] = [2.5 \times 10^{-5}][2.5 \times 10^{-5}] = 6.3 \times 10^{-10}$

Q is greater than K_{sp}, $(6.3 \times 10^{-10} > 1.8 \times 10^{-10})$

If $Q > K_{sp}$, then precipitate forms; solution is supersaturated.

Both ions will be reduced in concentration to arrive finally at the equilibrium curve (when $Q = K_{sp}$).

EXERCISE 17.7 PLOT OF EQUILIBRIUM CONSTANT FOR AgCl
Using the graph in Figure 17.2, verify that a solution having a concentration of Ag^+ of 2.5×10^{-5} *M* and Cl^- of 1.0×10^{-5} *M* cannot be at equilibrium. Does this mean that AgCl will precipitate or that more AgCl(s) will dissolve?

K_{sp} AND PRECIPITATIONS

With some knowledge of the reaction quotient, we can decide (a) if a precipitate will form when the ion concentrations are known and (b) what concentrations of ions are required to begin the precipitation of an insoluble salt.

EXAMPLE 17.5

DECIDING WHETHER A PRECIPITATE WILL FORM

Suppose the concentration of aqueous nickel(II) ion in a solution is 1.5×10^{-6} *M*. If enough Na_2CO_3 is added to make the solution 6.0×10^{-4} *M* in the carbonate ion, CO_3^{2-}, will precipitation of nickel(II) carbonate occur? Will it occur if the CO_3^{2-} concentration is raised by a factor of 100? The K_{sp} of $NiCO_3$ is 6.6×10^{-9}.

Solution The insoluble salt $NiCO_3$ dissolves according to the balanced equation

$$NiCO_3(s) \rightleftharpoons Ni^{2+}(aq) + CO_3^{2-}(aq)$$

and the K_{sp} expression is $K_{sp} = [Ni^{2+}][CO_3^{2-}]$. In the first of the cases above, Q is

$$Q = [Ni^{2+}][CO_3^{2-}] = (1.5 \times 10^{-6})(6.0 \times 10^{-4}) = 9.0 \times 10^{-10}$$

and you see that it is *less than* the K_{sp}. Therefore, precipitation of $NiCO_3$ does *not* occur.

If the carbonate concentration is increased by a factor of 100, it becomes 6.0×10^{-2} *M*. Q is now 9.0×10^{-8}, a value larger than K_{sp}. Thus, precipitation occurs, and it will continue until the Ni^{2+} and CO_3^{2-} concentrations have declined so that their product is equal to K_{sp}.

EXERCISE 17.8 WILL PRECIPITATION OCCUR?
If the concentration of strontium ion is 2.5×10^{-4} *M*, will precipitation of $SrSO_4$ occur when enough Na_2SO_4 is added to make the solution 2.5×10^{-4} *M* in sodium sulfate? K_{sp} for $SrSO_4 = 2.8 \times 10^{-7}$.

Now that you know how to decide *if* a precipitate will form when the concentration of each ion involved is known, let us turn to the problem of deciding *how much* of the precipitating agent is required to begin the precipitation of an ion at a given concentration level.

EXAMPLE 17.6

PRECIPITATION OF AN INSOLUBLE SALT

If the concentration of Ni^{2+} ion in water is 0.010 *M*, (a) what is the minimum concentration of S^{2-} necessary to begin precipitating NiS, and (b) what will be the concentration of Ni^{2+} when the S^{2-} concentration reaches 0.00010 *M*? K_{sp} for NiS is 3.0×10^{-21}.

Solution As usual, it is essential to begin with a balanced equation for the equilibrium that will exist when NiS(s) is precipitated,

$$NiS(s) \rightleftharpoons Ni^{2+}(aq) + S^{2-}(aq)$$

and the equilibrium constant expression?

$$K_{sp} = [Ni^{2+}][S^{2-}] = 3.0 \times 10^{-21}$$

The K_{sp} expression tells us that the product of the nickel(II) and sulfide ion concentrations cannot exceed 3.0×10^{-21}. If that happens, then precipitation of NiS will occur. This means that, if the nickel ion concentration is known, as in question (a), we can obtain the sulfide ion concentration that satisfies the K_{sp} expression.

$$K_{sp} = 3.0 \times 10^{-21} = (0.010)[S^{2-}]$$

$$[S^{2-}] = \frac{K_{sp}}{[Ni^{2+}]} = \frac{3.0 \times 10^{-21}}{0.010} = 3.0 \times 10^{-19}\ M$$

If you were to increase the sulfide ion concentration slightly, to a value *greater than* 3.0×10^{-19} *M*, precipitation of NiS will begin because $Q = [Ni^{2+}][S^{2-}]$ becomes greater than K_{sp}.

(b) If sulfide ion is added until it reaches 0.00010 *M*, the maximum amount of nickel that can be in solution can be obtained from the K_{sp} expression.

$$[Ni^{2+}] = \frac{K_{sp}}{[S^{2-}]} = \frac{3.0 \times 10^{-21}}{0.00010} = 3.0 \times 10^{-17}\ M$$

This means that the nickel(II) ion has been almost completely removed from the solution by the time the sulfide ion concentration reaches the very low value of 10^{-4} *M*.

EXERCISE 17.9 K_{sp} AND PRECIPITATIONS

If the concentration of Ba^{2+} in water is 1.0×10^{-3} *M*, what concentration of SO_4^{2-} is necessary to just begin precipitating $BaSO_4(s)$? K_{sp} for $BaSO_4$ is 1.1×10^{-10}.

EXAMPLE 17.7

K_{sp} AND PRECIPITATIONS

Mercury(I) chloride, Hg_2Cl_2, is commonly called "calomel." (As a mixture with proteins, it has been used as a laxative.) It is insoluble, having a K_{sp} of 1.1×10^{-18}. If you have 100. mL of a solution having a 0.0010 *M* concentration of Hg_2^{2+}, how many grams of NaCl must be added to begin precipitation of Hg_2Cl_2?

Solution The equilibrium reaction for calomel is

$$Hg_2Cl_2(s) \rightleftharpoons Hg_2^{2+}(aq) + 2\ Cl^-(aq)$$

and the K_{sp} expression is

$$K_{sp} = [Hg_2^{2+}][Cl^-]^2$$

(Notice in the equation above that the Hg(I) cation is diatomic.)

To answer the question, you need to know what concentration of Cl^- can exist in equilibrium with 0.0010 M Hg_2^{2+} ion. This means we must solve the K_{sp} expression for $[Cl^-]$.

$$[Cl^-] = \sqrt{\frac{K_{sp}}{[Hg_2^{2+}]}} = \sqrt{\frac{1.1 \times 10^{-18}}{0.0010}} = 3.3 \times 10^{-8}\ M$$

Our calculation shows that precipitation will begin when $[Cl^-]$ just exceeds the very small value of 3.3×10^{-8} M. To find out how many grams of NaCl are required, you need the molar mass of NaCl (58.45 g/mol), and you can then perform the following calculation:

$$\left(\frac{3.3 \times 10^{-8} \text{ moles } Cl^-}{\text{L of solution}}\right) 0.100 \text{ L of solution} \left(\frac{1 \text{ mol NaCl}}{1 \text{ mol } Cl^-}\right)\left(\frac{58.45 \text{ g}}{1 \text{ mol}}\right)$$
$$= 1.9 \times 10^{-7} \text{ g NaCl}$$

Thus, precipitation of Hg_2Cl_2 begins when the amount of NaCl added is slightly in excess of 1.9×10^{-7} g of NaCl.

EXERCISE 17.10 PRECIPITATION OF $BaCO_3$

If you have 100. mL of a 0.0010 M solution of CO_3^{2-} ion, how many grams of $BaCl_2$ must be added to begin precipitation of $BaCO_3$? How many grams of barium chloride, $BaCl_2$, must be added to reduce the carbonate ion concentration to 1.0×10^{-6} M?

EXERCISE 17.11 PRECIPITATION REACTION

If you mix 100. mL of a 0.0010 M solution of Na_2CO_3 with 100. mL of 0.0010 M $BaCl_2$, will a precipitate of $BaCO_3$ form?

17.6 SOLUBILITY AND THE COMMON ION EFFECT

Using the methods of Section 17.4, you can estimate that the solubility of AgCl in pure water at 25°C is 1.3×10^{-5} M. If instead you tried to dissolve AgCl in a sample of seawater, you would find that its solubility is considerably lower due to the presence of Cl^-, an ion **common** to AgCl and the NaCl of seawater. The reason for the lower solubility can be explained by **Le Chatelier's principle**. Recall that this principle states that a change in any of the factors that determine the equilibrium conditions of a system will cause the system to change in such a manner as to offset the effect of the change. Let us see how that applies here.

Le Chatelier's principle was first applied to chemical equilibria in Chapter 14. See also the "common ion effect" in Chapter 16.

If we write an equation for the dissolving of AgCl,

$$AgCl(s) \rightleftharpoons Ag^+(aq) + Cl^-(aq)$$
$$K_{sp} = [Ag^+][Cl^-]$$

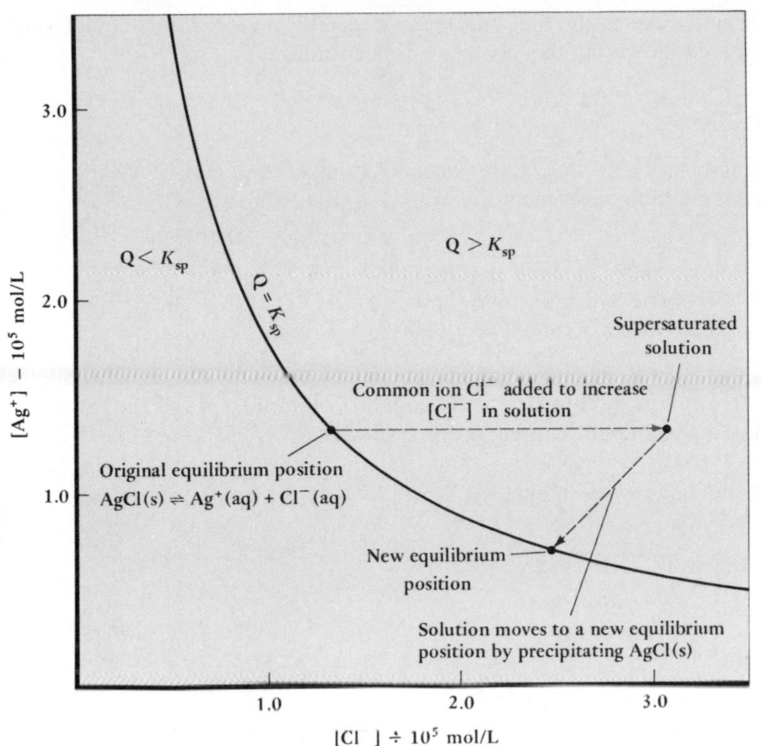

FIGURE 17.3

The common ion effect. Excess Cl^- is added to a solution containing AgCl(s) and equal concentrations of Ag^+ and Cl^-. The system returns to equilibrium along the red line. (The slope of this line is 1, since one Ag^+ ion is removed from solution for every Cl^- ion removed.)

we see that the concentrations of Ag^+ and Cl^- in equilibrium with AgCl(s) must be equal in pure water. However, if *extra* chloride ion is added to a solution of solid AgCl in equilibrium with its ions, the Cl^- concentration is now too large to satisfy the K_{sp} expression, and the situation might be as depicted in Figure 17.3. To return to equilibrium, both $[Ag^+]$ and $[Cl^-]$ must decline, an effect accomplished by precipitating some AgCl(s). Notice, however, that we have arrived at a new equilibrium position; the concentrations of Ag^+ and Cl^- are not the same as those before we added extra Cl^-.

The common ion effect is further illustrated for silver acetate in Figure 17.4, and a more quantitative analysis is done for AgCl in the next example.

The "common ion effect" is the basis of ion separations in qualitative analysis (Section 17.7).

EXAMPLE 17.8

THE COMMON ION EFFECT

The solubility of AgCl in pure water is $1.3 \times 10^{-5}\ M$. What is its solubility in seawater where the Cl^- concentration is 0.55 M?

Solution The solubility of silver chloride can be expressed by the silver ion concentration, $[Ag^+]$. As usual, we assign the variable S to this concentration.

$$\text{Solubility of AgCl defined by } [Ag^+] = S$$

In *pure* water, S is equal to both $[Ag^+]$ and $[Cl^-]$. However, in *seawater* containing the common ion Cl^-, S is equal only to $[Ag^+]$, and it must have a smaller value

FIGURE 17.4

The common ion effect. The tube at the left contains a saturated solution of silver acetate, $AgC_2H_3O_2$. When 1 *M* $AgNO_3$ is added to the tube, the equilibrium

$$AgC_2H_3O_2(s) \rightleftharpoons Ag^+(aq) + C_2H_3O_2^-(aq)$$

shifts to the left, as evidenced by the tube at the right, where more solid silver acetate has formed. (Charles D. Winters)

than in pure water due to the fact that the excess Cl^- ion forces the equilibrium to the left, thereby lowering the silver ion concentration.

$$AgCl(s) \rightleftharpoons Ag^+(aq) + Cl^-(aq)$$
$$\longleftarrow \text{Excess } Cl^- \text{ forces equilibrium left} \longrightarrow$$

The following table shows the concentrations of Ag^+ and Cl^- when equilibrium is attained in the presence of extra Cl^-.

	$[Ag^+]$	$[Cl^-]$
Before AgCl begins to dissolve (*M*)	0	0.55
After some AgCl dissolves and equilibrium is established (*M*)	S	$(S + 0.55)$

Some AgCl dissolves in the presence of Cl^- ion and produces ion concentrations of S mol/L. However, there was already some Cl^- ion present, so the total Cl^- concentration is the amount coming from AgCl (equals S) *plus* what was already there (equals 0.55 *M*).

Now the K_{sp} expression can be written, using the equilibrium concentrations from the table,

$$K_{sp} = [Ag^+][Cl^-] = (S)(S + 0.55)$$

and rearranged to

$$S^2 + 0.55S - K_{sp} = 0$$

This equation is a quadratic of the form $ax^2 + bx + c = 0$ (here $a = 1$, $b = 0.55$ and $c = -K_{sp}$). The equation can be solved for S (x) using the technique outlined in Appendix A. However, here we can make an *approximation* that will give us an answer that is *very* close to the value of S obtained from the quadratic formula. The value of S is small; it equals 1.3×10^{-5} *M* without the common ion, and it will be even smaller with the common ion. This means that $S + 0.55$ *M* is very nearly equal to 0.55 *M* (since $0.000013 + 0.55 \cong 0.55$). Thus, we now have

$$K_{sp} = 1.8 \times 10^{-10} \cong (S)(0.55\ M)$$
$$S = [Ag^+] \cong 3.3 \times 10^{-10}\ M$$

(If S is obtained by solving the full expression using the quadratic formula, we find it to be 3.25×10^{-10}. Since we have been using two significant figures, the approximate and more "accurate" results are identical.)

Notice the problem-solving strategy here. We assume Cl^- is already present in solution from the seawater. Then we add AgCl and assume *additional* Cl^- enters the solution as this salt dissolves. Thus, $[Cl^-]$ at equilibrium is the sum of the concentrations of the ion from two sources. You will see many more such "common ion" problems, and the strategy used to solve them will be the same. See page 611.

As predicted by Le Chatelier's principle, the solubility of AgCl in the presence of added Cl^-, an ion common to the equilibrium, is clearly less (3.3×10^{-10} *M*) than in pure water (1.3×10^{-5} *M*).

As a final step, let us check the approximation we made. To do this, we substitute the approximate value of S into the exact expression $K_{sp} = (S)(S + 0.55)$. Then, if the product $(S)(S + 0.55)$ agrees with the given value of K_{sp}, the approximation is valid.

$$K_{sp} = (S)(S + 0.55) = (3.3 \times 10^{-10})(3.3 \times 10^{-10} + 0.55)$$
$$= 1.8 \times 10^{-10}$$

Here the calculated and given K_{sp} values are the same, so the approximation was valid. If the values are different by more than 5%, the solubility should be calculated without making an approximation.

EXERCISE 17.12 THE COMMON ION EFFECT

Calculate the solubility of $BaSO_4$ (a) in pure water and (b) in the presence of 0.010 *M* $Ba(NO_3)_2$. K_{sp} for $BaSO_4$ is 1.1×10^{-10}.

There are two important facts to be learned from Example 17.8 above.

(a) The solubility of AgCl in the presence of a common ion was reduced by a factor of about 10^5, in accordance with Le Chatelier's principle.
(b) We made the approximation that the amount of common ion added to the solution is very large in comparison with the amount of that ion coming from the insoluble salt, and this allows us to simplify our calculations. This is *almost* always the case, but you should always check the approximation.

EXAMPLE 17.9

THE COMMON ION EFFECT

Calculate the solubility of silver chromate, Ag_2CrO_4, at 25°C in (a) pure water and in (b) the presence of 0.0050 M K_2CrO_4 solution. At 25°C, $K_{sp} = 9.0 \times 10^{-12}$.

Solution The balanced equation defining silver chromate solubility is

$$Ag_2CrO_4(s) \rightleftharpoons 2\ Ag^+(aq) + CrO_4^{2-}(aq)$$

and the K_{sp} expression is

$$K_{sp} = [Ag^+]^2[CrO_4^{2-}]$$

(a) *Solubility in pure water.* Because the stoichiometric ratio of moles of CrO_4^{2-} in solution to moles of Ag_2CrO_4 dissolved is 1:1, we define the solubility of this salt in terms of $[CrO_4^{2-}] = S$.

	$[Ag^+]$	$[CrO_4^{2-}]$
Before Ag_2CrO_4 begins to dissolve	0	0
After equilibrium is achieved (M)	$2S$	S

This means that $[Ag^+]$ must equal $2S$, so the K_{sp} expression can be written

$$9.0 \times 10^{-12} = (2S)^2(S)$$
$$= 4S^3$$
$$S^3 = 2.3 \times 10^{-12}$$

Solving the expression above, we obtain

$$S = \text{solubility of } Ag_2CrO_4 \text{ in pure water} = 1.3 \times 10^{-4}\ M$$

Thus, the ion concentrations at equilibrium in pure water are

$$[CrO_4^{2-}] = 1.3 \times 10^{-4}\ M \quad \text{and} \quad [Ag^+] = 2S = 2.6 \times 10^{-4}\ M$$

Do not be confused about the use of $(2S)^2$, where the factor 2 appears both as the multiplier of S and the exponent of $(2S)$. The reason for this is that, since $[CrO_4^{2-}]$ is defined as S, the stoichiometry demands that $[Ag^+]$ be defined as $2S$; this is substituted into the K_{sp} expression in place of $[Ag^+]$. The K_{sp} expression then demands that $[Ag^+]$ or the equivalent expression $2S$ be squared.

(b) *Solubility in chromate solution.* In the presence of excess chromate ion, we again say that S represents the solubility of the salt.

	$[Ag^+]$	$[CrO_4^{2-}]$
Before Ag_2CrO_4 begins to dissolve (M)	0	0.0050
After some Ag_2CrO_4 has dissolved and equilibrium has been established (M)	$2S$	$(0.0050 + S)$

Substituting the equilibrium amounts into the K_{sp} expression, we have

$$9.0 \times 10^{-12} = (2S)^2(0.0050 + S)$$
$$9.0 \times 10^{-12} \cong (2S)^2(0.0050)$$

Again, we make the approximation that $0.0050 + S \cong 0.0050$, and proceed to solve the resulting equation for S.

$$S = [CrO_4^{2-}] \text{ from } Ag_2CrO_4 = 2.1 \times 10^{-5}\ M$$

$$[Ag^{2+}] = 2S = 4.2 \times 10^{-5}\ M$$

The silver ion concentration is less than its value in pure water ($2.6 \times 10^{-4}\ M$), and you again see the results of adding a common ion.

As a final step, check the approximation. Substitute the calculated S back into the expression $K_{sp} = (2S)^2(0.0050 + S)$ to see if K_{sp} calculated from the expression agrees with the given K_{sp}.

$$K_{sp} = [2(2.1 \times 10^{-5})]^2(0.0050 + 0.000021) = 8.9 \times 10^{-12}$$

Here you see that the calculated K_{sp} is only about 1% different than the given K_{sp}, a difference probably smaller than experimental error. The approximation is valid.

EXERCISE 17.13 THE COMMON ION EFFECT

Calculate the solubility of $Zn(CN)_2$ at 25°C (a) in pure water and (b) in the presence of 0.010 M KCN. K_{sp} at 25°C for $Zn(CN)_2$ is 8.0×10^{-12}.

17.7 SOLUBILITY, ION SEPARATIONS, AND QUALITATIVE ANALYSIS

In many courses in introductory chemistry a portion of the laboratory work is devoted to the qualitative analysis of aqueous solutions, the identification of anions and metal cations. The purpose of such laboratory work is (a) to introduce you to some basic chemistry of various metal ions and (b) to illustrate how the principles of chemical equilibria can be applied.

Assume you have an aqueous solution that contains some or all the following metal ions: Ag^+, Pb^{2+}, Cd^{2+}, and Ni^{2+}. Your objective is to separate the ions from one another so that each type of ion ends up in a separate test tube; the presence or absence of the ion can then be established. As a first step in this process, you want to find one reagent that will form a precipitate with one or more of the ions and leave the others in solution.

Looking over the list of solubility products in Appendix I, you notice that all of the ions in solution form very insoluble sulfides (Ag_2S, PbS, CdS, and NiS). However, only two of them form insoluble chlorides, AgCl and $PbCl_2$. Thus, your magic reagent could be aqueous HCl, which will cause precipitates to form with two of the ions while the other two ions remain in solution (Figure 17.5).

COMPOUND	K_{sp} AT 25°C
Ag_2S	1.0×10^{-49}
PbS	8.4×10^{-28}
CdS	3.6×10^{-29}
NiS	3.0×10^{-21}
AgCl	1.8×10^{-10}
$PbCl_2$	1.7×10^{-5}

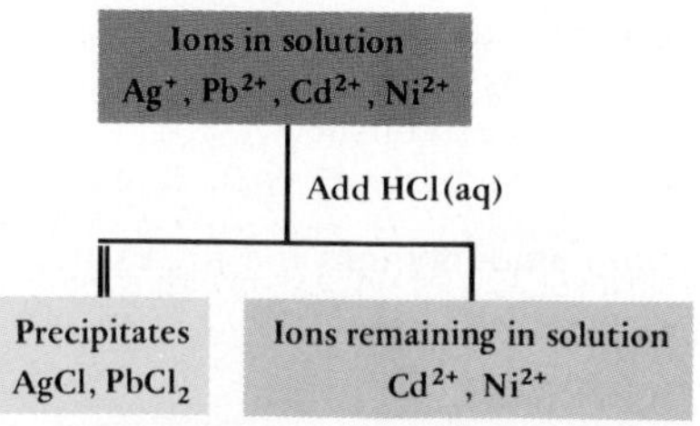

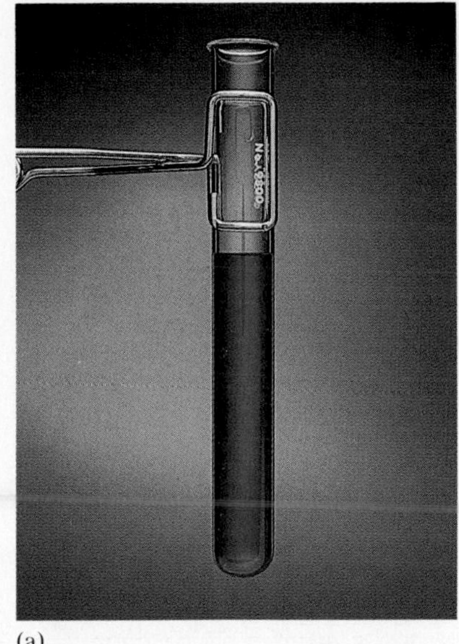
(a)

(b)

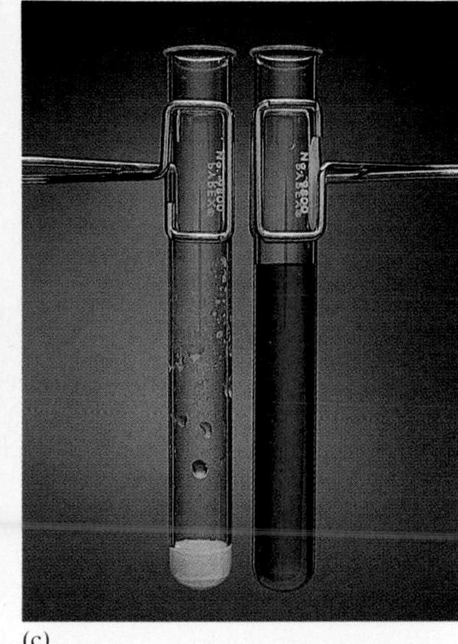
(c)

FIGURE 17.5
Ion separations by solubility difference. (a) The solution contains nitrate salts of Ag^+, Pb^{2+}, Cd^{2+}, and Ni^{2+}. (The Ni^{2+} ion in water is light green, while the others are colorless.) (b) Aqueous HCl is added in an amount sufficient to precipitate completely AgCl and $PbCl_2$ (both white solids). (c) The green solution, now containing only Ni^{2+} and Cd^{2+}, is poured into another test tube, leaving AgCl and $PbCl_2$ in the first test tube. (Charles D. Winters)

Now you are left with the task of separating AgCl and $PbCl_2$ from one another and Cd^{2+} and Ni^{2+} from one another. The separation of AgCl and $PbCl_2$ is not difficult, since $PbCl_2$ dissolves in hot water, while AgCl remains insoluble (Figure 17.5).

The separation of Cd^{2+} and Ni^{2+} can be done by selective precipitation with sulfide ion, S^{2-}. Ordinarily this is done in the presence of H^+ ion (Section 17.10), but in the example that follows we wish only to ask a question that is common in chemistry. As a precipitating reagent is added to a mixture of two ions, which ion precipitates first? Just before the second ion begins to precipitate as an insoluble salt, how much of the first ion remains in solution?

EXAMPLE 17.10

SEPARATION OF TWO IONS BY DIFFERENCE IN SOLUBILITY

Assume you have a solution that is 0.020 *M* in both Cd^{2+} and Ni^{2+}. When you add sulfide ions, CdS and NiS will precipitate. (a) Which precipitates first? (b) Just before the second ion begins to precipitate as an insoluble sulfide, what is the concentration in solution of the ion that precipitated first?

$$K_{sp} \text{ for CdS} = 3.6 \times 10^{-29} \qquad K_{sp} \text{ for NiS} = 3.0 \times 10^{-21}$$

Solution The key to answering part (a) is to realize that *the substance that precipitates first is the one whose K_{sp} is first exceeded.* Therefore, let us calculate the sulfide ion concentration, $[S^{2-}]$, necessary to satisfy the solubility product expression for each metal sulfide. These amounts of S^{2-} will bring each metal salt to "incipient" precipitation.

$$\text{For CdS, the needed } [S^{2-}] = \frac{K_{sp}}{[Cd^{2+}]} = \frac{3.6 \times 10^{-29}}{0.020} = 1.8 \times 10^{-27}\ M$$

$$\text{For NiS, the needed } [S^{2-}] = \frac{K_{sp}}{[Ni^{2+}]} = \frac{3.0 \times 10^{-21}}{0.020} = 1.5 \times 10^{-19}\ M$$

Clearly, a much smaller sulfide ion concentration is needed to begin precipitating CdS(s) than to begin forming NiS(s). CdS precipitates before NiS(s).

Part (b) can now be rephrased more explicitly. Just before NiS begins to precipitate, how much Cd^{2+} remains in solution? We have already found that $[S^{2-}]$ needs to be slightly in excess of $1.5 \times 10^{-19}\ M$ for NiS to begin to precipitate. Therefore, let us calculate the cadmium ion concentration that can exist when $[S^{2-}] = 1.5 \times 10^{-19}\ M$.

$$[Cd^{2+}] \text{ just before NiS begins to precipitate} = \frac{K_{sp} \text{ of CdS}}{[S^{2-}]} = \frac{3.6 \times 10^{-29}}{1.5 \times 10^{-19}} = 2.4 \times 10^{-10}\ M$$

We now know that just before NiS begins precipitating, the Cd^{2+} ion concentration has decreased from 0.020 M to $2.4 \times 10^{-10}\ M$, a change of about 10 million! This means that we can separate Cd^{2+} and Ni^{2+} in aqueous solution by careful control of the sulfide ion concentration. In Section 17.10, we shall outline how this control can be achieved.

EXERCISE 17.14 ION SEPARATION BY SOLUBILITY DIFFERENCES

Under the right circumstances, Ag^+ can be separated from Pb^{2+} by the difference in the solubilities of their chloride salts, AgCl and $PbCl_2$.

(a) If you begin with both metal ions having a concentration of 0.0010 M, which ion precipitates first as an insoluble chloride on adding HCl? (b) What is the concentration of the metal ion that precipitates first just before the second metal chloride begins to precipitate?

EXERCISE 17.15 SCHEMES FOR ION SEPARATION

The cations of each pair given below appear together in one solution. You may add only one reagent to precipitate one cation and *not* the other. Consult the solubility product table in Appendix I and tell whether you would use Cl^-, S^{2-}, or OH^- as the precipitating ion in each case.

(a) Ag^+ and Bi^{3+}

(b) Fe^{2+} and K^+

17.8 SIMULTANEOUS EQUILIBRIA

It is possible that when another reagent is added to a saturated solution of an insoluble salt the salt will be transformed into another, even less soluble salt. Such situations represent one example of **simultaneous equilibria**.

Consider two important minerals: gypsum ($CaSO_4 \cdot 2\ H_2O$) and anhydrite ($CaSO_4$). Gypsum is slightly less soluble than anhydrite, because K_{sp} for gypsum is smaller than that for anhydrite. This means that anhydrite will change spontaneously into gypsum under the right conditions.

Gypsum $CaSO_4 \cdot 2\ H_2O$ $K_{sp} = 2.4 \times 10^{-5}$

Anhydrite $CaSO_4$ $K_{sp} = 4.2 \times 10^{-5}$

The observation that anhydrite takes on water to become gypsum can be proved by deriving the equilibrium constant, K_{net}, for the process

$$CaSO_4(s) + 2\ H_2O(\ell) \rightleftharpoons CaSO_4 \cdot 2\ H_2O(s) \qquad K_{net}$$

This reaction is just the sum of two reactions whose equilibrium constants are known from above.

$$CaSO_4(s) \rightleftharpoons Ca^{2+}(aq) + SO_4^{-}(aq) \qquad K_1 = 4.2 \times 10^{-5}$$

$$Ca^{2+}(aq) + SO_4^{-}(aq) + 2\ H_2O(\ell) \rightleftharpoons CaSO_4 \cdot 2\ H_2O \qquad K_2 = \frac{1}{2.4 \times 10^{-5}}$$

The manipulation of equilibrium constants and their values is described in Section 14.2. Recall that, when a reaction is reversed, $K_{new} = 1/K_{old}$.

The equilibrium constant for the overall reaction, the addition of water to anhydrite, is the *product* of the equilibrium constants for the summed reactions. That is, $K_{net} = K_1K_2 = 1.75$. This is a small equilibrium constant, but it is greater than 1 and indicates that the reaction should proceed from left to right.

Simultaneous equilibria are quite common, and you will see more examples in the sections that follow.

EXERCISE 17.16 SIMULTANEOUS EQUILIBRIA

Silver forms many insoluble salts. Which is more insoluble, AgCl or AgBr? If you add sufficient bromide ion to an aqueous suspension of AgCl(s), can AgCl be transformed into AgBr? To answer this, derive the equilibrium constant for

$$AgCl(s) + Br^-(aq) \rightleftharpoons AgBr(s) + Cl^-(aq)$$

17.9 SOLUBILITY AND COMPLEX IONS

In Chapter 15 we described reactions of Lewis acids and Lewis bases. When a metal ion, a Lewis acid, reacts with a Lewis base, a complex ion can form. Since such ions are often soluble in water, their formation can be used to dissolve otherwise insoluble salts. For example, if sufficient aqueous ammonia is added to insoluble silver chloride, the latter can be dissolved in the form of the complex ion $[Ag(NH_3)_2]^+$ (see the Special Section on Silver Chemistry). The overall reaction is

The extent of a Lewis acid–base reaction is given by the formation constant, $K_{formation}$, as described in Section 15.7.

$$AgCl(s) + 2\ NH_3(aq) \longrightarrow [Ag(NH_3)_2]^+(aq) + Cl^-(aq)$$

and you may recognize this as the sum of two reactions whose equilibrium constants are known:

$$AgCl(s) \rightleftharpoons Ag^+(aq) + Cl^-(aq) \qquad K_{sp} = 1.8 \times 10^{-10}$$

$$Ag^+(aq) + 2\,NH_3(aq) \rightleftharpoons [Ag(NH_3)_2]^+(aq) \qquad K_{formation} = 1.6 \times 10^7$$

That is, there are two simultaneous equilibria occurring in this solution. As described in Section 17.8, the equilibrium constant for the overall reaction is the product of the constants of the added reactions. Thus, for silver chloride plus ammonia,

$$K = K_{sp} \times K_{formation} = 2.9 \times 10^{-3} = \frac{\{[Ag(NH_3)_2{}^+]\}[Cl^-]}{[NH_3]^2}$$

The equilibrium constant for dissolving AgCl in ammonia is not large, but if the concentration of ammonia is sufficiently high, this means that $\{[Ag(NH_3)_2^+]\}$ and $[Cl^-]$ must be high, and AgCl will dissolve.

EXAMPLE 17.11

DISSOLVING PRECIPITATES USING COMPLEX ION FORMATION

How many moles of ammonia must be added to dissolve 0.050 mole of AgCl suspended in 1.0 L of water?

Solution As shown in the text above, the reaction that occurs is

$$AgCl(s) + 2\ NH_3(aq) \longrightarrow [Ag(NH_3)_2]^+(aq) + Cl^-(aq)$$

If 0.050 mole of AgCl is to be dissolved, this means that 0.050 mol/L of the complex ion $[Ag(NH_3)_2]^+$ will be formed. The $[Cl^-]$ must also be 0.050 *M*. If these are the concentrations that we wish to have at equilibrium, what must the concentration of ammonia be to satisfy this?

$$K = 2.9 \times 10^{-3} = \frac{\{[Ag(NH_3)_2^+]\}[Cl^-]}{[NH_3]^2} = \frac{(0.050)(0.050)}{[NH_3]^2}$$

Solving for $[NH_3]$, we find that it is 0.93 *M*. Therefore, to dissolve AgCl, enough NH_3 must be added to form the complex with the silver ion (2 × 0.050 mole or 0.10 mole required) and then to bring the concentration up to 0.93 *M*. Thus, the total required is

$$\text{Total } NH_3 = 0.10\ M + 0.93\ M = 1.03\ M$$

EXERCISE 17.17 USING FORMATION CONSTANTS
Will 100. mL of 4.0 *M* aqueous NH_3 completely dissolve 0.010 mole of AgCl suspended in 1.0 L of pure water?

17.10 ACID–BASE AND PRECIPITATION EQUILIBRIA OF PRACTICAL SIGNIFICANCE

SOLUBILITY OF SALTS IN WATER AND ACIDS

The next time you are tempted to wash a supposedly insoluble salt down the laboratory drain, stop and consider the consequences. Many metal ions such as lead, chromium, and mercury are toxic in the environment. Even if you throw away an insoluble salt, its solubility in water may be greater than you would think, largely owing to the possibility of hydrolysis of the anion of the salt.

Lead sulfide, PbS, is found in nature as the mineral galena. The solubility of any salt depends on a host of factors: temperature, impurities, and so on. However, we keep things as simple as possible by considering only two reactions for lead sulfide:

(a) the solubility of PbS(s) in water

$$PbS(s) \rightleftharpoons Pb^{2+}(aq) + S^{2-}(aq) \qquad K_{sp} = 8.4 \times 10^{-28}$$

(b) the hydrolysis of $S^{2-}(aq)$

$$S^{2-}(aq) + H_2O \rightleftharpoons HS^-(aq) + OH^-(aq) \qquad K_b = 0.077$$

This means that the overall process is

$$PbS(s) + H_2O(\ell) \rightleftharpoons Pb^{2+}(aq) + HS^-(aq) + OH^-(aq)$$

with an equilibrium constant of $K_{total} = K_{sp} \times K_b = 6.5 \times 10^{-29}$. The first reaction produces a tiny concentration of sulfide. The second reaction, however, removes S^{2-} ion from solution by hydrolysis, a reaction with a relatively large equilibrium constant. According to Le Chatelier's principle, more PbS will therefore dissolve. Indeed, it can be calculated using K_{total} that, in an equilibrium system composed of these reactions, the lead ion concentration is approximately 10^{-10} M. Sulfide ion hydrolysis thus leads to an increase in environmental lead concentration by a factor of about 10,000 over the solubility of PbS calculated simply from K_{sp}!

See Study Question 96 at the end of the chapter.

The lead sulfide example leads to the general observation that *any salt containing an anion that is the conjugate base of a weak acid will dissolve in water to a greater extent than given by* K_{sp}. This means that salts of phosphate, acetate, carbonate, and cyanide ions will be affected, since all of these ions can undergo the general reaction

$$X^-(aq) + H_2O(\ell) \rightleftharpoons HX(aq) + OH^-(aq)$$

in aqueous solution. This reaction, however, leads to another useful, general conclusion. If a strong acid is added to a water-insoluble salt such as ZnS or $CaCO_3$, OH^- ion from X^- hydrolysis is removed (by formation of water from $H^+ + OH^-$). This shifts the X^- hydrolysis reaction further to the right; the weak acid HX is formed and the salt dissolves. For example, calcium carbonate will dissolve readily when hydrochloric acid is added, and the main reactions occurring are

$$CaCO_3(s) \rightleftharpoons Ca^{2+}(aq) + CO_3^{2-}(aq)$$
$$K = K_{sp} = 3.8 \times 10^{-9}$$

$$H^+ \text{ (from HCl)} + CO_3^{2-}(aq) \rightleftharpoons HCO_3^-(aq)$$
$$K = 1/K_{a2} \text{ for } H_2CO_3 = 1/(4.8 \times 10^{-11})$$

$$H^+ \text{ (from HCl)} + HCO_3^-(aq) \rightleftharpoons H_2CO_3(aq)$$
$$K = 1/K_{a1} \text{ for } H_2CO_3 = 1/(4.2 \times 10^{-7})$$

$$CaCO_3(s) + 2\,H^+(aq) \rightleftharpoons Ca^{2+}(aq) + H_2CO_3(aq)$$
$$K_{total} = K_{sp}\,(1/K_{a1})(1/K_{a2}) = 1.9 \times 10^8$$

Now the equilibrium constant (K_{total}) is enormous, and the reaction goes essentially to completion. This is especially the case for carbonate salts, since H_2CO_3 rapidly reverts to CO_2 gas and water.

$$H_2CO_3(aq) \rightleftharpoons CO_2(g) + H_2O(\ell) \qquad K \cong 10^5$$

The CO_2 bubbles out of the solution, and the equilibrium is moved further to the right.

Carbonates are generally soluble in strong acids, and so are many metal sulfides

$$ZnS(s) + 2\,H^+(aq) \rightleftharpoons Zn^{2+}(aq) + H_2S(aq)$$

and metal hydroxides.

$$Mg(OH)_2(s) + 2\ H^+(aq) \rightleftharpoons Mg^{2+}(aq) + 2\ H_2O(\ell)$$

The only exceptions are sulfides of mercury, copper, cadmium, and a few other metals to be discussed in Section 17.10. In general, *insoluble inorganic salts containing basic anions tend to be soluble in solutions of strong acids*. In contrast, the salt is *not* soluble in strong acid if the anion is the conjugate base of a strong acid. For example, AgCl is not soluble in strong acid

$$AgCl(s) \rightleftharpoons Ag^+(aq) + Cl^-(aq) \qquad K = K_{sp} = 1.8 \times 10^{-10}$$

$$Cl^-(aq) + H^+(aq) \longrightarrow HCl(aq) \qquad K << 1$$

because Cl^- is a *very* weak base (Table 15.2), and HCl is a strong acid; there is no significant reaction of Cl^- with H^+. You can see that this same conclusion would apply to salts of SO_4^{2-} and NO_3^- as well as Cl^-.

PREPARING SALTS BY ACID–BASE AND PRECIPITATION REACTIONS

You learned in Chapter 3 that salts can be prepared by the reaction of a strong acid with a strong base. The reason such reactions work is the formation of water, the very weak conjugate acid of OH^- and the weak conjugate base of H_3O^+. You now know that strong acid/weak base and weak acid/strong base reactions occur for much the same reason. All of these reactions can be used to prepare salts. But let us expand on those kinds of reactions.

Let us prepare nickel nitrate, starting with the commonly available salt nickel chloride (Figure 17.6). Both of these salts are soluble in water, so simply adding nitrate ion, in some form, to aqueous $NiCl_2$ will not

(a)

(b)

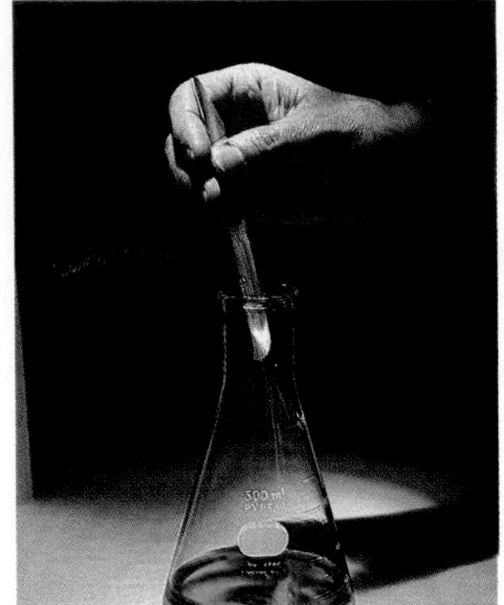

(c)

(d)

(e)

FIGURE 17.6

Preparation of $Ni(NO_3)_2$ from $NiCl_2$. (a) $NiCl_2$, on the watch glass, is dissolved in water, and then (b) Na_2CO_3 is added to precipitate insoluble nickel carbonate. (c) After filtering to collect the $NiCO_3$, 6 *M* HNO_3 is added to an aqueous suspension of $NiCO_3$. (d) It dissolves completely. (A sample of $NiCO_3$ is on the watch glass.) The reaction occurring is $NiCO_3(s) + 2\ HNO_3(aq) \rightarrow Ni(NO_3)_2(aq) + CO_2(g) + H_2O(\ell)$. (e) If the aqueous solution of $Ni(NO_3)_2$ is evaporated, crystals of the hydrated salt $NiNO_3 \cdot 6\ H_2O$ are isolated. (Charles D. Winters)

produce $Ni(NO_3)_2$. Therefore, we first convert $NiCl_2$ into an insoluble material using an anion that is the conjugate base of a weak acid. A good choice is CO_3^{2-} (from Na_2CO_3).

$$NiCl_2(aq) + CO_3^{2-}(aq) \longrightarrow NiCO_3(s) + 2\ Cl^-(aq)$$

Nickel carbonate is insoluble and can be separated from solution by filtration and washed with distilled water to remove contaminants. Next, the strong acid nitric acid is added to the pure $NiCO_3$ to give the desired salt, $Ni(NO_3)_2$, plus water and CO_2.

$$2\ HNO_3(aq) + NiCO_3(s) \longrightarrow Ni(NO_3)_2(aq) + H_2O(\ell) + CO_2(g)$$

As soon as the solid nickel carbonate has dissolved, we stop adding nitric acid and now have a solution containing only nickel nitrate. If the solution is evaporated to dryness, we can obtain the desired salt as a solid hexahydrate, $Ni(NO_3)_2 \cdot 6\ H_2O$.

Using approaches similar to the one outlined above, you can prepare many inorganic salts. In general, it is best to begin with a hydroxide, soluble or insoluble, or a carbonate and add a strong acid containing the desired anion (H_2SO_4, HCl, HNO_3, $HClO_4$). All of these reactions have a great driving force, the formation of water or CO_2 gas.

EXERCISE 17.18 PREPARATION OF SALTS

Outline a method of preparing copper(II) chloride beginning with copper(II) sulfate.

SEPARATION OF IONS BY SELECTIVE PRECIPITATION

Some insoluble salts can be dissolved by adding an acid (Figure 17.7), and such reactions can be used to distinguish between metal ions in solution or to separate two or more metal ions from one another. As an example, let us say you have a solution that may contain Cl^- or PO_4^{3-} ions. To help you decide which ion is present, you are given aqueous $AgNO_3$ and HNO_3. Appendix I shows that both chloride and phosphate form insoluble salts with silver ion, AgCl and Ag_3PO_4, respectively. How-

(a)

(b)

FIGURE 17.7

Dissolving precipitates in acid. (a) A precipitate of AgCl (white) and Ag_3PO_4 (yellow). (b) Adding a strong acid dissolves Ag_3PO_4 and leaves insoluble AgCl. (Charles D. Winters)

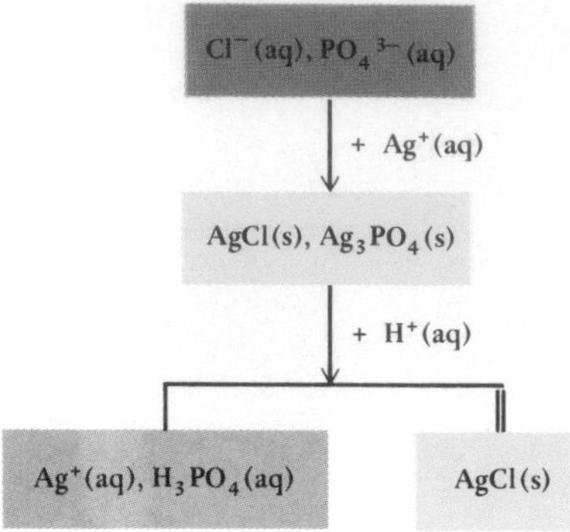

ever, the phosphate ion is the conjugate base of a weak acid, while chloride is the conjugate base of a strong acid. This means that Ag_3PO_4 can dissolve in excess strong acid, while AgCl cannot.

$$Ag_3PO_4(s) + 3\,H^+(aq) \rightleftharpoons 3\,Ag^+(aq) + H_3PO_4(aq)$$

$$AgCl(s) + H^+(aq) \longrightarrow \text{no reaction}$$

Many metal sulfides are insoluble in water, although some are more soluble than others (Table 17.3). This fact can be used to assist in the separation of a group of metal ions from one another and in their ultimate identification.

Taking the metal ions in Table 17.3, we can see that the first three sulfides are much less soluble than the last three. Therefore, if you have a solution of all six ions, it is convenient to add S^{2-} ion in such a way that only HgS, CuS, and CdS precipitate and leave Co^{2+}, Zn^{2+}, and Mn^{2+} in solution. This can be accomplished *if* $[S^{2-}]$ is kept larger than 3.6×10^{-26} M but less than 5.9×10^{-18} M.

To see how the sulfide ion concentration can be adjusted to the desired value, let us look again at the weak acid H_2S.

$$H_2S(aq) \rightleftharpoons H^+(aq) + HS^-(aq) \qquad K_1 = 1.0 \times 10^{-7}$$
$$HS^-(aq) \rightleftharpoons H^+(aq) + S^{2-}(aq) \qquad K_2 = 1.3 \times 10^{-13}$$
$$H_2S(aq) \rightleftharpoons 2\,H^+(aq) + S^{2-}(aq) \qquad K_{total} = K_1K_2 = 1.3 \times 10^{-20}$$

The sulfide ion concentration in aqueous solution, where $[H_2S]$ and $[H^+]$ are known, can be found from the rearranged equilibrium constant expression.

$$[S^{2-}] = \frac{[H_2S]}{[H^+]^2} K_{total}$$

From Le Chatelier's principle and the common ion effect, we can predict that raising $[H^+]$ will cause $[S^{2-}]$ to decrease. Conversely, adding OH^- will consume H^+ and raise the sulfide ion concentration. Therefore, to set $[S^{2-}]$ at a given value, we only need to adjust the H^+ ion concentration.

Let us say we want to separate Cd^{2+} and Zn^{2+}, both 0.0010 M, from one another. Precipitation of CdS requires a lower S^{2-} ion concentration (3.6×10^{-26} M) than precipitation of ZnS (1.1×10^{-18} M). Therefore, if we can adjust $[S^{2-}]$ to a value slightly less than 1.1×10^{-18} M, say 1.0×10^{-19} M, we can precipitate the maximum amount of Cd^{2+} and leave Zn^{2+} ion in solution.

TABLE 17.3 Metal Sulfides and Their Solubility

METAL (=MS) SULFIDE	K_{sp}	$[S^{2-}]$ REQUIRED TO BEGIN PRECIPITATION OF MS FROM 0.001 M M^{2+}
HgS	3.0×10^{-53}	3.0×10^{-50}
CuS	8.7×10^{-36}	8.7×10^{-33}
CdS	3.6×10^{-29}	3.6×10^{-26}
ZnS	1.1×10^{-21}	1.1×10^{-18}
CoS	5.9×10^{-21}	5.9×10^{-18}
MnS	5.1×10^{-15}	5.1×10^{-12}

Now, the concentration of H_2S in a saturated aqueous solution at 1 atmosphere pressure is 0.10 *M*, so a S^{2-} ion concentration of 1.0×10^{-19} would be achieved when

$$[H^+]^2 = \frac{[H_2S]}{[S^{2-}]} K_{total} = \frac{(0.10)}{1.0 \times 10^{-19}} \times 1.3 \times 10^{-20}$$

$$[H^+] = 0.11\ M \qquad \text{or} \qquad pH = 0.94$$

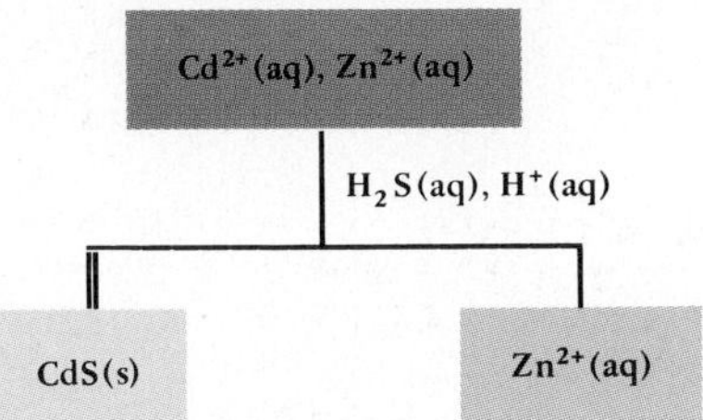

This H^+ ion concentration is easily achieved, so Cd^{2+}, and the Hg^{2+} and Cu^{2+} that form less soluble sulfides than Cd^{2+}, can be separated from Zn^{2+} and the other ions.

EXERCISE 17.19 ION SEPARATION

You have an aqueous solution that may contain CO_3^{2-} or SO_4^{2-}. Show how you can use aqueous $BaCl_2$ and HCl to decide which ion is in solution.

EXERCISE 17.20 ION SEPARATION

You have an aqueous solution of Co^{2+} and Mn^{2+}; both are 0.0010 *M*. Approximately what H^+ ion concentration would allow you to precipitate the maximum amount of CoS and leave Mn^{2+} in solution when H_2S is added?

SUMMARY

When an insoluble compound is placed in water, the concentrations of the ions will increase until the salt achieves its maximum **solubility** at the particular temperature; at this point the solution is **saturated,** and a dynamic equilibrium is achieved between the ions in solution and the solid salt. Under these conditions, the equilibrium constant expression for an insoluble salt such as zinc phosphate

$$Zn_3(PO_4)_2(s) \rightleftharpoons 3\ Zn^{2+}(aq) + 2\ PO_4^{3-}(aq)$$

can be written as

$$K_{sp} = 9.1 \times 10^{-33} = [Zn^{2+}]^3[PO_4^{3-}]^2$$

The constant is labeled with a subscript "sp" to indicate that it is the **solubility product constant** (Section 17.2). The name comes from the fact that the concentrations of the cations and anions are measures of the aqueous *solubility* of the compound. Solubility constants are typically in the range 10^{-3} to 10^{-50}, although some are much smaller.

K_{sp} values can be determined by various laboratory measurements of salt solubility (Section 17.3). Many such values have been tabulated in Appendix I, and we can use them to derive the solubility of a salt in pure water (Section 17.4).

When the reaction quotient (Q) is less than K_{sp}, the solution is said to be **undersaturated**, and additional solid can dissolve (Section 17.5). When equilibrium is achieved, and Q is equal to K_{sp}, the solution is **saturated**, and there is no further *net* dissolution of solid. However, if Q is larger than K_{sp}, the solution is **supersaturated** and some solid must precipitate from solution until saturation is achieved.

In addition to molar solubility, K_{sp} values can be used to determine whether a precipitate will form if an anion is mixed with a particular cation and what the salt solubility can be in the presence of an ion common to the salt (Sections 17.6 and 17.7).

It is possible that when another reagent is added to a saturated solution of an insoluble salt the salt will be transformed into another, even less soluble salt or a soluble complex ion (Sections 17.8 and 17.9). Such situations are examples of **simultaneous equilibria**, and a net equilibrium constant can be derived.

There are many acid–base and precipitation reactions of practical significance, and a few are described in Section 17.10. One reaction of importance is increased salt solubility in water due to hydrolysis of the anion. Also described are (a) the solubility in strong acids of salts having an anion that is the conjugate base of a weak acid, (b) the preparation of salts by acid–base reactions, and (c) the separation of ions using acid–base reactions.

STUDY QUESTIONS

REVIEW QUESTIONS

1. Explain why [AgCl] does not appear in the K_{sp} expression for

   $AgCl(s) \rightleftharpoons Ag^+(aq) + Cl^-(aq)$

2. What is a "reaction quotient" and how does it differ from an equilibrium constant? Use the equilibrium

   $CaCO_3(s) \rightleftharpoons Ca^{2+}(aq) + CO_3^{2-}(aq)$

   as an example in your discussion.
3. In Appendix I, find two salts that have K_{sp} values less than 1×10^{-40} and (a) write balanced equations to show the equilibria existing when the compounds dissolve in water and (b) write the K_{sp} expression for each compound.
4. Explain the terms saturated, undersaturated, and supersaturated. Use an equilibrium such as

   $NiS(s) \rightleftharpoons Ni^{2+}(aq) + S^{2-}(aq)$

   and its K_{sp} expression in your discussion.
5. What is the "common ion effect"? Use the equilibrium

   $Fe(OH)_2(s) \rightleftharpoons Fe^{2+}(aq) + 2\ OH^-(aq)$

   in your discussion.
6. Explain how Le Chatelier's principle operates when some excess cyanide ion, CN^-, is added to AgCN in equilibrium with its ions.

   $AgCN(s) \rightleftharpoons Ag^+(aq) + CN^-(aq)$

   What happens when excess AgCN(s) is added to a beaker of water already containing solid AgCN in equilibrium with its ions? Will the concentrations of Ag^+ and CN^- increase, decrease, or not change?
7. Explain why the solubility of AgCN can be greater in water than is calculated from the K_{sp} of the salt.
8. Explain why $BaCO_3$ can dissolve in strong acid but $BaSO_4$ cannot.

SOLUBILITY GUIDELINES

9. Name two insoluble salts of each of the following ions:
   (a) the chloride ion (c) the zinc(II) ion
   (b) the sulfide ion (d) the iron(II) ion
10. Using the table of solubility guidelines (Table 3.1), predict whether each of the following is insoluble or soluble in water.
    (a) $(NH_4)_2S$ (e) FeS
    (b) $ZnCO_3$ (f) $BaSO_4$
    (c) $Mg(OH)_2$ (g) Hg_2Cl_2
    (d) Na_2CO_3
11. If Na_2CO_3 is added to an aqueous solution of $AgNO_3$, can a precipitate occur? If so, what solid will form? Write a balanced equation for the reaction.
12. If Na_2S is added to an aqueous solution of $CuSO_4$, can precipitation occur? If so, what solid will form? Write a balanced equation for the reaction.
13. Each pair of salts below is soluble in water. State whether a precipitate can form when you add one salt to the other in aqueous solution. If so, give the formula of the precipitate.
    (a) NaCl + $AgNO_3$
    (b) KOH + $MgCl_2$
    (c) KOH + NaCl
    (d) KCl + $Pb(NO_3)_2$
    (e) $(NH_4)_2S$ + $Ni(NO_3)_2$

WRITING SOLUBILITY PRODUCT CONSTANT EXPRESSIONS

14. For each of the following salts, (a) write a balanced equation showing the equilibrium occurring when the salt is added to water and (b) write the K_{sp} expression. Give the value of K_{sp} for each salt.
    (a) ZnS (c) $NiCO_3$
    (b) SnI_2 (d) Ag_2SO_4

15. For each of the following salts, (a) write a balanced equation showing the equilibrium occurring when the salt is added to water and (b) write the K_{sp} expression. Give the value of K_{sp} for each salt.
(a) Bi_2S_3 (c) BaF_2
(b) $Ni(OH)_2$ (d) $Al(OH)_3$

16. If the K_{sp} expression for magnesium arsenate is $[Mg^{2+}]^3[AsO_4^{3-}]^2$, what is the formula of the salt?

17. If the K_{sp} expression for cadmium hexacyanoferrate is $[Cd^{2+}][Fe(CN)_6^{2-}]$, what is the formula of the salt?

DETERMINING K_{sp}

18. When 100. mg of solid CoS is added to 1.00 L of water, the cobalt(II) ion concentration in the aqueous solution of Co^{2+} and S^{2-} in equilibrium with CoS is 7.7×10^{-11} *M*. What is the K_{sp} for CoS?

19. At 20°C, a saturated aqueous solution of silver acetate, $AgC_2H_3O_2$, contains 1.0 g dissolved in 100.0 mL of solution. Calculate the K_{sp} for silver acetate.

$$AgC_2H_3O_2(s) \rightleftharpoons Ag^+(aq) + C_2H_3O_2^-(aq)$$

20. $Ca(OH)_2$ dissolves in water to the extent of 0.93 g per liter. What is the K_{sp} of $Ca(OH)_2$?

21. Barium fluoride, 2.75 g, was placed in 1.00 L of pure water at 25°C. After some BaF_2 had dissolved, and equilibrium had been established, the fluoride ion concentration was found to be 0.0150 *M*. What is the K_{sp} of barium fluoride?

22. At 25°C, 34.9 mg of Ag_2CO_3 will dissolve in 1.0 L of pure water. What is the solubility product constant for this salt?

23. At 25°C, 7.4 mg of $Zn(CN)_2$ will dissolve in 0.50 L of pure water. What is the solubility product constant for this salt?

SALT SOLUBILITY FROM K_{sp}

24. Estimate the solubility of silver cyanide in (a) moles per liter and (b) grams per liter of pure water.

25. What is the molar concentration of Au^+ in a saturated solution of AuI in pure water at 25°C?

26. In a saturated, aqueous solution of gold(III) iodide, how many milligrams of gold(III) ion are there per liter of solution? What is the I^- ion concentration in mg/L?

27. Estimate the solubility of lead bromide in terms of (a) moles per liter and (b) grams per liter of pure water.

28. In a saturated solution of silver carbonate, Ag_2CO_3, how many milligrams of Ag^+ are there in 1.0×10^2 mL of solution at 25°C?

29. Although it is a violent poison if swallowed, mercury(II) cyanide, $Hg(CN)_2$, has been used as a topical (skin) antiseptic.
(a) What is the molar solubility of this salt in pure water?
(b) How many milligrams of the salt dissolve per liter of pure water?
(c) How many milliliters of water are required to dissolve 1.0 g of the salt?

30. Limestone, $CaCO_3$, can exist in two mineral forms: calcite and aragonite. In pure water, the former has a K_{sp} of 3.8×10^{-9}, whereas K_{sp} of the latter is 6.0×10^{-9}. Which is the more soluble in pure water?

31. Using K_{sp} values, decide which compound in each of the following pairs is the more soluble.
(a) AgBr or AgSCN (d) ZnS or Bi_2S_3
(b) $SrCO_3$ or $SrSO_4$ (e) Ag_2S or SnS_2
(c) MgF_2 or CaF_2 (f) $Ca_3(PO_4)_2$ or AgI

32. Rank the following insoluble compounds in order of increasing solubility in water: BaF_2, $BaCO_3$, and Ag_2CO_3.

33. List the following in order of increasing solubility in water: AgI, HgS, PbI_2, $PbSO_4$, and NH_4NO_3.

34. Barium sulfate is very insoluble in water and is opaque to x-rays. Thus, if you drink a "barium cocktail" it does not dissolve in your stomach and intestines, and its progress through the digestive organs can be followed by x-ray analysis. If you place 1.00×10^2 mg of $BaSO_4$ in 1.0 L of water, approximately how many milligrams of $BaSO_4$ are dissolved at 25°C?

35. If you place the amounts given below in pure water, will all of the salt dissolve before equilibrium can be established, or will some salt remain undissolved?
(a) 5.0 milligrams of AgCl in 1.0 L of pure water
(b) 5.0 milligrams of $NiCO_3$ in 1.0 L of pure water
(c) 5.0 milligrams of MgF_2 in 125 mL of pure water
(d) 0.50 g of CaF_2 in 100. mL of pure water

PRECIPITATIONS

36. If the concentration of the nickel(II) ion in 100 mL of pure water is 0.0024 *M*, will precipitation of $NiCO_3$ occur when the concentration of the CO_3^{2-} ion is (a) 1.0×10^{-6} *M* or (b) 100 times greater?

37. To make up an unknown solution for the students in your laboratory, you dissolve 1.0×10^{-5} moles of $AgNO_3$ in 1.0 L of water. Unfortunately, you used a liter of tapwater instead of distilled water. If the chloride ion concentration in the tapwater is 2.0×10^{-4} *M*, will your error be revealed immediately by the formation of a white precipitate of AgCl?

38. Assume you have 100. mL of a solution that has a Pb^{2+} concentration of 0.0012 *M*.
(a) If enough of a soluble chloride-containing salt is added so that the Cl^- concentration is 0.010 *M*, will $PbCl_2$ precipitate?
(b) If 100. mL of 0.10 *M* HCl is added, will $PbCl_2$ precipitate?
(c) If 1.20 g of solid NaCl is added, will $PbCl_2$ precipitate?

39. If the concentration of Zn^{2+} in 10. mL of pure water is 1.6×10^{-4} *M*, will precipitation of zinc hydroxide occur when 4.0 mg of NaOH is added?

40. The mineral fluorite is CaF_2. If you have 500. mL of a solution that contains Ca^{2+} ion with a concentration of 0.0021 *M*, how many milligrams of KF must be added to the solution to just begin precipitation of CaF_2?

41. Fluoridation of city water supplies (to aid in the preventation of dental caries) may add about 1 part per million of F^- ion. This means the F^- ion concentration is about 5×10^{-5} *M*. Some water supplies are also "hard," that is, they contain alkaline earth cations such as Ca^{2+}. Under such circumstances, calcium ion can combine with the fluoride to produce the insoluble salt CaF_2. Will a precipitate of CaF_2 form when hard water, in which $[Ca^{2+}] = 2 \times 10^{-4}$ *M*, is fluoridated?

42. In each instance below, tell how many grams of the second reagent are needed to begin precipitation of an insoluble salt of the first-named metal ion. For example, how many grams of Na_2S are necessary to begin precipitation of PbS?
(a) 1.0 L of 0.0012 *M* $Pb(NO_3)_2$ + Na_2S
(b) 1.0 L of 1.0×10^{-6} *M* $CoCl_2$ + NaOH
(c) 125 mL of 3.2×10^{-4} *M* $BaCl_2$ + $Na_2SO_4 \cdot 10\ H_2O$
(d) 1.5 mL of 0.014 *M* $CaCl_2$ + Na_2HPO_4

43. If you mix 50. mL of 0.0012 *M* $BaCl_2$ with 25 mL of 1.0×10^{-6} *M* H_2SO_4 will a precipitate of $BaSO_4$ form?

44. Will a precipitate of $Mg(OH)_2$ form when 25.0 mL of 0.010 *M* NaOH is combined with 75.0 mL of a 0.10 *M* solution of magnesium chloride?

45. Lead chloride has the relatively large K_{sp} of 1.7×10^{-5}. If you mix 10. mL of 0.0010 *M* $Pb(NO_3)_2$ with 5.0 mL of 0.015 *M* HCl, will a precipitate of $PbCl_2$ form?

46. You can test for the presence of silver ion in aqueous solution by adding a soluble chloride salt. A white precipitate of AgCl would indicate silver ion. If you add one drop (0.05 mL) of 1.0 *M* NH_4Cl to 4.0 mL of a solution suspected to contain Ag^+, what is the maximum concentration of Ag^+ that can be present in the original 4.0 mL without a precipitate of AgCl forming?

47. Magnesium is extracted from seawater by precipitating it as $Mg(OH)_2$. If the concentration of Mg^{2+} in seawater is 1350 mg per liter, what must the OH^- concentration be in order to precipitate $Mg(OH)_2$?

48. A sample of hard water contains about 2.0×10^{-3} *M* Ca^{2+}. The water is to be fluoridated (see Study Question 41). Calculate the maximum concentration of F^- that can be present in the water without precipitation of CaF_2.

GRAPHICAL INTERPRETATION OF K_{sp}

49. If a point on the graph in Figure 17.2 is at $[Ag^+] = 3.0 \times 10^{-5}$ *M* and $[Cl^-] = 0.75 \times 10^{-5}$ *M*, is the solution saturated, undersaturated, or supersaturated? If the system is not at equilibrium, in which direction does the reaction proceed?

50. A silver chloride solution is represented by a point on the graph in Figure 17.2 where $[Ag^+] = 0.5 \times 10^{-5}$ *M* and $[Cl^-] = 3.0 \times 10^{-5}$ *M*. Is the solution supersaturated, saturated, or undersaturated?

51. Draw a graph similar to Figure 17.2 for $BaSO_4$.
(a) Use the plot to estimate the concentrations of Ba^{2+} and SO_4^{2-} when equilibrium has been established by dissolving some solid $BaSO_4$.
(b) If 50.0 mg of $BaSO_4$ are placed in 1.00 L of pure water, how many milligrams of the compound remain undissolved when equilibrium is established?
(c) Use the plot to estimate the concentration of SO_4^{2-} when equilibrium has been established with $[Ba^{2+}] = 8 \times 10^{-5}$ *M*.

ORDER OF PRECIPITATION

52. The cations Co^{2+}, Mn^{2+}, and Ni^{2+} can all be precipitated as very insoluble sulfides, CoS, MnS, and NiS. If you add a soluble sulfide to a solution containing these ions, all with a concentration of 0.1 *M*, which metal is precipitated first?

53. HI is added slowly to a solution 0.10 *M* in each of the following ions: Pb^{2+}, Ag^+, and Hg_2^{2+}. The insoluble salts PbI_2, AgI, and Hg_2I_2 will eventually form. In what order do these salts precipitate?

54. You often work with salts of Fe^{3+}, Pb^{2+}, and Al^{3+} in the laboratory, and all are found in nature and are important economically. If you have a solution containing these three ions, each at a concentration of 0.1 *M*, what is the order in which their hydroxides precipitate?

55. Alkaline earth metal ions can be precipitated as insoluble carbonates. If you have a solution of Mg^{2+}, Ca^{2+}, Sr^{2+}, and Ba^{2+} ions, all with the same concentration, what is the order in which their carbonates are precipitated as sodium carbonate is slowly added?

56. The alkaline earth metal cations Mg^{2+}, Ca^{2+}, and Ba^{2+} can be precipitated as their fluoride salts. If you have a solution that is 0.015 *M* in each of these ions, and you slowly add KF, in what order do you expect to see the fluoride salts precipitated?

COMMON ION EFFECT

57. Calculate the molar solubility of silver thiocyanate, AgSCN, in pure water and in 0.010 *M* NaSCN.

58. Calculate the molar solubility of silver carbonate, Ag_2CO_3, in pure water and after 0.15 g of Na_2CO_3 has been added to 225 mL of a saturated silver carbonate solution.

59. What is the solubility, in grams per mL, of BaF_2, in (a) pure water and in (b) water containing 5.0 mg per milliliter of KF?

60. Calculate the solubility, in milligrams per milliliter, of silver phosphate, Ag_3PO_4, in (a) pure water and in (b) 0.020 *M* $AgNO_3$.

61. Calculate the molar solubility of BiI_3 in (a) pure water and in (b) 0.020 *M* sodium iodide.

62. The solubility product for silver iodate, $AgIO_3$, is 1.0×10^{-8}. If 0.10 g of solid $AgIO_3$ is added to a 0.020 *M* solution of KIO_3, what are the concentrations of Ag^+, IO_3^-, and K^+ at equilibrium?

63. The alkaline earth metal ions can be precipitated from aqueous solution by addition of fluoride ion. (a) If you have a 0.015 *M* solution of Ba^{2+}, what is the minimum concentration of F^- necessary to begin precipitation of BaF_2? (b) If you started with 100. mL of the 0.015 *M* Ba^{2+}-containing solution, how many milligrams of NaF would be required to just begin precipitation? (c) After adding 0.50 g of NaF to 100. mL of 0.015 *M* Ba^{2+} solution, what is the concentration of Ba^{2+} remaining in solution?

64. Assume you have 1.0 L of a solution of Ba^{2+}(aq) and SO_4^{2-}(aq) in equilibrium at 25°C with solid $BaSO_4$. Estimate the concentrations of Ba^{2+}(aq) and SO_4^{2-}(aq) after adding 2.8 mg of Na_2SO_4.

SEPARATIONS

65. The ions Mn^{2+} and Co^{2+} can be separated in the qualitative analysis scheme in the laboratory by the difference in the solubility of their sulfides, MnS and CoS. If you have a 0.10 *M* solution of Mn^{2+} and Co^{2+}, CoS will begin to precipitate first as S^{2-} is slowly added to the solution. What is the concentration of Co^{2+} that remains in solution when MnS begins to precipitate?

66. To separate Ca^{2+} from Mg^{2+}, ammonium oxalate, $(NH_4)_2C_2O_4$, is added to a solution that is 0.020 *M* in both metal ions. If the concentration of the oxalate ion is adjusted properly, the metal oxalates can be precipitated separately.
    (a) What concentration of oxalate ion, $C_2O_4^{2-}$, will precipitate the maximum amount of Ca^{2+} without precipitating Mg^{2+}?
    (b) What concentration of Ca^{2+} remains when Mg^{2+} just begins to precipitate?

67. A solution contains Ca^{2+} and Pb^{2+} ions, both at a concentration of 0.010 *M*. You wish to separate the two ions from each other as completely as possible by precipitating one but not the other using aqueous Na_2SO_4 as the precipitating reagent.
    (a) Which will precipitate first as sodium sulfate is added, $CaSO_4$ or $PbSO_4$?
    (b) Just before the more soluble salt begins to precipitate, what molarity of the first metal ion remains in solution?

68. A solution contains 0.10 *M* I^- and 0.10 *M* CO_3^{2-}. (a) If solid $Pb(NO_3)_2$ is slowly added to the solution, which salt will precipitate first, PbI_2 or $PbCO_3$? (b) What will be the concentration of the first anion that precipitates (I^- or CO_3^{2-}) when the second or more soluble salt begins to precipitate?

69. Lead and zinc compounds are often found together in nature, since the solubilities of their salts are similar. For example, the K_{sp} of $Pb(OH)_2$ is 2.8×10^{-16} and of $Zn(OH)_2$ is 4.5×10^{-17}.
    (a) If Pb^{2+} and Zn^{2+} are present in ground water, each with a concentration of 1.0×10^{-6} *M*, and if the ground water is 1.0×10^{-6} *M* in OH^- ion, will either $Pb(OH)_2$ or $Zn(OH)_2$ precipitate?
    (b) What is the maximum concentration of OH^- that can exist in ground water that will allow you to precipitate one of the metal hydroxides and leave the other one in solution?

70. Each pair of ions below is found together in aqueous solution. Using the table of solubility products in Appendix I, devise a way of separating these ions by precipitating one of them as an insoluble salt and leaving the other in solution.
    (a) Ba^{2+} and Na^+
    (b) Ca^{2+} and Zn^{2+}
    (c) Bi^{3+} and Cd^{2+}
    (d) Cu^{2+} and Ag^+
    (e) Pb^{2+} and Sn^{2+}

SIMULTANEOUS EQUILIBRIA AND COMPLEX IONS

71. Solid AgBr is dissolved when excess thiosulfate ion, $S_2O_3^{2-}$, is added to give a water-soluble complex ion.

$$AgBr(s) + 2\ S_2O_3^{2-}(aq) \rightleftharpoons [Ag(S_2O_3)_2]^{3-}(aq) + Br^-(aq)$$

Show that this equation is the sum of two other equations, one for the dissolving of AgBr to give its ions and the other for the formation of the $[Ag(S_2O_3)_2]^{3-}$ ion from its ions. Calculate *K* for the overall reaction.

72. Using the appropriate K_{sp} and $K_{formation}$ constants in the Appendices, calculate the equilibrium constant for the reaction

$$AgI(s) + 2\ NH_3(aq) \rightleftharpoons [Ag(NH_3)_2]^+(aq) + I^-(aq)$$

73. If I^- ion is added to a saturated solution of AgCl will AgI form? What is the equilibrium constant for this process?

74. Can zinc sulfide, ZnS, be transformed extensively into zinc cyanide? Calculate the equilibrium constant for the reaction

$$ZnS(s) + 2\ CN^-(aq) \rightleftharpoons S^{2-}(aq) + Zn(CN)_2(s)$$

75. Will 5.0 mL of 2.5 *M* NH_3 dissolve 0.0001 moles of AgCl? 0.0005 moles of AgBr? (Both salts are suspended in 1.0 L of pure water.)

ACID–BASE REACTIONS FOR DISSOLVING SOLIDS AND PREPARING SALTS

76. Explain why $CaCO_3$ is soluble in aqueous HCl but $CaSO_4$ is not.

77. Which of the following barium salts should be soluble in strong acid: $Ba(OH)_2$, $BaSO_4$, and $BaCO_3$?

78. Explain why some metal sulfides can dissolve in a strong acid such as HCl.

79. Explain how you can prepare pure $CaSO_4$ starting with limestone, $CaCO_3$.

80. Describe the method of preparation of $NiSO_4$ from $NiCO_3$.

81. How can you prepare pure magnesium nitrate beginning with pure magnesium chloride?

82. Describe a preparation of iron(III) perchlorate, $Fe(ClO_4)_3$, from iron(III) chloride.

83. Beginning with magnesium carbonate, describe how to prepare (a) magnesium sulfate, (b) magnesium bromide, (c) magnesium oxalate (MgC_2O_4), (d) magnesium perchlorate, and (e) magnesium fluoride.

84. You have an aqueous solution that may contain CO_3^{2-} or Cl^-. If you have available aqueous $AgNO_3$ and HCl, tell *two* ways you can decide which ion the solution contains.

85. You have an aqueous solution of Zn^{2+} and Pb^{2+}, both 0.0010 *M*. Both form insoluble precipitates with S^{2-} ion. Approximately what H^+ ion concentration will allow you to precipitate the maximum amount of one ion while leaving the other in solution?

86. Some solid sulfides can dissolve in strong acid.

$$MS(s) + 2\,H^+(aq) \longrightarrow M^{2+}(aq) + H_2S(aq) \qquad K = K_{dissolve}$$

This reaction can be considered the sum of two other reactions:

$$MS(s) \rightleftharpoons M^{2+}(aq) + S^{2-}(aq) \qquad K = K_{sp}$$

$$S^{2-}(aq) + 2\,H^+(aq) \rightleftharpoons H_2S(aq) \qquad K = 1/1.3 \times 10^{-20} = 7.7 \times 10^{19}$$

Therefore, we can write the expression

$$K_{dissolve} = K_{sp}\,(7.7 \times 10^{19}) = \frac{[M^{2+}][H_2S]}{[H^+]^2}$$

This means we can find the equilibrium concentration of M^{2+} when $[H^+]$ is known and if we assume $[H_2S]$ is 0.10 *M* in a saturated solution.

(a) Calculate $K_{dissolve}$ for ZnS.

(b) Calculate $[Zn^{2+}]$ when the H^+ ion concentration is 0.30 *M*.

87. Study Question 86 outlines a method for calculating the solubility of metal sulfides in strong acid. Use this approach to compare the solubility of CuS and ZnS in 1.0 *M* HCl.

GENERAL PROBLEMS

88. AgCl(s) is in equilibrium with $Ag^+(aq)$ and $Cl^-(aq)$. If you have 1.0 L of this saturated solution, and you add 1.0 mg of solid NaCl, what are the new concentrations of $[Ag^+]$ and $[Cl^-]$ when equilibrium is reestablished? How many milligrams of AgCl will be precipitated by adding the extra NaCl?

89. You have 1.0 L of pure water at 25°C and add 25 mg of $Zn(CN)_2$.

(a) Does all of the zinc cyanide dissolve, or does some solid remain when equilibrium is achieved? What are the concentrations of Zn^{2+} and CN^-? What happens to the Zn^{2+} and CN^- concentrations when another 25 mg of $Zn(CN)_2$ is added?

(b) If you add another liter of water to the original mixture of 25 mg of solid and 1.0 L of water, will any zinc cyanide remain undissolved? What are the concentrations of Zn^{2+} and CN^-?

(c) If 100. mg of KCN is added to the original mixture, how many milligrams of solid $Zn(CN)_2$ are there now in the beaker?

90. $Zn(OH)_2$ is a relatively insoluble base. A saturated solution has a pH of 8.65. Calculate K_{sp} for $Zn(OH)_2$.

91. A saturated solution of $Ce(OH)_3$ has a pH of 9.00. (a) What are the Ce^{3+} and OH^- concentrations? (b) What is the K_{sp} of $Ce(OH)_3$?

92. Calcite and aragonite are two forms of $CaCO_3$. In pure water, their respective K_{sp} values are 3.8×10^{-9} and 6.0×10^{-9}. Will calcite spontaneously transform into aragonite in pure water? What is the equilibrium constant for this transformation?

93. If 0.581 g of $Mg(OH)_2$ is added to 1.00 L of water, will it all dissolve? If the solution is buffered at a pH of 5.00, will it all dissolve?

94. The Ca^{2+} ion in hard water is often precipitated as $CaCO_3$ by adding soda ash, Na_2CO_3. If the calcium ion concentration in hard water is 0.010 *M*, and if Na_2CO_3 is added until the carbonate ion concentration is 0.050 *M*, what percentage of the calcium ion has been removed from the water? (You may neglect hydrolysis of the carbonate ion.)

95. Photographic film is coated with crystals of AgBr suspended in gelatin. Light exposure leads to reduction of some of the silver ions to metallic silver. Unexposed AgBr is dissolved with sodium thiosulfate in the fixing step.

$$AgBr(s) + 2\,S_2O_3^{2-}(aq) \longrightarrow [Ag(S_2O_3)_2]^{3-}(aq) + Br^-(aq)$$

(a) Using the appropriate K_{sp} and $K_{formation}$ values in Appendices I and J, calculate the equilibrium constant for the dissolving process.
(b) If you want to dissolve 1.0 g of AgBr in 1.0 L of solution, how many moles of $Na_2S_2O_3$ must be added?

96. The solubility of some metal salts is affected by hydrolysis of the anion. For PbS, the reaction

$$PbS(s) + H_2O(\ell) \rightleftharpoons Pb^{2+}(aq) + HS^-(aq) + OH^-(aq)$$

$$K_{total} = [Pb^{2+}][HS^-][OH^-] = 6.5 \times 10^{-29}$$

produces a lead(II) ion concentration of 3.9×10^{-10} *M*. (This can be calculated when $[HS^-]$ and $[OH^-]$ are known.)

(a) If sulfide ion hydrolysis leads to a concentration of HS^- of 4.1×10^{-10} at 25°C (with no Pb^{2+} present), what is the concentration of OH^- in the solution?
(b) Assuming that the values of $[HS^-]$ and $[OH^-]$ calculated in part (a) are always observed when a metal sulfide is placed in water, calculate the solubility of CdS assuming sulfide ion hydrolysis.
(c) Compare the result in part (b) with the solubility calculated without assuming sulfide hydrolysis (from K_{sp} alone).

SPECIAL SECTION: CARBON DIOXIDE EQUILIBRIA: OF OCEANS AND EGGS

Our biosphere is a complicated mixture of carbon compounds, some being created, some transformed, and others decomposed at any one time (Figure 17.8). The compound that links these processes and their materials or products is carbon dioxide, which is produced in the biological process of respiration and consumed by photosynthetic organisms. Although CO_2 constitutes only about 0.0325% of the atmosphere, a recent estimate is that the earth's atmosphere contains 700 billion tons of carbon in the form of the gas.

Our atmosphere normally contains about 0.0003 atm of CO_2, and, when this is in equilibrium with CO_2 dissolved in water, the concentration of aqueous H_2CO_3 is about 10^{-5} M. Because of this solubility, the oceans are thought to hold roughly 60 times 700 billion tons in the form of (a) dissolved CO_2 or carbonic acid, (b) one of the dissociation products of H_2CO_3, (c) solid metal carbonates such as $CaCO_3$ and $MgCO_3$, and (d) organic matter.

$$H_2CO_3(aq) \rightleftharpoons H^+(aq) + HCO_3^-(aq) \quad K_1 = 4.2 \times 10^{-7}$$

$$HCO_3^-(aq) \rightleftharpoons H^+(aq) + CO_3^{2-}(aq) \quad K_2 = 4.8 \times 10^{-11}$$

Because H_2CO_3 is a weak acid, solutions of CO_2 in pure water are slightly acidic, and, since K_2 is so much smaller than K_1, we can estimate the pH from the first equilibrium reaction. Using $[H_2CO_3] \cong 10^{-5}$, we can calculate that $[H^+]$ is about 2.2×10^{-6} M and that the pH is therefore

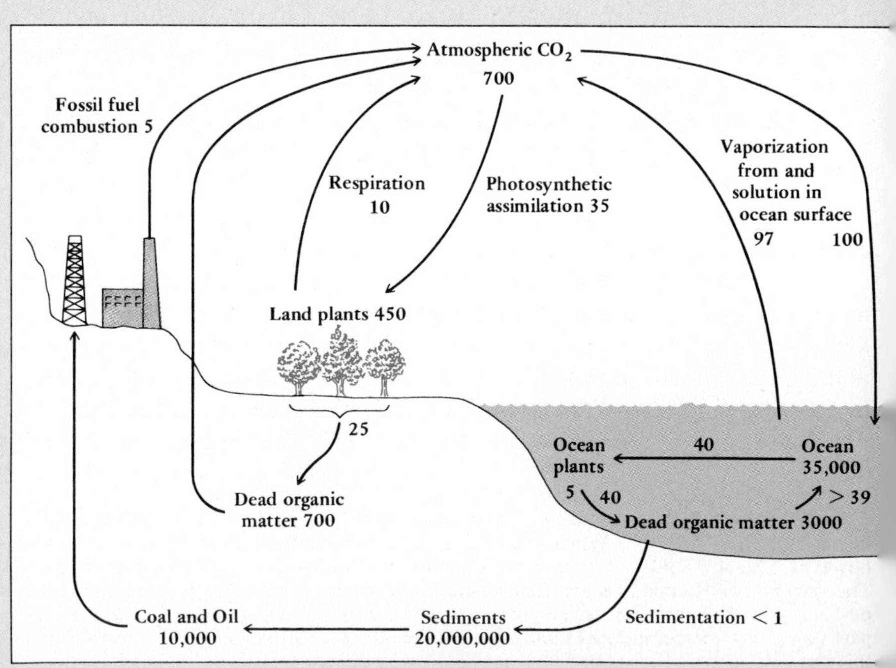

FIGURE 17.8
The carbon cycle in the biosphere. The numbers represent quantities of C (in units of 10^9 tons) in a particular reservoir or flowing from one reservoir to another.

about 5.6 (assuming $[H_2CO_3] = 10^{-5}$ *M*). This means that the rain that falls even in a nonpolluted environment should be slightly acidic.

The oceans of the earth contain enormous quantities of calcium carbonate produced by various sea creatures. This means there is the additional equilibrium of calcium carbonate to consider.

$$CaCO_3(s) \rightleftharpoons Ca^{2+}(aq) + CO_3^{2-}(aq) \qquad K_{sp} = 4.8 \times 10^{-9}$$

When this is included, seawater saturated with carbon dioxide is found to have a pH of 8.2 ± 0.2. The slightly alkaline character of seawater arises from the fact that the carbonate ion from the mineral calcite is the conjugate base of a weak acid, the bicarbonate ion, and so carbonate hydrolysis produces OH^-.

$$CO_3^{2-}(aq) + H_2O(\ell) \longrightarrow HCO_3^-(aq) + OH^-(aq)$$
$$K_b = 2.3 \times 10^{-8}$$

The most important aspect of the carbonate equilibrium system is that it buffers the pH of seawater. The addition of acids by undersea volcanic activity or other natural processes is countered primarily by carbonic acid formation (and subsequent loss of carbon dioxide to the atmosphere)

$$HCO_3^-(aq) + H^+(aq) \rightleftharpoons H_2CO_3(aq)$$

and secondarily by bicarbonate formation.

$$H^+(aq) + CO_3^{2-}(aq) \rightleftharpoons HCO_3^-(aq)$$

On the other hand, an increase in hydroxide ion concentration is counteracted by the reaction

$$OH^-(aq) + HCO_3^-(aq) \rightleftharpoons H_2O(\ell) + CO_3^{2-}(aq)$$

This reaction of course leads to an increase in carbonate concentration. If the seawater is above the saturation limit of calcium carbonate, limestone is precipitated (Figures 17.9 and 17.10).

Carbon dioxide is being generated in ever increasing amounts, in part due to the increase in the population of the earth, in part to the clearing of forests (and thus to less use of CO_2 in photosynthesis), and in part to increased combustion of fossil fuels. Indeed, there is fear that this could lead to a global warming trend caused by the "greenhouse effect."*

Fortunately, the amount of CO_2 in the atmosphere is not increasing as rapidly as might be expected, largely because the ocean is a great CO_2 sink. As the partial pressure of CO_2 increases, CO_2 solubility increases, and it is estimated that the sea has absorbed roughly half of the increase in CO_2. Although this should lead in turn to an increase in hydrogen ion concentration, it can be controlled through reaction with carbonate ion.

The ocean, a sink for atmospheric CO_2 (Charles D. Winters).

*The atmosphere is transparent to sunlight in the visible region (and at some wavelengths in the ultraviolet as well). When the earth's surface absorbs and reradiates this energy, however, it is in the infrared part of the spectrum, for which CO_2 is an excellent absorber. The energy thus trapped in the atmosphere could cause a warming of the entire earth, perhaps enough to melt part of the polar ice caps and raise the level of the oceans and turn part of the now temperate zones into deserts. Scientists are still debating the probability of this outcome.

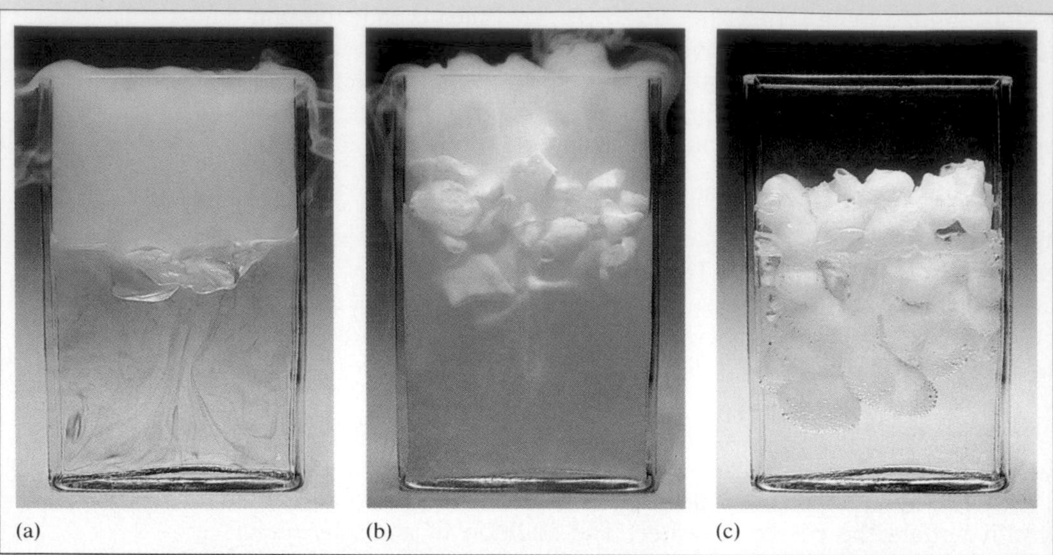

FIGURE 17.9

The system $CO_2/Ca(OH)_2(aq)$. (a) Dry ice (solid CO_2) is added to saturated $Ca(OH)_2(aq)$. (The bromthymol blue indicator shows the solution is basic.) The cloudiness in the solution indicates $CaCO_3(s)$ is just beginning to be precipitated.

$$CO_2(g) + H_2O(\ell) \rightleftharpoons H_2CO_3(aq)$$
$$H_2CO_3(aq) + Ca(OH)_2(aq) \rightleftharpoons CaCO_3(s) + 2\,H_2O(\ell)$$

(b) As the concentration of dissolved CO_2 builds up, more $CaCO_3$ precipitates. (c) When the concentration of dissolved CO_2 is large enough (the indicator now shows the solution is acidic), $CaCO_3$ dissolves according to the equilibrium

$$CO_2(aq) + CaCO_3(s) + H_2O(\ell) \rightleftharpoons Ca^{2+}(aq) + 2\,HCO_3^-(aq)$$

It is just this reaction that leads to the dissolution of limestone in CO_2–saturated ground water. Reversal of this reaction then leads to reprecipitation of limestone in caves (Figure 14.1). (Charles D. Winters)

Furthermore, as the upper layers of seawater containing dissolved carbon dioxide are mixed with the lower layers in contact with carbonate-containing sediments, hydrogen ion can be removed by reactions such as

$$CaCO_3(s) + H^+(aq) \rightleftharpoons HCO_3^-(aq) + Ca^{2+}(aq)$$

The problem is that ocean mixing is relatively slow, requiring times on the order of 1000 years.

There is a problem in agriculture that is closely related to absorption of CO_2 by the sea. Like dogs, chickens pant when they are hot. This is not an obscure fact meant to inject a little humor into your reading. Rather, it has serious consequences. The problem is that the chicken loses CO_2 at a more rapid rate when panting. Lowering the partial pressure of CO_2 in the chicken means, according to Le Chatelier, that the carbonate equilibrium is shifted away from CO_3^{2-} and toward CO_2. Less CO_3^{2-} is thus available for the formation of $CaCO_3$ shells, and the egg shells produced by chickens in hot weather are often fragile. The solution? During warm weather, chickens are given carbonated drinking water. (Rumors that chickens are demanding Perrier water are unfounded!)

FIGURE 17.10

The Redwall limestone formation (Grand Canyon, Arizona), a remnant of an ancient sea. (James Cowlin)

SPECIAL SECTION: SILVER CHEMISTRY

Starting with soluble silver nitrate, $AgNO_3$, various reagents are added to form progressively less soluble silver salts. For NH_3 and $S_2O_3^{2-}$, the formation constants of their Ag^+ complexes are large enough that the insoluble salts AgCl and AgBr, respectively, can be dissolved. For a discussion of the equilibria involved [especially the reaction $Ag_2CO_3(s) + 2\ OH^-(aq) \rightarrow Ag_2O(s) + CO_3^{2-}(aq)$, where $K = 2 \times 10^4$], see B. Shakhashiri, *Chemical Demonstrations,* Volume 1, University of Wisconsin Press, 1983, page 307.

18

The Spontaneity of Chemical Reactions: Entropy and Free Energy

$K(s) + H_2O(\ell) \rightarrow KOH(aq) + \frac{1}{2} H_2(g)$; $\Delta H° = -196.5$ kJ; $\Delta S° = 22.9$ J/K; $\Delta G° = -203.4$ kJ (Charles D. Winters)

CHAPTER OUTLINE

Some chemical and physical processes will take place by themselves, given enough time. If you stretch a rubber band, it will snap back spontaneously. If you put a spoonful of sugar in your coffee, it will dissolve and the molecules will distribute themselves evenly throughout the liquid. Because of chemical reactions in our bodies, we grow older whether we like it or not. Diseases like cancer can arise spontaneously from chemistry gone astray, in spite of our best efforts to live a healthy life.

To control chemical reactions or to halt a cancer, we must understand why some chemical reactions are spontaneous and others are not. We must understand how to predict when a reaction will be spontaneous and how to control the spontaneity if necessary. This chapter has the limited objective of working out a method of predicting the spontaneity of reactions.

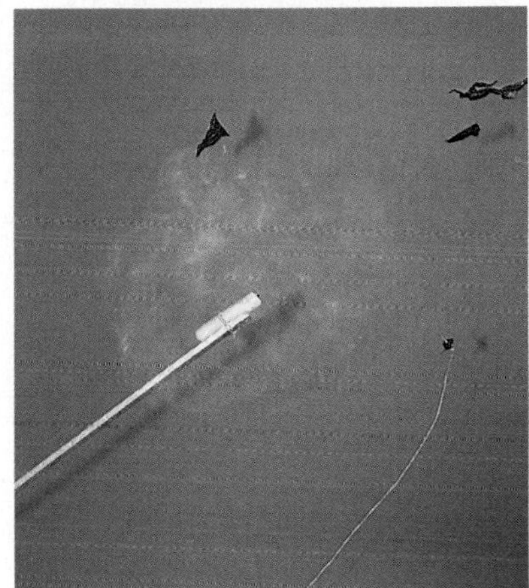

FIGURE 18.1
A balloon, filled with hydrogen gas, explodes after being ignited with a candle. The spontaneous reaction occurring is $2\ H_2(g) + O_2(g) \rightarrow 2\ H_2O(g)$. (Charles D. Winters)

18.1 SPONTANEOUS REACTIONS AND SPEED: THERMODYNAMICS AND KINETICS

A **spontaneous chemical reaction** is one that achieves chemical equilibrium by reacting from left to right as written, if only there is enough time (Figure 18.1). Thus, hydrogen will react spontaneously with oxygen to give water.

$$\xrightarrow{\text{Spontaneous}}$$
$$2\ H_2(g) + O_2(g) \longrightarrow 2\ H_2O(\ell)$$

A *nonspontaneous* reaction, in spite of the misleading name, also tends to occur, but this time it occurs spontaneously from right to left, in a direction opposite from the way the equation is written. The dissociation of N_2O_4 molecules is nonspontaneous,

$$\xrightarrow{\text{Nonspontaneous}}$$
$$O_2N{-}NO_2(g) \rightleftharpoons 2\ NO_2(g)$$
$$\xleftarrow{\text{Spontaneous}}$$

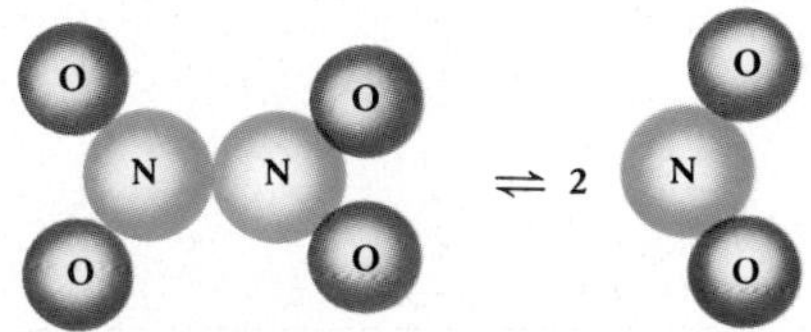

but NO_2 molecules will spontaneously revert to N_2O_4. Only when a reaction has reached equilibrium is there no tendency to move spontaneously in one direction or the other; at equilibrium there is no net change.

You live everyday with spontaneous chemical and physical transformations. The oxidation of H_2 is a spontaneous reaction, but it is rapid only if you ignite the mixture (Figure 18.1). The mere presence of oxygen is not enough, since H_2 can stay in contact with air a long time if you are careful not to set off the reaction. This means there is a difference between the speed of a reaction and its spontaneity. **Thermodynamics**, the subject of this chapter, is the science of heat and energy transfer, and it will help us predict whether a reaction can occur given enough time, that is, whether it is spontaneous. However, *thermodynamics can tell us nothing at all about the speed of the reaction*. The study of the rates of reactions, and why some are fast but others slow, is called **kinetics**, and it was the topic of Chapter 13.

Beginning with Chapter 5, we have repeatedly used ideas of energy differences and energy transfer. In Chapter 5 we described the energy involved in chemical reactions, in Chapter 7 the concern was with energy on the atomic level, and in Chapters 11 and 12 we learned about the energy involved in changes in physical state. In this chapter we are going to define more completely the energy changes that cause some chemical reactions and physical transformations to be spontaneous regardless of their speed.

You should briefly review the following concepts in Chapter 5: the idea of systems and their surroundings, the first law of thermodynamics, standard states, and Hess's Law.

18.2 ENERGY AND SPONTANEITY

One of the first experimental observations you made in your life was that some spontaneous processes occur with loss of energy: If you drop a glass bottle from your hand, it falls spontaneously to the ground (and can break into a thousand pieces); a stone or snowball rolls downhill; rain, snow, and hail fall from the sky. In every instance, the object falls or

moves because gravity exerts a force on it. If the object is free to move, this force results in a conversion of potential to kinetic energy for the object, a form of energy converted to work (breaking of the bottle) or heat when the object is stopped.

When gasoline is ignited, it burns spontaneously in oxygen to give CO_2 and H_2O, and equations such as that for the combustion of octane

$$2\ C_8H_{18}(g) + 25\ O_2(g) \longrightarrow 16\ CO_2(g) + 18\ H_2O(g) + \text{heat}$$

can be written. The first thing you may notice about this reaction is that heat is evolved. The reaction is exothermic because 16 moles of CO_2 and 18 moles of H_2O possess less energy than do 25 moles of O_2 and 2 moles of octane (at 298K). The energy of the chemical system has decreased.

In all these examples of spontaneous processes, the system moved to a position of lower energy, but is it a general law of nature that in order to be spontaneous a process must move in the direction of lower energy? This certainly is a tempting conclusion, and it is valid for strictly mechanical systems, that is, for falling rain and rocks. However, for systems in general, it is incomplete, and exceptions are not hard to find. The evaporation of liquid water is a spontaneous process, but one in which the system moves to a state of higher energy. Dissolving solutes in solvents is often endothermic (Chapter 12), and some chemical reactions are endothermic.

An experiment you can try is the spontaneous motion made by a stretched rubber band. Take a thick rubber band and hold it gently against your upper lip. When you stretch it to its limit, you will notice that it has very slightly warmed (so stretching the rubber band is an *exo*thermic process). Holding it stretched, let the heat dissipate, and then, while still holding the ends of the rubber band, let it spontaneously contract. You will notice that it has cooled! The spontaneous contraction of a rubber band is an *endo*thermic process.

Where does this leave us? Spontaneous processes are indeed often exothermic, but they can also be endothermic. Energy is not the only determining factor for spontaneity. We are missing something.

18.3 SPONTANEITY, DISORDER, AND ENTROPY

Look again at an endothermic change such as the melting of ice or the evaporation of water to water vapor under conditions where the change occurs spontaneously. Ice is a highly organized structure but, as it melts, the structure becomes less ordered. In liquid water and, finally, in water vapor the molecules are widely separated and free to roam (Figure 18.2). In the case of octane combustion, a total of 27 molecules in the gas phase were converted to 34 gas molecules. Although no mass was lost, rearrangement of atoms resulted in a greater number of molecules. When salt or sugar dissolves in water, a regular orderly arrangement of ions or molecules in the crystal becomes a random jumble of ions or molecules in the solvent. In every case, a few molecules gave many more, or a compact material flew apart into many pieces. Order became disorder. Regularity became chaos. *A spontaneous process creates disorder.*

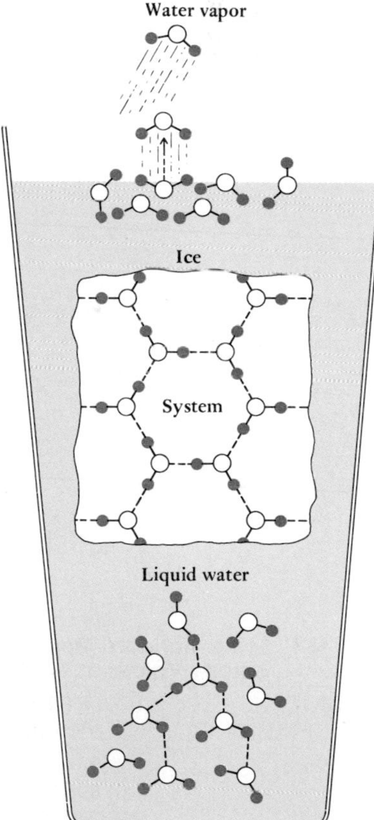

FIGURE 18.2
Increasing disorder in an endothermic, spontaneous process, the melting of ice to liquid water and the evaporation of the liquid.

BOLTZMANN'S TOMBSTONE

Ludwig Boltzmann (1844–1906) was an Austrian mathematician who did much of the work on the kinetic theory of gases and who gave us a useful interpretation of entropy. Engraved on his tombstone in Vienna is his equation relating entropy and probability (where k is a constant in energy units), $S = k \ln W$. That is, Boltzmann proposed that the entropy of a system is related to W, the number of ways that atoms or molecules can be arranged in a given state, always keeping their sum total energy fixed. If there are but a few ways to arrange the atoms of a substance, then the entropy is low. On the other hand, the entropy is high if there are many possible ways to arrange things ($W >> 1$). Thus, the entropy of a crystalline solid, where the atoms, ions or molecules are located at fixed points in space, will be low. On the other hand, the molecules of a gas are in constant, random motion, and the entropy of a system is higher. Low entropy is associated with order, and high entropy is associated with disorder, randomness, or chaos.

Clearly, two factors should be considered when trying to predict the spontaneity of an event. A decrease in energy or enthalpy is one determinant, and the creation of disorder or chaos is another. Sometimes these effects reinforce one another, but at other times they are opposing, and the final outcome is determined by their relative magnitudes. In the case of water evaporation, for example, the increase in molecular disorder must more than offset the increase in energy.

The thermodynamic function called **entropy** is a measure of the disorder of a system. Entropy is the invention of one of the great scientists of the 19th century, Rudolf Clausius, and it was he who gave it the symbol we use today, ***S***. Entropy came from consideration of the usual tools of the thermodynamic trade, heat and temperature, and if you make some careful measurements in the laboratory you can derive a value for the entropy of a substance (as described in Sections 18.5 and 18.6).

Entropy, like enthalpy, is a *state function*, and the quantity of entropy of a system depends only on the temperature and pressure of the system. Values of the entropy for a few substances at 298K are collected in Table 18.1, and many more are listed in Appendix L. Notice that the units of entropy are joules/kelvin · mole (J/K · mol). The superscript "°" indicates that all are "standard state entropies," that is, the entropy of the compound or element in its most stable form at a pressure of 1 atm; for substances in solution, the concentration is 1 m.

TABLE 18.1 Some Standard Molar Entropy Values at 298K

COMPOUND OR ELEMENT	ENTROPY, $S°$ (J/K · mol)
C(graphite)	5.74
C(g)	158.096
CO_2(gas)	213.74
Ca(s)	41.42
$CaCO_3$(s) (as calcite)	92.9
O_2(g)	205.138
$H_2O(\ell)$	69.91
H_2O(gas)	188.825

One of the most important aspects of the entropy values in Table 18.1 is that there are entries for elements in their standard states. You will recall that, in contrast, standard enthalpies of formation of the elements are defined as zero. You may have noticed further that the entropies given are not entropies of formation (there is no subscript "$_f$"). To understand these observations, we must delve a bit deeper into the nature of entropy.

Unlike enthalpies, it is possible to measure the entropy of any *pure* material at any temperature (Sections 18.5 and 18.6). To understand this you need to visualize a system at the absolute zero of temperature. At 0K any system will consist of atoms, ions, or molecules that can be perfectly ordered with respect to each other in a solid.* Since entropy has been equated with disorder, this means the entropy of a pure, *perfectly ordered* crystalline substance is 0 at 0K. This statement is known as the **third law of thermodynamics**. Because of this, we have an absolute starting place for the measurement of entropy, and the *absolute entropy* of a substance can then be determined at any temperature higher than 0K.†

Many times in this text you have seen that the enthalpy change for a process can be calculated from the difference between the enthalpies of the initial and final states.

$$\Delta H^\circ_{process} = H^\circ(\text{final}) - H^\circ(\text{initial})$$

Likewise, it is possible to calculate the entropy change in a process.

$$\Delta S^\circ_{process} = S^\circ(\text{final}) - S^\circ(\text{initial})$$

For example, the evaporation of one mole of liquid water to water vapor at 25°C occurs with an entropy change of

$$\Delta S^\circ_{vaporization} = S^\circ[H_2O(g)] - S^\circ[H_2O(\ell)]$$

$$\Delta S^\circ_{vaporization} = 188.8 \text{ J/K} \cdot \text{mol} - 69.9 \text{ J/K} \cdot \text{mol} = 118.9 \text{ J/K} \cdot \text{mol}$$

It is important to recognize that the entropy has *increased*: ΔS° is positive. Since we have equated entropy and randomness or disorder, this is a reasonable result: molecules in the gas phase do indeed have more freedom to move about than do molecules in the liquid state.

18.4 ENTROPY, ELEMENTS, AND COMPOUNDS

A common rule of thumb when dealing with entropy is that the greater the disorder or randomness in a system, the larger the entropy. In this short section we want to examine some processes in which entropy changes are observed, since it will help you later in the prediction of reaction spontaneity.

A. *The entropy of a substance increases as it changes from solid to liquid to gas.*

$$\underset{S^\circ = 69.9 \text{ J/K}\cdot\text{mol}}{H_2O(\ell)} \xrightarrow{\Delta S^\circ = +118.9 \text{ J/K}} \underset{188.8 \text{ J/K}\cdot\text{mol}}{H_2O(g)}$$

B. *When pure solids or liquids dissolve in a solvent, the entropy of the substance increases.*

---

*Even at absolute zero there is some residual, internal motion as atoms within molecules vibrate against each other and as the molecules vibrate against one another. This residual motion is a quantum effect associated with small particles. It is analogous to the electron in the H atom, which still has motion even in its lowest energy state.

†You may be wondering why it is not possible to say that all molecules have zero enthalpy at zero kelvin, just as $S = 0$ at this temperature. For the answer, refer to the previous footnote. All molecules of the same kind have some residual, internal (vibrational) energy, but it is different for different elements and compounds. This means that you cannot say that the enthalpies of all systems at 0 K are the same, let alone that H = 0.

$$\underset{S^\circ = 151.1 \text{ J/K}\cdot\text{mol}}{NH_4NO_3(s)} \xrightarrow{\Delta S^\circ = +108.7 \text{ J/K}} \underset{259.8 \text{ J/K}\cdot\text{mol}}{NH_4NO_3(aq)}$$

In Chapter 12 you learned that, when ammonium nitrate dissolves in water, the process is quite endothermic. However, the entropy certainly increases because the system has gone from an ordered crystalline solid to a jumble of ions in aqueous solution, and it is this enormous entropy increase (as you will see in Section 18.6) that allows the process to occur.

C. *When a gas dissolves in a solvent, its entropy declines.*

$$\underset{S^\circ = 186.9 \text{ J/K}\cdot\text{mol}}{HCl(g)} \xrightarrow{\Delta S^\circ = -130.4 \text{ J/K}} \underset{56.5 \text{ J/K}\cdot\text{mol}}{HCl(aq)}$$

A gas is clearly the most disordered state. When the substance dissolves, there are intermolecular interactions that impose some order, and the entropy declines.

D. *Entropy generally increases with increasing molecular complexity.* There are several aspects to the relation between molecular complexity and entropy. For example, the standard entropy increases as the number of ions per formula unit in an ionic solid increases.

	$KCl(s)$	$CaCl_2(s)$	$GaCl_3(s)$
S° (J/K · mol)	83	105	142

Similarly, entropy increases as the number of atoms increases in a related series of molecules. In the following series of straight-chain hydrocarbons (all C atoms are bound into a nonbranching chain), notice that the entropy increases by about 40 J/K · mol every time another CH_2 group is added to the chain.

COMPOUND	STRUCTURAL FORMULA	S° (J/K · mol) FOR GASEOUS COMPOUND
Methane	CH_4	186.3
Ethane	$H_3C{-}CH_3$	229.6
Propane	$H_3C{-}CH_2{-}CH_3$	269.9
n-Butane	$H_3C{-}CH_2{-}CH_2{-}CH_3$	310.0
n-Octane	$H_3C{-}CH_2{-}CH_2{-}CH_2$ (bonded to) $H_3C{-}CH_2{-}CH_2{-}CH_2$	463.6

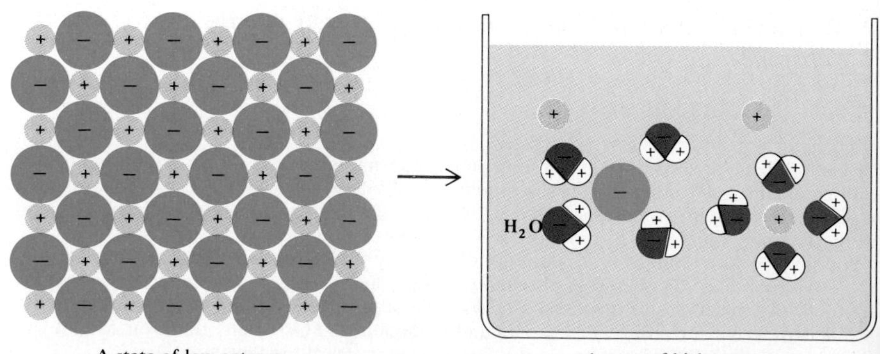

A state of low entropy

A state of high entropy

Solid | Gas

						H_2 130.7	He 126.2
Li 29.1	Be 9.5	B 5.86	C diamond 2.38 graphite 5.74	N_2 191.6	O_2 205.1	F_2 202.8	Ne 146.3
Na 51.2	Mg 32.7	Al 28.3	Si 18.8	P (white P) 41.1	S 31.8	Cl_2 223.1	Ar 154.8
K 64.2	Ca 41.4	Ga 40.88	Ge 31.1	As (gray As) 35.1	Se (black Se) 42.4	Br_2 152.2 (liquid Br_2)	Kr 164.1
Rb 76.8	Sr 52.3	In 57.8	Sn (white tin) 51.6	Sb 45.7	Te 49.8	I_2 116.1	Xe 169.7
Cs 85.2	Ba 62.8	Tl 64.2	Pb 64.8	Bi 56.7	Po	At	Rn 176.1

FIGURE 18.3
Entropy values, in J/K · mol, for some of the elements at 25°C.

A corollary to this is that entropy increases when going from a three-dimensional network solid to one that is only two dimensional. The best examples are of course diamond and graphite, whose very low entropies reflect the high degree of order. More entropy values are given in the

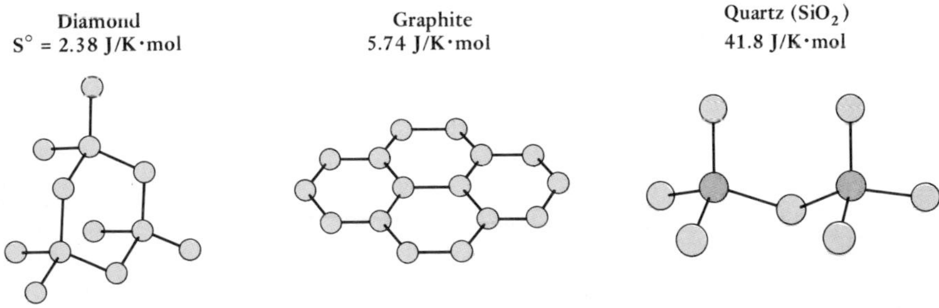

abbreviated periodic table in Figure 18.3. In general, elements that have a high degree of covalent bonding in the solid state (boron, carbon, phosphorus, sulfur) have relatively low standard entropies. Metals have values ranging from about 10 J/K · mol to 80 J/K · mol, while the gaseous elements have the highest values of all.

EXERCISE 18.1 PREDICTING SIGNS OF ENTROPY CHANGES AND RELATIVE ENTROPY VALUES

(1) For each of the following changes in state, predict the sign of the entropy change.
(a) $Hg(\ell) \longrightarrow Hg(g)$. (b) $I_2(g) \longrightarrow I_2(s)$. (c) $AgNO_3(s) \longrightarrow 1\ m\ AgNO_3(aq)$

(2) In each pair of compounds, tell which should have the higher value for $S°$.
(a) $CH_3OH(\ell)$ or $CH_3CH_2OH(\ell)$. (b) KI(s) or $AlI_3(s)$.

18.5 ENTROPY AND THE SECOND LAW

The **second law of thermodynamics** states that the combined entropy of a system and its surroundings always increases in a spontaneous process. This statement helps us to accomplish two objectives: (1) to find experimental methods for determining the entropy of a substance and (2) to be able to predict the spontaneity of chemical and physical processes. We shall turn to the first objective, since it will lead directly to an answer to the spontaneity problem.

THE SECOND LAW

There are two kinds of systems: equilibrium and non-equilibrium. The difference between them is that non-equilibrium systems *spontaneously* change to achieve equilibrium. Our interest in non-equilibrium systems is that we would like to know in which direction they will change to get to equilibrium and how much useful work we can get out of them as they change. The concepts of entropy and enthalpy are crucial to these goals.

In this text we will be interested in systems that exchange heat with their surroundings. Changes in the entropies of the system and surroundings always accompany heat flow. First, let us examine the special case of systems at equilibrium. If q_{sys} is the amount of heat that flows into a system at equilibrium, the system seeks a new equilibrium and undergoes an entropy increase.

$$\Delta S_{sys} = \frac{q_{sys}}{T} \qquad \text{for processes with the } \textbf{system at equilibrium} \qquad (18.1)$$

In Chapter 11 we described phase changes, processes that occur at, or very close to, equilibrium. For example, ice can be in equilibrium with liquid water at 0°C. If heat is added, the ice will be transformed to liquid. The entropy of the system certainly increases in the process, since the disorder of the system increases, and Equation 18.1 can be applied to find ΔS_{sys}.

Having described ΔS_{sys}, what about the change in entropy of the surroundings of the system, ΔS_{surr}? It is an important fact that Equation 18.1 applies to the surroundings, whether the system is at equilibrium or is ready to make a spontaneous change. Since the surroundings of a system can be considered the rest of the universe, heat flow to or from the surroundings never significantly disturbs the equilibrium within the surroundings. Therefore, we can say

$$\Delta S_{surr} = \frac{q_{surr}}{T} \qquad \text{for } \textbf{all processes} \qquad (18.2)$$

An important consequence of Equation 18.2 is that the more heat given off by a system the greater the disordering of the surroundings as they absorb that heat ($q_{surr} > 0$). Now, what does the *second law of thermodynamics* have to do with this? In quantitative form, the second law is

$$\Delta S_{universe} = \Delta S_{sys} + \Delta S_{surr} > 0 \qquad \text{for a } \textbf{spontaneous process}$$

which means $\Delta S_{sys} > -\Delta S_{surr}$ and $\Delta S_{sys} > -q_{surr}/T$ for a spontaneous process. Therefore,

$$\Delta S_{sys} > \frac{q_{sys}}{T} \quad \text{for a } \textbf{spontaneous process} \qquad (18.3)$$

If heat flows out of the surroundings (q_{surr}), it must flow *into* the system (q_{sys}); $q_{sys} = -q_{surr}$.

Equation 18.3 makes an important distinction between equilibrium and non-equilibrium systems: spontaneously changing systems experience greater disorder (18.3) than they would if the change were made under equilibrium conditions (18.1).

An important point for your understanding is that ΔS_{sys} and ΔS_{surr} cannot both be negative for a spontaneous process; according to the second law, entropy must increase somewhere. Both ΔS_{sys} and ΔS_{surr} can be positive, or one can be negative and the other positive, as long as the positive one is larger than the negative one. Notice especially that the system entropy change does not have to be positive in a spontaneous process, as long as enough heat is given off by the system ($q_{sys} < 0$) to make ΔS_{surr} large and positive. There is more on this point in Sections 18.6 and 18.8. Before turning to the use of the criterion of entropy changes to define the spontaneity of a process, we shall first see how entropy values are determined.

DETERMINING ENTROPY CHANGES FOR PHASE CHANGES

The entropy change for the system in an equilibrium process is given by the heat absorbed by the system, q, divided by the temperature in kelvins. In Chapter 11 you learned that the heat added at constant pressure to change a liquid to a vapor, for example, is just the enthalpy of vaporization, ΔH_{vap}. Therefore, we can find ΔS for a liquid → vapor change from $\Delta H_{vap}/T$, and ΔS for solid → liquid is just $\Delta H_{fusion}/T$.

EXAMPLE 18.1

THE ENTROPY CHANGE OF A PHASE CHANGE

The enthalpy of vaporization of benzene (C_6H_6) is 30.8 kJ/mol at the boiling point of 80.1°C. Calculate the entropy change for benzene going from (a) liquid to vapor and (b) vapor to liquid at 80.1°C.

Solution The transition from liquid to vapor is an endothermic process, so the enthalpy of vaporization is a positive quantity. Since entropy is usually given in units of J/K · mol, we convert ΔH_{vap} to joules and divide by the absolute temperature at which the transition occurs.

$$\Delta S^\circ = \frac{\Delta H_{vap}}{T} = \frac{3.08 \times 10^4 \text{ J/mol}}{353\text{K}} = +87.3 \text{ J/K} \cdot \text{mol}$$

The entropy change is +87.3 J/K · mol for the liquid to vapor phase change. The sign of ΔS° is positive, as expected for a change from a liquid to molecules moving about randomly in the vapor phase.

When considering the change in phase from vapor to liquid, we recall from Chapter 11 that heat is *evolved* on condensing a gas. Thus, the process is *exothermic*; the enthalpy change is −30.8 kJ/mol. The change in entropy is

$$S^\circ = \frac{-\Delta H_{vap}}{T} = \frac{-3.08 \times 10^4 \text{ J/mol}}{353\text{K}} = -87.3 \text{ J/K} \cdot \text{mol}$$

a negative quantity, reflecting the *decrease in disorder* as intermolecular bonding occurs to cause the vapor to form the more orderly liquid state.

EXAMPLE 18.2

THE ENTROPY OF A COMPOUND

The entropy, $S°$, of liquid ethyl alcohol (C_2H_5OH) (Appendix L) is known to be 161 J/K · mol. If the enthalpy of vaporization of the alcohol is 36.4 kJ/mol at 25°C, what is the value of the absolute entropy, $S°$, for ethyl alcohol vapor in equilibrium with the liquid at 25°C?

Solution The entropy change for the phase transition at 25°C is

$$\Delta S° = \frac{36.4 \times 10^3 \text{ J/mol}}{298\text{K}} = 122 \text{ J/K} \cdot \text{mol}$$

This tells us that the vapor contains more entropy than the liquid by 122 J/K · mol at 25°C. Therefore, we can write

$$\Delta S° = S°_{\text{vapor}} - S°_{\text{liquid}}$$

$$122 \text{ J/K} \cdot \text{mol} = S°_{\text{vapor}} - 161 \text{ J/K} \cdot \text{mol}$$

$$\text{and so } S°_{\text{vapor}} = 283 \text{ J/K} \cdot \text{mol.}$$

EXERCISE 18.2 ENTROPY CHANGES

Liquid carbon tetrachloride (CCl_4) has an entropy, $S°$, of 216.4 J/K · mol at 25°C. If its enthalpy of vaporization at 25°C is 27.7 kJ/mol, calculate the standard entropy, $S°$, of $CCl_4(g)$ at 25°C.

DETERMINING AND USING STANDARD ENTROPIES

We can determine experimentally ΔS for a phase change, but what about finding the *absolute entropy* for a substance? Since we know from the third law of thermodynamics that pure, perfectly crystalline substances have $S = 0$ at $T = 0$K, all one has to do is measure the increase in entropy from absolute zero to a particular temperature. The necessary calculations are illustrated graphically in Figure 18.4 for cyclopropane, C_3H_6. Here you see that the entropy of the solid compound has been found to increase from 0 to 66.8 J/K · mol as the temperature increased from 0K to the melting point (145.54K). At this temperature the solid changes to a liquid with a large jump in entropy, as expected for a phase change ($\Delta S° = 37.4$ J/K · mol). Once in the liquid state, the entropy once again increases slowly with temperature (by another 38.4 J/K · mol) up to the boiling point (240.3K) where it undergoes another very large increase (83.5 J/K · mol) as the liquid becomes a vapor. The total entropy change from 0K to a gas at 240.3K is thus 226.1 J/K · mol, and, since $S = 0$ at 0K, this means that $S°$ for $C_3H_6(g)$ is 226 J/K · mol at 240.3K.

Considerable experimental information is needed to derive an entropy value, but many have been determined. Some values have already been mentioned in Section 18.4 and others are given in Appendix L.

ENTROPY CHANGES OF REACTIONS The enthalpy change for a reaction can be calculated by subtracting the enthalpies of formation of the reactant molecules from those of the product molecules

$$\Delta H°_{\text{reaction}} = \Sigma m \Delta H_f°(\text{products}) - \Sigma n \Delta H_f°(\text{reactants})$$

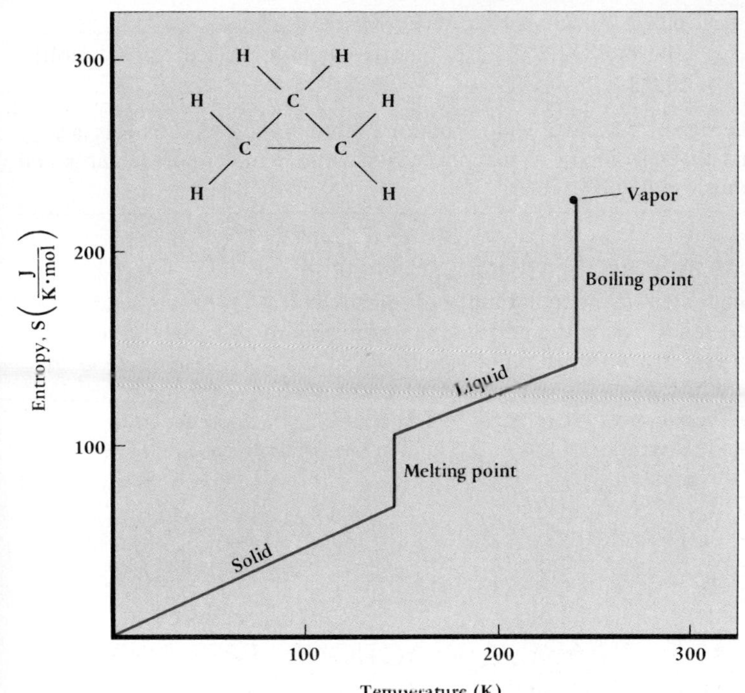

FIGURE 18.4

The entropy of cyclopropane, C_3H_6, as a function of temperature. The entropy of a given phase rises slowly with temperature, but a sharp jump occurs when there is a phase change.

where m and n are the number of moles of products and reactants, respectively.

In a similar manner, the entropy change for a chemical reaction can be calculated.

$$\Delta S^\circ_{reaction} = \Sigma m S^\circ(\text{products}) - \Sigma n S^\circ(\text{reactants})$$

This was applied to phase changes in Sections 18.3 and 18.4, and the following example and exercise show how to find ΔS° for a reaction.

EXAMPLE 18.3

CALCULATING THE CHANGE IN ENTROPY, ΔS°, UPON FORMATION OF A COMPOUND

Nitrogen dioxide, NO_2, is the red-brown gas that you can sometimes see in bottles of nitric acid and in polluted air. Determine the standard entropy change, ΔS°, involved in the formation of NO_2.

Solution As in any problem involving a chemical reaction, first write the balanced equation. Since we are describing the formation of NO_2, this means we begin with the elements in their standard state.

$$\tfrac{1}{2} N_2(g) + O_2(g) \longrightarrow NO_2(g)$$

Next, subtract the sum of the entropies of the reactants from that of the product (Appendix L), paying careful attention to scale each entropy by the number of moles of reagent involved.

$$\Delta S^\circ_{rxn} = (1 \text{ mol } NO_2)(240.0 \text{ J/K} \cdot \text{mol}) - [(1 \text{ mol } O_2)(205.1 \text{ J/K} \cdot \text{mol}) + (\tfrac{1}{2} \text{ mol } N_2)(191.6 \text{ J/K} \cdot \text{mol})] = -60.9 \text{ J/K}$$

Notice that the sign of the entropy of formation for NO_2 is negative. This is largely due to the fact that the chemical reaction began with $1\frac{1}{2}$ moles of reactants and ended with only 1 mole of product.

EXERCISE 18.3 CALCULATING ΔS°

The active ingredient in a popular antiacid remedy is $CaCO_3$ (more familiar as chalk or limestone). Using the entropy values in Figure 18.3 and Table 18.1, calculate the entropy of formation of 1 mole of $CaCO_3$(s). (Use graphite as the standard state of carbon.) Is the sign of the entropy change for the formation of calcium carbonate positive or negative? Account for the increase or decrease in entropy in the formation of this compound in terms of your notion of disorder in chemical systems.

18.6 ENTROPY AND CHEMICAL REACTIONS: FREE ENERGY

The major goal of this chapter is to develop a method for predicting whether a reaction is spontaneous, that is, whether it will occur by itself given sufficient time. With our knowledge of enthalpy and entropy in chemical systems, this goal can now be realized.

Many spontaneous systems are exothermic, but a release of heat by the system alone is not a guarantee of spontaneity. *The combined entropy of the system and entropy of the surroundings must increase.* However, since these two determinants cannot always be counted on to work in the same direction, we need a way to put them together into a single thermodynamic quantity that includes both effects and measures the balance between them. This can be done by referring again to the second law of thermodynamics. Previously in Equation 18.3, this was expressed as (where ΔS here and below is for the system)

The second law of thermodynamics (p. 672) states that $\Delta S_{universe} = \Delta S_{sys} + \Delta S_{surr} > 0$.

Unless otherwise noted, *all* ΔS, ΔH, and ΔG values here and on subsequent pages are for the *system.*

$$\Delta S > \frac{q}{T} \qquad \text{for a spontaneous process}$$

From Chapter 5, you know that the heat involved in a process at constant pressure is the enthalpy change, ΔH. Thus, for a system at constant P and T,

$$\Delta S > \frac{q}{T} = \frac{\Delta H}{T} \qquad \text{for the system in a \textit{spontaneous process at constant T and P}}$$

This expression tells us that ΔS must be greater than $\Delta H/T$ for the system at constant P and T for a spontaneous process. If ΔS is subtracted from both sides of the last expression,

$$0 > \frac{\Delta H}{T} - \Delta S \qquad \text{for a \textit{spontaneous process at constant T and P}}$$

and then both sides are multiplied by T, we can write it in a useful alternative form.

$0 > \Delta H - T\Delta S$ for a *spontaneous process at constant T and P*

This form of the expression tells us that the difference between ΔH and $T\Delta S$ must be less than zero (a negative number) for a spontaneous process! This conclusion is important enough that the function $\Delta H - T\Delta S$ is combined into one thermodynamic state function called the **Gibbs free energy, *G***, which is defined

The compound word "free energy" is useful because it reminds you that enthalpy change (energy) and entropy change ($T\Delta S$, energy of disorder or freedom) are both involved in G.

Gibbs free energy: $G = H - TS$

At constant temperature, the change in the Gibbs free energy is

$$\Delta G = \Delta H - T\Delta S \quad (18.4)$$

Energy free to do work = total energy change − temperature × entropy change (18.5)

ΔG has units of energy since both ΔH and the $T\Delta S$ term are energy terms [since $T\Delta S$ = (K)(J/K · mol) = J/mol].

The first term of the Gibbs equation (Equation 18.4) reflects the heat energy transferred between system and surroundings, and the $T\Delta S$ term reflects the energy consumed or evolved in ordering or disordering the system. Thus, ΔG is the maximum energy that is "free" to be extracted from the system as it passes from its initial state to its final state at a temperature T. It is this that will allow us finally to tell whether a process will be spontaneous. The **criteria of spontaneity** can be summarized briefly as follows:

J.W. Gibbs (AIP, Niels Bohr Library) The symbol G for ($H - TS$) was chosen to honor J. Willard Gibbs (1839–1903) who pioneered and developed the concept of free energy. Gibbs received a Ph.D. from Yale University in 1863 and was a faculty member there until his death.

(a) If ΔG is negative (<0), a process is spontaneous. (The system passes on excess energy to the surroundings as the process occurs.)
(b) If ΔG is positive (>0), a process is *not* spontaneous under the specified conditions. (If the system is to change, the surroundings must give up energy to the system.)
(c) If $\Delta G = 0$, the process is at equilibrium. (An input of energy from the surroundings is required for the system to change.)

Before applying these criteria, let us look in more detail at the ΔG function itself.

VALUES FOR THE STANDARD FREE ENERGY OF FORMATION, ΔG_f°

Just as the thermodynamic functions ΔH° and S° have been tabulated for substances under a set of standard conditions, so too have *standard free energy* values (Table 18.2 and Appendix L). These values represent the free energy change observed when elements in their standard states form a compound in its standard state at 298K. Hence, they are labeled by the symbol ΔG_f°.

Enthalpy and entropy changes can be calculated for chemical reactions using values of ΔH_f° and S° for the substances in the reaction, and the same can be done to find ΔG_{rxn}° as illustrated in the following example.

TABLE 18.2 Standard Molar Free Energies of Formation for Some Substances at 298K

ELEMENT OR COMPOUND	ΔG_f°(kJ/mol)
C(graphite)	0
C(diamond)	2.90
CO(g)	−137.2
CO_2(g)	−394.4
CH_4(g)	−50.72
H_2O(g)	−228.572
H_2O(ℓ)	−237.129
NH_3(g)	−16.45

EXAMPLE 18.4

CALCULATING ΔG°_{rxn}

Calculate the standard free energy for the following reaction at 298K:

$$C(graphite) + 2\,H_2(g) \longrightarrow CH_4(g)$$

Solution The following values for ΔH°_f and S° are provided in Appendix L.

	C(graphite)	$H_2(g)$	$CH_4(g)$
ΔH°_f(kJ/mol)	0	0	−74.81
S°(J/K · mol)	5.74	130.7	186.3

From these values, we can find both ΔH° and ΔS° for the reaction:

$$\begin{aligned}\Delta H^\circ_{rxn} &= \Delta H^\circ_f[CH_4(g)] - \{\Delta H^\circ_f[C(gr)] + 2\,\Delta H^\circ_f[H_2(g)]\}\\ &= -74.81\text{ kJ} - (0 + 0)\\ &= -74.81\text{ kJ}\end{aligned}$$

$$\begin{aligned}\Delta S^\circ_{rxn} &= S^\circ[CH_4(g)] - \{S^\circ[C(gr)] + 2\,S^\circ[H_2(g)]\}\\ &= 186.3\text{ J/K}\cdot\text{mol} - [(5.74\text{ J/K}\cdot\text{mol}) + 2(130.7\text{ J/K}\cdot\text{mol})]\\ &= -80.84\text{ J/K}\end{aligned}$$

The enthalpy and entropy changes for the reaction can now be combined in the Gibbs equation to give the standard free energy for the reaction.

$$\begin{aligned}\Delta G^\circ_{rxn} &= \Delta H^\circ_{rxn} - T\,\Delta S^\circ_{rxn}\\ &= -74.81\text{ kJ} - (298\text{K})(-80.84\text{ J/K})(1.000\text{ kJ}/1000.\text{ J})\\ &= -74.81\text{ kJ} - (-24.09\text{ kJ})\\ &= -50.72\text{ kJ}\end{aligned}$$

Notice that this value is consistent with that listed in Table 18.2.

EXERCISE 18.4 CALCULATING ΔG°_{rxn}

Using values of ΔH°_f and S°, calculate the free energy change for the following reaction occurring under standard conditions at 25°C:

$$\tfrac{1}{2}\,N_2(g) + \tfrac{3}{2}\,H_2(g) \longrightarrow NH_3(g)$$

Elements in their standard states have ΔG°_f values of 0.

In Example 18.4 the calculated standard free energy change was that for the formation of $CH_4(g)$ in its standard state from the elements in their standard states. Therefore, the calculated ΔG°_{rxn} can be identified as the *standard free energy of formation* of CH_4, $\Delta G^\circ_f[CH_4(g)]$ at 298K. It is in fact standard free energies of formation that are tabulated in Table 18.2 and in Appendix L. Be sure to notice that standard free energies of formation of elements in their standard states are 0, just as are values of ΔH°_f. Just as ΔH°_{rxn} can be calculated from standard enthalpies of formation, the free energy change for a reaction can be found from values of ΔG°_f by the general equation

$$\Delta G^\circ_{rxn} = \Sigma m\Delta G^\circ_f(\text{products}) - \Sigma n\Delta G^\circ_f(\text{reactants}) \qquad (18.6)$$

where m and n are the number of moles of a given product or reactant respectively.

EXAMPLE 18.5

CALCULATING ΔG°_{rxn} FROM ΔG°_f

Calculate the free energy change for the combustion of methane, $CH_4(g)$, from the standard free energies of formation of the products and reactants.

Solution We first write a balanced equation for the reaction and then find values of ΔG°_f for each reactant and product.

$$\begin{array}{lcccc} & CH_4(g) & + \; 2\,O_2(g) & \longrightarrow \; 2\,H_2O(\ell) & + \; CO_2(g) \\ \Delta G^\circ_f(\text{kJ/mol}) & -50.7 & 0 & -237.1 & -394.4 \end{array}$$

$$\begin{aligned} \Delta G^\circ_{rxn} &= 2\Delta G^\circ_f[H_2O(\ell)] + \Delta G^\circ_f[CO_2(g)] - \{\Delta G^\circ_f[CH_4(g)] + 2\Delta G^\circ_f[O_2(g)]\} \\ &= 2\text{ mol}(-237.1\text{ kJ/mol}) + (1\text{ mol})(-394.4\text{ kJ/mol}) - \\ &\qquad [(1\text{ mol})(-50.7\text{ kJ/mol}) + (2\text{ mol})(0)] \\ &= -817.9\text{ kJ} \end{aligned}$$

Be sure to notice that free energy of formation values are given for 1 mole. Therefore, each value must be multiplied by the number of moles involved.

EXERCISE 18.5 STANDARD FREE ENERGIES OF FORMATION, ΔG°_f

(a) Write a balanced chemical equation depicting the formation of gaseous carbon dioxide (CO_2) from its elements.
(b) What is the free energy of formation of 1.00 mole of gaseous CO_2?
(c) What is the standard free energy change for the reaction when 2.5 moles of $CO_2(g)$ are formed from the elements?

EXERCISE 18.6 CALCULATING ΔG°_{rxn} FROM ΔG°_f

Calculate the standard free energy change for the combustion of 1 mole of benzene, $C_6H_6(\ell)$, to give $CO_2(g)$ and $H_2O(\ell)$.

ΔG° AS A CRITERION FOR SPONTANEITY

We can now turn to the important question of the use of ΔG° values in determining reaction spontaneity.

It is easy to prove that if there is "free" energy released in a process, then it will be spontaneous. That is, *the sign of ΔG must be negative for a process to be spontaneous*. The process will occur spontaneously if the energy of the system declines and its entropy increases. In this case, ΔH is negative and ΔS is positive, so $\Delta H - T\Delta S$ *must* give a negative ΔG. Of course, it is also possible that the enthalpy and entropy changes will work against each other, or that neither favor spontaneity of the process. All of these possibilities are summarized in Figure 18.5 and in Table 18.3.

Remember that the ΔS, ΔH, and ΔG values described are for the *system*.

Consider beginning at point A in Figure 18.5; this represents a process that has a negative ΔH. If the entropy increases, the term $(T\Delta S)$ contributes a negative quantity, and you would follow line 1 to more negative values of ΔG as the entropy increases. Processes that behave this way are spontaneous under all conditions. If, on the other hand, the

FIGURE 18.5
Changes in ΔG as a function of the sign and magnitude of ΔS for an exothermic reaction ($-\Delta H$, point A) or an endothermic reaction ($+\Delta H$, point B).

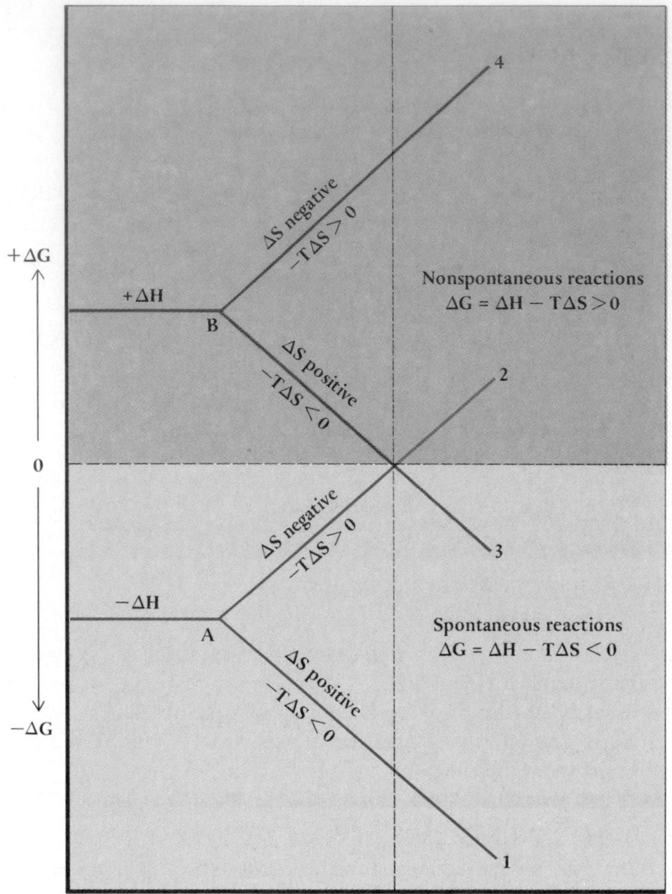

It is entirely possible that the entropy change is so negative that a positive $-T\Delta S$ term outweighs a negative ΔH, and ΔG_{rxn} becomes positive; that is, the reaction becomes nonspontaneous (end of line 2).

entropy decreases, the reaction can be spontaneous only if ΔH is large and negative and outweighs the positive ($T\Delta S$) term, as shown by the early portion of line 2. Such a reaction is called *enthalpy driven*.

Now consider a reaction in which the enthalpy increases (point B). The reaction can still be spontaneous if ΔS is very large and positive or

TABLE 18.3 Reaction Spontaneity as a Function of ΔH and ΔS*

ENTHALPY CHANGE	ΔH	SYSTEM ENTROPY CHANGE	ΔS	SIGN OF ΔG	PROCESS SPONTANEITY
Line 1 exothermic	−	Increasing	+	−	Yes
Line 2 exothermic	−	Decreasing	−	− or +	Depends on T and on relative size of ΔH and ΔS
Line 3 endothermic	+	Increasing	+	+ or −	Depends on T and on relative size of ΔH and ΔS
Line 4 endothermic	+	Decreasing	−	+	No

*Line numbers refer to Figure 18.5.

the temperature is high [$T\Delta S$ is positive and large; line 3], and we would say that the process is *entropy driven*. Finally, the case given by line 4, with an increase in enthalpy and a decrease in entropy, represents a process that cannot be spontaneous under any conditions.

What does it mean when ΔG is positive instead of negative? This is a question best answered on exploring the relation between ΔG and the equilibrium constant K for a reaction in Section 18.7. From Figure 18.5, however, you can conclude that there are two conditions under which a nonspontaneous reaction can be made spontaneous. For reactions following line 3 in Figure 18.4, a higher temperature is required. On the other hand, for a reaction given by line 2, a lower temperature is required.

FREE ENERGY CHANGES FOR REACTIONS

Now let us examine a few actual reactions to see how the criterion of reaction spontaneity works in practice.

THE COMBUSTION OF CARBON:

$\Delta H°$ Negative and $\Delta S°$ Positive → $\Delta G°$ Negative at All T

	C(graphite)	+ O_2(g)	⟶ CO_2(g)	OVERALL
$\Delta H_f°$ (kJ/mol)	0	0	−393.5	−393.5
$S°$ (J/K · mol)	5.74	205.1	213.7	+2.9
$\Delta G_f°$ (kJ/mol)	0	0	−394.4	−394.4

The free energy change for this reaction can be calculated in either of two ways. First, you can add up the $\Delta G_f°$ values for the products and subtract the sum of those for the reactants.

$$G°_{rxn} = \Sigma m\Delta G_f°(\text{products}) - \Sigma n\Delta G_f°(\text{reactants})$$
$$= -394.4 \text{ kJ} - [0 + 0] = -394.4 \text{ kJ}$$

Alternatively, you can calculate the enthalpy and entropy changes for the reaction, $\Delta H°_{rxn}$ and $\Delta S°_{rxn}$, and combine them using the Gibbs equation.

$$\Delta G°_{rxn} = \Delta H°_{rxn} - T\Delta S°_{rxn}$$
$$= -393.5 \text{ kJ} - (298 \text{ K})(+2.9 \text{ J/K} \cdot \text{mol})(1 \text{ kJ}/1000 \text{ J})$$
$$= -394.4 \text{ kJ}$$

Both the enthalpy and entropy changes contribute to making this reaction spontaneous. Not only does the reaction evolve a large amount of heat, but the disordering, going from a solid and gas to a gaseous product, contributes to a small extent to the spontaneity. Although the reaction liberates 393.5 kJ of heat energy, the positive entropy change means that the energy "free" to do some useful work is actually somewhat higher, that is, 394.4 kJ. Reactions such as this one would correspond to line 1 in Figure 18.5.

THE COMBUSTION OF METHANE:

$\Delta H°$ and $\Delta S°$ Both Negative → $\Delta G°$ with Variable Sign Methane burns spontaneously in oxygen to give liquid water and carbon dioxide.

	$CH_4(g)$ +	$2\ O_2(g)$ ⟶	$2\ H_2O(\ell)$ +	$CO_2(g)$	OVERALL
ΔH_f° (kJ/mol)	−74.8	0	2(−285.8)	−393.5	−890.3
S° (J/K · mol)	186.3	2(205.1)	2(69.91)	213.7	−243.0
ΔG_f° (kJ/mol)	−50.7	0	2(−237.1)	−394.4	−817.9

The free energy change, ΔG°, is a large negative number, clearly indicating the reaction is spontaneous under standard conditions, something you already knew. However, although the reaction generates 890 kJ of heat energy, only 818 kJ are available and capable of being used for some useful purpose. That is, 72 kJ of energy ($T\Delta S^\circ$) is not "free," having been used to create some order. Three molecules of reactants gave three of products, but two molecules of the products are in the more ordered liquid state. Thus, the combustion of methane, as in most combustion reactions, is enthalpy driven and corresponds to line 2 in Figure 18.5.

DISSOLVING NH_4NO_3 IN WATER:

ΔH° and ΔS° Both Positive → ΔG° with Variable Sign A situation in which both ΔH° and ΔS° are positive corresponds to line 3 in Figure 18.5. The only way that the reaction can be spontaneous is for $T\Delta S^\circ$ to be large enough that ΔG° is negative. This is true for many salts dissolving in water. Since the salt begins as a highly ordered solid with low entropy, dissolving it in water gives a jumble of ions with a high entropy.

	$NH_4NO_3(s)$ ⟶	$NH_4NO_3(aq, 1\ M)$	OVERALL
ΔH_f° (kJ/mol)	−365.6	−339.9	25.7
S° (J/K · mol)	151.1	259.8	108.7
ΔG_f° (kJ/mol)	−183.9	−190.6	−6.7

The enthalpy and entropy changes for the reaction clearly show that the reaction is entropy driven. That is, although the process is endothermic (and you would feel an obvious cooling effect if you held your hand on a beaker containing dissolving ammonium nitrate), this is outweighed by the large increase in entropy, and the reaction is spontaneous under standard conditions. The disordering of the system (solid + solvent) when the solid dissolves is a very potent driving force.

Entropy is the "force" that ideally drives the mixing of any two substances, liquid or gas, with each other. Chapter 12 described the formation of ideal solutions, ones in which the forces between solute molecules and between solvent molecules are the same as between solute and solvent. Since energy changes *cannot* be involved in forming an ideal solution, it is the increase in entropy experienced by the solute and solvent on mixing with one another that provides the driving force.

EXERCISE 18.7 CALCULATING ΔG° FOR A REACTION

Using free energies of formation in Appendix L, calculate ΔG° for (a) the formation of $CaCO_3(s)$ from $CO_2(g)$ and $CaO(s)$ and (b) for the decomposition of $CaCO_3$ to give $CaO(s)$ and $CO_2(g)$. Which reaction is spontaneous under standard conditions?

FREE ENERGY AND TEMPERATURE

If a reaction has a positive enthalpy change, the only way that it can be spontaneous is if $-T\Delta S°$ is large enough to outweigh $\Delta H°$. This can happen in two ways: the entropy change can be positive and quite large (as is the case in dissolving ammonium nitrate and many other ionic solids in water) or the entropy change can be positive and the temperature high. The latter is in fact one of the reasons that reactions are often carried out at high temperatures. Let us consider an example of this case.

Our economy is based in large measure on the production of iron, and we can think of at least three different ways to reduce iron(III) oxide to metallic iron. One way to do this is to heat the iron(III) oxide and hope that the oxide simply decomposes to oxygen and iron.

$$Fe_2O_3(s) \longrightarrow 2\ Fe(s) + \tfrac{3}{2}\ O_2(g)$$

We can obtain some idea of the feasibility of this by calculating $\Delta G°$. Using the data in Appendix L, we find that $\Delta H°$ for the reaction is +824.2 kJ and $\Delta S°$ is +275 J/K. Enthalpy says "no" to spontaneity and entropy says "yes." Unfortunately, $\Delta H°_{rxn}$ is so positive that it cannot be outweighed by $-T\Delta S°$ at any reasonable temperature, and $\Delta G°_{rxn}$ at room temperature is +742 kJ!

Another way to reduce iron(III) oxide is the so-called **thermite reaction**.

$$Fe_2O_3(s) + 2\ Al(s) \longrightarrow 2\ Fe(s) + Al_2O_3(s)$$

$$\Delta H°_{rxn} = -852\text{ kJ} \qquad \Delta S°_{rxn} = -39\text{ J/K} \qquad \Delta G°_{rxn} = -840\text{ kJ}$$

This is a spectacularly spontaneous process (Figure 18.6). The unfavorable $\Delta S°_{rxn}$ at 298K is offset by a very large and negative enthalpy change. That means that a large amount of heat is produced, so much in fact that the products are raised to the melting point of iron (1530°C), and white hot, molten metal streams out of the crucible. The reaction has been applied to welding procedures, but it is unfortunately not practical for the pro-

(a)

(b)

(c)

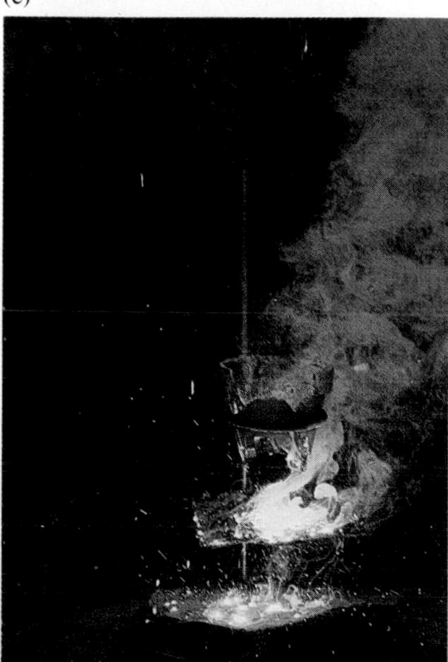

FIGURE 18.6
The thermite reaction. After starting the reaction with a fuse of burning magnesium wire (a), iron(III) oxide reacts with aluminum powder to give aluminum oxide and iron. The Fe_2O_3/Al reaction (b) generates so much heat that the iron is produced in the molten state. It has dropped out of the clay pot, which originally contained the reactants, and burned through a sheet of iron placed under the pot (c). (Charles D. Winters)

duction of iron on a large scale. The cost of producing the aluminum to use as the reducing agent is much larger than the value of the iron produced.

In practice, Fe_2O_3 is not reduced directly by C. Instead, the carbon is burned to give CO, which then acts as the reducing agent. See Chapter 25 for more information.

The usual method of reducing iron(III) oxide to iron metal is to use a much cheaper reducing agent, carbon. The overall process can be represented by

$$2\ Fe_2O_3(s) + 3\ C(s) \longrightarrow 4\ Fe(s) + 3\ CO_2(g)$$

$$\Delta H^\circ_{rxn} = +467.9\ kJ \qquad \Delta S^\circ_{rxn} = +560.3\ J/K \qquad \Delta G^\circ_{rxn} = +300.9\ kJ$$

Even though the entropy is large and positive, the $-T\Delta S^\circ$ term is not large enough to outweigh the very unfavorable (positive) enthalpy change at 25°C. Why then is this process used in industry? Precisely because the large entropy change allows the reaction to become spontaneous at higher temperatures. Let us calculate the temperature, T, at which ΔG°_{rxn} is no longer positive, that is, the temperature at which it is zero.

$$\Delta G^\circ_{rxn} = \Delta H^\circ_{rxn} - T\Delta S^\circ_{rxn}$$

$$0 = +467.9\ kJ - T(0.5603\ kJ)$$

$$T = 835K \text{ or } 562°C$$

The free energy change for the reaction becomes zero at 562°C, and at any higher temperature it will be negative. Thus, spontaneity can be achieved by raising the temperature to a point easily reached in an industrial furnace.*

EXERCISE 18.8 TEMPERATURE AND FREE ENERGY

Is the reduction of magnesia, MgO, with carbon a spontaneous process at 25°C? If not, at what temperature does it become spontaneous?

$$MgO(s) + C(graphite) \longrightarrow Mg(s) + CO(g)$$

18.7 THERMODYNAMICS AND THE EQUILIBRIUM CONSTANT

You know now that when ΔG° for a reaction is negative, the reaction is predicted to be spontaneous. But does this mean that reactants are converted *completely* to products? Does the fact that a positive ΔG° signals a nonspontaneous reaction also mean that the reaction will not proceed at all? The answer to both of these questions is *no*, and we can understand why by developing a relationship between ΔG° and the equilibrium constant K for a reaction.

To develop the relationship between ΔG°_{rxn} and K we return to the *reaction quotient*, Q, first introduced in Chapter 14. For the general re

*To calculate ΔG at a much higher temperature, we used the enthalpy and entropy value appropriate for a temperature of 25°C. This is not entirely correct, because ΔH and ΔS ar temperature dependent. However, their dependency is *much* smaller than that of ΔG (a long as the temperature range does not include a phase change). Thus, our calculatio provides a good estimate of the point at which the reaction becomes spontaneous.

action of A and B giving products C and D

$$aA + bB \rightleftharpoons cC + dD$$

we defined the reaction quotient as

$$\text{where } Q = \frac{[C]^c[D]^d}{[A]^a[B]^b}$$

You learned in Chapter 14 that if $Q = K$ (or $Q/K = 1$) then the system is at equilibrium. However, if $Q < K$ (or $Q/K < 1$), the reaction will proceed spontaneously to the right as written until equilibrium is reached. Conversely, if $Q > K$ (or $Q/K > 1$), then the reaction proceeds spontaneously to the left, counter to the direction in which it is written. Now, let us summarize these relationships and compare them with conclusions from thermodynamics in the following table.

	ΔG	Q/K	$\ln(Q/K)$
Spontaneous reaction	less than 0	less than 1	less than 0
Nonspontaneous reaction	more than 0	more than 1	more than 0
Reaction at equilibrium	= 0	= 1	= 0

What is important is that there is a *direct* relationship between ΔG for a reaction and the *logarithm* of Q/K, $\Delta G = c \ln(Q/K)$. From experiment, it has been found that the proportionality constant c is RT where $R =$ 8.314 J/K · mol and T is the absolute temperature. Therefore,

$$\Delta G = RT \ln(Q/K)$$

and expanding the equation we have

$$\Delta G = RT \ln Q - RT \ln K$$

If we now set up a reaction with the rather peculiar arrangement that all components in Q are at unit concentration, then $Q = 1$ and $\ln Q = 0$. This means that the free energy change *when going from this special situation to equilibrium* is $-RT \ln K$, and it is given a special symbol, $\Delta G°$, and name, the standard free energy change. Therefore, we can write

$$\mathbf{\Delta G^\circ_{rxn} = -RT \ln K} \quad (18.7)$$

For reactions involving gases, Equation 18.7 always gives K_p.

This relationship now shows clearly that a free energy change occurs whether or not a reaction is spontaneous. An equilibrium constant greater than 1 means ΔG°_{rxn} must be negative. Conversely, when K is less than 1 (and so $\ln K$ is negative), ΔG°_{rxn} is positive. And finally, when $K = 1$, then $\ln K = 0$, and ΔG°_{rxn} is zero.

Two applications of Equation 18.7 are the calculation of equilibrium constants from the values of ΔG°_f of reactants and products and the evaluation of ΔG°_{rxn} from an experimental determination of K. These uses are illustrated in Examples 18.6 and 18.7, and the information that can be extracted from enthalpy, entropy, and free energy data is thoroughly explored in Example 18.8.

EXAMPLE 18.6

CALCULATING K FROM ΔG°_{rxn}

The standard free energy change for the reaction

$$N_2(g) + 3\ H_2(g) \longrightarrow 2\ NH_3(g)$$

is -32.9 kJ. Calculate the equilibrium constant for this reaction at 25°C.

Solution In this case, we only need to substitute the appropriate values into Equation 18.7, taking care that the units of ΔG°_{rxn} are the same as those of RT.

$$\begin{aligned} \Delta G^\circ_{rxn} &= -RT \ln K_p \\ (-32.9\ \text{kJ/mol})(1000\ \text{J/kJ}) &= -(8.314\ \text{J/K} \cdot \text{mol})(298\text{K})\ln K_p \\ \ln K_p &= 13.28 \\ K_p &= e^{13.28} \\ K_p &= 5.9 \times 10^5 \end{aligned}$$

The equilibrium constant has a very large value, which means that the equilibrium position lies very far to the product side at room temperature. (To find K_p using a calculator, enter 13.28 and then strike the key labeled "e^x" or "inv(erse) ln x." See Appendix A for more information.)

EXAMPLE 18.7

CALCULATING ΔG°_{rxn} FROM K_p

Calculate ΔG°_{rxn} for the conversion of oxygen to ozone at 25°C from $K_p = 2.47 \times 10^{-29}$.

$$\tfrac{3}{2}\ O_2(g) \longrightarrow O_3(g)$$

Solution To find ΔG°_{rxn} we substitute K_p into Equation 18.6.

$$\begin{aligned} \Delta G^\circ_{rxn} &= -RT \ln K_p = -(8.314\ \text{J/K} \cdot \text{mol})(298\ \text{K})\ln(2.47 \times 10^{-29}) \\ &= 163000\ \text{J/mol} \\ &= 163\ \text{kJ/mol} \end{aligned}$$

From Appendix L you can verify that the free energy change for the reaction should indeed be 163 kJ/mol.

EXAMPLE 18.8

USES OF ΔH°_f, S°, ΔG°_f AND THE CALCULATION OF K

In Chapter 3 you studied in a qualitative way the driving forces of chemical reactions. One of these is the formation of a gaseous product that can escape when the reaction is open to the surroundings. Let us look at such a reaction from a quantitative, thermodynamic point of view.

$$MgCO_3(s) \longrightarrow MgO(s) + CO_2(g)$$

(a) Is the reaction spontaneous at room temperature?
(b) Is the reaction driven by enthalpy, entropy, or both?
(c) What is the value of K at 25°C?

(d) At what temperature is $K = 1$?
(e) Does high temperature make the reaction more or less spontaneous?

Solution To answer the first two questions, you need to know ΔG°_{rxn} and its sign, in addition to ΔH°_{rxn} and ΔS°_{rxn}. ΔG°_{rxn} can of course be calculated using the Gibbs equation

$$\Delta G^\circ_{rxn} = \Delta H^\circ_{rxn} - T\Delta S^\circ_{rxn}$$

Since we need to know ΔH°_{rxn} and ΔS°_{rxn} to answer question (b), we shall first calculate these values from ΔH°_f and S° for the reactants and products. The enthalpy change for the reaction can be calculated as we first outlined in Chapter 5.

$$\Delta H^\circ_{rxn} = \Delta H^\circ_f[CO_2(g)] + \Delta H^\circ_f[MgO(s)] - \Delta H^\circ_f[MgCO_3(s)]$$

Using data from Appendix L, and ΔG°_f $[MgCO_3(s)] = -1095.8$ kJ/mol

$$\begin{aligned}\Delta H^\circ_{rxn} &= (1\text{ mol})(-393.5\text{ kJ/mol}) + (1\text{ mol})(-601.7\text{ kJ/mol}) \\ &\quad - (1\text{ mol})(-1095.8\text{ kJ/mol}) \\ &= 100.6\text{ kJ}\end{aligned}$$

The positive enthalpy change indicates that enthalpy opposes reaction spontaneity under standard conditions. Therefore, for question (b), you now know that the reaction is *not* enthalpy driven.

Next, the entropy change for the reaction can be calculated from values in Appendix L and the Bureau of Standards.

$$\begin{aligned}\Delta S^\circ_{rxn} &= S^\circ[CO_2(g)] + S^\circ[MgO(s)] - S^\circ[MgCO_3(s)] \\ &= (1\text{ mol})(213.6\text{ J/K}\cdot\text{mol}) + (1\text{ mol})(26.9\text{ J/K}\cdot\text{mol}) \\ &\quad - (1\text{ mol})(65.7\text{ J/K}\cdot\text{mol}) \\ &= 174.8\text{ J/K (or 0.175 kJ/K)}\end{aligned}$$

The entropy change for the reaction is positive, which means that spontaneity can be achieved if the temperature is right. Referring again to question (b), you now know that the reaction can be entropy driven.

Next, we combine ΔH°_{rxn} and ΔS°_{rxn} to find ΔG°_{rxn}; we can thus discover whether enthalpy or entropy wins out in the competition and whether the reaction is spontaneous at room temperature (25°C).

$$\begin{aligned}\Delta G^\circ_{rxn} &= \Delta H^\circ_{rxn} - T\Delta S^\circ_{rxn} \\ &= 100.6\text{ kJ} - (298\text{ K})(0.175\text{ J/K}) \\ &= 48.5\text{ kJ}\end{aligned}$$

Now you know that $MgCO_3(s)$ does not spontaneously generate 1 atm of CO_2 (its standard state) at room temperature; the enthalpy difference between products and reactants is simply too high to be offset by the increase in entropy on CO_2 evolution at 25°C.

Having found ΔG°_{rxn}, you can now calculate K_p at room temperature from

$$\Delta G^\circ_{rxn} = -RT\ln K_p$$

This is to remind you that K in Equation 18.7 is K_p, the equilibrium constant in terms of partial pressures.

or

$$\ln K_p = -\frac{\Delta G^\circ_{rxn}}{RT} = -\frac{48{,}500\text{ J}}{(8.314\text{ J/K}\cdot\text{mol})(298\text{ K})} = -19.6$$

The unit of mol in the ln K_p expression can be ignored. It actually cancels as a result of ΔG° being calculated for reaction per mole of material.

Since ln K_p is -19.6, this means that $K_p = e^{-19.6} = 3.2 \times 10^{-9}$.

$$K_p = P_{CO_2} = 2.9 \times 10^{-9}$$

K_p is extremely small, because ΔG°_{rxn} has such a large positive value.

The next question is (d): at what temperature is $K = 1$? From the preceding analysis, you see that ΔS°_{rxn} drives the reaction. This means that, at a sufficiently high temperature, the negative value of $-T\Delta S^\circ$ could become large enough to outweigh the inhibiting effect of a positive ΔH°, and ΔG°_{rxn} would become negative. A balance between ΔH° and $-T\Delta S^\circ$ is reached when $\Delta G^\circ = 0$, that is, when $K = 1$. What is the temperature at which this is achieved? When $\Delta G^\circ = 0$, we can write $\Delta H^\circ - T\Delta S^\circ = 0$. This means that the break even temperature is

$$T = \frac{\Delta H^\circ}{\Delta S^\circ} = \frac{100.8 \text{ kJ}}{0.175 \text{ kJ/K}} = 576\text{K (or 303°C)}$$

At approximately 300°C the equilibrium constant is 1, but at still higher temperatures, the $-T\Delta S^\circ$ term increasingly dominates the ΔH° term, and the yield increases because K increases. This means the answer to question (e) is yes, and this is true for any reaction with $\Delta S^\circ_{rxn} > 0$ (see Figure 18.5).

EXERCISE 18.9 CALCULATING *K* FROM FREE ENERGY OF REACTION

Calculate K_p at 298K from the value of ΔG°_{rxn} for each of the following reactions.

(a) $S(s) + O_2(g) \longrightarrow SO_2(g)$

(b) $CaCO_3(s) \longrightarrow CaO(s) + CO_2(g)$

18.8 THERMODYNAMICS, PERPETUAL MOTION, AND THE UNIVERSE

With this chapter, we have now brought together the **three laws of thermodynamics.**

Some cynic long ago paraphrased the first two of these laws into simpler statements:

First law: You can't win!
Second law: You can't break even either!

First law: The total energy of the universe is a constant.

Second law: The total entropy of the universe is always increasing.

Third law: The entropy of every pure, perfectly crystalline substance at absolute zero is zero.

Neither of the first two laws has ever been or can be proven. It is just that there has never been a single, concrete example showing otherwise. No less a scientist than Albert Einstein has remarked that thermodynamic theory ". . . is the only physical theory of universe content [which], within the framework of applicability of its basic concepts, will never be overthrown."

Einstein's statement does not mean that people have not tried (and are continuing to try) to disprove the laws of thermodynamics. Someone is always claiming to have invented a machine that performs useful work without expending energy—a perpetual motion machine. Although such a machine was actually granted a patent recently by the U.S. Patent Office (presumably the examiner had not had a course in thermodynamics), no workable perpetual motion machine has ever been demonstrated; the laws of thermodynamics are safe.

An interesting discussion of attempts to build perpetual motion machines, and the relation of such machines to the laws of thermodynamics, is given by J. Tierney, *Science 83*, May, 1983, p. 31.

It is rather easy to understand the first law; the conservation of energy is not difficult to observe. (And this is the downfall of most so-called "perpetual motion" machines.) It is the second law that presents problems in understanding. After all, you have seen several cases where the entropy declined, but the process was still spontaneous. However, what you must realize is that we have focused only on the *system*, and not on the total entropy change of the system plus surroundings, that is, of the universe. A brief example will illustrate this point. The formation of water from its elements evolves considerable heat energy, and the system (the reaction mixture) declines in entropy.

$$H_2(g) + \tfrac{1}{2}\,O_2(g) \longrightarrow H_2O(g)$$
$$\Delta H^\circ_f = -242\text{ kJ} \qquad \Delta S^\circ_{rxn} = -44.5\text{ J/K}$$

But what is the effect on the outside world of the 242 kJ of heat energy? It can be harnessed to do some work, to warm your home, or to make electricity. That is, it is released to the surroundings, where its effect is to increase the entropy of the surroundings (Equation 18.2). In this case, $\Delta H^\circ_{surr} = +242$ kJ/mol, the sign being positive because heat has flowed *from* the reaction (the system) *into* its surroundings. This means that the entropy change of the surroundings is

$$\Delta S^\circ_{surr} = \frac{+242\text{ kJ}}{298\text{K}} = +812\text{ J/K}$$

Now, what is the total entropy change in the universe due to the combustion of 1 mole of $H_2(g)$? It is the sum of that for the system and its surroundings.

$$\begin{aligned}\Delta S^\circ_{universe} &= \Delta S^\circ_{rxn} + \Delta S^\circ_{surr}\\ &= (-44.4\text{ J/K}) + (+812\text{ J/K})\\ &= +768\text{ J/K}\end{aligned}$$

The entropy of the universe increases when some hydrogen and oxygen are burned in a spontaneous process.

There is an interesting lesson to be drawn from this calculation of $\Delta S^\circ_{universe}$. The idea of adding the entropy changes for system and surroundings amounts to finding a quantity related to the free energy change for the system. You can see this if you rearrange the Gibbs equation slightly.

$$\Delta G_{system} = \Delta H_{system} - T\Delta S_{system}$$

$$-\frac{\Delta G_{sys}}{T} = -\frac{\Delta H_{sys}}{T} + \Delta S_{sys}$$

Since the law of the conservation of energy states that $-\Delta H_{system} = \Delta H_{surroundings}$, you can change this equation to read

$$-\frac{\Delta G_{sys}}{T} = \frac{\Delta H_{surr}}{T} + \Delta S_{sys}$$

Since $\Delta H_{surr}/T = \Delta S_{surr}$

$$-\frac{\Delta G_{sys}}{T} = \Delta S_{surr} + \Delta S_{sys} = \Delta S_{universe}$$

This means that if the entropy change of the universe is positive, the free energy change for the system must be negative, exactly the criterion you have been using to predict a spontaneous change.

The fact that the entropy of the universe is always increasing agrees with some current ideas about the origin of the universe and its constant expansion (Special Section, Chapter 8). But how do we explain the existence of highly organized plants and animals? Since a living creature survives only as long as its reactions are spontaneous, why does it continue to exist as an organized entity? The answer is that the earth and its plants and animals are not closed thermodynamic systems isolated from their surroundings. The continual flow of energy from the sun provides the means for the photosynthesis of compounds of high free energy such as glucose and amino acids from CO_2, H_2O, and other nutrients. Consumption of these compounds by animals fuels their existence, although only about 10% to 15% of the energy available is actually transferred at each step along the food chain. Along this chain, waste products of low free energy and high entropy such as CO_2 and H_2O are discarded to the surroundings. These can be recycled by the further input of energy, but a net increase in entropy and disorder over time is an inescapable conclusion of the second law of thermodynamics.

SUMMARY

A **spontaneous chemical reaction** is one that tends to occur from left to right as written, given enough time, and it is found that the creation of disorder favors spontaneity (Sections 18.2 and 18.3). **Entropy**, ***S***, a measure of the disorder or randomness in a system, is a state function that has units of J/K · mol (Section 18.4). The **third law of thermodynamics** states that the entropy of a pure, perfectly crystalline substance is zero at 0K. This means that, unlike enthalpy, there is an absolute starting place for measuring the **absolute entropy** of a substance, so the entropy possessed by a substance can be determined at some temperature other than 0K. The **standard state entropy**, $S°$, is that measured for a mole of the substance at a pressure of 1 atm (solutions have a concentration of 1 *m*).

The **second law of thermodynamics** states that the combined entropy of a system and its surroundings always increase in a spontaneous process (Section 18.5). From this law, we can state that

$$\Delta S > \frac{q}{T} \text{ for a spontaneous process}$$

$$\Delta S = \frac{q}{T} \text{ for an equilibrium process}$$

where q is the heat involved in a process at temperature T. Since $q = \Delta H$ for a reaction at constant T and P, this means that ΔS can be determined for an equilibrium phase change from the enthalpy change and temperature of the transition [$\Delta S = \Delta H/T$] (Section 18.5).

Values of $S°$ have been measured for many substances (Section 18.5) and can be used to determine the entropy change for a reaction by

$$\Delta S^\circ_{rxn} = \Sigma m S^\circ(\text{products}) - \Sigma n S^\circ(\text{reactants})$$

where m and n are the number of moles of products or reactants, respectively.

The **Gibbs function**, G, also a state function, has been derived from the second law and is defined by the **Gibbs equation** (Section 18.6).

$$\Delta G = \Delta H - T\Delta S \qquad (18.4)$$

Using ΔH_{rxn} and ΔS_{rxn}, ΔG_{rxn} can be determined. When all substances are in their standard states, the standard free energy change ΔG° is derived. If the reaction involved is that for the formation of a compound from the elements, with the products and reactants in their standard states, then ΔG°_{rxn} is the standard free energy of formation, ΔG°_f. Using values of ΔG°_f for all substances in a reaction (with $\Delta G^\circ_f = 0$ for elements in their standard states), ΔG°_{rxn} can be calculated from the expression

$$\Delta G^\circ_{rxn} = \Sigma\, m\, \Delta G^\circ_f(\text{products}) - \Sigma\, n\, \Delta G^\circ_f(\text{reactants})$$

Based on Gibbs's equation, the *criteria for spontaneity of a reaction* are

(a) If ΔG is negative (a value < 0), a process is spontaneous.
(b) If ΔG is positive (a value > 0), a process is not spontaneous under the specified conditions.
(c) If $\Delta G = 0$, the process is at equilibrium.

The use of these criteria is outlined in Section 18.6 (see Table 18.3 and Figure 18.5) and detailed examples are described.

If a reaction has a positive enthalpy change, the only way the reaction can be spontaneous is if $-T\Delta S$ is larger than ΔH. This can occur if (a) ΔS is positive and large (an increase in disorder) or (b) if ΔS is positive and small and the temperature T is high (Section 18.6).

The relation between ΔG°_{rxn} and the equilibrium constant of a reaction, K, is given by

$$\Delta G^\circ_{rxn} = -RT \ln K \qquad (18.7)$$

where R is 8.314 J/K · mol and T is the absolute temperature. Thus, reactions for which ΔG°_{rxn} is negative are predicted to be spontaneous and to have high concentrations of products at equilibrium. When ΔG°_{rxn} is positive, the reaction is predicted to be nonspontaneous at that temperature and to have low concentrations of products at equilibrium.

STUDY QUESTIONS

Many questions require thermodynamic data. If the required data are not given in the question, consult Appendix L.

REVIEW QUESTIONS

1. Give the three laws of thermodynamics.

2. What is meant by a "spontaneous chemical reaction?"

3. Criticize the following statements:
(a) The entropy increases in all spontaneous reactions.
(b) All spontaneous processes are exothermic.
(c) A reaction with a negative free energy change, $-\Delta G$, is predicted to be spontaneous with rapid transformation of reactants to products.

(d) In general, endothermic reactions are not spontaneous.

4. Tell if each of the following statements is true or false. If false, rewrite the statement to make it true.
   (a) The entropy of a substance increases when going from the liquid to the vapor state.
   (b) In general, an exothermic reaction in which the number of moles of products exceeds the number of moles of reactants, and in which all reactants and products are in the gas phase, will be thermodynamically spontaneous.
   (c) Reactions with a positive ΔH°_{rxn} and a positive ΔS°_{rxn} can never be spontaneous.
   (d) Reactions with $-\Delta G^\circ_{rxn}$ always have an equilibrium constant greater than 1.
   (e) When the equilibrium constant of a reaction is less than 1, then ΔG°_{rxn} is negative.
5. Explain why the entropy of NaCl increases on dissolving the solid in water {$S°$[NaCl(s)] = 72.1 J/K · mol and $S°$[NaCl(aq)] = 115.5 J/K · mol}.

ENTROPY

6. For each of the following, specify which has the higher entropy.
   (a) A sample of dry ice (solid CO_2) at −78°C or CO_2 vapor at 0°C.
   (b) Sugar, as a solid or after dissolving in a cup of coffee.
   (c) A chunk of metallic sodium or a flask of argon at room temperature.
   (d) Two connected flasks of equal volume, one containing pure ammonia and the other pure nitrogen, or both gases mixed and occupying the flasks equally.
7. In each pair of compounds, tell which should have the higher entropy:
   (a) KCl(s) or $AlCl_3$(s)
   (b) CH_3I(ℓ) or CH_3CH_2I(ℓ)
   (c) NH_4Cl(s) or NH_4Cl(aq)
8. In each of the following pairs, decide which compound has the higher entropy at 298K:
   (a) NaCl or $MgCl_2$
   (b) Cl_2(g) or P_4(g)
   (c) CH_3NH_2 or $(CH_3)_2NH$
   (d) Au(s) or Hg(ℓ)
9. Calculate entropy changes for each of the following changes of state and comment on the sign of the change:
   (a) C(graphite) ⟶ C(diamond)
   (b) Na(g) ⟶ Na(s)
   (c) Hg(ℓ) ⟶ Hg(g) ($S°$ for Hg(g) = 175 J/K · mol)
   (d) NH_4Cl(s) ⟶ NH_4Cl(aq)
10. The entropy values for several straight-chain hydrocarbons are given in Section 18.4 Using these values, estimate the entropy, $S°$, of pentane, C_5H_{12} ($H_3C—CH_2—CH_2—CH_2—CH_3$).
11. What trend is observed in the entropies of the rare gases in Figure 18.3? Can you rationalize this trend? Is it observed elsewhere in the periodic table?
12. In general, $S°$ values increase as the elements of a periodic group increase in atomic mass (Figure 18.3). Why is the standard state entropy of Br_2 greater than that of I_2?

DETERMINATION OF ENTROPY

13. The enthalpy of vaporization of liquid diethyl ether, $(C_2H_5)_2O$, is 26.0 kJ/mol at the boiling point of 35.0°C. Calculate ΔS for (a) liquid to vapor and (b) vapor to liquid at 35.0°C.
14. The standard enthalpies of formation of liquid and vapor methyl iodide, CH_3I, are −15.5 kJ/mol and +13.0 kJ/mol, respectively. (a) Calculate the enthalpy of vaporization at 25°C. (b) If the liquid has a standard entropy, $S°$, of 163 J/K · mol, what is the value of $S°$ for CH_3I(g)?
15. Using $S°$ for liquid and vapor $CHCl_3$, estimate the enthalpy of vaporization of the liquid at 25°C. Compare your estimate with a calculation of the enthalpy using enthalpies of formation of liquid and vapor $CHCl_3$.
16. The standard enthalpies of formation of liquid and vapor BCl_3 are −427.2 kJ/mol and −403.76 kJ/mol, respectively. If $S°[BCl_3(g)]$ is 290 J/K · mol, what is $S°$ for BCl_3(ℓ) at 25°C?

ENTROPY OF REACTIONS

17. Using the following reaction, calculate the standard molar entropy change for the formation of gaseous propane at 25°C.

    3 C(graphite) + 4 H_2(g) ⟶ C_3H_8(g)
18. Calculate the standard entropy change for the formation of one mole of each of the following compounds from its elements at 25°C:
    (a) H_2O(ℓ) (d) Al_2O_3(s)
    (b) MgO(s) (e) Fe_2O_3(s)
    (c) CaS(s) (f) $PbCl_2$(s)
19. Calculate the standard molar entropy change for the formation of each of the following compounds from its elements at 25°C:
    (a) ICl(g) (c) $CaCO_3$(s)
    (b) $COCl_2$(g) (d) $CHCl_3$(g)
20. Calculate the standard entropy change for each of the following reactions at 25°C:
    (a) Mg(s) + ½ O_2(g) ⟶ MgO(s)
    (b) Pb(s) + Cl_2(g) ⟶ $PbCl_2$(s)
    (c) 2 ICl(g) ⟶ I_2(s) + Cl_2(g)
    (d) C_2H_5OH(ℓ) + 3 O_2(g) ⟶ 2 CO_2(g) + 3 H_2O(ℓ)
    (e) NaCl(s) ⟶ NaCl(aq)
21. Using the concept of entropy in terms of disorder or randomness, account for the signs of the calculated values of $\Delta S°$ in Study Question 20.

22. Sodium reacts violently with water according to the equation

$$Na(s) + H_2O(\ell) \longrightarrow NaOH(aq) + \tfrac{1}{2} H_2(g) + \text{heat}$$

Without doing calculations, predict the signs of $\Delta H°$ and $\Delta S°$ for this reaction. Verify your prediction with a calculation.

23. Calculate the entropy change involved in formation of 1.0 mole of each of the following gaseous hydrocarbons under standard conditions. (Use graphite as the standard state of carbon.)

(a) H—C≡C—H (ethyne or acetylene)

(b) $H_2C{=}CH_2$ (ethene or ethylene)

(c) $H_3C{-}CH_3$ (ethane)

What trend do you see in these values? Does $\Delta S°$ increase or decrease on adding H atoms? Can you think of some reasons for the observed trend?

FREE ENERGY

24. For each of the following processes, give the algebraic sign of $\Delta H°$, $\Delta S°$, and $\Delta G°$. (No calculations are necessary. Use your common sense.)
(a) The splitting of liquid water to give gaseous oxygen and hydrogen.
(b) Dissolving a small amount of NH_4Cl in water. (The solution becomes quite cold in the process.)
(c) The explosion of dynamite (where liquid nitroglycerin gives gaseous products such as water, CO_2, and others; much heat is evolved.)

25. Using values of $\Delta H_f°$ and $S°$, calculate $\Delta G°$ for the following reactions:
(a) $Pb(s) + Cl_2(g) \longrightarrow PbCl_2(s)$
(b) $Mg(s) + \tfrac{1}{2} O_2(g) \longrightarrow MgO(s)$
(c) $NH_3(g) + HCl(g) \longrightarrow NH_4Cl(s)$
(d) $Br_2(\ell) + 3\,F_2(g) \longrightarrow 2\,BrF_3(g)$

26. Which of the values of $\Delta G°$ calculated in Study Question 25 correspond to standard free energies of formation? Do your calculations agree with values of $\Delta G_f°$ listed in Appendix L?

27. Using $\Delta H_f°$ and $S°$ values, calculate the standard molar free energy of formation, $\Delta G_f°$, for each of the following compounds:
(a) $CS_2(g)$
(b) $N_2H_4(\ell)$
(c) $COCl_2(g)$
(d) $Mg(OH)_2(s)$
Compare your calculated value of $\Delta G_f°$ with that listed in Appendix L.

28. Write a balanced equation depicting the formation of 1 mole of $Fe_2O_3(s)$ from its elements. What is the free energy of formation of 1.00 mole of Fe_2O_3? What is the standard free energy change when 454 g of Fe_2O_3 is formed from its elements?

29. Using values of $\Delta G_f°$, calculate $\Delta G°_{rxn}$ for each of the following reactions:
(a) $Ca(s) + Cl_2(g) \longrightarrow CaCl_2(s)$
(b) $HgO(s) \longrightarrow Hg(\ell) + \tfrac{1}{2} O_2(g)$
(c) $HgS(s) + O_2(g) \longrightarrow Hg(\ell) + SO_2(g)$
(d) $NH_3(g) + 2\,O_2(g) \longrightarrow HNO_3(\ell) + H_2O(\ell)$

30. Yeast can produce ethyl alcohol by the fermentation of glucose, the basis for the production of most alcoholic beverages.

$$C_6H_{12}O_6(s) \longrightarrow 2\,C_2H_5OH(\ell) + 2\,CO_2(g)$$

Calculate $\Delta H°$, $\Delta S°$, and $\Delta G°$ for the reaction. In addition to the thermodynamic values in Appendix L, you will need the following for $C_6H_{12}O_6(s)$: $\Delta H°$ (kJ/mol) = −1260.0; $S°$(J/K · mol) = 288.9; $\Delta G°$ (kJ/mol) = −919.2.

31. There has been great interest in the splitting of water into its elements, since the $H_2(g)$ thereby produced could be used as fuel. Calculate $\Delta H°$, $\Delta S°$, and $\Delta G°$ for the water-splitting reaction at 25°C.

$$H_2O(g) \longrightarrow H_2(g) + \tfrac{1}{2} O_2(g)$$

32. Calculate $\Delta G°_{rxn}$ for each of the following reactions:
(a) $4\,NH_3(g) + 5\,O_2(g) \longrightarrow 4\,NO(g) + 6\,H_2O(\ell)$
(b) $2\,H_2S(g) + 3\,O_2(g) \longrightarrow 2\,H_2O(g) + 2\,SO_2(g)$
(c) $SiCl_4(g) + 2\,Mg(s) \longrightarrow 2\,MgCl_2(s) + Si(s)$

33. A value for $\Delta G_f°$ is not available in Appendix L for the first compound in each of the following reactions. Given $\Delta G°_{rxn}$, and data in Appendix L, calculate the standard molar free energy of formation for that compound.
(a) $\Delta G°_{rxn}$ = +48.2 kJ for
$MgCO_3(s) \longrightarrow MgO(s) + CO_2(g)$
(b) $\Delta G°_{rxn}$ = +20.5 kJ for
$NOCl(g) \longrightarrow NO(g) + \tfrac{1}{2} Cl_2(g)$
(c) $\Delta G°_{rxn}$ = −272.8 kJ for
$TiCl_2(s) + Cl_2(g) \longrightarrow TiCl_4(\ell)$
(d) $\Delta G°_{rxn}$ = −1830 kJ for
$2\,CH_2CO(g) + 4\,O_2(g) \longrightarrow 4\,CO_2(g) + 2\,H_2O(g)$

34. Hydrogenation, the addition of hydrogen to an organic compound, is a reaction of some industrial importance. Determine $\Delta H°$, $\Delta S°$, and $\Delta G°$ at 25°C for the hydrogenation of octene, C_8H_{16}, to give octane, C_8H_{18}.

$$C_8H_{16}(g) + H_2(g) \longrightarrow C_8H_{18}(g)$$

The following information is necessary:

COMPOUND	$\Delta H_f°$(kJ/mol)	$S°$(J/K · mol)
Octene	−82.93	462.8
Octane	−208.45	463.6

35. If gaseous hydrogen can be produced cheaply, it can be burned directly as a fuel or converted to another fuel, methane for example.

$$3\ H_2(g) + CO(g) \longrightarrow CH_4(g) + H_2O(g)$$

Calculate $\Delta H°$, $\Delta S°$, and $\Delta G°$. Is this reaction predicted to be spontaneous under standard conditions?

36. In a (fictitious) court case, a woman sued a chemical company for damages. She claimed the company produced ozone, O_3, which reacted with water in the air to produce hydrogen peroxide. The hydrogen peroxide in turn bleached her beautiful black hair to a red. You are her lawyer. Is the following reaction thermodynamically feasible? Is it spontaneous under standard conditions?

$$O_3(g) + H_2O(g) \longrightarrow O_2(g) + H_2O_2(\ell)$$

Prove your answer with appropriate calculations.

37. Elemental boron, in the form of thin fibers, can be made by reducing a boron halide with H_2:

$$BCl_3(g) + \tfrac{3}{2}\ H_2(g) \longrightarrow B(s) + 3\ HCl(g)$$

The standard heat of formation of $BCl_3(g)$ is -403.8 kJ/mol and its entropy, $S°$, is 290 J/K · mol. The entropy, $S°$, for B(s) is 5.86 J/K · mol. Calculate $\Delta H°$, $\Delta S°$, and $\Delta G°$ for this reaction. Is it predicted to be spontaneous under standard conditions? If spontaneous, is it enthalpy driven or entropy driven? If not spontaneous, in your view why is it not?

38. Wood alcohol (methyl alcohol) is a valuable starting material in the chemical industry. Although it was originally made from the destructive distillation of wood (hence its name), it is now made by the reaction below.

$$CO(g) + 2\ H_2(g) \longrightarrow CH_3OH(\ell)$$

Using thermodynamic calculations, predict the spontaneity of this reaction under standard conditions. The thermodynamic functions for $CH_3OH(\ell)$ are $\Delta H_f° = -238.7$ kJ/mol, $\Delta G_f° = -166.3$ kJ/mol, and $S° = 126.8$ J/K · mol.

THERMODYNAMICS AND EQUILIBRIUM CONSTANTS

39. The formation of NO(g) from its elements

$$N_2(g) + O_2(g) \longrightarrow 2\ NO(g)$$

has a standard free energy change, $\Delta G°$, of $+86.57$ kJ per mole of NO formed at 25°C. Calculate K_p. Are reactants or the product favored at this temperature?

40. The equilibrium constant for the *n*-butane $\rightleftarrows$ isobutane equilibrium is 2.5 at 25°C. Calculate $\Delta G°$ for the reaction in kJ/mol.

41. The dissociation of gaseous chlorine to Cl atoms has an equilibrium constant of 0.106 at 1800K. Calculate the free energy change for this reaction.

$$Cl_2(g) \rightleftharpoons 2\ Cl(g)$$

42. The equilibrium constant, K_p, for $N_2O_4(g) \rightleftarrows 2\ NO_2(g)$ is 0.11 atm at 25°C. Calculate $\Delta G°$ from this constant and compare your calculated value with that determined from the $\Delta G_f°$ values in Appendix L.

43. Water reacts with $SiCl_4$ to produce SiO_2 and HCl.

$$SiCl_4(g) + 2\ H_2O(\ell) \longrightarrow SiO_2(s) + 4\ HCl(aq)$$

(a) Using the data in Appendix L, calculate $\Delta G°$ for the reaction. Is the reaction predicted to be spontaneous under standard conditions?

(b) From $\Delta G°_{rxn}$ calculate K_p. Are products or reactants favored at 25°C?

44. Most metal oxides can be reduced with hydrogen to the pure metal. (Although such reactions work well, it is an expensive method and not often used.) The reduction of iron(II) oxide

$$FeO(s) + H_2(g) \longrightarrow Fe(s) + H_2O(g)$$

has an equilibrium constant of 0.422 at 700°C. Estimate $\Delta G°$.

GENERAL QUESTIONS

45. Benzene, C_6H_6, is made in quantities ranging from 8 to 10 billion pounds a year. It is used as a starting material for many other compounds and as a solvent. (Although it is an excellent solvent for many purposes, it has recently been found to be a carcinogen, and its use is now restricted.) One compound that can be made from it is cyclohexene, C_6H_{12}.

$$C_6H_6(g) + 3\ H_2(g) \longrightarrow C_6H_{12}(g)$$

$$\Delta H°_{rxn} = -206.1\ \text{kJ} \qquad \Delta S°_{rxn} = -363.1\ \text{J/K}$$

(a) Is this reaction predicted to be spontaneous under standard conditions at 25°C? Is the reaction entropy driven or enthalpy driven?

(b) Calculate $\Delta G_f°$ for benzene vapor [$C_6H_6(g)$] from $\Delta H°$ and $\Delta S°$ for the reaction above, together with the free energy of formation of cyclohexane, $C_6H_{12}(g)$, which is $\Delta G_f° = 31.76$ kJ/mol.

46. Iodine, I_2, dissolves readily in carbon tetrachloride with an enthalpy change that is effectively zero. What is the sign of ΔG for the reaction below? Is the solution process entropy driven or enthalpy driven? Explain briefly.

$$I_2(s) \xrightarrow[CCl_4]{} I_2(\text{in } CCl_4 \text{ solution})$$

47. Animals, except those that are deep divers such as the whales, operate under conditions of constant pressure. The oxidation of glucose, $C_6H_{12}O_6$, is a source of energy for their nervous and muscular activity. If one mole of glucose is oxidized according to the following overall equation, how much free energy is available under standard conditions at 25°C?

$$C_6H_{12}O_6(s) + 6\ O_2(g) \longrightarrow 6\ CO_2(g) + 6\ H_2O(\ell)$$

For glucose, $\Delta H_f^\circ = -1260$ kJ and $S^\circ = 288.9$ J/K · mol. (The biological reactions are equivalent to this reaction, but they involve many intermediate steps.)

48. If a 70. kg man climbs 3.0 meters vertically, he needs to do 21 kJ of work. Refer to Study Question 47 and calculate the approximate amount of glucose the man's body would have to burn to climb Mt. McKinley, the highest mountain in North America. The mountain is 20,300 feet high. (1 meter = 3.28 feet)

49. A crucial reaction for the production of synthetic fuels is the conversion of coal into H_2 by steam.

$C(s) + H_2O(g) \longrightarrow CO(g) + H_2(g)$

(a) Calculate ΔG° for this reaction at 25°C.
(b) Calculate K_p for the reaction at 25°C.
(c) Is the reaction predicted to be spontaneous under standard conditions? If not, at what temperature will it become so?
(d) Calculate the temperature at which the equilibrium constant for the reaction is 1.0×10^{-4}.

50. Calculate ΔG° for the decomposition of sulfur trioxide to sulfur dioxide and oxygen.

$2\ SO_3(g) \longrightarrow 2\ SO_2(g) + O_2(g)$

(a) Is the reaction spontaneous under standard conditions at 25°C?
(b) If the reaction is not spontaneous at 25°C, is there a temperature at which it will become so?
(c) What is the equilibrium constant for the reaction at 1500°C?
(d) What is the equilibrium constant for the oxidation of SO_2 to give SO_3 at 25°C?

51. Wood alcohol (methyl alcohol) is relatively inexpensive to produce. Much consideration has been given to using it as a precursor to other fuels such as methane, which could be obtained by the decomposition of the alcohol. (See Study Question 38 for the necessary thermodynamic functions for methyl alcohol.)

$CH_3OH(\ell) \longrightarrow CH_4(g) + \frac{1}{2}\ O_2(g)$

(a) What are the sign and magnitude of the entropy change for the reaction? Does the sign of ΔS° agree with your expectations? Why or why not?
(b) Is the reaction spontaneous under standard conditions? Use thermodynamic calculations to prove your answer.
(c) At what temperature does the reaction become spontaneous?

52. Consider the formation of NO(g) from its elements.

$N_2(g) + O_2(g) \longrightarrow 2\ NO(g)$

(a) Use the free energy data in Appendix L to calculate K_p at 25°C.
(b) Assume that ΔH°_{rxn} and ΔS°_{rxn} are nearly constant with temperature and calculate ΔG°_{rxn} at 700°C. Estimate K_p at 700°C from the new value of ΔG°_{rxn}.
(c) Using K_p for 700°C, calculate the equilibrium partial pressures of the three gases if you mix 1.00 atm each of N_2 and O_2.

53. As noted in Study Question 31, there has been a great deal of interest in the splitting of water into hydrogen and oxygen. The free energy of the reaction itself is so positive there is no hope of causing the reaction to occur without coupling it to another process. For example, it has been proposed that the reaction can be promoted by first reacting silver with water to produce hydrogen and silver oxide.

$2\ Ag(s) + H_2O(g) \longrightarrow Ag_2O(s) + H_2(g)$

and then decomposing the silver oxide at high temperature to give oxygen and recover silver.

$Ag_2O(s) \longrightarrow 2\ Ag(s) + \frac{1}{2}\ O_2(g)$

To solve this problem, you will need the following information in addition to that found in Appendix L: for $Ag_2O(s)$ $\Delta H_f^\circ = -31.0$ kJ/mol, $S^\circ = 121$ J/K · mol. (For more information on reactions of this type see E. L. King, *J. Chem. Educ.*, 58: 975, 1981.)

(a) Calculate ΔH°, ΔS°, and ΔG° for both reactions at 25°C.
(b) Write a balanced equation for the overall reaction which occurs on summing these two equations.
(c) What is the enthalpy change, ΔH°, for the two reactions added together?
(d) What is ΔG° for the two reactions combined? Is the overall process spontaneous under standard conditions?
(e) At what temperature is the second reaction above spontaneous?

54. Photosynthetic bacteria carry out the synthesis of high free energy compounds such as glucose from CO_2 and H_2O using light as the source of energy.

$6\ CO_2(g) + 6\ H_2O(\ell) \longrightarrow C_6H_{12}O_6(s) + 6\ O_2(g)$

The free energy change is +2870 kJ/mol of glucose. However, in the deep ocean where there is no light, this same synthesis can apparently be done by bacteria using hydrogen sulfide as the energy source. Show that, by adding the following reaction

$H_2S(g) + \frac{1}{2}\ O_2(g) \longrightarrow H_2O(\ell) + S(s)$

to the glucose synthesis reaction above, sufficient free energy will be produced so that the overall process

$$24\ H_2S(g) + 6\ CO_2(g) + 6\ O_2(g) \longrightarrow C_6H_{12}O_6(s) + 18\ H_2O(\ell) + 24\ S(s)$$

is spontaneous. (ΔG_f° for glucose is -919.2 kJ/mol.) (For further information see S. Krishnamurthy, *J. Chem. Educ.*, 58: 981, 1981.)

19

Oxidation–Reduction Reactions

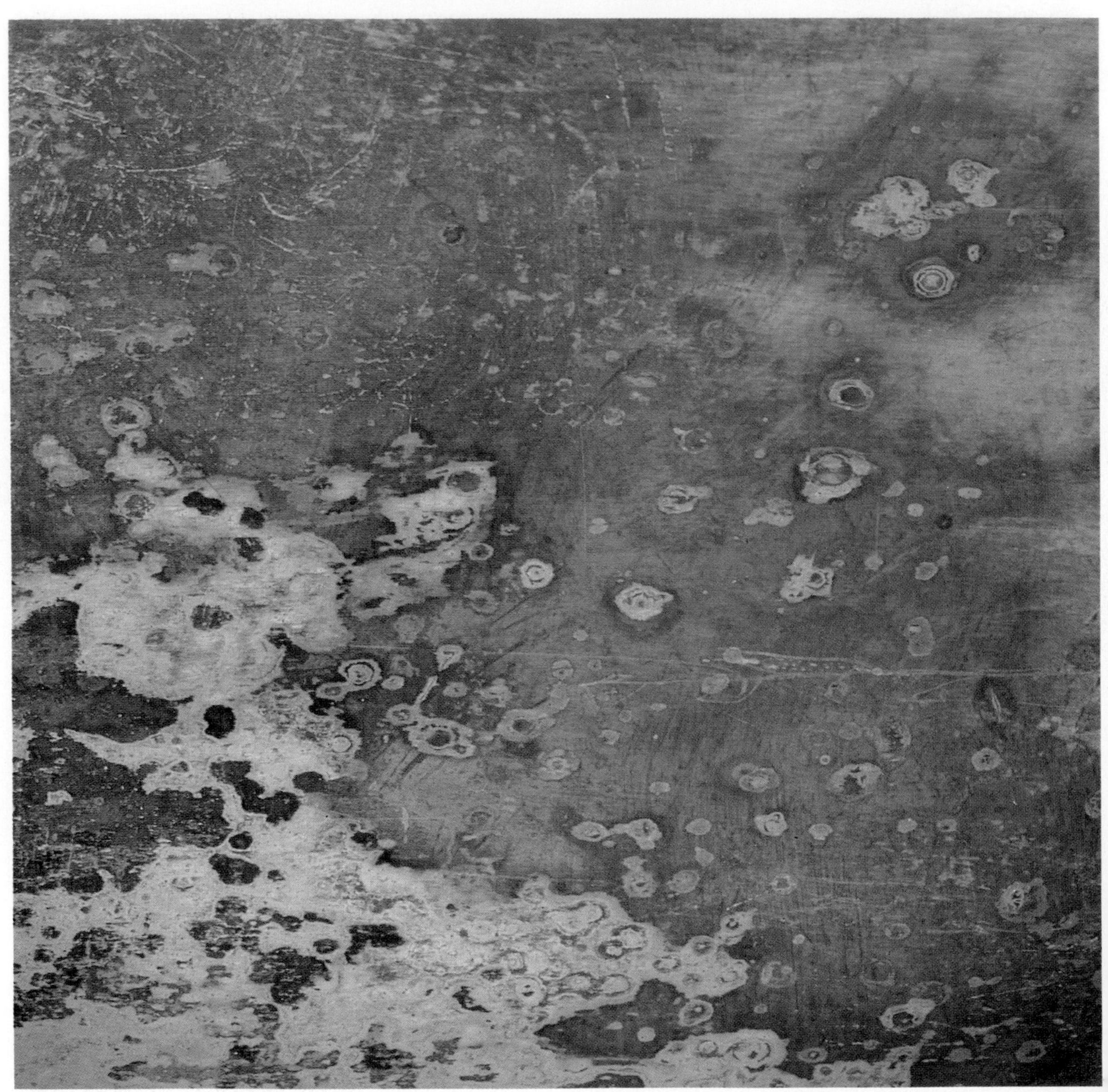

Tarnish on silver, the result of oxidation. (R. Megna, Fundamental Photographs)

CHAPTER OUTLINE

You learned in Chapter 3 that the transfer of an electron from one compound to another results in the oxidation of the electron donor and reduction of the electron acceptor.

Loss of electrons (oxidized)

$$Zn(s) + Cu^{2+}(aq) \longrightarrow Zn^{2+}(aq) + Cu(s)$$

Gain of electrons (reduced)

Hence, reactions involving electron transfer are often called oxidation–reduction or redox reactions. Our goal in this chapter is to investigate some of the practical consequences of electron transfer.

Electron transfer reactions are so numerous that you experience such reactions and their consequences daily. Let's look at a few examples.

(a) **Corrosion** occurs by oxidation–reduction reactions. In some areas salt is spread on the roads to melt the snow and ice. Although accidents are prevented, the salt accelerates the formation of rust on car bodies, bridges, and other steel structures. Near the ocean, the salt suspended in the air assists in corroding metal objects. In each of these cases the corrosion process is central to the oxidation of the metal (usually iron or aluminum) to its oxide (Figure 19.1).

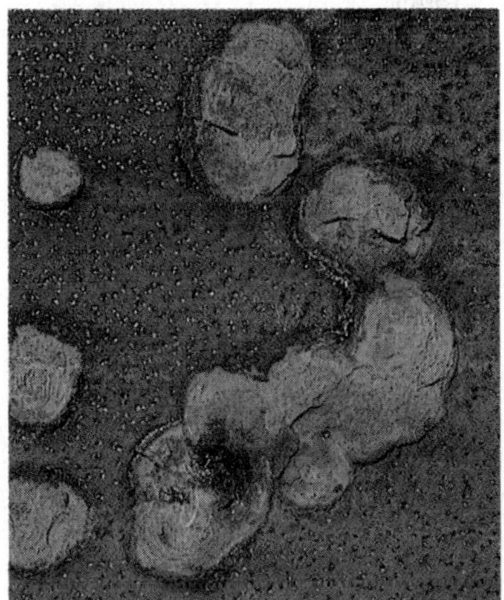

FIGURE 19.1
Rust on a shipwreck, from an oxidation–reduction reaction involving iron, water, and oxygen. See Section 19.8. (Fundamental Photographs)

(b) Many **biological processes** depend on the transfer of electrons. For example, when you breathe air, oxygen is converted ultimately to water and carbon dioxide. The oxidation number of the oxygen atom in the product molecules (H_2O and CO_2) is -2, so electrons must be transferred to O_2 to cause its reduction. Where did the electrons come from? At least in the final step they are transferred to O_2 from hemoglobin, a large molecule containing Fe^{2+}, and the Fe^{2+} is oxidized to Fe^{3+} in the process. Other biological oxidation–reduction processes include the conversion of water to O_2 in green plants by photosynthesis, and the conversion of atmospheric N_2 to usable nitrogen, such as NH_4^+, by bacteria.

(c) The electrical power to start your car comes from a battery. A **battery** is an electrochemical device that produces a current or flow of electrons at a constant voltage as a result of an oxidation–reduction reaction. Batteries will be described in some detail later.

A battery is also sometimes called a **voltaic cell** or a **galvanic cell**. These names derive from Alessandro Volta (1745–1827), who studied such cells, and Luigi Galvani (1737–1798) who made early studies of animal electricity and electricity produced chemically.

(d) Many electron transfer processes are important in **industry**. For example, metals such as sodium, aluminum, and copper are commercially prepared or purified by the direct application of electricity,

$$Na^+ + e^- \longrightarrow Na(s)$$
$$Cu^{2+} + 2e^- \longrightarrow Cu(s)$$

while other metals, such as iron, are prepared by using chemical reducing agents.

$$Fe_2O_3(s) + 3\ CO(g) \longrightarrow 2\ Fe(s) + 3\ CO_2(g)$$

Chemists have only recently begun to understand the way in which an electron is transferred from one site to another. Moreover, the process of corrosion and its prevention, the construction of more powerful batteries, and the plating of metals using electricity are now better understood. The general subject of oxidation–reduction reactions is fascinating since it applies to so many problems of practical interest.

Before turning to your study of electron transfer reactions, you should review Chapter 3. There you learned to recognize and balance redox equations and to understand the following terms:

The following photos of redox reactions are found earlier in the book:
Al/Cu^{2+} (Figure 1.8)
Mg/O_2 (Figure 3.1)
Al/Br_2 (Figure 3.3)
Zn/I_2 (Figure 3.2)
Mg/HCl (Figure 3.4)
Fe/O_2 (Figure 3.10)
Cu/HNO_3 (Figure 3.11)
S/O_2 (Figure 3.12)
Zn/Cu^{2+} (Figure 3.18)

Oxidation: An atom, ion, or molecule releases electrons and is oxidized. The oxidation number of an element increases.

Reduction: An atom, ion, or molecule accepts electrons and is reduced. The oxidation number of an element decreases.

Reducing agent: The atom, ion, or molecule providing electrons.

Oxidizing agent: The atom, ion, or molecule accepting electrons from the reducing agent.

19.1 CHEMICAL CHANGE LEADING TO ELECTRIC CURRENT

The type of reaction capable of producing an electrical current, or flow of electrons, is illustrated in Figure 19.2. A piece of zinc is partially immersed in an aqueous solution of copper sulfate. After a time, the blue color of the aqueous Cu^{2+} ion begins to fade, the edges of the zinc plate are eaten away, and copper begins "to plate out" or form a covering on

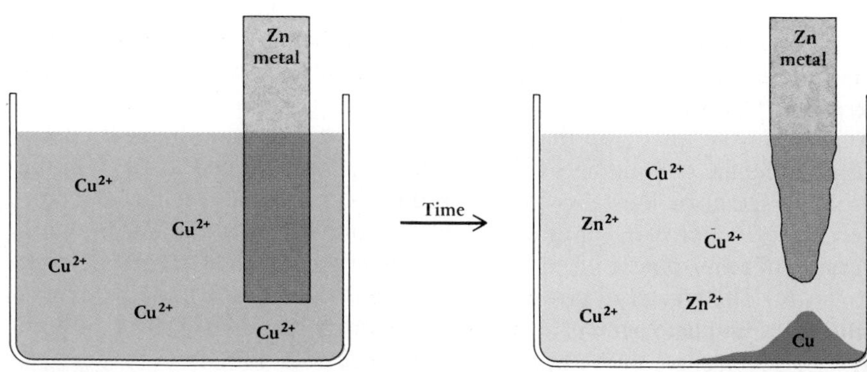

FIGURE 19.2
An oxidation–reduction reaction. A Zn plate is gradually "eaten away" as zinc reduces aqueous copper(II) ions to copper metal. The net chemical reaction is

$$Zn(s) + Cu^{2+}(aq) \longrightarrow Zn^{2+}(aq) + Cu(s)$$

See also Figures 3.18 and 19.3.

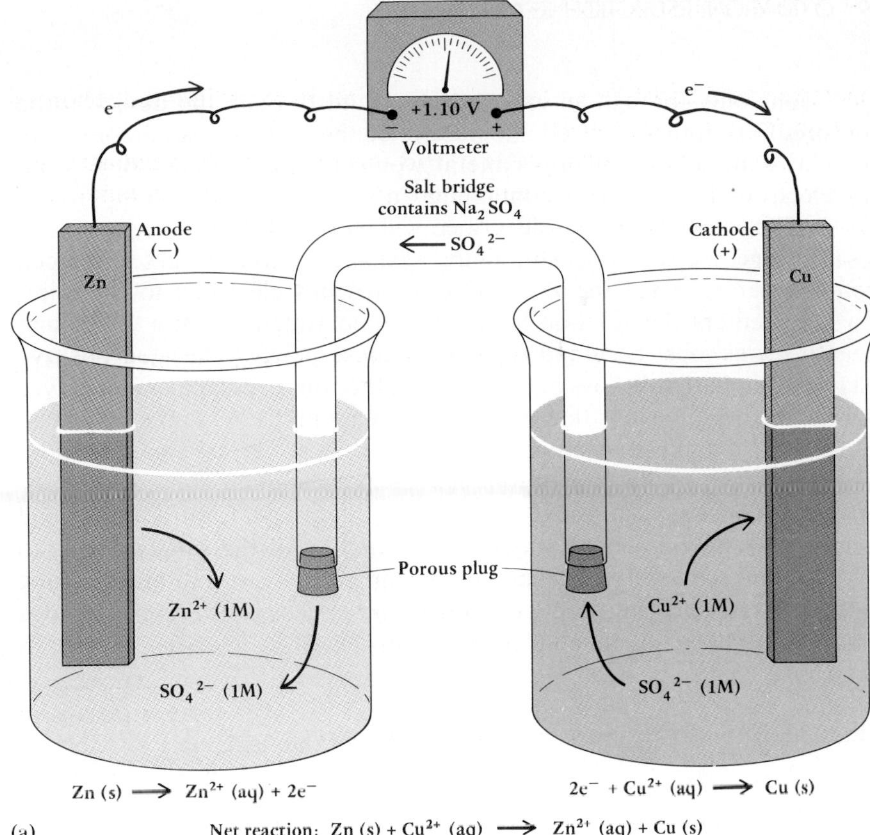

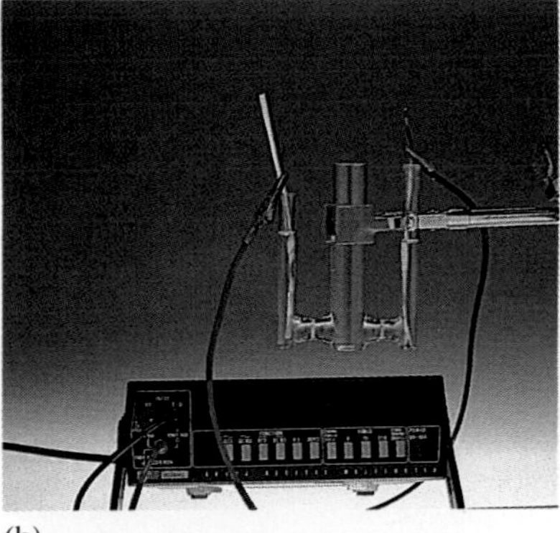

(b)

FIGURE 19.3

A galvanic or voltaic cell using Cu(s)/Cu^{2+}(aq) and Zn(s)/Zn^{2+}(aq) half-cells. (a) A voltage of 1.10 V is generated if the cell is set up under the conditions shown. Electrons flow through the external wire from the Zn electrode (anode) to the Cu electrode (cathode). A salt bridge provides a connection between the half-cells for ion flow; thus, SO_4^{2-} ions flow from the copper to the zinc half-cell. (b) An actual cell operating under nearly standard conditions. The negative Zn electrode is at the left, and the positive Cu electrode is at the right. The compartments are separated from one another by porous glass disks. The center compartment is the salt bridge and contains Na_2SO_4. (Charles D. Winters)

the zinc strip. After still more time, the zinc strip disappears, copper is piled up on the bottom of the container, and the color of the copper ion fades still more (see Figure 3.18). What is happening?

The fate of the Cu^{2+} ion in the beaker in Figure 19.2 is probably obvious from the fact that the blue color of the aqueous ion has faded and a copper cover has appeared on the zinc strip: the copper ion has been reduced to the metal.

$$Cu^{2+}(aq) + 2\,e^- \longrightarrow Cu(s)$$

But what about the zinc? Our observation was that it disappeared, so it must have been the source of the electrons that caused reduction of Cu^{2+}; that is, the zinc was the reducing agent and formed aqueous Zn^{2+}.

$$Zn(s) \longrightarrow Zn^{2+}(aq) + 2\,e^-$$

The *net chemical reaction* occurring in the beaker, therefore, was the *spontaneous* reduction of Cu^{2+} and the simultaneous *spontaneous* oxidation of Zn(s).

$$Cu^{2+}(aq) + Zn(s) \longrightarrow Cu(s) + Zn^{2+}(aq)$$

The reaction above is interesting, but it could not be used as a source of current in the physical arrangement we have described. The electrons provided by the zinc moved directly to the aqueous Cu^{2+} ions on contact. In order to use the Cu^{2+}/Zn reaction as the basis of a battery, the zinc must be placed in a container separate from the Cu^{2+} ions (Figure 19.3). Electrons can then pass from the zinc **electrode** to the solution of

An electrode conducts electrons into and out of a solution. It is most often a metal plate or wire or a piece of graphite.

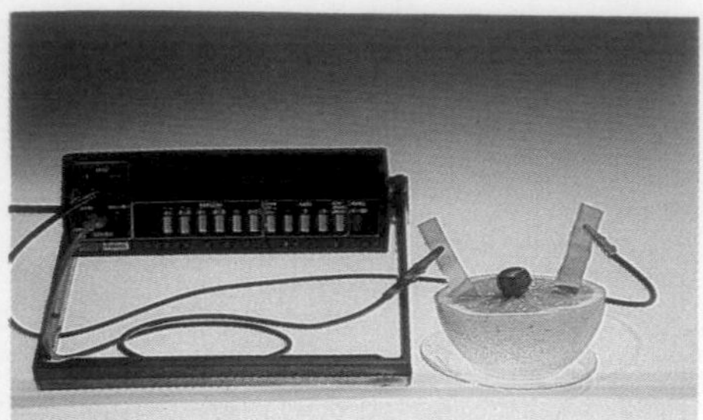

A galvanic cell made by inserting zinc and copper electrodes in a grapefruit. A voltage of over 0.9 V is obtained. (The water and citric acid of the fruit allow for ion conduction between electrodes.) (Charles D. Winters)

Cu^{2+} ions only through an external wire (and through the device to be powered) to another electrode (a piece of copper in this case) dipping into the Cu^{2+}(aq) solution. Copper metal would then be "plated out" onto the electrode in the beaker containing Cu^{2+}.

The arrangement just described will work *only* if we have provided a **salt bridge**, a device for maintaining a balance of ion charges in the cell compartments. When the Zn electrode provides electrons to the wire, Zn^{2+}(aq) enters the solution in the Zn compartment (Figure 19.3), and negative ions must be found to balance these newly generated positive charges. Similarly, the loss of Cu^{2+} ions in the copper compartment leaves behind the negative ions that were associated with Cu^{2+}. Some way must be found so that these negative ions, now in excess, can leave this solution. Thus, to achieve a balance of ion charges in each compartment, the negative ion concentration must decrease in the copper compartment and increase in the zinc compartment. The function of the salt bridge is to allow anions to pass freely from the compartment where cations are being lost to the compartment where cations are being generated. The salt bridge could be simply an aqueous solution of Na_2SO_4, allowing SO_4^{2-} to transfer.*

You can keep track of the movement of electrons and ions in an electrochemical cell this way: The movement of negative charges (both electrons and anions) is in a "circle." As shown in Figure 19.3, electrons move through the external circuit from reducing agent (Zn) to oxidizing agent (Cu^{2+}); the negative ions complete the circle through the salt bridge (from Cu^{2+} to Zn^{2+}).

$$SO_4^{2-} \text{ in } Cu^{2+} \text{ solution} \longrightarrow SO_4^{2-} \text{ in salt bridge} \longrightarrow SO_4^{2-} \text{ in } Zn^{2+} \text{ solution}$$

All batteries operate in a similar fashion. The oxidation–reduction reaction must be spontaneous. There must be an external circuit to allow the electrons to perform useful work, and there must be a salt bridge to allow ion flow between the battery compartments.

There is some additional terminology peculiar to electrochemical cells that you must know. *Oxidation always occurs at the anode* of an electrochemical cell and *reduction always occurs at the cathode*. Furthermore, since the electrons are being produced at the *anode* of a battery, this terminal *bears a negative charge*. This means the other terminal in a battery, the *cathode*, *bears a positive charge*. To depict cells in an abbreviated way, the following notation is often used.

$$\overbrace{\text{M(electrode)}|\text{M}^+\text{(solution)}|}^{\text{anode compartment}}\ \overbrace{|\text{N}^+\text{(solution)}|\text{N(electrode)}}^{\text{cathode compartment}}$$

where M and N are general terms for ions and the single line (|) indicates a phase difference between a solid electrode and the solution. The double line (||) is a notation for the salt bridge connecting the anode and cathode. The notation tells us that the cell reaction is

$$M + N^+ \longrightarrow M^+ + N$$

*This is not to imply that the cations in the cell do not move as well. Since Zn^{2+} ions are being generated in the anode compartment and Cu^{2+} ions are being consumed in the cathode compartment, the Na^+ ions in the salt bridge can also migrate toward the copper compartment to maintain charge balance.

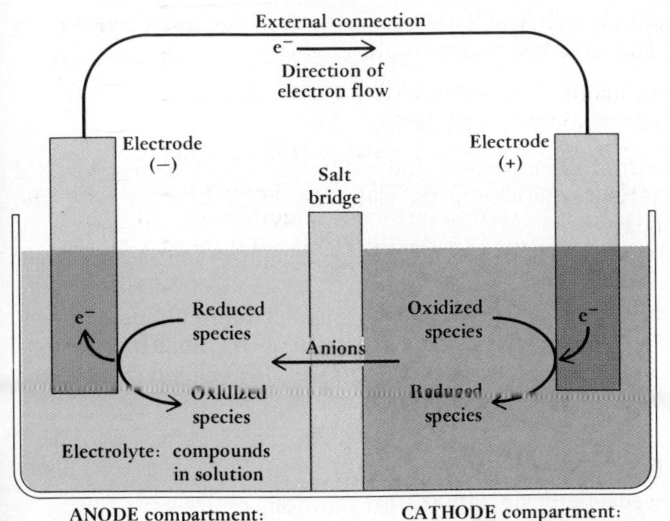

FIGURE 19.4
Summary of the terminology used in voltaic cells.

since $M \rightarrow M^+ + e^-$ occurs at the anode, and $N^+ + e^- \rightarrow N$ occurs at the cathode. This means the copper/zinc cell described above would be given by

$$Zn|Zn^{2+}(aq)||Cu^{2+}(aq)|Cu$$

Important terms are summarized in Figure 19.4.

To remember that *A*node and *O*xidation are paired as are *C*athode and *R*eduction, note the alphabetic orders:

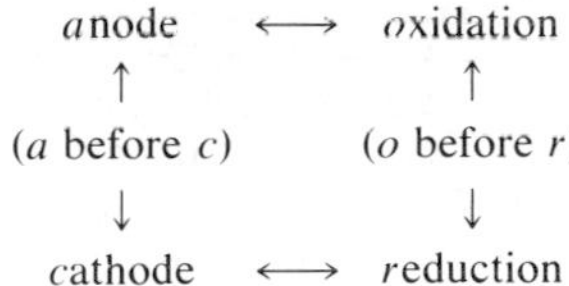

EXAMPLE 19.1

ELECTROCHEMICAL CELLS

A simple voltaic cell has been assembled with Ni(s) and $Ni(NO_3)_2$(aq) in one compartment and Cd(s) and $Cd(NO_3)_2$(aq) in the other. An external wire connects the two electrodes, and a salt bridge containing $NaNO_3$ connects the two solutions. The net reaction is

$$Ni^{2+}(aq) + Cd(s) \longrightarrow Ni(s) + Cd^{2+}(aq)$$

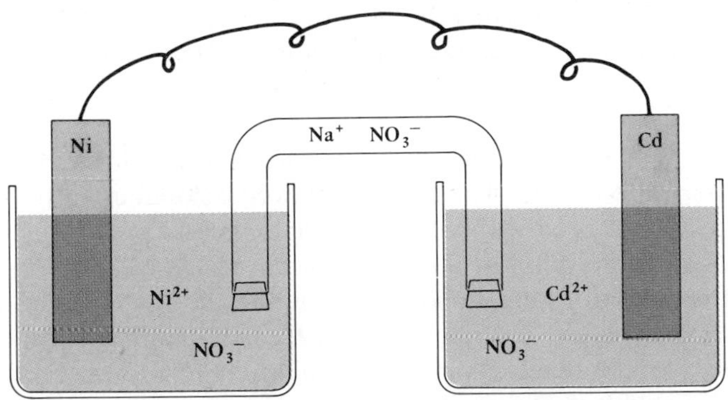

What half reaction occurs at each electrode? Which is the anode, and which is the cathode? What is the direction of electron flow in the external wire and of anion flow in the salt bridge?

Solution The net equation tells you that Cd(s) is the reducing agent and Ni^{2+} is the oxidizing agent. Thus, the half reactions are

Oxidation at the anode: $Cd(s) \longrightarrow Cd^{2+}(aq) + 2e^-$
Reduction at the cathode: $Ni^{2+}(aq) + 2\ e^- \longrightarrow Ni(s)$

Electrons flow from their source (the oxidation of cadmium at the Cd electrode or anode) through the wire to the electrode where they are used (the Ni electrode or cathode).

Since Cd^{2+} ions are being formed in the anode compartment, anions must move into that compartment from the salt bridge. The Ni^{2+} concentration in the cathode compartment is being depleted, so anions move out of this compartment into the salt bridge. The "circle" of flow of negative charge is complete: electrons flow from Cd to Ni and anions from Ni to Cd. Finally, the cell would have the notation $Cd|Cd^{2+}(aq)||Ni^{2+}(aq)|Ni$.

EXERCISE 19.1 ELECTROCHEMICAL CELLS

A voltaic cell has been assembled with the net reaction

$$Ni(s) + 2\ Ag^+(aq) \longrightarrow Ni^{2+}(aq) + 2\ Ag(s)$$

Give the half-reactions for this redox process, indicate whether each is an oxidation or reduction, and tell which happens at the anode and which at the cathode. What is the direction of electron flow in an external wire connecting the two electrodes? If a salt bridge connecting the cell compartments contains KNO_3, what is the direction of flow of the nitrate ions? Show the abbreviated notation for the cell.

19.2 ELECTROCHEMICAL CELLS AND VOLTAGE

Electrons generated at the site of oxidation, the anode, of a battery are thought to be "driven" or "pushed" toward the cathode by an **electromotive force** or **emf**. This force is due to the difference in (potential) energy of an electron at the two electrodes, and it is this energy in which we are interested. If the energy difference for a coulomb of charge is 1 joule, we say that the **cell potential** or **cell voltage**, E_{cell}, is 1 volt. (Though a misnomer, common practice is to call this potential the "emf.")

$$1\ \text{volt (V)} = 1\ \text{joule/coulomb} = 1\ \text{J/C}$$

The voltage is a readily measured characteristic of an electrochemical cell, and its value can be predicted under well defined conditions. When the voltage is measured with all of the reactants and products present as pure solids or in solution at a concentration of 1.0 M, or with gases at 1.0 atm, the conditions are **standard conditions** and the measured voltage is the **standard voltage**, $E°$. This voltage, $E°$, is a quantitative measure of the tendency of the reactants in their standard states to proceed to products in their standard states. As an example of a cell at standard conditions, consider the cell in Figure 19.3. This cell will deliver exactly 1.10 volts at 25°C.

Unless specified otherwise, all values of $E°$ are given at 25°C.

$E°$ AND $\Delta G°$

The standard cell potential is clearly a measure of reaction spontaneity, but so is the standard free energy change for a reaction, $\Delta G°$. The exact relation between them is

$$\Delta G° = -nFE°$$

$$F = \text{Faraday constant} = 9.65 \times 10^4 \text{ joule/volt} \cdot \text{mol} \qquad (19.1)$$

n is the number of electrons transferred between oxidizing and reducing agents in a balanced redox reaction, and F is the Faraday constant (named in honor of Michael Faraday, the first person to investigate quantitatively the relation between chemistry and electricity; see Chapter 3).

The reaction of Zn(s) and Cu^{2+}(aq) produces a current, so you readily conclude that it is a spontaneous reaction as written.

$$Zn(s) + Cu^{2+}(aq, 1\ M) \longrightarrow Zn^{2+}(aq, 1\ M) + Cu(s) \qquad E° = +1.10\ V$$

Since spontaneous processes have a negative free energy change, $-\Delta G°$ (Chapter 18), this is the reason for the negative sign in the relation between free energy and cell voltage. Thus, *all spontaneous electron transfer reactions will have a positive $E°$.*

If the direction of a reaction is reversed, the sign of $\Delta G°$ and so the sign of $E°$ are reversed. Thus, if we write the equation for the reduction of Zn^{2+} by Cu,

$$Cu(s) + Zn^{2+}(aq, 1\ M) \longrightarrow Cu^{2+}(aq, 1\ M) + Zn(s) \qquad E° = -1.10\ V$$

this reaction is nonspontaneous and must have a positive $\Delta G°$ and a negative $E°$. *Nonspontaneous reactions have a negative $E°$.*

When a reaction is reversed, the magnitudes of $\Delta G°$ and $E°$ remain the same but their signs are reversed [(+ → −) or (− → +)].

The relation between the voltage of an electrochemical cell and the free energy for the reaction is important. After looking into the matter of cell voltages in more detail, we shall return to the thermodynamic relationship later in this section and in Section 19.3. For the moment, the calculation of $\Delta G°$ from $E°$ is illustrated in Example 19.2.

EXAMPLE 19.2

$\Delta G°$ AND $E°$

The Cu^{2+}(aq)/Zn(s) reaction above has a standard cell voltage $E°$ of +1.10 V at 25°C. Calculate $\Delta G°$ for the reaction.

$$Cu^{2+}(aq) + Zn(s) \longrightarrow Cu(s) + Zn^{2+}(aq)$$

Solution To obtain $\Delta G°$ we need only to substitute into Equation 19.1.

$$\Delta G° = -(2.00 \text{ mol electrons transferred})\left(\frac{9.65 \times 10^4\ J}{V \cdot mol}\right)(1.10\ V)\left(\frac{1\ kJ}{10^3\ J}\right)$$

$$= -212\ kJ$$

EXERCISE 19.2 USING THE $\Delta G°/E°$ RELATIONSHIP
The following reaction has an $E°$ value of -0.76 V. Calculate $\Delta G°$ and tell whether the reaction is spontaneous as written.

$$Zn^{2+}(aq) + H_2(g) \longrightarrow Zn(s) + 2\,H^+(aq)$$

CALCULATING THE VOLTAGE, $E°$, OF AN ELECTROCHEMICAL CELL

As you learned in Chapter 3, an oxidation–reduction reaction is the sum of two half-reactions, one for oxidation and the other for reduction. For the $Zn(s)/Cu^{2+}$ reaction we have been considering, this means the overall process is the sum of the reactions

Oxidation occurs; reducing agent, Zn:	$Zn(s) \longrightarrow Zn^{2+}(aq) + 2e^-$
Reduction occurs; oxidizing agent, Cu^{2+}:	$Cu^{2+}(aq) + 2e^- \longrightarrow Cu(s)$
Net oxidation–reduction reaction:	$Cu^{2+}(aq) + Zn(s) \longrightarrow Zn^{2+}(aq) + Cu(s)$

Since redox reactions are the sum of two half-reactions, it would be very helpful if we could assign a voltage to each half-reaction. Summing the half-reaction voltages would then give the cell voltage. The problem is that voltages for isolated half reactions cannot be obtained directly, because the voltage measures the potential energy *difference* for electrons in *two different* chemical environments. However, we can always measure the electrical potential for any half reaction in combination with some other, standard half reaction. Indeed, it has been decided by the chemistry community that the **standard half-cell** reaction against which all others are measured is the **standard hydrogen electrode** (S.H.E.)

$$2\,H^+(aq,\ 1\ M) + 2e^- \longrightarrow H_2(g,\ 1\ atm)$$

and a half-cell potential of 0.00 V has been *assigned* to this half reaction. (This value has no physical meaning in itself, just as the half reaction alone has no meaning.) To measure the voltage for any half-cell, we make that half reaction (under standard conditions) one side of an electrochemical cell and the H_2/H^+ half reaction the other side. H_2 is a reducing agent and $H^+(aq)$ is an oxidizing agent. Thus, when the hydrogen electrode is paired with another half-cell in an electrochemical cell, the H_2/H^+ cell reaction can be either an oxidation or reduction. If the other half-cell contains a better reducing agent than H_2, then H^+ is reduced to H_2.

H^+ reduced: $\quad 2\,H^+(aq) + 2\,e^- \longrightarrow H_2(g) \qquad E° = 0.00$ V

If the other half-cell contains a better oxidizing agent than H^+, then H_2 is oxidized.

H_2 oxidized: $\quad H_2(g) \longrightarrow 2\,H^+(aq) + 2\,e^- \qquad E° = 0.00$ V

In either direction, the $H_2(g)/H^+(aq)$ half-cell has a potential of 0.00 V. The measured voltage of the electrochemical cell is then *assigned* as the voltage of the half-cell being studied. To demonstrate the strategy for determining half-cell potentials, consider the following examples.

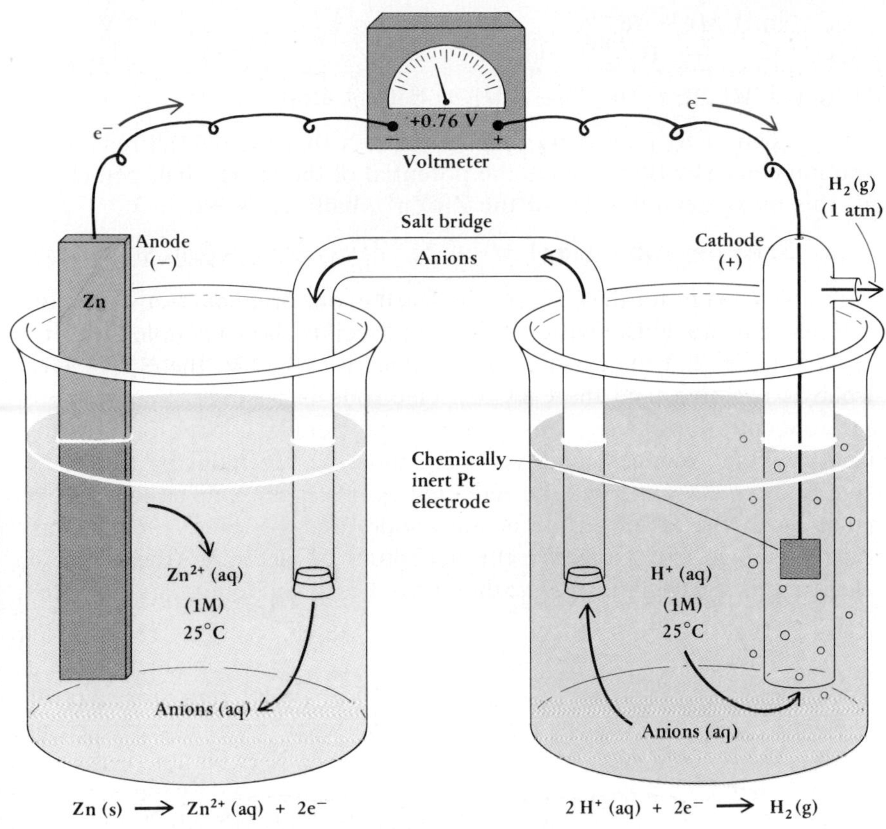

FIGURE 19.5
An electrochemical cell using $Zn(s)/Zn^{2+}(aq)$ and $H_2(g)/H^+(aq)$ half-cells. A voltage of +0.76 V is generated when the cell is set up under the conditions shown at 25°C. Electrons flow from the Zn electrode (anode) to the H_2/H^+ electrode (cathode) to produce $Zn^{2+}(aq)$ and H_2 gas. Zinc is the reducing agent and $H^+(aq)$ is the oxidizing agent.

In Figure 19.5, we have diagrammed a cell in which one compartment contains the H_2/H^+ reaction mixture, and the other compartment has a zinc electrode dipping into a solution of 1 *M* Zn^{2+}. As usual, the compartments are connected by an external wire (for electron flow) and a salt bridge (for ion flow).

When a redox reaction involves an ion or compound that cannot be made into solid electrodes, a chemically inert conductor of electricity can be used. Platinum or gold are frequently used as such inert electrodes.

The measured voltage of an electrochemical cell is always positive. The device used to measure voltages, a voltmeter, is designed to give a positive voltage only when the positive terminal (+) of the voltmeter is connected to the positive electrode and the negative terminal (−) to the negative electrode. In this way we not only measure the voltage but we find the sign of each of the electrodes. Thus, when a voltmeter is attached to the voltaic cell shown in Figure 19.5, it is found that the H_2/H^+ electrode is positive, the zinc electrode is negative, and the measured voltage is $E° = +0.76$ V. Although both Zn and H_2 can potentially function as reducing agents here, the observation that the Zn electrode is negative means that Zn is the source of electrons: metallic zinc is a better reducing agent than H_2 gas. Therefore, it is observed that the zinc electrode dissolves owing to oxidation of metallic zinc, and H_2 gas bubbles form owing to reduction of H^+ to H_2.

In Figure 19.3, the green wire is attached to the negative terminal, the Zn electrode.

Anode, oxidation:	$Zn(s) \longrightarrow Zn^{2+}(aq, 1\ M) + 2e^-$	$E° = ?$ V
Cathode, reduction:	$2\ H^+(aq, 1\ M) + 2e^- \longrightarrow H_2(g, 1\ atm)$	$E° = 0.00$ V
Net cell reaction:	$Zn(s) + 2\ H^+(aq, 1\ M) \longrightarrow Zn^{2+}(aq, 1\ M) + H_2(g, 1\ atm)$	$E°_{cell} = +0.76$ V

The voltage produced by an electrochemical cell is the sum of the voltages of the oxidizing half reaction and the reducing half reaction.

The + sign of $E°_{cell}$ correctly reflects the fact that the overall reaction is spontaneous as written. Since the potential of the H_2/H^+ half-cell is 0.00 V, this must mean that $E°$ for the Zn/Zn^{2+} half-cell is +0.76 V.

$$Zn(s) \longrightarrow Zn^{2+}(aq, 1\ M) + 2e^- \qquad E° = +0.76\ V$$

What is $E°$ for the Cu/Cu^{2+} half-cell in our original electrochemical cell in Figure 19.3? In Figure 19.6 this half-cell is shown coupled with the H_2/H^+ half-cell. The measured cell voltage is +0.34 V, the H_2/H^+ electrode is negative, and the Cu/Cu^{2+} electrode is positive. Furthermore, the concentration of Cu^{2+} ions declines, and metallic copper forms. Thus, Cu^{2+} is being reduced, and hydrogen must be the reducing agent, the source of electrons. Since the reducing agent H_2 is oxidized to H^+, this must mean the H_2 electrode is the anode (and is negatively charged). Further, since Cu^{2+} ions are the acceptors of electrons (the oxidizing agent), this electrode is the cathode (and is positively charged). Most

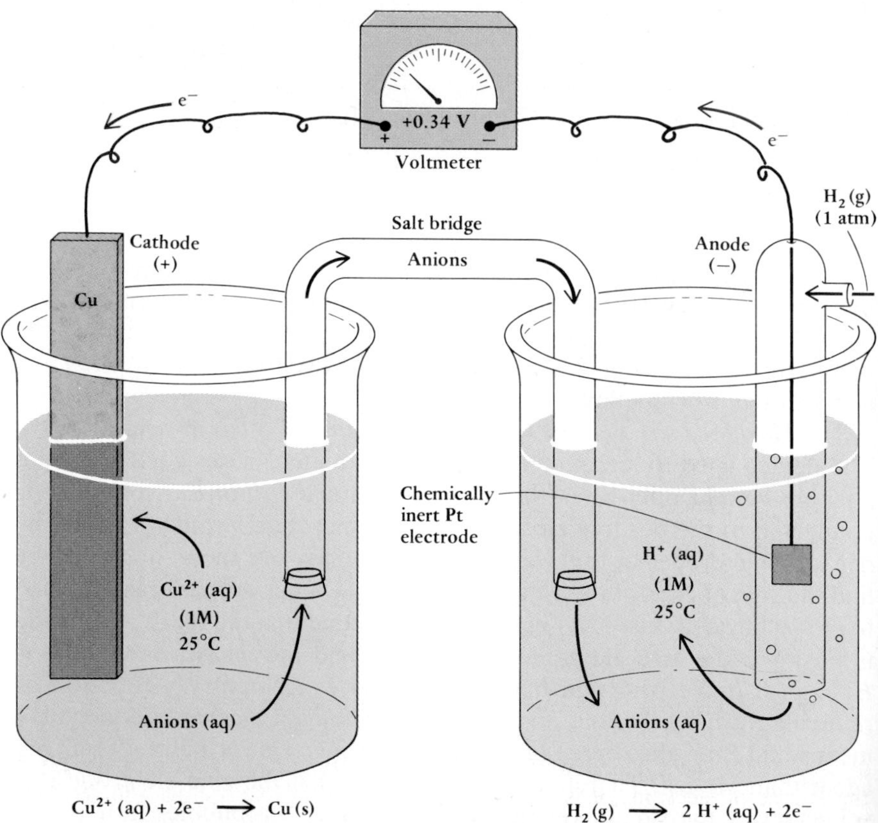

FIGURE 19.6

An electrochemical cell using $Cu(s)/Cu^{2+}(aq)$ and $H_2(g)/H^+(aq)$ half-cells. A voltage of +0.34 V is generated when the cell is set up as pictured. Electrons flow from the H_2/H^+ electrode (the anode) to the copper electrode (cathode) to produce copper and H^+ ions. $H_2(g)$ is the reducing agent and $Cu^{2+}(aq)$ is the oxidizing agent.

importantly, we now see that $H_2(g)$ is a better reducing agent than Cu(s), and the appropriate half reactions and net reaction are

Anode, oxidation:	$H_2(g, 1\ atm) \longrightarrow 2\ H^+(aq, 1\ M) + 2e^-$	$E° = 0.00\ V$
Cathode, reduction:	$Cu^{2+}(aq, 1\ M) + 2e^- \longrightarrow Cu(s)$	$E° = ?\ V$
Net cell reaction:	$H_2(g, 1\ atm) + Cu^{2+}(aq, 1\ M) \longrightarrow Cu(s) + 2\ H^+(aq, 1\ M)$	$E°_{cell} = +0.34\ V$

The half-cell voltage for $Cu^{2+}(aq, 1\ M) + 2e^- \rightarrow Cu(s)$ must be $+0.34$ V at 25°C.

The voltage of a cell where Zn(s) reduces $Cu^{2+}(aq)$ to Cu(s) can now be calculated, since we have $E°$ values for the half reactions involved.

Anode, oxidation:	$Zn(s) \longrightarrow Zn^{2+}(aq, 1\ M) + 2e^-$	$E° = +0.76\ V$
Cathode, reduction:	$Cu^{2+}(aq, 1\ M) + 2e^- \longrightarrow Cu(s)$	$E° = +0.34\ V$
Net cell reaction:	$Zn(s) + Cu^{2+}(aq, 1\ M) \longrightarrow Zn^{2+}(aq, 1\ M) + Cu(s)$	$E°_{cell} = +1.10\ V$

This is an important result. We have independently measured two half-cell voltages against the same standard H_2/H^+ half-cell, and we now find that these two half-cell voltages can be used to obtain the voltage of a new reaction. This was our original objective. Now we could use a similar technique, with any of the Cu/Cu^{2+}, Zn/Zn^{2+}, or H_2/H^+ cells as reference cells, to determine $E°$ values for hundreds of other possible half-cells (see Example 19.3).

EXAMPLE 19.3

DETERMINING A HALF-REACTION POTENTIAL

The cell illustrated below generates a voltage of $E° = +0.51$ V under standard conditions at 25°C, and the net cell reaction is

$$Zn(s) + Ni^{2+}(aq, 1\ M) \longrightarrow Zn^{2+}(aq, 1\ M) + Ni(s)$$

Tell which electrode is the anode and which is the cathode, give the signs of the electrodes, and calculate the half-cell potential for $Ni^{2+}(aq) \rightarrow Ni(s)$.

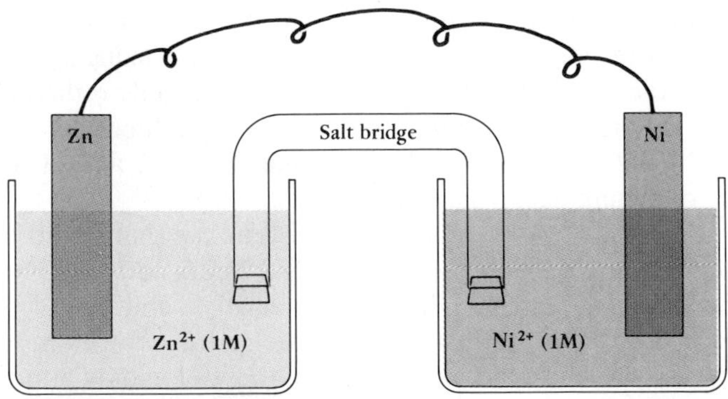

Solution The electrode at which oxidation occurs is the anode (and, as this is the source of electrons, it is negative). Since Zn(s) is oxidized to $Zn^{2+}(aq)$, the Zn electrode is the anode. Nickel(II) ions are reduced at the Ni electrode, so Ni metal is the cathode.

Since the overall cell voltage is known, and the voltage for the $Zn(s)/Zn^{2+}$(aq, 1 *M*) half-cell is known, $E°$ for Ni^{2+}(aq, 1 *M*) → Ni(s) can be calculated.

Anode, oxidation:	$Zn(s) \longrightarrow Zn^{2+}(aq) + 2e^-$	$E° = +0.76$ V
Cathode, reduction:	$Ni^{2+}(aq) + 2e^- \longrightarrow Ni(s)$	$E° = ?$
Net reaction	$Zn(s) + Ni^{2+}(aq) \longrightarrow Zn^{2+}(aq) + Ni(s)$	$E°_{cell} = +0.51$ V

At 25°C, the value of $E°$ for the Ni^{2+}(aq, 1 *M*) + $2e^-$ → Ni(s) half reaction is −0.25 V.

EXERCISE 19.3 HALF-REACTION POTENTIALS

Given that the reduction of aqueous copper(II) with iron has an $E°$ value of +0.78 V, what is $E°$ for the half-cell Fe(s) → Fe^{2+}(aq, 1 *M*) + $2e^-$?

$$Fe(s) + Cu^{2+}(aq, 1\ M) \longrightarrow Fe^{2+}(aq, 1\ M) + Cu(s)$$

USING STANDARD CELL POTENTIALS

In Example 19.3, we found that $E°$ for the reduction of Ni^{2+} to metallic nickel is −0.25 V. What does this mean? This tells us that if Ni^{2+} is used as an oxidizing agent, coupled with H_2 as the reducing agent,

$Ni^{2+}(aq, 1\ M) + 2e^- \longrightarrow Ni(s)$	$E° = -0.25$ V
$H_2(g) \longrightarrow 2\ H^+(aq, 1\ M) + 2e^-$	$E° = 0.00$ V
$Ni^{2+}(aq, 1\ M) + H_2(g) \longrightarrow Ni(s) + 2\ H^+(aq, 1\ M)$	$E°_{cell} = -0.25$ V

the reaction is *not spontaneous* under standard conditions; $\Delta G°$ is positive so $E°_{cell}$ is negative. From thermodynamics, you know that if a reaction is not spontaneous in the direction written ($\Delta G°$ is positive), the *reverse* reaction is spontaneous ($\Delta G°$ is negative). Therefore, the reaction

$$Ni(s) + 2\ H^+(aq, 1\ M) \longrightarrow Ni^{2+}(aq, 1\ M) + H_2(g, 1\ atm) \qquad E°_{cell} = +0.25\ V$$

is spontaneous, and $E°_{cell}$ for the reaction is positive. Just as acid–base reactions always move in the direction of the weaker acid–base pair (Section 15.2), redox reactions move toward the weaker oxidizing agent/reducing agent pair. In the Ni/H^+ reaction, there are two possible reducing agents, Ni(s) and H_2. Since nickel metal spontaneously reduces hydrogen ion, nickel metal must be a better reducing agent than H_2, and H^+(aq) must be a better oxidizing agent than Ni^{2+}(aq).

Four elements have been described thus far that could act as reducing agents: Zn, Cu, H_2, and Ni. What are their relative abilities to act in this manner? Conversely, what are the relative abilities of their ions (Zn^{2+}, Cu^{2+}, H^+, and Ni^{2+}) to act as electron acceptors or oxidizing agents? If we write half reactions involving these elements and their ions in the form

$$\text{Oxidized form} + ne^- \longrightarrow \text{reduced form}$$

and list them in order of descending $E°$ values, we will have placed the oxidizing agents in descending order of their ability to attract electrons.

Increasing strength as oxidizing agent ↑		Increasing strength as reducing agent ↓		
	Cu^{2+}(aq, 1 *M*) + 2e$^-$ ⟶ Cu(s)		$E° = +0.34$ V	$\Delta G° = -65.6$ kJ
	2 H^+(aq, 1 *M*) + 2e$^-$ ⟶ H_2(g, 1 atm)		$E° = 0.00$ V	$\Delta G° = 0.0$ kJ
	Ni^{2+}(aq, 1 *M*) + 2e$^-$ ⟶ Ni(s)		$E° = -0.25$ V	$\Delta G° = +48.3$ kJ
	Zn^{2+}(aq, 1 *M*) + 2e$^-$ ⟶ Zn(s)		$E° = -0.76$ V	$\Delta G° = +146.7$ kJ

The value of $E°$ becomes more negative, and $\Delta G°$ becomes more positive, down the series. This means that Cu^{2+}(aq) is the best oxidizing agent (most electron-attracting ion) of the substances on the left; that is, Cu^{2+} shows the greatest tendency to be reduced. Conversely, Zn^{2+} is the worst oxidizing agent; it is the least electron-attracting ion. Of the substances on the right, Zn(s) is the best reducing agent (best electron donor), since $E°$ for the half reaction

$$Zn(s) \longrightarrow Zn^{2+}(aq), 1\ M) + 2e^- \qquad E° = +0.76\ V \qquad \Delta G° = -146.7\ kJ$$

has the most positive value. By the same reasoning, Cu is the worst reducing agent.

The table of half reaction potentials above tells us that, at standard conditions, the following reactions are spontaneous:

Increasing oxidizing ability ↑		Increasing reducing ability ↓	
	Cu^{2+} can oxidize H_2, Ni, and Zn		H_2 can reduce Cu^{2+}
	H^+ can oxidize Ni and Zn		Ni can reduce H^+ and Cu^{2+}
	Ni^{2+} can oxidize Zn		Zn can reduce Ni^{2+}, H^+, and Cu^{2+}

Each of these reactions will have a positive $E°_{cell}$ and a negative $\Delta G°$. Many have been described in the text above, and the reduction of Cu^{2+}(aq) with Ni further illustrates the point.

Reducing agent	Ni(s) ⟶ Ni^{2+}(aq) + 2e$^-$	$E° = +0.25$ V
Oxidizing agent	Cu^{2+}(aq) + 2e$^-$ ⟶ Cu(s)	$E° = +0.34$ V
Overall reaction	Ni(s) + Cu^{2+}(aq) ⟶ Ni^{2+}(aq) + Cu(s)	$E°_{cell} = +0.59$ V

The spontaneity of this reaction is confirmed by the positive $E°$ (and negative $\Delta G°$) for the reaction.

We have just created a small portion of a **Table of Standard Reduction Potentials** (Table 19.1). Important points concerning this table are

Many more values of $E°$ are given in Appendix K.

1. The $E°$ values are for the reaction written in the form "oxidized form + electrons → reduced form."
2. When writing the reaction as "reduced form → oxidized form + electrons," the sign of $E°$ is reversed. Thus,

$$2\ F^-(aq, 1\ M) \longrightarrow 2e^- + F_2(g, 1\ atm) \qquad E° = -2.87\ V$$

TABLE 19.1 Standard Reduction Potentials in Aqueous Solution at 25°C*

REDUCTION HALF REACTION			$E°$ (V)
$F_2(g) + 2\ e^-$	$\longrightarrow$	$2\ F^-$	+2.87
$H_2O_2(aq) + 2\ H^+(aq) + 2\ e^-$	$\longrightarrow$	$2\ H_2O(\ell)$	+1.77
$PbO_2(s) + SO_4^{2-}(aq) + 4\ H^+(aq) + 2\ e^-$	$\longrightarrow$	$PbSO_4(s) + 2\ H_2O(\ell)$	+1.685
$Au^{3+}(aq) + 3\ e^-$	$\longrightarrow$	$Au(s)$	+1.50
$Cl_2(g) + 2\ e^-$	$\longrightarrow$	$2\ Cl^-(aq)$	+1.360
$O_2(g) + 4\ H^+(aq) + 4\ e^-$	$\longrightarrow$	$2\ H_2O(\ell)$	+1.229
$Br_2(\ell) + 2\ e^-$	$\longrightarrow$	$2\ Br^-(aq)$	+1.08
$Hg^{2+}(aq) + 2\ e^-$	$\longrightarrow$	$Hg(\ell)$	+0.855
$Ag^+(aq) + e^-$	$\longrightarrow$	$Ag(s)$	+0.80
$Hg_2^{2+}(aq) + 2\ e^-$	$\longrightarrow$	$2\ Hg(\ell)$	+0.789
$Fe^{3+}(aq) + e^-$	$\longrightarrow$	$Fe^{2+}(aq)$	+0.771
$O_2(g) + 2\ H_2O(\ell) + 4\ e^-$	$\longrightarrow$	$4\ OH^-(aq)$	+0.40
$Cu^{2+}(aq) + 2\ e^-$	$\longrightarrow$	$Cu(s)$	+0.337
$Sn^{4+}(aq) + 2\ e^-$	$\longrightarrow$	$Sn^{2+}(aq)$	+0.15
$2\ H^+(aq) + 2\ e^-$	$\longrightarrow$	$H_2(g)$ REFERENCE	0.00
$Sn^{2+}(aq) + 2\ e^-$	$\longrightarrow$	$Sn(s)$	−0.14
$Ni^{2+}(aq) + 2\ e^-$	$\longrightarrow$	$Ni(s)$	−0.25
$PbSO_4(s) + 2\ e^-$	$\longrightarrow$	$Pb(s) + SO_4^{2-}(aq)$	−0.356
$Cd^{2+}(aq) + 2\ e^-$	$\longrightarrow$	$Cd(s)$	−0.40
$Fe^{2+}(aq) + 2\ e^-$	$\longrightarrow$	$Fe(s)$	−0.44
$Zn^{2+}(aq) + 2\ e^-$	$\longrightarrow$	$Zn(s)$	−0.763
$2\ H_2O(\ell) + 2\ e^-$	$\longrightarrow$	$H_2(g) + 2\ OH^-(aq)$	−0.8277
$Al^{3+}(aq) + 3\ e^-$	$\longrightarrow$	$Al(s)$	−1.66
$Mg^{2+}(aq) + 2\ e^-$	$\longrightarrow$	$Mg(s)$	−2.37
$Na^+(aq) + e^-$	$\longrightarrow$	$Na(s)$	−2.714
$Li^+(aq) + e^-$	$\longrightarrow$	$Li(s)$	−3.045

Increasing strength of oxidizing agents (↑, left side)

Increasing strength of reducing agents (↓, right side)

*In volts (V) versus the standard hydrogen electrode (S.H.E.).

3. The more positive the value of $E°$, the better the oxidizing ability (the greater the tendency to be reduced) of the ion or compound. Thus, $F_2(g)$ is the best oxidizing agent in the table,

$$F_2(g,\ 1\ atm) + 2e^- \longrightarrow 2\ F^-(aq,\ 1\ M) \qquad E° = +2.87\ V$$

since the element has the most positive $E°$ of all the compounds or ions on the left. The ion at the bottom of the table, $Li^+(aq)$, is the poorest oxidizing agent, since its $E°$ is the most negative. Conversely, F^- is the weakest reducing agent and Li is the strongest reducing agent. Under standard conditions, the *oxidizing agents* (ions, elements, and compounds at the left) *increase in strength from the bottom to the top of the table*. The *reducing agents* (the ions, elements, and compounds at the right) *increase in strength from the top to the bottom*.

4. All the half reactions listed are reversible. A given substance can act as the anode or cathode, depending on conditions. For example, $H^+(aq)$ is reduced to $H_2(g)$ in Figure 19.5, whereas $H_2(g)$ is oxidized to $H^+(aq)$ in Figure 19.6.

5. Under standard conditions, any substance on the left in this table (the reduced form) will spontaneously oxidize any substance lower than it on the right (the reduced form). This point was explored with the abbreviated table above.

The same kind of information about redox reactions is conveyed in Table 19.1 as that about acid–base reactions in Table 15.4.

6. The algebraic sign of the half reaction voltage is the sign of the electrode when it is attached to a H_2/H^+ standard half-cell. This is illustrated in Example 19.3.
7. Electrode potentials are *intensive* properties (see page 11). Therefore, changing the stoichiometric coefficients for a half reaction does *not* change the value of $E°$. For example, the reduction of Fe^{3+} has an $E°$ of +0.771 V whether the reaction is written as

$$Fe^{3+}(aq, 1\ M) + e^- \longrightarrow Fe^{2+}(aq, 1\ M) \qquad E° = +0.771\ V$$

or as

$$2\ Fe^{3+}(aq, 1\ M) + 2e^- \longrightarrow 2\ Fe^{2+}(aq, 1\ M) \qquad E° = +0.771\ V$$

Recall that "volt" is defined as "energy" per "unit charge." Multiplying a reaction by some number causes *both* the energy *and* the charge to be multiplied by that number. Thus, the ratio energy/charge = volt does not change.

EXAMPLE 19.4

PREDICTING REACTION SPONTANEITY AND $E°$

For each of the two reactions below, predict whether or not the reaction is spontaneous, under standard conditions, in the direction in which it is written. Calculate $E°_{cell}$ for each.

(a) $2\ Al(s) + 3\ Sn^{4+}(aq) \longrightarrow 2\ Al^{3+}(aq) + 3\ Sn^{2+}(aq)$

(b) $Cd^{2+}(aq) + Cu(s) \longrightarrow Cd(s) + Cu^{2+}(aq)$

Solution *For reaction (a)*:

$$2\ [Al(s) \longrightarrow Al^{3+}(aq) + 3\ e^-] \qquad E° = +1.66\ V$$
$$3\ [Sn^{4+}(aq) + 2\ e^- \longrightarrow Sn^{2+}(aq)] \qquad E° = +0.15\ V$$
$$2\ Al(s) + 3\ Sn^{4+}(aq) \longrightarrow 2\ Al^{3+}(aq) + 3\ Sn^{2+}(aq) \qquad E°_{cell} = +1.81\ V$$

This reaction is predicted to be spontaneous as written (and thus to have a positive $E°$ and negative $\Delta G°$), because as indicated in Table 19.1, Al(s) is a stronger reducing agent than Sn^{2+}, and Sn^{4+} is a stronger oxidizing agent than Al^{3+}.

For reaction (b):

$$Cd^{2+}(aq) + 2\ e^- \longrightarrow Cd(s) \qquad E° = -0.40\ V$$
$$Cu(s) \longrightarrow Cu^{2+}(aq) + 2\ e^- \qquad E° = -0.34\ V$$
$$Cd^{2+}(aq) + Cu(s) \longrightarrow Cu^{2+}(aq) + Cd(s) \qquad E°_{cell} = -0.74\ V$$

The data in Table 19.1 indicate that Cu is a poorer reducing agent than Cd, and Cd^{2+} is a poorer oxidizing agent than Cu^{2+}. Therefore, the reaction is not spontaneous in the direction written, and the negative sign of $E°_{cell}$ confirms this conclusion. If an electrochemical cell is assembled from a Cu/Cu^{2+} half-cell and a Cd/Cd^{2+} half-cell, the observed reaction will be the opposite of that given in question (b).

$$Cu^{2+}(aq) + Cd(s) \longrightarrow Cd^{2+}(aq) + Cu(s) \qquad E° = +0.74\ V$$

EXERCISE 19.4 PREDICTING REACTION SPONTANEITY AND $E°$

For each of the following reactions at standard conditions, decide if it occurs spontaneously in the direction written and calculate $E°$.

(a) $Zn(s) + Sn^{2+}(aq) \longrightarrow Sn(s) + Zn^{2+}(aq)$

(b) $Br_2(aq) + 2\ Cl^-(aq) \longrightarrow 2\ Br^-(aq) + Cl_2(aq)$

EXAMPLE 19.5

CONSTRUCTING AN ELECTROCHEMICAL CELL

Using the half reactions $Fe(s)/Fe^{2+}(aq)$ and $Cu(s)/Cu^{2+}(aq)$, construct an electrochemical cell and predict its standard voltage.

Solution As a first step, decide which is the reducing agent and which is the oxidizing agent. The data in Table 19.1 show that Fe(s) is a better reducing agent than Cu, and Cu^{2+} is a better oxidizing agent than Fe^{2+}. Therefore, Fe will reduce $Cu^{2+}(aq)$. This means the half reactions and the overall reaction must be

Oxidation:	$Fe(s) \longrightarrow Fe^{2+}(aq) + 2e^-$	$E° = +0.44$ V
Reduction:	$Cu^{2+}(aq) + 2e^- \longrightarrow Cu(s)$	$E° = +0.34$ V
Net reaction:	$Fe(s) + Cu^{2+}(aq) \longrightarrow Fe^{2+}(aq) + Cu(s)$	$E°_{cell} = +0.78$ V

To construct a cell that operates under standard conditions, we shall have a solid iron electrode dipping into a solution of 1.0 *M* $Fe^{2+}(aq)$. Another compartment will contain a solid copper electrode in a solution that is 1.0 *M* in $Cu^{2+}(aq)$. An external wire connects the two electrodes, and a salt bridge will allow for ion flow. (Since virtually all salts are soluble as nitrates, we shall use $Cu(NO_3)_2$ and $Fe(NO_3)_2$ in the cell compartments, and $NaNO_3$ in the salt bridge.)

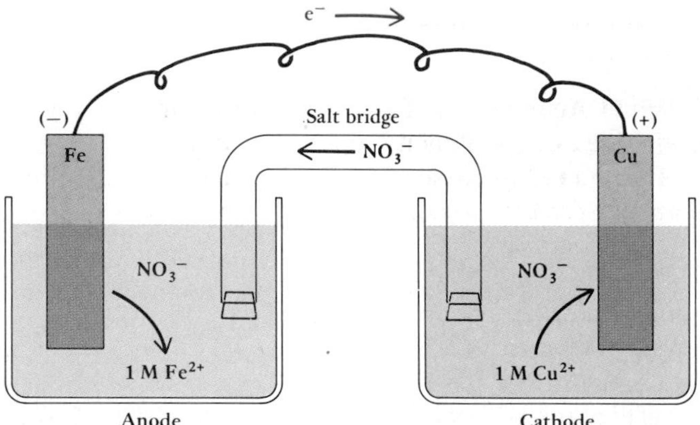

When the cell is assembled, electrons will flow from the Fe(s) anode to the Cu(s) cathode, and anions (NO_3^- in this case) will flow from the Cu^{2+} compartment to the Fe(s) electrode compartment.

EXERCISE 19.5 SETTING UP AN ELECTROCHEMICAL CELL

Draw a diagram of a working battery using the half-cells $Zn(s)/Zn^{2+}(aq)$ and $Al(s)/Al^{3+}(aq)$. Decide first on the net spontaneous reaction and predict its $E°$ value. Show the direction of electron flow in the external wire and the direction of ion flow in the salt bridge. Tell which compartment is the anode and which is the cathode.

OVERVOLTAGE

Batteries (and the electrolysis cells you will see below) that involve aqueous ions and compounds exhibit an effect called **overvoltage**, which arises especially for reactions that produce $H_2(g)$ and $O_2(g)$. The $E°$ values in

$E°$ AND DISCOMFORT FROM DENTAL FILLINGS

The most common filling material for tooth cavities is "dental amalgam," a solid solution of tin and silver in mercury. Dental amalgams are generally inert and cause few if any health problems. However, if you bite a piece of aluminum foil from a gum or candy wrapper with a filled tooth, your tooth nerves will complain! The sensation is caused by the nerves detecting the electron flow from an electrochemical reaction.

Dental amalgam consists of three phases, having compositions approximating Ag_2Hg_3, Ag_3Sn, and Sn_xHg (where x is 7 to 9). All three of these phases may undergo electrochemical reactions; for example,

$$3\,Hg_2^{2+}(aq) + 4\,Ag(s) + 6e^- \longrightarrow 2\,Ag_2Hg_3(s) \qquad E° = +0.85\text{ V}$$

$$Sn^{2+}(aq) + 3\,Ag(s) + 2e^- \longrightarrow Ag_3Sn(s) \qquad E° = -0.05\text{ V}$$

Aluminum is a much better reducing agent than any of the solid solutions above ($E°$ for $Al(s) \rightarrow Al^{3+}(aq) + 3e^-$ is $+1.66$ V). Therefore, if a piece of aluminum comes into contact with a dental filling, the saliva and gum tissue act as a salt bridge, and a short circuit results. The minute current produces a jolt of pain.

Table 19.1 (and Appendix K) for half-reactions involving H_2 and O_2 are based on the idea of spontaneity alone, without reference to the kinetic question of how rapidly these substances react. It is often the case that the mechanisms for their reactions prevent the reactions from occurring rapidly at the spontaneous $E°$ values. In such cases, the reactions occur at observable rates with voltages (signified by E without a superscript °) that are quite different from the spontaneous values listed in Table 19.1. The difference between the observed E and $E°$ is called the *overvoltage*, and it can amount to about -0.6 V for $H_2(g)$ and $O_2(g)$ evolution. Because of overvoltage, other half reactions, which are ideally less favorable (spontaneous), may actually occur in preference to reactions involving $H_2(g)$ and $O_2(g)$. This effect is another good example of kinetic control of reaction products, rather than thermodynamic control; its effect is to make the H_2 and O_2 reactions appear less spontaneous.

EXAMPLE 19.6

EFFECT OF OVERVOLTAGE ON REACTION SPONTANEITY

Using the $E°$ values in Table 19.1, list the following half reactions in decreasing order of spontaneity, most spontaneous first.

(a) $Al^{3+}(aq,\ 1\ M) + 3e^- \longrightarrow Al(s)$

(b) $2\,H_2O(\ell) + 2e^- \longrightarrow H_2(g,\ 1\text{ atm}) + 2\,OH^-(aq,\ 1\ M)$

(c) $Zn^{2+}(aq,\ 1\ M) + 2e^- \longrightarrow Zn(s)$

(d) $Hg(\ell) \longrightarrow Hg^{2+}(aq,\ 1\ M) + 2e^-$

(e) $2\,Cl^-(aq,\ 1\ M) \longrightarrow Cl_2(g,\ 1\text{ atm}) + 2e^-$

After making this list, repeat the process, but take into account that reaction (b) can have an overvoltage of -0.6 V.

Solution First, we shall refer to Table 19.1 and list the $E°$ values for each reaction above *as written*.

(a) -1.66 V (d) -0.85 V
(b) -0.83 V (e) -1.36 V
(c) -0.76 V

Now we can arrange the reactions in order of decreasing spontaneity, listing the reaction with the most positive (least negative) voltage first:

(c) > (b) > (d) > (e) > (a)

Now the effect of overvoltage can be taken into account. The overvoltage *inhibits* reaction (b) by -0.6 V. Therefore, the true voltage for (b) is

$$E \text{ for reaction (b)} = -0.83 - 0.6 = -1.4 \text{ V}$$

Due to the overvoltage, the new list is

(c) > (d) > (e) > (b) > (a)

In practice, overvoltage causes the oxidations of Hg and Cl^-(aq) to appear to be more spontaneous than the reduction of H_2O to H_2(g).

When base is added to aqueous Cr^{3+} (left), the solution turns green (middle). When an oxidizing agent (H_2O_2) is added, Cr^{3+}(aq) is oxidized to the yellow chromate ion (CrO_4^{2-}) (right). (Charles D. Winters)

SOME USES OF REDOX POTENTIALS IN THE INTRODUCTORY CHEMISTRY LABORATORY

Many of the laboratory experiments commonly done in the introductory course involve oxidation–reduction reactions, so it is interesting to see what voltages could be produced in such reactions and whether they are indeed predicted to be spontaneous under standard conditions.

One such reaction is the quantitative oxidation of an iron(II) salt by titration with the permanganate ion in acid solution.

$$5[Fe^{2+}(aq) \longrightarrow Fe^{3+}(aq) + e^-] \qquad E° = -0.77 \text{ V}$$
$$5e^- + 8\,H^+(aq) + MnO_4^-(aq) \longrightarrow Mn^{2+}(aq) + 4\,H_2O(\ell) \qquad E° = +1.52 \text{ V}$$
$$5\,Fe^{2+}(aq) + 8\,H^+(aq) + MnO_4^-(aq) \longrightarrow 5\,Fe^{3+}(aq) + Mn^{2+}(aq) + 4\,H_2O(\ell) \qquad E°_{cell} = +0.75 \text{ V}$$

The overall reaction is certainly spontaneous, and it proceeds essentially to completion. As a result, it is often used for the analysis of iron-containing samples.

If your laboratory program includes the qualitative analysis of solutions for the presence or absence of metal ions, you may have seen many examples of reactions involving electron transfer. An example already used in this chapter is the reduction of tin(IV) ion with aluminum (page 711). Yet another reaction is the hydrogen peroxide oxidation of chromium(III) hydroxide to the chromate ion, CrO_4^{2-}, in basic solution. The data in Appendix K predict these reactions to be spontaneous under standard conditions.

$$2\,[Cr(OH)_4^-(aq) + 4\,OH^-(aq) \longrightarrow CrO_4^{2-}(aq) + 4\,H_2O(\ell) + 3e^-] \qquad E° = +0.12 \text{ V}$$
$$3\,[H_2O_2(aq) + 2e^- \longrightarrow 2\,OH^-(aq)] \qquad E° = +0.88 \text{ V}$$
$$2\,Cr(OH)_4^-(aq) + 3\,H_2O_2(aq) + 2\,OH^-(aq) \longrightarrow 2\,CrO_4^{2-}(aq) + 8\,H_2O(\ell) \qquad E°_{cell} = +1.00 \text{ V}$$

Watch for other such reactions in your laboratory and see if they are predicted to be spontaneous based on the sign of $E°$.

EXERCISE 19.6 LABORATORY EXPERIMENTS AND $E°$

One method of determining the concentration of iron(II) in solution is to oxidize it quantitatively to iron(III) with the dichromate ion, $Cr_2O_7^{2-}$, in acid solution. Using the half-reactions in Appendix K, determine $E°$ for the overall reaction.

19.3 VOLTAIC CELLS AT NONSTANDARD CONDITIONS

In the previous section you learned how to predict the voltage produced by a spontaneous chemical reaction proceeding under standard conditions. Reactions in the real world are not always so accommodating; the concentrations of the ions are usually not 1 *M*. Even if the cell started out with all dissolved species at 1 *M* concentration, these would change as the reaction progressed, reactants decreasing in concentration and products increasing. Thus, we have to define the potential of the cell under nonstandard conditions.

THE NERNST EQUATION

The free energy for an oxidation–reduction reaction under standard conditions ($\Delta G°$) is proportional to the cell potential at these conditions (Equation 19.1). It should make sense that the free energy under some set of nonstandard conditions (ΔG without the superscript °) is proportional to the potential (E without the superscript °) under these same conditions, that is, $\Delta G = -nFE$. Standard conditions are just a special case of this more general relation.

Now we can relate the standard potential $E°$ to the nonstandard potential E using another equation from thermodynamics. In Chapter 18 we considered the general reaction

$$aA + bB \longrightarrow cC + dD$$

and wrote the equation

$$\Delta G = RT \ln Q - RT \ln K$$

where

$$Q\text{, the reaction quotient} = \frac{[C]^c[D]^d}{[A]^a[B]^b}$$

Refer to Section 18.7. Recall that Q has the same form as K_{eq}, but the concentrations are those of the ions in an actual electrochemical cell, and the cell is not at equilibrium if a net chemical reaction is occurring to produce a current and voltage.

Substituting for ΔG and using the fact that $-RT\ln K = \Delta G° = -nFE°$, we have

$$-nFE = -nFE° + RT \ln Q$$

which rearranges to

$$E = E° - (RT/nF)\ln Q$$

where R is the gas constant (8.3144 J/K · mol), F is the Faraday constant (9.65 × 10^4 J/V · mol), and n is the number of electrons transferred between oxidizing and reducing agents in a balanced redox reaction. When T is 298K (and when the logarithmic term is given as a base-10 log), we can write the **Nernst equation**.

$$E = E^\circ - \frac{0.0592}{n} \log Q \qquad (19.2)$$

This equation allows us to find the voltage produced by a simple cell under nonstandard conditions or to find the concentration of a reactant or product ion by measuring the voltage produced by a cell.

EXAMPLE 19.7

USING THE NERNST EQUATION

Determine the cell voltage for

$$Fe(s) + Cd^{2+}(aq) \longrightarrow Fe^{2+}(aq) + Cd(s)$$

when (a) $[Fe^{2+}] = 0.10\ M$ and $[Cd^{2+}] = 1.0\ M$ and (b) when $[Fe^{2+}] = 1.0\ M$ and $[Cd^{2+}] = 0.010\ M$.

Solution When the conditions are those in (a), we first calculate E°,

$$\begin{array}{ll} Fe(s) \longrightarrow Fe^{2+}(aq) + 2e^- & E^\circ = +0.44\ V \\ Cd^{2+}(aq) + 2e^- \longrightarrow Cd(s) & E^\circ = -0.40\ V \\ \hline Fe(s) + Cd^{2+}(aq) \longrightarrow Fe^{2+}(aq) + Cd(s) & E^\circ = +0.04\ V \end{array}$$

and then substitute into the Nernst equation to find E.

$$E = E^\circ - \frac{0.0592}{n} \log \frac{[Fe^{2+}]}{[Cd^{2+}]}$$

As in chemical equilibrium expressions (Chapter 14), the "concentration" of solid does not enter into the expression. Now, using $n = 2$ (the number of mole[s] of electrons transferred), and the ion concentrations given above,

$$E = +0.04\ V - \frac{0.0592}{2} \log \frac{(0.10)}{1.0} = +0.07\ V$$

Notice that the voltage is now larger than E°, so the tendency for Fe to transf[er] electrons to Cd^{2+} is even greater than under standard conditions.

For the conditions in part (b), the first step again is to calculate E°. Sin[ce] this is the same as in (a) above, we can proceed to set up and solve the Nern[st] equation for the new conditions.

$$E = +0.04\ V - \frac{0.0592}{2} \log \frac{1.0}{0.010} = -0.02\ V$$

Now we can tell from the equation that the cell potential is negative. This mea[ns] that the reaction does not proceed spontaneously in the direction in which it w[as] originally written, but in the opposite direction; iron(II) ion is reduced to metal iron by transfer of electrons from cadmium when $[Fe^{2+}] > [Cd^{2+}]$.

$$Fe^{2+}(aq) + Cd(s) \longrightarrow Fe(s) + Cd^{2+}(aq)$$

Walter Nernst (1864–1941) was a German physicist and chemist known especially for his work related to the third law of thermodynamics. (Francis Simon, AIP Niels Bohr Library)

EXERCISE 19.7 USING THE NERNST EQUATION

The aqueous tin(II) ion, Sn^{2+}, is commonly used in the laboratory as a reducing agent. However, its solutions are not stable in air, as the ion is readily oxidized by the oxygen in air to Sn^{4+}. What is the potential for the O_2 oxidation of 0.10 *M* Sn^{2+}(aq) dissolved in 0.10 *M* HCl when $P(O_2) = 0.20$ atm and Sn^{4+}(aq) has a concentration of 1.0×10^{-6} *M*?

$E°$ AND THE EQUILIBRIUM CONSTANT

We saw in Example 19.7 that the cell voltage, and even the reaction direction, can change when the concentrations of reactants and products change. Thus, as reactants are converted to products in a spontaneous reaction, the value of E_{cell} must decline from its initial positive value to eventually reach zero. A *potential of zero* means that *no net reaction* is occurring; it is an indication that *the cell has reached equilibrium*. Thus, when $E = 0$, the Q term in the Nernst equation is now equivalent to the equilibrium constant, K. So, when equilibrium has been attained, we can rewrite the Nernst equation as

$$E = 0 = E° - \frac{0.0592}{n} \log K$$

and so

$$\log K = \frac{nE°}{0.0592} \tag{19.3}$$

This is an extremely useful equation, since it tells you that the equilibrium constant for a reaction can be obtained from a calculation or measurement of $E°$.

EXAMPLE 19.8

THE NERNST EQUATION AND K

Calculate the equilibrium constant for the reaction

$$\text{Fe(s)} + \text{Cd}^{2+}\text{(aq)} \longrightarrow \text{Fe}^{2+}\text{(aq)} + \text{Cd(s)}$$

Solution The first step is to calculate $E°$. In this case, +0.0400 V was found in Example 19.7, so we use this $E°$ in the Nernst equation under equilibrium conditions.

$$\log K = \frac{(2.00 \text{ mol})(0.0400 \text{ V})}{0.0592 \text{ V} \cdot \text{mol}}$$

$$\log K = 1.35$$

$$K = 22$$

An interesting sidelight to this problem is the evaluation of the concentrations of Fe^{2+} and Cd^{2+} when the cell has reached equilibrium after starting at standard conditions. You can begin by writing the equilibrium expression.

$$K = 22 = \frac{[\text{Fe}^{2+}]}{[\text{Cd}^{2+}]}$$

The concentrations at standard conditions are 1.0 *M*. The amount of Cd^{2+} lost in the reaction and the amount of Fe^{2+} produced are both set equal to x.

$$K = 22 = \frac{1.0 + x}{1.0 - x}$$

Solving, we find $x = 0.91\ M$. Therefore, the equilibrium concentrations are

$$[Fe^{2+}] = 1.0 + x = 1.9\ M \qquad \text{and} \qquad [Cd^{2+}] = 1.0 - x = 0.1\ M$$

EXAMPLE 19.9

$E°$ AND K_{sp}

The table of reduction potentials can be used to estimate the equilibrium constant for an electron transfer reaction, as in the example above, as well as other types of reactions that are not redox reactions. For example, let us determine *K* for the reaction

$$AgBr(s) \longrightarrow Ag^+(aq) + Br^-(aq)$$

Solution In solving this problem, the trick is to find and combine two half-cell reactions that will produce the desired net equation. In this case, we can find a half reaction for the reduction of silver in AgBr (1) and a half reaction for the oxidation of silver to silver ion (2).

(1)	$AgBr(s) + e^- \longrightarrow Ag(s) + Br^-(aq)$	$E° = +0.10\ V$
(2)	$Ag(s) \longrightarrow Ag^+(aq) + e^-$	$E° = -0.80\ V$
(Net)	$AgBr(s) \longrightarrow Ag^+(aq) + Br^-(aq)$	$E° = -0.70\ V$

Now that $E°$ for the desired reaction is known, *K* can be calculated.

$$\log K = \frac{nE°}{0.0592} = \frac{(1.0\ \text{mol})(-0.70\ \text{V})}{0.0592\ \text{V} \cdot \text{mol}}$$

$$\log K = -12$$
$$K = 1 \times 10^{-12}$$

EXERCISE 19.8 $E°$ AND EQUILIBRIUM CONSTANTS

A disproportionation reaction is one in which identical molecules or ions react so that some molecules are oxidized and others are reduced. Calculate the equilibrium constant for the disproportionation of the mercury(I) ion. (*Hint:* both half reactions have $Hg_2^{2+}(aq)$ as the reactant.)

$$Hg_2^{2+}(aq) \longrightarrow Hg(\ell) + Hg^{2+}(aq)$$

If you make a 0.10 *M* solution of $Hg_2(NO_3)_2$, what are the equilibrium concentrations of Hg_2^{2+} and Hg^{2+}?

19.4 COMMON BATTERIES AND STORAGE CELLS

The voltaic cells we have described to this point can produce a useful voltage, but the voltage will decline rapidly as the reactant concentrations decline. For this reason, there has been great interest over the years in the design of usable batteries, voltaic cells that deliver current at a constant voltage.

SOME COMMON BATTERIES

THE COMMON DRY CELL The LeClanché or dry cell (Figure 19.7) is commonly used in flashlights, portable radios, and toys. The battery contains a carbon rod electrode inserted into a moist paste of NH_4Cl, $ZnCl_2$, and MnO_2 in a zinc can that serves as the anode.

$$\text{Anode reaction (oxidation):} \quad Zn(s) \longrightarrow Zn^{2+} + 2\,e^-$$

The electrons produced reduce the ammonium ion at a carbon cathode.

$$\text{Cathode reaction (reduction): } 2\,NH_4^+ + 2\,e^- \longrightarrow 2\,NH_3(g) + H_2(g)$$

The products of the cathode reaction are gases and would cause the sealed dry cell to explode if they were not removed by further reactions. This is the reason the battery contains manganese(IV) oxide, an oxidizing agent that consumes the hydrogen.

$$2\,MnO_2(s) + H_2(g) \longrightarrow Mn_2O_3(s) + H_2O(\ell)$$

and the NH_3 is taken up by zinc(II) ions.

$$Zn^{2+}(aq) + 2\,NH_3(g) + 2\,Cl^-(aq) \longrightarrow Zn(NH_3)_2Cl_2(s)$$

All these reactions lead to the net process below and produce a voltage of 1.5 V.

$$2\,MnO_2(s) + 2\,NH_4Cl(s) + Zn(s) \longrightarrow Zn(NH_3)_2Cl_2(s) + H_2O(\ell) + Mn_2O_3(s)$$

Unfortunately, there are at least two disadvantages to this battery. If current is drawn from the battery rapidly, the gaseous products cannot be consumed rapidly enough; the voltage drops as a result, although it is restored after standing. Furthermore, there is a spontaneous but slow direct reaction between the zinc electrode and ammonium ion that leads to further deterioration, and the battery has a poor "shelf life."

THE ALKALINE BATTERY The somewhat more expensive "alkaline" battery has come into use (Figure 19.8) because it avoids some of the problems of the common dry cell. An alkaline battery produces 1.54 V, and a key reaction is again the oxidation of zinc, this time under alkaline or basic conditions. The oxidation or anode reaction is

$$Zn(s) + 2\,OH^- \longrightarrow ZnO(s) + H_2O(\ell) + 2e^-$$

and the electrons produced by the anode are consumed by manganese(IV) oxide through the reduction or cathode half reaction.

$$2\,MnO_2(s) + H_2O(\ell) + 2e^- \longrightarrow Mn_2O_3(s) + 2\,OH^-$$

In contrast to the LeClanché battery, no gases are formed in the alkaline battery, and there is no decline in voltage under high current loads.

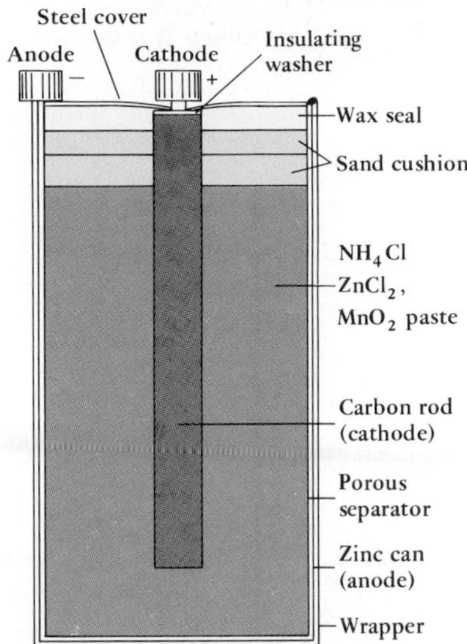

FIGURE 19.7
The LeClanché dry cell. The cell consists of a zinc anode (the battery container), a graphite cathode, and an electrolyte consisting of a moist paste of MnO_2 and NH_4Cl. The voltage developed by the cell is +1.5 V.

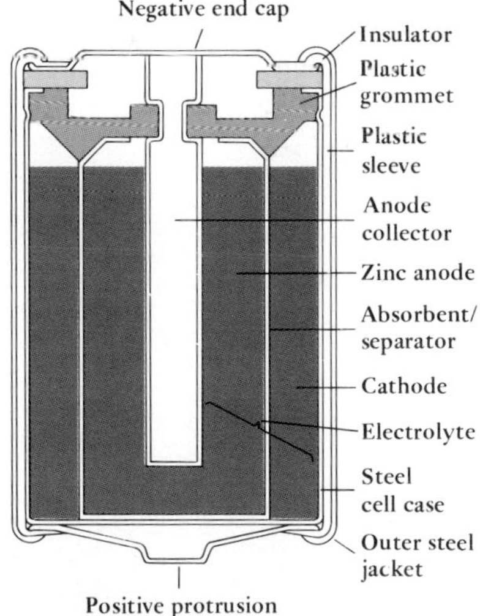

FIGURE 19.8
The alkaline battery. Like the LeClanché cell, this battery consists of a zinc anode, but the container is a steel case serving as a conductor of electrons to the oxidizing agent MnO_2. The battery produces 1.54 V.

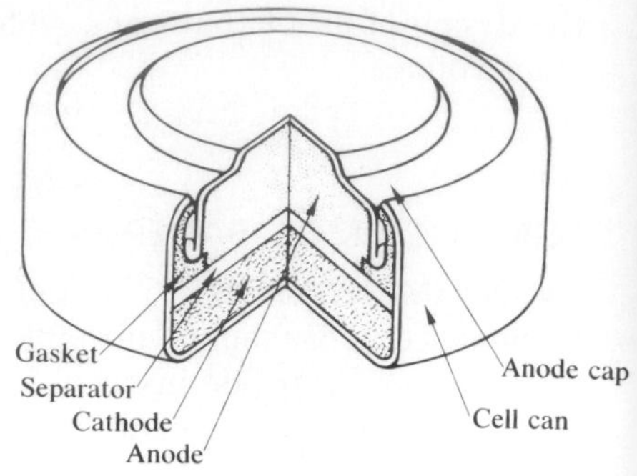

FIGURE 19.9
The mercury battery. (Eveready Batteries)

THE MERCURY BATTERY A close relative of the alkaline battery is the mercury cell (Figure 19.9), the type of battery used in calculators, cameras, watches, heart pacemakers, and other devices in which a small cell is required. As in the two previous batteries, the anode (the reducing agent) is metallic zinc, but the cathode (oxidizing agent) is mercury(II) oxide. These materials are tightly compacted powders separated by a moist paste of HgO containing some sodium or potassium hydroxide. Moistened paper serves as the "salt bridge." The battery produces a voltage of 1.35 V by the reactions

Anode (oxidation):

$$Zn(s) + 2\ OH^-(aq) \longrightarrow ZnO(s) + H_2O(\ell) + 2e^-$$

Cathode (reduction):

$$HgO(s) + H_2O(\ell) + 2e^- \longrightarrow Hg(\ell) + 2\ OH^-(aq)$$

These batteries are widely used, but, because they contain mercury, they can lead to some environmental problems. Mercury and its compounds are poisonous, so mercury cells should be reprocessed to recover the metal when the battery is no longer useful.

STORAGE BATTERIES

The LeClanché cell, the alkaline battery, and the mercury cell all will no longer produce a current when the chemicals inside have reached equilibrium conditions. They must be discarded. In contrast, **storage batteries** can be recharged. That is, the original reactant concentrations can be restored by reversing the net cell reaction by using an external source of the electrical energy. An automobile battery is perhaps the best example. That battery is used to supply the energy to the engine starter, but once the engine is running, the battery is recharged by current from the alternator, a mechanical device used to generate electricity.

THE LEAD STORAGE BATTERY There are two types of electrodes in a lead storage battery (Figure 19.10), one made of porous lead and the other of compressed, insoluble lead(IV) oxide. The electrodes, arranged alternately in a stack and immersed in moderately concentrated sulfuric acid are separated by thin fiberglass sheets. When the cell acts as a supplier of electrical energy, the lead electrode is oxidized to insoluble lead(II sulfate that adheres to the surface of the electrode.

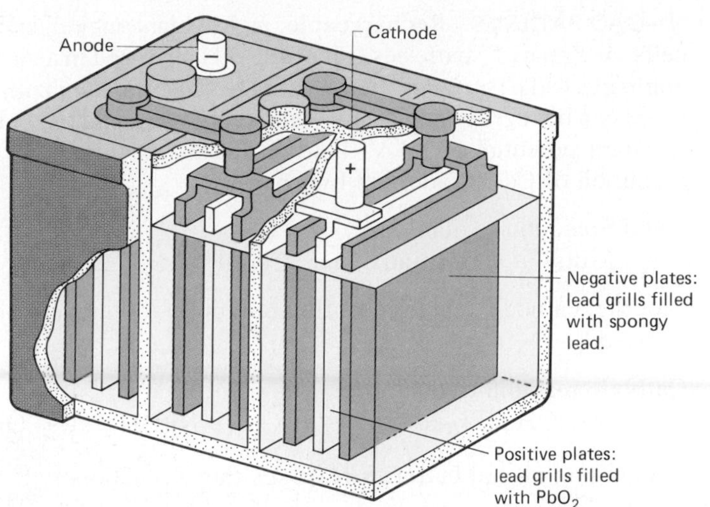

FIGURE 19.10
The lead storage battery.

Anode reaction (oxidation):

$$Pb(s) + SO_4^{2-}(aq) \longrightarrow PbSO_4(s) + 2e^-$$

The electrons move through the external circuit to the PbO_2 electrode where they cause reduction of the lead(IV) oxide.

Cathode reaction (reduction):

$$PbO_2(s) + 4\ H^+(aq) + SO_4^{2-}(aq) + 2e^- \longrightarrow PbSO_4(s) + 2\ H_2O(\ell)$$

The net result of using the cell to supply electrical energy is to coat both electrodes with an adhering film of white lead(II) sulfate and to consume the sulfuric acid.

Net process:

$$Pb(s) + PbO_2(s) + 2\ H_2SO_4(aq) \longrightarrow 2\ PbSO_4(s) + 2\ H_2O(\ell)$$

A lead storage battery can usually be recharged by supplying electrical energy to reverse the net process above. The film of lead(II) sulfate clinging to the electrodes is converted back to metallic lead and lead(IV) oxide at their respective electrodes, and sulfuric acid is regenerated.

Many attempts have been made to improve upon lead storage cells, because they are large and heavy and produce a relatively low power for their mass. Nonetheless, they do produce 2 V and an enormously large initial current. This is a combination that has been hard to surpass, and they remain in widespread use.

EXERCISE 19.9 $E°$ FOR THE LEAD STORAGE BATTERY

Using the half reaction potentials listed in Table 19.1, calculate $E°$, the standard potential, for the reaction in the lead storage battery. How does $E°$ compare with the actual value of the voltage produced by the battery?

FIGURE 19.11

The nickel–cadmium or "ni–cad" battery. (Eveready Battery Company)

NI–CAD BATTERIES Rechargeable, lightweight nickel–cadmium alkaline cells, or "ni–cad" batteries (Figure 19.11), are used in a variety of cordless appliances. These have the advantage that the oxidizing and reducing agent can be regenerated easily when recharged, and they produce a nearly constant potential of 1.4 V. During discharge, the anode reaction is the oxidation of Cd to cadmium hydroxide,

Anode reaction (oxidation):

$$Cd(s) + 2\,OH^-(aq) \longrightarrow Cd(OH)_2(s) + 2e^-$$

and the cathode reaction is the reduction of nickel ion to nickel(II) hydroxide.

Cathode reaction (reduction):

$$NiOOH(s) + H_2O(\ell) + e^- \longrightarrow Ni(OH)_2(s) + OH^-(aq)$$

Recharging ni–cad batteries reverses these reactions.

FUEL CELLS

Storage batteries contain the chemicals necessary for a reversible electrochemical reaction. When the reaction proceeds in the spontaneous direction, a current is produced until the chemicals reach equilibrium. Subsequent reversal of the spontaneous reaction by an external current source restores the original reactant concentrations. A **fuel cell** is also an electrochemical device, but, in contrast to a storage battery, it does not involve a reversible reaction in the practical sense; the reactants are continually supplied from an external reservoir.

The best known fuel cell is the hydrogen–oxygen cell (Figure 19.12), used in the Gemini, Apollo, and Space Shuttle programs. The net cell reaction is the very simple oxidation of hydrogen with oxygen to give water. Rather than allowing these gases to react directly and produce energy in the form of heat, they are made to react in such a way that the energy produced can be tapped by an electrical device. A stream of $H_2(g)$ is pumped onto the anode of the cell, and pure oxygen is directed to the cathode. Since the cell contains concentrated KOH, the reactions occur under basic conditions.

Anode (oxidation of H_2): $2\,H_2(g) + 4\,OH^-(aq) \longrightarrow 4\,H_2O(g) + 4e^-$

Cathode (reduction of O_2): $O_2(g) + 2\,H_2O(\ell) + 4e^- \longrightarrow 4\,OH^-(aq)$

The product water is swept out of the cell as a vapor in the hydrogen stream. Such a cell normally operates at approximately 70°C to 140°C and delivers about 0.9 V. (The cell heating indicates, by the way, that not all of the energy available has been tapped as electrical energy, so the efficiency is about 95%.)

EXERCISE 9.10 VOLTAGE PRODUCED BY A FUEL CELL

Hydrazine, N_2H_4, can be used as the reducing agent in a fuel cell.

$$N_2H_4(aq) + O_2(g) \longrightarrow N_2(g) + 2\,H_2O(\ell)$$

If $\Delta G°$ for this reaction is -607 kJ, calculate the theoretical voltage for the reaction from the relation between $\Delta G°$ and $E°$ (Equation 19.1).

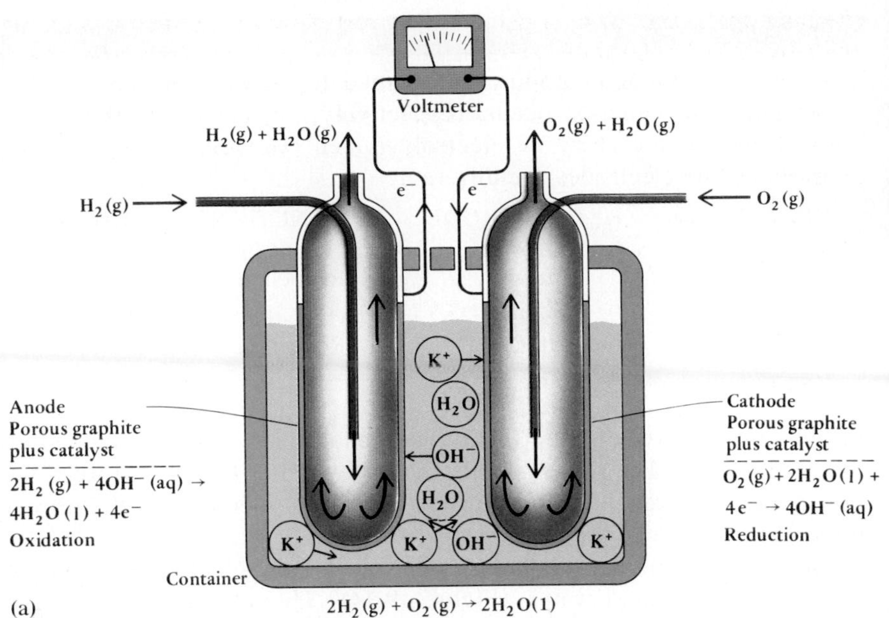

(b)

FIGURE 19.12
Fuel cells. (a) Schematic of a H_2/O_2 fuel cell. (b) A 4.5-megawatt fuel cell power plant in Tokyo. Unlike the fuel cells used in space vehicles, which have an alkaline (KOH) electrolyte, larger plants such as the one in this photo use phosphoric acid as the electrolyte. (Johnson-Matthey)

19.5 ELECTROLYSIS: CHEMICAL CHANGE FROM ELECTRICAL ENERGY

Thus far we have described the production of electric current from chemical reactions. Equally important, however, is the opposite process, **electrolysis**, the use of an electric current to bring about chemical change.

Let us consider first the manufacture of sodium and chlorine by electrolysis. In a very simple fashion, a cell having a pair of inert electrodes dipping into a bath of *molten* NaCl is depicted in Figure 19.13. (The fact that the NaCl is in the liquid state means the Na^+ and Cl^- ions are free to move in the melt.) The cell has been attached to a source of electric current, for example, a battery. The external voltage source acts like an "electron pump," and electrons flow from this source onto one of the electrodes, thereby giving it a negative charge. Sodium ions are attracted to this negative electrode and are reduced when electrons from the electrode are accepted, making it the cathode. The battery simultaneously draws electrons from the other electrode, giving it a positive electrical charge. Chloride ions are attracted to this electrode and surrender electrons. Since oxidation has occurred, this is the anode. The following reactions have thus occurred in molten NaCl:

Anode reaction (oxidation):	$2\ Cl^- \longrightarrow Cl_2(g) + 2\ e^-$
Cathode reaction (reduction):	$Na^+ + e^- \longrightarrow Na(s)$
Net reaction	$2\ Cl^- + 2\ Na^+ \longrightarrow 2\ Na(s) + Cl_2(g)$

The half-cell voltages listed in Table 19.1 apply only to aqueous solution. If we use these to *estimate* the voltage for the reaction above, we obtain a value of about −4 V. The reaction clearly does not occur spontaneously

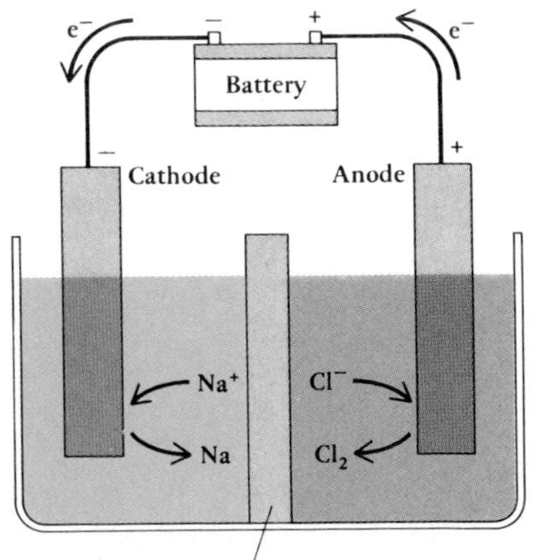

FIGURE 19.13
Electrolysis of molten sodium chloride.

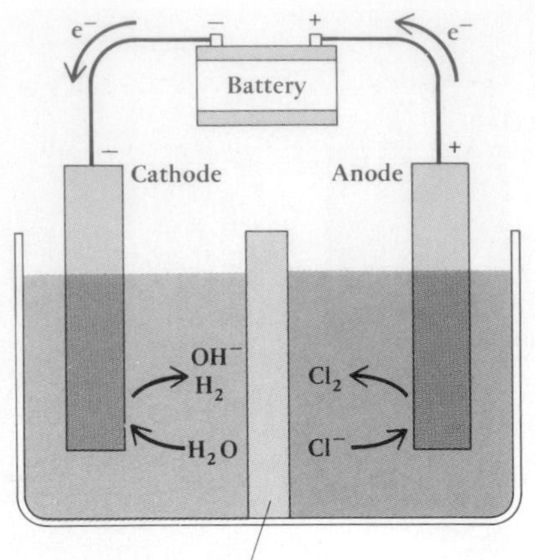

FIGURE 19.14

The electrolysis of aqueous sodium chloride.

ELECTROCHEMICAL TERMS

The terms anode and cathode always refer to the electrodes at which oxidation and reduction occur, respectively, no matter whether you are discussing a battery or electrolysis cell. However, the charges carried by the electrodes are different.

TYPE OF CELL	ELECTRODE	FUNCTION	CHARGE
Battery	Anode	Oxidation	−
	Cathode	Reduction	+
Electrolysis	Anode	Oxidation	+
	Cathode	Reduction	−

in the direction written, and this is the reason that we have attached an external battery. The battery, with a voltage greater than 4 V, forces the nonspontaneous reaction to occur by "pumping" electrons in the proper direction.

What if you use an *aqueous solution* of sodium chloride instead of a molten salt (without water) (Figure 19.14)? The left-hand electrode in the cell is again the cathode, the site of the reduction half reaction, and the right-hand one is the anode, the site of the oxidation half reaction. Electrons are pumped onto the cathode by an external voltage source. With water now present, we must ask whether it is still Na^+ and Cl^- ions that are reduced and oxidized, respectively, or whether water might be involved. To answer this question, we can compare $E°$ values to determine the relative spontaneities of the half reactions. For reduction, we have the possibilities

$$Na^+(aq) + e^- \longrightarrow Na(s) \qquad E° = -2.71 \text{ V}$$

$$2\,H_2O(\ell) + 2\,e^- \longrightarrow H_2(g) + 2\,OH^-(aq) \qquad E = -0.42 \text{ V}$$

The Nernst equation (Equation 19.2) has been used to convert $E°$ (−0.83 V) for the water half reaction, which assumes $[OH^-] = 1$ *M*, to E for $[OH^-] = 10^{-7}$ *M* in pure water. The correction is

$$-(0.0592/2)\log[OH^-]^2 = +0.41 \text{ V}$$

Even allowing for the effect of overvoltage for H_2 evolution at the electrode, it is clear that H_2O will be reduced to $H_2(g)$, rather than $Na^+(aq)$ to Na(s). That is, the H_2O/H_2 half reaction has a less negative potential than the $Na^+(aq)/Na(s)$ half reaction.

For the oxidation processes possible in aqueous sodium chloride, we need to compare

$$2\,Cl^-(aq) \longrightarrow Cl_2(g) + 2\,e^- \qquad E° = -1.36 \text{ V}$$

and

$$2\,H_2O(\ell) \longrightarrow O_2(g) + 4\,H^+(aq) + 4\,e^- \qquad E = -0.82 \text{ V}$$

[where we have corrected $E°$ for water oxidation to E (for $[H^+] = 10^{-7}$ *M*)]. Remembering that the overvoltage for $O_2(g)$ evolution is high (−0.6 V) and makes water oxidation *much* more difficult, we conclude that it is $Cl^-(aq)$ that is oxidized. Therefore, the reactions occurring in the aqueous sodium chloride cell are those with the least negative potentials.

Anode (oxidation):	$2\,Cl^-(aq) \longrightarrow Cl_2(g) + 2\,e^-$	$E° = -1.36$ V
Cathode (reduction):	$2\,H_2O(\ell) + 2\,e^- \longrightarrow H_2(g) + 2\,OH^-(aq)$	$E = -0.42$ V
Net reaction	$2\,Cl^-(aq) + 2\,H_2O(\ell) \longrightarrow H_2(g) + OH^-(aq) + Cl_2(g)$	$E_{cell} = -1.78$ V

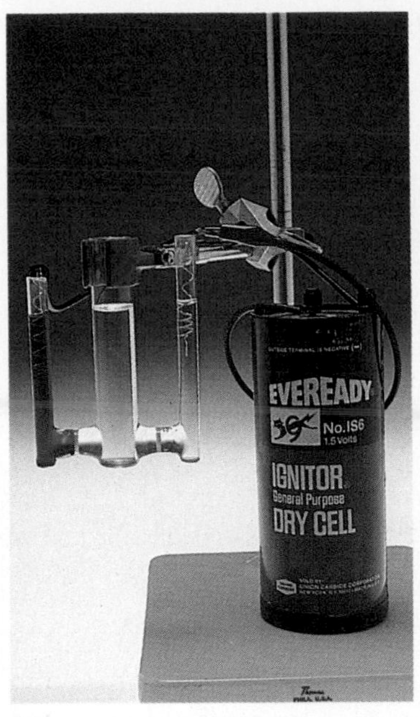

(a)

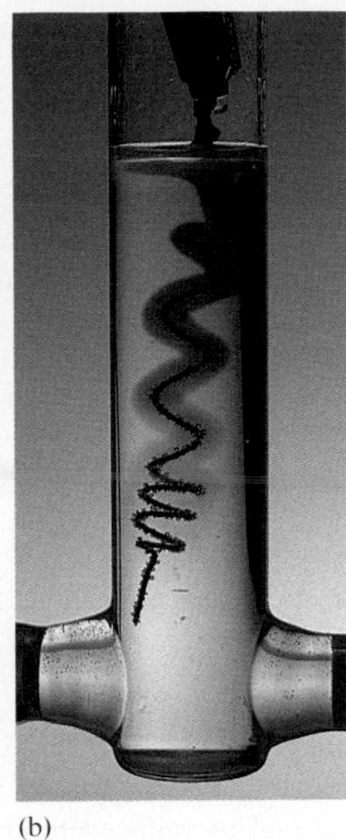

(b)

FIGURE 19.15

The electrolysis of aqueous potassium iodide. Aqueous KI is contained in all three compartments of the cell, and both electrodes are platinum. (a) At the positive electrode or anode (right), the I^- ion is oxidized to iodine, which gives the solution a brown color.

$$2I^-(aq) \longrightarrow I_2(aq) + 2e^-$$

At the negative electrode or cathode (left), water is reduced, and presence of the OH^- ion is indicated by the red color of phenolphthalein.

$$2\,H_2O(\ell) + 2\,e^- \longrightarrow H_2(g) + 2\,OH^-(aq)$$

(b) In a close-up of the cathode, bubbles of H_2 and evidence of OH^- are clearly seen. (Charles D. Winters)

Next, let us change the sodium chloride solution above to an aqueous solution of some other metal halide like $CuCl_2$ or KI (Figure 19.15). As before, consult Table 19.1, and, considering all possible reactions, find the oxidation and reduction reactions that require the least potential. Again, for $CuCl_2(aq)$, the only two possible oxidations are those of $Cl^-(aq)$ to $Cl_2(g)$ and H_2O to $O_2(g)$. As before, we expect the former. Unlike the aqueous sodium chloride cell, however, $Cu^{2+}(aq)$ is much more easily reduced ($E^\circ = +0.34$ V) than water ($E = -0.42$ V at $[OH^-] = 10^{-7}\,M$). Therefore, the products of the electrolysis of the aqueous $CuCl_2$ solution should be $Cl_2(g)$ and Cu(s).

Anode (oxidation):	$2\,Cl^-(aq) \longrightarrow Cl_2(g) + 2\,e^-$	$E^\circ = -1.36$ V
Cathode (reduction):	$Cu^{2+}(aq) + 2\,e^- \longrightarrow Cu(s)$	$E^\circ = +0.34$ V
Net reaction:	$Cu^{2+}(aq) + 2\,Cl^-(aq) \longrightarrow Cu(s) + Cl_2(g)$	$E^\circ_{cell} = -1.02$ V

Again, this process is of obvious commercial importance. The copper that is used in wiring, in coins, and for other purposes is purified by electrolysis (Chapter 25).

From the discussion above, we can arrive at a useful general principle. If you pass an electric current through a solution, the electrode reactions will be most likely those requiring the least potential to overcome their nonspontaneity. In water, this means that *a substance will be reduced if it has a reduction potential less negative than about −0.42 V*, the potential for the reduction of pure water. A check of Table 19.1 (and Appendix K) shows that this includes commercially useful metals such as Pt, Cu, Ag, Au, and Cd. Indeed, the electrolysis method is used to coat or "plate" other materials with these metals. *If a substance has a reduction potential more negative than −0.42 V, then only water is reduced.* Substances falling into this category include Na, K, Mg, and Al.

The electrolyses of molten NaCl and of aqueous sodium chloride are both commercial processes. (Section 19.7).

To produce these metals requires methods other than the reduction of their ions in aqueous solution.

EXAMPLE 19.10

ELECTROLYSIS OF AQUEOUS NaOH

Predict the result of passing an electric current into an aqueous solution of NaOH.

Solution First, list all of the species in solution. In this case, they will be Na^+, OH^-, and H_2O. Then, using Table 19.1, decide which of the species in solution can be oxidized and which reduced and note the potential of each possible reaction.

Reductions:

$$Na^+(aq) + e^- \longrightarrow Na(s) \qquad E° = -2.71\ V$$

$$2\ H_2O(\ell) + 2\ e^- \longrightarrow H_2(g) + 2\ OH^-(aq) \qquad E° = -0.83\ V$$

Oxidations:

$$4\ OH^-(aq) \longrightarrow O_2(g) + 2\ H_2O(\ell) + 4\ e^- \qquad E° = -0.40\ V$$

$$2\ H_2O(\ell) \longrightarrow O_2(g) + 4\ H^+(aq) + 4\ e^- \qquad E° = -1.23\ V$$

It is evident that water will be reduced to H_2 at the cathode (even considering the problem of overvoltage), and OH^- will be oxidized to O_2 at the anode. The net cell reaction is $2\ H_2O(\ell) \rightarrow 2\ H_2(g) + O_2(g)$ and the potential under standard conditions is -1.23 V (see Chapter 20).

We should note that this is again a process of some commercial importance, since it produces both hydrogen and oxygen, both useful fuels in rockets and for industrial purposes.

EXERCISE 19.11 ELECTROLYSIS OF SALTS
Predict the results of passing an electric current into each of the following solutions: (a) molten NaBr, (b) aqueous NaBr, and (c) aqueous $SnCl_2$.

19.6 ELECTRICAL ENERGY

Metallic sodium is produced at the cathode in the electrolysis of NaCl, the reaction being

$$Na^+ + e^- \longrightarrow Na(s)$$

One mole of electrons is required to produce one mole of sodium from one mole of sodium ions. In contrast, two moles of electrons are required to produce one mole of copper.

$$Cu^{2+}(aq) + 2e^- \longrightarrow Cu(s)$$

It follows from this that, if you could measure the number of moles of electrons flowing through the electrolysis cell, you would know the number of moles of sodium or copper produced. Conversely, if you knew the amount of sodium or copper produced, you could calculate the number of moles of electrons used.

The moles of electrons consumed or produced in a redox reaction are usually obtained by measuring the current flowing in the external electrical circuit in a given time. The **current** flowing in an electrical circuit is the amount of charge (in units of coulombs) passed in unit time, and the usual unit for current is **amperes**.

$$\text{Current} = \text{electrical charge passing/time}$$
$$\text{I(amps)} = \text{coulombs/seconds}$$

Michael Faraday first explored the quantitative aspects of electricity. In his honor scientists have defined the **Faraday** as the charge carried by one mole of electrons. This quantity can be calculated, since the charge on one electron is known (see Chapters 2 and 7), and you know that there is Avogadro's number of electrons in a mole.

Charge on 1 mole of electrons
$= (1.6022 \times 10^{-19}$ coulombs/electron$)(6.022045 \times 10^{23}$ electrons/mole$)$
$= 9.6485 \times 10^{4}$ coulombs/mole
$= 1$ Faraday or $1\ F$

The number, **9.65×10^4 coulomb/mole of electrons**, is called the **Faraday constant** and is often abbreviated *F*. From the relation between charge, current, and time, you can obtain the charge by multiplying the current (in amperes) by the time (in seconds) during which that current passed. These quantities are easily measured with modern instruments. Knowing the charge, and using the Faraday constant as a conversion factor, you can then obtain the moles of electrons that passed through the cell.

EXAMPLE 19.11

USING THE FARADAY CONSTANT

Assume that 1.50 amperes of current flow through a solution containing silver ions for 15.0 minutes. The voltage is such that silver is deposited at the cathode. How many grams of silver metal are deposited?

$$Ag^{+}(aq) + e^{-} \longrightarrow Ag(s)$$

Solution Our objective is to obtain the mass of silver, so we should aim our sequence of calculations to obtain moles of silver. The half reaction tells you that, if 1 mole of electrons is passed, then 1 mole of silver is obtained. Thus, we must find the charge passed, because this will give us moles of electrons. The charge comes from the relation between current and time, the two factors given by the experiment. Thus, the calculation sequence here is

$$\text{Time} \times \text{current} \longrightarrow \text{charge} \longrightarrow \text{moles } e^{-} \longrightarrow \text{moles Ag} \longrightarrow \text{mass Ag}$$

(a) Calculate the number of coulombs of charge passed in 15.0 minutes.

$$\begin{aligned}\text{Charge(coulombs)} &= \text{amps} \times \text{time(sec)}\\ &= 1.50\ \text{amps}(15.0\ \text{min})(60.0\ \text{s/min})\\ &= 1.35 \times 10^{3}\ \text{coulombs}\end{aligned}$$

(b) Calculate the number of moles of electrons, that is, the number of Faradays of electricity.

$$1.35 \times 10^3 \text{ coulombs} \left(\frac{1 \text{ mole e}^-}{9.65 \times 10^4 \text{ coulombs}}\right) = 1.40 \times 10^{-2} \text{ moles e}^-$$
$$= 1.40 \times 10^{-2} \text{ Faradays}$$

(c) Calculate the moles of Ag^+ and then the mass of silver deposited. When passing 1.40×10^{-2} moles of electrons through the cell, 1.40×10^{-2} moles of Ag(s) must have been formed. From this, we obtain the mass of silver deposited on the electrode.

$$1.40 \times 10^{-2} \text{ mol Ag} \left(\frac{108 \text{ g}}{1 \text{ mol}}\right) = 1.51 \text{ g Ag}$$

EXAMPLE 19.12

USING THE FARADAY CONSTANT

One of the half reactions occurring in the lead storage battery (Section 19.4) is

$$Pb(s) + SO_4^{2-}(aq) \longrightarrow PbSO_4(s) + 2\,e^-$$

If the battery delivers 1.50 amperes, and if its Pb electrode contains 1.00 pound of lead, how long can current flow before the lead in the electrode is used up?

Solution Here we want to know time, and we begin with information on mass (i.e., moles) and current. This is just the reverse of the previous problem. That is, our conversion sequence here should be

$$\text{Mass} \longrightarrow \text{moles} \longrightarrow \text{charge} \longrightarrow \text{time}$$

(a) Calculate the number of moles of lead. Since 1.00 lb. of Pb = 454 g,

$$454 \text{ g} \left(\frac{1 \text{ mol}}{207.2 \text{ g}}\right) = 2.19 \text{ mol Pb}$$

(b) Calculate the number of moles of electrons produced by the available lead.

$$2.19 \text{ mol Pb} \left(\frac{2 \text{ mol e}^-}{1 \text{ mol Pb}}\right) = 4.38 \text{ mol e}^- = 4.38 \text{ Faradays}$$

(c) Calculate the number of coulombs of charge carried by the moles of electrons produced.

$$4.38 \text{ mol e}^- \left(\frac{9.65 \times 10^4 \text{ coulombs}}{1 \text{ mol e}^-}\right) = 4.23 \times 10^5 \text{ coulombs}$$

(d) Using the charge produced, and its relation to current and time, calculate time.

$$4.23 \times 10^5 \text{ coulombs} = 1.50 \text{ amps} \times \text{time(secs)}$$
$$\text{Time} = 2.82 \times 10^5 \text{ seconds (or 78.3 hours)}$$

EXERCISE 19.12 USING THE FARADAY CONSTANT

In the commercial production of sodium by electrolysis (see Section 19.7), the cell operates at 7.0 V and a current of 25×10^3 amps. How many grams of sodium can be produced in one hour?

The appliances in your home are often rated in watts, a unit that tells you the *rate* of energy consumption or production; that is, the **watt** has units of energy/time and is defined as

$$1 \text{ watt} = 1 \text{ joule/second}$$

It is related to electrical units by the fact that a volt is defined as 1 joule/coulomb (Section 19.2). The joule is a relatively small unit of energy, so it is more common to talk about power, the rate of energy consumption or production, in terms of *kilowatt-hours* (kwh).

$$1 \text{ kwh} = (1000 \text{ watts})(1 \text{ hr}) \left(\frac{3600 \text{ s}}{\text{hr}}\right) \left(\frac{1 \text{ J/s}}{\text{watt}}\right) = 3.60 \times 10^6 \text{ J}$$

Energy production is obviously important in the design of batteries and power plants, just as energy consumption is important in the production of metals by electrolysis in a chemical plant. The application of energy calculations in a chemical plant is illustrated in Example 19.13.

EXAMPLE 19.13

ENERGY EXPENDITURE IN AN ELECTROCHEMICAL PROCESS

Hydrogen has great promise as a fuel in our economy, and, as illustrated in Example 19.10, it can be produced by electrolysis of water (Figure 20.2). Electrolyzing a solution of sulfuric acid will lead to the production of $H_2(g)$ at the cathode and $O_2(g)$ at the anode. If the minimum voltage required is 1.24 V (no overvoltage is considered), calculate the number of kilowatt-hours of energy required to produce 1.00 kilogram of $H_2(g)$.

Solution In this problem the conversion sequence is

$$\text{Mass} \longrightarrow \text{moles} \longrightarrow \text{charge} \longrightarrow \text{energy}$$

(a) Calculate the number of coulombs required. One mole of $H_2(g)$ will require 2 moles of electrons.

$$2 \text{ H}^+(aq) + 2e^- \longrightarrow H_2(g)$$

$$1.00 \times 10^3 \text{ g} \left(\frac{1 \text{ mol } H_2}{2.02 \text{ g } H_2}\right) = 4.95 \times 10^2 \text{ moles of } H_2$$

$$4.95 \times 10^2 \text{ mol } H_2 \left(\frac{2 \text{ mol } e^-}{1 \text{ mol } H_2}\right) = 9.90 \times 10^2 \text{ moles } e^-$$

$$9.90 \times 10^2 \text{ mol } e^- \left(\frac{9.65 \times 10^4 \text{ coulombs}}{1 \text{ mol } e^-}\right) = 9.55 \times 10^7 \text{ coulombs}$$

(b) Calculate the number of joules of energy required.

$$\text{Number of joules} = \text{coulombs} \times \text{volts} = (9.55 \times 10^7 \text{ coulombs})(1.24 \text{ V}) = 1.18 \times 10^8 \text{ joules}$$

(c) Calculate the number of kilowatt-hours.

$$\text{kwh} = 1.18 \times 10^8 \text{ J} \left(\frac{1 \text{ kwh}}{3.60 \times 10^6 \text{ J}}\right) = 32.9 \text{ kilowatt-hours}$$

EXERCISE 19.13 ENERGY EXPENDITURE IN SODIUM PRODUCTION

Using the information from Exercise 19.12, calculate the kilowatt-hours of energy used per hour in the production of sodium by electrolysis.

19.7 THE COMMERCIAL PREPARATION OF CHEMICALS BY ELECTROCHEMICAL METHODS

ALUMINUM PRODUCTION

Aluminum is the third most abundant element in the earth's crust and has many important uses in our economy. You probably know it best in its use in the kitchen as a food wrapper, a use demonstrating its excellent formability. Just as importantly, aluminum has a low density and excellent corrosion resistance; the latter property comes from the fact that a transparent, chemically inert film of aluminum oxide clings tightly to the metal's surface. It is these properties that have led to the many other uses of aluminum in aircraft parts, ladders, and automobile parts, for example.

As with some other metals of commercial importance, the history of the development of a practical method for aluminum production is interesting. Aluminum was originally made in the 19th century by reducing $AlCl_3$ with sodium,

The aluminum that caps the Washington Monument was made by the sodium-reduction method.

$$3\ Na(s) + AlCl_3(s) \longrightarrow Al(s) + 3\ NaCl(s)$$

but only at a very high cost. It was therefore considered a precious metal, chiefly used in jewelry. In fact, in the 1855 Exposition in Paris (France), some of the first aluminum metal produced was exhibited along with the crown jewels of France. Napoleon III saw its possibilities for military use, however, and commissioned studies on improving its production. The French had a ready source of aluminum-containing ore (called bauxite, since it came from the tiny medieval fortress town of Les Baux in southern France), and in 1886, a 23 year old Frenchman, Paul Heroult, conceived the electrochemical method in use today. In an interesting coincidence, an American, Charles Hall, who was also 23 years old, announced his invention of the identical process in the same year. Hence, the commercial process is now known as the Hall–Heroult process.

Charles Martin Hall was only 22 years old in 1886 when he discovered the electrolytic process for extracting aluminum from Al_2O_3 in a woodshed behind his family's home in Oberlin, Ohio. Following his discovery, Hall went on to found the company that eventually became Aluminum Corporation of America. He died a multimillionaire in 1914.

The essential features of the Hall–Heroult process are illustrated in Figure 19.16. The aluminum-containing ore, chiefly in the form of Al_2O_3, is mixed with cryolite, Na_3AlF_6, at about 980°C and electrolyzed. Aluminum is produced at the cathode and oxygen at the anode, both electrodes being graphite. The cells operate at the very low voltage of 4.0 to 5.5 V, but at a current of 50,000 to 150,000 amps. Each pound of aluminum produced requires 6 to 8 kilowatt-hours of energy, exclusive of that required to maintain the heat of the furnace. This is why there has been so much interest in recycling soft drink cans and other aluminum objects. The recycled metal can be purified and made into new objects at a fraction of the cost of making aluminum from the ore.

EXERCISE 19.14 ENERGY AND POWER IN ALUMINUM PRODUCTION

Given that an aluminum electrolysis cell operates at 5.0 V and a current of 1.0×10^5 amps, calculate the power required to produce 1 ton (2.0×10^3 pounds) of aluminum.

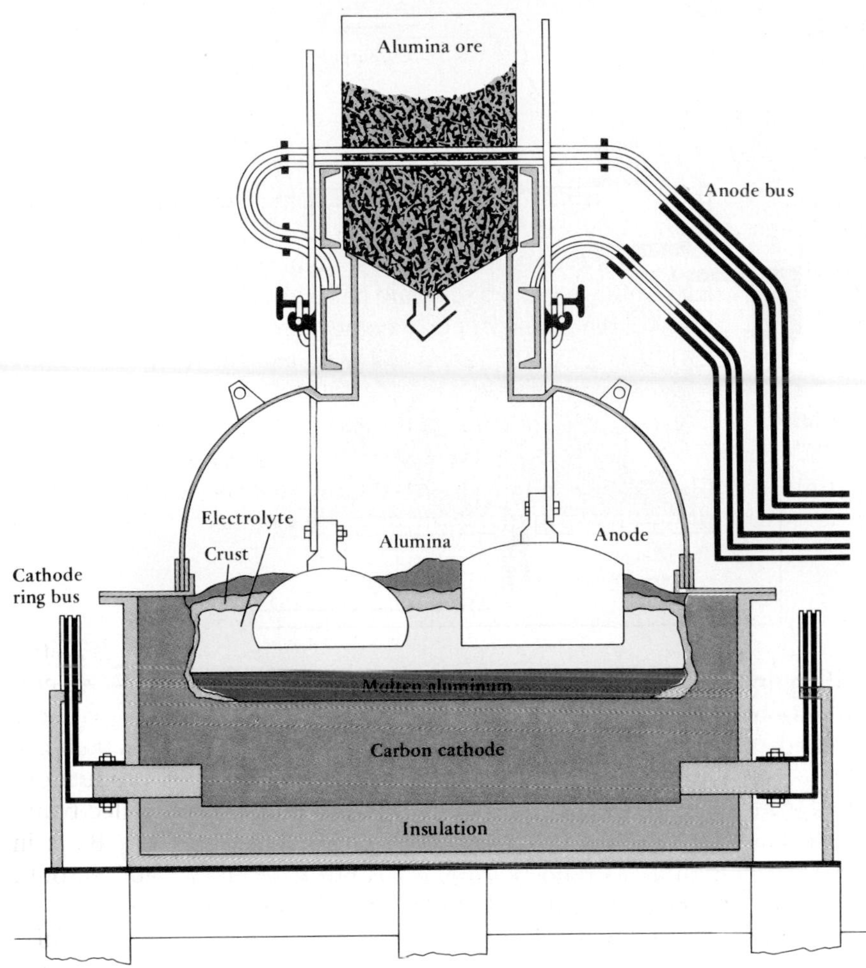

FIGURE 19.16

Industrial production of aluminum by electrolysis. The aluminum-containing ore, essentially Al_2O_3, is mixed with cryolite (Na_3AlF_6) to give a mixture that melts at a lower temperature than Al_2O_3 alone. The aluminum-containing materials are reduced at the graphite cathode to give molten aluminum. Oxygen is produced at the carbon anode; the gas reacts slowly with the carbon to give CO_2, leading to eventual loss of the electrode.

SODIUM PRODUCTION

The English chemist Sir Humphrey Davy first isolated sodium in 1807 by the electrolysis of molten sodium hydroxide. However, the element was a laboratory curiosity until, in 1824, it was found to reduce aluminum chloride to metallic aluminum (see above). The possibility of producing commercially useful aluminum led to considerable interest in manufacturing sodium. By 1886 a practical method for sodium manufacture had been devised (the reduction of NaOH with carbon), but in the same year Hall and Heroult invented the electrolytic method for aluminum production. For those working on sodium production, the unfortunate consequence of the Hall–Heroult invention was to remove the primary market for sodium, at least temporarily. However, other uses developed, and by 1921 the Downs process for sodium manufacture allowed the production of several million pounds each year. A less expensive process for producing a metal leads to more new uses, and sodium soon found a market in the manufacture of tetraethyllead, a compound added to gasoline in small concentrations to raise the octane rating. We now know that this

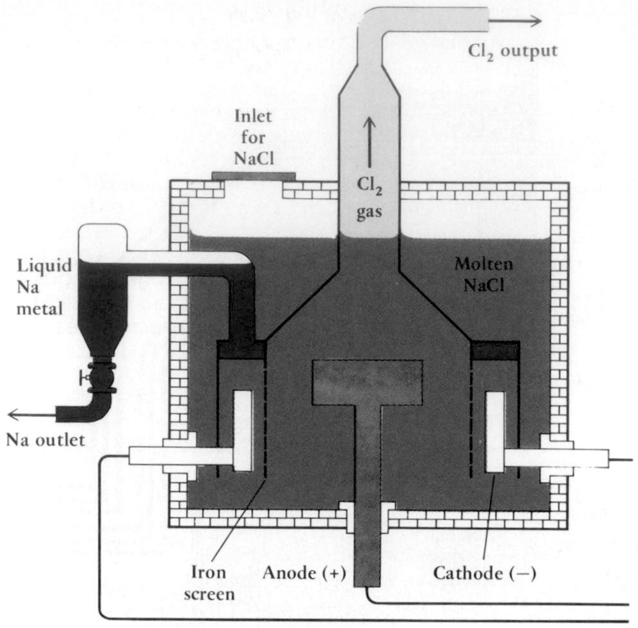

FIGURE 19.17

The Downs cell for the electrolysis of molten NaCl. The circular iron cathode is separated from the graphite anode by an iron screen. Since the cell operates at about 600°C, the sodium is produced in the molten state. The liquid metal has a low density, so it floats and can be drawn off. The chlorine gas bubbles out of the anode compartment and is collected.

lead compound has a great environmental impact; its use is being rapidly phased out, and so, as a result, is this use of sodium. However, sodium still finds commercial application in making alloys and in street lights.

Sodium-containing lights are effective because electrically excited sodium atoms emit a bright yellow light that is useful for street lighting in foggy areas. (See photograph beginning Chapter 20.)

The Downs cell for sodium manufacture by the electrolysis of molten NaCl (Figure 19.17) operates at 7 to 8 volts and currents of 25,000 to 40,000 amps. The cell is filled with a mixture of dry NaCl and $CaCl_2$ approximately in a 1:3 ratio. Pure NaCl is not used because its melting point is 800°C. As you learned in Chapter 12, however, the melting point of a solvent can be lowered by adding a solute; in this case addition of $CaCl_2$ to NaCl gives a mixture melting at approximately 600°C.

The sodium in the Downs cell is produced at a copper or iron cathode that surrounds a circular graphite anode. Immediately over the cathode there is an inverted trough in which the low density, molten sodium collects (melting point of pure sodium = 97.83°C), and the by-product, gaseous Cl_2, is collected in an inverted cone that extends through the molten salt mixture and almost to the level of the anode.

THE MANUFACTURE OF CHLORINE AND SODIUM HYDROXIDE

Chlorine is used to treat water and sewage and in the production of organic chemicals such as pesticides and vinyl chloride, the building block of plastics called PVC's (polyvinyl chlorides). In 1984 chlorine was ninth on the list of chemicals produced in largest amount, about 21.5 billion pounds having been manufactured. Almost all Cl_2 is made by electrolysis, with 95% coming from the electrolysis of NaCl brine, that is, aqueous NaCl. (Part of the remainder comes from the electrolysis of molten NaCl as described above.)

A simplified version of one type of cell used for Cl_2 production, the mercury cell, is illustrated in Figure 19.18. As predicted in Section

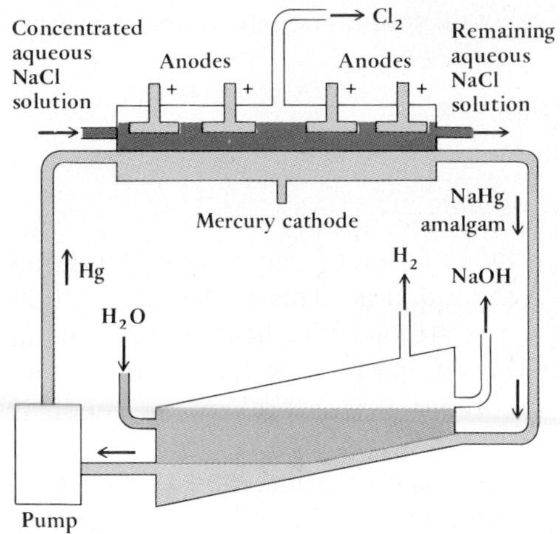

FIGURE 19.18

A *mercury cell* for the electrolysis of aqueous NaCl. Chlorine is produced at anodes immersed in the aqueous NaCl, and sodium is produced at the mercury cathode. The sodium dissolves in the liquid mercury to form a liquid "amalgam," which is pumped out of the electrolysis cell into another chamber where the Na/Hg amalgam is stirred with water to give H_2 gas and aqueous sodium hydroxide. An unfortunate aspect of these cells is that waste products from them have been responsible for putting thousands of pounds of mercury into the environment. Therefore, mercury cells are being replaced by new types of cells.

19.5, Cl^- is oxidized at the graphite anode to give Cl_2. However, the reaction at the cathode in this cell is rather different from that predicted from the values in Table 19.1. Sodium can form a liquid solution with mercury called an *amalgam*.

$$Na^+ + e^- + Hg \longrightarrow \underset{\text{sodium amalgam}}{Na(Hg)}$$

and the voltage required for this is less than that for the reduction of water to H_2. Thus, Na^+ is reduced to Na(Hg), and this liquid solution flows out of the cell. When the Na(Hg) solution is vigorously stirred with water, the sodium reacts to give NaOH and H_2.

$$2\ Na(Hg) + 2\ H_2O(\ell) \longrightarrow 2\ NaOH(aq) + H_2(g) + Hg(\ell)$$

In this manner, electrolysis of brine can yield two pure, commercially useful gases as well as pure NaOH. The latter is also a very important industrial chemical, 20.5 billion pounds having been produced in 1983. Most of it is used to make various other chemicals, but a very large amount is used in the paper industry.

19.8 CORROSION: AN EXAMPLE OF ELECTRON TRANSFER

Corrosion is the deterioration of metals, usually with loss of metal to a solution in some form, by an oxidation–reduction reaction. The corrosion of iron is the conversion of the metal to red-brown rust, hydrated iron(III) oxide [$Fe_2O_3 \cdot H_2O$] and other products. The significance of this is that 25% of the annual steel production in the United States is estimated to be for the replacement of material lost to corrosion.

Metals generally are found in the earth as their oxides or similar compounds; in the presence of oxygen, metal oxides are thermodynamically more stable than the pure metals. Thus, all refined metals (except perhaps silver, gold, and platinum) will spontaneously revert to some oxidized state. For example,

$$4\ Fe(s) + 3\ O_2(g) \longrightarrow 2\ Fe_2O_3(s)$$
$$4\ Cu(s) + O_2(g) \longrightarrow 2\ Cu_2O(s)$$
$$4\ Al(s) + 3\ O_2(g) \longrightarrow 2\ Al_2O_3(s)$$

The oxide film that forms on the surface of a metal can be quite hard and can protect the metal against further oxidation. This is the reason that medieval makers of armor and swords so carefully heated these metal objects; a microscopically thin coating of metal oxide formed to protect the beauty of the object. However, if this film is broken, the metal is exposed to outside elements, and corrosion can begin.

For corrosion to occur at the surface of a metal, there must be anodic areas where the metal can be oxidized to metal ions as electrons are produced,

$$\text{Anode reaction:} \quad M(s) \longrightarrow M^{n+} + ne^-$$

and cathodic areas where the electrons are consumed by any or all of several possible half reactions, such as

$$\text{Cathode reactions:} \quad 2\ H^+(aq) + 2\ e^- \longrightarrow H_2(g)$$
$$2\ H_2O(g) + 2\ e^- \longrightarrow 2\ OH^-(aq) + H_2(g)$$
$$O_2(g) + 2\ H_2O(\ell) + 4e^- \longrightarrow 4\ OH^-(aq)$$

Anodic areas occur at cracks in the oxide coating, at boundaries between phases, or around impurities. The cathodic areas occur at the metal oxide coating, at less reactive metallic impurity sites, or around other metal compounds such as sulfides.

The other requirements for corrosion are an electrical connection between the anode and cathode and an electrolyte with which both anode and cathode are in contact. Both requirements are easily fulfilled, as seen in Figures 19.19 and 19.20.

If the relative rates of the anodic and cathodic corrosion reactions could be measured independently, the anodic reaction would be found to be the faster. When the two reactions are coupled to each other as in a

FIGURE 19.19

The corrosion of iron in an aqueous environment with oxygen present.

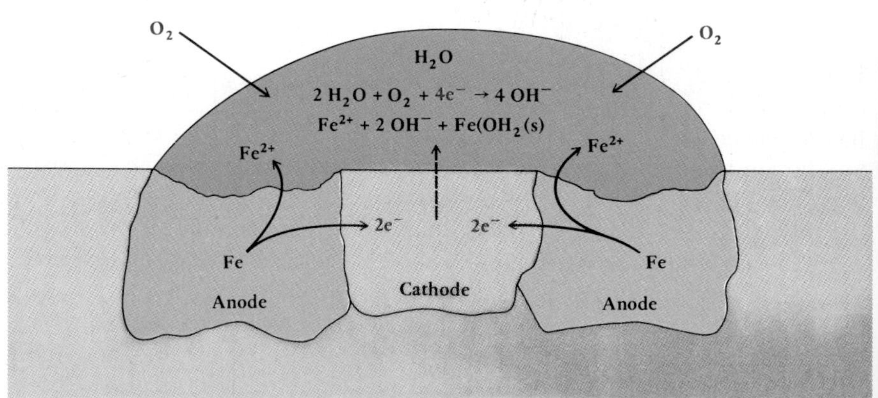

corroding metal, however, the overall rate can only be that of the slower process. In general, this means that corrosion is controlled by the relative rate of the cathodic process. This fact helps to explain the basic chemistry of corroding systems and how to prevent corrosion.

If the cathode reaction is very slow, the electrons produced by the anode reaction have nowhere to go, so the oxidation must slow down.

In the corrosion of iron, the anodic reaction is clearly the oxidation of iron. However, as noted just above, there are three possible cathodic reactions; which of these is fastest depends, in part, on the acidity of the surrounding solution and the amount of oxygen present. When little or no oxygen is present (the case when a piece of iron is buried in a soil such as moist clay), hydrogen ion and water are reduced. As indicated in the reactions above, $H_2(g)$ and hydroxide ion are the products. Since iron(II) hydroxide is relatively insoluble, this can precipitate on the metal surface and so inhibit the further formation of Fe^{2+} at the anodic site.

Anode:	$Fe(s) \longrightarrow Fe^{2+}(aq) + 2\,e^-$
Cathode:	$2\,H_2O(\ell) + 2\,e^- \longrightarrow 2\,OH^-(aq) + H_2(g)$
Net reaction:	$Fe(s) + 2\,H_2O(\ell) \longrightarrow \underbrace{Fe^{2+}(aq) + 2\,OH^-(aq)}_{\downarrow\ Fe(OH)_2(s)} + H_2(g)$

The $Fe(OH)_2$ coating, and the slowness of H_2O reduction, mean that corrosion under oxygen-free conditions is slow.

If both water and $O_2(g)$ are present, the chemistry of iron corrosion is somewhat different, and the corrosion reaction is about 100 times faster than without oxygen.

Anode:	$2[Fe(s) \longrightarrow Fe^{2+}(aq) + 2\,e^-]$
Cathode:	$O_2(g) + 2\,H_2O(\ell) + 4\,e^- \longrightarrow 4\,OH^-(aq)$
Net reaction:	$2\,Fe(s) + O_2(g) + 2\,H_2O(\ell) \longrightarrow 2\,Fe(OH)_2(s)$

If $O_2(g)$ is not freely available, further oxidation of the iron(II) hydroxide is limited to the formation of magnetic iron oxide (which can be thought of as a mixed oxide of Fe_2O_3 and FeO).

$$6\,Fe(OH)_2(s) + O_2(g) \longrightarrow \underset{\text{green hydrated magnetite}}{2\,Fe_3O_4 \cdot H_2O(s)} + 4\,H_2O(\ell)$$

$$Fe_3O_4 \cdot H_2O(s) \longrightarrow H_2O(\ell) + \underset{\text{black magnetite}}{Fe_3O_4(s)}$$

It is the black magnetite that you find coating an iron object that has corroded by resting in a moist soil. On the other hand, if the iron object has free access to $O_2(g)$ and H_2O, as in the open or in flowing water, red-brown iron(III) oxide will form (Figure 19.1).

$$4\,Fe(OH)_2(s) + O_2(g) \longrightarrow 2\,H_2O(\ell) + \underset{\text{red brown}}{2\,Fe_2O_3 \cdot H_2O(s)}$$

This is the familiar rust you see on cars and buildings, and the substance that colors the water red in some mountain streams or in your home.

Other substances in air and water can assist in the corrosion process. Chlorides, from sea air or from salt spread on roadways in the winter,

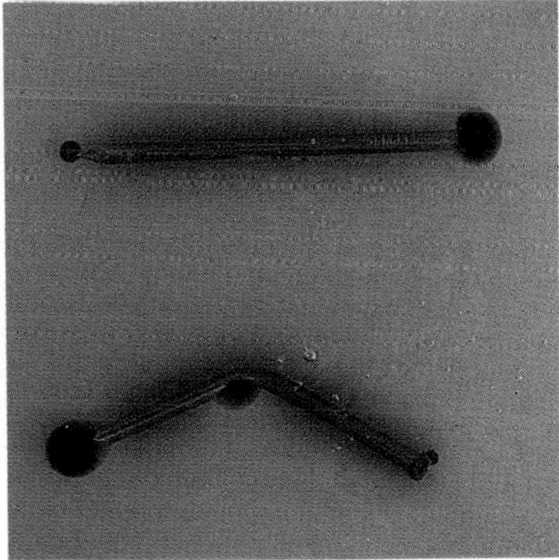

FIGURE 19.20

Corroding iron nails. Two nails were placed in an agar gel, which also contained phenolphthalein and the $Fe(CN)_6^{3-}$ ion. The nails began to corrode and gave Fe^{2+} ions at the tip or where the nail is bent. That these points are the anode is indicated by the blue-green color of Prussian blue, a complex of Fe^{2+} and $Fe(CN)_6^{3-}$. The remainder of the nail is the cathode, since water reduction occurs to give OH^- ion, the presence of which is detected by phenolphthalein. (See Figure 19.15.) (Charles D. Winters)

are notorious. Since the chloride ion is relatively small, it can diffuse into and through a protective metal oxide coating. Metal chlorides, which are more soluble than metal oxides and hydroxides, can then form. These chloride salts leach back through the oxide coating with the metal ion, and a path is now open for further attack on the underlying metal by oxygen and water. This is the reason that you often see small pits on the surface of a corroded metal.

Sulfur dioxide is a notorious air pollutant formed in the combustion of oil and coal in power plants and homes. It is 1300 times more soluble in water than is O_2, and, after further oxidation, it can form sulfuric acid solutions—the reason for so-called "acid rain." It is these solutions that lead to the greatly accelerated deterioration of buildings and art objects in a polluted environment. The following redox reactions are believed to be responsible for the corrosion of iron objects when SO_2 is present.

$$Fe(s) + SO_2(g) + O_2(g) \longrightarrow FeSO_4(s)$$

$$4\ FeSO_4(s) + O_2(g) + 6\ H_2O(\ell) \longrightarrow 2\ Fe_2O_3 \cdot H_2O(s) + 4\ H_2SO_4(aq)$$

$$4\ H_2SO_4(aq) + 4\ Fe(s) + 2\ O_2(g) \longrightarrow 4\ FeSO_4(s) + 4\ H_2O(\ell)$$

Once these reactions have begun, the sulfuric acid that forms is difficult to remove. In fact, even if an iron object is carefully cleaned, corrosion will continue as long as sulfates are present.

How can you stop a metal object from corroding? There are many methods, some more effective than others, but none totally successful. The general approaches are (a) to inhibit the anodic process, (b) to inhibit the cathodic process, or (c) to do both. The more usual method is **anodic inhibition**, attempting directly to limit or prevent the oxidation half reaction by painting the metal surface or by allowing a thin film of oxide to form. More recently developed methods are illustrated by the following reaction:

$$2\ Fe(s) + 2\ Na_2CrO_4(aq) + 2\ H_2O(\ell) \longrightarrow Fe_2O_3(s) + Cr_2O_3(s) + 4\ NaOH(aq)$$

An iron surface is oxidized by a Cr(VI) salt to give Fe(III) and Cr(III) oxides. These form a coating impervious to O_2 and water, and further atmospheric oxidation is inhibited.

There are several other ways to inhibit metal oxidation, and one of these is to force the metal to become the cathode, instead of the anode, in an electrochemical cell. Hence, this is called **cathodic protection**. This is usually done by attaching another, more readily oxidized metal. The best example of this is **galvanized iron**, iron that has been coated with a thin film of zinc (Figure 19.21). $E°$ for zinc oxidation is considerably more positive than $E°$ for iron oxidation (Table 19.1), so the zinc metal film is oxidized before any of the iron. Thus, the zinc coating forms what is called a **sacrificial anode**. Another reason for using a zinc coating on iron is that, when the zinc is corroded, $Zn(OH)_2$ forms on the surface. This metal hydroxide is even less soluble than $Fe(OH)_2$, so the insoluble hydroxide film further slows corrosion.

The corrosion of the hulls of ships in sea water is a major problem. One way to prevent it is to attach small blocks of platinum-coated titanium

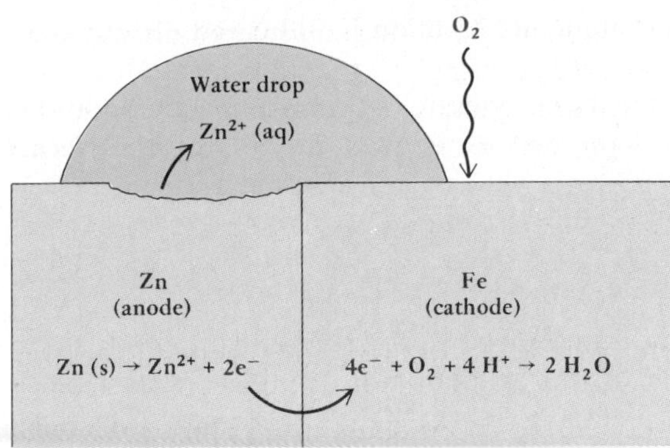

FIGURE 19.21

Cathodic protection of an iron-containing object. The iron is coated with a film of zinc, a metal more easily oxidized than iron. Therefore, the zinc acts as an anode, and forces iron to become the cathode.

to the hull (Figure 19.22). Platinum is a rather inert metal, not subject to corrosion under the usual environmental conditions. By attaching the block to the ship's hull and then to a battery, the block can be made the anode of an electrolysis cell. This forces the ship's hull to be the cathode, and oxidation of the iron in the steel hull is prevented. The same technique is being used in oil wells, for example, to prevent the corrosion of the steel pipes deep in the ground.

SUMMARY

A spontaneous oxidation–reduction or electron transfer reaction can be the basis of an electrochemical device called a **voltaic cell**. A **battery** is a voltaic cell that can produce a direct electric current at a constant voltage. In a voltaic cell, the oxidation half-reaction occurs at the **anode**, which supplies electrons to the external circuit (and bears a negative charge). The **cathode** receives electrons from the external circuit and is therefore the site of the reduction half reaction (and bears a positive charge). Charge balance inside the cell is maintained by a flow of ions across a **salt bridge** separating the anode and cathode compartments (Section 19.1).

The voltage of an electrochemical cell operating at standard conditions (all dissolved species at 1.0 *M*, gases at 1 atm) is $E°$, a quantity directly related to the standard free energy of the reaction by

$$\Delta G° = -nFE° \qquad (19.1)$$

where n is the moles of electrons transferred and F is the **Faraday constant** (9.65×10^4 J/V · mol) (Section 19.2). Thus, a positive cell potential indicates a spontaneous reaction. Each half reaction can be assigned a potential relative to the 0.00 V assigned to the half reaction $2\ H^+(1\ M, aq) + 2\ e^- \rightarrow H_2(1\ atm, g)$, the **standard hydrogen electrode (S.H.E.)** at 25°C. Thus, species having $E°$ for reduction more negative than 0.00 V are poorer oxidizing agents than H^+, while those having more positive potentials than 0.00 V are better oxidizing agents than H^+. ($E°$ values for a variety of reduction half reactions are collected in Table 19.1 and in Appendix K.) An oxidizing half reaction and a reducing half reaction can be summed to obtain a net redox reaction. Similarly, the sum of the $E°$ values of the two half reactions is the potential observed for the voltaic

FIGURE 19.22

Cathodic protection of a ship's hull. Platinum-coated titanium blocks (the four strips in line with the propeller shaft) are attached to the hull. The strips are made the anode of an electrolysis cell, and so the hull is forced to become the cathode. Since reduction always occurs at a cathode, oxidation of the iron in the ship's hull is prevented.

cell. If E°_{cell} is positive, a spontaneous reaction is indicated (at standard conditions).

When an electrochemical cell operates at nonstandard conditions (Section 19.3), the voltage of the cell, E, is given at 25°C by the **Nernst equation**

$$E = E^\circ - \frac{0.0592}{n} \log Q \qquad (19.2)$$

where Q = reaction quotient = $[C]^c[D]^d/[A]^a[B]^b$ for a general reaction $aA + bB \rightarrow cC + dD$.

As the reactants are converted to products in an electrochemical cell, concentrations change, and E is given by Equation 19.2. Eventually, equilibrium is reached, and E goes to 0. At this point, $Q = K_{eq}$, and as shown in Section 19.3,

$$\log K = \frac{nE^\circ}{0.0592} \qquad (19.3)$$

Common **batteries** (the LeClanché dry cell, alkaline batteries, mercury battery) are described in Section 19.4. **Storage batteries** (the lead battery and ni–cad batteries) can be recharged by using an external voltage source to reverse the discharge reaction. **Fuel cells** are voltaic cells in which reactants are continuously added or which can be restored after discharge by physically adding new reactants.

Electrolysis is the use of electrical energy to produce chemical change (Section 19.5). Given a choice of electrode reactions, those which occur are those requiring the least potential to overcome their nonspontaneity. In water, this generally means that substances having reduction potentials more positive than about −0.42 V can be reduced. Thus, in aqueous NaCl, water and not Na^+ is reduced at the cathode. At the anode, the **overvoltage** (Section 19.2) for $O_2(g)$ production makes oxygen production unfavorable, and $Cl_2(g)$ is produced instead. Commercially, Al, Na, $Cl_2(g)$, and NaOH are among the chemicals produced electrochemically (Section 19.7).

Electrical current (in amps) is the number of coulombs that are passed per second. Since the charge on one mole of electrons is 1 Faraday (9.65×10^4 coulombs), the number of coulombs of charge that flow in an electrochemical cell can be found by measuring time and current. The quantity of material consumed or produced is related to time and current through the Faraday constant. The energy expended in an electrochemical process (Section 19.6) can be determined from the relationship joule = 1 volt · coulomb, and the power expended (rate of energy use or production) in watts is joules per second.

Corrosion reactions (Section 19.8) are examples of electron transfers of great consequence in our environment. Most metals are thermodynamically unstable with respect to oxidation, and corrosion is the oxidative deterioration of these metals. Metals can be protected from corrosion by inhibiting access of the oxidant (**anodic protection**) or by attaching some other, more easily oxidized substance (a **sacrificial anode**; **cathodic protection**).

STUDY QUESTIONS

REVIEW QUESTIONS

1. In each of the following reactions, tell which substance is oxidized and which is reduced. Tell which is the oxidizing agent and which is the reducing agent.
(a) $2\ Al(s) + 3\ Cl_2(g) \longrightarrow 2\ AlCl_3(s)$
(b) $8\ H^+(aq) + MnO_4^-(aq) + 5\ Fe^{2+}(aq) \longrightarrow 5\ Fe^{3+}(aq) + Mn^{2+}(aq) + 4\ H_2O(\ell)$
(c) $FeS(s) + 3\ NO_3^-(aq) + 4\ H^+(aq) \longrightarrow 3\ NO(g) + SO_4^{2-}(aq) + Fe^{3+}(aq) + 2\ H_2O(\ell)$
2. Explain the function of a salt bridge in an electrochemical cell.
3. Tell if each of the following statements is true or false. If false, rewrite it to make it a correct statement.
(a) Oxidation always occurs at the anode of an electrochemical cell.
(b) The anode of a battery is the site of reduction and is negative.
(c) Standard conditions for electrochemical cells are a concentration of 1.0 *M* for dissolved species and 1 atm pressure for gases.
(d) The potential of a cell does not change with temperature.
(e) All spontaneous oxidation–reduction reactions have an E°_{cell} with a negative sign.
4. Tell which phrase (a through d) best completes the sentence. A spontaneous oxidation–reduction reaction has (a) a positive ΔG° and a positive E°, (b) a negative ΔG° and a positive E°, (c) a positive ΔG° and a negative E°, or (d) a negative ΔG° and a negative E°.
5. Tell which phrase (a through c) best completes the sentence. In Table 19.1, Zn(s) can (a) oxidize Fe(s) and Cd(s), (b) reduce Al^{3+} and Mg^{2+}, or (c) reduce Cd^{2+} and Ag^+.
6. Tell if each of the following statements is true or false. If false, rewrite it to make it a correct statement.
(a) The value of an electrode potential changes when the half reaction is multiplied by a factor. That is, E° for ($Li^+ + e^- \rightarrow Li$) is different than that for ($2\ Li^+ + 2\ e^- \rightarrow 2\ Li$).
(b) Al is the strongest reducing agent listed in Table 19.1.
(c) The equilibrium constant for an oxidation–reduction can be calculated using the Nernst equation.
(d) Changing the concentrations of dissolved substances does not change the potential observed for an electrochemical cell.
7. What are the advantages and disadvantages of lead storage batteries?
8. How does a fuel cell differ from a battery?
9. Explain why the products of electrolysis of molten NaCl differ from those obtained from aqueous NaCl.
10. Describe the electrochemical method for the manufacture of Cl_2 and NaOH.
11. What is the difference between anodic and cathodic protection against corrosion? Explain how each works.

CELLS AND CELL POTENTIALS

12. Copper can reduce silver(I) to metallic silver, a reaction that could in principle be used as a battery.
$Cu(s) + 2\ Ag^+(aq) \longrightarrow Cu^{2+}(aq) + 2\ Ag(s)$
(a) Write the half reactions involved.
(b) Which half reaction is an oxidation and which is a reduction? Which half reaction occurs in the anode compartment and which in the cathode compartment?
(c) Write the abbreviated notation for the cell.
(d) If a strip of copper is used as the electrode in the battery, does it have a positive or a negative sign?
13. Suppose the half-cells $Cu(s)/Cu^{2+}(aq)$ and $Sn(s)/Sn^{2+}(aq)$ are to be used as the basis of a battery. If the sign of the copper electrode is found to be positive and that of the tin electrode negative, write the half reactions that occur in each half-cell. Tell which is the oxidation and which is the reduction. Decide which half reaction occurs at the anode and which at the cathode. Write the abbreviated notation for the cell.
14. If the standard voltage of the reaction of Cu(s) with $Ag^+(aq)$ in Study Question 12 is +0.46 V, what is the standard free energy change, ΔG°, for the reaction?
15. What is the standard potential, E°, of the redox reaction
$H_2(g) + I_2(aq) \longrightarrow 2\ HI(aq)$
if the free energy, ΔG°, is −103 kJ? Is the reaction spontaneous as written? What are the half reactions involved?
16. What are the values of E° for the following reactions? Are the reactions spontaneous as written?
(a) $2\ H^-(aq) + Zn^{2+}(aq) \longrightarrow H_2(g) + Zn(s)$
(b) $Zn^{2+}(aq) + Ni(s) \longrightarrow Zn(s) + Ni^{2+}(aq)$
17. Balance each of the following *unbalanced* equations. Next, calculate values of the standard potential, E°, for each reaction, and tell whether each is or is not spontaneous as written. (Half reaction potentials are found in Appendix K.)
(a) $Sn^{2+}(aq) + Ag(s) \longrightarrow Sn(s) + Ag^+(aq)$
(b) $Zn(s) + Sn^{4+}(aq) \longrightarrow Sn^{2+}(aq) + Zn^{2+}(aq)$
(c) $I_2(s) + Br^-(aq) \longrightarrow I^-(aq) + Br_2(\ell)$
(d) $Ce^{4+}(aq) + Cl^-(aq) \longrightarrow Ce^{3+}(aq) + Cl_2(g)$
(e) $Cu(s) + NO_3^-(aq) + H^+(aq) \longrightarrow Cu^{2+}(aq) + NO(g) + H_2O(\ell)$
18. Consider the half reactions in the table on p. 740.

HALF REACTION	$E°$(V)
$Ce^{4+}(aq) + e^- \longrightarrow Ce^{3+}(aq)$	+1.61
$Ag^+(aq) + e^- \longrightarrow Ag(s)$	+0.80
$Hg_2^{2+}(aq) + 2e^- \longrightarrow 2\,Hg(\ell)$	+0.79
$Sn^{2+}(aq) + 2e^- \longrightarrow Sn(s)$	−0.14
$Ni^{2+}(aq) + 2e^- \longrightarrow Ni(s)$	−0.25
$Al^{3+}(aq) + 3e^- \longrightarrow Al(s)$	−1.66

(a) Which is the weakest oxidizing agent in the list?
(b) Which is the strongest oxidizing agent?
(c) Which is the strongest reducing agent?
(d) Which is the weakest reducing agent?
(e) Will Sn(s) reduce $Ag^+(aq)$ to Ag(s)?
(f) Will $Hg(\ell)$ reduce $Sn^{2+}(aq)$ to Sn(s)?
(g) Name the ions which can be reduced by Sn(s).
(h) What metals can be oxidized by $Ag^+(aq)$?

19. Use the table of half reaction potentials in Study Question 18 to answer the following questions:
(a) What reaction leads to the maximum positive standard potential?
(b) If the $Ni(s)/Ni^{2+}(aq)$ half-cell is combined with the $Hg(s)/Hg_2^{2+}(aq)$ half-cell, write the spontaneous reaction which occurs. What will be its standard potential?
(c) Write the spontaneous reaction that occurs when the half reactions $Ag(s)/Ag^+(aq)$ and $Ce^{4+}(aq)/Ce^{3+}(aq)$ are combined. What is the value of $E°$ for this reaction?

20. Four metals, A, B, C, and D, exhibit the following properties:
(a) Only A and C react with 1.0 *M* hydrochloric acid to give $H_2(g)$.
(b) When C is added to solutions of the ions of the other metals, metallic B, D, and A are formed.
(c) Metal D reduces B^{n+} to give metallic B and D^{n+}.
Based on the information above, arrange the four metals in order of increasing ability to act as reducing agents.

21. Assume that you assemble an electrochemical cell based on the half reactions $Zn(s)/Zn^{2+}(aq)$ and $Ag(s)/Ag^+(aq)$.
(a) Write the reaction that occurs spontaneously.
(b) Which electrode is the anode and which is the cathode? Write the abbreviated notation for the cell.
(c) Diagram the components of the cell.
(d) If you used a strip of zinc as an electrode, is it the anode or cathode?
(e) Do electrons flow from the Zn electrode to the Ag electrode, or vice versa?
(f) If a salt bridge containing $NaNO_3(aq)$ connects the two half-cells, in which direction do the nitrate ions move, from the zinc to the silver compartment, or vice versa?

22. An electrochemical cell uses Al(s) and $Al^{3+}(aq)$ in one compartment and Ag(s) and $Ag^+(aq)$ in the other.
(a) Write a balanced equation for the reaction that will occur spontaneously in this cell, and calculate $E°$.
(b) Which is the better reducing agent, Ag or Al?
(c) Which is the anode and which is the cathode? Write the abbreviated notation for the cell.
(d) Sodium nitrate is in the salt bridge connecting the two half-cells. In which direction do the nitrate ions flow, from Al to Ag or Ag to Al?

23. Given the reaction

$$2\,Cr(s) + 3\,Fe^{2+}(aq) \longrightarrow 2\,Cr^{3+}(aq) + 3\,Fe(s)$$

(a) Calculate $E°$ and $\Delta G°$.
(b) Sketch a complete picture of an electrochemical cell that might be constructed to carry out the reaction. Include in your diagram the following labels: anode; cathode; the sign of each electrode [Cr(s) and Fe(s)]; the direction of electron flow in an external wire; and the direction of anion flow in a salt bridge (containing $NaNO_3(aq)$).

24. An electrochemical cell is assembled using the half-cells $Pb^{2+}(aq)/Pb(s)$ and $Ni(s)/Ni^{2+}(aq)$.
(a) Which is the better reducing agent, Pb or Ni?
(b) Write the spontaneous reaction which can occur and calculate $E°$.
(c) Which electrode is the cathode, Pb or Ni?
(d) What is the direction of electron flow in an external wire, from Pb to Ni or Ni to Pb?
(e) If $NaNO_3(aq)$ is placed in a salt bridge connecting the two half-cells, in which direction do the anions flow?

25. Which of the following aqueous metal ions can be reduced by $H_2(g)$: (a) Sn^{2+}; (b) Hg^{2+}; (c) Sn^{4+}.

26. You are told to assemble an electrochemical cell with one half-cell being $Cl^-/Cl_2(g)$. The other half-cell could be $Al(s)/Al^{3+}(aq)$, $Mg(s)/Mg^{2+}$, or $Zn(s)/Zn^{2+}(aq)$. Which of the metal/metal ion combinations would you choose to produce the largest possible positive $E°$? Write a balanced equation for the reaction you have chosen.

27. A simple electrochemical cell has been constructed using a zinc electrode in a solution of 1.0 *M* $Zn(NO_3)_2$ and a tin electrode in a 1.0 *M* solution of $Sn(NO_3)_2$.
(a) Write a balanced equation for the reaction which occurs spontaneously.
(b) What is the value of $E°$?
(c) Do electrons move in an external wire from Sn to Zn or from Zn to Sn?
(d) Which is the anode compartment and which is the cathode compartment?
(e) What is the sign of the Sn electrode?

28. The reaction of Zn(s) and Cl_2(g) has been suggested as the basis of a new battery.
(a) If the product is zinc chloride, write a balanced equation describing this reaction.
(b) If the reaction occurs under standard conditions, what does this mean in terms of gas pressure, temperature, and concentration of water soluble species?
(c) What is $E°$ for the reaction?
(d) If fluorine, F_2(g), were used instead of Cl_2(g), what would the value of $E°$ be?

29. Calculate $E°$ for the disproportionation (self oxidation–reduction) of the copper(I) ion.

$$2\ Cu^+(aq) \longrightarrow Cu(s) + Cu^{2+}(aq)$$

30. Calculate $E°$ for the reaction Hg(s) + S(s) → HgS(s). Is the reaction spontaneous? Calculate $\Delta G°$.

CELLS UNDER NONSTANDARD CONDITIONS AND EQUILIBRIUM CONSTANTS

31. Calculate the voltage delivered by an electrochemical cell using the following reaction:

$$2\ Fe^{3+}(aq) + 2\ I^-(aq) \longrightarrow 2\ Fe^{2+}(aq) + I_2(s)$$

All dissolved species are at a concentration of 0.1 *M*.

32. An electrochemical cell is constructed of one half-cell in which a silver wire dips into an aqueous solution of $AgNO_3$. The other half-cell consists of an inert platinum wire in an aqueous solution of Fe^{2+} and Fe^{3+}.
(a) What is the reaction occurring when this cell operates spontaneously?
(b) When the dissolved metal ions have a concentration of 1.0 *M*, what is $E°$?
(c) If $[Ag^+(aq)]$ = 0.1 *M* but $[Fe^{2+}(aq)]$ and $[Fe^{3+}(aq)]$ remain 1.0 *M*, what is the value of *E*? Is the cell reaction still that in part (a) above? If not, what is the net reaction when the cell operates spontaneously?

33. Calculate the equilibrium constant, *K*, for the reaction in Study Question 31.

34. Calculate equilibrium constants for the reactions
(a) Zn^{2+}(aq, 1 *M*) + Ni(s) → Zn(s) + Ni^{2+}(aq, 1 *M*)
(b) I_2(aq) + 2 Br^-(aq) → 2 I^-(aq) + Br_2(aq)
(c) Cu(s) + 2 Ag^+(aq) → Cu^{2+}(aq) + 2 Ag(s)

35. The corrosion of iron in the presence of water and O_2(g) gives $Fe(OH)_2$(s) (see Section 19.8). What is $E°$ for this reaction? If the pressure of oxygen is only 0.2 atm (as it is in air), what is the value of *E* under these conditions? Calculate the equilibrium constant, *K*, for the process.

36. Determine $E°$ for $PbSO_4(s) \rightarrow Pb^{2+}(aq) + SO_4^{2-}(aq)$. Calculate the equilibrium constant, *K*, for the reaction. [*Hint:* To solve this problem, first find a half reaction for the reduction of $PbSO_4$ in Appendix K. When you do, you will notice that one product is Pb(s). Since the final, net reaction must have Pb^{2+}(aq) as a product, the next step is to find an oxidation half reaction that carries Pb(s) to Pb^{2+}. Adding these together will give $E°$ for the net process.]

37. Determine the equilibrium constant for the dissociation of $AgCl(s) \rightarrow Ag^+(aq) + Cl^-(aq)$. (For a hint on how to do this problem, see Study Question 36.)

ELECTRICAL ENERGY AND POWER

38. A current of 2.50 amps is passed through an aqueous solution of $CuSO_4$ for 2.00 hours. What mass of copper is deposited at the cathode?

39. The basic reaction occurring in the cell in which Al_2O_3 and aluminum salts are electrolyzed is

$$Al^{3+} + 3\ e^- \longrightarrow Al(s)$$

If the cell operates at 5.0 V and 1.0×10^5 amps, how many grams of aluminum metal can be produced in an 8.0 hour day?

40. The vanadium(II) ion can be produced by electrolysis of a vanadium(III) salt in solution. How long must you carry out an electrolysis if you wish to convert completely 1.0 L of 0.010 *M* V^{3+}(aq) to V^{2+}(aq) if the current is 0.268 amps?

41. The reactions occurring in a lead storage battery are given in Section 19.4. A typical battery might be rated "50 ampere-hours," which means that it has the capacity to deliver 50. amps for 1.0 hour or 1.0 amp for 50. hours. If it does deliver 1.0 amp for 50. hours, how many grams of lead would be used up to accomplish this?

42. Suppose you use Cu(s)/Cu^{2+}(aq) and Sn(s)/Sn^{2+}(aq) half-cells as the basis of an electrochemical cell (Study Question 13). If the cell starts at standard conditions, and 0.400 amp of current flows for 48.0 hours, what are the concentrations of each of the dissolved species at this point? What is the potential of the cell after 48.0 hours? (Assume 1.00 L of solution.)

43. A lead storage battery operates at 12.0 V and is rated at "100 ampere-hours." The latter means that it can deliver 1.0 amp for 100 hours. What is the power rating of the battery in watts?

44. A hydrogen/oxygen fuel cell operates on the simple reaction

$$H_2(g) + \tfrac{1}{2}\ O_2(g) \longrightarrow H_2O(\ell)$$

If the cell is designed to produce 1.5 amps of current, and if the H_2(g) fuel is contained in a 1.0-L tank at 200. atm pressure (at 25°C), how long can the fuel cell operate before the $H_2(g)$ runs out? (We assume there is an unlimited supply of O_2.)

45. It has been demonstrated that an effective battery can be built using the reaction between Al metal and O_2 from the air. If the Al anode of this battery consists of a 3-ounce piece of aluminum (84 grams), how many hours can the battery produce 1.0 amp of electricity (assuming unlimited O_2).

46. A proposed automobile battery involves the reaction of Zn(s) and $Cl_2(g)$ to give $ZnCl_2$. If you want such a battery to operate for 10. hours and deliver 1.5 amps of current, what is the minimum mass of zinc that the anode must contain?

47. Batteries are listed in a popular mail order catalog by their "cranking power." This is the amount of current that the battery can produce for 30 seconds, and a typical value is 450 amps. How many coulombs flow through the battery in 30. seconds? If this is a lead storage battery, how much lead, Pb(s), is consumed in 30. seconds?

ELECTROLYSIS

48. Current is passed through a solution containing $Ag^+(aq)$. How much silver was there in the solution if all the silver was removed completely as Ag metal by electrolysis for 14.5 min with a current of 1.10 mA (1 mA = 10^{-3} amp)?

49. An old method of measuring the current flowing in a circuit was to use a "silver coulometer." The current passed first through a solution of $Ag^+(aq)$. The amount of metallic silver deposited was weighed, and, if the time during which the current flowed was noted, the current could be calculated.
(a) If 0.052 g of Ag was deposited during a 450-second experiment, what was the current flowing in the circuit?
(b) If the current flowing through the $Ag^+(aq)$ then flowed into a cell containing gold in the form of $AuCl_4^-(aq)$, how much gold was deposited in this electrolysis cell?
(c) At what electrode was the gold deposited, the anode or the cathode?

50. A solution of 5.0 mg of an organometallic compound, $C_{10}H_{10}Fe$, was electrolyzed at an anode for 15.0 minutes with an average current of 0.00288 amps. How many electrons per molecule of the iron compound are transferred? Was the compound oxidized or reduced? Write a half reaction showing the electrochemical reaction of the compound.

51. An aqueous solution of KI is placed in a beaker with two inert platinum electrodes. When the cell is attached to an external battery, electrolysis occurs.
(a) Write the half reaction occurring at the anode. What is its sign?
(b) Write the half reaction occurring at the cathode. What is its sign?
(c) If 0.05 amps flow through the cell for 5.0 hours, how many grams of each product are formed?

52. For each of the following solutions, tell what happens at the anode and at the cathode on electrolysis.
(a) KBr(aq)
(b) KI(aq)
(c) NaF(molten)
(d) NaF(aq)
(e) $NiCl_2(aq)$

53. Electrolysis of brine leads to gaseous chlorine, Cl_2. About 20 billion pounds were manufactured in 1982. How much energy in kilowatt-hours must have been used to produce this amount of chlorine from NaCl? Modern mercury cells operate at about 4.6 V and 3.0×10^5 amps.

54. Electrolysis of molten NaCl is done in a Downs cell operating at 7.0 volts and 4.0×10^4 amps. How much Na(s) and $Cl_2(g)$ can be produced in one day in such a cell? What is the energy consumption in kilowatt-hours? (Assume 100% efficiency.)

GENERAL

55. Ni–cad batteries are rechargeable and are commonly used in cordless appliances. Although such batteries actually function under basic conditions, imagine an electrochemical cell using the setup below.

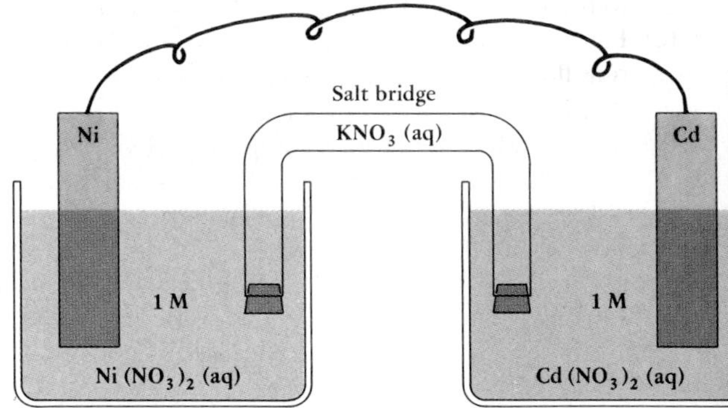

(a) Write a balanced equation depicting the reaction occurring in the cell above.
(b) What is oxidized? What is reduced? What is the reducing agent and what is the oxidizing agent?
(c) Which is the anode, Ni or Cd? What is the sign of the Cd electrode?
(d) Write the abbreviated notation for the cell.
(e) What is the direction of electron flow in the external wire?
(f) If the salt bridge contains KNO_3, toward which compartment will the NO_3^- ions migrate?
(g) Calculate the equilibrium constant for the cell.
(h) If the concentration of Cd^{2+} is reduced to 0.010 *M*, and $[Ni^{2+}]$ = 1.0 *M*, what is the potentia

produced by the cell? Is it still spontaneous in the direction written in part (a)?

(i) If 0.050 amps are drawn from the battery, how long can it last if you began with 1.0 L of each of the solutions and each was initially 1.0 *M* in dissolved species? The electrodes weighed 50. g each at the beginning.

56. Tell what happens when you bubble $O_2(g)$ through an acidic aqueous solution of each of the following:
(a) $FeBr_2$ (b) $CuBr_2$ (c) PbI_2

57. Living organisms derive energy from the oxidation of food, typified by glucose.

$$C_6H_{12}O_6(aq) + 6\ O_2(g) \longrightarrow 6\ CO_2(g) + 6\ H_2O(\ell)$$

Electrons in this redox process are transferred from glucose to oxygen in a series of at least 25 reactions. It is interesting to calculate the total daily current flow in a typical organism and the rate of energy expenditure (power). (See T.P. Chirpich, *J. Chem. Educ.*, *52*:99, 1975.)

(a) The molar heat of combustion of glucose is −2800 kJ. If you are on a typical daily diet of 2400 Calories (2.4×10^3 kcal; 1 cal = 4.184 J), how many moles of glucose must be consumed in one day? How many moles of O_2 must be consumed in the oxidation of the glucose?

(b) How many moles of electrons must be supplied to reduce the amount of O_2 calculated in part (a)?

(c) From your answer in part (b), calculate the current flowing in your body from the combustion of glucose per second.

(d) If the average standard potential in the electron transport chain is 1.0 V, what is the rate of energy expenditure in watts?

58. An expensive but lighter alternative to the lead storage battery is the silver–zinc battery. Zinc is the reducing agent and silver oxide is the oxidizing agent.

$$Ag_2O(s) + Zn(s) + H_2O \longrightarrow Zn(OH)_2(s) + 2\ Ag(s)$$

The electrolyte is 40% KOH, and silver/silver oxide electrodes are separated from zinc/zinc hydroxide electrodes by a plastic sheet permeable to hydroxide ion. Under normal operating conditions, the battery has a potential of 1.59 V. (R.C. Plumb, *J. Chem. Educ.*, *50*:857, 1973.)

(a) How much energy per gram can be produced by the silver–zinc battery? (Add up the weight of all reactants, and divide that into the energy produced.)

(b) How much energy per gram of reactants can be obtained from the standard lead storage battery with a voltage of 2.0 V? (see Section 19.4 for the net reaction).

(c) Which battery produces more energy per gram, the silver–zinc battery or the lead storage battery?

59. Copper can be obtained from a solution of copper(II) by reducing the ion with scrap iron. How many tons of scrap iron (95% Fe) will it take to recover all of the copper from 8.0×10^6 liters of mine drainage that is 1.0×10^{-3} *M* in Cu^{2+}? (1 ton = 2000 lbs; 1 lb = 454 g)

60. Hydrazine, N_2H_4, has been proposed as the basis of a fuel cell. The reactions are

$$N_2H_4(aq) + 4\ OH^-(aq) \longrightarrow N_2(g) + 4\ H_2O(\ell) + 4\ e^-$$

$$O_2(g) + 2\ H_2O(\ell) + 4\ e^- \longrightarrow 4\ OH^-(aq)$$

(a) Which reaction occurs at the anode and which at the cathode?

(b) What is the net cell reaction?

(c) If the cell is to produce 0.50 amps of current for 50. hours, how many grams of hydrazine (N_2H_4) must be present?

(d) How many grams of O_2 must be available to react with the amount of N_2H_4 determined in (c)?

PART FIVE

THE CHEMISTRY OF THE ELEMENTS AND THEIR COMPOUNDS

Fireworks are based on salts of many different metals. (Comstock, Inc.)

PART FIVE: PREFACE

The goal of this section is to lay out, in a reasonably systematic way, something of the chemistry of the elements, especially those of economic and biochemical importance. In this short space we can only give a very brief introduction to the subject. For this reason, we plan to dwell on the chemistry you may find interesting and useful in everyday life, rather than dig into every nook and cranny of all 108 elements!

We shall first look into the chemistry of periodic groups 1A–4A (Chapters 20 and 21). These groups are dominated by metallic elements, although we shall begin to encounter a few nonmetals and metalloids. Elements of groups 5A–8 are almost entirely nonmetals, and we shall take them up next (Chapter 22). Since carbon forms such a huge number of compounds, which are generally called *organic,* we shall look at this subject as separate chapters (Chapters 22 and 23).

The transition metals, which include the lanthanides and actinides, make up a huge portion of the known elements, and their chemistry is best considered separately (Chapter 25).

As a final topic in this section, we want to give you a very brief introduction to radiochemistry (Chapter 26), a subject of great interest in these days of controversy over nuclear warfare and nuclear power.

20

The Chemistry of Hydrogen and the *s*-Block Elements

Sodium salts in a flame emit a bright yellow light. (Fundamental Photographs)

CHAPTER OUTLINE

20.1 THE CHEMISTRY OF HYDROGEN
20.2 DIAGONAL RELATIONSHIPS
20.3 THE ALKALI METALS: GROUP 1A
20.4 THE ALKALINE EARTH METALS: GROUP 2A
SPECIAL SECTION: THE CHEMISTRY OF FIREWORKS

1A	2A
1 H 1.0079	
3 Li 6.941	4 Be 9.01218
11 Na 22.9898	12 Mg 24.305
19 K 39.0983	20 Ca 40.08
37 Rb 85.4678	38 Sr 87.62
55 Cs 132.905	56 Ba 137.33
87 Fr (223)	88 Ra 226.025

Hydrogen was the first chemical element formed in the moment of creation, and it is now the most abundant element in the universe. All other elements were formed from it, and virtually every element, with the exception of the rare gases, combines chemically with hydrogen. Indeed, hydrogen-containing compounds are of such importance that you have seen many of them in this book and will be introduced to many more.

The elements of Groups 1A and 2A of the periodic table are all metals and all have valence shell electron configurations with one or two electrons, respectively. They are clearly important to all of us. Human blood and sea water are buffered NaCl solutions. Ions of four metals in these groups (Na^+, K^+, Mg^{2+}, and Ca^{2+}) constitute 99% of the positive ions in the human body. Eight of the fifty chemicals produced in largest amounts in the United States are based on metal ions of Groups 1A and 2A. Thus, in this chapter you will not only see some compounds of great commercial importance but you will also be treated to a journey through salt and emeralds, a treatment for manic depression, and a metal that could melt in your hand.

20.1 THE CHEMISTRY OF HYDROGEN

In spite of the great role played by hydrogen in the creation of the universe and the other chemical elements, it was not until 1766 that Cavendish, an English chemist, actually isolated gaseous H_2. Unfortunately, he did not recognize it as a new substance; rather, he called it "flammable air." Shortly thereafter, however, the gas was found to form water on combustion in air and was given its name, hydrogen, which means "water former."

Little use was made of H_2 until the middle of the 19th century when **coal gas** came into use. When soft coal is heated in a sealed vessel, a gas is given off that contains about 20% hydrogen plus a number of lightweight carbon–hydrogen compounds. It was found that this gas could be used for cooking and lighting, and the solid residue from the process, coke, could be used as a special purpose fuel, specifically in the reduction of iron ore to metallic iron in a blast furnace.

Since coal gas was so useful, new methods were sought for its production. It was then found that water injected into a bed of red hot coke would produce a gas that was a combustible mixture of H_2 and CO a mixture that came to be known as **water gas** or **synthesis gas**.

$$C(s) + H_2O(g) \longrightarrow \underbrace{H_2(g) + CO(g)}_{\textbf{water gas}}$$

Water gas burns cleanly, and it can be handled readily. However, it has the serious drawback that the amount of heat produced is only about half that from the combustion of coal gas, and the flame is nearly invisible and produces almost no light. Furthermore, since carbon monoxide is poisonous, water gas is very dangerous. Despite its hazards, water gas was used as a cooking gas to some extent until about 1950, but only after adding other material to make the flame luminous and a malodorous compound to detect leaks. Water gas has been in use almost 100 years, and there is a greatly renewed interest in using it to manufacture hydrocarbons.

A modern plant for the production of synthesis gas. Coal, oxygen, and steam are heated under pressure to produce a mixture of gases. In this particular plant, located in South Africa, the mixture consists of 57% hydrogen plus carbon monoxide, 9.4% methane, 32% carbon dioxide, 0.7% hydrogen sulfide (from sulfur in the coal), 0.3% nitrogen, and 0.5% hydrocarbons. After purification, a mixture of hydrogen, carbon monoxide, and methane is used to manufacture various hydrocarbons. (Floor Engineers, Inc.)

In 1783, Jacques Charles first used hydrogen to fill a balloon large enough to float above the French countryside (Chapter 6), a method used in World War I to float observation balloons. The *Graf Zeppelin*, a passenger-carrying dirigible built in Germany in 1928, was filled with hydrogen. She carried over 13,000 people between Germany and the United States until 1937, when she was replaced by the infamous *Hindenburg*. The *Hindenburg* was designed to be filled with helium. However, World War II was approaching, and the United States, which has much of the world's supply of helium, would not sell the gas to Germany; the *Hindenburg* had to use H_2. Unfortunately, when landing in Lakehurst, New Jersey in May, 1939, the *Hindenburg* exploded and burned; of the 62 people on board, only about half escaped uninjured (Figure 20.1). The result was that H_2 gas gained a reputation as a very dangerous substance. Actually, it is as safe to handle as any other fuel, as evidenced by the large quantities used in rockets today.

CHEMICAL AND PHYSICAL PROPERTIES OF HYDROGEN

Hydrogen has three isotopes, two of them stable (protium and deuterium) and one radioactive (tritium).

ISOTOPES OF HYDROGEN

ISOTOPE MASS	SYMBOL	NAME
1.0078 (amu)	^1H (H)	Hydrogen (protium)
2.0141	^2H (D)	Deuterium
3.0160	^3H (T)	Tritium

Of the three isotopes, only H and D are found in measurable quantities in nature (H = 99.985% and D = 0.015%). In contrast, tritium, which is produced by cosmic ray bombardment of water in the atmosphere, is found to the extent of 1 atom per 10^{18} of ordinary hydrogen. Tritium is radioactive, half of a sample of the element disappearing in 12.26 years by the loss of an electron (often called a beta particle) and formation of an unstable isotope of helium (Chapter 26).

$$\underset{\text{tritium}}{{}^3_1H} \longrightarrow \underset{\text{helium isotope}}{{}^3_2He} + \underset{\text{electron (beta particle)}}{{}^{\;\,0}_{-1}e}$$

FIGURE 20.1

The disastrous end of the hydrogen-filled dirigible *Hindenburg*, May, 1939. The accident occurred on landing in Lakehurst, New Jersey. (United Press International)

Deuterium compounds generally have higher melting and boiling points than the analogous protium compounds.

	H_2O	D_2O
Freezing point (K)	273.15	276.97
Boiling point (K)	373.15	374.57

Deuterium compounds have been thoroughly studied in the laboratory. One important observation is that, since D has twice the atomic mass of H, deuterium compounds react slightly more slowly than similar protium-containing compounds, and this leads to a way to produce D_2O or "heavy water." Hydrogen can be produced, albeit expensively, by electrolysis (Chapter 19 and Figure 20.2).

$$H_2O(\ell) + \text{electrical energy} \longrightarrow H_2(g) + \tfrac{1}{2}\,O_2(g)$$

In any sample of natural water there is always a small concentration of D_2O. When electrolyzed, H_2 is liberated about six times more rapidly than D_2. This means that as the electrolysis proceeds the liquid remaining is increasingly D_2O, and nearly pure D_2O can be obtained eventually. Heavy water is very valuable as a moderator in some nuclear power reactions, such as those used in southeastern Canada.

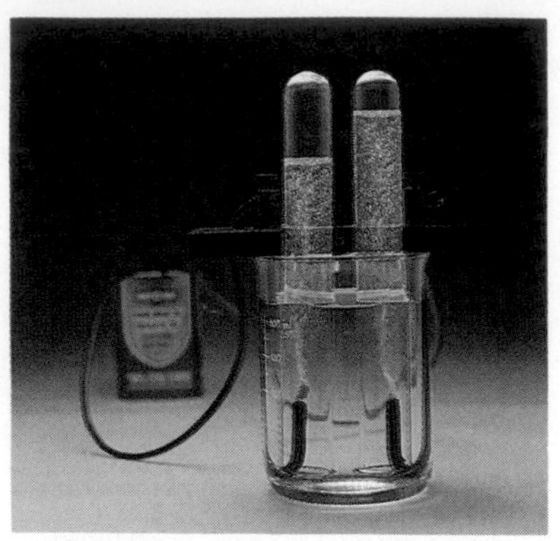

FIGURE 20.2

Electrolysis of a dilute aqueous sulfuric acid solution gives H_2 and O_2. Notice that the volume ratio is 2 (H_2, left) to 1 (O_2, right), as expected. (Charles D. Winters)

Hydrogen will enter into chemical combination with virtually every element, the rare gases being the only exception. Although hydrogen-containing compounds seem to come in a bewildering variety, there are only three different types.

(a) *Anionic hydrides* form when H_2 interacts with metals of low electronegativity such as Na, Li, and Ca.

$$Na(s) + \tfrac{1}{2}\,H_2(g) \longrightarrow NaH(s)$$

$$Ca(s) + H_2(g) \longrightarrow CaH_2(s)$$

Here the metal formally has a positive oxidation number, while H is in the form of the hydride ion, H^-.

(b) *Covalent hydrides* are formed with electronegative elements such as the nonmetals C, N, O, and F (Figure 20.3). Here the formal oxidation number of the H atom is +1.

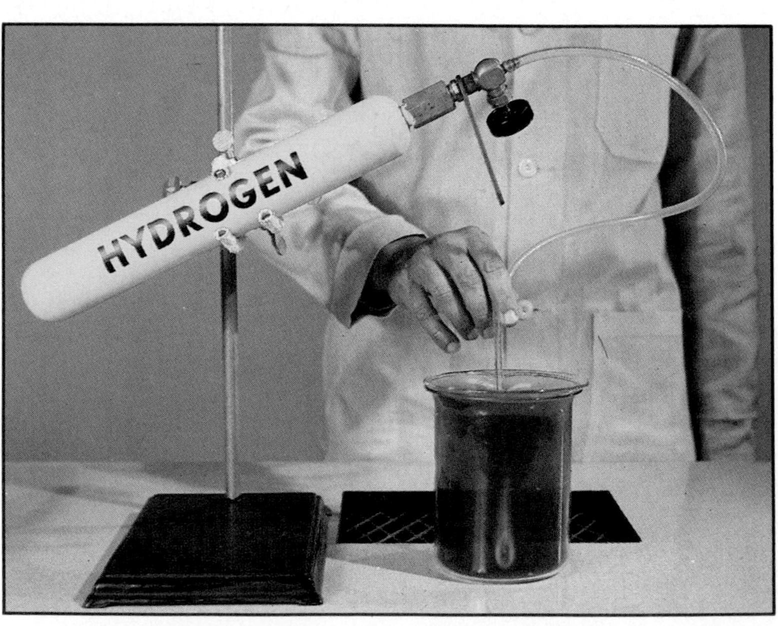

FIGURE 20.3

Hydrogen burns in an atmosphere of bromine to give hydrogen bromide, HBr. (CHEM Study Films)

$$N_2(g) + 3\ H_2(g) \longrightarrow 2\ NH_3(g)$$

$$F_2(g) + H_2(g) \longrightarrow 2\ HF(g)$$

(c) *Interstitial metallic hydrides*. Hydrogen is a very small molecule, and so can be absorbed, apparently as H atoms, into the holes or interstices in the crystal lattice of a metal (Figure 20.4). For example, when a piece of palladium metal is used as an electrode for the electrolysis of water, the metal can soak up a thousand times its volume of H_2 (at STP). However, the H_2 can be driven out of the metal by heating. This is why interstitial hydrides have been very thoroughly examined as a possible way to store H_2 for later use, the same as a sponge can store water.

THE SYNTHESIS OF HYDROGEN

Virtually all the hydrogen now produced in the world is used immediately in another process, so it is difficult to estimate how much is manufactured. However, we can estimate that approximately 300 billion STP-liters of the gas are produced worldwide in a year, and most of this is made from hydrocarbons such as natural gas, petroleum, or coal.

In today's technology, the method used to produce the largest quantities of H_2 is the catalytic *steam reformation of hydrocarbons* in a two-step process. The most desirable starting material is natural gas or methane, CH_4. In the first step, CH_4 reacts with steam at high temperature.

$$CH_4(g) + H_2O(g) + 205\ kJ \longrightarrow 3\ H_2(g) + CO(g)$$

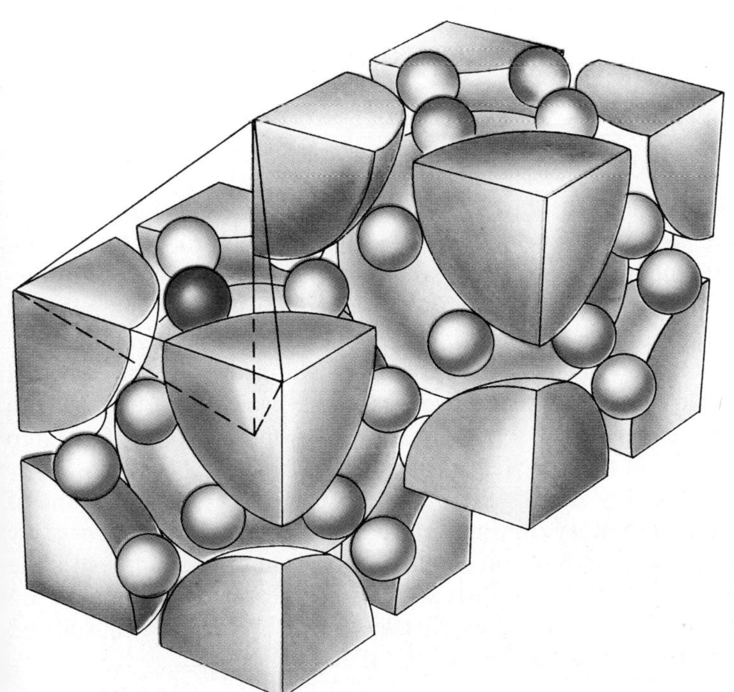

FIGURE 20.4

An interstitial metal hydride. The large spheres are niobium atoms packed in a body-centered cubic crystal lattice (Chapter 11). Hydrogen atoms (the small spheres) occupy the (tetrahedral) holes or interstices of the lattice. (See *Scientific American*, February, 1980, p. 118.)

The reaction is rapid in the 900°C to 1000°C range and goes nearly t completion. At a lower temperature, 400°C to 500°C, the CO is used a a reducing agent to extract still more hydrogen from water. In this process the "water gas shift" reaction, poisonous CO is converted to CO_2, nontoxic compound.

$$H_2O(g) + CO(g) \longrightarrow H_2(g) + CO_2(g) + 42 \text{ kJ}$$

The unwanted CO_2 is removed by carbonate formation, for example.

$$CO_2(g) + CaO(s) \longrightarrow CaCO_3(s)$$

Hydrocarbons other than methane can also be used as a starting material but methane has the highest H to C ratio (4:1). (Petroleum has a H:C ratio of about 2:1, while coal is about 1:1.) Therefore, methane give more total hydrogen per gram than any other hydrocarbon, and the rati of by-product CO_2 to H_2 is much lower. No matter what method is chose there is an environmental problem: CO_2 or $CaCO_3$ are produced in ver large quantities and must be disposed of in some manner.

The 19th century science fiction writer Jules Verne, author of *20,000 Leagues Under the Sea*, wrote that "Water will someday be employed as a fuel.... Water will be the coal of our future."

The electrolysis of water is perhaps the cleanest method of H production (Figure 20.2), and high-purity O_2 is a valuable by-product However, electrical energy is quite expensive, so only in places wher electricity is very cheap is electrolysis of water possible.

Water can also be split into hydrogen and oxygen using heat energy Almost 10,000 approaches to this have been suggested, all characterize by the use of a sequence of reactions such as

$$CaBr_2 + H_2O \xrightarrow{750°C} 2\,HBr + CaO$$

$$Hg + 2\,HBr \xrightarrow{100°C} HgBr_2 + H_2$$

$$HgBr_2 + CaO \xrightarrow{25°C} HgO + CaBr_2$$

$$HgO \xrightarrow{500°C} Hg + \tfrac{1}{2}\,O_2$$

$$\text{Net:} \quad H_2O(\ell) \longrightarrow H_2(g) + \tfrac{1}{2}\,O_2(g)$$

There are clearly major problems with hydrogen-generating cycles suc as that above. First, many require very high temperatures to operat efficiently, and the only economical source of these temperatures at pres ent seems to be nuclear reactors, with all the environmental and politica problems thereby posed. Second, if a hydrogen-generating cycle is to b successful, each step must be of high yield. The reason for this is tha the overall yield of any multistep process is the product of the yields i the individual steps. (A 90% yield in each step means that a four-ste process will have an overall yield of only 66%.)

There are many times when a chemist needs a small amount o hydrogen in the laboratory, and there are a number of *simple laborator methods* for making hydrogen on a small scale. One of the simplest is th reaction of a metal with an acid [reaction (a) in Table 20.1; Figures 3. and 6.8]. Indeed, in 1783 Jacques Charles used the reaction of sulfuri acid with iron to produce the hydrogen for his balloon.

TABLE 20.1 Methods for Preparing H_2 in the Laboratory

(a) Acid + metal ⟶ metal salt + H_2

$$Mg(s) + 2\ HCl(aq) \longrightarrow MgCl_2(aq) + H_2(g)$$

(b) Metal + H_2O or base ⟶ metal hydroxide or oxide + H_2

$$2\ Na(s) + 2\ H_2O(\ell) \longrightarrow 2\ NaOH(aq) + H_2(g)$$
$$2\ Fe(s) + 3\ H_2O(\ell) \longrightarrow Fe_2O_3(s) + 3\ H_2(g)$$
$$2\ Al(s) + 2\ KOH(aq) + 6\ H_2O(\ell) \longrightarrow 2\ KAl(OH)_4(aq) + 3\ H_2(g)$$

(c) Metal hydride + H_2O ⟶ metal hydroxide + H_2.

$$CaH_2(s) + 2\ H_2O(\ell) \longrightarrow Ca(OH)_2(aq) + 2\ H_2(g)$$

The reaction of Al with NaOH [reaction (b) in Table 20.1; Figure 6.9] is impractical due to the cost of materials. However, during World War II it was used to obtain H_2 to inflate small balloons for weather observation and to raise radio antennas. Metallic aluminum was plentiful, since it came from damaged and scrapped aircraft parts. Finally, reaction (c) in Table 20.1 is very useful, and this is perhaps the most efficient way to synthesize H_2 in the laboratory (Figure 15.1). It is also useful for removing traces of water from liquid compounds that do not react with hydrides, so metal hydrides are widely used for this purpose.

SOME USES OF HYDROGEN

The largest use of H_2 gas is in the production of ammonia, NH_3, by the *Haber process*.

$$N_2(g) + 3\ H_2(g) \longrightarrow 2\ NH_3(g)$$

Approximately 15 million tons of this fundamental chemical are made each year in the United States. As described in Chapter 22, NH_3 is used directly as a fertilizer or is converted to other fertilizers such as ammonium nitrate or urea $[(H_2N)_2CO]$.

The second largest use of hydrogen is in the manufacture of methyl alcohol or methanol, CH_3OH, by the reaction of H_2 with carbon monoxide.

$$2\ H_2(g) + CO(g) \xrightarrow{\text{catalyst}} CH_3OH(\ell)$$

Almost 7 billion gallons of this colorless liquid were produced in 1983. The compound, which can cause blindness if consumed internally, is increasingly used as an additive in nonleaded gasoline and is even being converted directly to gasoline in a special process invented recently. It can itself also be used as a fuel, since combustion of 1 mole gives 676 kJ.

$$2\ CH_3OH(\ell) + 3\ O_2(g) \longrightarrow 4\ H_2O(g) + 2\ CO_2(g) + \text{heat}$$

Aside from these uses, large quantities of methanol are used to make various adhesives, plastics, and synthetic fibers.

Still another large use of hydrogen is known to every lover of peanut butter. Peanut butter contains "hydrogenated (or saturated) vegetable

FIGURE 20.5

A prototype car run by hydrogen combustion in a slightly modified combustion engine. The hydrogen is stored in the form of an iron–titanium hydride, and heat from the engine exhaust is used to release H_2 from the hydride. (Mercedes-Benz)

oils,'' which means that H_2 has been added to a vegetable oil to change its properties. The essence of this *hydrogenation* reaction is illustrated by the following equation in which the simple compound ethylene, which contains a C=C double bond, has been converted to ethane by hydrogenation, the addition of H_2 to the double bond.

$$\underset{\text{ethylene}}{H_2C{=}CH_2}(g) + H_2(g) \longrightarrow \underset{\text{ethane}}{H{-}CH_2{-}CH_2{-}H}(g)$$

Fats contain, in varying proportions, organic esters with C=C double bonds. In general, the greater the number of double bonds the lower the melting point of the fat, so some fats are liquids at room temperature. The value of hydrogenation is that the oil is *hardened* to a solid with a consistency comparable to that of lard or butter. This is the basis of a large industry; cheap oils from cottonseed, corn, or soybeans are converted to cooking fats (e.g., Crisco or Spry) or margarine by this process.

HYDROGEN AS A FUEL

There is a growing awareness that reserves of coal and oil are limited. Nuclear power has recently posed severe political and environmental problems, and, even if these problems are solved, it is not useful for propelling cars, planes, or trains. Thus, there has been a search for alternatives, and one that has many attractive features is the ''hydrogen economy.''

See Figure 18.1, the explosive reaction of H_2 with O_2.

There are two major barriers to the goal of a hydrogen economy. The first is to find an inexpensive method for the synthesis of H_2, a problem discussed above. The second is a means of convenient storage. The space program has demonstrated that hydrogen can be stored relatively easily and safely as a liquid. However, this may not be convenient in a home or car, so an alternative is to store H_2 in the form of an interstitial hydride (Figure 20.4). The hydrogen is released as the solid is heated, and the gas can then be used in a slightly modified internal combustion engine (Figure 20.5).

The most visible use of H_2 as a fuel is in the space shuttle. The cigar-shaped tank strapped onto the shuttle contains 385,265 gallons of liquid H_2 and 143,351 gallons of liquid O_2 on lift-off (Figure 20.6). Just a few minutes into the flight, the fuel is exhausted, the hydrogen and oxygen having been converted completely to water.

On board the space shuttle is another example of hydrogen as a fuel. The H_2/O_2 fuel cell (Section 19.4) is routinely used as a power source

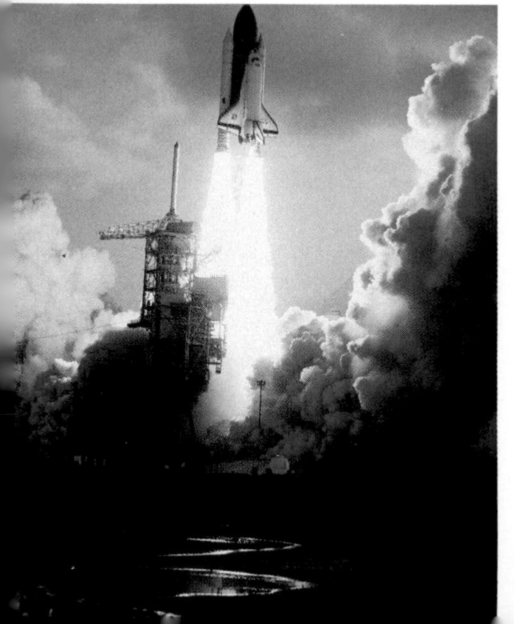

FIGURE 20.6

H_2 is used as a fuel in the space shuttle. The cigar-shaped tank contains 385,265 gallons of liquid H_2 and 143,351 gallons of liquid O_2 on lift-off. (NASA)

on space craft. Unfortunately, such fuel cells are too expensive to be used in more ordinary, consumer-oriented applications.

Using the methods in Chapter 5, you can calculate that the combustion of H_2 to give water vapor will give about 121 kJ per gram of H burned. While this seems like a large quantity of energy, the fusion of hydrogen nuclei to give helium, the reaction used by our sun (or in a hydrogen bomb), gives a million times more energy per gram of H. This is why scientists are trying to devise ways of carrying out the hydrogen "fusion" process on earth (Chapter 26). A small amount of hydrogen could provide an enormous amount of energy for consumers. A hydrogen economy holds great promise, but it poses many problems. We are doubtless many years away from its realization.

EXERCISE 20.1 THE CHEMISTRY OF HYDROGEN

Give balanced equations for four methods of preparing H_2. Give three uses of H_2.

The exercises in this and the chapters that follow are review questions meant to help you organize the chemistry of the elements.

20.2 DIAGONAL RELATIONSHIPS

When moving from hydrogen into the main groups of the periodic table, we find that the chemistry of the first element in a group is often markedly different from that of the second element. Beryllium hydroxide is amphoteric, while the hydroxides of the other alkaline earth elements live up to their name: they form basic solutions in water. Boron is a metalloid and forms volatile hydrides, while the other Group 3A elements are metallic in character and form more nearly ionic hydrides. Carbon is the basis for literally millions of compounds with C—C and C—H bonds, while silicon forms relatively few with Si—Si bonds. A closer look at the chemistry of the first element in a group reveals that it behaves more like the second element in the following group. That is, lithium and magnesium are rather alike, as are beryllium and aluminum, and boron and silicon share some similarities (except for oxidation number).

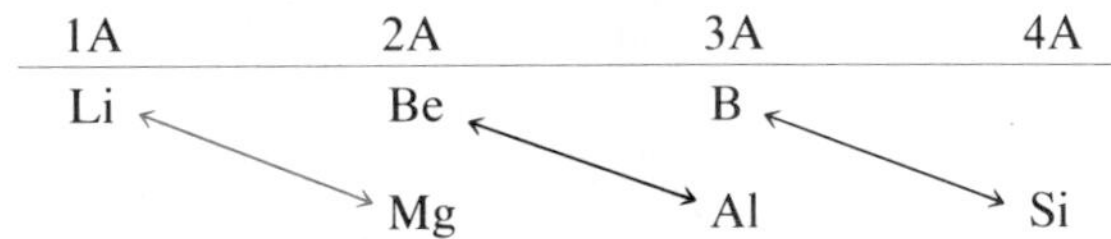

This has come to be known as the **diagonal relationship**.

The underlying cause of the diagonal relationship is that diagonally related ions have similar effective nuclear charges. For example, although the radius of the Mg^{2+} ion (65 pm) is somewhat larger than that of Li^+ (60 pm), the Mg^{2+} ion has a larger nuclear charge. Thus, the ability of a Mg^{2+} ion to attract a negative ion or dipole is approximately the same as that of Li^+.

In the text that follows, you will see examples of the diagonal relationship in the six elements above in particular, so knowledge of one element helps you to know something about another element. But be careful! Don't push the relationship unreasonably. The oxygen and hydrogen compounds of boron and silicon, for example, do have similarities, but there are some great differences as well.

1A
1 H 1.0079
3 Li 6.941
11 Na 22.9898
19 K 39.0983
37 Rb 85.4678
55 Cs 132.905
87 Fr (223)

20.3 THE ALKALI METALS: GROUP 1A

The elements of Group 1A (Table 20.2) are characterized by an electron configuration of [core]ns^1, and all occur as +1 ions. None are found in nature as free elements, since all combine rapidly and completely with virtually all the nonmetals. They are called the alkali metals, because they react with water to form hydroxides, which are all soluble bases, historically known as alkalies.

COMMON MINERALS AND RECOVERY OF THE METAL

Although sodium and potassium are comparatively abundant in nature, the other elements are more rare. The lightest element, lithium, is found in trace amounts in natural waters, soils, and rocks throughout the world. Not surprisingly, the chief mineral form of lithium is an aluminosilicate,

TABLE 20.2 Group 1A: The s^1 Elements (Alkali Metals)

	LITHIUM	SODIUM	POTASSIUM	RUBIDIUM	CESIUM	FRANCIUM
Symbol	Li	Na	K	Rb	Cs	Fr
Atomic number	3	11	19	37	55	87
Atomic weight	6.941	22.99	39.10	85.47	132.9	(223)*
Valence e^-	$2s^1$	$3s^1$	$4s^1$	$5s^1$	$6s^1$	$7s^1$
mp, °C	186	97.5	63.65	38.89	28.5	27
bp, °C	1326	889	774	688	690	677
d, g/cm^3	0.534	0.971	0.862	1.53	1.87	—
Atomic radius, pm	152	186	231	244	262	—
Ion radius, pm	60	95	133	148	169	—
Pauling EN	1.0	0.9	0.8	0.8	0.7	0.7
Standard reduction potential	−3.05	−2.71	−2.92	−2.93	−2.92	—
Oxidation numbers	+1	+1	+1	+1	+1	—
Ionization energy*, kJ	519	498	418	402	377	—
Heat of vaporization, kJ	159	107	89	81	76	—
Discoverer	Arfvedson	Davy	Davy	Bunsen and Kirchhoff	Bunsen and Kirchhoff	Perey
Date of discovery	1817	1807	1807	1861	1860	1939
rpw* pure O_2	Li_2O	Na_2O, Na_2O_2	K_2O_2, KO_2	RbO_2	CsO_2	—
rpw H_2O	LiOH	NaOH	KOH	RbOH	CsOH	—
rpw N_2	Li_3N	None	None	None	None	—
rpw halogens	LiX	NaX	KX	RbX	CsX	—
rpw H_2	Li^+H^-	Na^+H^-	K^+H^-	Rb^+H^-	Cs^+H^-	—
Flame color	Bright red	Yellow	Violet	Purple	Blue	—
Mohs hardness	0.6	0.4	0.5	0.3	ul*	—
Crystal structure	Cubic bc*	Cubic bc	Cubic bc	Cubic bc	Cubic bc	—

*Atomic weights in parentheses are those of the most stable isotope. All energies and heats are in kJ per mole. The letters rpw stand for "reaction product with"; bc = body-centered; ul = usually liquid. Electrode potentials are in volts.
(Adapted from E.G. Rochow, *Modern Descriptive Chemistry*, W.B. Saunders, 1977.)

$LiAlSi_2O_6$. Lithium can be recovered from the mineral by first heating it with sodium carbonate (soda ash) to form lithium carbonate.

$$Na_2CO_3(aq) + 2\ LiAlSi_2O_6(s) \longrightarrow Li_2CO_3(s) + 2\ NaAlSi_2O_6(s)$$

To get the poorly soluble lithium carbonate into solution so that it can be separated from the insoluble silicates, CO_2 is bubbled into the mixture, forming lithium bicarbonate.

$$Li_2CO_3(s) + CO_2(g) + H_2O(\ell) \longrightarrow 2\ LiHCO_3(aq)$$

This is similar to the reaction that occurs when CO_2-laden groundwater encounters a bed of limestone ($CaCO_3$) and dissolves the limestone as $Ca(HCO_3)_2$.

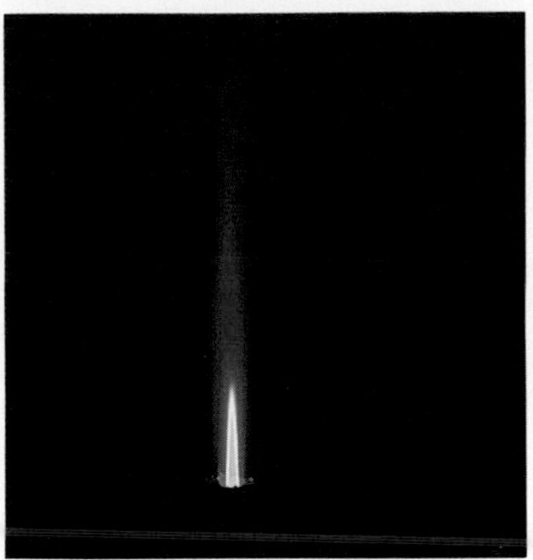

When salts of the alkali metals are placed in a flame, light of characteristic colors is emitted. In this photograph, lithium bromide emits red light, as do all lithium salts. Sodium salts emit yellow light (see chapter opening photograph), while potassium salts emit a violet light (see photograph opening Chapter 18). (Fundamental Photographs)

Within the earth's crust, sodium and potassium are about equally abundant (2.83% and 2.59%, respectively). However, sea water contains 2.8% NaCl, but only 0.08% KCl. Why the great difference? Sodium and potassium salts are both water soluble. Since they have about the same relative abundance on land, why didn't the rain dissolve Na- and K-containing minerals over the centuries and carry them down to the sea, to appear in the same proportions as on land? The answer lies in the fact that potassium is an important factor in plant growth. Much of the potassium appearing in groundwater from dissolved minerals is taken up preferentially by plants, while the sodium ion continues on to the sea. Most plants contain 4 to 6 times as much combined potassium as sodium, so potassium compounds find important use as fertilizers.

A reasonable amount of salt is essential in the diet of humans and animals, because many biological functions are controlled by concentrations of sodium and chloride ions. Animals will travel great distances to a "salt lick," and farmers and ranchers often place large blocks of salt in the fields for cattle to lick. Indeed, salt has been so important for so long that it has influenced us in ways you may not have thought about. When we are paid for work done, we can be paid a "salary." The word comes from the Latin word *salarium*, which originally meant "salt money" or money paid to Roman soldiers for guarding salt deposits. And we still talk of "salting away" some money for a rainy day, a term related to the practice of preserving meat by salting it.

Although sodium chloride is the most common form of sodium in nature, where it is found as rock salt or halite in salt lakes or underground deposits (Figure 20.7), there are other important sodium-containing minerals: sodium borate (borax), sodium carbonate (soda or trona), sodium nitrate (Chile saltpeter), and sodium sulfate (mirabilite).

Potassium deposits have also come from the crystallization of minerals from ancient seas, and the largest of these found thus far is in Saskatchewan, Canada; the deposit is thought to contain 10 billion tons of KCl.

Metallic sodium is produced by electrolysis of a molten NaCl (Chapter 19). Although potassium could be similarly made, there are many problems with the method, not the least of which is that the molten metal is soluble in molten KCl, and the two cannot be separated. Instead, potassium is produced by allowing sodium vapor to react with molten KCl

FIGURE 20.7
A dry salt lake (Bonneville Salt Flat, Utah). (P. and L. Hermann, Tom Stack & Associates)

$$Na(g) + KCl(\ell) \longrightarrow K(g) + NaCl(\ell)$$

This reaction is a good example of the importance of understanding chemical equilibria. It has an equilibrium constant less than 1 ($\Delta G^\circ > 0$), so the reactant side is favored. However, since potassium vapor is continually removed, we can tell from Le Chatelier's principle that more product will be formed as the system attempts to come to equilibrium, and KCl can eventually be converted completely to potassium metal.

Rubidium and cesium are also prepared by reaction of their chlorides with sodium.

Francium is found only in trace amounts, and all its isotopes are radioactive. Indeed, its longest-lived isotope (^{223}Fr) has a half-life of only 22 minutes, making it rather difficult to carry out any extensive chemical experiments.

THE GROUP 1A ELEMENTS: THEIR GENERAL PROPERTIES AND USES

All the uses of the alkali metals themselves stem from their great ability to act as reducing agents. One of the best known reactions of the alkali metals is that with water. If you drop a small piece of sodium into water, it will react with great vigor, the heat of reaction sometimes igniting the hydrogen that is produced (Figure 20.8)

$$2\ Na(s) + 2\ H_2O(\ell) \longrightarrow 2\ NaOH(aq) + H_2(g)$$

Lithium is more docile, but potassium, rubidium, and cesium will react with great violence.

See photo beginning Chapter 18 for the reaction of K and H_2O.

Ammonia is certainly not considered as strong a Brønsted acid as water, and so it differs from water on reaction with sodium. When sodium is put in liquid ammonia (a reaction that must be done at temperatures below −33°C, the boiling point of ammonia), a beautiful deep blue solution is formed. Sodium, being a powerful reducing agent, loses an electron, but the electron is "solvated" by the ammonia, just as an anion can be bound up by a hydrogen-bonding solvent.

$$Na(s) \xrightarrow[\text{in liquid } NH_3]{} Na^+(\text{in } NH_3) + e^-(\text{in } NH_3)$$

This solution has great reducing power, and it is widely used for this purpose.

Until quite recently, the largest use of sodium as a reducing agent was in the synthesis of tetraethyllead.

$$4\ Na(s) + 4\ C_2H_5Cl(\ell) + Pb(s) \longrightarrow 4\ NaCl(s) + Pb(C_2H_5)_4(\ell)$$

The lead compound can be added to gasoline to raise the octane rating. However, due to severe pollution problems, leaded gasolines are being phased out, at least in the United States. Sodium is now a product in search of a market. Some of it, for example, is used to reduce $TiCl_4$ to metallic titanium, which is then used as a strong, lightweight, but expensive structural material.

$$TiCl_4(\ell) + 4\ Na(s) \longrightarrow Ti(s) + 4\ NaCl(s)$$

In principle, sodium can be used to reduce many metal halides to metal and sodium halide. However, since sodium is made by electrolysis and electrical power is expensive, the sodium is costly. Producing metals by sodium reduction is therefore an expensive process and is used only in special cases.

FIGURE 20.8

Sodium reacting with water. The vigorous reaction gives H_2 and NaOH. An indicator (phenolphthalein) in the water has turned red in the presence of the NaOH. CAUTION! This is a potentially explosive reaction and should only be done with the proper safety precautions. (Charles D. Winters)

Not only can the Group 1A metals reduce water and ammonia, but they most assuredly react with oxygen, one of the most powerful oxidizing agents. What is fascinating about this reaction, however, is that the different metals produce such different products. Lithium is the *only* alkali metal to give a simple oxide when reacting with O_2.

$$4\ Li(s) + O_2(g) \longrightarrow 2\ Li_2O(s)$$

On the other hand, sodium produces sodium **peroxide**, a yellowish-white solid.

$$2\ Na(s) + O_2(g) \longrightarrow Na_2O_2(s)$$

The peroxide ion is O_2^{2-}, essentially an oxygen molecule that has been reduced by two electrons. This is a commercially useful material, because it hydrolyzes to produce hydrogen peroxide, H_2O_2.

Recall the discussion of the molecular orbital structure of O_2 and its ions in Chapter 10.

$$Na_2O_2(s) + 2\ H_2O(\ell) \longrightarrow 2\ NaOH(aq) + H_2O_2(aq)$$

Because of the great oxidizing power of sodium peroxide, it is widely used in the paper and textile industries as a bleaching agent.

Potassium, rubidium, and cesium also reduce O_2, but they all form **superoxides**, MO_2, where the alkali metal ion is associated with the O_2^- ion.

As explained in Chapter 10, the superoxide ion, O_2^-, is paramagnetic.

$$K(s) + O_2(g) \longrightarrow KO_2(s)$$

The most important commercial use of KO_2 is as a source of oxygen in an emergency breathing apparatus. The so-called "oxygen masks" are designed so that the CO_2 and water vapor in the exhaled breath of the wearer reacts with superoxide to provide oxygen.

The products listed for alkali metals + O_2 are the major products. In addition, Na forms some Na_2O and K forms some K_2O and K_2O_2.

$$4\ KO_2(s) + 4\ CO_2(g) + 2\ H_2O(g) \longrightarrow 4\ KHCO_3(s) + 3\ O_2(g)$$

It is clear that the uses of the Group 1A elements as metals depend on their reducing ability. The data in Table 20.2 show that Li is the best reducing agent in the group, whereas Na is the poorest, and the remainder are comparable in their ability.* This lack of a clear-cut trend is curious, so it is useful to analyze the relative reducing ability of these metals to understand their behavior.

Analysis of $E°$ is a thermodynamic problem, and, as in any thermodynamic problem, we can break the process $M(s) \rightarrow M^+(aq) + e^-$ into a series of steps

$$\begin{array}{ccc} M(g) & \xrightarrow{IE\text{ (ionization energy)}} & M^+(g) + e^- \\ \uparrow \Delta H_{sub} & & \downarrow \Delta H_{hyd} \\ M(s) & \xrightarrow[\Delta H_{net}]{} & M^+(aq) + e^- \end{array}$$

from which we write

*The $E°$ values found in Table 20.2 are given by convention as reduction potentials; that is, the values are for the reaction $M^+(aq) + e^- \rightarrow M(s)$. Therefore, the value for $Li(s) \rightarrow Li^+(aq) + e^-$ is +3.05 V, for example.

$$\Delta H_{net} = \Delta H_{sub} + IE + \Delta H_{hyd} = IE + (\Delta H_{sub} + \Delta H_{hyd})$$

In this case, we imagine the solid metal sublimes (sub) to form the vapor, an electron is removed from a gaseous atom, and then the gaseous ion is hydrated (hyd). The first two steps require energy, but the third is exothermic. You can see now that the element that is the best reducing agent should have the most negative (or least positive) value of ΔH_{net}, and the sum ($\Delta H_{sub} + \Delta H_{hyd}$) is responsible for the difference between reducing ability and ionization energy.

Since the enthalpy changes for all of these steps are known for all of the Group 1A elements, we can easily calculate ΔH for $M(s) \rightarrow M^+(aq) + e^-$. Before doing the numerical analysis, however, there are two important points to be made. First, $E°$ for an electrochemical half reaction is directly related to $\Delta G°$, and not to the $\Delta H°$, for the reaction. Fortunately, $\Delta G°$ is largely determined by $\Delta H°$ in this case, so it is legitimate to ascribe variations in $E°$ to variations in $\Delta H°$. Second, we are calculating $\Delta H°$ for a *hypothetical* half reaction, not an actual reaction. Therefore, the numbers derived are not the absolute values but are only relative values that allow us to compare relative reducing ability.

The values of ΔH for each of the steps, for each of the Group 1A elements, are given in Table 20.3, as is the calculated value of ΔH for the element acting as a reducing agent. According to these calculations, Li is the best reducing agent, Na the poorest, and the others all have about the same ability, exactly in agreement with experimentally determined $E°$ values (see Table 20.3). And now we can see why. The small lithium ion has a high effective nuclear charge. Thus, it forms very strong bonds with water and so has the most negative heat of hydration. In reality it is this great release of energy when solvating $Li^+(g)$ that offsets its relatively high ionization energy and that makes lithium the best reducing agent. *The hydration energy plays a very large role in determining $E°$.*

The very small size of Li^+ is important in determining many of its properties and in setting it apart in many ways from the other elements of the group. You will see this effect again.

A useful lesson to be learned from this analysis of alkali metal $E°$ values is that a physical property such as $E°$ is not determined solely by one underlying property of the element; rather, it is an amalgamation of several such properties, interacting in a complex way. Indeed, it is likely that many periodic trends are really a complex interplay of several underlying properties.

SOME IMPORTANT GROUP 1A COMPOUNDS

THE CHLOR–ALKALI INDUSTRY The "chlor–alkali" industry is the biggest user and producer of alkali metal compounds, and the main building block is sodium chloride. Chief among the products of the industry are chlorine

The chemistry of chlorine is described in Chapter 22.

TABLE 20.3 Enthalpy Changes for Electrode Half Reactions*

	Li	Na	K	Rb	Cs
Ionization energy	519	498	418	402	377
ΔH_{sub}	159	107	89	81	76
ΔH_{hyd}	−506	−397	−318	−289	−259
$(\Delta H_{sub} + \Delta H_{hyd})$	−347	−290	−229	−208	−183
$\Delta H[M(s) \longrightarrow M^+(aq) + e^-]$	172	208	189	194	194

*Energies are all in kJ/mol. sub = sublimation; hyd = hydration.

TABLE 20.4 Chemicals Produced Directly or Indirectly by the Chlor–Alkali Industry (1984)

COMPOUND	AMOUNT (IN BILLIONS OF POUNDS)	RANK (AMONG TOP 50 CHEMICALS PRODUCED)
Sodium hydroxide (caustic soda)	22.45	7
Chlorine	21.45	9
Sodium carbonate (soda ash)	17.02	10
Calcium chloride	2.10	40
Sodium sulfate	1.74	44
Sodium silicate	1.50	46
Sodium triphosphate	1.33	49

and sodium hydroxide from the electrolysis of brine, aqueous sodium chloride (Chapter 19),

$$2\ NaCl(aq) + 2\ H_2O(\ell) \xrightarrow[\text{electrolysis}]{} 2\ NaOH(aq) + H_2(g) + Cl_2(g)$$

To give you some idea of the size and importance of the industry, consider the amounts of the chemicals produced in 1984 (Table 20.4).

Sodium hydroxide is normally sold as a concentrated aqueous solution called "caustic soda." Approximately half of it is used in the manufacture of other chemicals (such as in the aluminum industry), and a large part of the remainder is used in the pulp and paper industry.

Sodium carbonate has been obtained, since prehistoric times, from naturally occurring deposits of $Na_2CO_3 \cdot 10\ H_2O$, commonly known as **washing soda**. (Anhydrous sodium carbonate is called **soda ash**.) More recently, however, a deposit of *trona*, $Na_2CO_3 \cdot NaHCO_3 \cdot 2\ H_2O$, estimated at 6×10^{10} tons has been discovered in Wyoming, and virtually all the sodium carbonate in the United States now comes from that source. Its largest use is in the manufacture of *glass*.

Until quite recently, the **Solvay process** (Figure 20.9) was the major source of sodium carbonate and sodium bicarbonate. This process has been superseded, particularly in the United States, by less costly mined soda ash, but it is still in use in other parts of the world. It is worth describing, not only because of its economic importance, but also because it illustrates many chemical principles.

The overall reaction of the Solvay process is a simple exchange,

$$\underset{\text{brine}}{2\ NaCl} + \underset{\text{limestone}}{CaCO_3} \longrightarrow \underset{\text{soda ash}}{Na_2CO_3} + CaCl_2$$

but the reaction will not occur directly. In practice, the first step is to convert limestone to CO_2 and lime, CaO, in a decomposition reaction (Chapter 3).

$$CaCO_3(s) \longrightarrow CaO(s) + CO_2(g)$$

and then to allow the CO_2 to react with aqueous ammonia.

$$CO_2(g) + NH_3(aq) + H_2O(\ell) \longrightarrow NH_4HCO_3(aq)$$

This reaction occurs because CO_2 produces the weak Brønsted acid H_2CO_3 in water, and this in turn reacts with the weak Brønsted base NH_3. Sodium

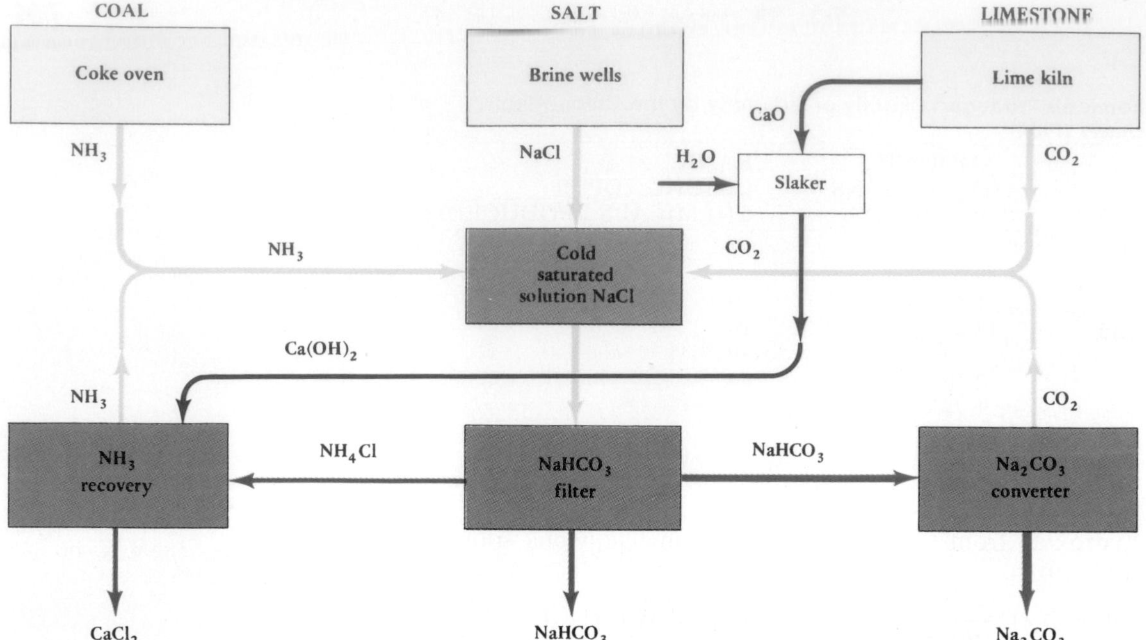

FIGURE 20.9

The Solvay process for the production of sodium carbonate and sodium bicarbonate. It is named after Ernest Solvay, an Englishman who first succeeded in getting the process to work on an economical basis in 1869. Although the process continues to be important in other countries, it is being phased out in the United States; here the mineral *trona* ($Na_2CO_3 \cdot NaHCO_3 \cdot H_2O$) now supplies most of our domestic needs.

chloride then undergoes an exchange reaction with the ammonium bicarbonate.

$$NaCl(aq) + NH_4HCO_3(aq) \longrightarrow NaHCO_3(s) + NH_4Cl(aq)$$

The product is sodium bicarbonate (or sodium hydrogen carbonate), a salt that is relatively insoluble in the reaction medium. The sodium bicarbonate is isolated and used directly for some purposes, but most of it is converted by heating to soda ash, sodium carbonate.

$$2\ NaHCO_3(s) \longrightarrow Na_2CO_3(s) + CO_2(g) + H_2O(g)$$

Ammonia is an expensive material, so it is not economic to waste it as NH_4Cl, a by-product of the reactions above; the NH_3 must be recovered to be used again. To do this, water is added to the CaO produced in the first step to give the base $Ca(OH)_2$, **slaked lime**.

$$\underset{\text{lime}}{CaO(s)} + H_2O(\ell) \longrightarrow \underset{\text{slaked lime}}{Ca(OH)_2(aq)}$$

The base reacts with the ammonium ion, a Brønsted acid, to regenerate NH_3,

$$NH_4^+(aq) + OH^-(aq) \longrightarrow NH_3(aq) + H_2O(\ell)$$

so the net reaction is

$$2\ NH_4Cl(aq) + CaO(s) \longrightarrow 2\ NH_3(aq) + H_2O(\ell) + CaCl_2(aq)$$

A big problem is that the calcium is left over as $CaCl_2$. Some is used in the winter on roads and sidewalks to melt ice and snow, but much of it remains to be disposed of in some manner.

The amount of sodium bicarbonate produced annually is approximately 2% of the amount of sodium carbonate produced. However, you are probably more aware of the bicarbonate, since it is commonly called

baking soda. Not only is this used in cooking, but it is also added in small amounts to table salt. Sodium chloride is often contaminated with small amounts of $MgCl_2$. The magnesium salt is hygroscopic; that is, it picks water up from the air and, in doing so, causes the NaCl to clump on damp days. Adding $NaHCO_3$ converts $MgCl_2$ to magnesium carbonate, a non-hygroscopic salt.

$$MgCl_2(s) + 2\ NaHCO_3(s) \longrightarrow MgCO_3(s) + 2\ NaCl(s) + H_2O(\ell) + CO_2(g)$$

OTHER ALKALI METAL SALTS Many other salts of the alkali metals are important to commerce, and a few are of interest here.

Sodium sulfate or "salt cake" is a product of an exchange reaction (Figure 20.10).

This is also a good method of preparing HCl gas in the laboratory. It is especially useful because H_2SO_4 is a good drying agent, so the HCl that evolves is quite dry.

$$2\ NaCl(s) + H_2SO_4(aq) \longrightarrow Na_2SO_4(s) + 2\ HCl(g)$$

It is inexpensive to produce in this manner, since both NaCl and sulfuric acid, a by-product of many industrial processes, are inexpensive. In addition, a commercially useful product, HCl, is also formed. Sodium sulfate itself is used mainly in the production of wood pulp.

Large deposits of *sodium nitrate*, $NaNO_3$, are found in Chile, hence the name "Chile saltpeter." It is thought that these deposits were formed by bacterial action on organisms in shallow seas. The initial product was ammonia that was subsequently oxidized to nitrate ion; combination with sea salt led to sodium nitrate. However, because nitrates in general, and the alkali metal nitrates in particular, are so water soluble, deposits are only found in areas of very low rainfall.

Sodium nitrate can be converted easily into KNO_3 by the exchange reaction,

$$NaNO_3(aq) + KCl(aq) \longrightarrow KNO_3(aq) + NaCl(s)$$

because, of the four salts in this reaction, NaCl is the least soluble in hot water. These nitrates are largely used for their oxidizing ability. For example, for centuries KNO_3 has been used as the oxidizer in gunpowder. A mixture of KNO_3, charcoal, and sulfur will spontaneously react when ignited according to the equations

$$2\ KNO_3(s) + 4\ C(s) \longrightarrow K_2CO_3(s) + 3\ CO(g) + N_2(g)$$

$$2\ KNO_3(s) + 2\ S(s) \longrightarrow K_2SO_4(s) + SO_2(g) + N_2(g)$$

Notice that both reactions produce gases, and it is these hot gases that propel a bullet from a gun or cause a firecracker to explode. (See the Special Section on the Chemistry of Fireworks at the end of this chapter.)

Over 1 billion pounds of *sodium tripolyphosphate*, $Na_5P_3O_{10}$, are produced annually in the United States to be used as a "builder" in household detergents. Not only does it give a basic solution on hydrolysis,

$$P_3O_{10}^{5-}(aq) + H_2O(\ell) \longrightarrow HP_3O_{10}^{4-}(aq) + OH^-(aq)$$

but the tripolyphosphate ion forms soluble complexes with the troublesome calcium ion found in hard water. However, there is a continuing controversy about the role of phosphates in water pollution. As a result,

FIGURE 20.10
When concentrated sulfuric acid (in the funnel at the top) drops onto solid NaCl in the flask, the products are Na_2SO_4 ("salt cake") and HCl gas. The HCl vapor exits through the side arm of the flask. When it passes over an indicator on the filter paper, the indicator turns bright red, its color in acid solution. (Charles D. Winters)

their use in the United States has declined, and some manufacturers are increasingly using sodium carbonate (washing soda) again.

Lithium carbonate, Li_2CO_3, has been used for more than 30 years as a treatment for manic depression. Manic depression involves alternating phases of depression and mania or over-excitement, mood swings occurring in some instances in periods of a few weeks to a year or so. Although the treatment of manic depression with lithium salts is well known, the mechanism of lithium action is not understood. However, a good guess is that it disturbs Mg^{2+} metabolism. In Section 20.2, it was noted that there is a "diagonal relationship" between the properties of Li and Mg, so it is possible that Li^+ takes the place of Mg^{2+} in some key process. The problem is that Mg^{2+} is important to so many biochemical processes that it may be difficult to pinpoint the one process accounting for manic depression.

All the alkali metals react with hydrogen when heated to produce salt-like hydrides such as *sodium hydride*, NaH.

$$2\ Na(s) + H_2(g) \longrightarrow 2\ NaH(s)$$

The chief chemical characteristic of alkali metal hydrides, MH, is that the hydride ion, H^-, is a strong Lewis and Brønsted base; as such, it reacts vigorously with acids such as water (Figure 15.1).

$$NaH(s) + H_2O(\ell) \longrightarrow NaOH(aq) + H_2(g)$$

The heat of this reaction is so great that the hydrogen can ignite. In research and industrial laboratories the hydrides are commonly used as drying agents for nonaqueous solvents and as reducing agents.

There are many important alkali metal compounds in commerce and biochemistry. Watch for more of them, in this book and in consumer products.

EXERCISE 20.2 THE CHEMISTRY OF GROUP 1A ELEMENTS

(a) Write balanced equations to show how elemental Na and K are prepared.
(b) Write balanced equations to illustrate two uses of sodium metal.
(c) Name three uses of sodium-containing compounds.
(d) Give the formulas for washing soda and baking soda.

FIGURE 20.11

Calcium oxide (lime), a white solid, reacts with water to produce $Ca(OH)_2$ (slaked lime). The test tube at the left contains the indicator phenolphthalein, which turns red in basic solution. The tube at the right contains only CaO in water. (Charles D. Winters)

20.4 THE ALKALINE EARTH METALS: GROUP 2A

The elements of the alkaline earth group (Table 20.5) are all characterized by electron configurations of the type [core]ns^2. As a result, all form compounds in the +2 oxidation state such as calcium oxide, CaO. With the exception of BeO, all Group 2A oxides hydrolyze to give a basic solution (Figure 20.11); hence the name "alkaline" in the name of this periodic group.

$$CaO(s) + H_2O(\ell) \longrightarrow Ca(OH)_2(s)$$
$$K_{sp} = 7.9 \times 10^{-6}$$

The "earth" part of the group name is left over from the days of alchemy. To medieval alchemists, any solid substance that did not melt

TABLE 20.5 Group 2A: The s^2 Elements (Alkaline-Earth Metals)

	BERYLLIUM	MAGNESIUM	CALCIUM	STRONTIUM	BARIUM	RADIUM
Symbol	Be	Mg	Ca	Sr	Ba	Ra
Atomic number	4	12	20	38	56	88
Atomic weight	9.012	24.31	40.08	87.62	137.33	226.03
Valence e^-	$2s^2$	$3s^2$	$4s^2$	$5s^2$	$6s^2$	$7s^2$
mp, °C	1283	650	845	770	725	700
bp, °C	2970	1120	1420	1380	1640	1140
d, g/cm³	1.85	1.74	1.55	2.60	3.51	5
	5.0	14.0	29.9	33.7	39.0	—
Atomic radius, pm	111	160	197	215	217	220
Ion radius, pm	31	65	99	113	135	—
Pauling EN	1.5	1.2	1.0	1.0	1.0	1.0
Standard reduction potential*	−1.85	−2.37	−2.87	−2.89	−2.90	−2.92
Oxidation numbers	+2	+2	+2	+2	+2	+2
Ionization energy*	900	736	590	548	502	—
Heat of vaporization	324	147	178	164	180	—
Discoverer*	Vauquelin	Bussy	Berzelius	Davy	Davy	Curie
Date of discovery	1798	1831	1808	1808	1808	1911
rpw* pure O_2	BeO	MgO	CaO	SrO, SrO_2	BaO_2	RaO
rpw H_2O	None	None	$Ca(OH)_2$	$Sr(OH)_2$	$Ba(OH)_2$	$Ra(OH)_2$
rpw N_2	None	Mg_3N_2	Ca_3N_2	Sr_3N_2	Ba_3N_2	Ra_3N_2
rpw halogens	BeX_2	MgX_2	CaX_2	SrX_2	BaX_2	RaX_2
rpw H_2	None	MgH_2	CaH_2	SrH_2	BaH_2	—
Flame color	—	—	Brick red	Crimson	Green	—
Mohs hardness	4	2.0	3	1.8	1.5	—
Crystal structure	Hexagonal	Hexagonal	Cubic fc*	Cubic fc	Cubic bc*	—

*rpw = reaction product with; fc = face-centered; bc = body-centered. "Discoverer" refers to first isolation; Mg, Ca, and Ba were known to ancients. All energies and heats are in kJ per mole; potentials are in volts.
(Adapted from E.G. Rochow, *Modern Descriptive Chemistry*, W.B. Saunders, 1977.)

and was not changed by fire into another substance was called an "earth." Compounds of Group 1A elements, such as Na_2CO_3 and NaOH, are certainly alkaline according to the experimental tests of the alchemists: the compounds all have a bitter taste and have the ability to neutralize acids. However, some Group 1A compounds melted in a fire or combined with the clay containers in which they were heated. On the other hand, since the melting point of CaO is 2572°C, a temperature well beyond the range of any ordinary fire, it was unchanged in fire. Since it also was alkaline, it was called an "alkaline earth," a phrase that came to be applied to the elements themselves in Group 2A.

The abundances of the elements vary widely (Table 20.6); just as in Group 1A, the lightest element is very rare and the heaviest element

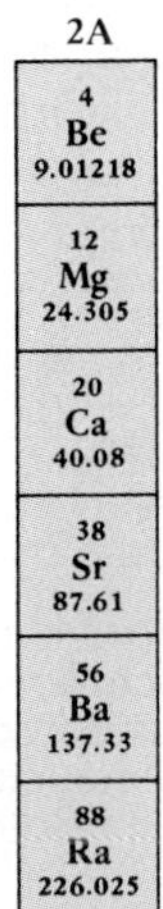

TABLE 20.6 Abundance of Group 2A Elements*

	IGNEOUS ROCKS	LUNAR ROCKS	OCEANS
Be	<6	—	—
Mg	18,700	42,000	1,310
Ca	29,600	103,000	420
Sr	340	166	8–12
Ba	650	210	0.03
Ra	9×10^{-7}	—	10^{-10}

*In parts per million.

is radioactive. The great abundance of calcium and magnesium led to their wide occurrence in plants and animals, and both elements form many compounds that are commercially important. It is on this chemistry that we want to focus most of our attention.

COMMON MINERALS OF THE ALKALINE EARTH METALS

Masses of beryl weighing more than a ton have been found in New Hampshire.

Like so many elements, beryllium is commonly found in the form of an aluminosilicate, in this case the mineral beryl, $3\ BeO \cdot Al_2O_3 \cdot 6\ SiO_2$. Although beryl is colorless when pure, it can be a brilliant green if some of the Al^{3+} ions are replaced by Cr^{3+} ions, and you know it as an emerald. If, on the other hand, it contains Fe^{2+} and Fe^{3+} as impurities, it is blue-green and is known as aquamarine, the birthstone for March.

Crystals of calcite, a form of $CaCO_3$. (Allen B. Smith, Tom Stack & Associates)

The great abundance of magnesium and calcium means that they are found in common minerals (Table 20.7). Strontium and barium can be found in relatively concentrated deposits, but radium is extremely widely distributed. The latter is always found in association with uranium, since it is a product of the radioactive disintegration of ^{238}U. It is long lived enough to isolate, but uranium ore only contains about 1 mg of Ra for every 3 kg of uranium.

The principal source of commercial calcium compounds is limestone, which occurs in immense sedimentary beds over large parts of the earth's surface. These are deposits formed from the fossilized remains of marine life and are mostly the *calcite* form of the compound. However, you often know calcite by other names. Limestone is generally contaminated with other metal ions such as those of iron, but *marble* is fairly pure calcite, formed in nature by recrystallization of $CaCO_3$ under pressure. High quality deposits are found in Italy and in the United States in Vermont, Georgia, and Colorado. Yet another form of $CaCO_3$ is *Iceland spar*, which forms large, clear crystals. Finally, *chalk* is a fine grained, powdery form that is found not only in classrooms but also in such spectacular deposits as the cliffs of the Grand Canyon (Figure 17.10).

Gypsum, or hydrated calcium sulfate, is extensively mined. Some of it is used in portland cement, but most is heated in large kilns to form "plaster of Paris," a process called *calcining*.

$CaSO_4 \cdot \frac{1}{2} H_2O$ is known as plaster of Paris because it was originally obtained from gypsum mined in Montmartre, a district of Paris.

$$\underset{\text{gypsum}}{CaSO_4 \cdot 2\ H_2O(s)} \longrightarrow \underset{\text{plaster of Paris}}{CaSO_4 \cdot \tfrac{1}{2}\ H_2O(s)} + \tfrac{3}{2}\ H_2O(g)$$

Dolomite, a mixture of magnesium and calcium carbonate, is a common mineral and has given its name to the mountainous region of northern Italy, the so-called Dolomite Alps.

TABLE 20.7 Common Minerals of the Group 2A Elements

COMMON NAME	FORMULA
Beryl	$3\ BeO \cdot Al_2O_3 \cdot 6\ SiO_2$
Magnesite	$MgCO_3$
Talc or soapstone	$3\ MgO \cdot 4\ SiO_2 \cdot H_2O$
Asbestos (chrysotile)	$3\ MgO \cdot 2\ SiO_2 \cdot 2\ H_2O$
Dolomite	$MgCO_3 \cdot CaCO_3$
Limestone (calcite or aragonite)	$CaCO_3$
Gypsum	$CaSO_4 \cdot 2\ H_2O$
Fluorspar	CaF_2
Fluorapatite	$CaF_2 \cdot Ca_3(PO_4)_2$
Barites	$BaSO_4$

If enough water is added to plaster of Paris to make a paste, it quickly hardens as it reverts to gypsum. The mixture also expands as it hardens, so it forms a sharp impression of anything molded in it. Virtually all calcined gypsum is used to make wallboard (sheet rock or plaster board), and the rest is used for industrial and building plasters. In fact, this use of gypsum is very old; there is evidence that the interiors of some of the great pyramids of Egypt were coated with gypsum plaster.

The mineral alabaster is nearly pure gypsum; it is dense and fine textured, but it is still soft enough that it can be worked with common tools, and it has been used since ancient times to make bowls, statues, and similar items.

Alabaster, a fine-grained variety of gypsum ($CaSO_4 \cdot 2\ H_2O$), has long been used for ornamental objects such as the large Egyptian urns in this photograph. However, the palace at Knossos in Crete, which is about 4000 years old, has a toilet seat carved of alabaster in the queen's quarters. (J.C. Brice)

RECOVERY AND USES OF THE ALKALINE EARTH METALS

In spite of its rarity, beryllium is a fascinating element, and its metallic form does have uses. It is as strong as steel, has a very high melting point (1283°C), and is very lightweight (the density of beryllium, 1.85 g/cm^3, is only one-fourth that of iron, 7.85 g/cm^3). The element itself is rather brittle, but it is widely used with other metals to make alloys. For example, adding about 2% Be to copper makes an alloy that is almost as good an electrical conductor as copper, but the material wears better than copper. Since it is also a strong material and does not create sparks when struck, it is used to make tools for areas where explosive gases might be present, for example.

The recovery of beryllium from beryl involves some complex chemistry, but the end product is BeF_2 or $BeCl_2$. Like all the elements of Groups 1A and 2A, the chloride can be reduced electrochemically to the metal.

$$BeCl_2(\text{as a melt with NaCl}) \xrightarrow[\text{electrical energy}]{} Be(s) + Cl_2(g)$$

Magnesium is also generally isolated in an electrolytic process, but the total chemistry of this industrial isolation process is quite interesting. Since magnesium is so abundant in the oceans, this has been one of the major sources of the metal (Figure 20.12). A very useful way to isolate the magnesium from the ocean is in the form of its relatively insoluble hydroxide (K_{sp} for $Mg(OH)_2 = 1.5 \times 10^{-11}$). To precipitate the hydroxide, the only thing needed is a ready supply of an inexpensive base, a need fulfilled nicely by sea shells, $CaCO_3$. As you first learned in Chapter 3, heating a metal carbonate generally leads to loss of CO_2 and formation of the metal oxide. Thus, calcium carbonate is converted to lime and then to calcium hydroxide, the source of hydroxide ion to precipitate $Mg(OH)_2$.

$$\underset{\text{sea shells}}{CaCO_3(s)} \longrightarrow \underset{\text{lime}}{CaO(s)} + CO_2(g)$$

$$CaO(s) + H_2O(\ell) \longrightarrow Ca(OH)_2(aq)$$

$$Mg^{2+}(aq) + Ca(OH)_2(aq) \longrightarrow Mg(OH)_2(s) + Ca^{2+}(aq)$$

The magnesium hydroxide can be isolated by filtration and then neutralized by another inexpensive chemical, hydrochloric acid.

$$Mg(OH)_2(s) + 2\ HCl(aq) \longrightarrow MgCl_2(aq) + 2\ H_2O(\ell)$$

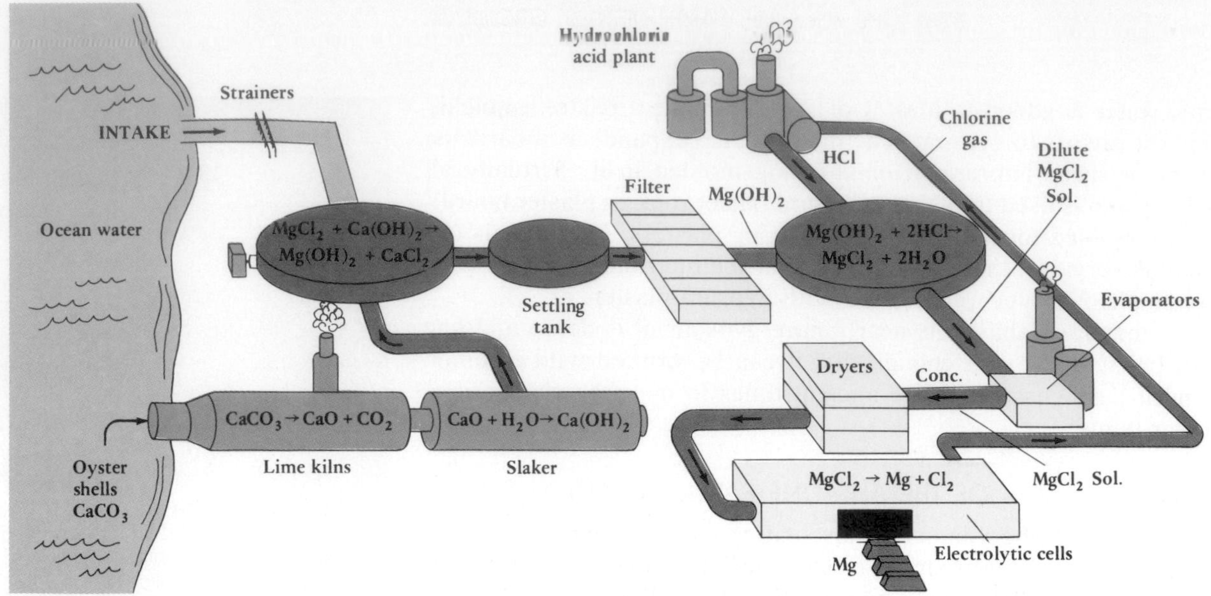

FIGURE 20.12
Diagram of an industrial plant for the production of magnesium from the magnesium ions present in seawater.

If the water is evaporated, solid hydrated magnesium chloride is left. After drying, it is melted at 708°C and then electrolyzed to give the metal and a commercially useful by-product, chlorine.

$$MgCl_2(melt) \xrightarrow[\text{electrical energy}]{} Mg(s) + Cl_2(g)$$

Several hundred thousand tons of magnesium are produced annually, and all its uses depend on its low density (1.74 g/cm^3) or high reactivity. In fact, most of the magnesium that is produced is used to make lightweight alloys. Most aluminum has about 5% magnesium added to it to improve its mechanical properties and to make it more resistant to corrosion under basic conditions. There are also alloys that have the reverse formulation, that is, more magnesium than aluminum. These alloys are used where a high strength-to-weight ratio is needed and where corrosion resistance is important. Much of it is used in aircraft and automotive parts and in lightweight tools, luggage trim, and so on.

The reactivity of magnesium leads to its use as a sacrificial anode to give cathodic corrosion protection for the hulls of ships and underground pipelines (see Chapter 19).

Calcium, strontium and barium can be prepared by electrolysis or by aluminum reduction, but only barium has some commercial uses.

Al can be used to reduce oxides of all but the most "active" metals. Such reactions depend on the very high heat of formation of Al_2O_3. See the thermite reaction on page 683.

$$3\ CaO(s) + 2\ Al(s) \longrightarrow Al_2O_3(s) + 3\ Ca(s)$$

In spite of the revolution in the production and use of semiconductor "chips," vacuum tubes are still needed for high-power electrical devices. In such tubes, the electrical circuitry is enclosed in a glass tube or envelope, which is evacuated. Since it is difficult to remove all traces of gases such as O_2 and N_2 from the tubes, a "getter" such as barium is placed in the tube. The function of the getter is to react with the residual gases, effectively cleaning the tube.

$$2\ Ba(s) + O_2(g) \longrightarrow 2\ BaO(s)$$
$$3\ Ba(s) + N_2(g) \longrightarrow Ba_3N_2(s)$$

Products are the metal oxide and the metal nitride, the latter an ionic compound based on the nitride ion, N^{3-}.

Elemental radium has no large-scale commercial uses, but its salts do. Radium bromide can be used, for example, in luminous materials for the faces of wristwatches and clocks. The radium compound is mixed with a phosphorescent material, and the latter glows because it is bombarded by alpha particles from the disintegration of the radium. Although only about one part of radium to 20,000 parts of the phosphor are used, radium is being replaced in this use by somewhat less dangerous radioactive materials such as polonium.

THE CHEMISTRY OF THE ALKALINE EARTH METALS

Because a sample of beryllium metal is coated with a chemically inert coating of oxide, it cannot be dissolved in water. Furthermore, it must be heated to 600°C in air before there is any visible change. In contrast, freshly cleaned magnesium will slowly evolve H_2 when heated gently in water.

$$Mg(s) + 2\ H_2O(\ell) \longrightarrow Mg(OH)_2(aq) + H_2(g)$$

This reaction was used to produce H_2 for weather balloons in World War II.

The remaining alkaline earth elements, on the other hand, must be protected from air and moisture because they react so rapidly.

The halides, oxides, and carbonates are the most important compounds of these elements, so we shall focus on them. Before beginning, however, we want to remind you that their chemistry is dominated by the +2 ions of the elements. Another dominating fact is their relatively high effective nuclear charge, Z^*. Each alkaline earth +2 ion is smaller than the +1 ion of the preceding alkali metal, so the +2 ions of Group 2A are much more electron attracting and are capable of forming stronger bonds. This is particularly true of beryllium, so its bonds to other elements can have a high degree of covalency. One example of this effect is the behavior of Be^{2+} in aqueous solution (Section 15.7). All of the alkaline earth +2 ions are hydrated in water, but Be^{2+} is involved in a complex series of equilibria beginning with the tetrahedrally coordinated ion, among them the following.

$$[Be(H_2O)_4]^{2+}(aq) \longrightarrow [Be(H_2O)_3(OH)]^{+}(aq) + H^{+}(aq)$$

$$[Be(H_2O)_4]^{2+}(aq) + [Be(H_2O)_3(OH)]^{+}(aq) \longrightarrow [(H_2O)_3Be{-}O{-}Be(H_2O)_3]^{2+}(aq) + H^{+}(aq) + H_2O(\ell)$$

The highly electron-attracting Be^{2+} ion weakens the O—H bond of coordinated water, and a proton is lost to the best available Lewis base, another water molecule.

$$\left[Be \longleftarrow :\!O\!:\!\!\begin{smallmatrix} H \\ H \end{smallmatrix}\right]^{2+}(aq) + H_2O(\ell) \longrightarrow \left[Be{-}\ddot{O}\!:\!\!\begin{smallmatrix} \\ H \end{smallmatrix}\right]^{+}(aq) + H_3O^{+}(aq)$$

Thus, aqueous Be^{2+} ion is acidic, just like Al^{3+} ion in water (Section 15.5). In another parallel with aluminum chemistry, both $Be(OH)_2$ and $Al(OH)_3$ are amphoteric. Beryllium hydroxide acts as a Lewis acid in forming $[Be(OH)_4]^{2-}$ with additional base,

$$Be(OH)_2(s) + 2\ OH^-(aq) \longrightarrow [Be(OH)_4]^{2-}(aq)$$

and it is a Brønsted base when it reacts with acid to form Be^{2+} and H_2O.

$$Be(OH)_2(s) + 2\ H^+(aq) \longrightarrow Be^{2+}(aq) + 2\ H_2O(\ell)$$

These parallels between Be^{2+} and Al^{3+} chemistry are good examples of a diagonal relationship (Section 20.2).

ALKALINE EARTH HALIDES The most important fluoride of the alkaline earth metals is CaF_2, fluorspar (Figure 11.20), but fluorapatite ($CaF_2 \cdot 3\ Ca_3(PO_4)_2$) is becoming increasingly important as a commercial source of fluorine. Although fluorspar is widely distributed, Europe and Mexico are the largest producers.

Almost half of the CaF_2 mined is used in the steel industry. It is added to the mixture of materials that are melted to make crude iron in a steel mill. The CaF_2 serves to remove some impurities, and it improves the separation of the molten metal from *slag*, the layer of silicate impurities and by-products that come from reducing iron ore to the metal (Chapter 25).

The steel industry is described more fully in Chapter 25 and the properties of HF in Chapter 22.

The other half of the fluorspar is used to manufacture hydrofluoric acid by reaction of the mineral with concentrated sulfuric acid.

$$CaF_2(s) + H_2SO_4(\ell) \longrightarrow 2\ HF(g) + CaSO_4(s)$$

HF is an extremely important material used in the aluminum industry to make cryolite, Na_3AlF_6 (see p. 731) and to make fluorocarbons, the Freons of air conditioners, and the Teflon of nonstick frying pans.

The apatites are collectively known as phosphate rock, and over 100 million tons are mined annually; the mines of Florida alone account for about one third of the world's output. Much of this rock is converted to phosphoric acid by reaction with sulfuric acid.

$$\underset{\text{fluorapatite}}{CaF_2 \cdot 3\ Ca_3(PO_4)_2(s)} + 10\ H_2SO_4(aq) \longrightarrow 10\ CaSO_4(s) + 6\ H_3PO_4(aq) + 2\ HF(g)$$

The phosphoric acid is used to manufacture a multitude of products (fertilizers and detergents), and its reaction products are found in baking powder, in frozen fish, and in many other food products (Chapter 22).

The Group 2A chlorides are interesting, albeit in different ways. The dichloride of beryllium is prepared by a type of reaction that can be used to prepare many other metal halides. That is, the oxide is reduced to metal with carbon, and the metal is oxidized with Cl_2 to the metal chloride.

$$BeO(s) + C(s) + Cl_2(g) \xrightarrow[600°C-800°C]{} BeCl_2(s) + CO(g)$$

The structure of $BeCl_2$ is interesting and unusual. Since Be in the dichloride apparently forms two bonds, the VSEPR theory predicts that the molecule is linear. Valence bond theory then suggests that the Be must be *sp* hybridized.

sp hybridized

$$:\ddot{\underset{..}{Cl}}—Be—\ddot{\underset{..}{Cl}}:$$

However, as there are two vacant valence orbitals located on the beryllium, this means that the atom does not adhere to the octet rule. As you saw in Chapter 15, the availability of empty orbitals means the Be can function as an acceptor of electron pairs, that is, as a Lewis acid. But what is the donor atom in a solid composed only of $BeCl_2$ molecules? In this case, the Cl atoms of $BeCl_2$ each have three unshared pairs of electrons, so $BeCl_2$ molecules can *associate* as pictured at right. A Cl atom from one molecule uses an electron pair to form a donor–acceptor or coordinate covalent bond with a Be center in a neighboring molecule. Notice that the Be in the center is now surrounded by four Cl atoms, so it is expected that the Cl's will lie at the corners of a tetrahedron. Valence bond theory rationalizes this structure by saying that the beryllium is sp^3 hybridized.

:Cl—Be—Cl:
:Cl—Be—Cl:
:Cl—Be—Cl:

sp³ hybridized Be atom

M-halogen-M bridges, such as the ones in $BeCl_2$, are common in chemistry, especially in aluminum and transition metal chemistry.

Large quantities of $CaCl_2$ are produced in the Solvay process (Section 20.3), but demand for it is low. Some is used as an additive in concrete mixes, some to melt snow and ice on roadways in the winter, and some as a heat transfer agent in refrigerating plants. Some is also used as a desiccant or drying agent, since the hydration of $CaCl_2$ to give the $CaCl_2 \cdot 6\ H_2O$ is highly exothermic (91 kJ/mol). Much of the $CaCl_2$, however, is simply discarded, and serious water pollution problems have resulted.

ALKALINE EARTH OXIDES AND CARBONATES Oxides and carbonates of the Group 2A elements are intimately linked, since all of the oxides are generally prepared by thermal decomposition of their respective carbonates (Chapter 3).

An exception to this is beryllium. It does not form a simple carbonate, nor does the element to which Be is diagonally related, Al.

$$CaCO_3(s) \xrightarrow[>850°C]{} CaO(s) + CO_2(g)$$

$$K_p = 1.0 \text{ at } 897°C$$

The ease of the decomposition process decreases with increasing atomic weight, that is, with increasing basicity of the metal oxide. The reaction is reversible, as you might have suspected given the importance of $CaCO_3$ and $MgCO_3$ in nature. All the oxides are white powders, and none is decomposed even at 3000°C!

Beryllium oxide does not react with water, while MgO (magnesia) reacts only slowly. Lime or CaO reacts more rapidly in a reaction called "slaking" (Figure 20.11). The heat of the reaction is so great that in the days of wooden sailing ships it was considered highly dangerous to carry lime as a cargo. If the ship should spring a leak, the lime would hydrolyze, and the heat was sufficient to ignite the wooden parts of the ship's hull.

Crystalline magnesite, MgO, is a useful material, because it is stable at high temperatures and is a good conductor of heat; however, it is a poor conductor of electricity. These properties make it useful as an insulator for the wires within electrical heating units, such as the ones in a home cooking range or space heater.

Because of their economic importance, calcium carbonate and calcium oxide are of special interest to us. As already mentioned, calcium carbonate can be decomposed to CaO or lime in one of the oldest chemical transformations used by humans. Until lime came to be used primarily as a chemical in this century, it had been used for several thousand years

Lime is third on the list of the top 50 chemicals produced in the United States (after sulfuric acid and nitrogen). About 32 billion pounds were produced in 1985. (See also Section 3.4.)

as a building material in the form of mortar (a lime, sand, and water paste) to secure stones to one another in building houses, walls, and roads. The Greeks used lime mortar in building the Temples of Apollo and Elis in 450 BC, and the Chinese used it in laying up the stones in the Great Wall. The Romans, however, perfected the use of lime mortar, and the fact that many of their constructions still stand is testament to their skill and the usefulness of lime. In 312 BC, the famous Appian Way, a Roman highway from Rome to Brindisi (a distance of about 350 miles) was begun, and lime mortar was used between several layers of its stones. Wherever the Romans went, they built magnificent structures. However, in central France they could find no deposits of limestone or marble, so they dug the powdery form of $CaCO_3$, chalk, from the ground. This has an added benefit for us today—the caves left in quarrying the rock are ideal places for making and aging champagne.

The usefulness of mortar depends on some simple chemistry. It consists of one part lime to three parts sand, with water added to make a thick paste. The first reaction that occurs, therefore, is the hydrolysis or "slaking" of the lime. When the mortar is placed between bricks or stone blocks, it slowly absorbs CO_2 from the air, and the slaked lime reverts to calcium carbonate.

$$Ca(OH)_2(s) + CO_2(g) \longrightarrow CaCO_3(s) + H_2O(\ell)$$

Although the sand in the mortar is chemically inert, the grains are bound together by the particles of calcium carbonate, and a hard material results.

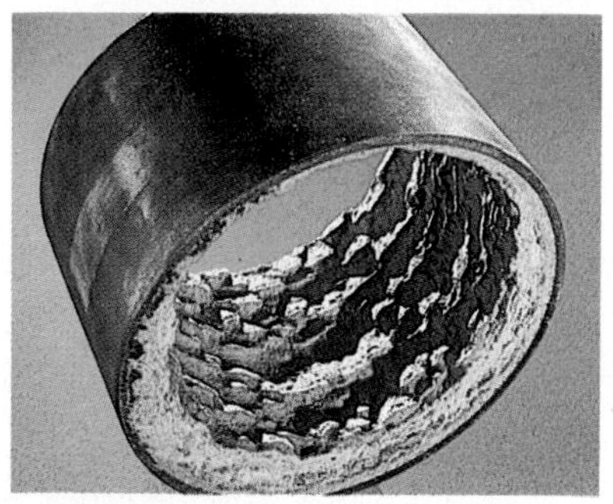

A section of a water pipe that was coated on the inside with $CaCO_3$ deposited from hard water. (Betz Laboratories)

Limestone is one of the oldest agricultural chemicals known. Yields of many crops can be greatly increased by spreading limestone on fields. The calcium carbonate neutralizes acidic compounds in the soil and supplies the essential nutrient Ca^{2+}; since magnesium carbonate also is often found in limestone, "liming" a field also supplies Mg^{2+}, another important plant nutrient.

The largest quantities of limestone and lime are now used in the chemicals industry rather than as mortar or in agriculture. About 45% of the lime used now is in steel making by the "basic oxygen process" (Chapter 25), and much of the rest is used to soften water by the "lime–soda" process. "Hard water" contains dissolved ions, chiefly Ca^{2+} and Mg^{2+}. If the water for a home or a city passes through limestone and if CO_2 is present, some limestone is dissolved,

$$CaCO_3(s) + H_2O(\ell) + CO_2(g) \rightleftharpoons Ca^{2+}(aq) + 2\ HCO_3^-(aq)$$

"Temporarily" hard water contains both metal ions and CO_2, but "permanently" hard water has only metal ions.

and this solution is hard water. As described in Chapter 14, this reaction is reversible. Therefore, if the water is heated, the solubility of CO_2 drops and from Le Chatelier's principle we see that the equilibrium shifts to the left. If this happens in a heating system or steam-generating plant, the walls of the hot water pipes can become coated or even blocked with limestone. If you have hard water in your house, you will notice a coating of calcium carbonate on the inside of cooking pots. Another problem is that common soaps are a mixture of sodium and potassium salts of organic acids with long carbon chains.

$$\underset{\text{soap}}{2\,[H_3C\text{—}(CH_2)_{16}\text{—}CO_2]Na(aq)} + Ca^{2+}(aq) \longrightarrow \underset{\text{soap scum}}{[H_3C\text{—}(CH_2)_{16}\text{—}CO_2]_2Ca(s)} + 2\,Na^{2+}(aq)$$

While these salts are soluble in water, those with Ca^{2+} and Mg^{2+} are not, and calcium and magnesium salts precipitate on clothes as a slimy, sticky residue.

There are several ways to remove Ca^{2+} (and Mg^{2+}) from water, that is, to *soften the water*. One of these is the "lime–soda process," and a large fraction of the 32 billion pounds of lime (CaO) and 17 billion pounds of soda (Na_2CO_3) produced annually in the United States is used for this purpose. The soda is added to precipitate Ca^{2+}, for example, as the carbonate.

$$\underset{\text{from hard water}}{Ca^{2+}(aq)} + \underset{\text{from soda}}{CO_3^{2-}(aq)} \rightleftharpoons CaCO_3(s)$$

Although it would seem to be nonsense to also add calcium oxide to hard water, the chemistry shows we come out ahead. Slaked lime from CaO is an extremely inexpensive source of OH^-,

$$\underset{\text{slaked lime}}{Ca(OH)_2(s)} \rightleftharpoons Ca^{2+}(aq) + 2\,OH^-(aq) \qquad K_{sp} = 7.9 \times 10^{-6}$$

and the base transforms the hydrogen carbonate ion in hard water into carbonate ion.

$$2\,OH^-(aq) + 2\,HCO_3^-(aq) \rightleftharpoons 2\,CO_3^{2-}(aq) + 2\,H_2O(\ell)$$

The carbonate ion then precipitates not only the calcium ion from the added lime but also the calcium and other ions present in the hard water. Looking carefully at the balanced equations, we notice that 1 mole of $Ca(OH)_2$ furnishes 2 moles of OH^- which lead to 2 moles of CO_3^{2-}; these in turn precipitate 2 moles of Ca^{2+}, one from the added lime and one from the hard water.

Finally, another enormous use of lime is in the production of **calcium carbide**. If lime is heated at 2000°C with coke, the product is a very interesting ionic compound containing the carbide ion, $:C{\equiv}C:^{2-}$.

$$CaO(s) + 3\,C(s) \xrightarrow{2000°C} \underset{\text{calcium carbide}}{CaC_2(s)} + CO(g)$$

Calcium carbide, a dirty yellow solid, is useful because the carbide ion is the salt of a weak acid, acetylene ($K_a \cong 10^{-20}$) (Figure 20.13). Hydrolysis of the anion, therefore, gives the acid.

$$\underset{\text{calcium carbide}}{CaC_2(s)} + 2\,H_2O(\ell) \longrightarrow Ca(OH)_2(aq) + \underset{\text{acetylene}}{H\text{—}C{\equiv}C\text{—}H(g)}$$

Acetylene made in this manner was once used as a major starting material for the manufacture of organic chemicals. Now that ethylene, $H_2C\text{—}CH_2$, can be made inexpensively, however, acetylene is no longer competitive.

FIGURE 20.13
Calcium carbide is a brown solid (right). When water is added (left), gaseous acetylene forms as indicated by the bubbles. The other product is $Ca(OH)_2$. The fact that the reaction evolves considerable heat is indicated by a wisp of steam rising from the reacting CaC_2. (See also Figure 9.4) (Charles D. Winters)

EXERCISE 20.3 THE CHEMISTRY OF CALCIUM OXIDE

Tell how lime, CaO, is produced, and give four uses of the chemical.

BIOLOGICAL EFFECTS OF ALKALINE EARTH COMPOUNDS

Your body contains many metal ions that serve important regulatory functions, and Na^+, K^+, Mg^{2+}, and Ca^{2+} ions constitute about 99% of the total. The graphs in Figure 20.14 give you some idea of the relative amounts of these ions in various body fluids as compared with sea water. The feature that you may notice immediately is the fact that K^+ and Mg^{2+} are by far the most important cations within cells, whereas the Na^+ concentration is much higher outside the cell. This difference is especially important to the operation of nerve cells, but the way that Na^+ is "pumped" out of cells and K^+ "pumped" in is one of the great mysteries of science. The difference in Ca^{2+} concentrations inside and outside cells is also

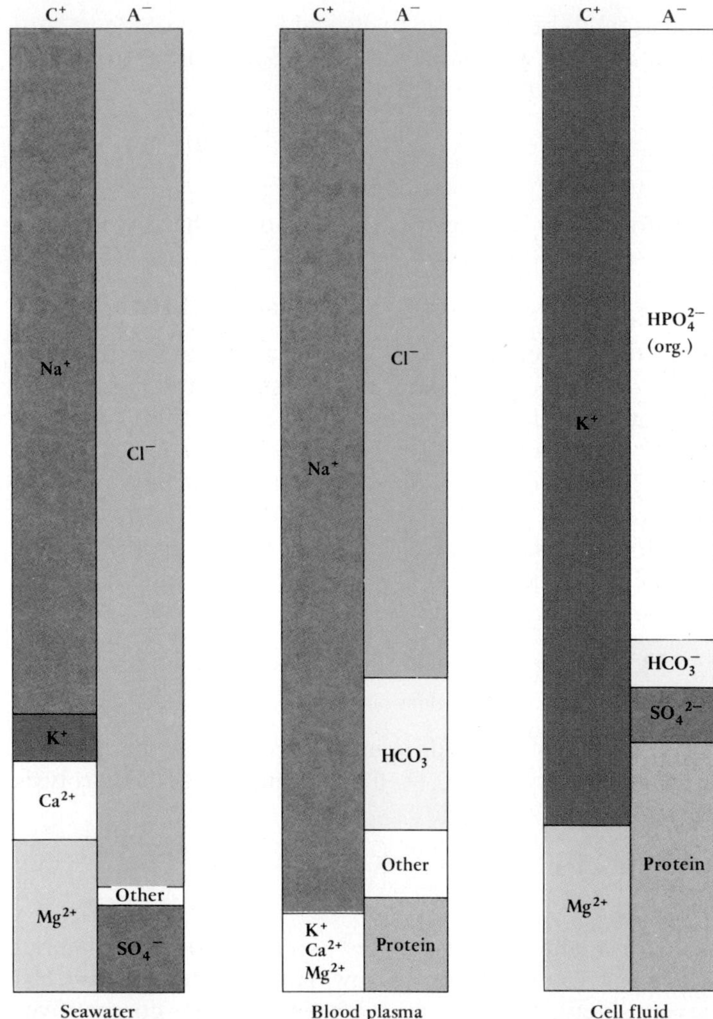

FIGURE 20.14

Relative ionic compositions of seawater and body fluids.

important. Many biochemical and physiological processes are triggered by entry of Ca^{2+} into a cell or release of Ca^{2+}. One such process is muscle contraction. Initiation of contraction results from the arrival of a nerve impulse at a motor nerve ending in a muscle fiber, and this causes Ca^{2+} to be released from a storage site. The released Ca^{2+} then interacts with a regulatory protein.

When you cut yourself, blood coagulates to prevent excessive bleeding. This complicated process occurs in a number of steps, almost all of which depend on Ca^{2+}.

Plants and animals derive energy from the oxidation of a sugar, glucose, with O_2. Plants are unique, though, in being able to synthesize glucose from CO_2 and H_2O using sunlight as the source of energy. The process is initiated by *chlorophyll*, a very large, magnesium-based molecule.

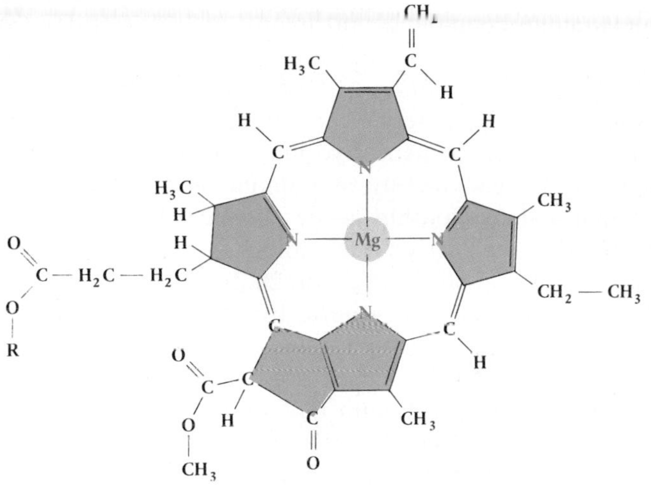

Although Ca^{2+} and Mg^{2+} are required by living systems, the other Group 2A elements are toxic. For example, soluble barium salts are toxic, so you may be concerned if you are told by a physician to drink a "barium cocktail" so that the condition of your digestive tract may be checked. The "cocktail," however, contains the very *insoluble* salt $BaSO_4$ with a K_{sp} value of 1.1×10^{-10}. With this information, you know the concentration of Ba^{2+} ions produced by the salt is extraordinarily small, so the cocktail is safe.

All simple beryllium-containing compounds are highly toxic and can be carcinogens. They *must* be handled with great care.

We have already mentioned the apatites, calcium- and phosphorus-containing minerals that are mined for their phosphorus content. However, they serve a most important biochemical function: hydroxyapatite, $Ca_5(PO_4)_3(OH)$, is the main component of tooth enamel. Cavities in your teeth are formed when acids decompose the weakly basic apatite coating.

The reason for drinking a slurry of barium sulfate is that the compound is opaque to x-rays. Thus, the path taken by the $BaSO_4$ through your digestive tract appears on the x-ray photograph.

$$Ca_5(PO_4)_3(OH)(s) + 4\ H^+(aq) \longrightarrow 5\ Ca^{2+}(aq) + 3\ HPO_4^{2-}(aq) + H_2O(\ell)$$

This can be prevented, however, by converting hydroxyapatite to a much more acid-resistant coating, fluorapatite.

$$\underset{\text{hydroxyapatite}}{Ca_5(PO_4)_3(OH)(s)} + F^-(aq) \longrightarrow \underset{\text{fluorapatite}}{Ca_5(PO_4)_3F(s)} + OH^-(aq)$$

The source of the fluoride ion can be stannous fluoride, sodium fluoride, or sodium monofluorophosphate (Na_2FPO_3, commonly known as MFP) in your toothpaste, or a soluble fluoride salt such as NaF in your water supply. Indeed, about half the population of the United States now drinks water that is fluoridated artificially or naturally.

SUMMARY

Hydrogen, the most abundant element in the universe, has three isotopes (protium, ^1H = H; deuterium, ^2H = D; and tritium, ^3H = T) (Section 20.1). There are measurable quantities of deuterium in nature, but only 1 atom in 10^{18} H atoms is radioactive tritium. There are three types of hydrogen-containing compounds: (a) anionic hydrides (NaH); (b) covalent hydrides (CH_4); and (c) interstitial hydrides. Hydrogen is produced industrially by *steam reformation of hydrocarbons* or *electrolysis*. In the laboratory it is made by reacting (a) a metal and acid, (b) a metal with water or a base, or (c) a metal hydride with water. The main uses of hydrogen are in the manufacture of ammonia and hydrocarbons.

The **diagonal relationship** is the chemical similarity between elements situated diagonally from one another in the periodic table (Li and Mg, Be and Al, and B and Si) (Section 20.2). For example, both Be and Al form inert oxides and amphoteric hydroxides. The reason for this similarity is that they have similar values of Z^*, so the abilities of each ion to attract another ion or a dipole are comparable.

The **alkali metals**, the Group 1A elements, are all metals and none exist free in nature (Section 20.3). Na and K are especially abundant, forming many common minerals, while the others are more rare. The heaviest element, Fr, is radioactive.

All the alkali metals can be recovered from one of their salts by electrolysis, and their uses depend on their ability to act as reducing agents. The hydration energy of the +1 ion plays a pivotal role in determining the potential of $M(s) \rightarrow M^+(aq) + e^-$. All react with O_2, but only lithium forms a normal oxide, Li_2O. Sodium gives sodium **peroxide**, Na_2O_2, while the remaining elements form **superoxides** such as KO_2 (a salt of the O_2^- ion).

The *chlor–alkali industry* consumes NaCl and produces billions of pounds of chemicals such as Cl_2, NaOH, and Na_2CO_3. The *Solvay process* was used until recently in the United States to manufacture Na_2CO_3 (and $NaHCO_3$).

The **alkaline earth elements** of Group 2A (Section 20.4) are all metals, and all form +2 ions. The lightest element, Be, is relatively rare, and the heaviest element, Ra, is radioactive. Magnesium and calcium are especially abundant and important. They form many common minerals, $CaCO_3$ or *limestone* being particularly common. Heating limestone gives CaO or *lime*, an oxide that reacts vigorously with water to give $Ca(OH)_2$ or *slaked lime*.

Recovery of the alkaline earth metals from their compounds can be done by electrolysis, the usual method for Mg. Other methods are possible, however, such as the aluminum reduction of CaO to give Ca and Al_2O_3.

All alkaline earth elements form halides of the type MX_2. Beryllium, however, is unique in that its halides are associated in the solid state through Be—X → Be bridges, a form of bonding found in aluminum chemistry as well.

Like aluminum, beryllium forms a very stable oxide and an amphoteric hydroxide. Since BeO oxide is so stable, this coating on a piece of Be metal means it does not react readily with water. The reaction of magnesium with water is also slow, while the other metals all react vigorously. All form normal oxides, and calcium oxide, or lime, is used in the chemical industry as a base.

The ions Na^+, K^+, Mg^{2+}, and Ca^{2+} are the most abundant metal ions in biochemical systems. Ca^{2+}, for example, is important in the processes of muscle contraction and blood coagulation and is a component of tooth enamel. Mg^{2+} is the metal ion in chlorophyll.

STUDY QUESTIONS

1. Write balanced equations for the following:
   (a) the reaction of potassium with hydrogen
   (b) the reaction of chlorine and hydrogen
   (c) the reaction of sulfur with hydrogen
   (d) the reaction of propane, C_3H_8, with steam
   (e) the reaction of potassium with water
2. One recently suggested method for the preparation of hydrogen (and oxygen) proceeds as follows: (a) Sulfuric acid and hydrogen iodide are formed from sulfur dioxide, water, and iodine. (b) The sulfuric acid from the first step is decomposed by heat to water, sulfur dioxide, and oxygen. (c) The hydrogen iodide from the first step is decomposed with heat to hydrogen and iodine. Write a balanced equation for each of these steps and show that their sum is the decomposition of water to hydrogen and oxygen. (*High Technology*, May 1983, p. 67.)
3. Compare the mass of H_2 expected for 100% reaction of steam (H_2O) with CH_4, with petroleum, and with coal. Calculate the mass of H_2 produced per gram of starting material. Use CH_2 and CH as representative formulas for petroleum and coal, respectively.
4. How much energy is required to convert 10.0 L of H_2 gas at 25.0°C and 3.50 atm to H atoms? When these H atoms recombine to H_2, how much heat energy is evolved?
5. The electrolysis of aqueous NaCl gives NaOH, Cl_2, and H_2.
   (a) Write a balanced equation for the process.
   (b) In 1983 in the United States, 20.46 billion pounds of NaOH and 19.92 billion pounds of Cl_2 were produced. Does the ratio of masses of NaOH and Cl_2 actually produced agree with the ratio of masses expected from the balanced equation? If not, what does this tell you about the way in which NaOH or Cl_2 are actually produced? Is the electrolysis of aqueous NaCl the only source of these chemicals?
6. To store 2.88 kg of gasoline with an energy equivalence of 1.43×10^8 J requires a volume of 4.1 L. In comparison, 1.0 kg of H_2 has this same energy equivalence. What volume is required if this quantity of H_2 is to be stored at 25°C and 1.0 atm of pressure?
7. Write complete electron configurations using the spectroscopic notation for francium and radium.
8. What is oxidized, and what is reduced, if an aqueous solution of KCl is electrolyzed under the same conditions as the electrolysis of aqueous NaCl? What happens when CsI is electrolyzed under these conditions?
9. Explain how magnesium acts as a sacrificial anode in providing cathodic protection against oxidative corrosion on a ship's hull.
10. Why are lithium salts often hydrated, whereas those of the other alkali metals are often anhydrous? (If you need help, see Chapter 11.)
11. How would you extinguish a sodium fire in the laboratory? What is the worst thing you could do?
12. Give three uses for Group 2A metals.
13. Beryllium fluoride is a linear molecule in the gas phase. What is the orbital hybridization of the central beryllium atom?
14. When Be^{2+} is placed in aqueous solution, it forms the hydrated ion $[Be(H_2O)_4]^{2+}$. Sketch the structure of ion showing the arrangement of water molecules about the beryllium ion. What is the orbital hybridization used by the central beryllium?
15. When magnesium burns in air, it forms both an oxide and a nitride. Write a balanced equation for the formation of the nitride.

16. Heating barium oxide in pure oxygen gives barium peroxide. Write a balanced equation for this process.
17. Barium peroxide is an excellent oxidizing agent. Reaction with iron gives iron(III) oxide and barium oxide as products. Write a balanced equation for this reaction.
18. Write balanced equations for the reaction of all of the Group 1A elements with O_2. Specify which metals form oxides, which ones form peroxides, and which ones form superoxides.
19. Magnesium is made in large amounts by electrolysis of molten $MgCl_2$ and by the reduction of calcined dolomite with ferrosilicon at 1430K. The basic reaction for the latter process is

$$MgO \cdot CaO(s) + Si(s) \longrightarrow Mg(g) + Ca_2SiO_4(s)$$

Balance this equation.
20. Complete and balance equations for the following reactions:
(a) $SrCO_3(s) + HCl(aq) \longrightarrow$
(b) $BaCO_3(s) + HNO_3(aq) \longrightarrow$
(c) $BaO(s) + Al(s) \longrightarrow$
(d) $LiH(s) + H_2O(\ell) \longrightarrow$
(e) $Na_2CO_3(aq) + H_2O(\ell) + CO_2(g) \longrightarrow$
21. Why is a solution of LiCl very slightly acidic? Write a balanced equation to demonstrate your answer.
22. Calcium forms a hydride, as do sodium and the other alkali metals. As it reacts readily with water, it is an excellent drying agent for organic solvents. (a) Write a balanced equation showing the formation of calcium hydride from calcium and H_2. (b) Write a balanced equation showing the reaction of calcium hydride with water.
23. $Ca(OH)_2$ has a K_{sp} of 7.9×10^{-6}, whereas that for $Mg(OH)_2$ is 1.5×10^{-11}. Calculate the equilibrium constant for the reaction,

$$Ca(OH)_2(s) + Mg^{2+}(aq) \longrightarrow Ca^{2+}(aq) + Mg(OH)_2(s)$$

and explain why this reaction can be used in the commercial isolation of magnesium from sea water.
24. (a) When 1000. kg of $MgCl_2$ is electrolyzed to produce magnesium, how many kilograms of metal are produced at the cathode? (b) What is produced at the anode? How many kilograms of the other product are produced? (c) What is the total number of Faradays of electricity used in the process? (d) One industrial process has an energy consumption of 8.4 kilowatt-hours per pound of Mg. How many joules are therefore required per mole? (e) How does this energy compare with the energy of the process $MgCl_2(s) \rightarrow Mg(s) + Cl_2(g)$?
25. One way to produce pure potassium is to use sodium to reduce KCl.

$$KCl(s) + Na(g) \longrightarrow NaCl(s) + K(g)$$

(a) Calculate the enthalpy change for this reaction using the following information:

lattice energy of KCl = 694 kJ/mol
lattice energy of NaCl = 768 kJ/mol
ionization energy of K = 418 kJ/mol
ionization energy of Na = 498 kJ/mol

(b) If ΔS for the reaction is zero, what is the approximate composition of the gas phase in equilibrium with the excess solids at 1000K?
26. Using data in Appendix L and that given below, calculate $\Delta G°$ values for the decomposition of MCO_3 to MO and CO_2 where M = Mg, Ca, and Ba. What is the relative tendency of these carbonates to decompose?

COMPOUND	$\Delta G_f°$(kJ/mol)
$MgCO_3$	−1012.1
$BaCO_3$	−1137.6
BaO	−525.1

27. Calculate the heat of the reaction of CaO(s) with aluminum.
28. Name three uses of limestone. Write a balanced equation for the reaction of limestone with CO_2 in water.
29. Explain what is meant by "hard water." Outline one method for softening hard water.
30. Give three biologically important functions of Group 1A and Group 2A metal ions.
31. CaO is used to remove SO_2 from power plant exhaust, since the two react to give solid $CaSO_3$. How many grams of SO_2 can be removed using 1000. kg of CaO?
32. Calcium fluoride can be used in the fluoridation of water supplies. (a) Calculate the solubility of CaF_2 in grams per liter. (b) If you wanted to achieve a fluoride ion concentration of 2.0×10^{-5} *M*, how many grams of CaF_2 would have to be used for 1 million liters of water?
33. Identify the lettered compounds in the following reaction scheme. When 1.00 g of a white solid A is strongly heated, you obtain another white solid, B, and a gas. (The gas exerts a pressure of 209 mmHg in a 450-mL flask at 25°C.) Bubbling the gas into a solution of $Ca(OH)_2$ gives another white solid, C. If the white solid B is added to water, the resulting solution turns red litmus paper blue. To the solution of B, you add dilute, aqueous HCl and evaporate to dryness to yield a white solid D. When D is placed in a bunsen flame, it colors the flame green. Finally, if the aqueous solution of B is treated with sulfuric acid, a white precipitate, E, forms.

SPECIAL SECTION: THE CHEMISTRY OF FIREWORKS

The development of black powder, a mixture of potassium nitrate, charcoal, and sulfur, has been one of society's most important developments. The discovery of this mixture, apparently well before AD 1000, is usually attributed to the Chinese, and it was they who began the development of pyrotechnic devices or fireworks. They found that enclosing the black powder in a sealed tube led to an explosion and a loud bang when ignited. When a small hole was made at the end of the tube, on the other hand, the hot, gaseous products rushed out of the opening and sky rockets were born.

Black powder made its way to Europe by the 1300s where it was used not only by the military but also for fireworks. By the time of the American Revolution, fireworks formulations and manufacturing methods had been worked out that are still in use today.

Manufacturing black powder is a hazardous business, and the only plant currently operating in the United States produces only about 500,000 pounds annually for the fireworks industry.

Before the 19th century, fireworks displays were nothing but noise and rockets, with little or no color. Occasionally, some extra charcoal or iron filings were added, and orange or gold effects were produced (Figure 3.10). During the 19th century, however, potassium perchlorate ($KClO_4$) and potassium chlorate ($KClO_3$) were first produced and used as oxidizing agents in fireworks, and copper, barium, and strontium salts were introduced to provide colors. Production of magnesium around 1860 and of aluminum slightly later led to fireworks devices that gave a brilliant white light.

A typical firework contains an oxidizer, fuel, binder, and a material for some special effect (Table 20.8 and Figure 20.15). You will notice that potassium salts are commonly used because sodium salts have two important drawbacks. They are hygroscopic, absorbing water from the air, and so do not remain dry in storage. Furthermore, when heated, sodium compounds give off an intense, yellow light that is so bright it can mask other colors.

In addition to KNO_3 as an oxidizer, $KClO_3$ and $KClO_4$ are also used, but not without problems. Mixtures of $KClO_3$ and sulfur are so dangerous that they were banned in England in 1875, but $KClO_4$ is safer. The problem is that it may become difficult to obtain. The only supplier of perchlorates in the United States makes ammonium perchlorate, the oxidizer used in the Space Shuttle solid-fuel booster rockets. Each shuttle launch requires about 1.5 million pounds of *ammonium* perchlorate, about twice the annual U.S. consumption of *potassium* perchlorate. Thus, frequent shuttle launchings could threaten the fireworks industry.

The parts of any fireworks display we remember best are the vivid colors and brilliant flashes. White light can be produced by oxidizing magnesium or aluminum metal at high temperatures (Figure 3.1), and the

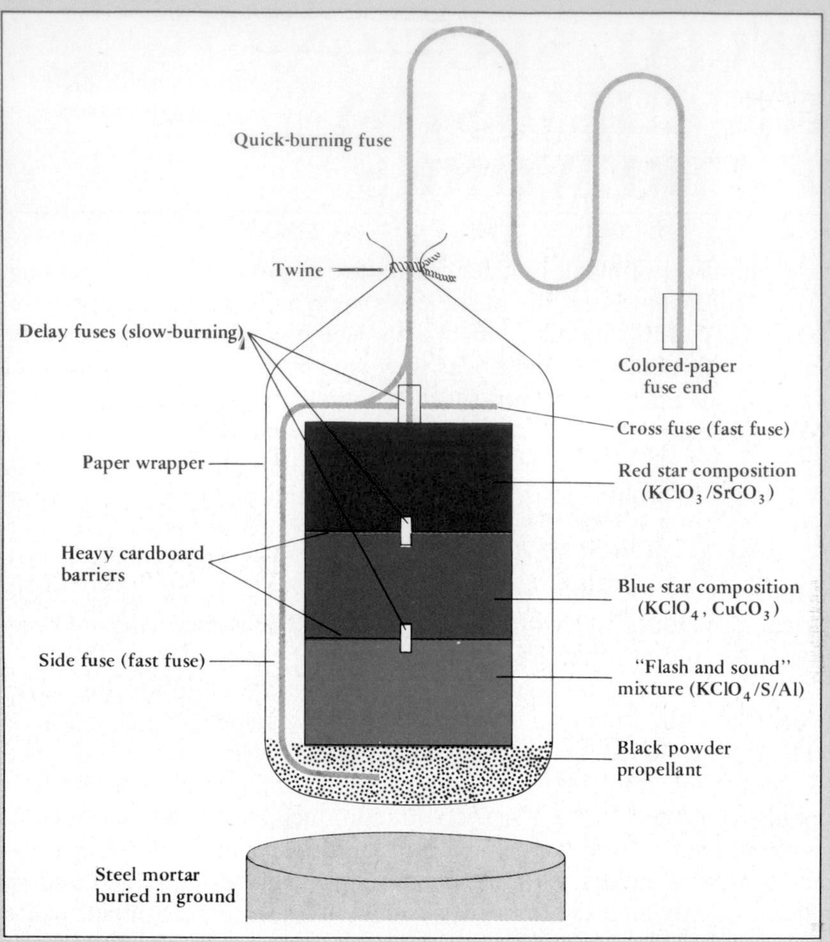

The construction of a "red/blue/report three-break aerial shell." The shell, 2 to 8 inches in diameter, is placed in a steel tube buried in the ground. When the fuse is ignited, it quickly burns to the delay fuse at the top of the red star mixture as well as burning to the black powder propellant at the bottom of the shell. The propellant ignites, sending the shell into the air. In the meantime, the slower or delay fuse is burning. If the timing is correct, the shell bursts high in the sky into a red star. This burst ignites a second delay fuse, leading to a blue burst, and finally to the "flash and sound mixture" that gives that wonderful, final booming report.

flashes you see at rock concerts or other stage productions are typically magnesium/potassium perchlorate mixtures.

Salts of all of the alkali metals and some of the alkaline earth metals emit light when heated or excited in some way. See Tables 20.2 and 20.5 and the chapter-opening photograph for the colors emitted by each element.

Yellow light is the easiest to produce, since sodium salts give an intense light with a wavelength of 598 nm. Fireworks mixtures usually contain sodium in the form of such nonhygroscopic compounds as cryolite, Na_3AlF_6. Strontium salts are most often used to produce a red light and green fireworks are produced by barium salts such as $Ba(NO_3)_2$.

When you next see a fireworks display, watch for ones that are blue. You will probably observe that there are relatively few aerial rockets that burst into blue, since this color is by far the most difficult to achieve. The best blue color comes from emission of light by copper(I) chloride (CuCl) at low temperatures.

TABLE 20.8 Chemicals Used in Modern Fireworks Formulations

OXIDIZERS		
Potassium nitrate		Barium nitrate
Potassium chlorate		Barium chlorate
Potassium perchlorate		Strontium nitrate
Ammonium perchlorate		
FUELS		
Aluminum	Charcoal	Dextrin
Magnesium	Sulfur	Red gum
Titanium	Antimony sulfide	Polyvinyl chloride
BINDERS		
Dextrin, red gum, synthetic polymers		
SPECIAL EFFECTS		
Red flame—strontium nitrate, strontium carbonate		
Green flame—barium nitrate, barium chlorate		
Blue flame—copper carbonate, copper sulfate, copper oxide, copper(I) chloride		
Yellow flame—sodium oxalate, cryolite (Na_3AlF_6)		
White flame—magnesium, aluminum metals		
Gold sparks—iron filings, charcoal		
White sparks—aluminum, magnesium, aluminum-magnesium alloy, titanium		
Whistle effect—potassium benzoate or sodium salicylate		
White smoke—potassium nitrate/sulfur mixture		
Colored smoke—potassium chlorate/sulfur/organic dye mixture		

21

Metals, Metalloids, and Nonmetals: Periodic Groups 3A and 4A

Crystalline tin (Kurt Nassau, AT&T Bell Laboratories)

21.1 THE ELEMENTS OF GROUP 3A
21.2 THE ELEMENTS OF GROUP 4A
SPECIAL SECTION: LEAD CHEMISTRY

CHAPTER OUTLINE

The elements of Groups 3A and 4A are a bridge between the metals of Groups 1A and 2A and the largely nonmetallic elements of Groups 5A through 8A. Thus, the Groups 3A and 4A exhibit a tremendous range of chemistry. They include one of the least dense metals and one of the most dense metals, the most abundant metal and some of the least abundant metals, volatile hydrides that burn or explode in air, diamonds, and sand. They have an equally wide range of uses, from detergents to semiconductors, and some have no significant commercial importance.

As we have done in the previous chapter, we shall emphasize the common minerals of the elements, the methods of obtaining the element itself, and compounds of commercial importance.

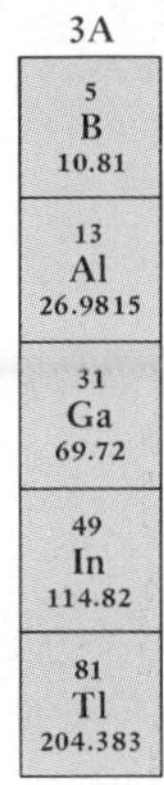

21.1 THE ELEMENTS OF GROUP 3A

The elements of Group 3A all have electron configurations of the type ns^2np^1 (Table 21.1). This means that each may lose these electrons to reach the +3 formal oxidation state, although the heavier elements, especially thallium (Tl), also form compounds in the +1 state.

TABLE 21.1 The p^1 Elements of Group 3A (the Boron and Aluminum Family)

	BORON	ALUMINUM	GALLIUM	INDIUM	THALLIUM
Symbol	B	Al	Ga	In	Tl
Atomic number	5	13	31	49	81
Atomic weight	10.81	26.98	69.72	114.82	204.38
Valence e^-	$2s^22p^1$	$3s^23p^1$	$4s^24p^1$	$5s^25p^1$	$6s^26p^1$
mp, °C	2300	660	29.78	156.6	303.5
bp, °C	2550	2327	2403	2000	1457
d, g/cm^3	2.34	2.70	5.91	7.31	11.85
Atomic radius, pm	88	143	122	162	171
Ion radius, pm		50	62	81	95
Pauling EN	2.0	1.5	1.7	1.6	1.6
Standard reduction potential	−0.90	−1.66	−0.56	−0.34	−0.33
Oxidation numbers	Covalent	+3	+1, +3	+1, +2, +3	+1, +3
Ionization energy*	801	577	579	560	589
Heat of vaporization*	562	326	277	243	182
Isolated by	Gay–Lussac	Wöhler	Boisbaudran	Reich	Crookes
Date of isolation	1808	1827	1875	1863	1861
rpw† pure O_2	B_2O_3, 1200°C	Al_2O_3, 800°C	Ga_2O_3, 1600°C	In_2O_3, 600°C	Tl_2O, 400°C
rpw H_2O	None	None	None	None	None
rpw N_2	BN, 1200°C	AlN, 740°C			None
rpw halogens	BX_3, 400°C	Al_2X_6, 200°C	Ga_2X_6	In_2X_6	TlX
Flame color	Grass green		Violet	Blue violet	Bright green

*All energies in kJ/mol.
†rpw = reaction product with.
(Adapted from Rochow, *Modern Descriptive Chemistry*, W.B. Saunders, 1977.)

TABLE 21.2 Relative Terrestial Abundance

IN PARTS PER MILLION			
Si	257,000	B	3
Al	75,000	Sn	2.1
C	800	Tl	1.8
Ga	15	Ge	1.5
Pb	13	In	1.8

The elements vary greatly in their relative abundances on earth. The first few elements of the periodic table, up to carbon, are very low in abundance (Chapter 8). Aluminum, however, is very abundant and is our most abundant metal (Table 21.2).

Except for boron, a metalloid, all the elements of Group 3A are metals. In Group 4A, carbon is clearly a nonmetal, silicon and germanium are metalloids, and tin and lead are metals. According to the *diagonal relationship* (Section 20.2), aluminum and beryllium should have similar chemistries, and boron and silicon should show some similarities. A preliminary comparison of boron and silicon chemistry and beryllium and aluminum chemistry indicates

(a) The oxide of boron, B_2O_3, and boric acid, $B(OH)_3$, are weakly acidic, just as are SiO_2 and its acid, orthosilicic acid (H_4SiO_4). In contrast, $Be(OH)_2$ and $Al(OH)_3$ are amphoteric, both dissolving in a strong base such as NaOH (Chapter 15).

(b) Boron–oxygen compounds, *borates*, are somewhat similar to silicon–oxygen compounds, the *silicates*.

(c) The halides of both boron and silicon (e.g., BCl_3 and $SiCl_4$) react vigorously with water. Aluminum halides are only partly hydrolyzed.

(d) The hydrides of boron and silicon are volatile, flammable, and readily hydrolyzed (with some exceptions). In contrast, aluminum hydride, AlH_3, is a colorless, nonvolatile solid that is extensively polymerized through Al—H—Al bonds.

There is clearly evidence for the diagonal relationship. However, the fact that boron and silicon have different numbers of valence shell electrons means that there will also be differences in their chemistries.

COMMON MINERALS AND RECOVERY OF THE ELEMENTS

The Group 3A elements do not occur free in nature. Rather, they are generally found as oxides in many locations around the earth. Although boron is very low in relative abundance, its common minerals are found in concentrated deposits, especially in California. Large deposits of *borax*, $Na_2B_4O_7 \cdot 10\ H_2O$, are currently being mined in the Mojave Desert near the town of Boron (Figure 21.1). Deposits of borate ores were originally mined in Death Valley in the late 1800s and were hauled out using the famous 20-mule teams.

FIGURE 21.1

An aerial view of the open pit borax mine of the U.S. Borax and Chemical Corporation in the Mojave Desert near Boron, California. (Rick McIntyre, Tom Stack & Associates)

Isolation of pure, elemental boron is extremely difficult and is done only in small quantities. Annual worldwide production is probably less than 10 tons. As is true of most metals and metalloids, it can be obtained by chemically or electrolytically reducing an oxide or halide. Magnesium, for example, has been used as the reducing agent for many years, but the product is a noncrystalline boron of low purity.

$$B_2O_3(s) + 3\ Mg(s) \longrightarrow 2\ B(s) + 3\ MgO(s)$$

Very pure boron can be made on the kilogram scale by reducing BBr_3 with H_2 on hot tantalum metal.

$$2\ BBr_3(g) + 3\ H_2(g) \xrightarrow{Ta} 2\ B(s) + 6\ HBr(g)$$

There are several allotropes of boron, elemental boron with different solid state structures, but all are characterized by having the *icosahedron* as one structural element (Figure 21.2). Partly as a result of extended, covalent bonding, boron is very hard, refractory (resistant to heat), and a nonconductor. In this regard, it is quite different from the other Group 3A elements; Al, Ga, In, and Tl are all relatively low melting, rather soft metals with high electrical conductivity.

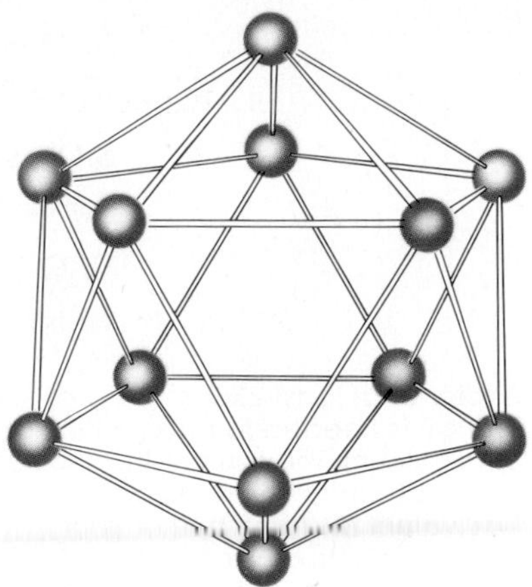

FIGURE 21.2
The icosahedron, a regular polyhedron having 20 faces. In the many different types of solid boron, the boron atoms are bound together in this polyhedron. In some cases, there are then bonds from one polyhedron to another.

If boric oxide is reduced with carbon, the product is boron carbide. (Recall that reduction of calcium oxide with carbon gives calcium carbide; Section 20.4)

$$2\ B_2O_3(s) + 4\ C(s) \longrightarrow B_4C(s) + 3\ CO_2(g)$$

The carbide is extraordinarily high melting (2250°C) and very hard. It also has a low density (2.5 g/cm^3), so one of its largest uses has been in bulletproof armor. Thousands of suits of this armor were used in the war in Vietnam.

Aluminum is found in varying amounts in nature as aluminosilicates, minerals such as clay that are based on aluminum, silicon, and oxygen (see Section 21.2). As these minerals are weathered, they gradually break down to various forms of hydrated aluminum oxide, $Al_2O_3 \cdot nH_2O$, and a mixture of these is called *bauxite*. Dehydrating the hydrated oxides leads finally to anhydrous Al_2O_3, commonly called *corundum*.

Recovery of aluminum from purified bauxite is done by electrolysis as outlined in Chapter 19. However, the chemistry for the purification of the bauxite is a good illustration of some aspects of Group 3A chemistry. Bauxite ore contains iron and silicon oxides as well as hydrated aluminum oxide. The aluminum oxide is purified by the *Bayer process* by making use of the amphoteric, basic, or acidic nature of the various oxides. Silica, SiO_2, is an acidic oxide, Al_2O_3 is amphoteric,* while Fe_2O_3 is a basic oxide. Therefore, the first two oxides will dissolve in a hot concentrated solution of caustic soda (NaOH).

An aluminum "pot line." Electrolytic reduction "pots" convert refined bauxite (aluminum oxide) to molten aluminum metal. (Atlantic Richfield)

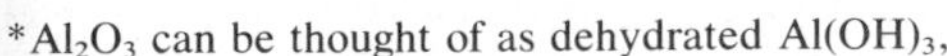

*Al_2O_3 can be thought of as dehydrated $Al(OH)_3$,

$$2\ Al(OH)_3(s) \longrightarrow Al_2O_3(s) + 3\ H_2O(\ell)$$

and the hydroxide is clearly amphoteric, dissolving in acid or in strong base.

$$\left.\begin{array}{c} Al_2O_3(s) \\ \text{amphoteric} \\ SiO_2\ (s) \\ \text{acidic} \\ Fe_2O_3(s) \\ \text{basic} \end{array}\right\} + 30\%\ NaOH \xrightarrow[190^{\circ}C]{} NaAl(OH)_4(aq) + Na_2Si(OH)_6(aq) + Fe_2O_3(s)$$

A total of at least 235 × 10^6 kJ of energy is needed to make 1000 kg of aluminum. This is more than 4700 kJ per 20-gram soft drink can, equivalent to about half of your daily food intake. Since only 4.5% as much energy is needed to recycle aluminum, this is why there is interest in reusing aluminum.

The insoluble materials are removed by filtration, and Al_2O_3 is precipitated by treating the solution with CO_2. Recall that CO_2 forms the weak acid H_2CO_3 in water, so Al_2O_3 is precipitated in an acid–base reaction,

$$\underset{\text{acid}}{H_2CO_3(aq)} + \underset{\text{base}}{2\ NaAl(OH)_4(aq)} \longrightarrow Na_2CO_3(aq) + Al_2O_3(s) + 5\ H_2O(\ell)$$

and the silicate ion remains in solution.

Metallic aluminum has thousands of uses as a structural material and in packaging. However, pure aluminum is rarely used, since it is soft and weak. What is more, it loses strength rapidly above 300°C. To strengthen the metal, and improve its properties, it is alloyed with small amounts of other metals. A typical alloy, for example, may contain about 4% copper with smaller amounts of silicon, magnesium, and manganese. A typical large passenger plane today may use more than 50 tons of such alloy. To make a softer, more corrosion resistant alloy for window frames, furniture, highway signs, and cooking utensils, however, only manganese may be included.

Much of the usefulness of aluminum comes from its corrosion resistance. This is due to the formation of a thin, tough, and transparent skin of oxide, Al_2O_3.

$$4\ Al(s) + 3\ O_2(g) \longrightarrow 2\ Al_2O_3(s) + 3351.4\ kJ$$

Partly because the heat of formation of the oxide is so large, the oxide layer is rapidly self-repairing. If you scratch a piece of aluminum it quickly forms a new layer of oxide that covers the damaged area.

Aluminum will dissolve in acids such as HCl, but not in nitric acid (Figure 21.3). The latter is a powerful oxidizing agent, so it oxidizes the surface of aluminum rapidly, and Al_2O_3 protects the metal from further attack. In fact, nitric acid is often shipped in aluminum tanker trucks.

FIGURE 21.3
Copper reacts vigorously with nitric acid (left) to give $Cu(NO_3)_2$ and NO_2, but aluminum (right) is unreactive. (Charles D. Winters)

The remaining elements of Group 3A, gallium, indium, and thallium, are very low in abundance, but gallium is rapidly gaining in commercial importance. All are found in small amounts associated with other minerals, and all are obtained as by-products in the recovery of other metals such as zinc and aluminum. As a result, these metals are expensive. Pure gallium, for example, costs about \$4 per gram, whereas one gram of aluminum of comparable purity costs only about 20 cents.

Gallium was one of the elements that was not known at the time Mendeleev developed his notion of the periodic table (Chapter 8), but he predicted its existence and properties, a fact that helped greatly in its discovery just a few years later. It is truly a remarkable element. It has the greatest liquid range of all known elements: it can melt in your hand (see Table 21.1), but it does not boil until the temperature reaches 2403°C! Finally, like water, gallium is one of the few known materials that *expands* upon freezing.

There have been repeated suggestions that Ga be used in high temperature thermometers, but so far this possibility has not been exploited.

The greatest use for gallium, and one that may continue to grow, is in the semiconductor gallium arsenide, GaAs (Figure 21.4). Integrated circuits based on GaAs have achieved operating speeds up to five times that of the fastest silicon chips currently available, and they will operate over a wider temperature range than silicon circuits. However, these advantages come at a price: arsenic is volatile and toxic, so GaAs is difficult to make.

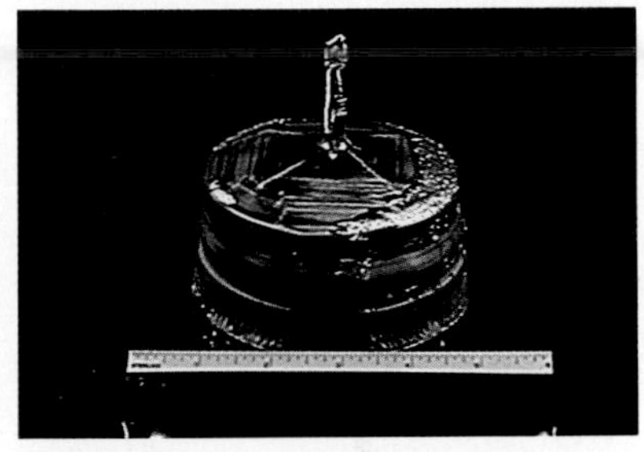

FIGURE 21.4
A GaAs semiconductor. (Harris Semiconductor Company)

EXERCISE 21.1 ALUMINUM CHEMISTRY

The Hall–Heroult electrolysis process for aluminum outlined in Chapter 19 uses a mixture of Al_2O_3 and Na_3AlF_6, cryolite. The mixture melts at only 960°C, whereas pure Al_2O_3 melts at more than 2000°C. Cryolite is made by the reaction

$$Al_2O_3(s) + HF(aq) + NaOH(aq) \longrightarrow Na_3AlF_6(s) + H_2O(\ell)$$

Balance the equation.

THE CHEMISTRY OF THE GROUP 3A ELEMENTS

Group 3A elements all form compounds in the +3 oxidation state. One of the most interesting aspects of this group, however, is that the +1 oxidation state is observed, and it becomes more important as the atomic number increases. Thus, there is evidence of the +1 oxidation state for Al^+ compounds, and with Ga^+ and In^+ compounds the +1 oxidation state exists but the compounds are unstable. In contrast, Tl^+ salts are as common as compounds of Tl^{3+}. In the +1 state, thallium compounds resemble those of silver (Group 1B) and of the Group 1A elements: TlCl is insoluble, just like AgCl, and TlOH is a strong base, just like NaOH.

The occurrence of an oxidation state that is 2 less than the group number is sometimes called the "inert pair" effect.

GROUP 3A OXYGEN COMPOUNDS Although many chemists have spent years researching boron compounds and have uncovered fascinating structures and reactions, the compounds making up the boron products industry are the chemically unspectacular oxides and oxyacids: boric acid, borax, and their simple derivatives. With an annual worldwide production in the range of two million tons, it is an important industry.

Borax, $Na_2B_4O_7 \cdot nH_2O$ (n = 5 or 10), is the most important boron–oxygen compound and is the form of the element most often found in nature. It has been used for centuries as a low-melting *flux* in metallurgy, because of the ability of molten borax to dissolve other metal oxides, thus cleaning the surfaces to be joined and permitting good metal-to-metal contact. As it is usually written, the formula of borax is misleading, since the salt really contains the ion $B_4O_5(OH)_4^{2-}$. Borax is thus better written as $Na_2[B_4O_5(OH)_4] \cdot 8\ H_2O$ (n = 10).

The Venetian adventurer Marco Polo (1254–1324?) brought borax back from the Far East, along with gunpowder and spaghetti.

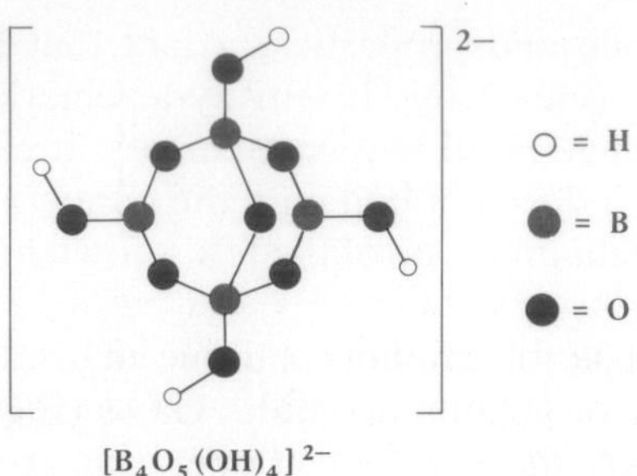

$[B_4O_5(OH)_4]^{2-}$

The structure of the borate ion illustrates several commonly observed phenomena. First, many minerals consist of MO_n groups that share oxygen atoms. Second, this fusion often takes the form of metal–oxygen ring systems. And third, it has both sp^2 and sp^3 hybridized boron atoms.

After refinement, borax can be treated with sulfuric acid and converted to boric acid, $B(OH)_3$ (Figure 21.5).

$$Na_2B_4O_7 \cdot 10\ H_2O(s) + H_2SO_4(aq) \longrightarrow 4\ B(OH)_3(aq) + Na_2SO_4(aq) + 5\ H_2O(\ell)$$

Boric acid is a Lewis acid that acts by accepting an hydroxyl ion from water.

$$H{-}\ddot{O}{:}\,(H) + B(OH)_3 \longrightarrow \left[B(OH)_4 \right]^- + H^+ \qquad K_a = 7.3 \times 10^{-10}$$

The second largest use of boric acid and borates is as a flame retardant for cellulose home insulation. Such insulation, which is mostly made from scrap paper, is inexpensive but flammable. To control its flammability, 5% to 10% of the weight of the insulation is boric acid.

Because of its weak acid properties and slight biological toxicity, the acid has been used for many years as an antiseptic. Furthermore, since the acid is so weak, this means that salts of borate ions, such as the $[B_4O_5(OH)_4]^{2-}$ ion in borax, are hydrolyzed in water to give basic solutions. For this reason, borax has also been used for years in soap and detergent systems.

Boric acid is dehydrated to boric oxide when heated strongly.

$$2\ B(OH)_3(s) \longrightarrow B_2O_3(s) + 3\ H_2O(g)$$

By far the largest use of borax and of boric oxide is in the manufacture of borosilicate glass. This type of glass is composed of about 76% SiO_2, 13% B_2O_3, and much smaller amounts of Al_2O_3 and Na_2O. The presence of boric oxide gives the glass a higher softening temperature, a better resistance to attack by acids, and makes it expand less on heating.

One common brand of borosilicate glass is Pyrex, the trademark of the Corning Glass Company. The source of sodium in glass is Na_2CO_3 (page 761).

The chemistry of the remaining Group 3A elements with oxygen has some similarities with boron, but also some great differences. In water, the Group 3A 3+ ions all undergo hydrolysis to give acidic solutions.

$$[M(H_2O)_6]^{3+}(aq) + H_2O(\ell) \rightleftharpoons [M(H_2O)_5(OH)]^{2+}(aq) + H_3O^+(aq)$$

ELEMENT	K_a
Al	7.9×10^{-6}
Ga	2.5×10^{-3}
In	2.0×10^{-4}
Tl	$\approx 7 \times 10^{-2}$

The ions with six water molecules exist only in very acidic solutions. As the pH is raised, the hydrated oxide [$Al_2O_3 \cdot 3\ H_2O = 2\ Al(OH)_3$] precipitates. Aluminum and gallium hydroxides are amphoteric, and both dissolve in excess base to give $[M(OH)_4]^-$.

$$Ga(OH)_3(s) + OH^-(aq) \longrightarrow [Ga(OH)_4]^-(aq)$$

As you will see below, these properties lead to many uses of aluminum salts in particular.

Aluminum oxide, Al_2O_3, which can be formed by dehydration of $Al(OH)_3$, is quite insoluble in water and generally resistant to chemical attack. In the crystalline form, aluminum oxide is known as *corundum*, and it is extraordinarily hard. This property has led to its use as the abrasive in grinding wheels, "sandpaper," and toothpaste.

When Al_2O_3 is contaminated with SiO_2 and iron oxides, it is known as *emery*.

Some gems are just impure aluminum oxide. Rubies, the beautiful red crystals prized as gems and as the materials used in some lasers, are Al_2O_3 contaminated with a small amount of Cr^{3+} in place of some Al^{3+} ions. Blue sapphires occur when Fe^{2+} and Ti^{4+} impurities are present in Al_2O_3. As you will learn in Chapter 25, it is just these transition metal "impurities" that give the minerals their beautiful colors and other desirable properties. Synthetic rubies were first made in 1902, and the worldwide capacity is now about 200,000 kg; much of this production is used for the jewel bearings in watches and instruments (Figure 21.6).

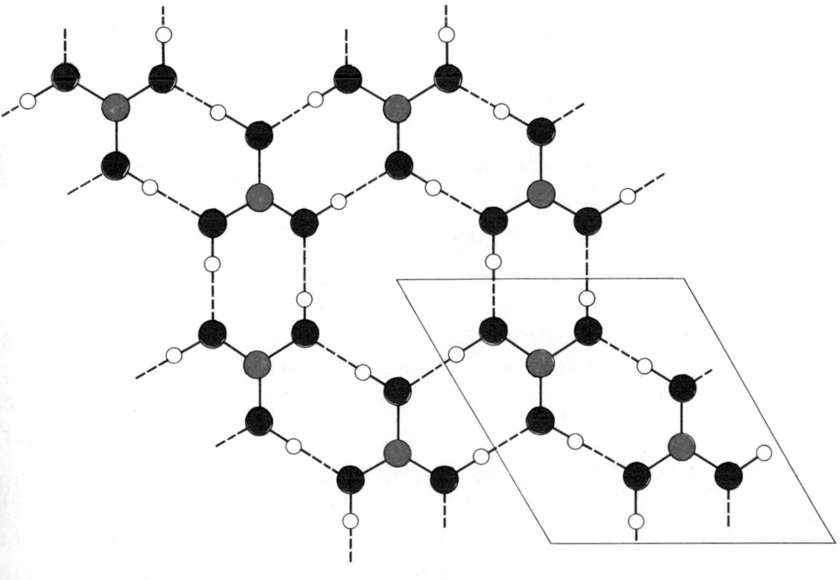

FIGURE 21.5

Solid boric acid is a good example of hydrogen bonding in the solid state. $B(OH)_3$ molecules are hydrogen bonded into six-membered rings in layers. The layers are 318 pm apart.

FIGURE 21.6
A synthetic ruby, a crystal of Al_2O_3 containing a small amount of Cr^{3+} in place of Al^{3+}. (Kurt Nassau)

If bauxite or clay is treated with sulfuric acid, the product is aluminum sulfate.

$$\underset{\text{clay}}{H_2Al_2(SiO_4)_2 \cdot H_2O(s)} + \underset{\text{sulfuric acid}}{3\ H_2SO_4(aq)} \longrightarrow \underset{\text{aluminum sulfate}}{Al_2(SO_4)_3(aq)} + 2\ H_4SiO_4(s) + H_2O(\ell)$$

This and the related compounds called *alums* [potassium alum is $KAl(SO_4)_2 \cdot 12\ H_2O$] are very useful. Large quantities of aluminum sulfate, for example, are used in the paper industry to make the product stronger and nonporous. Alums are also important in water treatment as "clarifiers." Remember that in any but *very* acidic media, Al^{3+} hydrolyzes ultimately to the hydrated oxide. When this occurs in water containing fine, suspended particles (mud!), the alumina surrounds the particles and precipitates them.

For similar reasons, aluminum sulfate or alums are used in dyeing cloth. The aluminum hydroxide is called a *mordant* in the textile industry, because it binds to both cloth and dye molecules, thereby serving to "fix" the dye to the cloth (Figure 21.7).

FIGURE 21.7
Aluminum hydroxide, $Al(OH)_3$, readily adsorbs dyes. At the left is a precipitate of pure $Al(OH)_3$, while the $Al(OH)_3$ at the right has adsorbed the red dye aluminon. (Charles D. Winters)

EXERCISE 21.2 BORON CHEMISTRY

The structure of boric acid is illustrated in Figure 21.5. (a) One unit cell is outlined in the figure. How many molecules of $B(OH)_3$ are contained in this unit cell? (b) What is the hybridization of the boron atom in the acid? (c) On the basis of the boron hybridization, explain why $B(OH)_3$ is open to attack by the Lewis base H_2O.

BORON AND ALUMINUM HALIDES None of the simple halides of Group 3A elements exists in nature for the simple reason that B—O, Al—O bonds, and so on, are so thermodynamically stable that oxygen compounds are found instead. Nonetheless, the simple halides have some interesting properties and are important as catalysts in the organic chemicals industry.

Boron trifluoride is a colorless gas made by heating boric oxide with calcium fluoride and sulfuric acid. The essence of the process is that the acid and CaF_2 combine to produce hydrofluoric acid

$$CaF_2(s) + H_2SO_4(\ell) \longrightarrow CaSO_4(s) + 2\ HF(g)$$

and the acid then reacts with boric oxide.

$$B_2O_3(s) + 6\ HF(g) \longrightarrow 2\ BF_3(g) + 3\ H_2O(\ell)$$

Since BF_3 does hydrolyze in water to some extent to give HF and boric acid, it is imperative that the water by-product be removed. Therefore, this process uses excess sulfuric acid, since the concentrated acid is an excellent drying agent.

Although the boron trihalides are gases or volatile liquids or solids (Figure 21.8), the aluminum halides are all solids. Boron trifluoride has a boiling point of $-100°C$, while AlF_3 is an ionic solid in which Al^{3+} ions are surrounded by an octahedron of F^- ions. In fact, cryolite (Na_3AlF_6), the mineral so important in the electrolytic recovery of aluminum, contains octahedral AlF_6^{3-} ions (see Exercise 21.1).

FIGURE 21.8
Liquid BBr_3 (left) and solid BI_3 (right). (Charles D. Winters)

Aluminum chloride is made by direct reaction of the metal and chlorine. The reaction generates so much heat that the metal melts and continues to "burn" in the stream of chlorine once the reaction is started (see Figure 3.3).

$$2\ Al(s) + 3\ Cl_2(g) \longrightarrow 2\ AlCl_3(s) + 1408.4\ kJ$$

Alternatively, $AlCl_3$ can be prepared by heating a mixture of aluminum oxide, carbon, and chlorine.

$$Al_2O_3(s) + 3\ C(s) + 3\ Cl_2(g) \longrightarrow 2\ AlCl_3(s) + 3\ CO(g)$$

This reaction can be seen as a two-step process. The carbon effectively reduces the metal oxide to metal and CO, and then chlorine reacts with the metal to produce the metal chloride. The solid trichloride is, like AlF_3, an ionic material. However, when it melts, the $AlCl_3$ units form Al_2Cl_6 molecules. Aluminum bromide and iodide are composed of Al_2X_6 molecules even in the solid state.

Crystals of alum, hydrated potassium aluminum sulfate [$KAl(SO_4)_2 \cdot 12H_2O$]. The alum in this crystal was prepared in an introductory chemistry laboratory by digesting aluminum cans in potassium hydroxide, followed by treatment with sulfuric acid. (Charles D. Winters)

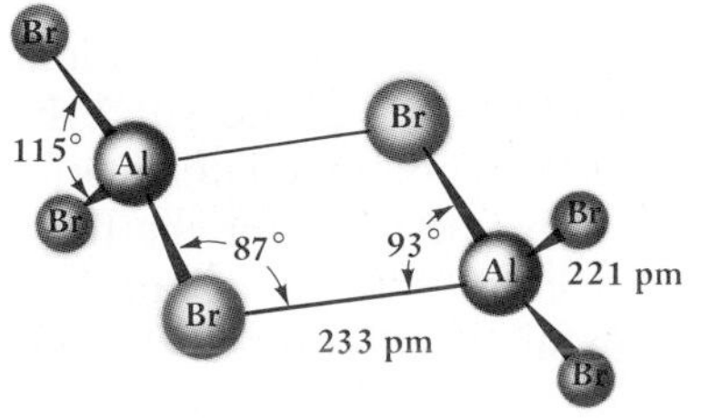

Aluminum chloride, bromide, and iodide all hydrolyze in water, partial hydrolysis giving hydroxyhalides, $Al(OH)_nX_{3-n}$, of various compositions. For example,

$$AlCl_3(s) + 2\ H_2O(\ell) \longrightarrow 2\ HCl(aq) + Al(OH)_2Cl(s)$$

Dehydration then leads to a solid material whose composition approximates the formula AlOCl. Such substances have some commercial value in antiperspirants and deodorants.

BORON HYDRIDES Although the boron hydrides were once proposed as lightweight, high-energy fuels for airplanes and rockets, they are without great commercial interest at the present time. Their importance lies in the role they have played in the development of theories of chemical structure and bonding.

Compounds that contain only an element and hydrogen are often named by adding the suffix -ane to the element name. Thus, there are boranes (BH), alanes (AlH), silanes (SiH), and germanes (GeH).

The boron hydrides are frequently called *boranes*. At the present time over twenty neutral boranes, of general formula $(BH)_pH_q$, are known. The simplest borane is *diborane*, B_2H_6, where p is 2 and q is 4.

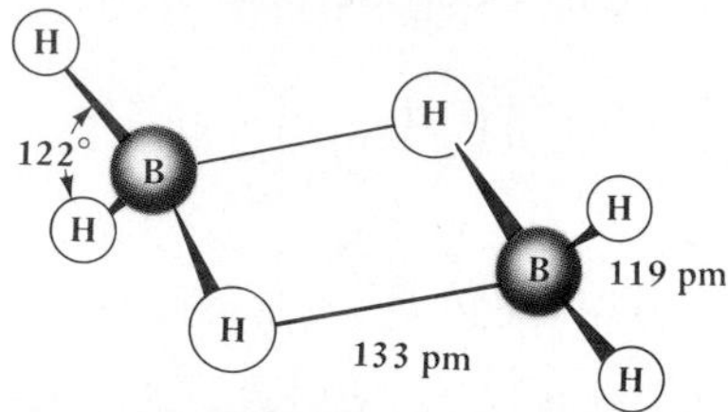

This is curious, since you might have expected BH_3 to be the simplest hydride, by analogy with boron trihalides. It is even more curious if you examine the structure above. Two boron atoms and six hydrogen atoms bring a total of twelve valence electrons to bind the molecule together. If you take each of the eight lines in the structure above as a two-electron bond, then sixteen electrons are required. Thus, the boron hydrides came to be called "electron deficient" molecules. But clearly the molecule cannot be electron deficient; if it did not have sufficient electrons to satisfy its bonding needs it would be quite unstable or nonexistent.

Many boranes can be viewed as fragments of the icosahedron used by elemental boron as its basic structural element (Figure 21.2).

There are several ways to solve the diborane bonding dilemma, and one of them can be outlined as follows. The boron atoms are surrounded tetrahedrally by H atoms, so we assume that the borons are sp^3 hybridized. The "outside" or terminal B—H bonds are assumed to be normal, two-electron bonds formed by overlap of the H 1*s* orbital with a boron sp^3 orbital. The four bonds of this type account for eight of the twelve electrons available, so four electrons remain to construct the two B—H—B bridges. Each boron has two sp^3 hybrid orbitals, and these extend into the bridging region as illustrated in Figure 21.9. Here a spherical hydrogen 1*s* orbital can overlap one sp^3 hybrid from each boron, creating a *three-center bond*. This three-center bridge bond can hold two electrons, so the two bridges account for the four electrons remaining. Although you have not seen such forms of bonding before this, they are not uncommon in the chemistry of boron and other elements.

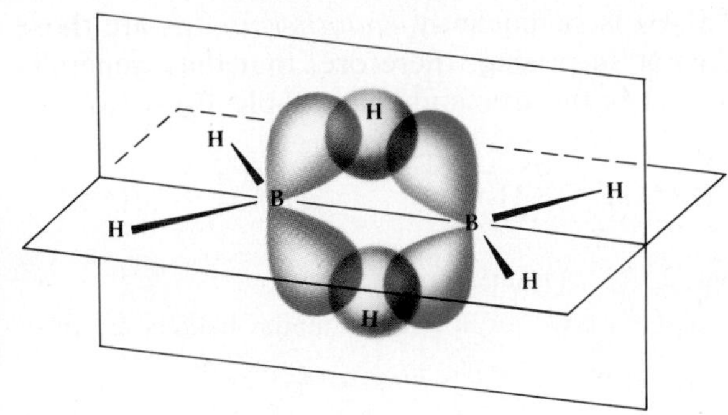

FIGURE 21.9

B—H—B bridge bonding in diborane (B_2H_6). After accounting for bonding to two "terminal" H atoms, two sp^3 orbitals remain on each boron atom. One such orbital from each boron may overlap a hydrogen 1*s* orbital in the bridge to give a three-center (three-orbital) bond occupied by two electrons.

The only B—H compounds being produced in ton quantities are the *borohydrides*. By far the most common of these is sodium borohydride, a white, crystalline, water-soluble solid. It is usually made by reacting sodium hydride with trimethylborate, a compound made from methanol and boric acid.

$$B(OH)_3(s) + 3\ CH_3OH(\ell) \xrightarrow{-3\ H_2O(\ell)}$$

$$\underset{\text{sodium hydride}}{4\ NaH(s)} + \underset{\text{trimethylborate}}{B(OCH_3)_3(\ell)} \longrightarrow \underset{\text{sodium borohydride}}{NaBH_4(s)} + 3\ NaOCH_3(s)$$

Over 2000 tons of $NaBH_4$ are produced annually, and one of its main uses is as a reducing agent. Since it readily reduces metal ions to the metal, the reaction can be used to plate metals onto surfaces without using electrodes and to remove potentially harmful metal ions in the effluent from a chemical or sewage treatment plant. It is also used to recover silver from the solutions used in developing photographic film (Figure 21.10).

Sodium borohydride is also the usual starting point for the synthesis of diborane, and an excellent approach is

$$2\ NaBH_4(s) + I_2(s) \longrightarrow B_2H_6(g) + 2\ NaI(s) + H_2(g)$$

$$B_2H_6\text{: bp, } -92.6°C,\ \Delta H_f^\circ = +35.6\ kJ/mol$$

FIGURE 21.10

Reducing Ag^+(aq) with sodium borohydride, $NaBH_4$. At the left, aqueous $NaBH_4$ is contained in a flask, while the dish holds $AgNO_3$(aq). When $NaBH_4$(aq) is added to $AgNO_3$(aq), silver ion is reduced to silver, the black solid seen in the picture at the right. (H_2 gas is evolved as well.) In the foreground is a "button" of pure silver (labeled with $NaBH_4$) made by the reaction illustrated. (Charles D. Winters)

The heat of formation of B_2H_6 is significantly *endothermic*, as are those of the other boranes. It is not surprising, therefore, that they generally burn in air, since the product is the tremendously stable B_2O_3 (ΔH°_f = −1254.5 kJ/mol).

$$B_2H_6(g) + 3\ O_2(g) \longrightarrow B_2O_3(s) + 3\ H_2O(g) \qquad \Delta H^\circ = -2016\ \text{kJ}$$

EXERCISE 21.3 BORON AND ALUMINUM HALIDES

Write balanced equations to show how boron and aluminum halides are obtained from their oxides.

21.2 THE ELEMENTS OF GROUP 4A

With Group 4A we move away from elements that are metals or metalloids and begin to see nonmetallic behavior (Table 21.3). Carbon, the lightest element of this group, is distinctly nonmetallic in its chemistry, but silicon is classed as a metalloid. Germanium begins the trend to metallic behavior for the heavier elements of the group.

All these elements are characterized by half-filled valence shells with two electrons in the *ns* orbital and two electrons in *np* orbitals (where *n* is the period number). The bonding in carbon compounds, like that of

TABLE 21.3 The p^2 Elements of Group 4A (the Carbon Family)

	CARBON	SILICON	GERMANIUM	TIN	LEAD
Symbol	C	Si	Ge	Sn	Pb
Atomic number	6	14	32	50	82
Atomic weight	12.011	28.086	72.59	118.71	207.19
Valence e^-	$2s^22p^2$	$3s^23p^2$	$4s^24p^2$	$5s^25p^2$	$6s^26p^2$
mp, °C	3570	1414	937	232†	328
bp, °C	Sublimes	2355	2830	2270	1750
d, g/cm^3	2.266*	2.33	5.32	7.30†	11.35
Atomic radius, pm	77	177	122	140	175
Ion radius (+2), pm			73	93	121
Pauling EN	2.5	1.8	1.9	1.8	1.8
Standard reduction potential (V)	+0.39	+0.10	−0.3	−0.15	−0.126
Oxidation numbers	−4 to +4	−4, +2, +4	−4, +2, +4	−4, +2, +4	+2, +4
Ionization energy‡	1090	782	762	704	714
Heat of vaporization‡	717	456	377	302	195
Isolated by	Antiquity	Berzelius	Winkler	Antiquity	Antiquity
Date of isolation	Antiquity	1824	1886	Antiquity	Antiquity
rpw§ pure O_2	CO, CO_2, 600°C	SiO_2, 1200°C	GeO_2, 1000°C	SnO_2, 800°C	PbO, 600°C
rpw H_2O	None	None	None	None	None
rpw N_2	None	Si_3N_4, 1400°C	None	Sn_3N_4, 2000°C	None
rpw halogens	CX_4, 800°C	SiX_4, 400°C	GeX_4, 400°C	SnX_2, 400°C	PbX_2, 400°C
rpw H_2	CH_4, 1000°C	None	None	None	None
Flame color	None	None	None	None	None

*Data are given for graphite.
†Data are given for white tin.
‡Energies given in kJ/mol.
§rpw = reaction product with.
Adapted from E.G. Rochow, *Modern Descriptive Chemistry*, W.B. Saunders, 1977.

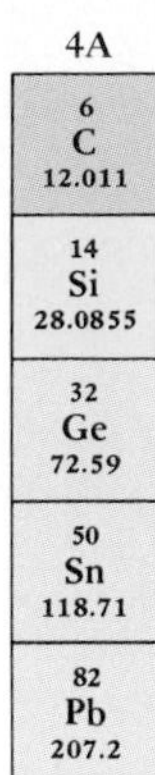

the second period elements in general, is largely covalent. This means that, depending on the relative electronegativity of the bonded element, the formal oxidation number can be ±2 or ±4. Because of generally lower ionization energies, elements of Group 4A of higher atomic number have only positive oxidation numbers of +2 and +4. An oxidation number of +2 for silicon and germanium is unusual, but it becomes more common for the heavier elements tin and especially lead. Recall that a similar trend to more stable, lower oxidation numbers was seen in Group 3A, and the trend will continue in Group 5A as well.

COMMON MINERALS AND RECOVERY OF THE ELEMENTS

Carbon is reasonably abundant on our planet and is certainly widely distributed. The cosmic abundance of carbon is six times that of silicon, and carbon-containing material, including diamonds, has been found in meteorites. In the earth's crust, the most abundant material not containing silicon is calcium carbonate, a substance already discussed many times in this text, and all of us are composed of a myriad of carbon-containing molecules.

In elementary form, carbon occurs as the **allotropes** diamond and graphite. Graphite is composed of layers of interconnected, planar, six-membered rings of sp^2 hybridized carbon atoms (Figure 11.30). The standard heats of formation are 0 for graphite, the thermodynamic standard state of carbon, and +1.895 kJ/mol for diamond, and the entropy of graphite (5.740 J/K · mol) is slightly greater than that of diamond (2.377 J/K · mol). This means that diamonds should spontaneously revert to graphite. To the considerable relief of diamond owners, the transformation involves such an enormous geometry change that the rate is negligibly slow, except at *very* high temperatures.

Allotropes are different forms of the same element that exist in the same physical state under the same conditions of pressure and temperature.

Graphite melts at the extraordinarily high temperature of 3570°C, so it is used to make crucibles for casting metals and to line electric furnaces. Since it is also a reasonably good conductor of electricity, it is used to make electrodes for industrial processes such as aluminum reduction (Figure 19.16). However, it is also soft and marks paper, so it is the "lead" in pencils. Finally, it is commonly used as a lubricant; the very weak interactions between the layers allow them to slide over one another. Very pure graphite is actually an abrasive, however, so it is thought that it is impurities, such as oxygen, between the layers that reduce layer-to-layer bonding enough to give rise to the lubricant property.

Worldwide consumption of graphite is on the order of a million tons annually.

Charcoal, carbon black, and common soot are other reasonably pure forms of microcrystalline graphite. Activated charcoal is made by heating charcoal in steam at about 1000°C, a process that removes volatile materials from the pores of the solid and makes it an excellent absorbent; it is widely used, for example, to purify molasses by absorbing impurities. Most of the carbon black produced is used as a reinforcing agent in rubber, hence we have black automobile tires and not the white or amber of pure rubber. Carbon black is also used as the pigment in printing inks.

Finally, there are new high-strength materials made by mixing pure graphite fibers with various man-made plastics. You may have seen them

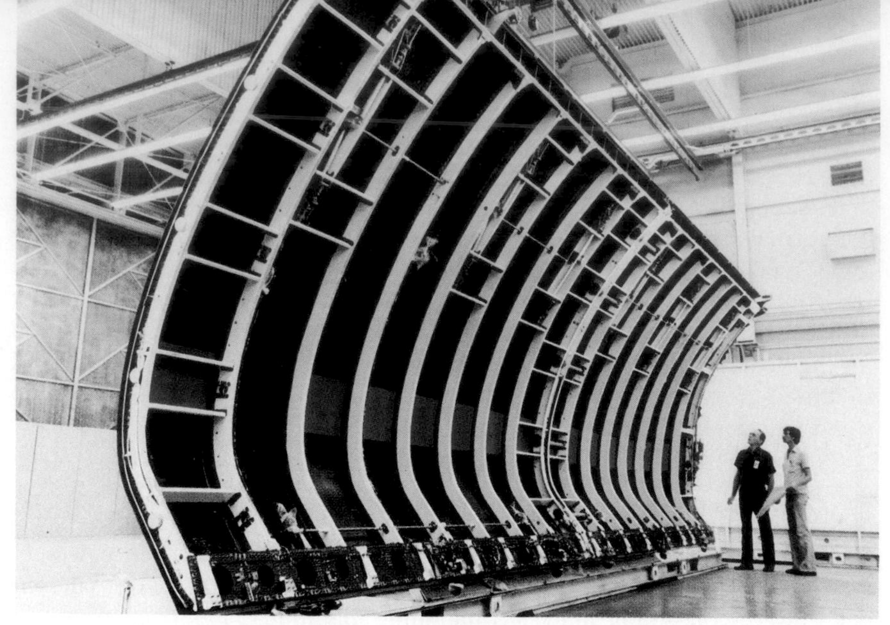

FIGURE 21.11
The Space Shuttle uses graphite–epoxy composites in several areas, such as the cargo door pictured here. (From *High Technology*, October, 1983, p. 63.)

in tennis rackets or golf clubs, but they are also used in less common places such as the payload bay doors of the Space Shuttle and the cockpits of race cars (Figure 21.11).

The world's largest known diamond, weighing 570 g, was found near Pretoria, South Africa, in 1905.

The diamond allotrope has fascinated man for hundreds of years. Diamonds are found around the world, but most come from South Africa. Approximately six tons of diamonds are used annually, mostly as industrial abrasives. Natural diamonds do not completely satisfy this demand, so some diamonds are made by subjecting pure graphite to very high pressures at high temperature (70,000 atm and 1800°C) (Figure 21.12). The transformation is possible because the density of diamond (3.514 g/cm^3) is much greater than that of graphite (2.266 g/cm^3).

The structure of a diamond (Figure 11.31) shows that it is an interconnected network of tetrahedral, sp^3 hybridized carbon atoms, and it is this interconnectedness that makes diamond so hard and relatively inert chemically. It is also important that diamond has the highest thermal conductivity of any known substance (about 5 times that of Cu); for this reason, diamond-tipped tools do not overheat when used for drilling and cutting.

FIGURE 21.12
Synthetic diamonds. (General Electric Company)

Silicon is the second most abundant element in the earth's crust, so its compounds have been important to the development of our society. Pottery, made of silicon-based natural materials, was made at least 6000 years ago in the Middle East, and sophisticated techniques were developed by the Chinese 5000 years ago. Our modern name for the element is derived from the Latin word for flint, a silicate mineral often used by prehistoric people to make knives and other tools. Today we are surrounded by silicon-containing materials: bricks, pottery, porcelain, lubricants, sealants, and computer chips and solar cells.

Reasonably pure silicon is made in large quantities by heating pure silica sand with purified coke to approximately 3000°C in an electric furnace.

$$SiO_2(s) + 2\ C(s) \longrightarrow Si(\text{liquid at } 3000°C) + 2\ CO(g)$$

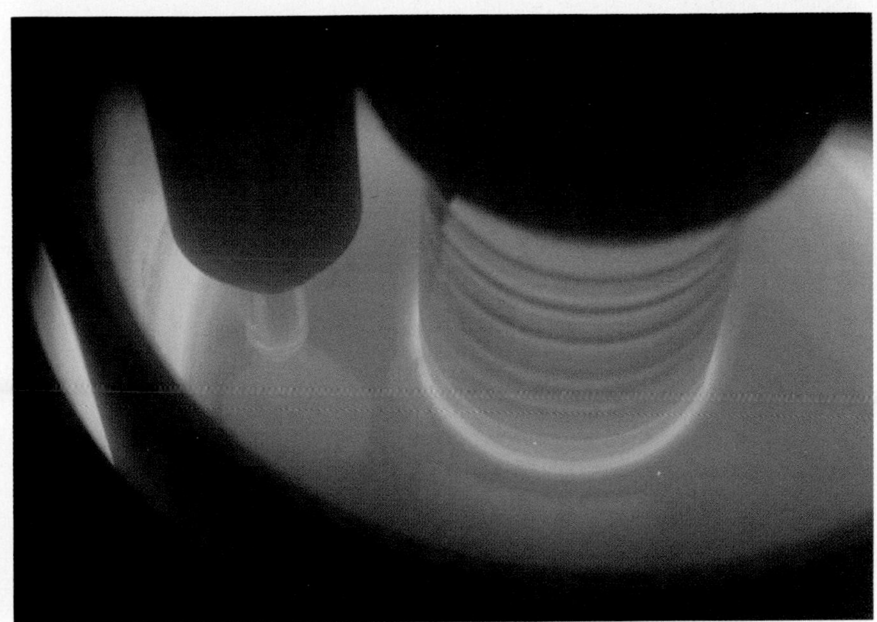

FIGURE 21.13

Zone refining silicon to produce a very pure form of the element. The method relies on the fact that impurities do not fit into the lattice of a pure crystal. One portion of the bar is melted with an induction heater surrounding the bar. As the heater moves on, that portion slowly crystallizes, leaving the impurities in the melt. When the heater has moved to the end of the bar, the impurities have collected at that point. (Great Western Silicon Company)

The molten silicon is drawn off the bottom of the furnace and allowed to cool to a shiny blue-gray solid. This can then be purified further for the electronics industry by first chlorinating the silicon to form silicon tetrachloride.

$$Si(s) + 2\ Cl_2(g) \longrightarrow SiCl_4(\ell,\ bp = 57.6°C)$$

The volatile tetrachloride is carefully purified and then reduced with very pure magnesium or zinc.

$$SiCl_4(g) + 2\ Mg(s) \longrightarrow 2\ MgCl_2(s) + Si(s)$$

Any magnesium chloride that may remain in the solid silicon is washed out with water. The silicon is then remelted and cast into bars that are finally purified by *zone refining* (Figure 21.13).

Germanium, tin, and lead are not abundant (Table 21.3), but they occur in workable mineral deposits. Tin and lead are especially easily recovered from their ores, and they have been used for hundreds of years. As evidence of their long history of use, their symbols, Sn and Pb, come from their Latin names, *stannum* and *plumbum* (Figure 21.14).

After the stone age, the earliest period of civilization, periods of mankind's development are often expressed in terms of copper, bronze, and iron. The copper that was first used was probably "native copper," naturally occurring metal. However, there is evidence that the Sumerians in their prehistoric home in southern Iran learned to obtain copper from its ores by heating these in the presence of wood. Since no attempt was made to purify the ore, there must have been times when a mixture of copper and tin-bearing ores was heated, and *bronze*, an alloy consisting

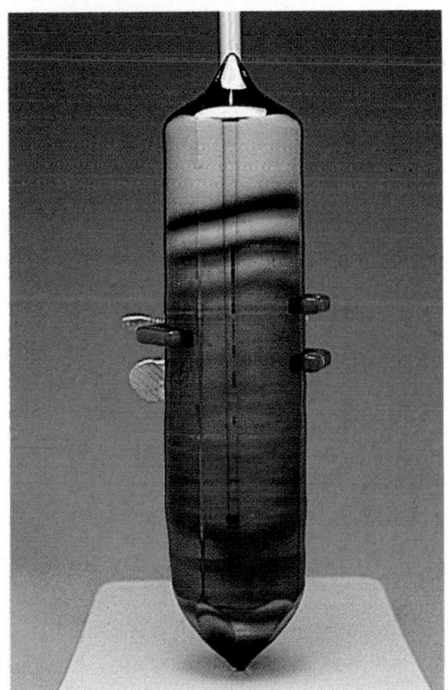

A solid cylinder of nearly pure silicon, approximately 10 inches long. (Charles D. Winters)

FIGURE 21.14
A lead crystal. (Dave Davidson, Tom Stack & Associates)

of about 20% tin and 80% copper, was obtained. Since bronze melts at a lower temperature than copper and is much harder, it was clearly a desirable material. In Egypt, bronzes have been found dating from as early as 2500 BC, and even older objects have been found in Mesopotamia.

By the time the Greek and Roman civilizations had developed, it was known that bronze could be obtained by mixing certain copper- and tin-containing ores from Cyprus and Britain. From Cyprus came copper-bearing ores (and the Latin name for the island gave us the name of this element). From Britain came *cassiterite*, SnO_2, an oxide that could be reduced easily to the metal with glowing charcoal.

$$SnO_2(s) + C(s) \longrightarrow Sn(s) + CO_2(g)$$

Tin is obtained today by the same process, and it is still widely used. Although most of the 210,000 tons of cassiterite mined today comes from such countries as Malaysia (25%) and Bolivia (14%), the United States is the largest consumer. Tin remains an expensive and fairly rare metal (about $4.00 per pound in 1986).

Tin cans are used increasingly for beer and soft drinks. In the United States, 27 billion of the 40 billion drink cans sold annually are tin plated, the remainder are aluminum cans.

Tin is very resistant to corrosion, so almost 40% of the metal produced is used to plate the surfaces of other metals, as in the tin-coated iron cans that you have always called "tin cans." The coating, typically 0.0004 to 0.025 mm thick, is applied by dipping the object in molten tin or by electroplating.

Since it is quite unreactive to air and water, tin is also used in the "float" process of making glass. Molten glass is poured onto large vats of molten tin; the metal has such a smooth surface that the glass, when cooled, is also smooth and does not need to be polished.

Lead was also known to ancient civilizations. The Hanging Gardens of Babylon were said to be floored with sheets of lead to retain moisture, and the Romans used lead extensively for water pipes and plumbing. One reason for the importance of lead in early civilizations is that its ore, a heavy black lead(II) sulfide called *galena*, was easily recognized. Just as importantly, the ore was readily reduced by heating with wood charcoal. If galena is first roasted in a limited amount of air, lead(II) oxide forms.

Roasting is a common step in recovery of metals from their ores. It consists of heating the ore, often a metal sulfide, in air to convert the sulfide to the oxide.

$$2\ MS(s) + 3\ O_2(g) \longrightarrow 2\ MO(s) + 2\ SO_2(g)$$

Unfortunately, if the SO_2 is vented to the atmosphere it contributes to air pollution and is one source of "acid rain." (See also Chapter 25.)

$$2\ PbS(s) + 3\ O_2(g) \longrightarrow 2\ PbO(s) + 2\ SO_2(g)$$

The carbon then reduces the oxide to the metal,

$$PbO(s) + C(s) \longrightarrow Pb(\ell) + CO(g)$$

as does the carbon monoxide produced in the first step.

$$PbO(s) + CO(g) \longrightarrow Pb(\ell) + CO_2(g)$$

Approximately one million tons of lead are produced in the United States annually. Approximately 60% of this is used as an alloy for battery plates (91% lead and 9% antimony).

Pewter is about 85% tin and 7% copper; the rest is antimony and bismuth. The solder used in electronic circuits averages 33% tin.

EXERCISE 21.4 GRAPHITE AND DIAMOND

Using the data in the text, prove that diamonds should spontaneously revert to graphite.

SPECIAL SECTION: LEAD CHEMISTRY

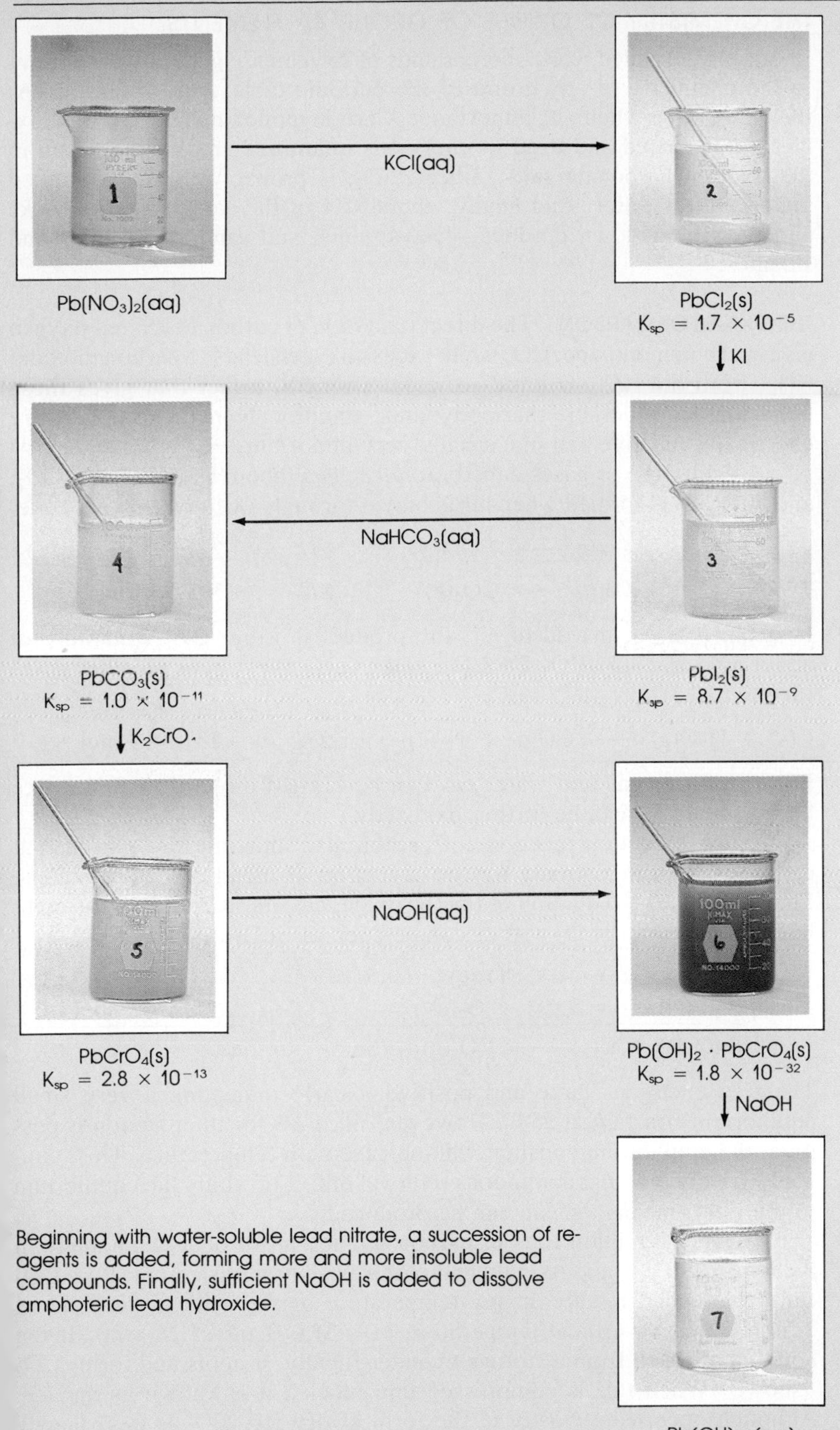

Beginning with water-soluble lead nitrate, a succession of reagents is added, forming more and more insoluble lead compounds. Finally, sufficient NaOH is added to dissolve amphoteric lead hydroxide.

THE CHEMISTRY OF OXIDES OF GROUP 4A ELEMENTS

In our oxygen-filled world, compounds of oxygen are generally important to the chemistry of any group of the periodic table, but, in Group 4A, they are of overwhelming importance. Carbon monoxide and carbon dioxide are produced and used in enormous quantities and often end up in the form of carbonate salts. Silica, SiO_2, is probably the most studied material after water. And finally, about 95% of the earth's crustal rocks and their breakdown products—clays, soils, and sands—are silica and silicates.

THE OXIDES OF CARBON The direct oxidation of carbon in limited oxygen gives carbon monoxide, CO, while excess oxygen leads to carbon dioxide, CO_2. Both have C—O bonds of great strength, a fact that gives these molecules considerable thermodynamic stability. Nonetheless, they are chemically reactive and industrially very important.

A mixture of gases called *producer gas* (about 25% CO, 4% CO_2, and 70% N_2) is formed when air is blown through a bed of glowing coke.

$$2\,C(s) + O_2(g) \longrightarrow 2\,CO(g) \qquad \Delta H^\circ = -221.0 \text{ kJ/mol of } O_2$$

$$C(s) + O_2(g) \longrightarrow CO_2(g) \qquad \Delta H^\circ = -393.5 \text{ kJ/mol}$$

If steam is used instead of air, the product is *water gas*, a mixture of about 50% H_2, 40% CO, 5% CO_2, and 5% N_2 plus traces of other gases (Section 20.1).

$$C(s) + H_2O(g) \longrightarrow CO(g) + H_2(g) \qquad \Delta H^\circ = +131.3 \text{ kJ/mol}$$

Both producer gas and water gas can be burned directly as fuel, since both CO and H_2 can be further oxidized.

The oxidation reactions of carbon also illustrate the reason that carbon is used industrially for the reduction of metal oxides to metals. For example, the reduction of tin(IV) oxide, cassiterite, couples the combustion of carbon in O_2 with decomposition of SnO_2 to Sn and O_2.

$$\begin{array}{ll} C(s) + O_2(g) \longrightarrow CO_2(g) & \Delta G^\circ = -393.5 \text{ kJ} \\ SnO_2(s) \longrightarrow Sn(s) + O_2(g) & \Delta G^\circ = +580.7 \text{ kJ} \\ \hline SnO_2(s) + C(s) \longrightarrow CO_2(g) + Sn(s) & \Delta G^\circ = +187.2 \text{ kJ} \end{array}$$

The free energy is large and positive, clearly indicating a very small equilibrium constant at 25°C. However, since ΔS for the reaction is positive, the equilibrium constant will increase with temperature. This, coupled with the fact that continuous removal of $CO_2(g)$ shifts the equilibrium to the right, means that tin can be produced.

Over 30 million tons of carbon dioxide are produced annually in the United States, and roughly half of it is used as a refrigerant. Pure CO_2 can be liquefied readily at any temperature between its triple point of −56.6°C and its critical temperature of +31.0°C (at 75.28 atm). If the liquid is released from a storage cylinder rapidly, it cools and forms CO_2 "snow." When this is compressed into blocks, it is known as *dry ice*. Although once used mostly in the form of dry ice, CO_2 is increasingly used now as a liquid refrigerant and as the propellant gas in aerosol cans.

Approximately 25% of the CO_2 manufactured is used to carbonate beverages. Over 400 bottles of carbonated beverage are produced per person in the United States per year. (See the special section of Chapter 17.)

Carbon monoxide, unlike CO_2, is a Lewis base, using the unshared electron pair of the carbon atom (:C≡O:) to form coordinate covalent bonds with some Lewis acids, the low-valent transition metals in particular. For example, CO readily interacts with nickel to give a nickel carbonyl.

$$Ni(s) + 4\,CO(g) \xrightleftharpoons{50°C} Ni(CO)_4(\ell)$$

This reaction is reversible, so heating the metal compound in a vacuum causes it to dissociate. The *Mond process* for purifying nickel takes advantage of this. Nickel oxide ore is first heated in the presence of water gas (H_2 + CO). The hydrogen reduces the metal oxide to impure metal, and the residual CO then reacts to give volatile $Ni(CO)_4$. When $Ni(CO)_4$ vapor is passed over a nickel surface heated to 230°C, the vapor decomposes to leave pure nickel and CO gas, which is then recycled.

$$\text{Impure ore containing } Ni^{2+} \xrightarrow[H_2 + CO]{} Ni(CO)_4(g)$$

$$Ni(CO)_4(g) \xrightarrow[\text{heat to 230°C}]{} Ni(s, \text{pure}) + 4\,CO(g)$$

Carbon monoxide is a toxic gas. Just as CO can interact with metallic nickel, it can also react with the iron of the hemoglobin in your blood. The function of hemoglobin is to activate O_2 to reduction, so, if you breathe CO, it displaces O_2 from the hemoglobin, and you suffocate. Exposure to even small concentrations (120 parts per million) can impair your abilities, and concentrations as high as 100 ppm are not unusual in tunnels, garages, and even the streets of large cities.

FIGURE 21.15
A pure quartz crystal (pure SiO_2). Quartz is one of the most common minerals on earth; it occurs in measurable quantities in almost every type of rock exposed at the earth's surface. Pure, colorless quartz was used as an ornamental material as early as the Stone Age, and by Roman times, it was known that a wedge of colorless quartz could be used to concentrate the sun's rays. Now quartz is used for its electrical properties in phonographs and other devices and for its optical properties to make special lenses and prisms. (Ward's Natural Science Establishment)

THE OXIDES OF SILICON Just as carbon forms CO_2, the simplest oxide of silicon is SiO_2, commonly called **silica**. Unlike CO_2, however, SiO_2 is not a simple molecule with Si=O double bonds. The energy of two Si=O double bonds is estimated to be much less than that of four Si—O single bonds, so silicon in SiO_2 and all other Si—O compounds is always surrounded tetrahedrally by four oxygen atoms. It is the interconnections between SiO_4 tetrahedra that lead to the tremendous variety of silicon–oxygen compounds observed.

More than 22 phases of pure silica, SiO_2, have been described, the most common being "low temperature" or α-quartz. It is the major constituent of many rocks such as granite and sandstone, and it occurs alone as pure rock crystal or in a variety of less pure forms (Figures 21.15 and 21.16).

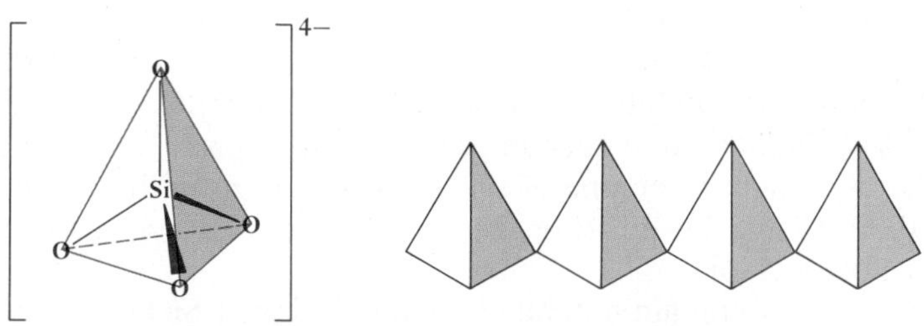

FIGURE 21.16
Amethyst is the most highly prized variety of quartz. Its color, due to Fe^{3+}-containing impurities in SiO_2, can range from pale lilac to a deep, royal purple. The name comes from the Greek "amethustos," meaning "not drunken." In ancient Greece it was believed that an amethyst wearer could never become intoxicated. (Ward's Natural Science Establishment)

The main crystalline modifications of SiO_2 consist of infinite arrays of SiO_4 tetrahedra sharing corners. In one form of quartz, the tetrahedra form chains, and the chains interlink. If α-quartz is heated to almost 1500°C, it transforms ultimately to another crystalline form, cristobalite. Elemental silicon has the diamond structure (Figure 11.31). If an oxygen atom is placed between each pair of silicon atoms in this lattice, the structure of cristobalite is generated (Figure 21.17).

The most important commercial application of crystalline quartz is to control the frequency of almost all radio and television transmissions. These and related applications use so much quartz that there is not enough natural material to fulfill the demand, so quartz is synthesized by the following procedure. Noncrystalline or vitreous quartz, made by melting pure silica sand, is more soluble in hot water than α-quartz. The vitreous quartz is placed in a steel "bomb" and dilute aqueous NaOH is added. A "seed" crystal is placed in the mixture, just as you place a seed crystal in a hot sugar solution to grow rock candy. When the mixture is heated above the critical temperature of water (above 400°C and 1700 atm) for some days, pure quartz crystallizes on the seed crystal.

Silica is resistant to attack by all acids except HF, but it does dissolve slowly in hot, molten NaOH or Na_2CO_3 (soda ash) to give Na_4SiO_4, the exact stoichiometry depending on the ratio of soda ash to silica.

$$2\ Na_2CO_3 + SiO_2 \longrightarrow Na_4SiO_4 + 2\ CO_2$$

When the fused glass has cooled, it is dissolved in hot water under pressure and any insoluble sand or glass is filtered off. The sodium silicate obtained from the solution is isolated by evaporation and is called *water glass*. Its biggest single use is in household and industrial detergents. The silicates maintain pH by their buffering ability and can degrade animal and vegetable fats and oils. They are also used in various adhesives and binders, especially for glueing corrugated cardboard boxes.

If the soluble sodium silicate is treated with acid, a gelatinous precipitate of noncrystalline or amorphous SiO_2 is formed. After washing and drying this so-called *silica gel*, the white residue is a very porous material with dozens of uses. Since it can absorb up to 40% of its own weight of water, you may know it as a drying agent. (Small packets of silica gel are often placed in packing boxes of merchandise during storage.) When stained with $(NH_4)_2CoCl_4$, it is a humidity detector, turning pink when hydrated, but remaining blue when dry. Finally, one other use of silicates is to clarify beer, that is, to remove minute particles that make the brew cloudy.

The **silicate minerals** are a world in themselves. The simplest silicates, *orthosilicates*, are based on discrete, tetrahedral $SiO_4{}^{4-}$ anions. The -4 charge of the anion can be balanced by metal ions: four M^+ ions, two M^{2+} ions, or a combination of ions. For example, the calcium orthosilicates, Ca_2SiO_4, are vital components of *Portland cement*. In *olivine*, the mineral thought to be one of the most important in the mantle of the earth, the ions are those of Mg, Fe, and Mn; the Fe^{2+} gives it its characteristic olive color.

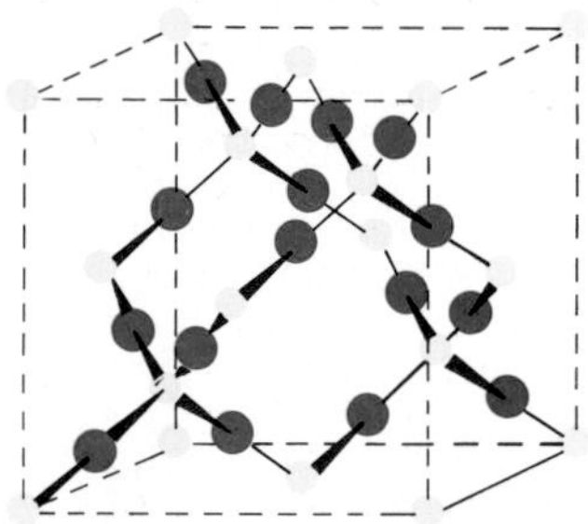

FIGURE 21.17
The structure of cristobalite, one of the solid forms of SiO_2. Notice the similarity to diamond (Figure 11.31): Each C—C bond in diamond is replaced by a Si—O—Si linkage in cristobalite.

Pyroxenes contain extended chains of linked SiO_4 tetrahedra. All are based on the *metasilicate* ion, $SiO_3{}^{2-}$, as in $Mg_2Si_2O_6$.

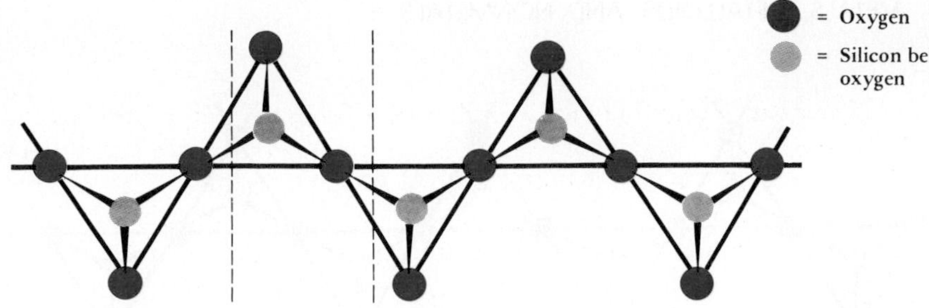

A chain of linked SiO_4 tetrahedra. The dotted lines enclose the repeating SiO_3 unit of a pyroxene. The view shown is from on top of the tetrahedra.

If two such chains are laid side by side, they may link together by sharing oxygen atoms in adjoining chains.

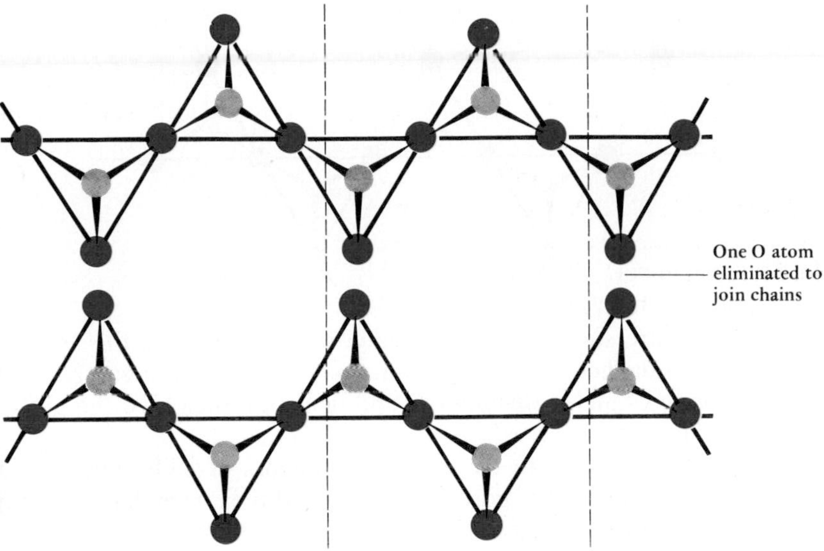

The result is an *amphibole*, of which the *asbestos* minerals are an excellent example. Because of their double-stranded chain structure, asbestos minerals are fibrous materials. Their best known characteristic is their very low thermal conductivity, which has led to their use in insulation and fireproofing.

The fibrous silicates called collectively *asbestos* have been used for thousands of years. Canada and the USSR are the largest producers of asbestos, with the United States mining only about 2% of the world's annual production of about 5 million tons. Asbestos-reinforced cement accounts for the greatest use of the mineral, but a large amount is consumed in making vinyl floor tiles, brake linings, automobile clutches, fire-fighting garments, and so on. Prolonged exposure to asbestos dust is now known to be very dangerous, and there is increasing concern about *asbestosis*, a nonmalignant scarring of lung tissue. Victims become short of breath and eventually struggle so hard to breathe that they die of heart failure. Unfortunately, it takes many years (typically 20 to 30 years) before the effects of asbestos inhalation show up, so the cases being discovered now came from exposure during and shortly after World War II.

If the linking of silicate chains is continued in two dimensions, sheets of SiO_4 tetrahedra result.

Asbestos is not a discrete chemical compound or even a single group of minerals. The term is an imprecise one that refers to several fibrous minerals valued for their high tensile strength and resistance to heat. All commercial varieties of asbestos are silicates and most belong to the amphibole group. Amphiboles are characterized by strips of cations (usually sodium, calcium, magnesium, iron, or aluminum) sandwiched between two strips of linked SiO_4 tetrahedra. (Particulate Mineralogy Unit, Avondale Research Center, U.S. Bureau of Mines)

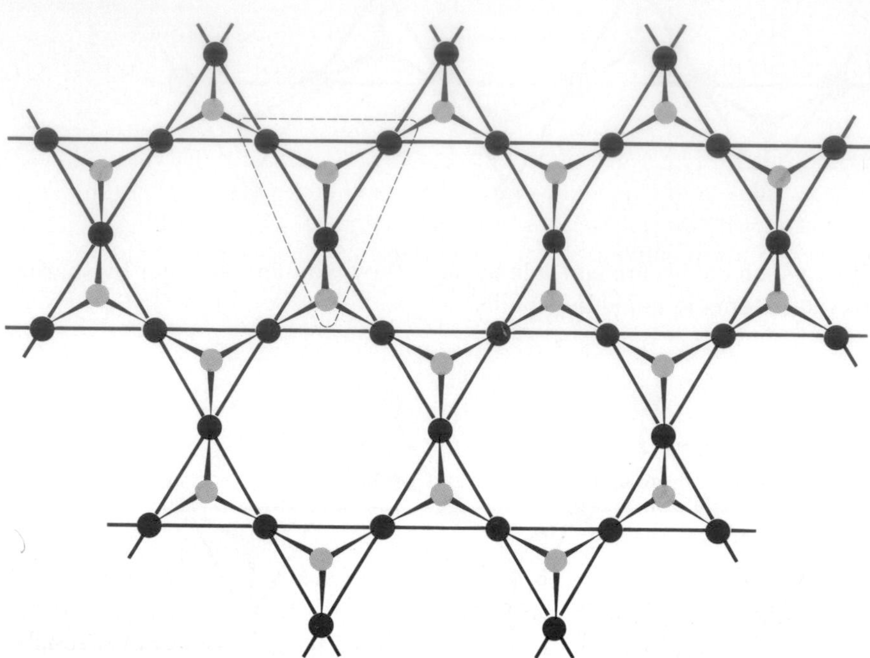

In this category are some of the most important minerals in the ancient and modern world: the clay minerals and mica. Sheets of mica are used in furnace windows and as insulation, and flecks of mica give the glitter to "metallic" paints.

The *clay minerals* are essential components of soils, and, since they are the raw material for pottery, bricks, and tiles, they have played a major role in the development of our civilization. Clays come from the weathering and decomposition of igneous rocks. Specifically, the aluminosilicate *kaolinite* comes from the weathering of feldspar, and an idealized version of this reaction would be

$$\underset{\text{feldspar}}{2\,[KAlSi_3O_8](s)} + CO_2(g) + 2\,H_2O(\ell) \longrightarrow \underset{\text{kaolinite}}{[Al_2(OH)_4Si_2O_5](s)} + 4\,SiO_2(s) + K_2CO_3(aq)$$

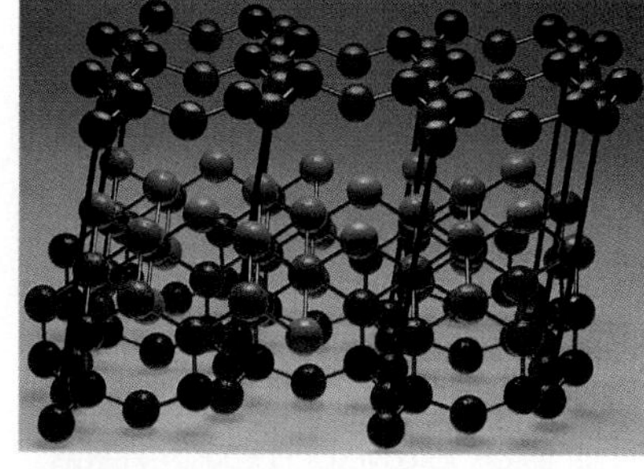

FIGURE 21.18
Kaolinite, an aluminosilicate. At the bottom is a network of silicon–oxygen rings. Each silicon (black) is surrounded tetrahedrally by oxygen atoms (red) to give rings consisting of 6 O atoms and 6 Si atoms. In addition, a layer of Al ions (silver) is attached through O atoms to Si—O rings, and the Al ions are bridged by OH ion (light green). The net result is a layered structure typical of clays in general.

Its structure consists of layers of SiO_4 tetrahedra linked into sheets, as in mica, but these sheets are interleaved with six-coordinate Al^{3+} ions surrounded octahedrally by O atoms of the silicon–oxygen sheets and OH^- ions (Figure 21.18).

China clay, or kaolin, is primarily kaolinite. It is practically free of iron, and so it is colorless and particularly valuable. World production in 1974 was approximately 17 million tons, with the predominant use in the United States being for paper filling and coating. Some was also used for china, crockery, and earthenware, but most of these articles are made from ball clay, an especially fine-grained clay that is mostly kaolinite but also contains mica and quartz. Approximately 800,000 tons of ball clay are used annually in the United States for tableware, porcelain, and wall and floor tiles.

A final category of silicates are those with three-dimensional structures, the *aluminosilicates*. This group includes *feldspars* (among the most common minerals; they make up about 60% of the earth's crust) and *zeolites* (see below). Both materials are composed of SiO_4 tetrahedra in which each oxygen is shared between two tetrahedra; however, in addition, some of the Si atoms have been replaced by Al atoms. Since the silicon atoms formally bear a +4 charge and are replaced by a +3 aluminum, other positive ions must be added for charge balance. Typically these are alkali and alkaline earth ions. For example, the synthetic zeolite "Linde A" has the formula $Na_{12}(Al_{12}Si_{12}O_{48}) \cdot 27\ H_2O$.

The structure of a zeolite is illustrated in Figure 21.19. The main feature of zeolite structures is that there are tunnels and cavities of regular shape. Since the diameter of the holes is typically 300 to 1000 pm, and since water molecules have an effective diameter of 265 pm, they can fit comfortably into the cavities of the zeolite and can be absorbed selectively from air or a solvent. This is why zeolites are excellent drying agents, and small amounts are sealed into multipane windows to keep the air dry between the panes. Zeolites have also been used as catalysts. Mobil Oil, for example, has patented a process in which a simple, one-carbon compound methyl alcohol, CH_3OH, forms gasoline in the presence of specially tailored zeolites. Finally, zeolites are used as water-softening agents in detergents, since the sodium ions of the zeolite can be exchanged for Ca^{2+} ions in hard water, effectively removing Ca^{2+} from the water.

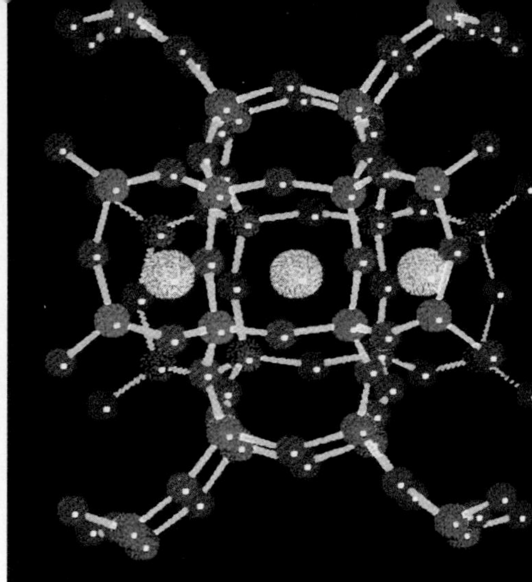

FIGURE 21.19

The structure of a zeolite. It is essentially built of SiO_4 tetrahedra. There are, however, occasional Al^{3+} ions in place of Si^{4+}. This means other positive ions, such as Na^+, are also present to balance the charge of the lattice. (The three pink spheres are xenon atoms placed in the lattice to give some idea of the size of the cavities.)

Zeolites make up as much as 25% of many detergents; 325 million pounds of zeolites were consumed in the United States in 1980, over 75% being used in detergents.

One current theory of the origin of life on earth is that zeolites and clay minerals provided an organizing influence so that simple molecules could form the precursors to biologically important molecules we know today.

TIN AND LEAD OXIDES Other Group 4A oxides are also economically important. The mineral cassiterite, SnO_2, is not only a source of tin, but thousands of tons are also used in the ceramics industry in glazes, enamels, and pigments. For example, yellow glazes are a mixture of SnO_2 and V_2O_5, and blue-grey colors are obtained with SnO_2 and Sb_2O_5.

Lead(II) oxide, PbO, is a red solid at room temperature (commonly known as *litharge*) or a yellow solid at higher temperatures. It is the most important inorganic compound of lead, and one large use is in the manufacture of leaded glass, a glass with a high lead content having a high refractive index and thus a special brilliance. The other major consumer of PbO is storage batteries (Section 19.4). The oxide is applied as a paste in sulfuric acid to the lead battery plates. The positive plates, the cathode, are activated by oxidizing PbO to the lead(IV) oxide, PbO_2, and the negative plates, the anode, are activated by reducing the oxide to Pb.

If PbO is heated in air, "red lead" is formed.

$$6\ PbO(s) + O_2(g) \longrightarrow 2\ Pb_3O_4(s)$$

This solid can best be formulated as a mixed Pb(II)/Pb(IV) oxide, $(2\ Pb^{2+})(Pb^{4+})(4\ O^{2-})$. It is widely used in rust-inhibiting paints and primers for iron and steel, but other lead salts are important in the same manner. The chromate $PbCrO_4$ is a brilliant yellow and is used in road markings, and a mixed carbonate, $2\ PbCO_3 \cdot Pb(OH)_2$, has traditionally been used as a white pigment. More recently, lead-containing paints have ceased to be used in houses, because of the toxicity of lead as described below.

Lead oxide chemistry, and the compounds such as tetraethyllead, $Pb(C_2H_5)_4$, raise the important issue of *lead poisoning*. Chronic exposure

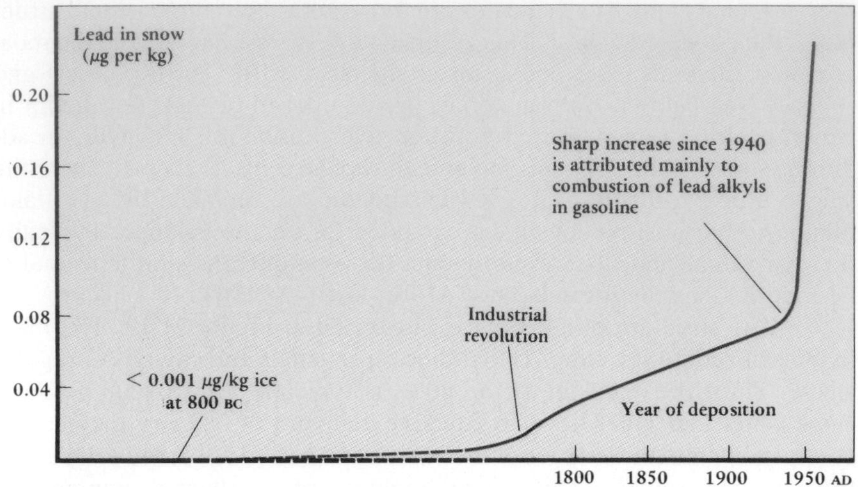

FIGURE 21.20
Environmental lead. The amount of lead introduced into the environment was measured by examining snow samples in northern Greenland.

Lead poisoning has been known for centuries, the first clear description coming from a Greek poet in the second century BC.

to lead compounds can lead to cholic, anemia, headaches, convulsions, and brain damage. The sources of lead in our environment and the solutions to the problem are still controversial. Tetraethyllead, however, is clearly one source. The recent use of this compound as a gasoline additive is thought to have contributed to the tremendous increase in environmental lead since about 1940 (Figure 21.20). Although the U.S. Environmental Protection Agency has mandated a gradual decline in lead use in gasoline (from 3 g/gallon in 1974 to 0.1 g/gallon presently), there are still other sources of lead in the environment. Since we know that lead has insidious effects, it seems wise to reduce its introduction into the environment in every possible way.

EXERCISE 21.5 SILICATES
Describe the main structural characteristics of orthosilicates, pyroxenes, mica, and zeolites.

THE HALIDES OF GROUP 4A ELEMENTS

All the Group 4A elements form chlorides of the type MCl_4, and all are liquids. In keeping with the general trend in Group 4A, lead(IV) chloride is not stable, however, decomposing to $PbCl_2$ and Cl_2 above 50°C.

$$PbCl_4(\ell) \xrightarrow{>50°C} PbCl_2(s) + Cl_2(g)$$

Carbon tetrachloride is well known as a common laboratory and industrial solvent. Almost 600 million pounds are made in the United States annually, usually by treating methane with Cl_2.

$$CH_4(g) + 4\ Cl_2(g) \longrightarrow CCl_4(\ell) + 4\ HCl(g)$$

Its use as a solvent is declining due to its toxicity, but increasingly it is used to make "Freons" such as $CFCl_3$, CF_2Cl_2, and CF_3Cl.

$$CCl_4(\ell) + HF(\ell) \xrightarrow{\text{catalyst} = SbFCl_4} \underset{\text{Freon-11}}{CFCl_3(g)} + HCl(g)$$

Freons are used as the working fluid in refrigerators and air conditioners, and, until recently, as the propellant gas in aerosol cans. However, in the past few years there has been increasing concern that the Freons released from these spray cans (by 1977 the annual production of spray cans in the United States was several billion) were working their way into the upper atmosphere where they could destroy the earth's protective shield of ozone (see Chapter 22). For this reason, this use of Freons has declined rapidly.

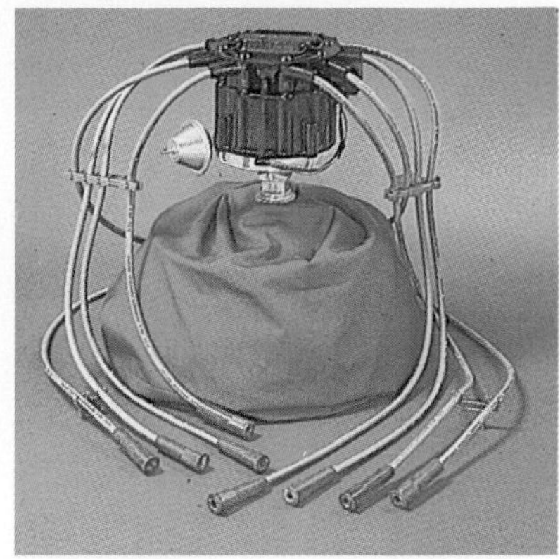

FIGURE 21.21
Silicone rubber is used extensively in automotive ignition systems because of its superior electrical properties and heat resistance. (Stauffer-Wacker Silicones Corp.)

You learned previously that silicon reacts readily with chlorine to produce $SiCl_4$. A similar reaction occurs between silicon and methyl chloride, CH_3Cl, but now two methyl groups (CH_3) and two Cl atoms are bound in a tetrahedral fashion to Si.

$$\text{Si(s)} + 2\,\text{CH}_3\text{Cl(g)} \xrightarrow[300^\circ\text{C}]{\text{Cu powder catalyst}} (\text{H}_3\text{C})_2\text{SiCl}_2(\ell)$$
(70% yield)

Unlike CCl_4 and other compounds with C—Cl bonds, halides based on other Group 4A elements are hydrolyzed to form M—O bonds. For example,

$$(\text{H}_3\text{C})_3\text{Si—Cl}(\ell) + \text{H}_2\text{O}(\ell) \longrightarrow (\text{H}_3\text{C})_3\text{Si—OH}(\ell) + \text{HCl(aq)}$$

As often happens with compounds having the M—O—H grouping, two such molecules come together and eliminate a molecule of water between them.

$$(\text{H}_3\text{C})_3\text{Si—O—H} + \text{H—O—Si(CH}_3)_3 \longrightarrow (\text{H}_3\text{C})_3\text{Si—O—Si(CH}_3)_3 + \text{H}_2\text{O}$$

For $(H_3C)_2SiCl_2$, with more than one Si—Cl bond, reaction with water can initially produce $(H_3C)_2Si(OH)_2$, and each Si—O—H linkage can interact with a neighbor until a long chain of Si—O—Si—O—Si links has been forged.

```
 CH3    CH3    CH3    CH3    CH3
  |      |      |      |      |
—Si—O—Si—O—Si—O—Si—O—Si—O—
  |      |      |      |      |
 CH3    CH3    CH3    CH3    CH3
```

This is a *polymer* called polydimethylsiloxane, a member of the *silicone* polymer family (Figure 21.21). Silicones are nontoxic and have good stability to heat, light, and oxygen; they are chemically inert and have valuable antistick and antifoam properties. Since they can be made in the form of oils, greases, and resins, or with rubber-like properties ("Silly Putty," for example), they are commercially useful. Approximately 300,000 tons are made worldwide annually and are used in a wide variety of products: as lubricants and the antistick material for peel-off labels, in lipstick, suntan lotion, and car polish, and as the antifoam substance in stomach remedies.

Polymers are discussed in more detail in Chapter 24.

Germanium, tin, and lead form halides of the type MCl_2 and MCl_4. Since the +2 oxidation state becomes more important as the element atomic weight increases, $GeCl_4$ is more stable than $GeCl_2$, but the reverse is true of the lead chlorides. At least two tin halides are used in considerable amounts. Tin(II) fluoride (stannous fluoride) is put into some toothpastes to aid in the prevention of dental caries. Tin(IV) chloride is sprayed onto freshly made glass bottles where it leaves a film of SnO_2 due to the hydrolysis of the chloride.

$$SnCl_4(\ell) + 2\ H_2O(\ell) \longrightarrow SnO_2(s) + 4\ HCl(g)$$

This transparent film of tin(IV) oxide toughens the surface of the glass and improves its resistance to abrasion.

EXERCISE 21.6 SILICON CHEMISTRY

Write a balanced equation for the reaction of $SiCl_4$ with water. Indicate the physical state of each reactant and product. Tell why you believe $SiCl_4$ reacts with water while CCl_4 does not.

THE HYDROGEN COMPOUNDS OF GROUP 4A

The general trends in Group 4A behavior are never more evident than when the elements are bonded to hydrogen. Carbon is rather electronegative, so, when H is bonded to carbon, the bond polarity is $C(-\delta)$—$H(+\delta)$. The other elements of Group 4A, however, become less electronegative with increasing atomic weight, and the negative end of the E—H bond shifts to the hydrogen atom. This means the hydrogen is increasingly "hydridic" or H^- in character.

The other trend in Group 4A chemistry is that element–hydrogen bond energies decrease in the order C > Si > Ge > Sn > Pb. Thus, CH_4 is a stable, gaseous compound, but SiH_4 is a stable gas only in the absence of air and water. Tin(IV) hydride (SnH_4), however, slowly decomposes to tin and H_2,

$$SnH_4(g) \longrightarrow Sn(s) + 2\ H_2(g)$$

and there is serious doubt that PbH_4 has ever been prepared.

Literally thousands of carbon–hydrogen compounds are known and some will be described in Chapter 24. The simplest of these compounds is methane, CH_4, the first in a series of compounds, the *alkanes*, whose general formula is C_nH_{2n+2}. One of the features of carbon chemistry in general is that there are bonds between carbon atoms, a property called *catenation*. Long, straight chains, branched chains, and rings (see graphite and diamond) of carbon can form. Except in the elements themselves, this ability declines severely as the atomic weight of the element increases. Silicon forms a series of *silanes*, Si_nH_{2n+2}, analogous to the alkanes. The simplest of these, silane, SiH_4, is made by treating $SiCl_4$ with a metal hydride in an exchange reaction.

Other alkanes are, for example, H_3C—CH_3 (ethane), H_3C—CH_2—CH_3 (propane), and H_3C—CH_2—CH_2—CH_3 (butane). All are gases at room temperature. There is about 1 ppm of CH_4 in the atmosphere, much of it produced in the digestion of food by animals.

$$SiCl_4(\ell) + 4\ NaH(s) \longrightarrow SiH_4(g) + 4\ NaCl(s)$$

Although the silanes up to H_3Si—SiH_2—SiH_3 can be made in good yield and longer chain silanes are known up to $n = 8$, they are increasingly unstable; only SiH_4 is stable indefinitely at room temperature. All are extremely reactive and spontaneously ignite in air.

)THER GROUP 4A COMPOUNDS

'arbon forms simple binary compounds called *carbides* with many metals nd metalloids. An especially important example of the latter is silicon arbide, SiC, a lustrous black solid commonly known as *carborundum.*

$$SiO_2(s) + 3\ C(s) \xrightarrow{2000°C} SiC(s) + 2\ CO(g)$$

ts crystal structure is similar to the diamond structure, with every other tom being silicon. Because of this similarity, SiC is harder than Al_2O_3 :orundum) but not quite as hard as diamond. Therefore, it has been used /idely as an inexpensive abrasive. More recently, however, industry has ıken advantage of the ability of silicon carbide to withstand very high :mperatures, and it is now being used to make parts for ceramic auto-ıobile engines. Such engines can run at higher temperatures and are ghter in weight than conventional iron and steel engines, leading to ignificant fuel savings (Figure 21.22).

Crystalline silicon carbide, SiC, also known as carborundum. (Charles D. Winters)

Metal carbides contain carbon as an anion, primarily C^{4-} or C_2^{2-}. n Section 20.4 we described the synthesis of calcium carbide, CaC_2, an ıdustrially important material that gives the hydrocarbon acetylene, IC≡CH, when hydrolyzed. Another industrially important reaction of alcium carbide is its ability to "fix" nitrogen from the air.

$$\underset{\text{:alcium :arbide}}{'aC_2(s)} + N_2(g) \xrightarrow{1000°C} \underset{\text{calcium cyanamide}}{CaNCN(s)} + C(s) \qquad \Delta H° = -296\ kJ/mol$$

he product is *calcium cyanamide*, in which the NCN^{2-} cyanamide anion . isoelectronic with CO_2. The salt hydrolyzes in the atmosphere to give ee cyanamide.

$$CaNCN(s) + CO_2(g) + H_2O(\ell) \longrightarrow CaCO_3(s) + \underset{\text{cyanamide}}{H_2NCN(s)}$$

ecause of this reaction, calcium cyanamide has been widely used as a :rtilizer, since it is one way to bring usable nitrogen into the soil.

Hydrogen cyanide, HCN, is a very weak Brønsted acid in aqueous olution and an extraordinarily poisonous gas (mp, −13.4°C; bp, 25.6°C). 'ormerly it was made by adding acid to cyanide salts such as NaCN, but ıost of the 500,000 tons produced annually are made by a more direct, atalyzed process.

$$CH_4(g) + NH_3(g) \xrightarrow[1200°C]{\text{Pt catalyst}} HCN(g) + 3\ H_2(g)$$

'he vast majority of the HCN is used in the polymer industry, but some used to make NaCN for extraction of metals from their ores. Cyanide ɔn, an anionic Lewis base, can form very stable complexes with metal ɔns such as Ag^+ and Au^+. For example, gold is often obtained by bub-ling air through a cyanide-containing alkaline suspension of its ores.

$$CN^-(aq) + 4\ Au(s) + 2\ H_2O(\ell) + O_2(g) \longrightarrow 4\ [Au(CN)_2]^-(aq) + 4\ OH^-(aq)$$

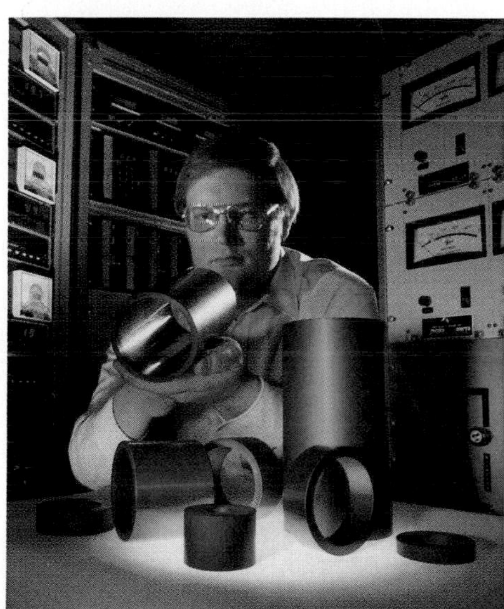

FIGURE 21.22

Silicon carbide, SiC, is commonly known as carborundum, and it has been widely used as an abrasive. More recently, it has been used to manufacture high-strength, heat-resistant ceramics. (General Electric Company)

SUMMARY

The elements of **Group 3A** (B, Al, Ga, In, and Tl) all have the valenc shell electron configuration ns^2np^1, and all form compounds in the + oxidation state. There is, however, an increasing tendency to form th +1 state for elements of higher atomic number.

None of the elements exists in nature uncombined. Boron form many compounds with oxygen, one of the most common being **borax** $Na_2B_4O_7 \cdot 10\ H_2O$. All form oxides of general formula M_2O_3, and alu minum commonly exists as a hydrated oxide **bauxite**, $Al_2O_3 \cdot nH_2O$ o anhydrous **corundum**, Al_2O_3.

Unlike the metals of Group 3A, boron does not form simple + ions. The hydrated +3 ions of the metals are all weak acids in aqueou solution.

In general, nonmetal or metalloid oxides are acidic, and meta oxides are amphoteric or basic. This is true in Group 3A, since the hy drated oxide of boron, $B(OH)_3$, **boric acid**, is a weak acid in water. I contrast $Al(OH)_3$ and $Ga(OH)_3$ are amphoteric, dissolving in both aci and strong base. This property plays a key role in the recovery of alu minum from bauxite in the **Bayer process**.

The elements are obtained by chemical or electrochemical reduc tion. Boron is a hard, covalently bonded insulator (a metalloid), wherea the other elements are metals, electrically conducting and relatively sof

All Group 3A elements form trihalides that are excellent Lewi acids. Unlike the boron trihalides, Al and Ga at least form halide-bridge dimers $X_2M(X)_2MX_2$, a structural type observed in other metal–haloge compounds.

Boron, but not the metals of Group 3A, forms a large series c hydrides of which the simplest is B_2H_6, **diborane**. The bonding in thi B—H—B bridged compound is found throughout chemistry; you shoul understand its formation.

Group 4A elements all have a ns^2np^2 configuration and all can hav an oxidation number of +4. Again, there is a tendency to lower oxidatio numbers (+2) with the heavier metals (Sn and Pb), and carbon is com monly found in the −4 state.

Carbon is a nonmetal, silicon a metalloid, and germanium, tin, an lead become progressively more metallic. Carbon exists in thousands c different compounds exhibiting two features not shared to any extent wit the other Group 4A elements: **catenation**, the ability to form long chain of the elements, and **multiple bonding**.

Carbon is found in nature in elemental form as the **allotropes graph ite** and **diamond**. In diamond, each carbon atom is tetrahedral, sp^3 hy bridized, and bonded to four other carbon atoms. In graphite, the carbon are arranged in sheets of interconnected, planar, six-membered rings; th atoms are sp^2 hybridized (Figures 11.30 and 11.31).

As expected from the diagonal relationship of boron and silicon the latter forms many oxides: silica (SiO_2) in its many forms (e.g., **quartz** silicates (Ca_2SiO_4, found in **Portland cement**), and **aluminosilicates** (cla minerals, zeolites). All are based on networks of interconnected SiO tetrahedra. The most important ore of tin is cassiterite, SnO_2. Litharg (PbO) and red lead (Pb_3O_4) are important oxides of lead.

Silicon and the Group 4A metals are obtained by chemical reduction, often with carbon.

All Group 4A elements form tetrahalides such as CCl_4, CF_2Cl_2 (a "Freon"), $SiCl_4$ (starting material to make silicon computer chips), and $SnCl_4$. All except CX_4 are susceptible to hydrolysis. Hydrolysis of $(CH_3)_2SiCl_2$ gives silicone polymers.

Carbon forms thousands of C—H containing compounds where the bond polarity is $C(\delta-)—H(+\delta)$. Since the remaining elements of Group 4A are less electronegative, their hydrogen compounds are called **hydrides** because the bond polarity is $M(\delta+)—H(-\delta)$.

Other important Group 4A compounds are silicon carbide, SiC **(carborundum)**, and HCN (hydrogen cyanide).

STUDY QUESTIONS

1. Complete and balance the following equations for reactions of Group 3A elements:
   (a) $Ga(s) + O_2(g) \rightarrow$
   (b) $In(s) + Br_2(\ell) \rightarrow$
   (c) $BBr_3(g) + H_2(g) \rightarrow$
   (d) $B(OH)_3(s) + CH_3OH(\ell) \rightarrow$
   (e) $BF_3(g) + H_2O(\ell) \rightarrow$
   (f) $B_2O_3(s) + HF(g) \rightarrow$
   (g) $NaBH_4(s) + I_2(s) \rightarrow$
   (h) $B_5H_9(g) + O_2(g) \rightarrow$
   (B_5H_9 is a borane like B_2H_6)
2. Complete and balance the following equations for reactions of Group 4A elements:
   (a) $SnCl_4(\ell) + H_2O(\ell) \rightarrow$
   (b) $Pb(s) + O_2(g) \rightarrow$
   (c) $Ge(s) + O_2(g) \rightarrow$
   (d) $Sn(s) + Cl_2(g) \rightarrow$
   (e) $GeCl_4(\ell) + NaH(s) \rightarrow$
   (f) $SiH_4(g) + O_2(g) \rightarrow$
3. Complete and balance the following equations for reactions of Group 4A elements:
   (a) $SiO_2(s) + C(s) \rightarrow$
   (b) $Si(s) + Cl_2(g) \rightarrow$
   (c) $PbS(s) + O_2(g) \rightarrow$
   (d) $PbO(s) + CO(g) \rightarrow$
   (e) $CCl_4(\ell) + HF(\ell) \rightarrow$
   (f) $Si(s) + CH_3Cl(g) \rightarrow$
4. Write complete, balanced equations for the preparation of each of the following compounds. Where possible, indicate the phase of each reactant and product.
   (a) diborane
   (b) indium trioxide
   (c) boron trifluoride
   (d) aluminum trichloride (starting with aluminum oxide)
   (e) germanium tetrachloride (starting with germanium(IV) oxide)
   (f) $Ni(CO)_4$ (starting with nickel(II) oxide)
   (g) carborundum
   (h) calcium carbide (starting with limestone)
   (i) cyanamide
5. Name at least one use for each of the following compounds:
   (a) cyanamide
   (b) hydrogen cyanide
   (c) boron carbide
   (d) aluminum sulfate
   (e) boric acid
   (f) sodium borohydride
   (g) carbon black
6. Draw a possible structure for the cyclic anion in the salt $K_3B_3O_6$ and the chain anion of $Ca_2B_2O_5$.
7. The boron trihalides (except BF_3) hydrolyze completely to boric acid and the corresponding HX acid.
   (a) Write a balanced equation depicting this hydrolysis reaction for BCl_3.
   (b) Calculate $\Delta H°$ for the hydrolysis of BCl_3. ($\Delta H_f°$: $BCl_3(g) = -408$ kJ/mol; $B(OH)_3(s) = -968.92$ kJ/mol)
8. When the boron hydrides burn in air, a tremendous amount of heat is evolved. Calculate the heat of combustion of B_5H_9 ($\Delta H_f° = 54$ kJ/mol) and compare it with that of B_2H_6 in the text. Calculate the heat of combustion of C_2H_6 (ethane) for comparison.
9. Diborane, B_2H_6, can be prepared by mixing $NaBH_4$ with I_2 in an oxidation–reduction reaction. What substance is reduced and what is oxidized in this reaction?
10. Diborane can be made on an industrial scale by reducing BF_3 with NaH in a metathesis reaction. Write a balanced equation depicting this process.
11. When BCl_3 is passed through an electric discharge, small amounts of B_2Cl_4 are isolated. Draw a possible Lewis structure of this highly reactive molecule. What is the hybridization of the boron atom? What is the shape of the molecule in the vicinity of the boron atom?
12. Sodium borohydride, $NaBH_4$, is used to reduce metal ions to metals. Write a balanced equation to depict

the reaction of $NaBH_4$ with $AgNO_3$ in water to give silver metal, H_2, boric acid, and $NaNO_3$ (see Figure 21.10).

13. Boron carbide is an abrasive and is chemically inert. Consequently it has many commercial uses. Boron carbide fibers can be grown by passing a mixture of BCl_3 and H_2 over hot graphite (C) fibers. The products are B_4C and HCl. Write a balanced equation for this process.
14. Suggest a reason for the fact that Al^{3+} forms an octahedral ion AlF_6^{3-} with fluoride ion but a tetrahedral ion, AlX_4^- with ions of the other, larger halides.
15. Even though the table of $E°$ values in Chapter 19 shows that metallic aluminum should be a good reducing agent, a typical sample of the metal is resistant to attack by air and water. Explain this observation.
16. In 1870 Mendeleev predicted the existence of a Group 3A element that he called "eka-aluminum." He stated, among other things, that its atomic weight would be about 69 and its density 5.9. The predicted element was found shortly thereafter. What element is "eka-aluminum" and how do its actual properties compare with those predicted?
17. The density of solid gallium at 29.6°C is 5.904 g/cm³, whereas that of liquid gallium at 29.8°C is 6.095. Would the liquid–solid equilibrium line in a phase diagram of gallium have a positive or negative slope? Will solid gallium float or sink in a bath of liquid gallium? Explain your choice briefly.
18. Aluminum dissolves readily in hot aqueous base (NaOH) to give the aluminate ion, $Al(OH)_4^-$, and H_2. Write a balanced equation for this reaction.
19. Plot the half reaction potential for $M(s) \rightarrow M^{3+}(aq) + 3e^-$ versus atomic number for the Group 3A elements (see Table 21.1). What trend or trends do you see? What do they tell you about the chemistry of Group 3A elements? About the stability of the +3 oxidation state?
20. Al_2O_3 is amphoteric, so it will dissolve when heated strongly or "fused" with an acidic oxide or a basic oxide.
    (a) Write a balanced equation for the reaction of alumina with silica, an acidic oxide, to give an aluminum metasilicate, $Al_2(SiO_3)_3$.
    (b) Write a balanced equation for the reaction of alumina with the basic oxide CaO to give a calcium aluminate, $Ca(AlO_2)_2$.
21. Gallium hydroxide, like aluminum hydroxide, is amphoteric. Write balanced equations showing how $Ga(OH)_3$ can dissolve in both HCl and NaOH.
22. Halides of Group 3A elements are excellent Lewis acids.
    (a) When a Lewis base such as Cl^- (why is this a Lewis base?) interacts with $AlCl_3$, the ion $AlCl_4^-$ is formed. What is the structure of this ion?
    (b) BF_3 will react with ammonia, $:NH_3$, to a give a Lewis acid–base complex or adduct. Draw the Lewis structure of the product. What is the geometry around the N atom? Around the B atom? What is the hybridization of each atom?
23. Diborane reacts readily with Lewis bases (L:) to give $L:BH_3$ complexes. The hydrogen-bridged structure of diborane is clearly dismantled in the reaction.
    (a) Write a balanced equation for the interaction of the Lewis base trimethylamine, $N(CH_3)_3$, with diborane.
    (b) What is the geometry around the N atom? Around the B atom in the Lewis acid–base adduct?
    (c) What is the hybridization used by N and B in the adduct?
24. Carbon exists in several allotropic forms, one of which is diamond. The size of diamonds is often measured in *carats* (1 carat = 0.200 g). If worldwide production of gem quality diamonds in 1974 was 12.52 million carats, how many grams were produced? What would be the volume of a single crystal of this mass?
25. When calcium carbide, CaC_2, is treated with water, it is hydrolyzed to $Ca(OH)_2$ and acetylene, H—C≡C—H.
    (a) Write a balanced equation for the hydrolysis of calcium carbide.
    (b) Draw an electron dot structure of the carbide ion, C_2^{2-}.
26. The metal carbide produced in largest amount is calcium carbide. This is often called an "acetylide" since it hydrolyzes to give acetylene. Other metal carbides are "methanides," because they give methane on hydrolysis. One example of the latter is Al_4C_3. Write a balanced equation showing the hydrolysis of aluminum carbide.
27. Calcium carbide has a face-centered structure closely resembling that of NaCl. If Ca^{2+} ions define the face-centered lattice, in what type of holes are the C_2^{2-} ions located? How many Ca^{2+} and C_2^{2-} ions are there per unit cell?
28. Carbon forms a series of simple halides of the type CX_4. If the boiling point of CF_4 is −128.5°C, do you anticipate the boiling point of CCl_4 will be be higher or lower? Why?
29. Carbon is widely used as an industrial reducing agent to reduce metal oxides to metals. Write balanced equations for the following:
    (a) $Cu_2O(s) + C(s) \rightarrow$
    (b) $TiO_2(s) + C(s) \rightarrow$
    (c) $FeO(s) + C(s) \rightarrow$

30. This question compares the reduction of two metal oxides with carbon.
(a) Calculate the free energies of reaction for the reduction of PbO and CaO with C to give the metal and $CO_2(g)$.
(b) Which of the two reductions would you consider more feasible on a commercial scale? You can answer by deciding which is more likely to be spontaneous at a reasonable temperature.

31. In the presence of light, CO reacts with Cl_2 to give *phosgene* (Cl_2CO). It is a deadly gas and was used in World War I as a chemical warfare agent. Sketch the electron dot structure of the molecule. What is the hybridization of the central atom?

32. The planet Venus is thought to have clouds of sulfuric acid. One mechanism for their formation involves the reaction of sulfur with CO to form the deadly gas carbonyl sulfide, OCS. Draw the electron dot picture of OCS.

33. Many compounds have been detected in interstellar space, and three quarters of those observed involve carbon. Carbon monoxide dominates, binding perhaps 10% of the interstellar carbon, but a number of the other compounds are based on the cyanide ion, a close relative of CO.
(a) Compare the electron dot structures of CO and CN^-.
(b) One interstellar compound is cyanamide, H_2NCN. Draw its electron dot structure.
(c) Sketch the molecular structure of cyanamide. What hybridization is utilized by the central N atom and by the C atom?

34. Worldwide production of silicon carbide, SiC, is several hundred thousand tons annually. If you want to make 100,000 tons of the carbide, how many tons of silica sand (SiO_2) will you have to use if 70% of the sand is converted to SiC?

35. Silicates are enormously important in our lithosphere.
(a) Name one mineral that is a simple orthosilicate.
(b) Name one mineral that is an amphibole.
(c) Name one mineral that is a sheet silicate.
(d) Give two examples of aluminosilicates.

36. If galena is heated with lead(II) oxide, the product is lead and SO_2. Write a balanced equation depicting this reaction.

37. 227,250 tons of tetraethyllead were burned during gasoline combustion in one year. How much lead(II) oxide was thereby introduced into the atmosphere? (One ton is equivalent to 2.00×10^3 pounds)

38. Thousands of tons of organotin compounds are used annually as stabilizers for PVC (polyvinyl chloride) plastics, the clear material used for bottles and jars for food packaging. Other such compounds are added to antifouling paint for the hulls of ships, and still others are used in agriculture to control fungal growth, for example. If dibutyltin dichloride, $(C_4H_9)_2SnCl_2$, is made by treating tin with butylchloride, C_4H_9Cl,

$$2\ C_4H_9Cl + Sn \longrightarrow (C_4H_9)_2SnCl_2$$

how many tons of each of the two reactants are necessary to make 1.0 ton of product if the yield is 40%? (One ton is equivalent to 2.00×10^3 pounds)

39. Liquid HCN is dangerously unstable with respect to trimer formation (i.e., formation of $[HCN]_3$).
(a) Propose a structure for the cyclic trimer, a six-membered ring of alternating C and N atoms.
(b) Estimate the energy of the trimerization reaction. Obtain bond energies from Table 9.4 and use the following additional values: $E_{C\equiv N} = 887$ kJ/mol, $E_{C=N} = 615$ kJ/mol, and $E_{C-N} = 305$ kJ/mol.

40. The cyanide ion forms stable complexes with many metal ions. One example is $[Ag(CN)_2]^-$. If the formation constant (Chapter 15) for the ion is 5.6×10^{18}, what is the concentration of $Ag^+(aq)$ in a 0.010 *M* solution of the silver–cyanide complex ion?

22

The Chemistry of the Nonmetals: Periodic Groups 5A Through 7A and the Rare Gases

Elemental sulfur at the rim of a volcano (David Cavagnaro)

CHAPTER OUTLINE

We come now to a region of the periodic table where virtually all of the elements are nonmetals. Recall that ionization energies, electron affinities, and, therefore, electronegativities generally increase when moving across the periodic table. As a result, not only is there an increasing tendency to form negative oxidation states but compounds of the nonmetals with oxygen are more generally acidic, in contrast to basic metal oxides.

In these groups we find elements that are important constituents of compounds essential to life (nitrogen, phosphorus, oxygen, and sulfur) and that are also the basis of the strongest aqueous acids. Their abundances range from the highest on earth to among the lowest and from among the most reactive to the most inert.

22.1 THE CHEMISTRY OF GROUP 5A ELEMENTS

Group 5A elements are characterized by their ns^2np^3 configuration with its half-filled np subshell (Table 22.1). This configuration gives the ele-

TABLE 22.1 Group 5A: The p^3 Elements (Nitrogen Through Bismuth)

	NITROGEN	PHOSPHORUS	ARSENIC	ANTIMONY	BISMUTH
Symbol	N	P	As	Sb	Bi
Atomic number	7	15	33	51	83
Atomic weight	14.007	30.974	74.922	121.75	208.98
Valence e	$2s^22p^3$	$3s^23p^3$	$4s^24p^3$	$5s^25p^3$	$6s^26p^3$
mp, °C	−210	44 (a)*	814 (b)*	631	271
bp, °C	−196	280 (a)	Sublimes (b)	1380	1560
d, g/cm³	1.25 g/L	1.83 (a)	5.73 (b)	6.69	9.75
Atomic volume	17.3	17	13.1	18.4	21.3
Atomic radius, pm	70	110	121	141	146
Ion radius (+5), pm	11	34	47	62	74
Pauling EN	3.0	2.1	2.1	1.9	1.8
Standard potential (c)*	+0.27	−0.06	−0.60	0.51	−0.8
Oxidation numbers	−3 to +5	−3 to +5	−3 to +5	−3 to +5	
Ionization energy*	1400	1013	946	830	704
Isolated by	Rutherford	Brandt	Albertus	Antiquity	Geoffroy
Date of isolation	1772	1669	1250	Antiquity	1753
rpw* pure O_2	NO_x, 1200°C	P_4O_6, P_4O_{10}	As_4O_6	Sb_2O_3, 500°C	Bi_2O_3, 700°C
rpw H_2O	None	None	None	None	None
rpw N_2		None	None	None	None
rpw halogens	(d)*	PX_3, PX_5	AsX_3, AsX_5	SbX_3, SbX_5	BiX_3
rpw H_2	NH_3, 350°C	PH_3, 300°C	(d)	(d)	(d)

*Energies and heats are in kJ/mol; a = data on the white form of phosphorus; b = data on gray arsenic, yellow waxy form has density 2.01 g/cm³; c = std. potential for hydride EH_3 to element E, in volts; d indicates indirect preparation not from elements; rpw = reaction product with.

Adapted from E.G. Rochow, *Modern Descriptive Chemistry*, W.B. Saunders, 1977.

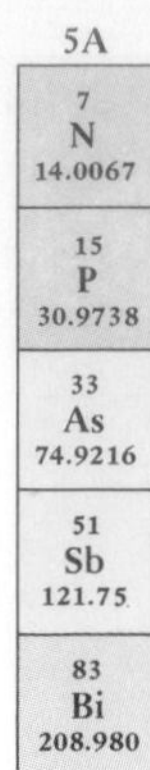

ments a tremendous range of chemistry, with oxidation numbers ranging from −3 (the *n*th shell completely full) to +3 and +5 (the shell partially or completely empty). Once again, as in Groups 3A and 4A, the most positive oxidation number is not as stable for the heavier elements, and arsenic and bismuth compounds with oxidation numbers of +5 are powerful oxidizing agents.

THE ELEMENTS AND THEIR RECOVERY

The relative abundances of the elements of a group give us some clue about the importance of an element and its possible commercial value. In Table 22.2, you see that phosphorus is the most abundant of the Group 5A elements in the earth's crust, with nitrogen only about as abundant as Ga (Table 21.2). On the other hand, nitrogen comprises 78.1% by volume (or 75.5% by weight) of the air around us.

Nitrogen and its compounds play a key role in our economy (Figure 22.1). Of the "top 14" chemicals produced by industry in 1984, five contain nitrogen.

CO and CN^- are also triply bonded and isoelectronic with N_2, but both are more reactive than N_2 due to their polarity.

N_2 is chemically rather inert for two reasons: the N atoms are bound with a triple bond of great energy (945.4 kJ/mol) and the molecule is nonpolar. If sufficient energy is used, however, it will react directly with many metals to give nitrides (N^{3-}); for example,

$$3\ Mg(s) + N_2(g) \longrightarrow \underset{\text{magnesium nitride}}{Mg_3N_2(s)}$$

Like oxygen, nitrogen forms bonds with virtually all the elements in the periodic table, but, in contrast to O_2, this does not usually occur by direct reaction of N_2 with the element.

Elemental nitrogen, N_2, is a very useful material. Because of its inertness, the largest quantity of gas is used to provide a nonoxidizing atmosphere for packaged foods and wine, for example, and to pressurize electric cables and telephone wires. Nitrogen is also easily converted to a liquid (bp, −196°C) that is convenient to handle, and about 10% of the N_2 produced is used in this form to freeze soft materials such as rubber so they can be ground to a powder, to preserve biological samples (e.g., blood and semen), and so on.

See the effect of liquid nitrogen in Figure 1.7.

Over 200 different **phosphorus**-containing minerals have been described, and all are *orthophosphates*, that is, they all contain the tetrahedral PO_4^{3-} anion or a derivative. By far the largest source of the element is the mineral family called the *apatites*, with the general formula $3\ Ca_3(PO_4)_2 \cdot CaX_2$ where X is commonly F, Cl, or OH. Thousands of

Mines in Florida produce almost one third of the world's phosphate minerals. Phosphates occur in all living things. Tooth enamel is nearly pure hydroxyapatite (X = OH), and the orthophosphate group is also part of DNA, RNA, and ATP, molecules of overwhelming biochemical importance.

TABLE 22.2 Relative Abundance of the Group 5A Elements in the Earth's Crust

ELEMENT	ABUNDANCE (IN PPM)	RANK AMONG ALL ELEMENTS
N	19	33
P	1120	11
As	1.8	51
Sb	0.2	62
Bi	0.008	69

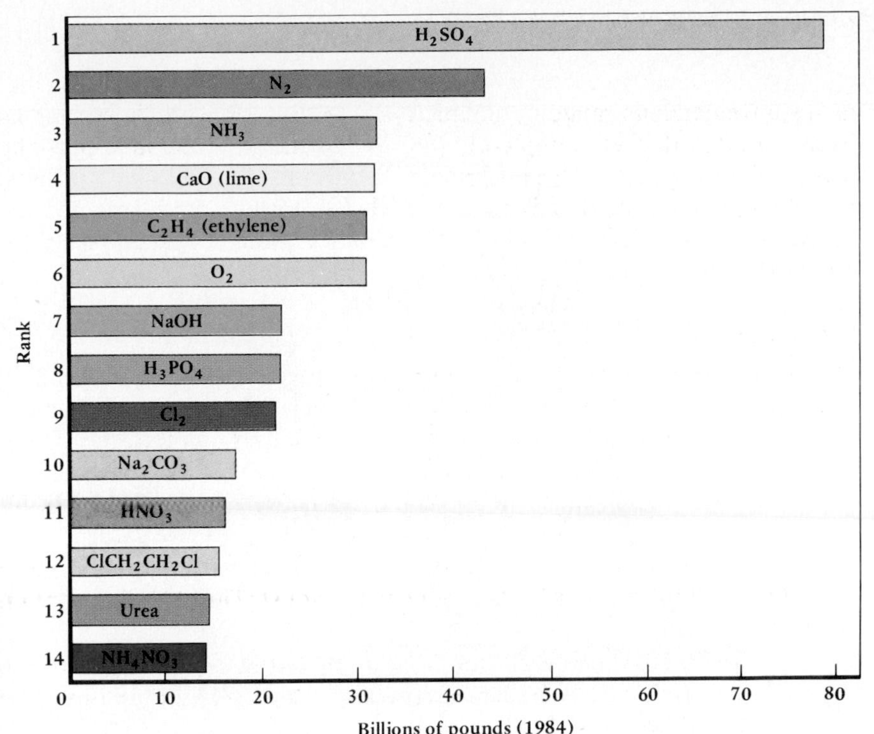

FIGURE 22.1
The 14 chemicals produced in largest amount in the United States annually. Data reflect 1984, but little change occurs from year to year.

tons of elemental phosphorus are produced each year by the reduction of apatite with carbon in the presence of silica (Figure 22.2).

$$\underset{\text{apatite}}{2\ Ca_3(PO_4)_2(s)} + 10\ C(s) + 6\ SiO_2(s) \longrightarrow \underset{\text{slag}}{6\ CaSiO_3(s)} + 10\ CO(g) + P_4(s)$$

P is unique in that it was first isolated from an animal source rather than a mineral. In 1669 an alchemist putrefied urine and then obtained P_4 when it was distilled.

Almost all of the solid P_4 produced is oxidized to pure phosphoric acid, and much of the remainder is used to make phosphorus sulfides for matches (Figure 22.3).

As you have seen, the chemistry of the lightest element in a group is often rather different from that of the heavier elements. The same relationship continues in Group 5A (and in 6A) in that nitrogen exists as a diatomic gas, but phosphorus, like B, C, Si, Ge, and Sn, is a covalently bonded solid with several allotropic forms. These forms are commonly called white, red, and black phosphorus for obvious reasons. *White phosphorus*, P_4, the most common allotrope, is a simple tetrahedron of atoms. It is insoluble in and unreactive with water, but it burns in air to give the oxide P_4O_{10} as described below. It is extremely toxic, and any contact should be avoided. Heating P_4 in the absence of air leads to another allotrope, *red phosphorus*, a polymer of P_4 units.

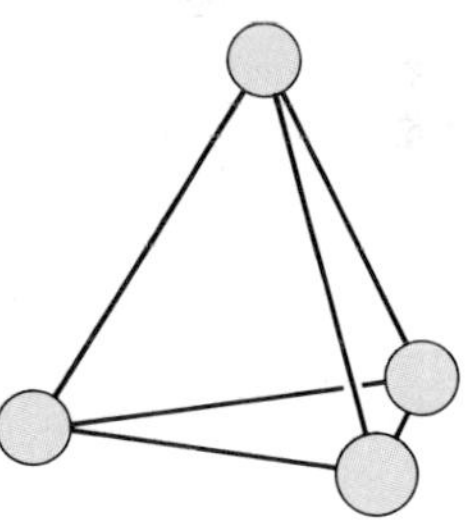

P_4, white phosphorus

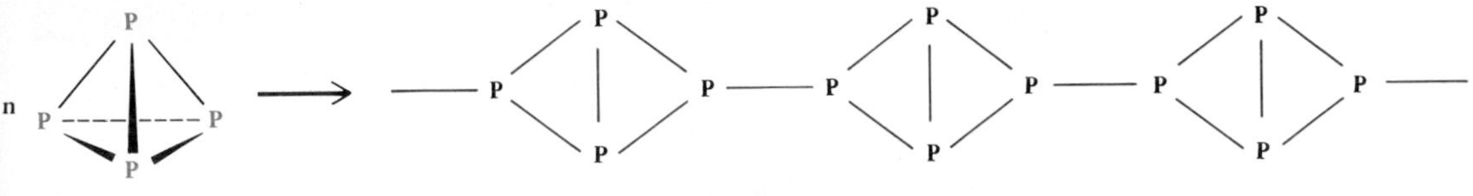

$(P_4)_n$, red phosphorus

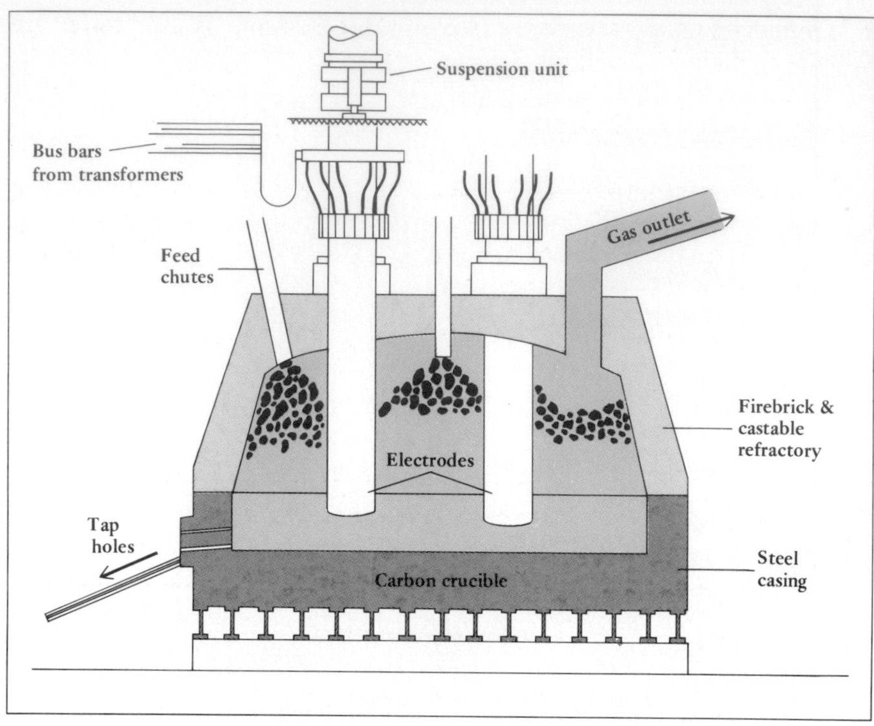

FIGURE 22.2

A furnace for phosphorus production. The "feed" is a mixture of $Ca_3(PO_4)_2$, SiO_2, and C. A mixture of P_4 gas, CO, and H_2 is driven off the top of the furnace, and molten slag (containing calcium silicate among other things) is tapped off the bottom at intervals.

The red and white allotropes of phosphorus. (Charles D. Winters)

Black phosphorus, an even more complicated polymeric form, is obtained by heating white P_4 under pressure. Both the red and black forms are much less reactive than the white allotrope.

Arsenic, antimony, and **bismuth** are among the oldest elements known, in spite of their low abundance. Although all form stable oxides, they all resemble the transition metals in forming very stable sulfides, and it is in this form that they are commonly found. Lemon yellow *orpiment*, As_2S_3, was used by physicians and assassins for hundreds of years, and black *stibnite*, Sb_2S_3, was used as a cosmetic hundreds of years ago. Like all of the elements you have seen thus far, these elements can be obtained

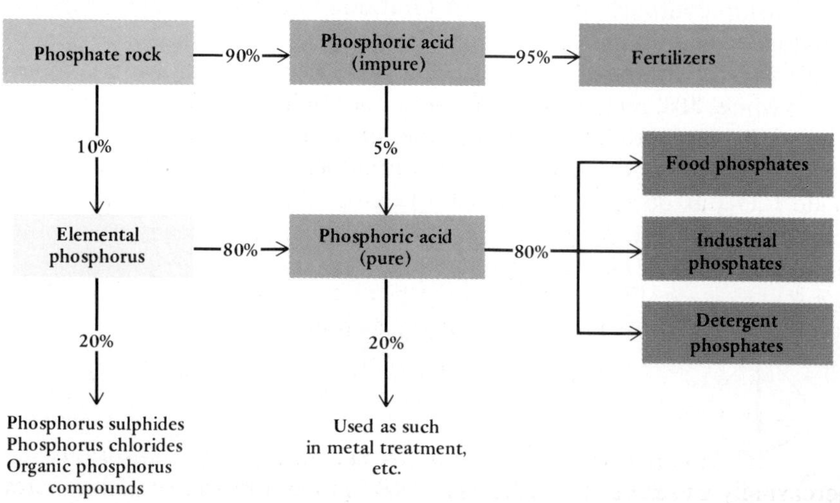

FIGURE 22.3

Uses of phosphate-containing rock.

by reducing a mineral source. Scrap iron, for example, is used to reduce stibnite to elemental antimony.

$$3\ Fe(s) + \underset{\text{stibnite}}{Sb_2S_3(s)} \longrightarrow 3\ FeS(s) + 2\ Sb(s)$$

A mineral containing yellow orpiment (As_2S_3) and red realgar (AsS). (Dave Davidson,Tom Stack & Associates)

A principal use of arsenic and antimony is in automobile batteries. The battery plates are lead, but their performance is greatly improved by adding 2.5% to 3% Sb and a trace of As. The U.S. consumption of antimony is about 12,000 tons per year.

EXERCISE 22.1 GROUP 5A ELEMENTS

(a) Name three uses of Group 5A elements or their compounds.
(b) Name the three allotropes of phosphorus.

COMPOUNDS OF GROUP 5A ELEMENTS WITH HYDROGEN

The elements of Group 5A form simple gaseous compounds with hydrogen of formula EH_3 (Table 22.3). Their names may be a bit puzzling at first. NH_3 is called ammonia, a common name that has come into general use. The other names, however, come from the old element name with the suffix "-ine" added.

TABLE 22.3 The EH_3 Series of Compounds of Group 5A

FORMULA	NAME	NORMAL BOILING POINT	ΔH_f°(kJ/mol)
NH_3	Ammonia	−34.5°C	−46.1
PH_3	Phosphine	−87.5°C	−9.6
AsH_3	Arsine	−62.4°	66.4
SbH_3	Stibine	−18.4°C	145.1
BiH_3	Bismuthine	Unstable	277.8

All the molecules have the same structural pair geometry and hence the same triangular pyramid molecular structure. The lone pair of electrons on the central atom means that all behave as Lewis bases.

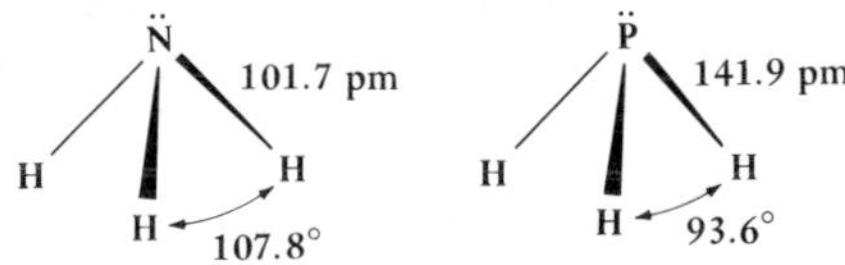

Ammonia, NH_3, is the most important of the EH_3 series (Figure 22.1), but the others are not without interest. Phosphine is a poisonous, highly reactive gas that has a faint garlic odor. In industry it is made by alkaline hydrolysis of white phosphorus.

$$P_4(s) + 3\ KOH(aq) + 3\ H_2O(\ell) \longrightarrow PH_3(g) + 3\ KH_2PO_2(aq)$$

One major use of PH_3 is to convert it to a salt that is a major ingredient in compounds used for flameproofing cotton cloth.

$$\underset{\text{phosphine}}{PH_3(g)} + \underset{\text{formaldehyde}}{4\ HCHO(aq)} + HCl(aq) \longrightarrow [P(CH_2OH)_4]Cl$$

The remaining EH_3 compounds are also exceedingly poisonous and increasingly unstable. Nonetheless, AsH_3 is used in the semiconductor industry in the manufacture of materials such as gallium arsenide, GaAs (Figure 21.4).

Before World War I, the majority of commercial nitrogen fertilizer came from deposits of Chile saltpeter ($NaNO_3$) and the excrement (guano) of bats and sea birds.

AMMONIA AND NITROGEN FIXATION Nitrogen is vital to life. Even though we are bathed in tons of nitrogen gas, it cannot be used by plants until it is "fixed," converted into biologically useful forms such as ammonia. Nitrogen fixation is done naturally by organisms such as blue-green algae and some field crops such as alfalfa and soybeans. Most plants cannot fix N_2, however, so "fixed" nitrogen must be supplied by an external source. This is particularly true of the new varieties of wheat, corn, and rice that were developed during the "green revolution" of the past several decades.

A feasible process for fixing nitrogen in the form of NH_3 was devised by Fritz Haber, a great chemist but a tragic figure in the history of chemistry. It should be possible to fix nitrogen in some usable form from simple molecules using any of several possible reactions, but Haber chose *direct synthesis of ammonia* from its elements.

The Haber Process

$$\tfrac{1}{2}\,N_2(g) + \tfrac{3}{2}\,H_2(g) \longrightarrow NH_3(g)$$

$$\Delta H^\circ = -46.1 \text{ kJ at } 25°C \qquad K = 8.3 \times 10^2$$

$$\Delta H^\circ = -55.6 \text{ kJ at } 450°C \qquad K \cong 6.5 \times 10^{-3}$$

At the outset, this seems a poor choice for several reasons. H_2 is available naturally only in combined form in water or hydrocarbons, for example. This means that some H-containing compounds must be destroyed first at the cost of considerable energy. Not only that, but this energy expense is completely wasted because the hydrogen of ammonia is oxidized to water and the nitrogen to nitrate by soil bacteria before ammonia can be taken up by the plants. The plant must then expend more energy, derived from photosynthesis, to reduce the nitrate back to ammonia. Nonetheless, because the Haber process has now been so well developed, the ammonia is so inexpensive (about $150 per ton) that it is used as a fertilizer.

The direct synthesis of ammonia is an equilibrium process that has been carefully fine-tuned by industry (Figure 22.4).

1. To achieve a reasonable rate, the temperature must be raised. However, notice that the equilibrium constant is smaller at higher temperature, so the yield of ammonia decreases.
2. The unfavorable equilibrium constant at higher temperature is partly countered by carrying the reaction out at higher total pressure. While this does not change the equilibrium constant, it does result in a higher percentage conversion of N_2 and H_2 to NH_3.
3. Finally, to give a reasonable reaction rate, a catalyst (Fe_3O_4 mixed with KOH, SiO_2, and Al_2O_3) is used. Since the catalyst is not efficient below about 400°C, this means the process is usually operated at about 450°C.

The Haber process is so efficient that the cost of ammonia is now almost entirely the cost of the hydrogen consumed in making it, generally from natural gas by *steam reforming* (Chapter 20).

Over 25% of the ammonia produced is used directly as a fertilizer, and the remainder is the starting material for other nitrogen-containing compounds. Either directly or indirectly, ammonia is the key to many industries: it goes into the starting materials for nylon production, household detergents, water purification, and the production of pharmaceuticals.

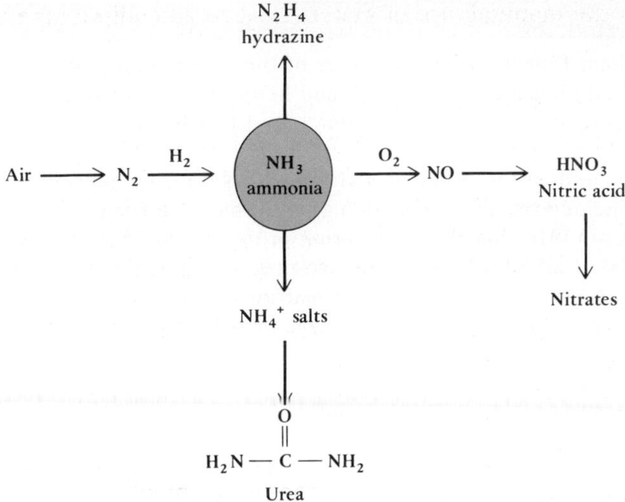

NITROGEN CATENATION AND HYDRAZINE All the second period, *p*-block elements readily form bonds to another of the same atom, a property called **catenation**. An atom of nitrogen can similarly form single, double, and triple bonds to another nitrogen, but catenation is very limited: the longest known N atom chain has only 8 atoms. An important, catenated nitrogen compound is also one of the simplest: N_2H_4, hydrazine.

You have also seen catenation in the chemistry of boron, carbon, and silicon (Chapter 21).

hydrazine

mp, 2.0°C; bp, 113.5°C

It is a colorless, fuming liquid with an ammonia-like odor. If you compare its melting and boiling points with those of NH_3 (mp, −77.8°C; bp, −34.5°C), it is clear that hydrogen bonding is important in liquid and solid hydrazine.

FIGURE 22.4

An industrial plant for the manufacture of ammonia. (M.W. Kellogg Company)

FRITZ HABER (1868–1934)

In 1898 William Ramsay (the discoverer of the rare gases) pointed out the depletion of fixed nitrogen in the world, and he predicted world disaster due to a "fixed N shortage" by mid-20th century. That this has not occurred is due to the work of Fritz Haber.

Haber studied the $N_2(g) + 3\ H_2(g) \rightarrow 2\ NH_3(g)$ equilibrium in the early 1900s and concluded that direct ammonia synthesis should be possible. However, it was not until 1914 that the engineering problems and catalyst question had been solved by Carl Bosch, and ammonia production began just in time for the start of World War I. Ammonia is the starting material to make nitric acid, a vital material in the manufacture of the explosives TNT and nitroglycerin. Therefore, this is thought to be the first large-scale use of a synthetic chemical for military purposes.

Haber's contract with the manufacturer of ammonia called for him to receive 1 pfennig per kilogram of ammonia, and he soon became not only famous but rich. Unfortunately, he joined the German Chemical Warfare Service at the start of World War I and became its director in 1916. The primary mission of the service was to develop gas warfare, and in 1915 he supervised the first use of Cl_2 at the battle of Ypres. Not only was this a tragedy of modern warfare but also to Haber personally. His wife pleaded with him to stop his work in this area, and, when he refused, she committed suicide.

In 1918 he was awarded the Nobel Prize for the ammonia synthesis, but the choice was criticized because of his role in chemical warfare.

After World War I, Haber did some of his best work, continuing on in thermodynamics (the Born–Haber cycle introduced in Chapter 11 was a product of this period). However, because he had a Jewish background, Haber left Germany in 1933, worked for a time in England, and died in Switzerland in 1934.

The Raschig process was used as an example of reaction mechanisms in Chapter 13. Even though the process was first introduced in 1907, the purpose of the gelatin is still not clear.

Approximately 20 million pounds of hydrazine are produced annually by the *Raschig process*, the oxidation of ammonia with alkaline sodium hypochlorite in the presence of gelatin.

$$2\ NH_3(aq) + NaOCl(aq) \xrightarrow[\text{gelatin}]{\text{aqueous alkali}} N_2H_4(aq) + NaCl(aq) + H_2O(\ell)$$

Hydrazine is widely used because it has three important properties: it is (a) a Lewis and Brønsted base, (b) a reducing agent, and (c) an energy-rich compound. As a base in water it has an equilibrium constant for the first ionization of 8.5×10^{-7},

$$N_2H_4(aq) + H_2O(\ell) \rightleftharpoons N_2H_5^+(aq) + OH^-(aq)$$

making it only slightly less basic than ammonia ($K_b = 1.8 \times 10^{-5}$). Its strong reducing ability is reflected in the $E°$ value in basic solution,

$$N_2H_4(aq) + 4\ OH^-(aq) \longrightarrow N_2(g) + 4\ H_2O(\ell) + 4e^- \quad E° = 1.16\ V$$

and this ability has been put to good use. For example, it is currently used to treat waste water from chemical plants to remove ions such as CrO_4^{2-} and thus prevent them from entering the environment.

An important new use of the compound is in the treatment of water boilers in large electrical generating plants. Oxygen dissolved in the water is a great problem in these plants, because the dissolved gas can oxidize the metal of the boiler and pipes and lead to corrosion. Hydrazine is thus added to reduce the oxygen to water.

$$N_2H_4(aq) + O_2(g) \longrightarrow N_2(g) + 2\ H_2O(\ell)$$

The important noncommercial use of hydrazine, and a spectacular example of its reducing ability, is in rocket fuels. It has been used on the Apollo missions to the moon and in the steering rockets of the Space Shuttle. The lunar lander rockets were powered by mixing liquid N_2O_4, an oxidizing agent, with a 1:1 mixture of methyl derivatives of hydrazine [$H_2N—NH(CH_3)$ and $H_2N—N(CH_3)_2$].

The landing on the moon required about 4.5 tons of the mixed hydrazines and 3 tons of N_2O_4. To take off again, only about one third of these amounts was used.

$$H_2N—N(CH_3)_2(\ell) + 2\ N_2O_4(\ell) \longrightarrow 3\ N_2(g) + 4\ H_2O(g) + 2\ CO_2(g)$$

These reactants are called *hypergolic*; that is, they ignite on contact. Thus, when the rocket is to be fired, the astronaut only needs to open valves and allow the chemicals to mix.

EXERCISE 22.2 NITROGEN FIXATION

The Haber process is carried out at a high temperature, in spite of the fact that the reaction is exothermic. Explain why a high temperature is necessary.

OXYGEN COMPOUNDS OF NITROGEN AND PHOSPHORUS

Nitrogen and phosphorus oxides bear some resemblance to their Group 4A neighbors in that nitrogen, a second period element, forms only simple binary oxides, whereas phosphorus, a third period element, forms more extended molecules and ions. In contrast to carbon, however, nitrogen is unique in forming *seven* binary oxides, all of them thermodynamically unstable with respect to decomposition to N_2 and O_2.

NITROGEN–OXYGEN COMPOUNDS Six of the seven known binary nitrogen–oxygen compounds are listed in Table 22.4. Some are important commercially and some must be explored because they are implicated in air pollution.

Dinitrogen oxide, N_2O, the oxide having nitrogen in the lowest oxidation state (+1) can be made by the careful decomposition of ammonium nitrate at 250°C.

$$NH_4NO_3(s) \xrightarrow{250°C} N_2O(g) + 2\ H_2O(g)$$

The gas is nontoxic, odorless, and tasteless. It can be used as an anesthetic in minor surgery and has come to be called "laughing gas" because of its effects. Since it is soluble in vegetable fats, its largest commercial use is as a propellent and aerating agent in cans of whipped cream.

Nitrogen oxide, NO, is the simplest, thermally stable, odd-electron molecule known. With a total of 11 valence electrons, both atoms cannot simultaneously have a completed octet. Therefore, since the nitrogen atom

TABLE 22.4 Some Oxides of Nitrogen

FORMULA	NAME	STRUCTURE	N OXIDATION NUMBER	DESCRIPTION
N_2O	Dinitrogen oxide	$:N{\equiv}N{-}\ddot{\underset{..}{O}}:$ linear	+1	Colorless gas (laughing gas)
NO	Nitrogen oxide	$:N{=}\ddot{O}:$	+2	Odd-electron molecule (paramagnetic)
N_2O_3	Dinitrogen trioxide	O=N–N(=O)–O planar	+3	Reversibly dissociates to NO and NO_2
NO_2	Nitrogen dioxide	O–N=O	+4	Brown, paramagnetic gas
N_2O_4	Dinitrogen tetroxide	$O_2N{-}NO_2$ planar	+4	Colorless liquid/gas dissociates to NO_2 (Fig. 14.5)
N_2O_5	Dinitrogen pentoxide	$O_2N{-}O{-}NO_2$	+5	Colorless ionic solid

is less electronegative than oxygen, 8 electrons surround the O atom and the odd electron resides closer to the N atom (Table 22.4).

As you will see below, NO is an intermediate in the synthesis of nitric acid by oxidation of ammonia. On a laboratory scale, however, it can be synthesized conveniently by using a mild reducing agent on a nitrogen oxide in the +3 oxidation state.

$$\underset{+3\text{ nitrogen}}{KNO_2(aq)} + \underset{\text{reducing agent}}{KI(aq)} + H_2SO_4(aq) \longrightarrow NO(g) + K_2SO_4(aq) + H_2O(\ell) + \tfrac{1}{2}\,I_2(aq)$$

Dinitrogen trioxide, a blue solid with no particular commercial value, can be isolated only at very low temperatures (mp, −100.1°C). As the temperature climbs the color fades and becomes greenish as the trioxide dissociates to brown NO_2 and colorless NO.

The N_2O_3 dissociation is a *disproportionation reaction.* The N atom in the trioxide has an oxidation number of +3, while it is either +4 (in NO_2) or +2 (in NO) in the products.

$$\underset{\text{blue}}{N_2O_3(\ell)} \rightleftharpoons \underset{\text{brown}}{NO_2(g)} + \underset{\text{colorless}}{NO(g)}$$

Nitrogen dioxide, NO_2, is the brown gas you see when a bottle of nitric acid is allowed to stand in the sunlight, and it is a culprit in air pollution.

$$2\ HNO_3(aq) \rightleftharpoons 2\ NO_2(g) + H_2O(\ell) + \tfrac{1}{2}\ O_2(g)$$

Like NO, the dioxide NO_2 is an odd-electron molecule. Since the odd electron again resides mainly on the nitrogen atom,

$$2\ \cdot NO_2 \rightleftharpoons O_2N{-}NO_2$$

brown gas — colorless gas

this means that two NO_2 molecules can combine to form $O_2N{-}NO_2$, dinitrogen tetroxide, with a N—N single bond. When N_2O_4 is frozen (mp, −11.2°C), the solid consists entirely of N_2O_4 molecules; as the solid melts and the temperature increases to the boiling point, dissociation to NO_2 begins to occur. At the boiling point (21.5°C), the gas phase consists of 15.9% NO_2 and is distinctly brown (Figure 11.1).

Dinitrogen tetroxide, an oxidizing agent in rocket fuels, is best made by condensing the NO_2 that comes from the thermal decomposition of $Pb(NO_3)_2$.

$$2\ Pb(NO_3)_2(s) \xrightarrow{\sim 400°C} 4\ NO_2(g) + 2\ PbO(s) + O_2(g)$$

It is also formed when NO reacts with oxygen,

$$\underset{\text{colorless}}{2\ NO(g) + O_2(g)} \longrightarrow \underset{\text{deep brown gas}}{2\ NO_2(g)} \rightleftharpoons \underset{\text{colorless solid at } -11.2°C}{N_2O_4(\ell)}$$

and it is this reaction that is so important in producing NO_2 in polluted air.

Nitrogen dioxide and N_2O_4 react with water to form nitric acid, so the moist gases are not only toxic but highly corrosive as well.

$$N_2O_4(g) + H_2O(\ell) \longrightarrow HNO_3(aq) + HNO_2(aq)$$

$$3\ HNO_2(aq) \longrightarrow HNO_3(aq) + 2\ NO(g) + H_2O(\ell)$$

The "true" anhydride of nitric acid, however, is the strange compound *dinitrogen pentoxide*, N_2O_5. This compound is made by chemically dehydrating nitric acid,

$$4\ HNO_3 + P_4O_{10} \longrightarrow 2\ N_2O_5 + 4\ HPO_3$$

and so nitric acid is regenerated if the pentoxide is put into water again. Although the structure is not well established, it is considered to be an ionic solid, consisting of loosely bonded linear NO_2^+ and planar trigonal NO_3^- ions.

NITROGEN OXYACIDS AND RELATED COMPOUNDS As with other nonmetal oxides, nitrogen oxides are acidic and interact with water to produce a number of acids. Several of these acids are unstable in the pure state and are known only in aqueous solution or as their salts. Only two are important enough for you to know about at this point.

Nitrous acid, HNO_2, has never been isolated as a pure compound, but it is known as a weak acid in aqueous solution, dissociating to its conjugate base, the *nitrite* ion.

$$\underset{\text{Nitrous acid}}{HNO_2(aq)} + H_2O(\ell) \overset{K_a = 4.5\ 10^{-4}}{\rightleftharpoons} \underset{\text{Nitrite ion}}{NO_2^-(aq)} + H_3O^+(aq)$$

H, O, N, O; 102.1; 110.7°; 143.3 pm; 117.7 pm

The acid is usually made by acidifying a cooled solution of a nitrite salt,

$$NaNO_2(aq) + HCl(aq) \longrightarrow NaCl(aq) + HNO_2(aq)$$

and salts of NO_2^- are made commercially by treating the acidic oxides NO and NO_2 with a base.

$$Na_2CO_3(aq) + NO(g) + NO_2(g) \longrightarrow 2\ NaNO_2(aq) + CO_2(g)$$

Sodium nitrite is mildly toxic to human beings, but it and the nitrate have been used since Roman times for curing meat. Nitrites can be added directly, or they can be produced by bacterial reduction of nitrates:

$$NO_3^-(aq) + 2\ H^+(aq) + 2e^- \longrightarrow NO_2^-(aq) + H_2O(\ell)$$

The nitrite ion is known to prevent the growth of microorganisms that lead to deadly botulisms, but there is now evidence that nitrites can be converted during digestion to compounds that are known to be carcinogens.

Nitric acid, HNO_3, has been known for centuries and is still one of the most important compounds in our economy (Figure 22.1). The oldest way to make the acid is to treat Chile saltpeter, $NaNO_3$, with sulfuric acid.

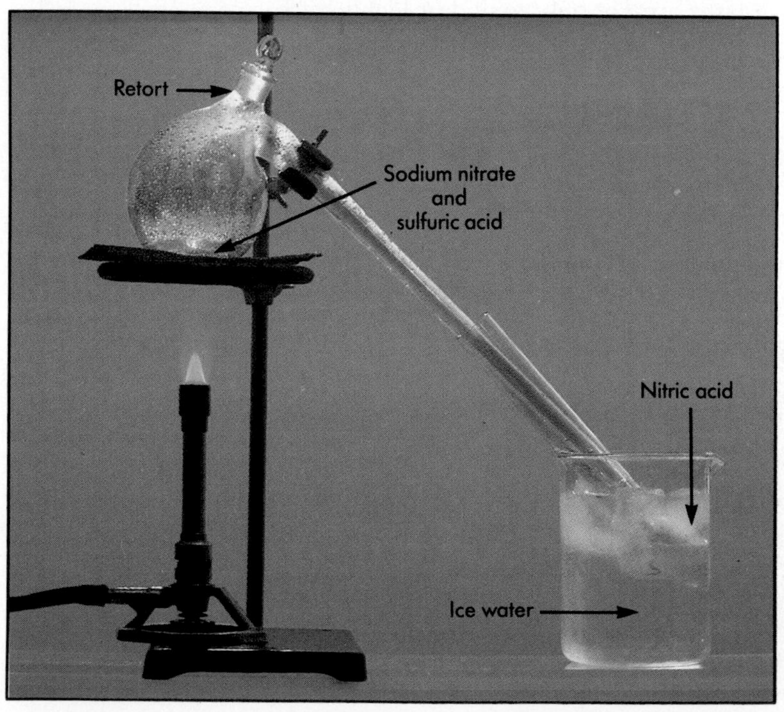

The preparation of nitric acid by the reaction of sulfuric acid and sodium nitrate. Pure HNO_3 is colorless, but some acid decomposes during the preparation to give the brown gas NO_2, and it is this gas that fills the apparatus and colors the product. (Yoav Levy)

$$2\ NaNO_3(s) + H_2SO_4(aq) \longrightarrow 2\ HNO_3(aq) + Na_2SO_4(s)$$

However, enormous quantities are now produced by the *Ostwald process*. The first step involves the controlled oxidation of ammonia over a Pt catalyst,

$$NH_3(g) + \tfrac{5}{4}\ O_2(g) \xrightarrow{Pt} NO(g) + \tfrac{3}{2}\ H_2O(g) \qquad \Delta H^\circ = -292.5\ kJ$$

followed by further oxidation of NO to NO_2.

$$NO(g) + \tfrac{1}{2}\ O_2(g) \longrightarrow NO_2(g) \qquad \Delta H^\circ = -56.8\ kJ$$

In a typical industrial plant, a mixture of air and 10% NH_3 is passed very rapidly over a Pt catalyst at 5 atm pressure and ~850°C (Figure 22.5). Roughly 96% of the ammonia is converted to NO_2, making this one of the most efficient industrial catalytic reactions. The final step is to absorb the NO_2 into water to give the acid and NO, the latter being recycled back into the process.

$$NO_2(g) + \tfrac{1}{3}\ H_2O(\ell) \longrightarrow \tfrac{2}{3}\ HNO_3(aq) + \tfrac{1}{3}\ NO(g) \qquad \Delta H^\circ = -23.3\ kJ$$

This means the overall reaction is

$$NH_3(g) + 2\ O_2(g) \longrightarrow HNO_3(aq) + H_2O(\ell) \qquad \Delta H^\circ = -412.7\ kJ$$

Nitric acid is produced as a concentrated aqueous solution, but careful procedures can convert this to the anhydrous acid (mp, −41.6°; bp, 82.6°C). At room temperature, HNO_3 is a colorless liquid with a pungent, choking odor. In the gas phase the molecule is planar and has a structure expected from VSEPR theory.

Roughly 20% of the ammonia produced every year is converted to nitric acid, which in turn finds many uses. By far the greatest amount of

FIGURE 22.5
The gold–palladium gauze catalyst for the oxidation of ammonia to nitric acid. (Johnston-Matthey)

FIGURE 22.6

The decomposition of NH_4NO_3. In the presence of H_2O, and Cl^- as catalyst, ammonium nitrate decomposes according to the equation

$$NH_4NO_3(s) \longrightarrow N_2O(g) + 2\ H_2O(g)$$

If zinc is present, the following oxidation can occur.

$$Zn(s) + NH_4NO_3(s) \longrightarrow N_2(g) + ZnO(s) + 2\ H_2O(g)$$

(Charles D. Winters)

the acid is turned into ammonium nitrate by neutralization of nitric acid with ammonia. Most of the NH_4NO_3 is consumed as fertilizer, but another use depends on the fact that it is thermally unstable and potentially explosive (Figure 22.6):

$$2\ NH_4NO_3(s) \xrightarrow{>300°C} 2\ N_2(g) + O_2(g) + 4\ H_2O(g)$$

$$NH_4NO_3(s) \xrightarrow{200\text{–}260°C} N_2O(g) + 2\ H_2O(g)$$

This decomposition is catalyzed by organic material, so a mixture of NH_4NO_3 and fuel oil is used as an explosive in mining operations.

Nitric acid is a powerful oxidizing agent, as the large, positive $E°$ values for the following half reactions illustrate.

$$NO_3^-(aq) + 4\ H^+(aq) + 3e^- \longrightarrow NO(g) + 2\ H_2O(\ell) \quad E° = +0.96\ V$$

$$NO_3^-(aq) + 2\ H^+(aq) + e^- \longrightarrow NO_2(g) + H_2O(\ell) \quad E° = +0.80\ V$$

Concentrated nitric acid will attack and oxidize almost all metals, but the exact nitrogen-containing product depends on the metal (the reducing agent) and acid concentration (Figure 21.3).

$$Cu(s) + \underbrace{4\ H^+(aq) + 2\ NO_3^-(aq)}_{\text{concentrated}} \longrightarrow Cu^{2+}(aq) + 2\ H_2O(\ell) + 2\ NO_2(g)$$

$$3\ Cu(s) + \underbrace{8\ H^+(aq) + 2\ NO_3^-(aq)}_{\text{dilute}} \longrightarrow 3\ Cu^{2+}(aq) + 4\ H_2O(\ell) + 2\ NO(g)$$

There are at least four metals that nitric acid will not attack—Au, Pt, Rh, and Ir—so these came to be known as the ''noble metals.'' However, the alchemists of the 14th century knew that if you mixed HNO_3 with HCl in a ratio of about 1:3, this *aqua regia* or ''kingly water'' would attack even the noblest of metals, as in this reaction with platinum.

$$3\ Pt(s) + 4\ NO_3^-(aq) + 18\ Cl^-(aq) + 22\ H^+(aq) \longrightarrow 3\ H_2PtCl_6(aq) + 4\ NO(g) + 8\ H_2O(\ell)$$

EXERCISE 22.3 NITROGEN OXIDES

(a) Explain why N_2O is ''linear,'' whereas NO_2 is ''bent.''
(b) Give three commercial uses of nitrogen oxides or oxyacids.

Phosphorus was named for the phosphorescent glow observed when P_4 oxidizes.

PHOSPHORUS OXIDES AND SULFIDES The most important compounds of phosphorus are those with oxygen, and there are at least six simple binary P–O compounds. All of them can be thought of as structurally derived from the P_4 tetrahedron of elemental phosphorus. For example, if P_4 is carefully oxidized, P_4O_6 is formed; an oxygen atom has been inserted into each of the P—P bonds of the tetrahedron.

221 pm

P_4 $\xrightarrow{O_2}$ P_4O_6 (165.6 pm, 127.0°, 99.5°) $\xrightarrow{O_2}$ P_4O_{10} (143 pm, 102°, 123°, 160 pm)

P_4O_6 $\xrightarrow{H_2O}$ H_3PO_3 Phosphorous acid

P_4O_{10} $\xrightarrow{H_2O}$ H_3PO_4 Phosphoric acid

The most common and important oxide of phosphorus is P_4O_{10}, a compound commonly called "phosphorus pentoxide" because its empirical formula is P_2O_5. It is formed as a fine, white powder when P_4 burns in air, and its structure is just an extension of that of P_4O_6. Each P atom in P_4O_{10} is surrounded tetrahedrally by O atoms. In this regard, phosphorus chemistry resembles silicon–oxygen chemistry, in which the most important structural element is the $\{SiO_4\}$ tetrahedron.

Other binary P–O compounds have formulas between P_4O_6 and P_4O_{10}. They are formed by starting with P_4O_6 and adding O successively to the P atom vertices.

Unlike nitrogen, phosphorus also forms an extensive series of compounds with sulfur that are structurally analogous with the P–O series, and the most stable of these is tetraphosphorus trisulfide, P_4S_3.

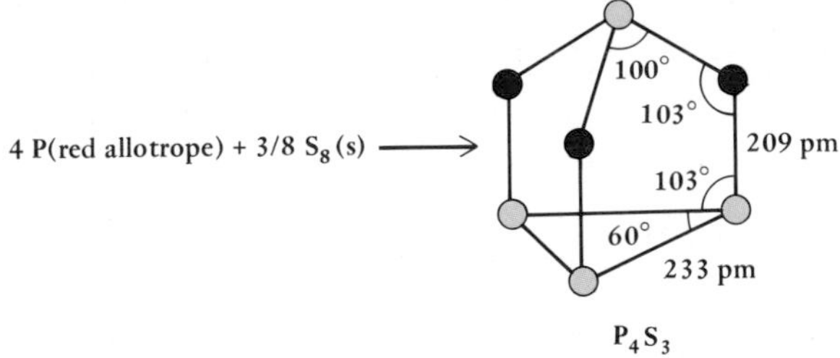

As in P–O chemistry, the basic structural element is again the P_4 tetrahedron, but S atoms are inserted into only three of the six P—S bonds. The principal use of P_4S_3 is in "strike anywhere" matches, the kind of match that will light if you rub it against a stone. The active ingredients are P_4S_3 and the powerful oxidant potassium chlorate, $KClO_3$, and it is the violent reaction between them that lights the match. (Other ingredients of such matches are ground glass, Fe_2O_3, ZnO, and glue). The "safety

The other common P—S compound is P_4S_{10}, structurally analogous to P_4O_{10}. Since P_4S_{10} is used as a starting material for compounds containing the P—S bond, world production exceeds 250,000 tons.

match" is now more common than the "strike anywhere" match. In this case the match head is predominantly $KClO_3$, and the material on the match book is red P (about 50%), Sb_2S_3, Fe_2O_3, and glue.

PHOSPHORUS OXYACIDS AND THEIR SALTS Only a few of the many phosphorus oxyacids are illustrated in Table 22.5. Indeed, there are so many types that some structural principles have been developed to help us organize and understand them.

Chemists typically draw a double bond for the PO terminal bond. This reduces the formal charge (page 320) on P and the terminal O atom. Although the P atom is now involved in five bonds, recall that P and other third period elements may have an expanded octet.

TABLE 22.5 Some Phosphorus Oxyacids

FORMULA	NAME	STRUCTURE
H_3PO_4	Orthophosphoric	O, P, HO, OH, OH
$H_4P_2O_7$	Diphosphoric (pyrophosphoric)	O, O, P, P, O, HO, OH, HO, OH
$(HPO_3)_3$	Metaphosphoric	O, O, O, P, P, HO, OH, O, O, P, HO, O
H_3PO_3	Phosphonic (phosphorous)	O, P, H, OH, OH
H_3PO_2	Phosphinic (hypophosphorous)	O, P, H, OH, H

(a) All the P atoms in the oxyacids and their anions are four-coordinate and tetrahedral. All have at least one P=O bond.

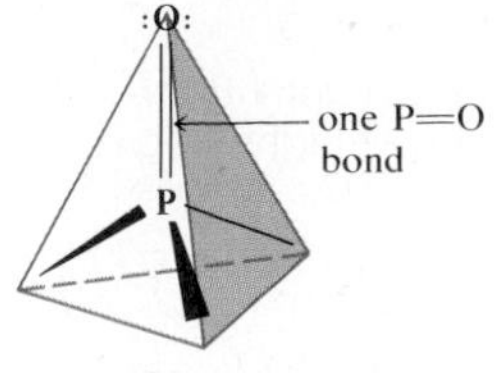

(b) All of the P atoms in the acids have at least one P—OH group, and this often occurs in the anions as well. In every case, the H is ionizable as H^+.

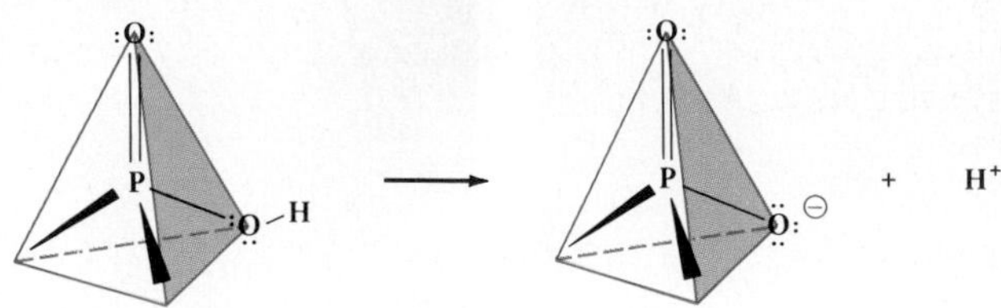

(c) Some oxyacids and anions have one or more P—H bonds. This hydrogen is not ionizable as H^+.

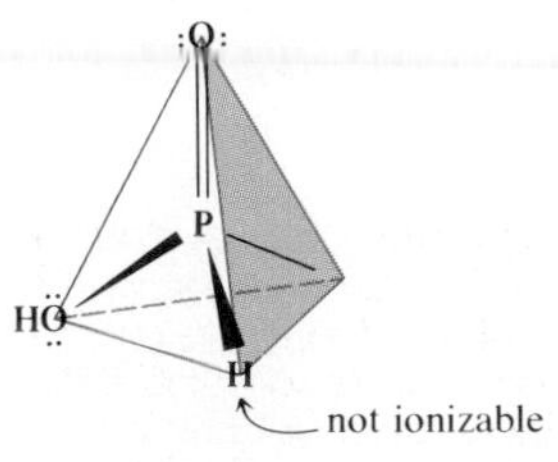

(d) Polymerization occurs by P—O—P or P—P bonding (catenation) to give both linear and cyclic species. Two P atoms have never been found to be joined by more than one P—O—P bridge.

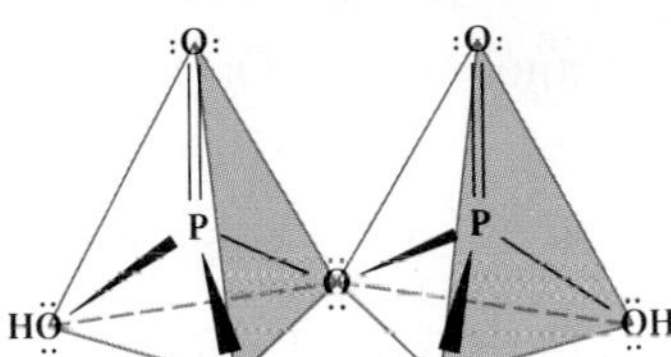

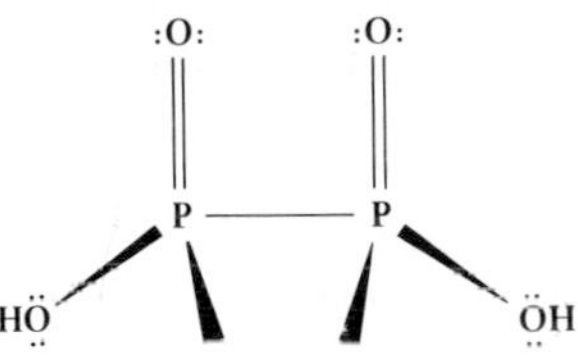

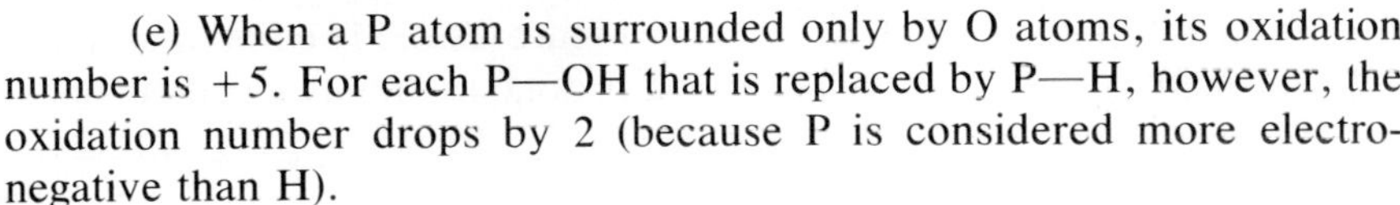

(e) When a P atom is surrounded only by O atoms, its oxidation number is +5. For each P—OH that is replaced by P—H, however, the oxidation number drops by 2 (because P is considered more electronegative than H).

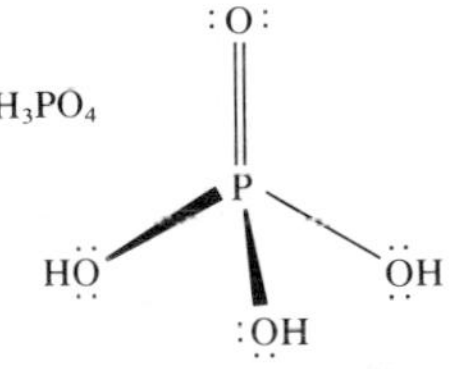

Nitrogen oxyacids are in general good oxidizing agents, while the P–O acids (except for H_3PO_4 where P has its maximum oxidation number of +5) are often good reducing agents. The few half reactions below illustrate the point.

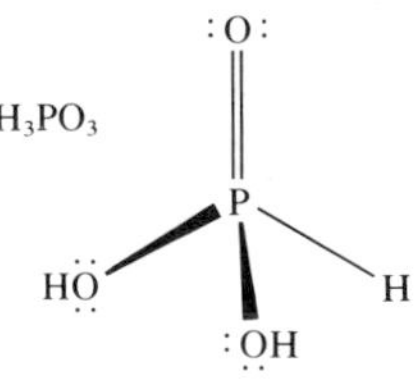

$$H_3PO_2(aq) + H^+(aq) + e^- \longrightarrow P + 2\ H_2O(\ell) \qquad E^\circ = -0.508\ V$$

$$H_3PO_3(aq) + 2\ H^+(aq) + 2e^- \longrightarrow H_3PO_2(aq) + H_2O(\ell) \qquad E^\circ = -0.499\ V$$

$$H_3PO_4(aq) + 2\ H^+(aq) + 2e^- \longrightarrow H_3PO_3(aq) + H_2O(\ell) \qquad E^\circ = -0.276\ V$$

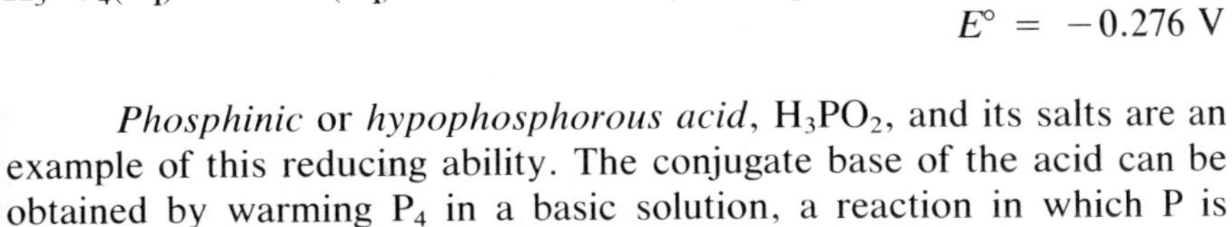

Phosphinic or *hypophosphorous acid*, H_3PO_2, and its salts are an example of this reducing ability. The conjugate base of the acid can be obtained by warming P_4 in a basic solution, a reaction in which P is oxidized to +1 in $H_2PO_2^-$ and reduced to −3 in PH_3.

$$P_4(s) + 3\ OH^-(aq) + 3\ H_2O(\ell) \longrightarrow 3\ H_2PO_2^-(aq) + PH_3(g)$$

The sodium salt of the $H_2PO_2^-$ ion has been increasingly used in industry as a reducing agent to plate metals onto plastic surfaces. Since plastic

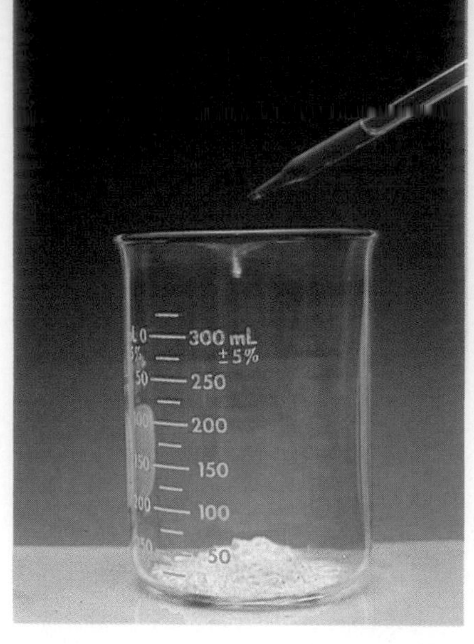

FIGURE 22.7

The white solid P_4O_{10} (left) reacts vigorously with water (right) to give phosphoric acid, H_3PO_4. (Charles D. Winters)

will not carry electricity, a plastic part cannot be made part of an electrochemical cell. Therefore, chemical reducing agents are used instead, and one of the best is a basic solution of NaH_2PO_2.

$$H_2PO_2^-(aq) + 3\ OH^-(aq) \longrightarrow HPO_3^{2-}(aq) + 2\ H_2O(\ell) + 2e^- \qquad E \simeq 1.6\ V$$

Orthophosphoric acid, H_3PO_4, and its derivatives are far more important industrially than the other P–O acids. Almost 11 million tons of phosphoric acid are made annually, and the acid currently sells for about $300 per ton.* A small fraction of the acid is made by the "thermal" or "furnace" method (Section 22.1 and Figure 22.2). The initial product, P_4, is burned to give P_4O_{10}, and, since it is done in the presence of steam, a concentrated aqueous solution of H_3PO_4 results (Figure 22.7).

$$P_4O_{10}(s) + 6\ H_2O(\ell) \longrightarrow 4\ H_3PO_4(aq)$$

H_3PO_4 is also made by the "gypsum method." Here $Ca_3(PO_4)_2$ is treated with sulfuric acid to give the acid and gypsum, $CaSO_4$. It was originally patented in 1842 as a way of making fertilizer from bones, a material high in phosphate.

Since this approach gives a pure product, it is used to make acid for use in food products in particular. When the pure acid is dilute it is nontoxic and is used to give the tart or sour taste to carbonated "soft drinks" such as various colas (~0.05% H_3PO_4, pH 2.3) or root beer (~0.01% H_3PO_4, pH 5.0).

A major use of phosphoric acid is to impart corrosion resistance to metal objects such as nuts and bolts, tools, and car-engine parts by plunging the object into a hot acid bath (Figure 22.3). Car bodies are similarly treated with phosphoric acid containing metal ions such as Zn^{2+}, and aluminum auto trim is "polished" by treating with the acid.

The behavior of H_3PO_4 as a typical polyprotic acid was described in Chapter 15.

All three protons of H_3PO_4 can be removed to give a series of phosphate salts such as NaH_2PO_4, Na_2HPO_4, and Na_3PO_4. In industry, the monosodium and disodium salts are produced using Na_2CO_3 as the base, but an excess of the more expensive base NaOH must be used to remove the final H^+ to give Na_3PO_4. Phosphates are used in such a variety

*Production figures and prices in the phosphoric acid industry are based on the amount of "contained P_4O_{10}." To convert to the equivalent amount of anhydrous H_3PO_4, multiply by 1.380.

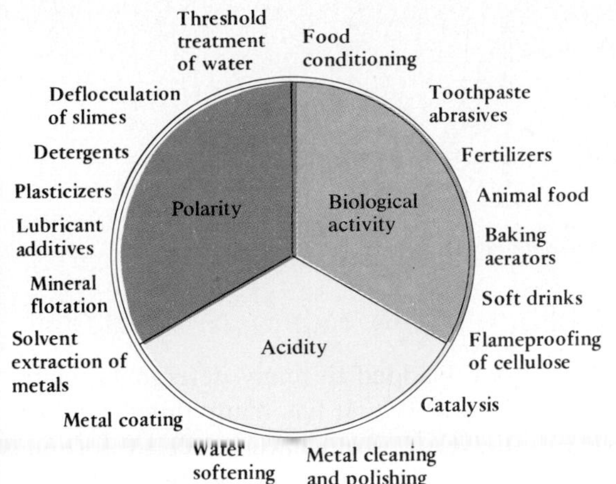

FIGURE 22.8

Uses of phosphoric acid and its derivatives. Some uses rely on the acidity of the compound, some on its polarity, and still others on its biological activity.

of ways in our modern economy that we can only mention a few (Figure 22.8).

Sodium phosphate (Na_3PO_4), for example, is used in scouring powders and paint strippers, since the ion PO_4^{3-} is a very strong base in water. Sodium monohydrogen phosphate, Na_2HPO_4, has a less basic anion (HPO_4^{2-}) than Na_3PO_4, so the former is widely used in food products. In 1916 J.L. Kraft patented a process using the salt in the manufacture of pasteurized process cheese. Thousands of tons of the phosphate are still used for this purpose, even though the function of the salt is still not fully understood. In addition, a small amount in pudding mixes enables the mix to gel in cold water, and the basic nature of HPO_4^{2-} raises the pH of cereals to provide "quick-cooking" breakfast cereal. (The OH^- from hydrolysis accelerates the breakdown of the cellulose material.)

Calcium phosphates are used in a broad spectrum of products. For example, the weak acid $Ca(H_2PO_4)_2 \cdot H_2O$ is used as the acid leavening agent in baking powder, since it reacts with $NaHCO_3$ to produce CO_2 (see page 569).

$$Ca(H_2PO_4)_2 \cdot H_2O(s) + 2\ NaHCO_3(aq) \longrightarrow 2\ CO_2(g) + 3\ H_2O(\ell) + Na_2Ca(HPO_4)_2(aq)$$

A typical baking powder contains 28% $NaHCO_3$, 10.7% $Ca(H_2PO_4)_2 \cdot H_2O$, 21.4% $NaAl(SO_4)_2$ (also a weak acid; see Section 15.5), and 39.9% starch. This same phosphate, but made by treating phosphate rock with sulfuric acid, is used as a fertilizer called *superphosphate*.

$$2\ Ca_5(PO_4)_3F(s) + 7\ H_2SO_4(aq) \longrightarrow 3\ Ca(H_2PO_4)_2(s) + 2\ HF(aq) + 7\ CaSO_4(aq)$$

Finally, the monohydrogen phosphate, $CaHPO_4$, also finds a large use as the abrasive and polishing agent in toothpaste.

When phosphates are heated to high temperature, they eliminate water in forming P—O—P links. For example, if $H_2PO_4^-$ and HPO_4^{2-} are heated to 450°C in a 1:2 ratio, the product is the tripolyphosphate ion.

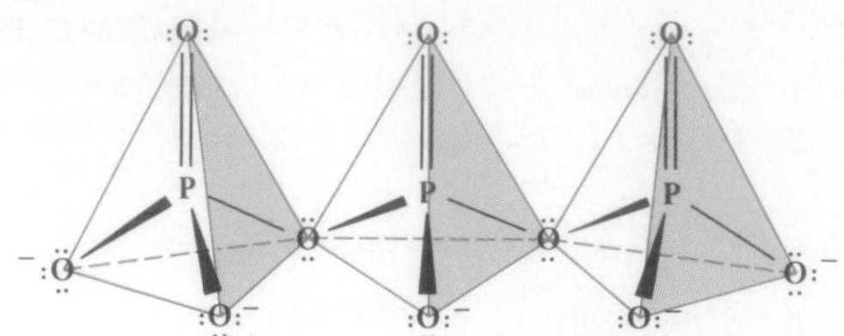

The tripolyphosphate ion, $P_3O_{10}^{5-}$

$$2\ Na_2HPO_4(s) + NaH_2PO_4(s) \xrightarrow{450°C} Na_5P_3O_{10}(s) + 2\ H_2O(g)$$

This is the "phosphate builder" that is added to many detergents, some containing as much as 25% to 40% $Na_5P_3O_{10}$. It has many functions, one of them being to remove Mg^{2+} and Ca^{2+} from hard water by forming complex ions, an action called *sequestering*. Another function is to provide a mildly basic solution through hydrolysis of the anion as in the reaction

$$P_3O_{10}^{5-}(aq) + H_2O(\ell) \longrightarrow HP_3O_{10}^{4-}(aq) + OH^-(aq)$$

Many thousands of tons of phosphates are used annually for this function around the world.

There is considerable controversy in the United States over the use of phosphates. Since inorganic phosphate is an essential nutrient for plant growth, introducing large quantities into natural waters can lead to *eutrophication* or overfertilization of the water. It can promote a massive growth of algae that in turn depletes the oxygen in water and kills off the fish life.

EXERCISE 22.4 PHOSPHORUS OXYACIDS

Hypophosphoric acid has the formula $H_4P_2O_6$. Its sodium salt can be made by reacting red phosphorus with sodium chlorite at room temperature.

$$2\ P(s) + 2\ NaClO_2 + 2\ H_2O(\ell) \longrightarrow Na_2H_2P_2O_6(aq) + 2\ HCl(aq)$$

Using the principles of phosphorus–oxygen chemistry, develop a plausible structure for the acid. What is the oxidation number of P in the acid?

OTHER OXIDES AND HALIDES OF GROUP 5A ELEMENTS Arsenic, antimony, and bismuth all form oxides with the general formula E_2O_3 where the oxidation number of the Group 5A element E is +3. The structure of As_4O_6 is analogous to that of P_4O_6, but Bi_2O_3 is ionic.

The more metallic nature of bismuth is clear from the chemistry of these three elements. In general, when descending one of the Groups 3A through 7A in the periodic table, the oxides go from acidic to amphoteric to basic. Thus, nitrogen and phosphorus oxides are decidedly acidic, arsenic and antimony oxides are acidic to amphoteric, and Bi_2O_3 is basic. This means that, while As_4O_6 and Sb_4O_6 are soluble in both acid and base, Bi_2O_3 is soluble only in acid.

$$\underset{\text{white arsenic}}{As_4O_6(s)} + 6\ H_2O(\ell) \xrightarrow{\text{slow}} 4\ \underset{\text{arsenious acid}}{H_3AsO_3(aq)}$$

$K_{1a} = 6 \times 10^{-10}$

$$As_4O_6(s) + 12\ NaOH(aq) \longrightarrow 4\ \underset{\text{sodium arsenite}}{Na_3AsO_3(aq)} + 6\ H_2O(\ell)$$

$$As_4O_6(s) + 12\ HCl(aq) \longrightarrow 4\ AsCl_3(aq) + 6\ H_2O(\ell)$$

$$Bi_2O_3(s) + 6\ HNO_3(aq) \longrightarrow 2\ Bi(NO_3)_3(aq) + 3\ H_2O(\ell)$$

As in Groups 3A and 4A, lower oxidation states are more important for the heavier elements. Oxidation numbers of +3 and +5 are common for arsenic, antimony can be forced to +5, but bismuth +5 is unstable. Heating As_4O_6 with concentrated nitric acid gives As_4O_{10}, which dissolves to form *arsenic acid*, analogous to orthophosphoric acid.

$$As_4O_{10}(s) + 6\ H_2O(\ell) \longrightarrow 4\ H_3AsO_4(aq)$$
$$K_{1a} = 2.5 \times 10^{-4}$$

It is not as strong an acid as orthophosphoric, but it is decidedly stronger than arsenious acid.

In general, acids of higher oxidation state oxides are more acidic than those of lower oxidation state oxides.

All arsenic compounds are poisonous, so they have been the favorites of writers of mystery novels for years. Arsenic can be detected in the stomach of a victim of poisoning by the *Marsh test*. An arsenic-containing solution is treated with zinc granules and hydrochloric acid, and arsenites or arsenates are reduced to the deadly gas AsH_3 (arsine).

$$Zn(s) + 2\ H^+(aq) \longrightarrow Zn^{2+}(aq) + H_2(g)$$
$$H_3AsO_4(aq) + 4\ H_2(g) \longrightarrow AsH_3(g) + 4\ H_2O(\ell)$$

The evolved gas is passed through a heated glass tube where it decomposes to the elements and leaves a shiny film of As on the glass.

$$2\ AsH_3(g) \longrightarrow 2\ As(s) + 3\ H_2(g)$$

As little as 0.5 milligrams of arsenic can be detected in this manner.

The *halogen compounds of Group 5A* also reflect the increasing metallic character of the elements when descending the group. While halogen compounds of nitrogen are generally unstable, the heavier elements form many different species with the halogens.

Fluorine is especially valuable in forming halides with Group 5A elements. NF_3, for example, is the most stable nitrogen trihalide,

$$4\ NH_3(g) + 3\ F_2 \xrightarrow{\text{copper catalyst}} 3\ NH_4F(s) + NF_3(g)$$
$$(\text{mp}, -206.8°\text{C};\ \text{bp}, -129.0°\text{C})$$

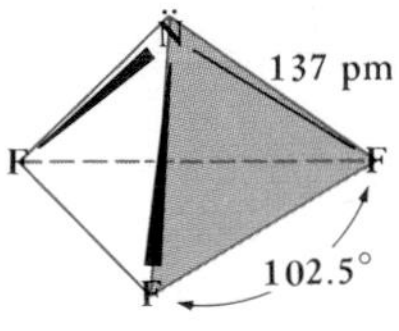

The gas is relatively unreactive in the sense that it is unaffected by water or dilute acids or bases. In contrast, NCl_3 and NBr_3 are highly unstable. The chloride was prepared in 1811 by P.L. Dulong who lost three fingers and an eye in the attempt. The brown solid nitrogen triiodide ($NI_3 \cdot NH_3$) can be handled when wet, but when dry a very light touch will cause it to explode violently (Figure 22.9).

In contrast to nitrogen, phosphorus and the remaining Group 5A elements all form well defined and stable halides in both the +3 and +5 oxidation states. Many are commercially available. Phosphorus trifluoride, for example, is a colorless gas at room temperature and has the pyramidal structure predicted by VSEPR theory.

$$2\ PCl_3(\ell) + 3\ CaF_2(s) \longrightarrow 3\ CaCl_2(s) + 2\ PF_3(g)$$

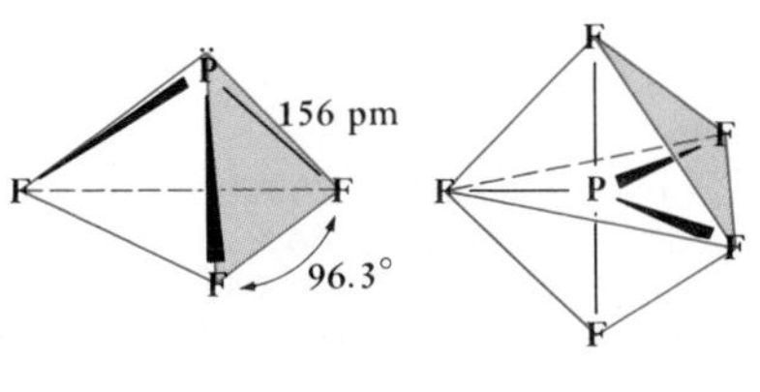

PF_3 PF_5

It is odorless and quite toxic because, like CO, it combines with the hemoglobin of your blood and leads to suffocation.

FIGURE 22.9

Nitrogen triiodide, NI_3. When dry, this dark brown compound is so sensitive that the touch of a small tree branch will set off an explosion. CAUTION! Do not perform this experiment yourself. (The filter paper and branch are blurred in the photo at the right because it was taken at the instant the compound exploded.) (Charles D. Winters)

Based on the number of valence electrons, VSEPR leads you to expect compounds such as PX_5 to be trigonal bipyramidal. PF_5 is indeed an excellent example of that structure,

The halogen chemistry of As, Sb, and Bi is a treasure trove of interesting compounds and structures. However, in practical terms their most important function is as fluorinating agents. For example, Freon-12 is made by exchanging Cl for F on carbon with an antimony chlorofluoride.

$$SbCl_5(s) + HF(\ell) \longrightarrow SbCl_4F(s) + HCl(g)$$

$$CCl_4(\ell) + 2\ SbCl_4F(s) \longrightarrow \underset{\text{Freon-12}}{CCl_2F_2(g)} + 2\ SbCl_5(s)$$

22.2 THE CHEMISTRY OF GROUP 6A ELEMENTS

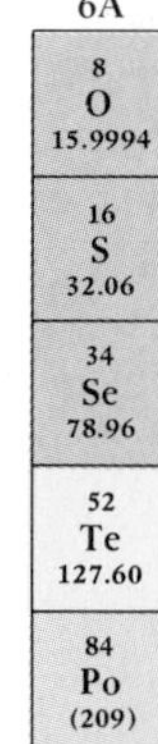

Group 6A, with elements of configuration ns^2np^4, consists almost completely of nonmetals (Table 22.6). All the 6A elements form compounds with metals or hydrogen in which the oxidation number is -2; oxidation numbers of $+2$, $+4$, and $+6$ occur when a Group 6A element is combined with more electronegative elements such as oxygen and the halogens.

THE ELEMENTS: THEIR RECOVERY AND USES

Oxygen is the most abundant element on our planet, occurring free as O_2 gas and combined in the form of water and oxides of many kinds. However, elemental oxygen, O_2, did not appear in the atmosphere of the earth until about 2 billion years ago, when it began to arise from photosynthesis occurring in the earliest green plants.

$$CO_2(g) + H_2O(\ell) + \text{light energy} \xrightarrow[\text{catalysts}]{\text{chlorophyll}} O_2(g) + \underset{\text{carbohydrates}}{\{CH_2O\}}$$

Oxygen makes up 23% by weight of the atmosphere, 46% of the lithosphere (the crustal rocks of the earth), and more than 85% of the hydrosphere, the lakes and oceans. Even on the moon, the rocks are 44.6% oxygen.

It is not yet known how and why the photosynthetic process began, nor is it completely clear why there is an excess of oxygen in the atmosphere. respiration by animals, the "combustion" of carbohydrates in what is effectively the reverse of the photosynthetic process, should consume the

TABLE 22.6 Group 6A: The p^4 Elements (Oxygen, Sulfur, and Their Congeners)

	OXYGEN	SULFUR	SELENIUM	TELLURIUM	POLONIUM
Symbol	O	S	Se	Te	Po
Atomic number	8	16	34	52	84
Atomic weight	15.999	32.064	78.96	127.60	(209)*
Valence e^-	$2s^22p^4$	$3s^23p^4$	$4s^24p^4$	$5s^25p^4$	$6s^26p^4$
mp, °C	−218	112	217	450	254
bp, °C	−183	444	685	990	962
d, g/cm³	1.43 g/liter	2.07 (a)*	4.79	6.24	9.32
Atomic volume	14.0	15.5	16.5	20.5	22.5
Atomic radius, pm	66	104	117	137	150
Ion radius (−2),	140	184	198	221	169
Pauling EN	3.5	2.5	2.4	2.1	1.9
Standard potential (V)	+1.229	+0.141	−0.40	−0.72	−1.0
Oxidation numbers	−1, −2	−2 to +6	−2 to +6	−2 to +6	
Ionization energy*	1312	1004	946	869	811
Isolated by	Priestley	Antiquity	Berzelius	Mueller	Curie
Date of isolation	1774	Antiquity	1817	1782	1898
rpw* pure O_2		SO_2, SO_3	SeO_2	TeO_2	
rpw H_2O	None	None	None	None	
rpw N_2	NO_x, 1200°C	(b)*	None	None	
rpw halogens		S_2X_2 to SX_6	SeX_2, SeX_4	TeX_2	
rpw H_2	H_2O, 500°C	H_2S, 400°C	H_2Se, 400°C	b	
Color of element	Pale blue	Bright yellow	Brick red (c)*	Brown (c)	

*Energies and heats are in kJ/mol; atomic weight in parentheses is that of the most stable isotope; a = data for orthorhombic S; potentials are for $O_2 + 4H^+ + 4e^- \rightarrow 2H_2O$ and the like; b indicates nitride or hydride formed only by indirect methods; rpw = reaction product with; c = color of nonmetallic form.
Adapted from E.G. Rochow, *Modern Descriptive Chemistry*, W.B. Saunders, 1977.

evolved oxygen. These are some of the mysteries of science to be solved by people with skill and imagination in chemistry, geology, and biology.

Molecular oxygen can be prepared in the laboratory by decomposing a salt of an oxyacid. A particularly convenient source is potassium chlorate, whose decomposition is catalyzed by transition metal oxides such as MnO_2.

$$2\ KClO_3(s) \xrightarrow[70-100°C]{MnO_2\ catalyst} 2\ KCl(s) + 3\ O_2(g)$$

The major industrial source of O_2, however, is fractional distillation of the gases of the atmosphere. Over 30 billion pounds of O_2 are produced in the United States every year (Figure 22.1), largely for use in the Bessemer process of steel making (Section 25.2).

Oxygen is a colorless, odorless gas at room temperature that is unique in being a paramagnetic molecule, even though it has an even number of electrons (see Chapter 10). Partly because of the unpaired electrons, the liquid (bp, −183.0°C at 1 atm) is blue, as is the solid (mp, −218.8°C).

O_2 is the most common allotrope of oxygen; *ozone* or O_3 is the other. It is a blue, diamagnetic gas with an odor so strong that it can be detected in concentrations as low as 0.01 ppm. Ozone is conveniently synthesized by passing O_2 through an electrical discharge, but an alternative preparation is to irradiate O_2 with ultraviolet light (Figure 7.9).

Some details of the evolution of our oxygen environment are described in "The Oxygen Cycle," an article by P. Cloud and A. Gibor in *Chemistry in the Environment*, pp. 31–41, Readings from *Scientific American*, W.H. Freeman, San Francisco, 1973.

A postage stamp honoring Joseph Priestley (1733–1804), who discovered oxygen in 1774. For religious reasons he fled his native England in 1794 and settled in Northumberland, Pennsylvania, where he continued his experiments in chemistry to the end of his life. (See also page 81.)

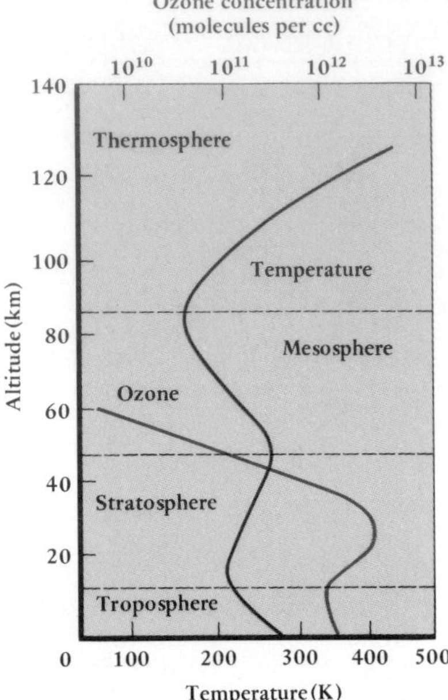

FIGURE 22.10

Ozone concentration (white curve) and temperature (blue curve) as a function of altitude. Ozone concentration is highest in the stratosphere. The absorption of ultraviolet radiation by ozone in this region leads to a permanent temperature inversion. That is, warm air masses are permanently located over cooler air masses, and little mixing occurs. Because of this, substances reaching the stratosphere tend to stay there.

O_3, Ozone

mp, −192.5°C; bp, −111.9°C
$\Delta G_f^\circ = +163.2$ kJ/mol

Geometry: 127.8 pm; 116.8°

Electron dot structure

$$O_2(g) \xrightarrow{\text{ultraviolet light}} 2\ O(g)$$

$$O(g) + O_2(g) \longrightarrow O_3(g) + 109\ \text{kJ}$$

The gas is unstable with respect to decomposition back to O_2, a reaction that is normally slow but can be speeded up by ultraviolet irradiation. These reactions occur in the stratosphere and serve not only to warm the earth's atmosphere but also to protect plants and animals from intense ultraviolet radiation. Radiation from the sun is sufficiently intense down to about 30 km above the earth's surface that ozone can be synthesized. Once made, the ozone itself is a good absorber of ultraviolet radiation in the very energetic 220 to 290 nm region, a process that helps to prevent damaging radiation from reaching the earth (Figure 22.10).*

Since ozone is hazardous to handle, it is usually generated where it is needed in amounts approaching perhaps a ton per day. Its primary use is to produce oxygen-containing compounds in industry, and, in Europe in particular, it is used to purify drinking water.

Sulfur is roughly twice as abundant as carbon in the crust of the earth. It can be found in the elemental form in nature, but only in certain concentrated deposits. Generally, it exists in the form of sulfur-containing compounds in natural gas and oil and as metal sulfide minerals, especially iron pyrites (FeS_2, commonly known as "fool's gold"). Indeed, there are many common sulfide-containing minerals; examples include cinnabar, HgS, and galena, PbS. Other sulfur-containing compounds in the environment are sulfates such as gypsum ($CaSO_4 \cdot 2\ H_2O$), and the sulfur oxides (SO_2 and SO_3), products of volcanic activity. Rarely are these sources worked for their sulfur content, however.

One major source of elemental sulfur (about 10 million tons per year) are the deposits along the Gulf coast of the United States and Mexico. These occur in the caprock over subterranean salt domes (Figure 22.11), typically at a depth of 150 to 750 m below the surface in layers perhaps 30 m thick. The theory is that the sulfur was formed by anaerobic ("without air") bacteria feeding on sedimentary sulfate deposits such as gypsum ($CaSO_4 \cdot 2\ H_2O$).

In the United States and in many other countries the major source of elemental sulfur is actually the H_2S and sulfur compounds found in natural gas and crude oil. Although the conversion is a complicated process, the net reaction is simply an oxidation of hydrogen sulfide.

$$2\ H_2S(g) + O_2(g) \longrightarrow \tfrac{1}{4}\ S_8(s) + 2\ H_2O(g)$$

*Unfortunately, there may be a problem with our ozone security blanket. A recent fea raised by some chemists is that the Freons used as propellents in spray cans have bee diffusing to the upper atmosphere where they interact with the ozone layer and lead to i partial destruction.

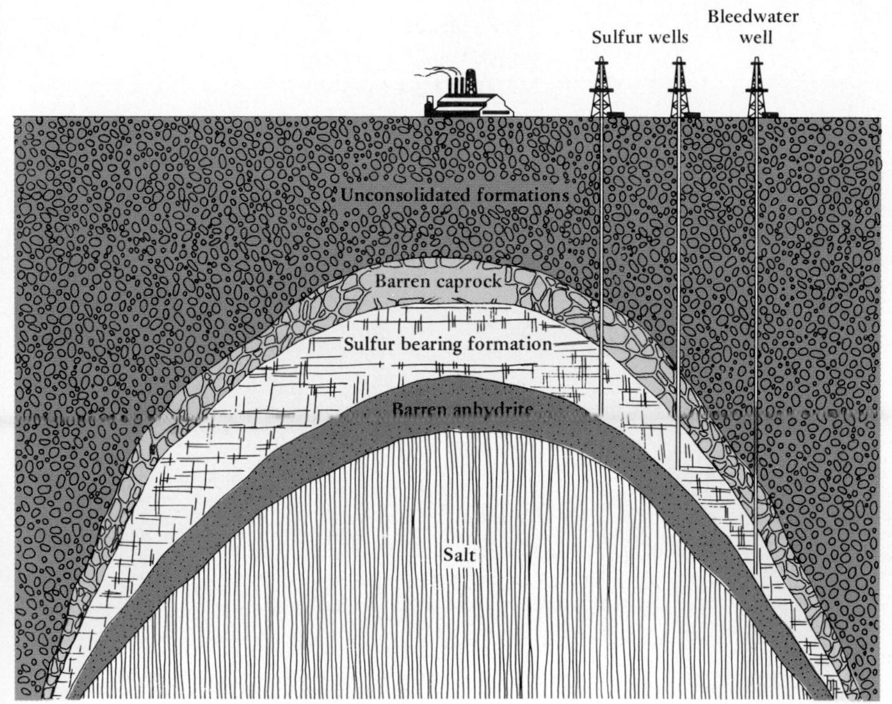

FIGURE 22.11

Diagram of the sulfur and caprock over a salt dome. (Anhydrite is anhydrous $CaSO_4$. It is thought that bacterial action on the sulfate resulted in the sulfur deposit.) Such sulfur deposits account for about 10% of the sulfur used in the United States annually. Recovery of sulfur from these deposits is done by a process devised by Herman Frasch about 1900. Superheated water (at 165°C) and then air are forced into the deposit; the sulfur is melted (mp, 119°C) and is forced to the surface as a frothy, yellow stream.

The largest use of sulfur by far is the production of sulfuric acid, the compound produced in largest quantity by the chemical industry (Figure 22.1). As indicated in Figure 22.12, sulfur and its compounds, usually sulfuric acid, are widely used in manufacturing, but only rarely appear in the final product.

Sulfur has perhaps more allotropes than any other element. These arise because there are many ways in which S—S catenation can occur. The most common and most stable allotrope is the yellow, orthorhombic

Two common sulfur-containing minerals. At the left is iron pyrite (FeS_2), commonly known as "fool's gold." At the right is gypsum, $CaSO_4 \cdot 2\ H_2O$. (Ward's Natural Science Establishment)

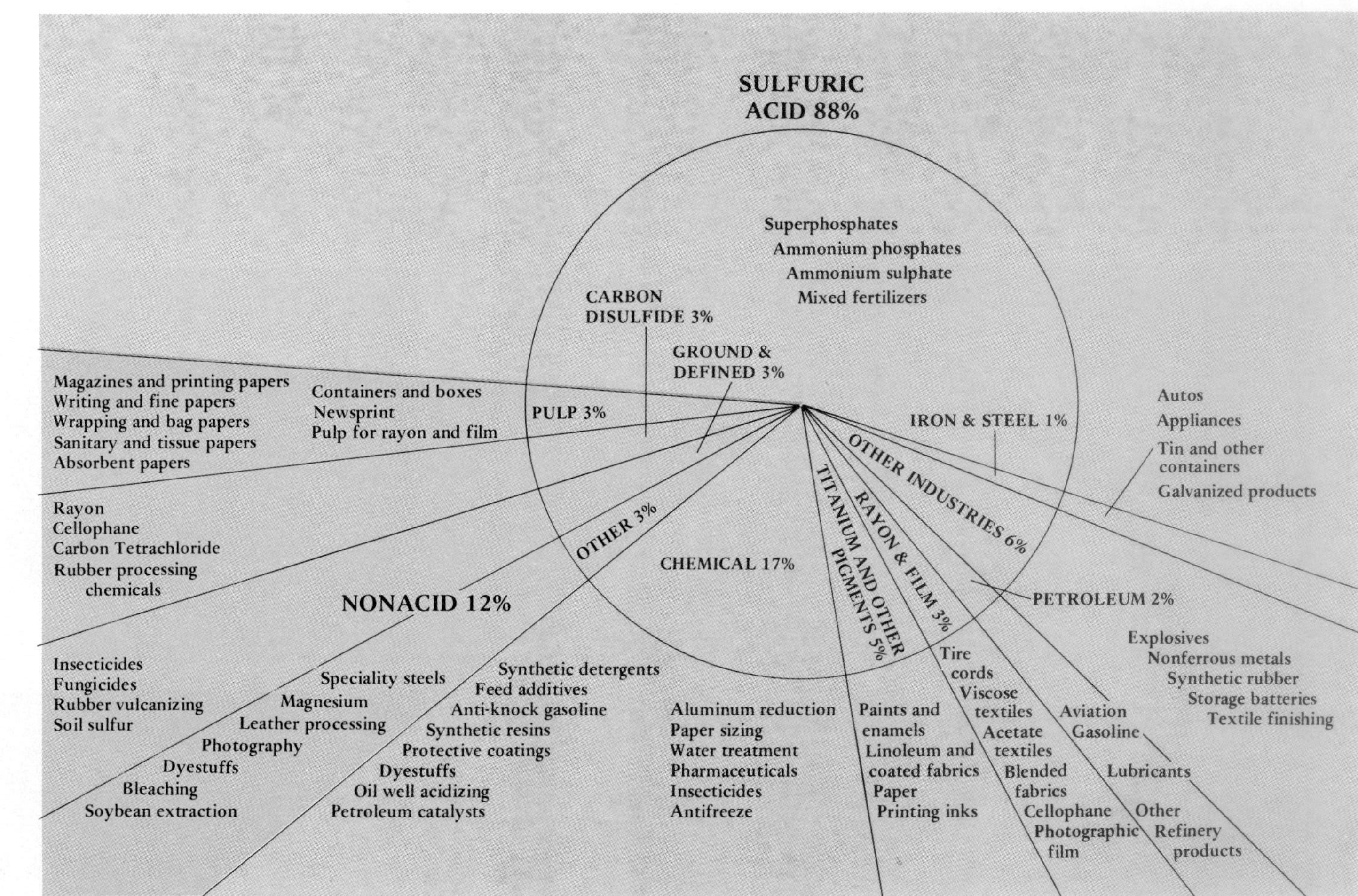

FIGURE 22.12
The uses of sulfur. About 88% of the elemental sulfur produced is converted to sulfuric acid, which in turn has myriad uses.

form, in which the sulfur atoms are arranged in a crown-shaped ring of eight atoms.

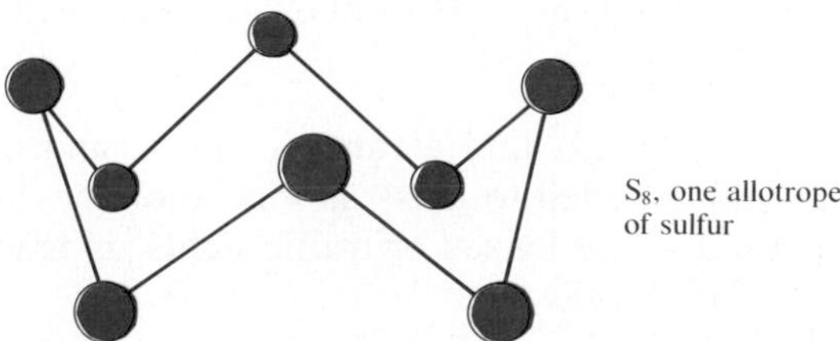

S_8, one allotrope of sulfur

Less stable are at least eight other allotropes having rings of 6 to 20 S atoms per ring and various allotropes having open chains of atoms (Figure 22.13).

Selenium and **tellurium** are comparatively rare, having abundances similar to those of silver and gold, respectively. Since their chemistry is so similar to that of sulfur, they are found in minerals associated with the sulfides of Cu, Ag, Fe, and As, and they are recovered as by-products of the industries devoted to these metals. For example, a selenite salt can be recovered from silver selenide by treating it with oxygen in a basic medium.

$$Ag_2Se(s) + Na_2CO_3(aq) + O_2(g) \longrightarrow 2\ Ag(s) + \underset{\text{sodium selinite}}{Na_2SeO_3(aq)} + CO_2(g)$$

(a)

(b)

FIGURE 22.13

The two common, lemon-yellow forms of sulfur are orthorhombic sulfur (mp, 12°C) and monoclinic sulfur (mp, 119°C), both consisting of S_8 rings. (a) When the sulfur is heated above about 150°C, the S_8 rings begin to break open, and the melt becomes viscous due to tangling of the chains. At the same time the color darkens because the chain ends have unpaired electrons. (b) When the viscous melt is poured into cool water, the sulfur slowly reverts to the orthorhombic form. (Charles D. Winters)

After the sodium selenite is converted to selenous acid, elemental selenium is recovered by reduction with SO_2.

$$\underset{\text{selenous acid}}{H_2SeO_3(aq)} + 2\ SO_2(g) + H_2O(\ell) \longrightarrow Se(s) + 2\ H_2SO_4(aq)$$

There have been some reports of an inverse relation between blood selenium and cancer incidence in humans. However, other studies clearly show high levels of Se cause cancer. It is not wise to use dietary selenium supplements sold in health food stores.

Selenium and its compounds find a range of uses, and one is in glassmaking. They are often added to glass as a decolorizer, but brilliant red glass, such as is used in the lenses of traffic lights, is made by adding a mixed sulfide/selenide, CaSSe.

The use of selenium you must be most familiar with is in "xerography." At the heart of most photocopying machines is an aluminum plate coated with a film of selenium. Light coming from the imaging lens selectively discharges a static electric charge in the selenium film, and the black "toner" sticks only to the areas that remain charged. A copy is made when the toner is transferred to a sheet of plain paper.

Finally, the heaviest element in Group 6A, **polonium**, is radioactive. It was discovered in 1898 by Marie Sklodowska Curie (1867–1934) and her husband Pierre Curie (1869–1906) (Chapter 26). The Curies painstakingly separated the elements in a large quantity of uranium-containing ore, pitchblende, and found the new elements radium and polonium. The latter was named in honor of Marie Curie's home country, Poland.

Xerography means "dry printing," or printing without ink. In such copiers the image of a document is first formed electrostatically on a selenium-coated drum of the type shown in this photograph. (Xerox Corporation)

EXERCISE 22.5 GROUP 6A ELEMENTS

Name three commercial uses of elements of Group 6A. What are the natural sources of these elements?

COMPOUNDS OF OXYGEN

In spite of the high bond energy of O_2 (498 kJ), the molecule reacts with virtually all the elements. Some elements such as W, Pt, and Au do not combine *directly* with O_2, but all of the elements—except the rare gases He, Ne, Ar, and possibly Kr—ultimately form oxo compounds. The reason for the great reactivity of O_2 is the stability of the resulting products.

Oxygen is of course an excellent oxidizing agent, as indicated by its $E°$ value in acidic media.

$$O_2(g) + 4\ H^+(aq) + 4e^- \longrightarrow 2\ H_2O(\ell)$$

$$E° = +1.229\ \text{V at 1}\ M\ H^+ \text{ and 1 atm } O_2$$

This is effectively the reaction that keeps all of us operating, since O_2 acts as an acceptor of H^+ and e^- in oxidative metabolism, the process that supplies the energy for all higher forms of life.

The ultimate product of O_2 reduction is the oxide ion, O^{2-}, in compounds of great thermodynamic stability. One way to systematize what is known about oxides is in terms of their acid–base behavior. In general, we can say that (a) oxides are *acidic* if combined with a nonmetal (e.g., CO_2, NO_2, P_4O_{10}, SO_3); (b) oxides are *basic* if combined with an electropositive element, a metal (e.g., Li_2O, CaO, Tl_2O, Bi_2O_3); (c) oxides are *amphoteric* if combined with a weakly electropositive element (e.g., BeO, Al_2O_3, ZnO); (d) oxides are *neutral* if they do not interact with water or aqueous acids and bases (e.g., CO, NO); (e) across a period, the oxides begin as basic compounds but become progressively more acidic, the middle ones being amphoteric. For example, basicity declines in the series Na_2O, MgO, Al_2O_3, SiO_2, P_4O_{10}, SO_3, and ClO_2; (f) acidity increases with oxidation number (As_2O_5 is more acidic than As_2O_3); and (g) in the main groups, oxide basicity increases down the group (BeO < MgO < CaO < SrO < BaO).

Not all binary oxygen compounds are oxides, wherein the oxygen is considered as O^{2-}. You learned in Group 1A chemistry that sodium and O_2 give a *peroxide*, Na_2O_2 (an O_2^{2-} salt), and potassium and O_2 give a *superoxide*, KO_2 (a salt of O_2^-).

Bond lengths in superoxide and peroxide ions were compared with O_2 in Chapter 10.

In Group 5A chemistry, the simplest hydrogen compound of nitrogen is ammonia, NH_3. Nitrogen can catenate or form N—N bonds to a limited extent, and the simplest compound of this type is hydrazine, $H_2N—NH_2$. In oxygen chemistry, the analogs are water and hydrogen peroxide, H_2O and HO—OH.

Hydrogen peroxide, H_2O_2, was first made in 1818 by the acidification of barium peroxide, still a convenient laboratory method.

$$BaO_2(s) + 2\,H^+(aq) \longrightarrow \underset{\substack{\text{mp, } -0.43^\circ\text{C; bp, } 150.2^\circ\text{C} \\ \Delta G_f^\circ = -118.0 \text{ kJ/mol}}}{H_2O_2(aq)} + Ba^{2+}(aq)$$

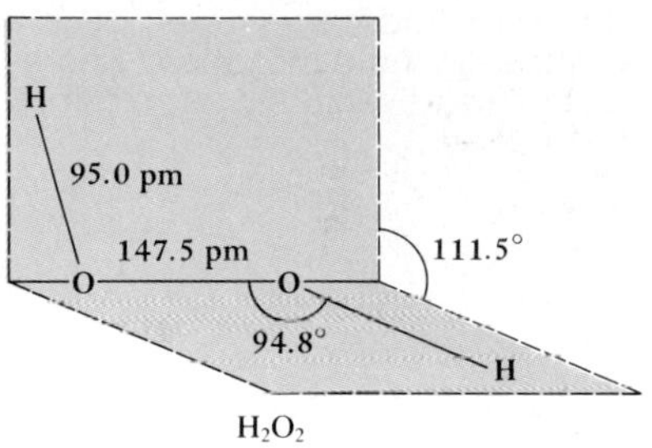

This method is not suitable for large scale production, however, so other methods such as the "isopropanol process" are employed.

$$\underset{\text{isopropanol}}{HO—\overset{\overset{CH_3}{|}}{\underset{\underset{CH_3}{|}}{C}}—H} + O_2 \xrightarrow[90\text{–}140^\circ\text{C}]{15\text{–}20 \text{ atm}} \underset{\text{acetone}}{O{=}\overset{\overset{CH_3}{|}}{\underset{\underset{CH_3}{|}}{C}}} + HOOH$$

Hydrogen peroxide mainly finds use as an oxidizing agent.

$$H_2O_2(aq) + 2\,H^+(aq) + 2e^- \longrightarrow 2\,H_2O(\ell) \qquad E^\circ = +1.77 \text{ V}$$

Thus, a significant fraction is used to bleach paper and textiles, and about 10% is consumed for environmental purposes: purification of drinking water, sewage treatment, and sterilization of milk containers.

A typical H_2O_2 plant can produce 15,000 tons of 100% H_2O_2 per year along with 30,000 tons of acetone, a valuable solvent.

Hydrogen peroxide is generally sold as 30% aqueous solutions for laboratory and industrial use. If you have handled such solutions, you know that they are always shipped in plastic containers, since traces of metal or alkali metal ions (dissolved from a glass container) can catalyze the explosive decomposition of the peroxide (Figure 13.15).

The catalyzed decomposition of H_2O_2 is used by the bombardier beetle to protect itself (Figure 13.1).

$$H_2O_2(aq) \longrightarrow H_2O(\ell) + \tfrac{1}{2}\,O_2(g) \qquad \Delta H^\circ = -98.2 \text{ kJ}$$

EXERCISE 22.6 OXIDE CHEMISTRY

Tell if each of the following oxides is acidic, basic, neutral, or amphoteric: (a) MgO, (b) CO_2, (c) Ga_2O_3, and (d) CO. Dichlorine heptoxide, Cl_2O_7, is the anhydride of perchloric acid. That is, adding water to the oxide gives the acid. Write a balanced equation for this reaction.

THE CHEMISTRY OF SULFUR

IONS OF SULFUR Just as O_2 can be reduced to the oxide (O^{2-}), peroxide (O_2^{2-}), and superoxide (O_2^-) ions, sulfur can be reduced. In the case of sulfur, however, many more ions are known, all the way from S^{2-}, the sulfide ion, to S_n^{2-}, where n can be 2 to 6. It can be predicted by the octet rule and confirmed by experiment that none of these ions is a ring of atoms; adding two electrons breaks down the ring structure of the S_n allotropes and gives S_n^{2-} chains.

The simplest anion of sulfur is sulfide, S^{2-}, the anion associated in nature with metals of great commercial interest. As illustrated in Figure 22.14, iron, copper, silver, and mercury, among many others, are commonly found as their sulfides. You recall from Chapter 17 that many metal sulfides are poorly soluble in water, so these compounds can accumulate in geologic zones and not be dispersed by ground water.

If there is poor pollution control during the roasting process, SO_2 escapes to the atmosphere where it can lead eventually to acid rain. See below.

The recovery of metals from their sulfide ores is usually done by heating or *roasting* the ore in air, a process that can have any of three outcomes. The most common of these is the conversion of the metal sulfide to an oxide, the sulfur appearing as SO_2.

$$2\ ZnS(s) + 3\ O_2(g) \longrightarrow 2\ ZnO(s) + 2\ SO_2(g)$$

Alternatively, the sulfide can be converted to a sulfate, which is often more water soluble than the sulfide. Finally, if the metal oxide product from roasting is less stable than SO_2, the free metal is obtained by sulfide roasting. Examples include Cu, Ag, Hg, and Pb. For example, if lead sulfide or galena is heated to form the oxide, and then mixed with fresh galena, lead and SO_2 result.

$$2\ PbO(s) + PbS(s) \longrightarrow 3\ Pb(s) + SO_2(g)$$

SULFUR–HYDROGEN COMPOUNDS Hydrogen sulfide, H_2S, is the simplest hydrogen–containing compound of sulfur, and its structure resembles that of water. However, unlike water, hydrogen sulfide is a gas under normal conditions (mp, −85.6°C; bp, −60.3°C), presumably due to very weak hydrogen bonding between molecules as compared with water (see Chapter 11). Furthermore, hydrogen sulfide is a deadly poison, comparable to hydrogen cyanide. Fortunately, the sulfide has a terrible odor and can be smelled in concentrations as low as 0.02 ppm. You must be careful with H_2S, however, because it has an anesthetic effect on the olfactory nerve, so your nose can lose its ability to detect H_2S. Death occurs at H_2S concentrations of 100 ppm.

H_2S dissolves to a small extent in water (~0.10 *M*) where it behaves as a polyprotic acid (Chapter 15).

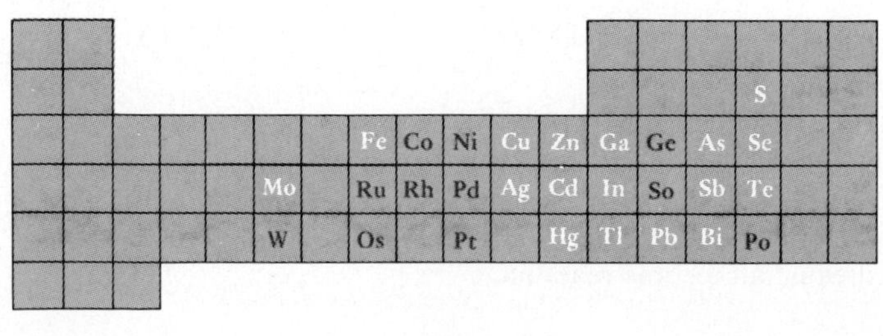

FIGURE 22.14

The elements that are most likely to occur as sulfides in nature are shown in white, while those in blue have a slightly lesser tendency to be found as sulfides.

SULFUR–OXYGEN COMPOUNDS At least thirteen oxides of sulfur are known to exist, but only the dioxide (SO_2) and trioxide (SO_3) are important. The former is produced on an enormous scale by the combustion of sulfur, by roasting sulfide ores (especially iron pyrites, FeS_2) in air or by the combustion of sulfur-containing coal and fuel oil. It has been estimated that about 200 million tons of sulfur are released into the atmosphere each year by human activities, primarily in the form of SO_2; this is more than half of the total emitted by all other natural sources of sulfur in the environment.

SO_2	SO_3
mp, −75.5°C	mp, 16.86°C
bp, −10.0°C	bp, 44.6°C
S–O 143.1 pm; O–S–O 119°	S–O 142 pm; O–S–O 120°

Sulfur dioxide is a colorless, toxic gas with a choking odor. It readily dissolves in water to give solutions of a species that we write as H_2SO_3 (sulfurous acid),

$$SO_2(g) + H_2O(\ell) \longrightarrow \text{“}H_2SO_3\text{”}(aq)$$

but this is an unstable molecule that has never been isolated. The ions SO_3^{2-} (sulfite) and HSO_3^- (hydrogen sulfite), however, are known in the form of salts of sulfurous acid.

The most important reaction of SO_2 is its oxidation to the trioxide.

$$SO_2(g) + \tfrac{1}{2}\,O_2(g) \longrightarrow SO_3(g) \qquad \Delta H^\circ = -98.9 \text{ kJ/mol}$$

Sulfur trioxide is extremely reactive and is very difficult to handle; as described below, it is almost always deliberately converted to sulfuric acid by reaction with water.

As in nitrogen chemistry there are numerous **sulfur oxyacids**. Few can be isolated, however, and most are known only in aqueous solution or as salts of the oxyacid anions. Several of the most important species are shown in Table 22.7.

The figure below shows the sulfur budget for the land-ocean-atmosphere (in units of 10^6 tons). Geothermal activity (especially volcanoes) releases large amounts of SO_2 into the atmosphere, along with smaller amounts of SO_3, H_2S, sulfates, and elemental sulfur. The most important sources of sulfur in the atmosphere are (a) the biological reduction of sulfur-containing compounds (much of this sulfur is released as H_2S) and (b) human activities such as power generation from coal and metal processing.

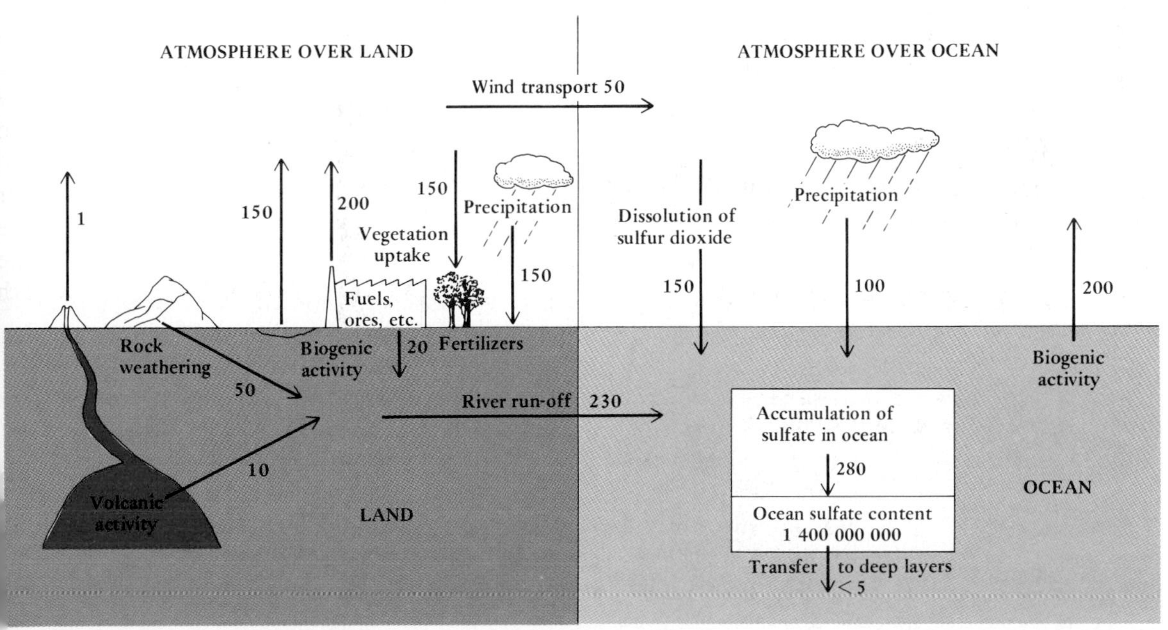

TABLE 22.7 Some Oxyacids of Sulfur

ACID	SULFUR OXIDATION NUMBER	SALTS
Sulfuric acid 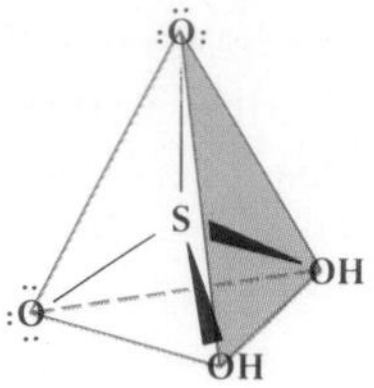	+6	SO_4^{2-}, sulfate HSO_4^-, hydrogen sulfate
Dissulfuric 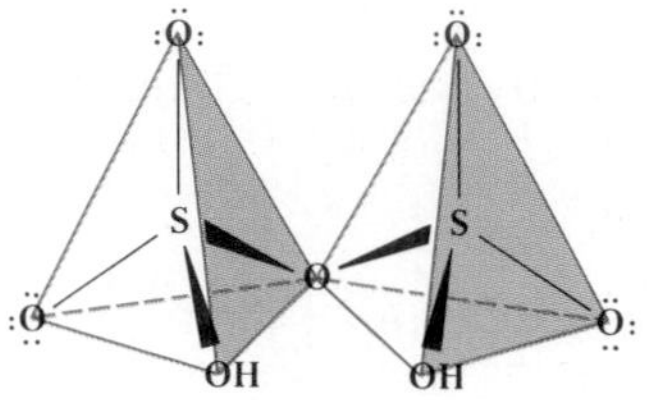	+6	$O_3SOSO_3^{2-}$, disulfate
Thiosulfuric 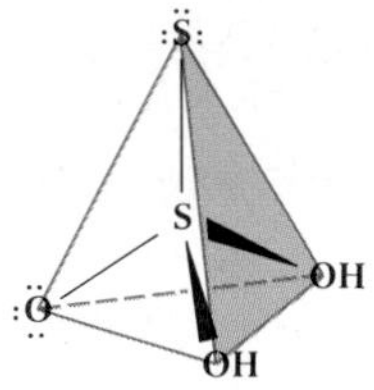	+4 and 0	SSO_3^{2-}, thiosulfate
Sulfurous 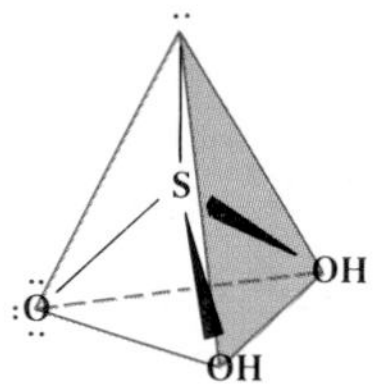	+4	SO_3^{2-}, sulfite HSO_3^-, hydrogen sulfite

Sulfuric acid production is taken by some economists as a reliable guide to a nation's industrial strength, because the acid is consumed in so many ways in a modern industrial society. The United States, for example, produces almost 40 million tons of the acid annually. It is manufactured in the highly efficient *contact process* from SO_2 produced by burning sulfur or by roasting sulfide ores. The first step is the oxidation of the SO_2 to SO_3, the reversible and exothermic process referred to above. According to LeChatelier's principle, the yield of SO_3 can be increased if the pressure is increased, if the concentration of O_2 is increased, or if SO_3 is removed from the reaction zone. An increase in temperature will increase the reaction rate, but, since the reaction is exothermic, it will also decrease the yield. This means a catalyst is required, V_2O_5 in this case, to increase the rate at a sufficiently low temperature that the yield is not affected greatly.

Following SO_3 formation in a sulfuric acid plant, the oxide is absorbed into a 98% aqueous solution of H_2SO_4, and water is added to maintain the concentration approximately at this level. The concentrated acid used in the laboratory is a solution that is 96% H_2SO_4.

In the United States, roughly 70% of the sulfuric acid is used to manufacture "superphosphate" fertilizer (p. 833), and smaller amounts are used, for example, in the conversion of ilmenite, a titanium-bearing ore, to TiO_2, which is used as the white pigment in paint, plastics, and paper.

The only other oxyacid of sulfur that is industrially useful is thiosulfuric acid. Just as H_2O and SO_3 form H_2SO_4, thiosulfuric acid is derived from H_2S and SO_3.

$$H_2S + SO_3 \xrightarrow[\text{absence of water}]{-78°C} H_2S_2O_3$$

The acid is not stable in the presence of water or above about 0°C. Instead, it is the salts of the thiosulfate ion, $S_2O_3^{2-}$, that are important. This ion, which resembles the sulfate ion except that one O atom is replaced by an S atom, is conveniently manufactured by boiling an aqueous solution of sodium sulfite with sulfur.

$$Na_2SO_3(aq) + \tfrac{1}{8} S_8(s) \xrightarrow{H_2O/100°C} Na_2S_2O_3(aq)$$

The hydrated salt, $Na_2S_2O_3 \cdot 5\ H_2O$ is commonly known as "hypo," and its aqueous solutions are used as the "fixer" in photography. Photographic films are coated with silver halide, and light falling on the film causes the reduction of the "exposed" halide to silver metal.

$$AgX(\text{on film surface}) + \text{light} \longrightarrow Ag(s) + \tfrac{1}{2} X_2(\text{on film surface})$$

After an exposed film is developed, the silver halide that was not touched by light is washed away with sodium thiosulfate solution, because silver ion and thiosulfate ion form a stable, water-soluble complex ion. (See Special Section: Silver Chemistry in Chapter 17.)

$$AgBr(s) + 2\ S_2O_3^{2-}(aq) \longrightarrow [Ag(S_2O_3)_2]^{3-}(aq) + Br^-(aq)$$

What remains on the film negative is the silver metal particles where light struck the coating. That is why "negatives" are dark where the image was bright.

EXERCISE 22.7 SULFUR CHEMISTRY

Sodium sulfide, Na_2S, is used extensively in the leather industry to remove hair from hides. Its synthesis is another example of the use of carbon as a reducing agent. Balance the equation for this reaction.

$$Na_2SO_4 + C \longrightarrow Na_2S + CO$$

The sulfide is readily oxidized to sodium thiosulfate; almost 50,000 tons of the latter are produced annually in the United States. Balance the equation for the oxidation.

$$Na_2S(aq) + O_2(g) + H_2O(\ell) \longrightarrow Na_2S_2O_3(aq) + NaOH(aq)$$

22.3 THE CHEMISTRY OF GROUP 7A ELEMENTS: THE HALOGENS

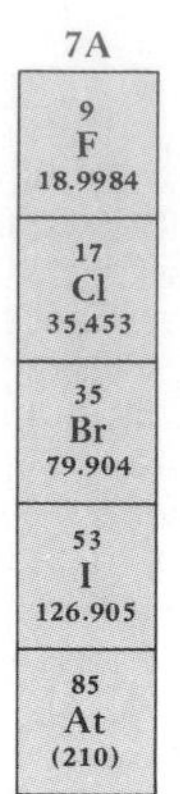

Compounds of the halogens (Table 22.8) have been known for thousands of years, and many have already been described in this book. Thus, in this section, we shall discuss only a few aspects of halogen chemistry: the diatomic, elemental halogens, the hydrogen halides, and halogen oxides and oxyacids. All are used in large quantities in commerce in ways with which you should be familiar.

THE ELEMENTS AND THEIR RECOVERY

Fluorspar, CaF_2, is one of the chief minerals containing **fluorine**. Since the mineral was originally used as a flux in metalworking, its name comes from the Latin meaning "to flow." In the 17th century, it was discovered that solid CaF_2 would emit light when heated, and the phenomenon was called *fluorescence*. Thus, when it was recognized in the early 1800s that a new element, F, was contained in fluorspar, A. Ampere suggested that the element be called *fluorine*.

CaF_2 is pictured in Figure 11.20.

Although fluorine was recognized as an element by 1812, it was not until 1886 that it was isolated in elemental form as a colorless gas by Moissan by the electrolysis of KF in anhydrous HF, still the only practical way to obtain diatomic F_2 (Figure 22.15). Experimentally, this is a difficult preparation because F_2 is such a powerful oxidizing agent that it corrodes the apparatus and will react violently with traces of grease or similar

TABLE 22.8 Group 7A: The p^5 Elements (the Halogen Family)

	FLUORINE	CHLORINE	BROMINE	IODINE	ASTATINE
Symbol	F	Cl	Br	I	At
Atomic number	9	17	35	53	85
Atomic weight	18.998	35.453	79.904	126.905	(210)*
Valence e^-	$2s^22p^5$	$3s^23p^5$	$4s^24p^5$	$6s^26p^5$	$7s^27p^5$
mp, °C	−220	−101	−7.3	114	
bp, °C	−188	−34	58.8	184	
d, g/cm³	1.81 g/L	3.21 g/L	3.12	4.94	
Atomic volume	12.8	14.5	23.5	25.7	
Atomic radius, pm	64	99	114	133	
Ion radius (−1),	136	181	195	216	
Pauling EN	4.0	3.0	2.8	2.5	2.1
Standard potential (a)*	+2.87	+1.36	+1.07	+0.615	+0.3
Oxidation numbers	−1	−1 to +7	−1 to +7	−1 to +7	
Ionization energy*	1680	1254	1139	1113	897
Isolated by	Moisson	Scheele	Balard	Courlois	CMS*
Date of isolation	1886	1774	1826	1811	1940
rpw* pure O_2	O_2F_2 (el)*	None (i)*	None (i)	None (i)	
rpw H_2O	HF, O_2, O_3	HCl, HOCl	HBr, HOBr	HI, HOI	
rpw N_2	None	None	None	None	
rpw H_2	HF	HCl	HBr	HI	
Color of element	Pale yellow	Yellow-green	Dark red	Black	

*Energies and heats are in kJ/mol; atomic weight in parentheses is that of the most stable isotope; rpw = reaction product with; a = standard electrode potential for reaction $X_2 + 2e^- \rightarrow 2X^-$ (reaction $F_2 + H_2 \rightarrow 2HF$ gives 3.06 V); CMS = Corson, McKenzie, and Segré; (i) indicates oxides, etc., made by indirect methods; el = electric discharge.
Adapted from E.G. Rochow, *Modern Descriptive Chemistry*, W.B. Saunders, 1977.

contaminants. Furthermore, F_2 and H_2, the products of the electrolysis, will recombine explosively, so they must be carefully separated. Still, a large plant can make 9 tons of liquid F_2 a day, and the annual U.S. and Canadian production is about 5000 tons.

Almost 80% of the fluorine produced is used for processing and reprocessing uranium for fueling nuclear power plants. For example, the oxide of uranium reacts with hydrofluoric acid to give uranium(IV) fluoride,

$$UO_2(s) + 4\ HF(aq) \longrightarrow UF_4(\text{green solid}) + 2\ H_2O(\ell)$$

and the latter is then oxidized to uranium(VI) hexafluoride by reaction with F_2.

$$UF_4(s) + F_2(g) \longrightarrow UF_6(\text{volatile white solid})$$

In the uranium fissioning process that provides the heat energy in a nuclear power plant, other elements are produced, among them plutonium. These elements and their oxides also react with fluorinating agents, but only uranium forms a fluoride that can be vaporized at a relatively low temperature. This means that UF_6 can be separated easily from the other elements, and the uranium can be recovered for reprocessing.

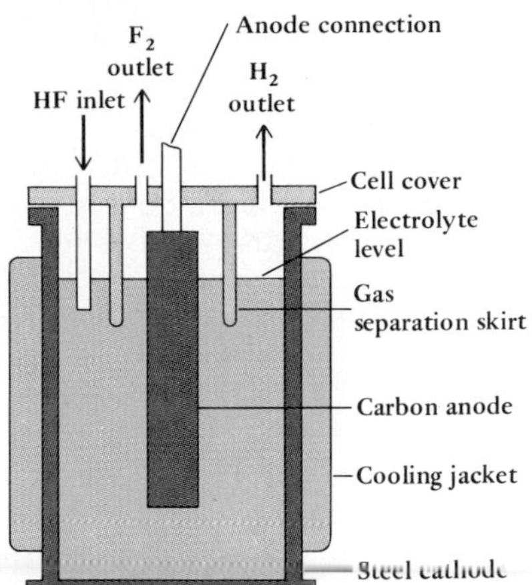

FIGURE 22.15

A cell for the electrolytic production of fluorine. Gaseous F_2 is produced at the carbon anode and H_2 is produced at the steel cathode.

Other major uses of fluorine and its compounds are in Teflon, the nonstick surface on cooking utensils, and the Freons, compounds such as CCl_2F_2 that are used as the "working fluid" in refrigerators and air conditioners. Furthermore, many municipal water supplies now contain fluoride ion, introduced by adding NaF; as explained on page 775, the ion is incorporated into tooth enamel, making it more resistant to decay. Finally, the cryolite (Na_3AlF_6) vital to the aluminum industry (pages 731 and 770) is now manufactured rather than mined.

Chlorine was the first of the halogens to be isolated in elemental form as a yellow-green gas. It was made by C.W. Scheele in 1774 by a reaction still used today as a laboratory source of Cl_2 (Figure 22.16).

$$2\ NaCl(s) + 2\ H_2SO_4(aq) + MnO_2(s) \xrightarrow{\text{heat}} Na_2SO_4(aq) + MnSO_4(aq) + 2\ H_2O(\ell) + Cl_2(g)$$

Today, of course, chlorine is made by electrolysis of brine (Chapter 19) in enormous quantities and ranks number 8 in production (Figure 22.1). Almost 70% of the Cl_2 manufactured is used for the production of organic chemicals such as vinyl chloride (that is then converted to PVC or polyvinyl chloride, a plastic used in many consumer items).

$$\underset{\text{ethylene}}{CH_2{=}CH_2(g)} + Cl_2(g) \longrightarrow \underset{\text{ethylene dichloride}}{H{-}CHCl{-}CHCl{-}H} \xrightarrow{\text{heat}} HCl + \underset{\text{vinyl chloride}}{H_2C{=}CHCl}$$

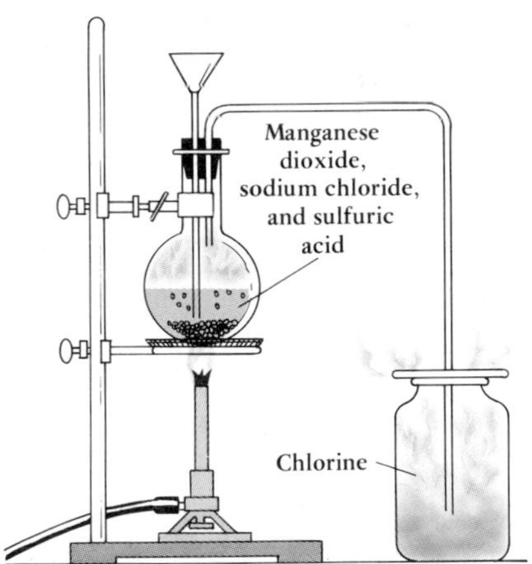

FIGURE 22.16

One laboratory method for preparing chlorine is to heat sodium chloride with sulfuric acid in the presence of manganese dioxide. CAUTION: This experiment must be performed in a hood.

One of the first properties that Scheele recognized about Cl_2 was its ability to bleach textiles and paper, and soon thereafter its ability to act as a disinfectant in water was also recognized. Today, these two uses account for about 20% of Cl_2 consumption. Finally, about 10% of the Cl_2 production goes into the manufacture of a wide range of inorganic chloride-containing compounds.

The structure of the indigo dye known to the Romans as Tyrian purple.

```
Br   H
  C—C
HC//  \\CH
  \   /
   C=C
  /   \
HN     C
  \   /  \\O
    C
    ||
    C
O// \ 
  C    NH
  \    /
   C=C
  /   \
HC     CH
  \\  //
   C—C
  H    Br
```

In Biblical times a beautiful purple dye, called Tyrian purple, was in great demand. Almost 12,000 small purple snails were required to prepare 1.5 g of the substance, now known to be a naturally occurring compound containing **bromine**. Red-brown, elemental bromine, Br_2, has an unpleasant and penetrating odor and is the only nonmetallic element that is a liquid under normal conditions (Figure 2.4).

All the halogens are oxidizing agents, but the ability declines as they become heavier (see the $E°$ values in Table 22.8) (Figure 22.17). That is,

$$\xleftarrow{\text{oxidizing ability}}$$

$$F_2 \quad > \quad Cl_2 \quad > \quad Br_2 \quad > \quad I_2$$
$$2.866\text{ V} \qquad 1.395\text{ V} \qquad 1.087\text{ V} \qquad 0.615\text{ V}$$
$$E° \text{ [for } X_2(aq) + 2e^- \longrightarrow 2\,X^-(aq)]$$

This means that Cl_2 will oxidize Br^- to Br_2 in aqueous solution.

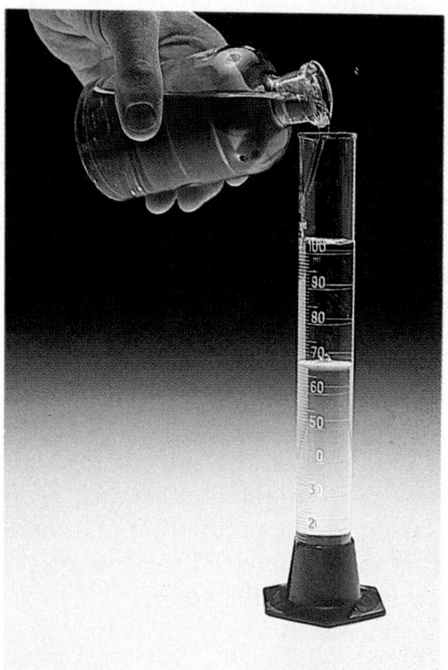

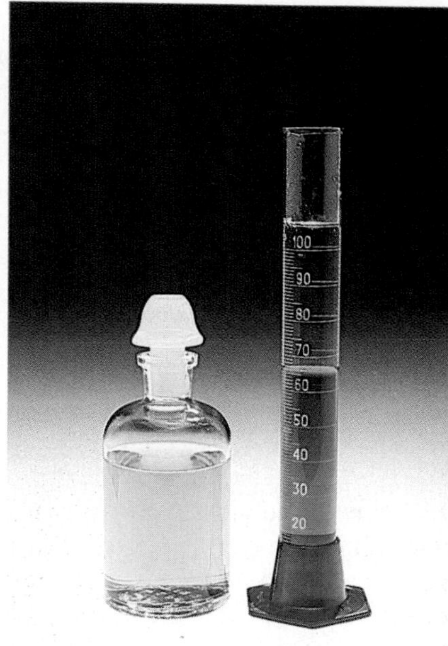

FIGURE 22.17

Formation of iodine in the reaction of bromine with an iodide salt.

$$Br_2(aq) + 2\,I^-(aq) \longrightarrow 2\,Br^-(aq) + I_2(\text{in } CCl_4)$$

At the left is a graduated cylinder containing CCl_4 (bottom layer) and NaI in water (top layer). When yellow-brown aqueous Br_2 is added and the reactants are mixed, reaction occurs. At the right, we now see that purple I_2 has collected in the CCl_4 layer. (The top aqueous layer now contains Br^- and some unreacted Br_2.) (See also Figure 12.4.) (Charles D. Winters)

$$\begin{array}{lll} Cl_2(aq) + 2e^- \longrightarrow 2\,Cl^-(aq) & E^\circ = & 1.395 \text{ V} \\ 2\,Br^-(aq) \longrightarrow Br_2(aq) + 2e^- & E^\circ = & -1.087 \text{ V} \\ \hline 2\,Br^-(aq) + Cl_2(aq) \longrightarrow 2\,Cl^-(aq) + Br_2(aq) & E^\circ = & 0.308 \text{ V} \end{array}$$

In fact, this equation represents the commercial method for preparing bromine, when NaBr is obtained from brine wells in Arkansas and Michigan and from the ocean.

Until recently the major use of Br_2 was in the manufacture of EDB, $C_2H_4Br_2$. The latter is added to leaded gasoline to remove lead after combustion and has been used as a general agricultural pesticide. EDB is now known to be a carcinogen.

Iodine, a lustrous, purple-black solid, is easily sublimed at room temperature and atmospheric pressure. The element was first isolated in 1811 from certain seaweeds and kelps, extracts of which had long been used in the treatment of goiter, the enlargement of the thyroid gland. It is now known that the thyroid gland produces a growth-regulating hormone (thyroxine) that contains iodine. Most of the table salt sold in the United States has 0.01% NaI added to provide the necessary iodine in the diet.

Thyroxine, the iodine-containing compound produced by the thyroid gland.

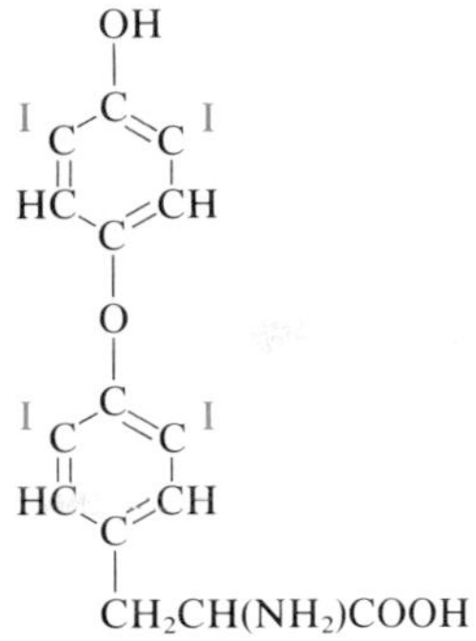

OH
I C C I
HC CH
C
O
I C C I
HC CH
C
$CH_2CH(NH_2)COOH$

The commercial preparation of iodine depends on the source of I^- and its concentration. One method is oxidation with Cl_2 to give I_2, the same as the Br_2 method. But another method is interesting, because it involves some simple chemistry you have already seen.

$$I^-(aq) \xrightarrow{AgNO_3} AgI(s) \xrightarrow{Fe} Ag(s) + FeI_2(aq) \xrightarrow{Cl_2} FeCl_3(aq) + I_2$$

(HNO_3 returns Ag(s) to the $AgNO_3$ step.)

The iodide ion is precipitated as the poorly soluble silver salt. This is then reduced with clean scrap iron to give silver and aqueous FeI_2. (Since silver is so expensive, it is recycled by oxidizing the metal with nitric acid.) The iodide ion is then oxidized to I_2 with Cl_2. A laboratory method for preparing I_2 is illustrated in Figure 22.18.

World production of iodine is much smaller than that of the other halogens, but it finds many uses. Most is consumed in the manufacture of organic chemicals, but smaller amounts are used in animal-feed supplements, in ink pigments, and in making AgI for high-speed photographic films.

All the isotopes of the heaviest halogen, **astatine**, are radioactive. None has a long lifetime, so the element and its compound occur in only trace amounts in nature.

GENERAL ASPECTS OF HALOGEN CHEMISTRY

Fluorine is the most reactive of all of the elements, forming compounds with all of the other elements with the exception of He, Ne, and Ar. In some cases, the elements combine directly in reactions that are so vigorous that they are explosive. This can be explained by at least two factors: the relatively weak F—F bond (low dissociation energy) compared with the other halogens, and the relatively strong bond formed between F and other elements. This general feature is illustrated by the bond energies in the table in the margin.

As you have already seen, F_2 is the best oxidizing agent of all of the halogens. This ability, combined with the small size of the fluorine atom, means that fluorine can form compounds with other elements where the other element has a high oxidation number, as in IF_7, PtF_6, AgF_2, UF_6, and XeF_4.

Bond Energies of Some Halogen Compounds (kJ/mol)

X	X—X	H—X	C—X (in CX_4)
F	159	574	456
Cl	243	428	327
Br	192	363	272
I	151	294	239

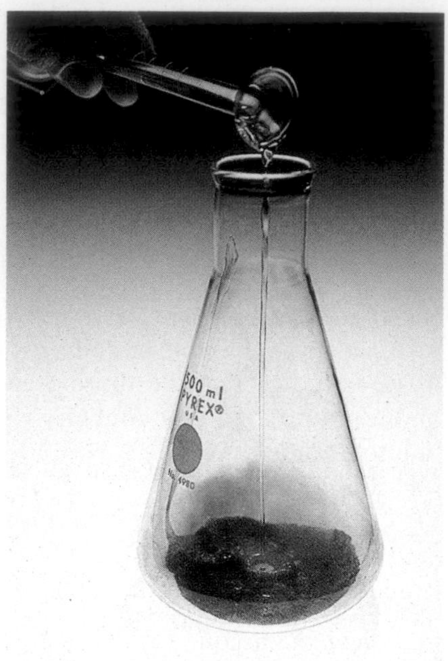

FIGURE 22.18

The preparation of iodine. A mixture of sodium iodide and manganese(IV) oxide was placed in a flask (photo at left). On adding concentrated sulfuric acid (photo at right), brown, gaseous I_2 was evolved.

$$2\ NaI(s) + 2\ H_2SO_4(aq) + MnO_2(s) \longrightarrow Na_2SO_4(aq) + MnSO_4(aq) + 2H_2O(\ell) + I_2(g)$$

(Charles D. Winters)

HYDROGEN HALIDES The pure, anhydrous compounds HX are commonly called hydrogen halides (Table 22.9). Aqueous solutions of them, however, are usually referred to as the "hydrohalic acids." Of the four possible HX compounds, HF and hydrofluoric acid are produced in substantial amounts, and HCl and hydrochloric acid are industrial chemicals of major importance; HBr and hydrobromic acid are made on a much smaller scale, and there is little demand for HI and hydriodic acid.

The estimated worldwide production of *hydrogen fluoride* is about 1 million tons annually, virtually all by the action of concentrated H_2SO_4 on fluorspar, CaF_2.

$$CaF_2(s) + H_2SO_4(\ell) \longrightarrow CaSO_4(s) + 2\ HF(g)$$

TABLE 22.9 The Hydrogen Halides

	HF	HCl	HBr	HI
mp(°C)	−83.4	−114.7	−88.6	−51.0
bp(°C)	19.5	−84.2	−67.1	−35.1
ΔH_f°(kJ/mol)	−271.1	−92.3	−36.4	26.5
ΔG_f°(kJ/mol)	−273.2	−95.3	−53.5	1.72
H—X bond energy(kJ/mol)	574.0	428.1	362.5	294.6
H—X bond length(pm)	91.7	127.4	141.4	160.9

The fluorspar must be quite pure and especially free of silica, since the latter reacts with HF to produce H_2SiF_6 in the following way.

$$SiO_2(s) + 4\ HF(g) \longrightarrow SiF_4 + 2\ H_2O \xrightarrow{\text{aq HF}} H_2SiF_6(aq)$$

This series of reactions allows HF to etch or frost glass, and designs on glass are often made this way. This is also the reason that HF is never shipped or used in glass containers.

Most of the HF produced is used to make *cryolite*, since the mineral is necessary for making aluminum (page 730), but is found in only small amounts in nature.

$$6\ HF(aq) + Al(OH)_3(s) + 3\ NaOH(aq) \longrightarrow \underset{\text{cryolite}}{Na_3AlF_6(s)} + 6\ H_2O(\ell)$$

Large amounts of HF also go into the production of the Freons, and smaller amounts are used to make NaF (for water fluoridation), SnF_2 (for use in fluoride-containing toothpaste), and to reprocess nuclear fuel.

Hydrogen chloride ranks 27th among industrial chemicals, world-wide production being on the order of 2.5 million tons annually. The classical method of making laboratory quantities of dry HCl gas is the *salt-cake* method (Figure 20.10).

$$2\ NaCl(s) + H_2SO_4(\ell) \longrightarrow Na_2SO_4(s) + 2\ HCl(g)$$

Now, however, it is usually made by burning hydrogen in chlorine (if the reactants are available cheaply) (Figure 9.3),

$$H_2(g) + Cl_2(g) \longrightarrow 2\ HCl(g)$$

Hydrogen also burns in a bromine atmosphere to give HBr (Figure 20.3, page 750).

or as a by-product of the organic chemicals industry.

All the hydrohalic acids are corrosive. However, hydrofluoric acid and HF are especially so, and, if you ever work with them, you must be very careful. Serious symptoms of an HF burn are slow to develop, but they are particularly painful. The F^- ion removes Ca^{2+} from tissues and precipitates it as insoluble CaF_2. It is thought that this leaves an excess of K^+ ions in the affected tissue, and nerve stimulation occurs.

The H—F bond is highly polar, so hydrogen bonding is important to the properties of the compound (Chapter 11). For example, gaseous HF is an equilibrium mixture of single HF molecules and rings consisting of six HF molecules, while HF in the solid state is a zigzag polymer held together by H . . . F hydrogen bonds (Figure 22.19).

HCl, HBr, and HI are not strongly hydrogen bonded in the gaseous state, in contrast to HF.

All the anhydrous hydrogen halides are excellent reagents for the halogenation of metals, metal and nonmetal oxides, and hydrides. For

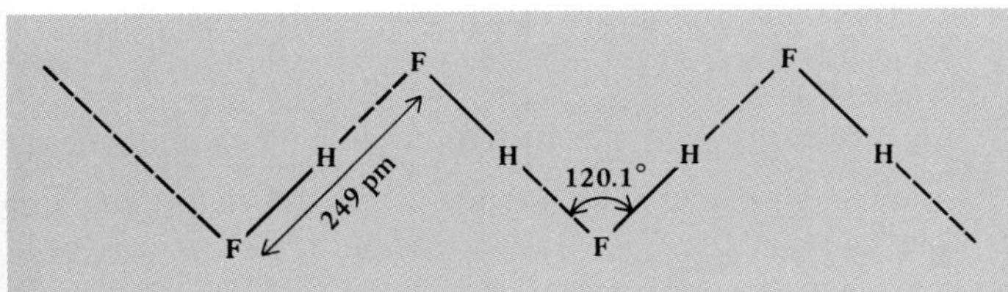

FIGURE 22.19
The structure of HF in the solid state. Hydrogen bonding accounts of a chain of HF molecules.

example,

$$Mg(s) + 2\ HCl(g) \longrightarrow MgCl_2(s) + H_2(g)$$

Many of these reactions are thermodynamically spontaneous but kinetically slow and must be speeded up by heating or by adding a catalyst. Some useful predictions can be made concerning the general reaction of a metal with a hydrogen halide.

$$M(s) + nHX(g) \longrightarrow MX_n + \tfrac{1}{2}\ H_2(g)$$

Since the standard free energies of formation of M and H_2 are zero, the net free energy for the reaction is

$$\Delta G^\circ_{net} = \Delta G^\circ_f(MX_n) - n\Delta G^\circ_f[HX(g)]$$

Consulting the table of HX properties above, we see that ΔG°_{net} will be negative (and the reaction thermodynamically spontaneous) if

$\Delta G^\circ_f(MF_n)$ is $< -273n$ kJ/mol for the fluoride

$\Delta G^\circ_f(MCl_n)$ is $< -95n$ kJ/mol for the chloride

$\Delta G^\circ_f(MBr_n)$ is $< -54n$ kJ/mol for the bromide

$\Delta G^\circ_f(MI_n)$ is $< \sim 2$ kJ/mol for the iodide

To make some predictions about halogenation reactions, you need only to consult tables of free energies of MX_n. It will then be clear that most metals will react spontaneously with most of the hydrogen halides. In the cases of the alkali metals, the alkaline earths, and Zn, Al, and the lanthanides, the reactions are quite exothermic. In other cases, HCl, HBr, and HI should react (for instance, with silver to produce AgCl, AgBr, and AgI), but HF will not. Silicon should react with all but HI, but in practice only HF reacts at room temperature; the other reactions are kinetically slow. Remember: when predicting reactivity, you must take both thermodynamics and kinetics into account.

EXERCISE 22.8 HALOGEN CHEMISTRY

(a) Name three uses each for fluorine and chlorine and their compounds.

(b) Is the reaction of Si(s) and $Cl_2(g)$ to give $SiCl_4(g)$ predicted to be thermodynamically spontaneous? ΔG°_f for $SiCl_4(g)$ is -617 kJ/mol.

HALOGEN OXYACIDS There are numerous oxyacids of the halogens, but only chlorine forms a complete series of such acids. They range from HOCl, with Cl in the formal $+1$ oxidation state, to $HClO_4$ where Cl is in the $+7$ state.

ACID	NAME	SALT	NAME
HOCl	Hypochlorous	OCl^-	Hypochlorite
HOClO	Chlorous	ClO_2^-	Chlorite
$HOClO_2$	Chloric	ClO_3^-	Chlorate
$HOClO_3$	Perchloric	ClO_4^-	Perchlorate

Hypohalous acids, HOX, are known for all of the halogens. For fluorine, the unstable HOF is the only known oxyacid, and it was only discovered a few years ago. In contrast, HOCl was discovered over two centuries ago in the original work on chlorine. This and the other hypohalous acids are made by the hydrolytic disproportionation of the halogen.

$$X_2(g) + H_2O(\ell) \rightleftharpoons H^+(aq) + X^-(aq) + HOX(aq)$$

Here, one halogen atom is oxidized (to HOX) and the other is reduced (to X^-). Generally, this equilibrium process is carried out in *cold* alkaline solutions so the H^+ ion is eliminated.

$$X_2(g) + 2\ OH^-(aq) \rightleftharpoons OX^-(aq) + X^-(aq) + H_2O(\ell)$$

When Cl_2 is treated with NaOH, the resulting alkaline solution is the "liquid bleach" used in home laundries, while $Ca(OCl)_2$ is the product when calcium hydroxide (slaked lime) is used. The latter compound is an easily handled form of a powerful oxidant, so it is sold for swimming pool sterilization.

Hypohalites, the OX^- ions, are formed when the halogens disproportionate in cold alkaline solution. If the solution is *hot*, however, disproportionation leads to the halogen in the +5 oxidation state, and the product is a halate, a salt of a halic acid.

$$3\ X_2(g) + 6\ OH^-(aq) \longrightarrow XO_3^-(aq) + 5\ X^-(aq) + 3\ H_2O(\ell)$$

Sodium and potassium chlorates are made in huge quantities to fulfill a variety of needs. The sodium salt, for example, is largely reduced to ClO_2 for bleaching paper pulp. Some of the remainder is oxidized to perchlorates (see below) or is used as a herbicide. And finally, some $NaClO_3$ is converted to the potassium chlorate, the preferred oxidizer in fireworks (page 779) and a crucial component in the head of safety matches.

Nearly 12,000 tons of $KClO_3$ are consumed annually in the United States to make matches.

The most stable halogen compounds are those with the halogen in the lowest (−1) or highest (+7) oxidation states. Consequently, perchlorates (ClO_4^-) are the most stable oxychlorine compounds, although they remain *powerful* oxidants. *Pure* perchloric acid, $HClO_4$, is a colorless liquid that will explode if shocked. It explosively oxidizes organic materials and even rapidly oxidizes silver and gold. More dilute aqueous solutions of the acid, however, have very little oxidizing power.

Industrially, perchlorates are produced exclusively by the electrolysis of aqueous sodium chlorate, $NaClO_3$. All other perchlorate salts, and perchloric acid, are made from the sodium salt. Perchlorate salts of most metals in the periodic table are known. Although many are relatively stable, others are dangerously unpredictable and *great care should be used when handling any perchlorate salt.* Ammonium perchlorate, for example, bursts into flame if heated above 200°C.

$$2\ NH_4ClO_4(s) \longrightarrow N_2(g) + Cl_2(g) + 2\ O_2(g) + 4\ H_2O(g)$$

This property of the ammonium salt is put to use as the oxidizer in the solid booster rockets for the Space Shuttle. The solid propellent in these rockets is largely NH_4ClO_4, the remainder being the reducing agent, powdered aluminum. Each shuttle launch requires about 700 tons of ammonium perchlorate, so more than half of the sodium perchlorate manufactured is converted to the ammonium salt. The process for doing this is a simple exchange reaction (Chapter 3)

$$NaClO_4(aq) + NH_4Cl(aq) \longrightarrow NaCl(aq) + NH_4ClO_4(s)$$

that takes advantage of the fact that ammonium perchlorate is less soluble in water than sodium perchlorate.

22.4 THE CHEMISTRY OF THE RARE GASES

At the right side of the periodic table is the group of elements that all exist as monatomic gases. Helium, the lightest element of the group, is certainly not rare: it is the second most abundant element in the universe after hydrogen. However, it is rare on earth, because the gravitational field of our planet is too small to retain this light gas. Altogether, the gases of the group make up about 1% of the earth's atmosphere, so we have come to call them the ''rare gases.'' Alternatively, they have been called the ''inert gases,'' but, since the discovery of xenon compounds in 1962, only the lighter elements are considered truly inert.

William Ramsay (1842–1919). With Lord Raleigh, Ramsay made the first discovery of a rare gas, argon (1894). In the next four years, Ramsay also discovered helium, neon, krypton, and xenon. (AIP Niels Bohr Library)

In the 1890s, Lord Raleigh measured the density of some common gases and found that a sample of N_2 from the air had a slightly different density than N_2 prepared from ammonia. When the N_2 was removed from the sample of ''atmospheric nitrogen'' by forming magnesium nitride,

$$3\,Mg(s) + N_2(g) \longrightarrow Mg_3N_2(s)$$

Raleigh and William Ramsay found a very small amount of a much denser, monatomic gas that they named *argon* from the Greek word for ''lazy,'' an appropriate name for a chemically inert gas. Soon thereafter Ramsay suggested that this element and the recently discovered element helium were members of an entirely new group in the periodic table.

Argon is the most abundant Group 8A element in the atmosphere, but the other rare gases are present in minute quantities (Table 22.10). About 250,000 tons of argon are recovered annually from air in the United States, and the gas is used primarily as an inert atmosphere for high temperature metallurgical processes. Small amounts of neon, krypton, and xenon are also recovered from air, and neon especially is used to fill the ''neon'' signs used in advertising. Xenon has an interesting potential medical use. It is readily soluble in blood and acts as an inhalation anesthetic in much the same way as dinitrogen oxide or ''laughing gas'' (N_2O).

TABLE 22.10 Composition of Dry Air

GAS	PERCENT BY VOLUME
N_2	78.03
O_2	20.99
Ar	0.93
CO_2	0.03
Ne	0.0015
H_2	0.0010
He	0.0005
Kr	0.0001
Xe	0.000008

Helium is usually obtained from natural gas wells in which it is found in concentrations up to 7% by volume. Among other things, helium is used for inert atmospheres and, along with oxygen, as a gas in deep-sea diving gases (see Chapter 6). Its normal boiling point is only 4.2K, so liquid helium is the coldest liquid refrigerant available and is used to study very low temperature phenomena.

The discovery of the first rare gas compound provides us with the lesson that the most interesting discoveries are often the result of lucky accidents and of being prepared to interpret an observation in a new way.

While studying the chemistry of PtF_6, Neil Bartlett noticed, quite accidentally, that exposing it to air led to a compound that he showed to be $[O_2^+][PtF_6^-]$. Just as importantly, Bartlett recognized that the ionization energy of O_2 (to form O_2^+) was comparable to that of Xe. Thus, he quickly proceeded to treat PtF_6 with Xe and discovered $Xe^+PtF_6^-$, the first rare gas compound. Very soon after this work, other xenon fluorides were discovered, and the field has been an active one since. Isolable compounds have been obtained only for Xe and Kr, but, were it not for the intense radioactivity of Rn, compounds of this element would surely be isolated. Virtually all the known compounds contain bonds to F or O, but a few exist with bonds to C, Cl, and N.

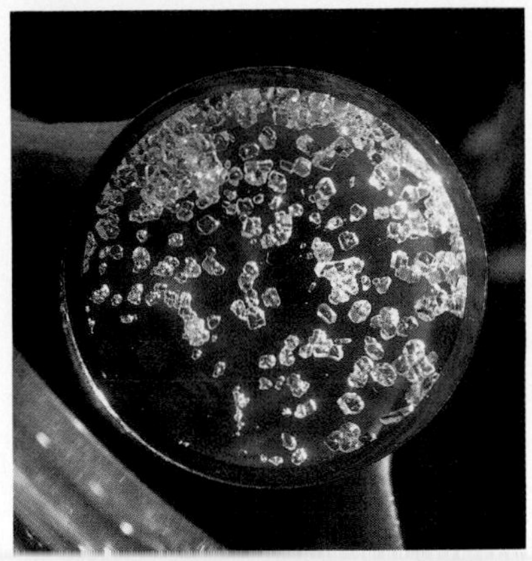
Crystals of xenon tetrafluoride, XeF_4. (Argonne National Laboratory)

Among the best known rare gas compounds are the first discovered, the simple xenon fluorides, XeF_2, XeF_4, and XeF_6. All are colorless, volatile solids that can be prepared by combining the elements under carefully controlled conditions. It is ironic, in view of the historical idea of the inertness of these elements, that XeF_2 can be obtained simply by exposing a mixture of Xe and F_2 to sunlight.

$$Xe(g) + F_2(g) \xrightarrow{\text{sunlight}} XeF_2(g)$$

All of the xenon fluorides are good fluorinating agents.

$$XeF_4(g) + 2\ SF_4(g) \longrightarrow 2\ SF_6(g) + Xe(g)$$

The difluoride is stable in water, but the tetrafluoride and hexafluoride are both hydrolyzed

$$XeF_4(aq) + 2\ H_2O(\ell) \longrightarrow \tfrac{1}{3}\ XeO_3(s) + \tfrac{2}{3}\ Xe(g) + \tfrac{1}{2}\ O_2(g) + 4\ HF(aq)$$

to the dangerously explosive compound XeO_3.

XeF_2	XeF_4
mp, 129°C	mp, 117.1°C
Xe—F = 200 pm	Xe—F = 195.2 pm
:F—Xe—F:	F, F / Xe / F, F
linear	square planar
Xe = sp^3d	Xe = sp^3d^2

EXERCISE 22.9 XENON CHEMISTRY

Using the ideas of VSEPR (Chapter 9), explain why XeF_2 has a linear molecular structure.

SUMMARY

The elements of **Group 5A** all have the configuration ns^2np^3, so oxidation numbers of the elements can range from −3 to +5. Only nitrogen is found uncombined in nature. Phosphorus is found in the form of *orthophosphates*, and the others (As, Sb, and Bi) occur as oxides and sulfides. Elemental phosphorus has three *allotropic forms* denoted by the color of the solid: white (or yellow), red, and black. White phosphorus is a tetrahedron of P atoms.

All Group 5A elements form hydrogen compounds of the type EH_3, of which *ammonia*, NH_3, is most important commercially. Ammonia is produced in the *Haber process*, the direct, catalyzed combination of N_2 and H_2.

Unlike the extensive *catenation* observed for boron and carbon, N—N bonding occurs only to a limited extent. The simplest N—N bonded compound, hydrazine (N_2H_4), is produced in the *Raschig process*. It is an excellent reducing agent.

Seven binary N—O compounds are known (Table 22.2), among them "laughing gas," N_2O, and the brown gas nitrogen dioxide, NO_2. The most common phosphorus oxide is P_4O_{10}, while the formula E_2O_3 is typical of As, Sb, and Bi.

As expected, nitrogen and phosphorus oxides are acidic, arsenic and antimony oxide are amphoteric, and bismuth oxide, Bi_2O_3, is basic. Nitric acid, HNO_3, is produced by the oxidation of ammonia in the *Ostwald process*, while orthophosphoric acid, H_3PO_4, is produced by hydrolysis of P_4O_{10}, the latter a product of oxidation of P_4. Orthophosphoric acid has many uses (Figure 22.8).

Group 6A elements, with the electron configuration ns^2np^6, are almost all nonmetals. Oxygen has two allotropic forms, O_2 and O_3 (ozone). Of the allotropes of sulfur, the most common is an eight-membered ring of sulfur atoms. Elemental sulfur is obtained from deep mines or by oxidation of hydrogen sulfide. Sulfur has many uses (Figure 22.12); one of the most important of these is its conversion to sulfuric acid, the chemical produced in largest amount in industrialized countries.

Oxygen forms compounds directly or indirectly with virtually all of the elements (except for some of the rare gases). Most are oxides; some *peroxides* (O_2^{2-}) and *superoxides* (O_2^-) are also known. The most important peroxide is H_2O_2, a powerful oxidizing agent.

Many metals readily form sulfides (Figure 22.14), compounds of great commercial interest. Other important sulfur compounds are hydrogen sulfide (H_2S), SO_2, and SO_3. Sulfur dioxide is the anhydride of sulfurous acid, H_2SO_3, and the trioxide is the anhydride of sulfuric acid, H_2SO_4.

The halogens of **Group 7A** have the electron configuration ns^2np^5 and all form the halide ion, X^- (oxidation number = −1). None of the elements occurs naturally. Both fluorine and chlorine are prepared by electrolysis. Bromine and iodine, however, are prepared by adding Cl_2 (an oxidizing agent) to an aqueous solution of Br^- or I^- (a reducing agent).

$$Cl_2(aq) + 2\,Br^-(aq) \longrightarrow 2\,Cl^-(aq) + Br_2(aq)$$

Fluorine, in the form of F_2, is the most reactive of all of the elements; it forms compounds with all of the elements except He, Ne, and Ar.

Among the best known halogen compounds are the hydrogen halides, HX. All are gases at room temperature and pressure, but strong hydrogen bonding gives HF a much higher boiling point than the others. A number of oxygen–halogen compounds are known, but among the most familiar are oxyacids such as HOCl (hypochlorous acid) and $HClO_4$ (perchloric acid).

The lighter rare gases of **Group 8A** are chemically unreactive, but compounds of the heavier ones are known. Xenon in particular forms chemical compounds with highly electronegative elements such as F and O.

STUDY QUESTIONS

1. Many chemical reactions or industrial processes bear the name of the person who discovered the process or who worked out the details. Write balanced equations and give the reaction conditions for the following named processes: Haber, Marsh, and Raschig. What is the "contact process"?
2. Complete and balance the following equations:
   (a) $Mg(s) + N_2(g) \rightarrow$
   (b) $P_4(s) + KOH(aq) + H_2O(\ell) \rightarrow$
   (c) $CH_4(g) + H_2O(g) \rightarrow$
   (d) $HNO_3(aq) + KOH(aq) \rightarrow$
   (e) $NH_3(aq) + OCl^-(aq) \rightarrow$
   (f) $NH_4NO_3(s) + heat \rightarrow$
   (g) $NaNO_2(aq) + HCl(aq) \rightarrow$
   (h) $NH_3(g) + O_2(g) \rightarrow$
   (i) $Cu(s) + HNO_3(aq) \rightarrow$
   (j) $P_4O_{10}(s) + H_2O(\ell) \rightarrow$
3. Write a balanced equation to show the preparation of each of the following:
   (a) $NO_2(g)$
   (b) $N_2O_5(s)$
   (c) $HNO_2(aq)$
   (d) $AsH_3(g)$
   (e) NF_3
   (f) CCl_2F_2 (Freon-12)
   (g) "laughing gas," dinitrogen oxide
   (h) hydrazine
4. A major use of hydrazine, N_2H_4, is in steam boilers in power plants. (a) The reaction of hydrazine with O_2 dissolved in water gives N_2 and water. Write a balanced equation for this reaction. (b) O_2 dissolves in water to the extent of 3.08 cm^3 (gas at STP) in 100. mL of water at 20°C. In order to destroy all of the dissolved O_2 in 3.00×10^4 L of water (enough to fill a small swimming pool), how many grams of N_2H_4 are needed?
5. Before hydrazine came into use to remove dissolved oxygen in the water in steam boilers, Na_2SO_3 was commonly used. Assuming the appropriate reaction is

   $$Na_2SO_3(aq) + \tfrac{1}{2}\, O_2(aq) \longrightarrow Na_2SO_4(aq)$$

   how many grams of Na_2SO_3 would be required to remove O_2 from 30,000 L of water as outlined in Study Question 4?
6. The various steps in the industrial synthesis of nitric acid are given on page 827. By adding up the heats of the steps, show that the heat of the overall reaction is −412.7 kJ.

   $$NH_3(g) + 2\, O_2(g) \longrightarrow HNO_3(aq) + H_2O(\ell)$$

7. A common analytical method for hydrazine involves its oxidation with iodate, IO_3^-. In the process hydrazine behaves as a four-electron reducing agent.

   $$N_2(g) + 5\, H^+(aq) + 4e^- \longrightarrow N_2H_5^+(aq) \qquad E^\circ = -0.23 \text{ V}$$

   Write the balanced equation for the reaction of hydrazine in acid solution ($N_2H_5^+$) with IO_3^- to give N_2 and I_2. Calculate E° for this reaction.
8. If the lunar lander on the moon missions used 4.5 tons of a hydrazine derivative, $H_2N{-}N(CH_3)_2$ (page 823), how many tons of N_2O_4 are required to react with it? How many tons of each of the products of this reaction are generated?
9. Unlike carbon, which can form extended chains of atoms, nitrogen can form chains of very limited length. Draw the Lewis electron dot structure of the azide ion, N_3^-.
10. The cyanide ion, CN^-, can be oxidized to the cyanate ion, NCO^-. Compounds of this ion are quite stable, in contrast to those of the fulminate ion, CNO^-. Mercury(II) fulminate, $Hg(CNO)_2$, for example, explodes when struck and is used in blasting caps. Show the Lewis electron dot structures for the cyanate and fulminate ions. (Show all resonance structures.) For each ion, calculate the formal charge on each atom in each resonance structure and decide which is the most important resonance structure. (See Section 9.4 for a discussion of formal charges.)
11. The boiling points of the EH_3 molecules of Group 5A are given in Table 22.3 on page 819. Explain the observed trend in these temperatures.
12. Dinitrogen trioxide, N_2O_3, has the structure shown below.

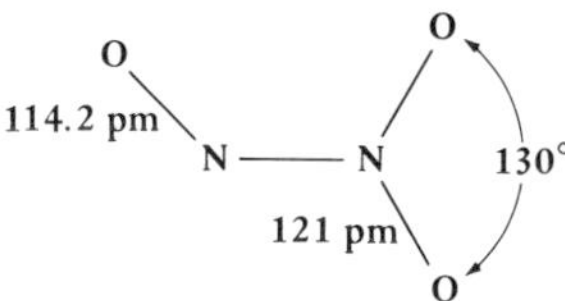

   The oxide is unstable, dissociating to NO and NO_2 in the gas phase at 25°C.

   $$N_2O_3(g) \rightleftharpoons NO(g) + NO_2(g)$$

   (a) Explain why one N—O bond distance in N_2O_3 is 114.2 pm, while the other two are shorter and nearly equal to one another.
   (b) For the dissociation reaction, $\Delta H^\circ = 40.5$ kJ/mol and $\Delta G^\circ = -1.59$ kJ/mol. Calculate ΔS° and K for the reaction, and ΔH_f° for N_2O_3.

13. The structure of HNO_3 is given on page 827. Explain why one NO bond is longer than the other two (which have the same bond length).
14. When trade journals discuss the amount of phosphoric acid produced or used annually, they always do so in terms of equivalent P_2O_5. Recall that P_2O_5 hydrolyzes to H_3PO_4, so the mass of anhydrous acid can be obtained by multiplying the given phosphorus content by the factor 1.380.
    (a) Show why the factor relating the mass of P_2O_5 to that of anhydrous H_3PO_4 is 1.380.
    (b) Estimated 1984 production of phosphoric acid (measured as P_2O_5) was 10.8 million tons. How many tons of H_3PO_4 does this represent?
    (c) If the prices are listed as $300/ton of P_2O_5, what is the price of 2.00×10^3 pounds (1.00 ton) of anhydrous H_3PO_4?
15. What is the approximate pH of a 0.010 *M* solution of Na_3PO_4?
16. Phosphonic acid (Table 22.5) has phosphorus in the formal oxidation state of +3. Draw the structure of diphosphonic acid, $H_4P_2O_5$. What is the maximum number of protons that this acid can dissociate in water?
17. $CaHPO_4$ is used as an abrasive in toothpaste. Write a balanced equation showing one possible preparation for this salt.
18. Sketch the structures of NH_3 and NF_3 and compare them on the following basis:
    (a) Each molecule has a dipole moment. In each molecule, toward which end does the negative portion of the dipole lie?
    (b) Give a possible explanation for the fact that the dipole moment of NH_3 is 1.47 D whereas that of NF_3 is only 0.234 D.
19. When base is added to a water suspension of P_4, several reactions are possible, and one of them produces the ion $H_2PO_2^-$ and the gas PH_3 (phosphine). Write a balanced equation to depict this reaction.
20. On page 831 it was noted that $H_2PO_2^-$ salts are excellent reducing agents in basic solution and can be used for the electrodeless plating of metals. Will the ion reduce a 1 *M* solution of a zinc salt to zinc metal in basic solution? What about 1 *M* Cr^{3+} to Cr? Write a balanced equation for each process and predict the net $E°$ for each reaction. (For the necessary potentials see Appendix K.)
21. Arsenic burns in air to give As_4O_6, a poisonous material with a garlic-like odor.
    (a) By analogy with phosphorus chemistry, draw an approximate structure for this molecule.
    (b) Again by analogy with phosphorus chemistry, predict the product of hydrolysis of As_4O_6. Draw an approximate structure of the product, and give the oxidation number of the arsenic in the product.
22. Elemental arsenic is attacked by dilute nitric acid to give H_2 and arsenious acid, H_3AsO_3. Write a balanced equation for this oxidation–reduction process.
23. Complete and balance the following equations:
    (a) $KClO_3(s) + \text{heat} \rightarrow$
    (b) $H_2S(g) + O_2(g) \rightarrow$
    (c) $Na(s) + O_2(g) \rightarrow$
    (d) $K(s) + O_2(g) \rightarrow$
    (e) $ZnS(s) + O_2(g) \rightarrow$
    (f) $SO_2(g) + H_2O(\ell) \rightarrow$
24. Write balanced equations for the preparation of each of the following:
    (a) hydrogen peroxide
    (b) potassium superoxide
    (c) sodium thiosulfate
    (d) sulfuric acid
25. Oxygen is an excellent oxidizing agent; $E°$ for the following reaction is 1.229 V.

$$O_2(g) + 4\,H^+(aq) + 4e^- \longrightarrow 2\,H_2O(\ell)$$

    (a) What is *E* when the pH is 1.00 (and $P_{oxy} = 1.0$ atm)?
    (b) What is *E* when the partial pressure of O_2 is the value normally found in air? (Use $P_{oxy} = 0.22$ atm and pH = 1.00.)
26. O_2 has a relatively high bond energy, but it is a reactive molecule. Account for its great reactivity.
27. The structure of H_2O_2, hydrogen peroxide, is not planar. Suggest a possible reason for this observation.
28. Hydrogen peroxide is a weak acid in aqueous solution. Write a balanced chemical equation showing how H_2O_2 can function as a Brønsted acid in water. If $K_a = 1.78 \times 10^{-12}$ (for the loss of one H^+), what is the pH of a 0.10 *M* aqueous solution of the peroxide?
29. The standard heat of formation of OF_2 gas is 24.5 kJ/mol. Calculate the average O—F bond energy.
30. Place the following oxides in order of increasing basicity: As_2O_3, Ga_2O_3, GeO_2, BrO_2, CaO, and K_2O.
31. Which is more basic, BeO or CaO? Write a balanced chemical equation for the hydrolysis of SrO.
32. In addition to the simple sulfide ion, S^{2-}, there are polysulfides, S_n^{2-}. (Such ions are chains of sulfur atoms, not rings.) Draw a Lewis electron dot structure of the S_3^{2-} ion.
33. Sulfur forms a range of compounds with fluorine. Draw Lewis electron structures for S_2F_2 (connectivity is FSSF), SF_2, SF_4, and SF_6. What is the formal oxidation number of sulfur in each of these compounds?
34. In the modern "contact process" for making sulfuric acid, sulfur is first burned to SO_2. Environmental restrictions allow no more than 0.3% of this SO_2 to be vented to the atmosphere.

(a) If enough sulfur is burned in a plant to produce 2.00×10^3 tons of pure, anhydrous H_2SO_4 per day, how much SO_2 is allowed to be exhausted to the atmosphere?

(b) One way to prevent even this much SO_2 from reaching the atmosphere is to "scrub" the exhaust gases with hydrated lime.

$$Ca(OH)_2 + SO_2 \longrightarrow CaSO_3 + H_2O$$

$$2\ CaSO_3 + O_2 \longrightarrow 2\ CaSO_4$$

How many tons of $Ca(OH)_2$ are needed to remove the SO_2 calculated in part (a)?

35. A sulfuric acid plant produces an enormous amount of heat. To keep costs as low as possible, much of this heat is used to make steam to generate electricity. Some of the electricity is used to run the plant, and the excess is sold to the local electrical utility. Three reactions are important in sulfuric acid production: (1) burning S to SO_2; (2) oxidation of SO_2 to SO_3; and (3) hydrolysis of SO_3.

$$SO_3(g) + H_2O \text{ (in 98\% } H_2SO_4) \longrightarrow H_2SO_4$$

If the heat of the third reaction is -130 kJ/mol, estimate the total heat produced per mole of H_2SO_4 produced. How much heat is produced per ton of H_2SO_4?

36. Complete and balance the following equations:
(a) $UF_4(s) + F_2(g) \rightarrow$
(b) $Br^-(aq) + Cl_2(aq) \rightarrow$
(c) $I^-(aq) + Br_2(aq) \rightarrow$
(d) $Br^-(aq) + AgNO_3(aq) \rightarrow$
(e) $Cl_2(g) + OH^-(aq) \rightarrow$

37. Write a balanced equation for the preparation of each of the following:
(a) HF(g)
(b) HCl(g)
(c) xenon tetrafluoride

38. If aqueous Br_2 is mixed with an aqueous solution of KI, what reaction will occur? Write a balanced chemical equation.

39. What is the value of $E°$ for the reaction between

40. $Cl_2(aq)$ and $I^-(aq)$.
(a) Name three mineral sources of fluoride ion.
(b) What is the usual commercial source of the iodide ion?

41. If an electrolytic cell for producing F_2 (Figure 22.15) operates at 5.00×10^3 amps (at 10.0 V), how many tons of F_2 can be produced per 24 hour day if the conversion of F^- to F_2 is assumed to be 100% efficient?

42. The annual U.S. and Canadian fluorine production is 5.00×10^3 tons (one ton is 2.00×10^3 pounds), and 70% of this is used to manufacture UF_6. How much uranium(VI) fluoride is manufactured every year (assume 100% efficiency in using the F_2)?

43. The F—F bond energy (page 851) is among the weakest of the halogens. Suggest a possible reason for this.

44. Suggest two reasons for the observation that F_2 is the most reactive of the halogens.

45. Halogens combine with one another to form *interhalogens* such as BrF_3. Sketch a possible molecular structure for this molecule and tell whether the observed F—Br—F bond angles will be less than or greater than ideal.

46. Why is the ClO_2 molecule paramagnetic? Sketch its electron dot structure.

47. Metal halides are soluble in water, the relative solubility order generally being $MF_n < MCl_n < MBr_n < MI_n$ (where M = K^+ or Ca^{2+}, for example). The melting points of these same halides increase in the opposite order. Suggest a common reason for these observations.

48. Halogens form polyhalide ions. Sketch Lewis electron dot structures, and molecular structures, for (a) I_3^-, (b) $BrCl_2^-$, (c) ClF_2^+.

49. The halogen oxides and oxyions are generally good oxidizing agents. For example, the bromate ion oxidizing half reaction has an $E°$ value of 1.495 V in acid solution.

$$BrO_3^-(aq) + 6\ H^+(aq) + 6e^- \longrightarrow Br^-(aq) + 3\ H_2O(\ell)$$

Can you oxidize aqueous 1 *M* Mn^{2+} to aqueous MnO_4^- with 1 *M* bromate ion?

50. The hypohalite ions, OX^-, are the salts of weak acids. Calculate the pH of a 0.10 *M* solution of NaOCl. What is the concentration of HOCl in this solution?

51. Ammonium perchlorate, NH_4ClO_4, is used as an oxidizer in launching the Space Shuttle. If one launch requires 700. tons of the salt (one ton is 9.08×10^5 grams), and the salt decomposes according to the equation on page 855, how many tons of water are produced? How many grams of O_2 are produced? If the O_2 produced is assumed to react with the powdered aluminum present in the rocket engine, how much Al is necessary to use up the O_2 and how much Al_2O_3 is formed in the Al/O_2 reaction?

52. Consider the reaction of a metal with a hydrogen halide (page 854).
(a) Which of the following metals are predicted to react spontaneously with HCl: Ba, Pb, Hg or Ti?
(b) Will HF react spontaneously with Ca?
(c) Will Ni react spontaneously with HF (ΔG_f° of $NiF_2(s)$ is -604 kJ/mol)?

(d) Will Ni react spontaneously with HBr (ΔG_f° of $NiBr_2(s)$ is -253.6 kJ/mol)?
(e) Will B react spontaneously with HI (ΔG_f° of $BI_3(g)$ is $+20.7$ kJ/mol)?

53. Of the rare gases, xenon is the only one that forms an extensive series of compounds.
(a) Sketch the Lewis electron dot structure of XeF_4. What is its molecular structure?
(b) Sketch a Lewis electron dot structure for the ion XeF_3^+. What is its molecular structure?

54. XeF_6 hydrolyzes to form $XeOF_4$ and HF.
(a) Write a balanced chemical equation for the hydrolysis reaction.
(b) Sketch the electron dot and molecular structure of $XeOF_4$. (*Hint:* You can think of this molecule as resulting from the addition of an O atom to XeF_4.)

55. XeO_3 is a powerful oxidizing agent in acid solution.

$$XeO_3(aq) + 6\,H^+(aq) + 6e^- \longrightarrow Xe(g) + 3\,H_2O \qquad E^\circ = 2.10\text{ V}$$

Is it capable of oxidizing Mn^{2+} to MnO_4^- under standard conditions? Can it oxidize NaF to F_2? (See Appendix K.)

23
Organic Chemistry

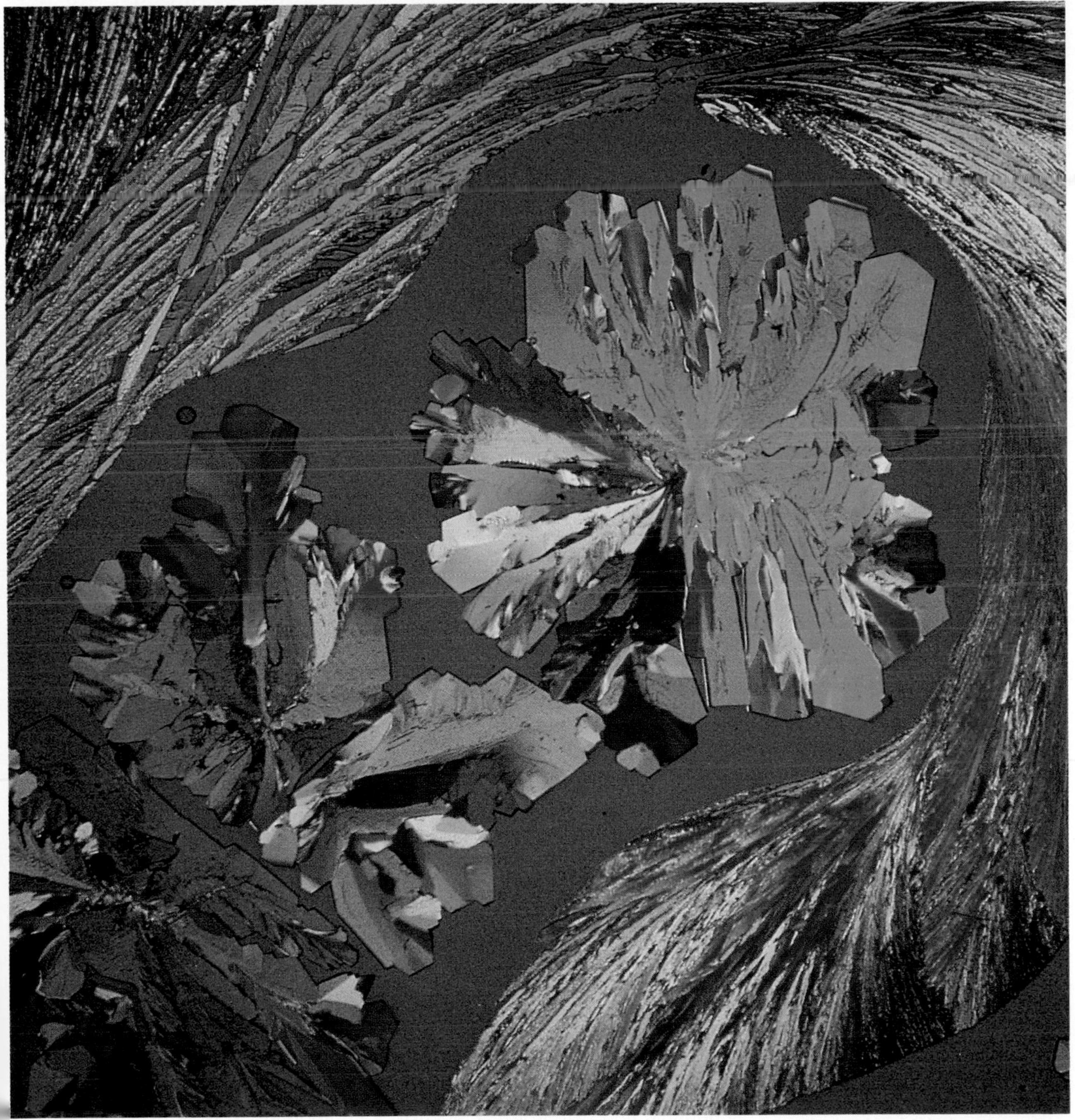

Polarized light micrograph of citric acid crystals (Jan Hinsch/Science Photo Library/Photo Researchers)

CHAPTER OUTLINE

Organic chemistry is the study of carbon compounds. Although it may seem surprising that the study of compounds of a single element requires a separate branch of chemistry, it is more reasonable when you realize that over 90% of all of the compounds ever prepared contain carbon.

Why are there so many compounds of carbon? It is largely the result of the unique ability of carbon atoms to bond together to form rings or chains of almost unlimited length and at the same time form bonds to atoms of a large number of other elements.

Until about the middle of the 19th century, it was generally believed that organic compounds were different from other chemical compounds in that organic compounds could only be formed by living organisms and that they contained a ''vital force'' somehow associated with the life process. This vital force theory was gradually abandoned after 1828, however, when Friedrich Wöhler found that urea, a compound present in human urine, could be synthesized from the purely inorganic salt ammonium cyanate.

$$\underset{\text{ammonium cyanate}}{NH_4OCN(aq)} + \text{heat} \longrightarrow \underset{\text{urea}}{H_2N{-}\overset{\overset{\displaystyle O}{\|}}{C}{-}NH_2}$$

Soon thereafter many other ''natural'' compounds such as methyl alcohol, ethyl alcohol, and acetic acid, as well as compounds not found in nature, were synthesized by other chemists.

The synthesis of new organic compounds has continued at an ever increasing pace since the middle of the 19th century. Of the organic compounds known today, only a few can be found in nature. These synthetic efforts have been directed to the production of (a) new compounds with properties superior to those of naturally occurring compounds, (b) compounds with properties that cannot be found in naturally occurring compounds, and (c) naturally occurring compounds in quantities far greater than can be obtained efficiently from natural sources.

FRIEDRICH WÖHLER (1800–1882)

It has been said that Friedrich Wöhler was the last great "all rounder" in chemistry. He began his work as a boy, preparing oxygen, extracting phosphorus, and isolating potassium in the bedroom of his home in Frankfurt, Germany. After his graduation in medicine in 1823, he went to work for the great chemist Berzelius in Stockholm and began a lifetime of contributions to both organic and inorganic chemistry. It may well be a chemical myth that Wöhler was the first to prepare urea from inorganic sources, but he had many admirers who claimed for him many doubtful priorities. When he died in 1882, his obituary said, "As a colleague, a teacher, and in every relation of private life, Wöhler was esteemed and beloved for his kindliness and geniality of his disposition, his modesty and uprightness." (*Platinum Metals Review*, 29:81, 1985.)

The development of new synthetic methods of production is a major area of organic chemistry. In addition, organic chemists study the mechanisms of organic reactions and develop methods for determining the structures and other properties of organic compounds.

23.1 SOME PRELIMINARIES: STRUCTURAL FORMULAS AND BONDING

STRUCTURAL FORMULAS The molecular formula of a compound is a listing of the elements of the molecule, usually in alphabetic order. Thus, the molecular formula of ethyl alcohol is C_2H_6O and that of sucrose is $C_{12}H_{22}O_{11}$. Unfortunately, these give us relatively little information. For example, dimethyl ether has the same molecular formula as ethyl alcohol, but the compounds function chemically in very different ways. As for sucrose, over 100 different compounds have the same formula! For this reason, chemists routinely use **structural formulas** to represent organic compounds. The differences between ethyl alcohol and dimethyl ether, for example, are clearly evident from their structural formulas.

$$\text{Structural formula:}\quad \begin{array}{ccccccc} & & \mathrm{H} & & \mathrm{H} & & \\ & & | & & | & & \\ \mathrm{H} & - & \mathrm{C} & - & \mathrm{C} & - & \mathrm{O}-\mathrm{H} \\ & & | & & | & & \\ & & \mathrm{H} & & \mathrm{H} & & \end{array} \qquad \begin{array}{ccccccccc} & & \mathrm{H} & & & & \mathrm{H} & & \\ & & | & & & & | & & \\ \mathrm{H} & - & \mathrm{C} & - & \mathrm{O} & - & \mathrm{C} & - & \mathrm{H} \\ & & | & & & & | & & \\ & & \mathrm{H} & & & & \mathrm{H} & & \end{array}$$

$$\text{Condensed formula:}\quad CH_3CH_2OH \qquad\qquad CH_3OCH_3$$

ethyl alcohol — dimethyl ether

For convenience, the lines representing single bonds between atoms are usually omitted, and condensed formulas are written.

BONDING The Pauling electronegativity of carbon (2.5) (Table 9.5) is roughly intermediate between that of the alkali metals and the halogens, and it is similar to that of elements to which carbon is often bonded (H, O, N, S, P, and the halogens). Thus, while the bonds between carbon and other elements are frequently polar, they are highly covalent.

23.2 THE ALKANES

Petroleum and natural gas are composed primarily of alkanes. Consequently, alkanes are important to us as a source of energy and as raw material in the industrial production of other organic compounds.

Hydrocarbons are compounds containing only carbon and hydrogen, an the first class of such compounds to be considered is the **alkanes**. I alkanes, any one carbon atom is bonded to four other atoms (either C c H). Since four sigma bonds are the maximum number of bonds carbo can form, alkanes are often called **saturated** compounds. Later, we sha study **unsaturated** hydrocarbons, compounds that contain double and tr ple C–C bonds.

The simplest alkane, **methane** (CH_4) is most familiar as the princip constituent of natural gas. As outlined in Chapters 9 and 10, the molecu has a tetrahedral structure with an sp^3 hybridized carbon atom.

Methane, CH_4

structural formula "ball and stick" model

As suggested by the VSEPR theory (Chapter 9), a similar tetrahedr arrangement is found for all carbon atoms in all organic compounds i which a carbon atom is attached to four other groups. All carbon atom in alkanes are sp^3 hybridized.

Ethane (C_2H_6) and propane (C_3H_8) are the two- and three-carbo alkanes, respectively. Like methane, they are gases at room temperatur Propane is the principal constituent of liquefied petroleum gas (LPG c bottled gas). You should notice particularly that the tetrahedral geometr of the central carbon atom in propane results in a nonlinear carbon chair

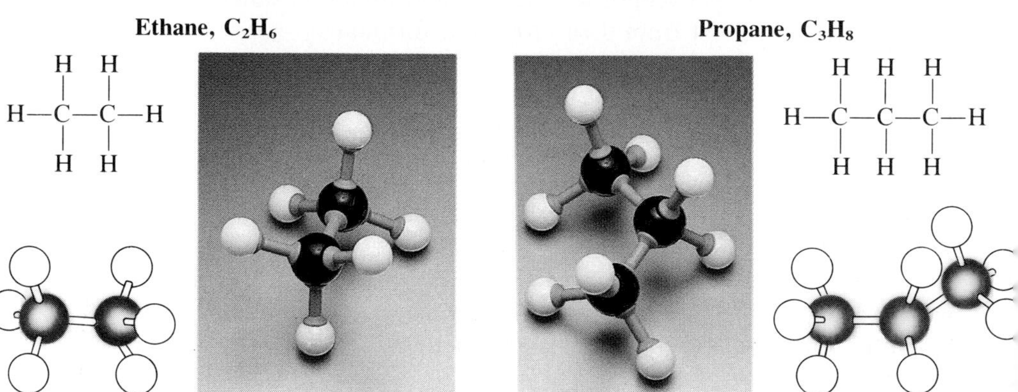
Ethane, C_2H_6 Propane, C_3H_8

This is always the case when three or more tetrahedral carbon atoms ar connected in a chain or ring.

It is often useful to refer to fragments of alkane structures tha result from the removal of a hydrogen from an alkane. These fragment are called **alkyl groups**. For example, removal of H from methane gives a methyl group,

```
     H                  H
     |       -H         |
 H—C—H   ———→   H—C—
     |                  |
     H                  H
 methane          methyl group
```

and ethane gives an **ethyl** group.

```
     H  H                 H  H
     |  |       -H        |  |
 H—C—C—H   ———→   H—C—C—
     |  |                 |  |
     H  H                 H  H
  ethane            ethyl group
```

Two different alkyl groups can be derived from propane,

```
            H   H   H
            |   |   |
        H—C3—C2—C1—H
            |   |   |
            H   H   H
             propane

 − H from C1                       − H from C2

     H  H  H                  H  H  H
     |  |  |                  |  |  |
 H—C—C—C—              H—C—C—C—H
     |  |  |                  |     |
     H  H  H                  H     H
  normal-propyl                isopropyl
    n-propyl
```

where the labels *n* and *iso* are explained just below. The names of alkyl groups come from the parent alkane by dropping "-ane" and adding "-yl."

STRUCTURAL ISOMERS

There are two alkanes with four carbon atoms, one having a straight chain and the other a branched chain.

n-butane, $CH_3CH_2CH_2CH_3$
mp, −138.3°C; bp, −0.50°C

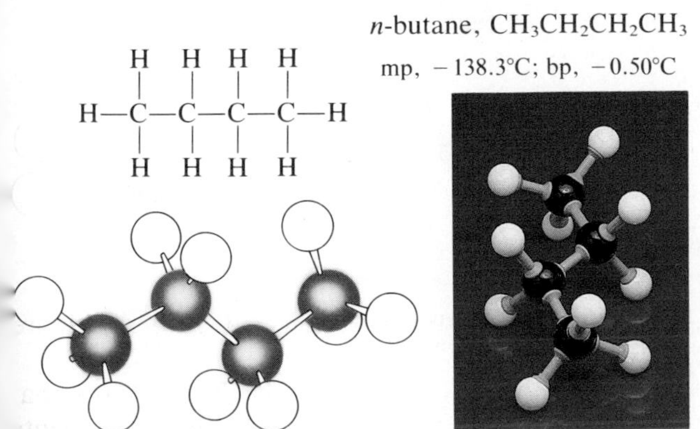

*iso*butane, CH_3CHCH_3 (with CH_3 on the central C)
mp, −159.6°C; bp, −11.7°C

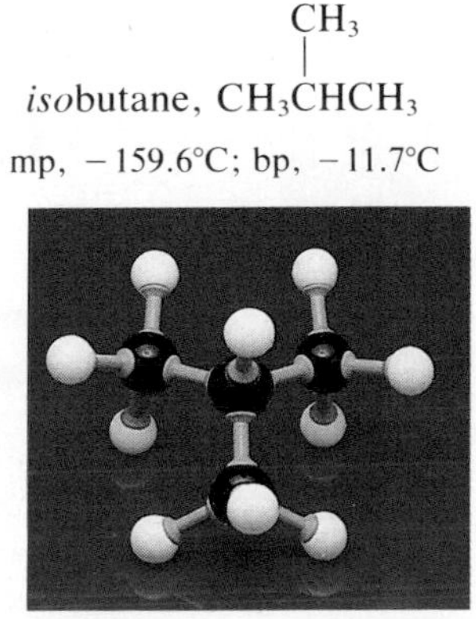

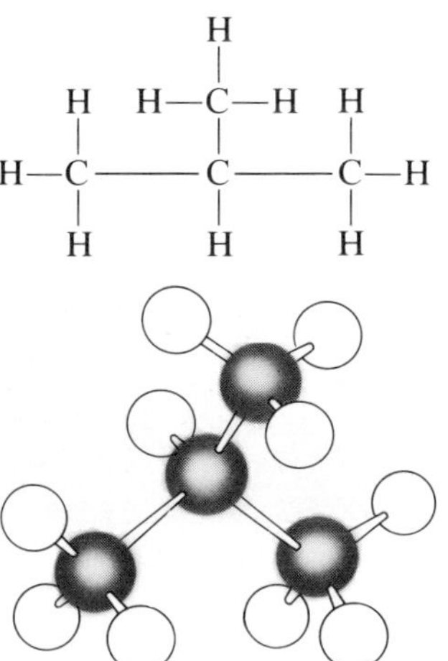

These two compounds are **isomers** of one another. They have the same molecular formulas, C_4H_{10}, but they have different atom arrangements. Thus, we call them more specifically **structural isomers**. The straight-chain isomer is called *n*-butane where the *n* stands for "normal," the term always applied to the straight-chain isomer of any hydrocarbon. The branched-chain *iso*mer is *iso*butane.

OTHER ALKANES

A very large number of carbon atoms can be bonded into a chain. Beyond butane, you encounter these and others.

HYDRO-CARBON	MOLECULAR FORMULA	STRUCTURAL FORMULA
n-Pentane	C_5H_{12}	$CH_3CH_2CH_2CH_2CH_3$
n-Hexane	C_6H_{14}	$CH_3CH_2CH_2CH_2CH_2CH_3$
n-Heptane	C_7H_{16}	$CH_3CH_2CH_2CH_2CH_2CH_2CH_3$
n-Octane	C_8H_{18}	$CH_3CH_2CH_2CH_2CH_2CH_2CH_2CH_3$
n-Nonane	C_9H_{20}	$CH_3CH_2CH_2CH_2CH_2CH_2CH_2CH_2CH_3$
n-Decane	$C_{10}H_{22}$	$CH_3CH_2CH_2CH_2CH_2CH_2CH_2CH_2CH_2CH_3$

To this point, we have written formulas for alkanes having one to ten carbons. It is important to notice that all of these compounds fit the general formula C_nH_{2n+2} where n is the number of carbon atoms. For example,

	n	$2n+2$
CH_4	1	4
C_5H_{12}	5	12
$C_{10}H_{22}$	10	22

On this basis, you can predict that an alkane with 15 carbon atoms would have the molecular formula $C_{15}H_{2(15)+2} = C_{15}H_{32}$.

With more complex alkanes, the number of structural isomers increases rapidly as the number of carbon atoms increases. For example, there are 3 structural isomers of pentane, 5 of hexane, 9 of heptane, 18 of octane, 75 of decane, and 366,319 structural isomers of $C_{20}H_{42}$!

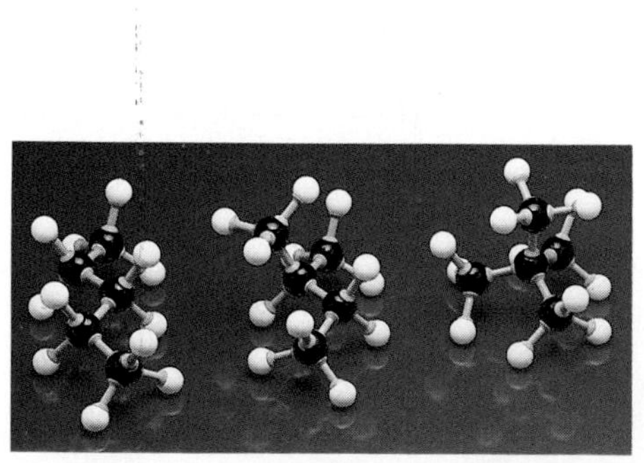
Pentane isomers. (Charles D. Winters)

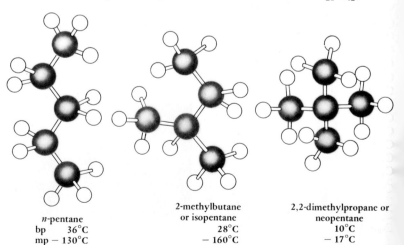

	n-pentane	2-methylbutane or isopentane	2,2-dimethylpropane or neopentane
bp	36°C	28°C	10°C
mp	− 130°C	− 160°C	− 17°C

EXERCISE 23.1 FORMULAS AND STRUCTURAL ISOMERS
(a) Draw the structural isomers of hexane, C_6H_{14}.
(b) Write the molecular formula for the alkane with 22 carbon atoms.

PHYSICAL PROPERTIES OF ALKANES

The physical properties of a few alkanes are listed in Table 23.1. Notice that alkanes having 1 to 4 carbon atoms are gases at room temperature, while those with 5 to about 16 carbons are liquids at room temperature and normal atmospheric pressure. Thus, natural gas, consisting mostly of methane, is transported within the United States as a gas in pipelines. Butane, although a gas at room temperature, has a relatively high critical temperature (Table 11.5) and so can be liquefied easily.

Camping stoves often use liquid butane; liquid alkanes, both normal and isomeric, are the principal components of various petroleum products such as gasoline, kerosene, fuel oil, and diesel oil (Section 23.13).

NAMING ALKANES

Common names and *systematic names* are used for alkanes in particular and for organic compounds in general. Many of the names we have used to this point are **common names** that simply came into use over the years. With the enormous number of organic compounds, however, it is apparent that it would be a problem learning many common names. Thus, we use *systematic names,* which can be derived from a simple set of rules.

The **systematic names** of alkanes are based on the number of carbon atoms in the *longest continuous* carbon chain. If that chain contains three carbons, the parent name is propane, if four carbons the parent name is butane, and so on. The remaining parts of the structure are treated as

After butane, the number of carbon atoms in the chain is given by the prefix to the systematic name: *pent* for 5, *hex* for 6, *hept* for 7, *oct* for 8, and so on.

TABLE 23.1 Some Normal Hydrocarbons (Alkanes)

MOLECULAR FORMULA	NAME	BOILING POINT (°C)	MELTING POINT (°C)	STATE AT ROOM TEMP.
CH_4	Methane	−161	−184	
C_2H_6	Ethane	−88	−183	Gas
C_3H_8	Propane	−42	−188	
C_4H_{10}	*n*-Butane	−0.5	−138	
C_5H_{12}	*n*-Pentane	36	−130	
C_6H_{14}	*n*-Hexane	69	−94	
C_7H_{16}	*n*-Heptane	98	−91	
C_8H_{18}	*n*-Octane	126	−57	
C_9H_{20}	*n*-Nonane	150	−54	
$C_{10}H_{22}$	*n*-Decane	174	−30	
$C_{11}H_{24}$	*n*-Undecane	194.5	−25.6	Liquid
$C_{12}H_{26}$	*n*-Dodecane	214.5	−9.6	
$C_{13}H_{28}$	*n*-Tridecane	234	−6.2	
$C_{14}H_{30}$	*n*-Tetradecane	252.5	+5.5	
$C_{15}H_{32}$	*n*-Pentadecane	270.5	10	
$C_{16}H_{34}$	*n*-Hexadecane	287.5	18	
$C_{17}H_{36}$	*n*-Heptadecane	303	22.5	
$C_{18}H_{38}$	*n*-Octadecane	317	28	
$C_{19}H_{40}$	*n*-Nonadecane	330	32	
$C_{20}H_{42}$	*n*-Eicosane	205 (at 15 torr)	36.7	Solid

substituents on the chain and are indicated as prefixes in the name. Numbers are used to indicate the positions of the substituents on the parent carbon chain.

The alkyl group derived from *iso*butane by removing H from the center C atom is called *tert*-butyl. *Tert* stands for *tertiary*, which means that the central C atom is bonded to three other C atoms.

*Iso*butane is the common name of one of the structural isomers of C_4H_{12}. The longest continuous chain of carbons is three atoms in length, so the systematic name is based on propane. Finally, since a methyl group appears on the second carbon, the correct name is 2-methylpropane.

```
      CH3
 1   2|  3
CH3—C—CH3
      |
      H
```

Common name: *iso*butane
Systematic name: 2-methylpropane

One of the isomers of pentane is 2-methylbutane. The parent chain, a four-carbon butane chain, is numbered beginning at the end nearer the substituent group. Therefore, the methyl group is indicated as being attached to carbon atom number 2.

```
      CH3
      |
CH3—CH—CH2—CH3
 1    2    3    4
```

Systematic name: 2-methylbutane

The presence of two or more identical substituent groups is indicated by the prefixes "di-, tri-, tetra-," and so on, and the position of each substituent is indicated by a number in the prefix. For example, both methyl groups in 2,2-dimethylbutane are attached to carbon atom 2 of a butane chain.

```
      CH3
 1   2|  3     4
CH3—C—CH2—CH3
      |
      CH3
```

Systematic name: 2,2-dimethylbutane

Be careful naming the next compound. The longest chain has five carbon atoms; the fact that they are not printed in a linear fashion is not important. Either methyl group at the right-hand end of the chain may be included in the *longest continuous* carbon chain, the pentane chain. There are three methyl substituents, so the name prefix is "*tri*methyl." Two of these methyl groups are attached to carbon atom 2 and one to carbon atom 4. By convention, the numbering begins at the end giving the lowest possible set of numbers. This means, therefore, that the name is 2,2,4-trimethylpentane (and not the alternative 2,4,4-trimethylpentane).

```
                    5
      CH3          CH3
 1   2|  3        4|
CH3—C—CH2—CH
      |             |
      CH3          CH3
```

Common name: *iso*octane
Systematic name: 2,2,4-trimethylpentane

The compound 2,2,4-trimethylpentane is the systematic name of a compound commonly called *iso*octane. It is used as a standard in assigning the octane ratings of various types of gasoline. The compound is arbitrarily assigned an octane number of 100, while *n*-hexane is assigned the rating of 0. The antiknock performance of a particular gasoline is compared with that of various mixtures of 2,2,4-trimethylpentane and hexane, and an octane number is assigned on that basis. As the octane numbers of the two standards suggest, branched-chain hydrocarbons have better antiknock properties than straight-chain hydrocarbons.

EXAMPLE 23.1

NAMING COMPOUNDS

Give the systematic name for each of the following compounds:

```
        CH3  CH3
        |    |
(a) CH3—CH—CH—CH2—CH3

                        CH3
                        |
(b) CH3—CH2—CH2—C—CH3
                        |
                        CH3

                CH2—CH3
                |
(c) CH3—CH2—CH—CH2—CH3
```

Solution (a) In this compound, the longest continuous carbon chain has five carbon atoms, so the parent name is pentane. There are two methyl groups, one at carbon 2 and the other at carbon 3, so the systematic name is 2,3-dimethylpentane.

(b) The longest continuous carbon chain has five atoms, so the parent name is pentane. The carbon atoms are numbered beginning at the end closest to the substituents. Therefore, the two methyl groups are attached to carbon atom 2, and the systematic name is 2,2-dimethylpentane.

(c) In this compound, the parent name is again pentane. In this case, however, the substituent is an ethyl group ($—CH_2CH_3$), so the systematic name is 3-ethylpentane.

As a final point, notice that these three compounds have the same molecular formula, C_7H_{16}, and so are structural isomers.

EXERCISE 23.2 NAMING COMPOUNDS

(a) The three compounds in Example 23.1 are all isomers with the molecular formula C_7H_{16}. What is the name of the parent or unbranched alkane with this formula?

(b) In Exercise 23.1, you drew the structural isomers of C_6H_{14}. Give the systematic name of each of the isomers.

23.3 FUNCTIONAL GROUPS AND COMMON CLASSES OF ORGANIC COMPOUNDS

Organic compounds are characterized by **functional groups,** which are structural fragments found in all members of a given class of compounds and which are centrally involved in the chemical reactions of the class. So, instead of having to learn the properties of an overwhelming number of individual compounds, we study the properties associated with about a dozen basic classes of organic compounds. First, we shall survey briefly the most common classes of organic compounds and their characteristic functional groups. After this survey, we shall then consider a *few* of the properties of each class in some detail.

There are several classes of hydrocarbons in addition to alkanes. The **alkenes** are characterized by the presence of a C=C double bond. Ethylene and propylene, the simplest alkenes, are compounds of major industrial importance, since they serve as the starting materials in the synthesis of a wide variety of compounds.

Alkenes are also referred to as olefins.

Alkene	$CH_2{=}CH_2$	$CH_3CH{=}CH_2$
Common name	Ethylene	Propylene
Systematic name	Ethene	Propene

```
H       H          H  H  H
 \     /           |  |  |
  C = C         H—C—C=C
 /     \           |     |
H       H          H     H
```

Ethene (left) and propene (right). (Charles D. Winters)

Alkenes and alkynes are also called **unsaturated** compounds. The latter name comes from the fact that each C atom of the multiple bond is bound to only two or three other C or H atoms; a *saturated* C atom bonds to four C or H atoms.

The **alkynes** are hydrocarbons having a carbon–carbon triple bond, and the simplest member of the class is *acetylene,* H—C≡C—H, a compound whose synthesis from calcium carbide was first mentioned in Chapter 21.

Aromatic compounds constitute yet another class of hydrocarbons, and the fundamental compound in this class is benzene. Aromatic compounds always contain one or more benzene rings or related structures.

Benzene, C_6H_6

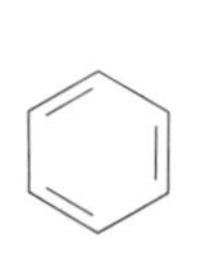

or

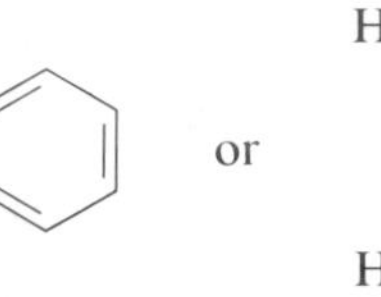

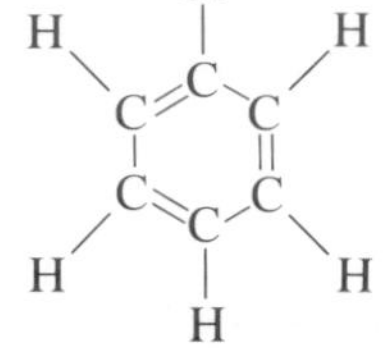

Naphthalene, $C_{10}H_8$

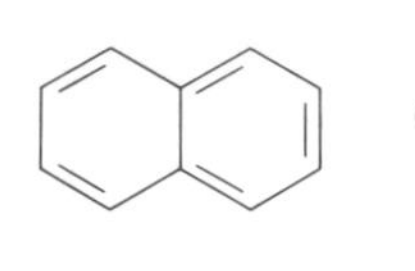

or

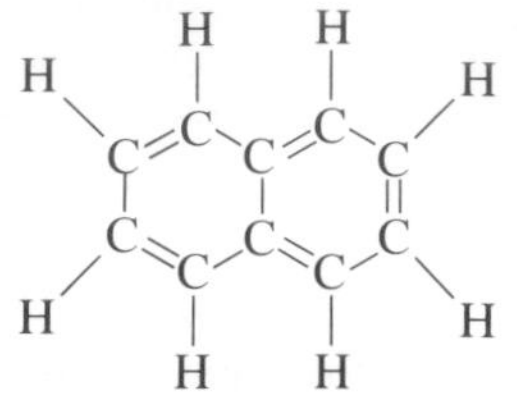

Bromoethane (Charles D. Winters)

If a hydrogen atom of an alkane, for example, is replaced by a halogen atom, the result is an **alkyl halide.** Two common members of this class are iodomethane and bromoethane.

Iodomethane, CH_3I Bromoethane, CH_3CH_2Br

$$\begin{array}{c} H \\ | \\ H—C—I \\ | \\ H \end{array} \qquad \begin{array}{c} H\ \ H \\ |\ \ \ | \\ H—C—C—Br \\ |\ \ \ | \\ H\ \ H \end{array}$$

Methyl alcohol (Charles D. Winters)

Alcohols are characterized by the presence of a hydroxyl group (—OH) covalently bonded to a saturated carbon atom. The simplest and most common alcohols are methyl and ethyl alcohol.

Methyl alcohol, CH_3OH Ethyl alcohol, C_2H_5OH

$$\begin{array}{c} H \\ | \\ H—C—O—H \\ | \\ H \end{array} \qquad \begin{array}{c} H\ \ H \\ |\ \ \ | \\ H—C—C—O—H \\ |\ \ \ | \\ H\ \ H \end{array}$$

Ethyl alcohol (Charles D. Winters)

Alcohols can be viewed as having been derived from water; one of the H atoms of H_2O is replaced by an nonaromatic group. Another class of compounds, **ethers**, is also based on water, but both H atoms have been replaced by organic groups.

Diethyl ether, $C_2H_5—O—C_2H_5$

$$\begin{array}{c} H\ \ H\ \ \ \ \ H\ \ H \\ |\ \ \ |\ \ \ \ \ \ |\ \ \ | \\ H—C—C—O—C—C—H \\ |\ \ \ |\ \ \ \ \ \ |\ \ \ | \\ H\ \ H\ \ \ \ \ H\ \ H \end{array}$$

There are several other classes of oxygen-containing organic compounds, in addition to alcohols and ethers. For example, organic acids are characterized by the **carboxyl** group and are often called **carboxylic acids**.

```
   O
   ‖
—C—O—H
```

carboxyl group

The two simplest members of the class are commonly called formic acid and acetic acid, and both are weak Brønsted acids in aqueous solution (Chapter 15).

Diethyl ether

Formic acid, HCOOH Acetic acid, H_3CCOOH

```
    O              H  O
    ‖              |  ‖
H—C—O—H         H—C—C—O—H
                   |
                   H
```

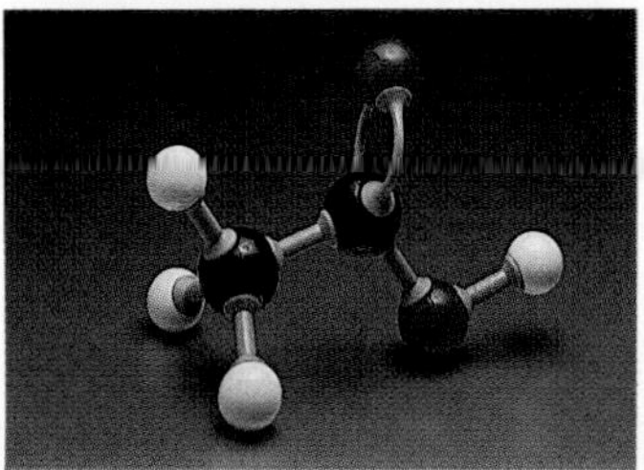

Acetic acid

Esters are structurally related to carboxylic acids. Esters differ in that they have a hydrocarbon group in place of the OH hydrogen atom of an acid. Thus, ethyl acetate can be viewed as derived from acetic acid; an ethyl group has replaced the OH hydrogen in this case.

Ethyl acetate, $CH_3COOCH_2CH_3$

```
   H  O     H  H
   |  ‖     |  |
H—C—C—O—C—C—H
   |        |  |
   H        H  H
```

Ethyl acetate (Charles D. Winters)

The grouping >C=O is called a **carbonyl group**, and it is the important functional group of the other major classes of oxygen-containing compounds, **aldehydes** and **ketones**. In aldehydes, the carbonyl carbon is bonded to at least one H atom; in ketones, the carbonyl group is bonded to two other carbon atoms. Formaldehyde and acetaldehyde are the simplest aldehydes, and acetone is the simplest ketone.

Formaldehyde, HCHO Acetaldehyde, H_3CCHO Acetone, H_3CCOCH_3

```
    O            H  O            H  O  H
    ‖            |  ‖            |  ‖  |
H—C—H         H—C—C—H         H—C—C—C—H
                 |               |     |
                 H               H     H
```

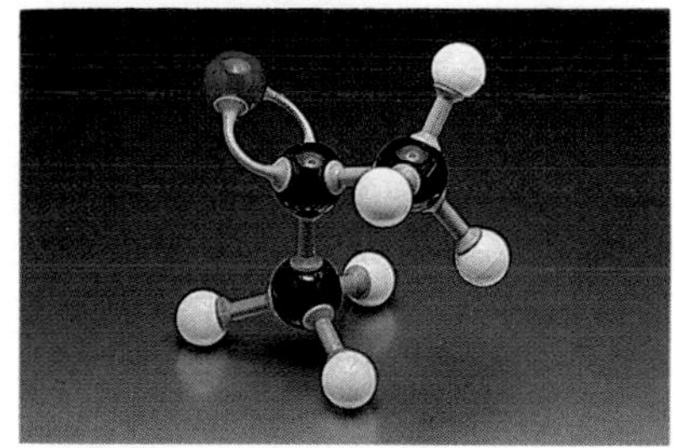

Acetone

Nitrogen is another common substituent atom in organic compounds, and **amines** represent one major class of nitrogen-containing compounds. These can be viewed as derivatives of ammonia where one or more of the NH hydrogen atoms is replaced by a hydrocarbon group.

Ethylamine, $H_3CCH_2NH_2$ Dimethylamine, $(H_3C)_2NH$

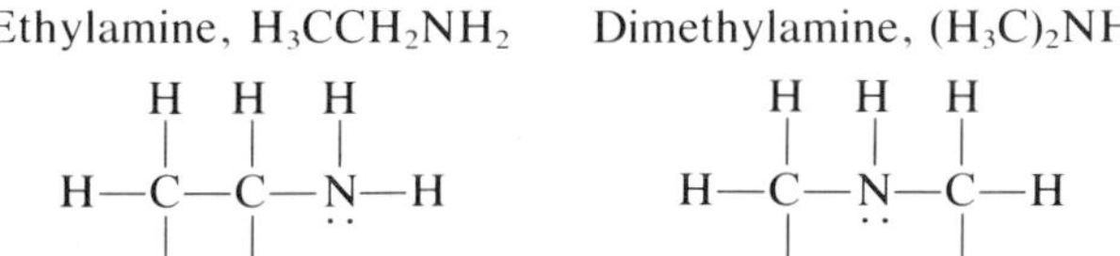

Ethylamine (Charles D. Winters)

With this brief survey of many of the major classes of organic compounds, we can turn to an expanded description of the important aspects of each type.

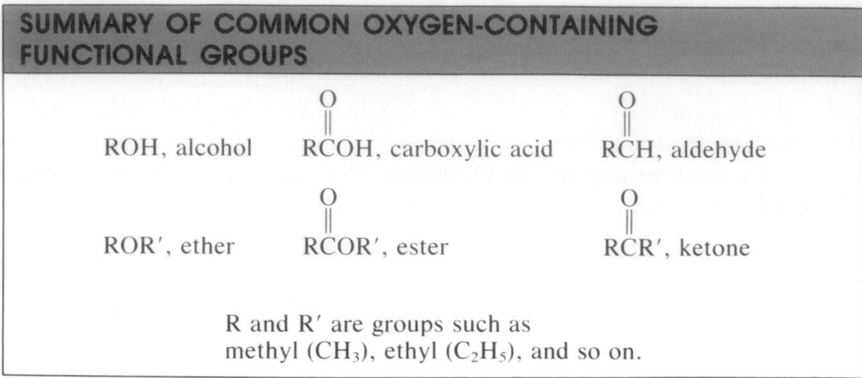

SUMMARY OF COMMON OXYGEN-CONTAINING FUNCTIONAL GROUPS

ROH, alcohol

$RC(=O)OH$, carboxylic acid

$RC(=O)H$, aldehyde

ROR′, ether

$RC(=O)OR'$, ester

$RC(=O)R'$, ketone

R and R′ are groups such as methyl (CH_3), ethyl (C_2H_5), and so on.

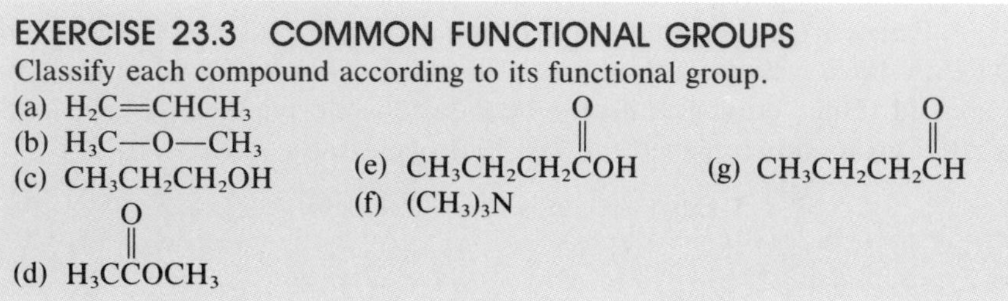

EXERCISE 23.3 COMMON FUNCTIONAL GROUPS

Classify each compound according to its functional group.

(a) $H_2C{=}CHCH_3$

(b) $H_3C{-}O{-}CH_3$

(c) $CH_3CH_2CH_2OH$

(d) $H_3CC(=O)OCH_3$

(e) $CH_3CH_2CH_2C(=O)OH$

(f) $(CH_3)_3N$

(g) $CH_3CH_2CH_2C(=O)H$

23.4 ALCOHOLS, ROH

It is common practice to write generalized structures using the letter **R** to represent alkyl or substituted alkyl groups and **Ar** to represent aromatic or substituted aromatic groups.

The general formula of alcohols, ROH, suggests a structural similarity to water (Table 23.2), and we shall see that some of the properties of alcohols resemble those of water.

Over 1 billion gallons of methyl alcohol, CH_3OH, are produced annually in the United States, and most of it is used to make formaldehyde and acetic acid, important components of a variety of polymers. In addition, methyl alcohol is used as a solvent and as an octane booster and anti-icing agent in gasoline.

Ethyl alcohol is probably the oldest as well as one of the most important synthetic chemicals. It is the "alcohol" of alcoholic beverages and is prepared for this purpose by fermentation of sugar from a wide variety of plant sources. For many years industrial alcohol, which is used

TABLE 23.2 Some Representative Alcohols

STRUCTURE	SYSTEMATIC NAME	COMMON NAME	Bp(°C)	Mp(°C)
CH_3OH	Methanol	Methyl alcohol	65	−97
CH_3CH_2OH	Ethanol	Ethyl alcohol	78	−115
$CH_3CH(OH)CH_3$	2-Propanol	Isopropyl alcohol	97	−126
$CH_3C(CH_3)(OH)CH_3$	2-Methyl-2-propanol	*tert*-Butyl alcohol	83	26

PATTERNS OF REACTIVITY

In the sections that follow, you will explore the characteristic reactions of organic compounds. To help you organize your reading, it is useful to see that many of these reactions fall into *four classes.*

Oxidation–reduction reactions are one major class of reactions. Virtually all organic compounds combust in oxygen,

$$CH_4(g) + 2\ O_2(g) \longrightarrow CO_2(g) + 2\ H_2O(g)$$

but many react with other oxidizing agents such as $KMnO_4$, $Na_2Cr_2O_7$, or even silver ion.

Reductions of certain organic compounds can be done with various metal hydrides such as lithium aluminum hydride, $LiAlH_4$.

$$CH_3C(=O)CH_3 + 2\ H^- \text{ (from } LiAlH_4) \longrightarrow CH_3CH(OH)CH_3 + 2e^-$$

Addition and elimination reactions are contrasting reaction classes. For example, a characteristic reaction of alkenes is the addition of halogens to the C=C bond.

$$H_2C{=}CH_2 + Br_2 \xrightarrow{\text{addition}} Br{-}CH_2{-}CH_2{-}Br$$

The converse reaction, elimination, is observed when an alcohol is induced to eliminate water to form an alkene.

$$H{-}CH_2{-}CH_2{-}OH \xrightarrow[H_2SO_4]{\text{elimination}} H_2C{=}CH_2 + HOH$$

Finally, there are many examples of **substitution reactions**. For example, a hydrogen atom of benzene can be substituted by bromine if an appropriate catalyst is used.

$$C_6H_6 \xrightarrow[FeBr_3]{Br_2} HBr + C_6H_5Br$$

as a solvent and as a starting material for the synthesis of other compounds, was also made by fermentation. However, in the last several decades, it became cheaper to make the alcohol from petroleum by-products, specifically by the catalyzed addition of water to ethylene.

$$H_2C{=}CH_2(g) + H_2O(g) \xrightarrow[\text{catalyst (phosphoric acid)}]{\text{addition}} H_3C{-}\overset{\displaystyle H}{\underset{\displaystyle H}{C}}{-}OH \quad (g)$$

NAMES AND CLASSES OF ALCOHOLS

Systematic names of alcohols are derived from the names of the corresponding alkanes by dropping the "-e" ending and adding "-ol." Thus, the systematic name of methyl alcohol is methanol; for ethyl alcohol it is ethanol. Where necessary, a numerical prefix is used to designate the position of the hydroxyl group (Table 23.2).

Alcohols can be classified into three categories based on the number of carbons bonded to the carbon bearing the hydroxyl group. *Primary alcohols* have one carbon and two hydrogen atoms attached to the carbon atom to which the OH group is attached. When two carbon atoms are attached to the OH carbon, the alcohol is called a *secondary alcohol*, while a *tertiary alcohol* has three carbon atoms attached to the OH carbon.

$$-\overset{|}{\underset{|}{C}}-\overset{\displaystyle H}{\underset{\displaystyle H}{C}}-O-H \qquad -\overset{|}{\underset{|}{C}}-\overset{\displaystyle H}{\underset{\displaystyle -C-}{C}}-O-H \qquad -\overset{|}{\underset{|}{C}}-\overset{\displaystyle -C-}{\underset{\displaystyle -C-}{C}}-O-H$$

primary secondary tertiary

> **EXERCISE 23.4 ALCOHOL NAMES AND STRUCTURES**
> (a) What are the systematic names for $CH_3CH_2CH_2OH$ and $CH_3CH(OH)CH_3$?
> (b) Draw structural formulas for 3-pentanol and 3-methyl-2-butanol.
> (c) Tell whether each alcohol in (b) is primary, secondary, or tertiary.

Reaction of sodium metal with ethyl alcohol to give sodium ethoxide and hydrogen. (Charles D. Winters)

SOME CHEMISTRY OF ALCOHOLS

In analogy with water, alcohols react with alkali metals to produce hydrogen. The other product, instead of *hydr*oxide ion, is called the *alk*oxide ion.

$$HOH(\ell) + Na(s) \longrightarrow \tfrac{1}{2}\,H_2(g) + Na^+(aq) + OH^-(aq)$$

$$ROH(\ell) + Na(s) \longrightarrow \tfrac{1}{2}\,H_2(g) + Na^+(\text{in alc}) + RO^-(\text{in alc})$$

Alcohols react with hydrogen bromide or hydrogen iodide in a *substitution reaction* to give the corresponding alkyl halides.

$$ROH + HX \xrightarrow[X\,=\,Br,\,I]{} RX + H_2O$$

Concentrated sulfuric acid has a great affinity for water. Thus when alcohols (with H and OH on *adjacent carbons*) are treated with this acid, the alcohol is dehydrated to give the corresponding alkene in an *elimination reaction*.

$$\underset{\text{alcohol}}{-\underset{\substack{|\\H}}{\overset{|}{C}}-\underset{\substack{|\\OH}}{\overset{|}{C}}-} \xrightarrow[180°C]{\text{conc. } H_2SO_4} \underset{\text{alkene}}{>C=C<} + H_2O$$

By modifying the conditions slightly, an *ether* can be obtained instead of an alkene. In this process, one mole of water is removed from *two* moles of alcohol.

$$\underset{\text{alcohol}}{2\ ROH} \xrightarrow[\text{heat}]{\text{conc. } H_2SO_4} \underset{\text{ether}}{R-O-R} + H_2O$$

Alcohols can be oxidized, but the products vary according to the type of alcohol involved and the oxidizing agent. Primary alcohols can be oxidized to aldehydes, which in turn can be oxidized to carboxylic acids.

$$\underset{\text{primary alcohol}}{RCH_2OH} \xrightarrow{\text{oxidizing agent}} \underset{\text{aldehyde}}{R-\overset{\overset{O}{\|}}{C}-H} \xrightarrow{\text{oxidizing agent}} \underset{\text{acid}}{R-\overset{\overset{O}{\|}}{C}-O-H}$$

Aldehydes are very useful reagents, but getting the reaction to stop at the aldehyde stage is a significant problem. With most oxidizing agents the subsequent oxidation of aldehydes to acids occurs more readily than the first step. Fortunately, a specialized reagent, PCC, has been developed that will oxidize primary alcohols and stop at the aldehyde stage.

Pyridinium chlorochromate, PCC, is a salt prepared from CrO_3, HCl, and pyridine, an aromatic amine.

$$\text{C}_5\text{H}_5\text{N}^+\text{—H} \quad CrO_3Cl^-$$

$$\underset{\text{primary alcohol}}{RCH_2OH} \xrightarrow{\text{PCC}} \underset{\text{aldehyde}}{R-\overset{\overset{O}{\|}}{C}-H}$$

Conversion of primary alcohols to carboxylic acids can be accomplished with a variety of oxidizing agents, including potassium permanganate and sodium dichromate.

$$\underset{\text{primary alcohol}}{RCH_2OH} \xrightarrow{KMnO_4 \text{ or } Na_2Cr_2O_7} \underset{\text{carboxylic acid}}{R-\overset{\overset{O}{\|}}{C}-O-H}$$

Secondary alcohols are readily oxidized, but ketones are the only product; no further oxidation can occur. Any of the oxidizing agents mentioned so far is effective.

$$\underset{\text{secondary alcohol}}{R-\underset{\substack{|\\H}}{\overset{\overset{O-H}{|}}{C}}-R'} \xrightarrow{PCC,\ KMnO_4, \text{ or } Na_2Cr_2O_7} \underset{\text{ketone}}{R-\overset{\overset{O}{\|}}{C}-R'}$$

In the oxidation of a primary or secondary alcohol, a hydrogen atom is removed from the carbon atom to which the OH group is bonded. Tertiary alcohols do not have the grouping >CH(OH), so they are not oxidized under normal conditions.

$$\begin{array}{c} \text{O—H} \\ | \\ \text{R—C—R}' \\ | \\ \text{R}'' \end{array} \xrightarrow{\text{oxidizing agents}} \text{no reaction}$$

tertiary alcohol

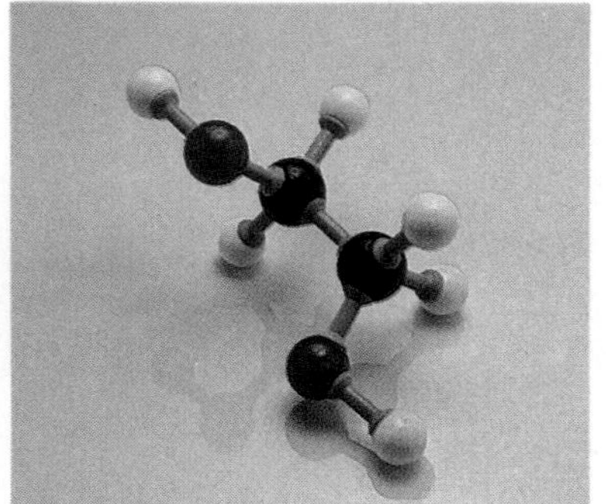
Ethylene glycol (Charles D. Winters)

Several important alcohols have more than one OH group. Two of the most familiar are glycerine and ethylene glycol.

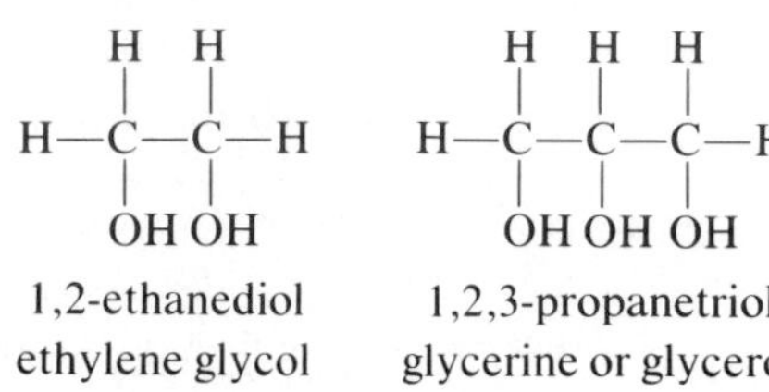

Systematic name:	1,2-ethanediol	1,2,3-propanetriol
Common name:	ethylene glycol	glycerine or glycerol

Ethylene glycol is familiar as automobile antifreeze. Glycerine is about 0.6 times as sweet as cane sugar and so is often used in confectioneries; its greatest use, however, is as a softener in soaps and lotions. When it reacts with nitric acid, it produces nitroglycerine,

$$\begin{array}{c} H_2C\text{—}OH \\ | \\ HC\text{—}OH \\ | \\ H_2C\text{—}OH \\ \text{glycerine} \end{array} + 3\ HNO_3 \longrightarrow \begin{array}{c} H_2C\text{—}O\text{—}NO_2 \\ | \\ HC\text{—}O\text{—}NO_2 \\ | \\ H_2C\text{—}O\text{—}NO_2 \\ \text{nitroglycerine} \end{array} + 3\ H_2O$$

the explosive component of dynamite. However, nitroglycerine is also used to treat a common heart condition called angina.

EXERCISE 23.5 REACTIONS OF ALCOHOLS

Complete each of the following reactions:

(a) $CH_3CH_2CH_2OH \xrightarrow{\text{PCC}}$

(b) $CH_3CH_2CH_2OH \xrightarrow{H_2SO_4/180°C}$

(c) $CH_3CH_2CH_2OH + HI \rightarrow$

23.5 ETHERS, ROR

Ethers, like alcohols, can be viewed as structural derivatives of water. However, in ethers, both H atoms of water are replaced by hydrocarbon groups. Because they lack the —OH group, ethers cannot hydrogen bond with one another. This causes their boiling points to be relatively low closer to those of alkanes of similar molecular weight. Compare, for ex ample, diethyl ether with pentane and 1-butanol, all compounds having similar molecular weights.

COMPOUND	FORMULA	MW	Bp, °C
Diethyl ether	$CH_3CH_2OCH_2CH_3$	74	35
Pentane	$CH_3CH_2CH_2CH_2CH_3$	72	36
1-Butanol	$CH_3CH_2CH_2CH_2OH$	74	118

Ethers are much less reactive than alcohols. For example, they are unaffected by treatment with oxidizing agents, reducing agents, bases, and most acids. Nonetheless, their relative inertness, but mild Lewis basicity, makes them ideal as solvents, and diethyl ether is widely used for this purpose.

Ethers do form peroxides on standing for long periods in air. As these peroxides are explosive, great care should be exercised in storing and using ethers.

23.6 UNSATURATED COMPOUNDS: ALKENES AND ALKYNES

Compounds with double and triple bonds between carbon atoms are often referred to as **unsaturated**, because the pi electrons of the multiple bond can be used to form additional sigma bonds.

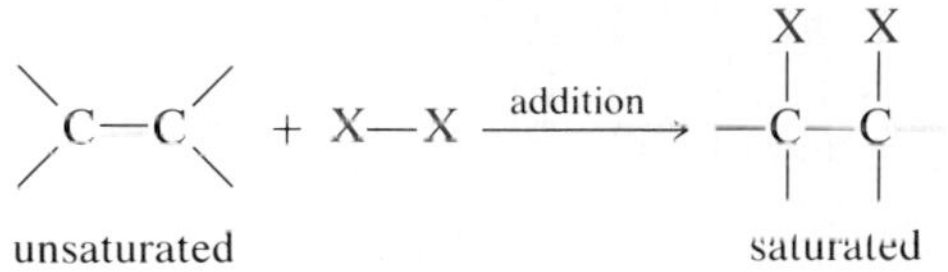

$$\underset{\text{unsaturated}}{>C{=}C<} + X{-}X \xrightarrow{\text{addition}} \underset{\text{saturated}}{-\overset{X}{\underset{|}{\overset{|}{C}}}-\overset{X}{\underset{|}{\overset{|}{C}}}-}$$

Thus, compounds of this class are reactive with respect to the formation of new bonds, and so they are often the building blocks used by the chemical industry to synthesize new compounds. As such, they represent a class of organic compounds of considerable economic importance.

ALKENES, $R_2C{=}CR_2$

NAMING The *systematic names* of compounds such as those listed in Table 23.3 are derived from the name of the corresponding alkane by dropping the "-ane" ending and adding "-ene" in its place. Where necessary, the position of the double bond is included as a prefix. To determine the number to be used, begin counting at the end of the carbon chain closest to the double bond.

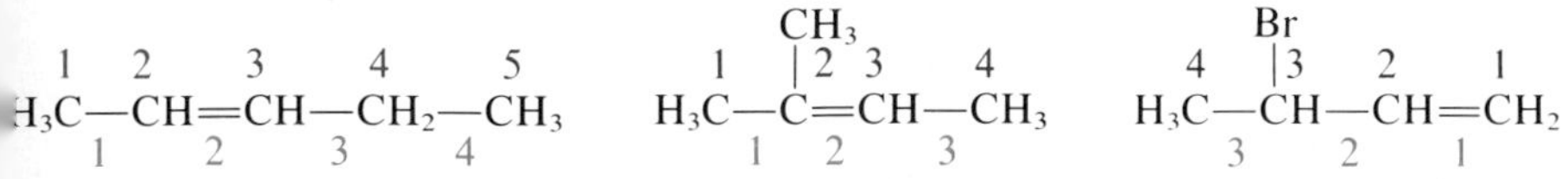

$$\underset{1\quad 2\quad 3\quad 4}{\overset{1\quad 2\quad 3\quad 4\quad 5}{H_3C{-}CH{=}CH{-}CH_2{-}CH_3}} \qquad \underset{1\quad 2\quad 3}{\overset{1\quad 2\quad 3\quad 4}{H_3C{-}\overset{CH_3}{\overset{|}{C}}{=}CH{-}CH_3}} \qquad \underset{3\quad 2\quad 1}{\overset{4\quad 3\quad 2\quad 1}{H_3C{-}\overset{Br}{\overset{|}{C}H}{-}CH{=}CH_2}}$$

2-pentene 2-methyl-2-butene 3-bromo-1-propene

BONDING, STRUCTURE, AND ISOMERISM We can tell from the VSEPR theory, and it is confirmed by experiment, that the substituents of a carbon atom doubly bonded to another carbon will be arranged as a planar triangle.

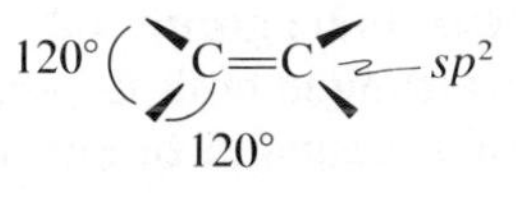

$$120^\circ\;>C{=}C<\;\leftarrow sp^2 \qquad 120^\circ$$

TABLE 23.3 Some Representative Alkenes

STRUCTURE	SYSTEMATIC NAME	COMMON NAME	Bp(°C)	Mp(°C)
$CH_2{=}CH_2$	Ethene	Ethylene	−104	−169
$CH_3CH{=}CH_2$	Propene	Propylene	−48	−185
$CH_3CH_2CH{=}CH_2$	1-Butene	—	−6	−185
$CH_3CH{=}CHCH_3$	2-Butene	—	See text	
CH_3 $CH_3C{=}CH_2$	Methylpropene	*Iso*-butylene	−7	−140
CH_3 $CH_3C{=}CCH_3$	2-Methyl-2-butene	—	39	−134
CH_3 $CH_2{=}CHC{=}CH_2$	2-Methyl-1,3-butadiene	Isoprene	34	−146

This means that the C atom must be considered sp^2 hybridized, and the pi bond can form using the unhybridized *p* orbitals, one on each carbon, as explained in Chapter 10.

One of the most important aspects of alkene chemistry is to recognize the effect of the pi bond on molecular structure. Ordinarily, when two carbon atoms are sigma bonded to one another, the groups at the ends of the bond rotate rapidly about the bond axis.

However, the effect of a carbon–carbon pi bond is to *prevent* rotation about the C=C bond axis. Rotation of the ends of the molecule can occur if the two C atoms are joined by *only* a sigma bond. This means that rotation can occur in alkenes ($R_2C{=}CR_2$) *only* if the pi bond is broken, and this would require considerable energy. Therefore, if a chlorine atom is attached at each carbon in ethylene, for example, there are *two* possible *permanent* arrangements.

Cis and *trans* isomers. (Charles D. Winters)

cis-1,2-dichloroethene (H, H on one side; Cl, Cl on the other)	*trans*-1,2-dichloroethene (H, Cl on top; Cl, H on bottom)
mp, −80°C; bp, 60°C	mp, −50°C; bp, 48°C

These two molecules have the same formula, but restricted rotation about the C—C bond causes them to have different structures and different properties. Hence, these are *isomers*, since they have the same molecular formula but the atoms are arranged in a different manner. The isomer with chlorine atoms on opposite sides of the bond is called the *tran* isomer and that with the chlorine atoms on the same side is called the *cis* isomer. This form of isomerism is often referred to as *cis–trans* isomerism.

EXERCISE 23.6 *CIS–TRANS* ISOMERS
Draw the *cis–trans* isomers for 2-pentene.

ALKENE CHEMISTRY Ethylene and propylene are prepared industrially on a large scale by *steam cracking* hydrocarbons found in natural gas and petroleum.

30.6 billion pounds of ethylene worth $6 billion were produced in 1985 in the United States, making it the sixth largest chemical produced. Propylene, $CH_3CH{=}CH_2$, is 12th on the list of chemicals produced in the United States (Figure 22.1). About 25% of the production is used to make the polymer polypropylene (Chapter 24). Polypropylene fibers are now widely used in carpets.

$$\underset{\text{ethane}}{C_2H_6(g)} \xrightarrow{\text{heat}} \underset{\text{ethylene}}{H_2C{=}CH_2(g)} + H_2(g)$$

Almost half of the ethylene produced is used to make the polymer *polyethylene* (see Chapter 24), but a large amount is used to make ethylene glycol, vinyl chloride (to make PVC plastics), styrene (to make polystyrene), and a wide variety of other products.

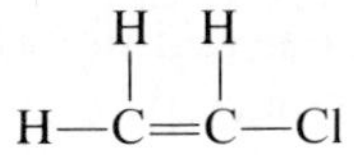

$$\underset{\substack{\text{chloroethene}\\\text{(vinyl chloride)}}}{H{-}\overset{H}{\overset{|}{C}}{=}\overset{H}{\overset{|}{C}}{-}Cl} \qquad \underset{\substack{\text{phenylethene}\\\text{(styrene)}}}{C_6H_5{-}CH{=}CH_2}$$

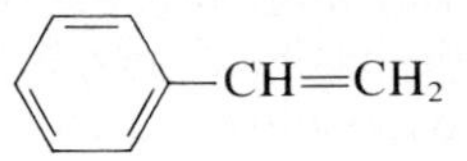

The **addition reaction** is the most common reaction of alkenes (and alkynes): reagents add to the carbon atoms of the C—C bond. If the reagent is H_2, the reaction is often called a **hydrogenation** and the product is an alkane. Specially prepared forms of metals such as platinum, palladium, nickel, and rhodium, among others, are used as catalysts for this reaction.

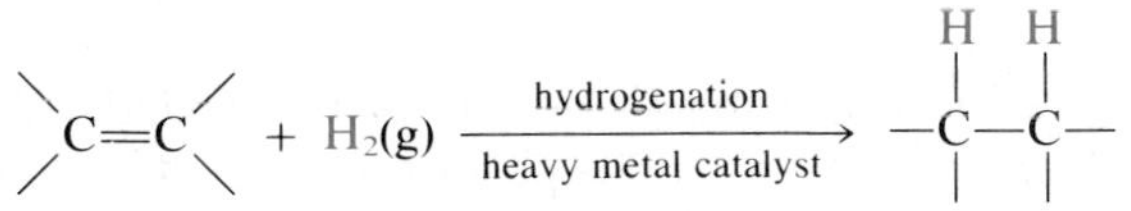

$$>C{=}C< + H_2(g) \xrightarrow[\text{heavy metal catalyst}]{\text{hydrogenation}} -\overset{H}{\overset{|}{\underset{|}{C}}}-\overset{H}{\overset{|}{\underset{|}{C}}}-$$

The halogens chlorine and bromine also add to alkenes to give 1,2-dihaloalkanes,

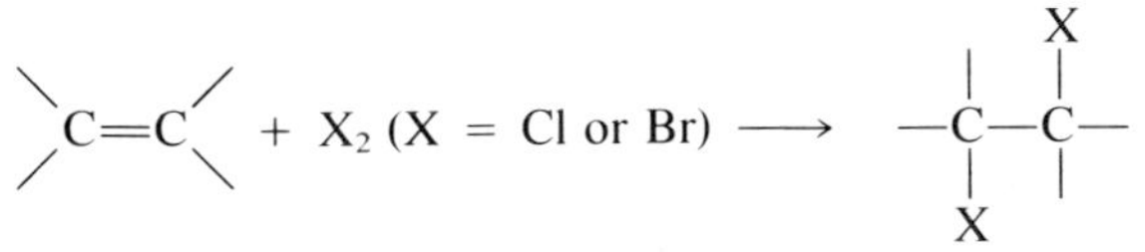

$$\underset{\text{alkene}}{>C{=}C<} + X_2\ (X = Cl \text{ or } Br) \longrightarrow \underset{\text{1,2-dihaloalkane}}{-\overset{|}{\underset{\underset{X}{|}}{C}}-\overset{X}{\overset{|}{\underset{|}{C}}}-}$$

and hydrogen halide (HCl, HBr, and HI) addition produces haloalkanes.

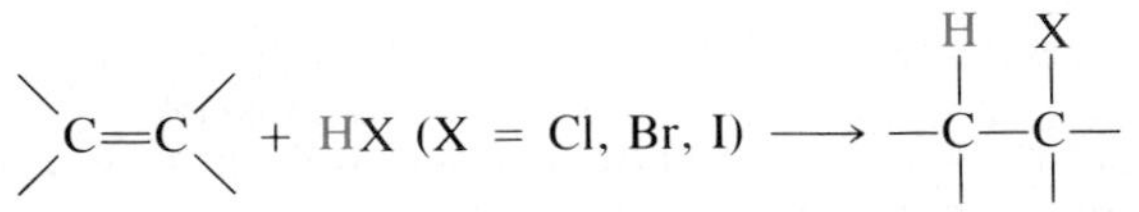

$$\underset{\text{alkene}}{>C{=}C<} + HX\ (X = Cl, Br, I) \longrightarrow \underset{\text{haloalkane}}{-\overset{H}{\overset{|}{\underset{|}{C}}}-\overset{X}{\overset{|}{\underset{|}{C}}}-}$$

When a hydrogen halide adds to propene, two products can be imagined: the halogen can end up attached either to carbon 1 or carbon 2.

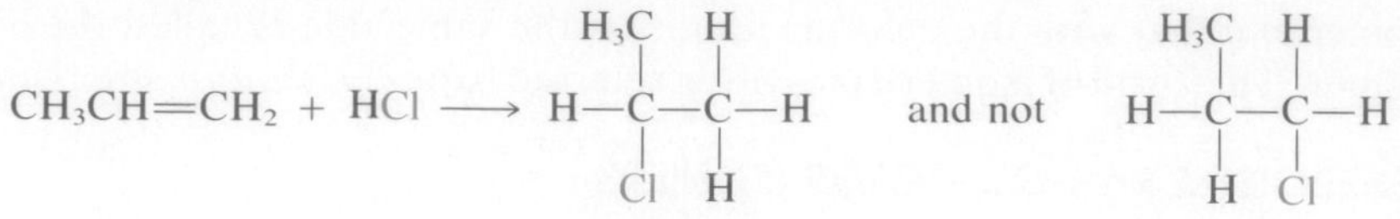

$$\underset{\text{propene}}{CH_3CH{=}CH_2} + HCl \longrightarrow \underset{\substack{\text{observed product}\\ \text{2-chloropropane}}}{\begin{array}{c} H_3C\quad H \\ |\qquad | \\ H{-}C{-}C{-}H \\ |\qquad | \\ Cl\quad H \end{array}} \quad \text{and not} \quad \underset{\substack{\text{not observed}\\ \text{1-chloropropane}}}{\begin{array}{c} H_3C\quad H \\ |\qquad | \\ H{-}C{-}C{-}H \\ |\qquad | \\ H\quad Cl \end{array}}$$

However, only 2-chloropropane is observed. Similarly, reaction of a hydrogen halide with methylpropene results only in 2-halo-2-methylpropane; no 1-halo product is formed.

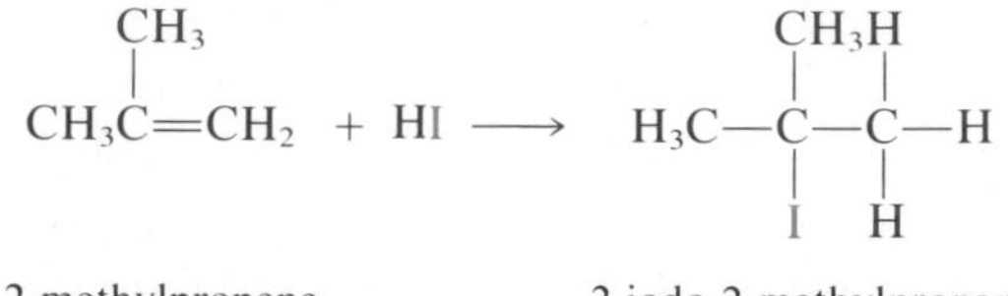

$$\underset{\text{2-methylpropene}}{\begin{array}{c} CH_3 \\ | \\ CH_3C{=}CH_2 \end{array}} + HI \longrightarrow \underset{\text{2-iodo-2-methylpropane}}{\begin{array}{c} CH_3H \\ |\quad | \\ H_3C{-}C{-}C{-}H \\ |\quad | \\ I\quad H \end{array}}$$

Markovnikov's rule can be stated succinctly as "Hydrogen goes where hydrogen is." It is an empirical observation that is understandable in terms of the properties of carbon-containing compounds. A course in organic chemistry will cover this topic.

During the 1860s the Russian chemist Vladimir Markovnikov examined the results of a large number of alkene addition reactions. In cases where two isomeric products might be expected, one usually predominated. He observed a pattern, now known as **Markovnikov's rule**, that can be stated as follows: When a reagent of the type HY is added to an unsymmetrical alkene, the H atom of the reagent becomes attached to the carbon that is already bonded to the greater number of hydrogen atoms. In the reaction of hydrogen chloride with propene, the H atom of HCl becomes attached to carbon 1, the carbon atom bearing two H atoms, and the Cl atom goes to carbon 2, which is bonded to only one H atom.

EXERCISE 23.7 ALKENE ADDITION REACTIONS

What products are expected from the reaction of HBr with (a) 1-butene and (b) 2-methyl-2-butene?

There are exceptions to Markovnikov's rule (called "anti-Markovnikov" additions) that can be explained in terms of modern bonding theory. The subject is studied in detail in courses in organic chemistry.

Alkenes react with water in the presence of acids to produce alcohols. Markovnikov's rule is followed in this *hydration reaction* when an unsymmetrical alkene is involved.

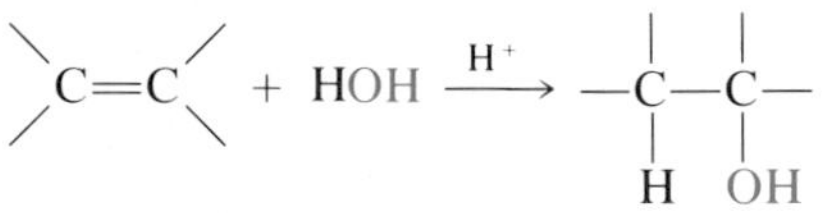

$$\underset{\text{alkene}}{\gt C{=}C \lt} + HOH \xrightarrow{H^+} \underset{\text{alcohol}}{\begin{array}{c} |\quad | \\ {-}C{-}C{-} \\ |\quad | \\ H\quad OH \end{array}}$$

$$\underset{\text{propene}}{CH_3CH{=}CH_2} + HOH \xrightarrow{H^+} \underset{\text{2-propanol}}{\begin{array}{c} H_3C\quad H \\ |\qquad | \\ H{-}C{-}C{-}H \\ |\qquad | \\ OH\quad H \end{array}}$$

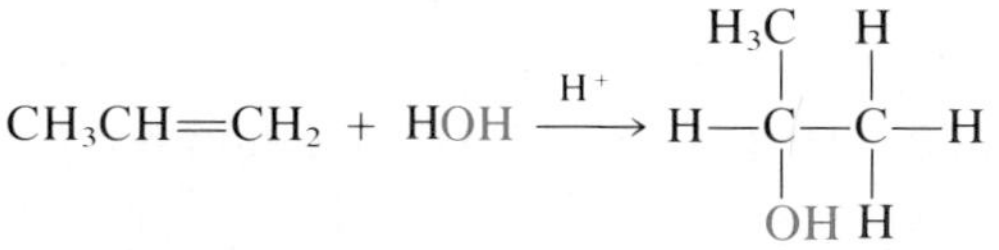

EXERCISE 23.8 ALKENE REACTIONS

What product is expected when water is added to 2-methyl-1-butene?

ALKYNES, RC≡CR

At one time most synthetic organic compounds were derived from coal, and acetylene (HC≡CH) was an important intermediate in industrial organic synthesis. However, the methods of preparing it

$$\text{(a) } CaO(s) + 3\,C(s) + \text{heat} \longrightarrow CO(g) + CaC_2(s)$$
$$CaC_2(s) + 2\,H_2O(\ell) \longrightarrow HC{\equiv}CH(g) + Ca(OH)_2(aq)$$

or

$$\text{(b) } 2\,CH_4(g) + \text{heat} \longrightarrow HC{\equiv}CH(g) + 3\,H_2(g)$$

involved high temperature processes and were therefore costly. When relatively inexpensive ethylene became available from petroleum and natural gas about 40 years ago, acetylene was almost completely replaced by the alkene.

The bonding in the linear acetylene molecule, with sp hybridized C atoms, was described in Chapter 10.

The systematic names of alkynes are formed similarly to alkenes: drop the "-ane" ending from the corresponding alkane and add "-yne." Thus, the systematic name of acetylene is ethyne. As is the case with many simple, widely used compounds the common name is almost always used. Where necessary the position of the triple bond and the location of substituent groups are indicated by numerical prefixes (Table 23.4).

Like alkenes, alkynes undergo addition reactions. An important difference between alkenes and alkynes, however, is that *two* molecules of reagent usually add across a triple bond because addition of the first H_2 gives an alkene.

$$\underset{\text{propyne}}{CH_3C{\equiv}CH} + 2\,H_2(g) \xrightarrow[\text{Pt catalyst}]{\text{hydrogenation}} \underset{\text{propane}}{CH_3CH_2CH_3}$$

$$CH_3C{\equiv}CH + 2\,Br_2 \xrightarrow[X_2 \text{ can be } Br_2 \text{ or } Cl_2]{\text{halogen addition}} CH_3CBr_2{-}CBr_2{-}H$$

1,1,2,2-tetrabromopropane

$$CH_3C{\equiv}CH + 2\,HCl \xrightarrow[\text{X in HX can be Cl, Br, I}]{\text{hydrogen halide addition}} CH_3CCl_2{-}CH_2{-}H$$

2,2-dichloropropane

TABLE 23.4 Some Simple Alkynes

STRUCTURE	SYSTEMATIC NAME	COMMON NAME	Bp, °C
H—C≡C—H	Ethyne	Acetylene	−75
$CH_3C{\equiv}CH$	Propyne	Methylacetylene	−23
$CH_3CH_2C{\equiv}CH$	1-Butyne	Ethylacetylene	9
$CH_3C{\equiv}CCH_3$	2-Butyne	Dimethylacetylene	27
$CH_3CH(CH_3){-}C{\equiv}CH$	3-Methyl-1-butyne	*iso*-Propylacetylene	40

Notice that the addition of hydrogen halides produces dihaloalkanes. In accordance with Markovnikov's rule, both H atoms are attached to the same carbon atom.

EXERCISE 23.9 ALKYNE NAMES AND FORMULAS
(a) Draw structural formulas for (1) 2-pentyne and (2) 4-methyl-1-pentyne.
(b) What is the name of the compound $ClCH_2C{\equiv}CH$?

EXERCISE 23.10 ALKYNE REACTIONS
Complete and balance the following reactions:
(a) $CH_3CHC{\equiv}CH + HBr \longrightarrow$ (with CH_3 on the second C)

(b) $CH_3CH_2C{\equiv}CH + Cl_2 \longrightarrow$

23.7 BENZENE AND AROMATIC COMPOUNDS

Benzene, C_6H_6, is about 15th on the list of the top 50 chemicals produced in the United States every year, and it is clearly the most important of the aromatics. Almost 10 billion pounds of the liquid were produced in 1985, for example, and almost all of that was converted into the types of products illustrated in Figure 23.1.

Benzene, toluene, the xylenes, and most naphthalene are obtained from petroleum. Heavier aromatics are derived from coal.

Benzene occupies an important place in the history of chemistry. Michael Faraday discovered the compound in 1825 as a by-product of illuminating gas, itself a product of heating coal. The formula of the compound suggested to 19th century chemists that it should be unsaturated, but, if viewed this way, its chemistry was perplexing. Whereas an alkene readily adds Br_2, and is oxidized by acidic potassium permanganate, benzene is not attacked under the same conditions. Higher temperatures or catalysts, however, do lead to reaction, but the products are those of substitution reactions and not addition or oxidation (Figure 23.1). The fact that only one monosubstitution product is ever obtained for a given

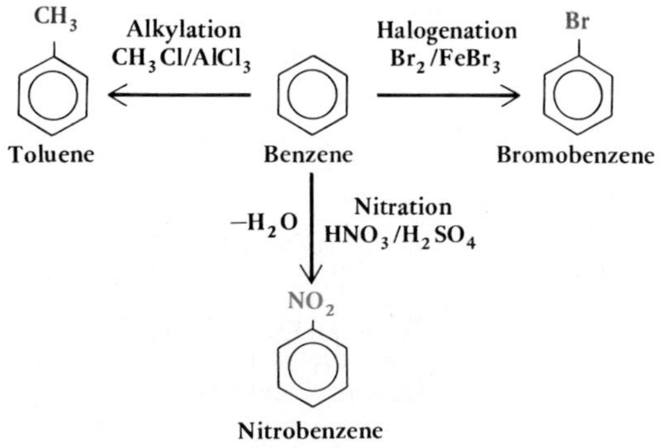

FIGURE 23.1
Some products formed from benzene.

reaction, along with other experimental observations, strongly suggested to August Kekulé (1829–1896) a *symmetrical ring structure* where all of the H atoms are equivalent.

Benzene, C_6H_2

Symmetrical molecule with all C and H atoms equivalent

It was Kekulé who then suggested (in 1865) that the molecule could be represented by some combination of two structures that we now call *resonance structures*.

benzene resonance structures

The bonding in benzene can be understood on the basis of the principles in Chapter 10. The geometry of atoms around each carbon atom is trigonal and planar, thereby implying sp^2 hybridization for each carbon. Using one of these hybrid orbitals for binding to an H atom and the other two orbitals for binding to adjacent carbons, an unhybridized p orbital remains on each carbon. Each of these p orbitals, perpendicular to the plane of the ring, contains an electron, and overlap of adjacent orbitals produces a set of *alternating* pi bonds.

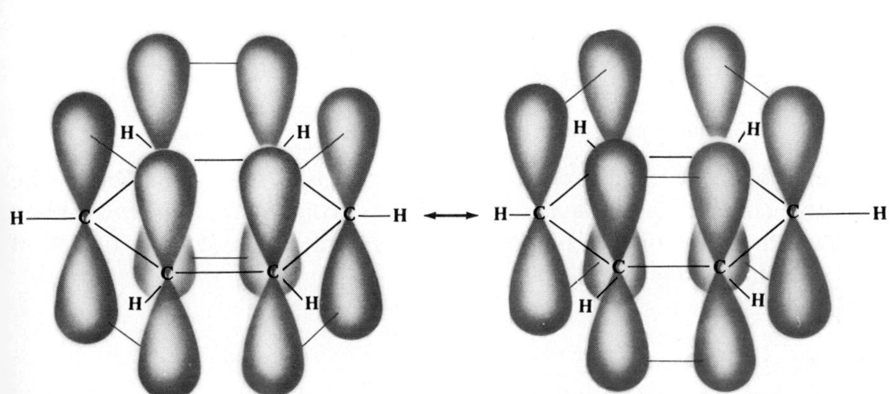

Unhybridized atomic p orbitals on carbon atoms overlap to form the multicenter π system

Experimental measurements clearly show that the actual structure of benzene does *not* have alternating single C—C and double C=C bonds. A typical C—C bond has a bond length of 154 pm and the typical C=C bond length is 134 pm. In benzene, all six carbon–carbon distances are the same, 139 pm. This is best explained by assuming that the six unhybridized p orbitals form six molecular orbitals, which delocalize the electrons evenly over the carbon framework. It is this delocalization of

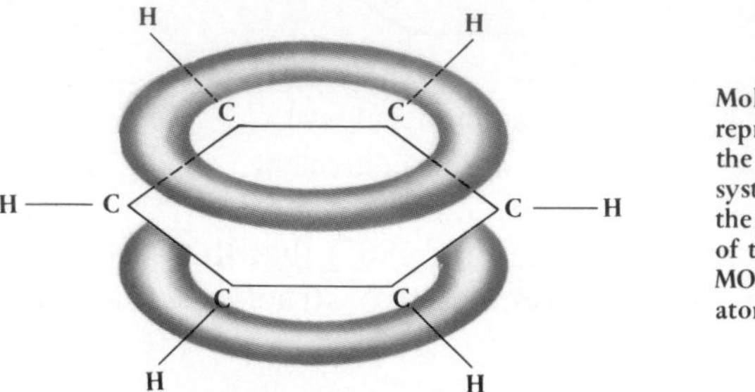

Molecular orbital representation of the delocalized π system formed by the combination of three π-bonding MO's on carbon atoms

electron density that accounts for six carbon–carbon bonds of equal length and leads to the extraordinary stability of aromatic systems. To convey this notion of delocalization, we often draw the benzene ring with a circle in the middle, as shown in the margin.

Condensed formula of benzene to convey electron delocalization

NAMING AROMATIC COMPOUNDS

Most aromatic compounds are named as derivatives of benzene.

Cl — chlorobenzene

NO_2 — nitrobenzene

but a few have their own, common names.

CH_3 — toluene

CH_3, CH_3 — *m*-xylene

$CH{=}CH_2$ — styrene

OH — phenol

When a benzene ring bears two substituents, as in the xylene above, the relative positions are indicated by the terms *ortho, meta, and para*. When writing the name of the compound, usually only the first letter of one of these terms is given.

X

ortho to X

meta to X

para to X

When two substituents are attached to adjacent carbons, they are said to be *ortho*.

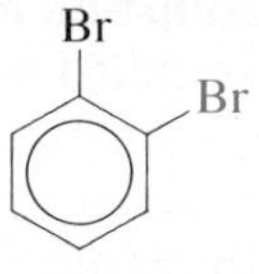

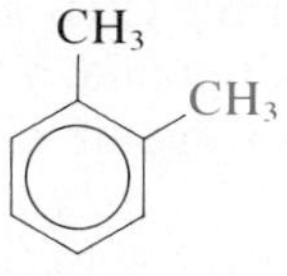

	Br / Br	CH_3 / CH_3
Systematic name:	*o*-dibromobenzene	*o*-dimethylbenzene
Common name:	----	*o*-xylene

If one carbon intervenes between the substituted carbon atoms, then the substituents are *meta*.

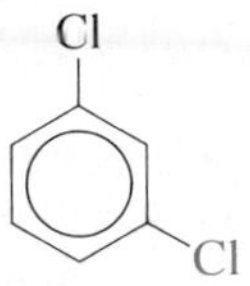

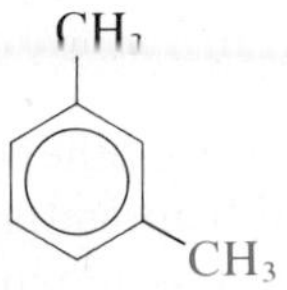

	Cl / Cl	CH_3 / CH_3
Systematic name:	*m*-dichlorobenzene	*m*-dimethylbenzene
Common name:	-----	*m*-xylene

Finally, if the substituents are separated by four carbon atoms, then they are *para* to one another.

When two different substituents are attached to the benzene ring, they are named in alphabetic order.

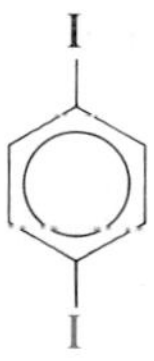

	I / I	NO_2 / OH
Systematic name:	*p*-diiodobenzene	*p*-hydroxynitrobenzene
Common name:	----	*p*-nitrophenol

If benzene has three or more substituents, the system just outlined does not work, and the substituent positions are numbered and are given the lowest possible set of numbers.

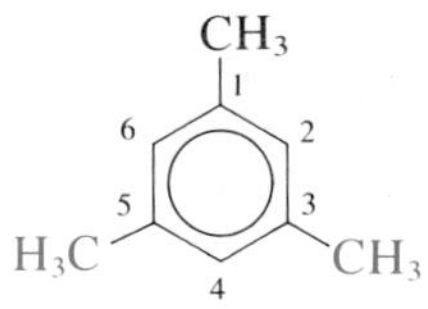

CH_3 (1); 2; 3 CH_3; 4; H_3C 5; 6 — Br; Br; Br

1,3,5-trimethylbenzene 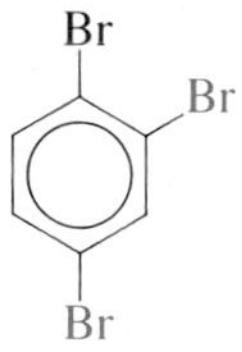 1,2,4-tribromobenzene

If the aromatic molecule is one of those with a common name, then the substituent is implicitly assumed to be in position 1.

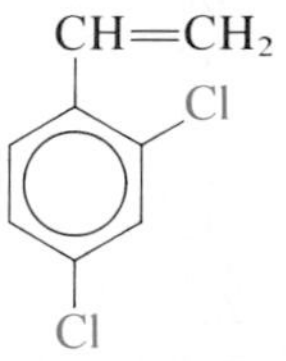

CH=CH_2; Cl; Cl — CH_3; O_2N; NO_2; NO_2

2,4-dichlorosytrene 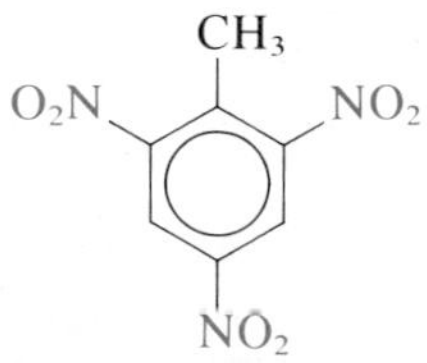 2,4,6-trinitrotoluene (TNT)

When the benzene ring has only a single substituent, and when that substituent is more than just an atom or a simple group, it is better to consider the C_6H_5 group as a substituent of an alkane, alkene, alkyne, or other compound. Then C_6H_5 is called a *phenyl* group, and names such as the following are given.

diphenylmethane 2-phenylpropane (cumene)

There are other aromatic compounds beside benzene, and they can be obtained from petroleum or coal tar (Table 23.5). Some of these compounds, such as naphthalene ($C_{10}H_8$, page 872), consist of two or more aromatic rings sharing ring edges to give even more extensive pi electron delocalization. One reason for the interest in compounds such as *benzopyrene* is they are powerful cancer-causing agents. If painted on the skin of mice, for example, such compounds produce skin tumors. Many such aromatics are found in the smoke from tobacco and from grilling meat on charcoal and are thought to induce colon cancer.

benzopyrene

THE CHEMISTRY OF BENZENE

As was mentioned at the beginning of this section, the chemistry of benzene does not resemble an alkene. Addition to a double bond in an aromatic hydrocarbon would disrupt the delocalization of electrons over the

TABLE 23.5 Compounds from Coal Tar

NAME	FORMULA	BOILING POINT (°C)	MELTING POINT (°C)	SOLUBILITY
Benzene	C_6H_6	80	+6	
Toluene	$C_6H_5CH_3$	111	−95	
o-Xylene	$C_6H_4(CH_3)_2$	144	−27	All
m-Xylene	$C_6H_4(CH_3)_2$	139	−54	insoluble
p-Xylene	$C_6H_4(CH_3)_2$	138	+13	in
Naphthalene	$C_{10}H_8$	218	+80	water
Anthracene	$C_{14}H_{10}$	342	+218	
Phenanthrene	$C_{14}H_{10}$	340	+101	

entire ring, the feature that adds so much stability to aromatic systems. Thus, the typical addition reactions of alkenes do *not* occur. Benzene and bromine, for example, do not react in the absence of catalysts, and the ring undergoes catalytic hydrogenation only very slowly and then only under extreme conditions.

$$C_6H_6 \text{ (benzene)} + 3\ H_2(g) \xrightarrow[\text{150°C, pressure}]{\text{Ni catalyst}} C_6H_{12} \text{ (cyclohexane)}$$

benzene　　　　cyclohexane

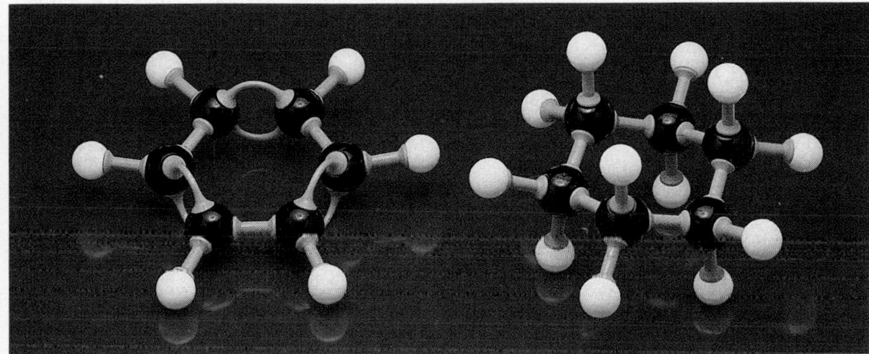

Benzene (left) and cyclohexane (right). (Charles D. Winters)

Instead of addition reactions, benzene and other aromatic compounds typically undergo *substitution reactions*. For example, one can halogenate (with a catalyst), nitrate, or alkylate the ring (Figure 23.1).

Cyclohexane, C_6H_2, is one of a class of cyclic alkanes of general formula C_nH_{2n}.

EXERCISE 23.11 NAMING AROMATIC COMPOUNDS

Give the systematic names for

$Br-C_6H_4-NO_2$　　　　$C_6H_4(Cl)(OH)$ (Cl and OH on adjacent ring carbons)

23.8 ALKYL HALIDES

Alkyl halides such as iodomethane (CH_3I) undergo a wide variety of reactions and are valuable intermediates in the synthesis of many other organic compounds (Table 23.6).

REACTIONS OF ALKYL HALIDES

An important reaction of alkyl halides is **elimination** of hydrogen halide, a reaction commonly called *dehydrohalogenation*. A hydrogen atom and a halogen atom are removed from *adjacent* carbon atoms by reacting the

TABLE 23.6 Some Simple Alkyl Halides

FORMULA	SYSTEMATIC NAME	COMMON NAME	Bp, °C
CH_3Cl	Chloromethane	Methyl chloride	−24
CH_3I	Iodomethane	Methyl iodide	41
CH_2Cl_2	Dichloromethane	Methylene chloride	40
$CHCl_3$	Trichloromethane	Chloroform	61
CH_3CH_2Cl	Chloroethane	Ethyl chloride	12
$CH_3CH(Br)CH_3$	2-Bromopropane	Isopropyl bromide	59
$CH_3—C(CH_3)_2—Br$	2-Bromo-2-methylpropane	*tert*-Butyl bromide*	72

**tert* stands for *tertiary*; it means that three methyl groups are attached to a single carbon.

alkyl halide with a strong base (potassium hydroxide) in ethanol as the solvent. The result is an alkene.

$$\underset{\substack{\text{alkyl halide}\\ X\,=\,Cl,\,Br,\,I}}{—\underset{H}{\overset{|}{\underset{|}{C}}}—\underset{X}{\overset{|}{\underset{|}{C}}}—} \xrightarrow[\text{KOH in ethanol}]{\text{elimination}} \underset{\text{alkene}}{>C{=}C<} + HX$$

In many cases, particularly with more complex alkyl halides, more than one isomeric alkene can be formed. Generally, the favored product has the fewest H atoms attached to the carbon atoms of the double bond.

$$CH_3CH_2\underset{Br}{\underset{|}{C}}HCH_3 \xrightarrow[\text{ethanol}]{\text{KOH}} \underset{\substack{\text{major product}\\ \text{2-butene}\\ \text{(both } cis \text{ and } trans)}}{CH_3CH{=}CHCH_3} + \underset{\substack{\text{minor product}\\ \text{1-butene}}}{CH_3CH_2CH{=}CH_2} + HBr$$

When aqueous NaOH is substituted for KOH in ethanol, alkyl halides follow a different reaction course. Now the reaction is a **substitution**, and OH^- replaces the halogen to give an alcohol.

$$\underset{\substack{\text{alkyl halide}\\ X\,=\,Cl,\,Br,\,\text{or } I}}{RX} \xrightarrow[\text{NaOH/H}_2\text{O}]{\text{substitution}} \underset{\text{alcohol}}{ROH} + NaX$$

Other negatively charged ions can substitute for the halogen as well. For example, the methoxide ion, prepared by reacting methanol with sodium, will react with 1-bromopropane to produce methyl propyl ether.

$$\underset{\substack{\text{alkyl halide}\\ X\,=\,Cl,\,Br,\,I}}{RX} + \underset{\substack{\text{sodium alkoxide}\\ \text{(prep'd from}\\ \text{corresponding alcohol)}}}{NaOR'} \xrightarrow{\text{substitution}} \underset{\text{ether}}{R—O—R'} + NaX$$

$$\underset{\text{sodium methoxide}}{NaOCH_3} + \underset{\text{1-bromopropane}}{CH_3CH_2CH_2Br} \longrightarrow \underset{\text{methyl propyl ether}}{CH_3CH_2CH_2{-}O{-}CH_3} + NaBr$$

One of the most important reactions of alkyl (and aromatic) halides is their reaction with metallic magnesium to give an organometallic compound known as a **Grignard reagent**. These reagents are extraordinarily versatile synthetic intermediates and are widely used. For example, iodomethane reacts with magnesium to produce methylmagnesium iodide.

$$\underset{\text{alkyl or aryl halide}}{RX} + Mg \longrightarrow \underset{\text{Grignard reagent}}{RMgX}$$

$$\underset{\text{iodomethane}}{CH_3I} + Mg \longrightarrow \underset{\text{methylmagnesium iodide}}{CH_3MgI}$$

Grignard reagents react with a wide range of Brønsted acids, including water, to produce the corresponding alkanes.

$$\underset{\text{Grignard reagent}}{RMgX} + H^+ \longrightarrow RH + MgX^+$$

For this reason, the presence of water must be avoided when Grignard reagents are prepared and when they are used in reactions.

Victor Grignard (1871–1935) was a French chemist. He received the Nobel Prize in Chemistry in 1912 for the discovery of the compounds we now call "Grignard reagents."

EXERCISE 23.12 REACTIONS OF ALKYL AND ARYL HALIDES

Complete the following reactions:

(a) $CH_3\underset{\underset{CH_3}{|}}{C}HCH_2CH_2I$ + KOH in ethanol $\longrightarrow$

(b) $CH_3CH_2{-}\underset{\underset{Cl}{|}}{C}H{-}CH_3$ + KOH in ethanol $\longrightarrow$

(c) $CH_3\overset{\overset{CH_3}{|}}{C}H{-}CH_2I$ + NaOH in water $\longrightarrow$

(d) $CH_3\overset{\overset{CH_3}{|}}{C}H{-}CH_2I + Mg \longrightarrow ? \xrightarrow{H_2O} ?$

23.9 CARBOXYLIC ACIDS AND ESTERS

Carboxylic acids are the end products of alcohol oxidation. A number are found in nature and have been known for many years. As a result, some of the familiar carboxylic acids are known by their common names.

The systematic names of carboxylic acids are easily derived: "-e" is dropped from the name of the corresponding alkane and "-oic" added followed by the word "acid" (as in "alkanoic acid").

The simplest common acid is formic acid (Table 23.7), the substance responsible for the sting of an ant. Therefore, the name of the acid comes from the Latin word (*formica*) for ant. Acetic acid gives the sour taste to vinegar, and the name comes from the Latin word for this substance. Butyric acid gives rancid butter its unpleasant odor, and the name is related to the Latin word for butter, *butyrum*. The names for caproic acid, as well as caprilic (C_8) and capric (C_{10}) acids are all derived from

TABLE 23.7 Some Simple Carboxylic Acids

STRUCTURE	COMMON NAME	SYSTEMATIC NAME	Bp, °C	K_a
$H{-}C({=}O){-}OH$	Formic acid	Methanoic acid	101	1.8×10^{-4}
$CH_3C({=}O)OH$	Acetic acid	Ethanoic acid	118	1.8×10^{-5}
$CH_3CH_2C({=}O)OH$	Propionic acid	Propanoic acid	141	1.3×10^{-5}
$CH_3(CH_2)_2C({=}O)OH$	Butyric acid	Butanoic acid	163	1.5×10^{-5}
$CH_3(CH_2)_3C({=}O)OH$	Caproic acid	Pentanoic acid	187	1.4×10^{-5}

the Latin word for goat, since these acids combine to give goats their characteristic odor. In general, the simpler carboxylic acids have unpleasant odors.

Acetic acid is a major chemical with almost 3 billion pounds produced in the United States annually. Most of the production is used ultimately in the manufacture of synthetic fibers of various kinds.

The only other acids produced in large quantity are several with two carboxylic acid groups.

$HOC({=}O)(CH_2)_4C({=}O)OH$ — adipic acid

$HO{-}C({=}O){-}C_6H_4{-}C({=}O){-}OH$ — terephthalic acid

$C_6H_4(C({=}O){-}OH)_2$ — phthalic acid

The three acids above are used to manufacture polymers of types described in Chapter 24. There are others, however, that you may have seen before (Table 23.8), since some occur in nature.

TABLE 23.8 Some Common Dicarboxylic Acids

FORMULA	COMMON NAME	NATURAL SOURCE
$HOOC{-}COOH$	Oxalic acid	Occurs in many plants and vegetables
$HOOC{-}CH_2{-}COOH$	Malonic acid	—
$HOOC{-}CH_2CH_2{-}COOH$	Succinic acid	Occurs in fossils, lichens; discovered by Agricola in 1546
$HOOC{-}CH_2CH_2CH_2{-}COOH$	Glutaric acid	In sugar beets

PROPERTIES OF CARBOXYLIC ACIDS

Carboxylic acids are polar and readily form hydrogen bonds. In the pure liquid state, a pair of carboxylic acid molecules is held together by two hydrogen bonds.

$$\text{R—C}\begin{matrix}\nearrow\!\!\!=\text{O---H—O}\searrow \\ \searrow\text{O—H---O}=\!\!\!\nearrow\end{matrix}\text{C—R}$$

Hydrogen-bonded dimer of acetic acid. (Charles D. Winters)

This hydrogen bonding results in relatively high boiling points for the acid, even higher than those of alcohols of comparable molar mass. For example, formic acid (46 g/mol) has a boiling point of 101°C, while ethanol (46 g/mol) has a boiling point of only 78°C. As a final comparison, you might note that propane, a nonpolar alkane with a comparable molar mass (44 g/mol) has a very low boiling point (−42°C).

All organic acids are weak Brønsted acids with K_a values in the vicinity of 10^{-5}. As you would expect, all react with bases to form salts.

Names of carboxylic acid salts are in two parts: (a) name of cation; (b) name of acid anion derived by dropping the "-ic" ending from acid name and adding "-ate." For example, CH_3COONa is sodium acetate.

$$\text{R}\overset{\text{O}}{\overset{\|}{\text{C}}}\text{OH(aq)} + \text{base(aq)} \longrightarrow \text{baseH}^+\text{(aq)} + \text{R}\overset{\text{O}}{\overset{\|}{\text{C}}}\text{O}^-\text{(aq)}$$

$$\text{C}_6\text{H}_5\text{—}\overset{\text{O}}{\overset{\|}{\text{C}}}\text{OH(aq)} + \text{KOH(aq)} \longrightarrow \text{H}_2\text{O} + \text{C}_6\text{H}_5\text{—}\overset{\text{O}}{\overset{\|}{\text{C}}}\text{OK(aq)}$$

benzoic acid — potassium benzoate

CHEMISTRY OF CARBOXYLIC ACIDS

One of the major reactions of primary alcohols is their oxidation to carboxylic acids, and the reverse of this is also important. An acid can be **reduced** to the corresponding alcohol with a metal hydride such as $LiAlH_4$, lithium aluminum hydride.

$$\underset{\text{carboxylic acid}}{\text{R}\overset{\text{O}}{\overset{\|}{\text{C}}}\text{OH}} \xrightarrow[\text{LiAlH}_4]{\text{reduction}} \underset{\text{primary alcohol}}{\text{RCH}_2\text{OH}}$$

Such reactions are often used in the synthesis of new alcohols.

Carboxylic acids react with alcohols to produce **esters**, as in the reaction of acetic acid with ethanol to give ethyl acetate.

The reaction of an acid and an alcohol is often called an *esterification.*

The two-part name of an ester is given by (a) the name of the alkyl group from the alcohol and (b) the name of the carboxylate group derived from the acid. In general, ester names are of the form "alkyl carboxylate."

$$\underset{\text{carboxylic acid}}{\text{R}\overset{\text{O}}{\overset{\|}{\text{C}}}\text{—(O—H)}} + \underset{\text{alcohol}}{\text{R}'\text{—O—(H)}} \xrightarrow{\text{H}^+} \underset{\text{ester}}{\text{R}\overset{\text{O}}{\overset{\|}{\text{C}}}\text{OR}'} + \text{H}_2\text{O}$$

$$\underset{\text{acetic acid}}{\text{CH}_3\overset{\text{O}}{\overset{\|}{\text{C}}}\text{OH}} + \underset{\text{ethanol}}{\text{CH}_3\text{CH}_2\text{OH}} \xrightarrow{\text{H}^+} \underset{\text{ethyl acetate}}{\text{CH}_3\overset{\text{O}}{\overset{\|}{\text{C}}}\text{OCH}_2\text{CH}_3} + \text{H}_2\text{O}$$

TABLE 23.9 Some Common Esters

NAME	FORMULA	ODOR OF
n-Butyl acetate	$CH_3COOC_4H_9$	Bananas
Ethyl butyrate	$C_3H_7COOC_2H_5$	Pineapples
n-Amyl butyrate	$C_3H_7COOC_5H_{11}$	Apricots
n-Octyl acetate	$CH_3COOC_8H_{17}$	Oranges
Isoamyl isovalerate	$C_4H_9COOC_5H_{11}$	Apples
Methyl salicylate	$C_6H_4(OH)(COOCH_3)$	Oil of wintergreen
Methyl anthranilate	$C_6H_4(NH_2)(COOCH_3)$	Grapes

Although small amounts of esters are used as fragrances, the ester group as a side group or as the main link in polymers is more important. Polymers such as cellulose acetate and various polyesters such as "Dacron" are described in Chapter 24.

Unlike the acids from which esters are derived, esters often have pleasant odors (Table 23.9). A typical example is methyl salicylate or "oil of wintergreen."

$$C_6H_4(OH)-\overset{O}{\overset{\|}{C}}-OCH_3$$

methyl salicylate

EXERCISE 23.13 NAMES OF ACIDS AND ESTERS

Give a name for each of the following:

(a) $CH_3(CH_2)_4\overset{O}{\overset{\|}{C}}OH$

(b) $O_2N-C_6H_4-\overset{O}{\overset{\|}{C}}OH$

(c) $CH_3\overset{O}{\overset{\|}{C}}OCH_2CH_2CH_3$

EXERCISE 23.14 REACTIONS OF ACIDS

Complete the following reactions:

(a) $CH_3\underset{}{\overset{CH_3}{\overset{|}{C}}}HCH_2\overset{O}{\overset{\|}{C}}OH + LiAlH_4 \longrightarrow$

(b) $C_6H_5-\overset{O}{\overset{\|}{C}}OH + CH_3OH$ (in the presence of acid) $\longrightarrow$

(c) $CH_3\underset{\underset{CH_3}{|}}{C}H\overset{O}{\overset{\|}{C}}OH + KOH \longrightarrow$

Methyl salicylate, oil of wintergreen. (Charles D. Winters)

As a class, esters are not very reactive. Perhaps their most important reaction is *hydrolysis* in the presence of a strong base to give the constituents of the ester: the alcohol and a salt of the acid from which the ester was formed.

$$\underset{\text{ester}}{RC(=O)—O—R'} + NaOH(aq) \xrightarrow{\text{heat}} \underset{\text{carboxylate salt}}{RC(=O)—O^-Na^+} + \underset{\text{alcohol}}{R'OH}$$

$$\underset{\text{ethyl acetate}}{CH_3C(=O)—O—CH_2CH_3} + NaOH \xrightarrow{\text{heat}} \underset{\text{sodium acetate}}{CH_3C(=O)—O^-Na^+} + \underset{\text{ethanol}}{CH_3CH_2OH}$$

The hydrolysis of esters is also called a **saponification** reaction, since the hydrolysis of a special type of esters, fats and oils, produces soaps. **Fats** (solids) and **oils** (liquids) are **triesters** of acids with 1,2,3-trihydroxypropane (glycerol) and have the general structural formula

$$\underset{\text{glycerol}}{CH_2(OH)—CH(OH)—CH_2(OH)} + \underset{\text{"fatty acid"}}{3\ RC(=O)OH} \longrightarrow \underset{\text{fat or oil}}{CH_2(O—C(=O)R)—CH(O—C(=O)R)—CH_2(O—C(=O)R)}$$

R can be the same or different groups within the same fat or oil, and the R groups can be saturated or unsaturated. The acid portions of fats almost always have an even number of carbon atoms (C_{16} or C_{18}), and the R group is usually saturated. In contrast, the R group in oils is usually unsaturated. Some acids that commonly occur in fats and oils are

SATURATED ACIDS

Butyric	C_4	$CH_3CH_2CH_2COOH$
Lauric	C_{12}	$CH_3(CH_2)_{10}COOH$
Myristic	C_{14}	$CH_3(CH_2)_{12}COOH$
Palmitic	C_{16}	$CH_3(CH_2)_{14}COOH$
Stearic	C_{18}	$CH_3(CH_2)_{16}COOH$

UNSATURATED ACIDS

Oleic	C_{18}	$CH_3(CH_2)_7CH{=}CH(CH_2)_7COOH$
Linolenic	C_{18}	$CH_3CH_2CH{=}CHCH_2CH{=}CHCH_2CH{=}CH(CH_2)_7COOH$

Waxes are closely related to fats and oils in that all are esters. Waxes, however, are esters fomed between a carboxylic acid and an alcohol with only one —OH group. For example, beeswax ($C_{15}H_{31}COOC_{30}H_{61}$) is an ester of myricyl alcohol, $C_{30}H_{61}OH$. (Charles D. Winters)

Some oils with which you may be familiar are olive oil and soybean oil. However, a bottle of "vegetable oil" in the grocery store indicates that it is "partially" hydrogenated soybean oil; hydrogen has been added to some of the double bonds in the "fatty acid" to improve the properties of the oil. On the other hand, peanut butter usually contains fully hydrogenated oil. There are also many examples of naturally occurring fats such as milk or butter fat.

Like any ester, fats and oils are hydrolyzed by strong bases to give glycerol and a salt of the "fatty acid." The acid salt is a **soap** and is the origin of the name of the process (saponification).

$$\begin{array}{l} CH_2-O-\overset{\overset{\displaystyle O}{\|}}{C}-(CH_2)_{16}CH_3 \\ | \\ CH-O-\overset{\overset{\displaystyle O}{\|}}{C}-(CH_2)_{16}CH_3 \\ | \\ CH_2-O-\overset{\overset{\displaystyle O}{\|}}{C}-(CH_2)_{16}CH_3 \end{array} + 3NaOH \xrightarrow{\Delta} \begin{array}{l} CH_2-OH \\ | \\ CH-OH \\ | \\ CH_2-OH \end{array} + NaO-\overset{\overset{\displaystyle O}{\|}}{C}-(CH_2)_{16}CH_3$$

glyceryl tristearate, a fat — glycerol — sodium stearate, a soap

23.10 CARBONYL COMPOUNDS: ALDEHYDES AND KETONES

This means that there is an unhybridized *p* orbital perpendicular to the trigonal plane, a *p* orbital that can be used to form the C=O pi bond (as explained in Chapter 10).

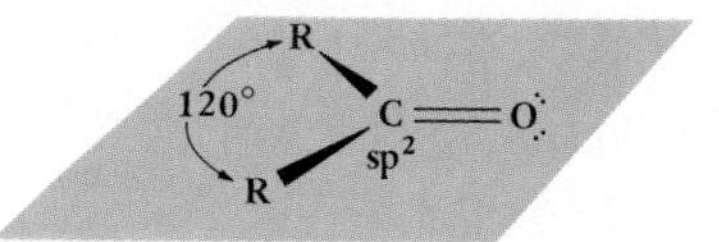

Aldehydes and ketones are characterized by the $\rangle C{=}O$ or carbonyl group. In both types of molecules, the carbon atom is the center of a trigonal, planar arrangement of atoms, so this central carbon atom is sp^2 hybridized.

Like esters, aldehydes and ketones can have pleasant odors and are often used as the basis of fragrances. Benzaldehyde, for example, is responsible for the characteristic odor of almonds, and citral is found in lemons and oranges and is used in perfumes. Muscone, a ketone, is the odoriferous component of musks; the compound is now synthesized, but it can be obtained from the musk glands of various animals.

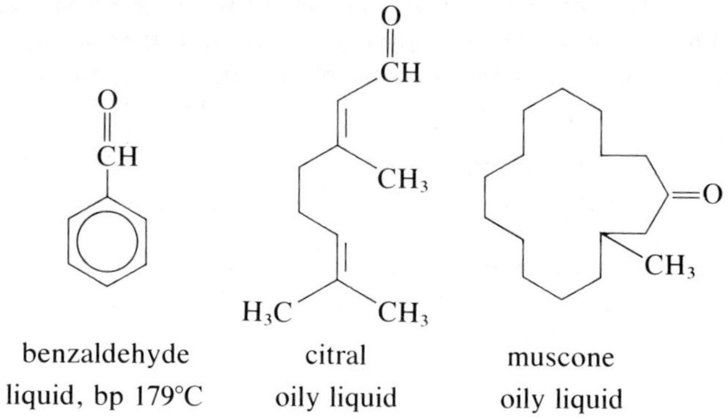

benzaldehyde, liquid, bp 179°C — citral, oily liquid — muscone, oily liquid

NAMING ALDEHYDES AND KETONES

As in other classes of organic compounds, simple aldehydes generally have common names. These are derived from the common name of the acid by dropping the "-ic" suffix and adding "aldehyde." Thus, we have formaldehyde and acetaldehyde, among others (Table 23.10). Systematic names of aldehydes are formed by dropping the final "-e" from the name of the parent alkane and adding "-al." Thus, formaldehyde, HCHO, should be named methanal.

The simplest possible ketone, $CH_3C(O)CH_3$, is called *acetone* (Table 23.10). Most other ketones are named systematically by replacing the

TABLE 23.10 Simple Aldehydes and Ketones

STRUCTURE	COMMON NAME	SYSTEMATIC NAME	BP, °C
$H{-}C({=}O){-}H$	Formaldehyde	Methanal	−21
$CH_3C({=}O)H$	Acetaldehyde	Ethanal	21
$CH_3C({=}O)CH_3$	Acetone	Propanone	56
$CH_3C({=}O)CH_2CH_3$	Methyl ethyl ketone	Butanone	80
$CH_3CH_2C({=}O)CH_2CH_3$	Diethyl ketone	3-Pentanone	102

"-e" suffix of the corresponding alkane with "-one." Thus, the general systematic name is "alkanone" (Table 23.10).

CHEMISTRY OF ALDEHYDES AND KETONES

As you learned from the chemistry of alcohols, aldehydes are intermediate in oxidation level between *primary* alcohols and carboxylic acids. Thus, aldehydes are easily oxidized to the corresponding carboxylic acid

$$\underset{\text{aldehyde}}{R{-}\overset{\overset{\displaystyle O}{\|}}{C}{-}H} \xrightarrow{KMnO_4 \text{ or } Na_2Cr_2O_7} \underset{\text{carboxylic acid}}{R{-}\overset{\overset{\displaystyle O}{\|}}{C}{-}OH}$$

or reduced to the corresponding alcohol. Common reducing agents are the metal hydrides sodium borohydride, $NaBH_4$, or lithium aluminum hydride, $LiAlH_4$.

$$\underset{\text{aldehyde}}{R{-}\overset{\overset{\displaystyle O}{\|}}{C}{-}H} \xrightarrow{NaBH_4 \text{ or } LiAlH_4} \underset{\text{primary alcohol}}{RCH_2OH}$$

In a similar manner, ketones can be reduced to *secondary* alcohols with metal hydride reducing agents.

$$\underset{\text{ketone}}{R{-}\overset{\overset{\displaystyle O}{\|}}{C}{-}R'} \xrightarrow{NaBH_4 \text{ or } LiAlH_4} \underset{\text{secondary alcohol}}{R{-}\underset{\underset{\displaystyle H}{|}}{\overset{\overset{\displaystyle O{-}H}{|}}{C}}{-}R'}$$

Alkyl and aryl halides react with magnesium to give Grignard reagents (Section 23.8). Reaction of these reagents with aldehydes and ketones is of great synthetic utility, since this represents a way to attach the R group of the Grignard reagent to another compound.

$$\underset{\text{aldehyde or ketone}}{-\overset{\overset{\displaystyle O}{\|}}{C}-} + \underset{\text{Grignard reagent}}{RMgX} \longrightarrow \left[R-\overset{|}{\underset{|}{C}}-O-MgX \right] \xrightarrow[H_2O]{H^+} \underset{\text{alcohol}}{R-\overset{|}{\underset{|}{C}}-OH}$$

The initial product of the reaction is the magnesium salt of the weakly acidic alcohol. However, on adding aqueous acid, the salt is converted in high yield to the alcohol itself.

Formaldehyde reacts with Grignard reagents to produce only primary alcohols.

$$\underset{\text{formaldehyde}}{H-\overset{\overset{\displaystyle O}{\|}}{C}-H} + \underset{\text{Grignard reagent}}{RMgX} \longrightarrow H-\overset{\overset{\displaystyle O-MgX}{|}}{\underset{\underset{\displaystyle R}{|}}{C}}-H \xrightarrow[H_2O]{H^+} \underset{\text{primary alcohol}}{H-\overset{\overset{\displaystyle O-H}{|}}{\underset{\underset{\displaystyle R}{|}}{C}}-H}$$

Thus, if you wish to synthesize a primary alcohol, you can begin with an alkyl halide with one less carbon atom than the desired alcohol. After making the Grignard reagent, treatment with formaldehyde leads to the desired product, often in good yield.

Careful examination of the reaction of formaldehyde above shows that an aldehyde other than formaldehyde will lead to *secondary* alcohols.

$$\underset{\text{aldehyde}}{R-\overset{\overset{\displaystyle O}{\|}}{C}-H} + \underset{\text{Grignard reagent}}{R'MgX} \longrightarrow R-\overset{\overset{\displaystyle O-MgX}{|}}{\underset{\underset{\displaystyle R'}{|}}{C}}-H \xrightarrow[H_2O]{H^+} \underset{\text{secondary alcohol}}{R-\overset{\overset{\displaystyle O-H}{|}}{\underset{\underset{\displaystyle R'}{|}}{C}}-H}$$

$$\underset{\text{propanal}}{CH_3CH_2\overset{\overset{\displaystyle O}{\|}}{C}H} + \underset{\text{methyl magnesium iodide}}{CH_3MgI} \longrightarrow CH_3CH_2-\overset{\overset{\displaystyle O-MgBr}{|}}{\underset{\underset{\displaystyle CH_3}{|}}{C}}-H \xrightarrow[H_2O]{H^+} \underset{\text{2-butanol}}{CH_3CH_2-\overset{\overset{\displaystyle O-H}{|}}{\underset{\underset{\displaystyle CH_3}{|}}{C}}-H}$$

In contrast, *ketones provide tertiary alcohols* on treatment with a Grignard reagent.

$$\underset{\text{ketone}}{R-\overset{\overset{\displaystyle O}{\|}}{C}-R'} + \underset{\text{Grignard reagent}}{R''MgX} \longrightarrow R-\overset{\overset{\displaystyle O-MgX}{|}}{\underset{\underset{\displaystyle R''}{|}}{C}}-R' \xrightarrow[H_2O]{H^+} \underset{\text{tertiary alcohol}}{R-\overset{\overset{\displaystyle OH}{|}}{\underset{\underset{\displaystyle R''}{|}}{C}}-R'}$$

$$\underset{\text{acetone}}{CH_3\overset{\overset{O}{\|}}{C}CH_3} + \underset{\text{ethyl magnesium bromide}}{CH_3CH_2MgBr} \longrightarrow CH_3-\overset{\overset{O-MgBr}{|}}{\underset{\underset{CH_2CH_3}{|}}{C}}-CH_3 \xrightarrow[H_2O]{H^+} \underset{\text{2-methyl-2-butanol}}{CH_3-\overset{\overset{O-H}{|}}{\underset{\underset{CH_2CH_3}{|}}{C}}-CH_3}$$

Carbon dioxide, O=C=O, can be considered as a simple carbonyl functional group, so it is not surprising that it also reacts with Grignard reagents. The reaction is very useful synthetically, since the ultimate product is a carboxylic acid.

$$\underset{\text{Grignard reagent}}{RMgX} + O=C=O \longrightarrow R-\overset{\overset{O}{\|}}{C}-OMgX \xrightarrow[H_2O]{H^+} \underset{\text{carboxylic acid}}{R-\overset{\overset{O}{\|}}{C}-OH}$$

$$\underset{\text{phenyl magnesium bromide}}{C_6H_5-MgBr} + O=C=O \longrightarrow C_6H_5-\overset{\overset{O}{\|}}{C}-OMgBr \xrightarrow[H_2O]{H^+} \underset{\text{benzoic acid}}{C_6H_5-\overset{\overset{O}{\|}}{C}-OH}$$

Aldehydes and ketones are particularly valuable compounds in organic chemistry, since they can be transformed into so many other compounds. These reactions can be summarized as follows:

	Reducing agent		**Grignard**	
methanol	⟵	formaldehyde	⟶	primary alcohol
primary alcohol	⟵	general aldehyde	⟶	secondary alcohol
secondary alcohol	⟵	ketone	⟶	tertiary alcohol

EXERCISE 23.15 STRUCTURES OF ALDEHYDES AND KETONES
Draw structural formulas of the following compounds:
(a) 2-pentanone (b) *m*-bromobenzaldehyde (c) butanal

EXERCISE 23.16 NAMES OF ALDEHYDES AND KETONES
Give systematic names for the following compounds:

(a) $CH_3CH_2CH_2\overset{\overset{O}{\|}}{C}CH_2CH_3$ (b) $CH_3CH_2CH_2CH_2\overset{\overset{O}{\|}}{C}H$

EXERCISE 23.17 REACTIONS OF ALDEHYDES AND KETONES
Complete the following reactions:

(a) $H_3C-CH_2-\overset{\overset{O}{\|}}{C}H + NaBH_4 \longrightarrow ?$

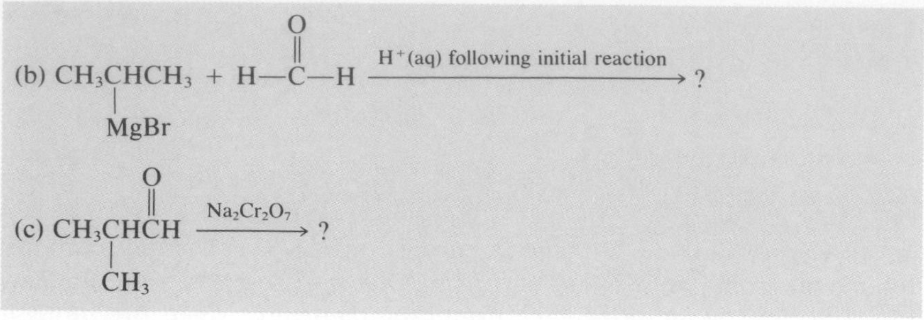

(b) $CH_3\underset{\displaystyle MgBr}{\underset{|}{C}}HCH_3 + H—\overset{\displaystyle O}{\overset{\|}{C}}—H \xrightarrow{H^+(aq)\text{ following initial reaction}} ?$

(c) $CH_3\underset{\displaystyle CH_3}{\underset{|}{C}}H\overset{\displaystyle O}{\overset{\|}{C}}H \xrightarrow{Na_2Cr_2O_7} ?$

23.11 SYNTHESIS OF ORGANIC COMPOUNDS

The purpose of the enormous organic chemicals and pharmaceuticals industries is to prepare new materials for consumers or to prepare existing compounds more efficiently. To design a route from a starting material to a final product, chemists piece together one or more reactions, often going through several steps to accomplish the synthesis. You have been introduced to one or two reactions of each class of organic compounds. Our goal in doing this was to place you in a position to do what chemists actually do: to use these reactions to design, at least on paper, the synthesis of a compound. We shall look at two simple examples, but you should be aware that there are many more reactions available to the synthetic chemist than we can possibly describe here. Such reactions are covered in courses on organic chemistry.

Suppose we wished to make 3-hexanone, and the only available organic starting material was 1-propanol. The way to approach such problems is usually to work backward from the desired compound. For example, we know that 3-hexanone can be made by the oxidation of 3-hexanol.

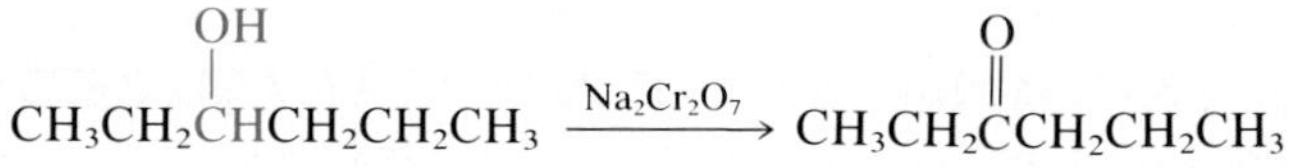

$$\underset{\substack{\text{3-hexanol}\\\text{(starting material)}}}{CH_3CH_2\overset{\displaystyle OH}{\overset{|}{C}}HCH_2CH_2CH_3} \xrightarrow{Na_2Cr_2O_7} \underset{\text{3-hexanone}}{CH_3CH_2\overset{\displaystyle O}{\overset{\|}{C}}CH_2CH_2CH_3}$$

The starting material for the reaction above is a secondary alcohol, and we know that secondary alcohols can be made by treating an aldehyde with the appropriate Grignard reagent. Thus, to obtain a six-carbon alcohol, we can use a three-carbon Grignard reagent and a three-carbon aldehyde.

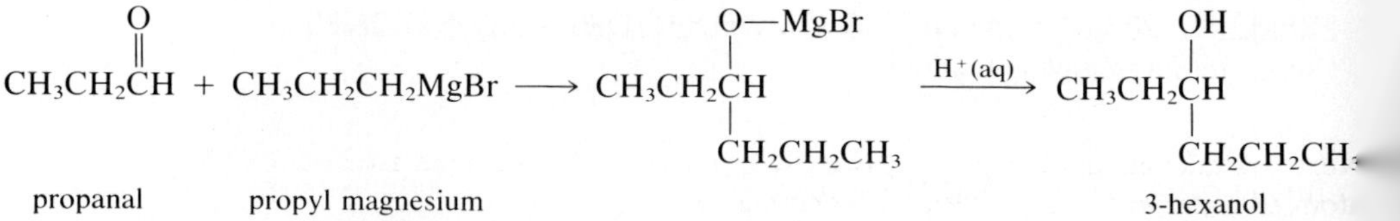

$$\underset{\text{propanal}}{CH_3CH_2\overset{\displaystyle O}{\overset{\|}{C}}H} + \underset{\substack{\text{propyl magnesium}\\\text{bromide}}}{CH_3CH_2CH_2MgBr} \longrightarrow CH_3CH_2\underset{\displaystyle CH_2CH_2CH_3}{\underset{|}{\overset{\displaystyle O—MgBr}{\overset{|}{C}}}}H \xrightarrow{H^+(aq)} \underset{\text{3-hexanol}}{CH_3CH_2\underset{\displaystyle CH_2CH_2CH_3}{\underset{|}{\overset{\displaystyle OH}{\overset{|}{C}}}}H}$$

To carry out the reaction above, we must have propanal and a Grignard reagent derived from 1-bromopropane. Both of these can be obtained from 1-propanol by the following reactions

$$\underset{\text{1-propanol}}{CH_3CH_2CH_2OH} \xrightarrow{\text{PCC}} \underset{\text{propanal}}{CH_3CH_2\overset{\overset{\Large O}{\|}}{C}H}$$

$$\underset{\text{1-propanol}}{CH_3CH_2CH_2OH} \xrightarrow[\text{HBr}]{} \underset{\text{1-bromopropane}}{CH_3CH_2CH_2Br}$$

Thus, it is possible to prepare 3-hexanone from 1-propanol in a series of reactions.

EXAMPLE 23.2

ORGANIC SYNTHESIS

Synthesize 2-methylpropene (commonly called isobutene) from 1-propanol and bromoethane.

Solution As in the text above, we begin by working backward from the desired product. Here we see it is possible to obtain 2-methylpropene by dehydration of 2-methyl-2-propanol (*tert*-butyl alcohol),

$$\underset{\textit{tert}\text{-butyl alcohol}}{CH_3-\overset{\overset{\large CH_3}{|}}{\underset{\underset{\large OH}{|}}{C}}-CH_3} \xrightarrow{H_2SO_4} \underset{\text{2-methylpropene}}{CH_3-\overset{\overset{\large CH_3}{|}}{C}=CH_2} + H_2O$$

and the required starting material for this step can be prepared from a ketone and a Grignard reagent. Specifically, the latter reaction must be between acetone and CH_3MgBr, where the bromomethane starting material is the source of the Grignard reagent.

$$\underset{\text{acetone}}{CH_3-\overset{\overset{\large O}{\|}}{C}-CH_3} + \underset{\text{methyl magnesium bromide}}{CH_3MgBr} \longrightarrow CH_3-\overset{\overset{\large O-MgBr}{|}}{\underset{\underset{\large CH_3}{|}}{C}}-CH_3 \xrightarrow{H^+(aq)} \underset{\textit{tert}\text{-butyl alcohol}}{CH_3-\overset{\overset{\large OH}{|}}{\underset{\underset{\large CH_3}{|}}{C}}-CH_3}$$

The acetone required in this step can be synthesized from 1-propanol in three steps. First, 1-propanol is dehydrated to propene.

$$\underset{\text{1-propanol}}{CH_3CH_2CH_2OH} \xrightarrow{H_2SO_4} \underset{\text{propene}}{CH_3CH{=}CH_2} + H_2O$$

Next, we take advantage of Markovnikov's rule and rehydrate the alkene, the H atom of water attaching itself to the end carbon.

$$CH_3CH{=}CH_2 + H_2O \longrightarrow CH_3\underset{}{\overset{\displaystyle OH}{\overset{|}{C}}}HCH_3$$

propene 2-propanol

Finally, oxidation of this secondary alcohol produces the required ketone, acetone.

$$CH_3\overset{\displaystyle OH}{\overset{|}{C}}HCH_3 \xrightarrow{KMnO_4} CH_3\overset{\displaystyle O}{\overset{\|}{C}}CH_3$$

2-propanol acetone

EXERCISE 23.18 ORGANIC SYNTHESIS

Using an appropriate organic halide as a starting material, outline a synthesis for

$$CH_3\underset{\displaystyle CH_3}{\underset{|}{C}}H\overset{\displaystyle O}{\overset{\|}{C}}OH$$

23.12 AMINES AND AMIDES

Amines are structural derivatives of ammonia with one or more of the three hydrogen atoms replaced by alkyl or aryl groups. They play a significant role in biochemistry and are important industrial reagents.

STRUCTURES, NAMES, AND GENERAL PROPERTIES

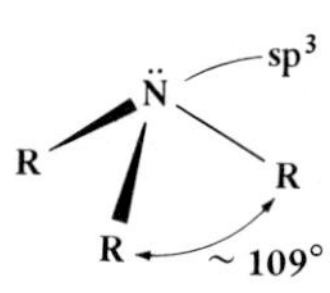

Ammonia is a pyramidal molecule due to the presence of the nitrogen lone pair. Therefore, all amines have similar structures based on an sp^3 hybridized N atom.

Amines are classed as *primary, secondary,* or *tertiary* according to the number of groups bonded to the N atom.

$$CH_3CH_2{-}\overset{\displaystyle H}{\overset{|}{\ddot{N}}}{-}H \qquad CH_3{-}\overset{\displaystyle H}{\overset{|}{\ddot{N}}}{-}CH_3 \qquad CH_3CH_2{-}\overset{\displaystyle CH_3}{\overset{|}{\ddot{N}}}{-}CH_3$$

ethylamine	dimethylamine	ethyldimethylamine
primary	secondary	tertiary

Notice that alkylamines are named simply by identifying the groups bound to the nitrogen, with the group names in alphabetic order; the compound name is completed with the word ''amine'' at the end.

The simplest aromatic amine is aniline, $C_6H_5NH_2$. By convention, aromatic amines are usually named as derivatives of aniline.

:NH_2 on benzene ring; :NH_2 on benzene ring with Cl at positions 2 and 4

aniline 2,4-dichloroaniline

Some of the oxygen-containing organic compounds we have examined thus far are very weak Lewis and Brønsted bases. However, amines are the only organic bases that can turn red litmus blue in aqueous solution.

$$RNH_2(aq) + H_2O(\ell) \rightleftharpoons RNH_3^+(aq) + OH^-(aq)$$

Aliphatic amines are slightly stronger bases than ammonia, but aromatic amines such as aniline are generally significantly weaker (Table 23.11).

TABLE 23.11 Ionization Constants for Some Amines

COMPOUND	K_b
Ammonia	1.8×10^{-5}
Methylamine	5.0×10^{-4}
Dimethylamine	7.4×10^{-4}
Trimethylamine	7.4×10^{-5}
Aniline	4.2×10^{-10}

Amines often have an offensive odor. In more familiar terms, amines are responsible for the odor of decaying fish. The "smell of death" of decaying flesh is due to two appropriately named amines, cadaverine and putrescine.

$$H_2N{-}CH_2CH_2CH_2CH_2{-}NH_2 \qquad H_2N{-}CH_2CH_2CH_2CH_2CH_2{-}NH_2$$

putrescine (1,4-butanediamine) cadaverine (1,5-pentanediamine)

REACTIONS OF AMINES

As bases, amines react with inorganic and organic acids to form ammonium salts.

$$CH_3NH_2(aq) + HCl(aq) \longrightarrow CH_3NH_3^+Cl^-(aq)$$

methylamine methylammonium chloride

Although only a few amines are soluble in water, their salts with inorganic acids are often water soluble. Conversely, the amines are soluble in hydrocarbon solvents, but the salts are not. For example, the amine procaine

$$H_2N{-}C_6H_4{-}\overset{O}{\overset{\|}{C}}CH_2CH_2N(CH_2CH_3)_2$$

procaine

is soluble in ethanol, chloroform, and benzene, but it dissolves in water only to the extent of 1 g in 200 mL. On the other hand, the HCl salt of procaine is soluble to the extent of 1 g in 1 mL of water. In the form of the HCl salt, the compound is sold as a local anesthetic and is most commonly known as Novocain.

AMIDES

Just as an ester can be viewed as derived from a carboxylic acid and an alcohol, an **amide** can be said to be derived from a carboxylic acid and an amine.

$$R\overset{O}{\overset{\|}{C}}{-}OH + H{-}NR_2 \longrightarrow HOH + R\overset{O}{\overset{\|}{C}}{-}NR_2$$

carboxylic acid amine amide

Although this reaction can be used in principle, amides are usually made from a primary or secondary amine and a carboxylic acid derivative, an acid chloride.

$$R'-\overset{\overset{\large O}{\|}}{C}-OH$$

$$\downarrow + SOCl_2$$

$$\underset{\text{amine}}{RNH_2} + \underset{\text{acid chloride}}{R'-\overset{\overset{\large O}{\|}}{C}-Cl} \longrightarrow \underset{\text{amide}}{R'-\overset{\overset{\large O}{\|}}{C}-N\begin{matrix}H\\ R\end{matrix}} + HCl$$

The amide grouping is important in some synthetic polymers and in many naturally occurring compounds, especially in proteins (Chapter 24). However, if you love camping and hiking, one example of an amide you may find especially interesting is *N,N*-diethyl-*m*-toluamide, a compound which is the active ingredient in most effective insect repellents.

$$m\text{-}CH_3C_6H_4-\overset{\overset{\large O}{\|}}{C}-N(CH_2CH_3)_2$$

N,N-diethyl-*m*-toluamide

EXERCISE 23.19 AMINES AND AMIDES

(a) Name the compound $(CH_3CH_2)_2NH$.

(b) Complete the following reaction

$$CH_3\overset{\overset{\large O}{\|}}{C}-Cl + (CH_3)_2NH \longrightarrow$$

23.13 PETROLEUM AND REFINERY PROCESSES

One barrel of petroleum equals 42 U.S. gallons

About 17 million barrels of petroleum are consumed each day in the United States. Approximately 8 million barrels are converted to gasoline, and another 8 million barrels are consumed primarily for other fuel needs, among them heating oil, diesel fuel, and jet engine fuel. The remaining one million barrels used daily provide the raw material for the production of organic chemicals and polymers.

Petroleum is a complex mixture of alkanes, alkenes, alkynes, aromatic hydrocarbons, and other compounds. To obtain useful materials the components must be separated as efficiently as possible. Since most of the materials are volatile, *fractional distillation* is used. The basic technique of separating volatile materials by distillation was described in Section 12.4. This same method on a much larger and more complex scale (Figure 23.2) is applied to the separation of the components of crude petroleum into fractions according to the temperature range in which they boil.

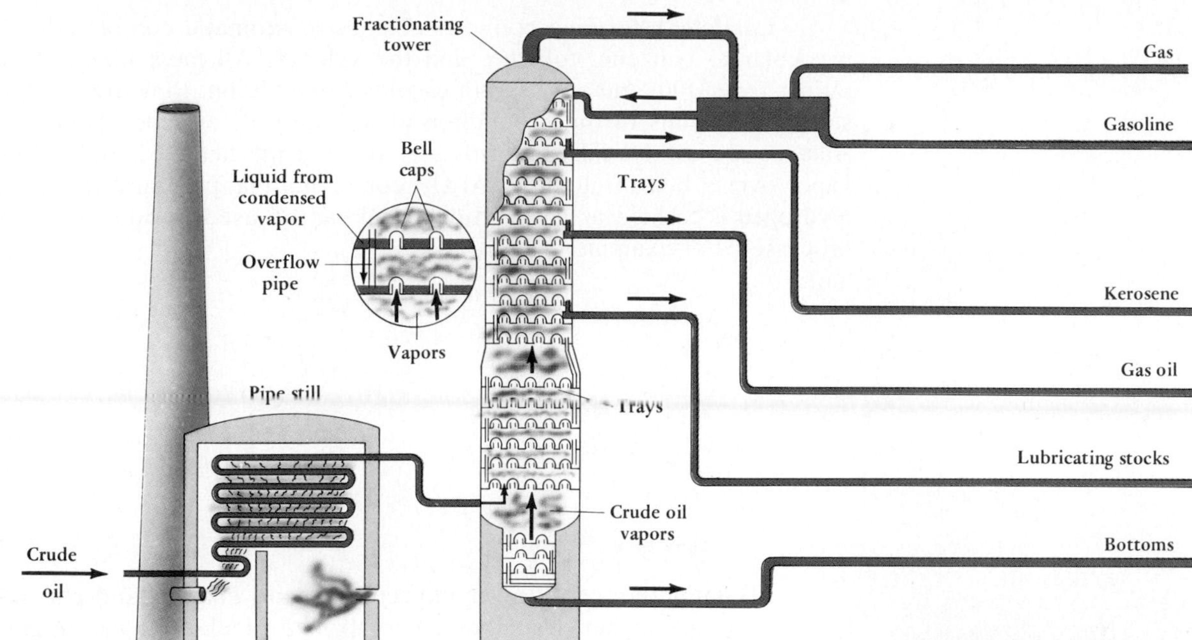

(a)

(b)

FIGURE 23.2

A diagram (a) and a photograph (b) of a modern petroleum refinery that uses distillation to separate volatile components from one another. (b, Exxon Company, USA)

In the United States the most valuable petroleum product is gasoline derived from the fractions boiling from about 30°C to about 200°C. However, the lower boiling material contains a high proportion of straight chain alkanes, and it is not suitable for use in modern higher performance engines; such hydrocarbons have a low octane rating.

One way to raise the octane rating of the lower boiling fractions of petroleum is to add various substances. Tetraethyllead, $(C_2H_5)_4Pb$, is one of the major additives, but this is being phased out because of environmental lead pollution and because the lead poisons the catalysts used in exhaust systems. Other currently used additives are methanol (CH_3OH) and

$$CH_3—O—\underset{CH_3}{\overset{CH_3}{\overset{|}{\underset{|}{C}}}}—CH_3 \qquad CH_3—\underset{CH_3}{\overset{CH_3}{\overset{|}{\underset{|}{C}}}}—OH$$

methyl *tert*-butyl ether *tert*-butyl alcohol

Several refinery processes are also used to raise the octane rating of petroleum to be used as gasoline. For example, **catalytic cracking** breaks down large alkanes from high boiling fractions into smaller, branched-chain alkanes more suitable for use in gasoline. This is done by passing the alkane vapor over a catalyst [such as silica–alumina (SiO_2–Al_2O_3)] at 450°C to 550°C.

Catalytic reforming converts alkanes to aromatic compounds, in particular to benzene, toluene, and the xylenes. All have high octane ratings (over 100) and are used in gasoline. In addition, they are used as starting materials in the production of a variety of organic chemicals. Alkanes are catalytically reformed by passing the heated hydrocarbon vapor over a bed of alumina (Al_2O_3) containing platinum and rhodium. Hydrogen is formed as a by-product and can be used in other refinery processes. For example,

$$C_6H_{12}(g) \xrightarrow{\text{catalytic reforming}} C_6H_6(g) + 3\ H_2(g)$$

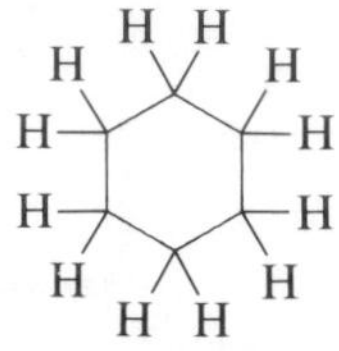

cyclohexane 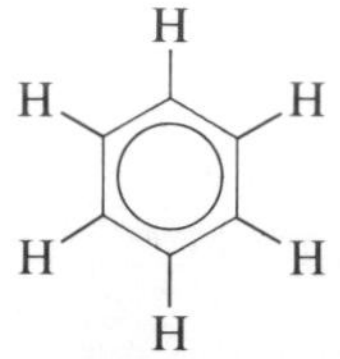   benzene

Alkylation converts lower molecular weight alkanes and alkenes (C_3 and C_4) into higher, branched-chain alkanes suitable for use in gasoline. Hydrofluoric acid and sulfuric acid are typically used as catalysts in this process.

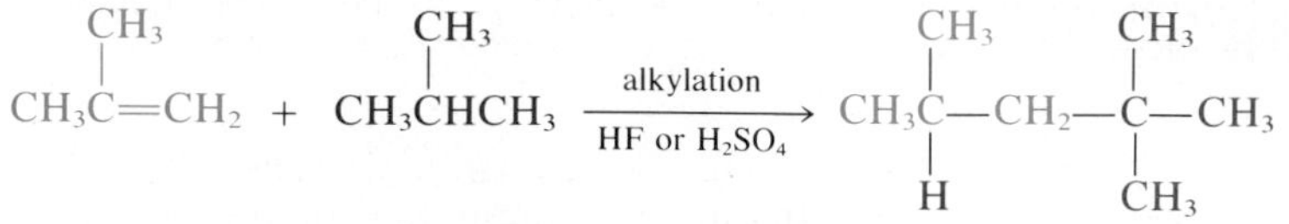

$$CH_3C(CH_3){=}CH_2 + CH_3CH(CH_3)CH_3 \xrightarrow[\text{HF or } H_2SO_4]{\text{alkylation}} CH_3CH(CH_3){-}CH_2{-}C(CH_3)_2{-}CH_3$$

methylpropene   methylpropane   2,2,4-trimethylpentane

The three refinery processes outlined above are used to augment and upgrade the lower boiling fraction of petroleum for use as gasoline. A fourth refinery process, **steam cracking** (page 752) can be used to convert ethane and propane in natural gas to ethylene and propylene, the building blocks of polyethylene and polypropylene.

EXERCISE 23.20 CATALYTIC REFORMING

Write a balanced equation for the catalytic reforming of heptane (C_7H_{16}) to toluene $C_6H_5CH_3$. How many moles of hydrogen are produced per mole of heptane converted?

SUMMARY

Organic chemistry is the study of carbon-containing compounds. Hydrocarbons contain only C and H, and **alkanes**, C_nH_{2n+2}, are one major class of such compounds. (They are often called **saturated** compounds.) For alkanes with 4 or more carbon atoms, **structural isomers** are possible. (Isomers have identical molecular formulas but different atom-to-atom connections.)

Compounds other than alkanes have various **functional groups**, structural fragments found in all members of a class of compounds. Functional groups are the site of the characteristic reactions of the class. The following is a summary of functional groups:

CLASS	FUNCTIONAL GROUP/STRUCTURAL FRAGMENT
Alkenes	C=C double bond
Alkynes	C≡C triple bond
Aromatic compounds	Based on benzene (C_6H_6) 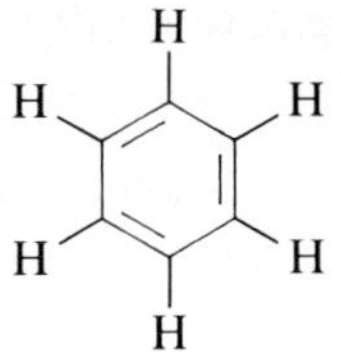
Alkyl halides	R—X where X is F, Cl, Br, or I*
Alcohols	R—OH
Ethers	R—O—R
Carboxylic acid	R—C(=O)—OH
Esters	R—C(=O)—O—R
Aldehydes	R—C(=O)—H (contains a carbonyl group)
Ketones	R—C(=O)—R (contains a carbonyl group)
Amines	R_3N (R is H or an organic group)
Amides	R—C(=O)—NR_2 (R is H or an organic group)

*R is an alkyl group (e.g., CH_3) or aryl group (e.g., C_6H_5).

Major forms of reactivity of organic compounds (Section 23.4) include **oxidation** and **reduction**, **addition** of small molecules such as H_2, X_2 (halogen), H_2O, or HX (where X is a halogen), **elimination** of H_2O, and **substitution** of one atom for another. Each functional group will participate in one or more of these reactions.

Alcohols can undergo substitution of —OH by —X (Br, Cl, or I), elimination of H_2O to give alkenes or ethers, and oxidation to aldehydes or ketones (Section 23.5).

Alkenes (and **alkynes**) (Section 23.7) are often called **unsaturated** because the pi electrons of the carbon–carbon multiple bond can be used to form additional sigma bonds. A C=C double bond prevents rotation about the carbon–carbon bond, so ***cis–trans*** **isomerism** is possible. Alkenes (and alkynes) add small molecules such as H_2 (hydrogenation), X_2 and HX (X is Cl, Br, I), and H_2O (hydration). For the addition of HX to a double bond, **Markovnikov's rule** is usually followed.

Benzene, C_6H_6, is the simplest **aromatic** compound (Section 23.8). The C_6 ring has six equivalent C—C bonds and there are six equivalent C—H bonds, an observation explained by invoking pi resonance structures. Substitution of H is the major form of reactivity of benzene and other aromatic compounds.

Alkyl halides (R—X where X is a halogen) (Section 23.9) participate in elimination reactions. H and X are lost from adjacent atoms to give an alkene; the reaction is called a **dehydrohalogenation**. Alkyl halides also form **Grignard reagents**, R—Mg—X, versatile synthetic reagents for adding R to another molecule.

Carboxylic acids ($R{-}\overset{\overset{O}{\|}}{C}{-}OH$) (Section 23.10) are Brønsted acids and form salts with bases. Acids can be reduced to alcohols. Reaction of an alcohol with an acid gives an **ester**, $R{-}\overset{\overset{O}{\|}}{C}{-}O{-}R$ (esterification). The resulting esters can be hydrolyzed back to acid and alcohol (saponification). **Fats** and **oils** are esters of a tri-alcohol, glycerol.

Carbonyl compounds include **aldehydes** and **ketones** (Section 23.11). Both contain the $\gt C{=}O$ functional group. Aldehydes can be oxidized to acids and reduced to primary alcohols. Ketones cannot be oxidized chemically, but they are reduced to secondary alcohols.

Many organic compounds can be synthesized by combining, in appropriate order, the functional group reactions described above (Section 23.12).

Amines, R_3N (Section 23.13) are classed as primary (two R groups are H), secondary (one R group is H), or tertiary (all R groups are organic fragments). Amines are Brønsted and Lewis bases. **Amides**, $R{-}\overset{\overset{O}{\|}}{C}{-}NR_2$, are made effectively from a carboxylic acid and a primary or secondary amine.

Petroleum is a complex mixture of alkanes, alkenes, alkynes, aromatic compounds, and others (Section 23.14). Refining petroleum into useful products involves many different processes (fractional distillation, catalytic cracking, catalytic reforming, alkylation, and steam cracking, among others).

STUDY QUESTIONS

ALKANES

1. What is the name of the straight-chain alkane with the formula C_8H_{18}?
2. Which of the following hydrocarbons is an alkane: (a) C_2H_4, (b) C_5H_{10}, (c) $C_{14}H_{30}$, (d) C_7H_8?
3. What is the name of alkyl group $CH_3CH_2CH_2CH_2{-}$?
4. What is the molecular formula for an alkane with 13 carbon atoms?
5. Draw four of the nine structural isomers of heptane, C_7H_{16}. Give the systematic name of each isomer you have drawn.
6. In terms of the intermolecular forces described in Chapter 11, explain the trend in boiling points in Table 23.1.
7. Give the systematic name of each of the following compounds:

(a) $CH_3\underset{\underset{CH_3}{|}}{C}H\overset{\overset{CH_3}{|}}{C}HCH_3$

(b) $CH_3CH_2CH_2\underset{\underset{CH_3}{|}}{C}HCH_3$

(c) $CH_3CH_2CH_2\underset{\underset{CH_3}{|}}{\overset{\overset{CH_3}{|}}{C}}CH_2CH_2CH$

(d) $CH_3CH_2CH_2\underset{\underset{CH_2CH_3}{|}}{C}HCH_3$

8. Draw the structure of each of the following compounds:
   (a) 2,3-dimethylpentane

(b) 2,4-dimethyloctane
(c) 3-ethylhexane
(d) 2-methyl-3-ethylhexane

9. Draw the structure of each of the following compounds:
(a) 2,2-dimethylhexane
(b) 3,3-diethylpentane
(c) 2-methyl-3-ethylhexane
(d) isobutane

FUNCTIONAL GROUPS

10. Classify each of the compounds below according to its functional group.
(a) $CH_3CH{=}CHCH_3$
(b) $CH_3{-}O{-}CH_2CH_3$
(c) $C_6H_5C{\equiv}CCH_3$
(d) $CH_3CH_2NH_2$

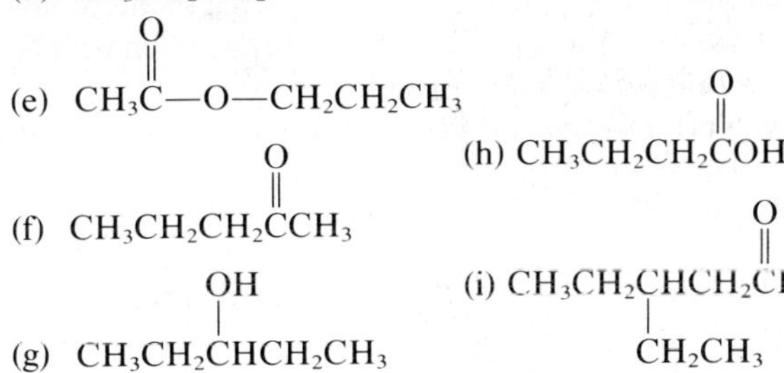

(e) $CH_3C(=O){-}O{-}CH_2CH_2CH_3$
(f) $CH_3CH_2CH_2C(=O)CH_3$
(g) $CH_3CH_2CH(OH)CH_2CH_3$
(h) $CH_3CH_2CH_2C(=O)OH$
(i) $CH_3CH_2CH(CH_2CH_3)CH_2C(=O)H$

11. Draw an electron dot structure of dimethyl ether, $H_3C{-}O{-}CH_3$. What is the approximate C—O—C bond angle? What is the hybridization of the O atom?

12. Draw an electron dot structure for acetaldehyde, $H_3C{-}C(=O){-}H$. What is the C—C—O bond angle? Specify the hybridization of each C atom.

13. Draw an electron dot structure for formic acid, $H{-}C(=O){-}O{-}H$. Specify all of the bond angles in the molecule. What is the hybridization of the C atom?

14. One functional group not mentioned in Section 23.3 is the nitrile or cyano group, $-C{\equiv}N$. Draw an electron dot structure for cyanomethane (common name, acetonitrile), $H_3C{-}C{\equiv}N$. What is the C—C—N angle? Specify the hybridization of the C atom in the cyano group.

ALCOHOLS

15. Give systematic names for the following alcohols:
(a) CH_3CH_2OH
(b) $H_3C{-}C(CH_3)_2{-}OH$
(c) $CH_3CH_2CH_2CH_2OH$
(d) $CH_3C(OH)(CH_3)CH_2CH_3$

16. Draw structural formulas for the following alcohols:
(a) 1-pentanol and 2-pentanol
(b) 2-methyl-1-hexanol
(c) 3,3-dimethyl-2-butanol

17. For each alcohol in Study Question 15, tell if it is primary, secondary, or tertiary.

18. Draw structural formulas for all of the alcohols of the formula $C_4H_{10}O$.

19. Draw structural formulas for all of the alcohols of the formula $C_5H_{12}O$.

20. Complete the following reactions:
(a) $CH_3CH_2CH_2CH_2OH + KMnO_4(aq) \rightarrow$
(b) $CH_3CH_2CH_2CH_2OH + PCC \rightarrow$
(c) $CH_3CH_2CH(OH)CH_3 + KMnO_4(aq) \longrightarrow$
(d) $CH_3CH_2CH(OH)CH_3 + HBr \longrightarrow$
(e) $CH_3CH_2CH_2CH_2OH + H_2SO_4/180°C \rightarrow$
(f) $CH_3CH_2OH + Na \rightarrow$

21. Which compound should have the lower boiling point: $CH_3CH_2CH_2OH$ or $H_3C{-}O{-}CH_2CH_3$? Explain your answer briefly.

ALKENES AND ALKYNES

22. What structural requirement is necessary for an alkene to have *cis* and *trans* isomers? Can *cis* and *trans* isomers exist for alkynes?

23. Draw the structures of the *cis–trans* isomers of
(a) 1,2-dichloroethene
(b) 3-methyl-3-hexene

24. Complete the following reactions:
(a) $CH_3CH_2CH{=}CH_2 + Br_2 \rightarrow$
(b) $CH_3CH_2CH{=}CH_2 + HBr \rightarrow$
(c) $CH_3CH_2CH{=}CH_2 + H_2 \rightarrow$
(d) $CH_3CH_2CH{=}CH_2 + H_2O \rightarrow$

25. What products are expected for the following reactions?
(a) $(H_3C)_2C{=}C(H)(CH_2CH_3) \xrightarrow{HBr}$
(b) $(H_3C)_2C{=}C(H)(CH_2CH_3) \xrightarrow{H_2}$

26. Give the systematic name of each of the products of the reactions in Study Question 20.
27. Starting with the appropriate alkene, tell how you could prepare each of the following in a single reaction step.
    (a) 1,2-dibromobutane (c) 2-butanol
    (b) 2-bromobutane
28. Starting with 1-butanol, show how one could prepare 2-butanol in two reaction steps.
29. Starting with 1-butanol, show how you could prepare 2-butanone in three reaction steps.
30. Draw structural formulas for (a) 3-hexyne, (b) 1-pentyne, and (c) 4-methyl-2-pentyne.
31. Complete and balance the following reactions:
    (a) $CH_3CH_2C{\equiv}CH + H_2 \rightarrow$
    (b) $CH_3CH_2C{\equiv}CCH_3 + HBr \rightarrow$
    (c) $CH_3CH_2C{\equiv}CCH_3 + Br_2 \rightarrow$

BENZENE AND AROMATIC COMPOUNDS

32. Draw structural formulas for the following compounds:
    (a) *o*-dichlorobenzene (alternatively called 1,2-dichlorobenzene)
    (b) *m*-chlorotoluene (alternatively called 3-chlorotoluene)
    (c) *p*-diethylbenzene
    (d) *p*-chlorophenol
    (e) *p*-chlorostyrene
33. Show how to prepare isopropylbenzene (common name, cumene) from benzene and the appropriate propyl derivative.
34. Show how to prepare chlorobenzene in a single step from benzene and other appropriate reagents.

ALKYL HALIDES

35. One of the typical reactions of an alkyl halide is a dehydrohalogenation. For this reaction to occur, what atoms must be present in a molecule and how must they be placed?
36. Name the three types of reactions of alkyl halides described in the text. Using 2-iodopropane, give an example of each reaction type.
37. Name the type of reaction needed to produce the products below from alkyl halides. In each case give the alkyl halide starting material.
    (a) $CH_3CH_2CH{=}CH_2$
    (b) $CH_3CH_2CH_2CH_2OH$
    (c) $CH_3CH_2CH_2CH_2{-}O{-}CH_3$ (Here the OCH_3 portion arises from methanol. What is the source of the remainder of the compound?)
    (d) $CH_3CH_2CH_2CH_2CH_3$
38. Tell how you could prepare di-*n*-propyl ether starting with 1-bromopropane.
39. Outline a synthesis of 2-deuteropropane from an alkyl iodide, D_2O, and other necessary reagents.

CARBOXYLIC ACIDS AND ESTERS

40. Give a name for
    (a) $CH_3CH_2CH_2CH_2CH_2COOH$.
    (b) $CH_3CH_2\overset{\overset{\large O}{\|}}{C}{-}O{-}CH_3$
    (c) $H_3C\overset{\overset{\large O}{\|}}{C}{-}O{-}CH_2CH_2CH_2CH_3$
    (d) 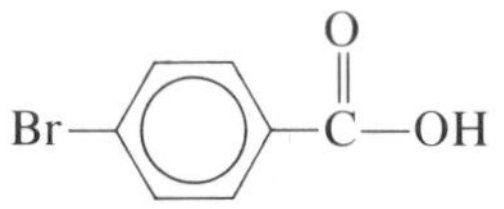 $Br{-}C_6H_4{-}\overset{\overset{\large O}{\|}}{C}{-}OH$ (para-substituted benzene ring)
41. Beginning with butanoic acid, what is the product of
    (a) reduction with $LiAlH_4$ (b) reaction with NaOH
    (c) reaction with ethanol
42. Tell how you would synthesize propyl propanoate beginning with propanoic acid.
43. Give the structure and name of the product of the reaction between benzoic acid and isopropyl alcohol.
44. Maleic acid is a *di*carboxylic acid prepared by the catalytic oxidation of benzene. Its formula is $C_4H_4O_4$.
    (a) Give the structural formula for the compound.
    (b) How many milliliters of 0.130 *M* NaOH are required to titrate 0.522 g of the acid (so that both carboxylic acid groups donate a proton)?
    (c) What is the product of the titration with NaOH?
45. What is the saponification reaction? Give an example of this important reaction.
46. Considering each ester below, give the structural formula of each product and its systematic or common name.
    (a) $CH_3\overset{\overset{\large O}{\|}}{C}{-}O{-}CH_2CH_2CH_3$ + NaOH (followed by acid) $\longrightarrow$
    (b) 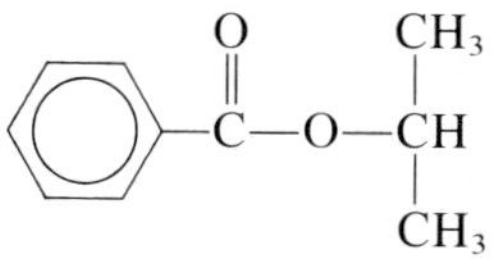 $C_6H_5{-}\overset{\overset{\large O}{\|}}{C}{-}O{-}CH(CH_3)_2$ + NaOH (followed by acid) $\longrightarrow$
47. Define fats and oils and give an example of each.

ALDEHYDES AND KETONES

48. Draw structural formulas for
    (a) 2-hexanone (b) pentanal (c) *m*-chlorobenzaldehyde
49. Give systematic names for

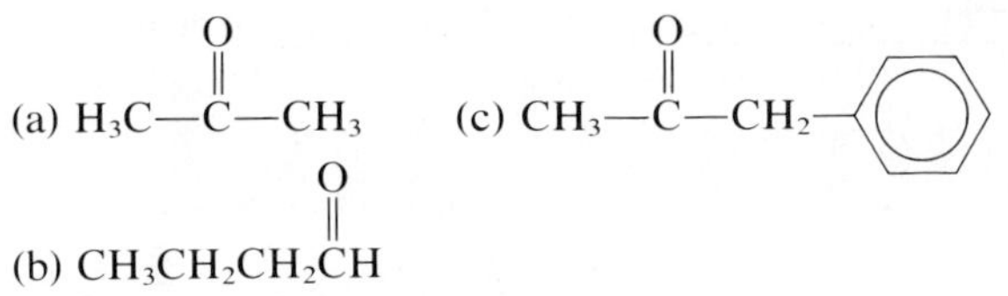

(a) $H_3C-\overset{O}{\overset{\|}{C}}-CH_3$ (c) $CH_3-\overset{O}{\overset{\|}{C}}-CH_2-C_6H_5$

(b) $CH_3CH_2CH_2\overset{O}{\overset{\|}{C}}H$

50. Give the structural formula and name of the product from each of the following reactions:
(a) propanal and $KMnO_4$
(b) propanal and $LiAlH_4$
(c) 2-butanone and $LiAlH_4$
(d) 2-butanone and $KMnO_4$
(e) butanal plus CH_3MgI followed by aqueous acid
(f) 2-butanone plus CH_3MgI followed by aqueous acid

51. If acetone is reduced with $LiAlH_4$, give the structural formula and name of the product. If this product is then treated with hot concentrated H_2SO_4, what are the structural formula and name of the product? Finally, if this second product is treated with HBr, what are the structural formula and name of the final product?

52. Tell how you could prepare 2-pentanol beginning with an appropriate ketone. Prepare this same alcohol beginning with an appropriate aldehyde and other reagents as necessary.

53. Tell how you can prepare 2-methyl-2-pentanol beginning with 2-pentanone and other appropriate reagents.

54. Give a synthesis for 1-butanol beginning with 1-propanol and other appropriate reagents.

SYNTHESIS

55. Outline a method of synthesizing 2-hexanol from acetaldehyde, 1-bromobutane, and any necessary inorganic reagents.

56. Tell how you could convert *tert*-butyl bromide (2-bromo-2-methylpropane) to 2,2-dimethylpropanol.

57. Show how you could prepare methyl ethyl ketone (commercially known as MEK) from ethanol and inorganic reagents. What is the systematic name of MEK?

58. Outline a procedure by which 3-methyl-3-pentanol can be synthesized from ethanol and inorganic reagents.

59. Outline a procedure to prepare 1,2-dibromo-2-methylpropane from 2-propanol and bromomethane.

AMINES AND AMIDES

60. Name the following amines.
(a) $(C_6H_5)_2NH$
(b) $(CH_3CH_2)_3N$
(c) 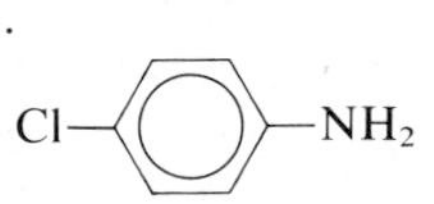 $Cl-C_6H_4-NH_2$

61. What is the pH of a 0.10 *M* aqueous solution of methylamine?

62. What is the pH of a 0.20 *M* aqueous solution of pyridine?

63. What is the product of the reaction between triethylamine and hydrogen bromide? Give its systematic name.

64. Give the structural formula for the product of the reaction of diethylamine with acetyl chloride.

65. Tri-*n*-propylamine does not react with acetyl chloride. Explain briefly why there is no reaction.

PETROLEUM REFINING

66. Name the various petroleum refinery processes and tell what each does.

GENERAL QUESTIONS

67. In addition to the structural isomerism of alkanes, and *cis–trans* isomerism of some alkenes, other types of isomers are found in organic chemistry. For example, dimethyl ether (CH_3-O-CH_3) and ethyl alcohol (CH_3CH_2OH) are isomers; they have the same molecular formulas but their chemical functionality is quite different.
(a) Draw all of the isomers possible for C_3H_8O. Give the systematic name of each and tell what class of compounds it fits.
(b) Draw the structural formula for an aldehyde and a ketone with the molecular formula C_4H_8O. Give the systematic name of each.

68. Draw the structural formula for each of the nine possible isomers of $C_4H_8Cl_2$. Name each compound.

69. Salicylic acid has both a carboxyl group and a hydroxyl group attached to a benzene ring. Consequently, it is both a carboxylic acid and a phenol. Describe syntheses for methyl salicylate (the fragrance in oil of wintergreen) and acetylsalicylic acid (aspirin) starting with salicylic acid.

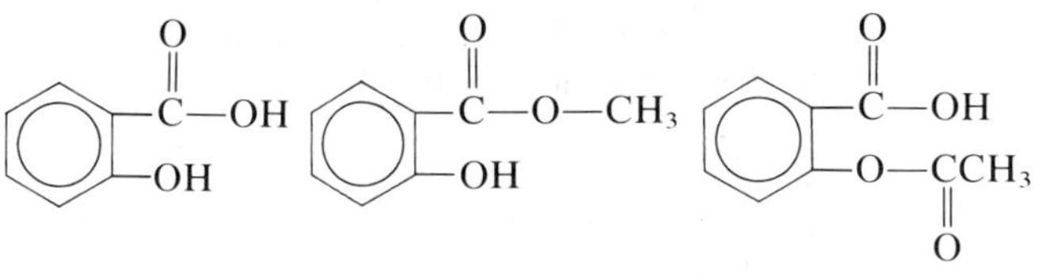

$C_6H_4(OH)-\overset{O}{\overset{\|}{C}}-OH$ $C_6H_4(OH)-\overset{O}{\overset{\|}{C}}-O-CH_3$ $C_6H_4(O-\underset{O}{\underset{\|}{C}}CH_3)-\overset{O}{\overset{\|}{C}}-OH$

salicylic acid   methyl salicylate   acetylsalicylic acid

70. Suggest ways to carry out the following multistep syntheses.
(a) $CH_3CH_2CH_2OH \rightarrow CH_3CH_2CH_2CH_2OH$
(b) 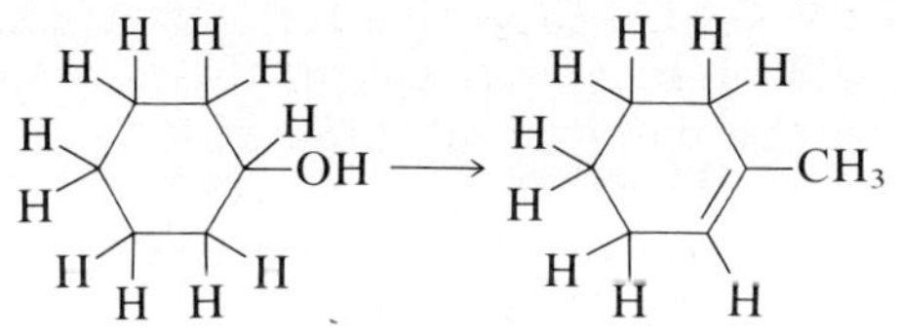 cyclohexanol (ring carbons with H's; one carbon bearing H and OH) $\longrightarrow$ 1-methylcyclohexene (ring with CH_3 on a double-bonded carbon, H's on the other carbons)

24

Polymers: Natural and Synthetic Macromolecules

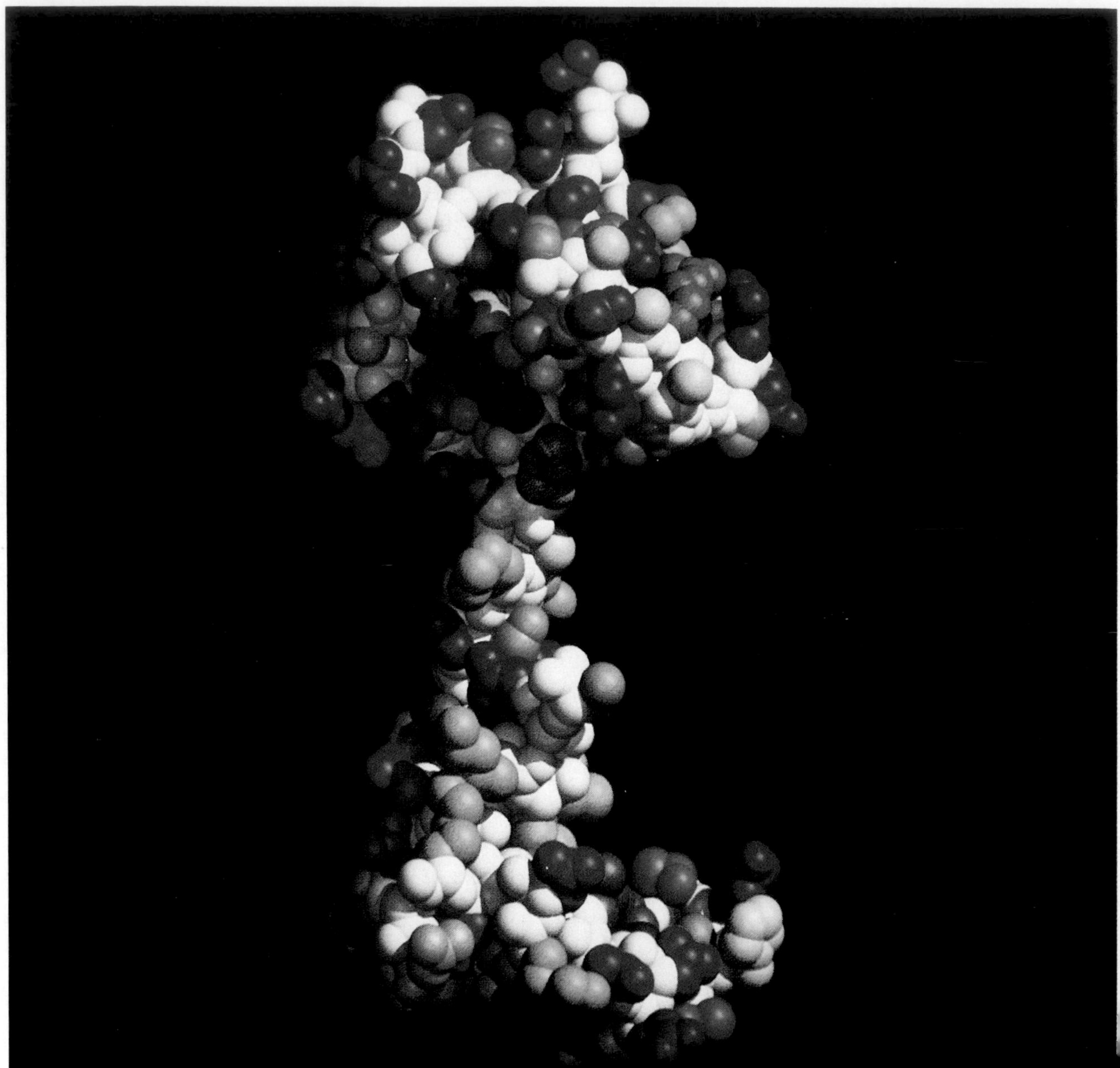

Computer-drawn model of calcium-binding protein (Dr. Osnat Herzberg and Dr. Michael N.G. James, University of Alberta)

CHAPTER OUTLINE

Approximately 80% of the organic chemical industry is devoted to the production of synthetic polymers. Phonograph records are polyvinyl chloride, the plastic squeeze bottle in your laboratory desk is polyethylene, and frying pans are coated with Teflon. You may wear clothes made of polyester, Dacron, or Orlon and walk in shoes made of a synthetic material. Synthetic polymers are so widely used that it should come as no surprise to learn that almost 180 pounds of such materials are made per person in the United States annually.

Materials such as polyethylene can have molecular weights approaching 50,000, and so we can refer to them as **macromolecules**. More generally, however, we call them **polymers**. The latter word comes from the Greek in which *poly* means "many" and *mer* means "unit." The small, individual units from which a polymer is constructed are called *monomers*. Thus, the monomer of polyethylene is ethylene,

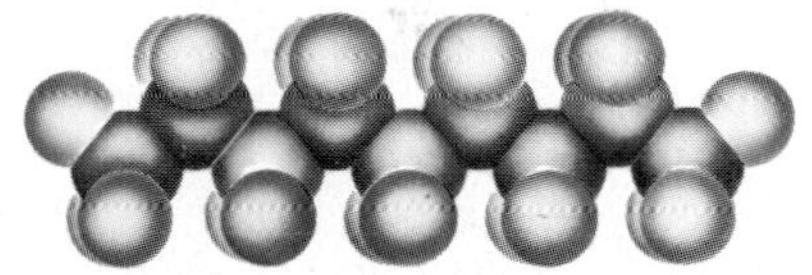

Portion of polyethylene polymer.

$$\underset{\substack{\text{monomer}\\\text{ethylene}}}{n\mathrm{CH_2{=}CH_2}} \xrightarrow{\text{addition}} \underset{\substack{\text{addition polymer}\\\text{polyethylene}}}{\mathrm{+CH_2{-}CH_2{-}CH_2{-}CH_2+}}$$

and that of a polypeptide or protein is an amino acid.

$$\underset{\substack{\text{monomer}\\\text{amino acid}}}{n\mathrm{H_2N{-}CH(R){-}COOH}} \xrightarrow{\text{condensation}} \underset{\substack{\text{condensation polymer}\\\text{polypeptide}}}{+\overset{\mathrm{H}}{\overset{|}{\mathrm{N}}}\mathrm{{-}CH(R){-}}\overset{\mathrm{O}}{\overset{\|}{\mathrm{C}}}\mathrm{{-}}\overset{\mathrm{H}}{\overset{|}{\mathrm{N}}}\mathrm{{-}CH(R){-}}\overset{\mathrm{O}}{\overset{\|}{\mathrm{C}}}+} + n\mathrm{H_2O}$$

Polyethylene is called an *addition polymer*, since the empirical formula of the polymer is the same as that of the monomer. In contrast, a polypeptide is a *condensation polymer*; the monomers have condensed by eliminating a small molecule, in this case water.

Polymers may be divided into two broad classes: *natural* and *synthetic*. The polypeptide or protein above is a natural polymer, as are cellulose and nucleic acids such as DNA. After examining some structural principles important to protein chemistry, we shall turn to these and other natural polymers. Purely synthetic polymers are the subject of the latter part of the chapter.

24.1 MOLECULAR STEREOCHEMISTRY: CHIRALITY

Isomerism is common in organic chemistry. Several molecules can have the same molecular formula, but the atoms can be arranged differently. So far you have seen two types of isomerism. In one, the atom-to-atom bonding sequences differ, and the molecules often function chemically in quite different ways. One example would be the isomers of C_2H_6O. These atoms can be bonded to give either ethanol or dimethyl ether. Since alcohols and ethers differ chemically, these might be called **functional isomers**.

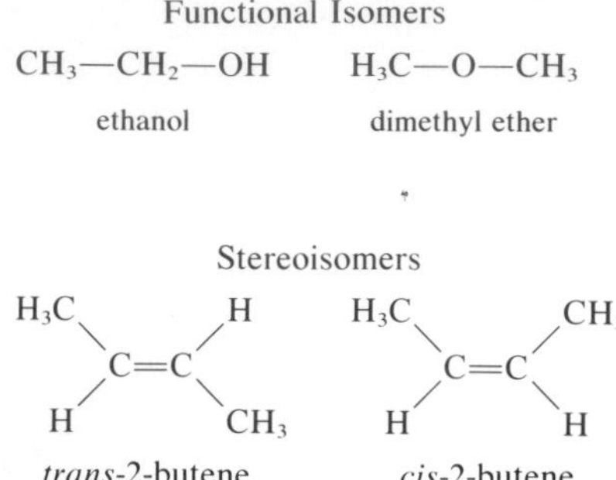

The other major type of isomerism is **stereoisomerism**. Here the isomers have the same atom-to-atom bonding sequences but the atoms differ in their arrangement in space. Two such isomers of 2-butene would be *trans*-2-butene and *cis*-2-butene. Recall from Chapter 23 that the double bond restricts rotation of one end of the molecule relative to the other, so both methyl groups can be on one "side" of the molecule (*cis*) or on opposite sides (*trans*).

Cis–trans isomerism is only one form of stereoisomerism. The other, extremely important in biochemistry, is illustrated by lactic acid, $C_3H_6O_3$. The central carbon atom of the molecule has four *different* groups bonded to it: $—CH_3$, —OH, —H, and —C(O)OH.

H
C
HO CH₃ COOH

lactic acid

As a result of the tetrahedral arrangement around this sp^3 hybridized atom, it is possible to have two different arrangements of the four groups. If a lactic acid molecule is placed so the C—H bond is vertical, as illustrated in Figure 24.1, one possible arrangement of the remaining groups would be that where —OH, $—CH_3$, and —C(O)OH are attached in a clockwise manner (isomer I). Alternatively, these groups can be attached in a counterclockwise fashion (isomer II).

To see further that the arrangements in Figure 24.1 are different, we place isomer I in front of a mirror (Figure 24.2a). Now you see that isomer II is the mirror image of isomer I. What is important, however, is that these mirror-image molecules cannot be superimposed on one another (Figure 24.2b). The best analogy to this situation is to look at your hands. If you hold your right hand up to a mirror, the image you see is that of your left hand (Figure 24.3). Your mirror-image hands cannot be superimposed on one another, and the mirror-image molecules of lactic acid cannot be superimposed on each other.

By "superimposable" we mean that one molecule can be placed on top of the other and the two will match *in all positions.*

Since the two forms of lactic acid in Figure 24.2 are nonsuperimposable mirror images, they are stereoisomers. Such stereoisomers are said to be **chiral** (pronounced chī′ral), and a chiral compound and its mirror image isomer are called **enantiomers**. The central carbon atom of lactic acid, around which there are four different atoms or groups, is

Both enantiomers of lactic acid are found in nature, one form in muscle tissue and the other in souring milk.

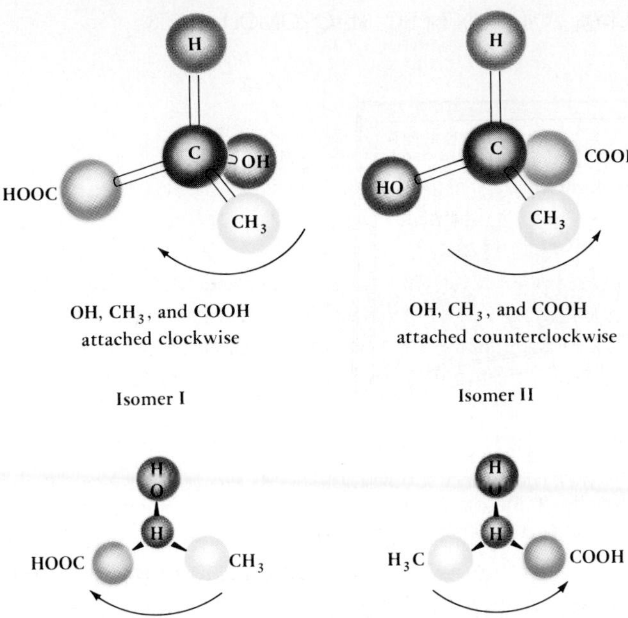

FIGURE 24.1

The two stereoisomers (enantiomers) of lactic acid. If you were to look down the molecule along the H—C bond, you would see that one of the isomers (I) has the other groups attached such that OH, CH_3, and COOH are arranged clockwise. In the other isomer (II), the arrangement is counterclockwise.

referred to as a **chiral center**. Any compound that has four different groups bonded to a carbon atom is chiral and can exist in two enantiomeric forms.

Although a pair of enantiomers may appear different on paper, they have the same physical properties such as melting point, boiling point, density, and solubility in various solvents. There is, however, one exception: when a beam of plane polarized light is passed through a solution of a pure enantiomer, the plane of polarization is twisted in one direction

The fact that enantiomers have identical physical properties means that it is difficult to separate one from the other in a mixture of the two.

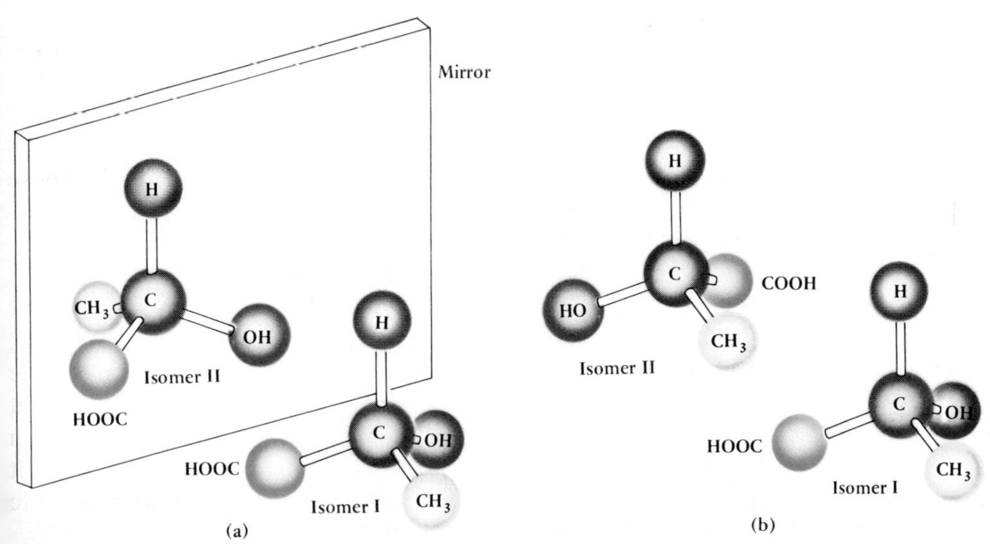

FIGURE 24.2

The stereoisomers of lactic acid. (a) Isomer I is placed in front of a mirror, and its mirror image is isomer II. (b) Isomers I and II are placed so that the grouping H—C—CH_3 is oriented the same way in each molecule. If isomer II were placed on top of isomer I, the C—H and C—CH_3 bonds would be superimposed, but the C—OH bond of II would rest on top of the C—COOH bond in I. The isomers are nonsuperimposable.

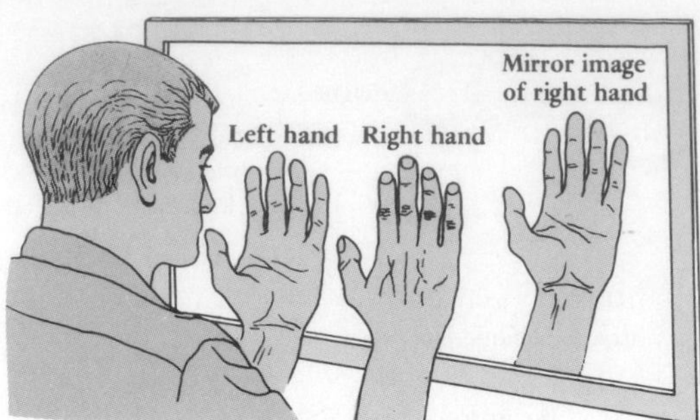

FIGURE 24.3

Mirror images. Each of your hands is a nonsuperimposable mirror image of the other hand. The mirror image of your right hand, for example, looks like your left hand. If you try to place one hand directly on the other, they will not line up completely; one of your hands cannot be superimposed on its mirror image.

or the other (Figure 24.4). For this reason, chiral compounds are sometimes referred to as **optical isomers** and are said to be **optically active**.

If the plane of polarization in Figure 24.4 is twisted to the right, the enantiomer is referred to as *dextrorotatory* (*d*) (from the Latin *dexter* meaning ''right''). The other enantiomer will differ in that it will rotate the plane of polarization by the same number of degrees to the left, and so the other isomer is referred to as *levorotatory* (*l*) (from the Latin *laevus* meaning ''left'').

EXERCISE 24.1 CHIRALITY

Amino acids are taken up in the section that follows. One amino acid is alanine and has the structure below.

$$H_2N—CH(CH_3)—COOH$$

Show the enantiomers of alanine and how they are mirror images of each other.

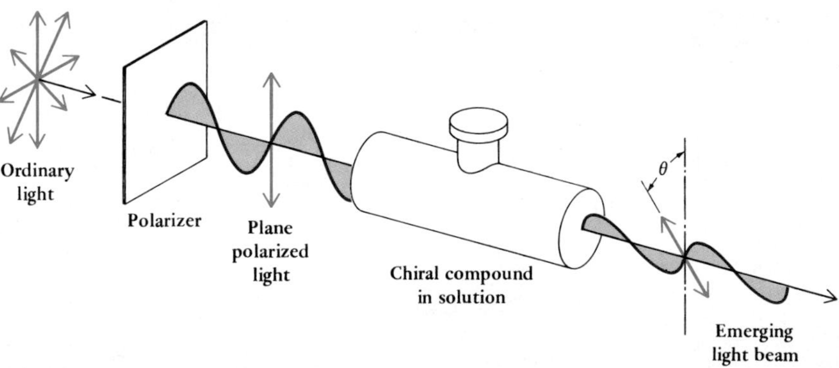

FIGURE 24.4

Rotation of the plane of polarization of light by a chiral compound. If light is passed through a polarizing filter (of the type found in some sunglasses), the light is polarized so that the electric field of the light beam lies in a given plane. When this light passes through a solution of a chiral compound, the plane of polarization is rotated. One enantiomer of a pair rotates the light beam to the left by θ degrees, and the other enantiomer rotates the beam by an equal number of degrees to the right.

24.2 AMINO ACIDS

All proteins are built of amino acid units, all of which have the same general formula.

$$\mathrm{R{-}\underset{\displaystyle NH_2}{\overset{\displaystyle H}{C}}{-}\overset{\displaystyle O}{\overset{\|}{C}}{-}O{-}H} \qquad \mathrm{CH_3{-}\underset{\displaystyle NH_2}{\overset{\displaystyle H}{C}}{-}\overset{\displaystyle O}{\overset{\|}{C}}{-}O{-}H}$$

general α-amino acid — the amino acid alanine

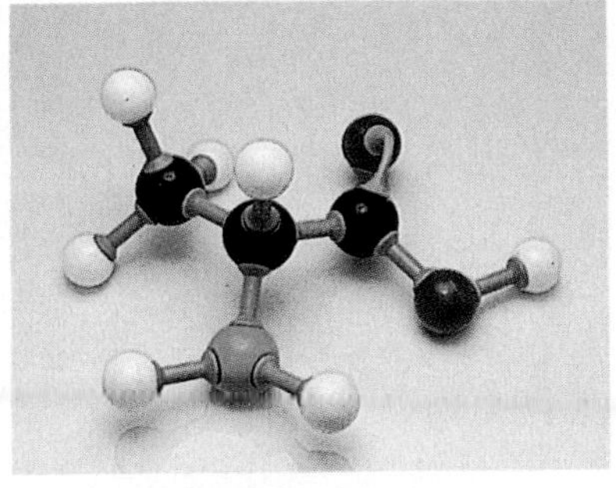

Alanine, one of the essential amino acids. (Charles D. Winters)

All amino acids are characterized by having an amino group (generally $-NH_2$) and a carboxylic acid group ($-COOH$). Almost all of the amino acids found in living systems are α-amino acids where the symbol α is the Greek letter *alpha*. This signifies that the amino group is attached to the first or alpha carbon atom beyond the $-COOH$ group.

In view of the fact that a large number of proteins exists and that they vary so widely in their properties, it is striking that they are derived from only about 20 amino acid building units. All 20 of the common amino acids are α-amino acids and, with the exception of glycine, the alpha carbon atom is a chiral center (Table 24.1) because four different groups

TABLE 24.1 Ten of the 20 α-amino acids

$CH_3-CH(NH_2)-COOH$	Alanine	$HC{=}C(-NH-CH{=}N-)-CH_2-CH(NH_2)-COOH$ (imidazole ring: HC=C, N, NH, C–H)	Histidine
$HOOC-CH_2-CH(NH_2)-COOH$	Aspartic Acid	$NH_2-CH_2-CH_2-CH_2-CH_2-CH(NH_2)-COOH$	Lysine
$HS-CH_2-CH(NH_2)-COOH$	Cysteine	$C_6H_5-CH_2-CH(NH_2)-COOH$	Phenylalanine
$HOOC-CH_2-CH_2-CH(NH_2)-COOH$	Glutamic Acid	$HO-CH_2-CH(NH_2)-COOH$	Serine
$H-CH(NH_2)-COOH$	Glycine	$(CH_3)_2CH-CH(NH_2)-COOH$	Valine

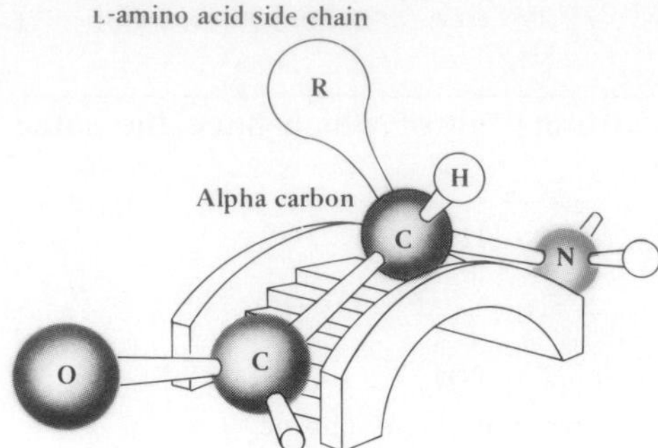

FIGURE 24.5
The chirality of α-amino acids. This figure represents a way to remember the two enantiomeric forms of an α-amino acid. When walking along the C—C—N bridge from the C—O end to the NH_2 end, the acid is in the L form if the R group is on the left. If R is on the right, it is a D acid. The L form is the one found predominantly in nature.

(H, R, NH_2, and COOH) are attached to the carbon. However, it is equally striking that only one of the enantiomeric forms of a given amino acid exists in nature. In virtually every case, living organisms use *only* one enantiomer, the so-called "L" form, a configuration defined in Figure 24.5.

Most amino acids dissolve reasonably well in water. For example, 25 g of glycine dissolve per 100 g of water at 25°C, and 3 g of phenylalanine dissolve under the same conditions. In every case, the following equilibrium exists in aqueous solution.

$$R{-}\overset{\displaystyle H}{\underset{\displaystyle NH_2}{\overset{|}{\underset{|}{C}}}}{-}\overset{\displaystyle O}{\overset{\|}{C}}{-}O{-}H \rightleftharpoons R{-}\overset{\displaystyle H}{\underset{\displaystyle NH_3^{+}}{\overset{|}{\underset{|}{C}}}}{-}\overset{\displaystyle O}{\overset{\|}{C}}{-}O^{-}$$

zwitterion

Since the amino group is more basic than the carboxylate group, the molecule exists primarily as a *dipolar ion* known as a **zwitterion**, a term taken from the German word *zwitter* meaning "double."

EXERCISE 24.2 AMINO ACID STRUCTURES
Draw the L form of the amino acid serine.

24.3 PROTEINS

Proteins are important in a variety of ways. As *enzymes* they serve as catalysts in biological synthesis and degradation reactions. *Hormones* serve a regulatory role, and as *antibodies* they protect us against disease. The protein *hemoglobin* carries oxygen from the lungs to the various parts of the body. Finally, proteins are the major constituents of cellular and intracellular membranes, skin, hair, muscle, and tendons.

Proteins can be divided into two classes: simple and conjugated. **Simple proteins** consist only of amino acids, and **conjugated proteins** may contain some other group in addition to the amino acids. Examples of conjugated proteins are hemoglobin and myoglobin where the iron-containing heme group (see below) is the site of oxygen binding.

To form a protein, the first step is a *condensation reaction* of two amino acids.

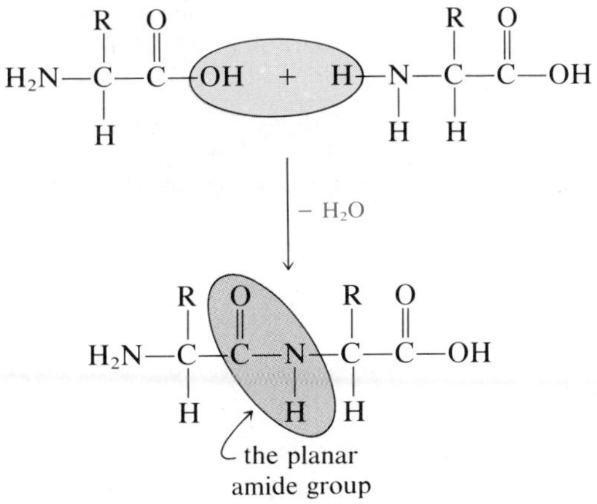

$$H_2N-\underset{H}{\overset{R}{C}}-\overset{O}{\overset{\|}{C}}-OH + H-\underset{H}{N}-\underset{H}{\overset{R}{C}}-\overset{O}{\overset{\|}{C}}-OH \xrightarrow{-H_2O} H_2N-\underset{H}{\overset{R}{C}}-\overset{O}{\overset{\|}{C}}-\underset{H}{N}-\underset{H}{\overset{R}{C}}-\overset{O}{\overset{\|}{C}}-OH$$

the planar amide group

One molecule of water is eliminated between the carboxylic acid of one amino acid and the amine group of another. The result is a **peptide** bond, a grouping called an amide in Chapter 23, and the molecule is a *dipeptide*. Either end of the dipeptide can react with another amino acid to give a tripeptide, which can in turn condense with another amino acid, and so on. Ultimately, many amino acids can be linked to give a *polypeptide* or protein (Figure 24.6). Thus, *a protein is a polymer of amino acids*.

Proteins occur in a variety of sizes. The common protein insulin has only 51 amino acid units in two linked chains. In contrast, human hemoglobin contains four protein chains, two identical ones having 141 amino acids and the other two, again identical, having 146. Considering all of the possible ways that 20 α-amino acids could be put together to form proteins, it is remarkable that so few proteins are in fact known (the human body is thought to contain only about 100,000 different proteins)

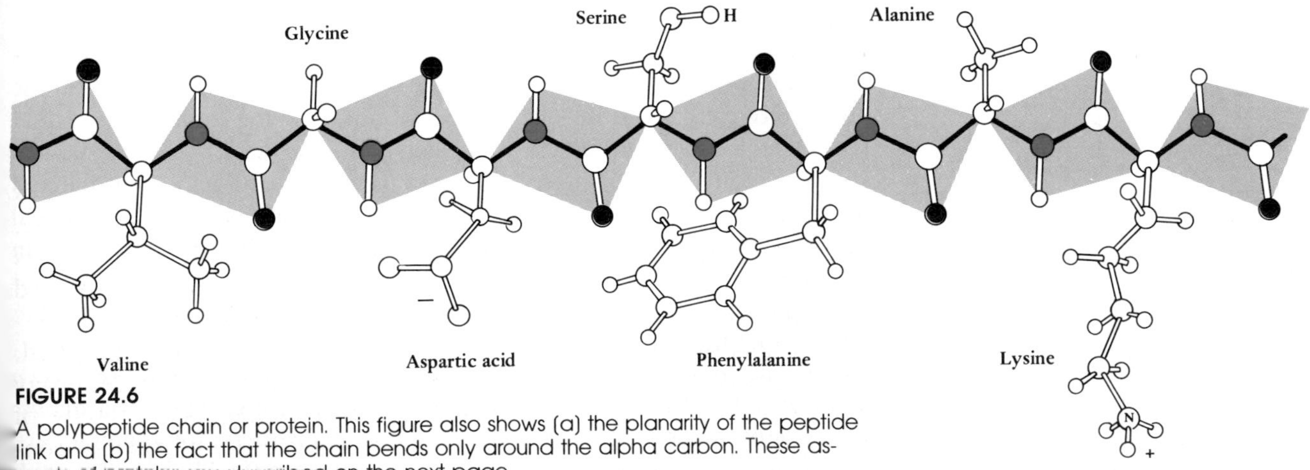

FIGURE 24.6

A polypeptide chain or protein. This figure also shows (a) the planarity of the peptide link and (b) the fact that the chain bends only around the alpha carbon. These aspects of proteins are described on the next page.

Nature can go awry and produce different proteins from a collection of amino acids. The genetic disease "sickle cell anemia" results when valine is substituted for glutamic acid at a specific point in two of the four hemoglobin chains.

and that cells produce generation after generation of identical proteins that serve a given function.

EXERCISE 24.3 PEPTIDES

(a) Any two amino acids can form two dipeptides. For example, an N—H of glycine can condense with the —COOH group of serine, or the serine N—H can condense with the —COOH of glycine. Draw structures of these two possibilities.
(b) Draw the structure of the tripeptide formed from glycine molecules.

PROTEIN STRUCTURE

When many amino acids are linked together in a particular sequence with a particular geometry, a specific protein is formed. To fully understand the functioning of the protein, we must know its *primary structure*, the sequence of amino acid units. Equally important, however, is the three-dimensional structure of a protein that arises from interactions between chains. These interactions are important because they ultimately affect the functioning of the protein, such as its ability to act as a biochemical catalyst or to bind oxygen.

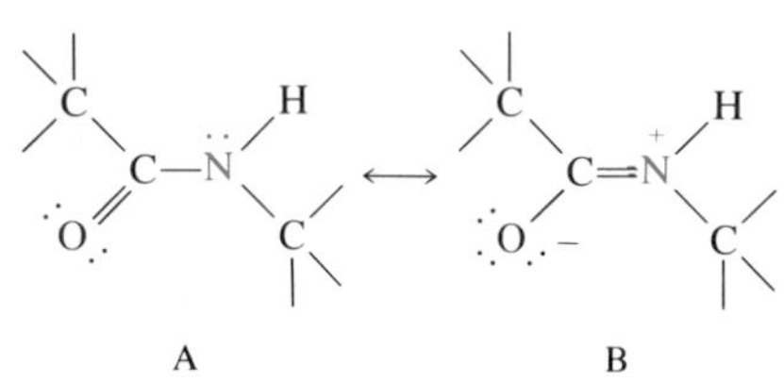

Once the primary structure of a protein is established, the next question is whether the protein chains are straight or folded up in a particular pattern. To see the possibilities, first look at the resonance structures of the peptide bond. Structure B contributes to the overall structure, so there is significant double bond character to the C—N bond. As you learned in Chapter 23, energy is required to rotate the groups at one end of a double bond relative to the other end, so the peptide linkage is locked in a flat configuration.

The flat peptide unit, with a slight negative charge on the oxygen atom and a slight positive charge on the N—H hydrogen, has two important consequences. First, the peptide chain can only bend around the alpha carbon atom of each amino acid unit (see Figure 24.6). Second, the polarities of the C—O and N—H bonds lead to hydrogen bonding between amino acid units (Figure 24.7).

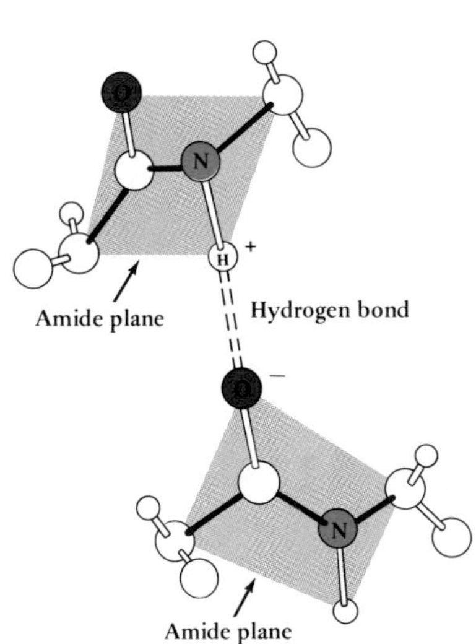

FIGURE 24.7
Hydrogen bonding between two amino acid units of a polypeptide chain. Notice also that the C—O and N—H are *trans,* a feature almost always observed in proteins.

Given that the peptide linkage is flat and that hydrogen bonding is possible, Linus Pauling demonstrated some years ago that there must be two most common *secondary structures* of protein molecules: the **α-helix** and the **β-pleated sheet**. The secondary structure adopted depends on the position of the C=O···H—N hydrogen bonds. If the hydrogen bonding is *intra*molecular (within the protein chain), the result is the α-helical structure. The protein shown in Figure 24.8 is wound as a right-handed helix, just as the threads of a screw or bolt are right handed. The helical structure is fixed by the fact that *all* of the C=O and N—H groups in peptide linkages participate in hydrogen bonding parallel to the axis of the helix; each C=O group binds to a N—H bond in a unit four residue away along the chain.

The α-helix is the basic structural unit of *fibrous proteins* known as α-keratins. These include wool, hair, skin, beaks, nails, and claws. Because of the helical protein chains, human hair fibers, for example, are stretchable and elastic to a small extent. Stretching the fibers will stretch

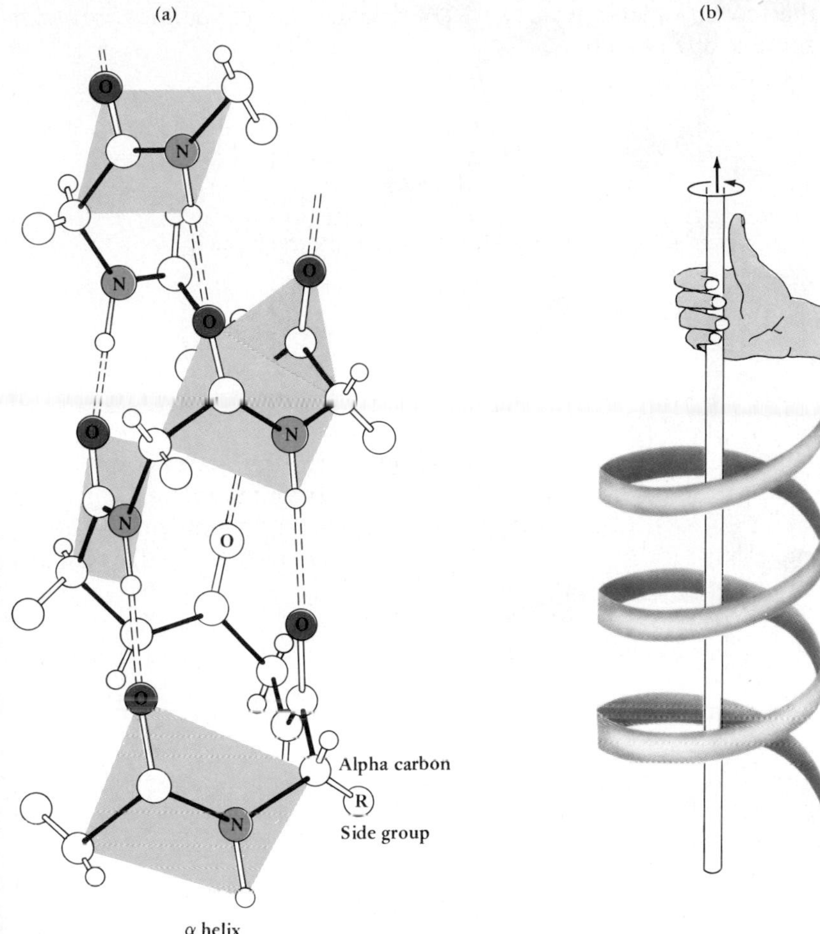

FIGURE 24.8
The helical structure of proteins. (a) Hydrogen bonding within a polypeptide chain leads to a spiral arrangement called an α-helix. (b) In proteins, the helix always has a right-handed arrangement. If you hold your right hand so that your thumb points along the axis of the helix, your fingers curl in the same direction as the curl of the spiral.

the relatively weak hydrogen bonds, but not the covalent bonds. If the hair fiber is not stretched to the point where the covalent bonds begin to break, the fiber will snap back to its original length. The fiber is elastic, because the hydrogen bonds can be re-formed.

In other fibrous proteins, such as those in silk, the C=O···H—N hydrogen bonding is *inter*molecular, one protein chain being bound to a neighboring chain (Figure 24.9). The polypeptide chain is fully extended, and hydrogen bonding between chains leads to a sheet-like structure. Sheets of proteins can then be stacked on top of one another like the pages of this book. This structure, called the **β-sheet structure**, means that silk and other insect fibers are not stretchable or elastic. Pulling on them would break covalent bonds or the many hydrogen bonds holding the individual protein strands in the sheet. However, just as you can bend the stack of pages in this book, so too can the protein stack be bent.

Not only are α-helix and β-sheet structures commonly used in fibrous proteins but they are also common in the *globular proteins*. Two globular proteins are hemoglobin, the protein that carries O_2 in the blood from the lungs to the tissues, and myoglobin, the O_2-storage protein in

Hemoglobin has four folded, helical protein chains, each having a heme group. The four chains are packed into a compact unit.

FIGURE 24.9

The β-sheet structure. In silk and other insect fibers, polypeptide chains are bound to one another by hydrogen bonds.

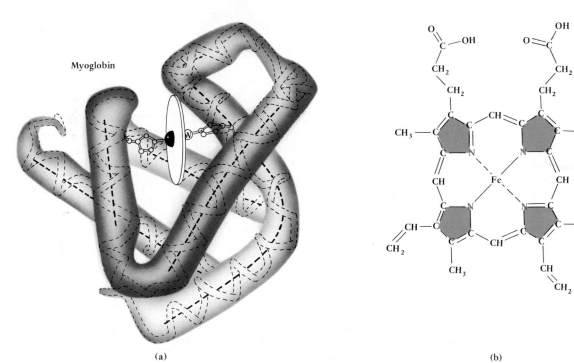

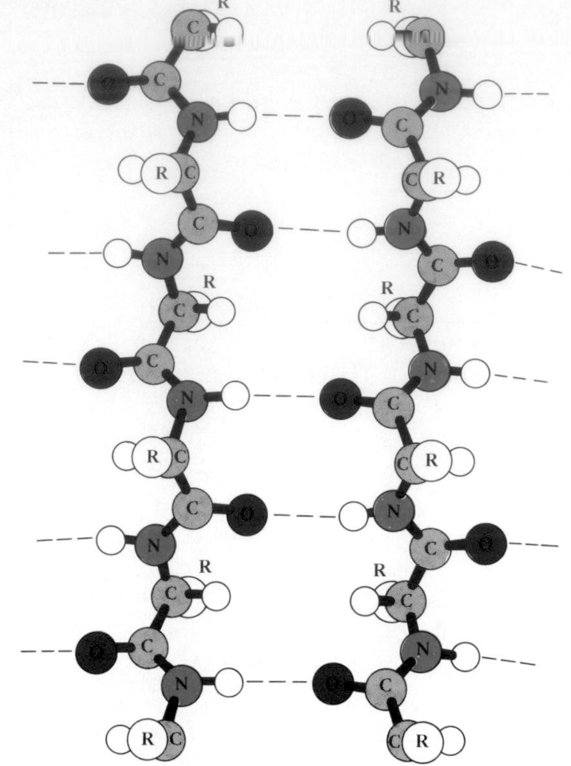

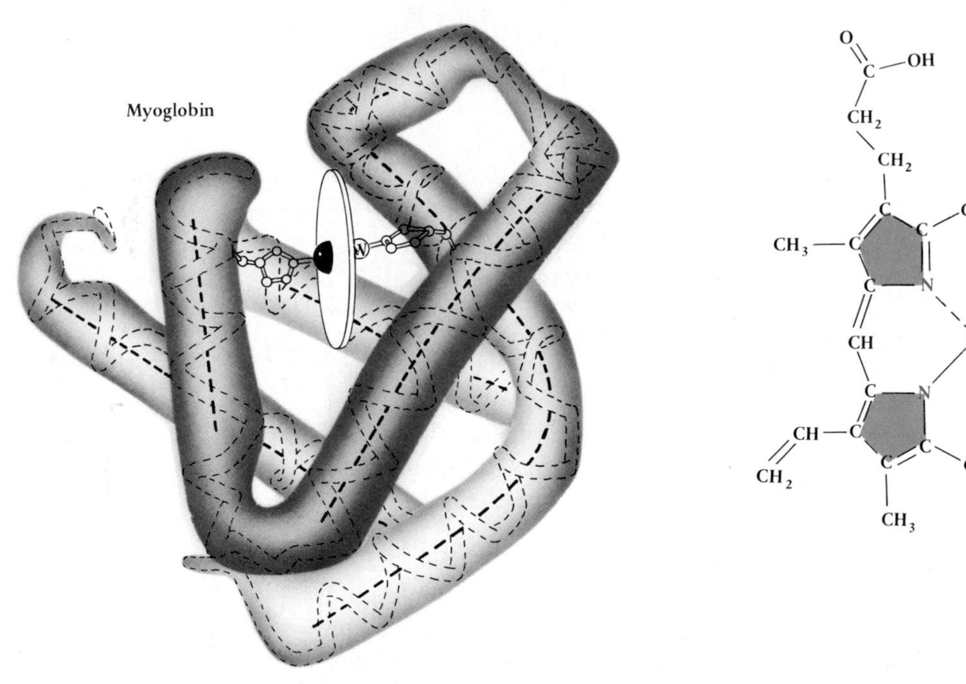

FIGURE 24.10

Myoglobin, the O_2-storage protein in muscles. (a) A continuous chain of 153 amino acid units encloses a heme group. The iron of the heme, shown as the black hemisphere, is the site of O_2-binding. When not bound to O_2, the metal binds water, designated by W in the figure. (b) The heme group, where an iron atom is enclosed within an organic group called a "porphin."

muscles (Figure 24.10a). Both are characterized by α-helical protein chains, and both contain the heme group (Figure 24.10b). It is this latter group that gives the red color characteristic of blood or red meat.

24.4 CARBOHYDRATES

The word "carbohydrate" literally means "hydrate of carbon." Thus, **carbohydrates** have the general formula $C_x(H_2O)_y$ in which x and y are integers. However, in spite of the fact that reaction of the carbohydrate sucrose with sulfuric acid produces carbon (Figure 24.11), this does not mean that sucrose is a simple combination of carbon and water. Rather, carbohydrates are all molecules that are polyhydroxyaldehydes or polyhydroxyketones and share some chemical properties with alcohols and aldehydes or ketones.

Although carbohydrates account for only about 1% of total body mass, they are centrally important in human biochemistry. Carbohydrates are formed in plants by the process of photosynthesis from carbon dioxide, water, and energy from the sun.

$$xCO_2 + yH_2O + \text{energy} \longrightarrow \underset{\text{carbohydrate}}{C_x(H_2O)_y} + xO_2$$

Not only do they serve as the ultimate source of much of the food for animals, but the carbohydrate cellulose is the main structural component of plants. Furthermore, carbohydrates are major components of wood,

FIGURE 24.11

Reaction of sucrose with concentrated sulfuric acid. Sucrose is a carbohydrate with a formula of $C_{12}(H_2O)_{11}$ or $C_{12}H_{22}O_{11}$. Sulfuric acid is an excellent dehydrating agent, so carbon is the final product. (Charles D. Winters)

cotton, paper, and many other important materials derived from natural sources. Finally, the nucleic acids (Section 24.5) contain the carbohydrate units in their repeating structure.

Carbohydrates are placed in one of three classes, depending on their molecular size. *Monosaccharides* are the simplest and are the building blocks of *disaccharides*, which contain two monosaccharide units, and of *polysaccharides*, which are polymers of monosaccharides.

The names of saccharides end in "-ose." The general name of a saccharide with n = 5 is a pentose, while one with n = 6 is a hexose.

SIMPLE SUGARS: MONOSACCHARIDES

Monosaccharides have one unit of water per carbon atom ($x = y$) and so have the general formula $(CH_2O)_n$ where n can be three to six or more. One of the important simpler sugars is *ribose*, a *pentose* with n = 5 and a formula of $(CH_2O)_5$ or $C_5H_{10}O_5$,

β-D-ribose

and another common monosaccharide is *glucose*, a *hexose* with a formula of $C_6H_{12}O_6$ (n = 6).

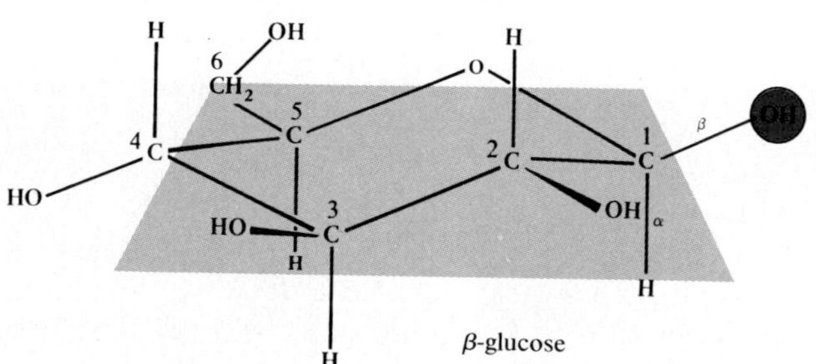

β-glucose

The structure above is drawn so that three of the ring carbon atoms lie in a plane. Notice that this allows us to discriminate *two types of groups* attached to the ring carbon atoms: (a) those above or below the plane, called **axial** groups, and (b) those that lie approximately in the plane, called **equatorial** groups. Here the axial groups are all H atoms, while the equatorial groups are —OH or —CH_2OH.

The position of the —OH group attached to the carbon labeled 1 is important to the functioning of the molecule, as you shall see below. If the group is in an equatorial position, as pictured above, the molecule is called β-glucose. On the other hand, if —OH is in the axial position on C-1, the molecule is labeled α-glucose.

An aqueous solution of pure glucose consists of an equilibrium mixture of the α and β isomers. They can interconvert by opening, and then reclosing, the ring.

Two other common hexoses, mannose and galactose, differ from glucose only in that one —OH group is moved to an axial position.

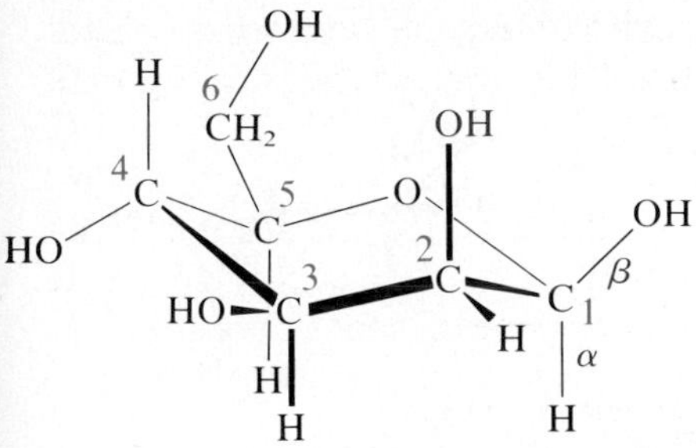

β-D-mannose 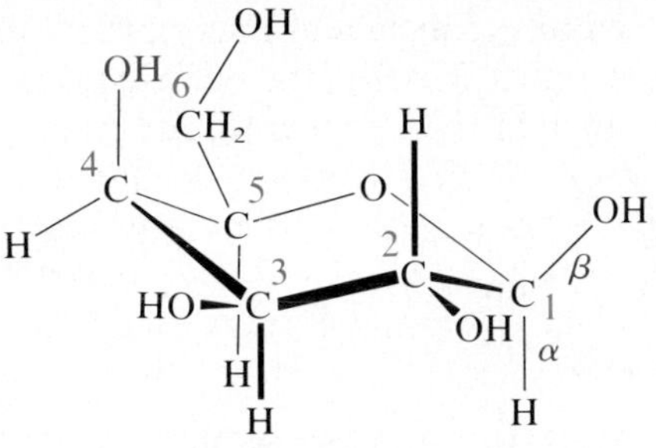 β-D-galactose 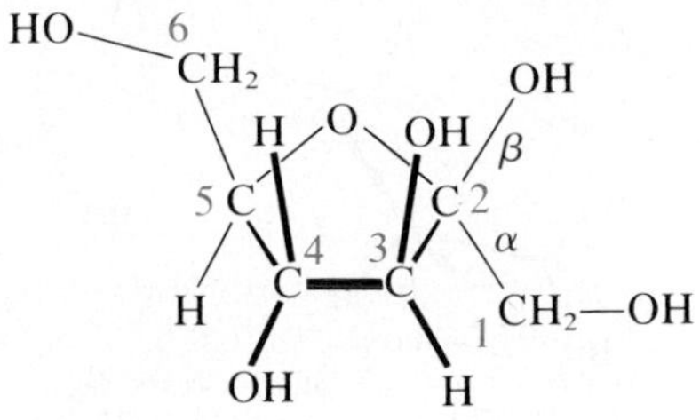 β-D-fructose

Fructose, or fruit sugar, is another six-carbon monosaccharide, but only four of the carbon atoms participate in the ring, so it is similar in this regard to ribose.

DISACCHARIDES

The monosaccharide units described above can condense, with the elimination of water, to give disaccharides.

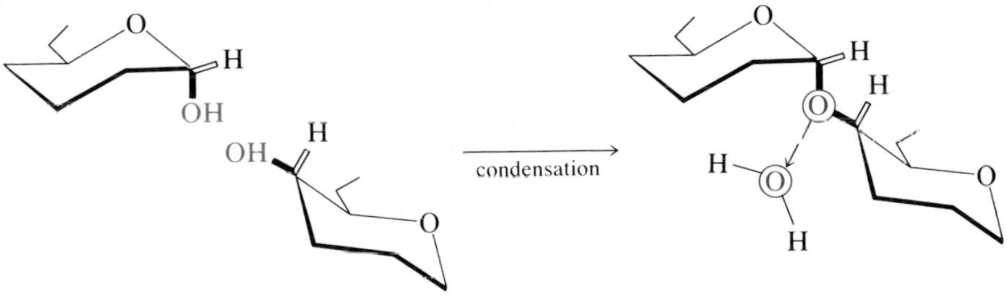

two monosaccharides —condensation→ disaccharide + H_2O

The structures of four common disaccharides are illustrated in Figure 24.12. In every case the condensation leads to the so-called **glucoside** linkage. When two hexoses are combined, the C—O—C linkage is between C-1 of one molecule and C-4 of the other molecule; such bonds are called "1,4-linkages."

Honey, a mixture of the monosaccharides glucose and fructose, has been used for centuries as a natural sweetener for foods. In contrast, sucrose, derived from sugar cane or sugar beets, is a disaccharide. Honey is a popular sweetener because cane sugar is not as sweet as a mixture of pure glucose and fructose. To convert cane sugar into glucose and fructose requires treatment with acid or with a natural enzyme or catalyst called "invertase." The sugar industry goes to considerable expense to convert sugar chemically into "dextrose" or "levulose," a mixture of glucose and fructose. Bees are much better chemists, however; they supply the necessary enzyme along with the sucrose.

Cane sugar first appeared in Southeast Asia around the 5th century BC, but it did not arrive in the western world until about the 7th century AD. However, not until the great sugar plantations were established in the New World in the 17th century was "stone honey," as the Chinese called cane sugar, readily available.

POLYSACCHARIDES: CELLULOSE AND STARCH

The three most important polysaccharides are **cellulose**, **starch**, and **glycogen**, all polymers of glucose. Cellulose is the most abundant organic compound on earth, and its purest natural form is cotton. This polysac-

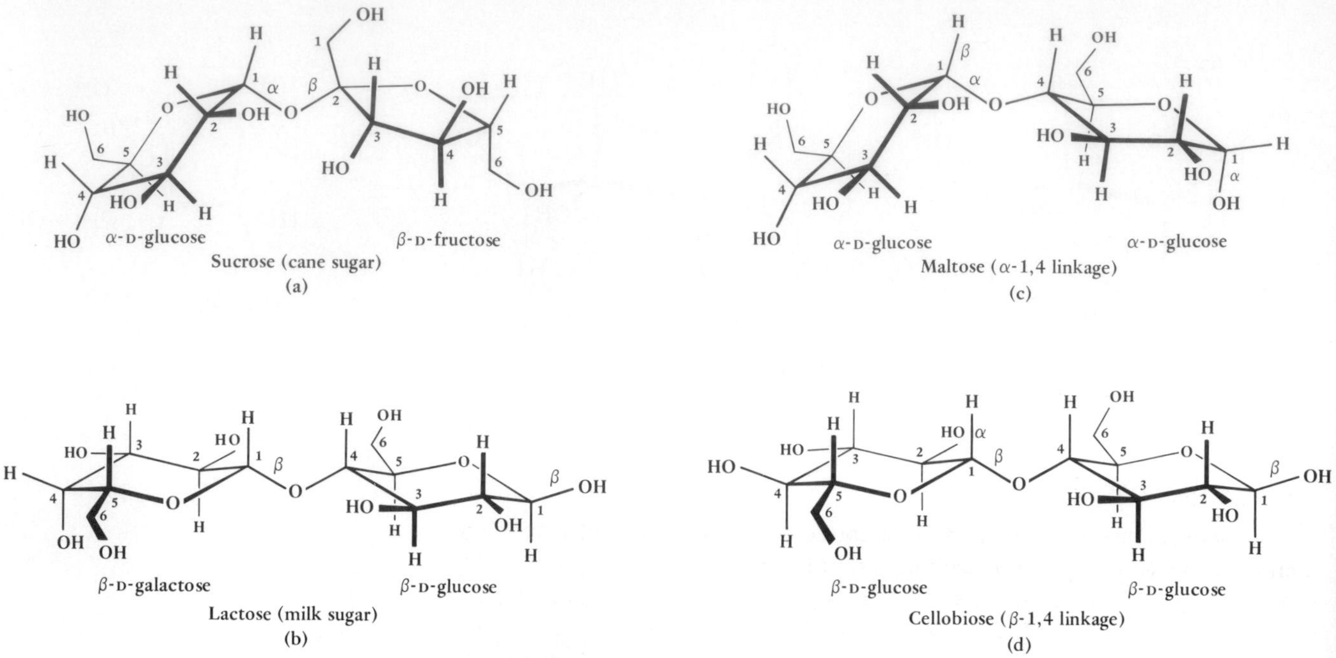

α-D-glucose β-D-fructose
Sucrose (cane sugar)
(a)

α-D-glucose α-D-glucose
Maltose (α-1,4 linkage)
(c)

β-D-galactose β-D-glucose
Lactose (milk sugar)
(b)

β-D-glucose β-D-glucose
Cellobiose (β-1,4 linkage)
(d)

FIGURE 24.12

Four common disaccharides, all combinations of two sugar monomers. (a) Sucrose is a combination of the α form of sucrose and the β form of fructose. (b) Lactose or milk sugar is formed from the β forms of galactose and glucose. Therefore, the linkage is called a β-1,4 glucoside linkage. (c) Maltose is a degradation product of starch, a glucose polymer with an α-1,4 linkage. (d) Cellobiose comes from breaking down the cellulose polymer and has the β-1,4 linkage characteristic of this substance.

charide is also found as the woody part of trees and the supporting material in plants and leaves.

Cellobiose, shown in Figure 24.12, is a fragment of **cellulose**, a *linear* polymer of β-glucose units. There are typically 300 to 3000 glucose molecules in the chain, so the molecular weight is between 50,000 and 500,000.

Since cellulose is so abundant, it would be advantageous if humans could use it for food. Unfortunately, we cannot digest it, because we lack the necessary enzyme or biochemical catalyst to chew up the β-1,4 bonds. On the other hand, termites, a few species of cockroaches, and ruminant mammals such as cows, sheep, goats, and camels do have the proper internal chemistry laboratory for this purpose.

Because we cannot digest cellulose, we must rely partly on **starch** as a source of glucose. Starch occurs in two forms, *amylose* and *amylopectin*. Both are polymers of glucose units bound through α-1,4 linkages (as in maltose, Figure 24.12), but amylopectin is the more common form. It usually has around 1000 units in a branched chain (whereas the amylose chain is unbranched). Branching occurs at intervals of 20 to 25 units and involves a glucoside linkage between C-6 of a glucose unit of the main chain and C-1 of the first glucose unit of the branch chain (Figure 24.13).

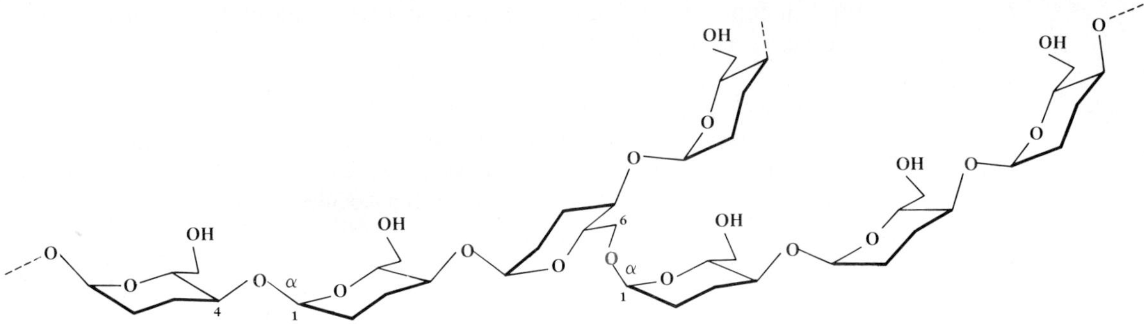

FIGURE 24.13
Amylopectin, a starch. The basic polymer forms by α-1,4 linkages between glucose units, but the chain branches using the —OH group of C-6.

Although we do not have the necessary enzyme to break down cellulose, we do have "amylase" in our saliva and pancreatic juices, the enzyme required to "crack" starch into glucose. You know this because, if you hold a piece of bread in your mouth for a few minutes, it begins to taste sweet, a signal that glucose and maltose are being produced. When the bread is swallowed, the acid in your stomach finishes what your saliva began, and the freed glucose can be absorbed into the bloodstream and carried off for eventual oxidation.

Starch is very attractive as a material for energy storage. The polymer is easily synthesized and only one type of chemical bond needs to be broken to give instant energy, glucose. On the other hand, fats and fatty acids (Chapter 23) require many steps for their synthesis and their metabolism is complex. The complexity of fat chemistry is offset by the fact that fats provide much more energy per gram than glucose (see Table 24.2). Thus, animals and other organisms that move around often use fat for energy storage, because less mass needs to be transported to provide a given energy. In contrast, an immobile plant has little need for a substance with a high energy-to-mass ratio, so plants choose the simplicity of carbohydrate chemistry.

To provide instant energy, however, animals do use one form of carbohydrate. They synthesize *glycogen*, a more highly branched form of amylopectin. This is stored in the liver and muscle tissues and is used

TABLE 24.2 Heat Energy Provided by Various Fuels

FUEL	HEAT OF COMBUSTION (kJ/g)
Hydrogen, $H_2(g)$	142
Methane, $CH_4(g)$	56
Octane (gasoline), $C_8H_{18}(g)$	48
Stearic acid (fatty acid), $C_{17}H_{35}COOH(s)$	40
Glucose, $C_6H_{12}O_6(s)$	15

for "instant" energy until the process of fat metabolism can take over and serve as the energy source.

EXERCISE 24.4 DISACCHARIDES

At the beginning of the section on disaccharides, the condensation of two monosaccharides is illustrated. What are these monosaccharides? What disaccharide is formed? Is an α- or a β-1,4 glycoside link formed?

24.5 ENERGY AND INFORMATION: NUCLEOTIDES AND NUCLEIC ACIDS

The nucleic acids store and transmit genetic information. *Deoxyribonucleic acid* (**DNA**) is the permanent repository of genetic information in the nucleus of the cell where it controls the synthesis of *ribonucleic acid* (**RNA**). It is the RNA that is involved in the biochemical machinery that transmits the genetic information and directs the synthesis of proteins. Close relatives of DNA and RNA are nucleotides such as **ATP** (adenosine triphosphate), much smaller molecules that are the chief short-term energy storage molecules for all life processes.

Nucleic acids and nucleotides are composed of three structural units: a sugar, a base, and one or more phosphates. The five organic bases that are the heart and soul of these systems are shown in the margin. Adenine (A) and guanine (G) are derivatives of the organic base purine ($C_5H_4N_4$), while thymine (T), cytosine (C), and uracil (U) are derivatives of pyrimidine ($C_4H_4N_2$). These bases are so important to any discussion of nucleic acid chemistry that they are often referred to only by the first letter of their name.

The five organic bases of nucleic acids and nucleotides

Purine Derivatives

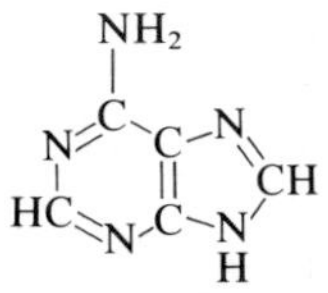

adenine (A) guanine (G)

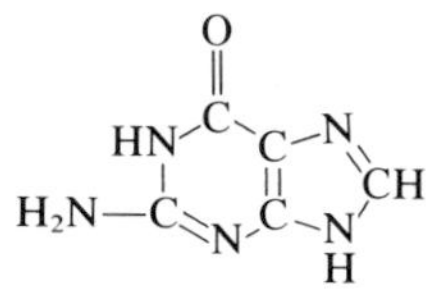

Pyrimidine Derivatives

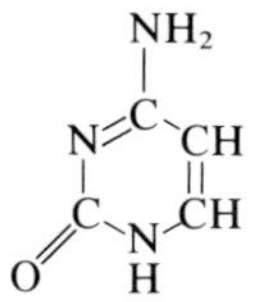

cytosine (C)

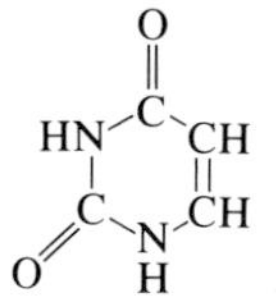

uracil (U) only in RNA

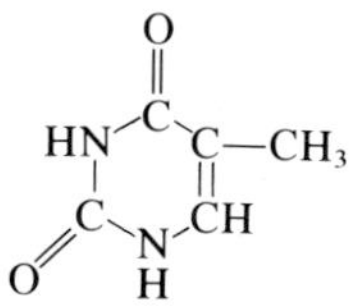

thymine (T) only in DNA

In RNA the sugar involved is *ribose*, whereas DNA is based on *deoxyribose*. The latter is also a pentose, but it has no —OH group on C-2.

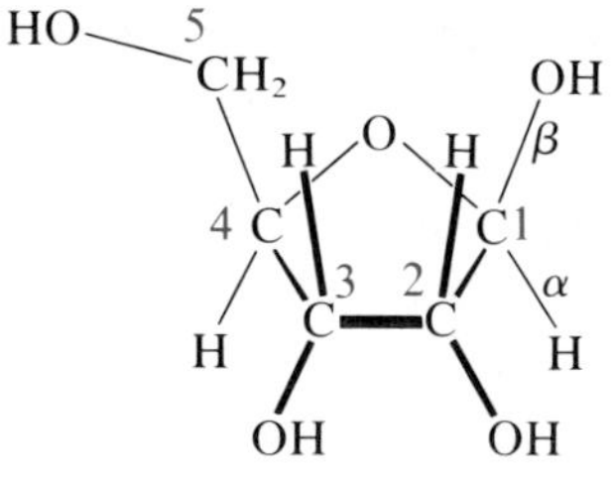

D-ribose, found in RNA 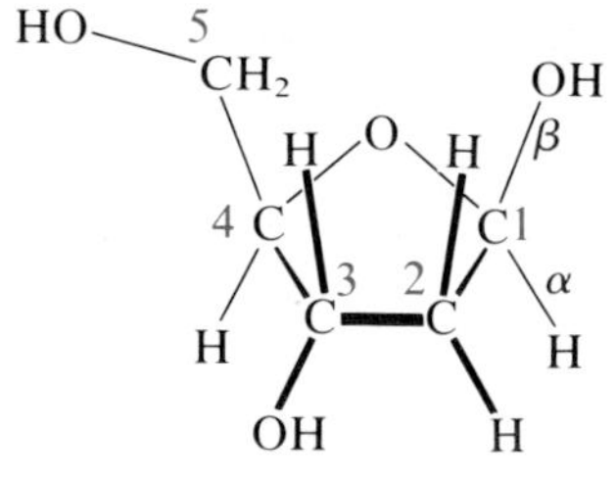 D-deoxyribose, found in DNA

When any one of the five bases is bound to C-1 of the carbohydrate ribose, the result is a *nucleoside* such as adenosine (Figure 24.14). These nucleosides can in turn form *esters* with the phosphate ion by the elimination of water between the HPO_4^{2-} ion and the —OH group at C-2, C-3, or C-5.

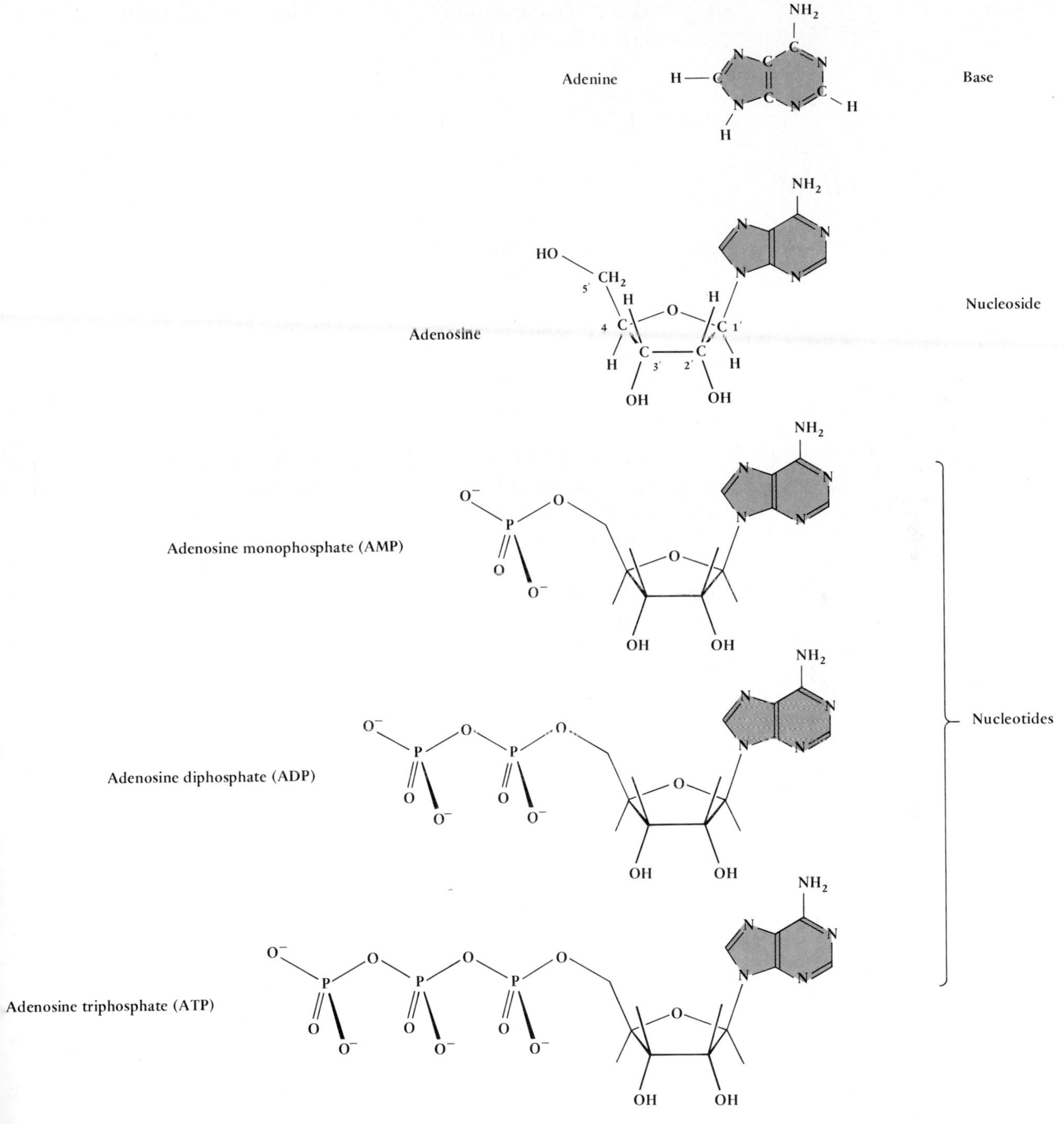

FIGURE 24.14

The formation of nucleosides and nucleotides. Adenine combines at C-1 of ribose to form the nucleoside adenosine. If the —OH group at C-5 condenses with phosphate, the ester AMP is formed. Addition of other phosphate linkages gives ADP and ATP.

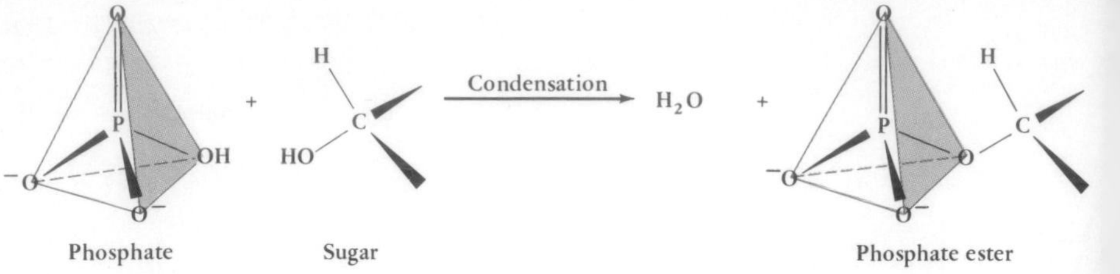

$$\text{Phosphate} + \text{Sugar} \xrightarrow{\text{Condensation}} H_2O + \text{Phosphate ester}$$

The phosphate esters are called *nucleotides*, and the most common are those involving the C-5 hydroxyl group. (Living organisms use one of these, the nucleotide ATP, as an energy storage system.) If the phosphate bound to C-5 forms an ester linkage to another ribose or deoxyribose, then a polymer is formed. The backbone of DNA (deoxyribonucleic acid) or RNA (ribonucleic acid) is just such a polymer of alternating phosphate and deoxyribose or ribose units (Figure 24.15), respectively, the phosphate bridging between C-5 of one sugar and C-3 of an adjacent sugar.

In Figure 24.15 the backbone of the DNA polymer is shown with only the positions of the bases indicated. It is these bases in which genetic information is coded. The sequence of the bases along the chain is specific to the new protein to be synthesized by the cell in which the DNA is located. Each three-base sequence in the DNA, called a *codon*, is the code for one of the amino acids that is to be assembled into a new protein. For example, if one three-base sequence on the DNA is A—G—C, this is the code for serine to be added to the protein being synthesized by the cell.

You know from making long-distance telephone calls that information can sometimes be lost. Losing genetic information in the course of protein synthesis can be disastrous, as exemplified by genetic disorders such as sickle cell anemia and Down's syndrome or mongolism. To solve this problem, each polymeric strand of DNA has a *complementary* partner or strand, and either partner is coded for the synthesis of a specific protein. James D. Watson and Francis H.C. Crick showed in 1953 that DNA is composed of two polymeric strands coiled into a *double helix* (Figures 24.16 and 24.17). The base units of each strand are pointed into the interior of the helix, and pairs of bases from the two strands are linked together by hydrogen bonds. The critical point of the Watson–Crick model is that *hydrogen bonding can occur only between specific bases*. A purine base of one strand is always paired with a pyrimidine base of the other strand: adenine is always paired with thymine and guanine is always paired with cytosine. Each base is the donor in one hydrogen bond and the acceptor in the other. Because of the sizes and geometries of the bases, the *only possible pairings in DNA are A with T and C with G* (Figure 24.16).

Due to the tetrahedral geometry around all of the atoms making up the polymeric backbone of DNA and the requirements of base pairing, the double strand must assume a helical shape about 2000 pm wide (Figure 24.17). Since base pairs occur every 340 pm along the helix and since there are 10 base pairs for every complete turn of the helix, each repeating unit is about 3400 pm long.

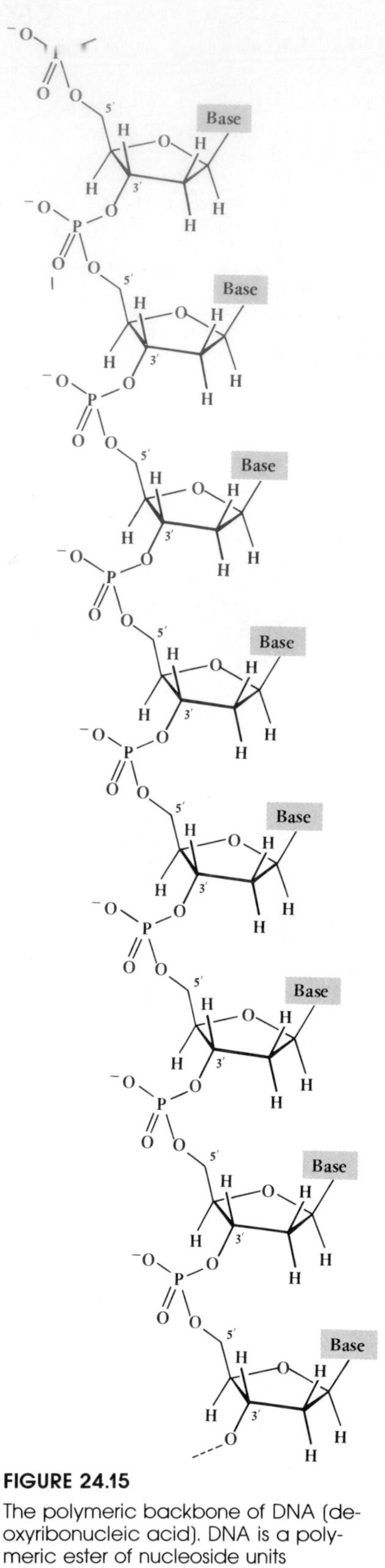

FIGURE 24.15
The polymeric backbone of DNA (deoxyribonucleic acid). DNA is a polymeric ester of nucleoside units bridged by phosphate groups.

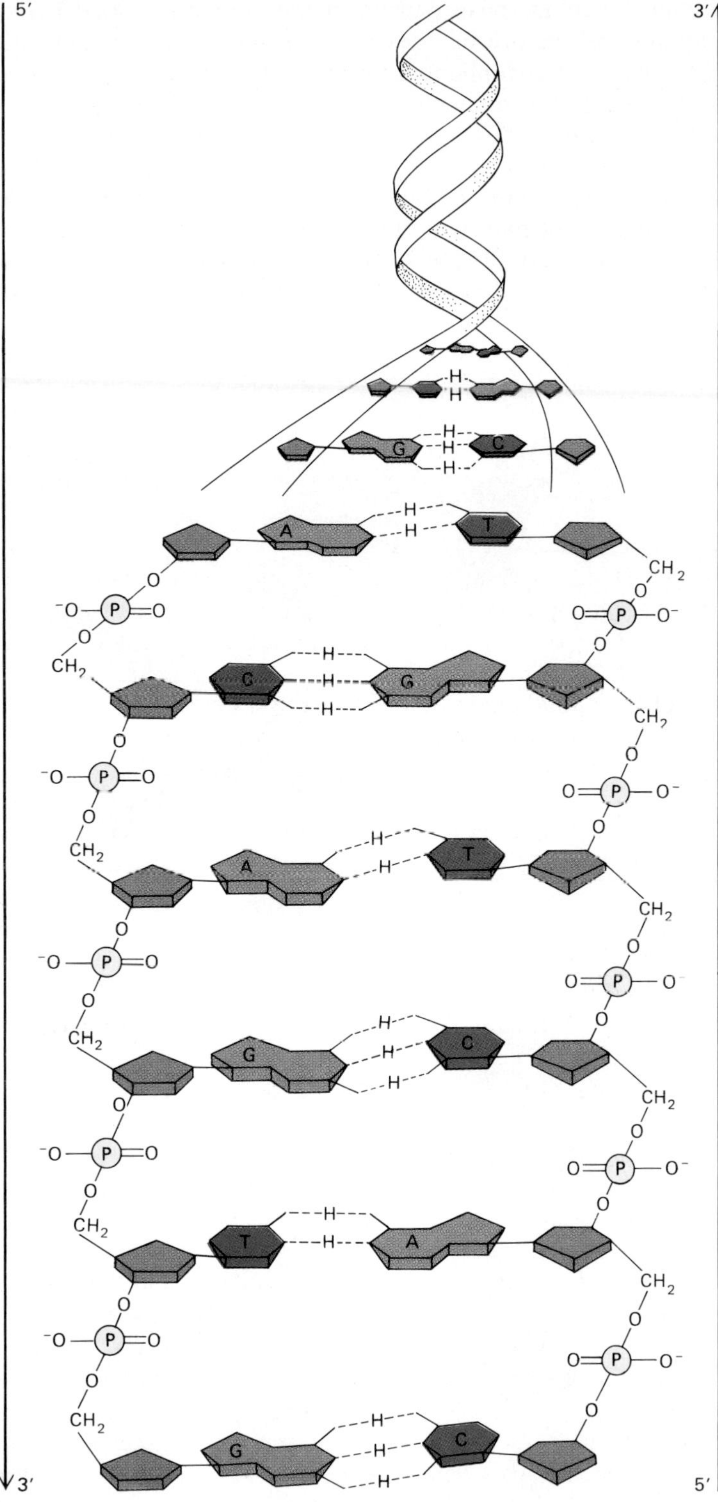

FIGURE 24.16

Structure of DNA. Notice that the two sugar-phosphate chains run in opposite directions (b). This permits complementary bases to pair by hydrogen bonding as illustrated in (a). Only certain base pairings can occur in DNA: A with T and G with C. Because of the tetrahedral geometry of the phosphate and carbon atoms of the backbone, the strands twist into a helical structure.

The complementary base pairing in the double-stranded helix is the "back-up system" nature has devised to insure accurate replication in new DNA. The DNA is replicated by unwinding the double helix and attaching a nucleoside phosphate to the exposed bases on each strand (Figure 24.18). Since each base in a given strand can only pair with its particular complement, this means that the newly formed "daughter" strand must be an exact copy of one of the "parent" strands. The net result is the formation of two new DNA double helices from the parent double helix. Each new double helix contains one parent strand and one new strand.

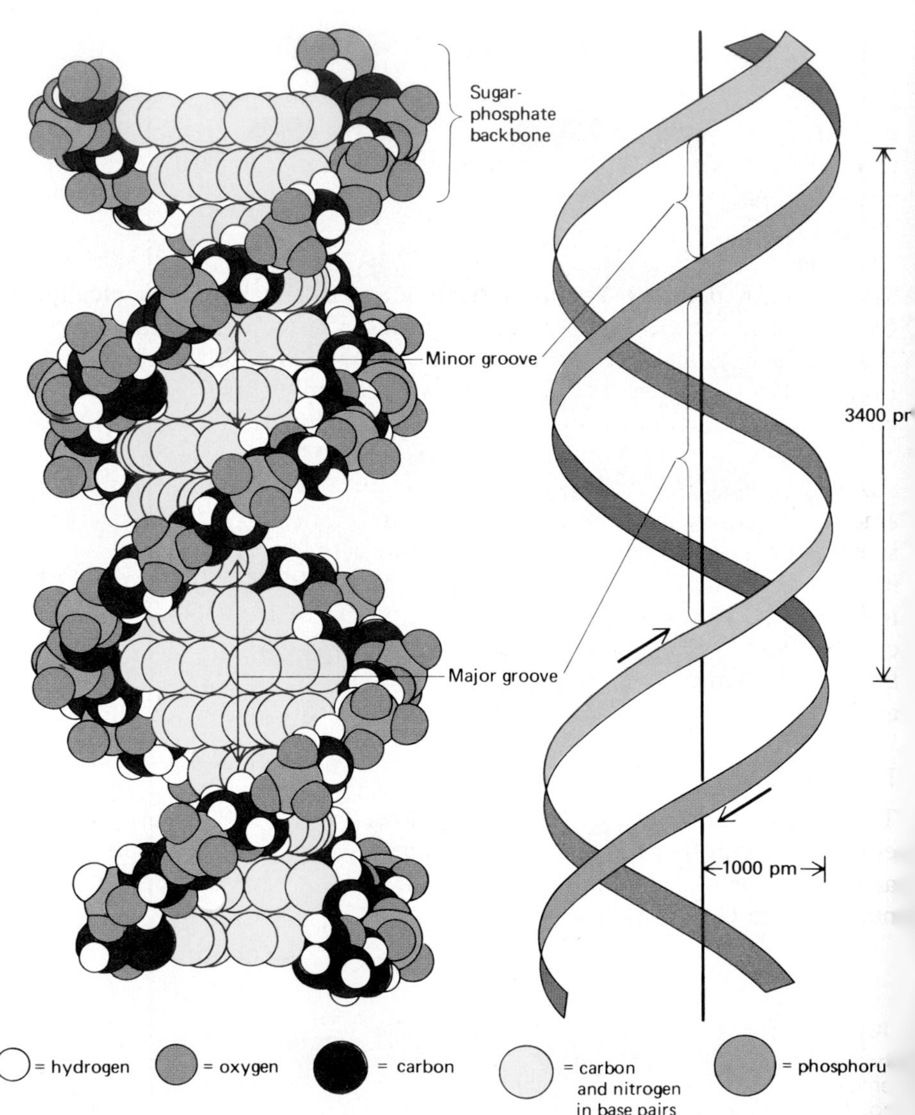

FIGURE 24.17

According to modern theory, most DNA exists in this double-helical form. The molecule actually has two backbones held together by base-pairing.

24.6 SYNTHETIC POLYMERS: PLASTICS, FIBERS, AND ELASTOMERS

It is almost certain that some of your clothing and many of the objects around you are synthetic, all products of chemistry. In spite of their familiarity, this has been a recent development. Synthetic resins such as *Bakelite*, fibers such as *rayon*, and plastics such as *celluloid* were made early in this century. However, the plethora of synthetic polymers has been available only since World War II. In 1983 over 36 pounds of man-made fibers and 140 pounds of plastics were produced for every person in the United States. By 1976, plastic outstripped steel as the nation's most widely used material, and we now use more plastic than steel, aluminum, and copper combined.

The polymers described in this section are organic polymers. There are, however, inorganic polymers such as the silicones described in Section 21.2.

THE TERMINOLOGY OF THE POLYMER INDUSTRY

Polymers are classified as **elastomers**, **plastics**, and **fibers** depending on their elasticity. As the name suggests, elastomers can be highly stretched, to over ten times their normal length and returned to their normal dimensions many times. Fibers, in contrast, are the least elastic, while plastics fall in between.

POLYMER TYPE	UPPER LIMIT OF EXTENSIBILITY (%)
Elastomer	100–1000
Plastic	20–100
Fiber	Less than 10

Polymers are also classified on the basis of ther response to heating. **Thermoplastic** polymers soften on heating and are unaltered chemically. On the other hand, **thermosetting** polymers degrade or decompose on heating. In general, linear polymers such as polyethylene and the others listed in Table 24.3 are thermoplastic, whereas polymers with one backbone linked to another (as in proteins) are thermosetting (Figure 24.19).

The reasons for the differences in polymer properties depend on molecular weight, extent of cross-linking between chains, and crystallinity among other things. For example, polymers begin to develop mechanical strength after about 50 monomer units have been bound together, and it increases up to about 500 units; after this, increases in chain length seem to have little effect on strength. If the polymer chains are cross-linked by covalent bonds or dispersion forces, the polymer is more likely to be a fiber. On the other hand, if the forces between chains are weak or there are bulky substituents along a chain, it is more likely to be an elastomer.

THERMOPLASTIC POLYMERS

There are five major thermoplastic polymers: low- and high-density polyethylene, polypropylene, polystyrene, and polyvinyl chloride (Tables 24.3 and 24.4). Almost 35 billion pounds of these polymers are produced annually in the United States.

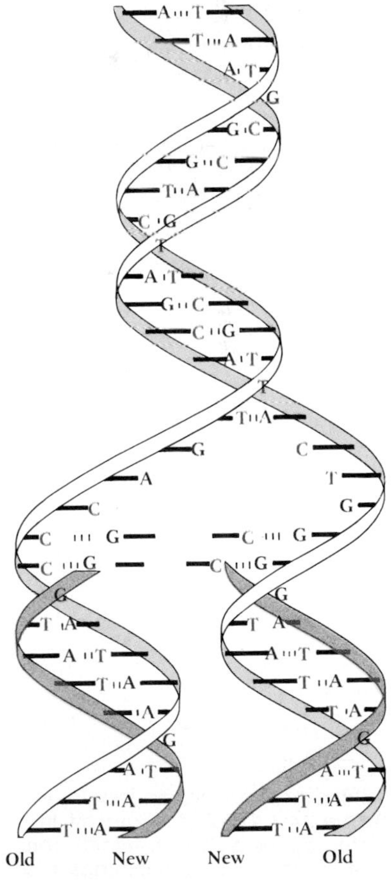

IGURE 24.18

he replication of DNA. The original helical coil is unwound and each strand acts as a emplate for the formation of a new strand. Owing to the requirements of base pairing, the new strand is an exact copy of one of the old strands.

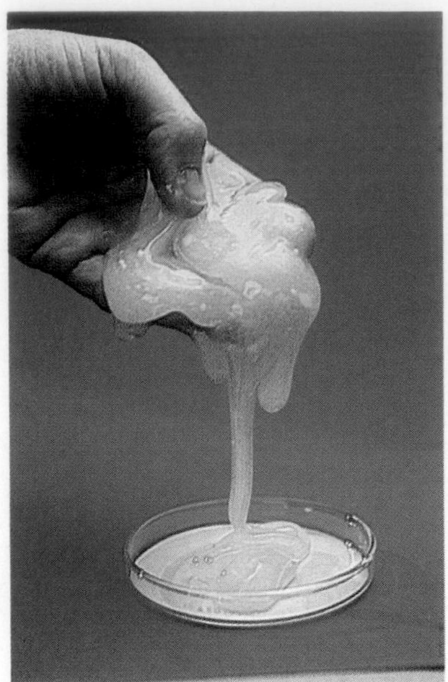

FIGURE 24.19
Polyvinyl alcohol, derived from vinyl alcohol (CH_2=CHOH), is a close relative of PVC and PVA in Table 24.3. It is used as a coating on grease-proof paper and to thicken foods. When an aqueous solution of polyvinyl alcohol is mixed with boric acid (or its sodium salt), an ester forms and leads to "cross-linking" of the polymer. The viscous mixture is popularly known as "slime."

$$2\ \text{~OH} + B(OH)_3 \longrightarrow \text{~O—B(—OH)—O~} + 2\,H_2O$$

The most common polymer is **polyethylene**, of which about 7 million tons are produced annually in the United States. *Low-density polyethylene* (LDPE) is a tough, waxy, mainly linear polymer; it has a molecular weight in the range of 10,000 to 40,000 and a density of 0.92 g/cm^3. *High density polyethylene* (HDPE) is an unbranched, more crystalline polymer with a density of 0.97 g/cm^3. HDPE has greater strength and rigidity, a higher softening temperature, and is better able to withstand boiling water than LDPE. However, since LDPE is more transparent than HDPE, the low-density form is used in the form of film for packaging.

The thermoplastics described so far are formed by polymerizing one type of olefin. However, many polymers are a combination of two different monomers and are referred to as **copolymers**. *Saran*, a polymeric film used to wrap food in your home and the supermarket, is an excellent example of a thermoplastic copolymer. It is prepared by mixing 1,1-dichloroethene and chloroethene (vinyl chloride) in a 4:1 ratio.

$$\underset{\text{1,1-dichloroethene}}{CCl_2{=}CH_2} + \underset{\text{vinyl chloride}}{CH_2{=}CHCl} \longrightarrow \text{Saran}$$

Polyurethanes are also thermoplastics. When an organic isocyanate is treated with an alcohol, the alcohol adds to the CN double bond to give a urethane.

$$\underset{\text{isocyanate}}{R{-}N{=}C{=}O} + \underset{\text{alcohol}}{R'{-}OH} \longrightarrow \underset{\text{a urethane}}{R{-}N(H){-}C({=}O){-}OR'}$$

With some imagination you can see that a diisocyante (OCN—R—NCO) and a dialcohol (HO—R'—OH) will lead to a polymer.

$$\underset{\text{polyurethane}}{-C({=}O){-}N(H){-}R{-}N(H){-}C({=}O){-}O{-}R'{-}O{-}C({=}O){-}N(H){-}R{-}N(H){-}C({=}O){-}O{-}R'{-}O-}$$

Polyurethanes are used as coatings when high impact and abrasion resistance are important, so you often find them on dance floors, gymnasium floors, and bowling alleys. If gas is bubbled into polyurethane as it forms, a foam is produced that is used in bedding and car upholstery. By modifying the R groups of the urethane system, the polyurethanes can be cross-linked and made into fibers that are used to make swim suits, ski wear, and other garments that must stretch.

THERMOSETTING POLYMERS

The polymers in the previous section and in Table 24.3 are thermoplastic. They are made from monomers that can form two bonds to neighboring molecules. Thermosetting polymers, on the other hand, are made from monomers that form more than two bonds to neighboring units, and the polymers are thus highly branched or cross-linked. For this reason, polymers of this type often do not melt when heated.

All the polymerization reactions in Table 24.3 require a catalyst. In some cases a *Ziegler–Natta* catalyst, a mixture of compounds such as $Al(C_2H_5)_3$ and $TiCl_3$, is used.

TABLE 24.3 Some Common Polymers Based on Ethylene

Monomer	Polymer	Name
$CH_2{=}CH_2$ → Ethylene	$-\overset{H}{\underset{H}{C}}-CH_2-\overset{H}{\underset{H}{C}}-CH_2-\overset{H}{\underset{H}{C}}-CH_2-$	Polyethylene
$\overset{Cl}{\underset{H}{C}}{=}CH_2$ → Vinyl chloride	$-\overset{Cl}{\underset{H}{C}}-CH_2-\overset{Cl}{\underset{H}{C}}-CH_2-\overset{Cl}{\underset{H}{C}}-CH_2-$	Polyvinyl chloride, PVC
$\overset{Cl}{\underset{Cl}{C}}{=}CH_2$ → Vinylidene chloride	$-\overset{Cl}{\underset{Cl}{C}}-CH_2-\overset{Cl}{\underset{Cl}{C}}-CH_2-\overset{Cl}{\underset{Cl}{C}}-CH_2-$	Polyvinylidene chloride
$\underset{C_6H_5}{CH}{=}CH_2$ → Styrene	$-\overset{H}{\underset{C_6H_5}{C}}-CH_2-\overset{H}{\underset{C_6H_5}{C}}-CH_2-\overset{H}{\underset{C_6H_5}{C}}-CH_2-$	Polystyrene
$\underset{C\equiv N}{CH}{=}CH_2$ → Acrylonitrile	$-\overset{H}{\underset{C\equiv N}{C}}-CH_2-\overset{H}{\underset{C\equiv N}{C}}-CH_2-\overset{H}{\underset{C\equiv N}{C}}-CH_2-$	Polyacrylonitrile (Orlon)
$\underset{CH_3}{CH}{=}CH_2$ → Propylene	$-\overset{H}{\underset{CH_3}{C}}-CH_2-\overset{H}{\underset{CH_3}{C}}-CH_2-\overset{H}{\underset{CH_3}{C}}-CH_2-$	Polypropylene
$\underset{OCCH_3(=O)}{CH}{=}CH_2$ → Vinyl acetate	$-\overset{H}{\underset{OCCH_3(=O)}{C}}-CH_2-\overset{H}{\underset{OCCH_3(=O)}{C}}-CH_2-\overset{H}{\underset{OCCH_3(=O)}{C}}-CH_2-$	Polyvinylacetate (PVA)
$CF_2{=}CF_2$ → Tetrafluoroethylene	$-\overset{F}{\underset{F}{C}}-CF_2-\overset{F}{\underset{F}{C}}-CF_2-\overset{F}{\underset{F}{C}}-CF_2-$	Polytetrafluoroethylene (Teflon)

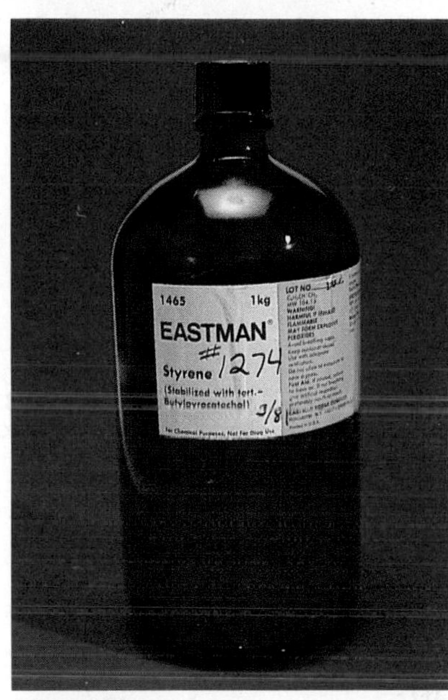

(a)

(b)

After storage for some time, the styrene in a bottle (a) was found to have polymerized to a clear block of polystyrene (b). (Charles D. Winters)

FIGURE 24.20

Phenol–formaldehyde polymer. When a mixture of phenol and formaldehyde in acetic acid is treated with concentrated HCl, polymer begins to grow. (Charles D. Winters)

The formation of thermosetting polymers is more complex than thermoplastics. One example of the former is the phenol–formaldehyde resin called *Bakelite* (Figure 24.20). When phenol and a concentrated aqueous solution of formaldehyde are mixed in the presence of HCl, compounds such as the following are formed.

$$\underset{\text{phenol}}{C_6H_5OH} + \underset{\text{formaldehyde}}{HCHO} \xrightarrow{HCl} o\text{-}HOC_6H_4CH_2OH \text{ and } p\text{-}HOC_6H_4CH_2OH$$

These products react with one another by **condensation** in much the same way as two amino acids or two sugars.

$$>C{-}CH_2{-}O{-}H + H{-}C< \longrightarrow >C{-}CH_2{-}C< + H_2O$$

Eventually, a highly cross-linked polymer is formed.

TABLE 24.4 Major Thermoplastic Polymers Produced in the United States in 1984

POLYMER	AMOUNT (billions of pounds)	USES
Low-density polyethylene	8.5	Film for packaging; molded products
Polyvinyl chloride	7.0	Molded pipes (about 40% of the usage of PVC); phonograph records; bottles
High-density polyethylene	6.0	Molded containers (e.g., 1-gal milk)
Polypropylene	4.8	Molded items (e.g., appliance parts, battery cases); fibers
Polystyrene	3.9	Packaging; more than 1 pound is used in each video cassette

ELASTOMERS: RUBBER

As we usually think of it, rubber is not a synthetic polymer; it is a natural substance. Nonetheless, it is useful to describe it briefly here because it is the best example of an elastomer and because attempts to duplicate its properties have heavily influenced the polymer industry.

The 18th century English chemist Joseph Priestley invented the name "rubber" because of the ability of the substance to erase pencil marks.

Natural rubber is a polymer of 2-methyl-1,3-butadiene (isoprene) and is obtained from the tree *Hevea brasiliensis*.

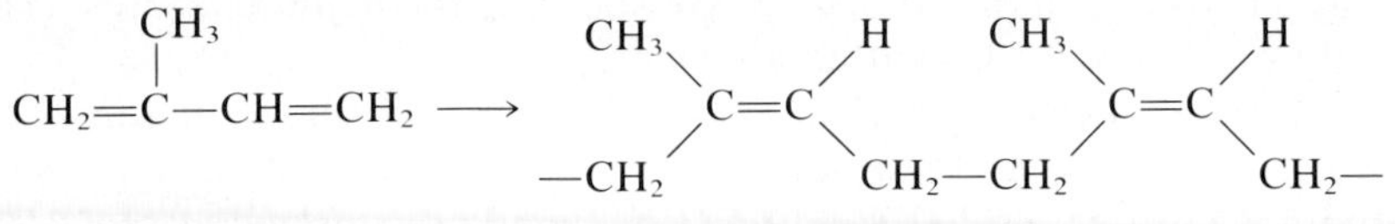

$$CH_2{=}C(CH_3){-}CH{=}CH_2 \longrightarrow {-}CH_2{-}C(CH_3){=}CH{-}CH_2{-}CH_2{-}C(CH_3){=}CH{-}CH_2{-}$$

isoprene — Hevea rubber

The substituent groups are *cis*, that is, they lie on the same side of the C=C double bond. The *trans* form is also found in nature, but it has come to be called *gutta-percha*.

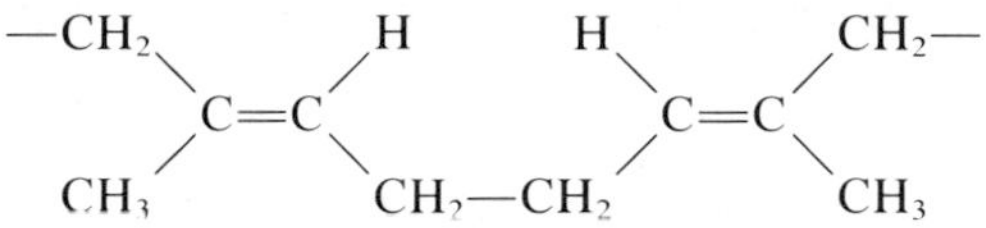

$${-}CH_2{-}C(CH_3){=}CH{-}CH_2{-}CH_2{-}CH{=}C(CH_3){-}CH_2{-}$$

gutta-percha rubber

Unlike Hevea rubber, gutta-percha is hard, brittle, and nonelastic. An apparently minor change in structure leads to a major change in properties.

Rubber was first introduced into Europe in 1740, but it remained a curiosity until 1823 when Charles Mackintosh invented a method of waterproofing cotton cloth with a solution of rubber. The mackintosh, as rain coats are even now sometimes called, became extremely popular, despite some major drawbacks: natural rubber is notably weak, and it is thermoplastic, soft and tacky when warm but brittle at lower temperature. The American inventor Charles Goodyear found a way around these difficulties with the *vulcanization* process. If rubber, natural or synthetic, is heated with sulfur, the polymer chains are linked with C—S—S—C bonds.

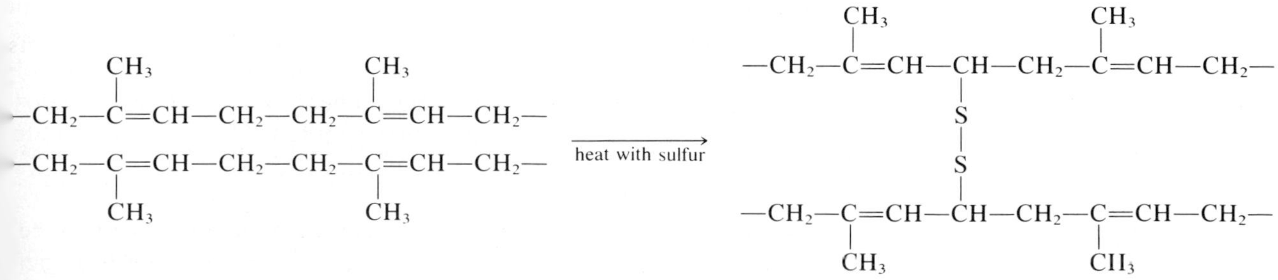

$$\begin{array}{l}{-}CH_2{-}C(CH_3){=}CH{-}CH_2{-}CH_2{-}C(CH_3){=}CH{-}CH_2{-}\\{-}CH_2{-}C(CH_3){=}CH{-}CH_2{-}CH_2{-}C(CH_3){=}CH{-}CH_2{-}\end{array} \xrightarrow{\text{heat with sulfur}} \begin{array}{l}{-}CH_2{-}C(CH_3){=}CH{-}CH(S{-}){-}CH_2{-}C(CH_3){=}CH{-}CH_2{-}\\ \qquad\qquad\qquad\quad |\\ \qquad\qquad\qquad\quad S\\ \qquad\qquad\qquad\quad |\\{-}CH_2{-}C(CH_3){=}CH{-}CH(S{-}){-}CH_2{-}C(CH_3){=}CH{-}CH_2{-}\end{array}$$

Cross-linking makes the rubber harder, stronger, and less thermoplastic. With Goodyear's invention, rubber became more useful, and the demand increased. Unfortunately, the only sources of supply of rubber in the 19th century were wild trees in the jungles of Brazil and the Congo. This problem was solved by an enterprising Englishman who smuggled seeds for rubber trees out of Brazil and planted them in large plantations in Sri Lanka and Malaysia. By 1939, almost 1,400,000 tons of rubber were consumed worldwide per year. World War II intervened, however, and cut off supplies of natural rubber to Europe and the United States. This brought on the hunt for synthetic rubber.

The obvious way to synthesize rubber is to duplicate nature. This is easier said than done, since only recently have catalysts been discovered that will do what a rubber tree does: polymerize isoprene so that all of the substituent groups are *cis*.

As alternatives to duplicating natural rubber, many different elastomers have been developed, all of which involve butadiene or some derivative of this olefin. For example, 2-chloro-1,3-butadiene (commonly called chloroprene) polymerizes to form *neoprene* rubber.

$$\underset{\text{chloroprene}}{n\mathrm{CH_2{=}CH{-}\underset{\substack{|\\ Cl}}{C}{=}CH_2}} \longrightarrow \underset{\text{neoprene}}{\mathrm{{-}CH_2{-}CH{=}\underset{\substack{|\\ Cl}}{C}{-}CH_2{-}CH_2{-}CH{=}\underset{\substack{|\\ Cl}}{C}{-}CH_2{-}}}$$

The latest high-tech resin, a so-called engineering plastic, is ABS resin. The letters stand for "acrylonitrile–butadiene–styrene." ABS resins are used in computer cases, electric tools, watch cases, and automobile parts.

The high chlorine content of neoprene gives the polymer resistance to heat and flames, as well as to oil and chemicals. It is thus used for hoses carrying gasoline, for protective gloves, and for balloons. About 6% of the synthetic rubber produced in the United States is polychloroprene.

Most synthetic elastomers, however, are copolymers of two different olefins. For example, the most important synthetic rubber in the United States market (about 45%) is SBR, or a styrene–butadiene rubber.

$$\underset{\text{butadiene}}{\mathrm{CH_2{=}CH{-}CH{=}CH_2}} + \underset{\text{styrene}}{\mathrm{\underset{\substack{|\\ C_6H_5}}{CH}{=}CH_2}} \longrightarrow \underset{\text{styrene–butadiene rubber (SBR)}}{\mathrm{{-}CH_2{-}CH{=}CH{-}CH_2{-}\underset{\substack{|\\ C_6H_5}}{CH}{-}CH_2{-}CH_2{-}CH{=}CH{-}CH_2{-}\underset{\substack{|\\ C_6H_5}}{CH}{-}CH}}$$

FIBERS

More than 9 billion pounds of fibrous polymers, some purely synthetic and others based on cellulose, are produced per year in the United States. Some of these are semisynthetic (partly synthetic) and others are purely synthetic fibers. The chief semisynthetic fibers are rayon and cellulose acetate, and there are three main groups of synthetic fibers: (a) polyamides (nylon); (b) polyesters (Dacron); and acrylics (Orlon).

SEMISYNTHETIC FIBERS Semisynthetic fibers are made by modifying cellulose in some way. In Section 24.4 you learned that cellulose is a linear polymer of glucose molecules. If this polymer is heated in sodium hydroxide with carbon disulfide, a so-called "viscose" solution is formed. If this solution is then forced through very small holes, continuous filaments are formed. When the filaments drop into a sulfuric acid bath, cellulose is regenerated from the viscose, and the fibers are then woven into cloth that we know as rayon.

Cellulose acetate, used in "wash and wear" fabrics, is made by converting some of the —OH substituents of cellulose into ester groups by reaction with acetic acid.

cellulose acetate

SYNTHETIC FIBERS **Nylon** was the first fiber to result from a deliberate search for a purely synthetic material, and it was greatly in demand from the beginning. As women's dresses became shorter about 70 years ago, silk stockings were very fashionable but also very expensive. Late in the 1930s Wallace H. Carothers of Du Pont discovered nylon-66, and it was soon found that nylon fibers could be woven into sheer hosiery. The first such stockings went on sale in October, 1939, and they were so popular they had to be rationed. Unfortunately, World War II caused the commercial use of nylon to cease until 1945. Today it is produced in many forms at an annual rate of more than 9 pounds per person in the United States.

Nylons are *polyamides*, condensation polymers formed from a diacid and a diamine (Figure 24.21). For example, adipic acid and hexamethylenediamine produce nylon-66, the 66 suffix coming from the fact that both the diacid and the diamine have 6 carbon atoms.

$$\underset{\text{adipic acid}}{HOOC(CH_2)_4COOH} + \underset{\text{hexamethylenediamine}}{H_2N(CH_2)_6NH_2} \xrightarrow{-H_2O}$$

$$-\overset{O}{\overset{\|}{C}}-(CH_2)_4-\overset{O}{\overset{\|}{C}}-\overset{H}{\overset{|}{N}}-(CH_2)_6-\overset{H}{\overset{|}{N}}-\overset{O}{\overset{\|}{C}}-(CH_2)_4-\overset{O}{\overset{\|}{C}}-\overset{H}{\overset{|}{N}}-(CH_2)_6-\overset{H}{\overset{|}{N}}-$$

amide link

nylon-66

About half of the nylon produced is used to make cords for tires. The remainder goes into ropes, cords, rugs, fish nets and lines, clothes, thread, hose, coats, dresses and on and on.

FIGURE 24.21
Nylon-66. Hexamethylenediamine is dissolved in water (bottom layer), and a derivative of adipic acid (adipyl chloride) is dissolved in hexane (top layer). The chemicals mix at the interface between the two layers to form nylon (which is being wound onto a stirring rod). (Charles D. Winters)

Polyesters are also condensation polymers used as fibers and are produced in even greater amounts than nylon. The most common one is sold in the United States as *Dacron*, a polymer formed from a dialcohol and an aromatic diacid.

$$\underset{\text{ethylene glycol}}{HO{-}CH_2CH_2{-}OH} + \underset{\text{terephthalic acid}}{HO{-}\overset{\overset{O}{\|}}{C}{-}C_6H_4{-}\overset{\overset{O}{\|}}{C}{-}OH} \xrightarrow{-H_2O} \underset{\substack{\text{polyethyleneterephthalate}\\ \text{Dacron}}}{{-}O{-}CH_2CH_2{-}O{-}\overset{\overset{O}{\|}}{C}{-}C_6H_4{-}\overset{\overset{O}{\|}}{C}{-}}$$

Polyester fibers are generally used today in a blend with cotton fibers for wash and wear garments.

Finally, **acrylics** are polymers based on cyanoethene or acrylonitrile (Table 24.3). Polyacrylonitrile fibers are in great demand because they are only about half as expensive as wool, can be woven into textiles such as *Orlon* that have a "wooly" feel, and are machine washable and do not shrink. Although only about $\frac{1}{4}$ as much polyacrylonitrile is produced as nylon, demand for the acrylic polymer seems certain to increase.

SUMMARY

Polymers are constructed of many small, individual units or **monomers**. The monomer of polyvinyl chloride is vinyl chloride (chloroethene),

$$\underset{\text{vinyl chloride}}{nCH_2{=}CHCl} \longrightarrow \underset{\text{polyvinyl chloride}}{{-}CH_2{-}\underset{\underset{Cl}{|}}{CH}{-}CH_2{-}\underset{\underset{Cl}{|}}{CH}{-}CH_2{-}\underset{\underset{Cl}{|}}{CH}{-}}$$

and the basic building blocks of proteins or polypeptides are amino acids.

$$\underset{\text{amino acid}}{nH_2N{-}CH(R){-}COOH} \xrightarrow{-nH_2O} \underset{\text{polypeptide}}{{-}\overset{\overset{H}{|}}{N}{-}CH(R){-}\overset{\overset{O}{\|}}{C}{-}\overset{\overset{H}{|}}{N}{-}CH(R){-}\overset{\overset{O}{\|}}{C}{-}}$$

Polyvinyl chloride is an **addition polymer**, because the empirical formula of the polymer is the same as the monomer. In contrast, a polypeptide is a **condensation polymer**, since the monomers come together or "condense" by eliminating a small molecule.

Polymers in this chapter were divided into two types: **natural** (proteins, carbohydrates, and nucleic acids) and **synthetic** (e.g., polyviny chloride, polyethylene, Teflon, polyesters).

Before discussing polymer chemistry, the subject of **stereoisomerism** was outlined in Section 24.1. Stereoisomers have the same atom-to atom bonding sequences but the atoms differ in their arrangement in space All of the α-amino acids (Section 24.2 and Table 24.1) but glycine exhibi a form of stereoisomerism called **chirality**. A compound is chiral if th compound and its mirror image are not superimposable. Such compounds are called **enantiomers**. If a beam of plane polarized light is passed through a solution of the compound, the plane of polarization is rotated. Thus, chiral compounds are often said to be **optically active** or to be **optical isomers**. A carbon atom will be a center of chirality if it is bound to four *different* groups.

Simple proteins consist only of amino acids, while **conjugated proteins** contain some other group in addition to the amino acids. When two amino acids condense, an amide linkage or **peptide bond** is formed.

$$H_2N-\underset{H}{\overset{R}{C}}-\overset{O}{\overset{\|}{C}}-OH + H_2N-\underset{H}{\overset{R'}{C}}-\overset{O}{\overset{\|}{C}}-OH \longrightarrow H_2N-\underset{H}{\overset{R}{C}}-\overset{O}{\overset{\|}{C}}-\overset{H}{N}-\underset{H}{\overset{R'}{C}}-\overset{O}{\overset{\|}{C}}-OH$$

amide or peptide linkage

The geometry and polarity of the peptide bond have a profound effect on protein structure.

The **primary structure** of a protein is determined by the sequence of amino acids. Due to hydrogen bonding between peptide units, the protein chain can have a **secondary structure**: an α-helix or β-sheet structure. **Fibrous** and **globular** proteins have the α-helix. Segments of a protein coil can also interact by hydrogen bonding and other mechanisms.

Carbohydrates have the formula $C_x(H_2O)_y$ and are divided into **monosaccharides**, **disaccharides**, and **polysaccharides** (Section 24.4). When monosaccharides condense to form disaccharides or polysaccharides, a C—O—C **glucoside** linkage is formed. **Cellulose**, **starch**, and **glycogen** are polymers of glucose. The first two differ in the geometry of the glucoside linkage, so cellulose is a linear polymer, whereas starch is a helical coil.

If one of five organic bases (Section 24.5) is bound to the monosaccharide ribose, a **nucleoside** results. If the nucleoside condenses with a phosphate, a **nucleotide** is the product. Linking nucleotides together gives the polymer **RNA**, **ribonucleic acid**. **DNA**, or **deoxyribonucleic acid**, results when deoxyribose is the sugar of the polymer backbone. Due to the fact that bases of DNA can hydrogen bond with one another, the polynucleic acid chains interact to form the double stranded α-helix of DNA.

Synthetic polymers (Section 24.6) are classified as **elastomers, plastics**, or **fibers** depending on their elasticity. On the basis of their response to heat, they are also classed as **thermoplastic** (unaltered chemically on heating) or **thermosetting** (degrade on heating). Most thermoplastic polymers are addition polymers, while thermosetting polymers are often condensation polymers. Both purely synthetic and semisynthetic fibers are made. Nylon, a condensation polymer of a diacid and diamine, is synthetic, but rayon is a cellulosic, that is, a modified form of cellulose.

STUDY QUESTIONS

STEREOCHEMISTRY

1. Define the term chirality using the molecule glyceraldehyde, the simplest carbohydrate. The structural formula of the compound is

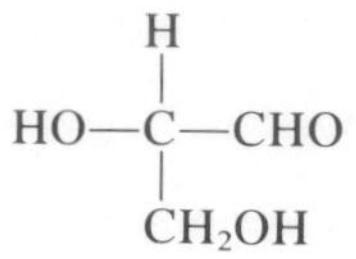

```
      H
      |
HO—C—CHO
      |
    CH2OH
```

2. Using drawings like those in Figure 24.2, prove that CH_2ClBr is not a chiral molecule. Explain your reasoning briefly.
3. Using drawings such as those in Figure 24.2, illustrate the enantiomers of the amino acid cysteine (Table 24.1).
4. For each of the following molecules, decide if the underlined carbon atom is or is not a chiral center.
   (a) $\underline{C}H_2Cl_2$
   (b) $C_6H_5—\underline{C}H(CH_3)—NH_2$
   (c) $H_2N—\underline{C}H_2—COOH$
   (d) $HOOC—\underline{C}H(OH)—CH_2—COOH$

AMINO ACIDS

5. Draw the D form of the amino acid serine.
6. Which one of the ten α-amino acids in Table 24.1 is not chiral? Explain briefly why it is not chiral.

PROTEINS

7. Draw the structures of the four dipeptides that can form if alanine and serine are mixed.
8. Draw the structure of one of the tripeptides that can form from a mixture of glycine, histidine, and cysteine.
9. Define the terms primary and secondary as applied to proteins.
10. Name three features of the peptide linkage that are important in controlling the secondary structures of proteins.
11. Give two examples of each type of protein: (a) fibrous, (b) globular.
12. Interactions between side chains (or R groups) of amino acids can influence the structures of proteins (and control what is known as the protein's tertiary structure). If you have a protein with a serine unit ($R = CH_2OH$) close to an aspartic acid unit ($R = CH_2COOH$), what type of interaction can occur between the R groups of these units?

CARBOHYDRATES

13. Show that glyceraldehyde, $CH(OH)(CHO)CH_2OH$, is a carbohydrate. (See its structure in Study Question 1.)
14. Consider the structure of ribose below. Which carbon atoms are chiral centers? Which are not?

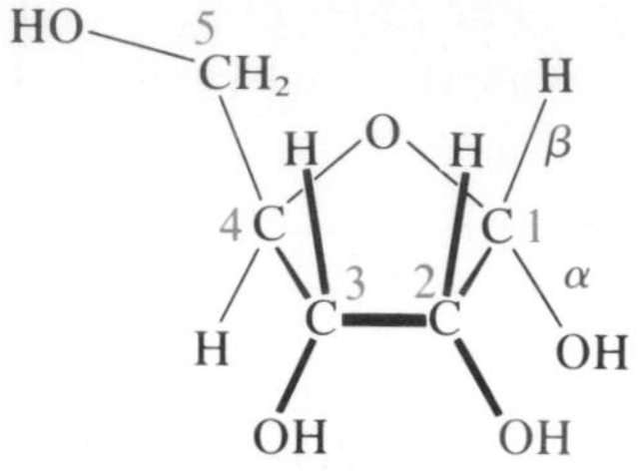

```
HO—5CH2          H
      \   H  O  H / β
      4C        C1
      / \3    2/ \ α
     H   C ——C    OH
        /     \
      OH       OH
```

α-D-ribose

15. Name two examples of each of the following: (a) monosaccharide, (b) disaccharide, and (c) polysaccharide.
16. Why can humans not use cellulose as an energy source?

NUCLEIC ACIDS

17. Describe the features of polynucleotides that lead to the double helix structure of DNA.
18. When β-D-ribose reacts with adenine, what two products form?
19. Explain why base pairing cannot occur between adenine and cytosine, for example.
20. How is genetic information contained in DNA?
21. The single-strand polynucleotide mRNA is made from DNA. The bases on mRNA will be bases complementary to those found on one strand of DNA. Assume mRNA is "copied" from segment B of the DNA illustrated below (where S is ribose and P is the phosphate link).

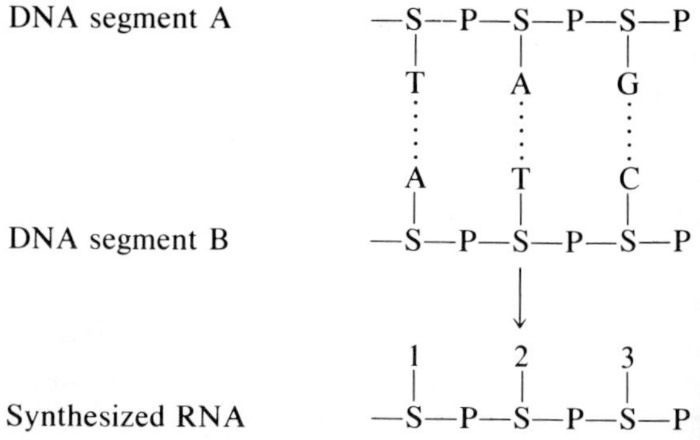

```
DNA segment A      —S—P—S—P—S—P
                    |    |    |
                    T    A    G
                    :    :    :
                    A    T    C
                    |    |    |
DNA segment B      —S—P—S—P—S—P
                         |
                         ↓
                    1    2    3
                    |    |    |
Synthesized RNA    —S—P—S—P—S—P
```

Identify bases 1, 2, and 3 in mRNA. [Note on p. 929 that the base uracil (U) takes the place of thymine (T) in RNA.]

SYNTHETIC POLYMERS

22. Name two synthetic polymers that you may have encountered today and the objects made from these polymers.

23. Polyvinyl acetate is the binder in water-based paints. Draw the structure of this polymer.

24. The squeeze bottles you use in your chemistry laboratory are polyethylene. If the empty bottle has a mass of 150 g, how many moles of ethylene (ethene) must have been used to make the bottle?

25. Polymers are classified in several ways, one depending on their elasticity and the other according to their response to heat. Briefly describe these classifications. Give an example of each class.

26. The two reactions by which polymers are formed are addition and condensation. Give an example of each kind.

27. Condensation reactions are important in forming disaccharides and polysaccharides, nucleosides, and polymers such as nylon-66. Illustrate the condensation reaction for each of these cases.

28. Define the term "copolymer." Give an example of a copolymer.

29. Methyl methacrylate has the structural formula below. When polymerized it is very transparent and is used where this property is appropriate. It is sold in the United States as "Lucite" or "Plexiglas." Draw a polymethyl methacrylate chain that is four monomer units in length.

$$H_2C{=}C(CH_3){-}C({=}O){-}O{-}CH_3 \quad = \text{methyl methacrylate}$$

30. In what way is nylon-66 comparable to proteins?

31. Tetrafluoroethylene is made by first reacting HF with chloroform followed by cracking the resultant difluorochloromethane.

$$CHCl_3 + 2\ HF \longrightarrow CHClF_2 + 2\ HCl$$

$$2\ CHClF_2 + \text{heat} \longrightarrow CF_2{=}CF_2 + 2\ HCl$$

$$CF_2{=}CF_2 + \text{peroxide catalyst} \longrightarrow \text{Teflon}$$

If you wished to make 1.0 kilogram of Teflon, what mass of chloroform and HF must you use to make the starting material, $CHClF_2$? (Although it is not realistic, assume each reaction step proceeds in 100% yield.)

32. A "styrofoam" drinking cup, made of polystyrene has a mass of 8.0 g. How many moles of styrene were used to make the cup?

33. Saran is a copolymer of 1,1-dichloroethene and 5% to 17% chloroethene (vinyl chloride). Draw a possible structure for the polymer called Saran.

34. In 1983 2.42 million pounds of nylon were made in the United States. How many grams each of adipic acid and hexamethylenediamine must have been combined to make this amount of nylon-66? (Recall that 1.00 pound = 454 g.)

25

The Transition Elements and Their Chemistry

Crystals of a manganese-containing mineral (rhodochrosite, $MnCO_3$) (Brian Parker, Tom Stack & Assoc)

CHAPTER OUTLINE

The large block of elements in the central portion of the table is a bridge between the *s*-block elements at the left of the table and the *p*-block metals, metalloids, and nonmetals on the right (Figure 25.1). The first three rows of these elements (Sc to Zn, Y to Cd, and La to Hg) are generally called the **transition elements** or **transition metals**. All these elements count *d* electrons among their valence electrons, and so we can call them the ***d*-block** elements. The elements fitting into the periodic table between La and Hf, and Ac and element 104, are the **inner transition elements**. Here *f* electrons are involved as valence electrons, so they can be called

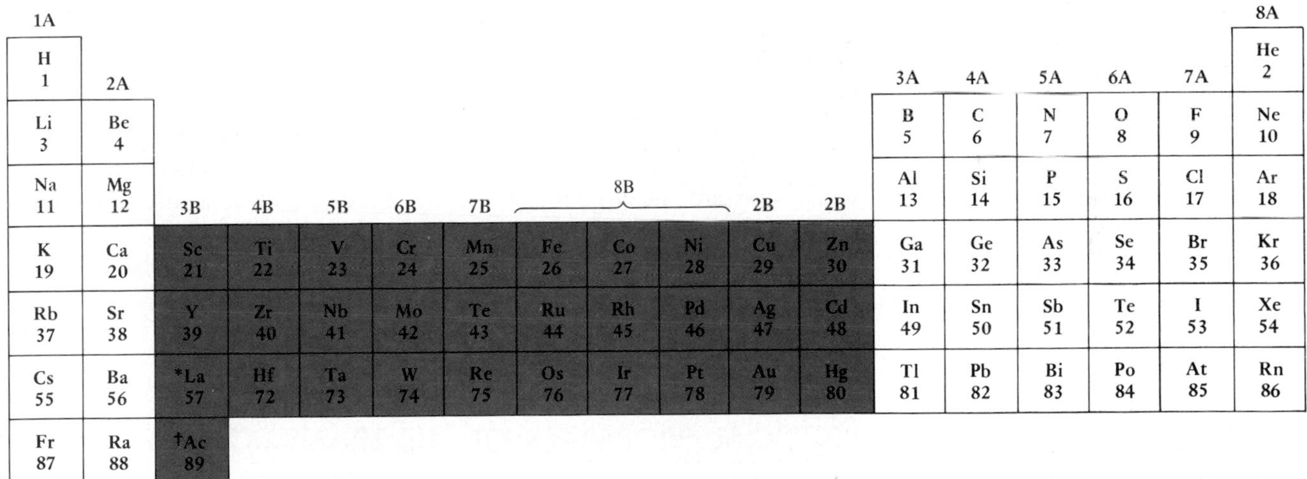

FIGURE 25.1
Periodic table showing the *d* block or transition elements (red) and the *f*-block elements, the lanthanides (blue) and the actinides (yellow).

For historical reasons, the lanthanide elements are sometimes called the "rare earth elements." However, many are actually not rare. For example, Ce is 5 times as abundant as Pb and only about one half as abundant as Cl.

***f*-block elements**. More specifically, the elements coming after lanthanum (La) and before hafnium (Hf) are called the **lanthanide elements**, and those following actinium (Ac), but coming before element 104, are the **actinide elements**. Although you may see all the elements referred to simply as "transition elements," we shall use this term only for the *d*-block elements.

The *d*-block elements in particular are important in living organisms, in certain industrial processes, and as materials. For example, cobalt is the crucial element in vitamin B_{12}, where its role is to act as a catalyst. Iron is the key element in biochemical oxidation–reduction processes, and its prominence in myoglobin is shown in Figure 24.10. Molybdenum and iron, together with sulfur, form the reactive portion of nitrogenase, a biological catalyst used by plants to convert atmospheric nitrogen into ammonia, and copper and zinc are important in other biological catalysts. As catalysts in nonbiological reactions, you have already seen iron used in the Haber process for ammonia, rhodium and platinum in automobile catalytic converters, and titanium in Ziegler–Natta catalysts for the synthesis of polypropylene.

For colors of transition metal compounds see the Special Sections on Iron and Nickel Chemistry and other figures in this chapter.

Many transition metals compounds are highly colored, which makes them useful as pigments in paints and dyes. Prussian blue, $Fe_4[Fe(CN)_6]_3$, is the "bluing agent" in laundry bleach and in engineering blueprints. When transition metal ions are present in crystalline silicates or alumina, the minerals become gems. For example, rubies contain Cr^{3+} substituted for some of the Al^{3+} ions in a crystal lattice of Al_2O_3; the transition metal ion gives the red color to the material and is the reason rubies can function as lasers. Blue glass is made by adding a small amount of a Co^{2+} salt, and the green patina on copper statues and roofs is an oxidized form of copper.

The most obvious use of transition elements is as the metals themselves (Figure 2.4). They are used in coins in many countries around the world and are the primary structural materials in cars and appliances. Approximately 700 million tons of raw steel, 8 million tons of copper, and 750,000 tons of nickel are produced annually on a worldwide basis to meet these needs.

25.1 PHYSICAL PROPERTIES OF THE TRANSITION METALS

To organize the chemistry of the *d*- and *f*-block elements, it is useful to know their electron configurations and their commonly observed oxidation numbers. You recall from Chapter 8 that all these elements have incomplete *d*- or *f*-subshells, so it is possible that many different positive oxidation numbers can result from loss of a varying number of electrons.

OXIDATION NUMBERS

Rutile, TiO_2. (Allen B. Smith, Tom Stack & Assoc)

Oxidation numbers of +2 and +3 are commonly observed for the *d*-block elements, but others are certainly possible (Figures 25.2 and 25.3). For example, common minerals of titanium are TiO_2 (rutile, see photograph)

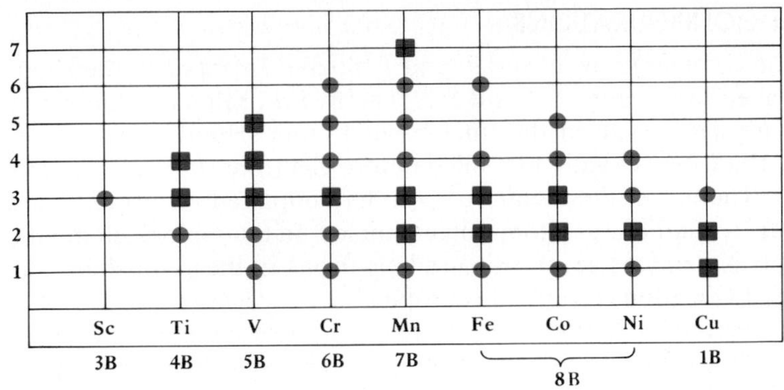

FIGURE 25.2
Oxidation numbers for the fourth period transition metals. More common ones are designated in red.

and ilmenite ($FeTiO_3$) in which titanium has an oxidation number of +4; the most common oxidation number for silver is +1, and copper often forms compounds of Cu^+. On the high end of the oxidation number scale, you have already encountered +7 in an ion such as MnO_4^- (permanganate). The maximum possible oxidation number is +8, but it is rarely seen; perhaps the best example is OsO_4, a powerful oxidizing agent but a very toxic compound.

The lanthanide elements are also commonly found with +2 and +3 oxidation numbers in their compounds, but +4 is important for some, including cerium. The actinide elements, however, generally have higher oxidation numbers such as +4 and even +6. For example, UO_2 is a common oxide of uranium, and a compound important in reprocessing uranium fuel for nuclear reactors is UF_6 (Section 22.3).

In main group metal chemistry, the highest possible oxidation number is equal to the group number, and these are generally observed (Na^+ and Al^{3+}, for example). In transition metal chemistry it is still true, in principle, that the maximum oxidation number is given by the group number. However, transition metal ions usually do not exhibit the maximum possible number in their compounds, especially for elements of the later groups (Figure 25.2). When high oxidation numbers are found, it is usually when the metal is combined with highly electronegative oxygen or fluorine.

FIGURE 25.3
Some common salts of chromium, illustrating oxidation numbers of +3 [in $Cr(NO_3)_3$(green) and $CrCl_3$(violet)] and +6 [in K_2CrO_4(yellow) and $K_2Cr_2O_7$(orange)]. (Charles D. Winters)

ELECTRON CONFIGURATIONS

See Table 8.4 for electron configurations of the *d*-block and *f*-block elements.

The electron configurations of the *d*- and *f*-block elements and their ions were described in Sections 8.4 and 8.7. The five *d* orbitals or the seven *f* orbitals are degenerate in the uncombined atom. Hund's rule means that these orbitals are exactly half filled before electrons are paired in any one orbital. Therefore, there can be up to five unpaired electrons in a *d* subshell; Mn^{2+} and Fe^{3+} are common examples. In the *f*-block elements, there can be as many as seven unpaired electrons in the *f* subshell, and the Eu^{2+} and Gd^{3+} ions are good examples.

The difference between paramagnetic and ferromagnetic materials is that domains of aligned magnets do not form in paramagnetic substances. Ferromagnets are thus "super-paramagnets."

Because there are unpaired electrons in most common oxidation states, ions of the *d*-block and *f*-block elements are typically **paramagnetic** (Section 8.1). The metals themselves can be **ferromagnetic**. Recall that electrons can act as subatomic bar magnets. If, in a solid metal, the atomic magnets of a group of atoms (called a *domain*) cooperate and orient their magnets in the same direction, the magnitude of the magnetic effect is much larger than paramagnetism. Only the metals of the iron, cobalt, and nickel subgroups exhibit this property. They are also unique in that, once the electron magnets are aligned by an external magnetic field, the metal is permanently magnetized. In such a case the magnetism can only be eliminated by heating or vibrating the metal to rearrange the electron spin domains.

EXERCISE 25.1 ELECTRON CONFIGURATIONS

Give electron configurations for (a) Ti^{3+}, (b) Fe^{2+}, (c) Cu^{+}, and (d) Gd^{3+}. Are any of the ions paramagnetic? If so, how many unpaired electrons does the ion have? (For a review, refer to Chapter 8.)

Another unique property of transition metal ions is that it is rare for the ion to follow the "octet rule" in forming compounds. The octet rule was developed for main group elements with four valence atomic orbitals that, when assigned lone and bond pairs, can account for no more than eight electrons. For transition metal ions, the minimum number of valence orbitals is five and, if the next higher *s* and *p* orbitals are counted as valence orbitals, the number is nine. Metals like Ni^{2+} ($3d^8$ electron configuration), Cu^{2+} ($3d^9$), and Zn^{2+}($3d^{10}$) meet or exceed the "octet rule" before they form their first bond to another atom. Consequently, transition metal ions routinely have expanded octets.

METAL ATOM RADII AND DENSITY

Metal radius is a measure of the size of transition atoms; it is found by determining the size of atoms from the solid state structure of the metal. See the discussion of periodic trends in atomic radii in Chapter 8.

The periodic trend in the radii of transition metal atoms is illustrated in Figure 25.4. Here you see that the sizes of the metal atoms change very little across a series, especially beginning at Group 5B (V, Nb, or Ta). The reason for this is that the atom size is determined by the radius of the *ns* orbital (n = 4, 5, or 6), which in every case has at least one electron. The variation in electrons occurs in the $(n - 1)d$ orbitals, which are somewhat smaller than the *ns* orbitals. As the number of electrons in these $(n - 1)d$ orbitals increases, they nearly completely screen the increase in nuclear charge from the 4*s* electrons. Consequently, the 4*s* electrons experience only a slightly increasing effective nuclear charge across the series, and element radii remain nearly constant. Finally, the

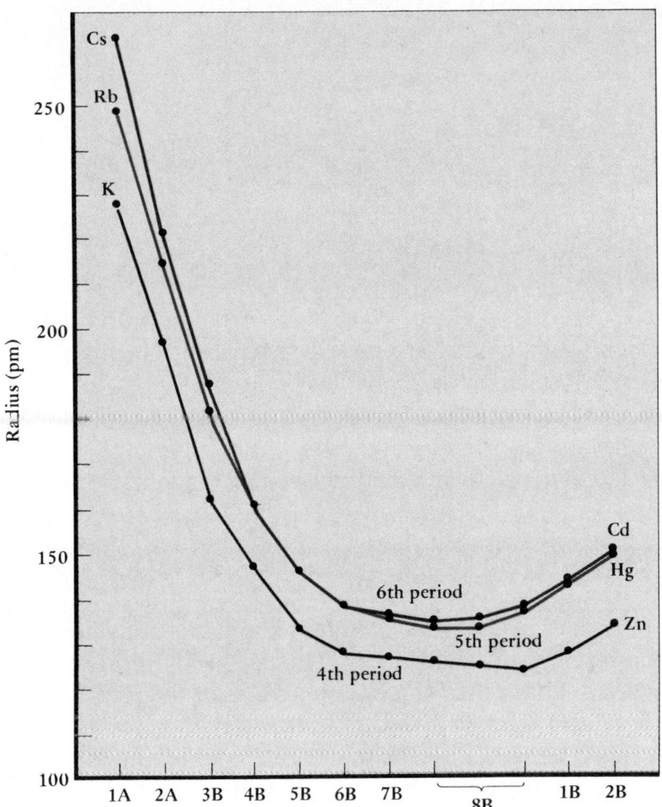

FIGURE 25.4
The radii of the transition metals as a function of their periodic group.

slight rise in radius at Groups 1B and 2B is due to increasing electron–electron repulsions as the *d* subshell is completed.

As a consequence of the variation in metal radius, the densities of the metals tend to increase and then decrease across a period (Figure 25.5). (The atom mass uniformly increases, but the atom volume $[4/3(\pi r^3)]$ decreases across a period.)

The contraction in metal radius across the fourth period transition metals has an effect on the radii of main group elements that follow (Table 25.1). Instead of the normal increase in size down a periodic group, we see that gallium is actually smaller than aluminum. In addition, contraction across the sixth period transition elements causes a smaller increase in radius from In to Tl than would be expected. Indeed, this points to another interesting observation: while the fifth period transition metals are larger than the fourth period metals, as expected, the sixth period transition elements have radii almost identical to those of the fifth period metals (Figure 25.4). The reason for this is that the sixth period transition elements (those filling 5*d* orbitals) follow the lanthanide series. The filling of 4*f* orbitals through the lanthanide elements causes a steady contraction in size, called the **lanthanide contraction**. The contraction in lanthanide radii carries over to the sixth period transition metals (beginning with Hf, radius = 159 pm), and the latter are smaller than expected.

One result of the lanthanide contraction is the great density of elements such as platinum and gold. The relatively small radii of sixth

TABLE 25.1
Radii of the Group 3A Elements

B	88 pm
Al	143 pm
Ga	122 pm
In	162 pm
Tl	171 pm

Radii of Some Lanthanide Elements
La, 106.1 pm
Gd, 93.8 pm
Lu, 84.8 pm

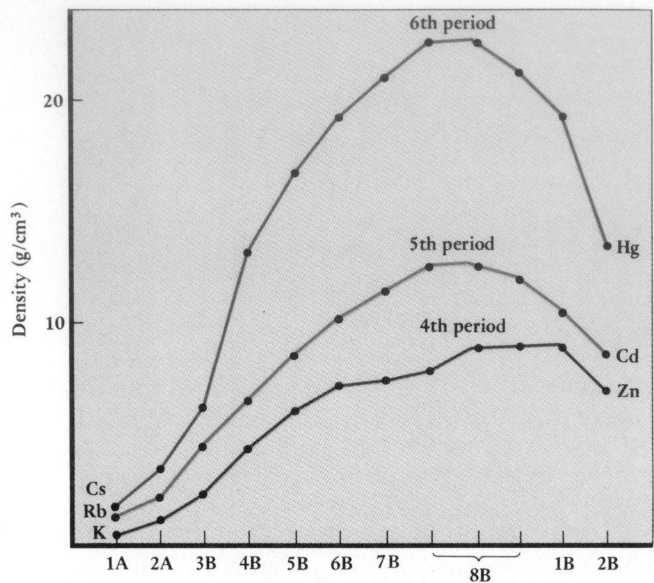

FIGURE 25.5
The density of the transition metals as a function of their periodic group.

Densities of Sixth Period Transition Metals (g/cm^3)
W, 19.5
Pt, 21.4
Hg, 13.5

period transition metals combined with fact that their mass is considerably larger than their counterparts in the fifth period cause sixth period metal densities to be very large. Platinum, for example, is one of the most dense metals known.

The similar radii of the transition metals and their ions have an important effect on their chemistry. For example, the nearly identical radii of fifth and sixth period transition elements leads to difficult problems of metal recovery. As described in "Platinum" (Special Section, Chapter 13), the metals Ru, Os, Rh, Ir, Pd, and Pt are called the "platinum group metals" because they occur together in nature. The reason for this is that their radii and chemistry are so similar that their minerals are similar and are found in the same geologic zones.

25.2 COMMERCIAL PRODUCTION OF TRANSITION METALS

Most metals are found in nature as oxides, sulfides, halides, carbonates or other salts (Figure 25.6). Many of the metal-containing mineral deposits are of little value, either because the deposit is impure or the metal is too difficult to separate from impurities. The relatively few minerals from which elements can be obtained profitably are called **ores**, and it is those that are listed in Figure 25.6.

Pyrometallurgy and hydrometallurgy are two methods of recovering metals from their ores. **Pyrometallurgy** is a high temperature method and is illustrated by iron production. **Hydrometallurgy** uses aqueous solutions at relatively low temperatures and is used to obtain many metals such as copper, zinc, tungsten, and gold.

Very few ores are chemically pure substances. They are usually mixtures of the desired mineral and large quantities of impurities such as sand and clay minerals called **gangue** (pronounced "gang"). Therefore a major step in obtaining the desired metal is to separate its mineral from the gangue. One way to do this is illustrated by the production of iron

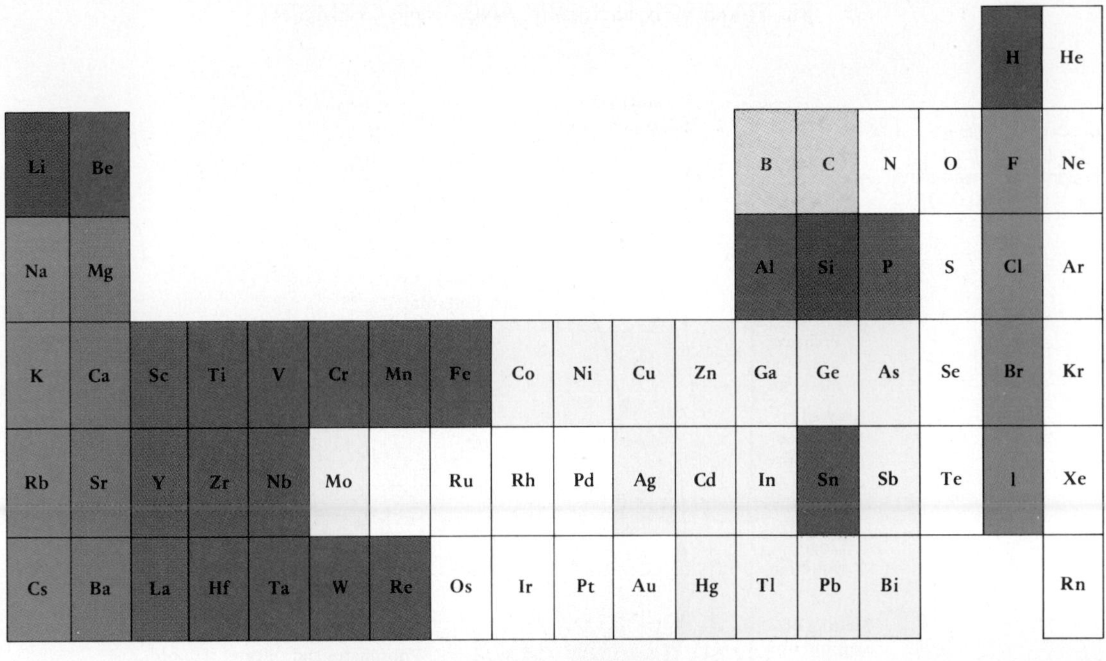

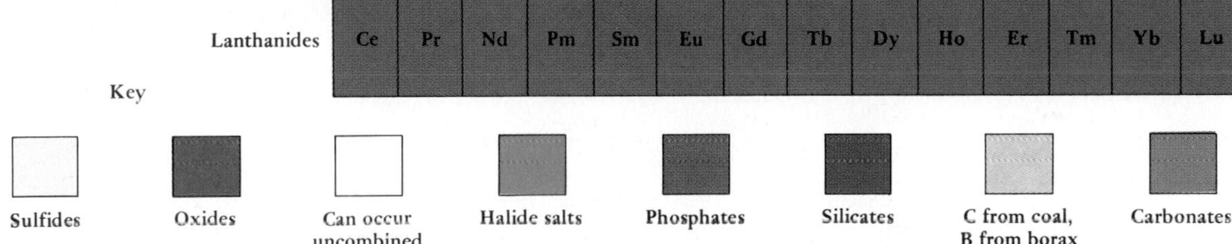

FIGURE 25.6

Sources of the elements. The *d*-block elements are found uncombined or as oxides, sulfides, and halides. The lanthanides occur predominantly as phosphate salts.

from its ores. Another method is **flotation**, used to purify copper-bearing ore, and a third approach is **chemical leaching**.

IRON PRODUCTION

Our economy depends on iron and steel. World production of iron ore in 1980 was 904 million tons, most coming from the USSR. Also in 1980, world production was 706 million tons of raw steel and 507 million tons of iron.

The production of iron from its ores involves oxidation–reduction reactions carried out in a blast furnace (Figure 25.7). The furnace is charged at the top with a mixture of ore (usually hematite, Fe_2O_3), coke, and

The production of 1 ton of pig iron requires about 1.7 tons of iron ore, 0.5 tons of coke, 0.25 tons of limestone, and 2 tons of air.

TABLE 25.2 Percentage of Steel and Pig Iron Production by Country in 1980

COUNTRY	RAW STEEL (%)*	PIG IRON (%)†
USSR	21.0	21.1
Japan	15.8	17.3
USA	14.4	11.3
West Germany	6.2	6.7
China	5.3	7.5
Other	37.3	36.1

*706 million tons total.
†507 million tons total.

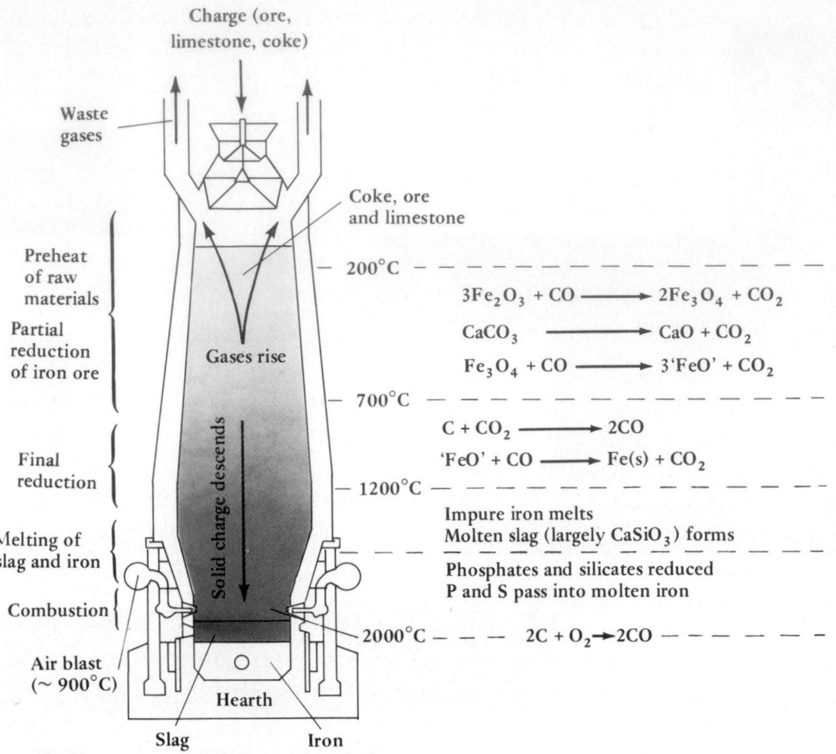

FIGURE 25.7
A blast furnace and some of the reactions that occur. The largest modern furnaces have hearths 14 meters in diameter and produce up to 10,000 tons of iron a day.

limestone, and a blast of hot air is forced in at the bottom. The coke burns with such an intense heat that the temperature at the bottom is almost 2000°C,

$$2\ C(s, coke) + O_2(g) \longrightarrow 2\ CO(g) + heat$$

and temperatures of about 200°C are attained at the top of the furnace. The main reaction of interest is the net reduction of iron oxide with carbon monoxide to give impure metal,

$$\underset{\text{hematite}}{Fe_2O_3(s)} + 3\ CO(g) \longrightarrow 2\ Fe(\ell) + 3\ CO_2(g)$$

but some of the iron oxide is also reduced by the coke.

$$Fe_2O_3(s) + 3\ C(s) \longrightarrow 2\ Fe(\ell) + 3\ CO(g)$$

Much of the carbon dioxide formed in the reduction process (and from heating limestone) is itself reduced on contact with unburned coke and produces more reducing agent.

$$CO_2(g) + C(s) \longrightarrow 2\ CO(g)$$

The molten iron flows down through the furnace and collects in a pool in the bottom where it can be tapped off through an opening in the side. When cooled, the impure iron is called "cast" or "pig iron." The material is brittle and soft, generally undesirable properties, in part due to the presence of impurities such as elemental carbon, phosphorus, and sulfur.

The latter two elements are formed from sulfate and phosphate salts in the reducing atmosphere of the furnace. For example, P_2O_5 can be reduced to elemental phosphorus by carbon in an endothermic reaction.

$$P_2O_5(s) + 5\ C(s) + \text{heat} \longrightarrow 2\ P(s) + 5\ CO(g)$$

Iron ores are contaminated with gangue, which includes various silicate minerals (containing SiO_4^{4-}, for example) and sand (SiO_2). These react with lime from the limestone to give calcium silicate.

Limestone is called a "flux" in steel making.

$$\underset{\text{gangue}}{SiO_2(s)} + \underset{\text{limestone}}{CaO(s)} \longrightarrow \underset{\text{slag}}{CaSiO_3(s)}$$

This is an acid–base reaction, since CaO is a basic oxide and SiO_2 is an acidic oxide. The salt from the reaction, calcium silicate, is less dense than molten iron, so the silicate floats on the iron as a separate layer. Other metal oxides that may be present dissolve in the silicate layer, and the mixture is called a "slag." At the interface or boundary between molten iron and slag, ionic impurities tend to move into the slag layer, while elements tend to concentrate in the iron. The floating slag layer is easily removed and frees the iron of many of the original impurities.

The impure pig iron coming from the bottom of the blast furnace is treated in a second step, involving oxidation, to remove the nonmetal impurities. Several technologies are used at this stage, but currently the most important is the **basic oxygen furnace** (BOF) (Figure 25.8). In this furnace, oxygen is blown into the molten pig iron to oxidize phosphorus, sulfur, and most of the excess carbon.

$$4\ P(s) + 5\ O_2(g) \longrightarrow 2\ P_2O_5(s)$$

$$C(s) + O_2(g) \longrightarrow CO_2(g)$$

$$S(s) + O_2(g) \longrightarrow SO_2(g)$$

All the oxides formed are acidic and react with basic oxides (such as CaO), which are added or are used to line the furnace. The products are calcium phosphate, calcium sulfate, and calcium carbonate. For example,

$$\underset{\text{acidic oxide}}{P_2O_5(s)} + \underset{\text{basic oxide}}{3\ CaO(s)} \longrightarrow \underset{\text{salt}}{Ca_3(PO_4)_2(s)}$$

These salts form a floating layer of slag, which can be poured off to free the more dense molten iron layer of impurities.

Most of the slag from steel making is used to make cement, but some is blown with air to form rock wool insulation.

Pig iron typically contains up to 4.5% carbon, 1.7% manganese, 0.3% phosphorus, 0.04% sulfur, and 1% silicon. The final oxidation step removes much of the P, S, and Si, and reduces the carbon content to about 1.3%. The result is ordinary *carbon steel*. Almost any degree of flexibility, hardness, strength, and malleability can be achieved in carbon steel by proper cooling, reheating, and tempering. The drawbacks to carbon steel, however, are that it corrodes easily and loses its properties when heated strongly.

During the processing of steel, other transition metals, such as chromium, manganese, and nickel, can be added to give alloys having specific physical, chemical, and mechanical properties. One type of *stain-*

FIGURE 25.8

A "basic oxygen furnace" in operation.

less steel, for example, contains up to 18% to 20% Cr and 8% to 12% Ni, but very little carbon. Such steels are very resistant to corrosion and are used in automobile trim, laboratory and kitchen ware, and so on. *Magnetic alloys* are solutions of iron and other elements that are permanently magnetic (and are used in loudspeaker magnets and the like) or that can be temporarily magnetized (and so are used in electric motors, generators, and transformers). Alnico V, as its name implies, contains five elements: 8% Al, 14% Ni, and 24% Co, as well as 51% Fe and 3% Cu.

COPPER PRODUCTION

The production of copper illustrates some other of the basic approaches of the metals industry.

The most abundant copper-bearing ore, chalcopyrite, $CuFeS_2$, can be enriched readily by flotation. Here the powdered ore is mixed with oil and agitated with soapy water in a large tank (Figure 25.9a). Compressed air is forced through the mixture, and the lightweight, oil-covered copper sulfide particles are carried to the top and float on the froth. The heavier gangue settles to the bottom of the tank, and the copper-laden froth is skimmed off.

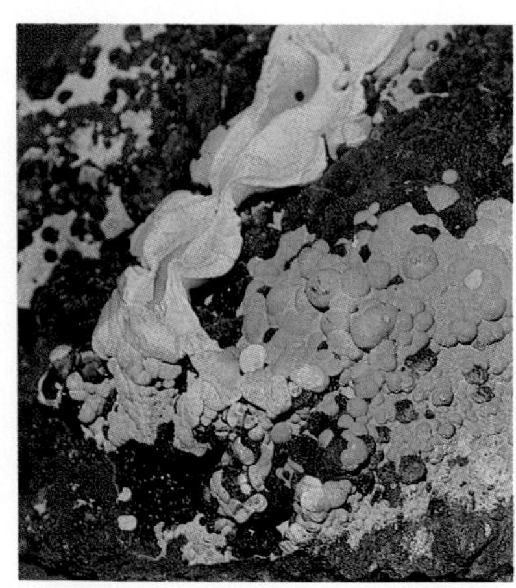

Copper-bearing minerals. The blue mineral is azurite [$2CuCO_3 \cdot Cu(OH)_2$], and the green mineral is malachite [$CuCO_3 \cdot Cu(OH)_2$]. (Brian Parker, Tom Stack & Assoc)

In the pyrometallurgy of copper, the enriched ore is **roasted** with enough air to convert the iron selectively to its oxide and leave copper as its sulfide.

$$2\ CuFeS_2(s) + 3\ O_2(g) \longrightarrow 2\ CuS(s) + 2\ FeO(s) + 2\ SO_2(g)$$

This mixture of copper sulfide and iron oxide is mixed with ground limestone, sand, and some fresh concentrated ore and then heated to 1100°C in a reverberatory furnace (Figure 25.9). As in the blast furnace, limestone and sand (SiO_2) form glassy calcium silicate, and this dissolves the iron oxide.

$$CaSiO_3(\ell) + FeO(s) + SiO_2(s) \longrightarrow 2\ (Fe,Ca)SiO_3(\ell)$$

calcium silicate slag

At the same time, excess sulfur in the ore reduces copper(II) sulfide, CuS, to copper(I) sulfide, Cu_2S, which melts and flows to the bottom of the furnace. Since the iron-containing slag is less dense than molten Cu_2S, the slag can be drawn off, thereby separating iron and copper.

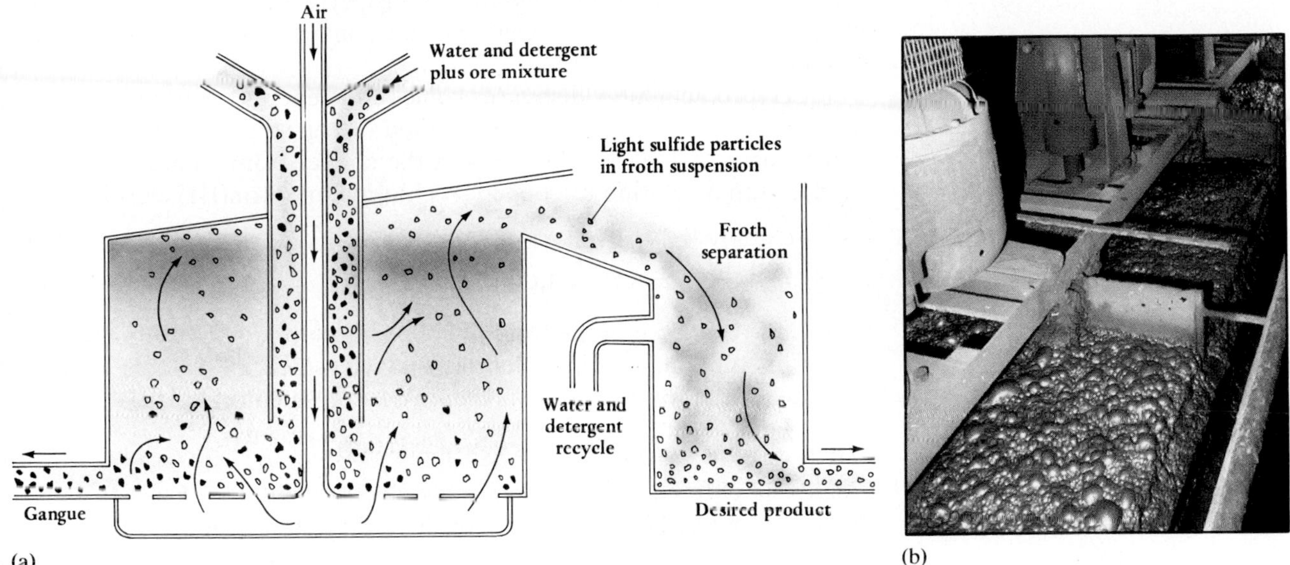

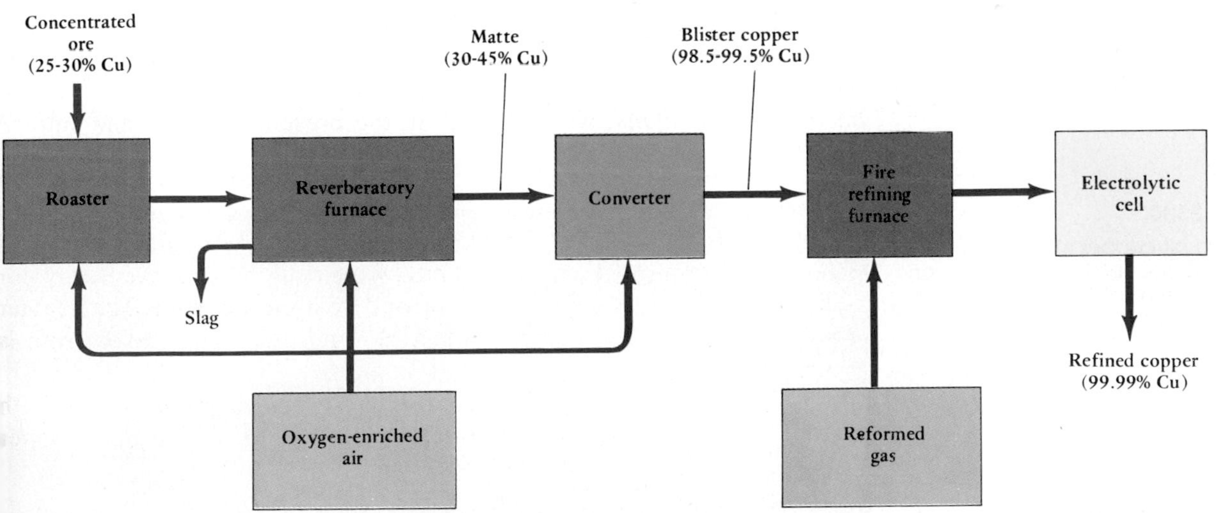

(c)

FIGURE 25.9

"Winning" copper from its ores. (a) The sulfide ores are enriched in the flotation process. The lighter metal sulfide particles are trapped in soapy bubbles and float on the water. The heavier gangue settles to the bottom. (b) A photograph shows the flotation process in operation. (c) After concentrating the ore by flotation, the copper is refined in a sequence of steps as outlined in the flow diagram and in the text.

The bottom Cu_2S layer, called *copper matte*, is tapped off and run into another furnace (the converter) where it is "blown" with air to oxidize the sulfide ions and produce impure copper metal. This is further refined in yet another furnace.

$$Cu_2S(\ell) + O_2(g) \longrightarrow 2\ Cu(\ell) + SO_2(g)$$

In the pyrometallurgical recovery of copper, each ton of copper produced also leads to 1.5 tons of iron silicate slag and 2 tons of SO_2. These by-products must be disposed of properly, not a simple task. However, one solution for SO_2 disposal is to convert it to economically important sulfuric acid (Section 22.2).

Hydrometallurgy decreases some of the energy costs and pollution problems of pyrometallurgy. One method of copper recovery now in use in Arizona is to *leach* or dissolve out the copper from its ore by treating the ore with a solution of copper(II) chloride and iron(III) chloride.

$$CuFeS_2(s) + 3\ CuCl_2(aq) \longrightarrow 4\ CuCl(s) + FeCl_2(aq) + 2\ S(s)$$

$$CuFeS_2(s) + 3\ FeCl_3(aq) \longrightarrow CuCl(s) + 4\ FeCl_2(aq) + 2\ S(s)$$

Copper is recovered in the form of copper(I) chloride. To keep the compound in solution, sodium chloride is added, since the soluble complex ion $[CuCl_2]^-$ is formed in the presence of excess chloride ion.

$$CuCl(s) + Cl^-(aq) \longrightarrow [CuCl_2]^-(aq)$$

Finally, impure copper is obtained by electrolyzing the $[CuCl_2]^-$ ion to the metal and $CuCl_2$; the latter is used to continue the leaching process.

$$2\ [CuCl_2]^-(aq) \longrightarrow Cu(s) + CuCl_2(aq) + 2\ Cl^-(aq)$$

Approximately 10% of the copper produced in the United States is actually obtained using bacteria. Acidified water is sprayed onto copper-mining wastes, which can contain low levels of copper (Figure 25.10). As the water trickles down through the crushed rock, the bacterium *Thiobacillus ferrooxidans*, which thrives in the presence of acid and sulfur,

FIGURE 25.10

An open pit copper mine near Bagdad, Arizona. Additional copper can be extracted from the waste rock by a bacterium. (James Cowlin)

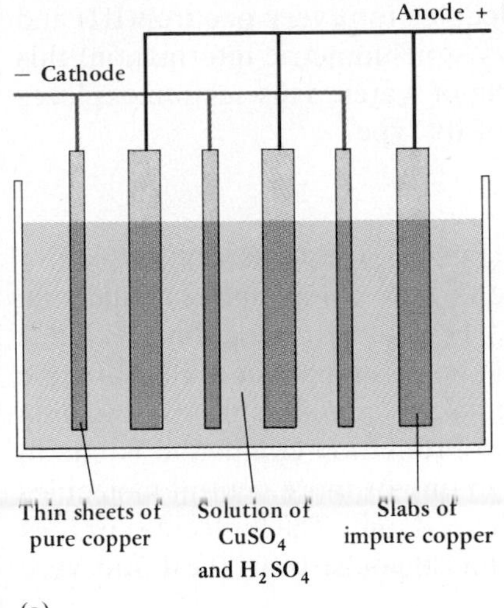

FIGURE 25.11
Electrolytic refining of copper. (a) A schematic diagram. Copper atoms from slabs of impure copper, called "blister copper," are made the anode of an electrolysis cell. The copper is oxidized and passes into the solution as Cu^{2+} ions. These ions migrate to the negative electrode, or cathode (made of thin sheets of pure copper), where they are reduced again to copper metal. (b) Photograph of electrolysis cells for refining copper.

breaks down the iron sulfides in the rock and converts iron(II) to iron(III).* The iron(III) ion in turn oxidizes the sulfide ion of copper sulfides, leaving copper(II) ion in the water. This copper-laden water is recovered at the bottom of the pile, and metallic copper is obtained by reduction with scrap iron.

$$Cu^{2+}(aq) + Fe(s) \longrightarrow Cu(s) + Fe^{2+}(aq)$$

Regardless of the method of copper recovery from its ores, the final step is purification by electrolysis. Thin sheets of pure copper metal and slabs of impure copper are immersed in a solution of $CuSO_4$ and H_2SO_4 (Figure 25.11). The pure copper sheets are made the cathode of an electrolysis cell, and the impure slabs are the anode. This means that copper ions are formed at the anode (oxidation occurs) and move into solution. The ions migrate to the cathode where they are reduced to pure copper.

25.3 COORDINATION COMPOUNDS

You are familiar with simple binary compounds of transition metals with ions such as O^{2-}, S^{2-}, and halides such as Cl^- and Br^-. For example, rutile or TiO_2 is used as a pigment in white paints, FeS_2 is a mineral commonly known as "fool's gold," and AgBr is the light-sensitive substance in black and white film. There is another closely related class of compounds in which neutral molecules such as water and ammonia are incorporated. Examples include $FeCl_3 \cdot 6\ H_2O$ and $CoCl_2 \cdot 6\ NH_3$. The notation "$\cdot\ 6\ H_2O$" has historically been used to show that the basic

A great deal of transition metal chemistry is centered on coordination compounds. See also the Special Sections on iron and nickel chemistry.

*You can read more about microbiological mining in (a) C.L. Brierley, *Scientific American*, 247:44, 1982; and in (b) J. Alper, *High Technology*, April, 1984, p. 32.

chemical unit consists of six water molecules for every one iron(III) and three chloride ions. Although it conveys stoichiometric information, this notation is not informative about the role of water. This section explores that role in this compound and others of its type.

COMPLEXES AND LIGANDS

It is more useful to write the formula $FeCl_3 \cdot 6\ H_2O$ as $[Fe(H_2O)_6]Cl_3$. The square brackets show that the cation in the compound is actually an Fe^{3+} ion bound to six water molecules, $[Fe(H_2O)_6]^{3+}$, and this +3 ion is associated with three Cl^- ions. We call such a compound a **coordination compound** or **coordination complex** because it contains the **coordination cation** or **complex ion** $[Fe(H_2O)_6]^{3+}$. A coordination compound is one in which the metal ion or atom is bonded to one or more neutral molecules or anions so as to define an integral structural unit. The molecules or ions bonded to the central metal ion are called **ligands**, from the Latin verb *ligare*, to bind.

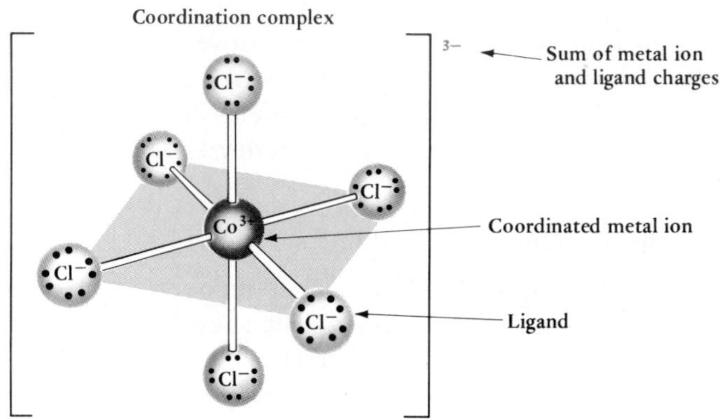

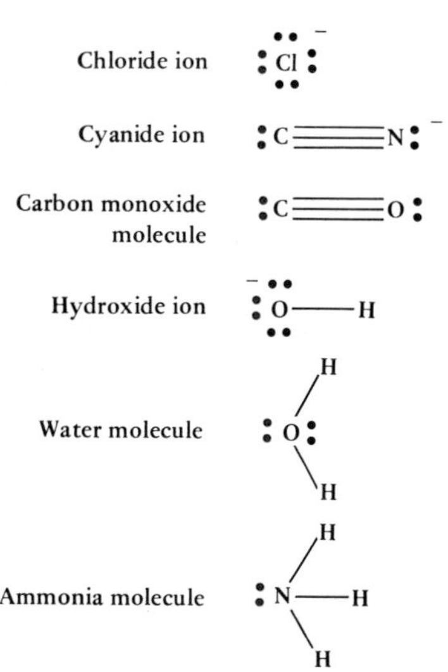

Ligands having one lone pair available for binding to an electron-pair acceptor.

Coordination complexes can occur as cations, anions, or neutral molecules (Figure 25.12). For example, $K_3[Fe(CN)_6]$ contains the anion $[Fe(CN)_6]^{3-}$, whereas $Cr(CO)_6$ consists of a chromium atom bonded to six carbon monoxide molecules.

Ligands are **Lewis bases** because they always bear at least one atom having a lone pair of electrons accessible for bonding to another atom. A positive metal ion can accept electron pairs from a Lewis base, so the metal ion is a ligand **Lewis acid** (see Section 16.7). As the ligand approaches a metal ion, a ligand lone pair forms a sigma bond with the metal through an interaction between the lone pair orbital and an empty orbital on the metal ion or atom.

$$M^{n+} + :NH_3 \longrightarrow [M{\leftarrow}NH_3]^{n+}$$

coordinate covalent bond

Some ligands have more than one Lewis base site. When these atoms are separated from each other by several intervening atoms, it is possible that all the lone pair atoms can bind to the metal to form a complex called a **chelate** (pronounced "key-late"). A good example of a chelating ligand is 1,2-diaminoethane ($H_2N—CH_2—CH_2—NH_2$), commonly called

FIGURE 25.12

Coordination complexes. Starting at the bottom and moving clockwise they are: $[Co(NH_3)_5H_2O]Cl_3$ (red), ligands are NH_3 and H_2O; $K_3[Fe(CN)_6]$ (red-orange), ligand is CN^-; $Cr(CO)_6$ (white), ligand is CO; $K_3[Fe(C_2O_4)_3]$ (green), ligand is $C_2O_4{}^{2-}$, oxalate ion; and $[Co(H_2N—CH_2—CH_2—NH_2)_3]I_3$ (yellow-orange), ligand is ethylenediamine. (Charles D. Winters)

ethylenediamine and abbreviated *en*. When binding to the metal ion, such ligands form a ring of atoms, of which the metal is one member. Numerous complexes are formed by *en*, and $[Co(en)_3]^{3+}$ is an excellent example (Figures 25.12 and 25.13). Other common ligands capable of binding through more than one atom are illustrated in Figure 25.14.

Ligands like H_2O and NH_3, with only a single Lewis base atom, are termed **monodentate**. The word "dentate" stems from the Latin word *dentis* for tooth, so NH_3 is a "one-toothed" ligand. This means en, phen,

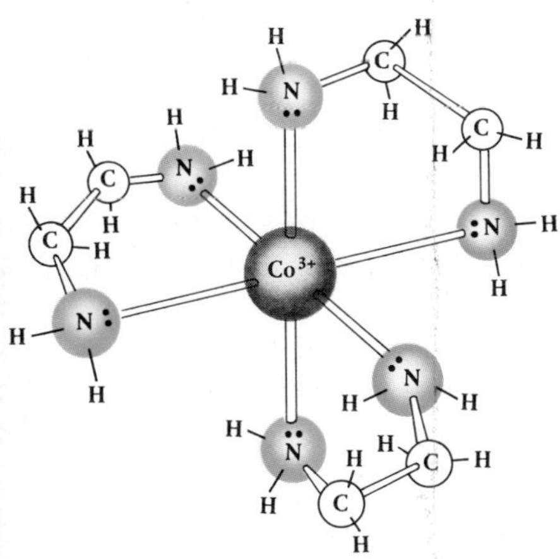

$[Co(ethylenediamine)_3]^{3+}$

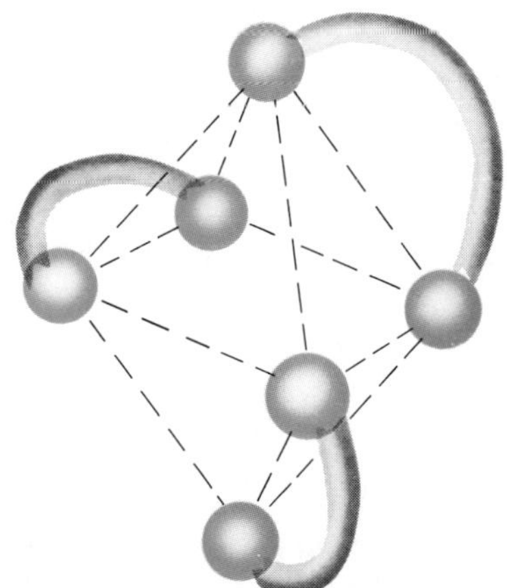

The arrangement of "two-toothed" (bidentate) ethylenediamine ligands on the octahedron of N atoms about Co^{3+}

FIGURE 25.13

The structure of $[Co(en)_3]^{3+}$, a complex ion formed from three bidentate, chelating ethylenediamine molecules and Co^{3+}.

FIGURE 25.14

Some common chelating ligands. The origin of the word chelate is the Greek word *chele* for claw.

Bidentate ligands

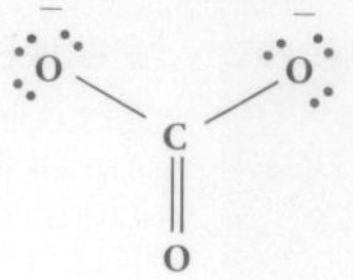

Carbonate ion

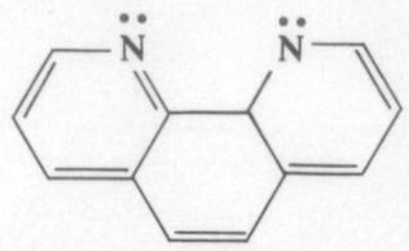

ortho-phenanthroline (phen)

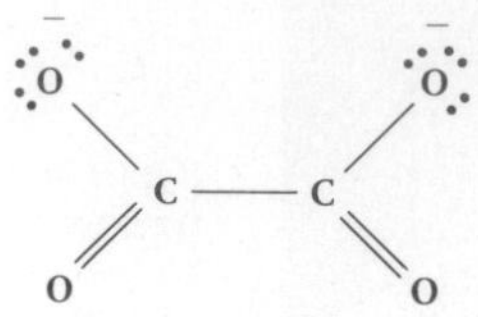

Oxalate ion (ox^{2-})

$H_2\ddot{N}—CH_2—CH_2—\ddot{N}H_2$

Ethylenediamine (en)

Tridentate ligand

$H_2\ddot{N}—CH_2—CH_2—\ddot{N}(H)—CH_2—CH_2—\ddot{N}H_2$

Diethylenetriamine (dien)

Sexadentate ligand

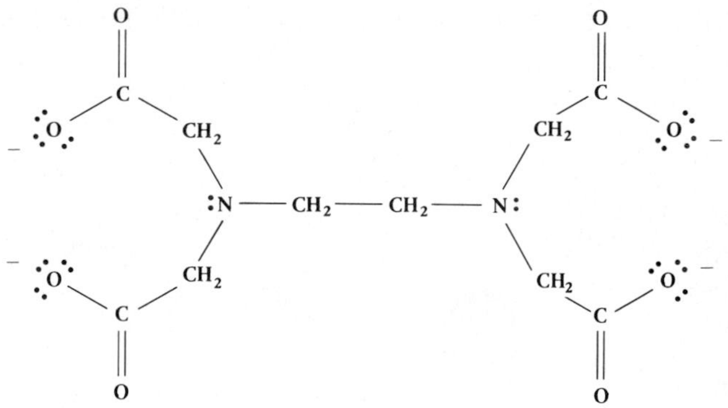

$EDTA^{4-}$, ethylenediaminetetraactate ion

and ox^{2-} are *bi*dentate ligands, or "two-toothed," while $EDTA^{4-}$ is *sexa*-dentate because it has six Lewis base sites.

Chelated complexes are important in everyday life. For example, one way to clean the rust out of water-cooled automobile engines and steam boilers is to add a solution of oxalic acid. Iron oxide dissolves in the presence of the acid to give the water-soluble iron oxalate complex and water.

$$Fe_2O_3(s) + 6\ H_2C_2O_4(aq) \longrightarrow 2\ [Fe(C_2O_4)_3]^{3-}(aq) + 3\ H_2O(\ell) + 6\ H^+(aq)$$

$EDTA^{4-}$ is an excellent chelating ligand; it encapsulates and firmly binds metal ions.

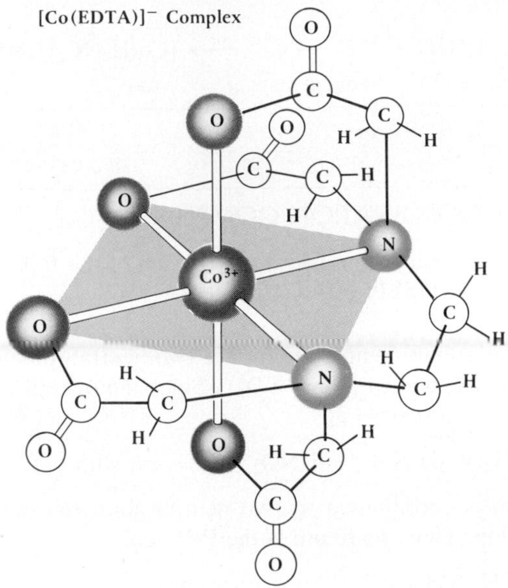

It is often added, for example, to commercial salad dressing to remove traces of metal ions from solution, since these metal ions can otherwise act as catalysts for the oxidation of the oils in the product; without $EDTA^{4-}$, the dressing would become rancid. Another use is in bathroom cleansers, where $EDTA^{4-}$ removes deposits of $CaCO_3$ and $MgCO_3$ left by hard water; the $EDTA^{4-}$ encapsulates Ca^{2+} or Mg^{2+} just like Co^{3+} above. And finally, you have already seen the chelating porphin ligand in hemoglobin and myoglobin (Chapter 24), and plant chlorophyll is a related compound.

It is useful to be able to predict the formula of a coordination complex, given the metal ion and ligands, and to predict the oxidation number of the coordinated metal ion. The following examples explore these questions.

Be aware that a bidentate ligand, for example, is not necessarily chelating. Some bidentate ligands can use both sites to bind one metal or they can be a *bridge* between two metals, using one site per metal.

EXAMPLE 25.1

COORDINATION COMPLEXES

Given the following combinations of metal ion and ligands, give the formula of the coordination complex where the metal ion is coordinated to six Lewis base sites. (a) One Ni^{2+} ion is bound to two water molecules and two bidentate oxalate ions. (b) One Co^{3+} ion is bound to one Cl^- ion, one ammonia molecule, and two bidentate ethylenediamine molecules.

Solution (a) In the nickel(II) complex, the ligands consist of two neutral molecules and two with -2 charges. When these are combined with $+2$ nickel, the net charge is -2.

$$\underset{}{Ni^{2+}} + \underset{\text{monodentate}}{2\,H_2O} + \underset{\text{bidentate}}{2\,C_2O_4^{2-}} \longrightarrow [Ni(C_2O_4)_2(H_2O)_2]^{2-}$$

(b) In the cobalt(III) complex there are two neutral en molecules, one neutral NH_3 molecule, and one Cl^- ion. When combined with Co^{3+}, the net charge is 2+.

$$\underset{}{Co^{3+}} + \underset{\text{bidentate}}{2\ H_2NC_2H_4NH_2} + \underset{\text{monodentate}}{NH_3} + Cl^- \longrightarrow [Co(H_2NC_2H_4NH_2)_2(NH_3)Cl]^{2+}$$

EXAMPLE 25.2

FORMULAS OF COORDINATION COMPLEXES

Give the oxidation number of the metal ion in each of the following complexes: (a) $[Co(en)_2(NO_2)_2]^+$ and (b) $Pt(NH_3)_2(C_2O_4)$.

Solution (a) In this cobalt complex there are two neutral, bidentate ethylenediamine molecules and two nitrite ions, NO_2^-. Since the overall charge on the ion is +1, the cobalt ion must be +3.

$$Co^{3+} + 2\ H_2NC_2H_4NH_2 + 2\ NO_2^- \longrightarrow \text{ion with } +1 \text{ charge}$$

(b) Platinum is coordinated to two neutral ammonia molecules and one bidentate oxalate ion. Thus, platinum is the Pt^{2+} ion.

$$Pt^{2+} + 2\ NH_3 + C_2O_4^{2-} \longrightarrow \text{charge of } 0$$

EXERCISE 25.2 FORMULAS OF COORDINATION COMPLEXES
Give the oxidation number of platinum in $Pt(NH_3)_2Cl_2$. What is the formula of a complex assembled from one Co^{3+}, five ammonia molecules, and one monodentate carbonate ion?

NAMING COORDINATION COMPOUNDS

Just as there are rules for naming simple inorganic and organic compounds, coordination compounds are named according to an established system. For example, the following compounds are named according to the rules outlined below.

COMPOUND	SYSTEMATIC NAME
$[Ni(H_2O)_6]SO_4$	Hexaaquonickel(II) sulfate
$[Cr(en)_2(CN)_2]Cl$	Dicyanobis(ethylenediamine)chromium(III) chloride
$K[Pt(NH_3)Cl_3]$	Potassium amminetrichloroplatinate(II)

As you read through the rules, notice how they apply to these examples.

1. In naming a coordination compound that is a salt, name the cation first and then the anion, as is usually done (Appendix B).
2. When giving the name of the complex ion or molecule, the ligands are named first, in alphabetical order, followed by the name of the metal.

   (a) If a ligand is an anion whose name ends in *-ite* or *-ate*, the final *e* is changed to *o* (as in sulfat*e* → sulfat*o* or nitrit*e* → nitrit*o*).

   (b) If the ligand is an anion whose name ends in *-ide*, the ending is changed to *o* (as in chlor*ide* → chlor*o* or cyan*ide* → cyan*o*).

   (c) If the ligand is a neutral molecule, its common name is used The important exceptions at this point are water, which is called *aquo* ammonia, which is called *ammine*, and CO, called *carbonyl*.

(d) When there is more than one of a particular monodentate ligand with a simple name, the number of ligands is designated by the appropriate prefix: *di*, *tri*, *tetra*, *penta*, or *hexa*. If the ligand name is complicated (whether monodentate or bidentate), the prefix changes to *bis*, *tris*, *tetrakis*, *pentakis*, or *hexakis*, followed by the ligand name in parentheses.

(e) If the complex ion is an anion, the suffix *-ate* is added to the metal name.

3. Following the name of the metal, the oxidation number of the metal is given in Roman numerals.

Complex ions can become more complicated than those described in this chapter, and even more rules of nomenclature must be applied. However, the brief set just outlined is sufficient for the vast majority of complexes.

EXAMPLE 25.3

NAMING COORDINATION COMPOUNDS

Name the following complexes: (a) $[Cu(NH_3)_4]SO_4$; (b) $K_2[CoCl_4]$; (c) $Co(phen)_2Cl_2$; and (d) $[Co(en)_2(H_2O)Cl]Cl_2$.

Solution (a) The sulfate ion has a −2 charge, so the complex ion has a +2 charge. Since NH_3 is a neutral molecule, this means the copper ion is Cu^{2+}. Therefore, the compound's name is tetraamminecopper(II) sulfate.

(b) There are two K^+ ions in this complex, so the complex ion has a −2 charge ($[CoCl_4]^{2-}$). Since there are four Cl^- ions in the complex ion, the cobalt center is Co^{2+}. Thus, the name of the compound is potassium tetrachlorocobaltate(II).

(c) This is a neutral complex. Since two Cl^- ions and two neutral phen ligands are bonded to a cobalt ion, the metal ion must be Co^{2+}. This means the compound name is dichlorobis(phenanthroline)cobalt(II).

(d) Here the complex ion has a +2 charge, since it is associated with two uncoordinated Cl^- ions. The cobalt ion must be Co^{3+} because it is bound to two neutral en ligands, one neutral water, and one Cl^-. Therefore, the name is aquochlorbis(ethylenediamine)cobalt(II) chloride.

EXERCISE 25.3 NAMING COORDINATION COMPOUNDS
Name the two compounds (a) $[Ru(phen)_2(H_2O)CN]Cl$ and $Pt(NH_3)_2Cl_2$.

25.4 STRUCTURES OF COORDINATION COMPOUNDS AND ISOMERS

Coordination compounds are characterized by integral structural units consisting of a metal ion or atom surrounded by a sheath of ligands. The number of metal-ligand sigma bonds on the metal is the **coordination number**. Although there are examples of coordination numbers of two through twelve, only four and six are very common, and two deserves mention.

COMMON COORDINATION NUMBERS AND GEOMETRIES

If we consider for a moment only complexes of M with the monodentate ligand L, common coordination complexes would have stoichiometries of ML_6, ML_4, and ML_2. Complexes of ML_2 are always linear, and the silver-containing ion that comes from dissolving AgCl in concentrated ammonia is a good example.

$$AgCl(s) + 2\,NH_3(aq) \longrightarrow \underset{\text{Linear ML}_2\text{ Complex}}{[H_3N{:}Ag{:}NH_3]^+(aq)} + Cl^-(aq)$$

If we consider only M—L bond pairs in coordination complexes, the VSEPR theory of Chapter 9 leads us to expect tetrahedral structures for ML_4 complexes. Indeed, this is observed for many molecules, such as $[CoCl_4]^{2-}$, $Ni(CO)_4$, and $[Zn(NH_3)_4]^{2+}$.

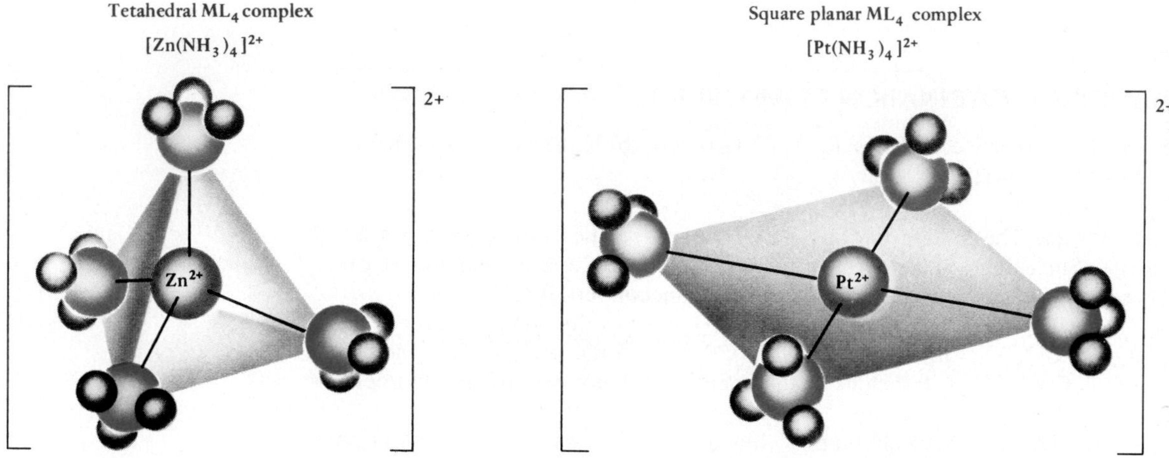

However, as we pointed out in Chapter 9, the VSEPR theory does not always apply to transition metal complexes, and, for ML_4 complexes, *square planar* geometry is observed more often than tetrahedral geometry. This is particularly true for ML_4 complexes of Pt^{2+} and Pd^{2+}, and sometimes Ni^{2+} and Cu^{2+}.

All of the complex ions or molecules in Figure 25.12 have the coordinating atoms at the corners of an octahedron.

With *very* rare exceptions, ML_6 complexes have the six ligands arranged at the corners of an octahedron, as in the $[CoCl_6]^{3-}$ ion pictured on page 958. Two views of an octahedron are given in Figure 25.15. The drawing at the left emphasizes that the regular octahedron is a figure with six corners and eight faces, all of which are equilateral triangles; the metal ion is at the center of the octahedron. On the right in Figure 25.15, the drawing shows that the ligands are located along three axes at right angles to one another (the x, y, and z axes), and the ligands are *equidistant from the metal atom or ion in the center*. Finally, the drawing at the right shows us we can consider an octahedron as a *square plane* of M and four ligand atoms L, with other ligand atoms above and below the plane.

ISOMERS

Given these three common stoichiometries and the four possible structural types, there remain various possibilities for the arrangement of the ligand about the metal ion or atom.

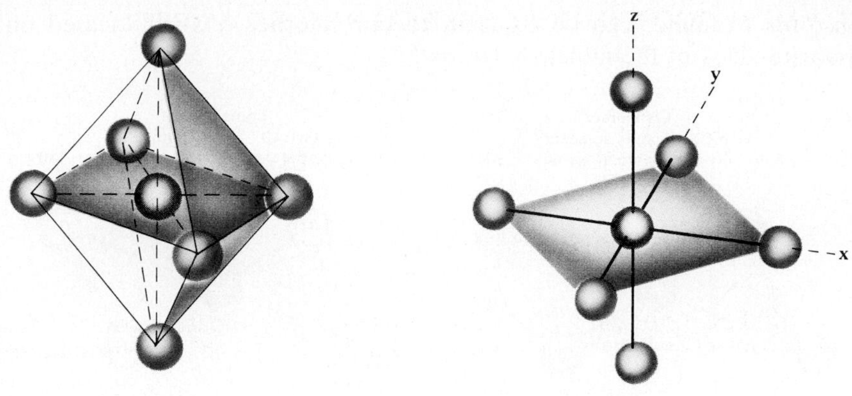

FIGURE 25.15
Two views of a regular octahedron.

Isomers are molecules with the same stoichiometry but different atomic arrangements. In Chapter 24 we defined **stereoisomers** as isomers with the same atom-to-atom connections but with different three-dimensional arrangements of those atoms. There we described the possibility of two arrangements, *cis* and *trans*, for a molecule such as 1,2-dichloroethene. We can refer to this as **geometrical isomerism**.

Some complexes can exist in both square planar and tetrahedral forms and so are isomers of one another.

```
Cl       Cl        Cl       H
  \     /            \     /
   C=C                C=C
  /     \            /     \
H        H         H        Cl
```

cis isomer   *trans* isomer

Optical isomerism arises when two compounds are *nonsuperimposable mirror images* of one another (Section 24.1).

Optical Isomers of Lactic Acid

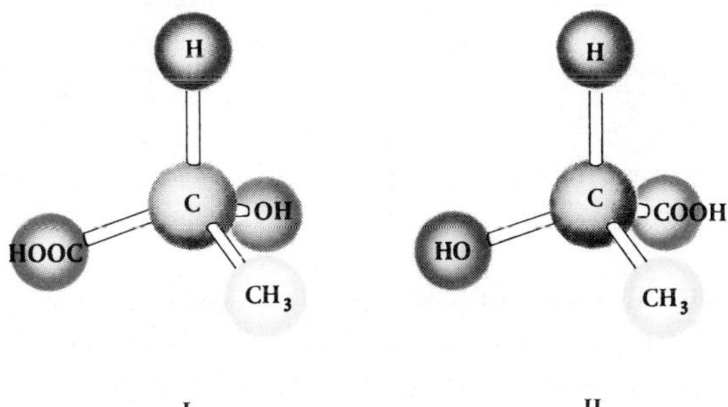

As outlined in Section 24.1, optical isomers differ from one another in that they rotate the plane of plane polarized light in opposite directions (Figure 24.4).

There are many different types of isomerism observed for coordination compounds of the transition metals. However, we shall describe only geometric and optical isomerism.

GEOMETRIC ISOMERISM Geometric isomers result when the atoms bonded *directly* to the metal are sequenced in a different order about the metal. The simplest example of geometrical isomerism occurs with square planar complexes, such as $Pt(NH_3)_2Cl_2$, with two identical ligands. This complex is formed from Pt^{2+}, two NH_3 molecules, and two Cl^- ions. The two Cl^-

In Chapter 1 you learned that $Pt(NH_3)_2Cl_2$ is used in cancer chemotherapy. It is interesting that only the *cis* isomer is physiologically active.

ions, for example, can be adjacent to one another (*cis*) or located on opposite sides of the molecule (*trans*).

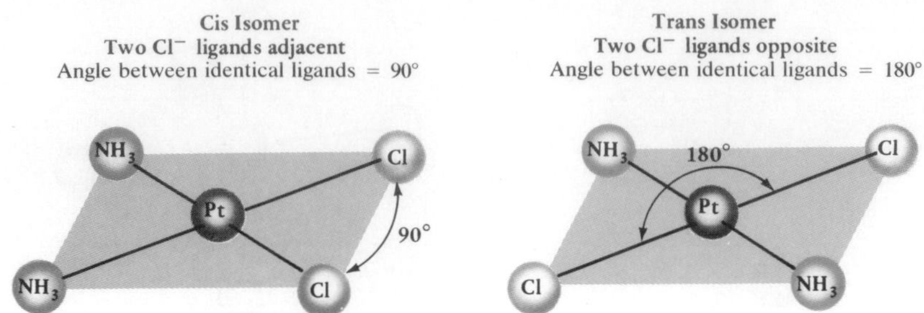

Now consider square planar $[Pt(NH_3)Cl_3]^-$, a complex with three like ligands.

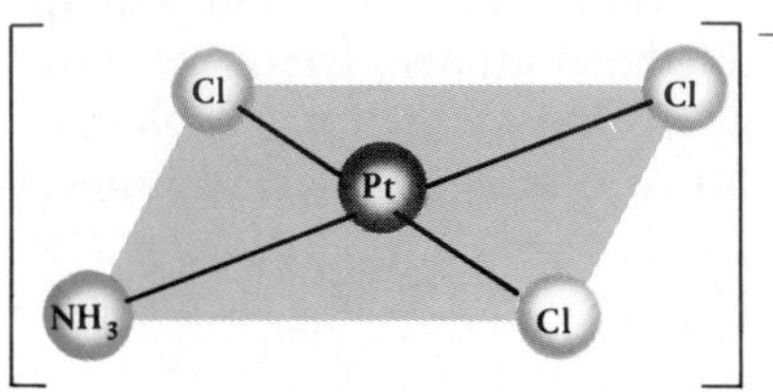

Cis–trans isomers are not possible here. No matter how you draw the molecule, there are always two Cl^- ions 180° apart and two Cl^- ions 90° apart. On the other hand, *cis* and *trans* isomer can be constructed for $Pt(NH_3)_2(Cl)(NO_2)$.

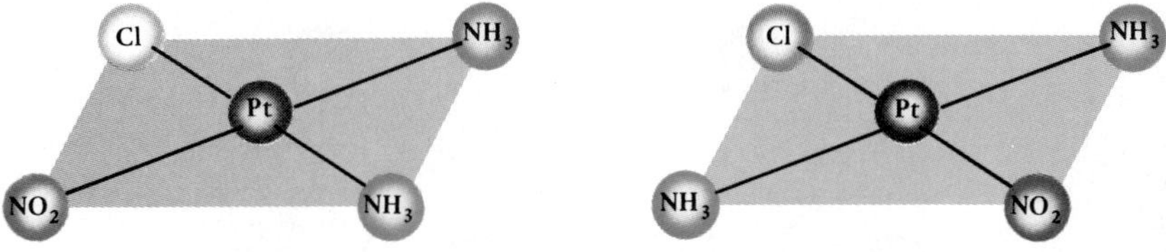

Four-coordinate complexes can have tetrahedral geometry. However, *cis–trans isomerism is not possible for tetrahedral complexes* because it is not possible for two ligands to be "across" the molecule from one another. All possible angles in a tetrahedron are 109°, so all ligands are effectively adjacent.

All angles in a tetrahedron are 109°, so *cis–trans* isomerism is not possible.

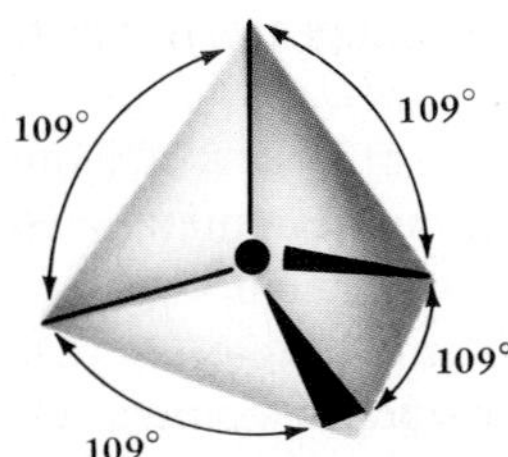

Geometrical isomerism is widely observed for octahedral complexes. As an example consider $[Co(H_2NC_2H_4NH_2)_2Cl_2]^+$, an octahedral complex ion with two different ligands. Here the two Cl^- ions can occupy adjacent or opposite positions of the octahedron of ligands,

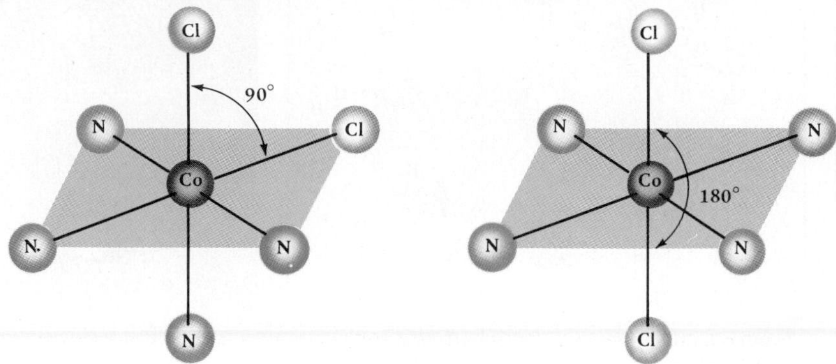

so the resulting isomers of $[Co(en)_2Cl_2]^+$ are illustrated in Figure 25.16. It is interesting that the different ways of connecting the ligands lead to different colors; the *cis* isomer is green, but the *trans* isomer is purple.

When three unique ligands are present in an octahedral complex, such as the NH_3 molecules or the Cl^- ions in $Cr(NH_3)_3Cl_3$, geometrical isomerism occurs. Here the three Cl^- ligands, for example, can all lie in the same plane as the metal, or they can be at the corners of an equilateral triangle.

The left isomer is often called the *mer* isomer, because the Cl^- ions are located on the *meridian* of the complex. The right isomer is called the *fac* isomer, since the three Cl^- ions are found on a *face* of the octahedron.

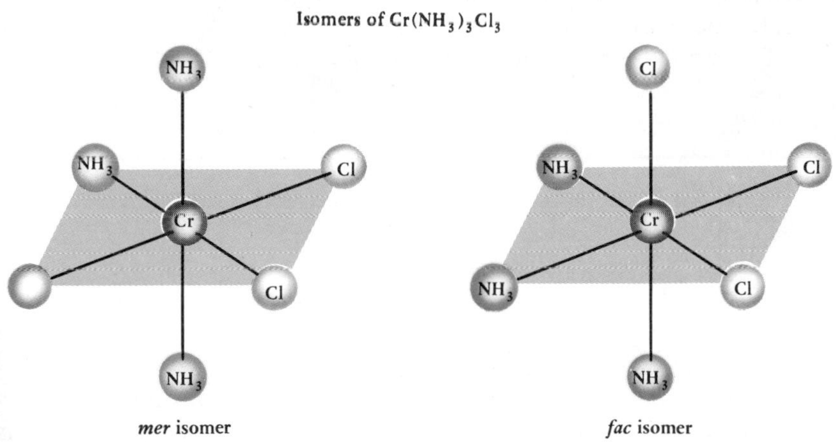

mer isomer

fac isomer

OPTICAL ISOMERISM Optical isomers have the same stoichiometry and he same atom-to-atom bonding sequence, but they differ in the details of the arrangement of atoms in space. Begin by considering tetrahedral ML_4. As explained in Section 24.1, if four different groups are bound to a central atom, the molecule can exist in two forms that are mirror images of one another, but where one form cannot be superimposed on the other. The lactic acid molecule described on page 965 is an example. In coordination chemistry there are no examples of stable complexes with a metal bonded tetrahedrally to four *different types* of ligands, so optical isomers of tetrahedral coordination complexes need not be considered further.

The situation is quite different with octahedral complexes, where optical isomerism can arise in a variety of ways. Only one of these ways

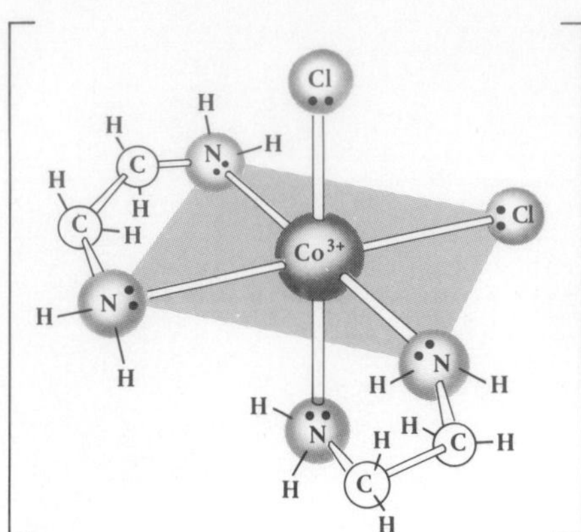

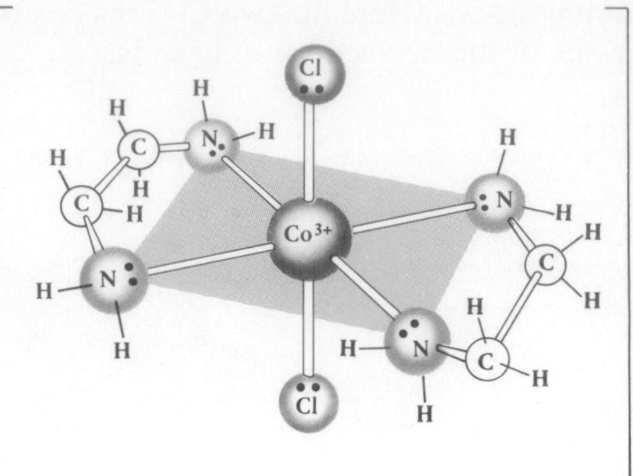

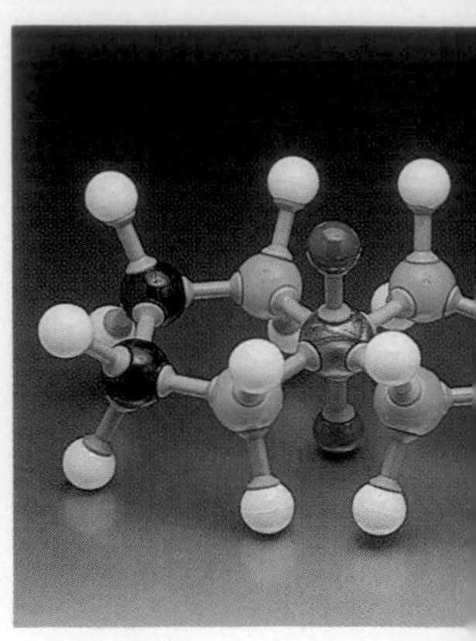

FIGURE 25.16

The geometrical isomers of $[Co(H_2N—C_2H_4—NH_2)_2Cl_2]^+$, with a photograph of a model of the *trans* isomer. (Charles D. Winters)

will be described here. When three bidentate ligands, such as en or phen, bind to a metal ion or atom, three metal-containing rings are formed, as in the $[Co(en)_3]^{3+}$ complex pictured in Figure 25.13. Two optical isomers result from the two ways of arranging these rings in space (Figure 25.17). Think of the complex as a three-bladed propeller. The blades can be arranged so that the propeller twists clockwise or counterclockwise. One arrangement is the mirror image of the other, and neither is superimposable on the other.

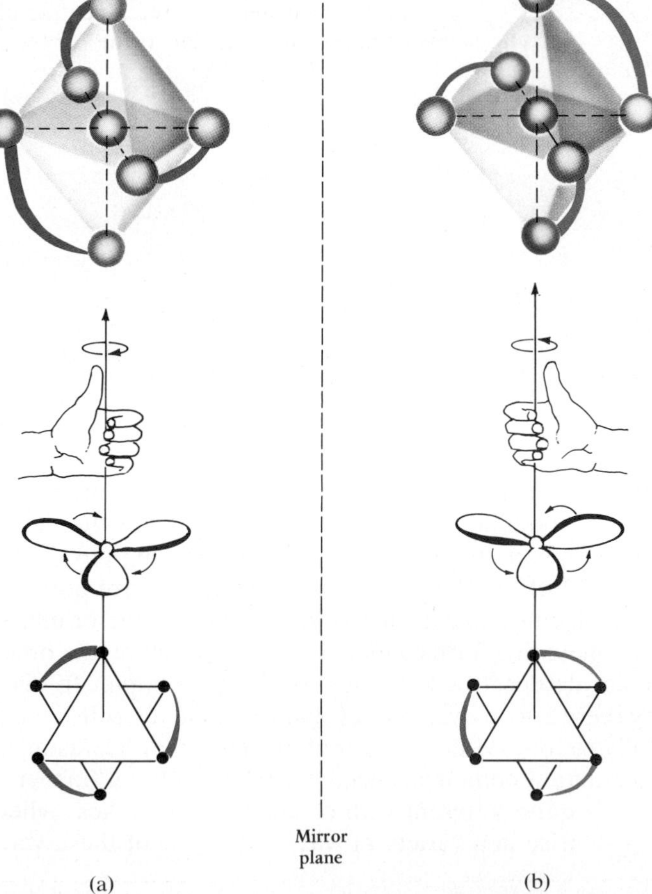

FIGURE 25.17

Optical isomerism in complexes of the type $M(bidentate)_3$. The three chelate rings are arranged so that the complex is essentially a three-bladed propeller. One of the propellers twists clockwise (a) and the other twists counterclockwise (b). The mirror images cannot be superimposed.

EXAMPLE 25.4

ISOMERISM

Which of the following complexes exhibit geometrical and/or optical isomerism?

(a) $[Pt(NH_3)_4Cl_2]^{2+}$ (b) $[Ru(phen)_3]Cl_2$ (c) $K_2[Pt(CN)_2Cl_2]$

Solution (a) This Pt^{4+} complex has an octahedral structure with two geometric isomers. It is similar to $[Co(en)_2Cl_2]^+$ in the text, one isomer with two Cl^- ions *cis* and the other with the two ions *trans*.

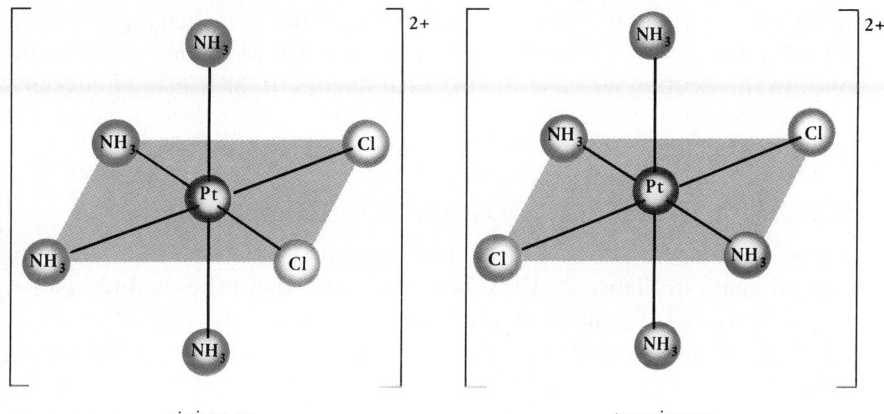

cis isomer *trans* isomer

(b) In $[Ru(phen)_3]^{2+}$, the Ru^{2+} ion is surrounded by three bidentate phen ligands (Figure 25.14). Therefore, this complex ion is similar to the $[Co(en)_3]^{3+}$ ion described in the text (Figure 25.17), an ion that can have optical isomers.

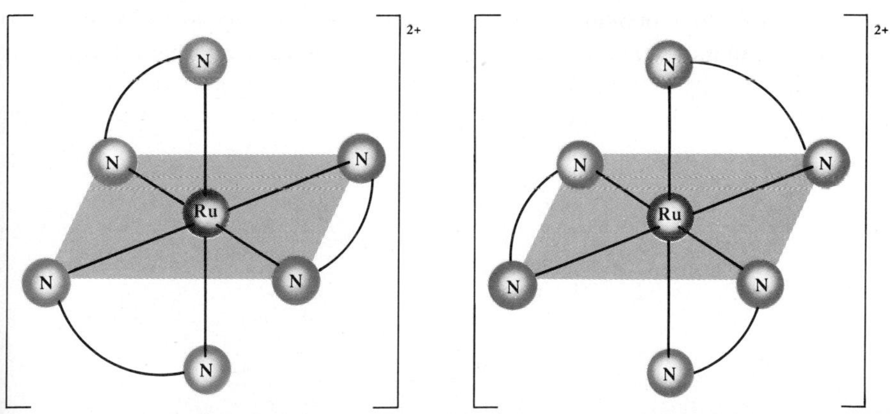

(c) This complex has two different types of ligands, in two pairs, attached to Pt^{2+}, and it is square planar. This means geometrical isomers are possible.

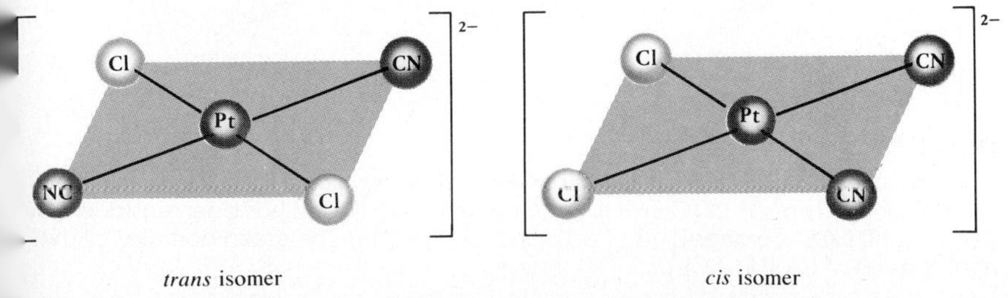

trans isomer *cis* isomer

EXERCISE 25.4 IDENTIFYING ISOMERS
What types of isomers are possible for (a) $[Co(en)_2(CN)_2]Br$ and (b) $[Cr(phen)_3]Cl_3$?

25.5 BONDING IN COORDINATION COMPOUNDS

The reactions, magnetic properties, and colors (Figure 25.18) of the complexes of a particular metal are sensitive to the attached ligands. To understand this, we shall have to delve into the nature of the metal–ligand bond. In this section we describe (a) what occurs to the *d*-orbital energies of a metal atom or ion as ligand lone pairs approach to form a bond and (b) the consequences of this for the color of complexes.

d-ORBITAL ENERGIES IN COORDINATION COMPOUNDS

Shapes of atomic *d* orbitals were first described in Chapter 7. They are pictured again in Figure 25.19, where the relation of orbitals and ligands is also illustrated. We have divided the five orbitals into two sets: the z^2 and $x^2–y^2$ orbitals in one set and the xy, xz, and yz orbitals in the other. Notice that the first two orbitals have their greatest amplitude along the cartesian axes. The orbitals of the other set, in contrast, have their greatest amplitude *between* the axes. This division of orbitals into two sets is reasonable because the ligands in square planar and octahedral complexes lie along the x, y, and z axes, with their lone pair orbitals pointing toward the metal atom or ion at the center of the coordinate system.

Molecular orbital theory and **crystal field theory** are two ways to approach metal–ligand bonding using metal *d* orbitals and ligand lone pair orbitals. As ligands approach the metal to form bonds, there are two effects: (a) orbital overlap and (b) repulsion between electrons in metal

FIGURE 25.18
The colors of the complexes of a given metal ion depend on the ligand. The yellow solid at the left is a salt of $[Co(NH_3)_6]^{3+}$. One ammonia ligand has been replaced by (left to right) NCS^- (orange), H_2O (red), and Cl^- (purple). The green complex at the right is a salt of $[Co(NH_3)_4Cl_2]^+$.

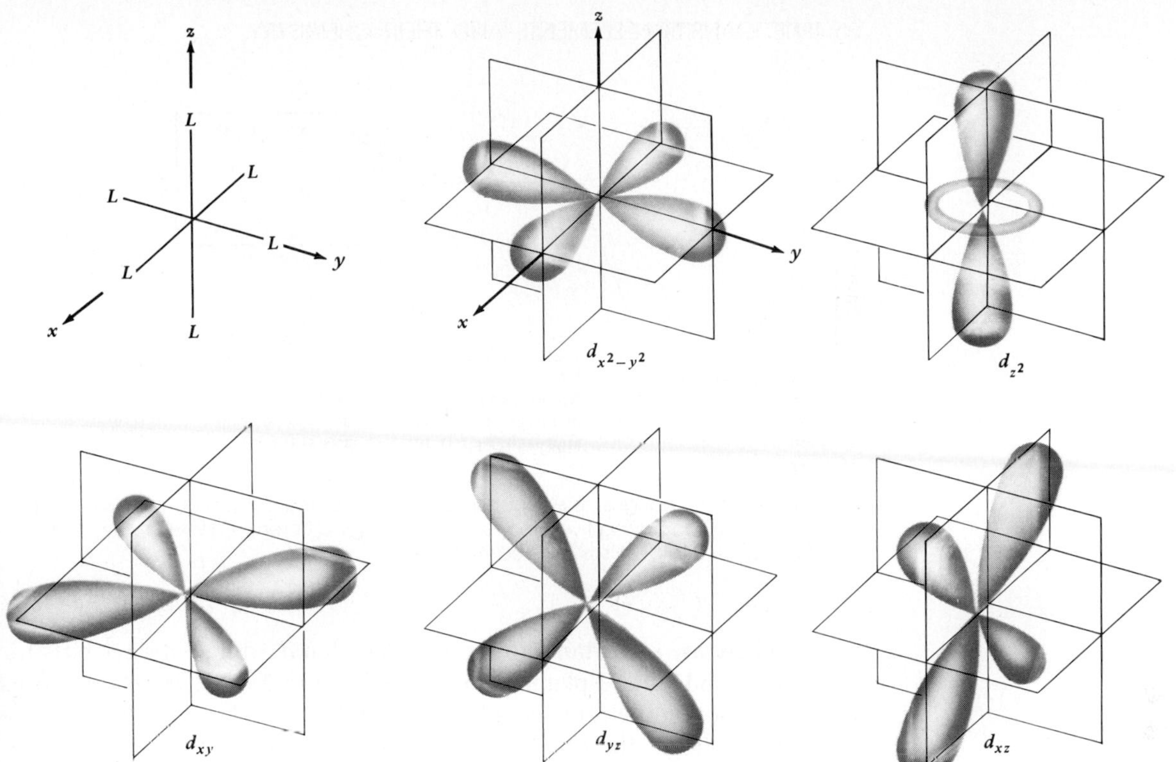

FIGURE 25.19
The five *d* orbitals and their relation to ligands on the *x*, *y*, and *z* axes.

and ligand. Molecular orbital theory takes into account both effects, while the crystal field model focuses exclusively on metal–ligand electron repulsion. The molecular orbital model assumes that metal and ligand are bound by molecular orbitals formed by metal–ligand atomic orbital overlap. In contrast, the crystal field model assumes the positive metal ion and negative ligand lone pair are *electrostatically* attracted. The name of the theory comes from the fact that the ligands are assumed to create an electrostatic "crystal field" that influences metal electrons. Both models ultimately produce the same *qualitative* result, and we shall focus on the crystal field approach.

The name "crystal field theory" comes from the original theory, which assumed a metal ion in a crystal surrounded by ligands. A similar theory is *ligand field theory*, but the latter allows for metal–ligand covalent bonding.

The *d* orbitals of a free atom or ion are degenerate, that is, they have the same energy. However, according to the crystal field model, there is a repulsion between metal electrons and ligand lone pairs as ligands approach the metal atom or ion, and the *d* electron orbitals rise in energy (Figure 25.20). The largest repulsion is felt by the z^2 and x^2–y^2 electrons, since these electrons are in orbitals pointing directly at the incoming ligand electron pairs. A somewhat smaller repulsive effect, however, is experienced between *xy*, *xz*, and *yz* metal electrons and ligand electrons. The difference in degree of repulsion means that there is a difference in the energy increase for the two sets of orbitals. The "splitting" in energy of the (z^2, x^2–y^2) and (*xy*, *xz*, *yz*) sets, given by the symbol Δ_0, depends on the metal and the ligands and can change as these change from one complex to another. As you shall soon see, this splitting of *d* orbitals into two sets can lead to the distinctive colors of transition metal complexes.

Δ_0 is sometimes called the "octahedral crystal field splitting parameter."

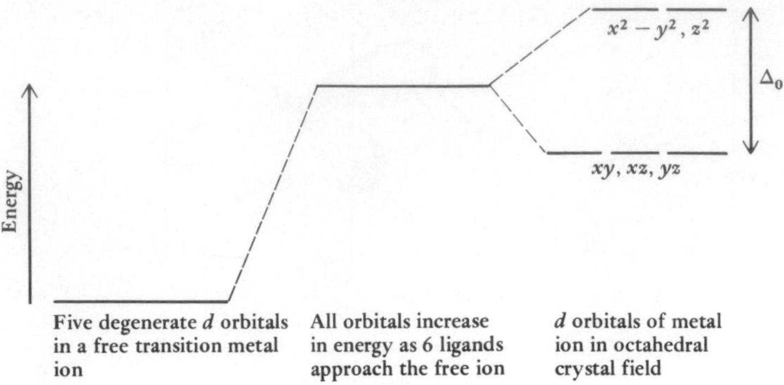

FIGURE 25.20

d Orbital energy changes as six ligands approach, at the corners of an octahedron, a transition metal ion having five electrons. The energy difference between the (z^2,x^2–y^2) and (*xy, xz, yz*) orbital sets is labeled Δ_o.

There are important differences between *d* orbital energy shifts for octahedral and square planar complexes (Figure 25.21). Both structures have in common the four ligands in the molecular plane, assumed to be the *xy* plane. Thus, the x^2–y^2 orbital is raised in energy by the same amount in ML_4 as in ML_6. In square planar ML_4, however, the upward shift of z^2 is smaller because only $\frac{1}{3}$ of the z^2 orbital (the "doughnut") is concentrated in the *xy* plane. Finally, there are no ligands in the *xz* or *yz* planes of square planar ML_4. This means electrons in *xz* and *yz* orbitals have an energy lower than electrons in an *xy* orbital.

ELECTRON CONFIGURATIONS OF METAL IONS OR ATOMS IN COMPLEXES AND MAGNETISM

The consequences of *d* orbital splitting in coordination complexes are far-reaching, affecting the color and magnetic behavior of complexes. To understand these properties, we must first understand how to assign electrons to the "split" orbital sets of square planar and octahedral complexes.

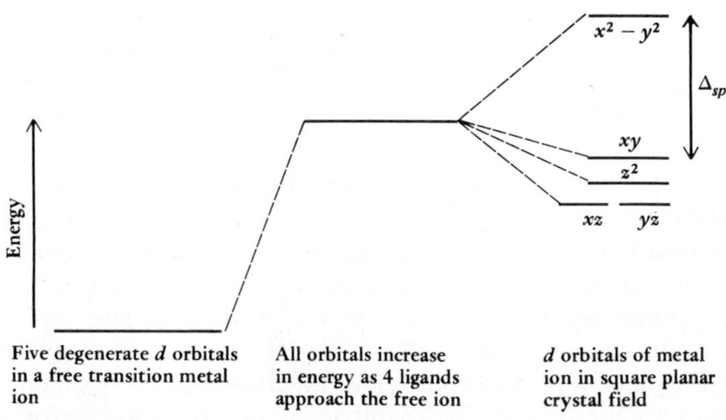

FIGURE 25.21

The splitting of *d* orbitals for a square planar complex. Here the energy difference between the x^2–y^2 orbital and the next highest energy orbital is given by the symbol Δ_{sp}

The Cr^{2+} ion has the electron configuration [Ar]$3d^4$. Hund's rule means that the five degenerate *d* orbitals of a free Cr^{2+} ion will have the following orbital diagram.

Electron Configuration of Cr^{2+}

[Ar] ↑ ↑ ↑ ↑ __

five degenerate $3d$ orbitals

When Cr^{2+} is bound in an octahedral complex, the *xy*, *xz*, and *yz* orbitals are degenerate and at a lower energy than the degenerate (z^2, $x^2–y^2$) set. Having two sets of orbitals means that two electron configurations are possible for four electrons. The first three electrons are assigned, one each, to the lower energy (*xy*, *xz*, *yz*) orbitals. The fourth electron, however, could be assigned to the higher energy (z^2, $x^2–y^2$) set (Configuration A) *or* pair up with an electron already in the lower energy set (Configuration B).

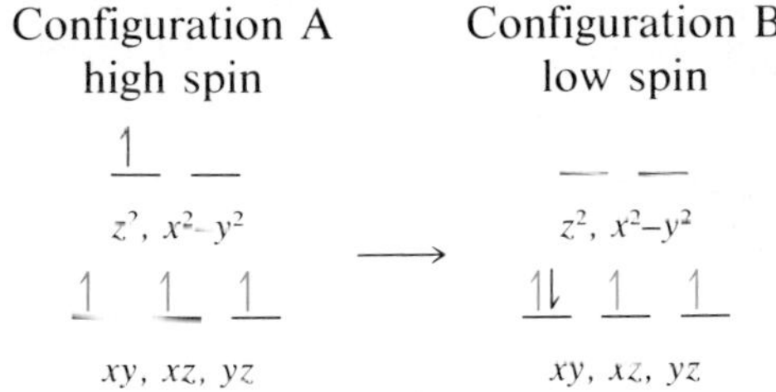

The first arrangement is commonly called **high spin** because it has the maximum number of unpaired electrons. In contrast, the second arrangement is called **low spin** because it has the minimum number of unpaired electrons.

The correct arrangement of the four Cr^{2+} electrons in a complex containing this metal ion depends on two factors. We can see what these are if we write an equation for the energy difference between the low and high spin configurations.

$$\Delta E = E_{\text{low spin}} - E_{\text{high spin}} = -(\Delta_0) + P$$

The quantity Δ_0 is the energy difference between the two orbital sets. If an electron moves from the higher energy set to the lower energy set, energy in an amount equal to Δ_0 is evolved, so Δ_0 is given a negative sign. However, to form the low spin configuration, an electron pair is formed, and, due to higher electron–electron repulsions in a pair, this *requires energy*. *P* is the energy of electron pairing, and it is given a positive sign. From this equation, we can come to the *general conclusion* that a low spin configuration will be observed in preference to high spin if energy is evolved on going from high spin to low spin, that is, if ΔE is negative. This will occur if Δ_0 is large or *P* is small. This conclusion is summarized for a simple two-orbital system in Figure 25.22.

There is a choice between high and low spin only for configurations d^4 through d^7 in octahedral complexes (Figure 25.23). In each case, the outcome is dictated by the relative size of Δ_0 and the electron pairing energy *P*. This means that, for a given metal ion with a constant value of

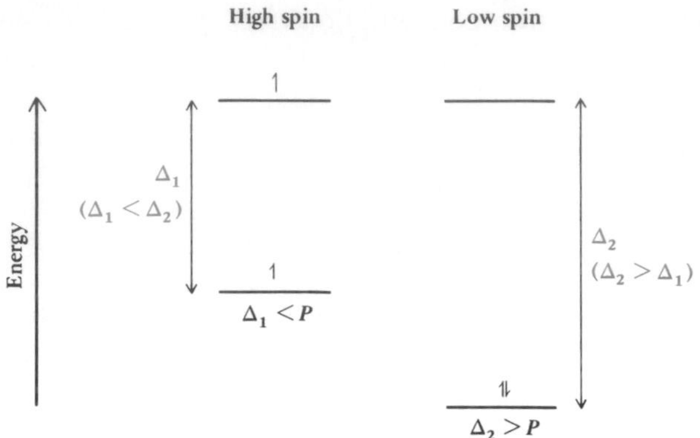

FIGURE 25.22
Dependence of high and low spin configurations on extent of orbital splitting (Δ) and electron pairing energy P.

P, whether the complex is high or low spin will depend on Δ_o. Thus, for Fe^{2+}, for example, the complex formed when the ion is placed in water, $[Fe(H_2O)_6]^{2+}$, is high spin, whereas $[Fe(CN)_6]^{4-}$ is low spin.

Electron Configuration for Fe^{2+} in an Octahedral Complex

High spin		Low spin	
↑ ↑ $x^2{-}y^2$, z^2	Δ_o (H_2O)	— — $x^2{-}y^2$, z^2	Δ_o (CN^-)
↑↓ ↑ ↑ xy, xz, yz		↑↓ ↑↓ ↑↓ xy, xz, yz	
high spin $[Fe(H_2O)_6]^{2+}$		low spin $[Fe(CN)_6]^{4-}$	

Ions with eight d electrons (Ni^{2+}, Pd^{2+}, and Pt^{2+}) are especially prone to form square planar complexes. The Ni^{2+} ion, for example, has the electron configuration

Electron Configuration of a Free Ni^{2+} Ion

[Ar] — — — — —

five degenerate $3d$ orbitals

In a square planar complex, however, the five degenerate orbitals form four sets of orbitals (Figure 25.21). With eight electrons to be accommodated, high and low spin configurations are possible.

Electron Configuration for Ni^{2+} in a Square Planar Complex

High spin		Low spin	
↑ $x^2{-}y^2$	Δ_{sp}	— $x^2{-}y^2$	Δ_{sp}
↑ xy		↑↓ xy	
↑↓ z^2		↑↓ z^2	
↑↓ ↑↓ xz, yz		↑↓ ↑↓ xz, yz	
high spin		low spin	

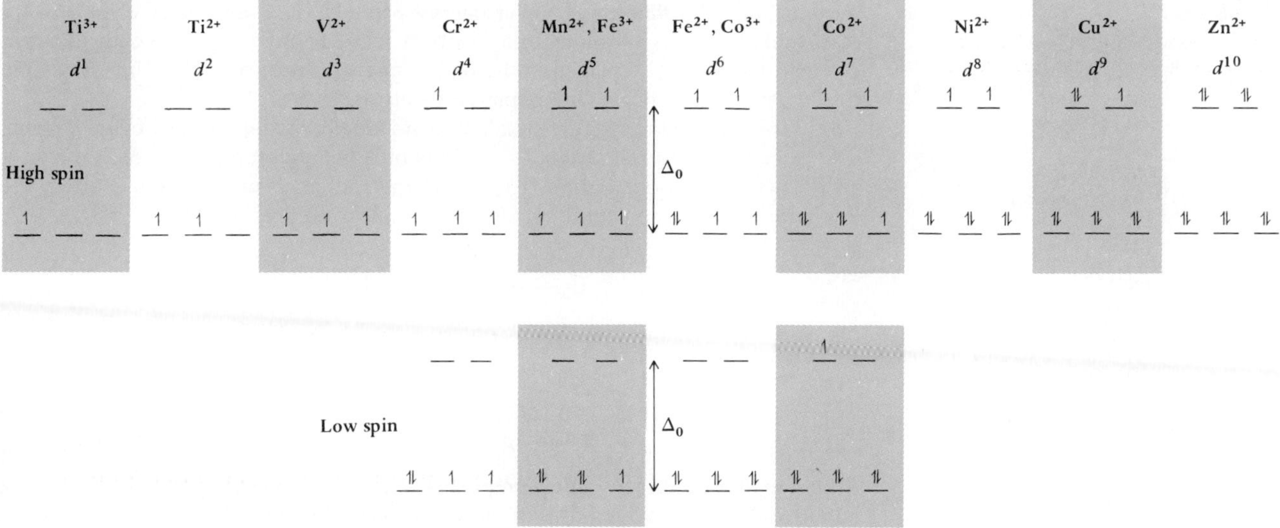

FIGURE 25.23
Electron configurations for metal ions with d^1 to d^{10} configuration in octahedral complexes. Only d^4 through d^7 configurations can have either high or low spin arrangements.

However, all known square planar Ni^{2+} complexes are low spin, so Δ_{sp} clearly outweighs the pairing energy, P.

Refer to Section 8.1 for a discussion of paramagnetism.

One can tell the difference between low and high spin complexes by determining the magnetic behavior of the substance. For example, a high spin, square planar Ni^{2+} complex has two unpaired electrons and is *paramagnetic*, whereas a low spin Ni^{2+} complex has no unpaired electrons and is *diamagnetic*. Similarly, $[Fe(H_2O)_6]^{2+}$ is paramagnetic to the extent of four unpaired electrons, whereas $[Fe(CN)_6]^{4-}$ is diamagnetic.

EXAMPLE 25.5

HIGH AND LOW SPIN COMPLEXES AND MAGNETISM

Write the low and high spin electron configurations for each of the complexes below and tell how many unpaired electrons are present in each.

(a) $[Co(NH_3)_6]^{2+}$ (b) $[Mn(CN)_6]^{4-}$

Solution (a) Since there are six ligands surrounding cobalt, we assume this is an octahedral complex. Further, since the NH_3 ligands are neutral molecules and since the overall charge on the complex is +2, this means the complex is based on Co^{2+}. Cobalt(II) ion has an electron configuration of $[Ar]3d^7$. Therefore, the metal ion configurations in the complex would be

high spin	low spin
__ __ (x^2-y^2, z^2)	__ __ (x^2-y^2, z^2)
Δ_0	Δ_0
⇅ ⇅ 1 (xy, xz, yz)	⇅ ⇅ 1 (xy, xz, yz)

To obtain the high spin configuration, first half fill all five orbitals, and then pair up any remaining electrons in the lower energy orbitals. This leads to the configuration above with three unpaired electrons. The low spin configuration is obtained by first half filling the three lower energy orbitals. The next three electrons are then added to these orbitals and electron pairs formed. The seventh and last electron must then be placed in the higher energy set of orbitals. Thus, low spin Co^{2+} is paramagnetic by one unpaired electron.

(b) The ligands in the octahedral manganese complex are cyanide ions, CN^-. Since the overall charge is -4, this means the complex is based on Mn^{2+}, an ion with the configuration $[Ar]3d^5$. Using the same procedure as above, the low and high spin configurations in the complex ion are

high spin	low spin
↑ ↑	— —
x^2-y^2, z^2	x^2-y^2, z^2
↑ ↑ ↑	↑↓ ↑↓ ↑
xy, xz, yz	xy, xz, yz

Both the high and low spin complexes are paramagnetic, but the number of unpaired electrons is very different. This difference can be measured by experiment, so it is possible to tell if the complex is low or high spin.

EXERCISE 25.5 LOW AND HIGH SPIN CONFIGURATIONS AND MAGNETISM

For each of the following complex ions, give the oxidation number of the metal ion, write the low and high spin configurations, give the the number of unpaired electrons in each state, and tell whether each is paramagnetic or diamagnetic.

(a) $[Ru(H_2O)_6]^{2+}$ (b) $[Ni(NH_3)_6]^{2+}$

25.6 THE COLORS OF COORDINATION COMPOUNDS

One of the most interesting properties of the transition elements is that their compounds are usually colored, whereas those of main group metals are almost always colorless (Figure 25.24). With an understanding of *d* orbital splitting, we can now explain the origin of the color.

FIGURE 25.24

Compounds of the transition elements are often colored, while those of main group elements are usually colorless. Pictured are aqueous solutions of the nitrate salts of (left to right) Fe^{3+}, Co^{2+}, Ni^{2+}, Cu^{2+}, and Zn^{2+}. (Charles D. Winters)

COLOR AND LIGHT ABSORPTION AND TRANSMISSION

In Chapter 7 you saw that white light covers a range of different colors, from red light of low energy and long wavelength to violet light of high energy and short wavelength.

You can remember the succession of colors from the acronym ROY G BIV.

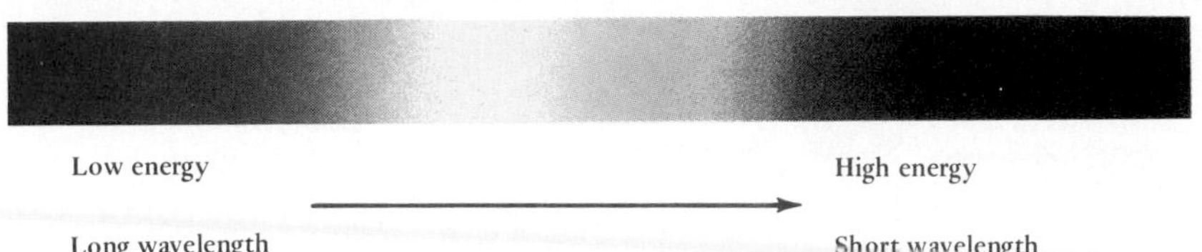

You perceive the color of transition metal complexes in solution by the color of the *transmitted* or nonabsorbed light. A solution of the compound absorbs light of one or more colors, and the color or colors not absorbed are perceived by your eyes (Figure 25.25).

The six components of white light consist of three primary colors (red, yellow, and blue) and three secondary colors (orange, green, and violet), which can be made by mixing the primary colors in pairs as follows:

$$\text{Red} + \text{yellow} \longrightarrow \text{orange} \qquad \text{yellow} + \text{blue} \longrightarrow \text{green}$$
$$\text{blue} + \text{red} \longrightarrow \text{violet}$$

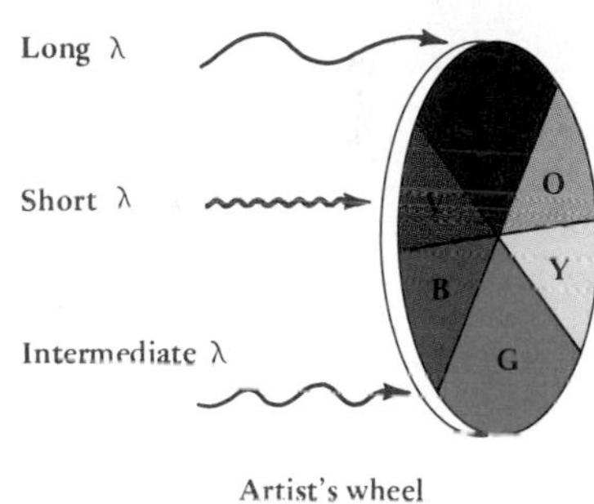

Artist's wheel

A convenient way to remember this is to arrange the colors circularly on an "artist's wheel." Colors on opposite sides of the wheel are complementary; they add together to give white light. If a solution absorbs light of one or more colors on one side of the wheel, you see the remaining colors. Thus, if you see green, this means one of two things happened. If all but green light is absorbed, only green light is transmitted (Figure 25.26a). Alternatively, if violet, red, and orange were absorbed, then blue,

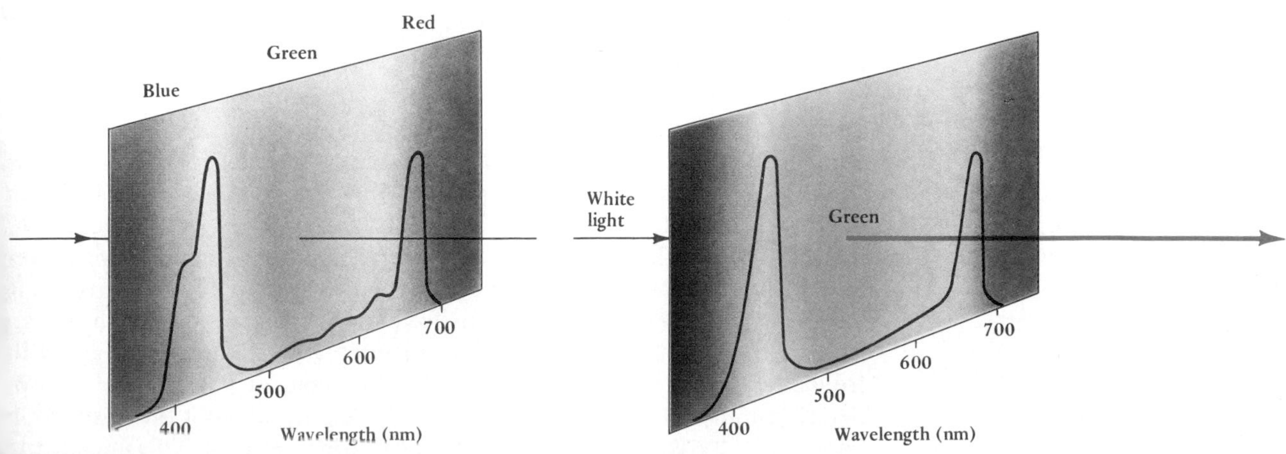

FIGURE 25.25
A compound dissolved in water can be green because the molecule of the compound absorbs blue and red light but not the wavelengths of the colors in between. The transmitted colors, chiefly yellow to green, are perceived by your eyes.

FIGURE 25.26
A sample absorbs light of one or more colors and transmits those on the opposite side of the "artist's color wheel."

(a) Sample absorbs all but green light. Green is perceived.

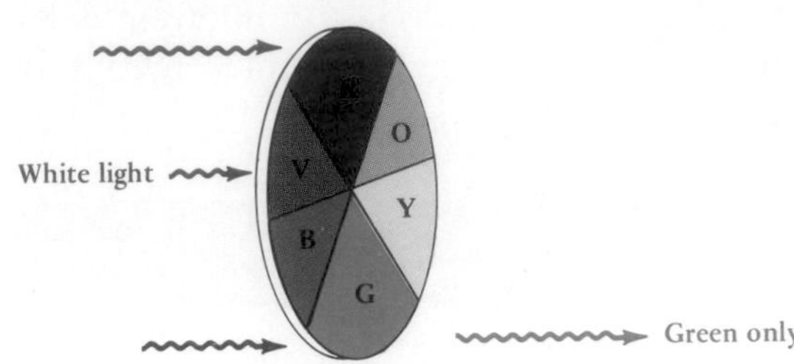

(b) Sample absorbs violet, red, and orange light. Blue, green, and yellow light are transmitted. Green light is perceived.

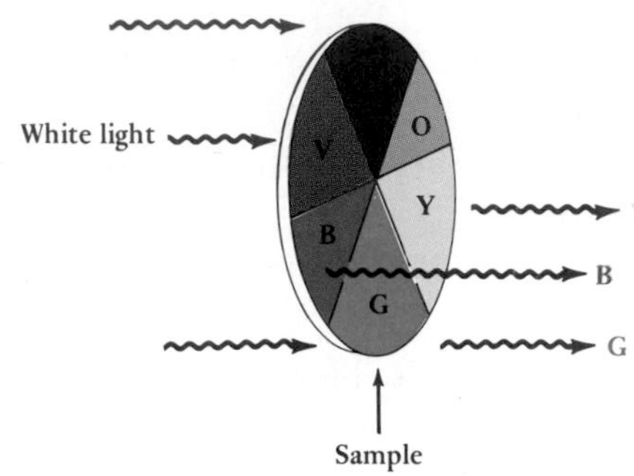

green, and yellow are transmitted, and you perceive the middle color of the three (Figure 25.26b).

Octahedral Cu^{2+} coordination compounds are typically blue. This means that blue light, and perhaps its flanking colors (violet and green) are passed on to your eyes, while orange (and perhaps red and yellow) are absorbed. This can be verified with a *spectrophotometer* (Figure 25.27). White light from a glowing filament is first passed through a device (a prism or diffraction grating) that disperses the photons of white light according to their frequencies. The instrument can select photons of each different frequency to pass through a solution of the compound to be studied. If photons of a given frequency are not absorbed, that light is unchanged in its intensity when it emerges from the sample. On the other hand, if photons of some frequency are absorbed, by a mechanism described below, light of that frequency emerges from the sample with a decreased intensity. By making a plot of the frequency or wavelength of the light against the intensity of light absorbed at that frequency or wavelength, we obtain an **absorption spectrum** of the sample. When there is an absorption of light in a certain frequency range, the plot shows an *absorption band*, and the variation in intensity of light absorbed with frequency is called a **spectrum**. In the spectrum of $[Cu(H_2O)_6]^{2+}$, copper(II) ion in water (Figure 25.27), you should notice that the absorption band spans the red-orange-yellow region of the spectrum. This means that violet, blue, and green light are transmitted, so the solution appears blue.

Transition metal complexes can absorb light because photons of the appropriate energy can excite the complex from its ground state to a

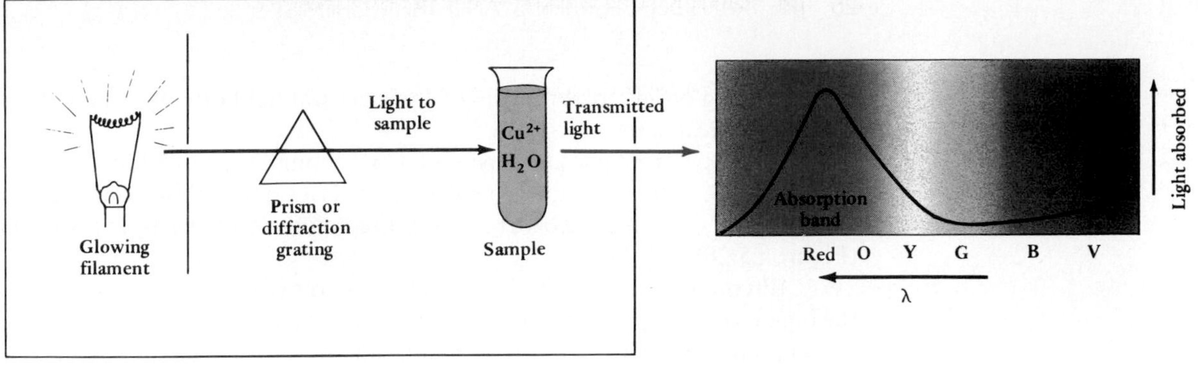

FIGURE 25.27
A spectrophotometer and the absorption spectrum of Cu^{2+} in water.

higher energy or excited state. Although the details of the process are beyond the scope of this book, in a case such as copper(II) ion in water you can imagine that one of the *d* electrons in a d_{xy}, d_{xz}, and d_{yz} orbital is excited to a $d_{x^2-y^2}$ or d_{z^2} orbital.

↑↓ ↑ (x^2-y^2, z^2) ⟶ ↑↓ ↑↓ (x^2-y^2, z^2)

Δ_0 + energy (Δ_0) (light absorbed)

↑↓ ↑↓ ↑↓ (xy, xz, yz) ⟶ ↑↓ ↑↓ ↑ (xy, xz, yz)

ground state Cu^{2+} excited state Cu^{2+}

The fact that high energy light is transmitted means that low energy light was absorbed. In turn, this means that the band gap, Δ_0, is relatively small for the Cu^{2+} ion in water.

THE SPECTROCHEMICAL SERIES OF LIGANDS

Experience with many coordination complexes has revealed that, for a given metal ion, some ligands lead to a small energy separation of the *d* orbitals, while others lead to a large separation. In the language of crystal field theory, some ligands create a small crystal field while others create a large field. What happens as ligands surrounding Co^{3+} are changed is shown in Figure 25.28. Fluoride ions, F^-, create a small field, so Δ_0 for the hexafluorocobaltate(III) ion is small. This means that only low energy light is required to induce an electron transition, and the complex ion absorbs in the red region of the spectrum and transmits green light. In contrast, the hexacyanocobaltate(III) ion absorbs in the high energy ultraviolet region owing to the very strong field of CN^- ligands; Δ_0 is large.

From thousands of experiments, it has been found that the separation or "splitting" of *d* orbitals increases in the following order:

Spectrochemical Series

$$Cl^- < F^- < H_2O < NH_3 < \text{en} < \text{phen} < CN^-$$

small orbital splitting
small Δ_0
weak field ligands

large orbital splitting
large Δ_0
strong field ligands

This ordering of ligands in terms of their crystal field effects is called the **spectrochemical series**.

The spectrochemical series of Co^{3+} complexes are illustrated in Figure 25.18. As ammonia is replaced by H_2O and halide ions, longer wavelengths of light are absorbed, and the colors of the complexes differ (Figure 25.28).

Predicting the possible color of a complex based on the nature of the ligands is risky, since the details of the spectroscopy of such complexes are complicated. However, it is clear that weak field ligands lead to a small splitting, so the complex will absorb relatively low energy photons. Like hexaaquocopper(II) ion or $[CoF_6]^{3-}$, such complexes *tend* to have colors at the blue end of the spectrum. Conversely, strong field ligands cause a large splitting, and their complexes *tend* to have colors at the red end of the spectrum. The red complex $[Fe(CN)_6]^{3-}$ shown in Figure 25.12 is a good example of this.

From the relative position of the ligand in the series, you can make some prediction about the compound's magnetic behavior. Recall that d^4,

FIGURE 25.28

The spectral properties of some cobalt(III) complexes.

Co^{3+} complex	Wavelength of light absorbed (nm)	Color of light absorbed	Color seen
$[CoF_6]^{3-}$	700	red	green
$[Co(H_2O)_6]^{3+}$	600	orange	blue
$[Co(NH_3)_5Cl]^{2+}$	535	yellow	purple
$[Co(NH_3)_5H_2O]^{3+}$	500	blue-green	red
$[Co(NH_3)_6]^{3+}$	475	blue	yellow-orange
$[Co(CN)_6]^{3-}$	310	ultraviolet	tail of absorption band in visible gives pale yellow

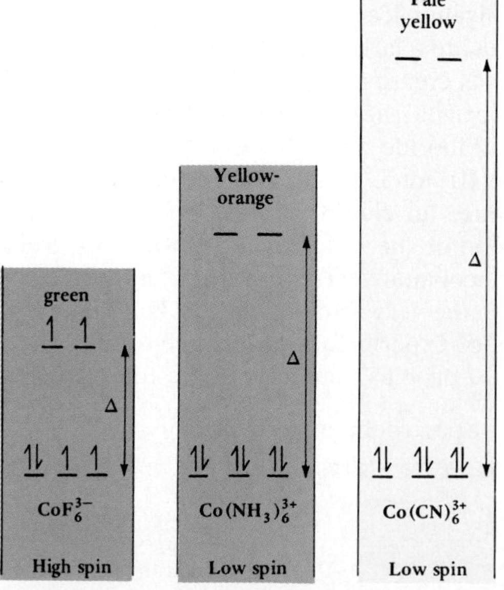

d^5, d^6, and d^7 complexes can be high or low spin, depending on the value of Δ_0 and the pairing energy (Figure 25.23). Thus, for a given metal ion, ligands near the left of the spectrochemical series (as given above) lead to a small Δ_0 and high spin complexes result. In contrast, ligands near the right end lead to a large Δ_0 and give low spin configurations. Notice that $[CoF_6]^{3-}$ is high spin, while $[Co(NH_3)_6]^{3+}$ and the others shown in Figure 25.28 are low spin.

EXAMPLE 25.6

SPECTROCHEMICAL SERIES

The iron(II) ion in water, $[Fe(H_2O)_6]^{2+}$, is light green. Do you expect the d^6 Fe^{2+} ion in this complex to have a high or low spin configuration?

Solution The green color implies that low energy photons are absorbed by the complex. The artist's color wheel shows that green can arise if the complex absorbs only low energy red photons. This means the *d* orbital splitting, given by Δ_0, is small. A relatively small value of Δ_0 means that the complex has a good chance of being a high spin complex, and this is verified experimentally.

High spin Fe^{2+}

↑ ↑
x^2-y^2, z^2 $\quad\Delta_0$
↑↓ ↑ ↑
xy, xz, yz

EXERCISE 25.6 THE SPECTROCHEMICAL SERIES

Which complex should absorb the lower energy visible light, hexaaquonickel(II) or hexaamminenickel(II)?

SUMMARY

The **transition elements** or **transition metals** are the elements from Sc to Zn, Y to Cd, and La to Hg. All have *d* electrons among their valence electrons and so are also called ***d*-block elements**. The **inner transition elements** or **f-block elements** are those between La and Hf (the **lanthanides**) and between Ac and element 104 (the **actinides**) (Figure 25.1).

Oxidation numbers of +2 and +3 are common for transition elements, but others are also observed (Section 25.1). Many transition metals, lanthanides, and actinides, as well as their ions, have unpaired electrons and so can be **paramagnetic**. Some metals and their alloys are **ferromagnetic**.

Periodic trends in metal atom radii of transition metals are illustrated in Figure 25.4. The near constancy of size for the elements of Groups 6B through 1B, and for elements of the fifth and sixth periods, is important to their chemistry. The great increase in metal density across the transition metals reflects not only the constancy of size but also the increase in atomic mass (Figure 25.5).

Pyrometallurgy, high temperature processes, and **hydrometallurgy**, techniques using water, are the two major methods for recovering metals from their ores. Ores are usually mixtures of the desired mineral and **gangue**, sand or clay minerals. To separate the desired metal or its compound, **flotation** or **chemical leaching** is used. **Iron** production in a **blast furnace** uses pyrometallurgy and depends primarily on the reduction of iron oxides (e.g., Fe_2O_3) with CO (generated by reacting coke, C, with O_2 in the blast furnace) (Section 25.2). Impure "pig" or "cast" iron is purified by oxidation of impurities in a **basic oxygen furnace**.

Production of **copper** uses both pyrometallurgy and hydrometallurgy (Section 25.2). After separation of copper ore from gangue by flotation, the ore is **roasted** in air to give impure copper. This is refined by electrolysis.

Transition metals and their ions form **coordination compounds** or **coordination complexes**. In a complex ion such as $[Fe(CN)_6]^{4-}$ the CN^- ions are called the **ligands** (Section 25.3). The CN^- ion, H_2O, NH_3, and other molecules or ions donating one electron pair to a metal ion are **monodentate**, and those with more than one such atom are **multidentate** (bidentate, tridentate, and so on) (Figure 25.14). Multidentate ligands that use more than one atom to bind to the metal are called **chelating** ligands.

The rules for systematically naming coordination compounds are given in Section 25.3.

The number of ligand-metal sigma bonds is the metal **coordination number**. Numbers of 2, 4, and 6 are common. Two-coordinate complexes are linear; four-coordinate ones are tetrahedral or square planar; and six-coordinate complexes are usually octahedral.

Isomers are compounds having the same stoichiometry but different atomic arrangements. As in organic chemistry (Chapters 23 and 24) there can be **geometric** and **optical** isomers (Section 25.4). Square planar complexes may exist as geometric isomers but not optical isomers. Octahedral complexes can exhibit both forms of isomerism.

Metal–ligand bonding can be explained using **molecular orbital** and **crystal field** theory, but this chapter focuses on the latter (Section 25.5). Crystal field theory assumes metal–ligand bonding arises from an electrostatic attraction between the metal cation and ligands. As ligands approach a metal ion, repulsion between ligand lone pairs and the metal electrons affects the metal *d* orbitals differently (Figures 25.20 and 25.21). For an octahedral complex, the x^2-y^2 and z^2 orbitals are affected more than the xy, xz, and yz orbitals by this "crystal field" effect. The energy difference between the sets of *d* orbitals is given by Δ_0. Because there are two sets of *d* orbitals in an octahedral complex (or four sets in a square planar complex), there are two ways of assigning electrons (Section 25.5). The configuration with the maximum number of unpaired electrons is called **high spin**, while a **low spin** configuration has the minimum number of unpaired electrons. The exact assignment depends on Δ and P, the electron pairing energy.

The energy difference between sets of *d* orbitals in complexes is comparable to the energy of visible light. Therefore, coordination complexes are usually colored (Section 25.6). The observed color depends in large part on Δ_0. From experiments, chemists have derived a **spectrochemical series**, an ordering of ligands according to the magnitude of Δ_0.

STUDY QUESTIONS

REVIEW QUESTIONS

1. What distinguishes the transition elements and the lanthanides and actinides?
2. What are the two major types of processes for the recovery of metals from their ores? Give an example of the use of each type of process.
3. What transition elements are commonly used in the manufacture of permanent magnets?
4. Define the following words or phrases: (a) transition elements, (b) coordination compound, (c) complex ion, (d) ligand, (e) chelate, (f) bidentate. Give an example to illustrate each word or phrase.
5. What are three common metal coordination numbers and what structure or structures are possible for each?
6. Distinguish an optical isomer from a geometric isomer. Give an example of each.
7. According to the crystal field model, what is the origin of the splitting of metal *d* orbitals into two sets in an octahedral complex?
8. What are the factors that determine if a complex will be high or low spin?

CONFIGURATIONS AND PHYSICAL PROPERTIES

9. The chromium compounds pictured in Figure 25.3 have two different oxidation numbers. Using an orbital box diagram, show the electron configuration of chromium with each of its oxidation numbers. Are either of these expected to be paramagnetic?
10. Give electron configurations for each of the following ions. Tell if each is paramagnetic or diamagnetic.
    (a) Y^{3+} (d) Pt^{4+}
    (b) Rh^{3+} (e) V^{2+}
    (c) Ce^{4+} (f) U^{4+}
11. Which element in each of the following pairs should be more dense? Explain each answer briefly.
    (a) Ti or Fe (c) Ti or Os
    (b) Ti or Zr (d) Zr or Hf

METALLURGY

12. The following equations represent various ways of obtaining transition metals from their compounds. Balance each equation.
    (a) $Cr_2O_3(s) + Al(s) \longrightarrow Al_2O_3(s) + Cr(s)$
    (b) $TiCl_4(\ell) + Mg(s) \longrightarrow Ti(s) + MgCl_2(s)$
    (c) $[Ag(CN)_2]^-(aq) + Zn(s) \longrightarrow Ag(s) + Zn^{2+}(aq) + CN^-(aq)$
13. In the first step in the recovery of copper, an ore such as chalcopyrite is roasted in air to give CuS. If you begin with one ton (908 kg) of the ore, how many tons of SO_2 are produced?
14. Titanium is the fourth most abundant metal. It is strong, lightweight, and corrosion resistant and thus is used in aircraft engines and chemical plants. To recover the titanium, ilmenite ($FeTiO_3$) is first treated with sulfuric acid to give $Ti(SO_4)_2$, and, after separating $FeSO_4$ and $Ti(SO_4)_2$, the latter is converted to TiO_2 in basic solution.

$$FeTiO_3(s) + 3\ H_2SO_4(\ell) \longrightarrow FeSO_4(aq) + Ti(SO_4)_2(aq) + 3\ H_2O(\ell)$$

$$Ti^{4+}(aq) + 4\ OH^-(aq) \longrightarrow TiO_2(s) + 2\ H_2O(\ell)$$

If you begin with 1.00 kg of ilmenite, how many liters of 18.0 *M* H_2SO_4 are required to react completely with the ilmenite, and how many kilograms of TiO_2 can theoretically be produced?

15. In the titanium process in the preceding question, ilmenite ore is leached with sulfuric acid. However, this leads to the significant environmental problem of disposal of the iron(II) sulfate (which, in its hydrated form, is commonly called "copperas"). To avoid this, it has been suggested that HCl be used to leach ilmenite so that the iron-containing product is $FeCl_2$. This can be treated with water and air to give commercially useful iron(III) oxide and regenerate HCl by the reaction

$$2\ FeCl_2(aq) + 2\ H_2O(\ell) + \tfrac{1}{2}\ O_2(g) \longrightarrow Fe_2O_3(s) + 4\ HCl(aq)$$

    (a) Write a balanced equation for the treatment of ilmenite with HCl to give iron(II) chloride, titanium(IV) oxide, and water.
    (b) If the equation written in part (a) is combined with the equation for oxidation of $FeCl_2$ to Fe_2O_3 above, is the HCl used in the first step recovered in the second step?
    (c) How many grams of iron(III) oxide can be obtained from one ton (908 kg) of ilmenite?
16. Worldwide production of nickel is 750,000 tons per year; most comes from Canada. The *Mond process* is one method for extracting the pure metal. Here the oxide is treated with water gas (H_2 + CO), and the H_2 reduces the oxide to the metal.

$$NiO(s) + H_2(g) \longrightarrow Ni(s) + H_2O(g)$$

At a temperature of about 50°C, and at atmospheric pressure, the impure metal reacts with the residual CO to give the volatile compound tetracarbonylnickel(0), $Ni(CO)_4$. If this compound is passed into another part of the reactor, and is heated to 250°C, it reverts to Ni and CO. The newly formed nickel is pure.

$$Ni(s) + 4\ CO(g) \underset{250^\circ C}{\overset{50^\circ C}{\rightleftharpoons}} Ni(CO)_4$$

If you wish to produce one ton (908 kg) of pure nickel, how much nickel oxide, H_2, and CO are required?

LIGANDS AND FORMULAS OF COMPLEXES

17. Which of the following ligands are expected to be monodentate and which are multidentate?

(a) CH_3NH_2
(b) Br^-
(c) en
(d) N_3^-
(e) $C_2O_4^{2-}$
(f) $H_3C—C≡N$
(g) phen

18. Draw the electron dot structure of the glycinate ion, $H_2N—CH_2—COO^-$. Show how this ion can bind to a metal ion using an oxygen atom and a nitrogen atom as electron pair donors. Is this a monodentate or bidentate ligand? Can it be a chelating ligand?

19. Give the oxidation number of the metal ion in each of the following compounds.
(a) $[Ni(NH_3)_6]SO_4$
(b) $[Co(NH_3)_4Cl_2]Cl$
(c) $K_4[Fe(CN)_6]$
(d) $Ni(en)_2Cl_2$

20. Give the formula of a compound formed from one Ni^{2+} ion, one ethylenediamine ligand, three ammonia molecules, and one water molecule. Is the complex neutral or is it charged? If charged, give the charge.

NAMING

21. Write formulas for the following ions or compounds.
(a) dichlorobis(ethylenediamine)nickel(II)
(b) potassium tetrachloroplatinate(II)
(c) diamminetriaquohydroxochromium(II) nitrate
(d) hexaammineiron(III) nitrate
(e) pentacarbonyliron(0)

22. Name the following ligands:
(a) OH^- (b) O^{2-} (c) I^- (d) $C_2O_4^{2-}$

23. Name the compounds pictured in Figure 25.12.

24. Name the ions or compounds in Examples 25.1 and 25.2.
(a) $[Ni(C_2O_4)_2(H_2O)_2]^{2-}$
(b) $[Co(en)_2(NH_3)Cl]^{2+}$
(c) $[Co(en)_2(NO_2)_2]^+$
(d) $Pt(NH_3)_2(C_2O_4)$

25. Give the name or formula for each ion or compound, as appropriate.
(a) hydroxopentaaquoiron(III) ion
(b) $K_2[Ni(CN)_4]$
(c) $K[Cr(C_2O_4)_2(H_2O)_2]$
(d) $[Cr(NH_3)_5SO_4]Cl$
(e) sodium tetrachloropalladate(II)

ISOMERS

26. Draw the geometric isomers of
(a) $Pd(NH_3)_4Cl_2$
(b) $Pt(NH_3)_2(NCS)(Br)$
(c) $Co(NH_3)_3(NO_2)_3$

27. Which of the following complexes can have geometrical isomers? If isomers are possible, draw the structures of the isomers and label them as cis or trans.
(a) $[Co(H_2O)_4Cl_2]^+$
(b) $[Pt(NH_3)Br_3]^-$
(c) $Co(H_2O)_3F_3$
(d) $[Co(en)_2(NH_3)Cl]^{2+}$

28. Four isomers are possible for $[Co(en)(NH_3)_2(H_2O)Cl]^{2+}$. (Two of the four are optical isomers and so have a nonsuperimposable mirror image.) Draw the structures of the four isomers.

MAGNETISM OF COORDINATION COMPLEXES

29. What *d* electron configurations can exhibit both high and low spin in square planar complexes? (In Figure 25.25 notice that the largest energy gap occurs between the *xy* and $x^2–y^2$ orbitals.)

30. Write high and low spin configurations for each of the ions below. Tell if each is diamagnetic or paramagnetic. Give the number of unpaired electrons for the paramagnetic cases.
(a) $[Fe(CN)_6]^{4-}$ (b) $[Co(NH_3)_6]^{3+}$ (c) $[Fe(H_2O)_6]^{3+}$

31. From experiment we know that $[CoF_6]^{3-}$ is paramagnetic and $[Co(NH_3)_6]^{3+}$ is diamagnetic. Using the crystal field model, write electron configurations for each ion. What can you conclude about the effect of the ligand on the magnitude of Δ_o?

COLOR

32. Arrange the following ligands in order of increasing crystal field splitting.
(a) CN^- (b) NH_3 (c) F^- (d) H_2O

33. The titanium(III) ion in water, $[Ti(H_2O)_6]^{3+}$, is red. What color of the visible spectrum is absorbed by the ion? (Its broad absorption band occurs at about 500 nm.)

34. The chromium(II) ion in water, $[Cr(H_2O)_6]^{2+}$, absorbs light with a wavelength of about 715 nm. What color is the solution?

GENERAL QUESTIONS

35. From the position of the ligand in the spectrochemical series, decide whether each ion below is high or low spin and whether each is paramagnetic or diamagnetic. If paramagnetic, give the number of unpaired electrons. Use the crystal field model to write the electron configuration of each ion.
(a) $[Fe(CN)_6]^{4-}$
(b) $[Cr(en)_3]^{3+}$
(c) $[MnF_6]^{4-}$
(d) $[Cu(phen)_3]^{2+}$

36. In $Pt(C_2O_4)(NH_3)_2$ the metal ion is surrounded by a square plane of coordinating atoms. Draw a structure for this molecule. Give the oxidation number of the platinum and the name of the compound. Specify whether each ligand is monodentate or bidentate.

37. A complex formed from a cobalt(III) ion, five ammonia molecules, a bromide ion, and a sulfate ion exists in two forms, one dark violet (A) and the other violet-red (B). The dark violet form (A) gives a precipitate with $BaCl_2$ but none with silver nitrate. Form B behaves in the opposite manner. This tells you that one form is $[Co(NH_3)_5Br]SO_4$ and the other is $[Co(NH_3)_5SO_4]Br$. Which is A and which one is B? (*Note*: only when ions such as Br^- or SO_4^{2-} are *not* directly coordinated to the metal ion can they form free ions in aqueous solution.)

38. A platinum-containing compound, known as Magnus's Green Salt, has the formula $[Pt(NH_3)_4][PtCl_4]$. Name the compound.

39. Titanium metal is valued for its relatively low density, high strength, and corrosion resistance. In the Kroll method, it is made by starting with a titanium-containing ore and using carbon and chlorine to give $TiCl_4$. The latter is then reduced with magnesium to give titanium.

$$\underset{\text{ilmenite}}{2\ FeTiO_3(s)} + 7\ Cl_2(g) + 6\ C(s) \longrightarrow 2\ TiCl_4(\ell) + 2\ FeCl_3(s) + 6\ CO(g)$$

$$TiCl_4(\ell) + 2\ Mg(s) \longrightarrow Ti(s) + 2\ MgCl_2(s)$$

Is the first reaction an oxidation–reduction or an acid–base reaction? If you have 25.0 kg of ilmenite, how many kilograms of Cl_2 and carbon are required for complete reaction? How many kilograms of titanium metal can be produced ultimately from the ilmenite?

40. Give the formula of a complex ion formed from one Pt^{2+} ion, one nitrite ion (NO_2^-), one chloride ion, and two ammonia molecules. Are isomers possible? If so, draw the structure of each isomer and tell what type of isomerism is observed. Name the ion.

41. 0.213 g of uranyl(VI) nitrate, $UO_2(NO_3)_2$, was dissolved in 20.0 mL of 1.0 M H_2SO_4 and shaken with Zn. The zinc is a reducing agent and reduces the uranyl ion, UO_2^{2+}, to an ion with a lower oxidation number, U^{n+} ($n < 6$). The uranium-containing solution, after reduction, was titrated with 0.0173 M $KMnO_4$. The potassium permanganate is an oxidizing agent, which oxidizes the uranium back to the +6 oxidation state. 12.47 mL of the potassium permanganate were required for titration to a permanent pink color of the equivalence point. Calculate the oxidation number of the uranium after the uranyl(VI) nitrate was reduced with zinc. Write a balanced, net ionic equation for the oxidation of U^{n+} (where you now know n) with MnO_4^- to give UO_2^{2+} and Mn^{2+} in acid solution.

42. Comment on the fact that, while an aqueous solution of cobalt(III) sulfate is diamagnetic, the solution becomes paramagnetic when a large excess of fluoride ion is added. (*Hint:* In aqueous solution the metal ion is coordinated with water, while excess F^- ion displaces some or all of the water in favor of fluoride ion coordination.)

43. Experiment shows that the compound $K_4[Cr(CN)_6]$ is paramagnetic and has two unpaired electrons. In contrast, the related complex $K_4[Cr(SCN)_6]$ is paramagnetic to the extent of four unpaired electrons. Account for these differences using the crystal field model. Where does the SCN^- ion occur in the spectrochemical series relative to CN^-?

44. In this question, we wish to explore the differences between metal coordination by monodentate and bidentate ligands.

$$Ni^{2+}(aq) + 6\ NH_3(aq) \longrightarrow [Ni(NH_3)_6]^{2+}(aq) \qquad K_{formation} = 10^8$$

$$Ni^{2+}(aq) + 3\ H_2N{-}CH_2{-}CH_2{-}NH_2 \longrightarrow [Ni(H_2NCH_2CH_2NH_2)_3]^{2+} \qquad K_{formation} = 10^{18}$$

The vast difference in K_{form} between these complexes, reflecting a large increase in stability of the chelated complex, is called the *chelate effect*. Recall that K is related to the standard free energy of the reaction by

$$\Delta G° = -RT \ln K$$

and that

$$\Delta G° = \Delta H° - T\,\Delta S°$$

Here we know from experiment that $\Delta H°$ for the NH_3 reaction is −109 kJ/mol, and $\Delta H°$ for the en reaction is −117 kJ/mol. Is the difference in $\Delta H°$ sufficient to account for the 10^{10} difference in K_{form}? Comment on the role of entropy in the second reaction compared with that in NH_3 reaction.

45. You have a sample of an alloy of copper and aluminum and wish to determine the weight percentage of each element in the mixture. A 2.1309-g sample was first dissolved in a mixture of HCl and HNO_3. The resulting solution was made basic with excess ammonia, and the $Al(OH)_3$ that precipitated was collected and dried in a furnace to give 3.8249 g of Al_2O_3. What is the weight percentage of Al and Cu in the alloy?

46. Three different compounds of chromium(III) with water and chloride ion have the same composition: 19.44% Cr, 39.83% Cl, and 40.60% H_2O. One of the compounds is violet and dissolves in water to give a complex ion with a +3 charge and three Cl^- ions. All of the chloride ions precipitate immediately as AgCl on adding $AgNO_3$. Draw the structure of the complex ion and name the compound.

47. It is usually observed that, for analogous complexes, their stability is in the order $Mn^{2+} < Fe^{2+} < Co^{2+} < Ni^{2+} < Cu^{2+} > Zn^{2+}$. (This order of ions is called the *Irving–Williams* series.) Look up the values of K_{form} for ammonia complexes of Co^{2+}, Ni^{2+}, Cu^{2+}, and Zn^{2+} in Appendix J and verify this statement.

48. The Co^{3+} ion in water, $[Co(H_2O)_6]^{3+}$, is a powerful oxidizing agent (see Appendix K). Is the ion in fact stable in water? (You can decide this by looking up the value of $E°$ for the oxidation of H_2O to O_2.) What is the effect on the oxidizing ability of Co^{3+} of adding ammonia to form $[Co(NH_3)_6]^{3+}$? Is the latter ion capable of oxidizing water?

SPECIAL SECTION: Iron Chemistry

When $Fe(NO_3)_3 \cdot 9\ H_2O$, a purple solid, dissolves in water, iron(III) enters the solution as hexaaquoiron(III), $[Fe(H_2O)_6]^{3+}$. This complex ion is acidic (see Chapter 15) due to a hydrolysis reaction, the product of which is slightly yellow.

Beaker 1:

$$[Fe(H_2O)_6]^{3+}(aq) \longrightarrow [Fe(H_2O)_5(OH)]^{2+}(aq) + H^+(aq)$$

When a strong base is added, orange-brown iron(III) hydroxide precipitates (beaker 2).

Beaker 2:

$$[Fe(H_2O)_6]^{3+}(aq) + 3\ OH^-(aq) \longrightarrow Fe(H_2O)_3(OH)_3(s) + 3\ H_2O(\ell)$$

However, if a strong acid (HNO_3) is added to the hydroxide or to the solution containing $[Fe(H_2O)_5(OH)]^{2+}$, the hydroxide is neutralized and the hydrolysis reaction is reversed. The nearly colorless hexaaquoiron(III) ion predominates (beaker 3).

Addition of Cl^- or SCN^- to $[Fe(H_2O)_6]^{3+}$ leads to replacement of one or more water ligands to give new complex ions with characteristic colors.

Beaker 4:

$$[Fe(H_2O)_6]^{3+}(aq) + Cl^-(aq) \longrightarrow [Fe(H_2O)_5Cl]^{2+}(aq) + H_2O(\ell)$$

Beaker 5:

$$[Fe(H_2O)_6]^{3+}(aq) + SCN^-(aq) \longrightarrow [Fe(H_2O)_5SCN]^{2+}(aq) + H_2O(\ell)$$

Adding a large quantity of F^- ion causes displacement of the thiocyanate ion and gives the colorless pentaaquofluoroiron(III) ion.

Beaker 6a:

$$[Fe(H_2O)_5SCN]^{2+}(aq) \longrightarrow [Fe(H_2O)_5F]^{2+}(aq) + SCN^-(aq)$$

Finally, addition of a few drops of $[Fe(CN)_6]^{4-}$ to either $[Fe(H_2O)_5F]^{2+}$ or $[Fe(H_2O)_6]^{3+}$ gives a deep blue precipitate of "Prussian" or "Turnbull's blue," $Fe_4[Fe(CN)_6]_3$ (beaker 6b). In all of these reactions it is important to notice the effect of ligand on the color of the complex.

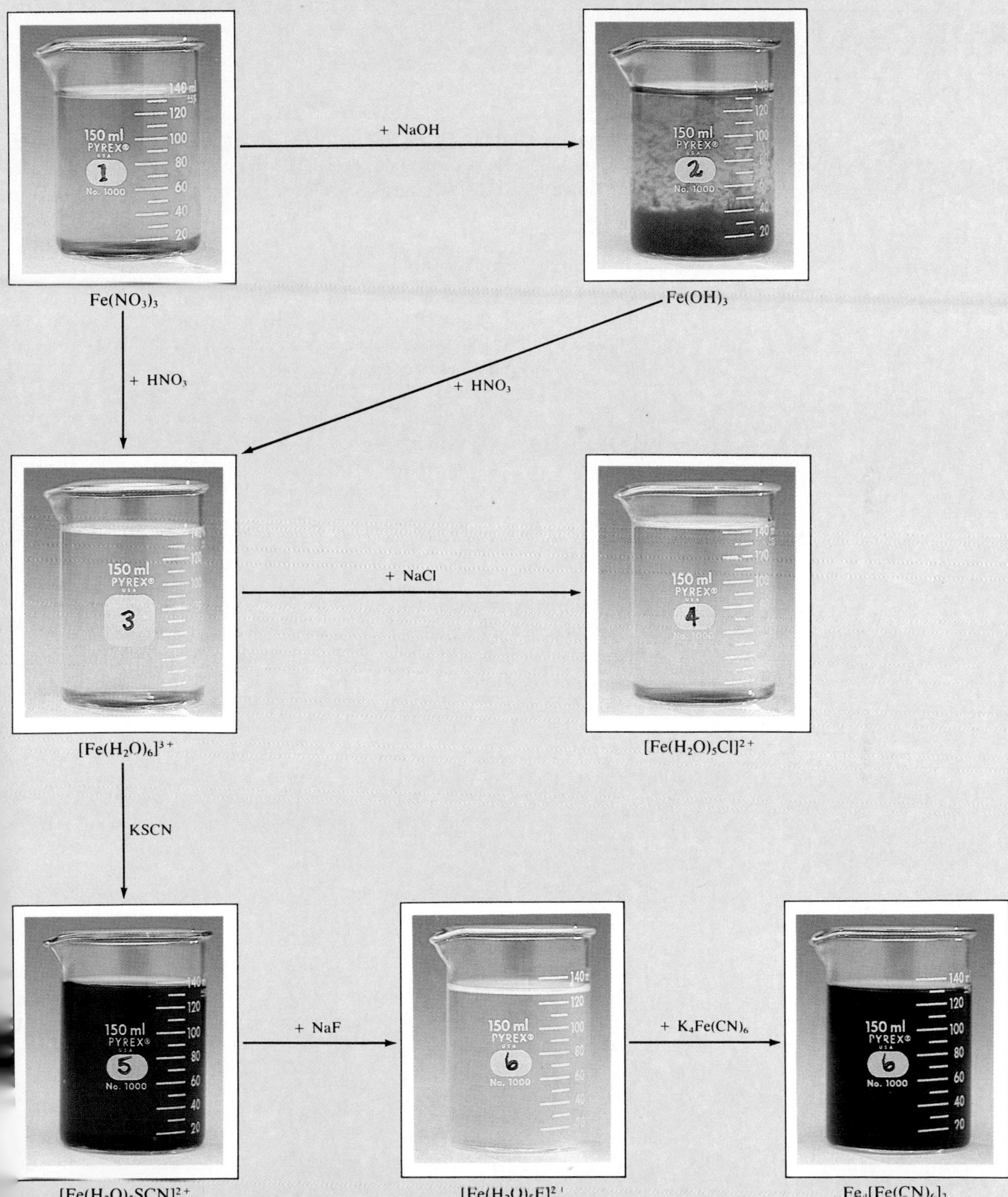
+ NaOH
$Fe(NO_3)_3$
$Fe(OH)_3$
+ HNO_3
+ HNO_3
+ NaCl
$[Fe(H_2O)_6]^{3+}$
$[Fe(H_2O)_5Cl]^{2+}$
KSCN
+ NaF
+ $K_4Fe(CN)_6$
$[Fe(H_2O)_5SCN]^{2+}$
$[Fe(H_2O)_5F]^{2+}$
$Fe_4[Fe(CN)_6]_3$

SPECIAL SECTION: Nickel Chemistry

When the green solid $NiSO_4 \cdot 6\ H_2O$ is dissolved in water a green solution of hexaaquonickel(II) results (beaker 1). If the monodentate ligand NH_3 is added, however, the blue complex ion hexaamminenickel(II) is formed.

Beaker 2:

$$\underset{\text{green}}{[Ni(H_2O)_6]^{2+}(aq)} + 6\ NH_3(aq) \longrightarrow \underset{\text{blue}}{[Ni(NH_3)_6]^{2+}(aq)} + 6\ H_2O(\ell)$$

Alternatively, if the bidentate ligand ethylenediamine is added to $[Ni(H_2O)_6]^{2+}$, pairs of water molecules are replaced successively to give complexes of varying shades of purple.

Beaker 3:

$$[Ni(H_2O)_6]^{2+}(aq) + H_2NC_2H_4NH_2(aq) \longrightarrow [Ni(H_2O)_4(H_2NC_2H_4NH_2)]^{2+}(aq) + 2\ H_2O(\ell)$$

Beaker 4:

$$[Ni(H_2O)_4(H_2NC_2H_4NH_2)]^{2+}(aq) + 2\ H_2NC_2H_4NH_2(aq) \longrightarrow [Ni(H_2NC_2H_4NH_2)_3]^{2+}(aq) + 4\ H_2O(\ell)$$

Finally, if dimethylglyoxime is added to the hexaaquonickel(II) ion, reaction produces a very insoluble strawberry red complex (beaker 5). (Notice that the solution is still green, indicating that some aqueous nickel(II) ion remains in solution).

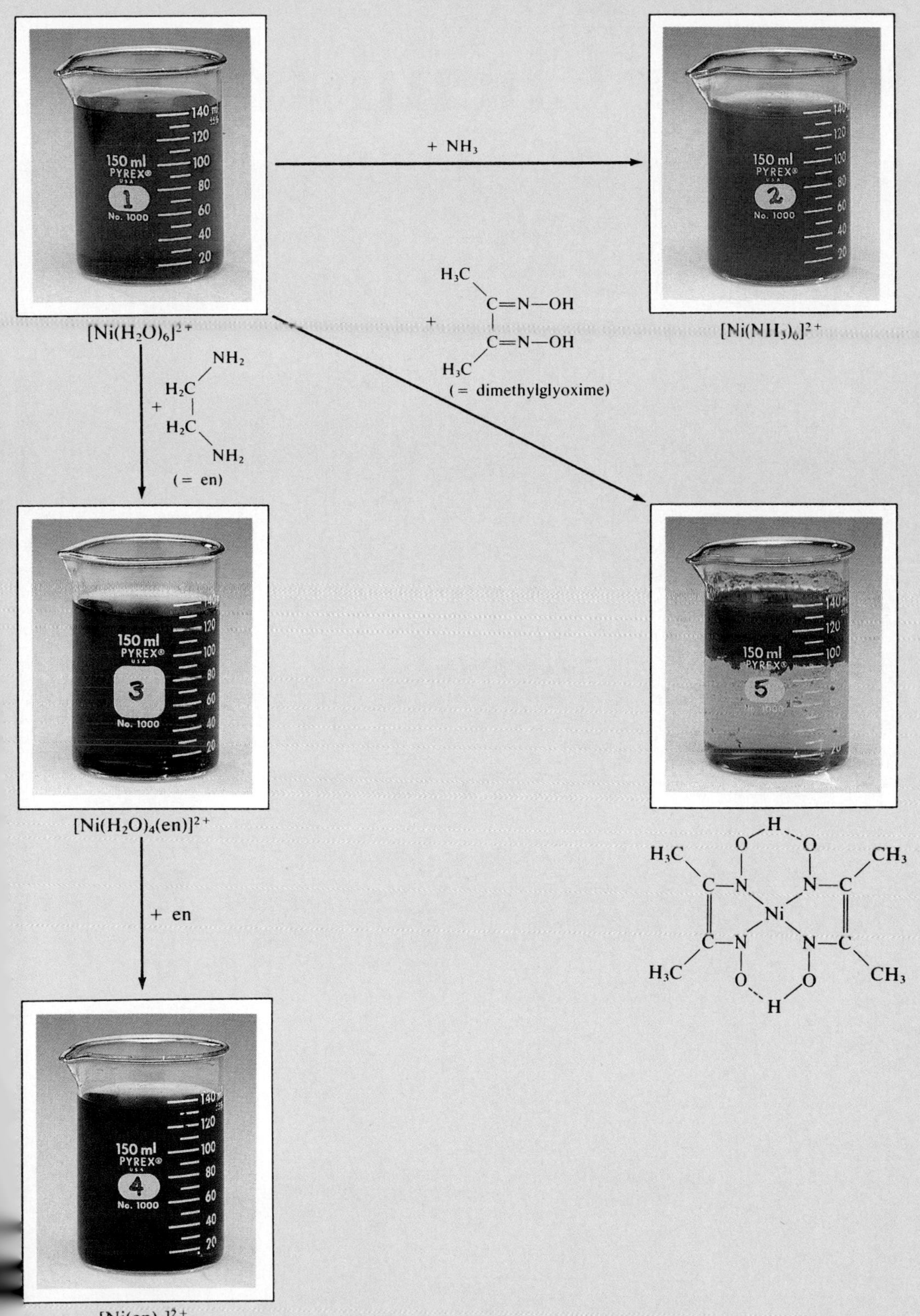

150 ml
PYREX
1
No. 1000
+ NH_3
150 ml
PYREX
2
No. 1000
$[Ni(H_2O)_6]^{2+}$
$[Ni(NH_3)_6]^{2+}$
+
H_3C
C=N—OH
C=N—OH
H_3C
(= dimethylglyoxime)
+
NH_2
H_2C
H_2C
NH_2
(= en)
150 ml
PYREX
3
No. 1000
150 ml
PYREX
5
$[Ni(H_2O)_4(en)]^{2+}$
+ en
H_3C
CH_3
Ni
H_3C
CH_3
150 ml
PYREX
4
No. 1000
$[Ni(en)_3]^{2+}$

26

Nuclear Chemistry

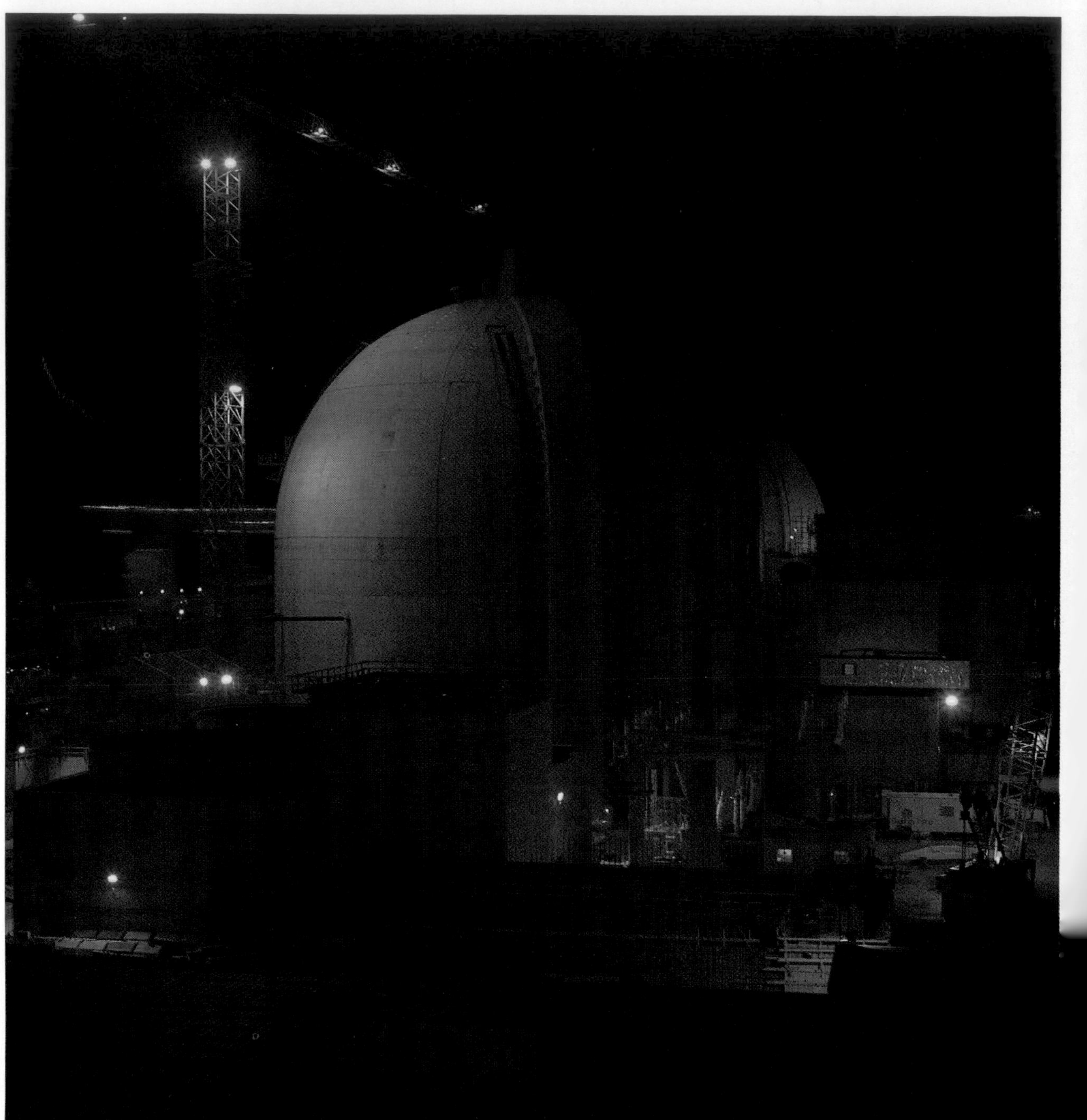

Reactor containment buildings of a nuclear power plant (Black Star)

CHAPTER OUTLINE

On August 2, 1939, as the world was on the brink of World War II, Albert Einstein sent a letter, which profoundly changed the course of history, to President Franklin D. Roosevelt. In this letter, Einstein called attention to work then being done on the physics of the atomic nucleus; he and others believed this work suggested the possibility that "uranium may be turned into a new and important source of energy . . . and [that it was] conceivable . . . that extremely powerful bombs of a new type may thus be constructed. . . ."

Powerful indeed! Einstein's letter was the beginning of the Manhattan Project, the project that led to the detonation of the first atomic bomb at 5:30 AM on July 16, 1945, in the desert of New Mexico. The rest of the world would learn the terrible truth of the power locked in the atomic nucleus a few weeks later, on August 6 and August 9, when the United States used atomic weapons against Japan.

As the physicist Freeman Dyson has put it, "the discovery of nuclear fission has gotten the world profoundly stuck: stuck in a buildup of nuclear weapons, stuck in outdated concepts of war and peace, and stuck in human nature. There are many benefits to come from this discovery, but we must find a way to get ourselves unstuck."

J. Robert Oppenheimer, the director of the atomic bomb project, is said to have recalled the following words from the sacred Hindu epic, Bhagavad–Gita, at the moment of the explosion of the first atomic bomb.

"If the radiance of a thousand suns
Were to burst at once upon the sky,
That would be like the splendor of
the Mighty One . . .
I am become Death,
The shatterer of worlds."

26.1 NATURE OF RADIOACTIVITY

Many minerals, called phosphors, glow for some time after exposure to sunlight or ultraviolet light. In 1896, Henri Becquerel was engaged in the study of this phenomenon, called phosphorescence, when he accidentally discovered radioactivity.

The glow of phosphorescence will darken a previously unexposed photographic film. Since Becquerel knew that x-rays, which had just been discovered, also produced a glow in phosphors, he set out to study a possible connection between phosphors and x-rays. One mineral that Becquerel knew to display phosphorescence was a uranium salt, so he placed the salt on a photographic film with the film wrapped in paper to prevent it from being exposed to sunlight. Much to his astonishment, he found that the phosphorescence emitted by the uranium salt *penetrated the paper*. This suggested to Becquerel the presence of x-radiation in the uranium radiation, and his initial assumption was that the x-radiation came from exposing the uranium salt to sunlight. To test this assumption, he

Marie and Pierre Curie. The Curies and H.A. Becquerel shared the Nobel prize in physics in 1903 for their research on radioactivity. Marie Curie received a second Nobel Prize, this in chemistry, in 1911 for the discovery of radium and polonium. The latter was named in honor of her homeland, Poland.

wrapped *both* the photographic plate *and* the uranium sample in paper. The uranium salt still led to darkening of the photographic plate. The sample clearly emitted some form of energy without absorption of light. This phenomenon was named **radioactivity**. Elements that naturally emit energy without the absorption of energy are said to be naturally radioactive.

On studying Becquerel's discovery further, it was observed that the mineral *pitchblende* exhibited greater activity than other uranium salts. Since this observation indicated the presence of a substance more radioactive than uranium, Pierre and Marie Curie carried out an analysis of the mineral. Using several tons of pitchblende, they isolated (in 1898) two new chemical elements: polonium, Po, 400 times more radioactive than uranium, and radium, Ra, a million times more radioactive than uranium.

The study of radioactivity excited many chemists and physicists around the turn of the century, and one who was involved was Ernest Rutherford, whom we first mentioned in Chapter 8. Using the type of apparatus shown in Figure 26.1, Rutherford and Paul Villard identified three types of radiation emitted by radioactive substances. They were labeled according to the first three letters of the Greek alphabet: **alpha**, $\boldsymbol{\alpha}$, **beta**, $\boldsymbol{\beta}$, and **gamma**, $\boldsymbol{\gamma}$. The alpha beam, attracted to the electrically negative plate, must be composed of positively charged particles, and further studies showed these particles to be *helium nuclei*, ${}^{4}_{2}He^{2+}$, ejected at high speeds from certain radioactive isotopes.

The β beam must be composed of negatively charged particles, since it is attracted to the electrically positive plate. Further study of this beam showed particles to have an electric charge and mass equal to those of an electron. Thus, *beta particles* are electrons ejected at high speeds from some radioactive nuclei.

Finally, the γ-ray beam is not affected by the electric field. The properties of γ-rays are similar to those of x-rays, except that γ-rays have higher energies than x-rays (see Figure 7.8).

The characteristics of the three types of radioactivity are summarized in Table 26.1. The heaviest particles, alpha particles, have limited penetrating power, as they can be stopped by several sheets of ordinary

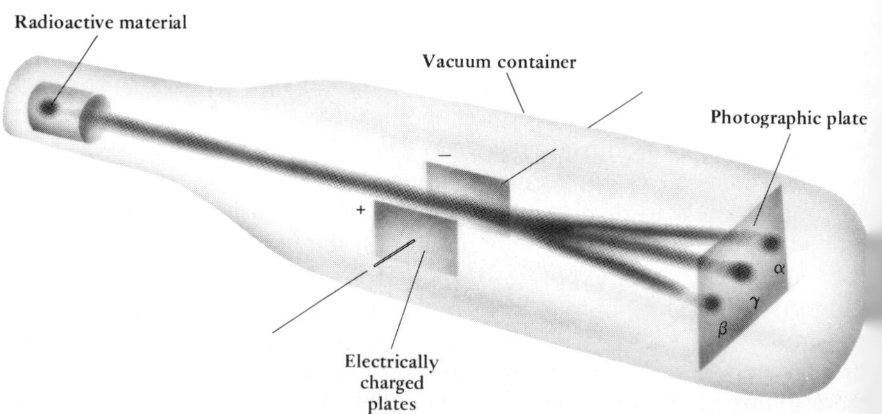

FIGURE 26.1

Behavior of α, β, and γ emissions in an electric field.

TABLE 26.1 Characteristics of α, β, and γ Emissions

NAME	SYMBOL	CHARGE	MASS (g/PARTICLE)	VELOCITY IN VACUUM (km/s)	RELATIVE ABILITY TO IONIZE AIR
Alpha	${}^{4}_{2}He^{2+}$, ${}^{4}_{2}\alpha$	+2	6.65×10^{-24}	$1.6–2.1 \times 10^{4}$	10,000
Beta	${}^{0}_{-1}e$, ${}^{0}_{-1}\beta$	−1	9.11×10^{-28}	2.7×10^{5}	100
Gamma	${}^{0}_{0}\gamma$, γ	0		3.0×10^{5}	1

paper or clothing. In contrast, beta particles are somewhat more penetrating than alpha particles; at least a $\frac{1}{8}$-inch piece of aluminum is necessary to stop beta particles. They will penetrate several millimeters of living tissue. Finally, gamma radiation is most penetrating, since it can pass completely through the human body. Thick layers of lead and concrete are required to minimize penetration.

All three types of radiation disrupt normal cellular processes in living organisms. Controlled exposure can be beneficial in destroying unwanted tissue, as in the radiation therapy used in treating some types of cancer. Low levels of x-radiation are used to photograph human tissue. However, the potential for serious radiation damage to humans is well known. The biological effects of the atomic bombs exploded at Hiroshima and Nagasaki, Japan, in 1945 have been well documented.

To quantify the biological effects of radiation, a unit called the *rem* (standing for *r*oentgen *e*quivalent *m*an) is used. It is defined as the amount of radiation required to produce a certain biological effect. Some effects of whole-body irradiation as a function of dose level are given in Table 26.2.

The level of radioactivity of a material is given in units of *curies*. One curie is the amount of material undergoing 3.7×10^{10} disintegrations per second.

Humans are constantly being exposed to natural and man-made background radiation (Figure 26.2). Sources include cosmic rays, radiation from naturally radioactive isotopes in rocks, the soil, water, air, and food, as well as radioactive isotopes naturally occurring in our bodies (^{40}K). In recent years, radioactive decay products from testing nuclear explosives in the atmosphere have added to the background. In addition, there are man-made sources of radiation such as x-ray generators, television, nuclear power plants, wastes from nuclear power plants and weapons manufacture, nuclear fuel reprocessing, and radioactive isotopes used for medical purposes.

For a discussion of the effects of radiation, and the history of the worst radioactive spill in North America, see *Science 84*, December, 1984, page 28.

Burning fossil fuels (coal and oil) releases naturally occurring radioactive isotopes into the atmosphere. This has added significantly to the background radiation in recent years.

TABLE 26.2 Effects of Whole-Body Radiation

DOSE LEVEL(rem)	EFFECTS
0–25	A dose around 25 rem may reduce the white blood cell count.
25–100	Nausea for about half those exposed, fatigue, changes in blood
100–200	Nausea, vomiting, fatigue, death possible, susceptible to infection (low white blood cell count)
200–400	Lethal dose for 50% of those exposed, especially in the absence of treatment. Bone marrow, spleen (blood-forming organs) damaged
> 600	Fatal, possibly even with treatment

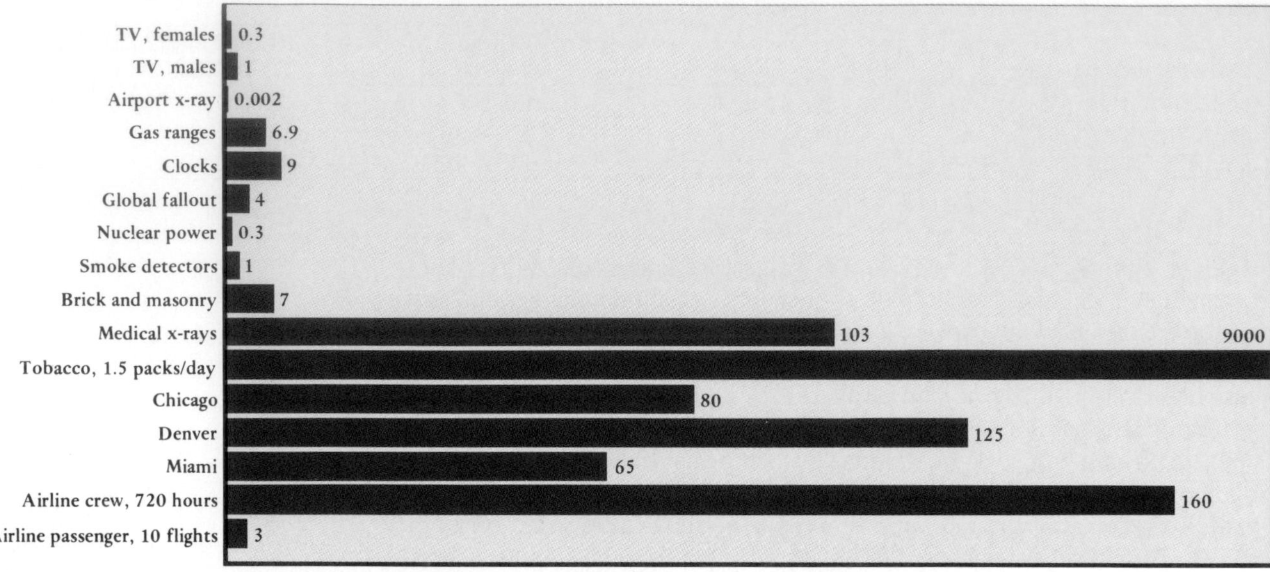

FIGURE 26.2
Everyday radiation.

26.2 NUCLEAR REACTIONS

Ernest Rutherford found that radium not only emits alpha particles but it produces the radioactive gas radon in the process. It was observations such as this that led Rutherford and Frederick Soddy, in 1902, to propose the theory that radioactivity is the result of a natural change of the isotope of one element into an isotope of a different element. Such changes, **nuclear reactions** or **transmutations**, generally involve a change in the atomic number and the mass number of the radioisotope. For example, the reaction studied by Rutherford can be written as

$$^{226}_{88}\text{Ra} \longrightarrow {}^{4}_{2}\text{He} + {}^{222}_{86}\text{Rn}$$

In this balanced equation the subscripts are the atomic numbers and the superscripts the mass numbers of the radioisotope or fragment.

The atoms in molecules and ions are rearranged in a chemical change; they are not created or destroyed. The number of atoms remains the same. Similarly in nuclear reactions the nuclear particles, or **nucleons**, can be rearranged but not created or destroyed. The essence of nuclear reactions is that one nucleon can change into a different nucleon. A proton and electron, for example, can change to a neutron or a neutron can change to a proton and electron, but the total number of nucleons remains the same. Therefore, *the sum of the mass numbers of reacting nuclei must equal the sum of the mass numbers of the nuclei produced*. Furthermore, to maintain electrical balance, *the sum of the atomic numbers of the products must equal the sum of the atomic numbers of the reactants*. These principles may be verified by the nuclear reaction above

$$^{226}_{88}\text{Ra} \longrightarrow {}^{4}_{2}\text{He} + {}^{222}_{86}\text{Rn}$$

$$\text{Mass number: } 226 \longrightarrow 4 + 222$$

$$\text{Atomic number: } 88 \longrightarrow 2 + 86$$

and the decay of an isotope of uranium to give an alpha particle and thorium.

$$^{234}_{92}\text{U} \longrightarrow {}^{4}_{2}\text{He} + {}^{230}_{90}\text{Th}$$

$$\text{Mass number: } 234 \longrightarrow 4 + 230$$

$$\text{Atomic number: } 92 \longrightarrow 2 + 90$$

EXERCISE 26.1 WRITING NUCLEAR REACTIONS

Write an equation showing the emission of an alpha particle by an isotope of neptunium, $^{227}_{93}\text{Np}$, to produce protactinium.

EXERCISE 26.2 NUCLEAR REACTIONS

Give the symbol, mass number, and atomic number of the product when each of the following isotopes transforms by alpha particle emission.

(a) $^{221}_{88}\text{Ra}$, (b) $^{257}_{103}\text{Lr}$, (c) $^{220}_{86}\text{Np}$

Emission of a beta particle is another way for a given isotope to be radioactive. For example, loss of a beta particle by uranium-235 is represented by

$$\underset{\text{uranium-235}}{^{235}_{92}\text{U}} \longrightarrow {}^{0}_{-1}\beta + \underset{\text{neptunium-235}}{^{235}_{93}\text{Np}}$$

Since a beta particle has a charge of −1, electrical balance makes the atomic number of the product greater than that of the reacting nucleus by one:

$$\text{U} \qquad e^- + \text{Np}$$

$$\text{Mass number } 235 \longrightarrow 0 + 235$$

$$\text{Atomic number } 92 \longrightarrow -1 + 93$$

How does a nucleus, composed only of protons and neutrons, eject an electron? It is generally accepted that a series of reactions is involved, but the net process is

$$\underset{\text{neutron}}{^{1}_{0}\text{n}} \longrightarrow \underset{\text{electron}}{^{0}_{-1}\beta} + \underset{\text{proton}}{^{1}_{1}\text{p}}$$

Notice that ejection of a beta particle always means that a new element is formed with an atomic number one unit greater than the present nucleus.

EXERCISE 26.3 BETA PARTICLE EMISSION

Write a balanced equation showing the emission of a beta particle by sulfur-35, $^{35}_{16}\text{S}$, to produce an isotope of chlorine.

EXERCISE 26.4 BETA PARTICLE EMISSION

Predict the product formed when each of the following radioactive isotopes emits a beta particle:

(a) $^{3}_{1}\text{H}$ (b) $^{60}_{26}\text{Fe}$ (c) $^{129}_{53}\text{I}$

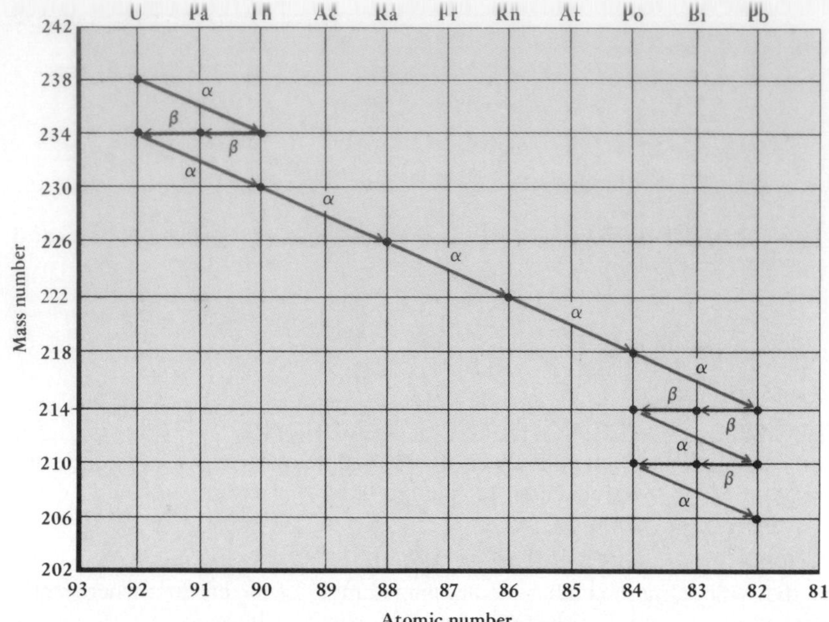

FIGURE 26.3

$^{238}_{92}U$ radioactive series. The radioisotopes emit either alpha or beta particles. $^{206}_{82}Pb$ is the "end product." (A few elements may emit both alpha and beta particles, but these are not shown for simplicity.)

In many cases, the emission of an alpha or beta particle results in the formation of an isotope that is also radioactive.* The new radioactive isotope may therefore undergo a number of successive transformations until a stable, nonradioactive isotope is finally produced. Such a series of reactions is called a **radioactive series**. One such series that begins with uranium-238 and ends with lead-206 is illustrated in Figure 26.3. The first step in the series is

$$^{238}_{92}U \longrightarrow {}^{4}_{2}He + {}^{234}_{90}Th$$

and the equation for the final step, the conversion of polonium-210 to lead-206, is

$$^{210}_{84}Po \longrightarrow {}^{206}_{82}Pb + {}^{4}_{2}He$$

EXERCISE 26.5 A RADIOACTIVE SERIES

The actinium series begins with uranium-235, $^{235}_{92}U$, and ends with lead-207, $^{207}_{82}Pb$. The first five steps in the series are α, β, α, α, β. Identify the radioactive isotope produced in each of these steps beginning with $^{235}_{92}U$.

In addition to radioactive decay by alpha, beta, or gamma radiation, other decay processes are observed. Some nuclei decay, for example, by emission of a **positron**, $_{+1}^{0}\beta$, or what is effectively a positively charged electron.† Positron emission by polonium-207 leads to the formation of bismuth-207.

---

*A nucleus formed as the result of an alpha or beta particle emission is generally in an excited state and so emits a γ-ray (photon). Photon emission by an excited state of the nucleus is analogous to the emission of a photon by an excited hydrogen atom (page 228).
†The positron was discovered in 1932 by Carl Anderson. It is sometimes called an "antielectron," one of a group of particles that have become known as "antimatter." "Matter" and "antimatter" have identical masses but opposite charge signs. Contact between a particle and its antiparticle always leads to mutual annihilation of both particles with production of high energy photons (γ-rays).

$$^{207}_{84}Po \longrightarrow {}^{0}_{+1}\beta + {}^{207}_{83}Bi$$

This process leads to *reduction* of the atomic number, as does a process called **K-capture**. This is the capture of an electron, usually from the lowest atomic energy level (the K or 1*s* level), by the nucleus.

$$^{7}_{4}Be + e^{-}(\text{orbital electron}) \longrightarrow {}^{7}_{3}Li$$

$$^{40}_{19}K + e^{-}(\text{orbital electron}) \longrightarrow {}^{40}_{18}Ar$$

EXERCISE 26.6 NUCLEAR REACTIONS

Balance the following nuclear reactions. Indicate the symbol, the mass number, and the atomic number of "?."

(a) $^{13}_{7}N \longrightarrow {}^{13}_{6}C + ?$

(b) $^{41}_{20}Ca + e^{-}(\text{orbital electron}) \longrightarrow ?$

(c) $^{90}_{38}Sr \longrightarrow {}^{90}_{39}Y + ?$

(d) $^{11}_{6}C \longrightarrow {}^{11}_{5}B + ?$

(e) $^{43}_{21}Sc \longrightarrow ? + {}^{1}_{1}H$

26.3 STABILITY OF ATOMIC NUCLEI

The fact that some nuclei are stable (nonradioactive), while others are unstable (radioactive), leads to a consideration of the factors that impart stability to the nucleus. For some insight into this question, it is interesting to ask first why there are not many more isotopes of the known elements.

WHY AREN'T THERE 15,862 ISOTOPES OF THE KNOWN ELEMENTS?

In its simplest and most abundant form, hydrogen has only one nuclear particle, the proton. The element has two well known isotopes: deuterium, with one proton and one neutron, and tritium (an unstable isotope), with one proton and two neutrons (Figure 26.4). Helium, the next element, has four nuclear particles, two protons and two neutrons, in its more stable form. Jumping to the end of the actinide series you come to element number 103, lawrencium. One isotope has an atomic weight of 257 and so has 154 neutrons. As you travel along the periodic table between He and Lr, you should observe that the atomic weight of an element (except for H) is always *at least* twice as large as the atomic number. The reason for this is that, except for hydrogen, every isotope of every element has a nucleus containing *at least one neutron for every proton*. Apparently the tremendous repulsive forces between protons in the nucleus are moderated by the presence of neutrons. The atomic nucleus is believed to have a regular structure similar to the "shell" arrangement of electrons. In fact, nuclear structure reveals what has come to be called "magic numbers." There are nuclear shells having a capacity of 2, 8, 20, 28, 50,

Except for H and ^{3}He the nucleus of every element contains at least one neutron for every proton.

$^{1}_{1}H$

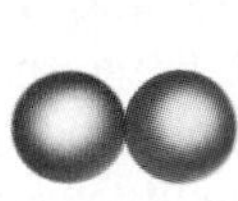

$^{2}_{1}H$ = D = Deuterium

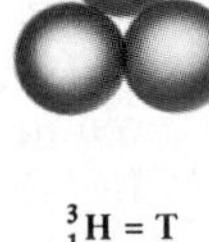

$^{3}_{1}H$ = T = Tritium

$^{3}_{2}He$

$^{4}_{2}He$

FIGURE 26.4

Isotopes of hydrogen and helium. The proton is a blue circle and the neutron is a red circle.

82, and 126 nucleons of one type (protons or neutrons), and a nucleus with a magic number of *either* protons *or* neutrons has filled shells and is generally likely to be stable. Nuclei with magic numbers of *both* protons and neutrons will be *especially* stable, and some examples are given in the table in the margin. There also seem to be other magic numbers corresponding to filled nuclear subshells of 6 (carbon), 14 (silicon), 16 (sulfur), and 26 (iron) protons *or* neutrons. As you know, it is just these elements that play such a crucial role in the universe as we know it.

ELEMENT	PROTONS (NO.)	NEUTRONS (NO.)
$^{4}_{2}He$	2	2
$^{16}_{8}O$	8	8
$^{40}_{20}Ca$	20	20
$^{208}_{82}Pb$	82	126

More insight on what makes the universe "tick" can be obtained from a plot of the atomic number of each element against the number of neutrons in its naturally occurring isotopes. Such a plot for the naturally occurring elements through bismuth ($Z = 83$), the heaviest stable isotope of which has 126 neutrons, is shown in Figure 26.5. In this plot there is a total of $83 \times 126 = 10{,}458$ possible combinations of protons and neutrons. (Considering elements to $Z = 103$ and $N = 154$, that is, to Lr, there would be 15,862 combinations.) However, it is clear from the information in Figure 26.5 that *there are only a very few stable combinations of protons and neutrons.*

There are several further points of interest shown in Figure 26.5.

(a) For the light elements up to calcium ($Z = 20$), the stable isotopes usually have an equal number of protons and neutrons. Examples include $^{12}_{6}C$, $^{16}_{8}O$, and $^{32}_{16}S$. Where this is not true, the stable isotope contains only one more neutron than the number of protons ($^{7}_{3}Li$ and $^{19}_{9}F$, for example).

(b) Beyond calcium there is an increasing surplus of neutrons. For the heavier elements, the repulsion between nuclear protons is larger, so more neutrons are needed for nuclear stability. For example, whereas one stable isotope of Fe has 26 protons and 30 neutrons, one of the stable isotopes of platinum has 78 protons and 117 neutrons.

(c) Above bismuth with 83 protons and 126 neutrons, all isotopes are radioactive. Beyond this point there is apparently no nuclear "super glue" sufficient to hold heavy nuclei together, and fission or splitting of the nucleus into smaller particles occurs. Furthermore, the rate of this nuclear splitting becomes greater the heavier the nucleus. For example, while half of a sample of $^{238}_{92}U$ disintegrates in 10^9 years, hal of $^{257}_{103}Lr$ is gone in 8 seconds.

(d) Included in Figure 26.5 is a portion of the curve (from Ca through Kr that is enlarged. This more detailed view shows some interesting fea tures. One is that elements of even atomic number have more isotope than those of odd atomic number. Furthermore, stable isotopes gen erally have even numbers of neutrons. For elements of odd atomi number, this means that the most stable isotope has an even numbe of neutrons. To emphasize this point, it is interesting to find that, c the more than 300 stable isotopes, roughly 200 have an even numbe of neutrons and an even number of protons. Only about 120 have a odd number of protons and an even number of neutrons, or vice vers Only 4 stable isotopes ($^{2}_{1}H$, $^{6}_{3}Li$, $^{10}_{5}B$, $^{14}_{7}N$) have both an odd number protons and neutrons! It is from observations of this type that t notion of magic numbers of nucleons arose.

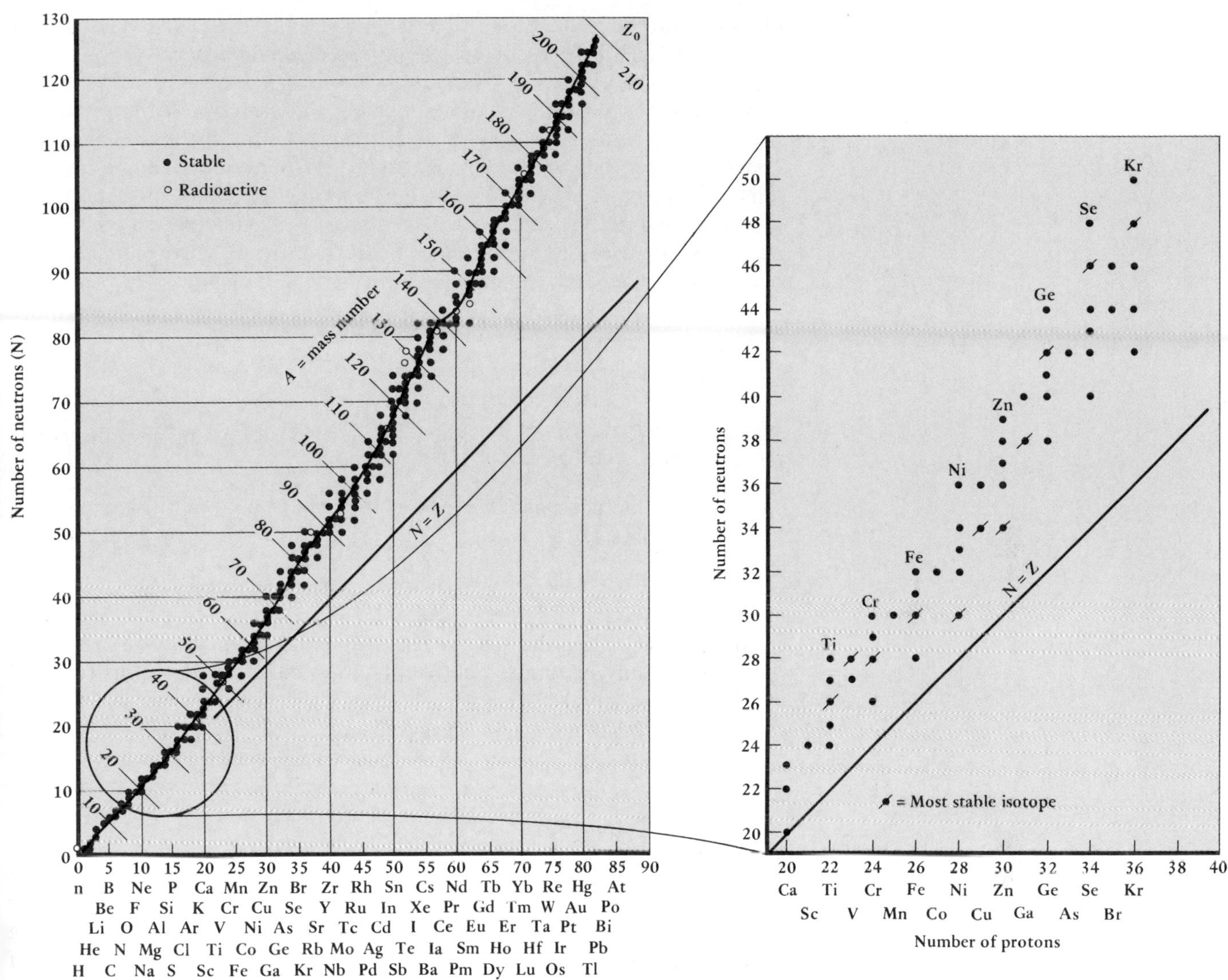

FIGURE 26.5
The naturally occurring isotopes of the elements from H (Z = 1) to Bi (Z = 83).

It is clear that, as protons and neutrons accumulate to form one element after another, a large number of isotopes of each element will not arise or are not stable. Only a limited number of isotopes of each element is possible.

BINDING ENERGY

In Chapter 9 you learned that the energy of a bond in a molecule is given by the energy required to separate a given pair of bonded atoms into gaseous atoms. Similarly, the energy required to separate a nucleus into its individual nucleons, or the energy evolved when these nucleons combine to form the nucleus, is the **binding energy** of a nucleus. The greater the binding energy of a nucleus, the greater its stability.

A measure of nuclear stability is the energy evolved when nucleons combine to form a nucleus. When nucleons coalesce to form a nucleus, the mass of the nucleus is *always* less than the sum of the masses of the constituent protons and neutrons. Since Einstein's equation, $E = mc^2$, tells us that mass can be converted to energy, it is this "missing mass" or its energy equivalent that is a measure of the binding energy. As an example, consider the formation of deuterium from hydrogen and a neutron: ${}^1_1H + {}^1_0n \rightarrow {}^2_1H$. The masses are

$$\begin{aligned} {}^1_1H &= 1.007825 \text{ g/mol} \\ {}^1_0n &= \underline{1.008665 \text{ g/mol}} \\ \text{Sum} &= 2.016490 \text{ g/mol} \end{aligned}$$

Since the actual mass of deuterium, ${}^2_1H$, is 2.01410 g/mol, the difference in mass, Δm, is

$$\begin{aligned} \Delta m &= \text{actual mass of product} - \text{sum of masses of reactants} \\ &= 2.01410 \text{ g/mol} - 2.016490 \text{ g/mol} \\ &= -0.00239 \text{ g/mol} \end{aligned}$$

The energy liberated can be calculated from $\Delta E = (\Delta m)c^2$ where Δm must be given in kg and c in m/s. (Recall that 1 J = 1 kg · m^2/s^2; Appendix D.)

$$\Delta E = \left(-2.39 \times 10^{-6} \frac{\text{kg}}{\text{mol}}\right)\left(3.00 \times 10^8 \frac{\text{m}}{\text{s}}\right)^2$$

$$\Delta E = -2.15 \times 10^{11} \frac{\text{J}}{\text{mol}}$$

For comparison, you find in Appendix L that ΔE for 2 H(g) $\rightarrow$ H_2(g) is only -435 kJ/mol.

The binding energy of other nuclei can be calculated in the same manner, and it is evident that the greater the energy released per nucleon, the greater the stability of the nucleus. In fact, if this is taken as a measure of relative nuclear stability, and it is plotted as a function of mass number, Figure 26.6 is obtained. The point of maximum stability occurs in the vicinity of ${}^{56}_{26}Fe$.

There are two interesting points to be made regarding the observation of the stability of iron. First, refer to Figure 8.18 and notice that although there is a decline in the relative abundance of the elements from carbon to lead, there is a pronounced maximum in the curve around iron. The binding energy explains this fact. Second, in terms of binding energy all other nuclei are unstable compared to the nuclei near iron. This means that very heavy nuclei may split or **fission**, with the release of tremendous quantities of energy, to give stable nuclei with atomic numbers near iron. Also, very light nuclei may undergo **fusion** exothermically to form heavier nuclei. Very heavy or very light nuclei are thermodynamically unstable with respect to nuclei in the neighborhood of ${}^{56}_{26}Fe$.

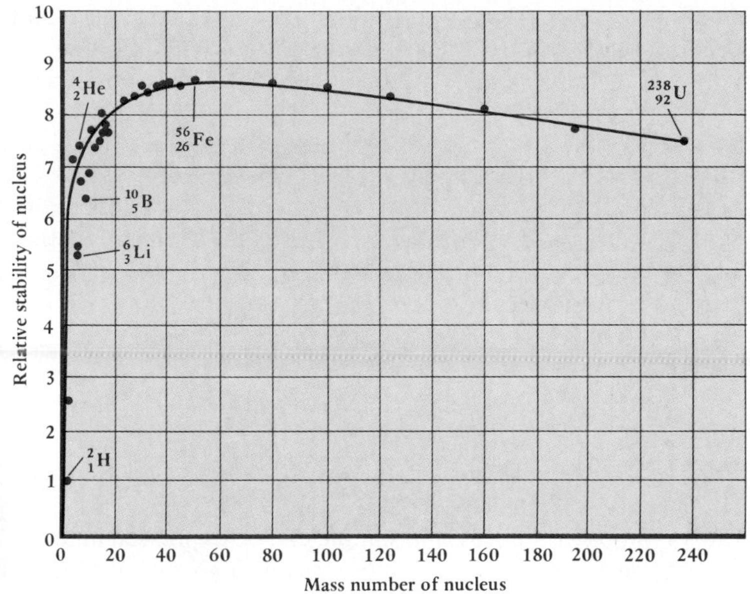

FIGURE 26.6

The relative stability of nuclei. (Based on the energy evolved in the formation of the nucleus divided by the number of nucleons.)

EXERCISE 26.7 BINDING ENERGY

Calculate the binding energy, in kJ/mol, for the formation of lithium-6.

$$3\ ^1_1\text{H} + 3\ ^1_0\text{n} \longrightarrow\ ^6_3\text{Li}$$

The necessary masses are $^1_1$H = 1.00783, $^1_0$n = 1.00867 g/mol, and $^6_3$Li = 6.01690 g/mol.

26.4 RATES OF DISINTEGRATION REACTIONS

You have just seen that radioactive nuclei have different stabilities and thus disintegrate at different rates. The lower the stability, the faster the disintegration rate. In this section, we wish to describe this relation in more detail and the uses that can be made of disintegration rates.

RATE OF RADIOACTIVE DECAY

The rate of nuclear decay or disintegration is often described in terms of **activity** (A), the number of disintegrations per unit time, a quantity proportional to the number of atoms present (N).

$$\text{Activity } (A) \propto \text{number of atoms present } (N)$$

The activity of a sample can be measured using a Geiger counter (Figure 26.7), and a measure of the number of radioactive atoms present is provided.

A chemical reaction where the rate is directly proportional to the quantity of material present is a *first order reaction*, so radioactive decay

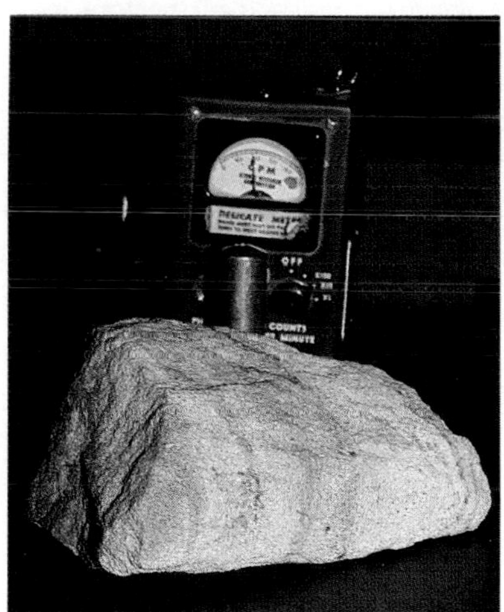

FIGURE 26.7

A Geiger counter with a sample of carnotite, a mineral containing uranium oxide. The Geiger counter was invented by Hans Geiger and Ernest Rutherford in 1908. A charged particle (such as a beta or alpha particle), when entering a gas-filled tube, ionizes the gas. These ions are attracted to electrically charged plates, giving rise to a "pulse," a momentary flow of electric current. The current is amplified and used to operate a counter. (Dave Davidson, Tom Stack & Associates)

For a review of the principles of chemical kinetics, and first order reactions in particular, see Chapter 13.

processes are all first order with respect to the radioactive substance, and the language and methods of chemical kinetics can be readily applied. Thus, the relation between activity and number of atoms present can be expressed as

$$A = kN$$

where k is the rate constant or *decay constant*. Since A = (change in number of atoms)/(change in time) = $\Delta N/\Delta t$, we can write an equation relating the number of atoms and time as

$$\ln\left(\frac{N}{N_0}\right) = -kt$$

If we express the equation in words, we have

$$\ln\left(\frac{\text{quantity after some elapsed time} = t}{\text{quantity at time} = 0}\right) = -(\text{rate constant})(\text{time}, t)$$

The rate of radioactive decay of a sample can be expressed in terms of the decay constant k. However, it is more usual to give it in terms of the **half-life**, $t_{1/2}$, the time required for one half of the sample to disintegrate. The faster the rate of decay, the larger the rate constant, and the shorter the half-life.

For nuclear disintegration reactions, the half-life is a constant, independent of temperature and of the amount of reactant or nuclei present.

As an example of the concept of half-life in radioactive decay, consider the disintegration of oxygen-15, $^{15}_{8}O$, by positron emission.

$$^{15}_{8}O \longrightarrow {}^{15}_{7}N + {}^{0}_{+1}\beta$$

The half-life of $^{15}_{8}O$ is 2.0 minutes. This means that one-half the quantity of $^{15}_{8}O$ present at any given time will disintegrate every 2.0 minutes. Thus, if we begin with 20 mg of $^{15}_{8}O$, 10 mg of the isotope remain after 2.0 minutes, 5.0 mg remain after 4.0 minutes (two half-lives), 2.5 mg remain after 6.0 minutes (three half-lives), and so on. The amounts of $^{15}_{8}O$ present at various times are illustrated in Figure 26.8 and Table 26.3. Half-lives of some other radioisotopes are given in Table 26.4.

EXAMPLE 26.1

DISINTEGRATION RATE AND HALF-LIFE

Tritium ($^{3}_{1}H$), a radioactive isotope of hydrogen, has a half-life of 12.3 years.

$$^{3}_{1}H \longrightarrow {}^{0}_{-1}\beta + {}^{3}_{2}He$$

If you begin with 1.5 mg of the isotope, how many milligrams remain after 49.2 years?

Solution First, we find the number of half-lives in the given time period of 49.2 years. Since the half-life is 12.3 years, the number of half-lives is

$$49.2 \text{ years}\left(\frac{1 \text{ half-life}}{12.3 \text{ years}}\right) = 4.00 \text{ half-lives}$$

This means that the initial quantity of 1.5 mg is reduced four times by $\frac{1}{2}$.

$$1.5 \times \tfrac{1}{2} \times \tfrac{1}{2} \times \tfrac{1}{2} \times \tfrac{1}{2} = 1.5 \times (\tfrac{1}{2})^4 = 1.5 \times \tfrac{1}{16} = 0.094 \text{ mg}$$

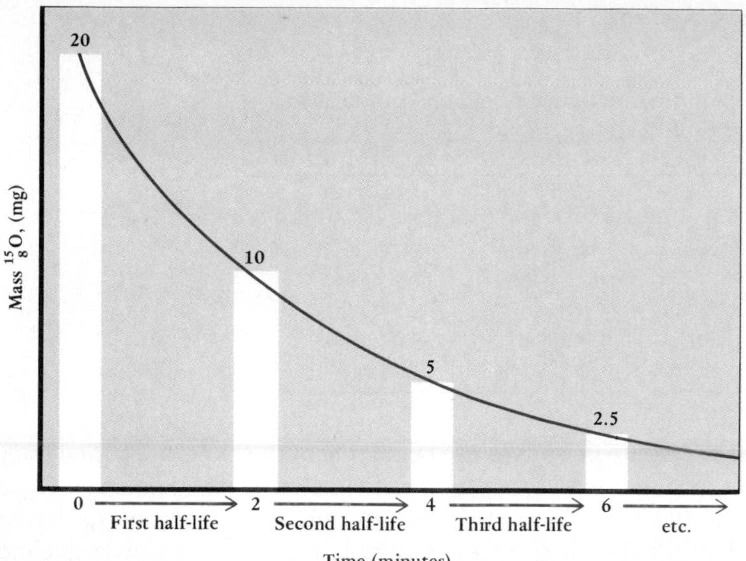

FIGURE 26.8

Decay of 20 mg of $^{15}_{8}O$. After each half-life period, the quantity present is reduced by half.

EXERCISE 26.8 RADIOACTIVITY AND HALF-LIVES

Strontium-90 ($^{90}_{38}Sr$) is a radioisotope ($t_{1/2}$ = 28 years) produced in atomic bomb explosions. Its long life and tendency to concentrate in bone marrow make it particularly dangerous to man and animals.

(a) If the isotope decays with loss of a beta particle, write a balanced equation showing the other product of the decay.

(b) A sample of $^{90}_{38}Sr$ emits 2000 beta particles per minute. How many half-lives and how many years are necessary to reduce the emission to 125 beta particles per minute?

The half-life is the time for half of the material present at $t = 0$ to have disappeared. Thus, time = $t_{1/2}$ when $N = \frac{1}{2}N_0$.

$$\ln\left[\frac{\frac{1}{2}N_0}{N_0}\right] = -kt_{1/2}$$

$$\ln \tfrac{1}{2} = -kt_{1/2} \qquad \text{or} \qquad \ln \tfrac{1}{2} = -\ln 2 = -kt_{1/2}$$

and the half-life and rate constant are related by the natural logarithm of 2.

$$t_{1/2} = \frac{0.693}{k}$$

In practical terms, this means that you can measure the activity or disintegrations over some elapsed time for some radioisotope. This gives

TABLE 26.3 Effect of the Number of Half-Lives on Initial Quantity of a Sample

NUMBER OF HALF-LIVES	FRACTION OF INITIAL QUANTITY REMAINING	QUANTITY REMAINING (mg)
	1	20.0 (*initial*)
	1/2	10.0
	1/4	5.00
	1/8	2.50
	1/16	1.25
	1/32	0.625

TABLE 26.4 Half-Lives of Some Common Radioactive Isotopes

ISOTOPE	DECAY PROCESS	HALF-LIFE ($t_{1/2}$)
$^{238}_{92}U$	$^{238}_{92}U \longrightarrow ^{234}_{90}Th + ^{4}_{2}He$	4.47×10^9 years
$^{3}_{1}H$	$^{3}_{1}H \longrightarrow ^{3}_{2}He + ^{0}_{-1}\beta$	12.3 years
$^{14}_{6}C$	$^{14}_{6}C \longrightarrow ^{14}_{7}N + ^{0}_{-1}\beta$	5.73×10^3 years
$^{32}_{15}P$	$^{32}_{15}P \longrightarrow ^{32}_{16}S + ^{0}_{-1}\beta$	14.3 days
$^{60}_{27}Co$	$^{60}_{27}Co \longrightarrow ^{60}_{28}Ni + ^{0}_{-1}\beta$	5.27 years
$^{131}_{53}I$	$^{131}_{53}I \longrightarrow ^{131}_{54}Xe + ^{0}_{-1}\beta$	8.04 days
$^{239}_{94}Pu$	$^{239}_{94}Pu \longrightarrow ^{235}_{92}U + ^{4}_{2}He$	2.44×10^4 years

you N_0 and N (disintegrations at the beginning of the experiment and after some time t) and the time. From this, you can derive k and thus $t_{1/2}$. Once $t_{1/2}$ is known, it may be used with the general rate expression to derive such useful information as the time required for a given isotope to decline to some level of radioactivity.

EXAMPLE 26.2

DETERMINATION OF HALF-LIFE

A sample of radon initially emitted 7.0×10^4 alpha particle disintegrations per second (dps). After 6.6 days, it emitted only 2.1×10^4 alpha particle dps. What is the half-life of this isotope of radon?

Solution Experiment has provided us with both N and N_0.

$$N = 2.1 \times 10^4 \text{ dps} \qquad N_0 = 7.0 \times 10^4 \text{ dps}$$

and the time (t = 6.6 days). Therefore, we can find the value of k.

$$\ln\left(\frac{2.1 \times 10^4}{7.0 \times 10^4}\right) = -k(6.6 \text{ days})$$

$$\ln(0.30) = -k(6.6 \text{ days})$$

$$k = -\frac{\ln(0.30)}{6.6 \text{ days}} = -\frac{(-1.20)}{6.6 \text{ days}} = 0.18 \text{ day}^{-1}$$

and from k we can obtain $t_{1/2}$.

$$t_{1/2} = \frac{0.693}{k} = \frac{0.693}{0.18 \text{ day}^{-1}} = 3.8 \text{ days}$$

EXAMPLE 26.3

TIME AND RADIOACTIVITY

Some high-level radioactive waste with a half-life, $t_{1/2}$, of 200. years is stored i underground tanks. What time is required to reduce an activity of 6.50×10 disintegrations per minute (dpm) to a fairly harmless activity of 3.00×10^{-3} dpm

Solution The data give you the initial quantity ($N_0 = 6.50 \times 10^{12}$ dpm) and th quantity after some elapsed time ($N = 3.00 \times 10^{-3}$ dpm). In order to find th elapsed time t, you must first find k from the half-life.

$$k = \frac{0.693}{t_{1/2}} = \frac{0.693}{200.\ \text{years}} = 0.00347\ \text{yr}^{-1}$$

With k known, the time t can be calculated.

$$\ln\left(\frac{3.00 \times 10^{-3}}{6.50 \times 10^{12}}\right) = -\ [0.00347\ \text{yr}^{-1}]\ t$$

$$-35.31 = -\ [0.00347\ \text{yr}^{-1}]\ t$$

$$t = \frac{-35.31}{-(0.00347)\ \text{yr}^{-1}}$$

$$t = 1.02 \times 10^4\ \text{years}$$

EXERCISE 26.9 RATE OF RADIOACTIVE DECAY

Gallium citrate, containing the radioactive isotope gallium-67, is used medically as a tumor-seeking agent. It has a half-life of 77.9 hours. How much time must elapse for a sample of gallium citrate to decay to 10.% of its original activity?

RADIOCHEMICAL DATING

Radiochemical dating is the determination of the age of rocks and fossils that date back to the origin of humans and the development of life in the very distant past. The significance of fossils, tools, and other artifacts of our ancestors discovered at several sites in Africa and Peru, for example, could not be realized until the ages of those objects were determined (Figure 26.9).

Willard Libby (1950) developed the technique of age determination using radioactive carbon-14 ($^{14}_{6}C$). Natural carbon contains small amounts of this isotope. The important observation about carbon-14 is that its activity in living plants and animals and in the air is approximately constant at about 14 disintegrations per minute per gram of carbon. We *assume* that this activity was about the same in ancient times as it is now. The activity remains fairly constant, because the activity lost through disintegration, $^{14}_{6}C \rightarrow {}^{0}_{-1}\beta + {}^{14}_{7}N$ ($t_{1/2} = 5.73 \times 10^3$ years), is balanced by the production of carbon-14 through the action of neutrons with $^{14}_{7}N$ in the atmosphere, $^{14}_{7}N + {}^{1}_{0}n \rightarrow {}^{14}_{6}C + {}^{1}_{1}H$. Plants absorb carbon dioxide from the atmosphere, converting it to food, and so incorporate the ^{14}C into living tissue. As long as the plant is alive, the carbon-14 activity remains constant. However, when the plant dies or is ingested by an animal, carbon-14 disintegration continues *without being replaced*; consequently, the activity decreases with passage of time. The smaller the activity the longer the period between the death of the plant and the present.

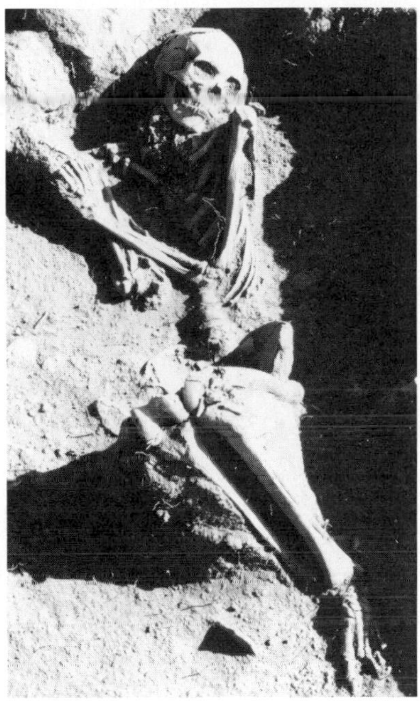

FIGURE 26.9

Radioisotopes can be used to date and study fossils. The skeleton of this 11th-century inhabitant of a South African village was puzzling to anthropologists. The man's skeleton was different from others found in the region, suggesting that he was not native to the area. However, when the skeleton was analyzed for carbon isotopes, the ratio of carbon isotopes was found to be similar to that of other skeletons located in the same village. Since different kinds of plants incorporate different proportions of isotopes as they grow, the similarity of the carbon isotope ratios in all the skeletons indicates that all the people in the village ate the same food. Thus, the scientists concluded that the man probably spent most of his life in the village where he was buried, after migrating there from some distant region. (N. J. van der Merwe, American Scientist, 70:596–606, 1982)

EXAMPLE 26.4

RADIOCHEMICAL DATING

The so-called Dead Sea Scrolls, Hebrew manuscripts of the books of the Old Testament, were found in 1947. The activity of carbon-14 in the linen wrappings of the book of Isaiah is about 11 disintegrations per minute per gram (d/min · g). Calculate the approximate age of the scrolls.

Solution The standard first order rate law is used

$$\ln\left(\frac{N}{N_0}\right) = -kt$$

where N is proportional to the activity at the present time (11 d/min · g) and N_0 is proportional to the activity of carbon-14 in the living material (14 d/min · g). In order to calculate the time elapsed since the linen wrappings were part of a living plant, we first need k, the rate constant. From the text you know that $t_{1/2}$ is 5.73×10^3, so

$$k = \frac{0.693}{t_{1/2}} = \frac{0.693}{5.73 \times 10^3 \text{ years}} = 1.21 \times 10^{-4} \text{ yr}^{-1}$$

Now everything is in place to calculate t.

$$\ln\left(\frac{11 \text{ d/min} \cdot \text{g}}{14 \text{ d/min} \cdot \text{g}}\right) = -[1.21 \times 10^{-4} \text{ yr}^{-1}]\, t$$

$$t = \frac{\ln 0.79}{-[1.21 \times 10^{-4} \text{ yr}^{-1}]}$$

$$= \frac{-0.24}{-[1.21 \times 10^{-4} \text{ yr}^{-1}]}$$

$$= 2.0 \times 10^3 \text{ years}$$

Therefore, the Scrolls are about 2000 years old.

EXERCISE 26.10 RADIOCHEMICAL DATING

A Japanese wooden temple guardian statue of the Kamakura period (*AD* 1185–1334) has a carbon-14 activity of 12.9 d/min · g). What is the age of the statue? In what year was the statue made? Initial quantity of carbon-14 = 14 d/min · g and $t_{1/2} = 5.73 \times 10^3$ years.

26.5 ARTIFICIAL TRANSMUTATIONS

Experiments over the years since Rutherford's discovery have showed that alpha particles will dislodge a proton from nearly all of the elements up to potassium, element 19.

In the course of his experiments, Rutherford found in 1919 that alpha particles ionize hydrogen gas, knocking off an electron. If nitrogen gas was used instead, he found that bombardment with alpha particles *also produced protons*. Quite correctly he concluded that the alpha particles had knocked a proton out of the nitrogen nucleus and that an isotope of another element had been produced. Nitrogen had undergone a *transmutation* to oxygen.

$$^4_2\text{He} + ^{14}_7\text{N} \longrightarrow ^{17}_8\text{O} + ^1_1\text{H}$$

Rutherford had proposed that protons and neutrons are the fundamental building blocks of nuclei. Although Rutherford's search for the neutron was not successful, it was found by James Chadwick in 1932 as a product of the alpha particle bombardment of beryllium.

$$^9_4\text{Be} + ^4_2\text{He} \longrightarrow ^{12}_6\text{C} + ^1_0\text{n}$$

Transforming one element into another by alpha particle bombardment has its limitations. Before a charged particle, such as the alpha particle, can be captured by a positively charged nucleus, the particle must have a sufficient kinetic energy to overcome the repulsive force developed as the positive particle approaches the positive nucleus. But the neutron is electrically neutral. Enrico Fermi (1934) reasoned, therefore, that a nucleus would not oppose its entry. Using this approach, practically all elements have since been transmuted, and a number of **transuranium elements** (elements beyond uranium) have been prepared. For example, uranium-238 forms neptunium-239 on neutron bombardment.

$$^{238}_{92}U + ^{1}_{0}n \longrightarrow ^{239}_{92}U \longrightarrow ^{239}_{93}Np + ^{0}_{-1}\beta$$

and the latter decays to plutonium.

$$^{239}_{93}Np \longrightarrow ^{0}_{-1}\beta + ^{239}_{94}Pu$$

FIGURE 26.10
Glenn Seaborg, recipient of the Nobel Prize in Chemistry in 1951, received his Ph.D. from the University of California in 1937. He has been a president of the American Chemical Society and director of the Lawrence Berkeley Laboratory.

During World War II other new transuranium elements were prepared, and by 1945 the list had extended to curium, element 96. All of the new elements were made by a team of scientists headed by Glenn Seaborg at the Lawrence Berkeley Laboratory in Berkeley, California (Figure 26.10). One member of Seaborg's team was Albert Ghiorso. In 1952 he had a hunch that small quantities of elements 99 and 100 could be found in the debris from the detonation of the first H bomb at Eniwetok in the South Pacific. Indeed, he found einsteinium (element 99) on a piece of filter paper that had been flown on an airplane through the blast zone, and he located fermium (element 100) in a garbage can containing coral from a nearby island.

Of the 108 elements presently known, only elements up to uranium exist in nature (except for Tc, Pm, At, and Fr). The transuranium elements are all synthetic. Up to element 101, mendelevium, all of the elements can be produced by bombarding a lighter nucleus with a small particle such as an alpha particle or neutron. Beyond 101 special techniques using heavier particles are required and are still being developed. For example, lawrencium is made by bombarding californium-252 with boron nuclei.

$$^{252}_{98}Cf + ^{10}_{5}B \longrightarrow ^{257}_{103}Lr + 5\,^{1}_{0}n$$

and the latest element to be discovered, 109, was made by firing iron nuclei at bismuth atoms.

Elements 104 through 107 and 109 (108 is not yet known) have not yet been named. The International Union of Pure and Applied Chemistry has suggested three-letter symbols, but there is by no means international agreement on this. Many have suggested simply naming the elements by their number.

EXERCISE 26.11 NUCLEAR TRANSMUTATION

Balance the following nuclear reactions, indicating the symbol, the mass number, and atomic number for "?."

(a) $^{13}_{6}C + ^{1}_{0}n \longrightarrow ^{4}_{2}He + ?$
(b) $^{14}_{7}N + ^{4}_{2}He \longrightarrow ^{1}_{0}n + ?$
(c) $^{253}_{99}Es + ^{4}_{2}He \longrightarrow ^{1}_{0}n + ?$

26.6 NUCLEAR FISSION

The discovery of nuclear fission began in 1934. That year Irène Curie, daughter of Pierre and Marie, and her husband, Frédéric Joliot, discovered "artificial" radioactivity. Before that time, all radioactive substances had

FIGURE 26.11

Enrico Fermi, the Italian physicist who first experimentally observed nuclear fission and who demonstrated a nuclear chain reaction.

been found in minerals and ores. Joliot and Curie, however, found that they could *create* radioactive elements by bombarding nonradioactive ones with alpha particles. Certain stable nuclei, apparently content to sit quietly forever, could be made unstable if they were forced to absorb additional subatomic particles. These force-fed atomic nuclei, in an agitated state, began spewing out little pieces of themselves, just as in a "naturally" radioactive element. Enrico Fermi (Figure 26.11), then working in Rome, decided that neutrons may work better than protons, because a nucleus may more readily absorb a neutral particle than a positively charged alpha particle. Thus, Fermi bombarded a uranium nucleus to see what would happen, assuming, as did others, that the result would be an element close in mass to uranium. However, in 1938 the radiochemists Otto Hahn and Fritz Strassman found some barium in the remnants of the bombarded uranium! Further work and interpretation by Lise Meitner, Otto Frisch, Niels Bohr, and Leo Szilard confirmed that a uranium-235 nucleus had captured a neutron to form uranium-236 and that this heavier nucleus had undergone fission; that is, the nucleus had split in two (Figure 26.12).

$$^{235}_{92}\text{U} + ^{1}_{0}\text{n} \longrightarrow ^{236}_{92}\text{U} \longrightarrow ^{141}_{56}\text{Ba} + ^{92}_{36}\text{Kr} + 3\ ^{1}_{0}\text{n}$$

$$\Delta E = -2 \times 10^{10}\ \text{kJ}$$

The fact that the fission reaction produces more neutrons than are required to begin the process is important. A single neutron begins a process producing 3 neutrons capable of inducing 3 more fission reactions, which release 9 neutrons to induce 9 more fissions from which 27 neutrons are obtained, and so on. Since the fission of uranium-236 is extremely rapid, this sequence of reactions can be an explosively rapid **chain reaction** as illustrated in Figure 26.13. If the amount of uranium-235 is small, so many neutrons escape (are not captured by ^{235}U nuclei) that the chain reaction cannot be sustained. In an atomic bomb, two small pieces of uranium, neither capable of sustaining a chain reaction, are brought together to form one piece capable of supporting a chain reaction, and an atomic explosion results.

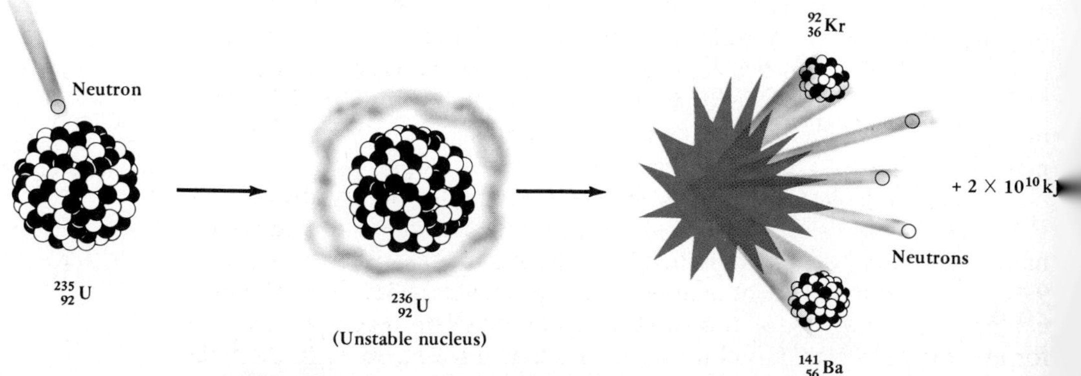

FIGURE 26.12

The fission of a $^{236}_{92}$U nucleus, induced by collision with a neutron. The electrical repulsion between protons rips the nucleus apart.

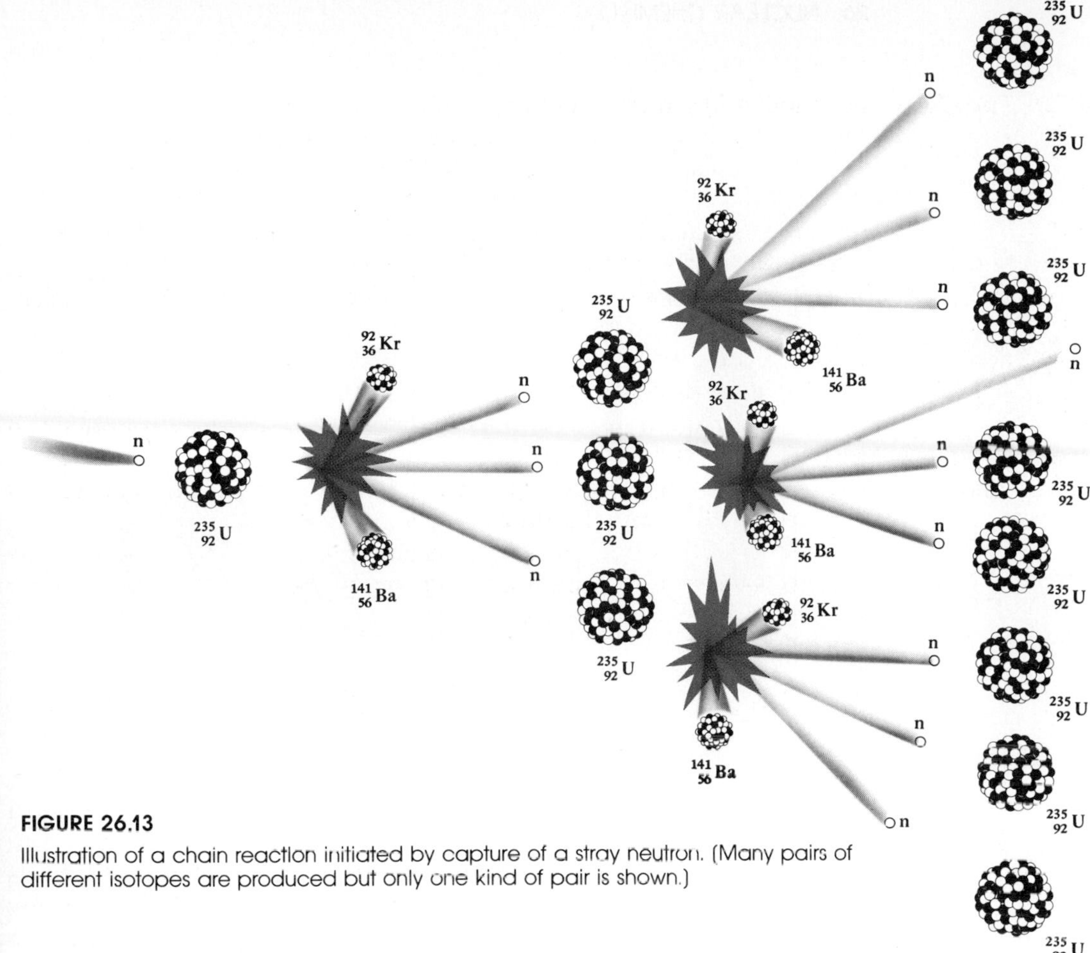

FIGURE 26.13

Illustration of a chain reaction initiated by capture of a stray neutron. (Many pairs of different isotopes are produced but only one kind of pair is shown.)

26.7 CONTROLLED NUCLEAR REACTIONS

Rather than allowing a nuclear chain reaction to run away explosively, it can be slowed down by limiting the number of neutrons available, and energy can be derived safely and used as the heat source in a nuclear power plant (Figure 26.14). In a **nuclear** or **atomic reactor**, the rate of fission is controlled by inserting cadmium rods or other "neutron absorbers" into the reactor. The rods absorb the neutrons and, by withdrawing or inserting the rods, the rate of the reaction can be increased or decreased. The rods are adjusted to make only about one neutron per fission available for the next fission reaction. If the number increases much above one, then the reaction is dangerously fast. If less than one neutron is produced, then the reaction effectively stops.

Not all nuclei can undergo fission on colliding with a neutron, but ^{235}U and ^{239}Pu are two nuclides where fission is possible. Unfortunately, natural uranium contains only 0.72% of the fissionable 235 isotope; over 99% of the natural element is nonfissionable uranium-238. Since the percentage of natural ^{235}U is too small to sustain a chain reaction, uranium for nuclear power plant fuel must be enriched. That is, some of the ^{238}U isotope is effectively discarded, thereby raising the concentration of ^{235}U. The most common way to do this is by *gaseous diffusion* as described in Section 6.7.

If the uranium-235 content of a sample is over 90%, it is considered of weapons quality.

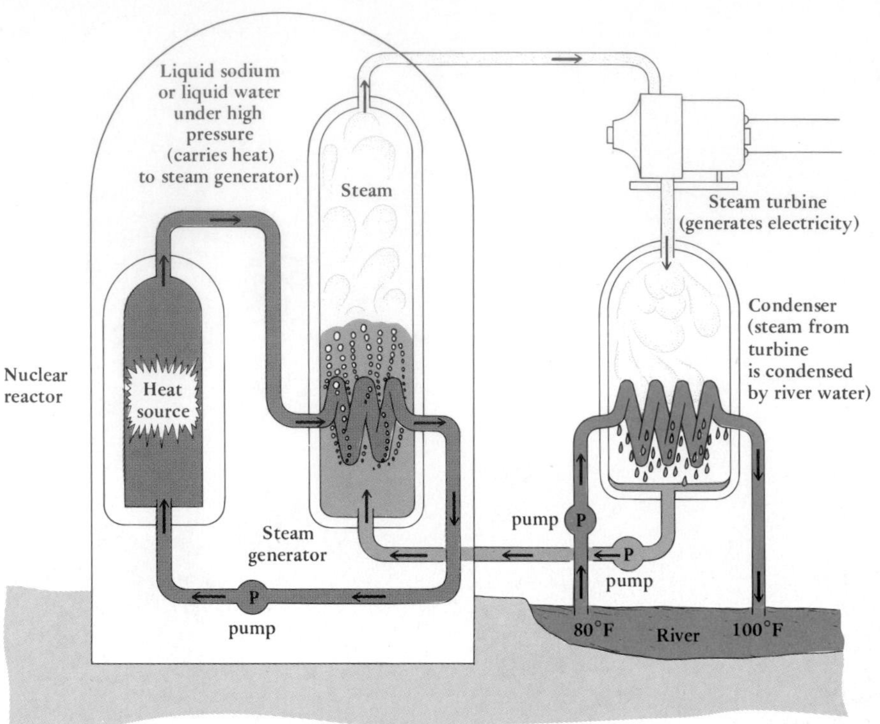

FIGURE 26.14

The essential parts of a nuclear power plant. Liquid water (or liquid sodium) is circulated through the reactor, where it is heated to about 325°C. This hot water (or liquid sodium) converts water in the steam generator to steam, which in turn drives a steam turbine. After passing through the turbine, the steam is converted back into liquid water and is recirculated through the steam generator. Enormous quantities of outside cooling water from rivers or lakes are necessary to condense the steam. These outside sources of water are returned to their source at appreciably higher temperatures than when they were taken into the condenser. This increase in temperature of the cooling water may cause a disturbance (sometimes called "thermal pollution") in the ecology of the lake or river.

If the United States and other countries continue to build fission reactors relying on ^{235}U as the fuel, there are predictions of a severe shortage of uranium ore before the year 2000. Therefore, there is great interest in (a) more efficient means of isotope separation and (b) breeder reactors.

Laser isotope separation is being intensively studied. To separate the uranium isotopes, for example, the uranium is converted to the volatile compound UF_6, just as in the gaseous diffusion method. A laser is tuned to excite (or introduce energy into) only the UF_6 based on ^{235}U.

$$^{235}UF_6 + {}^{238}UF_6 + \text{energy from laser} \longrightarrow {}^{235}UF_6(\text{excited state}) + {}^{238}UF_6(\text{ground state})$$

Since laser isotope separation is so efficient, there is concern that the technique will further the spread of nuclear weapons.

The excited UF_6 is then removed by a chemical process affecting only the excited state.

Another solution to the uranium shortage is to use **breeder reactors** that manufacture more fuel than they consume. One type of breeder i

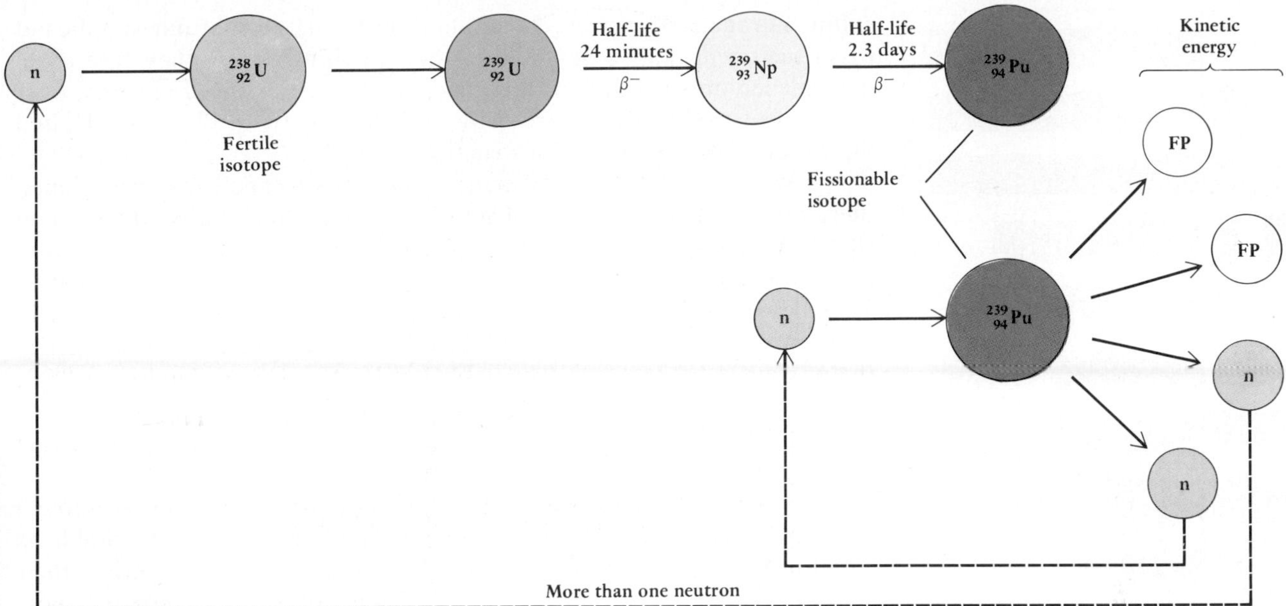

FIGURE 26.15

A "uranium cycle" for breeding fuel in a "fast breeder reactor." An atom of ^{238}U absorbs a neutron and emits a beta particle to become neptunium, which then undergoes further beta decay to become fissionable ^{239}Pu. When an atom of ^{239}Pu absorbs a neutron, it can fission to release energy, two fission products (FP), and at least two neutrons. One of the two neutrons is required to continue the chain reaction, but the other is available to continue the creation of ^{239}Pu from ^{238}U. A breeder reactor doubles its original fuel inventory in about 30 years.

based on ^{238}U, the more abundant isotope of uranium. This isotope absorbs a neutron which leads ultimately to plutonium-239 (Figure 26.15), a fissionable nuclide. Some of the neutrons from the fission of the plutonium lead to the fission of other ^{239}Pu nuclei and some lead to the formation of more plutonium. However, there are two significant problems with breeder reactors. First, plutonium is in demand for atomic weapons manufacture, since plutonium-based atomic weapons are smaller (smaller critical mass needed) and more effective than uranium-based weapons. Second, plutonium is radiologically very toxic; current occupational limits allow a worker to inhale no more than about 0.2 millionths of a gram over a lifetime. Thus, very great care must be used in handling the material and in securing it from capture by terrorist organizations.

It is estimated that the United States now has approximately 70 tons of reactor plutonium containing 35 to 40 tons of plutonium-239, enough to make 7000 to 8000 bombs.

At this point, it is worthwhile to consider briefly the controversy surrounding the use of nuclear power plants. Their proponents regard nuclear power to be an essential part of an advancing, technologically dependent society, independent of foreign energy sources. The health of our economy and our standard of living are dependent on inexpensive, reliable, and safe sources of energy. Some argue that the production of electric power is barely keeping pace with demand, so nuclear power plants must be built to meet the demand. Nuclear power plants are capable of supplying these demands, and they are also a source of "clean" energy in that they do not pollute the atmosphere with ash, smoke, or oxides of

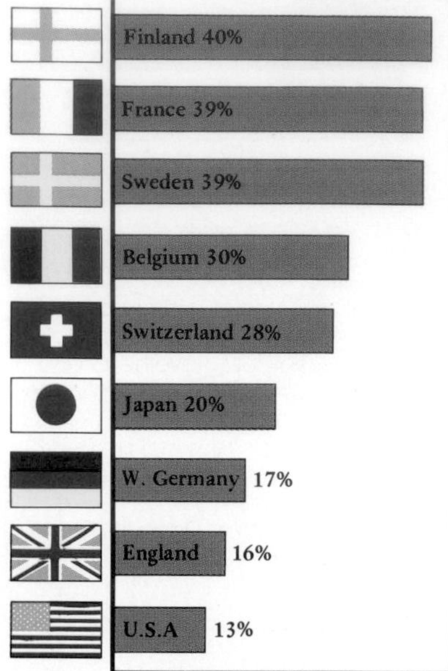

The share of electricity generated by nuclear power plants in 1982.

sulfur, nitrogen, or carbon. In addition, they ensure that our supplies of fossil fuels will not be depleted in the near future, and they free us of dependence on such fuels from other countries. For these reasons, over 340 nuclear power plants are in use worldwide (with 85 in the United States) and 200 are under construction.

While there are clearly many advantages to nuclear power plants, there are also disadvantages. There is evidence that radioactive tritium (^3_1H) is formed in the cooling water taken from and discharged into rivers and lakes, and some radioactive gases are emitted from nuclear plants. These levels of radioactivity are below the levels to which we are naturally exposed, however, and the United States Environmental Protection Agency (EPA) claims that they are acceptable levels. As suggested in Figure 26.2, we are always exposed to radiation originating from natural sources. Moreover, coal-burning power plants also release radioactive gases owing to naturally radioactive substances trapped in coal.

Another problem is presented by reactor fission products. Although some are put to various uses (see Section 26.9), many are not suitable as a fuel or for other purpose. Since they are often highly radioactive, their disposal poses an enormous environmental problem. Perhaps the most reasonable suggestion is that radioactive wastes can be converted to a glassy material having a volume of about 2 m^3 per reactor per year; this relatively small volume of material can then be stored underground in geological formations like salt deposits, known to be stable for hundreds of millions of years.

One of the greatest fears is a catastrophic accident leading to a large release of radioactivity. The United States apparently came close to having such an accident in the incident at the Three Mile Island plant in Pennsylvania on March 28, 1979 (Figure 26.16). The cooling system failed, and water gushed out of the reactor. The loss of cooling water allowed the core temperature to climb from its normal temperature of about 650°F to at least 4000°F, and the upper portion of the core began to melt. The zirconium alloy tubes containing the uranium oxide fuel also began to react with steam at around 2000°F, and the reaction produced hydrogen.

Metals commonly react with water, sometimes even at lower temperatures, to produce H_2 and the metal oxide.

$$Zr(s) + 2\ H_2O(\ell) \longrightarrow ZrO_2(s) + 2\ H_2(g)$$

This hydrogen accumulated in the upper part of the reactor containment vessel, along with other radioactive gases such as Kr and Xe. Since there was danger of explosion from the hydrogen, the gas mixture was vented to the atmosphere.* Within a week of the accident, normal cooling water circulation was restored, and the core began to cool. However, the plant

*It is estimated that about 2.5 million curies of radioactive xenon and krypton gas were released. To keep this in perspective, it is important to compare this with the almost 3 million curies of Rn gas (and other radioactive elements) released by the explosion of the Mt. St. Helens volcano on May 18, 1980. Not only did the volcano emit considerably more radiation than Three Mile Island, but radon is considered 1000 times more hazardous to health than xenon on a per curie basis due to the decay products of radon. Finally, it is important to note that xenon and krypton are chemically inert gases, and so do not readily enter the biosphere.

(a)

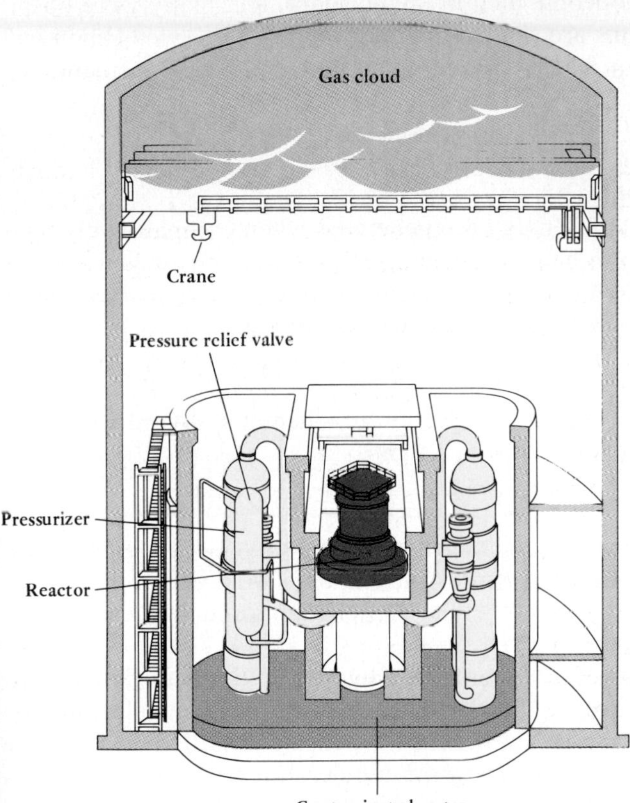

(b)

FIGURE 26.16

The Three Mile Island nuclear power plant after the reactor accident on March 28, 1979. (a) The arrow shows the still intact reactor containment building. (b) A drawing of the reactor containment building. The accident started when a valve, which keeps the pressure balanced in the reactor's cooling system, stuck open. This allowed contaminated cooling water to spill out of the reactor into the basement. Further, this loss of cooling water allowed the reactor to overheat, and chemical reactions occurred between residual water and metals in the reactor. The gas cloud consisted of H_2 from these reactions, as well as water vapor and other gases such as Kr and Xe. (See T. Dworetzky, *Discover*, October, 1984, p. 29.)

was closed, and cleanup is still proceeding and will continue until about 1988 or so.

Unfortunately, another major reactor accident occurred in April 1986. On April 28, scientists in Sweden noticed an appreciable increase in traces of radioactive gases in the atmosphere. From the nature of the isotopes, they concluded these isotopes could only have come from a significant reactor accident, and they traced the source to a place 700 miles away in the Soviet Union. Apparently, as early as April 25, 1986, one of four 1000-megawatt power reactors in Chernobyl, a town near Kiev in the Soviet Union, exploded and caught fire. These reactors are different from almost all reactors in the United States in that the Soviet reactors are encased in 1000 tons of graphite, a moderator of neutrons. Further-

"Chernobyl is a reminder of the two great fears of the age. One is the fear we will destroy ourselves. It comes and goes. But the other fear is always with us. It's what keeps the factories going—the fear of a civilization addicted to abundance, willing to risk anything except running out." Thomas Powers, *Discover*, p. 34, June, 1986.

more, Soviet reactors are not enclosed in a building designed to contain radioactive materials in the event of an accident. Nuclear scientists have speculated that cooling water stopped flowing through the graphite casing. The reactor heat turned the water remaining in the cooling system to superheated steam, which reacted with the zirconium cladding on the fuel rods, just as in the Three Mile Island accident. Explosion of the H_2 produced in the reaction, and the heat of the runaway reactor, may then have started a fire in the graphite reactor casing.

Because the reactor was not enclosed in a containment building, radioactive materials dispersed to the atmosphere and the surrounding countryside. Several people died in the accident, and many were hospitalized. All indications are that a small portion of the Soviet Union will not be inhabitable for many years to come because of this accident.

26.8 NUCLEAR FUSION

Nuclear fusion is the basis of the hydrogen bomb. A fission or atomic bomb provides the temperatures required for the fusion of hydrogen or of other light nuclei.

Tremendous amounts of energy are generated when comparatively light nuclei combine to form heavier nuclei. Such a reaction is called nuclear **fusion**, and an excellent example is the combination of two deuterium (or heavy hydrogen) nuclei to give an isotope of helium.

$$^{2}_{1}\text{H} + {}^{2}_{1}\text{H} \longrightarrow {}^{3}_{2}\text{He} + {}^{1}_{0}\text{n} \qquad \Delta E = -2.9 \times 10^{8}\ \text{kJ}$$

As outlined in the Special Section to Chapter 8, hydrogen fusion to give helium on the sun and other stars

$$4\ {}^{1}_{1}\text{H} \longrightarrow {}^{4}_{2}\text{He} + 2\ {}^{\ 0}_{+1}\beta \qquad \Delta E = -2.5 \times 10^{9}\ \text{kJ}$$

is our source of energy and the elements of the universe. A temperature of 10^6 to 10^7°C, the estimated temperature of the sun, is required to bring the positively charged nuclei together with enough force to overcome the repelling electrical force.

Rather than unleash uncontrolled hydrogen fusion to destroy our civilization, it is the dream of many that fusion can be controlled and the vast energy emitted can be harnessed for peaceful use. However, there are enormous problems producing the temperatures required in a controlled environment, and it has not yet been achieved. Of course, even if this problem is solved, there will be others. For example, there will be waste products produced in a fusion reactor, and we shall again have the problem of their disposal.

26.9 USES OF RADIOACTIVITY

Radioactivity has found many uses in medicine, in chemical research as a "tracer" or marker in chemical reactions, in chemical analysis, and in the preservation of food (Fig. 26.17).

Although radioactivity may induce cancer in living organisms, radiation has also been used successfully to treat cancer. Cancerous cells are more sensitive to radiation than normal cells, so the physician attempts

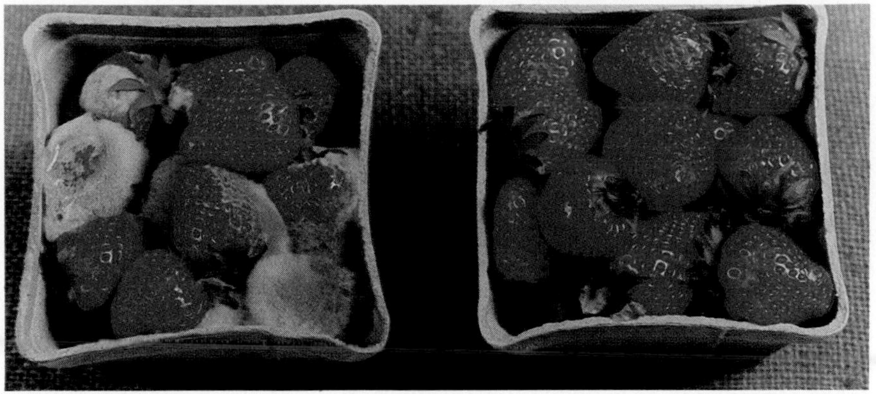

FIGURE 26.17
Strawberries irradiated with gamma rays from radioactive isotopes are still fresh after 15 days storage at 4°C (right), while those not irradiated are moldy (left). Preserving food by irradiation works by killing microorganisms and insects on food. There are clearly many benefits to food irradiation. For example, the sprouting of potatoes and onions after harvesting is inhibited, the parasite that causes trichinosis is inactivated in pork, and irradiated chicken can pose less risk from salmonella. (International Atomic Energy Agency)

to find a dosage that destroys malignant cells but not normal cells. Generally the cancerous cells are irradiated with gamma radiation in one of three ways: (a) from an external source, usually cobalt-60, directed into the body so that it passes through the growth; (b) by inserting a radioisotope directly into cancerous tissue; or (c) by inserting the radiation source in an accessible body cavity near the cancer.

One of the most widespread applications of radioisotopes is their use as tracers. Since the chemical properties of the isotopes of an element are identical,* a radioisotope may be used to "tag" an element without changing its chemical properties. Since the radioisotope is easily detected with a Geiger counter, the element may be followed or "traced" through a series of reactions. For example, radioactive iodine-131 may be used in the diagnosis and treatment of thyroid gland disorders. The thyroid gland, located in the neck, affects the rate of growth and metabolism. Disorders of the thyroid are associated with the rate at which iodine-131 appears in the gland. Thus, a patient is given a solution tagged with NaI* (where I* is ^{131}I), and a detector placed near the thyroid measures the uptake of iodine by the gland.

Reactions of many biochemicals important to metabolism, such as fats, sugars, amino acids, hormones, and drugs, can be followed in the body by tagging these compounds with carbon-14 or tritium ($^{3}_{1}H$). In a recent study of the metabolism of tetrahydrocannibinol (the active component of marijuana), the drug was tagged with radioactive atoms and then administered intravenously. Using a detector, it was found that the compound persisted in the bloodstream for more than three days, and the

Significant differences, resulting from mass differences, are found in the physical properties nd reaction rates for the lighter elements, particularly the isotopes of hydrogen.

products of metabolism of the compound were excreted in the urine for more than eight days.

The ban on the use of certain pesticides such as DDT has accelerated the search for other methods for the control of destructive insects. An application of radioactivity is based on the fact that less than lethal doses of radioactivity can cause sterility in male insects. The release of sterile insects into the environment means that the majority of the eggs produced by the female do not hatch, and the insect population is greatly reduced. Such a technique has been used successfully in the past few years to control the Mediterranean fruit fly in California and Florida.

The intensity of radiation depends on the quantity of the material through which it passes. Because of this, radiation is commonly used to gauge the thickness of industrial products such as metal sheets, plastic films, and cigarettes. The wear-resistant properties of moving parts of an automobile engine, for example, are easily measured by using radioactive parts and determining the amount of radioactivity transferred to the lubricating oil. In a similar way, the wear resistance of automobile tires and floor wax can be determined.

For one use of activation analysis see K.A. Rahn and D.H. Lowenthal, "Pollution Aerosol in the Northeast." *Science*, 228:275, 1985.

Neutron activation analysis allows us to analyze for elements in complex biological systems, archeological samples, and meteorites, for example, *without* destruction of the sample. When a sample is exposed to neutrons, each constituent element forms a specific radioisotope that can be identified by its characteristic radiation and half-life. The mass of the element present can then be determined from the activity of the radioisotope derived from it; the greater the activity, the greater the quantity of the element present. The sensitivity of neutron activation analysis depends on the intensity of the neutron beam used, the ability of the isotopes of a particular element to absorb neutrons, and the energy of the decay process. It is possible to detect as little as 10^{-10} grams of sodium, copper, or tungsten by this method, but other elements are somewhat less sensitive. An example of the use of activation analysis is the very thorough study of airborne pollution in the Northeast. The elements Se, Zn, Sb, Mn, V, As, In, and S as sulfate were monitored in the air in two locations over a period of months.

Americium, an artificial element, is widely used in smoke detectors in the home and in hotels.

The uses of radioisotopes are important in medicine, industry, and scientific research. We have given only a brief glimpse into their current uses, and their future applications are limited only by our imaginations.

SUMMARY

Elements that naturally emit energy without the absorption of energy are said to be **radioactive** (Section 26.1). Three types of radiation are emitted: **alpha particles**, helium nuclei ($^{4}_{2}\alpha$, $^{4}_{2}He$); **beta particles**, electrons ($^{0}_{-1}\beta$, $^{0}_{-1}e$), and **gamma rays** (γ).

Nuclear reactions or **transmutations** generally involve a change in the atomic number and mass of a radioisotope (Section 26.2). For example,

$$^{238}_{92}U + ^{1}_{0}n \longrightarrow ^{239}_{94}Pu + 2\ ^{0}_{-1}\beta$$

when uranium-238 absorbs a neutron ($^{1}_{0}n$), it is transformed into plutonium-239 and two beta particles. The sum of the mass numbers of the interacting particles (238 + 1) equals the sum of the mass numbers of the products

(239 + 0). Similarly, the sum of atomic numbers of interacting particles and products must be the same [92 + 0 = 94 + 2(−1)]. A radioactive nucleus may undergo a series of reactions called a **radioactive series** (Figure 26.4).

In addition to emission of alpha, beta, or gamma particles, two other modes of decay are emission of a **positron** (a positive electron, $_{+1}^{0}\beta$) and K-capture. The latter occurs when the nucleus of an atom absorbs one of its own electrons (usually a 1*s* or K electron).

All nuclei contain at least one neutron for every proton, and heavier nuclei have n/p ratios greater than 1 (Section 26.3 and Figure 26.5). There is evidence that the nucleus has a "shell" structure, just as electrons are arranged in energy shells. Nuclear shells have a capacity of 2, 8, 20, 28, 50, 82, and 126 nucleons of one type (protons or neutrons); these numbers are called nuclear "magic numbers." Nuclei with a magic number of either protons or neutrons are likely to be stable, and those with magic numbers of both types of nucleons are especially stable.

The energy evolved when nucleons combine to form a nucleus is the **binding energy**. The greater the binding energy, the greater the nuclear stability (Figure 26.6). A maximum in the curve of binding energy occurs near iron. Thus, all other nuclei are thermodynamically unstable with respect to nuclei near iron. Lighter nuclei can undergo **fusion** to form heavier nuclei, and heavier nuclei can undergo **fission** to form lighter nuclei nearer iron in atomic number and mass.

Nuclear disintegrations occur at different rates (Section 26.4). All follow a first order rate law where the rate of disintegration or **activity** (A) is proportional to the number of nuclei present (N). Thus, $A = kN$, where k is the rate constant. Solving this first order rate law as in Chapter 13, we find that $\ln (N/N_0) = -kt$, where t is the time elapsed on going from N_0 nuclei to N nuclei. The half-life of a particular ratioactive isotope ($t_{1/2}$) is the time required to go from N_0 to $\frac{1}{2}N_0$. (The greater the half-life, the more stable the nucleus.) As in Chapter 13, $t_{1/2} = 0.693/k$. These relations allow you to (a) find the half-life of an isotope from activity measurements, (b) find the time elapsed between measurements, and (c) find the quantity of material remaining after a certain time.

Uranium is the heaviest element occurring in nature in recoverable amounts. The **transuranium elements**, those beyond uranium, are all produced by forcing a lighter nucleus to absorb an appropriate particle (Section 26.5).

When certain isotopes of some of the actinide elements are struck by neutrons, they undergo fission and produce more neutrons (Sections 26.6 and 26.7). These neutrons begin a **chain reaction** and lead to the release of enormous quantities of energy (Figures 26.12 and 26.13). Such reactions are the basis of nuclear power plants and atomic bombs.

Nuclear fusion, the combination of lighter elements to form a heavier one, is also being explored as a source of energy for power generation (Section 26.8).

Radioisotopes have found a broad range of uses in, for example, the treatment of cancer by radiation, as "tags" or "tracers" to follow the fate of an element in a chemical reaction and in chemical analysis (Section 26.9).

STUDY QUESTIONS

GENERAL

1. Define or illustrate (a) α particle, (b) β particle, (c) γ-ray, (d) positron, (e) binding energy, (f) radioactive series, (g) half-life, (h) fission, (i) fusion, and (j) radioactive tracer.
2. Name three people who made important contributions to nuclear chemistry and indicate their contributions.

NUCLEAR REACTIONS

3. Balance the following nuclear reactions. Write the mass number and atomic number for "?," as well as the element symbol where possible.
   (a) $^{54}_{26}Fe + ^{4}_{2}He \longrightarrow 2\ ^{1}_{1}H + ?$
   (b) $^{27}_{13}Al + ^{4}_{2}He \longrightarrow ^{30}_{15}P + ?$
   (c) $^{32}_{16}S + ^{1}_{0}n \longrightarrow ^{1}_{1}H + ?$
   (d) $^{96}_{42}Mo + ^{2}_{1}H \longrightarrow ^{1}_{0}n + ?$
   (e) $^{98}_{42}Mo + ^{1}_{0}n \longrightarrow ^{99}_{43}Tc + ?$
   (f) $^{13}_{6}C + ? \longrightarrow ^{14}_{6}C$
   (g) $^{40}_{18}Ar + ? \longrightarrow ^{43}_{19}K + ^{1}_{1}H$
4. Balance the following nuclear reactions. Write the mass number and atomic number for "?," as well as the element symbol where possible.
   (a) $^{9}_{4}Be + ? \longrightarrow ^{6}_{3}Li + ^{4}_{2}He$
   (b) $? + ^{1}_{0}n \longrightarrow ^{24}_{11}Na + ^{4}_{2}He$
   (c) $^{40}_{20}Ca + ? \longrightarrow ^{40}_{19}K + ^{1}_{1}H$
   (d) $^{241}_{95}Am + ^{4}_{2}He \longrightarrow ^{243}_{97}Bk + ?$
   (e) $^{246}_{96}Cm + ^{12}_{6}C \longrightarrow 4\ ^{1}_{0}n + ?$
   (f) $^{238}_{92}U + ? \longrightarrow ^{249}_{100}Fm + 5\ ^{1}_{0}n$
   (g) $^{250}_{98}Cf + ^{11}_{5}B \longrightarrow 4\ ^{1}_{0}n + ?$
   (h) $^{53}_{24}Cr + ^{4}_{2}He \longrightarrow ? + ^{56}_{26}Fe$
5. Balance the following nuclear reactions. Write the mass number and atomic number for ?, as well as the element symbol where possible.
   (a) $^{104}_{47}Ag \longrightarrow ^{104}_{48}Cd + ?$
   (b) $^{87}_{36}Kr \longrightarrow ^{0}_{-1}\beta + ?$
   (c) $^{231}_{91}Pa \longrightarrow ^{227}_{89}Ac + ?$
   (d) $^{230}_{90}Th \longrightarrow ^{4}_{2}He + ?$
   (e) $^{82}_{35}Br \longrightarrow ^{82}_{36}Kr + ?$
   (f) $? \longrightarrow ^{24}_{12}Mg + ^{0}_{-1}\beta$
   (g) $^{212}_{84}Po \longrightarrow ^{208}_{82}Pb + ?$
   (h) $^{122}_{53}I \longrightarrow ^{122}_{54}Xe + ?$
   (i) $? \longrightarrow ^{23}_{11}Na + ^{0}_{-1}\beta$
6. Balance the following nuclear reactions. Write the mass number and atomic number for "?," as well as the element symbol where possible.
   (a) $^{19}_{10}Ne \longrightarrow ^{0}_{+1}\beta + ?$
   (b) $^{56}_{26}Fe \longrightarrow ^{0}_{-1}\beta + ?$
   (c) $^{40}_{19}K \longrightarrow ^{0}_{-1}\beta + ?$
   (d) $^{37}_{18}Ar + e^-$ (orbital electron) $\longrightarrow ?$
   (e) $^{56}_{26}Fe + e^-$ (orbital electron) $\longrightarrow ?$
   (f) $^{26}_{13}Al \longrightarrow ^{25}_{12}Mg + ?$
   (g) $^{137}_{53}I \longrightarrow ^{1}_{0}n + ?$
   (h) $^{22}_{11}Na \longrightarrow ^{1}_{1}H + ?$
7. The radioactive series that begins with $^{235}_{92}U$ and ends with $^{207}_{82}Pb$ undergoes the sequence of emissions: α, β, α, β, α, α, α, α, β, β, α. Identify the radioisotope produced in each step.
8. Given the following reaction sequence, state the atomic numbers and element symbols of W, X, Y, and Z.

$$_{92}Pb + _{24}Cr \longrightarrow W$$
$$_{98}Cf + _{8}O \longrightarrow W$$
$$W \longrightarrow \alpha + X$$
$$X \longrightarrow \alpha + Y$$
$$Y \longrightarrow \alpha + Z$$

NUCLEAR STABILITY

9. Which isotope in each of the following pairs should be the more stable?
   (a) $^{12}_{6}C$ or $^{13}_{6}C$; (b) $^{15}_{7}N$ or $^{14}_{7}N$; (c) $^{28}_{14}Si$ or $^{29}_{14}Si$
10. Calculate the binding energy in kJ per mole of P for the formation of $^{30}_{15}P$,

$$15\ ^{1}_{1}H + 15\ ^{1}_{0}n \longrightarrow ^{30}_{15}P$$

and for the formation of $^{31}_{15}P$.

$$15\ ^{1}_{1}H + 16\ ^{1}_{0}n \longrightarrow ^{31}_{15}P$$

Which is the more stable isotope? (The required masses are $^{1}_{1}H$ = 1.00783; $^{1}_{0}n$ = 1.00867; $^{30}_{15}P$ = 29.9880; and $^{31}_{15}P$ = 30.97376.)

11. Calculate the energy released or absorbed in kJ/mol for the following reactions:
   (a) $^{59}_{27}Co + ^{2}_{1}H \longrightarrow ^{60}_{27}Co + ^{1}_{1}H$
   (b) $^{230}_{90}Th \longrightarrow ^{4}_{2}He + ^{226}_{88}Ra$
   The required masses are: $^{59}_{27}Co$ = 58.9332; $^{2}_{1}H$ = 2.01410; $^{60}_{27}Co$ = 59.9529; $^{1}_{1}H$ = 1.00783; $^{230}_{90}Th$ = 230.0332; $^{4}_{2}He$ = 4.00260; and $^{226}_{88}Ra$ = 226.02544.

RATES OF DISINTEGRATION REACTIONS

12. Cobalt-60 was involved in the worst radioactive accident in North America (page 933). The half-life of the isotope is 5.3 years. Starting with 10.0 mg of ^{60}Co, how much will remain after 21.2 years? Approximately how much will remain after a century?
13. Gallium-67 is used in the medical diagnosis of tumors. If you ingest a compound containing 0.15 mg of this isotope, how many milligrams will remain in your body after 13 days? (The half-life of ^{67}Ga is 77.9 hours.)
14. A radioisotope disintegrates at the rate of 6400 counts per minute. 6.00 hours later, the disintegration rate is 1600 counts per minute. What is the half-life of the isotope?
15. The activity of a sample of ^{90}Sr is 6000 disintegrations per second (dps). After 87 years, the activity is 750 dps. Calculate the half-life of ^{90}Sr.
16. Strontium-90 is produced in nuclear explosions. Because strontium can be ingested and replace calcium in bone, radioactive ^{90}Sr is a dangerous isotope. As-

sume the quantity of strontium-90 in the bones of an individual in 1985 was about 80 disintegrations per seconds (dps). How many half-lives and how many years are necessary to reduce the activity to the natural background level of 2.5 dps? In what year will the activity be 2.5 dps? (The half-life of strontium-90 is 29 years.)

17. Radioisotopes of iodine are widely used in medicine. For example, iodine-131 (half-life = 8.05 days) is used to treat thyroid cancer. If you ingest a sample of NaI containing ^{131}I, how much time is required for the isotope to fall to 5.0% of its original activity?

18. A piece of charred bone found in the ruins of an American Indian village has a ^{14}C to ^{12}C ratio of 0.72 times that found in living organisms. Calculate the age of the bone fragment.

19. A sample of wood from a Thracian chariot found in an excavation in Bulgaria has a ^{14}C activity of 11.2 disintegrations per minute per gram. Estimate the age of the chariot and the year it was made.

20. The oldest known fossil cells form a biological cluster found in South Africa. The fossil has been dated by the reaction $^{87}Rb \rightarrow {}^{87}Sr + {}^{0}_{-1}\beta$, $t_{1/2} = 4.7 \times 10^{10}$ years. The ratio of the present quantity to the original quantity of ^{87}Rb is 0.951. Calculate the age of the fossil cells.

NUCLEAR TRANSMUTATIONS

21. There are two isotopes of americium both with half-lives sufficiently long to allow the handling of massive quantities. ^{241}Am, for example, has a half-life of 248 years as an alpha emitter. Thus, it is used in gauging and in smoke detectors. The isotope is formed from ^{239}Pu by absorption of two neutrons followed by emission of a beta particle. Write a balanced nuclear equation for this process.

22. Americium-240 is made by bombarding a plutonium-239 atom with an alpha particle. In addition to ^{240}Am, the products are a proton and two neutrons. Write a balanced equation for this process.

23. To synthesize the heavier transuranium elements, one must bombard a lighter nucleus with a relatively large particle. If you know the product is to be californium-246 (plus four neutrons), what particle would you force uranium-238 to absorb?

24. Balance the following reactions used for synthesis of transuranium elements.

(a) ${}^{238}_{92}U + {}^{14}_{7}N \longrightarrow ? + 5\,{}^{1}_{0}n$

(b) ${}^{238}_{92}U + ? \longrightarrow {}^{249}_{100}Fm + 5\,{}^{1}_{0}n$

(c) ${}^{253}_{99}Es + ? \longrightarrow {}^{256}_{101}Md + {}^{1}_{0}n$

(d) ${}^{246}_{96}Cm + ? \longrightarrow {}^{254}_{102}No + 4\,{}^{1}_{0}n$

(e) ${}^{252}_{98}Cf + ? \longrightarrow {}^{257}_{103}Lr + 5\,{}^{1}_{0}n$

25. The element with the highest known atomic number is 109. It is thought that still heavier elements are possible, especially with $Z = 114$ and $N = 184$. To this end, serious attempts have been made to force calcium-48 and curium-248 to merge. What would be the atomic number of the element formed?

NUCLEAR FISSION AND POWER

26. On December 2, 1942, the first man-made self-sustaining nuclear fission chain reactor was operated by Enrico Fermi under the University of Chicago Stadium. In June, 1972, natural fission reactors, operating billions of years ago, were discovered in Oklo, Gabon. At present, natural uranium contains 0.72% ^{235}U. How many years ago did natural uranium contain 3.0% ^{235}U, sufficient to sustain a natural reactor? ($t_{1/2}$ for ^{235}U is 7.04×10^8 years.)

27. A good grade of coal averages an energy output of 2.6×10^7 kJ/ton. Fission of one mole of ^{235}U releases 2.1×10^{10} kJ. Find the coal tonnage needed to produce the same energy as one pound of ^{235}U. (See Appendix D for conversion factors.)

28. Thorium-232 is a possible fuel for breeder reactors, since it ultimately produces fissionable uranium-233 on neutron bombardment. When ^{233}U fissions, it gives some fission products plus two neutrons, one of which can lead to fission of more ^{233}U and the other of which can bombard ^{232}Th to begin the breeding of more ^{233}U. When ^{232}Th swallows a neutron, it produces an unstable thorium isotope which emits a beta particle to give an isotope of another element. This element in turn emits another beta particle to give ^{233}U. Draw a diagram similar to Figure 26.15 to illustrate the operation of the "thorium breeder cycle."

USES OF RADIOISOTOPES

29. Acetic acid reacts with methyl alcohol, CH_3OH, by eliminating a molecule of H_2O to form methyl acetate. Explain how you would use the isotope ^{18}O to show whether the oxygen atom in the water product comes from the —OH of the acid or the —OH of the alcohol.

$$CH_3{-}\overset{\overset{\displaystyle O}{\|}}{C}{-}OH + CH_3OH \longrightarrow CH_3{-}\overset{\overset{\displaystyle O}{\|}}{C}{-}O{-}CH_3 + H_2O$$

30. In order to measure the volume of the blood system of an animal, the following experiment was done. A 1.0-mL sample of an aqueous solution containing 2.0×10^6 disintegrations per second (dps) of tritium was injected into the bloodstream. After time was allowed for complete circulatory mixing, a 1.0-mL blood sample was withdrawn and found to have an activity of 1.5×10^4 dps. What was the volume of the circulatory system? (The half-life of tritium is 12.3 years, so this experiment assumes that only a negligible amount of tritium has decayed in the time of the experiment.)

GENERAL QUESTIONS

31. In studies near Boulder, Colorado, 6.4×10^{17} atoms of radioactive ^{210}Po have been detected per gram of pine needles. Calculate the number of atoms, moles, and grams of ^{210}Po deposited on 1.0 kg of pine needles.

32. One of the problems in the accident at the Three Mile nuclear plant was that the zirconium-clad fuel rods reacted at high temperature with water (in the form of steam) to form H_2. (a) If the zirconium was converted to ZrO_2, write the balanced equation for the hydrogen-producing reaction. (b) If 5.0×10^2 kg of zirconium reacted, how many moles of H_2 could be produced? (c) Assuming the H_2 was released into a building with a volume of 1.5×10^6 m^3, what is the partial pressure of the H_2 at 25°C?

33. Since radioactive decay is a first order process, the rate law can be written

Activity $= kN$

where N is the number of atoms of the isotope present in the sample. Calculate the alpha activity in disintegrations per minute for a 0.0010-g sample of ^{226}Ra ($t_{1/2}$ = 1620 years).

34. Plutonium-239 has a half-life of 24,300 years. What is the alpha activity of a 1.0 mg sample of ^{239}Pu in disintegrations per minute? (See Question 33 for further information.)

35. If a sample of ^{147}Pm has an activity of 1.0×10^7 disintegrations per second, what is the amount of ^{147}Pm in the sample? $t_{1/2}$ for ^{147}Pm is 2.64 years. (See Question 33 for further information.)

36. The age of the earth can be estimated using naturally occurring radioactive isotopes. For example, ^{238}U has a half-life of 4.5×10^9 years, whereas that of ^{235}U is 7.1×10^8 years. The isotopic abundance of ^{238}U is 99.28%, and that of ^{235}U is 0.72%. If one *assumes* that the two isotopes began with the same relative abundance on formation of the earth, and if there has been no separation of isotopes in some manner, then the abundance ratio reflects the shorter half-life of ^{235}U.

From the rate law for radioactive decay, you can write

$$^{235}U = {}^{235}U_0 e^{-kt} \quad \text{and} \quad {}^{238}U = {}^{238}U_0 e^{-k't}$$

where ^{235}U is the amount of the 235 isotope present now, and $^{235}U_0$ is that present at the beginning; the same applies to the 238 isotope. k is the rate constant for decay of ^{235}U and k' is that for decay of ^{238}U. t is the time elapsed from the beginning of the earth to the present. Now, taking the ratio of these two expressions, we have

$$\frac{^{238}U}{^{235}U} = \frac{^{238}U_0 e^{-k't}}{^{235}U_0 e^{-kt}}$$

or, since $^{235}U_0$ and $^{238}U_0$ are *assumed* to be equal, we have

$$\frac{^{238}U}{^{235}U} = \frac{e^{-k't}}{e^{-kt}} = \frac{99.28}{0.72} = 138$$

Solving this, we find that

$$\ln 138 = -(k' - k)t$$

Since you can calculate both k and k', you can now solve for t and obtain an upper limit for the age of the earth. Why is the calculated age an "upper limit" estimate?

APPENDICES

A. Some Mathematical Operations
B. Naming Inorganic Compounds
C. Four-Place Table of Logarithms
D. Common Units, Equivalences, and Conversion Factors
E. Physical Constants
F. Some Physical Constants for Water and a Few Common Substances
G. Ionization Constants for Weak Acids at 25°C
H. Ionization Constants for Weak Bases at 25°C
I. Solubility Product Constants for Some Inorganic Compounds at 25°C
J. Formation Constants for Some Complex Ions in Aqueous Solution
K. Standard Reduction Potentials in Aqueous Solution at 25°C
L. Selected Thermodynamic Values
M. Answers to Exercises in the Text
N. Answers to Selected Study Questions

APPENDIX A

Some Mathematical Operations

The mathematical skills required in this introductory course are basic skills in algebra and a knowledge of (a) exponential or scientific notation, (b) logarithms, and (c) quadratic equations. This appendix reviews each of the last three topics.

A.1 ELECTRONIC CALCULATORS

The advent of inexpensive electronic calculators a few years ago has made calculations in introductory chemistry much more straightforward. You are well advised to purchase a calculator that has the capability of performing calculations in scientific notation, has both base-10 and natural logarithms, and is capable of raising any number to any power and of finding any root of any number. In the discussion below, we shall point out in general how these functions of your calculator can be used.

Although electronic calculators have greatly simplified calculations, they have also forced us to focus again on significant figures. A calculator easily handles 8 or more significant figures, but real laboratory data is never known to this accuracy. Therefore, you are urged to review Section 1.5 on handling numbers.

A.2 EXPONENTIAL OR SCIENTIFIC NOTATION

In exponential or scientific notation, a number is expressed as a product of two numbers: $N \times 10^n$. The first number, N, is the so-called *digit term* and is a number between 1 and 10. The second number, 10^n, the *exponential term*, is some integer power of 10. For example, 1234 would be written in scientific notation as 1.234×10^3 or 1.234 multiplied by 10 three times.

$$1234 = 1.234 \times 10^1 \times 10^1 \times 10^1 = 1.234 \times 10^3$$

Conversely, a number less than 1, such as 0.01234, would be written as 1.234×10^{-2}. This notation tells us that 1.234 should be divided twice by 10 in order to obtain 0.01234.

$$0.01234 = \frac{1.234}{10^1 \times 10^1} = 1.234 \times 10^{-1} \times 10^{-1} = 1.234 \times 10^{-2}$$

Some other examples of scientific notation are

$$10000 = 1 \times 10^4 \qquad 12345 = 1.2345 \times 10^4$$
$$1000 = 1 \times 10^3 \qquad 1234 = 1.234 \times 10^3$$

$100 = 1 \times 10^2$	$123 = 1.23 \times 10^2$
$10 = 1 \times 10^1$	$12 = 1.2 \times 10^1$
$1 = 1 \times 10^0$	(any number to the zero power $= 1$)
$1/10 = 1 \times 10^{-1}$	$0.12 = 1.2 \times 10^{-1}$
$1/100 = 1 \times 10^{-2}$	$0.012 = 1.2 \times 10^{-2}$
$1/1000 = 1 \times 10^{-3}$	$0.0012 = 1.2 \times 10^{-3}$
$1/10000 = 1 \times 10^{-4}$	$0.00012 = 1.2 \times 10^{-4}$

When converting a number to scientific notation, notice that the exponent n is positive if the number is greater than 1 and negative if the number is less than 1. The value of n is the number of places by which the decimal was shifted to obtain the number in scientific notation.

$$1\ 2\ 3\ 4\ 5. = 1.2345 \times 10^4$$

Decimal shifted 4 places to the left. Therefore, n is positive and equal to 4.

$$0.0\ 0\ 1\ 2 = 1.2 \times 10^{-3}$$

Decimal shifted 3 places to the right. Therefore, n is negative and equal to 3.

If you wish to convert a number in scientific notation to the usual form, the procedure above is simply reversed.

$$6\ .\ 2\ 7\ 3 \times 10^2 = 627.3$$

Decimal point moved 2 places to the right, since n is positive and equal to 2.

$$0\ 0\ 6.273 \times 10^{-3} = 0.006273$$

Decimal point shifted 3 places to the left, since n is negative and equal to 3.

EXERCISE APP-1 SCIENTIFIC NOTATION

Convert the following numbers to scientific notation:

(a) 6273
(b) 0.0627
(c) 12144
(d) 0.00000178

Convert the following numbers in scientific notation to the usual form:

(e) 5.12×10^2
(f) 6.02×10^4
(g) 2.58×10^{-2}
(h) 1.89×10^{-5}

There are two final points to be made concerning scientific notation. First, if you are used to working on a computer, you may be in the habit of writing a number such as 1.23×10^3 as 1.23E3 or 6.45×10^{-5} as 6.45E-5. Second, some electronic calculators allow you to convert numbers readily to the scientific notation. If you have such a calculator, you can change a number shown in the usual form to scientific notation simply by pressing the EE or EXP key and then the "=" key.

1. ADDING AND SUBTRACTING NUMBERS

When adding or subtracting two numbers, they must first be converted to the same powers of ten. The digit terms are then added or subtracted as appropriate.

$$(1.234 \times 10^{-3}) + (5.623 \times 10^{-2}) = (0.1234 \times 10^{-2}) + (5.623 \times 10^{-2}) = 5.746 \times 10^{-2}$$

$$(6.52 \times 10^{2}) - (1.56 \times 10^{3}) = (6.52 \times 10^{2}) - (15.6 \times 10^{2}) = -9.1 \times 10^{2}$$

2. MULTIPLICATION

The digit terms are multiplied in the usual manner, and the exponents are added algebraically. The result is expressed with a digit term with only one nonzero digit to the left of the decimal.

$$(1.23 \times 10^{3})(7.60 \times 10^{2}) = (1.23)(7.60) \times 10^{3+2} = 9.35 \times 10^{5}$$

$$(6.02 \times 10^{23})(2.32 \times 10^{-2}) = (6.02)(2.32) \times 10^{23-2} = 13.966 \times 10^{21} = 1.40 \times 10^{22} \text{ (answer in 3 significant figures)}$$

3. DIVISION

The digit terms are divided in the usual manner, and the exponents are subtracted algebraically. The quotient is written with one nonzero digit to the left of the decimal in the digit term.

$$\frac{7.60 \times 10^{3}}{1.23 \times 10^{2}} = \frac{7.60}{1.23} \times 10^{3-2} = 6.18 \times 10^{1}$$

$$\frac{6.02 \times 10^{23}}{9.10 \times 10^{-2}} = \frac{6.02}{9.10} \times 10^{(23)-(-2)} = 0.662 \times 10^{25} = 6.62 \times 10^{24}$$

4. POWERS OF EXPONENTIALS

When raising a number in exponential notation to a power, treat the digit term in the usual manner. The exponent is then multiplied by the number indicating the power.

$$(1.25 \times 10^{3})^2 = (1.25)^2 \times 10^{3\times2} = 1.5625 \times 10^{6} = 1.56 \times 10^{6}$$

$$(5.6 \times 10^{-10})^3 = (5.6)^3 \times 10^{(-10)\times3} = 175.6 \times 10^{-30} = 1.8 \times 10^{-28}$$

The calculator instructions apply to such common calculators as those made by Texas Instruments, Casio, and Sharp. If you have another brand, consult the instruction booklet that came with the calculator.

Electronic calculators usually have two methods of raising a number to a power. To square a number, enter the number and then press the "x^2" key. To raise a number to any power, use the "y^x" key. For example, to raise 1.42×10^2 to the 4th power,

(a) enter 1.42×10^2
(b) press "y^x"

(c) enter 4 (this should appear on the display)
(d) press "=" and 4.0659×10^8 will appear on the display.

As a final step, express the number in the correct number of significant figures (4.07×10^8) in this case.

5. ROOTS OF EXPONENTIALS

Unless you use an electronic calculator, the number must first be put into a form where the exponential is exactly divisible by the root. The root of the digit term is found in the usual way, and the exponent is divided by the desired root.

$$\sqrt{3.6 \times 10^7} = \sqrt{36 \times 10^6} = \sqrt{36} \times \sqrt{10^6} = 6.0 \times 10^3$$
$$\sqrt[3]{2.1 \times 10^{-7}} = \sqrt[3]{210 \times 10^{-9}} = \sqrt[3]{210} \times \sqrt[3]{10^{-9}} = 5.9 \times 10^{-3}$$

To take a square root on an electronic calculator, enter the number and then press the "$\sqrt{x}$" key. To find a higher root of a number, such as the 4th root of 5.6×10^{-10},

(a) enter the number.
(b) press the "$\sqrt[x]{y}$" key. (On most calculators, the sequence you actually use is to press "2ndF" and then "$\sqrt[x]{y}$." Alternatively, you press "INV" and then "y^x.")
(c) enter the desired root, 4 in this case.
(d) press "=". The answer here is 4.8646×10^{-3} or 4.9×10^{-3}.

A general procedure for finding any root is to use the "y^x" key. For a square root, x is 0.5 (or ½), whereas it is 0.33 (or ⅓) for a cube root, 0.25 (or ¼) for a 4th root, and so on.

EXERCISE APP-2 USING NUMBERS IN SCIENTIFIC NOTATION

Perform the indicated operations below.
(a) $(1.56 \times 10^{-2}) + (6.27 \times 10^{-3})$
(b) $(7.60 \times 10^2) \times (2.5 \times 10^{-2})$
(c) $(1.5 \times 10^{-2})/(8.91 \times 10^2)$
(d) $(6.02 \times 10^5)^3$
(e) $\sqrt[2]{4.83 \times 10^{-3})} = (4.83 \times 10^{-3})^{1/2}$
(f) $\sqrt[3]{1.68 \times 10^5} = (1.68 \times 10^5)^{1/3}$
(g) $(1.8 \times 10^{-4})^2/(3.20 \times 10^{-3})$

A.3 LOGARITHMS

There are two types of logarithms used in this text: (a) common logarithms (abbreviated log) whose base is 10 and (b) natural logarithms (abbreviated ln) whose base is e (= 2.71828).

$$\log x = n \text{ where } x = 10^n$$
$$\ln x = m \text{ where } x = e^m$$

Most equations in chemistry and physics were developed in natural or base e logarithms and this practice is followed in this text. The relation between log and ln is

$$\ln x = 2.303 \log x$$

Aside from the different bases of the two logarithms, they are used in the same manner. What follows is largely a description of the use of common logarithms.

A common logarithm is the power to which you must raise 10 to obtain the number. For example, the log of 100 is 2, since you must raise 10 to the second power to obtain 100. Other examples are

$$\begin{aligned}
\log 1000 &= \log (10^3) = 3 \\
\log 10 &= \log (10^1) = 1 \\
\log 1 &= \log (10^0) = 0 \\
\log 1/10 &= \log (10^{-1}) = -1 \\
\log 1/10000 &= \log (10^{-4}) = -4
\end{aligned}$$

To obtain the common logarithm of a number other than a simple power of 10, you must resort to a log table (Appendix C) or an electronic calculator. For example,

$$\begin{aligned}
\log 2.10 &= 0.3222, \text{ which means that } 10^{0.3222} = 2.10 \\
\log 5.16 &= 0.7126, \text{ which means that } 10^{0.7126} = 5.16 \\
\log 3.125 &= 0.49485, \text{ which means that } 10^{0.49485} = 3.125
\end{aligned}$$

To check this on your calculator, enter the number and then press the "log" key. When using Appendix C, the logs of the first two numbers above can be read directly from the table. The log of the third number (3.125), however, must be interpolated. That is, 3.125 is midway between 3.12 and 3.13, so the log is midway between 0.4942 and 0.4955.

To obtain the natural logarithm, ln, of the numbers above, use a calculator having this function. Enter each number and press "ln."

$$\begin{aligned}
\ln 2.10 &= 0.7419, \text{ which means that } e^{0.7419} = 2.10 \\
\ln 5.16 &= 1.6409, \text{ which means that } e^{1.6409} = 5.16
\end{aligned}$$

To find the common logarithm of a number greater than 10 or less than 1 with the log table in Appendix C, first express the number in scientific notation. Then find the log of each part of the number and add the logs. For example,

$$\begin{aligned}
\log 241 &= \log (2.41 \times 10^2) = \log 2.41 + \log 10^2 \\
&= 0.382 + 2 = 2.382
\end{aligned}$$

$$\begin{aligned}
\log 0.00573 &= \log (5.73 \times 10^{-3}) = \log 5.73 + \log 10^{-3} \\
&= 0.758 + (-3) = -2.242
\end{aligned}$$

Logarithms and Nomenclature
The number to the left of the decimal is called the *characteristic* and the number to the right of the decimal is the *mantissa* of the logarithm.

SIGNIFICANT FIGURES AND LOGARITHMS Notice that the mantissa has as many significant figures as the number whose log was found. (So that you could more clearly see the result obtained with a calculator or a table this rule was not strictly followed until the last two examples.)

OBTAINING ANTILOGARITHMS If you are given the logarithm of a number, and find the number from it, you have obtained the "antilogarithm" or "antilog" of the number. There are two common procedures used by electronic calculators to do this:

PROCEDURE A	PROCEDURE B
(a) enter the log or ln	(a) enter the log or ln
(b) press 2ndF	(b) press INV
(c) press 10^x or e^x	(c) press log or ln x

Test one or the other of these procedures with the following examples:

1. Find the number whose log is 5.234.

   Recall that $\log x = n$ where $x = 10^n$. In this case $n = 5.234$. Enter that number in your calculator and find the value of 10^n, the antilog. In this case,

   $$10^{5.234} = 10^{0.234} \times 10^5 = 1.71 \times 10^5$$

   Notice that the characteristic (5) sets the decimal point; it is the power of 10 in the exponential form. The mantissa (0.234) gives the value of the number x. Thus, if you use Appendix C to find x, you need only look up 0.234 in the table and see that it corresponds to 1.71.
2. Find the number whose log is -3.456.

   $$10^{-3.456} = 10^{0.544} \times 10^{-4} = 3.50 \times 10^{-4}$$

Notice here that -3.456 must be expressed as the sum of -4 and $+0.544$.

MATHEMATICAL OPERATIONS USING LOGARITHMS Because logarithms are exponents, operations involving them follow the same rules as the use of exponents. Thus, multiplying two numbers can be done by adding logarithms.

$$\log xy = \log x + \log y$$

For example, we multiply 563 by 125 by adding their logarithms and finding the anti-logarithm of the result.

$$\begin{aligned} \log 563 &= 2.751 \\ \log 125 &= \underline{2.097} \\ \log xy &= 4.848 \end{aligned}$$

$$xy = 10^{4.848} = 10^4 \times 10^{0.848} = 7.05 \times 10^4$$

One number (x) can be divided by another (y) by subtraction of their logarithms.

$$\log \frac{x}{y} = \log x - \log y$$

For example, to divide 125 by 742,

$$\begin{aligned} \log 125 &= 2.097 \\ -\log 742 &= \underline{2.870} \\ \log x/y &= -0.773 \end{aligned}$$

$$x/y = 10^{-0.773} = 10^{0.227} \times 10^{-1} = 1.69 \times 10^{-1}$$

Similarly, powers and roots of numbers can be found using logarithms.

$$\log x^y = y(\log x)$$

$$\log \sqrt[y]{x} = \log x^{1/y} = \frac{1}{y} \log x$$

As an example, find the fourth power of 5.23. We first find the log of 5.23 and then multiply it by 4. The result, 2.874, is the log of the answer. Therefore, we find the antilog of 2.874.

$$(5.23)^4 = ?$$
$$\log (5.23)^4 = 4 \log 5.23 = 4\ (0.719) = 2.874$$
$$(5.23)^4 = 10^{2.874} = 748$$

As another example, find the fifth root of 1.89×10^{-9}.

$$\sqrt[5]{1.89 \times 10^{-9}} = (1.89 \times 10^{-9})^{1/5} = ?$$

$$\log (1.89 \times 10^{-9})^{1/5} = \frac{1}{5} \log (1.89 \times 10^{-9}) = \frac{1}{5}(-8.724) = -1.745$$

The answer is the antilog of -1.745.

$$(1.89 \times 10^{-9})^{1/5} = 10^{-1.745} = 1.80 \times 10^{-2}$$

EXERCISE APP-3 LOGARITHMS

Give common logs of
(a) 1.29 (b) 596 (c) 0.436
Give natural logarithms of
(d) 2.36 (e) 45.6 (f) 0.00253
Find the number corresponding to each of the following common logs:
(g) 0.895 (h) 1.408 (i) −2.249
Find the number corresponding to each of the following natural logarithms:
(j) 0.940 (k) 4.476 (l) −4.234

A.4 QUADRATIC EQUATIONS

Algebraic equations of the form $ax^2 + bx + c = 0$ are called **quadratic equations**. The coefficients a, b, and c may be either positive or negative. The two roots of the equation may be found using the *quadratic formula*.

$$x = \frac{-b \pm \sqrt{b^2 - 4ac}}{2a}$$

As an example, solve the equation $5x^2 - 3x - 2 = 0$. Here $a = 5$, $b = -3$, and $c = -2$. Therefore,

$$x = \frac{3 \pm \sqrt{(-3)^2 - 4(5)(-2)}}{2(5)}$$

$$= \frac{3 \pm \sqrt{9 - (-40)}}{10} = \frac{3 \pm \sqrt{49}}{10} = \frac{3 \pm 7}{10}$$

$$x = 1 \text{ and } -0.4$$

How do you know which of the two roots is the correct answer? You have to decide in each case which root has physical significance. However, it is *usually* true in this course that negative values are not significant.

When you have solved a quadratic expression, you should always check your values by substitution into the original equation. In the example above, we find that $5(1)^2 - 3(1) - 2 = 0$ and that $5(-0.4)^2 - 3(-0.4) - 2 = 0$.

EXERCISE APP-4 QUADRATIC EQUATIONS

Find the roots of the equation $3.6 = \dfrac{x^2}{1.5 - x}$

APPENDIX B

Naming Inorganic Compounds

It is important to know how to name common inorganic compounds. Here we describe the names of simple ionic compounds such as those introduced in Chapter 2. Additionally, names of oxides and other compounds of nonmetals are introduced, as are the names of common acids. Both of the latter types of compounds are described in Chapter 3.

A. IONIC COMPOUNDS

Names of ionic compounds are derived from the ions of which they are composed, the positive ion name being given first followed by the name of the negative ion.

NAMING POSITIVE IONS Positive ions are named by the following rules:

(a) For a **monatomic positive ion**, that is, a metal cation, the name is simply the name of the metal from which it is derived.

K^+ is potassium Mg^{2+} is magnesium Al^{3+} is aluminum

(b) There are cases, especially in the transition series, where a metal can form more than one positive ion. The most common practice today is to use the **Stock system**, a system in which the oxidation number of the ion is designated with a Roman numeral in parentheses immediately following the name of the ion. For example,

Ti^{2+} is titanium(II) and Ti^{4+} is titanium(IV)
Co^{2+} is cobalt(II) and Co^{3+} is cobalt(III)
Sn^{2+} is tin(II) and Sn^{4+} is tin(IV)

Although not used in this book, you should be aware of an older system of naming. For example, in this system Sn^{2+} would be named *stannous* and Sn^{4+} would be *stannic*. In general, one takes the stem of the Latin name of the element and adds *-ous* or *-ic* to designate the lower or higher oxidation number, respectively. Other examples are

Fe^{2+} is ferrous ion and Fe^{3+} is ferric ion
Cu^+ is cuprous ion and Cu^{2+} is cupric ion

Finally, two common inorganic positive ions are NH_4^+ (ammonium ion) and Hg_2^{2+} (mercury(I) or mercurous ion).

NAMING NEGATIVE IONS Two types of negative ions must be considered: those having only one atom and those having several atoms.

(a) **Monatomic** or single-atom negative ions are named by adding

"-ide" to the stem of the name of the nonmetal element from which it is derived. See Table 2.4 for examples.

(b) Negative **polyatomic** or many-atom ions are common, especially ones containing oxygen (oxyanions). The names of some of the most common of these ions are given in Table 2.5. Most of these names must simply be learned. However, there are some important guidelines. Consider, for example, NO_3^- and NO_2^- and SO_4^{2-} and SO_3^{2-}.

NO_3^-: nitr*ate* SO_4^{2-}: sulf*ate*
NO_2^-: nitr*ite* SO_3^{2-}: sulf*ite*

The ion with the greater number of oxygen atoms is given the suffix *-ate*, while the ion with the smaller number of oxygen atoms has the suffix *-ite*. For a series of oxyanions with more than two members, the ion with the smallest number of oxygen atoms has the prefix *hypo-* and the suffix *-ite*. The ion with the largest number of oxygen atoms has the prefix *per-* and the suffix *-ate*. The chlorine-containing oxyanions are the most common examples.

ClO^- = *hypo*chlor*ite*
ClO_2^- = chlor*ite*
ClO_3^- = chlor*ate*
ClO_4^- = *per*chlor*ate*

Oxyanions that contain hydrogen can be named simply by prefixing the name with the word "hydrogen." Notice that the number of such hydrogen atoms is also indicated by *di-*, *tri-*, and so on.

ION	SYSTEMATIC NAME	COMMON NAME
HCO_3^-	Hydrogen carbonate	Bicarbonate
HSO_4^-	Hydrogen sulfate	Bisulfate
HSO_3^-	Hydrogen sulfite	Bisulfite
HPO_4^{2-}	Hydrogen phosphate	
$H_2PO_4^-$	*Di*hydrogen phosphate	

Many of these hydrogen-containing oxyanions have common names that are so often used that you must know them. For example, $NaHCO_3$ is very commonly called sodium bicarbonate (although you may know it as "baking soda").

NAMING IONIC COMPOUNDS Some examples were given in Table 2.6, and others include the following:

IONIC COMPOUND	NAME
$CaCl_2$	Calcium chloride
$NaHSO_4$	Sodium hydrogen sulfate
KH_2PO_4	Potassium dihydrogen phosphate
$(NH_4)_2CO_3$	Ammonium carbonate
$TiCl_2$	Titanium(II) chloride
$TiCl_4$	Titanium(IV) chloride
$Mg(OH)_2$	Magnesium hydroxide

B. NAMING BINARY COMPOUNDS OF THE NONMETALS

When two nonmetallic elements combine to form a binary (or "two-element") compound, the result is not an ionic compound but a "covalent" one (as seen in Chapters 9 and 10).

Hydrogen forms binary compounds of the type H_xX_y with all of the nonmetals (except the rare gases). Except in a few cases (see below), the H atom is written first in the formula and is named first. The other nonmetal is named as if it were a negative ion. For example,

COMPOUND	NAME
HF	Hydrogen fluoride
HCl	Hydrogen chloride
H_2S	Hydrogen sulfide

Virtually all the binary, nonmetal compounds you will see have at least one element from Groups 6A or 7A. This element is always listed second in the formula and is named second. The number of such elements in the molecule is designated with a prefix such as "di-, tri-, tetra-, penta-," and so on. For example,

COMPOUND	NAME
NF_3	Nitrogen trifluoride
NO_2	Nitrogen dioxide
N_2O	Dinitrogen oxide
PCl_3	Phosphorus trichloride
PF_5	Phosphorus pentafluoride
SF_6	Sulfur hexafluoride
S_2F_{10}	Disulfur decafluoride
N_2O_4	Dinitrogen tetroxide*

*The name should be tetr*ao*xide, but the *a* is dropped so that the name sounds better.

Many of the binary compounds of nonmetals were discovered years ago and have names so common that they continue to be used. These names must simply be learned.

COMPOUND	COMMON NAME
H_2O	Water
H_2O_2	Hydrogen peroxide
NH_3*	Ammonia
N_2H_4*	Hydrazine
PH_3*	Phosphine
AsH_3*	Arsine
SbH_3*	Stibine
NO	Nitric oxide
N_2O	Nitrous oxide

*The formulas of Group 5A hydrogen compounds have, by tradition, the Group 5A element given first in the formula.

C. NAMING OXYACIDS

The names of the most common acids containing oxygen are listed in Table 3.2. Acids that have a name ending in *-ic* contain an *-ate* anion. On the other hand, if the acid has an *-ous* suffix, then the anion name ends in *-ite*.

ANION AND NAME	ACID AND NAME
NO_3^- nitrate	HNO_3 nitric
NO_2^- nitrite	HNO_2 nitrous
SO_4^{2-} sulfate	H_2SO_4 sulfuric
SO_3^{2-} sulfite	H_2SO_3 sulfurous
CO_3^{2-} carbonate	H_2CO_3 carbonic
ClO_3^- chlorate	$HClO_3$ chloric
ClO_4^- perchlorate	$HClO_4$ perchloric
PO_4^{3-} phosphate	H_3PO_4 phosphoric

EXERCISE APP-5 NAMING COMPOUNDS

1. Name each of the following ionic compounds:
   (a) $CaSO_4$ (c) VCl_3 (e) $NaClO_4$
   (b) $SrCO_3$ (d) VCl_4 (f) $K_2Cr_2O_7$
2. Name each of the following binary, nonmetal compounds:
   (a) BCl_3 (d) P_2O_5
   (b) XeF_2 (e) ClF_3
   (c) SO_2 (f) N_2F_4
3. Give the formula of each of the following ionic compounds.
   (a) barium oxide (d) iron(III) perchlorate
   (b) vanadium(III) oxide (e) potassium chlorate
   (c) zinc chloride (f) calcium hypochlorite
4. Give the formula of each of the following binary, nonmetal compounds.
   (a) carbon dioxide (d) boron trifluoride
   (b) phosphorus triiodide (e) dioxygen difluoride
   (c) sulfur dichloride (f) xenon trioxide

APPENDIX C

Four-Place Table of Logarithms

	0	1	2	3	4	5	6	7	8	9
1.0	.0000	.0043	.0086	.0128	.0170	.0212	.0253	.0294	.0334	.0374
1.1	.0414	.0453	.0492	.0531	.0569	.0607	.0645	.0682	.0719	.0755
1.2	.0792	.0828	.0864	.0899	.0934	.0969	.1004	.1038	.1072	.1106
1.3	.1139	.1173	.1206	.1239	.1271	.1303	.1335	.1367	.1399	.1430
1.4	.1461	.1492	.1523	.1553	.1584	.1614	.1644	.1673	.1703	.1732
1.5	.1761	.1790	.1818	.1847	.1875	.1903	.1931	.1959	.1987	.2014
1.6	.2041	.2068	.2095	.2122	.2148	.2175	.2201	.2227	.2253	.2279
1.7	.2304	.2330	.2355	.2380	.2405	.2430	.2455	.2480	.2504	.2529
1.8	.2553	.2577	.2601	.2625	.2648	.2672	.2695	.2718	.2742	.2765
1.9	.2788	.2810	.2833	.2856	.2878	.2900	.2923	.2945	.2967	.2989
2.0	.3010	.3032	.3054	.3075	.3096	.3118	.3139	.3160	.3181	.3201
2.1	.3222	.3243	.3263	.3284	.3304	.3324	.3345	.3365	.3385	.3404
2.2	.3424	.3444	.3464	.3483	.3502	.3522	.3541	.3560	.3579	.3598
2.3	.3617	.3636	.3655	.3674	.3692	.3711	.3729	.3747	.3766	.3784
2.4	.3802	.3820	.3838	.3856	.3874	.3892	.3909	.3927	.3945	.3962
2.5	.3979	.3997	.4014	.4031	.4048	.4065	.4082	.4099	.4116	.4133
2.6	.4150	.4166	.4183	.4200	.4216	.4232	.4249	.4265	.4281	.4298
2.7	.4314	.4330	.4346	.4362	.4378	.4393	.4409	.4425	.4440	.4456
2.8	.4472	.4487	.4502	.4518	.4533	.4548	.4564	.4579	.4594	.4609
2.9	.4624	.4639	.4654	.4669	.4683	.4698	.4713	.4728	.4742	.4757
3.0	.4771	.4786	.4800	.4814	.4829	.4843	.4857	.4871	.4886	.4900
3.1	.4914	.4928	.4942	.4955	.4969	.4983	.4997	.5011	.5024	.5038
3.2	.5051	.5065	.5079	.5092	.5105	.5119	.5132	.5145	.5159	.5172
3.3	.5185	.5198	.5211	.5224	.5237	.5250	.5263	.5276	.5289	.5302
3.4	.5315	.5328	.5340	.5353	.5366	.5378	.5391	.5403	.5416	.5428
3.5	.5441	.5453	.5465	.5478	.5490	.5502	.5514	.5527	.5539	.5551
3.6	.5563	.5575	.5587	.5599	.5611	.5623	.5635	.5647	.5658	.5670
3.7	.5682	.5694	.5705	.5717	.5729	.5740	.5752	.5763	.5775	.5786
3.8	.5798	.5809	.5821	.5832	.5843	.5855	.5866	.5877	.5888	.5899
3.9	.5911	.5922	.5933	.5944	.5955	.5966	.5977	.5988	.5999	.6010
4.0	.6021	.6031	.6042	.6053	.6064	.6075	.6085	.6096	.6107	.6117
4.1	.6128	.6138	.6149	.6160	.6170	.6180	.6191	.6201	.6212	.6222
4.2	.6232	.6243	.6253	.6263	.6274	.6284	.6294	.6304	.6314	.6325
4.3	.6335	.6345	.6355	.6365	.6375	.6385	.6395	.6405	.6415	.6425
4.4	.6435	.6444	.6454	.6464	.6474	.6484	.6493	.6503	.6513	.6522
4.5	.6532	.6542	.6551	.6561	.6571	.6580	.6590	.6599	.6609	.6618
4.6	.6628	.6637	.6646	.6656	.6665	.6675	.6684	.6693	.6702	.6712
4.7	.6721	.6730	.6739	.6749	.6758	.6767	.6776	.6785	.6794	.6803
4.8	.6812	.6821	.6830	.6839	.6848	.6857	.6866	.6875	.6884	.6893
4.9	.6902	.6911	.6920	.6928	.6937	.6946	.6955	.6964	.6972	.6981
5.0	.6990	.6998	.7007	.7016	.7024	.7033	.7042	.7050	.7059	.7067
5.1	.7076	.7084	.7093	.7101	.7110	.7118	.7126	.7135	.7143	.7152
5.2	.7160	.7168	.7177	.7185	.7193	.7202	.7210	.7218	.7226	.7235
5.3	.7243	.7251	.7259	.7267	.7275	.7284	.7292	.7300	.7308	.7316
5.4	.7324	.7332	.7340	.7348	.7356	.7364	.7372	.7380	.7388	.7396
5.5	.7404	.7412	.7419	.7427	.7435	.7443	.7451	.7459	.7466	.7474
5.6	.7482	.7490	.7497	.7505	.7513	.7520	.7528	.7536	.7543	.7551

APPENDIX C (*Continued*)

Four-Place Table of Logarithms

	0	1	2	3	4	5	6	7	8	9
5.7	.7559	.7566	.7574	.7582	.7589	.7597	.7604	.7612	.7619	.7627
5.8	.7634	.7642	.7649	.7657	.7664	.7672	.7679	.7686	.7694	.7701
5.9	.7709	.7716	.7723	.7731	.7738	.7745	.7752	.7760	.7767	.7774
6.0	.7782	.7789	.7796	.7803	.7810	.7818	.7825	.7832	.7839	.7846
6.1	.7853	.7860	.7868	.7875	.7882	.7889	.7896	.7903	.7910	.7917
6.2	.7924	.7931	.7938	.7945	.7952	.7959	.7966	.7973	.7980	.7987
6.3	.7993	.8000	.8007	.8014	.8021	.8028	.8035	.8041	.8048	.8055
6.4	.8062	.8069	.8075	.8082	.8089	.8096	.8102	.8109	.8116	.8122
6.5	.8129	.8136	.8142	.8149	.8156	.8162	.8169	.8176	.8182	.8189
6.6	.8195	.8202	.8209	.8215	.8222	.8228	.8235	.8241	.8248	.8254
6.7	.8261	.8267	.8274	.8280	.8287	.8293	.8299	.8306	.8312	.8319
6.8	.8325	.8331	.8338	.8344	.8351	.8357	.8363	.8370	.8376	.8382
6.9	.8388	.8395	.8401	.8407	.8414	.8420	.8426	.8432	.8439	.8445
7.0	.8451	.8457	.8463	.8470	.8476	.8482	.8488	.8494	.8500	.8506
7.1	.8513	.8519	.8525	.8531	.8537	.8543	.8549	.8555	.8561	.8567
7.2	.8573	.8579	.8585	.8591	.8597	.8603	.8609	.8615	.8621	.8627
7.3	.8633	.8639	.8645	.8651	.8657	.8663	.8669	.8675	.8681	.8686
7.4	.8692	.8698	.8704	.8710	.8716	.8722	.8727	.8733	.8739	.8745
7.5	.8751	.8756	.8762	.8768	.8774	.8779	.8785	.8791	.8797	.8802
7.6	.8808	.8814	.8820	.8825	.8831	.8837	.8842	.8848	.8854	.8859
7.7	.8865	.8871	.8876	.8882	.8887	.8893	.8899	.8904	.8910	.8915
7.8	.8921	.8927	.8932	.8938	.8943	.8949	.8954	.8960	.8965	.8971
7.9	.8976	.8982	.8987	.8993	.8998	.9004	.9009	.9015	.9020	.9026
8.0	.9031	.9036	.9042	.9047	.9053	.9058	.9063	.9069	.9074	.9079
8.1	.9085	.9090	.9096	.9101	.9106	.9112	.9117	.9122	.9128	.9133
8.2	.9138	.9143	.9149	.9154	.9159	.9165	.9170	.9175	.9180	.9186
8.3	.9191	.9196	.9201	.9206	.9212	.9217	.9222	.9227	.9232	.9238
8.4	.9243	.9248	.9253	.9258	.9263	.9269	.9274	.9279	.9284	.9289
8.5	.9294	.9299	.9304	.9309	.9315	.9320	.9325	.9330	.9335	.9340
8.6	.9345	.9350	.9355	.9360	.9365	.9370	.9375	.9380	.9385	.9390
8.7	.9395	.9400	.9405	.9410	.9415	.9420	.9425	.9430	.9435	.9440
8.8	.9445	.9450	.9455	.9460	.9465	.9469	.9474	.9479	.9484	.9489
8.9	.9494	.9499	.9504	.9509	.9513	.9518	.9523	.9528	.9533	.9538
9.0	.9542	.9547	.9552	.9557	.9562	.9566	.9571	.9576	.9581	.9586
9.1	.9590	.9595	.9600	.9605	.9609	.9614	.9619	.9624	.9628	.9633
9.2	.9638	.9643	.9647	.9652	.9657	.9661	.9666	.9671	.9675	.9680
9.3	.9685	.9689	.9694	.9699	.9703	.9708	.9713	.9717	.9722	.9727
9.4	.9731	.9736	.9741	.9745	.9750	.9754	.9759	.9763	.9768	.9773
9.5	.9777	.9782	.9786	.9791	.9795	.9800	.9805	.9809	.9814	.9818
9.6	.9823	.9827	.9832	.9836	.9841	.9845	.9850	.9854	.9859	.9863
9.7	.9868	.9872	.9877	.9881	.9886	.9890	.9894	.9899	.9903	.9908
9.8	.9912	.9917	.9921	.9926	.9930	.9934	.9939	.9943	.9948	.9952
9.9	.9956	.9961	.9965	.9969	.9974	.9978	.9983	.9987	.9991	.9996

APPENDIX D

Common Units, Equivalences, and Conversion Factors

FUNDAMENTAL UNITS OF THE SI SYSTEM

The metric system was begun by the French National Assembly in 1790 and has undergone many modifications. The International System of Units or *Système International* (SI), which represents an extension of the metric system, was adopted by the 11th General Conference of Weights and Measures in 1960. It is constructed from seven base units, each of which represents a particular physical quantity (Table I).

TABLE I SI Fundamental Units

PHYSICAL QUANTITY	NAME OF UNIT	SYMBOL
Length	meter	m
Mass	kilogram	kg
Time	second	s
Temperature	kelvin	K
Amount of substance	mole	mol
Electric current	ampere	A
Luminous intensity	candela	cd

The first five units listed in Table I are particularly useful in general chemistry. They are defined as follows.

1. The *meter* was redefined in 1960 to be equal to 1,650,763.73 wavelengths of a certain line in the emission spectrum of krypton-86.
2. The *kilogram* represents the mass of a platinum-iridium block kept at the International Bureau of Weights and Measures at Sèvres, France.
3. The *second* was redefined in 1967 as the duration of 9,192,631,770 periods of a certain line in the microwave spectrum of cesium-133.
4. The *kelvin* is 1/273.16 of the temperature interval between absolute zero and the triple point of water.
5. The *mole* is the amount of substance that contains as many entities as there are atoms in exactly 0.012 kg of carbon-12 (12 g of ^{12}C atoms).

PREFIXES USED WITH TRADITIONAL METRIC UNITS AND SI UNITS

Decimal fractions and multiples of metric and SI units are designated by using the prefixes listed in Table II. Those most commonly used in general chemistry are in italics.

TABLE II Traditional Metric and SI Prefixes

FACTOR	PREFIX	SYMBOL	FACTOR	PREFIX	SYMBOL
10^{12}	tera	T	10^{-1}	*deci*	d
10^{9}	giga	G	10^{-2}	*centi*	c
10^{6}	mega	M	10^{-3}	*milli*	m
10^{3}	*kilo*	k	10^{-6}	micro	μ
10^{2}	hecto	h	10^{-9}	*nano*	n
10^{1}	deka	da	10^{-12}	pico	p
			10^{-15}	femto	f
			10^{-18}	atto	a

DERIVED SI UNITS

In the International System of Units, all physical quantities are represented by appropriate combinations of the base units listed in Table I. A list of the derived units frequently used in general chemistry is given in Table III.

TABLE III Derived SI Units

PHYSICAL QUANTITY	NAME OF UNIT	SYMBOL	DEFINITION
Area	square meter	m^2	
Volume	cubic meter	m^3	
Density	kilogram per cubic meter	kg/m^3	
Force	newton	N	$kg\ m/s^2$
Pressure	pascal	Pa	N/m^2
Energy	joule	J	$kg\ m^2/s^2$
Electric charge	coulomb	C	A s
Electric potential difference	volt	V	J/(A s)

Common Units of Mass and Weight

1 POUND = 453.59 GRAMS

1 pound = 453.59 grams = 0.45359 kilogram
1 kilogram = 1000 grams = 2.205 pounds
1 gram = 10 decigrams = 100 centigrams = 1000 milligrams
1 gram = 6.022×10^{23} atomic mass units
1 atomic mass unit = 1.6606×10^{-24} gram
1 short ton = 2000 pounds = 907.2 kilograms
1 long ton = 2240 pounds
1 metric tonne = 1000 kilograms = 2205 pounds

Common Units of Length

1 INCH = 2.54 CENTIMETERS (EXACTLY)

1 mile = 5280 feet = 1.609 kilometers
1 yard = 36 inches = 0.9144 meter
1 meter = 100 centimeters = 39.37 inches = 3.281 feet = 1.094 yards
1 kilometer = 1000 meters = 1094 yards = 0.6215 mile
1 Ångstrom = 1.0×10^{-8} centimeter = 0.10 nanometer = 100 picometers
= 1.0×10^{-10} meter = 3.937×10^{-9} inch

Common Units of Volume

1 QUART = 0.9463 LITER
1 LITER = 1.056 QUARTS

1 liter = 1 cubic decimeter = 1000 cubic centimeters = 0.001 cubic meter
1 milliliter = 1 cubic centimeter = 0.001 liter = 1.056×10^{-3} quart
1 cubic foot = 28.316 liters = 29.902 quarts = 7.475 gallons

Common Units of Force* and Pressure

1 atmosphere = 760 millimeters of mercury = 1.013×10^5 pascals
= 14.70 pounds per square inch
1 bar = 10^5 pascals
1 torr = 1 millimeter of mercury
1 pascal = 1 kg/m s^2 = 1 N/m^2

*Force: 1 newton (N) = 1 kg m/s^2, i.e., the force that when applied for 1 second gives a 1 kilogram mass a velocity of 1 meter per second.

Common Units of Energy

1 JOULE = 1×10^7 ERGS

1 thermochemical calorie* = 4.184 joules = 4.184×10^7 ergs
= 4.129×10^{-2} liter-atmospheres
= 2.612×10^{19} electron volts
1 erg = 1×10^{-7} joule = 2.3901×10^{-8} calorie
1 electron volt = 1.6022×10^{-19} joule = 1.6022×10^{-12} erg = 96.487 kJ/mol†
1 liter-atmosphere = 24.217 calories = 101.32 joules = 1.0132×10^9 ergs
1 British thermal unit = 1055.06 joules = 1.05506×10^{10} ergs = 252.2 calories

*The amount of heat required to raise the temperature of one gram of water from 14.5°C to 15.5°C.

†Note that the other units in this line are per particle and must be multiplied by 6.022×10^{23} to be strictly comparable.

APPENDIX E

Physical Constants

QUANTITY	SYMBOL	TRADITIONAL UNITS	SI UNITS
Acceleration of gravity	g	980.6 cm/s	9.806 m/s
Atomic mass unit (1/12 the mass of ^{12}C atom)	amu or u	1.6606×10^{-24} g	1.6606×10^{-27} kg
Avogadro's number	N	6.022×10^{23} particles/mol	6.022×10^{23} particles/mol
Bohr radius	a_0	0.52918 Å 5.2918×10^{-9} cm	5.2918×10^{-11} m
Boltzmann constant	k	1.3807×10^{-16} erg/K	1.3807×10^{-23} J/K
Charge-to-mass ratio of electron	e/m	1.7588×10^{8} coulomb/g	1.7588×10^{11} C/kg
Electronic charge	e	1.6022×10^{-19} coulomb 4.8033×10^{-10} esu	1.6022×10^{-19} C
Electron rest mass	m_e	9.1095×10^{-28} g 0.00054859 amu	9.1095×10^{-31} kg
Faraday constant	F	96,487 coulombs/mol e^- 23.06 kcal/volt mol e^-	96,487 C/mol e^- 96,487 J/V mol e^-
Gas constant	R	$0.08206 \dfrac{\text{L atm}}{\text{mol K}}$ $1.987 \dfrac{\text{cal}}{\text{mol K}}$	$8.3145 \dfrac{\text{Pa dm}^3}{\text{mol K}}$ 8.3145 J/mol K
Molar volume (STP)	V_m	22.414 L/mol	22.414×10^{-3} m^3/mol 22.414 dm^3/mol
Neutron rest mass	m_n	1.67495×10^{-24} g 1.008665 amu	1.67495×10^{-27} kg
Planck's constant	h	6.6262×10^{-27} erg s	6.6262×10^{-34} J s
Proton rest mass	m_p	1.6726×10^{-24} g 1.007277 amu	1.6726×10^{-27} kg
Rydberg constant	R_∞	3.289×10^{15} cycles/s 2.1799×10^{-11} erg	1.0974×10^{7} m^{-1} 2.1799×10^{-18} J
Velocity of light (in a vacuum)	c	2.9979×10^{10} cm/s (186,281 miles/second)	2.9979×10^{8} m/s

$\pi = 3.1416$
$e = 2.7183$
$\ln X = 2.303 \log X$
$2.303\ R = 4.576$ cal/mol·K $= 19.15$ J/mol·K
$2.303\ RT$ (at 25°C) $= 1364$ cal/mol $= 5709$ J/mol

APPENDIX F

Some Physical Constants for Water and a Few Common Substances

Vapor Pressure of Water at Various Temperatures

TEMPERATURE °C	VAPOR PRESSURE torr	TEMPERATURE °C	VAPOR PRESSURE torr	TEMPERATURE °C	VAPOR PRESSURE torr	TEMPERATURE °C	VAPOR PRESSURE torr
−10	2.1	21	18.7	51	97.2	81	369.7
−9	2.3	22	19.8	52	102.1	82	384.9
−8	2.5	23	21.1	53	107.2	83	400.6
−7	2.7	24	22.4	54	112.5	84	416.8
−6	2.9	25	23.8	55	118.0	85	433.6
−5	3.2	26	25.2	56	123.8	86	450.9
−4	3.4	27	26.7	57	129.8	87	468.7
−3	3.7	28	28.3	58	136.1	88	487.1
−2	4.0	29	30.0	59	142.6	89	506.1
−1	4.3	30	31.8	60	149.4	90	525.8
0	4.6	31	33.7	61	156.4	91	546.1
1	4.9	32	35.7	62	163.8	92	567.0
2	5.3	33	37.7	63	171.4	93	588.6
3	5.7	34	39.9	64	179.3	94	610.9
4	6.1	35	42.2	65	187.5	95	633.9
5	6.5	36	44.6	66	196.1	96	657.6
6	7.0	37	47.1	67	205.0	97	682.1
7	7.5	38	49.7	68	214.2	98	707.3
8	8.0	39	52.4	69	223.7	99	733.2
9	8.6	40	55.3	70	233.7	100	760.0
10	9.2	41	58.3	71	243.9	101	787.6
11	9.8	42	61.5	72	254.6	102	815.9
12	10.5	43	64.8	73	265.7	103	845.1
13	11.2	44	68.3	74	277.2	104	875.1
14	12.0	45	71.9	75	289.1	105	906.1
15	12.8	46	75.7	76	301.4	106	937.9
16	13.6	47	79.6	77	314.1	107	970.6
17	14.5	48	83.7	78	327.3	108	1004.4
18	15.5	49	88.0	79	341.0	109	1038.9
19	16.5	50	92.5	80	355.1	110	1074.6
20	17.5						

Some Physical Constants for Water and a Few Common Substances

Specific Heats and Heat Capacities for Some Common Substances

SUBSTANCE	SPECIFIC HEAT cal/g·°K	SPECIFIC HEAT J/g·°K	MOLAR HEAT CAPACITY cal/mol·°K	MOLAR HEAT CAPACITY J/mol·°K
Al (s)	0.215	0.902	5.81	24.3
Ca (s)	0.156	0.653	6.25	26.2
Cu (s)	0.092	0.385	5.85	24.5
Fe (s)	0.106	0.451	5.92	24.8
Hg (*l*)	0.0331	0.138	6.62	27.7
H_2O (s), ice	0.500	2.06	9.00	37.7
H_2O (*l*), water	1.00	4.18	18.0	75.3
H_2O (*g*), steam	0.484	2.03	8.71	36.4
C_6H_6 (*l*), benzene	0.415	1.74	32.4	136
C_6H_6 (*g*), benzene	0.249	1.04	19.5	81.6
C_2H_5OH (*l*), ethanol	0.587	2.46	27.0	113
C_2H_5OH (*g*), ethanol	0.228	0.954	10.5	420
$(C_2H_5)_2O$ (*l*), diethyl ether	0.893	3.74	41.1	172
$(C_2H_5)_2O$ (*g*), diethyl ether	0.561	2.35	25.8	108

Heats of Transformation and Transformation Temperatures of Several Substances

SUBSTANCE	MP °C	HEAT OF FUSION cal/g	HEAT OF FUSION J/g	ΔH_{fus} kcal/mol	ΔH_{fus} kJ/mol	BP °C	HEAT OF VAPORIZATION cal/g	HEAT OF VAPORIZATION J/g	ΔH_{vap} kcal/mol	ΔH_{vap} kJ/mol
Al	658	94.5	395	2.54	10.6	2467	2515	10520	67.9	284
Ca	851	55.7	233	2.23	9.33	1487	963	4030	38.6	162
Cu	1083	49.0	205	3.11	13.0	2595	1146	4790	72.8	305
H_2O	0.0	79.8	333	1.44	6.02	100	540	2260	9.73	40.7
Fe	1530	63.7	267	3.56	14.9	2735	1515	6340	84.6	354
Hg	−39	2.7	11	5.57	23.3	357	69.8	292	14.0	58.6
CH_4	−182	14.0	58.6	0.22	0.92	−164	—	—	—	—
C_2H_5OH	−117	26.1	109	1.20	5.02	78.0	204	855	9.39	39.3
C_6H_6	5.48	30.4	127	2.37	9.92	80.1	94.3	395	7.36	30.8
$(C_2H_5)_2O$	−116	23.4	97.9	1.83	7.66	35	83.9	351	6.21	26.0

APPENDIX G

Ionization Constants for Weak Acids at 25°C

ACID	FORMULA AND IONIZATION EQUATION	K_a
Acetic	$CH_3COOH \rightleftharpoons H^+ + CH_3COO^-$	1.8×10^{-5}
Arsenic	$H_3AsO_4 \rightleftharpoons H^+ + H_2AsO_4^-$	$K_1 = 2.5 \times 10^{-4}$
	$H_2AsO_4^- \rightleftharpoons H^+ + HAsO_4^{2-}$	$K_2 = 5.6 \times 10^{-8}$
	$HAsO_4^{2-} \rightleftharpoons H^+ + AsO_4^{3-}$	$K_3 = 3.0 \times 10^{-13}$
Arsenous	$H_3AsO_3 \rightleftharpoons H^+ + H_2AsO_3^-$	$K_1 = 6.0 \times 10^{-10}$
	$H_2AsO_3^- \rightleftharpoons H^+ + HAsO_3^{2-}$	$K_2 = 3.0 \times 10^{-14}$
Benzoic	$C_6H_5COOH \rightleftharpoons H^+ + C_6H_5COO^-$	6.3×10^{-5}
Boric	$H_3BO_3 \rightleftharpoons H^+ + H_2BO_3^-$	$K_1 = 7.3 \times 10^{-10}$
	$H_2BO_3^- \rightleftharpoons H^+ + HBO_3^{2-}$	$K_2 = 1.8 \times 10^{-13}$
	$HBO_3^{2-} \rightleftharpoons H^+ + BO_3^{3-}$	$K_3 = 1.6 \times 10^{-14}$
Carbonic	$H_2CO_3 \rightleftharpoons H^+ + HCO_3^-$	$K_1 = 4.2 \times 10^{-7}$
	$HCO_3^- \rightleftharpoons H^+ + CO_3^{2-}$	$K_2 = 4.8 \times 10^{-11}$
Citric	$H_3C_6H_5O_7 \rightleftharpoons H^+ + H_2C_6H_5O_7^-$	$K_1 = 7.4 \times 10^{-3}$
	$H_2C_6H_5O_7^- \rightleftharpoons H^+ + HC_6H_5O_7^{2-}$	$K_2 = 1.7 \times 10^{-5}$
	$HC_6H_5O_7^{2-} \rightleftharpoons H^+ + C_6H_5O_7^{3-}$	$K_3 = 4.0 \times 10^{-7}$
Cyanic	$HOCN \rightleftharpoons H^+ + OCN^-$	3.5×10^{-4}
Formic	$HCOOH \rightleftharpoons H^+ + HCOO^-$	1.8×10^{-4}
Hydrazoic	$HN_3 \rightleftharpoons H^+ + N_3^-$	1.9×10^{-5}
Hydrocyanic	$HCN \rightleftharpoons H^+ + CN^-$	4.0×10^{-10}
Hydrofluoric	$HF \rightleftharpoons H^+ + F^-$	7.2×10^{-4}
Hydrogen peroxide	$H_2O_2 \rightleftharpoons H^+ + HO_2^-$	2.4×10^{-12}
Hydrosulfuric	$H_2S \rightleftharpoons H^+ + HS^-$	$K_1 = 1.0 \times 10^{-7}$
	$HS^- \rightleftharpoons H^+ + S^{2-}$	$K_2 = 1.3 \times 10^{-13}$
Hypobromous	$HOBr \rightleftharpoons H^+ + OBr^-$	2.5×10^{-9}
Hypochlorous	$HOCl \rightleftharpoons H^+ + OCl^-$	3.5×10^{-8}
Nitrous	$HNO_2 \rightleftharpoons H^+ + NO_2^-$	4.5×10^{-4}
Oxalic	$H_2C_2O_4 \rightleftharpoons H^+ + HC_2O_4^-$	$K_1 = 5.9 \times 10^{-2}$
	$HC_2O_4^- \rightleftharpoons H^+ + C_2O_4^{2-}$	$K_2 = 6.4 \times 10^{-5}$
Phenol	$HC_6H_5O \rightleftharpoons H^+ + C_6H_5O^-$	1.3×10^{-10}
Phosphoric	$H_3PO_4 \rightleftharpoons H^+ + H_2PO_4^-$	$K_1 = 7.5 \times 10^{-3}$
	$H_2PO_4^- \rightleftharpoons H^+ + HPO_4^{2-}$	$K_2 = 6.2 \times 10^{-8}$
	$HPO_4^{2-} \rightleftharpoons H^+ + PO_4^{3-}$	$K_3 = 3.6 \times 10^{-13}$
Phosphorous	$H_3PO_3 \rightleftharpoons H^+ + H_2PO_3^-$	$K_1 = 1.6 \times 10^{-2}$
	$H_2PO_3^- \rightleftharpoons H^+ + HPO_3^{2-}$	$K_2 = 7.0 \times 10^{-7}$
Selenic	$H_2SeO_4 \rightleftharpoons H^+ + HSeO_4^-$	$K_1 =$ very large
	$HSeO_4^- \rightleftharpoons H^+ + SeO_4^{2-}$	$K_2 = 1.2 \times 10^{-2}$
Selenous	$H_2SeO_3 \rightleftharpoons H^+ + HSeO_3^-$	$K_1 = 2.7 \times 10^{-3}$
	$HSeO_3^- \rightleftharpoons H^+ + SeO_3^{2-}$	$K_2 = 2.5 \times 10^{-7}$
Sulfuric	$H_2SO_4 \rightleftharpoons H^+ + HSO_4^-$	$K_1 =$ very large
	$HSO_4^- \rightleftharpoons H^+ + SO_4^{2-}$	$K_2 = 1.2 \times 10^{-2}$
Sulfurous	$H_2SO_3 \rightleftharpoons H^+ + HSO_3^-$	$K_1 = 1.2 \times 10^{-2}$
	$HSO_3^- \rightleftharpoons H^+ + SO_3^{2-}$	$K_2 = 6.2 \times 10^{-8}$
Tellurous	$H_2TeO_3 \rightleftharpoons H^+ + HTeO_3^-$	$K_1 = 2 \times 10^{-3}$
	$HTeO_3^- \rightleftharpoons H^+ + TeO_3^{2-}$	$K_2 = 1 \times 10^{-8}$

APPENDIX H

Ionization Constants for Weak Bases at 25°C

BASE	FORMULA AND IONIZATION EQUATION	K_b
Ammonia	$NH_3 + H_2O \rightleftharpoons NH_4^+ + OH^-$	1.8×10^{-5}
Aniline	$C_6H_5NH_2 + H_2O \rightleftharpoons C_6H_5NH_3^+ + OH^-$	4.2×10^{-10}
Dimethylamine	$(CH_3)_2NH + H_2O \rightleftharpoons (CH_3)_2NH_2^+ + OH^-$	7.4×10^{-4}
Ethylenediamine	$(CH_2)_2(NH_2)_2 + H_2O \rightleftharpoons (CH_2)_2(NH_2)_2H^+ + OH^-$	$K_1 = 8.5 \times 10^{-5}$
	$(CH_2)_2(NH_2)_2H^+ + H_2O \rightleftharpoons (CH_2)_2(NH_2)_2H_2^{2+} + OH^-$	$K_2 = 2.7 \times 10^{-8}$
Hydrazine	$N_2H_4 + H_2O \rightleftharpoons N_2H_5^+ + OH^-$	$K_1 = 8.5 \times 10^{-7}$
	$N_2H_5^+ + H_2O \rightleftharpoons N_2H_6^{2+} + OH^-$	$K_2 = 8.9 \times 10^{-16}$
Hydroxylamine	$NH_2OH + H_2O \rightleftharpoons NH_3OH^+ + OH^-$	6.6×10^{-9}
Methylamine	$CH_3NH_2 + H_2O \rightleftharpoons CH_3NH_3^+ + OH^-$	5.0×10^{-4}
Pyridine	$C_5H_5N + H_2O \rightleftharpoons C_5H_5NH^+ + OH^-$	1.5×10^{-9}
Trimethylamine	$(CH_3)_3N + H_2O \rightleftharpoons (CH_3)_3NH^+ + OH^-$	7.4×10^{-5}

APPENDIX I

Solubility Product Constants for Some Inorganic Compounds at 25°C

SUBSTANCE	K_{sp}
Aluminum compounds	
$AlAsO_4$	1.6×10^{-16}
$Al(OH)_3$	1.9×10^{-33}
$AlPO_4$	1.3×10^{-20}
Antimony compounds	
Sb_2S_3	1.6×10^{-93}
Barium compounds	
$Ba_3(AsO_4)_2$	1.1×10^{-13}
$BaCO_3$	8.1×10^{-9}
$BaC_2O_4 \cdot 2H_2O$*	1.1×10^{-7}
$BaCrO_4$	2.0×10^{-10}
BaF_2	1.7×10^{-6}
$Ba(OH)_2 \cdot 8H_2O$*	5.0×10^{-3}
$Ba_3(PO_4)_2$	1.3×10^{-29}
$BaSeO_4$	2.8×10^{-11}
$BaSO_3$	8.0×10^{-7}
$BaSO_4$	1.1×10^{-10}
Bismuth compounds	
$BiOCl$	7.0×10^{-9}
$BiO(OH)$	1.0×10^{-12}
$Bi(OH)_3$	3.2×10^{-40}
BiI_3	8.1×10^{-19}
$BiPO_4$	1.3×10^{-23}
Bi_2S_3	1.6×10^{-72}
Cadmium compounds	
$Cd_3(AsO_4)_2$	2.2×10^{-32}
$CdCO_3$	2.5×10^{-14}
$Cd(CN)_2$	1.0×10^{-8}
$Cd_2[Fe(CN)_6]$	3.2×10^{-17}
$Cd(OH)_2$	1.2×10^{-14}
CdS	3.6×10^{-29}
Calcium compounds	
$Ca_3(AsO_4)_2$	6.8×10^{-19}
$CaCO_3$	3.8×10^{-9}
$CaCrO_4$	7.1×10^{-4}
$CaC_2O_4 \cdot H_2O$*	2.3×10^{-9}
CaF_2	3.9×10^{-11}
$Ca(OH)_2$	7.9×10^{-6}
$CaHPO_4$	2.7×10^{-7}
$Ca(H_2PO_4)_2$	1.0×10^{-3}
$Ca_3(PO_4)_2$	1.0×10^{-25}
$CaSO_3 \cdot 2H_2O$*	1.3×10^{-8}
$CaSO_4 \cdot 2H_2O$*	2.4×10^{-5}
Chromium compounds	
$CrAsO_4$	7.8×10^{-21}
$Cr(OH)_3$	6.7×10^{-31}
$CrPO_4$	2.4×10^{-23}
Cobalt compounds	
$Co_3(AsO_4)_2$	7.6×10^{-29}
$CoCO_3$	8.0×10^{-13}
$Co(OH)_2$	2.5×10^{-16}
CoS (α)	5.9×10^{-21}
$Co(OH)_3$	4.0×10^{-45}
Co_2S_3	2.6×10^{-124}
Copper compounds	
$CuBr$	5.3×10^{-9}
$CuCl$	1.9×10^{-7}
$CuCN$	3.2×10^{-20}
Cu_2O ($Cu^+ + OH^-$)†	1.0×10^{-14}
CuI	5.1×10^{-12}
Cu_2S	1.6×10^{-48}
$CuSCN$	1.6×10^{-11}
$Cu_3(AsO_4)_2$	7.6×10^{-36}
$CuCO_3$	2.5×10^{-10}
$Cu_2[Fe(CN)_6]$	1.3×10^{-16}
$Cu(OH)_2$	1.6×10^{-19}
CuS	8.7×10^{-36}
Gold compounds	
$AuBr$	5.0×10^{-17}
$AuCl$	2.0×10^{-13}
AuI	1.6×10^{-23}
$AuBr_3$	4.0×10^{-36}
$AuCl_3$	3.2×10^{-25}
$Au(OH)_3$	1×10^{-53}
AuI_3	1.0×10^{-46}
Iron compounds	
$FeCO_3$	3.5×10^{-11}
$Fe(OH)_2$	7.9×10^{-15}
FeS	4.9×10^{-18}
$Fe_4[Fe(CN)_6]_3$	3.0×10^{-41}
$Fe(OH)_3$	6.3×10^{-38}
Fe_2S_3	1.4×10^{-88}
Lead compounds	
$Pb_3(AsO_4)_2$	4.1×10^{-36}
$PbBr_2$	6.3×10^{-6}

APPENDIX I (*Continued*)

Solubility Product Constants for Some Inorganic Compounds at 25°C

SUBSTANCE	K_{sp}
$PbCO_3$	1.5×10^{-13}
$PbCl_2$	1.7×10^{-5}
$PbCrO_4$	1.8×10^{-14}
PbF_2	3.7×10^{-8}
$Pb(OH)_2$	2.8×10^{-16}
PbI_2	8.7×10^{-9}
$Pb_3(PO_4)_2$	3.0×10^{-44}
$PbSeO_4$	1.5×10^{-7}
$PbSO_4$	1.8×10^{-8}
PbS	8.4×10^{-28}
Magnesium compounds	
$Mg_3(AsO_4)_2$	2.1×10^{-20}
$MgCO_3 \cdot 3H_2O$*	4.0×10^{-5}
MgC_2O_4	8.6×10^{-5}
MgF_2	6.4×10^{-9}
$Mg(OH)_2$	1.5×10^{-11}
$MgNH_4PO_4$	2.5×10^{-12}
Manganese compounds	
$Mn_3(AsO_4)_2$	1.9×10^{-11}
$MnCO_3$	1.8×10^{-11}
$Mn(OH)_2$	4.6×10^{-14}
MnS	5.1×10^{-15}
$Mn(OH)_3$	$\sim 1 \times 10^{-36}$
Mercury compounds	
Hg_2Br_2	1.3×10^{-22}
Hg_2CO_3	8.9×10^{-17}
Hg_2Cl_2	1.1×10^{-18}
Hg_2CrO_4	5.0×10^{-9}
Hg_2I_2	4.5×10^{-29}
$Hg_2O \cdot H_2O$ ($Hg_2^{2+} + 2OH^-$)*†	1.6×10^{-23}
Hg_2SO_4	6.8×10^{-7}
Hg_2S	5.8×10^{-44}
$Hg(CN)_2$	3.0×10^{-23}
$Hg(OH)_2$	2.5×10^{-26}
HgI_2	4.0×10^{-29}
HgS	3.0×10^{-53}
Nickel compounds	
$Ni_3(AsO_4)_2$	1.9×10^{-26}
$NiCO_3$	6.6×10^{-9}
$Ni(CN)_2$	3.0×10^{-23}
$Ni(OH)_2$	2.8×10^{-16}
NiS (α)	3.0×10^{-21}
NiS (β)	1.0×10^{-26}
NiS (γ)	2.0×10^{-28}
Silver compounds	
Ag_3AsO_4	1.1×10^{-20}
$AgBr$	3.3×10^{-13}
Ag_2CO_3	8.1×10^{-12}
$AgCl$	1.8×10^{-10}
Ag_2CrO_4	9.0×10^{-12}
$AgCN$	1.2×10^{-16}
$Ag_4[Fe(CN)_6]$	1.6×10^{-41}
Ag_2O ($Ag^+ + OH^-$)†	2.0×10^{-8}
AgI	1.5×10^{-16}
Ag_3PO_4	1.3×10^{-20}
Ag_2SO_3	1.5×10^{-14}
Ag_2SO_4	1.7×10^{-5}
Ag_2S	1.0×10^{-49}
$AgSCN$	1.0×10^{-12}
Strontium compounds	
$Sr_3(AsO_4)_2$	1.3×10^{-18}
$SrCO_3$	9.4×10^{-10}
$SrC_2O_4 \cdot 2H_2O$*	5.6×10^{-8}
$SrCrO_4$	3.6×10^{-5}
$Sr(OH)_2 \cdot 8H_2O$*	3.2×10^{-4}
$Sr_3(PO_4)_2$	1.0×10^{-31}
$SrSO_3$	4.0×10^{-8}
$SrSO_4$	2.8×10^{-7}
Tin compounds	
$Sn(OH)_2$	2.0×10^{-26}
SnI_2	1.0×10^{-4}
SnS	1.0×10^{-28}
$Sn(OH)_4$	1×10^{-57}
SnS_2	1×10^{-70}
Zinc compounds	
$Zn_3(AsO_4)_2$	1.1×10^{-27}
$ZnCO_3$	1.5×10^{-11}
$Zn(CN)_2$	8.0×10^{-12}
$Zn_3[Fe(CN)_6]$	4.1×10^{-16}
$Zn(OH)_2$	4.5×10^{-17}
$Zn_3(PO_4)_2$	9.1×10^{-33}
ZnS	1.1×10^{-21}

*Since $[H_2O]$ does not appear in equilibrium constants for equilibria in aqueous solution in general, it does *not* appear in the K_{sp} expressions for hydrated solids.

†Very small amounts of oxides dissolve in water to give the ions indicated in parentheses. Solid hydroxides are unstable and decompose to oxides as rapidly as they are formed.

APPENDIX J

Formation Constants for Some Complex Ions in Aqueous Solution

FORMATION EQUILIBRIUM	*K*
$Ag^+ + 2Br^- \rightleftharpoons [AgBr_2]^-$	1.3×10^7
$Ag^+ + 2Cl^- \rightleftharpoons [AgCl_2]^-$	2.5×10^5
$Ag^+ + 2CN^- \rightleftharpoons [Ag(CN)_2]^-$	5.6×10^{18}
$Ag^+ + 2S_2O_3^{2-} \rightleftharpoons [Ag(S_2O_3)_2]^{3-}$	2.0×10^{13}
$Ag^+ + 2NH_3 \rightleftharpoons [Ag(NH_3)_2]^+$	1.6×10^7
$Al^{3+} + 6F^- \rightleftharpoons [AlF_6]^{3-}$	5.0×10^{23}
$Al^{3+} + 4OH^- \rightleftharpoons [Al(OH)_4]^-$	7.7×10^{33}
$Au^+ + 2CN^- \rightleftharpoons [Au(CN)_2]^-$	2.0×10^{38}
$Cd^{2+} + 4CN^- \rightleftharpoons [Cd(CN)_4]^{2-}$	1.3×10^{17}
$Cd^{2+} + 4Cl^- \rightleftharpoons [CdCl_4]^{2-}$	1.0×10^4
$Cd^{2+} + 4NH_3 \rightleftharpoons [Cd(NH_3)_4]^{2+}$	1.0×10^7
$Co^{2+} + 6NH_3 \rightleftharpoons [Co(NH_3)_6]^{2+}$	7.7×10^4
$Cu^+ + 2CN^- \rightleftharpoons [Cu(CN)_2]^-$	1.0×10^{16}
$Cu^+ + 2Cl^- \rightleftharpoons [CuCl_2]^-$	1.0×10^5
$Cu^{2+} + 4NH_3 \rightleftharpoons [Cu(NH_3)_4]^{2+}$	1.2×10^{12}
$Fe^{2+} + 6CN^- \rightleftharpoons [Fe(CN)_6]^{4-}$	7.7×10^{36}
$Hg^{2+} + 4Cl^- \rightleftharpoons [HgCl_4]^{2-}$	1.2×10^{15}
$Ni^{2+} + 4CN^- \rightleftharpoons [Ni(CN)_4]^{2-}$	1.0×10^{31}
$Ni^{2+} + 6NH_3 \rightleftharpoons [Ni(NH_3)_6]^{2+}$	5.6×10^8
$Zn^{2+} + 4OH^- \rightleftharpoons [Zn(OH)_4]^{2-}$	2.9×10^{15}
$Zn^{2+} + 4NH_3 \rightleftharpoons [Zn(NH_3)_4]^{2+}$	2.9×10^9

APPENDIX K

Standard Reduction Potentials in Aqueous Solution at 25°C

ACIDIC SOLUTION	STANDARD REDUCTION POTENTIAL, E^0 (volts)
F_2 (g) + 2e $\longrightarrow$ $2F^-$ (aq)	2.87
Co^{3+} (aq) + e^- $\longrightarrow$ Co^{2+} (aq)	1.82
Pb^{4+} (aq) + $2e^-$ $\longrightarrow$ Pb^{2+} (aq)	1.8
H_2O_2 (aq) + $2H^+$ (aq) + $2e^-$ $\longrightarrow$ $2H_2O$	1.77
NiO_2 (s) + $4H^+$ (aq) + $2e^-$ $\longrightarrow$ Ni^{2+} (aq) + $2H_2O$	1.7
PbO_2 (s) + SO_4^{2-} (aq) + $4H^+$ (aq) + $2e^-$ $\longrightarrow$ $PbSO_4$ (s) + $2H_2O$	1.685
Au^+ (aq) + e^- $\longrightarrow$ Au (s)	1.68
2HClO (aq) + $2H^+$ (aq) + $2e^-$ $\longrightarrow$ Cl_2 (g) + $2H_2O$	1.63
Ce^{4+} (aq) + e^- $\longrightarrow$ Ce^{3+} (aq)	1.61
$NaBiO_3$ (s) + $6H^+$ (aq) + $2e^-$ $\longrightarrow$ Bi^{3+} (aq) + Na^+ (aq) + $3H_2O$	~1.6
MnO_4^- (aq) + $8H^+$ (aq) + $5e^-$ $\longrightarrow$ Mn^{2+} (aq) + $4H_2O$	1.51
Au^{3+} (aq) + $3e^-$ $\longrightarrow$ Au (s)	1.50
ClO_3^- (aq) + $6H^+$ (aq) + $5e^-$ $\longrightarrow$ $\frac{1}{2}Cl_2$ (g) + $3H_2O$	1.47
BrO_3^- (aq) + $6H^+$ (aq) + $6e^-$ $\longrightarrow$ Br^- (aq) + $3H_2O$	1.44
Cl_2 (g) + $2e^-$ $\longrightarrow$ $2Cl^-$ (aq)	1.360
$Cr_2O_7^{2-}$ (aq) + $14H^+$ (aq) + $6e^-$ $\longrightarrow$ $2Cr^{3+}$ (aq) + $7H_2O$	1.33
$N_2H_5^+$ (aq) + $3H^+$ (aq) + $2e^-$ $\longrightarrow$ $2NH_4^+$ (aq)	1.24
MnO_2 (s) + $4H^+$ (aq) + $2e^-$ $\longrightarrow$ Mn^{2+} (aq) + $2H_2O$	1.23
O_2 (g) + $4H^+$ (aq) + $4e^-$ $\longrightarrow$ $2H_2O$	1.229
Pt^{2+} (aq) + $2e^-$ $\longrightarrow$ Pt (s)	1.2
IO_3^- (aq) + $6H^+$ (aq) + $5e^-$ $\longrightarrow$ $\frac{1}{2}I_2$ (aq) + $3H_2O$	1.195
ClO_4^- (aq) + $2H^+$ (aq) + $2e^-$ $\longrightarrow$ ClO_3^- (aq) + H_2O	1.19
Br_2 (ℓ) + $2e^-$ $\longrightarrow$ $2Br^-$ (aq)	1.08
$AuCl_4^-$ (aq) + $3e^-$ $\longrightarrow$ Au (s) + $4Cl^-$ (aq)	1.00
Pd^{2+} (aq) + $2e^-$ $\longrightarrow$ Pd (s)	0.987
NO_3^- (aq) + $4H^+$ (aq) + $3e^-$ $\longrightarrow$ NO (g) + $2H_2O$	0.96
NO_3^- (aq) + $3H^+$ (aq) + $2e^-$ $\longrightarrow$ HNO_2 (aq) + H_2O	0.94
$2Hg^{2+}$ (aq) + $2e^-$ $\longrightarrow$ Hg_2^{2+} (aq)	0.920
Hg^{2+} (aq) + $2e^-$ $\longrightarrow$ Hg (ℓ)	0.855
Ag^+ (aq) + e^- $\longrightarrow$ Ag (s)	0.7994
Hg_2^{2+} (aq) + $2e^-$ $\longrightarrow$ 2Hg (ℓ)	0.789
Fe^{3+} (aq) + e^- $\longrightarrow$ Fe^{2+} (aq)	0.771
$SbCl_6^-$ (aq) + $2e^-$ $\longrightarrow$ $SbCl_4^-$ (aq) + $2Cl^-$ (aq)	0.75
$[PtCl_4]^{2-}$ (aq) + $2e^-$ $\longrightarrow$ Pt (s) + $4Cl^-$ (aq)	0.73
O_2 (g) + $2H^+$ (aq) + $2e^-$ $\longrightarrow$ H_2O_2 (aq)	0.682
$[PtCl_6]^{2-}$ (aq) + $2e^-$ $\longrightarrow$ $[PtCl_4]^{2-}$ (aq) + $2Cl^-$ (aq)	0.68
H_3AsO_4 (aq) + $2H^+$ (aq) + $2e^-$ $\longrightarrow$ H_3AsO_3 (aq) + H_2O	0.58
I_2 (s) + $2e^-$ $\longrightarrow$ $2I^-$ (aq)	0.535
TeO_2 (s) + $4H^+$ (aq) + $4e^-$ $\longrightarrow$ Te (s) + $2H_2O$	0.529
Cu^+ (aq) + e^- $\longrightarrow$ Cu (s)	0.521
$[RhCl_6]^{3-}$ (aq) + $3e^-$ $\longrightarrow$ Rh (s) + $6Cl^-$ (aq)	0.44
Cu^{2+} (aq) + $2e^-$ $\longrightarrow$ Cu (s)	0.337
$HgCl_2$ (s) + $2e^-$ $\longrightarrow$ 2Hg (ℓ) + $2Cl^-$ (aq)	0.27
AgCl (s) + e^- $\longrightarrow$ Ag (s) + Cl^- (aq)	0.222
SO_4^{2-} (aq) + $4H^+$ (aq) + $2e^-$ $\longrightarrow$ SO_2 (g) + $2H_2O$	0.20
SO_4^{2-} (aq) + $4H^+$ (aq) + $2e^-$ $\longrightarrow$ H_2SO_3 (aq) + H_2O	0.17

Standard Reduction Potentials in Aqueous Solution at 25°C

ACIDIC SOLUTION	STANDARD REDUCTION POTENTIAL, E^0 (volts)
Cu^{2+} (aq) + e^- ⟶ Cu^+ (aq)	0.153
Sn^{4+} (aq) + $2e^-$ ⟶ Sn^{2+} (aq)	0.15
S (s) + $2H^+$ (aq) + $2e^-$ ⟶ H_2S (aq)	0.14
AgBr (s) + e^- ⟶ Ag (s) + Br^- (aq)	0.0713
$2H^+$ (aq) + $2e^-$ ⟶ H_2 (g) (reference electrode)	0.0000
N_2O (g) + $6H^+$ (aq) + H_2O + $4e^-$ ⟶ $2NH_3OH^+$ (aq)	−0.05
Pb^{2+} (aq) + $2e^-$ ⟶ Pb (s)	−0.126
Sn^{2+} (aq) + $2e^-$ ⟶ Sn (s)	−0.14
AgI (s) + e^- ⟶ Ag (s) + I^- (aq)	−0.15
$[SnF_6]^{2-}$ (aq) + $4e^-$ ⟶ Sn (s) + $6F^-$ (aq)	−0.25
Ni^{2+} (aq) + $2e^-$ ⟶ Ni (s)	−0.25
Co^{2+} (aq) + $2e^-$ ⟶ Co (s)	−0.28
Tl^+ (aq) + e^- ⟶ Tl (s)	−0.34
$PbSO_4$ (s) + $2e^-$ ⟶ Pb (s) + SO_4^{2-} (aq)	−0.356
Se (s) + $2H^+$ (aq) + $2e^-$ ⟶ H_2Se (aq)	−0.40
Cd^{2+} (aq) + $2e^-$ ⟶ Cd (s)	−0.403
Cr^{3+} (aq) + e^- ⟶ Cr^{2+} (aq)	−0.41
Fe^{2+} (aq) + $2e^-$ ⟶ Fe (s)	−0.44
$2CO_2$ (g) + $2H^+$ (aq) + $2e^-$ ⟶ $(COOH)_2$ (aq)	−0.49
Ga^{3+} (aq) + $3e^-$ ⟶ Ga (s)	−0.53
HgS (s) + $2H^+$ (aq) + $2e^-$ ⟶ Hg (ℓ) + H_2S (g)	−0.72
Cr^{3+} (aq) + $3e^-$ ⟶ Cr (s)	−0.74
Zn^{2+} (aq) + $2e^-$ ⟶ Zn (s)	−0.763
Cr^{2+} (aq) + $2e^-$ ⟶ Cr (s)	−0.91
FeS (s) + $2e^-$ ⟶ Fe (s) + S^{2-} (aq)	−1.01
Mn^{2+} (aq) + $2e^-$ ⟶ Mn (s)	−1.18
V^{2+} (aq) + $2e^-$ ⟶ V (s)	−1.18
CdS (s) + $2e^-$ ⟶ Cd (s) + S^{2-} (aq)	−1.21
ZnS (s) + $2e^-$ ⟶ Zn (s) + S^{2-} (aq)	−1.44
Zr^{4+} (aq) + $4e^-$ ⟶ Zr (s)	−1.53
Al^{3+} (aq) + $3e^-$ ⟶ Al (s)	−1.66
H_2 (g) + $2e^-$ ⟶ $2H^-$ (aq)	−2.25
Mg^{2+} (aq) + $2e^-$ ⟶ Mg (s)	−2.37
Na^+ (aq) + e^- ⟶ Na (s)	−2.714
Ca^{2+} (aq) + $2e^-$ ⟶ Ca (s)	−2.87
Sr^{2+} (aq) + $2e^-$ ⟶ Sr (s)	−2.89
Ba^{2+} (aq) + $2e^-$ ⟶ Ba (s)	−2.90
Rb^+ (aq) + e^- ⟶ Rb (s)	−2.925
K^+ (aq) + e^- ⟶ K (s)	−2.925
Li^+ (aq) + e^- ⟶ Li (s)	−3.045

APPENDIX K (*Continued*)

Standard Reduction Potentials in Aqueous Solution at 25°C

BASIC SOLUTION	STANDARD REDUCTION POTENTIAL, E^0 (volts)
ClO^- (aq) + H_2O + $2e^-$ ⟶ Cl^- (aq) + $2OH^-$ (aq)	0.89
OOH^- (aq) + H_2O + $2e^-$ ⟶ $3OH^-$ (aq)	0.88
$2NH_2OH$ (aq) + $2e^-$ ⟶ N_2H_4 (aq) + $2OH^-$ (aq)	0.74
ClO_3^- (aq) + $3H_2O$ + $6e^-$ ⟶ Cl^- (aq) + $6OH^-$ (aq)	0.62
MnO_4^- (aq) + $2H_2O$ + $3e^-$ ⟶ MnO_2 (s) + $4OH^-$ (aq)	0.588
MnO_4^- (aq) + e^- ⟶ MnO_4^{2-} (aq)	0.564
NiO_2 (s) + $2H_2O$ + $2e^-$ ⟶ $Ni(OH)_2$ (s) + $2OH^-$ (aq)	0.49
Ag_2CrO_4 (s) + $2e^-$ ⟶ $2Ag$ (s) + CrO_4^{2-} (aq)	0.446
O_2 (g) + $2H_2O$ + $4e^-$ ⟶ $4OH^-$ (aq)	0.40
ClO_4^- (aq) + H_2O + $2e^-$ ⟶ ClO_3^- (aq) + $2OH^-$ (aq)	0.36
Ag_2O (s) + H_2O + $2e^-$ ⟶ $2Ag$ (s) + $2OH^-$ (aq)	0.34
$2NO_2^-$ (aq) + $3H_2O$ + $4e^-$ ⟶ N_2O (g) + $6OH^-$ (aq)	0.15
N_2H_4 (aq) + $2H_2O$ + $2e^-$ ⟶ $2NH_3$ (aq) + $2OH^-$ (aq)	0.10
$[Co(NH_3)_6]^{3+}$ (aq) + e^- ⟶ $[Co(NH_3)_6]^{2+}$ (aq)	0.10
HgO (s) + H_2O + $2e^-$ ⟶ Hg (ℓ) + $2OH^-$ (aq)	0.0984
O_2 (g) + H_2O + $2e^-$ ⟶ OOH^- (aq) + OH^- (aq)	0.076
NO_3^- (aq) + H_2O + $2e^-$ ⟶ NO_2^- (aq) + $2OH^-$ (aq)	0.01
MnO_2 (s) + $2H_2O$ + $2e^-$ ⟶ $Mn(OH)_2$ (s) + $2OH^-$ (aq)	−0.05
CrO_4^{2-} (aq) + $4H_2O$ + $3e^-$ ⟶ $Cr(OH)_3$ (s) + $5OH^-$ (aq)	−0.12
$Cu(OH)_2$ (s) + $2e^-$ ⟶ Cu (s) + $2OH^-$ (aq)	−0.36
S (s) + $2e^-$ ⟶ S^{2-} (aq)	−0.48
$Fe(OH)_3$ (s) + e^- ⟶ $Fe(OH)_2$ (s) + OH^- (aq)	−0.56
$2H_2O$ + $2e^-$ ⟶ H_2 (g) + $2OH^-$ (aq)	−0.8277
$2NO_3^-$ (aq) + $2H_2O$ + $2e^-$ ⟶ N_2O_4 (g) + $4OH^-$ (aq)	−0.85
$Fe(OH)_2$ (s) + $2e^-$ ⟶ Fe (s) + $2OH^-$ (aq)	−0.877
SO_4^{2-} (aq) + H_2O + $2e^-$ ⟶ SO_3^{2-} (aq) + $2OH^-$ (aq)	−0.93
N_2 (g) + $4H_2O$ + $4e^-$ ⟶ N_2H_4 (aq) + $4OH^-$ (aq)	−1.15
$[Zn(OH)_4]^{2-}$ (aq) + $2e^-$ ⟶ Zn (s) + $4OH^-$ (aq)	−1.22
$Zn(OH)_2$ (s) + $2e^-$ ⟶ Zn (s) + $2OH^-$ (aq)	−1.245
$[Zn(CN)_4]^{2-}$ (aq) + $2e^-$ ⟶ Zn (s) + $4CN^-$ (aq)	−1.26
$Cr(OH)_3$ (s) + $3e^-$ ⟶ Cr (s) + $3OH^-$ (aq)	−1.30
SiO_3^{2-} (aq) + $3H_2O$ + $4e^-$ ⟶ Si (s) + $6OH^-$ (aq)	−1.70

APPENDIX L
Selected Thermodynamic Values*

SPECIES	ΔH_f°(298.15K) kJ/mol	S°(298.15K) J/K·mol	ΔG_f°(298.15K) kJ/mol
Aluminum			
Al(s)	0	28.3	0
$AlCl_3$(s)	−704.2	110.67	−628.8
Al_2O_3(s)	−1675.7	50.92	−1582.3
Barium			
$BaCl_2$(s)	−858.6	123.68	−810.4
BaO(s)	−553.5	70.42	−525.1
$BaSO_4$(s)	−1473.2	132.2	−1362.2
Beryllium			
Be(s)	0	9.5	0
$Be(OH)_2$	−902.5	51.9	−815.0
Bromine			
Br(g)	111.884	175.022	82.396
$Br_2(\ell)$	0	152.2	0
Br_2(g)	−30.907	245.463	3.110
BrF_3(g)	−255.60	292.53	−229.43
HBr(g)	−36.40	198.695	−53.45
Calcium			
Ca(s)	0	41.42	0
Ca(g)	178.2	158.884	144.3
Ca^{2+}(g)	1925.90	—	—
CaC_2(s)	−59.8	69.96	−64.9
$CaCO_3$(s; calcite)	−1206.92	92.9	−1128.79
$CaCl_2$(s)	−795.8	104.6	−748.1
CaF_2(s)	−1219.6	68.87	−1167.3
CaH_2(s)	−186.2	42.	−147.2
CaO(s)	−635.09	39.75	−604.03
CaS(s)	−482.4	56.5	−477.4
$Ca(OH)_2$(s)	−986.09	83.39	−898.49
$Ca(OH)_2$(aq)	−1002.82	−74.5	−868.07
$CaSO_4$(s)	−1434.11	106.7	−1321.79
Carbon			
C(s, graphite)	0	5.740	0
C(s, diamond)	1.895	2.377	2.900
C(g)	716.682	158.096	671.257
$CCl_4(\ell)$	−135.44	216.40	−65.21
CCl_4(g)	−102.9	309.85	−60.59
$CHCl_3$(liq)	−134.47	201.7	−73.66
$CHCl_3$(g)	−103.14	295.71	−70.34
CH_4(g, methane)	−74.81	186.264	−50.72
C_2H_2(g, ethyne)	226.73	200.94	209.20
C_2H_4(g, ethene)	52.26	219.56	68.15
C_2H_6(g, ethane)	−84.68	229.60	−32.82
C_3H_8(g, propane)	−103.8	269.9	−23.49
$C_6H_6(\ell$, benzene)	49.03	172.8	124.5
$CH_3OH(\ell$, methanol)	−238.66	126.8	−166.27
CH_3OH(g, methanol)	−200.66	239.81	−161.96
$C_2H_5OH(\ell$, ethanol)	−277.69	160.7	−174.78
C_2H_5OH(g, ethanol)	−235.10	282.70	−168.49
CO(g)	−110.525	197.674	−137.168

APPENDIX L (*Continued*)

Selected Thermodynamic Values*

SPECIES	ΔH_f°(298.15K) kJ/mol	S°(298.15K) J/K·mol	ΔG_f°(298.15K) kJ/mol
CO_2(g)	−393.509	213.74	−394.359
CS_2(g)	117.36	237.84	67.12
$COCl_2$(g)	−218.8	283.53	−204.6
Cesium			
Cs(s)	0	85.23	0
Cs^+(g)	457.964	—	—
CsCl(s)	−443.04	101.17	−414.53
Chlorine			
Cl(g)	121.679	165.198	105.680
Cl^-(g)	−233.13	—	—
Cl_2(g)	0	223.066	0
HCl(g)	−92.307	186.908	−95.299
HCl(aq)	−167.159	56.5	−131.228
Chromium			
Cr(s)	0	23.77	0
Cr_2O_3(s)	−1139.7	81.2	−1058.1
$CrCl_3$(s)	−556.5	123.0	−486.1
Copper			
Cu(s)	0	33.150	0
CuO(s)	−157.3	42.63	−129.7
$CuCl_2$(s)	−220.1	108.07	−175.7
Fluorine			
F_2(g)	0	202.78	0
F(g)	78.99	158.754	61.91
F^-(g)	−255.39	—	—
F^-(aq)	−332.63	−13.8	−278.79
HF(g)	−271.1	173.779	−273.2
HF(aq)	−332.63	−13.8	−278.79
Hydrogen			
H_2(g)	0	130.684	0
H(g)	217.965	114.713	203.247
H^+(g)	1536.202	—	—
$H_2O(\ell)$	−285.830	69.91	−237.129
H_2O(g)	−241.818	188.825	−228.572
$H_2O_2(\ell)$	−187.78	109.6	−120.35
Iodine			
I_2(s)	0	116.135	0
I_2(g)	62.438	260.69	19.327
I(g)	106.838	180.791	70.250
I^-(g)	−197.	—	—
ICl(g)	17.78	247.551	−5.46
Iron			
Fe(s)	0	27.78	0
FeO(s)	−272.	—	—
Fe_2O_3(s, hematite)	−824.2	87.40	−742.2
Fe_3O_4(s, magnetite)	−1118.4	146.4	−1015.4
$FeCl_2$(s)	−341.79	117.95	−302.30
$FeCl_3$(s)	−399.49	142.3	−344.00
FeS_2(s, pyrite)	−178.2	52.93	−166.9
$Fe(CO)_5(\ell)$	−774.0	338.1	−705.3

Selected Thermodynamic Values*

SPECIES	ΔH_f°(298.15K) kJ/mol	S°(298.15K) J/K·mol	ΔG_f°(298.15K) kJ/mol
Lead			
$Pb(s)$	0	64.81	0
$PbCl_2(s)$	−359.41	136.0	−314.10
PbO(s, yellow)	−217.32	68.70	−187.89
$PbS(s)$	−100.4	91.2	−98.7
Lithium			
$Li(s)$	0	29.12	0
$Li^+(g)$	685.783	—	—
$LiOH(s)$	−484.93	42.80	−438.95
$LiOH(aq)$	−508.48	2.80	−450.58
$LiCl(s)$	−408.701	59.33	−384.37
Magnesium			
$Mg(s)$	0	32.68	0
$MgCl_2(s)$	−641.32	89.62	−591.79
$MgO(s)$	−601.70	26.94	−569.43
$Mg(OH)_2(s)$	−924.54	63.18	−833.51
$MgS(s)$	−346.0	50.33	−341.8
Mercury			
$Hg(\ell)$	0	76.02	0
$HgCl_2(s)$	−224.3	146.0	−178.6
HgO(s, red)	−90.83	70.29	−58.539
HgS(s, red)	−58.2	82.4	−50.6
Nickel			
$Ni(s)$	0	29.87	0
$NiO(s)$	−239.7	37.99	−211.7
$NiCl_2(s)$	−305.332	97.65	−259.032
Nitrogen			
$N_2(g)$	0	191.61	0
$N(g)$	472.704	153.298	455.563
$NH_3(g)$	−46.11	192.45	−16.45
$N_2H_4(\ell)$	50.63	121.21	149.34
$NH_4Cl(s)$	−314.43	94.6	−202.87
$NH_4Cl(aq)$	−299.66	169.9	−210.52
$NH_4NO_3(s)$	−365.56	151.08	−183.87
$NH_4NO_3(aq)$	−339.87	259.8	−190.56
$NO(g)$	90.25	210.76	86.55
$NO_2(g)$	33.18	240.06	51.31
$N_2O(g)$	82.05	219.85	104.20
$N_2O_4(g)$	9.16	304.29	97.89
$NOCl(g)$	51.71	261.69	66.08
$HNO_3(\ell)$	−174.10	155.60	−80.71
$HNO_3(g)$	−135.06	266.38	−74.72
$HNO_3(aq)$	−207.36	146.4	−111.25
Oxygen			
$O_2(g)$	0	205.138	0
$O(g)$	249.170	161.055	231.731
$O_3(g)$	142.7	238.93	163.2
Phosphorus			
P_4(s, white)	0	164.36	0
P_4(s, red)	−70.4	91.2	−48.4

APPENDIX L (*Continued*)

Selected Thermodynamic Values*

SPECIES	ΔH_f°(298.15K) kJ/mol	S°(298.15K) J/K·mol	ΔG_f°(298.15K) kJ/mol
P(g)	314.64	163.193	278.25
PH_3(g)	5.4	310.23	13.4
PCl_3(g)	−287.0	311.78	−267.8
P_4O_{10}(s)	−2984.0	228.86	−2697.7
H_3PO_4(s)	−1279.0	110.5	−1119.1
Potassium			
K(s)	0	64.18	0
KCl(s)	−436.747	82.59	−409.14
$KClO_3$(s)	−397.73	143.1	−296.25
KI(s)	−327.90	106.32	−324.892
KOH(s)	−424.764	78.9	−379.08
KOH(aq)	−482.37	91.6	−440.50
Silicon			
Si(s)	0	18.83	0
$SiBr_4$(ℓ)	−457.3	277.8	−443.9
SiC(s)	−65.3	16.61	−62.8
$SiCl_4$(g)	−657.01	330.73	−616.98
SiH_4(g)	34.3	204.62	56.9
SiF_4(g)	−1614.94	282.49	−1572.65
SiO_2(s, quartz)	−910.94	41.84	−856.64
Silver			
Ag(s)	0	42.55	0
Ag_2O(s)	−31.05	121.3	−11.20
AgCl(s)	−127.068	96.2	−109.789
$AgNO_3$(s)	−124.39	140.92	−33.41
Sodium			
Na(s)	0	51.21	0
Na(g)	107.32	153.712	76.761
Na^+(g)	609.358	—	—
NaBr(s)	−361.062	86.82	−348.983
NaCl(s)	−411.153	72.13	−384.138
NaCl(g)	−176.65	229.81	−196.66
NaCl(aq)	−407.27	115.5	−393.133
NaOH(s)	−425.609	64.455	−379.494
NaOH(aq)	−470.114	48.1	−419.150
Na_2CO_3(s)	−1130.68	134.98	−1044.44
Sulfur			
S(s, rhombic)	0	31.80	0
S(g)	278.805	167.821	238.250
S_2Cl_2(g)	−18.4	331.5	−31.8
SF_6(g)	1209.	291.82	−1105.3
H_2S(g)	−20.63	205.79	−33.56
SO_2(g)	−296.830	248.22	−300.194
SO_3(g)	−395.72	256.76	−371.06
$SOCl_2$(g)	−212.5	309.77	−198.3
H_2SO_4(ℓ)	−813.989	156.904	−690.003
H_2SO_4(aq)	−909.27	20.1	−744.53
Tin			
Sn(s, white)	0	51.55	0
Sn(s, gray)	−2.09	44.14	0.13

APPENDIX L (*Continued*)

Selected Thermodynamic Values*

SPECIES	ΔH_f°(298.15K) kJ/mol	S°(298.15K) J/K·mol	ΔG_f°(298.15K) kJ/mol
$SnCl_4(\ell)$	−511.3	258.6	−440.1
$SnCl_4(g)$	−471.5	365.8	−432.2
$SnO_2(s)$	−580.7	52.3	−519.6
Titanium			
Ti(s)	0	30.63	0
$TiCl_4(\ell)$	−804.2	252.34	−737.2
$TiCl_4(g)$	−763.2	354.9	−726.7
TiO_2	−939.7	49.92	−884.5
Zinc			
Zn(s)	0	41.63	0
$ZnCl_2(s)$	−415.05	111.46	−369.398
ZnO(s)	−348.28	43.64	−318.30
ZnS(s, sphalerite)	−205.98	57.7	−201.29

*Taken from "The NBS Tables of Chemical Thermodynamic Properties," 1982.

APPENDIX M

Answers to Exercises

CHAPTER 1

1. 2.5 cm(1 meter//100 cm) = 0.025 m
   2.5 cm(10 mm/1 cm) = 25 mm
2. (a) 750 mL (1 L/1000 mL) = 0.750 L
   (b) There are 4 quarts in 1 gallon. Therefore, 2.0 quarts = 0.50 gal; (0.50 gal)(3.785 L/1 gal) = 1.9 L
3. (a) (453.6 g/lb)(3.00 lb)(1 kg/1000 g) = 1.36 kg
   (b) (0.4536 kg/lb)(0.500 lb)(1.00×10^6 mg/kg) = 2.27×10^5 mg
   (c) (1.00 lb/0.4536 kg)(4.00 kg) = 8.82 lb
4. 1.0 kg = 1.0×10^3 g
   1.0×10^3 g(1 cm^3/0.00112 g) = 8.93×10^5 cm^3
5. t°C = (5°C/9°F)(110°F − 32°F) = 43°C
6. (a) Temp in kelvin degrees = 25°C + 273 = 298 K
   (b) 77 K − 273 = −196°C
   −196°C = (5°C/9°F)(t°F − 32°F)
   t°F = −321°F
7. (a) 12.63 = 4 significant figures; 0.063 = 2 significant figures
   (b) 12.63 + 0.063 = 12.69; the answer can have only 2 places past the decimal point.
   (c) 12.63 × 0.063 = 0.80; the answer is limited to 2 significant figures.
8. (a) 100. cm^3 (13.6 g/cm^3) = 1360 g
   1360 g (1 lb/453.6 g) = 3.00 lb
   (b) A layer 2.0 mm thick is 0.20 cm thick.
   Area of puddle = volume/thickness
   = 100. cm^3/0.20 cm = 5.0×10^2 cm^2
   5.0×10^2 cm^2 (1.0 in/2.54 cm)^2 = 78 in^2

CHAPTER 2

1. (a) Li = 3; Ar = 18; Cr = 24; Ag = 47; Ra = 88
   (b) Li = lithium; Ar = argon; Cr = chromium; Ag = silver; Ra = radium
   (c) beryllium = Be; chlorine = Cl; manganese = Mn; antimony = Sb; platinum = Pt; and plutonium = Pu
2. (a) Mg has 12 protons and 12 neutrons. Therefore, A = 24.
   (b) Cu has 29 protons and 34 neutrons. A = 63.
   (c) Hg has 80 protons and 120 neutrons. A = 200.
3. (a) Iron-57 is ${}^{57}_{26}Fe$ and tin-119 is ${}^{119}_{50}Sn$.
   (b) Platinum-195 has 78 protons and 78 electrons. Since the mass number is 195, there are 195 − 78 = 117 neutrons.
4. 35.45 = (0.7577)(34.96885) + (0.2423)(36.96590)
5. (a) Manganese atomic weight = 54.9380 amu; mass of 1.00 mole = 54.9 g.
   (b) 6.022×10^{23} atoms of Si is one mole or 28.09 g.
   (c) 196.97 g is the molar mass of gold. Therefore, there must be 6.022×10^{23} atoms in that mass.
6. (a) 2.5 mol Al (27 g Al/mol) = 67 g
   (b) (1.00 lb)(454 g/lb)(1 mol Pb/207 g) = 2.19 mol Pb
7. 50.0 g U(1 mol U/238 g) = 0.210 mol U
   0.210 mol (6.022×10^{23} atoms/mol) = 1.27×10^{23} atoms U
8. (a) (1.01 g H/mol)(1 mol H/6.022×10^{23} atoms) = 1.67×10^{-24} g H/atom
   (b) $\dfrac{3.33 \times 10^{-22} \text{ g for 1 Hg atom}}{1.67 \times 10^{-24} \text{ g for 1 H atom}} = 199$
   The ratio of 1 Hg atom to 1 H atom must be the same as the ratio of one mole of Hg atoms to one mole of H atoms. Therefore, once the mass of an atom of one element is known, the mass of an atom of another element can be determined using the ratio of atomic masses in the periodic table.
9. (a) N_2H_4; (b) $C_2H_2O_4$; (c) $C_{10}H_{10}Fe$
10. (a) K^+; (b) Se^{2-}; (c) Be^{2+}; (d) V^{2+} or V^{3+}; (e) Co^{2+} or Co^{3+}; Cs^+
11. (a) 1 Na^+ ion and one F^- ion, sodium fluoride; (b) 1 Na^+ ion and 1 $C_2H_3O_2^-$ ion, sodium acetate; (c) 1 Cu^{2+} ion and 2 NO_3^- ions, copper(II) nitrate.
12. (a) $NH_4^+ + NO_3^- \rightarrow NH_4NO_3$, ammonium nitrate
    (b) $Co^{2+} + SO_4^{2-} \rightarrow CoSO_4$, cobalt(II) sulfate
    (c) $Ni^{2+} + 2\ CN^- \rightarrow Ni(CN)_2$, nickel(II) cyanide
13. (a) Limestone, $CaCO_3$
    40.08 g for Ca + 12.01 g for C + 3(16.00 g for 0) = 100.1 g/mol
    (b) Caffeine, $C_8H_{10}N_4O_2$
    8(12.0 g) + 10(1.01 g) + 4(14.0 g) + 2(16.0 g) = 194 g/mol
14. (a) (1.00×10^3 g)(1 mol/100.1 g) = 9.99 mol
    (b) (2.50×10^{-3} mol)(194 g/mol) = 0.485 g caffeine
15. (a) (0.070 g)(1 mol/294/g) = 2.4×10^{-4} mol aspartame
    (b) (2.4×10^{-4} mol)(6.02×10^{23} molecules/mol) = 1.4×10^{20} molecules

(c) (294 g/mol)(1 mol/6.022 × 10^{23} molecules) = 4.88 × 10^{-22} g/molecule

16. (a) %Na = (23.0/58.4)100 = 39.4% Na
%Cl = (35.5/58.4)100 = 60.7% Cl
(b) %C = (12.0/16.0)100 = 75.0% C
%H = (4.00/16.0)100 = 25.0% H
(c) %N = [(2 × 14.0)/132)]100 = 21.2% N
%H = [(12 × 1.01)/132]100 = 6.10% H
%S = (32.1/132)100 = 24.3% S
%O = [(4 × 16.0)/132] = 48.4% O
(d) %C = [(8 × 12.0)/194)100 = 49.5% C
%H = [(10 × 1.0)/194)100 = 5.2% H
%N = [(4 × 14.0)/194)100 = 28.9% N
%O = [(2 × 16.0)/194)100 = 16.5% O
(These add up to 100.1% owing to rounding of results.)

17. 78.3 g B(1 mol B/10.8 g) = 7.25 mol B
21.7 g H(1 mol H/1.01 g) = 21.5 mol H
Mole ratio of H to B = 21.5/7.25 = 2.97 H/1.00 B.
The empirical formula is BH_3.
The molecular weight is twice as large as the empirical formula weight (13.8 amu/formula unit). Therefore, the molecular formula is 2 × BH_3 = B_2H_6.

18. 1.542 g CO_2 (12.01 g C/44.01 g CO_2) = 0.4208 g C
0.315 g H_2O (2.02 g H/18.02 g H_2O) = 0.0353 g H
The weight of iron in the 0.652 g-sample =
0.652 g − (0.421 g C + 0.0353 g H) = 0.196 g Fe
From the weight of each element, the weight percent composition of ferrocene is: 64.5% C, 5.41% H, and 30.1% Fe. The empirical formula is $C_{10}H_{10}Fe$, and the molar mass is 186 g/mol.

19. 2.145 g − 1.130 g = 1.015 g H_2O lost on heating (= 0.05634 mol H_2O)
Moles of $Na_2B_4O_7$ = 1.130 g(1 mol $Na_2B_4O_7$/201.2 g) = 0.005616 mol
Mole ratio = 0.05634 mol H_2O/0.005616 mol $Na_2B_4O_7$ = 10.03/1.00.
The formula of borax is $Na_2B_4O_7 .10H_2O$.

20. (a) 0.560 g $BaSO_4$ (137.3 g Ba/233.4 g $BaSO_4$) = 0.329 g Ba; 32.2% Ba
(b) 0.329 g Ba (208.2 g $BaCl_2$/137.3 g Ba) = 0.499 g $BaCl_2$

CHAPTER 3

1. (a) 4 Fe(s) + 3 O_2(g) → 2 Fe_2O_3(s)
(b) CH_4(g) + 2 O_2(g) → CO_2(g) + 2 H_2O(g)
(c) 2 B_4H_{10}(l) + 11 O_2(g) → 4 B_2O_3(s) + 10 H_2O(g)
(d) CO(g) + 2 H_2(g) → CH_3OH(ℓ)
(e) 2 C_8H_{18}(l) + 25 O_2(g) → 16 CO_2(g) + 18 H_2O(ℓ)

2. (a) NaBr, soluble, Na^+, Br^-
(b) $BaSO_4$, insoluble
(c) K_2CO_3, soluble, 2 K^+, CO_3^{2-}
(d) AgCl, insoluble
(e) $(NH_4)_2SO_4$, soluble, 2 NH_4^+, SO_4^{2-}

3. (a) KNO_3; (b) NiS; (c) $NiNO_3$

4. (a) Balanced complete ionic equation:

$$Ba^{2+}(aq) + 2Cl^-(aq) + 2Na^+(aq) + SO_4^{2-}(aq) \longrightarrow BaSO_4(s) + 2Na^+(aq) + 2Cl^-(aq)$$

Net ionic equation:

$$Ba^{2+}(aq) + SO_4^{2-}(aq) \longrightarrow BaSO_4(s)$$

(b) Balanced complete ionic equation:

$$2\ NH_4^+(aq) + S^{2-}(aq) + Cd^{2+}(aq) + 2\ NO_3^-(aq) \longrightarrow CdS(s) + 2\ NH_4^+(aq) + 2\ NO_3^-(aq)$$

Net ionic equation:

$$Cd^{2+}(aq) + S^{2-}(aq) \longrightarrow CdS(s)$$

(c) Balanced complete ionic equation:

$$Pb^{2+}(aq) + 2NO_3^-(aq) + 2K^+(aq) + 2Cl^-(aq) \longrightarrow PbCl_2(s) + 2K^+(aq) + 2NO_3^-(aq)$$

Net ionic equation:

$$Pb^{2+}(aq) + 2\ Cl^-(aq) \longrightarrow PbCl_2(s)$$

5. (a) 4 Li(s) + O_2(g) → $2Li_2O$(s)
(b) 4 V(s) + 3 O_2(g) → 2 V_2O_3(s)
(c) 4 Ga(s) + 3 O_2 → 2 Ga_2O_3(s)

6. (a) Ge(s) + O_2(g) → GeO_2(s)
(b) 4 Cr(s) + 3 O_2(g) → 2 Cr_2O_3(s)

7. (a) $AgNO_3$(aq) + LiCl(aq) → AgCl(s) + $LiNO_3$(aq)
(b) $NiCl_2$(aq) + Na_2S(aq) → 2NaCl(aq) + NiS(s)

8. H_2SO_4(aq) + 2 KOH(aq) → K_2SO_4(aq) + 2 H_2O(ℓ)

9. (a) H_2SO_4(aq) + 2 CsOH(aq) → Cs_2SO_4(aq) + 2 H_2O(ℓ), Acid-base reaction
(b) $MgCl_2$(aq) + Na_2CO_3(aq) → $MgCO_3$(s) + 2 NaCl(aq), Precipitation reaction
(c) 2 HNO_3(aq) + $Ca(OH)_2$(aq) → $Ca(NO_3)_2$(aq) + 2 H_2O(ℓ), Acid-base reaction
(d) $CdCl_2$(aq) + Na_2S(aq) → 2 NaCl(aq) + CdS(s), Precipitation reaction

10. (a) NaOH(aq) + HCl(aq) → NaCl(aq) + H_2O(ℓ)
(b) KOH(aq) + HNO_3(aq) → KNO_3(aq) + H_2O(ℓ)
(c) $Ba(OH)_2$(s) + 2 HNO_3(aq) → $Ba(NO_3)_2$(aq) + 2 H_2O(ℓ)
(e) $FeCl_2$(aq) + $(NH_4)_2S$(aq) → 2 NH_4Cl(aq) + FeS(s)

11. $MgCO_3$(s) + heat → MgO(s) + CO_2(g)
Na_2CO_3(s) + 2 HNO_3(aq) → 2 $NaNO_3$(aq) + H_2O(ℓ) + CO_2(g)

12. (a) Mg = +2 and O = −2
(b) P = −3 and H = +1
(c) Al = +3 and H = −1

(d) H = +1 and S = −2
(e) Cl = +5 and O = −2
(f) S = +2 and O = −2
(g) Na = +1 and Cl = −1
(h) H = +1, N = +5, and O = −2
(i) Zn = +2 and S = −2

13. (a) S oxidation number changes from 0 to +4, and the oxidation number of oxygen changes from 0 in O_2 to −2 in the product. S is oxidized and O_2 is reduced.
(b) Fe oxidation number changes from 0 to +2, while that of Cl changes from 0 to −1. Iron is oxidized and Cl is reduced.
(c) No elements change in oxidation number. This is not an oxidation-reduction reaction; it is an acid-base reaction.
(d) S changes in oxidation number from −2 in ZnS to +6 in the SO_4^{2-} ion in $ZnSO_4$. N changes from +5 in HNO_3 to +2 in NO. Thus, ZnS is oxidized to $ZnSO_4$ and HNO_3 is reduced to NO.

14. $2\,[Cr^{2+}(aq) \longrightarrow Cr^{3+}(aq) + e^-]$
$2\,e^- + I_2(aq) \longrightarrow 2\,I^-(aq)$

$2\,Cr^{2+}(aq) + I_2(aq) \longrightarrow 2\,Cr^{3+}(aq) + 2\,I^-(aq)$

(a) I_2, oxidizing agent (gains electrons); (b) Cr^{2+}, reducing agent (loses electrons); (c) I_2, substance reduced (oxidation number declines); (d) Cr^{2+}, substance oxidized (oxidation number increases).

15. $Cu(s) \longrightarrow Cu^{2+}(aq) + 2\,e^-$
$2\,[e^- + 2\,H^+(aq) + NO_3^-(aq) \longrightarrow NO_2(g) + H_2O(\ell)]$

$Cu(s) + 4\,H^+(aq) + 2\,NO_3^-(aq) \longrightarrow Cu^{2+}(aq) + 2\,NO_2(g) + 2\,H_2O(\ell)$

Cu is the reducing agent and so is oxidized. NO_3^- is the oxidizing agent (as it often is in acid solution) and so is reduced.

16. $2\,[Cr(s) + 3\,OH^-(aq) \longrightarrow Cr(OH)_3(s) + 3\,e^-]$
$3\,[ClO_4^-(aq) + H_2O(\ell) + 2\,e^- \longrightarrow ClO_3^-(aq) + 2\,OH^-(aq)]$

$2\,Cr(s) + 3\,ClO_4^-(aq) + 3\,H_2O(\ell) \longrightarrow 2\,Cr(OH)_3(s) + 3\,ClO_3^-(aq)$

Cr is the reducing agent and so is oxidized. ClO_4^- is the oxidizing agent (as it often is in acid solution) and so is reduced.

17. (a) $2\,S_2O_3^{2-}(aq) \rightarrow S_4O_6^{2-}(aq) + 2\,e^-$
Oxidation; $S_2O_3^{2-}$ is a reducing agent, and it is oxidized.
(b) $NO_3^-(aq) + 4\,H^+(aq) + 3\,e^- \rightarrow NO(g) + 2\,H_2O$
Reduction; NO_3^- is an oxidizing agent, and is itself reduced.
(c) $C_2H_5OH(aq) \rightarrow C_2H_4O(aq) + 2\,H^+(aq) + 2\,e^-$
Oxidation: C_2H_5OH (ethyl alcohol) is a reducing agent and is oxidized.
(d) $PH_3(aq) + 4\,H_2O \rightarrow H_3PO_4(aq) + 8\,H^+(aq) + 8\,e^-$
Oxidation; PH_3 is a reducing agent and is oxidized.
(e) $ClO^-(aq) + H_2O + 2\,e^- \rightarrow Cl^-(aq) + 2\,OH^-(aq)$
Reduction; ClO^- is an oxidizing agent, and is itself reduced.

18. (a) $2\,S_2O_3^{2-}(aq) \longrightarrow S_4O_6^{2-}(aq) + 2\,e^-$ (red. ag.)
$I_2(aq) + 2\,e^- \longrightarrow 2\,I^-(aq)$ (ox. ag.)

$2\,S_2O_3^{2-}(aq) + I_2(aq) \longrightarrow S_4O_6^{2-}(aq) + 2\,I^-(aq)$

(b) $3\,[Cu(s) \longrightarrow Cu^{2+}(aq) + 2\,e^-]$ (red. ag.)
$2\,[NO_3^-(aq) + 4\,H^+(aq) + 3\,e^- \longrightarrow NO(g) + 2\,H_2O]$ (ox. ag.)

$3\,Cu(s) + 2\,NO_3^-(aq) + 8\,H^+(aq) \longrightarrow 2NO(g) + 4\,H_2O + 3\,Cu^{2+}$

(c) $5\,[Fe^{2+}(aq) \longrightarrow Fe^{3+}(aq) + e^-]$ (red.ag.)
$MnO_4^-(aq) + 8\,H^+(aq) + 5\,e^- \longrightarrow Mn^{2+}(aq) + 4\,H_2O$ (ox. ag.)

$5\,Fe^{2+}(aq) + MnO_4^-(aq) + 8\,H^+(aq) \longrightarrow Mn^{2+}(aq) + 5\,Fe^{3+} + 4\,H_2O$

(d) $2\,[Cr(OH)_3(s) + 5\,OH^- \longrightarrow CrO_4^{2-}(aq) + 4\,H_2O + 3\,e^-]$ (red. ag.)
$3\,[2\,ClO^-(aq) + 2\,H_2O + 2\,e^- \longrightarrow Cl_2(aq) + 4\,OH^-(aq)]$ (ox. ag.)

$2\,Cr(OH)_3(s) + 6\,ClO^-(aq) \longrightarrow 2\,CrO_4^{2-}(aq) + 2\,H_2O(\ell) + 3\,Cl_2(aq) + 2\,OH^-(aq)$

CHAPTER 4

1. 454 g S_8(1 mol S_8/257 g) = 1.77 mol S_8
1.77 mol S_8 (8 mol O_2/1 mol S_8) = 14.2 mol O_2 required
14.2 mol O_2(32.0 g/1 mol O_2) = 453 g O_2 required
1.77 mol S_8 (8 mol SO_2/1 mol S_8) = 14.2 mol SO_2 produced
14.2 mol SO_2(64.1 g/mol) = 908 g SO_2 produced

2. 1.05 g Fe (1 mol/55.8 g) = 0.0188 mol Fe
Balanced equation: $4\,Fe(s) + 3\,O_2(g) \rightarrow 2\,Fe_2O_3(s)$
0.0188 mol Fe (2 mol Fe_2O_3/4 mol Fe) = 0.00940 mol Fe_2O_3 produced
0.00940 mol Fe_2O_3 (160. g/mol) = 1.50 g Fe_2O_3
0.0188 mol Fe (3 mol O_2/4 mol Fe) = 0.0141 mol O_2 required
0.0141 mol O_2 (32.0 g/mol) = 0.451 g O_2 required

3. 2.70 g Al = 0.100 mol and 4.05 g Cl_2 = 0.0571 mol
Actual mole ratio = (0.100 mol Al/0.0571 mol Cl_2) = 1.75 mol Al/1 mol Cl_2
Required mole ratio = 2 mol Al/3 mol Cl_2
= 0.67 mol Al/1 mol Cl_2
This means that Al is present in excess, so Cl_2 is the limiting reagent.
The yield of product is calculated from the limiting reagent.
0.0571 mol Cl_2 (2 mol $AlCl_3$/3 mol Cl_2) = 0.0381 mol $AlCl_3$
0.0381 mol $AlCl_3$ (133 g/mol) = 5.08 g $AlCl_3$ expected.

4. 18.9 g $NaBH_4$(1mol/37.8 g) = 0.500 mol
0.500 mol(2 mol B_2H_6/3 mol $NaBH_4$) = 0.333 mol B_2H_6
0.333 mol B_2H_6(27.7 g/mol) = 9.22 g
% yield of B_2H_6 = (7.50 g/9.22g)100 = 81.3%

5. 26.3 g $NaHCO_3$(1 mol/84.0 g) = 0.313 mol $NaHCO_3$
0.313 mol/0.200 L = 1.57 *M* = [$NaHCO_3$]

6. (a) HCl dissociates in water to give H^+ and Cl^-.

$HCl(aq) \longrightarrow H^+(aq) + Cl^+(aq)$

Therefore, 1 *M* HCl means that [H^+] = [Cl^-] = 1 *M*
Total ion concentration = 2 *M*

(b) Ammonium sulfate dissociates according to the equation

$(NH_4)_2SO_4(aq) \longrightarrow 2\ NH_4^+(aq) + SO_4^{2-}(aq)$

Therefore, [NH_4^+] = 2 × 0.5 *M* = 1.0 *M* and [SO_4^{2-}] = 0.5 *M*
Total ion concentration = 1.5 *M*

7. Moles of $KMnO_4$ required = (0.500 L)(0.0200 mol/L) = 0.0100 mol $KMnO_4$

0.0100 mol $KMnO_4$(158 g/mol) = 1.58 g

Place 1.58 g $KMnO_4$ in a 500. mL volumetric flask, add water slowly, shaking to ensure that the $KMnO_4$ dissolves. Fill with water to the mark on the flask.

8. If you require 300. mL of 1.00 M NaOH, this means you need

(0.300 L)(1.00 mol/L) = 0.300 mol NaOH

This amount of NaOH can be taken from the more concentrated solution. The volume of more concentrated NaOH required is

(0.300 mol NaOH required)(1.00 L/3.00 mol NaOH) = 0.100 L or 100. mL

Therefore, take 100. mL of 3.00 *M* NaOH and carefully add 200. mL water to bring the total volume of the diluted solution to 300. mL.

9. Balanced equation: 2 HCl(aq) + Na_2CO_3(aq) → $H_2O(\ell)$ + CO_2(g) + 2 NaCl(aq)
0.100 L Na_2CO_3(1.50 mol/L) = 0.150 mol Na_2CO_3
0.150 mol Na_2CO_3(2 mol NaCl/1 mol Na_2CO_3)
= 0.300 mol NaCl produced
0.300 mol NaCl(58.5 g/mol) = 17.6 g NaCl
0.150 mol Na_2CO_3(1 mol CO_2/1 mol Na_2CO_3)
= 0.150 mol CO_2
0.150 mol CO_2(44.0 g/mol) = 6.60 g CO_2
Concentration of NaCl = [NaCl] =
0.300 mol/0.500 L = 0.600 *M*
(Note that the total solution volume is the sum of the volumes of the reacting solutions.)

10. See balanced equation in previous exercise.
(0.0250 L)(0.750 mol/L) = 0.0188 mol HCl
0.0188 mol HCl(1 mol Na_2CO_3/2 mol HCl) = 0.00938 mol Na_2CO_3
0.00938 mol Na_2CO_3(106 g/mol) = 0.994 g Na_2CO_3
0.0188 mol HCl(2 mol NaCl/2 mol HCl) = 0.0188 mol NaCl
0.0188 mol NaCl(58.5 g/mol) = 1.10 g NaCl

11. (0.02833 L)(0.953 mol/L) = 0.0270 mol NaOH
0.0270 mol NaOH(1 mol $HC_2H_3O_2$/1 mol NaOH) = 0.0270 mol $HC_2H_3O_2$
0.0270 mol $HC_2H_3O_2$(60.1 g/mol) = 1.62 g $HC_2H_3O_2$
0.0270 mol $HC_2H_3O_2$/0.0250 L = 1.08 *M*

12. Balanced equation: HCl(aq) + NaOH(aq) → NaCl(aq) + $H_2O(\ell)$
Moles HCl = (0.03256 L)(0.100 mol/L) = 0.00326 mol HCl
0.00326 mol HCl(1 mol NaOH/1 mol HCl) = 0.00326 mol NaOH
0.00326 mol NaOH/0.0250 L = 0.130 *M* = [NaOH]

13. Moles $KMnO_4$ = (0.02434 L)(0.0200 mol/L) = 0.000487 mol $KMnO_4$
0.000487 mol $KMnO_4$(5 mol Fe^{2+}/1 mol $KMnO_4$) = 0.00243 mol Fe^{2+}
0.00243 mol Fe^{2+}(55.8 g/mol) = 0.136 g Fe
Weight percent = (0.136 g Fe/1.026 g sample)100 = 13.2% Fe

CHAPTER 5

1. The relation between mass, kinetic energy, and velocity is KE = $(\frac{1}{2})mv^2$.
m = 9.11×10^{-31} kg and KE = 1.60×10^{-18} J
v = $(2KE/m)^{1/2}$ =
[$2 \times 1.60 \times 10^{-18}$ kg·m²/s²]/[9.11×10^{-31} kg]$^{1/2}$
= 1.87×10^6 m/s
1.87×10^6 m/s (3600 s/hr)(1 km/1000 m)(0.621 miles/km) = 4.18×10^6 mph

2. 150 Calories (1000 calories/Calorie) = 1.50×10^5 cal
1.50×10^5 cal (4.184 J/cal) = 6.28×10^5 J
3. Heat input = 24.1 kJ or 24.1×10^3 J
Heat input = 24.1×10^3 J =
(0.902 J/g-deg)(250.g)$(t_{\text{final}} - t_{\text{initial}})$
Solving this expression for t_{initial} = 37.0°C gives
t_{final} = 144°C.
4. The text indicates that 243 kJ are required to decompose 1.0 mol or 18 g of H_2O to H_2 and O_2. This relation provides the necessary conversion factor. 12.5 g H_2O (243 kJ/18.0 g H_2O) = 169 kJ required
5. $3\ [H_2(g) + \frac{1}{2}O_2(g) \longrightarrow H_2O(\ell)]$
$2\ [C(\text{graphite}) + O_2(g) \longrightarrow CO_2(g)]$
$C_2H_5OH(\text{liq}) \longrightarrow \frac{1}{2}O_2(g) + 3\ H_2(g) + 2\ C(\text{graph})$

$C_2H_5OH(l) + 3\ O_2(g) \longrightarrow 2\ CO_2(g) + 3\ H_2O(\ell)$

3 [ΔH = −285.8 kJ]
2 [ΔH = −393.5 kJ]
ΔH = + 277.7 kJ

ΔH = −1366.7kJ

6. (a) $Ag(s) + \frac{1}{2}Cl_2(g) \rightarrow AgCl(s)$ $\Delta H_f^\circ = -127.0$ kJ
(b) $C(\text{graphite}) + 2\ H_2(g) + \frac{1}{2}O_2(g) \rightarrow CH_3OH(\ell)$
$\Delta H_f^\circ = -238.7$ kJ
7. Balanced equation $C_3H_8(g) + 5\ O_2(g) \rightarrow$
$3\ CO_2(g) + 4\ H_2O(\ell)$

$\Delta H^\circ_{\text{reaction}} = \Delta H^\circ_{\text{combustion}}$
$= 3\,\Delta H_f^\circ[CO_2(g)] + 4\,\Delta H_f^\circ[H_2O(\ell)] - \Delta H_f^\circ[C_3H_8(g)]$
= 3 mol(−393.5 kJ/mol)
+ 4 mol(−285.8 kJ/mol)
− 1 mol(−103.8 kJ/mol)
= −2219.9 kJ

8. Change in temperature = 27.32 deg − 25.00 deg = 2.32 deg

Heat input to calorimeter water
= (1.50×10^3 g)(4.184 J/g-deg)(2.32 deg)
= 14.6×10^3 J
Heat input to calorimeter bomb
= (837 J/deg)(2.32 deg)
= 1.94×10^3J

Total heat evolved by 1.00 g of sucrose = (−14.6 kJ) + (−1.94 kJ) = −16.5 kJ
Heat of combustion per mole =
(−16.5 kJ/g) (342 g/mol) = −5640 kJ/mol

CHAPTER 6

1. 75 kPa (0.740 atm) > 0.60 atm > 350. mm Hg (0.500 atm) > 300. torr (0.395 atm)

2. P_1 = 55 mm Hg and V_1 = 125 mL
P_2 = 78 mm Hg and V_2 = ?
$V_2 = P_1V_1/P_2$ = 88 mL
3. V_1 = 4.5 L and T_1 = 298 K
V_2 = ? and T_2 = 263 K
$V_2 = V_1(T_2/T_1)$ = 4.0 L
4. 22.4 L CH_4 (2 L O_2/1 L CH_4) = 44.8 L O_2 required
44.8 L of H_2O and 22.4 L of CO_2 are produced.
5. n = 1300 mol, P = 750/760 = 0.99 atm, T = 293 K, R = 0.082 L·atm/K·mol.
$V = nRT/P = 3.2 \times 10^4$ L
[Using $V = 3.2 \times 10^7$ cm³, the radius r = 200 cm]
6. P_1 = 150 atm, T_1 = 303 K, V_1 = 20. L
P_2 = 755/760 = 0.99 atm, T_2 = 295 K, V_2 = ?
$V_2 = (P_1/P_2)(T_2/T_1) = 2.9 \times 10^3$ L
2.9×10^3 L(1 balloon/5.0 L) = 590 balloons
7. M = 29.0 g/mol, P = 745/760 = 0.980 atm, T = 295 K, R = 0.0821 L·atm/K·mol
$d = PM/RT$ = 1.17 g/L
8. P = 0.737 atm, V = 0.125 L, T = 296.2 K, R = 0.0821 L·atm/K·mol
$n = PV/RT = 3.79 \times 10^{-3}$ mol
molar mass = 0.105 g/3.83×10^{-3} mol = 27.7 g/mol
9. 180 g (1 mol N_2H_4/32.0 g) = 5.6 mol N_2H_4
5.6 mol N_2H_4 (1 mol O_2/1 mol N_2H_4) = 5.6 mol O_2
$V(O_2) = nRT/P$ = 140 L when n = 5.6 mol, R = 0.082 L·atm/K·mol, T = 294 K, and P = 0.99 atm.
10. For N_2: P_1 = 1.0 atm, V_1 = 1.0 L, P_2 = ?, V_2 = 4.0 L
This gives $P_2(N_2)$ = 0.25 atm

For H_2: P_1 = 1.0 atm, V_1 = 3.0 L, P_2 = ?, V_2 = 4.0 L
This gives $P_2(H_2)$ = 0.75 atm

$P_{\text{total}} = P_2(N_2) + P_2(H_2)$ = 1.0 atm
11. P = 742 mm Hg − vapor pressure of H_2O = (742 − 20) mm Hg = 722 mm Hg.
Therefore, P is 0.950 atm (= 722 mm Hg).
V = 0.352 L, T = 295 K, R = 0.0821 L·atm/K·mol
$n = PV/RT$ = 0.0138 mol N_2 = 0.387 g
12. Using equation 6.17 with M = 4.00 g/mol (or 4.00×10^{-3} kg/mol), T = 298 K, R = 8.314 J/K.mol, one obtains a root mean square speed of 1360 m/s. Note that this same result could be obtained from the results of Example 6.15 and equations 6.19 and 6.20.

$$\frac{\text{rms for He}}{\text{rms for } O_2} = \sqrt{\frac{M(O_2)}{M(\text{He})}} \quad \text{or}$$

$$\frac{\text{rms for He}}{482 \text{ m/s}} = \sqrt{\frac{32 \text{ g/mol}}{4.0 \text{ g/mol}}}$$

Therefore, rms for He = 1360 m/s.

13. When the mass of gas molecules is increased, the average speed decreases. This same result can be achieved by lowering the temperature.

14. Assume that each sample of gas has a mass of 1.00 g. The molar mass of CH_4 is 16.0 g/mol. Therefore,

$$\frac{\text{rate for } CH_4}{\text{rate for unk}} = \frac{1.00 \text{ g}/1.50 \text{ min}}{1.00 \text{ g}/4.73 \text{ min}} = \sqrt{\frac{M}{16.0}}$$

M = 159 g/mol

15. For n = 10.0 moles, V = 1.00 L, T = 298 K, R = 0.0821 L·atm/K·mol

(a) $P = nRT/V$ = 245 atm

(b) Van der Waals's equation with a = 0.034 atm·L²/mol² and b = 0.0237 L/mol
P = 321 atm

(c) Plot of ψ^2 versus r for a 3p orbital.

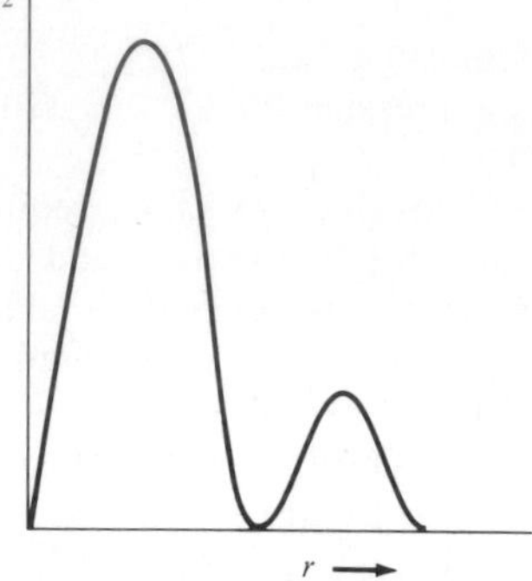

CHAPTER 7

1. 104.6 MHz = $104.6 \times 10^6 \text{ s}^{-1}$
$\lambda = c/\nu = 2.998 \times 10^8 \text{ m/s}(1\text{s}/104.6 \times 10^6) = 2.866$ m

2. (a) Highest frequency, violet; (b) lowest frequency, red; (c) microwave frequency is greater than used for FM radio.

3. (a) λ = 10 cm;
(b) λ = 6 cm;
(c) 2 waves; 5 nodes

4. (a) To calculate the wavelength (where $\lambda = h/mv$) you must first calculate the velocity of the neutron.

$v = [2E/m]^{1/2} =$
$[2(6.21 \times 10^{-21} \text{ kg·m}^2/\text{s}^2)/(1.675 \times 10^{-27} \text{ kg})]^{1/2}$
v = 2723 m/s

(b) $\lambda = h/mv = (6.6262 \times 10^{-34} \text{ kg·m}^2/\text{s}^2)(\text{s})/(1.675 \times 10^{-27} \text{ kg})(2723 \text{ m/s})$
$\lambda = 1.45 \times 10^{-10}$ m

5. (a) ℓ = 0 and 1
(b) m_ℓ = −1, 0, +1; subshell is p
(c) When ℓ = 2, then the subshell is a d.
(d) When the subshell label is s, then ℓ is 0, and m_ℓ is 0.
(e) When the subshell label is p, then there are 3 orbitals in the subshell.
(f) For an f subshell, there are 7 values of m_ℓ, and there are 7 orbitals within the subshell.

6. (a)

Orbital Label	n	ℓ
6s	6	0
4p	4	1
5d	5	2
4f	4	3

(b) 2p has 1 planar node (= ℓ) and 0 spherical nodes ($n - \ell - 1 = 2 - 1 - 1 = 0$).

CHAPTER 8

Note: for simplicity, electron configurations given by the orbital box notation will use only a line instead of a box to describe an orbital. Thus, the configuration of H is depicted as

H: $\underset{1s}{\underline{1}}$

1. (a) An electron is assigned first to 3s, before 3d, because $n + \ell = 3 + 0 = 3$ for 3s, whereas it is 5 (= 3 + 2) for 3d.
(b) 6s ($n + \ell = 6 + 0 = 6$) before 5d (5 + 2 = 7).
(c) 5s ($n + \ell$ = 5) before 4f (4 + 3 = 7).

2. (a) Chlorine, Cl.
(b) Sulfur, S: $1s^22s^22p^63s^23p^4$

⇅	⇅	⇅	⇅	⇅	⇅	⇅	↿	↿
1s	2s	2p	2p	2p	3s	3p	3p	3p

(c) $n = 3$, $\ell = 1$, $m_\ell = +1$, $m_s = +\frac{1}{2}$.

3. Check Table 8.4.

4. (a) H radius = 37 pm, O radius = 66 pm, and S radius = 104 pm. H—O = (37 + 66)pm = 103 pm; H—S = (37 + 104)pm = 141 pm
(b) Br radius = ½ (Br—Br) = ½(228 pm) = 114 pm. Br—Cl = 114 pm + 99 pm = 213 pm

5. (a) Radius: C < Si < Al
(b) Ionization energy: Al < Si < C
(c) Si

6. V^{2+}

[Ar]	↿	↿	↿			
	3d	3d	3d	3d	3d	4s

Paramagnetic; 3 unpaired electrons

V^{3+}

[Ar] $\frac{1}{3d}$ $\frac{1}{3d}$ $\frac{\ }{3d}$ $\frac{\ }{3d}$ $\frac{\ }{3d}$ $\frac{\ }{4s}$

Paramagnetic; 2 unpaired electrons

Co^{3+}

[Ar] $\frac{\uparrow\downarrow}{3d}$ $\frac{1}{3d}$ $\frac{1}{3d}$ $\frac{1}{3d}$ $\frac{1}{3d}$ $\frac{\ }{4s}$

Paramagnetic; 4 unpaired electrons

7. Ion sizes: $N^{3-} > O^{2-} > F^-$. This is an isoelectronic series. All ions have 10 electrons. However, the effective nuclear charge is smallest at N^{3-} and largest at F^-.
8. (a) F^- $1s^22s^22p^6$ = [Ne], rare gas configuration
   (b) S^{2-} $1s^22s^22p^63s^23p^6$ = [Ar], rare gas configuration
   (c) In^+ [Kr]$5s^24d^{10}$, pseudo-rare gas configuration

CHAPTER 9

1. a, d, e, and h are consistent with the octet rule.
2.

Molecule	Atom in Question	Number of Sigma Bonds	Number of Pi Bonds
BH_3	B	3	0
CH_4	C	4	0
NH_3	N	3	0
H_2O	O	2	0
HF	F	1	0
BF_3	B	3	0
BF_3	F	1	0
C_2H_4	C	3	1
N_2	N	1	2
SCl_2	S	2	0
SCl_2	Cl	1	0
ClF	Cl or F	1	0

3. (a) :C≡O: C and O both have 1 sigma bond and 2 pi bonds.
   (b) :N≡O:$^+$ NO^+ has the same number of valence electrons as CO. Both N and O have 1 sigma bond and 2 pi bonds.
   (c) H—C≡N: C has 2 sigma bonds and 2 pi bonds, and N has 1 sigma bond and 2 pi bonds.
   (d) H—S—H S has 2 sigma bonds.
4. SF_4 ClF_3 PCl_5

5. Resonance structures for SO_2

Resonance structures for NO_3^-

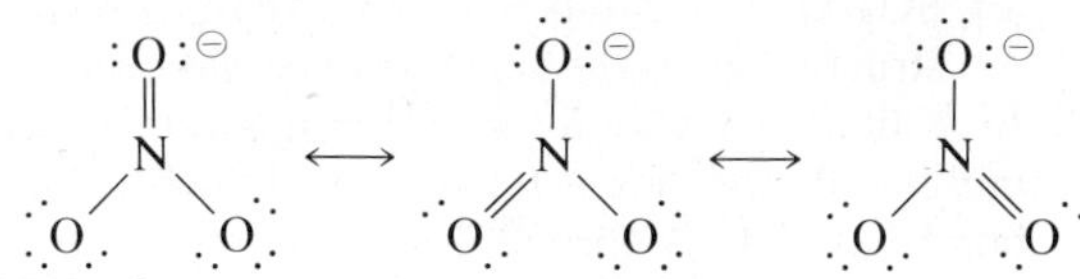

6. (a) Decreasing CN bond distance (bond order): C—N(1) > C=N(2) > C≡N(3).
   (b) The NO bond order in NO_2^- is 1.5. Therefore, the observed bond length (124 pm) should be approximately midway between an N—O single bond (136 pm) and and N=O double bond (115 pm).

7. $\Delta H_f^\circ = [D_{N-N} + 2D_{F-F}] - [4D_{N-F} + D_{N-N}]$
   $= [(946\ kJ) + 2(159\ kJ)] - [4(272\ kJ) + (159\ kJ)]$
   $= 17$ kJ
8. Heat of formation of $CCl_4(g)$:

C(graphite) ⟶ C(g)	ΔH° = +717 kJ
2 Cl_2(g) ⟶ 4 Cl(g)	ΔH° = $2D_{Cl-Cl}$ = 2(243 kJ)
C(g) + 4 Cl(g) ⟶ CCl_4(g)	ΔH° = $-4D_{C-Cl}$ = −4(330 kJ)
C(graphite) + 2 Cl_2(g) → CCl_4(g)	ΔH° = −117 kJ

9. Heat of formation of HCO_2H = −378.6 kJ/mol = $= \Delta H^\circ_{vap}$ [C(graph) → C(g)] + $D_{O=O} + D_{H-H} - [D_{C-H} + D_{C=O} + D_{C-O} + D_{O-H}]$
   = [(+717 kJ) + (498 kJ) + (436 kJ)] − [414 kJ + $D_{C=O}$ + 351 kJ + 464 kJ]
   $D_{C=O}$ = 801 kJ/mol
10. (a) H—F ($\Delta\chi$ = 4.0 − 2.1 = 1.9) is more polar than H—I ($\Delta\chi$ = 2.5 − 2.1 = 0.4). In both bonds H is positive and the halogen negative.
    (b) B—F ($\Delta\chi$ = 4.0 − 2.0 = 2.0) is more polar than B—C ($\Delta\chi$ = 2.5 − 2.0 = 0.5). In both bonds B is positive and F or C is negative.
    (c) C—Si is more polar ($\Delta\chi$ = 2.5 − 1.8 = 0.7) than C—S ($\Delta\chi$ = 2.5 − 2.5 = 0.0, a nonpolar bond). In C—Si the C is more electronegative than Si and so is the negative end of the bond.

11. In each of the following the first-named element is more electronegative and "sets" the oxidation number of the other element:
(a) SF_4: F = −1 and S = +4
(b) CO_3^{2-}: O = −2 and C = +4
(c) ClO_3^-: O = −2 and Cl = +5
(d) PF_3: F = −1 and P = +3
(e) SO_3: O = −2 and S = +6. Any one resonance structure gives the same set of oxidation numbers.

12. In *A* all atoms have a formal charge of 0. However, in *B*, the B atom has −1 and the F of the F═B double bond has +1. A formal charge of +1 on F is not reasonable, since F is so electronegative.

13.

Molecule	Structural-Pair Geometry
H_2O	tetrahedral (2 sigma and 2 lone pairs around O)
NO_2^+	linear (2 sigma and 2 pi pairs around N)
SF_4	trigonal bipyramidal (4 sigma and 1 lone pair around S)

14.

Molecule	Structural-Pair Geometry	Molecular Geometry
CS_2	linear	linear
H_2S	tetrahedral	bent
PO_4^{3-}	tetrahedral	tetrahedral

15. ICl_2^-: structural-pair geometry is trigonal bipyramidal and molecular geometry is linear.

:Cl:⁻ — I — :Cl:

16. S═C═S The S—C—S bond angle in CS_2 is 180° (the molecule has a linear molecular geometry).

H—S—H Like H_2O, the H_2S molecule has a tetrahedral structural-pair geometry and a bent molecular geometry. The ideal H—S—H angle is 109.5°, but the presence of the lone pairs on S will force the H atoms somewhat closer together, and the angle will be slightly smaller than 109°.

17. $BFCl_2$ is polar with the B—F side more negative.

NH₂Cl (molecular geometry is pyramidal) is polar with the direction of the dipole indicated below.

SCl_2 (molecular geometry is bent) is polar with the negative end of the dipole toward the Cl atoms.

CHAPTER 10

1. The structural-pair geometry around S in SCl_2 is tetrahedral (and the molecular geometry is bent). The tetrahedral pair geometry means the S is sp^3 hybridized.

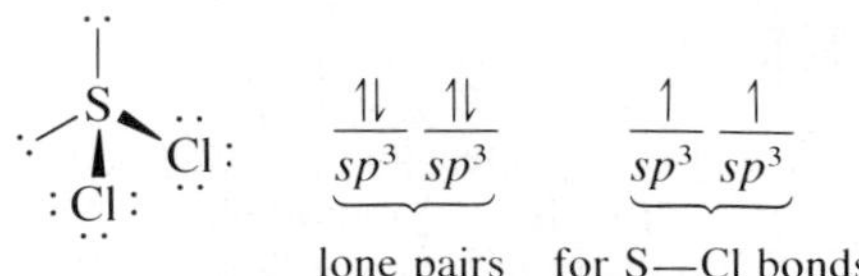

2. The structural-pair geometry around Xe is octahedral, so the hybridization of the atom is sp^3d^2.

lone pairs: two sp^3d^2 (↿⇂ ↿⇂); for 4 Xe—F sigma bonds: four sp^3d^2 (1 1 1 1)

3. The bonding in N_2 is identical to that in CO (one sigma bond, two pi bonds, and one lone pair on each atom), except that both atoms are of course N. Each N atom can be described as *sp* hybridized. The assignments of the 5 valence electrons of each N and their roles are as follows:

sp (↿⇂): lone pair; *sp* (1): N—N sigma bond; 2*p* 2*p* (1 1): for the two N—N pi bonds

4. H—C(H)(H)—C≡N:

(a) Bond angles
H—C—H = 109°; H—C—C = 109°; C—C—N = 180°

(b) Atom hybridizations
CH_3 carbon has a tetrahedral structural-pair geometry; sp^3 hybridized. CN carbon has a linear structural-pair geometry; sp hybridized. N atom has a linear structural-pair geometry (triple bond and lone pair are 180° apart); sp hybridized.

5. The electron dot structure of NO_2^- shows the N atom has a trigonal planar structural-pair geometry, so the N atom is sp^2 hybridized.

⇅ 1 1	1
sp^2 sp^2 sp^2	$2p$
for the 2 N—O sigma bonds and the N lone pair	for the N—O pi bond

One of the 3 sp^2 hybrid orbitals is assigned the N lone pair, while the other two orbitals are each assigned an electron to be utilized in N—O sigma bonding. The N—O pi bond is then formed using the electron assigned to the unhybridized N $2p$ orbital.

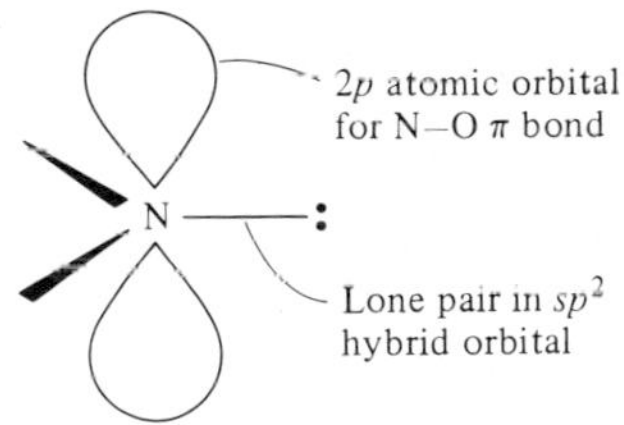

6. H_2^-: $(\sigma_{1s})^2(\sigma^*_{1s})^1$
The bond order is $\frac{1}{2}$, so the ion should exist under the right conditions,

7. Be_2^+: $(\sigma_{1s})^2(\sigma^*_{1s})^2(\sigma_{2s})^2(\sigma^*_{2s})^1$
The net bond order is $\frac{1}{2}$, so, unlike the neutral Be_2 molecule, the positive ion should exist under the right conditions.

8. O_2^+: [core electrons]$(\sigma_{2s})^2(\sigma^*_{2s})^2(\pi_{2p})^4(\sigma_{2p})^2(\pi^*_{2p})^1$
The net bond order is 2.5, a higher bond order than in O_2 and thus a stronger bond. The ion is paramagnetic to the extent of one unpaired electron.

CHAPTER 11

1. (a) LiF, because the ion-ion distance is shorter than in KI. (b) MgO, because there are +2 and −2 charges on the ions, whereas there are only +1 and −1 charges for NaCl.

2. The order of hydration energies is $Cs^+ < Na^+ < Mg^{2+}$. Mg^{2+} is the smallest ion and the ion with the highest charge. Since the force of attraction between an ion and a given dipole is greater for smaller ions, and for ions of higher charge, Mg^{2+} has the highest energy. The energy of hydration is smaller for Cs^+ than for Na^+ since the former ion is the larger of the two, while both have the same charge.

3. (a) N_2 interactions occur by induced dipole-induced dipole forces, the weakest of intermolecular forces. (b) Dipole-induced dipole forces. H_2O is a polar molecule, so it can more readily induce a dipole in CO_2. As a consequence, CO_2 dissolves in H_2O.

4. $(1.00 \times 10^3 \text{ g})(1 \text{ mol}/32.0 \text{ g})(37.6 \text{ kJ/mol}) = 1180 \text{ kJ}$

5. Enthalpy of vaporization
$$= \Delta H_f^\circ[CCl_4(g)] - \Delta H_f^\circ[CCl_4(\ell)]$$
$$= -102.9 \text{ kJ} - (-135.4 \text{ kJ}) = 32.5 \text{ kJ}$$

6. Ethane molecules (C_2H_6) are attracted to one another through very weak induced dipole forces, whereas NH_3 molecules can interact through dipole-dipole forces (and weak hydrogen bonds). Thus, because forces between C_2H_6 molecules are weaker than between NH_3 molecules, less energy is required to vaporize C_2H_6 than NH_3, and the boiling point of the C_2H_6 is lower than that of NH_3.

7. $P = nRT/V =$
$(0.028 \text{ mol})(0.0821 \text{ L·atm/K·mol})(333 \text{ K})/5.0 \text{ L}$
$= 0.15 \text{ atm} = 120 \text{ mm Hg}$

Appendix F gives a vapor pressure of H_2O at 60°C of 149 mm Hg. The calculated pressure of water in the flask is smaller than this, so all of the 0.50 g of water evaporates. With 2.0 g, however, the calculated pressure (460 mm Hg) is much larger than the vapor pressure of water at 60°C, so only enough water can evaporate to give an equilibrium pressure of 149 mm Hg.

8. $$\ln\frac{534}{57} = \frac{\Delta H_{vap}}{8.314 \times 10^{-3} \text{ J/K·mol}}\left[\frac{1}{250.4} - \frac{1}{298.2}\right]$$
$\Delta H_{vap} = 29.0 \text{ kJ}$

9. (8 corner Cl^- ions)($\frac{1}{8}$ per corner) = 1 net Cl^- ion in the unit cell. Since there is 1 Cs^+ ion in the center of the unit cell, the formula of the salt must be CsCl.

10. Cube edge = 2(radius of K^+) + 2(radius of Cl^-)
= 2(133 pm) + 2(181 pm) = 628 pm
Volume = $(\text{edge})^3 = 2.48 \times 10^8 \text{ pm}^3$ or $2.48 \times 10^{-22} \text{ cm}^3$

11. There are 4 KCl's in one unit cell. The mass of these is (4 KCl/unit cell)(1 mol/6.022 × 10^{23} KCl)(74.6 g/mol) = 4.95 × 10^{-22} g/unit cell

$$\text{density} = \text{mass/volume} = \frac{4.95 \times 10^{-22}\ \text{g/unit cell}}{2.48 \times 10^{-22}\ \text{cm}^3} = 2.00\ \text{g/cm}^3$$

12. $\Delta H_f^\circ[\text{KCl(s)}] = \Delta H_f^\circ[\text{K(g)}]$ + Ion. Eneg. K(g) + 1/2 D[Cl_2(g)] + Elec. Aff. Cl + ΔH_{cryst}
= 90.14 kJ + 415 kJ + 1/2(243.3 kJ) + (−348.5 kJ) + (−708 kJ)
= −430 kJ/mol

13. Heat evolved by NH_3 = (17.0 g)(4.70 J/g·K)[−43.0 − (−33.0)](K) = −799 J
Heat evolved by liquid water = (18.0 g)(4.184 J/g·K)(−10.0 K) = −753 J

14. Liquid CO_2 should exist at a temperature of −20°C and a pressure of 12 atm.

CHAPTER 12

1. 1.0 × 10^3 g glycol = 16 mol; 4.0 kg water = 220 mol H_2O
m = 16 mol/4.0 kg = 4.0 molal

$$X_{\text{glycol}} = \frac{16\ \text{mol glycol}}{16\ \text{mol glycol} + 220\ \text{mol H}_2\text{O}} = 0.068$$

2. For AgCl: Heat of solution = 916 kJ/mol + (−851 kJ/mol) = +65 kJ/mol
For RbF: Heat of solution = 776 kJ/mol + (−792 kJ/mol) = −16 kJ/mol
AgCl, an insoluble salt, has a large, positive heat of solution, whereas the soluble salt RbF has a negative heat of solution.

3. Heat of solution =
ΔH_f°[Solution] − ΔH_f°[Solid]
= −339.9 kJ/mol − (−365.6 kJ/mol) = +25.7 kJ/mol

4. m (CO_2) = (4.44 × 10^{-5} molal/mm Hg)(253 mm Hg) = 1.12 × 10^{-2} m

5. Solution consists of sucrose (= 0.0292 mol) and water (= 12.5 mol).

$$X_{\text{water}} = \frac{12.5\ \text{mol H}_2\text{O}}{12.5\ \text{mol H}_2\text{O} + 0.0292\ \text{mol sucrose}} = 0.998$$

P_{water} = (0.998)(526) = 525 mm Hg

Even with 10.0 g of sugar, the vapor pressure of hot water has changed little.

6. Solution consists of 0.00200 mol nitroglycerin and 1.28 mol benzene.
$X_{\text{nitro}} = 1.56 \times 10^{-3}$
$P = X_{\text{nitro}}P^\circ_{\text{benzene}} = (1.56 \times 10^{-3})(95\ \text{mm Hg}) = 0.15\ \text{mm Hg}$

7. Δt = 80.31°C − 80.10°C = 0.21°C

$$m = \frac{\Delta t}{K_{bp}} = \frac{0.21°\text{C}}{2.53\ \text{deg/molal}} = 0.083\ \text{mol/kg}$$

(0.083 mol/kg)(0.100 kg) = 0.0083 mol

$$\text{molar mass} = \frac{1.25\ \text{g}}{0.0083\ \text{mol}} = 150\ \text{g/mol}$$ (actual mass of $C_8H_8O_3$ = 152 g/mol)

8. Δt = 5.10°C − 5.50°C = −0.40°C

$$m = \frac{-0.40°\text{C}}{-4.90\ \text{deg/molal}} = 0.082\ \text{mol/kg}$$

0.082 mol/kg (0.100 kg) = 0.0082 mol

$$\text{molar mass} = \frac{1.25\ \text{g}}{0.0082\ \text{mol}} = 150\ \text{g/mol}$$

9. 866 g of NaCl (14.8 mol) give a 2.69 m solution in 5.5 kg water. More importantly, the total concentration of Na^+ and Cl^- ions is 2 × 2.69 m = 5.38 m.
Δt_{bp} = (5.38 molal)(+0.512 deg/molal) = 2.76°C
Boiling point = 100.00°C + 2.76°C = 102.76°C

10. (a) 25.0 g of NaCl in 0.500 kg water = 0.855 m
Total molality of ions (Na^+ + Cl^-) = 2 × 0.855 = 1.71 m
Δt_{fp} = 1.71 m(−1.86 deg/molal) = −3.18°C
(b) 25.0 g of $CaCl_2$ in 0.500 kg water = 0.450 m
Total molality of ions (Ca^{2+} + 2 Cl^-) = 1.35 m
Δt_{fp} = 1.35 m(−1.86 deg/molal) = −2.51°C

11. $$M = \Pi/RT = \frac{(26.57\ \text{mm Hg}/760.\ \text{mm Hg atm}^{-1})}{(0.0821\ \text{L·atm/K·mol})(298\ \text{K})} = 1.43 \times 10^{-3}\ \text{mol/L}$$

(1.43 × 10^{-3} mol/L)(0.0100 L) = 1.43 × 10^{-5} mol

$$\text{Molar mass} = \frac{7.68 \times 10^{-3}\ \text{g}}{1.43 \times 10^{-5}\ \text{mol}} = 537\ \text{g/mol}$$

The actual mass of beta-carotene ($C_{40}H_{56}$) is 536.85 g/mol.

CHAPTER 13

1. −1/2(Δ[NOCl]/Δt) = +1/2(Δ[NO]/Δt) = +(Δ[Cl_2]/Δt)

2. (a) m = 1. The initial rates and initial concentrations are directly proportional. For example, as the concentration is doubled, the rate is doubled.
(b) Taking the data from Experiment 1, we have

$$k = \frac{\text{Rate}}{[\text{reactant}]} = \frac{1.3 \times 10^{-7}\ \text{mol/(L·min)}}{1.0 \times 10^{-3}\ \text{mol/L}} = 1.3 \times 10^{-4}/\text{min}$$

3. ln ([cis]/[cis]$_0$) = $-kt$ = $-(2.22/\text{hr})(19\ \text{min})(1\ \text{hr}/60\ \text{min}) = -0.70$
[cis]/[cis]$_0$ = 0.50 so [cis] = (0.50)(0.010 M) = 0.0050 M

4. $\frac{1}{[\text{HI}]} - \frac{1}{[\text{HI}]_0} = kt$
When $[\text{HI}]_0 = 0.010\ M$, $k = 30./M\cdot\text{min}$, and $t = 10.$ min, [HI] = 0.0025 M.

5. (a) $t_{1/2} = \frac{0.693}{5.40 \times 10^{-2}/\text{hr}} = 12.8\ \text{hr}$
(b) 51.2 hr = 4.00 half-lives. After 4.00 half-lives the fraction remaining is 1/16.
(c) ln (fraction remaining) = $-kt = -(5.40 \times 10^{-2}/\text{hr})(18\ \text{hr}) = -0.97$; fraction remaining = $[A]/[A]_0 = 0.38$

6.

Time (s)	$[A]M$	ln$[A]$	$1/[A]$
0	0.10	−2.30	10.
30	0.074	−2.60	13.5
60	0.055	−2.90	18.2
90	0.041	−3.20	24.4

If ln$[A]$ is plotted vs. time, a straight line is obtained, whereas a curved line is observed for $1/[A]$ vs. time. From the ln$[A]$ vs. time plot we find a slope of −0.01. Since k = −slope, k = 0.01.

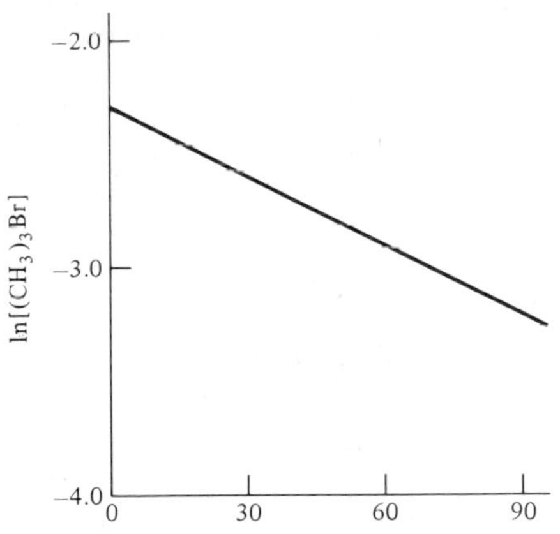

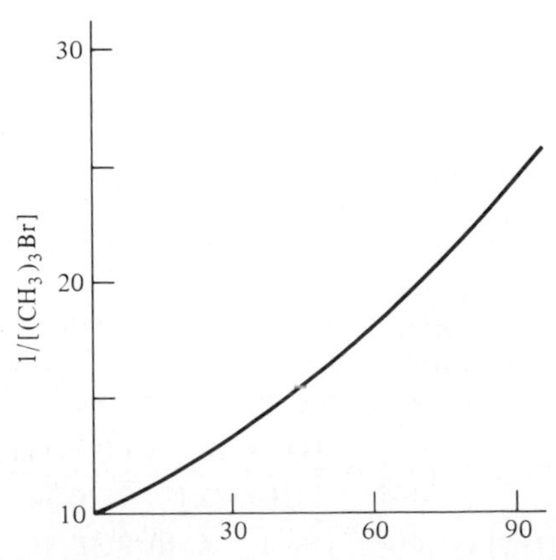

7. Substituting into equation 13.5, we have
$$\ln \frac{9.839}{5.076} = -\frac{E^*}{8.31 \times 10^{-3}\ \text{kJ/K}\cdot\text{mol}}\left[\frac{1}{330.} - \frac{1}{300.}\right]$$
$E^* = 18.1$ kJ

8. (a) Rate = $k[\text{Ni(CO)}_4]$; (b) Rate = $k[O_3]$; (c) Rate = $k[NO_2][NO_3]$

9. (a) Bimolecular; two species are involved. (b) Rate = k[Cl][CO]

10. No. If the chemical equation represented the single, elementary step, then the rate law would have been Rate = $k[NO_2]$[CO].

11. (a) Step 2, the slow step. (b) Rate = $k[NH_2Cl][NH_3]$

12. Rate law for the slow step: Rate = k_2[Br]$[H_2]$
Since the concentration of the intermediate Br cannot appear in the rate law, we eliminate it by using the equilibrium constant expression for the equilibrium step 1.

$$K = \frac{[\text{Br}]^2}{[\text{Br}_2]} \text{ which gives } [\text{Br}] = K^{1/2}[\text{Br}_2]^{1/2}$$

Substituting this into the rate law for the slow step:
Rate = $k_2\{K^{1/2}[\text{Br}_2]^{1/2}\}[H_2]$
Taking $k_2K^{1/2}$ as the constant k, we have the experimentally observed rate law.

CHAPTER 14

1. (a) $K = \frac{[PCl_3][Cl_2]}{[PCl_5]}$ (b) $K = \frac{[H_2]}{[HCl]}$

(c) $K = [Cu^{2+}][OH^-]^2$ (d) $K = \frac{[Cu^{2+}][NH_3]^4}{\{[Cu(NH_3)_4]^{2+}\}}$

2. (a) $K_p' = \frac{[N_2O_4]}{[NO_2]^2} = \frac{1}{K_p} = \frac{1}{6.75} = 0.148$

$$K_p'' = \frac{[NO_2]}{[N_2O_4]^{1/2}} = (K_p)^{1/2} = (6.75)^{1/2} = 2.60$$

(b) $K_p' = 1/K_p = 1/6.6 \times 10^{11} = 1.5 \times 10^{-12}$

3. Concentration of Ag^+ in the AgCl beaker (1.3×10^{-5} M) is greater than in the AgI beaker (1.2×10^{-8}).

4. n = 2 moles gaseous products − 4 moles gaseous reactants = −2

$K_c = K_p/(RT)^n = K_p/(RT)^{-2} = K_p(RT)^2$

$K_c = (6.0 \times 10^5)[(0.082\ \text{L·atm/K·mol})(298\ \text{K})]^2 = 3.6 \times 10^8$

The units are accounted for in these calculations. K_p was effectively in units appropriate to gases (atm^{-2} in this case). The units in which K_c is obtained are then M^{-2}, appropriate for this equilibrium constant in concentration units.

5. When [iso] = 2.0, then $[n]$ = 0.80.
$K = [\text{iso}]/[n] = 2.0/0.80 = 2.5$

6. $$Q_p = \frac{[NO]^2}{[N_2][O_2]} = \frac{(4.2 \times 10^{-3})^2}{(0.50)(0.25)} = 1.4 \times 10^{-4}$$

Since $Q_p < K_p$, more product is produced by the reaction.

7.

	$[PCl_5]$	$[PCl_3]$	$[Cl_2]$
initial P (atm)	1.50	0	0
change in P (atm)	−1.34	+1.34	+1.34
final P (atm)	0.16	1.34	1.34

$$K_p = \frac{[PCl_3][Cl_2]}{[PCl_5]} = \frac{(1.34)^2}{0.16} = 11$$

8.

	$[PCl_5]$	$[PCl_3]$	$[Cl_2]$
initial P (atm)	3.00	0	0
change in P (atm)	$-x$	$+x$	$+x$
final P (atm)	$3.00 - x$	x	x

$$K_p = \frac{[PCl_3][Cl_2]}{[PCl_5]} = \frac{x^2}{3.00 - x} = 11.2$$

$11.2(3.00 - x) = x^2$, and on rearrangement we have $x^2 + 11.2x - 33.6 = 0$

Solving this quadratic equation, $x = [PCl_3] = [Cl_2]$ = 2.46 atm. This means $[PCl_5]$ = 0.54 atm.

9. $K_c = [Mn^{2+}][S^{2-}]$. Using $K_c = 5.1 \times 10^{-15}$ and $[S^{2-}] = 2.0 \times 10^{-9}\ M$, $[Mn^{2+}] = 2.6 \times 10^{-6}\ M$.

10. (a) $[NH_4^+]$ increases with increasing temperature. (b) $[N_2]$ increases as the temperature declines.

11.

	$[n]$	[iso]
initial conc. (M)	0.20	0.50
conc. after adding 2.0 M more iso (M)	0.20	2.0 + 0.50
change on proceeding to equilibrium (M)	$+x$	$-x$
conc. at equilibrium (M)	$0.20 + x$	$2.50 - x$

$K = [\text{iso}]/[n] = (2.50 - x)/(0.20 + x)$. Solving for x gives $x = 0.57\ M$.

Therefore, the new equilibrium concentrations are [iso] = 1.93 M and $[n]$ = 0.77 M.

12. (a) Added H_2 shifts the equilibrium right, and added NH_3 shifts it left.
(b) Increasing the volume decreases all concentrations. The equilibrium shifts left, toward the side with the greater number of molecules.

CHAPTER 15

1. (a) HCN is a Brønsted acid and CN^- is its conjugate base. NH_3 is a Brønsted base and NH_4^+ is its conjugate acid.
(b) HS^- is the conjugate base of H_2S.
(c) HNO_3 is the conjugate acid of NO_3^-.

2. (a) NH_4^+ is a weaker acid than HSO_4^- (and SO_4^{2-} is a weaker base than NH_3), so the equilibrium lies to the right.
(b) HCO_3^- is a weaker acid than H_2S (and HS^- is a weaker base than CO_3^{2-}), so the equilibrium lies to the left.

3. The Cl^- and Na^+ ions do not enter the problem, since they are the conjugates of a strong acid and a strong base, respectively. The reaction that may occur is

$$NH_4^+(aq) + SO_4^{2-}(aq) \rightleftharpoons NH_3(aq) + HSO_4^-(aq)$$

However, since NH_4^+ is a weaker acid than HSO_4^- (and SO_4^{2-} is a weaker base than NH_3), the equilibrium lies well to the left.

4. 0.073 g of HCl is 2.0×10^{-3} mol. Therefore, [HCl] = $[H^+]$ = 2.0×10^{-3} mol/0.50 L = $4.0 \times 10^{-3}\ M$. From $K_w = 1.0 \times 10^{-14} = [H^+][OH^-]$, we have $[OH^-] = 2.5 \times 10^{-12}\ M$.

5. $[H^+] = 10^{-pH} = 10^{-7.30} = 5.0 \times 10^{-8}\ M$. From $K_w = 1.0 \times 10^{-14} = [H^+][OH^-]$, we have $[OH^-] = 2.0 \times 10^{-7}\ M$.

6. This question considers the equilibrium propH(aq) ⇌ H^+(aq) + $prop^-$(aq). From the pH we obtain $[H^+]$ and $[prop^-]$ at equilibrium.

$[H^+] = [prop^-] = 10^{-pH} = 10^{-2.94} = 1.1 \times 10^{-3}\ M$

This means that [propH] = $0.10 - 1.1 \times 10^{-3} \cong 0.10\ M$.

$$K_a = \frac{[H^+][prop^-]}{[propH]} = \frac{(1.1 \times 10^{-3})^2}{(0.10)} = 1.3 \times 10^{-5}$$

7. This question considers the reaction HOAc(aq) ⇌ H^+(aq) + OAc^-(aq)

	[HOAc]	$[H^+]$	$[OAc^-]$
Initial concentration (M)	0.10	10^{-7}	0
Change in concentration	$-x$	$+x$	$+x$
Concentration at equilibrium (M)	$0.10 - x$	$10^{-7} + x$	x
Equilibrium concentrations after making approximations (M)	0.10	x	x

$$K_a = 1.8 \times 10^{-5} = \frac{[H^+][OAc^-]}{[HOAc]} = \frac{x^2}{(0.10)}$$

$x = [H^+] = [OAc^-] = 1.3 \times 10^{-3}\ M$

8. (a) $NH_3(aq) + H_2O(\ell) \rightleftharpoons NH_4^+(aq) + OH^-(aq)$

	$[NH_3]$	$[OH^-]$	$[NH_4^+]$
Initial concentration (M)	0.010	10^{-7}	O
Change in concentration	$-x$	$+x$	$+x$
Concentration at equilibrium (M)	$0.010 - x$	$10^{-7} + x$	x
Equilibrium concentrations after making approximations (M)	0.010	x	x

$$K_b = \frac{[OH^-][NH_4^+]}{[NH_3]} = \frac{x^2}{0.010} = 1.8 \times 10^{-5}$$

$x = [OH^-] = [NH_4^+] = 4.2 \times 10^{-4}\ M$. This gives a pOH of 3.37 or a pH of 10.63.

(b) $CN^-(aq) + H_2O(\ell) \rightleftharpoons HCN(aq) + OH^-(aq)$

Solving for equilibrium concentrations in the same manner as (a) above, we have

$$K_b = \frac{[OH^-][HCN]}{[CN^-]} = \frac{x^2}{0.010} = 2.5 \times 10^{-5}$$

$x = [OH^-] = [HCN] = 5.0 \times 10^{-4}\ M$. This gives a pOH of 3.30 or a pH of 10.70.

9. (a) Na^+ and Cl^- both lead to a neutral solution; pH is 7. (b) Fe^{2+} ions hydrolyze to give an acidic solution. Chloride ion, Cl^-, has no effect on pH. A solution of $FeCl_2$ should have a pH less than 7. (c) The ammonium ion, NH_4^+, is a weak acid, while the nitrate ion leads to a neutral solution. Thus, pH of NH_4NO_3(aq) is less than 7. (d) The Na^+ ion does not hydrolyze, but the bicarbonate ion, HCO_3^-, is a weak base (K_b for HCO_3^- is greater than K_a for HCO_3^-). The solution should be slightly basic.

10. The pH of the solution is determined by the first equilibrium process:

$H_2C_2O_4(aq) \rightleftharpoons H^+(aq) + HC_2O_4^-(aq)$

$K_{a1} = 5.9 \times 10^{-2}$

If $[H_2C_2O_4]$ is $(0.10 - x)M$, and $[H^+] = [HC_2O_4^-] = xM$, then

$$K_{a1} = \frac{[H^+][HC_2O_4^-]}{[H_2C_2O_4]} = \frac{x^2}{0.10 - x} = 5.9 \times 10^{-2}$$

Since 100K is not less than the original concentration of the acid, this must be solved by the quadratic formula. From the expanded equation

$0.059(0.10) - 0.059x = x^2$

$x^2 + 0.059x - 0.0059 = 0$

we find $x = [H^+] = 0.053\ M$. This means the pH is 1.28.

The concentration of $[C_2O_4^{2-}]$, the ion from the second equilibrium process,

$HC_2O_4^-(aq) \rightleftharpoons H^+(aq) + C_2O_4^{2-}(aq)$

is equivalent to the value of K_{a2}, which is 6.4×10^{-5} M in this case.

11. (a) H_2SO_4 has more O atoms attached to the S than H_2SO_3, and so H_2SO_4 is the stronger acid (K_{a1} for H_2SO_4 = very large, while K_{a1} for H_2SO_3 = 1.2×10^{-2}).

(b) H_3AsO_4 (arsenic acid) is stronger than H_3AsO_3 (arsenous acid) for the same reasons as in (a).

12. (a) $:PH_3$ is a Lewis base; there is a lone pair of electrons on the P atom to be shared with another atom. (b) BCl_3, like BF_3 in the text, is a Lewis acid. (c) H_2S, like water, is a Lewis base due to the lone pairs of electrons on the S atom. (d) SF_4 has a lone pair of electrons on the S atom and can, in principle, act as a Lewis base.

13. When the solutions of Cu^{2+} and NH_3 are mixed, the volume is doubled. This means the concentrations of Cu^{2+} and NH_3 are half of their value in the original solutions. Thus, we have the values in the table below.

	$[Cu^{2+}]$	$[NH_3]$	$\{[Cu(NH_3)_4^{2+}\}$
Concentrations after mixing solutions but before reaction (M)	0.00100	1.50	0
Concentrations after reaction to form $[Cu(NH_3)_4^{2+}$ (M)	0	$1.50 - 4(0.00100)$	0.00100
Change on proceeding to equilibrium (M)	$+x$	$+4x$	$-x$
Equilibrium concentrations (M)	x	$1.50 - 0.00400 + 4x$	$0.00100 - x$
Equilibrium concentrations after approximations (M)	x	1.50	0.00100

For the reaction $Cu^{2+}(aq) + 4\ NH_3(aq) \rightleftharpoons [Cu(NH_3]_4^{2+}(aq)$

$$K = 6.8 \times 10^{12} = \frac{\{[Cu(NH_3]_4^{2+}\}}{[Cu^{2+}][NH_3]^4} = \frac{0.00100}{(x)(1.50)^4}$$

$x = [Cu^{2+}]$ at equilibrium $= 2.9 \times 10^{-17}\ M$

CHAPTER 16

1. $NaOH(aq) + C_6H_5COOH(aq) \rightleftharpoons NaC_6H_5COO(aq) + H_2O(\ell)$

0.976 g of benzoic acid $= 8.00 \times 10^{-3}$ mol

8.00×10^{-3} mol acid (1 mol NaOH/1 mol acid)(1.00 L NaOH/0.100 mol NaOH) $= 0.0800$ L (or 80 mL) of NaOH

8.00×10^{-3} mol benzoic acid gives 8.00×10^{-3} mol benzoate ion. Its concentration is

$[C_6H_5COO^-] = 8.00 \times 10^{-3}$ mol/0.0800 L $= 0.100\ M$

The $C_6H_5COO^-$ ion is a weak base, reacting in water according to the equation

$$C_6H_5COO^-(aq) + H_2O(\ell) \rightleftharpoons C_6H_5COOH(aq) + OH^-(aq)$$

$$K_b = 1.6 \times 10^{-10} = \frac{[OH^-][C_6H_5COOH]}{[C_6H_5COO^-]} = \frac{x^2}{0.100}$$

$x = [OH^-] = [C_6H_5COO^-] = 4.0 \times 10^{-6}\ M$

pOH $= 5.40$ and pH $= 8.60$

2. Moles of HCl $= (0.050\text{ L})(0.20\ M) = 0.010$ mol
Moles of aniline $= 0.93$ g(1 mol/93.1 g) $= 0.010$ mol
Since 1 mol of aniline requires 1 mol of HCl, and since they were present initially in equal molar quantities, the reaction completely consumes the acid and base. The solution after reaction contains 0.010 mol of $C_6H_5NH_3^+$ in 0.050 L of solution, so its concentration is 0.20 M. It is the conjugate acid of the weak base aniline, so

$$C_6H_5NH_3^+(aq) \rightleftharpoons C_6H_4NH_2(aq) + H^+(aq)$$

$$K_a = 2.4 \times 10^{-5} = \frac{[H^+][C_6H_5NH_2]}{[C_6H_5NH_3^+]} = \frac{x^2}{0.20}$$

$x = [H^+] = 2.2 \times 10^{-3}\ M$, which gives a pH of 2.66.

3. Equal numbers of moles of the acid (HOAc) and base (pyridine) were mixed, so reaction was complete to produce OAc^-, a weak base, and $C_5H_5NH^+$, a weak acid.

K_b for $OAc^- = 5.6 \times 10^{-10}$ and K_a for $C_5H_5NH^+ = 6.7 \times 10^{-6}$

The acid is stronger than the base here, so the solution containing the two ions is slightly acidic. (The pH is about 5; see Study Question 15.110.)

4. (a) pH of 0.30 M formic acid

$$K_a = 1.8 \times 10^{-4} = \frac{[H^+][HCOO^-]}{[HCOOH]} = \frac{x^2}{0.30}$$

$x = [H^+] = 7.4 \times 10^{-3}\ M$, which gives a pH of 2.13

(b) pH of 0.30 M formic acid + 0.10 M NaCHOO

	[HCOOH]	[H⁺]	[HCOO⁻]
Before acid ionization (M)	0.30	0	0.10
Change	$-x$	$+x$	$+x$
At equilibrium (M)	$0.30 - x$	x	$0.10 + x$
After approximations (M)	0.30	x	0.10

$$K_a = 1.8 \times 10^{-4} = \frac{[H^+][HCOO^-]}{[HCOOH]} = \frac{(x)(0.10)}{0.30}$$

$x = [H^+] = 5.4 \times 10^{-4}\ M$; pH $= 3.27$

5. K_a for formic acid is 1.8×10^{-4}

(a) Find pH of buffer before adding HCl.

$$[H^+] \text{ before adding excess HCl} = \frac{[\text{acid}]}{[\text{salt}]} K_a = \frac{0.50}{0.70}(1.8 \times 10^{-4})$$

$[H^+] = 1.3 \times 10^{-4}\ M$, which gives pH $= 3.89$

(b) Now 10. mL of 1.0 M HCl (= 0.010 mol) is added. The reaction occurring is

$$H^+(aq) + HCOO^-(aq) \longrightarrow HCOOH(aq)$$

and the amounts and concentrations at this stage are as follows:

	[H⁺] from HCl	[HCOO⁻] from buffer	[HCOOH] from buffer
Before reaction (mol)	0.010	0.35	0.25
Change when HCl reaction occurs (mol)	−0.010	−0.010	+0.010
After HCl reaction (mol)	0	0.34	0.26
After reaction (moles in 0.510 L total volume)	0	0.67	0.51

To find the $[H^+]$ at this stage, we use the chemical equation for the dissociation of formic acid, HCOOH.

$$HCOOH(aq) \rightleftharpoons H^+(aq) + HCOO^-(aq)$$

$$[H^+] \text{ after adding excess HCl} = \frac{[\text{acid}]}{[\text{salt}]} K_a = \frac{0.51}{0.67}(1.8 \times 10^{-4}) = 1.4 \times 10^{-4}\ M$$

pH = 3.86

Only a *very* small change in pH occurred on adding a concentrated acid.

6. (a) $pK_a = -\log K_a = 10.251$

(b) Molar mass of $NaHCO_3$ is 84.0 g/mol and that of Na_2CO_3 is 106 g/mol. This means $[NaHCO_3] = 0.179\ M$ and $[Na_2CO_3] = 0.170\ M$.

$$pH = pK_a + \log \frac{[CO_3^{2-}]}{[HCO_3^-]}$$

$$= 10.251 + \log \frac{0.170}{0.179} = 10.228$$

7. A pH of 5.00 corresponds to $[H^+] = 1.0 \times 10^{-5}\ M$. This means that

$$1.0 \times 10^{-5} = [H^+] = \frac{[HOAc]}{[OAc^-]}(1.8 \times 10^{-5})$$

The ratio $[HOAc]/[OAc^-]$ must be 1/1.8 to achieve the correct H^+ concentration. Therefore, 1.0 mol of HOAc is mixed with 1.8 mol of a salt of OAc^- (say NaOAc) in some amount of water. (The volume of water is not critical; only the relative amounts of acid and salt are important.)

8. (a) Using the expression in the Example to find $[H^+]$ before the equivalence point, we have

$$[H^+] = \frac{0.00500 \text{ mol HCl} - (0.0400 \text{ L NaOH})(0.100\ M)}{0.0500 \text{ L HCl} + 0.0400 \text{ L NaOH}} = 0.0111\ M$$

pH = 1.954

(b) After 60.0 mL of base have been added, the HCl has been completely consumed, and we have added 10.0 mL of base in excess of the equivalence point.

$$[OH^-] = \frac{\text{moles excess base}}{\text{total volume}} = \frac{(0.0100 \text{ L})(0.100\ M)}{0.050 \text{ L acid} + 0.060 \text{ L NaOH}} = 9.1 \times 10^{-3}\ M$$

pOH = 2.04, which gives a pH of 11.96.

(c) When 49.9 mL of NaOH has been added, we are still 0.1 mL short of the equivalence point. Therefore, $[H^+]$ is calculated as in part (a). This gives $[H^+] = 1.00 \times 10^{-4}\ M$, or a pH of 4.000. Even this close to the equivalence point, the solution is still relatively acidic.

9. The chemical equation for the reaction occurring here is $NaOH(aq) + HOAc(aq) \rightarrow NaOAc(aq) + H_2O(\ell)$

(a) 50.0 mL of 0.100 *M* NaOH are required to neutralize the acid completely. Therefore, to neutralize only 20.0%, only 20.0% of 50.0 mL or 10.0 mL of NaOH are required. Concentrations of compounds in solution at this point are:

(i) 80.0% of the HOAc remains and is contained in 60.0 mL of solution.

$$[HOAc] = \frac{\text{Amount of HOAc remaining } (= 80.0\% \text{ of original})}{0.0500 \text{ L HOAc originally} + 0.0100 \text{ L NaOH added}}$$

$$= \frac{(0.0500 \text{ L HOAc})(0.100\ M)(0.800)}{0.0600 \text{ L}}$$

$$= 6.67 \times 10^{-2}\ M$$

(ii) 20.0% of the HOAc has been converted to NaOAc, and the salt is contained in 60.0 mL of solution.

$$[OAc^-] = \frac{(0.0500 \text{ L HOAc})(0.100\ M)(0.200)}{0.0600 \text{ L}}$$

$$= 1.67 \times 10^{-2}\ M$$

(iii) This is a buffer solution, and $[H^+]$ is calculated in the usual manner.

$$[H^+] = \frac{[HOAc]}{[OAc^-]}K_a = \frac{6.67 \times 10^{-2}}{1.67 \times 10^{-2}}(1.8 \times 10^{-5})$$

$[H^+] = 7.2 \times 10^{-5}$, which gives a pH of 4.14.

(b) After 55.0 mL of NaOH has been added, we have added 5.00 mL of NaOH in excess of the equivalence point. Therefore,

$$[OH^-] = \frac{(0.0050 \text{ L})(0.100\ M)}{0.105 \text{ L}} = 4.8 \times 10^{-3}\ M$$

pOH = 2.32, which gives a pH of 11.68.

10. The equivalence point occurs at pH = 5.21. A suitable indicator might be methyl orange, which is yellow in a basic solution but red to red-orange in an acidic solution. The color change occurs around pH 4.5–5.0. Other choices of indicator would include bromcresol green, bromphenol blue, and methyl red.

CHAPTER 17

1. (a) AgBr, insoluble due to Ag^+ and Br^-; (b) K_2CrO_4, soluble (due to K^+); (c) $SrCO_3$, insoluble (carbonates are not usually soluble); (d) KF, soluble (due to K^+); (e) $NH_4C_2H_3O_2$ (ammonium acetate), soluble (due to NH_4^+); (f) $Ba(NO_3)_2$, soluble (due to NO_3^-)

2. (a) $NaCl(aq) + AgNO_3(aq) \rightarrow AgCl(s) + NaNO_3(aq)$

(b) $2\ KI(aq) + Pb(NO_3)_2(aq) \rightarrow PbI_2(s) + 2\ KNO_3(aq)$

(c) No precipitate forms (possible products $Cu(NO_3)_2$ and $MgCl_2$ are both water-soluble).

3. (a) $K_{sp} = [Cd^{2+}][S^{2-}] = 3.6 \times 10^{-29}$
(b) $K_{sp} = [Bi^{3+}][I^-]^3 = 8.1 \times 10^{-19}$
(c) $K_{sp} = [Ag^+]^2[CO_3^{2-}] = 8.1 \times 10^{-12}$

4. (a) $K_{sp} = [Ag^+][SCN^-] = 1.0 \times 10^{-12}$ since $[Ag^+] = [SCN^-] = 1.0 \times 10^{-6}$ *M*.
(b) $K_{sp} = [Ag^+]^2[S^{2-}] = 9.8 \times 10^{-50}$ since $[Ag^+] = 5.8 \times 10^{-17}$ *M* and $[S^{2-}] = 1/2(5.8 \times 10^{-17}\ M)$

5. (a) Solubility of CdS $= [Cd^{2+}] = [S^{2-}] = \sqrt{K_{sp}} = \sqrt{3.6 \times 10^{-29}} = 6.0 \times 10^{-15}$ mol/L
(b) Solubility of $Mg(OH)_2 = [Mg^{2+}] = S$. Therefore, $[OH^-] = 2S$. $K_{sp} = [Mg^{2+}][OH^-]^2 = (S)(2S)^2$. $S = 1.6 \times 10^{-4}\ M$

6. (a) AgCl, (b) ZnS, (c) $MgCO_3$, (d) $Ca(OH)_2$, (e) FeS

7. $Q = [Ag^+][Cl^-] = 2.5 \times 10^{-10}$. Not at equilibrium since $Q > K_{sp}$. More AgCl will precipitate.

8. $Q = [Sr^{2+}][SO_4^{2-}] = (2.5 \times 10^{-4})(2.5 \times 10^{-4}) = 6.3 \times 10^{-8}$. No precipitate forms, since $Q < K_{sp}$.

9. $[SO_4^{2-}] = K_{sp}/[Ba^{2+}] = (1.1 \times 10^{-10})/(1.0 \times 10^{-3}) = 1.1 \times 10^{-7}\ M$

10. (a) $[Ba^{2+}]$ required for equilibrium $= K_{sp}/[CO_3^{2-}] = (8.1 \times 10^{-9})/(0.0010) = 8.1 \times 10^{-6}\ M$.

$$0.100\ \text{L}\left(\frac{8.1 \times 10^{-6}\ \text{mol Ba}^{2+}}{\text{L}}\right) \times \left(\frac{1\ \text{mol BaCl}_2}{1\ \text{mol Ba}^{2+}}\right) \times \left(\frac{208.2\ \text{g}}{\text{mol}}\right) = 1.7 \times 10^{-4}\ \text{g BaCl}_2$$

(b) $[Ba^{2+}]$ required $= 8.1 \times 10^{-3}$ *M*. Solving for mass of $BaCl_2$ as above gives 0.17 g.

11. Total volume of solution after mixing is 200. mL. Therefore, the initial concentrations of CO_3^{2-} and Ba^{2+} (before precipitation) will be *half* of those in the original 100. mL solutions. This means $Q = [Ba^{2+}][CO_3^{2-}] = (5.0 \times 10^{-4})(5.0 \times 10^{-4}) = 2.5 \times 10^{-7}$. Since $Q > K_{sp}$ precipitation will occur.

12. (a) In pure water $S = [Ba^{2+}] = \sqrt{K_{sp}} = \sqrt{1.1 \times 10^{-10}} = 1.0 \times 10^{-5}\ M$.
(b) When $[Ba^{2+}] = 0.010\ M$, $S = [SO_4^{2-}] = K_{sp}/[Ba^{2+}] = 1.1 \times 10^{-8}\ M$.

13. $Zn(CN)_2(s) \rightleftharpoons Zn^{2+}(aq) + 2\ CN^-(aq)$
(a) In pure water solubility $= [Zn^{2+}] = S$ and so $[CN^-] = 2S$. This means $K_{sp} = (S)(2S)^2$ and so $S = 1.3 \times 10^{-4}\ M$.
(b) When initial concentration of $CN^- = 0.010\ M$, this means the K_{sp} expression is $K_{sp} = (S)(0.010 + 2S)^2$ where $S = [Zn^{2+}]$. Assuming $2S$ is very small compared with 0.010 *M*, $S = 8.0 \times 10^{-8}\ M$.

14. (a) AgCl precipitates before $PbCl_2$.
$[Cl^-]$ required by 0.0010 *M* $Ag^+ = K_{sp}/(0.0010) = 1.8 \times 10^{-7}\ M$
$[Cl^-]$ required by 0.0010 *M* $Pb^{2+} = \sqrt{K_{sp}/(0.0010)} = 0.13\ M$
(b) When $[Cl^-] = 0.13\ M$, $[Ag^+] = (K_{sp}$ of AgCl$)/0.13 = 1.4 \times 10^{-9}\ M$

15. (a) Use HCl to give insoluble AgCl. $BiCl_3$ is soluble in strong acid.
(b) Use S^{2-} (to give insoluble FeS) or OH^- (to give insoluble $Fe(OH)_2$). K^+ gives soluble salts with both S^{2-} and OH^-.

16. $AgCl(s) \rightleftharpoons Ag^+(aq) + Cl^-(aq)$
$K_{sp} = 1.8 \times 10^{-10}$
$Br^-(aq) + Ag^+(aq) \rightleftharpoons AgBr(s)$
$K = 1/K_{sp} = 1/3.3 \times 10^{-13}$
$AgCl(s) + Br^-(aq) \rightleftharpoons AgBr(s) + Cl^-(aq)$
$K_{net} = 5.5 \times 10^2$

17. 0.010 mol AgCl will give 0.010 *M* $[Ag(NH_3)_2]^+$ and 0.010 *M* Cl^- when dissolved completely in NH_3. Solving for $[NH_3]$ as in Example 17.11, $[NH_3] = 0.19\ M$. The quantity of NH_3 available $= (0.100\ \text{L})(4.0\ M) = 0.40$ mol. Only 0.020 mol of NH_3 is required to form 0.010 M $[Ag(NH_3)_2]^+$ and to achieve a concentration of 0.19 *M*. Therefore, there is sufficient NH_3 to dissolve the AgCl completely.

18. $CuSO_4(aq) + Na_2CO_3(aq) \rightarrow Na_2SO_4(aq) + CuCO_3(s)$
$CuCO_3(s) + 2\ HCl(aq) \rightarrow CuCl_2(aq) + H_2O(\ell) + CO_2(g)$

19. Ba^{2+} forms insoluble salts with both anions ($BaCO_3$ and $BaSO_4$). However, only $BaCO_3$ will dissolve in HCl.

20. CoS is less soluble than MnS; less S^{2-} is required to precipitate CoS than MnS when both metal ions are 0.0010 *M* (see Table 17.3). To precipitate MnS, Table 17.3 shows that $[S^{2-}]$ must be 5.1×10^{-12} *M*. Therefore, if one comes up to *slightly* less than this concentration, then the maximum amount of CoS is precipitated while leaving Mn^{2+} in solution. The strategy is to find the $[H^+]$ required to achieve just the right S^{2-} concentration.($= 5.1 \times 10^{-12}\ M$)
$[H^+]^2 = \{[H_2S]/[S^{2-}]\}(1.3 \times 10^{-20})$
$[H^+] = 1.6 \times 10^{-5}\ M$ (pH = 4.80)

CHAPTER 18

1. (1a) positive; (1b) negative; (1c) positive. (2a) CH_3CH_2OH; (2b) AlI_3.

2. $\Delta S^\circ_{vap} = (27.7 \times 10^3\ \text{J/mol})/298\ \text{K} = 93.0$ J/K·mol
$\Delta S^\circ_{vap} = S^\circ_{vap} - S^\circ_{liq}$
93.0 J/K·mol $= S^\circ_{vap} - 216.4$ J/K·mol
$S^\circ_{vap} = 309.4$ J/K·mol

3. $\Delta S^\circ = S^\circ[CaCO(s)] - S^\circ[Ca(s)] - S^\circ[C(graph)]$
$- 3/2\ S^\circ[O_2(g)]$
$= -262.0$ J/K
When $CaCO_3(s)$ is formed from two other solids (Ca and C) and a gas (O_2), the entropy must decline as the gas is converted to a solid product.

4. $\Delta H^\circ_{rxn} = \Delta H^\circ_f[NH_3(g)] - 1/2\ \Delta H^\circ_f[N_2(g)] - 3/2\ \Delta H^\circ_f[H_2(g)]$
$= -46.11$ kJ/mol $- 1/2(O) - 3/2(O)$
$\Delta S^\circ_{rxn} = S^\circ[NH_3(g)] - 1/2\ S^\circ[N_2(g)] - 3/2\ S^\circ[H_2(g)]$
$= -99.38$ J/mol
$\Delta G^\circ_{rxn} = \Delta H^\circ_{rxn} - T\,\Delta S^\circ_{rxn} = -46.11$ kJ $- (298\text{ K})(-99.38\text{ J/K})(1\text{ kJ}/10^3\text{ J})$
$= -16.5$ kJ

5. (a) C(graphite) + $O_2(g) \rightarrow CO_2(g)$
(b) -394.359 kJ/mol
(c) 2.5 mol(-394.359 kJ/mol) $= -990$ kJ

6. $C_6H_6(\ell) + 15/2\ O_2(g) \rightarrow 6\ CO_2(g) + 3\ H_2O(\ell)$
$\Delta G^\circ_{rxn} = 3\ \Delta G^\circ_f[H_2O(\ell)] + 6\ \Delta G^\circ_f[CO_2(g)] - \{\Delta G^\circ_f[C_6H_6(\ell)] + 15/2\ \Delta G^\circ_f[O_2(g)]\}$
$= 3$ mol(-237.129 kJ/mol) $+ 6$ mol(-394.359 kJ/mol) $-$ [1 mol(125.5 kJ/mol) + 15/2 (O)]
$\Delta G^\circ_{rxn} = -3203.0$ kJ

7. (a) $CaO(s) + CO_2(g) \rightarrow CaCO_3(s)$
$\Delta G^\circ_{rxn} = \Delta G^\circ_f[CaCO_3(s)] - \{\Delta G^\circ_f[CaO(s)] + \Delta G^\circ_f[CO_2(g)]\}$
$= 1$ mol(-1128.79 kJ/mol) $- \{1$ mol(-604.03 kJ/mol) $+ 1$ mol(-394.36 kJ/mol)$\}$
$\Delta G^\circ_{rxn} = -130.4$ kJ
(b) $CaCO_3(s) \rightarrow CaO(s) + CO_2(g)$
ΔG° for the decomposition $= -\Delta G^\circ$ for the formation $= +130.4$ kJ

8. $\Delta H^\circ_{rxn} = +491.18$ kJ and $\Delta S^\circ_{rxn} = 197.67$ J/K
$T = \Delta H^\circ_{rxn}/\Delta S^\circ_{rxn} = 491\text{ kJ}/(0.198\text{ kJ/K}) = 2480$ K (or about 2200°C)

9. (a) $\Delta G^\circ_{rxn} = \Delta G^\circ_f[SO_2(g)] = -300.194$ kJ/mol
Solving for $\ln K_p$ as in 18.7, $\ln K_p = 1.21 \times 10^2$, and so $K_p = 4.18 \times 10^{52}$
(b) $\Delta G^\circ_{rxn} = +130.4$ kJ (from Exercise 18.7)
$\ln K_p = -52.6$, and so $K_p = 1.39 \times 10^{-23}$

CHAPTER 19

1. Oxidation, at anode: $Ni(s) \rightarrow Ni^{2+}(aq) + 2e^-$
Reduction, at cathode: $e^- + Ag^+(aq) \rightarrow Ag(s)$
Electrons flow from anode (Ni) to cathode (Ag), while NO_3^- ions flow from the cathode compartment (containing a declining concentration of Ag^+) to the anode compartment (containing an increasing concentration of Ni^{2+}). Abbreviated notation: $Ni(s)|Ni^{2+}(aq)||Ag^+(aq)|Ag(s)$

2. $\Delta G^\circ = -(2.00\text{ mol }e^-)(9.65 \times 10^4\text{ J/V.mol})(-0.76\text{ V})(1.0\text{ kJ}/1.0 \times 10^3\text{ J})$
$= +150$ kJ

3. $Fe \rightarrow Fe^{2+} + 2e^-$ $E^\circ = +0.44\ V$
$2e^- + Cu^{2+} \rightarrow Cu$ $E^\circ = +0.34\ V$
$Fe + Cu^{2+} \rightarrow Fe^{2+} + Cu$ $E^\circ = +0.78\ V$

4. (a) $Zn \rightarrow Zn^{2+} + 2e^-$
$2e^- + Sn^{2+} \rightarrow Sn$
$Zn + Sn^{2+} \rightarrow Zn^{2+} + Sn$
$E^\circ = +0.763\ V$
$E^\circ = -0.14\ V$
$E^\circ = +0.62\ V$ (spontaneous)
(b) $Br_2 + 2e^- \rightarrow 2\ Br^-$
$2\ Cl^- \rightarrow Cl_2 + 2e^-$
$Br_2 + 2\ Cl^- \rightarrow 2\ Br^- + Cl_2$
$E^\circ = +1.08\ V$
$E^\circ = -1.36\ V$
$E^\circ = -0.28\ V$ (not spontaneous)

5. $2\ Al(s) \rightarrow 2\ Al^{3+}(aq) + 6e^-$
$6e^- + 3\ Zn^{2+}(aq) \rightarrow 3\ Zn(s)$
$2\ Al(s) + 3\ Zn^{2+}(aq) \rightarrow 2\ Al^{3+}(aq) + 3\ Zn(s)$
$E^\circ = +1.66\ V$
$E^\circ = -0.76\ V$
$E^\circ = +0.90\ V$

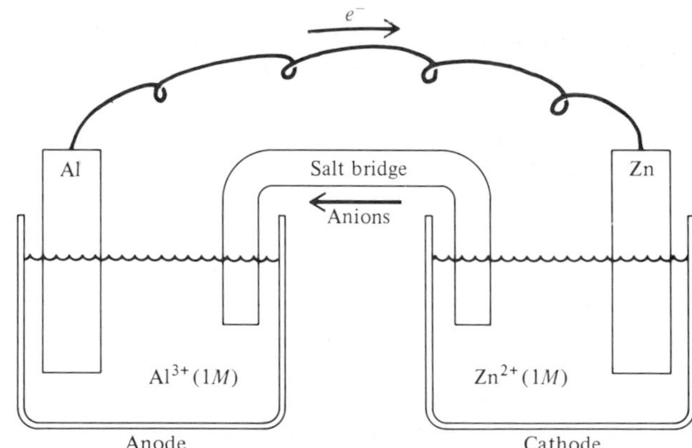

Electrons flow through external wire from Al (anode) to Zn (cathode). Anions flow through the salt bridge from the Zn^{2+}-containing compartment (cathode) to the Al^{3+}-containing compartment (anode).

6. $Cr_2O_7^{2-}(aq) + 14\ H^+(aq) + 6e^- \rightarrow 2\ Cr^{3+}(aq) + 7\ H_2O(\ell)$
$6\ Fe^{2+}(aq) \rightarrow 6\ Fe^{3+}(aq) + 6e^-$

$Cr_2O_7^{2-}(aq) + 14\ H^+(aq) + 6\ Fe^{2+}(aq) \rightarrow 2\ Cr^{3+}(aq) + 7\ H_2O(\ell) + 6\ Fe^{3+}(aq)$

$E° = +1.33\ V$
$E° = -0.77\ V$

$E° = +0.56\ V$

7. $2\ Sn^{2+}(aq) \rightarrow 2\ Sn^{4+}(aq) + 4e^-$
$O_2(g) + 4\ H^+(aq) + 4e^- \rightarrow 2\ H_2O(\ell)$

$2\ Sn^{2+}(aq) + O_2(g) + 4\ H^+(aq) \rightarrow 2\ Sn^{4+}(aq) + 2\ H_2O(\ell)$

$E° = -0.15\ V$
$E° = +1.23\ V$

$E° = +1.08\ V$

$$E = E° - \frac{0.059}{n} \log \frac{[Sn^{4+}]^2}{[H^+]^4[Sn^{2+}]^2P(O_2)}$$

$$E = 1.08 - \frac{0.059}{4} \log \frac{1.0 \times 10^{-6}}{(0.10)^4(0.10)^2(0.20)}$$

$E = 1.16\ V$

8. $1/2\ Hg_2^{2+}(aq) + e^- \rightarrow Hg(\ell)$ $E° = +0.789\ V$
$1/2\ Hg_2^{2+}(aq) \rightarrow Hg^{2+}(aq) + e^-$ $E° = -0.920\ V$

$Hg_2^{2+}(aq) \rightarrow Hg^{2+}(aq) + Hg(\ell)$ $E° = -0.131\ V$

$$\log K = \frac{nE°}{0.0592} = \frac{(1\ \text{mol}\ e^-)(-0.131\ V)}{0.0592\ V\cdot\text{mol}\ e^-} = -2.21$$

$K = 6.1 \times 10^{-3}$
$[Hg_2^{2+}] = 0.10\ M$ and $[Hg^{2+}] = 6.1 \times 10^{-4}\ M$

9. $Pb(s) + SO_4^{2-}(aq) \rightarrow PbSO_4(s) + 2e^-$
$PbO_2(s) + 4\ H^+(aq) + SO_4^{2-}(aq) + 2e^- \rightarrow PbSO_4(s) + 2\ H_2O(\ell)$

$Pb(s) + PbO_2(s) + 2\ H_2SO_4(aq) \rightarrow 2\ PbSO_4(s) + 2\ H_2O(\ell)$

$E° = +0.356\ V$
$E° = +1.685\ V$

$E° = +2.041\ V$

10. $\Delta G° = -nFE°$
$-6.07 \times 10^3\ J = -(4\ \text{mol}\ e^-)(9.65 \times 10^4\ J/V\cdot\text{mol})E°$
$E° = 1.57\ V$

11. (a) NaBr(molten) + electricity $\rightarrow$ Na(s) + $Br_2(\ell)$
(b) NaBr(aq) + electricity $\rightarrow$ $OH^-(aq)$ and $H_2(g)$ from water + $Br_2(aq)$ from $Br^-(aq)$.
(c) $SnCl_2(aq) \rightarrow Sn(s) + Cl_2(aq)$

12. $(25 \times 10^3\ \text{amps})(3600\ \text{s/hr}) = 9.0 \times 10^7\ C$
$(9.0 \times 10^7\ C)(1\ \text{mol}\ e^-/96500\ C) = 9.3 \times 10^2\ \text{mol}\ e^-/\text{hr}$.
This can produce 9.3×10^2 mol Na/hr.
$(9.3 \times 10^2\ \text{mol Na/hr})(23\ \text{g/mol}) = 2.2 \times 10^4$ g Na/hr (= 47.2 lb)

13. $(9.0 \times 10^7\ C)(7.0\ V) = 6.3 \times 10^8\ J$
$(6.3 \times 10^8\ J)(1\ \text{kwh}/3.60 \times 10^6\ J) = 180$ kwh

14. $(2.0 \times 10^3\ \text{lb})(454\ \text{g/lb})(1\ \text{mol}/27.0\ \text{g}) = 3.4 \times 10^4$ mol Al
$(3.4 \times 10^4\ \text{mol Al})(3\ \text{mol}\ e^-/1\ \text{mol Al}) = 1.0 \times 10^5$ mol e^- required
$(1.0 \times 10^5\ \text{mol}\ e^-)(96500\ C/\text{mol}\ e^-) = 9.7 \times 10^9\ C$
$(9.7 \times 10^9\ C)(5.0\ V) = 4.9 \times 10^{10}\ J$
$(4.9 \times 10^{10}\ J)(1\ \text{kwh}/3.60 \times 10^6\ J) = 1.4 \times 10^4$ kwh

CHAPTER 20

The Exercises in Chapter 20 are review questions. All answers are found directly in the text.

CHAPTER 21

1. $Al_2O_3(s) + 12\ HF(aq) + 6\ NaOH(aq) \rightarrow 2\ Na_3AlF_6(s) + 9\ H_2O(\ell)$
2. (a) 2 $B(OH)_3$ per unit cell. (b) B is sp^2 hybridized. (c) The B atom has an unoccupied *p* orbital, which is open to attack by a Lewis base such as water.
3. (a) $B_2O_3(s) + 6\ HF(g) \rightarrow 2\ BF_3(g) + 3\ H_2O(\ell)$
(b) $Al_2O_3(s) + 3\ C(s) + 3\ Cl_2(g) \rightarrow 2\ AlCl_3(s) + 3\ CO(g)$
4. $\Delta G°$ for diamond $\rightarrow$ graphite is -2.9 kJ, predicting a spontaneous reaction.
5. (a) Orthosilicates: contain tetrahedral SiO_4^{4-} ions. (b) Pyroxenes: extended chains of SiO_4 units. (c) Mica: sheets of linked SiO_4 units. (d) Zeolite: three-dimensional network of linked SiO_4 units.
6. $SiCl_4(\ell) + 2\ H_2O(\ell) \rightarrow SiO_2(s) + 4\ HCl(aq)$
Silicon has unoccupied 3*d* orbitals that are open to attack by a Lewis base such as water. The C atom in CCl_4, for example, has no such unoccupied orbitals.

CHAPTER 22

Answers are not given to exercises that ask simply to look up information in the text.

2. High temperature is required in the Haber process to raise the reaction rate.
3. (a) N_2O has a linear structural-pair geometry, whereas NO_2 has a planar trigonal structural-pair geometry.

$:\!N\!=\!N\!=\!O\!:$ $\quad$ $O\!-\!\dot{N}\!=\!O$

4. The oxidation number of P is +4.

$$\left[H{-}O{-}\underset{\displaystyle O}{\overset{\displaystyle O}{\overset{|}{\underset{|}{P}}}}{-}\underset{\displaystyle O}{\overset{\displaystyle O}{\overset{|}{\underset{|}{P}}}}{-}O{-}H \right]^{2-}$$

6. (a) MgO, basic; CO_2, acidic; Ga_2O_3, amphoteric; CO, neutral.
(b) $Cl_2O_7(s) + H_2O(\ell) \rightarrow 2\ HClO_4(aq)$

7. (a) $Na_2SO_4(s) + 4\ C(s) \rightarrow Na_2S(s) + 4\ CO(g)$
(b) $2\ Na_2S(aq) + 2\ O_2(g) + H_2O(\ell) \rightarrow Na_2S_2O_3(aq) + 2\ NaOH(aq)$

8. (b) The reaction is predicted to be spontaneous, since ΔG° is a negative quantity.

9. The structural-pair geometry of XeF_2 is trigonal bipyramidal. Since lone pairs occupy a larger volume of space than bond pairs, the three lone pairs occupy the trigonal plane, and the F atoms must be placed at the peaks of the pyramids (see page 329).

$$\underset{\displaystyle :\ddot{F}:}{\overset{\displaystyle :\ddot{F}:}{\overset{|}{\underset{|}{:\!\!\blacktriangleleft Xe{-}:}}}}$$

CHAPTER 23

1. (a,1) $CH_3CH_2CH_2CH_2CH_2CH_3$;

(a,2) $CH_3\underset{\displaystyle CH_3}{\underset{|}{C}}HCH_2CH_2CH_3$;

(a,3) $CH_3CH_2\underset{\displaystyle CH_3}{\underset{|}{C}}HCH_2CH_3$;

(a,4) $CH_3\overset{\displaystyle CH_3}{\overset{|}{\underset{\displaystyle CH_3}{\underset{|}{C}}}}CH_2CH_3$;

(a,5) $CH_3\underset{\displaystyle CH_3}{\underset{|}{C}}H\overset{\displaystyle CH_3}{\overset{|}{C}}HCH_3$

(b) $C_{22}H_{46}$

2. (a) heptane; (b) (a,1) hexane; (a,2) 2-methylpentane; (a,3) 3-methylhexane; (a,4) 2,2-dimethylbutane; (a,5) 2,3-dimethylbutane

3. (a) alkene; (b) ether; (c) alcohol; (d) ester; (e) carboxylic acid; (f) amine; (g) aldehyde

4. (a) 1-propanol and 2-propanol

(b) $CH_3CH_2\underset{\displaystyle OH}{\underset{|}{C}}HCH_2CH_3$ and $CH_3\underset{\displaystyle OH}{\underset{|}{C}}H\overset{\displaystyle CH_3}{\overset{|}{C}}HCH_3$

(c) Both are secondary alcohols, since —OH is bound to a carbon atom that is in turn bound to two other C atoms.

5. The products of these reactions are: (a) $CH_3CH_2\overset{\displaystyle O}{\overset{\|}{C}}H$; (b) $CH_3CH{=}CH_2$; (c) $CH_3CH_2CH_2I$.

6.

$$\begin{matrix} H & & H \\ & C{=}C & \\ H_3C & & CH_2CH_3 \end{matrix} \qquad \begin{matrix} H & & CH_2CH_3 \\ & C{=}C & \\ H_3C & & H \end{matrix}$$

cis-2-pentene   *trans*-2-pentene

7. (a) $CH_3CH_2\overset{\displaystyle Br}{\overset{|}{C}}HCH_3$ (b) $CH_3\overset{\displaystyle Br}{\overset{|}{\underset{\displaystyle CH_3}{\underset{|}{C}}}}CH_2CH_3$

8. $CH_3\overset{\displaystyle OH}{\overset{|}{\underset{\displaystyle CH_3}{\underset{|}{C}}}}CH_2CH_3$

9. (a,1) $CH_3C{\equiv}CCH_2CH_3$; (a,2) $CH_3\underset{\displaystyle CH_3}{\underset{|}{C}}HCH_2C{\equiv}CH$;
(b) 3-chloropropyne

10. (a) $CH_3\underset{\displaystyle CH_3}{\underset{|}{C}}HC{\equiv}CH + 2\ HBr \longrightarrow CH_3\underset{\displaystyle CH_3}{\underset{|}{C}}H{-}\overset{\displaystyle Br}{\overset{|}{\underset{\displaystyle Br}{\underset{|}{C}}}}{-}CH_3$

(b) $CH_3CH_2C{\equiv}CH + 2\ Cl_2 \longrightarrow CH_3CH_2\overset{\displaystyle Cl}{\overset{|}{\underset{\displaystyle Cl}{\underset{|}{C}}}}{-}\overset{\displaystyle Cl}{\overset{|}{\underset{\displaystyle Cl}{\underset{|}{C}}}}H$

11. (a) p-bromonitrobenzene; (b) o-chlorophenol

12. (a) $CH_3CHCH{=}CH_2$ (with CH_3 on the second carbon); (b) $CH_3CH{=}CHCH_3$ (*cis* and *trans*) and $CH_3CH_2CH{=}CH_2$;

(c) CH_3CHCH_2OH (with CH_3 on the second carbon);

(d) CH_3CHCH_2MgI, CH_3CHCH_2OH (each with CH_3 on the second carbon)

13. (a) hexanoic acid; (b) p-nitrobenzoic acid; (c) propyl ethanoate (systematic name) or propyl acetate (common name)

14. (a) $CH_3CHCH_2CH_2OH$ (with CH_3 on the second carbon); (b) C_6H_5–$\overset{O}{\overset{\|}{C}}OCH_3$;

(c) $CH_3CH\overset{O}{\overset{\|}{C}}O^-K^+ + H_2O$ (with CH_3 on the second carbon)

15. (a) $CH_3\overset{O}{\overset{\|}{C}}CH_2CH_2CH_2$; (b) C_6H_4Br–$\overset{O}{\overset{\|}{C}}H$ (Br in the meta position);

(c) $CH_3CH_2CH_2\overset{O}{\overset{\|}{C}}H$

16. (a) 3-hexanone; (b) pentanol

17. (a) $CH_3CH_2CH_2OH$

(b) CH_3CHCH_2OH (with CH_3 on the second carbon) (where CH_3CHCH_2OMgBr (with CH_3 on the second carbon) is an intermediate)

(c) $CH_3CH\overset{O}{\overset{\|}{C}}OH$ (with CH_3 on the second carbon)

18. CH_3CHCH_3 (with Br on the second carbon) $\xrightarrow{Mg}$ CH_3CHCH_3 (with MgBr on the second carbon) $\xrightarrow{H\overset{O}{\overset{\|}{C}}H}$ CH_3CHCH_2OMgBr (with CH_3 on the second carbon) $\xrightarrow{H^+/H_2O}$ CH_3CHCH_2OH (with CH_3 on the second carbon) $\xrightarrow{PCC}$ $CH_3CH\overset{O}{\overset{\|}{C}}H$ (with CH_3 on the second carbon)

19. (a) diethylamine; (b) $CH_3\overset{O}{\overset{\|}{C}}N(CH_3)_2$

20. $CH_3CH_2CH_2CH_2CH_2CH_2CH_3 \rightarrow C_6H_5{-}CH_3 + 4\ H_2$

Four moles of H_2 are produced per mole of heptane converted.

CHAPTER 24

1. Enantiomers of alanine.

H above C, with H_2N, CH_3, COOH; and H above C, with HOOC, CH_3, NH_2

2. L form of serine

H above C, with H_2N, COOH, CH_2OH

3. (a) $H_2N{-}\underset{CH_2OH}{\overset{H}{C}}{-}\overset{O}{\overset{\|}{C}}{-}\overset{H}{N}{-}\underset{H}{\overset{H}{C}}{-}\overset{O}{\overset{\|}{C}}{-}O{-}H$

serine-glycine

(b) $H_2N{-}\underset{H}{\overset{H}{C}}{-}\overset{O}{\overset{\|}{C}}{-}\overset{H}{N}{-}\underset{CH_2OH}{\overset{H}{C}}{-}\overset{O}{\overset{\|}{C}}{-}O{-}H$

glycine-serine

(c) $H_2N{-}\underset{H}{\overset{H}{C}}{-}\overset{O}{\overset{\|}{C}}{-}\underset{H}{N}{-}\underset{H}{\overset{H}{C}}{-}\overset{O}{\overset{\|}{C}}{-}\underset{H}{N}{-}\underset{H}{\overset{H}{C}}{-}\overset{O}{\overset{\|}{C}}{-}O{-}H$

4. (a) glucose; (b) maltose; (c) α-1,4 glycoside link

CHAPTER 25

1. (a) Ti^{3+}: $[Ar]3d^1$, paramagnetic
   (b) Fe^{2+}: $[Ar]3d^6$, paramagnetic
   (c) Cu^+: $[Ar]3d^{10}$, diamagnetic
   (d) Gd^{3+}: $[Xe]4f^7$, paramagnetic
2. (a) The platinum in $Pt(NH_3)_2Cl_2$ is Pt^{2+}. The complex has no charge, since Pt^{2+} is combined with two negative ligands (Cl^-) and two neutral ligands (NH_3).
   (b) $[Co(NH_3)_5CO_3]^+$
3. (a) aquocyanobis(phenanthroline)ruthenium(II) chloride
   (b) diamminedichloroplatinum(II)
4. (a) *cis-trans* isomers (see the figure of $[Co(en)_2Cl_2]^+$ on page 968).
   (b) optical isomers (see Figure 25.17).
5. (a) $[Ru(H_2O)_6]^{2+}$. The Ru^{2+} ion has 6 $4d$ electrons. Its high spin configuration is paramagnetic.

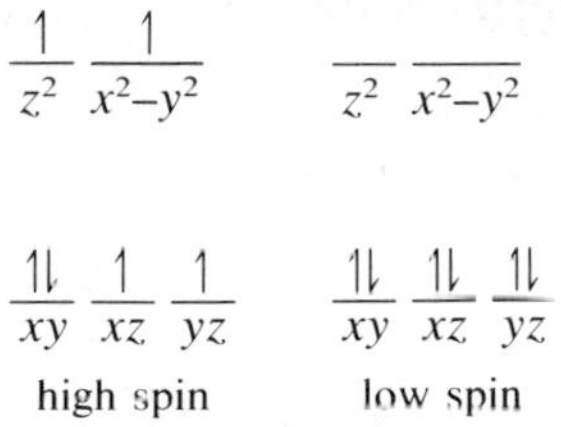

high spin	low spin
$\underline{\uparrow}$ $\underline{\uparrow}$ z^2 $x^2{-}y^2$	$\underline{\ \ }$ $\underline{\ \ }$ z^2 $x^2{-}y^2$
$\underline{\uparrow\downarrow}$ $\underline{\uparrow}$ $\underline{\uparrow}$ xy xz yz	$\underline{\uparrow\downarrow}$ $\underline{\uparrow\downarrow}$ $\underline{\uparrow\downarrow}$ xy xz yz

   (b) $[Ni(NH_3)_6]^{2+}$. The Ni^{2+} ion has 8 $3d$ electrons. Here the low and high spin configurations are the same. The complex is paramagnetic.

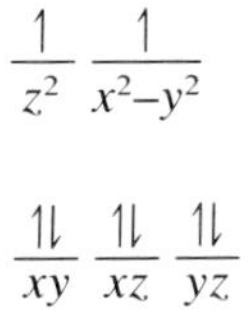

$\underline{\uparrow}$ $\underline{\uparrow}$
z^2 $x^2{-}y^2$

$\underline{\uparrow\downarrow}$ $\underline{\uparrow\downarrow}$ $\underline{\uparrow\downarrow}$
xy xz yz

6. Since both complexes are based on the same metal ion, Ni^{2+}, we can predict that the water-containing complex requires less energy to promote an electron. Water is lower in the spectrochemical series than ammonia.

CHAPTER 26

1. ${}^{227}_{93}Np \rightarrow {}^{4}_{2}He + {}^{223}_{91}Pa$
2. ${}^{221}_{88}Ra \rightarrow {}^{4}_{2}He + {}^{217}_{86}Rn$
   ${}^{257}_{103}Lr \rightarrow {}^{4}_{2}He + {}^{253}_{101}Md$
   ${}^{220}_{86}Np \rightarrow {}^{4}_{2}He + {}^{216}_{84}Po$
3. ${}^{35}_{16}S \rightarrow {}^{0}_{-1}e + {}^{35}_{17}Cl$
4. ${}^{3}_{1}H \rightarrow {}^{0}_{-1}e + {}^{3}_{2}H$
   ${}^{60}_{26}Fe \rightarrow {}^{0}_{-1}e + {}^{60}_{27}Co$
   ${}^{129}_{53}I \rightarrow {}^{0}_{-1}e + {}^{129}_{54}Xe$
5. ${}^{235}_{92}U \rightarrow {}^{4}_{2}He + {}^{231}_{90}Th$
   ${}^{231}_{90}Th \rightarrow {}^{0}_{-1}e + {}^{231}_{91}Pa$
   ${}^{231}_{91}Pa \rightarrow {}^{4}_{2}He + {}^{227}_{89}Ac$
   ${}^{227}_{89}Ac \rightarrow {}^{4}_{2}He + {}^{223}_{87}Fr$
   ${}^{223}_{87}Fr \rightarrow {}^{0}_{-1}e + {}^{223}_{88}Ra$
6. positron emission: ${}^{13}_{7}N \rightarrow {}^{13}_{6}C + {}^{0}_{+1}e$
   K-capture: ${}^{41}_{20}Ca + {}^{0}_{-1}e \rightarrow {}^{41}_{19}K$
   β-emission: ${}^{90}_{38}Sr \rightarrow {}^{0}_{-1}e + {}^{90}_{39}Y$
   positron emission: ${}^{11}_{6}C \rightarrow {}^{11}_{5}B + {}^{0}_{+1}e$
   proton emission: ${}^{43}_{21}Sc \rightarrow {}^{1}_{1}H + {}^{42}_{20}Ca$
7. Mass defect $= \Delta m = -0.03260$ g/mol

   $\Delta E = (-32.60 \times 10^{-6}\ \text{kg/mol})(3.00 \times 10^{8}\ \text{m/s})^2$
   $= -2.93 \times 10^{12}$ J/mol
8. (a) ${}^{90}_{38}Sr \rightarrow {}^{0}_{-1}e + {}^{90}_{39}Y$
   (b) Disintegrations are reduced to 1000 after 1 half-life, to 500 after 2 half-lives, to 250 after 3 half-lives, and to 125 after 4 half-lives. Thus, a total of $4 \times 28 = 112$ years is required.
9. $k = 0.693/t_{1/2} = 8.90 \times 10^{-3}\ \text{hr}^{-1}$
   $\ln(0.10/1.0) = -kt = -(8.90 \times 10^{-3}\ \text{hr}^{-1})t$
   $t = 259$ hr
10. $\ln(12.9/14.0) = -(1.21 \times 10^{-4}\ \text{yr}^{-1})t$
    $t = 676$ years
    $1986 - 676 = 1310$ AD
11. (a) ${}^{13}_{6}C + {}^{1}_{0}n \rightarrow {}^{4}_{2}He + {}^{10}_{4}Be$
    (b) ${}^{14}_{7}N + {}^{4}_{2}He \rightarrow {}^{1}_{0}n + {}^{17}_{9}F$
    (c) ${}^{253}_{99}Es + {}^{4}_{2}He \rightarrow {}^{1}_{0}n + {}^{256}_{101}Md$

APPENDIX N

Answers to Designated Study Questions

CHAPTER 1

1. (a) physical; (b) chemical; (c) chemical; (d) physical; and (e) physical
2. (a) solid; (b) gas; (c) solid; (d) liquid
3. Heterogeneous. The iron chips can be removed with a magnet; they are attracted to the magnet but the sand is not.
5.

	Qualitative	Quantitative
(a)	purple solid	1.25 g
(b)	silvery magnesium floats on oil	0.025 g
(c)	blue copper sulfate colorless ammonia	25 mL volumes

7. 1.9×10^2 mm, 0.19 m, 7.5 in
10. 22.5 miles per hour
13. 310 miles
14. 89 km/hr; 81 ft/s
15. 5.3 cm^2; 5.3×10^{-4} m^2; 0.82 in^2
18. 1.50×10^3 L; 1.50 m^3
20. 2.7×10^4 g; 27 kg
22. 0.00563 kg; 5630 mg
25. 800. cm^3; 0.800 L; 8.00×10^{-4} m^3; 0.800 dm^3
26. 1.97 L
28. (a) 5.63 Å; (b) 0.178 nm^3; 1.78×10^{-22} cm^3
30. 2 ounces = 57 g; protein = 14%; carbohydrate = 74%; fat = 1.8%

31. 3.51×10^3 kg/m^3
34. (a) 2.79 cm^3; (b) $6.71
36. (a) 272 cm^3; (b) No, the volume of the ice exceeds the volume of the container.
39. 0.57 cm^3

40. 77°F; 298 K
41.

	°F	°C	K
(a)	57	14	287
(b)	99	37	310
(c)	−40	−40	233
(d)	−321	−196	77
(e)	140	60.	333
(f)	1000	537.8	811.0

44. Your normal body temperature is 98.6°F or 37°C; therefore, the gallium should melt in your hand.

47. 9.6 g; only one place past the decimal is allowed.
48. 6.6×10^4 cm^3; one of the dimensions has only two digits, so the answer is limited to two as well.
49. 2.1×10^3 cm^3; two significant figures
51. 0.0743; three significant figures

52. 3.473×10^7 tons; 3.150×10^{10} kg; 3.150×10^{13} g
54. 6.2 cm^3; 0.74 g/cm^3
57. 0.018 mm thick
58. 3.3×10^7 m^2; 13 miles^2

CHAPTER 2

5. Element 25 is manganese, Mn.
6. (a) A = 9 for Be; (b) A = 48 for Ti; (c) A = 70 for Ga.
8. (a) $^{23}_{11}$Na; (b) $^{39}_{18}$Ar; (c) $^{208}_{82}$Pb
10.

(a)	$^{45}_{21}$Sc	$^{33}_{16}$S	$^{17}_{8}$O	$^{56}_{25}$Mn
(b)	21	16	8	25
(c)	24	17	9	31
(d)	21	16	8	25

12. a, c, and e
13. 69.75
15. 28.1
18. Rn, radon, 222 amu.
21. ^{121}Sb = 57.5% and ^{123}Sb = 42.5%
23. (a) 1.9998 mol Cu; (b) 0.499 mol Ca; (c) 0.6208 mol Al; (d) 3.1×10^{-4} mol K; (e) 2.1×10^{-5} mol Am
25. 45.8 g S = 1.43 mol; 64.9 g C = 5.40 mol
28. (a) 5.6 g Fe; (b) 64.9 g Si; (c) 0.028 g C; (d) 12 g Na; (e) 1.19×10^3 g Au
32. (a) 0.159 mol Pt; (b) $5000 buys 478.5 g or 2.453 mol of Pt.
34. 0.0698 mol of Pt
35. (a) 1.43×10^5 mol; (b) 100. cm or 1 meter on side
37. 9.08×10^7 g; 1.43×10^6 mol copper
38. (a) 6.0×10^{22} atoms Fe
    (b) 1.39×10^{24} atoms Si
    (c) 1.4×10^{21} atoms C
    (d) 3.3×10^{23} atoms Na
    (e) 3.63×10^{24} atoms Au

40. Cu atom = 1.06×10^{-22} g. (b) W atom = 3.05×10^{-22} g

43. 3.8 g of Cu (= 0.060 mol or 3.6×10^{22} atoms) and 1.3 g Ni (= 0.022 mol or 1.3×10^{22} atoms).

46. The volume of the cube of 8 silicon atoms is 1.602×10^{-22} cm^3. Using the density, the mass of this cube is found to be 3.731×10^{-22} g. Since there are 8 atoms in the cube, the mass of one atom is 4.664×10^{-23} g. Dividing the atomic mass by the mass of one atom, we find Avogadro's number.

50. (a) TiO_2 (c) AlC_3H_9
(b) B_4H_{10} (d) $C_6H_8O_6$

51. (a) KI (e) MgS
(b) CaO (f) Na_2CO_3
(c) NaBr (g) NH_4NO_3
(d) Li_2O (h) CsOH

53. (a) $NaC_2H_3O_2$ (e) $TiBr_4$
(b) $AgClO_4$ (f) UF_4
(c) $KClO_3$ (g) $Ca(OCl)_2$
(d) $CaHPO_4$ (h) $NaNO_2$

54. (a) sodium hydrogen carbonate (Na^+, HCO_3^-) (or sodium bicarbonate)
(b) calcium phosphate (3 Ca^{2+}, 2 PO_4^{3-})
(c) ammonium bromide (NH_4^+, Br^-)
(d) potassium perchlorate (K^+, ClO_4^-)
(e) sodium cyanide (Na^+, CN^-)
(f) copper(II) sulfate (Cu^{2+}, SO_4^{2-})
(g) potassium permanganate (K^+, MnO_4^-)
(h) zinc oxide (Zn^{2+}, O^{2-})

56. Na_2CO_3, sodium carbonate; NaI, sodium iodide; $NaNO_3$, sodium nitrate; $SrCO_3$, strontium carbonate; SrI_2, strontium iodide; $Sr(NO_3)_2$, strontium nitrate; $(NH_4)_2CO_3$, ammonium carbonate; NH_4I, ammonium iodide; NH_4NO_3, ammonium nitrate

58. (a) 159.7 (d) 197.9
(b) 67.81 (e) 176.1
(c) 44.02

59. (a) 122.2 (d) 90.18
(b) 227.1 (e) 324.4
(c) 300.1

60. (a) 0.0312 mol (d) 0.00406 mol
(b) 0.0101 mol (e) 0.00599 mol
(c) 0.0125 mol

62. (a) 3.2×10^{-4} mol; (b) 3.8×10^{20} F^- ions and 1.9×10^{20} Sn^{2+} ions.

63. 1.4×10^{21} molecules

65. 1 molecule of N_2 < 1 penny < 1 mole B_2H_6 < 6×10^{23} molecules CO < 3×10^{23} molecules C_4H_{10}

66. (a) sodium hydrogen carbonate (or sodium bicarbonate)
(b) 10.8 mol $NaHCO_3$
(c) 32.4 mol O; 519 g oxygen

67. (a) $C_5H_8NNaO_4$; mol wt = 169.1 g/mol
(b) 0.0118 mol; 7.12×10^{21} molecules

68. (a) 537.5 g/mol
(b) 4.7×10^{-4} mol
(c) 199 g of beryl required

69. $20.46 per mole of ferrocene

70. (a) 0.00180 mol aspirin; 0.02267 mol $NaHCO_3$; 0.00521 mol citric acid; (b) 1.08×10^{21} molecules aspirin

72. 5.51×10^{10} mol benzene

73. 7.71×10^{23} molecules; 227 g chlorine

74. 0.00166 mol of "active ingredient"; 346 mg Bi.

76. (a) $C_6H_8O_6$
(b) 3.4×10^{-4} mol vitamin C
(c) 5.68×10^{-3} mol vitamin C
(d) 2.05×10^{22} atoms of O

79. (a) C_2H_6; 30.1 g/mol; 79.9% C and 20.1% H
(b) $C_2H_4O_2$; 60.1 g/mol; 40.0% C, 6.71% H, 53.3% O
(c) $C_2H_3NO_5$; 121 g/mol; 19.8% C, 2.50% H, 11.6% N, and 66.1% O
(d) $C_4H_{10}NO_3PS$; 183 g/mol; 26.2% C, 5.50% H, 7.65% N, 26.2% O, 16.9% P, 17.5% S

80. (a) 53.1 g/mol; (b) 67.9% C, 5.69% H, and 26.4% N

82. (Molar mass = 406.9 g/mol) 38.37% C, 1.49% H, 52.28% Cl, and 7.86% O

83. (Molar mass = 345.7 g/mol) (a) 55.59% C, 4.37% H, 30.76% Cl, and 9.26% O; (b) 510 g for 150-pound person

84. $C_4H_4O_4$

86. B_5H_7

88. Empirical formula = CH_2O; molecular formula = $C_2H_4O_2$

89. SF_4Cl_2

90. Nicotine: Empirical formula = C_5H_7N; molecular formula = $C_{10}H_{14}N_2$

91. Cacodyl: Empirical formula = C_2H_6As; molecular formula = $C_4H_{12}As_2$

92. Cadaverine = $C_5H_{14}N_2$

94. Fluorocarbonyl hypofluorite: empirical and molecular formulas = CO_2F_2

96. MMT = $C_9H_7MnO_3$

98. Empirical formula = $Mn(CO)_5$; molecular formula = $Mn_2(CO)_{10}$

100. Anthracene: empirical formula = C_7H_5; molecular formula = $C_{14}H_{10}$

102. B_5H_9

104. Empirical formula = $Fe(CO)_4$; molecular formula = $Fe_3(CO)_{12}$

105. Vinyl chloride: 38.4% C and 56.8% Cl; empirical and molecular formulas are C_2H_3Cl.

106. $x = 3$

110. 24.2% Ba

112. 20.0% Ni

114. (a) 8.13% water; (b) 55.1% $BaCl_2 \cdot 2H_2O$; (c) 60.6% Ba
116. 6.94 g M/mol; M = Li
118. Cu_2S
121. 0.74
122. 0.45 g Cu_2O

CHAPTER 3

6. (a) $4\ Al(s) + 3\ O_2(g) \longrightarrow 2\ Al_2O_3(s)$
(b) $N_2(g) + 3\ H_2(g) \longrightarrow 2\ NH_3(g)$
(c) $2\ C_6H_6(\ell) + 15\ O_2(g) \longrightarrow 6\ H_2O(g) + 12\ CO_2(g)$
8. (a) $UO_2(s) + 4\ HF(\ell) \longrightarrow UF_4(s) + 2\ H_2O(\ell)$
(b) $B_2O_3(s) + 6\ HF(\ell) \longrightarrow 2\ BF_3(g) + 3\ H_2O(\ell)$
(c) $BF_3(g) + 3\ H_2O(\ell) \longrightarrow 3\ HF(\ell) + H_3BO_3(s)$
10. (a) $Na_2O_2(s) + 2\ H_2O(\ell) \longrightarrow 2\ NaOH(aq) + H_2O_2(aq)$
(b) $4\ PH_3(g) + 8\ O_2(g) \longrightarrow P_4O_{10}(s) + 6\ H_2O(g)$
(c) $2\ C_2H_3Cl(\ell) + 5\ O_2(g) \longrightarrow 4\ CO_2(g) + 2\ H_2O(g) + 2\ HCl(g)$
12. (a) $H_2NCl(aq) + 2\ NH_3(g) \longrightarrow NH_4Cl(aq) + N_2H_4(aq)$
(b) $(CH_3)_2N_2H_2(\ell) + 2\ N_2O_4(\ell) \longrightarrow 3\ N_2(g) + 4\ H_2O(g) + 2\ CO_2(g)$
(c) $CaC_2(s) + 2\ H_2O(\ell) \longrightarrow Ca(OH)_2(s) + C_2H_2(g)$
14. (a) $CuCl_2$; (b) $AgNO_3$; (c) all; (d) NaCl, $BaCl_2$
15. (a) $NaC_2H_3O_2$; (b) CuS; (c) NaOH; (d) AgCl
16. (a) Na^+, I^-; (b) 2 K^+, SO_4^{2-}; (c) K^+, HSO_4^-; (d) Na^+, CN^-
18. (a) soluble, Ba^{2+} and 2 Cl^-;
(b) soluble, Cr^{2+} and 2 NO_3^-;
(c) soluble, Pb^{2+} and 2 NO_3^-;
(d) insoluble
20. Soluble compound: use Cl^-, Br^-, NO_3^-, and $C_2H_3O_2^-$ (acetate).
Insoluble compound: use S^{2-} or CO_3^{2-}, for example.
25. The complete, balanced equation is given first followed by the net ionic equation.
(a) $Zn + 2\ HCl \longrightarrow H_2 + ZnCl_2$
$Zn(s) + 2\ H^+(aq) \longrightarrow H_2(g) + Zn^{2+}(aq)$
(b) $Mg(OH)_2 + 2\ HCl \longrightarrow MgCl_2 + 2\ H_2O$
$Mg(OH)_2(s) + 2\ H^+(aq) \longrightarrow Mg^{2+}(aq) + 2\ H_2O$
(c) $2\ HNO_3 + Ca(OH)_2 \longrightarrow Ca(NO_3)_2 + 2\ H_2O$
$2\ H^+(aq) + Ca(OH)_2(s) \longrightarrow Ca^{2+}(aq) + 2\ H_2O$
(d) $4\ HCl + MnO_2 \longrightarrow MnCl_2 + Cl_2 + 2\ H_2O$
$4\ H^+(aq) + 2\ Cl^-(aq) + MnO_2(s) \longrightarrow Mn^{2+}(aq) + Cl_2(g) + 2\ H_2O(\ell)$
28. (a) $2\ Mg(s) + O_2(g) \longrightarrow 2\ MgO(s)$
(b) $2\ Ca(s) + O_2(g) \longrightarrow 2\ CaO(s)$
(c) $4\ In(s) + 3\ O_2(g) \longrightarrow 2\ In_2O_3(s)$
30. (a) $CH_4(g) + 2\ O_2(g) \longrightarrow CO_2(g) + 2\ H_2O(g)$
(b) $2\ C_8H_{18}(\ell) + 25\ O_2(g) \longrightarrow 16\ CO_2(g) + 18\ H_2O(g)$
(c) $C_2H_5OH(\ell) + 3\ O_2(g) \longrightarrow 2\ CO_2(g) + 3\ H_2O(g)$
32. (a) $2\ C(s) + O_2(g) \longrightarrow 2\ CO(g)$
(b) $2\ Ni(s) + O_2(g) \longrightarrow 2\ NiO(s)$
(c) $4\ Cr(s) + 3\ O_2(g) \longrightarrow 2\ Cr_2O_3(s)$
34. (a) Acid-base reaction:
$HCl(aq) + KOH(aq) \longrightarrow H_2O(\ell) + KCl(aq)$
(b) Precipitation reaction:
$AgNO_3(aq) + KCl(aq) \longrightarrow AgCl(s) + KNO_3(aq)$
(c) Acid-base reaction:
$H_2SO_4(aq) + 2\ NaOH(aq) \longrightarrow Na_2SO_4(aq) + 2\ H_2O(\ell)$
36. (a) Acid-base reaction:
$Al(OH)_3(s) + 3\ HNO_3(aq) \longrightarrow Al(NO_3)_3(aq) + 3\ H_2O(\ell)$
(b) Precipitation reaction:
$BaCl_2(aq) + H_2SO_4(aq) \longrightarrow BaSO_4(s) + 2\ HCl(aq)$
(c) Precipitation reaction:
$MnCl_2(aq) + (NH_4)_2S(aq) \longrightarrow MnS(s) + 2\ NH_4Cl(aq)$
38. (a) $NaOH(aq) + HNO_3(aq) \longrightarrow NaNO_3(aq) + H_2O(\ell)$
(b) $KOH(aq) + HCl(aq) \longrightarrow KCl(aq) + H_2O(\ell)$
(c) $3\ KOH(aq) + H_3PO_4(aq) \longrightarrow K_3PO_4(aq) + 3\ H_2O(\ell)$
(d) $2\ CsOH(aq) + H_2SO_4(aq) \longrightarrow Cs_2SO_4(aq) + 2\ H_2O(\ell)$
40. In each case we begin with soluble salts containing the required cation and anion.
(a) $AgNO_3(aq) + KBr(aq) \longrightarrow AgBr(s) + KNO_3(aq)$
(b) $Na_2CO_3(aq) + CaCl_2(aq) \longrightarrow CaCO_3(s) + 2\ NaCl(aq)$
(c) $NiCl_2(aq) + Na_2S(aq) \longrightarrow NiS(s) + 2\ NaCl(aq)$
(d) $BaCl_2(aq) + Na_2SO_4(aq) \longrightarrow BaSO_4(s) + 2\ NaCl(aq)$
(e) $Zn(NO_3)_2(aq) + (NH_4)_2S(aq) \longrightarrow ZnS(s) + 2\ NH_4NO_3(aq)$
43. (a) Salts are most easily obtained by reacting an acid with a base. Here we use hydrochloric acid with sodium hydroxide. Evaporating the solution will give the water-soluble NaCl.
$HCl(aq) + NaOH(aq) \longrightarrow NaCl(aq) + H_2O(\ell)$
Another approach is the following reaction. Sodium chloride can be separated by filtering away the barium sulfate and evaporating the solution to obtain the soluble NaCl.
$BaCl_2(aq) + Na_2SO_4(aq) \longrightarrow 2\ NaCl(aq) + BaSO_4(s$
(b) Lead chloride is insoluble, so one can use an exchange reaction starting with soluble salts.
$Pb(NO_3)_2(aq) + 2\ KCl(aq) \longrightarrow PbCl_2(s) + 2\ KNO_3(a$

Again, reaction of a metal carbonate with acid is also an effective way to prepare salts.

$PbCO_3(s) + 2\ HCl(aq) \longrightarrow PbCl_2(s) + CO_2(g) + H_2O(\ell)$

(c) Barium sulfate is an insoluble salt. Begin with a soluble barium salt and a soluble sulfate salt (or sulfuric acid).

$BaCl_2(aq) + Na_2SO_4(aq) \longrightarrow 2\ NaCl(aq) + BaSO_4(aq)$

(d) Silver iodide, an insoluble salt, can only be prepared by a precipitation reaction starting with soluble silver and iodide salts.

$AgNO_3(aq) + KI(aq) \longrightarrow AgI(s) + KNO_3(aq)$

44. (a) $CaCO_3(s) + 2\ HNO_3(aq) \longrightarrow Ca(NO_3)_2(aq) + H_2O(\ell) + CO_2(g)$
(b) $SrCO_3(s) + heat \longrightarrow SrO(s) + CO_2(g)$
(c) $K_2CO_3(aq) + H_2SO_4(aq) \longrightarrow K_2SO_4(aq) + H_2O(\ell) + CO_2(g)$
(d) $CuCO_3(s) + 2\ HClO_4(aq) \longrightarrow Cu(ClO_4)_2(aq) + H_2O(\ell) + CO_2(g)$

46. (a) $4\ Li(s) + O_2(g) \longrightarrow 2\ Li_2O(s)$
(b) $Li_2CO_3(s) + heat \longrightarrow Li_2O(s) + CO_2(g)$

47. (a) $Li_2CO_3(s) + 2\ HClO_4(aq) \longrightarrow 2\ LiClO_4(aq) + H_2O(\ell) + CO_2(g)$
(b) $NiCO_3(s) + 2\ HCl(aq) \longrightarrow NiCl_2(aq) + H_2O(\ell) + CO_2(g)$
(c) $CuCO_3(s) + heat \longrightarrow CuO(s) + CO_2(g)$

48. (a) Br = +1, O = −2
(b) C = +3, O = −2
(c) I = 0
(d) I = +5, O = −2
(e) H = +1, Cl = +7, O = −2

50. (a) Xe = +6, O = −2
(b) N = −2, H = +1
(c) C = −8/3, H = +1
(d) C = −4/7, H = +1, O = −2

51. (a) Mg is oxidized; O_2 is reduced
(b) C in C_2H_4 is oxidized (from −2 to +4); O_2 is reduced
(c) Si is oxidized; Cl_2 is reduced

53. Reaction	**Reactant function**	**Overall reaction**
(a) $Cr \longrightarrow Cr^{3+} + 3e^-$	reducing agent	oxidation
(b) $Fe^{3+} + e^- \longrightarrow Fe^{2+}$	oxidizing agent	reduction
(c) $AsH_3 \longrightarrow As + 3\ H^+ + 3e^-$	reducing agent	oxidation
(d) $6\ H^+ + 3e^- + VO_3^- \longrightarrow V^{2+} + 3\ H_2O$	oxidizing agent	reduction

56. Reaction	**Reactant function**	**Overall reaction**
(a) $HOI + H^+ + 2e^- \longrightarrow I^- + H_2O$	oxidizing agent	reduction
(b) $NO + H_2O \longrightarrow HNO_2 + H^+ + e^-$	reducing agent	oxidation
(c) $C_6H_5CH_3 + 2\ H_2O \longrightarrow C_6H_5COOH + 6\ H^+ + 6e^-$	reducing agent	oxidation
(d) $SO_4^{2-} + 4\ H^+ + 2e^- \longrightarrow H_2SO_3 + H_2O$	oxidizing agent	reduction

58. Reaction	**Reactant function**	**Overall reaction**
(a) $CrO_2^- + 4\ OH^- \longrightarrow CrO_4^{2-} + 2\ H_2O + 3e^-$	reducing agent	oxidation
(b) $Br_2 + 12\ OH^- \longrightarrow 2\ BrO_3^- + 6\ H_2O + 10e^-$	reducing agent	oxidation
(c) $Ni(OH)_2 + 2\ OH^- \longrightarrow NiO_2 + 2\ H_2O + 2e^-$	reducing agent	oxidation

59. (a) $Cl_2 + 2e^- \longrightarrow 2\ Cl^-$
$2\ Br^- \longrightarrow Br_2 + 2e^-$
$Cl_2 + 2\ Br^- \longrightarrow 2\ Cl^- + Br_2$

(b) $Sn \longrightarrow Sn^{2+} + 2e^-$
$2\ H^+ + 2e^- \longrightarrow H_2$
$Sn + 2\ H^+ \longrightarrow Sn^{2+} + H_2$

(c) $2[Al \longrightarrow Al^{3+} + 3e^-]$
$3[Sn^{4+} + 2e^- \longrightarrow Sn^{2+}]$
$2\ Al + 3\ Sn^{4+} \longrightarrow 2\ Al^{3+} + 3\ Sn^{2+}$

(d) $Zn \longrightarrow Zn^{2+} + 2e^-$
$2[VO^{2+} + 2\ H^+ + e^- \longrightarrow V^{3+} + H_2O]$
$Zn + 2\ VO^{2+} + 4\ H^+ \longrightarrow Zn^{2+} + 2\ V^{3+} + 2\ H_2O$

61. (a) $2[Ag^+ + e^- \longrightarrow Ag]$
$HCHO + H_2O \longrightarrow HCOOH + 2\,H^+ + 2e^-$
$2\,Ag^+ + HCHO + H_2O \longrightarrow 2\,Ag + HCOOH + 2\,H^+$

(b) $2[MnO_4^- + 8\,H^+ + 5e^- \longrightarrow Mn^{2+} + 4\,H_2O]$
$5[C_2H_5OH \longrightarrow C_2H_4O + 2\,H^+ + 2e^-]$
$5C_2H_5OH + 2\,MnO_4^- + 6\,H^+ \longrightarrow 5\,C_2H_4O + 2\,Mn^+ + 8\,H_2O$

(c) $Cr_2O_7^{2-} + 14\,H^+ + 6e^- \longrightarrow 2\,Cr^{3+} + 7\,H_2O$
$3[H_2S \longrightarrow S + 2\,H^+ + 2e^-]$
$3\,H_2S + Cr_2O_7^{2-} + 8\,H^+ \longrightarrow 3\,S + 2\,Cr^{3+} + 7\,H_2O$

(d) $3[Zn \longrightarrow Zn^{2+} + 2e^-]$
$2[VO_3^- + 6\,H^+ + 3e^- \longrightarrow V^{2+} + 3\,H_2O]$
$3\,Zn + 2\,VO_3^- + 12\,H^+ \longrightarrow 3\,Zn^{2+} + 2\,V^{2+} + 6\,H_2O$

63. (a) $IO_3^- + 6\,H^+ + 6e^- \longrightarrow I^- + 3\,H_2O$
$3[HSO_3^- + H_2O \longrightarrow SO_4^{2-} + 3\,H^+ + 2e^-]$
$IO_3^- + 3\,HSO_3^- \longrightarrow I^- + 3\,SO_4^{2-} + 3\,H^+$

(b) $OsO_4 + 8\,H^+ + 4e^- \longrightarrow Os^{4+} + 4\,H_2O$
$2[C_4H_8(OH)_2 \longrightarrow C_4H_8O_2 + 2\,H^+ + 2e^-]$
$OsO_4 + 2\,C_4H_8(OH)_2 + 4\,H^+ \longrightarrow Os^{4+} + 2\,C_4H_8O_2 + 4\,H_2O$

(c) $5[U^{4+} + 2\,H_2O \longrightarrow UO_2^+ + 4\,H^+ + e^-]$
$MnO_4^- + 8\,H^+ + 5e^- \longrightarrow Mn^{2+} + 4\,H_2O$
$5\,U^{4+} + MnO_4^- + 6\,H_2O \longrightarrow 5\,UO_2^+ + Mn^{2+} + 12\,H^+$

(d) $3[CuS \longrightarrow Cu^{2+} + S + 2e^-]$
$2[NO_3^- + 4\,H^+ + 3e^- \longrightarrow NO + 2\,H_2O]$
$3\,CuS + 2\,NO_3^- + 8\,H^+ \longrightarrow 3\,Cu^{2+} + 3\,S + 2\,NO + 4\,H_2O$

65. (a) $Zn + 2\,OH^- \longrightarrow Zn(OH)_2 + 2e^-$
$ClO^- + H_2O + 2e^- \longrightarrow Cl^- + 2\,OH^-$
$Zn + ClO^- + H_2O \longrightarrow Cl^- + Zn(OH)_2$

(b) $5[Br_2 + 2e^- \longrightarrow 2\,Br^-]$
$Br_2 + 12\,OH^- \longrightarrow 2\,BrO_3^- + 6\,H_2O + 10e^-$
$6\,Br_2 + 12\,OH^- \longrightarrow 10\,Br^- + 2\,BrO_3^- + 6\,H_2O$
Or dividing the final coefficients by 2,
$3\,Br_2 + 6\,OH^- \longrightarrow 5\,Br^- + BrO_3^- + 3\,H_2O$

(c) $3[ClO^- + H_2O + 2e^- \longrightarrow Cl^- + 2\,OH^-]$
$2[CrO_2^- + 4\,OH^- \longrightarrow CrO_4^{2-} + 2\,H_2O + 3e^-]$
$3\,ClO^- + 2\,CrO_2^- + 2\,OH^- \longrightarrow 3\,Cl^- + 2\,CrO_4^{2-} + H_2O$

CHAPTER 4

1. 4.3 g $AgNO_3$; 2.1 g $NaNO_3$; 4.7 g AgBr

3. 2.13 g O_2

5. 17.3 g H_2; 606 g Cl_2; 684 g NaOH

6. (a) $2\,NaCl(s) + H_2SO_4(aq) \rightarrow 2\,HCl(g) + Na_2SO_4(aq)$
(b) 17 g H_2SO_4
(c) 45%
(d) 11 g Na_2SO_4

7. (a) 3.6 mol CO required; 1.0×10^2 g
(b) 170 g CO
(c) 39.9 kg Fe_2O_3
(d) 71.5%

9. (a) 1.05×10^4 tons N_2; 5.99×10^3 tons O_2; 1.35×10^4 tons H_2O
(b) 1.00×10^2 g NH_4NO_3
(c) 63%

12. 6.00 mol H_2O; 160. g $CaCN_2$
13. (a) 0.14 g H_2O_2; (b) 0.30 g $PbSO_4$
15. Formula of oxide is Sc_2O_3; 0.557 g H_2O
16. 9.961 g
18. 1.49 g of NH_3 is present in excess; 7.66 g N_2F_4; 8.84 g HF
20. (a) O_2 tank emptied first; (b) 62.9 kg B_5H_9 remains; 81.1 kg H_2O formed.
21. (a) 1.3 g H_2 are in excess; (b) 85.2 g CH_3OH theoretically obtainable.
24. 0.158 *M*; 250 mL
25. (a) 1.7 g; (b) 0.01 g; (c) 2.1 g; (d) 0.055 g
27. (a) 0.092 *M*; (b) 0.23 *M*; (c) 0.083 *M*; (d) 0.27 *M*; (e) 0.931 *M*
28. (a) $[Na^+] = [Br^-] = 1.0\ M$; (b) $[NH_4^+] = 1.5\ M$ and $[PO_4^{3-}] = 0.50\ M$; (c) $[Zn^{2+}] = 0.10\ M$ and $[Cl^-] = 0.20\ M$; (d) $[K^+] = [ClO_4^-] = 0.10\ M$
31. 0.040 *M*
32. 1.2 g $MgCl_2$
33. 230 g $CuSO_4{\cdot}5H_2O$
35. 1.5 g H_2; 60. g NaOH
37. 0.00313 mol Cl_2; 0.222 g Cl_2; 0.272 g MnO_2
39. 2.1×10^7 g Cl_2
40. (a) 0.0354 *M*; (b) 14.2 moles O_2
42. 250. mL HCl
44. 0.0292 moles HCl required; 265 mL HCl
45. 0.112 *M*
48. 38.5 mL NaOH
50. 0.0749 *M* in succinic acid; 0.841 g KOH required. Balanced equation:

$$C_4H_6O_4(aq) + 2\ KOH(aq) \longrightarrow K_2C_4H_4O_4(aq) + 2\ H_2O(\ell)$$

52. $[KMnO_4] = 0.01650\ M$
54. 198 mL $KMnO_4$
56. 96.8% $Na_2S_2O_3$
58. 8.20% Cr
60. 31.7% Pb
62. 120. mL $Na_2S_2O_3$
63. (a) 0.109 g MnO_2; (b) 125 mL $S_2O_3^{2-}$; (c) 31.6 g $Na_2S_2O_3$

CHAPTER 5

5. (a) System, propane; surroundings, laboratory. Heat flows from system to surroundings; process is exothermic with respect to the system.
(b) System, water drop; surroundings, skin and atmosphere. Process is endothermic with respect to heat flow into the system from the surroundings.
(c) System, chemicals mixed in a flask; surroundings, flask, bench, laboratory. Process is exothermic; heat flows from system into surroundings.

6. Temperature of room does not depend on how it was achieved.
10. (a) 6008 J/mol; (b) 333.4 J/g; (c) When water changes to ice, 1436 cal/mol of heat energy are evolved.
11. 11.2 kJ
12. 5.68×10^7 kJ
13. 45 g of ice must melt.
15. (a) 24.3 J/mol·K; (b) 25.2 J/mol·K; (c) 132 J/mol·K
18. Balanced equation: $2\ Fe(s) + 3/2\ O_2(g) \rightarrow Fe_2O_3(s)$; 206 kJ of heat energy evolved.
19. 1370 kJ/mol
21. 56.6 g
23. 2.21×10^7 kJ
24. The process evolves 1280 kJ per kilogram of PbS.
26. 3.03×10^9 kJ
27. ΔH for $C{-}N(g) \rightarrow C(gas) + N(g) = +774$ kJ
28. (a) Exothermic;
(b) $6\ C(graphite) + 6\ H_2(g) + 3\ O_2(g) \rightarrow C_6H_{12}O_6(s) + 1260$ kJ
30. (a) endothermic; (b) +890.3 kJ
32. $\Delta H = -98.8$ kJ
34. -48.3 kJ per gram of C_2H_2
35. (a) -478.4 kJ; (b) -69.1 kJ; (c) -905.2 kJ
38. $\Delta H_f^\circ[C_2H_4Cl_2(g)] = -64$ kJ
40. (a) $\Delta H_f^\circ[fat] = -7070$ kJ/mol
(b) 1.72×10^4 kJ; 4100 Calories
41. $\Delta H_f^\circ[B_2H_6(g)] = -57.4$ kJ/mol
43. $\Delta H_f^\circ[MnO_2(s)] = -504$ kJ/mol
44. Heat absorbed by water = 4460 J; heat absorbed by bomb = 2160 J; total heat evolved by reaction = 6.62 kJ
46. $\Delta H_f^\circ[H_2CO(g)] = -109$ kJ
48. (a) Heat of reaction = -36.0 kJ/mol
(b) Heat evolved = -11800 kJ
50. Heat absorbed by water = $-$heat given off by iron bar = 53.6 kJ
Initial temperature = 331°C
52. $\Delta H_f^\circ[B_5H_9(liq)] = 80.9$ kJ/mol

CHAPTER 6

5. Avogadro's law: The volume of a gas, at a given temperature and pressure, is directly proportional to the quantity of the gas. 2 moles of H_2 at STP has a volume of 2×22.4 L (=44.8 L); 1.0 mol O_2 required or one-half the volume of H_2 (=22.4 L) and 2.0 mol $H_2O(g)$ produced (or the same volume as the volume of H_2 used).
6. Dalton's law: The total pressure exerted by a mixture of gases is the sum of the partial pressures of the individual gases in the mixture. $X(O_2) = 0.22$ and $X(N_2) = 0.78$. $P(O_2) = 164$ mm Hg.

12. (a) 0.97 atm; (b) 950. mm Hg; (c) 19 mm Hg; (d) 0.021 atm; (e) 542 torr; (f) 45 mm Hg; (g) 99 kPa; (h) 51 kPa
15. 56.3 mm Hg; 0.0741 atm; 56.3 torr; 7.50 kPa
17. 128 mm Hg
19. 23.1 mL
21. 0.50 L O_2
22. (a) 10.4 L O_2 required;
(b) 10.4 L H_2O produced.
23. 0.193 atm (147 mm Hg)
25. 2.6 L
26. Largest number in (d) and smallest number in (c).
28. 1.27 atm
30. 0.90 g
32. 501 mL
35. B_2H_6
36. 61.7 g/mol
38. 3.7×10^{-4} g/L
40. 119 g/mol
42. 9.00 g
44. H_2S
46. (a) H_2; (b) He
48. (a) C_2H_2 (B); (b) Flasks A and B have same number of molecules (to 2 significant figures).
50. 1.44 g $Ni(CO)_4$
52. 17 atm O_2 (required by stoichiometry)
53. 3.3 g O_2
56. P_{He} = 87 mm Hg, P_{Ar} = 140 mm Hg; P_{total} = 230 mm Hg
58. 0.0946 mm Hg
60. $P(O_2)$ = 7.5 mm Hg; $P(B_2H_6)$ = 2.5 mm Hg
62. 2.0 L at 500°C and 0.78 L at 25°C
64. (a) 29 g/mol;
(b) $X(N_2)$ = 0.75 and $X(O_2)$ = 0.25
66. 67.5% $KClO_3$
67. (a) Average kinetic energy is the same, since temperature is the same.
(b) Root mean square speed of H_2 is greater than that of CO_2 (H_2 is less massive).
(c) Twice as many molecules in B as in A.
70. Root mean square speed for Xe = 239 m/s. Rms for He is 5.75 times greater than rms for Xe, since $[M_{Xe}/M_{He}]^{1/2}$ = 5.75.
73. Xe < C_2H_6 < CO < He
76. Molar mass uranium fluoride = 354 g/mol (for UF_6)
78. Using a = 6.49 atm·L²/mol² and b = 0.0562 L/mol in van der Waals's equation, P = 29.5 atm. Using the ideal gas law, P = 49.3 atm.
80. C_2H_7N
81. 86.0% $NaNO_2$
84. (a) O_2 < B_2H_6 < H_2O; 15 atm B_2H_6 required.
85. XeF_4
89. (a) Empirical formula, B_2H_3; (b) molecular formula, B_8H_{12}
91. 1.24 g Al

CHAPTER 7

17.

ℓ value	Orbital Type
3	*f*
0	*s*
1	*p*
2	*d*

19. 0.8×10^{-19} coulombs
21. 2100 atoms
22. 3.9×10^6 m; 2500 miles
24. 6.00×10^{14} s^{-1}; 3.98×10^{-19} J/photon; 2.39×10^5 J/mol
26. 5.090×10^{14} s^{-1}; 2.031×10^5 J/mol
28. radar < microwave < red light < ultraviolet < gamma rays
30. 5×10^3 s (1.5 hours)
32. 5.44×10^{-7} m or 544 nm; visible region
35. (a) 6 lines;
(b) from n = 4 to n = 1;
(c) n = 4 to n = 3
36. (a), (b), and (d)
38. (a) Energy emitted for n = 4 to n = 1 is −0.94Rhc, whereas it is −0.21Rhc for n = 5 to n = 2. The former transition leads to a greater emission of energy than the latter.
(b) 97.2 nm, ultraviolet
40. 4.02×10^{13} s^{-1}; 7460 nm
42. 2.4×10^{-10} m (0.24 nm)
44. 9.9×10^{-7} m (990 nm)
46. (a) When n = 3, ℓ can be 0, 1, and 2
(b) When ℓ = 3 then m_ℓ = −3, −2, −1, 0, +1, +2, +3. An ℓ of 3 designates *f* atomic orbitals (of which there are 7).
(c) For a 4*s* orbital n = 4, ℓ = 0, and m_ℓ = 0.
(d) For a 5*f* orbital, n = 5, ℓ must be 3, and m_ℓ has any integer value from −3 to +3 including 0.
(e) 4*d*
(f) 36 orbitals (= n^2); 6 subshells (= n); subshell labels: *s, p, d, f, g, and h*
(g) For a *g* subshell, ℓ = 4, there are 9 values of m_ℓ (−4 . . . 0 . . . +4) corresponding to 9 orbitals.
47. 4*p* electron: n = 4, ℓ must be 1, and m_ℓ can be −1, 0, or +1.
50. (a) 5;
(b) 25;
(c) 0 (no electrons may have this set of quantum numbers where n = ℓ);
(d) 1 (*p* orbital);
(e) 0 (no electrons may have this set of quantum numbers where ℓ is less than m_ℓ);
(f) 3 (*p* orbitals);
(g) 11.

52. (a) A 4s orbital has 3 spherical nodes and 0 planar nodes.

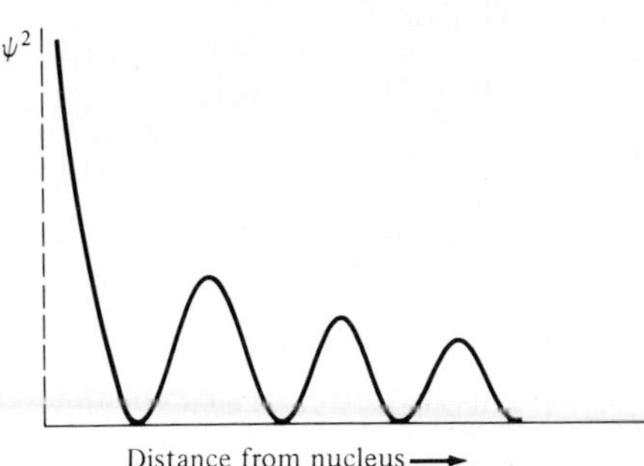

55. (a) 2d (n = 2 and ℓ = 2 is not possible because ℓ can be no larger than n − 1); (b) 3f (n = 3 and ℓ = 3; same reason as (a) above).

56. 1s < 2s = 2p < 3s = 3p = 3d < 4s

57.

Orbital	n	ℓ	m_ℓ
2p	2	1	−1, 0, +1
3d	3	2	−2, −1, 0, 1, 2
4f	4	3	−3, −2, −1, 0, 1, 2, 3

60. 5f orbitals have 1 spherical node and 3 planar nodes.

64. (a) n = 3 and ℓ = 1, orbital is labeled 3p; (b) n = 3 and ℓ = 2, orbital designation is 3d.

CHAPTER 8

Note: For simplicity, electron configurations given by the orbital box notation will use only a line instead of a box. For example, the configuration of H is depicted as

	↑
H:	1s

25. F and Cl have the configurations [rare gas]ns^2np^5 where n = 2 for F and n = 3 for Cl. The orbital box notation is

	↑↓	↑↓	↑↓	↑
[rare gas]	ns	np	np	np

For Br and I, 10 d electrons are included, so their configurations are [rare gas]$(n-1)^{10}\,ns^2np^5$ (n = 4 for Br and 5 for I).

	↑↓	↑↓	↑↓	↑↓	↑↓	↑↓	↑↓	↑↓	↑
[rare gas]	$(n-1)d$					ns	np	np	np

26. (a) Na^+

↑↓	↑↓	↑↓	↑↓	↑↓				
1s	2s	2p	2p	2p	3s	3p	3p	3p

(b) Al^{3+}, same as Na^+

(c) Cl^-

↑↓	↑↓	↑↓	↑↓	↑↓	↑↓	↑↓	↑↓	↑↓
1s	2s	2p	2p	2p	3s	3p	3p	3p

33. Element 109: [Rn]$5f^{14}6d^77s^2$. Cobalt is in the same group.

34. (a) Ti

	↑	↑				↑↓
[Ar]	3d	3d	3d	3d	3d	4s

(b) Ti^{2+}

	↑	↑				
[Ar]	3d	3d	3d	3d	3d	4s

(c) Ti^{4+}

[Ar]	3d	3d	3d	3d	3d	4s

36. Pt^{2+}

	↑↓	↑↓	↑↓	↑	↑	
[Ar] ($4f^{14}$)	5d	5d	5d	5d	5d	6s

38. In^{3+}: [Kr]$4d^{10}5s^05p^0$

40. Element 113 should be in Group 3A (with B, Al, Ga, In, Tl): [Rn]$5f^{14}6d^{10}7s^27p^1$

41. Beryllium, Be:

↑↓	↑↓
1s	2s

Electron	n	ℓ	m_ℓ	m_s
1	1	0	0	+1/2
2	1	0	0	−1/2
3	2	0	0	+1/2
4	2	0	0	−1/2

43. See question 34 for the configuration of Ti.

Electron	n	ℓ	m_ℓ	m_s
1	3	2	+2	+1/2
2	3	2	+1	+1/2
3	4	0	0	+1/2
4	4	0	0	−1/2

45. Total number of electrons is 19 (and it is K). (a) atomic number = 19; (b) 7 *s* electrons (2 in 1*s*, 2 in 2*s*, 2 in 3*s*, and 1 in 4*s*). (c) 12 *p* electrons (6 in 2*p* and 6 in 3*p*). (d) No *d* electrons.

46. (a) 6; (b) 50; (c) None. n and ℓ cannot have the same value; (d) 2; (e) 10; (f) None. m_ℓ can only be 0 when ℓ is 0.

48. (a) ℓ cannot be larger than $n - 1$.
(b) m_s can only be + or − 1/2.

50. 6 electrons can be assigned to the three 7*p* orbitals.

52. 2 electrons assigned to the single 7*s* orbital.

54. Cu, copper.

56. Set (c).

58. Possible sets of quantum numbers for electron assigned to a 4*p* orbital.

Electron	n	ℓ	m_ℓ	m_s
1	4	1	+1	±1/2
2	4	1	0	±1/2
3	4	1	−1	±1/2

60. C—Cl = 176 pm; Si—Cl = 216 pm; Ge—Cl = 221 pm: Sn—Cl = 239 pm; Pb—Cl = 274 pm.

62. C (77 pm) < B (88 pm) < Al (143 pm) < Na (186 pm) < K (231 pm)

64. Isoelectronic series (all have 18 electrons): $S^{2-} > Cl^- > Ar > K^+ > Ca^{2+}$

66. 1st IE of K < 1st IE of Li < 1st IE of Be < 2nd IE of Be < 2nd IE of Na

68. Na. The second electron removed is from an inner shell.

71. (c)

73. (a) K; (b) C; (c) K < Li < C < N

75. (a) S < O < F. IE increases moving up a group (S < O) and on moving across a period (O < F).
(b) Oxygen. IE increases on moving up a periodic group.
(c) Cl. EA generally increases on moving to the right in a period and on moving up in a group.
(d) O^{2-}. This ion has the largest electron/proton ratio and so the smallest effective nuclear charge.

77. Ti^{2+} has 2 unpaired (3*d*) electrons. Co^{3+} has 4 unpaired (3*d*) electrons. Both ions are paramagnetic.

79. (a) Only Zn^{2+} is diamagnetic. (b) Mn^{2+} has 5 unpaired *d* electrons.

80. Cr^{2+} and Fe^{2+} each have 4 unpaired *d* electrons.

82. Group 2A [also some elements of Group 4A (Ge, Sn, Pb)].

84. In^{4+}: Too much energy required to lose 4th electron (from 4*d* orbitals).
Fe^{6+}: Energy cost in creating ions of such high charge is too great.
Sn^{5+}: Same reason as In^{4+}.

85. (a) sulfur; (b) radium; (c) nitrogen; (d) Ru^{2+}; (e) At, astatine; (f) copper; (g) gadolinium, Gd.

87. (a) metal (Ca); (b) nonmetal (Br); (c) B (Br); (d) B (Br); (e) B (Br).

88. (a) $[Rn]5f^{14}6d^{10}7s^27p^1$; (b) Group 3A: boron, aluminum, gallium, indium, thallium; (c) Et_2O_3 and $EtCl_3$ (compare with Al_2O_3 and $AlCl_3$).

CHAPTER 9

1. Li, 1; Sc, 3; Zn, 2; Si, 4; and Cl, 7.

6. Odd-electron molecules; NO_2, NO_3.

8. (a) and (c) are identical and are resonance structures of (b). There are 1.5 CO bonds, on average.

9. C—H = bond order of 1; C—C = bond order of 3. Acetylene has 3 sigma bonds (2 C—H and 1 C—C) and 2 pi bonds (between the C atoms).

11. C—F (141 pm) < C—O (143 pm) < C—N (147 pm) < C—C (154 pm) < C—B

26. NH_3 has a dipole moment. N is the negative end of the dipole and the three H atoms are the positive side.

27. (a) N, 5, 5A (f) C, 4, 4A
(b) B, 3, 3A (g) Cl, 7, 7A
(c) S, 6, 6A (h) P, 5, 5A
(d) Na, 1, 1A (i) Ne, 8, 8A
(e) Mg, 2, 2A

31. Only elements beyond the second period of the periodic table (Li–Ne) can have expanded valence. Therefore, P, Cl, Se, and Sn can bind to as many as five or six other atoms.

33. (a) $H_2C{=}\ddot{O}{:}$ (two H atoms bonded to C, C=O)

(b) $H—\ddot{\underset{..}{O}}—\ddot{\underset{..}{Cl}}{:}$

(c) $:\ddot{\underset{..}{F}}—\ddot{N}—\ddot{\underset{..}{F}}:$ with $:\ddot{\underset{..}{F}}:$ bonded below N

(d) $:\ddot{O}:^{2-}$ bonded to $\ddot{S}$, which is bonded to two $:\ddot{\underset{..}{O}}:$ atoms

(e) $:\ddot{\underset{..}{O}}—\ddot{Cl}$ bonded to two $:\ddot{O}:$ atoms ($^-$)

(f) $:\ddot{F}:^-$ above; $:\ddot{\underset{..}{F}}—B—\ddot{\underset{..}{F}}:$; $:\underset{..}{F}:$ below

(g) $\ddot{O}{=}N{=}\ddot{O}^+$

(h) $:\ddot{O}:^{3-}$ above; $:\ddot{\underset{..}{O}}—P—\ddot{\underset{..}{O}}:$; $:\underset{..}{O}:$ below

35. (a) SO_2

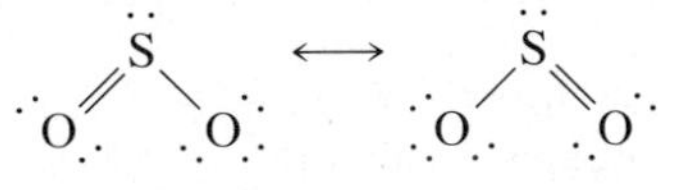

$O{=}S{-}O \longleftrightarrow O{-}S{=}O$

SO_3

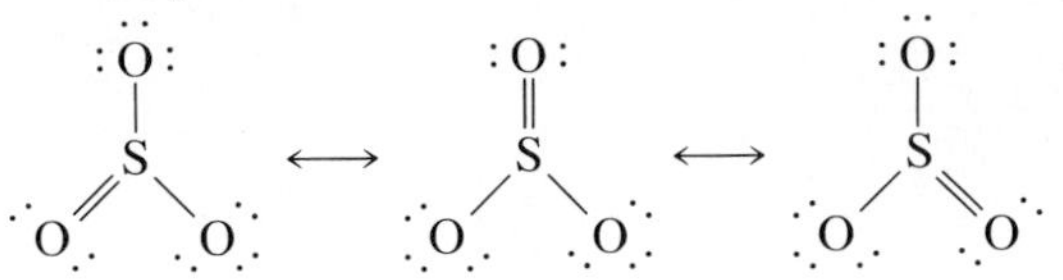

$\longleftrightarrow \quad \longleftrightarrow$

(b) HNO_3

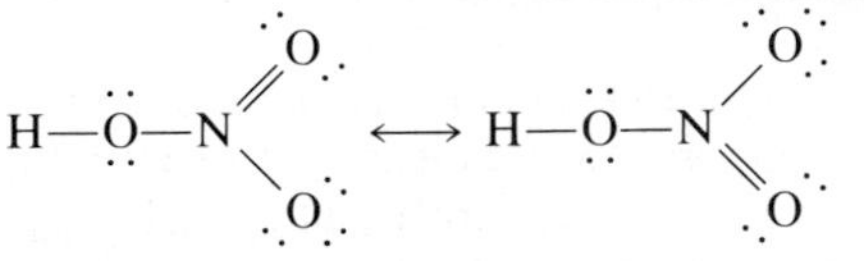

$H{-}O{-}N(=O)(-O) \longleftrightarrow H{-}O{-}N(-O)(=O)$

(c) $H_3C{-}N(=O)(-O) \longleftrightarrow H_3C{-}N(-O)(=O)$

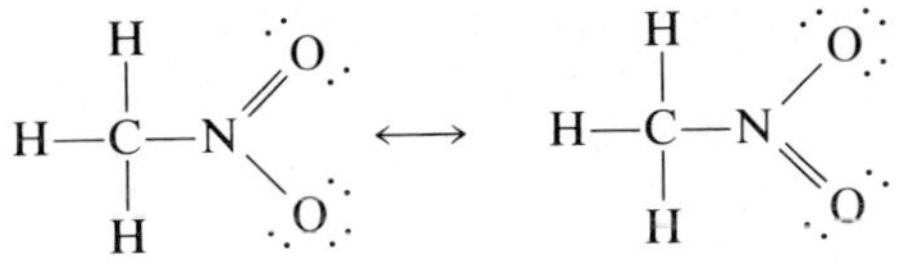

(d) $O{=}CH{-}NH_2 \longleftrightarrow O{-}CH{=}NH_2$

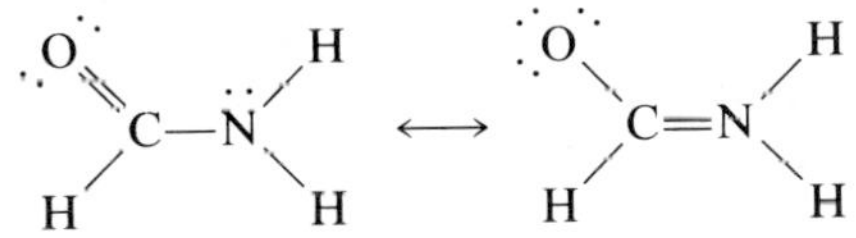

37. BrF_3 $\quad I_3^-$ $\quad XeO_2F_2$

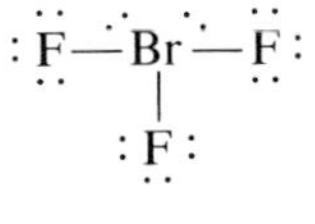

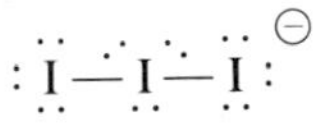

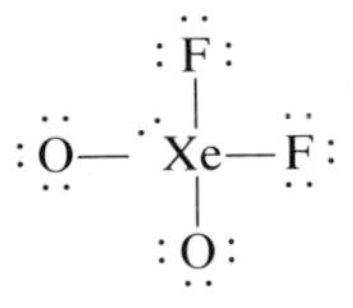

38.

Molecule	Sigma Bonds	Pi Bonds	Bond Orders
H_2CO	3	1	CH = 1; CO = 2
SO_3^{2-}	3	0	SO = 1
NO_2^+	2	2	NO = 2
CN^-	1	2	CN = 3
H_3CCN	5	2	CH = 1; C—C = 1, CN = 3
SO_2	2	1	SO = 1.5
SO_3	3	1	SO = 1.33

40. NO bond order in NO_2^- is 1.5, while it is 2 in NO_2^+. This means the NO bond in NO_2^+ should be shorter (= 110 pm) than in NO_2^- (124 pm).

41. (a) B—Cl; (b) C—O; (c) P—O; (d) P—O

43. (a) NH_3, −40. kJ/mol; (b) H_2O, −243 kJ/mol; (c) H_2O_2, −132 kJ/mol.

44. Heat of formation of C_2H_4 = +39 kJ/mol. The heat of formation of acetylene (C_2H_2) in Example 19.5 was found as +205 kJ/mol. The formation of 2 additional C—H bonds in C_2H_4, relative to C_2H_2, causes C_2H_4 to have a less positive heat of formation (and partly offsets the formation of one more pi bond in acetylene).

46. D_{H-I} = 298 kJ/mol. The trend in H—X bond energies is to lower values with increasing halogen atomic number.

48. Heat of combustion = −658 kJ/mol

50. Heat of formation of glycine = −311 kJ/mol

52. D_{O-F} = 195 kJ/mol

54. Heat of formation = −87 kJ/mol

56. Heat of reaction = −48 kJ

59. (a) C—O is more polar than C—N

(b) N—O is more polar than P—S

(c) P—N is more polar than P—H (which is nonpolar, since P and H have the same electronegativity)

(d) B—I is more polar than B—H

61. F—F < O—F < N—F < H—F < Be—F

63. (a) H_2S: H = +1 and S = −2

(b) CH_4: C = −4 and H = +1

C_2H_4: C = −2 and H = +1

(c) H_2CO: H = +1, O = −2, and C = 0

(d) ClF_3: F = −1 and Cl = +3

(e) XeF_2: F = −1 and Xe = +2

(f) ICl_2^+: Cl = −1 and I = +3

(g) OF_2: O = +2 and F = −1

64. (a) H_2O: H and O are both 0 (zero)

(b) CH_4: H and C are both zero.

(c) NO_2^+: O = zero and N = +1

(d) OCl^-: O = −1 and Cl = 0

(e) HOF: H, O, and F are all zero.

(f) XeF_2: Xe and F are both zero.

(g) ICl_2^-: Cl = 0 and I = −1

66. (a) $:N{=}N{=}O:^- \longleftrightarrow :N{\equiv}N{-}O:^-$

$\longleftrightarrow :N{-}N{\equiv}O:^-$

(b) −1 +1 0, $\quad$ 0 +1 −1, $\quad$ −2 +1 +1

(c) The middle structure, with a −1 charge on O, is the most reasonable. The smallest contribution to the overall electronic structure is the structure at the right.

68. (a) :C̈=N=Ö:⁻ ⟷ :C≡N—Ö:⁻

⟷ :C̈—N≡O:⁻

(b) −2 +1 0, −1 +1 −1, −3 +1 +1

(c) The middle structure is the most reasonable, since it places the smallest possible charges on the atoms.

(d) The middle structure above, although the most reasonable of the three, still has a +1 charge on the relatively electronegative N atom and a negative charge on the least electronegative atom, C.

71. (a) CO_2, 2 pairs (each double bond counting as "one" structural pair); (b) $CO_3{}^{2-}$, 3 pairs; (c) NH_3, 4 pairs; (d) CH_4, 4 pairs; (e) PF_5, 5 pairs; (f) SF_6, 6 pairs.

74.

Molecule	Structural-Pair Geometry	Molecular Shape
BH_3	trigonal planar	trigonal planar
NH_2Cl	tetrahedral	pyramidal
$SnCl_3{}^-$	tetrahedral	pyramidal
$SnCl_4$	tetrahedral	tetrahedral
$ClF_2{}^+$	tetrahedral	bent
SF_2	tetrahedral	bent
BeF_2	linear	linear

75.

Molecule	Structural-Pair Geometry	Molecular Shape
ClF_3	trigonal bipyramidal	T-shape
$ClF_4{}^-$	octahedral	square planar
ClF_5	octahedral	square pyramidal
SF_4	trigonal bipyramidal	"seesaw"
PF_5	trigonal bipyramidal	trigonal bipyramidal
$PF_6{}^-$	octahedral	octahedral
$SiF_6{}^{2-}$	octahedral	octahedral
XeF_4	octahedral	square planar

77. The $NO_2{}^+$ ion is linear (O—N—O angle = 180°) whereas the $NO_2{}^-$ ion is bent (trigonal planar structural pair geometry) with an O—N—O angle closer to 120°.

79. (a) 120°

(b) 120°

(c) 1 = 109° and 2 = 120°

(d) 1 = 109° and 2 = 180°

(e) 1 and 2 = 109°

(f) 109°

(g) 1 = 109° and 2 = 120°

(h) SeF_4 (i) SOF_4 (j) $XeOF_4$

120° 90° 120° 90° 90° 90° 90°

80. F—Cl—F angle in $ClF_2{}^+$ is about 109°.

~109°

In $ClF_2{}^-$ it is most likely that the F atoms are in axial positions, so the F—Cl—F angle is 180°.

82. (a) N_2O, polar N=N=O

(b) CS_2, a linear molecule, is nonpolar.

(c) PCl_3, polar

(d) XeF_2, a linear molecule, is nonpolar.

:F̈—Xe—F̈:

84. (a) FBH_2, polar

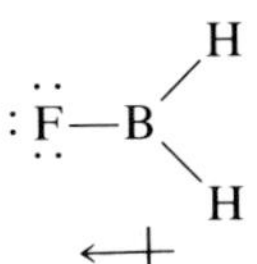

(b) PH_3, nonpolar (P and H have the same electronegativity)

(c) XeF_4, a square planar molecule, is nonpolar.

86. (a) 10 sigma bonds; (b) 2 pi bonds; (c) 1 = 109°, 2 = 120°, 3 = 109°, 4 = 120°.

89. (a) Heat of reaction = −205.9 kJ

(b) Heat of hydrogenation = −384 kJ

(c) Assuming the double bonds are fixed in benzene means that much more heat is given off on going to cyclohexane. The "fixed bond" molecule is less stable (by the difference between −384 kJ and −206 kJ or 178 kJ) than benzene molecules where resonance is taken into account (as in the calculation in part (a) above).

CHAPTER 10

5. 6 (a set of 6 hybrid orbitals, each labeled sp^3d^2).

16. H_2S, hydrogen sulfide, H—S̈—H
    (a) structural-pair geometry, tetrahedral
    (b) molecular geometry, bent
    (c) S atom hybridization, sp^3

18.

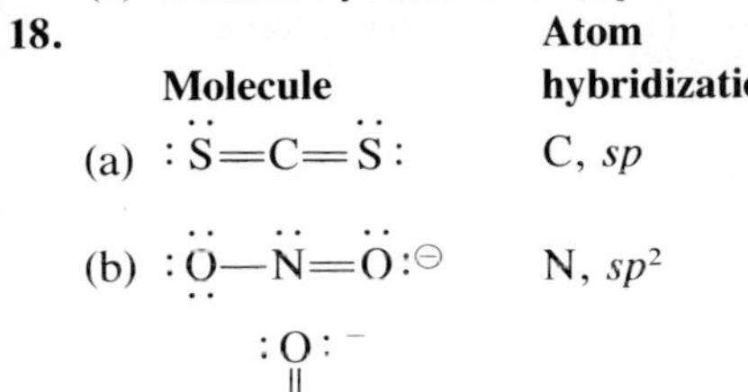

	Molecule	Atom hybridization
(a)	:S̈=C=S̈:	C, sp
(b)	:Ö—N̈=Ö:⊖	N, sp^2
(c)	:O:⁻ ‖ N (with two :Ö: atoms)	N, sp^2
(d)	H⁻ | H—B—H | H	B, sp^3

20. (a) :N̈=N=Ö:
    (b) structural pair geometry, linear; molecular geometry, linear
    (c) N atom hybridization, sp

22. XeF_2 has 8 + 2(7) = 22 valence electrons

:F̈—Xe—F̈:

There are five electron pairs (2 bond pairs and 3 lone pairs) around the Xe atom. Therefore, the pairs are at the corners of a trigonal bipyramid, and the Xe atom hybridization is sp^3d.

24.

Ion	ClF_2^+	ClF_2^-
Dot structure	:F̈—C̈l—F̈:⊕	:F̈—Cl—F̈:⁻
Structural pair geometry	tetrahedral	trigonal bipyramid
Molecular geometry	bent	linear
Cl hybrid orbitals	sp^3	sp^3d

26. :Ö=C=Ö:
The central atom is sp hybridized. This leaves two unhybridized p orbitals on the C atom to be used in pi bonding.

C atom: $\frac{\uparrow\downarrow}{2s}\ \frac{\uparrow}{2p}\ \frac{\uparrow}{2p}\ \frac{\ }{2p} \xrightarrow{\text{hybridize}} \frac{\uparrow}{sp}\ \frac{\uparrow}{sp}\ \frac{\uparrow}{2p}\ \frac{\uparrow}{2p}$

The C atom sp hybrid orbitals are used to form sigma bonds with the O atoms,

and the unhybridized $2p$ orbitals are used to form pi bonds to the O atoms. One $2p$ C atom orbital is illustrated below, forming one of the C=O pi bonds.

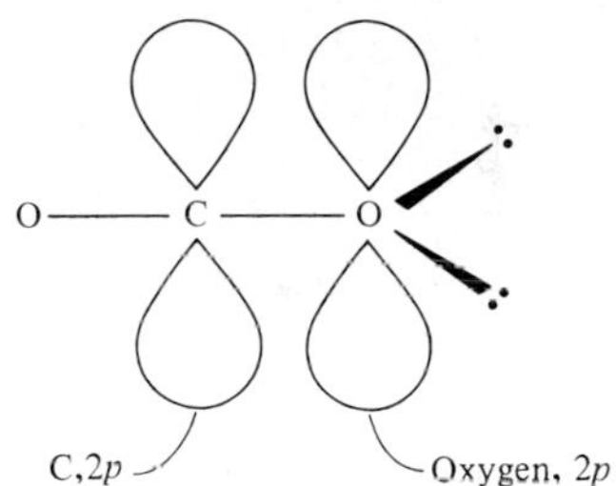

28. Dot structure:

:O:
‖
:C̈l—C—C̈l:

(a) The C atom is sp^2 hybridized, each of these hybrid atomic orbitals being used to form a sigma bond (one to O and two to the Cl atoms). The unhybridized $2p$ orbital is used to form a CO pi bond as in CO_2 in Study Question 26.

C atom: $\underbrace{\frac{\uparrow}{sp^2}\ \frac{\uparrow}{sp^2}\ \frac{\uparrow}{sp^2}}_{\text{for sigma bond formation}}\quad \underbrace{\frac{\uparrow}{p}}_{\text{for pi bond formation}}$

30. One of the resonance structures of the NO_3^- ion is

:O:⁻
‖
N
:Ö ⁄ ∖ Ö:

The geometry of the N sigma bonds is trigonal planar, so the N atom hybridization is sp^2.

N atom: $\frac{\uparrow\downarrow}{2s}\ \frac{1}{2p}\ \frac{1}{2p}\ \frac{1}{2p}\ \xrightarrow{\text{hybridize}}\ \frac{\uparrow\downarrow}{sp^2}\ \frac{1}{sp^2}\ \frac{1}{sp^2}\ \frac{1}{2p}$

This means there is one, unhybridized $2p$ orbital remaining on the N atom.

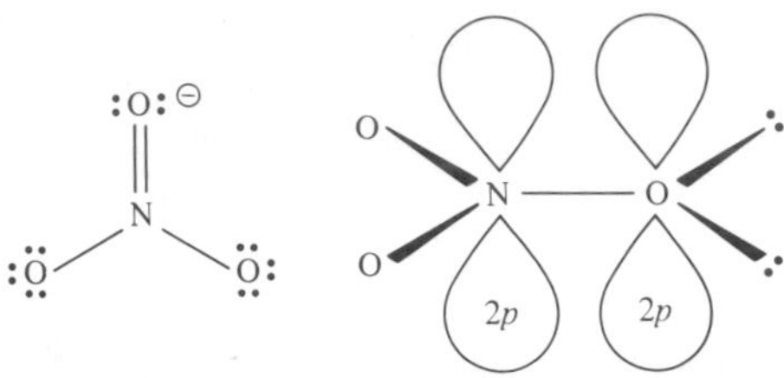

This orbital can be used to form a pi bond with one of the O atoms.

32. H_2^+, $(\sigma_{1s})^1(\sigma^*_{1s})^0$. Bond order = 1/2. The H—H bond in H_2^+ is weaker than in H_2 (where the bond order is 1).

34. (a) Paramagnetic molecules: Only B_2 and O_2 are paramagnetic (that is, have one or more unpaired electrons).
(b) Bond order of 1: Li_2, B_2, F_2
(c) Bond order of 2: C_2 and O_2
(d) Highest bond order: N_2

36. NO, Ne_2^+, CN

38. (a) NO is paramagnetic, since it has an odd number of electrons (a total of 11 valence electrons). It must have 1 unpaired electron.
(b) The highest-energy molecular orbital assigned to an electron is π^*_{2p}.
(c) Bond order = 2.5
(d) When NO ionizes to form NO^+ the electron in π^*_{2p} is lost. This strengthens the N—O bond.

42. (a) The CH_3 carbon atom is sp^3 hybridized, and the C═N carbon atom, sp^2 hybridized. (b) The C—N—O bond angle is approximately 120°.

44. (a) 5 pi bonds and 21 sigma bonds. (b) A = 120°, B = 109°, C = 109°. (c) C-1, sp^2; C-2, sp^2; O-3, sp^3.

46. (a) 17 sigma bonds and 2 pi bonds. (b) N-1, sp^3; C-2, sp^3; C-3, sp^2; N-4, sp^2. (c) A = 109°; B = 120; C = 109°.

48. Electron dot structure of O_2^{2-} [:Ö—Ö:$^{2-}$] shows a net of one sigma bond, and a bond order of 1. The MO picture leads to the same result.

O_2^{2-}: $(\sigma_{2s})^2(\sigma^*_{2s})^2(\pi_{2p})^4(\sigma_{2p})^2(\pi^*_{2p})^4$

Both approaches indicate a molecule with no paramagnetism.

50. The electron added to form O_3^- from O_3 is assigned to a pi antibonding orbital, so the O—O pi bonds are affected. To form O_3^+ from O_3, an electron assigned to a pi nonbonding orbital is removed. The pi bonding in the ion is not affected relative to the neutral molecule.

52. (a) Sb in SbF_5 is sp^3d hybridized. In $[SbF_6]^-$ the Sb is sp^3d^2 hybridized. (b) H—F̈—H$^+$. (c) Bent (structural-pair geometry is tetrahedral). (d) sp^3.

53. (a) C atoms are sp^2 hybridized in the reactant and sp^3 hybridized in the product. (b) P atom goes from sp^3 hybrid to sp^3d hybrid. (c) Xe goes from sp^3d hybrid to sp^3d^2 hybrid. (d) Sn is sp^3 hybridized in $SnCl_4$, but it is sp^3d^2 hybridized in $SnCl_6^{2-}$.

56. (a) Orbital assignment for NH_4^+ is the same as for CH_4 (since they have the same number of valence electrons). (b) The N—H bonds are lengthened because the added electron is assigned to an antibonding orbital. The N—H bonds in NH_4 have an order of 7/8.

59. (a) sp^2; (b) sp^3; (c) sp^2; (d) sp^3

CHAPTER 11

17. (a) ion-dipole; (b1) hydrogen bonds; (b2) induced dipole-induced dipole; (b3) ion-ion; (b4) dipole-dipole (and weak hydrogen bonds)

18. Ne < CH_4 < CO < KI < $MgSO_4$; Ne, CH_4, and CO exist as gases at room temperature and normal pressures.

19. Increase. The force of attraction between ions depends directly on ion charge and inversely on ion-ion distance. If the charges are constant, but the distance decreases, then the ion-ion attractive force increases. This should result in an increase in melting point.

22. C_2H_5—Ö: . . . H—Ö:; hydrogen bonds (each O bonded to an H below)

24. (a) O_2, induced dipole forces; (b) SO_2, dipole-dipole forces; (c) HF, strong hydrogen bonds; (d) GeH_4, induced dipole forces

26. (a) induced dipole forces; (b) dipole-dipole (and weak hydrogen bonds); (c) dipole-dipole forces; (d) induced dipole forces

28. (a) Li_2SO_4; (b) $CaCl_2$; (c) $Mg(NO_3)_2$; (d) $NiCl_2$

31. (a) increase; (b) decrease; (c) no change

34. 181 kJ

36. (a) 30.9 kJ/mol; (b) 714.8 kJ/mol; (c) 31.3 kJ/mol; (d) 39.8 kJ/mol

38. (a) About 150 mm Hg (Appendix F gives 149.4 mm Hg); (b) about 95°C (Appendix F suggests a temper ature between 93° and 94°C); (c) at 70°C the vapor pressure of water (about 225 mm Hg) is only about half that of ethyl alcohol (about 475 mm Hg); (d) about 400 mm Hg

41. (a) Approximately 600 mm Hg (the vapor pressure of diethyl ether at 30°C). (b) Additional ether vapor will condense to liquid. (c) Gas pressure (= vapor pressure of ether) at 0°C is about 200 mm Hg.

43. Molar heat of vaporization = 38.6 kJ/mol; estimated boiling point = 402 K (= 129°C)(actual bp = 125.6°C).

45. All the ethyl alcohol evaporates.

47. Concave

49. $CaCl_2$ cannot have the NaCl structure. The latter must have a 1:1 ratio of anion to cation. $CaCl_2$ requires a ratio of 2 anions to 1 cation.

51. (a) Volume of unit cell = 2.85×10^{-22} cm^3; (b) 2.82 g/cm^3 (literature value is 2.80 g/cm^3).

52. 3.3 g/cm^3

53. (a) 4; (b) LiH can have the face-centered cubic unit cell.

55. (a) 2 atoms per unit cell; (b) 2.36×10^{-23} cm^3; (c) 7.86 g/cm^3

57. 8 C atoms per unit cell; (a) 4.55×10^{-23} cm^3; (b) 3.57×10^{-8} cm

58. MgO consists of ions with +2 and −2 charges, so the ion-ion forces are much stronger than in NaF (where the ion charges are +1 and −1).

60. (a) CsF has a higher melting point (682°C) than CsI (621°C). (b) KBr has a higher melting point (730°C) than RbBr (682°C).

62. 679.4 kJ/mol

65. 5.01 kJ

67. (a) 1.97 kJ evolved; (b) 1.97 kJ required

69. 1.6×10^7 J

71. (a) vapor; (b) liquid; (c) about −117°C; (d) 0.15 atm; (e) solid is more dense than liquid; S-L equilibrium line has a positive slope.

74. 6.9×10^{15} J

75. Estimated bp = 76°C (actual, 76.7°C)

77. (a) 79.4 kJ; (b) vapor pressure at 0°F is estimated to be 368 mm Hg. Since this is much lower than ordinary barometric pressure, there is not sufficient vapor pressure to force the vapor out of the tank of the stove so that the gas can be burned.

78. Assuming a radius for the circles of 1.00 cm, the area covered by circles *inside* the square in A (of area 4.00 cm^2) is 3.14 cm^2. Thus, in A, 78.5% of the area is covered. In B, the area of the triangle is 1.73 cm^2, while the area covered by circles is 1.57 cm^2. This means that 90.8% of the surface is covered.

80. (a) Cs^+—I^- distance = 0.385 nm [= $(1/2)(3)^{1/2}$(edge)]; (b) Cs^+ radius = 0.169 nm.

81. Vapor pressure at 25°C = 0.0021 mm Hg

83. $MgSO_4$

86. 2710 kJ/mol (approximately 4 times the NaF value). The ion radii are approximately the same, but the charges on Mg and O are both 2.

CHAPTER 12

8. 1 mol of NaCl provides 2 mol of ions, whereas 1 mol of $CaCl_2$ provides 3 mol of ions. Therefore, in equal quantities of water, $CaCl_2$ will have 3/2 or 1.5 times the effect on freezing point, boiling point, and osmotic pressure.

11. (a) 13.5%; (b) 0.0458; (c) 2.66 *m*

14.

Compound	Molality	Weight Percent	Mole fraction
NaCl	0.25	*1.4*	4.5×10^{-3}
C_2H_5OH	*1.1*	5.0	*0.020*
$C_{12}H_{22}O_{11}$	0.10	*3.3*	*0.0018*
NH_3	*25.2*	30.0	*0.312*
CH_3COOH	*0.0083*	0.050	1.5×10^{-4}

16. (a) 194 molal; (b) 17.8 molar

19. (a) 0.162; (b) 30.0%; (c) 9.97 molar

21. (a) 0.024 *m* in $CaCl_2$; (b) 0.072 *m* in ions

26. 104 g/100. g solvent

27. (a) positive; (b) +34.9 kJ/mol

29. −299.7 kJ/mol

31. −296 kJ/mol

34. 6.7×10^{-5} *m*

36. 691 mm Hg

38. 101 g/mol

42. 100.256°C

43. 62.5°C

45. Actual solubility of NaCl at 100°C is 39.1 g/100. g H_2O. This gives a boiling point of 103.43°C.

47. 980 g

49. 940 g

51. Empirical formula, BF_2; molecular formula, B_2F_4.

53. First equation is correct. 9.41 g $NaHSO_3$ = 0.0905 mol. From Δt, molality of ions = 2 × 0.09 = 0.18 *m*.

57. 0.296 mol/L

59. 59 mm Hg; 13% toluene and 87% benzene

61. 410. mm Hg

63. 288 mm Hg

65. (a) 125 g; (b) 40 g

67. 27 mm Hg

69. 0.20 *m* Na_2SO_4 (lowest vapor pressure) < 0.50 *m* sugar < 0.20 *m* KBr < 0.35 *m* $C_2H_4(OH)_2$ (highest vapor pressure)

70. (a) $C_{10}H_{12}O$; (b) 148 g/mol; (c) Same as empirical formula

74. The empirical and molecular formulas are $C_{18}H_{24}Cr$.

CHAPTER 13

5. Reaction is second order in A and zero order in B. Total order is 2 + 0 = 2.
10. After 5 half-life periods the fraction remaining = $(1/2)^5$ = 1/32.
11. 1/[reactant] vs. time gives a straight line for a process second order in reactant. If the process is first order in reactant, then ln[reactant] vs. time is a straight line.
18. (a) true; (b) true; (c) false, ". . . the rate declines;" (d) true.
21. (a) $-1/2(\Delta[O_3]/\Delta t) = +1/3(\Delta[O_2]/\Delta t)$
    (b) $-1/2(\Delta[HOF]/\Delta t) = +1/2(\Delta[HF]/\Delta t) = (\Delta[O_2]/\Delta t)$
    (c) $-(\Delta[N_2]/\Delta t) = -1/3(\Delta[H_2]/\Delta t) = +1/2(\Delta[NH_3]/\Delta t)$
23. (a) 0 to 10 s, $\Delta[A]/\Delta t = -0.0167$ mol/(L.s); 10 to 20 s, $\Delta[A]/\Delta t = -0.0119$ mol/(L.s); 20 to 30 s, $\Delta[A]/\Delta t = -0.0089$ mol/(L.s). (b) $\Delta[B]/\Delta t = 2(-\Delta[A]/\Delta t) = 0.0238$ mol/(L.s).
25. (a) Rate for 1000 s to 2000 s = $\Delta[C_4H_6]/\Delta t = -2.3 \times 10^{-6}$ mol(L.s)
    (b) $\frac{1}{2} - \Delta[C_4H_6]/\Delta t = \Delta[C_8H_{12}]/\Delta t = 1.1 \times 10^{-6}$ mol/(L.s)
    (c) Rate for first 1000 s = $\Delta[C_4H_6]/\Delta t = -3.2 \times 10^{-6}$
26. (a) Rate = $k[NO_2]^2[CO]^o$
    (b) If $[NO_2]$ is halved, then the rate drops to 1/4 (or 1/2 squared) of its original value.
28. Rate = $k[B]_o^2$.
30. (a) Rate = $k[A]_o[B]_o^2$. (b) Overall order = 3. (c) Unlikely to occur in a single step, since the rate law implies it must then be termolecular. This type of elementary step is improbable.
32. (a) CH_2O must appear twice as fast as O_3 disappears (since 2 moles of CH_2O are formed for each mole of O_3 consumed); $-(\Delta[O_3]/\Delta t) = +\frac{1}{2}(\Delta[CH_2O]/\Delta t)$.
    (b) Rate = $k[O_3]_o[C_2H_4]_o$; k = 2.0×10^3 L/mol·s
    (c) Reaction is first order in both reactants.
34. (a) First order in O_2 and second order in NO; (b) overall order = 3; (c) rate of appearance of NO_2 = $k[O_2]_o[NO]_o^2$; (d) rate = 8.8×10^{-4} M/s.
36. t = 13 hr (one half-life)
38. (a) 1400 s; (b) 4600 s
40. (a) 4.5×10^3 s; (b) 4.0×10^4 s
42. (a) k = 0.017/min; (b) 0.00075 M (after 3 half-lives); (c) 240 min
43. (a) 0.648/hr; (b) fraction remaining after 3 half-lives = 1/8 = 0.125; (c) rate = 0.0810 M/hr; (d) 7.11 hr
45. (a) P(HOF) = 50 mm Hg and P(total) = 125 mm Hg; (b) P(HOF) = 35 mm Hg and P(total) = 133 mm Hg
46. (a) First order in N_2O_5; (b) rate = $k[N_2O_5]$
49. (a) The plot of ln[A] vs. t is curved, whereas 1/[A] vs. t is a straight line. Therefore, the reaction is second order in A.
    (b) k = 0.0444 /min-M
    (c) $t_{1/2}$ = 585 min
    (d) Rate = $k[A][B]$
51. Exothermic (by 26 kJ/mol)
53. 1.0×10^2 kJ
55. 85 kJ
56. 0.52 L/(mol.s)
58.

	Rate Law	**Molecularity**
(a)	Rate = $k[NO][NO_3]$	bimolecular
(b)	Rate = $k[Cl][H_2]$	bimolecular
(c)	Rate = $k[(CH_3)_3CBr]$	unimolecular

60. (a) Step 2 is rate-determining, as it is the slow step.
    (b) Rate law for step 2 is Rate = $k[O_3][O]$
    (c) Step 1 is unimolecular and step 2 is bimolecular.
62. (a) Rate = $k[A][B]^2$; (b) $k = 7.4 \times 10^{-4}/M^2$.s
65. Mechanism 1 would lead to a rate law of Rate = $[H_2][NO]$, an equation not in agreement with experiment. Mechanism 2 gives the rate law Rate = $kK[NO]^2[H_2]$ (where k is the rate constant for the second step and K is the equilibrium constant for the reaction in step 1). Mechanism 2 agrees with experiment.
69. Statements (a), (d), and (e) are true.
74. 7.5 times faster

CHAPTER 14

2. (a) False. The magnitude of the equilibrium constant is dependent on the temperature.
   (b) True.
   (c) False. The equilibrium constant for a reaction is the reciprocal of the constant for the reverse reaction.
   (d) True.
   (e) True.
6. More B is formed from A.
8. (a) $K = \dfrac{[H_2O]^2[O_2]}{[H_2O_2]^2}$ (b) $K = \dfrac{[PCl_5]}{[PCl_3][Cl_2]}$
   (c) $K = [CO]^2$; (d) $K = [H_2S]/[H_2]$; (e) $K = [O_3]^2/[O_2]^3$
   (f) $K = \dfrac{[H_2O]^2}{[SiH_4][O_2]^2}$; (g) $K = 1/[O_2]^{1/2}[SO_2]$;
   (h) $K = [SO_2]^2/[O_2]^3$
10. (a) K_c = 1.6; (b) products favored ($K > 1$); (c) $K_{reverse}$ = 0.63
12. (a) K = 870; (b) $K_p = K_c$
14. K_3 = 0.14 and $K_4 = 1/K_3$ = 7.3
16. $K_{net} = K_1K_2K_3$
18. (b) Slope = 8.33 = reciprocal of K for hex $\rightleftharpoons$ pent equilibrium; (c) when [hex] = 4.5, then value of [pent] from the graph is 0.54, the same as that calculated from K.

20. (a) $K_c = 1.6$; (b) $[H_2O] = [CO] = 11.2$ moles
22. (a) $K_c = 0.035$; (b) original $[H_2O] = 5.5\ M$
24. $K_c = K_p = 0.0156$
26. Q (= 0.5) is greater than K. Therefore, NO and Br_2 are consumed and produce more NOBr.
28. (a) Not at equilibrium; Q_c (= 1.67) > K_c
(b) Since $Q_c > K_c$, some NO_2 is converted to NO and O_2.
30. $[Zn^{2+}] = 3.3 \times 10^{-11}\ M$
32. $[Zn^{2+}] = [CO_3^{2-}] = 3.9 \times 10^{-6}$
33. $K_p = 2.31 \times 10^{-4}$
35. $K_c = 1.08 \times 10^7$; $K_p = 1.35 \times 10^5$
37. 4.4 g CO_2
39. $P(C_2N_2) = P(H_2) = 0.481$ atm and $P(HCN) = 24.0$ atm
41. $[NH_3] = 1.49$, $[H_2] = 0.459$, $[N_2] = 0.153$
43. (a) shifts left; (b) shifts left; (c) shifts right; (d) shifts right
45. (a) $[n] = 1.1\ M$ and [iso] = 2.9 M; (b) $[n] = 1.1\ M$ and [iso] = 2.9 M
47. Final pressure of CO_2 is greater than 0.05 atm. Adding CO_2 must lead to a decline in $P(NH_3)$.
49. $K_c = 4.1$, so $[PCl_5] = 0.070\ M$
50. (a) 82% conversion to dimer; (b) less dimer is present at higher temperatures.
52. (a) $K_p = K_c = 4.0 \times 10^{-4}$; (b) Q_c (= 2.8×10^{-4}) < K_c, so the system is not at equilibrium; (c) More NO is produced from N_2 and O_2; that is, the reaction shifts to the right. (d) $[N_2] = [O_2] = 0.25\ M$ and $[NO] = 0.0050\ M$.
54. (a) The complex with NH_3; (b) $P(NH_3) = P(BH_3) = 2.1$ atm, $P(H_3N{-}BH_3) = 0.96$ atm; total pressure = 5.2 atm. (c) 69% of $H_3N{-}BH_3$ has dissociated; (d) $P(BH_3) = P[(CH_3)_3N] = 0.99$ atm and $P[(CH_3)_3N{-}BH_3] = 2.07$ atm, so the % dissociation is 32%.
57. (a) Q_c (= 2.1×10^5) < K_c so the system is not at equilibrium. More products (O_2 and NO_2) will form; that is, the reaction proceeds to the right. (b) Using enthalpies of formation, the enthalpy of reaction is found to be −199.8 kJ. Since the reaction is exothermic, an increase in temperature will shift the equilibrium away from products; their concentrations will decrease.

CHAPTER 15

6. NH_4^+, acid; NH_3, conjugate base. CN^-, base; HCN, conjugate acid.
9. $[Ni(H_2O)_6]^{2+}(aq) \rightleftharpoons [Ni(H_2O)_5OH]^+(aq) + H^+(aq)$

$$K = \frac{\{[Ni(H_2O)_5OH]^+\}[H^+]}{\{[Ni(H_2O)_6]^{2+}\}}$$

15. $K_2CO_3(s) \xrightarrow{H_2O} 2\ K^+(aq) + CO_3^{2-}(aq)$
Basic solution by carbonate ion hydrolysis:
$CO_3^{2-}(aq) + H_2O(\ell) \rightleftharpoons HCO_3^-(aq) + OH^-(aq)$
17. Acid: $HPO_4^{2-}(aq) \rightleftharpoons H^+(aq) + PO_4^{3-}(aq)$
Base: $HPO_4^{2-}(aq) + H_2O(\ell) \rightleftharpoons H_2PO_4^-(aq) + OH^-(aq)$
18. $H^-(aq) + H_2O(\ell) \rightarrow H_2(g) + OH^-(aq)$; solution is basic.
20. (a) $NaOH(aq) + H_2SO_4(aq) \rightarrow NaHSO_4(aq) + H_2O(\ell)$
(b) $2\ NaOH(aq) + H_2SO_4(aq) \rightarrow Na_2SO_4(aq) + 2\ H_2O(\ell)$
21. (a) CN^-, cyanide; (b) SO_4^{2-}, sulfate; (c) F^-, fluoride; (d) NO_2^-, nitrite; (e) NH_2^-, amide.
23. (a) Br^-, base; HBr, conjugate acid. (b) $[Al(H_2O)_6]^{3+}$, acid; $[Al(H_2O)_5OH]^{2+}$, conjugate base. (c) H_3PO_4, acid; $H_2PO_4^-$, conjugate base. (d) CH_3COO^-, base; CH_3COOH, conjugate acid.
25.

	Acid	Conjugate Base	Base	Conjugate Acid
(a)	HCOOH	$HCOO^-$	H_2O	H_3O^+
(b)	H_2S	HS^-	NH_3	NH_4^+
(c)	HSO_4^-	SO_4^{2-}	OH^-	H_2O

27. $NH_4^+(aq) + H_2PO_4^-(aq) \rightleftharpoons NH_3(aq) + H_3PO_4(aq)$
H_3PO_4 is a stronger acid than NH_4^+ (and NH_3 is a stronger base than $H_2PO_4^-$), so the equilibrium lies predominantly to the left.
30. (a) HF, strongest acid; HS^-, weakest acid; (b) F^-; (c) HF; (d) HS
32. (a) NH_3, strongest base; C_5H_5N, weakest base; (b) $C_5H_5NH^+$; (c) C_5H_5N has strongest conjugate acid; NH_3 has weakest conjugate acid.
34. SO_4^{2-} is the weakest base because, according to Table 15.4, it has the strongest conjugate acid.
36. (a) right; (b) right; (c) left; (d) left
38. $[H^+] = 4.0 \times 10^{-4}$, acidic
40.

	pOH	$[OH^-]$
(a)	13.00	1.0×10^{-13}
(b)	3.5	3.2×10^{-4}
(c)	9.11	7.7×10^{-10}
(d)	4.75	1.8×10^{-5}
(e)	6.83	1.5×10^{-7}
(f)	6.65	2.2×10^{-7}
(g)	8.75	1.8×10^{-9}
(h)	12.40	4.0×10^{-13}

42. pH = 6.80
44. phenolphthalein, red; alizarin, red to blue
45. (a) $[H^+] = 1.59 \times 10^{-4}\ M$; (b) moderately weak
47. $[H^+] = [OAc^-] = 1.9 \times 10^{-3}\ M$; $[HOAc] = 0.20\ M$
49. $[H^+] = 1.5 \times 10^{-6}\ M$; pH = 5.83
51. pH of nicotinic acid (= 3.43) > pH barbituric acid (= 3.00)

52. (a) Increasing strength: HOBr < valeric < glutaric. (b) Increasing pH: glutaric < valeric < HOBr

54. Quadratic equation is required (100K = 0.13 > $[HA]_o$). $[H^+] = 3.0 \times 10^{-3}$ *M* and pH = 2.52

56. $[OH^-] = [NH_4^+] = 1.6 \times 10^{-3}$ *M* and $[NH_3]$ = 0.15 *M*; pH = 11.22

57. $[OH^-] = 1.1 \times 10^{-2}$ *M*; pH = 12.05 and pOH = 1.95

59. pH = 9.26 and pOH = 4.74

61. $K_a = 1.4 \times 10^{-3}$

63. $K_a = 1.6 \times 10^{-5}$

65. [HF] at equilibrium = 3.5×10^{-2} *M* and [HF] originally present = 4.0×10^{-2} *M*. At equilibrium $[H^+] = [F^-] = 5.0 \times 10^{-3}$ *M*.

67. (b) sodium benzoate and (c) Na_2HPO_4

69. (a) $AlCl_3$, pH < 7; (b) NH_4Br, < 7; (c) Na_2S, > 7; (d) NaF, > 7; (e) NaH_2PO_4, < 7; (f) $NaNO_3$, neutral; (g) $KClO_4$, neutral

71. $[H^+] = 3.5 \times 10^{-9}$ *M*

73. $[H^+] = 1.2 \times 10^{-9}$ *M* and $[OH^-] = 8.4 \times 10^{-6}$ *M*; pH = 8.93

75. $K_a = 1.3 \times 10^{-5}$

78. (a) pH = 4.00; (b) $[H_2S]$ decreases and $[S^{2-}]$ increases

79. $K_{total} = 3.8 \times 10^{-6}$

80. (a) $[OH^-] = [N_2H_5^+] = 9.2 \times 10^{-5}$ *M*; $[N_2H_6^{2+}] = K_{b2} = 8.9 \times 10^{-16}$ *M*; (b) pH = 9.96

83. (a) pH = 1.17; (b) $[SO_3^{2-}] = K_{a2} = 6.2 \times 10^{-8}$ *M*

84. (a) $H_3PO_4 > H_3PO_3$; (b) $H_2SeO_4 > H_2SeO_3$

86. strongest, ClOH; weakest, LiOH

88. (a) acid; (b) base; (c) base; (d) base; (e) acid

90. SbF_5 is a Lewis acid

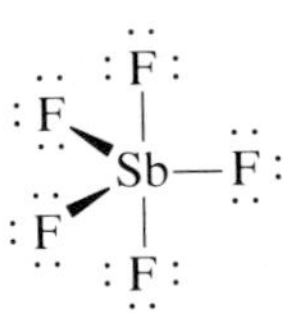

92. Lewis base (electron pair donor)

94. Lewis base

95. CN^-

97. $[Cd^{2+}] = 1.2 \times 10^{-6}$ *M*; $[Cl^-] = 0.96$ *M*; $\{[CdCl_4^{2-}]\}$ = 0.010 *M*

98. $[Ag^+] = 7.6 \times 10^{-8}$ *M*

100. 400. mL

102. Increasing pH: HCl < $HC_2H_3O_2$ < NaCl < $NaC_2H_3O_2$ < NH_3 < KOH

104. NaOH and $Ca(OH)_2$ can be described as Arrhenius bases, while CO_3^{2-}, CN^-, and $C_2H_5O^-$ are best described as proton acceptors, that is, Brønsted bases.

105. $K_a = 3 \times 10^{-5}$

107. pH = 8.64

108. $P_{total} = P(BF_3) + P[(C_2H_5)_2O] + P[(C_2H_5)_2O\!:\!BF_3]$ = 0.13 atm + 0.13 atm + 0.090 atm = 0.35 atm

CHAPTER 16

2. (a) acidic, pH < 7; (b) neutral, pH = 7; (c) basic, pH > 7

6. $[H^+] = 1.9 \times 10^{-9}$ *M*; pH = 8.72

8. (a) pH > 7; (b) pH < 7; (c) pH = 7; (d) pH > 7; (e) pH < 7

10. $[Na^+] = [C_6H_5O^-] = 3.47 \times 10^{-2}$ *M*; $[OH^-] = 1.63 \times 10^{-3}$ *M* and $[H^+] = 6.12 \times 10^{-12}$ *M*; pH = 11.213

12. pH = 3.56

14. (a) pH = 2.78; (b) 83 g $NaC_2H_3O_2$

16. (a) pH decreases (by about 0.1); (b) pH decreases (to about 9); (c) no effect on pH; (d) pH decreases (to about 7)

18. 4.8 g NH_4Cl

20. pH = 3.90. pH does not change on diluting a buffer solution.

23. 1.0×10^2 g $NaNO_2$

24. (a) pH = 9.26; (b) dilution has no effect on pH.

25. (a) pH = 9.56; (b) pH = 9.54

27. 0.082 mol of strong base would be required to increase the pH by 1.00 unit, for example.

29. pH = 5.23

32. (a) pH of original solution = 11.15; (b) after 5.00 mL HCl, pH = 9.79; (c) after 11.0 mL HCl, half-neutralization point, pH = 9.26; (d) after 15.0 mL HCl, pH = 8.92; (e) after 20.0 mL HCl, pH = 8.26; (f) after 22.0 mL HCl, equivalence point, pH = 5.28; (g) after 25.0 mL HCl, pH = 2.18. A suitable indicator would be methyl red.

34. (a) 7.48×10^{-2} *M*; (b) pH = 11.37; (c) pH at half-neutralization point = 9.87; (d) 37.4 mL of acid are required to neutralize the amine; (e) pH at equivalence point = 5.62; (f) bromcresol green would be a suitable indicator.

35. The table below gives the approximate pH at the equivalence point, assuming that the weak acid or base at that point is 0.10 *M*. (For a calculation of the necessary K_a or K_b values see page 581.)

	Weak Acid or Base Produced in Titration	Approximate pH	Indicator
(a)	$C_5H_5NH^+$	3.1	thymol blue
(b)	$HCOO^-$ (formate)	8.4	phenol red
(c)	NH_4^+, $HCOO^-$	7	bromthymol b
(d)	$N_2H_5^+$	4.5	bromphenol bl
(e)	HCO_3^-	9.7	phenolphthale

38. $[Na^+] = [OH^-] = 0.040\ M$; $[NH_3] = 0.20\ M$; $[NH_4^+] = 9.0 \times 10^{-5}\ M$
40. Lactate ion
42. 11%
44. (a) HB is a stronger acid than HA. (b) A^- is a stronger base than B^-.
45. 4.40
49. 1×10^{-11}

CHAPTER 17

10. (a) soluble, (b) insoluble, (c) insoluble, (d) soluble, (e) insoluble, (f) insoluble, (g) insoluble
12. $CuSO_4(aq) + Na_2S(aq) \rightarrow Na_2SO_4(aq) + CuS(s)$
16. $Mg_3(AsO_4)_2$
18. 5.9×10^{-21}
20. 7.9×10^{-6}
22. 8.1×10^{-12}
24. (a) $1.1 \times 10^{-8}\ M$; (b) 1.5×10^{-6} g/L
26. 2.7×10^{-7} mg Au^{3+}/L; $[I^-] = 5.3 \times 10^{-7}$ mg/L.
28. 2.7 mg
30. Aragonite
32. $BaCO_3$ (solubility $= S = 9.0 \times 10^{-5}\ M) < Ag_2CO_3$ $(S = 1.3 \times 10^{-4}\ M) < BaF_2$ $(S = 7.5 \times 10^{-3}\ M)$
34. 2.4 mg
36. (a) $Q < K_{sp}$, no precipitate; (b) $Q > K_{sp}$, $NiCO_3$ precipitates
38. (a) and (b) no precipitate forms because $Q < K_{sp}$. (c) $Q > K_{sp}$, so precipitate of $PbCl_2$ forms.
40. 4.1 mg
42. (a) 5.5×10^{-23} g Na_2S; (b) 6.3×10^{-3} g NaOH; (c) 1.4×10^{-5} g $Na_2SO_4.10H_2O$; (d) 4.1×10^{-6} g Na_2HPO_4.
44. $Q = 4.7 \times 10^{-7} > K_{sp}$ of $Mg(OH)_2$. Precipitation occurs.
46. $1.5 \times 10^{-8}\ M\ Ag^+$
48. $1.4 \times 10^{-4}\ M\ F^-$
50. $Q = [Ag^+][Cl^-] = (0.50 \times 10^{-5})(3.0 \times 10^{-5}) = 1.5 \times 10^{-10} < K_{sp}$; No precipitation; undersaturated
52. NiS
54. First $Fe(OH)_3$, then $Al(OH)_3$, and $Pb(OH)_2$ last
56. CaF_2 first, then MgF_2, and BaF_2 last
57. (a) $1.0 \times 10^{-6}\ M$; (b) $1.0 \times 10^{-10}\ M$
59. (a) 1.3 g/L; (b) 0.040 g/L
61. (a) $1.3 \times 10^{-5}\ M$; (b) $1.0 \times 10^{-13}\ M$
63. (a) $[F^-] = 1.1 \times 10^{-2}\ M$; 45 mg NaF; (c) $[Ba^{2+}] = 1.2 \times 10^{-4}\ M$
65. $1.2 \times 10^{-7}\ M$
67. (a) $PbSO_4$; (b) $[Pb^{2+}]$ remaining $= 7.5 \times 10^{-6}\ M$
69. (a) Neither $Pb(OH)_2$ nor $Zn(OH)_2$ precipitates ($Q < K_{sp}$ for both);
(b) $1.7 \times 10^{-5}\ M$
71. $AgBr(s) \rightleftharpoons Ag^+(aq) + Br^-(aq)$
$Ag^+(aq) + 2\ S_2O_3^{2-}(aq) \rightleftharpoons [Ag(S_2O_3)_2]^{3-}(aq)$
$AgBr(s) + 2\ S_2O_3^{2-}(aq) \rightleftharpoons [Ag(S_2O_3)_2]^{3-}(aq) + Br^-(aq)$
$K_{net} = K_{sp}K_{form} = 6.6$
74. $K_{net} = 1.4 \times 10^{-10}$ for $ZnS(s) + 2\ CN^-(aq) \rightleftharpoons Zn(CN)_2(s) + S^{2-}(aq)$; K_{net} is too small for an appreciable amount of ZnS to dissolve.
75. 0.0001 moles of AgCl will dissolve but 0.0005 moles of AgBr will not.
77. $Ba(OH)_2$, $BaCO_3$
79. Add H_2SO_4 to $CaCO_3$ to form $CaSO_4$, CO_2, and H_2O.
82. Add NaOH to $FeCl_3$ to form insoluble $Fe(OH)_3$ (see Special Section on Iron Chemistry in Chapter 25). Next, add perchloric acid, $HClO_4$, to the insoluble $Fe(OH)_3$ to give $Fe(ClO_4)_3$ and H_2O.
84. (a) Add HCl. If CO_3^{2-} is present, then CO_2 bubbles out of solution.
(b) Add $AgNO_3$ to precipitate either AgCl or $AgCO_3$. If $AgCO_3$ has formed, the precipitate will dissolve in exccess HCl. In contrast, AgCl does not dissolve in HCl.
86. (a) $K_{dissolve} = 8.5 \times 10^{-?}$; (b) $[Zn^{2+}] = 7.6 \times 10^{-2}\ M$
88. 1.9 mg AgCl dissolves in pure water; after adding NaCl, 1.1 mg dissolves. Therefore, 0.8 mg of AgCl precipitates on adding NaCl.
90. 4.5×10^{-17}
93. (a) No; only 9.1 mg of $Mg(OH)_2$ can dissolve in pure water. (b) Yes; 8.75 g $Mg(OH)_2$ can dissolve per liter when pH is buffered at 5.

CHAPTER 18

6. (a) CO_2 vapor; (b) dissolved sugar; (c) flask of argon gas; (d) mixed gases
8. (a) $MgCl_2$; (b) P_4; (c) $(CH_3)_2NH$; (d) $Hg(\ell)$
9. (a) −3.363 J/K; (b) −102.50 J/K; (c) 99 J/K; (d) 75.3 J/K
10. Adding one CH_2 increases entropy about 40 J/K.mol. Therefore, S° for C_4H_{10} is about 350 J/K.mol.
13. (a) +84.4 J/K; (b) −84.4 J/K
15. (a) 28.0 kJ/mol; (b) 31.3 kJ/mol
17. −270.1 J/K
20. (a) −108.31 J/K; (b) −151.9 J/K; (c) −155.89 J/K; (d) −138.9 J/K; (e) 43.4 J/K
23. (a) 58.78 J/K·mol; (b) −53.29 J/K·mol; (c) −179.93 J/K·mol. As more H_2 is used, the entropy change becomes more negative because more gas molecules are being used to produce only one molecule of gaseous product.

24. (a) Splitting of water: $H_2O(liq) \longrightarrow H_2(g) + \frac{1}{2} O_2(g)$ A liquid is producing 1.5 moles of gas. Therefore, the entropy change is positive (+163 J/K; Question 18a). Energy is required to split water, so ΔH is positive. (Here the energy required is the opposite of the enthalpy of formation, that is, +285.83 kJ/mol.) ΔG° is positive.
(b) Dissolving NH_4Cl: $NH_4Cl(s) \longrightarrow NH_4Cl(aq)$ A regular solid gives ions, randomly moving in solution. Thus, the entropy change is positive. Since the solution becomes cold, this means the process is endothermic; it has a positive enthalpy change. ΔG° is negative for spontaneous process.
(c) Liquid nitroglycerin ⟶ gases The explosion gives off gas and heat. Therefore, the entropy change is positive, but the enthalpy change is negative. ΔG° is negative.

25. (a) $\Delta H^\circ_{rxn} = -359.41$ kJ and $\Delta S^\circ_{rxn} = -151.9$ J/K; $\Delta G^\circ = -314.1$ kJ
(b) $\Delta H^\circ_{rxn} = -601.70$ kJ and $\Delta S^\circ_{rxn} = -108.31$ J/K; $\Delta G^\circ = -569.42$ kJ
(c) $\Delta H^\circ_{rxn} = -176.01$ kJ and $\Delta S^\circ_{rxn} = -284.8$ J/K; $\Delta G^\circ = -91.15$ kJ
(d) $\Delta H^\circ_{rxn} = -511.20$ kJ and $\Delta S^\circ_{rxn} = -175.51$ J/K; $\Delta G^\circ = -458.9$ kJ

28. 2 Fe(s) + 3/2 $O_2(g) \rightarrow Fe_2O_3(s)$. $\Delta G^\circ_f[Fe_2O_3(s)] = -742.2$ kJ/mol or -2110 kJ for 454 g

30. (a) $\Delta H^\circ = -82.4$ kJ; (b) $\Delta S^\circ = 460.0$ J/K; (c) $\Delta G^\circ = -219.1$ kJ

32. (a) -1010.8 kJ; (b) -990.41 kJ; (c) -566.60 kJ

34. (a) $\Delta H^\circ = -125.52$ kJ; (b) $\Delta S^\circ = -129.9$ J/K; (c) $\Delta G^\circ = -86.81$ kJ

37. (a) $\Delta H^\circ = 126.9$ kJ;
(b) $\Delta S^\circ = 81$ J/K;
(c) $\Delta G^\circ = +102.8$ kJ. The reaction is not spontaneous at room temperature; the enthalpy change is too positive to be outweighed by $-T\Delta S$.

39. $K_p = 4.5 \times 10^{-31}$. Reactants favored.

41. $\Delta G^\circ = +33.6$ kJ

43. (a) $\Delta G^\circ = -290.31$ kJ and the reaction is spontaneous; (b) $K_p = 7.74 \times 10^{50}$

45. (a) $\Delta G^\circ = -97.9$ kJ. The reaction is spontaneous and is enthalpy-driven.
(b) $\Delta G^\circ_f[C_6H_6(g)] = 129.7$ kJ/mol

47. $\Delta G^\circ_{rxn} = -2870$ kJ

49. (a) $\Delta G^\circ_{rxn} = 91.4$ kJ; (b) 9.5×10^{-17}; (c) not spontaneous at 298 K; a temperature of 981 K is required to make it spontaneous; (d) temperature = 1190 K (920°C).

51. (a) The entropy change (162.0 J/K) is positive as expected for a reaction that converts 1 mole of liquid to 1.5 moles of gas. (b) The reaction is not spontaneous at 298 K ($\Delta G^\circ_{rxn} = +115.6$ kJ). (c) It becomes spontaneous at temperatures over 1012 K (739°C).

53. (a) For Ag + H_2O: $\Delta H^\circ = +210.8$ kJ, $\Delta S^\circ = -22$ J/K, and $\Delta G^\circ = 217.4$ kJ. For decomposition of Ag_2O: $\Delta H^\circ = 31.0$ kJ, $\Delta S^\circ = 67$ J/K, and $\Delta G^\circ = 11.1$ kJ
(b) $H_2O(g) \longrightarrow H_2(g) + 1/2\ O_2(g)$
(c) $\Delta H^\circ = 241.8$ kJ
(d) $\Delta G^\circ = 228.5$ kJ; overall reaction is not spontaneous.
(e) 464 K

CHAPTER 19

1. (a) Al, oxidized, reducing agent; Cl_2, reduced, oxidizing agent
(b) Fe^{2+}, oxidized, reducing agent, MnO_4^-, reduced, oxidizing agent
(c) FeS, oxidized, reducing agent; NO_3^-, reduced, oxidizing agent

5. (c), reduce Cd^{2+} and Ag^+

12. (a,b) $Cu(s) \longrightarrow Cu^{2+}(aq) + 2e^-$; oxidation, at anode
$Ag^+(aq) + e^- \longrightarrow Ag(s)$; reduction, at cathode
(c) The copper electrode is the site of electron generation and so is the negative electrode.

14. $\Delta G^\circ = -89$ kJ

16. (a) $E^\circ = +1.49$ V, spontaneous; (b) $E^\circ = -0.51$ V, not spontaneous

18. (a) Al^{3+} is the weakest oxidizing agent.
(b) Ce^{4+} is the strongest oxidizing agent.
(c) Al(s) is the strongest reducing agent.
(d) Ce^{3+} is the weakest reducing agent.
(e) Sn(s) is lower in the table than Ag. Therefore, Sn(s) will reduce Ag^+ to Ag(s).
(f) Sn(s) is lower in the table than Hg(liq). Therefore, Hg(liq) cannot reduce Sn^{2+} to Sn(s).
(g) Sn(s) can reduce Ag^+, Hg_2^{2+}, and Ce^{4+}.
(h) Ag^+ can oxidize Sn, Ni, and Al.

20. (a) A or C + $H^+(aq) \longrightarrow A^{n+}(aq)$ or $C^{n+}(aq) + H_2$ This means A and C are better reducing agents than H_2.
(b) C + B^{n+} (or either D^+ or A^+) $\longrightarrow C^{n+}$ + B (or either D or A) This means C is a better reducing agent than A, B, or D.
(c) D + $B^{n+} \longrightarrow$ B + D^{n+} D is a better reducing agent than B.
The results above mean that the metals are in the following order of reducing ability: $B < D < H_2 < A < C$

22. (a) $Al(s) + 3\ Ag^+(aq) \longrightarrow Al^{3+}(aq) + 3\ Ag(s)$, $E^\circ = 2.46$ V

(b) Al

(c) Al is the anode and Ag the cathode; $Al(s)|Al^{3+}||Ag^+(aq)|Ag(s)$

(d) NO_3^- migrate from the Ag^+-containing compartment to the Al^{3+}-containing compartment.

24. (a) Ni is a better reducing agent than Pb.

(b) $Ni(s) + Pb^{2+}(aq) \longrightarrow Ni^{2+}(aq) + Pb(s)$; $E^\circ = +0.12\ V$

(c) Cathode = Pb

(d) Electrons flow from Ni to Pb in the external circuit.

(e) $NO_3^-(aq)$ ions flow from the Pb^{2+}/Pb compartment to the Ni^{2+}/Ni compartment.

26. To maximize E°, react the oxidizing agent Cl_2 with the most powerful reducing agent from the list. In this case it is Mg.

$Mg(s) + Cl_2(g) \longrightarrow Mg^{2+}(aq) + 2\ Cl^-(aq)$

$E^\circ = 3.73\ V$

28. (a) $Zn(s) + Cl_2(g) \longrightarrow ZnCl_2(s)$

(b) $P(Cl_2) = 1.0$ atm, $[ZnCl_2] = 1.0\ M$, temperature usually specified as 25°C.

(c) $E^\circ = 2.12\ V$

(d) $E^\circ = 3.63\ V$

30. $E^\circ = +0.86$ V, reaction spontaneous. $\Delta G^\circ = -170$ kJ.

32. (a) $Ag^+(aq) + Fe^{2+}(aq) \longrightarrow Fe^{3+}(aq) + Ag(s)$

(b) $E^\circ = +0.03\ V$

(c) The cell reaction is now the *reverse* of that in part (a).

34. (a) $E^\circ = -0.513\ V$, $K = 4.7 \times 10^{-18}$; (b) $E^\circ = -0.55\ V$, $K = 3.9 \times 10^{-19}$; (c) $E^\circ = +0.462\ V$, $K = 4.0 \times 10^{15}$.

36. $E^\circ = -0.230\ V$, $K = 1.7 \times 10^{-8}$

38. 5.93 g Cu

40. 3.60×10^3 s (1.00 hr)

42. (a) $[Sn^{2+}] = 1.36\ M$ and $[Cu^{2+}] = 0.642\ M$; (b) $E = +0.47\ V$

44. 8.2 mol of H_2 (which react by $H_2 \rightarrow 2\ H^+ + 2e^-$) will last 2.4×10^6 s or 660 hours.

46. The reaction at the anode is $Zn(s) \rightarrow Zn^{2+}(aq) + 2e^-$, and 18 g of Zn are required.

48. 9.92×10^{-6} mol Ag^+

50. (a) 2.68×10^{-5} mol e^- transferred by 2.69×10^{-5} mol of compound. Therefore, 1 mol e^- transferred per mol $C_{10}H_{10}Fe$.

(b) The electrolysis occurred at an anode. Therefore, it was an oxidation. The balanced equation is $C_{10}H_{10}Fe \rightarrow C_{10}H_{10}Fe^+ + e^-$

52. (a) Anode (oxidation of bromide):

$2\ Br^-(aq) \longrightarrow Br_2(aq) + 2e^-$

Cathode (reduction of water):

$2\ H_2O(\ell) + 2e^- \longrightarrow H_2(g) + 2\ OH^-(aq)$

(b) Anode (oxidation of I^-):

$2\ I^-(aq) \longrightarrow I_2(aq) + 2e^-$

Cathode (reduction of water):

$2\ H_2O(\ell) + 2e^- \longrightarrow H_2(g) + 2\ OH^-(aq)$

(c) Anode (oxidation of fluoride):

$2\ F^- \longrightarrow F_2(g) + 2e^-$

Cathode (reduction of sodium):

$Na^+ + e^- \longrightarrow Na(s)$

(d) Anode (oxidation of water):

$2\ H_2O(\ell) \longrightarrow O_2(g) + 4\ H^+(aq) + 4e^-$

(This occurs because it is easier than the oxidation of F^-.) Cathode (reduction of water):

$2\ H_2O(\ell) + 2e^- \longrightarrow H_2(g) + 2\ OH^-(aq)$

(e) Anode (oxidation of chloride):

$2\ Cl^-(aq) \longrightarrow Cl_2(g) + 2e^-$

Cathode (reduction of nickel ion):

$Ni^{2+}(aq) + 2e^- \longrightarrow Ni(s)$

54. 6700 kwh; 8.2×10^5 g Na; 1.3×10^6 g Cl_2.

55. (a) $Cd(s) + Ni^{2+}(aq) \longrightarrow Cd^{2+}(aq) + Ni(s)$

(b) Cd is oxidized and is the reducing agent. $Ni^{2+}(aq)$ reduced and is the oxidizing agent.

(c) Oxidation occurs at the (negative) Cd anode.

(d) $Cd(s)|Cd^{2+}(aq)||Ni^{2+}(aq)|Ni(s)$

(e) Electron flow is from the Cd electrode to the Ni electrode.

(f) NO_3^- ions migrate from the Ni/Ni^{2+} compartment to the Cd/Cd^{2+} compartment.

(g) $K = 1.2 \times 10^5$ (for $E^\circ = +0.15$ V).

(h) $E = +0.21\ V$; cell reaction is still spontaneous in the original direction.

(i) The "limiting reagent" with respect to the length of time the battery can operate is Cd. It will last 1.7×10^6 s (= 480 hours).

57. (a) 3.6 mol glucose require 22 mol O_2.

(b) 22 mol O_2 require 88 mol e^- ($4e^- + O_2 \rightarrow 2\ O^{2-}$)

(c) 8.5×10^6 C/day or 98 C/s

(d) 98 J/s or 98 watts

60. (a) $N_2H_4 + 4\ OH^-$ is an oxidation; occurs at the anode. $O_2 + 2\ H_2O$ is a reduction; occurs at the cathode.

(b) $N_2H_4(aq) + O_2(g) \longrightarrow N_2(g) + 2\ H_2O(\ell)$

(c) 7.5 g N_2H_4

(d) 7.5 g O_2

CHAPTER 20

1. (a) $2\ K(s) + H_2(g) \longrightarrow 2\ KH(s)$

(b) $Cl_2(g) + H_2(g) \longrightarrow 2\ HCl(g)$
(c) $S_8(s) + 8\ H_2(g) \longrightarrow 8\ H_2S(g)$
(d) $C_3H_8(g) + 3\ H_2O(g) \longrightarrow 3\ CO(g) + 7\ H_2(g)$
(e) $2\ K(s) + 2\ H_2O(\ell) \longrightarrow 2\ KOH(aq) + H_2(g)$

3. (a) 0.38 g H_2/g CH_4; (b) 0.29 g H_2/g CH_2; (c) 0.23 g H_2/g CH

5. (a) $2\ NaCl(aq) + 2\ H_2O(\ell) \longrightarrow 2\ NaOH(aq) + H_2(g) + Cl_2(g)$
(b) Electrolysis of NaCl should give 0.89 g Cl_2/1.0 g NaOH. Actual production is 0.97 g Cl_2/1.0 g NaOH. The reason that more Cl_2 is actually produced than expected from NaCl electrolysis is that some Cl_2 comes from the HCl and metals industries.

6. 1.2×10^4 L (or a cubic container about 90 inches on a side).

8. (a) $2\ KCl(aq) + 2\ H_2O(\ell) \longrightarrow 2\ KOH(aq) + H_2(g) + Cl_2(g)$. $Cl^-(aq)$ is oxidized to Cl_2 and H_2O is reduced to H_2.
(b) $2\ CsI(aq) + 2\ H_2O(\ell) \longrightarrow 2\ CsOH(aq) + H_2(g) + I_2(aq)$

13. Be is *sp* hybridized.

14. Be is sp^3 hybridized.

H_2O | Be $^{2+}$ — H_2O, H_2O, OH_2

16. $2\ BaO(s) + O_2(g) \longrightarrow 2\ BaO_2(s)$

18. (a) $4\ Li(s) + O_2(g) \longrightarrow 2\ Li_2O(s)$
(b) $2\ Na(s) + O_2(g) \longrightarrow Na_2O_2(s)$ (sodium peroxide). Some Na_2O forms as well.
(c) $K(s) + O_2(g) \longrightarrow KO_2(s)$ (potassium superoxide). Some K_2O and K_2O_2 form as well.

20. (a) $SrCO_3(s) + 2\ HCl(aq) \longrightarrow SrCl_2(aq) + H_2O(\ell) + CO_2(g)$
(b) $BaCO_3(s) + 2\ HNO_3(aq) \longrightarrow Ba(NO_3)_2(aq) + H_2O(\ell) + CO_2(g)$
(c) $3\ BaO(s) + 2\ Al(s) \longrightarrow Al_2O_3(s) + 3\ Ba(s)$
(d) $LiH(s) + H_2O(\ell) \longrightarrow LiOH(aq) + H_2(g)$
(e) $Na_2CO_3(aq) + H_2O(\ell) + CO_2(g) \longrightarrow 2\ NaHCO_3(aq)$

24. (a) 255.3 kg Mg; (b) 744.7 kg $Cl_2(g)$; (c) 2.100×10^4 Faradays; (d) 1600 kJ/mol Mg; (e) 641 kg/mol Mg

25. (a) $\Delta H° = 6$ kJ; (b) assuming that $\Delta H° = \Delta G°$, then $K = 0.5 = P_K/P_{Na}$. This means there is about half as many moles of K at equilibrium as moles of Na.

27. $\Delta H° = 229.6$ kJ

31. 1.142×10^6 g SO_2 (or about 1.3 tons)

33. A = $BaCO_3$ (1.00 g = 0.00507 mol, which gives 0.00507 mol CO_2 on heating. This is the amount calculated from PVT data); B = BaO (a basic oxide); C = $CaCO_3$; D = $BaCl_2$; E = $BaSO_4$.

CHAPTER 21

1. (a) $4\ Ga(s) + 3\ O_2(g) \longrightarrow 2\ Ga_2O_3(s)$
(b) $2\ In(s) + 3\ Br_2(\ell) \longrightarrow 2\ InBr_3(s)$
(c) $2\ BBr_3(g) + 3\ H_2(g) \longrightarrow 2\ B(s) + 6\ HBr(g)$ (requires high temperature)
(d) $B(OH)_3(s) + 3\ CH_3OH(\ell) \longrightarrow B(OCH_3)_3(\ell) + 3\ H_2O(\ell)$
(e) $BF_3(g) + 3\ H_2O(\ell) \longrightarrow B(OH)_3(s) + 3\ HF(g)$
(f) $B_2O_3(s) + 6\ HF(g) \longrightarrow 2\ BF_3(g) + 3\ H_2O(g)$
(g) $2\ NaBH_4(s) + I_2(s) \longrightarrow B_2H_6(g) + 2\ NaI(s) + H_2(g)$
(h) $2\ B_5H_9(g) + 12\ O_2(g) \longrightarrow 5\ B_2O_3(s) + 9\ H_2O(\ell)$

3. (a) $SiO_2(s) + 3\ C(s) \longrightarrow SiC(s) + 2\ CO(g)$
(b) $Si(s) + 2\ Cl_2(g) \longrightarrow SiCl_4(\ell)$
(c) $2\ PbS(s) + 3\ O_2(g) \longrightarrow 2\ PbO(s) + 2\ SO_2(g)$
(d) $PbO(s) + CO(g) \longrightarrow Pb(s) + CO_2(g)$
(e) $CCl_4(\ell) + HF(\ell) \longrightarrow CFCl_3(g) + HCl(g)$
(f) $Si(s) + 2\ CH_3Cl(g) \longrightarrow (H_3C)_2SiCl_2(\ell)$

5. (a) Cyanamide: used as a fertilizer (and also in the plastics industry)
(b) HCN; used in the polymer industry; some used in metals extraction
(c) boron carbide: used in bulletproof armor
(d) $Al_2(SO_4)_3$: used in the paper industry and in water treatment
(e) $B(OH)_3$: used in the manufacture of glass
(f) $NaBH_4$: reducing agent
(g) carbon black: used in rubber tires and printing inks

7. (a) $BCl_3(g) + 3\ H_2O(\ell) \longrightarrow B(OH)_3(s) + 3\ HCl(aq)$
(b) $\Delta H°_{rxn} = -205$ kJ

9. $2\ NaBH_4(s) + I_2(s) \rightarrow B_2H_6(g) + 2\ NaI(s) + H_2(g)$. I_2 is reduced to I^- and H^- (in BH_4^-) is oxidized to H_2.

11. B_2Cl_4 has two, sp^2 hybridized B atoms. The geometry about these atoms is planar and triangular.

:Cl̈ Cl̈: — B—B — :Cl̈ Cl̈:

13. $4\ BCl_3(g) + 6\ H_2(g) + C(s) \longrightarrow B_4C(s) + 12\ HCl(g)$

17. Solid gallium has a lower density than liquid gallium and so will float on the liquid. The S/L line in the phase diagram should have a negative slope (as in water).

18. $2\ Al(s) + 2\ OH^-(aq) + 6\ H_2O(\ell) \longrightarrow 2\ Al(OH)_4^-(aq) + 3\ H_2(g)$

20. (a) $Al_2O_3(s) + 3\ SiO_2(s) \longrightarrow Al_2(SiO_3)_3(s)$
(b) $CaO(s) + Al_2O_3(s) \longrightarrow Ca(AlO_2)_2(s)$

22. (a) tetrahedral
(b) The geometry about both N and B in $F_3B{-}NH_3$ is tetrahedral. Both atoms are sp^3 hybridized.

```
      ..
     :F:  H
      |   |
 ..   |   |
:F—B—N—H
 ..   |   |
      |   |
     :F:  H
      ..
```

See the structure of the complex on page 590.

24. (a) 2.50×10^6 g; (b) 7.13×10^5 cm^3.

25. (a) $CaC_2(s) + 2\ H_2O(\ell) \longrightarrow Ca(OH)_2(s) + C_2H_2(g)$. See Figure 9.4 (page 313).
(b) $:C{\equiv}C:^{2-}$

27. (a) Octahedral holes; see Figure 11.28; (b) 4 Ca^{2+} and 4 C_2^{2-} ions per unit cell

29. (a) $Cu_2O(s) + C(s) \longrightarrow 2\ Cu(s) + CO(g)$
(b) $TiO_2(s) + 2\ C(s) \longrightarrow Ti(s) + 2\ CO(g)$
(c) $FeO(s) + C(s) \longrightarrow Fe(s) + CO(g)$

31. C atom hybridization is sp^2.

34. 2×10^5 tons of SiO_2 required

37. 156,810 tons PbO

40. $[Ag^+] = 7.6 \times 10^{-8}\ M$

CHAPTER 22

2. (a) $3\ Mg(s) + N_2(g) \rightarrow Mg_3N_2(s)$
(b) $P_4(s) + 3\ KOH(aq) + 3\ H_2O(\ell) \rightarrow PH_3(g) + 3\ KH_2PO_2(aq)$
(c) $CH_4(g) + H_2O(g) \rightarrow CO(g) + 3\ H_2(g)$ (high temperature)
(d) $HNO_3(aq) + KOH(aq) \rightarrow KNO_3(aq) + H_2O(\ell)$ (acid-base reaction)
(e) $2\ NH_3(aq) + OCl^-(aq) \rightarrow N_2H_4(aq) + Cl^-(aq) + H_2O(\ell)$ (the Raschig process; done in the presence of aqueous alkali and gelatin)
(f) $NH_4NO_3(s) + \text{heat} \rightarrow N_2O(g) + 2\ H_2O(g)$
(g) $NaNO_2(aq) + HCl(aq) \rightarrow NaCl(aq) + HNO_2(aq)$ (acid-base reaction)
(h) $4\ NH_3(g) + 5\ O_2(g) \rightarrow 4\ NO(g) + 6\ H_2O(g)$ (catalyzed by Pt)
(i) $3\ Cu(s) + 8\ HNO_3(aq, \text{dilute}) \rightarrow 3\ Cu(NO_3)_2(aq) + 4\ H_2O(\ell) + 2\ NO(g)$
(j) $P_4O_{10}(s) + 6\ H_2O(\ell) \rightarrow 4\ H_3PO_4(aq)$

4. (a) $N_2H_4(aq) + O_2(g) \rightarrow N_2(g) + H_2O(\ell)$
(b) 1.32×10^3 g N_2H_4

7. (a) $4\ IO_3^-(aq) + 5\ N_2H_5^+(aq) \rightarrow 2\ I_2(aq) + 5\ N_2(g) + 12\ H_2O(\ell) + H^+(aq)$
(b) $E° = 1.42\ V$

8. (a) 14 tons N_2O_4 required; (b) 6.3 tons N_2, 5.4 tons H_2O, and 6.6 tons CO_2 produced

12. (a) The electron dot structure of N_2O_3 shows that the 114.2 pm bond is a N=O double bond (bond order = 2). The longer NO bonds have a bond order of 1.5.
(b) $\Delta S° = 141$ J/K; $K = 1.90$; and $\Delta H_f°[N_2O_3(s)] = 82.9$ kJ/mol.

14. (a) $P_2O_5(s) + 3\ H_2O(\ell) \rightarrow 2\ H_3PO_4(aq)$
1 mole of P_2O_5 (142 g) give 2 moles of H_3PO_4 (196 g). Therefore, 1.380 g of H_3PO_4 are produced for every 1.000 g of P_2O_5 used.
(b) 14.9 million tons of H_3PO_4.
(c) \$217 per ton H_3PO_4

15. pH = 11.89

17. $Ca(OH)_2(s) + H_3PO_4(aq) \rightarrow CaHPO_4(s) + 2\ H_2O(\ell)$

19. $P_4(s) + 3\ OH^-(aq) + 3\ H_2O(\ell) \rightarrow 3\ H_2PO_2^-(aq) + PH_3(g)$

22. $2\ As(s) + 6\ H_2O(\ell) \rightarrow 2\ H_3AsO_3(aq) + 3\ H_2(g)$

23. (a) $2\ KClO_3(s) + \text{heat (and catalyst)} \rightarrow 2\ KCl(s) + 3\ O_2(g)$
(b) $2\ H_2S(g) + 3\ O_2(g) \rightarrow 2\ SO_2(g) + 2\ H_2O(g)$. The SO_2 produced in this reaction further oxidizes H_2S:

$$16\ H_2S(g) + 8\ SO_2(g) \rightarrow 16\ H_2O(\ell) + 3\ S_8(s).$$

(c) $2\ Na(s) + O_2(g) \rightarrow Na_2O_2(s)$
(d) $K(s) + O_2(g) \rightarrow KO_2(s)$
(e) $2\ ZnS(s) + 3\ O_2(g) \rightarrow 2\ ZnO(s) + 2\ SO_2(g)$
(f) $SO_2(g) + H_2O(\ell) \rightarrow H_2SO_3(aq)$

25. (a) 1.170 V; (b) 1.160 V

28. (a) $H_2O_2(aq) \rightarrow H^+(aq) + HO_2^-(aq)$;
(b) pH = 6.37

30. $BrO_2 < As_2O_3 < GeO_2 < Ga_2O_3 < CaO < K_2O$.

31. (a) CaO is more basic; (b) $SrO(s) + H_2O(\ell) \rightarrow Sr(OH)_2(aq)$

34. (a) 3.92 tons SO_2; (b) 4.53 tons $Ca(OH)_2$ required

36. (a) $UF_4(s) + F_2(g) \rightarrow UF_6(g)$
(b) $2\ Br^-(aq) + Cl_2(aq) \rightarrow Br_2(aq) + 2\ Cl^-(aq)$
(c) $I^-(aq) + Br_2(aq) \rightarrow I_2(aq) + 2\ Br^-(aq)$
(d) $Br^-(aq) + AgNO_3(aq) \rightarrow AgBr(s) + NO_3^-(aq)$
(e) $Cl_2(aq) + 2\ OH^-(aq) \rightarrow Cl^-(aq) + OCl^-(aq) + H_2O(\ell)$

38. I^- oxidized to I_2 and Br_2 reduced to Br^-: $2\ I^-(aq) + Br_2(aq) \rightarrow I_2(aq) + 2\ Br^-(aq)$

41. 0.0937 tons F_2

46. The molecule has an odd number of valence electrons (19).

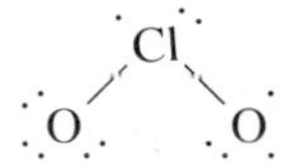

48. (a) I_3^- (b) $BrCl_2^-$ (c) ClF_2^+

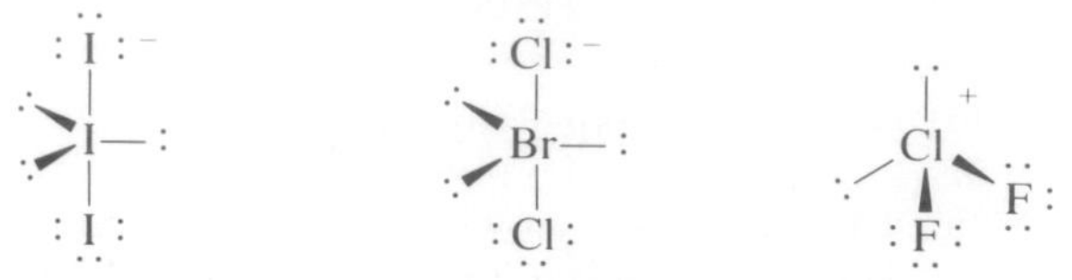

49. Not spontaneous; $E°$ for the net reaction is -0.02 V.
51. (a) 215 tons H_2O; (b) 1.74×10^8 g O_2 produced. (c) 215 tons of Al are required and 407 tons of Al_2O_3 are produced.
55. (a) yes, $E° = 0.59$ V; (b) no, $E° = -0.77$ V.

CHAPTER 23

1. octane (systematic) or *n*-octane (common)
2. (c)
4. $C_{13}H_{28}$
7. (a) 2,3-dimethylbutane; (b) 2-methylpentane; (c) 4,4-dimethylbutane; (d) 3-methylhexane
10. (a) alkene; (b) ether; (c) alkyne; (d) amine; (e) ester; (f) ketone; (g) alcohol; (h) carboxylic acid; (i) aldehyde
14. $H-\overset{H}{\underset{H}{C}}-C\equiv N:$, C—C—N bond angle is 180°; C atom of CN is *sp* hybridized.
15. (a) ethanol; (b) 2-methyl-2-propanol; (c) 1-butanol; (d) 2-methyl-2-butanol
17. (a) primary; (b) tertiary; (c) primary; (d) tertiary
20. (a) $CH_3CH_2CH_2\overset{O}{\overset{\|}{C}}OH$;

(b) $CH_3CH_2CH_2\overset{O}{\overset{\|}{C}}H$;

(c) $CH_3CH_2\overset{O}{\overset{\|}{C}}CH_3$;

(d) $CH_3CH_2\overset{Br}{\overset{|}{C}}HCH_3$

(e) $CH_3CH_2CH_2CH_2OCH_2CH_2CH_2CH_3$;
(f) $CH_3CH_2O^-Na^+$
22. There must be two *different* atoms or groups at each of the double bond carbon atoms. *Cis-trans* isomerism cannot exist with alkynes.

24. (a) $CH_3CH_2\underset{Br}{\underset{|}{C}}HCH_2Br$;

(b) $CH_3CH_2\underset{Br}{\underset{|}{C}}HCH_3$;

(c) $CH_3CH_2CH_2CH_3$;

(d) $CH_3CH_2\underset{OH}{\underset{|}{C}}HCH_3$

25. (a) $CH_3\overset{O}{\overset{\|}{C}}CH_3$ and $CH_3CH_2\overset{O}{\overset{\|}{C}}OH$

(b) $CH_3\overset{Br}{\underset{CH_3}{C}}CH_2CH_2CH_3$

(c) $CH_3\underset{CH_3}{\underset{|}{C}}HCH_2CH_2CH_3$

27. (a) $CH_3CH{=}CHCH_3 + Br_2 \rightarrow CH_3\underset{Br}{\underset{|}{C}}H-\underset{Br}{\underset{|}{C}}HCH_3$

(b) $CH_3CH_2CH{=}CH_2$ (or $CH_3CH{=}CHCH_3$) + $HBr \rightarrow CH_3CH_2\underset{Br}{\underset{|}{C}}HCH_3$

(c) $CH_3CH_2CH{=}CH_2$ (or $CH_3CH{=}CHCH_3$) + H_2O (in the presence of H^+) $\rightarrow CH_3CH_2\underset{OH}{\underset{|}{C}}HCH_3$

28. $CH_3CH_2CH_2CH_2OH \xrightarrow{H_2SO_4} CH_3CH_2CH{=}CH_2 \xrightarrow{H_2O/H^+} CH_3CH_2\underset{OH}{\underset{|}{C}}HCH_3$

33. Benzene + $CH_3\underset{Br}{\underset{|}{C}}HCH_3 \xrightarrow{AlCl_3} C_6H_5-\underset{CH_3}{\underset{|}{C}}HCH_3$

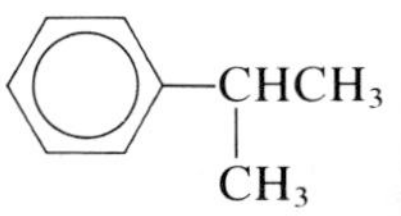

35. An H atom and a halogen atom on adjacent carbon atoms
37. (a) elimination, $CH_3CH_2CH_2CH_2X$ or $CH_3CH_2\underset{X}{\underset{|}{C}}HCH_3$

(b) substitution, $CH_3CH_2CH_2CH_2X$
(c) substitution, $CH_3CH_2CH_2CH_2X$
(d) Grignard, 1,2, or 3-halopentane

38. $CH_3CH_2CH_2Br \xrightarrow{NaOH/H_2O} CH_3CH_2CH_2OH \xrightarrow{Na} CH_3CH_2CH_2O^-Na^+ \xrightarrow{CH_3CH_2CH_2Br} CH_3CH_2CH_2{-}O{-}CH_2CH_2CH_3$

40. (a) hexanoic acid; (b) methyl propanoate; (c) butyl ethanoate; (d) p-bromobenzoic acid.

44. (a) $HO\overset{O}{\overset{\|}{C}}CH{=}CH\overset{O}{\overset{\|}{C}}OH$;

(b) 69.2 mL;

(c) $2Na^+(^-O\overset{O}{\overset{\|}{C}}CH{=}CH\overset{O}{\overset{\|}{C}}O^-)$

49. (a) propanone; (b) butanal; (c) 1-phenylpropanone

51. $CH_3\overset{O}{\overset{\|}{C}}CH_3$ (acetone) $\xrightarrow{LiAlH_4}$ $CH_3\overset{OH}{\overset{|}{C}}HCH_3$ (2-propanol) $\xrightarrow{conc.\ H_2SO_4}$

$CH_3CH{=}CH_2$ (propene) $\xrightarrow{HBr}$ $CH_3\overset{Br}{\overset{|}{C}}HCH_3$ (2-bromopropane)

53. $CH_3CH_2CH_2\overset{O}{\overset{\|}{C}}CH_3 + CH_3MgI(\text{from } CH_3I + Mg) \longrightarrow$

$CH_3CH_2CH_2\overset{OMgI}{\overset{|}{C}}CH_3$ (with CH_3 on C) $\xrightarrow{H^+/H_2O}$ $CH_3CH_2CH_2\overset{OH}{\overset{|}{C}}CH_3$ (with CH_3 on C)

55. $CH_3CH_2CH_2CH_2Br$ (1-bromobutane) $\xrightarrow{Mg}$

$CH_3CH_2CH_2CH_2MgBr \xrightarrow{CH_3\overset{O}{\overset{\|}{C}}H}$

$CH_3\overset{OMgBr}{\overset{|}{C}}HCH_2CH_2CH_2CH_3 \xrightarrow{H^+/H_2O}$

$CH_3\overset{OH}{\overset{|}{C}}HCH_2CH_2CH_2CH_3$

57. (a) $CH_3CH_2OH \xrightarrow{PCC} CH_3\overset{O}{\overset{\|}{C}}H$

(b) $CH_3CH_2OH \xrightarrow{HBr} CH_3CH_2Br \xrightarrow{Mg} CH_3CH_2MgBr$

(c) $CH_3\overset{O}{\overset{\|}{C}}H$ (from step a)

$+ CH_3CH_2MgBr$ (from step b) $\longrightarrow$

$CH_3\overset{OMgBr}{\overset{|}{C}}HCH_2CH_3 \xrightarrow{H^+/H_2O} CH_3\overset{OH}{\overset{|}{C}}HCH_2CH_3 \xrightarrow{Na_2Cr_2O_7}$

$CH_3\overset{O}{\overset{\|}{C}}CH_2CH_3$

methyl ethyl ketone (MEK) or butanone

59. (a) $CH_3Br \xrightarrow{Mg} CH_3MgBr$

(b) $CH_3\overset{OH}{\overset{|}{C}}HCH_3 \xrightarrow{Na_2Cr_2O_7} CH_3\overset{O}{\overset{\|}{C}}CH_3$

(c) CH_3MgBr (from step a) $+ CH_3\overset{O}{\overset{\|}{C}}CH_3$

(from step b) $\rightarrow$ $CH_3\overset{OMgBr}{\overset{|}{C}}CH_3$ (with CH_3 on C) $\xrightarrow{H^+/H_2O}$

$CH_3\overset{OH}{\overset{|}{C}}CH_3$ (with CH_3 on C) $\xrightarrow{conc.\ H_2SO_4}$ $CH_3C{=}CH_2$ (with CH_3 on C)

$\xrightarrow{Br_2}$ $CH_3\overset{Br}{\overset{|}{C}}CH_2Br$ (with CH_3 on C)

61. pH = 11.85

63. $(CH_3CH_2)_3N + HBr \rightarrow (CH_3CH_2)_3NH^+Br^-$
triethylammonium bromide

67. (a) $CH_3CH_2CH_2OH$, 1-propanol (primary alcohol); $CH_3\overset{OH}{\overset{|}{C}}HCH_3$, 2-propanol (secondary alcohol); $CH_3CH_2{-}O{-}CH_3$, ethyl methyl ether (ether).

(b) $CH_3CH_2CH_2\overset{O}{\overset{\|}{C}}H$, butanal (aldehyde); $CH_3CH_2\overset{O}{\overset{\|}{C}}CH_3$, butanone (ketone)

69. 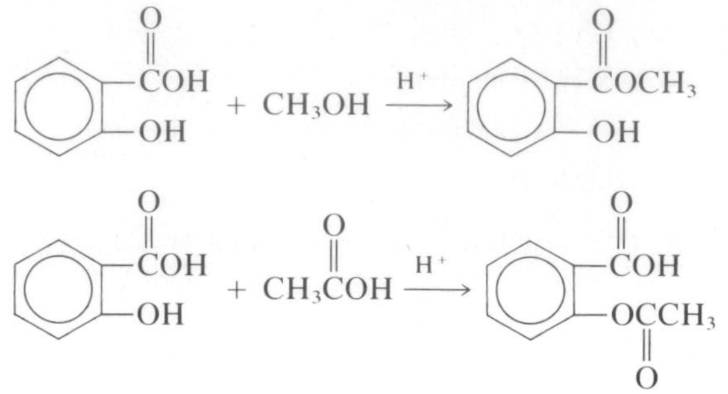 $C_6H_4(OH)COOH + CH_3OH \xrightarrow{H^+} C_6H_4(OH)COOCH_3$

$C_6H_4(OH)COOH + CH_3COOH \xrightarrow{H^+} C_6H_4(OCOCH_3)COOH$

23. $-\!\!(CH_2-CH(OCOCH_3)-CH_2-CH(OCOCH_3)-CH_2-CH(OCOCH_3))\!\!-$

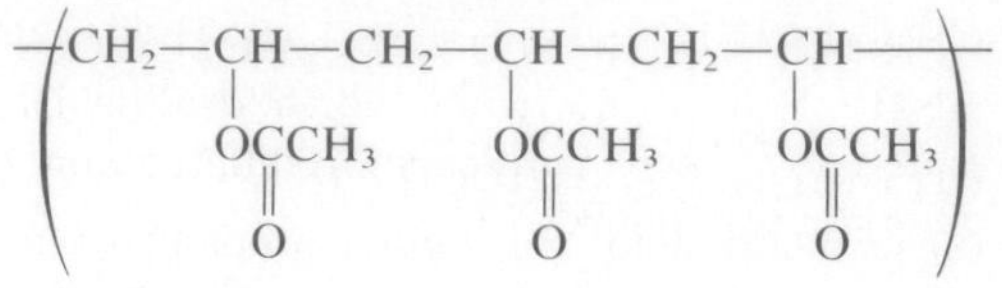

24. 5.4 moles of C_2H_4

26. Polyethylene is formed by addition polymerization.

$$n\ CH_2{=}CH_2 \longrightarrow -\!(CH_2CH_2)_n\!-$$

Nylon-66 is a condensation polymer.

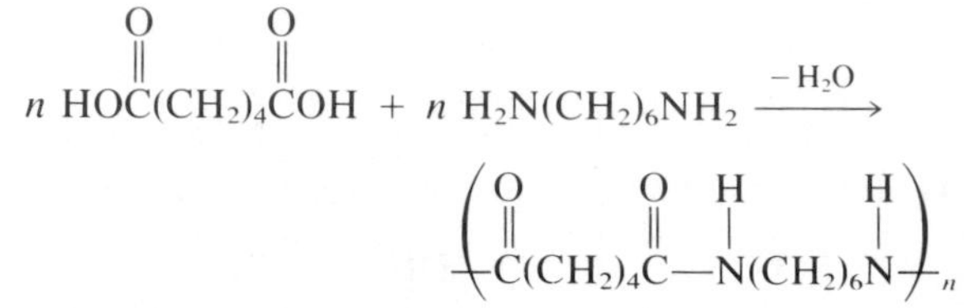

$$n\ HOC(O)(CH_2)_4C(O)OH + n\ H_2N(CH_2)_6NH_2 \xrightarrow{-H_2O} -\!(C(O)(CH_2)_4C(O)-NH(CH_2)_6NH)_n\!-$$

31. 2.4 kg of $CHCl_3$ and 8.0×10^2 g of HF

32. 0.077 mol styrene

34. 7.09×10^8 g of adipic acid and 5.64×10^8 g of hexamethylenediamine

CHAPTER 24

4. (a) Not a chiral center; (b) yes; (c) no; (d) yes

6. Glycine. The central C atom does not have four different groups bonded to it.

7. (a) 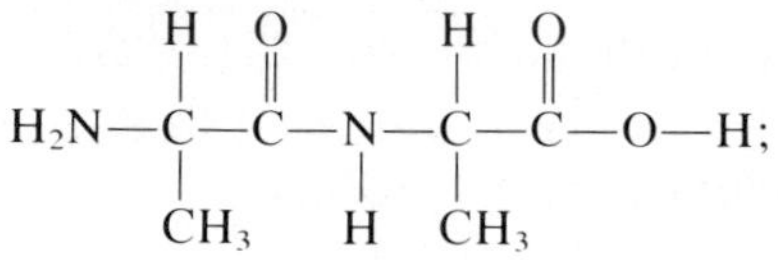 $H_2N-CH(CH_3)-C(O)-NH-CH(CH_3)-C(O)-O-H$;

alanine-alanine

(b) 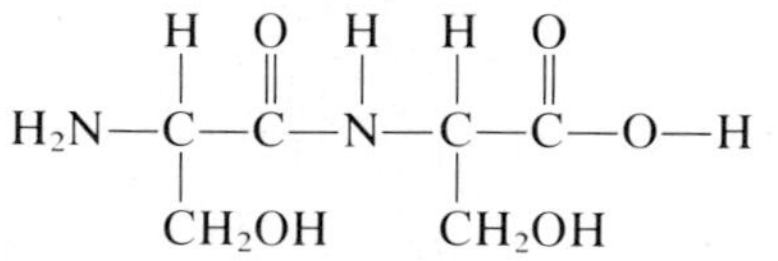 $H_2N-CH(CH_2OH)-C(O)-NH-CH(CH_2OH)-C(O)-O-H$

serine-serine

(c) 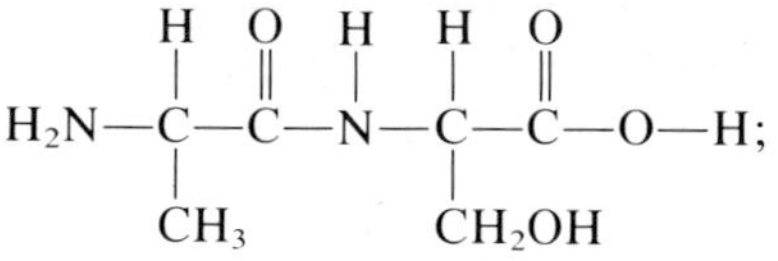 $H_2N-CH(CH_3)-C(O)-NH-CH(CH_2OH)-C(O)-O-H$;

alanine-serine

(d) 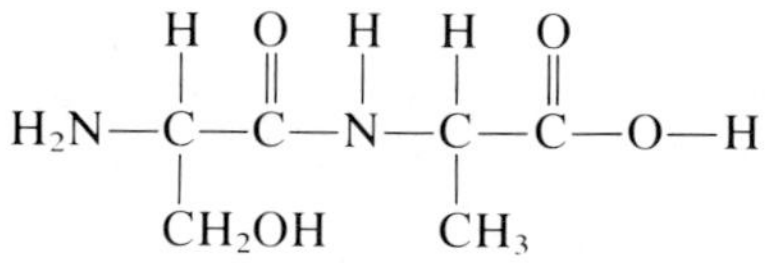 $H_2N-CH(CH_2OH)-C(O)-NH-CH(CH_3)-C(O)-O-H$

serine-alanine

12. Hydrogen bonding

13. It is a dihydroxylaldehyde. It molecular formula is $C_3H_6O_3$ and fits the $C_x(H_2O)_y$ general formula for carbohydrates (here x and y are both 3).

14. Carbon atoms 1, 2, 3 and 4.

18. The nucleoside adenosine and water

21. 1 = *U*, 2 = *A*, and 3 = *G*

CHAPTER 25

10. (a) Y^{3+}: Kr core remains after removing the 5*s* and 4*d* electrons of the atom to form the diamagnetic ion.
 (b) Rh^{3+}: $[Kr]4d^6$, paramagnetic.
 (c) Ce^{4+}: Xe core remains. The two 6*s*, one 5*d*, and one 4*f* electrons have been removed from the atom to form the diamagnetic ion.
 (d) Pt^{4+}: $[Xe]4f^{14}5d^6$, paramagnetic
 (e) V^{2+}: $[Ar]3d^3$, paramagnetic
 (f) U^{4+}: $[Rn]5f^2$, paramagnetic

13. $2\ CuFeS_2(s) + 3\ O_2(g) \rightarrow 2\ CuS(s) + 2\ FeO(s) + 2\ SO_2(g)$
 1.0 ton of $CuFeS_2$ gives 0.35 tons of SO_2.

15. (a) $FeTiO_3(s) + 2\ HCl(aq) \rightarrow FeCl_2(aq) + TiO_2(s) + H_2O(\ell)$
 (b) The HCl can be recovered completely (in theory).
 (c) 4.78×10^5 g

17. (a) CH_3NH_2, monodentate (lone pair of electrons on N); (b) monodentate; (c) ethylenediamine, bidentate (d) N_3^-, monodentate; (e) $C_2O_4^{2-}$, oxalate ion, bidentate; (f) acetonitrile, monodentate (lone pair of electrons on N); (g) bidentate

21. (a) $Ni(en)_2Cl_2$; (b) $K[PtCl_4]$; (c) $[Cr(NH_3)_2(H_2O)_3OH]NO_3$; (d) $[Fe(NH_3)_6](NO_3)_3$; (e) $Fe(CO)_5$

25. (a) $[Fe(H_2O)_5OH]^{2+}$; (b) potassium tetracyanonickelate(II); (c) potassium diaquobis(oxalato)chromate(III); (d) pentaamminesulfatochromium(III) chloride; (e) $Na_2[PdCl_4]$.

26. (a) $Pd(NH_3)_4Cl_2$

H_3N Cl NH_3 / Pd / H_3N Cl NH_3 — H_3N Cl Cl / Pd / H_3N NH_3 NH_3

(b) $Pt(NH_3)_2(NCS)(Br)$

H_3N NCS / Pt / Br NH_3 — H_3N NCS / Pt / H_3N Br

(c) $Co(NH_3)_3(NO_2)_3$

H_3N NO_3 NO_3 / Co / H_3N NH_3 NO_3 — H_3N NH_3 NO_3 / Co / O_3N NH_3 NO_3

28. $[Co(en)(NH_3)_2(H_2O)Cl]^{2+}$, en is depicted by N—N.

N NH_3 Cl / Co / N NH_3 OH_2 $]^{2+}$ — N Cl NH_3 / Co / N H_2O NH_3 $]^{2+}$

N NH_3 NH_3 / Co / N H_2O Cl $]^{2+}$ — N NH_3 NH_3 / Co / N Cl OH_2 $]^{2+}$

30. (a) $[Fe(CN)_6]^{4-}$. Ion is based on Fe^{2+}, which has 6 $3d$ electrons. If high spin, the complex would be paramagnetic by 4 unpaired electrons. (The ion is actually low spin.)

high spin	low spin
1 1 (z^2 x^2-y^2)	— — (z^2 x^2-y^2)
1l 1 1 (xy xz yz)	1l 1l 1l (xy xz yz)

(b) $[Co(NH_3)_6]^{3+}$. Ion is based on Co^{3+}, which has 6 $3d$ electrons. Therefore, the configurations would be the same as in part (a) above. (As in (a) above, the ion is actually low spin.)

(c) $[Fe(H_2O)_6]^{3+}$. The ion is based on Fe^{3+}, which has 5 $3d$ electrons. Here both the low and high spin configurations would be paramagnetic.

high spin	low spin
1 1 (z^2 x^2-y^2)	— — (z^2 x^2-y^2)
1 1 1 (xy xz yz)	1l 1l 1 (xy xz yz)

33. Blue

35. (a) CN^- creates a very high crystal or ligand field. Complex is low spin and not paramagnetic. (See question 30, part a.)

(b) The bidentate ligand en creates a strong ligand field. However, here we have d^3 Cr^{3+} ion, so only one configuration is possible.

— — (z^2 x^2-y^2)

1 1 1 (xy xz yz)

(c) The F^- ion creates the weakest ligand field. Therefore, $[MnF_6]^{4-}$, based on a d^5 Mn^{2+} ion, is likely to be high spin.

1 1 (z^2 x^2-y^2)

1 1 1 (xy xz yz)

(d) The phen ligand creates a very strong ligand field, and its complexes are usually low spin. This complex is based on a d^9 Cu^{2+} ion, so there is no difference between low and high spin.

1l 1 (z^2 x^2-y^2)

1l 1l 1l (xy xz yz)

37. A = $[Co(NH_3)_5Br]SO_4$ and B = $[Co(NH_3)_5SO_4]Br$

39. (a) oxidation-reduction; (b) 40.9 kg Cl_2 and 5.93 kg C; (c) 7.89 kg Ti

41. (a) U^{4+}; (b) $5\ U^{4+} + 2\ H_2O + 2\ MnO_4^- \rightarrow 5\ UO_2^{2+} + 4\ H^+ + 2\ Mn^{2+}$

43. Both complexes are based on Cr^{2+}, an ion with four $3d$ electrons.

high spin	low spin
1 — (z^2 x^2-y^2)	— — (z^2 x^2-y^2)
1 1 1 (xy xz yz)	1l 1 1 (xy xz yz)

Both complexes are paramagnetic. However, the high spin complex is expected to have four unpaired electrons, while the low spin ion will have only two.

45. 95.00% Al and 5.00% Cu

46. $[Cr(H_2O)_6]Cl_3$, hexaaquochromium(III) chloride

CHAPTER 26

3. (a) $^{56}_{26}Fe$; (b) $^{1}_{0}n$; (c) $^{32}_{15}P$; (d) $^{97}_{43}Tc$; (e) $^{0}_{-1}e$; (f) $^{1}_{0}n$; (g) $^{4}_{2}He$

5. (a) $^{0}_{-1}e$; (b) $^{87}_{37}Rb$; (c) $^{4}_{2}He$; (d) $^{226}_{88}Ra$; (e) $^{0}_{-1}e$; (f) $^{24}_{11}Na$; (g) $^{4}_{2}He$; (h) $^{0}_{-1}e$; (i) $^{23}_{10}Ne$

7. $^{235}_{92}U \xrightarrow{-\alpha} {}^{231}_{90}Th \xrightarrow{-\beta} {}^{231}_{91}Pa \xrightarrow{-\alpha} {}^{227}_{89}Ac \xrightarrow{-\beta}$

$^{227}_{90}Th \xrightarrow{-\alpha} {}^{223}_{88}Ra \xrightarrow{-\alpha} {}^{219}_{86}Rn \xrightarrow{-\alpha} {}^{215}_{84}Po \xrightarrow{-\alpha}$

$^{211}_{82}Pb \xrightarrow{-\beta} {}^{211}_{83}Bi \xrightarrow{-\beta} {}^{211}_{84}Po \xrightarrow{-\alpha} {}^{207}_{82}Pb$

9. (a) $^{12}_{6}C$; (b) $^{14}_{7}N$; (c) $^{28}_{14}Si$

10. $^{30}_{15}P$, -2.34×10^{13} J/mol; $^{31}_{15}P$, -2.54×10^{13} J/mol; ^{31}P is the more stable.

12. (a) 0.63 mg; (b) 2.0×10^{-5} mg

14. $t_{1/2} = 3.00$ hr

16. (a) 5 half-lives; (b) 2130 AD

18. 2700 years

20. 3.4×10^{9} years

22. $^{239}_{94}Pu + {}^{4}_{2}He \longrightarrow {}^{240}_{95}Am + {}^{1}_{1}H + 2\, {}^{1}_{0}n$

24. (a) $^{247}_{99}Es$; (b) $^{16}_{8}O$; (c) $^{4}_{2}He$; (d) $^{12}_{6}C$; (e) $^{10}_{5}B$

26. 1.45×10^{9} years

28. $^{232}_{90}Th \longrightarrow {}^{233}_{90}Th \longrightarrow {}^{0}_{-1}e + {}^{233}_{91}Pa$

$^{1}_{0}n$ + fission products $\longleftarrow {}^{233}_{92}U + {}^{0}_{-1}e$

30. 130 mL

32. (a) $Zr(s) + 2\, H_2O(g) \longrightarrow ZrO_2(s) + 2\, H_2(g)$
(b) 1.1×10^{4} mol H_2
(c) 1.8×10^{-4} atm

33. 1.14×10^{15} yr^{-1} (or 3.6×10^{7} s^{-1})

35. 1.2×10^{15} atoms (or 2.9×10^{-7} g Pm)

INDEX/GLOSSARY

Boldface page numbers indicate pages on which the index item is defined; italicized page numbers indicate pages containing illustrations. Glossary terms, printed in boldface, are defined here as well as in the text. Some terms used generally in the text are defined here without giving specific page references.

The following illustrations are based upon drawings by Irving Geis:

Figure 7.5 (page 220)

Figure 11.36 (page 416)

Figure 11.4 (page 384)

Figure 24.5 (page 918)

Figure 24.6 (page 919)

Figure 24.8 (page 921)

Figure 24.10 (page 922)

Unnumbered figure (page 924)

Unnumbered figure (page 925)

Figure 24.12 (page 926)

Figure 24.13 (page 927)

Figure 24.14 (page 929)

Figure 24.15 (page 930)

Figure 24.16a (page 931)

Unnumbered figure (page 958)

Unnumbered figure (page 958)

Figure 25.13 (page 959)

Unnumbered figure (page 961)

Figure 25.25 (page 977)

Figure 25.28 (page 980)

Physical and Chemical Constants

Avogadro's number	$N = 6.022045 \times 10^{23}$ /mol
Electronic charge	$e = 1.6022 \times 10^{-19}$ C
Faraday's constant	$F = Ne = 9.6485 \times 10^{4}$ C/mol electrons
Gas constant	$R = 8.3144$ J/K.mol
	$= 0.082057$ L.atm/K.mol
Pi	$\pi = 3.1415926536$
Planck's constant	$h = 6.6262 \times 10^{-34}$ J.s
Speed of light (in a vacuum)	$c = 2.997925 \times 10^{8}$ m/s

Useful Conversion Factors and Relationships

LENGTH

SI unit: Meter (m)

1 kilometer = 1000 meters
= 0.62137 mile
1 meter = 100 centimeters
1 centimeter = 10 millimeters
1 nanometer = 1×10^{-9} meter
1 picometer = 1×10^{-12} meter
1 inch = 2.54 centimeters (exactly)
1 Ångstrom = 1×10^{-10} meter

MASS

SI unit: Kilogram (kg)

1 kilogram = 1000 grams
1 gram = 1000 milligrams
1 pound = 453.6 grams = 16 ounces
1 ton = 2000 pounds

VOLUME

SI unit: Cubic meter (m^3)

1 liter (L) = 1×10^{-3} m^3
= 1000 cm^3
= 1.056710 quarts
1 gallon = 4 quarts

PRESSURE

SI unit: Pascal (Pa)

1 pascal = 1 N/m^2
= 1 $kg/m^1.s^2$
1 atmosphere = 101.325 kilopascals
= 760 mm Hg = 760 torr
= 14.70 lb/in^2

ENERGY

SI unit: Joule (J)

1 joule = 1 $kg.m^2/s^2$
= 0.23901 calorie
= 1 C × 1 V
1 calorie = 4.184 joules
1 electron-volt = 96.485 kJ/mol

TEMPERATURE

SI unit: Kelvin (K)

0 K = −273.15°C
K = °C + 273.15
?°C = (5°C/9°F)(°F − 32°F)
?°F = (9°F/5°C)°C + 32°F